ANNUAL REVIEW OF BIOCHEMISTRY

EDITORIAL COMMITTEE (1992)

ANNUAL REVIEW OF BIOCHEMISTRY

VOLUME 61, 1992

CHARLES C. RICHARDSON, *Editor*
Harvard Medical School

JOHN N. ABELSON, *Associate Editor*
California Institute of Technology

ALTON MEISTER, *Associate Editor*
Cornell University Medical College

CHRISTOPHER T. WALSH, *Associate Editor*
Harvard Medical School

ANNUAL REVIEWS INC.   4139 EL CAMINO WAY   P.O. Box 10139   PALO ALTO, CALIFORNIA 94303-0897

ANNUAL REVIEWS INC.
Palo Alto, California, USA

International Standard Serial Number: 0066–4154
International Standard Book Number: 0–8243–0861–1
Library of Congress Catalog Card Number: 32-25093

Annual Review and publication titles are registered trademarks of Annual Reviews Inc.

⊗ The paper used in this publication meets the minimum requirements of
American National Standard for Information Sciences—Permanence of Paper
for Printed Library Materials, ANZI Z39.48-1984.

Annual Reviews Inc. and the Editors of its publications assume no responsibility for the
statements expressed by the contributors to this *Review*.

Typesetting by Kachina Typesetting Inc., Tempe, Arizona; John Olson, President;
Janis Hoffman, Typesetting Coordinator; and by the Annual Reviews Inc. Editorial Staff

PRINTED AND BOUND IN THE UNITED STATES OF AMERICA

Annual Review of Biochemistry
Volume 61, 1992

CONTENTS

SOME RELATED ARTICLES IN OTHER *ANNUAL REVIEWS*

From the *Annual Review of Biophysics and Biophysical Structure,* Volume 21 (1992)

The Mechanism of α-Helix Formation by Peptides, J. M. Scholtz and R. L. Baldwin
Protein Folding in the Cell: The Role of Molecular Chaperones Hsp70 and Hsp60, F. U. Hartl, J. Martin, and W. Neupert
The Single-Nucleotide Addition Cycle in Transcription: A Biophysical and Biochemical Perspective, D. A. Erie, T. D. Yager, and P. H. von Hippel
Structure and Mechanism of Alkaline Phosphatase, J. E. Coleman
NMR Structure Determination in Solution: A Critique and Comparison with X-Ray Crystallography, G. Wagner, S. G. Hyberts, and T. F. Havel

From the *Annual Review of Cell Biology,* Volume 8 (1992)

Cadherins, B. Geiger
Structure Framework for the Protein Kinase Family, S. S. Taylor, D. R. Knighton, J. Zheng, L. F. Ten Eyck, J. M. Sowadski
Chromatin Structure and Transcription, R. Kornberg
Retroviral Reverse Transcription and Integration: Progress and Problems, J. M. Whitcomb and S. H. Hughes
Regulation of Translation in Eukaryotic Systems, M. Kozak

From the *Annual Review of Genetics,* Volume 25 (1991)

Import of Proteins into Mitochondria, B. Glick and G. Schatz
Mechanisms and Biological Effects of Mismatch Repair, P. Modrich
Regulation of Bacterial Oxidative Stress Genes, B. Demple

From the *Annual Review of Immunology,* Volume 10 (1992)

Antigen Receptors on B Lymphocytes, M. Reth
V(D)J Recombination: Molecular Biology and Regulation, D. G. Schatz, M. A. Oettinger, M. S. Schlissel

From the *Annual Review of Medicine,* Volume 43 (1992)

The Molecular Basis of Colon Cancer, A. K. Rustgi and D. K. Podolsky
Biological Regulation of Factor VIII Activity, R. J. Kaufman

ERRATA

On the following four pages, we reprint the two tables in last year's chapter by Lena Kjellén and Ulf Lindahl, titled Proteoglycans: Structures and Interactions. Table 1, which was on pp. 452–54, had several misaligned or missing brackets, and poor spacing; Table 2, on p. 459, had one misaligned bracket. We apologize to the authors for not catching these errors earlier, and hope that no one was mislead by them.

Table 1 Properties and functions of proteoglycans

Source/designation	Core protein size (kDa)[a]	GAG chains	Proposed functions/characteristics	Refs.
Extracellular				
Family a				
Cartilage/aggrecan	208–221[b]	>100 CS, 20–30 KS		54–63
Fibroblasts/versican	265[b]	12–15 CS	Aggregated by HA.	64–67
Embryonic mesenchyme/PG-M	550	CS	Provide mechanical support; fixed negative charge	68–70
Aorta		CS/DS	Regulate cell migration and cell aggregation?	71–74
Aortic endothelial cells				75
Aortic smooth muscle cells	>100–550			76
Embryonic skeletal muscle				77
Developing bone				78
Glial cells				79
Nucleus pulposus				80
Sclera				81
Tendon				82
Brain	75–420	CS	Role in development?	83–85
Neurons/CAT-301	580	CS	Stabilize synaptic structure?	86, 87
Endothelial cells	>200	DS	Role in endothelial regeneration?	88
Reicherts membrane	160	13–22 CS	Stabilize basement membranes?	89, 89a, 255, 269
Family b				
Connective tissues/decorin	36[b]	1 CS/DS	Modulate collagen fibrillogenesis; regulate cell growth.	90–102
Connective tissues/biglycan	38[b]	2 CS/DS		94, 97, 102–104
Connective tissues/fibromodulin	41[b]	1 KS		105–108
Osteosarcoma cells, fibroblasts/PG 100	106	1 CS	Modulate collagen fibrillogenesis.	108a

	Cartilage, vitreous humor/Collagen type IX	68[b]	1 CS	Binds to collagen fibrils.	53, 109, 110
	Cornea, other tissues	25, 37–51	KS	Organize collagen fibrils, required for corneal transparency.	111–116
	Hepatocytes	28	1–2 CS	Urinary trypsin inhibitor	116a
Family c	EHS-tumor	400–450	2–3 HS	⎫ Modulate assembly of basement membranes.	117–122
	Fibroblasts	250–400	HS		67, 123–125
	Colon carcinoma cells	240–400	10–15 HS		19a, 126, 127
	Basement membranes	150–400	HS	⎬ Provide filtration barrier.	128–131
	Endothelial cells	350	HS		132–134
	Mammary epithelial cells	?	HS	⎭	135

Cell surface

Family d	Mammary epithelial cell/syndecan	31[b]	1–2 CS/ 1–2 HS	Role in morphogenesis; link cytoskeleton to extracellular matrix	16, 136–141
	Fibroblasts/fibroglycan	48(90)	HS		142, 143
	Liver	35, 44, 49, 77	HS		144–146a
	Brain	55	CS/HS		147–149
	Fibroblasts	125, 35	HS		67, 142, 150, 151
	Fibroblasts	90, 52	CS/DS		67, 152, 153
	Neurons	280	CS	Mediate cell adhesion. Role in morphogenesis?	154–156
	Marine ray electric organ/TAP-1	180	≈20 CS		157
	Melanoma cells	250	3–12 CS	Modulate assembly of matrix.	158–160
	Hepatocyte cell line	?	HS		161
	Ovarian granulosa cells	80	2–4 HS		162
	Schwann cells	?	HS, CS?	⎫ Phosphatidyl-inositol linked PGs.	163
	Central nervous system	50, 59	HS		85
	Fibroblasts	64	HS	⎭	163a

Table 1 (*Continued*)

Source/designation	Core protein size (kDa)[a]	GAG chains	Proposed functions/characteristics	Refs.
Fibroblasts/betaglycan	100–120	CS/HS	Provide reservoir for TGF-β	164, 165
Endothelial cells/thrombomodulin	58–60[b]	1 CS	Regulate blood coagulation.	17, 166–169
Lymphocytes/CD44, homing receptor	37[b]	CS	Mediate cell adhesion.	10, 170–174
Fibroblasts/transferrin receptor	2 × 90[b]	4–6 HS	Mediate uptake of transferrin.	29, 67
B-cells, antigen presenting cells/invariant chain PG	31[b]	CS	Role in antigen presentation.	52, 175, 176
Intracellular				
Chromaffin granules/Chromogranin A PG	48[b]	1 CS	Part-time PGs.	177, 178
Family e { Yolk sac carcinoma cells[c] / Mast cells / Basophilic leukemia cells / Platelets / NK-cells / Macrophage-like cells / Eosinophils } /serglycin	17–19[b]	CS, DS, HS, heparin	Store and modulate activity of granular proteases. Prevent blood coagulation (heparin).	179–195
Chromaffin granules	14	1–2 CS/DS		177

[a] If not otherwise indicated, M_r was estimated from the migration of core proteins in SDS-PAGE after removal of GAG chains with endoglycosidases.

[b] M_r calculated from amino acid/cDNA sequence data.

[c] In these cells, serglycin is primarily located extracellularly.

Table 2 Heparin-binding proteins—functional role of interaction[a]

Protein	Proposed functional role of interaction	Refs.
Enzymes		
Mast cell proteinases	Provide intracellular storage of enzymes in inactive form; modulate enzyme activity	190, 225–227
Lipoprotein lipase	Anchor enzyme at cell surface; regulate enzyme expression	228–238
Coagulation enzymes	Facilitate enzyme inactivation by antithrombin	239–244
Elastase	Promote enzyme inhibition	245
Extracellular superoxide dismutase	Anchor enzyme at cell surface or in extracellular matrix	246–249
Serine protease inhibitors		
Antithrombin	Promote inactivation of coagulation enzymes at vascular endothelium	250–260
Heparin cofactor II	Promote inactivation of thrombin	261–264
Protease nexins	Inactivate serine proteases in the extracellular matrix	265–266
Extracellular matrix proteins		
Vitronectin	Modulate coagulation and complement processes	283–284
Fibronectin		15, 101, 150, 276–282
Laminin	Mediate/modulate cell-substrate interactions	285–287
Thrombospondin		288–292
Collagens		293, 294
Growth factors		
Fibroblast growth factors		295–301
Schwann cell growth factor	Sequester (protect) growth factors in the extracellular matrix	302
Retinal survival factor		303
Hemopoietic growth factors		303a, b
Other proteins		
Apolipoproteins	Mediate lipoprotein uptake by cells; role in atherosclerosis	267–275
Platelet factor 4	Neutralize biological activity of heparin/HS	304, 305
N-CAM	Promote nerve cell adhesion	306–308
Viral coat protein	Mediate binding of virus (HIV; Herpes) to cells	309
Transcription factors?	Modulate gene expression	310, 311

[a] The ligand-binding GAG in vivo is likely to be HS (or in some cases, a galactosaminoglycan) rather than heparin; see the text. For recent reviews on GAG-protein interactions see Cardin & Weintraub (313), Jackson et al (7), and Lane (311a).

Eugene P. Kennedy

Annu. Rev. Biochem. 1992. 61:1–28

SAILING TO BYZANTIUM

Eugene P. Kennedy

Department of Biological Chemistry and Molecular Pharmacology, Harvard Medical School, Boston, Massachusetts 02115

KEY WORDS: mitochondria, protein phosphorylation, phospholipid biosynthesis, phospholipid translocation, membrane-derived oligosaccharides

CONTENTS

EARLY EDUCATION

I was born on September 4, 1919, in Chicago, the fourth of five children of Catherine Frawley Kennedy and Michael Kennedy. My parents were Irish immigrants who had been born within a few miles of each other in County Clare, near Miltown Malbay, but as was not infrequently the case with such couples, had not met before coming to America. My mother told me later that I had been named Eugene, not an uncommon name, after the great socialist

1

0066-4154/92/0701-0001$02.00

Eugene V. Debs. This was not a political statement on her part; she simply liked the sound of the name.

My father was a motorman on the streetcars of the Chicago Surface Lines. With five children, and a motorman's salary of about $35 per week in the 1930s, our family was distinctly on the lower end of the lower middle-class. Although plagued by a chronic shortage of money, we were never really poor, as urban poverty is now known. My father's education had stopped at about the fifth grade; my mother had somewhat more schooling, with even a smattering of French. Both urged their children to strive for better education and better opportunities than they had had.

I began my education in Catholic parochial schools. I did not at first easily learn to read, but rather as many children do, tried to memorize the simple text of the primer rather than puzzling out the words. We had boxes of letters that we were told to arrange into words, and I distinctly remember the shock of recognition when it dawned on me that the letter "y" at the end of a word was pronounced "eee." Somehow that seemed the key to all of phonetics, and I was soon reading easily.

Not long thereafter, I was introduced to the local branch of the public library, a store-front operation on North Avenue, by my brother Joseph, six years older, and in this as in other ways a strong influence on my early life. He also lent me his "intermediate" library card so that I was not confined to the children's section in borrowing books. I became a voracious, indiscriminate reader. While still in grammar school, I happily read my way through a ton of junk, such stuff as the Fu Manchu novels of Sax Rohmer, the detective stories of S. S. Van Dine, all the available works of Zane Grey, James Oliver Curwood, and on a somewhat higher level, Jack London.

In 1933, having completed grammar school, I entered St. Philip High School, taught by priests of the Servite order, a few of whom made a lasting impression on me. The principal of the school was Father Wolfe, a fierce disciplinarian, who was held in awe and some terror by the students, not only from fear of corporal punishment, which he was not slow to administer, but also from a kind of intensity that seemed to simmer just below the surface of his personality.

My teacher of English was Father Peter Doherty. Sandy-haired, rather pale, he was soft-spoken, but given to outbursts of rage at infractions of order or defective preparation of an assigned task, so that even the most indifferent or refractory students took care not to come to class unprepared as they might in other courses. His instruction stressed fundamental English grammar, which was reviewed meticulously from start to finish, and expository writing. He was a regular reader of the *New Yorker*, then considered rather sophisticated fare, and was a strong advocate of the simple, clear prose for which the style of that magazine's E. B. White became celebrated. When I became editor of the school newspaper, which he supervised, I saw a little more of him. As a year-end treat for the staff he took us to dinner in a downtown restaurant, followed by a play, *Three Men on a Horse*, the first I had ever seen. It was a bit strange to encounter him outside the context of the school in a more relaxed mood. Father Doherty was rumored to be a poet of considerable powers, but I never saw any of his verse. I had the feeling that he lived as much removed from his fellow Servites as from the students.

In 1937, I entered De Paul University, to which I had been awarded a scholarship. Located in Chicago, not far from Lincoln Park, it was known then as now principally for its basketball team. Nearly all of the students were residents of Chicago and continued to live at home, as I did, of economic necessity.

I enrolled as a chemistry major, although my strongest bent was toward literature. At that time, the Department of Chemistry was very small, with a staff of only three or four, headed by Malcolm Dull, who was, however a first-rate teacher of organic chemistry. Later I was to regret that, diverted by other pursuits including the editorship of the university newspaper, I gave only a divided attention to the courses in chemistry.

Events unfolding on the international scene during my college years, 1937–1941, were surely the most dismal of this century, perhaps of any century. A child of the Great Depression, I was an enthusiastic Roosevelt Democrat, and followed international politics closely, particularly of course after the beginning of World War II. It is hard now to convey the nightmare quality of May, 1940, as one watched, helpless, what appeared to be the inexorable triumph of Hitlerism in Europe.

After graduation in June 1941, I was uneasily aware that my training in chemistry had suffered from the diversity of my other interests. I decided nevertheless to enroll as candidate for the PhD degree in organic chemistry at the University of Chicago, which in that simpler time I did simply by writing a letter of application, together with a transcript of college credits. I was accepted also by mail, without the formality of a interview, which was perhaps just as well from the point of view of my candidacy. Of course, the Department of Chemistry was making no commitment whatever to me since I would be paying full tuition. If as a graduate student I sank, there was no loss to them; if I swam, so much the better.

THE PLASMA FRACTIONATION PROJECT

It was now necessary to find a full-time job. The expenses of college had been modest, because I had had a full scholarship and lived at home, but had nevertheless been a strain on the very limited resources of my family. Through an employment agency, I applied for a job in the chemical research department of Armour and Company, one of the giant meat-packers in the world-famous stockyards that dominated the South Side of Chicago at that time. I was interviewed by Lemuel C. Curlin, a biochemist who had only recently received his PhD degree from the University of Chicago.

"What do you know about proteins?" he asked. I fleetingly considered assembling the few pitiful scraps of what I knew about proteins into some kind of a response, but fortunately opted for candor, and said firmly that I knew nothing about proteins. Curlin took this in good part, perhaps concluding that my answer indicated some sophistication rather than dismal ignorance as was the fact. He said that not much was known about proteins in any case, which was certainly true in 1941, and without further interrogation went on to outline the project for which he was assembling a team.

The laboratory of Edward J. Cohn in the Harvard Medical School, of which John Edsall has given a valuable account (1), was then one of the few great centers for the study of proteins. By 1941, many foresaw that the United States would soon be forced into war. The Cohn laboratory was diverted to an all-out effort to meet the anticipated need for blood plasma and plasma substitutes by developing new methods for the fractionation of plasma. Because serum albumin is the most abundant protein in plasma, it makes a major contribution to the maintenance of normal blood volume, a problem of special importance to patients in shock. The strategy therefore was to fractionate large volumes of plasma to isolate the serum albumin in purified, stable form to administer as needed under battlefield conditions. The fractionation procedure was further designed to make available the other valuable proteins of plasma, notably the gamma globulins and fibrinogen.

It was also hypothesized that bovine serum albumin might be an effective plasma substitute for administration to human subjects, based on the theory that, if absolutely pure and completely free from highly antigenic globulins, it so closely resembled human serum albumin that it would not provoke a dangerous immune response. This was the reason for the entry of the Armour laboratories into the picture. Through its pharmaceutical division, Armour had much experience in the fractionation of extracts of animal tissues on a large scale, and therefore undertook the large-scale fractionation of bovine blood with the aim of preparing bovine serum albumin of sufficient purity for adminstration to human subjects.

Curlin hired me as a member of his "plasma gang" at a salary of $40 per week, which I thought quite generous, because entry-level chemists with a B.Sc. degree at that time usually received $25–30. The job provided a jolting introduction to reality. The laboratories were located in the stockyards, a roaring bustling place. To reach the laboratories from Ashland Avenue one passed the loading-dock of one of the nation's largest knackers, specialists in converting the carcasses of animals condemned for use as food into soap and fertilizer. On a summer day, the smell was staggering but was rivalled by the sight of scattered parts of animals, often the heads of horses and cattle, which had spilled onto the dock as they were unloaded from freight cars. When I see the paintings of Georgia O'Keeffe, depicting the spare, clean skulls of animals dried in the pure desert sun of New Mexico, I sometimes think that there can hardly be a more striking example of the transforming power of art.

The work that we underlings carried out on the plasma project was not intellectually demanding but required patience and care. The fractionation procedures, rigidly stipulated by the Cohn laboratory in Boston, were based on control of temperature, pH, ionic strength, and ethanol concentration. Much of the work was done at -5 °C in a large cold-room equipped with large Sharples continuous centrifuges and glass-lined tanks.

In September of 1941, I began graduate studies in chemistry at the University of Chicago. Work on the plasma project was now being pushed forward in shifts around the clock, and my new schedule found me working from midnight to 8:00 AM in the cold-rooms at Armour, followed by morning classes at the University of Chicago, not far distant. The folly of attempting such a schedule while I was acutely aware of the need for extensive review of my background in chemistry is apparent, but the very magnitude of the challenge lent it some of the glamor to which the young are so susceptible. I managed to get through the courses in organic chemistry which were prerequisite for the qualifying examination before beginning research, but I really enjoyed only one course, that in organic qualitative analysis. The instructor was a young chemist named Frank Westheimer. The star pupil was Dan Koshland, and the teaching assistant was Aaron Novick, who was given to cryptic hints as to the identity of the compounds we were required to identify. All three of course became leaders in biochemistry and molecular biology in their later careers, and our paths fortunately were to cross again and again.

Our entry into the war, followed by the almost unrelieved series of military disasters in 1942 that brought the United States and its allies to the very brink of defeat, greatly increased the urgency of the plasma project. Despite sporadic encouraging results, by the end of 1942, hope had almost faded that bovine serum albumin would provide an effective treatment for human patients in shock. The Red Cross was now engaged in collecting blood from human volunteers on a vast scale, and it was decided that Armour would open a new facility in Fort Worth, Texas, where it already had extensive meat-packing operations, for the fractionation of human blood from donors in a large catchment area of the Southwest.

I was one of the team sent to Fort Worth in October, 1943 to begin this new effort. Fortunately, I was soon joined by Adelaide Majewski, who had been a classmate in college, and we were married in Fort Worth on October 27, 1943. Our family was later to be completed by the births of Lisa (1950), now a lawyer and married to the writer Mark Helprin; Sheila (1957), now an architect and married to Frano Violich, also an architect; and Katherine (1960), now a lawyer married to Matthew Diller, also a lawyer.

The brand-new laboratory at Fort Worth was equipped with huge cold-rooms, containing glass-lined tanks with capacities up to a thousand liters, and was expressly designed for fractionation of plasma by the Cohn procedures. As in Chicago, we worked in three shifts around the clock, and my task was to supervise the work of one such shift, staffed mostly by women recruited from the local labor force, and presumed to have some minimum training in chemistry. Some proved to have only a very limited grasp of the fundamentals of chemistry and required very close supervision, but were highly motivated and conscientious.

As in Chicago, the work offered little opportunity for real research. I was, however, given some scope in the design of large-scale procedures for the production of fibrin foam, sponge-like materials designed for use by neuro-surgeons in the treatment of massive head-wounds to control bleeding and to promote healing. Never an admirer of high technology, I managed to do this with a minimum of expense. I was particularly pleased with an apparatus that I designed for the large-scale lyophilization of fibrin foams, which had been assembled in the local Armour shop for a few hundred dollars, equipped with a thermocouple device that I assembled myself to follow the internal tempera-ture of the foams, and thus their moisture content. When the chief engineer on a inspection visit from headquarters in Chicago saw this make-shift apparatus, his face plainly revealed what he thought of it, although he could hardly object since it was working quite well. I later surmised that part of his displeasure arose from the fact that I had spent only a very small fraction of the funds that he had earlier estimated would be needed.

In the spring of 1945, World War II was clearly approaching its end, and large amounts of human plasma proteins had been stockpiled. I returned to Chicago, where I began work on a quasi-independent project on the crystallization of enzymes (particularly pepsin) in which the company was interested.

GRADUATE STUDIES: MITOCHONDRIAL FUNCTION

With the generous support of my supervisors, James Lesh and Jules Porsche, I was now able to arrange a much more flexible schedule to pursue my studies at the University of Chicago. I immediately transferred from the Department of Chemistry to the Department of Biochemistry. Experience on the plasma project, lowly as it was, had offered some glimpse of the fascination of biochemistry.

The Department of Biochemistry at the University of Chicago at that time was headed by Earl Alison Evans Jr., who in 1942 at the age of 32 had succeeded F. C. Koch as chairman. Evans had assembled a brilliant group of colleagues, including Konrad Bloch, Herbert Anker, Albert Lehninger, Birgit Vennesland, Frank Putnam, Hans Gaffron, James Moulder, and until his untimely death in a mountaineering accident, John Speck. Like Bloch and Anker, Evans was a product of Hans Clarke's great Department of Biochemistry at Columbia University, one of the earliest centers for the application of the isotope tracer technique, which was soon to transform biochemistry. Evans later worked with Hans Krebs on the newly formulated Krebs tricarboxylic acid cycle. He had become well known for his studies on the fixation of carbon dioxide in animal tissues. While continuing work along these lines, his laboratory was making a transition to the study of the replication of bacteriophage, thus revealing in the 1940s a remarkable insight into the importance of an approach that was to prove to be one of the precursors of molecular biology, a discipline then still slouching toward Bethlehem to be born. Evans, a widely cultured and urbane man, rather formal in speech and manner, was held in some awe by students in the Department.

It is not easy to convey the heady sense of excitement and adventure that caught up those entering biochemistry at that time. The feeling was that of an intellectual gold rush. Everyone was interested in and more or less knowledgeable about the work of everyone else. At Chicago, Konrad Bloch had begun the series of studies on the biosynthesis of steroids that was to win him a Nobel prize. Herbert Anker, whose critical insight and remarkable technical abilities have perhaps never been sufficiently recognized, was also working on the biosynthesis of lipids but was further attempting to develop approaches to probe the biosynthesis of proteins, an enterprise in which he was soon to be joined by Howard Green. Bloch was also interested in protein synthesis, and began the studies that led to his elucidation of the biosynthesis of the tripeptide glutathione, perhaps as a kind of warm-up for the main event. Lehninger, deeply immersed in his studies of oxidative phosphorylation and fatty acid oxidation, had very wide interests and regarded the problem of photosynthesis, then under vigorous attack at Chicago by Hans Gaffron and James Franck, and later by Birgit Vennesland, as one of the central problems of biology.

Graduate students in the Department worked for the most part in a single large laboratory on the third floor of the Abbott building rather than in the laboratories of their mentors. They were a lively group, including at this time Irving Zabin, Eugene Goldwasser, Robert Johnston, Lloyd Kozloff, and Ray Koppleman. Morris Friedkin, another graduate student, was a teaching assistant in the introductory biochemistry course that I took. His abilities and personality, marked by a kind of sunny integrity, greatly impressed me. He and his wife Bert, and Adelaide and I became fast friends, and have remained so for almost five decades. Friedkin had also been involved in war work, on the development of penicillin production, but was a year or so ahead of me in graduate studies and had already begun a research project in Lehninger's laboratory.

In 1947, when I was ready to begin research for a dissertation, upon Friedkin's recommendation, I went to Lehninger to discuss the possibility of working under his direction. With staggering naiveté, I suggested to him that the proper approach would be to purify the various enzymes undoubtedly involved in fatty acid oxidation and crystallize them. He agreed that this would be desirable, but went on to point out rather gently that fatty acid oxidation had not yet been demonstrated in a soluble extract from which individual enzymes might be isolated. To reach that stage, it would first be necessary to discover the nature of the energy-requiring activation or "sparking" of fatty acid oxidation and the special dependence of the process on particulate structures. He agreed, however, to take me on as a graduate student to work on this project, perhaps in part because he felt that experience in the fractionation of proteins might in fact be useful at a later stage in the research.

Lehninger was only a few years older than Friedkin and I, his only students. Although he held an appointment in the Department of Biochemistry, his laboratory was in the Department of Surgery, to which he had been recruited from the University of Wisconsin by Charles Huggins at the recommendation of Earl Evans. This was the beginning of the research enterprise fostered by Huggins that was to grow rapidly and culminate in the establishment of the Ben May Laboratory for Cancer Research. Huggins, already a famed cancer researcher, and a brilliant, idiosyncratic and sardonic man, was a formidable figure to a neophyte entering his domain. He possessed and cherished a semi-micro balance, the only one available, which was kept not in the laboratory but in his own office. It was a bit of a test to knock, enter, and weigh out a reagent under his watchful eye.

Lehninger had set Friedkin to work on the problem of oxidative phosphorylation, then one of the most baffling in all of biochemistry. Friedkin was able to show clearly that the reoxidation of reduced carriers such as NADH was linked to the phosphorylation of ADP to ATP, an important conceptual advance (2). To do so, he used $^{32}P_i$ as a tracer, and found that it was converted not only to ATP, but also further metabolized in this cell-free preparation, with the formation of labeled phospholipid, nucleic acid, and phosphoprotein (3). Friedkin had uncovered a marvelous set of leads for explorations in biosynthesis, but some time was to pass before they were explored.

On June 30, 1947 I began work on the problem of fatty acid oxidation. The central approach to intermediary metabolism in that day was through the measurement of oxygen uptake or carbon dioxide production in a highly sensitive manometric procedure brought to perfection by Otto Warburg. The Warburg apparatus was then the centerpiece of nearly every laboratory. It is now as defunct as the dodo, its passing yet another reminder of the rapid obsolescence of experimental techniques in biochemistry. Our enzyme source was rat liver, and because of the great lability of the systems under study, it was necessary to prepare fresh enzyme every day. Killing the rats was a necessary but odious task, to which my experience collecting blood in a slaughterhouse had by no means inured me. Although Lehninger had become allergic to rats, and had already begun to suffer from the asthma that was so serious an affliction in his last years, on that first day, he prepared the enzyme and taught me the art of Warburg manometry.

Lehninger had a very clear idea of the central importance of structure in the biological processes we were studying, and long before methods became available for the isolation of functionally as well morphologically intact mitochondria, he was convinced that mitochondria were in fact the organelles in which the oxidative and energy-conserving systems were localized. He had read and also set me to reading much of the old literature on mitochondria. One of the clues guiding Lehninger's thinking arose from his observation that both fatty acid oxidation and oxidative phosphorylation were inhibited in strikingly parallel fashion by exposure of our particulate enzyme preparations to hypotonic buffers. The activity could be preserved by adding to such buffers either salts such as KCl or iso-osmotic amounts of sucrose. My first project in Lehninger's laboratory (4) was a detailed study of these effects, which led us to surmise that fatty acid oxidation, oxidative phosphorylation, and the accompanying reactions of the Krebs tricarboxylic acid cycle must all be taking place in a single organelle, bounded by a membrane impermeable to certain solutes. Although our particulate preparations were still quite crude, we were convinced that this organelle was the mitochondrion.

Palade and his collaborators in the meantime were developing the methods for the separation and identification of cell organelles that were to prove so important in linking biochemistry to cell biology, work which was later to be recognized by a Nobel prize. They described a procedure for the isolation of purified mitochondria by differential centrifugation in 0.88 M sucrose (5). I immediately tested mitochondria isolated by this method and obtained convincing evidence that oxidative phosphorylation, fatty acid oxidation, and the reactions of the Krebs tricarboxylic acid cycle are indeed localized in mitochondria (6, 7).

I also worked hard on the problem of fatty acid oxidation. Previous work by Leloir as well as by Lehninger showed that the initiation of fatty acid oxidation requires metabolic energy, presumably in the form of ATP. It was further known on the basis of evidence dating from the classical studies of Knoop that degradation of fatty acids via beta-oxidation led to the formation of a "two-carbon unit" or "active acetate," which could either be oxidized via the Krebs tricarboxylic acid cycle, or yield acetoacetate, long known to be a characteristic end-product of fat metabolism under certain conditions. A shrewd suspicion was also spreading that Lipmann's coenzyme A must be involved, but the pure cofactor was not available and there was still much uncertainty about its structure.

My studies consisted principally of a detailed examination of the conditions regulating the relative yield of acetoacetate and of carbon dioxide during the oxidation of fatty acids of varying chain-length (8).

WORK WITH H. A. BARKER AT BERKELEY

During my last year of graduate study, I obtained a fellowship from the Nutrition Foundation upon the recommendation of Lehninger. This enabled me to apply for a leave of absence from Armour, which was generously granted, to work full time on research for the dissertation. Up to this time, I had not very seriously considered the possibility of an academic career, but Lehninger now suggested that I undertake a year of postdoctoral study, which entailed resignation from Armour and considerable uncertainty for the future. He recommended that I apply to the laboratory of H. A. Barker, at the University of California in Berkeley. He greatly admired Barker, whom (I believe) he had not yet met, and said: "Whatever he touches turns to gold," a judgment later amply confirmed. Barker and his brilliant student Earl Stadtman had just set the world of enzymology on its ear with their discovery of an extraordinary enzyme preparation derived from cells of the obligate anaerobe *Clostridium kluyveri*, an organism discovered by Barker.

Stadtman and Barker discovered that soluble extracts of dried autolyzed cells of *C. kluyveri* when shaken in air would oxidize fatty acids of chain-length up to eight carbon atoms to a mixture of acetyl phosphate and acetate (9). Acetyl phosphate had been found by Lipmann (10) to be a form of "active acetate" in bacteria, but (disappointingly) not in animal tissues. The same enzyme preparation, when shaken under an atmosphere of hydrogen, would resynthesize short-chain fatty acids from acetyl phosphate and acetate! To those toiling away at comparable problems with feeble and unstable preparations from animal tissues, this seemed almost too good to be true, like the legendary Shmoo in the comic strip, which when fried, tastes like chicken and when broiled, tastes like steak . . .

I was accepted by Barker for a year's study, and my wife and I drove west from Chicago in a new automobile, picked up from a dealer in Chicago to be delivered to another dealer in San Francisco. This was a common mode of shipping new automobiles to the West Coast in those days, and since even the cost of gasoline was repaid by the dealer, it represented the cheapest mode of travel, an important consideration for us. The drive itself was a memorable experience, through parts of the country we had never visited, and we found Lake Tahoe, then almost pristine and now so sadly overdeveloped, and the mountains around it especially attractive.

Arrived in Barker's laboratory, not much larger than Lehninger's, I found him to be a mild-mannered, soft-spoken man, who with all his great abilities was simple, direct, and friendly. Barker and his wife Margaret were warm and generous in their hospitality to Adelaide and to me, as they were to others who worked in his laboratory. I later found that Barker rarely appeared at national meetings, avoided service on committees, and was completely devoid of the arts by which some workers in science think it necessary to attract notice. It is pleasant to observe that these traits have not prevented the wide recognition of his work.

Barker asked me to study the oxidation of butyrate in extracts of *C. kluyveri* in greater detail. In Stadtman's experiments, the oxidation of fatty acids had been completely dependent upon the addition of inorganic phosphate. With the extracts I studied, however, the oxidation of butyrate proceeded in the absence of added phosphate at about one-third the maximum rate. This might perhaps be thought to reflect the presence of low levels of phosphate in the enzyme preparation itself, but I also found that the product of oxidation of butyrate in the absence of phosphate was acetoacetate, not acetyl phosphate (11). This was interesting because Stadtman had shown that neither free acetoacetate nor β-hydroxybutyrate were intermediates in the oxidation of butyrate. This led us to formulate a scheme, which also fit much that was known about β-oxidation in animal tissues, in which a derivative of acetoacetate (designated $CH_3COCH_2CO - X$) was the true intermediate. In the presence of phosphate, it could be converted very rapidly to acetyl phosphate. In the absence of phosphate, it could be hydrolyzed more slowly and irreversibly to free acetocetate. We suggested that X represented CoA, but definitive evidence on this point was tantalizingly elusive.

In 1949, Berkeley was already a great center for the biological sciences. Through Barker, I became acquainted with his colleagues Michael Doudoroff and Zev Hassid, who had worked with him on the study of sucrose phosphorylase, which had won them a prize awarded for a successful synthesis of sucrose, which was indeed catalyzed by their enzyme, although not alas on a scale that could compete with the processing of sugar cane. Roger Stanier, like Barker a representative of the van Niel school of microbiology, was also an occasional visitor to the laboratory. Paul Stumpf had laboratory space directly adjacent to Barker. He had just bought a new house perched on a hill at the edge of Berkeley. When the winter rains began, the hillside began to slide, and for a time, each morning began with an appraisal of the ominous weather reports and the progress of the erosion. The house finally had to be abandoned, but fortunately Paul did not have to bear the entire financial loss.

During this time, my wife Adelaide and I lived in San Francisco where she had a job on the *San Francisco Chronicle*, working in the so-called morgue, where prewritten obituary notices of notable people were filed, ready for publication as needed. I commuted to Berkeley each day on the interurban trains that then linked Berkeley and San Francisco via the Bay Bridge. San Francisco was experiencing a mild economic downturn after the boom period of war. We were able to rent an apartment atop Russian hill for $60 per month. Restaurants in the largely Italian area around us offered delicious meals for one or two dollars. We enjoyed the city thoroughly.

At this point, quite out of the blue, a letter arrived from William Mansfield Clark, head of the Department of Physiological Chemistry at Johns Hopkins Medical School, offering me a job as instructor in his department. In those innocent days, such an offer could be made without an interview, and without elaborate screening by committees. The work on mitochondrial function had attracted some attention, and presumably I had also been recommended to him by someone at Chicago. If I were indeed to undertake an academic career, this seemed to be its start. I sought Barker's advice, however, and was surprised to hear him suggest that I decline the offer and go instead to Lipmann's laboratory for another year of postdoctoral work, suggesting that it would not be difficult to find another job at the end of that time. There was a strong link between Lipmann and Barker, who at one time had worked in Lipmann's laboratory, and had recommended Stadtman to him. Lipmann was already recognized as one of the great figures in biochemistry, and I gratefully accepted Barker's good offices in arranging for me to work with Lipmann.

WITH FRITZ LIPMANN AT THE MASSACHUSETTS GENERAL HOSPITAL

In 1950, when I joined Lipmann's laboratory, Earl and Thressa Stadtman had just departed to take up positions at the National Institutes of Health (NIH). The group then included David Novelli, John Gregory, Morris Soodak, Harold Klein, Charles Du Toit, and Lipmann's research assistant Ruth Flynn. We were crammed into a tiny laboratory in the Massachusetts General Hospital next to the famous Ether Dome, the scene of the first (or perhaps only almost the first) use of diethyl ether as an anesthetic. In the course of the year, we were to move into much more spacious, even elegant, quarters in a new research building.

With abundant hair just becoming a little gray, and usually wearing a soft bow tie, Lipmann presented a figure closer to the stereotype of the artist than of the scientist. He spoke softly, and his sentences often trailed off into the distance. He thought not only in terms of the essential technical details of an experimental approach but also deeply about the fundamental biology underlying a problem. He was originally trained as a physician, and I believe the breadth of this training helped give him the feeling that nothing in biology was beyond his range. Again and again, he proved ready to tackle new problems, no matter how far removed from previous work in his laboratory.

Lipmann had a great respect for the chemical approach to biology, and perhaps regretted his own rather limited background in chemistry of which he made nevertheless quite effective use. He discovered the role of acetyl phosphate as a product of the oxidation of pyruvate and an "energy-rich" precursor of ATP in bacterial extracts, not by attempting the isolation of small amounts of the labile compound from the bacterial enzyme preparations, but by synthesizing acetyl phosphate, which he then added to the enzyme preparation and showed that it was a precursor of ATP (10). I imitated this strategy in later research of my own.

Lipmann's manner toward those who worked in his laboratory was rather formal. He was friendly, but a little aloof. He inspired nevertheless not only loyalty and admiration but also lasting affection in those who worked under his direction.

Loomis & Lipmann (12) had discovered that 2,4-dinitrophenol prevented the phosphorylation of ADP to form ATP and at the same time stimulated the rate of oxygen uptake of cell-free preparations containing mitochondria. It was soon learned that a large number of other compounds, some not obviously related in structure to 2,4-dinitrophenol, also had this uncoupling effect, which Lipmann rightly regarded as fundamental to an understanding of oxidative phosphorylation. The uncoupling phenomenon was later to be brilliantly explained by Mitchell (13) as an important part of his formulation of the chemiosmotic theory of oxidative phosphorylation.

Lipmann had noted with interest that di-iodophenols were uncoupling agents similar to dinitrophenols, and had the idea that the hormonal action of thyroxine, with its di-iodophenol residues, might be due to its action as an uncoupling agent. Indeed, poisoning of human subjects with 2,4-dinitrophenol induces heightened metabolic rate and wasting of tissue somewhat similar to that seen in hyperthyroidism. Further, there was some evidence that treatment of mitochondria with very high levels of thyroxine in vitro sometimes brought about uncoupling. Lipmann therefore suggested that Du Toit and I look for an uncoupling action of thyroxine in intact animals and in isolated mitochondria, respectively. We could find no evidence to support Lipmann's idea. Du Toit's experiments I thought had special force. Mitochondria from the tissues of rats made severely thyrotoxic by the administration of huge doses of thyroxine were not significantly uncoupled, nor did tissue slices from such animals incorporate amino acid into protein at a lower rate. Lipmann readily accepted these negative findings, but was characteristically unwilling to discard completely an idea that he felt had appealing elements. He would concede the validity of our data, and then at the end of a long discussion would murmur: "But still . . . but still . . ."

Lipmann also had the notion that there might be two distinct pathways of electron transport in mitochondria, a pathway coupled to phosphorylation, and another for uncoupled oxidation. Britton Chance was the master of the spectroscopic methods that might perhaps detect differences in the putative two pathways. Arrangements were made for me to go to Chance's laboratory in Philadelphia to test this idea. I believe that I was the first person in Chance's laboratory to prepare mitochondria, later to be so important a subject of his research. Certainly I had to borrow most of the needed equipment from other laboratories. The spectral studies were done under the tutelage of Chance's colleague Lucille Smith. The spectrophotometer was in an improvised dark-room not much larger than a closet, closed off by heavy, dark curtains. It was midsummer in Philadelphia, without air-conditioning, and sweat dripped from my face as we made the readings. We could find no evidence for spectral differences between intact and uncoupled mitochondria. Years later, when Lucille Smith introduced me in a seminar at Dartmouth, I startled the audience by informing them that she and I had once spent a week together in a small room in Philadelphia.

Although Lipmann was very interested in the work on oxidative phosphorylation, the principal focus of the laboratory was on the isolation and characterization of coenzyme A, on which progress was discouragingly slow. We were convinced that "active acetate," of central importance to intermediary metabolism, was an adduct of some kind of acetate and CoA, but every effort to isolate it was unsuccessful. Because acetyl phosphate, known to be active in bacterial systems, is quite labile, we surmised that acetyl CoA, whatever its structure might be, was even more labile, and this supposed lability seemed to stand in the way of most schemes we tried to design to isolate it.

Pursuing a habit of wide, although rather haphazard, reading I came one day upon an article in *Angewande Chemie* (14) from the laboratory of Feodor Lynen, then unknown to me. He and his student Ernestine Reichert reported definitive evidence for an essential sulfhydryl residue on CoA, and had isolated acetyl CoA and proved it to be a thioester! I brought the article at once to Lipmann, who apparently had no previous notice of this work. Although the Munich laboratory had obviously stolen some of his thunder, Lipmann was generous in his praise of the work. He was particularly impressed by the fact that in isolating acetyl CoA from yeast, they had begun by boiling the yeast. So much for the hyperlability that we had so much feared! We should have realized, Lipmann pointed out, that an intermediate that plays such varied roles must have a certain stability.

RETURN TO CHICAGO: THE PHOSPHORYLATION OF PROTEINS

In 1951, after a year in Lipmann's laboratory, I returned to the University of Chicago with a joint appointment in the Department of Biochemistry and in the newly organized Ben May Laboratory for Cancer Research, founded by Charles Huggins with the financial support of Mr. Ben May of Mobile Alabama. Lehninger, who was a leading member of the new Ben May research group, was about to depart for a sabbatical year in England, and made his laboratory as well as his research assistant Sylvia Smith available to me during his absence, while space that I would later occupy was to be readied on another floor. While on sabbatical leave, however, he accepted an invitation to succeed William Mansfield Clark as De La Mar Professor and head of the Department of Physiological Chemistry at the Johns Hopkins Medical School, and I inherited his facilities on a more permanent basis.

Charles Huggins had gathered into the Ben May a number of talented young people, including Paul Russell, Paul Talalay, Guy Williams-Ashman, Jean Sice, and Elwood Jensen. The atmosphere among these juniors was easy and friendly, and enthusiasm for research was high. Huggins, in his 90th year and still active at the time of this writing, was then barely 50 but was already "the old man," around whom legends were beginning to gather. He was master of a deadpan delivery, which added effect to occasional outrageous pronouncements on subjects sacred or profane. Himself completely committed to research, he was caustic in his appraisal of others he thought to be less so, and particularly scathing in his dissection of a colleague in another department who was said to spend some time each day on his stamp collection. Huggins was helpful and generous to the young scientists in the Ben May, and strongly fostered their careers. In a more Puckish mood, however, he had a trick of treating someone with mockingly exaggerated courtesy and flattery so overblown that it made its object squirm. It was more pleasant for the junior staff to be invited to drink tea with him, one on one. Tea would be brewed by him in his laboratory, directly connected to his office, in beakers, accompanied by sugar in paper packets thriftily saved from airline dinners. He would inquire about one's research, which he clearly followed with great interest, and tell about his own, which had two notable themes: the hormonal regulation of tumor growth (for which he was to receive a Nobel prize in 1966) and the induction of tumors in experimental animals by highly specific carcinogens. Although he had many irons in the fire, as head of the Ben May and professor of surgery with continuing involvement in the care of patients, often critically ill with cancer of the breast, he continued always to do research with his own hands. Like many other surgeons, he was an early riser, and had a great part of a day's work accomplished by the time we later arrivals appeared. He always worked on Sunday morning, and perhaps took pleasure in the absence of weekday bustle. Because we all lived within an easy walk of the laboratory in the Hyde Park district of Chicago, many of the rest of us also turned up on Sunday, although less regularly, and it was felt that this was a good time to approach Huggins with some special request.

Following some of the leads that had emerged from Friedkin's work, I began a study of the incorporation of $^{32}P_i$ into phosphoproteins catalyzed by cell-free, particulate preparations from liver and from tumors. Some credence was still given to the doctrine of Otto Warburg that the energy metabolism of tumors was largely anaerobic. To test this idea, Williams-Ashman and I (15) carried out a study of oxidative phosphorylation in particulate preparations from highly malignant tumors, and found them to be almost as active as those from normal tissues. Further, tumor preparations also actively incorporated $^{32}P_i$ into lipid, nucleic acid, and phosphoprotein. The specific activity of the phosphoprotein fraction was an order of magnitude higher than that of the other fractions, which I took to be clear proof of the biological importance of the phosphorylation of proteins. I undertook a further study, and found that only phosphoserine could be identified amongst the labeled products of the partial acid hydrolysis of phosphoproteins derived from Ehrlic ascites cells (16).

George Burnett and I then searched for an enzyme that would catalyze the phosphorylation of proteins, and found that a soluble enzyme could be extracted from acetone powder extracts of our particulate enzyme preparations, and partially purified by ammonium sulfate fractionation (17). The enzyme was quite specific for casein, and we further found that casein stripped of its phosphate by treatment with alkali under relatively mild conditions was inactive, although its primary structure must have remained largely intact, suggesting that specific conformations as well as sequences were recognized by the enzyme. This was the first demonstration of a protein kinase. In the discussion we stated: "The physiological significance of this new type of enzymatic action still remains to be discovered. The great biological reactivity of the phosphoproteins of a wide variety of tissues, particularly tumors, as evidenced by the very high rate of renewal of phosphoprotein phosphorus appears to point to some role of importance." I dropped the study of protein kinases, and like the base Indian, cast a pearl away, else richer than all his tribe.

THE BIOSYNTHESIS OF PHOSPHOLIPIDS IN ANIMAL TISSUES

A brief account of the origins of the work on phospholipid biosynthesis has been given elsewhere (18). My interest from the start was centered on the mechanism of formation of the phosphodiester bond of phosphatidylcholine because of the hope that it might shed light on the formation of the phosphodiester bond in nucleic acids. Arthur Kornberg had the same idea, and for a time his studies on phospholipid biosynthesis ran parallel to my own.

The incorporation of ^{32}P into phospholipid in our preparations was greatly stimulated by the addition of glycerol, an observation that promptly led to the finding that sn-3-glycerophosphate was an intermediate in the process (19), which in turn led to the discovery of the enzyme glycerokinase (20). The principal labeled phospholipid formed in this system was phosphatidic acid. The important studies of Kornberg & Pricer (21) established that acyl CoA was the activated form of fatty acid for the acylation of glycerophosphate.

An essential first step in the biosyntheses carried out by our type of enzyme preparation was the conversion of labeled phosphate to ATP via oxidative phosphorylation, and this of course required mitochondria, leading to the mistaken impression that the subsequent biosyntheses were also mitochondrial. Only later did it become clear that microsomes, derived from endoplasmic reticulum, and also present in these enzyme preparations, were a major, although not the sole, site of phospholipid biosynthesis.

In 1952, I began to study the origins of the phosphodiester bond of phosphatidylcholine in investigations with labeled choline, and immediately fell into error, from which I was later fortunately able to retrieve myself. I found that free choline, but not phosphocholine, was converted to lipid in an enzymic reaction dependent upon the generation of ATP via oxidative phosphorylation (22). The lipid product cochromatographed with phosphatidylcholine on columns of alumina, and was alkali-labile, and therefore not sphingomyelin, the only other lipid form of choline known at that time. I therefore concluded, wrongly as it turned out, that the labeled product was phosphatidylcholine. Only later did I establish (23) that the labeled product was in fact a long-chain fatty acyl ester of choline, the exact biological function of which remains uncertain.

In the meantime, Kornberg & Pricer (24) had reported experiments in which phosphocholine, doubly labeled with ^{32}P and ^{14}C, was converted to phosphatidylcholine with little alteration of the isotope ratio, in a reaction that required added ATP. Their formulation of the synthesis of phosphatidylcholine was:

Phosphatidic acid + [^{32}P]phosphocholine →
[^{32}P]phosphatidylcholine + P$_i$ 1.

This was the state of affairs in July 1954, when Samuel Weiss, my first postdoctoral fellow, arrived on the scene. Sam proved to be a gifted researcher, as was to be amply confirmed by his later brilliant studies leading to the discovery of DNA-directed RNA polymerase in animal tissues and in *Micrococcus luteus*. We decided to make a detailed examination of the difference between my previous studies of choline incorporation and those of Kornberg & Pricer. ATP in our system was generated from added AMP via oxidative phosphorylation. Sam first carefully checked to make sure that oxidative phosphorylation was in fact generating ATP and that under these conditions phosphocholine could not be converted to phospholipid. When ATP was added from the bottle, however, phosphocholine was indeed converted to phosphatidylcholine, confirming the findings of Kornberg & Pricer.

ATP was needed at very high levels for the incorporation of phosphocholine at maximum rates, giving rise to our suspicion that it might contain some more active impurity. Crystalline ATP had just come on the market, and we were able to obtain some through the generosity of Drs. S. A. Morrel, S. Lipton, and A. Frieden of the Pabst Laboratories. It was quite inactive, making it certain that some impurity in the amorphous ATP was the true cofactor. At this point, we considered fractionating the amorphous ATP to isolate the active substance, but decided first to test other nucleoside triphosphates, then rare substances. CTP proved to be the active triphosphate.

We formulated many schemes that might account for the participation of CTP. By analogy with the uridine coenzymes then only recently discovered by Leloir, the intermediary formation of cytidine diphosphate choline seemed an attractive candidate. Just at this time, again through rather random reading, I came upon a paper from Khorana's laboratory (25) describing an elegant new method for the synthesis of ATP by the carbodiimide method, which offered a new approach to the synthesis of substituted pyrophosphates. It occurred to me that it might be applied to the synthesis of asymmetical, di-substituted pyrophosphates such as cytidine diphosphate choline. At this stage, we had no evidence whatever that CDP-choline was in fact involved, and the synthesis of the reagent, dicyclohexylcarbodiimide, required several steps that might be time-consuming for a chemist of my limited skills. On the other hand, I was encouraged by our success thus far in arriving at the truth by conjecture, and both ^{32}P-phosphocholine and CMP needed for the synthesis were already available.

I was deeply involved at a critical point in my first attempt to prepare dicyclohexylcarbodiimide when I received an urgent telephone call. It was my wife Adelaide's mother, babysitting with our daughter Lisa, and she informed me that the house was on fire! I lived in an apartment in a house on Drexel avenue, owned by the University, only a few hundred yards from the laboratory. I raced down the street, lab coat flying, only to find that Adelaide had just returned. She and her mother and Lisa were on the sidewalk outside the house. Fire trucks were arriving, and the small blaze in the basement caused by an overheated furnace was soon extinguished. After hasty greetings to my family, I raced back to the laboratory and finished the preparation without further incident.

CDP-choline, synthesized by the carbodiimide method (26), proved to be the key intermediate in the biosynthesis of phosphatidylcholine that we had been seeking (27). We immediately synthesized CDP-ethanolamine, although in very poor yield, and showed that it was similarly the precursor of phosphatidylethanolamine. A little later, Henry Paulus synthesized CDP-diacylglycerol by a similar approach (28), and showed that it is the activated form of phosphatidic acid and the essential intermediate needed for the synthesis of phosphatidylinositol, as Agronoff (29) had earlier suggested. In each case, the coenzyme was synthesized before it was isolated from nature. As a happy consequence of this small venture into nucleotide chemistry, I became acquainted with Gobind Khorana, whose procedures I had adapted, and whose friendship I have valued now for many years.

The pace now quickened, and a rather complete picture of the biosynthesis of phosphatidylcholine and of phosphatidylethanolamine began to emerge (Figure 1). An important piece of the puzzle was the function of a specific phosphatase (30) that catalyzed the dephosphorylation of phosphatidic acid, the precursor of all other glycerophosphatides, to generate sn-1,2-diacylglycerol, which Sam Weiss showed was the precursor of triacylglycerol (31) as well as of glycerophosphatides. Shortly thereafter, Weiss left the laboratory to work for a time with Fritz Lipmann, to whom I had enthusiastically recommended him. He was to return to the University of Chicago in 1959 to begin his pioneering studies on RNA synthesis.

In 1959, while I was on sabbatical leave at Oxford, I was invited to succeed Baird Hastings as Hamilton Kuhn professor and head of the Department of Biological Chemistry at the Harvard Medical School, an offer which, attractive as it was, I accepted only after considerable soul-searching. I felt strong affection for the University of Chicago, which when I became a student in 1941 could claim with some plausibility to be the nation's leading university, and I realized how much I owed to the support not only of Lehninger and of Huggins but also of Evans, to whose Department of Biochemistry I had moved in 1956.

The move to Harvard was considerably eased by the kind agreement of Louise Fencil Borkenhagen, my colleague at Chicago, to come with me for a few months to help set up the laboratory and to transplant to it the research we had been pursuing on the biosynthesis of phosphatidylserine, which stubbornly resisted our efforts to fit it into the same pattern as that established for phosphatidylcholine and phosphatidylethanolamine. Our search for a serine kinase, expected by analogy with choline and ethanolamine kinases, had instead led to the discovery of a specific phosphatase that converts

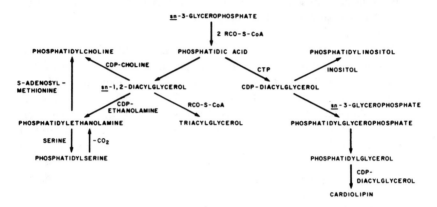

Figure 1   Biosynthesis of phosphatidylcholine and phosphatidylethanolamine

phosphoserine to serine, the final step in the biosynthesis of serine. The phosphatase is inhibited by serine, which it readily equilibrates with phosphoserine, simulating the action of a kinase (32). We could find no evidence for a requirement of cytidine nucleotides for the conversion of serine to phosphatidylserine, which we found to occur instead by an exchange of free serine with the ethanolamine moiety of pre-formed phosphatidylethanolamine. By now thoroughly convinced that even a modest investment in organic synthesis might pay big dividends in enzymology of this kind, we synthesized phosphatidylserine labeled with ^{14}C in the serine carboxyl group. With this substrate, the missing link, phosphatidylserine decarboxylase, was discovered (33) that enabled us to formulate a phospholipid cycle, driven by the decarboxylation reaction, in which free serine may be converted to free ethanolamine.

It was possible by 1961 to formulate (34) a rather detailed picture (Figure 1) of the pathways of biosynthesis of the principal glycerophosphatides and of triacylglycerol in animal tissues. It has stood the test of time.

There has been some confusion in textbooks and elsewhere as to which are de novo and which are scavenger reactions in schemes such as Figure 1. The term de novo denotes those pathways that lead to a net, new formation of glycerophosphodiester bonds in phospholipids. In Figure 1, the stepwise methylation of phosphatidylethanolamine to phosphatidylcholine, discovered by Bremer & Greenberg (35), is highly important because it provides the only known pathway for the synthesis of choline. In mammalian tissues, however, it occurs at a significant rate only in liver. In all other tissues, phosphatidylcholine is synthesized almost entirely from CDP-choline. In animal tissues, so far as is known, the sole reaction leading to the synthesis of phosphatidylserine is the exchange reaction depicted in Figure 1. Therefore both phosphatidylserine and phosphatidylethanolamine are entirely derived from CDP-ethanolamine in all animal tissues. In yeast, on the other hand, phosphatidylserine can be synthesized from CDP-diacylglycerol and serine. Its decarboxylation to phosphatidylethanolamine, which is then methylated to phosphatidylcholine, provides a de novo pathway that does not require CDP-choline or CDP-ethanolamine. In yeast, but not in mammalian tissues, the CDP-choline and CDP-ethanolamine reactions may be regarded as a scavenger pathway for the re-utilization of choline and ethanolamine, respectively.

It seems likely that this difference in metabolic pathways reflects the fact that in the course of evolution, mammals have become almost entirely dependent upon choline in the diet as a precursor of phospholipid and of acetylcholine, and most tissues have lost the capacity to synthesize it at a significant rate.

MEMBRANE BIOGENESIS AND FUNCTION IN BACTERIA

In 1963, while continuing to work on some aspects of lipid biosynthesis in mammal tissues, I turned to the study of the membrane biogenesis and function in *Escherichia coli*. At this point, I was joined by Marilyn Rumley, whose contributions were to prove so important to the life and work of the laboratory. We have now been friends and collaborators for almost 30 years.

The rapidly growing body of information on the genetics of *E. coli* made it extremely attractive for the study of membrane function as well as the biosynthesis of membrane lipids. The very rapid rate of growth of bacterial cells also makes them a rich source of biosynthetic enzymes. In contrast, the extraction and purification of lipid biosynthetic enzymes from animal tissues was frustratingly difficult, a situation that was to persist for decades.

The pathways for the synthesis of the principal membrane phospholipids of *E. coli* have been reviewed elsewhere (36), and are shown in abbreviated form in Figure 2. CDP-diacylglycerol derived, as in animal tissues, from a reaction of CTP with phosphatidic acid, plays a central role. The reactions of Figure 2 are also found in many other bacteria and in yeast. As a result of the work of many laboratories as well as our own, nearly every enzyme involved in phospholipid bioysnthesis in *E. coli* has been purified to homogeneity. The structural genes for these enzymes have been identified and many have been cloned. More is now known about the enzymology and genetic regulation of the biosynthesis of phospholipids in *E. coli* than in any other organism.

Those who contributed to this work in our laboratory during the 1960s included Julian Kanfer, Ying-ying Chang, Ronald Pieringer, and James Carter, to be followed in the 1970s by William Wickner, William Dowhan,

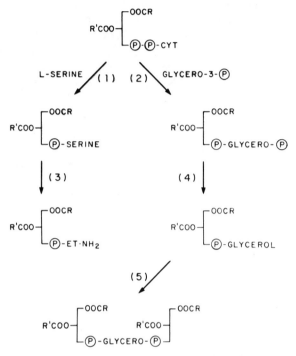

Figure 2   Pathways for the synthesis of the principal membrane phospholipids of *E. coli*

Edward Hawrot, Christopher Raetz, Keith Langley, Martin Snider, Gayle Schneider, Richard Tyhach, Michel Satre, and Carlos Hirschberg.

PHOSPHOLIPID TRANSLOCATION AND THE BIOGENESIS OF ORGANELLES

In contrast to the much-studied translocation and targetting of proteins to destinations in the cell, little is known about translocation of membrane phospholipids, which obviously must be coordinated with that of proteins in the synthesis of organelle membranes. Studies with Wilgram (37) and Dennis (38) on the localization of specific steps in the biosynthesis of phospholipids in liver and in *Tetrahymena* showed the need for specific mechanisms for the translocation of phospholipids to mitochondria from sites of synthesis in the endoplasmic reticulum. It is widely assumed that soluble "phospholipid transfer proteins," which have been shown to bind phospholipids and interchange them between different membranes in the test tube, carry out this function in vivo. The experiments of Yaffe (39), which revealed that the specificity of the phospholipid transfer proteins of BHK cells does not correspond to the specificity of translocation within the cell, cast doubt on this dogma. Recent work in Dowhan's laboratory has led to the isolation of the structural gene for one such "phospholipid transfer protein" in yeast (40), an essential step towards a definitive test of its function.

James Rothman devised an experimental approach to measure the "flip-flop" or transverse movement of newly synthesized phosphatidylethanolamine from inner to outer face of the membrane in cells of *Bacillus megaterium*, and found that its rate is orders of magnitude greater than in model lipid bilayers (41). Further investigation by Keith Langley showed that this rapid movement is independent of sources of metabolic energy (42). Despite much work in a number of laboratories, the fundamental mechanisms involved in phospholipid "flip-flop" remain obscure.

LACTOSE PERMEASE AND THE DOUBLE LABEL EXPERIMENT

In 1965, Hiroshi Nikaido, then working in Herman Kalckar's laboratory at the Massachusetts General Hospital, reported that the function of the lactose permease in *E. coli* caused a "phospholipid effect" revealed by an increased rate of incorporation of $^{32}P_i$ into phospholipid (43). This phenomenon appeared to be fundamentally similar to the effects discovered in animal tissues in the now classic studies of the Hokins (44).

Alvin Tarlov, a young physician from the Department of Medicine of the University of Chicago, where he was later to become chairman, had just arrived for a period of postdoctoral work. I found him to be an able experimentalist, so that I half-seriously advised him that he was too smart to end as a professor of medicine and he should switch completely to biochemistry, a suggestion that he wisely ignored. Tarlov, like others who came to the laboratory, was surprised at the simplicity of the equipment, arising partly from my own adherence to the wax-and-string school of experiment, and partly from the vigilance of Phyllis Elfman, my secretary for more than 30 years, who screened requested expenditures with the parsimony of a New England Yankee. Tarlov was particularly annoyed that he was forced to scrounge in other laboratories for crushed ice. He finally won a hard-fought campaign to get an ice-maker, which for some years after his departure was known as the Alvin Tarlov Memorial Ice Machine.

Tarlov confirmed Nikaido's findings that the operation of the permease increased the labeling of phospholipids from $^{32}P_i$. In a more detailed analysis, however, Tarlov showed that the effect was not specific for phosphatidic acid; all of the phospholipids showed increased ^{32}P-labeling. Further, a similar effect could be noted with labeled glycerol, or with labeled serine (45). The cells in these experiments were suspended in a medium containing no source of utilizable nitrogen. Under these conditions, the rate of phospholipid biosynthesis is very severely reduced, presumably in part at least because of "stringent" regulation. The initiation of lactose transport causes a transient release from this regulation by a mechanism that is not known but does not involve a specific "phospholipid effect."

If the mechanism of lactose transport did not involve a phospholipid effect, what was the mechanism? It is now hard to credit, but at that time the role of proteins in membrane transport was by no means certain. To test the idea that the specific recognition of a β-galactoside such as lactose by the transport system must involve a protein, Fred Fox and I first tried affinity labeling. I prepared the diazonium derivative of p-aminophenyl-β-galactoside. Fox tested it and found that it was indeed an irreversible inhibitor of the permease, but also found that other diazonium compounds unrelated to β-galactoside structure were equally effective. They were not affinity but sulfhydryl reagents. Kepes (46) had already found that the lactose permease was inhibited by treatment with sulfhydryl reagents.

Some years earlier, an elegant paper by Cohen et al (47) had caught my eye. These workers showed that the irreversible inhibition of acetylcholine esterase by diisopropylfluorophosphate may be prevented by saturating the enzyme with substrate, and exploited this finding to develop a method for the specific labeling of the enzyme with ^{32}P-diisopropylfluorophosphate. Driving back home to Brookline from Woods Hole one Sunday evening, I recalled this paper, and it suggested another approach to the specific labeling of the sugar-binding component of the lactose permease.

As Fred and I finally worked it out, the experiment involved two forms of the sulfhydryl reagent N-ethylmaleimide (NEM), one labeled with ^{3}H and the other with ^{14}C. Two cultures of cells of E. coli were prepared, one induced and therefore containing the lactose permease, and the other uninduced. In stage 1, these cultures were incubated with unlabeled NEM in the presence of saturating amounts of thiodigalactoside, a specific substrate for the lactose permease that protects it from reaction with NEM. The great bulk (but not all) of the proteins accessible to NEM were therefore blocked in this step, except the lactose permease, which was protected by the substrate thiodigalactoside, which was then removed by washing the cells.

In the second stage, induced cells, containing the permease protein, were incubated with ^{14}C-NEM in the absence of thiodigalactoside, resulting in the labeling of the permease protein with ^{14}C-NEM. Other "background" proteins that had not been completely blocked in the first stage also become labeled.

The uninduced cells were treated in identical fashion with ^{3}H-NEM. Because the permease was not present in these cells, only the "background" proteins became labeled with tritium. When portions of the two types of cells were mixed, all "background" proteins contained both tritium and ^{14}C, but the permease protein contained only ^{14}C. Its presence in partially purified fractions extracted from the mixed cells was therefore revealed by an increase in the ratio of ^{14}C/^{3}H.

This double-label approach led to the identification of a membrane protein component of the lactose permease (48), which later was identified by electrophoresis in gels containing sodium dodecyl sulfate as a protein apparently of molecular weight 30,000 (49). When the sequence of the protein finally became known from the sequence of its gene (50), its size was found to be 46,500; its behavior on gels, like that of some other membrane proteins, is quite anomalous.

Our procedure specifically labeled the lactose permease by a step that also inactivated it. We now began a series of attempts to extract and purify the permease in functional form. Despite vigorous efforts by students and postdoctoral fellows whose later careers proved their ingenuity, we were completely unsuccessful. It was small consolation for this great disappointment that the efforts of other laboratories for almost 20 years were also unsuccessful. It remained for Michael Newman in Thomas Wilson's laboratory (51) to solve this problem, which opened a new era in the study of the lactose permease. Aided by recombinant DNA technology, the analysis of structure and function of the permease is now proceeding at a level of resolution and sophistication undreamt of in our early studies.

PERIPLASMIC GLUCANS AND CELL SIGNALLING IN BACTERIA

Early pulse-chase experiments with ^{32}P$_i$ revealed a continuous turnover of the headgroups of phospholipids in *E. coli* (52). Later work by van Golde (53) showed that this turnover is the result of the transfer of phosphoglycerol and phosphoethanolamine residues from membrane phospholipids to a novel type of cell-constituent, termed membrane-derived oligosaccharides (MDO). These are periplasmic glucans, closely related to the cyclic glucans of the Rhizobiaceae.

Both in *E. coli* and in the Rhizobiaceae the synthesis of periplasmic glucans is a striking instance of osmotic adaptation, being 10–20 times higher in cells growing in medium of low osmolarity than in similar medium of high osmolarity (54,55). Periplasmic glucans also have important although poorly understood functions in cell signalling. Fiedler & Rotering (56) reported that mutants of *E. coli* defective in the production of MDO have greatly impaired motility, an increased production of capsular polysaccharide, and a decreased production of outer membrane protein OmpF when grown in medium of low osmolarity. These are all properties expected during growth in high osmolarity, consistent with a model in which the presence of MDO in the periplasm is needed to signal to certain osmoregulatory systems that the osmolarity of the medium is low, but much further work is needed to understand the real basis of these effects.

Howard Schulman helped to outline the framework of the biosynthesis of MDO, followed by later studies of Daniel Goldberg, Barbara Jackson, and Jean-Pierre Bohin. Audrey Weissborn (57) discovered a membrane-bound glucosyltransferase and found that it also requires a soluble factor, which Helene Therisod (58) identified as acyl carrier protein. This completely unexpected role of acyl carrier protein is presently under intensive scrutiny, as is the requirement for polyprenol phosphate in this surprisingly complex system.

The work of Karen Miller (59) strongly underscored the fundamental similarity between the MDO of *E. coli* and the cyclic glucans of *Rhizobium* and *Agrobacterium*. In Rhizobia, cyclic glucans appear to play a part in the complex interchange of signals between plant and bacterium that leads to the mutual recognition needed for symbiosis. Mutations that block the synthesis of glucans also prevent the formation of normal root nodules.

Stimulated by the discovery of the unusual role of acyl carrier protein in MDO synthesis, we recently turned to an examination of the acyl carrier proteins of *Rhizobium*. Mark Platt isolated the constitutive acyl carrier protein responsible for the synthesis of cell lipids needed for growth (60); Otto Geiger and Herman Spaink isolated the inducible acyl carrier protein, product of the *nodF* gene, which functions in cell signalling (61). The two proved to be distinctly different. In an exciting recent development, Spaink and his collaborators (62) have isolated rhizobial signalling compounds containing a novel fatty acid for the synthesis of which the NodF acyl carrier protein is needed.

Because membranes of every type of living cell are freely permeable to water, osmotic regulation is a fundamental problem in biology as is cell signalling in symbiotic nitrogen fixation. How does the cell detect the osmolarity of the medium in which it is growing? Is there a hierarchy of regulation of osmotic adaptation in which one system derives information from another? Are periplasmic glucans important in these processes? What is their role in symbiotic nitrogen fixation? What is the significance of the involvement of membrane phospholipids in their biosynthesis?

Fortunately, no problem is ever really solved; it simply opens to reveal another.

SAILING TO BYZANTIUM

An aged man is but a paltry thing,
A tattered coat upon a stick, unless
Soul clap its hands and sing, and louder sing
For every tatter in its mortal dress,
Nor is there singing school but studying
Monuments of its own magnificence;
And therefore I have sailed the seas and come
To the holy city of Byzantium.

Yeats

What are these monuments of the soul's magnificence that we must study to transcend age and physical decay? Yeats meant the monuments of art, but surely there is another Byzantium to which we may sail and it contains the monuments of science. For those to whom fame is the spur, that last infirmity of noble minds, to labor on the monuments of art seems to hold a distinct advantage. Mozart, Beethoven, Yeats himself produced works that seem forever stamped with their personalities, like thumbprints on pottery. In contrast, scientists are interchangeable, with the the rarest exceptions, perhaps none. The towering figure of Otto Warburg dominated biochemistry during the first half of this century. Today it is not easy to find a graduate student who can give a meaningful synopsis of his work.

The anonymity that is the fate of nearly every scientist as the work of one generation blends almost without a trace into that of the next is a small price to pay for its unending progress, the great long-march of human reason. In his Nobel address, Fritz Lipmann said of scientists: "Their purpose often may be none but just to push back a little the limits of our comprehension. Their findings mostly have to be expressed in a scientific language that is understood only by a few. We feel, nevertheless, that the drive and urge to explore nature in all its facets is one of the most important functions of humanity."

To feel that one has contributed to this splendid enterprise, on however small a scale, is reward enough for labor at the end of a day.

Literature Cited

1. Edsall, J. T. 1971. *Annu. Rev. Biochem.* 40:1–28
2. Friedkin, M., Lehninger, A. L. 1949. *J. Biol. Chem.* 178:611–23
3. Friedkin, M., Lehninger, A. L. 1949. *J. Biol. Chem.* 177:775–88
4. Lehninger, A. L., Kennedy, E. P. 1948. *J. Biol. Chem.* 173:753–71
5. Hogeboom, G. H., Schneider, W. C., Palade, G. E. 1947. *Proc. Soc. Exp. Biol. Med.* 65:320–21
6. Kennedy, E. P., Lehninger, A. L. 1949. *J. Biol. Chem.* 179:957–72
7. Kennedy, E. P., Lehninger, A. L. 1948. *J. Biol. Chem.* 172:847–48
8. Kennedy, E. P., Lehninger, A. L. 1950. *J. Biol. Chem.* 185:275–85
9. Stadtman, E. R., Barker, H. A. 1949. *J. Biol. Chem.* 180:1095–115
10. Lipmann, F. 1939. *Nature* 144:381
11. Kennedy, E. P., Barker, H. A. 1951. *J. Biol. Chem.* 191:419–38

12. Loomis, W. F., Lipmann, F. 1948. *J. Biol. Chem.* 173:807–8
13. Mitchell, P. 1961. *Nature* 191:144–48
14. Lynen, F., Reichert, E. 1951. *Angew. Chem.* 63:47–48
15. Williams-Ashman, H. G., Kennedy, E. P. 1951. *Cancer Res.* 12:415–21
16. Kennedy, E. P., Smith, S. W. 1954. *J. Biol. Chem.* 207:153–63
17. Burnett, G., Kennedy, E. P. 1954. *J. Biol. Chem.* 211:969–80
18. Kennedy, E. P. 1989. *Phosphatidylcholine Metabolism*, ed. D. Vance, pp. 1–8. Boca Raton: CRC Press
19. Kennedy, E. P. 1953. *J. Biol. Chem.* 201:399–412
20. Bublitz, C., Kennedy, E. P. 1954. *J. Biol. Chem.* 211:951–61
21. Kornberg, A., Pricer, W. E. Jr. 1952. *J. Am. Chem. Soc.* 74:1617
22. Kennedy, E. P. 1953. *J. Am. Chem. Soc.* 75:249
23. Kennedy, E. P. 1956. *Can. J. Biochem. Physiol.* 34:334–47
24. Kornberg, A., Pricer, W. E. Jr. 1952. *Fed. Proc.* 11:242
25. Khorana, H. G. 1954. *J. Am. Chem. Soc.* 76:3517–22
26. Kennedy, E. P. 1956. *J. Biol. Chem.* 222:185–91
27. Kennedy, E. P., Weiss, S. B. 1956. *J. Biol. Chem.* 222:193–214
28. Paulus, H., Kennedy, E. P. 1959. *J. Am. Chem. Soc.* 81:4436
29. Agronoff, B. W., Bradley, R. M., Brady, R. O. 1958. *J. Biol. Chem.* 233:1077–83
30. Smith, S. W., Weiss, S. B., Kennedy, E. P. 1957. *J. Biol. Chem.* 228:915–22
31. Weiss, S. B., Kennedy, E. P. 1956. *J. Am. Chem. Soc.* 78:3550
32. Borkenhagen, L. F., Kennedy, E. P. 1958. *Biochim. Biophys. Acta* 28:222–23
33. Borkenhagen, L. F., Kennedy, E. P., Fielding, L. 1961. *J. Biol. Chem.* 236:28–29
34. Kennedy, E. P. 1961. *Fed. Proc.* 20:934–40
35. Bremer, J., Greenberg, D. M. 1961. *Biochim. Biophys. Acta* 46:205–16
36. Kennedy, E. P. 1986. *Lipids and Membranes Past, Present and Future*, ed. J. A. F. op den Kamp, B. Roelofsen, K. W. A. Wirtz, pp. 171–206. Amsterdam: Elsevier
37. Wilgram, G. F., Kennedy, E. P. 1963. *J. Biol. Chem.* 238:2615–19
38. Dennis, E. A., Kennedy, E. P. 1972. *J. Lipid Res.* 13:263–67
39. Yaffe, M. P., Kennedy, E. P. 1983. *Biochemistry* 22:1497–507
40. Cleves, A. E., McGee, T. P., Whitters, E. A., Champion, K. M., Aitken, J. R., et al. 1991. *Cell* 64:789–800
41. Rothman, J. E., Kennedy, E. P. 1977. *Proc. Natl. Acad. Sci. USA* 74:1821–25
42. Langley, K., Kennedy, E. P. 1979. *Proc. Natl. Acad. Sci. USA* 76:6245–49
43. Nikaido, H. 1962. *Biochem. Biophys. Res. Commun.* 9:486–92
44. Hokin, L. E., Hokin, M. R. 1953. *J. Biol. Chem.* 203:967–77
45. Tarlov, A. R., Kennedy, E. P. 1965. *J. Biol. Chem.* 240:49–53
46. Kepes, A. 1960. *Biochim. Biophys. Acta* 40:70–84
47. Cohen, J. A., Oosterbaan, R. A., Warringa, M. G. P. J., Jansz, H. S. 1955. *Disc. Faraday Soc.* 20:114–19
48. Fox, C. F., Kennedy, E. P. 1965. *Proc. Natl. Acad. Sci. USA* 54:891–99
49. Jones, T. H. D., Kennedy, E. P. 1969. *J. Biol. Chem.* 244:5981–87
50. Buchel, D. E., Gronenborn, B., Muller-Hill, B. 1980. *Nature* 283:541–45
51. Newman, M. J., Foster, D. L., Wilson, T. H., Kaback, H. R. 1981. *J. Biol. Chem.* 256:11804–8
52. Kanfer, J., Kennedy, E. P. 1963. *J. Biol. Chem.* 238:2919–22
53. Van Golde, L. M. G., Schulman, H., Kennedy, E. P. 1973. *Proc. Natl. Acad. Sci. USA* 70:1368–72
54. Kennedy, E. P. 1982. *Proc. Natl. Acad. Sci. USA* 79:1092–95
55. Miller, K. J., Kennedy, E. P., Reinhold, V. N. 1986. *Science* 231:48–51
56. Fiedler, W., Rotering, H. 1988. *J. Biol. Chem.* 263:14684–89
57. Weissborn, A. C., Kennedy, E. P. 1984. *J. Biol. Chem.* 259:12644–51
58. Therisod, H., Weissborn, A. C., Kennedy, E. P. 1986. *Proc. Natl. Acad. Sci. USA* 83:7326–40
59. Miller, K. J., Reinhold, V. N., Weissborn, A. C., Kennedy, E. P. 1987. *Biochim. Biophys. Acta* 901:112–18
60. Platt, M., Miller, K. J., Lane, W. S., Kennedy, E. P. 1990. *J. Bacteriol.* 172:5440–44
61. Geiger, O., Spaink, H. P., Kennedy, E. P. 1991. *J. Bacteriol.* 173:2872–78
62. Spaink, H. P., Sheeley, D. M., van Brussel, A. A. N., Glushka, J., York, W. S., et al. 1991. *Nature.* 354:125–30

Annu. Rev. Biochem. 1992. 61:29–54

CATALYTIC ANTIBODIES

Stephen J. Benkovic

Department of Chemistry, The Pennsylvania State University, 152 Davey Laboratory, University Park, Pennsylvania 16802

KEY WORDS: abzymes, catalytic antibodies, transition state analogs

CONTENTS

INTRODUCTION

We have long been fascinated by the intricate stereospecificities and astonishing turnover numbers of nature's catalysts, the enzymes. There exists a voluminous literature describing ingenious experiments and methodologies to delineate the structural basis of the molecular recognition inherent in enzyme-substrate specificity, the origins of the enzyme's catalytic power, and the chemical mechanisms by which the substrate's transformation to product by the enzyme is achieved. With at least partial understanding of these features of enzymic catalysis has come the desire to imitate or improve, leading on one hand to the construction of macrocylic molecules of various types capable of

29

0066-4154/92/0701-0029$02.00

substrate discrimination and possessing at present modest catalytic power—the field of supermolecular chemistry (1–3)—and on the other hand to the modification of existing enzymes through genetic or chemical means to alter their substrate specificity without loss of catalytic efficiency—the field of protein engineering (4, 5).

A third approach draws on the remarkable capacity of the immune system to generate—in vast numbers—immunoglobulins that possess high affinity and unmatched structural specificity towards virtually any molecule (6–9) (some 10^8 different antibody molecules are available in the primary response, a number further expanded by somatic mutation). The experimental challenge is clear: how to harness and to select, from the enormous numbers and diversity of the immunological response, molecular frameworks in the form of antibodies, inherently able not only to bind a given substrate but also to catalyze a preselected chemical transformation of it. Thus one might bypass for the time-being the thorny unsolved problems that bedevil the design of de novo catalysts: the fit of substrate to catalyst, the flexibility of the catalyst to facilitate a step-wise process, the location and strength of complementary noncovalent bonding interactions, etc. To set the stage, I begin with an abbreviated, historical perspective; a brief description of the physical and structural properties of antibodies; and a limited survey of the reaction types now known to be catalyzed by antibodies. More complete discussions can be found in other reviews (10–15). My main intent, however, is to examine the information available on the mechanisms of action of catalytic antibodies in order to draw parallels and contrasts to those of enzymes: How do catalytic antibodies function kinetically? For a particular chemical transformation will the same reaction mechanism be favored by both enzyme and catalytic antibody? Will catalytic antibodies use less complex kinetic sequences owing to their selection by a single immunogen? How high are their turnover numbers, and how can they be improved? These are some of the queries that owing to the rapid development of the field can now be partially answered.

HISTORICAL PERSPECTIVE

The concept of inducing antibodies that would possess, in addition to their exquisite ligand specificity, catalytic potential has its roots in the seminal contributions of Pauling (16, 17). His proposal that the ability of an enzyme to speed up a chemical reaction stemmed from the "complementarity of the enzyme's active site structure to the activated complex" shifted the research focus from concerns about substrate-enzyme fit to a means for defining the structural requirements for binding the transition state. Since a transition state by definition has a negligible lifetime, evidence was sought for transition-state stabilization in the tighter binding of inhibitors whose structures mimicked those of the presumed transition state relative to the weaker binding of substrate (18, 19). Many examples of such high-affinity transition-state inhibitors are now documented (20–22). At the risk of being pedantic, stabilization of the transition state alone is necessary but not sufficient to give catalysis (23, 24), which requires differential binding of substrate and transition state. As we see later, the use of transition-state analogs to induce catalytic antibodies successfully stems not only from this differential binding but other factors as well. Nevertheless, it was apparent that such inhibitors would furnish a convenient starting-point for creating catalytic antibodies (25, 26) or, even possibly by immunizing with the substrate, if the challenge was repeatedly presented to the immunological system (27). Earlier attempts (28–30), however, were thwarted by the use of polyclonal rather than monoclonal antibodies (31), and by the need for better transition-state mimics. Many earlier experiments also lacked adequate controls, for example the necessary demonstration that the immunogen acted as a competitive inhibitor of the catalysis, so that the attribution of catalytic properties to the antibody in this case is equivocal (30).

PERTINENT CHARACTERISTICS OF ANTIBODIES

Antibodies are large proteins assembled in a disulfide cross-linked four-chain structure. The major serum antibody, IgG, consists of two identical heavy chains of molecular weight approximately 50,000 and two identical light chains of molecular weight 25,000 (32). Sequence comparison of monoclonal IgG proteins indicates that the carboxy-terminal half of the light chain and roughly three quarters of the heavy chain from the carboxy end show little sequence variation (33, 34). The antigen-combining site of the molecule is in the first 100 amino acids of the amino-terminal regions of both light and heavy chains, referred to as V_L and V_H domains, which show considerable sequence variability. Within these variable regions are short stretches of extreme amino acid sequence variation—three such regions in both the heavy and the light chains—associated with antigen recognition and designated as complementarity-determining regions. Proteolytic cleavage of the molecule on the carboxy-terminal side of the interstrand disulfide linkage connecting the light and heavy chains generates two F_{ab} molecules, each containing an antigen-binding region.

Crystallographic studies of F_{ab} fragments, reviewed most recently by Davies et al (36), reveal that the *immunological fold* consists of two twisted stacked β-sheets (37), a structural motif characterizing the V_H and V_L domains. One sheet has four and the other three antiparallel β-strands related by a pseudo two-fold axis. These strands are joined at their ends by the six loops of the complementarity-determining regions, creating a key β-barrel fold that can tolerate sequence and conformational changes in the loop region (35). On the basis of comparative studies of known antibody structures, Chothia et al have argued that there is a small repertoire of main-chain conformations— "canonical structures"—for at least five of the six variable regions of antibodies (38) whose conformation is determined by a few key residues. The area of interaction between the antigen and antibody may be relatively flat and extensive for protein antigen binding to an antibody (700–750 Å^2) (39–41), whereas in the case of small organic molecules the binding may occur by way of clefts whose volumes are in excess of 600 Å^3 (35). For small organic molecules such as fluorescein the dissociation constant of the antibody-antigen complex ranges from 10^{-4} to 10^{-14} M^{-1} (42), which if totally coupled to drive a chemical transformation would provide a free energy change up to 20 kcal M^{-1}, sufficient to promote most reactions in aqueous solution. The binding of antigen does not result in a global conformational change in the antibody. Rather the union is accommodated by conformational adjustments in the specific amino acid side-chains that improve the weakly binding interactions that involve hydrogen bonds, van der Waal, and electrostatic forces.

REACTION TYPES CATALYZED BY ANTIBODIES

There are now ca. 50 reactions that have been catalyzed by antibodies (10). I have selected several representative examples that are instructive as to the present scope of reaction types and that also serve as my basis for discussion of mechanism. Figures 1–10 depict these reactions along with the inducing antigen to the left.

Antibodies As "Free Energy Traps"

Conceptually, the reactions most susceptible to antibody catalysis would be those originally viewed as "no-mechanism" reactions because their transition states evinced little polar or radical character. Broadly speaking, these are pericyclic processes that involve bond formation through a symmetry-allowed synchronous reorganization involving the highest occupied and lowest unoccupied molecular orbitals (56). These transformations include electrocyclic reactions, sigmatropic rearrangements, and cycloadditions (57), and generally are not reactions found to be catalyzed by enzymes.

The Claisen rearrangement and Diels-Alder addition described in Figures 1 and 2 are two such examples. The former is a 3,3-sigmatropic rearrangement of chorismate to prephenate in which formation of a carbon-carbon bond is accompanied by breaking of a carbon-oxygen bond. In this case there is a biological counterpart; the same reaction is catalyzed by chorismate mutase from *Escherichia coli* (58). The nonenzymatic reaction has been demon-

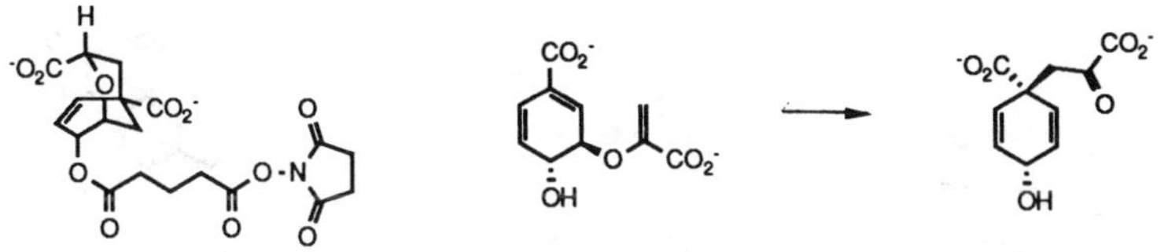

Figure 1 Claisen rearrangement (49–51).

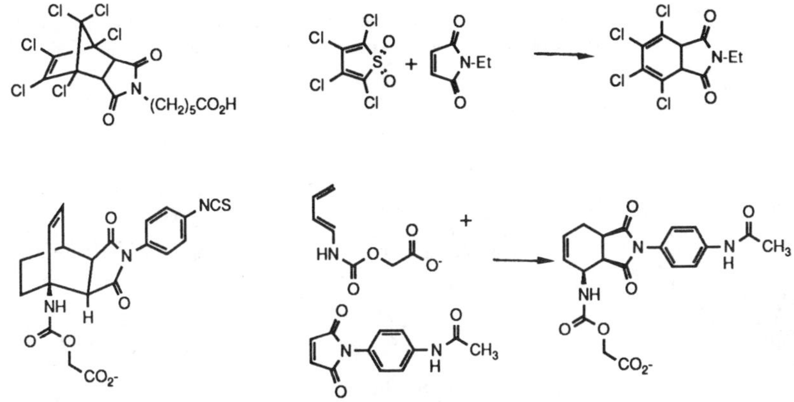

Figure 2 (*Top & bottom*) Diels-Alder condensation (52, 53).

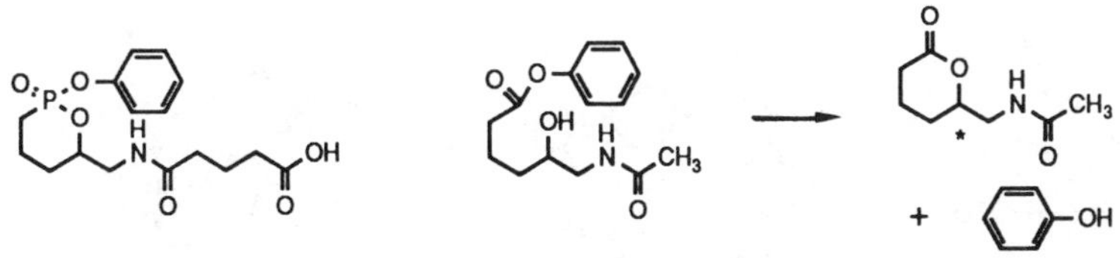

Figure 3 Lactonization (the same antibody catalyzes lactonization and amide bond formation) (46).

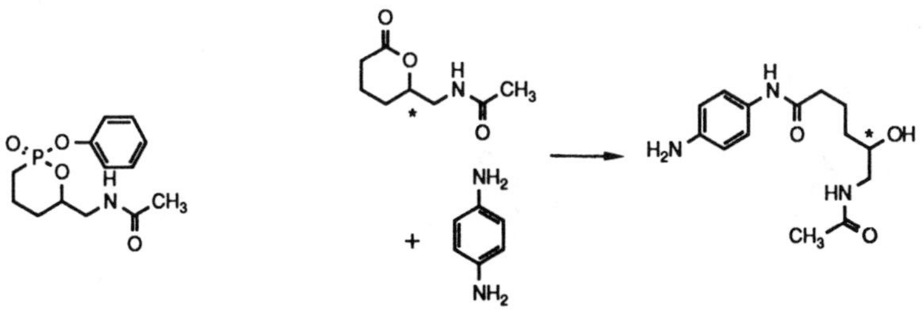

Figure 4 Amide bond formation (47).

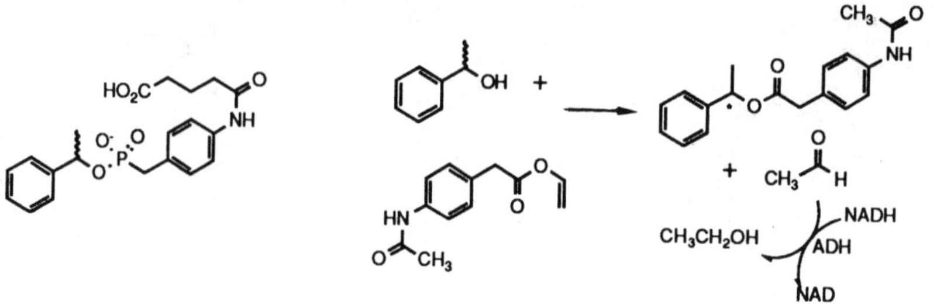

Figure 5 Transesterification (the same set of antibodies catalyze either the stereospecific hydrolysis of the R- or S- sec-phenethyl esters or the respective transesterification) (44).

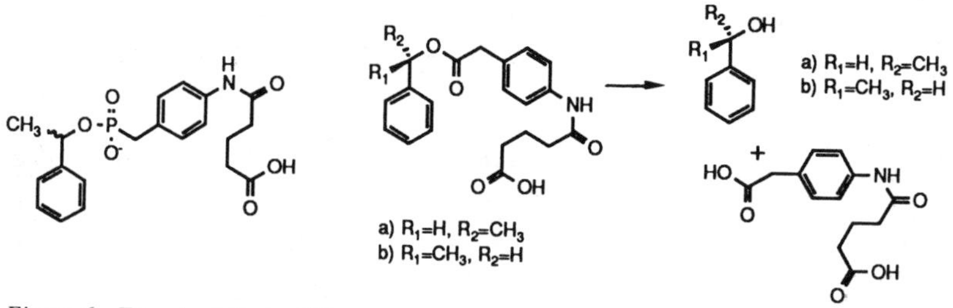

Figure 6 Ester hydrolysis (43).

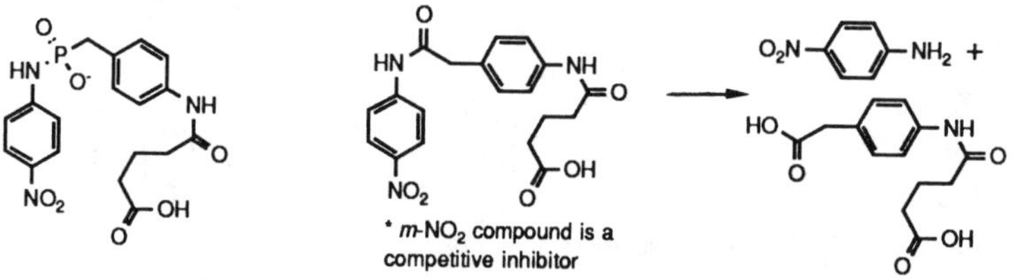

Figure 7 Amide hydrolysis (the same antibody catalyzes ester hydrolysis when the leaving alcohol is one of a series of *p*-substituted phenols) (45).

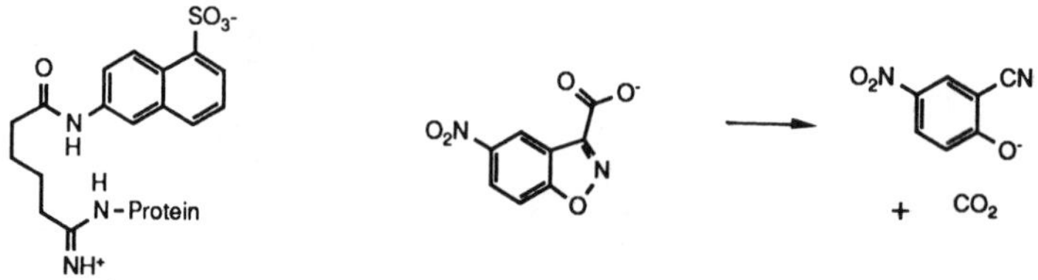

Figure 8 Decarboxylation (55).

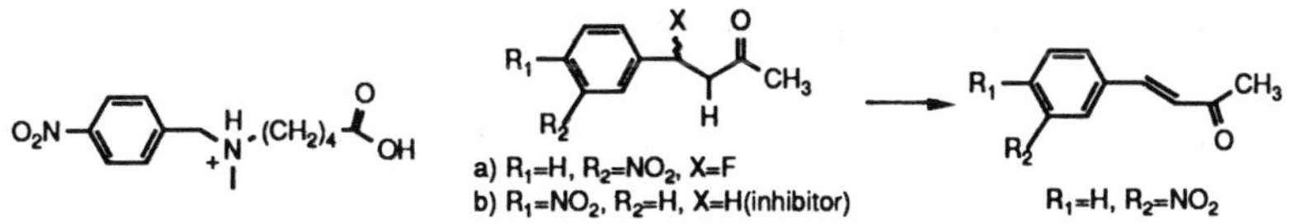

Figure 9   β-Elimination (48).

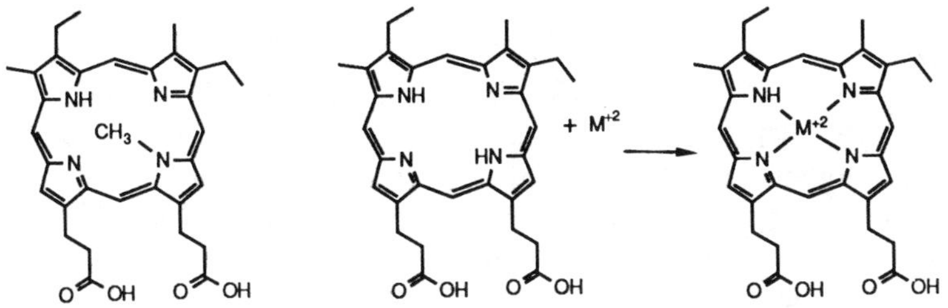

Figure 10   Metallation (54).

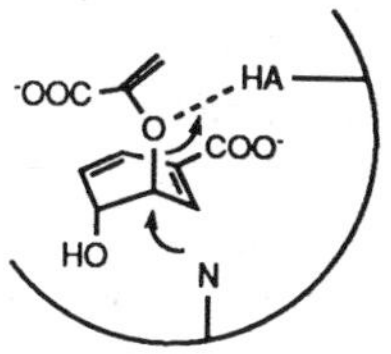

Figure 11   Transition state for Claisen rearrangement.

strated to occur through an asymmetric chairlike transition state in which the carbon-oxygen bond is nearly broken, whereas carbon-carbon bond formation has lagged substantially behind (59–61). On the other hand, although the *E. coli* enzyme-catalyzed reaction may proceed through a covalently linked intermediate based on measurements of secondary tritium (C-4 and C-5) and deuterium solvent isotope effects, the key transition state (Figure 11) is also thought to be a chairlike species (62, 63).

Consequently, the inducing antigen depicted in Figure 1 is a potent inhibitor of the enzyme with a K_i of 0.12 μM (64), since it closely mimics the putative bicyclic structure of the chairlike transition state shown in Figure 11. As is the case for all transition-state mimics, it is not perfect since it does not reflect either the asynchronous bond breakage and formation or the planarity of the carboxylate substituent.

This mimic was used independently by two groups to produce two antibodies that catalyzed the rearrangement of chorismate to prephenate (50, 51). Values of k_{cat} were 2.7 min^{-1} and 0.072 min^{-1} (K_Ms were 260 μM and 51

μM), illustrating the very important principle that the use of an identical immunogen does not perforce dictate isolation of an antibody with identical catalytic parameters because of the immense number of potentially catalytic antibodies. For the more reactive antibody there was no measurable deuterium solvent isotope effect, although like the mutase enzyme it catalyzed the rearrangement of the O-methyl ether, thus ruling out mechanisms involving 4-hydroxyl loss or its participation in oxirinium formation (62). The antibody-catalyzed process is also stereospecific, kinetically resolving the ($\pm$) racemate of chorismate (49).

There are various means of estimating the rate acceleration achieved by restricting the conformational freedom of substrate molecules. Since the results are superficially discordant, one should view them as merely establishing a range for reference. One method employs quantum mechanical calculations of the entropy loss (presuming $\Delta H^{\ddagger}$ is fixed) to determine the effect on rate, i.e. the rate constant is proportional to $\Delta S^{\ddagger}/R$. For a reaction model featuring the conversion of succinic acid to its anhydride, an entropy loss of ca. 15 eu results from the disappearance of two degrees of internal rotation and a low-entropy asymmetrical and symmetrical bending mode (65). This translates to a rate advantage of ca. 40 [7.5 entropy units (eu)] per rotation. A similar calculation based on cyclopentadiene dimerization provides a value of ca. 10 as the rate advantage for freezing out of a single bond rotation (66). On the other hand, experimental observations that compare various intramolecular cyclizations furnish estimates of 10^2-10^4-fold rate enhancements (67, 68) associated with a single frozen rotation in the absence of steric strain or electronic effects. The discrepancy between theory and experiment is not as grave as it might first appear; the inclusion of the loss-of-bending motions in the calculation provides entropy losses up to 14 eu, or a rate factor of 10^3 for a single rotation (69). Consequently, the restriction of chorismate by either antibody into a more reactive pseudo diaxial chair conformation with appropriate alignment of reacting bonds (a point we return to later) alone may be sufficient to rationalize the observed rate accelerations of up to 10^4 over background, a factor only ca. 10^2-fold less than the enzyme's.

The Diels-Alder reaction provides a second example of a transformation proceeding through a highly ordered, entropy disfavored, transition state with little charge development (70, 71), which must more closely resemble the product than the conjugated diene and olefin starting materials. However, inducing antibodies to the product itself would produce binding sites subject to severe product inhibition. Two strategies have been created to circumvent this undesirable property: (*a*) catalyze the formation of an initial, unstable bicyclic intermediate that subsequently rearranges or, as in Figure 2 (*top*) extrudes SO_2, to yield the ultimate product; or (*b*) incorporate into the transition-state analog a molecular constraint that restricts the analog to a

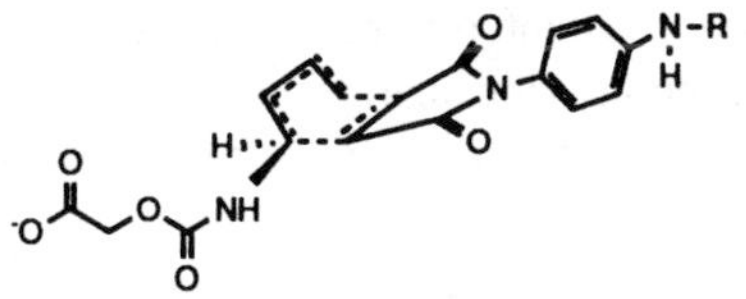

Figure 12   Transition state for Diels-Alder reaction.

higher-energy conformational state than the product. In Figure 2 (*bottom*) the ethano bridge locks the cyclohexene ring of the hapten into a conformation that resembles the pericyclic transition state for the Diels-Alder reaction (Figure 12) but that corresponds to a less favored boat conformation of the product. Both strategies led to antibodies that promoted multiple turnovers. It is worth noting that the antibodies used to catalyze the reaction of tetrachlorothiophene dioxide with *N*-ethylmaleimide had to first be protected from the reactivity of the former reagent with lysine by reductive methylation, a protocol that may be necessary with other substrates capable of protein modification.

From analyses and experimental comparisons along the lines discussed earlier, various estimates have been projected for the advantage of converting a solution bimolecular process to an intramolecular one through formation of a catalyst bisubstrate complex. If we use the same quantum mechanical calculation, the outcome depends on whether the complexed state is *loose*—internal and rotational degrees of freedom of the complex offsetting the loss of translational entropy—or *tight*—internal degrees of freedom converted to bending motions. For a loose complex the entropy loss is only 4.1 eu, corresponding to ca. 8-fold rate acceleration (65), similar to a value reached earlier by analysis of the combination of two bromine atoms (72, 73). For a tight complex the loss of entropy is an additional 35 eu, equivalent to an overall rate acceleration of 10^8 M, similar to that obtained from the analysis of the cyclopentadiene dimerization (66). (The standard state is 1 M in reactants, 25°C.)

These calculations presume that the orientation of the reacting groups is favorable for reaction. There is little theoretical or experimental support for the hypothesis that the orientation of the reactive groups need be confined to only a very narrow spatial range (73); in fact theoretically there is little entropy to be gained by restriction of vibrational amplitudes (65, 74), and experimentally the orientation of reactive moieties can deviate by angular displacements of $>10°$ with no effect on reaction rates (75). (Of course, gross misalignment of reactive moieties—pointing in opposite directions—would not be accommodated.)

From experimental comparisons of intramolecular and intermolecular reactions operating by the same mechanism, an effective molarity parameter has been devised to quantify intramolecularity. The ratio k_{intra}/k_{inter} where k represents the first- and second-order rate constant for the respective processes ranges from < 1 M to $> 10^{10}$ M, reflecting ring size, reaction type, etc (76). For our purposes, a value of 10^8 M represents an upper limit for a tight, strain-free complex.

The antibodies that catalyze the Diels-Alder reactions exhibit effective molarities of > 110 M (Figure 2, *top*) and 0.35 M or 335 M relative to buffer or acetonitrile solvent (Figure 2, *bottom*). For both antibodies the binding of the final product molecule was 100–1000-fold less than the hapten molecule, further confirming the validity of the design strategy. It will be of considerable interest to determine the mechanistic characteristics of these antibody-catalyzed transformations, since there is no known enzyme counterpart.

The intramolecular cyclization reaction (Figure 3) presents an additional opportunity, namely for general acid-base catalysis, as well as a need to reduce the rotational entropy of the substrate in order to maximize the rate of lactonization. The ring closure in the absence of antibody is specific-base catalyzed, consistent with nucleophilic attack by alkoxide ion generated from the hydroxy ester. The pH-rate profiles for k_{cat} and k_{cat}/K_M for the antibody-catalyzed process likewise are first-order in hydroxide ion (pH 7–10), providing no evidence for the dissociation of a participating binding-site residue. The rate acceleration is 790, suggesting that values of 10^2–10^4 may be typical for uncatalyzed intramolecular rearrangements within the antibody-binding pocket. Most importantly, this antibody catalyzed the enantioselective cyclization of one stereoisomer of the racemic hydroxy ester with $>94\%$ enantiomeric excess as determined by ^{1}H NMR in the presence of chiral shift reagents. This early example demonstrated the feasibility of catalytic antibodies for chemical transformations that require precise stereochemical control.

Two examples that fit into this class of reactions should be mentioned in passing, namely the photochemically allowed dimerization of substituted olefins (77) and the photo-induced cleavage of thymidine dimers (78). Both take advantage of the ubiquitous decoration of antibody-binding sites with tryptophans and use the indole side-chain as a photo sensitizer. In the case of the cleavage process, the rate rivaled that of the *E. coli* DNA photolyase (k_{cat} of 1.2 min^{-1}, antibody vs k_{cat} of 3.4 min^{-1}, enzyme). More examples of catalytic antibody–induced excited state chemistry should be pursued.

Bimolecular Reactions

Two general classifications of kinetic sequences, distinguishable by steady-state kinetics, describe bimolecular enzyme-catalyzed reactions: *sequential* and *ping-pong*. For the sequential process, the chemistry of bond formation and cleavage takes place within the bisubstrate enzyme ternary complex; in contrast, for the ping-pong sequence, more molecular finesse is required, with the enzyme acting to shuttle through covalent attachment an isolated fragment of one substrate for chemical union with the second. Of course within the sequential pathways there are variations imposed by whether the binding of either substrate is independent of the other, is at kinetic equilibrium relative to turnover, etc. Two bimolecular antibody-catalyzed reactions now have been examined in sufficient detail to permit their unequivocal classification.

The initial velocity of the enantioselective reaction of 1,4-phenylenediamine with the lactone shown in Figure 4 was studied over a range of substrate concentrations for both the lactone and 1,4-phenylenediamine substrates. The resulting Lineweaver-Burke plots showed an intersection pattern consistent with a random equilibrium process. Since the value of K_i for the phosphonate hapten was also accurately measured from its competitive inhibition of the cyclization reaction along with the respective K_Ms of lactone and p-phenylenediamine, one could estimate the extent to which the factor of 10^8 M noted above might be captured in the antibody reaction from a reaction cycle (Figure 13) based on transition-state theory (79).

The values of K_A, K_L, and K_T correspond to the K_M (1,4-phenylenediamine), K_M (lactone), and K_i (phosphonate hapten) in so far as the immunogen resembles the transition state for the reaction. The more favorable binding of the phosphonate (75 nM) relative to the two substrates (1.2 mM for the diamine and 4.9 mM for the lactone) provides an advantage of 155 M for the bimolecular reaction and 1000 for the previously discussed cyclization reaction. The observed ratio of k_{ab}/k_N (equivalent to k_{cat}/k_{uncat}) is 16 M and 790, in fair agreement with the theoretical estimates, which ignore the inadequacies of the transition-state mimic. Whereas the weaker-than-expected binding of transition-state analogs to enzymes has been attributed to either unfavorable steric interactions (a form of destabilization) or failure to form binding-site contacts (a form of unproductive binding) for antibodies, the transition-state analog acts as both model and mold. Thus the question of

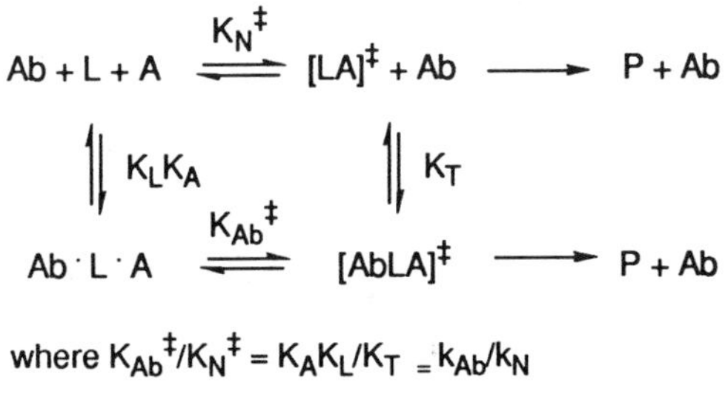

$$Ab + L + A \xrightleftharpoons{K_N^\ddagger} [LA]^\ddagger + Ab \longrightarrow P + Ab$$

$$\Big\Updownarrow K_L K_A \qquad\qquad \Big\Updownarrow K_T$$

$$Ab \cdot L \cdot A \xrightleftharpoons{K_{Ab}^\ddagger} [AbLA]^\ddagger \longrightarrow P + Ab$$

$$\text{where } K_{Ab}^\ddagger/K_N^\ddagger = K_A K_L/K_T = k_{Ab}/k_N$$

Figure 13 Reaction cycle to calculate the rate acceleration of the antibody-catalyzed reaction from the relative affinity of the antibody for the transition state versus the reactant (47).

fit is much less important than the accurate mimicry of the stereochemical and electronic characteristics of the transition state for the reaction in question.

A compilation (86) of 18 K_M/K_i ratios, where K_M is the Michaelis constant for substrate dissociation and K_i is the dissociation constant for the phosphonate inhibitor for predominantly antibody-catalyzed ester hydrolysis (80–85), showed a linear though scattered correlation (slope = 0.86, R^2 = 0.80) with the ratio of k_{ab}/k_N. The values of k_{ab}/k_N range from 30 to 10^6 M and K_M/K_i from 4 to 3×10^4 M, (H_2O at 1 M), with the former values roughly within one or two orders of magnitude of the latter (86). Since it is unlikely that the ratio of K_M/K_i will exceed 10^9 M (although this is more than respectable), it is clear that rate-enhancements above this limit probably arise from the active participation of amino acid side-chains or added cofactors within the antibody-binding site. Positive deviations from this relationship, as we shall see, indeed are suggestive of more complex mechanistic processes.

The transesterification reaction (Figure 5) is formally a bimolecular substitution reaction that like the aforementioned aminolysis reaction must compete with the hydrolysis of the ester. Antibodies that previously had been induced with racemic phosphonate to catalyze the hydrolysis of the esters of R- and S- sec-phenethyl alcohols fell into two classes: a group that acted on the R- and a second set that acted on the S-substrates. One antibody, designated 21H3, which was an S-specific esterase, also was found to catalyze the transesterification reaction in a mixture of water and 10% dimethyl sulfoxide through the two-step process shown in Figure 14.

Remarkably, this antibody uses a ping-pong kinetic sequence, which is not limited to the vinyl ester but includes substituted benzylic esters as well. Compelling evidence adduced for the ping-pong sequence includes: (a) double reciprocal initial velocity plots that are a family of parallel lines; (b) inhibition patterns in which the ester product is a noncompetitive inhibitor of the alcohol substrate or the first alcohol product released competitively inhibits addition of the second alcohol; (c) the rapid, stoichiometric reaction of 21H3 with the p-nitrophenylester; and (d) the labeling of the antibody with a ^{14}C radiolabel located in the acyl portion of the ester.

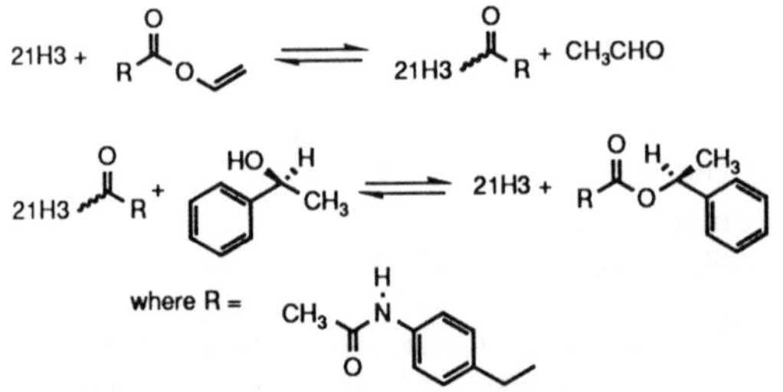

Figure 14   Ping-pong kinetic sequence for transesterification reaction.

A measurement of the reaction rate in the absence of antibody was complicated by hydrolytic cleavage of the ester at pHs at which sufficient alkoxide ion could be generated for observable transesterification. The comparison of k_N and k_{ab} is further compromised by the fact that the mechanisms in the two cases are different, but with these caveats, a minimal estimate of 10^5–10^6 M was obtained from the antibody-catalyzed rate for transesterification of the vinyl ester with sec-phenethyl alcohol.

A common feature ascribed to protein conformation flexibility and often associated with enzymic catalysis is the induced fit of substrate to stabilize a rare, reactive conformer. The presence of sec-phenethyl alcohol provides a route for transesterification but also increases the esterase activity 10-fold (J. Stewart, unpublished results). Similarly, surrogates devoid of an accepting hydroxyl, e.g. sec-phenethyl chloride and bromide, increase the rate of hydrolysis (V/K) of the vinyl ester by the antibody by a factor of 10. The simple induced-fit interpretation (88) is that in the absence of the surrogate, the antibody has either a conformation in which bound ester or the acyl intermediate is less accessible to water, or key transition states in which the acylation/deacylation pathway is less stabilized.

The induction of an antibody capable of a ping-pong kinetic sequence would not at first glance appear to be implicit in the hapten design. But recall that these antibodies were produced in response to a monoesterified phosphonic acid derivative (Figure 6) with the intent to elicit esterase activity. If transesterification activity were the initial target, then the hapten should have been a diesterified phosphonate with either similar or dissimilar alcohol moieties. This would provide sufficient space in a programmed binding site for the simultaneous binding of both ester and alcohol to the antibody. One might argue, then, that in order for the transesterification reaction to be catalyzed by the present set of antibodies, it must follow a ping-pong sequence since the acceptor alcohol must occupy the same volume in space as the alcohol portion of the ester. What is remarkable then is the induction of a nucleophile at the binding site of the antibody in proper juxtaposition, with an appropriate pK_a, and with a balanced acylation/deacylation reactivity to catalyze the transesterification reaction at such high efficiency.

Hydrolysis

The antibody, 43C9, which catalyzes the hydrolysis of the p-nitroanilide (Figure 7) and a series of related aromatic esters (pNO_2, pCl, pCH_3, pCH_3CO), has been the subject of an extensive number of experiments to establish its mechanism of action (89). The steady-state Michaelis-Menten parameters, k_{cat}/K_M and k_{cat}, as a function of pH for both the p-nitroanilide and ester substrates, are exhibited in Figure 15. At high pH the observed k_{cat} exceeds the ratio of K_i/K_M by >100-fold. The data can be fit either to a reaction mechanism involving the titration of a group whose dissociation (pK_a

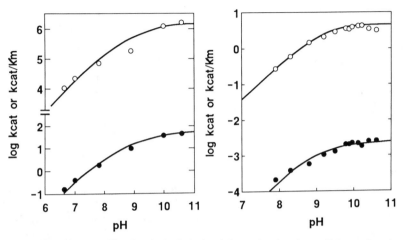

Figure 15 pH-rate profiles for the hydrolysis of the *p*-nitrophenylester (*left panel*) and *p*-nitroanilide (*right panel*) by antibody, 43C9 (see Figure 7) (89).

≈ 9.0) promotes substrate hydrolysis, or to a mechanism that features a change in the rate-limiting step around an intermediate antibody-bound species due to changes in pH.

Support for the latter sequence stems from several lines of evidence. Firstly, that a true pK_a is not being observed in the pH-rate profiles is in accord with the observations that (*a*) the plateau rate encountered in k_{cat} at pH > 9.0 for hydrolysis of the *p*-nitroester is a consequence of the slow dissociation of the *p*-nitrophenol product; (*b*) the rate constants for binding of *p*-nitrophenol as well as the dissociation constant K_i (for the *p*-nitroanilide as a competitive inhibitor of ester hydrolysis) are pH independent over the range of measurement; (*c*) slow dissociation of *p*-nitrophenol should perturb the pK_a of a group in common to both profiles by ca. 0.7 pH units for the ester relative to the anilide (90). Secondly, measurement of the pH-k_{cat} and k_{cat}/K_M profiles for both the hydrolysis of the anilide and the *p*-Cl ester in deuterium oxide revealed no deuterium solvent isotope effect on k_{cat} or k_{cat}/K_M at pH > 9.0 but a substantial effect ($k_2^{H_2O}/k_2^{D_2O} = 2$–4) on k_{cat} and k_{cat}/K_M at pH < 9.0 (91, 92). These observations rule out a simple general base-catalyzed hydrolytic mechanism involving a single dissociable group, and are only in accord with two different rate processes defining the apparent pK_a.

Evidence for the presence of an acyl intermediate sought by both stopped flow and rapid quench methods did not reveal any accumulation above undetectable steady-state levels (89, 92). However, the hydrolysis of a series of aromatic esters by 43C9 gave k_{cat} values above pH 9.0, which were strongly dependent on the electronic nature of the *p*-substituent. The Hammett ρ value is 2.3, consistent with values of 2–3 observed for attack on acyl esters

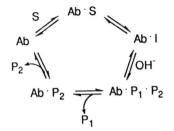

Figure 16   Kinetic reaction sequence for ester and anilide hydrolysis catalyzed by 43C9 (89).

by nitrogen nucleophiles (93) but not values of 1.0–1.2 and 0.5–0.7 found for nucleophilic attack by oxyanions or general base catalysis (94). This observation, coupled with a lack of ^{18}O exchange from $H_2^{18}O$ into unreacted anilide in the presence of antibody, also rules out the accumulation of a symmetrical tetrahedral intermediate.

The weight of the data then favor the reaction sequence given in Figure 16, where S is either anilide or ester substrate, I is the acylated antibody plus P_2, P_1 is the acid product, and P_2 the phenol or aniline. The pH-rate profiles arise from a change in rate-limiting step, from product release (ester, k_{cat}) or acylation (anilide, k_{cat}/K_M) at high pH, to deacylation (k_{cat} and k_{cat}/K_M) at low pH. From the primary sequence of the antibody, 43C9, the most likely candidate for the nucleophile is a histidine located in the L3 loop of the light chain. However, the appearance of a large deuterium solvent isotope effect on the deacylation leg is not in accord with attack by hydroxide on an acyl intermediate, but suggests that a second basic group on the antibody may be acting to catalyze deacylation. Its pK_a, like that of the nucleophile, is outside the range of the pH-rate profiles.

The relative importance of the various kinetic steps in antibody turnover can be more easily visualized through the aid of a free energy ($\Delta G^{\ddagger}$) reaction-coordinate diagram (Figure 17). There are several striking features in the two profiles. The first is the general unevenness in the $\Delta G^{\ddagger}$ barriers for the various internal antibody-substrate complexes and their respective transition states. A more optimal situation would balance the differences in $\Delta G^{\ddagger}$ between the external and internal ground states as well as their respective transition states—as is the case for enzymes operating at optimal flux—so that no single $\Delta G^{\ddagger}$ barrier would be egregious (95). A second is the high stability of the $Ab·P_1·P_2$ product complex relative to the respective uncomplexed substrates. It is this tight product complex that ultimately limits the rate of ester hydrolysis (k_{cat}) and also is responsible for the lack of Ab·I accumulation. The distortion introduced by tight product binding has been remedied by changing the nature of the *p*-substituent, e.g. *p*NO₂ to *p*Cl, but with a drop in the acylation rate (92). A third feature is the increased kinetic

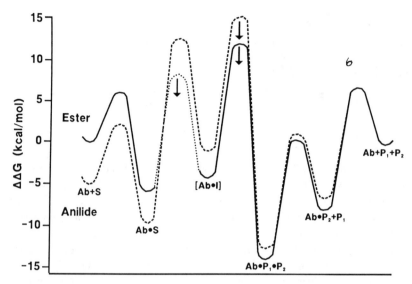

Figure 17   Free energy reaction-coordinate diagram for the reaction sequence in Figure 16 (89). Vertical arrows indicate transition states for which $\Delta\Delta G$ is a maximal estimate.

barrier for formation of Ab·I from the amide relative to the ester substrate that limits k_{cat}. The rate of deacylation, on the other hand, can be increased by raising the pH.

Despite its limitations, 43C9 is a potent, multistep catalyst with k_{cat} values at pH > 9.0 that are comparable to chymotrypsin, whose kinetic sequence it closely resembles (98, 99). Both use a series of steps around an acyl intermediate. Rate constants for the acylation of the enzyme by *N*-acetyl-L-Phe-*p*-nitrophenylester and *N*-acetyl L-tyr-*p*-nitroanilide are 23,700 and 0.05 s^{-1}, some 130- to 25-fold faster than the corresponding steps for the antibody, albeit the latter are at pH 9.0, and in the case of the amide, 12°C higher in temperature (96, 97). However, since deacylation is rate limiting for chymotrypsin-catalyzed ester hydrolysis, k_{cat} is 77 s^{-1} compared to 40 s^{-1} for 43C9. Thus, like 21H3, an antibody with potent catalytic properties uses a multistep pathway featuring the active participation of amino acid side-chains.

IMPROVING THE ODDS

The antibody-catalyzed reactions described earlier drew heavily on the advantages offered by monoclonal hybridoma development. Typically 30–50 monoclonal antibodies were selected as potential catalysts owing to their high affinity for the hapten conjugated to carrier protein in ELISA assays. High affinity for the transition-state analog is no guarantee of catalytic activity; the antibody affinity may be a consequence of primary binding interactions to residues on the hapten not central to mimicking transition-state structure. Furthermore, this number of antibodies is conservatively less than 1% of the 10^4 specific binders estimated to be produced in the response. Consequently, as might be anticipated, the screening of ca. 1200 monoclonals produced only ca. 10 antibodies capable of catalyzing the decarboxylation reaction in Figure 8.

Several approaches have been enlisted to improve the odds for creating catalytic antibodies, including: (*a*) cloning of the immunological response into *E. coli;* (*b*) hapten design that programs the introduction of suitably juxtaposed catalytic residues or modulates the hydrophilic/hydrophobic character of the antibody-binding site; (*c*) genetic and chemical modification of the antibody-combining site; and (*d*) introduction of cofactors. These strategies are discussed in turn.

Cloning

In that individual F_{ab} molecules behave as whole antibodies in terms of antigen recognition with retention of the specificity and affinity of the parent, the route to exploring the rich diversity of the antibody system lies in the expression of the repertoires of heavy and light chains followed by their combination in vitro. The construction of bacteriophage libraries (100) used PCR amplification of mRNA from immunized spleen cells or hybridomas (101, 102) to produce separate heavy and light chain libraries followed by their recombination at asymmetric restriction sites present in each vector. Typically combinatorial libraries of ca. one million clones overexpressing light and heavy chain fragments as functional F_{ab}s can be constructed, and unlike monoclonals can then be expressed and assembled in *E. coli* (103, 104).

Such libraries capture the diversity of the immunological response while retaining the recognition and affinity properties of the parent monoclonals. An analysis of 18 clones selected at random from one such heavy chain library showed at least seven different subgroup classifications based on amino acid framework sequences (105). The diversity can be expanded by extending the number of primers used in the PCR amplification or by shuffling the library (106). Antigen screening of heavy, light, and combinatorial chain libraries constructed from mRNA isolated from spleens of mice immunized with the transition-state analog shown in Figure 7 showed that the separate heavy and light chain libraries did not contain clones that bound antigen, implying that heavy chain binding of antigen is not general (107). On the other hand, within one F_{ab} library ca. 100 out of 10^6 clones were identified as originally binding antigen with $<0.1 \ \mu M$ affinity.

One criticism of the random combinatorial library assembled from immunized mice is that the chances of recovering the original pairs of V genes is low and therefore the highest-affinity antibodies will be lost (108). The argument is based on the presumption that there is equal representation of heavy and light chain genes so that even a random combination of heavy and light chain genes would necessitate the screening of $>10^6$ clones to recover the original pairings. However, it is more likely that mRNA for sequences expressed by stimulated B cells can be expected to predominate over those of unstimulated cells because of higher levels of expression. Indeed frequencies for functional heavy chains of 1 in 50 have been observed in mouse anti-hemagglutinin (109) and human antitetanus toxoid libraries (M. A. Persson, D. R. Burton, unpublished results). Moreover, unique combinations of heavy and light chains are not required for antigen binding. Shuffling of the heavy and light chain genes from binding clones for the antigen in Figure 7 generated a fivefold higher frequency of antigen-binding clones with retained affinity than anticipated if the original heavy-light chain combinations were unique. This promiscuity probably results from heavy chains from a given clonal family pairing with different light chains from a complementary family and vice-versa. However, this promiscuity does not extend to heavy-light chain combinations with chains derived from immunized and nonimmunized mice (106). In short, chain shuffling is an effective route to optimization (i.e. maximization) of the pairing of light and heavy chains for antigen recognition and possibly catalysis.

Two issues are the focus of considerable research activity to implement the use of the combinatorial libraries: (a) their rapid screening by methods other than radiolabelled probing of filter lifts and (b) improved means of antibody expression. One method of accessing such libraries is to link the recognition element and the instructions for its production. Such linkage has been accomplished for peptide libraries, proteins, and a single-chain antibody (111–114). The method depends upon the assembly of the antibody molecule in the periplasm in concert with phage morphogenesis so that the F_{ab} is expressed as a fusion protein on the phage surface (115). The multivalency of the phage particle when the F_{ab} is coexpressed with the CpVIII phage coat protein allows its rapid screening through precipitation of the particle upon addition of hapten-carrier protein conjugates. Thus a much wider range of affinities ($<10\ \mu M$) may be accessed. A second version of this method exhibits the F_{ab} molecules only on the head of phage by using a heavy chain–gene III fusion, which can be readily screened for high-affinity antibodies by various affinity chromatographic methods (116). Although these methods do not reveal whether the captured antibody is catalytically active, they do greatly expand the number of antibodies facilely screened for hapten binding. It is obvious that since these particles retain their infectivity, they might be used to complement auxotrophic bacteria or yeast as a means of selection. The expression in yeast of a catalytically active F_{ab} with chorismate mutase represents a step in that direction (117).

The expression of antibodies or antibody fragments in *E. coli* has confronted the problem of folding the two chains to the native, functional state. A favored method for F_{ab} as well as single-chain F_V fragments is functional expression by secretion into the periplasmic space with its diminished protease activity accompanied by disulfide bond formation and correct folding (118, 119). The demonstration that the single-chain F_V version of the antibody, 43C9, exhibits the same catalytic parameters as the parent monoclonal, coupled with its potential for large-scale expression (120), is valuable for structural and mutagenesis studies. The high-level production of functional immunoglobulin heterodimers in baculovirus is also a welcome development (121).

Hapten Design

Since many chemical reactions are accelerated by general acid-base catalysis, the induction of desirable, appropriate functionalities has been attempted by a "bait and switch" process (122, 123). Here, an additional structural element, for example a charged residue, is present in the immunogen but not in the substrate so that the complementary functionality induced in the antibody may assume a different role with the substrate molecule. A pertinent example is given in Figure 9, in which the presence of the positively charged ammonium group is located to induce a complementary carboxylate, which with bound substrate might abstract the proton in the β-elimination reaction. Charge complementarity that induces catalytically important amino acids in the antibody-combining site also has been demonstrated in ester hydrolysis. In this case the transition-state analog featured a positive or negative point charge adjacent to a neutral secondary hydroxylic group that successfully served as a representation of the acyl-transfer transition state (124, 125).

The appearance and identity of a catalytic amino acid side-chain for β-elimination in the antibody's combining site was probed by determining the pH dependence of k_{cat} and also by chemical modification. The pH-rate profile revealed that k_{cat} is increased by the dissociation of a single titratable group of $pK_a = 6.2$, a finding within the range of the pK_a of carboxyl groups at enzyme active sites. Specific modification of carboxylate residues by the reagent, diazoacetamide, diminished the catalytic activity by ca. 70%, but in the presence of a competitive inhibitor, *p*-nitrobenzyl acetone, 80% of the catalytic activity was retained. Incubation with the trans-epoxide (Figure 18) inactivated the antibody in a time-dependent manner to less than 5% of its original activity, but treatment with hydroxylamine restored it. All these results are consistent with the presence of carboxylate residue within the binding site of the antibody. That this residue is acting as a general base

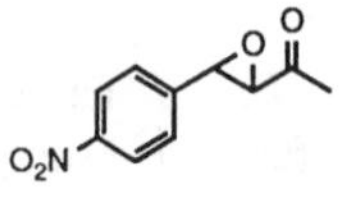

Figure 18 Epoxide inhibitor of antibody-catalyzed β-elimination (see Figure 9).

catalyst is strongly supported by the observation of a kinetic isotope effect on k_{cat} of 2.4 for the deuterated vs protium substrate, consistent with an E2 elimination mechanism. Relative to acetate, this antibody has an effective molarity of ca. 45 M.

The "bait and switch" strategy need not be confined to charge complementarity. One can imagine the induction of hydrogen bonds in response to the presence of neutral basic residues on the antigen, or the creation of a hydrophobic pocket in response to a large patch of nonpolar surface on the antigen. The ability of antibodies to catalyze the decarboxylation reaction shown in Figure 8 is in large measure due to the hydrophobic nature of their binding sites, which destabilizes the ground state of the charged substrate and which is relieved upon carbon dioxide formation (55). Furthermore, the "switch" may be made with elements of substrate structure more removed from the reaction center. The replacement of the pNO_2 substituent in the aromatic ester series catalyzed by 43C9 effectively improves the rate for phenolate product release so that it no longer governs turnover. Similarly, the replacement of the quinaldine heterocycle in an inducing immunogen with napthyl or phenyl in the substrate confers esterase activity on the antibodies (126).

Chemical and Genetic Modification

Can catalytic residues be introduced and positioned into the antibody-combining site so that they function efficiently? Precedent exists for two methods by which this may be achieved: (*a*) selective chemical derivitization of the combining site, permitting the incorporation of natural or synthetic catalytic groups (127) and (*b*) site-specific mutagenesis (4) for amino acid replacement.

Chemical modification of the F_{ab} fragment of MOPC315, an antibody that binds substituted 2,4-dinitrophenyl ligands, led to the introduction of a butyl-thiol unit ($-(CH_2)_4SH$) attached to Lys 52 known to be proximal to the binding site (128). This thiol then acted as a chemical handle for the introduction of imidazole via 4-mercaptomethylimidazole, forming a disulfide linkage (129). The modified antibody was then reacted with a series of coumarin esters in which the distance between the carboxyl function and the 2,4-dinitrophenyl moiety was varied (Figure 19). The rate of ester hydrolysis depended on the value of n, being maximal at $n = 2$. Although the pH-rate profile for this ester exhibited a pK_a of 7.5, the rate acceleration in terms of

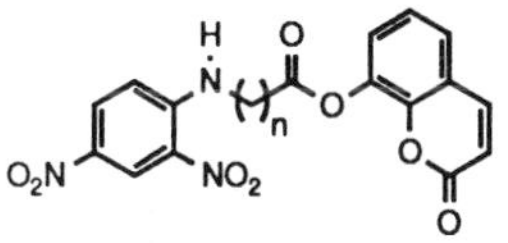

Figure 19   Variable coumarin substrates.

effective molarity was only ca. 0.05 M. It is not known yet whether the imidazole acts as a nucleophilic or general acid-base catalyst.

The low effective molarities for the semisynthetic antibodies undoubtedly reflect large populations of unproductive conformations derived from the large degrees of freedom introduced by the tether that attaches the potential nucleophile to the antibody. This may be overcome by reducing tether flexibility, although catalyst optimization will largely be empirical. The thiol handle, however, provides a means of introducing powerful synthetic, catalytic groups such as metal ion chelates, which can function as Lewis acids or chemical oxidants.

Site-directed mutagenesis provides a means for evaluating the contributions of various amino acid residues to hapten binding (130) and for improving the catalytic power of the antibody. Most of the effort to date has been directed at antibody F_{VS} (cross-linked by glutaldehyde) and F_{ab} fragments derived from McPC603, MOPE167, T15, and S107, a class of homologous antibodies that bind phosphoryl choline mono- and diesters (131–134). All have been shown to catalyze the hydrolysis of choline carbonates. The structure of one, McPC603, has been solved by X-ray crystallography (135), permitting the identification of side-chains important in phosphoryl choline binding now verified by the effect of their mutagenesis on binding of antigen. Mutagenesis of comparable residues on S107 but using the change in the catalytic activity of the antibody towards hydrolysis of p-nitrophenyl choline carbonate demonstrated the modest importance of electrostatic stabilization (k_{cat} decreased 20-fold upon replacement of Arg 52 that contacts the phosphate group), but the apparent unimportance of hydrogen bonding in the anionic transition state (136). The ca. 50-fold increase in k_{cat} brought about by the introduction of a His 33 proximal to the bound substrate is a consequence of its participation in turnover as either a general base or electrophilic catalyst (137). The low catalytic activity of this mutant exemplifies the nettlesome problem of optimal placement, which often punctures attempts to alter enzyme activity and specificity by mutagenesis methods.

Cofactors

The union of cofactors with antibodies to broaden their catalytic versatility is only in its initial stages, but includes several examples featuring redox cofactors. Antibodies induced to the oxidized form of riboflavin lowered the reduction potential of the flavin by 136 mV (ϵ_M −206 to −342 mV) (138), permitting the reduction of a Saframine T dye. Since a bent conformation is favored by the reduced flavin, whereas the oxidized form is planar, the fit to the antibody apparently stabilizes the oxidized species. Similarly, antibodies induced to a distorted porphyrin macrocycle resulting from N-alkylation (Figure 10) facilitated the metal ion chelation of Zn^{2+} and Cu^{2+} to the planar porphyrin (139). At 1 mM Zn^{2+} and Cu^{2+} (nonsaturating levels), the rate acceleration is 2600- and 1700-fold over background. Relative to ferrochelatase, the antibody-catalyzed process is only 10-fold slower, but the binding of metal ion by the enzyme exhibits a $K_M = 32$ μM whereas for the antibody $K_M > 1$ mM.

Metal ions have also now been successfully incorporated into single-chain antibodies by remodeling, through site-specific mutagenesis, the light chain of antibodies to create potential Zn^{2+} binding sites (140, 141). These designer light chains can then be crossed with members of heavy-chain libraries with appropriate substrate specificity as a general strategy for developing catalytic metal cofactor sites. At present a preexisting F_{ab} fragment has been genetically manipulated and shown to bind Zn^{2+} and Cu^{2+} but does not perform a catalytic formation (141). Alternatively, the metal function can be introduced from solution as a complex into an antibody whose combining site was formed to bind the metal ion complex and substrate in juxtaposition. Such antibodies were the first to cleave peptide linkages with sequence specificity (142).

CONCLUSION

The activities and stereospecificities of catalytic antibodies can, for certain cases, rival those of comparable enzymes. Rate accelerations for catalytic antibodies up to 10^6 over the uncatalyzed reaction have been measured; however, enzymes such as adenosine deaminase, alkaline phosphatase, and triosephosphate isomerase exhibit rate accelerations from 10^9 to 10^{12} fold (143). It is unlikely that the difference between the antibody affinity for the transition-state mimic and the actual substrate would approach this ratio; furthermore, given the analog's imperfections, it will be necessary to have present appropriate bonding and nonbonding interactions furnished by combining-site side-chains or cofactors. The examination of various structures of enzyme-substrate complexes reveals multiple bonding networks between numerous residues within the active site and portions of the substrate all poised to promote the chemical transformation (144). Although there is an entropic cost, the catalytic gain is substantial; thus bait/switch hapten design, heavy/light chain shuffling, mutagenesis, and cofactor introduction will be important for improving turnover. Not to be overlooked is that unique stereospecific catalysts can now be sought for reactions not known to be enzyme catalyzed.

We may not have anticipated that catalytic antibodies would parallel the relevant enzyme in the use of multistep kinetic sequences for substrate processing because of their induction by a single transition-state analog. In retrospect, however, the opportunistic molding of the combining site by the immunogen provides myriad chances for the creation of such a kinetic sequence. It would appear that the mechanistic nuances classified by enzymologists simply follow from the tuning of the chemical potential of such a binding site. The evolution of this process and the attendant insights into the origins of biological catalysis represent another major opportunity created by this field.

ACKNOWLEDGMENTS

This work was supported in part by a grant from SmithKline Beecham. The author gratefully acknowledges the suggestions of Richard Lerner and Peter Schultz.

Literature Cited

1. Cram, D. J. 1988. *Angew. Chem. Int. Ed. Engl.* 27:1009–20
2. Lehn, J. M. 1988. *Angew. Chem. Int. Ed. Engl.* 27:89–112
3. Rebek, J. Jr. 1990. *Acc. Chem. Res.* 23:399–404
4. Johnson, K. A., Benkovic, S. J. 1990. *Enzymes* 19:159–211
5. Clarke, A. R., Atkinson, T., Holbrook, J. J. 1989. *Trends Biochem. Sci.* 14:145–48
6. Pressman, D., Grossberg, A. 1968. *The Structural Basis of Antibody Specificity.* New York: Benjamin
7. Hood, L. E., Weissman, I. L., Wood, W. B. 1978. *Immunology.* Menlo Park, Calif: Benjamin/Cummings
8. Alt, F. W., Blackwell, T. K., Yancopoulos, G. D. 1987. *Science* 238:1079–87
9. Rajewsky, K., Förster, I., Cumano, A. 1987. *Science* 238:1088–94
10. Lerner, R. A., Benkovic, S. J., Schultz, P. G. 1991. *Science* 252:659–67
11. Schultz, P. G. 1989. *Angew. Chem. Int. Ed. Engl.* 28:1283–95
12. Lerner, R. A., Benkovic, S. J. 1988. *BioEssays* 9:107–12
13. Lerner, R. A., Benkovic, S. J. 1990. *Chemtracts-Org. Chem.,* pp. 1–36
14. Hilvert, D. 1990. *Biomimetic Polymers,* ed. G. Gebelein, pp. 95–114. New York: Plenum
15. Green, B. S. 1989. *Monoclonal Antibodies: Production and Application,* pp. 359–93. New York: Liss
16. Pauling, L. 1946. *Chem. Eng. News* 24:1375–77
17. Pauling, L. 1948. *Am. Sci.* 36:51–58
18. Wolfenden, R. 1969. *Nature* 223:704–5
19. Wolfenden, R. 1972. *Acc. Chem. Res.* 5:10–18
20. Wolfenden, R., Frick, L. 1987. *Enzyme Mechanisms,* ed. M. I. Page, A. Williams, pp. 97–122. London: Royal Soc. London
21. Bartlett, P. A., Marlowe, C. K. 1987. *Biochemistry* 26:8553–61
22. Lolis, E., Petsko, G. A. 1990. *Annu. Rev. Biochem.* 59:597–630
23. Kraut, J. 1988. *Science* 242:533–40
24. Jencks, W. P. 1987. *Cold Spring Harbor Symp. Quant. Biol.* 52:65–73
25. Jencks, W. P. 1969. *Catalysis in Chemistry and Enzymology,* p. 288. New York: McGraw-Hill
26. Pauling, L. 1990. *J. Natl. Insts. Health Res.* 2:63
27. Woolley, D. W. 1952. *A Study of Antimetabolites,* p. 87. New York: Wiley
28. Slobin, L. I. 1966. *Biochemistry* 5:2836–44
29. Kohen, F., Kim, J.-B., Barnard, G., Lindner, H. R. 1980. *Biochim. Biophys. Acta* 629:328–37

30. Raso, V., Stollar, B. D. 1975. *Biochemistry* 14:591–99
31. Köhler, G., Milstein, C. 1975. *Nature* 256:495–97
32. Edelman, G. M., Cunningham, B. A., Gall, W. E., Gottlieb, P. D., Rutishauser, U., Waxdal, M. J. 1969. *Proc. Natl. Acad. Sci. USA* 63:78–85
33. Kabat, E. A., Wu, T. T., Bilofsky, H., Reid-Miller, M., Perry, H. 1983. *Sequences of Protein of Immunological Interest.* Washington, DC: US Dept. Health Hum. Serv. NIH
34. Burton, D. R. 1987. *Molecular Genetics of Immunoglobulins,* ed. F. Calabi, M. S. Neuberger, pp. 1–50. Netherlands: Elsevier
35. Getzoff, E. D., Tainer, J. A., Lerner, R. A., Geysen, H. M. 1988. *Adv. Immunol.* 43:1–98
36. Davies, D. R., Padlan, E. A., Sheriff, S. 1990. *Annu. Rev. Biochem.* 59:439–73
37. Poljak, R. J., Amzel, L. M., Avey, H. P., Chen, B. L., Phizackerley, R. P., Saul, F. 1973. *Proc. Natl. Acad. Sci. USA* 70:3305–10
38. Chothia, C., Lesk, A. M., Tramontano, A., Levitt, M., Smith-Gill, S. J., et al. 1989. *Nature* 342:877–83
39. Amit, A. G., Mariuzza, R. A., Phillips, S. E. V., Poljak, R. J. 1986. *Science* 233:747–53
40. Sheriff, S., Silverton, E. W., Padlan, E. A., Cohen, G. H., Smith-Gill, S. J., et al. 1987. *Proc. Natl. Acad. Sci. USA* 84:8075–79
41. Colman, P. M., Laver, W. G., Varghese, J. N., Baker, A. T., Tulloch, P. A., et al. 1987. *Nature* 326:358–63
42. Watt, R. M., Herron, J. N., Voss, E. W. 1980. *Mol. Immunol.* 17:1237–43
43. Janda, K. D., Benkovic, S. J., Lerner, R. A. 1989. *Science* 244:437–40
44. Wirsching, P., Ashley, J. A., Benkovic, S. J., Lerner, R. A. 1991. *Science* 252:680–85
45. Janda, K., Schloeder, D., Benkovic, S. J., Lerner, R. A. 1988. *Science* 241:1188–91
46. Napper, A. D., Benkovic, S. J., Tramontano, A., Lerner, R. A. 1987. *Science* 237:1041–43
47. Benkovic, S. J., Napper, A. D., Lerner, R. A. 1988. *Proc. Natl. Acad. Sci. USA* 85:5355–58
48. Shokat, K. M., Leumann, C. J., Sugasawara, R., Schultz, P. G. 1989. *Nature* 338:269–71
49. Hilvert, D., Nared, K. D. 1988. *J. Am. Chem. Soc.* 110:5593–94
50. Hilvert, D., Carpenter, S. H., Nared, K. D., Auditor, M. T. M. 1988. *Proc. Natl. Acad. Sci. USA* 85:4953–55
51. Jackson, D. Y., Jacobs, J. W., Sugasawara, R., Reich, S. H., Bartlett, P. A., Schultz, P. G. 1988. *J. Am. Chem. Soc.* 110:4841–42
52. Hilvert, D. H., Hill, K. W., Nared, K. D., Auditor, M. T. M. 1989. *J. Am. Chem. Soc.* 111:9261–62
53. Braisted, A., Schultz, P. G. 1990. *J. Am. Chem. Soc.* 112:7430–31
54. Cochran, A. G., Schultz, P. G. 1990. *Science* 249:781–83
55. Lewis, C., Kramer, T., Robinson, S., Hilvert, D. 1991. *Science* 253:1019–21
56. Woodward, R. B., Hoffmann, R. 1969. *Angew. Chem. Int. Ed. Engl.* 8:781–853
57. Bushby, R. J. 1989. *Annu. Rep. R. Soc. Chem.* 86(Sect. B):45–85
58. Weiss, V., Edwards, J. M. 1980. *The Biosynthesis of Aromatic Amino Compounds,* pp. 134–84. New York: Wiley
59. Copley, S. D., Knowles, J. R. 1985. *J. Am. Chem. Soc.* 107:5306–8
60. Gajewski, J. J., Jurayj, J., Kimbrough, D., Gande, M. E., Ganem, B., Carpenter, B. K. 1987. *J. Am. Chem. Soc.* 109:1170–86
61. Coates, R. M., Rogers, B. D., Hobbs, S. J., Peck, D. R., Curran, D. P. 1987. *J. Am. Chem. Soc.* 109:1160–70
62. Guilford, W. J., Copley, S. D., Knowles, J. R. 1987. *J. Am. Chem. Soc.* 109:5013–19
63. Sogo, S. G., Widlanski, T. S., Hoare, J. H., Grimshaw, C. E., Berchtold, G. A., Knowles, J. R. 1984. *J. Am. Chem. Soc.* 106:2701–3
64. Bartlett, P. A., Nakagawa, Y., Johnson, C. R., Reich, S. H., Luis, A. 1988. *J. Org. Chem.* 53:3195–210
65. Hackney, D. D. 1990. *Enzymes* 19:1–36
66. Page, M. I., Jencks, W. P. 1971. *Proc. Natl. Acad. Sci. USA* 68:1678–83
67. Bruice, T. C., Pandit, U. K. 1960. *J. Am. Chem. Soc.* 82:5858–65
68. Menger, F. M. 1985. *Acc. Chem. Res.* 18:128–34
69. Jencks, W. P., Page, M. I. 1974. *Biochem. Biophys. Res. Commun.* 57:887–92
70. Sauer, J., Sustman, R. 1980. *Angew. Chem. Int. Ed. Engl.* 19:779–807
71. Brown, F., Houk, K. N. 1984. *Tetrahedron Lett.* 25:4609–12
72. Dafforn, A., Koshland, D. E. 1971. *Proc. Natl. Acad. Sci. USA* 68:2463–67
73. Dafforn, A., Koshland, D. E. 1973. *Biochem. Biophys. Res. Commun.* 52:779–85
74. Bruice, T. C., Brown, A., Harris, D. O.

1971. *Proc. Natl. Acad. Sci. USA* 68:658–61
75. Menger, F. M., Glass, L. E. 1980. *J. Am. Chem. Soc.* 102:5404–6
76. Kirby, A. J. 1980. *Adv. Phys. Org. Chem.* 17:183–278
77. Balan, A., Doctor, B. P., Green, B. S., Torten, M., Ziffer, H. 1988. *J. Chem. Soc.-Chem. Commun.*, pp. 106–8
78. Cochran, A. G., Sugasawara, R., Schultz, P. G. 1988. *J. Am. Chem. Soc.* 110:7888–90
79. Jencks, W. P. 1975. *Adv. Enzymol. Relat. Areas Mol. Biol.* 43:219–410
80. Pollack, S. J., Jacobs, J. W., Schultz, P. G. 1986. *Science* 234:1570–73
81. Tramontano, A., Janda, K. D., Lerner, R. A. 1986. *Science* 234:1566–70
82. Tramontano, A., Ammann, A. A., Lerner, R. A. 1988. *J. Am. Chem. Soc.* 110:2282–86
83. Pollack, S. J., Hsiun, P., Schultz, P. G. 1989. *J. Am. Chem. Soc.* 111:5961–62
84. Jacobs, J., Schultz, P. G., Sugasawara, R., Powell, M. 1987. *J. Am. Chem. Soc.* 109:2174–76
85. Jacobs, J. W. 1989. *Catalytic antibodies*. PhD thesis. Dept. Chem. Univ. Calif., Berkeley
86. Jacobs, J. W. 1991. *Bio-Technology* 9:258–62
87. Deleted in proof
88. Ray, W. J. Jr., Long, J. W., Owens, J. D. 1976. *Biochemistry* 15:4006–17
89. Benkovic, S. J., Adams, J. A., Borders, C. L., Janda, K. D., Lerner, R. A. 1990. *Science* 250:1135–39
90. Cleland, W. W. 1977. *Adv. Enzymol.* 45:273–387
91. Janda, K. D., Ashley, J. A., Jones, T. M., McLeod, D. A., Schloeder, D. M., et al. 1991. *J. Am. Chem. Soc.* 113:291–97
92. Gibbs, R. A., Janda, K. M., Benkovic, P. A., Lerner, R. A., Benkovic, S. J. 1992. *J. Am. Chem. Soc.* In press
93. Bruice, T. C., Mayahi, M. F. 1960. *J. Am. Chem. Soc.* 82:3067–71
94. Kirsch, J. F., Clewell, W., Simon, A. 1968. *J. Org. Chem.* 33:127–32
95. Albery, W. J., Knowles, J. R. 1976. *Biochemistry* 15:5631–40
96. Zerner, B., Bond, R. P. M., Bender, M. L. 1964. *J. Am. Chem. Soc.* 86:3674–79
97. Ingami, T., Patchornik, A., York, S. S. 1969. *J. Biochem.* 65:809–19
98. Bender, M. L., Kezdy, F. J. 1964. *J. Am. Chem. Soc.* 86:3704–14
99. Bender, M. L., Kezdy, F. J., Gunter, C. R. 1964. *J. Am. Chem. Soc.* 86:3714–21
100. Huse, W. D., Sastry, L., Iverson, S. A.,

Kang, A. S., Alting-Mees, M., et al. 1989. *Science* 246:1275–81
101. Larrick, J. W., Danielsson, L., Brenner, C. A., Wallace, E. F., Abrahamson, M., et al. 1989. *Bio-Technology* 7:934–38
102. Orland, R., Gussow, D. H., Jones, P. T., Winter, G. 1989. *Proc. Natl. Acad. Sci. USA* 86:3833–37
103. Better, M., Chang, C. P., Robinson, R., Horwitz, A. H. 1988. *Science* 240:1041–43
104. Skerra, A., Pluckthun, A. 1988. *Science* 240:1038–41
105. Sastry, L., Alting-Mees, M., Huse, W. D., Short, J. M., Sorge, J. A., et al. 1989. *Proc. Natl. Acad. Sci. USA* 86:5728–32
106. Kang, A. S., Jones, T. M., Burton, D. R. 1991. *Proc. Natl. Acad. Sci. USA.* 88:11120–23
107. Ward, E. S., Gussow, D., Griffiths, A. D., Jones, P. T., Winter, G. 1989. *Nature* 341:544–46
108. Winter, G., Milstein, C. 1991. *Nature* 349:293–99
109. Caton, A. J., Koprowski, H. 1990. *Proc. Natl. Acad. Sci. USA* 87:6450–54
110. Deleted in proof
111. Parmley, S. F., Smith, G. P. 1988. *Gene* 73:305–18
112. Scott, J. K., Smith, G. P. 1990. *Science* 249:386–90
113. Devlin, J. J., Panganiban, L. C., Devlin, P. E. 1990. *Science* 249:404–6
114. McCaffarty, J., Griffiths, A. D., Winter, G., Chiswell, D. J. 1990. *Nature* 348:552–54
115. Kang, A. S., Barbas, C. F., Janda, K. D., Benkovic, S. J., Lerner, R. A. 1991. *Proc. Natl. Acad. Sci. USA* 88:4363–66
116. Barbas, C., Khan, A., Lerner, R. A., Benkovic, S. J. 1991. *Proc. Natl. Acad. USA.* 88:7978–82
117. Bowdish, K., Tang, Y., Hicks, J. B., Hilvert, D. 1991. *J. Biol. Chem.* 266:11901–8
118. Pluckthun, A. 1991. *Bio-Technology* 9:545–51
119. Skerra, A., Pfitzinger, I., Pluckthun, A. 1991. *Bio-Technology* 9:273–78
120. Gibbs, R. A., Posner, B. A., Filpula, D. R., Dodd, S. W., Finkelman, M. A., et al. 1991. *Proc. Natl. Acad. Sci. USA* 88:4001–4
121. Hasemann, C. A., Capra, J. D. 1990. *Proc. Natl. Acad. Sci. USA* 87:3942–46
122. Shokat, K. M., Schultz, P. G. 1991. *Catalytic Antibodies. Ciba Found. Symp.* 159:118–34
123. Janda, K. D., Weinhouse, M. I.,

Schloeder, D. M., Lerner, R. A., Benkovic, S. J. 1990. *J. Am. Chem. Soc.* 112:1274–75

124. Janda, K. D., Weinhouse, M. I., Danon, T., Pacelli, K. A., Schloeder, D. M. 1991. *J. Am. Chem. Soc.* 113:5427–34

125. Shokat, K. M., Ko, M. K., Scanlan, T. S., Kochersperger, L., Yonkovich, S., et al. 1990. *Angew. Chem. Int. Ed. Engl.* 29:1296–303

126. Janda, K. D., Benkovic, S. J., McLeod, D. A., Schloeder, D. M., Lerner, R. A. 1991. *Tetrehedron* 47:2503–6

127. Kaiser, E. T., Lawrence, D. S. 1984. *Science* 226:505–11

128. Pollack, S. J., Nakayama, G. R., Schultz, P. G. 1988. *Science* 242:1038–40

129. Pollack, S. J., Schultz, P. G. 1989. *J. Am. Chem. Soc.* 111:1929–31

130. Roberts, S., Cheetham, J. C., Rees, A. R. 1987. *Nature* 328:731–34

131. Perlmutter, R. M., Crews, S. T., Douglas, R., Sorensen, G., Johnson, N., et al. 1984. *Adv. Immunol.* 35:1–37

132. Pollack, S. J., Jacobs, J. W., Schultz, P. G. 1986. *Science* 234:1570–73

133. Jackson, D. Y., Prudent, J. R., Baldwin, E. P., Schultz, P. G. 1991. *Proc. Natl. Acad. Sci. USA* 88:58–62

134. Pluckthun, A., Stadlmuller, J. 1991. *Catalytic Antibodies. Ciba Found. Symp.* 159:103–17

135. Satow, Y., Cohen, G. H., Padlan, E. A., Davies, D. R. 1986. *J. Mol. Biol.* 190:593–604

136. Glockshuber, R., Stadlmuller, J., Pluckthun, A. 1991. *Biochemistry* 30:3049–54

137. Baldwin, E., Schultz, P. G. 1989. *Science* 245:1104–7

138. Shokat, K. M., Leumann, C. J., Sugasawara, R., Schultz, P. G. 1988. *Angew. Chem. Int. Ed. Engl.* 27:1172–74

139. Cochran, A. G., Schultz, P. G. 1990. *Science* 249:781–83

140. Roberts, V. A., Iverson, B. L., Iverson, S. A., Benkovic, S. J., Lerner, R. A., et al. 1990. *Proc. Natl. Acad. Sci. USA* 87:6654–58

141. Iverson, B. L., Iverson, S. A., Roberts, V. A., Getzoff, E. D., Tainer, J. A., et al. 1990. *Science* 249:659–62

142. Iverson, B. L., Lerner, R. A. 1989. *Science* 243:1184–88

143. Frick, L., MacNeela, J. P., Wolfenden, R. 1987. *Bioorg. Chem.* 15:100–8

144. Hansen, D. E., Raines, R. T. 1990. *J. Chem. Ed.* 67:483–89

Annu. Rev. Biochem. 1992. 61:55–85

STRUCTURE AND FUNCTION OF SIMIAN VIRUS 40 LARGE TUMOR ANTIGEN

Ellen Fanning

Institut für Biochemie, Universität München, D 8000 München, Federal Republic of Germany

Rolf Knippers

Fakultät für Biologie, Universität Konstanz, D 7750 Konstanz, Federal Republic of Germany

KEY WORDS: specific DNA binding, DNA helicase, protein phosphorylation, replication transcriptional activation, transformation

CONTENTS

PERSPECTIVES AND SUMMARY

The interactions of animal viruses with their host cells open an avenue to understanding the key pathways controlling cell metabolism, growth, and differentiation, as viruses not only depend on the cellular environment for their own multiplication, but often alter cellular processes to promote their own growth. Viruses thus encode regulatory proteins that have evolved to subvert cellular control processes, thereby ensuring optimal conditions for viral development and multiplication. The biochemical and genetic study of viral regulatory proteins and the mechanisms through which they perturb cellular control processes have thus been particularly useful in identifying important cellular regulatory proteins and elucidating their functions.

Simian Virus 40 (SV40), a simple virus with a small DNA genome, encodes one major multifunctional regulatory protein, the large tumor antigen (T antigen), which is responsible for both the control of viral infection and the required alterations of cellular processes.

SV40 T antigen contains 708 amino acids with multiple posttranslational modifications. It has several intrinsic biochemical activities required for production of viral progeny. It regulates the timing of the infection cycle, repressing transcription of its own gene, initiating viral DNA replication, and stimulating the expression of viral capsid proteins. T antigen stimulates cell proliferation, as measured by induction of DNA synthesis in quiescent cells, immortalization of primary cells, cell transformation, and induction of tumors in animals. T antigen binds to a number of interesting cellular proteins, such as the tumor suppressor p53, the retinoblastoma suppressor gene product, and DNA polymerase α-primase. It also stimulates transcription of cellular genes. The multiple functions of T antigen raise the question of how one polypeptide chain can accomplish so much, and moreover, do so in a temporally controlled fashion.

This extraordinary protein has been well-studied since its discovery as a novel antigen in SV40 virus–induced tumors of laboratory animals (1). Its structure and function were comprehensively reviewed in 1983 (2) and in 1989 (3). Specialized aspects of its function in control of viral DNA replication and transformation have also been reviewed (4–14). However, the recent accumulation of biochemical and genetic data on T antigen has led to a clearer view of its remarkably complex structure and function, and it is our goal in this review to make this information accessible to those outside the field who may find it useful for comparison with other viral and cellular regulatory proteins.

BACKGROUND

SV40 infects a variety of cultured mammalian cells, but initiates a productive infection cycle only in monkey cells, its natural host, and in cells of other primate species. Thousands of progeny virus particles per infected monkey cell are eventually produced. Cells of rodents and other mammalian species are nonpermissive and are abortively infected by SV40. Infected nonpermissive cells can be immortalized, and transformed to make possible anchorage-independent growth. In suitable animals, these cells may induce tumors.

The SV40 genome is a closed circular double-stranded DNA molecule of 5243 base pairs (bp) (15, 16). The genome associates with histones to form a chromatin structure with 24–28 nucleosomes, frequently referred to as the SV40 minichromosome (17–19).

Functionally, the genome consists of two genetic regions of about equal size, oriented in opposite directions. The two divergent promoters reside within a ca. 300-bp section, the genomic control region, which also includes the viral origin of replication. The early genetic region is transcribed soon after infection, resulting in a pre-mRNA, which is then differentially spliced to give mRNAs for the two early proteins, the *small t antigen* and the *large T antigen*. The late genetic region mainly encodes structural proteins, the building blocks of progeny viral shells.

Large T antigen, the subject of this review, serves as a regulatory protein determining the course of events in productively infected as well as in abortively infected cells. The small t antigen appears to regulate protein phosphatase 2A (20, 20a, 21, 21a). It is dispensable for virus multiplication, at least under laboratory conditions, but positively affects the ability of SV40 to transform nonpermissive cells (22 and references therein). Small t antigen is not considered further in this review.

In permissive monkey cells, large T antigen initiates viral DNA replication and regulates early and late viral gene transcription. In addition, T antigen induces cellular DNA replication and activates a number of cellular genes, preparing the cell for a massive multiplication of the infecting virus.

In nonpermissive cells, viral DNA is not replicated, and late viral genes are not expressed. Instead the infecting viral DNA molecule may integrate into the host genome. Integration occurs without obvious preferences for particular sites on either the host or the viral DNA. T antigen, encoded by the integrated early viral genes, is essential for the establishment and maintenance of the transformed phenotype.

A subset of SV40-transformed cells is able to grow as tumors in susceptible animals. The immune systems of animals bearing tumors of SV40-transformed cells produce antibodies against the early viral proteins, hence their traditional designation as tumor antigens or T antigens. Another name for T antigen is *A protein* since the protein is encoded by a gene in the viral complementation group A (23).

The ability of large T antigen to perform a variety of functions in replication, transcriptional regulation, and transformation, described briefly in this introduction, is the reason that T antigen has attracted and continues to attract the attention of molecular biologists of all persuasions. A complete list of references on the subject of SV40 T antigen includes more than a thousand papers. Since we cannot cover this vast amount of information within the limits of this review, we rather intend to concentrate on more recent developments.

BIOCHEMICAL PROPERTIES

T antigen is composed of 708 amino acids with a calculated M_r of ca. 82, 500. A considerable fraction of T antigen in acutely infected as well as in transformed cells occurs in dimeric, tetrameric, hexameric, or larger complexes. T antigen also associates with a variety of cellular proteins.

Furthermore, T antigen can be modified posttranslationally by phosphorylation of certain threonine and serine side chains (24–26), amino-terminal acetylation (27), O-glycosylation (28), acylation (29), adenylylation (30), poly(ADP-ribosyl)-ation (31), and covalent coupling to RNA (32). Except for phosphorylation, discussed later in this review, little is known about the physiological consequences of the modification reactions.

More than 90–95% of T antigen in infected and transformed cells is located in the nucleus, where it occurs as a free nucleoplasmic protein or bound to chromatin or the nuclear matrix (33). The remaining 5–10% of intracellular T antigen is found in the cytoplasm and associated with plasma membrane structures. A minor subfraction (0.1–0.2%) is integrated into the plasma membrane and partially exposed on the surface of the infected or transformed cell (8, 34, 35). This fraction of T antigen is modified by palmitoylation, which probably serves to anchor the protein in the plasma membrane (29).

Even though the total amount of T antigen per infected or transformed cell can be substantial (10–50 $\mu g/10^8$ infected cells, or, roughly, 5–10 $\times$ 10^5 molecules/cell), conventional column purification schemes have usually not been very successful for the isolation of T antigen. This is probably because of the high degree of heterogeneity caused by aggregation and posttranslational modifications, preventing the formation of homogeneous chromatographic peaks. These problems were alleviated by the development of immunoaffinity purification procedures (36, 37), which yield highly concentrated and biochemically active T antigen preparations largely free of contaminating cellular proteins. The yields of T antigen can be increased by a factor of 5–10 using an adenovirus vector that contains the SV40 early coding region controlled by the major late adenovirus promoter (38). In recent years, homogeneous T antigen preparations of high concentration have been obtained using specifically engineered bacteria (39, 40) or baculoviruses (41–43). The baculovirus system offers several advantages over bacteria for an efficient expression of T antigen, as the protein is more likely to assume its native conformation and acquires several functionally important phosphate groups (41–44).

DNA Binding

T antigen binds to DNA of any sequence and conformation, but has a somewhat higher affinity for single-stranded than for double-stranded DNA (45–47). In fact, the binding of T antigen to double-stranded DNA of random base sequence is generally weak and can best be observed in the absence of salt and at pH values below 7. At higher pH values and in the presence of 25–50 mM NaCl, T antigen binds to double-stranded DNA sequences containing the pentanucleotide G(A)AGGC, the *recognition pentanucleotide* (48). But not all GAGGC pentanucleotides permit T antigen binding with equal stability, and adjacent DNA sequences also help to stabilize T antigen–DNA interaction. A major stabilizing factor is a second nearby GAGGC, but the size and the base sequence of the spacer between adjacent pentanucleotides clearly affect the stability of the T antigen–DNA complex (48–52).

DNA regions with these sequence characteristics are not uncommon in the cellular genome (53, 54). However, the prototypes of these specific T antigen–binding sites are found in the genomic control region of the SV40 genome with its two divergent promoter/enhancer elements and the origin of viral DNA replication (Figure 1).

Most T antigen extracted from lytically infected cells binds to a viral DNA sequence commonly referred to as *binding site I* (Figure 1). The important elements of this site are two GAGGC elements separated by an AT-rich spacer of seven bp (55). T antigen binds to this site with an affinity that has been estimated to be at least 1000-fold higher than its affinity to random double-stranded DNA. This is reflected in the stability of the complex, which remains intact at 150 mM NaCl and pH 7.5–8 while the binding of T antigen to nonspecific DNA sequences or to sequences with a single GAGGC element is completely suppressed (55).

One T antigen molecule binds to each of the two recognition pentanucleotides in binding site I (at low temperature and in the absence of ATP) (56). Methylation protection and other experiments have shown that T antigen forms a specific contact with each of the guanine residues in the recognition pentanucleotides. These contacts are mainly formed in the major groove of the double helix (52). However, the oligo(dA-dT) spacer is an essential and integral element of the binding site (55) and may favor specific protein-protein contacts between bound T antigens, thereby stabilizing the interaction of T antigen with site I DNA (55, 57). Thus, T antigen is thought be wrapped around the GAGGC-spacer-GAGGC sequence, forming a complex that extends to 5–7 bp on either side of the 17-bp core element (52, 55–57).

The main physiological function of T antigen binding to site I appears to be the autoregulation of early gene transcription, since mutations in binding site I cause an overexpression of the T antigen gene (58), and T antigen bound to this site represses the transcription of the adjacent early coding region (59, 60).

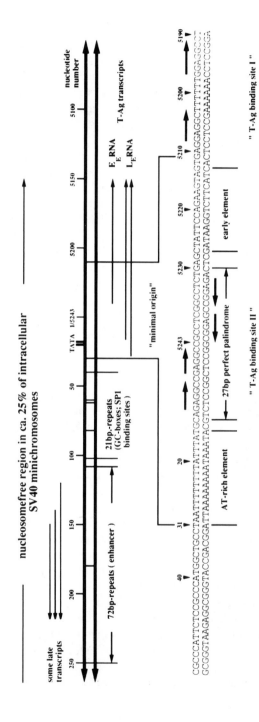

Figure 1 The SV40 genomic control region. *Upper part.* Some landmarks in the nucleosome-free region of the SV40 minichromosome. (*i*) The enhancer, the GC-box-elements, and the TATA-box of the early promoter directing the synthesis of T antigen mRNA. Immediate early (E$_E$) start sites are used before and late early (E$_L$) start sites are used after sufficient amounts of T antigen have accumulated in the cell. (*ii*) The heterogeneous late transcriptional start sites. (*iii*) The minimal origin of replication (see text for details). *Lower part.* The nucleotide sequence of the minimal origin of replication. The three origin "domains" are indicated. The recognition pentanucleotides in T antigen–binding sites I and II are shown by arrows. This figure is reproduced from (9) with permission.

The second strong T antigen–binding site in the SV40 genomic control region has a composition distinctly different from that of binding site I. *Binding site II* contains four GAGGC elements, organized as two pairs in opposite directions, forming a perfect palindrome of 27 bp (Figure 1).

Site II is the central section or domain of the "minimal" viral origin of replication, a regulatory element of 64 bp which, in addition to site II, includes an imperfect palindrome on the early side and a stretch of 17 AT bp on the late side (Figure 1). All three domains of the viral origin are essential for function; most nucleotide exchanges or deletions in these regions drastically reduce replication, whereas mutations between these domains have much less effect (61–64).

The primary role of the GAGGC pentanucleotides in site II is to bind T antigen and to position it in the proper location and orientation for its subsequent function in DNA replication.

At low temperature (<30°C) and in the absence of ATP, T antigen protects against nuclease attack a region of about 35 bp centered about binding site II (48, 50). Under these conditions a tetrameric T antigen is the largest structure found on the origin (56). This complex is unable to activate the origin for subsequent replication.

However, at 37°C and in the presence of ATP, a 10-fold enhancement of origin binding is observed, with an enlargement of the protected region to 70 bp, covering the entire origin of replication (from SV40 nucleotide 5205 to 33; see Figure 1) (65–68). Apparently, ATP induces a conformational change in T antigen that allows the protein to form specific protein-protein contacts, which stabilize its binding to the origin region. ATP hydrolysis is not required for this reaction, as ADP and a number of nonhydrolyzable ATP analogues also lead to enhanced origin binding (65, 66).

As convincingly shown by scanning transmission electron microscopy (STEM), which measures the mass distribution in the protein-DNA complex, T antigen, in the presence of ATP, assembles as a two-lobed structure, each lobe containing a T antigen hexamer, centered at the origin (68).

The assembly of the double-hexameric complex requires an intact zinc-finger region (amino acid residues 302–320; see Figure 2) and probably occurs in steps, as free monomeric or dimeric T antigen appears to be activated or conformationally altered by bound ATP before it assembles as hexamers to form the bilobed double-hexameric complex (69–71). The STEM data suggest that DNA-bound T antigen hexamers surround the double helix, in agreement with DNase I protection and methylation interference data (68).

As analyzed by a variety of chemical and enzymatic probes, the bound T antigen bilobe changes the double-helical structure in the origin sequence. T antigen causes the separation of a ca. 8-bp purine-pyrimidine tract in the early imperfect inverted repeat and untwists the 17-bp late AT-rich element (72–75).

Thus, the assembly of the T antigen double-hexamer complex at binding site II differs clearly from the association of T antigen with site I, in keeping with the different biological functions of DNA binding at these sites.

Interestingly, the T antigen–origin complex assembles in a manner distinctly different from that of the initiator-origin nucleoprotein complexes in some well-studied prokaryotic systems: the SV40 origin sequence is surrounded by T antigen molecules, whereas the bacterial or bacteriophage λ origin sequences are wrapped about a complex of their respective initiator proteins (76, 77).

Unwinding of DNA Double Strands

T antigen binds one mole of ATP per mole of polypeptide chain (78). In the presence of magnesium ions, bound ATP is rapidly hydrolyzed. The ATPase activity is stimulated by single-stranded DNA, most effectively by poly-(dT) (79). This property is a hallmark of DNA helicases. A DNA unwinding function of T antigen was proposed as early as 1980 (80). However, this idea was soon discarded (81), mainly because the lack of homogeneous T antigen preparations and of effective in vitro assays made a critical test of its function impossible.

The possibility of an intrinsic ATP-driven DNA unwinding activity of T antigen was reinvestigated after it had been shown that T antigen is tightly associated with replication forks in replicating SV40 minichromosomes (82, 83), and that certain T antigen–specific monoclonal antibodies inhibit the movement of replication forks in replicating SV40 DNA in vitro (84–86). The role that T antigen plays at replication forks became obvious when highly purified T antigen was shown to displace complementary strand sequences hydrogen bonded to circular single-stranded phage DNA in a reaction that requires magnesium salts and the hydrolysis of ATP or dATP (87).

T antigen, probably in its active, hexameric form (88), uses as an entry site the junction between single-stranded and double-stranded DNA in these partially double-stranded substrates and then moves in the 3' to 5' direction on the DNA strand to which it is bound (89–91). T antigen is able to unwind double-stranded regions of several thousand base pairs at an in vitro rate of a few hundred base pairs per minute (89).

T antigen is an unusual DNA helicase, distinguished from the classical bacterial and from the known eukaryotic DNA helicases (92) by its ability to unwind purely double-stranded DNA starting from internal sites (90, 93–96). This property of T antigen can be most clearly demonstrated using either linear or closed circular double-stranded DNA carrying the viral origin of replication. As described above, ATP, but not ATP hydrolysis, is required for the proper alignment and orientation of double-hexameric T antigen complexes at the origin. The DNA double helix in these complexes is partially unwound (72), but a complete strand separation occurs only when ATP is hydrolyzed (92–96). Once unwound, the origin region no longer serves as a specific binding substrate, and T antigen functions as a DNA helicase on the emerging single-stranded DNA independently of sequence-specific DNA binding (97). Recent electron microscopic investigation of active unwinding complexes suggests that the double-hexameric T antigen complex does not separate to give two independently diverging hexameric helicase units. It rather appears that the DNA is reeled through a stationary double-hexameric complex of T antigen (88).

Single-strand-specific DNA binding (SSB) proteins are essential cofactors in the double-strand unwinding reaction. These proteins are needed to prevent the reannealing of separated DNA complementary strands. Furthermore, with closed circular DNA molecules as substrates, a DNA topoisomerase must be present to relieve the torsional strain generated when a topologically closed DNA molecule is unwound (93–96).

Even though DNA double strands carrying the viral origin of replication are the preferred substrates, T antigen is also able to catalyze the unwinding of linear or circular double-stranded DNA of any sequence (90, 98). Since T antigen binding to nonspecific DNA is weak (see above), the reaction must be performed at low salt concentrations (90) and at relatively high concentrations of T antigen to increase the probability of formation of properly oriented hexameric complexes at two adjacent sites (98).

In summarizing this and the preceding section, we wish to emphasize the dual role of ATP in the DNA binding/unwinding function of T antigen. Nonhydrolyzed ATP, bound to the single nucleotide-binding site of the protein, has a *regulatory role,* inducing a conformational change suitable for the assembly of a functional initiation complex on DNA substrates. The functional complex is a hexamer when a partially single-stranded entry site exists on suitable substrates, or it is a double-hexameric bilobe when the substrate is an intact DNA duplex. In addition, ATP has a *catalytic role,* being hydrolyzed by the T antigen–associated ATPase either in an idling reaction (in the absence of DNA duplex regions) or to provide the energy needed for the separation of complementary DNA strands and for the movement of T antigen relative to the DNA strand to which it is bound.

Unwinding of RNA Double Strands

With ATP as a cofactor, T antigen also unwinds partially double-stranded DNA-RNA hybrid molecules provided that the overhanging single strand is DNA. With cofactors other than ATP, T antigen also unwinds partially double-stranded RNA-DNA hybrids with overhanging RNA, as well as partially double-stranded RNA substrates. In these latter cases, ATP does not serve as an energy source, and the unwinding reaction depends on non-ATP nucleotides such as GTP, UTP, or CTP (99). This curious property could arise from a conformational change induced by the bound single-stranded nucleic acid. According to this model, the conformational change induced by bound DNA would cause an optimal placement of ATP relative to the active center of the ATPase domain, whereas single-stranded RNA fails to induce this conformational change, and non-ATP nucleotides are then utilized as an energy source to drive the separation of complementary RNA strands (100).

While the T antigen–associated DNA helicase performs a well-defined and vital function for the viral replication process, the physiological role of its RNA helicase activity has yet to be identified. Like some of the cellular RNA helicases (101), it could be involved in splicing or posttranscriptional gene regulation, for example, by eliminating inhibitory RNA secondary structures.

DOMAIN STRUCTURE

T antigen has so far defied all attempts to determine its structure directly by X-ray crystallography, despite serious efforts in several laboratories to crystallize the protein. Nevertheless, studies of T antigen peptides generated by partial proteolysis of the native protein indicate that T antigen is composed of several protease-resistant structural domains (102–105). The amino-terminal 130-amino-acid fragment, internal fragments containing approximately residues 131 to 371, and a carboxy-terminal fragment were identified. Early studies established that the extreme carboxy-terminus of T antigen was highly susceptible to proteolysis during extraction of the protein (24, 106). The protease-resistant fragments were correlated in several cases with biochemical activities of T antigen, including sequence-specific binding to the SV40 control region, ATPase, and binding to the p53 protein, allowing the tentative assignment of functional domains (Figure 2). Partial proteolysis of DNA-bound T antigen generated a novel peptide extending from approximately residue 140 to 281 that retained DNA-binding activity (104), providing support for a DNA-binding domain in this region of the protein. However, the difficulty of obtaining pure peptides from T antigen digests left these conclusions still open to question.

Genetic analysis has also been employed to identify functional domains of T antigen. Deletion analysis demonstrated clearly that the DNA-binding domain of T antigen must lie within the amino-terminal 272 amino acids (40, 107–109). Finer deletion analysis and expression of the peptides in bacteria confirmed that residues 131 to 259 or 131 to 246 were sufficient for sequence-specific binding to SV40 DNA sites I and II (40, 110–112) (Figure 2). The fact that a peptide comprising this DNA-binding domain accumulated as a

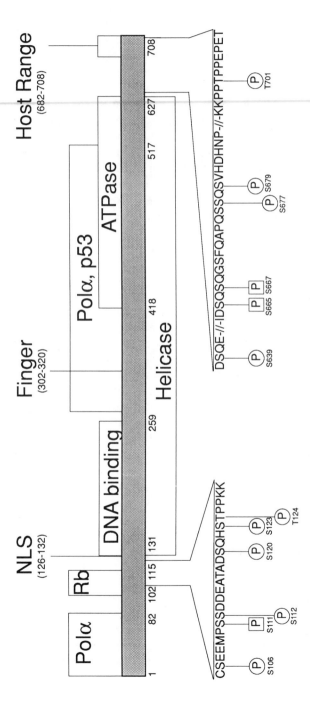

Figure 2   Functional domains of SV40 T antigen. The minimal regions of the protein that retain activity in binding to DNA polymerase α-primase (pol α), to the products of the retinoblastoma and p53 tumor suppressor genes (*Rb, p53*), and to the control region of SV40 DNA (DNA binding) are discussed in detail in the text. Cellular protein p107 also binds to the Rb region (252a). NLS, nuclear location signal; Zn finger, potential metal-binding region; host range, a function required in some cells for virion production but not viral DNA replication. In the two clusters of phosphorylation sites, phosphorylated residues found in T antigen from mammalian cells are indicated by a circled P; those found so far only in T antigen overexpressed in insect cells are indicated by a boxed P (174, 175).

stable active protein of the expected size in bacteria argues for the existence of a stably folded structural domain in this region of the full-length T antigen.

Numerous point mutations, small deletions, and insertions have been generated in attempts to characterize the DNA-binding domain further (113–116). Loss-of-function mutations obtained with this approach have often been interpreted to indicate that mutated residues are required as part of the DNA-binding domain. In most cases, this interpretation has in fact been confirmed by deletion analyses. However, in several cases, mutations outside the DNA-binding domain do cause loss of DNA-binding activity (40, 69, 107), indicating that results obtained using point mutations must be interpreted with caution. Moreover, the binding specificity of the minimal DNA-binding domain can be modulated by sequences outside this domain. For example, a peptide comprising residues 1–371 binds efficiently to site I but not to site II (112), and mutations at some phosphorylation sites also alter the binding-site specificity (117) (see below).

The ATPase domain of T antigen has been localized by several methods to the carboxy-terminal half of the protein (Figure 2). Covalent affinity labeling with ATP and with ATP analogues, coupled with examination of tryptic peptides, demonstrated only one ATP-binding site in T antigen, localized between residues 418 and 528 (78, 118, 119). A computer homology search indicated a potential nucleotide-binding fold in this region of T antigen (120). Partial proteolysis also indicates that the ATPase and helicase activity correlate with the integrity of a region between amino acid residues 131 and 708 (105). A point mutation at residue 522 causes loss of ATPase function (121). Amino acids beyond residue 627 may be deleted without loss of ATPase function (122–124). Finally, a monoclonal antibody, Pab 204, whose epitope maps between residues 453 to 469 of T antigen, specifically inhibits the ATPase and helicase activities (85, 108, 124). However, it has not yet been unequivocally demonstrated that the region between amino acids 418 to 627 is sufficient by itself for ATP binding and ATPase activity, or that the region between amino acids 131 to 627 is sufficient for helicase activity.

Several defined regions of T antigen interact specifically with cellular proteins. The retinoblastoma suppressor protein–binding domain was identified by loss of function in mutant T antigens (113, 114, 125), but has also been confirmed using a synthetic peptide comprising residues 102 to 115 (126), exchange of domains with adenovirus E1A protein (127), and by deletion analysis (22).

The p53-binding domain has been localized to residues overlapping the ATPase domain (Figure 2). A 53-bound proteolytic fragment comprising residues 131 to 517 was generated by partial digestion of T antigen–p53 complexes (129), and a bacterial fusion protein bearing T antigen residues 272 to 708 is sufficient to bind p53 in vitro (108). Point mutations at residue

402 reduce the p53-binding of T antigen without apparent loss of virus viability (130), suggesting that the p53-binding domain may be only a small region within residues 272 to 517.

DNA polymerase α-primase, a cellular DNA replication protein required for SV40 replication (11, 12), binds specifically to two regions of T antigen (Figure 2). Monoclonal antibodies against T antigen whose epitopes have been mapped within the ATPase domain block T antigen binding to the polymerase in crude cell extracts (131–134) and to purified polymerase-primase (135, 136). Moreover, p53 competes with polymerase α-primase for binding to T antigen (133, 137–139). In contrast, the 180-kDa subunit of purified DNA polymerase α-primase was shown to bind specifically to a T antigen fragment comprising residues 1 to 82, as well as to the full-length protein (135). Initially, these results appeared to be in disagreement, but more recent experiments have shown that purified polymerase can also bind to T antigen residues 83 to 708 (A. K. Arthur, I. Gilbert, I. Moarefi, M. K. Bradley, E. Fanning, unpublished data). Since T antigen residues 83 to 259 do not bind to the polymerase (unpublished), the data suggest the existence of two independent polymerase α-primase binding sites in T antigen (Figure 2).

An additional site, responsible for protein-protein interaction, has been identified, the zinc-finger motif at amino acids 302–320 (Figure 2). Point mutations at defined sites within the zinc-finger region affect viral viability (69) and sequence-specific DNA-binding activity (40, 112). As described above, these mutations also affect the hexamer formation required for effective T antigen–DNA interaction (70).

Two additional functionally important regions of T antigen have been identified, the nuclear location signal and the extreme carboxy-terminal region. The nuclear location signal consists of a short basic region from residue 126 to 132. This region is both necessary and sufficient to localize a normally cytoplasmic protein to the nucleus (140–142). Remarkably, a synthetic peptide containing this sequence is sufficient, when covalently cross-linked to any of several non-nuclear proteins, to direct them to the nucleus (143).

The carboxy-terminus of T antigen (residues 621 to 708) harbors a poorly understood function required for late SV40 gene expression in certain host cells in culture. The same region serves to overcome a block that prevents adenovirus multiplication in monkey cells. The carboxy-terminal region is therefore known as the *host range* or *adenovirus helper* functional domain of T antigen. This region is not required for either viral DNA replication or cell transformation activity of T antigen, although mutations in this region may affect these activities (117, 144–149). Remarkably, the host range function of this region was retained when it was expressed as part of the SV40 capsid protein VP1, demonstrating that this function is genetically independent of the other functions of T antigen (145, 148). However, the mechanism by which the host range domain facilitates expression of SV40 capsid proteins remains poorly understood (150, 151).

T ANTIGEN IN VIRAL DNA REPLICATION

SV40 has become a very useful model for the study of mammalian DNA replication. The SV40 genome is replicated by the host replication machinery in conjunction with a single viral protein, T antigen. Replication of the viral genome begins at the well-defined origin, and, like chromosomal DNA replication, proceeds in a bidirectional and semidiscontinuous manner. The usefulness of this simple system became obvious when a cell-free replication assay was developed. This assay has facilitated the identification and purification of several cellular proteins which, in combination with T antigen, are sufficient to reconstitute bidirectional, origin-dependent DNA replication (reviewed in 11, 12, 152, 153).

Initiation of SV40 DNA replication proceeds in discrete steps, each of which is characterized by defined biochemical parameters. As a first step, T antigen binds as a double hexamer to specific sequences within the origin and catalyzes the local unwinding of the SV40 genome. As described above, this reaction requires ATP, the activity of a DNA topoisomerase, and a single-strand-specific DNA binding (SSB) protein. In a biochemical unwinding assay, prokaryotic SSB proteins are sufficient to prevent the reannealing of the separated complementary strands. However, for the initiation of a replication cycle, a mammalian SSB protein, variously termed replication factor A (RF-A), replication protein A (RP-A), or human SSB protein, is necessary (154–157). The unwound origin, carrying T antigen and RP-A, has been described as the *preinitiation complex*.

In a second step, the preinitiation complex is converted to the *initiation complex* by recruitment of the DNA polymerase α-primase. This step is determined by specific protein-protein contacts between T antigen, RP-A, and DNA polymerase α-primase (132, 134, 135, 136, 158). The high specificity of these protein-protein contacts is exemplified by the fact that T antigen forms functional complexes only with the polymerase α from primate sources, and that polymerase α-primase preparations from other mammalian cells are not active in SV40 DNA replication (159, 160). The inability of rodent polymerase α-primase to associate properly with SV40 T antigen is thought to be responsible, at least in part, for the inability of rodent cells to support SV40 multiplication.

The most probable function of the specific contacts between T antigen and the polymerase α-primase is to position the enzyme in an orientation suitable for the ensuing DNA synthesis events. These events begin with the synthesis of short ribonucleotide primers, which are then extended into short DNA chains by the polymerase α activity (161–163).

After formation of the initiation complex, T antigen functions as a DNA helicase to unwind the two strands of the prefork parental DNA. According to biochemical studies, T antigen does so by translocating in the 3' to 5' direction along each parental strand, thus generating two potential replication forks (85, 89, 91, 164). During replicative chain elongation, T antigen probably remains in association with the DNA polymerase α-primase complex (165, 166).

A close association of a helicase and a primase function appears to be a characteristic feature of the design of prokaryotic replication machineries. Particularly clear examples are found in the bacteriophage T4 and the T7 replication systems (167, 168).

In fact, it is interesting to compare the known prokaryotic and the SV40 replication systems (12). For example, in *Escherichia coli* three polypeptides are required for the activation of the origin: (*a*) the DnaA-protein binds to and melts the origin DNA; (*b*) the DnaB-helicase, in conjunction with (*c*) the DnaC-protein, further unwinds the DNA strands as a prerequisite for the establishment of replication forks (169, 170). These functions are combined on one polypeptide chain in the SV40 system. The advantage of the viral system may be a higher rate of initiation events.

REGULATION BY PHOSPHORYLATION

SV40 T antigen is a phosphoprotein (24, 171). The phosphorylated residues, serine (Ser) and threonine (Thr), cluster in two regions of the primary sequence, one at the aminoterminus and the other at the carboxyterminus (25, 26, 171–175) (Figure 2).

The level of phosphorylation of T antigen from mammalian cells is not identical in all molecules; i.e. a minor subclass of underphosphorylated T antigen is observed in infected monkey cells, and to a lesser extent in transformed cells (177–180). An underphosphorylated subclass of T antigen, either isolated from infected cells or generated by partial enzymatic dephosphorylation, has a higher specific activity of sequence-specific DNA binding and of SV40 DNA replication activity (39, 47, 176, 178, 180–185). Chromatin-bound T antigen prepared from infected cells is also underphosphorylated and has a higher SV40 DNA-binding activity than nucleoplasmic or nuclear-matrix bound T antigen (180, 181, 186).

Which residues of T antigen down-regulate SV40 DNA binding activity and DNA replication activity when they are phosphorylated? Several approaches have been taken to address this question. Enzymatic dephosphorylation of T antigen with alkaline phosphatase enhances DNA-binding activity and DNA replication (176, 182–185). Alkaline phosphatase removes most of the phosphates from Ser, but not from Thr (176, 182, 184), suggesting that down-regulation results from modification on Ser. In contrast, treatment of T antigen with acid phosphatase to dephosphorylate Thr 124 as well as most Ser residues results in loss of sequence-specific binding activity (182), raising the question of whether the phosphorylation of Thr 124 might enhance DNA-binding activity, while phosphorylation of Ser, or perhaps certain Ser residues, diminishes DNA-binding activity. Better resolution of the Ser residues involved in down-regulation was obtained using the catalytic subunit of protein phosphatase 2A (PP2A) purified from human cell extracts as a stimulatory factor for initiation of SV40 DNA replication. Limited concentrations of PP2A dephosphorylated Ser 120 and Ser 123 preferentially and Ser 677 and Ser 679 to a lesser extent (187, 188). Dephosphorylation correlated well with a gain of T antigen binding to site II and replication activity.

PP2A may fulfill a similar role in the infected cell. Extracts prepared from cells in the G1 phase of the cell cycle support SV40 DNA replication in vitro only poorly in comparison with S phase extracts (187). Supplementation of G1 extracts with PP2A stimulates SV40 replication to a level comparable to that observed with S phase extracts (187). Thus, PP2A may be involved in activating SV40 replication in the S phase.

The enzymatic dephosphorylation studies agree well with the properties of a series of mutant T antigens carrying alanine (Ala) in place of Ser (117). The mutant at residue 679 replicated in vivo better than wild-type and the site II binding activity of mutant T antigen was greater than wild-type, suggesting that phosphorylation at Ser 679 contributes to down-regulation. Mutations at Ser 120 and 123 enhanced site II binding activity but, for unknown reasons, these mutants replicated poorly in vivo (117). Mutations at most other phosphorylation sites had little or no effect on the replication functions of T antigen.

There was, however, one important exception: the substitution of Thr 124 by Ala led to a complete loss of DNA replication activity in vivo and in vitro as well as to a severely defective site II DNA-binding activity in vitro (117). This result supports the view that phosphorylation of Thr 124 is required for DNA replication activity. This idea was tested directly in a cell-free SV40 replication assay by using bacterially expressed, i.e. unphosphorylated, T antigen. Treatment with the mammalian p34[cdc2] kinase specifically modified Thr 124 in vitro (44, 189), and the replication activity of phosphorylated T antigen was greatly enhanced (189), confirming that phosphorylation at Thr 124 is required for replication activity.

In summary, differential phosphorylation of T antigen at specific Ser and Thr residues controls SV40 DNA replication activity both positively and negatively. Much of this control is mediated by alteration of the site II DNA-binding activity, but additional more complex mechanisms, such as origin-unwinding, may also be involved (190). However, the phosphorylation pattern of T antigen does not appear to affect its ATPase or DNA helicase activities (117, 176, 182, 184).

What role does the differential phosphorylation of T antigen play in the infected cell, and which cellular enzymes regulate the phosphorylation state of T antigen? In early studies, the kinetics of phosphate turnover in T antigen were observed to follow two different rates, both more rapid than the turnover of T antigen protein (191). Subsequent studies demonstrated that the turnover of Thr phosphate was rather slow, whereas the turnover of Ser phosphate was somewhat faster on residues 112 and 639, and fastest on residues 120, 123, 677, and 679 (192, 193). The data are consistent with a role for differential phosphorylation in the regulation of T antigen function, but do not specify how the regulation is manifested. Several speculative models for control of viral DNA replication based on differential phosphorylation of T antigen have been presented (14, 44, 190).

The question of which protein kinases modify T antigen has been addressed primarily by treatment of un- or under-phosphorylated T antigen (expressed using bacterial vectors or recombinant baculovirus) with purified cellular kinases. A number of protein kinases were unable to phosphorylate T antigen, but casein kinase II phosphorylated Ser 106 and Ser 112 in vitro (194, 195). Purified casein kinase I modified Ser 123 and several carboxyterminal Ser residues (194). Thus, these enzymes are possible candidates for in vivo phosphorylation. As mentioned above, the mammalian p34^{cdc2} kinase specifically phosphorylates T antigen on Thr 124 (44, 189). The unusually slow phosphorylation reaction with T antigen as substrate compared to histone H1 justifies some concern about whether the p34^{cdc2} is involved in this reaction in vivo (196). One possible solution to this conundrum postulates an S-phase-specific form of the cdc2 kinase; indeed, a cdc2-related kinase called RF-S has been purified from S-phase human cells and shown to stimulate SV40 DNA replication in vitro, though it is not thought to act by modifying T antigen (197). Further work is required to resolve this question.

T antigen was recently found to be a good substrate for a novel protein kinase whose activity is stimulated by relaxed double-stranded DNA (DNA-PK) (175, 198, 199). The enzyme prefers to phosphorylate Ser or Thr followed by glutamine and usually preceded by an acidic residue (175, 199). This sequence or a related one occurs at several T antigen phosphorylation sites, and indeed DNA-PK can phosphorylate some of these sites in vitro (175). Thus, DNA-PK is another candidate for a T antigen kinase in vivo. It is tempting to speculate that DNA-PK may preferentially modify T antigen bound to relaxed replicating SV40 DNA, thereby down-regulating its replication activity and preventing it from initiating a new round of replication.

T ANTIGEN AND TRANSCRIPTIONAL REGULATION

T antigen regulates the transcription of the SV40 early and late genes. T antigen is also able to trans-activate a number of other viral and cellular promoters. One region of T antigen responsible for transcriptional activation resides in the amino-terminal hydrophilic domain of 85 amino acids, but the trans-activating function of this domain depends critically on its orientation with respect to the overall conformation of the protein (200).

SV40 Gene Transcription

The SV40 early promoter is composed of three important regions: a TATA box, which directs the site of transcription initiation, GC-box elements within the 21-bp-repeat region, serving as binding sites for the general transcription factor SP1, and upstream enhancer regions within the 72-bp repeats (Figure 1). Cellular factors necessary for the expression of the early promoter appear to be nearly ubiquitous, since the promoter is used in a wide variety of cells and in different mammalian species (reviewed in 201).

At low levels, T antigen appears to stimulate the transcription of SV40 early genes (202, 203), whereas at higher concentrations, T antigen acts to repress early mRNA synthesis. As mentioned in a previous section, in vitro assays show that T antigen, bound to its specific binding sites in the viral genomic control region, autoregulates its synthesis by sterically interfering with RNA polymerase binding or progression (59, 60, 204). In contrast, repression of early gene transcription in vivo appears to be only partially dependent on T antigen binding to DNA. Autoregulation of early gene expression in vivo may depend on an additional mechanism involving the specific interaction of T antigen with the enhancer-binding protein AP2. The binding of AP2 to a specific site in the early promoter/enhancer region is essential for the full expression of the early SV40 transcription unit, but T antigen–AP2 complexes are unable to bind to the enhancer site, and their formation causes a reduction of early gene transcription (205).

A major shift in the expression of SV40 genes occurs with the start of SV40 DNA replication in permissive cells. Early mRNAs are initiated at sites further upstream of the original transcription initiation site (Figure 1), and late mRNAs are synthesized at maximal rates.

The impact of DNA replication on late SV40 gene expression appears to be twofold: it increases the gene copy number, and activates the late promoter. This latter effect could be accomplished by changes in the organization or structure of the viral chromatin to increase its accessibility for essential transcription factors (206–208).

The late promoter does not contain a TATA-box element, which may explain why late mRNAs are initiated at heterogeneous transcriptional start sites (between SV40 nucleotides 120 and 325; see Figure 1). The late promoter depends for full activity on a complex combination of DNA elements, including parts of the GC-rich 21-bp-repeat region and parts of the 72-bp-repeat enhancer region (209–212).

T antigen has a decisive function in trans-activation of late genes, even in the absence of DNA replication. In fact, the late SV40 promoter, linked to a reporter gene, is efficiently activated in monkey CV1 cells or human HeLa cells when cotransfected with a T antigen–expressing plasmid (213). T antigen does not appear to act by binding to a promoter or enhancer element, as no T antigen was found in SV40 late transcription complexes (208), and T antigen mutants that cannot be transported to the nucleus activate late gene transcription (214). It has therefore been concluded that T antigen in the cytoplasm may modify the quantity or the activity of cellular factors that mediate the activation of the early and late SV40 genes. Indeed, even though the cellular factors have not yet been isolated, it was clearly shown that crude extracts from permissive cells contain a cellular protein whose DNA-binding characteristics are modified by T antigen in infected cells (215). This factor binds to an essential late promoter element, a specific site within the Oct/SPH region of the SV40 72-bp-repeat enhancer (216). The factor has the binding properties of a known cellular transcription protein, TEF-1 (217).

It will be very interesting to elucidate the biochemical mechanism by which T antigen affects the activity of this transcription factor, as this may reveal an as-yet-unrecognized, additional function of T antigen.

Trans-activation of Cellular Genes

T antigen stimulates the expression of not only a variety of cellular protein-encoding genes, but also genes encoding the large rRNA species as well as genes for some small RNAs (218–223, 223a). In addition, T antigen positively regulates some viral promoters such as the long terminal repeat promoter of Rous Sarcoma Virus and the E3 promoter of adenovirus (213).

In most cases, no T antigen–binding site was detected in the promoter-enhancer region of these genes, excluding a direct effect of T antigen in promoter-activation. However, there may be exceptions, and at least one transcription unit has been identified where a T antigen–binding site could be part of an enhancer element (224, 225). Indeed, the role of T antigen in the stimulation of cellular gene activity is probably complex and must not be identical for all genes.

Recently, a striking increase in the amount of the general transcription factor SP1 was detected in SV40-infected cells (226), and that may at least partially account for the activation of genes containing binding sites for SP1 in their promoters. But, in general, it is assumed that trans-activation by T antigen occurs through modification of the amounts or the activities of enhancer-binding proteins as described above for the activation of the late SV40 promoter. However, the nature of these transcription factors is still unknown. A possible exception may be the transcription factor TFIIIC, which is present in higher amounts in some SV40-transformed murine cells (227). An activation of this factor is responsible for the increased transcription by RNA polymerase III of the repetitive B1 and B2 elements in transformed murine cells (221, 228).

In addition to activating transcription factors, T antigen may regulate transcription by attenuation (229). Furthermore, T antigen may be involved in the maturation or translation of mRNA. This latter possibility is suggested by the finding that a significant fraction of T antigen is found in association with ribonucleoprotein particles (230). In fact, after extraction, some T antigen is closely associated with RNA fragments, of which a considerable fraction possesses a snap-back hairpin structure (231, 232). It has been suggested, moreover, that the major function of the T antigen–associated RNA helicase activity is somehow involved in translational regulation by resolving double-stranded regions in certain mRNA species (99).

Clearly, much has yet to be learned about the mechanisms by which T antigen affects cellular gene expression. In fact, T antigen not only activates gene expression, but also is able to down-regulate the activity of some genes, perhaps by binding and inactivating the transcription factor AP2 (205). A more recent example of down-regulation of gene expression reveals that T antigen binds to and inactivates the trans-acting protein pX of hepatitis B virus; conversely, the pX protein inhibits the trans-activating function of T antigen (233). Additional mechanisms for down-regulation of cellular gene expression by T antigen may also exist.

T ANTIGEN IN TRANSFORMATION

It is well established that T antigen can change the growth properties of cultured rodent cells (1). SV40 DNA does not replicate in nonpermissive rodent cells, and the viral DNA may integrate into the host genome. If the early coding region remains intact after integration, expression of T antigen can take place, which is essential and sufficient to initiate the change of the cellular phenotype. Typically, following virus infection or transfection of SV40 DNA, dense foci of morphologically altered cells appear on monolayers of unaltered cells. When transferred to semisolid media (0.3% agarose or 1.3% methylcellulose), cells derived from such foci are frequently able to grow in an anchorage-independent manner. These cells are said to be fully transformed if they are able to form tumors after injection in susceptible animals (reviewed in 234).

The analysis of specific T antigen mutants has so far not led to a clear picture of whether the same functions of T antigen are required for the immortalization of primary cells and the transformation of established cells. For lack of space, it is not possible to review these data (235–243, 243a).

The fraction of cells converted to a transformed state is rather low after infection with virus particles or after transfection of protein-free SV40 DNA. This problem was overcome by using recombinant retrovirus vectors, carrying the SV40 early coding region. Because of the high frequency of integration of these retroviral recombinants, all infected cells receive an integrated T antigen gene and, most significantly, all infected cells become transformed, providing strong evidence that the presence of intact T antigen suffices to transform primary and established cells fully (244–247).

The same conclusion can be drawn from an entirely different approach involving the generation of transgenic mice carrying the early SV40 coding region in all cells (247a). In combination with its natural promoter-enhancer, the T antigen–coding gene is expressed in the cells of the choroid plexus of the brain. The T antigen–expressing cells become hyperplastic, and some of them develop into tumors (248). Moreover, the expression of T antigen can be targeted to certain cell types by linking the early SV40 coding region to the promoter-enhancer of tissue-specific genes. For example, the 5' flanking region of the insulin gene leads to an expression of T antigen in the β cells of the pancreas, whereas the 5' region of the elastase gene leads to its expression in the acinar cells of the pancreas (249, 250). T antigen expression in these different cell types causes cellular dysplasia and increased cell growth, and some of these cells develop tumors that eventually kill the host animal. These experiments leave little doubt that it is the expression of T antigen that induces abnormal growth properties, not only in cultured cells but also in intact animals.

Analyses of a number of T antigen mutants, both deletion and nucleotide-exchange mutants, have clearly shown that the specific DNA-binding properties of T antigen are not important for transformation (114, 121). In fact, two regions outside of the DNA-binding domain (see Figure 2) are essential for transformation.

Most of the amino-terminal 130 amino acids are required for the transformation of virtually all rodent cell lines examined. This portion of the protein is sufficient to transform some selected cell lines (for example, the C3H10T1/2 cell line). It also cooperates with the Ha-*ras* oncogene product for full transformation of other cells. However, in the absence of the Ha-*ras* product, and for the transformation of most other primary and established cells, an additional, carboxy-terminal portion of T antigen (roughly localized between amino acid residues 400 and 600) is necessary (251, 252).

Each of the two functional domains contains a binding site for one or more cellular proteins (see Figure 2). The amino-terminal domain specifically interacts with the product of the retinoblastoma susceptibility (RB) gene and a related cellular protein known as p107 (252a), and the central domain interacts with the cellular p53 protein. It is widely believed that these protein-protein interactions play decisive roles in the transformation process, because most mutants of T antigen that are unable to form a complex with one or the other of these proteins are at least partially defective for transformation.

Interaction with the Product of the RB Gene

A region within the amino-terminal part of T antigen, amino acid residues 105 to 115, is of particular interest. The integrity of this region is required for full transformation (113, 126, 253). Interestingly, this region shares considerable primary and secondary structure similarities with a region in the E1A oncogene product of transforming adenoviruses. As in corresponding mutants of T antigen (126), mutations in the conserved E1A region impair the ability of its product to cooperate with the *ras* oncogene product for transformation (254, 255); and a region including residues 108–118 of T antigen can functionally substitute for the corresponding E1A region in a chimeric protein (127). Most importantly, both the 105-to-115 region of T antigen and the homologous part of the E1A protein are required for the specific association of these proteins with a cellular protein, $p110^{Rb}$, the product of the RB gene (125, 126, 256).

The RB gene is considered to be the prototype of a class of tumor suppressor genes (257). Inactivation of the RB gene has been found not only in retinoblastomas but also in a variety of other more common human carcinomas. The RB gene, introduced into cultured tumor cells, causes slower growth rates and the loss of tumorigenicity in nude mice, suggesting that the RB gene product can influence neoplastic properties, and that its inactivation plays an essential role in transformation and oncogenesis (reviewed in 258; 259, 260). The RB gene product is a phosphoprotein that is un- or under-phosphorylated in stationary and early G1 phase cells and becomes hyper-phosphorylated later during the progression through the cell cycle (261–263, 263a). Thus, either hyperphosphorylation or binding to T antigen (or the adenovirus E1A protein) appears to neutralize the growth-arresting properties of $p110^{Rb}$.

It has been proposed recently that $p110^{Rb}$ may be (part of) an inhibitor of the transcription factor E2F (also known as DRTF). Experiments suggest that this factor, when associated with $p110^{Rb}$, is inactive, and that either T antigen or the E1A protein dissociates the complex by binding to $p110^{Rb}$ and releasing an active form of the transcription factor (264–266).

Binding sites for E2F are present in the promoter-enhancer regions of genes known to be involved in cell-cycle regulation, such as *fos* and *myc,* or of genes encoding functions required for DNA synthesis (267, 268). These observations suggest a relatively simple model: growth-regulating genes are silent in resting cells because E2F (and, possibly, related transcription factors) are inhibited by bound p110^{Rb} alone or as a multiprotein complex; T antigen dissociates this complex by sequestering p110^{Rb}, thereby activating transcription factor E2F, which then turns on growth-promoting genes as a prerequisite for cellular transformation.

We note that the amino-terminal segment of T antigen is also able to induce cellular DNA synthesis in resting cells (220). It would therefore not be surprising if T antigen could induce cellular DNA synthesis by activating a pathway that starts with the dissociation of a complex between p110^{Rb} and a transcription factor. On the other hand, it has been clearly shown that the T antigen–mediated activation of the SV40 late promoter or of the retroviral long terminal repeat promoter (see above) does not depend on T antigen–p110^{Rb} dissociation (200), suggesting that not all genes in SV40-infected cells are activated by the same mechanism.

Interaction with the p53 Protein

The p53 phosphoprotein was the first cellular protein found to interact specifically with T antigen in infected and in transformed cells. Originally, the *p53* gene was classified as an oncogene because its product was found to immortalize primary cells in culture and to cooperate with the *ras* oncogene product to transform primary cells to tumorigenicity. However, it was shown later that only mutated *p53* was active as an oncogene. In fact, wild-type *p53* has no transforming activity, and, in sufficient quantities, it can even suppress the ability of other oncogenes to transform cells. And, when introduced into tumor cells, *p53* can inhibit uncontrolled proliferation. The significance of these experimental results became obvious when it was found that in a large number of human cancers, the *p53* gene is either lost or altered by mutation. Thus, the wild-type *p53* gene must now be classified as a tumor-suppressor gene that can be converted into an oncogene by mutation (reviews in 269–273).

These observations suggest a mechanism for the transforming function of T antigen. By forming specific protein-protein contacts with p53, T antigen could neutralize the growth-suppressing activity of p53, leading to unregulated DNA replication and cell division.

Unfortunately, at the time of writing this review, it is not possible to describe clearly the function of p53 in more biochemical terms. The effects of bound p53 on the functions of T antigen may give important clues. Bound p53 from permissive monkey cells stimulates the ATPase and helicase functions of T antigen (274, 275). Significantly, however, bound wild-type, but not mutant, p53 from nonpermissive murine cells blocks the replicative activity of T antigen in vivo and in vitro (137, 139, 276). Murine p53 inhibits the T antigen–associated helicase activity (139, 277), and competes with DNA polymerase α-primase for the association with T antigen, thus preventing an important step in the formation of replication initiation complexes (133). Moreover, p53 binds to specific sites within the promoter-enhancer region, close to the viral origin of replication (278), possibly also interfering with the formation of a productive replication initiation complex.

p53 is a DNA-binding protein with high affinity for specific cellular DNA sequences that could well be elements of cellular replication start sites (279). It has therefore been hypothesized that p53 prevents the assembly of active replication complexes at cellular origins as it does at the viral origin (279). However, p53 has also been shown to interact with the $p34^{cdc2}$ cell-cycle-regulating kinase (280) and to possess a strong trans-activating activity (281–283). These activities could also be somehow required for the function of p53 as an inhibitor of uncontrolled cell proliferation. Whatever the precise biochemical functions of p53 may be, it is probably safe to conclude that T antigen, by interacting with p53, releases a brake imposed by p53 on the proliferative potential of cells.

Tumorigenicity

As described in the preceding sections, T antigen is able to drive cells into uncontrolled proliferation by inhibiting at least two known growth-arrest functions. How is this related to tumorigenicity? We may recall that, in transgenic mice, most cells expressing T antigen are stimulated to hyperplastic growth, but, in all cases, only a few cells are transformed into tumor cells, and tumors do not develop until the mice are 3–7 months of age. This implies that T antigen is necessary, but not sufficient for tumorigenesis.

Additional events are required, but their nature is presently not known. It may be important, though, that T antigen causes a large variety of chromosomal aberrations in transformed cultured cells (284, 285). Indeed, T antigen is known to induce recombinational events in transformed and transfected cells (286, 287, 287a). One possible explanation for these reactions is that T antigen molecules bound to physically separated regions of double-stranded DNA can form protein-protein contacts, leading to DNA loops when T antigen resides on two sites of the same DNA strand, or to the joining of different DNA strands when T antigen is bound to two separated DNA molecules (288). The association of distinct DNA sections may facilitate recombination processes as the basis for the observed chromosomal aberrations. But the consequences of chromosomal changes for the pathway to full tumorigenicity remain to be explored.

CONCLUSION

It appears to be unlikely that T antigen has evolved to transform nonpermissive cells to immortalization and tumorigenicity, because this is a dead end with respect to virus multiplication. It is more likely that these functions serve special and important purposes in the normal infection cycle in permissive monkey cells.

SV40 depends on the host cell machinery for the replication of its genome, and essential components of the replication apparatus may not be present in sufficient amounts in resting cells. With the inactivation of negative growth-regulating proteins, such as $p110^{Rb}$ and p53, T antigen forces the infected cell to express its growth-regulated genes, including genes that encode enzymes required for nucleotide metabolism and for DNA synthesis. Thus, T antigen prepares the intracellular milieu for the replication of viral DNA.

The broader significance of the process, outlined above, is illustrated by the fact that other DNA tumor viruses have evolved very similar mechanisms. We have already mentioned that the E1A proteins of transforming adenoviruses are able to bind to (256) and to neutralize the function of the $p110^{Rb}$ protein (264–266); a second adenovirus protein, E1B, binds to and probably inactivates the p53 protein (289). The highly transforming human papilloma viruses types 16 and 18 also encode two proteins that bind to these cellular proteins. The papilloma virus E6 protein binds to p53 (290) and appears to promote its proteolytic breakdown (291); the E7 protein interacts with $p110^{Rb}$ (292). Thus, the SV40-encoded T antigen combines the binding sites for $p110^{Rb}$ and for p53 on one polypeptide chain, whereas the other two DNA viruses encode two proteins, binding to one or the other product of a cellular growth suppressor gene. But, in principle, all three DNA viruses have evolved to use very similar strategies for their propagation.

ACKNOWLEDGMENTS

We thank the following for communicating results prior to publication: C. W. Anderson, M. Bradley, C. Cole, W. Deppert, T. J. Kelly, and H. Stahl. We thank I. Moarefi for designing Figure 2, and S. Dehde, E. Rohrer, and S. Karstenmüller for their help with the preparation of the manuscript.

Literature Cited

1. Tooze, J., ed. 1980. *Molecular Biology of Tumor Viruses*, Part 2. *DNA Tumor Viruses*. Cold Spring Harbor, NY: Cold Spring Harbor Lab.
2. Rigby, P. W. J., Lane, D. 1983. *Adv. Viral Oncol.* 3:31–57
3. Levine, A. J. 1989. In *Molecular Biology of Chromosome Function*, ed. K. W. Adolph, pp. 71–96. New York: Springer-Verlag
4. Butel, J. S., Jarvis, D. L. 1986. *Biochim. Biophys. Acta* 865:171–95
5. De Pamphilis, M. L., Bradley, M. K. 1986. In *The Papovaviridiae*, Vol. 1,

The Polyoma Viruses, ed. N. Salzman, pp. 99–246. New York: Plenum
6. Fried, M., Prives, C. 1986. *Cancer Cells* 4:1–16
7. Monier, R. 1986. See Ref. 5, pp. 247–94
8. Butel, J. S. 1986. *Cancer Surv.* 5:343–65
9. Stahl, H., Knippers, R. 1987. *Biochim. Biophys. Acta* 910:1–10
10. Livingston, D. M., Bradley, M. K. 1987. *Mol. Biol. Med.* 4:63–80
11. Challberg, M. D., Kelly, T. J. 1989. *Annu. Rev. Biochem.* 58:671–717

12. Stillman, B. 1989. *Annu. Rev. Cell Biol.* 5:197–245
13. Borowiec, J. A., Dean, F. B., Bullock, P. A., Hurwitz, J. 1990. *Cell* 60:181–84
14. Prives, C. 1990. *Cell* 61:735–38
15. Fiers, W., Contreras, R., Haegeman, G., Rogiers, R., van der Voorde, A., et al. 1978. *Nature* 273:113–20
16. Reddy, V. B., Thimmappaya, B., Dhar, R., Subramanian, K. N., Zain, B. S., et al. 1978. *Science* 200:494–502
17. Varshavsky, A. J., Sundin, O. M., Bohn, M. 1979. *Cell* 16:453–66
18. Jakobovits, E. G., Bratosin, S., Aloni, Y. 1980. *Nature* 285:263–65
19. Sogo, J. M., Stahl, H., Koller, T., Knippers, R. 1986. *J. Mol. Biol.* 189:189–204
20. Pallas, D. C., Shahrik, L. K., Martin, B. L., Jaspers, S., Miller, T. B., et al. 1990. *Cell* 60:167–76
20a. Walter, I. G., Ruediger, R., Slaughter, C., Mumby, M. 1990. *Proc. Natl. Acad. Sci. USA* 87:2521–25
21. Yang, S. I., Lickteig, R. L., Estes, R., Rundell, K., Walter, G., Mumby, M. C. 1991. *Mol. Cell. Biol.* 11:1988–95
21a. Scheidtmann, K. H., Mumby, M. C., Rundell, K., Walter, G. 1991. *Mol. Cell. Biol.* 11:1996–2003
22. Montano, X., Millikan, R., Milhaven, J., Newsome, D., Ludlow, J. W., et al. 1990. *Proc. Natl. Acad. Sci. USA* 87:7448–52
23. Tegtmeyer, P. 1980. See Ref. 1, pp. 297–337
24. Tegtmeyer, P., Rundell, K., Collins, J. K. 1977. *J. Virol.* 21:647–57
25. Scheidtmann, K. H., Kaiser, A., Carbone, A., Walter, G. 1981. *J. Virol.* 38:59–69
26. Scheidtmann, K. H., Echle, B., Walter, G. 1982. *J. Virol.* 44:116–33
27. Paucha, E., Mellor, A., Harvey, R., Smith, A. E., Hewick, R., Waterfield, M. D. 1978. *Proc. Natl. Acad. Sci. USA* 75:2165–69
28. Jarvis, D. L., Butel, J. S. 1985. *Virology* 141:173–89
29. Klockmann, U., Deppert, W. 1983. *EMBO J.* 2:1151–57
30. Bradley, M. K., Hudson, J., Villanueva, M. S., Livingston, D. M. 1984. *Proc. Natl. Acad. Sci. USA* 81:6574–78
31. Goldman, N., Brown, M., Khoury, G. 1981. *Cell* 24:567–72
32. Carroll, R. B., Samad, A., Mann, A., Harper, J., Anderson, C. W. 1988. *Oncogene* 2:437–44
33. Staufenbiel, M., Deppert, W. 1983. *Cell* 33:173–81
34. Walser, A., Rinke, Y., Deppert, W. 1989. *J. Virol.* 63:3926–33
35. Rinke, Y., Deppert, W. 1989. *Virology* 170:424–32
36. Simanis, V., Lane, D. P. 1985. *Virology* 144:776–85
37. Dixon, R. A. F., Nathans, D. 1985. *J. Virol.* 53:1001–4
38. Stillman, B. W., Gluzman, Y. 1985. *Mol. Cell. Biol.* 5:2051–60
39. Mohr, I. J., Gluzman, Y., Fairman, M. P., Strauss, M., McVey, D., et al. 1989. *Proc. Natl. Acad. Sci. USA* 86:6479–83
40. Arthur, A. K., Höss, A., Fanning, E. 1988. *J. Virol.* 62:1999–2006
41. Lanford, R. E. 1988. *Virology* 167:72–81
42. Murphy, C. I., Weiner, B., Bikel, I., Piwnica-Worms, H., Bradley, M. K., Livingston, D. M. 1988. *J. Virol.* 62:2951–59
43. O'Reilly, D., Miller, L. K. 1988. *J. Virol.* 62:3109–19
44. Höss, A., Moarefi, I., Scheidtmann, K. H., Cisek, L. J., Corden, J. L., et al. 1990. *J. Virol.* 64:4799–807
45. Spillman, T., Giacherio, D., Hager, L. P. 1979. *J. Biol. Chem.* 254:3100–4
46. Oren, M., Winocour, E., Prives, C. 1980. *Proc. Natl. Acad. Sci. USA* 77:220–24
47. Dorn, A., Brauer, D., Otto, B., Fanning, E., Knippers, R. 1982. *Eur. J. Biochem.* 128:53–62
48. Tjian, R. 1978. *Cell* 13:165–79
49. Shalloway, D., Kleinberger, T., Livingston, D. M. 1980. *Cell* 20:411–22
50. DeLucia, A. L., Lewton, B. A., Tjian, R., Tegtmeyer, P. 1983. *J. Virol.* 46:143–50
51. Wright, P. J., DeLucia, A. L., Tegtmeyer, P. 1984. *Mol. Cell. Biol.* 12:2631–38
52. Jones, K. A., Tjian, R. 1984. *Cell* 36:155–62
53. Pollwein, P., Wagner, S., Knippers, R. 1987. *Nucleic Acids Res.* 15:9741–59
54. Gruss, C., Wetzel, E., Baack, M., Mock, U., Knippers, R. 1988. *Virology* 167:349–60
55. Ryder, K., Vakalopoulou, E., Mertz, R., Mastrangelo, I., Hough, P., et al. 1985. *Cell* 42:539–48
56. Mastrangelo, I. A., Hough, P. V. C., Wilson, V. G., Wall, J. S., Hainfeld, J. E., Tegtmeyer, P. 1985. *Proc. Natl. Acad. Sci. USA* 82:3626–30
57. Ryder, K., Silver, S., DeLucia, A. L., Fanning, E., Tegtmeyer, P. 1986. *Cell* 44:719–25
58. DiMaio, D., Nathans, D. 1982. *J. Mol. Biol.* 156:531–48
59. Hansen, U., Tenen, D. G., Livingston,

D. M., Sharp, P. A. 1981. *Cell* 27:603–12
60. Rio, D., Robbins, A., Myers, R., Tjian, R. 1980. *Proc. Natl. Acad. Sci. USA* 77:5706–10
61. Deb, S., DeLucia, A. L., Baur, C. P., Koff, A., Tegtmeyer, P. 1986. *Mol. Cell. Biol.* 6:1663–70
62. Li, J. J., Peden, K. W. C., Dixon, R. A. F., Kelly, T. 1986. *Mol. Cell. Biol.* 6:1117–28
63. Deb, S., Tsui, S., Koff, A., DeLucia, A. L., Parsons, R., Tegtmeyer, P. 1987. *J. Virol.* 61:2143–49
64. Dean, F. B., Borowiec, J. A., Ishimi, Y., Deb, S., Tegtmeyer, P., Hurwitz, J. 1987. *Proc. Natl. Acad. Sci. USA* 84:8267–71
65. Borowiec, J. A., Hurwitz, J. 1988. *Proc. Natl. Acad. Sci. USA* 85:64–68
66. Deb, S. P., Tegtmeyer, P. 1987. *J. Virol.* 61:3649–54
67. Dean, F. B., Dodson, M., Echols, H., Hurwitz, J. 1987. *Proc. Natl. Acad. Sci. USA* 84:8981–85
68. Mastrangelo, I. A., Hough, P. V. C., Wall, J. S., Dodson, M., Dean, F. B., Hurwitz, J. 1989. *Nature* 338:658–62
69. Loeber, G., Parsons, R., Tegtmeyer, P. 1989. *J. Virol.* 63:94–100
70. Loeber, G., Stenger, J. E., Ray, S., Parsons, R. E., Anderson, M. E., Tegtmeyer, P. 1991. *J. Virol.* 65:3167–74
71. Parsons, R. E., Stenger, J. E., Ray, S., Welker, R., Anderson, M. E., Tegtmeyer, P. 1991. *J. Virol.* 65:2798–806
72. Borowiec, J. A., Hurwitz, J. 1988. *EMBO J.* 7:3149–58
73. Parsons, R., Anderson, M. E., Tegtmeyer, P. 1990. *J. Virol.* 64:509–18
74. Roberts, J. M. 1989. *Proc. Natl. Acad. Sci. USA* 86:3939–43
75. Borowiec, J. A., Dean, F. B., Hurwitz, J. 1991. *J. Virol.* 65:1228–35
76. Echols, H. 1989. *Science* 233:1050–56
77. Bramhill, D., Kornberg, A. 1988. *Cell* 54:915–19
78. Bradley, M. K. 1990. *J. Virol.* 64:4939–47
79. Giacherio, D., Hager, L. P. 1979. *J. Biol. Chem.* 254:8113–20
80. Giacherio, D., Hager, L. P. 1980. *J. Biol. Chem.* 255:8963–66
81. Myers, R. M., Kligman, M., Tjian, R. 1981. *J. Biol. Chem.* 256:10156–60
82. Stahl, H., Knippers, R. 1983. *J. Virol.* 47:65–76
83. Tack, L. C., Proctor, G. N. 1987. *J. Biol. Chem.* 262:6339–49
84. Stahl, H., Dröge, P., Zentgraf, H., Knippers, R. 1985. *J. Virol.* 54:473–82
85. Wiekowski, M., Dröge, P., Stahl, H. 1987. *J. Virol.* 61:411–18

86. Dröge, P., Sogo, J. M., Stahl, H. 1985. *EMBO J.* 4:3241–46
87. Stahl, H., Dröge, P., Knippers, R. 1986. *EMBO J.* 5:1939–44
88. Wessel, R., Schweizer, J., Stahl, H. 1992. *J. Virol.* 62:In press
89. Wiekowski, M., Schwarz, M. W., Stahl, H. 1988. *J. Biol. Chem.* 263:436–42
90. Stahl, H., Scheffner, M., Wiekowski, M., Knippers, R. 1988. *Cancer Cells* 6:105–12
91. Goetz, G. S., Dean, F. B., Hurwitz, J., Matson, S. W. 1988. *J. Biol. Chem.* 263:383–92
92. Matson, S. W., Kaiser-Rogers, K. A. 1990. *Annu. Rev. Biochem.* 59:289–329
93. Wold, M. S., Li, J. J., Kelly, T. J. 1987. *Proc. Natl. Acad. Sci. USA* 84:3643–47
94. Dean, F. B., Bullock, P., Murakami, Y., Wobbe, C. R., Weissbach, L., Hurwitz, J. 1987. *Proc. Natl. Acad. Sci. USA* 84:16–20
95. Dodson, M., Dean, F. B., Bullock, P., Echols, H., Hurwitz, J. 1987. *Science* 238:964–67
96. Parsons, R., Anderson, M. E., Tegtmeyer, P. 1990. *J. Virol.* 64:509–18
97. Auborn, K. J., Markowitz, R. B., Wang, E., Yu, Y. T., Prives, C. 1988. *J. Virol.* 63:912–18
98. Scheffner, M., Wessel, R., Stahl, H. 1989. *Nucleic Acids Res.* 17:93–106
99. Scheffner, M., Knippers, R., Stahl, H. 1989. *Cell* 57:955–63
100. Scheffner, M., Knippers, R., Stahl, H. 1991. *Eur. J. Biochem.* 195:49–54
101. Wassarman, D. A., Steitz, J. A. 1991. *Nature* 349:463–64
102. Schwyzer, M., Weil, R., Frank, G., Zuber, H. 1980. *J. Biol. Chem.* 255:5627–34
103. Simmons, D. T. 1986. *J. Virol.* 57:776–85
104. Simmons, D. T. 1988. *Proc. Natl. Acad. Sci. USA* 85:2086–90
105. Wun-Kim, K., Simmons, D. T. 1990. *J. Virol.* 64:2014–20
106. Smith, A. E., Smith, R., Paucha, E. 1978. *J. Virol.* 28:140–53
107. Clark, R., Peden, K., Pipas, J., Nathans, D., Tjian, R. 1983. *Mol. Cell. Biol.* 3:220–28
108. Mole, S. E., Gannon, J. V., Ford, J. J., Lane, D. P. 1987. *Philos. Trans. R. Soc. London Ser. B* 317:455–69
109. Strauss, M., Argani, P., Mohr, I. J., Gluzman, Y. 1987. *J. Virol.* 61:3326–30
110. McVey, D., Strauss, M., Gluzman, Y. 1989. *Mol. Cell. Biol.* 9:5525–36
111. Fanning, E., Schneider, J., Arthur, A.,

K., Höss, A., Moarefi, I., Modrow, S. 1989. *Curr. Top. Microbiol. Immunol.* 144:9–19

112. Höss, A., Moarefi, I. F., Fanning, E., Arthur, A. K. 1990. *J. Virol.* 64:6291–96

113. Kalderon, D., Smith, A. E. 1984. *Virology* 139:109–37

114. Paucha, E., Kalderon, D., Harvey, R., Smith, A. E. 1986. *J. Virol.* 57:50–64

115. Simmons, D. T., Loeber, G., Tegtmeyer, P. 1990. *J. Virol.* 64:1973–83

116. Simmons, D. T., Wun-Kim, K., Young, W. 1990. *J. Virol.* 64:4858–65

117. Schneider, J., Fanning, E. 1988. *J. Virol.* 62:1598–605

118. Clertant, P., Cuzin, F. 1982. *J. Biol. Chem.* 257:6300–05

119. Clertant, P., Gaudray, P., May, E., Cuzin, F. 1984. *J. Biol. Chem.* 259: 15196–203

120. Bradley, M. K., Smith, T. F., Lathrop, R. H., Livingston, D. M., Webster, T. A. 1987. *Proc. Natl. Acad. Sci. USA* 84:4026–30

121. Manos, M., Gluzman, Y. 1985. *J. Virol.* 52:120–27

122. Cole, C. N., Tornow, J., Clark, R., Tjian, R. 1986. *J. Virol.* 57:539–46

123. Pipas, J. M., Peden, K. W. C., Nathans, D. 1983. *Mol. Cell. Biol.* 3:204–13

124. Clark, R., Lane, D. P., Tijan, R. 1981. *J. Biol. Chem.* 256:11854–58

125. DeCaprio, J. A., Ludlow, J. W., Figge, J., Shaw, J.-Y., Huang, C.-M., et al. 1988. *Cell* 54:275–83

126. DeCaprio, J. A., Ludlow, J. W., Lynch, D., Furukawa, Y., Griffin, J., et al. 1989. *Cell* 58:1085–89

127. Moran, E. 1988. *Nature* 334:168–70

128. Deleted in proof

129. Schmieg, F. I., Simmons, D. 1988. *Virology* 164:132–40

130. Lin, J.-Y., Simmons, D. T. 1991. *J. Virol.* 65:2066–72

131. Smale, S. T., Tjian, R. 1986. *Mol. Cell. Biol.* 6:4077–87

132. Gough, G., Gannon, J. V., Lane, D. P. 1988. *Cancer Cells* 6:153–58

133. Gannon, J. V., Lane, D. P. 1987. *Nature* 329:456–58

134. Gannon, J. V., Lane, D. P. 1990. *New Biol.* 2:84–92

135. Dornreiter, I., Höss, A., Arthur, A. K., Fanning, E. 1990. *EMBO J.* 9:3329–36

136. Collins, K. L., Kelly, T. J. 1991. *Mol. Cell. Biol.* 11:2108–15

137. Braithwaite, A. W., Stürzbecher, H., Addison, C., Palmer, C., Rudge, K., Jenkins, J. R. 1987. *Nature* 329:458–60

138. Stürzbecher, H. W., Brain, R.,

Maimets, T., Addison, C., Rudge, K., Jenkins, J. R. 1988. *Oncogene* 3:405–13

139. Wang, E. H., Friedman, P. N., Prives, C. 1989. *Cell* 57:379–92

140. Lanford, R. E., Butel, J. S. 1984. *Cell* 37:801–13

141. Kalderon, D., Richardson, W. D., Markham, A. F., Smith, A. E. 1984. *Nature* 311:33–38

142. Kalderon, D., Roberts, B. L., Richardson, W. D., Smith, A. E. 1984. *Cell* 39:499–509

143. Lanford, R. E., Kanda, P., Kennedy, R. C. 1986. *Cell* 46:575–82

144. Lewis, E. D., Chen, S., Kumar, A., Blanck, G., Pollack, R. E., Manley, J. L. 1983. *Proc. Natl. Acad. Sci. USA* 80:7065–69

145. Tornow, J., Polvino-Bodnar, M., Santangelo, G., Cole, C. N. 1985. *J. Virol.* 53:415–24

146. Pipas, J. M. 1985. *J. Virol.* 54:569–75

147. Schneider, J., Schindewolf, C., van Zee, K., Fanning, E. 1988. *Cell* 54:117–25

148. Tornow, J., Cole, C. N. 1983. *Proc. Natl. Acad. Sci. USA* 80:6312–16

149. Cole, C. N., Tornow, J., Clark, R., Tjian, R. 1986. *J. Virol.* 57:539–46

150. Khalili, K., Brasy, J., Pipas, J. M., Spence, S. L., Sadofsky, M., Khoury, G. 1988. *Proc. Natl. Acad. Sci. USA* 85:354–58

151. Stacy, T., Chamberlain, M., Cole, C. N. 1989. *J. Virol.* 63:5208–15

152. Kelly, T. J., Wold, M. S., Li, J. J. 1988. *Adv. Virus Res.* 34:1–42

153. Hurwitz, J., Dean, F. B., Kwong, A. D., Lee, S. H. 1990. *J. Biol. Chem.* 265:18043–46

154. Fairman, M. P., Stillman, B. 1988. *EMBO J.* 7:1211–18

155. Wold, M. S., Weinberg, D. H., Virshup, D. M., Li, J. J., Kelly, T. J. 1989. *J. Biol. Chem.* 264:2801–9

156. Bullock, P. A., Seo, Y. S., Hurwitz, J. 1989. *Proc. Natl. Acad. Sci. USA* 86:3944–48

157. Kenny, M. K., Lee, S. H., Hurwitz, J. 1989. *Proc. Natl. Acad. Sci. USA* 86:9757–61

158. Dornreiter, I., Erdile, L. F., Gilbert, I. U., von Winkler, D., Kelly, T. J., Fanning, E. 1992. *EMBO J.* 11:In press

159. Murakami, Y., Wobbe, C. R., Weissbach, L., Dean, F. B., Hurwitz, J. 1986. *Proc. Natl. Acad. Sci. USA* 83:2869–73

160. Bennett, E. R., Naujokas, M., Hassell, J. A. 1989. *J. Virol.* 63:5371–85

161. Hay, R. T., DePamphilis, M. L. 1982. *Cell* 26:767–79

162. Taljanidisz, J., Decker, R. S., Guo, Z.

S., DePamphilis, M. L., Sarkar, N. 1987. *Nucleic Acids Res.* 19:7877–88

163. Bullock, P. A., Seo, Y. S., Hurwitz, J. 1991. *Mol. Cell. Biol.* 11:2350–61
164. Reynisdottir, I., O'Reilly, D. R., Miller, L. K., Prives, C. 1990. *J. Virol.* 64:6234–45
165. Tsurimoto, T., Melendy, T., Stillman, B. 1990. *Nature* 346:534–39
166. Weinberg, D. H., Collins, K. L., Simancek, P., Russo, A., Wold, M. S., et al. 1990. *Proc. Natl. Acad. Sci. USA* 87:8692–96
167. Huber, H. E., Bernstein, J., Nakai, H., Tabor, S., Richardson, C. C. 1988. *Cancer Cells* 6:11–17
168. Alberts, B. M. 1987. *Philos. Trans. R. Soc. London Ser. B* 317:395–420
169. Kornberg, A. 1982. *1982 Supplement to DNA Replication.* New York: Freeman
170. Bramhill, D., Kornberg, A. 1988. *Cell* 52:743–55
171. Walter, G., Flory, J. 1979. *Cold Spring Harbor Symp. Quant. Biol.* 44:165–69
172. Kress, M., Resche-Rigon, M., Feunteun, J. 1982. *J. Virol.* 43:761–71
173. VanRoy, F., Fransen, L., Fiers, W. 1983. *J. Virol.* 45:315–31
174. Scheidtmann, K. H., Buck, M., Schneider, J., Kalderon, D., Fanning, E., Smith, A. E. 1991. *J. Virol.* 65:1479–90
175. Chen, Y.-R., Lees-Miller, S., Tegtmeyer, P., Anderson, C. W. 1991. *J. Virol.* 65:5131–40
176. Grässer, F. A., Mann, K., Walter, G. 1987. *J. Virol.* 61:3373–80
177. Fanning, E., Novak, B., Burger, C. 1981. *J. Virol.* 37:92–102
178. Fanning, E., Westphal, K. H., Brauer, D., Cörlin, D. 1982. *EMBO J.* 1:1023–28
179. Greenspan, D. S., Carroll, R. B. 1981. *Proc. Natl. Acad. Sci. USA* 84:105–9
180. Scheidtmann, K. H., Hardung, M., Echle, B., Walter, G. 1984. *J. Virol.* 50:1–12
181. Gidoni, D., Scheller, A., Barnet, B., Hantzopoulos, P., Oren, M., Prives, C. 1982. *J. Virol.* 42:456–66
182. Klausing, K., Scheidtmann, K. H., Baumann, E., Knippers, R. 1988. *J. Virol.* 62:1258–65
183. Klausing, K., Scheffner, M., Scheidtmann, K. H., Stahl, H., Knippers, R. 1989. *Biochemistry* 28:2238–44
184. Mohr, I. J., Stillman, B., Gluzman, Y. 1987. *EMBO J.* 6:153–60
185. Simmons, D. T., Chou, W., Rodgers, K. 1986. *J. Virol.* 60:888–94
186. Schirmbeck, R., Deppert, W. 1989. *J. Virol.* 63:2308–16
187. Virshup, D. M., Kauffman, M. G., Kelly, T. J. 1989. *EMBO J.* 8:3891–98
188. Scheidtmann, K. H., Virshup, D. M., Kelly, T. J. 1991. *J. Virol.* 65:2098–101
189. McVey, D., Brizuela, L., Mohr, I., Marshak, D. R., Gluzman, Y. 1989. *Nature* 341:503–7
190. Erdile, L. F., Collins, K. L., Russo, A., Simancek, P., Small, D., et al. 1991. *Cold Spring Harbor Symp. Quant. Biol.* 57:In press
191. Edwards, C. A. F., Khoury, G., Martin, R. G. 1979. *J. Virol.* 29:753–62
192. VanRoy, F., Fransen, L., Fiers, W. 1983. *J. Virol.* 45:442–46
193. Scheidtmann, K. H. 1986. *Virology* 150:85–95
194. Grässer, F. A., Scheidtmann, K. H., Tuazon, P. T., Traugh, J. A., Walter, G. 1988. *Virology* 165:13–22
195. Carroll, D., Santoro, N., Marshak, D. R. 1988. *Cold Spring Harbor Symp. Quant. Biol.* 53:91–95
196. Moreno, S., Nurse, P. 1990. *Cell* 61:549–51
197. D'Urso, G., Marraccino, R. L., Marshak, D. R., Roberts, J. M. 1990. *Science* 250:786–91
198. Carter, T., Vancurova, I., Sun, I., Lou, W., DeLeon, S. 1990. *Mol. Cell. Biol.* 10:6460–71
199. Lees-Miller, S. P., Chen, Y.-R., Anderson, C. W. 1990. *Mol. Cell. Biol.* 10:6472–81
200. Zhu, J., Rice, P. W., Chamberlain, M., Cole, C. N. 1991. *J. Virol.* 65:2778–90
201. Jones, N. C., Rigby, P. W. J., Ziff, E. B. 1988. *Genes Dev.* 2:267–81
202. Wildeman, A. G. 1989. *Proc. Natl. Acad. Sci. USA* 86:2123–27
203. Kelly, J. J., Munholland, J. M., Wildeman, A. G. 1989. *J. Virol.* 63:383–91
204. Rio, D. C., Tjian, R. 1983. *Cell* 32:1227–40
205. Mitchell, P. J., Wang, C., Tjian, R. 1987. *Cell* 50:847–61
206. Contreras, R., Gheysen, D., Knowland, J., van der Voorde, A., Fiers, W. 1982. *Nature* 300:500–5
207. Hartzell, S. W., Byrne, B. J., Subramanian, K. N. 1984. *Proc. Natl. Acad. Sci. USA* 81:6335–39
208. Tack, L. C., Beard, P. 1985. *J. Virol.* 54:207–18
209. Keller, J. M., Alwine, J. C. 1985. *Mol. Cell. Biol.* 5:1859–69
210. May, E., Omilli, F., Ernoult-Lange, E., Zenke, M., Chambon, P. 1987. *Nucleic Acids Res.* 15:2445–61
211. Gong, S. S., Subramanian, K. N. 1988. *Virology* 163:481–93
212. Dynan, W. S., Chervitz, S. A. 1989. *J. Virol.* 63:1420–27

213. Alwine, J. C. 1985. *Mol. Cell. Biol.* 5:1034–42
214. Pannuti, A., Pascucci, A., La Mantia, G., Fischer-Fantuzzi, L., Vesco, C., Lania, L. 1987. *J. Virol.* 61:1296–99
215. Gallo, G. J., Gruda, M. C., Manupello, J. R., Alwine, J. C. 1990. *J. Virol.* 64:173–84
216. Gruda, M. C., Alwine, J. C. 1991. *J. Virol.* 65:3553–58
217. Davidson, I., Xiao, J. H., Rosales, R., Staub, A., Chambon, P. 1988. *Cell* 54:931–42
218. Schutzbank, T., Robinson, R., Oren, M., Levine, A. J. 1982. *Cell* 30:481–90
219. Scott, M. R. D., Westphal, K. H., Rigby, P. W. J. 1983. *Cell* 34:557–67
220. Soprano, K. J., Galanti, N., Jonak, G. J., McKercher, S., Pipas, J. M., et al. 1983. *Mol. Cell. Biol.* 3:214–19
221. Carey, M. F., Singh, K., Botchan, M., Cozzarelli, N. R. 1986. *Mol. Cell. Biol.* 6:3068–76
222. Segawa, K., Yamaguchi, N. 1987. *Mol. Cell. Biol.* 7:556–59
223. Hiscott, J., Wong, A., Alper, D., Xanthoudakis, S. 1988. *Mol. Cell. Biol.* 8:3397–405
223a. Taylor, I. C. A., Solomon, W., Weiner, B., Paucha, E., Bradley, M., Kingston, R. E. 1989. *J. Biol. Chem.* 264:16160–64
224. Wagner, S., Knippers, R. 1990. *Oncogene* 5:353–59
225. Wagner, S., Cullmann, G., Knippers, R. 1991. *J. Virol.* 65:3259–67
226. Saffer, J. D., Jackson, S. P., Thurston, S. J. 1990. *Genes Dev.* 4:659–66
227. White, R. J., Stott, D., Rigby, P. W. J. 1990. *EMBO J.* 9:3713–21
228. Singh, K., Carey, M., Saragosti, S., Botchan, M. 1985. *Nature* 314:553–56
229. Hay, N., Skolnik, H., Aloni, Y. 1982. *Cell* 29:183–93
230. Michel, M. R., Schwyzer, M. 1982. *Eur. J. Biochem.* 129:25–32
231. Khandjian, E. W., Loche, M., Darlix, J. L., Türler, H., Weil, R. 1982. *Proc. Natl. Acad. Sci. USA* 79:1139–43
232. Darlix, J. L., Khandjian, E. W., Weil, R. 1984. *Proc. Natl. Acad. Sci. USA* 81:5425–29
233. Seto, E., Yen, T. S. B. 1991. *J. Virol.* 65:2351–56
234. Martin, R. G. 1981. *Adv. Cancer Res.* 34:1–68
235. Tevethia, M. J. 1984. *Virology* 137:414–21
236. Sugano, S., Yamaguchi, N. 1984. *J. Virol.* 52:884–91
237. Peden, K. W., Pipas, J. M. 1985. *J. Virol.* 55:1–9
238. Asselin, C., Bastin, M. 1985. *J. Virol.* 56:958–88
239. Lanford, R. E., Long, C. W., Butel, J. S. 1985. *Mol. Cell. Biol.* 5:1043–50
240. Jat, P. S., Sharp, P. A. 1986. *J. Virol.* 59:746–50
241. Rutila, J. E., Imperiale, M. J., Brockmann, W. W. 1986. *J. Virol.* 58:526–35
242. Sompayrac, L. M., Danna, K. J. 1988. *Virology* 163:391–96
243. Chen, S., Paucha, E. 1990. *J. Virol.* 64:3350–57
243a. Thompson, D. L., Kalderon, D., Smith, A. E., Tevethia, M. J. 1990. *Virology* 178:15–34
244. Kriegler, M., Perez, C. F., Hardy, C., Botchan, M. 1984. *Cell* 38:483–91
245. Brown, M., McCormack, M., Zinn, K. G., Farrel, M. P., Bikel, I., Livingston, D. M. 1986. *J. Virol.* 60:290–93
246. Jat, P. S., Cepko, C. L., Mulligan, R. C., Sharp, P. A. 1986. *Mol. Cell. Biol.* 6:1204–17
247. Jat, P. S., Sharp, P. A. 1989. *Mol. Cell. Biol.* 9:1672–81
247a. Hanahan, D. 1989. *Science* 246:1265–75
248. Brinster, R. L., Chen, H. Y., Messing, A., Van Dyke, T., Levine, A. J., Palmiter, R. D. 1984. *Cell* 37:367–79
249. Hanahan, D. 1985. *Nature* 315:115–22
250. Ornitz, D. M., Palmiter, R. D., Hammer, R. E., Brinster, R. L., Swift, G. H., McDonald, R. J. 1985. *Nature* 313:115–22
251. Srinivasan, A., Peden, K. W. C., Pipas, J. M. 1989. *J. Virol.* 63:5459–63
252. Peden, K. W. C., Sinivasan, A., Farber, J. M., Pipas, J. M. 1989. *Virology* 168:13–21
252a. Ewen, M. E., Xing, Y., Lawrence, J. B., Livingston, D. M. 1991. *Cell* 66:1155–64
253. Figge, J., Webster, T., Smith, T. F., Paucha, E. 1988. *J. Virol.* 62:1814–18
254. Moran, B., Zerler, B. 1988. *Mol. Cell. Biol.* 8:1756–64
255. Velcich, A., Ziff, E. 1988. *Mol. Cell. Biol.* 8:2177–83
256. Whyte, P., Buchkovich, K. J., Horowitz, J. M., Friend, S. H., Raybuck, M., et al. 1988. *Nature* 334:124–29
257. Sager, R. 1989. *Science* 246:1406–12
258. Marshall, C. J. 1991. *Cell* 64:313–26
259. Huang, H. J. S., Yee, J. K., Shew, J. Y., Chen, P. L., Bookstein, R., et al. 1988. *Science* 242:1563–66
260. Bookstein, R., Shew, J. Y., Chen, P. L., Scully, P., Lee, W. H. 1990. *Science* 247:712–15
261. Chen, P. L., Scully, P., Shew, J. Y.,

Wang, J. Y. J., Lee, W. H. 1989. *Cell* 58:1193–98

262. Buchkovich, K., Duffy, L. A., Harlow, E. 1989. *Cell* 58:1097–105

263. Ludlow, J. W., DeCaprio, J. A., Huang, C. M., Lee, W. H., Paucha, E., Livingston, D. M. 1989. *Cell* 58:57–65

263a. Ludlow, J. W., Shon, J., Pipas, J. M., Livingston, D. M., DeCaprio, J. A. 1990. *Cell* 60:387–96

264. Chellappan, S. P., Hiebert, S., Mudryj, M., Horowitz, J. M., Nevins, J. R. 1991. *Cell* 65:1053–61

265. Bagchi, S., Weinmann, R., Raychaudhuri, P. 1991. *Cell* 65:1063–82

266. Bandara, L. R., Adamczewski, J. P., Hunt, T., LaThangue, N. B. 1991. *Nature* 352:249–51

267. Chittenden, T., Livingston, D. M., Kaelin, W. G. 1991. *Cell* 65:1073–82

268. Mudryj, M., Hiebert, S. W., Nevins, J. R. 1990. *EMBO J.* 7:2179–84

269. Oren, M. 1985. *Biochim. Biophys. Acta* 823:67–78

270. Jenkins, J. R., Stürzbecher, H. W. 1988. In *The Oncogene Handbook*, ed. E. P. Reddy, A. M. Skalka, T. Curran, pp. 403–32. New York: Elsevier Science

271. Lane, D. P., Benchimol, S. 1990. *Genes Dev.* 4:1–9

272. Levine, A. J. 1990. *Virology* 177:419–28

273. Levine, A. J., Momand, J., Finlay, C. A. 1991. *Nature* 351:453–55

274. Tack, L. C., Wright, J. H., Gurney, E. G. 1988. *J. Virol.* 62:1028–37

275. Tack, L. C., Wright, J. H., Deb, S. P., Tegtmeyer, P. 1989. *J. Virol.* 63:1310–17

276. Stürzbecher, K. W., Braithwaite, A. W., Addison, C., Palmer, C., Rudge, K., et al. 1988. *Cancer Cells* 6:159–63

277. Kienzle, H., Baack, M., Knippers, R. 1989. *Eur. J. Biochem.* 184:181–86

278. Bargonetti, J., Friedman, P. N., Kern, S. E., Vogelstein, B., Prives, C. 1991. *Cell* 65:1083–91

279. Kern, S. E., Kinzler, K. W., Bruskin, A., Jaroz, D., Frieman, P., et al. 1991. *Science* 252:1708–11

280. Stürzbecher, H. W., Maimets, T., Chumakov, R., Brain, R., Addison, C., et al. 1990. *Oncogene* 5:795–803

281. Fields, S., Jang, S. K. 1990. *Science* 249:1046–49

282. Raycroft, L., Wu, H., Lozano, G. 1990. *Science* 249:1049–51

283. O'Rourke, R. W., Miller, C. W., Kato, G. J., Simon, K. J., Chen, D. L., et al. 1990. *Oncogene* 5:1829–32

284. Ray, F. A., Peabody, D. S., Cooper, J. L., Cram, L. S., Kraemer, P. M. 1990. *J. Cell. Biochem.* 42:13–31

285. Stewart, N., Bacchetti, S. 1991. *Virology* 180:49–57

286. Gurney, T., Gurney, E. G. 1989. *J. Virol.* 63:165–74

287. Stary, A., James, M. R., Sarasin, A. 1989. *J. Virol.* 63:3837–43

287a. St. Onge, L., Bouchard, L., Laurent, S., Bastin, M. 1990. *J. Virol.* 64:2958–66

288. Schiedner, G., Wessel, R., Scheffner, M., Stahl, H. 1990. *EMBO J.* 9:2937–43

289. Sarnow, P., Ho, Y. S., Williams, J., Levine, A. J. 1982. *Cell* 28:387–94

290. Werness, B. A., Levine, A. J., Howley, P. M. 1990. *Science* 248:76–79

291. Scheffner, M., Werness, B. A., Huibregtse, J. M., Levine, A. J., Howley, P. M. 1990. *Cell* 63:1129–36

292. Dyson, N., Howley, P. M., Munger, K., Harlow, E. 1989. *Science* 243:934–36

Annu. Rev. Biochem. 1992. 61:87–111

BIOCHEMICAL INSIGHTS DERIVED FROM INSECT DIVERSITY

John H. Law*, José M. C. Ribeiro⁺, and Michael A. Wells*

Center for Insect Science and Departments of Biochemistry* and Entomology⁺, University of Arizona, Tucson, Arizona 85721

KEY WORDS: blood feeding, digestion and absorption, lipid transport and storage, storage proteins, insect egg

CONTENTS

INTRODUCTION AND PERSPECTIVES

Insects are usually regarded as foes of humankind: bearers of human, plant, and animal diseases or fierce competitors for food and other natural resources. Occasionally, their roles as pollinators or producers of honey and silk are

87

0066-4154/92/0701-0087$02.00

recalled, as an afterthought. However, insects have much potential as objects of research, the results of which can be highly beneficial to humankind. Not only are they established sources of important biologically active materials, but they also share with us much biology, often in a simpler and experimentally more accessible form. Moreover, the tremendous diversity of insect species represents a great, untapped resource of biochemistry, which, if more fully exploited, could uncover novel solutions to problems shared with other organisms.

First, a few words about insects to orient the reader to their biology: imagine an experimental animal that increases in size by severalfold in a few days; or one that can be raised in large numbers in a small space on an inexpensive diet; or one in which parts of the body can be isolated and maintained in a viable condition for several days without any specialized equipment; or one in which genetics are well understood and readily manipulated. Among the insects, one can find all of these useful properties, and more.

In most of their biochemistry and physiology, insects resemble mammals closely. Most basic metabolic pathways and their control are similar; neurons and muscles operate on the same principles; and endocrine control is exerted by like mechanisms, although hormone structures differ. Digestive processes are often parallel, and transport of nutrients is basically the same in insects and mammals. Some notable differences between insects and vertebrates concern respiration, blood circulation and hemostasis, and immunity. Insects do not carry oxygen in the blood (hemolymph), and lack either heme or copper proteins often used for this purpose in other species. Instead, oxygen is transported by a tubular (tracheolar) system to the cellular level. Hemolymph is enclosed by basement membranes that surround all tissues; there are no vessels for its transport. Pulsatory organs aid in moving hemolymph through the extremities, and a dorsal aorta (sometimes called a heart) moves hemolymph from posterior to anterior. A wound in the body wall can result in significant loss of hemolymph. An effective clotting mechanism is present, but it appears to be very different from that of mammals and is not well studied. Insects lack an antibody system similar to that of mammals, but many insects respond to bacterial or parasitic invasion by the induction of several peptides and proteins that are lethal to the invader (1). Alternatively, melanization reactions, involving a proteolytic cascade leading to active phenoloxidase, encapsulate the invader (2). Cells in the hemolymph (hemocytes) are in lower concentration than in mammalian blood (especially since erythrocytes are lacking). Hemocytes perform phagocytosis (2), are involved in hemolymph clotting (2), and may also be involved in the encapsulation reactions mentioned above, but in general their functions are not well understood.

A specialized tissue characteristic of insects is the fat body. Dispersed throughout the abdomen, this tissue has many of the functions of both liver and adipose tissue in vertebrates. It has a versatile synthetic capacity and is responsible for production of most of the hemolymph proteins. It can store fat, glycogen, waste material, and specialized proteins, and even house microscopic symbionts. Its cell membranes are major sites for metabolite interchange. It generally produces most of the components that make up the insect egg yolk.

Insects are divided into two groups on the basis of their life histories, the hemimetabola, in which adults differ from juveniles in that they are reproductively mature, and the holometabola, in which the juveniles (larvae) are completely unlike the adults. Holometabola, insects with a complete metamorphosis, have divided their lives into two distinct phases for two distinct purposes. Larvae feed, grow, and store nutrients; adults carry out reproduction and dispersal, and may or may not feed. Complete metamorphosis offers the investigator an unusual opportunity to study cell death and tissue remodeling within a living organism.

Studies of insects have provided insights that have contributed to an understanding of mammalian systems. The effects of the insect steroid hormone, ecdysone, led Karlson (3) to postulate that hormones have a direct action on gene expression, a principle amply verified in mammals. Roth & Porter (4) observed coated pits and vesicles in mosquito oocytes and postulated that these were involved in endocytosis of yolk proteins. Later investigations have revealed the general importance of coated membrane–mediated endocytosis. The studies of Berridge (5) on fly salivary glands led to the discovery of the phosphoinositide second-messenger system. Extensive studies of the molecular genetics of the fruit fly *Drosophila melanogaster* have contributed to many areas of contemporary biochemistry, and cell and developmental biology, e.g. homeotic genes (6). The use of P element–mediated transformation in *D. melanogaster* to create mutant proteins is a very powerful tool that is still underexploited for biochemical studies.

One of the essentially untapped resources of insect biochemistry is the tremendous variety of insect species, which have been evolving and diversifying for hundreds of millions of years. One area in which this rich diversity can be put to good use is in studies of protein evolution. Comparative sequence analyses of the same protein from several insect species and/or between insects and vertebrates may identify highly conserved regions that are essential for the biological activity of the protein. For example, when the cDNA sequence for a cytochrome P450 of unknown specificity from the cockroach *Blaberus discoidalis* was compared to the sequences of known cytochrome P450s, it was possible to determine the family to which the cockroach P450 belonged based on the conservation of a 13-residue sequence in the active site of the protein (7). Likewise, such diversity holds promise of a treasury of new materials with unique properties that will have practical importance.

The purpose of this review is to introduce non-insect biochemists to the potential usefulness of insects as experimental systems—it is not intended to be a comprehensive review of insect biochemistry. The topics covered reflect the interests of the authors and most of the examples are taken from work in our laboratories. One area we do not discuss is insect endocrinology, an active focus of research in several laboratories. The isolation of the ecdysone receptor from *D. melanogaster* S3 cells (7a), the cloning of the genes from two of the early ecdysone-induced gene puffs from *D. melanogaster,* and the identification of their products as DNA-binding proteins and possibly the ecdysone receptor (7b–7d) are quite exciting. These results open the way for the use of the power of *D. melanogaster* genetics in studies of hormone-regulated gene regulation. Perhaps insects will return to center stage in this area of research, a position they occupied some 30 years ago. More in-depth coverage can be found in the comprehensive treatise edited by Kerkut & Gilbert (8), or in articles appearing in recent issues of *Annual Review of Entomology* or *Advances in Insect Physiology.*

BLOOD FEEDING

Insects have adapted to exploit virtually every organic resource on the planet. In many respects, this resource utilization is welcome, for insects recycle material more efficiently than people, and remove vast quantities of detritus that would otherwise overwhelm us. Different feeding styles have led to unusual designs for mouthparts and alimentary tracts. Materials indigestible for mammals (e.g. wool and cellulose) are no problem for an insect (sometimes aided by symbionts) that has filled a specific resource niche. Some interesting symbioses have developed that maximize utilization of a food source. For example, aphids process many times their body weight in plant phloem in order to assimilate sufficient quantities of amino acids, which occur in low concentrations in phloem, and in the process defecate large amounts of sugar syrup, because the carbohydrate content of phloem exceeds their ability to use it. Another group of insects, the ants, utilize this excreted sugar, even to the point of husbanding the aphids.

The practice of blood feeding evolved independently in many different arthropods (insects and ticks), as well as in annelids (leeches), and vertebrates (bats). Even within a group of related insects (e.g. the Diptera), blood feeding has arisen independently several times. From such a diverse background, these animals evolved different solutions to the common problem of avoiding host hemostatic mechanisms that prevent blood loss. Because hemostasis is a complex and redundant physiological phenomenon that involves blood clotting, platelet aggregation, and vasoconstriction (Figure 1), blood-sucking arthropods, in their struggle to survive, have become extremely good pharmacologists, with the means to neutralize or manipulate host hemostasis to their advantage. Here we review the pharmacological substances produced by these resourceful animals, and not the allergic reactions resulting from this interaction, which often leads to a negative human attitude regarding insects.

Blood coagulation occurs through two pathways, the intrinsic and the extrinsic, both of which, through independent mechanisms, lead to the conversion of Factor X to Factor Xa. Factor Xa converts prothrombin to thrombin, the enzyme that produces fibrin from fibrinogen (9). The intrinsic pathway is triggered by the exposure of blood to subendothelial components, such as collagen. This can activate plasma Factor XII, leading to a number of reactions ultimately activating Factor X. The extrinsic pathway is triggered by the release of tissue thromboplastin from injured cells, which also ultimately leads to activation of Factor X. Platelets can aggregate by at least three pathways, induced by ADP, thrombin, or collagen (10). Although each of these mediators may potentiate the action of the others, each independently can give a full aggregatory response if the proper dose is present. Finally, vasoconstriction of venules and arterioles can be promoted by serotonin and thromboxane A2 released by aggregating platelets (11), or other inflammatory

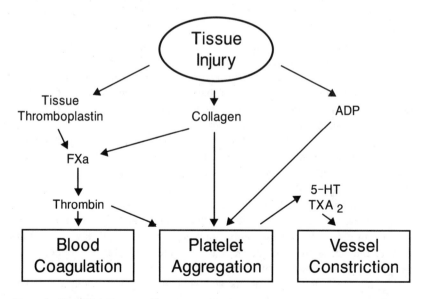

Figure 1 Simplified diagram of hemostasis. The figure shows how tissue injury can activate blood coagulation, platelet aggregation, and vessel constriction.

cells, such as neutrophils and mast cells. Accordingly, blood-sucking arthropods have evolved an array of different anticoagulants, antiplatelet substances, and vasodilators (summarized in Table 1).

Anticoagulants

Anticoagulants were first described in the salivary glands of blood-sucking arthropods 70 years ago, and the diversity in the mode of action of different salivary homogenates on the clotting cascade has been well documented (12–19). More recently, an anticoagulant peptide with anti-Xa properties from the tick *Ornithodorus moubata* has been sequenced (20–23). The black fly *Simulium vitatum* also contains a low-molecular-weight protein with anti-Xa activity (23), and potent antithrombin activity resides in the salivary glands of the tsetse fly, *Glossina morsitans* (24, 25). The antithrombin, hirudin, from the leech *Hirudus medicinalis* has been known for many years, and a recombinant form of this peptide has been made (26). Anticlotting peptides from other leeches have also been characterized or sequenced (27–33). Pharmaceutical companies are presently engaged in the characterization and cloning of insect-derived anticoagulants, which may represent alternatives to heparin in antithrombotic therapies (34).

Antiplatelet Factors

Platelet aggregation represents the most immediate reaction against blood loss, and is sufficient to arrest blood flow from small vessels in the absence of a functional blood-clotting system (35). Not surprisingly, most hematophagous arthropods studied to date have abundant antiplatelet activities in their saliva or salivary gland homogenates (36). Among the antiplatelet compounds, the enzyme ATP-diphosphohydrolase (apyrase), which hydrolyzes ATP and ADP (or other nucleotides), has been commonly found (36). Additionally, ticks have prostaglandins that have antiplatelet activity, but these compounds will be discussed in another section.

One of the most relevant physiological triggers of platelet aggregation is ADP, released by injured cells into the extracellular space that normally contains less than 10^{-7} M nucleotides. Salivary gland homogenates of tsetse flies (25), mosquitoes (37), sand flies (38), black flies (39; E. W. Cupp, M. S. Cupp, unpublished), fleas (40), triatomine bugs (41), and ticks (42, 43) have elevated levels of apyrase activity, when compared to non-blood-sucking insects, which presumably function to reduce ADP levels and thus prevent platelet aggregation at the site of feeding. The kinetic properties of salivary apyrases indicate two general types, some being activated by both Ca^{2+} and Mg^{2+} ions, and others activated only by Ca^{2+} ions (43). The remarkable finding of high levels of apyrase activity in the salivary glands of most blood-sucking arthropods suggested that the enzyme, whose normal

Table 1 The strategies used by arthropods to inhibit hemostasis in vertebrates

Activity	Insect	Species	Mode of action	Reference
Anticoagulant	Ticks	*Ornithodorus moubata*	Anti-Factor Xa	20–22
		Dermacentor andersoni	Anti-Factor V and VII	19
	Bugs	*Rhodinus prolixus*	Anti-Factor VII	13, 14
	Tsetse	*Glossina morsitans*	Antithrombin	24, 25
	Black fly	*Simulium vittatum*	Anti-Factor Xa	23
Antiplatelet	Ticks	*Ornithodorus moubata*	Apyrase	43
		Ixodes dammini	Apyrase, PGI_2 (?)	42, 51
	Bugs	*Rhodinus prolixus*	Apyrase, NO	60
	Tsetse	*Glossina morsitans*	Apyrase	25
	Fleas	*Oropsylla bacchi*	Apyrase	40
		Orchopea howardii	Apyrase	40
		Xenopsylla cheopis	Apyrase	40
	Mosquitoes	*Anopheles freeborni*	Apyrase	133
		Anopheles stephensi	Apyrase	133
		Aedes aegypti	Apyrase	37
	Sand flies	*Lutzomyia longipalpis*	Apyrase	38
		Phlebotomus argentipes	Apyrase	39
		Phlebotomus papatasi	Apyrase	39
		Phlebotomus perniciosus	Apyrase	39
Vasodilator	Ticks	*Ixodes dammini*	PGE_2	42
		Boophilus microplus	$PGE_{2\alpha}$	49, 50
		Amblyomma americanum	PGE_2, $PGF_{2\alpha}$	J. M. C. Ribeiro et al, unpublished
	Bugs	*Rhodinus prolixus*	NO compound	60
	Mosquitoes	*Aedes aegypti*	Tachykinin	J. M. C. Ribeiro, unpublished
		Anopheles albimanus	Unidentified	J. M. C. Ribeiro, unpublished
	Sand flies	*Lutzomyia longipalpis*	Maxidilan (EIF)	55, 56

function might be to make irreversible the glycosylation reactions that produce UDP or ADP as a product (44), had become a secretory product during adaptation to the blood-sucking mode of feeding. Support for this suggestion is the finding of elevated apyrase activities in plant tissues involved in carbohydrate synthesis, an activity that produces ADP or UDP (45). This hypothesis about the normal role of apyrase remains to be tested, but may prove useful in identifying a function for the intracellular apyrases found in most eukaryotic cells. Extracellular vertebrate apyrases, as proposed by Burnstock (46), may be related to the turnover of nucleotides released by purinergic nerves or injured cells. Indeed, most ecto-ATPases of vertebrate cells have been characterized as apyrases, and some of them have been recently purified and their genes cloned (47). Interestingly, an endothelium apyrase from bovine aorta was purified to homogeneity (48). Are these enzymes related to endothelial protection against platelet aggregation? Does a decrease in their activity lead to aggregation, thrombosis, and arteriosclerosis? Studies of the similarity between invertebrate salivary apyrases and vertebrate ecto-ATPases may provide further insights into this aspect of hemostatic physiology and pathology.

Vasodilators

Salivary vasodilators are widely found in blood-sucking arthropods, and are thought to help feeding by counteracting the vasoconstrictors released by aggregating platelets, or to increase blood flow in the skin and thus the supply of food to the feeding insect. In contrast to the uniform presence of apyrase in salivary glands, the great diversity of salivary vasodilators is becoming apparent. Thus, prostaglandins, peptides, and even nitrovasodilators have been found in the saliva or salivary gland homogenates of blood-sucking arthropods. Because very few arthropods have been screened for such biological activity, it is expected that many powerful compounds await discovery.

PROSTAGLANDINS   Prostaglandins have been found in the saliva or salivary gland homogenates of the ticks *Boophilus microplus* (49, 50), *Ixodes dammini* (42, 51), *Hyalomma excavatum* (52), and *Amblyomma americanum* (J. M. C. Ribeiro, P. H. Evans, A. V. Schleger, A. A. Fawcett, unpublished). In the first three, evidence for the presence of PGE_2 was obtained by thin-layer chromatography, bioassay, or radioimmunoassay. Recently, both PGE_2 and $PGF_{2\alpha}$ were identified in the saliva of *A. americanum* by HPLC/bioassay and by GC-mass spectrometry (J. M. C. Ribeiro, P. H. Evans, J. L. MacSwain, J. R. Sauer, unpublished). Concentrations of salivary prostaglandins are of pharmacological significance, ranging from 100 to 500 ng/ml of pilocarpine-induced tick saliva (42, 49, 50; J. M. C. Ribeiro, P. H. Evans, A. V. Schleger, A. A. Fawcett, unpublished). These substances are potent inducers of skin vasodilation, and were proposed to help the tick by increasing the amount of blood at the feeding site (53). Salivary prostaglandins may also be involved in the generalized lymphocyte suppression seen in tick-infested animals (42, 54). Tick salivary glands may be a good source of invertebrate cyclo-oxygenase and prostaglandin synthetases for studies of the properties of these enzymes, which are potential targets for drugs to control ectoparasites.

PEPTIDES   The salivary gland of the sand fly *Lutzomyia longipalpis* contains a novel vasodilatory peptide that has been characterized as the most potent and persistent vasodilator known (55, 56). Recently, pharmacological characterization of an endothelium-dependent vasodilator from the salivary glands of the mosquito *Aedes aegypti* indicated the presence of a peptide of the tachykinin family (J. M. C. Ribeiro, unpublished). Currently, vasodilators from black flies and other mosquito species are under investigation.

NITROVASODILATORS   Since 1980, it has been known that many vasodilators, including acetylcholine, act on endothelial cells to cause the release of a very labile substance that relaxed vascular smooth-muscle cells (57). This uncharacterized compound became known as endothelium-derived relaxation factor (EDRF). In the past three years, nitric oxide (NO), or a parent molecule that releases NO, has been shown to be EDRF (58, 59). Nitric oxide, which readily permeates membranes, acts on the smooth-muscle cell by activating cytoplasmic guanylate cyclase after binding to the heme moiety of the enzyme. Interestingly, homogenates of the salivary gland of the triatomine bug *Rhodinus prolixus* contain a nitrovasodilator (60). It is thought that this vasodilator is a 16-kDa hemoprotein that binds NO and serves as a storage form of NO. The metabolism of NO in the salivary glands of this bug, as well as the biophysical interaction of NO with a heme protein that seems designed to interact with NO, will probably yield interesting insights in this new area of biochemical and pharmacological research.

DIGESTION AND ABSORPTION

Insects use many different sources of food and exhibit an enormous diversity in the morphology and biochemistry of the gut (61). The gut represents a major interface between the insect and its environment, and an understanding of the function of the gut is essential to evaluating methods of insect control that act through the gut, e.g. the use of genetically engineered plants. The gut can be divided into three parts: the cuticle-lined foregut, often including a crop; the midgut, in which most digestion and absorption occur; and the cuticle-lined hindgut, which serves to absorb some material, especially water. The midgut is the focus of the remainder of this discussion. It is composed primarily of a single layer of epithelial cells, interspersed with endocrine cells [whose function is beginning to be explored (62)], innervation from the stomatogastric nervous system, some muscle, and an abundant tracheal supply, which brings oxygen to the tissue. The environment within the midgut is as varied as its structure, ranging from the highly alkaline midgut of moths and butterflies (pH > 11), to the nearly neutral pH of the cockroach midgut, to the slightly acidic midgut of beetles.

The insect gut is distinguished by the presence of a peritrophic matrix, which surrounds the bolus of food (63). The peritrophic matrix is a chitin-protein complex made either in the anterior end of the midgut or by the midgut epithelial cells. The peritrophic matrix may serve as a mechanical barrier to protect the midgut epithelium from abrasive plant material, and as a barrier to exclude microorganisms and noxious macromolecules in the diet from interacting with the midgut cells. The matrix also serves to partition the midgut lumen into endoperitrophic and exoperitrophic spaces (64). Often there is a countercurrent flow of water between the endo- and exo-peritrophic spaces, in which fluid moves from anterior to posterior in the endoperitrophic space and in the opposite direction in the exoperitrophic space. This countercurrent flow increases the efficiencies of digestion and absorption, because digestive enzymes and their hydrolysis products, when they cross the peritrophic matrix, are swept anteriorly, which increases the effective absorptive area and also permits digestive enzymes to re-enter the endoperitrophic space and carry out further digestion, instead of being excreted (64).

Although our understanding of the biochemistry of the insect gut is rudimentary at present, a number of interesting observations suggest that further study of this tissue offers rich opportunities for increasing our understanding of the evolution of digestive processes.

Digestive Enzymes

Many insects with neutral to alkaline midgut pHs have a complement of digestive enzymes that are similar to those of vertebrates: trypsin and chymotrypsin act as endoproteases, and aminopeptidase and carboxypeptidase act as exopeptidases. In insects that have an acidic midgut pH, protein digestion is carried out by cathepsins, enzymes usually associated with lysosomes (65).

Recently, cDNA sequences of trypsin from the moth *Manduca sexta* (A. M. Peterson, M. A. Wells, unpublished) and the mosquito *A. aegypti* (66), as well as genomic sequences for putative trypsins from *D. melanogaster* (67), have shown that the primary translation products are most likely pre-proenzymes, which contain a signal sequence for secretion, and an activation peptide, which must be cleaved in order to produce the mature active enzyme. The cleavage of the activation peptide is probably tryptic, because in the three sequences known, there is an arginine preceding the amino terminus of the mature enzyme. In the blood-feeding stable fly *Stomoxys calcitrans*, trypsin-like enzymes are stored in zymogen granules in a specialized region of the midgut (68). At present the enzyme(s) involved in zymogen activation is unknown.

In *M. sexta*, both the tryptic and chymotryptic activities have high pH optima (about 11), but show inhibition reactions typical for serine proteases: trypsin is inhibited by DFP, PMSF, TLCK, and soybean trypsin inhibitor; chymotrypsin is inhibited by DFP, PMSF, TPCK, and chymostatin (A. M. Peterson, M. A. Wells, unpublished). Indeed the sequences of the trypsins noted above show the presence of the Ser-His-Asp catalytic triad. Why these "typical" serine proteases have such high pH optima is unknown at present.

Little is known about the cathepsin-like enzymes present in insects with acidic midgut pHs. It is of considerable interest to understand if the midgut secretory system has been modified to permit secretion of lysosomal enzymes into the lumen of the gut.

Ticks have a unique digestive system (69). The midgut lumen serves merely as a reservoir, and proteolysis occurs intracellularly. The proteins in the blood meal are taken into the midgut cells by both fluid-phase and clathrin-mediated endocytosis. After a transient appearance in endosomes, the proteins are digested in secondary lysosomes. When all of the digestible material has been hydrolyzed, a residual body is formed. When the absorptive/digestive cell is filled with residual bodies, it is sloughed into the lumen and replaced by a new absorptive/digestive cell. Whether there are specific receptors associated with the clathrin-mediated endocytosis is unknown.

Models of Tissue Maturation

The midgut of the mosquito is a good model system for the study of gut maturation. In the first 3–4 days of adult life, the female mosquito feeds only on nectar. During this period, the midgut consists of an immature epithelium in which the endoplasmic reticulum is present as condensed whorls, and there is little or no production of digestive enzymes (71). Females require a blood meal to produce eggs. The blood meal triggers a remarkable series of events in the midgut within 2–4 hours (71). The stretching of the midgut after consumption of the blood meal may induce the numerous endocrine cells (71) in the gut to secrete peptides (62), and these substances in turn may be responsible for initiating maturation of the midgut. Whatever the signal, within about 2–4 hours after the blood meal, the cells of the midgut begin to synthesize and secrete digestive enzymes. This process involves the synthesis of rough endoplasmic reticulum and Golgi and transcriptional activation of a number of genes, including those for the digestive enzymes. Within 36–48 hours the blood meal has been completely digested and absorbed, a process of "dedifferentiation" begins, and by 72 hours the gut has returned to its initial stage. When another blood meal is taken, the whole process is repeated.

While midgut morphology is relatively well characterized (71), to date, little molecular work has been carried out in this system. A cDNA for a trypsin induced by the blood meal has been sequenced (66) and used to isolate a genomic clone (C. Barillas-Mury, M. A. Wells, unpublished). Because of the rapidity with which maturation occurs and the fact that it can be initiated by allowing the mosquito to take a blood meal, we suggest that this is an excellent model system for studies of tissue maturation deserving of extensive analysis. The tick midgut also merits investigation, because its epithelium is stimulated to mature by a blood meal (69).

Ion and Nutrient Absorption

One of the unique aspects of the insect midgut is that secondary active transport is driven by a ouabain-insensitive potassium pump, instead of the usual Na^+/K^+-ATPase found in vertebrate cells (61, 70). The active step in K^+ secretion occurs across the apical membrane of specialized cells in the midgut (globlet cells) (61) via an electrogenic K^+/nH^+ antiporter (70), which is energized by a H^+-driven ATPase (70). The physiological roles of the high lumenal potassium concentration are to drive active transport of amino acids, whereas sugar transport across the midgut is passive (61); alkalinization of the midgut (pH > 11) (61); and regulation of hemolymph K^+ levels (61). There are undoubtedly many other unique ion and nutrient transport systems yet to be discovered in the insect midgut; only a few have been investigated so far.

The absorption of sterols, which is still poorly understood in vertebrates, might be studied to advantage in insects, which do not synthesize sterols and therefore require them in their diets. Insects may also prove worthwhile in the study of carotenoid absorption, because there are several mutants in the silkworm *Bombyx mori* in which carotenoid absorption is affected (72). Numerous questions concerning nutrient and micronutrient absorption in insects remain essentially unexplored, e.g. how is iron absorbed from the lepidopteran midgut with a pH near 12, which should lead to formation of insoluble iron hydroxides?

Fatty Acid–Binding Proteins

Although the mechanism has not been studied extensively, fatty acid absorption in the insect midgut is efficient (73). Potentially important components of fatty acid absorption are the fatty acid–binding proteins (FABPs), which have been recently identified in the midgut of *M. sexta*, from which two proteins (MFB1 and MFB2) have been isolated and their cDNA sequences determined (74). A FABP has also been isolated from the flight muscle of the migratory locust *Locusta migratoria* (75). The sequences of the *M. sexta* midgut FABPs

(Figure 2) place them in the superfamily of cytoplasmic hydrophobic ligand-binding proteins (76). This superfamily contains low-molecular-weight proteins (14–16 kDa) that bind hydrophobic ligands, such as fatty acids, fatty acyl CoAs, sterols, and retinol (76). A number of roles have been proposed for FABPs, including facilitating fatty acid uptake by cells, targeting fatty acids to organelles and to specific metabolic pathways, altering the activity of enzymes involved in fatty acid metabolism, and protecting cellular proteins and membranes from the detergent effects of fatty acids or their CoA derivatives. In spite of considerable effort, the physiological role(s) played by these proteins remains obscure. The presence of FABPs in insects and their similarity to vertebrate proteins show that FABPs are more ancient than previously supposed. It is interesting to note that MFB2 has a Lys instead of an Arg at the site where the carboxyl group of the fatty acid binds to the protein. This may account for the fact that MFB2 binds fatty acids more tightly than either MFB1 or the vertebrate FABPs. No other member of the FABP superfamily has a Lys in this position; they contain either Arg or a nonbasic residue.

TRANSPORT AND STORAGE

Lipoproteins

LIPOPROTEIN METABOLISM   The lipoprotein is a universal vehicle for efficient transport of hydrophobic materials through an aqueous medium. Lipoproteins probably exist in all animals. Only in the past 10 years have the lipoprotein systems of insects received intensive study (77–79). Although on first inspection, the insect systems appear very different from those of humans, a closer examination reveals similarities. Insect lipoproteins can function to transport lipids from the gut to storage tissues or to rapidly growing tissues. They can also move lipids from the site of synthesis to sites of utilization, or from storage sites to active muscles or developing oocytes. These functions have parallels in vertebrates. The compounds transported may differ, however. Diacylglycerol is the major form in which dietary fatty acids are transported in the hemolymph of most insects, whereas triacylglycerol serves this purpose in mammalian blood. Sterols are usually transported by insect lipoproteins as free sterols, instead of sterol esters, as in vertebrate lipoproteins. Hydrocarbons are abundant in insect lipoproteins because insects require large quantities of hydrocarbons to coat their exoskeleton and prevent water loss.

A major lipoprotein found in most insects has been given the name *lipophorin* (Lp) (80). It is generally a particle of about 700 kDa composed of about 40% lipid, with diacylglycerol (DG) and phospholipids (PL) pre-

```
MFB1  AYLGKVYKFDREENFDGFLKSIGLSEEQVQKYLQYKPSSQLVKEGDKYKYISVSDGTKETVFESGVETDDVVQGGLPIKTTYTVDG NTVTQVVN  SAQGSATFKREYNGDELKVTITSSEMDGVAYRYYKA
MFB2  SYLGKVYSLVKQENFDGFLKSAGLSDDKIQALVSDKPTQKMEANGDSYSITSTGIGGERTVSFKSGVEFDDVIGAGESVKSMYTVDG NVVTHVVK  GDAGVATFKKEYNGDDLVVTITSSNMDGVARRYYKA
RFBL  NFSGK YQVQSQENFEPFKAMGLPEDLIQKGKDIKGVSEIVHEGKKVKLTITYGSKVIHNEFTLGEECELETMTGEKVKAVVKMEGDNKMVTTFK  GIKSVTEFNGDTITNTMTLG  DIVYKRVSGRI
HFBL  SFSGK YQLQSQENFEAFMKAIGLPEELIQKGKDIKGVSEIVQNGKHFKFTITAGSKVIQNEFTVGEECELETMTGEKVKTVVQLEGDNKLVTTFK  NIKSVTELNGDIITNTMTLG  DIVFKRISKRI
RFBI  AFDGT MKVYRNEYEKFMEKMGINVVKRKLGAHDNLKLTITQEGNKFTVKESSNFRNIDVVFELGVDFAYSLADGTELTGTLTMEG NKLVGKFKRVQNGKELIAVREISGNELIQTYYE  GVEAKRIFKK
HFBI  AFDST MKVDRSENYDKFMEKMGVNIVKRKLAHDNLKLTITQEGNKFTVKESSAFRNIEVVFELGVTFNYNLADGTELRGTWSLEG NKLIGKFKRTDMGNELNTVREIIGDELVQTYYE  GVEAKRIFKD
HFBH  DAFLGT MKLVDSKNFDDYMKSLGVGFATRQVASMTKPTTIIEKNGDILTLKTHSTFKNTEISFKLGVEFDETTADDRKVKSIVTLDG GKLVHLQKW DGQETTKVRELIDGKLILTLTHG TAVSTRTYEKEA
RFBH  ADAFVGT MKLVDSKNFDDYMKSLGVGFATRQVASMTKPTTIIEKNGDTITIKTHSTFKNTEISFQLGVEFDEVTADDRKVKSVVTLDG GKLVHVQKW DGQETTLTRELSDGKLILTLTHG NVVSTRTYEKEA
```

Figure 2 Alignment of members of the FABP family. The amino acid sequences were aligned by the method of Feng & Doolittle (134). Percent identities are derived from the generated alignments. Identical residues are marked by *; residues involved in fatty acid binding as derived from the X-ray structure of rat FABP-intestine (RFBI) (135) are indicated by ↑. Sequences are: *M. sexta* FABP 1 (MFB1); *M. sexta* FABP 2 (MFB2); rat FABP-liver (RFBL); human FABP-liver (HFBL); rat FABP-intestine (RFBI); human FABP-intestine (HFBI); human FABP-heart (HFBH); rat FABP-heart (RFBH).

dominating, and about 60% protein. Because lipophorin is a high-density lipoprotein (HDL), it is referred to as high-density lipophorin, or HDLp (81). There are two apoproteins, one (apoLp-I) of about 240 kDa and the second (apoLp-II) of about 80 kDa. Both apoproteins are usually glycosylated by mannose-rich oligosaccharides (82).

Whereas vertebrate lipoproteins are produced by both the liver and the intestine, lipophorin is not synthesized in the insect gut: only in fat body tissue is there appreciable synthesis (83). Lipophorin, more like vertebrate HDL than low-density lipoprotein (LDL), acts as a reusable lipid shuttle, deriving lipids from and delivering lipids to tissues without cellular retention and degradation of the protein components (73, 77, 84). For example, in larval M. $sexta$, lipophorin produced by the fat body consists of apoLp-I and apoLp-II plus about 10% PL; it is a very high-density lipoprotein (VHDLp) (83). At the midgut it loads dietary fat in the form of DG to become a typical HDLp (83). DG is delivered to fat body and other tissues, regenerating VHDLp (73, 83). Immunocytochemical observations on locust flight muscle detected apolipoproteins in the intercellular spaces, but not intracellularly (85). Although this indicates that HDLp is not taken up by the cell, it is not possible to eliminate a process in which HDLp is transiently internalized by the cell, delivers DG, and then is re-secreted; a similar process exists for vertebrate transferrin (86).

APOLIPOPROTEINS   During long-distance flight in adults of several insect species, triacylglycerol stores in the fat body are converted to DG, which is carried by lipophorin to the flight muscle, where the fatty acids are oxidized to provide the energy required for flight (79). This uptake of DG by lipophorin results in a much larger and less dense lipoprotein, low-density lipophorin (LDLp). LDLp is stabilized by an abundant hemolymph protein, apolipophorin-III (apoLp-III, $M_r = 18,000$), which binds to lipophorin and greatly increases its capacity to carry DG (87). The X-ray structure of apoLp-III from $L.$ $migratoria$ has been reported (88). This represents the first structural determination of an apolipoprotein from any source. The overall molecular architecture consists of five long alpha helices connected by short loops. The helices are distinctly amphiphilic with the hydrophobic residues pointing toward the center of the protein and the hydrophilic side-chains pointing toward the solvent. In order for the amphiphilic helices to play a role in the binding of the protein to a lipid surface, there must be a structural reorganization, which exposes the hydrophobic interior to the lipid surface. We have proposed a model in which the protein binds to the lipid surface via a relatively nonpolar loop, and then spreads on the surface in such a way as to cause the hydrophobic side-chains of the helices to come in contact with the lipid surface, while the charged and polar residues remain in contact with water (88–90). In this way, the overall helical motif of the protein is maintained, and the reorganization requires only the movement of the helices relative to each other around "hinges" located in connecting loops between helices. Recently, the X-ray structure of the receptor-binding domain of the vertebrate apolipoprotein, apoE, has been reported (91). Although the structure determined does not represent the entire protein molecule, it is strikingly similar to apoLp-III in that it consists of four long amphiphilic alpha helices in which the hydrophobic residues point toward the interior. Based on these two examples, it appears that the insect apolipoprotein is indeed an excellent model for vertebrate apolipoproteins.

What are the lipid-binding motifs in apolipoproteins? One approach to this problem is to compare the amino acid sequences of apoLp-III from a variety of insect species, because apoLp-IIIs from different sources seem to have the same function (92). We have determined the amino acid sequence of apoLp-III from *M. sexta* (93), *L. migratoria* (94), and *Acheta domesticus* (A. F. Smith, M. R. Kanost, M. A. Wells, unpublished). Surprisingly, there is little sequence identity among apoLp-III from these three species (Figure 3), in spite of the fact that they function identically. In searching for ways to compare these sequences for common structural elements, we found that a simple helical wheel representation (for an example see Figure 4) shows that all three sequences contain five amphiphilic helices. Thus, the *L. migratoria* structure can form the basis for models for the other two proteins, and preliminary molecular-replacement studies using the *L. migratoria* structure and the amino acid sequence of the *M. sexta* and *A. domesticus* apoLp-IIIs show that reasonable models for the structures of the latter two proteins can be constructed. Thus, although amino acid sequence identity is low, one can still use these divergent sequences to learn something important about the conservation of structure of apoLp-III and, because apoLp-III and apoE have common structural features, about apolipoprotein structure in general.

LIPOPROTEIN RECEPTORS   Whether lipophorin is loaded and unloaded at the cell surface or taken up by endocytosis and recycled, we would expect to find receptors on the cells. The receptors should have tissue-specific properties that would enable loading or unloading of lipid. A lipophorin receptor has been purified and characterized from larval *M. sexta* fat body and another has been partially purified and characterized from the midgut (95; K. Tsuchida, M. A. Wells, unpublished). The fat-body receptor has a native molecular mass of 120,000 and has an absolute requirement for Ca^{2+}; receptor binding is inhibited by suramin; and the K_d for HDLp is 4.1×10^{-8} M. The fat-body receptor shows a 10-fold higher affinity for DG-rich lipophorin than for DG-poor lipophorin. The midgut receptor has a molecular mass of about 140,000 and does not require metal ions for binding activity. The midgut

```
                                  *                                    *
ACHE   DAGTTGADFNSLFEAAQRHFQN LTATIQNALPSQ  EEVRTQLQTHAQTFANNLQAAATQFNEKAAELSGDAQTAVRQAAQQLEQQVSNLRQQFP
MAND   DAPAGGNAFEEM EKHAKEFQKTFSEQFNSLVNSKNTQDFNKALKDGSDSVLQQLSAFSSSLQGAISDANGKAKEALEQARQNVEKTAEELRKAHP
LOCU   DAAGHVNIAEAVQQLNHTIVN AAHELHETLGLPTPDEALNLLTEQANAFKTKIAEVTTSLKQEAEKHQGSVAEQLNRFARNLNNSIHDAATSAQ
       !-------- helix 1 --------!  !----------- helix 2 -----------!  !---- helix 3 ---!


                   *            *                     *             *     *
ACHE   DGAQAA    DKLKASIESALAEVQ  EAVQPHADAVAESLKTAARTAVQQATVITNQVQQSVQQAANAH
MAND   DVEKEANAFKDKLQAAVQTTVQESQKLAKEVASNMEETNKKLAPKIKQAYDDFVKHAEEVQKKLHEAATKQ
LOCU   PADQL     NSLQSALTNVGHQWQT  SQPRPSVAQEAWAPVQSALQEAAEKTKEAAANLQNSIQSAVQKPAN
       !----------- helix 4-----------!  !------- helix 4 --------!
```

Percent identity:

	ACHE	MAND	LOCU
ACHE		23.0	13.6
MAND	23.0		13.3
LOCU	13.6	13.3	

Figure 3   Alignment of members of the apoLp-III family. The amino acid sequences were aligned by the progressive method of Feng & Doolittle (134). Percent identities are derived from the generated alignments. Identical residues are marked by *. Helix 1–5 refer to amphiphilic helices found in the X-ray structure of *L. migratoria* apoLp-III (88).

receptor has a 20-fold higher affinity for DG-poor lipophorin than for DG-rich lipophorin.

The properties of these two receptors, coupled with a theoretical study of the structure and stability of lipophorins (96), allow the development of a model for DG transport and storage in feeding larvae. As the first component of the model, we propose that the movement of DG from HDLp to the fat body is driven by the conversion of DG to triacylglycerol within the fat-body cells. Based on the studies of lipophorin stability (96), the depletion of DG from HDLp will lead to an unstable particle, which contains an excess of surface components. This DG-poor lipophorin can regain its stability by either shedding surface components, an unlikely process, or by taking on additional DG. Therefore, as the second component of the model, we propose that DG movement from the midgut to DG-poor lipophorin is driven by the instability of the DG-poor lipophorin, and the elimination of this instability by the uptake of DG. The third component of the model depends on the higher affinity of the fat-body receptor for DG-rich lipophorin and the higher affinity of the midgut receptor for DG-poor lipophorin. Thus, DG-rich lipophorin will bind to the fat-body receptor and after delivery of DG to the fat body will be converted to DG-poor lipophorin, which dissociates from the receptor. The DG-poor lipophorin will bind to the midgut receptor, which facilitates delivery of DG from the midgut epithelial cells to the lipophorin. When the lipophorin is loaded with DG, it dissociates from the midgut receptor and returns to the fat body to complete the cycle. The actual mechanism whereby DG moves between lipophorin and cells is unknown and represents a challenging problem in physical biochemistry.

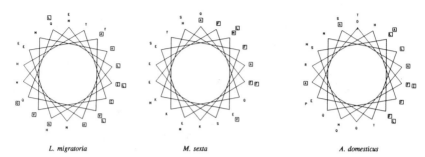

L. migratoria   M. sexta   A. domesticus

Figure 4   Helical wheel representations of amphiphilic helices in apoLp-IIIs. Depicted are helix 1 from *L. migratoria* apoLp-III (residues 7–32), and putative helices 1 from *M. sexta* (residues 8–33) and *A. domesticus* (residues 8–33) apoLp-IIIs. Hydrophobic residues are boxed.

FLIGHT   In the case of adult insects that have an elevated lipid transport function, e.g. those that engage in sustained flight, HDLp can be converted to a low-density form, LDLp, which has a high lipid-carrying capacity. LDLp is about twice the size of HDLp, and it functions to deliver DG to active flight muscles and the developing oocyte. Release of a neurosecretory peptide, the adipokinetic hormone, into the hemolymph at the onset of flight stimulates the fat-body cells to convert triacylglycerol to DG (79), which is loaded onto HDLp through the agency of a lipid transfer particle (LTP) (97). Up to 14 molecules of apoLp-III also associate with lipophorin, presumably to stabilize the expanded hydrophobic surface (90).

The lipid transfer particle is very large (greater than 800 kDa) and has a most unusual morphology (98, 99). In *M. sexta* it consists of three apoproteins and about 14% of lipids (100). LTP mediates the transfer of DG from the fat body to lipoproteins, but not the reverse process (97). It probably has other functions as well, and in vitro, it can promote equilibration of DG and PL between lipoproteins (100). Insect LTP has been used to alter the lipid composition of human lipoproteins (101, 102).

At the flight-muscle cell, DG is processed by a lipoprotein lipase in the muscle cell membrane (103). As DG is removed from LDLp, the size of the particle decreases, apoLp-III dissociates, and HDLp is regenerated (104). This is similar to the transfer of triacylglycerol from the mammalian chylomicron to the adipose cell, but whereas the latter process results in storage of fat, the insect flight muscle utilizes the fatty acids of DG to power its highly active energy metabolism.

Usually, lipid reserves power flight in insects that engage in sustained flights or do not feed as adults. Other insects generally use sugar reserves for flight. The principal sugar of insect blood is not glucose, but trehalose, a nonreducing disaccharide made from glucose in the fat body. A small group of insects, represented by the tsetse fly *Glossina* and the Colorado potato beetle *Leptinotarsa decemlineata*, uses proline as the principal fuel for flight (105). These insects have essentially partitioned the citric acid cycle so that half occurs in the fat body, converting acetyl CoA from fatty acid oxidation plus pyruvate to α-ketoglutarate, and half in the flight muscle, where α-ketoglutarate is converted to pyruvate. The transport between the tissues is accomplished by alanine coming from the flight muscle to the fat body and proline coming from the fat body to the flight muscle. In the fat body, α-ketoglutarate is transaminated to glutamate with alanine as a partner, then cyclized to proline. In the flight muscle, proline is converted to glutamate, which is transaminated to α-ketoglutarate, while pyruvate is converted to alanine. Conversion of one equivalent of α-ketoglutarate to pyruvate in the flight muscle provides nine equivalents of ATP for powering muscle.

Iron Transport and Storage

Iron is essential to all organisms that carry out oxidative metabolic processes. Although they generally lack hemoglobin as an oxygen carrier, insects require cytochromes and iron-sulfur proteins in their tissues, especially the highly active flight muscles. Thus there is a need both to store dietary iron for future development and to transport it between tissues.

Structures resembling mammalian ferritin have been observed in insect tissues for many years, but only recently has insect ferritin been isolated and characterized. From *M. sexta*, a 490-kDa ferritin with subunits of 24 kDa and 26 kDa was isolated (106). From another lepidopteran, the butterfly *Calpodes ethlius*, a somewhat larger ferritin (660 kDa) with larger subunits (24 kDa and 31 kDa) was described (107). The *C. ethlius* ferritin was shown to be abundant in the midgut and hemolymph (107), and later apoferritin was identified in the fat body (108). Translation of midgut poly A^+ RNA in vitro, either without or with microsomes, suggested that the 24-kDa subunit was not glycosylated but could be converted to the 28–30-kDa subunit by glycosylation (109). The ferritin in the tissues of *C. ethlius* is not cytoplasmic, but is found in the vacuolar system and is abundant in the hemolymph. In mammals, ferritin is predominately in the cytoplasm, but micromolar amounts occur in the serum (110), and this serum ferritin is glycosylated (111). Little is known of the origin, nature, and function of vertebrate serum ferritin, and perhaps a study of insect ferritin will provide insights into this rare vertebrate form. Some insects have cytoplasmic ferritin (112), and thus it may be possible to develop good insect models for studying either cytoplasmic or exported ferritin.

An iron-transport protein was isolated from hemolymph of *M. sexta* and shown to deliver iron to the fat body (106). Although this protein had a molecular size (about 80 kDa) similar to that of vertebrate transferrins, it bound only one, rather than two, iron atoms, and was not considered to be a transferrin (106). Further investigation, however, showed that this insect protein had significant sequence similarity to vertebrate transferrins, including two similar domains, but that the putative iron-binding ligands were present only in the N-terminal domain and were missing from the C-terminal domain (113). Whether this unusual structural feature is general in insects or is an isolated anomaly cannot be determined until the structures of other insect transferrins are known.

We have no information about the functions of transferrin and ferritin in insects at the cellular level. Are there receptors that allow for endocytosis of or iron delivery by these proteins? Is there a reciprocal translational control of ferritin and transferrin receptor synthesis in insects as there is in vertebrates (114)? These and other aspects of insect metabolism need to be explored.

Storage Proteins

In late larval life, many insects synthesize in the fat body and secrete into the hemolymph large amounts of 500-kDa hexameric proteins that have been called storage proteins (115). These proteins are often utilized during development of the adult, and therefore are considered to be depots of amino acids for future protein synthesis. Just prior to or during the pupal stage, many of these proteins are taken up by the fat body and stored in large crystalline granules (115). They may also be taken up by the fat body for degradation. In some moths, the uptake of storage proteins depends upon the insect steroid hormone 20-hydroxyecdysone. The storage proteins are composed of six subunits of approximately 72–80 kDa, which may or may not be identical. In Lepidoptera, such as *M. sexta* and *B. mori*, there are two well-characterized classes of storage proteins: the arylphorins and the methionine-rich proteins (115). Recently, cDNA and gene sequences for arylphorin (116) and the methionine-rich protein (117; X.-Y. Wang, M. A. Wells, unpublished) from *M. sexta*, arylphorin and the methionine-rich protein from *B. mori* (118, 119), and the cDNA sequence for a storage protein from *Trichoplusia ni* (120) have been reported. The cDNA-derived amino acid sequences have established that several families of storage proteins exist and that the insect storage proteins have homology to the hemocyanins of other arthropods (horseshoe crab, spider, and lobster) (116, 120). Arthropod hemocyanins are copper-containing proteins that transport oxygen in the hemolymph (121). The hemocyanins are hexamers or multihexamers with subunits of ~75 kDa. Thus, at least from a structural perspective, the storage proteins and the hemocyanins have similar subunit sizes and form hexamers. The storage proteins are almost as similar to the various hemocyanins as are the various hemocyanins to each other.

A 3.2-Å structure has been reported for the *Panulirus interruptus* (spiny lobster) hemocyanin (122). When we examined the aligned sequences of this hemocyanin and the insect storage proteins, we noted that those regions of the proteins that contained the highest concentration of identical residues or conservative replacements corresponded to the portions of the hemocyanin protein that form the subunit contacts between the hexamers. These residues would be expected to be conserved because of structural requirements for interactions along the interface between subunits. Thus, it appears reasonable that the hexameric insect storage proteins and the hexameric arthropod hemocyanins may have a common three-dimensional structure. If this is the case, it may be that these two classes of proteins have evolved from a common ancestor. These observations suggest that a hexameric structure of blood proteins may be a common feature in arthropods, although at this time there is no evidence to suggest why a hexameric protein is advantageous. To date, hexameric blood proteins have not been reported outside the phylum Arthropoda.

Construction of the Insect Egg

As mentioned earlier, insect oocytes gave early clues to the nature of endocytosis of proteins. Indeed, the oocyte is a locus of prodigious endocytotic activity: an insect oocyte can double in size in six hours, principally by accretion of protein. The insect endocytotic systems have been largely neglected, while the characterization of the mammalian LDL receptor (123) has received considerable attention. A vitellogenic receptor from the membrane of locust oocytes has now been isolated and characterized (124). Ligand blotting of oocyte membrane proteins from other insect species has identified proteins of 200 kDa (125) and 205 kDa (126) that bind vitellogenin. The locust receptor is a 156-kDa protein with an acidic PI, and suramin and trypan blue (polyanionic compounds) inhibit binding of vitellogenin, presumably because they mask a cationic site on the ligand that normally interacts with the receptor. Calcium ion is required for binding. The vitellogenin receptor shares these properties with the human LDL receptor, which, however, is smaller, 130 kDa (123), and the fat body lipophorin receptor (95). It seems, therefore, that the endocytotic vitellogenin receptor of the insect oocyte resembles the human LDL receptor. However, the events subsequent to ligand dissociation and receptor recycling are very different in the two cases. The LDL receptor directs its ligand to the lysosomal compartment, where it is hydrolyzed, whereas the vitellogenin receptor directs its ligand to a storage compartment, the yolk body.

In some insect species, proteins other than vitellogenin are taken into the egg by receptor-mediated endocytosis, but we know nothing about the receptors involved (127). In *M. sexta*, HDLp is taken up by oocytes and stripped of much of its diacylglycerol to produce a VHDLp that is also stored in the yolk body (128). This represents an exception to the generalization that lipophorin is not taken up by cells. The insect egg also contains a large store of fat in the form of lipid droplets, which may account for 35% of the weight of the egg (84). In *M. sexta*, most of the egg lipid is derived from fat body stores transported to the egg by the DG-rich LDLp. This lipoprotein can deliver lipid to the egg without the apoproteins being entrapped in the cell (84). How this is accomplished is by no means clear.

CONCLUDING REMARKS

We have attempted to provide the reader with some glimpses of the diverse nature of insect biochemistry. Clearly there is much more to be explored and much new and useful information to be discovered. We hope that the reader will be inclined to look with favor on these remarkable creatures whose planet we share, and perhaps even to invite them into the laboratory.

ACKNOWLEDGMENTS

We thank Carolina Barillas-Mury, Elizabeth Bernays, René Feyereisen, Don Frohlich, Henry Hagedorn, John Hildebrand, Peter Kulakosky, Fernando Noriega, Ann Peterson, Alan Smith, José Soulages, Rik Van Antwerpen, Miranda Van Heusden, Joy Winzerling, and Rolf Ziegler for reading the manuscript and offering helpful suggestions. Our work was supported by NIH grants GM 29238 (JHL), AI 18694 (JMCR), HL 39116 (MAW), AI29434 (MAW), and HL42322 (MAW).

Literature Cited

1. Boman, H. G. 1991. *Cell* 65:205–7
2. Kanost, M. R., Kawooya, J. K., Law, J. H., Ryan, R. O., Van Heusden, M. C., Ziegler, R. 1990. *Adv. Insect Physiol.* 22:299–396
3. Karlson, P. 1963. *Perspect. Biol. Med.* 6:203–14
4. Roth, T. F., Porter, K. R. 1964. *J. Cell Biol.* 20:313–32
5. Berridge, M. J. 1983. *Biochem. J.* 212:849–58
6. Manley, J. L., Levine, M. S. 1985. *Cell* 43:1–2
7. Bradfield, J. Y., Lee, Y.-H., Keeley, L. L. 1991. *Proc. Natl. Acad. Sci. USA* 88:4558–62
7a. Luo, Y., Amin, J., Voellmy, R. 1991. *Mol. Cell. Biol.* 11:3660–75
7b. Burtis, K. C., Thummel, C. S., Karim, F. D., Hogness, D. S. 1990. *Cell* 61:85–99
7c. Seagraves, W. A., Hogness, D. S. 1990. *Genes Dev.* 4:204–19
7d. Thummel, C. S., Burtis, K. C., Hogness, D. S. 1990. *Cell* 61: 101–11
8. Kerkut, G., Gilbert, L., eds. 1985. *Comprehensive Insect Physiology, Biochemistry and Pharmacology.* Oxford: Pergamon
9. Sundsomo, J. S., Fair, D. S. 1983. *Springer Semin. Immunopathol.* 6:231–57
10. Leung, L., Nachman, R. 1986. *Annu. Rev. Med.* 37:179–86
11. Vargaftig, B. B., Chignard, M., Benveniste, J. 1981. *Biochem. Pharmacol.* 30:263–71
12. Cornwall, J. W., Patton, W. S. 1914. *Indian J. Med. Res.* 2:569
13. Hellmann, K., Hawkins, R. I. 1964. *Nature* 201:1008–9
14. Hellmann, K., Hawkins, R. I. 1965. *Nature* 207:265–67
15. Hawkins, R. I., Hellmann, K. 1966. *Br. J. Haematol.* 12:86–91

16. Hellmann, K., Hawkins, R. I. 1966. *Br. J. Haematol.* 12:376–84
17. Hawkins, R. I. 1966. *Nature* 212:738–39
18. Hellman, K., Hawkins, R. I. 1967. *Thromb. Diath. Haemorrh.* 18:617–25
19. Gordon, J. R., Allen, J. R. 1991. *J. Parasitol.* 77:167–70
20. Waxman, L., Smith, D. E., Arcuri, K. E., Vlasuk, G. P. 1990. *Science* 248:593–96
21. Jordan, S. P., Waxman, L., Smith, D. E., Vlasuk, G. P. 1990. *Biochemistry* 29:11095–100
22. Vlasuk, G. P., Ramjit, D., Fujita, T., Dunwiddie, C. T., Nutt, E. M., et al. 1991. *Thromb. Haemost.* 65:257–62
23. Jacobs, J. W., Cupp, E. W., Sardana, M., Friedman, P. A. 1990. *Thromb. Haemost.* 64:235–38
24. Parker, K. R., Mant, M. J. 1979. *Thromb. Haemost.* 42:743–51
25. Mant, M. J., Parker, K. R. 1981. *Br. J. Haematol.* 48:601–8
26. Benatti, L., Scacheri, E., Bishop, D. H., Sarmientos, P. 1991. *Gene* 101: 255–60
27. Budzynski, A. Z., Olexa, S. A., Brizuela, B. S., Sawyer, R. T., Stent, G. S. 1981. *Proc. Soc. Exp. Biol. Med.* 168:266–75
28. Nutt, E. M., Jain, D., Lenny, A. B., Schaffer, L., Siegl, P. K., Dunwiddie, C. T. 1991. *Arch. Biochem. Biophys.* 285:37–44
29. Seymour, J. L., Henzel, W. J., Nevins, B., Stults, J. T., Lazarus, R. A. 1990. *J. Biol. Chem.* 265:10143–47
30. Swadesh, J. K., Huang, I. Y., Budzynski, A. Z. 1990. *J. Chromatogr.* 502:359–69
31. Blankenship, D. T., Brankamp, R. G., Manley, G. D., Cardin, A. D. 1990. *Biochem. Biophys. Res. Commun.* 166:1384–89
32. Condra, C., Nutt, E., Petroski, C. J., Simpson, E., Friedman, P. A., Jacobs, J. W. 1989. *Thromb. Haemost.* 61:437–41
33. Han, J. H., Law, S. W., Keller, P. M., Kniskern, P. J., Silberklang, M., et al. 1989. *Gene* 75:47–57
34. Walenga, J. M., Bakhos, M., Messmore, H. L., Fareed, J., Pifarre, R. 1991. *Ann. Thorac. Surg.* 51:271–77
35. Mustard, J. F., Packham, M. A. 1977. *Br. Med. Bull.* 33:187–91
36. Ribeiro, J. M. C. 1987. *Annu. Rev. Entomol.* 32:463–78
37. Ribeiro, J. M. C., Sarkis, J. J. F., Rossignol, P. A., Spielman, A. 1984. *Comp. Biochem. Physiol. B* 79: 81–86
38. Ribeiro, J. M. C., Rossignol, P. A.,

Spielman, A. 1986. *Comp. Biochem. Physiol. A* 83:683–86
39. Ribeiro, J. M. C., Modi, G. B., Tesh, R. B. 1989. *Insect Biochem.* 19:409–12
40. Ribeiro, J. M. C., Vaughan, J. A., Azad, A. F. 1990. *Comp. Biochem. Physiol. B* 95:215–18
41. Sarkis, J. J. F., Guimaraes, J. A., Ribeiro, J. M. C. 1986. *Biochem. J.* 233:885–91
42. Ribeiro, J. M. C., Makoul, G., Levine, J., Robinson, D., Spielman, A. 1985. *J. Exp. Med.* 161:332–44
43. Ribeiro, J. M. C., Endris, T. M., Endris, R. 1991. *Comp. Biochem. Physiol.* In press
44. Sadler, E., Beyer, T. A., Oppenheimer, C. L., Paulson, J. C., Prieels, J.-P., et al. 1982. *Methods Enzymol.* 83:458–514
45. Preiss, J. 1982. *Annu. Rev. Plant Physiol.* 33:431–54
46. Burnstock, G. 1972. *Pharmacol. Rev.* 24:509–81
47. Lin, S. H. 1990. *Ann. NY Acad. Sci.* 603:394–99
48. Yagi, K., Arai, Y., Kato, N., Hirota, K., Miura, Y. 1989. *Eur. J. Biochem.* 180:509–13
49. Dickinson, R. G., O'Hagan, J. E., Schotz, M., Binnington, K. C., Hegarty, M. P. 1976. *Aust. J. Exp. Biol. Med. Sci.* 54:475–86
50. Higgs, G. A., Vane, J. R., Hart, R. J., Porter, C., Wilson, R. G. 1976. *Bull. Entomol. Res.* 66:665–70
51. Ribeiro, J. M. C., Makoul, G., Robinson, D. 1988. *J. Parisol.* 74:1068–69
52. Shemesh, M., Hadani, A., Shklar, A., Shore, L. S., Meleguir, F. 1979. *Bull. Ent. Res.* 69:381–85
53. Kemp, D. H., Hales, J. R. S., Schleger, A. V., Fawcett, A. A. 1983. *Experientia* 39:725–27
54. Wikel, S. K. 1982. *Ann. Trop. Med. Parasitol.* 76:627–32
55. Ribeiro, J. M. C., Vachereau, A., Modi, G. B., Tesh, R. B. 1989. *Science* 243:212–14
56. Lerner, E. A., Ribeiro, J. M. C., Nelson, R. J., Lerner, M. R. 1991. *J. Biol. Chem.* 266:11234–36
57. Furchgott, R. F. 1990. *Acta Physiol. Scand.* 139:257–70
58. Ignarro, L. J., Buga, G. M., Wood, K. S., Byrns, R. E., Chaudhuri, G. 1987. *Proc. Natl. Acad. Sci. USA* 84:9265–70
59. Moncada, S., Palmer, R. M., Higgs, E. A. 1988. *Hypertension* 12:365–72
60. Ribeiro, J. M. C., Gonzales, R., Marinotti, O. 1990. *Br. J. Pharmacol.* 101:932–36
61. Dow, J. A. T. 1986. *Adv. Insect Physiol.* 19:187–328

62. Brown, M. R., Crim, J. W., Lea, A. O. 1986. *Tissue Cell* 18:419–28
63. Richards, A. G., Richards, P. A. 1977. *Annu. Rev. Entomol.* 22:219–40
64. Terra, W. R. 1990. *Annu. Rev. Entomol.* 35:181–200
65. Murdock, L. L., Brookhart, G., Dunn, P. E., Foard, D. E., Kelley, S., et al. 1987. *Comp. Biochem. Physiol. B* 87: 783–87
66. Barillas-Mury, C., Graf, R., Hagedorn, H. H., Wells, M. A. 1991. *Insect Biochem.* In press
67. Davis, C. A., Riddle, D. C., Higgins, M. J., Holden, J. J. A., White, B. N. 1985. *Nucleic Acids Res.* 13:6605–19
68. Moffatt, M. R., Lehane, M. J. 1990. *Insect Biochem.* 20:719–23
69. Coons, L. B., Rosell-Davis, R., Tarnowski, B. I. 1986. In *Morphology, Physiology, and Behavioral Biology of Ticks*, ed. J. R. Sauer, J. A. Hair, pp. 249–79. Chichester: Horwood
70. Wieczirek, H., Putzenlechner, M., Zeiske, W., Klein, U. 1991. *J. Biol. Chem.* 266:15340–47
71. Billingsley, P. F. 1990. *Annu. Rev. Entomol.* 35:219–48
72. Doira, H. 1978. In *The Silkworm: An Important Laboratory Tool*, ed. Y. Tazima, pp. 54–81. Tokyo: Kodansha
73. Tsuchida, K., Wells, M. A. 1988. *Insect Biochem.* 18:263–68
74. Smith, A. F., Tsuchida, K., Hanneman, E., Suzuki, T. C., Wells, M. A. 1991. *J. Biol. Chem.* In press
75. Haunerland, N. H., Chisholm, J. M. 1991. *Biochim. Biophys. Acta* 1047: 233–38
76. Sweetser, D. A., Heuckeroth, R. O., Gordon, J. I. 1987. *Annu. Rev. Nutr.* 7:337–59
77. Chino, H. 1985. In *Comprehensive Insect Physiology, Biochemistry and Pharmacology*, ed. G. Kerkut, L. Gilbert, Vol. 10, pp. 115–35. Oxford: Pergamon
78. Shapiro, J. P., Law, J. H., Wells, M. A. 1988. *Annu. Rev. Entomol.* 33:297–318
79. Beenakkers, A. M. Th., Van der Horst, D. J., Van Marrewijk, W. J. A. 1985. *Prog. Lipid Res.* 24:19–67
80. Chino, H., Downer, R. G. H., Wyatt, G. R., Gilbert, L. I. 1981. *Insect Biochem.* 1:491
81. Beenakkers, A. M. Th., Chino, H., Law, J. H. 1988. *Insect Biochem.* 11:1–2
82. Nagao, E., Takahashi, N., Chino, H. 1987. *Insect Biochem.* 17:531–38
83. Prasad, S. V., Fernando-Warnakulasuriya, G. J. P., Sumida, M., Law, J. H., Wells, M. A. 1986. *J. Biol. Chem.* 261:17174–76
84. Kawooya, J. K., Law, J. H. 1988. *J. Biol. Chem.* 263:8748–53
85. Van Antwerpen, R., Linnemans, W. A. M., Van der Horst, D. J., Beenakkers, A. M. Th. 1988. *Cell Tissue Res.* 252:661–68
86. Kuhn, L. C., Schulman, H. M., Ponka, P. 1990. In *Iron Transport and Storage*, ed. P. Ponka, H. Schulman, R. Woodworth, pp. 150–91. Boca Raton: CRC
87. Wheeler, C. H., Goldsworthy, G. J. 1983. *J. Insect Physiol.* 29:349–54
88. Breiter, D. R., Kanost, M. R., Benning, M. M., Law, J. H., Wells, M. A., et al. 1991. *Biochemistry* 30:603–8
89. Kawooya, J. K., Meredith, S. C., Wells, M. A., Kézdy, F. J., Law, J. H. 1986. *J. Biol. Chem.* 261:13588–91
90. Wells, M. A., Ryan, R. O., Kawooya, J. K., Law, J. H. 1987. *J. Biol. Chem.* 262:4172–76
91. Wilson, W. M., Wardell, M. R., Weisgraber, K. H., Mahley, R. W., Agard, D. A. 1991. *Science* 252:1817–22
92. Van der Horst, D. J., Ryan, R. O., Van Heusden, M. C., Schulz, T. K. F., Van Doorn, J. M., et al. 1988. *J. Biol. Chem.* 263:2027–33
93. Cole, K. D., Fernando-Warnakulasuriya, G. J. P., Boguski, M. S., Freeman, M., Gordon, J. I., et al. 1987. *J. Biol. Chem.* 262:11794–800
94. Kanost, M. R., Boguski, M. S., Freeman, M., Gordon, J. I., Wyatt, G. R., Wells, M. A. 1988. *J. Biol. Chem.* 263:10568–73
95. Tsuchida, K., Wells, M. A. 1990. *J. Biol. Chem.* 265:5761–67
96. Soulages, J. L., Brenner, R. R. 1991. *J. Lipid Res.* 32:407–15
97. Van Heusden, M. C., Law, J. H. 1989. *J. Biol. Chem.* 264:17287–92
98. Ryan, R. O., Hicks, L. D., Kay, C. M. 1990. *FEBS Lett.* 267:305–10
99. Ryan, R. O., Howe, A., Scraba, D. G. 1990. *J. Lipid Res.* 31:871–79
100. Ryan, R. O., Senthilathipan, K. R., Wells, M. A., Law, J. H. 1988. *J. Biol. Chem.* 263:14140–45
101. Ryan, R. O., Wessler, A. N., Ando, S., Price, H. M., Yokoyama, S. 1990. *J. Biol. Chem.* 265:10551–55
102. Silver, E. T., Scraba, D. G., Ryan, R. O. 1990. *J. Biol. Chem.* 265:22487–92
103. Van Heusden, M. C., Van der Horst, D. J., Van Doorn, J. M., Wes, J., Beenakkers, A. M. Th. 1986. *Insect Biochem.* 16:517–23
104. Van Heusden, M. C., Van der Horst, D.

J., Voshol, J., Beenakkers, A. M. Th. 1987. *Insect Biochem.* 17:771–76

105. Bursell, E. 1981. In *Energy Metabolism in Insects*, ed. R. Downer, pp. 135–54. New York: Plenum

106. Huebers, H. A., Huebers, E., Finch, C. A., Webb, B. A., Truman, J. W., et al. 1988. *J. Comp. Physiol. B* 158:291–300

107. Nichol, H. K., Locke, M. 1989. *Insect Biochem.* 19:587–602

108. Locke, M., Ketola-Pirie, C., Leung, H., Nichol, H. 1991. *J. Insect Physiol.* 37:297–309

109. Ketola-Pirie, C. A. 1990. *Biochem. Cell Biol.* 68:1005–11

110. Monro, H. N., Linder, M. C. 1978. *Physiol. Rev.* 58:317–96

111. Theil, E. C. 1990. *Adv. Enzymol.* 63:421–49

112. Nichol, H., Locke, M. 1990. *Tissue Cell* 22:767–77

113. Bartfeld, N. S., Law, J. H. 1990. *J. Biol. Chem.* 265:21684–91

114. Theil, E. C. 1990. *J. Biol. Chem.* 265:4771–74

115. Telfer, W. H., Kunkel, J. G. 1991. *Annu. Rev. Entomol.* 36:205–28

116. Willott, E., Wang, X-Y., Wells, M. A. 1989. *J. Biol. Chem.* 264:19052–59

117. Corpuz, L. M., Choi, H., Muthukrishnan, S., Kramer, K. J. 1991. *Insect Biochem.* 21:265–76

118. Fujii, T., Sakurai, H., Izumi, S., Tomino, S. 1989. *J. Biol. Chem.* 264:11020–25

119. Sakurai, H., Fujii, T., Izumi, S., Tomi-no, S. 1988. *Nucleic Acids Res.* 16:7717–18

120. Jones, G., Brown, N., Manczak, M., Hiremath, S., Kafatos, F. C. 1990. *J. Biol. Chem.* 265:8596–602

121. Linzen, B., Soeter, N. M., Riggs, A. F., Schneider, H. J., Schartau, W., et al. 1985. *Science* 229:519–24

122. Gaykema, W. P. J., Hol, W. G. J., Vereijken, J. M., Soeter, N. M., Bak, H. J., et al. 1984. *Nature* 309:23–29

123. Brown, M. S., Goldstein, J. L. 1986. *Science* 232:34–47

124. Roehrkasten, A., Ferenz, H.-J., Buschmann-Gebhardt, B., Hafer, J. 1989. *Arch. Insect Biochem. Physiol.* 10:141–49

125. Indrasith, L. S., Kindle, H., Lanzrein, B. 1990. *Arch. Insect Biochem. Physiol.* 15:1–16

126. Raikel, A. S., Dhadialla, T. S. 1992. *Annu. Rev. Entomol.* 37:217–51

127. Law, J. H. 1990. *Adv. Invertebrate Reprod.* 5:97–102

128. Kawooya, J. K., Osir, E. O., Law, J. H. 1988. *J. Biol. Chem.* 263:8740–47

129. Deleted in proof

130. Deleted in proof

131. Deleted in proof

132. Deleted in proof

133. Riberio, J. M. C., Rossignol, P. A., Spielman, A. 1985. *J. Insect Physiol.* 31:689–92

134. Feng, D.-F., Doolittle, R. F. 1987. *J. Mol. Evol.* 25:351–60

135. Sacchettini, J. C., Gordon, J. I., Banaszak, L. J. 1988. *J. Biol. Chem.* 263: 5815–19

Annu. Rev. Biochem. 1992. 61:113–29

TELOMERASES

E. H. Blackburn

Department of Microbiology and Immunology, University of California, San Francisco, California, 94143-0414

KEY WORDS: telomere synthesis, telomerase RNA, *Tetrahymena*, reverse transcriptase

CONTENTS

Perspectives

The DNA at the ends of the linear eukaryotic chromosomes usually consists of tandemly repeated simple sequences. In general, the presence of this simple-sequence telomeric DNA at the chromosomal termini is essential for chromosome stability. Telomeric DNA is unusual in that one strand is synthesized by a ribonucleoprotein enzyme, telomerase, which is distinct from the conventional DNA replication machinery. The telomeric DNA sequence is specified by copying an RNA template sequence within the RNA moiety of telomerase. Telomerases appear to be widespread among eukaryotes, as judged, first, by the evolutionary conservation of telomeric DNA structure and in vivo behavior among eukaryotes, and, more directly, by the identification of telomerase activities from diverse eukaryotes. This review examines the specialized mechanism of telomeric DNA synthesis by telomerase. The structure and function of telomeres and telomeric DNA, and the role of

113

0066-4154/92/0701-0113$02.00

telomerase in vivo, have been the subjects of previous reviews and are not covered in detail here. The reader is referred to (1–4) for such reviews.

Telomerase is a DNA polymerase that can be classified as a reverse transcriptase, because its mechanism of action involves the copying of an RNA template into DNA. It is an unusual reverse transcriptase, however, because unlike the conventional, purely protein reverse transcriptases found in systems ranging from retroviruses to prokaryotes, it is a ribonucleoprotein which contains its own RNA template as an integral part of the enzyme. The RNA component of telomerases from several species has been identified, sequenced, and a secondary structure model proposed, and the mechanism of telomerase has been investigated both in vitro and in vivo. The protein component(s) of telomerases have to date resisted unambiguous biochemical identification. Thus the relationship of telomerase to conventional reverse transcriptases remains to be elucidated. In particular, whether the RNA component of telomerase plays roles in the action of telomerase besides being the template for telomeric DNA synthesis is a question of considerable interest, from the standpoints of both enzyme (protein or RNA) mechanisms and the evolution of telomerase.

Telomeric DNA Sequences

A given eukaryotic species has a characteristic telomeric DNA sequence, although the same sequence may occur in more than one species. Telomeric repeats can be regular, consisting of perfect tandem repeats of a fixed repeat unit sequence, or irregular, consisting of length or sequence variations of a basic repeat unit (1). However, regardless of whether the repeat units are perfect or irregular, one strand of the tandemly repeated simple telomeric DNA sequences is characterized by containing clusters of G residues, giving the two telomeric DNA strands a composition bias such that there is a G-rich and a C-rich strand. These strands have an invariable orientation with respect to the chromosome end: the G-rich strand is found at the 3' end of each chromosomal DNA strand, and at each end this G-rich strand protrudes 12–16 nucleotides beyond the complementary C-rich strand, at least in the various species in which it has been possible to analyze it (5, 6). Examples of G-rich strand repeat units include the regular sequences TTGGGG, found in the ciliates *Tetrahymena* and *Glaucoma*, TTTTGGGG, in the ciliates *Euplotes* and *Oxytricha*, and TTAGGG, found in humans and other vertebrates as well as some protozoans and molds. Examples of irregular repeats include TG_{1-3} and GGG(G/T)TT, characteristic of the yeast *Saccharomyces cerevisiae* and the ciliate *Paramecium*, respectively (1). (Unless specified otherwise, these and other nucleic acid sequences in this review are written in the usual 5' to 3' direction.)

The Telomerase Reaction

The telomerase of a given species synthesizes the G-rich strand DNA sequence characteristic of its species. Telomerase requires a DNA primer, to which telomeric repeats are added by polymerization in the usual 5' to 3' direction, using the appropriate deoxynucleoside triphosphates (dNTPs) as substrates (7). As described below, the most efficient primers in vitro consist of a few repeats of a single-stranded G-rich telomeric or telomere-like DNA sequence. These correspond to the G-rich protruding strand of natural telomeres (5, 6). Blunt-ended fully duplex DNAs are not used as primers (M. Lee, E. Blackburn, unpublished work). A typical example of an in vitro reaction carried out by the telomerase from *Tetrahymena thermophila*, in which the primer is the DNA oligonucleotide GGGGGGTTGGTT, can be depicted thus:

$$\underline{\text{GGGGTTGGGGTT}} + 4n \text{ dGTP} + 2n \text{ TTP} \rightarrow \text{GGGGTTGGGGTT}$$
$$(\text{GGGGTT})_n + 6n \text{ pyrophosphate.} \qquad\qquad 1.$$

The primer supplied to the reaction is the underlined sequence. Other than those incorporated into the telomeric sequence, no other deoxynucleoside (or ribonucleoside) triphosphates, or nucleotide cofactors, are required for the reaction (7), and under typical in vitro conditions n can be in the hundreds (8).

RECOGNITION OF THE PRIMER 3' END SEQUENCE   An aspect of the telomerase reaction that was important for determining its mechanism is the recognition of the 3' end of the primer, such that the appropriate next nucleotides are added to complete a telomeric repeat unit. It was found that primers with, for example, a TTGGG, a GGTT, or TTGG 3' end were extended by the *Tetrahymena* telomerase to produce respectively TTGGGGTTGGGGTT.., GGTTGGGGTTGG.., and TTGGGGTTGGGG..; i.e. in each case the next appropriate nucleotides were added to complete some permutation of a perfect TTGGGG repeat that included the 3' end of the primer (8, 9).

Analogous results were obtained subsequently with the telomerase activities isolated from the ciliates *Oxytricha* and *Euplotes* (10, 11), and from human tissue culture cells (12, 13). For example, when supplied with a primer with a GGTT 3' end, the telomerase activities from *Euplotes* or *Oxytricha* produce the reaction product GGTTTT(GGGGTTTT)$_n$. Note that whereas this GGTT 3' primer is extended by the *Tetrahymena* telomerase with nucleotides beginning with four G residues, the *Euplotes* or *Oxytricha* activity extends the same primer first by two T residues, which completes a T_4G_4 repeat unit that includes the primer 3' end sequence. Similarly, the human telomerase activity, which synthesizes AGGGTT repeats, was found to extend such a primer to produce <u>GGTT</u>AGGG<u>GTT</u>AGGG, the first added nucleotides thus completing a <u>TT</u>AGGG telomeric repeat unit (12, 13).

The basis of the 3' end recognition of the primer by telomerase became clear when it was found that the *Tetrahymena* telomerase contained an RNA moiety that had within it the template sequence for telomeric G-rich strand synthesis.

The RNA Moiety of Telomerase

PRIMARY AND SECONDARY STRUCTURE   The *Tetrahymena* telomerase has an apparent molecular size between 200 and 500 kDa, as judged by its elution position from a gel filtration column under various salt conditions (9). The first indication that a nucleic acid was an intrinsic part of telomerase and was essential for its enzymic activity came from the finding that partially purified telomerase activity in *Tetrahymena* extracts was sensitive to nuclease activities as well as to protease or heat. From experiments in which the enzyme was pretreated with either micrococcal nuclease or RNase A, the nuclease inactivated, and then telomerase activity assayed, it was deduced that an RNA was required for activity (9). The telomerase activities in *Oxytricha*, *Euplotes*, and human extracts were found to be similarly nuclease sensitive (10–12).

The RNA moiety of the *Tetrahymena* telomerase was initially identified by its co-chromatography with telomerase activity through several column chromatographic fractionations (9). The sequence of this 159-nucleotide RNA species was determined, and was found to include the sequence 3' AACC-CCAAC 5' (14). As described below, this sequence was shown to act as the template for TTGGGG repeat synthesis by telomerase. A prediction made from this work was that the telomerase RNAs from other species would each have a template sequence corresponding to its species-specific telomeric repeat sequence. This prediction has been borne out for the telomerase RNAs of the ciliates *Euplotes* (which has T_4G_4 telomeric repeats), *Glaucoma*, and five other *Tetrahymena* species (which have T_2G_4 telomeric repeats) (15, 16). The *Euplotes* telomerase RNA was first identified by its sensitivity to cleavage by RNase H directed by a DNA oligonucleotide containing T_4G_4 repeats (15). The telomerase RNAs of the *Tetrahymena* spp. and the related ciliate *Glaucoma* were initially identified by cross-hybridization of their single-copy genes to the *Tetrahymena thermophila* gene (15). These ciliate RNAs are the only telomerase RNA sequences that have been identified. By analogy and extension of the model for telomere synthesis deduced from the ciliate telomerases (see below), a template sequence has been suggested for the human telomerase RNA (13), although this RNA has not yet been identified.

The *Tetrahymena* telomerase RNA, like a variety of other small RNAs found in ribonucleoprotein complexes, is an RNA polymerase III transcript.

This was deduced from its lack of a 5' cap, the presence in its gene of a run of T residues encompassing the position of the 3' end of the RNA (14), and, most informatively, its sensitivity to α-amanitin in run-off transcription experiments with isolated nuclei (17). All the ciliate telomerase RNA genes have a pair of highly conserved upstream sequences, and no conserved Box A consensus sequence, suggesting that, like U6 RNA, their RNA polymerase III–mediated transcription is regulated by upstream cis-acting elements (15, 16). The regulation of telomerase RNA transcription has not been studied. However, when overexpressed in vivo, the steady-state level of accumulated telomerase RNA was unchanged from the normal situation, even though the transcription rate of the overexpressed genes, as judged by nuclear run-off experiments, was very high. These findings suggested that excess telomerase RNA transcribed but not assembled into the telomerase RNA complex was degraded (17).

In each sequenced ciliate RNA, the predicted telomeric C-rich putative template sequence is present, as shown in Figure 1 (14–16). The *Euplotes* telomerase RNA is 192 nucleotides in length [recent work has shown that the A residue thought to be upstream of this RNA (16), whose length was previously estimated to be 191 nucleotides, is the 5' nucleotide of this RNA; D. Shippen-Lentz, E. Blackburn, unpublished work]. This RNA contains the 15-nucleotide sequence 3' CCAAAACCCCAAAAC 5', complementary to nearly two full T_4G_4 repeats. However, as described below, only the 12 5' nucleotides of this sequence act as the template (16). The *Tetrahymena* spp. and *Glaucoma* RNAs, all ~160 nucleotides in length, contain a 22-nucleotide sequence that is absolutely conserved among this group of species and which includes the putative template sequence (Figure 1). Apart from the template, and a five-nucleotide sequence 5' to the template which is conserved between *Euplotes* and the distantly related *Tetrahymena* spp. group (see Figure 1), the primary sequences of the telomerase RNAs are highly divergent overall (14–16). This divergence prevents a reliable alignment from being made between the *Euplotes* and *Tetrahymena* RNAs, apart from the template domain. Among the *Tetrahymena* group of RNAs the maximum divergence is ~35%, allowing them to be aligned with each other and compensatory base-pair changes in putative secondary structures to be identified. Such comparison of the sequences of these RNAs allowed a phylogenetically supported secondary structure model to be deduced. Despite their high degree of primary sequence divergence, these RNAs have a strikingly well-conserved secondary structure (15).

THE TEMPLATE FUNCTION OF TELOMERASE RNA   The first evidence indicating that the 3' AACCCCAAC 5' sequence in the *Tetrahymena* telomerase RNA acts as the template for synthesis of the complementary

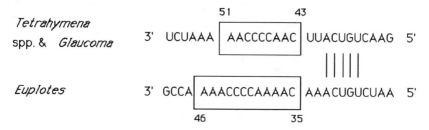

Figure 1 Sequences of the template domain (boxed) and surrounding region of telomerase RNAs from *Tetrahymena* spp. and *Glaucoma* (top) and *Euplotes* (bottom). Nucleotide positions of the template in each RNA sequence are indicated by numbers (14–16).

telomeric TTGGGG repeats came from experiments designed to test the effect on telomerase activity of treatment with RNase H in the presence of DNA oligonucleotides complementary to the putative telomerase RNA sequence (14). Since RNase H will cleave RNA where it is base-paired to a complementary DNA sequence, the loss of telomerase activity in such an experiment indicated the corresponding RNA sequence was required for telomerase. It was found that RNase H cleavage directed by a DNA oligonucleotide whose 3' end was complementary to the 3'AACCCCAAC 5' sequence in the telomerase RNA specifically inactivated telomerase. Furthermore, this oligonucleotide competed with a $(TTGGGG)_4$ primer for telomerase, suggesting it was blocking access of the primer to a site necessary for telomerase activity. A DNA oligonucleotide whose 3' end was complementary to the adjacent sequence, but ending one nucleotide from the 3' end of the putative template sequence, was also extended by GGGGTT repeat addition, suggesting that base-pairing of a DNA oligonucleotide to the telomerase RNA in the vicinity of the template could allow the oligonucleotide to be utilized as a primer. Based on these results, and on the results with 3' primer recognition described above, it was proposed that the 3'AACCCCAAC 5' sequence in the *Tetrahymena* telomerase RNA is the template sequence, and a model was proposed for the telomerase mechanism (14). This model, shown in Figure 2 and described below, was verified in the experiments described next.

A prediction of the model shown in Figure 2 is that altering the template RNA sequence should result in synthesis of telomeric DNA with the corresponding altered sequence. This prediction was confirmed by experiments done in vivo (17). The template sequence of the cloned *Tetrahymena thermophila* telomerase RNA gene was mutated by site-directed mutagenesis to produce three different templates: one with an additional C residue, converting the CCCC sequence to CCCCC, which was predicted to specify the synthesis of G_5T_2 repeats, one with the A at position 44 (see Figure 1) substituted by G (predicting G_4TC repeats), and a third template sequence

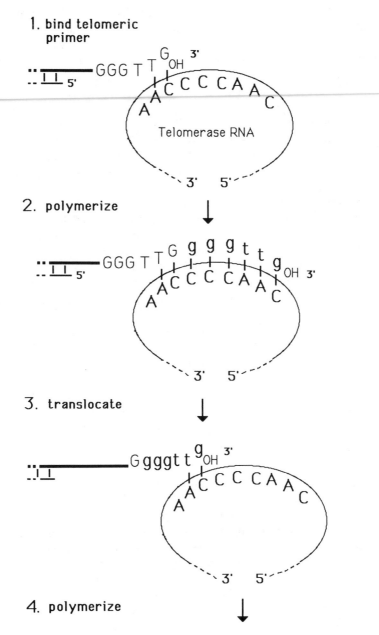

Figure 2 Synthesis of telomeric DNA by the ribonucleoprotein enzyme telomerase from *Tetrahymena*. See text and (14, 16) for explanation.

with the C at position 48 substituted by T (predicting GAG_2T_2 repeats). Each mutated telomerase RNA gene was overexpressed in *Tetrahymena* cells. The first two each resulted in synthesis in vivo of telomeres with the expected mutant telomeric repeat sequences. The C-to-T substitution at position 48, in contrast, resulted in telomere shortening, and none of the expected mutated repeat sequence was detectable in these cells (17). These results are intriguing because they suggest the possibility that this C residue plays a special role in the action of telomerase. This possibility is also interesting in light of a possible base modification of C residues 46, 47, and 48. This was deduced from the resistance of the polynucleotide backbone to cleavage with RNases at these positions, although these positions showed normal sensitivity to alkaline hydrolysis (14).

Independent evidence for the templating function of the RNA moiety of telomerase was obtained by in vitro experiments with the enzyme from *Euplotes* (16). As mentioned above, this enzyme synthesizes T_4G_4 repeats (10). A series of DNA oligonucleotides complementary to the region of the *Euplotes* telomerase RNA that is 3' of the putative templating domain (see Figure 1), and whose 3' ends extended towards, or varying numbers of nucleotides into, the template region, were tested for their ability to prime telomerase (16). In a manner reminiscent of the situation with the *Tetrahymena* enzyme, with which an oligonucleotide complementary to the region 3' of its template was able to prime repeat addition (14), these complementary oligonucleotides competed with a telomeric G-rich DNA primer, suggesting that they interacted with a similar region of the *Euplotes* telomerase. By determining which positions of the primer 3' ends allowed priming of telomeric repeat addition, the functional template domain of the *Euplotes* RNA was demarcated as shown in the boxed sequence in Figure 1 (16).

The Mechanism of Telomeric DNA Synthesis by Telomerase

Figure 2 shows the current model for the mechanism of telomerase (14, 16). The example shown is the *Tetrahymena* telomerase, for which in vivo proof for the template function exists, as described above. However, the 3' boundary of the template region has not been unequivocally deduced by, for example, the methods used to delineate the corresponding border in the *Euplotes* enzyme (14, 16). Otherwise, the model has been confirmed by findings made with the *Tetrahymena* and *Euplotes* enzymes in vitro and in vivo (14, 16, 17), and is supported by results obtained with telomerase activities from *Oxytricha* and human cells (11–13). In this model for synthesis of telomeric DNA:

1. The 3' nucleotides of the terminal chromosomal G-rich overhang in vivo, or of a single-stranded DNA primer oligonucleotide in vitro (thick line;

shown arbitrarily as a TTG 3'end), base-pair with the telomere-comple-
mentary sequence in the telomerase RNA.
2. The chromosomal end is extended by polymerization of dGTP and dTTP
   using the RNA as a template, resulting in the addition of six telomeric
   nucleotides.
3. The extended DNA terminus unpairs from its RNA template and is
   repositioned on the 3' portion of the template, becoming available for
   another round of elongation by telomerase.

Several points about this model merit further consideration.

DNA POLYMERIZATION   A round of telomeric repeat synthesis involves a
six-nucleotide polymerization cycle in the case of *Tetrahymena* and human
telomerases, and an eight-nucleotide cycle in the case of the *Euplotes* and
Oxytricha enzymes. Inherent in this model is the idea that telomerase has
conformational flexibility, since the polymerization active site and the template region of the RNA, which is physically associated with the telomerase
protein(s) in the telomerase ribonucleoprotein complex, must move in relation
to each other as each template position is copied (16). Thus, each polymerization step is expected to be distinct, as at each position the primer-template will
have a particular spatial relation to the telomerase active site. Indeed in vitro,
telomerases have characteristic patterns of pausing points along the template,
which can be different for different telomerases (9–12, 14, 16). For example,
under identical reaction conditions (including dNTP concentration), the
telomerases of *Oxytricha* and *Euplotes* predominantly pause respectively at
the fourth T and the fourth G (corresponding to position 40 on the template;
see Figure 1) during TTTTGGGG repeat synthesis (D. Shippen-Lentz, E.
Blackburn, unpublished work).

Results obtained with a mutated telomerase RNA gene of *Tetrahymena* in
vivo support the idea that the spatial relationship of the template with the rest
of the telomerase ribonucleoprotein is important (18). The *Tetrahymena*
telomerase normally synthesizes perfect T_2G_4 hexanucleotide repeats. When
the template sequence was expanded with an extra C residue as described
above, in addition to synthesis of telomeres with the expected G_5T_2 repeats,
under certain circumstances irregular repeats, with the sequence $G_{6-8}T_2$, were
synthesized in vivo. From this result it was proposed that expanding the
template promoted slippage of the active site on the run of C residues (18).

TRANSLOCATION   One of the most striking features of the telomerase reaction is that it involves not only copying of an internal template, but also an
efficient translocation event which occurs after the last [5' most] residue of
the template has been copied into DNA (see Figure 2). The efficiency of the
translocation step has been deduced from studies showing that telomerase is

processive, at least in vitro (8, 10, 19) (M. Lee, E. Blackburn, unpublished results) (but see below for a discussion of the in vivo situation). In vitro, telomerase will initiate synthesis on a telomeric sequence DNA primer and, in the presence of an excess of the same primer, or a high concentration of a challenging primer added to the reaction, continue to elongate the first primer for up to hundreds of nucleotides before dissociation (19) (M. Lee, E. Blackburn, unpublished results). Because of this high processivity, the rate of elongation of an individual primer molecule before it dissociates can be determined by measuring the length of the elongation products of short reactions fractionated on DNA sequencing gels. Depending on reaction conditions, this growth rate, which is determined by the rates of the translocation as well as all six polymerization steps, has been determined to be ~30 to 70 nucleotides polymerized per minute for the *Tetrahymena* enzyme (19) (M. Lee, E. Blackburn, unpublished results). Similar rates have been estimated for the *Euplotes* activity (D. Shippen-Lentz, E. Blackburn, unpublished results). Despite this processivity, under certain conditions the translocation step is marked by a significant pause in both the *Tetrahymena* and *Euplotes* telomerase elongation reactions (14, 16). Detailed analyses of the kinetics and products of the *Tetrahymena* telomerase reaction with various DNA primers suggest when dissociation of the elongating primer does occur, it is most likely to be at the translocation step (M. Lee, E. Blackburn, unpublished results). This finding is reminiscent of the findings with *Escherichia coli* RNA polymerase, in which termination (transcript-template dissociation) is mechanistically closely linked to pausing (20).

The translocation position clearly does not correspond to the 5' end of the telomerase RNA—i.e. it is not determined by "run-off" from the RNA template. In the *Tetrahymena* telomerase RNA the translocation position (position 43) is followed by two U residues (14), so it could be argued that this telomerase, which normally only utilizes dGTP and TTP, cannot incorporate dATP efficiently, and hence polymerization is prevented from continuing along this RNA sequence. [It should be noted, however, that when the RNA template was mutated to include an rG residue, this telomerase used dCTP to synthesize G_4TC repeats in vivo (17).] However, the corresponding 5' end of the C- and A-containing template in the *Euplotes* telomerase RNA (position 35) is followed by three more upstream A residues, then a C residue, but none of these are copied (16). Thus there must exist in telomerase mechanisms to (*a*) prevent copying of nucleotides upstream of the template domain, and (*b*) promote both dissociation of the newly elongated primer from the template and its repositioning on the 3' portion of the template, preparatory to another round of template copying. These mechanisms could be coupled or distinct.

Models of how the 5' end of the template sequence is defined fall into two classes. The first involves simple steric hindrance: it can be proposed that the telomerase RNA sequence 5' to the template is blocked by tightly bound telomerase protein, so that further polymerization is prevented. Consistent with this idea, experiments determining the accessibility of oligonucleotides complementary to the *Tetrahymena* telomerase RNA to various regions of this RNA indicated that only the region including the template was accessible (14). The rest of the RNA is presumably buried by protein in the ribonucleo-protein complex. But simple steric hindrance does not explain the efficiency with which the elongated primer is dissociated from the template in order to be repositioned. Implicit in the model for the mechanism of telomerase shown in Figure 2, and in the above discussion, is the idea that a DNA-RNA helix builds up along the template. As shown in Figure 2, up to nine Watson-Crick base-pairs of an RNA-DNA helix could exist at the end of a cycle of templated synthesis on the RNA template of the *Tetrahymena* telomerase. With the *Euplotes* telomerase, from results obtained in vitro using primers complementary to the region 3' of the template and extending into the template, it was deduced that a minimum of 11 base-pairs must be dissociated after the initial round of elongation of such a primer, and eight base-pairs must be dissociated in each subsequent elongation round, as each GGGGTTTT repeat is added (16). Is such a long helix in fact made? If so, unwinding it is expected to require energy. The possibility that a DNA-RNA helicase activity is either intrinsic to, or associated with, telomerase needs to be investigated. Although ATP is not required for telomerase activity, the possibility has not been excluded that the same dNTPs incorporated into the product are also used by telomerase in an energy-requiring step.

In an alternative type of model, as polymerization proceeds along the template, steric strain within the RNP is built up. This strain could prevent further polymerization once the 5' limit of the template is reached. The energy for building up such strain could come from the polymerization of the dNTPs and the formation of the DNA-RNA helix as the template is copied. It was further proposed that as a strained RNA structure builds up, the strain could be relieved by dissociation of the DNA-RNA helix when the translocation point is reached (15). In the secondary structure model for telomerase RNA, the template is located in an RNA loop whose ends are confined by base-pairing (15). It was proposed that the strain from building a DNA-RNA helix could result from distortion of this constrained loop (15). However, cleavage of the telomerase RNA sequence downstream from, but not including, the template by oligonucleotide-directed RNase H cleavage did not inactivate telomerase (14). This cleaved sequence is located within the template loop in the secondary structure model (15). Its cleavage would therefore be expected to prevent build-up of steric strain and, hence, translocation, if translocation depended on such strain. This result is still consistent with a model in which steric strain is built up because the RNA is constrained by other interactions, such as RNA-protein interactions, with the rest of the telomerase particle.

Processivity of Telomerase In Vivo

As discussed above, telomerase is a highly processive enzyme in vitro. However, the sequence analysis of telomeres bearing two telomeric repeat unit variants strongly suggests that this is not the case in vivo. A *Tetrahymena* telomerase RNA gene with a mutated template specifying mutant G_5T_2 repeats (17, 18) was expressed in *Tetrahymena* cells along with the wild-type gene, which specifies G_4T_2 repeat synthesis. Cloned telomeres from these cells were found to be made up of a highly interspersed mixture of both mutant G_5T_2 and wild-type G_4T_2 repeats (18). Previous results had indicated that the two repeat sequences result from synthesis by different telomerase ribonucleoprotein molecules carrying RNAs with different template sequences (17). Hence the interspersion of the different repeat units provides information about the processivity of telomerase in vivo. The number of tandem repeats of the same sequence was most frequently one to four, suggesting that telomerase acts in a mostly distributive fashion in vivo (18). Thus, unlike the situation in vitro, telomerase in vivo may synthesize as little as one telomeric repeat following binding to a primer, and then dissociate so that the next repeat can be added by another telomerase molecule with a different template RNA. *Tetrahymena* telomeres in vivo appear to be largely if not completely complexed with telomere-binding proteins (21), and these could compete efficiently with telomerase for the newly elongated telomeric terminus after each round of template copying. In support of this idea, in vitro experiments show that competition occurs between the purified telomere-binding protein of *Oxytricha*, which binds the G-rich overhanging end of the telomere (22–24), and the *Tetrahymena* telomerase (D. Shippen-Lentz, E. Blackburn, C. Price, unpublished results). It has been proposed that the extended G-rich strand synthesized by telomerase is copied by lagging-strand synthesis in vivo (25). Polymerase-primase could therefore also compete for this G-rich strand and curtail telomerase elongation. Alternatively, the difference in processivity in vitro and in vivo could reflect a difference in the intrinsic reaction mode of the enzyme under the two different kinds of conditions.

In some species, the wild-type telomeres are mixtures of two repeat unit sequences of identical length, but with two alternative nucleotides in a given position in the repeat. In both *Paramecium* [G_4T_2 and G_3T_3 repeats (1)] and *Plasmodium* [AG_3T_3 and AG_3T_2C repeats; (1)] the distribution of different repeats appears random, and a histogram of the interspersion of different repeat types in the published telomeric sequences of these organisms is qualitatively very similar to that obtained in the *Tetrahymena* cells transformed with a mutated telomerase RNA gene (18). Results obtained in *Tetrahymena* suggest that *Paramecium* and *Plasmodium* contain two telomerase RNAs, with two alternative nucleotides at one position in the template sequence. Specifically, in *Plasmodium*, one telomerase RNA would have an rG and the other an rA at the position where either a dC or a dT is incorporated in the two repeat types, AG_3T_2C or AG_3T_3. The *Paramecium* telomerase RNAs would have either an rC or rA in the template position specifying the alternative dG or dT in the G_4T_2 and G_3T_3 repeats.

Primer Specificity of Telomerase

PRIMER SPECIFICITY IN VITRO   Telomeric DNA sequences are characterized by a highly conserved G-richness, in the form of clusters of G residues, on the strand that is extended by telomerase (1). This conservation could result from (as yet undefined) specific structural requirements for the functioning of telomeric DNA, which are uniquely met by clusters of G or C residues on one or the other strand. One such function could be some aspect of telomere synthesis by telomerase, since this mechanism appears to be highly conserved (1–3). The only hints of a mechanistic reason for the necessity of telomerase to synthesize G-rich DNA are the observations that mutating the C residue at position 48 in the template of the *Tetrahymena* telomerase RNA appears to inhibit telomere synthesis in vivo (17), and this particular C residue was one of the three C residues which are apparently base-modified (14). These findings suggest the possibility that these template C residues play a special base-specific role in the telomerase reaction, for which another base cannot substitute, and that the synthesis of G-rich telomeric DNA is simply the consequence of the requirement for these C residues in the template. However, currently no definitive data exist to refute or support this possibility.

A feature of telomerases that appears to be conserved is their efficiency of utilization of G-rich DNA as a primer in vitro. This property was proposed to underlie the recognition of G-rich but heterologous telomeric sequences that occurs in vivo [reviewed in (1–3)]. Again, this recognition could be a secondary result of the other possible reasons for G-rich telomeric DNA mentioned above, or it could reflect a more fundamental aspect of the telomerase enzyme. Therefore it is of interest to define which aspects of the telomerase reaction are influenced by G-richness of the primer. Recognition of G residues at the 3' end sequence of the primer can be understood based on their base-pairing the RNA template sequence (Figure 2). However, extensive analysis also indicates that additional sequence features of primers, besides the 3' end sequence, determine their efficiency of utilization by telomerase (7–9, 16); M. Lee, E. Blackburn, unpublished results). Interestingly, although in vitro G-rich DNA sequences of sufficient length or concentration can form respectively intra- or intermolecular structures stabilized by "G-quartets" (26–29), the G-quartet form of DNA is strongly disfavored as a primer by the *Oxytricha* telomerase (30), and apparently also by the *Tetrahymena* enzyme (31; M. Lee, E. Blackburn, unpublished results).

The initial characterization of the telomerase activity of *Tetrahymena* (7, 9) suggested that G-rich primers were utilized by telomerase much more efficiently than non-G-rich primers of comparable length. For the *Tetrahymena* and *Euplotes* activities this was true even for long primers lacking G-richness except for a ..GGG or a ..GG 3' end (7, 9, 16), which would enable them to base-pair appropriately with the RNA template sequence. However, the results were obtained with crude telomerase preparations contaminated with nuclease and other unknown DNA-binding activities, making it difficult to attribute differences in primer utilization unambiguously to telomerase. These comparisons were made using high (micromolar) primer concentrations, and differences in binding versus k_{cat} could not be distinguished. With the further purification of telomerases from *Tetrahymena* and human cells, some of these problems have been minimized and more valid comparisons between primers can be made. Partial purification has led to increased telomerase activities, and allowed priming activity of short but template-complementary oligonucleotides such as the hexanucleotide TTGGGG, previously thought to be unable to act as a primer, to be detected (M. Lee, E. Blackburn, unpublished results). However, kinetic analysis shows that much higher primer concentrations are required for comparable extents of product formation than with longer G-rich primers (M. Lee, E. Blackburn, unpublished results). These results are consistent with a higher affinity for the latter primers. Using extensively purified *Tetrahymena* telomerase, this has been confirmed directly in primer competition assays (M. Lee, E. Blackburn, unpublished results). For single-stranded primers with the same length and 3' end sequence, and which are unable to form G-quartet structures under the assay conditions used, primers that have a G-rich sequence 5' to the template-complementary region at their 3' end are recognized much more efficiently by the *Tetrahymena* telomerase than non-G-rich primers (M. Lee, E. Blackburn, unpublished results). Whether this is true for the telomerases of other species is currently being debated, although the weight of current evidence is not inconsistent with this conclusion for these telomerases as well.

Results with different telomerases have often been difficult to compare because experiments have been carried out in different ways. However, using both the *Tetrahymena* and (unpurified) *Oxytricha* activities, the activity of various DNA primers has been compared in "abortive initiation" assays, in which polymerization is confined to one or a few nucleotides by including only one dNTP. With the *Oxytricha* enzyme, the k_{cat}/K_m ratio (which measures substrate specificity of the enzyme) of T_4G_4 is higher than that for $(T_4G_4)_2$ (9). From this result the authors concluded that only base-pairing with the template was important for primer utilization, and G-richness did not increase primer utilization. However, as the k_{cat} for the abortive initiation reaction can be dominated by dissociation of the elongated primer product, these results are not informative about the more normal processive reaction in which large numbers of nucleotides are added. Contrasting results were obtained with the extensively purified *Tetrahymena* enzyme (M. Lee, E. Blackburn, unpublished results). With this telomerase, in the abortive initiation reaction, k_{cat}/K_m is higher for $G_4T_2G_4$ than for T_6G_4 or $TA_3T_2G_4$. More significantly, with the extensively purified *Tetrahymena* enzyme, in the processive reactions k_{cat}/k_m is higher for $G_4T_2G_4$ than for T_6G_4, T_2G_4, or $TA_3T_2G_4$. Hence in all these experiments the *Tetrahymena* enzyme exhibited a clear specificity for a G-rich oligonucleotide, compared with other primers of the same length and with the same (or more extensive) base-pairing possible with the template.

In a study with partially purified human telomerase activity, a somewhat different conclusion was reached: that interaction with the template was more important than the G-rich nature of the rest of the primer in determining its priming activity (13). However, in contrast to the *Tetrahymena* telomerase studies, these assays were carried out in very long reactions. Results with the *Tetrahymena* enzyme have shown that in long reactions, measurement of the extent of the reaction is complicated by dissociation and re-initiation events (M. Lee, E. Blackburn, unpublished results). Therefore the conclusions made with the different enzymes as to primer specificity have not yet been made under comparable conditions, so the question of species variation in the relative importance of different aspects of primer recognition remains open.

HEALING OF BROKEN CHROMOSOMES BY TELOMERASE IN VIVO   All the available results with telomerases from different species are consistent in showing that some combination of template pairing and G-richness, and possibly other, as yet poorly defined, structural features of the primer, are important for its recognition by telomerase in vitro. Oligonucleotides consisting of the sequence of the position to which telomeric repeats are added to a healed human chromosome have been tested in vitro and shown to act as primers for the human and *Tetrahymena* telomerases (13, 31). However, this sequence had a G-rich portion, and in this respect was not unlike telomeric DNA sequences. In contrast, much indirect evidence indicates that under some circumstances in vivo, telomeric repeats are added to broken chromosome ends lacking any semblance of telomeric sequences (2). Direct in vivo evidence showing that this addition is accomplished by telomerase has been obtained in *Tetrahymena* (18). These experiments showed that, at least in one developmental stage, in vivo telomerase adds telomeric repeats onto nontelomeric sequences lacking G-rich sequences or the ability to pair at their 3' end with the telomerase RNA. This conclusion came from experiments in which a *Tetrahymena* telomerase RNA gene with a mutated template specifying mutant G_5T_2 repeats was overexpressed in *Tetrahymena* cells, as described above. The mutant *Tetrahymena* telomerase RNA was used to monitor telomerase action in vivo, during the particular developmental stage in which a developmentally programmed chromosome fragmentation process occurs in this organism. Telomerase was shown to heal the nontelomeric ends generated by the chromosome fragmentation process, by adding the first telomeric repeats de novo onto these ends (18).

Concluding Remarks and Future Directions

Telomerase is a ribonucleoprotein polymerase whose specialized mechanism involves copying an internal RNA template sequence into DNA. Many interesting questions about this mechanism remain unanswered. Whether the RNA moiety of telomerase collaborates with the protein moiety of the enzyme in catalysis or substrate binding, in addition to the template interaction with the DNA primer, or has catalytic activity in the absence of protein, or simply supplies the template sequence, is not yet clear. The protein component(s) of the telomerase ribonucleoprotein have not been biochemically identified. Reconstitution of telomerase in vitro therefore awaits this identification. The mechanisms that accomplish the efficient translocation and resulting high processivity characteristic of the in vitro reaction of telomerase are yet to be determined.

ACKNOWLEDGMENTS

I thank Margaret Lee for critical reading of the manuscript. Support was provided by grants GM26259 and 32565 from the National Institutes of Health to E.H.B.

Literature Cited

1. Blackburn, E. H. 1990. *Science* 249: 489–90
2. Blackburn, E. H. 1991. *Nature* 350: 569–73
3. Blackburn, E. H., Szostak, J. W. 1984. *Annu. Rev. Biochem.* 53:163–94
4. Zakian, V. A. 1989. *Annu. Rev. Genet.* 23:579–604
5. Klobutcher, L. A., Swanton, M. T., Donini, P., Prescott, D. M. 1981. *Proc. Natl. Acad. Sci. USA* 78:3015–19
6. Henderson, E. R., Blackburn, E. H. 1989. *Mol. Cell. Biol.* 9:345–48
7. Greider, C. W., Blackburn, E. H. 1985. *Cell* 43:405–13
8. Blackburn, E. H., Greider, C. W., Henderson, E., Lee, M. S., Shampay, J., Shippen-Lentz, D. 1989. *Genome* 31: 553–60
9. Greider, C. W., Blackburn, E. H. 1987. *Cell* 51:887–98
10. Shippen-Lentz, D., Blackburn, E. H. 1989. *Mol. Cell. Biol.* 9:2761–64
11. Zahler, A. M., Prescott, D. M. 1988. *Nucleic Acids Res.* 16:6953–72
12. Morin, G. B. 1989. *Cell* 59:521–29

13. Morin, G. 1991. *Nature* 353:454–56
14. Greider, C. W., Blackburn, E. H. 1989. *Nature* 337:331–37
15. Romero, D., Blackburn, E. 1991. *Cell* 67:343–53
16. Shippen-Lentz, D., Blackburn, E. H. 1990. *Science* 247:546–52
17. Yu, G.-L., Bradley, J. D., Attardi, L. D., Blackburn, E. H. 1990. *Nature* 344:126–32
18. Yu, G.-L., Blackburn, E. H. 1991. *Cell* 67:823–32
19. Grieder, C. W. 1991. *Mol. Cell. Biol.* 11:4572–80
20. Fisher, R. F., Yanofsky, C. 1983. *J. Biol. Chem.* 258:8146–50
21. Budarf, M., Blackburn, E. H. 1986. *J. Biol. Chem.* 261:363–69
22. Gottschling, D. E., Zakian, V. A. 1986. *Cell* 47:195–205
23. Price, C. M., Cech, T. R. 1989. *Biochemistry* 28:769–74
24. Raghuraman, M. K., Dunn, C. J., Hicke, B. J., Cech, T. R. 1989. *Nucleic Acids Res.* 17:4235–53
25. Shampay, J., Szostak, J. W., Blackburn, E. H. 1984. *Nature* 310:154–57
26. Henderson, E. R., Moore, M., Malcolm, B. A. 1990. *Biochemistry* 29:732–37
27. Sen, D., Gilbert, W. 1990. *Nature* 344:410–14
28. Sundquist, W. I., Klug, A. 1989. *Nature* 342:825–29
29. Williamson, J. R., Raghuraman, M. K., Cech, T. R. 1990. *Cell* 59:871–80
30. Zahler, A., Williamson, J. R., Cech, T. R., Prescott, D. M. 1991. *Nature* 350:718–20
31. Harrington, L. A., Greider, C. W. 1991. *Nature* 353:451–54

Annu. Rev. Biochem. 1992. 61:131–56

POLYMERASE CHAIN REACTION STRATEGY

Norman Arnheim

Molecular Biology Section, University of Southern California, Los Angeles, California 90089-1340

Henry Erlich

Human Genetics Department, Roche Molecular Systems, 1400 53rd Street, Emeryville, California 94608

CONTENTS

131

0066-4154/92/0701-0131$02.00

INTRODUCTION

About 20 years ago, recombinant DNA technology was introduced as a tool for the biological sciences. Molecular cloning has allowed the study of the structure of individual genes of living organisms. This method depends on the replication of the DNA of plasmids or other vectors during cell division of microorganisms. In 1984, a team of scientists at Cetus Corporation developed a DNA amplification procedure based on an in vitro rather than an in vivo process. Known as the polymerase chain reaction (PCR; 1–3), this method can produce large amounts of a specific DNA fragment from a complex DNA template in a simple enzymatic reaction. Cell-free gene amplification by PCR can simplify many of the standard procedures for cloning, analyzing, and modifying nucleic acids. PCR is characterized by the three Ss: selectivity, sensitivity, and speed. Virtually pure DNA fragments from complex genomes can be obtained in a matter of hours rather than the weeks or months traditional cloning requires. The method utilizes a DNA polymerase and two oligonucleotide primers to synthesize a specific DNA fragment from a single-stranded template sequence. The amount of starting material needed for PCR can be as little as a single molecule rather than the usual millions of molecules required for standard cloning and molecular biological analysis. Although purified DNA is used in many applications, it is not required for PCR, and crude cell lysates also provide excellent templates. The DNA need not even be intact, in contrast to the requirements of other standard molecular biological procedures, as long as some molecules exist that contain sequences complementary to both primers (see below). The speed and sensitivity of PCR have been widely recognized by scientists in both medicine and basic biology, and the method has been applied to problems that a few years ago were thought to be inaccessible to molecular analysis.

THE BASIC METHOD

The principle of the method is shown in Figure 1. To amplify a specific DNA segment by PCR it is not necessary to know the nucleotide sequence of the target DNA. Ideally two small stretches of known unique sequence that flank the target are used to design two oligonucleotide primers. The length of the primers (usually ≥ 20 bases) must be sufficient to overcome the statistical likelihood that their sequence would occur randomly in the overwhelmingly large number of nontarget DNA sequences in the sample. PCR is carried out in a series of cycles. Each cycle begins with a denaturation step to render the target nucleic acid single-stranded. This is followed by an annealing step during which the primers anneal to their complementary sequences so that their 3' hydroxyl ends face the target. Finally each primer is extended through the target region by the action of DNA polymerase. These three-step cycles are repeated over and over until a sufficient amount of product is produced. A critical requirement is that the extension products of each primer extend far enough through the target region to include the sequences of the other flanking primer. In this way, each extension product made in one cycle can serve as a template for extension in the next cycle. This results in an exponential increase in PCR product as a function of cycle number. After the first few cycles the major product is a DNA fragment that is exactly equal in length to the sum of the lengths of the two primers and the intervening target DNA.

The earliest PCR experiments (1, 2) utilized the Klenow fragment of *Escherichia coli* DNA polymerase I at a temperature of 37 °C and often produced incompletely pure target product as judged by gel electrophoresis. The isolation of a heat-resistant DNA polymerase from *Thermus aquaticus* (Taq) allows primer annealing and extension to be carried out at an elevated temperature (3), thereby reducing mismatched annealing to nontarget sequences. This added selectivity results in the production of large amounts of virtually pure target DNA. Beginning with 1 μg of human DNA, which contains about 300,000 copies of each unique sequence, 25 cycles of PCR can generate up to several μg (a few picomoles) of a specific product several hundred base pairs in length.

Another important advantage of Taq polymerase is that it escapes inactivation during each cycle, unlike the Klenow enzyme, which had to be added after every denaturation step. This has allowed automation of PCR using machines that have controlled heating and cooling capability. A number of thermocyclers are commercially available at relatively low cost. This development has been a significant factor in the rapid application of this technology by the scientific community. Since the first report in 1985, more than 5000 scientific papers have been published using PCR (G. Mcgregor, personal communication). The large number of publications of course makes it impossible for us to review all of the important contributions to the development and application of PCR technology.

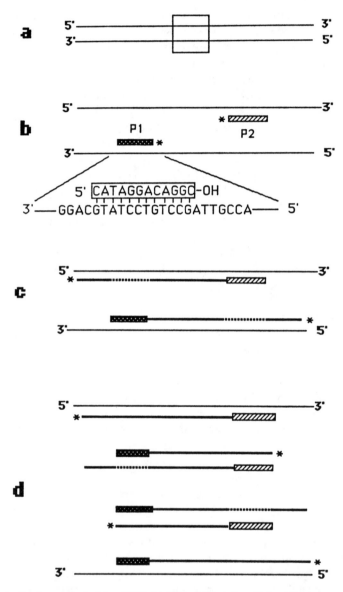

Figure 1 Fundamental principle of PCR. (*a*) DNA double helix with boxed target segment. (*b*) Hybridization of PCR primers to opposite strands of the region flanking the target. The 3' end of each primer is indicated by an *. The base-base interactions of one primer (boxed) with the region flanking the target is shown below. (*c*) Result of extending each primer with a DNA polymerase. The dotted region of each extension product indicates the portion that is complementary to the other primer. (*d*) PCR products after the material in (*c*) was subjected to another round of amplification. Reprinted with permission of the American Institute of Biological Sciences (139).

BIOCHEMISTRY OF THE REACTION

PCR Specificity

In a typical mammalian genome (with a complexity on the order of 10^9 nucleotides), a 1000-base-pair target would represent only one millionth of the DNA. The annealing of the primers to other sequences besides the target in such complex templates could result in extension and eventually amplification of nontarget sequences.

Specificity is achieved by designing primers flanking the target that are of sufficient length so that their sequence is virtually unique in the genome. The specificity of the interaction of the primer with the desired template versus nontarget DNA is temperature and salt concentration dependent, and appropriate conditions must be determined empirically. The conditions of the reaction must also be compatible with full activity of the polymerase.

The Taq polymerase has measurable activity at room temperature and at almost all of the temperatures up to the DNA denaturation temperature (4). The activity at low temperatures allows extension to take place on primer-template complexes that are not perfectly matched along their whole length. For optimal specificity, the highest annealing temperature possible should be used to reduce nonspecific primer extension. Nonspecific extensions themselves are not necessarily a serious problem. However, once an extension product is present in the reaction, there is a finite chance that it could serve as a template for nonspecific extension by another primer. This would create a product that has primer sequences at each end and therefore could be amplified along with the desired target; if the nonspecific product is smaller than the target, it may be amplified more efficiently.

It is the usual practice to set up the reaction at room temperature and to begin it with a 92–96 °C denaturation step. It has been suggested that even while the samples are being prepared primer extension by the Taq DNA polymerase could occur. At room temperature there would be little specificity to primer-template interactions. Experiments have shown that some of the nonspecific amplification products can be eliminated under so-called "hot start" conditions (5, 6). This approach keeps the sample at a temperature greater than the calculated annealing temperature for the specific primer before the reaction is started.

Limiting amplification to the desired target can also be enhanced using a nesting strategy that involves two different rounds of PCR. After amplification with one set of primers, a small aliquot is taken and amplified in a second round with either two new primers that are internal to those used in the first round (nesting; 2) or one new internal primer and one of the original primers (heminesting; 7). This strategy works because, unlike the desired target, any nontarget sequences that were amplified in the first round cannot be further amplified with the internal target-specific primer(s) used in the second round. Recently a new approach to the heminesting strategy has been developed in which it is not necessary to carry out two distinct rounds in separate tubes. Following the first round in which all three primers are present, the thermal cycing conditions are changed so as to preferentially enhance the annealing of the nested primer (8).

Details of the Reaction

PCR is quite complex even though there are a limited number of reagents used. In addition to a genomic DNA sample usually containing less than 1 amol of specific target sequence, the 25–100 μliter volume includes 20 nmol of each of the four deoxynucleoside triphosphates (dATP, dCTP, dGTP, and TTP), 10 to 100 pmol of each primer, the appropriate salts and buffers, and DNA polymerase. The nucleotide concentration must be sufficient to saturate the enzyme, but not so low or unbalanced as to promote misincorporation (see below). The primer concentration must be high enough to anneal rapidly to the single-stranded target and, in later stages of the reaction, faster than target-target reassociation. Temperature control and timing are also important. Denaturation must be efficient, but the temperature must not be too high or held for too long a period, because the Taq polymerase, although heat-resistant, is not indefinitely stable (4). The temperature used for annealing must maximize specific primer annealing and polymerase elongation but not sacrifice yield by reducing primer-template hybridization.

One of the advantages of PCR is its speed. The Taq polymerase can extend sequences up to 1000 bases in length in less than a minute (4). With short PCR cycle times, rapid temperature equilibration of the sample is necessary. The reaction mixture is usually overlaid with mineral oil to prevent evaporation, thereby contributing to rapid thermal equilibration and eliminating a concentration of reagents during the course of the reaction. A newly designed thermocycler is capable of very rapid temperature change, and because the whole sample tube including the cap is heated, mineral oil is not required to prevent evaporation (9). In general, using 20-nucleotide-length primer sequences with a 50% GC content, denaturation at 92–96 °C for 30–60 seconds, annealing at 55–60 °C for 30 seconds, and extension at 72 °C for 1 minute is satisfactory for targets less than 500 bp. It is often found that a simple two-step cycle (95 °C denaturation; 60 °C annealing and extension) also gives excellent results.

If every template molecule in the sample is completely extended at each cycle then the amplification efficiency is 100%. In practice, efficiencies can vary quite significantly throughout the course of amplification. For example, the amount of product produced at each cycle eventually levels off. This plateau can be explained by two phenomena. As the concentration of double-stranded product reaches high levels, competition increases between annealing of template (PCR product) to primer and reannealing of the complementary template strands. Secondly, the amount of enzyme is finite and eventually there is not enough to extend all of the primer-template complexes in the allotted time. The efficiency may also depend upon the amount of original target and on properties of the target that are not well understood such as the likelihood of secondary structure in the single-stranded template. In general, long targets are less efficiently amplified than short ones, but there are exceptions to this generalization, which could be due to the actual sequences involved. Finally, some primers seem to work better than others even when the target sequence is the same, but the rules that govern this phenomenon are not yet known. The addition of reagents that tend to reduce secondary structure formation can in some circumstances improve amplification (10, 11).

The average efficiency of a series of PCR cycles can be described by the following equation: $N = n (1 + E)^c$, where N = the final amount of product, n = the initial amount of target, E = the efficiency of amplification, and c = the number of PCR cycles. Because of the plateau effect, the measurement of the efficiency of PCR in its exponential phase must be carried out when only small amounts of product are present. The reported average efficiencies are usually greater than 0.60 and can be as high as 0.95 (3, 12–14).

In general, PCR has been used as a tool rather than being the object of detailed investigations of its complex reaction dynamics. Trying to understand the nature of the structure of the nucleic acids, the protein–nucleic acid interactions, and the fundamental enzymology and structure of the thermostable polymerases could lead to important advances in this technology.

PROPERTIES OF THERMOSTABLE POLYMERASES

As discussed above, the introduction of a thermostable DNA polymerase from *Thermus aquaticus* (Taq polymerase) into the PCR greatly simplified the PCR protocol and allowed the development of simple thermal cycling instruments to automate the reaction. It also dramatically increased the specificity and yield of the PCR by allowing primer annealing and extension to be carried out at higher temperatures. The Taq DNA polymerase isolated from the cloned *Thermus aquaticus* gene expressed in *E. coli* (15) has a specific activity of 200,000 units/mg and an inferred molecular weight of 93,910 (4). It has a temperature optimum of 75–80 °C, depending on the DNA template. Under appropriate conditions, it is highly processive and has been reported to have an extension rate of >60 nucleotides per second at 70 °C using M13 phage

DNA as template (16). At lower temperatures, the extension rate and the extent of processivity are both reduced.

Taq polymerase lacks a 3'–5' exonuclease activity or "proofreading" function but does contain a 5'–3' exonuclease activity during polymerization (16–19). A primer annealed to a single-stranded template will be degraded from its 5' end by polymerase extension of a second primer annealed to the template 5' of the first primer. This activity has recently been developed into an assay for detecting the presence of specific DNA templates in a sample (20). Genetically engineered variants of the Taq enzyme lack this activity but carry out polymerization (8).

Under some conditions, the Taq DNA polymerase is capable of adding a single nontemplate-directed nucleotide (preferentially dATP) to a blunt-ended duplex DNA fragment (21). This activity, common to other polymerases that lack the 3'–5' proofreading exonuclease, can pose problems for blunt-ended cloning of PCR products. This activity may also be involved in the formation of the so-called "primer-dimer," a double-stranded PCR artifact consisting of the two primers and their complementary sequences (22) and, occasionally, additional bases inserted between the primers. The formation of "primer-dimer" probably occurs when one primer is extended using the other primer as a template, and would be expected to be dependent on primer sequence, primer concentration, and enzyme concentration. Primer-dimer usually is seen in PCR experiments involving high cycle numbers, such as those necessitated in amplifying rare targets. When primers are inadvertently designed with partially complementary 3'-termini, primer-dimer formation appears to be more frequent. The amount of primer-dimer formed is inversely proportional to the amount of the target PCR product because of enzyme competition between the various DNA molecules being replicated. Efficient primer-dimer amplification reduces the sensitivity of specific target analysis. The initial formation of these products can occur at ambient temperatures; "hot start" (see above) eliminates this phase of the thermal profile and, consequently, can reduce the accumulation of "primer-dimer."

Recently, a variety of thermostable DNA polymerases with different properties have been isolated from other bacteria. One, from the thermoacidophilic archebacterium *Sulfolobus acidocaldarius,* has been shown to carry out polymerization at 100 °C (23–25), which could facilitate the amplification of regions of high secondary structure and enhance specificity. In the case of Taq polymerase, the enzymatic incorporation of modified bases such as 7-Aza-dGTP has proved useful in the amplification of sequences with secondary structures in GC-rich regions (26). Some of the new thermostable polymerases may allow the efficient amplification of larger PCR products (E. Rose, personal communication). The introduction of thermostable accessory proteins may also prove helpful in increasing the processivity of polymerases during PCR and allow the amplification of longer products.

The search for new thermostable polymerases has resulted in the discovery of one with reverse transcriptase activity (27). A polymerase from *Thermus thermophillus* can reverse transcribe RNA efficiently in the presence of $MnCl_2$. The DNA polymerase activity of the same enzyme can be stimulated by chelating the $MnCl_2$ and adding $MgCl_2$, allowing cDNA synthesis and PCR amplification to be carried out in a single-enzyme, single-tube reaction. The thermostability of the enzyme could minimize the effect of secondary structure in the RNA template and allow efficient cDNA synthesis at high temperatures. The topic of cDNA amplification is further discussed below.

Finally, polymerases from *Thermoplasma acidophilum, Thermococcus litoralis,* and *Methanobacterium thermoautotrophicum* have been reported to have 3'–5' exonuclease activities (28–30). The consequences for misincorporation during polymerization are discussed below.

PRACTICAL CONSIDERATIONS

While PCR provides a very powerful tool for molecular analysis, the uninitiated investigator should be aware of some aspects of the procedure that should be considered in the design and interpretation of experiments.

Misincorporation

The rate of nucleotide misincorporation during the PCR has been estimated in various studies by determining the frequency of misincorporated nucleotides in the sequence of cloned PCR products and calculating an average rate per cycle (3, 31, 32). The misincorporation rate would be expected to reflect not only the properties of the polymerase itself (e.g. presence or absence of 3'–5' exonuclease proofreading activity), but also the reaction conditions. Biochemical fidelity measurements of "nonproofreading polymerases" (e.g. AMV reverse transcriptase and *Drosophila melanogaster* DNA polymerase α) revealed that the nucleotide misincorporation is dependent upon the dNTP concentration (33). The initial estimate for nucleotide misincorporation (1.7×10^{-4} nucleotides polymerized per cycle) by Taq polymerase during PCR (3) used 1.5 mM of each dNTP, 10 mM $MgCl_2$, and a 37 °C annealing temperature to amplify a 272-base-pair fragment from the *HLA-DPB1* gene. More recent studies (31, 32) gave an average nucleotide misincorporation rate of 5×10^{-6} per cycle. These studies used lower dNTP and $MgCl_2$ concentrations (200 μM of each dNTP and 1.5 mM $MgCl_2$) as well as higher annealing temperatures (54–55 °C). In general, the rate of misincorporation as well as that of extension from a 3' mispaired primer may be reduced by minimizing the annealing/extension time, maximizing the annealing temperature, and by minimizing the dNTP and $MgCl_2$ concentrations (34). However, reaction conditions that maximize fidelity may reduce PCR efficiency. Also, although the calculations yield an "average" misincorporation rate per cycle, the fidelity of thermostable polymerases may vary during the PCR (4). The misincorporation rate of thermostable polymerases with $3'-5'$ exonuclease proofreading activity is likely to be lower. A comparison of the fidelity of the Taq and *T. litoralis* polymerases shows that the latter enzyme, which has proofreading activity, may misincorporate at 25% of the rate of the Taq enzyme (30).

For most PCR applications, such as direct sequencing (see below) or characterization by hybridization probes or restriction enzyme digestions, it is the population of amplified products that are analyzed and therefore, rare misincorporated nucleotides do not pose any problem in the interpretation of experiments. The actual fraction of amplified molecules that contain a mutation is related to the number of bases in each target molecule, the rate of misincorporation per base per cycle, and the number of cycles. A mathematically precise theory for estimating the frequency of molecules without any misincorporations has recently been put forward (35). Using the most recent average estimate of misincorporation (8.5×10^{-6}; 31, 32), only 3% of the molecules of 200-base-pair PCR product will contain a misincorporated base somewhere in the sequence after 30 cycles. Of course most of the molecules with misincorporated nucleotides would not contain the same mutation. Additional estimates of misincorporation are required since the rates might vary for the particular sequence studied and the actual PCR conditions used.

While misincorporations appear to have relatively little effect on analysis by most methods, the determinations of individual cloned sequences derived from PCR products of course reveal such rare errors. It is advisable either to sequence multiple clones derived from a single PCR or to clone the products from several independent amplification reactions to identify any potential errors.

Hybrid PCR Products

When the sample being amplified is heterozygous (two different alleles for the target locus) or when many different but related sequences are amplified with the same primers (e.g. a multigene family), in vitro recombination or template-strand switching is theoretically possible.

Although a very rare event, PCR hybrid sequences can be formed when a primer that has been partially extended on one of the templates is annealed and extended on a similar template in a subsequent cycle (3, 36–38). For this so-called "jumping" to occur, the concentration of partially extended PCR primers must be quite high. The occurrence of very rare hybrid PCR products amplified from genomic DNA was initially discussed in Saiki et al (3). The model experiments that studied this phenomenon with plasmid DNA, however, used high concentrations of template, thereby increasing the likelihood that the levels of partially extended primers would be sufficient to promote jumping (36–38). The likelihood of partial primer extension is also dependent on the distance between the primers, secondary structure, the time allowed for polymerase extension, the processivity of the polymerase, and the extent to which the original target DNA sequences contained single strand breaks. Again, most methods of analysis characterize the population of PCR products, not rare individual hybrid products. In general, the formation of hybrid molecules can be minimized if the minimum number of PCR cycles required for product detection are carried out.

Under some circumstances jumping PCR may be beneficial (36). Assume a sample of DNA is highly degraded and no completely intact target molecules exist. If the DNA contained a set of partially overlapping fragments from the target, then repeated jumping could eventually produce a full-length PCR product.

Preferential Amplification

Under some circumstances preferential amplification of one allele relative to the other can occur in heterozygous samples. One possible mechanism for differential priming is that one of the two alleles may have a rare DNA sequence polymorphism in the primer-binding site. Another potential mechanism is that one allele could have a higher GC content, and therefore require a higher denaturation temperature to generate functional PCR template than does the other allele. In the case of length polymorphisms detected as PCR products of different size, the smaller allele can, under some circumstances, be amplified preferentially. In addition, one allele in a heterozygous sample can be amplified preferentially due to stochastic fluctuation when sampling a very low number of target molecules.

Contamination

Contamination of sample tubes by PCR product derived from earlier experiments can cause serious problems. Amplified DNA from a single sample can contain several picomoles of PCR product. Even one millionth of that product could have disastrous consequences if it contaminated a 1-microgram sample of genomic DNA containing 300,000 copies of a unique sequence. When carrying out PCR experiments, aerosol formation and other possibilities of contamination must be minimized. Physically separating the areas in the laboratory where PCR reactions are set up from those where the PCR products are analyzed can be of significant help. A set of pipettes that are dedicated to setting up PCR reactions is also warranted. Reagent solutions should be aliquoted. Micropipette tips with built-in filters can also be useful. A number of other precautions in the technical aspects of carrying out PCR are discussed in Kwok & Higuchi (39).

Recently, several protocols for eliminating contaminating DNA sequences carried over from previous reactions have been developed. UV irradiation of the reaction tube contents can damage any contaminating sequences before the DNA template and polymerase are added (40). Another approach uses dUTP instead of dTTP for PCR (41; J. J. Sninsky et al, in preparation). PCR products therefore are susceptible to degradation by the enzyme uracil N-glycosylase. Before beginning a PCR experiment, a sample is treated with the glycosylase, which will destroy any carried-over PCR product but will not degrade the desired genomic DNA template. Following heat inactivation of the glycosylase, PCR is carried out with dUTP. In a still different approach to the contamination problem, a photochemical reagent is added before the reaction and activated after PCR is completed. The reagent will cross-link the two strands of the PCR product and render them unamplifiable (42).

GENE AND cDNA AMPLIFICATION

Cloning PCR Product

PCR produces large amounts of product derived from most if not all of the copies of the target present in the original sample, but individual product molecules can be cloned by recombinant techniques. Cloning can be simplified if the primers used for PCR are constructed so that sequences containing a restriction endonuclease cleavage site are appended to their 5' end (43). Following restriction enzyme digestion, the PCR product has "sticky ends," which can be ligated to an appropriately digested cloning vector. PCR product can also be cloned following blunt-end ligation, although nontemplate-directed nucleotide addition (see above) can lead to lower efficiency.

Recently, a strategy for introducing "sticky ends" into the product and vector by appropriately designed PCR primers has eliminated the need for restriction enzyme digestion and in vitro ligation (44).

Single-Sided DNA Amplification

INVERSE PCR   The basic method of PCR requires that DNA sequence information be available on both sides of the target sequence. Inverse PCR allows amplification of targets where precise DNA sequence information is available on only one side (45–47). If two primers are designed in the known region so that, unlike conventional PCR, they face away from each other, the flanking sequences can be amplified if a DNA fragment containing the known and flanking sequences is first circularized by ligation following restriction enzyme digestion.

LIGATION-MEDIATED PCR   Inverse PCR can be difficult to carry out, primarily because self-ligation of the ends of linear DNA fragments in complex genomes is rather inefficient. A different approach to single-sided amplification involves the addition of a known sequence, in the form of a linker, to one end of a duplex DNA fragment whose other end has sufficient sequence information available to allow a primer to be made (48–50). The linker is added by DNA ligation under normal conditions. One elegantly designed primer for this purpose (48),

5'   OH-GCGGTGACCCGGGAGATCTGAATTC-OH   3'
3'   OH - CTAGACTTAAG-OH   5',

has the following important attributes. One end is a blunt-ended duplex, while the other is single stranded. Neither of the 5' ends is phosphorylated. Thus, the linkers cannot ligate to each other and only the long strand will be ligated to the blunt end of a genomic DNA fragment. The short strand is AT rich compared to the long strand and thus, after ligation, an appropriate temperature can be found that will prevent the two linker strands from annealing. PCR can be carried out using the long linker sequence as one primer and a genomic sequence from the other end as the other primer. Sequences designed to be used as primer sites can also be added to the 3' end of a target sequence by enzymatic extension using terminal transferase (see below).

Whole Genome PCR

Another application of the linker ligation strategy allows amplification of every DNA sequence present in a complex mixture. While this approach lacks specificity, it is useful when starting out with a very small amount of a complex DNA and large amounts are desired for some purpose. For example, DNA libraries have been constructed starting from specific segments of chromosomes dissected from whole chromosomes on a microscope slide (51, 52). After dissection, the DNA is digested with a restriction enzyme and the fragments ligated to a plasmid. Plasmid sequences flanking the insertion site are used for PCR primers (51). Whole-genome PCR can also be applied to the analysis of protein-DNA interactions (see below).

cDNA Amplification

GENERAL APPROACHES   Amplification of cDNAs where sequence information permits primer design at both flanking regions is straightforward. Total cDNA can be made with a reverse transcriptase (RT) using standard procedures including oligo dT or random hexanucleotide primers (53). Amplification of a specific cDNA with the appropriate message-specific primers is then carried out (54). Whenever possible, primers used for amplifying cDNAs of RNAs should not be positioned within the same exon. This strategy will minimize amplification from small amounts of contaminating DNA (55) since, if the primers lie within the same exon, the size of the PCR product produced from genomic DNA would be indistinguishable from the product expected from the cDNA target. If a nonprocessed RNA is the PCR target, RNase-free DNase I digestion should be employed to destroy possible contaminating DNA sequences and confirmed with appropriate controls, including samples amplified without a prior reverse transcription step. The fact that primers can be designed for any pair of exons led to an early application of cDNA PCR in the analysis of RNA processing (56).

mRNA QUANTITATION   Several approaches using PCR to quantitate the amount of specific mRNAs have been published. The final amount of PCR product depends not only on the initial amount of mRNA but also on the efficiency of reverse transcription and PCR. Two basic approaches have been used. In one, two different cDNAs are amplified and, from the ratio obtained, absolute levels of one can be calculated if already known for the other, assuming that the two cDNA targets are amplified at similar efficiencies (14). An alternative approach (57) involves spiking each sample with a known amount of a control DNA sequence. The control sequence is designed to be as similar in sequence as possible to the cDNA target so that they will have identical amplification efficiencies using the same pair of primers. The target and control can differ by as little as a single nucleotide substitution that alters a restriction enzyme site. Since each unknown sample receives a known amount of the control sequence, the absolute amount of the target can be determined by comparing the levels of the control and target products produced in the same PCR reaction after restriction enzyme digestion. While this method should be accurate for quantitating cDNA levels, it does not take the efficiency of reverse transcription into consideration. However, Wang et al (58) spike the sample with an RNA transcript from a plasmid containing a modified target DNA. A known amount of the RNA transcribed from the plasmid is added to the sample as a control sequence.

SINGLE-SIDED cDNA AMPLIFICATION   In some cases the available sequence information may not permit both flanking primers to be designed for cDNA amplifiation. Two examples of this are T cell receptor and immunoglobulin mRNAs where a 3' constant region suitable for primer design is associated with 5' variable regions of unknown sequence. The linker ligation strategy cannot be used because the product of reverse transcription using a primer from the 3' end of the message, although blunt-ended, is an RNA-DNA hybrid and therefore not a substrate for DNA ligase. PCR of RT copies of specific RNAs where only one primer sequence is available, however, can be carried out using "single-sided" or "anchor" PCR (59–61). Assume that specific sequence information for primer construction is only available at the 3' end of the messenger RNA. Following reverse transcription, the 3' end of the cDNA is modified by addition of nucleotides with terminal deoxynucleotidyl transferase. The modified cDNA product containing, for example, a string of dGs at the 5' end (60) is a substrate for second strand synthesis with an "anchor" primer containing, in addition to a poly-C region at its 3' end, a unique sequence at its 5' end. PCR is carried out with a primer containing the unique sequence portion of the "anchor" and the original message-specific primer from the 3' end. If specific sequence information is known only at the 5' end of the RNA, an oligo dT containing anchor primer can be used for the reverse transcription step at the 3' poly A tail.

cDNA LIBRARIES   In analogy with whole genome PCR, it may be advantageous in some circumstances to amplify all of the cDNAs derived from an RNA preparation. For example one may want to make a cDNA library from a very small amount of starting material. In this case a primer specific for the poly A tract and one specific for whatever nucleotide was added to the 3' end of the cDNA by terminal deoxynucleotidyl transferase can be used for amplification (62–64).

Enriching Gene or cDNA Libraries using PCR

Methods of using PCR to enrich a cDNA (65–67) or genomic DNA (68) sample for specific classes of sequences have recently been devised. Although these enrichment strategies are primarily based on standard molecular biological practice, PCR enables very small amounts of starting material to be used. Conventionally produced double-stranded cDNA can be amplified after ligation of a specific linker to both ends (65). In this case single-stranded cDNA from a mouse plasmacytoma cell line was first subjected to subtractive hybridization with RNA from a B cell lymphoma line. Following second strand synthesis and linker ligation to both ends, the selected cDNA was amplified. Low-abundance cDNAs specific to the plasmacytoma line were eventually cloned. PCR can be used as an approach to construct a "normalized" cDNA library (66–68). The advantage of such a library is that all of the cDNAs would be present at about the same concentration, allowing cDNAs of very rare messages to be cloned with the same efficiency as highly abundant ones. In one strategy (67) cDNAs produced by conventional methods using random hexanucleotide primers are first cloned into a bacteriophage library and the inserts are amplified using PCR primers from the vector. The PCR products are then denatured, allowed to reassociate for various time periods, and the remaining single-stranded PCR product purified. Using an appropriate reassociation time decreases the amount of highly abundant cDNA sequences present in the single-stranded fraction relative to the rare cDNA sequences. Following PCR using the single-stranded sequences as template, the product is recloned into a vector. These methods could allow cloning of very rare but developmentally important cDNAs.

Amplification with Degenerate Primers

Although perfect complementarity is usually desirable in the design of PCR primers to maximize specificity, it is not absolutely required for successful PCR. Primer-template annealing under less stringent conditions allows successful PCR, although the products may be less homogeneous. Sometimes this can provide a distinct advantage. For example, sequence information from a known gene can be used to design primers to amplify a heretofore undiscovered but related gene. If the related gene has some regions of amino acid sequence identity with the known gene, a mixture of primers based on the sequence of the known gene can be used for PCR in a manner similar to the standard practice of using mixed oligonucleotide probes for screening cDNA libraries by hybridization. For PCR, two primer mixtures, one for each flanking sequence, are required. Each primer mixture needs to be composed of all of the different nucleotide sequences that can code for the same amino acid sequence and therefore is expected to contain at least one primer perfectly complementary to the coding sequence of the related gene as well as a number of primers differing only by one or a few bases. Primer mixtures containing more than 200,000 different sequences have been used successfully (69). This approach is especially useful in the search for members of gene families with similar physiological functions. Based upon protein sequence data, new members of a number of protein families have been discovered (61, 70–74).

At the extreme, PCR makes it possible to clone specific cDNAs even in the absence of any direct nucleic acid sequence information. Using only protein sequence data, degenerate primers can be successfully utilized (75).

Degenerate primers have also been used as an approach to detecting uncharacterized viruses related to viruses that are already known. Mack & Sninsky (76) used a Hepadnavirus model system to demonstrate the utility of this method. Using a similar procedure, sequences related to retroviral reverse transcriptases in human DNA have been detected (77–79).

ALLELE-SPECIFIC PCR

The basis for many of the PCR modifications mentioned above is that single base-pair mismatches between primer and template do not necessarily prevent the production of the desired PCR product. However, under some circumstances, even a single mismatch between primer and template can prevent amplification. In fact, there are a number of situations in which such specificity is desirable. For example, alternate forms of a single gene (alleles) that differ by a single substitution can exist in a population, and amplification of only one may be desired. Allele-specific PCR relies on the fact that a base-pair mismatch at the 3' end of a primer not only reduces the efficiency of extension by the polymerase, but can also reduce the thermal stability of the primer-template complex (80–83). The extension of a primer mismatched at the 3' end by a DNA polymerase lacking a 3' to 5' exonuclease activity can be dramatically reduced by lowering the total dNTP concentration by almost a factor of 100 (7, 83). This effect of dNTP concentration is expected based on the fundamental enzymology of mismatch extension (84). The use of lower dNTP levels, however, imposes some constraints on overall amplification efficiency. Recent studies on the effects of mismatches at and near the 3' end of PCR primers provide important information for the design of allele-specific primers (85).

PCR SENSITIVITY

The sensitivity of PCR is based primarily on two factors: (*a*) the number of target molecules present in the sample and (*b*) the complexity of the nontarget sequences. The detection of specific PCR product amplified from a unique sequence gene present in one or a few cells requires a great many cycles. As a result, many more opportunities exist for nontarget amplification and primer-dimer formation, both of which ultimately reduce sensitivity because of competition for available PCR reagents (e.g. enzyme). "Hot start" is one strategy that could minimize these artifacts; a nesting strategy is another.

Amplification of a single-copy gene in a single haploid cell can produce virtually pure PCR product as detected by gel electrophoresis (7). The analysis of single sperm for genetic mapping indicates that the probability of amplifying the desired unique sequence to a detectable level ranges between 70% and 95% (86–88). Single-cell PCR has also been applied to the analysis of oocytes and polar bodies in preimplantation genetic disease diagnosis (see 88), and characterization of immunoglobulin gene rearrangements in memory B cells (89).

PCR also makes it possible to study RNAs present in one or a few cells. PCR of cDNA has already been used to search for the presence of specific messenger RNAs in the earliest stages of mouse embryogenesis (90). Messenger RNAs of various growth factors from small numbers of macrophages isolated from wounds have also been examined in an effort to better understand the healing process (91). An interesting observation from studies on cDNA amplification is that some cells not expected to transcribe certain genes may have very low levels of so-called "ectopic" or "illegitimate" transcripts (92, 93). This finding can be of practical use in the analysis of mRNAs normally found in tissues that are not readily accessible. For example, mRNA for the muscle- and brain-specific protein dystrophin, which is altered in Duchenne muscular dystrophy, can be amplified from more readily accessible white blood cells. Using this source of DNA from patients with the disease, many new dystrophin mutations have been discovered (94).

A problem similar to single-cell analysis exists when the number of target molecules is greater than one but the ratio of target DNA sequences to total DNA is significantly less than that typical for a unique sequence gene in a mammalian genome. For example, detecting the presence of a few viral genomes in a sample containing millions of cells poses similar sensitivity problems and calls for many of the same solutions described above (95).

SEQUENCING PCR PRODUCT

A number of different strategies have been devised for directly sequencing PCR product (96–100). Double-stranded DNA product can be sequenced immediately after removing unincorporated dNTPs and primers by standard methods and has the advantage that the effect of misincorporations is eliminated. A modification of the standard PCR protocol, called "asymmetric" PCR, has been introduced to facilitate direct sequencing (101). In this procedure unequal amounts of the two primers are used. One primer is eventually used up, and the remaining primer continues to be extended each cycle so that an excess of one of the strands of the PCR product is generated. Additional methods have been developed to produce single-stranded template (102, 103). Although standard dideoxy dNTP sequencing methods work well, a chemical degradation method for directly determining the sequence of amplified DNA fragments that have incorporated phosphorothioate nucleotides during PCR is also possible (104). Carrying out PCR in the presence of low levels of dideoxy dNTPs has also been reported as a shortcut to DNA sequencing (105, 105a).

In diploid organisms where allelic differences might exist, the PCR product of a single gene could be heterogeneous with respect to DNA sequence. Direct sequencing of double-stranded PCR product from heterozygotes has been shown to be able to identify those positions at which a polymorphism exists (100). For some loci, however, alleles differ by multiple substitutions. It is impossible to know, therefore, which of the alternative bases at each polymorphic position are associated with which alleles by sequencing pooled PCR product. A number of different strategies to overcome this problem have been discussed by Gyllensten (106).

GENOMIC FOOTPRINTING

The ability of PCR to increase the concentration of rare DNA fragments in a complex mixture has been adapted to the analysis of DNA-protein interactions in living cells (genomic footprinting). The goal of this procedure is to determine the exact contacts between DNA bases and specific proteins in the living cell, in contrast to DNA footprinting, which determines these interactions in vitro. The PCR approach devised by Mueller & Wold in 1989 (48) has a number of steps in common with the original genomic footprinting method (107, 108), but is more rapid and sensitive. Cells are treated with the methylating reagent dimethyl sulfate (DMS), which reacts with guanosines that are not protected by bound protein. DNA is isolated from the cells and treated with piperidine, which induces single-strand breaks in the phosphodiester bond at the position of all modified guanosines. The DNA is denatured, annealed to a specific primer downstream of the potential protein-binding site, and extended upstream using a DNA polymerase. The extension products are expected to terminate at positions of piperidine cleavage. The double-stranded fragments are ligated to a linker as described in the section on LIGATION-MEDIATED PCR. After denaturation, PCR is initiated using a second downstream gene-specific primer and the linker fragment. The PCR products span the region from the gene-specific primer to the end of the linker attached at each piperidine cleavage site. Finally, a third radioactivity labeled primer is used to extend the PCR fragments and the products are run on a sequencing gel. Compared to a naked DNA sample, the absence of a fragment in a DNA sample from DMS-treated cells indicates the position of a guanosine that was protected from methylation by bound protein. Ligation-mediated PCR has also been shown to be effective in producing in vivo footprints using DNase I in place of DMS (109).

GENOMIC SEQUENCING

Genomic sequencing by PCR (110) is a modification of the procedure originally developed by Church & Gilbert (111), and uses the chemistry of purine and pyrimidine modification and cleavage developed for DNA sequencing (112). The application of PCR to this procedure has greatly reduced the amount of material and time and effort required. DNA is subjected to the chemical cleavage reactions followed by denaturation, primer extension, linker-ligation, and amplification using the methods described above for genomic footprinting. In the final step, the fragments are run on a DNA sequencing gel and, following electroblotting to a nylon membrane, hybridized to a probe specific for the gene under investigation. Since the presence of 5-methyl cytosine in the original chromosomal DNA will result in a gap in the C ladder on the sequencing gel, the position of every 5-methyl cytosine in the sequence can be determined. The method has been successfully applied to the analysis of the in vivo methylation state of several genes (110, 113, 114). Previous methods of determining the methylation state relied on Southern blotting and the use of restriction enzymes, which limited the analysis of 5-methyl cytosines to specific restriction enzyme sites.

Finally, a highly novel application of the genomic sequencing strategy has been developed to analyze sequence-specific DNA damage in vivo (115). When cells grown in culture are exposed to UV irradiation, pyrimidine (6–4) pyrimidone photoproducts are formed. DNA is isolated from treated cells and strand breakage initiated at the sites of photodamage by heating in piperidine. These lesions can be studied at specific sites in any gene using specific PCR primers as described above for the genomic sequencing and footprinting strategies. The authors also propose that the repair of this damage in living cells could be studied using the same methodology. This protocol could provide very important information concerning DNA repair mechanisms.

MUTATION DETECTION AND ANALYSIS

PCR can play a major role in discovering the molecular basis for new mutations, since it is much simpler to analyze PCR-amplified segments from organisms carrying specific mutants rather than having to clone each gene fragment using recombinant techniques.

Gel Electrophoretic Methods

Denaturing gradient gel electrophoresis (DGGE) was introduced several years ago and makes use of the fact that two double-stranded DNA fragments that differ only by a single nucleotide substitution can have different thermal denaturation profiles (116, 117). Although this difference is subtle, it can be detected by running the two DNA samples in adjacent lanes of a polyacrylamide gel that has a gradient of a DNA denaturant along the axis of DNA migration. When the fragments reach the position in the gel where the concentration of denaturant is sufficient to cause DNA melting, their mobility is severely retarded. If one fragment has a lower melting temperature than the other it will not migrate as far. Although DGGE was originally developed for the analysis of cloned DNA segments, PCR products are easily adapted to this method (118). A recent innovation of the method involved attaching a GC-rich stretch ("clamp") to one of the PCR primers (119). Having a very stable structure at one end of the PCR product enhances the ability to detect all possible single-base substitutions.

Two other new approaches to mutation detection by electrophoresis have been developed. One, thermal-gradient gel electrophoresis, makes use of a temperature as well as chemical gradient within the gel (120). Another method is based upon being able to detect subtle conformational differences due to a nucleotide substitution by analyzing the electrophoretic mobility of PCR products that are rendered single stranded (single strand conformational polymorphisms; SSCP) and run on nondenaturing polyacrylamide gels (121, 122).

Methods Involving Chemical Cleavage

Another approach to mutation analysis is based upon the detection of base-pair mismatches in heteroduplex DNA using either chemical or enzymatic methods. PCR product from a wild-type allele, when denatured and annealed to the product of a mutant allele, will form double-stranded DNA containing a base-pair mismatch. In one method, hydroxylamine is used to chemically modify mispaired Cs or osmium tetroxide for mispaired Ts (123–125). Following piperidine cleavage, the position of the mismatch can be determined by sizing single-stranded DNA fragments on a gel. This method has, for example, been recently applied successfully to the analysis of Factor IX mutations associated with hemophilia B (126).

DNA AND RNA SELECTION TECHNIQUES

Nucleic Acid–Protein Interactions

In the past several years, there has been an explosion in the detection of DNA sequence motifs involved in the binding of transcription factors. The finding of similar motifs in different gene segments has suggested that these proteins could be involved in regulating the expression of many genes or gene families. In some cases the proteins themselves have been cloned. Once cloned, the challenge is to identify what other genes have the same binding motifs. Unfortunately, this can not be done at the level of nucleic acid hybridization, because the length of the critical binding sequences is too short to allow specificity. An alternative approach based on PCR, devised by Kinzler & Vogelstein (127), was demonstrated in an analysis of the transcription factor TFIIIA, which is involved in the regulation of 5S RNA gene transcription. Their strategy began with the shearing of total genomic DNA and ligation of "catch linkers" to each fragment. Following exposure to purified TFIIIA protein, any DNA bound was immunoprecipitated with anti-TFIIIA antibodies. The DNA extracted from the immunoprecipitate was amplified by PCR using primer sequences on the ligated linkers. Following additional rounds of protein binding, immunoprecipitation, and amplification, the DNA fragments were cloned and characterized. The selected sequences showed a significant enrichment of those with the TFIIIA-binding motif. This technique was subsequently used to detect human genomic sequences that bound a zinc finger protein, which is amplified in a number of different tumors (128). This approach could contribute significantly to our understanding of how specific transcription factors can coordinate the expression of unlinked genes.

PCR is required for the above approach to studying DNA-protein interactions, because the level of nonspecific binding can be quite large and the concentration of true DNA-binding sites so low. Without PCR, multiple rounds of selection would result in extremely low yields of bound DNA fragments, which would be substantially contaminated. Because PCR can increase the absolute amount of precipitated DNA after a round of selection, even sequences with relatively small binding constants could eventually be enriched and characterized.

Similar principles have also been applied to the problem of understanding the strict nucleotide sequence requirements for the binding of a protein to a DNA motif. Traditionally, this information has come from mutagenesis experiments in which specific base substitutions in small DNA fragments have been made and analyzed by their effects on protein-DNA binding using the gel retardation and DNase I footprinting procedures (129–131). These experiments are laborious, and each synthetic oligonucleotide is usually studied individually. Three groups have independently devised a PCR-based method for studying this problem, which has a number of advantages (132–134). A synthetic double-stranded oligonucleotide is made so that it contains at each end a specific sequence that can be used for amplification. Internal to these primer sequences is a stretch of n nucleotides, which are chosen at random so that the collection of molecules contains 4^n different sequences. Protein is incubated with the oligonucleotide, subjected to electrophoresis, and the retarded band used as a substrate for PCR. The gel retardation step enriches for those sequences that are capable of binding the protein and the PCR step produces enough DNA so that additional rounds of selection and finally characterization can be carried out.

RNA-protein interactions can also be studied using the same basic strategy (135). To create the pool of RNAs used for selection, a DNA template containing a randomized sequence is transcribed by a bacteriophage RNA polymerase. Following each round of selection, the RNAs are reverse transcribed, amplified by PCR, transcribed by T7 polymerase, and subjected to the next round of selection.

Nucleic Acid–Ligand Interactions

The study of nucleic acid binding is not limited to the interactions with proteins. Szostak and collegues (136, 137) developed a method to select RNAs that specifically interact with column-bound organic dyes that mimic metabolic cofactors. They independently developed a strategy for producing the random pool of RNA sequences and amplifying those that are selected on the column.

The developers of these PCR-based selection techniques visualize a number of additional important applications. Selection of RNA species that bind transition-state analogs could lead to the production of novel ribozymes (136). The isolation of RNAs that could inhibit a step in the life cycle of an infectious agent could have therapeutic potential (135). Finally, it may be possible to select for RNAs that code for peptides or proteins with desired functions (135). This could theoretically be accomplished by creating a pool of mRNA variants, precipitating those RNAs that are on ribosomes followed by selecting for those nascent polypeptides that have the function desired.

CONCLUSION

The speed, specificity, and sensitivity of PCR, along with the ease with which it can be carried out and its versatility, make it ideally suited for application to many problems in biology. Many recent reviews and books indicate the extent to which PCR has had an impact on fields in addition to molecular biology, including human genetics, immunology, forensic science, evolutionary biology, and ecology and population biology (8, 88, 138–145). In addition to allowing the application of molecular biological procedures to scientific disciplines that had not exploited these techniques, other applications of PCR have produced information that was impossible to obtain previously. Among these applications include studies on RNA and DNA sequences present in individual cells (86–89, 91) retrospective medical studies on DNA in pathological samples stored for decades as formalin-fixed paraffin-embedded tissues without losing the histological information contained therein (146–148), and analysis of DNA from tissues of extinct plants and animals (149–151). Using PCR in the selection of nucleic acid and protein sequences with novel functions is also on the horizon. Only time will tell what new areas of biology may be influenced by the further development and application of this technique.

Literature Cited

1. Saiki, R., Scharf, S., Faloona, F., Mullis, K. B., Horn, G. T., Erlich, H. A., Arnheim, N. 1985. *Science* 230:1350–54
2. Mullis, K. B., Faloona, F. A. 1987. *Methods Enzymol.* 155:335–50
3. Saiki, R. K., Gelfand, D. H., Stoffel, S., Scharf, S., Higuchi, R., et al. 1988. *Science* 239:487–91
4. Gelfand, D. H., White, T. J. 1990. In *PCR Protocols, A Guide to Methods and Applications,* ed. M. A. Innis, D. H. Gelfand, J. J. Sninsky, T. J. White, pp. 129–41. New York: Academic
5. Faloona, F., Weiss, S., Ferre, F., Mullis, K. 1990. *6th Int. Conf. AIDS.* Abstr.
6. D'Aquila, R. T., Bechtel, L. J., Videler, J. A., Eron, J. J., Gorczyca, P., Kaplan, J. C. 1991. *Nucleic Acids Res.* 19:3749
7. Li, H., Cui, X., Arnheim, N. 1991. *Proc. Natl. Acad. Sci. USA* 87:4580–84

8. Erlich, H. A., Gelfand, D., Sninsky, J. 1991. *Science* 252:1643–51
9. Haff, L., Atwood, J. G., Dicesare, J., Katz, E., Picozza, E., et al. 1991. *Biotechniques* 10:102–12
10. Chamberlain, J. S., Gibbs, R. A., Ranier, J. E., Caskey, C. T. 1990. See Ref. 141, pp. 272–81
11. Smith, K. T., Long, C. M., Bowman, B., Manos, M. M. 1990. *Amplifications,* Sept: pp. 16–17. Perkin-Elmer Corp.
12. Lubin, M., Elashoff, J. D., Wang, S-J., Rotter, J., Toyoda, H. 1991. *Mol. Cell. Probes* 5:307–17
13. Syvanen, A-C., Bengtstrom, M., Tenhunen, J., Soderlund, H. 1988. *Nucleic Acids Res.* 16:11327–38
14. Chelly, J., Kaplan, J. C., Maire, P., Gautron, S., Kahn, A. 1988. *Nature* 333:858–60
15. Lawyer, F. C., Stoffel, S., Saiki, R. K.,

Myambo, K., Drummond, R., Gelfand, D. H. 1989. *J. Biol. Chem.* 264:6427–37

16. Innis, M. A., Myambo, K. B., Gelfand, D. H., Brow, M. A. D. 1988. *Proc. Natl. Acad. Sci. USA* 85:9436–40
17. Gelfand, D. H. 1989. See Ref. 142, pp. 17–22
18. Longley, M. J., Bennett, S. E., Mosbaugh, D. W. 1990. *Nucleic Acids Res.* 18:7317–24
19. Tindall, K. R., Kunkel, T. A. 1988. *Biochemistry* 27:6008–13
20. Holland, P. M., Abramson, R. D., Watson, R., Gelfand, D. H. 1991. *Proc. Natl. Acad. Sci. USA* 88:7276–80
21. Clark, J. M. 1988. *Nucleic Acids Res.* 16:9677–86
22. Watson, R. 1989. *Amplifications,* May: pp. 5–6. Perkin-Elmer Corp.
23. Klimczak, L. J., Grummt, F., Burger, K. J. 1985. *Nucleic Acids Res.* 13:5269–82
24. Elie, C., Salhi, S., Rossignol, J-M., Forterre, P., deRecondo, A-M. 1988. *Biochem. Biophys. Acta* 951:261–67
25. Salhi, S., Elie, C., Forterre, P., deRecondo, A-M., Rossignol, J-M. 1989. *J. Mol. Biol.* 209:635–44
26. McConlogue, L., Brow, M. A. D., Innis, M. A. 1988. *Nucleic Acids Res.* 16:9869
27. Myers, T. W., Gelfand, D. H. 1991. *Biochemistry* 30:7661–66
28. Klimczak, L. J., Grummt, F., Burger, K. J. 1986. *Biochemistry* 25:4850–55
29. Hamal, A., Forterre, P., Elie, C. 1990. *Eur. J. Biochem.* 190:517–21
30. Cariello, N. F., Swenberg, J. A., Skopek, T. R. 1991. *Nucleic Acids Res.* 19:4193–98
31. Goodenow, M., Huet, T., Saurin, W., Kwok, S., Sninsky, J., Wain-Hobson, S. 1989. *J. Acquired Immunol. Defic. Syndr.* 2:344–52
32. Fucharoen, S., Fucharoen, G., Fucharoen, P., Fukumaki, Y. 1989. *J. Biol. Chem.* 264:7780–83
33. Mendelman, L. V., Boosalis, M. S., Petruska, J., Goodman, M. F. 1990. *J. Biol. Chem.* 264:14415–23
34. Eckert, K. T., Kunkel, T. A. 1990. *Nucleic Acids Res.* 18:3734–44
35. Krawzak, M., Reiss, J., Schmidtke, J., Rosler, U. 1989. *Nucleic Acids Res.* 17:2197–201
36. Paabo, S., Irwin, D. M., Wilson, A. C. 1990. *J. Biol. Chem.* 265:4718–21
37. Meyerhans, A., Vartanian, J-P., Wain-Hobson, S. 1990. *Nucleic Acids Res.* 18:1687–91
38. Marton, A., Delbecchi, L., Bourgaux, P. 1991. *Nucleic Acids Res.* 19:2423–26

39. Kwok, S., Higuchi, R. 1989. *Nature* 239:237–38
40. Sarkar, G., Sommer, S. 1990. *Nature* 343:27
41. Longo, M. C., Berninger, M. S., Harley, J. L. 1990. *Gene* 93:125
42. Cimino, G. D., Metchette, K. C., Tessman, J. W., Hearst, J. E., Isaacs, S. 1991. *Nucleic Acids Res.* 19:99–108
43. Scharf, S. J., Horn, G. T., Erlich, H. A. 1986. *Science* 233:1076–78
44. Aslanidis, C., de Jong, P. J. 1990. *Nucleic Acids Res.* 18:6069
45. Silver, J., Keerikatte, V. 1989. *J. Cell. Biochem.* Abstr. WH239, Suppl. 13E
46. Triglia, T., Peterson, M. G., Kemp, D. J. 1988. *Nucleic Acids Res.* 16:8186
47. Ochman, H., Gerber, A. S., Hartl, D. L. 1988. *Genetics* 120:621–23
48. Mueller, P. R., Wold, B. 1989. *Science* 246:780–86
49. Ko, M. S. H., Ko, S. B. H., Takahashi, N., Abe, K. 1990. *Nucleic Acids Res.* 18:4293–94
50. Shyamala, V., Ames, G. F. 1989. *J. Cell. Biochem.* SBE:306
51. Ludecke, H-J., Senger, G., Claussen, U., Horsthemke, B. 1989. *Nature* 338:348–50
52. Wesley, C. S., Ben, M., Kreitman, M., Hagag, N., Eanes, W. F. 1990. *Nucleic Acids Res.* 18:599–603
53. Maniatis, T., Fritsch, E. F., Sambrook, J. 1982. *Molecular Cloning, A Laboratory Manual.* New York: Cold Spring Harbor Lab.
54. Todd, J. A., Bell, J. I., McDevitt, H. O. 1987. *Nature* 329:599–604
55. Kawasaki, E. S., Wang, A. M. 1989. See Ref. 142, pp. 89–97
56. Powell, L. M., Wallis, S. C., Pease, R. J., Edwards, Y. H., Knott, T. J., Scott, J. 1987. *Cell* 50:831–40
57. Gilliland, G., Perrin, S., Blanchard, K., Bunn, H. F. 1990. *Proc. Natl. Acad. Sci. USA* 87:2725–29
58. Wang, A. M., Doyle, M. V., Mark, D. F. 1989. *Proc. Natl. Acad. Sci. USA* 86:9717–21
59. Frohman, M. A., Dush, M. K., Martin, G. R. 1988. *Proc. Natl. Acad. Sci. USA* 85:8998–9002
60. Loh, E. Y., Elliott, J. F., Cwirla, S., Lanier, L. L., Davis, M. M. 1989. *Science* 243:217–20
61. O'Hara, O., Dorit, R. L., Gilbert, W. 1989. *Proc. Natl. Acad. Sci. USA* 86:5673–77
62. Belyavsky, A., Vinogradova, T., Rajewsky, K. 1989. *Nucleic Acids Res.* 17:2919–32
63. Brunet, J-F., Shapiro, E., Foster, S. A.,

Kandel, E. R., Iino, Y. 1991. *Science* 252:856–59

64. Welsh, J., Liu, J-P., Efstratiadis, A. 1990. *Genet. Anal. Techn. Appl.* 7:5–17

65. Timblin, C., Battey, J., Kuehl, W. M. 1990. *Nucleic Acids Res.* 18:1587–93

66. Ko, M. S. H. 1990. *Nucleic Acids Res.* 18:5705–11

67. Patanjali, S. R., Parimoo, S., Weissman, S. M. 1991. *Proc. Natl. Acad. Sci. USA* 88:1943–47

68. Wieland, I., Bolger, G., Asouline, G., Wigler, M. 1990. *Proc. Natl. Acad. Sci. USA* 87:2720–24

69. Gould, S. J., Subramani, S., Scheffler, I. E. 1989. *Proc. Natl. Acad. Sci. USA* 86:1934–38

70. Kamb, A., Weir, M., Rudy, B., Varmus, H., Kenyon, C. 1989. *Proc. Natl. Acad. Sci. USA* 86:4372–76

71. Libert, F., Parmentier, M., Lefort, A., Dinsart, C., Van Sande, J., et al. 1989. *Science* 244:569–72

72. Gautam, N., Baetscher, M., Aebersold, R., Simon, M. I. 1989. *Science* 244:971–74

73. Wilks, A. F. 1989. *Proc. Natl. Acad. Sci. USA* 86:1603–7

74. Wilkie, T. M., Simon, M. I. 1991. See Ref. 143, pp. 32–41

75. Lee, C. C., Wu, X., Gibbs, R. A., Cook, R. G., Muzny, D. M., Caskey, C. T. 1988. *Science* 239:1288–91

76. Mack, D. H., Sninsky, J. J. 1988. *Proc. Natl. Acad. Sci. USA* 85:6977–81

77. Shih, A., Misra, R., Rush, M. G. 1989. *J. Virol.* 63:64–75

78. Nunberg, J. H., Wright, D. K., Cole, G. E., Petrovskis, E. A., Post, L. E., et al. 1989. *J. Virol.* 63:3240–49

79. Greenberg, S. J., Erlich, G. D., Abbott, M. A., Hurwitz, B. J., Waldmann, T. A., Poiesz, B. J. 1989. *Proc. Natl. Acad. Sci. USA* 86:2878–82

80. Wu, D. Y., Ugozzoli, L., Pal, B. K., Wallace, R. B. 1989. *Proc. Natl. Acad. Sci. USA* 86:2757–60

81. Newton, C. R., Graham, A., Heptinstall, L. E., Powell, S. J., Summers, C., et al. 1989. *Nucleic Acids Res.* 17:2503–16

82. Gibbs, R. A., Nguyen, P-N., Caskey, C. T. 1989. *Nucleic Acids Res.* 17:2437–48

83. Ehlen, T., Dubeau, L. 1989. *Biochem. Biophys. Res. Commun.* 160:441–47

84. Petruska, J., Goodman, M. F., Boosalis, M. S., Sowers, L. C., Chaejoon, C., Tinoco, I. 1988. *Proc. Natl. Acad. Sci. USA* 85:6252–56

85. Kwok, S., Kellogg, D. E., McKinney, N., Spasic, D., Goda, L., et al. 1990. *Nucleic Acids Res.* 18:999–1005

86. Cui, X., Li, H., Goradia, T. M., Lange, K., Kazazian, H. H. Jr., et al. 1989. *Proc. Natl. Acad. Sci. USA* 86:9389–93

87. Goradia, T. M., Stanton, V. P. Jr., Cui, X., Aburatani, H., Li, H., et al. 1991. *Genomics* 10:748–55

88. Arnheim, N., Li, H., Cui, X. 1990. *Genomics* 8:415–19

89. McHeyzer-Williams, M. G., Nossal, G. J. V., Lalor, P. A. 1991. *Nature* 350:502–5

90. Rappolee, D. A., Brenner, C. A., Schultz, R., Mark, D., Werb, Z. 1988. *Science* 241:1823–25

91. Rappolee, D. A., Mark, D., Banda, M. J., Werb, Z. 1988. *Science* 241:708–12

92. Chelly, J., Concordet, J-P., Kaplan, J-C., Kahn, A. 1989. *Proc. Natl. Acad. Sci. USA* 86:2617–21

93. Sarkar, G., Sommer, S. S. 1989. *Science* 244:331–34

94. Roberts, R. G., Barby, T. F. M., Manners, E., Bobrow, M., Bentley, D. R. 1991. *Am. J. Hum. Genet.* 49:298–310

95. Sninsky, J. 1991. In *Viral Hepatitis and Liver Disease*, ed. F. B. Hollinger, S. M. Lemon, H. S. Margolis, pp. 799–805. Baltimore: Williams and Wilkins

96. Wrischnik, L. A., Higuchi, R. G., Stoneking, M., Erlich, H. A., Arnheim, N., Wilson, A. C. 1987. *Nucleic Acids Res.* 15(2):529–42

97. Stoflet, E. S., Koeberl, D. D., Sarkar, G., Sommer, S. S. 1988. *Science* 239:491–94

98. Engelke, D. R., Hoener, P. A., Collins, F. S. 1988. *Proc. Natl. Acad. Sci. USA* 85:544–48

99. McMahon, G., Davis, E., Wogan, G. N. 1987. *Proc. Natl. Acad. Sci. USA* 84:494–78

100. Wong, C., Dowling, C. E., Saiki, R. K., Higuchi, R. G., Erlich, H. A., Kazazian, H. H. 1987. *Nature* 330:384–86

101. Gyllensten, U. B., Erlich, H. A. 1988. *Proc. Natl. Acad. Sci. USA* 85:7652–56

102. Hornes, E., Hultman, T., Moks, T., Uhlen, M. 1990. *Biotechniques* 9:730

103. Higuchi, R., Ochman, H. 1989. *Nucleic Acids Res.* 17:5865

104. Nakamaye, K. L., Gish, G., Eckstein, F., Vosberg, H. P. 1988. *Nucleic Acids Res.* 16:9947–59

105. Ruano, G., Kidd, K. K. 1991. *Proc. Natl. Acad. Sci. USA* 88:2815

105a. Lee, J-S. 1991. *DNA Cell Biol.* 10:67–73

106. Gyllensten, U. B. 1989. See Ref. 142, pp. 45–60

107. Ephrussi, A., Church, G. M., Tonegawa, S., Gilbert, W. 1985. *Science* 227:134–40

108. Church, G. M., Ephrussi, A., Gilbert, W., Tonegawa, S. 1985. *Nature* 313:798–801
109. Pfeifer, G. P., Riggs, A. D. 1991. *Genes Dev.* 5:1102–13
110. Pfeifer, G. P., Steigerwald, S. D., Mueller, P. R., Wold, B., Riggs, A. D. 1989. *Science* 246:810–13
111. Church, G. M., Gilbert, W. 1984. *Proc. Natl. Acad. Sci. USA* 81:1991–95
112. Maxam, A. M., Gilbert, W. 1980. *Methods Enzymol.* 65:499–560
113. Pfeifer, G. P., Steigerwald, S. D., Hansen, R. S., Gartler, S. M., Riggs, A. D. 1990. *Proc. Natl. Acad. Sci. USA* 87:8252–56
114. Rideout, W. M., Coetzee, G. A., Olumi, A. F., Jones, P. A. 1990. *Science* 249:1288–90
115. Pfeifer, G. P., Drouin, R., Riggs, A. D., Holmquist, G. P. 1991. *Proc. Natl. Acad. Sci. USA* 88:1374–78
116. Fischer, S. G., Lerman, L. S. 1983. *Proc. Natl. Acad. Sci. USA* 80:1579–83
117. Myers, R. M., Maniatis, T. 1986. *Cold Spring Harbor Symp. Quant. Biol.* 51:275–84
118. Keohavong, P., Kat, A. G., Cariello, N. F., Thilly, W. G. 1988. *DNA* 7(1):63–70
119. Sheffield, V. C., Cox, D. R., Lerman, L. S., Myers, R. M. 1989. *Proc. Natl. Acad. Sci. USA* 86:232–36
120. Wartell, R. M., Hosseini, S. H., Moran, C. J. Jr. 1990. *Nucleic Acids Res.* 18:2699–705
121. Orita, M., Iwahana, H., Kanazawa, H., Hayashi, K., Sekiya, T. 1989. *Proc. Natl. Acad. Sci. USA* 86:2766–70
122. Orita, M., Suzuki, Y., Sekiya, T., Hayashi, K. 1989. *Genomics* 5:874–79
123. Johnston, B. H., Rich, A. 1985. *Cell* 42:713–24
124. McClellan, J. A., Palecek, E., Lilley, D. M. J. 1986. *Nucleic Acids Res.* 14:9291–309
125. Cotton, R. G. H., Rodrigues, N. R., Campbell, R. D. 1988. *Proc. Natl. Acad. Sci. USA* 85:4397–401
126. Montandon, A. J., Green, P. M., Giannelli, F., Bentley, D. R. 1989. *Nucleic Acids Res.* 17:3347–58
127. Kinzler, K. W., Vogelstein, B. 1989. *Nucleic Acids Res.* 17:3645–53
128. Kinzler, K. W., Vogelstein, B. 1990. *Mol. Cell. Biol.* 10:634–42
129. Galas, D. J., Schmitz, A. 1978. *Nucleic Acids Res.* 5:3157–70
130. Fried, M., Crothers, D. M. 1981. *Nucleic Acids Res.* 9:6505–25
131. Garner, M. M., Revzin, A. 1981. *Nucleic Acids Res.* 9:3047–60
132. Blackwell, T. K., Weintraub, H. 1990. *Science* 250:1104–10
133. Thiesen, H-J., Bach, C. 1990. *Nucleic Acids Res.* 18:3203–9
134. Mavrothalassitis, G., Beal, G., Papas, T. S. 1990. *DNA Cell Biol.* 9:783–88
135. Tuerk, C., Gold, L. 1990. *Science* 249:505–10
136. Ellington, A. D., Szostak, J. W. 1990. *Nature* 346:818–22
137. Green, R., Ellington, A. D., Bartel, D. P., Szostak, J. W. 1991. See Ref. 143, pp. 75–86
138. Arnheim, N., Levenson, C. H. 1990. *Chem. Eng. News* 68:36–47
139. Arnheim, N., White, T., Rainey, W. 1990. *BioScience* 40:174–82
140. Erlich, H. A. 1989. *J. Clin. Immunol.* 9:437–47
141. Innis, M. A., Gelfand, D. H., Sninsky, J. J., White, T. J., eds. 1990. *PCR Protocols, A Guide to Methods and Applications*. New York: Academic. 482 pp.
142. Erlich, H. A., ed. 1989. *PCR Technology: Principles and Applications for DNA Amplification*. New York: Stockton. 246 pp.
143. Arnheim, N., ed. 1991. *Polymerase Chain Reaction. Methods: A Companion to Methods in Enzymology*, Vol. 2. 86 pp.
144. Reiss, J., Cooper, D. N. 1990. *Hum. Genet.* 85:1–8
145. White, T., Madej, R., Persing, D. H. 1991. *Advances in Clinical Chemistry*, ed. H. E. Spiegel. In press
145a. White, T., Arnheim, N., Erlich, H. A. 1989. *Trends Genet.* 5:185–89
146. Almoguera, C., Shibata, D., Forrester, K., Martin, J., Arnheim, N., Perucho, M. 1988. *Cell* 53:549–54
147. Shibata, D., Martin, W. J., Arnheim, N. 1988. *Cancer Res.* 48:4564–66
148. Smit, V. T., Boot, A. J., Smits, A. A., Fleuren, G. J., Cornelisse, C. J., Bos, J. L. 1988. *Nucleic Acids Res.* 16:7773–82
149. Paabo, S., Higuchi, R. G., Wilson, A. C. 1989. *J. Biol. Chem.* 264:9709–12
150. Thomas, R. H., Schaffner, W., Wilson, A. C., Paabo, S. 1989. *Nature* 340:465–67
151. Golenberg, E. M., Giannasi, D. E., Clegg, M. T., Smiley, C. J., Durbin, M. et al. 1990. *Nature* 344:656–58

Annu. Rev. Biochem. 1992. 61:157–97

BIOCHEMISTRY OF PEROXISOMES[1]

H. van den Bosch

Centre for Biomembranes and Lipid Enzymology, University of Utrecht, 3584 CH Utrecht, The Netherlands

R. B. H. Schutgens and R. J. A. Wanders

Department of Pediatrics, Academic Medical Centre, University of Amsterdam, 1105 AZ Amsterdam, The Netherlands

J. M. Tager*

E. C. Slater Institute for Biochemical Research, Academic Medical Centre, University of Amsterdam, 1105 AZ Amsterdam, The Netherlands

KEY WORDS: comparison of microbodies, morphology of peroxisomes, peroxisomal disorders, assembly of peroxisomes, integral membrane proteins

CONTENTS

*Present address: Institute of Medical Biochemistry and Chemistry, University of Bari, 70124 Bari, Italy

[1]Abbreviations used: DHAP, dihydroxyacetone phosphate; AT, acyltransferase; VLCFA, very long-chain fatty acid; DHCA, dihydroxy-5-β-cholestanoic acid; THCA, trihydroxy-5-β-cholestanoic acid; AGT, alanine:glyoxylate aminotransferase; HMG-CoA, 3-hydroxy-3-methylglutaryl-CoA; SCP-2, sterol carrier protein 2; nsLTP, nonspecific lipid transfer protein; CHO, Chinese hamster ovary; NALD, neonatal adrenoleukodystrophy; IRD, infantile Refsum disease; RCDP, rhizomelic chondrodysplasia punctata; X-ALD, X-linked adrenoleukodystrophy; IMP, integral membrane protein; ER, endoplasmic reticulum

INTRODUCTION

Compartmentation of biochemical pathways and processes is a most characteristic feature of eukaryotic cellular metabolism. Apart from various kinds of transient endocytotic vesicles and multivesicular bodies, which appear to have ever-changing protein compositions depending on their exact location in endocytotic and related pathways, the peroxisome was the last "true" subcellular organelle structure to be detected. Originally discovered in the early 1950s in mouse kidney cells by electron microscopy and classified as a microbody because of the lack of knowledge regarding its associated biochemistry, the organelle became a biochemical entity named *peroxisome* through the pioneering studies of De Duve and his colleagues (1, 2).

The early history of the mammalian peroxisome and its relationship with microbody structures in other kingdoms has been reviewed extensively (2–8). Reviews concentrating on microbodies in plants (4), protozoa (9, 10), and fungi (11) are available. It is now well-documented that microbodies from these various sources contain different arrays of enzymes. More than a decade ago Tolbert (5) presented a comprehensive summary in this series of the enzyme content of mammalian peroxisomes and their counterpart from the plant kingdom, *glyoxysomes*. There have been many new developments since that time. The involvement of peroxisomes in ether lipid biosynthesis has been further substantiated, and evidence for a role in cholesterol and dolichol metabolism has been presented. The recognition of a new class of inheritable diseases in humans caused by the deficiency of peroxisomes or single peroxisomal enzymes (reviewed in 12–16) has provided new insights into the function of processes taking place in peroxisomes or has assigned processes to peroxisomes that were previously thought to take place in other organelles. Major breakthroughs in establishing the site of synthesis of peroxisomal proteins, their import into the organelle, and the targeting sequences involved have been reported (6, 7, 17 for reviews). As a consequence of these studies, the prevailing opinion of a decade ago that peroxisomes are formed by budding from the ER (5) has had to be abandoned and is now replaced by a model according to which peroxisomes grow by posttranslational import of proteins into existing peroxisomes and new peroxisomes are formed by fission (6, 18). Complementation analysis of cells from patients with the Zellweger syndrome and related disorders of peroxisome biogenesis has indicated that at least six gene products are involved in peroxisome assembly. This review concentrates on these new developments.

CHARACTERIZATION OF PEROXISOMES

Nomenclature and Comparison of Microbodies

Microbodies are bounded by a single membrane and generally have a diameter of 0.1 to 1.0 μm and an equilibrium density of 1.21 to 1.25 g/cm^3. They are widespread in nature, and the number of enzymes that can be found in microbodies under various conditions has extended to such an extent that it is no longer meaningful to attempt to tabulate all of them. Table 1 presents a brief summary of microbody nomenclature and characteristic microbody processes. This summary illustrates both the relationships among microbodies and the fact that none of them contains all characteristic microbody processes.

The mammalian peroxisome was originally defined as an organelle containing at least one flavinoxidase producing H_2O_2 and catalase (2). The term *microperoxisome* reflects the observation that in many tissues the size of organelles showing a positive cytochemical reaction for catalase is smaller than that of peroxisomes in liver and kidney. Initially, this heterogeneity was the cause of considerable confusion in the localization of enzymes to (micro)peroxisomes after differential centrifugation of homogenates. It is now generally believed that the prefix "micro-" reflects a difference in size only rather than enzyme content and function, and the term *peroxisome* will be used indiscriminately in this review.

An organelle in germinating castor beans containing the five enzymes of the glyoxylate cycle received the name "glyoxysome" (19). It was then found to have a microbody-like morphology and to contain oxidases, catalase, and a β-oxidation system (3–5). On the basis of these properties, *glyoxysomes* are considered to be the plant kingdom variant of mammalian peroxisomes. Upon the transition from germinating seeds to seedlings with green leaves, the glyoxysomes develop the capacity for a photorespiration process in which glyoxylate is produced from glycolate and converted by a transaminase to

Table 1 Main microbody processes

| Nomenclature | Mammals | Plants | | Yeast | Protozoa | |
		seeds	leaves		Tetrahymena	Trypanosoma
	(Micro) peroxisomes	Glyoxysomes	Peroxisomes	Peroxisomes	Peroxisomes	Glycosomes
Enzyme system						
Oxidases (H$_2$O$_2$)	+	+	+	+	+	−*
Catalase	+	+	+	+	+	±
β-oxidation	+	+	+	+	+	+
Glyoxylate cycle	−	+	−	+	+	−*
Ether lipid synthesis	+	−	−	−		+
Photorespiration	−	−	+	−	−	−
Glycolysis	−	−	−	−	−	+
References	3, 6	3–6	3–6, 20	3, 5, 11, 22	3, 5, 6, 9	6, 9, 10, 18

Symbols: +, present; −, absent; ±, present in some but not all species; blank, not yet reported; *, present in some species but apparently not located in the microbodies.

glycine (3, 5). Initially it was reported (5) that leaf peroxisomes lacked the β-oxidation system present in glyoxysomes, but more recent data indicate that fatty acid β-oxidation is a general function of higher plant peroxisomes. In fact, microbodies are now regarded as the sole location in plant cells where fatty acid β-oxidation takes place (20). Ether-linked phospholipids have been reported in higher plants and their seeds, but only trace quantities are present, generally amounting to less than 1% of the total phospholipids (21). There have been no reports on the presence in plant glyoxysomes/peroxisomes of acyl-CoA:DHAP-AT, alkyl-DHAP synthase, and NADPH:alkyl-DHAP oxidoreductase, three enzymes required for the synthesis of ether phospholipids in mammalian peroxisomes.

Yeast microbodies contain a complete β-oxidation system, several H_2O_2-producing oxidases, and catalase, thus characterizing them as peroxisomes. In addition, the two glyoxylate cycle enzymes, isocitrate lyase and malate synthase, are present when the cells are grown on C_2-compounds like acetate and ethanol or on compounds that are degraded to C_2-compounds like alkanes and fatty acids (22). The content of other enzymes in yeast microbodies is strongly dependent on the carbon and nitrogen sources used for growth. With the exception of *Pullularia pullulans*, in which choline- and ethanolamine phospholipids have 12% and 32% plasmalogens, respectively (23), most common yeasts like *Saccharomyces cerevisiae* and *Hansenula polymorpha* have at best trace amounts of plasmalogens (21, 23). In addition, the specific peroxisomal DHAP-AT involved in ether lipid biosynthesis in mammalian tissues could not be demonstrated in yeast. Mutants of *S. cerevisiae* defective in *sn*-glycerol-3-phosphate AT showed a simultaneous loss of DHAP-AT activity, indicating a common structural gene for both activities (24).

A new function for microbodies in some fungi was recently suggested when the localization of enzymes involved in penicillin biosynthesis was studied in *Penicillium chrysogenum*. Immunogold cytochemistry and cell fractionation provided strong evidence that the acyltransferase, which catalyzes the final step in penicillin biosynthesis, is located in organelles with a diameter of 0.2 to 0.8 μm that most probably represent microbodies (25).

The microbody of the aerobic protozoa *Tetrahymena pyriformis* played an important role in establishing the phylogenetic relationship between peroxisomes and glyoxysomes (3). Like the yeast peroxisome, it contains the two glyoxylate cycle enzymes, isocitrate lyase and malate synthase, H_2O_2-producing oxidases, catalase, and a complete β-oxidation system. *T. pyriformis* strains do not contain plasmalogens, but are very rich in alkyl-acyl glycerophospholipids (20, 26). Neither the mechanism nor the localization of the responsible enzymes has been resolved. It should be remembered that the mere presence of ether-linked phospholipids cannot be taken as evidence for the operation of the DHAP pathway for ether lipid synthesis. In fact, experiments using [2-^{3}H]- and [2-^{14}C]glycerol revealed that DHAP did not serve as an intermediate in the synthesis of plasmalogens in anaerobic bacteria and protozoa (27).

In *Trypanosomatides* an organelle resembling peroxisomes in morphology, but lacking peroxisomal marker enzymes like H_2O_2-producing oxidases and catalase, was termed "glycosome" because it harbored seven enzymes of glycolysis (28). Enzymes of other pathways including ether lipid synthesis and β-oxidation are also present in *Trypanosoma brucei* or closely related organisms (7, 10, 18, 29). The β-oxidation system in these and other catalase-free microbodies may use acyl-CoA dehydrogenase rather than acyl-CoA oxidase for the initial desaturation of acyl-CoA (18). The glycosome of *T. brucei* contains a specific DHAP-AT that does not use glycerol-3-phosphate, alkyl/acyl-DHAP oxidoreductase, and acyl-CoA reductase for the formation of long-chain alcohols (29, 30). This strongly suggests that the glycosome of *T. brucei* shares the DHAP-pathway of ether lipid synthesis with the mammalian peroxisome, although the actual formation of glycero-ether bonds by alkyl-DHAP synthase has not yet been shown.

All of the above microbodies are surrounded by a single membrane and do not contain DNA. An apparently unrelated microbody in *Trichomonatides* received the name "hydrogenosome" because of its capacity to transfer electrons from pyruvate oxidation to H^+ under anaerobic conditions (9). Hydrogenosomes share with other microbodies a high equilibrium density of 1.24 to 1.26 g/cm^3 but differ in characteristic other properties, i.e. they have been reported to contain DNA (9) and appear to be bounded by a double membrane (31). For these reasons they are not included in Table 1.

Morphology and Induction of Peroxisomes

In liver and kidney, peroxisomes appear as round or slightly oval structures with a diameter of 0.3 to 1.0 μm. In liver they are conspicuous in the parenchymal cells where they occupy 2.4% of the cellular volume (32). Hepatic peroxisomes often contain crystalloid cores believed to represent urate oxidase aggregates. This structure is not present in human tissues, which lack urate oxidase (12). Further support for this notion was provided by protein A-gold immunochemistry, demonstrating that the crystalloid cores in bovine kidney peroxisomes were the exclusive site of urate oxidase localization (33). By contrast, rat kidney peroxisomes lack urate oxidase and do not contain a crystalloid core (33, 34). Marginal plates, another common type of crystalline inclusion in renal peroxisomes, were recently shown to be composed mainly of the B isozyme of L-α-hydroxyacid oxidase (35).

The detection and identification of peroxisomes are facilitated by using 3,3'-diaminobenzidine staining employing the peroxidative activity of catalase at alkaline pH. In most mammalian tissues other than liver and kidney, catalase-positive particles are smaller, with diameters of 0.05 to 0.2 μm, and devoid of a crystalloid core (36, 37). Peroxisomes are abundant in tissues active in lipid metabolism such as liver, sebaceous glands, and brown fat. In nervous tissue of the rat, they are especially enriched in myelin-producing oligodendrocytes and most abundant in the oligodendrocyte processes adjacent to the growing myelin sheets at day 17 postnatal when myelin formation peaks (38, 39).

A close spatial relationship between peroxisomes and sites of lipid synthesis, such as ER membranes, has frequently been observed (37, 40–42). Initially, this close association was misinterpreted as a luminal continuity between both organelles and contributed to the view that peroxisomes were formed by budding from the ER. (6). This view was revised, in part because of more recent experiments employing serial sectioning electron microscopy (41, 43). Such experiments demonstrated instead clusters of interconnected peroxisomes that form a peroxisomal reticulum in mouse liver and in regenerating rat liver. These elongated tubular structures provide an explanation for the variation in size and shape of catalase-positive particles in thin sections.

Morphological studies using catalase cytochemistry in conjunction with more specific immunocytochemistry have contributed to our understanding of peroxisome biogenesis. Fahimi and coworkers have used specific antibodies against peroxisomal matrix and membrane proteins to study the gradual reappearance of these proteins after partial hepatectomy. The synthesis of membrane proteins preceded that of matrix proteins, suggesting that peroxisome biogenesis is initiated by the synthesis of new membranes (44). Additional evidence for this was obtained from peroxisome proliferation in rat liver (45) and yeast (46). In rats treated with a hypocholesterolemic drug, double-membrane loops resembling smooth ER were seen in the proximity of some peroxisomes. Serial sectioning revealed that these loops were continuous with the peroxisomal membranes and that they reacted with antibodies against unique peroxisomal membrane proteins. These loops may serve as peroxisomal membrane extensions prior to import of peroxisomal matrix proteins.

In methylotrophic yeasts, peroxisomes are massively induced upon growth in media containing methanol. In *H. polymorpha*, single peroxisomes mature to one large organelle before up to 18 smaller organelles are split off, which may occupy more than 80% of the cell volume (47, 48). Similarly, *Candida boidinii* contained two to five small peroxisomes when cultured in glucose media (46). A study of their proliferation upon a shift to methanol-containing media indicated that several elongated peroxisomes developed from a pre-existing organelle. These incorporated catalase and a 47-kDa membrane protein, but not yet alcohol oxidase, dihydroxy acetone synthase, and a

20-kDa membrane protein. The latter three proteins were imported during further proliferation when up to 30 clustered peroxisomes were formed by division. These findings and those obtained after peroxisome proliferation in liver (44, 45) provide strong morphological evidence that newly formed peroxisomes arise by growth and division of pre-existing peroxisomes. *S. cerevisiae* long seemed to form an exception to the general observation that peroxisomes in yeasts can be induced by changes in growth media. Peroxisome proliferation in this organism was recently shown to occur upon growth in media containing oleic acid. Glucose-grown cells generally contained one to four small peroxisomes. After the shift to oleic acid media, these developed into up to 14 mature peroxisomes with diameters of up to 0.5 μm (49).

Hypolipidemic drugs, herbicides, and phthalate ester plasticizers induce proliferation of peroxisomes in hepatic parenchymal cells of rodents (50, 51). Additional morphometric studies reported a doubling in the number and average size of peroxisomes in mouse hepatocytes after four days of clofibrate feeding (52), resulting in an increase of the cytoplasmic area occupied by peroxisomes from 1.8% to 7.6%. A similar increase, from 1.4% to 5.4%, was observed in rat liver after two weeks of plasticizer feeding (53). Proliferation can also be induced in vitro in cultured rat hepatocytes to cause the cytoplasmic volume of peroxisomes to increase from less than 1% to 9–12% (54). No increases in cultures derived from monkey and human were observed (55). Since peroxisome proliferation by hypolipidemic drugs has been implicated in liver carcinogenesis (50, 51), the absence of or limited proliferation in primates may suggest the absence of tumorigenic activity in these species. Reddy and coworkers (50, 51) suggest that the carcinogenicity of peroxisome proliferators is attributable to the increased activity of peroxisomal β-oxidation resulting in overproduction of H_2O_2 and possible oxidative damage to DNA. In this respect, it is interesting to note that a new hypolipidemic drug, 4-(2-[4-(chlorocinnamyl)piperazin-1-yl]ethyl)benzoic acid, was recently shown (56) to induce marked proliferation of peroxisomes in rat liver without the significant elevation of the β-oxidation capacity that generally accompanies proliferation. The induction of β-oxidation usually reaches a maximum after two weeks of treatment (57). The activity versus time profile is nearly identical for the three individual β-oxidation enzymes, i.e. acyl-CoA oxidase, bifunctional enzyme, and thiolase, and is mainly due to an increased rate of synthesis from the enhanced mRNA levels (51, 54, 57, 59). In comparison to the approximately 20-fold increase in the levels of mRNA for peroxisomal β-oxidation enzymes in rat liver, the increases in kidney, small intestine, and heart were small, varying from 2- to 4-fold, and no increases were observed in nine other extrahepatic tissues (59). Even in liver, other peroxisomal enzymes such as catalase, urate-, α-hydroxyacid-, and D-amino-acid oxidase exhibit only marginal increases, remain unchanged, or are reduced (57–60). Some, but not all, membrane proteins are inducible (see section PEROXISOMAL MEMBRANES).

The mechanism of action of the proliferators remains largely to be clarified. It has been suggested that a carboxylate function, which is either present or can be formed by oxidation, and which can be activated to a CoA ester, forms a common motif of otherwise structurally nonrelated proliferators (61, 62). The detection by affinity chromatography of a peroxisome proliferator-binding protein in rat liver (63), which turned out to be identical with the 72-kDa member of the heat-shock protein family (64), suggested that these proliferators could act by a mechanism similar to that of steroid hormones to activate transcription of peroxisomal β-oxidation genes. Following this reasoning, Isseman & Green (65) recently cloned a novel member of the steroid hormone receptor family of ligand-activated transcription factors that is activated by a diverse class of peroxisome proliferators. The mouse gene coded for a 52-kDa protein and could be expressed in liver, kidney, and heart in a manner that mirrored the tissue-specific effects of peroxisome proliferators. These findings suggest that this receptor directly mediates the effect of such proliferators, although further ligand-binding studies are needed. The relationship with the 72-kDa peroxisome proliferator-binding protein from rat liver remains at present unclear.

SELECTED METABOLIC PATHWAYS

Peroxisomal Oxidation and Respiration

The first peroxisomal function discovered by de Duve and coworkers (2) was its simple respiratory pathway. The pathway is based upon the formation of hydrogen peroxide by a collection of oxidases and the decomposition of the H_2O_2 by catalase. These reactions may be responsible for as much as 20% of oxygen consumption in rat liver (2). Among the oxidases identified are: urate oxidase, D-aminoacid oxidase, L-α-hydroxyacid oxidase A and B (2), acyl-CoA oxidase (66), glutaryl-CoA oxidase (67), polyamine oxidase (68), pipecolic acid oxidase (69, 70), oxalate oxidase (71), trihydroxycholestanoyl-CoA oxidase (72–74), and pristanoyl-CoA oxidase (75). The decomposition of H_2O_2 may occur catalytically by reaction of two molecules of H_2O_2 to form H_2O and O_2 or, alternatively, peroxidatically. Both reactions are catalyzed by catalase. Catalase accepts a range of peroxidation substrates such as ethanol, methanol, nitrites, quinones, and formate (76). Until recently it was generally considered that ethanol is oxidized predominantly via cytosolic alcohol dehydrogenase. However, as shown by Handler & Thurman (77, 78), ethanol elimination via catalase may be significant if substrates such as fatty acids are available to generate H_2O_2 inside the peroxisome.

Fatty Acid β-Oxidation

Cooper & Beevers (79) were the first to show that glyoxysomes are capable of catalyzing the β-oxidation of fatty acids. Inspired by these studies, Lazarow & de Duve (80) demonstrated the presence of a fatty acid β-oxidation system in rat liver peroxisomes. Subsequent studies have shown that nonmitochondrial β-oxidation is widely distributed in nature, much more so in fact than β-oxidation in mitochondria, which is mainly restricted to the animal kingdom. Nonmitochondrial β-oxidation has been detected in animals, plants, and eukaryotic microorganisms (81).

ENZYMIC ORGANIZATION   Since acyl-CoAs rather than free fatty acids are the true substrates for β-oxidation, conversion to CoA-esters is the first obligatory step. This is brought about by a variety of acyl-CoA synthetases, which differ with regard to their substrate specificity and subcellular localization. Long-chain fatty acid activation can occur via a long-chain acyl-CoA synthetase, which is located at the cytosolic face of the peroxisomal membrane (82). An identical enzyme is present on the outer mitochondrial membrane and the membrane of the ER (83). Recently, evidence has been brought forward suggesting that liver peroxisomes contain a second acyl-CoA synthetase, which preferentially activates very long-chain fatty acids (VLCFA) such as tetracosanoic (C24:0) and hexacosanoic (C26:0) acid (84–86). A similar activity was detected in the ER but not in mitochondria (85, 86). There is disagreement with regard to the topological localization of the VLCFA-activating enzyme in the peroxisomal membrane. Whereas Lazo et al (87) reported that the active site faces the peroxisomal interior, Lageweg et al (88) found a cytosolic orientation. Studies on the human genetic disorder X-ALD indicated a deficiency of the peroxisomal VLCFA-CoA synthetase but normal activity of the ER enzyme (89, 90). This suggests an obligatory coupling between the activation of VLCFA on the peroxisomal membrane and the subsequent β-oxidation of VLCFA within this organelle. Elegant studies by Numa and coworkers (91) have shown a similar type of compartmentation of the metabolism of oleic acid in the yeast *Candida lipolytica*. However, it is not a general principle of fatty acid degradation in peroxisomes that activation and subsequent β-oxidation are coupled processes. Activation of di- and trihydroxycholestanoic acid (92, 93), dicarboxylic acids (94), and prostaglandins (95) occurs at the ER.

Following activation, the acyl-CoAs must be carried across the peroxisomal membrane via a mechanism that is as yet not well understood. Once inside the peroxisome, CoA-esters are subjected to oxidation by acyl-CoA oxidases. Like the mitochondrial acyl-CoA dehydrogenases, these enzymes contain FAD as a prosthetic group. Reoxidation of enzyme-bound $FADH_2$, however, occurs via direct interaction with molecular oxygen, yielding H_2O_2

(66). Acyl-CoA oxidases have been purified from a number of sources, including rat liver, cucumber seeds, and different yeast species, and possess very similar subunit sizes (96, for review). In rat liver, two mRNAs for acyl-CoA oxidase were identified, which are produced by alternative splicing of the primary transcript of a single acyl-CoA oxidase gene (97). Comparison of the acyl-CoA oxidase sequences from rat (97) and different yeast species (98–100) shows significant homology. Recently, evidence was reported for the existence of two acyl-CoA oxidases in rat liver peroxisomes, one of which is clofibrate-inducible and the other not (72). Whether these acyl-CoA oxidases are the products of the two mRNAs identified by Miyazawa et al (97, 101) is not yet known. Apart from these acyl-CoA oxidases, there is a separate oxidase catalyzing the first step in trihydroxycholestanoyl-CoA oxidation (72, 73). The latter enzyme is expressed in liver only and is not induced by clofibrate (72).

The next two reactions in peroxisomal β-oxidation are catalyzed by multifunctional proteins, which, apart from enoyl-CoA hydratase and L-3-hydroxyacyl-CoA dehydrogenase activities, may also contain other enzyme activities. A trifunctional β-oxidation protein harboring 3-hydroxyacyl-CoA epimerase as a third enzyme activity was found in *Candida tropicalis*, *C. lipolytica*, and *Neurospora crassa*, whereas a bifunctional protein was found in *S. cerevisiae* (96). The rat liver enzyme, long known as a bifunctional protein (102), is in fact a trifunctional protein having Δ^3-*cis*-Δ^2-*trans*-enoyl-CoA isomerase as third inherent activity (103). Molecular cloning of the cDNA for rat liver trifunctional protein has revealed strong sequence similarity with both mitochondrial hydratase and mitochondrial 3-hydroxyacyl-CoA dehydrogenase (104, 105), suggesting that the gene has originally been produced by gene fusion.

Finally, the last reaction in peroxisomal β-oxidation is catalyzed by 3-oxoacyl-CoA thiolase (106). Peroxisomal thiolases have been purified from a number of sources including rat liver, cucumber seed glyoxysomes, and several yeast species (96). In contrast to the mitochondrial and cytosolic thiolases, which are tetramers, peroxisomal thiolases are dimeric with a subunit size of 40–50 kDa (96). In rat liver there are two peroxisomal thiolase species differing with respect to molecular weight and inducibility by clofibrate. These two thiolases are encoded for by different genes, which show strong sequence homology (107–109). Apart from the enzymes directly involved in β-oxidation, peroxisomes also contain the auxiliary enzymes required for the degradation of unsaturated fatty acids, i.e. 2,4-dienoyl-CoA reductase (110) and Δ^3-*cis*-Δ^2-*trans*-enoyl-CoA isomerase (111). Recently, Schulz and coworkers (112) identified a novel enzyme in peroxisomes, D-3-hydroxyacyl-CoA dehydratase, which may function in the β-oxidation of unsaturated fatty acids and/or D-hydroxy fatty acids.

Peroxisomes also possess medium-chain carnitine acyltransferase activity and carnitine acetyltransferase activity (113–115). Farrell & Bieber (116) suggested that these enzymes are involved in the transport of the products of peroxisomal β-oxidation (acetyl-CoA, propionyl-CoA, medium-chain acyl-CoA esters) to the mitochondria as carnitine esters. Importantly, the peroxisomal medium-chain carnitine acyltransferase is inhibited by malonyl-CoA (117), just like the outer mitochondrial carnitine palmitoyltransferase (118) and the microsomal medium-chain carnitine acyltransferase (119), suggesting that it may be subject to regulation.

PHYSIOLOGICAL FUNCTION OF THE PEROXISOMAL β-OXIDATION SYSTEM   The peroxisomal β-oxidation system is not just a functional duplicate of the mitochondrial system. It is now clear that peroxisomes are involved in the degradation of a distinct set of compounds.

With regard to long-chain saturated fatty acids, it is generally agreed that at least in rat liver, oxidation takes place predominantly in mitochondria, with estimates varying from 70% to more than 95% (120, 121). The initial steps involved in the β-oxidation of VLCFA proceed in peroxisomes rather than in mitochondria (122, 123). This might well be due to the fact that mitochondria lack the capability to activate VLCFA (85, 86). Recently, it has been argued that although oxidation of VLCFA may be peroxisomal in rat liver and human skin fibroblasts, it may be mitochondrial in brain (124).

Following the discovery of a fatty acid β-oxidation system in rat liver peroxisomes (80), Pedersen & Gustafsson (125) reinvestigated the subcellular site of synthesis of cholic acid, which is formed from $3\alpha,7\alpha,12\alpha$-THCA via β-oxidation. These studies (125, 126) showed that peroxisomes are probably the sole site of THCA oxidation, a conclusion strengthened by the finding that THCA degradation is deficient in patients lacking peroxisomes (127).

Like mitochondria, peroxisomes possess 2,4-dienoyl-CoA reductase activity and Δ^3-cis-Δ^2-trans-enoyl-CoA isomerase activity, thus enabling them to handle the double bonds in unsaturated fatty acids. Studies (128) on the β-oxidation of fatty acids with chain lengths of 20 and 22 carbon atoms and possessing from zero to six double bonds have shown that in general the peroxisomal β-oxidation system shows a preference for the more polyunsaturated fatty acids. No quantitative analysis has been made in intact cells to evaluate the contribution of peroxisomes and mitochondria to overall β-oxidation of these polyunsaturated fatty acids. However, recent studies using peroxisome-deficient cell lines indicated that the oxidation of erucic acid (C22:1 (n-9)), adrenic acid (C22:4 (n-6)), arachidonic acid (C20:4 (n-6)), and tetracosatetraenoic acid (C24:4 (n-6)) is initiated in peroxisomes and not in mitochondria (129, 130). Also, the β-oxidation of branched-chain fatty acids, such as pristanic acid, appears to occur in peroxisomes (131). In addition, peroxisomes have been implicated in the β-oxidation of dicarboxylic acids (132), prostaglandins (95), and xenobiotics with an acyl side-chain (133). It is as yet uncertain, however, whether or not peroxisomes are obligatory in the β-oxidation of these compounds (134, for review).

Glyoxylate Metabolism

The glyoxylate cycle, discovered by Kornberg & Krebs (135), provides a mechanism for the net conversion of acetyl units to carbohydrate in microorganisms and germinating fat-bearing seeds (5). This conversion is brought about by malate synthase and isocitrate lyase acting together with enzymes of the citric acid cycle. The glyoxylate cycle is generally considered to be absent from animal tissues. Goodman and coworkers, however, have reported that isocitrate lyase and malate synthase are present in several vertebrate tissues, including adipose tissue of the hibernating bear (136) and rat liver (137). The activities found by these authors are surprisingly high. While a role for the glyoxylate cycle could be envisaged in the former case, this pathway is unlikely to be of quantitative significance in rat liver in view of the fact that gluconeogenesis from C_3 compounds in liver and kidney is essential for glucose homeostasis in mammals.

In mammals glyoxylate can be metabolized by conversion to glycine in a reaction catalyzed by AGT, an enzyme detected only in liver and to a lesser extent in kidney (138–144). In this way glyoxylate enters the gluconeogenic pathway (138–141, 144, 145). This is an essential metabolic pathway in humans, as indicated by the fact that an inherited deficiency of the enzyme leads to primary hyperoxaluria type I (140, 141, 145, 146), a serious disease caused by massive production of oxalate from glyoxylate (see section PER-OXISOMAL DISORDERS). The subcellular distribution of AGT varies with species and tissues. The enzyme in liver is exclusively present in peroxisomes in humans (145), baboons (147), rabbits (147), and guinea pigs (139). A peroxisomal location is also found in guinea pig kidney (142), but not in rabbit kidney, where the enzyme is found nearly exclusively in mitochondria (142), as is the case in dog and cat liver (139, 147). Despite the fact that the localization of the enzyme in dogs differs from that in humans, primary hyperoxaluria type I has also been encountered in dogs (143).

AGT in marmoset (147) and rat and mouse liver (139, 148–151) is present both in peroxisomes and in mitochondria. The peroxisomal and mitochondrial forms have identical physicochemical and enzymological characteristics and are immunologically indistinguishable (139, 152). Indeed, recent molecular genetic studies (152) have shown that the two forms are products of a single gene with the mRNAs being transcribed from different initiation sites (see section ASSEMBLY OF PEROXISOMES). In rats, the peroxisomal enzyme is induced by feeding with clofibrate (139) or plasticizer (149), whereas the mitochondrial form is induced by glucagon (139, 148–151, 153, 154).

It should be noted that AGT is identical to the enzyme referred to by Ichiyama and coworkers (152) as serine:pyruvate aminotransferase (139, 150, 153), and that both activities are deficient in primary hyperoxaluria type I (155). In fact, rat liver AGT has a rather wide substrate specificity (139, 148).

Cholesterol and Dolichol Metabolism

Mammalian peroxisomes have recently been implicated as a site of cholesterol biosynthesis. Immunoelectronmicroscopy (156) and subcellular fractionation studies demonstrated that substantial amounts of HMG-CoA reductase could be detected in peroxisomes (157, 158). HMG-CoA reductase in the ER is considered to be the key regulatory enzyme of isoprenoid biosynthesis. The molecular masses of the peroxisomal and the ER enzyme proteins are apparently identical (158), even though the ER enzyme is a glycosylated transmembrane protein (159) whereas the peroxisomal enzyme is a soluble protein found in the matrix space of the organelle. Highly purified peroxisomes contain a thiolase activity (160) that is able to catalyze the initial step in cholesterol synthesis, i.e. the condensation of acetyl-CoA units to acetoacetyl-CoA, and are able to convert mevalonic acid to cholesterol in the presence of a cytosolic fraction in vitro (161). The presence of individual enzymes of cholesterol biosynthesis in rat liver peroxisomes was also reported by Appelkvist et al (162). The physiological and quantitative significance of this putative peroxisomal pathway for cholesterol synthesis is not yet clear.

Peroxisomes in liver play an essential role in the biosynthesis of bile acids. The conversion of the intermediate $3\alpha,7\alpha,12\alpha$-THCA to cholic acid is a peroxisomal process (126) involving oxidative cleavage of the side-chain with release of propionyl-CoA (163).

It is of interest that other proteins thought to be involved in cholesterol utilization are also present in peroxisomes. Especially the subcellular localization of sterol carrier protein 2 (SCP-2; or nonspecific lipid transfer protein, nsLTP) has been studied in detail (164–167). SCP-2 is a nonenzymic protein of 13.5 kDa involved in the conversion of lanosterol to cholesterol, the esterification of cholesterol, and the synthesis of bile acids. It is also involved in the intracellular transfer of cholesterol, which is required in pregnenolone production in adrenal (168) and ovarian (169) tissues. Immunoelectronmicroscopic studies in rat liver demonstrated that SCP-2 is present inside mitochondria and associated with the ER and the cytosol, but the largest concentration of SCP-2 is inside the peroxisomes. The dependence of SCP-2 on the peroxisomal compartment is also illustrated by the greatly reduced amount of protein in liver homogenates of Zellweger patients (170) and in CHO mutants deficient in peroxisomes (171). Cloning and expression of cDNAs encoding rat liver (172–174) and human liver (175) SCP-2 indicated that the protein is synthesized in both species as a 143-amino-acid long pre-SCP-2, including a 20-residue N-terminal leader sequence (see section ASSEMBLY OF PEROXISOMES). Molecular cloning (173, 174, 176) also detected much larger cDNAs derived from the same gene that encoded a 59-kDa protein containing the sequence of pre-SCP-2 as its C-terminal part. This protein has consistently been found in peroxisomes using antibodies against SCP-2 (164–166, 170, 171). As pointed out by Mori et al (174), the presence of this protein in peroxisomes makes it unlikely that it is processed to SCP-2, and its function needs to be studied further. Notably, apart from containing the SCP-2 sequence, the 59-kDa protein has significant sequence homology with mitochondrial and peroxisomal 3-oxoacyl-CoA thiolases (174, 176). The implications of these observations are presently not clear, but suggest that the 59-kDa protein may somehow be related to β-oxidation. In this respect, it is worth noting that a peroxisomal nsLTP from *C. tropicalis*, showing 33% homology with nsLTP from rat liver, was exclusively expressed in cells grown on oleate, also suggesting a role in the β-oxidation of fatty acids (177).

The initial steps in the biosynthetic pathway leading to cholesterol are also involved in dolichol biosynthesis. Based on in vitro and in vivo experiments, Appelkvist and Kalén (178) concluded that dolichol is synthesized in both the ER and peroxisomes.

Ether Lipid Synthesis

LOCALIZATION AND TOPOLOGY   The pathway of glycero-ether bond formation was elucidated two decades ago (179, 180 for reviews), and starts with the acylation of DHAP. The acyl-DHAP can be reduced by NADPH to 1-acyl-*sn*-glycerol-3-phosphate, the first lipidic intermediate in the biosynthesis of triglycerides and diacyl phosphoglycerides, compounds which can also be formed through direct acylation of glycerol-3-phosphate. However, the main role of the DHAP pathway appears to be in the synthesis of ether lipids in which acyl-DHAP functions as an obligatory intermediate. Alkyl-DHAP synthase, the enzyme responsible for glycero-ether bond formation, replaces the acyl moiety in acyl-DHAP by a long-chain alcohol (Figure 1). A recent claim (181) that this replacement could also take place in the nonphosphorylated acyl-dihydroxyacetone appeared to be based on erroneous product identification (182).

The initial confusion about the subcellular localization of DHAP-AT, i.e. in the crude mitochondrial fraction of liver and in the microsomal fraction of several extrahepatic organs, was nicely resolved when Hajra and coworkers showed that the enzyme was in fact mainly localized in peroxisomes (183, 184). In guinea pig liver also, alkyl-DHAP synthase was found to be nearly

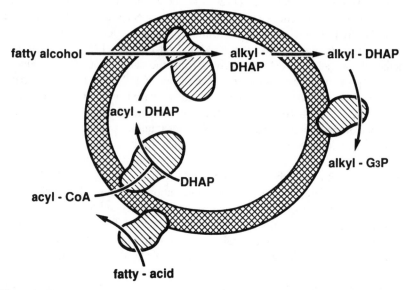

Figure 1   Sidedness of peroxisomal enzymes required for complex ether lipid synthesis. DHAP, dihydroxyacetone phosphate; G3P, glycerol-3-phosphate.

exclusively present in this organelle, which also contained acyl/alkyl-DHAP reductase (184). The presence of the DHAP-pathway enzymes in peroxisomes suggested an important role of this organelle in glycero-ether lipid synthesis. This importance only became fully appreciated after we demonstrated that tissues from Zellweger patients, known to lack peroxisomes (185), were severely deficient in ether-linked phosphoglycerides (186), and that fibroblasts from such patients were impaired in plasmalogen biosynthesis (187). This latter condition appeared to be caused by a deficiency in both DHAP-AT (188–190) and alkyl-DHAP synthase (189–191). The essential role of peroxisomes in ether lipid synthesis was also established in CHO mutants defective in peroxisome biogenesis (192, 193). In all cases the deficient ether lipid synthesis could be corrected by supplementation of the cells with alkylglycerol (187, 194, 195), which convincingly demonstrated that the deficiency in ether lipids in these mutant cells was solely caused by impaired formation of the glycero-ether bond in peroxisomes, and indicated that ether lipid synthesis cannot take place anywhere else in the cell to any appreciable extent.

The peroxisomal contribution to complex ether lipid biosynthesis is outlined in Figure 1. Acyl-CoA can be provided by an acyl-CoA ligase, the active site of which faces the cytoplasm (82, 88). By contrast, DHAP-AT and alkyl-DHAP synthase are not sensitive to proteolysis unless detergent is used to disrupt the membrane (179, 196), suggesting a matrix-oriented localization of the active sites of these enzymes. For DHAP-AT this conclusion was recently supported by experiments with selectively permeabilized fibroblasts. Under these conditions DHAP-AT showed latency that was not observed in sonicated cells. This latency could be overcome by addition of ATP, suggesting that ATP is required for the transport of at least one of the substrates of DHAP-AT over the peroxisomal membrane to reach the active site of this enzyme at the inner aspect of the membrane (197).

It has been suggested that a fatty alcohol–producing acyl-CoA reductase is present in rat brain peroxisomes (198), but it is generally believed that this enzyme is located in microsomes (199). The fact that this membrane-associated enzyme retains near normal activity in ZS fibroblasts (200) would be in line with an extraperoxisomal localization. Once alkyl-DHAP has been formed, presumably at the inner aspect of the peroxisomal membrane, it has to be translocated for reduction by acyl/alkyl-DHAP reductase located at the cytoplasmic side of either peroxisomes (179) or ER (201). Acyl-DHAP and alkyl-DHAP generated in isolated liver peroxisomes transverse the membrane within minutes to become available at the outer surface for extraction with bovine serum albumin (202). The next step in complex ether lipid synthesis concerns the acylation of the sn-2-hydroxyl. This and further enzymes in the pathway are not found in peroxisomes (203), and the process is completed in the ER. It is not known whether alkyl-DHAP, alkyl-glycerol-3-phosphate, or both will leave the peroxisome, nor whether the transport is protein-mediated or involves monomer diffusion. It is clear, however, that the function of peroxisomes is to provide ether-linked intermediates for the final steps of ether lipid biosynthesis in the ER.

Peroxisome proliferation in liver by clofibrate or plasticizer results in a 2- to 3-fold increase in total and specific activity of DHAP-AT (204–206) and a less than 2-fold increase in acyl/alkyl-DHAP reductase (205, 207). However, in purified peroxisomes the specific activity of DHAP-AT exhibited no significant increase under conditions causing an at least 10-fold increase in fatty acid β-oxidation specific activity (206). In homogenates the specific activity of alkyl-DHAP synthase was hardly affected by proliferation (205, 206), and a slight decrease in purified peroxisomes was observed (206). In spite of the increase in total activity of the enzymes per liver cell, the plasmalogen content did not change appreciably upon proliferation (206).

ENZYMOLOGY   Neither DHAP-AT nor alkyl-DHAP synthase have been obtained in homogeneous form. A partially purified preparation of DHAP-AT from guinea pig liver peroxisomes was devoid of glycerol-3-phosphate acyltransferase activity. The biosynthetic reaction is reversible as shown by the formation of palmitoyl-[^{3}H]CoA from palmitoyl-DHAP and [^{3}H]CoA. In addition, the enzyme catalyzes the CoA-dependent exchange of the acyl group between acyl-DHAP and DHAP (208).

The mechanism of the unique reaction catalyzed by alkyl-DHAP synthase has received considerable attention (209–214). The use of partially purified enzyme preparations, devoid of contaminating acyl hydrolases, has been particularly rewarding (211–213). Characteristic features in the formation of the glycero-ether bond are the loss of the pro-R hydrogen at the esterified C-atom in acyl-DHAP and the stoichiometric replacement of the acyl group by a fatty alcohol donating the oxygen for the ether linkage (210). These features predicted an unusual ester cleavage between the C-atom of DHAP and oxygen rather than the usual cleavage between oxygen and the carbonyl carbon of the fatty acid. This exceptional cleavage mechanism was confirmed experimentally using palmitoyl-[^{18}O]DHAP (213, 214).

Acyl/alkyl-DHAP reductase has recently been obtained in homogenous form as a 60-kDa protein after 370-fold purification from guinea pig liver peroxisomes (215). The enzyme reduces acyl-DHAP and alkyl-DHAP with nearly equal K_M and V_{max} values, and exhibits a large preference for NADPH over NADH with K_M values of 20 μM and 1.7 mM, respectively.

PEROXISOMAL DISORDERS

Introduction

In humans the vital importance of peroxisomes received particular emphasis by the recognition of a number of inherited diseases involving either a defect in the biogenesis of peroxisomes or a deficiency of one (or more) peroxisomal enzymes (13–16, 134). So far 12 different peroxisomal disorders have been described, most of them lethal diseases.

In 1973 Goldfischer and coworkers (185) showed that morphologically distinguishable peroxisomes are absent in liver and renal tubule cells of patients suffering from the cerebro-hepato-renal (Zellweger) syndrome. This finding was later confirmed by other investigators, who also showed that the deficiency of peroxisomes is manifested in all tissues and cells studied so far, including cultured skin fibroblasts (15). The absence of peroxisomes in the Zellweger syndrome is accompanied by the following biochemical abnormalities (see 13–16, 134 for reviews). Firstly, the biosynthesis of ether phospholipids is impaired, resulting in a deficiency of plasmalogens in all tissues. Secondly, the β-oxidation of fatty acids is impaired due to a deficiency of all peroxisomal β-oxidation enzyme proteins. This impairment results both in an accumulation of very long-chain (>C22) fatty acids in tissues and blood and of intermediates in the biosynthesis of bile acids in blood. Schram et al (216) have shown that in Zellweger fibroblasts the peroxisomal β-oxidation enzyme proteins are synthesized normally but become rapidly degraded thereafter. Thirdly, there is a defective degradation of pipecolic acid due to a deficiency of pipecolic acid oxidase (69). Fourthly, the degradation of phytanic acid is impaired due to a deficiency at the level of phytanic acid α-oxidation (15). Fifthly, several peroxisomal matrix enzymes like catalase, D-aminoacid oxidase, L-α-hydroxyacid oxidase, and AGT are localized in the cytosol instead of being particle-bound (15). Recent studies by several groups have shown that some of the peroxisomal integral membrane proteins are present in fibroblasts from Zellweger patients in unusual, largely empty membrane structures called peroxisomal "ghosts" (217–219, sections ASSEMBLY OF PEROXISOMES and PEROXISOMAL MEMBRANES).

Classification

The genetic diseases in humans involving peroxisomes can tentatively be classified into three groups depending on the degree to which peroxisomal functions are deficient. The first category consists of disorders like the Zellweger syndrome with a generalized impairment of peroxisomal functions, and also comprises two clinically somewhat milder disorders, the neonatal type of adrenoleukodystrophy (NALD) and the infantile type of Refsum disease (IRD). There may be heterogeneity with respect to the degree of peroxisome deficiency. In cultured skin fibroblasts from some Zellweger and IRD patients (220) and in intestinal epithelia from several Zellweger and NALD patients (221), the presence of a limited number of peroxisomes has been demonstrated by ultrastructural enzyme cytochemistry. A remarkable genetic heterogeneity was found within this category by complementation after somatic cell fusion of fibroblasts. Complementation was assessed by measuring DHAP-AT activity (222), particle-bound catalase (222), de novo plasmalogen biosynthesis (223), phytanic acid oxidase activity (224), and peroxisomal β-oxidation activity (225). The results indicate that at least six different gene products play a role in peroxisome biogenesis.

A second category comprises two diseases, the rhizomelic type of chondrodysplasia punctata (RCDP) and Zellweger-like syndrome. A limited number of peroxisomal functions is impaired in these disorders (226, 227). In RCDP the de novo plasmalogen biosynthesis is deficient, phytanic acid oxidation is impaired, and the peroxisomal thiolase is present only as a 44-kDa precursor protein. Most of this precursor thiolase is extraperoxisomal (228). However, no accumulation of very long-chain fatty acids or intermediates of bile acid biosynthesis is found, indicating that the peroxisomal β-oxidation capacity is sufficient to prevent accumulation of substrates. Complementation studies indicate that RCDP is distinct from Zellweger syndrome, NALD, and IRD (222–224). In Zellweger-like syndrome, a very rare disorder, peroxisomes appear to be present, but the peroxisomal β-oxidation enzyme proteins were found to be deficient upon immunoblotting, there was deficiency of DHAP-AT, and the content of plasmalogens in tissues and cultured cells was low (229).

The third category includes a larger group of genetic disorders in which only one peroxisomal enzyme is deficient and in which normal peroxisomes can be found in tissues and cultured cells in most patients. The most commonly encountered disorder in this category is X-linked adrenoleukodystrophy (X-ALD) and its phenotypic variants (230), in which very long-chain fatty acids (VLCFA) accumulate in tissues, plasma, and cultured cells. A key observation was that in X-ALD fibroblasts, the oxidation of lignoceroyl-CoA was unimpaired, in contrast to the oxidation of free lignoceric acid (231). Subsequent studies confirmed that the biochemical defect in X-ALD is at the level of the peroxisomal VLCFA-CoA synthetase (89, 90).

Patients have recently been described with either a deficiency of acyl-CoA oxidase (232), the bifunctional protein (233), or peroxisomal thiolase (234). In some patients the deficiency was established by the absence of immunoreactive protein in immunoblots (233, 235). In others, immunoreactive protein is present, and the mutation presumably affects the catalytic activity of the enzyme (236). In patients with a deficiency of the bifunctional protein or of peroxisomal thiolase, not only VLCFA, but also the bile acid intermediates DHCA and THCA accumulate. In contrast, bile acid intermediates do not accumulate in patients with an acyl-CoA oxidase deficiency, which is in accordance with the recent finding of separate oxidases for THCA-CoA and VLCFA-CoA (73). Patients with an accumulation of THCA and DHCA but not of VLCFA have also been described, presumably reflecting a defect in THCA-CoA oxidase or THCA-CoA synthetase (237, 238).

Hyperoxaluria type I can also be classified as a peroxisomal disorder of the third category. It has recently been shown that the defect in this disorder is a deficiency of the enzyme AGT (239, 240), which is a peroxisomal enzyme in humans (145). Interestingly, some patients have a trafficking defect resulting in diversion of AGT to the mitochondria (239, section ASSEMBLY OF PEROXISOMES).

Finally, acatalasemia is a condition in which the peroxisomal enzyme catalase is deficient (241). Individuals with a deficiency of catalase usually do not show severe clinical abnormalities.

ASSEMBLY OF PEROXISOMES

Introductory Remarks and Genes for Peroxisomal Proteins

It is now well-established that peroxisomes, like mitochondria and chloroplasts, arise by growth and division of pre-existing peroxisomes (6, 18), rather than by budding from the ER (242). They may transiently or permanently be interconnected to form a "peroxisomal reticulum" (see 6) containing elements resembling smooth ER (43, 45, 243, 244) (see also section *Morphology and Induction of Peroxisomes*).

Unlike mitochondria and chloroplasts, peroxisomes contain no DNA (6, 18, 245), so that all peroxisomal proteins must be coded for by nuclear genes. During the past few years, the genes for a large number of peroxisomal proteins, including mammalian peroxisomal β-oxidation enzymes (97, 105, 108, 109, 246), have been cloned and characterized. No features common to all of these genes are as yet apparent, except perhaps for the presence of GC boxes in the promoter region.

In the rat a single acyl-CoA oxidase gene is present, which gives rise to two different mRNAs by differential splicing of the primary transcript (97). These two mRNAs may correspond to the two fatty acyl-CoA oxidases identified in rat liver by Schepers et al (72), one of which is inducible by clofibrate in rodent liver. Only one gene for enoyl-CoA hydratase:3-hydroxyacyl-CoA dehydrogenase has been identified in the rat (105). This gene is inducible, and there is no evidence for the presence of more than one mRNA or more than one protein. Finally, two genes for peroxisomal 3-oxoacyl-CoA thiolase are present in the rat, one of which is inducible by clofibrate (108, 109). Thus, entirely different mechanisms are responsible for the induction by clofibrate of individual β-oxidation enzymes, e.g. (*a*) enhanced synthesis of a specific mRNA formed by alternative splicing of the transcript of a single gene (acyl-CoA oxidase), or (*b*) enhanced transcription of a separate gene (3-oxoacyl-CoA thiolase), or (*c*) enhanced transcription of a single gene (enoyl-CoA hydratase:3-hydroxyacyl-CoA dehydrogenase).

Only a single gene for peroxisomal 3-oxoacyl-CoA thiolase appears to be present in humans (246), and only one fatty acid acyl-CoA oxidase, with properties similar to those of the noninducible isoform in rat liver (247). This may be related to the fact that compounds like clofibrate do not induce peroxisomal β-oxidation enzymes in primates (248). In this respect it will be of interest to see if the lack of response in primates correlates with an absence of the 70-kDa peroxisome proliferator–binding protein in rat liver described by Alvarez et al (64) or of the 52.4-kDa peroxisomal proliferator–activated receptor in mouse liver recently identified by Isseman & Green (65, see also section *Morphology and Induction of Peroxisomes*).

Biosynthesis and Import of Peroxisomal Proteins

GENERAL CHARACTERISTICS OF BIOSYNTHESIS AND IMPORT   All peroxisomal proteins investigated so far, including soluble matrix proteins (6, 18), core proteins (6), and integral membrane proteins (249–252), are synthesized on free ribosomes and are imported posttranslationally into peroxisomes with half-lives in the cytosol of 1–15 min (6). Almost all are synthesized in their final size (6, 18). Exceptions are discussed below.

The criteria required for demonstrating import of peroxisomal proteins, both in vivo and in vitro, have been summarized recently (18). Convincing evidence has been presented for the import by isolated peroxisomes of a number of peroxisomal proteins synthesized in vitro (6, 18, 253, 254). Import is time-dependent, occurs at 25 °C or 37 °C but not at 0 °C, and requires ATP. The question of whether protein import into peroxisomes, like that into mitochondria, requires a membrane potential is still unresolved (18). There is some evidence for a requirement for cytosolic factors (unfoldases?) for protein import into peroxisomes (253).

A TOPOGENIC SIGNAL FOR PROTEIN IMPORT INTO PEROXISOMES: THE SKL MOTIF   In 1987 Subramani and coworkers (255) showed that when animal cells are transfected with cDNA for luciferase, a peroxisomal enzyme occurring in the lantern organ of the firefly, the protein is synthesized and is found in the peroxisomes. They subsequently showed that the targeting signal for directing luciferase to the peroxisomes resides in the C-terminus of the protein, and identified the C-terminal tripeptide -Ser-Lys-Leu (SKL) as the peroxisome targeting signal not only in luciferase but also in other peroxisomal proteins (256–258). Gene fusion experiments indicated that reporter proteins with this sequence fused to their C-terminus are directed to the peroxisomes (257, 258). Certain conservative substitutions are permitted in the first two amino acids (but not in the third) without loss of targeting activity, and the consensus sequence could be defined as -(Ser/Ala/Cys)-(Lys/Arg/His)-Leu (258).

The SKL motif is present at the C-terminus of a number of peroxisomal proteins from several different organisms (17, 256–259). Antibodies raised to a synthetic peptide containing SKL at the C-terminus react with 15 to 20 protein bands in an immunoblot of rat-liver peroxisomal proteins (260). It has now been demonstrated that the SKL motif is required and is usually sufficient to direct a large number of peroxisomal proteins to the peroxisomes (254–259). Furthermore, this targeting signal has been conserved during evolution and functions in insects, mammals, plants, and different yeasts (254–258, 261, 262). In some proteins deletion of certain sequences leads to diminished import even though the C-terminal SKL is intact (254, 258). This may be due to a change in the conformation of the protein, leading to diminished exposure of the SKL motif.

In some peroxisomal proteins the SKL motif (or a variant of it) is not present at the C-terminus but at an internal position in the protein. In human catalase, for instance, which contains 526 amino acids, the sequence -Ser-His-Leu is present at residues 516–518 (263). Gould et al (257) have shown that the C-terminal 27 amino acids of catalase acted as a peroxisome targeting signal when fused to a reporter protein, whereas two longer peptides containing the C-terminal 27 amino acids did not. The fact that addition of one or two amino acids to the C-terminus abolishes peroxisome targeting activity (258) suggests that an internal SKL should not function as a peroxisome targeting signal.

EVIDENCE FOR OTHER PEROXISOME TARGETING SIGNALS   The fact that the SKL motif or one of its variants is not present in all peroxisomal proteins indicates that such proteins must use some other peroxisome targeting signal. In a recent paper (264), Subramani's group report that the peroxisome targeting signal in the precursor of peroxisomal 3-oxoacyl-CoA thiolase is located in the N-terminal peptide that is removed when the precursor is converted to the mature form. Removal of the N-terminal peptide takes place, at least in part, after import (265, 266).

The targeting of glycolytic enzymes to the glycosome, a microbody-like organelle in the Trypanosomatidae, appears to involve different signals from those utilized for peroxisome biogenesis in other eukaryotes (30, 267). However, proteins carrying the peroxisomal targeting SKL-motif have recently been identified in glycosomes (267a).

DIFFERENCES AMONG SPECIES IN THE TARGETING OF ALANINE:GLYOXY-LATE AMINOTRANSFERASE   As indicated in the section on *Glyoxylate Metabolism*, AGT is found exclusively in peroxisomes in humans and some other species, nearly exclusively in mitochondria in the dog and cat, and in both organelles in rodents. Recent molecular genetic studies (152, 268–271) have led to some clarification of the molecular genetic mechanisms underlying these differences.

Two mRNAs for AGT are found in rat liver (268, 269). These mRNAs arise from a single gene (152). Transcription from two initiation sites, 66 nucleotides apart in exon 1, generates the two mRNAs; the larger one contains a 5'-terminal sequence coding for a 22-amino-acid N-terminal extension, which functions as a cleavable mitochondrial targeting signal (152). Generation of the larger mRNA is under control of glucagon (268). Both the mRNAs contain the nucleotide sequence coding for an internal SKL sequence (152), which may act as a peroxisome targeting signal (but see subsection A TOPOGENIC SIGNAL FOR PROTEIN IMPORT INTO PEROXISOMES: THE SKL MOTIF). Apparently the mitochondrial targeting signal is dominant in the protein derived from the larger mRNA.

Human liver contains a single mRNA for AGT in which the untranslated sequence upstream of the first AUG codon shows considerable similarity to the sequence encoding the N-terminal 22-amino-acid peptide of the rat mitochondrial enzyme (270, 271). In fact, a single base substitution (A in humans for G in the rat) has led to the loss in humans of the AUG initiation codon which, in the rat, allows translation of a protein with the N-terminal extension (271).

Danpure and coworkers (239) have found that in certain patients with primary hyperoxaluria type I, the deficiency of AGT activity is accompanied by an aberrant localization of the enzyme, the residual activity being present in the mitochondria rather than in the peroxisomes. It is not clear at present whether the aberrant routing is due to loss of the peroxisomal targeting signal, appearance of a putative mitochondrial targeting signal, or both. All such patients investigated so far have at least one allele with point mutations giving rise to amino acid substitutions at positions 11, 170, and 340 (272). The aberrant protein in mitochondria from the patients has the same M_r as that of AGT from control subjects (239), whereas in rat liver there is a difference in size of about 2 kDa between the precursor and mature form of mitochondrial AGT (251). Thus the mitochondrial localization of AGT in the patients is apparently not accompanied by detectable cleavage of a targeting signal.

PEROXISOMAL   PROTEINS   SYNTHESIZED   AS   LARGER   PRECURSORS
Whereas most peroxisomal proteins are synthesized in their final size (see 6, 18), there are some exceptions. Peroxisomal 3-oxoacyl-CoA thiolase in mammals is synthesized as a 44-kDa precursor that is converted to a 41-kDa mature enzyme (265, 273, 274), and mammalian nsLTP is synthesized as a 14.5-kDa protein whereas the mature protein is about 1.5-kDa smaller (275, 276). In both cases processing does not occur in mutant cells defective in peroxisome biogenesis (171, 216, 277). Comparison of the amino acid sequence of human and rat peroxisomal 3-oxoacyl-CoA thiolases and nsLTP indicates that there is a striking homology in the region of the peptide bond cleaved during conversion of the precursors to the mature proteins (Table 2). For instance, proline is present at the P-2 position in all cases. This homology suggests that a single protease, located in peroxisomes, is responsible for the processing of both proteins (see also Ref. 174).

Table 2 Amino acid sequence in the region of the peptide bond cleaved during the processing of the precursor to the mature forms of rat 3-oxoacyl-CoA thiolases A and B, human 3-oxoacyl-CoA thiolase, and human and rat nonspecific lipid transport protein (nsLTP)

Protein	Amino acid sequence[a]	Refs.
Rat thiolase A	-Gln-Ala-Ala-Pro-Cys⁎Ser-Ala-Thr-	108, 109
Rat thiolase B	-Gln-Ala-Ala-Pro-Cys⁎Ser-Ala-Gly-	107, 109
Human thiolase	-Gln-Ala-Ala-Pro-Cys⁎Leu-Ser-Gly-	282, 283
Rat nsLTP	-Ser-Ala-Ala-Pro-Thr⁎Ser-Ser-Ala-	172–174, 278
Human nsLTP	-Glu-Ala-Val-Pro-Thr⁎Ser-Ser-Ala-	175

[a] The (putative) cleavage site is indicated by*.

As indicated in the subsection EVIDENCE FOR OTHER PEROXISOME TARGETING SIGNALS, the N-terminal peptide of 3-oxoacyl-CoA thiolase appears to contain the signal for directing this protein to peroxisomes. It will be of interest to see if the N-terminal peptide in the precursor of nsLTP plays any role in targeting of this protein to the peroxisomes. Pre-nsLTP does, however, contain a C-terminal -Ala-Lys-Leu tripeptide (172–175, 278), which is a consensus peroxisome targeting signal.

A third peroxisomal protein in mammals that behaves aberrantly is acyl-CoA oxidase. This enzyme is synthesized as a 72-kDa precursor, which is slowly converted to two subunits of 52- and 20-kDa, both of which remain associated with the enzyme (279). Again, cleavage of the precursor is impaired in mutant cells lacking peroxisomes (216).

Finally, the isoform of malate dehydrogenase found in glyoxysomes in germinating cotyledons of pumpkin and watermelon is synthesized as a larger precursor with a cleavable presequence (280, 281).

RECEPTOR AND ACCESSORY PROTEINS REQUIRED FOR IMPORT   Import of proteins into peroxisomes requires at least two receptors, one recognizing proteins with a C-terminal SKL motif and the other recognizing the targeting signal in the N-terminal region of the precursor of 3-oxoacyl-CoA thiolase and perhaps nsLTP. There is as yet no information available about the nature and properties of those putative receptors. Furthermore, translocation of proteins across the peroxisomal membranes, like that across the mitochondrial membranes (see 284 for a recent review) can be expected to require the participation of accessory proteins such as unfoldases, channel-forming proteins, and chaperonins (see subsection MUTANT MAMMALIAN CELLS DEFECTIVE IN PEROXISOME BIOGENESIS).

In patients with the disease RCDP, the amount of 3-oxoacyl-CoA thiolase protein in the peroxisome is reduced and processing of the precursor is impaired (228, 285–287) (see section PEROXISOMAL DISORDERS). The primary lesion in this disease could be a mutation in a gene coding for a receptor required for the import into peroxisomes of a limited number of proteins, including 3-oxoacyl-CoA thiolase and a putative prothiolase protease.

Peroxisome Biogenesis

TRANSFORMATION OF GLYOXYSOMES INTO LEAF PEROXISOMES IN HIGHER PLANTS   During germination of fat-storing seeds, biogenesis of glyoxysomes occurs (see section *Glyoxylate Metabolism*). Upon illumination of the cotelydons the glyoxysomes are converted into peroxisomes characteristic of green leaves. The following changes occur (288–290). Glyoxysome-specific enzymes, like malate synthase, are broken down in the transition-state organelles present in greening cotyledons (291). Concomitantly, light induces an increase in the level of mRNA for leaf peroxisome–specific enzymes like glycolate oxidase (292) and in the level of these enzymes in the transition-state organelles (288, 289). Malate synthase synthesized in vitro and imported into isolated microbodies is stable in glyoxysomes and leaf peroxisomes (291). Thus the enzyme system involved in the degradation of glyoxysomal enzymes is only transiently expressed.

MUTANT MAMMALIAN CELLS DEFECTIVE IN PEROXISOME BIOGENESIS: CORRECTION OF THE DEFECT BY SOMATIC CELL FUSION   Mutant cell lines defective in peroxisome biogenesis are available from patients with the Zellweger syndrome and related disorders (see section PEROXISOMAL DISORD-ERS) and have in addition been isolated from mutagenized CHO cells (192, 193, 293). Fusion of complementary cell lines leads to correction of peroxi-some biogenesis in the heterokaryons as assessed by restoration of per-oxisomal functions like DHAP-AT activity (222), plasmalogen biosynthesis (223), oxidation of VLCFA (225), or processing of the precursor of per-oxisomal 3-oxoacyl-CoA thiolase to the mature enzyme (293). Complementa-tion analysis following somatic cell fusion indicates that the human cell lines can be divided into at least six complementation groups (222, 223, 225). Thus peroxisome biogenesis requires the participation of at least six gene products.

Peroxisome biogenesis in the heterokaryons can also be assessed by measuring the rate of incorporation of catalase, originally present in the cytosol in the parental cell lines, into newly assembled peroxisomes (222, 294). After fusion of some combinations of human cell lines, the assembly of peroxisomes measured by this criterion was found to be very rapid (complete within 3 h; see Figure 2) and insensitive to cycloheximide, so that the components required for peroxisome assembly must have been present in the parental cell lines (294). Uptake of catalase into peroxisomes in the heteroka-ryons is inhibited by aminotriazole (295), presumably because this irrevers-ible inhibitor of catalase prevents or retards unfolding of the enzyme.

EVIDENCE FOR THE PRESENCE OF PEROXISOMAL GHOSTS IN MUTANT MAMMALIAN CELLS DEFECTIVE IN PEROXISOME BIOGENESIS   It has been concluded that the lesion in diseases of peroxisome biogenesis must involve the import of peroxisomal proteins rather than the assembly of peroxisomal membranes (6, 15). This conclusion is based in part on the fact that per-oxisomal integral membrane proteins can be detected by means of im-munoblotting or immunocytochemistry in liver and fibroblasts from patients with diseases of peroxisome assembly (219, 296–299). The proteins are associated with membranes in large, vesicular structures sometimes devoid of content, which have been referred to as peroxisomal ghosts (219, 298). However, most of these structures also contain lysosomal enzymes and must therefore represent autophagocytosed material in secondary lysosomes (300).

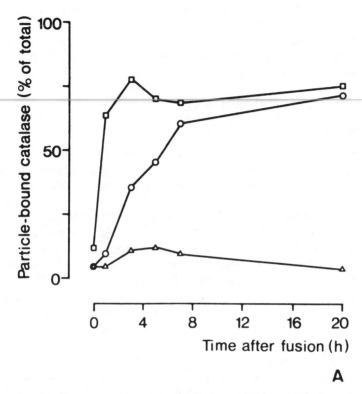

A

Figure 2 Kinetics of complementation in heterokaryons obtained after fusion of complementary cell lines deficient in peroxisomes. Complementation was assessed by determination of the activity of particle-bound catalase in the heterokaryons by the digitonin permeabilization technique. Polyethyleneglycol was used for fusion of: □, W78/515 (complementation group 2) × GOM85AD (group 3); ○, BOV84AD (group 2) x GOM85AD (group 3); △, self-fusion of GOM85AD cells. Reproduced from Ref. 297 with permission.

Nevertheless, in the somatic cell fusion experiments described in the previous section, the assembly of peroxisomes presumably requires the presence of some kind of scaffolding on which assembly of peroxisomal membranes can occur. Furthermore, it is possible that lack of complementation after fusion of two cell lines results from a lack of a scaffolding. Evidence for the absence of such a scaffolding in a CHO mutant has been presented (301).

IDENTIFICATION AND CHARACTERIZATION OF GENES AND GENE PRODUCTS REQUIRED FOR PEROXISOME BIOGENESIS   Fujiki and coworkers (302) have recently succeeded in restoring peroxisome biogenesis in a CHO mutant by transfecting the cells with rat-liver cDNA, and have identified the factor responsible for correction as a 35-kDa peroxisomal membrane protein. The function of the protein, which the authors refer to as peroxisome assembly factor PAF-1, is as yet unknown. An analogous study has been carried out by Kunau and coworkers (303) with one of the *S. cerevisiae* peroxisome assembly mutants, *pas1*, previously isolated by this group (304). The *pas1* gene product is a hydrophilic 117-kDa protein with a putative ATP-binding site. The findings of these two groups represent major achievements.

PEROXISOMAL MEMBRANES

Membrane Properties

Studies on peroxisomal membranes have greatly benefitted from the development by Lazarow and coworkers (305) of a one-step sodium carbonate extraction of purified organelles. This procedure effectively releases soluble and peripheral membrane proteins and yields a preparation consisting of membrane lipids and integral membrane proteins (IMPs). The total organelle proteins are generally recovered either in the soluble or in the membrane fraction. However, their enzymatic activity is frequently lost upon carbonate treatment (306). In such cases, their distribution can be established by an alternative technique involving Triton X-114 phase separation (306).

Studies on the osmotic behavior of isolated peroxisomes and the lack of latency of urate-, D-aminoacid-, and L-α-hydroxyacid oxidase indicated that sucrose and the substrates for these three peroxisomal oxidases easily permeate through the membrane of the isolated organelles (2). Direct permeability measurements with seemingly intact rat liver peroxisomes revealed them to be permeable to small solutes including glucose, sucrose, urea, acetate, methanol, uric acid, and the cofactors for fatty acid oxidation, i.e. NAD^+, CoA, ATP, and carnitine (307, 308). Evidence was obtained from reconstitution experiments with peroxisomal IMPs in phospholipid vesicles that the high permeability of the peroxisomal membrane is caused by the presence of a pore-forming protein. The authors suggested (308) that a 22-kDa IMP that is unique to peroxisomal membranes (305) is the most likely candidate for the pore-forming activity. It was also suggested that peroxisomal metabolism is not regulated at the level of membrane transport and that the concentrations of unbound substrates and cofactors are of the same order of magnitude in the peroxisomal matrix and the cytoplasm (308). This concept is hard to reconcile with the existence of a proton gradient across the peroxisomal membrane as observed in several yeasts by ^{31}P NMR measurements (309) and immunocytochemistry (310). A proton-translocating ATPase associated with the cytosolic face of the peroxisomal membrane of yeasts (311) was held responsible for the acidic internal pH of 5.8 to 6.0 of peroxisomes (309). The presence of an ATPase in yeast peroxisomes has recently been challenged, however (312). The existence of a pH gradient across peroxisomal membranes in organisms other than yeasts has not yet been established. However, the association of a cytosol-oriented ATPase activity with rat liver peroxisomes (313) and evidence for the involvement of an ATPase in the transport of at least one of the substrates for DHAP-AT across the peroxisomal membrane (197) have been reported. These results are all in line with the notion that the leakiness of isolated peroxisomes represents an isolation artifact.

Integral Mammalian Peroxisomal Membrane Proteins

Table 3 lists the IMPs of known molecular weights reported to be present in mammalian peroxisomes. The 76-kDa peroxisomal long-chain acyl-CoA synthetase in rat liver appears to be identical to the microsomal and mitochondrial enzyme (83, 314).

The 69/70-kDa protein has been detected in peroxisomal membranes from rat liver (305, 315, 316) and human liver (317) and fibroblasts (217, 298). It was also found in liver and fibroblasts from Zellweger patients (217, 298). The 69/70-kDa protein in rat liver peroxisomes extends into the cytoplasm, and evidence has been presented that 68-, 41/42-, and 28-kDa membrane proteins are derived from the 69/70-kDa polypeptide by the action of endogenous proteases during isolation of peroxisomes (315, 316). Masters and coworkers (318, 319) have reported that mouse liver peroxisomes contain one IMP with an M_r of 68,000, and another with an M_r of 70,000. The latter appeared to be identical to the bifunctional protein, which is generally not regarded as an IMP. It is known to be poorly soluble and to remain partly associated with the membrane fraction even after carbonate extraction (320). A cDNA for the 69/70-kDa peroxisomal IMP from rat liver has now been cloned and sequenced (321). Its C-terminal region shows strong sequence homology to a superfamily of ATP-binding proteins, many of which are involved in ATP-driven transport of small molecules or proteins through membranes (321). The ATP-binding domain of the peroxisomal membrane protein was localized to the part extending into the cytoplasm, in line with a proposed role in ATP-driven transport into the organelle.

Other peroxisomal IMPs for which a functional role has been proposed include 22-kDa and 35-kDa polypeptides. The 22-kDa protein has been detected in peroxisomes from rat (305, 315, 316, 322) and human (217) liver and human fibroblasts (217, 298), and is also found in liver and peroxisomal ghosts in fibroblasts from Zellweger patients (217, 298). As discussed in the previous section, this 22-kDa protein is believed to be involved in the permeability of peroxisomal membranes to small molecules (308). The protein consists 44% of hydrophobic amino acids, and appears to be embedded deeply in the membrane, in that proteolysis of intact peroxisomes produced a slightly smaller polypeptide of 21 kDa (249). Quite recently, the cloning and characterization of a cDNA encoding a unique peroxisomal membrane protein

Table 3 Integral mammalian peroxisomal membrane proteins

Subunit size (kDa)	Source	Organelle	Accessibility to added proteases	Induction by proliferators in rodent liver	Proposed function	Refs.
76	RL	P, Mt, Mc	accessible	++	acyl-CoA synthetase	83, 314
69/70	RL, HL, HF, ZSF	P, Pgh	accessible	++	ATP-binding, transport	217, 298, 305, 315–317, 321
57	RL	P	accessible	+		323
53	HL, HF, ZSF	P, Pgh	?			217, 298
36	RL	P, Mt	embedded	±		316, 322
35	CHO		accessible		assembly	302
26	RL	P	?	++		315
22	RL, HL, HF, ZSF	P, Pgh	accessible, embedded	±	pore-forming	217, 249, 298, 305, 308, 315, 316, 322
15	RL	P, Mt, Mc	?			305, 322

Abbreviations: RL, rat liver; HL, human liver; HF, human fibroblasts; ZSF, Zellweger syndrome fibroblasts; CHO, Chinese hamster ovary; P, peroxisome; Mt, mitochondria; Mc, microsomes; Pgh, peroxisomal ghosts; ?, not reported of 35 kDa that restores the biogenesis of peroxisomes and complements the defect of peroxisomal functions in a peroxisome-deficient CHO mutant was reported (302). This protein, denoted PAF-1, contains two putative hydrophobic membrane-spanning regions and is highly accessible to proteinase K, suggesting that a considerable part is exposed on the outer surface of peroxisomes.

Peroxisomal IMPs of as-yet-unknown function include a 57-kDa protein in rat liver (323), a 53-kDa protein present in liver, and fibroblasts from both controls and Zellweger patients (217, 298), and 36-kDa (316, 322), 26-kDa (315) and 15-kDa (305, 322) proteins in rat liver. Of the IMPs listed in Table 3, those of 69/70, 57, 53, 35, 26, and 22 kDa are exclusively present in peroxisomes.

The 76-, 69/70-, 26-, and to a lesser extent the 57-kDa protein increase in rodent liver upon administration of peroxisome proliferators, whereas the 36- and 22-kDa proteins are much less affected, thus causing a change in membrane polypeptide composition upon proliferation. The 69/70-, 36-, 26-, and 22-kDa polypeptides were all shown to be synthesized as the mature proteins on free polysomes (249–251).

Other proteins located to peroxisomal membranes include DHAP-AT, alkyl-DHAP synthase, and very long-chain acyl-CoA synthetase, but the polypeptides responsible for these activities have yet to be identified.

Integral Plant Glyoxysomal Membrane Proteins

In comparison to mammalian peroxisomes, less work has been reported on the IMP composition of glyoxysomes. In carbonate pellets of glyoxysomes isolated from Castor bean endosperm, 22 bands were detected by SDS-PAGE (324). Remarkably, 12 of these were reported to be glycoproteins. However, these results have not been confirmed by others and have been the subject of some controversy (325, 326). A possible relationship with IMPs from mammalian peroxisomes has not been investigated. It is noteworthy that the abundant, presumably pore-forming, 22-kDa protein from mammalian peroxisomes does not seem to be present in glyoxysomes. In this context it may be relevant to note also that isolated glyoxysomes do not seem to be freely permeable to small molecules as evidenced by latencies for malate synthase, hydroxyacyl-CoA dehydrogenase, and malate dehydrogenase of 98, 85, and 71%, respectively (327).

Integral Yeast Peroxisomal Membrane Proteins

Research on the IMPs of yeast peroxisomes is essentially confined to *C. tropicalis* and *C. boidinii*. Four major proteins of 44, 34, 29, and 24 kDa were found in *C. tropicalis* peroxisomal membranes, of which the 24-kDa entity was induced by growth on oleic acid (328). This pattern is significantly different from that observed in *C. boidinii*, where the most prominent bands consisted of 120-, 100-, 47-, 32-, 31-, and 20-kDa polypeptides (329). None of these appear to be glycosylated. The 20-kDa protein and its mRNA could only be detected after a shift from glucose to methanol as C-source (330, 331). Immunoelectronmicroscopy detected a predominant labeling over the peroxisomal membrane, but a significant fraction of the signal appeared in the matrix as well (330). This and the fact that a large proportion of the 20-kDa protein could be removed from the membrane by carbonate treatment (329) suggested that it may be a peripheral membrane– or a matrix protein. In agreement with this notion, no obvious membrane-spanning region could be predicted from the base sequence of the isolated gene (330). In contrast to the 20-kDa protein, the 31/32- and 47-kDa proteins represented abundant components of the peroxisomal membrane in *C. boidinii*, regardless of whether oleic acid, D-alanine, or methanol was used to induce peroxisome proliferation (331). The 47-kDa protein could also be observed by immunoelectronmicroscopy in the small organelles that are present in cells grown on glucose (46). The gene encoding this protein has been isolated, and its sequence predicts two membrane-spanning segments (332). It appears to be a novel protein with no sequence similarity to other proteins. The abundant presence of this protein regardless of the peroxisome inducer utilized suggests that it may be related to the structure, assembly, or some other general function of peroxisomes in yeast. However, using anti-47-kDa protein monoclonal antibodies, a homologue of this protein could not be detected in *S. cerevisiae* (46).

CONCLUSIONS AND FUTURE PROSPECTS

During the past decade the peroxisome has come of age. It is now recognized as a full-fledged intracellular compartment fulfilling essential functions in almost all eukaryotic organisms. The peroxisomal β-oxidation system, discovered by Lazarow & De Duve in 1976 and subsequently characterized in detail by Hashimoto's group, still occupies a central position in peroxisomal metabolism, particularly in mammals and other higher eukaryotes. The unique presence of specific accessory proteins enables the peroxisome to catabolize a variety of compounds that cannot be oxidized via the mitochondrial β-oxidation system.

Peroxisomes are highly versatile organelles. They contain enzymes that are essential for various catabolic and anabolic pathways, including the biosynthesis of ether phospholipids. Some of these enzymes are expressed constitutively, while others can be induced under appropriate conditions by mechanisms that are only now beginning to be understood. Undoubtedly many peroxisomal functions remain to be discovered.

Several metabolic pathways require participation of peroxisomal enzymes acting in consert with enzymes in other intracellular compartments. Little is known at present about the regulation of such pathways or about the shuttling of intermediates between compartments.

Strong evidence has now been brought forward indicating that the peroxisomal membrane in the intact cell is, like other biological membranes, not freely permeable to metabolites. The following questions arise. Is the transport of metabolites mediated by specific translocators or does it occur through gated pores? What is the role of ATP in metabolite transport? Is there a proton gradient in peroxisomes from higher organisms as there appears to be in those from yeasts? How can one reconcile a rather low pH in the peroxisomal matrix with a rather high pH optimum of many peroxisomal enzymes?

Considerable progress has recently been made in understanding the mechanisms involved in peroxisome biogenesis. Peroxisomal proteins are synthesized on free ribosomes and incorporated posttranslationally into preexisting peroxisomes, which grow and divide. In vivo and in vitro systems for studying protein import into peroxisomes have been devised, and the requirements for import are beginning to be understood. A major advance has been the discovery by Subramani and coworkers of a peroxisome targeting signal consisting of the consensus tripeptide sequence -(Ser/Ala/Cys)-(Lys/Arg/His)-Leu situated at the C-terminus of many peroxisomal proteins. Information on the receptor reacting with this targeting signal, which has been highly conserved during evolution, should be forthcoming in the near future.

There is evidence for at least one additional peroxisome targeting signal, present in the cleavable presequence of peroxisomal 3-oxoacyl-CoA thiolase. This is one of the few peroxisomal proteins synthesized as a larger precursor. Another is nonspecific lipid transport protein. The presequences of the two proteins show remarkable homology, the significance of which should be resolved in the near future.

Research on the biochemistry of peroxisomes has received a considerable stimulus from the discovery of genetic diseases in humans in which peroxisome biogenesis is impaired. These diseases fall into at least six complementation groups. Identification of the genes that are mutated in the different complementation groups and characterization of their gene products can be expected in the near future.

Two further experimental approaches have begun to yield important results pertaining to the biogenesis of peroxisomes. Firstly, Kunau and coworkers have discovered that peroxisome proliferation can be induced in *Saccharomyces cerevisiae*, a classic eukaryotic microorganism for genetic studies, by culture of the cells on oleate, and have used an oleate minus mutant to identify a gene product required for peroxisome biogenesis. Secondly, Fujiki and coworkers have utilized a procedure developed by Raetz and coworkers to generate mutants in peroxisome biogenesis in CHO cells. Transfection studies with one of these cell lines have enabled them to identify a membrane protein required for peroxisome biogenesis. The identification of further gene products required for the assembly of peroxisomes should follow rapidly, and the next step will be to ascertain what their functions are and whether they have been conserved during evolution or not.

This review is of necessity incomplete. For lack of space we have had to limit the number of topics and the coverage given to some topics. We have chosen to concentrate on the biochemistry of mammalian peroxisomes and have therefore had to omit many excellent papers concerning peroxisomes from other organisms. We hope the topics we have omitted will be the subject of future, more specialized reviews.

ACKNOWLEDGMENTS

Research in the authors' laboratories has been supported by grants from the Netherlands Organization for Scientific Research (NWO), Section for Medical Sciences, and from the Princess Beatrix Fund.

Literature Cited

1. Baudhuin, P., Beaufay, H., De Duve, C. 1965. *J. Cell Biol.* 26:219–43
2. De Duve, C., Baudhuin, P. 1966. *Physiol. Rev.* 46:323–57
3. De Duve, C. 1983. *Sci. Am.* 248:52–62
4. Beevers, H. 1979. *Annu. Rev. Plant Physiol.* 30:159–93
5. Tolbert, N. E. 1981. *Annu. Rev. Biochem.* 50:133–57
6. Lazarow, P. B., Fujiki, Y. 1985. *Annu. Rev. Cell Biol.* 1:489–530
7. Borst, P. 1986. *Biochim. Biophys. Acta* 866:179–203
8. Borst. P. 1983. *Trends Biochem. Sci.* 8:269–72
9. Müller, M. 1975. *Annu. Rev. Microbiol.* 29:467–83
10. Opperdoes, F. R. 1987. *Annu. Rev. Microbiol.* 41:127–51
11. Veenhuis, M., Van Dijken, J. P., Harder, W. 1983. *Adv. Microbiol. Physiol.* 24:1–82
12. Goldfischer, S., Reddy, J. K. 1984. *Int. Rev. Exp. Pathol.* 26:45–84
13. Schutgens, R. B. H., Heymans, H. S. A., Wanders, R. J. A., van den Bosch, H., Tager, J. M. 1986. *Eur. J. Pediatr.* 144:430–40
14. Moser, H. W. 1987. *Dev. Neurosci.* 9:1–18
15. Lazarow, P. B., Moser, H. W. 1989. In *The Metabolic Basis of Inherited Disease*, ed. C. R. Scriver, A. L. Beaudet, W. S. Sly, D. Valle, pp. 1479–1509. New York:McGraw-Hill. 6th ed.
16. Wanders, R. J. A., Heymans, H. S. A., Schutgens, R. B. H., Barth, P. G., van den Bosch, H., et al. 1988. *J. Neurol. Sci.* 88:1–39
17. Osumi, T., Fujiki, Y. 1990. *BioEssays* 12:217–22
18. Borst, P. 1989. *Biochim. Biophys. Acta* 1008:1–13
19. Breidenbach, R. W., Beevers, H. 1967. *Biochem. Biophys. Res. Commun* 27: 462–69
20. Gerhardt, B. 1987. *Methods Enzymol.* 148:516–25
21. Horrocks, L. A., Sharma, M. 1982. *New Comprehensive Biochemistry*, ed. J. N. Hawthorne, G. B. Ansell, 4:51–93. Amsterdam: Elsevier Biomedical
22. Veenhuis, M., Harder, W. 1987. *Peroxisomes in Biology and Medicine*, ed. H. D. Fahimi, H. Sies, pp. 436–58. Heidelberg: Springer-Verlag
23. Goni, F. M., Dominguez, J. B., Uruburu, F. 1978. *Chem. Phys. Lipids* 22: 79–81
24. Tillman, T. S., Bell, R. M. 1986. *J. Biol. Chem.* 261:9144–49
25. Müller, W. H., van der Krift, T. P., Krouwer, A. J. J., Wösten, H. A. B., van der Voort, L. H. M., et al. 1991. *EMBO J.* 10:489–95
26. Fukushima, H., Watanabe, T., Nozawa,

Y. 1976. *Biochim. Biophys. Acta* 436: 249–59
27. Prins, R. A., van Golde, L. M. G. 1976. *FEBS Lett.* 63:107–11
28. Opperdoes, F. R., Borst, P. 1977. *FEBS Lett.* 80:360–64
29. Opperdoes, F. R. 1990. *Biochem. Soc. Trans.* 18:729–31
30. Opperdoes, F. R. 1988. *Trends Biochem. Sci.* 13:255–60
31. Honigberg, B. M., Volkmann, D., Entzeroth, R., Scholtyseck, E. 1984. *J. Protozool.* 31:116–31
32. Blouin, A., Bolender, R. P., Weibel, E. 1977. *J. Cell Biol.* 72:441–55
33. Usuda, N., Usman, M. I., Reddy, M. K., Hashimoto, T., Reddy, J. K., et al. 1988. *J. Histochem. Cytochem.* 36:253–58
34. Beard, M. E. 1990. *J. Histochem. Cytochem.* 38:1377–82
35. Zaar, K., Völkl, A., Fahimi, H. D. 1991. *J. Cell Biol.* 113:113–21
36. Hruban, Z., Vigil, E. L., Slesers, A., Hopkins, E. 1972. *Lab. Invest.* 27:184–91
37. Böck, P., Kramar, R., Pavelka, M. 1980. *Cell Biol. Monogr.* 7:1–239
38. Holzmann, E. 1982. *Ann. NY Acad. Sci.* 386:523–25
39. Adamo, A. M., Aloise, P. A., Pasquini, J. M. 1986. *Int. J. Dev. Neurosci.* 4:513–17
40. Zaar, K., Gorgas, K. 1985. *Eur. J. Cell Biol.* 38:322–27
41. Gorgas, K. 1987. See Ref. 22, pp. 3–17
42. Zaar, K., Völkl, A., Fahimi, H. D. 1987. *Biochim. Biophys. Acta* 897:135–42
43. Yamamoto, K., Fahimi, H. D. 1987. *J. Cell Biol.* 105:713–22
44. Luers, G., Beier, K., Hashimoto, T., Fahimi, H. D., Völkl, A. 1990. *Eur. J. Cell Biol.* 52:175–84
45. Baumgart, E., Völkl, A., Hashimoto, T., Fahimi, H. D. 1989. *J. Cell Biol.* 108:2221–32
46. Veenhuis, M., Goodman, J. M. 1990. *J. Cell Sci.* 96:583–90
47. Veenhuis, M., Keizer, I., Harder, W. 1979. *Arch. Microbiol.* 120:167–75
48. Veenhuis, M., Harder, W., van Dijken, J. P., Mayer, F. 1981. *Mol. Cell Biol.* 1:949–57
49. Veenhuis, M., Mateblowski, M., Kunau, W. H., Harder, W. 1987. *Yeast* 3:77–84
50. Reddy, J. K., Lalwani, N. D. 1983. *Crit. Rev. Toxicol.* 12:1–58
51. Reddy, J. K. 1990. *Biochem. Soc. Trans.* 18:92–94
52. Meyer, J., Afzelius, B. A. 1989. *J. Ultrastr. Mol. Struct. Res.* 102:87–94
53. Ganning, A. E., Brunk, U., Dallner, G. 1983. *Biochim. Biophys. Acta* 763:72–82
54. Thangada, S., Alvares, K., Mangino, M., Usman, M. I., Rao, M. S., et al. 1989. *FEBS Lett.* 250:205–10
55. Blaauwboer, B. J., van Holsteijn, C. W. M., Bleumink, R., Mennes, W. C., van Pelt, F. N. A. M., et al. 1990. *Biochem. Pharmacol.* 40:521–28
56. Baumgart, E., Völkl, A., Pill, J., Fahimi, H. D. 1990. *FEBS Lett.* 264:5–9
57. Osumi, T., Hashimoto, T. 1984. *Trends Biochem. Sci.* 9:317–19
58. Beier, K., Völkl, A., Hashimoto, T., Fahimi, H. D. 1988. *Eur. J. Cell Biol.* 46:383–93
59. Nemali, M. R., Usuda, N., Reddy, M. K., Oyasu, K., Hashimoto, T., et al. 1988. *Cancer Res.* 48:5316–24
60. Leighton, F., Coloma, L., Koenig, C. 1975. *J. Cell Biol.* 67:281–309
61. Bronfman, M., Amigo, L., Morales, M. N. 1986. *Biochem. J.* 239:781–84
62. Aarsland, A., Berge, R. K., Bremer, J., Aarsaether, N. 1990. *Biochim. Biophys. Acta* 1033:176–83
63. Lalwani, N. D., Alvares, K., Reddy, M. K., Reddy, M. N., Parikh, I., et al. 1987. *Proc. Natl. Acad. Sci. USA* 84:5242–46
64. Alvares, K., Carrillo, A., Yuan, P. M., Kawano, H., Morimoto, R. I., et al. 1990. *Proc. Natl. Acad. Sci. USA* 87:5293–97
65. Isseman, I., Green, S. 1990. *Nature* 347:645–50
66. Osumi, T., Hashimoto, T. 1978. *Biochem. Biophys. Res. Commun.* 83:479–85
67. Vamecq, J., Van Hoof, F. 1984. *Biochem. J.* 221:203–11
68. Höltta, E. 1977. *Biochemistry* 16:91–100
69. Wanders, R. J. A., Romeyn, G. J., Schutgens, R. B. H., Tager, J. M. 1989. *Biochem. Biophys. Res. Commun.* 164:550–55
70. Mihalik, S. J., Rhead, W. J. 1989. *J. Biol. Chem.* 264:2509–17
71. Beard, M. E., Baker, R., Conomos, P., Pugatch, D., Holzman, E. 1985. *J. Histochem. Cytochem.* 33:460–64
72. Schepers, L., Van Veldhoven, P. P., Casteels, M., Eyssen, H. J., Mannaerts, G. P. 1990. *J. Biol. Chem.* 265:5242–46
73. Casteels, M., Schepers, L., Van Eldere, J., Eyssen, H., Mannaerts, G. P. 1988. *J. Biol. Chem.* 263:4654–61
74. Pedersen, J. I., Hvattum, E., Flatabo, T., Björkhem, I. 1988. *Biochem. Int.* 17:163–69
75. Wanders, R. J. A., Ten Brink, H. J.,

Van Roermund, C. W. T., Schutgens, R. B. H., Tager, J. M., Jakobs, C. 1990. *Biochem. Biophys. Res. Commun.* 172:490–95

76. Chance, B., Sies, H., Boveris, A. 1979. *Physiol. Rev.* 59:527–605

77. Handler, J. A., Thurman, R. G. 1988. *Eur. J. Biochem.* 176:477–84

78. Handler, J. A., Thurman, R. G. 1990. *J. Biol. Chem.* 265:1510–15

79. Cooper, T. G., Beevers, H. 1969. *J. Biol. Chem.* 244:3514–20

80. Lazarow, P. B., De Duve, C. 1976. *Proc. Natl. Acad. Sci. USA* 73:2043–46

81. Schulz, H. 1991. *Biochim. Biophys. Acta* 1081:109–20

82. Mannaerts, G. P., Van Veldhoven, P., Van Broekhoven, A., Vandebroek, G., De Beer, L. J. 1982. *Biochem. J.* 204:17–23

83. Miyazawa, S., Hashimoto, T., Yokota, S. 1985. *J. Biochem.* 98:723–33

84. Bhushan, A., Singh, R. P., Singh, I. 1986. *Arch. Biochem. Biophys.* 246: 374–80

85. Wanders, R. J. A., Van Roermund, C. W. T., Van Wijland, M. J. A., Heikoop, J., Schutgens, R. B. H., et al. 1987. *J. Clin. Invest.* 80:1778–83

86. Singh, H., Derwas, N., Poulos, A. 1987. *Arch. Biochem. Biophys.* 259: 382–90

87. Lazo, O., Contreras, M., Singh, I. 1990. *Biochemistry* 29:3981–86

88. Lageweg, W., Tager, J. M., Wanders, R. J. A. 1991. *Biochem. J.* 276:53–56

89. Wanders, R. J. A., Van Roermund, C. W. T., Van Wijland, M. J. A., Schutgens, R. B. H., Van den Bosch, H., et al. 1988. *Biochem. Biophys. Res. Commun.* 153:618–24

90. Lazo, O., Contreras, M., Hashmi, M., Stanley, W., Irazu, C., et al. 1988. *Proc. Natl. Acad. Sci. USA* 85:7647–51

91. Numa, S. 1981. *Trends Biochem. Sci.* 6:113–15

92. Schepers, L., Casteels, M., Verheyden, K., Parmentier, G., Asselberghs, S., et al. 1989. *Biochem. J.* 257:221–29

93. Prydz, K., Kase, B. F., Björkhem, I., Pedersen, J. I. 1988. *J. Lipid Res.* 29:997–1004

94. Vamecq, J., De Hoffmann, E., Van Hoof, F. 1985. *Biochem. J.* 230:683–93

95. Schepers, L., Casteels, M., Vamecq, J., Parmentier, G., Van Veldhoven, P. P., et al. 1988. *J. Biol. Chem.* 263:2724–31

96. Kunau, W. H., Kionka, C., Ledebur, A., Mateblowski, M., Moreno de la Garza, M., et al. 1987. See Ref. 22, pp. 128–40

97. Osumi, T., Ishii, N., Miyazawa, S., Hashimoto, T. 1987. *J. Biol. Chem.* 262:8138–43

98. Okazaki, K., Tan, H., Fukui, S., Kubota, I., Kamiryo, T. 1987. *Gene* 58:37–44

99. Okazaki, K., Takechi, T., Kambara, N., Fukui, S., Kubota, I., et al. 1986. *Proc. Natl. Acad. Sci. USA* 83:1232–36

100. Hill, D. E., Boulay, R., Rogers, H. 1988. *Nucleic Acids Res.* 16:365–66

101. Miyazawa, S., Hayashi, H., Hijikata, M., Ishii, N., Furuta, S., et al. 1987. *J. Biol. Chem.* 262:8131–37

102. Furuta, S., Miyazawa, S., Osumi, T., Hashimoto, T., Ui, N. 1980. *J. Biochem.* 88:1059–70

103. Palosaari, P. M., Hiltunen, J. K. 1990. *J. Biol. Chem.* 265:2446–49

104. Osumi, T., Ishii, N., Hijikata, M., Kamijo, K., Ozasa, H., et al. 1985. *J. Biol. Chem.* 260:8905–10

105. Ishii, N., Hijikata, M., Osumi, T., Hashimoto, T. 1987. *J. Biol. Chem.* 262:8144–50

106. Miyazawa, S., Furuta, S., Osumi, T., Hashimoto, T., Ui, N. 1981. *J. Biochem.* 90:511–19

107. Hijikata, M., Ishii, N., Kagamiyama, H., Osumi, T., Hashimoto, T. 1987. *J. Biol. Chem.* 262:8151–58

108. Hijikata, M., Wen, J. K., Osumi, T., Hashimoto, T. 1990. *J. Biol. Chem.* 265:4600–6

109. Bodnar, A. G., Rachubinski, R. A. 1990. *Gene* 91:193–99

110. Dommes, V., Baumgart, C., Kunau, W. H. 1981. *J. Biol. Chem.* 256:8259–62

111. Kärki, T., Hakkola, E., Hassinen, I. E., Hiltunen, J. K. 1987. *FEBS Lett.* 215:228–32

112. Li, J., Smeland, T. E., Schulz, H. 1990. *J. Biol. Chem.* 265:13629–34

113. Markwell, M. A. K., McGroarty, E. J., Bieber, L. L., Tolbert, N. E. 1973. *J. Biol. Chem.* 248:3426–32

114. Markwell, M. A. K., Tolbert, N. E., Bieber, L. L. 1976. *Arch. Biochem. Biophys.* 176:479–88

115. Miyazawa, S., Ozasa, H., Furuta, S., Osumi, T., Hashimoto, T. 1983. *J. Biochem.* 93:439–51

116. Farrell, S. O., Bieber, L. L. 1983. *Arch. Biochem. Biophys.* 222:123–32

117. Derrick, J. P., Ramsay, R. R. 1989. *Biochem. J.* 262:801–6

118. McGarry, J. D., Leatherman, G. F., Foster, D. W. 1978. *J. Biol. Chem.* 253:4128–36

119. Lilly, K., Bugaisky, G. E., Umeda, P. K., Bieber, L. L. 1990. *Arch. Biochem. Biophys.* 280:167–74

120. Mannaerts, G. P., De Beer, L. J., Thomas, J., De Schepper, P. J. 1979. *J. Biol. Chem.* 254:4585–95
121. Kondrup, J., Lazarow, P. B. 1985. *Biochim. Biophys. Acta* 835:147–53
122. Kawamura, N., Moser, H. W., Kishimoto, Y. 1981. *Biochem. Biophys. Res. Commun.* 99:1216–25
123. Singh, I., Moser, A. B., Goldfischer, S., Moser, H. W. 1984. *Proc. Natl. Acad. Sci. USA* 81:4203–7
124. Singh, H., Usher, S., Poulos, A. 1989. *J. Neurochem.* 53:1711–18
125. Pedersen, J. I., Gustafsson, J. 1980. *FEBS Lett.* 121:345–48
126. Kase, B. F., Björkhem, I., Pedersen, J. I. 1983. *J. Lipid Res.* 24:1560–67
127. Kase, B. F., Pedersen, J. I., Strandvick, B., Björkhem, I. 1985. *J. Clin. Invest.* 76:2393–402
128. Hovik, R., Osmundsen, H. 1987. *Biochem. J.* 182:779–88
129. Christensen, E., Hagve, T. A., Christopherson, B. O. 1988. *Biochim. Biophys. Acta* 959:95–99
130. Street, J. M., Johnson, D. W., Singh, H., Poulos, A. 1989. *Biochem. J.* 260:647–55
131. Singh, H., Usher, S., Johnson, D., Poulos, A. 1990. *J. Lipid Res.* 31:217–25
132. Kolvraa, S., Gregersen, N. 1986. *Biochim. Biophys. Acta* 876:515–25
133. Yamada, J., Itoh, S., Horie, S., Watanabe, T., Suga, T. 1986. *Biochem. Pharmacol.* 35:4363–68
134. Wanders, R. J. A., Van Roermund, C. W. T., Schutgens, R. B. H., Barth, P. G., Heymans, H. S. A., et al. 1990. *J. Inher. Metab. Dis.* 13:4–36
135. Kornberg, H. L., Krebs, H. A. 1957. *Nature* 179:988–91
136. Davis, W. L., Goodman, D. B. P., Crawford, L. A., Cooper, O. J., Matthews, J. L. 1990. *Biochim. Biophys. Acta* 1051:276–78
137. Davis, W. L., Matthews, J. L., Goodman, D. B. P. 1989. *FASEB J.* 3:1651–55
138. Rowsell, E. V., Snell, K., Carnie, J., Rowsell, K. V. 1972. *Biochem. J.* 127:155–65
139. Noguchi, T. 1987. See Ref. 22, pp. 235–43
140. Hillman, R. E. 1989. See Ref. 15, pp. 933–44
141. Danpure, C. 1989. *J. Inher. Metab. Dis.* 12:210–24
142. Hayashi, S., Noguchi, T. 1990. *Biochem. Biophys. Res. Commun.* 166:1467–70
143. Danpure, C. J., Jennings, P. R., Jansen, J. H. 1991. *Biochim. Biophys. Acta* 1096:134–38
144. Rowsell, E. V., Snell, K., Carnie, J. A., Al-Tai, A. H. 1969. *Biochem. J.* 115:1071–73
145. Noguchi, T., Takada, Y. 1979. *Arch. Biochem. Biophys.* 196:645–47
146. Thompson, J. S., Richardson, K. E. 1967. *J. Biol. Chem.* 242:3614–19
147. Danpure, C. J., Guttridge, K. M., Fryer, P., Jennings, P. R., Allsop, J., et al. 1990. *J. Cell Sci.* 97:669–78
148. Noguchi, T., Fujiwara, S. 1988. *J. Biol. Chem.* 263:182–86
149. Yokata, S. 1986. *Histochemistry* 85:145–55
150. Oda, T., Ichiyama, A., Miura, S., Mori, M., Takibana, M. 1981. *Biochem. Biophys. Res. Commun.* 102:568–73
151. Oda, T., Yanagisawa, M., Ichiyama, A. 1982. *J. Biochem.* 91:219–32
152. Oda, T., Funai, T., Ichiyama, A. 1990. *J. Biol. Chem.* 265:7513–19
153. Noguchi, T., Okuno, E., Takada, Y., Minatogawa, Y., Okai, K., et al. 1978. *Biochem. J.* 169:113–22
154. Fukushima, M., Aihara, Y., Ichiyama, A. 1978. *J. Biol. Chem.* 253:1187–94
155. Danpure, C. J., Jennings, P. R. 1986. *Biochem. Soc. Trans.* 14:1059–60
156. Keller, G. A., Barton, M. C., Shapiro, D. J., Singer, S. J. 1985. *Proc. Natl. Acad. Sci. USA* 82:770–74
157. Keller, G. A., Pazirandeh, M., Krisans, S. K. 1986. *J. Cell Biol.* 103:875–86
158. Krisans, S. K., Pazirandeh, M., Keller, G. A. 1987. See Ref. 25, pp. 40–52
159. Liscum, L., Finer-Moore, J., Stroud, R. M., Luskey, K. L., Brown, M. S., et al. 1985. *J. Biol. Chem.* 260:522–30
160. Thompson, S. L., Krisans, S. K. 1990. *J. Biol. Chem.* 265:5731–35
161. Thompson, S. L., Burrows, R., Laub, R. J., Krisans, S. K. 1988. *J. Biol. Chem.* 262:17420–25
162. Appelkvist, E.-L., Reinhart, M., Fischer, R., Billheimer, J. 1990. *Arch. Biochem. Biophys.* 282:318–25
163. Hagey, L. R., Krisans, S. K. 1982. *Biochem. Biophys. Res. Commun.* 107:834–41
164. Van der Krift, T. P., Leunissen, J., Teerlink, T., Van Heusden, G. P. H., Verkleij, A. J., et al. 1985. *Biochim. Biophys. Acta* 812:387–92
165. Tsuneoka, M., Yamamoto, A., Fujiki, Y., Tashiro, Y. 1988. *J. Biochem.* 104:560–64
166. Keller, G. A., Scallen, T. J., Clarke, D., Maher, P. A., Krisans, S. K. 1989. *J. Cell Biol.* 108:1353–61

167. Van Amerongen, A., Van Noort, M., Van Beckhoven, J. R. C. M., Rommerts, F. F. G., Orly, J., et al. 1989. *Biochim. Biophys. Acta* 1001:243–48
168. Chanderbhan, R., Noland, B. J., Scallen, T. J., Vahouny, G. V. 1982. *J. Biol. Chem.* 257:8928–34
169. Chanderbhan, R., Tasaka, T., Strauss, J. F., Irwin, D., Noland, B. J., et al. 1983. *Biochem. Biophys. Res. Commun.* 177:702–9
170. Van Amerongen, A., Helms, J. B., Van der Krift, T. P., Schutgens, R. B. H., Wirtz, K. W. A. 1987. *Biochim. Biophys. Acta* 919:149–55
171. Van Heusden, G. P. H., Bos, K., Raetz, C. R. H., Wirtz, K. W. A. 1990. *J. Biol. Chem.* 265:4105–10
172. Ossendorp, B. C., Van Heusden, G. P. H., Wirtz, K. W. A. 1990. *Biochem. Biophys. Res. Commun.* 168:631–36
173. Seedorf, U., Assmann, G. 1991. *J. Biol. Chem.* 266:630–36
174. Mori, T., Tsukamoto, T., Mori, H., Tashiro, Y., Fujiki, Y. 1991. *Proc. Natl. Acad. Sci. USA* 88:4338–42
175. Yamamoto, R., Kallen, C. B., Babalola, G. O., Rennert, H., Billheimer, J. T., et al. 1991. *Proc. Natl. Acad. Sci. USA* 88:463–67
176. Ossendorp, B. C., Van Heusden, G. P. H., De Beer, A. L. J., Bos, K., Schouten, G. L., et al. 1991. *Eur. J. Biochem.* 201:233–39
177. Tan, H., Okazaki, K., Kubota, I., Kamiryo, T., Utiyama, H. 1990. *Eur. J. Biochem.* 190:107–12
178. Appelkvist, E.-L., Kalén, A. 1989. *Eur. J. Biochem.* 185:503–9
179. Hajra, A. K., Ghosh, M. K., Webber, K. O., Datta, N. S. 1986. In *Enzymes of Lipid Metabolism II*, ed. L. Freysz, H. Dreyfus, R. Massarelli, S. Gatt, pp. 199–207. New York: Plenum
180. Snyder, F., Lee, T-C., Wykle, R. L. 1985. In *The Enzymes of Biological Membranes*, ed. A. N. Martonosi, Vol. 2, pp. 1–58. New York: Plenum. 2nd. ed.
181. Rabert, U., Völkl, A., Debuch, H. 1987. *Biol. Chem. Hoppe-Seyler* 367:215–22
182. Hardeman, D., van den Bosch, H. 1991. *Biochim. Biophys. Acta* 1081:285–92
183. Hajra, A. K., Burke, C. L., Jones, C. L. 1979. *J. Biol. Chem.* 254:10896–900
184. Hajra, A. K., Bishop, J. E. 1982. *Ann. NY Acad. Sci.* 386:170–82
185. Goldfischer, S., Moore, C. L., Johnson, A. B., Spiro, A. J., Valsamis, M. P., et al. 1973. *Science* 182:62–64
186. Heymans, H. S. A., Schutgens, R. B. H., Tan, R., van den Bosch, H., Borst, P. 1983. *Nature* 306:69–70
187. Schrakamp, G., Schutgens, R. B. H., Wanders, R. J. A., Heymans, H. S. A., Tager, J. M., et al. 1985. *Biochim. Biophys. Acta* 833:170–74
188. Schutgens, R. B. H., Romeyn, G. J., Wanders, R. J. A., van den Bosch, H., Schrakamp, G., et al. 1984. *Biochem. Biophys. Res. Commun.* 120:179–84
189. Datta, N. S., Wilson, G. N., Hajra, A. K. 1984. *New Eng. J. Med.* 311:1080–83
190. Singh, H., Usher, S., Poulos, A. 1989. *Arch. Biochem. Biophys.* 268:676–86
191. Schrakamp, G., Rosenboom, C. F. J., Schutgens, R. B. H., Wanders, R. J. A., Heymans, H. S. A., et al. 1985. *J. Lipid Res.* 26:867–73
192. Zoeller, R. A., Raetz, C. R. H. 1986. *Proc. Natl. Acad. Sci. USA* 83:5170–74
193. Zoeller, R. A., Allen, L.-A. H., Santos, M. J., Lazarow, P. B., Hashimoto, T., et al. 1989. *J. Biol. Chem.* 264:21872–78
194. Schrakamp, G., Schalkwijk, C. G., Schutgens, R. B. H., Wanders, R. J. A., Tager, J. M., et al. 1988. *J. Lipid Res.* 29:325–34
195. Zoeller, R. A., Morand, O. H., Raetz, C. R. H. 1988. *J. Biol. Chem.* 263:11590–96
196. Hardeman, D., van den Bosch, H. 1988. *Biochim. Biophys. Acta* 963:1–9
197. Wolvetang, E. J., Tager, J. M., Wanders, R. J. A. 1990. *Biochem. Biophys. Res. Commun.* 170:1135–43
198. Bishop, J. E., Hajra, A. K. 1978. *J. Neurochem.* 30:643–47
199. Riendeau, D., Meighen, E. 1985. *Experientia* 41:707–13
200. Webber, K. O., Datta, N. S., Hajra, A. K. 1987. *Arch. Biochem. Biophys.* 254:611–20
201. Bell, R. M., Ballas, L. M., Coleman, R. A. 1981. *J. Lipid Res.* 22:391–403
202. Hardeman, D., van den Bosch, H. 1989. *Biochim. Biophys. Acta* 1006:1–8
203. Ballas, L. M., Lazarow, P. B., Bell, R. M. 1984. *Biochim. Biophys. Acta* 795:297–300
204. Burke, C. L., Hajra, A. K. 1980. *Biochem. Int.* 1:312–18
205. Horie, S., Utsumi, K., Suga, T. 1990. *Biochim. Biophys. Acta* 1042:294–300
206. Hardeman, D., Zomer, H. W. M., Schutgens, R. B. H., Tager, J. M., van den Bosch, H. 1990. *Int. J. Biochem.* 22:1413–18
207. Ghosh, A. K., Hajra, A. K. 1986. *Arch. Biochem. Biophys.* 245:523–30

208. Jones, C. L., Hajra, A. K. 1983. *Arch. Biochem. Biophys.* 226:155–65
209. Davis, P. A., Hajra, A. K. 1981. *Arch. Biochem. Biophys.* 211:20–29
210. Snyder, F., Rainey, W. J., Blank, M. L., Christie, W. H. 1970. *J. Biol. Chem.* 245:5453–56
211. Brown, A. J., Snyder, F. 1982. *J. Biol. Chem.* 257:8835–39
212. Brown, A. J., Snyder, F. 1983. *J. Biol. Chem.* 258:4184–89
213. Brown, A. J., Glish, G. L., McBay, E. H., Snyder, F. 1985. *Biochemistry* 24:8012–16
214. Friedberg, S. J., Weintraub, S. T., Peterson, D., Satsangi, N. 1987. *Biochem. Biophys. Res. Commun.* 145:1177–84
215. Datta, S. C., Ghosh, M. K., Hajra, A. K. 1990. *J. Biol. Chem.* 265:8268–74
216. Schram, A. W., Strijland, A., Hashimoto, T., Wanders, R. J. A., Schutgens, R. B. H., et al. 1986. *Proc. Natl. Acad. Sci. USA* 83:6156–58
217. Santos, M. J., Imanaka, T., Shio, H., Lazarow, P. B. 1988. *J. Biol. Chem.* 263:10502–9
218. Gärtner, J., Chen, W. W., Kelley, R. I., Mihalik, S. J., Moser, H. W. 1991. *Pediatr. Res.* 29:141–46
219. Wiemer, E. A. C., Brul, S., Just, W. W., van Driel, R., Brouwer-Kelder, E., et al. 1989. *Eur. J. Cell. Biol.* 50:407–17
220. Arias, J. A., Moser, A. B., Godfischer, S. L. 1985. *J. Cell Biol.* 100:1789–92
221. Black, V. H., Cornacchia, L. III. 1986. *Am. J. Anat.* 177:107–18
222. Brul, S., Westerveld, A., Strijland, A., Wanders, R. J. A., Schram, A. W., et al. 1988. *J. Clin. Invest.* 81:1710–15
223. Roscher, A. A., Hoefler, S., Hoefler, G., Paschke, E., Paltauf, F., et al. 1989. *Pediatr. Res.* 26:67–72
224. Poll-Thé, B.-T., Skjeldal, O. H., Stokke, O., Poulos, A., Demaugre, F., et al. 1989. *Hum. Genet.* 81:175–81
225. McGuinness, M. C., Moser, A. B., Moser, H. W., Watkins, P. A. 1990. *Biochem. Biophys. Res. Commun.* 172:364–69
226. Schutgens, R. B. H., Heymans, H. S. A., Wanders, R. J. A., Oorthuys, J. W. E., Tager, J. M., et al. 1988. *Adv. Clin. Enzymol.* 6:57–65
227. Hoefler, G., Hoefler, S., Watkins, P. A., Chen, W. W., Moser, A., et al. 1988. *J. Pediatr.* 112:726–33
228. Heikoop, J. C., van Roermund, C. W. T., Just, W. W., Ofman, R., Schutgens, R. B. H., et al. 1990. *J. Clin. Invest.* 86:126–30
229. Suzuki, Y., Shimozawa, N., Orii, T., Igarashi, N., Kono, N., et al. 1988. *J. Pediatr.* 113:841–45
230. Moser, H. W., Moser, A. B. 1989. See Ref. 15, pp. 1511–32
231. Hashmi, M., Stanley, W., Singh, I. 1986. *FEBS Lett.* 196:247–50
232. Poll-Thé, B. T., Roels, F., Ogier, H., Scotto, J., Vamecq, J., et al. 1988. *Am. J. Hum. Genet.* 42:422–34
233. Naidu, S., Hoefler, G., Watkins, P. A. 1988. *Neurology* 38:1100–07
234. Goldfischer, S., Collins, J., Rapin, I., Neumann, P., Neglia, W., et al. 1986. *J. Pediatr.* 108:25–32
235. Schram, A. W., Goldfischer, S., van Roermund, C. W. T., Brouwer-Kelder, E. M., Collins, J., et al. 1987. *Proc. Natl. Acad. Sci. USA* 84:2494–96
236. Wanders, R. J. A., van Roermund, C. W. T., Schelen, A., Schutgens, R. B. H., Tager, J. M., et al. 1990. *J. Inher. Met. Dis.* 13:375–79
237. Przyrembel, H., Wanders, R. J. A., van Roermund, C. W. T., Schutgens, R. B. H., Mannaerts, G. P., et al. 1990. *J. Inher. Met. Dis.* 13:367–70
238. Christensen, E., van Eldere, J., Brandt, N. J., Schutgens, R. B. H., Wanders, R. J. A., et al. 1990. *J. Inher. Met. Dis.* 13:363–66
239. Danpure, C. J., Cooper, P. J., Wise, P. J., Jennings, P. R. 1989. *J. Cell. Biol.* 108:1345–52
240. Wanders, R. J. A., Ruiter, J., van Roermund, C. W. T., Schutgens, R. B. H., Ofman, R., et al. 1990. *Clin. Chim. Acta* 189:139–44
241. Eaton, J. W. 1989. See Ref. 15, pp. 1551–61
242. Novikoff, A. B., Shin, W.-Y. 1964. *J. Microscopie.* 3:187–206
243. Gorgas, K. 1982. *Ann. NY Acad. Sci.* 386:519–22
244. Gorgas, K. 1985. *Anat. Embryol.* 172:21–32
245. Kamiryo, T., Abe, M., Okazaki, K., Kato, S., Shimamoto, N. 1982. *J. Bacteriol.* 152:269–74
246. Bout, A., Franse, M. M., Collins, J., Blonden, L., Tager, J. M., et al. 1991. *Biochim. Biophys. Acta.* 1090:43–51
247. Casteels, M., Schepers, L., van Veldhoven, P., Eyssen, H. J., Mannaerts, G. P. 1990. *J. Lipid Res.* 31:1865–72
248. Holloway, B. R., Thorp, J. M., Smith, G. D., Peters, T. J. 1982. *Ann. NY Acad. Sci.* 386:453–55
249. Fujiki, Y., Rachubinski, R. A., Lazarow, P. B. 1984. *Proc. Natl. Acad. Sci. USA* 81:7127–31
250. Köster, A., Heisig, M., Heinrich, P. C.,

Just, W. W. 1986. *Biochem. Biophys. Res. Commun.* 137:626–32

251. Suzuki, Y., Orri, T., Takiguchi, M., Mori, M., Hijikata, M., et al. 1987. *J. Biochem.* 101:491–96

252. Just, W. W., Hartl, F-U. 1987. See Ref. 22, pp. 235–43

253. Imanaka, T., Small, G. M., Lazarow, P. B. 1987. *J. Cell Biol.* 105:2915–22

254. Miyazawa, S., Osumi, T., Hashimoto, T., Ohno, K., Miura, S., et al. 1989. *Mol. Cell. Biol.* 9:83–91

255. Keller, G. A., Gould, S. J., De Luca, M., Subramani, S. 1987. *Proc. Natl. Acad. Sci. USA* 84:3264–68

256. Gould, S. J., Keller, G. A., Subramani, S. 1987. *J. Cell Biol.* 105:2923–31

257. Gould, S. J., Keller, G. A., Subramani, S. 1988. *J. Cell Biol.* 107:897–905

258. Gould, S. J., Keller, G. A., Hosken, N., Wilkinson, J., Subramani, S. 1989. *J. Cell Biol.* 108:1657–64

259. Motojima, K., Goto, S. 1989. *Biochim. Biophys. Acta* 1008:116–18

260. Gould, S. J., Krisans, S., Keller, G. A., Subramani, S. 1990. *J. Cell Biol.* 110:27–34

261. Gould, S. J., Keller, G. A., Schneider, M., Howell, S. H., Garrard, L. J., et al. 1990. *EMBO J.* 9:85–90

262. Distel, B., Gould, S. J., Voorn-Brouwer, T., van den Berg, M., Tabak, H. F., et al. 1991. *New Biol.* Submitted

263. Bell, G. I., Najarian, R. C., Mullenbach, G. T., Hallewell, R. A. 1986. *Nucleic Acids Res.* 14:5561–62

264. Swinkels, B. W., Gould, S. J., Bodnar, A. G., Rachubinski, R. A., Subramani, S. 1991. *EMBO J.* 10:3255–62

265. Miura, S., Mori, M., Takiguchi, M., Tatibana, M., Furuta, S., et al. 1984. *J. Biol. Chem.* 259:6397–402

266. Wiemer, E. A. C., Brul, S., Bout, A., Strijland, A., Heikoop, J. C., et al. 1989. In *Organelles of Eukaryotic Cells: Molecular Structure and Interactions*, ed. A. Azzi, F. Guerrieri, S. Papa, J. M. Tager, pp. 27–46. New York:Plenum

267. Swinkels, B. W., Evers, R., Borst, P. 1988. *EMBO J.* 7:1159–65

267a. Keller, G. A., Krisans, S., Goula, S. J., Sommer, J. M., Wang, C. C., et al. 1991. *J. Cell Biol.* 114:893–904

268. Oda, T., Kitamura, N., Nakanishi, S., Ichiyama, A. 1985. *Eur. J. Biochem.* 150:415–21

269. Oda, T., Miyajima, H., Suzuki, Y., Ichiyama, A. 1987. *Eur. J. Biochem.* 168:537–42

270. Nishiyama, K., Berstein, G., Oda, T., Ichiyama, A. 1990. *Eur. J. Biochem.* 194:9–18

271. Takada, Y., Kaneko, N., Esumi, H., Purdue, P. E., Danpure, C. J. 1990. *Biochem. J.* 268:517–20

272. Purdue, P. E., Takada, Y., Danpure, C. J. 1990. *J. Cell Biol.* 111:2341–51

273. Furuta, S., Hashimoto, T., Miura, S., Mori, M., Tatibana, M. 1982. *Biochem. Biophys. Res. Commun.* 105:639–46

274. Fujiki, Y., Rachubinski, R. A., Mortensen, R. M., Lazarow, P. B. 1985. *Biochem. J.* 226:697–704

275. Trzeciak, W. H., Simpson, E. R., Scallen, T. J., Vahouny, G. V., Waterman, M. R. 1987. *J. Biol. Chem.* 262:3713–17

276. Fujiki, Y., Tsuneoka, M., Tashiro, Y. 1989. *J. Biochem.* 106:1126–31

277. Suzuki, Y., Yamaguchi, S., Orii, T., Tsuneoka, M., Tashiro, Y. 1990. *Cell Struct. Funct.* 15:301–08

278. Billheimer, J. T., Strehl, L. L., Davis, G. L., Strauss, J. F., Davis, L. G. 1990. *DNA Cell Biol.* 9:156–65

279. Osumi, T., Hashimoto, T., Ui, N. 1980. *J. Biochem.* 87:1735–46

280. Yamaguchi, J., Mori, H., Mishimura, M. 1987. *FEBS Lett.* 213:329–32

281. Gietl, C. 1990. *Proc. Natl. Acad. Sci. USA* 87:5773–77

282. Bout, A., Teunissen, Y., Hashimoto, T., Benne, R., Tager, J. M. 1988. *Nucleic Acids Res.* 16:10369

283. Fairbairn, L. J., Tanner, M. J. A. 1989. *Nucleic Acids Res.* 17:3588

284. Pfanner, N., Söllner, T., Neupert, W. 1991. *Trends Biochem. Sci.* 16:63–66

285. Singh, I., Lazo, O., Contreras, M., Stanley, W., Hashimoto, T. 1991. *Arch. Biochem. Biophys.* 286:277–83

286. Heikoop, J. C., van den Berg, M., Strijland, A., Weijers, P. J., Schutgens, R. B. H., et al. 1991. *Biochim. Biophys. Acta* 1097:62–70

287. Balfe, A., Hoefler, G., Chen, W. W., Watkins, P. A. 1990. *Pediatr. Res.* 27:304–10

288. Titus, D. E., Becker, W. M. 1985. *J. Cell Biol.* 101:1288–99

289. Nishimura, M., Yamaguchi, T., Mori, H., Akazawa, T., Yokota, S. 1986. *Plant Physiol.* 80:313–16

290. Hock, B., Gietl, C., Sautter, C. 1987. See Ref. 22, pp. 417–25

291. Mori, H., Nishimura, M. 1989. *FEBS Lett.* 244:163–66

292. Gerdes, H. H., Kindl, H. 1988. *Biochim. Biophys. Acta* 949:195–205

293. Tsukamoto, T., Yokota, S., Fujiki, Y. 1990. *J. Cell Biol.* 110:651–60

294. Brul, S., Wiemer, E. A. C., Westerveld, A., Strijland, A., Wanders, R. J.

A., et al. 1988. *Biochem. Biophys. Res. Commun.* 152:1083–89
295. Middelkoop, E., Strijland, A., Tager, J. M. 1991. *FEBS Lett.* 279:79–82
296. Lazarow, P. B., Fujiki, Y., Small, G. M., Watkins, P., Moser, H. 1986. *Proc. Natl. Acad. Sci. USA* 83:9193–96
297. Suzuki, Y., Shimozawa, N., Orii, T., Aikawa, J., Tada, K., et al. 1987. *J. Inherit. Metab. Dis.* 10:297–300
298. Santos, M. J., Imanaka, T., Shio, H., Small, G. M., Lazarow, P. B. 1988. *Science* 234:1536–38
299. Suzuki, Y., Shimozawa, N., Orii, T., Hashimoto, T. 1989. *Pediatr. Res.* 26: 150–53
300. Heikoop, J. C., van den Berg, M., Strijland, A., Weijers, P. J., Just, W. W., et al. 1992. *Eur. J. Cell Biol.* In press
301. Allen, L.-A. H., Morand, O. H., Raetz, C. R. H. 1989. *Proc. Natl. Acad. Sci. USA* 86:7012–16
302. Tsukamoto, T., Miura, S., Fujiki, Y. 1991. *Nature* 350:77–81
303. Erdmann, R., Wiebel, F. F., Flessau, A., Rytka, J., Beyer, A., et al. 1991. *Cell* 64:499–510
304. Erdmann, R., Veenhuis, M., Mertens, D., Kunau, W.-H. 1989. *Proc. Natl. Acad. Sci. USA* 86:5419–23
305. Fujiki, Y., Fowler, S., Shio, H., Hubbard, L., Lazarow, P. B. 1982. *J. Cell Biol.* 93:103–10
306. Hardeman, D., Versantvoort, C., van den Brink, J. M., van den Bosch, H. 1990. *Biochim. Biophys. Acta* 1027: 149–54
307. Van Veldhoven, P., De Beer, L. J., Mannaerts, G. P. 1983. *Biochem. J.* 210:685–93
308. Van Veldhoven, P. P., Just, W. W., Mannaerts, G. P. 1987. *J. Biol. Chem.* 262:4310–18
309. Nicolay, K., Veenhuis, M., Douma, A. C., Harder, W. 1987. *Arch. Microbiol.* 147:37–41
310. Waterham, H. R., Keizer-Gunnik, I., Goodman, J. M., Harder, W., Veenhuis, M. 1990. *FEBS Lett.* 262:17–19
311. Douma, A. C., Veenhuis, M., Sulter, G. J., Harder, W. 1987. *Arch. Microbiol.* 147:42–47
312. Whitney, A. B., Bellion, E. 1991. *Biochim. Biophys. Acta* 1058:345–55

313. Del Valle, R., Soto, U., Necochea, C., Leighton, F. 1988. *Biochem. Biophys. Res. Commun.* 156:1353–59
314. Tanaka, T., Hosaka, K., Hoshimaru, M., Numa, S. 1979. *Eur. J. Biochem.* 98:165–72
315. Hashimoto, T., Kuwabara, T., Usuda, N., Nagata, T. 1986. *J. Biochem.* 100:301–10
316. Hartl, F.-U., Just, W. W. 1987. *Arch. Biochem. Biophys.* 255:109–19
317. Small, G. M., Santos, M. J., Imanaka, T., Poulos, A., Danks, D. M., et al. 1988. *J. Inherit. Metab. Dis.* 11:358–71
318. Chen, N., Crane, D. I., Masters, C. J. 1988. *Biochim. Biophys. Acta* 945:135–44
319. Crane, D. I., Chen, N., Masters, C. J. 1989. *Biochim. Biophys. Res. Commun.* 160:503–8
320. Alexson, S. E. H., Fujiki, Y., Shio, H., Lazarow, P. B. 1985. *J. Cell Biol.* 101:294–305
321. Kamijo, K., Taketani, S., Yokota, S., Osumi, T., Hashimoto, T. 1990. *J. Biol. Chem.* 265:4534–40
322. Hartl, F.-U., Just, W. W., Köster, A., Schimassek, H. 1985. *Arch. Biochem. Biophys.* 237:124–34
323. Imanaka, T., Lazarow, P. B., Takano, T. 1991. *Biochim. Biophys. Acta* 1062:264–70
324. Beevers, H., Gonzáles, E. 1987. *Methods Enzymol.* 148:526–32
325. Kruse, C., Kindl, H. 1982. *Ann. NY Acad. Sci.* 386:499–501
326. Völkl, A., Lazarow, P. B. 1982. *Ann. NY Acad. Sci.* 386:504–6
327. Donaldson, R. P., Tully, R. E., Young, O. A., Beevers, H. 1981. *Plant Physiol.* 67:21–26
328. Nuttley, W. M., Bodnar, A. G., Mangroo, D., Rachubinski, R. A. 1990. *J. Cell Sci.* 95:463–70
329. Goodman, J. M., Maher, J., Silver, P. A., Pacifico, A., Sanders, D. 1986. *J. Biol. Chem.* 261:3464–68
330. Garrard, L. J., Goodman, J. M. 1989. *J. Biol. Chem.* 264:13929–37
331. Goodman, J. M., Trapp, S. B., Hwang, H., Veenhuis, M. 1990. *J. Cell Sci.* 96:193–204
332. McCammon, M. T., Dowds, C. A., Orth, K., Moomaw, C. R., Slaughter, C. A., et al. 1990. *J. Biol. Chem.* 265:20098–105

Annu. Rev. Biochem. 1992. 61:199–223

DNA LOOPING

Robert Schleif

Biology Department, Johns Hopkins University, 34th and Charles Streets, Baltimore, Maryland 21218

KEY WORDS: gene regulation, protein-DNA interactions, cooperativity

CONTENTS

199

0066-4154/92/0701-0199$02.00

INTRODUCTION

DNA looping is generated by a protein or complex of proteins that simultaneously binds to two different sites on a DNA molecule. Consequently, the intervening tens to thousands of base pairs of DNA loop out. This seemingly simple phenomenon is central in the regulation of many biochemical transactions involving DNA. The most prominent current examples of looping are in the regulation of the expression of prokaryotic and eukaryotic genes, regulation of site-specific recombination, and in the regulation of DNA replication.

WHY NATURE USES DNA LOOPING

Multiple proteins can be required to regulate properly the activity of a complex on DNA. For example, transcription of a gene could require the simultaneous presence of four different conditions, the status of each of which must be transmitted by a protein sensor to the initiation complex. In order that the protein sensors, which we normally call regulatory proteins, confine their activities to the correct genes, they bind to specific sequences located near the genes to be regulated. There is space for only two or three proteins to bind to DNA alongside an initiation complex. DNA looping permits additional proteins to bind in the vicinity of the complex and to interact with it. Multiple proteins could also assist or interfere with the interactions necessary for looping. Thus, a rich diversity of regulation systems can be expected to utilize DNA looping.

DNA looping also generates cooperativity in the binding of a looping protein to the two DNA sites involved. When a protein has bound to one of the two DNA sites, it is held near the other DNA site because the furthest the two sites can separate is the length of the DNA between the two. This tethering of one site to another increases the concentration of the protein near the second site and generates cooperativity in the binding to the two sites—whenever one site is bound by the protein, the other tends also to be bound.

The cooperativity resulting from DNA looping permits proteins to saturate their binding sites even though the protein concentration is well below the dissociation constant of the protein from an individual site. Thus, the microscopic dissociation of the protein from a single site can still occur, but overall, the protein remains at the two sites involved. Another way to view this phenomenon is that the protein does not come free of the DNA and diffuse away from the vicinity of the binding sites unless it lets go of both DNA-binding sites simultaneously. As a result, comparatively low concentrations of proteins are required to saturate DNA-binding sites involved with looping, and the sites need not be constructed to possess particularly high affinities for the proteins.

Why would nature want to saturate DNA sites while using low concentrations of protein, and without building the DNA sites with particularly high affinities? One answer seems to be that a large number of regulatory proteins, perhaps 50,000 in a eukaryotic cell, must simultaneously be present in the nucleus. Therefore, the concentration of any single one must be low, perhaps 10^{-10} M. While this concentration of protein could saturate sites with dissociation constants of 10^{-11} M, binding many proteins to DNA with such a high affinity could well interfere with other DNA processes such as replication and recombination. DNA looping takes care of this problem through the cooperativity. The DNA sites involved can have individual dissociation constants for the protein much lower, say 10^{-9} M, but by virtue of looping, pairs of such sites would still be nearly fully occupied. Thus, by virtue of looping, the objectives of low concentrations of the proteins, binding sites that do not weld the proteins to the DNA, and nearly complete occupancy of the binding sites can all be simultaneously achieved.

At one extreme DNA looping is generated by one protein with two DNA-binding sites. At the other extreme, two different proteins bind the two different sites, and because they are then held near one another by the DNA, they bind to each other and form a loop. Therefore, we can expect DNA looping to be generated by a single protein, by an oligomeric protein, or by two proteins that only associate at high concentrations or in the presence of DNA containing their DNA-binding sites.

BRIEF HISTORY AND OVERVIEW

DNA looping was an unanticipated discovery. In fact, it was sufficiently unexpected that a number of clues to its existence were overlooked in the years preceding its discovery. When it was finally proposed and presented with strong supporting evidence, the idea was sufficiently alien that it was not readily accepted for publication. Once reasonably solid data did get published, however, acceptance of looping came rapidly because it aided understanding of several widely observed phenomena.

The clue that motivated the experiments that ultimately uncovered DNA looping was an observation made by Englesberg on the arabinose operon in *Escherichia coli* (1). He obtained genetic evidence for a site lying upstream of the *araBAD* promoter that acted to depress activity of the promoter. A long series of studies followed that were aimed at determining the mechanism by which repression in the *ara* system could operate from the upstream site. Ultimately, it was found that this was accomplished by protein bound to the upstream site as well as to a site beside the promoter 210 base pairs downstream (2–9). Subsequently, DNA looping has been shown to function in a wide variety of other systems, both prokaryotic and eukaryotic.

The objective of this review is to give the reader an understanding of how and why looping occurs, and what it has been found to accomplish. First, the various assays that have been used to detect and measure looping are described. These different methods give an idea of the richness and diversity of the looping phenomena, as they range from detecting changes in the conformation of the DNA involved in looping to measuring changes in the regulation of genes in growing cells. As the heart of looping is the physical chemistry, an overview of the physical basis for looping is given. Then, a number of important biological systems are described in which looping plays a prominent role. This review does not contain an exhaustive list of references relevant to DNA looping; the available electronic bibliographic databases can provide this for the interested reader.

ASSAYS OF LOOPING

Helical-Twist Experiments in vivo and in vitro

A variety of assays have been developed to detect and quantitate DNA looping. The first of these, used with the intention of demonstrating looping, was the helical-twist experiment in the arabinose operon (4). In this type of experiment the two sites involved in looping are rotated around the axis of the cylindrical DNA with respect to one another by the insertion or deletion of differing amounts of DNA between the sites (Figure 1, *Top*). If the two sites are positioned for easiest looping, insertion of five base pairs between them rotates one site with respect to the other half a turn around the DNA. As a result, the formation of a loop now requires not only bending the DNA to bring the sites near one another, but also twisting of the DNA to bring the sites to the correct face of the DNA. Because DNA resists torsional stress, such a positioning increases the energetic costs of looping, thereby decreasing its frequency of occurrence. Of course, the amount of interference with looping depends upon the size, shape, and flexibility of the proteins involved, the distance separating the binding sites, and the energies available for looping. The most convincing helical-twist experiments measure the responses to large numbers of spacing changes and show multiple cycles of oscillations as the sites are positioned alternatingly to favor and disfavor looping (Figure 1).

The conceptual basis for helical-twist experiments comes from DNA cyclization experiments first done by Shore, Langowski, & Baldwin (10). For DNA several hundred base pairs in length, the cyclization rate of linear molecules to covalently closed circles in the presence of DNA ligase depended not only on the total length of the DNA, but also on the relative phases of the short single-stranded ends. Misphasing the ends by half a helical turn dramatically decreased the cyclization rate.

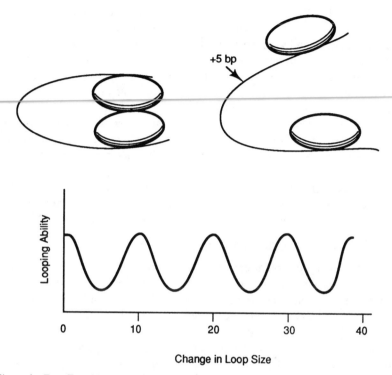

Figure 1   Top. Two DNA-binding sites on the same face of the DNA favor DNA looping. Insertion of five base pairs between the sites rotates one of the sites half a turn around the DNA axis so that now the proteins are on the opposite sides of the DNA and looping is disfavored. *Bottom.* The DNA looping abilities of multiple spacings between the protein-binding sites oscillates with a period equal to the helical repeat of the DNA.

Electrophoresis of Looped Small Circles

The binding of proteins to sites located on small supercoiled DNA molecules of about 400 base pairs, as well as electrophoresis of the complexes, has proven to be a particularly useful tool for studying DNA looping (11). The power of the assay results from the fact that weakly looping systems can be studied. This results from the fact that the supercoiling places the sites in very close proximity even in the absence of the protein. As long as the sites are located on the proper face of the DNA, little of the protein-DNA or protein-protein binding energy needs to be expended to bend the DNA or position the sites so that the protein can simultaneously contact both.

The assay built upon gel retardation experiments in which the binding of a protein to DNA can be detected by performing electrophoresis under conditions where the protein remains bound during the electrophoretic run. Normally in these experiments the protein-DNA complex migrates more slowly than the DNA molecules free of protein. On small supercoiled circles such retardation can be seen when the DNA contains only a single binding site for a protein. When looping between two sites on the circle occurs, however, the complex may migrate faster than either the free DNA or the DNA with protein bound to a single site (11). The reason for the mobility increase when two operators are present must be that looping between the two sites holds the DNA in a highly compact structure.

Although the dependence on both protein-binding sites for the existence of the altered mobility species suggests DNA looping, additional properties strengthen this conclusion. First, the complex is much more stable when two binding sites are present than when only a single site is present (11, 12). That is, its dissociation rate is much reduced when the two sites are present. Second, the increased stability and faster mobility properties show helical-twist effects when one site is rotated to the opposite face of the DNA.

Generation of Sandwich (Crossjoined) Structures

In DNA looping, a protein or complex of proteins binds to two DNA sites. There is no fundamental reason that the same two sites could not be on two different DNA molecules. Then, the DNA molecules would be held together by the protein(s), and a sandwich structure would be formed. The detection of such sandwiches is easy in the gel retardation assay in which proteins remain bound to DNA during the electrophoresis. Unfortunately, such sandwiches are hard to form. As described above, one reason looping occurs is because the two DNA sites are held in the vicinity of one another by the intervening DNA. This greatly reduces the entropic costs of loop formation. When the two sites are on different molecules, the entropic costs of forming the sandwich are high. Therefore many looping systems lack sufficient binding energy to form sandwiches, although their presence has occasionally been used to support conclusions that a system loops.

Altered Structure of DNA in Loops

Looped DNA has to be bent. When the loop is small enough, the bending will be great enough to affect substantially the widths of the major and minor grooves of the DNA. These width changes can be revealed by DNaseI digestion since the activity of this enzyme is sensitive to the width of the minor groove (13, 14). The regions facing inward will be so compressed that the enzyme will have little activity there, but portions of the regions facing outward will be suitably spaced for rapid attack by the enzyme, and will be readily cleaved. In some cases then, the pattern of sites that have gained and lost DNaseI sensitivity reflects the helical repeat of the DNA and shows periodicities of 10 or 11 base pairs.

A dramatic example of this phenomenon is an artificial looping situation involving lambda phage repressor (15). Normally the repressor dimers bind to DNA alongside one another and also interact with one another via protein-protein interactions. These protein-protein interactions are sufficiently strong that if the binding sites are moved 55 base pairs apart, the proteins will still interact with one another and in doing so, will form a DNA loop. The region between the binding sites shows areas of strongly protected and enhanced DNase sensitivity.

Looping can alter not only the conformation of DNA lying between the two protein-binding sites, but it can also alter DNA beyond the looped region. The holding of the two protein-binding sites near one another directs the segments of DNA coming into the loop and leaving the loop to lie in similar directions. This constraint tends to hold the ends of the DNA near one another. Such a proximity increase can increase the rate at which the ends will ligate together when DNA ligase is added (16). Such ligation rate effects have been observed in a looping system comprising a DNA replication protein from plasmid RP6 (16).

Electron Microscopic Observation of Looping

DNA looping formed by protein binding to two sites separated by more than about 50 base pairs should be readily observable by electron microscopy. Such looping has been observed in a number of systems (17–21). Although seeing is close to believing, the labor of obtaining statistically valid data has greatly limited the use of microscopy in looping studies. Careful control experiments are also necessary in this type of experiment because random aggregation of protein that has bound to the correct sites can generate a loop, but one without biological significance.

Cooperativity from a Distance

As described earlier, the binding of a bivalent protein to one site on the DNA increases the concentration of the protein's second DNA-binding site to nearby DNA sequences. Simultaneous binding to a second site generates looping. The affinity of the second site for the protein can be so weak and the concentration of the protein so low that the site is occupied only by virtue of looping. Another way to describe the phenomenon is that binding to the two DNA sites shows cooperativity. The presence of one binding site increases the binding to a second site. Chemists call this phenomenon the chelate effect, and its origins can equally well be discussed in terms of the decreased entropy. The existence of cooperativity has been used as an argument for DNA looping in a number of systems. The cooperativity can be observed physiologically, as for example, in the *lac* operon where a mutant *lac* operator binds repressor so weakly that only minimal repression is observed (22). The introduction of a second operator in the vicinity increases the repressor occupancy of the first operator.

Tethering

DNA looping is an attractive means by which a protein bound at one site can interact with a DNA site or protein bound to a DNA site located within a few hundred or few thousand base pairs. Instead of looping, some have suggested that two nearby sites might interact by direct communication along the DNA separating the sites through crawling, threading, or conformational telegraphic mechanisms. The two basic options of looping vs DNA-mediated communication can be distinguished by tethering two sites near one another without providing a direct DNA connection between the two. In this case, communication between the sites should be blocked if normally they communicate by mechanisms directly utilizing the intervening DNA, but communication between the sites will not be blocked if they normally utilize DNA-looping mechanisms.

Two ways of fastening sites near one another have been used. These are (*a*) use of interlocked DNA circles, called catenanes, and (*b*) linking DNA by a protein bridge. Interlocked circles have been used in vitro with the *gln* system to demonstrate that stimulation of transcription by NtrC was via looping. The NtrC-binding site was placed on one circle and the promoter on the other (23). The promoter was activated by NtrC despite being bound to a different DNA molecule than the promoter. Analogous experiments were performed with the enhancer and promoter for ribosomal RNA synthesis in *Xenopus*, although there the interlocked circles were injected into oocytes for assay of enhancement (24). In these experiments the topological interlocking was removed in vivo within 30 minutes, but the enhancement persisted for more than 18 hours. This implies that a stable initiation stimulating complex was set up early on. DNA molecules can also be linked by a protein bridge of avidin or streptavidin binding to biotinylated ends of DNA molecules (25). In this case too, the enhancer worked fine.

SOME PHYSICAL CHEMISTRY OF DNA LOOPING

General Considerations

Too many effects complicate in vivo DNA-looping situations for us to be able to perform meaningful calculations on looping probabilities in growing cells. For some cases, however, we can estimate whether looping will be more or less favored by certain changes in the system. In looping, the protein-binding sites must approach within a distance equal to the size of the looping protein and the orientation of the DNA at the sites must be constrained as well. Therefore DNA looping is closely related to ring closure of DNA molecules, and a number of results from physical chemistry on ring closure are applicable to or at least provide insight into this aspect of looping, particularly looping in vitro. Below are listed a few of the factors that clearly affect looping. I also discuss a few of the calculations related to looping with the objectives of seeing the consequences of changes in DNA length, lateral stiffness, and torsional stiffness.

LENGTH AND LATERAL STIFFNESS OF THE DNA   Two sites can only interact if the DNA is sufficiently flexible and the sites are sufficiently far apart that the DNA can bend to position the two sites the proper distance apart in three dimensions. The stiffer the DNA, the more difficult it is to form small loops.

A measure of the stiffness of polymers is their persistence length. Roughly this can be thought of, within a factor of two, as the lengths of inflexible rods connected by flexible joints that approximate the statistical behavior of the polymer. Therefore, bending the DNA into a circle of size near the persistence length would be very difficult because of the stiffness of the DNA.

The formal definition of persistence length comes from a consideration of the statistics of polymers. Consider a collection of polymers with one end at the origin. The average of the locations of the other end will also lie at the origin. Now consider orienting each of the polymers so that the initial direction of the polymer is along the x axis. Then, on average, the other ends of the polymers will be displaced some distance in the x direction. The amount of this displacement as the polymer length approaches infinity is defined as the persistence length.

TORSIONAL STIFFNESS OF THE DNA   If two sites are misoriented around the DNA, twisting of the DNA between the sites may be required to orient the sites properly for looping. On the other hand, if the sites are already properly oriented, torsional stiffness of the DNA holds the sites in the correct orientation and increases the ease of loop formation.

FLEXIBILITY OF THE LOOPING PROTEINS   If the looping protein is inflexible, then the DNA must assume exactly the correct conformation for looping to occur. A highly flexible protein would tolerate a large collection of DNA conformations and would loop more easily. Such a protein would not require the DNA to bend in the limit of small loops, ones in which the protein can span the distance between the two DNA sites.

THE GLOBAL STRUCTURE OF SUPERCOILED DNA   Points well separated from one another on a circle of 5000 base pairs are brought close to one another by supercoiling of the circle. In general the effective concentration of one point near another is increased by supercoiling. This concentration increase facilitates loop formation, but the magnitude of the increase decreases as the separation between sites becomes very large.

ADDITIONAL SECOND-ORDER EFFECTS   Proteins not involved with a particular looping reaction could affect the reaction. For example, a competing looped structure could hold the DNA such that the sites required for the first looping reaction could not occur. Alternatively, a protein could bind within a region that normally would be included in a loop and prevent loop formation if the length of the DNA is insufficient to reach around the interloper.

DNA kinks or bends specified by sequence could help or hinder formation of small loops by altering the amount of bend required to loop. They could also affect formation of larger loops by altering the conformations available to the DNA and therefore alter the looping probability.

A Zeroth Order Approximation, A Perfectly Flexible and Twistable String

How might the possibility of looping vary with DNA length if we consider the role of the intervening DNA as merely to hold the two sites in the vicinity of one another and not to let them wander independently of one another anywhere in the cell? Consider first the problem of estimating the average distance separating closest molecules in a solution of molarity M. Let us approximate the situation by taking the molecules to be located on the lattice points of a cubic lattice. By taking a one molar solution to contain Avagadro's number of molecules in 1000 cm^3, the spacing between adjacent lattice points turns out to be approximately $12/M^{1/3}$ Å. For a protein present at 10^{-8} M in cells, which is about 10 molecules per bacterial cell and about 10,000 per eukaryotic cell, the spacing is then about 6×10^4 Å. That is, the average spacing between such molecules is about 6×10^4 Å. If a pair of the protein molecules is forced closer together, the effective concentration of one of the proteins in the presence of the other is raised above 10^{-8} M. Consider that the protein is bound to a site on DNA. What is its effective concentration for binding to a second site d Å away on the DNA? Since the other DNA site is confined to the vicinity of the first site by the length d Å of the "string," we can invert the above formula to obtain an effective concentration of M = $(12/d)^3$. If d were 300 Å, which might correspond to about 100 base pairs of spacing, the effective concentration is about 3×10^{-5} M, which is a substantial increase from 10^{-8} M. This means that holding a bivalent protein at one site increases its effective concentration at the other DNA-binding site 300 base pairs away.

Including Lateral Stiffness

In the limit of small circles where the size of the circle approaches the persistence length of the DNA, circle formation becomes energetically more and more expensive as the circle size decreases. The persistence length of double-stranded DNA in physiological buffers is about 450 Å. In the other extreme, as the DNA becomes longer and longer, the concentration of one end in the presence of the other falls off, but rather slowly, and in this limit, can be calculated (26). Due to the random-walk nature of the path of long DNA, this concentration falls as $(3/2\pi nl^2)^{-3/2}$, where l is the step length in the random walk, or interbase distance, and n is the number of steps, or bases.

In the above calculation, only the presence of the two ends in the same volume element is calculated, independent of the directions from which the two ends approach one another. For some ring-closure problems and some looping problems the DNA segments should be oriented properly. A more complicated calculation by Yamakawa & Stockmayer (27) has considered this case. In such a situation we expect that the concentration of one end in the presence of the other will be low for short DNA, for which the necessary bending is difficult, will rise to a maximum for some intermediate length of DNA, and will then fall as $(\text{length})^{-3/2}$ as in the random-walk situation, due to the reducing probability that the two ends will occupy the same volume. The quantitative calculations bear this out (27).

Including Torsional Stiffness

In the example just considered, the torsional stiffness in the connecting string can increase or decrease the effective concentration of one site in the vicinity of the other. If the sites were properly oriented, then each collision would be a productive collision and functionally the effective concentration would be increased. The converse is true if the sites are misoriented. The torsional stiffness of DNA is such that it costs about 2 kcal/mole to twist a stretch of 200 base pairs half a turn (28). Of course, the precise values depend upon the nucleotide sequence and the buffer conditions. This energetic cost is inversely proportional to the length of the DNA separating the sites. Since the typical energies involved in protein-DNA binding are 10 to 15 kcal/mole, torsional stiffness can be expected to play an important role for looping involving sites closer together than about 500 base pairs. As mentioned earlier, if the binding energies are substantially greater or if the protein is unusually flexible or extended, such helical-twist effects can be greatly reduced.

Including Both Lateral and Torsional Stiffness

The consequences to ring closure of both lateral and torsional stiffness in the DNA have been considered by Shimada & Yamakawa (29). Their calculation shows that the probability of ring closure for DNA of less than about 500 base pairs long oscillates with period of 10.5 base pairs while following roughly the same shape of ring-closure curve as is followed when twisting is ignored. This is the behavior expected. When the ends are in phase with one another, the rate of ring closure is high, and when they are out of phase with each other, twisting of the DNA is required in addition to bringing the ends together, and the rate is reduced.

Monte Carlo Calculations

The ring-closure problem for a laterally and torsionally stiff polymer is just about as complex a problem as can be solved analytically. For some of the biological problems, however, we will need to consider the effects of natural bends in the DNA, the consequences of other proteins bound to the DNA, and of supercoiling of the DNA. In these cases, Monte Carlo calculations are the best computational approach. In such approaches one computes over and over, millions of times, the behavior of a system when particular randomly chosen values are applied to each of the variables that describe the system.

For example, in a Monte Carlo calculation of DNA conformation, the variables would be the various angles describing the path of the phosphodiester backbone, the angles describing the sugar conformation, and the angle of the base-sugar bond. The values assigned to the random variables are weighted in accordance with the distribution that occurs in nature. For example, if there is complete freedom of rotation about a particular bond, then any value between 0 and 360 degrees will be randomly chosen for the angle. On the other hand, if the bond is constrained to lie near 10 degrees, most of the time the angles will be chosen to lie near 10 degrees. The probability of deviating an amount δ from 10 degrees would be appropriately weighted.

In performing a Monte Carlo calculation of the conformation of DNA, the path of a DNA molecule is then computed for a set of the randomly chosen angles. This is done over and over, and the fraction of conformations that forms loops is obtained. This can be converted to an effective concentration of one end in the presence of the other. Levene & Crothers (30) did this ignoring the torsion angle, and later Hagerman included computation of the torsion angle (31). When applied to situations similar to those for which the analytical solutions of Shimada & Yamakawa apply, the results closely agree, suggesting that both the analytical and numerical approaches are correct (or most unlikely, both are incorrect).

LOOPING IN GENE REGULATION

This section describes and surveys a number of the most carefully studied DNA looping systems. Both an historical and a mechanistic perspective is taken in these descriptions. The systems illustrate the diversity of the uses to which nature has put DNA looping and show the generality of looping.

Prokaryotic Systems

ARA   Englesberg found a deletion entering the *araCBAD* operon from up-stream of the *araC* gene and ending before the *araB* gene had unusual properties (1) (Figure 2). When AraC protein was provided in *trans*, the deletion did not impair normal inducibility of the $arap_{BAD}$ promoter, but it did alter the uninduced, or basal level of the promoter in the presence of AraC protein. In the absence of arabinose, the uninduced level of the *araBAD* genes was about 10 times normal. That is, the deletion appeared to remove a site through which AraC protein acted to repress its own activation of the $arap_{BAD}$ promoter. Presumably, in the absence of arabinose, a small fraction of AraC would be in the inducing state, and would weakly activate transcription of the $arap_{BAD}$ promoter if the site required for repression by AraC were absent. Englesberg's deletion would remove this site.

At about the time the *ara* operon was being probed genetically, biochemi-cal studies were unraveling the *lac* operon. There repression occurred through steric hindrance in the binding of repressor or RNA polymerase. Therefore the possibility that a repressor could function from a position upstream of all the sites that are required for induction, as the genetic data in the *ara* system indicated, was not easily understandable. This led to an extensive study of the phenomenon of repression from upstream. Standard genetics reproduced the upstream repression phenomenon (2, 4). Subsequently, locating the AraC protein–binding sites on the DNA led to the mistaken conclusion that repres-sion was mediated by the $araO_1$ site (32). This is located immediately adjacent to the initiation complex (Figure 2). A protein bound in such a position can be imagined to be capable of generating repression by a mech-anism involving direct side-by-side touching of proteins all in a row along the DNA; from AraC protein bound at $araO_1$ contacting the cyclic AMP receptor protein contacting AraC protein bound at *araI* finally contacting RNA polymerase bound at the promoter.

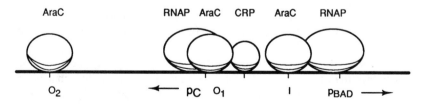

Figure 2   The regulatory region of the L-arabinose operon of *Escherichia coli*. The promoter p_C serves the *araC* gene, which begins just to the left of $araO_2$. The $araO_1$ site helps regulate p_C. The *araI* site activates transcription from p_{BAD}, and the *araBAD* genes lie just to the right of p_{BAD}.

Genetic engineering techniques eventually permitted more accurate mapping of the upstream site required for repression of the arabinose operon. It was found not to be $araO_1$, but another site that lies more than 200 base pairs upstream from the induction complex at the promoter (4, 6, 7), called $araO_2$. That is, a deletion ending as far as 200 base pairs upstream from p_{BAD} eliminated the self-repression by AraC protein. Clearly some signal was being sent from $araO_2$ to the initiation complex at p_{BAD}, and the domino mechanism of direct protein-to-protein interaction for transmitting a signal from one site to another by a continuous row of proteins on the DNA was impossible. Amongst the possibilities for transmission of the signal was DNA looping whereby AraC protein bound at the $araO_2$ site would contact AraC protein bound at the $araI$ site at p_{BAD}.

Helical-twist experiments were used to demonstrate that indeed, looping was the mechanism of interaction between $araO_2$ and $araI$ (4). Inserting five base pairs at an irrelevant site between $araO_2$ and $araI$ effectively rotates these sites one half a turn of the DNA helix with respect to one another. If previously the sites faced one another and looping required bringing proteins bound to these sites into contact, the insertion of five bases required not only bringing the proteins together, but also twisting the DNA half a helical turn. The energetic costs of such a twist are about 2 kcal/mole, which in the arabinose operon were sufficient to interfere substantially with looping. Inserting 10 base pairs at the same site restored looping. The only reasonable mechanism compatible with these observations is DNA looping.

One virtue of the arabinose system was the relative simplicity of utilizing genetics to study looping. One might expect that mutations in both the DNA and in the AraC protein would interfere with looping. Both types have been found and studied. Repression-negative mutations in the DNA were confined to the two AraC protein–binding sites, $araO_2$ and $araI$ (7). Those in $araO_2$ reduced the binding affinity to this site about 30-fold. Those in $araI$ were more complicated, and originally appeared to decrease the ability of AraC bound at $araI$ to repress without affecting the binding ability of AraC. More accurate binding measurements have shown they slightly increase the binding affinity of AraC protein and they may be constitutive in nature (R. Lobell and R. Schleif, unpublished). Mutations in AraC have also been found that either increase or decrease repression; their molecular mechanism is not yet understood.

The $araO_2$ site involved in DNA looping binds AraC protein about 50-fold less tightly than the $araI$ site. This is sufficiently weak binding that in the absence of looping to $araI$, the $araO_2$ site is not significantly occupied by AraC protein. The cooperativity in binding between $araO_2$ and $araI$ as measured with an in vivo footprinting system developed for the purpose provided good supporting data for the existence of looping in the ara system (7, 8).

Extensive helical-twist experiments have been done on the *ara* system (9). About 75 different spacings have been generated and the degree of repression in each measured. Sufficient data is available that by eye or by Fourier transform there is no question that the periodicity in the ability to repress expression oscillates with a period of a little over 11 base pairs, and is clearly not 10.5 base pairs per turn. The dramatic difference in helical repeat from the oft-measured value of 10.5 base pairs per turn of linear DNA in vitro must result from the linking-number deficit of DNA in vivo. DNA extracted from bacteria has a linking-number deficit that is commonly measured as supercoiling. In reality, the linking-number deficit can be partitioned between unwinding the DNA, which increases the number of base pairs per turn, and supercoiling. Clever measurements show that in vivo the partitioning between the two is roughly half and half (33). At the physiological linking-number deficit of -0.06 and such a 50-50 partitioning, the helical repeat of the DNA would be about 10.8 base pairs per turn. Therefore there remains 0.3 or so of base pairs per turn unaccounted for. It will be interesting when sufficiently accurate measurements have been made on other systems to see their helical repeat values.

AraC protein is a dimer in solution and binds to linear DNA as a dimer (34, 35). Therefore it was expected that looping in the *ara* system would involve the binding of a dimer at *araI* to a dimer bound to *araO₂*, and it was a shock to discover that, in fact, just a single dimer of AraC protein generates the loops (36). When arabinose is added, the loop opens, contacts to *araO₂* are lost, and DNA contacts are made only at *araI*. This behavior raises the question of what is it that makes the loop open and AraC protein now contact a larger stretch of DNA at *araI*? One attractive mechanism that could generate this behavior is subunit orientation or position. The subunits could be located in the absence of arabinose such that it is easier for them to contact two well-separated sites on the DNA via looping than two adjacent sites on the DNA. That is, the subunits could normally be misoriented. The addition of arabinose could alter the orientation such that the subunits become well positioned for binding to two adjacent sites on the DNA. This mechanism then could easily control looping, and could be extended to other systems in which looping is regulated.

LAC   Several early hints for the existence of looping in the *lac* system were not pursued. When, however, the powerful tools of the *lac* operon were then turned to investigating looping, a variety of wonderful experiments were done.

Amongst the hints for looping in the *lac* system were experiments with chimeric *lac* repressor-β-galactosidase molecules, the existence of multiple pseudo-operators, and the behavior of operator mutants. In 1977 Kania & Müller-Hill reported the DNA-binding abilities of tetrameric hybrids of β-galactosidase consisting of the normal β-galactosidase subunit and β-galactosidase subunits containing the N-terminal 60–80 amino acids of *lac* repressor (37). The various types of hybrids were partially separated from one another and their binding to linear DNA was investigated with the filter-binding assay in which uncomplexed DNA freely passes through a filter, but the protein bound on the DNA retards the DNA on the filter. Although significant cross-contamination existed in samples with differing numbers of DNA-binding domains, and there could be questions about proper positioning of DNA-binding domains or conformational effects on tetramers containing two or four of the repressor domains, a tighter DNA binding was seen in the fractions thought to contain molecules with four repressor domains than in the fractions enriched in hybrids containing two repressor domains. The authors suggested that these results could result from the binding of repressor to two operators on the DNA simultaneously. The authors did not propose that the effect occurs in vivo, and this work was apparently not pursued between 1977 and 1986. In retrospect, it is difficult to understand any higher binding affinity seen by the tetrameric molecule because subsequent studies have shown that supercoiling is required for the natural operators to be able to engage in looping (38, 39).

Cooperativity in repressor binding to *lac* operators was also an overlooked clue to the existence of DNA looping. For some years, the behavior of operator mutants in the *lac* operon was most puzzling. Point mutations in the operator seemed to raise the basal level of expression of the operon at most only 20-fold, yet study of the binding affinities to the operators in vitro and consideration of the in vivo concentrations of *lac* repressor suggested that the mutations should have had a much greater effect on repression (40). Apparently DNA looping ameliorated the effects of the operator-constitutive mutations. When wild-type *lac* operator was placed various distances from a mutant operator, it was apparent that the presence of the wild-type operator aided the binding of repressor to the mutant operator. By this time the concept of looping as a mechanism in gene regulation was accepted and the data was interpreted as possibly resulting from looping (41). Similarly, placing the *lac* operators upstream of the *lac* promoter gave no repression and placing an operator downstream gave moderate repression, but both operators together generated strong repression (42). Whether the repression results from looping, helping the repressor bind to the downstream site and thereby serve as a blockade to an active RNA polymerase, or whether the loop poses a steric hindrance to the binding of the polymerase is not yet known.

An elegant experiment showed that looping in the *lac* system is a natural part of the operon's regulation. This made use of the two pseudo-operators found near the *lac* regulatory region. Since 1974 it had been known that in addition to the primary *lac* operator located just downstream from the transcription initiation site, two additional weaker binding operators existed in the *lac* operon (43, 44). For more than 10 years these were thought merely to be remnants of evolution. To investigate a role of the downstream operator O_2, which is located 481 base pairs downstream from the primary operator, eight bases in *lac* O_2 were altered in such a way as not to change the amino acids encoded by this region of the DNA (45). Thus the β-galactosidase remained unaltered, but O_2 was so greatly altered that *lac* repressor no longer could bind. These changes were found to decrease the repression of the operon about fivefold. That is, they raised the uninduced or basal level fivefold, but had almost no effect on the induced activity of the promoter. This result indicates that the second *lac* operator does play a role in repressing the operon (46). Apparently DNA looping frequently occurs in vivo between the main *lac* operator and the downstream pseudo-operator.

The second pseudo-operator is located 83 base pairs upstream from O_1, at the end of *lacI*. It also apparently plays a role in normal regulation of the operon, for its destruction raises the basal level of *lacZ* about threefold (46).

Nearly the full set of experiments that can demonstrate looping have now been applied to the *lac* operon, and with dramatic success (17). In vitro and in vivo footprinting has been used to show cooperativity in the binding of repressor to the main *lac* operator and the two pseudo-operators O_2 and O_3 located 83 and 481 base pairs from the operator (47, 48).

DNA-binding studies of *lac* repressor-operator interactions have been performed with the filter-binding assay. The dissociation rate of repressor from such DNA is much reduced if the DNA is supercoiled and if it contains one of the pseudo-operators in addition to the operator at the promoter (38, 39).

In addition to using supercoiled DNA and the natural *lac* operators, looping in *lac* has also been studied by increasing the energies available for looping by using perfectly symmetric *lac* operators. Repressor binds to such operators about 100 times as tightly as it binds to the wild-type operator. With such operators, linear DNA could be forced to loop and at high repressor concentrations, "crossjoined" or sandwich structures could also be formed (17). The structures could be detected by gel retardation or observed by electron microscopy. They were identified as looped structures by their enhanced stability and their absence when the two operators were rotated half a turn with respect to one another. Additionally, the presence of these loops on the DNA generated the characteristic alternating pattern of DNaseI-hypersensitive and hyposensitive sites on the DNA between the two operators.

In an effort to apply principles of statistical mechanics to looping in vivo in the *lac* operon, Mossing & Record examined operator spacings of 118, 185, and 283 base pairs and concluded that supercoiled DNA in vivo possessed the same stiffness and therefore experienced the same difficulty in bending sharply to form small loops as would be predicted from the hydrodynamically determined stiffness of linear DNA (49). Later it was learned that the inability of the closest spaced operators, 118 base pairs, to loop in their experiment apparently resulted not from stiffness of the DNA, but from the fact that they were on opposite faces of the DNA as the helical repeat of DNA in vivo is not 10.5 base pairs per turn as it is on linear DNA in vitro. This they found in a subsequent helical-twist experiment in which the repression abilities of 12 different spacing mutations were clearly inconsistent with an helical repeat in vivo of 10.5 base pairs per turn (50). These authors have also considered the decrease in looping that occurs as loops become particularly large. This should occur when entropic effects might begin to become more important than the gains from not having to bend the DNA so sharply (50). This problem is particularly difficult, however, for in supercoiled DNA distal sites are forced into close contact with one another due to the wrapping, and simple polymer statistics should not be applied to the problem. The *deo* system is an example where operator separations of up to 4000 base pairs showed only a small distance effect (51). An additional difficulty in attempting to calculate looping is the unknown flexibility of the protein. Undoubtedly some proteins possess significant flexibility and their bending will reduce the amount of bending required by the DNA to form a loop.

DEO   The protein products of the *deo* operon in *Escherichia coli* permit the cells to catabolize nucleosides. At least four enzymes are involved in the catabolism and three proteins are involved in the regulation. The synthesis of *deoC*, *deoA*, *deoB*, and *deoD* message initiates from two promoters separated by 599 base pairs and lying upstream of the four *deo* genes (52). Messenger synthesis is regulated by three proteins, the cyclic AMP receptor protein, which activates transcription, and the CytR and DeoR repressor proteins. In the absence of CytR protein, DeoR can repress expression of the Deo proteins about 60-fold (53). This repression requires three operator sites, one over-lapping each of the -10 regions of the two promoters, and a third site lying an additional 270 base pairs upstream of the first two (54). The integrity of all three operators is required for maximal repression, suggesting that a double DNA loop might form. Indeed, such is possible, for the molecular weight of DeoR protein based on gel exclusion chromatography indicates that the protein is an octamer, and therefore it could bind three, or possibly four of the 16-base-pair palindromic binding sites simultaneously (55). Double looped structures can be formed in vitro and observed by electron microscopy (18).

The potential energies available for looping of the *deo* system must be large, for loops will form on linear DNA, and particularly large loops can form and repress in vivo. When only two operators are present, one of which overlaps the -10 region of a promoter, repression falls from 23-fold to

3.6-fold to 3.0-fold as the spacing between the two operators increases from 567 base pairs to 3700 base pairs to 4685 base pairs (51).

OTHER LOOPING SYSTEMS   The *gal* operon of *Escherichia coli* is subject to a complex control as galactose can be catabolized when it is available, but also must always be present to fulfill needs for cell wall synthesis. Two promoters separated by about 10 base pairs serve the *galETK* genes. The activity of one is stimulated by cyclic AMP receptor protein, whereas the other is repressed by cyclic AMP receptor protein. The activity of both is repressed by GalR repressor, which binds both at positions -60 and at +55 with respect to the start of transcription (56). Instead of repressing via steric hindrance in which repressor bound to either of the operators blocks the binding of RNA polymerase or cyclic AMP receptor protein, a DNA loop formed by repressor bound at both sites generates a loop that represses the promoter. Most likely the binding rate of RNA polymerase to such a sterically constrained promoter is much reduced from the nonlooped state. An elegant experiment showing that the repression was generated by looping was the finding that substitution of either binding site for the *lac* repressor binding site left the system poorly repressed, but when both of the *gal* operators were changed to *lac* operators, repression was strong when *lac* repressor was present (57). As expected, if a mutant *lac* repressor unable to form tetramers and therefore unable to loop was used, repression was poor (20).

The genes involved in the uptake and metabolism of *N*-acetylglucosamine are controlled by the NagC protein, a repressor that loops. This may be a particularly good system for physical studies as short linear DNA fragments can be looped by the NagC protein (58).

Activator Systems

A complex system utilizing a two-component signal-transducing system as well as protein phosphorylation regulates synthesis from the genes involved in nitrogen metabolism in *Escherichia coli* (59). Transcription of the *gln* genes does not use the major form of the RNA polymerase that contains the σ^{70} subunit. Instead the promoters require RNA polymerase containing σ^{54}, a subunit encoded by the *rpoN* gene whose other aliases are *glnF* and *ntrA* (60). The major difference between transcription from genes using the σ^{54} polymerase and those using σ^{70} is that transcription from the former requires an additional activating protein. Transcription of the *gln* genes requires either the NtrC protein or the NifA protein, which bind to DNA one or two hundred base pairs away from the promoter (61) and form a loop in the DNA as they interact with RNA polymerase.

Although no plausible mechanism except DNA looping could explain the action-at-a-distance effects shown by the activating protein NtrC in the transcription of *glnALG* genes, a nice looping confirmation was performed using linked or concatenated DNA circles as described earlier (23). The electron microscopic evidence for looping in the *nif* system is rather inconclusive as only 5% of the DNA molecules with bound protein were observed to be looped (19). Undoubtedly NtrC and bound RNA polymerase can interact in the samples prepared for microscopy, but random aggregation of protein would produce the same result.

The regulatory proteins NtrC and NifA regulate nitrogen fixation genes called *nif* in *Klebsella pneumoniae*, a close relative of *Escherichia coli*. Helical-twist experiments in which which 5, 11, 15, and 21 base pairs were added to the natural spacing of about 150 base pairs between the NtrC-binding site and the promoter for *nifLA* showed a cyclic dependence (62). Additional evidence for DNA looping in *nif* genes is that even if the natural spacing of one hundred or so base pairs from the promoter to the NifA-binding site were increased to as many as several thousand, activation by NifA could still be detected (63).

Eukaryotic Enhancers in General

Eukaryotic enhancers activate and regulate transcription initiating at promoters located hundreds, thousands, and sometimes tens of thousands of base pairs away. Enhancers can be moved about and turned end for end, usually with only a small effect on their stimulatory powers (63a, 64a). Enhancers were first recognized in animal viruses, and initially were hypothesized as being entry points for RNA polymerase molecules that would then drift to the promoter to start transcription (65).

One test of the drifting model is to place a blockade between the enhancer and promoter. Ptashne & Brent did this with the *gal4* system in yeast and found strong interference with promoter activity (65). This they interpreted as evidence in favor of the bind and drift model. After finding evidence for looping in vitro with the lambda repressor system, looping seemed more plausible and the blockade was interpreted as interfering with looping instead (66).

Relatively few definitive experiments have been done to prove that proteins bound to enhancers generate loops as they interact with other parts of the transcription apparatus at the promoter. Nonetheless, the prevailing opinion seems to be that most enhancers use a looping mechanism. This certainly seems the most plausible for enhancers with large separations between them and their promoters. Further, DNA looping is almost the only way that transvection (67) in *Drosophila* could work (68). There an enhancer on one chromosome activates a promoter on a different but homologous chromosome.

A few helical-twist experiments have been done with eukaryotic enhancers. Those that showed some helical effects were the late promoter of SV40 (69), and the heat-shock promoter of *Drosophila* (70). Other promoters fail to show helical-twist effects (71–74). These negative findings can be the result of several factors. The energies available for loop formation may greatly exceed the energetic costs of torsionally twisting the DNA. Another reason may be that the proteins involved are sufficiently flexible that they can accommodate substantial misorientations of their binding sites. This is highly likely, for many of the eukaryotic enhancer binding proteins seem to consist of loosely connected domains that can freely be interchanged without loss of activity.

As discussed earlier, tethering experiments support DNA looping. They have been done with the ribosomal promoter of *Xenopus* and the viral SV40 promoter (24, 25). Electron microscopy of enhancer binding proteins binding together and to two sites on the DNA suggest, but hardly prove, the existence of DNA looping.

Looping in Genetic Recombination

Because the DNA sites participating in site-specific genetic recombination must be close to one another at the time of strand exchange, proteins bound to these sites bind to each other. Consequently DNA looping, or its equivalent if the sites are on two different DNA molecules, occurs. These types of reactions are found in the *hin* inversion system of *Salmonella*, the *gin* and *pin* inversion systems of phages Mu and P1, resolution reactions of transposons, the transposition system of Mu, and the integration-excision system of lambdoid phages (75–77).

The resolvase protein encoded by transposons such as Tn3 and Mu phage catalyzes an exchange reaction between two *res* sites in the same orientation on the same DNA molecule (78). When the sites are oppositely oriented there is no reaction or a greatly reduced reaction. Initially this fact was taken as evidence for a DNA-threading reaction by resolvase from one *res* site to the other (78). By this means the recombinase could determine the relative orientation of the two sites. It did not seem possible that the two sites could communicate their relative orientations by any other means. Closely reasoned topological experiments (79) followed by an elegant linked-circle experiment with the Mu phage transposition system however, changed thinking on this issue (80). DNA molecules with two properly oriented Mu *att* sites that were capable of recombining were used in this experiment. The DNA substrates for this experiment also contained lambda phage *attB* and *attP* sites oriented such (Figure 3) that the integration reaction between them would either invert one of the Mu *att* sites or separate the two *att* sites onto two separate DNA molecules. Once the DNA was supercoiled and capable of supporting a reaction between the *att* sites, the lambda phage integrase reaction between

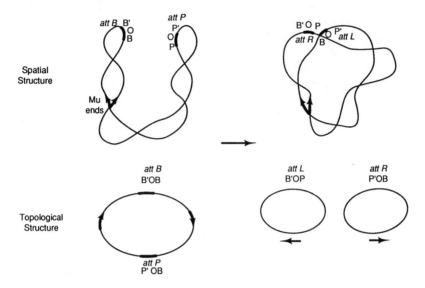

Figure 3   The spatial structure and the topological structure of the constructs used to demon-strate that neither tracking nor threading is used by the Mu phage integration system to monitor the relative orientation of the Mu ends of the DNA. Once the DNA is supercoiled, treating with lambda phage Int protein promotes recombination between *attB* and *attP* to generate *attL* and *attR*, which are then located on two topologically separate but spatially intertwined DNA molecules. Site-specific recombination promoted by Mu phage proteins still occurs between the two Mu phage ends.

the sites, which either inverted an *att* site or placed the two sites on two different molecules, had no effect on the reaction. The explanation for these results is that the two *att* sites are placed in close proximity and in the proper spatial orientation via the supercoiling of the DNA. If this supercoiling persists while the DNA molecule is cut and rejoined at sites elsewhere, the reaction between the *att* sites is unaltered. The cutting and rejoining by the lambda *int* system can separate the sites or even invert them topologically. In summary, what is important for interactions between two *att* sites is their local spatial orientation not their global topological structure.

The lambda phage integration-excision structure itself can also be thought of as a loop. The complex of Int, IHF, and Xis proteins at the phage attachment region is sharply bent by the proteins and appears to be nearly looped itself. Of course the Int protein participates in the strand scission and rejoining reactions, but it also helps hold the DNA looped by binding both at the crossover region as well as 125 and 150 base pairs away. Although six dimers of proteins, including three dimers of IHF, are present in the structure, they appear not to require direct protein-protein interaction. The cyclic AMP receptor protein bends DNA about as much as IHF protein (81–83), and it can replace IHF protein in the intasome. Also, a sequence of DNA possessing oriented runs of As and Ts so that it naturally bends about 100 degrees can replace one of the IHF proteins (83).

Looping in Regulation of DNA Replication

DNA replication is a highly important process, and it is likely that as discussed earlier, proteins that bind to DNA and that interact are able to form loops in the DNA. As a collection of proteins binds to the origin of replication of phage lambda and *E. coli*, it seems reasonable that the activity of most origins will require binding by multiple proteins. Such a situation opens the possibility of DNA looping, and indeed looping between two lambda replication origins that had been placed on the same DNA molecule mediated by the lambda O protein has been observed (84). It remains to be seen whether this type of looping is of biological significance. It is clear that the replication-initiator protein for plasmid R6K binds weakly to an active origin, but by virtue of DNA looping from a collection of much tighter binding sites, the weaker site is occupied by the initiator (85). This is a case where the cooperativity provided by looping helps transfer proteins to a DNA site.

SUMMARY

DNA looping is widely used in nature. It is well documented in the regulation of prokaryotic and eukaryotic gene expression, DNA replication, and site-specific DNA recombination. Undoubtedly looping also functions in other protein-DNA transactions such as repair and chromosome segregation. While the underlying physical chemistry of DNA looping is common to all systems, the precise biochemical details of looping and the utilization of looping by different systems varies widely. Looping appears to have been chosen by nature in such a wide variety of contexts because it solves problems both of binding and of geometry. The cooperativity inherent in binding a protein to multiple sites on DNA facilitates high occupancy of DNA sites by low concentrations of proteins. DNA looping permits a sizeable number of DNA-binding proteins to interact with one of their number, for example RNA polymerase. Finally, DNA looping may simplify evolution by not requiring a precise spacing between a protein's binding site and a second site on the DNA.

ACKNOWLEDGMENTS

I thank NIH for research support and Pieter Wensink and Susan Egan for comments on the manuscript.

222   SCHLEIF

Literature Cited

1. Englesberg, E., Squires, C., Meronk, F. 1969. *Proc. Natl. Acad. Sci. USA* 62:1100–107
2. Schleif, R., Lis, J. 1972. *Proc. Natl. Acad. Sci. USA* 69:3479–84
3. Lis, J., Schleif, R. 1975. *J. Mol. Biol.* 95:395–407
4. Dunn, T., Hahn, S., Ogden, S., Schleif, R. 1984. *Proc. Natl. Acad. Sci. USA* 81:5017–20
4a. Schleif, R., Lis, J. 1975. *J. Mol. Biol.* 95:417–31
5. Hahn, S., Dunn, T., Schleif, R. 1984. *J. Mol. Biol.* 180:61–72
6. Dunn, T., Schleif, R. 1984. *J. Mol. Biol.* 180:201–4
7. Martin, K., Huo, L., Schleif, R. 1986. *Proc. Natl. Acad. Sci. USA* 83:3654–58
8. Huo, L., Martin, K., Schleif, R. 1988. *Proc. Natl. Acad. Sci. USA* 85:5444–48
9. Lee, D. H., Schleif, R. F. 1989. *Proc. Natl. Acad. Sci. USA* 86:476–80
10. Shore, D., Langowski, J., Baldwin, R. L. 1981. *Proc. Natl. Acad. Sci. USA* 78:4833–37
11. Krämer, H., Amouyal, M., Nordheim, A., Müller-Hill, B. 1988. *EMBO J.* 7:547–56
12. Eismann, E. R., Müller-Hill, B. 1990. *J. Mol. Biol.* 213:763–75
13. Drew, H. R. 1984. *J. Mol. Biol.* 176:535–57
14. Drew, H. R., Travers, A. A. 1984. *Cell* 37:491–502
15. Hochschild, A., Ptashne, M. 1986. *Cell* 44:681–87
16. Mukherjee, S., Erickson, H., Bastia, D. 1988. *Proc. Natl. Acad. Sci. USA* 85:6287–91
17. Krämer, H., Niemoller, M., Amouyal, M., Revet, B., vonWilcken-Bergmann, B., et al. 1987. *EMBO J.* 6:1481–91
18. Amouyal, M., Mortensen, L., Buc, H., Hammer, K. 1989. *Cell* 58:545–51
19. Su, W., Porter, S., Kustu, S., Echols, H. 1990. *Proc. Natl. Acad. Sci. USA* 87:5504–8
20. Mandal, N., Su, W., Haber, R., Adhya, S., Echols, H. 1990. *Genes Dev.* 4:410–18
21. Théveny, B., Bailly, A., Rauch, C., Rauch, M., Delain, E., Milgram, E. 1987. *Nature* 329:79–81
22. Mossing, M. C., Record, M. T. Jr. 1986. *J. Mol. Biol.* 186:295–305
23. Wedel, A., Weiss, D. S., Popham, D., Droge, P., Kustu, S. 1990. *Science* 248:486–90
24. Dunaway, M., Droge, P. 1989. *Nature* 341:657–59
25. Muller, H. P., Sogo, J. M., Schaffner, W. 1989. *Cell* 58:767–77
26. Jacobson, H., Stockmayer, W. H. 1950. *J. Chem. Phys.* 18:1600–6
27. Yamakawa, H., Stockmayer, H. W. 1972. *J. Chem. Phys.* 57:2843–87
28. Shore, D., Baldwin, R. L. 1983. *J. Mol. Biol.* 170:957–81
29. Shimada, J., Yamakawa, H. 1984. *Macromolecules* 17:689–98
30. Levene, S. D., Crothers, D. M. 1986. *J. Mol. Biol.* 189:61–72
31. Hagerman, P. J., Rawadevi, V. A. 1990. *J. Mol. Biol.* 212:351–62
32. Ogden, S., Haggerty, D., Stoner, C., Kolodrubetz, D., Schleif, R. 1980. *Proc. Natl. Acad. Sci. USA* 77:3346–50
33. Bliska, J., Cozzarelli, N. 1987. *J. Mol. Biol.* 194:205–18
34. Steffen, D., Schleif, R. 1977. *Mol. Gen. Genet.* 157:341–44
35. Hendrickson, W., Schleif, R. 1985. *Proc. Natl. Acad. Sci. USA* 82:3129–33
36. Lobell, R. B., Schleif, R. F. 1990. *Science* 250:528–32
37. Kania, J., Müller-Hill, B. 1977. *Eur. J. Biochem.* 79:381–86
38. Whitson, P. A., Hsieh, W. T., Wells, R. D., Matthews, K. S. 1987. *J. Biol. Chem.* 262:14592–99
39. Whitson, P. A., Hsieh, W. T., Wells, R. D., Matthews, K. S. 1987. *J. Biol. Chem.* 262:4943–46
40. Smith, T., Sadler, J. 1971. *J. Mol. Biol.* 59:273–305
41. Mossing, M. C., Record, M. T. Jr. 1985. *J. Mol. Biol.* 186:295–305
42. Besse, M., von Wilcken-Bergmann, B., Müller-Hill, B. 1986. *EMBO J.* 5:1377–81
43. Reznikoff, W. S., Winter, R. B., Hurley, C. K. 1974. *Proc. Natl. Acad. Sci. USA* 71:2314–18
44. Gilbert, W., Maxam, A. 1975. *Protein-Ligand Interactions*, pp. 192–210. Berlin: deGruyter
45. Eismann, E., von Wilcken-Bergmann, B., Müller-Hill, B. 1987. *J. Mol. Biol.* 195:949–52
46. Oehler, S., Eismann, E. R., Krämer, H., Müller-Hill, B. 1990. *EMBO J.* 9:973–79
47. Sasse-Dwight, S., Gralla, J. D. 1988. *J. Mol. Biol.* 202:107–19
48. Borowiec, J. A., Zhang, L., Sasse-Dwight, S., Gralla, J. D. 1987. *J. Mol. Biol.* 196:101–11
49. Mossing, M. C., Record, M. T. Jr. 1986. *Science* 233:889–92

50. Bellomy, G. R., Mossing, M. C., Record, M. T. Jr. 1988. *Biochemistry* 27:3900–6
51. Dandanell, G., Valentin-Hansen, P., Larsen, J., Hammer, K. 1987. *Nature* 325:823–26
52. Valentin-Hansen, P., Aiba, H., Schümperli, D. 1982. *EMBO J.* 1:317–22
53. Dandanell, G., Hammer, K. 1985. *EMBO J.* 4:3333–38
54. Valentin-Hansen, P., Albrecht, B., Larsen, J. 1986. *EMBO J.* 5:2015–21
55. Mortensen, L., Dandanell, G., Hammer, K. 1989. *EMBO J.* 8:325–31
56. Majumdar, A., Adhya, S. 1984. *Proc. Natl. Acad. Sci. USA* 81:6100–4
57. Haber, R., Adhya, S. 1988. *Proc. Natl. Acad. Sci. USA* 85:9683–87
58. Plumbridge, J., Kolb, A. 1991. *J. Mol. Biol.* 217:661–79
59. Ninfa, A. J., Magasanik, B. 1986. *Proc. Natl. Acad. Sci. USA* 83:5909–13
60. Hirschman, J., Wong, P. K., Sei, K., Keener, J., Kustu, S. 1985. *Proc. Natl. Acad. Sci. USA* 82:7525–29
61. Reitzer, L. J., Magasanik, B. 1986. *Cell* 45:785–92
62. Minchin, S. D., Austin, S., Dixon, R. A. 1989. *EMBO J.* 8:3491–99
63. Buck, M., Miller, S., Drummond, M., Dixon, R. 1986. *Nature* 320:374–78
63a. Banerji, J., Rusconi, S., Schaffner, W. 1981. *Cell* 27:299–308
64a. Moreau, P., Hen, R., Wasylyk, B., Everett, R., Gaub, M. P., Chambon, P. 1981. *Nucleic Acids Res.* 9:6047–68
65. Brent, R., Ptashne, M. 1984. *Nature* 312:612–15
66. Ptsahne, M. 1988. *Nature* 335:683–89
67. Judd, B. H. 1988. *Cell* 53:841–43
68. Geyer, P. K., Green, M. M., Corces, V. G. 1990. *EMBO J.* 9:2247–56
69. Takahashi, K., Vigneron, M., Matthes, H., Wildeman, A., Zenke, M., Chambon, P. 1986. *Nature* 319:121–26
70. Cohen, R., Meselson, M. 1988. *Nature* 332:856–58
71. Chodosh, L. A., Carthew, R. W., Morgan, J. G., Crabtree, G. R., Sharp, P. A. 1987. *Science* 238:684–88
72. Wirth, T., Staudt, L., Baltimore, D. 1987. *Nature* 329:174–78
73. Ruden, D. M., Ma, J., Ptashne, M. 1988. *Proc. Natl. Acad. Sci. USA* 85: 4262–66
74. Wu, L., Berk, A. 1988. *Genes Dev.* 2:403–11
75. Plasterk, R. H. A., Iimer, T. A. M., van de Putte, P. 1983. *Virology* 127:24–36
76. Scott, T., Simon, M. 1982. *Mol. Gen. Genet.* 188:313–21
77. Enomoto, M., Oosawa, K., Momota, H. 1983. *J. Bacteriol.* 156:663–68
78. Krasnow, M., Cozzarelli, N. 1983. *Cell* 32:1312–24
79. Benjamin, H., Matzuk, M., Krasnow, M., Cozzarelli, N. 1985. *Cell* 40:147–58
80. Craigie, R., Mizuuchi, K. 1986. *Cell* 45:793–800
81. Robertson, C. A., Nash, H. A. 1988. *J. Biol. Chem.* 263:3554–57
82. Liu-Johnson, H. N., Gartenberg, M. R., Crothers, D. M. 1986. *Cell* 47:995–1005
83. Thompson, J. F., Landy, A. 1988. *Nucleic Acids Res.* 16:9687–705
84. Schnos, M., Zahn, K., Blattner, F. R., Inman, R. B. 1989. *Virology* 168:370–77
85. Mukherjee, S., Erickson, H., Bastia, D. 1988. *Cell* 52:375–83

Annu. Rev. Biochem. 1992. 61:225–50
Copyright © 1992 by Annual Reviews Inc. All rights reserved

INOSITOL PHOSPHATE BIOCHEMISTRY

Philip W. Majerus

Division of Hematology-Oncology, Washington University School of Medicine, St. Louis, Missouri 63110

KEY WORDS: messengers, signal transduction, inositol phospholipids

CONTENTS

PERSPECTIVES

An enormous literature has followed the last review of inositol phosphate metabolism in *Annual Review of Biochemistry* five years ago (1). At that time,

225

Berridge proposed a scheme in which phospholipase C (PLC) hydrolyzed phosphatidylinositol 4,5-bisphosphate [PtdIns(4,5)P$_2$] to yield inositol 1,4,5-trisphosphate [Ins(1,4,5)P$_3$] and diacylglycerol. However, the multiple sources of diacylglycerol (2), as well as the many intermediates, enzymes, and routes for formation of inositol phosphates (3), indicate that this signal-transducing pathway is much more complex than described previously. Many new reactions, enzymes, and inositol phosphate–binding proteins have been discovered, and molecular genetic approaches have begun to further define the structures and functions of the proteins involved. Rather than catalog all of the new information, I present selected topics that have yielded new insights for future directions of research. I do not discuss calcium ion mobilization in response to Ins(1,4,5)P$_3$ as this topic has been reviewed elsewhere (4–7). The main subjects to be discussed are (*a*) mechanisms for triggering phosphatidyl-inositol turnover and inositol phosphate production; (*b*) characterization of enzymes that metabolize inositol phosphates; (*c*) pathways for metabolism of inositol pentaphosphates and inositol hexaphosphates; (*d*) inositol phosphate–binding proteins, especially the Ins(1,4,5)P$_3$-binding protein; and (*e*) the newly discovered 3-phosphate-containing inositol phospholipids.

CONVENTIONAL PHOSPHATIDYLINOSITOLS

Inositol-containing phospholipids are ubiquitous components of eukaryotic cells and constitute 2–8% of the total phospholipids. There are three major *myo*-inositol containing lipids: phosphatidylinositol (PtdIns), phosphatidyl-inositol (4) phosphate [PtdIns(4)P], and PtdIns(4,5)P$_2$. PtdIns accounts for more than 80% of the total phosphatidylinositols (8, 9).

The pathway for biosynthesis of PtdIns in the endoplasmic reticulum was elaborated in the early 1960s (10, 11). PtdIns(4)P and PtdIns(4,5)P$_2$ are synthesized sequentially by PtdIns 4-kinase and PtdIns(4)P 5-kinase (12). PtdIns 4-kinase is membrane-associated in most tissues (13–16), and has been purified to homogeneity from bovine uterus (17), A431 cells (18), porcine liver (19), and rat brain (20). The properties of the purified enzymes are very similar.

PtdIns(4)P 5-kinase has been found both in the soluble and particulate fractions of cell homogenates (15, 21, 22). A 53-kDa form of this enzyme has been purified from human erythrocyte membranes (23). There appear to be two distinct forms of PtdIns(4)P 5-kinase in erythrocytes (24). Type II PtdIns(4)P 5-kinase is distributed in both membrane and cytosol and has a molecular weight of 53,000. A second form, Type I PtdIns(4)P 5-kinase, is found only in the membrane fraction of erythrocytes and though not yet purified to homogeneity, appears distinct from the previously characterized 53-kDa Type II enzyme. Antibodies against the cytosolic Type II enzyme neither inhibit the activity of Type I enzyme nor detect Type I enzyme by immunoblotting.

The reason for multiple isoforms and the physiologic functions of each remain to be determined. Functional characterization of these kinases shows that the Type I kinase has a lower K_m for PtdIns(4)P than does the Type II enzyme. Additionally the Type I PtdIns(4)P 5-kinase phosphorylates PtdIns(4)P in isolated erythrocyte membranes, whereas the Type II PtdIns(4)P 5-kinase has no activity towards membrane PtdIns(4)P.

PHOSPHOLIPASE C

PtdIns-specific PLC enzymes catalyze the hydrolysis of phosphatidylinositols to produce messenger molecules. These enzymes hydrolyze PtdIns, PtdIns(4)P, and PtdIns(4,5)P$_2$ to yield six water-soluble inositol phosphates corresponding to cyclic 1:2 phosphates and 1 phosphates of inositol mono, bis, and tris phosphates (Figure 1). There are multiple PLC enzymes in mammalian tissues as deduced from direct protein isolation and molecular cloning studies (25, 26). Nine isoforms of PLC identified to date have been categorized into four groups, designated α, β, γ, and δ (6).

None of the PLC isoforms contains a membrane-spanning sequence with the possible exception of PLCα. The phospholipid substrates for these enzymes are all in membrane bilayers; therefore PLC must bind to membranes before hydrolyzing phosphatidylinositols. Reversible membrane binding of PLC is an attractive hypothesis to explain changes in PtdIns turnover upon stimulation of cells by agonists, although such has not been demonstrated experimentally.

ACTIVATION OF PHOSPHOLIPASE C BY $G_\alpha Q$

Numerous studies have demonstrated that GTP and its nonhydrolyzable analogues stimulate PtdIns turnover in permeabilized cells, crude membrane fractions, or partially purified enzyme preparations (28). This has led to speculation that a guanine nucleotide–binding protein (G protein) is involved in coupling between receptors and PLC.

Recently a new class of G protein, designated Gq, that activates the PLCβ1 isoform of PLC has been discovered. Pang & Sternweis (29) isolated $G_\alpha q$ from rat brain by a novel strategy utilizing affinity chromatography on columns containing $\beta\gamma$ subunits. Detergent extracts of brain membranes were allowed to bind to the affinity column in the presence of GDP. When the extract was passed very slowly over the column, the α subunits of the G proteins dissociated from endogenous $\beta\gamma$ and bound to the matrix $\beta\gamma$ subunits. The α subunits of the G proteins were then eluted from the column by

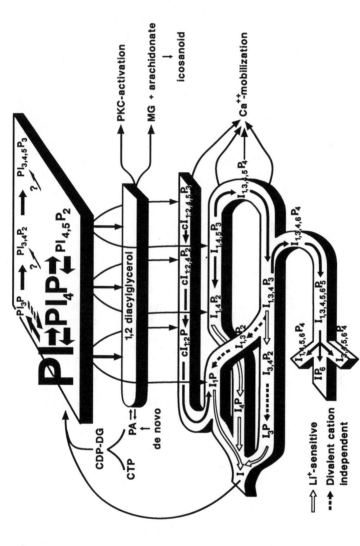

Figure 1 Pathway for inositol phosphate metabolism. In the top level, PI indicates phosphatidylinositol; PI$_4$P, phosphatidylinositol 4-phosphate; PI$_3$P, phosphatidylinositol 3-phosphate; PI$_{3,4}$P$_2$, phosphatidylinositol 3,4-bisphosphate; PI$_{3,4,5}$P$_3$, phosphatidylinositol 3,4,5-trisphosphate; and PI$_{4,5}$P$_2$, phosphatidylinositol 4,5-bisphosphate. On the bottom level, I, inositol; P, phosphate. The numbers preceding P refer to positions of phosphates on the inositol ring, and those following P, the number of phosphate groups. PA, phosphatidic acid; CDP-DG, cytidine diphosphate diacylglycerol; PKC, protein kinase C; MG, 2 monoacylglycerol.

addition of the nonhydrolyzable GTP analog, GTPγS. G$_\alpha$q remained bound to the column under these conditions because it exchanges GDP for GTPγS very slowly, but eluted with aluminum fluoride. Aluminum fluoride binds to GDP on G$_\alpha$q and thereby mimics GTP, causing dissociation of G$_\alpha$q from the column. Through use of this method, two novel α subunits were obtained, G$_\alpha$q and G$_\alpha$11. Limited sequence information indicated that these proteins correspond to two G protein α subunits that were cloned by Strathmann & Simon (30) using a strategy based on the polymerase chain reaction (PCR). Taylor et al have also isolated G$_\alpha$q from bovine liver by a different strategy (31). These workers treated liver membranes with GTPγS overnight prior to solubilization with detergents and in this way identified a fraction that could be separated from PLC and that stimulated PLC activity. The isolated G protein was shown to be Gq by immunoblotting with antipeptide antibodies specific for Gq.

The G$_\alpha$q subunit specifically activates PLCβ1 and has no stimulatory effect on other isoforms of PLC tested, including PLCγ and PLCδ (32). Taylor et al demonstrated a 30-fold stimulation of PLCβ1 by G$_\alpha$q from approximately 10 μmol/min/mg protein to almost 300 μmol/min/mg protein. The activation was independent of calcium ion concentration. In similar studies, Smrcka et al (33) found that a preparation containing PLCβ1 was markedly stimulated by Gq. However, in this case Gq appeared to lower the affinity of the enzyme for calcium ions. At 0.1 μM free calcium, marked stimulation occurred ($\sim$30-fold), whereas at 20 μM calcium only a threefold stimulation occurred. These studies identify at least one isoform of PLC that is activated by a traditional G protein mechanism. Whether other PLC isoforms are similarly activated by other types of G proteins remains to be elucidated.

ACTIVATION OF PLCγ BY TYROSINE PHOSPHORYLATION

Whereas the direct activation PLCβ_1 by Gq-type G proteins has been convincingly demonstrated, the mode of activation of PLCγ by tyrosine protein kinase receptors is not clear. PLCγ has been shown to be phosphorylated on tyrosine residues in cells responding to platelet-derived growth factor (PDGF) and epidermal growth factor (EGF) (34–39). In cells containing EGF receptors and PLC-β, -δ, and -γ, only PLCγ is phosphorylated in response to EGF stimulation of intact cells (34). Although it is clear that PLCγ is a substrate for receptor tyrosine protein kinases, an effect of the phosphorylation on the activity of the enzyme has not been demonstrated. In fact, the magnitude of phosphorylation of PLCγ on tyrosine is variable. For example, in A431 cells, which have large numbers of EGF receptors, up to 50% of PLCγ may be phosphorylated (40); in PDGF-stimulated Balb 3T3 cells less than 5% of PLCγ complexes with receptors and most is not phosphorylated, and much of the tyrosine-phosphorylated PLCγ is found in the cytosol (41). Kim et al (42) have demonstrated that EGF receptor kinase phosphorylates PLCγ stoichiometrically in vitro on tyrosines 472, 771, 783, and 1254. However, phosphorylation had no effect on PLC activity as measured using PtdIns(4,5)P$_2$ as substrate at a variety of calcium ion concentrations. In contrast, Nishibe et al (43) immunoprecipitated PLCγ from homogenates of A431 cells, phosphorylated it in vitro with EGF receptor kinase, and found approximately threefold increased PLC activity in the immunoprecipitates. They showed that treatment of these precipitates with a protein tyrosine phosphatase (T-cell phosphatase) both decreased tyrosine phosphate and reduced PLC activity by 50%. However, Goldschmidt-Clermont et al (44) found that PLCγ phosphorylated by the EGF receptor had only 25% increased enzyme activity using PtdIns(4,5)P$_2$ as substrate. Factors present in the immunoprecipitates (43) may account for the differences in the observed results.

Kim et al (45) have transfected mutant forms of PLCγ with substitution of phenylalanine for tyrosine into NIH 3T3 cells and studied the effects on PLC activity measured in cell homogenates versus inositol phosphate production in response to stimulation of intact cells with PDGF. Thus the former monitors the effects of tyrosine phosphorylation on enzyme activity, whereas the latter measures the effect of tyrosine phosphorylation on the coupling of phosphatidylinositol turnover to PDGF stimulation in intact cells. When either tyrosine 783 or 1254 was mutated to phenylalanine, cellular phosphatidylinositol turnover in response to PDGF was either totally ablated (Tyr783) or attenuated (Tyr1254). Homogenates from both of these mutants expressed the same PLC activity in vitro as those from cells transfected with native PLCγ. This study shows that phosphorylation of PLCγ is necessary for increased phosphatidylinositol turnover in cells stimulated with PDGF. The authors (45) speculate that tyrosine phosphorylation may reverse some inhibitory factor not apparent in enzyme assay of homogenates.

One cytosolic protein that may affect PLCγ activity is the actin-binding protein profilin. Goldschmidt-Clermont (44) showed that profilin binds PtdIns(4,5)P$_2$, inhibiting the ability of the lipid to be hydrolyzed by unphosphorylated PLCγ. In the presence of 50 μM profilin, PtdIns(4,5)P$_2$ hydrolysis by PLCγ was reduced fivefold. However, when PLCγ was phosphorylated by the EGF receptor, no inhibition was observed. Interestingly, profilin did not inhibit the action of PLCβ on PtdIns(4,5)P$_2$. One problem with this study is that the assay conditions used involved substrate vesicles containing inhibitory phospholipids, including phosphatidylserine and/or phosphatidylcholine. Thus, PLC activity in the absence of profilin was only $\sim$0.1% of that obtained under optimal conditions.

Another compound that may affect PLCγ activity is glucosphingolipid. Shayman et al (46) have shown that reduction of cellular glucosphingolipids with an inhibitor of glucosylceramide synthatase, threo-1-phenyl-2-decanoyl-amino-3-morpholino-1-propanol, increases phosphatidylinositol turnover in response to bradykinin or GTγS in MDCK cells. However, it is not known which PLC isozymes are present in MDCK cells.

INOSITOL POLYPHOSPHATE 5-PHOSPHATASE

There are multiple isoforms of inositol polyphosphate 5-phosphatase enzymes distributed between soluble and particulate fractions of various tissues. The 5-phosphatase acts on three substrates (Figure 1), namely Ins(1,4,5)P$_3$, Ins-(1,3,4,5)P$_4$ and cIns(1:2,4,5)P$_3$ (3).

Two forms of cytosolic 5-phosphatase have been isolated from human platelets (47, 48). Type I enzyme has an apparent molecular weight of 45,000 and requires Mg^{2+}. The K_m of Type I 5-phosphatase for Ins(1,3,4,5)P$_4$ is 0.5 μM, much lower than that for Ins(1,4,5)P$_3$ (K_m = 7.5 μM). The V_{max} for Ins(1,3,4,5)P$_4$ is only 2% of that obtained using Ins(1,4,5)P$_3$ as substrate (49). Consistent with these findings, low concentrations of Ins(1,3,4,5)P$_4$ (1 μM) inhibit the breakdown of either Ins(1,4,5)P$_3$ or cIns(1:2,4,5)P$_3$ in homogenates. This implies that cellular levels of InsP$_3$ metabolites may be sustained under conditions where Ins(1,3,4,5)P$_4$ is formed. Type I 5-phosphatase from human platelets is phosphorylated by protein kinase C (50), and this phosphorylation is associated with increased 5-phosphatase activity (50–52). Thus, protein kinase C action tends to reduce InsP$_3$ levels and terminate signalling.

Type II 5-phosphatase from human platelets hydrolyzes the same substrates as the Type I enzyme, although the kinetic parameters are different (K_m values for Ins(1,4,5)P$_3$ = 24 μM, and for Ins(1,3,4,5)P$_4$ = 7.5 μM) (48). The Type II enzyme has a molecular weight of 75,000 and is not phosphorylated by protein kinase C. Recently, a cDNA encoding human platelet Type II 5-phosphatase has been isolated by screening a placental λgt11 cDNA library for β-galactosidase fusion proteins that bind to Ins(1,3,4,5)P$_4$ (53). The sequences derived from the expression clone were used to screen human erythroleukemia and megakaryocyte cell cDNA libraries. The composite cDNA isolated consists of 2381 bp and predicts a protein that includes the amino-terminal amino acid sequence (19 residues) obtained from the platelet 75-kDa Type II 5-phosphatase. The cDNA predicts that the mature enzyme contains 635 amino acids (M_r = 72,891). The recombinant protein expressed in Cos-7 cells has the same intrinsic 5-phosphatase activity as the platelet 5-phosphatase as determined by immunoblotting.

All of the 5' cDNA sequences isolated to date maintain an open reading frame 5' to the predicted amino-terminal sequence of the platelet Type II enzyme. Thus it is possible that the platelet 5-phosphatase is formed by proteolytic processing of a larger precursor. The full length of the transcript based on Northern blotting is 4.4 kilobases. Antibodies directed against recombinant Type II 5-phosphatase immunoprecipitate Type II enzyme from homogenates of platelets and do not cross-react with Type I enzyme, further supporting the idea that two distinct isoenzymes are present.

There is no sequence homology between Type II 5-phosphatase and other cloned proteins of the phosphatidylinositol pathway, including PLC, Ins-$(1,4,5)P_3$ 3-kinase, inositol monophosphate phosphatase, inositol cyclic 1:2 phosphate 2-phosphohydrolase, inositol polyphosphate 1-phosphatase, or the Ins$(1,4,5)P_3$-binding protein. In addition no homologies were found when comparisons were made to sequences in the Genbank data base. Sequence analysis of the predicted protein identified several potential phosphorylation consensus peptide sequences (54) for cGMP-dependent protein kinase, pro-line-dependent kinase, casein kinase I, casein kinase II, mammary gland casein kinase, and glycogen synthetase kinase-3. Consistent with the lack of phosphorylation of Type II 5-phosphatase by protein kinase C in vitro, there are no putative protein kinase C phosphorylation sites.

5-phosphatase enzymes have been partially characterized in several other tissues. The characteristics of these enzymes suggest that additional isoforms of 5-phosphatase are very likely present. Thus molecular weights in excess of 100,000 have been noted as well as a near inability to hydrolyze Ins-$(1,3,4,5)P_4$ by some isoforms. In particular, two distinct 5-phosphatase enzymes have been identified in rat brain (55, 56) and bovine brain (57). 5-phosphatase activities have also been characterized in extracts from macrophages (58), pancreas (59), erythrocytes (60), and liver (61–63). In hepatocytes, it appears that the particulate 5-phosphatase may be assymetrically localized between canilicular and sinusoidal membranes (64).

INS(1,4,5)P_3 3-KINASE

Another enzyme that utilizes Ins$(1,4,5)P_3$ as a substrate is Ins$(1,4,5)P_3$ 3-kinase, a widely distributed soluble enzyme that converts Ins$(1,4,5)P_3$ to Ins$(1,3,4,5)P_4$ in the presence of Mg^{2+} and ATP (7). The 3-kinase enzyme phosphorylates Ins$(1,4,5)P_3$ but not cIns$(1:2,4,5)P_3$ (49, 65). The Ins-$(1,4,5)P_3$ 3-kinase has low K_m for Ins$(1,4,5)P_3$ (0.2 to 1.5 μM) (66–69). Thus the 3-kinase may compete effectively with the inositol polyphosphate 5-phosphatases for Ins$(1,4,5)P_3$, since the K_m for the latter enzymes are higher (7–25 μM) (48, 50, 64).

The native enzyme is composed of two catalytic subunits of 53 kDa plus calmodulin. Physiological concentrations of Ca^{2+} stimulate Ins$(1,4,5)P_3$ 3-kinase activity (67–70) via the calmodulin subunit (68, 71–74). Ins$(1,4,5)P_3$ 3-kinase may also be regulated by phosphorylation. Addition of phorbol esters to intact platelets or to a malignant T-cell line causes a twofold increase in 3-kinase activity (52, 75). In hepatocytes, the enzyme is activated by stimulation of protein kinase A and protein kinase C together (76). In contrast, when isolated $Ins(1,4,5)P_3$ 3-kinase is phosphorylated in vitro with protein kinase C, it is inhibited (77, 78). In vitro phosphorylation of the isolated enzyme by cAMP-dependent protein kinase increased the apparent enzyme activity (77). The site of phosphorylation by protein kinase A was serine 109, whereas multiple sites of phosphorylation by protein kinase C were observed including serine 109 and serine 175.

Results of other studies suggest additional mechanisms for regulation of $Ins(1,4,5)P_3$ 3-kinase. Cellular $Ins(1,4,5)P_3$ 3-kinase activity was increased 6- to 8-fold in cytosolic extracts prepared from vSRC-transformed cells (79). Striking developmental changes in the activity of 3-kinase occur in rat brain, where enzyme activity increases 15-fold from birth to 13 weeks of age (80).

$Ins(1,4,5)P_3$ 3-kinase has been isolated from rat and bovine brain (69, 81, 82). The enzyme is susceptible to proteolysis by calpain, which apparently accounts for the finding of multiple molecular-weight forms ranging from 32 to 53 kDa. A cDNA encoding $Ins(1,4,5)P_3$ 3-kinase has been isolated from a rat liver cDNA library and predicts a protein of 449 amino acids with a molecular weight of 49,853 (83). The protein has no sequence relationship to other enzymes of inositol phosphate metabolism. There are four short sequences from different regions of the molecule that are 20 to 50% identical to sequences in the $Ins(1,4,5)P_3$-binding protein.

NIH 3T3 cells and CCL39 hamster fibroblasts stably transfected with the 3-kinase cDNA overexpress $InsP_3$ 3-kinase 16- to 18-fold (84). These cells display lower levels of $Ins(1,4,5)P_3$ and higher levels of $Ins(1,3,4,5)P_4$ after stimulation with thrombin, PDGF, or bombesin, consistent with increased kinase activity. Blunted calcium mobilization was observed, presumably due to lower levels of $Ins(1,4,5)P_3$. The previously proposed role for $Ins(1,3,4,5)P_4$ in calcium mobilization (85) is not supported by this study, since in the 3-kinase-transfected cells the elevated $Ins(1,3,4,5)P_4$ levels had no effect in promoting calcium ion mobilization. Further study of the physiology of these cells will be required to uncover a function for $Ins(1,3,4,5)P_4$ or other "downstream" metabolites elevated in these cells.

INOSITOL POLYPHOSPHATE 1-PHOSPHATASE

Inhorn et al (86) discovered a phosphatase in calf brain homogenates that hydrolyzes $Ins(1,3,4)P_3$ and $Ins(1,4)P_2$ to $Ins(3,4)P_2$ and $Ins(4)P$, respectively. The enzyme was designated inositol polyphosphate 1-phosphatase, although it does not hydrolyze other 1-phosphate-containing inositol polyphosphates. Subsequently, others have found this enzyme in brain and liver (87). The enzyme has been isolated from calf brain and is a monomeric protein of 44 kDa (88, 89). The K_m for $Ins(1,4)P_2$ is 4 to 5 μM and that for $Ins(1,3,4)P_3$ is 20 μM. The enzyme requires magnesium ions and is inhibited at physiological calcium ion concentrations (88, 90). Rabbit polyclonal anti-sera against 1-phosphatase inhibits hydrolysis (greater than 95%) of Ins-$(1,4)P_2$ in extracts of bovine tissues, suggesting that 1-phosphatase accounts for the majority of $Ins(1,4)P_2$ hydrolysis (88). Lithium ions inhibit both $Ins(1,3,4)P_3$ and $Ins(1,4)P_2$ hydrolysis uncompetitively with a K_i of 0.3 mM and 6.0 mM, respectively (88, 90). The differences in K_i for lithium are consistent with the uncompetitive mode of inhibition, i.e. the inhibitor reacts only with the enzyme-substrate complex (88–91). The inhibitory potency of lithium for $Ins(1,3,4)P_3$ hydrolysis suggests that inhibition at this step may account for the pharmacologic action of lithium ions used to treat psychiatric disorders, since lithium levels in patients are approximately 1 mM (92).

Bovine brain inositol polyphosphate 1-phosphatase has been digested with cyanogen bromide, and trypsin and peptide fragments were isolated and sequenced (93). The sequence was used to design degenerate oligonucleotide primers that were used to isolate a 1572-bp cDNA with an open reading frame of 400 amino acids, corresponding to a predicted molecular weight of 43,980. The predicted 1-phosphatase amino acid sequence is not similar to any other sequences in Genbank with the exception of slight homology to inositol monophosphate phosphatase (93). In one region of the molecule, 9 of 20 amino acids are identical and in another segment 6 of 15 amino acids are identical. It is of interest that these similar regions are also conserved in three proteins found in *Escherichia coli*. These proteins have pleiotropic effects when mutated but have no other known functions (94). It is possible that they are signalling molecules of *E. coli*.

The cDNA for the 1-phosphatase has been expressed in *E. coli*. The recombinant protein has the same specific activity, affinity for substrates, and inhibition by lithium ions as native inositol polyphosphate 1-phosphatase. These results suggest that the activity of the native protein does not require posttranslational modifications.

The cDNA for bovine brain inositol polyphosphate 1-phosphatase was used to screen a human umbilical vein endothelial cell cDNA library to obtain a 1.75-kilobase cDNA encoding human inositol polyphosphate 1-phosphatase (J. D. York, P. W. Majerus, unpublished observations). The predicted amino acid sequence of the human 1-phosphatase is 82% identical to that of the bovine brain clone.

INOSITOL POLYPHOSPHATE 4-PHOSPHATASE

Inositol polyphosphate 4-phosphatase converts $Ins(1,3,4)P_3$ to $Ins(1,3)P_2$ and $Ins(3,4)P_2$ to $Ins(3)P$. The enzyme does not require metal ions, is not inhibited by Li^+ (95), and does not hydrolyze the 4-phosphate from other inositol polyphosphates (90, 95, 96). Bansal et al have purified this enzyme 3400-fold from calf brain–soluble extract (97). The isolated enzyme has an apparent molecular mass of 110 kDa as determined by gel filtration. On SDS polyacrylamide gel electrophoresis the protein migrates at 105 kDa, suggesting that it is monomeric.

The inositol polyphosphate 4-phosphatase has apparent K_m values of 40 and 25 μM for Ins(1,3,4)P$_3$ and Ins(3,4)P$_2$, respectively. The maximal velocities for these two substrates are 15 to 20 μM of product/minute/mg protein. Ins(1,3,4)P$_3$ is a competitive inhibitor of Ins(3,4)P$_2$ hydrolysis with an apparent K_i of 27 μM, implying that the same active site is involved in hydrolysis of both substrates. The ratio of inositol polyphosphate 1-phosphatase to inositol polyphosphate 4-phosphatase activity in hydrolyzing their common substrate Ins(1,3,4)P$_3$ varies among tissues. Only 5 to 20% of Ins(1,3,4)P$_3$ is utilized by inositol polyphosphate 4-phosphatase in various tissues, except in brain, where approximately 70% is metabolized by 4-phosphatase. These observations suggest that Ins(1,3)P$_2$ or a further metabolite plays a role in neuronal function.

INOSITOL POLYPHOSPHATE 3-PHOSPHATASE

Inositol polyphosphate 3-phosphatase catalyzes the hydrolysis of the 3-position phosphate bond of inositol 1,3-bisphosphate [Ins(1,3)P$_2$] to form inositol 1-phosphate and inorganic phosphate (95). Two isoforms of this enzyme designated Types I and II have been isolated from rat brain (98). The Type I 3-phosphatase consists of a protein doublet that migrates with a molecular mass of 65,000 upon SDS polyacrylamide gel electrophoresis. The molecular weight of this isoform upon size exclusion chromatography is 110,000, suggesting that the native enzyme is a dimer. The Type II enzyme consists of equal amounts of a molecular weight 65,000 doublet and a molecular weight 78,000 band upon SDS polyacrylamide gel electrophoresis. This isoform has a molecular weight of 147,000 upon size exclusion chromatography, indicating that it is a heterodimer. The 65,000-Da subunits of the two forms of 3-phosphatase appear to be the same based on comigration on SDS polyacrylamide gels and peptide maps generated with *Staphylococcus aureus* protease V8 and trypsin. The peptide map of the 78,000-Da subunit was different from that of the 65,000-Da subunits. The Type II 3-phosphatase catalyzes the hydrolysis of Ins(1,3)P$_2$ with a catalytic efficiency 1/19th of that measured for the Type I enzyme, suggesting that the presence of the 78-kDa subunit decreases activity. Both isoforms of 3-phosphatase enzyme also hydrolyze phosphatidylinositol 3-phosphate [PtdIns(3)P] to form phosphatidylinositol and inorganic phosphate. The Type II enzyme is more active than Type I enzyme in hydrolyzing Ptd Ins(3)P. Thus far this is the only example in inositol phosphate metabolism of an enzyme that hydrolyzes both a lipid and corresponding water-soluble substrate.

INOSITOL MONOPHOSPHATASE

The other enzyme of inositol phosphate metabolism that is inhibited by Li^+ in vitro is inositol monophosphatase. This enzyme was initially designated as inositol 1-phosphatase, but since it hydrolyzes phosphate groups of all inositol monophosphates with the exception of inositol 2-phosphate, its name has been changed to inositol monophosphate phosphatase (90, 99). Inositol monophosphate phosphatase requires Mg^{2+} for activity and other divalent ions, i.e. Ca^{2+} and Mn^{2+}, are competitive inhibitors (100). Hallcher & Sherman (100) demonstrated that Li^+ inhibits inositol monophosphate phosphatase uncompetitively with an apparent K_i of 0.8 μM; thus this enzyme is another potential target for the action of lithium as a therapeutic agent. Inositol monophosphate phosphatase has been purified to homogeneity from bovine and rat brain (101, 102). The enzyme ($M_r = 55,000$) is a dimer of identical subunits (100–102). A cDNA encoding inositol monophosphatase was recently isolated and predicts a protein of 277 amino acids. The recombinant protein has been expressed in *E. coli* and found to have the same activity and inhibition by Li^+ as native monophosphatase (103). The protein is not homologous to any other proteins in the Genbank data base with the exception of inositol polyphosphate 1-phosphatase and the bacterial proteins mentioned above. Modification of monophosphatase by the arginine-specific reagent phenylglyoxal results in enzyme inactivation, suggesting that an arginine is in the active site of this enzyme. One mole of arginine is modified per mole of protein in the process of enzyme inactivation (104).

INOSITOL (1:2 CYCLIC) PHOSPHATE 2-PHOSPHOHYDROLASE AND INOSITOL CYCLIC PHOSPHATES

The metabolism of inositol cyclic phosphates is distinct from that of the more widely studied noncyclic counterparts. They are produced by PLC and subsequently metabolized by distinct enzymes (Figure 1). The existence of inositol cyclic phosphates in vivo has been documented in a wide variety of cells and tissues (105–115). It is not clear what determines the ratio of cyclic versus noncyclic phosphates produced upon stimulation of cells with various agonists. In general, prolonged stimulation of cells or tissues tends to produce higher levels of cyclic phosphates, in part because these metabolites are metabolized slowly (109, 112). For example, inositol polyphosphate 5-phosphatase hydrolyzes $cIns(1:2,4,5)P_3$ very poorly compared to $Ins(1,4,5)P_3$ (100- to 200-fold less), which may explain the accumulation of cyclic trisphosphate in cells (116). A 4-phosphatase in bovine brain hydrolyzes $cIns(1:2,4)P_2$ and requires magnesium ions (V. S. Bansal, P. W. Majerus, unpublished observations). This phosphatase differs from inositol polyphosphate 4-phosphatase.

Inositol cyclic (1:2) phosphate 2-phosphohydrolase (cyclic hydrolase) is the only enzyme known to hydrolyze the cyclic phosphate bond, and this enzyme only utilizes inositol cyclic (1:2) phosphate [cIns(1:2)P] as a substrate. Since all inositol cyclic phosphates are finally metabolized by this enzyme, it is likely that it regulates cellular levels of these molecules. Both Zn^{2+} and Ins(2)P are potent inhibitors, while Mn^{2+} and acidic phospholipids stimulate the enzyme (117–119). Cyclic hydrolase also catalyzes the hydrolysis of a second substrate glycerophosphoinositol to form glycerol and inositol 1-phosphate (118). As in the case of hydrolysis of inositol cyclic phosphates (117), the glycerophosphoryl derivatives with additional monoester phosphate groups in the 4 or 4 and 5 positions are not substrates for cyclic hydrolase.

Cyclic hydrolase has been isolated from human placenta and characterized (117, 118) and shown to be identical to lipocortin III (119). Lipocortin III is one of eight related proteins that bind lipids and calcium ions (120–122) but have no previously known physiological functions (123, 124), and have recently been renamed *annexins*. In order to investigate possible functions for inositol cyclic phosphates, Ross et al have transfected a cDNA encoding cyclic hydrolase into 3T3 cells (125). 3T3 cells that overexpress cyclic hydrolase (3–10-fold) have decreased cellular levels of cIns(1:2)P, indicating that alterations in the cellular levels of the enzyme directly alter cellular cIns(1:2)P levels. Cells with increased cyclic hydrolase activity grow to a lower density at confluence (4-fold) than untransfected cells or cells transfected with vector alone. Similar results were obtained when the growth properties of several different cell types with varying endogenous cyclic hydrolase activity (100-fold range) were examined. Those cells with high levels of cyclic hydrolase had correspondingly low cIns(1:2)P levels and grew to a lower density at confluence than cells expressing low levels of enzyme (4-fold range) (125). Cyclic hydrolase levels have also been found to be lower within a given cell line during active cell growth than when cells are confluent. These studies suggest that cyclic hydrolase is antiproliferative and could be an example of an antioncogene. Whether cyclic hydrolase can reverse or prevent the transformed phenotype remains to be determined.

While cyclic hydrolase is important in determining cellular cIns(1:2)P levels, another way to vary this metabolite is through alterations in PLC activity. Camilli et al have shown that an essential virulence factor in the intracellular pathogen *Listeria monocytogenes* is a PLC enzyme (126). This enzyme produces only cIns(1:2)P and diacylglycerol from PtdIns. Thus it is possible that overproduction of either or both of these metabolites is required to produce disease.

OTHER ENZYMES OF INOSITOL PHOSPHATE METABOLISM

Several other enzymes involved in $InsP_4$ metabolism have been partially characterized. $Ins(1,3,4,6)P_4$ is formed by a soluble 6-kinase that metabolizes $Ins(1,3,4)P_3$ (127–130). The 6-kinase is widely distributed and is not affected by either calcium or lithium ions (129). A 3-phosphatase that converts $Ins(1,3,4,5)P_4$ to $Ins(1,4,5)P_3$ has been described in cell extracts (131–135). It has been suggested that this enzyme may actually be involved in $InsP_5$ and $InsP_6$ metabolism, since these substances are potent inhibitors of this 3-phosphatase (64).

INOSITOL PENTAPHOSPHATE AND HEXAPHOSPHATE METABOLISM

The metabolism of inositol phosphates containing 5 and 6 phosphate esters is not fully understood. Further, it appears that there are differences among organisms in the particular isomers of $InsP_5$ present. There are six possible $InsP_5$ isomers, most of which have been found in various plant species and *Dictyostelium* (136). In animal cells, the major $InsP_5$ isomer is Ins-$(1,3,4,5,6)P_5$, although other isomers including traces of $(D,L)Ins(1,2,4,5,6)P_5$ and $(D,L)Ins(1,2,3,4,5)P_5$ have been found in HL60 cells and NG115 401L-C3 cells (136, 137). In animal cells, the major pathway for $InsP_5$ synthesis appears to be that outlined in Figure 1. Thus, $Ins(1,3,4,5)P_4$ is isomerized to $Ins(1,3,4,6)P_4$ via sequential action of inositol polyphosphate 5-phosphatase and $Ins(1,3,4)P_3$ 6-kinase. $Ins(1,3,4,6)P_4$ is converted to Ins-$(1,3,4,5,6)P_5$ by a 5-kinase (128, 130, 139). $Ins(1,3,4,5,6)P_5$ is further metabolized to three compounds: $InsP_6$, $Ins(3,4,5,6)P_4$, and $Ins(1,4,5,6)P_4$ (the latter two compounds are an enantiomeric pair and thus not separable by most conventional means) (127–130, 138, 140).

The mass of $InsP_5$ and $InsP_6$ in HL60 cells has been measured as 35 μM and 50 μM, respectively (137). These levels are approximately 100-fold greater than those of other inositol phosphate metabolites. In HL60 cells stimulated with formyl-methionyl-leucyl-phenylalanine (FMLP), the levels of $InsP_5$ and $InsP_6$ rise within two minutes to 50 μM and 60 μM, respectively. Thus, these compounds appear to undergo accelerated synthesis in response to agonists. In studies of $InsP_5$ and $InsP_6$ turnover, as measured by changes in $[^3H]$ inositol incorporation into these compounds in response to agonists, conflicting results have been obtained, suggesting increased, decreased, or no change in turnover. Much of the variation in these results may be accounted for by the fact that isotope incorporation can reflect either changes in mass, specific activity, or a combination of the two. Many investigators assume that labelling of tissue-culture cells with [^{3}H]inositol for 48 to 72 hours yields steady-state labelling wherein all metabolites of inositol have the same specific activity. This appears not to be the case in some instances and is never proven without mass measurements.

In HL60 cells labelled with [^{3}H]Inositol for 48 hours (137), the specific activity of Ins(1,3,4,5)P$_4$, Ins(1,3,4,6)P$_4$, and InsP$_5$ was similar, yet upon stimulation with FMLP there was a rapid rise in specific activity of these compounds, thus indicating that they were not in equilibrium with precursor inositol phosphate metabolites. The specific activity of (D,L)Ins(3,4,5,6)P$_4$ was approximately half that of InsP$_5$ in unstimulated HL60 cells, implying that this metabolite turns over very slowly. Upon stimulation with FMLP, labelling of (D,L)Ins(3,4,5,6)P$_4$ increased rapidly, suggesting accelerated turnover of this metabolite. Similar results were found in bombesin-stimulated AR4-2J cells (140).

The pathway for metabolism of these compounds in avian erythrocytes appears to be quite different. Stephens & Downes have analyzed InsP$_4$ and InsP$_5$ metabolism by labelling erythrocytes with [^{3}H]Inositol and ^{32}PO$_4$ (139). They analyzed the incorporation of ^{32}PO$_4$ into various monoester phosphates of Ins(1,3,4,5,6)P$_5$ and Ins(3,4,5,6)P$_4$ in a protocol similar to that used by Cunningham et al (141) to determine pathways of 3-phosphate containing inositol phospholipid metabolism (see below). In this analysis the most highly labelled phosphate is the last added to the molecule. They found that the 1-position phosphate of Ins(1,3,4,5,6)P$_5$ is most highly labelled, suggesting that its precursor is Ins(3,4,5,6)P$_4$ rather than Ins(1,3,4,6)P$_4$ as shown in Figure 1. In fact, the overall scheme they propose is Ins(4,6)P$_2$ → Ins(3,4,6)P$_3$ → Ins(3,4,5,6)P$_4$ → Ins(1,3,4,5,6)P$_5$. This pathway predominates despite the fact that avian erythrocytes have the enzymes to form both Ins(1,3,4,5)P$_4$ and Ins(1,3,4,6)P$_4$. It will be interesting to analyze the InsP$_4$ and InsP$_5$ isomers of animal cells by this same strategy.

The functions of InsP$_4$, InsP$_5$, and InsP$_6$ in signalling remain a mystery. InsP$_5$ and InsP$_6$ have been proposed to be extracellular agonists that stimulate neuronal excitability and reduce heart rate and blood pressure (142). It is unclear, however, how these compounds could exit cells. InsP$_5$ does reduce the affinity of hemoglobin for oxygen in erythrocytes of birds (143). Proteins that bind InsP$_4$, InsP$_5$, and InsP$_6$ have been identified recently, as described below.

INOSITOL PHOSPHATE–BINDING PROTEINS

The action of Ins(1,4,5)P$_3$ in calcium ion mobilization has been validated through identification and isolation of an Ins(1,4,5)P$_3$-binding protein that serves as a calcium channel across various membranes (144, 145). The isolated Ins(1,4,5)P$_3$-binding protein has a subunit polypeptide chain of 260 kDa upon SDS polyacrylamide gel electrophoresis, but a molecular weight of approximately 1 million upon size exclusion chromatography, thus implying that the native molecule is a tetramer (144). Cross-linking studies of the isolated protein also suggest a tetramer (146). The binding protein had a K_d for Ins(1,4,5)P$_3$ of ~10 nM. Other inositol phosphates displace bound Ins-(1,4,5)P$_3$ poorly (1000-fold higher concentration).

The Ins(1,4,5)P$_3$-binding protein has been reconstituted into lipid vesicles and shown to directly mediate the transfer of calcium across the lipid bilayer (147). Photo-affinity labelling and direct binding studies have shown that the protein contains an ATP-binding site and further that ATP affects the flux of calcium (146, 148).

The distribution of the Ins(1,4,5)P$_3$-binding protein in brain is highly enriched in purkinje cells, where 10 to 100 times greater concentration of this protein is present than in other cells (148, 149). Mikoshiba and coworkers noted that the enrichment in purkinje cells and other properties of the Ins-(1,4,5)P$_3$-binding protein suggested that it was likely to be similar or identical to a protein that they had previously characterized as abundant in purkinje cells and nearly lacking in cerebella of mice with various mutations in which purkinje cells are deficient or absent (150). These workers utilized monoclonal antibodies to isolate a cDNA encoding the Ins(1,4,5)P$_3$-binding protein (151). The full-length transcript is approximately 10 kilobases and encodes a protein of 2749 amino acids with a molecular weight of 313,000. The predicted protein contains an ATP-binding motif and several putative transmembrane domains near the carboxyl terminus of the molecule. The majority of the protein is thought to reside in the cytoplasm of the cell with the carboxyl terminus in the lumen of the endoplasmic reticulum (i.e. outside the cell).

These authors concluded that there might be as many as seven membrane-spanning domains contained within the molecule. Based on studies with monoclonal antibodies, they concluded that the number of membrane-spanning domains was odd and therefore the amino terminus and carboxyl terminus were on opposite sides of the membrane. They and others also noted a striking homology between the Ins(1,4,5)P$_3$-binding protein and the ryanodine receptor (152). The homology is particularly strong in the carboxyl-terminal membrane-spanning domains. One puzzling difference is that the ryanodine receptor has an even number of membrane-spanning domains and the carboxyl terminus is on the cytoplasmic face of the membrane. Both the Ins(1,4,5)P$_3$-binding protein and ryanodine receptor appear to be tetramers of similar structure as determined by electron microscopy, and both are involved in the movement of calcium across cell membranes.

A partial cDNA encoding the purkinje cell Ins(1,4,5)P$_3$-binding protein had been isolated previously by obtaining cDNA clones encoding proteins uniquely expressed in mouse purkinje cells. This was achieved by using a library of normal mouse cerebellar cDNA subtracted with cDNA sequences from the cerebella of mice with mutations causing purkinje cell degeneration (153, 154). It is interesting that the predicted mass of the monomeric Ins-(1,4,5)P_3-binding protein is 313 kDa (plus some carbohydrates since the molecule is a glycoprotein), but by SDS polyacrylamide gel electrophoresis the molecular weight appears to be 260 kDa (155, 156). Recently a cDNA for rat brain Ins(1,4,5)P_3-binding protein was reported (157).

Several different forms of Ins(1,4,5)P_3-binding protein cDNA have been identified (158). Nakagawa et al found five different subtypes in various tissues during mouse brain development that they postulate arise by alternatively spliced exons. They used an S1 nuclease protection assay of RNA from different tissues and from mouse brain at different times after birth. They found that a "short-form" RNA with a deletion of 15 amino acids predominates in adult brain and accounts for 20–80% of the mRNA in other adult tissues. Danoff et al (159) found the same short form in rat brain by a PCR assay, but curiously found the opposite developmental correlation in that the short form was in fetal brain while the long form predominates in adult tissues. The significance of the different forms is unknown, and the lack of similarity of distribution in closely related species may suggest that the variant forms are not functionally important. Alternatively, the PCR assay (159) may not accurately reflect quantities of mRNA in various tissues.

While most of the physiological studies have suggested that calcium release is primarily from endoplasmic reticulum, localization of the Ins(1,4,5)P_3-binding protein by immunohistochemical methods suggests a potentially wider distribution of calcium-mobilizing sites. Ross et al (155), utilizing immunoelectron microscopy of rat cerebellar sections, have shown that the Ins(1,4,5)P_3-binding protein is distributed over much of the endoplasmic reticulum, but is also present over most of the nuclear membrane and in some areas of the cis golgi. This localization has suggested a possible nuclear site for calcium mobilization, which might mediate cellular responses involving cell growth or proliferation. Recently Ins(1,4,5)P_3 was shown to mediate calcium release from isolated nuclei (160, 161).

Evidence for other inositol phosphate–binding proteins has been obtained. A distinct Ins(1,3,4,5)P_4-binding protein ($K_d = 5$ nM) has been solubilized and partially purified from porcine and rat cerebellar membranes (162–165). Donie et al have purified a porcine cerebellar Ins(1,3,4,5)P_4-binding protein approximately 20,000-fold (165). The purified protein has a mass of 42,000 daltons as estimated by SDS polyacrylamide gel electrophoresis. It binds 0.2 mol of Ins(1,3,4,5)P_4/mol of protein. The Ins(1,3,4,5)P_4-binding protein shows an approximately 100 times greater affinity for Ins(1,3,4,5)P_4 than for Ins(1,4,5)P_3, the converse of that observed for the Ins(1,4,5)P_3-binding protein.

Theibert et al (164) recently isolated two Ins(1,3,4,5)P$_4$-binding proteins from rat brain that show similar binding specificities to that found by Donie & Reiser (165); however, the molecular weights were quite different (182,000 and a dimer of 174,000 and 84,000). The specificity of the various InsP$_4$-binding proteins for particular InsP$_4$ isomers has not been defined, and some preparations bind InsP$_5$ quite well.

Specific binding proteins for InsP$_6$ have also been identified (166, 167). InsP$_6$ binds to isolated cerebellar membranes with an apparent K_d of 60 nM. Displacement by InsP$_5$ required approximately 1 μM. Binding sites were found in rat brain membrane fractions from mitochondria, myelin, and synaptosomes and represented at least 1% of membrane protein (167). An apparently different InsP$_6$-binding protein has been purified from rat brain membranes by affinity chromatography on an Ins(1,3,4,5)P$_4$ derivative affinity column (164). It differs in that it binds InsP$_5$ almost as well as InsP$_6$. InsP$_6$-binding sites have also been found in brain membranes by autoradiographic staining of brain sections using [^{3}H]InsP$_6$ (166). It was determined that InsP$_6$-binding sites were particularly enriched in cerebellum, choroid plexis, and in the nucleus of the solitary tract. This distribution mimics that of the regions of brain mediating pharmacologic activities of InsP$_6$ (142). Definition of the functions of these proteins awaits further study.

PHOSPHATIDYLINOSITOL 3-PHOSPHATE PATHWAY

Phosphatidylinositols containing phosphate esters in the 3 position of inositol represent a recently discovered pathway of phosphatidylinositol metabolism (168). These include PtdIns(3)P, phosphatidylinositol 3,4-bisphosphate [PtdIns(3,4)P$_2$], and phosphatidylinositol 3,4,5-trisphosphate [PtdIns(3,4,5)P$_3$] (see Figure 1). PtdIns(3)P is present in cells under all conditions and serves as a precursor for the polyphosphorylated 3-phosphate-containing phosphatidylinositols (141). The last two mentioned compounds are formed transiently in many cells in response to agonists and growth factors (169). Whitman et al (170) first identified PtdIns(3)P in transformed fibroblasts and proposed that the 3-phosphate-containing lipids serve a function in controlling cell proliferation. However, they are also found in normal cells and in nongrowing cells such as neutrophils, platelets, and brain, indicating that they serve a function other than, or in addition to, stimulating cell proliferation (reviewed in reference 26).

The 3-phosphate-containing phosphatidylinositols are present in very small amounts in cells. In fact the mass of these lipids has not been measured in any system. Based on incorporation of trace radiolabelled inositol or phosphate, they represent only 1 to 2% of the levels of the previously known non-3-phosphate-containing phosphatidylinositol polyphosphates (171–177). The 3-phosphate-containing lipids are not hydrolyzed by PLC enzymes (178,

179). Given this lack of reactivity and the small amounts of these compounds, it is unlikely that they serve as precursors of water-soluble messenger molecules. The water-soluble inositol phosphates that would be derived from these lipids via PLC action, namely $Ins(1,3)P_2$, $Ins(1,3,4)P_3$, and $Ins(1,3,4,5)P_4$, are all present in cells and are formed by other reactions as described above. The levels of these metabolites in stimulated cells, based on estimates from radiolabelling, are 10 to 50 times greater than the estimated levels of 3-phosphate-containing phosphatidylinositols. Therefore, it is likely that the 3-phosphate-containing lipids either function as lipids per se or they are converted to other as-yet-unidentified compounds.

Many agonists that evoke classical PtdIns turnover also stimulate the production of 3-phosphate-containing phosphatidylinositols. In quiescent NIH 3T3 cells (169) or unstimulated platelets (175, 176), it is possible to label PtdIns(3)P with [^{3}H] inositol or $^{32}PO_4$, indicating that there is turnover or synthesis of PtdIns(3)P in a basal state. After stimulation of platelets with thrombin, neutrophils with FMLP (174), or NIH 3T3 cells with PDGF, there is a rapid increase in labelling of both $PtdIns(3,4)P_2$ and $PtdIns(3,4,5)P_3$, suggesting that the formation of the polyphosphorylated lipids is the functionally significant event. A question that remains to be addressed is whether all agonists that stimulate PtdIns turnover activate both 3-phosphate- and PLC-mediated inositol lipid turnover.

The pathway for formation of $PtdIns(3,4)P_2$ and $PtdIns(3,4,5)P_3$ is uncertain. Carpenter et al (180) have isolated a PtdIns 3-kinase that phosphorylates PtdIns(4)P and $PtdIns(4,5)P_2$ in vitro. However, labelling of intact cells suggests a different pathway. Cunningham et al (141) labelled human platelets with $^{32}PO_4$ for brief periods and stimulated them with thrombin. The relative specific activities of the phosphate group at each position of the 3-phosphate-containing phosphatidylinositols were determined by degrading the deacylated and deglycerated compounds with specific inositol phosphate phosphatases. As labelling reaches a steady state, all phosphates are equally labelled. Prior to a steady state, the order of phosphate addition is determined by the relative specific activity of each phosphate. In this analysis, the last phosphate added reaches steady state first and therefore is identified by demonstrating that it has the highest specific activity. These experiments clearly indicated that the major pathway in thrombin-stimulated platelets is PtdIns $\rightarrow$ PtdIns(3)P $\rightarrow$ $PtdIns(3,4)P_2$ $\rightarrow$ $PtdIns(3,4,5)P_3$.

This pathway for synthesis was also found in NIH 3T3 cells (181) by determining the specific activity of each phosphate in $PtdIns(3,4,5)P_3$ after brief PDGF stimulation of $^{32}PO_4$-labelled cells. The phosphate in position 5 was most highly labelled, followed by 4, then 3, and finally 1. Further support for this pathway comes from the discovery of a PtdIns(3)P 4-kinase in homogenates of platelets, erythrocytes, and human erythroleukemia cells that forms $PtdIns(3,4)P_2$ from PtdIns(3)P (182). Stephens et al (183) studied the pathway for PtdIns(3,4,5)P$_3$ synthesis in FMLP-stimulated neutrophils. They labelled cells with [^{3}H] inositol and ^{32}PO$_4$ and analyzed the relative specific activity of ^{32}PO$_4$ in each position of PtdIns(3,4,5)P$_3$. They found that the 3-position was most highly labelled and thus suggested that PtdIns(3,4,5)P$_3$ arises from PtdIns(4,5)P$_2$. It is difficult to reconcile this study with those of Cunningham et al (141, 181). The main differences in procedure are that Stephens et al used crude extracts as sources of the phosphatases to degrade the compounds and used a different cell type. Discounting possible errors because of the crude enzymes, it is possible that 3-phosphate-containing lipids may be synthesized by different routes under varied conditions in different cells.

PtdIns 3-kinase has recently been isolated from bovine thymus (184). Two discrete types of 3-kinase enzyme were isolated. Type I is a monomeric protein of 110 kDa, while Type II is a heterodimeric form with subunits of 110 kDa and 85 kDa. Peptide mapping analysis indicated that the 110-kDa proteins were the same, while the 85-kDa protein was different. This suggests that the enzyme catalytic unit is the 110-kDa protein and that the 85-kDa subunit is a regulatory unit. The specific activity of Type I enzyme was fivefold greater than that of the Type II PtdIns 3-kinase, indicating that the monomer is an active type and the heterodimer is less active. The PtdIns(3)P 3-phosphatase enzymes described above are analogous, since they show varied activity in that the Type I enzyme is less active than the heterodimeric Type II enzyme. The Type II PtdIns 3-kinase has also been isolated from rat liver (180).

Recently, a cDNA encoding the 85-kDa regulatory subunit has been obtained by three different laboratories (185–187). The 85-kDa regulatory subunit of PtdIns 3-kinase is phosphorylated on both tyrosine and serine (180). This protein forms complexes with receptors that stimulate PtdIns(3)P synthesis in a manner similar to that described above for PLCγ (188–193). PtdIns 3-kinase forms complexes with wild-type but not a mutant form of middle T antigen. The mutant, which encodes phenylalanine in place of tyrosine at the major site of phosphorylation by pp60^{c-src}, retains the ability to form complexes with *src,* and activates its tyrosine kinase. The incidence and spectrum of tumors induced by cells with this mutant middle T antigen is reduced, implying that tyrosine phosphorylation is needed for recruitment of PtdIns 3-kinase into a complex that is in turn important for oncogenesis (194). The significance of these complexes in activating PtdIns 3-phosphate-containing phospholipid turnover is as yet uncertain.

An additional uncertainty is the messenger role that could be served by the 3-phosphate-containing inositol phospholipids. PtdIns(4,5)P$_2$ has been shown to promote actin polymerization in vitro by causing dissociation of gelsolin-actin complexes and association of profilin-actin complexes (195–198). Profilin inhibits actin polymerization by binding G-actin, while gelsolin severs actin filaments. PtdIns(4,5)P$_2$ binds profilin with high affinity ($K_d < 0.1$ μM) with a stoichiometry of 5 moles PtdIns(4,5)P$_2$/mole profilin (44, 199). PtdIns(4,5)P$_2$ also binds to gelsolin. When platelets are stimulated by thrombin, there is rapid polymerization of actin, causing them to extend pseudopods and spread on surfaces. These events occur at a time when PtdIns(4,5)P$_2$ levels fall transiently and then return to basal levels. This pattern of change is not consistent with the proposed role for PtdIns(4,5)P$_2$ in binding profilin to promote actin polymerization. Dadabay et al (200) found no correlation between changes in PtdIns(4,5,)P$_2$ and actin polymerization in EGF-stimulated A431 cells. In contrast, PtdIns(3,4,5)P$_3$ levels do rise rapidly upon thrombin stimulation, at least as measured by incorporation of ^{32}PO$_4$ (175). Perhaps PtdIns(3,4,5)P$_3$ binds profilin or gelsolin to serve this proposed function. Eberle et al (201) have shown that the time course of actin polymerization and subsequent depolymerization in neutrophils in response to FMLP stimulation parallels the rise and fall of PtdIns(3,4,5)P$_3$ in these same cells. When ligand binding is reversed with an antagonist, F-actin rapidly depolymerizes and ^{32}P-labelled PtdIns(3,4,5)P$_3$ disappears. If these patterns of labelling actually reflect changes in mass, then PtdIns(3,4,5)P$_3$ may control transient actin polymerization. Mass quantities of PtdIns(3,4,5)P$_3$ are needed to determine whether this molecule binds profilin as well or better than PtdIns(4,5)P$_2$. It is possible that proliferative responses to growth factors require changes in actin polymerization and associations between actin and cellular cytoskeletal elements (202). Future studies along the lines of those described for PLC should provide a much clearer picture of the role of the 3-phosphate-containing inositol lipids.

ACKNOWLEDGMENTS

I am very grateful to Theo Ross and Linda Pike for their help in writing this article. I thank Kevin Caldwell, Anne Bennett Jefferson, John York, and Kim Aiken for reading the manuscript and for suggestions of ways to clarify it.

    This research was supported by Grants HL 14147 (Specialized Center for Research in Thrombosis), HL 16634, and Training Grant HL 07088 from the National Institutes of Health and Monsanto.

Literature Cited

1. Berridge, M. J. 1987. *Annu. Rev. Biochem.* 56:59–193
2. Pelech, S. L., Vance, D. E. 1989. *Trends Biochem. Sci.* 14:28–30
3. Majerus, P. W., Connolly, T. M., Bansal, V. S., Inhorn, R. C., Ross, T. S., Lips, D. L. 1988. *J. Biol. Chem.* 263:3051–54
4. Rana, R. S., Hokin, L. E. 1990. *Physiol. Rev.* 70:115–61
5. Berridge, M. J., Irvine, R. F. 1989. *Nature* 341:97–205
6. Putney, J. W., Takemura, H., Hughes, A. R., Horstman, D. A., Thastrup, O. 1989. *FASEB J.* 3:1899–905
7. Irvine, R. F., Moor, R. M., Pollock, W.

K., Smith, P. M., Wreggett, K. A. 1988. *Philos. Trans. R. Soc. London Ser. B* 320:281–98

8. Majerus, P. W., Neufeld, E. J., Wilson, D. B. 1984. *Cell* 37:701–3

9. Majerus, P. W., Connolly, T. M., Deckmyn, H., Ross, T. S., Bross, T. E., et al. 1986. *Science* 234:1519–26

10. Agranoff, B. W., Bradley, R. M., Brady, R. O. 1958. *J. Biol. Chem.* 233:1077–83

11. Paulus, H., Kennedy, E. P. 1960. *J. Biol. Chem.* 235:1303–11

12. Brockerhoff, H., Ballou, C. E. 1962. *J. Biol. Chem.* 237:1764–66

13. Smith, C. D., Wells, W. W. 1983. *J. Biol. Chem.* 258:9368–73

14. Campbell, C. R., Fishman, J. B., Fine, R. E. 1985. *J. Biol. Chem.* 260:10948–51

15. Hawthorne, J. N. 1983. *Biosci. Rep.* 3:887–904

16. Kurosawa, M., Parker, C. W. 1986. *J. Immunol.* 136:616–22

17. Porter, F. D., Li, Y. S., Deuel, T. F. 1988. *J. Biol. Chem.* 263:8989–95

18. Walker, D. H., Dougherty, N., Pike, L. J. 1988. *Biochemistry* 27:6504–11

19. Hou, W. M., Zhang, Z. L., Tai, H. H. 1988. *Biochim. Biophys. Acta* 959:67–75

20. Yamakawa, A., Takenawa, T. 1988. *J. Biol. Chem.* 263:17555–60

21. Lundberg, G. A., Jergil, B., Sundler, R. 1986. *Eur. J. Biochem.* 161:257–62

22. Vandongen, C. J., Kok, J. W., Schrama, L. H., Oestreicher, A. B., Gispen, W. H. 1986. *Biochem. J.* 253:859–64

23. Ling, L. E., Schulz, J. T., Cantley, L. C. 1989. *J. Biol. Chem.* 264:5080–88

24. Bazenet, C. E., Ruano, A. R., Brockman, J. L., Anderson, R. A. 1990. *J. Biol. Chem.* 265:18012–22

25. Bansal, V. S., Majerus, P. W. 1990. *Annu. Rev. Cell Biol.* 6:41–67

26. Majerus, P. W., Ross, T. S., Cunningham, T. W., Caldwell, K. K., Jefferson, A. B., Bansal, V. S. 1990. *Cell* 63:459–65

27. Deleted in proof

28. Fain, J. N., Wallace, M. A., Wojcikiewicz, R. J. H. 1988. *FASEB J.* 2:2569–74

29. Pang, I. H., Sternweis, P. C. 1990. *J. Biol. Chem.* 265:18707–12

30. Strathmann, M., Simon, M. I. 1990. *Proc. Natl. Acad. Sci. USA* 87:9113–17

31. Taylor, S. J., Smith, J. A., Exton, J. H. 1990. *J. Biol. Chem.* 265:17150–56

32. Taylor, S. J., Chae, H. Z., Rhee, S. G., Exton, J. H. 1991. *Nature* 350:516–18

33. Smrcka, A. V., Hepler, J. R., Brown, K. O., Sternweis, P. C. 1990. *Science* 251:804–7

34. Nishibe, S., Wahl, M. I., Rhee, S. G., Carpenter, G. 1989. *J. Biol. Chem.* 264:10335–38

35. Wahl, M. I., Nishibe, S., Suh, P. G., Rhee, S. G., Carpenter, G. 1989. *Proc. Natl. Acad. Sci. USA* 86:1568–72

36. Margolis, B., Rhee, S. G., Felder, S., Mervic, M., Lyall, R., et al. 1989. *Cell* 57:1101–7

37. Downing, J. R., Margolis, B. L., Zilberstein, A., Ashmun, R. A., Ullrich, A., et al. 1989. *EMBO J.* 8:3345–50

38. Meisenhelder, J., Suh, P. G., Rhee, S. G., Hunter, T. 1989. *Cell* 57:1109–22

39. Kumjian, D. A., Wahl, M. I., Rhee, S. G., Daniel, T. O. 1989. *Proc. Natl. Acad. Sci. USA* 86:8232–36

40. Todderud, G., Wahl, M. I., Rhee, S. G., Carpenter, G. 1990. *Science* 249:296–98

41. Kumjian, D. A., Branstein, A., Rhee, S. G., Daniel, T. O. 1991. *J. Biol. Chem.* 266:3973–80

42. Kim, J. W., Sim, S. S., Kim, U. H., Nishibe, S., Wahl, M. I., et al. 1990. *J. Biol. Chem.* 265:3940–43

43. Nishibe, S., Wahl, M. I., Hernandez-Sotomayor, S. M. T., Tonks, N. K., Rhee, S. G., Carpenter, G. 1990. *Science* 250:1253–56

44. Goldschmidt-Clermont, P. J., Kim, J. W., Machesky, L. M., Rhee, S. G., Pollard, T. D. 1991. *Science* 251:1231–33

45. Kim, H. K., Kim, J. W., Zilberstein, A., Margolis, B., Kim, J. G., Schlessinger, J., Rhee, S. G. 1991. *Cell* 65:435–41

46. Shayman, J. A., Mahdiyoun, S., Deshmukh, G., Barcelon, F., Inokuchi, J. I., Radin, N. S. 1990. *J. Biol. Chem.* 265:12135–38

47. Connolly, T. M., Bross, T. E., Majerus, P. W. 1985. *J. Biol. Chem.* 260:7868–74

48. Mitchell, C. A., Connolly, T. M., Majerus, P. W. 1989. *J. Biol. Chem.* 264:8873–77

49. Connolly, T. M., Bansal, V. S., Bross, T. E., Irvine, R. F., Majerus, P. W. 1987. *J. Biol. Chem.* 262:2146–49

50. Connolly, T. M., Lawing, W. J., Majerus, P. W. 1986. *Cell* 46:951–58

51. Molina y Vedia, L. M., Lapetina, E. G. 1986. *J. Biol. Chem.* 261:10493–95

52. King, W. G., Rittenhouse, S. E. 1989. *J. Biol. Chem.* 264:6070–74

53. Ross, T. S., Jefferson, A. B., Mitchell,

C. A., Majerus, P. W. 1991. *J. Biol. Chem.* 266:20283–89
54. Kemp, B. E., Pearson, R. B. 1990. *Trends Biochem. Sci.* 15:342–46
55. Hansen, C. A., Johanson, R. A., Williamson, M. T., Williamson, J. R. 1987. *J. Biol. Chem.* 262:17319–26
56. Takimoto, K., Okada, M., Nakagawa, H. 1989. *J. Biol. Chem.* 106:684–90
57. Erneux, C., Lemos, M., Verjans, B., Vanderhaeghen, P., Delvaux, A., Dumont, J. E. 1989. *Eur. J. Biochem.* 181:317–22
58. Kukita, M., Hirata, M., Koga, T. 1986. *Biochim. Biophys. Acta* 885:121–28
59. Rana, R. S., Sekar, M. C., Hokin, L. E., MacDonald, M. J. 1986. *J. Biol. Chem.* 261:5237–40
60. Downes, C. P., Mussat, M. C., Michell, R. H. 1982. *Biochem. J.* 203:169–77
61. Storey, D. J., Shears, S. B., Kirk, C. J., Michell, R. H. 1984. *Nature* 312:374–76
62. Seyfred, M. A., Farrell, L. E., Wells, W. W. 1984. *J. Biol. Chem.* 259:13204–8
63. Joseph, S. K., Williams, R. J. 1985. *FEBS Lett.* 180:150–54
64. Shears, S. B. 1991. *Pharmacol. Ther.* 49:79–104
65. Morris, A. J., Murray, K. J., England, P. J., Downes, C. P., Michell, R. H. 1988. *Biochem. J.* 251:157–63
66. Irvine, R. F., Letcher, A. J., Heslop, J. P., Berridge, M. J. 1986. *Nature* 320:631–34
67. Biden, T. J., Wollheim, C. B. 1986. *J. Biol. Chem.* 261:11931–34
68. Ryu, S. H., Lee, S. Y., Lee, K. Y., Rhee, S. G. 1987. *FASEB J.* 1:388–93
69. Johanson, R. A., Hansen, C. A., Williamson, J. R. 1988. *J. Biol. Chem.* 263:7465–71
70. Daniel, J. L., Dangelmaier, C. A., Smith, J. B. 1988. *Biochem. J.* 253:789–94
71. Biden, T. J., Peter-Riesch, B., Schlegel, W., Wollheim, C. B. 1987. *J. Biol. Chem.* 262:3567–71
72. Yamaguchi, K., Hirata, M., Kuriyama, H. 1987. *Biochem. J.* 244:787–91
73. Kimura, Y., Hirata, M., Yamaguchi, K., Koga, T. 1987. *Arch. Biochem. Biophys.* 257:363–69
74. Morris, A. J., Downes, C. P., Hardin, T. K., Michell, R. H. 1987. *Biochem. J.* 248:489–93
75. Imboden, J. B., Pattison, G. 1987. *J. Clin. Invest.* 79:1538–41
76. Biden, T. J., Altin, J. G., Karajalainen,
A., Bygrave, F. L. 1988. *Biochem. J.* 256:697–701
77. Sim, S. S., Kim, J. W., Rhee, S. G. 1990. *J. Biol Chem.* 265:10367–72
78. Lin, A. N., Barnes, S., Wallace, R. W. 1990. *Biochem. Biophys. Res. Commun.* 170:1371–76
79. Johnson, R. M., Wasilenko, W. J., Mattingly, R. R., Weber, M. J., Garrison, J. C. 1989. *Science* 246:121–24
80. Moon, K. H., Lee, S. Y., Rhee, S. G. 1989. *Biochem. Biophys. Res. Commun.* 164:370–74
81. Lee, S. Y., Sim, S. S., Kim, J. W., Moon, K. H., Kim, J. H., Rhee, S. G. 1990. *J. Biol. Chem.* 265:9434–40
82. Takazawa, K., Passareiro, H., Dumont, J. E., Ereneux, C. 1989. *Biochem. J.* 261:483–88
83. Choi, K. Y., Kim, H. K., Lee, S. Y., Moon, K. H., Sim, S. S., et al. 1990. *Science* 248:64–66
84. Bala, T., Sim, S. S., Iida, T., Choi, K. Y., Catt, K. J., Rhee, S. G. 1991. *J. Biol. Chem.* 266:24719–26
85. Irvine, R. F., Moore, M. 1986. *Biochem. J.* 240:917–20
86. Inhorn, R. C., Bansal, V. S., Majerus, P. W. 1987. *Proc. Natl. Acad. Sci. USA* 84:2170–74
87. Shears, S. B. 1989. *Biochem. J.* 260:313–24
88. Inhorn, R. C., Majerus, P. W. 1988. *J. Biol. Chem.* 263:14559–65
89. Gee, N. S., Reid, G. G., Jackson, R. G., Barnaby, R. J., Ragan, C. I. 1988. *Biochem. J.* 253:777–82
90. Inhorn, R. C., Majerus, P. W. 1987. *J. Biol. Chem.* 262:15946–52
91. Hansen, C. A., Inubushi, T., Williamson, M. T., Williamson, J. R. 1988. *Biochim. Biophys. Acta* 1001:134–44
92. Berridge, M. J., Downes, C. P., Hanley, M. R. 1989. *Cell* 59:411–19
93. York, J. D., Majerus, P. W. 1990. *Proc. Natl. Acad. Sci. USA* 87:9548–52
94. Newald, A. F., York, J. D., Majerus, P. W. 1991. *FEBS Lett.* 294:16–18
95. Bansal, V. S., Inhorn, R. C., Majerus, P. W. 1987. *J. Biol. Chem.* 262:9444–47
96. Howell, S., Barnaby, R. J., Rowe, T., Ragan, C. I., Gee, N. S. 1989. *Eur. J. Biochem.* 183:169–72
97. Bansal, V. S., Caldwell, K. K., Majerus, P. W. 1990. *J. Biol. Chem.* 265:1806–11
98. Caldwell, K. K., Lips, D. L., Bansal, V. S., Majerus, P. W. 1991. *J. Biol. Chem.* 266:18378–86
99. Ackermann, K. E., Gish, B. G., Hon-

char, M. P., Sherman, W. R. 1987. *Biochem. J.* 242:517–24

100. Hallcher, L. M., Sherman, W. R. 1980. *J. Biol. Chem.* 255:10896–901

101. Takimoto, K., Okada, M., Matsuda, Y., Nakagawa, H. 1985. *J. Biochem.* 98:363–70

102. Gee, N. S., Ragan, C. I., Watling, K. J., Aspley, S., Jackson, R. G., et al. 1988. *Biochem. J.* 249:883–89

103. Diehl, R. E., Whiting, P., Potter, J., Gee, N., Ragan, C. I., et al. 1990. *J. Biol. Chem.* 265:5946–49

104. Jackson, R. G., Gee, N. S., Ragan, C. I. 1989. *Biochem. J.* 264:419–22

105. Koch, M. A., Diringer, H. 1974. *Biochem. Biophys. Res. Commun.* 58:361–67

106. Binder, H., Weber, P. C., Siess, W. 1985. *Anal. Biochem.* 148:220–27

107. Dixon, J. F., Hokin, L. E. 1985. *J. Biol. Chem.* 260:16068–71

108. Dixon, J. F., Hokin, L. E. 1987. *Biochem. Biophys. Res. Commun.* 149:1208–13

109. Dixon, J. F., Hokin, L. E. 1987. *J. Biol. Chem.* 262:13892–95

110. Shayman, J. A., Auchus, R. J., Morrison, A. R. 1986. *Biochim. Biophys. Acta* 886:171–75

111. Ishii, H., Connolly, T. M., Bross, T. E., Majerus, P. W. 1986. *Proc. Natl. Acad. Sci. USA* 83:6397–400

112. Sekar, M. C., Dixon, J. F., Hokin, L. E. 1987. *J. Biol. Chem.* 262:340–44

113. Graham, R. A., Meyer, R. A., Szwergold, B. S., Brown, T. R. 1987. *J. Biol. Chem.* 262:35–37

114. Hughes, A. R., Takemura, H., Putney, J. W. Jr. 1988. *J. Biol. Chem.* 263:10314–19

115. Tarver, A. P., King, W. G., Rittenhouse, S. E. 1987. *J. Biol. Chem.* 262:17268–71

116. Connolly, T. M., Wilson, D. B., Bross, T. E., Majerus, P. W. 1986. *J. Biol. Chem.* 261:122–26

117. Ross, T. S., Majerus, P. W. 1986. *J. Biol. Chem.* 261:11119–23

118. Ross, T. S., Majerus, P. W. 1991. *J. Biol. Chem.* 266:851–56

119. Ross, T. S., Tait, J. F., Majerus, P. W. 1990. *Science* 248:605–7

120. Tait, J. F., Sakata, M., McMullen, B. A., Miao, C. H., Funakoshi, T., et al. 1988. *Biochemistry* 27:6268–76

121. Pepinsky, R. B., Tizard, R., Mattaliano, R. J., Sinclair, L. K., Miller, G. T., et al. 1988. *J. Biol. Chem.* 263:10799–811

122. Kaetzel, M. A., Hazarika, P., Dedman, J. R. 1989. *J. Biol. Chem.* 264:14463–70

123. Crompton, M. R., Moss, S. E., Crumpton, M. J. 1988. *Cell* 55:1–3

124. Haigler, H. T., Fitch, J. M., Schlaepfer, D. D. 1989. *Trends Biochem. Sci.* 14:48–51

125. Ross, T. S., Whiteley, B., Graham, R. A., Majerus, P. W. 1991. *J. Biol. Chem.* 266:9086–92

126. Camilli, A., Goldfine, H., Portnoy, D. A. 1991. *J. Exp. Med.* 173:751–54

127. Balla, T., Guillemette, G., Baukal, A. J., Catt, K. J. 1987. *J. Biol. Chem.* 262:9952–55

128. Stephens, L. R., Hawkins, P. T., Barker, C. J., Downes, C. P. 1988. *Biochem. J.* 253:721–23

129. Hansen, C. A., Dahl, S. V., Huddell, B., Williamson, J. R. 1988. *FEBS Lett.* 236:53–56

130. Shears, S. B. 1989. *J. Biol. Chem.* 264:19879–86

131. Cunha-Melo, J. R., Dean, N. M., Ali, H., Beavan, M. A. 1988. *J. Biol. Chem.* 263:14245–50

132. Doughney, C., McPherson, M. A., Dormer, R. L. 1988. *Biochem. J.* 251:927–29

133. Hoer, D., Kwiatkowski, A., Seib, C., Rosenthal, W., Schultz, G., Oberdisse, E. 1988. *Biochem. Biophys. Res. Commun.* 154:668–75

134. Hodgson, M. E., Shears, S. B. 1990. *Biochem. J.* 267:831–34

135. McIntosh, R. P., McIntosh, J. E. A. 1990. *Biochem. J.* 268:141–45

136. Stephens, L. R., Hawkins, P. T., Stanley, A. F., Moore, T., Poyner, D. R., et al. 1991. *Biochem. J.* 275:485–99

137. Pittet, D., Schlegel, W., Lew, D. P., Monod, A., Mayr, G. W. 1989. *J. Biol. Chem.* 264:18489–93

138. Balla, T., Hunyady, L., Baukal, A. J., Catt, K. J. 1989. *J. Biol. Chem.* 264:9386–90

139. Stephens, L. R., Downes, C. P. 1990. *Biochem. J.* 265:435–52

140. Menniti, F. S., Oliver, K. G., Nogimori, K., Obie, J. F., Shears, S. B., Putney, J. W. 1990. *J. Biol. Chem.* 265:11167–76

141. Cunningham, T. W., Lips, D. L., Bansal, V. S., Caldwell, K. K., Mitchell, C. A., Majerus, P. W. 1990. *J. Biol. Chem.* 265:21676–83

142. Vallejo, M., Jackson, P., Lightman, S., Hanley, M. R. 1987. *Nature* 330:656–58

143. Isaacks, R. E., Harkens, D. R. 1980. *Am. Zool.* 20:115–29

144. Supattapone, S., Worley, P. F., Baraban, J. M., Snyder, S. H. 1988. *J. Biol. Chem.* 263:1530–34

145. Ferris, C. D., Huganir, R. L., Supatta-

pone, S., Snyder, S. H. 1989. *Nature* 342:87–89
146. Maeda, N., Kawasaki, T., Nakade, S., Yokota, N., Taguchi, T., et al. 1991. *J. Biol. Chem.* 266:1109–16
147. Ferris, C. D., Huganir, R. L., Snyder, S. H. 1990. *Proc. Natl. Acad. Sci. USA* 87:2147–51
148. Worley, P. F., Baraban, J. M., Colvin, J. S., Snyder, S. H. 1987. *Nature* 325:159–61
149. Nunn, D. L., Potter, B. V., Taylor, C. W. 1990. *Biochem. J.* 265:393–98
150. Mikoshiba, K., Okano, H., Tsukada, Y. 1985. *Dev. Neurosci.* 7:179–87
151. Furuichi, T., Yoshikawa, S., Miyawaki, A., Wada, K., Maeda, N., Mikoshiba, K. 1989. *Nature* 342:32–38
152. Takeshima, H., Nishimura, S., Matsumoto, T., Ishida, H., Kangawa, K., et al. 1989. *Nature* 339:439–45
153. Nordquist, D. T., Kozak, C. A., Orr, H. T. 1989. *J. Neurosci.* 8:4780–89
154. Mignery, G. A., Sudhof, T. C., Takei, K., De Camilli, P. 1989. *Nature* 342:192–95
155. Ross, C. A., Meldolesi, J., Milner, T. A., Satoh, T., Supattapone, S., Snyder, S. H. 1989. *Nature* 339:468–70
156. Maeda, N., Niinobe, M., Mikoshiba, K. 1990. *EMBO J.* 9:61–67
157. Mignery, G. A., Newton, C. L., Archer, B. T., Sudhof, T. C. 1990. *J. Biol. Chem.* 265:12679–85
158. Nakagawa, T., Okano, H., Furuichi, T., Aruga, J., Mikoshiba, K. 1991. *Proc. Natl. Acad. Sci. USA* 88:6244–48
159. Danoff, S. K., Ferris, C. D., Donath, C., Fischer, G. A., Minemitsu, S., et al. 1991. *Proc. Natl. Acad. Sci. USA* 88:2951–55
160. Malviya, A. N., Rogue, P., Vincendon, G. 1990. *Proc. Natl. Acad. Sci. USA* 87:9270–74
161. Nicotera, P., Orenius, S., Nilsson, T., Derggren, P. O. 1990. *Proc. Natl. Acad. Sci. USA* 87:6858–62
162. Donie, F., Hulser, E., Reiser, G. 1990. *FEBS Lett.* 268:194–98
163. Theibert, A. B., Supattapone, S., Ferris, C. D., Danoff, S. K., Evans, R. K., Snyder, S. H. 1990. *Biochem. J.* 267:441–45
164. Theibert, A. B., Estevez, V. A., Ferris, C. D., Danoff, S. K., Barrow, R. K., et al. 1991. *Proc. Natl. Acad. Sci. USA* 88:3165–69
165. Donie, F., Reiser, G. 1991. *Biochem. J.* 275:453–57
166. Hawkins, P. T., Reynolds, D. J., Poyner, D. R., Hanley, M. R. 1990. *Biochem. Biophys. Res. Commun.* 167:819–27
167. Nicoletti, F., Bruno, V., Cavallar, S., Copani, A., Sortino, M. A., Canono, P. L. 1990. *Mol. Pharmacol.* 37:689–93
168. Carpenter, C. L., Cantley, L. C. 1990. *Biochemistry* 29:11147–56
169. Serunian, L. A., Auger, K. R., Roberts, T., Cantley, L. C. 1990. *J. Virol.* 64:4718–25
170. Whitman, M., Downes, C. P., Keeler, M., Keller, T., Cantley, L. 1988. *Nature* 332:644–46
171. Varticovski, L., Drucker, B., Morrison, D., Cantley, L., Roberts, T. 1989. *Nature* 342:699–702
172. Auger, K. R., Serunian, L. A., Soltoff, S. P., Libby, P., Cantley, L. C. 1989. *Cell* 57:167–75
173. Lips, D. L., Majerus, P. W., Gorga, F. R., Young, A. T., Benjamin, T. L. 1989. *J. Biol. Chem.* 264:8759–63
174. Traynor-Kaplan, A. E., Thompson, B. L., Harris, A. L., Taylor, P., Omann, G. M., Sklar, L. A. 1989. *J. Biol. Chem.* 264:15668–73
175. Kucera, G. L., Rittenhouse, S. E. 1990. *J. Biol. Chem.* 265:5345–48
176. Sultan, C., Breton, M., Mauco, G., Grondin, P., Plantavid, M., Chap, H. 1990. *Biochem. J.* 269:831–34
177. Stephens, L. R., Hawkins, P. T., Downes, C. P. 1989. *Biochem. J.* 259:267–76
178. Lips, D. L., Majerus, P. W. 1989. *J. Biol. Chem.* 264:19911–15
179. Serunian, L. A., Haber, M. T., Fukui, T., Kim, J. W., Rhee, S. G., et al. 1989. *J. Biol. Chem.* 264:17809–15
180. Carpenter, C. L., Duckworth, B. C., Auger, K. R., Cohen, B., Schafferhausen, B. S., Cantley, L. C. 1990. *J. Biol. Chem.* 265:19704–11
181. Cunningham, T. W., Majerus, P. W. 1991. *Biochem. Biophys. Res. Commun.* 175:568–76
182. Yamamoto, K., Graziani, A., Carpenter, C., Cantley, L. C., Lapetina, E. G. 1990. *J. Biol. Chem.* 265:22086–89
183. Stephens, L. R., Hughes, K. T., Irvine, R. F. 1991. *Nature* 351:33–39
184. Shibasaki, F., Homma, Y., Takenawa, T. 1991. *J. Biol. Chem.* 266:8108–14
185. Otsu, M., Hiles, I., Gout, I., Fry, J., Ruiz-Larrea, F., Panayotou, G., et al. 1991. *Cell* 65:91–104
186. Escobedo, J. A., Navankasattusas, S., Kavanaugh, W. M., Milfay, D., Fried, V. A., Williams, L. T. 1991. *Cell* 65:75–82
187. Skolnik, E. Y., Margolis, B., Mohammadi, M., Lowenstein, E., Fischer, R., et al. 1991. *Cell* 65:83–90

188. Whitman, M., Downes, C. P., Keeler, M., Keller, T., Cantley, L. 1988. *Nature* 332:644–46
189. Fukui, Y., Kornbluth, S., Jong, S.-M., Wang, L.-H., Hanafusa, H. 1989. *Oncogene Res.* 4:283–92
190. Auger, K. R., Serunian, L. A., Soltoff, S. P., Libby, P., Cantley, L. C. 1989. *Cell* 57:167–75
191. Varticovski, L., Drucker, B., Morrison, D., Cantley, L., Roberts, T. 1989. *Nature* 342:699–702
192. Coughlin, S. R., Escobedo, J. A., Williams, L. T. 1989. *Science* 243:1191–94
193. Ullrich, A., Schlessinger, J. 1990. *Cell* 61:203–12
194. Talmage, D. A., Freund, R., Young, A. T., Dah, J., Dawe, C. J., Benjamin, T. L. 1989. *Cell* 59:55–65
195. Larsson, H., Lindberg, U. 1988. *Biochim. Biophys. Acta* 953:95–105
196. Lassing, I., Lindberg, U. 1985. *Nature* 314:472–74
197. Lassing, I., Lindberg, U. 1988. *J. Cell. Biochem.* 37:255–67
198. Janmey, P. A., Stossel, T. P. 1989. *J. Biol. Chem.* 264:4825–31
199. Goldschmidt-Clermont, P. J., Machesky, L. M., Baldassare, J. J., Pollard, T. D. 1990. *Science* 247:1575–78
200. Dadabay, C. Y., Patton, E., Cooper, J. A., Pike, L. J. 1991. *J. Cell Biol.* 112:1151–56
201. Eberle, M., Traynor-Kaplan, A. E., Sklar, L. A., Norgauer, J. 1990. *J. Biol. Chem.* 265:16725–28
202. Bockus, B. J., Stiles, C. D. 1984. *Exp. Cell Res.* 153:186–97

Annu. Rev. Biochem. 1992. 61:251–81

MAMMALIAN DNA LIGASES

Tomas Lindahl and Deborah E. Barnes

Imperial Cancer Research Fund, Clare Hall Laboratories, South Mimms, Hertfordshire, EN6 3LD, United Kingdom

KEY WORDS: DNA replication, DNA repair, recombination, DNA-protein interactions, biochemistry of disease

CONTENTS

PERSPECTIVES

Recent successful demonstrations of DNA replication (1), excision repair (2), and recombinational repair (3) in soluble extracts of mammalian cells have stimulated research aimed at a thorough characterization of the enzymes that catalyze these processes. The ultimate goal of such biochemical investigations

0066-4154/92/0701-0251$02.00

is to correctly reproduce, with purified proteins, the events and pathways that allow the regulated synthesis and maintenance of cellular DNA. The general approach is the same as that initially employed for characterization of DNA replication functions in *Escherichia coli* and its bacteriophages, that is, identification, purification, and characterization of relevant proteins by in vitro complementation assays. In addition to the discovery of novel mammalian factors required for DNA replication and repair, such projects have stimulated further investigation of the roles of those enzymes previously known to be active in DNA synthesis.

The DNA ligases are a case in point. After the discovery of this enzyme activity in the mid-1960s and an intense research effort in several laboratories over the next 5–10 years, the basic reaction mechanism and functions of microbial DNA ligases were clarified. General interest then waned and, in spite of the widespread use of DNA ligases as essential reagents in recombinant DNA research, very little biochemical work on their properties appeared. In the past few years the situation has changed. The identification of several distinct DNA ligases in mammalian cell nuclei, and the occurrence of DNA-joining defects in certain individuals with both clinical symptoms of immune deficiencies and cellular hypersensitivity to DNA-damaging agents, have encouraged detailed studies of DNA ligases. These projects can now be extended by construction of transgenic mouse models of the relevant human disorders.

Two potentially fruitful research topics are still in their infancies. Firstly, the biochemical investigation of DNA ligases in the genetically well-characterized budding and fission yeasts, which should provide excellent model systems for functional assignments, remains to be performed. Secondly, future identification of specific chemical inhibitors of the different mammalian DNA ligases should help to clarify the relative roles of these enzymes. Studies with inhibitors, although admittedly only yielding somewhat preliminary and tentative results, nevertheless have been very important for the assignments of specific cellular functions to different DNA polymerases, RNA polymerases, and DNA topoisomerases, as well as for manipulation of their activities. If, for example, different DNA ligases were responsible for DNA joining during homologous recombination than during illicit recombination, selective inhibition of the latter process would obviously be helpful in gene correction experiments. Furthermore, as is already the case for DNA topoisomerases, specific enzyme inhibitors that prevent DNA sealing may turn out to be clinically useful as cytotoxic drugs.

REACTION MECHANISM

The mechanism for joining of DNA strand interruptions by DNA ligases has been elucidated (4) and is described in standard textbooks of biochemistry.

The reaction is initiated by the formation of a covalent enzyme-adenylate complex. Mammalian and viral DNA ligases employ ATP as cofactor, whereas bacterial DNA ligases use NAD to generate the adenylyl group. The ATP is cleaved to AMP and pyrophosphate with the adenylyl residue linked by a phosphoramidate bond to the ε-amino group of a specific lysine residue at the active site of the protein (5, 6). Several enzymes that interact with nucleotide substrates, including DNA ligases, contain an unusually reactive lysine residue in their active site, which can form a Schiff base with pyridoxal phosphate. In consequence, the activity of mammalian DNA ligase I (and presumably other DNA ligases) is inhibited in vitro by pyridoxal phosphate (6). Detection of the enzyme-adenylate complex as a radioactively labelled intermediate by SDS-polyacrylamide gel electrophoresis and autoradiography provides a convenient method for the assay of DNA ligases in partially purified protein fractions. ATP-dependent DNA ligases can employ analogues such as dATP, but the anomalous complexes formed with some ligases may function poorly in subsequent steps of the DNA-joining reaction (7). The activated AMP residue of the DNA ligase–adenylate intermediate is transferred to the 5' phosphate terminus of a single strand break in double-stranded DNA to generate a covalent DNA-AMP complex with a 5'–5' phosphoanhydride bond. This reaction intermediate has also been isolated for microbial and mammalian DNA ligases, but is more short-lived than the adenylylated enzyme. In the final step of DNA ligation, unadenylylated DNA ligase is required for the generation of a phosphodiester bond and catalyzes displacement of the AMP residue through attack by the adjacent 3'-hydroxyl group on the adenylylated site.

Two related enzymes also employ this reaction mechanism of enzyme-adenylate formation, followed by transfer of the AMP residue to activate a 5'-phosphate residue of a nucleic acid substrate prior to joining. One of these is bacteriophage T4 RNA ligase, the product of T4 gene 63, which can catalyze the ligation of single-stranded ribo-oligonucleotides and deoxyribo-oligonucleotides with 5'-terminal phosphates and 3'-terminal hydroxyl groups (8). In contrast to DNA ligases, RNA ligases show no requirement for a complementary template strand in the nucleic acid substrate. The other enzyme is yeast tRNA ligase, an enzyme required for splicing of tRNA precursors. The locations of the adenylylated lysine residue within the sequences of the T4 RNA ligase and yeast tRNA ligase have been defined (9, 10). Recently, the active site for enzyme-adenylate formation of mammalian DNA ligase I was located by isolation and sequencing of a radioactively labelled tryptic peptide-AMP complex from bovine DNA ligase I, and by identification of the same peptide within the predicted reading frame of the human cDNA encoding this enzyme (6). The tryptic peptide contained a single internal lysine residue, protected against proteolysis by adenylylation. Since several DNA ligase genes have been cloned and sequenced, it is possible to predict the locations of the active site regions in all these enzymes by partial peptide sequence homology with the mammalian enzyme (Figure 1). The distance between the postulated adenylylation site and the carboxyl terminus of these enzymes is also practically invariant between the ATP-dependent DNA ligases.

The amino acid sequences of human DNA ligase I and the DNA ligase of the fission yeast, *Schizosaccharomyces pombe,* are virtually identical in the region of the active site, although the two enzymes show only 44% overall identity (11, 12). By comparison with DNA ligases from more distantly related organisms, several strongly conserved residues may be identified. The active lysine residue in DNA ligases is bracketed by a hydrophobic amino acid residue on each side, and the sequence E-KYDG-R is common to enzymes from such different sources as mammalian cells, yeasts, vaccinia virus, and bacteriophage T7. The T4 RNA ligase shows so little homology with DNA ligases that its sequence could not be used to predict the active sites in DNA ligases (9), but even in this case certain features are retained, such as the conserved glycine residue at position +3.

The mechanistic analogy between DNA ligases and RNA ligases has been confirmed and extended by recent site-specific mutagenesis experiments with T4 RNA ligase (9) and human DNA ligase I (13). As expected, replacement of the AMP-binding reactive lysine at the active site by another amino acid causes total loss of enzyme activity. In DNA ligase I, substitutions of the

		-7	-6	-5	-4	-3	-2	-1		+1	+2	+3	+4	+5	+6	+7	+8	+9	
Human	561	A	A	F	T	C	E	Y	K	Y	D	G	Q	R	A	Q	I	H	577
Sch.pombe	409	A	A	F	T	C	E	Y	K	Y	D	G	E	R	A	Q	V	H	425
S.cerevisiae	412	E	T	F	T	S	E	Y	K	Y	D	G	E	R	A	Q	V	H	428
Vaccinia	224	S	G	M	F	A	E	V	K	Y	D	G	E	R	V	Q	V	H	240
T7	27	G	Y	L	I	A	E	I	K	Y	D	G	V	R	G	N	I	C	43
T3	27	G	Y	L	I	A	D	C	K	Y	D	G	V	R	G	N	I	V	43
T4	152	F	P	A	F	A	Q	L	K	A	D	G	A	R	C	F	A	E	168
E.coli	108	V	T	W	C	C	E	L	K	L	D	G	L	A	V	S	I	L	124
T.thermophilus	111	F	A	Y	T	V	E	H	K	V	D	G	L	S	V	N	L	Y	127
T4 RNA ligase	92	D	V	D	Y	I	L	T	K	E	D	G	S	L	V	S	T	Y	108
tRNA ligase	107	G	P	Y	D	V	T	I	K	A	N	G	C	I	I	F	I	S	123

Figure 1 Active site regions for enzyme-adenylate formation in DNA and RNA ligases. Sequences are aligned by homology with the active site peptide of mammalian DNA ligase I. The reactive lysine residue within the mammalian enzyme and the RNA ligases, and that predicted in the other proteins, is shown in outline. The peptide sequences for the *Thermus thermophilus* DNA ligase and yeast tRNA ligase are from Lauer et al (142) and Xu et al (10), respectively; references to other sequences are in Tomkinson et al (6).

strongly conserved residues at positions -2, $+3$, and $+5$ also lead to very strong reduction or total loss of ability to form an enzyme-adenylate complex. Heaphy et al (9) found that substitution of the Asp residue at $+2$ in T4 RNA ligase still allowed enzyme-adenylate formation, but that the subsequent joining reaction did not occur. Similar results have been obtained for mammalian DNA ligase I, indicating the existence of shared structural and mechanistic features at the active sites of these two enzymes, although they have no detectable overall sequence homology. An antibody directed against the active site peptide of mammalian DNA ligase I does not bind to or inhibit the native enzyme (A. E. Tomkinson, T. Lindahl, unpublished data), indicating that this relatively hydrophobic sequence is protected within a groove or cleft of the protein.

DNA LIGASE I

Structure

Peptide sequences from the homogeneous bovine enzyme were used to synthesize corresponding oligonucleotide probes, and a partial cDNA sequence was isolated from a human cDNA library. A full-length human cDNA encoding DNA ligase I was subsequently obtained by functional complementation of a *Saccharomyces cerevisiae cdc9* temperature-sensitive DNA ligase mutant (11). The full-length cDNA encodes a 102-kDa protein of 919 amino acid residues. There is no marked sequence homology to other known proteins except for microbial DNA ligases. The active site lysine residue is located at position 568. The intact enzyme has been purified to apparent homogeneity from calf thymus (14, 15), and has also been highly purified from regenerating rat liver (16) and human lymphoblastoid cells (15). DNA ligase I accounts for the majority of the total DNA ligase activity in calf thymus extracts.

There is considerable confusion in the literature about the size of DNA ligase I. This is due to at least four complications:

1. The enzyme migrates anomalously slowly on SDS-polyacrylamide gels, corresponding to a 125–130-kDa protein (14–17). This retardation is largely due to a high proline content in the protein (15), and also to the presence of a small number of phosphoserine residues (18). Expression of the human cDNA encoding DNA ligase I in *S. cerevisiae* results in the production of a protein with the same anomalously slow migration rate during electrophoresis (11).

2. Mammalian DNA ligase I has a markedly asymmetric structure with a frictional ratio of 1.9 (14, 15) and resembles the *E. coli* and T4 DNA ligases in this regard. Consequently, the purified calf thymus enzyme sediments more slowly than expected for a 102-kDa protein on sucrose gradient centrifugation, whereas it elutes earlier than expected on gel filtration. Combination of

the data (19) for the sedimentation coefficient and the Stokes radius determined by gel filtration, nevertheless, allows for determination of an accurate native molecular weight by the Svedberg equation, within 5% of the value determined from the cDNA sequence (15). Thus, estimation of the size of the enzyme by hydrodynamic methods is more accurate than by SDS-polyacrylamide gel electrophoresis. On the other hand, use of gel filtration data by themselves to estimate a molecular mass for a hypothetical globular protein leads to a gross overestimate. This error, rather than enzyme dimerization, explains several reports of 200–240 kDa forms of DNA ligase I in solution.

3. Several antisera against purified DNA ligase I cross-react by immunoblotting with a 200-kDa protein present in crude cell extracts (20, 21). Moreover, a protein of ~200-kDa is one of several proteins in crude cell extracts that can be adenylylated (20, 22). It is not known if these two 200-kDa proteins are the same, and in any case the 200-kDa protein fraction has no DNA ligase activity (22). Nevertheless, it has been proposed by several authors that DNA ligase I might initially be synthesized as a 200-kDa protein that would be proteolytically processed to the active enzyme (17, 20–23). The cDNA sequencing of DNA ligase I and determination of the size of the mRNA (11) invalidate this model. Moreover, a longer cDNA clone has been obtained, which extends the 5' end of the published sequence (11) by 89 basepairs (bp) and places an in-frame stop codon 99 bp upstream of the initial methionine, confirming that this was correctly assigned (23a). We have noted that the 200-kDa protein cross-reacting with antisera against DNA ligase I is readily detectable by immunoblotting if bovine serum albumin is used to block nonspecific protein binding to nitrocellulose filters, but is barely seen if a more effective blocking agent such as nonfat milk is employed; the authentic DNA ligase I signal is not suppressed by either blocking agent (C. Prigent, T. Lindahl, unpublished data). These data indicate that the 200-kDa protein is not directly related to DNA ligase I but could have a common epitope.

4. DNA ligase I is highly sensitive to partial proteolysis, and it is difficult to avoid such degradation during enzyme purification. In calf thymus extracts, and partially purified enzyme fractions, DNA ligase I is attacked slowly by an endogenous protease activity that removes the N-terminal 216 amino acids to generate an active fragment of 703 amino acids. This fragment shows the same DNA-joining activity as the intact enzyme in standard DNA ligase assays, indicating that the N-terminal region of the protein is not required for DNA ligation activity in vitro. The same result is obtained when intact, highly purified DNA ligase I is treated with low concentrations of subtilisin (15) instead of being exposed to proteolysis in cell extracts. The large C-terminal fragment of DNA ligase I is relatively resistant to further proteolysis and apparently represents the catalytic domain of the enzyme.

Furthermore, a truncated human cDNA encoding this domain is able to complement DNA ligase mutants of *S. cerevisiae* (11) and *E. coli* (13). The nonessential N-terminal region contains a large proportion of proline and hydrophilic amino acids, in particular stretches of glutamic acid residues. It shows no detectable homology to the shorter N-terminal regions of the related DNA ligases from *Sch. pombe* and *S. cerevisiae*, although the three proteins are about 50% homologous in their C-terminal regions.

Because of the ready conversion of the intact form of DNA ligase I to an active proteolytic fragment, several studies on DNA ligase I have been concerned with the properties of the active fragment rather than the intact enzyme. This was the case for the first report on a mammalian DNA ligase activity (24) as well as for several more recent investigations (25–29). Other early studies reported on both the intact form and the active fragment of DNA ligase I (30). When mammalian cells growing in tissue culture are lysed instantaneously with hot SDS, followed by immunoblotting analysis with antibodies against DNA ligase I, the intact protein is the only one detected. Thus, the active fragment representing the catalytic domain does not seem to be present in vivo (31).

Deletion of terminal regions from the DNA ligase I cDNA, followed by functional expression in *E. coli*, has provided more precise information on the domain structure of DNA ligase I. The *E. coli* conditional lethal mutant strain SG251, which was originally isolated as a temperature-sensitive DNA replication mutant, and a transductant strain GR501 carrying the same mutation, have a heat-labile NAD-dependent DNA ligase and are unable to join Okazaki fragments at 42 °C (32). The physiological defect can be complemented by the catalytic domain of DNA ligase I, encoded by a fragment of the human cDNA expressed as a lacZ fusion protein (13). A fragment lacking the N-terminal 249 amino acids of DNA ligase I functions well as a DNA ligase in this heterologous in vivo system. This active fragment is 33 amino acids shorter than the catalytic domain generated by proteolysis of DNA ligase I in cell extracts. However, reduction of the size of the domain from 670 to 635 amino acids by a further N-terminal deletion causes complete loss of DNA ligation activity both in vivo and in vitro.

In the C-terminal region of the enzyme, human DNA ligase I shows homology with the DNA ligases of both *Sch. pombe* and *S. cerevisiae*, except that the human and *Sch. pombe* enzymes are 14 amino acids longer at optimal sequence alignment (11). There is little homology between the two proteins in this "tail" sequence (11). This short C-terminal sequence can be removed from human DNA ligase I without functional impairment, but deletion of 8 additional residues at the C-terminus causes loss of activity (13). The existence of an evolutionarily conserved C-terminal catalytic domain joined to an N-terminal region that is nonessential for enzyme activity in vitro is not unique to DNA ligase I, but has also been observed for several other mammalian nuclear enzymes acting on DNA including DNA topoisomerase I (33) and the DNA methyltransferase that converts cytosine residues to 5-methylcytosine (34). The role of the N-terminal region of DNA ligase I remains unclear, although it contains sequences that could function as nuclear translocation signals and stretches of acidic residues that might mediate interactions with chromatin or the nuclear scaffold. The N-terminal region might also be employed in species-specific protein-protein interactions during DNA replication. A cDNA subclone expressing this fragment in vivo might have a dominant negative effect and could elucidate the function of this hydrophilic region of the enzyme.

Gene Structure and Chromosome Mapping

The cloning of the human DNA ligase I cDNA allows for an analysis of the gene encoding this enzyme (11, 35). Southern hybridization analysis of EcoRI-digested genomic DNA with the cDNA probe identified a discrete number of hybridizing bands and gave no indication for the existence of multiple related genes or pseudogenes. The size of the gene is estimated to be 45–50 kilobases (kb), and sequence analysis of genomic clones has shown that the gene contains 27 introns (23a). A single mRNA of 3.2 kb is transcribed from the gene.

DNA hybridization with material from rodent-human somatic cell hybrids retaining subsets of human chromosomes showed that the DNA ligase I gene is located on chromosome 19. Four overlapping positive cosmids forming a contig have been identified in an arrayed human chromosome 19–specific cosmid library (35). More detailed mapping by in situ hybridization demonstrated that the *LIG1* gene is localized to the long arm of the chromosome, at 19q13.2–13.3. Since three different genes active in DNA repair, *ERCC1* (36), *ERCC2* (37), and *XRCC1* (38), have been mapped previously to this region of chromosome 19, a more detailed study was performed using two-color fluorescence in situ hybridization to position *LIG1* relative to *ERCC1*. The in situ hybridization results, together with genetic and physical data, give an order for the four genes on the long arm of this chromosome of: centromere ——-*XRCC1* ——*ERCC2* ——*ERCC1* ——*LIG1* ——-telomere.

A defect in the *XRCC1* gene leads to cellular sensitivity to ethyl methanesulfonate and ionizing radiation, and a reduced capacity to rejoin DNA single-strand breaks (38). The putative *XRCC1* gene product is a 69.5-kDa protein, which is clearly different from DNA ligase I. Furthermore, peptide sequences derived from purified DNA ligase II do not appear in the predicted XRCC1 protein sequence, indicating that the *XRCC1* gene does not encode DNA ligase II. The relationship, if any, of DNA ligase III to XRCC1 remains to be established.

Catalytic Properties

DNA ligase I requires Mg^{2+} and ATP for activity. GTP and NAD are inactive as cofactors. The K_M for ATP is low, 0.5–1 μM (6, 14, 39). This property of the enzyme facilitates assays of the enzyme by measuring DNA ligase–AMP complex formation. The reaction is readily reversed in the presence of pyrophosphate. The active proteolytic fragment of DNA ligase I, which represents the catalytic domain of the enzyme, retains the low K_M for ATP. The DNA-adenylate reaction intermediate in the joining process catalyzed by DNA ligase I has also been isolated and characterized (40). Thus, the general reaction mechanism for this mammalian enzyme appears identical to that established for microbial DNA ligases. DNA ligase I effectively seals single-strand breaks in DNA and joins restriction enzyme DNA fragments with staggered ends. The enzyme is also able to catalyze blunt-end joining of DNA (41). As for other DNA ligases, the latter reaction is stimulated greatly by the macromolecular crowding conditions inflicted by the presence of 10–20% polyethylene glycol in reaction mixtures (42). DNA ligase I is much more effective at blunt-end joining than mammalian DNA ligases II and III (41, 42a), but is less efficient in this regard than T4 DNA ligase. Similarly to microbial DNA ligases and *Drosophila* DNA ligase I (43), mammalian DNA ligase I can act at low efficiency as a topoisomerase, relaxing supercoiled DNA in an AMP-dependent reversal of the last step of the ligation reaction followed by re-ligation (44).

DNA ligase I can join oligo (dT) molecules hydrogen-bonded to poly (dA), but the enzyme differs from T4 DNA ligase in being unable to ligate oligo (dT) with a poly (rA) complementary strand (30, 41). The response with several other polynucleotide-oligonucleotide combinations is variable, but DNA ligase I can join oligo (rA) molecules hydrogen-bonded to poly (dT) (45). The lack of activity with an oligo (dT)·poly (rA) hybrid structure has turned out to be a helpful property of the enzyme, because DNA ligases II and III are able to join this particular substrate. Thus, the latter DNA ligase activities can be specifically followed during enzyme purification even in the presence of an excess of the intact form and active fragments of DNA ligase I (16, 41, 45). However, the sensitivity of the oligo (dT)·poly (rA) substrate to degradation by RNaseH makes this method unsuitable for cell extracts or crude enzyme fractions.

DNA Ligase I is a Phosphoprotein

Labelling of tissue culture cells with ^{32}P orthophosphate followed by immunoprecipitation of DNA ligase I indicated that the enzyme contained radioactive material. Addition of homogeneous calf thymus DNA ligase I to the cell extracts completely suppressed the immunoprecipitation of radioactive material, and established that the cellular DNA ligase I was ^{32}P-labelled.

Approximately one-third of the radioactive material could be released by exposure of the immune complexes to pyrophosphate and was due to the presence of activated enzyme-adenylate reaction intermediates in the cell. However, two-thirds of the radioactive material in DNA ligase I was refractory to release by pyrophosphate but could be converted to an acid-soluble form by treatment with *E. coli* alkaline phosphatase or potato acid phosphatase. Amino acid analysis by gel electrophoresis with appropriate markers demonstrated the presence of phosphoserine, whereas phosphothreonine and phosphotyrosine were not detected (18). Phosphatase treatment of purified calf thymus DNA ligase I caused a small shift in the migration of the enzyme on SDS-polyacrylamide gel electrophoresis. The corresponding molecular mass estimates by this method were 125 kDa for the phosphorylated enzyme isolated from tissue, and 115 kDa for the dephosphorylated enzyme.

Many different cellular proteins are phosphorylated, so these results are by themselves not surprising. However, the phosphorylation of DNA ligase I is of interest because it affects the catalytic activity of the enzyme. Dephosphorylation of DNA ligase I with *E. coli* alkaline phosphatase or potato acid phosphatase caused a drastic reduction in enzyme activity. In the converse experiment, additional phosphorylation of purified calf thymus DNA ligase I with the ubiquitous protein kinase, casein kinase II [CKII; (46)] using either ATP or GTP as cofactor caused an approximately threefold increase in the ability of DNA ligase I to generate the enzyme-adenylate intermediate. Thus, DNA ligase I is activated by protein phosphorylation (18). The reaction is most conveniently followed by using GTP rather than ATP as cofactor for CKII, since this avoids simultaneous adenylylation of the DNA ligase. In this way it was shown that formation and accumulation of the enzyme-AMP complex only occurs with the phosphorylated form of DNA ligase I. Several protein kinases were investigated, including the cAMP-dependent protein kinase A, protein kinase C, and purified p34cdc2 protein kinase, but CKII was the only kinase tested that activated DNA ligase I in vitro. Heparin is a potent inhibitor of CKII and also prevented the activation of DNA ligase I by CKII. In the predicted amino acid sequence of DNA ligase I there are at least three potential phosphorylation sites for CKII. Two of these are located in the N-terminal region of the enzyme outside the catalytic domain, whereas the third occurs within an evolutionarily conserved region just over a hundred amino acid residues from the C-terminus of the enzyme.

Expression of the entire 102-kDa human DNA ligase I in a DNA ligase-deficient *E. coli* strain yields no complementation at 42 °C, whereas a subclone expressing the catalytic domain of the enzyme allows efficient physiological complementation (13). The 102-kDa DNA ligase expressed in *E. coli* did not detectably form the enzyme-adenylate complex on incubation with ATP. However, this form of DNA ligase I was converted to an active enzyme by pretreatment with CKII and GTP (18). The data indicate that phosphorylation of one or both target serine residues in the N-terminal region of the enzyme is required to avoid suppression of the activity of the catalytic domain. This could provide a mechanism for regulation of DNA ligase I activity during the cell cycle, especially as the long half-life of the protein in vivo (31) would make its regulation at the mRNA level ineffectual.

Subcellular Localization and Functional Properties

Immunofluorescence studies with affinity-purified polyclonal antibodies and monoclonal antibodies against DNA ligase I have established that the enzyme is localized to cell nuclei (31). As also observed for DNA polymerase α (47), DNA ligase I leaches out of cell nuclei during isolation in isotonic sucrose by standard methods. Both DNA polymerase α and DNA ligase I exhibited granular nuclear staining and seem to be excluded from the nucleoli. Such a localization pattern coincides with that of apparent DNA replication complexes (48, 49). Moreover, during Herpes simplex 1 (HSV 1) infection of cells, certain host DNA replication factors are redistributed within nuclei and recruited to viral replication factories, whereas this is not the case for several other nuclear proteins. DNA ligase I is one of the recruited host proteins, as determined by immunofluorescence (50). Since HSV 1, as well as other herpesviruses, does not encode a DNA ligase, it is hardly surprising that the virus employs a host enzyme for replication, and it seems likely that DNA ligase I serves in this function. Adenovirus infection also depends on reprogramming of the host replication machinery, and it is intriguing that the viral regulatory E1A proteins bind to an unidentified host protein of the same size as DNA ligase I (130 kDa by SDS-polyacrylamide gel electrophoresis; 51) as well as modulators of cell growth, such as the retinoblastoma protein (52) and cyclin A (53). However, immunoprecipitation experiments with E1A and DNA ligase I antibodies are inconclusive at this point (C. Prigent and D. Barnes, unpublished). DNA ligase I appears to account for the ligation of Okazaki fragments during SV40 DNA synthesis in a cell-free system (54). In this connection, Malkas et al (55) have isolated a 21S protein complex from HeLa cells, which comprises several replication enzymes and can replicate SV40 DNA in the presence of large T antigen. The enzyme complex contains DNA polymerase α/primase, the DNA polymerase δ–associated factor PCNA, DNA topoisomerase I, an RNaseH, a DNA-dependent ATPase, and DNA ligase I, as well as a number of additional polypeptides (E. F. Baril, personal communication).

The half-life of DNA ligase I in growing tissue-culture cells is about 7 h as determined by pulse-chase experiments (31). The enzyme is present at a much higher level in proliferating than in nongrowing cells, both when measured by activity assays and by immunochemical protein determinations. In contrast,

DNA ligases II and III are present in similar amounts in proliferating and nongrowing cells (14, 16, 56, 57). DNA ligase I is also able to correct the DNA replication defect in conditional lethal DNA ligase mutants of *S. cerevisiae* (11) and *E. coli* (13). Taken together, these results strongly imply a role for DNA ligase I in DNA replication.

Direct evidence for this concept has been obtained recently by the characterization of the human cell line 46BR, described in detail below, as a mutant line with a structural defect in DNA ligase I on the amino acid sequence level (58). This cell line exhibits a strong delay in the rate of joining of Okazaki fragments at 37 °C (59, 60, 60a). Interestingly, not only is the 46BR cell line defective in DNA replication, but the cells are also hypersensitive to several DNA-damaging agents, implying a defect in DNA repair (60–63). These agents include ultraviolet light, simple alkylating agents, and ionizing radiation. The DNA damage induced by these agents is chemically different and is corrected by separate excision-repair processes (64). DNA ligase I apparently completes all these repair pathways, which are largely carried out by different sets of enzymes. In agreement with these data, it has been found that the human DNA ligase I cDNA corrects the defective phenotype of a *S. cerevisiae cdc9* mutant both with regard to replication and several forms of repair (11; L. H. Johnston, personal communication). In conclusion, it would appear that DNA ligase I is the key enzyme for joining Okazaki fragments during lagging-strand DNA synthesis in mammalian cells and also for completion of DNA excision-repair processes.

Activators and Inhibitors

Several reports have appeared on cellular factors that modulate the activity of DNA ligase I. Preliminary observations on a heat-stable protein activator of DNA ligase activity in extracts of the human heteroploid cell line EUE indicated that the stimulatory factor(s) was either very heterogeneous in size or aggregated easily (30, 65). However, the factor could not be replaced with bovine serum albumin and caused a four- to five-fold promotion of DNA ligase I activity. Similarly, the human fibroblast line GM1492, but not several related fibroblast lines, expresses substantial amounts of a heat-stable protease-sensitive factor, which stimulates the activity of calf thymus DNA ligase I (66). The active factor has been partially purified by single-stranded DNA cellulose chromatography.

Induction of differentiation in mouse erythroleukemia cells by growth in the presence of dimethyl sulfoxide causes a fourfold decrease in DNA ligase activity in cell extracts. This reduction in DNA ligase activity may be associated with a decreased replication rate in differentiated cells, as observed in the *Drosophila* and *Xenopus* systems. However, the high levels of heme accumulated in the differentiating erythroleukemia cells may also contribute to suppress DNA ligase activity (67). In a separate study, a protein inhibitor of DNA ligase I has been extensively purified from HeLa cell extracts (67a). The inhibitor is heat-labile and sensitive to proteases. The specificity of this inhibitor is of interest; it interferes effectively with DNA joining catalyzed by mammalian DNA ligase I but not enzyme-AMP formation, and it has no effect on either mammalian DNA ligase II or T4 DNA ligase. This inhibitor is a protein of about 65 kDa as estimated by gel filtration, and it appears to form a protein-protein complex with DNA ligase I. Thus, the inhibitor initially copurifies with DNA ligase I, but the two proteins can be separated by Mono S ion-exchange chromatography. The purified inhibitor exhibits no detectable phosphatase, ATPase, DNase, or poly ADP-ribose polymerase activity. A modulating effect of poly ADP-ribosylation of DNA ligase II has been reported but is controversial (68, 69). Further studies on the significance and binding specificity of the cellular factors that affect DNA ligase activity, and cloning of cDNAs encoding putative protein activators and inhibitors, should clarify their possible roles in modulating DNA ligase activity in vivo.

In a recent survey of ATP analogues, arabinosyl-2-fluoro-ATP was found to inhibit DNA ligase I (69a). Thus, the ligase may be a target in tumor cells treated with the anticancer drug arabinosyl-2-fluoroadenine.

DNA LIGASE II

A DNA ligase distinct from DNA ligase I is more firmly associated with cell nuclei and is only extracted into soluble form by salt-containing buffers (27, 39, 70). The enzyme has been purified from calf thymus as a 68–72-kDa protein of apparent homogeneity, estimated by SDS-polyacrylamide gel electrophoresis, by three different groups (27, 45, 71). The protein has a blocked N-terminal residue. The enzyme is present in this form both in crude cell extracts and in purified form, and there is no detectable proteolytic processing or generation of active fragments during isolation (45, 71). In common with other DNA ligases, this protein exhibits a high frictional ratio, reflecting an apparently asymmetric structure. DNA ligase II is generally found at a lower level of activity than DNA ligase I in mammalian cells and tissues. Moreover, the enzyme is a labile protein, which is rapidly inactivated at 42 °C. This intrinsic lability causes problems during enzyme purification, although some stabilization of the purified active enzyme can be achieved by addition of a neutral detergent such as Tween 20 to 0.2% (45) or a carrier protein such as bovine serum albumin (71). Several groups have documented their inability to detect or purify DNA ligase II. The various reasons for these technical difficulties have been discussed elsewhere (72); possible explanations include irreversible denaturation of the enzyme by exposure to SDS prior to attempted enzyme assays (17), failure to extract the enzyme from nuclei, or loss in the first batch purification step.

The properties of highly purified DNA ligase II initially indicated that the enzyme had been purified to homogeneity from calf thymus. A single narrow protein band was observed on SDS-polyacrylamide gel electrophoresis, and a single amino-terminal sequence was detected by protein sequencing. However, isoelectric focusing followed by SDS-polyacrylamide gel electrophoresis demonstrated the existence of two proteins of slightly different pI present in similar amounts in the preparation (45). Separation of the two proteins by chromato-focusing was inefficient. Sequencing of several peptides from the preparation established that the purified DNA ligase II was heavily contaminated with the abundant cellular structural protein ezrin, also called cytovillin (73, 74). Ezrin/cytovillin contributed the single nonblocked N-terminal sequence of the preparation. DNA ligase II and ezrin/cytovillin are of the same size, and these two proteins also share the unusual property of adsorbing very tenaciously to hydroxylapatite, which consequently is a key purification step for either protein (45, 75, 76). Probably the easiest remedy for this copurification problem is to isolate DNA ligase II from cell nuclei rather than from whole cell extracts (27). The presence of variable amounts of ezrin/cytovillin in DNA ligase II preparations is unlikely to affect the catalytic properties of the DNA ligase, however, and the ezrin/cytovillin may even serve as a carrier protein instead of externally added bovine serum albumin. Since the sequence of ezrin/cytovillin is known, it is easy to deduce which peptides have been derived from this structural protein, and which ones from DNA ligase II, during amino acid sequencing.

The sequence of DNA ligase II appears to be quite different from that of DNA ligase I. Several peptides derived from DNA ligase II (45) show no detectable sequence homology with the predicted coding sequence of the DNA ligase I cDNA (11). The two enzymes show no serological cross-reactivity, that is, neutralizing rabbit antibodies against DNA ligase I do not bind or inhibit DNA ligase II (56), and neutralizing antibodies against DNA ligase II similarly do not inhibit DNA ligase I (71). Furthermore, an antibody directed against a peptide strongly conserved between DNA ligase I, the *Sch. pombe cdc17*+-encoded DNA ligase, the *S. cerevisiae CDC9* gene product, and vaccinia virus–encoded DNA ligase (15) readily detects these enzymes by immunoblotting but fails to cross-react with DNA ligase II (45, 77). A recent report of a fingerprinting study indicated that DNA ligase II may have a long sequence around its active site that seems identical to that of DNA ligase I (29), but it is presently unclear whether the peptide analyzed was really derived from the partially purified DNA ligase II or from a contaminating fragment of DNA ligase I. Subtilisin digestion of homogeneous DNA ligase I does not generate a fragment the size of DNA ligase II (15).

DNA ligase II resembles other eukaryotic DNA ligases in requiring ATP as cofactor, but the enzyme differs from DNA ligase I in having a much higher

K_M for ATP. Enzyme preparations from different sources consistently exhibit K_M values of 10^{-5}–10^{-4}M in different laboratories. DNA ligase II forms an enzyme-adenylate complex and apparently acts by the same mechanism as other DNA ligases (45, 71), but complex formation is inefficient and difficult to detect at low ATP concentrations due to the high K_M for the cofactor. The enzyme can catalyze the formation of phosphodiester bonds with an oligo (dT)·poly (rA) substrate, but not with an oligo (rA)·poly (dT) substrate, so it differs completely from DNA ligase I in this regard (41, 45). The cellular function of DNA ligase II remains unknown. Whereas DNA ligase I activity is induced in regenerating rat liver, as expected for an enzyme involved in DNA replication, DNA ligase II levels are unaltered in normal versus regenerating liver (57, 70).

DNA LIGASE III

Two different DNA ligases are present in calf thymus cell extracts that can join the oligo (dT)·poly (rA) substrate refractory to DNA ligase I. The recently detected enzyme, which is larger than DNA ligase II and apparently unrelated to that protein, has been named DNA ligase III (45). An enzyme with the properties of DNA ligase III was partially purified from rat liver cell extracts by Elder & Rossignol (16), who speculated that the activity might represent a more native form of DNA ligase II. However, DNA ligase II is not derived by processing from a larger precursor protein, and partial digestion of purified DNA ligase III with reagent proteases does not generate an active fragment of the size of DNA ligase II (45).

DNA ligase III is a minor DNA ligase activity in calf thymus, and was not detected for a long time because much of the biochemical work on DNA ligases in various laboratories employed this particular source for large-scale enzyme purification. In other cells and tissues, the level of DNA ligase III can approach that of DNA ligase I (58). DNA ligase III elutes before DNA ligase I on gel filtration, and the molecular mass of the active enzyme determined by hydrodynamic methods is about 147 kDa. The most purified fractions of the calf thymus enzyme show only two peptide bands by silver staining after SDS-polyacrylamide gel electrophoresis, corresponding to 100 kDa and 46 kDa (45). The partially purified rat liver enzyme appears to have the same structure (16). The 100-kDa polypeptide contains the AMP residue of a DNA ligase–adenylate complex. These data indicate that DNA ligase III may be a heterodimer comprising two subunits, with the active site for enzyme-AMP formation present in the larger subunit. All other DNA ligases investigated to date have a monomeric structure, although dimerization or aggregation of DNA ligase I has been observed under certain conditions (78, 79). For this reason, it cannot yet be concluded whether the 46-kDa peptide is an essential and integral part of DNA ligase III, or if it represents a separate protein that binds tenaciously to the 100-kDa DNA ligase component and may be involved in related cellular functions.

DNA ligase III resembles DNA ligase I, and differs from DNA ligase II, in binding only weakly to hydroxylapatite and having a low K_M, ~$2\mu M$, for ATP. However, polyclonal antisera against DNA ligase I or the evolutionarily conserved peptide sequence in the C-terminal region of DNA ligase I failed to detect DNA ligase III by immunoblotting (16, 45). These data indicate that DNA ligases I and III are not closely related. DNA ligase III repairs single-strand breaks in DNA efficiently, but is unable to perform either blunt-end joining or AMP-dependent relaxation of supercoiled DNA (42a). The substrate specificity of DNA ligase III differs from that of both DNA ligase I and II in that the enzyme can join both the oligo (dT)·poly (rA) and oligo (rA)·poly (dT) hybrid substrates. However, all three enzymes effectively ligate single-strand breaks with 3'-hydroxyl and 5'-phosphate termini in double-stranded DNA or the synthetic oligo (dT)·poly (dA) substrate (45).

In common with DNA ligase II, DNA ligase III is not induced during rat liver regeneration (16). Furthermore, the presence of the enzyme in the postmicrosomal supernatant of rat liver cells extracted with an isotonic sucrose buffer indicates that the enzyme has a similar subcellular distribution to DNA ligase I, which leaches out of nuclei under such conditions. In contrast, DNA ligase II is not extracted from nuclei with isotonic sucrose (27, 39, 70).

The definition of precise functions for DNA ligases II and III remains a challenging problem. The essential requirements for DNA ligase activity in joining of Okazaki fragments during DNA replication and completion of DNA repair pathways are most likely fulfilled by DNA ligase I. Consequently, it seems probable that DNA ligases II and III are involved in more specialized and less obvious functions. These are at present a matter of speculation. However, there is no shortage of interesting putative roles for the enzymes. For example, which DNA ligase joins larger discrete DNA replication intermediates (80), or adjacent replicons? Which enzyme is responsible for the joining of unusual replication intermediates and DNA structures occurring at telomeres (81) and perhaps at centromeres? The presence of stretches of single-stranded DNA up to 10^5 nucleotides long in gently prepared DNA from human lymphoid cells in S phase (82) indicates that asymmetric DNA replication occurs in addition to simultaneous leading- and lagging-strand synthesis—which enzymes are involved in such replication? Furthermore, does the observed increase in DNA ligase II activity during meiotic prophase indicate a specific role for the enzyme in completing meiotic recombination reactions (27)? Is there a requirement in cells for DNA ligases that can act on the DNA·RNA hybrid structures occurring during transcription? Similarly, is a DNA ligase required to seal breaks in unusual DNA

Table 1 Properties of the three mammalian DNA ligases

	I	II	III
Molecular mass estimated by SDS-PAGE	125 kDa	72 kDa	100 kDa (associated with a 46-kDa polypeptide)
by cDNA sequence	102 kDa	—	—
Chromosomal localization	19q13.2–13.3	—	—
Ligation of oligo(dT) · poly(dA)	Yes	Yes	Yes
oligo(dT) · poly(rA)	No	Yes	Yes
oligo(rA) · poly(dT)	Yes	No	Yes
K_m for ATP	Low	High	Low
Adsorption to hydroxylapatite	Weak	Strong	Weak
Recognition by DNA ligase I–specific antisera	Yes	No	No
Subcellular localization	Nucleus	Nucleus	Nucleus/cytoplasm?
Induction upon cell proliferation	Yes	No	No
Proportion of DNA ligase activity in calf thymus extract	~85%	5–10%	5–10%

structures such as Z-DNA or H-DNA? Finally, joining of nonhomologous DNA molecules with protruding single-strand ends occurs efficiently in vertebrate cell extracts and is believed to be important during illicit recombination (83–85), but the DNA ligase(s) responsible for this reaction has not yet been identified. On a less speculative note, it should be mentioned that neither DNA ligase I, II, nor III is exclusively a mitochondrial enzyme (16, 27, 31; A. E. Tomkinson, T. Lindahl, unpublished data), but it is not known which of these three activities (if any) accounts for the circularization of mitochondrial DNA. A comparison of the properties of DNA ligases I, II, and III is shown in Table 1.

DNA LIGASES IN OTHER EUKARYOTIC SYSTEMS

Joining of Okazaki fragments during DNA replication is an essential cellular function, and ATP-dependent DNA ligases have been found to be ubiquitous in eukaryotic cells. Early work documenting the presence of a DNA ligase in, for example, plant cells has been reviewed (39). Recent investigations have focused on well-characterized organisms that allow genetic or cell cycle analysis.

Schizosaccharomyces pombe *and* Saccharomyces cerevisiae *DNA Ligases*

The distantly related fission yeast and budding yeast each have an essential gene encoding a DNA ligase, *cdc17+* in *Sch. pombe* (86) and *CDC9* in *S. cerevisiae* (87). Conditional lethal DNA ligase mutants exhibit a phenotype typical of a cell division cycle (cdc) defect, and arrest in medial nuclear division at the restrictive temperature. Mutant cells contain an anomalously heat-labile DNA ligase activity and accumulate Okazaki fragments at 37 °C (87). *Sch. pombe cdc17* and *S. cerevisiae cdc9* mutants are also hypersensitive to ultraviolet light, ionizing radiation, and alkylating agents, indicating a general defect in DNA excision-repair in addition to the replication defect. The *S. cerevisiae CDC9* and *Sch. pombe cdc17+* genes have been cloned by functional complementation of a conditional lethal yeast DNA ligase mutant (88, 89), and a similar approach was later employed for the cloning of a complete human cDNA encoding DNA ligase I in *S. cerevisiae* (11). A comparison of the coding sequences (12) shows that the two yeast genes are related to each other and to human DNA ligase I (see Figure 1), indicating that they perform similar cellular functions. In particular, these three DNA ligases show strong sequence conservation towards their C-termini, whereas there is little or no sequence homology close to the N-termini. The highly conserved regions toward the C-termini of the DNA ligase polypeptides may define essential enzyme functions.

The *Sch. pombe cdc17+* gene encodes an 86-kDa protein (12, 90). Although the enzyme has not been highly purified, it can be detected as an 85–86-kDa enzyme-adenylate complex in crude cell extracts and partially purified protein fractions. Similar results have been obtained for the *S. cerevisiae CDC9* gene product. The 87-kDa primary translation product is readily degraded to an active 78-kDa fragment by intrinsic proteolysis in yeast cell extracts. Thus, the yeast DNA ligases seem to share a property of human DNA ligase I in having an N-terminal region that is not required for DNA ligase activity in vitro. However, the size of this apparently nonessential N-terminal region in yeast DNA ligases is only 1/3–1/2 of that of mammalian DNA ligase I.

In *S. cerevisiae*, *CDC9* and several other genes involved in DNA synthesis are periodically expressed at the G1/S boundary (91). However, such periodic transcription is not apparent for the *cdc17+* gene of *Sch. pombe* (92). The coordinate regulation of the relevant genes in *S. cerevisiae* is mediated by binding of a specific transcription factor to an ACGCGT hexamer sequence occurring in the promoters of these genes (93). This factor may be a key regulatory element in the control of S phase entry in *S. cerevisiae*.

Transcription of the DNA ligase genes of *S. cerevisiae* and *Sch. pombe* is strongly induced by exposure to DNA-damaging agents (94, 95). However, dramatic changes in transcript levels are not matched by sizeable variations in the corresponding protein levels. DNA damage induces at most a 50% increase in DNA ligase activity in *S. cerevisiae*. This is probably due to the high level and long half-life of the enzyme in growing yeast cells. The increased transcription of the DNA ligase gene in response to DNA damage may be of greater importance in cells that have been in stationary phase for very long time periods (94).

The elegant and detailed molecular genetic studies of the major DNA ligase in yeast have not yet been extended by enzymological investigations and large-scale protein purification in an attempt to search for minor DNA ligase activities in these organisms. The question of whether yeast cells, like mammalian cells, contain more than one DNA ligase thus remains unresolved. Recent fractionation studies on yeast cell extracts have identified a second distinct DNA ligase activity that resembles the mammalian DNA ligases II and III in its ability to join an oligo (dT)·poly (rA) substrate (A. E. Tomkinson, unpublished data). In this regard, it is interesting that a DNA ligase–defective *cdc17* mutant of *Sch. pombe* with impaired DNA replication, repair, and mitotic recombination exhibits unaffected meiotic recombination (96). Higashitani et al (27) have proposed that mammalian DNA ligase II, rather than DNA ligase I, could serve to catalyze the final step in meiotic recombination. The yeast system offers the possibility of putting this speculative proposal of a defined function for DNA ligase II on a firm experimental footing.

Drosophila melanogaster *DNA Ligases*

Two distinct DNA ligases have been detected in *Drosophila*. DNA ligase I from *D. melanogaster* has been extensively characterized with regard to its mechanism of action (43, 97, 98). This enzyme, and DNA ligase I from bovine thymus, are the two most studied eukaryotic DNA ligases. The *Drosophila* DNA ligase I has been purified to apparent homogeneity from embryos. In common with DNA ligase I from other sources, it has a low K_M (1.6 μM) for ATP, is unable to join oligo (dT)·poly (rA), and performs blunt-end joining of DNA in the presence of polyethylene glycol. In addition, *Drosophila* DNA ligase I can ligate an oligo (rA)·poly (dT) hybrid substrate (43), as later also observed for mammalian DNA ligase I (45), but is unable to join oligo (rA)·poly (rU). Consequently, the substrate specificity of the *Drosophila* enzyme appears to be the same as for mammalian DNA ligase I, but different from that of DNA ligases II and III. Detailed mechanistic studies have established that *Drosophila* DNA ligase I acts by precisely the same mechanism as that established for *E. coli* and bacteriophage T4 DNA ligases.

Drosophila DNA ligase I is isolated as an 83–86-kDa protein, as estimated by SDS-polyacrylamide gel electrophoresis, which is very susceptible to proteolysis and can be degraded by a variety of reagent proteases in vitro to an active fragment of 75 kDa, presumably by removal of N-terminal sequences. The 83–86-kDa form exhibits a high frictional ratio, indicating an asymmetric shape of the protein. Extensive proteolysis also generates a smaller 64-kDa form, which retains the ability to generate an enzyme-adenylate complex but can no longer catalyze the DNA-joining reaction. Such partially active proteolytic fragments have been observed upon digestion of *E. coli* DNA ligase (99) but have not been seen during limited proteolysis of mammalian DNA ligase I (15). These results indicate that *Drosophila* DNA ligase I is of a similar size to the corresponding *Sch. pombe* enzyme, although it cannot be excluded at present that the primary translation product is as large as the mammalian or *Xenopus* enzyme but is very rapidly cleaved to an 83–86-kDa active fragment in crude cell extracts. Cloning and sequencing of the *Drosophila* cDNA should resolve this problem. Alternatively, antibodies recognizing *Drosophila* DNA ligase I could be employed to define the size of the DNA ligase polypeptide by immunoblotting after gel electrophoresis of cell extracts prepared by instant lysis with hot SDS. Such detergent cell lysates were useful to define the correct size of the primary translation product of mammalian DNA ligase I (31).

A second DNA ligase with properties similar to those of mammalian DNA ligase II has been partially purified from *Drosophila* pupae and embryos (98, 100). In common with the mammalian enzyme, *Drosophila* DNA ligase II is not extracted from cell nuclei with an isotonic sucrose buffer of low ionic strength, and this accounts for an early unsuccessful attempt to detect the enzyme (97). However, *Drosophila* DNA ligase II is readily extracted from disrupted cells or isolated cell nuclei with 1 M NaCl, and the two *Drosophila* ligases can then be separated conveniently by phosphocellulose chromatography (98). The purified *Drosophila* DNA ligase II enzyme has a low affinity for the ATP cofactor and exhibits a K_M 10-fold higher than that of *Drosophila* DNA ligase I. Furthermore, the purified DNA ligase II can catalyze joining of an oligo (dT)·poly (rA) substrate. Two different inactivating monoclonal antibodies directed against *Drosophila* DNA ligase II have been isolated, and neither antibody binds or neutralizes *Drosophila* DNA ligase I (100). The *Drosophila* DNA ligase II is a 70-kDa protein, as defined both by immunoblotting and by analysis of a radioactive enzyme-AMP complex. In all these properties, the enzyme resembles mammalian DNA ligase II. Interestingly, a small amount of an apparent "larger form of DNA ligase II" has also been observed during fractionation of *Drosophila* extracts (98, 100), hinting at the possibility that a third DNA ligase might be present in this organism.

Drosophila DNA ligase I is present at a high level in unfertilized eggs and early developing embryos, but at a 10-fold lower level in larvae, pupae, and adults (43, 98, 100). The DNA ligase I activity correlates well with changes in DNA replication during development, since a very high replication rate is characteristic of early embryos. In complete contrast, the level of DNA ligase II activity does not change significantly between different developmental stages. For this reason, the enzyme has been purified from pupae, where DNA ligase I levels are low. The availability of highly purified enzymes and appropriate antibodies provides a good starting point for attempts to define the functions of the different DNA ligases in *Drosophila*.

Xenopus laevis *DNA Ligase*

The dominant DNA ligase activity in oocytes and eggs of the clawed toad, *Xenopus laevis,* is due to a large enzyme of highly asymmetric shape, which is detected by antisera against mammalian DNA ligase I (101). The *Xenopus* DNA ligase cannot join an oligo (dT)·poly (rA) hybrid substrate, but catalyzes blunt-end joining of DNA. Similarly to DNA ligase I of *Drosophila*, the *Xenopus* enzyme is present at a 10-fold lower level in adult tissue than in eggs. The *Xenopus* DNA ligase has been purified to near homogeneity, and a specific antiserum against the enzyme obtained. However, the occurrence of further DNA ligases in this organism has not been investigated.

Molecular size determinations have been performed by SDS-polyacrylamide gel electrophoresis, either by analysis of a radioactive enzyme-AMP complex, or by immunoblotting with antibodies against the *Xenopus* enzyme or bovine DNA ligase I. The results indicate that the *Xenopus* DNA ligase is a 180-kDa protein, which can be rapidly degraded by proteolysis in vitro to active fragments of 130 kDa and then 76 kDa. These data show that the *Xenopus* enzyme has a 76-kDa catalytic domain, which is very similar in size to that present in mammalian DNA ligase I, *Drosophila* DNA ligase I, and the yeast DNA ligases. However, the native *Xenopus* enzyme is the largest DNA ligase observed to date. It seems likely that it has an even longer nonessential N-terminal region than mammalian DNA ligase I. Even if the *Xenopus* DNA ligase I shares the property of mammalian DNA ligase I of migrating anomalously slowly in SDS-polyacrylamide gels, it is still clearly larger than the bovine enzyme. DNA ligases have been partially purified from two other amphibians, *Pleurodeles* and axolotl (102). These are also large proteins of 180 kDa and 160 kDa, respectively, which are detected by an antibody directed against the *Xenopus* DNA ligase. In all three amphibians, the DNA ligase is found at a high level in oocytes, unfertilized eggs, and embryos. A 25-kDa *Xenopus* DNA-binding protein that strongly promotes concatenation of linear DNA molecules by the DNA ligase has been identified (102a). Early reports of different size classes of DNA ligase before and after fertilization of amphibian eggs, as determined by sucrose gradient centrifugation of extracts (103, 104), probably reflected different levels of intrinsic proteolysis (101, 102), which converts the 180-kDa DNA ligase to the 76-kDa form. A second DNA ligase, if demonstrated in *Xenopus*, might be expected to occur at a similar level throughout development in analogy with *Drosophila* DNA ligase II. Injection of ultraviolet- or X-irradiated plasmid DNA into nuclei of *Xenopus* oocytes leads to rapid DNA repair with conversion of nicked circles to a covalently closed form (105, 106). It seems likely that the abundant DNA ligase accounts for this joining process.

Vaccinia Virus DNA Ligase

Human tumor viruses that encode a DNA polymerase, such as herpesviruses, adenoviruses, and retroviruses, do not produce a virus-specific DNA ligase. Thus, the complete DNA sequence of Epstein-Barr virus (107) does not contain an open reading frame with the active site motif (Figure 1) of a DNA ligase. Early investigations of vaccinia virus–infected cells demonstrated a 10-fold increase in DNA ligase activity in cell extracts (108), but failed to distinguish a virus-induced enzyme from host cell DNA ligase activity (109). However, recent studies (110–112) have clearly established that vaccinia virus encodes a DNA ligase. Since poxviruses, including vaccinia virus, have large genomes and replicate in the host cell cytoplasm, it is not surprising that they encode several different enzymes of DNA metabolism. The DNA ligase–encoding gene is transcribed early during infection. A restriction enzyme fragment carrying the relevant gene has been isolated and codes for a 63-kDa protein. The sequence of the vaccinia virus DNA ligase shows about 30% identity with those of the *Sch. pombe* and *S. cerevisiae* DNA ligases. The vaccinia enzyme is also distantly related to mammalian DNA ligase I, has a similar active site sequence (Figure 1), and contains a highly conserved peptide towards the C-terminus, which is also present in the C-terminal regions of mammalian DNA ligase I and the yeast DNA ligases (Figure 2). The vaccinia DNA ligase has been shown to form an enzyme-adenylate complex of about 61 kDa on incubation with ATP, in good agreement with the molecular mass estimate of the protein from the open reading frame. The adenylate residue is released from the enzyme-AMP complex on incubation with nicked DNA or with pyrophosphate, as expected for a DNA ligase. Moreover, the vaccinia enzyme catalyzes ligation of strand interruptions in an oligo (dT)·poly (dA) substrate.

The vaccinia virus–encoded DNA ligase is nonessential for viral DNA replication and growth on several types of host cells (112, 113). Possibly,

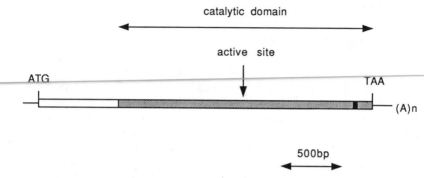

Figure 2 Schematic representation of the human DNA ligase I cDNA. Landmark features encoded by the cDNA are highlighted. The coding sequence is represented by an open box, the catalytic domain is shaded, and the position of a 16-amino-acid peptide, which is highly conserved in human DNA ligase I, as well as in the yeast and vaccinia DNA ligases (15), is marked by a solid bar.

vaccinia virus replicates its DNA in an asymmetric fashion without requirement for a DNA ligase (114), or traces of a host DNA ligase are recruited to the cytoplasmic viral replication factories. Recent studies have shown that a virus DNA ligase–deficient deletion mutant is anomalously sensitive to ultraviolet light and bleomycin during infection, indicating a role for the enzyme in viral DNA repair, and the mutant strain also exhibits attenuated virulence in vivo (77).

HUMAN CELL LINES DEFICIENT IN DNA LIGATION

Several human syndromes with autosomal recessive inheritance confer severe immunodeficiency and a greatly increased cancer frequency, while at the cellular level they are characterized by chromosome instability and hypersensitivity to DNA-damaging agents. These include (*a*) ataxia-telangiectasia (McKusick nr. 208900) and its two clinical variants (McKusick nrs. 208910 and 208920), and the closely related Nijmegen breakage syndrome (McKusick nr. 251260); (*b*) Fanconi's anemia (two genetic complementation groups, McKusick nrs. 227650 and 227660) and several variants (McKusick nrs. 227700, 227800, 227810, 227850), and the closely related Blackfan-Diamond anemia (McKusick nr. 205900); (*c*) Bloom's syndrome (McKusick nr. 210900). Direct measurements of enzyme activities in fractionated extracts of representative cell lines did not reveal any alteration of DNA ligases in ataxia-telangiectasia or Fanconi's anemia (115). However, Bloom's syndrome cells exhibit an apparent DNA ligation deficiency in vitro and in vivo (28, 78, 79, 115–117).

The DNA Ligase I–Defective Cell Line 46BR

A cell line with a DNA ligation deficiency was established from a young Irish female patient with clinical symptoms resembling Bloom's syndrome, that is, stunted growth, sun sensitivity, and severe immunodeficiency with recurrent infections. The patient died at age 19, apparently as a consequence of a slow-growing abdominal lymphoma in combination with an acute pulmonary infection. A fibroblast cell strain, 46BR, and an SV40-transformed line 46BR.1G1 with a closely similar phenotype, have been established in culture. 46BR cells are hypersensitive to a wide range of DNA-damaging agents including alkylating agents such as methyl methanesulfonate and dimethyl sulfate, ionizing radiation, and ultraviolet light. The cells are also hypersensitive to the poly ADP-ribose polymerase inhibitor 3-aminobenzamide, which interferes with DNA excision-repair. In contrast, ataxia-telangiectasia cells and Fanconi's anemia cells appear uniquely sensitive to ionizing radiation as opposed to DNA cross-linking agents and are normally resistant to methyl methanesulfonate and ultraviolet light. 46BR cells show a slightly but significantly elevated frequency of spontaneous sister chromatid exchange, and are hypersensitive to induction of such events by DNA-damaging agents. However, the spontaneous level of sister chromatid exchange is lower than that observed in Bloom's syndrome. This has been a main reason for considering 46BR as different from Bloom's syndrome, especially since measurements of sister chromatid exchange frequency are employed in the clinical diagnosis of Bloom's syndrome (59–63).

46BR fibroblasts, as well as the SV40-transformed subline, exhibit strongly delayed joining of Okazaki fragments during DNA replication (59, 60, 60a). Radioactive labeling of newly synthesized DNA in vivo with a 10-minute pulse of [^{3}H]thymidine, followed by DNA size determination by alkaline sucrose gradient centrifugation, showed that about one-half of the newly synthesized DNA remained as Okazaki fragments, whereas the great majority of newly replicated DNA in control cells was already in high-molecular-weight form. The delay in DNA joining observed in 46BR cells was only transient, because the newly synthesized DNA could be converted to a high-molecular-weight form during a 30-min chase period after the radioactive labeling. Furthermore, the rate of joining of DNA strand breaks occurring as a consequence of exposure of 46BR and 46BR.1G1 cells to dimethyl sulfate or ultraviolet light was markedly slower than that characteristic of control cells. These observations led to the proposal that 46BR cells might be defective in a DNA ligase (59, 63, 118).

Direct measurements of enzyme activities in cell extracts of 46BR.1G1, which grows better than 46BR in tissue culture, suggested similar levels of DNA-joining activity in standard in vitro DNA ligase assays for DNA ligases

I and II. However, DNA ligase I exhibited an unusual tendency to aggregate on size fractionation (60). Recent experiments have indicated that DNA ligase III activity is normal in 46BR.1G1 cells (58). Further investigations of the properties of DNA ligase I from 46BR.1G1 cells demonstrated a distinct anomaly, in that the amount of enzyme-adenylate complex formed on incubation with ATP is greatly reduced; only traces of the complex can be detected (58). In this regard, the DNA ligase I of 46BR.1G1 cells resembles the defective DNA ligase present in the conditional lethal *E. coli* mutant SG251 when grown at the permissive temperature; DNA-joining activity seems normal, but DNA ligase-AMP formation cannot be detected (32).

In a search for mutations in the structural gene encoding DNA ligase I in 46BR cells, the entire coding sequence was analyzed by direct DNA sequencing of cDNA amplified by the polymerase chain reaction (PCR) (58). In general, the coding sequence of the DNA ligase I gene is strongly conserved in the human population. Seven different DNA ligase I cDNA sequences from individuals of Anglo-Saxon, Ashkenazi Jewish, or Japanese origin did not reveal any amino acid replacements; the only polymorphisms observed were third-letter C/T variations in the codons for residues Gly377 and Asp802. However, diploid 46BR cells exhibited two sequence alterations in 50% of the cDNA. Both mutations represented CpG-to-TpG changes, a transition commonly found in human genetic disease (119) that may be ascribed to hydrolytic spontaneous deamination of a 5-methylcytosine residue. One of the mutations in the 46BR DNA ligase I cDNA would affect the active site region of the enzyme, leading to a replacement of the strongly conserved Glu566 residue (position −2 in Figure 1) with a Lys residue. Site-specific mutagenesis experiments (13) indicate that this alteration leads to an almost complete loss of enzyme activity. DNA ligase I protein synthesized in 46BR cells with the Glu566 → Lys alteration would be expected to be enzymatically inactive. The other mutation in 46BR DNA causes a replacement of Arg771, which is located in a region of the sequence showing strong evolutionary conservation, with a Trp residue. PCR amplification of appropriate sections of genomic DNA from 46BR cells and DNA sequencing established that both the mutations first seen in cDNA sequences are present in the cellular genome. Furthermore, genomic DNA from the mother of the patient contains the Arg771 → Trp mutation but not the Glu566 → Lys mutation. Thus, the two mutations seen in diploid 46BR cells are carried on different alleles, as also directly confirmed by cloning individual cDNAs, and one is maternally inherited. In the SV40-transformed aneuploid subline 46BR.1G1, selected as having the same hypersensitivity to DNA-damaging agents, gene rearrangements have occurred so that the line is homozygous for the Arg771 → Trp mutation but has the normal DNA sequence in the active site region. Consequently, the Arg771 → Trp mutation in DNA ligase I would appear to account for the malfunctioning but partly active enzyme present in 46BR cells that allows cell proliferation.

Bloom's Syndrome

Several cases of Bloom's syndrome form a single genetic complementation group (120). Cells of Bloom's syndrome origin exhibit delayed joining of DNA replication intermediates (80, 121, 122). However, the replication intermediates that accumulate in vivo are not Okazaki fragments but larger intermediates of 20 kb. The rate of DNA replication is reduced, in particular in early S phase, and replication fork displacement rates are 55–65% of normal (123–125). After transfection of Bloom's syndrome cells with a linearized shuttle vector plasmid, recircularization by end-joining was reduced and error-prone compared to that observed in control cells, indicating a DNA ligation deficiency (117). Bloom's syndrome cells are moderately hypersensitive to many different types of DNA-damaging agents, but differ from 46BR cells in that a strong delay in joining of DNA strand breaks caused by dimethyl sulfate treatment does not occur (60).

DNA polymerases appear to be normal in Bloom's syndrome (126). In contrast, partially purified DNA ligase I from Bloom's syndrome cells exhibits reduced activity and altered fractionation properties, whereas DNA ligase II seems normal (28, 78, 79, 115, 116). In more detailed studies on the nucleic acid sequence level, PCR amplification and DNA sequencing of DNA ligase I cDNA from several different Bloom's syndrome cell lines has not revealed mutations within the coding sequence (58, 126a). The molecular explanation for the reduced DNA ligase activity in cell extracts of Bloom's syndrome origin remains to be firmly defined, but the following possibilities could be considered:

1. Measurements of DNA ligase activities in fractionated extracts of Bloom's syndrome cells were performed before DNA ligase III was identified (45). Thus, size fractionation of crude cell extracts to separate DNA ligases I and II yielded a high-molecular-weight active protein fraction of partially purified DNA ligase I, which in fact contained both DNA ligase I and DNA ligase III. Cloning and sequencing of cDNA sequences representative of DNA ligase III should now be performed, to investigate whether this activity is defective in Bloom's syndrome.

2. The activity of DNA ligase I is dependent on the phosphorylation state of the enzyme (18). The phosphorylation patterns and other posttranslational modifications of DNA ligase I from normal cells as compared to Bloom's syndrome cells need to be defined.

3. Several cellular protein factors that modulate the activity of DNA ligase I have been described. These include a ~65-kDa inhibitor that initially copurifies with DNA ligase I (67a) but is not present in the homogeneous calf

thymus enzyme (15). By comparison, purified DNA ligase III contains two different polypeptides, the larger of which carries the active site for adenylylation (16, 45). The DNA ligase I–associated protein factor has not yet been studied in Bloom's syndrome cells.

4. The alteration of DNA ligase I activity in Bloom's syndrome cell extracts could be the indirect consequence of a more general defect in cellular metabolism. For example, it has been proposed that Bloom's syndrome cells might be defective in detoxification of active oxygen species (127). The evidence for this is quite indirect, but includes an elevated cellular level of superoxide dismutase (128). The synthesis of a number of proteins seems to be abnormally regulated in Bloom's syndrome cells, including uracil-DNA glycosylase (129) and c-myc protein (130). However, the overproduction of c-myc protein may be related to the demonstrated induction of c-myc protein synthesis in response to the introduction of chain breaks in DNA (131).

The resolution of these alternatives has a bearing on the nomenclature and classification of the important DNA ligase I–defective 46BR cell line. If Bloom's syndrome is due to a mutation specifically affecting the molecular activity of a DNA ligase, 46BR could be considered a variant, or second genetic complementation group, of the syndrome. On the other hand, if the basic defect in Bloom's syndrome affects several cellular functions in addition to DNA ligation, 46BR might be regarded as the first representative of a novel human syndrome.

Acute Lymphoblastic Leukemia

Rusquet et al (132) reported that extracts of leukemic cells from patients with T-cell acute lymphoblastic leukemia had very low or undetectable levels of DNA ligase activity. These studies remain unconfirmed; a T-cell line representative of the disease, Jurkat/J-6, had normal levels of DNA ligase activities (133).

Concluding Statement

The biochemical defects in the inherited human syndromes associated with hypersensitivity to DNA-damaging agents remain largely unknown. Work is in progress in many laboratories to clone and characterize cDNA sequences that might complement the defects in diseases such as ataxia-telangiectasia, Fanconi's anemia, Cockayne's syndrome, and Werner's syndrome, and variants of these syndromes. With regard to the intensely studied and genetically complex disease, xeroderma pigmentosum, cDNA sequences for three of at least seven genes required for enzymatic incision at pyrimidine dimers in ultraviolet-irradiated DNA have been cloned. The sequences contain structural motifs indicating the presence of zinc fingers in the XP-A protein, and putative DNA helicase domains in the XP-B and XP-D proteins (37, 134–

136). However, the only protein purified to date is the XP-A protein, which binds specifically to ultraviolet-irradiated DNA (137). The isolated XP-A protein has no enzymatic activity by itself and presumably forms part of an incision complex. Thus, the single presently documented case within this group of inherited syndromes, in which a DNA sequence alteration can be correlated directly with a biochemical defect in an enzyme of known function, is represented by the individual from whom the 46BR cell line was established. The IgA and IgG deficiencies in this patient may be explained by inadequate DNA ligation during the recombinational processes in which V and C genes are joined, and antibody variability is induced by V gene processing. The hypersensitivity to DNA-damaging agents is undoubtedly due to inadequate DNA ligation during DNA excision repair. Interest in this general area has been furthered by the discovery that cells from severely immunodeficient mice homozygous for the *scid* mutation on chromosome 16 and defective in V(D)J [variable (diversity) joining] recombination are also hypersensitive to DNA damage introduced by ionizing radiation (138–141). Further investigation of cell lines derived from immunodeficient human individuals who are also hypersensitive to DNA-damaging agents promises new insights into the correlation between enzymes active in DNA metabolism and distinct forms of inherited disease.

Literature Cited

1. Li, J. J., Kelly, T. J. 1984. *Proc. Natl. Acad. Sci. USA* 81:6973–77
2. Wood, R. D., Robins, P., Lindahl, T. 1988. *Cell* 53:97–106
3. Jessberger, R., Berg, P. 1991. *Mol. Cell. Biol.* 11:445–57
4. Lehman, I. R. 1974. *Science* 186:790–97
5. Gumport, R. I., Lehman, I. R. 1971. *Proc. Natl. Acad. Sci. USA* 68:2559–63
6. Tomkinson, A. E., Totty, N. F., Ginsburg, M., Lindahl, T. 1991. *Proc. Natl. Acad. Sci. USA* 88:400–4
7. Montecucco, A., Lestingi, M., Pedrali-Noy, G., Spadari, S., Ciarrocchi, G. 1990. *Biochem. J.* 271:265–68
8. Uhlenbeck, O. C., Gumport, R. I. 1982. *Enzymes* 15B:31–58
9. Heaphy, S., Singh, M., Gait, M. J. 1987. *Biochemistry* 26:1688–96
10. Xu, Q., Teplow, D., Lee, T. D., Abelson, J. 1990. *Biochemistry* 29:6132–38
11. Barnes, D. E., Johnston, L. H., Kodama, K., Tomkinson, A. E., Lasko, D. D., Lindahl, T. 1990. *Proc. Natl. Acad. Sci. USA* 87:6679–83
12. Barker, D. G., White, J. H. M., Johnston, L. H. 1987. *Eur. J. Biochem.* 162:659–67
13. Kodama, K., Barnes, D. E., Lindahl, T. 1991. *Nucleic Acids Res.* 19: 6093–99
14. Teraoka, H., Tsukada, K. 1982. *J. Biol. Chem.* 257:4758–63
15. Tomkinson, A. E., Lasko, D. D., Daly, G., Lindahl, T. 1990. *J. Biol. Chem.* 265:12611–17
16. Elder, R. H., Rossignol, J. M. 1990. *Biochemistry* 29:6009–17
17. Mezzina, M., Rossignol, J. M., Philippe, M., Izzo, R., Bertazzoni, U., Sarasin, A. 1987. *Eur. J. Biochem.* 162:325–32
18. Prigent, C., Lasko, D. D., Kodama, K. I., Woodgett, J., Lindahl, T. 1992. Submitted
19. Siegel, L. M., Monty, K. J. 1966. *Biochim. Biophys. Acta* 112:346–62
20. Teraoka, H., Tsukada, K. 1985. *J. Biol. Chem.* 260:2937–40
21. Teraoka, H., Tsukada, K. 1986. *Biochim. Biophys. Acta* 873:297–303
22. Prigent, C., Aoufouchi, S., Philippe, M. 1990. *Biochem. Biophys. Res. Commun.* 169:888–95
23. Mezzina, M., Sarasin, A., Politi, N., Bertazzoni, U. 1984. *Nucleic Acids Res.* 12:5109–22
23a. Noguiez, P., Barnes, D. E., Mohren-

weiser, H. W., Lindahl, T. 1992. Submitted
24. Lindahl, T., Edelman, G. M. 1968. *Proc. Natl. Acad. Sci. USA* 61:680–87
25. Zimmerman, S. B., Levin, C. J. 1975. *J. Biol. Chem.* 250:149–55
26. Bhat, R., Grossman, L. 1986. *Arch. Biochem. Biophys.* 244:801–12
27. Higashitani, A., Tabata, S., Endo, H., Hotta, Y. 1990. *Cell Struct. Funct.* 15:67–72
28. Chan, J. Y. H., Becker, F. F. 1988. *J. Biol. Chem.* 263:18231–35
29. Yang, S. W., Becker, F. F., Chan, J. Y. H. 1990. *J. Biol. Chem.* 265:18130–34
30. Pedrali-Noy, G. C. F., Spadari, S., Ciarrocchi, G., Pedrini, A. M., Falaschi, A. 1973. *Eur. J. Biochem.* 39:343–51
31. Lasko, D. D., Tomkinson, A. E., Lindahl, T. 1990. *J. Biol. Chem.* 265:12618–22
32. Dermody, J. J., Robinson, G. T., Sternglanz, R. 1979. *J. Bacteriol.* 139:701–4
33. d'Arpa, P., Machlia, P. S., Ratrie, H., Rothfield, N. F., Cleveland, D. W., Earnshaw, W. C. 1988. *Proc. Natl. Acad. Sci. USA* 85:2543–47
34. Bestor, T., Laudano, A., Mattaliano, R., Ingram, V. 1988. *J. Mol. Biol.* 203:971–83
35. Barnes, D. E., Kodama, K., Tynan, K., Trask, B. J., de Jong, P. J., et al. 1992. *Genomics* 12:164–66
36. van Duin, M., deWit, J., Odijk, H., Westerveld, A., Yasui, A., et al. 1986. *Cell* 44:913–23
37. Weber, C. A., Salazar, E. P., Stewart, S. A., Thompson, L. H. 1990. *EMBO J.* 9:1437–47
38. Thompson, L. H., Brookman, K. W., Jones, N. J., Allen, S. A., Carrano, A. V. 1990. *Mol. Cell. Biol.* 10:6160–71
39. Soderhall, S., Lindahl, T. 1976. *FEBS Lett.* 67:1–8
40. Soderhall, S. 1975. *Eur. J. Biochem.* 51:129–36
41. Arrand, J. E., Willis, A. E., Goldsmith, I., Lindahl, T. 1986. *J. Biol. Chem.* 261:9079–82
42. Zimmerman, S. B., Pfeiffer, B. H. 1983. *Proc. Natl. Acad. Sci. USA* 80:5852–56
42a. Elder, R. H., Montecucco, A., Ciarrocchi, G., Rossignol, J. M. 1992. *Eur. J. Biochem.* 203:53–58
43. Rabin, B. A., Chase, J. W. 1987. *J. Biol. Chem.* 262:14105–11
44. Montecucco, A., Fontana, M., Focher, F., Lestingi, M., Spadari, S., Ciarrocchi, G. 1991. *Nucleic Acids Res.* 19:1067–72
45. Tomkinson, A. E., Roberts, E., Daly,

G., Totty, N. F., Lindahl, T. 1991. *J. Biol. Chem.* 266:21728–35
46. Pinna, L. A. 1990. *Biochim. Biophys. Acta* 1054:267–84
47. Bensch, K. G., Tanaka, S., Hu, S.-Z., Wang, T. S.-F., Korn, D. 1982. *J. Biol. Chem.* 257:8391–96
48. Nakamura, H., Morita, T., Sato, C. 1986. *Exp. Cell Res.* 165:291–97
49. Nakayasu, H., Berezney, R. 1989. *J. Cell. Biol.* 108:1–11
50. Wilcock, D., Lane, D. P. 1991. *Nature* 349:429–31
51. Harlow, E., Whyte, P., Franza, B. R., Schley, C. 1986. *Mol. Cell Biol.* 6:1579–89
52. Whyte, P., Buchkovich, K. J., Horowitz, J. M., Friend, S. H., Raybuck, M., et al. 1988. *Nature* 334:124–29
53. Pines, J., Hunter, T. 1990. *Nature* 346:760–63
54. Goulian, M., Richards, S. H., Heard, C. J., Bigsby, B. M. 1990. *J. Biol. Chem.* 265:18461–71
55. Malkas, L. H., Hickey, R. J., Li, C., Pedersen, N., Baril, E. F. 1990. *Biochemistry* 29:6362–74
56. Soderhall, S., Lindahl, T. 1975. *J. Biol. Chem.* 250:8438–44
57. Soderhall, S. 1976. *Nature* 260:640–42
58. Barnes, D. E., Tomkinson, A. E., Lehmann, A. R., Webster, D., Lindahl, T. 1992. *Cell.* In press
59. Henderson, L. M., Arlett, C. F., Harcourt, S. A., Lehmann, A. R., Broughton, B. C. 1985. *Proc. Natl. Acad. Sci. USA* 82:2044–48
60. Lehmann, A. R., Willis, A. E., Broughton, B. C., James, M. R., Steingrimsdottir, H., et al. 1988. *Cancer Res.* 48:6343–47
60a. Lönn, U., Lönn, S., Nylen, U., Winblad, G. 1989. *Carcinogenesis* 10:981–85
61. Webster, D., Arlett, C. F., Harcourt, S. A., Teo, I. A., Henderson, L. 1982. In *Ataxia-telangiectasia—A Cellular and Molecular Link between Cancer, Neuropathology, and Immune Deficiency*, ed. B. A. Bridges, D. G. Harnden, pp. 379–86. New York: Wiley
62. Teo, I. A., Arlett, C. F., Harcourt, S. A., Priestley, A., Broughton, B. C. 1983. *Mutat. Res.* 107:371–86
63. Teo, I. A., Broughton, B. C., Day, R. S., James, M. R., Karran, P., Mayne, L. V., Lehmann, A. R. 1983. *Carcinogenesis* 4:559–64
64. Sancar, A., Sancar, G. B. 1988. *Annu. Rev. Biochem.* 57:29–67
65. Spadari, S., Ciarrocchi, G., Falaschi, A. 1971. *Eur. J. Biochem.* 22:75–78
66. Kenne, K., Ljungquist, S. 1988. *Eur. J. Biochem.* 174:465–70

67. Scher, B. M., Scher, W., Waxman, S. 1988. *Cancer Res.* 48:6278–84
67a. Yang, S. W., Becker, F. F., Chan, J. Y. H. 1992. *Proc. Natl. Acad. Sci. USA.* In press
68. Creissen, D., Shall, S. 1982. *Nature* 296:271–72
69. Ueda, K., Hayaishi, O. 1985. *Annu. Rev. Biochem.* 54:73–100
69a. Yang, S. W., Huang, P., Plunkett, W., Baker, F. F., Chan, J. Y. H. 1992. *J. Biol. Chem.* In press
70. Chan, J. Y. H., Becker, F. F. 1985. *Carcinogenesis* 6:1275–77
71. Teraoka, H., Sumikawa, T., Tsukada, K. 1986. *J. Biol. Chem.* 261:6888–92
72. Lasko, D. D., Tomkinson, A. E., Lindahl, T. 1990. *Mutat. Res.* 236:277–87
73. Gould, K. L., Bretscher, A., Esch, F. S., Hunter, T. 1989. *EMBO J.* 8:4133–42
74. Turunen, O., Winqvist, R., Pakkanen, R., Grzeschik, K.-H., Wahlstrom, T., Vaheri, A. 1989. *J. Biol. Chem.* 264: 16727–32
75. Bretscher, A. 1986. *Methods Enzymol.* 134:24–37
76. Bretscher, A. 1989. *J. Cell. Biol.* 108: 921–30
77. Kerr, S. M., Johnston, L. H., Odell, M., Duncan, S. A., Law, K. M., Smith, G. L. 1991. *EMBO J.* 10:4343–50
78. Chan, J. Y. H., Becker, F. F., German, J., Ray, J. H. 1987. *Nature* 325:357–59
79. Willis, A. E., Weksberg, R., Tomlinson, S., Lindahl, T. 1987. *Proc. Natl. Acad. Sci. USA* 84:8016–20
80. Lönn, U., Lönn, S., Nylen, U., Winblad, G., German, J. 1990. *Cancer Res.* 50:3141–45
81. Blackburn, E. H. 1991. *Nature* 350: 569–73
82. Bjursell, G., Gussander, E., Lindahl, T. 1979. *Nature* 280:420–23
83. Roth, D. B., Wilson, J. H. 1986. *Mol. Cell. Biol.* 6:4295–304
84. Thode, S., Schäfer, A., Pfeiffer, P., Vielmetter, W. 1990. *Cell* 60:921–28
85. North, P., Ganesh, A., Thacker, J. 1990. *Nucleic Acids Res.* 18:6205–10
86. Nasmyth, K. A. 1977. *Cell* 12:1109–20
87. Johnston, L. H., Nasmyth, K. A. 1978. *Nature* 274:891–93
88. Barker, D. G., Johnston, L. H. 1983. *Eur. J. Biochem.* 134:315–19
89. Johnston, L. H., Barker, D. G., Nurse, P. 1986. *Gene* 41:321–25
90. Banks, G. R., Barker, D. G. 1985. *Biochim. Biophys. Acta* 826:180–85
91. Johnston, L. H. 1990. *Curr. Opin. Cell Biol.* 2:274–79
92. White, J. H. M., Barker, D. G., Nurse,

P., Johnston, L. H. 1986. *EMBO J.* 5:1705–9
93. Lowndes, N. F., Johnson, A. L., Johnston, L. H. 1991. *Nature* 350:247–50
94. Johnson, A. L., Barker, D. G., Johnston, L. H. 1986. *Curr. Genet.* 11:107–12
95. Peterson, T. A., Prakash, L., Prakash, S., Osley, M. A., Reed, S. I. 1985. *Mol. Cell. Biol.* 5:226–35
96. Sipiczki, M., Grossenbacher-Grunder, A.-M., Bódi, Z. 1990. *Mol. Gen. Genet.* 220:307–13
97. Rabin, B. A., Hawley, R. S., Chase, J. W. 1986. *J. Biol. Chem.* 261:10637–45
98. Takahashi, M., Senshu, M. 1987. *FEBS Lett.* 213:345–52
99. Panasenko, S. M., Modrich, P., Lehman, I. R. 1976. *J. Biol. Chem.* 251:3432–35
100. Takahashi, M., Tomizawa, K. 1990. *Eur. J. Biochem.* 192:735–40
101. Hardy, S., Aoufouchi, S., Thiebaud, P., Prigent, C. 1991. *Nucleic Acids Res.* 19:701–5
102. Aoufouchi, S., Hardy, S., Prigent, C., Philippe, M., Thiebaud, P. 1991. *Nucleic Acids Res.* 19:4395–98
102a. Bayne, M. L., Alexander, R. F., Benbow, R. M. 1984. *J. Mol. Biol.* 172:87–108
103. Signoret, J., David, J. C. 1986. *Int. Rev. Cytol.* 103:249–79
104. Signoret, J., David, J. C., Lefresne, J., Jouillon, C. 1983. *Proc. Natl. Acad. Sci. USA* 80:3368–71
105. Legerski, R. J., Penkala, J. E., Peterson, C. A., Wright, D. A. 1987. *Mol. Cell. Biol.* 7:4317–23
106. Sweigert, S. E., Carroll, D. 1990. *Mol. Cell. Biol.* 10:5849–56
107. Baer, R., Bankier, A. T., Biggin, M. D., Deininger, P. L., Farrell, P. J., et al. 1984. *Nature* 310:201–11
108. Sambrook, J., Shatkin, A. J. 1969. *J. Virol.* 4:719–26
109. Spadari, S. 1976. *Nucleic Acids Res.* 3:2155–67
110. Kerr, S. M., Smith, G. L. 1989. *Nucleic Acids Res.* 17:9039–50
111. Smith, G. L., Chan, Y. S., Kerr, S. M. 1989. *Nucleic Acids Res.* 17:9051–62
112. Colinas, R. J., Goebel, S. J., Davis, S. W., Johnson, G. P., Norton, E. K., Paoletti, E. 1990. *Virology* 179:267–75
113. Kerr, S. M., Smith, G. L. 1991. *Virology* 180:625–32
114. Moss, B. 1990. *Annu. Rev. Biochem.* 59:661–88
115. Willis, A. E., Lindahl, T. 1987. *Nature* 325:355–57
116. Kurihara, T., Teraoka, H., Inoue, M.,

Takebe, H., Tatsumi, K. 1991. *Jpn. J. Cancer Res.* 82:51–57
117. Rünger, T. M., Kraemer, K. H. 1989. *EMBO J.* 8:1419–425
118. Squires, S., Johnson, R. T. 1983. *Carcinogenesis* 4:565–72
119. Cooper, D. N., Youssoufian, H. 1988. *Hum. Genet.* 78:151–55
120. Weksberg, R., Smith, C., Anson-Cartwright, L., Maloney, K. 1988. *Am. J. Hum. Genet.* 42:816–24
121. Giannelli, F., Benson, P. F., Pawsey, S. A., Polani, P. E. 1977. *Nature* 265:466–69
122. Ockey, C. H., Saffhill, R. 1986. *Carcinogenesis* 7:53–57
123. Hand, R., German, J. 1975. *Proc. Natl. Acad. Sci. USA* 72:758–62
124. Kapp, L. N. 1982. *Biochim. Biophys. Acta* 696:226–27
125. Fujikawa-Yamamoto, K., Odashima, S., Kurihara, T., Murakami, F. 1987. *Cell Tissue Kinet.* 20:69–76
126. Spanos, A., Holliday, R., German, J. 1986. *Hum. Genet.* 73:119–22
126a. Petrini, J. H. J., Huwiler, K. G., Weaver, D. T. 1991. *Proc. Natl. Acad. Sci. USA* 88:7615–19
127. Hirschi, M., Netrawali, M. S., Remsen, J. F., Cerutti, P. A. 1981. *Cancer Res.* 41:2003–7
128. Nicotera, T. M., Notaro, J., Notaro, S., Schumer, J., Sandberg, A. A. 1989. *Cancer Res.* 49:5239–43
129. Vollberg, T. M., Seal, G., Sirover, M. A. 1987. *Carcinogenesis* 8:1725–29
130. Sullivan, N. F., Willis, A. E., Moore, J. P., Lindahl, T. 1989. *Oncogene* 4:1509–11
131. Sullivan, N. F., Willis, A. E. 1989. *Oncogene* 4:1497–502
132. Rusquet, R. M., Feon, S. A., David, J. C. 1988. *Cancer Res.* 48:4038–44
133. Lindahl, T., Willis, A. E., Lasko, D. D., Tomkinson, A. 1989. In *DNA Repair Mechanisms and their Biological Implications in Mammalian Cells,* ed. M. Lambert, J. Laval, pp. 429–38. New York:Plenum
134. Tanaka, K., Miura, N., Satokata, I., Miyamoto, I., Yoshida, M. C., et al. 1990. *Nature* 348:73–76
135. Weeda, G., van Ham, R. C. A., Vermeulen, W., Bootsma, D., van der Eb, A. J., Hoeijmakers, J. H. J. 1990. *Cell* 62:777–91
136. Fleiter, W. L., McDaniel, L. D., Johns, D., Friedberg, E. C., Schulz, R. A. 1992. *Proc. Natl. Acad. Sci. USA* 89:261–65
137. Robins, P., Jones, C. J., Biggerstaff, M., Lindahl, T., Wood, R. D. 1991. *EMBO J.* 10:3913–21
138. Fulop, G. M., Phillips, R. A. 1990. *Nature* 347:479–82
139. Biedermann, K. A., Sun, J., Giaccia, A. J., Tosto, L. M., Brown, J. M. 1991. *Proc. Natl. Acad. Sci. USA* 88:1394–97
140. Hendrickson, E. A., Qin, X.-Q., Bump, E. A., Schatz, D. G., Oettinger, M., Weaver, D. T. 1991. *Proc. Natl. Acad. Sci. USA* 88:4061–65
141. Hendrickson, E. A., Lin, V. F., Weaver, D. T. 1991. *Mol. Cell. Biol.* 11:3155–62
142. Lauer, G., Rudd, E. A., McKay, D. L., Ally, A., Ally, D., Backman, K. C. 1991. *J. Bacteriol.* 173:5047–53

Annu. Rev. Biochem. 1992. 61:283–306

CHROMOSOME AND PLASMID PARTITION IN *ESCHERICHIA COLI*

Sota Hiraga

Department of Molecular Genetics, Institute for Medical Genetics, Kumamoto University Medical School, Kumamoto 862, Japan

KEY WORDS: chromosome partition, plasmid partition, MukB protein, ATPase, filamentous protein polymer

CONTENTS

INTRODUCTION

In bacteria, chromosomal DNA is replicated and daughter chromosome molecules are accurately partitioned into daughter cells prior to cell division.

0066-4154/92/0701-0283$02.00

Daughter cells almost always receive at least one copy of the chromosome; anucleate (chromosome-less) cells account for less than 0.03% of the population in growing cultures of wild-type strains of *Escherichia coli*. Before septation at midcell, daughter chromosome molecules are separated from each other and located at the cell quarter positions (1/4 and 3/4 cell lengths), where the daughter chromosomes appear more compact than replicating chromosomes.

Jacob et al (1) proposed a model in which the partitioning of replicated chromosomes and molecules of F plasmid (the classical *E. coli* sex factor) into daughter cells involves a physical connection between a daughter DNA strand and part of the cell envelope, and in which the driving force for segregation of the two daughter chromosomes and of F plasmids must be provided by insertion of new cell envelope material at midcell, i.e. between two attachment sites. According to one version of this model, the chromosome replication origin *oriC* or its flanking region is the specific site that is attached to the cell envelope, and daughter chromosomes move to opposite directions by elongation of the envelope, in concert with the progress of chromosome replication. However, experimental results showed that growth of the cell wall takes place by random insertion over the entire surface during most of the *E. coli* cell cycle, although during septum formation peptidoglycan precursors are inserted preferentially at midcell (2–4).

Although biochemical data showed temporary association of the bacterial membrane with the *oriC* regions of the bacterial chromosome and with *oriC* plasmids (minichromosomes) (for example, 5–7), it is doubtful that the membrane attachment of the *oriC* region actually plays an essential role in partition of the chromosome and of *oriC* plasmids to daughter cells in vivo. In the absence of selective pressure, *oriC* plasmids are unstably maintained in a population of actively dividing bacterial cells, and are lost rapidly from the bacterial population. They do not have a partition mechanism, but are partitioned essentially at random into daughter cells. By contrast, an *oriC* plasmid carrying the *sopA, sopB,* and *sopC* genes (*parA, parB,* and *parC*) of the F plasmid is stably maintained in spite of the low copy number of plasmid molecules per cell (8). Therefore it is unlikely that *oriC* and/or its flanking regions play an essential role in chromosome partitioning.

The two strands of the double-stranded chromosomal DNA in a newborn cell can be distinguished by the fact that one strand (the *younger* strand) was formed during the most recent round of replication and the other (the *older* strand) was formed during some earlier round of replication. Upon replication, the younger and older chromosomes separate from one another to eventually reside within the complementary new daughter cells. The directions in which the new chromosomes are partitioned can also be easily discriminated by the fact that every cell has a younger polar cap, formed at the previous division, and an older polar cap formed during some earlier division. Does the cell distinguish between the two daughter chromosomes in conjunction with the direction of partition? To answer the question, studies have been made of the partitioning of younger and older daughter chromosomes into daughter cells of *E. coli* during division, and it has been shown that there is a preferental segregation of the older chromosome towards the cell with the older polar cap (for a review, see 9). Helmstetter & Leonard (10) propose that the cell envelope contains a large number of sites capable of binding to the chromosomal replication origin, *oriC*, that a polymerizing DNA strand becomes attached to one of the sites at initiation of a round of replication, and that the attachment sites are distributed throughout the actively growing cell envelope, i. e. lateral envelope and septum, but not in the existing cell poles. This asymmetric distribution of *oriC* attachment sites could account for the experimentally observed nonrandom segregation of the older daughter chromosome with cell strain and growth rate. The multisite attachment concept could also account for at least some of the instability with which minichromosomes are maintained (9).

It has been found that the "positioning" of daughter chromosome molecules at the cell quarter positions requires postreplicational protein synthesis (11, 12; for a review, see 13). When protein synthesis is inhibited by starvation for amino acids or by the addition of an inhibitor of protein synthesis, a replicating chromosome completes its replication and the resulting daughter chromosomes remain at the midcell, close to each other. When protein synthesis resumes by the addition of the amino acids or by removal of the inhibitor, the daughter chromosomes can rapidly move from midcell to the cell quarter positions before a detectable increase in cell length occurs (12). This suggests that daughter chromosomes are transported by an unknown mechanism, but not by elongation of the cell envelope itself. The positioning of daughter chromosomes from midcell to cell quarter positions after resumption of protein synthesis is achieved even in the presence of an inhibitor of DNA gyrase, nalidixic acid (50 μg/ml), or novobiocin (100 μg/ml). Moreover, it was found that the positioning after resumption of protein synthesis is achieved even when the initiation of chromosome replication is inhibited, using a temperature-sensitive *dnaC* mutant, which is defective in initiation of chromosome replication at a nonpermissive temperature. Thus the reinitiation of chromosome replication is not required for the positioning of daughter chromosomes. The results described above strongly indicate that *E. coli* has a mechanism, albeit unknown, for positioning daughter chromosomes. This mechanism may be promoted by events that are dependent on postreplicational protein synthesis (12). The events may also promote septum formation, causing the cell to enter the D (division) period from the C period (for a review, see 14). Individual chromosomes are fixed in position within growing cells both before and during replication, but they are rapidly moved apart by a fixed distance (unit length) immediately after replication has been completed (15).

Understanding how daughter chromosomes are positioned at cell quarter positions before cell division will provide insights into the fundamental process of partition. To analyze the molecular mechanism of chromosome positioning, isolation and characterization of mutants defective in this mechanism are advantageous. Techniques have been developed by us to isolate mutants that produce, upon cell division and at a non-negligible frequency, one anucleate daughter cell of normal cell size and one nucleate daughter cell carrying two copies of daughter chromosomes (16). Many mutants have been isolated by using these techniques. Our recent results indicate that a newly discovered gene, *mukB*, is involved in chromosome positioning at the cell quarter positions. The amino acid sequence deduced from the nucleotide sequence of the cloned *mukB* gene and the properties of the purified MukB protein suggest that the MukB protein is the first candidate found for a force-generating (mechanochemical) enzyme in bacteria (17, 18). The mechanism of chromosome partition is discussed in this review. In addition, findings on partition mechanisms of plasmids are described.

ANALYSIS OF GENES INVOLVED IN CHROMOSOME PARTITION

Two Categories of Mechanisms

Two categories of mechanisms are involved in the overall process from termination of chromosome replication to partition of the daughter chromosomes into daughter cells upon cell division. The first category includes decatenation of chromosome catenanes, resolution of chromosome dimers or oligomers, and other topological events preparatory to the separation of daughter chromosomes (for a review, see 19). The second category consists of mechanisms for the active positioning of daughter chromosomes at cell quarter sites (12).

Bacterial par Mutants Defective in Topoisomerases

Conditionally lethal *par* mutants of *E. coli* are capable of DNA synthesis and septation for some time after transfer to a nonpermissive temperature, but have lost control over the spatial location of the septa. During incubation at nonpermissive temperature, these mutants form elongated cells with a large conglomerated nucleoid or nucleoids. This kind of abnormal cell formation has been called the "Par phenotype." In some cases, abnormal cell division occurs near the ends of elongated cells, producing anucleate cells. Some *par* mutants have been found to be defective in the genes coding for topoisomerases. The *parA* mutant carries at least two temperature-sensitive mutations; one of them is probably located in the *gyrB* gene encoding the B subunit of DNA gyrase and may be responsible for the Par phenotype (20). The *parD* mutant carries an *amber* mutation in the *gyrA* gene encoding the A subunit of DNA gyrase, and the formation of elongated cells is probably due to the *gyrA* mutation rather than to another mutation present in the mutant and assigned to a separate gene called *parD* (21, 22). In the temperature-sensitive *gyrB* mutant, which is defective in the B subunit of DNA gyrase, a catenane of replicated daughter chromosomal DNA molecules is produced at nonpermissive temperature (23, 24). Therefore, the daughter chromosomes cannot be physically separated.

The *parC* mutation is located at 65 min on the chromosome map (25). The nucleotide sequence of the *parC* gene was determined (26). The *parC* gene encodes a 81.2-kDa protein, and the deduced amino acid sequence of the protein has homology to that of the A subunit of DNA gyrase. Another newly discovered gene, *parE*, which is located about 5 kilobases (kb) upstream from the *parC* gene and codes for a ca. 70-kDa protein, was sequenced, and the deduced amino acid sequence shows that the gene product (66.7 kDa) has homology to the gyrase B subunit. Mutants of this gene show the typical Par phenotype at nonpermissive temperature; thus the gene was named *parE*. Relaxation activity of supercoiled plasmid molecules is enhanced in vitro in the combined crude cell lysates prepared from the ParC and ParE overproducers. The *parC* and *parE* genes code for the subunits of a newly discovered topoisomerase, named topoisomerase IV. The complex of ParC and ParE proteins causes a decrease of superhelicity as does eukaryotic type II topoisomerase (26), whereas bacterial DNA gyrase causes an increase of superhelicity. A *topA* mutation causing a defect in topoisomerase I can be compensated in vivo by increasing the gene dosage of both *parC* and *parE* (26). ParC protein is specifically associated with the inner membrane, suggesting that topoisomerase IV is a membrane-bound enzyme (25; J. Kato and H. Ikeda, unpublished information).

The *parC* and *parE* genes of *Salmonella typhimurium* encode homologues of the *E. coli* ParC and ParE proteins, respectively (27). The *parE* gene is about 5 kb upstream of the *parC* gene, and a third gene, *parF*, is just downstream of the *parC* gene. The DNA sequence of the *S. typhimurium* *parF* gene was determined and could encode a protein with a hydrophobic amino terminus (28). It was speculated that the ParF protein interacts with ParC and ParE to anchor these proteins to the membrane.

Some other *par* mutants exhibit an inhibition of cell division, probably by the SOS response. For example, the *parB* mutant is an allele of the *dnaG* gene, which encodes a primase essential for DNA replication, and the Par phenotype observed at nonpermissive temperature reflects the inhibition of cell division associated with perturbed DNA synthesis (29, 30). The *pcsA* mutant (31), which shows a cold-induced Par phenotype, induces the RecA protein at nonpermissive temperature, suggesting that the Par phenotype of the *pcsA* mutant also reflects the inhibition of cell division by SOS-associated division inhibitors. Therefore these mutations causing SOS-associated cell division inhibition do not appear to belong to the category of mutations that specifically affect partition mechanisms.

The *dif* (deletion-induced filamentation) site, which lies within the terminal region of the *E. coli* chromosome, may act to resolve dimeric chromosomes produced by sister chromatid exchange. The *dif* site is a *cis*-acting, *recA*-independent recombination site. Deletion of the *dif* site causes the Dif phenotype, characterized by formation of a subpopulation of filamentous cells with abnormal nucleoids and induction of the SOS repair system (32 see also 33, 34). This finding shows that site-specific resolution of chromosome dimers is also required for chromosome partition.

Three new classes of cell cycle mutants showing increased ploidy were identified by selecting for cells that grow in the presence of camphor vapors (35, 36). The mutants, named *mbr* (moth ball resistant), map to four loci on the *E. coli* chromosome: *mbrA* maps to 68 min, *mbrB* to 88.5 min, *mbrC* to 89.5 min, and *mbrD*, which may be allelic to *rpoB*, to 90 min. Based on several tests, *mbrA* may define a novel checkpoint that couples DNA replication to cell division. The *mbrB* mutant appears to be defective in the synchrony of initiation of DNA replication. The *mbrC* and *mbrD* mutants are most likely defective in chromosome partitioning, although in precisely what way remains to be determined.

We have isolated mutants characterized by the production, in rich media at 37 °C, of a subpopulation of filamentous cells with abnormal nucleoids distributed irregularly along the cell. One of these mutants was demonstrated to have a defect in the *ruvB* gene located at 41 min on the *E. coli* chromosome (J. Feng, K. Yamanaka, S. Hiraga, unpublished results). The *ruvB* and *ruvA* genes constitute an operon, which belongs to the SOS regulon (37). The gene products are involved in DNA repair and recombination. Purified RuvB protein shows weak ATPase activity and binds ADP more strongly than ATP (37). The SOS repair system may be spontaneously induced in rich media in our mutant. Thus it appears that the primary defect in mutants of this type is not in chromosome partition.

muk *Mutants: Mutants that Produce Chromosome-less Cells*

If the positioning of daughter chromosomes at the cell quarter positions is controlled by gene products, mutants defective in the relevant genes would be expected generally to be nonlethal. Cell division in these mutants would occur normally, and replicated daughter chromosomes would be appropriately decatenated and folded. However, the daughter chromosomes would tend to be located close together in the position where the chromosome had been replicated, because of the defect in the active positioning of daughter chromosomes at cell quarter positions. Therefore, one daughter cell carrying two copies of the chromosome and one daughter cell carrying no chromosome would be spontaneously produced by cell division in the growing cell population at a non-negligible frequency as shown in Figure 1 (16; for a review, see 13). In a daughter cell that received two copies of the chromosome, the initiation of chromosome replication should be inhibited for one generation until the next cell division.

We have developed techniques to isolate mutants that produce anucleate cells during cell division, and have indeed isolated mutants defective in chromosome positioning (16). One such isolation procedure makes use of a *lacZ*-bearing plasmid that is restrained from runaway replication and from full *lacZ* expression by copy control and *lacI* repressor genes situated on the bacterial chromosome. Mutants that produce anucleate cells form blue colonies on agar plates containing X-gal, because anucleate cells lack both restraints on β-galactosidase synthesis from the *lacZ* genes. Mutants so obtained we designate *muk* (from the Japanese *mukaku*, meaning "anucleate").

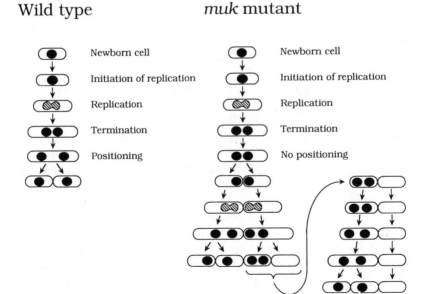

Figure 1 Chromosome partitioning in the *E. coli* wild-type strain and in a *muk* mutant defective in "positioning" of daughter chromosomes at cell quarter sites (16).

Among these *muk* mutants, *mukA1*, located at 66 min on the *E. coli* chromosome, was selected for detailed analysis of its properties (16). This mutant shows all the expected properties of the hypothetical mutant described above. The *mukA1* mutant produces normal-sized anucleate cells as about 3% of the population in an exponentially growing culture in a minimum-salt medium containing glucose. Fully replicated daughter chromosomes tend to be located close to each other. The intracellular location of the chromosome is broader in the mutant than in the parental wild-type strain. Abnormal pairs of one anucleate cell and one nucleate cell having two copies of chromosomal DNA can be observed at low frequency in a culture of the mutant. A wild-type chromosomal DNA segment that can complement the *mukA1* mutation was cloned and sequenced (16, 38). Results obtained indicated that the *mukA* gene is identical to the *tolC* gene, which codes for an outer membrane protein with a signal peptide. The *tolC* mutants show pleiotropic properties, tolerance to colicin E1 protein, slow growth, hypersensitivity to sodium dodecyl sulfate and drugs, and abnormal expression of outer membrane proteins OmpF and OmpC (for a review, see 39). In addition, TolC protein is specifically required for hemolysin secretion and might be a component of the secretion apparatus allowing a specific interaction between the inner and outer membranes (40). The molecular mechanism of the TolC (MukA) protein in chromosome positioning is still unknown.

A mutation belonging to another linkage group, *mukB106*, is located at 21 min on the *E. coli* chromosome (17). The *mukB* gene has been found to code for a large protein that has a surprising predicted secondary structure and interesting properties in vitro as described below.

THE *mukB* GENE AND ITS PRODUCT

Properties of the mukB *Mutants*

The original *mukB106* mutant grows nearly normally but spontaneously produces anucleate cells of normal size (about 5% of the population) during growth at 22 °C. The mutant shows temperature-sensitive growth and forms minute colonies at 42 °C in rich medium. A *mukB*-disrupted null mutant is very temperature sensitive and cannot form colonies at 42 °C in rich medium. The null mutant grows nearly normally at 22 °C, but spontaneously produces anucleate cells of normal cell size. In addition to anucleate cells there are pairs of cells: one that is anucleate with about 1 unit of cell mass and one that is nucleate of 1–2 units of cell mass (see Figure 1). There is also another type of cell pair: one nucleate cell with about two copies of chromosomal DNA and one cell having small amounts (5–20%) of chromosomal DNA at the cell pole near the division site, suggesting a guillotine effect by septal closure on a chromosome. When these *muk* mutants grown at 22 °C are transferred to

42 °C, nucleate cells elongate and nucleoids become irregularly distributed along the elongated cells within 3–4 h (17). During prolonged incubation at 42 °C, the elongated cells gradually divide, resulting in normal-sized anucleate cells and 40–60% as many normal-sized nucleate cells carrying large quantities of DNA (18). These results suggest that the MukB protein is required for chromosome positioning at 22 °C through 42 °C. In the *mukB* mutants, cell division is partially inhibited and delayed at 42 °C.

It has been demonstrated that the production of anucleate cells in the *mukB* mutants is not due to inhibition of DNA replication. Flow cytometric analysis of the *mukB106* mutant showed that the control of initiation of chromosome replication is normal, and all replication origins within any individual cell initiate simultaneously, even in the population at intermediate temperature containing some elongated cells (17).

Predicted Secondary Structure of MukB Protein

A wild-type *E. coli* DNA segment that complements the *mukB106* mutation was cloned and sequenced (17). The *mukB* gene codes for a protein of 176,826 daltons consisting of 1543 amino acids. The gene is the largest described so far in *E. coli*. Computer analysis of the deduced amino acid sequence showed that the MukB protein is hydrophilic with no highly hydrophobic regions, such as are typical of the membrane domains of integral membrane proteins.

MukB protein was predicted to have five domains showing distinguishable secondary structure (Figure 2). The amino-terminal domain I (amino acid residues 1–338) was predicted to be globular and to include a nucleotide-binding consensus sequence. Domains II (residues 339–665) and IV (residues 934–1116) were predicted to be highly α-helical and to show an extended coiled-coil structure. These α-helical domains have seven-residue repeats, consistent with the ability to form a coiled-coil. The two α-helical domains were found to exhibit a heptapeptide repeat motif *a b c d e f g;* hydrophobic residues were enriched at positions *a* and *d,* and a periodicity of negatively charged residues (Asp and Glu) and positively charged residues (Lys, Arg, and His) was observed in these α-helical regions. These results are consistent with an α-helical coiled-coil conformation (41–44) in these regions. Domain III (residues 666–933) was predicted to be globular. The carboxyl-terminal region of domain V (residues 1116–1534) was also predicted to be globular.

A subregion (residues 1364–1525) of domain V is rich in cysteine and the positively charged residues arginine and lysine. Within the carboxyl-terminal region of domain V are three putative "zinc finger"–like structures. Loops may be formed around the central atom of Zn (or another metal). Four cysteine residues provide the linkers between consecutive loops in structures I and II. Three cysteine residues and one histidine residue provide the linkers in

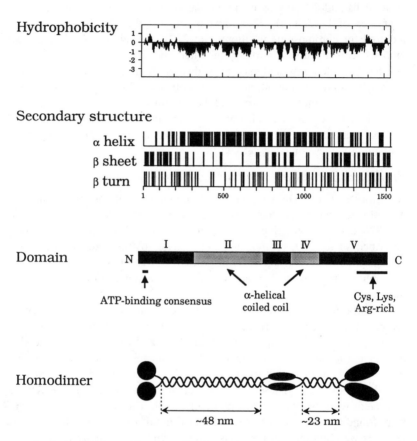

Molecular weight : 177 kDa (1534 amino acids)

Figure 2   Hydrophobicity profile and predicted secondary structure of MukB protein (17).

structure III. The loops are rich in positively charged residues and hydrophobic residues, like loops of previously described typical zinc fingers (45, 46).
Most of the negatively charged residues are located on the outside of these
zinc finger–like structures of the MukB protein. The carboxyl-terminal region
is presumably involved in specific interactions with other proteins and/or
DNA.

MukB's predicted secondary structure suggested that the protein is a good
candidate for a force-generating enzyme able to move chromosomes from
midcell toward the cell quarter positions. Two MukB protein molecules
presumably form a homodimer in the extended coiled-coil regions, similar to

that formed by myosin heavy chain (42) and kinesin heavy chain (44) in eukaryotic cells. The head domain containing a nucleotide-binding consensus sequence would be expected to act as a "motor" domain, as do the heads of other force-generating enzymes (for a review, see 47) such as myosin heavy chain and kinesin heavy chain. The coiled-coil regions of MukB protein have weak homology with the coiled-coil region of myosin heavy chain. The first globular domain of MukB protein has been found to have weak homology with rat dynamin D100 (48), but not with the amino-terminal motor domains of myosin heavy chain (42) and kinesin heavy chain (44).

Casaregola et al (49) described an *E. coli* protein of approximately 180 kDa that cross-reacted with a monoclonal antibody raised against myosin heavy chain encoded by the *MYO1* gene of the yeast *Saccharomyces cerevisiae*. The gene encoding the *E. coli* Hmp1, myosin-like protein (49, 50) has been shown to be identical to the *ams* gene (B. Holland and V. Norris, personal information). From sequence data, the *ams* gene encodes an approximately 120-kDa protein, which migrates aberrantly in the molecular weight range 180,000–200,000. The protein consists of at least two distinct domains, an amino-terminal domain with putative ATP- and membrane-binding regions and a carboxyl-terminal domain containing an Arg-, Glu-rich region with homology to mammalian snRNPs involved in RNA processing. In Ams this Arg-, Glu-rich region, which may be specifically involved in RNA splicing in the mammalian protein, is flanked by highly repetitive Val-, Ala-, Pro-rich sequences whose function is unknown. Mutations in *ams* have previously been shown to cause defects in mRNA stability and RNA processing. Recent evidence indicates that the Ams/Hmp1 protein is identical with ribonuclease E, an enzyme essential for processing of 5S rRNA and general mRNA decay (51).

Reeder & Schleif (52) have found that a long open reading frame, which is located upstream of *araJ* and codes for a protein having an extended α-helical coiled-coil in the carboxyl region, is identical to the *sbcC* gene (53).

Properties of MukB Protein in vitro

We purified the MukB protein from a sonicated cell lysate of a MukB-overproducing strain by centrifugation at $100,000 \times g$, precipitation with ammonium sulfate, gel filtration through Sephacryl S-400, and chromatography on DEAE Sephacel, and then on DNA-cellulose (18). The purified MukB protein showed ATP/GTP binding activity in the presence of $ZnCl_2$, but not in the presence of $MgCl_2$ or $CaCl_2$. The MukB protein adsorbed to a calf thymus DNA cellulose column (18). It remains for further experiments to determine the nucleotide sequence specificity and localized superhelicity of the DNA that is involved in the interaction between MukB protein and the chromosomal DNA.

Filamentous Protein Polymers in E. coli

We described previously a working hypothesis of chromosome positioning in *E. coli* to search for putative filaments of protein polymers interacting with MukB protein (58; Figure 3). It is interesting to search for these putative filaments, and to analyze interactions of these filaments, MukB protein, and specific membrane sites, for example, periseptal annuli (59–60c).

It has been suggested that cytoskeletonlike machinery may be involved in the process of growth and division of bacterial cells (54, 55a). There have been several reports of polypeptides that form long curvilinear filaments (5–6 nm wide) in *E. coli* (55b, 55c, 55d). The polypeptide fractions were obtained by reversible aggregation, under the same conditions in which eukaryotic actin is polymerized primarily through use of potassium. The fraction exhibited a distinct peak at the molecular weight of 45,000 upon electrophoresis on acrylamide gel in the presence of sodium dodecyl sulfate (55c). Beck et al (55e) repeated the procedures and showed that the actinlike component of 45 kDa was identical to elongation factor Tu.

Tomioka (56) reported that filaments and sheets of proteins are assembled in the soluble fraction of *E. coli* cell extract in the presence of 100 to 600 mM KCl. The cytoplasmic concentration of K^+ ion is 100 to 600 mM (57). The filaments appeared to have a diameter of about 6 nm under an electron microscope, and the sheets appeared to be an association product of the filaments. This filamentous assembly is depolymerized into soluble form in a buffer of low ionic strength. Polymerization and depolymerization could be repeated several times. The filament-forming proteins were purified, and two proteins with apparent molecular weights of 100,000 and 90,000 were found, which were named "100K" and "90K" protein, respectively. A filamentous assembly, which resembled that found in the cell extract, was reconstructed in

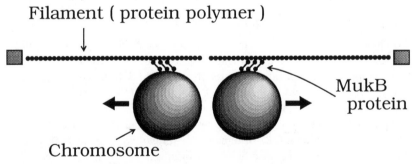

Figure 3   A working hypothesis of the molecular mechanism of chromosome positioning at cell quarter sites in *E. coli* (58).

a mixture of the two proteins by addition of 100 mM KCl. Antiserum to the 100K protein recognized the sheeted assembly formed in the cell extract, and thus the 100K protein appears to be component of it. A protein with apparent molecular weight 55,000 was copurified with the two proteins and might be associated to them. The 100K protein formed sheets and filaments with a different appearance upon addition of 100 mM KCl, and the 90K protein formed small particles under the same conditions. Subcellular fractionation and protease treatment of spheroplasts suggested that these proteins existed in the cytoplasm (56).

The MukB-containing fraction purified by Sephacryl S-400 gel filtration contained three proteins with apparent molecular weights of 100,000, 90,000, and 55,000, and other proteins in minor amounts, besides the 177-kDa MukB protein (18). The 100-, 90-, and 55-kDa proteins were also found in the corresponding fractions of the *mukB*-disrupted null mutant and also of an F$^-$ Pil$^-$ Hag$^-$ mutant, which did not have F-pili, I-pili, and flagella on the cell surface. We found that two distinct 100-kDa proteins existed in the fraction, and that one cross-reacted with the anti-100K rabbit antiserum obtained from Shigeo Tomioka, but not the other (18). This indicates that the former protein is identical with Tomioka's 100K protein. We identified recently the former with pyruvate dehydrogenase (99.6 kDa; the *aceE* gene), based on amino acid sequence of the amino-terminal region. We also identified the latter protein with alcohol dehydrogenase (95.9 kDa; the *adhE* gene). The latter protein was found to form filamentous polymers with widths of 10 nm (18). The 55-kDa protein in the MukB-containing fraction was identified with dihydrolipoamide dehydrogenase (50.6 kDa; the *lpd* gene), based on amino acid sequence of the amino-terminal region (18). There has been as yet no experimental result indicating functional interaction between the MukB protein and these proteins.

PARTITION MECHANISM OF PLASMIDS

Plasmid-Encoded Proteins and the cis-Acting Region of F and P1 Plasmids Essential for Partition

The stable maintenance in growing cultures of low copy-number plasmids such as the F plasmid implies that partition is not random during cell division; each daughter cell must receive at least one plasmid copy. Mini-F plasmids derived from the original F plasmid have been extensively analyzed for mechanisms of stable maintenance (for reviews, see 61–64). Mini-F plasmids encode the *trans*-acting *sopA* and *sopB* genes and the *cis*-acting *sopC* region essential for partitioning (8). The *sop* system, when associated with other replicons, for example, *oriC* plasmids, can stabilize them as well. In the *sopC* region, there are 12 direct repeats of a 43-basepair (bp) motif without any spacer regions, and one pair of 7-bp inverted repeats exists within each of the direct repeats (65, 66) as shown in Figure 4. Multicopy pBR322 derivatives carrying a segment containing the *sopC* region cause instability of a coexisting mini-F plasmid or an *oriC* plasmid carrying the *sopABC* segment (8). Thus the *sopC* region exerts incompatibility (termed IncD).

The purified SopA protein (43.7 kDa) binds to four repeated sequences

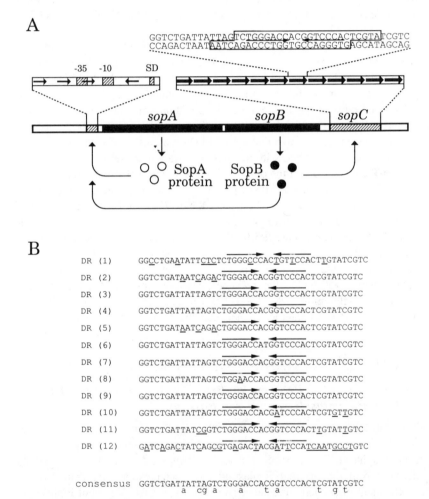

Figure 4 The *sop* region of F plasmid. *A*. Disposition of the *sopA* and *sopB* genes and the *cis*-acting *sopC/incD* region. SopA and SopB proteins act as autorepressor of the *sopAB* operon. The box including inverted repeats represents a SopB-binding site (67). *B*. Nucleotide sequences of the 43-bp direct repeats (DR) of the *sopC* region. The repeating units are aligned in the 5' to 3' direction (65).

(CTTTGG) located in the promoter-operator region of the *sopAB* operon (67). The binding activity of SopA is enhanced by the addition of SopB protein (35.4 kDa). SopB protein itself does not bind to these repeated sequences (67). These in vitro results are consistent with evidence that SopA and SopB act cooperatively as an autorepressor of the *sopAB* operon in vivo (H. Mori, unpublished data). SopB protein by itself also binds to the *sopC* region (67–69b). SopB protein recognizes the inverted repeats of 7 bp in the 12 direct repeats of the 43-bp nucleotide motif (67). SopA protein does not affect the SopB-binding activity to the *sopC* DNA segment (67). Bacterial cells harboring multicopy plasmids carrying the *sopB* gene, but not the *sopA* gene, produce excess SopB protein, causing inhibition of stable inheritance of a coexisting mini-F plasmid and of an *oriC* plasmid carrying the *sopABC* segment. This inhibition was termed IncG incompatibility (67, 69a). This kind of inhibition suggests that the SopB protein must be kept at appropriate concentrations to ensure stable maintenance of the mini-F plasmid (69a).

When the cell lysates were centrifuged in the presence of Mg^{2+}, SopB was found in the pellet where the cell membrane was recovered (69b). Further fractionation by centrifugation through a sucrose density gradient revealed that SopB in the cell lysates sedimented faster than the isolated SopB in the presence of Mg^{2+}, indicating that the protein made a complex with host factors. However, the sedimentation profile of the complex was not identical with that of the membrane. Therefore, there is no evidence at present that SopB is associated with the membrane (K. Nagai, unpublished information). This finding calls attention to the caution that must be exerted in the evaluation of "membrane fractions," for example, associations of membrane with partition protein, membrane with plasmid DNA (70), membrane with replication origins, etc.

The prophage of bacteriophages P1 and P7 are also maintained as unit copy plasmids (reviewed in 71), and yet are lost in less than $1/10^5$ cell division events (72, 73). The regions of P1 and P7 responsible for plasmid partition encode the *trans*-acting *parA* and *parB* genes and the *cis*-acting *incB/parS* region. P1 and P7 possess major regions of similarity (92% with respect to the P1 genome) (74). Homologies in deduced polypeptide sequences have been found between F SopA and P1 ParA, and between F SopB and P1 ParB (65, 75). The *incB* region (90 bp) of P1 includes the *parS* site, which is the region required in *cis* for partition when the P1 ParA and ParB proteins are supplied in *trans*. A 22-bp sequence possesses centromere activity and can exert incompatibility, albeit with reduced efficiencies (76; Figure 5). The *parS* site lies in a region consisting of a 13-bp palindrome and an adjacent AT-rich sequence. The 84-bp region of nucleotide positions 4303–4386 is sufficient to exert full IncB incompatibility (75, 77, 78). The P1 *parB* gene encodes a specific DNA-binding protein. The ParB protein binds to the palindromic region and again at sequences some 30 bp away. A heptamer repeat found in both regions is implicated in ParB binding (78; Figure 5). The spacer sequence is a specific binding site for the integration host factor (IHF) protein. ParB and IHF form a complex cooperatively on the 90-bp site (79–82). As IHF binding is thought to promote extreme DNA bending (84), it is likely that the *incB* region bends and forms a folded structure in the presence of IHF and ParB (79, 82, 83). The synthesis of P1 ParA and ParB is tightly autoregulated (85).

Broad host-range plasmid RK2 is capable of transfer between, and stable inheritance in, almost all Gram-negative bacterial species (for a review, see 86). It is 60 kilobases in size and is estimated to exist at two to seven copies per chromosome equivalent. During a search for functions of genes belonging to the autogenously regulated *korA-korB* operon, the RK2 *incC* polypeptide (87) was found to have homology to the SopA protein of plasmid F and to the ParA protein of prophage P1 (88). All three proteins are of similar size (RK2 IncC, 364 amino acids; F SopA, 388 amino acids; P1 ParA, 398 amino acids). In addition, comparison of the predicted polypeptide sequence of KorB protein (89, 90) with those of F SopB and P1 ParB shows homologies. Alignment of the IncC, ParA, and SopA proteins identifies two major regions of conservation. They contain sequences with strong similarity to the consensus for a type I nucleotide-binding motif, specifically that for ATP (91), although certain of the residues are not conserved, including the first and last glycine residues in region I. Comparison with the GTP-binding consensus (92) revealed no convincing fit that was conserved in all the partitioning proteins. In addition, a partitioning protein encoded by the *Agrobacterium tumefacieus* plasmid pTAR (93) and the *orf5* gene product of a cryptic plasmid from *Chlamydia trachomatis* (94) show significant similarity to the ParA protein family. A data base search also revealed homology between ParA family members and the *E. coli* cell-division inhibitor protein, MinD, which (in conjunction with MinC) can inhibit septation at all potential division sites (95). It was proposed that ParA may phosphorylate ParB and induce

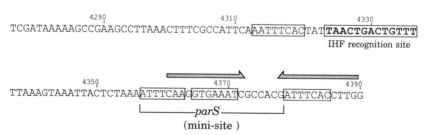

Figure 5 The *cis*-acting *incB/parS* region of P1 plasmid. Boxes mark heptamer sequences thought to be recognition sites for ParB binding (78).

a conformational change that allows separation of the paired plasmids equally to each side of the septation plane (88). Moreover, C. M. Thomas (unpublished information) revealed significant similarlity between parts of the P1 ParA protein and the ATP binding and hydrolysis site of ArsA, an ATPase involved in arsenate/arsenite resistance in *E. coli*; this comparison suggested that P1 ParA might be an ATPase itself.

Purified ParA protein does possess a weak ATPase activity in vitro; the reaction requires Mg^{2+}, and converts ATP to ADP and phosphoric acid. The ATPase activity is slightly stimulated by double-stranded DNA independent of the sequence, and strongly stimulated by purified P1 ParB protein (96). A *parA* mutant that has a base substitution causing the change Lys122 to Gln fails to support partition in a complementation assay. ParA possesses a site-specific DNA-binding activity in the presence of ATP; it protects a large (about 80-bp) region that overlaps the putative *par* operon promoter. This suggests that the ATPase may be involved in regulation of expression of the *par* operon.

Similarly, in F plasmid, specific binding to ATP and moderate ATPase activity has been detected with isolated SopA (K. Nagai, unpublished information). The ATPase requires Mg^{2+}, and the activity is significantly stimulated by single- or double-stranded DNA. The ATPase activity is further stimulated when double-stranded DNA and isolated SopB are simultaneously added to the reaction mixture, although no stimulation was caused by the addition of SopB alone. The stimulatory effect of DNA fragments with or without the *sopC* region does not differ significantly. The results suggest that partitioning of F plasmid DNA is supported by energy that is supplied by the ATPase activity of SopA in combination with SopB and DNA.

The *sop*-defective mutant mini-F plasmids differ in degree of supercoiling from the sop^+ parental mini-F plasmid, as judged by electrophoresis on chloroquine agarose gels of circular plasmid DNA isolated from cells. A mutation within the *sopB* gene or a deletion of the whole *sopC* region leads to increased superhelicity, whereas a mutation within the *sopA* gene leads to decreased superhelicity. Providing SopA and SopB proteins in *trans*, in vivo, to a pBR322 plasmid carrying the *sopC* region, results in decreased plasmid superhelicity; no such change is observed with pBR322 itself (D. Biek and J. Strings, unpublished information). These alterations in superhelicity are most easily explained by right-handed wrapping of the *sopC* DNA around the SopB protein in vivo. One right-handed wrapping would result in one localized positive turn in the *sopC* region. Topoisomerase action in vivo on the SopB-bound plasmid, followed by removal of SopB protein during plasmid isolation, would be observed as a net decrease in DNA superhelicity. A number of positive turns could be introduced by binding of SopB protein to all 12 SopB-binding sites that are present in the *sopC* region. Large amounts of SopB protein are produced in *sopA* mutants in which autorepression of the *sopAB* operon is defective, thus leading to the apparent decreased superhelicity of *sopA* mutant plasmids.

As described above, the *cis*-acting regions of F and P1 exert incompatibility against coexisting homologous plasmids. "Dicentric" plasmids containing both the P1 and F partition system are stably maintained, showing that the presence of two *cis*-acting partition sites does not cause interference of one system by the other during the orderly distribution of copies (97). There is some difficulty in reconciling the stability of dicentric plasmids with the proposed models for plasmid partition and for the incompatibility exerted by the *cis*-acting partition site (for reviews, see 71, 98, 99).

The P1 *par* system, when associated with other replicons, can stabilize them as well. In this sense the P1 partitioning system is replicon independent. The act of replication might nevertheless be obligatory at one of the steps of the partition process. For example, the passage of a replication fork through a plasmid partitioning site might be required in the initial pairing of plasmids for partition. This possibility was invoked to explain the stability of "dicentric" plasmids (97). Alternatively, replication might be required at a terminal stage in partitioning to release individual plasmids from membrane elements that had served to separate them (71). However, Treptow et al (100) have shown that the P1 *par* system acting on nonreplicating extrachromosomal lambda prophages that bear a *parS* element can delay the appearance of cells cured of prophage, in conformity with computer simulations. This result demonstrates that partitioning directed by the *par* system of P1 is replication independent.

To analyze host functions in plasmid partition, bacterial mutants that do not support stable maintenance of mini-F plasmids were isolated; the host partition (*hop*) mutants were classified into five linkage groups on the *E. coli* chromosome (101). *hopA* mutants carry mutations in the *gyrB* gene, coding for the B subunit of DNA gyrase. In these *gyrB(hopA)* mutants, the superhelicity of plasmid DNA was reduced, whereas SopB protein production and IncG incompatibility were increased. It is suggested that altered expression of the *sop* genes, which is due to relaxation of the mini-F plasmid DNA, causes both defective partitioning of the mini-F plasmids and increased IncG incompatibility in the *gyrB(hopA)* mutants (102). The *hopE* mutants have mutations in the *recD* gene coding for the D subunit of the RecBCD enzyme (exonuclease V). A large amount of linear multimer plasmid DNA accumulates in the mutants. To account for the mechanisms of linear multimer formation, models involving rolling circle replication of plasmid DNA have been proposed (103). The *hopD* mutants have mutations in the *hupB* gene coding for one of the subunits of the heterodimeric HU protein (104). Mini-F and mini-P1 plasmids could not be introduced into the Δ*hupA*-Δ*hupB* double deletion mutant (104, 105). The mini-P1 plasmid was slightly unstable in a *himA-hip* mutant lacking both moieties of IHF, whereas the mini-F plasmid was stable (105). The *hopD* mutants presumably affect plasmid replication rather than plasmid partition. Many other host mutants in which the F plasmid and mini-F plasmids are lost rapidly in a population in the absence of selective pressure have been described; however, these mutants are defective in plasmid replication rather than plasmid partition (106–109). No bacterial mutant that is defective in a gene encoding a protein showing direct involvement in the partition system of F or P1, but not affecting bacterial cell growth, has been described so far.

Keasling et al (110), contesting results of Leonard & Helmstetter (111), have presented evidence that the F plasmid replicates in a cell-cycle-specific manner. Their analysis of the replication of F plasmid at different growth rates suggests that the ratio of cell size at initiation of plasmid replication to the number of F plasmid origins is constant, with a pattern of regulation of F plasmid replication similar in this respect to that of the chromosome origin. However, if F plasmid does replicate at a specific initiation mass, it would appear to be a different initiation mass from that of the bacterial chromosome and *oriC* plasmids.

Partition Genes of the IncFII Plasmids R1 and NR1

Plasmids R1 (112) and NR1 (113, 114), which belong to the IncFII incompatibility group, have been analyzed for the DNA segment essential for plasmid partition. The stability locus (*par* or *stb*) of either R1 or NR1 encodes two *trans*-acting proteins, StbA (36,000 daltons) and StbB (13,000 daltons). The *stbA* and *stbB* genes constitute an operon. StbB protein may be an autorepressor of the *stbAB* operon (115; A. Tabuchi, unpublished information). The promoter proximal region of the operon, upstream from *stbA*, acts as a *cis*-acting site for plasmid stabilization. Unstable deletion mutants that retained this promoter proximal region but had deleted both *stbA* and *stbB* were stabilized in *trans* by plasmids that could supply StbA and StbB. In contrast, deletion mutants that had lost the *stbAB* promoter region were not complemented in *trans*, indicating that this region contained an essential *cis*-acting site. The *cis*-acting site of R1 exerts partition-mediated incompatibility (112), whereas the *cis*-acting site of NR1 does not express incompatibility (114). That is, plasmid clones that contained only the essential *cis*-acting promoter proximal region of NR1 did not cause destabilization of NR1-derived *stb*$^+$ plasmids in *trans*. The sequences around the putative *cis*-acting sites are identical, although there are differences elsewhere in the R1 and NR1 *par/stb* regions. This mysterious phenomenon has been discussed (for a review, see 99). The riddle of this phenomenon will be solved as data on molecular mechanisms of plasmid partitioning are accumulated.

In the case of the pTAR plasmid of *Agrobacterium tumefaciens*, only a single open reading frame (*parA*) encoding a 23.5-kDa protein exists within the *par* locus (1259 bp); a protein (approximately 25 kDa) encoded by the *par* locus is made in vitro in the *E. coli* and *A. tumefaciens* protein synthesis systems (80). A 248-bp region immediately upstream from the open reading frame, containing an array of 12 seven-base-pair palindromic repeats each of which are separated by exactly 10 bp of A+T-rich (75%) sequence, not only serves to provide the promoter but is also involved in *parA* autoregulation. In addition, this region is responsible for plasmid-associated incompatibility within the Inc Ag-1 group and functions as the *cis*-acting recognition site with which ParA interacts to bring about partitioning (93).

The cis-Acting par Site of pSC101 Plasmid

The pSC101 plasmid encodes a *cis*-acting locus termed *par*, which ensures the stable inheritance of this moderately low-copy-number plasmid in a population of actively dividing bacterial cells in the absence of selective pressure (116). *par*-defective plasmids are lost rapidly from the bacterial population in the absence of selection. The *par* locus (375 bp) of pSC101 does not include a protein-coding sequence, and this locus does not exert incompatibility against homologous plasmids (116), in contrast to the *cis*-acting loci of F and P1. The pSC101 *par* locus is adjacent to the replication origin of the plasmid and not required for plasmid replication. It appears to enhance replication capabilities (117, 118) and to have an effect on plasmid copy number (118). pSC101 derivatives lacking the *par* locus have decreased overall superhelical density as compared with DNA of wild-type pSC101 (119). pSC101 derivatives lacking the *par* locus are stabilized in *E. coli* by *topA* gene mutations, which increase negative DNA supercoiling. Conversely, mutations in DNA gyrase and DNA gyrase inhibitors increase the rate of loss of *par*-defective pSC101 derivatives (119). Possibly relevant to these observations is the finding that a strong binding site for DNA gyrase exists within the *par* region (120). The stability of *par*-deleted pSC101 derivatives is restored by introducing certain adventitious bacterial promoters onto the plasmid. A promoter-associated overall increase in negative superhelicity of plasmid DNA was observed; stabilized inheritance appeared to be dependent on localized rather than generalized supercoiling (121).

Plasmids defective in the *par* function, such as *sop*-deficient mini-F derivatives and *oriC* and p15A plasmids, are also stabilized in *E. coli* by *topA* gene mutations, which increase negative DNA supercoiling (119). The molecular mechanism of the stabilization by *topA* mutations is still unknown.

Partition of a Mini-F Plasmid in the mukB-Disrupted Null Mutant

Let us return to the molecular mechanisms of action of the plasmid-encoding partition proteins of unit copy number plasmids such as F and P1. How do the *par* (*sop*) gene products and the *cis*-acting specific *par* (*sop*) site perform exact partitioning of plasmid molecules into the two daughter cells? Are MukB protein and cytoplasmic filaments also involved in plasmid partitioning?

Our recent results (122) show that a partition-proficient mini-F plasmid derivative lacking determinants of a postsegregational killing system (*ccdA* and *ccdB*; Refs. 123, 124; for a review, see 64) is stably maintained at 22 °C in the *mukB* null mutant, which is defective in chromosome partitioning into the two daughter cells, in spite of the plasmid copy number being low in the null mutant as it is in the *muk*+ strain. This indicates that MukB protein is not necessary for partition of the mini-F plasmid. Anucleate cells were purified from a culture of the *mukB* null mutant harboring the mini-F plasmid, and DNA was extracted from the anucleate cell sample and analyzed for the amount of mini-F plasmid DNA and chromosome DNA by DNA-DNA hybridization. The results indicated that the plasmid was partitioned not only into nucleate daughter cells, but also into anucleate daughter cells, independently of host chromosomes, in the *mukB* null mutant (122). Kline & Miller (125) described previously evidence that the F plasmid DNA could be recovered from a folded bacterial chromosome fraction, and pointed out the possibility that plasmids associate with specific chromosomal loci noncovalently and partition with the chromosome into daughter cells. However, our findings described above show that association of the chromosome with the F plasmid does not seem to be required for partitioning of the plasmid (122).

CONCLUDING REMARKS

Our recent work reveals that replicated daughter chromosomes in *E. coli* are positioned at cell quarter sites by a mechanism that depends on the function of MukB protein. This has opened at last the heavy lid of the large black box of chromosome partitioning in bacteria. It is an interesting problem whether *E. coli* cells, like eukaryotic cells, have a genetically controlled mitotic apparatus acting to position daughter chromosomes at cell quarter sites. In addition to a genetic analysis of the genes encoding filament proteins, searches for specific MukB-binding site(s) on the *E. coli* chromosome acting as a centromere are important undertakings if we are to learn the molecular mechanism of chromosome partition. We have demonstrated that MukB protein is not required for partition of the mini-F plasmid, and that the mini-F plasmid is partitioned into daughter cells independently of the bacterial chromosome. Do cytoplasmic filaments have a role in the partition of mini-F plasmid and other plasmids? How do the plasmid-encoded partition proteins function in plasmid partitioning? What is the nature of the postreplicational protein synthesis involved in chromosome positioning at cell quarter sites? What trigger events of the cell cycle are involved in promoting septum formation? Answers to these questions will require genetic and biochemical analyses. The isolation and characterization of conditional mutants defective in the relevant events are certain to be fruitful.

ACKNOWLEDGMENTS

The author wishes to thank Dr. Michael Yarmolinsky for critical reading of the manuscript and valuable comments. The author also thanks Drs. Teru Ogura and Hironori Niki for their helpful discussion. Ms. Chiyome Ichinose provided assistance in preparing this manuscript. The author thanks all colleagues who provided reprints, preprints, and unpublished information. Work from author's laboratory has been supported by grants from the Ministry of Education, Science and Culture of Japan.

Literature Cited

1. Jacob, F., Brenner, S., Cuzin, F. 1963. *Cold Spring Harbor Symp. Quant. Biol.* 28:329–48
2. Woldringh, C. L., Huls, P., Pas, E., Brakenhoff, G. J., Nanninga, N. 1987. *J. Gen. Microbiol.* 133:575–86
3. Woldringh, C. L., Mulder, E., Valkenburg, J. A. C., Wientjes, F. B., Zaritsky, A., Nanninga, N. 1990. *Res. Microbiol.* 141:39–49
4. Nanninga, N., Wientjes, F. B., de Jonge, B. L. M., Woldringh, C. L. 1990. *Res. Microbiol.* 141:103–18
5. Ogden, G. B., Pratt, M. J., Schaechter, M. 1988. *Cell* 54:127–35
6. Schaechter, M., Polaczek, P., Gallegos, R. 1991. *Res. Microbiol.* 142:151–54
7. Kataoka, T., Gayama, S., Takahashi, K., Wachi, M., Yamasaki, M., Nagai, K. 1991. *Res. Microbiol.* 142:155–59
8. Ogura, T., Hiraga, S. 1983. *Cell* 32: 351–60
9. Helmstetter, C. E., Leonard, A. C. 1990. *Res. Microbiol.* 141:30–39
10. Helmstetter, C. E., Leonard, A. C. 1987. *J. Mol. Biol.* 197:195–204
11. Donachie, W. D., Begg, K. J. 1989. *J. Bacteriol.* 171:5405–9
12. Hiraga, S., Ogura, T., Niki, H., Ichinose, C., Mori, H. 1990. *J. Bacteriol.* 172:31–39
13. Hiraga, S. 1990. *Res. Microbiol.* 141:50–56
14. Cooper, S. 1990. *Res. Microbiol.* 141:17–29
15. Begg, K. J., Donachie, W. D. 1991. *New Biol.* 3:475–86
16. Hiraga, S., Niki, H., Ogura, T., Ichinose, C., Mori, H., et al. 1989. *J. Bacteriol.* 171:1496–505
17. Niki, H., Jaffé, A., Imamura, R.,

Ogura, T., Hiraga, S. 1991. *EMBO J.* 10:183–93
18. Niki, H., Imamura, R., Yamanaka, K., Ogura, T., Hiraga, S. 1991. *14th Conf. Mol. Biol. Soc. Jpn.* Fukuoka
19. Drlica, K. 1987. *In Escherichia coli and Salmonella typhimurium*, ed. F. C. Neidhardt, J. L. Ingraham, K. B. Low, B. Magasauik, M. Schaechter, H. E. Umbarger, 1:91–103. Washington, DC: Am. Soc. Microbiol. 806 pp.
20. Kato, J., Nishimura, Y., Suzuki, H. 1989. *Mol. Gen. Genet.* 217:178–81
21. Hussain, K., Begg, K. J., Salmond, G. P. C., Donachie, W. D. 1987. *Mol. Microbiol.* 1:73–81
22. Hussain, K., Elliott, E. J., Salmond, G. P. C. 1987. *Mol. Microbiol.* 1:259–73
23. Steck, T. R., Drlica, K. 1984. *Cell* 36:1081–88
24. Orr, E. N., Fairweather, F., Holland, I. B., Pritchard, R. H. 1979. *Mol. Gen. Genet.* 177:103–12
25. Kato, J., Nishimura, Y., Yamada, M., Suzuki, H., Hirota, Y. 1988. *J. Bacteriol.* 170:3967–77
26. Kato, J., Nishimura, Y., Imamura, R., Niki, H., Hiraga, S., Suzuki, H. 1990. *Cell* 63:393–404
27. Schmid, M. B. 1990. *J. Bacteriol.* 172:5416–24
28. Luttinger, A. L., Springer, A. L., Schmid, M. B. 1991. *New Biol.* 3:687–97
29. Norris, V., Alliotte, T., Jaffé, A., D'Ari, R. 1986. *J. Bacteriol.* 168:494–504
30. Grompe, M., Versalovic, J., Koeuth, T., Lupski, J. 1991. *J. Bacteriol.* 173:1268–78
31. Kudo, T., Nagai, K., Tamura, G. 1976. *Agric. Biol. Chem.* 41:607–8

32. Kuempel, P. L., Henson, J. M., Dircks, L., Tecklenburg, M., Lim, D. F. 1991. *New Biol.* 3:799–811
33. Clerget, M. 1991. *New Biol.* 3:780–88
34. Blakely, G., Colloms, S., May, G., Burke, M., Sherratt, D. 1991. *New Biol.* 3:789–98
35. Trun, N. J., Gottesman, S. 1990. *Genes Dev.* 4:2036–47
36. Trun, N. J., Gottesman, S. 1991. *Res. Microbiol.* 142:195–200
37. Iwasaki, H., Shiba, T., Makino, K., Nakata, A., Shinagawa, H. 1989. *J. Bacteriol.* 171:5276–80
38. Niki, H., Imamura, R., Ogura, T., Hiraga, S. 1990. *Nucleic Acids Res.* 18:45
39. Webster, R. E. 1991. *Mol. Microbiol.* 5:1005–11
40. Wandersman, C., Delepelaire, P. 1990. *Proc. Natl. Acad. Sci. USA* 87:4776–80
41. McLachlan, A. D., Karn, J. 1983. *J. Mol. Biol.* 164:1829–36
42. Molina, M. J., Kroppe, K. E., Gulick, J., Robbins, J. 1978. *J. Biol. Chem.* 262:6478–88
43. Kagawa, H., Gengyo, K., McLachlan, A. D., Brenner, S., Karn, J. 1989. *J. Mol. Biol.* 207:311–33
44. Yang, J., Laymon, R. A., Goldstein, L. S. B. 1989. *Cell* 56:879–89
45. Miller, J., McLachlan, A. D., Klug, A. 1985. *EMBO J.* 4:1609–14
46. Klug, A., Rhodes, D. 1987. *Trends Biochem. Sci.* 12:464–69
47. Vale, R. D., Goldstein, L. S. B. 1990. *Cell* 60:883–85
48. Obar, R. A., Collins, C. A., Hammarback, J. A., Shpetner, H. S., Vallee, R. B. 1990. *Nature* 347:256–61
49. Casaregola, S., Norris, V., Goldberg, M., Holland, I. B. 1990. *Mol. Microbiol.* 4:505–11
50. Casaregola, S., Chen, M., Bouquin, N., Norris, V., Jacq, A., et al. 1991. *Res. Microbiol.* 142:201–7
51. Babitzke, P., Kushner, S. R. 1991. *Proc. Natl. Acad. Sci. USA* 88:1–5
52. Reeder, T., Schleif, R. 1991. *J. Bacteriol.* 173:7765–71
53. Naom, I. S., Morton, S. J., Leach, D. R. F., Lloyd, R. G. 1989. *Nucleic Acids Res.* 17:8033–44
54. Lederberg, J., Clair, St. J. 1958. *J. Bacteriol.* 75:143–60
55a. Eda, T., Kanda, Y., Mori, C., Kimura, S. 1977. *J. Bacteriol.* 132:1024–26
55b. Minkoff, L., Damadian, R. 1976. *J. Bacteriol.* 125:353–65
55c. Nakamura, K., Watanabe, S. 1978. *J. Biochem.* 83:1459–70
55d. Nakamura, K., Takahashi, K., Watanabe, S. 1978. *J. Biochem.* 84:1453–58
55e. Beck, B. D., Arscott, P. G., Jacobson, A. 1978. *Proc. Natl. Acad. Sci. USA* 75:1250–54
56. Tomioka, S. 1991. *Studies on the formation of the cell surface structure in* Escherichia coli. PhD thesis. Univ. of Tokyo. (In Japanese)
57. Epstein, W., Schultz, S. G. 1965. *J. Gen. Phys.* 49:221–34
58. Hiraga, S., Niki, H., Imamura, R., Ogura, T., Yamanaka, K., et al. 1991. *Res. Microbiol.* 142:189–94
59. MacAlister, T. J., MacDonald, B., Rothfield, L. I. 1983. *Proc. Natl. Acad. Sci. USA* 80:1372–76
60a. Rothfield, L. I., Cook, W. R. 1988. *Microbiol. Sci.* 5:182–85
60b. Ishidate, K., Creeger, E. S., Zrike, J., Deb, S., Glauner, B., et al. 1986. *J. Biol. Chem.* 261:428–43
60c. Joseleau-Petit, D., Kepes, F., Peutat, L., D'Ari, R., Rothfield, L. I. 1990. *J. Bacteriol.* 172:6573–75
61. Lane, H. E. D. 1981. *Plasmid* 5:100–26
62. Kline, B. C. 1985. *Plasmid* 14:1–16
63. Hiraga, S., Ogura, T., Mori, H., Tanaka, M. 1985. In *Plasmids in Bacteria*, ed. D. R. Helinski, S. N. Cohen, D. B. Clewell, D. A. Jackson, A. Hollaender, pp. 469–87. New York:Plenum. 995 pp.
64. Hiraga, S. 1986. In *Advances in Biophysics*, ed. M. Kotani, pp. 91–103. Tokyo/Limerick: Jpn. Sci. Soc. Press/ Elsevier. 296 pp.
65. Mori, H., Kondo, A., Ohshima, A., Ogura, T., Hiraga, S. 1986. *J. Mol. Biol.* 192:1–15
66. Helsberg, M., Eichenlaub, R. 1986. *J. Bacteriol.* 165:1043–45
67. Mori, H., Mori, Y., Ichinose, C., Niki, H., Ogura, T., et al. 1989. *J. Biol. Chem.* 264:15535–41
68. Hayakawa, Y., Murotsu, T., Matsubara, K. 1985. *J. Bacteriol.* 163:349–54
69a. Kusukawa, N., Mori, H., Kondo, A., Hiraga, S. 1987. *Mol. Gen. Genet.* 208:365–72
69b. Watanabe, E., Inamoto, S., Lee, M.-H., Kim, S. U., Ogura, T., et al. 1989. *Mol. Gen. Genet.* 218:431–36
70. Gustafsson, P., Wolf-Watz, H., Lind, L., Lohansson, K.-E., Nordström, K. 1983. *EMBO J.* 2:27–32
71. Yarmolinsky, M. B., Sternberg, N. 1988. In *The Bacteriophages,* ed. R. Calender, pp. 291–438. New York: Plenum
72. Rosner, J. L. 1972. *Virology* 48:679–89
73. Wandersman, C., Yarmolinsky, M. 1977. *Virology* 77:386–400
74. Iida, S., Arber, W. 1979. *Mol. Gen. Genet.* 173:249–61

75. Abeles, A. L., Friedman, S. A., Austin, S. J. 1985. *J. Mol. Biol.* 185:261–72
76. Martin, K. A., Davis, M. A., Austin, S. 1991. *J. Bacteriol.* 173:3630–34
77. Austin, S., Abeles, A. 1983. *J. Mol. Biol.* 169:353–72
78. Martin, K. A., Friedman, S. A., Austin, S. J. 1987. *Proc. Natl. Acad. Sci. USA* 84:8544–47
79. Davis, M. A., Austin, S. J. 1988. *EMBO J.* 7:1881–88
80. Davis, M. A., Martin, K. A., Austin, S. J. 1990. *EMBO J.* 9:991–98
81. Funnell, B. E. 1988. *J. Bacteriol.* 170:954–60
82. Funnell, B. E. 1988. *Proc. Natl. Acad. Sci. USA* 85:6657–61
83. Funnell, B. E. 1991. *J. Biol. Chem.* 266:14328–37
84. Yang, C. C., Nash, H. A. 1989. *Cell* 57:869–80
85. Friedman, S. A., Austin, S. J. 1988. *Plasmid* 19:103–12
86. Thomas, C. M. 1989. In *Promiscuous Plasmids of Gram Negative Bacteria,* ed. C. M. Thomas, pp. 1–25. London: Academic
87. Thomas, C. M., Smith, C. A. 1986. *Nucleic Acids Res.* 14:4453–69
88. Motallebi-Veshareh, M., Rouch, D. A., Thomas, C. M. 1990. *Mol. Microbiol.* 4:1455–63
89. Theophilus, B. D. M., Thomas, C. M. 1987. *Nucleic Acids Res.* 15:7443–50
90. Kornacki, J. A., Balderes, P. J., Figurski, D. H. 1987. *J. Mol. Biol.* 198:211–22
91. Walker, J. E., Saraste, M., Runswick, M. J., Gay, N. J. 1982. *EMBO J.* 1:945–51
92. Dever, T. E., Glynias, M. J., Merrick, W. C. 1987. *Proc. Natl. Acad. Sci. USA* 84:1814–18
93. Gallie, D. R., Kado, C. I. 1987. *J. Mol. Biol.* 193:465–78
94. Commanducci, M., Ricci, S., Ratti, G. 1988. *Mol. Microbiol.* 2:531–38
95. de Boer, P. A. J., Crossley, R. E., Rothfield, L. I. 1989. *Cell* 56:641–49
96. Davis, M., Martin, K. A., Austin, S. J. 1992. Submitted
97. Austin, S. J. 1988. *Plasmid* 20:1–9
98. Nordström, K., Austin, S. J. 1989. *Annu. Rev. Genet.* 23:37–69
99. Austin, S., Nordström, K. 1990. *Cell* 60:351–54
100. Treptow, N., Yarmolinsky, M., Rosenfeld, R. 1990. *EMBO Workshop on the Bacterial Cell Cycle, Collonges-la-Rouge*
101. Niki, H., Ichinose, C., Ogura, T., Mori, H., Morita, M., et al. 1988. *J. Bacteriol.* 170:5272–78
102. Ogura, T., Niki, H., Mori, H., Morita, M., Hasegawa, M., et al. 1990. *J. Bacteriol.* 172:1562–68
103. Niki, H., Ogura, T., Hiraga, S. 1990. *Mol. Gen. Genet.* 224:1–9
104. Ogura, T., Niki, H., Kano, Y., Imamoto, F., Hiraga, S. 1990. *Mol. Gen. Genet.* 220:197–203
105. Wada, M., Kohno, K., Imamoto, F., Kano, Y. 1988. *Gene* 70:393–97
106. Wada, C., Hiraga, S., Yura, T. 1976. *J. Mol. Biol.* 108:25–41
107. Wada, C., Imai, M., Yura, T. 1987. *Proc. Natl. Acad. Sci. USA* 84:8849–53
108. Ezaki, B., Ogura, T., Mori, H., Niki, H., Hiraga, S. 1989. *Mol. Gen. Genet.* 218:183–89
109. Ezaki, B., Mori, H., Ogura, T., Hiraga, S. 1990. *Mol. Gen. Genet.* 223:361–68
110. Keasling, J. D., Palsson, B. O., Cooper, S. 1991. *J. Bacteriol.* 173:2673–80
111. Leonard, A. C., Helmstetter, C. E. 1988. *J. Bacteriol.* 170:1380–83
112. Gerdes, K., Molin, S. 1986. *J. Mol. Biol.* 190:269–79
113. Tabuchi, A., Min, Y.-n., Kim, C. K., Fan, Y.-l., Womble, D. D., Rownd, R. H. 1988. *J. Mol. Biol.* 202:511–25
114. Min, Y.-n., Tabuchi, A., Fan, Y.-l., Womble, D. D., Rownd, R. H. 1988. *J. Mol. Biol.* 204:345–56
115. Min, Y.-n., Tabuchi, A., Womble, D., Rownd, R. H. 1991. *J. Bacteriol.* 173:2378–84
116. Meacock, P. A., Cohen, S. N. 1980. *Cell* 20:529–42
117. Tucker, W. T., Miller, C. A., Cohen, S. N. 1984. *Cell* 38:191–201
118. Manen, D., Goebel, T., Caro, L. 1990. *Mol. Microbiol.* 4:1839–46
119. Miller, C. A., Beaucage, S. L., Cohen, S. N. 1990. *Cell* 62:127–33
120. Wahle, E., Kornberg, A. 1988. *EMBO J.* 7:1889–95
121. Beaucage, S. L., Miller, C. A., Cohen, S. N. 1991. *EMBO J.* 10:2583–88
122. Ezaki, B., Ogura, T., Niki, H., Hiraga, S. 1991. *J. Bacteriol.* 173:6643–46
123. Jaffé, A., Ogura, T., Hiraga, S. 1985. *J. Bacteriol.* 163:841–49
124. Hiraga, S., Jaffé, A., Ogura, T., Mori, H., Takahashi, H. 1986. *J. Bacteriol.* 166:100–4
125. Kline, B. C., Miller, J. R. 1975. *J. Bacteriol.* 121:165–72

Annu. Rev. Biochem. 1992. 61:307–30

STRUCTURE AND FUNCTION OF THE MANNOSE 6-PHOSPHATE/INSULINLIKE GROWTH FACTOR II RECEPTORS

Stuart Kornfeld

Department of Medicine, Washington University School of Medicine, St. Louis, Missouri 63110

KEY WORDS: mannose 6-phosphate receptors, insulinlike growth factor II receptor, lysosomes, lysosomal enzyme

CONTENTS

0066-4154/92/0701-0307$02.00

PERSPECTIVES AND SUMMARY

The discovery in 1987 that the cation-independent mannose 6-phosphate receptor and the insulinlike growth factor II receptor are the same protein raised the interesting possibility that this receptor functions in two distinct biologic processes, i.e. protein trafficking and transmembrane signal transduction. Cell transfection experiments have very recently provided direct evidence that the Man-6-P/IGF-II receptor mediates the transport of newly synthesized acid hydrolases to lysosomes. This function is shared with a second mannose 6-phosphate receptor that does not bind IGF-II. The cloning of cDNAs for these receptors has provided insights into their structures and has revealed that they are related proteins. The determinants on the receptors that mediate Man-6-P binding and direct their intracellular routing are being defined. A growing body of evidence indicates that the Man-6-P/IGF-II receptor also functions in transmembrane signal transduction. However, this property appears to be confined to certain species and cell types. This receptor may have additional roles in the clearance and activation of other growth factors.

This review focuses on recent findings concerning the structure, function, and cellular distribution of the Man-6-P/IGF-II receptors. This subject was last reviewed in 1989 (1, 2). The reader is referred to these reviews as well as to several other recent reviews (3–7) for a more complete summary of the earlier work in this area.

INTRODUCTION

Lysosomes are acidic cytoplasmic vacuoles that contain many hydrolytic enzymes that function in the degradation of internalized and endogenous macromolecules. Although the acid hydrolases themselves are relatively long-lived proteins, they do turn over and must be replaced. Additionally, dividing cells must be able to synthesize new lysosomes. To meet these needs, higher eukaryotes have developed an elaborate system for targeting newly synthesized acid hydrolases to lysosomes. With a few exceptions, the acid hydrolases are soluble glycoproteins that are synthesized in the rough endoplasmic reticulum where they undergo cotranslational glycosylation of selected asparagine residues. These early steps in the biosynthetic pathway are shared with secretory glycoproteins, and the two classes of proteins are transported together in vesicular carriers to the Golgi where they are sorted from one another for delivery to their final destinations. This sorting process is accomplished by the phosphomannosyl recognition system. Through the concerted action of two enzymes, the acid hydrolases selectively acquire phosphomannosyl residues, which serve as high-affinity ligands for binding to the Man-6-P receptors (MPRs) in the Golgi. This step results in the physical separation of the acid hydrolases from the proteins destined for secretion. The ligand-receptor complexes are then gathered into clathrin-coated pits and bud from the Golgi in coated vesicles that transport the acid hydrolase-receptor complexes to an acidified endosomal (prelysosomal) compartment. The low pH of this compartment induces dissociation of the ligand, which is then packaged into a lysosome. The receptor either returns to the Golgi to repeat this process or moves to the plasma membrane where it functions to internalize exogenous lysosomal enzymes and IGF-II or, in some instances, to mediate a transmembrane signaling event upon the binding of IGF-II.

RECEPTOR STRUCTURE

Primary Structure and Genomic Organization

Two distinct MPRs have been isolated, characterized, and their cDNAs cloned. One is a type I integral membrane glycoprotein with a M_r of 275,000. This receptor binds Man-6-P-containing ligands independent of divalent cations, and therefore has been called the cation-independent (CI)MPR. However, since this receptor also binds IGF-II, it will be referred to as the Man-6-P/IGF-II receptor. The other receptor is also a type I integral membrane glycoprotein with a subunit M_r of approximately 46,000. The bovine and murine forms of this receptor require divalent cations for optimal ligand binding, so this receptor is referred to as the cation-dependent (CD)MPR (8–11). However, the human and porcine forms of the receptor have little or no requirement for divalent cations (12–14). cDNAs for the Man-6-P/IGF-II receptor have been cloned from bovine (15, 16), human (17, 18), and rat (19) sources, while cDNAs for the CD-MPR have been cloned from bovine (20), human (21), and murine (11, 22) sources. The amino acid sequences deduced from these cDNAs have provided insights into the structures of the receptors and have revealed that the two receptors are related proteins.

The bovine Man-6-P/IGF-II receptor consists of four structural domains: a 44-residue amino-terminal signal sequence, a 2269-residue extracytoplasmic domain, a single 23-residue transmembrane region, and a 163-residue carboxyl-terminal cytoplasmic domain. The extracytoplasmic domain contains 19 potential Asn-linked glycosylation sites, a number of which are utilized, yielding a mature receptor of 275–300 kDa. Glycosylation is not required for the receptor to bind IGF-II (23). The extracytoplasmic domain also has a repetitive structure consisting of 15 contiguous repeating segments of approximately 147 amino acids each. Each repeating segment shares sequence identities with all the other repeats, with the percent of identical residues ranging from 16 to 38%. In addition, there are many conservative amino acid substitutions. The location of the cysteine residues is highly conserved among the repeating segments. One unusual feature is that the 13th repeat from the amino terminus contains a 43-residue insertion that is similar to sequences found in fibronectin, factor XII, and a bovine seminal fluid protein (24). This segment forms part of a collagen-binding domain in fibronectin, but its function in the Man-6-P/IGF-II receptor is unknown. This is the only sequence in the receptor that is similar to sequences in other known proteins.

The cytoplasmic domain of the receptor contains four regions with sequences that are known to be potential substrates for various protein kinases, including protein kinase C, cAMP-dependent protein kinase, and casein kinases I and II (19). These sequences are highly conserved among the bovine, human, and rat receptors, whereas the intervening sequences are quite degenerate (91% identity versus 38% for the intervening sequences). The receptor is known to be phosphorylated at a number of these sites (25–27). Overall, the amino acid sequences of the bovine and the human receptors are 80% identical.

The Man-6-P/IGF-II receptor from embryonic bovine tracheal cells and embryonic human skin fibroblasts has been shown to contain covalently bound palmitic acid, most likely in an amide-linkage (28). The functional significance of this modification is unknown.

The bovine CD-MPR also consists of four structural domains: a 28-residue amino-terminal signal sequence, a 159-residue extracytoplasmic domain, a single 25-residue transmembrane region, and a 67-residue carboxyl-terminal cytoplasmic domain. This receptor contains five potential Asn-linked glycosylation sites, four of which are used (20, 29). Alignment of the bovine, human, and mouse sequences reveals that the mature proteins are 93–95% identical, with the cytoplasmic domains being 100% identical. The cytoplasmic domain contains a single casein kinase II site. When these sequences are compared to the sequences of the Man-6-P/IGF-II receptor, it is evident that the entire extracytoplasmic domain of the CD-MPR is similar to each of the repeating units of the Man-6-P/IGF-II receptor, with sequence identities ranging from 14 to 28%. This suggests that the two receptors may be derived from a common ancestor, with the Man-6-P/IGF-II receptor arising from multiple duplications of a single ancestral gene. In contrast to this homology, there are no sequence similarities among the signal sequences, transmembrane regions, and the cytoplasmic domains of the two receptors.

The gene for the human CD-MPR has been mapped to chromosome 12 (21), and the genomic structure has been determined to consist of seven exons (110–1573 basepairs) spanning more than 12 kilobases (30). Exon 1 contains 5' untranslated sequence, exon 2 encodes the signal sequence and the beginning of the luminal domain, and exons 3–5 encode the remainder of the luminal domain. Exons 5 and 6 encode the transmembrane domain, and exons

6 and 7 encode the cytoplasmic domain. Interestingly, the cysteines thought to be involved in disulfide pairs are located on different exons. This indicates that the intron/exon borders of the CD-MPR gene do not reflect the protein domains. The genomic structure of the Man-6-P/IGF-II receptor has not been elucidated as yet, but the gene has been localized to the long arm of human chromosome 6, region 6q25→q27, and mouse chromosome 17, region A-C (31). Thus the two genes for the human Man-6-P receptors are on different chromosomes.

Oligomeric Structure

The oligomeric structure of the CD-MPR has been analyzed in several laboratories (10, 32–36). These reports indicate that detergent-solubilized receptor can exist as a monomer, dimer, or tetramer depending on the experimental conditions. The formation of dimeric and tetrameric forms is favored by low temperature ($<16°C$), neutral pH, the presence of Man-6-P, and high protein concentration, whereas monomer formation is favored by higher temperatures, low pH, and low receptor concentration (35). The kinetics of dissociation and reassociation are relatively rapid at 37°C, raising the possibility that the receptor may undergo these transitions as it cycles to different cellular compartments. In the studies that utilized the human receptor, the monomeric form did not bind to a phosphomannan-Sepharose affinity column (35). However, the monomeric form of the bovine receptor did bind to a pentamannosephosphate-affinity column when Mn^{2+} was present (10). Further evidence that the monomeric form of the bovine receptor can fold into an independent ligand-binding unit has come from studies of a truncated form of the receptor that lacks the transmembrane and cytoplasmic domains. This receptor, which behaves as a soluble monomer in solution, is capable of binding to a pentamannosephosphate-affinity column (33). Expression of a truncated form of the human CD-MPR gave rise to a soluble dimer under the conditions tested (36). These discrepancies may reflect species variation or possibly differences in the affinity column used to assess receptor-binding ability.

The relevance of these findings to the state of the receptor in cellular membranes has yet to be established. Several studies have shown that the CD-MPR exists primarily as a dimer in the membrane as analyzed by chemical cross-linking agents (10, 32, 33). However, Waheed et al (35) observed monomeric, dimeric, and tetrameric forms of the receptor in baby hamster kidney cells overexpressing the CD-MPR. These authors speculate that changes in the quaternary structure of the receptor during recycling may influence the biologic behavior of this molecule.

The quaternary structure of the Man-6-P/IGF-II receptor has not been analyzed as extensively, but hydrodynamic measurements are consistent with it being a monomer (37), while chemical cross-linking experiments indicate that it may be an oligomer (38).

Ligand-Binding Properties

Equilibrium dialysis experiments indicate that the CD-MPR binds one mole of the monovalent ligand Man-6-P and 0.5 mole of a diphosphorylated high-mannose oligosaccharide per monomeric subunit (39, 39a). Consequently each dimer would have two Man-6-P-binding sites, both of which can be occupied by a single oligosaccharide containing two Man-6-P residues. The Man-6-P/IGF-II receptor, on the other hand, binds two moles of Man-6-P or one mole of a diphosphorylated oligosaccharide per monomer (39a, 40). This suggests that only two of the 15 repeating segments of the receptor may function in the binding of Man-6-P. In this regard, two proteolytic fragments of the receptor corresponding to repeating units 1–3 and 7–11 have been shown to bind to Man-6-P-containing ligands (41). Therefore, these two regions of the receptor are likely to contain the two functional Man-6-P-binding domains.

Little is known about the actual binding sites for Man-6-P on either receptor. Wendland et al (42) mutated each of the five histidine and the eight arginine residues of the mature CD-MPR and found that His131 and Arg137 are essential for ligand binding. These investigators demonstrated that non-glycosylated CD-MPR is stable and binds ligands with high affinity, whereas mutants with replacement of any of the six luminal cysteine residues with glycine residues are completely inactive (29, 43). This latter finding is consistent with the observation that newly synthesized CD-MPR polypeptides lack ligand binding and only acquire this property after formation of in-tramolecular disulfide bonds (44, 45).

While both receptors bind Man-6-P with essentially the same affinity (7–8 $\times 10^{-6}$ M), the Man-6-P/IGF-II receptor binds a diphosphorylated oligosac-charide with a substantially higher affinity than does the CD-MPR (2×10^{-9} M vs 2×10^{-7} M, respectively) (39, 39a, 40). Since oligosaccharides with two phosphomonoesters bind to the MPRs with an affinity similar to that observed for lysosomal enzymes, the high-affinity binding of lysosomal enzymes can be explained by a two-site model in which two phosphoman-nosyl residues on the lysosomal enzyme interact with the receptor (40). This divalent interaction could either be mediated by two phosphomannosyl re-sidues on the same oligosaccharide of the lysosomal enzyme or by phospho-mannosyl residues located on different oligosaccharides. The latter interaction could result in an even better fit, resulting in higher-affinity binding. The secretion of the lysosomal enzyme cathepsin L by transformed mouse cells is of particular relevance in this regard. Dong et al (46) reported that NIH 3T3 cells transformed with the Kirsten sarcoma virus synthesize 25-fold more cathepsin L than nontransformed cells and secrete 94% of this enzyme while retaining most of the other lysosomal enzymes. The secreted cathepsin L, which contains a single oligosaccharide with two phosphomonoester moieties, bound to a MPR-affinity column with a 10-fold lower affinity than that observed with other lysosomal enzymes. Furthermore, cathepsin L synthesized by Chinese hamster ovary cells bound to the receptor with higher affinity than mouse cathepsin L, and upon analysis was found to contain two oligosaccharides rather than the single one present on the mouse enzyme (46), with each oligosaccharide having two phosphomannosyl residues (47). These data are consistent with the view that individual phosphomannosyl residues located on different oligosaccharides interact with the receptor with higher affinity than do two phosphomannosyl residues present on the same oligosaccharide. On the other hand, if the membrane form of the Man-6-P/IGF-II receptor proves to be a dimer (or the CD-MPR, a tetramer), then a four-site model for high-affinity ligand binding would be indicated and provide an explanation for the observed findings with cathepsin L. An alternative explanation for the poor binding of mouse cathepsin L to the Man-6-P/IGF-II receptor has been put forward by Lazzarino & Gabel (48). These investigators found that intact procathepsin L, either native or in a reduced and alkylated state, has a poor affinity for the receptor, whereas phosphorylated oligosaccharides released from the enzyme bind to the receptor with high affinity. These results indicate that the polypeptide portion of cathepsin L contains determinants that inhibit binding of the phosphorylated oligosaccharide to the MPR. One possibility is that the negatively charged phosphomannosyl groups of the oligosaccharide interact ionically with positively charged lysine or arginine side chains of the protein backbone, and thereby are impaired in their interaction with the MPR.

The bovine, human, and rat Man-6-P/IGF-II receptors all bind IGF-II, a nonglycosylated polypeptide, with high affinity (19, 49–53). In contrast, the chicken and frog receptors lack the high-affinity IGF-II-binding site (54, 55) as does the CD-MPR (49, 52). The stoichiometry of IGF-II binding to the bovine receptor has been determined to be one mole of ligand bound per polypeptide (49). Several studies have shown that the binding sites for Man-6-P and IGF-II on the receptor are distinct, and that the receptor can bind both ligands simultaneously (19, 49–52). However, lysosomal enzymes, in contrast to Man-6-P, do impair IGF-II binding, and IGF-II can inhibit lysosomal enzyme binding (19, 52, 56, 57). The significance of these inhibitory effects is unclear, since overexpression of IGF-II in NIH 3T3 cells does not impair the sorting of newly synthesized lysosomal enzymes or the uptake of exogenous arylsulfatase A (58).

RECEPTOR FUNCTION

Role in Lysosomal Enzyme Sorting and Endocytosis

There is considerable evidence that the Man-6-P/IGF-II receptor functions in the sorting of newly synthesized lysosomal enzymes and in the endocytosis of extracellular lysosomal enzymes, whereas the CD-MPR only participates in lysosomal enzyme sorting under physiologic conditions. Thus, cultured cells that either lack endogenous Man-6-P/IGF-II receptor (59) or are depleted of this receptor with specific antibodies (4, 60) secrete 60–70% of their newly synthesized lysosomal enzymes and do not endocytose extracellular lysosomal enzymes. Transfection of the receptor-deficient cell lines with Man-6-P/IGF-II receptor cDNA results in receptor expression and correction of the defects in lysosomal enzyme sorting and uptake (61, 62). The residual sorting found in the Man-6-P/IGF-II receptor-deficient cells appears to be mediated by the CD-MPR, since treatment of the cells with anti-CD-MPR antiserum results in increased secretion of the acid hydrolases (60). When such cells are transfected with CD-MPR cDNA to achieve 5–20 times the level of endogenous receptor expression, the sorting of lysosomal enzymes increases to about 50% (11, 14). However, endocytosis of exogenous β-glucuronidase is only observed in cells expressing 40–50 times the level of the endogenous CD-MPR, and even at these high levels of expression the efficiency of ligand uptake is only 1–2% of that mediated by physiologic levels of the Man-6-P/IGF-II receptor. The inability of the CD-MPR to function efficiently in endocytosis reflects its poor binding of ligand at the surface cell rather than its failure to recycle to the plasma membrane (11, 38, 60, 63).

When the CD-MPR is overexpressed in baby hamster kidney (BHK) and mouse L cells that contain normal levels of endogenous Man-6-P/IGF-II receptor, the sorting of lysosomal enzymes decreases from 90–95% to 50% (64). This effect requires the expression of high levels of CD-MPR that have the ability to bind ligand and to recycle. Cotransfection with Man-6-P/IGF-II receptor cDNA restores lysosomal enzyme sorting to the normal level. Chao et al (64) have proposed that the two receptors compete for lysosomal enzyme binding in the Golgi, with the ligands that bind to the CD-MPR being delivered to a site (early endosome or plasma membrane) from which they can be released from the cell. This interpretation would explain the inefficient sorting of lysosomal enzymes by Man-6-P/IGF-II receptor-deficient cells transfected with the CD-MPR cDNA.

Role in Signal Transduction

The discovery by Morgan et al (17) that the CI-MPR and the IGF-II receptor are the same protein immediately raised the fascinating possibility that this receptor may function in two diverse biologic processes, i.e. protein trafficking and transmembrane signaling. As discussed in the preceding section, the evidence for a role in protein trafficking is well established. However, it has been more difficult to establish a role in signal transduction because IGF-II also binds to the IGF-I receptor and the insulin receptor, both members of the tyrosine kinase family of receptors known to transmit signals across the plasma membrane (5, 65–67). In fact, it has been shown that some effects of IGF-II on various cell types are likely to be mediated by its binding to the IGF-I receptor or the insulin receptor or possibly yet another receptor (68–71). A clear example of IGF-II acting through a receptor other than the Man-6-P/IGF-II receptor is seen with chicken embryo fibroblasts. This cell type responds to IGF-II with enhanced protein synthesis and stimulation of cell division (72), even though the chicken CI-MPR does not bind IGF-II (54, 55). Nevertheless, a number of reports indicate that IGF-II does mediate responses through its own receptor. These responses include the stimulation of glycogen synthesis in rat hepatoma cells (73), the stimulation of amino acid uptake in human myoblasts (74), the promotion of cell proliferation in K562 cells (75), the stimulation of Na^+/H^+ exchange and inositol trisphosphate production in canine kidney proximal tubular cells (76–79), the growth and development of rat metanephroi (80), and the stimulation of Ca^{2+} influx and DNA synthesis in BALB/c 3T3 cells (81, 82). In two of these studies it was shown that antibodies to the Man-6-P/IGF-II receptor mimic the action of IGF-II (73, 81), while in another study Man-6-P potentiated the stimulatory effect of IGF-II (78). In the study utilizing K562 erythroleukemia cells, it was shown that the target cells lack the IGF-I receptor, and their growth response to insulin and IGF-I was extremely poor relative to their response to IGF-II (75).

Nishimoto and his colleagues have begun to elucidate the mechanism by which the Man-6-P/IGF-II receptor may transduce a signal across the plasma membrane. These workers found that IGF-II does not stimulate the opening of calcium channels or DNA synthesis in quiescent BALB/c 3T3 cells, but if the cells are first made competent by pretreatment with platelet-derived growth factor (PDGF) and then primed by a brief incubation with epidermal growth factor (EGF), IGF-II induces a twofold, sustained increase in calcium influx rate as well as an increase in [^{3}H]thymidine incorporation into DNA (82). These biologic responses could also be induced by anti-Man-6-P/IGF-II receptor antibodies (81). The stimulatory effects of IGF-II on calcium influx were completely abolished by pretreatment of the cells with pertussis toxin, suggesting that the IGF-II effects were mediated by a mechanism involving coupling to a G protein. Evidence that this is the case was obtained by demonstrating that IGF-II brings about the direct coupling of $G_{i-2\alpha}$, a G_i protein with a 40-kDa α subunit, with purified Man-6-P/IGF-II receptor reconstituted in phospholipid vesicles (83, 84). $G_{i-2\alpha}$ is a member of the membrane-associated oligomeric G proteins, which transduce receptor-mediated signals to intracellular effectors (85). Interestingly, Man-6-P and Man-6-P-containing β-glucuronidase did not stimulate GTPγS binding to G_i or activate its GTPase, but these agents completely inhibited the IGF-II-induced G_i protein activation (86). These results indicate that the Man-6-P/IGF-II receptor has two distinct signaling functions that positively or negatively regulate the activity of $G_{i-2\alpha}$ in response to the binding of IGF-II or Man-6-P.

When a synthetic peptide corresponding to residues 122–135 of the cytoplasmic tail of the human Man-6-P/IGF-II receptor was incubated with $G_{i-2\alpha}$, it activated the G protein in a manner similar to that observed with the intact receptor plus IGF-II (87). This region of the cytoplasmic tail was chosen for study since it shares the structural characteristics of mastoparan, a small peptide from wasp venom that mimics G-coupled receptors by directly activating G proteins (88). Peptides of similar size corresponding to other regions of the cytoplasmic tail had no biologic effects on these assays. The results are consistent with residues 122–135 of the cytoplasmic tail having a role in the $G_{i-2\alpha}$-activating function of the receptor. The coupling of the receptor to $G_{i-2\alpha}$ appears to be a regulatory step in the signal transduction process. Okamoto et al (89) have found that the failure of IGF-II to trigger the signaling pathway in quiescent BALB/c 3T3 cells is due to uncoupling of the receptor and $G_{i-2\alpha}$. This coupling could be restored by incubating the cells with the combination of PDGF and EGF or by transfection with Kirsten sarcoma virus bearing the *v-Ki-ras* gene. However, it is not known whether *ras p21* directly or indirectly acts on receptor-$G_{i-2\alpha}$ coupling. Okamoto et al (89) have proposed that Man-6-P/IGF-II receptor-$G_{i-2\alpha}$ uncoupling may be one mechanism that precludes quiescent cells from responding to IGF-II.

Sakano et al (89a) have taken a different approach to address the question of whether the biologic effects of IGF-II are mediated via the Man-6-P/IGF-II receptor or the IGF-I receptor. These investigators prepared IGF-II mutants that bind specifically to one or the other receptor. Thus, their Class I mutants with Tyr27 → Leu or Val43 → Leu substitutions bind with normal high affinity to the Man-6-P/IGF-II receptor and with extremely low affinity to the IGF-I receptor. In contrast, the Class II mutants with Phe48, Arg49, and Ser50 substituted with Thr, Ser and Ile residues, respectively, or Ala54Leu55 substituted with Arg residues, bind to the IGF-I receptor with high affinity but very poorly to the Man-6-P/IGF-II receptor. These results show that the two receptors interact with different domains of IGF-II. Similar findings have been reported by others (89b, 89c, 89d). When the IGF-II mutants of Sakano et al (89a) were tested for their activity in two biologic systems (stimulation of DNA synthesis in BALB/c 3T3 cells and glycogen synthesis in Hep G2 cells), the induction of a biologic response correlated with high-affinity binding to the IGF-I receptor and not with binding to the Man-6-P/IGF-II receptor. These results differ from the previous reports indicating that the biologic responses of these cells to IGF-II is mediated by the Man-6-P/IGF-II receptor (73, 81). The Class I and Class II mutants of IGF-II appear to be promising reagents for analyzing the biologically important targets of this hormone.

Role in Hormone Clearance and Activation

The Man-6-P/IGF-II receptor is known to bind and internalize IGF-II at the cell surface, resulting in the lysosomal degradation of this ligand (52, 90). In this manner the receptor may serve to clear IGF-II from the circulation. Binding of other hormones to the Man-6-P/IGF-II receptor at the cell surface may result in their activation. This appears to be the case with transforming growth factor-β1 (TGF-β1) precursor, the proform of a hormone that has multiple effects on cell growth and differentiation. This molecule has been found to contain Man-6-P residues and to bind to the Man-6-P/IGF-II receptor (91, 92). Dennis & Rifkin (93) reported that Man-6-P and anti-Man-6-P/IGF-II receptor antibodies inhibit the activation of TGF-β1 precursor by bovine aortic endothelial cell/bovine smooth muscle cell co-cultures, suggesting that binding to the Man-6-P/IGF-II receptor is required for latent TGF-β1 activation. It is not known whether this activation occurs at the cell surface or following internalization into acidified endosomes. Another growth factor that acquires Man-6-P residues is proliferin, a prolactin-related protein postulated to be an autocrine growth factor (94). Its binding to the Man-6-P/IGF-II receptor could result in its activation in endosomes or its degradation in lysosomes. Porcine thyroglobulin is a Man-6-P-containing glycoprotein that is secreted by thyroid follicle cells and then recaptured for degradation in lysosomes (95). The Man-6-P/IGF-II receptor may participate in its recapture and degradation.

Role in Extracellular Matrix Degradation

Wang and colleagues have presented evidence that MPRs located at the cell surface contain bound acid hydrolases that degrade cell surface and substratum-attached proteoglycans (96, 97). These investigators found that the addition of Man-6-P to human fibroblast cultures inhibited the turnover of extracellular ^{35}S-labeled proteoglycans. When the experiment was performed with fibroblasts from patients with I-cell disease, which lack acid hydrolases carrying the Man-6-P marker, no difference was seen in the turnover of the proteoglycans with or without added Man-6-P. However, the addition of acid hydrolases derived from normal human fibroblasts to ^{35}S-labeled I cells resulted in enhanced turnover of the proteoglycans, and this effect was inhibited by Man-6-P. These data suggest that the Man-6-P/IGF-II receptor molecules at the cell surface serve to anchor acid hydrolases and thereby allow them to degrade pericellular and extracellular proteoglycans more effectively. If the acid hydrolases were acting in solution rather than while anchored at the cell surface, then the addition of Man-6-P, which releases the hydrolases from the surface receptors, should have enhanced rather than impaired the degradation of the proteoglycans.

RECEPTOR DISTRIBUTION AND TRAFFICKING

Subcellular Localization

The earlier studies on the distribution and trafficking of the MPRs have been reviewed (1), so this information will be briefly summarized and the focus will be on the new contributions in this area. There is general agreement that a single pool of MPRs cycle constitutively among the Golgi, endosomes, and the plasma membrane. Most of the biochemical evidence suggests that the major site for lysosomal enzyme sorting is the last Golgi compartment, variously referred to as the trans Golgi network (TGN), the trans Golgi reticulum, the trans tubular network, and GERL (Golgi endoplasmic reticulum, lysosome) (98). In some cell types, such as pancreatic, hepatic, and epididymal cells, the Man-6-P/IGF-II receptor has been found by immunolocalization techniques to be most concentrated in the early (cis) Golgi, raising the possibility that lysosomal enzymes may bind to the receptor in that compartment and either pass through the Golgi as a complex or exit the Golgi at the cis side of the stack (99, 100). In most cell types, this receptor is found primarily in the TGN, with lesser amounts detected throughout the Golgi stack (100–102, 102a).

The immunolocalization studies have also revealed that at steady-state, most of the Man-6-P/IGF-II receptor is present in one or more populations of endosomes with very low, or undetectable amounts in structures identified as lysosomes (101–106). In normal rat kidney (NRK) cells, it has been estimated that 90% of the receptor is located in a late endosomal/prelysosomal compartment, with the rest being distributed over the plasma membrane, early endosomes, and the trans Golgi network (107). The intracellular distribution of the CD-MPR has been examined in this same cell type (108). At steady state, this receptor is concentrated in the Golgi complex, mainly in middle and trans cisternae, whereas in cells treated with weak bases, the receptor redistributes to endosomes that also contain the Man-6-P/IGF-II receptor. The CD-MPR is also concentrated in the Golgi in Clone 9 hepatocytes and in exocrine pancreatic cells, whereas it is localized primarily in endosomes in U937 monocytes, mouse macrophages, and proximal tubule cells (108, 109). In Madin-Darby Canine Kidney (MDCK) cells, which are polarized epithelial cells, the surface distribution of the Man-6-P/IGF-II receptor is exclusively basolateral (110). The CD-MPR could not be detected on the surface of this cell type.

The high level of receptor in endosomes, along with low or undetectable levels in lysosomes, has been interpreted to indicate that newly synthesized lysosomal enzymes are delivered to acidic prelysosomal/endosomal compartments rather than to lysosomes (1). The low pH of the endosomal compartment would cause the ligand-receptor complex to dissociate, and the released lysosomal enzymes could then be packaged into lysosomes while the MPRs could recycle either back to the Golgi and/or to the plasma membrane. It is currently unclear whether the Golgi-derived vesicles containing lysosomal enzyme-receptor complexes fuse with early or late endosomes or with both types of endosomes (110a). If delivery to early endosomes does occur, the ligand-MPR complex would presumably not dissociate until the complex migrates to late endosomes. This is because significant dissociation of ligand from either receptor requires pH <5.5, which is the pH of late endosomes, but not early endosomes (111).

Regulation of Receptor Trafficking

Several agents have been shown to alter the cellular distribution of the MPRs. In human fibroblasts, a rapid and transient redistribution of both MPRs from internal pools to the cell surface is induced by IGF-I, IGF-II, EGF, and phorbol myristate acetate (PMA) (112–114). This redistribution is associated with a 2–3-fold increase in the binding and uptake of exogenous lysosome enzymes. However, the sorting of newly synthesized lysosomal enzymes in the Golgi is not affected, because only 5–10% of the receptor is normally present on the cell surface, so a 2–3-fold increase will have only a minor effect on the intracellular pool. The effect of IGF-I and IGF-II on receptor redistribution varies with the cell type. In Hep G_2 cells and rat C6 glial cells, IGF-II does not alter the number of cell-surface receptors, and in the latter cell type, it inhibits the uptake of the lysosomal enzyme β-galactosidase by impairing binding of this ligand to the cell-surface receptors (56, 113). And in rat microvascular endothelial cells, IGF-II decreases the surface expression of the receptor, whereas PMA increases receptor number (115). The IGF-II effect is associated with decreased receptor phosphorylation, whereas the PMA effect is associated with a stimulation of receptor phosphorylation.

Prence et al (116) found that a short treatment of NIH 3T3 mouse fibroblasts with PDGF induces a redistribution of Man-6-P/IGF-II receptors to the cell surface. This treatment also causes an increase in the secretion of cathepsin L without altering the rate of synthesis of that protein or the secretion of other lysosomal enzymes. The authors proposed that the hypersecretion of cathepsin L is a consequence of the PDGF-induced changes in the distribution of the receptor, such that the Golgi concentration of receptors becomes limiting and the low-affinity cathepsin L is secreted rather than transported to lysosomes.

The most striking effects on Man-6-P/IGF-II receptor distribution have been observed in rat adipocytes and H-35 hepatoma cells, in both of which insulin causes a major redistribution of receptors from internal membranes to the cell surface (117–120). This effect is associated with an overall decrease in phosphorylation of the receptor molecules present in the plasma membrane (26, 121, 122). However, Corvera et al (122) found that a plasma membrane subfraction that is enriched in clathrin contains a highly phosphorylated form of the receptor. Based on these data, it has been proposed that insulin-mediated dephosphorylation of the receptor impairs its ability to concentrate in clathrin-coated regions of the membrane, thereby decreasing its rate of internalization and increasing its steady-state level at the cell surface.

Structural Determinants of Receptor Trafficking

The signals on the MPRs needed for rapid endocytosis from the cell surface and efficient lysosomal enzyme sorting in the Golgi have been localized to the cytoplasmic tails of these proteins (62, 123, 124). The approach for identifying these signals has been to transfect receptor-deficient cells with normal receptor cDNA or cDNAs mutated in the cytoplasmic domain and then test for the ability of the cells to sort and endocytose lysosomal enzymes. In some instances the rate of internalization of receptors from the plasma membrane was measured directly (124). These studies have revealed that the sequence Tyr-Lys-Tyr-Ser-Lys-Val, representing residues 24–29 of the cytoplasmic tail, serves as the internalization signal for the Man-6-P/IGF-II receptor (123). Alanine scanning of this sequence identified Tyr26 and Val29 as the most important residues for rapid internalization, with Tyr24 and Lys28 also contributing to the signal. The tyrosines could be substituted with phenylalanines with no loss of activity, indicating the requirement for an aromatic residue in these positions rather than tyrosine specifically. Mutant receptors containing leucine, isoleucine, methionine, or phenylalanine at position 29 in place of the critical valine were internalized at the same rate as the wild-type receptor, whereas replacement of this residue with glycine, arginine, or glutamate resulted in drastic reductions in the rate of receptor internalization (M. Jadot, W. Canfield, S. Kornfeld, manuscript in preparation). These results suggest that it is necessary to have a bulky hydrophobic/aromatic residue in the last position of the signal.

The CD-MPR contains at least two separate signals for rapid internalization in its 67-amino-acid cytoplasmic domain (124). One signal includes Phe13 and 18, while the other includes Tyr45. The Phe-containing signal appears to consist of six amino acids (Phe-Pro-His-Leu-Ala-Phe), while the Tyr-containing signal is postulated to consist of four amino acids (Tyr-Arg-Gly-

Val). The Tyr45-containing signal is less potent than the Phe13Phe18 signal.

The requirement for a tyrosine residue as a component of the internalization signal has been demonstrated in a number of instances (125–131). In a few cases, it has been shown that the tyrosine must be in the proper context relative to the surrounding amino acids to be functional (125, 128, 130). A comparison of the MPR internalization sequences to the sequences neighboring the critical tyrosines in the cytoplasmic tails of other proteins known to undergo rapid internalization is shown in Table 1. This alignment reveals some common elements. In four instances the internalization signal appears to consist of six residues with a critical aromatic residue in the first position and a bulky hydrophobic/aromatic residue in the last position. The common features of the other seven sequences are the presence of a four-amino-acid motif with a tyrosine in the first position and a bulky hydrophobic/aromatic residue in the fourth position. In one of these examples (transferrin receptor), the phenylalanine in the fourth position has been shown to be essential for rapid internalization (128). Furthermore, while the Man-6-P/IGF-II receptor contains a six-amino-acid signal, the mutant containing only the four amino acids Tyr-Ser-Lys-Val is rapidly internalized (123). Taken together, these data indicate that the signal for rapid internalization is a general motif rather than a specific sequence, with the essential elements being an aromatic residue in the amino-terminal position separated from a bulky hydrophobic/aromatic residue in the carboxyl-terminal position by two amino acids (123, 132).

Collawn et al (132) substituted the six-residue internalization signal of the Man-6-P/IGF-II receptor for the four-residue signal of the transferrin receptor, a type II membrane protein, and found that it promoted the rapid

Table 1 Cytoplasmic internalization signals of transmembrane receptors and proteins[a]

Bovine CI-MPR (123)	N V S **Y K Y S K V**	N K E E E	
Human LDL-receptor (125)	S I N	**F D N P V Y**	Q K T T E
Bovine CD-MPR[b] (124)	G M E	**F P H L A F**	W Q D L G
Bovine CD-MPR[b] (124)	N V P A A	**Y R G V**	G D D Q L
Human acid phosphatase (126)	A E P P G	**Y R H V**	A D G Q D
Human LAMP-1 (127)	R S H A G	**Y Q T I**	
Human transferrin-receptor[c] (128)	G E P L S	**Y T R F**	S L A R Q
Rabbit poly-Ig-receptor (129)	E A D L A	**Y S A F**	L L Q S N
Asialoglycoprotein receptor[c] (131)	M T K E	**Y Q D L**	Q H L D N

[a] Amino acids shown to be important for endocytosis are in boldface type.
[b] The CD-MPR has two internalization signals.
[c] This receptor is a type II membrane protein.

internalization of the transferrin receptor. In the converse experiment, Jadot et al found that the four-residue internalization signal of the transferrin receptor effectively replaced the internalization signal of the Man-6-P/IGF-II receptor (M. Jadot, W. Canfield, S. Kornfeld, manuscript in preparation). These results are consistent with the internalization signals being interchangeable, self-determined structural motifs that function in the setting of either a type I or type II membrane protein.

Collawn et al (128, 132) have noted that the internalization sequences of the Man-6-P/IGF-II, transferrin, and LDL receptors have a propensity to be in tight turns in proteins of known structure. On this basis, they suggested that these internalization sequences adopt tight turn configuations, which could account for the accessibility of critical residues in both four-residue and six-residue signals. They suggest that the two additional residues in the six-residue internalization sequences may place an aromatic side chain near the beginning of the tight turn and compensate for the absence of an aromatic side chain at the first position of the turn (as in the LDL receptor). Additional evidence that the Asn-Pro-Val-Tyr internalization sequence of the LDL receptor forms a β-turn structure has come from the nuclear magnetic resonance (NMR) analysis of nonapeptides containing this sequence (133).

The signal required for efficient sorting of lysosomal enzymes in the Golgi is more complex. Mutant Man-6-P/IGF-II receptors with 40 and 89 residues deleted from the carboxyl terminus of the 163-amino-acid cytoplasmic tail are impaired in sorting of newly synthesized lysosomal enzymes even though they function normally in endocytosis (62). In fact, deletion of the carboxyl-terminal four amino acids of the cytoplasmic tail (Leu-Leu-His-Val) gives the identical phenotype (K. Johnson, S. Kornfeld, manuscript in preparation). A similar result was obtained with the CD-MPR, in which deletion of the outer five amino acids of the cytoplasmic tail (His-Leu-Leu-Pro-Met) gives rise to a mutant receptor that is impaired in lysosomal enzyme sorting, but not in internalization at the plasma membrane (K. Johnson, S. Kornfeld, manuscript in preparation). All of these mutant receptors cycle from the plasma membrane to the Golgi with wild-type kinetics, indicating that their failure to sort lysosomal enzymes efficiently is due to an inability to enter the Golgi-clathrin-coated pits. Since the sorting function of the mutant Man-6-P/IGF-II receptors is decreased, but not abolished, it is possible that its cytoplasmic tail contains two (or more) signals that are required for efficient sorting, and removal of one of these leads to impairment of this function. Alternatively, the primary sequence required for efficient sorting may still be present on the mutant receptors, but be unable to function properly when the cytoplasmic tail is truncated.

Pearse and colleagues have provided some insight into how the MPRs may be concentrated in clathrin-coated pits (134, 135). These workers have analyzed the interaction of the Man-6-P/IGF-II receptor with coat proteins termed *adaptors* (136, 137). The adaptor proteins in the plasma-membrane coated pits consist of a heterodimer of 100-kDa polypeptides (HA-II adaptins) plus two smaller polypeptides of 50 kDa and 17 kDa. The Golgi clathrin-coated pits contain related, but distinct 100-kDa polypeptides (HA-I adaptins) as well as two different associated polypeptides of 47 kDa and 20 kDa (138, 139). Glickman et al (135) demonstrated that an immobilized fusion protein containing the cytoplasmic portion of the Man-6-P/IGF-II receptor bound both HA-II and HA-I adaptins. The binding of the HA-II adaptins, but not the HA-I adaptins, was dependent on the presence of the two tyrosines in the cytoplasmic domain of the receptor. The cytoplasmic portion of several other receptors also bound to the HA-II adaptins, but not to the HA-I adaptins (134). These data suggest that the plasma membrane–associated HA-II adaptins interact with the tyrosine-containing portion of the Man-6-P/IGF-II receptor cytoplasmic tail to allow its concentration in coated pits and rapid endocytosis, whereas the Golgi-associated HA-I adaptors interact with other determinants on the cytoplasmic tail to allow diversion to the prelysosomal compartment.

Méresse et al (27) have demonstrated that the 47-kDa subunit of the HA-I adaptor complex exhibits properties similar to casein kinase II and is capable of phosphorylating Ser85 and 156 of the cytoplasmic tail of the Man-6-P/IGF-II receptor. This result raises the possibility that the Man-6-P/IGF-II receptor forms an initial complex with the HA-I adaptors that is then stabilized by a subsequent phosphorylation event.

While the cytoplasmic tails of the MPRs are clearly involved in cellular trafficking, the extracellular and transmembrane domains of these receptors may influence their rate of movement between compartments. Dintzis & Pfeffer (140) constructed a chimeric receptor comprising human EGF receptor extracellular and transmembrane domains joined to the bovine Man-6-P/IGF-II receptor cytoplasmic domain. The expressed receptor was stable, bound EGF with high affinity and was efficiently endocytosed and recycled back to the cell surface. However, at steady state, more than 85% of the chimeric receptor molecules were located at the cell surface, whereas essentially all of the endogenous Man-6-P/IGF-II receptor molecules were intracellular. These data suggest that the extracellular and/or transmembrane portions of the Man-6-P/IGF-II receptor may contain an "endosome-retention" signal which, in some way, retains the native receptor within endosomal compartments. An alternative possibility is that the EGF receptor extracellular/transmembrane domains contain a dominant signal for rapid recycling to the cell surface.

Reconstitution of Receptor Trafficking in Cell Extracts

Goda & Pfeffer (141, 142) have developed a cell-free system that efficiently mediates the vesicular transport of the Man-6-P/IGF-II receptor from endosomes to the compartments that contain sialyltransferase, presumably the trans Golgi and trans Golgi network. This in vitro transport system requires GTP hydrolysis and cytoplasmic components, including an N-ethylmaleimide-sensitive protein of 50–100 kDa, which is distinct from NSF, an NEM-sensitive protein that facilitates the transport of proteins from the endoplasmic reticulum to the Golgi complex and between Golgi cisternae (143). Antibodies to clathrin that inhibit endocytosis do not impair the recycling of the Man-6-P/IGF-II receptor from endosomes to the Golgi, suggesting that this step in trafficking does not involve clathrin (144). These results show that it is possible to reconstitute at least one step in the vesicular movement of the MPRs, and indicate that this may be a fruitful approach for elucidating the components of the transport machinery.

DEVELOPMENTAL REGULATION AND TISSUE-SPECIFIC EXPRESSION

The Man-6-P/IGF-II receptor is developmentally regulated in the rat and mouse and subject to variable expression in different tissues (94, 145–147). The levels of receptor are highest in 16- and 20-day old fetal tissues, and decline dramatically during the postnatal period (145). In fetal heart, the concentration of the Man-6-P/IGF-II receptor reaches 1.7% of total protein at 16 days of gestation and falls to 0.07% of the total protein by the 20th postnatal day. The levels of receptor in other fetal tissues range from 1% (placenta) to 0.1% (brain) of total protein, and in all instances these levels decrease 5–10-fold during postnatal life. These changes in receptor protein concentration in the various rat tissues correlate with changes in Man-6-P/IGF-II receptor mRNA content, which is high in fetal tissues and rapidly decreases after birth (147, 148). There is also a direct association between the levels of the receptor in tissues and the presence of a soluble, truncated form of the receptor in the plasma. Kiess et al have shown that there are high levels of receptor in fetal and neonatal plasma, with a dramatic decline between days 20 and 40 of postnatal development (149). The plasma form of the receptor is truncated in the carboxyl-terminal domain, and appears to arise by proteolysis at the cell surface, leading to release of the extracytoplasmic domain into the circulation (150–152). This may represent a major mechanism for the turnover of the Man-6-P/IGF-II receptor in these settings (152).

The regulation of MPR expression has also been studied in cells grown in tissue culture. In mouse C2 muscle cells undergoing terminal differentiation, Man-6-P/IGF-II receptor mRNA levels increase by more than 10-fold, whereas CD-MPR mRNA levels remain constant (153). The kinetics of accumulation of Man-6-P/IGF-II receptor mRNA and protein in C2 cells correlate closely with those of IGF-II in the same cells (154). In contrast, there is only a small, transient increase in mRNA levels for several lysosomal enzymes during the muscle differentiation. These findings suggest a role for the Man-6-P/IGF-II receptor in C2 cell muscle formation that could involve signaling via IGF-II. The regulation of the Man-6-P/IGF-II receptor level differs in rat hepatoma cell lines compared to primary rat hepatocytes (155). Three hepatoma lines express 5- to 15-fold more surface receptor than normal hepatocytes at confluence. However, the normal hepatocytes showed a 6-fold increase in receptor level in sparse cultures, whereas the level of receptor in the hepatoma lines was constant regardless of cell density. There is also increased expression of Man-6-P/IGF-II receptor in regenerating rat liver (156).

The high levels of Man-6-P/IGF-II receptor expression in fetal rat tissues are accompanied by high levels of IGF-II expression, which also decline postnatally (157–162). This pattern of IGF-II expression has led to speculation that it may have a role in fetal growth and development. Direct evidence for this has been obtained by DeChiara et al (163, 164), who found that mouse neonates that are homozygous mutants for a disrupted IGF-II gene are small but normally proportioned. Interestingly, neonates that receive the disrupted gene from their father are indistinguishable in appearance from the homozygous mutants, whereas neonates that receive the disrupted gene from their mother are phenotypically normal. This difference in phenotype is due to the fact that the paternal allele is widely expressed in embryos, whereas the maternal allele is silent except in the choroid plexus and leptomeninges. These results demonstrate that IGF-II is required for normal embryonic growth and that the IGF-II gene is subject to tissue-specific parental imprinting.

Remarkably, Barlow et al (165) have found that the Man-6-P/IGF-II receptor gene is also imprinted in mouse embryos, but in this case the gene is transcribed from the maternal chromosome rather than the paternal chromosome, the site of the active IGF-II gene locus. These investigators studied the *Tme* deletion (T-associated maternal effect) of the mouse. Embryos that inherit a deletion of this locus from their mother die at day 15 of gestation, but if the deletion is paternally transmitted, the offspring are viable (166–168). The lethality of the embryos is not associated with growth deficiency. Of the four genes that have been mapped to the region of the *Tme* deletion, only the gene for the Man-6-P/IGF-II receptor is imprinted, being transcribed only from the maternal gene.

DeChiara et al (164) have suggested that the reciprocal imprinting of IGF-II and the Man-6-P/IGF-II receptor may be a coincidence because the mutant phenotypes are so disparate. They argue that if the ligand and the receptor were related in terms of developmental function, the phenotypes would have been similar, regardless of whether or not the Man-6-P/IGF-II receptor gene is proven to be the *Tme* gene. Haig & Graham (169) offer a different view. They suggest that it is no coincidence that IGF-II and its receptor are oppositely imprinted. They postulate that the major function of the maternally determined Man-6-P/IGF-II receptor in the embryo is not signal transduction, but rather the capture and degradation of IGF-II produced by the paternal genome, thereby preventing the IGF-II from binding to the IGF-I receptor and stimulating fetal growth. They hypothesize that the clearance of IGF-II may control and balance the growth of the offspring and ensure the survival of more of the litter while protecting the mother against potentially excessive demands on her metabolic resources. Thus, in this evolutionary trade-off, the benefits of the maternal expression of the Man-6-P/IGF-II receptor presumably outweigh the disadvantages of smaller offspring. One prediction of this theory is that genes that affect embryonic growth are not imprinted in oviparous taxa because offspring cannot influence how much yolk their eggs receive (170). In this regard it is of interest that the CI-MPRs of chickens and *Xenopus* do not bind IGF-II (54, 55). While this hypothesis is intriguing, it does not explain why the loss of the Man-6-P/IGF-II receptor gene in the *Tme* deletion results in embryonic death when the predicted outcome would be the formation of large offspring. One possibility is that the lethal phenotype arises from an impairment in another function of this receptor (e.g. lysosomal enzyme sorting). It is also possible that the *Tme* deletion is associated with the loss of yet another maternally imprinted gene that is essential for fetal development.

ACKNOWLEDGMENTS

The author would like to thank Deborah Sinak for her great help in the preparation of this manuscript. The research done in the author's laboratory has been supported by United States Public Health Service Grant CA 08759 and a Monsanto/Washington University Biomedical Research Grant.

Literature Cited

1. Kornfeld, S., Mellman, I. 1989. *Annu. Rev. Cell Biol.* 5:483–525
2. Dahms, N. M., Lobel, P., Kornfeld, S. 1989. *J. Biol. Chem.* 264:12115–18
3. von Figura, K., Hasilik, A. 1986. *Annu. Rev. Biochem.* 55:167–93
4. Nolan, C. M., Sly, W. S. 1987. *Adv. Exp. Med. Biol.* 225:199–212
5. Roth, R. A. 1988. *Science* 239:1269–71
6. Pfeffer, S. R. 1988. *J. Membr. Biol.* 103:7–16
7. Robbins, A. 1988. *Protein Transfer and Organelle Biogenesis*, pp. 463–520. Orlando: Academic
8. Hoflack, B., Kornfeld, S. 1985. *J. Biol. Chem.* 260:12008–14
9. Distler, J. J., Patel, R., Jourdian, G. W. 1987. *Anal. Biochem.* 166:65–71
10. Li, M., Distler, J. J., Jourdian, G. W. 1990. *Arch. Biochem. Biophys.* 283:150–57
11. Ma, Z. M., Grubb, J. H., Sly, W. S. 1991. *J. Biol. Chem.* 266:10589–95
12. Junghans, U., Waheed, A., von Figura, K. 1988. *FEBS Lett.* 237:81–84
13. Baba, T., Watanabe, K., Arai, Y. 1988. *Carbohydr. Res.* 177:153–61
14. Watanabe, H., Grubb, J. H., Sly, W. S. 1990. *Proc. Natl. Acad. Sci. USA* 87:8036–40
15. Lobel, P., Dahms, N. M., Breitmeyer, J., Chirgwin, J. M., Kornfeld, S. 1987.

Proc. Natl. Acad. Sci. USA 84:2233–37

16. Lobel, P., Dahms, N. M., Kornfeld, S. 1988. *J. Biol. Chem.* 263:2563–70

17. Morgan, D. O., Edman, J. C., Standring, D. N., Fried, V. A., Smith, M. C., et al. 1987. *Nature* 329:301–7

18. Oshima, A., Nolan, C. M., Kyle, J. W., Grubb, J. H., Sly, W. S. 1988. *J. Biol. Chem.* 263:2553–62

19. MacDonald, R. G., Pfeffer, S. R., Coussens, L., Tepper, M. A., Brocklebank, C. M., et al. 1988. *Science* 239:1134–37

20. Dahms, N. M., Lobel, P., Breitmeyer, J., Chirgwin, J. M., Kornfeld, S. 1987. *Cell* 50:181–92

21. Pohlmann, R., Nagel, G., Schmidt, B., Stein, M., Lorkowski, G., et al. 1987. *Proc. Natl. Acad. Sci. USA* 84:5575–79

22. Koster, A., Nagel, G., von Figura, K., Pohlmann, R. 1991. *Biol. Chem. Hoppe-Seyler* 372:297–300

23. Kiess, W., Greenstein, L. A., Lee, L., Thomas, C., Nissley, S. P. 1991. *Mol. Endocrinol.* 5:281–91

24. Hynes, R. 1985. *Annu. Rev. Cell Biol.* 1:67–90

25. Sahagian, G. G., Neufeld, E. F. 1983. *J. Biol. Chem.* 258:7121–28

26. Corvera, S., Czech, M. P. 1985. *Proc. Natl. Acad. Sci. USA* 82:7314–18

27. Méresse, S., Ludwig, T., Frank, R., Hoflack, B. 1990. *J. Biol. Chem.* 265:18833–42

28. Westcott, K. R., Rome, L. H. 1988. *J. Cell. Biochem.* 38:23–33

29. Wendland, M., Waheed, A., Schmidt, B., Hille, A., Nagel, G., et al. 1991. *J. Biol. Chem.* 266:4598–604

30. Klier, H.-J., von Figura, K., Pohlmann, R. 1991. *Eur. J. Biochem.* 197:23–28

31. Laureys, G., Barton, D. E., Ullrich, A., Francke, U. 1988. *Genomics* 3:224–29

32. Stein, M., Meyer, H. E., Hasilik, A., von Figura, K. 1987. *Biol. Chem. Hoppe-Seyler* 368:927–36

33. Dahms, N. M., Kornfeld, S. 1989. *J. Biol. Chem.* 264:11458–67

34. Waheed, A., von Figura, K. 1990. *Eur. J. Biochem.* 193:47–54

35. Waheed, A., Hille, A., Junghans, U., von Figura, K. 1990. *Biochemistry* 29:2449–55

36. Wendland, M., Hille, A., Nagel, G., Waheed, A., von Figura, K., et al. 1989. *Biochem. J.* 260:201–6

37. Perdue, J. F., Chan, J. K., Thibault, C., Radaj, P., Mills, B., et al. 1983. *J. Biol. Chem.* 258:7800–11

38. Stein, M., Braulke, T., Krentler, C., Hasilik, A., von Figura, K. 1987. *Biol. Chem. Hoppe-Seyler* 368:937–47

39. Tong, P. Y., Kornfeld, S. 1989. *J. Biol. Chem.* 264:7970–75

39a. Distler, J. J., Guo, J., Jourdian, G. W., Srivastava, O. P., Hindsgaul, O. 1991. *J. Biol. Chem.* 266:21687–92

40. Tong, P. Y., Gregory, W., Kornfeld, S. 1989. *J. Biol. Chem.* 264:7962–69

41. Westlund, B., Dahms, N. M., Kornfeld, S. 1991. *J. Biol. Chem.* 266:23233–39

42. Wendland, M., Waheed, A., von Figura, K., Pohlmann, R. 1991. *J. Biol. Chem.* 266:2917–23

43. Wendland, M., von Figura, K., Pohlmann, R. 1991. *J. Biol. Chem.* 266:7132–36

44. Hille, A., Waheed, A., von Figura, K. 1989. *J. Biol. Chem.* 264:13460–67

45. Hille, A., Waheed, A., von Figura, K. 1990. *J. Cell Biol.* 110:963–72

46. Dong, J., Prence, E. M., Sahagian, G. G. 1989. *J. Biol. Chem.* 264:7377–83

47. Dong, J., Sahagian, G. G. 1990. *J. Biol. Chem.* 265:4210–17

48. Lazzarino, D., Gabel, C. A. 1990. *J. Biol. Chem.* 265:11864–71

49. Tong, P. Y., Tollefsen, S. E., Kornfeld, S. 1988. *J. Biol. Chem.* 263:2585–88

50. Roth, R. A., Stover, C., Hari, J., Morgan, D. O., Smith, M. C., et al. 1987. *Biochem. Biophys. Res. Commun.* 149:600–6

51. Waheed, A., Braulke, T., Junghans, U., von Figura, K. 1988. *Biochem. Biophys. Res. Commun.* 152:1248–54

52. Kiess, W., Blickenstaff, G. D., Sklar, M. M., Thomas, C. L., Nissley, S. P., et al. 1988. *J. Biol. Chem.* 263:9339–44

53. Nolan, C. M., Kyle, J. W., Watanabe, H., Sly, W. S. 1990. *Cell Regul.* 1:197–213

54. Canfield, W. M., Kornfeld, S. 1989. *J. Biol. Chem.* 264:7100–3

55. Clairmont, K. B., Czech, M. P. 1989. *J. Biol. Chem.* 264:16390–92

56. Kiess, W., Thomas, C. L., Greenstein, L. A., Lee, L., Sklar, M. M., et al. 1989. *J. Biol. Chem.* 264:4710–14

57. Kiess, W., Thomas, C. L., Sklar, M. M., Nissley, S. P. 1990. *Eur. J. Biochem.* 190:71–77

58. Braulke, T., Bresciani, R., Buergisser, D. M., von Figura, K. 1991. *Biochem. Biophys. Res. Commun.* 179:108–15

59. Gabel, C. A., Goldberg, D. E., Kornfeld, S. 1983. *Proc. Natl. Acad. Sci. USA* 80:775–79

60. Stein, M., Zijderhand-Bleekemolen, J. E., Geuze, H., Hasilik, A., von Figura, K. 1987. *EMBO J.* 6:2677–81

61. Kyle, J. W., Nolan, C. M., Oshima, A., Sly, W. S. 1988. *J. Biol. Chem.* 263:16230–35

62. Lobel, P., Fujimoto, K., Ye, R. D.,

Griffiths, G., Kornfeld, S. 1989. *Cell* 57:787–96
63. Duncan, J. R., Kornfeld, S. 1988. *J. Cell Biol.* 106:617–28
64. Chao, H. H., Waheed, A., Pohlmann, R., Hille, A., von Figura, K. 1990. *EMBO J.* 9:3507–13
65. Rechler, M. M., Nissley, S. P. 1985. *Annu. Rev. Physiol.* 47:425–62
66. Froesch, E. R., Schmid, C., Schwander, J., Zapf, J. 1985. *Annu. Rev. Physiol.* 47:443–76
67. Czech, M. P. 1989. *Cell* 59:235–38
68. Mottola, C., Czech, M. P. 1984. *J. Biol. Chem.* 259:12705–13
69. Conover, C. A., Misra, P., Hintz, R. L., Rosenfeld, R. G. 1986. *Biochem. Biophys. Res. Commun.* 139:501–8
70. Kiess, W., Haskell, J. F., Lee, L., Greenstein, L. A., Miller, B. E., et al. 1987. *J. Biol. Chem.* 262:12745–51
71. Sinha, M. K., Buchanan, C., Raineri-Maldonado, C., Khazanie, P., Atkinson, S., et al. 1990. *Am. J. Physiol.* 258:E534–E542
72. Rechler, M. M., Zapf, J., Nissley, S. P., Froesch, E. R., Moses, A. C., et al. 1980. *Endocrinology* 107:1451–59
73. Hari, J., Pierce, S. B., Morgan, D. O., Sara, V., Smith, M. C., et al. 1987. *EMBO J.* 6:3367–71
74. Shimizu, M., Webster, C., Morgan, D. O., Blau, H. M., Roth, R. 1986. *Am. J. Physiol.* 251:E611–E615
75. Tally, M., Li, C. H., Hall, K. 1987. *Biochem. Biophys. Res. Commun.* 148:811–16
76. Mellas, J., Gavin, I. J. R., Hammerman, M. R. 1986. *J. Biol. Chem.* 261:14437–42
77. Rogers, S. A., Hammerman, M. R. 1988. *Proc. Natl. Acad. Sci. USA* 85:4037–41
78. Rogers, S. A., Hammerman, M. R. 1989. *J. Biol. Chem.* 264:4273–76
79. Rogers, S. A., Purchio, A. F., Hammerman, M. R. 1990. *J. Biol. Chem.* 265:9722–27
80. Rogers, S. A., Ryan, G., Hammerman, M. R. 1991. *J. Cell Biol.* 113:1447–53
81. Kojima, I., Nishimoto, I., Iiri, T., Ogata, E., Rosenfeld, R. 1988. *Biochem. Biophys. Res. Commun.* 154:9–19
82. Nishimoto, I., Hata, Y., Ogata, E., Kojima, I. 1987. *J. Biol. Chem.* 262:12120–26
83. Nishimoto, I., Murayama, Y., Katada, T., Ui, M., Ogata, E. 1989. *J. Biol. Chem.* 264:14029–38
84. Okamoto, T., Nishimoto, I., Murayama, Y., Ohkuni, Y., Ogata, E. 1990. *Biochem. Biophys. Res. Commun.* 168:1201–10

85. Gilman, A. G. 1987. *Annu. Rev. Biochem.* 56:615–49
86. Murayama, Y., Okamoto, T., Ogata, E., Asano, T., Iiri, T., et al. 1990. *J. Biol. Chem.* 265:17456–62
87. Okamoto, T., Katada, T., Murayama, Y., Ui, M., Ogata, E., et al. 1990. *Cell* 62:709–17
88. Higashijima, T., Uzu, S., Nakajima, T., Ross, E. M. 1988. *J. Biol. Chem.* 263:6491–94
89. Okamoto, T., Asano, T., Harada, S.-I., Ogata, E., Nishimoto, I. 1991. *J. Biol. Chem.* 266:1085–91
89a. Sakano, K., Enjoh, T., Numata, F., Fujiwara, H., Marumoto, Y., et al. 1991. *J. Biol. Chem.* 266:20626–35
89b. Cascieri, M. A., Chicchi, G. G., Applebaum, J., Green, B. G., Hayes, N. S., Bayne, M. L. 1989. *J. Biol. Chem.* 264:2199–202
89c. Bürgisser, D. M., Roth, B. V., Giger, R., Lüthi, C., Weigl, S., Zarn, J., Humbel, R. E. 1991. *J. Biol. Chem.* 266:1029–33
89d. Beukers, M. W., Oh, Y., Zhang, H., Ling, N., Rosenfeld, R. G. 1991. *Endocrinology* 128:1201–3
90. Oka, Y., Rozek, L. M., Czech, M. P. 1985. *J. Biol. Chem.* 260:9435–42
91. Purchio, A. F., Cooper, J. A., Brunner, A. M., Lioubin, M. N., Gentry, L. E., et al. 1988. *J. Biol. Chem.* 263:14211–15
92. Kovacina, K. S., Steele-Perkins, G., Purchio, A. F., Lioubin, M., Miyazono, K., et al. 1989. *Biochem. Biophys. Res. Commun.* 160:393–403
93. Dennis, P. A., Rifkin, D. B. 1991. *Proc. Natl. Acad. Sci. USA* 88:580–84
94. Lee, S. J., Nathans, D. 1988. *J. Biol. Chem.* 263:3521–27
95. Herzog, V., Neümuller, W., Holzmann, B. 1987. *EMBO J.* 6:555–60
96. Roff, C. F., Wozniak, R. W., Blenis, J., Wang, J. L. 1983. *Exp. Cell Res.* 144:333–44
97. Brauker, J. H., Roff, C. F., Wang, J. L. 1986. *Exp. Cell Res.* 164:115–26
98. Griffiths, G., Simons, K. 1986. *Science* 234:438–43
99. Brown, W. J., Farquhar, M. G. 1984. *Cell* 36:295–307
100. Brown, W. J., Farquhar, M. G. 1987. *Proc. Natl. Acad. Sci. USA* 84:9001–5
101. Geuze, H. J., Slot, J. W., Strous, G. J., Hasilik, A., von Figura, K. 1985. *J. Cell Biol.* 101:2253–62
102. Griffiths, G., Hoflack, B., Simons, K., Mellman, I., Kornfeld, S. 1988. *Cell* 52:329–41
102a. Brown, W. J. 1990. *Eur. J. Cell Biol.* 51:201–10

103. Willingham, M. C., Pastan, I. H., Sahagian, G. G., Jourdian, G. W., Neufeld, E. F. 1981. *Proc. Natl. Acad. Sci. USA* 78:6967–71
104. Brown, W. J., Goodhouse, J., Farquhar, M. G. 1986. *J. Cell Biol.* 103:1235–47
105. Geuze, H. J., Stoorvogel, W., Strous, G. J., Slot, J. W., Bleekemolen, J. E., et al. 1988. *J. Cell Biol.* 107:2491–501
106. Woods, J. W., Goodhouse, J., Farquhar, M. G. 1989. *Eur. J. Cell Biol.* 50:132–43
107. Griffiths, G., Matteoni, R., Back, R., Hoflack, B. 1990. *J. Cell Sci.* 95:441–61
108. Matovcik, L. M., Goodhouse, J., Farquhar, M. G. 1990. *Eur. J. Cell Biol.* 53:203–11
109. Bleekemolen, J. E., Stein, M., von Figura, K., Slot, J. W., Geuze, H. J. 1988. *Eur. J. Cell Biol.* 47:366–72
110. Prydz, K., Brandli, A. W., Bomsel, M., Simons, K. 1990. *J. Biol. Chem.* 265:12629–35
110a. Wood, S. A., Park, J. E., Brown, W. J. 1991. *Cell* 67:591–600
111. Schmid, S., Fuchs, R., Kieban, M., Helenius, A., Mellman, I. 1989. *J. Cell Biol.* 108:1291–300
112. Braulke, T., Tippmer, S., Neher, E., von Figura, K. 1989. *EMBO J.* 8:681–86
113. Braulke, T., Tippmer, S., Chao, H.-J., von Figura, K. 1990. *J. Biol. Chem.* 265:6650–55
114. Damke, H., von Figura, K., Braulke, T. 1991. *Biochem. J.* In press
115. Hu, K. Q., Backer, J. M., Sahagian, G., Feener, E. P., King, G. L. 1990. *J. Biol. Chem.* 265:13864–70
116. Prence, E. M., Dong, J., Sahagian, G. G. 1990. *J. Cell Biol.* 110:319–26
117. Oppenheimer, C. L., Pessin, J. E., Massagué, J., Gitomer, W., Czech, M. P. 1983. *J. Biol. Chem.* 258:4824–30
118. Wardzala, L. J., Simpson, I. A., Rechler, M. M., Cushman, S. W. 1984. *J. Biol. Chem.* 259:8378–83
119. Oka, Y., Mottola, C., Oppenheimer, C. L., Czech, M. P. 1984. *Proc. Natl. Acad. Sci. USA* 81:4028–32
120. Appell, K. C., Simpson, I. A., Cushman, S. W. 1988. *J. Biol. Chem.* 263:10824–29
121. Corvera, S., Roach, P. J., DePaoli-Roach, A. A., Czech, M. P. 1988. *J. Biol. Chem.* 263:3116–22
122. Corvera, S., Folander, K., Clairmont, K. B., Czech, M. P. 1988. *Proc. Natl. Acad. Sci. USA* 85:7567–71
123. Canfield, W. M., Johnson, K. F., Ye,

R. D., Gregory, W., Kornfeld, S. 1991. *J. Biol. Chem.* 266:5682–88
124. Johnson, K. F., Chan, W., Kornfeld, S. 1990. *Proc. Natl. Acad. Sci. USA* 87:10010–14
125. Chen, W.-J., Goldstein, J. L., Brown, M. S. 1990. *J. Biol. Chem.* 265:3116–23
126. Peters, C., Braun, M., Weber, B., Wendland, M., Schmidt, B., et al. 1990. *EMBO J.* 9:3497–506
127. Williams, M. A., Fukuda, M. 1990. *J. Cell Biol.* 111:955–66
128. Collawn, J. F., Stangel, M., Kuhn, L. A., Esekogwu, V., Jing, S. Q., et al. 1990. *Cell* 63:1061–72
129. Breitfeld, P. P., Casanova, J. E., McKinnon, W. C., Mostov, K. E. 1990. *J. Biol. Chem.* 265:13750–57
130. Ktistakis, N. T., Thomas, D., Roth, M. G. 1990. *J. Cell Biol.* 111:1393–407
131. Fuhrer, C., Geffen, I., Spiess, M. 1991. *J. Cell Biol.* 114:423–31
132. Collawn, J. F., Kuhn, L. A., Liu, L.-F. S., Tainer, J. A., Trowbridge, I. S. 1991. *EMBO J.* 10:3247–53
133. Bansal, A., Gierasch, L. M. 1991. *J. Cell. Biochem. Suppl.* 15G:171 (Abstr.)
134. Pearse, B. M. F. 1988. *EMBO J.* 7:3331–36
135. Glickman, J. N., Conibear, E., Pearse, B. M. F. 1989. *EMBO J.* 8:1041–47
136. Keen, J. H. 1990. *Annu. Rev. Biochem.* 59:415–38
137. Pearse, B. M. F., Robinson, M. S. 1990. *Annu. Rev. Cell Biol.* 6:151–71
138. Ahle, S., Mann, A., Eichelsbacher, U., Ungewickell, E. 1988. *EMBO J.* 7:919–29
139. Schröder, S., Ungewickell, E. 1991. *J. Biol. Chem.* 266:7910–18
140. Dintzis, S. M., Pfeffer, S. R. 1990. *EMBO J.* 9:77–84
141. Goda, Y., Pfeffer, S. R. 1988. *Cell* 55:309–20
142. Goda, Y., Pfeffer, S. R. 1991. *J. Cell Biol.* 112:823–31
143. Block, M. R., Glick, B. S., Wilcox, C. A., Wieland, F. T., Rothman, J. E. 1988. *Proc. Natl. Acad. Sci. USA* 85:7852–56
144. Draper, R. K., Goda, Y., Brodsky, F. M., Pfeffer, S. R. 1990. *Science* 248:1539–41
145. Sklar, M. M., Kiess, W., Thomas, C. L., Nissley, S. P. 1989. *J. Biol. Chem.* 264:16733–38
146. Alexandrides, T., Moses, A. C., Smith, R. J. 1989. *Endocrinology* 124:1064–76
147. Senior, P. V., Byrne, S., Brammar, W. J., Beck, F. 1990. *Development* 109:67–73
148. Ballesteros, M., Scott, C. D., Baxter,

R. C. 1990. *Biochem. Biophys. Res. Commun.* 172:775–79

149. Kiess, W., Greenstein, L. A., White, R. M., Lee, L., Rechler, M. M., et al. 1987. *Proc. Natl. Acad. Sci. USA* 84:7720–24

150. MacDonald, R. G., Tepper, M. A., Clairmont, K. B., Perregaux, S. B., Czech, M. P. 1989. *J. Biol. Chem.* 264:3256–61

151. Causin, C., Waheed, A., Braulke, T., Junghans, U., Maly, P., et al. 1988. *Biochem. J.* 252:795–99

152. Clairmont, K. B., Czech, M. P. 1991. *J. Biol. Chem.* 266:12131–34

153. Szebenyi, G., Rotwein, P. 1991. *J. Biol. Chem.* 266:5534–39

154. Tollefsen, S. E., Sadow, J. L., Rotwein, P. 1989. *Proc. Natl. Acad. Sci. USA* 86:1543–47

155. Scott, C. D., Taylor, J. E., Baxter, R. C. 1988. *Biochem. Biophys. Res. Commun.* 151:815–21

156. Scott, C. D., Ballesteros, M., Baxter, R. C. 1990. *Endocrinology* 127:2210–16

157. Moses, A. C., Nissley, S. P., Short, P. A., Rechler, M. M., White, R. M., et al. 1980. *Proc. Natl. Acad. Sci. USA* 77:3649–53

158. Adams, S. O., Nissley, S. P., Hanwerger, S., Rechler, M. M. 1983. *Nature* 302:150–53

159. Lund, P. K., Moats-Staats, B. M., Hynes, M. A., Simmons, J. G., Jansen, M., et al. 1986. *J. Biol. Chem.* 261:14539–44

160. Brown, A. L., Graham, D. E., Nissley, S. P., Hill, D. J., Strain, A. J., et al. 1986. *J. Biol. Chem.* 261:13144–50

161. Beck, F., Samani, N. J., Penschow, J. D., Thorley, B., Tregear, G. W., et al. 1987. *Development* 101:175–84

162. Romanus, J. A., Yang, Y. W.-H., Adams, S. O., Sofair, A. N., Tseng, L. Y.-H., et al. 1988. *Endocrinology* 122:709–16

163. DeChiara, T. M., Efstratiadis, A., Robertson, E. J. 1990. *Nature* 345:78–80

164. DeChiara, T. M., Robertson, E. J., Efstratiadis, A. 1991. *Cell* 64:849–59

165. Barlow, D. P., Stöger, R., Herrmann, B. G., Saito, K., Schweifer, N. 1991. *Nature* 349:84–87

166. Johnson, D. R. 1974. *Genetics* 76:795–805

167. Johnson, D. R. 1975. *Genet. Res.* 24:207–13

168. McGrath, J., Solter, D. 1984. *Nature* 308:550–51

169. Haig, D., Graham, C. 1991. *Cell* 64:1045–46

170. Haig, D., Westoby, M. 1989. *Am. Nat.* 134:147–55

Annu. Rev. Biochem. 1992. 61:331–54

RECOMBINANT TOXINS AS NOVEL THERAPEUTIC AGENTS[1]

Ira Pastan, Vijay Chaudhary, and David J. FitzGerald

Laboratory of Molecular Biology, Division of Cancer Biology, Diagnosis and Centers, National Cancer Institute, National Institutes of Health, 9000 Rockville Pike, Bethesda, MD 20829

KEY WORDS: immunotoxin, cancer, ricin, *Pseudomonas* exotoxin, diphtheria toxin

CONTENTS

PERSPECTIVE

Bacteria and plants produce powerful toxins, which may confer survival advantage in hostile environments or provide sources of nutrition. These

---

toxins kill many types of animal cells. Because of their potency, they have been coupled to antibodies and growth factors in order to kill target cells with specific surface properties. These agents are now being evaluated for treatment of cancer, autoimmune diseases, and chronic infectious diseases (1–4). DNA sequences encoding *Pseudomonas* exotoxin, diphtheria toxin, ricin, and several other toxins have been cloned and expressed in *Escherichia coli,* where these toxins can be produced in large amounts because they do not kill bacterial cells. Initially, toxins produced in bacteria were chemically coupled to antibodies to make classical immunotoxins. A significant advance has been the preparation of recombinant chimeric toxins by fusing cell-targeting genes to modified toxin genes (5, 6). The approaches used to design, express, and purify chimeric toxins, their properties, and their mechanisms of action are major areas of focus of this review.

Directly targeting cytotoxic agents to cancer cells is an old idea, but it has recently become a popular approach to therapy because modern biological techniques have identified molecules on the surfaces of cancer cells that can be used as targets for this type of therapy. These target proteins include oncoproteins that cause the cells to grow abnormally, differentiation antigens that are present on cancers derived from organs with dispensable functions, and other antigens commonly found on cancer cells, which may represent fetal antigens inappropriately expressed in these cancers. Because bacterial and plant toxins are very active cell-killing agents, one approach to targeted therapy is to attach one of these toxins to an antibody or a growth factor that preferentially binds to a cancer cell. Many toxins have been used for this purpose, but three of these, *Pseudomonas* exotoxin A (PE), ricin, and diphtheria toxin (DT) have been most intensively employed. Initially, toxins purified from natural sources and chemically attached to antibodies to make immunotoxins had to be chemically modified to reduce binding to toxin receptors in normal cells and allow preferential binding to the targets on the cancer cells. Several recent reviews describe the construction and activity of immunotoxins produced by the chemical linkage of toxins to antibodies (7–10). Recently, genes encoding these three toxins have been cloned (11–13), and expressed in *E. coli,* and novel molecules with more desirable biological properties produced (14–17). In this review, we discuss how these toxins act, how they can be genetically modified to produce more active and specific molecules, the results of animal experiments, and prospects for the future.

PSEUDOMONAS EXOTOXIN

Structure and Function

Although studies on the mechanism of action of DT and ricin were initiated before those on PE, there is currently much more information available about how PE acts than about the other toxins. PE is a single-chain toxin secreted by *Pseudomonas aeruginosa*. It kills cells by catalyzing the irreversible ADP-ribosylation and subsequent inactivation of elongation factor 2 (EF-2) (18). NAD^+ is an essential cofactor in this reaction:

$$\overset{\text{PE}}{NAD^+ + EF\text{-}2 \rightleftharpoons ADP \text{ ribose-EF-2 and Nicotinamide} + H^+}$$ 1.

The ADP-ribosylation site in EF-2 is a modified histidine residue, diphthamide, located at position 715 (19). DT catalyzes the identical ADP ribosylation reaction (20). It has been known for many years that PE must be endocytosed in order to kill a target cell, but essential information on the mechanism of cell killing has been absent (21, 22). A major advance in understanding how PE kills target cells came from the crystallization of PE and the elucidation of its three-dimensional structure (23). As shown schematically in Figure 1, PE consists of three major domains termed Ia (amino acids 1–252), II (aa 253–364), and III (aa 400–613). Domain Ib (amino acids 365–399) is a minor domain. Using the X-ray structure of PE as a guide, different portions of the PE gene were used to produce each of the domains in *E. coli* in order to ascertain their functions (15). These studies showed that domain Ia was responsible for the binding of PE to target cells, domain II was necessary for the translocation of the ADP-ribosylating activity into the cytosol where EF-2 is located, and domain III contained the ADP ribosylation activity. Since this initial study, a variety of genetic and biochemical investigations have increased our understanding of the mechanism of action of PE. The results of these studies are summarized in Table 1 and form the basis of Figure 2.

Domain Ia is the cell-binding domain. Binding and cytotoxicity are abolished or greatly diminished by mutating Lys57 to Glu or Gly, or by inserting a dipeptide at position 57 (24, 24a). Point mutations have been introduced into many other positions in domain Ia without large effects on cytotoxicity (24).

Domain II has an important role in translocation; it is a small domain that contains six alpha helices. Between the A and B helix, there is a loop containing three arginine residues, which lie on the surface of the molecule. After PE enters a cell, a proteolytic cleavage occurs near Arg279, which generates a 28-kDa amino-terminal fragment and a 37-kDa carboxyl-terminal fragment (25); the latter contains the ADP-ribosylating activity of PE. A disulfide bond (Cys265–287) that connects these two fragments is then broken, probably by reduction, and the 37-kDa fragment is translocated across the membrane into the cytosol (25). Mutations that change Arg279, 276, or 274 to Gly or Lys have been found to eliminate the cytotoxicity of PE

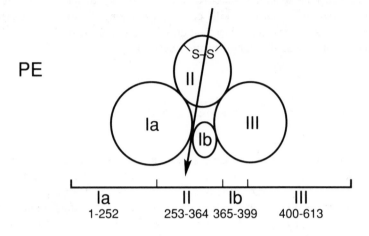

RICIN

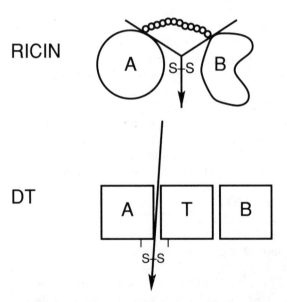

DT

Figure 1 Structure and processing of PE, ricin, and DT. Before reaching its intracellular target, each toxin must be processed to produce an active fragment, which translocates to the cell cytosol and arrests protein synthesis.

After cellular entry, several processing events are required to convert PE to an active toxin. First, the toxin is unfolded. It is the proteolytically cleaved between residues 279 and 280, and the disulfide bond that links residues 265 and 287 is reduced (see Figure 2). These events precede translocation. The arrow shows two processing steps: one cleaves the protein backbone; the other reduces the disulfide bond.

Table 1 Location of important mutations in PE

Domain	Mutation	Effect
I	Lys57→Glu	Inactivate
	Δ1–252	Inactivate
II	Arg279→Gly	Inactivate
	Arg276→Gly	Inactivate
	Cys265, 287→Ser	10-fold less active
	Δ346–365	No effect
	Δ343–345	Inactivate
Ib	Δ365–381	No effect
III	ΔGlu553	Inactivate
	ΔHis485–492	Inactivate
	His426	Inactivate
	REDLK→KDEL	Activate

(26). Biochemical studies with two of these, PE Gly276 and PE Gly279, have shown that these mutant molecules are resistant to cleavage by a cellular protease that carries out the processing of PE, and the 37-kDa fragment is not generated (25). Mutations that change Cys265 and 287 to Ser or Ala have been made and found to decrease cytotoxic activity about 10-fold (27, 28); the mechanism of this effect is not yet known. Finally, a portion of the E helix of domain II (amino acids 346–364) can be deleted without loss of activity, but larger deletions are very deleterious (29).

Domain Ib, a small domain extending from amino acids 365 to 399, is situated between domains Ia and III in the three-dimensional structure. A large part of this domain can be deleted (amino acids 365–381) without loss of activity; the function of domain Ib is unknown.

Domain III (amino acids 400–613) has two functions. It exhibits ADP-ribosylating activity and has a C-terminal sequence that directs endocytosed toxin into the endoplasmic reticulum. Deletion mapping shows that amino acids 400–600 are needed for the ADP-ribosylating activity (30, 31). Other

←

Ricin is produced in castor beans as preproricin. Proricin is generated by removal of a leader sequence at the N-terminus. Full toxin potential is realized after a 12-amino-acid linker peptide is excised (arrow). This excision occurs within the beans and leaves the A and B subunits attached by a disulfide bond. Reduction of the disulfide bond takes place within target cells (arrow).

DT is produced as a single-chain polypeptide. Like the other toxins, it must be proteolytically cleaved and reduced. Proteolysis can occur either extracellularly (e.g. in the growth medium) or within target cells, where reduction also occurs (arrow).

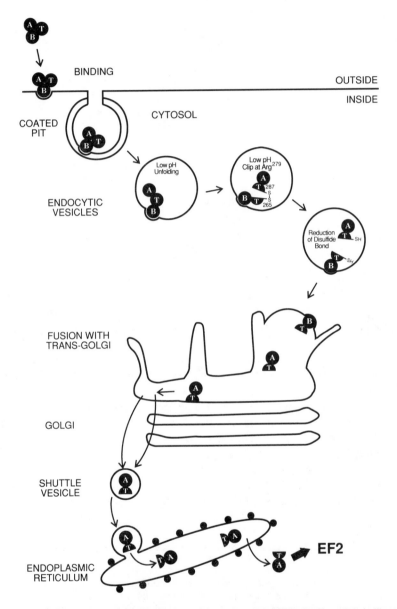

Figure 2 Model of endocytosis, processing, and translocation of PE. B, T, and A indicate binding domain (Ia), translocation domain (II), and ADP ribosylation domain (III), respectively. PE binds to a cell-surface receptor. The toxin-receptor complex is then internalized via coated pits into the endocytic vesicles where the low-pH environment causes toxin unfolding and facilitates proteolysis; the cleaved toxin is then reduced. The 37-kDa carboxyl-terminal fragment is

mutations that inactivate ADP-ribosylating activity are at positions 426 (32), 553 (33), and by deletion of amino acids 485–492 (31a). Deletion of the last several amino acids of PE also abolishes its cytotoxicity but has no effect on ADP-ribosylation activity (31). The residues at the carboxyl end of PE are Arg-Glu-Asp-Leu-Lys. Deletion of Lys613 has no effect, but removal of Leu612 or additional amino acids greatly reduces cytotoxicity (31). Biochemical experiments have shown that the 37-kDa fragment that is produced from this class of mutants is unable to reach the cytosol if the carboxyl end is mutated (25). Because REDL resembles the sequence KDEL, which functions to retain newly synthesized proteins in the endoplasmic reticulum (34), PE ending in KDEL was made and found to be fully active; in fact, it was somewhat more active than PE itself. A large number of other mutations have now been made at the carboxyl terminus of PE, and have shown that there is good correlation between carboxyl-terminal sequences that give an active PE molecule and sequences that are efficient in retaining proteins in the endoplasmic reticulum (35). It is of interest that two other toxins also end in REDL or KDEL; these are the A chains of cholera and *E. coli* heat labile toxin (36, 37).

The similarity in the sequence requirements for the cytotoxic action of PE, and for the retention of proteins in the endoplasmic reticulum, gives rise to the model of PE action shown in Figure 2. After binding, PE enters the cell by endocytosis and is exposed to an acidic pH within an early endocytic vesicle (endosome) that unfolds the toxin and activates a cellular protease (38–40). The protease cleaves PE between Arg279 and Gly280 (M. Ogata, I. Pastan, D. FitzGerald, unpublished results). The 37-kDa fragment that is produced by proteolytic cleavage and subsequent reduction of the disulfide bond (Cys265–287) is transferred by membrane fusion from endocytic vesicles to the *trans*-golgi and from there to a compartment that returns KDEL-tagged proteins to the endoplasmic reticulum (Figure 2). Once in the endoplasmic reticulum, the 37-kDa fragment may use pre-existing protein transport complexes or pores to gain access to the cytosol. This model, although highly speculative, is consistent with all of the data produced from genetic and biochemical studies. In particular, it does not require a portion of the PE molecule, such as domain Ia, to function as a translocation pore. Because domain Ia can be replaced by ligands that target the remainder of PE to different cellular receptors (see below), domain Ia is not necessary for the translocation process.

---

transported to the Golgi apparatus and ultimately reaches the endoplasmic reticulum via a shuttle vesicle. Specific residues (REDLK) at the C-terminus mediate retention in the endoplasmic reticulum. Translocation to the cytosol may occur in a reverse direction through translocation complexes or pores ordinarily used for translocating cellular proteins into the endoplasmic reticulum. Although no precedent exists for a pathway that transports proteins from the cell surface to the endoplasmic reticulum, it is possible that some lipids do so (155).

PE40

Deletion of domain Ia of PE (aa 1–252) produces a 40-kDa protein (PE40) that has extremely low cellular and animal cytotoxicity because it cannot bind to cellular receptors (17, 24). PE40 has been used to make conventional immunotoxins by chemically coupling it to antibodies (17, 41, 42). PE40 has also been used to make recombinant chimeric toxins by fusing DNA fragments encoding growth factors, antibody-combining sites, or other recognition elements to the PE40 gene [(6) and see below]. The gene encoding such a chimeric toxin is expressed in *E. coli,* from which source the chimeric toxin is purified.

Immunotoxins

PE or PE40 is attached to an antibody using a thioether linkage. To do this, the antibody is treated with iminothiolane, and the toxin with succinimidyl-4-(N-maleimidomethyl)cyclohexane-1-carboxylate (SMCC), as described in detail elsewhere (41, 43). The derivatized antibody and toxin are then mixed together, and the thioether-linked conjugate produced. PE40 is the preferred form of the toxin because of its low animal toxicity. To facilitate coupling to antibodies, an extra lysine residue has been added at the amino end of PE40 (41). Immunotoxins containing PE40 have been made with several antibodies, and have been shown to be potent cell-killing agents in tissue culture and to cause complete or substantial regression of subcutaneous human tumors growing in immunodeficient mice [HB21(41), B3 (42), or C242 (42a)]. But immunotoxins made with PE40 have no activity when the antigen-antibody complex on target cells is not internalized [OVB3 (17, 43), MRK16 (44)]. If PE is used in place of PE40, the interaction of domain Ia of PE with the cellular PE receptor will promote internalization of an immunotoxin that becomes bound to the target cell through the antibody interaction with a cell-surface antigen. With such immunotoxins, the therapeutic window is smaller because the binding of PE to the PE receptor, present on almost all normal cells, causes increased nonspecific toxicity to the animal. Nevertheless, therapeutic effects in animals have been obtained (42, 43, 45, 46).

Recombinant Toxins

A list of the chimeric toxins made by fusing the PE40 gene to cDNAs encoding different targeting molecules is shown in Table 2. It is important that the recognition element be placed at the amino end of the molecule because additions at the carboxyl end of domain III place the important REDLK sequence within the protein where it cannot function properly.

TGFα-PE40   Domain Ia, which has a molecular weight of 26,000, has been replaced by molecules ranging in molecular weight from 6000 for TGFα to

Table 2 Structure and activity of recombinant toxins derived from PE

Name	Structure	Target and cell lines	Activity in mice or rats
PE	(I)(II)(III)	most cells	liver toxicity
PE40	(II)(III)	—	
TGFα-PE40	(TGFa)(II)(III)	EGF receptor epidermoid carcinomas adenocarcinomas glioblastomas smooth muscle cells	epidermoid carcinomas
IL2-PE40	(IL2)(II)(III)	IL2 receptor leukemia	lymphoma, allograft rejection, autoimmune arthritis, uveitis, encephalitis
IL4-PE40	(IL4)(II)(III)	IL4 receptor	
IL6-PE40	(IL6)(II)(III)	IL6 receptor myelomas hepatomas prostate	hepatoma
acidic FGF-PE40	(aFGF)(II)(III)	many	
IGF1-PE40	(IGF1)(II)(III)	many	
CD4(178)-PE40	(CD4)(II)(III)	HIV infected	
AntiTac(Fv)-PE40	()()(II)(III)	human IL2 receptor leukemia	
B3(Fv)-PE38KDEL	()()(II)(III)	many carcinomas	epidermoid carcinoma, adenocarcinoma
PE-Barnase	(I)(II)(III*)[Bar]		

III*INACTIVE
ENZYME DOMAIN

25,000 for single-chain antigen binding proteins (single chain antibodies). However, there is no reason to believe that the size of the ligand is important as long as the specificity of binding is preserved. Transforming growth factor alpha (TGFα) targets PE40 to cells with epidermal growth factor (EGF) receptors (27, 47, 48). Although many normal cells contain some EGF receptors, tumor cells often have extremely large numbers of receptors resulting from amplification and overexpression of the EGF receptor gene (49–51). When given systemically, TGFα-PE40 has been found to cause regression of subcutaneous tumors in mice (an epidermoid carcinoma and a prostate carcinoma), showing that a therapeutic window exists in animals for this chimeric toxin (52, 53). Another approach with agents such as TGFα-PE40 is regional therapy. Many cancers, including bladder cancer, are routinely treated regionally with conventional chemotherapeutic drugs. Bladder cancer is an attractive candidate for TGFα-PE40 therapy because all bladder cancer cell lines have EGF receptors and tumor progression is associated with overexpression of EGF receptors (54). Clinical trials of TGFα-PE40 in human bladder cancer will be initiated soon using a TGFα-PE40 analog (TP40) produced by Merck and Co. in which the four cysteine residues in domains II and Ib are converted to alanines to prevent the incorrect disulfide bond formation that can occur when recombinant proteins are purified from bacteria (27, 52). Another use for the molecule is the elimination of the rapidly proliferating smooth muscle cells associated with restenosis of coronary arteries following angioplasty (52a).

IL2-PE40   IL2-PE40 is a chimeric toxin designed to kill cells with IL2 receptors (55). IL2 receptors are not present on normal cells in significant numbers. However, they appear on T cells when activated by exposure to antigen and IL2, and are also present on cells of patients with adult T cell leukemia (ATL) and some other lymphoid malignancies (56). IL2-PE40 was found to be very cytotoxic to activated mouse and rat cells and to have therapeutic effects in rodent models of several human diseases (Table 2). It was shown to cause regression of a mouse lymphoma (57), to permit survival of cardiac allografts in mice (58), and to suppress the development of autoimmune arthritis (59, 60), uveitis, and encephalitis (61). However, IL2-PE40 was not found to be very cytotoxic to human cells taken directly from patients with adult T cell leukemia or to activated human T cells (62). Therefore, alternative ways of targeting cells with IL2 receptors were initiated. One of these was the use of the variable regions of an antibody to the IL2 receptor (in a single-chain form) to deliver the toxin to IL2 receptor–bearing cells (see below). This single-chain immunotoxin, anti-Tac(Fv)-PE40, is very cytotoxic to cells from patients with ATL and to activated human T cells (63, 64). The basis of the discrepant activities of IL2-PE40 and anti-Tac(Fv)-PE40 is not known, but could be related to the fact the IL2-PE40 binds to both subunits of the IL2 receptor, whereas the latter only binds to the p55 subunit and as a consequence there is a difference in the intracellular processing or trafficking of the two molecules (Figure 2).

IL6-PE40   The IL6 receptor was initially identified on activated B cells but subsequently has been found on human liver and several other cell types (65). IL6-PE40 is cytotoxic to many human myeloma and hepatoma cell lines, which express high levels of IL6 receptors, as well as to several other carcinomas in which IL6 receptors are inappropriately expressed (66–70). In addition to PE40, other mutant forms of PE have been attached to IL6 to try to produce even more active cytotoxic agents, perhaps by altering the intracellular processing or trafficking of the chimeric toxin. IL6-PE66^{4Glu} is one of these. In IL6-PE66^{4Glu}, IL6 is fused to the whole PE molecule, which contains glutamate substitutions for basic amino acids located at positions 57, 246, 247, and 249. IL6-PE66^{4Glu} is considerably more active than IL6-PE40, and selectively kills hepatoma cells with as few as 600 receptors (69). In animal experiments, both IL6-PE40 and IL6-PE66^{4Glu} were found to inhibit the growth of a hepatoma expressing about 15,000 IL6 receptors per cell (70).

SINGLE-CHAIN IMMUNOTOXINS   Antibodies are large molecules in which the antigen recognition site, consisting of the light and heavy chain variable regions, is connected to a large constant region that has several different effector functions. The smallest antibody fragment that can bind antigen is the Fv region (fragment variable), which consists of two 110-amino-acid fragments from the amino-terminus of both the chains. The two fragments are not linked together by a disulfide bond and therefore form unstable complexes, which are stabilized by connecting them together with a linking peptide of about 15 amino acids in order to form a single-chain antigen-binding protein or single-chain antibody (71–73). These small molecules have high affinity for antigen, and have been used to target toxin to cells by fusing PE40 to their carboxyl end. Several single-chain immunotoxins have been made with PE40 (74–77). These include anti-Tac(Fv)-PE40 (described above), a single-chain immunotoxin directed at the human transferrin receptor [anti-TFR(Fv)-PE40], and B3(Fv)-PE40, a single-chain immunotoxin directed to an antigen found on many human carcinomas. A variant of the latter, B3(Fv)-PE38KDEL, which has a deletion of aa 365–384 encompassing the disulfide bond in domain Ib, and an altered carboxyl terminus (KDEL) to increase cytotoxicity (35), caused complete regression of human carcinomas growing in mice (77).

CD4-PE40   Although cytotoxic therapy is obviously useful in cancer, it also has a role in the treatment of other chronic illnesses, including autoimmune diseases and allograft rejection (discussed above), as well as in chronic viral infections such as AIDS and viral hepatitis. In AIDS, the human immunodeficiency virus (HIV) slowly spreads from infected to uninfected cells until the immune system is disabled. One approach to the treatment of this disease is to eliminate HIV-infected cells by targeting cytotoxic agents to viral proteins located on the surface of the virus-producing cells. One such protein is gp120, a viral coat protein that interacts strongly with CD4 (78). Accordingly, a portion of the CD4 molecule (aa 1–178) containing the gp120-binding site was attached to PE40 to make CD4-PE40 (79). This chimeric toxin binds to, and kills, cells in which HIV is replicating and producing gp120. Furthermore, in combination with the antiviral agent AZT, CD4-PE40 has been found to completely eliminate HIV from infected cultures (80). CD4-PE40 is currently being produced by the Upjohn Company and has entered clinical trials.

TRANSLOCATION OF FOREIGN PROTEINS BY *PSEUDOMONAS* EXOTOXIN   Because PE, DT, and ricin are able to cross cellular membranes, it should be possible to use the translocating function of a toxin to carry foreign antigens into the cytosol of target cells. To determine if this is possible, barnase, a small ribonuclease, was inserted into a nontoxic form of PE, $PE^{\Delta 553}$ (see Table 2), near its carboxyl terminus, leaving the last 10 amino acids, including REDLK, intact. The RNA of cells treated with $PE^{\Delta 533}$-barnase is degraded, showing that barnase is translocated to the cytosol where it hydrolyzes the cell's RNA (81). More recently, amino acids 413–606 of PE have been deleted and replaced with barnase to make an even more active molecule (T. I. Prior, D. J. FitzGerald, I. Pastan, unpublished results). The fact that the ribonuclease activity of this molecule is translocated to the cytosol firmly establishes that domain II contains the translocating activity of PE. Domain Ib is another site in PE where foreign peptides have been introduced so that they can be translocated to the cytosol (82). It should now be possible to use the translocating activity of PE to "inject" other peptides, such as antigens not ordinarily presented by the MHC Class I system, into cells. This approach could be used to generate types of cytotoxic T cells with novel specificities.

Expression Vector

Several vectors, which employ different regulated promoters, are available for the expression of genetically engineered proteins in *E. coli*. Studier and colleagues (83) have used elements of the bacteriophage T7 to construct a very efficient expression system. It is composed of a plasmid containing the phage promoter with appropriate restriction sites for insertion of DNA fragments. The expression vector used to produce chimeric proteins containing

PE or DT is a modification of the one described by Studier and coworkers (83). The chimeric toxin DNA sequences are inserted next to a T7 promoter along with a ribosome-binding site and an appropriately placed initiation codon (Figure 3). The vector contains a T7 transcription terminator which prevents readthrough. The phage f1 origin of replication allows the production of single-stranded DNA for use in site-directed mutagenesis. For expression purposes, the plasmids are transformed in *E. coli* BL21(λDE3), which carries as a λ-lysogen the late T7 RNA polymerase gene under *lac* promoter–operator control. Expression during normal growth is repressed because of the presence of the *lac* repressor. Protein expression is induced by the addition of IPTG, which inactivates the *lac* repressor allowing the synthesis of T7 RNA polymerase, which then transcribes DNA sequences next to the T7 promoter, and the desired recombinant protein is produced. Some vectors carry the Outer membrane protein A (OmpA) signal sequence preceding the coding sequence so the toxin can be secreted into the periplasm. Once the plasmid is transformed into BL21(λDE3), there is sometimes very low expression of the fusion protein due to incomplete repression of the *lac* promoter, which may be lethal to the host. Hence, the plasmids for genetic manipulation are propagated in *E. coli* hosts that do not carry a T7 RNA polymerase gene. In addition, other cloning vectors and *E. coli* hosts have been developed to overcome the problem of incomplete repression before adding IPTG (83).

Purification of Recombinant Toxins

Using the T7-based expression vector, PE and PE40 are produced as soluble proteins, which are secreted in a concentrated form into the periplasmic space. Maximal expression of PE in the periplasm requires the OmpA signal sequence (84), but for PE40 or LysPE40 this signal sequence is not required (84). Chimeric proteins that remain within the cell accumulate within inclusion bodies and can constitute up to 50% of the cellular protein. Highly purified chimeric toxins are prepared by dissolving the inclusion bodies in a

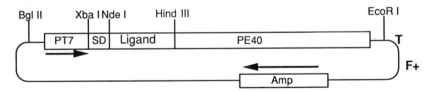

Figure 3 Expression vector for PE40 chimeric proteins. Sequences encoding ligands can be inserted as NdeI-HindIII fragment. PE40 starts from codon 253 of PE. The backbone of the vector is derived from pBluescript II KS(+) (Stratagene). PT7, T7 promoter; SD, Shine-Delgarno sequence; T, T7 transcription terminator; F+, f1 filamentous phage origin of replication; Amp, ampicillin resistance gene for antibiotic selection.

strong denaturant (7M guanidinium-HC1), renaturing the recombinant protein by rapid dilution, with purification achieved by successive anion and size-exclusion chromatography (66, 85).

RICIN

Ribosome-inactivating proteins are a family of toxic proteins produced by plants that inactivate eukaryotic ribosomes. Similar to PE and DT, these toxins have catalytic properties, and only a small number of molecules need to reach the cytosol in order to inactivate enough ribosomes to kill the cell. Plant toxins of this sort can be divided into two groups: those composed of two subunits and those having only one. When added to cells, those with two subunits, such as ricin, are cytotoxic since they have both a cell-binding domain and an enzymatic ribosome-inactivating domain. Those proteins that have enzymatic activity but no binding subunit, such as saporin, are nontoxic to intact cells. However, when saporin (or a similar single-subunit toxin) is chemically attached to an appropriate cell-binding antibody or growth factor, toxicity for intact cells can be achieved (86, 87).

Ricin has been the most popular choice for use as the toxin portion of immunotoxins (because of the limited scope of this review, other plant toxins are not discussed). Ricin has been targeted to cancer cells and other disease-causing cells using either monoclonal antibodies or various growth factors (1, 2). Ricin immunotoxins are currently being evaluated for use in experimental therapy in patients with lymphoma (88), graft verses host disease (89), and rheumatoid arthritis.

Structure and Function

The structural gene of ricin has been cloned and sequenced (13), and recombinant versions of the A subunit (14, 90, 91) and B subunit (92–94) have been expressed in bacteria, yeast, mammalian cells, and *Xenopus* oocytes. The native protein has been crystalized and a three-dimensional structure obtained with high resolution (2.5 Å) (95, 96).

The structural gene for ricin encodes a protein called preproricin. The removal of a leader sequence from the N-terminus results in proricin, which can bind galactose but which is nontoxic because its enzyme activity remains cryptic (97). Only the proteolytic excision of a 12-amino-acid joining peptide linking the cell-binding domain to the enzymatic domain renders the protein active as a toxin molecule (Figure 1) (98). The other notable posttranslational modification is the glycosylation of both subunits.

Traditionally ricin has been called a two-chain toxin. However, molecular studies have shown that the toxin is encoded from one gene and the mature toxin arises from the processing of a single polypeptide chain. Therefore, the term *subunit* is used here to describe the two functional units of the toxin. In the mature toxin, the A and B subunits are joined together by a disulfide bridge. Linkage via the disulfide bridge appears essential for toxic activity (99). The A subunit is at the N-terminus and has N-glycosidase activity, which is responsible for the inactivation of protein synthesis by causing the release of adenine 4324 from the 28S rRNA found in the 60S subunit of eukaryotic ribosomes (100). The B subunit, at the C-terminus, has two galactose-binding domains. It is not yet clear if these domains are functionally identical. Evidence from both genetic and structural studies suggests that the two binding domains arose from a gene duplication event (96). Each domain is composed of three ancient galactose-binding units, only one of which is active. The B subunit attaches to the surface of eukaryotic cells through the galactose-binding domains. This attachment is blocked by the addition of excess galactose-containing disaccharides or polysaccharides such as lactose.

Ricin binds to many different galactose-containing glycoproteins. It is not clear what fraction of cell-bound molecules are internalized or which remain "fixed" on the cell surface. Ricin uptake proceeds with a T 1/2 of about 180 min (101), but since cell-binding sites are heterogeneous some toxin molecules can enter very rapidly (102).

Ricin enters cells via coated pits and endocytic vesicles (103) (Figure 2). Once inside the cell, some portion of the Golgi apparatus, perhaps the *trans*-golgi system, appears to play a major role in mediating toxin entry into the cytosol (104–106). However, in contrast to PE and DT, a low-pH environment within the endosome is not required. In fact, agents that prevent formation of the pH gradient and inhibit the toxicity of DT and PE actually increase the toxicity of ricin (103). After ricin passes through the endosome, the A and B subunits must be separated by a process involving reduction of the interchain disulfide bond (Figure 1), and possibly also by a conformational change in the toxin. Only the A subunit (or a portion thereof) reaches the cytosol.

The process of translocation of the A subunit into the cytosol is poorly understood. Clearly, the A subunit itself has sufficient translocating activity to reach the cytosol. However, there are numerous reports indicating that the presence of the B subunit enhances toxicity (102). Whether this enhancement is a true increase in translocation activity by the B subunit, or merely reflects some kind of stabilization function, is unclear.

Using both the three-dimensional structure of ricin and sequence homologies with other plant toxins, residues have been identified that are thought to play key roles in toxin activity (107). Confirmation of their importance has been possible in some instances by the construction of mutant proteins. In the A subunit, a number of amino acid residues in a prominent cleft may participate in N-glycosidic bond hydrolysis. These include Asn78, Tyr80,

Tyr123, Glu177, Arg180, and Asn209, which are conserved among a series of ribosome-inactivating toxins. For example, mutation of Glu177 to Gln reduces the enzymatic activity by almost 200-fold (108). However, when Glu177 was changed to Ala there was very little change in activity, probably because a nearby Glu, at position 208, can substitute for it. A positive charge at position 180 was found necessary for both solubility of the protein and for enzyme activity (109).

Immunotoxin Construction

Two strategies are used to make ricin-containing immunotoxins. One is to reduce the disulfide bond that links the two subunits, and then cross-link the A subunit to a monoclonal antibody or growth factor (110). Since some antibodies do not internalize or are inappropriately routed during intracellular trafficking, some A subunit immunotoxins are not cytotoxic. For these a second approach has been devised. Whole ricin is chemically modified to reduce its binding to galactose or is reacted with a glycopeptide in order to block the galactose-binding sites (8, 111). These two procedures both generate a modified version of ricin that lacks the majority of the nonspecific cytotoxic activity but retains other functions, including translocation and ribosome inactivation.

The A subunit is naturally glycosylated by the addition of mannose-containing sugars. In animals, this directs the A subunit, and immunotoxins derived from it, to the liver, where undesirable side-effects occur. To avoid this problem, Thorpe and colleagues used chemical methods to alter these carbohydrate residues (112). Recently, a recombinant form of the A subunit has been produced in E. coli (14). This eliminates the need to separate the A and B subunits. In addition, this recombinant A subunit is not glycosylated, making it an attractive candidate for producing immunotoxins.

Immunotoxins composed of ricin or its A subunit, and monoclonal antibodies, are constructed using chemical cross-linking reagents (10, 113). Whole ricin immunotoxins are often constructed using thioether cross-linkers, while A subunit immunotoxins must use cross-linkers that employ a disulfide bond (113). More than 10 heterobifunctional agents are available to make such disulfide linkages (9, 114), the selection of which depends in large part on the desired stability of the immunotoxin. The optimal situation is one in which the immunotoxin remains stable and intact in serum, but dissociates within the target cell.

An alternative to using chemical cross-linkers is a gene fusion approach similar to that used to make PE and DT fusion proteins. However, recombinant ricin-based molecules have been difficult to produce, probably because the A chain of the plant toxins must be attached to the cell recognition domain by a disulfide bond, and disulfide-linked subunits are difficult to produce in bacteria. Recently Lord and colleagues introduced protease cleavage sites that may allow the production of ricin-based gene fusion proteins that retain cytotoxic activity (90).

Immunotoxin Activity

Many excellent reviews have already described the activities and properties of immunotoxins made with ricin and other plant toxins (1, 102, 115–117). A primary goal has been to produce agents that kill cancer cells. But ricin-containing immunotoxins have also been used to eliminate selected populations of lymphocytes and as a tool in neurobiology to kill specific neurons (118). In addition, a ricin-based conjugate of CD4 has been found to kill HIV-infected cells (119, 120).

Vitetta, Uhr, and colleagues as well as Thorpe and colleagues have produced ricin A chain coupled to antibodies to B-cell-specific antigens, and shown such chains to cause complete regression of B cell lymphomas in mice (121). Others have described antitumor activity of ricin A chain immunotoxins in solid or ascites tumor models (122, 123).

Several ricin-containing immunotoxins have been developed and approved for human trials, and two kinds of cancer-related trials have been conducted. The first involves the ex vivo treatment of harvested bone marrow to eliminate contaminating tumor cells prior to re-infusion in patients undergoing autologous bone marrow transplantation (124). A variety of antibodies, linked to ricin or ricin A chain, have been used for this purpose. The second kind of trial involves the parenteral administration of immunotoxins (125–128). A recent clinical trial indicates that an immunotoxin made with ricin A subunit has activity against certain lymphomas (88).

DIPHTHERIA TOXIN

Structure and Function

Diphtheria toxin (DT) is a single-chain polypeptide of 535 residues produced by *Corynebacterium diphtheria* carrying a lysogenic beta phage (11). DT is secreted into the extracellular fluid where it is processed by extracellular proteases to yield an N-terminal A fragment (M_r 21,000) and a C-terminal B fragment (M_r 37,000); the two fragments are held together by a disulfide bond (Figure 1). After binding to its cellular receptor, DT, like PE, is internalized through coated pits into endocytic vesicles (129, 130) (Figure 2). The low pH present in these vesicles initiates the translocation process whereby the disulfide bond is broken, and the A fragment is translocated into the cytosol. Whether DT translocation occurs from an early endocytic compartment or further along the endocytic pathway is not yet known. Despite the fact that DT has been subjected to intensive study, its three-dimensional structure has not been solved. Therefore, much of our information about the structure-function relationships of DT is derived from studies of mutant proteins that are no longer cytotoxic, but still cross-react with antibodies; these mutant toxins therefore are called crms (cross-reactive materials). The A fragment of DT ADP ribosylates EF-2 in a manner identical to PE (described above). The glutamic acid residue at position 148 is at the NAD^+-binding site (131), and corresponds to the glutamic acid at position 553 of PE (132). DT binds to its receptor on the cell surface of target cells through the carboxyl end of the B fragment. Cell binding is abolished, or greatly diminished, by deletion of a 15-kDa or a 6-kDa fragment from the C-terminus (16, 133, 134), by a point mutation at position 508 (crm 103), by point mutations at positions 390 and 525 (crm 107) (135), or by chemical cleavage and release of C-terminal residues (136). The N-terminal portion of the B fragment (amino acids 220 to 373) contains many hydrophobic residues, and this region has been postulated to be involved in the translocation of the active moiety of DT (A fragment) across the membrane into the cell. It has been suggested that the B fragment forms a pore, which allows translocation of the A fragment. The importance of the hydrophobic domains in the cytotoxic action of DT was confirmed by attaching several different A-chain fragments containing varying amounts of the hydrophobic region to a monoclonal antibody and showing that the cytotoxic activity of the immunotoxin was related to the amount of hydrophobic domain present (137).

Olsnes and coworkers have developed an artifical system to study the entry of DT into cells, which they believe mimics the process that normally occurs in endocytic vesicles (138). They, as well as Draper & Simon, observed translocation of the A fragment when cells with surface-bound, proteolytically (trypsin) nicked toxin were exposed to acidic medium (139, 140). Trypsin treatment generates several fragments that contain ADP-ribosylation activity owing to cleavage vicinal to the arginine residues at positions 190, 192, and 193. Only the DT fragment bearing a single arginine (at position 190) appeared to be translocation competent (141). In support of a model in which DT is cleaved at or near Arg193 is the finding that substitution of a glycine for arginine 193 reduced cellular toxicity by 1000-fold (150).

Immunotoxins

Because most humans are immunized against DT, as many as 80% of the people in the industrialized countries would have antibodies to DT. Therefore, there has been somewhat less interest in making immunotoxins using DT. The most active immunotoxins have been made by coupling antibodies to either full-length DT-containing point mutations in the binding region, or to truncated DT molecules that have lost their binding activity but retain translocation and ADP-ribosylation functions (137). These immunotoxins have high activity in tissue culture but have not been extensively studied in animals for their possible antitumor effects.

Recombinant Toxins

In order to direct DT to cells, ligands have been placed at the carboxyl end of DT to replace the cell-binding domain. As indicated in Table 3, both alpha MSH and IL2 have been used to prepare chimeric toxins (142–144). In addition, a single-chain immunotoxin has been made with DT using the variable regions of an antibody to the p55 subunit of the interleukin 2 receptor to direct DT to IL2 receptor–bearing cells (74).

Chimeric toxins containing DT are very toxic to cells bearing appropriate receptors, and DT-IL2 has been shown to have activity in several models of immunological diseases. Murphy, Strom, and colleagues have shown that a chimeric toxin termed DAB-IL2 can be used to prevent allograft rejection (145, 146), and inhibit delayed type hypersensitivity (147). DAB-IL2 also kills leukemic cells from some patients with Adult T-cell leukemia, and kills T cells infected with HIV 1 (148, 149). These chimeric proteins have also been useful in studying the intracellular processing of DT (150, 151).

PROSPECTS AND PROBLEMS

The studies summarized in this review show that it is possible to target several different toxins to specific cells, and to selectively kill cells in tissue culture and in animal models. A number of major problems, however, need to be solved. Conventional immunotoxins made by chemical methods have long circulation times in the blood (typical half-lives of 4–8 hours); however, they

Table 3 Structure and activity of recombinant toxins derived from DT

Name	Structure	Target and cells	Activity in mice or rats
DT-alpha-MSH	A T B (MSH)	alpha-MSH receptor	
DAB-IL2	A T B (IL2)	IL2 receptor leukemia HIV infected	allograft rejection DTH
DT-antiTac(Fv)	A T B ⋈	human IL2 receptor leukemia	
DT-antiTFR(Fv)	A T B ⋈	human TFR receptor	

appear to penetrate slowly into solid tumors, probably because of their large size (152). Recombinant toxins made from growth factors or single-chain antibodies are much smaller and penetrate tissues more rapidly. However, these molecules have very short survival times in the plasma, with half-lives ranging from 10–30 minutes. Their short half-life is probably a consequence of several factors. Their small size permits filtration by the kidney. They contain proteolytically sensitive sites so they may be degraded rapidly by circulating or cell-bound proteases, and newly exposed surface residues may cause nonspecific uptake by liver cells or other cell types. Toxins are very immunogenic molecules, and a strong primary antibody response usually develops within 8–10 days. However, in patients who have received extensive chemotherapy, the immune system is often severely compromised and antibody responses are poor. In normal individuals, the response is brisk and for continued or repeated therapy, immunosuppression will be required. In a recent study in mice, the drug 17-deoxyspergualin has been found to completely suppress the antibody response to PE in mice (153). Other immunosuppressive agents such as cyclosporine A, which are known to suppress the development of antibodies and human immunoglobulin in people, may be useful in suppressing the immune response to toxins (154). Finally, all the immunotoxins produced to date have dose-limiting toxicities. With ricin-based immunotoxins, patients develop a capillary leak syndrome (88, 125). With PE, the dose-limiting toxicity is damage to the liver. No information of the limiting toxicity of DT conjugates in humans is available. Using recombinant DNA techniques, it should be possible to produce recombinant immunotoxins with more desirable properties.

CONCLUSION

The experimental data reviewed here show that recombinant toxins can be constructed that are able to kill cells responsible for cancer, AIDS, and other human diseases. Many problems remain to be solved. But the recombinant DNA approach described here is both powerful enough and flexible enough to make these new types of therapeutic agents useful as drugs for treating human diseases for which no treatments currently exist.

ACKNOWLEDGMENTS

The authors acknowledge the many valuable contributions to the development of recombinant toxins made by members of the Laboratory of Molecular Biology both past and present. We also thank our colleagues who have shared with us reprints, preprints, and unpublished data.

Literature Cited

1. Vitetta, E. S., Fulton, R. J., May, R. D., Till, M., Uhr, J. W. 1987. *Science* 238:1098–104
2. Vitetta, E. S. 1990. *J. Clin. Immunol.* 10:15S–18S
3. FitzGerald, D., Pastan, I. 1989. *J. Natl. Cancer Inst.* 81:1455–63
4. Pastan, I., Willingham, M. C., FitzGerald, D. J. 1986. *Cell* 47:641–48
5. Murphy, J. R. 1988. *Cancer Treat. Res.* 37:123–40
6. Pastan, I., FitzGerald, D. 1989. *J. Biol. Chem.* 264:15157–60
7. FitzGerald, D. J. 1987. *Methods Enzymol.* 151:139–45
8. Thorpe, P. E., Ross, W. C., Brown, A. N., Myers, C. D., Cumber, A. J., et al. 1984. *Eur. J. Biochem.* 140:63–71
9. Thorpe, P. E., Wallace, P. M., Knowles, P. P., Relf, M. G., Brown, A. N., et al. 1987. *Cancer Res.* 47:5924–31
10. Cumber, A. J., Forrester, J. A., Foxwell, B. M., Ross, W. C., Thorpe, P. E. 1985. *Methods Enzymol.* 112:207–25
11. Greenfield, L., Bjorn, M. J., Horn, G., Fong, D., Buck, G. A., et al al. 1983. *Proc. Natl. Acad. Sci. USA* 80:6853–57
12. Gray, G. L., Smith, D. H., Baldridge, J. S., Harkins, R. N., Vasil, M. L., et al. 1984. *Proc. Natl. Acad. Sci. USA* 81:2645–49
13. Lamb, F. I., Roberts, L. M., Lord, J. M. 1985. *Eur. J. Biochem.* 148:265–70
14. Piatak, M., Lane, J. A., Laird, W., Bjorn, M. J., Wang, A., Williams, M. 1988. *J. Biol. Chem.* 263:4837–43
15. Hwang, J., FitzGerald, D. J., Adhya, S., Pastan, I. 1987. *Cell* 48:129–36
16. Colombatti, M., Greenfield, L., Youle, R. J. 1986. *J. Biol. Chem.* 261:3030–35
17. Kondo, T., FitzGerald, D., Chaudhary, V. K., Adhya, S., Pastan, I. 1988. *J. Biol. Chem.* 263:9470–75
18. Iglewski, B. H., Kabat, D. 1975. 72:2284–88
19. Omura, F., Kohno, K., Uchida, T. 1989. *Eur. J. Biochem.* 180:1–8
20. Collier, R. J. 1988. *Cancer Treat. Res.* 37:25–35
21. Morris, R. E., Saelinger, C. B. 1986. *Infect. Immun.* 52:445–53
22. FitzGerald, D., Morris, R. E., Saelinger, C. B. 1980. *Cell* 21:867–73
23. Allured, V. S., Collier, R. J., Carroll, S. F., McKay, D. B. 1986. *Proc. Natl. Acad. Sci. USA* 83:1320–24
24. Jinno, Y., Chaudhary, V. K., Kondo, T., Adhya, S., FitzGerald, D. J., Pastan, I. 1988. *J. Biol. Chem.* 263:13203–7
24a. Chaudry, G. J., Wilson, R. B., Draper, R. K., Clowes, R. C. 1989. *J. Biol. Chem.* 264:15151–56
25. Ogata, M., Chaudhary, V. K., Pastan, I., FitzGerald, D. J. 1990. *J. Biol. Chem.* 265:20678–85
26. Jinno, Y., Ogata, M., Chaudhary, V. K., Willingham, M. C., Adhya, S., et al. 1989. *J. Biol. Chem.* 264:15953–59
27. Edwards, G. M., DeFeo-Jones, D., Tai, J. Y., Vuocolo, G. A., Patrick, D. R., et al. 1989. *Mol. Cell. Biol.* 9:2860–67
28. Madshus, I. H., Collier, R. J. 1989. *Infect. Immun.* 57:1873–78
29. Siegall, C. B., Ogata, M., Pastan, I., FitzGerald, D. J. 1991. *Biochemistry* 30:7154–59
30. Siegall, C. B., Chaudhary, V. K., FitzGerald, D. J., Pastan, I. 1989. *J. Biol. Chem.* 264:14256–61
31. Chaudhary, V. K., Jinno, Y., FitzGerald, D., Pastan, I. 1990. *Proc. Natl. Acad. Sci. USA* 87:308–12
31a. Brinkmann, U., Pai, L. H., FitzGerald, D. J., Pastan, I. 1992. *Proc. Natl. Acad. Sci. USA*. In press
32. Wozniak, D. J., Hsu, L.-Y., Galloway, D. R. 1988. *Proc. Natl. Acad. Sci. USA* 85:8880–84
33. Lukac, M., Pier, G. B., Collier, R. J. 1988. *Infect. Immun.* 56:3095–98
34. Munro, S., Pelham, H. R. 1987. *Cell* 48:899–907
35. Seetharam, S., Chaudhary, V. K., FitzGerald, D., Pastan, I. 1991. *J. Biol. Chem.* In press
36. Spicer, E. K., Noble, L. A. 1982. *J. Biol. Chem.* 257:5716–21
37. Mekalanos, J. J., Swarth, D. J., Pearson, G. D. N., Harford, N., Groyne, F., deWilde, M. 1983. *Nature* 306:551–57
38. Sandvig, K., Moskaug, J. O. 1987. 245:899–901
39. Jiang, J. X., London, E. 1990. *J. Biol. Chem.* 265:8636–41
40. Idziorek, T., FitzGerald, D., Pastan, I. 1990. *Infect. Immun.* 58:1415–20
41. Batra, J. K., Jinno, Y., Chaudhary, V. K., Kondo, T., Willingham, M. C., et al. 1989. *Proc. Natl. Acad. Sci. USA* 86:8545–49
42. Pai, L. H., Batra, J. K., FitzGerald, D. J., Willingham, M. C., Pastan, I. 1991. *Proc. Natl. Acad. Sci. USA* 88:3358–62
42a. Debinski, W., Karlsson, B., Lindholm, L., Siegall, C. B., Willingham, M. C., et al. 1992. *J. Clin. Invest.* In press
43. FitzGerald, D., Idziorek, T., Batra, J.

K., Willingham, M. C., Pastan, I. 1990. *Bioconj. Chem.* 1:264–68
44. FitzGerald, D. J., Willingham, M. C., Cardarelli, C. O., Hamada, H., Tsuruo, T., et al. 1987. *Proc. Natl. Acad. Sci. USA* 84:4288–92
45. FitzGerald, D. J., Willingham, M. C., Pastan, I. 1986. *Proc. Natl. Acad. Sci. USA* 83:6627–30
46. Willingham, M. C., FitzGerald, D. J., Pastan, I. 1987. *Proc. Natl. Acad. Sci. USA* 84:2474–78
47. Siegall, C. B., Xu, Y. H., Chaudhary, V. K., Adhya, S., Fitzgerald, D., Pastan, I. 1989. *FASEB J.* 3:2647–52
48. Chaudhary, V. K., FitzGerald, D. J., Adhya, S., Pastan, I. 1987. *Proc. Natl. Acad. Sci. USA* 84:4538–42
49. Hall, W. A., Merrill, M. J., Walbridge, S., Youle, R. J. 1990. *J. Neurosurg.* 72:641–46
50. Hendler, F. J., Ozanne, B. W. 1984. *J. Clin. Invest.* 74:647–51
51. Scambia, G., Panici, P. B., Battaglia, F., Ferrandina, G., Almadori, G., et al. 1991. *Cancer* 67:1347–51
52. Heimbrook, D. C., Stirdivant, S. M., Ahern, J. D., Balishin, N. L., Patrick, D. R., et al. 1990. *Proc. Natl. Acad. Sci. USA* 87:4697–701
52a. Epstein, S. E., Siegall, C. B., Biro, S., Fu, Y.-M. FitzGerald, D., Pastan, I. 1991. *Circulation Res.* 84:778–87
53. Pai, L. H., Gallo, M. G., FitzGerald, D. J., Pastan, I. 1991. *Cancer Res.* 51:2808–12
54. Lau, J. L. T., Fowler, J. E. J., Ghosh, L. 1988. *J. Urol* 139:170–75
55. Lorberboum-Galski, H., FitzGerald, D., Chaudhary, V., Adhya, S., Pastan, I. 1988. *Proc. Natl. Acad. Sci. USA* 85:1922–26
56. Waldmann, T. A. 1990. *J. Am. Med. Assoc.* 263:272–74
57. Kozak, R. W., Lorberboum, G. H., Jones, L., Puri, R. K., Willingham, M. C., et al. 1990. *J. Immunol.* 145:2766–71
58. Lorberboum-Galski, H., Barrett, L. V., Kirkman, R. L., Ogata, M., Willingham, M. C., et al. 1989. *Proc. Natl. Acad. Sci. USA* 86:1008–12
59. Case, J. P., Lorberboum-Galski, H., Lafyatis, R., FitzGerald, D., Wilder, R. L., Pastan, I. 1989. *Proc. Natl. Acad. Sci. USA* 86:287–91
60. Lorberboum-Galski, H., Lafyatis, R., Case, J. P., FitzGerald, D. J., Wilder, R. L., Pastan, I. 1991. *Immunopharmacology* 13:305–15
61. Rose, J. W., Lorberboum-Galski, H., FitzGerald, D., McCarron, R., Hill, K.

E., et al. 1991. *J. Neuroimmunol.* 32:209–17
62. Lorberboum-Galski, H., Garsia, R. J., Gately, M., Brown, P. S., Clark, R. E., et al. 1990. *J. Biol. Chem.* 265:16311–17
63. Chaudhary, V. K., Gallo, M. G., FitzGerald, D. J., Pastan, I. 1990. *Proc. Natl. Acad. Sci. USA* 87:9491–94
64. Kreitman, R. J., Chaudhary, V. K., Waldmann, T., Willingham, M. C., FitzGerald, D. J., Pastan, I. 1990. *Proc. Natl. Acad. Sci. USA* 87:8291–95
65. Bauer, J., Lengyel, G., Bauer, T. M., Acs, G., Gerok, W. 1989. *FEBS Lett.* 249:27–30
66. Siegall, C. B., Chaudhary, V. K., Fitz-Gerald, D. J., Pastan, I. 1988. *Proc. Natl. Acad. Sci. USA* 85:9738–42
67. Siegall, C. B., FitzGerald, D. J., Pastan, I. 1990. *Curr. Top. Microbiol. Immunol.* 166:63–69
68. Siegall, C. B., Schwab, G., Nordan, R. P., FitzGerald, D. J., Pastan, I. 1990. *Cancer Res.* 50:7786–88
69. Siegall, C. B., FitzGerald, D. J., Pastan, I. 1990. *J. Biol. Chem.* 265:16318–23
70. Siegall, C. B., Kreitman, R. J., Fitz-Gerald, D. J., Pastan, I. 1991. *Cancer Res.* 51:2831–36
71. Bird, R. E., Hardman, K. D., Jacobson, J. W., Johnson, S., Kaufman, B. M., et al. 1988. *Science* 242:423–26
72. Huston, J. S., Levinson, D., Mudgett-Hunter, M., Tai, M. S., Novotny, J., et al. 1988. *Proc. Natl. Acad. Sci. USA* 85:5879–83
73. Skerra, A., Pluckthun, A. 1988. *Science* 240:1038–1041
74. Batra, J. K., FitzGerald, D. J., Chaudhary, V. K., Pastan, I. 1991. *Mol. Cell Biol.* 11:2200–5
75. Chaudhary, V. K., Queen, C., Junghans, R. P., Waldmann, T. A., Fitz-Gerald, D. J., Pastan, I. 1989. *Nature* 339:394–97
76. Chaudhary, V. K., Batra, J. K., Gallo, M. G., Willingham, M. C., FitzGerald, D. J., Pastan, I. 1990. *Proc. Natl. Acad. Sci. USA* 87:1066–70
77. Brinkmann, U., Pai, L. H., FitzGerald, D. J., Willingham, M. C., Pastan, I. 1991. *Proc. Natl. Acad. Sci. USA.* In press
78. Berger, E. A., Fuerst, T. R., Moss, B. 1988. *Proc. Natl. Acad. Sci. USA* 85:2357–61
79. Chaudhary, V. K., Mizukami, T., Fuerst, T. R., FitzGerald, D. J., Moss, B., et al. 1988. *Nature* 335:369–72
80. Ashorn, P., Moss, B., Weinstein, J. N.,

Chaudhary, V. K., FitzGerald, D. J., et al. 1990. *Proc. Natl. Acad. Sci. USA* 87:8889–93

81. Prior, T. I., FitzGerald, D. J., Pastan, I. 1991. *Cell* 64:1017–23

82. Debinski, W., Siegall, C. B., Fitzgerald, D., Pastan, I. 1991. *Mol. Cell. Biol.* 11:1751–53

83. Studier, F. W., Moffatt, B. A. 1986. *J. Mol. Biol.* 189:113–30

84. Chaudhary, V. K., Xu, Y. H., FitzGerald, D., Adhya, S., Pastan, I. 1988. *Proc. Natl. Acad. Sci. USA* 85:2939–43

85. Bailon, P., Weber, D. V., Gately, M., Smart, J. E., Lorberboum-Galski, H., et al. 1988. *Bio-Technology* 6:1326–29

86. French, R. R., Courtenay, A. E., Ingamells, S., Stevenson, G. T., Glennie, M. J. 1991. *Cancer Res.* 51:2353–61

87. Lappi, D. A., Martineau, D., Baird, A. 1989. *Biochem. Biophys. Res. Commun.* 160:917–23

88. Vitetta, E. S., Stone, M., Amlot, P., Fay, J., May, R., et al. 1991. *Cancer Res.* In press

89. Vallera, D. A., Uckun, F. M. 1990. *Prog. Clin. Biol. Res.* 333:191–204

90. O'Hare, M., Brown, A. N., Hussain, K., Gebhardt, A., Watson, G., et al. 1990. *FEBS Lett.* 273:200–4

91. Frankel, A., Schlossman, D., Welsh, P., Hertler, A., Withers, D., Johnston, S. 1989. *Mol. Cell. Biol.* 9:415–20

92. Vitetta, E. S., Yen, N. 1990. *Biochim. Biophys. Acta* 1049:151–57

93. Richardson, P. T., Gilmartin, P., Colman, A., Roberts, L. M., Lord, J. M. 1988. *Bio-Technology* 6:565–70

94. Richardson, P. T., Roberts, L. M., Gould, J. H., Lord, J. M. 1988. *Biochim. Biophys. Acta* 950:385–94

95. Katzin, B. J., Collins, E. J., Robertus, J. D. 1991. *Proteins: Struct. Funct. Genet.* 10:251–59

96. Rutenber, E., Robertus, J. D. 1991. *Proteins: Struct. Funct. Genet.* 10:260–69

97. Richardson, P. T., Westby, M., Roberts, L. M., Gould, J. H., Colman, A., Lord, J. M. 1989. *FEBS Lett.* 255:15–20

98. Harley, S. M., Lord, J. M. 1985. *Plant Sci.* 41:111–16

99. Wright, H. T., Robertus, J. D. 1987. *Arch. Biochem. Biophys.* 256:280–84

100. Endo, Y., Mitsui, K., Motizuki, M., Tsurugi, K. 1987. *J. Biol. Chem.* 262:5908–12

101. Sandvig, K., Olsnes, S., Pihl, A. 1976. *J. Biol. Chem.* 251:3977–84

102. Esworthy, R. S., Neville, D. J. 1984. *J. Biol. Chem.* 259:11496–504

103. Olsnes, S., Sandvig, K. 1988. *Cancer Treat. Res.* 37:39–73

104. van Deurs, B., Sandvig, K., Petersen, O. W., Olsnes, S., Simons, K., Griffiths, G. 1988. *J. Cell. Biol.* 106:253–67

105. Colombatti, M., Johnson, V. G., Skopicki, H. A., Fendley, B., Lewis, M. S., Youle, R. J. 1987. *J. Immunol.* 138:3339–44

106. Manske, J. M., Buchsbaum, D. J., Vallera, D. A. 1989. *J. Immunol.* 142:1755–66

107. Ready, M. P., Katzin, B. J., Robertus, J. D. 1988. *Proteins: Struct. Funct. Genet.* 3:53–59

108. Ready, M. P., Kim, Y., Robertus, J. D. 1991. *Proteins: Struct. Funct. Genet.* 10:270–78

109. Frankel, A., Welsh, P., Richardson, J., Robertus, J. D. 1990. *Mol. Cell. Biol.* 10:6257–63

110. Blythman, H. E., Casellas, P., Gros, O., Gros, P., Jansen, F. K., et al. 1981. *Nature* 290:145–46

111. Lambert, J. M., McIntyre, G., Gauthier, M. N., Zullo, D., Rao, V., et al. 1991. *Biochemistry* 30:3234–47

112. Thorpe, P. E., Wallace, P. M., Knowles, P. P., Relf, M. G., Brown, A. N., et al. 1988. *Cancer Res.* 48:6396–403

113. Marsh, J. W., Srinivasachar, K., Neville, D. J. 1988. *Cancer Treat. Rev.* 37:213–37

114. Goff, D. A., Carroll, S. F. 1990. *Bioconj. Chem.* 1:381–86

115. Ramakrishnan, S., Bjorn, M. J., Houston, L. L. 1989. *Cancer Res.* 49:613–17

116. Griffin, T. W., Childs, L. R., FitzGerald, D. J., Levin, L. V. 1987. *J. Natl. Cancer Inst.* 79:679–85

117. Marsh, J. W., Neville, D. J. 1986. *Biochemistry* 25:4461–67

118. Wiley, R. G., Oeltmann, T. N. 1986. *J. Neurosci. Methods* 17:43–53

119. Till, M. A., Zolla, P. S., Gorny, M. K., Patton, J. S., Uhr, J. W., Vitetta, E. S. 1989. *Proc. Natl. Acad. Sci. USA* 86:1987–91

120. Till, M. A., Ghetie, V., May, R. D., Auerbach, P. C., Zolla, P. S., et al. 1990. *J. Acquired Immune Defic. Syndr.* 3:609–14

121. Fulton, R. J., Uhr, J. W., Vitetta, E. S. 1988. *Cancer Res.* 48:2626–31

122. Griffin, T. W., Richardson, C., Houston, L. L., LePage, D., Bogden, A., Raso, V. 1987. *Cancer Res.* 47:4266–70

123. FitzGerald, D., Bjorn, M. J., Ferris, R. J., Winkelhake, J. L., Frankel, A. E., et al. 1987. *Cancer Res.* 47:1407–10

354   PASTAN ET AL

124. Uckun, F. M., Kersey, J. H., Vallera, D. A., Ledbetter, J. A., Weisdorf, D., et al. 1990. *Blood* 76:1723–33
125. Byers, V. S., Rodvien, R., Grant, K., Durrant, L. G., Hudson, K. H., et al. 1989. *Cancer Res.* 49:6153–60
126. Gould, B. J., Borowitz, M. J., Groves, E. S., Carter, P. W., Anthony, D., et al. 1989. *J. Natl. Cancer Inst.* 81:775–81
127. Hertler, A. A., Schlossman, D. M., Borowitz, M. J., Laurent, G., Jansen, F. K., et al. 1988. *J. Biol. Response Modif.* 7:97–113
128. Weiner, L. M., O'Dwyer, J., Kitson, J., Comis, R. L., Frankel, A. E., et al. 1989. *Cancer Res.* 49:4062–67
129. Keen, J. H., Maxfield, F. R., Hardegree, M. C., Habig, W. H. 1982. *Proc. Natl. Acad. Sci. USA* 79:2912–16
130. Morris, R. E., Gerstein, A. S., Bonventre, P. F., Saelinger, C. B. 1985. *Infect. Immun.* 50:721–27
131. Carroll, S. F., Collier, R. J. 1984. *Proc. Natl. Acad. Sci. USA* 81:3307–11
132. Carroll, S. F., Collier, R. J. 1988. *Mol. Microbiol.* 2:293–96
133. Pappenheimer, A. M. J. 1977. *Annu. Rev. Biochem.* 46:69–94
134. Giannini, G., Rappuoli, R., Ratti, G. 1984. *Nucleic Acids Res.* 12:4063–69
135. Greenfield, L., Johnson, V. G., Youle, R. J. 1987. *Science* 238:536–39
136. Myers, D. A., Villemez, C. L. 1988. *J. Biol. Chem.* 263:17122–27
137. Johnson, V. G., Wilson, D., Greenfield, L., Youle, R. J. 1988. *J. Biol. Chem.* 263:1295–300
138. Moskaug, J. O., Sandvig, K., Olsnes, S. 1988. *J. Biol. Chem.* 263:2518–25
139. Draper, R. K., Simon, M. I. 1980. *J. Cell. Biol.* 87:849–54
140. Sandvig, K., Olsnes, S. 1980. *J. Cell. Biol.* 87:828–32
141. Moskaug, J. O., Sletten, K., Sandvig, K., Olsnes, S. 1989. *J. Biol. Chem.* 264:15709–13
142. Murphy, J. R., Bishai, W., Borowski, M., Miyanohara, A., Boyd, J., Nagle, S. 1986. *Proc. Natl. Acad. Sci. USA* 83:8258–62
143. Williams, D. P., Parker, K., Bacha, P., Bishai, W., Borowski, M., et al. 1987. *Protein Eng.* 1:493–98
144. Wen, Z. L., Tao, X., Lakkis, F., Kiyokawa, T., Murphy, J. R. 1991. *J. Biol. Chem.* 266:12289–93
145. Kirkman, R. L., Bacha, P., Barrett, L. V., Forte, S., Murphy, J. R., Strom, T. B. 1989. *Transplantation* 47:327–30
146. Pankewycz, O., Mackie, J., Hassarjian, R., Murphy, J. R., Strom, T. B., Kelley, V. E. 1989. *Trnasplantation* 47:318–22
147. Kelley, V. E., Bacha, P., Pankewycz, O., Nichols, J. C., Murphy, J. R., Strom, T. B. 1988. *Proc. Natl. Acad. Sci. USA* 85:3980–84
148. Kiyokawa, T., Shirono, K., Hattori, T., Nishimura, H., Yamaguchi, K., et al. 1989. *Cancer Res.* 49:4042–46
149. Finberg, R. W., Wahl, S. M., Allen, J. B., Soman, G., Strom, T. B., et al. 1991. *Science* 252:1703–5
150. Williams, D. P., Wen, Z., Watson, R. S., Boyd, J., Strom, T. B., Murphy, J. R. 1990. *J. Biol. Chem.* 265:20673–77
151. Williams, D. P., Snider, C. E., Strom, T. B., Murphy, J. R. 1990. *J. Biol. Chem.* 265:11885–89
152. Jain, R. K. 1989. *J. Natl. Can. Inst.* 81:570–76
153. Pai, L. H., FitzGerald, D. J., Tepper, M., Schacter, B., Spitalny, G., Pastan, I. 1990. *Cancer Res.* 50:7750–53
154. Feutren, G. 1989. *Curr. Opin. Immunol.* 2:239–45
155. Pagano, R. E. 1990. *Biochem. Soc. Trans.* 18:361–66

Annu. Rev. Biochem. 1992. 61:355–86

PROTEIN ISOPRENYLATION AND METHYLATION AT CARBOXYL-TERMINAL CYSTEINE RESIDUES

Steven Clarke

Department of Chemistry and Biochemistry and the Molecular Biology Institute, University of California, Los Angeles, 90024-1569

KEY WORDS: farnesylation, geranylgeranylation, carboxyl methyltransferases, proteolysis, membrane proteins

CONTENTS

0066-4154/92/0701-0355$02.00

PERSPECTIVES AND SUMMARY

Eukaryotic polypeptides synthesized with one of at least three types of cysteine-containing carboxyl-terminal sequences are candidates for posttranslational modification reactions including isoprenylation, proteolytic cleavage, and methyl esterification. The major class of these proteins is formed by those that have precursor C-terminal sequences of -CXXX, in which the penultimate residue is generally aliphatic. Polypeptides with a terminal residue such as leucine are initially modified by a cytosolic isoprenyltransferase that transfers the C_{20} group from geranylgeranyl pyrophosphate to the cysteine sulfur atom. Alternatively, polypeptides synthesized with terminal residues such as serine, alanine, methionine, or glutamine are modified by a distinct cytosolic enzyme that utilizes farnesyl pyrophosphate to add the C_{15} isoprenyl group to the cysteine side chain. Following this step, the three terminal amino acids are cleaved by a cytosolic or membrane-bound protease. Finally, a membrane-bound enzyme catalyzes the S-adenosylmethionine-dependent addition of a methyl group to the newly exposed α-carboxyl group of the cysteine residue. The net effect of these modifications is the creation of a C-terminal S-isoprenyl cysteine α-methyl ester (Figure 1). A second type of processing occurs in at least one protein with a C-terminal -CXC precursor sequence. Here both cysteine residues are modified by geranylgeranylation and the carboxy-terminus is methyl esterified. Finally, proteins with certain C-terminal -CC precursor sequences can be modified by the addition of geranylgeranyl group to one or both of the cysteine residues.

Polypeptides containing these precursor sequence motifs include the ras proteins and many of the other small G-proteins, the γ-subunits of the large G-proteins, some of the nuclear lamins, the retinal cGMP phosphodiesterase, and several fungal mating pheromones. Many of these protein species function in signal transduction processes across the plasma membrane or in the control of cell division. At present, the specific utility to the cell of each of the C-terminal modification reactions is not clear. Possibly, the increased hydrophobicity of the C-terminus can lead to interactions with the membrane bilayer that result in membrane association of these proteins. Alternatively, the isoprenyl and methyl groups may be specific targets for binding by other membrane (or cytosolic) "receptor" proteins, leading to a specific alignment of protein partners in signalling pathways or in structural arrays.

Recent reviews in this area include those of Glomset et al (1), Maltese (2), and Sinensky & Lutz (2a), on protein isoprenylation, Rine & Kim (3) and Der & Cox (4) on ras isoprenylation and membrane association, Gibbs (5) on ras C-terminal processing enzymology, and Barten & O'Dea (6) on protein carboxyl methylation.

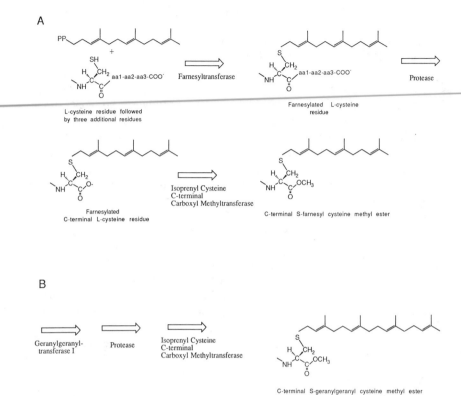

Figure 1 Biosynthetic pathways and structures of isoprenylated and methylated eukaryotic proteins with the -CXXX C-terminal motif. *A*. Formation of C-terminal farnesyl cysteine methyl esters from polypeptide precursors containing the -CXX(S/A/M/Q) sequence. Cofactors include *trans,trans*-farnesyl pyrophosphate as the isoprenyl donor and *S*-adenosylmethionine as the methyl donor. *B*. Formation of C-terminal geranylgeranyl cysteine methyl esters from precursors containing the -CXXL motif. All-*trans*-geranylgeranyl pyrophosphate is the isoprenyl donor.

C-TERMINAL -CXXX SEQUENCES CAN SPECIFY MODIFICATION BY ISOPRENYLATION, PROTEOLYSIS, AND METHYLATION

The recent discovery of the widespread occurrence of these interrelated protein modification reactions resulted from the recognition of the common features of several seemingly unrelated biological systems. Results obtained from studies of topics as diverse as the chemical nature of fungal peptide mating factors, the metabolism of mevalonic acid in mammalian cells, the functional role of eukaryotic protein carboxyl methylation reactions, and the posttranslational processing of *ras* oncogene products all contributed to establishing the generality of these novel posttranslational modifications.

Evidence for an isoprenylated C-terminal cysteine residue was first presented in 1979 for Rhodotorucine *A*, a peptide that mediates mating tube formation in the heterobasidiomyceteous yeast *Rhodosporidium toruloides*. The mature factor contains 11 amino acids including a C-terminal *S*-farnesyl cysteine residue (7, 8). Between 1979 and 1984 peptidyl sex hormones of the genus *Tremella*, a group of basidiomyceteous jelly fungi, were also found to terminate in a C-terminal cysteine residue that is *S*-isoprenylated by either an oxidized farnesyl group [*T. brasiliensis A* factors Ia and Ib (9) and *T. mesenterica A*-10 factor (10, 11)] or by an unmodified farnesyl group [*T. mesenterica* a-13 factor (11)]. Significantly, the α-carboxyl groups of the *A*-10 and *A*(Ia) factors were found to be further modified by methyl esterification (9–11).

Independently, work published in 1984 on the metabolic fate of radiolabelled mevalonic acid in a mouse cell line suggested that mammalian proteins might also be modified by isoprenylation reactions (12). Mevalonic acid is the precursor of farnesyl pyrophosphate and other isoprenoid intermediates in cholesterol, dolichol, and ubiquinone biosynthesis pathways (13). Although clear evidence for the covalent attachment of the isoprenyl group to protein was demonstrated, the chemical nature of the isoprenoid moiety and the type of linkage was not established (12). A functional role for this reaction was suggested in 1985 when the inhibition of DNA replication in cultured mammalian cells starved for mevalonate was correlated with defects in the synthesis of isoprenylated proteins (14). The close relationship between this type of protein modification and the isoprenylation seen in the fungal peptides was initially clouded, however, by the 1986 proposal that the mevalonate-derived additions to mouse kidney proteins were more similar to hydroxyl-linked C_{45} isoprenoid and C_{95} dolichol groups than to the sulfhydryl-linked C_{15} isoprenyl groups of the mating pheromones (15).

A crucial link that led to establishing the similarity of the reactions that modify the fungal mating peptides and mammalian cellular proteins was found in 1986 with the discovery of a gene in the ascomycetic yeast *Saccharomyces cerevisiae* that was required for the posttranslational processing of both the RAS proteins and the **a**-mating pheromone [named *RAM* for ras and mating (16)]. Analysis of the deduced amino acid sequences for the genes specifying yeast RAS proteins and the precursor polypeptide for **a**-factor revealed no similarity except at the C-terminus, where both sequences contained a cysteine residue followed by two aliphatic residues and the C-terminal residue. It was proposed that this Cys-Ali-Ali-Xaa sequence signalled the common modification to both polypeptides by the *RAM* gene product (16). Since such sequences were also found for a variety of ras proteins from higher eukaryotes, as well as the rho small G-proteins and the γ-subunit of bovine transducin, it was suggested the RAM-catalyzed modification occurred more widely in nature (16). Subsequent work demonstrated that mutations in this gene affected the processing of RAS proteins at a step prior to a palmitoylation reaction (17). Although the possibility that the defect in processing involved the loss of a farnesylation reaction was suggested based on similarities with the *Tremella* mating factors (17), the lack of a definitive structure of either the *S. cerevisiae* RAS proteins or the **a**-mating factor hampered direct comparisons of these systems. Further complicating the situation were a number of misleading clues including the proposals that mature **a**-factor might contain a palmitoylated cysteine residue followed by the last three amino acids (18), that the conserved cysteine residue on mammalian ras proteins was modified by a palmitoyl group (19), and that the C-terminal three residues of both mammalian ras proteins (19) and transducin γ-subunit (20) were not proteolytically cleaved.

The next piece of the puzzle was provided by research in a different area. By 1985, studies of protein modification by carboxyl methylation reactions had indicated that there were at least two major classes of methyltransferases with different biological functions (21). The first type of enzyme, limited to chemotactic bacteria, resulted in the modification of the side chains of glutamate residues in membrane-bound chemoreceptors (22, 23). The second type of enzyme, found widely distributed in both prokaryotic and eukaryotic cells, recognizes carboxyl groups on altered aspartyl residues, including the side chains of racemized D-aspartyl residues (24) and the α-carboxyl group of isomerized L-isoaspartyl residues (25, 26). There was, however, evidence for eukaryotic protein carboxyl methylation reactions that could not be readily assigned to either of these classes of enzymes. These reactions included those occurring in bovine retinal rods that resulted in the modification of the α-subunit of the cGMP phosphodiesterase (27) and those that resulted in the modification of nuclear lamin B (28). In 1988, with the realization that carboxyl methylation occurred on the C-terminus of *Tremella* mating factors, as well as the availability of cDNA-derived deduced amino acid sequences for both the lamin and cGMP phosphodiesterase polypeptides that showed C-terminal -CXXX sequences, it was possible to hypothesize that both the ras proteins and the latter polypeptides might contain C-terminal methylated structures similar to those seen in the fungal peptide pheromones (29). This idea was supported by the demonstration that the Ha-ras protein in rat embryo fibroblasts was indeed carboxyl methylated in a fashion consistent with a C-terminal localization (29). Using the common theme of methylation in the fungal mating factors, nuclear lamins, cGMP phosphodiesterase, and ras proteins, all of which shared -CXXX terminal sequences and at least two of which were known to be lipidated, it was then possible to propose that the

-CXXX sequence motif identified earlier (16) coded in some way for linked lipidation, proteolytic, and methylation reactions (29). However, the existing evidence that the ras proteins appeared to be palmitoylated on the conserved cysteine and the mating factors were either palmitoylated or isoprenylated still obscured the emerging picture.

The last crucial pieces of the puzzle were provided in 1988 when the *S. cerevisiae* **a**-factor was shown to contain a C-terminal farnesyl cysteine methyl ester (30) and in 1989 with the discovery that all of the mammalian ras proteins were in fact isoprenylated on the conserved cysteine residue and that palmitoylation occurred on only a subgroup of ras proteins at upstream cysteine residues (31). The last-mentioned study made it clear that the palmitoylation reaction is not an essential part of the modification reactions, and that the common modifications of all of these -CXXX-containing polypeptides are isoprenylation, proteolytic cleavage, and methylation.

While these developments were taking place, work continued on the analysis of mammalian isoprenylated proteins. Studies of mevalonate-labelled proteins in a mouse neuroblastoma cell line in 1987 demonstrated isoprenylated polypeptides of 66 kDa in the nuclear lamina/matrix fraction, of 45 kDa in both the soluble and particulate fractions, and 53 kDa, 22–26 kDa, and 17 kDa in the cytosolic fraction, suggesting that many different proteins are modified in this way (32). This work led the way to the first identification of an isoprenylated protein in 1988 when it was shown that the 66-kDa modified nuclear protein in mouse 3T3 cells and Chinese hamster ovary cells is lamin B (33, 34). In 1990, the identification of the major mammalian protein-bound isoprenyl groups as *trans,trans*-farnesyl and all-*trans*-geranylgeranyl (35, 36), as well as the specific determination between 1989 and 1990 of the farnesylation of nuclear lamin B (37) and mammalian ras (38), the farnesylation and methylation of yeast RAS (39), and the geranylgeranylation and methylation of the γ-subunit of the mammalian large G-proteins (40, 41) completed the outline of the picture. In the past year, rapid advances have been made in the identification of many other similarly modified proteins, in the characterization of the enzymes responsible for these modifications, as well as in the discovery of new C-terminal -CC and -CXC motifs for isoprenylation and methylation (42–44).

PROTEINS MODIFIED BY ISOPRENYLATION, PROTEOLYSIS, AND METHYLATION

Table 1 lists proteins known to date to be modified by C-terminal cysteinyl *S*-isoprenylation and α-carboxyl methylation, as well as those proteins containing the -CXXX motif that have been shown not to be modified. It is important to note that the experimental criteria to establish these modifications vary considerably among the different studies. For example, the incorporation of radiolabelled mevalonate into protein products has in some cases been the only evidence for isoprenylation. In other work, specific chemical assignments of isoprenyl groups have been based on data ranging from the chromatography of a single isoprenyl derivative in a single solvent system to data from more complete analyses using multiple derivatives and chromatography systems (39, 73) as well as advanced NMR and mass spectral techniques (30, 47, 74). One powerful new tool is electrospray mass spectroscopy, in which molecular weights of intact polypeptides can be determined with enough accuracy to confirm or exclude a specific postulated C-terminal structure (47). The ability to isolate proteolytic fragments that include the intact methyl-esterified S-isoprenylated cysteine residue (or derivatives) provides good evidence for C-terminal localization of this modification (40, 57, 59, 70, 75). Although no internally isoprenylated cysteine residues have been observed to date in the -CXXX proteins, the isolation of an isoprene derivative itself or a peptidyl isoprene derivative lacking the methyl ester would be also consistent with such a nonterminal localization. Finally, it should be noted that structural assignments of C-terminal modifications have been made in several cases utilizing heterologous systems. For example, endogenous isoprenyltransferases in a rabbit reticulocyte translation system can modify the in vitro translation products of mRNA species from various species (42, 56, 64, 65, 68). Alternatively, mammalian (52) and amphibian (72) cDNAs can be expressed and posttranslationally modified by enzymes in insect cells. Although the overall chemistry of modification appears to be relatively well conserved across evolutionary lines, it is possible that results obtained in such systems do not precisely reflect the chemistry of intact cells.

A number of farnesylated and geranylgeranylated proteins have now been described in eukaryotes (Table 1). Recent studies on the isoprenylation of the entire complement of polypeptides from several mammalian cell lines separated by dodecyl sulfate gel electrophoresis suggest that species of 66–72 kDa, 53–55 kDa, and 41–46 kDa are predominantly farnesylated, whereas those of 21–28 kDa are predominantly geranylgeranylated (76). The latter class includes both the farnesylated ras proteins and the large group of geranylgeranylated small G-proteins, while the former class includes the farnesylated nuclear lamin species. In most eukaryotic cells and tissues, the total number of molecules of geranylgeranylated proteins is greater than the total number of molecules of farnesylated proteins (35, 36, 76a).

Some indirect evidence has been presented for protein modification by isoprenoids distinct from the farnesyl or geranylgeranyl groups. It has been suggested that proteins can be isoprenylated with C_{45} groups (15) and C_{95}-dolichol (15) or dolichol phosphate groups (77), although the potential target proteins of such modifications have not been found and a conclusive chemical

Table 1 Isoprenylated and methylated proteins containing -CXXX, -CXC, or -CC C-termini

Protein	DNA-encoded carboxyl-terminal sequence	Known modifications	References
Fungal mating pheromones			
S. cerevisiae **a**-factor	DPACVIA	*trans,trans*-farnesyl-Cys Me ester	30
Tremella brasiliensis (A-9291-I)		*trans,trans*-OH-farnesyl-Cys Me ester	9
Tremella mesenterica (A-10)		*trans,trans*-OH-farnesyl-Cys Me ester	10, 11
R. toruloides Rhodotorucine A	RNGCTVA	C-terminal *trans,trans*-farnesyl-Cys	7, 8, 45
Ras proteins			
human/mouse H-ras	SCKCVLS	farnesyl-Cys Me ester	29, 31, 46
human K-ras-4B	KTKCVIM	farnesyl-Cys Me ester	31, 38, 47
human N-ras	GLPCVVM	isoprenyl Me ester	31, 48
S. cerevisiae RAS1	GGCCIIC	Me ester	49
S. cerevisiae RAS2	GGCCIIS	farnesyl-Cys Me ester	39, 49 – 51, 51a
Ras-related small G-proteins			
human rap1A/Krev-1/smg p21A	KKSCLLL	geranylgeranyl-Cys Me ester[a]	52
human rap1B/smg p21B	KSSCQLL	all-*trans*-geranylgeranyl-Cys Me ester	53
human rap2	CSACNIQ	isoprenyl	54
human rac1	KRKCLLL	geranylgeranyl[b,c]	55, 56
human rac2	KRACSLL	geranylgeranyl[b,c]	55, 56
human ralA	RERCCIL	geranylgeranyl[b]	56
bovine brain G25K	KRKCCIF[d] SRRCVLL[d]	all-*trans*-geranylgeranyl-Cys Me ester	57
human rab5	RNQCCSN	geranylgeranyl[b]	42
bovine smooth muscle rhoA	KSGCLVL	all-*trans*-geranylgeranyl-Cys Me ester	58
rat rab1B	SGGCC	geranylgeranyl[b]	42
		geranylgeranyl[a]	44
human rab2	GGGCC	geranylgeranyl[b]	42
bovine rab3A (smg p25A)	HQDCAC	all-*trans*-di-geranylgeranyl-Cys Me ester	43
human rab3A (smg p25A)	HQDCAC	geranylgeranyl[a]	44

Heterotrimeric (large) G-proteins			
bovine brain G-protein (γ-subunit)	KFFCAIL	all-*trans*-geranylgeranyl-Cys Me ester	40, 59
rat PC12 cell G-protein (γ-subunit)		geranylgeranyl	41
bovine transducin (γ-subunit)	KGGCVIS	*trans,trans*-farnesyl-Cys Me ester	60, 61
bovine transducin (α-subunit)	LKDCGLF	not modified	62
human/bovine/rat/mouse $G_i(1)$, $G_i(2)$ (α-subunit)	LKDCGLF	not modified	63–66
S. cerevisiae STE18 (γ-subunit)	SVCCTLM	isoprenyl	67
Nuclear lamin proteins			
chicken prelamin A	PQGCSIM	isoprenyl[b]	68
human prelamin A	PQNCSIM	isoprenyl	34
Chinese hamster ovary lamin B	NKSCAIM[f]	isoprenyl	34
chicken lamin B1	ERSCVVM	isoprenyl[b]	68
chicken lamin B2	SRGCLVM	isoprenyl Me ester[e]	68, 69
human lamin B	NRSCAIM	*trans,trans*-farnesyl-Cys	37
Additional proteins			
bovine retinal cGMP phosphodiesterase			
α-subunit	SKSCCVQ	C-terminal Cys Me ester	70, 70a
β-subunit	SSTCRIL	Me ester	70a
rat retinal cGMP phosphodiesterase			
α-subunit	SKSCCIQ[g]	farnesyl-Cys Me ester	71a
β-subunit	SSTCCIL[g]	geranylgeranyl-Cys Me ester	71a
Xenopus xlcaax-1	CQCCVVM	isoprenylated[b]	72

[a] Protein expressed in insect Sf9 cells.
[b] Modification determined in an in vitro rabbit reticulocyte protein synthesis system primed by the cDNA.
[c] Protein expressed in simian COS cells.
[d] Sequences determined for two human cDNA clones.
[e] Protein expressed in mouse cells or in an in vitro rabbit reticulocyte protein synthesis system.
[f] Sequence determined for mouse.
[g] Sequence is given of the mouse proteins (71).

identification has not been made (see also 78). It has also been reported that a 26-kDa protein in Chinese hamster ovary cells is modified by isopentenyladenine (79). Since all of these isoprenyl groups are derived from products of the mevalonic acid biosynthetic pathway, direct chemical analysis is essential in distinguishing these forms from the more intensely studied C_{15} farnesyl and C_{20} geranylgeranyl modifications. The biosynthetic origin of the terminal hydroxyl group of the oxidized farnesyl fungal mating peptides (9–11) has not been investigated; such a modified isoprenyl group has not been found to date in known isoprenylated proteins.

Isoprenylation, proteolysis, and methylation reactions appear to be linked in almost all cases. No examples have been found of mature isoprenylated proteins in nature with the -CXXX motif that retain the three terminal amino acids. On the other hand, evidence has been presented that methylation may not always stoichiometrically follow the cleavage reaction. For example, methylated forms of Rhodotorucine A have not been described (7, 8). It has been suggested that geranylgeranylated proteins may not be methylated (80), although the presence of stoichiometric levels of methyl groups on several specific proteins (40, 43, 52, 57, 58) indicates that this is not generally the case. The methyl ester linkage is labile to proteolytic treatments such as pronase digestion (39) as well as to spontaneous hydrolysis (see below), potentially resulting in the loss of methyl groups during analysis.

Although isoprenylation reactions have been shown to occur in many mammalian cell types, no evidence for farnesyl pyrophosphate synthesis (81) or C-terminal methylation (82) has been found in mature human erythrocytes. There is also no evidence to date for C-terminal protein isoprenylation, proteolysis, or methylation reactions in prokaryotic cells, with the possible exception of archaebacteria (76a). Specific assays have shown the absence of the C-terminal protein methyltransferase (49) and farnesyltransferase (109, 120) in *Escherichia coli* as well as the absence of protein-bound farnesyl and geranylgeranyl groups (76a). The absence of such modifications in this bacterium allows for the synthesis of unmodified precursor polypeptides as substrates for the isoprenylation reaction (see below).

Reversibility of Protein Isoprenylation and Methylation

Since reversible modification reactions, such as the phosphorylation and dephosphorylation of serine, threonine, or tyrosine residues, can modulate protein function, it is tempting to speculate that the ability to remove isoprenyl or methyl groups from C-terminal cysteine residues may be important to the cell. Although an initial study did not detect loss of isoprenyl groups from mammalian ras proteins (31), more work is needed to determine whether this is also the case for other proteins. The thioether linkage itself appears to be chemically stable under physiological conditions. In model compounds, however, oxidation of a thioether to the corresponding sulfoxide permits a [2,3]-sigmatropic rearrangement to give a sulfenic acid ester (83) that can be cleaved to produce a free sulfhydryl on the cysteine residue. Interestingly, enzymatic activities are known that catalyze the oxidation of thioethers (84). If an unmodified C-terminal cysteine residue is regenerated by this or an equivalent pathway from a mature isoprenylated protein in cells, its reisoprenylation may depend on an enzyme distinct from the isoprenyltransferases that require -CXXX sequences (see below).

It is also not clear whether or not the methylation reaction is physiologically reversible. No turnover of the methyl group was detected on the mammalian N-ras protein over at least a 2 h period in one study (48), but there has been a preliminary report suggesting the possibility of slow enzymatic cleavage of methylated transducin in rod outer segment membranes (85). The fact that nonspecific exopeptidases such as carboxypeptidase Y can also cleave protein C-terminal methyl ester linkages (50, 132a) complicates the interpretation of the latter type of study. In any case, the specificity of the C-terminal methyltransferase for an S-isoprenylated C-terminal cysteine residue (49, 82) should make it possible for a demethylated protein to be readily remethylated. That many isoprenylated proteins can be methylated to at least some extent in in vitro assays (59, 70, 75, 86–88) suggests that either the methylation reaction is not complete in vivo or that spontaneous or enzymatic hydrolysis has occurred, either in the cell or during the sample workup. Further studies are needed to examine these possibilities.

Modification of Proteins with the -CXXX Motif

The first examples of C-terminal polypeptide sequences shown to be modified contained two aliphatic residues following the cysteine residue (16, 29), and the designation -CaaX, where "a" refers to the aliphatic residue, has been frequently used to describe this motif. (The designation -CAAX has also been employed, but such usage creates confusion because the upper case A is the one-letter code for an alanine residue.) Although this motif has largely held for the penultimate residue, further examples of isoprenylated proteins have shown that the specificity for the residue adjacent to the cysteine is not as great. For example, in the isoprenylated polypeptides described in Table 1, all but one of the species has an isoleucine, leucine, or valine residue in the penultimate position. (The exception is the human rab5 protein with a serine residue at this site.) In this same group of proteins, the residues on the carboxyl side of the conserved cysteine include not only the aliphatic residues alanine, isoleucine, leucine, and valine, but the polar residues threonine, serine, cysteine, asparagine, glutamine, and arginine. Similarly, a variety of residues are present at the C-terminal position of the polypeptide precursors of the isoprenylated proteins. The results shown in Table 1 confirm the initial suggestion that whether a -CXXX-containing protein is farnesylated or geranylgeranylated may depend largely upon the nature of the C-terminal residue (53). Precursor polypeptides containing C-terminal serine, alanine, glutamine, or methionine residue are farnesylated, whereas those containing a leucine [or perhaps a phenylalanine (57) or asparagine (42)] residue are geranylgeranylated (see also Ref. 88a). In vitro studies on the specificity of the farnesyl and geranylgeranyl protein transferases discussed below suggest that the four terminal residues are the crucial sequence elements in the recognition of proteins for isoprenylation.

Based on these patterns, it is possible to scan the protein data bases for cDNA-derived amino acid sequences to predict additional proteins subject to isoprenylation, proteolytic cleavage, and methyl esterification at the C-terminus. Table 2 lists examples of proteins, excluding the numerous examples of small G-proteins (39, 96–98), which have not been demonstrated to contain these modifications but which have a cysteine four residues from the C-terminus and an aliphatic residue in the penultimate position (39). It will be interesting to see which of the proteins in Table 2 are in fact modified at their C-terminus, and also whether any of the much larger group of proteins containing a cysteine residue in the fourth position from the end can be modified.

In at least one protein, nuclear lamin A, a -CXXX motif specifies isoprenylation in a precursor species that is eventually proteolytically cleaved upstream to uncover a new unmodified C-terminus in the mature protein (99, 100). A similar situation may exist for the β-subunit of the retinal cGMP phosphodiesterase (71).

There are now two examples of eukaryotic proteins with the -CXXX motif that are not isoprenylated at the conserved cysteine residue. These are two α-subunits of heterotrimeric G-proteins with -CGLF terminal sequences (Table 1; cf. Refs. 63–66). When the C-terminal sequence of G_i was changed from -CGLF to -CVLS in proteins expressed in simian COS cells, mevalonate label could be incorporated (66), suggesting the significance of the glycine and the phenylalanine residues.

Modification of Proteins with the -CC or -CXC Motif

A class of ras-related small G-proteins are localized to membrane elements of the secretory pathway and include the mammalian rab proteins and the yeast YPT1 and SEC4 proteins (96–98,101,102). At least some of these protein species, however, do not contain the -CXXX C-terminal sequences described above, and yet do not appear to have other sequence elements, such as hydrophobic domains, which would lead to their membrane association. Recently, it has been shown that two of these proteins, the rat rab1B and

Table 2 Candidates for C-terminal isoprenylation, cleavage, and methylation

Protein	DNA-encoded C-terminal sequence	Comments	References
human brain 2',3'-cyclic nucleotide 3' phosphodiesterase	LQSCTII	present in myelin membranes	89
rabbit skeletal muscle phosphorylase kinase α-subunit	HSICAMQ	cytosolic enzyme	90
human extracellular superoxide dismutase	ESECKAA	binds to epithelial cell surfaces	91
mouse (2'-5')oligo(A) synthetase	NWTCILL		92
human (2'-5')oligo(A) synthetase E18	DWTCTIL		93
human leukemia antigen CALLA	EKKCRVW	cell surface glycoprotein	94
human/rat gap junction protein	SDRCSAC		95
rat kidney guanylate cyclase	SAFCVVL		95a

human rab2 proteins, can be isoprenylated when translated in a rabbit reticulocyte system (42, 44), and a yeast analog of the former species, the YPT1 protein, can be enzymatically modified in vitro with labelled geranylgeranyl pyrophosphate and a partially purified bovine brain transferase (103). Analyses of the mammalian isoprenylated products were consistent with geranylgeranyl groups, although the precise chemistry was not established (42, 44). The report that isoprenylation can occur at a reduced level when either of the YPT1 cysteine residues are replaced suggests that isoprenylation is possible at either site (103). Additionally, mutations to the *YPT1* gene such that one of the terminal cysteine residues was deleted resulted in a functional protein, whereas the loss of both cysteine residues gave a cytosolic and inactive protein (104). Interestingly, the addition of three C-terminal residues to generate a ras-like -CXXX motif generated a functional protein (105). Further studies will be required to pinpoint the isoprenylation sites and to ascertain whether isoprenylation is followed by methyl esterification.

At least one member of the rab class of small G-proteins contains a third type of modifiable cysteine-containing C-terminus, the -CXC motif (106). This protein, designated smg25A or rab3a, contains a geranylgeranyl group on each of the two cysteine residues as well as a terminal methyl ester (43, see also 44).

PROTEIN FARNESYLTRANSFERASES

An enzymatic activity was first detected in rat brain cytosol that catalyzed the incorporation of farnesyl pyrophosphate into protein acceptors containing -CXXX C-termini (107, 108), and this enzyme has now been found in a number of other mammalian cells and tissues (109–111) as well as in frog oocytes (112). The purified rat brain protein appears to be a heterodimer of approximately 100 kDa consisting of two polypeptides, a 49-kDa α-subunit and a 46-kDa β-subunit (107, 108, 113–115). The β-subunit appears to recognize the -CXXX acceptor-polypeptide (114), while the α-subunit may recognize the isoprenyl donor farnesyl pyrophosphate (113, 114).

Parallel work in the yeast *S. cerevisiae* has provided genetic evidence for two genes that specify distinct subunits of a cytosolic protein farnesyltransferase. The *RAM1* gene, originally described as *RAM* (16) or *DPR1* (17, 116), has been sequenced and shown to encode of polypeptide of 43 kDa (117). Although other proposals have been made for the function of this protein (16, 117), it has now been shown that extracts from cells deficient in this gene are defective in the in vitro farnesylation of **a**-mating pheromone precursor (118), RAS1 constructs (103), and the RAS2 protein (119). Comparison of the deduced amino acid sequence of the rat brain β-subunit (115) with that of the RAM1 protein (117) showed identical residues at 37% of the positions. A second unlinked gene, *RAM2,* encoding a 38-kDa protein, has recently been cloned (120) and identified as essential for farnesyltransferase activity on a-factor pheromone and the RAS proteins (103,119,120). The amino acid sequence of the *RAM2* product (120) shows homology with that of the α-subunit of the mammalian farnesyltransferase (121, 124). When the *RAM1* and *RAM2* genes were expressed singly in *E. coli,* no farnesyltransferase activity was detected, while a mixture of extracts from cells expressing both genes formed an active enzyme (120). These results suggest that the yeast farnesyltransferase contains subunits of both the *RAM1* and *RAM2* gene products, corresponding respectively to the β- and α-subunits of the mammalian enzyme.

At least two homologs of the *RAM1* product have been described in yeast that appear to be specificity subunits of protein geranylgeranyltransferases (see below). Interestingly, as the α-subunit of the mammalian farnesyltransferase has been suggested on immunological grounds to be a common subunit of the geranylgeranyltransferase active on -CXXL containing proteins (113), the *RAM2* product may also be shared with other yeast protein isoprenyltransferases (see below).

The sequence requirements of the isoprenyl acceptor protein for the farnesyltransferase have now been studied in a number of laboratories (103, 111, 121–123). In general, the results from these studies are consistent with the specificity deduced from the analysis of proteins known to be farnesylated in vivo and in vitro (Table 1, see discussion above). In all cases, acceptor species were found to have an absolute requirement for a cysteine residue in the fourth position from the C-terminus, but a variety of residues in the last three positions were recognized by the enzyme. There appears to be little or no specific recognition of sequences to the amino side of the modifiable cysteine residue (31). In fact, it has recently been shown that tetrapeptides such as CVIM (123) or CVLS (103) are themselves farnesyl-accepting substrates for the mammalian enzyme.

The specificity of the mammalian farnesyltransferase for the terminal three amino acids has been tested both by examining the inhibitory action of synthetic peptides on protein farnesylation and by modifying the sequences of the acceptor proteins themselves for in vitro analysis. In an initial study, the competitive inhibition of a group of 42 tetrapeptides modeled after CVIM, the C-terminus of the K-ras protein, was determined using the rat brain enzyme with bacterially produced human Ha-ras proteins as the acceptor protein (122). Half inhibition at concentrations of 1 μM or less was found for peptides containing valine, isoleucine, leucine, phenylalanine, and tyrosine in the penultimate position. The first three of these residues were found almost universally in isoprenylated proteins described to date (Table 1), whereas the last two aromatic residues were recently shown to block farnesyltransferase activity when present in this position (123). In the terminal position, inhibition was found for methionine, serine, and phenylalanine residues. The latter result was somewhat surprising, because other data suggest that terminal phenylalanine residues are specifically recognized by geranylgeranyltransferases (53, 103). Finally, a variety of residues adjacent to the conserved cysteine residue resulted in effective inhibition.

This type of specificity can be compared to that determined for a partially purified enzyme from bovine brain using a large number of bacterially produced yeast RAS1 constructs with various C-terminal tetrads, based on -CVLX and -CAIX sequences and representing all 20 possible C-terminal residues (103). The results obtained were consistent with the known farnesylation of proteins containing terminal alanine, serine, methionine, and glutamine residues (Table 1), and also showed that terminal glycine, alanine, valine, phenylalanine, threonine, histidine, asparagine, and cysteine residues may also be permissive for farnesylation (103). Finally, the specificity of the farnesyltransferase in rabbit reticulocytes was probed recently with the expression of modified G25K small G-protein. Farnesylation was detected with sequences terminating in serine, alanine, methionine, and glutamine, whereas little or no activity was detected when leucine, isoleucine, valine, glutamate, or lysine was in the terminal position (H. K. Yamane, R. A. Cerione, and B. K.-K. Fung, unpublished).

The ability of the farnesyltransferases to recognize sequences found to be apparently solely geranylgeranylated in vivo (103, 111) suggests that the relative concentrations of the two isoprenyl donors and the relative concentrations of the acceptor proteins may also help direct isoprenylation in intact cells.

PROTEIN GERANYLGERANYLTRANSFERASES

In contrast to the single type of protein farnesyltransferase described above, clear evidence has now been obtained for multiple types of protein geranylgeranyltransferases. The most intensively studied enzyme (designated GGT-I) is active on -CXXX-containing substrates (103, 111, 113, 125–128) and is probably responsible for the isoprenylation of the γ-subunit of the large G-proteins and several of the small G-proteins (Table 1). A second enzymatic activity (GGT-II) has been detected in bovine brain cytosol based on its ability to geranylgeranylate the yeast YPT1 protein (103). This activity is probably responsible for the modification of not only YPT1 in yeast but also of other -CC-terminating proteins such as yeast SEC4 and mammalian rab1B and rab2 small G-proteins. The isoprenyl-accepting protein specificities determined for these two enzymes (103, 111) suggest that a third type of enzyme (GGT-III) may be found that catalyzes the geranylgeranylation of the -CXC-terminating proteins (43, 44).

Recently, an enzyme activity has been partially purified from bovine brain that catalyzes the geranylgeranylation of the -CAC-terminating rab3a protein (129). It is not yet clear whether this activity is identical to the GGT-II activity described above (103) or whether it represents a distinct GGT-III. In the simplest case, a single enzyme could modify both -CC- and -CXC-containing proteins. On the other hand, the apparent failure of both enzymes to recognize peptide substrates (103, 129) suggests the possibility that not only are the GGT-II and GGT-III enzymes distinct, but that there may be multiple enzymes in each class that isoprenylate only a subgroup of -CC- or -CXC-containing proteins.

Analysis of yeast mutants has provided genetic insight into the structure of the protein geranylgeranyltransferases. Amino acid sequence similarities between the deduced sequence of the *RAM1* gene product (117) and the *CDC43/CAL1* gene product (130, 131) and the *BET2* gene product (132) suggest that the latter two may also be components of distinct isoprenyltransferases. Genetic analysis of the *CDC43/CAL1* locus suggests that its product may be essential for the modification of the *CDC42* gene product. The CDC42 protein is homologous to the geranylgeranylated mammalian G25K protein (57). In an in vitro geranylgeranyltransferase assay using a glutathione-*S*-transferase construct containing a -CIIL C-terminus as an acceptor, no activity was found in cytosolic extracts of cells lacking the *CDC43* gene, whereas normal activity was seen in extracts of cells lacking the *RAM1* gene (127). These studies suggest that the *CDC43* gene product may be a specificity component of the GGT-I. Evidence for a second yeast geranylgeranyltransferase has been provided by analysis of the *BET2* gene whose product is required for membrane attachment of the -CC-terminating proteins YPT1 and SEC4. This gene product shows 34% sequence identity to the *RAM1* gene product (132), suggesting that it may be a subunit of the yeast GGT-II.

Based on the homology of the *RAM2* protein product with the α-subunit of mammalian protein farnesyltransferase (121, 124) and the evidence that the latter subunit may also be present in the mammalian GGT-I (113), it would seem reasonable that the *RAM2* product may complex in a similar fashion with the *CDC43* and *BET2* proteins to make up the yeast geranylgeranyltransferases types I and II respectively. However, while at least one mutation in the *RAM2* gene in yeast completely abolishes farnesyltransferase activity, this mutation only partially reduces GGT-I activity and does not reduce GGT-II activity at all (103). These results suggest that the *RAM2* gene may not be an essential component of either enzyme, although it is possible that the specific *ram2* mutation investigated only affects its catalytic function when complexed to the RAM1 protein. Additionally, the lethality of *ram2*, but not *ram1*, mutations in *S. cerevisiae* indicates that the RAM2 product has some function in addition to its role as a subunit of the farnesyltransferase (120). Further work needs to be done to resolve this issue.

The mammalian GGT-I has been partially purified from bovine and rat brain cytosol and was shown to be partially resolved from the farnesyltransferase activity on anion exchange chromatography (103, 111, 113, 125). Further purification resulted in preparations of enzymes that were apparently completely resolved from each other (111) as well as from a geranylgeranyltransferase active on the -CC-containing YPT1 protein (103). As described above, immunological evidence has been presented that one subunit of the rat brain enzyme is the same (or very similar to) the α-subunit of the farnesyltransferase (113). Similar cytosolic activities have been found in rat liver, kidney, and heart (127, 128).

The specificity of the GGT-I active on -CXXX-terminating polypeptides has been tested in several systems. The crucial elements of recognition appear to be the cysteine and the C-terminal residue. As described above, it appears that a leucine residue is present at the C-terminus of the precursor protein in almost all of the known geranylgeranylated proteins of this type (Table 1), and all of the isoprenyl-accepting substrates for the in vitro assays have been based on various -CXXL sequences. For example, reduced or no activity is found when a serine residue is substituted for a terminal leucine residue with the yeast (111) or the bovine brain (103, 111) GGT-I. In an extensive study of isoprenylation of the RAS2-linked -CXXX tetrad by the bovine brain GGT-I, the highest activity was seen with a leucine at the C-terminal position, although some activity was seen with phenylalanine, valine, isoleucine, or surprisingly, methionine substituents (103). Similar results were found for the rabbit reticulocyte system (124). Although the sequence requirements for the residues between the cysteine and the terminal residue have not been mapped as extensively as for the farnesyltransferase, it appears that the GGT-I may be much more selective (103).

Finally, the specificities of the protein farnesyltransferase and geranylgeranyltransferase type I for isoprenyl donors and acceptors do not appear to be absolute in vitro, and significant cross-reactivity is seen (103, 111, 114, 120). The clearest evidence for this comes from studies in which both geranylgeranyl pyrophosphate and farnesyl pyrophosphate were used as isoprenyl donors with either the bovine brain farnesyltransferase or the GGT-I (103, 111). These studies suggest that geranylgeranyl pyrophosphate is a poor substrate for the farnesyltransferase and visa versa. Crossover was most apparent at high isoprenyl-acceptor concentrations; this result suggests that caution should be used in interpreting results of isoprenoid analyses of overexpressed proteins (111).

Results obtained so far with both the farnesyltransferase and the geranylgeranyltransferase active on -CXXX proteins suggest that the sequences of the C-terminal tetrad contain all of the information needed for recognition. A very different picture is seen with the geranylgeranyltransferases active on

-CC-terminating (103) and -CXC-terminating (129) proteins. Here, peptides containing the C-terminal sequences of their respective substrate proteins are not recognized by the enzyme. It thus appears that specific sequences or conformations of the protein itself are required for recognition.

C-TERMINAL PROTEASES

Evidence has recently been obtained for at least three enzymatic activities in the soluble and membrane fractions of yeast *S. cerevisiae* that can remove the three terminal amino acids of farnesylated -CXXX peptide substrates (132a,b). These activities are detected using a coupled assay with the STE14 C-terminal methyltransferase, an enzyme that is active on substrates having an isoprenylated C-terminal cysteine residue (49, 82). One activity found in the soluble fraction has a native molecular weight of about 110,000 and appears to be a novel carboxypeptidase that catalyzes the processive removal of free amino acids from both farnesylated and unfarnesylated peptides (132a). This enzyme is sensitive to inhibition by *o*-phenanthroline (132a,b). A second activity is membrane bound and is relatively insensitive to inhibition by *o*-phenanthroline (132a,b). This latter activity is distinct from the membrane-bound STE14 methyltransferase described below (132a), and appears to catalyze the removal of an intact C-terminal tripeptide from farnesylated substrates (132b). Both of these activities appear to be distinct from previously described yeast proteases (133). Finally, carboxypeptidase Y itself shows activity in this assay; because this enzyme is located in the yeast vacuole, this activity would presumably not occur under normal physiological conditions (132a).

In rat liver, a membrane-bound activity has been found that also appears to catalyze the removal of the C-terminal tripeptide from farnesylated -CXXX peptides (132b), while no evidence has been found for soluble activities comparable to those found in yeast (R. C. Stephenson and S. Clarke, unpublished).

C-TERMINAL METHYLTRANSFERASES

The first evidence for an enzymatic activity that catalyzes the *S*-adenosylmethionine-dependent methyl esterification at the C-terminus of isoprenylated proteins was provided with the incorporation of radiolabelled methyl groups from [*methyl*-^{3}H]methionine into the human Ha-ras polypeptide expressed in rat embryo fibroblasts (29). The stoichiometry of methyl incorporation (about 0.7 mol methyl esters/mol ras polypeptide) as well as the relative hydrolytic stability of the ester linkage was consistent with a C-terminal localization and was inconsistent with a methylation reaction catalyzed by the only previously established type of eukaryotic protein carboxyl methyltransferase, the L-isoaspartyl/D-aspartyl protein methyltransferase (21, 134).

In vitro evidence for the C-terminal methyltransferase was first obtained by analyzing the products of isolated bovine retinal rod outer segment membranes incubated with the methyl donor S-adenosyl-[*methyl*-³H]methionine. When the [³H]methylated 23–29-kDa membrane polypeptides were oxidized with performic acid and digested with a mixture of proteases, cysteic acid [³H]methyl ester was obtained (75). The isolation of this product revealed that methylation occurs on the α-carboxyl group of a C-terminal cysteine residue. Cysteic acid [³H]methyl ester was later identified from in vitro methylated preparations of the small G-protein G25K (86), the γ-subunit of the heterotrimeric G-proteins (40, 59), and the α-subunit of the retinal cGMP phosphodiesterase (70).

The specific requirements for methylation of the C-terminus were determined using a model peptide substrate in an in vitro reaction (82). A peptide with a sequence similar to that of the C-terminus of a *Drosophila* ras protein (LARYKC) was incubated with S-adenosyl-[*methyl*-¹⁴C]methionine with subcellular fractions of rat tissues including liver microsomal membranes. Little or no methylation was seen with the latter preparation on the unmodified peptide, while the S-farnesylated or S-geranylgeranylated peptides were excellent substrates (82). The S-farnesyl-LARYKC peptide was found to inhibit the carboxyl methylation of bovine retinal rod outer segment membrane proteins, demonstrating that the same enzyme can methylate both peptides and proteins. The requirement for an isoprenylated substrate demonstrated that the methylation step followed the isoprenylation and proteolytic steps. Recent studies have shown that the bulk of the enzyme activity is localized to the endoplasmic reticulum in rat liver (134a).

The availability of an in vitro C-terminal methyltransferase assay as well as mutants of *S. cerevisiae* lacking **a**-mating factor activity made it possible to screen the mutant collection for cells that are defective in the methyltransferase activity. It was found that membrane extracts of strains containing the **a**-specific *ste14* mutation completely lacked C-terminal methyltransferase activity (135). As with the mammalian enzyme, no activity was seen in cytoplasmic extracts and no activity was seen with the membrane fraction using the non-isoprenylated peptide. The suggestion that *STE14* is the structural gene for the C-terminal isoprenyl cysteine methyltransferase (135) was confirmed by expressing the *STE14* gene product as a fusion protein in *E. coli,* an organism lacking endogenous C-terminal methyltransferase activity (49). Importantly, it has been demonstrated that the STE14 methyltransferase is responsible for the modification of not only **a**-factor, but of the RAS1 and RAS2 proteins as well (49).

The methyl-accepting substrate specificity of the mammalian membrane-bound activity has been studied in some detail. Using a rat liver microsomal enzyme and the synthetic methyl-accepting peptide LARYKC, both the C_{15} S-farnesylated and C_{20} S-geranylgeranylated peptides were found to be excellent substrates, with K_m values of 2.2 μM and 10.9 μM, respectively (82). On the other hand, the C_{10} S-geranyl peptide was much more poorly recognized (K_m = 389 μM), as were C_8, C_{10}, C_{13}, and C_{15} S-n-alkylated peptides (K_m values from 480 μM to 1760 μM). The latter results suggest that both the length and the nature of the isoprenyl group on the cysteine residue are important to its recognition (82). The fact that this peptide sequence can be recognized by a yeast enzyme normally active on polypeptides with distinct C-terminal sequences suggested that the crucial part of the recognition was the isoprenylated cysteine residue (49, 135).

This concept is supported by the recent observations that N-acetyl farnesyl cysteine is a good substrate for this enzyme in bovine brain microsomes (Refs. 136, 136a; K_m about 20 μM), bovine retinal rod outer segment membranes (Refs. 85, 137; K_m value is 23 μM), and yeast membranes (49). It was subsequently found that S-farnesylthiopropionic acid, in which a hydrogen atom replaces the CH_3CONH- group of N-acetyl farnesyl cysteine, has a K_m of 14 μM with the bovine retinal membrane enzyme (137). Modifications including oxidation of the thioether to the sulfoxide, saturation of the farnesyl group, or the addition or subtraction of a methylene group between the sulfur atom and the carboxylic acid group, all resulted in the loss of methyl-accepting activity (137). It should be realized that the enzyme source for the mammalian activity is a membrane fraction, and it is possible that there are multiple enzymes that may catalyze these reactions. On the other hand, in the yeast S. cerevisiae it is clear that the only enzyme activity present is the STE14 gene product and that it can catalyze the methylation of both farnesylated and geranylgeranylated substrates (49, 135). Taken together, these results suggest that the high-affinity recognition requirement for the membrane-bound enzyme in intact cells is simply the presence of a C_{15} or C_{20} S-isoprenoid group on any terminal cysteine residue. Thus, it is possible that the membrane-bound enzyme in yeast, and perhaps in mammalian cells as well, catalyzes methyl ester formation on not only farnesylated and geranylgeranylated products of the -CXXX family, but the geranylgeranylated products of the -CC and -CXC families as well.

Additional protein carboxyl methyltransferases active on G-proteins may also be present in mammalian cells. For example, an activity has been described in several cells and tissues that catalyzes the methyl esterification of 23-kDa membrane proteins (138). This activity is stimulated by GTP or its nonhydrolyzable analogs, suggesting that its target methyl-accepting proteins may be small G-proteins. Although some activity is present in the membrane fraction alone, the addition of cytosolic extracts increases the activity markedly (138). These results can be compared to recently obtained data on the the in vitro carboxyl methylation of membrane-bound 23-kDa rap1 proteins in human platelet membranes incubated with a cytosolic fraction from the same cells (139). In this study, no methylation was observed with the membrane fraction alone, although a similar stimulation of methylation by GTPγS was found. Significantly, this activity was reported to be completely inhibited by N-acetyl farnesyl cysteine, a competitive inhibitor of the membrane-bound C-terminal methyltransferase (49, 136, 136a). Further work is needed to determine whether the cytosolic factor is an activator of a membrane-bound methyltransferase or substrate protein, or whether it represents a novel type of methyltransferase.

FUNCTIONAL ASPECTS OF ISOPRENYLATION AND METHYLATION REACTIONS

Role of Modification in Membrane Attachment

One of the most frequently suggested functional roles of protein isoprenylation and methylation reactions follows from the fact that proteins so modified have more hydrophobic C-termini than their unmodified precursors, and thus might be "anchored" by the insertion of the C-termini into membranes. Evidence has been obtained from many systems to correlate membrane attachment with modification (see 2–4 for reviews). The isoprenoid group, however, has a branched structure, which may disrupt a membrane bilayer composed of fatty acyl chains. This suggests that protein lipidation with a fatty acyl group such as a C_{14} myristoyl or C_{16} palmitoyl (140, 141) might be a superior way to simply anchor a protein into the bilayer. On the other hand, the apparent localization of isoprenylated and methylated proteins to specific cell membranes (plasma, Golgi, nuclear, etc), as well as the cytosolic localization of some modified polypeptides, suggests that more specific interactions with membrane and/or cytosolic proteins may occur (2–4).

The ras protein system has been investigated in the most detail. Here the situation is complicated by the presence of a membrane-associating sequence of lysine residues in the K-ras proteins (142) or by palmitoylation sites at upstream cysteine residues in the Ha-ras and N-ras proteins (31). When either of these elements is removed by mutation, membrane binding is reduced (31, 142). Evidence that the isoprenylation of the cysteine residue, the removal of the terminal three amino acids, and the carboxyl methylation reaction are all important for the membrane association of a K-ras polypeptide synthesized in an in vitro translation system has been recently presented (143). Although no membrane binding was found for an unmodified form (in which a serine residue replaces the conserved cysteine), 20% was found for the isoprenylated uncleaved form, 40% for the isoprenylated and cleaved form, and 60–80% for the completely modified methylated and isoprenylated form (143). From these results, it is clear that isoprenylation and methylation reactions can assist in membrane attachment, but are not in themselves sufficient for tight membrane association, which is dependent upon either an upstream palmitoylation site or a polybasic domain (30, 142).

A similar situation is found for some of the other small G-proteins. For example, in platelets the geranylgeranylated rap1B protein can be found after cell disruption in the membrane, cytosolic, or cytoskeletal fraction, depending upon whether the cells are aggregated or thrombin-activated (144). The G25K protein can also have different degrees of membrane association. Limited tryptic digestion of in vitro methylated G25K protein reveals that the C-terminal 1-kDa fragment, later demonstrated to include a geranylgeranyl cysteine methyl ester (57), contains a membrane attachment site (86). On the other hand, a recent study suggests that this protein expressed in murine erythroleukemia cells is largely cytosolic (145). Even in this case, however, blocking isoprenoid synthesis with mevinolin increased the fraction of the protein in the cytosol (145). Finally, the rab3A protein, shown to contain two geranylgeranyl groups and a C-terminal methyl ester on its -CAC C-terminus (43, 44), is found both in the cytosol and membrane fractions of rat brain (146). It has been suggested that its interaction with the membrane depends on a specific regulatory protein and is not an intrinsic property of the protein (147). It is also possible that there are differences in the chemical modifications of the soluble and membrane-bound forms (148, 148a). Further work in each of these systems will be needed to correlate the specific C-terminal modifications with subcellular localizations. It will be especially interesting to see whether the isoprenylation or methylation reactions are reversible (see above).

In the nuclear lamin system, the -CXXX motif functions in conjunction with a basic nuclear localization signal to target these proteins to the nuclear membrane (149, 150). Certain mutated forms of human lamin A (149) or chicken lamin B2 (69), in which the conserved cysteine is replaced with another residue or in which the terminal three or four residues are deleted, can be expressed in Chinese hamster ovary (149) or mouse cells (69), but are not isoprenylated or methylated. These proteins are imported into the nucleus, but are not specifically membrane associated. On the other hand, an intact -CXXX sequence does not lead to membrane association in the presence of changes elsewhere in the polypeptide chain (149), suggesting that the C-terminal sequence is not sufficient in itself for membrane localization.

Additional examples of a correlation of membrane attachment and isoprenylation include the case of the retinal cGMP phosphodiesterase, in which mild trypsin treatment removes a 1-kDa fragment from the C-terminus and results in the loss of membrane binding (70). Expression of the β and γ subunits of the large G-proteins in simian COS cells shows that the replacement of the conserved cysteine residue blocks membrane attachment (151).

When all of the results above are taken together, the conclusion emerges that isoprenylation and methylation of a C-terminal cysteine residue on a protein are not sufficient in themselves to allow a tight association with membranes, as would be expected of other types of membrane anchors that utilize fatty acyl chains (140, 141, 152–155). If this is the case, what is the role of the isoprenyl and methyl groups? One clue can be taken from studies on the interactions of various isoprenyl-containing small molecules with membranes. There is a body of evidence suggesting that nonprotein isoprenoid groups can interact not only with the hydrophobic part of the membrane bilayer but with specific membrane proteins as well (Table 3). For instance, the C_{20} isoprenoid retinal cofactor in bacteriorhodopsin is not exposed to the membrane bilayer, but is held entirely within the α-helical strands of protein (162). Similarly, the C_{45} isoprenoid moiety of the quinone component in bacterial photosynthetic reaction centers appears from three-dimensional structure determination to interact with both the membrane protein α-helices and the bilayer lipid (160). Less is known about the membrane interactions of bacteriochlorophyll molecules (geranylgeranyl side chains), bacterial and higher plant cell chlorophyll molecules [saturated geranylgeranyl (phytyl) side chains], ubiquinone (C_{30} to C_{50} isoprenyl chains), and dolichols (C_{95} isoprenyl chains). At least at some point, each of these molecules needs to interact with specific proteins to carry out its physiological functions, although it is unclear to what extent the isoprenylated groups (as opposed to other parts of these molecules) specifically participate in these interactions.

These considerations suggest that the modified C-terminus in isoprenylated proteins may interact with specific "receptor" proteins. Such interactions may align species for efficient information transfer or may mutually modulate their activities. If the interaction of isoprenylated fungal mating factors with G-protein-linked receptors (156) is similar to that of epinephrine with β-adrenergic receptors (164), this would also suggest specific isoprenylated polypeptide-protein interactions. It is interesting to note that many of the odorant molecules that interact with large G-protein-linked receptors are isoprenoid (159), as are the insect juvenile hormones that interact with receptors controlling development (165). It is also interesting to note that the latter hormones are methyl esterified (165).

Role of Modification in Fungal Mating Factor Activity

Neither the farnesyl nor the methyl ester modifications of these pheromones seem to be absolutely essential for their biological activity, although they appear to be important recognition elements for the membrane receptor

Table 3 Protein-isoprenoid interactions

Protein	Ligand	Comments	References
G-protein linked receptors			
yeast STE3 receptor	a-factor	farnesylated peptide ligand	156
rhodopsin (eukaryotic)	retinal	C_{20} isoprenoid ligand–amino acid interactions	157
olfactory receptors	various odorants	many ligands have isoprenoid structures	158, 159
Electron transport components			
photosynthetic reaction center (bacterial)	quinone	isoprenyl tail contacts with membrane helices	160
NADH dehydrogenase (mitochondrial)	ubiquinone		161
glucose dehydrogenase (aerobic bacterial)	ubiquinone		161
Other proteins			
bacteriorhodopsin	retinal	isoprenoid contacts with membrane helices	162
glycosyltransferase (yeast)	dolichol	sequence motif for isoprenoid binding	163

proteins on cells of the opposite mating type. For example, an *S. cerevisiae* **a**-factor derivative with a methyl group replacing the farnesyl group has about one-fifth of the biological activity of the native material (21). Similar results were also obtained with nonfarnesylated, *S*-prenyl, *S*-geranyl, *S*-benzyl, and *S*-hexadecanyl **a**-factor derivatives, in which relative activities ranging from 0.1 to 50% were observed (166, 167). In *T. brasiliensis,* both the methylated and unmethylated sex factors are found in the culture medium, with a predominance of the unmethylated species, and the activity of the methylated species is about 200 times greater than that of the unmethylated species (9). Similarly, demethylated *S. cerevisiae* **a**-mating factor gives 0.5 to 6.5% of normal activity (49, 167). Interestingly, the farnesylated (but unmethylated) mating factor *A* of the yeast *R. toruloides* appears to be fully active (168).

Requirement for Mammalian Cell Cycle Control

Mevalonic acid, the biosynthetic precursor to farnesyl and geranylgeranyl pyrophosphate, is required for DNA replication in mammalian cells (reviewed in Ref. 2). Inhibitors of mevalonate formation, such as mevinolin, lead to the growth arrest and the accumulation of cells in late G1 phase; this block is reversed with mevalonic acid but not with its product cholesterol (169). These inhibitors reduce the cellular production of isoprenylated metabolites, including farnesyl pyrophosphate and geranylgeranyl pyrophosphate, the substrates for the protein isoprenyltransferases. Thus, it was suggested that the requirement for mevalonic acid is based on the cell's need for isoprenylated proteins. The complete complement of isoprenylated polypeptides are synthesized in all stages of the cell cycle in murine erythroleukemia (170) and human hepatoma HepG2 cells (170a). There are at least two obvious protein targets for such effects: the nuclear lamins A and B, and the small and large G-proteins.

Isoprenoid modification of K-ras proteins appears to be essential for the processes that lead to the transformation of mouse NIH 3T3 cells in culture, although some controversy exists on whether ras membrane binding is needed (142, 171). Nevertheless, transformation can be restored to cells expressing a Ha-ras protein that lacks the isoprenylated cysteine residue when an N-terminal myristoylation signal is added (172). Furthermore, when the conserved cysteine residue of yeast RAS proteins is changed to a serine residue, thus preventing the isoprenylation reaction, there is still some activity of the RAS proteins (173). The latter results suggest that isoprenylation may not always be essential for ras activity.

Evidence for the role of carboxyl methylation reactions in cell cycle control has been presented in a study of nuclear lamin B methylation in Chinese hamster ovary cells (28, 174). Treatment of intact cells with 3 mM 5'-methylthioadenosine inhibits methylation of lamin B proteins from 40 to 64% and prevents reformation of the nuclear membrane. Although it seems most likely that the site of methylation is at a C-terminal isoprenylated cysteine residue (29), recent results indicate that 5'-methylthioadenosine is not a potent inhibitor of rat microsomal C-terminal methyltransferase in vitro (82). On the other hand, this metabolite has been reported to completely inhibit the methyl esterification of ras proteins by a dog pancreatic microsomal preparation, presumably on the C-terminus, at 3 mM in vitro (143).

In the yeast *S. cerevisiae,* the availability of cells lacking the STE14 C-terminal methyltransferase (49, 135) makes it possible to examine the effect of the methylation reaction on protein function. These cells, unable to methylate either **a**-factor or the RAS1 and RAS2 proteins, are sterile, but are viable and appear to have a normal ras activity (49). However, studies of the biosynthesis of the RAS2 protein revealed that there was a kinetic delay in the maturation and membrane localization of this protein in cells lacking the C-terminal methyltransferase (49). At least seven proteins in *S. cerevisiae* have either been shown to be isoprenylated and methylated at the C-terminus or are candidates for these reactions, including BUD1/RSR1, a candidate small G-protein required for bud site selection in growing cells (175).

Signalling Reactions

The fact that so many of the isoprenylated proteins, including the large and small G-proteins (176–180) and the cGMP phosphodiesterase (70), are involved as second-messengers suggests that a role of these modifications is to facilitate these functions. For example, evidence has been presented that the modified C-terminus of two small G-proteins is essential for their interaction with a GDP/GTP exchange protein (181, 181a). In the visual system, isoprenoid groups are involved in multiple steps, from the interaction of retinal with opsin to the activation of farnesylated transduction to the activation of farnesylated phosphodiesterase (60, 61, 70). It has recently been shown that inhibition of farnesyl pyrophosphate and geranylgeranyl pyrophosphate synthesis by mevinolin can inhibit signalling in two addtional systems including the lipopolysaccharide-induced activation of kappa gene expression in mouse pre-B cells (181b) and the IgE-induced actin assembly and inflammatory mediator secretion in a rat mast cell model (181c). In both of these cases, the inhibitory effect occurs at an early step that can be bypassed by agents such as phorbol esters that act later in the signal transduction pathway. Interestingly, both 5'-methylthioadenosine and mevinolin inhibited membrane protein methylation in the pre-B cell system (181b).

The discovery that protein carboxyl methylation reactions are an important part of the bacterial chemotactic sensing systems (reviewed in 21, 23) led to the search for similar reactions in mammalian cell chemotaxis. It was initially reported that chemotactic formylmethionyl peptides transiently increased carboxyl methylation of rabbit neutrophil proteins (182). The irreproducibility of these effects (183), as well as the difficulty in identifying specific methyl-transferases or methyl-accepting proteins, led to a lessening of interest in this area (6, 21). However, when the finding that formylpeptide receptor action was mediated by a large G-protein (184) was coupled to the knowledge of the C-terminal methyl esterification of a γ-subunit of this protein (40), the possibility arose that transient methylation of the G-protein may reflect receptor binding and can serve to modulate the chemotactic response. Preliminary studies have indicated that treatment of mouse peritoneal macrophage cells with N-acetyl farnesyl cysteine, a cell-permeable in vitro competitive inhibitor of the C-terminal methyltransferase, inhibits their chemotaxic response (136a). This suggests that methylation may be a reversible reaction that can modulate chemotaxis in mammalian cells as well as bacterial cells.

ACKNOWLEDGMENTS

Work in the author's laboratory was supported by a grant from the National Institutes of Health (GM 26020). I wish to thank all of my colleagues for their helpful discussions and for providing unpublished materials, including Bernard Fung, Harvey Yamane, Susan Michaelis, Mike Gelb, Scott Powers, Channing Der, Matt Ashby, Christine Hrycyna, Robert Stephenson, and Hongying Xie.

Literature Cited

1. Glomset, J. A., Gelb, M. H., Farnsworth, C. C. 1990. *Trends Biochem. Sci.* 15:139–42
2. Maltese, W. A. 1990. *FASEB J.* 4:3319–28
2a. Sinensky, M., Lutz, R. J. 1992. *BioEssays* 14:25–31
3. Rine, J., Kim, S.-H. 1990. *New. Biol.* 2:219–26
4. Der, C. J., Cox, A. D. 1991. *Cancer Cells* 3:1–11
5. Gibbs, J. B. 1991. *Cell* 65:1–4
6. Barten, D. M., O'Dea, R. F. 1990. *Life Sci.* 47:181–94
7. Kamiya, Y., Sakurai, A., Tamura, S., Takahashi, N., Tsuchiya, E., et al. 1979. *Agric. Biol. Chem.* 43:363–69
8. Kamiya, Y., Sakurai, A., Tamura, S., Takahasi, N. 1979. *Agric. Biol. Chem.* 43:1049–53
9. Ishibashi, Y., Sakagami, Y., Isogai, A., Suzuki, A. 1984. *Biochemistry* 23:1399–404
10. Sakagami, Y., Isogai, A., Suzuki, A., Tamura, S., Kitada, C., et al. 1979. *Agric. Biol. Chem.* 43:2643–45
11. Sakagami, Y., Yoshida, M., Isogai, A., Suzuki, A. 1981. *Science* 212:1525–27
12. Schmidt, R. A., Schneider, C. J.,

Glomset, J. A. 1984. *J. Biol. Chem.* 259: 10175–80
13. Goldstein, J. L., Brown, M. S. 1990. *Nature* 343:425–30
14. Sinensky, M., Logel, J. 1985. *Proc. Natl. Acad. Sci. USA* 82:3257–61
15. Bruenger, E., Rilling, H. C. 1986. *Biochem. Biophys. Res. Commun.* 139:209–14
16. Powers, S., Michaelis, S., Broek, D., Santa Anna-A., S., Field, J., et al. 1986. *Cell* 47:413–22
17. Fujiyama, A., Matsumoto, K., Tamanoi, F. 1987. *EMBO J.* 6:223–28
18. Becker, J. M., Marcus, S., Kundu, B., Shenbagamurthi, P., Naider, F. 1987. *Mol. Cell. Biol.* 7:4122–24
19. Chen, Z.-Q., Ulsh, L. S., DuBois, G., Shih, T. Y. 1985. *J. Virol.* 56:607–12
20. Hurley, J. B., Fong, H. K. W., Teplow, D. B., Dreyer, W. J., Simon, M. I. 1984. *Proc. Natl. Acad. Sci. USA* 81:6948–52
21. Clarke, S. 1985. *Annu. Rev. Biochem.* 54:479–506
22. Terwilliger, T. C., Koshland, D. E. Jr. 1984. *J. Biol. Chem.* 259:7719–25
23. Stock, J. B., Lukat, G. S., Stock, A. M.

1991. *Annu. Rev. Biophys. Biophys. Chem.* 20:109–36
24. McFadden, P. N., Clarke, S. 1982. *Proc. Natl. Acad. Sci. USA* 79:2460–64
25. Murray, E. D. Jr., Clarke, S. 1984. *J. Biol. Chem.* 259:10722–32
26. Aswad, D. W. 1984. *J. Biol. Chem.* 259: 10714–21
27. Swanson, R. J., Applebury, M. L. 1983. *J. Biol. Chem.* 258:10599-605
28. Chelsky, D., Olson, J. F., Koshland, D. E. Jr. 1987. *J. Biol. Chem.* 262:4303–9
29. Clarke, S., Vogel, J. P., Deschenes, R. J., Stock, J. 1988. *Proc. Natl. Acad. Sci. USA* 85:4643–47
30. Anderegg, R. J., Betz, R., Carr, S. A., Crabb, J. W., Duntze, W. 1988. *J. Biol. Chem.* 263:18236–40
31. Hancock, J. F., Magee, A. I., Childs, J. E., Marshall, C. J. 1989. *Cell* 57:1167–77
32. Maltese, W. A., Sheridan, K. M. 1987. *J. Cell. Physiol.* 133:471–81
33. Wolda, S. L., Glomset, J. A. 1988. *J. Biol. Chem.* 263:5977–6000
34. Beck, L. A., Hosick, T. J., Sinensky, M. 1988. *J. Cell Biol.* 107:1307–16
35. Farnsworth, C. C., Gelb, M. H., Glomset, J. A. 1990. *Science* 47:320–22
36. Rilling, H. C., Breunger, E., Epstein, W. W., Crain, P. F. 1990. *Science* 247:318–20
37. Farnsworth, C. C., Wolda, S. L., Gelb, M. H., Glomset, J. A. 1989. *J. Biol. Chem.* 264:20422–29
38. Casey, P. J., Solski, P. A., Der, C. J., Buss, J. E. 1989. *Proc. Natl. Acad. Sci. USA* 86:8323–27
39. Stimmel, J. B., Deschenes, R. J., Volker, C., Stock, J., Clarke, S. 1990. *Biochemistry* 29:9651–59
40. Yamane, H. K., Farnsworth, C. C., Xie, H., Howald, W., Fung, B. K.-K., et al. 1990. *Proc. Natl. Acad. Sci. USA* 87:5868–72
41. Mumby, S. M., Casey, P. J., Gilman, A. G., Gutowski, S., Sternweis, P. C. 1990. *Proc. Natl. Acad. Sci. USA* 87:5873–77
42. Kinsella, B. T., Maltese, W. A. 1991. *J. Biol. Chem.* 266:8540–44
43. Farnsworth, C. C., Kawata, M., Yoshida, Y., Takai, Y., Gelb, M. H., Glomset, J. A. 1991. *Proc. Natl. Acad. Sci. USA* 88:6196–200
44. Khosravi-Far, R., Lutz, R. J., Cox, A. D., Conroy, L., Bourne, J. R., et al. 1991. *Proc. Natl. Acad. Sci. USA* 88:6264–68
45. Akada, R., Minomi, K., Kai, J., Yamashita, I., Miyakawa, T., et al. 1989. *Mol. Cell. Biol.* 9:3491–98
46. Lowe, P. N., Sydenham, M., Page, M. J. 1990. *Oncogene* 5:1045–48
47. Page, M. J., Aitken, A., Cooper, D. J., Magee, A. I., Lowe, P. N. 1990. *Methods: Companion Methods Enzymol.* 1:221–30
48. Gutierrez, L., Magee, A. I., Marshall, C. J., Hancock, J. F. 1989. *EMBO J.* 8:1093–98
49. Hrycyna, C., Sapperstein, S., Clarke, S., Michaelis, S. 1991. *EMBO J.* 10:1699–709
50. Deschenes, R. J., Stimmel, J. B., Clarke, S., Stock, J., Broach, J. R. 1989. *J. Biol. Chem.* 264:11865–73
51. Fujiyama, A., Tamanoi, F. 1990. *J. Biol. Chem.* 265:3362–68
51a. Fujiyama, A., Tsunasawa, S., Tamanoi, F., Sakiyama, F. 1991. *J. Biol. Chem.* 266:17926–31
52. Buss, J. E., Quilliam, L. A., Kato, K., Casey, P. J., Solski, P. A., et al. 1991. *Mol. Cell. Biol.* 11:1523–30
53. Kawata, M., Farnsworth, C. C., Yoshida, Y., Gelb, M. H., Glomset, J. A., et al. 1990. *Proc. Natl. Acad. Sci. USA* 87:8960–64
54. Winegar, D. A., Vedia, L. M., Lapetina, E. G. 1991. *J. Biol. Chem.* 266:4381–86
55. Didsbury, J. R., Uhing, R. J., Snyderman, R. 1990. *Biochem. Biophys. Res. Commun.* 171:804–12
56. Kinsella, B. T., Erdman, R. A., Maltese, W. A. 1991. *J. Biol. Chem.* 266:9786–94
57. Yamane, H. K., Farnsworth, C. C., Xie, H., Evans, T., Howald, W. N., et al. 1991. *Proc. Natl. Acad. Sci. USA* 88:286–90
58. Katayama, M., Kawata, M., Yoshida, Y., Horiuchi, H., Yamamoto, T., et al. 1991. *J. Biol. Chem.* 266:12639–45
59. Fung, B. K.-K., Yamane, H. K., Ota, I. M., Clarke, S. 1990. *FEBS Lett.* 260:313–17
60. Fukada, Y., Takao, T., Ohguro, H., Yoshizawa, T., Akino, T., et al. 1990. *Nature* 346:658–60
61. Lai, R. K., Perez-Sala, D., Canada, F. J., Rando, R. R. 1990. *Proc. Natl. Acad. Sci.* 87:7673–77
62. West, R. E. Jr., Moss, J., Vaughan, M., Liu, T., Liu, T.-Y. 1985. *J. Biol. Chem.* 260:14428–30
63. Itoh, H., Kozasa, T., Nagata, S., Nakamura, S., Katada, T., et al. 1986. *Proc. Natl. Acad. Sci. USA* 83:3778–80
64. Maltese, W. A., Robishaw, J. D. 1990. *J. Biol. Chem.* 265:18071–74
65. Sanford, J., Codina, J., Birnbaumer, L. 1991. *J. Biol. Chem.* 266:9570–79
66. Jones, T. L. Z., Spiegel, A. M. 1990. *J. Biol. Chem.* 265:19389–92

67. Finegold, A. A., Schafer, W. R., Rine, J., Whiteway, M., Tamanoi, F. 1990. *Science* 249:165–69
68. Vorburger, K., Kitten, G. T., Nigg, E. A. 1989. *EMBO J.* 8:4007–13
69. Kitten, G. T., Nigg, E. A. 1991. *J. Cell Biol.* 113:13–23
70. Ong, O. C., Ota, I. M., Clarke, S., Fung, B. K.-K. 1989. *Proc. Natl. Acad. Sci. USA* 86:9238–42
70a. Catty, P., Deterre, P. 1991. *Eur. J. Biochem.* 199:262–99
71. Baehr, W., Champagne, M. S., Lee, A. K., Pittler, S. J. 1991. *FEBS Lett.* 278:107–14
71a. Anant, J. S., Ong, O. C., Xie, H., Clarke, S., O'Brien, P. J., et al. 1992. *J. Biol. Chem.* 267:687–90
72. Kloc, M., Reddy, B., Crawford, S., Etkin, L. D. 1991. *J. Biol. Chem.* 266:8206–12
73. Xie, H., Yamane, H. K., Stephenson, R. C., Ong, O. C., Fung, B. K.-K., Clarke, S. 1990. *Methods: Companion Methods Enzymol.* 1:276–82
74. Farnsworth, C. C., Casey, P. J., Howald, W. N., Glomset, J. A., Gelb, M. H. 1990. *Methods: Companion Methods Enzymol.* 1:231–40
75. Ota, I. M., Clarke, S. 1989. *J. Biol. Chem.* 264:12879–84
76. Reese, J. H., Maltese, W. A. 1991. *Mol. Cell. Biochem.* 104:109–16
76a. Epstein, W. W., Lever, D., Leining, L. M., Bruenger, E., Rilling, H. C. 1991. *Proc. Natl. Acad. Sci. USA* 88:9668–70
77. Thelin, A., Low, P., Chojnacki, T., Dallner, G. 1991. *Eur. J. Biochem.* 195:755–61
78. Rilling, H. C., Breunger, E., Epstein, W. W., Kandutsch, A. A. 1989. *Biochem. Biophys. Res. Commun.* 163:143–48
79. Faust, J. R., Dice, J. F. 1991. *J. Biol. Chem.* 266:9961–70
80. Epstein, W. W., Lever, D. C., Rilling, H. C. 1990. *Proc. Natl. Acad. Sci. USA* 87:7352–54
81. Mbaya, B., Rigomier, D., Edorh, G. G., Karst, F., Schrevel, J. 1990. *Biochem. Biophys. Res. Commun.* 173:849–54
82. Stephenson, R. C., Clarke, S. 1990. *J. Biol. Chem.* 265:16248–54
83. Braverman, S. 1988. In *The Chemistry of Sulphones and Sulphoxides,* ed. S. Patai, Z. Rappoport, C. J. M. Stirling, pp. 717–57. New York:Wiley
84. Cashman, J. R., Olsen, L. D., Bornheim, L. M. 1990. *Chem. Res. Toxicol.* 3:344–49
85. Perez-Sala, D., Tan, E. W., Canada, F.

J., Rando, R. R. 1991. *Proc. Natl. Acad. Sci. USA* 88:3043–46
86. Yamane, H. K., Fung, B. K.-K. 1989. *J. Biol. Chem.* 264:20100–5
87. Backlund, P. S. Jr., Simonds, W. F., Spiegel, A. M. 1990. *J. Biol. Chem.* 265:15572–76
88. Halkai, R., Kloog, Y. 1990. *Biochem. Pharmacol.* 40:1365–72
88a. Kinsella, B. T., Erdman, R. A., Maltese, W. A. 1991. *Proc. Natl. Acad. Sci. USA* 88:89334–38
89. Kurihara, T., Takahashi, Y., Nishiyama, A., Kumanishi, T. 1988. *Biophys. Res. Commun.* 152:837–42
90. Zander, N. F., Meyer, H. E., Hoffmann-Posorske, E., Crabb, J. W., Heilmeyer, L. M. G. Jr., et al. 1988. *Proc. Natl. Acad. Sci. USA* 85:2929–33
91. Hjalmarsson, K., Marklund, S. L., Engström, A., Edlund, T. 1987. *Proc. Natl. Acad. Sci. USA* 84:6340–44
92. Ichii, Y., Fukunaga, R., Shiojiri, S., Sokawa, Y. 1986. *Nucleic Acids Res.* 14:10117
93. Benech, P., Mory, Y., Revel, M., Chebath, J. 1985. *EMBO J.* 4:2249-56
94. Shipp, M. A., Richardson, N. E., Sayre, P. H., Brown, N. R., Masteller, E. L., et al. 1988. *Proc. Natl. Acad. Sci. USA* 85:4819–23
95. Kumar, N. M., Gilula, N. B. 1986. *J. Cell Biol.* 103:767–76
95a. Yuen, P. S. T., Potter, L. R., Garbers, D. L. 1990. *Biochemistry* 29:10872–78
96. Balch, W. E. 1990. *Trends Biochem. Sci.* 15:473–77
97. Downward, J. 1990. *Trends Biochem. Sci.* 15:469–572
98. Hall, A. 1991. *Science* 249:635–40
99. Beck, L. A., Hosick, T. J., Sinensky, M. 1990. *J. Cell Biol.* 110:1489–99
100. Weber, K., Plessmann, U., Traub, P. 1989. *FEBS Lett.* 257:411–14
101. Walworth, N. C., Goud, B., Kabcenell, A. K., Novick, P. J. 1989. *EMBO J.* 8:1685–93
102. Baker, D., Wuestehube, L., Schekman, R., Botstein, D., Segev, N. 1990. *Proc. Natl. Acad. Sci. USA* 87:355–59
103. Moores, S. L., Schaber, M. D., Mosser, S. D., Rands, E., O'Hara, M. B., et al. 1991. *J. Biol. Chem.* 266:14603–10
104. Molenaar, C. M. T., Prange, R., Gallwitz, D. 1988. *EMBO J.* 7:971–76
105. Chavrier, P., Vingron, M., Sander, C., Simons, K., Zerial, M. 1990. *Mol. Cell. Biol.* 10:6578–85
106. Zahraoui, A., Touchot, N., Chardin, P., Tavitian, A. 1989. *J. Biol. Chem.* 264:12394–401

107. Reiss, Y., Goldstein, J. L., Seabra, M. C., Casey, P. J., Brown, M. S. 1990. *Cell* 62:81–88

108. Reiss, Y., Seabra, M. C., Goldstein, J. L., Brown, M. S. 1990. *Methods: Companion Methods Enzymol.* 1:241–45

109. Manne, V., Roberts, D., Tobin, A., O'Rourke, E., DeVirgilio, M., et al. 1990. *Proc. Natl. Acad. Sci. USA* 87: 7541–45

110. Schaber, M. S., O'Hara, M. B., Farsky, V. M., Mosser, S. D., Bergstrom, J. D., et al. 1990. *J. Biol. Chem.* 265: 14701–4

111. Yokoyama, K., Goodwin, G. W., Ghomashchi, F., Glomset, J. A., Gelb, M. H. 1991. *Proc. Natl. Acad. Sci. USA* 88:5302–6

112. Kim, R., Rine, J., Kim, S.-H. 1990. *Mol. Cell. Biol.* 10:5945–49

113. Seabra, M. C., Reiss, Y., Casey, P. J., Brown, M. S., Goldstein, J. L. 1991. *Cell* 65:429–34

114. Reiss, Y., Seabra, M. C., Armstrong, S. A., Slaughter, C. A., Goldstein, J. L., et al. 1991. *J. Biol. Chem.* 266: 10672–77

115. Chen, W.-J., Andres, D. A., Goldstein, J. L., Russell, D. W., Brown, M. S. 1991. *Cell* 66:327–34

116. Tamanoi, F., Hsueh, E. C., Goodman, L. E., Cobitz, A. R., Detrick, R. J., et al. 1988. *J. Cell. Biochem.* 36:261–73

117. Goodman, L. E., Perou, C. M., Fujiyama, A., Tamanoi, F. 1988. *Yeast* 4:271–81

118. Schafer, W. R., Trueblood, C. E., Yang, C.-C., Mayer, M. P., Rosenberg, S., et al. 1990. *Science* 249:1133–39

119. Goodman, L. E., Judd, S. R., Farnsworth, C. C., Powers, S., Gelb, M. H., et al. 1990. *Proc. Natl. Acad. Sci. USA* 87:9665–69

120. He, B., Chen, P., Chen, S.-Y., Vancura, K. L., Michaelis, S., et al. 1991. *Proc. Natl. Acad. Sci. USA.* 88:11373–77

121. Kohl, N. E., Diehl, R. E., Schaber, M. D., Rands, E., Soderman, D. D., et al. 1991. *J. Biol. Chem.* 266: 18884–88

122. Reiss, Y., Stradley, S. J., Gierasch, L. M., Brown, M. S., Goldstein, J. L. 1991. *Proc. Natl. Acad. Sci. USA* 88:732–36

123. Goldstein, J. L., Brown, M. S., Stradley, S. J., Reiss, Y., Gierasch, L. M. 1991. *J. Biol. Chem.* 266:15575–78

124. Chen, W.-J., Andres, D. A., Goldstein, J. L., Brown, M. S. 1991. *Proc. Natl. Acad. Sci.* USA 88:11368–72

125. Yoshida, Y., Kawata, M., Katayama, M., Horiuchi, H., Kita, Y., et al. 1991. *Biochem. Biophys. Res. Commun.* 175: 720–28

126. Joly, A., Popjak, G., Edwards, P. A. 1991. *J. Biol. Chem.* 266:13495–98

127. Finegold, A. A., Johnson, D. I., Farnsworth, C. C., Gelb, M. H., Judd, S. R., et al. 1991. *Proc. Natl. Acad. Sci. USA* 88:4448–52

128. Casey, P. J., Thissen, J. A., Moomaw, J. F. 1991. *Proc. Natl. Acad. Sci. USA* 88:8631–35

129. Horiuchi, H., Kawata, M., Katayama, M., Yoshida, Y., Musha, T., et al. 1991. *J. Biol. Chem.* 266:16981–84

130. Johnson, D. I., O'Brien, J. M., Jacobs, C. W. 1991. *Gene* 98:149–50

131. Ohya, Y., Goebl, M., Goodman, L. E., Petersen-Bjorn, S., Friesen, J. D., et al. 1991. *J. Biol. Chem.* 266:12356–60

132. Rossi, G., Jiang, Y., Newman, A. P., Ferro-Novick, S. 1991. *Nature* 351: 158–61

132a. Hrycyna, C. A., Clarke, S. 1992. *J. Biol. Chem.* 267:In press

132b. Ashby, M. N., King, D. S., Rine, J. 1992. *Proc. Natl. Acad. Sci. USA.* 89:In press

133. Jones, E. W. 1991. *J. Biol. Chem.* 266:7963–66

134. Stephenson, R. C., Clarke, S. 1989. *J. Biol. Chem.* 264:6164–70

134a. Stephenson, R. C., Clarke, S. 1991. *FASEB J.* 5:1181

135. Hrycyna, C., Clarke, S. 1990. *Mol. Cell. Biol.* 10:5071–76

136. Volker, C., Miller, R. A., Stock, J. B. 1990. *Methods: Companion Methods Enzymol.* 1:283–87

136a. Volker, C., Miller, R. A., McCleary, W. R., Rao, A., Poenie, M., et al. 1991. *J. Biol. Chem.* 266:21515–22

137. Tan, E. W., Perez-Sala, D., Canada, F. J., Rando, R. R. 1991. *J. Biol. Chem.* 266:10719–22

138. Backlund, P. S. Jr., Aksamit, R. R. 1988. *J. Biol. Chem.* 263:15864-67

139. Huzoor-Akbar, Winegar, D. A., Lapetina, E. G. 1991. *J. Biol. Chem.* 266:4387–91

140. Schmidt, M. F. G. 1989. *Biochim. Biophys. Acta* 988:411–26

141. Grand, R. J. A. 1989. *Biochem. J.* 258:625–38

142. Hancock, J. F., Paterson, H., Marshall, C. J. 1990. *Cell* 63:133–39

143. Hancock, J. F., Cadwallader, K., Marshall, C. J. 1991. *EMBO J.* 10:641–46

144. Fisher, T. H., Gatling, M. N., Lacal, J.-C., White, G. C. II. 1990. *J. Biol. Chem.* 265:19405–8

145. Maltese, W. A., Sheridan, K. M. 1990. *J. Biol. Chem.* 265:17883–90

146. Burstein, E., Macara, I. G. 1989. *Mol. Cell. Biol.* 9:4807–11
147. Araki, S., Kikuchi, A., Hata, Y., Isomura, M., Takai, Y. 1990. *J. Biol. Chem.* 265:13007–15
148. von Mollard, G. F., Mignery, G. A., Baumert, M., Perin, M. S., Hanson, T. J., et al. 1990. *Proc. Natl. Acad. Sci. USA* 87:1988–92
148a. Johnston, P. A., Archer, B. T. III, Robinson, K., Mignery, G. A., Jahn, R., et al. 1991. *Neuron* 7:101–9
149. Holtz, D., Tanaka, R. A., Hartwig, J., McKeon, F. 1989. *Cell* 59:969–77
150. Krohne, G., Waizenegger, I., Hoger, T. H. 1989. *J. Cell Biol.* 109:2003–11
151. Simonds, W. F., Butrynski, J. E., Gautam, N., Unson, C. G., Spiegel, A. M. 1991. *J. Biol. Chem.* 266:5363–66
152. Gordon, J. L., Duronio, R. J., Rudnick, D. A., Adams, S. P., Gokel, G. W. 1991. *J. Biol. Chem.* 266:8647–50
153. Olson, E. N. 1988. *Prog. Lipid. Res.* 27:177–97
154. Low, M. G. 1989. *FASEB J.* 3:1600–8
155. Sefton, B. M., Buss, J. E. 1987. *J. Cell Biol.* 104:1449–53
156. Hagen, D. C., McCaffrey, G., Sprague, G. F. Jr. 1986. *Proc. Natl. Acad. Sci. USA* 83:1418–22
157. Nakayama, T. A., Khorana, H. G. 1991. *J. Biol. Chem.* 266:4269–75
158. Buck, L., Axel, R. 1991. *Cell* 65:175–87
159. Sklar, P. B., Anholt, R. R. H., Snyder, S. H. 1986. *J. Biol. Chem.* 261:15538–43
160. Allen, J. P., Feher, G., Yeates, T. O., Komiya, H., Rees, D. C. 1987. *Proc. Natl. Acad. Sci. USA* 84:5730–34
161. Friedrich, T., Strohdeicher, M., Hofhaus, G., Preis, D., Sahm, H., et al. 1990. *FEBS Lett.* 265:37–40
162. Henderson, R., Baldwin, J. M., Ceska, T. A., Zemlin, F., Beckmann, E., et al. 1990. *J. Mol. Biol.* 213:899–929
163. Albright, C. F., Orlean, P., Robbins, P. W. 1989. *Proc. Natl. Acad. Sci. USA* 86:7366–69
164. Strader, C. D., Sigal, I. S., Candelore, M. R., Rands, E., Hill, W. S., et al. 1988. *J. Biol. Chem.* 263:10267–71
165. Gilbert, L. I., Bollenbacher, W. E., Granger, N. A. 1980. *Annu. Rev. Physiol.* 42:493–510
166. Ewenson, A., Marcus, S., Becker, J. M., Naider, F. 1990. *Int. J. Peptide Protein Res.* 35:241–48
167. Marcus, S., Caldwell, G. A., Miller, D., Xue, C.-B., Naider, F., et al. 1991. *Mol. Cell. Biol.* 11:3603–12
168. Miyakawa, T., Tachikawa, T., Jeong, Y. K., Tsuchiya, E., Fukui, S. 1987. *Biochem. Biophys. Res. Commun.* 143:893–900
169. Quesney-Huneeus, V., Wiley, M. H., Siperstein, M. D. 1979. *Proc. Natl. Acad. Sci. USA* 76:5056–60
170. Maltese, W. A., Sheridan, K. M. 1988. *J. Biol. Chem.* 263:10104–10
170a. Sepp-Lorenzino, L., Rao, S., Coleman, P. S. 1991. *Eur. J. Biochem.* 200:579–90
171. Jackson, J. H., Cochrane, C. G., Bourne, J. R., Solski, P. A., Buss, J. E., et al. 1990. *Proc. Natl. Acad. Sci. USA* 87:3042–46
172. Buss, J. E., Solski, P. A., Schaeffer, J. P., MacDonald, M. J., Der, C. J. 1989. *Science* 243:1600–3
173. Deschenes, R. J., Broach, J. R. 1987. *Mol. Cell. Biol.* 7:2344–51
174. Chelsky, D., Sobotka, C., O'Neill, C. L. 1989. *J. Biol. Chem.* 264:7637–43
175. Chant, J., Herskowitz, I. 1991. *Cell* 65:1203–12
176. Taylor, C. W. 1990. *Biochem. J.* 272:1–13
177. Lochrie, M. A., Simon, M. I. 1988. *Biochemistry* 27:4957–65
178. Freissmuth, M., Casey, P. J., Gilman, A. G. 1989. *FASEB J.* 3:2125-31
179. Neer, E. J., Clapham, D. E. 1988. *Nature* 333:129–34
180. Bourne, H. R., Sanders, D. A., McCormick, F. 1991. *Nature* 349:117–27
181. Mizuno, T., Kaibuchi, K., Yamamoto, T., Kawamura, M., Sakoda, T., et al. 1991. *Proc. Natl. Acad. Sci. USA* 88:6442–46
181a. Shirataki, H., Kaibuchi, K., Hiroyoshi, M., Isomura, M., Araki, S., et al. 1991. *J. Biol. Chem.* 266:20672–77
181b. Law, R. E., Stimmel, J. B., Damore, M. A., Carter, C., Clarke, S., et al. 1992. *Mol. Cell. Biol.* 12:103–11
181c. Deanin, G. G., Cutts, J. L., Pfeiffer, J. R., Oliver, J. M. 1991. *J. Immunol.* 146:3528–35
182. O'Dea, R. F., Viveros, O. H., Axelrod, J., Aswanikumar, S., Schiffmann, E., et al. 1978. *Nature* 272:462–64
183. Venkatasubramanian, K., Hirata, F., Gagnon, C., Corcoran, B. A., O'Dea, R. F., et al. 1980. *Mol. Immunol.* 17:201–7
184. Polakis, P. G., Uhing, R. J., Snyderman, R. 1988. *J. Biol. Chem.* 263:4969–76

Annu. Rev. Biochem. 1992. 61:387–418

CONSTRAINED PEPTIDES: MODELS OF BIOACTIVE PEPTIDES AND PROTEIN SUBSTRUCTURES

Josep Rizo and Lila M. Gierasch

Department of Pharmacology, University of Texas Southwestern Medical Center, 5323 Harry Hines Boulevard, Dallas, Texas 75235-9041

KEY WORDS: peptide conformation, protein folding, cyclic peptides, nuclear magnetic resonance, molecular dynamics

CONTENTS

PERSPECTIVES AND SUMMARY

The great diversity of biological roles played by peptides and proteins is correlated with the immense number of possibilities that exist for their

387

0066-4154/92/0701-0387$02.00

primary sequences and three-dimensional structures. Hence, knowledge of the active conformation of a given polypeptide is a major step towards understanding its biological function. Although a wealth of biophysical techniques exist to study polypeptide conformation, a major obstacle in the study of small peptides is their intrinsic flexibility. In solution, small linear peptides are usually visiting a large number of structures that are almost equivalent energetically and, although slightly preferred conformations may be more populated in some cases, they are generally highly dependent on the environment. Paradoxically, despite the fact that the number of possible conformations increases enormously with molecular size, proteins are much more prone to adopt well-defined structures, and much information on the influence of amino acid sequence on polypeptide conformation has been obtained from the statistical analysis of known protein structures (see, for instance, Refs. 1–3). The basis for this behavior is twofold. First, the primary driving force for adoption of a compact conformation in a protein, which is generally accepted to be the sequestration of hydrophobic surface area, is expected to be less effective in directing peptide conformational preferences. Second, the intramolecular interactions that maintain a polypeptide chain in a specific folded structure are weak, and only a large number of such interactions can compensate for the conformational entropy lost when the polypeptide adopts a single major conformation.

The number of conformational possibilities in peptides and proteins can be reduced by introducing constraints. Indeed, nature uses several such constraints to reduce the flexibility of polypeptide chains, including the incorporation of cyclic amino acid residues (proline) and the formation of macrocycles through disulfide bonds. In these ways, particular conformations of peptides and proteins are stabilized by reducing the entropy cost upon folding into such conformations. Using the power of synthetic organic chemistry, peptide chemists have learned to constrain the conformational degrees of freedom in polypeptides with a wide variety of procedures, and the number of reported biochemical and biophysical studies based on constrained peptides has increased exponentially during the past decade. Correspondingly, the variety of applications of these studies has also expanded. Model constrained peptides have been used to study in detail the conformational preferences of natural and modified amino acid residues, affording new insight into their tendencies to adopt particular secondary structures. Constrained analogs of biologically active peptides have been designed and synthesized to induce specific structural features and to define the conformational requirements for binding to the corresponding receptor(s), leading to the development of potent agonists and antagonists. Constraints on peptides corresponding to specific protein sequences have been introduced to study the intrinsic conformational properties of such sequences and to elucidate the mechanisms of protein folding. Recently, constrained peptides have also been used in the synthesis of de novo designed proteins. One can expect that in the near future these and new applications will be in growing development, and that the same concepts will be applied to obtain constrained proteins.

Scope of the Review

It would be impossible to cover in this review all the literature on constrained peptides. We focus on the conformational aspects, and pay particular attention to the most recent developments. First, we discuss some general considerations about the methods used for the conformational analysis of constrained peptides, and give an overview of the different types of constraints that have been reported, with their structural consequences. Next, we illustrate how the constraint concept can be applied in the rational design of agonists and antagonists of biologically active peptides. Finally, the use of constrained peptide models to understand the rules that govern protein folding is treated.

A number of reviews on the topics addressed here and related subjects have appeared in recent years. Methods to define solution conformations for small linear peptides have been reviewed by Dyson & Wright (4). Toniolo has reviewed the use of short-range cyclizations to obtain conformationally restricted peptides (5). Spatola reviewed in 1983 the introduction of chemical modifications in the peptide chain (6), and Aubry & Marraud have recently summarized the conformational implications of some of these modifications (7). The structural and functional roles of turns in peptides and proteins were comprehensively reviewed by Rose, Gierasch, & Smith in 1985 (8). We have recently reviewed the use of cyclic pentapeptides (9) and other constrained peptides as models for reverse turns (10). The review on conformational constraints in biologically active peptides by Hruby in 1982 is a classic in which the principles underlying this strategy in drug design are clearly explained (11). More recent reviews on the design of peptide drugs, with particular emphasis on the use of conformational constraints, have been authored by Hruby et al (12–15), Marshall & Motoc (16), Kessler (17), Schiller (18), and Fauchère (19). The design of peptidase inhibitors has been reviewed by Rich (20). Methods for the prediction of the receptor-bound conformations of small peptides have been reviewed by Milner-White (21). Current methodologies for the design of peptides and proteins have been reviewed by DeGrado (22), and the use of templates to assemble proteins of de novo design has been reviewed by Mutter & Vuilleumier (23). Discussions on the synthetic methods used to obtain constrained peptides and on the biological assays used to test analogs of natural peptides can be found in many of the reviews cited above. A whole volume of the series *The Peptides* has been dedicated to the application of physical methods to the study of peptide conformation (24). Good perspectives on the problem of protein folding can be obtained from the review by Kim & Baldwin (25) and the book edited by King & Gierasch (26).

Conformational Analysis of Constrained Peptides

A variety of physical techniques have been used in the conformational analysis of peptides (24). In crystals, X-ray diffraction can yield very accurate three-dimensional structures (27), but complementary information can also be obtained by solid state nuclear magnetic resonance (NMR) (28). Methods used to study polypeptide conformation in solution include infrared (29), Raman (30), and fluorescence (31) spectroscopies, but those in most widespread use are circular dichroism (32) and NMR. NMR offers very detailed structural information and, indeed, the increase in the number of studies on peptide conformation reported in the literature in recent years has been associated with the explosion of multidimensional NMR techniques (33–35) and computational methods such as distance geometry (36), molecular dynamics (37, 38), simulated annealing (39), and variable target function algorithms (40). Most of these methods are used in combination with NMR data to obtain three-dimensional structures of target compounds, but techniques based on computation of the molecular energy (e.g. molecular dynamics) have also been extensively used in conformational studies with no experimental input. The application of all these techniques to study peptide conformation is well documented (24, 27–40), and here we want to emphasize considerations that have received less attention in the literature and are particularly important in the analysis of constrained peptides and in analog design.

Constrained peptides are designed and studied with the hope that they will adopt well-defined conformations. Although the use of constraints can highly reduce the number of conformational degrees of freedom in a peptide, a number of motions are still likely to exist. These motions may include librations around peptide bonds, rotations of side chains, and also more global conformational rearrangements. The time scale of most NMR measurements is in general slower than that of these motions and, consequently, some of the parameters measured may correspond to averaged values inconsistent with reasonable structures. In many studies, a set of torsional and interproton distance restraints deduced from NMR data is used in combination with distance geometry or molecular dynamics to obtain structures "compatible" with the experimental data, and the flexibility of the molecule is assessed from the degree of similarity between these structures. This approach is very powerful, but can lead to unrealistic structures unless the restraints used are carefully evaluated. Fortunately, there is an increasing number of reports in which the possibility of conformational averaging in fairly constrained peptides is considered explicitly (9, 10, 41–45). In addition, Ernst and coworkers have recently published a search algorithm that allows the systematic interpretation of NMR data in terms of multiple conformations (46), and van Gunsteren and coworkers have developed a molecular dynamics approach with time-averaged NMR restraints (47). These techniques will certainly aid in the assessment of the flexibility of the target compounds in given conditions, and further information can be obtained by performing experimental studies in different environments (e.g. using different solvents and comparing with the crystal structure if available).

Molecular dynamics is particularly useful to find fast motions that could account for averaged measurements and to detect restraints that produce excessive strain energy (45). However, an evaluation of the energy terms that contribute to the strain is indispensable. As an example, in a recent study of a cyclic decapeptide antagonist of gonadotropin-releasing hormone, we found that structures with NMR restraints had energies more than 30 kcal/mol above that of the lowest energy conformation found in vacuo without restraints. However, the low energy of the unrestrained minimum was in large part due to favorable nonbonding interactions involving aromatic side chains, which were lost when these side chains were forced onto the preferred conformers observed by NMR (45). Parallel results have been obtained by Kessler and coworkers (48). These observations have important and general consequences: wherever a molecule has some flexibility, folded conformations that maximize nonbonding interactions will be highly favored in vacuo. The large magnitude of such interactions has not been sufficiently recognized in many in vacuo modeling studies carried out with no experimental restraints and using energy thresholds in the selection of acceptable conformations. With the power of the present generation of computers, it is possible to include solvent in the calculations to obtain better energetic analyses and to refine the structures obtained in vacuo (45, 49). In the near future, screening of experimental restraints, systematic energy breakdown, and analysis of solvated systems will be in more general use in conformational studies of peptides.

TYPES OF CONSTRAINTS

Reverse turns, α helices, and β sheets are the three main classes of regular secondary structure in peptides and proteins that one may want to introduce into a polypeptide chain through the use of constraints. Relatively few examples of constraints used to elicit α-helix or β-sheet conformations in peptides have been described, and, as we will see through the examples given in this review, a large proportion of the constraints that have been reported were designed to force or stabilize reverse-turn conformations. This is due, in part, to the fact that a minimum number of residues is usually necessary to form stable helix or sheet conformations and, hence, turns are more amenable to be induced in small model peptides. Most importantly, a good part of the research on constrained peptides has been devoted to the development of analogs of biologically active peptides, and reverse turns are ideal sites for receptor recognition because they present side chains in a highly accessible arrangement around a compact folding of the peptide backbone (8). On the other hand, reverse turns are also important structural features in proteins, accounting for 25–30% of the residues in proteins of known structure (50, 51), and model constrained peptides containing turns have been widely used to study the rules that govern protein folding (10). The most common way to introduce constraints into polypeptides is through permanent chemical altera-tions involving covalent bonds, but some examples of noncovalent constraints such as metal ion binding have been described. According to the number of degrees of freedom directly affected, the constraints can be classified into two broad categories: local constraints and global constraints.

Local Constraints

A good number of chemical modifications of the peptide backbone or the side chains have been introduced to modulate local conformational properties of peptides (6), including peptide bond surrogates and cyclic as well as sterically constrained amino acid residues. Perhaps the simplest way to modify the local stereochemical properties of the peptide backbone is to introduce D-amino acid residues. Pioneering work by Venkatachalam (52) predicted the turn preferences of dipeptide sequences with different chiralities, and his con-clusions have been largely supported experimentally with model protected dipeptides and cyclic peptides (10). Compared to homochiral sequences, heterochiral (LD or DL) dyads have an increased tendency to form β turns, which can be correlated with the frequent appearance of Gly residues in protein β turns (51). These tendencies have been extensively applied in the design of peptide analogs and protein folding models, introducing D-amino acids (often replacing Gly) to stabilize β-turn conformations (see below).

The following groups are among the peptide bond surrogates that have been reported: CO-O (depsi) (53), CS-NH (thiated) (54), NH-CO (retro) (55), CH_2-NH (reduced) (56), CH_2-S (57), CH_2-SO (57), CO-CH_2 (58), CH=CH (59), and CH_2-CH_2 (60). Rather than acting as real constraints, these amide bond substitutes are modifications that alter to a greater or lesser extent the conformational behavior of the peptide backbone. In some cases, they in-crease its flexibility. Hence, peptide bond surrogates are ideal tools in drug design to introduce subtle changes, and sometimes the necessary flexibility, to fine-tune the binding properties of natural peptide analogs. In addition, these modifications can be used to improve enzymatic stability and to modulate solubility and in vivo transport characteristics of the analogs. Several detailed studies on the conformational consequences of various amide bond surrogates, with particular attention to their possible participation in β- and γ-turn structures, have been reported recently (7, 61–73). The tetrazoyl ring system with 1,5 substitutions has been proposed as a mimic of a *cis* peptide bond (74), and its geometry has been studied by X-ray diffraction (69, 75, 76).

The most common steric constraints used to narrow down the conformational space available to a polypeptide chain are alkylations in the N and Cα positions. N- and Cα-methyl substitutions were already studied in the early 1970s by Marshall, and used to obtain active analogs of angiotensin II (77). N-methylation restricts the conformations of both the residue bearing the modification and the preceding residue, and studies carried out on model dipeptides by Marraud and coworkers have shown the tendency of N-methylated residues to participate in different turn conformations, depending on the chirality of the sequence and the *trans* or *cis* conformation of the tertiary amide bond (7). A large decrease in the accessible conformational space, restricted to the right-handed and left-handed helical regions, occurs in the Cα-methylated derivative of alanine, α-aminoisobutyric acid (Aib) (77). Hence, the Aib residue has been extensively used to promote helical conformations (78–82), but it can also be accommodated in β-turns (8). Interestingly, Cα,Cα-diethylglycine residues adopt maximally extended conformations (with ϕ and ψ near 180°) (83). This shows the versatility of Cα-dialkylated amino acid residues to control conformational preferences, and there has recently been increased interest in such residues, in particular those in which the two Cα substituents are linked by cyclization (14, 62, 69, 83–86). Several residues with steric constraints in the side chains have been used in analog design (14), including penicillamine (β,β-dimethylcysteine) as a frequently used example (87). An alternative way to constrain the backbone and the side chain orientation at the same time is the utilization of α,β-unsaturated amino acids, which can be incorporated into reverse turns (7, 22, 88, 89).

Short-range cyclizations offer the possibility to establish well-defined constraints on particular backbone or side chain torsion angles. The amino acid proline represents a natural example of this type of constraint. The cyclic structure constrains the ϕ dihedral angle to values near $-60°$, while, as an N-alkylated amino acid, proline forces the preceding residue to adopt an extended conformation. These characteristics result in a high tendency for proline residues to participate in reverse-turn structures, and many of the model peptides that have been used to study reverse turns contain proline (10). A classic example of a synthetic constraint containing a small cycle is the S-γ-lactam introduced by Freidinger et al to stabilize a β-turn conformation in gonadotropin-releasing hormone (90). The S-γ-lactam links the $C\alpha$ of one residue to the N of the next residue, which constrains the ψ torsion angle of the first residue to about $-120°$ and, at the same time, imposes steric constraints in ϕ of the second residue (see Ref. 91 for a detailed conformational analysis). A large number of other short-range cyclic constraints, including some bicyclic structures, has been described, and a comprehensive review on such constraints has been presented recently by Toniolo (5). Cyclic constraints that force well-defined side-chain orientations have received increased interest recently (92–96).

Global Constraints

The most general way to introduce a global constraint into a peptide chain and to affect drastically its overall conformation is the formation of a covalent bond between distant parts in the sequence. These constraints affect all the degrees of freedom within the cycle formed, which should thus adopt a more defined conformation than in the parent linear form. In addition, the increased rigidity of the cycle can induce preferred conformations in the linear parts of the molecule. Cyclization is commonly performed by forming a link between the two backbone termini, between two side chains, or between one of the termini and a side chain. Alternatively, backbone atoms other than the termini can be linked or cyclic dimeric structures can be obtained by introducing two bonds between different peptide chains. Formation of amide or disulfide bonds is the simplest method to introduce cyclic constraints. In general, changes in the overall direction of the peptide chain are necessarily introduced by cyclic constraints and, hence, cyclic peptides are ideal models to study reverse-turn preferences in different amino acid residues and interactions that stabilize turns in peptides and proteins.

The literature on cyclic peptide models for reverse turns is abundant and has been extensively reviewed (8–10). In the next section we discuss several cases in which a cyclic constraint was introduced to stabilize turns in biologically active peptides, and in the last section we give some examples of cyclic peptides that were used as models for protein β turns. Some examples in which cyclization was used to stabilize α-helix and β-sheet conformations are also presented in the next sections. Cyclic peptides have also been used as models for antibody recognition (97) and for ion transport (98, 99), and polycyclic peptides have been proposed as novel cavitands (100–102). Finally, it is important to note that a new dimension in research with constrained peptides has been recently added with the increasing number of reports describing preformed peptide, pseudopeptide, and nonpeptide structures that can be included in polypeptide chains to mimic or induce particular conformational features. Much of the work in this area has been devoted to the development of structures that can act as β-turn mimics (see Ref. 103 for a review), but templates to induce α-helix or β-sheet conformations, as well as to orient peptide chains in a predetermined arrangement in de novo designed proteins, are receiving increased interest (see last section).

BIOLOGICALLY ACTIVE PEPTIDES

Many small peptides are key factors in the regulation of a wide variety of biological functions, acting as hormones, neurotransmitters, or inhibitors. Major efforts have thus been dedicated to the development of agonists or antagonists of these peptides that could be used as drugs with high specificity and low toxicity. A general problem in the rational design of analogs based on structural grounds is that many of the receptors for these peptides have not been characterized. Although the number of known receptor sequences will increase rapidly in the near future through application of molecular biology techniques, determination of their three-dimensional structures will still lag behind, in particular because most of them are membrane-bound proteins. Much of the research in the design of peptide drugs has been based on the synthesis of a large number of analogs with different substitutions and deletions, to determine the steric and hydrophobicity/hydrophilicity characteristics, as well as the minimum number of residues, necessary for biological activity. Although this strategy can be very informative, it is often difficult to rationalize the results. The different activities observed can be due to changes in the conformational properties of the analogs, rather than to the intended changes in polarity or bulkiness. On the other hand, different conformations may present equivalent side-chain orientations, and different binding modes may also exist. Moreover, the conformational analysis of the analogs and the native peptide is usually hampered by their high flexibility. The synthesis and study of constrained analogs has thus emerged as a very powerful approach in peptide drug design.

Ideally, a constraint that locks-in the active conformation of a peptide should result in an increased affinity, as there is little conformational entropy loss upon binding to the receptor. Furthermore, many biologically active peptides interact with multiple receptors, and constraining particular structural features can lead to drugs with high receptor specificity and free from undesired side effects. In addition, constrained peptides can offer increased enzymatic stability and improved transport properties. Possible structural features that may be important for the biological activity of a given peptide can be proposed from conformational analysis of the peptide, from results obtained with deletion-substitution studies, and from known conformational propensities of residues in the sequence. An iterative process can thus be started, using constraints to test and refine structural hypotheses. The conformational and topological requirements for agonist or antagonist activity, and for specificity when diverse receptors are involved, can be analyzed by modifying the constraints and using various biological assays.

Although the final evaluation of a peptide analog needs to be based on in vivo biological assays, it is particularly important in the rationalization of the design process to test binding affinities with sensitive in vitro assays, free from proteolysis interference. A potential drawback of the use of constraints is that some flexibility may be necessary for transduction of the biological response. In these cases, constrained peptides are more likely to behave as antagonists, rather than agonists, of the native peptide, but analysis of their conformation and flexibility may still yield information on requirements for agonist activity. In conformational studies of constrained peptides, it is also important to analyze their flexibility to assess if the structures observed are likely to be relevant for biological activity.

The caution necessary in the evaluation of a conformational model has been well illustrated recently with the discovery that cyclosporin A, an immunosuppressive cyclic undecapeptide, adopts similar conformations in solution and in crystals, with a *cis* peptide bond between residues 9 and 10 (104), yet it binds to cyclophilin with this peptide bond in a *trans* conformation (105). In the absence of direct information on the active conformation of a given peptide (which is the usual situation in peptide drug design), the biological relevance of a conformational model for a given peptide can be supported if the introduction of a constraint uniquely compatible with this model leads to an active analog.

The literature on constrained analogs of naturally occurring peptides has been exhaustively reviewed, and the principles outlined above have been extensively discussed (11–22). Many of the studies reported in the literature still rely on ill-defined structural hypotheses and do not include detailed conformational analyses of the analogs described. This tendency is mainly due to the interdisciplinary character of this research area, which involves the use of organic chemistry methods to synthesize the desired analogs, biological assays to test them, and physical methods to perform their conformational analysis. The use of constraints in analog design is most powerful when the design process is rationalized in terms of well-defined conformational and topological hypotheses. Here we try to illustrate different aspects of this heuristic process with some examples in which design based on conformational grounds has led to potent constrained analogs of diverse natural peptides. With the examples chosen, we would also like to show how the complexity of these studies increases when they involve several peptides with overlapping activities and multiple receptors that bind the same or related peptides. The reviews cited above include many other apt examples of successful design of biologically active constrained peptides. Excellent reviews with particular focus on the enkephalins or on somatostatin, the peptides that have been most extensively studied using this approach, have appeared recently (17, 18, 22, 106–108).

Gonadotropin-Releasing Hormone

Mammalian gonadotropin-releasing hormone (GnRH) is a hypothalamic decapeptide, pGlu1-His2-Trp3-Ser4-Tyr5-Gly6-Leu7-Arg8-Pro9-Gly10-NH$_2$, that acts in the pituitary gland to stimulate the release of luteinizing hormone and follicle-stimulating hormone, which in turn regulate ovulation and spermatogenesis in the gonads (109, 110). Hence, intense research has been directed to obtaining GnRH analogs with potential use as nonsteroidal contraceptives or as fertility agents (111), and more than 3000 derivatives have been obtained so far. The development of cyclic GnRH antagonists (112) exemplifies how a tentative structural hypothesis can be pursued through the use of constraints and how conformational analysis of the constrained analogs by NMR and molecular dynamics can lead to highly potent derivatives.

As can be expected for a linear peptide of its size, GnRH is largely unstructured in solution (113). However, several folded conformations of GnRH were suggested from empirical energy calculations, including the formation of a Type II' β turn around Gly6-Leu7 (114). This conformation could be favored by the presence of Gly in position $i+1$ of the turn, as Gly can adopt the torsion angles characteristic of a D-amino acid, which are required for a residue in this position of the turn (8). The possible relevance of a Gly6-Leu7 β-turn conformation for the biological activity of GnRH was supported by the increased potency of GnRH analogs with D-residues in position 6 (115). N-methylation of Leu7, which should be compatible with the turn conformation, produced a further increase in activity (116). Strong support for the hypothesis was obtained when Freidinger et al prepared a potent GnRH agonist in which the Type II' β-turn conformation was constrained with a S-γ-lactam linking Gly6 Cα and Leu7 N (90).

An approach between the N- and C-termini should be favored if a β turn around residues 6–7 is formed in the active conformation of GnRH, which suggests the possibility of cyclizing the molecule to stabilize the turn (Figure 1). Among the first cyclic GnRH analogs that were synthesized, some antagonist activity was found in an analog with an amide bond bridging residues 1 and 10: cyclo(1-10) [Δ^{3}Pro1,D-4ClPhe, D-Trp3,D-Trp6, NMeLeu7, βAla10] GnRH (117). Indeed, only one well-defined conformational family was found for this peptide using molecular dynamics simulations and energy minimizations (112, 118, 119). This structure, which includes a Type II' β turn with residues 6–7 in the corner positions, was later confirmed by two-dimensional NMR in solution (120). An interesting feature observed in the conformation of the cyclo(1-10) antagonist was the proximity of residues 4 and 9 (Figure 1), which suggested the possibility of bridging these two

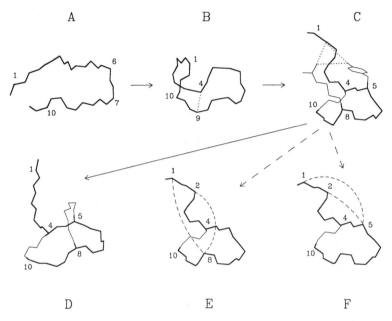

Figure 1   Development of highly potent, constrained antagonists of GnRH. *A:* GnRH with a
β-turn structure around residues 6–7 and the remainder of the molecule in a random conformation; the chain reversal, proposed to be important for biological activity, would make
approach of the N- and C-termini more probable. *B:* A cyclo(1-10) bridged GnRH analog is a
weak antagonist, and proximity between residues 4 and 9 is observed in its preferred conformations. *C:* A cyclo(4-10) bridged analog, designed to refine the putative binding conformation of the cyclo(1-10) bridged analog, is a potent GnRH antagonist; close N-terminus/Tyr5/Arg8
contacts suggest additional bridging possibilities. *D:* A bicyclo(4-10,5-8) analog is also a potent
antagonist and presents a similar conformation to the cyclo(4-10) analog. *E,F:*Bicyclic analogs
with residues 1 or 2 bridged to residues 5 or 8 are new targets to design even more potent GnRH
antagonists. Only the backbone (thick lines), and bridges and some side chains (both shown as
thin solid lines) are represented; dotted lines show close contacts that suggest additional constraints, and the numbers indicate Cα positions to help locating the termini and the residues of
interest.

positions. Hence, GnRH analogs with a bridge between residues 4 and 9 were
synthesized and found to have antagonist activities similar to the parent
cyclo(1-10) analog (121). These results were consistent with a retention of the
structure of the cyclo(1-10) antagonist, but it appeared that a slight modification was needed to fine-tune the binding conformation and obtain higher
activity. Based on previous results obtained in the agonist series, a (4-10)
bridge seemed attractive for this purpose, and several cyclo(4-10) GnRH
analogs were synthesized (112, 122). Some of these compounds showed a
very high antagonist activity and were almost as potent as the strongest linear
GnRH antagonists.

A conformational study of one of the most potent cyclo(4-10) antagonists, cyclo(4-10) [Ac-Δ^3Pro1,D-4FPhe2, D-Trp3,Asp4,D-2Nal6,Dpr10]GnRH, has been performed recently using NMR and restrained molecular dynamics and has shown that the β turn around residues 6–7 is also present in the preferred conformations of this compound (44, 45). Close contacts between the side chains of Tyr5 and Arg8, and between these side chains and the N-terminus, are observed in many of the restrained structures accessible to this GnRH analog, suggesting new constraints that could be introduced to obtain bicyclic GnRH antagonists (Figure 1). Bicyclic (4-10,5-8) analogs have already been synthesized and one of them, bicyclo (4-10,5-8) [Ac-D-2Nal1,D-4ClPhe2,D-Trp3,Asp4,Glu5,D-Arg6,Lys8,Dpr10]GnRH, is equipotent to the parent cyclo(4-10) antagonist (123) and has a similar conformational behavior (124) (Figure 1). The tail formed by residues 1–3 is relatively flexible, as it is in the cyclo(4-10) analog, and appears to interact frequently with the (5-8) bridge. Hence, bicyclic or tricyclic analogs with a bridge between the N-terminus and residue 5 or 8 (Figure 1) appear to be the likely targets to obtain more constrained, and perhaps more potent, GnRH antagonists.

α-Melanotropin

Melanin release in the skin of many vertebrates is regulated by the linear peptide α-melanocyte-stimulating hormone (α-MSH), also known as α-melanotropin (125). This tridecapeptide, Ac-Ser1-Tyr2-Ser3-Met4-Glu5-His6-Phe7-Arg8-Trp9-Gly10-Lys11-Pro12-Val13-NH$_2$, is synthesized in the pituitary gland and also appears to have several neurophysiological effects (126). The research on α-MSH analogs offers a good example of how the activity of a given peptide in different receptors can be modulated by diverse conformational features. Much of this research has been focused on the pigmentary activity of α-MSH, with the aim to obtain compounds with increased activity and prolonged duration of action. A linear analog with such properties in several bioassays, [Nle4,D-Phe7]α-MSH, was obtained after the observation that heat-alkali-treated α-MSH had prolonged activity and underwent substantial racemization in position 7 (127). On the basis of theoretical considerations and the activity enhancement produced by the D-Phe7 substitution, a turn conformation within residues 5–9 was proposed to be important for melanotropic activity. This hypothesis led to the synthesis of a cyclic analog, cyclo(4-10)[Cys4-Cys10]α-MSH, with extremely high activity in a frog skin assay, although the action of this compound had shorter duration than[Nle4,D-Phe7]α-MSH (128, 129). Both the linear and the cyclic analogs are substantially more potent in the frog skin assay than in a lizard skin assay, but only the linear compound is superpotent in a mouse melanoma tyrosinase assay, indicating that different features of the peptides are important for activity at different receptors (128–131).

Further research in the design of constrained α-MSH analogs has been directed to obtaining long-lasting compounds with high potency in the lizard skin and mammalian tyrosinase assays, and potential clinical use in the treatment of pigmentary disorders and melanoma cancer (131). Molecular dynamics simulations of α-MSH and [Nle4,D-Phe7]α-MSH yielded some low-energy structures with turn conformations within residues 6–9 and a definite amphiphilicity: polar and hydrophobic side chains were displayed on different sides of the structures (131). Although the side chains of Glu5 and Lys11 were not close enough to form a salt bridge, it was observed that such an interaction could easily be formed if the Lys side chain was located in position 10. A series of peptides with lactam bridges between residues 5 and 10 were obtained, and one of these compounds, cyclo(5-10) Ac-[Nle4,Asp5,D-Phe7,Lys10]α-MSH$_{4-10}$-NH$_2$, had very high and prolonged activity in both the lizard skin and melanoma tyrosinase bioassays, while it was equipotent to α-MSH in the frog skin assay (131, 132). Therefore, the cyclic (4-10) and (5-10) α-MSH analogs offer the opportunity to analyze additional details of the structural and topological requirements for activity in the different α-MSH receptors, and further conformational studies of these analogs should provide new insights on these requirements (132).

Oxytocin and Vasopressin

Oxytocin, cyclo(1-6) H-Cys1-Tyr2-Ile3-Gln4-Asn5-Cys6-Pro7-Leu8-Gly9-NH$_2$ (OT), is a neurophyseal hormone that mediates uterine contraction and milk ejection, and it was the first peptide hormone to be isolated, characterized, and synthesized by chemical methods (133). Closely related to this peptide are the vasoconstrictive and antidiuretic hormones Lys- and Arg-vasopressin, cyclo(1-6) H-Cys1-Tyr2-Phe3-Gln4-Asn5-Cys6-Pro7-(Lys8 or Arg8)-Gly9-NH$_2$ (LVP or AVP). Because of their cyclic nature, these peptide hormones have been the subject of extensive analog design based on structural hypotheses (11), and many of their analogs have been analyzed by theoretical and spectroscopic methods. The research on constrained oxytocin and vasopressin analogs illustrates how similar conformational elements can be required for the action of closely related peptides at different receptors and emphasizes the fact that not only different structural requirements, but also dynamic factors can govern the agonist or antagonist activity of analogs of natural peptides.

Despite the disulfide bond constraint, oxytocin and vasopressin are still highly flexible, as shown by NMR (134, 135), molecular dynamics simulations (136), and X-ray crystallography (137); thus, additional constraints can be used to modulate their activity and specificity. A "cooperative" conformational model based on NMR results predicted a proximity between the side chains in positions 5 and 8, and turn conformations for residues 3–4 and 7–8, as important structural features for the antidiuretic activity of vasopressins (138). Based on this model, a bicyclic analog with a 5-8 bridge, cyclo(1-6,5-8) [Mpa1,Phe2,Asp5]LVP, was synthesized and resulted in an antidiuretic antagonist (139). Although more potent monocyclic antidiuretic antagonists had been found earlier, all of these compounds were also potent antagonists of the pressor activity of vasopressin, which was not the case for the bicyclic compound (139).

In the oxytocin series, a "cooperative" model based on NMR and structure-activity studies was also proposed by Walter, with turns in residues 3–4 and 7–8, and proximity between the side chains of Tyr2 and Asn5 (140). The complementary "dynamic" model proposed by Meraldi et al predicted that conformational flexibility was necessary for agonist activity (87), and later Hruby suggested that different modes of interaction with the receptor exist for oxytocin agonists and antagonists (141). Analysis of several oxytocin antagonists with a penicillamine substitution in position 1 has shown that the double methylation on the β carbon of this residue produces a higher rigidity in the molecule (87, 142), giving support to the idea that conformational flexibility may be necessary for signal transduction. The solution of the crystal structure of deamino-oxytocin (137) has shown that, without the steric constraints on the β carbon, more than one conformation is adopted by the disulfide bridge even in the solid state. This X-ray structure has been used as a model to design a bicyclic analog, cyclo(1-6,4-8) [Mpa1,Glu4,Lys8]OT, which is one of the most potent oxytocin antagonists in the uterine receptor (143). The increased rigidity of this compound is supported by NMR results, and the observation that the parent monocyclic analog lacking the lactam bridge acts as a weak agonist supports the correlation between rigidity and antagonist activity (143). Indeed, the antidiuretic antagonist activity of the bicyclic vasopressin analog cited above could also be due to increased rigidity.

The mobility of the side chains can also be an important factor for agonist activity in oxytocin analogs, which is emphasized by the inhibitory effects observed in analogs with constrained side chains in positions 2 and 8 (93, 144). Interestingly, inhibition of the vasopressor response to vasopressin is also produced by one of these oxytocin analogs, containing a cycloleucine residue in position 8 (144). It is also important to point out that a new class of cyclic hexapeptide oxytocin antagonists have been recently isolated from *Streptomyces silvensis* (145). These molecules contain a proline and two noncoded, cyclic amino acid residues, which should certainly restrict their conformation. Substitution studies on these hexapeptides have led to potent oxytocin antagonists with high receptor selectivity and enhanced solubility (146, 147).

Tachykinins

The family of neuropeptides termed tachykinins includes several 10–12-residue long peptides with the common C-terminal sequence Phe-Xaa-Gly-Leu-Met-NH$_2$. These peptides have several peripheral actions such as contraction of some smooth muscles, salivation, hypotension, and vasodilatation, and they appear to interact with three different types of receptors, referred to as NK1, NK2, and NK3 (148). The tachykinins isolated in mammals are substance P (SP), neurokinin A (NKA), and neurokinin B (NKB), which act preferentially on the NK1, NK2, and NK3 receptors, respectively:

SP   H-Arg1-Pro2-Lys3-Pro4-Gln5-Gln6-Phe7-Phe8-Gly9-Leu10-Met11-NH$_2$
NKA   H-His1-Lys2-Thr3-Asp4-Ser5-Phe6-Val7-Gly8-Leu9 - Met10-NH$_2$
NKB   H-Asp1-Met2-His3-Asp4-Phe5-Phe6-Val7-Gly8-Leu9 - Met10-NH$_2$

Among the nonmammalian tachykinins, the most well studied are eledoisin (octopod), and physalaemin and kassinin (both amphibian) (148). The research on tachykinins epitomizes the complexity involved in the study of systems comprising several peptides that interact with a variety of receptors. Despite this complexity, the development of constrained tachykinin analogs based on conformational grounds has already yielded several specific derivatives.

Various tachykinins and analogs have been studied by NMR in different solvents (149–153) in order to define structural features that could be important for their biological activities and that could lead to constrained analogs with high receptor specificity. The structure of substance P, the most studied tachykinin, was found to be flexible and sensitive to the environment. An α-helical conformation in residues 4–8 was proposed for this peptide in methanol solution (149). This conformation could be favored by the helix-nucleating properties of Pro (see next section), while Gly could act as a helix breaker (154). Physalaemin, which also interacts preferentially with the NK1 receptor, also appears to adopt the helical conformation, in this case stabilized by an Asp3-Lys6 salt bridge (150). In contrast, no helicity was found in neurokinin A, which has lower NK1 potency (155). Several cyclic SP analogs with 3-6, 4-7, or 5-8 bridges were synthesized and cyclo(3-6) [Cys3,6]SP was found to have increased NK3 and decreased NK2 activities, while it was almost equipotent to SP in the NK1 receptor (156). A Tyr-for-Phe substitution in position 8 of this analog led to a further increase in NK1 and NK3 potencies, and NMR analysis of the Tyr8 analog indicated a β-turn conformation for residues 4–5 that initiates a helix in residues 6–8. High NK1 specificity was obtained by introducing Pro substitutions in the C-terminus of the Tyr analog, to yield cyclo(3-6) [Cys3,6,Tyr8,Pro(9 or 10)]SP (148). On the other hand, introduction of the Cys-Cys bridge in the analogous positions of neurokinin B, yielding cyclo(2-5) [Cys2,5]NKB, increased its specificity for the NK3 receptor (156). It should be noted that [Pro9 or 10]SP and [Pro-7]NKB also have high NK1 and NK3 specificities, respectively (148). It seems clear from these results that, apart from the short helix structure in the core, certain conformational features in the C-terminus of tachykinins also influence their activity and specificity, and several studies have tried to identify these features.

NMR analysis of [Pro7]NKB has led to the proposal of a turn structure within residues 6–9 that brings the N- and C-termini into proximity (151), but no cyclic compound linking the two termini has been described to our knowledge. The specific NK1 agonist Ac-[Arg6,Pro9]SP$_{6-11}$ has also been studied by NMR, and a Type I β-turn conformation with Pro9-Leu10 in the corner positions has been suggested (153). Based on the conformational model for this peptide, a cyclic analog of [Arg8]SP$_{6-11}$, with a $CO(CH_2)_4CONH(CH_2)_3$ bridge joining the N-terminus with Gly9 N, was designed and found to have a similar NK1 selectivity to the parent compound (157), but it is not clear how this constraint affects the conformation of the three last residues in the sequence. On the other hand, NMR studies suggested an interaction between the C-terminal amide and the side chains of Gln5 and Gln6 in the preferred conformations of substance P in methanol (149), but low or no activity was found in SP analogs with 5-9, 5-10, or 5-11 Cys-Cys bridges (148, 158). In another study, introduction of a R-γ-lactam in the Gly-Leu sequence of several tachykinin analogs produced NK3-specific compounds, while inactive analogs were obtained with the turn-inducing S-γ-lactam (159). It appears that conformational rigidity in the C-terminus increases the receptor specificity of tachykinins, and the cyclic constraints in the core of tachykinin analogs may also increase specificity by influencing the conformation of the C-terminus. Further research in this area should shed more light on what conformational features in the C-terminus of tachykinins are important for interaction with each receptor.

MIMICS OF PIECES OF PROTEINS

Acquiring a full understanding of the interactions that lead proteins to adopt particular three-dimensional structures, and of the pathways that proteins follow to fold into such structures, is one of the major goals of current research in biochemistry and biophysics. As we mentioned earlier, substantial knowledge of the factors that influence the preferred conformations of polypeptides has been gained through analysis of protein structures themselves. But, at the same time that we can learn about peptide conformation by studying proteins, we can also learn about proteins using peptide models. While cooperative effects are more amenable to study in protein domains or subdomains, very detailed studies of the individual factors that govern protein folding can be performed on lower-molecular-weight compounds. Perhaps the

main advantage of the use of peptide models to study protein folding is that
their sequences can be designed almost at will to ask specific questions about
intrinsic conformational properties, about particular interactions, and about
the energetics of these interactions, excised from the context of the overall
protein structure. By isolating these features in small molecules, their possible
roles in the mechanism of protein folding and in the stabilization of protein
structures can be analyzed. Such features may play active roles or may be
induced by other structural elements. Conformationally biased protein segments may direct the pathways of protein folding but not remain reflected in
the folded structures. Most peptide models used in these studies contain
constraints to force or to stabilize well-defined conformations. Rather than
giving an exhaustive account on the research in this area, we want to illustrate
with some examples how peptide models have been used to mimic pieces of
proteins with particular secondary and even tertiary structures, and how these
studies have given new insights into the rules that govern protein folding.

β Turns

Constrained peptides have been used extensively to study reverse turns (8–
10). Here, we consider a few cases where the main goal was to model β turns
occurring in specific sequences of proteins. The first example we want to
discuss is the cyclic hexapeptide cyclo(Gly1-Ile2-Leu3-Gln4-Pro5-D-
Tyr(Bzl)6), which was designed to model the Type I β turn adopted by the
sequence Gly54-Ile55-Leu56-Gln57 of lysozyme (160). Although β turns are
intrinsically polar structures that commonly occur at the surface of proteins
(8, 161), this particular turn is buried in a hydrophobic environment in the
interior of the protein. In order to satisfy the hydrogen-bonding needs of this
structural feature, a buried water is intimately associated with it.

The question arose from this observation whether an isolated β turn in a
hydrophobic environment would interact specifically with a water molecule.
To address this question, the model peptide was designed to mimic the buried
turn of lysozyme. The design was based on the known tendency of cyclic
hexapeptides to adopt two-β-turn conformations (162, 163). The Pro-D-
Tyr(Bzl) dyad was added to the lysozyme turn sequence as a strong determinant of a Type II β turn that could induce a chain reversal in the other side of
the cycle. The peptide was studied by NMR and was found to adopt the
predicted conformation, with Pro5-D-Tyr(Bzl)6 and Ile2-Leu3 in the corner
positions of Type II and Type I β turns, respectively, and the Gly1 and Gln4
residues in extended conformations (160). It should be noted that all the
residues in the Gly-Ile-Leu-Gln sequence have lower than average propensities to occur in their respective positions in protein Type I β turns (51). The
design of the cyclic hexapeptide showed that, despite these low propensities,
the tetrapeptide sequence could be induced with the appropriate constraints to

adopt the same conformation observed in the protein. This design strategy also provided a suitable model for well-defined water interaction studies, currently under way (A. Roby and M. Paulaitis, unpublished results). This example was one of the first to show that molecular modeling based on structural grounds could be used to study isolated pieces of proteins in their native conformations (160).

Similar principles were used in the design of the cyclic hexapeptide cyclo(Asn1-Pro2-Ala3-Leu4-Pro5-Gly6) to study the roles of asparagine in β-turn formation (10). Here, the sequence Asn-Pro-Ala-Leu corresponds to residues 148–151 of dihydrofolate reductase (DHFR), which form a Type I β turn on the surface of the protein (Figure 2), and Pro-Gly was expected to adopt a Type II β-turn conformation on the other side of the cycle. Statistics from proteins of known structure have shown that Asn and Pro are the residues with the highest probabilities of appearing in the i and $i+1$ positions, respectively, of Type I β turns (51). The occurrence of Asn in the i position seems to be associated with a stabilizing interaction between the Asn side chain CO and the NH of residue $i+2$, which is often observed in proteins (2). Such a hydrogen-bonding pattern occurs in the DHFR 148-151 turn. The cyclic hexapeptide model was shown, by NMR, to adopt the predicted two-β-turn conformation, including a hydrogen bond between the Asn1 side chain CO and Ala3 NH (10). Molecular dynamics analysis, combined with measured interproton distances, indicated that this interaction occurs with the Asn1 side chain in at least two different rotamers in the peptide model, and that interactions between this side chain and Leu4 NH were also likely to occur frequently (Figure 2). This study showed that the Asn side chain can form hydrogen-bonding interactions with the amide protons of the succeeding residues ($i+2$ and $i+3$) in the sequence in a variety of ways. In addition to stabilizing turns in native protein structures, these persistent interactions could play an active role in protein folding by increasing the population of turn conformations in the flexible chains of unfolded proteins, in particular when Asn is followed by Pro (10).

A different application of the design of protein β-turn models has been demonstrated with the development of a small pseudopeptide that mimics the binding and functional properties of monoclonal antibody (MAb) 87.92.6, which binds to reovirus type 3 cellular receptors (164). Constrained analogs of a heptadecapeptide, corresponding to a complementarity-determining region of a hypervariable domain light chain of this anti-idiotypic/antireceptor antibody, were synthesized in an attempt to model the active conformation of the heptadecapeptide (97). Cys-Cys bridges were introduced in different positions in order to optimize the binding conformation, and a cyclic analog including the sequence Cys9-Ile10-Tyr11-Ser12-Gly13-Ser14-Thr15-Cys16 was found to have an affinity for the receptor more than 40-fold larger than

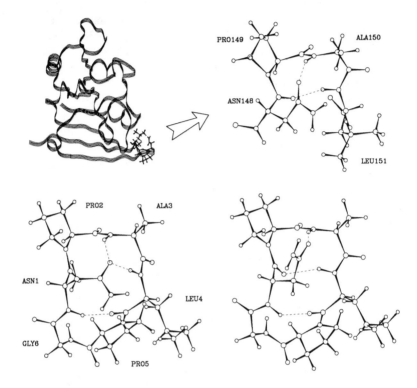

Figure 2 A cyclic hexapeptide model for a Type I β turn of dihydrofolate reductase. A ribbon diagram of the protein showing the location of the turn is shown at the *upper left,* and an expansion of the turn region as it occurs in the X-ray structure of the protein is shown on the *right. Below* are shown two conformations of the model cyclic hexapeptide cyclo(Asn-Pro-Ala-Leu-Pro-Gly), obtained by NMR and molecular dynamics analysis. Hydrogen bonds are represented by dashed lines. Note that an interaction between the Asn side chain CO and the Ala NH can be formed for different conformations of this side chain, due to its favorable orientation when Asn is in the *i* position of the turn.

that of the linear peptide (97). This and other cyclic analogs were analyzed by energy-minimization techniques, and the binding model deduced was used to design a pseudopeptide containing a cyclic mimic of a Gly-Ser β turn that joins a Tyr-Ser dipeptide and another Ser residue (164). This compound was shown to be functionally analogous to MAb 87.92.6, and to have increased solubility and enzymatic stability compared to the peptidic analogs. The small size and synthetic nature of this type of compound makes them especially interesting for potential therapeutic use, as they are expected to have low immunogenicity and good transport properties in vivo (164).

β Sheets

The formation of β sheets in small peptides often leads to aggregation. Hence, it is not surprising that model peptides have been used much more frequently to learn about the propensities of different amino acid residues to adopt β-turn or α-helix conformations than to study β sheets. Here, we would like to comment on some simple systems that have been used to model β-sheet conformations, and to discuss two cases where model peptides corresponding to pieces of proteins adopted β-hairpin conformations.

The simplest examples of an antiparallel β-sheet arrangement are cyclic hexapeptides such as those described above, where two transannular hydrogen bonds are formed between the two residues joining the β turns. Other simple and less distorted β-sheet models reported by Balaram, Karle, and coworkers are the hexapeptide with a disulfide bridge cyclo(1-6) Boc-Cys-Val-Aib-Ala-Leu-Cys-NHMe, and the antiparallel dimer with two disulfide bridges of the tripeptide Boc-Cys-Ala-Cys-NHMe. The first was shown, both by NMR (165) and by X-ray diffraction (166), to form a β hairpin with Aib-Ala in the corner positions and three interstrand hydrogen bonds. The tripeptide dimer also adopts, in solution (167) and in crystals (168), an antiparallel β-sheet conformation with four interchain hydrogen bonds. Hence, these systems offer a good opportunity to gain insight into the tendencies of different residues to participate in antiparallel β-sheet conformations, for example by introducing substitutions in the Val and Leu positions of the hexapeptide or in the middle residue of the tripeptide dimer. Cyclic decapeptides can adopt sheet conformations with four transannular hydrogen bonds, as is the case of gramicidin S (169), and could also be used for such studies. Very promising in this area are the functionalized 2,8-acylamidoepindolidiones that have been proposed by Kemp and coworkers as templates to nucleate both parallel and antiparallel β sheets (170).

Induction of a β-hairpin conformation with a disulfide bridge between N- and C-terminal Cys residues has been used successfully in the design of peptides mimicking the flap of human renin. The design of these peptides was based on a three-dimensional model of human renin built by analogy with the X-ray structure of other aspartyl proteases (171, 172). The flap is a protruding β-hairpin loop that belongs to a strongly antigenic region of human renin (172). According to the proposed model, the flap was assigned to residues 77–93 of human renin, and preliminary studies of a heptadecapeptide corresponding to this region suggested that the chain reversal could occur in the Tyr83-Ser84-Thr85-Gly86 sequence (173). To test the hypothesis, a decapeptide and a tetradecapeptide were synthesized, both with disulfide-bridged Cys residues at the N- and C-termini, and part of the flap sequence centered at the putative β turn. For comparison, linear peptides with alanines instead of cysteines were also studied. Both cyclic peptides were shown to form β-hairpin structures and were recognized by human antirenin antibodies, while the linear peptides were less structured and did not bind to the anti-protein antibodies (173, 174). These results indicated that the cyclic peptides adopted a very similar conformation to that expected for the renin flap and that such a conformation was necessary for antibody recognition. Further support for this conclusion was obtained with the observation that an analogous cyclic peptide with intermediate length (12 residues) did not adopt a full β-hairpin conformation and was substantially less antigenic than the 10- and 14-residue cyclic peptides (175). The lower tendency of the dodecapeptide to form a β hairpin is most likely due to the fact that only every other pair of residues added to the strand afford two new hydrogen bonds to stabilize the sheet.

An interesting approach, also involving Cys-Cys bridges in synthetic peptides, was used to study temperature-sensitive folding (*tsf*) defects in mutants of the P22 bacteriophage tailspike protein (176). These mutants fold properly at the permissive temperature, but form irreversible aggregates at the nonpermissive temperature (177). The point mutations causing these *tsf* defects occurred in sequences of the native protein with high turn propensities, and there was evidence that the structure of the protein is largely β. Based on these observations, it was suggested that the formation of critical cross-β patterns was hindered in the *tsf* mutants, in favor of the aggregation process. Synthetic deca- and dodecapeptides corresponding to a region containing one of these mutations, but flanked by Cys residues, were studied to test the hypothesis (176). The peptides contained Pro-Gly (native) or Pro-Arg (*tsf* mutant) dyads in the center of the sequence. NMR indicated that the reduced forms of all the peptides were largely unstructured in solution. However, the use of a thiol-disulfide exchange assay showed that the Pro-Gly peptides had a 3–5-fold higher tendency to form an intramolecular disulfide bridge than the Pro-Arg analogs. All the oxidized peptides adopted a Type II β-turn conformation around Pro-Gly or Pro-Arg, and there were also clear indications of the formation of a β hairpin. These results indicated that the mutant sequence can adopt the same conformation as the native, as in the proteins grown at permissive temperature, but it has a lower tendency to do so. These results gave support to the idea that a small difference in a local conformational propensity can be critical for the folding of a 666-residue-long protein (176).

α Helices

Among the regular secondary structures found in proteins, the most studied with medium-sized model peptides is the α helix (see Ref. 22 for a review), but very few examples have been described of covalent constraints that induce or stabilize α-helix conformations. As we mentioned earlier, the sterically constrained Aib residue is a strong helix former, and has often been used to nucleate both α- and 3_{10}-helices (78–82). Several other constraints to stabilize α helix have been based on $i/i+4$ side chain proximity. Lactam bridges between Asp and Lys residues in $i/i+4$ positions have been shown to stabilize an α-helical conformation in human growth hormone-releasing factor (GRF) analogs (178–180). Whereas [Ala15]GRF$_{1-29}$-NH$_2$ adopts 25% α-helix structure in water, an analog with Asp8/Lys12 and Lys21/Asp25 bridges, cyclo(8–12,21-25) [Asp8,Ala15]GRF$_{1-29}$-NH$_2$, is 45% helical in aqueous solution, and N-terminal substitutions in this analog led to a potent compound with highly increased resistance to degradation (179, 180). Peptides with the general formula H-(Lys-Leu-Xxx-Glu-Leu-Lys-Yyy)$_n$-OH, where $n=2–3$, Xxx=Lys or Orn, Yyy=Asp or Glu, have been used recently to find the length of the lactam bridge that offers optimal helix stabilization (181). The Lys/Asp pair was found to give maximum helicity (69% in water for $n=3$), but, interestingly, the isomeric Orn/Glu pair had a destabilizing effect that could be due to unfavorable electrostatic interactions with the backbone. An example of a noncovalent constraint to stabilize secondary structure, also involving $i/i+4$ side chain proximity in α helices, is the increased helicity observed in peptides with Cys(i)/His($i+4$) and His(i)/His($i+4$) pairs upon bidentate binding to metal ions (182). Without the need of external agents, ionic interactions between the side chains of Glu and Lys residues in positions i and $i+4$ can also stabilize α-helix formation, as has been elegantly shown by Marqusee & Baldwin (183).

In addition to the use of side-chain bridges, the formation of α helix can also be favored by nucleation with appropriate templates. Based on the nucleation/propagation model of Zimm &Bragg (184), no entropy would be lost in the initiation step if an alpha helix were started on a rigid template with three carbonyl groups in the correct orientation, and the amount of helix would only depend on the helix propagation tendency of the residues in the sequence. A helix template with a hydrazone link substituting an amide-amide $(i,i+4)$ hydrogen bond has been described (185), and preliminary results on its helix-promoting potential are encouraging (186). Noting that proline residues are often used by nature to initiate helices (3, 154), Kemp and coworkers have designed another helix-nucleating template consisting of an Ac-Pro1-Pro2 dipeptide constrained by a S-CH$_2$ bridge between Pro1 Cγ and Pro2 Cδ (187–189). The template can easily be incorporated at the N-terminus of any sequence, and adopts three major conformations in solution, one of them acting as an α-helical initiation site (Figure 3). Changes in the populations of the different conformations can be detected by NMR. As the helix-nucleating form is stabilized by an increased helical character in the sequence attached to the template, the population of this conformation can be correlated with α-helicity in the sequence. The values of the helix propagation parameter s for alanine obtained with peptides conjugated to this template are

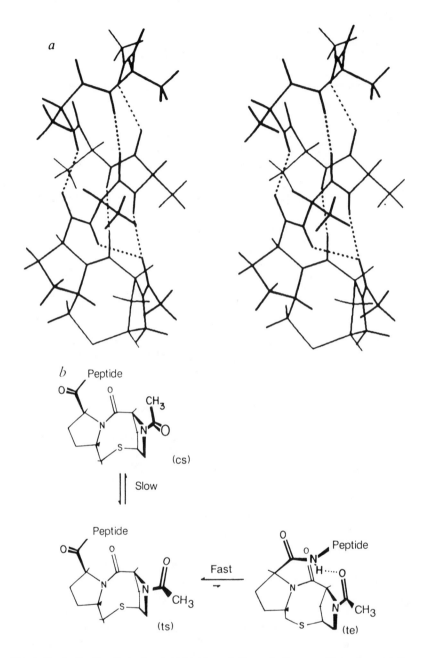

Figure 3 A diproline derivative template for α-helix nucleation. *a:* Stereodiagram of a penta-alanine conjugate of the template, with the peptide forming an α helix initiated at the template. *b:* Major conformations of the template observed in solution. The conformation designated (te) is helix-nucleating and its population increases with the helicity of the attached peptide. Reprinted by permission from *Nature* Vol. 352, pp. 451–54. Copyright © 1991 Macmillan Magazines Ltd.

close to 1 (187), in agreement with results calculated from host-guest experiments (190, 191), but substantially smaller than values obtained with alanine-based helical peptides (192). Further experimentation is needed to determine the extent to which either method can separate propagation from initiation contributions to α-helicity, but it is certain that α-helical templates can be highly useful tools for this research area.

Larger Domains, Active Site Models, and Entire Proteins

The constraints examined thus far have been introduced in relatively small peptides, and in most cases involved the use of covalent bonds to force or stabilize the better-known secondary structures. Constraints can also be applied to induce other secondary structure elements, and to stabilize tertiary structure in larger polypeptide chains that can span protein subdomains, protein domains, or even the entire protein. We would like to end this review with a brief overview of this exciting research area. The examples given below emphasize the fact that, although noncovalent interactions cannot produce permanent constraints, they are essential factors in the design of larger polypeptides.

Regions of proteins that do not fall into the categories of α helix, β sheets, or β turns are usually assigned to random coil. However, this assignment generally arises from the difficulty of identifying other nonperiodic structural motifs. For example, Leszczynski & Rose have shown that many protein loops (Ω-loops) are compact structures that could constitute independent folding units (193). Such structures are good targets for analysis with constrained peptides, and studies in this direction have been started very recently by at least two groups, yielding promising results (194, 195).

A striking example of how a piece of a protein can be isolated and still adopt a similar structure to that found in the native protein is the PαPβ analog corresponding to residues 20–33 (Pβ) and 43–58 (Pα) of bovine pancreatic trypsin inhibitor, described by Oas & Kim (196). These two peptides form β-sheet (Pβ) and α-helix (Pα) conformations in the protein and are the major parts of one of its hydrophobic cores. PαPβ folds into nativelike structure upon formation of an interchain disulfide bond between Cys30 and Cys51, a constraint that appears to be crucial to initiate the proper folding of the protein (197, 198).

Another remarkable case of native folding has been provided by a study of synthetic peptides corresponding to the calcium-binding sites III and IV of troponin-C, which are arranged in a symmetric way in the carboxy-terminal domain of the protein. A synthetic 34-residue peptide representing site III was shown to be mainly unstructured in aqueous solution, but adopted nativelike structure upon addition of calcium (199, 200). The structure consists of a symmetric dimer with an interchain association analogous to that observed between sites III and IV in the protein. Similar behavior has been observed for a 39-residue fragment corresponding to site IV (201). This example illustrates how specific noncovalent interactions can be essential for the stability of native three-dimensional structures in peptide models. The ability of some protein fragments to mimic interactions of the entire protein, whether these fragments are structured in bulk solution or not, has been demonstrated in a number of cases, including, for example, the recognition of linear peptide epitopes by antibodies (202), the activity of a peptide corresponding to a regulatory domain of protein kinase C (203), and the control of a voltage-sensitive ion channel by an isolated N-terminal peptide (204).

Constraints have been used in the de novo design of several proteins, mostly to place different peptide chains in the desired orientations (what could be called interchain constraints). The term *template-assembled synthetic proteins (TASP)* has been introduced by Mutter to refer to synthetic proteins formed by several amphiphilic peptides that are assembled with a predefined orientation on a template molecule (23, 205, 206). Among others, a protein with two domains, a four-helix bundle, and a β barrel, has been designed recently using this concept, and circular dichroism and infrared studies indicate spectroscopic properties consistent with the model proposed for the folding of this protein (207). A related approach has been used by Stewart and coworkers to build a protein with four amphiphilic helices tied together on a branched tripeptide unit (208). This protein was designed to bear a binding site analogous to chymotrypsin, and it has been shown to have chymotrypsin-like esterase activity. A number of de novo designed proteins have been obtained without the use of a preformed template, but disulfide bonds or short peptide links with high probability to form turns or loops have been introduced in most cases to ensure the proper localization of the chains to be assembled (209–215). It is important to underline that, whether templates, disulfide bonds, or peptide loops are used to assemble the proteins, the major driving forces to form tertiary structures are noncovalent interactions, namely hydrophobic collapse of amphiphilic chains and salt bridges that stabilize secondary structures. Covalent constraints are thus primarily introduced to maintain the tertiary arrangement.

Until recently, research on peptides and proteins has been for the most part divided into two different fields. We will gain considerably in the future by bridging these two fields to enhance our understanding of each, and several examples discussed in this review already point in this direction. The design of new proteins is perhaps the area that has made most use of principles and knowledge from the two fields. The rational design of peptide drugs will benefit from our growing understanding of the rules that govern polypeptide structure and folding. It is likely that a substantial part of such understanding will come from studies of model constrained peptides. The great power of the present methodology of peptide synthesis may allow, in the near future, the preparation of a variety of constrained analogs of natural proteins, which will be extremely useful tools in the study of protein folding.

ACKNOWLEDGMENTS

We would like to thank Arnie Hagler and Sam Landry for useful comments on the manuscript, and the many authors cited in this review for sending us reprints and preprints of their work. Ongoing research in our laboratory in the area of constrained peptides is supported by grants from the NIH (GM27616) and the Robert A. Welch Foundation.

Literature Cited

1. Chou, P. Y., Fasman, G. D. 1978. *Adv. Enzymol.* 47:45–148
2. Baker, E. N., Hubbard, R. E. 1984. *Prog. Biophys. Mol. Biol.* 44:97–179
3. Presta, L. G., Rose, G. D. 1988. *Science* 240:1632–41
4. Dyson, H. J., Wright, P. E. 1991. *Annu. Rev. Biophys. Chem.* 20:519–38
5. Toniolo, C. 1990. *Int. J. Peptide Protein Res.* 35:287–300
6. Spatola, A. F. 1983. *Chemistry and Biochemistry of Amino Acids, Peptides and Proteins,* ed. B. Weinstein, 7:267–357. New York:Dekker
7. Aubry, A., Marraud, M. 1989. *Biopolymers* 28:109–22
8. Rose, G. D., Gierasch, L. M., Smith, J. A. 1985. *Adv. Protein Chem.* 37:1–109
9. Stradley, S. J., Rizo, J., Bruch, M. D., Stroup, A. N., Gierasch, L. M. 1990. *Biopolymers* 29:263–87
10. Rizo, J., Dhingra, M., Gierasch, L. M. 1991. *Molecular Conformation and Biological Interactions,* ed. P. Balaram, S. Ramaseshan, pp. 469–96. Bangalore, India: Indian Acad. Sci. Publ.
11. Hruby, V. J. 1982. *Life Sci.* 31:189–99
12. Hruby, V. J., Sharma, S. D. 1991. *Curr. Opin. Biotechnol.* 2:599–605
13. Hruby, V. J., Kazmierski, W., Kawasaki, A. M., Matsunaga, T. O. 1990. *Peptide Pharmaceuticals,* ed. D. J. Ward, pp. 135–84. Milton Keynes: Open Univ. Press
14. Hruby, V. J., Al-Obeidi, F., Kazmierski, W. 1990. *Biochem. J.* 268:249–62
15. Hruby, V. J. 1990. *Comprehensive Medicinal Chemistry,* ed. C. Hansch, P. G. Sammes, J. B. Taylor, J. C. Emmett, 3:797–804. Oxford: Pergamon
16. Marshall, G. R., Motoc, I. 1986. *Molecular Graphics and Drug Design,* ed. A. S. V. Burgen, G. C. K. Roberts, M. S. Tute, pp. 115–56. Amsterdam: Elsevier Sci. Publ.
17. Kessler, H. 1990. *Trends in Drug Research,* ed. V. Claassen, pp. 73–84. Amsterdam: Elsevier Sci. Publ.
18. Schiller, P. W. 1991. *Medicinal Chemistry for the 21st Century,* ed. C. G. Wermuth. Oxford: IUPAC Publ./ Blackwell Publ. In press
19. Fauchère, J. L. 1986. *Adv. Drug Res.* 15:29–69
20. Rich, D. H. 1990. *Comprehensive Medicinal Chemistry,* ed. C. Hansch, P. G. Sammes, J. B. Taylor, 2:391–441. Oxford: Pergamon
21. Milner-White, E. J. 1989. *Trends Pharmacol. Sci.* 10:70–74
22. DeGrado, W. F. 1988. *Adv. Protein Chem.* 39:51–124
23. Mutter, M., Vuilleumier, S. 1989. *Angew. Chem. Int. Ed. Engl.* 28:535–54
24. Hruby, V. J., Udenfriend, S., Meienhofer, J., eds. 1985. *The Peptides: Analysis, Synthesis, Biology,* Vol. 7. Orlando: Academic. 495 pp.
25. Kim, P. S., Baldwin, R. L. 1990. *Annu. Rev. Biochem.* 59:631–60
26. Gierasch, L. M., King, J., eds. 1990. *Protein Folding: Deciphering the Second Half of the Genetic Code.* Washington, DC: AAAS Publ. 334 pp.
27. Karle, I. L. 1981. *The Peptides: Analysis, Synthesis, Biology,* ed. E. Gross, J. Meienhofer, 4:1–54. New York: Academic
28. Opella, S. J., Gierasch, L. M. 1985. See Ref. 24, pp. 405–36
29. Krimm, S. 1983. *Biopolymers* 22:217–25
30. Seaton, B. 1986. *Spectrochim. Acta* 42A:227–32
31. Schiller, P. W. 1985. See Ref. 24, pp. 115–64
32. Woody, R. W. 1985. See Ref. 24, pp. 15–114
33. Ernst, R. R., Bodenhausen, G., Wokaun, B. 1987. *Principles of Nuclear*

Magnetic Resonance in One and Two Dimensions. Oxford: Clarendon. 610 pp.

34. Wüthrich, K. 1986. *NMR of Proteins and Nucleic Acids.* New York: Wiley. 292 pp.

35. Clore, G. M., Gronenborn, A. M. 1991. *Science* 252:1390–99

36. Crippen, G. M., Havel, T. F. 1988. *Distance Geometry and Molecular Conformation.* New York: Wiley

37. Karplus, M., McCammon, J. A. 1981. *CRC Crit. Rev. Biochem.* 9:293–349

38. Hagler, A. T. 1985. See Ref. 24, pp. 213–99

39. Nilges, M., Clore, G. M., Gronenborn, A. M. 1988. *FEBS Lett.* 229:129–36

40. Braun, W., Go, N. 1985. *J. Mol. Biol.* 186:611–26

41. Kessler, H., Griesinger, C., Lautz, J., Müller, A., van Gunsteren, W. F., Berensden, H. J. C. 1988. *J. Am. Chem. Soc.* 110:3393–96

42. Kessler, H., Bats, J. W., Lautz, J., Müller, A. 1989. *Liebigs Ann. Chem.,* pp. 913–28

43. Constantine, K. L., De Marco, A., Madrid, M., Brooks, C. L. III, Llinás, M. 1990. *Biopolymers* 30:239–56

44. Rizo, J., Koerber, S. C., Bienstock, R. J., Rivier, J. E., Hagler, A. T., Gierasch, L. M. 1992. *J. Am. Chem. Soc.* In press

45. Rizo, J., Koerber, S. C., Bienstock, R. J., Rivier, J. E., Gierasch, L. M., Hagler, A. T. 1992. *J. Am. Chem. Soc.* In press

46. Brüschweiler, R., Blackledge, M., Ernst, R. R. 1991. *J. Biomol. NMR* 1:3–11

47. Torda, A. E., Scheek, R. M., van Gunsteren, W. F. 1990. *J. Mol. Biol.* 214:223–35

48. Kessler, H., Bats, J. W., Griesinger, C., Koll, S., Will, M., Wagner, K. 1988. *J. Am. Chem. Soc.* 110:1033–49

49. Lautz, J., Kessler, H., van Gunsteren, W. F., Weber, H.-P., Wenger, R. M. 1990. *Biopolymers* 29:1669–87

50. Chou, P. Y., Fasman, G.D. 1977. *J. Mol. Biol.* 115:135–75

51. Wilmot, C. M., Thornton, J. M. 1988. *J. Mol. Biol.* 203:221–32

52. Venkatachalam, C. M. 1968. *Biopolymers* 6:1425–36

53. Mathias, L. J., Fuller, W. D., Nissen, D., Goodman, M. 1978. *Macromolecules* 11:534–39

54. Jones, W. C. Jr., Nestor, J. J. Jr., du Vigneaud, V. J. 1973. *J. Am. Chem. Soc.* 95:5677–79

55. Goodman, M., Chorev, M. 1979. *Acc. Chem. Res.* 12:1–7

56. Roeske, R. W., Weitl, F. L., Prasad, K.

U., Thompson, R. M. 1976. *J. Org. Chem.* 41:1260–61

57. Spatola, A. F., Agarwal, N. S., Bettag, A. L., Yankeelov, J. A. Jr., Bowers, C. Y., Vale, W. W. 1980. *Biochem. Biophys. Res. Commun.* 97:1014–23

58. Almquist, R. G., Chao, W.-R., Ellis, M. E., Johnson, H. L. 1980. *J. Med. Chem.* 23:1392–98

59. Cox, M. T., Gormley, J. J., Hayward, C. F., Petter, N. N. 1980. *J. Chem. Soc. Chem. Commun.,* pp. 800–2

60. Hudson, D., Sharpe, R., Szelke, M. 1980. *Int. J. Peptide Protein Res.* 15:122–29

61. Spatola, A. F., Rockwell, A. L., Gierasch, L. M. 1983. *Biopolymers* 22:147–51

62. Goodman, M. 1985. *Biopolymers* 24:137–55

63. Hassan, B., Goodman, M. 1986. *Biochemistry* 25:7596–606

64. Mammi, N. J., Goodman, M. 1986. *Biochemistry* 25:7607–14

65. Spatola, A. F., Anwer, M. K., Rockwell, A. L., Gierasch, L. M. 1986. *J. Am. Chem. Soc.* 108:825–31

66. Van der Elst, P., Elseviers, M., De Cock, E., Van Marsenille, M., Tourwé, D., Van Binst, G. 1986. *Int. J. Peptide Protein Res.* 27:633–42

67. Zanotti, G., Toniolo, C., Owen, T. J., Spatola, A. F. 1988. *Acta Cryst.* C44:1576–79

68. Toniolo, C., Valle, G., Crisma, M., Kaltenbronn, J. S., Repine, J. T., et al. 1989. *Peptide Res.* 2:332–37

69. Toniolo, C. 1989. *Biopolymers* 28:247–57

70. Sherman, D. B., Spatola, A. F. 1990. *J. Am. Chem. Soc.* 112:433–41

71. Anwer, M. K., Sherman, D. B., Spatola, A. F. 1990. *Int. J. Peptide Protein Res.* 36:392–99

72. Mierke, D. F., Pattaroni, C., Delaet, N., Toy, A., Goodman, M., et al. 1990. *Int. J. Peptide Protein Res.* 36:418–32

73. Michel, A. G., Lajoie, G., Hassani, C. A. 1990. *Int. J. Peptide Protein Res.* 36:489–98

74. Marshall, G. R., Humblet, C., Van Opdenbosch, N., Zabrocki, J. 1981. *Peptides: Synthesis, Structure, Function,* ed. D. H. Rich, E. Gross, pp. 669–72. Rockford, Ill: Pierce Chem. Co.

75. Zabrocki, J., Smith, G. D., Dunbar, J. B. Jr., Iijima, H., Marshall, G. R. 1988. *J. Am. Chem. Soc.* 110:5875–80

76. Smith, G. D., Zabrocki, J., Flak, T. A., Marshall, G. R. 1991. *Int. J. Peptide Protein Res.* 37:191–97

77. Marshall, G. R. 1971. *Intra-science Chemistry Reports,* ed. N. Kharasch,

pp. 305–16. New York:Gordon & Breach
78. Karle, I. L., Balaram, P. 1990. *Biochemistry* 29:6747–56
79. Karle, I. L., Flippen-Anderson, J. L., Uma, K., Balaram, P. 1990. *Curr. Sci.* 59:875–85
80. Karle, I. L., Flippen-Anderson, J. L., Sukumar, M., Uma, K., Balaram, P. 1991. *J. Am. Chem. Soc.* 113:3952–56
81. Toniolo, C., Benedetti, E. 1991. *Trends Biochem. Sci.* 16:350–53
82. Pavone, V., Di Blasio, B., Santini, A., Benedetti, E., Pedone, C., et al. 1991. *J. Mol. Biol.* 214:633–35
83. Toniolo, C., Benedetti, E. 1991. *Macromolecules* 24:4004–9
84. Mapelli, C., Stammer, C. H., Lok, S., Mierke, D. F., Goodman, M. 1988. *Int. J. Peptide Protein Res.* 32:484–95
85. Ogawa, T., Kodama, H., Yoshioka, K., Shimohigashi, Y. 1990. *Peptide Res.* 3:35–41
86. Valle, G., Crisma, M., Toniolo, C., Polinelli, S., Boesten, W. H. J., et al. 1991. *Int. J. Peptide Protein Res.* 37: 521–27
87. Meraldi, J.-P., Hruby, V. J., Brewster, A. I. R. 1977. *Proc. Natl. Acad. Sci. USA* 74:1373–77
88. Bach, A. C. II, Gierasch, L. M. 1986. *Biopolymers* 25S:175–91
89. Aubry, A., Pietrzynski, G., Rzeszotarska, B., Boussard, G., Marraud, M. 1991. *Int. J. Peptide Protein Res.* 37: 39–45
90. Freidinger, R. M., Veber, D. F., Perlow, D. S., Brooks, J. R., Saperstein, R. 1980. *Science* 210:656–58
91. Paul, P. K. C., Burney, P. A., Campbell, M. M., Osguthorpe, D. J. 1990. *J. Comput.-Aided Mol. Design* 4:239–53
92. Kazmierski, W., Wire, W. S., Lui, G. K., Knapp, R. J., Shook, J. E., et al. 1988. *J. Med. Chem.* 31:2170–77
93. Lebl, M., Hill, P., Kazmierski, W., Kárászová, L., Slaninová, J., Friĉ, I., Hruby, V. J. 1990. *Int. J. Peptide Protein Res.* 36:321–30
94. Kazmierski, W. M., Yamamura, H. I., Hruby, V. J. 1991. *J. Am. Chem. Soc.* 113:2275–83
95. Holladay, M. W., Lin, C. W., May, C. S., Garvey, D. S., Witte, D. G., et al. 1991. *J. Med. Chem.* 34:455–57
96. Schiller, P. W., Weltrowska, G., Nguyen, T. M.-D., Lemieux, C., Chung, N. N., et al. 1991. *J. Med. Chem.* 34:3125–32
97. Williams, W. V., Kieber-Emmons, T., VonFeldt, J., Greene, M. I., Weiner, D. B. 1991. *J. Biol. Chem.* 266:5182–90
98. Ozeki, E., Kimura, S., Imanishi, Y.

1989. *Int. J. Peptide Protein Res.* 34: 14–20
99. Heitz, F., Kaddari, F., Heitz, A., Raniriseheno, H., Lazaro, R. 1989. *Int. J. Peptide Protein Res.* 34:387–93
100. Zanotti, G. C., Campbell, B. E., Easwaran, K. R. K., Blout, E. R. 1988. *Int. J. Peptide Protein Res.* 32:527–35
101. Bailey, P. D., Carter, S. R., Clarke, D. G. W., Crofts, G. A. 1991. *Peptides 1990*, ed. E. Giralt, D. Andreu, pp. 225–26. Leiden, The Netherlands: ESCOM
102. Barbato, G., D'Auria, G., Paolillo, L., Zanotti, G. 1991. *Int. J. Peptide Protein Res.* 37:388–98
103. Ball, J. B., Alewood, P. F. 1990. *J. Mol. Recognit.* 3:55–64
104. Loosli, H.-R., Kessler, H., Oschkinat, H., Weber, H.-P., Petcher, T. J., Widmer, A. 1985. *Helv. Chim. Acta* 68: 682–704
105. Fesik, S. W., Gampe, R. T. Jr., Holzman, T. F., Egan, D. A., Edalji, R., et al. 1990. *Science* 250:1406–9
106. Hruby, V. J., Pettitt, B. M. 1989. *Computer-Aided Drug Design*, ed. T. J. Perun, C. L. Propst, pp. 405–60. New York: Marcel-Dekker
107. Hruby, V. J. 1989. *New Methods in Drug Research*, ed. A. Makriyannis, 3:43–60. Prous Sci. Publ.
108. Kessler, H., Haupt, A., Will, M. 1989. See Ref. 106, pp. 461–84
109. Matsuo, H., Baba, Y., Nair, R. M. G., Arimura, A., Schally, A. V. 1971. *Biochem. Biophys. Res. Commun.* 43: 1334–39
110. Burgus, R., Butcher, M., Amoss, M., Ling, N., Monahan, M., et al. 1972. *Proc. Natl. Acad. Sci. USA* 69:278–82
111. Karten, M. J., Rivier, J. E. 1986. *Endocr. Rev.* 7:44–66
112. Struthers, R. S., Tanaka, G., Koerber, S. C., Solmajer, T., Baniak, E. L., et al. 1990. *Proteins* 8:295–304
113. Chary, K. V. R., Srivastava, S., Hosur, R. V., Roy, K. B., Govil, G. 1986. *Eur. J. Biochem.* 158:323–32
114. Momany, F. A. 1976. *J. Am. Chem. Soc.* 98:2990–96
115. Monahan, M. W., Amoss, M. S., Anderson, H. A., Vale, W. 1973. *Biochemistry* 12:4616–20
116. Ling, N., Vale, W. 1975. *Biochem. Biophys. Res. Commun.* 63:801–6
117. Rivier, J., Rivier, C., Perrin, M., Porter, J., Vale, W. 1981. *LHRH Peptides as Female and Male Contraceptives*, ed. G. I. Zatuchni, J. D. Shelton, J. J. Sciarra, pp. 13–23. Philadelphia: Harper & Row
118. Struthers, R. S., Rivier, J., Hagler, A.

T. 1984. *Conformationally Directed Drug Design*, ed. J. A. Vida, M. Gordon, pp. 239–61. Washington, DC: Am. Chem. Soc.

119. Struthers, R. S., Rivier, J., Hagler, A. T. 1985. *Ann. NY Acad. Sci.* 439:81–96

120. Baniak, E. L. II, Rivier, J. E., Struthers, R. S., Hagler, A. T., Gierasch, L. M. 1987. *Biochemistry* 26:2642–56

121. Rivier, J., Varga, J., Porter, J., Perrin, M., Hass, Y., et al. 1986. *Peptides: Structure and Function*, ed. C. M. Deber, V. J. Hruby, K. D. Kopple, pp. 541–44. Rockford, Ill: Pierce Chem. Co.

122. Rivier, J., Kupryszewski, G., Varga, J., Porter, J., Rivier, C., et al. 1988. *J. Med. Chem.* 31:677–82

123. Rivier, J., Rivier, C., Vale, W., Koerber, S., Corrigan, A., et al. 1990. *Peptides: Chemistry, Structure and Biology*, ed. E. J. Rivier, G. R. Marshall, pp. 33–37. Leiden, The Netherlands: ESCOM

124. Bienstock, R. J., Koerber, S. C., Rizo, J., Rivier, J. E., Hagler, A. T., Gierasch, L. M. 1992. In preparation

125. Hadley, M. E. 1989. *The Melanotropic Peptides*, Vols. I, II, and III. Boca Raton, Fla: CRC Press

126. O'Donohue, T. L., Jacobowitz, D. M. 1980. *Polypeptide Hormones*, ed. R. F. Beers, E. G. Bassett, pp. 203–22. New York: Raven

127. Sawyer, T. K., Sanfilippo, P. J., Hruby, V. J., Engel, M. H., Heward, C. B., et al. 1980. *Proc. Natl. Acad. Sci. USA* 77:5754–58

128. Sawyer, T. K., Hruby, V. J., Darman, P. S., Hadley, M. E. 1982. *Proc. Natl. Acad. Sci. USA* 79:1751–55

129. Knittel, J. J., Sawyer, T. K., Hruby, V. J., Hadley, M. E. 1983. *J. Med. Chem.* 26:125–29

130. Marwan, M. M., Malek, Z. A. A., Kreutzfeld, K. L., Hadley, M. E., Wilkes, B. C., et al. 1985. *Mol. Cell. Endocrinol.* 41:171–77

131. Al-Obeidi, F., Hadley, M. E., Pettitt, B. M., Hruby, V. J. 1989. *J. Am. Chem. Soc.* 111:3413–16

132. Al-Obeidi, F., Castrucci, A. M., Hadley, M. E., Hruby, V. J. 1989. *J. Med. Chem.* 32:2555–61

133. duVigneaud, V., Ressler, C., Swan, J. M., Roberts, C. W., Katsoyannis, P. G. 1954. *J. Am. Chem. Soc.* 76:3115–21

134. Brewster, A. I. R., Hruby, V. J., Glasel, J. A., Tonelli, A. E. 1973. *Biochemistry* 12:5294–304

135. Brewster, A. I. R., Hruby, V. J. 1973. *Proc. Natl. Acad. Sci. USA* 70:3806–9

136. Hagler, A. T., Osguthorpe, D. J., Dauber-Osguthorpe, P., Hempel, J. 1985. *Science* 227:1309–15

137. Wood, S. P., Tickle, I. J., Treharne, A. M., Pitts, J. E., Mascarenhas, Y., et al. 1986. *Science* 232:633–36

138. Walter, R., Smith, C. W., Mehta, P. K., Boonjarern, S., Arruda, J. A. L., Kurtzman, N. A. 1977. *Disturbances in Body Fluid Osmolality*, ed. T. E. Andreoli, J. J. Grantham, F. C. Rector, pp. 1–36. Bethesda: Am. Physiol. Soc.

139. Skala, G., Smith, C. W., Taylor, C. J., Ludens, J. H. 1984. *Science* 226:443–45

140. Walter, R. 1977. *Fed. Proc.* 36:1872–78

141. Hruby, V. J. 1987. *Trends Pharmacol. Sci.* 8:336–39

142. Hruby, V. J., Mosberg, H. I. 1982. *Peptides* 3:329–36

143. Hill, P. S., Smith, D. D., Slaninova, J., Hruby, V. J. 1990. *J. Am. Chem. Soc.* 112:3110–13

144. Frič, I., Hlavacek, J., Rockway, T. W., Chan, W. Y., Hruby, V. J. 1990. *J. Protein Chem.* 9:9–15

145. Pettibone, D. J., Clineschmidt, B. V., Anderson, P. S., Freidinger, R. M., Lundell, G. F., et al. 1989. *Endocrinology* 125:217–22

146. Freidinger, R. M., Williams, P. D., Tung, R. D., Bock, M. G., Pettibone, D. J., et al. 1990. *J. Med. Chem.* 33:1843–45

147. Bock, M. G., DiPardo, R. M., Williams, P. D., Pettibone, D. J., Clineschmidt, B. V., et al. 1990. *J. Med. Chem.* 33:2321–23

148. Lavielle, S., Chassaing, G., Ploux, O., Loeuillet, D., Besseyre, J., et al. 1988. *Biochem. Pharmacol.* 37:41–49

149. Chassaing, G., Convert, O., Lavielle, S. 1986. *Eur. J. Biochem.* 154:77–85

150. Chassaing, G., Convert, O., Lavielle, S. 1986. *Biochim. Biophys. Acta* 873:397–404

151. Loeuillet, D., Convert, O., Lavielle, S., Chassaing, G. 1989. *Int. J. Peptide Protein Res.* 33:171–80

152. Hölzemann, G., Pachler, K. G. R. 1989. *Int. J. Peptide Protein Res.* 34:139–47

153. Levian-Teitelbaum, D., Kolodny, N., Chorev, M., Selinger, Z., Gilon, C. 1989. *Biopolymers* 28:51–64

154. Richardson, J. S., Richardson, D. C. 1988. *Science* 240:1648–52

155. Chassaing, G., Convert, O., Lavielle, S. 1987. *Peptides 1986*, ed. D. Theodopoulos, pp. 303–6. Berlin:de Gruyter

156. Ploux, O., Lavielle, S., Chassaing, G., Julien, S., Marquet, A., et al. 1987. *Proc. Natl. Acad. Sci. USA* 84:8095–99

157. Gilon, C., Halle, D., Chorev, M.,

Selinger, Z., Goldshmith, R., Byk, G. 1991. See Ref. 101, pp. 404–6

158. Theodoropoulos, D., Poulos, C., Gatos, D., Cordopatis, P., Escher, E., et al. 1985. *J. Med. Chem.* 28:1536–39

159. Cascieri, M. A., Chicchi, G. G., Freidinger, R. M., Colton, C. D., Perlow, D. S., et al. 1985. *Mol. Pharmacol.* 29:34–38

160. Gierasch, L. M., Rockwell, A. L., Thompson, K. F., Briggs, M. S. 1985. *Biopolymers* 24:117–35

161. Kuntz, I. D. 1972. *J. Am. Chem. Soc.* 94:8568–72

162. Schwyzer, R., Sieber, P., Gorup, B. 1958. *Chimia* 12:90–91

163. Gierasch, L. M., Deber, C. M., Madison, V., Niu, C., Blout, E. R. 1981. *Biochemistry* 20:4730–38

164. Saragovi, H. U., Fitzpatrick, D., Raktabutr, A., Nakanishi, H., Kahn, M., Greene, M. I. 1991. *Science* 253:792–95

165. Kishore, R., Raghothama, S., Balaram, P. 1987. *Biopolymers* 26:873–91

166. Karle, I. L., Kishore, R., Raghothama, S., Balaram, P. 1988. *J. Am. Chem. Soc.* 110:1958–63

167. Kishore, R., Kumar, A., Balaram, P. 1985. *J. Am. Chem. Soc.* 107:8019–23

168. Karle, I. L., Flippen-Anderson, J. L., Kishore, R., Balaram, P. 1989. *Int. J. Peptide Protein Res.* 34:37–41

169. Hodgkin, D. C., Oughton, B. M. 1957. *Biochem. J.* 65:752–56

170. Kemp, D. S., Bowen, B. R., Muendel, C. C. 1990. *J. Org. Chem.* 55:4650–57

171. Blundell, T., Sibanda, B. L., Pearl, L. 1983. *Nature* 304:273–75

172. Bouhnik, J., Galen, F.-X., Menard, J., Corvol, P., Seyer, R., et al. 1987. *J. Biol. Chem.* 262:2913–18

173. Liu, C. F., Fehrentz, J. A., Heitz, A., Le Nguyen, D., Castro, B., et al. 1988. *Tetrahedron* 44:675–83

174. Fehrentz, J. A., Heitz, A., Seyer, R., Fulcrand, P., Devilliers, R., et al. 1988. *Biochemistry* 27:4071–78

175. Fehrentz, J. A., Heitz, A., Liu, C. F., Castro, B., Heitz, F., et al. 1989. *Second Forum on Peptides*, ed. A. Aubry, M. Marraud, B. Vitoux, 174:543–46. Colloq. INSERM/ Libbey Eurotext

176. Stroup, A. N., Gierasch, L. M. 1990. *Biochemistry* 29:9765–71

177. King, J., Fane, B., Haase-Pettingell, C., Mitraki, A., Villafane, R., Yu, M.-H. 1990. See Ref. 26, pp. 225–40

178. Felix, A. M., Heimer, E. P., Wang, C.-T., Lambros, T. J., Fournier, A., et al. 1988. *Int. J. Peptide Protein Res.* 32:441–54

179. Felix, A. M., Wang, C.-T., Heimer, E.

P., Campbell, R. M., Madison, V. S., et al. 1990. See Ref. 123, pp. 227–28

180. Madison, V. S., Fry, D. C., Greeley, D. N., Toome, V., Wegrzynski, B. B., et al. 1990. See Ref. 123, pp. 575–77

181. Gulyás, J., Profit, A. A., Gulyás, E. S., Ósapay, G., Taylor, J. W. 1992. *Peptides: Chemistry and Biology*, ed. J. A. Smith, J. E. Rivier. Leiden, The Netherlands: ESCOM. In press

182. Ghadiri, M. R., Choi, C. 1990. *J. Am. Chem. Soc.* 112:1630–32

183. Marqusee, S., Baldwin, R. L. 1987. *Proc. Natl. Acad. Sci. USA* 84:8898–902

184. Zimm, B. H., Bragg, J. K. 1959. *J. Chem. Phys.* 31:526–35

185. Satterthwait, A. C., Chiang, L.-C., Arrhenius, T., Cabezas, E., Zavala, F., et al. 1990. *Bull. WHO* 68S:17–25

186. Chiang, L., Cabezas, E., Noar, B., Arrhenius, T., Lerner, R. A., Satterthwait, A. C. 1991. See Ref. 101, pp. 465–67

187. Kemp, D. S., Boyd, J. G., Muendel, C. C. 1991. *Nature* 352:451–54

188. Kemp, D. S., Curran, T. P., Boyd, J. G., Allen, T. J. 1991. *J. Org. Chem.* 56:6683–97

189. Kemp, D. S., Curran, T. P., Davis, W. M., Boyd, J. G., Muendel, C. 1991. *J. Org. Chem.* 56:6672–82

190. Sueki, M., Lee, S., Powers, S. P., Denton, J. B., Konishi, Y., Scheraga, H. A. 1984. *Macromolecules* 17:148–55

191. Vasquez, M., Pincus, M. R., Scheraga, H. A. 1987. *Biopolymers* 26:351–71

192. Marqusee, S., Robbins, V. H., Baldwin, R. L. 1989. *Proc. Natl. Acad. Sci. USA* 86:5286–90

193. Leszczynski, J. F., Rose, G. D. 1986. *Science* 234:849–55

194. Barnett, J. K., Briggs, M. S. 1991. *J. Cell. Biochem.* 15G. *Keystone Symp. Mol. Biol. Abstr. R304*

195. Madison, V. S., Fry, D. C., Olson, G. L., Sarabu, R., Danho, W. 1991. *The Peptide-Protein Bridge*, Toronto, Canada. Abstr. Tu-PM-3

196. Oas, T. G., Kim, P. S. 1988. *Nature* 336:42–48

197. Creighton, T. E. 1978. *Prog. Biophys. Mol. Biol.* 33:231–97

198. Weissman, J. S., Kim, P. S. 1991. *Science* 253:1386–93

199. Shaw, G. S., Hodges, R. S., Sykes, B. D. 1990. *Science* 249:280–83

200. Shaw, G. S., Golden, L. F., Hodges, R. S., Sykes, B. D. 1991. *J. Am. Chem. Soc.* 113:5557–63

201. Kay, L. E., Forman-Kay, J. D., McCubbin, W. D., Kay, C. M. 1991. *Biochemistry* 30:4323–33

202. Dyson, H. J., Lerner, R. A., Wright, P. E. 1988. *Annu. Rev. Biophys. Biophys. Chem.* 17:305–24
203. House, C., Kemp, B. E. 1987. *Science* 238:1726–28
204. Zagotta, W. N., Hoshi, T., Aldrich, R. W. 1990. *Science* 250:568–71
205. Mutter, M. 1988. *Trends Biochem. Sci.* 13:260–65
206. Mutter, M., Tuchscherer, G. G., Miller, C., Altmann, K.-H., Carey, R. I., et al. 1992. *J. Am. Chem. Soc.* 114:1463–70
207. Mutter, M., Hersperger, R., Gubernator, K., Müller, K. 1989. *Proteins* 5:13–21
208. Hahn, K. W., Klis, W. A., Stewart, J. M. 1990. *Science* 248:1544–47
209. Hodges, R. S., Semchuk, P. D., Taneja, A. K., Kay, C. M., Parker, J. M. R., Mant, C. T. 1988. *Peptide Res.* 1:19–30
210. Ho, S. P., DeGrado, W. F. 1987. *J. Am. Chem. Soc.* 109:6751–58
211. Regan, L., DeGrado, W. F. 1988. *Science* 241:976–78
212. DeGrado, W. F., Wasserman, Z. R., Lear, J. D. 1989. *Science* 243:622–28
213. Richardson, J. S., Richardson, D. C. 1989. *Trends Biochem. Sci.* 14:304–9
214. Hecht, M. H., Richardson, J. S., Richardson, D. C., Ogden, R. C. 1990. *Science* 249:884–91
215. Kaumaya, P. T. P., Berndt, K. D., Heidorn, D. B., Trewhella, J., Kezdy, F. J., Goldberg, E. 1990. *Biochemistry* 29:13–23

Annu. Rev. Biochem. 1992. 61:419–40

THE BIOCHEMISTRY OF 3'-END CLEAVAGE AND POLYADENYLATION OF MESSENGER RNA PRECURSORS

Elmar Wahle and Walter Keller

Department of Cell Biology, Biozentrum, University of Basel, Klingelbergstrasse 70, CH-4056 Basel, Switzerland

KEY WORDS: gene expression, RNA processing, Poly(A) tails

CONTENTS

SUMMARY AND PERSPECTIVES

Poly(A) tails of mRNA are nearly ubiquitous. In animal cells, the only mRNAs known to lack poly(A) are those encoding the major histones (1–3). Poly(A) tails also occur in plants and in yeast (4) as well as in mitochondria (5, 6). Bacteria have a poly(A) polymerase (7), and short poly(A) tails have been reported to be attached to some of their mRNAs (8).

The enzyme that synthesizes poly(A) tails, poly(A) polymerase, was discovered as early as 1960 (9; for a historical review, see 10), but its role did not become clear until 10 years later when the poly(A) tails were found (11–13). Even before cloning and sequencing methods were available, hybridization

419

experiments confirmed that the poly(A) tails are not encoded in the DNA
(14). Differential inhibition of hnRNA and poly(A) synthesis supported the
idea that poly(A) was added after transcription by the template-independent
poly(A) polymerase (15, 16). Analyses of viral mRNA synthesis in infected
cells first demonstrated that transcription proceeds past the polyadenylation
site. This suggested that the 3'-end receiving poly(A) is generated by nuclease
action and not by transcription termination (17, 18). In animal cells,
endonucleolytic cleavage of the primary transcript at the poly(A) site, followed by the addition of the poly(A) tail, is now firmly established as the
pathway of 3'-end formation for polyadenylated mRNAs.

Poly(A) tails have been known for 20 years; still, their purpose in the cell is
not well understood. Evidence for a role in the control of mRNA stability and
in translation is emerging (19–21), but biochemical details are lacking, and
additional functions such as in the export of RNA from the nucleus (22)
cannot be ruled out. These questions are not treated here. The primary subject
of this review is the mechanism of 3'-end formation in animal cells, the only
polyadenylation pathway that has been studied in some detail. RNA sequences governing this process have been defined, several of the proteins
catalyzing cleavage and polyadenylation have been purified to homogeneity,
and the first cDNAs encoding them have been cloned. An exciting new field
is the regulation of poly(A) tail length, which is involved in the control of
translation. We also review what is known about 3'-end formation in yeast
and in plants. The regulation of mRNA production by alternative
polyadenylation is discussed only as it pertains to the mechanism of 3'-end
formation. Related reviews have been published (23–28).

POLYADENYLATION IN ANIMAL CELLS

Polyadenylation Signals

As mentioned above, 3'-end processing consists of endonucleolytic cleavage
of the primary transcript followed by polyadenylation of the upstream RNA
fragment. The cleavage event is directed by two sequence elements in the
RNA, one upstream and one downstream of the cleavage site.

THE AAUAAA SEQUENCE   Comparison of the nucleotide sequences preceding the polyadenylation sites in several messenger RNAs initially revealed the
presence of the hexanucleotide sequence AAUAAA in all of them (29).
Numerous mutagenesis experiments as well as the analysis of naturally
occurring mutations have confirmed that this sequence, typically located
10–35 nucleotides upstream of the poly(A) site, is essential for 3'-end formation of polyadenylated RNA (22, 30-37). Early experiments demonstrated
that, although mutations in AAUAAA strongly reduce 3'-end formation, all
detectable RNA is polyadenylated (22, 31). This was interpreted as evidence

that the hexanucleotide sequence is required only for cleavage but not for polyadenylation. However, subsequent use of an in vitro polyadenylation system that was independent of prior cleavage (see below) unequivocally demonstrated that AAUAAA is essential for polyadenylation as well (35). The initial finding that all the mutant RNA that was cleaved was also polyadenylated thus does not indicate AAUAAA-independent poly(A) synthesis. Rather, the result reflects the tight coupling between cleavage and polyadenylation. Mutations in AAUAAA affect the formation of the protein-RNA complex that catalyzes both reactions (see below); once a complex has managed to assemble even on a mutant substrate, both cleavage and polyadenylation are carried out.

Systematic mutagenesis of all the nucleotides in AAUAAA revealed that changes in the second A, in particular to a U, are relatively well tolerated (36, 37). This agrees with analyses of natural 3'-end sequences: while most polyadenylated mRNAs have a perfect hexanucleotide sequence, AUUAAA is by far the most frequent variant (36, 37).

Some mRNAs do not contain a recognizable AAUAAA element, i.e. no sequence differing in less than two nucleotides from the consensus. Interestingly, many, if not all, of these cases involve alternative polyadenylation: of the four major mRNAs derived from the mouse dihydrofolate reductase gene, one lacks AAUAAA by the definition given above whereas two others have unusual variants of the signal (38, 39). Likewise, one of two transcripts of human cyclin D (40) and a sperm-specific c-abl mRNA in the mouse (41) lack AAUAAA. The most significant group on this list are spermatid-specific transcripts of chicken histone genes. In somatic cells, these genes are transcribed and processed into mRNAs devoid of poly(A) tails, as is typical for histones. In spermatids, however, the RNA sequences signalling the normal histone-specific 3'-end processing pathway are ignored, and poly(A) tails are added to the RNA. Most of these histone mRNAs have no perfect copy of the AAUAAA sequence, and several have no sequence differing by less than two nucleotides (42). The polyadenylation pathway of any of the RNAs lacking AAUAAA is unknown. The association of unusual or missing hexanucleotide signals with alternative, in several cases tissue-specific, processing suggests that there might be a variant polyadenylation pathway for these RNAs.

THE DOWNSTREAM ELEMENTS   In addition to the hexanucleotide sequence upstream of the poly(A) addition site, sequences downstream are required for 3'-end cleavage. Mutagenesis studies in vivo and in vitro have defined these sequences with varying degrees of precision in a number of genes within approximately 50 nucleotides of the cleavage site (43–62). So far, no consensus sequence has emerged, and, in fact, there seem to be two different types of signals.

Experiments in which synthetic oligonucleotides were used to restore the activity of a polyadenylation site inactivated by the deletion of the downstream element (53, 56) identified a GU-rich and a U-rich element. The polyadenylation site of the α-globin gene contains both, and they augment each other (56). The GU-rich element resembles a proposed consensus, YGUGUUYY (52). Point mutations introduced into the GU-rich sequence (53) are not entirely consistent with this consensus. However, in the eight-nucleotide downstream element of the SV40 early polyadenylation site, the conversion of either of two particular G residues to a U improves the efficiency of 3'-end processing threefold, whereas the replacement of a different G by a U reduces it fourfold (53). Simultaneous alterations in two nucleotides of the element can abolish processing activity. These data argue that the GU-element is a specific sequence and not just characterized by its GU content. In contrast, point mutations in the U-rich element have little effect on 3'-end processing (53).

The activity of the downstream elements depends on their proximity to the polyadenylation site (53, 63). Two studies (64, 65) found that deletions downstream of the poly(A) site affected the position but not the efficiency of cleavage. Since the downstream element is poorly defined and possibly redundant, these results might be explained by a failure to eliminate the critical element(s). Alternatively, sequences upstream of AAUAAA might to some extent substitute for downstream sequences, just as they seem to be able to rescue unusual variants of the AAUAAA sequence (see below). RNA modification-interference experiments equivalent to saturation mutagenesis demonstrated that the only nucleotides stringently required for 3'-end processing are those in the AAUAAA sequence (66). This finding is in agreement with the poor conservation and redundancy of the downstream elements. The observation that, in contrast to the highly conserved AAUAAA sequence, the downstream elements are variable, has led to the suggestion that these may determine the varying efficiencies of poly(A) sites (45). In fact, there is now some experimental evidence for this (see below).

THE CLEAVAGE SITE   Little is known about the selection of the precise cleavage site. The distance between AAUAAA and the cleavage site is variable, and the sequence surrounding the latter is not highly conserved. A comparison of cDNA and genomic sequences shows that in 70% of all RNAs, the first nucleotide of the poly(A) tail could be either template-encoded or added by poly(A) polymerase (37). In the two cases analyzed in detail (see below), and thus probably in most others, it is template-encoded. Although the nucleotide at the cleavage site thus seems to be usually an A, this A is not strictly required (37). The nucleotide preceding the terminal A is a C in 59% of all genes analyzed (37). However, it is evident that the cleavage machinery does not simply select the first CA downstream from the recognition sequence. Alterations of the cleavage site itself as well as mutations downstream have been reported to alter the accuracy of cleavage (30, 37, 57, 64, 65).

UPSTREAM SEQUENCES AND THE INFLUENCE OF PROMOTERS Chemical synthesis of a "polyadenylation cassette" suggests that the combination of AAUAAA and downstream sequences is sufficient to cause cleavage and polyadenylation of a mRNA precursor (67). However, additional sequence elements influencing 3'-end processing have been identified upstream of AAUAAA in some genes. These include the L1 polyadenylation site of the adenovirus major late transcription unit (68) and the SV40 late polyadenylation site (69). The mechanism of action of an upstream sequence has to some extent been elucidated for a normal cellular gene, the doublesex (*dsx*) gene of *Drosophila*. *dsx* is part of a hierarchy of genes, largely regulated by alternative RNA processing, that determine sexual differentiation (70). Genetic analyses indicated that female-specific splicing and polyadenylation of *dsx* are induced by the transformer-2 (*tra-2*) gene product. In vitro, the Tra-2 protein binds to a sixfold repeated 13mer sequence located upstream of the female-specific polyadenylation site of *dsx*. This binding site is independently required for both sex-specific polyadenylation and splicing. Therefore, the Tra-2 protein acts as a positive regulator of polyadenylation (71). It is not yet clear in which way the protein interacts with the polyadenylation machinery. Note that this interpretation of *tra-2* action on the *dsx* polyadenylation site has been challenged (71a).

Upstream elements may be common among retroviruses and related elements. Their mechanism of replication requires terminally redundant RNA sequences. As a consequence, polyadenylation signals are found both at the 5'-end and at the 3'-end of the RNA. For the production of mRNA, the signal at the 5'-end must be ignored, and the one at the 3'-end must be used. The exclusive use of the second polyadenylation site often depends on signals that are present in the unique sequences upstream of this site. The AAUAAA present at the 5'-end lacks such upstream signals. The signals have been identified by deletions, but so far not further defined (72–76). In a hepatitis B virus, the dependence of 3'-end processing on the upstream regions is partially relieved by the conversion of the unusual hexanucleotide sequence, UAUAAA, to the regular AAUAAA (73).

In other cases, the retroviral dilemma of poly(A) site usage has been solved in a different way: proximity to the promoter inhibits the use of the 5' polyadenylation sites of some viruses (77–79). This inhibition by a nearby promoter was not observed for normal poly(A) sites of cellular genes. In the case of HIV 1, the inhibition has been attributed to the downstream part of the polyadenylation signal (79). The two mechanisms that secure the exclusive use of the second polyadenylation site, dependence on unique upstream sequences and inhibition of the first site by the promoter, may operate simultaneously, as both phenomena have been described for HIV 1 (74, 76, 79).

Another solution to the same problem is used by the human T cell leukemia virus (HTLV-1). In this virus, the AAUAAA sequence is located in the U3 sequences, which are unique for the second polyadenylation site. The hexanucleotide is separated by 250 nucleotides from the polyadenylation site. The intervening nucleotides form a very stable secondary structure that brings polyadenylation signal and polyadenylation site in close proximity (80). Similar mechanisms may be considered for cellular genes that lack obvious AAUAAA signals within the usual distance from the polyadenylation site (see above).

Mutations in polyadenylation signals result in a severely reduced concentration of mRNA, demonstrating that 3'-end formation is essential for the production of mRNA. This is most convincingly shown by the fact that certain types of thalassemias, human diseases that stem from a lack of globin synthesis, can be caused by polyadenylation defects (32, 34). Also, conversion of an imperfect hexanucleotide sequence (UAUAAA) to AAUAAA increased the production of mRNA in a hepatitis B virus (see above), and the steady-state mRNA level in transient expression experiments varied depending on the use of the SV40 late or SV40 early polyadenylation site (69). These data suggest that cells or viruses may use variations in polyadenylation signals to control the amount of mRNA produced. As an example, herpesvirus infection induces the activity of a factor that stimulates the use of a particular late viral poly(A) site (81). The identity of this factor and its mode of action have not been determined.

3'-End Processing in vitro

PROCESSING IN NUCLEAR EXTRACTS   Correct 3'-end cleavage followed by polyadenylation can be observed in nuclear extract from HeLa cells (82, 83). The in vitro reaction depends on the AAUAAA sequence as well as on the downstream element, and proceeds either with RNA synthesized by RNA polymerase II in the extract or with exogenous RNA made in vitro and added to the nuclear extract. Cleavage and polyadenylation are closely coupled as in vivo; cleaved RNA lacking poly(A) cannot be detected under normal circumstances. However, when ATP is replaced by an analogue containing a nonhydrolyzable α-β bond, the RNA is cleaved but not polyadenylated. Chain-terminating ATP-analogues also lead to the accumulation of the cleavage product, extended by a single nucleotide (83). Finally, chelation of divalent cations by EDTA allows cleavage but prevents polyadenylation.

EDTA also retards the degradation of the downstream RNA fragment. Detection of the downstream fragment confirmed the proposed endonucleolytic cleavage at the poly(A) site (83). Both RNA fragments have been analyzed in detail (84, 85). The upstream fragment ends in a 3'-hydroxyl at the site of poly(A) addition. In both cases analyzed, the 3'-terminal nucleotide of this fragment is an adenosine. The downstream fragment is heterogeneous, consisting of a family of molecules with 5'-phosphates. These RNA molecules differ in the position of their 5'-termini, the longest starting with the first nucleotide after the polyadenylation site. The easiest interpretation of these results is that the RNA is cleaved in an endonucleolytic manner precisely at the polyadenylation site, and the downstream fragment is then degraded exonucleolytically from the 5' end. However, imprecise cutting downstream of the poly(A) site with exonucleolytic trimming of the upstream fragment cannot be excluded. This question should be resolved by the analysis of cleavage products in a reconstituted system.

Just as cleavage of the RNA can occur under conditions that prevent polyadenylation, polyadenylation can occur without prior cleavage. This reaction depends on the presence of an AAUAAA sequence and is usually carried out with so-called "precleaved" RNA that ends near the polyadenylation site (35). There is no strict requirement for a particular distance between AAUAAA and the 3'-end; ends at least 400 nucleotides downstream of the polyadenylation signal can be elongated (33).

The factors carrying out the 3'-processing reactions have been partially or completely purified by resolution and reconstitution of the in vitro system. Specific polyadenylation of precleaved RNA is the simpler reaction and is discussed first.

RECONSTITUTION FROM PURIFIED COMPONENTS: POLY(A) POLYMERASE   AAUAAA-dependent polyadenylation of precleaved RNA requires two factors: poly(A) polymerase and a specificity factor (86–91). Poly(A) polymerase active in this reaction has been purified from calf thymus to near homogeneity (92). The purified enzyme exists in several chromatographically and electrophoretically distinct forms with molecular weights of 57,000 or 60,000, and behaves as a monomer in solution. Assay of the enzyme during purification was facilitated by its ability to polyadenylate nonspecifically any RNA in the absence of other factors. This nonspecific reaction is much more efficient when Mn^{2+} is substituted for Mg^{2+}, due to a higher affinity of the enzyme for the 3'-terminus of the RNA (92). Even under physiological conditions, i.e. in the presence of Mg^{2+}, the polymerase by itself has no primer specificity except for a slight preference for either poly(A) or a 3'-terminal adenylate residue. This specificity is in agreement with the presence of an A at the cleavage site of most transcripts (see above). Most importantly, poly(A) polymerase itself does not recognize the AAUAAA signal. Elongation of a poly(A) primer proceeds distributively and linearly without any lag phase. ATP is the only nucleotide that is polymerized to any significant extent (92).

cDNAs encoding bovine poly(A) polymerase have been isolated with probes derived from amino acid sequences of the purified protein. One clone encodes a protein with a molecular weight of 83,000 (93, 94). Its identity was confirmed by the expression in *Escherichia coli* of a protein of the expected molecular weight that copurified with specific and nonspecific polyadenylation activities (93). The difference between the molecular weight of this protein and that of the enzyme purified from calf thymus suggests partial proteolytic degradation of the latter. A C-terminal domain of 20 kDa thus seems to be dispensable for specific and nonspecific polyadenylation. This has been confirmed by deletions introduced into another cDNA clone. This clone encodes a protein of 77 kDa, differing in its C-terminal sequence from the 83-kDA protein. The 77-kDa protein, expressed in reticulocyte lysates, is also functional in specific and nonspecific polyadenylation assays. A deletion derivative lacking 150 amino acids from the C-terminus retains its activity (94). The N-terminus of poly(A) polymerase contains a possible match to the ribonucleoprotein particle (RNP) consensus sequence of RNA-binding proteins (93, 94). A similarity (94) to a "sequence module" found in template-dependent polymerases is of uncertain significance (see below).

Surprisingly, additional mRNAs related to poly(A) polymerase exist. A class of cDNA clones has been found that encodes a protein of 43 kDa almost identical to the N-terminal half of the 83-kDa poly(A) polymerase (93). High stringency Northern blots show two mRNAs near 4.5 kilobases (kb), which probably encode the 83- and 77-kDa proteins described above, an inabundant RNA near 2.4 kb, which may correspond to the 43-kDa cDNA clone, and another RNA of 1.3 kb, for which no cDNA clone has been found so far (93).

A considerable amount of work on poly(A) polymerase was done before the specific AAUAAA-dependent polyadenylation reaction was known (95, 96). Almost all of the enzymes purified in this earlier work had specific activities not exceeding 10^5 pmol AMP incorporated/min/mg in nonspecific poly(A) synthesis (95, 96). In contrast, the specific activities of the two preparations from calf thymus (92, 97), the identity and purity of which has now been confirmed by cloning, was higher than 10^7 pmol AMP/min/mg. The purity of other preparations is thus uncertain and results obtained with them may have to be reevaluated. Lack of purity may explain nonlinear kinetics of earlier preparations and their variable responses to divalent cations (95). The existence of several different types of poly(A) polymerases has often been claimed, based on the observation of different chromatographic peaks or the purification from the same source of enzymes differing in their properties (95). These results may now to some extent be attributed to partial proteolysis and incomplete purification. So far, all poly(A) polymerases tested, including chromatographically distinct fractions purified from one cell type or tissue, were functional in AAUAAA-dependent polyadenylation (92, 98, 99) and are, therefore, not likely to be fundamentally different. On the other hand, the existence of different poly(A) polymerase-related polypeptides is also suggested by cDNA cloning and Northern analysis (see above). The function of these proteins remains to be determined.

Poly(A) polymerase has also been purified from Vaccinia virus (100). The viral enzyme is a heterodimer with a catalytic subunit of 55 kDa and another subunit of 35 kDa. The genes for both polypeptides have been cloned (101). Surprisingly, there is no obvious sequence relationship between the enzymes from Vaccinia and from mammalian cells (see above) or yeast (see below). A requirement of the viral poly(A) polymerase for specific sequences in the RNA primer or for an interaction with other factors is not known.

RECONSTITUTION FROM PURIFIED COMPONENTS: CLEAVAGE AND POLY-ADENYLATION SPECIFICITY FACTOR   In the presence of Mg^{2+}, the mammalian cellular poly(A) polymerase is barely active on its own. An additional factor stimulates the polymerase to elongate specifically those RNAs that carry the sequence AAUAAA close to their 3'-ends (86–91). Since this factor is also required for the cleavage reaction, it has been termed Cleavage and Polyadenylation Factor (CPF; 86). What appears to be the same factor, as judged by chromatographic properties and function in the assay, has been described as Specificity Factor (SF; 89) or Polyadenylation Factor 2 (PF2; 91). In agreement with others (J. L. Manley, personal communication; J. R. Nevins, personal communication), we propose to call this factor Cleavage- and Polyadenylation Specificity Factor, CPSF.

CPSF has been purified to near homogeneity from calf thymus and from HeLa cells; preparations from both sources have the same properties (102). CPSF has a native molecular weight near 500,000. It consists of four polypeptides of 160, 100, 73, and 30 kDa, which are present in approximately equimolar amounts and comigrate in column chromatography and sedimentation (102). Immunoprecipitation of a partially purified preparation from HeLa cells with serum from a systemic lupus erythematosus patient revealed polypeptides of 170, 130, 100, 74, and 42 kDa (103). This polypeptide pattern is similar to that of the purified preparation (102), except for the presence of the 130-kDa polypeptide and the molecular weight estimate for the smallest polypeptide.

In gel retardation experiments, purified CPSF in the absence of any other factor specifically binds to RNAs containing the recognition sequence AAUAAA (102, 104, 105), confirming earlier experiments with partially purified fractions (91, 106). CPSF binding is abolished by point mutations in AAUAAA. RNA modification-interference experiments showed that all six bases within AAUAAA but no other nucleotide are required for binding (104). RNAs as short as 10 nucleotides are bound specifically and the sequence of the few nucleotides outside AAUAAA is irrelevant (106). CPSF thus seems to recognize only the AAUAAA sequence but no secondary structure, unlike many other RNA-binding proteins, which depend on the presence of a hairpin or other secondary structure in their binding sites (107). CPSF appears to interact with certain sugar residues as well as with the bases:

replacement of the ribo-UMP in AAUAAA with deoxyribo-UMP abolishes CPSF binding (106), and methylation of the 2'-hydroxyl group of any of several riboses in AAUAAA has the same effect (105). Upon UV-irradiation of CPSF-RNA complexes, two polypeptides of 160 and 35 kDa become covalently attached to the RNA (104). This suggests that the largest and the smallest subunits of CPSF make close contacts with the RNA. Probably CPSF also contacts poly(A) polymerase, thereby increasing the affinity of the enzyme for AAUAAA-containing RNAs. This suggestion has not yet been tested experimentally.

Previous evidence suggested the involvement of a small nuclear RNP (snRNP) in 3'-end processing of pre-mRNA. The U11 snRNP (108) was initially found to copurify with CPSF (86). However, a separation of CPSF from U11 was later achieved (89, 102). A pure preparation of CPSF contained the polypeptide subunits in amounts that could account for the amount of substrate RNA bound in gel retardation assays. U11 RNA, in contrast, was present at a more than 10-fold lower concentration (102). Other RNAs were also not found in significant quantities. Specific antisera (109) did not detect any of the common snRNP proteins in purified CPSF (102). A role of U11 or any other RNA in CPSF activity is thus unlikely. Other major RNA processing reactions, splicing and 3'-end processing of histone mRNAs, depend on the action of snRNPs (110, 111); so far, this has not been demonstrated for 3'-end cleavage and polyadenylation.

RECONSTITUTION FROM PURIFIED COMPONENTS: CLEAVAGE FACTORS AAUAAA is required for both polyadenylation and endonucleolytic cleavage of the pre-mRNA. Likewise, the AAUAAA-binding factor, CPSF, must be present not only for polyadenylation but also for the cleavage reaction (86, 89, 91). For the cleavage of at least some poly(A) sites, poly(A) polymerase is also necessary (87–89, 91). The presence of poly(A) polymerase in the complex that catalyzes the endonucleolytic cleavage may explain the tight coupling of cleavage and poly(A) addition. Endonucleolytic cleavage also requires, in addition to poly(A) polymerase and CPSF, cleavage factors which are specific for this reaction; they are dispensable for poly(A) addition. Up to three cleavage factors have been separated from HeLa cell nuclear extracts (86, 89, 91). The identity of the endonuclease cleaving the RNA is obscure.

Purified Cleavage Stimulation Factor (CStF; 89, 112) consists of three polypeptides of 77, 64, and 50 kDa, which form a tight complex. Monoclonal antibodies against the 64- and 50-kDa subunits inhibit the cleavage reaction, deplete purified fractions of CStF activity, and precipitate the heterotrimer (112). In nuclear extract, a protein of 64 kDa can be specifically cross-linked to substrate RNA undergoing 3'-end processing (113, 114). This polypeptide is probably the 64-kDa subunit of CStF, as it can be precipitated by the monoclonal antibodies against this cleavage factor (112). UV-cross-linking of the 64-kDa subunit of purified CStF to substrate RNA depends on the simultaneous presence of CPSF (115). How CStF and CPSF interact is unknown. Cleavage Factor 1 (CF1), purified by a different group (91, 103), has subunits of 76, 64, and 48 kDa. Its 64-kDa subunit can also be cross-linked to RNA (103). Therefore, CF1 is very likely identical with CStF.

Little is known about the other cleavage factors; none has been purified to homogeneity so far. A factor binding to a downstream polyadenylation element has been identified by UV-cross-linking and partially purified (116). It is un-known whether it is identical to any of the factors identified by functional assays. An involvement in 3'-end processing has not yet been demonstrated.

COMPLEX FORMATION   Native gel electrophoresis and sedimentation ex-periments demonstrated the formation of large nucleoprotein complexes on RNAs serving as cleavage/polyadenylation substrates in nuclear extract. The formation of these complexes is rapid, ATP dependent, and heparin resistant. Complexes form only on RNAs containing polyadenylation signals, and protect these sequences against various nucleases. Three different complexes containing precursor RNA, cleaved product, and polyadenylated product have been distinguished (117–124).

AAUAAA is the only strongly conserved sequence required for 3'-end processing, and the AAUAAA-binding factor, CPSF, appears to be the only 3'-end processing factor that by itself can form a complex with the RNA stable enough to be detectable in gel retardation experiments (see above). Thus, the interaction between the hexanucleotide sequence and CPSF is very likely the basis for the formation of 3'-end processing complexes. Binding of CPSF to AAUAAA is ATP independent (W. Keller, unpublished data). The role of ATP in complex formation has not been defined. The association of CPSF and its binding site is not very stable; it is easily challenged by AAUAAA-containing competitor RNA (91, 104). The cleavage factor CF1 (probably identical with CStF; see above) has been reported to stabilize the interaction between CPSF and the substrate RNA. This stabilization depends on the presence of the downstream RNA sequences (91, 103).

As the downstream sequences are the variable element among those that direct 3'-end processing, it has been proposed that they determine the variable efficiencies of poly(A) sites. In fact, a comparison of two downstream element mutants with the wild-type sequence revealed a qualitative correla-tion between their processing efficiencies and their abilities to form com-plexes stable to a challenge by a competing substrate (125). These data are consistent with a model in which the initial complex between AAUAAA and CPSF is converted into a "committed" complex by the interaction with CF1, and the stability of the committed complex determines the strength of the poly(A) site (125). The proposed role of CF1 is also consistent with UV-cross-linking of its 64-kDa subunit to RNA. This cross-linking is stimulated by the downstream sequences (103). UV cross-linking of the 64-kDa subunit of CStF (see above) is in agreement with these data and supports the identity of the two factors. However, binding of CF1/CStF to a particular RNA sequence remains to be demonstrated directly.

Whether poly(A) polymerase or an additional cleavage factor are stable components of the processing complex is unknown.

Reaction mechanisms and requirements of cleavage and polyadenylation beyond factors and RNA sequences have hardly been investigated. One group has claimed a dependence of cleavage in crude extract on the 5'-cap structure of the RNA substrate (50, 126), while others have denied such a requirement (83, 127). ATP requirements have also been investigated only in crude extract. ATP analogues with a nonhydrolyzable β-γ bond support both cleavage and polyadenylation, whereas analogues with a nonhydrolyzable α-β bond allow cleavage but prevent polyadenylation. No reaction at all is observed in the absence of ATP (83). This may be explained by the ATP requirement for complex formation, but additional roles of ATP are possible. Also in crude extract, a stimulation of 3'-end processing by the presence of an upstream intron has been reported (128).

POLY(A)-BINDING PROTEIN   Poly(A) synthesis in nuclear extracts proceeds in two phases: the first phase, addition of the first 10 AMP residues, is slow, whereas further elongation up to 200 is rapid. Only the addition of the first 10 nucleotides depends on the AAUAAA sequence, whereas the elongation phase is independent of AAUAAA and requires instead the oligo(A) tail added in the first phase (129). In contrast, polyadenylation by purified poly(A) polymerase and CPSF proceeds at a slow and uniform rate, and an oligo(A) tail does not relieve the AAUAAA dependence. An additional protein restores biphasic kinetics to the purified system (130). This protein, purified as a stimulatory factor for poly(A) synthesis, binds specifically to poly(A). Partial amino acid sequences of the prominent 49-kDa protein that copurifies with the activity identified the RNP consensus sequences typical for RNA-binding proteins (131), but no other relationship to the previously described 70-kDa poly(A)-binding protein (132–134). The stimulatory factor was thus termed poly(A)-binding protein II, PAB II. An oligo(A) tail of 10–11 residues is sufficient for binding of PAB II, and the same length of oligo(A) promotes PAB II–dependent elongation by purified poly(A) polymerase in the absence of CPSF and its binding site, AAUAAA. PAB II thus acts as a second specificity factor: while CPSF mediates AAUAAA-dependent polyadenylation, PAB II is responsible for oligo(A)-dependent polyadenylation (130). Oligo(A) extension is most efficient when both CPSF and PAB II are present. As a consequence, the growth of the poly(A) tail in the presence of both factors is biphasic as in the nuclear extract; the addition of the first 10 nucleotides, assisted by CPSF, is slow, and further extension, aided by both CPSF and PAB II, is rapid. It is not yet known whether PAB II affects the processivity of poly(A) polymerase.

Several properties of polyadenylation complexes in crude extracts (see above) change either abruptly after the addition of 10 As or more gradually during the elongation of the tail (135). These changes may be due to PAB II, suggesting that, beyond a mere binding to the growing tail, this protein may induce more profound alterations in the structure of the polyadenylation complex.

In nuclear extract, poly(A) tails are limited to the same length found in vivo, 200–250 nucleotides (129). The chain growth rate in the reconstituted system decreases drastically at this length only if PAB II is present, suggesting that this protein is involved in the control of poly(A) tail length (130). However, in the reconstituted reaction, the tails do grow longer than 250 nucleotides, albeit slowly. An additional factor inhibiting this elongation may thus be required.

Biphasic elongation kinetics were first shown with the purified heterodimeric poly(A) polymerase of Vaccinia virus (136). This behavior depends on the presence of the 39-kDa subunit; it is not observed when the catalytic 55-kDa subunit acts on its own. The 39-kDa subunit is a poly(A)-binding protein (101). The mechanism of the mammalian cellular poly(A) polymerase and the Vaccinia enzyme may, therefore, be similar with respect to the recognition of poly(A).

Relationship Between Polyadenylation and Termination

3'-end processing of mRNA appears to be independent of transcription termination. It can be performed in vitro on pre-made RNA substrates and can happen while transcription is still in progress (17). However, experiments with several genes have suggested a dependence of transcription termination on 3'-end processing (137–140). Mutations interfering with polyadenylation also inhibit transcription termination, and insertion of a polyadenylation site can induce termination. The biochemical mechanism of this coupling is unknown. It has been proposed that the uncapped RNA 5'-end generated by the endonucleolytic cleavage at the poly(A) site serves as an entry point for a termination factor that catches up with RNA polymerase II stalled at some pause site downstream (140). So far, this has remained speculation. A fragment from the *gypsy* transposon of *Drosophila* induces the use of upstream poly(A) sites. Evidence suggesting that the effect of this sequence is due to a protein binding to DNA rather than RNA has led to the speculation that an influence on transcription termination may be involved (141).

Regulated Polyadenylation

Many transcription units contain multiple polyadenylation sites, and the choice between them can determine which gene product is made. Polyadenylation sites apparently differ in their intrinsic strength, i.e. in the frequency with which they are used (see above). Beyond this, little is known about the rules that govern choices between alternative sites, and the question is not discussed here. However, there are also situations in which not the initial processing event but the extent of poly(A) elongation is regulated. A spectacular example is the vasopressin mRNA, which undergoes variations in poly(A) tail length not only in response to vasopressin induction (142), but also in a circadian rhythm, in parallel with the protein product. Long poly(A) tails are always associated with the expression of the hormone (143).

The most widespread regulated changes in poly(A) tail length occur during oocyte maturation and early embryogenesis of many animal species, including *Xenopus*, mouse, and the marine clam *Spisula*. Gene expression at these early stages of development is regulated to a large extent at the step of translation rather than transcription. The tightly controlled onset of translation of particular mRNAs during oocyte and early embryonal development is almost always associated with an equally precisely controlled elongation of the poly(A) tails. Other mRNAs are translated only at early stages and later lose their poly(A) tails at the same time that their translation ceases (144–148).

In contrast to regular polyadenylation, the developmentally regulated elongation of oligo(A) tails occurs in the cytoplasm (149, 150). The process can be observed in *Xenopus* oocytes induced to mature in vitro by progesterone treatment, it occurs on synthetic RNAs injected into such oocytes (149, 151), and can be faithfully reproduced in extracts prepared from oocytes (150, 152). The cyclin-dependent *CDC2*-kinase, which plays a key role in the regulation of the cell cycle, has been implicated in the induction of maturation-dependent polyadenylation (153). RNAs receive poly(A) if they contain the AAUAAA signal as well as a second sequence related to UUUUUAU (149–152). RNAs lacking these signals lose their poly(A) tails (154, 155). Extension of the poly(A) tail appears to be essential for the activation of translation: The so-called G10 RNA receives a long poly(A) tail and is translated during oocyte maturation. Injected synthetic G10 RNA is also polyadenylated and found on polysomes. However, if the 3'-end of the RNA is blocked in vitro by the addition of 3'-deoxy AMP so that it cannot be elongated, the RNA is not found on polysomes (151).

Regulated polyadenylation during mouse oocyte development has been studied in some detail for the mRNA encoding tissue plasminogen activator. This RNA is stored in an inactive state in primary oocytes. During oocyte maturation, the message acquires a long poly(A) tail and is translated. A not precisely defined sequence in the 3'-untranslated region of the RNA is responsible for the regulation of both polyadenylation and translation (156–158). As in the case of the *Xenopus* G10 RNA, polyadenylation and translation can be observed with injected RNA. A 3'-end blocked with 3'-dAMP prevents elongation and translation. Remarkably, the addition of a long poly(A) tail in vitro prior to injection is sufficient to induce translation in primary oocytes, which normally neither polyadenylate nor translate this RNA (158).

Other examples of regulated polyadenylation in *Drosophila* (159) and *Caenorhabditis elegans* (160) seem to differ in details from the examples discussed above, but confirm that translational regulation by poly(A) may be a widespread phenomenon.

POLYADENYLATION IN YEAST

Sequences Required

Only about 50% of all mRNAs of *Saccharomyces cerevisiae* carry an AAUAAA sequence close to their 3'-ends (161). Mutation or deletion of this sequence has no or at least not a very strong influence on the efficiency of 3'-end formation, nor does it change the site of polyadenylation (161–163). 3'-end formation of yeast mRNA must thus use sequences and possibly mechanisms different from those used in metazoa.

Several RNA (or DNA) sequences have been suggested to be involved in 3'-end formation in yeast mRNAs. The tripartite signal TAG.....TATGT (or TAGT)....TTT with variable lengths of T- or AT-rich sequence between the three parts was identified by a comparison of a number of mRNA 3'-end sequences with the *cyc1-512* mutation. This mutation is a 38-basepair (bp) deletion that completely blocks 3'-end formation of the *CYC1* message at the normal site (164). Subsequent work reduced the proposed signal to a bipartite motif, eliminating the group of Ts, but otherwise confirmed its involvement in 3'-end formation. Revertants that restored 3'-end formation in the *cyc1-512* mutation were isolated. Two out of six revertants had formed the sequence TAG...TATGTA (165). Point mutations affecting 3'-end formation have been isolated in the *ADH2* gene (161), and again some but not all of them are located in a sequence TAG...TATG. Sequences closely corresponding to the bipartite consensus are also found in 3'-end formation signals of the Ty transposon (166), of several genes in the 2-μm plasmid (167), and of the *ARO4, TRP1,* and *TRP4* genes (168). Simultaneous mutation of five nucleotides in the bipartite motif found in the *ARO4* sequence strongly reduces 3'-end formation. However, an insertion of just the consensus sequence into a test site does not suffice to induce 3'-end formation (168).

Unexpected evidence for the importance of TAG...TATGTA came from the observation that the 3'-end formation signal of the cauliflower mosaic virus (see below) functions in yeast (163). Deletion mutagenesis of the plant virus DNA fragment identified several regions influencing 3'-end formation in yeast. These are located between 180 bp upstream and 10 bp downstream of the poly(A) site. The most important region contains the sequence TAG-TATGTA 50 nucleotides upstream of the polyadenylation site. Point mutations in this motif reduce 3'-end formation more than fivefold. A conversion to a bipartite signal by the insertion of ATAT between TAG and TATGTA increases the efficiency of 3'-end formation (163).

A second class of sequences inducing 3'-end formation does not contain the bipartite motif. Several AT-rich sequences have been proposed instead. The motif TAAATAAA (169) has received little experimental support. Another potential signal for 3'-end formation, TTTTTATA, was found by deletion analysis of a *Drosophila* gene that complemented a yeast *ade8* mutation (162, 170). Four of the six revertants of the *cyc1-512* deletion discussed above had acquired the sequence TATATA or TACATA (165). A segment of 26 nucleotides involved in 3'-end formation of the *GAL7* gene is again extremely AT rich, and no similarity to the bipartite motif is apparent (171). A stretch of $(AT)_7$ contained in it, however, is not sufficient for 3'-end formation. 3'-end formation signals derived from the *GCN4, PHO5,* and *ADH1* genes all contain the sequence TTTTTAT, but a mutation of this sequence in the *PHO5* and *ARO4* genes has no effect, and its introduction into a test site does not lead to 3'-end formation(168).

Most 3'-end signals tested are functional only in one orientation (161, 171–174). Those that do function in either orientation carry similar sequences on both strands (168). In addition to the motifs discussed so far, which lie upstream of the polyadenylation site, sequences at the site itself and/or downstream of it also influence 3'-end formation (161, 162, 165, 171). A possible role for RNA secondary structure has been discussed (161).

In summary, sequences required for 3'-end formation in *Saccharomyces cerevisiae* are certainly more degenerate than the highly conserved AAUAAA of metazoa. The existence of at least two independent signals, the bipartite motif and some very AT-rich sequence, as yet undefined, is likely. The simultaneous presence of both sequences might also explain why extensive linker scanning mutagenesis of the *CYC1* 3'-end did not result in any mutation affecting 3'-end formation (174).

Mechanism

Not only are the sequences directing 3'-end formation in yeast mRNAs poorly understood, the mechanism is being debated as well. RNA extending beyond the polyadenylation site has never been reported to occur in vivo. This could be explained either by rapid processing of the primary transcript or by transcription termination at the poly(A) site. Two indirect in vivo assays have been used that measure specifically transcription termination as opposed to RNA processing. One assay uses the interference of transcription with centromere function to identify terminators as those sequences that are able to rescue centromeres. The other assay measures topological changes in a plasmid induced by transcription. Since the topological changes are related to the length of the transcription unit, terminators can be identified. In both assays, the sequence defined by the *cyc1-512* deletion (see above) qualifies as a transcription terminator (173, 175). Circumstantial evidence favoring 3'-end formation by transcription termination was also obtained by others (168).

However, when transcripts of several different genes, extending beyond the normal polyadenylation sites, are made in vitro and incubated in a crude extract of yeast, endonucleolytic cleavage at or near the natural poly(A) sites, followed by polyadenylation, is observed. Since processing of a *CYC1* transcript is abolished by the *cyc1-512* deletion, this sequence qualifies as a processing signal in this assay (176–178). The apparent contradiction may be resolved by the assumption that transcription termination occurs close to the site of poly(A) addition and that the endonucleolytic cleavage removes only a minor piece of the transcript. So far, no mutations have been reported that specifically affect either termination or processing. The possibility that the two different classes of 3'-end formation signals discussed above reflect two different mechanisms has not been ruled out.

Saccharomyces cerevisiae poly(A) polymerase has been purified to homogeneity (179). The enzyme has a molecular weight of 64,000, is highly specific for ATP, and elongates primers in a distributive fashion. All RNAs tested can serve as primers, and the enzyme has no preference for the authentic *CYC1* 3'-end as compared to several other RNAs. In contrast, polyadenylation in the crude extract discussed above is specific for the correct 3'-end: neither the precursor RNA nor the downstream cleavage product are extended (176, 177). This indicates that the polymerase must cooperate with a specificity factor. The enzyme is, however, apparently unable to recognize mammalian CPSF, since in AAUAAA-dependent polyadenylation the mammalian poly(A) polymerase cannot be replaced by its counterpart from yeast (179). A gene, *PAP1*, encoding poly(A) polymerase has been isolated and its identity confirmed by the expression of an active protein in *Escherichia coli*. Disruption of the gene is lethal (180). The size of the open reading frame in *PAP1* is in good agreement with that of poly(A) polymerase purified either from yeast or from the overproducing *E. coli* strain. The first 395 amino acids of the yeast enzyme are 47% identical with a corresponding domain in the bovine poly(A) polymerase (see above). This domain includes the RNP motifs and the amino acids that have been claimed to be similar to a polymerase consensus (see above). While the former can be found in the yeast sequence, the latter are not well conserved. The significance of the "polymerase module" thus remains to be confirmed by mutagenesis.

Two temperature-sensitive mutations, *rna14* and *rna15*, lead to a net loss of poly(A) tails at the restrictive temperature (181). The RNA15 protein is an RNA-binding protein as judged from its predicted amino acid sequence. Other than that, the biochemical functions of the two proteins are unknown. Also, it is not clear whether in these mutants the rate of mRNA deadenylation is normal and the net loss is due to a defect in the de novo synthesis of poly(A) tails, or whether the mutations accelerate deadenylation. A mutation in *PAB1*, the gene encoding the 70-kDa poly(A)-binding protein, leads to a lengthening of the poly(A) tails. Apparently the poly(A)-binding protein is required for cytoplasmic poly(A) shortening (182).

POLYADENYLATION IN PLANTS

Polyadenylation signals in plants have received little attention until recently. In contrast to mammalian mRNAs, which usually have well-defined 3'-ends, plant mRNAs generally have heterogeneous 3'-ends, spread over a region of up to 150 nucleotides (183). Mammalian polyadenylation signals function inefficiently and inaccurately in plants, suggesting that the process of 3'-end processing may be related but is not identical (184). A comparison of plant mRNA 3'-ends did not reveal any sequence comparable in conservation to the mammalian AAUAAA. Perfect copies of this sequence are found upstream of the polyadenylation site in less than one half of all plant genes examined (185). A point mutation in the AAUAAA sequence of the cauliflower mosaic virus inhibits, and a total deletion of the motif abolishes, 3'-end formation (186). Somewhat less severe effects of an AAUAAA mutation have been reported for a different gene (187). Thus, AAUAAA may be involved in 3'-end formation in plants, but the sequence requirement is certainly less stringent than in animal cells.

Deletion analyses have also identified multiple essential sequences extending more than 100 nucleotides upstream of the poly(A) site (186–189). Sequences downstream of the poly(A) site have been found to be necessary for 3'-end formation in the ribulose-1,5-bisphosphate carboxylase gene of pea (188) and in the octopine synthase gene of the Ti plasmid (189) but not in cauliflower mosaic virus (186). As mentioned before, the viral poly(A) signal functions in *Saccharomyces cerevisiae* (163) and also in *Schizosaccharomyces pombe* (190). RNA 3'-ends have been mapped to the same sites in yeast and in plant cells (163). Whether this reflects a true conservation of the mechanism of 3'-end formation between yeast and plants is not clear; deletion analyses have shown that different motifs in the viral DNA fragment are involved in polyadenylation in the two organisms (163, 186).

The mechanism of 3'-end formation in plants is unknown, and so are the enzymes catalyzing the reaction, with the exception of poly(A) polymerase (95).

CONCLUDING REMARKS

Considerable progress has been made towards understanding the pathway of 3'-end cleavage and polyadenylation in animal cells: sequences directing the process have been defined, and polyadenylation can be reconstituted from purified components. Purification of at least one additional cleavage factor is required for a reconstitution of the entire reaction. Purified components will permit detailed studies of the interactions of processing factors with the pre-mRNA substrates and with each other. Cloning of several processing factors has been achieved or is in progress, and this will facilitate biochemical studies by protein overproduction and mutagenesis. Finally, an understanding of the mechanism of 3'-end formation will be the basis for investigations concerning the choice between alternative polyadenylation sites, regulated polyadenylation, and the connection between polyadenylation and transcription termination. With the combined application of biochemical assays and genetic techniques, rapid progress may also be anticipated in the analysis of 3'-end formation in yeast.

ACKNOWLEDGMENTS

We are grateful to Thomas Hohn for information on plant poly(A) sites, to the colleagues who sent us preprints, and to the members of our laboratory who read the manuscript. Work in our laboratory was supported by grants from the Kantons of Basel and the Schweizerischer Nationalfonds.

Literature Cited

1. Adesnik, M., Darnell, J. E. 1972. *J. Mol. Biol.* 67:397–406
2. Adesnik, M., Salditt, M., Thomas, W., Darnell, J. E. 1972. *J. Mol. Biol.* 71:21–30
3. Greenberg, J. R., Perry, R. P. 1972. *J. Mol. Biol.* 72:91–98
4. McLaughlin, C. S., Warner, J. R., Edmonds, M., Nakazato, H., Vaughan, M. H. 1973. *J. Biol. Chem.* 248:1466–71
5. Hirsch, M., Penman, S. 1973. *J. Mol. Biol.* 80:379–91
6. Ojala, D., Montoya, J., Attardi, G. 1981. *Nature* 290:470–74
7. Sippel, A. E. 1973. *Eur. J. Biochem.* 37:31–40
8. Karnik, P., Taljanidisz, J., Sasvari-Szekely, M., Sarkar, N. 1987. *J. Mol. Biol.* 196:347–354
9. Edmonds, M., Abrams, R. 1960. *J. Biol. Chem.* 235:1142–49
10. Edmonds, M., Winters, M. A. 1976. *Prog. Nucleic Acid Res. Mol. Biol.* 17:149–79
11. Edmonds, M., Caramela, M. G. 1969. *J. Biol. Chem.* 244:1314–24
12. Edmonds, M., Vaughan, M. H., Nakazato, H. 1971. *Proc. Natl. Acad. Sci. USA* 68:1336–40
13. Molloy, G. R., Sporn, M. B., Kelley, D. E., Perry, R. P. 1972. *Biochemistry* 11:3256–60
14. Philipson, L., Wall, R., Glickman, G., Darnell, J. E. 1971. *Proc. Natl. Acad. Sci. USA* 68:2806–9
15. Penman, S., Rosbash, M., Penman, M. 1970. *Proc. Natl. Acad. Sci. USA* 67:1878–85
16. Darnell, J. E., Jelinek, W. R., Molloy, G. R. 1973. *Science* 181:1215–21
17. Nevins, J. R., Darnell, J. E. 1978. *Cell* 15:1477–93
18. Ford, J. P., Hsu, M.-T. 1978. *J. Virol.* 28:795–801
19. Brawerman, G. 1981. *Crit. Rev. Biochem.* 10:1–38
20. Atwater, J. A., Wisdom, R., Verma, I. M. 1990. *Annu. Rev. Genet.* 24:519–41
21. Jackson, R. J., Standart, N. 1990. *Cell* 62:15–24
22. Wickens, M., Stephenson, P. 1984. *Science* 226:1045–51
23. Birnstiel, M. L., Busslinger, M., Strub, K. 1985. *Cell* 41:349–59
24. Manley, J. L. 1988. *Biochim. Biophys. Acta* 950:1–12
25. Wickens, M. 1990. *Trends Biochem. Sci.* 15:277–81
26. Wickens, M. 1990. *Trends Biochem. Sci.* 15:320–24
27. Wahle, E. 1992. *Bioessays.* In press
28. Proudfoot, N. 1991. *Cell* 64:671–74
29. Proudfoot, N. J., Brownlee, G. G. 1976. *Nature* 263:211–14
30. Fitzgerald, M., Shenk, T. 1981. *Cell* 24:251–60
31. Montell, C., Fisher, E. F., Caruthers, M. H., Berk, A. J. 1983. *Nature* 305:600–5
32. Higgs, D. R., Goodbourn, S. E. Y., Lamb, J., Clegg, J. B., Weatherall, D. J., et al. 1983. *Nature* 306:398–400

33. Manley, J. L., Yu, H., Ryner, L. 1985. *Mol. Cell. Biol.* 5:373–79
34. Orkin, S. H., Cheng, T.-C., Antonarakis, S. E., Kazazian, H. H. 1985. *EMBO J.* 4:453–56
35. Zarkower, D., Stephenson, P., Sheets, M., Wickens, M. 1986. *Mol. Cell. Biol.* 6:2317–23
36. Wilusz, J., Pettine, S. M., Shenk, T. 1989. *Nucleic Acids Res.* 17:3899–908
37. Sheets, M. D., Ogg, S. C., Wickens, M. P. 1990. *Nucleic Acids Res.* 18: 5799–805
38. Setzer, D. R., McGrogan, M., Nunberg, J. H., Schimke, R. T. 1980. *Cell* 22:361–70
39. Hook, A. G., Kellems, R. E. 1988. *J. Biol. Chem.* 263:2337–43
40. Yue, X., Connolly, T., Futcher, B., Beach, D. 1991. *Cell* 65:691–99
41. Meijer, D., Hermans, A., von Lindern, M., van Agthoven, T., de Klein, A., et al. 1987. *EMBO J.* 6:4041–48
42. Challoner, P. B., Moss, S. B., Groudine, M. 1989. *Mol. Cell. Biol.* 9:902–13
43. Simonsen, C. C., Levinson, A. D. 1983. *Mol. Cell. Biol.* 3:2250–58
44. Gil, A., Proudfoot, N. J. 1984. *Nature* 312:473–74
45. McDevitt, M. A., Imperiale, M. J., Ali, H., Nevins, J. R. 1984. *Cell* 37:993–99
46. Sadofsky, M., Alwine, J. C. 1984. *Mol. Cell. Biol.* 4:1460–68
47. Conway, L., Wickens, M. 1985. *Proc. Natl. Acad. Sci. USA* 82:3949–53
48. Cole, C. N., Stacy, T. P. 1985. *Mol. Cell. Biol.* 5:2104–13
49. Sadofsky, M., Connelly, S., Manley, J. L., Alwine, J. C. 1985. *Mol. Cell. Biol.* 5:2713–19
50. Hart, R. P., McDevitt, M. A., Nevins, J. R. 1985. *Cell* 43:677–83
51. Hart, R. P., McDevitt, M. A., Ali, H., Nevins, J. R. 1985. *Mol. Cell. Biol.* 5:2975–83
52. McLauchlan, J., Gaffney, D., Whitton, J. L., Clements, J. B. 1985. *Nucleic Acids Res.* 13:1347–68
53. McDevitt, M. A., Hart, R. P., Wong, W. W., Nevins, J. R. 1986. *EMBO J.* 5:2907–13
54. Zhang, F., Denome, R. M., Cole, C. N. 1986. *Mol. Cell. Biol.* 6:4611–23
55. Kessler, M. M., Beckendorf, R. C., Westhafer, M. A., Nordstrom, J. L. 1986. *Nucleic Acids Res.* 14:4939–52
56. Gil, A., Proudfoot, N. J. 1987. *Cell* 49:399–406
57. Prochownik, E. V., Smith, M. J., Markham, A. 1987. *J. Biol. Chem.* 262:9004–10
58. Gimmi, E. R., Soprano, K. J., Rosenberg, M., Reff, M. E. 1988. *Nucleic Acids Res.* 16:8977–97
59. Zarkower, D., Wickens, M. 1988. *J. Biol. Chem.* 263:5780–88
60. Green, T. L., Hart, R. P. 1988. *Mol. Cell. Biol.* 8:1839–41
61. Böhnlein, S., Hauber, J., Cullen, B. R. 1989. *Mol. Cell. Biol.* 63:421–24
62. Ryner, L. C., Takagaki, Y., Manley, J. L. 1989. *Mol. Cell. Biol.* 9:1759–71
63. Heath, C. V., Denome, R. M., Cole, C. N. 1990. *J. Biol Chem.* 265:9098–104
64. Woychik, R. P., Lyons, R. H., Post, L., Rottman, F. M. 1984. *Proc. Natl. Acad. Sci. USA* 81:3944–48
65. Mason, P. J., Elkington, J. A., Lloyd, M. M., Jones, M. B., Williams, J. G. 1986. *Cell* 46:263–70
66. Conway, L., Wickens, M. 1987. *EMBO J.* 6:4177–84
67. Levitt, N., Briggs, D., Gil, A., Proudfoot, N.J. 1989. *Genes Dev.* 3:1019–25
68. DeZazzo, J. D., Imperiale, M. J. 1989. *Mol. Cell. Biol.* 9:4951–61
69. Carswell, S., Alwine, J. C. 1989. *Mol. Cell. Biol.* 9:4248–58
70. Baker, B. S. 1989. *Nature* 340:521–24
71. Hedley, M. L., Maniatis, T. 1991. *Cell* 65:579–86
71a. Ryner, L. C., Baker, B. S. 1991. *Genes Dev.* 5:2071–85
72. Dougherty, J. P., Temin, H. M. 1987. *Proc. Natl. Acad. Sci. USA* 84:1197–201
73. Russnak, R., Ganem, D. 1990. *Genes Dev.* 4:764–76
74. Valsamakis, A., Zeichner, S., Carswell, S., Alwine, J. C. 1991. *Proc. Natl. Acad. Sci. USA* 88:2108–12
75. Kurkulos, M., Weinberg, J. M., Pepling, M. E., Mount, S. M. 1991. *Proc. Natl. Acad. Sci. USA* 88:3038–42
76. DeZazzo, J. D., Kilpatrick, J. E., Imperiale, M. J. 1991. *Mol. Cell. Biol.* 11:1624–30
77. Iwasaki, K., Temin, H. M. 1990. *Genes Dev.* 4:2299–307
78. Sanfacon, H., Hohn, T. 1990. *Nature* 346:81–84
79. Weichs an der Glon, C., Monks, J., Proudfoot, N. J. 1991. *Genes Dev.* 5:244–53
80. Ahmed, Y. F., Gilmartin, G. M., Hanly, S. M., Nevins, J. R., Greene, W. C. 1991. *Cell* 64:727–37
81. McLauchlan, J., Simpson, S., Clements, J. B. 1989. *Cell* 59:1093–105
82. Moore, C. L., Sharp, P. A. 1984. *Cell* 36:581–91
83. Moore, C. L., Sharp, P. A. 1985. *Cell* 41:845–55
84. Moore, C. L., Skolnik-David, H., Sharp, P. A. 1986. *EMBO J.* 5:1929–38
85. Sheets, M. D., Stephenson, P., Wickens, M. P. 1987. *Mol. Cell. Biol.* 7:1518–29

86. Christofori, G., Keller, W. 1988. *Cell* 54:875–89
87. Christofori, G., Keller, W. 1989. *Mol. Cell. Biol.* 9:193–203
88. Takagaki, Y., Ryner, L. C., Manley, J. L. 1988. *Cell* 52:731–42
89. Takagaki, Y., Ryner, L. C., Manley, J. L. 1989. *Genes Dev.* 3:1711–24
90. McDevitt, M. A., Gilmartin, G. M., Reeves, W. H., Nevins, J. R. 1988. *Genes Dev.* 2:588–97
91. Gilmartin, G. M., Nevins, J. R. 1989. *Genes Dev.* 3:2180–89
92. Wahle, E. 1991. *J. Biol. Chem.* 266:3131–39
93. Wahle, E., Martin, G., Schilz, E., Keller, W. 1991. *EMBO J.* 10:4251–57
94. Raabe, T., Bollum, F. J., Manley, J. L. 1991. *Nature* 353:229–34
95. Edmonds, M. 1982. *Enzymes,* 15(Part B):217–44
96. Edmonds, M. 1990. *Methods Enzymol.* 181:161–70
97. Tsiapalis, C. M., Dorson, J. W., Bollum, F. J. 1975. *J. Biol. Chem.* 250:4486–96
98. Ryner, L. C., Takagaki, Y., Manley, J. L. 1989. *Mol. Cell. Biol.* 9:4229–38
99. Bardwell, V. J., Zarkower, D., Edmonds, M., Wickens, M. 1990. *Mol. Cell. Biol.* 10:846–49
100. Moss, B., Rosenblum, E. N., Gershowitz, A. 1975. *J. Biol. Chem.* 250: 4722–29
101. Gershon, P. D., Ahn, B.-Y., Garfield, M., Moss, B. 1991. *Cell* 66:1269–78
102. Bienroth, S., Wahle, E., Suter-Crazzolara, C., Keller, W. 1991. *J. Biol. Chem.* 266:19768–76
103. Gilmartin, G. M., Nevins, J. R. 1991. *Mol. Cell. Biol.* 11:2432–38
104. Keller, W., Bienroth, S., Lang, K. M., Christofori, G. 1991. *EMBO J.* 10: 4241–49
105. Bardwell, V. J., Wickens, M., Bienroth, S., Keller, W., Sproat, B. S., et al. 1991. *Cell* 65:125–33
106. Wigley, P. L., Sheets, M. D., Zarkower, D. A., Whitmer, M. E., Wickens, M. 1990. *Mol. Cell. Biol.* 10:1705–13
107. Whitherell, G. W., Gott, J. M., Uhlenbeck, O. C. 1991. *Prog. Nucleic Acids Res. Mol. Biol.* 40:185–220
108. Krämer, A. 1987. *Proc. Natl. Acad. Sci. USA* 84:8408–12
109. Lerner, E. A., Lerner, M. R., Janeway, C. A., Steitz, J. A. 1981. *Proc. Natl. Acad. Sci. USA* 78:2737–41
110. Steitz, J. A., Black, D. L., Gerke, V., Parker, K. A., Krämer, A., et al. 1988. *Structure and Function of Major and Minor Small Nuclear Ribonucleoprotein Particles,* ed. M. L. Birnstiel, pp. 115–54. Berlin:Springer
111. Birnstiel, M. L., Schaufele, F. J. 1988. See Ref. 110, pp. 155–82
112. Takagaki, Y., Manley, J. L., MacDonald, C. C., Wilusz, J., Shenk, T. 1990. *Genes Dev.* 4:2112–20
113. Wilusz, J., Shenk, T. 1988. *Cell* 52: 221–28
114. Moore, C. L., Chen, J., Whoriskey, J. 1988. *EMBO J.* 7:3159–69
115. Wilusz, J., Shenk, T., Takagaki, Y., Manley, J. L. 1990. *Mol. Cell. Biol.* 10:1244–48
116. Qian, Z., Wilusz, J. 1991. *Mol. Cell. Biol.* 11:5312–20
117. Zarkower, D., Wickens, M. 1987. *EMBO J.* 6:177–86
118. Zarkower, D., Wickens, M. 1987. *EMBO J.* 6:4185–92
119. Humphrey, T., Christofori, G., Lucijanic, V., Keller, W. 1987. *EMBO J.* 6:4159–68
120. Skolnik-David, H., Moore, C. L., Sharp, P. A. 1987. *Genes Dev.* 1:672–82
121. Zhang, F., Cole, C. N. 1987. *Mol. Cell. Biol.* 7:3277–86
122. Moore, C. L., Skolnik-David, H., Sharp, P. A. 1988. *Mol. Cell. Biol.* 8:226–33
123. Stefano, J. E., Adams, D. E. 1988. *Mol. Cell. Biol.* 8:2052–62
124. McLauchlan, J., Moore, C. L., Simpson, S., Clements, J. B. 1988. *Nucleic Acids Res.* 16:5323–44
125. Weiss, E. A., Gilmartin, G. M., Nevins, J. R. 1991. *EMBO J.* 10:215–19
126. Gilmartin, G. M., McDevitt, M. A., Nevins, J. R. 1988. *Genes Dev.* 2:578–87
127. Ryner, L. C., Manley, J. L. 1987. *Mol. Cell. Biol.* 7:495–503
128. Niwa, M., Rose, S. D., Berget, S. M. 1990. *Genes Dev.* 4:1552–59
129. Sheets, M. D., Wickens, M. 1989. *Genes Dev.* 3:1401–12
130. Wahle, E. 1991. *Cell* 66:759–68
131. Kenan, D. J., Query, C. C., Keene, J. D. 1991. *Trends Biochem. Sci.* 16:214–20
132. Sachs, A. B., Bond, M. W., Kornberg, R. D. 1986. *Cell* 45:827–35
133. Adam, S. A., Nakagawa, T., Swanson, M. S., Woodruff, T. K., Dreyfuss, G. 1986. *Mol. Cell. Biol.* 6:2932–43
134. Grange, T., de Sa, C. M., Oddos, J., Pictet, R. 1987. *Nucleic Acids. Res.* 15:4771–87
135. Bardwell, V. J., Wickens, M. 1990. *Mol. Cell. Biol.* 10:295–302
136. Shuman, S., Moss, B. 1988. *J. Biol. Chem.* 263:8405–12
137. Whitelaw, E., Proudfoot, N. 1986. *EMBO J.* 5:2915–22
138. Logan, J., Falck-Pedersen, E., Darnell,

J. E., Shenk, T. 1987. *Proc. Natl. Acad. Sci. USA* 84:8306–10
139. Lanoix, J., Acheson, N. H. 1988. *EMBO J.* 7:2515–22
140. Connelly, S., Manley, J. L. 1988. *Genes Dev.* 2:440–52
141. Dorsett, D. 1990. *Proc. Natl. Acad. Sci. USA* 87:4373–77
142. Carrazana, E. J., Pasieka, K. B., Majzoub, J. A. 1988. *Mol. Cell. Biol.* 8:2267–74
143. Robinson, B. G., Frim, D. M., Schwartz, W. J., Majzoub, J. A. 1988. *Science* 241:342–44
144. Rosenthal, E. T., Tansey, T. R., Ruderman, J. V. 1983. *J. Mol. Biol.* 166:309–27
145. Dworkin, M. B., Dworkin-Rastl, E. 1985. *Dev. Biol.* 112:451–57
146. Rosenthal, E. T., Ruderman, J. V. 1987. *Dev. Biol.* 121:237–46
147. Paynton, B. V., Rempel, R., Bachvarova, R. 1988. *Dev. Biol.* 129:304–14
148. Hyman, L. E., Wormington, W. M. 1988. *Genes Dev.* 2:598–605
149. Fox, C. A., Sheets, M. D., Wickens, M. P. 1989. *Genes Dev.* 3:2151–62
150. Paris, J., Richter, J. D. 1990. *Mol. Cell. Biol.* 10:5634–45
151. McGrew, L. L., Dworkin-Rastl, E., Dworkin, M. B., Richter, J. D. 1989. *Genes Dev.* 3:803–15
152. McGrew, L. L., Richter, J. D. 1990. *EMBO J.* 9:3743–51
153. Paris, J., Swenson, K., Piwnica-Worms, H., Richter, J. D. 1991. *Genes Dev.* 5:1697–708
154. Varnum, S. M., Wormington, W. M. 1990. *Genes Dev.* 4:2278–86
155. Fox, C. A., Wickens, M. 1990. *Genes Dev.* 4:2287–98
156. Huarte, J., Belin, D., Vassalli, A., Strickland, S., Vassalli, J.-D. 1987. *Genes Dev.* 1:1201–11
157. Strickland, S., Huarte, J., Belin, D., Vassalli, A., Rickles, R. J., et al. 1988. *Science* 241:680–84
158. Vassalli, J.-D., Huarte, J., Belin, D., Gubler, P., Vassalli, A., et al. 1989. *Genes Dev.* 3:2163–71
159. Schäfer, M., Kuhn, R., Bosse, F., Schäfer, U. 1990. *EMBO J.* 9:4519–25
160. Ahringer, J., Kimble, J. 1991. *Nature* 349:346–48
161. Hyman, L. E., Seiler, S. H., Whoriskey, J., Moore, C. L. 1991. *Mol Cell. Biol.* 11:2004–12
162. Henikoff, S., Kelly, J. D., Cohen, E. H. 1983. *Cell* 33:607–14
163. Irniger, S., Sanfacon, H., Egli, C. M., Braus, G. H. 1992. Submitted for publication
164. Zaret, K. S., Sherman, F. 1982. *Cell* 28:563–73

165. Russo, P., Li, W.-Z., Hampsey, D. M., Zaret, K. S., Sherman, F. 1991. *EMBO J.* 10:563–71
166. Yu, K., Elder, R. T. 1989. *Mol. Cell. Biol.* 9:2431–44
167. Sutton, A., Broach, J. R. 1985. *Mol. Cell. Biol.* 5:2770–80
168. Irniger, S., Egli, C. M., Braus, G. H. 1991. *Mol. Cell. Biol.* 11:3060–69
169. Bennetzen, J. L., Hall, B. D. 1982. *J. Biol. Chem.* 257:3018–25
170. Henikoff, S., Cohen, E. H. 1984. *Mol. Cell. Biol.* 4:1515–20
171. Abe, A., Hiraoka, Y., Fukasawa, T. 1990. *EMBO J.* 9:3691–97
172. Ruohola, H., Baker, S. M., Parker, R., Platt, T. 1988. *Proc. Natl. Acad. Sci. USA* 85:5041–45
173. Russo, P., Sherman, F. 1989. *Proc. Natl. Acad. Sci. USA* 86:8348–52
174. Osborne, B. I., Guarente, L. 1989. *Proc. Natl. Acad. Sci. USA* 86:4097–101
175. Osborne, B. I., Guarente, L. 1988. *Genes Dev.* 2:766–72
176. Butler, J. S., Platt, T. 1988. *Science* 242:1270–74
177. Butler, J. S., Sadhale, P. P., Platt, T. 1990. *Mol. Cell. Biol.* 10:2599–605
178. Sadhale, P. P., Sapolsky, R., Davis, R. W., Butler, J. S., Platt, T. 1991. *Nucleic Acids Res.* 19:3683–98
179. Lingner, J., Radtke, I., Wahle, E., Keller, W. 1991. *J. Biol. Chem.* 266:8741–46
180. Lingner, J., Kellermann, J., Keller, W. 1991. *Nature* 354:496–98
181. Minvielle-Sebastia, L., Winsor, B., Bonneaud, N., Lacroute, F. 1991. *Mol. Cell. Biol.* 11:3075–87
182. Sachs, A. B., Davis, R. W. 1989. *Cell* 58:857–67
183. Dean, C., Tamaki, S., Dunsmuir, P., Favreau, M., Katayama, C., et al. 1986. *Nucleic Acids Res.* 14:2229–40
184. Hunt, A. G., Chu, N. M., Odell, J. T., Nagy, F., Chua, N.-H. 1987. *Plant Mol. Biol.* 8:23–35
185. Joshi, C. P. 1987. *Nucleic Acids Res.* 15:9627–40
186. Sanfacon, H., Brodmann, P., Hohn, T. 1991. *Genes Dev.* 5:141–49
187. Mogen, B. D., MacDonald, M. H., Graybosch, R., Hunt, A. G. 1990. *Plant Cell* 2:1261–72
188. Hunt, A. G., MacDonald, M. H. 1989. *Plant Mol. Biol.* 13:125–38
189. MacDonald, M. H., Mogen, B. D., Hunt, A. G. 1991. *Nucleic Acids Res.* 19:5575–81
190. Hirt, H., Kögel, M., Murbacher, T., Heberle-Bors, E. 1990. *Curr. Genet.* 17:473–79

Annu. Rev. Biochem. 1992. 61:441–70

ANIMAL CELL CYCLES AND THEIR CONTROL

Chris Norbury and Paul Nurse

ICRF Cell Cycle Group, Microbiology Unit, Department of Biochemistry, University of Oxford, Oxford OX1 3QU, United Kingdom

KEY WORDS: *cdc2,* protein kinase, phosphorylation, mitosis, cyclin

CONTENTS

PERSPECTIVES AND SUMMARY

In its simplest form, the animal cell cycle consists of a round of chromosomal DNA replication in S phase followed by segregation of the replicated chromosomes into two daughter nuclei during M phase. In most animal cells, *gap*

0066-4154/92/0701-0441$02.00

phases termed G1 and G2 are introduced between M and S and between S and M, respectively. During these gap phases, although not necessarily throughout them, information is integrated in order to determine the readiness of a cell to enter either S or M. The molecular events that underlie these crucial cell-cycle decisions have only recently begun to become apparent. Thus until a few years ago, accounts of cell cycle biochemistry in animals were largely limited to descriptions of aspects of the mechanisms of DNA replication and mitosis per se. Since Pardee and coworkers reviewed this field (1) in these pages in 1978, our understanding of the molecular basis of cell cycle control has been substantially advanced. In this review we discuss these advances, concentrating on molecules for which there is functional, rather than merely correlative, evidence for involvement in regulation of progression through the animal cell cycle. On this basis we choose not to review the "mechanical" aspects of DNA replication, mitosis, and cytokinesis, instead concentrating on the regulatory events in G1 that lead to commitment to enter S phase, and those in G2 that determine the timing of entry into mitosis. The latter is without question the area in which the most progress has been made towards a molecular description of a cell-cycle control mechanism.

It has become apparent that a central and rate-limiting function in the transition from G2 into M is performed by a protein kinase that is highly conserved in its primary amino acid sequence and completely conserved in its M-phase-promoting activity throughout the eukaryotes (2). This activity was first described biochemically as maturation-promoting factor (MPF), a protein fraction able to promote M phase events in amphibian eggs (3–5). It is now known that MPF activity depends on two protein species present in equimolar amounts in purified MPF preparations (6). The first component is the protein kinase catalytic subunit $p34^{cdc2}$ (7, 8), independently identified genetically in yeast as having a function required for the onset of M phase (9, 10). The second component of MPF is a B-cyclin protein (11, 12), the biochemical role of which remains unclear, though potential functions include specification of substrates or subcellular localization of the MPF protein kinase heterodimer. The activity of $p34^{cdc2}$ is itself subject to regulation by phosphorylation (13–19). The predicted ATP-binding region of the protein is phosphorylated during interphase (17–19), resulting in inactivation of the pre-MPF complex until M phase onset (17, 19). Additional threonine phosphorylation may stabilize association of the $p34^{cdc2}$ and B-cyclin subunits, thus potentially activating the protein kinase complex (19, 20).

The M phase state is characterized in part by the appearance of newly phosphorylated and hyperphosphorylated proteins (21). The $p34^{cdc2}$/MPF complex is capable of phosphorylating a range of proteins in vitro (22–37), but the extent to which these substrates are phosphorylated in vivo by $p34^{cdc2}$ and the physiological relevance for cell cycle progression has not been established in many cases. Nonetheless, the phosphorylation of nuclear lamins to promote nuclear disassembly (25, 31, 38–40), and the phosphorylation of vimentin (41) and caldesmon (30), substrates potentially involved in cytoskeletal rearrangement, are reasonably well-substantiated roles for p34^{cdc2} at M phase onset. While protein phosphorylation is clearly of fundamental importance for the initiation of M phase, protein phosphatases also have important roles to play, not only in the activation of the pre-MPF complex, but also in removing MPF-driven phosphorylation to restore the interphase state on completion of M phase.

In some biological systems, cell-cycle delay and long-term arrest in G2 are well-documented phenomena (42–45), but much of the variation among tissues in cell cycle duration is due to variability in the length of G1. Growth factor and protein synthesis requirements for commitment to enter S suggest that this commitment process must occur in late G1, typically one to two hours before the initiation of DNA replication itself (for reviews see 1, 46). Compared to the recent advances relating to controls over entry into M phase, progress towards a molecular understanding of the controls in G1 determining commitment to enter S phase has been slow. Whereas the G2/M transition is characterized by dramatic intracellular changes that can be monitored both microscopically and by the use of in vitro assays, the changes in late G1 that commit a cell to enter S phase are by no means as obvious.

An approach taken by some investigators has been to reconstitute in vitro the components required for the initiation of DNA replication (47; reviewed in 48), but the relevance of this approach to the late G1 commitment point has yet to be clearly established. One interesting possibility, suggested initially by analogy with yeast systems (49–51), is that p34^{cdc2} is required for the G1 commitment process in animal cells, as well as for M phase onset. Investigation of this aspect of animal cell cycle control is complicated by the profusion of p34^{cdc2}-related proteins present in higher eukaryotic species (52–56). Still less well understood are the relationship between cell growth and cell cycle control, and the ways in which negative controls over proliferation act to block both cell growth and cell cycle progression.

Purely biochemical strategies, notably the purification and characterization of MPF, have led to major advances in the field of cell cycle control, but the contributions of genetical approaches are also of great importance, particularly for establishing the functional significance of particular gene products. Animal gene functions required for cell cycle progression can be identified through corresponding conditional mutant animal cell lines defective in such functions (57–66). An alternative approach has been to identify and isolate animal genes on the basis of their ability to complement yeast mutants defective in particular cell cycle functions (56, 67, 68). In taking this route, many of the experimental difficulties associated with mammalian cell systems are avoided. The combination of these genetical strategies with biochemical and physiological approaches is leading to a unified view of the ways in which animal cell cycles are controlled.

THE ONSET OF M PHASE IN ANIMAL CELLS

Many cell types in culture have G2 phases of relatively constant duration, although contrary to earlier arguments in the literature, this should not be taken as evidence against regulation of the animal cell cycle in G2. The transition from G2 into M phase requires the higher eukaryotic cell to initiate, in a highly coordinated manner, complex processes including nuclear envelope breakdown (69–71), chromatin condensation (72, 73), and reorganization of the cytoskeleton (74–77). The onset of M phase is only normally permitted after completion of a round of DNA replication earlier in the cycle (reviewed in 78, 79), further demonstrating the existence of controls operative at the G2/M boundary.

Experiments using cytoplasmic transfer into amphibian oocytes (3–5) and somatic cell fusion (80) first provided direct evidence that a dominantly acting intracellular factor is responsible for the establishment of the M phase state. The universal nature of this factor in all eukaryotic species has become apparent during the past three years (2).

Maturation-Promoting Factor and Growth-Associated Kinase

Comparatively straightforward assays for the complex events of M phase have facilitated the biochemical pursuit of the key molecules that trigger M phase onset. These assays were initially based on microinjection of material into immature oocytes arrested in meiotic prophase (4, 5). Injected material was tested for its capacity to cause meiotic maturation, that is to say entry into meiotic metaphase, without requiring further protein synthesis (81, 82). Maturation was scored by microscopic observation of the breakdown of the oocyte nuclear membrane, the germinal vesicle. Such maturation-promoting factor (MPF) was detected in metaphase-arrested eggs and in extracts of somatic cells from distantly related eukaryotes (83–91). During meiotic and mitotic cell cycles, MPF-specific activity was seen to oscillate, peaking in M phase (84, 85, 92). Similar oscillation was reported for a protein kinase activity called variously growth-associated kinase or M phase histone H1 kinase (93–100).

MPF induces M phase changes in extracts prepared from activated eggs in interphase of the first embryonic cell cycle, specifically changes in microtubule dynamics (101), including the induction of mitotic spindle formation (102), chromosome condensation (102, 103), inhibition of vesicular fusion (104), and nuclear envelope breakdown (102). The latter was used as an in vitro assay for MPF activity, facilitating the purification of MPF greater than 3000-fold from eggs of *Xenopus laevis* (6). Highly purified *Xenopus* MPF was shown to contain two protein species detectable by silver staining; one of 45 kDa and one of 32 kDa (6). This material was shown to have protein kinase activity (6), and MPF copurified with the M phase–specific histone H1 kinase from starfish (14), clearly implicating this protein kinase activity in the events leading to initiation of M phase.

$p34^{cdc2}$: the Catalytic Subunit of MPF

A parallel, primarily genetical line of research established that a 34-kDa protein kinase that is rate limiting for entry into mitosis in yeast is functionally conserved in human cells. This protein kinase is the product of the fission yeast *cdc2* gene (10, 13, 105), whose functional counterpart in budding yeast is *CDC28* (49, 50, 106). In both yeasts the gene product $p34^{cdc2/28}$ is a protein serine/threonine kinase (13, 105, 107, 108), and the gene function is required both for entry into M (10, 109) and for traversal of the G1 "start" control point (49–51), at which the cell becomes committed to S phase entry. In fission yeast, $p34^{cdc2}$ was implicated in determining the timing of M phase onset by the discovery of gain-of-function *cdc2* mutants that advance cells into mitosis (10, 110), thus shortening G2. Functional complementation of a temperature-sensitive *cdc2* mutant was used to clone a human gene encoding a 34-kDa protein kinase capable of performing both the G1 and G2 roles of $p34^{cdc2}$ (67), demonstrating the conservation of this function through evolution of the higher eukaryotes. The human *cdc2* gene, termed *CDC2Hs* (for *Homo sapiens*), encodes a protein that displays 63% identity in its amino acid sequence with fission yeast $p34^{cdc2}$ (67). Functional *cdc2* homologues were subsequently isolated from mouse (33) and chicken (111), as well as from fruit flies (54) and plants (112–114), and antibodies raised against conserved $p34^{cdc2}$ peptides recognized proteins of similar size in all eukaryotic species tested (67, 115, 116).

The biochemical and genetic approaches to understanding the control of M phase initiation were mutually reinforced when it was demonstrated that the 32-kDa component of MPF is a $p34^{cdc2}$-related polypeptide (7, 8). Thus a $p34^{cdc2}$ protein kinase activity, identical to the growth-associated or M phase–specific H1 histone kinase (14, 100) and the enzymic basis for MPF activity, appears to be responsible for governing entry into M phase in all eukaryotes. This conclusion was further substantiated by experiments in which microinjection of antibodies against $p34^{cdc2}$ was found to block entry into mitosis in fibroblasts (117), and by the identification of a mouse temperature-sensitive cell line that has a thermolabile $p34^{cdc2}$ protein (66), and that consequently becomes arrested in late G2 at the restrictive temperature (63). In evolution, the appearance of the *cdc2* function probably occurred close to the emergence of the eukaryotes (118), and may indeed have been essential for the evolution of some of the distinguishing features of eukaryotic cells, such as nuclear compartmentation and spindle formation.

Cyclin B: the Second Essential Component of MPF

The cyclins were first described as proteins periodically synthesized during early embryonic development of marine invertebrates (119). Cyclin levels were seen to peak at each M phase, with two types, A and B, being distinguished in surf clams by their different gel mobilities and the slightly earlier appearance and disappearance of the A type (119, 120). Molecular cloning of cyclin genes from a variety of eukaryotes confirmed the distinction between the A and B type cyclins at the amino acid level, though the limited similarity among cyclins from different species was restricted to an internal region of approximately 150 amino acid residues (120–129), and also identified the 56-kDa product of the *cdc13* gene as a B type cyclin in fission yeast (130–133). This discovery provided a further example of the strength of combining biochemical and genetic strategies, as in this case the purely genetic analysis of yeast mutants had already demonstrated that $p56^{cdc13}$ and $p34^{cdc2}$ interact closely at M phase onset (132, 134). This information, combined with the description of cyclin behavior in the early embryo, set the stage for the discovery that the 45-kDa component of purified MPF is a B cyclin (11, 12).

No enzymic activity has been attributed to the cyclin B moiety, while $p34^{cdc2}$ has protein kinase activity that appears to be essential for the initiation of M phase (66, 135). Functional B cyclin protein is required for entry into M phase (126, 134, 136, 137), and association with a cyclin is generally regarded as essential for activity of $p34^{cdc2}$ (138, 139). It has therefore been concluded that $p34^{cdc2}$ is the catalytic subunit of the MPF heterodimer, and that cyclin B has an accessory function that is not catalytic. This function remains uncertain, though two possibilities, which are not mutually exclusive, have been proposed. The first is that cyclin B determines the subcellular location of $p34^{cdc2}$, while the second is that cyclin B determines the substrate specificity of the MPF complex in vivo (129, 140, 141). As M phase is normally transient, there must be mechanisms to ensure that MPF is inactivated once its function is completed. This inactivation is achieved in part at least by proteolytic degradation of the cyclin B subunit, probably leading to loss of $p34^{cdc2}$ protein kinase activity. A truncated cyclin lacking N-terminal sequences required for proteolysis caused egg extracts to enter and become blocked in M phase (142). The sequences deleted from this truncated protein were able to confer ubiquitin-mediated proteolysis on a marker protein to which they were fused (143), suggesting that cyclin degradation may occur by the ubiquitin pathway. The means by which this degradation is triggered only after MPF has executed its function are unknown.

The function of cyclin A is not clearly understood (144). Exogenous cyclin A mRNA caused entry into M phase when injected into *Xenopus* oocytes (120), suggesting a potential cell cycle involvement, and cyclin A proteins are capable of forming complexes with p34^{cdc2} that have protein kinase activity that peaks earlier in the cell cycle than that of p34^{cdc2}/cyclin B complexes (128, 129, 140). In *Drosophila* the cyclin A gene was shown to be essential, but analysis of a cyclin A homozygous deletion mutant did not indicate a function for cyclin A in a specific cell cycle phase (123, 124). The cyclin A mutant was able to complete the first 15 embryonic cell cycles, presumably using the maternal store of cyclin A protein, before completing the following cycle with very low levels of cyclin A, becoming arrested before reaching mitosis 16 (123, 124). In mammalian cells and *Xenopus* eggs, cyclin A binds to a 32–33 kDa protein that is structurally related to, but distinct from, p34^{cdc2}. Immunodepletion of this 32-kDa protein, termed cdc2B or cdk2, from *Xenopus* egg extracts significantly reduced the capacity of the extracts to replicate added DNA (145), demonstrating a possible role for the 32-kDa species in S phase (see below). Inhibition of cyclin resynthesis in this system did not inhibit DNA replication, however, presumably suggesting that the cdk2/cyclin A complex is not normally important for S phase entry (145). Cyclin A is also capable of forming complexes with the cellular transcription factors E2F and DRFT1 (146, 147) and also with the E1a protein of adenovirus (128). It is not yet known whether any of these complexes participates directly in any aspect of the cell cycle.

Regulation of p34^{cdc2} Activity by Phosphorylation

Accumulation of cyclin B during S and G2 is a gradual process, but the onset of mitosis is in contrast an abrupt and dramatic change. This contrast has been explained in two general ways, taking into account the essential role for cyclin B in MPF formation and M phase entry. In the first model, cyclin B must accumulate to a particular threshold level, which determines when MPF activity appears (137). In the second model, sufficient cyclin B to support M phase entry is synthesized relatively early in the cycle, and a pool of inactive p34^{cdc2}/cyclin B heterodimers, or "pre-MPF," accumulates before being rapidly and dramatically activated to effect the G2/M transition (2, 139, 148, 149). This second model is supported by a number of experimental observations. Firstly, while cyclin B was found to be the only protein whose synthesis de novo in *Xenopus* was required for an interphase egg extract to enter an M phase state (137), this protein synthesis requirement is completed well before transition into M phase (126, 150–152). Though a threshold level of cyclin B could be said to be required, this level is normally surpassed well before entry into M phase. Further evidence for accumulation of a pool of inactive pre-MPF comes from the observation of "autocatalytic" MPF amplification. Injection of small amounts of MPF activity into immature oocytes led to the rapid generation of a much larger total activity of MPF without the need for protein synthesis (81, 153).

Pre-MPF accumulation and subsequent activation appear to be achieved by changes in the phosphorylation state of p34^{cdc2}. Here again studies of the fission yeast have provided a key to understanding the controls operative in higher eukaryotic species. Fission yeast p34^{cdc2} was found to be phosphorylated at two sites: Tyr15 and Thr167 (17, 20). In yeast cells entering mitosis, Tyr15 was found to be abruptly dephosphorylated (17), suggesting that this dephosphorylation might be important for activation of p34^{cdc2} protein kinase activity. This interpretation was supported by the observed behavior of a cdc2 mutant in which Tyr15 was replaced by a phenylalanine residue, which can no longer be phosphorylated. This cdc2 mutation appeared to cause yeast cells to enter mitosis prematurely, suggesting that Tyr15 phosphorylation normally functions to delay M phase entry by preventing p34^{cdc2} activation (17). As Tyr15 lies within the region of p34^{cdc2} identified by analogy with other nucleotide-binding proteins as the ATP-binding fold, phosphorylation at this position may inhibit the protein kinase by influencing ATP binding directly, though this has yet to be tested.

The protein kinases responsible for Tyr15 phosphorylation have yet to be identified unambiguously, but possible candidates are the products of the fission yeast wee1 and mik1 genes (154, 155). The activity of these proteins is essential to prevent the premature mitosis phenotype seen in the cdc2 Phe15 mutant (155). Both appear to be protein kinases on the basis of their amino acid sequences, and the p107^{wee1} protein has protein serine/tyrosine kinase activity in vitro (156, 157). Activation of p34^{cdc2} in fission yeast normally depends upon the activity of the cdc25 gene product, p80^{cdc25} (135, 158, 159), which appears to be a novel protein phosphatase directly responsible for Tyr15 dephosphorylation (160–163). Cyclic accumulation of p80^{cdc25} and its corresponding mRNA is thought to play a part in regulating this gene function (164, 165). This regulation may play a role in linking completion of DNA replication to onset of the subsequent M phase, as this cell cycle dependence is lost when the cdc25 gene function is deregulated (166).

This regulatory organization has been largely conserved in animal cells. The region of p34^{cdc2} spanning Tyr15 is identical in its amino acid sequence from yeast to mammals (67), and Tyr15 is also the sole site of tyrosine phosphorylation of p34^{cdc2} in animals (18, 19). Additional phosphorylation on the adjacent residue, Thr14, also suppresses p34^{cdc2} activity in interphase (19), so that phosphate must be removed from both Thr14 and Tyr15 in order to activate the pre-MPF complex. Dephosphorylation of these residues is probably mediated by animal proteins homologous to p80^{cdc25}, which may have a general role in determining the timing of the onset of mitosis in postembryonic animal cell cycles. Such functional homologues have been cloned from humans (167), as well as fruit flies (68, 168) and budding yeast (169), either by complementation of a fission yeast *cdc25* mutant or on the basis of conserved amino acid sequences, which are limited largely to the carboxy-terminal 100 amino acid residues. The *cdc25* homologue of *Drosophila, string,* was independently identified as an essential gene required for entry into mitosis in the blastoderm, immediately after the activation of zygotic gene expression (168). Ectopic expression of this gene was found to be capable of determining the timing of mitosis during postblastoderm development, when the rapid semisynchronous nuclear division cycles of early embryogenesis are replaced by cycles that differ in their G2 duration in a cell type–specific manner (43). In this system, commitment to the subsequent cell cycle appeared to occur in G2, as entry into mitosis in response to a pulse of *string* expression also led to completion of a round of DNA replication in the next cell cycle.

Inhibitor studies demonstrated that pre-MPF activation is due to a protein phosphatase distinct from the classical types 1, 2A, 2B, or 2C (170), supporting the idea that a novel $p80^{cdc25}$-related species is responsible. This phosphatase was shown to be able to dephosphorylate both of the electrophoretically retarded forms of $p34^{cdc2}$ (161, 162), which result from Thr14 and Tyr15 phosphorylation (19). The mechanism of MPF amplification that results in rapid conversion of the pre-MPF pool to the active state without the need for further protein synthesis (81, 153) has yet to be fully elucidated, but it seems likely that generation of a small amount of active MPF feeds back positively to one or more of the pathways controlling $p34^{cdc2}$ phosphorylation and dephosphorylation at Thr14 and Tyr15, leading to the rapid loss of phosphate from these positions.

Phosphorylation of $p34^{cdc2}$ at Thr14 and Tyr15 appears to occur after association between $p34^{cdc2}$ and cyclin B (19, 139). Thus the mutant $p34^{cdc2}$ Asp161, which exhibited reduced cyclin binding, did not become phosphorylated on these residues (19), while the double mutant Ala14 Phe15, though it cannot be phosphorylated in its ATP-binding region, could still bind cyclin B (19). The failure of monomeric $p34^{cdc2}$ to become phosphorylated on Thr14 and Tyr15 (139) could imply that $p34^{cdc2}$ undergoes a conformational change on binding cyclin B, resulting in the exposure of residues 14 and 15 to protein kinase or kinases as yet uncharacterized, but perhaps analogues of the *wee1* and *mik1* gene products of fission yeast. It is noteworthy that $p107^{wee1}$ was not able to phosphorylate purified $p34^{cdc2}$ in vitro (156), suggesting that if $p34^{cdc2}$ is the target in vivo, it must be presented in a specific conformation, perhaps as a result of cyclin B binding. In line with this interpretation, $p34^{cdc2}$ was phosphorylated on Tyr15 when coexpressed with $p107^{wee1}$ in a baculovirus/insect cell system, but became more heavily phosphorylated at this position when cyclin B was coexpressed in the same cells (157). Regardless of the mechanism, phosphorylation of Thr14 and Tyr15 immediately after assembly of the pre-MPF heterodimer facilitates its accumulation in an inactive, though potentially activatable, form.

Additional threonine phosphorylation of $p34^{cdc2}$ may be important for the regulation of MPF assembly. Phosphothreonine detected in vertebrate $p34^{cdc2}$ after [^{32}P] phosphate labelling in vivo has been attributed to phosphorylation at residue 161 (18, 19), analogous to Thr167, the sole reported site of threonine phosphorylation in fission yeast $p34^{cdc2}$ (20). Mutation of this threonine to a nonphosphorylatable residue resulted in much reduced cyclin B binding and consequent loss of $p34^{cdc2}$ protein kinase activity (19, 20). Interestingly, the additional threonine phosphorylation appeared, like phosphorylation on Thr14 and Tyr15, to depend upon the presence of cyclin B (139). Perhaps the simplest interpretation of these results is that phosphorylation of $p34^{cdc2}$ at Thr161 stabilizes an otherwise transient association between $p34^{cdc2}$ and cyclin B (Figure 1). Although $p34^{cdc2}$ is phosphorylated at Thr14 and Tyr15 during interphase in the first embryonic mitotic cycle and in cells derived from fully developed animals, Tyr15 was not seen to be phosphorylated in mitotic cycles 2 to 12 of the rapidly cleaving *Xenopus* embryo (171). If changes in $p34^{cdc2}$ phosphorylation are in part responsible for the cyclic appearance of MPF in these rapid cycles, it is possible that these changes occur at Thr14 or Thr161, or at both positions. Alternatively, periodic accumulation and degradation of cyclin B could be directly responsible for determining the timing of MPF appearance and disappearance in these cycles.

Substrates of $p34^{cdc2}$ and their Significance for M Phase Events

An increasingly large number of proteins have been found to serve as substrates for the $p34^{cdc2}$/cyclin B complex when tested in vitro (Table 1). The targets for phosphorylation are serine or threonine residues immediately N-terminal to proline, frequently where the flanking amino acid residues are basic. In many cases the sites of phosphorylation in vitro have been shown to be phosphorylated in vivo, but on its own this cannot be taken as watertight evidence for their phosphorylation in vivo by $p34^{cdc2}$ (22). This is in part because $p34^{cdc2}$ is only one of a family of closely related protein kinases present in higher eukaryotes (53, 54, 56, 128), and members of this family may well have similar substrate specificities. Candidate substrates for MPF are more likely to be physiologically significant if the [Ser/Thr]-Pro sites become phosphorylated or hyperphosphorylated in M phase, and if a functional consequence of this phosphorylation for an M phase process can be identified (22).

Using these criteria, the best-qualified candidate substrates so far described for $p34^{cdc2}$ in animal cells are nuclear lamins (25, 31, 38–40), vimentin (41),

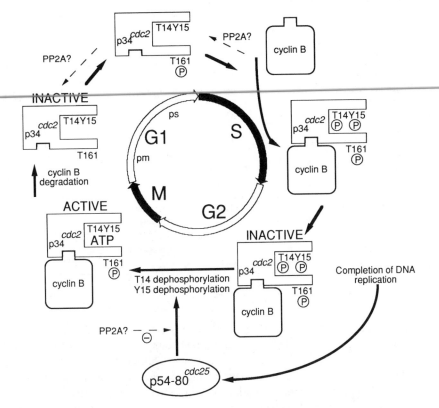

Figure 1 Regulation of p34^{cdc2} activity in the animal cell cycle. Cyclin B association with p34^{cdc2} occurs during interphase, possibly regulated by Thr161 phosphorylation of p34^{cdc2}. Sufficient cyclin B can be accumulated early in the cell cycle for subsequent protein kinase activation, though the molecular basis of the cyclin B function is not understood. Phosphorylation of Thr14 and Tyr15 in the ATP-binding site of p34^{cdc2} prevents protein kinase activation for the remainder of interphase. This phosphorylation may be promoted indirectly by phosphatase 2A/INH (protein fraction INH is described in the text—see section on *Cell Cycle Roles for Phosphoprotein Phosphatases*); alternatively, phosphatase 2A could inhibit kinase activation by direct dephosphorylation of Thr161 (see text). Dephosphorylation of Thr14 and Tyr15 by p54–80^{cdc25} brings about protein kinase activation, but this cannot usually occur unless DNA replication is completed.

and caldesmon (30). Lamins A, B, and C are intermediate filament proteins (172) that polymerize to generate the nuclear lamina, a structure that underlies the nuclear membrane in interphase (71, 173, 174), while vimentin is a constituent of the intermediate filament network that extends throughout the cytoplasm in higher eukaryotic cells (77). Caldesmon is a nonstructural component of cytoplasmic microfilaments that is thought to inhibit the actomyosin ATPase of these filaments (175, 176).

Table 1 Candidate substrates for the p34^{cdc2} protein kinase in animal cells

Substrate	Same site(s) in vivo as in vitro?	Possible role[a]	Reference(s)
Nuclear lamins	yes (M)[b]	Nuclear lamina disassembly	(25, 31, 38, 39)
Vimentin	yes (M)	Intermediate filament disassembly	(41)
Caldesmon	yes (M)	Microfilament contraction	(30)
Histone H1	yes (M)	(Chromosome condensation)	(100)
pp60^{c-src}	yes (M)	(Cytoskeletal rearrangements)	(27, 35)
NO38, nucleolin	yes (M)	(Nucleolar reorganization)	(26, 28)
SV40 T antigen	yes	(DNA replication)	(34)
c-Abl	yes (M)	Unknown	(29)
p105^{Rb}	yes	Unknown	(23, 260)
p53	yes	Unknown	(32)
RNA polymerase II	unknown	(Transcription inhibition)	(33)
EF-1γ	unknown	(Translation inhibition)	(24)
Cyclin B	unknown	(Regulation of p34^{cdc2})	(6, 122)
Myosin light chain	unknown	(Contractile ring activation)	(261)
Casein kinase II	unknown	unknown	(36)

[a] Roles proposed but as yet largely speculative are indicated in parentheses
[b] Substrates hyperphosphorylated in vivo during mitosis are denoted (M)

Lamins, vimentin and caldesmon are newly phosphorylated at mitosis in vivo (71, 176, 177) and by p34^{cdc2} in vitro at some of the same sites (25, 30, 41). Lamin hyperphosphorylation at these and additional sites in vivo coincides with disassembly of the nuclear lamina and nuclear envelope breakdown (25, 38), and furthermore mutation of two of the phosphorylation sites of lamin A to nonphosphorylatable residues generated a mutant lamin protein that prevented nuclear disassembly (39). Phosphorylation of a chicken lamin B by fission yeast extracts depended on active p34^{cdc2} (40), and treatment of isolated chicken nuclei with purified p34^{cdc2}/cyclin B caused depolymerization of the lamina, while other protein kinase activities such as protein kinase A or C had no such effect on nuclear structure (25). The sufficiency of the p34^{cdc2}/cyclin B kinase alone to cause complete lamina disassembly has been questioned, however (31), and it is possible that additional activities are required in vivo. Nonetheless a clear link has been established between activation of p34^{cdc2} and nuclear envelope breakdown, a major feature of M phase in higher eukaryotes.

Reorganization of the cytoskeleton at M phase may result in part from phosphorylation of caldesmon and vimentin. Like lamins, these proteins are subject to mitosis-specific phosphorylation at a number of sites, a subset of which can be phosphorylated by p34^{cdc2} in vitro (30, 41). A highly enriched preparation of one of two endogenous vimentin kinases was found to contain p34^{cdc2}, and highly purified p34^{cdc2}/cyclin B caused depolymerization of vimentin intermediate filaments in vitro (41). Similarly, phosphorylation in vitro by immunoprecipitated p34^{cdc2} reduced the affinity of caldesmon for actin and calmodulin (30). Dissociation of caldesmon from actin is thought to promote contraction of microfilaments, vital to contractile ring function during cytokinesis and possibly also involved in the "rounding up" characteristic of mitosis in many animal cell types.

In addition to the putative in vivo substrates described above, a number of proteins have been found to act as substrates for p34^{cdc2} in vitro (Table 1). These include histone H1, widely used in assays for M phase p34^{cdc2} protein kinase activity, and the products of a number of viral and cellular oncogenes and tumor suppressor genes. For these candidate substrates, evidence suggestive of a cell cycle function is either incomplete or absent, and in many cases phosphorylation in vivo by p34^{cdc2} has yet to be distinguished from phosphorylation by a related protein kinase of similar substrate specificity (22).

It is noteworthy that the best-qualified candidate substrates for p34^{cdc2} are either structural components or accessory proteins of subcellular features that undergo large-scale reorganization during M phase. This suggests that a major function of p34^{cdc2} in mitosis may be as a "workhorse," directly responsible for effecting some of the intracellular changes that characterize the G2/M transition. In such cases there would be no opportunity for further amplification of the signal generated by p34^{cdc2} activation, perhaps explaining the evolution of a mechanism for the rapid activation of a large pool of pre-MPF discussed above. However, although the p34^{cdc2}/cyclin B heterodimer is able to initiate M phase events, it is likely to be responsible only for a subset of M phase–specific phosphorylations. This implies that in addition to the "workhorse" functions alluded to above, there are other functions of MPF, which lead either directly or indirectly to the activation of other protein kinases, potentially generating amplification or diversification of the M phase signal through protein kinase cascades (36, 178). The in vitro effects of MPF on microtubule behavior mimic the interphase-metaphase transition in vivo (101), but this effect may be indirect, as purified *Xenopus* microtubule-associated protein (MAP) kinase caused similar behavior in vitro (179). As the *Xenopus* MAP kinase was found to be phosphorylated and activated in M phase, it is possible that a protein kinase cascade beginning with p34^{cdc2} and including MAP kinase is responsible for the reorganization of the microtubular network at M phase onset.

Dependence of Mitosis on Completion of DNA Replication

The interphase and M phase states are so mutually exclusive that entry into M phase must be very highly regulated. If this were not so, potentially damaging partial M phase states such as premature chromatin condensation might be generated in interphase cells. In general, blocking DNA replication inhibits entry into mitosis (79), and breakdown of this normal cell cycle order has been reported only in mutant cell lines (59, 180), in cells treated with caffeine (181) or okadaic acid (182), or in early embryos in which DNA replication is blocked (183–185). The appearance of mitotic features under these unusual circumstances can be taken as evidence that, in the normal course of events, M phase onset is delayed until cellular mechanisms deem it appropriate. Mechanisms such as this that delay onset of a cell cycle process until an earlier process has been completed have been termed "checkpoints." One such mechanism appears able to sense lack of completion of chromosomal DNA replication, such that M phase onset is normally delayed in response to agents that prolong S phase (79).

Although early embryos provide one of the few examples of an animal cell system in which inhibition of DNA replication fails to block entry into M phase, this dependence can be imposed upon activated egg extracts if they are provided with a sufficiently high concentration of unreplicated DNA (186). Using this system, it was shown that incomplete DNA replication blocks activation of pre-MPF, probably by preventing dephosphorylation of p34^{cdc2} at Tyr15 and possibly also Thr14 (19, 186), and this blockage of MPF activation was overcome by the addition of the protein product of the *Drosophila cdc25* homologue, confirming analogous studies in yeast. The fission yeast mutant *cdc2* Phe15 appeared to enter mitosis without necessarily first completing S phase (187), and was independent of the *cdc25* gene function (17) that seems to form part of a pathway linking M phase onset to the completion of S phase (166). Overproduction of p80^{cdc25} caused fission yeast cells arrested in S phase to undergo premature mitosis.

The blockage of MPF activation by unreplicated DNA in a *Xenopus* egg extract was overcome by addition of caffeine (186), a drug previously shown to induce M phase events in cultured baby hamster kidney (BHK) cells arrested in S phase (181). Similar results were obtained in BHK cells using okadaic acid, an inhibitor of phosphoprotein phosphatases (182), or by shifting the temperature-sensitive BHK cell line tsBN2 (59) to its restrictive temperature (188). In these mammalian systems, mitotic features did not appear unless the cells had entered S phase before addition of the drug or temperature shift, and acquisition of the capacity eventually to execute M phase events correlated with the first phosphorylation of p34^{cdc2} on Tyr15 (188). Incubation of the tsBN2 mutant at its restrictive temperature leads to degradation of the thermolabile product of the *RCC1* gene (188), a DNA-binding protein that prevents premature chromosome condensation by an as-yet-unknown mechanism (189). This protein is also required for progression through the G1 phase, where it could perform a different function.

Degradation of the *RCC1* product from the G1/S boundary onwards, like caffeine treatment, activates pre-MPF by a mechanism that requires protein synthesis and presumably therefore acts through one or more intermediates (181, 188). Nonetheless, these systems give some insight into the pathways through which M phase onset can be delayed in response to incomplete DNA replication.

Meiotic Maturation and Cytostatic Factor

Mature vertebrate eggs arrested in meiotic metaphase contain a cytoplasmic factor distinct from MPF that is capable of inducing metaphase arrest when injected into early embryos (5). This cytostatic factor (CSF) appears to act by stabilizing MPF and thus preventing exit from M phase. A number of lines of evidence indicate that the product of the proto-oncogene *c-mos* may act as a CSF (190). The pattern of expression of *c-mos* closely follows the appearance of CSF activity, and indeed the *c-mos* protein product $p39^{mos}$ was detectable only during oocyte maturation. Microinjection of antisense *c-mos* oligonucleotides into oocytes prevented both $p39^{mos}$ expression and meiotic maturation (191). Though this suggested a role for $p39^{mos}$ in promoting maturation rather than stabilizing M phase, injection of *c-mos* mRNA into two-cell embryos resulted in metaphase arrest, and CSF activity was removed from egg cytoplasm when it was immunodepleted of $p39^{mos}$ (190).

Like CSF, $p39^{mos}$ is calcium-sensitive both in vivo and in vitro, being selectively degraded by a calcium-dependent cysteine protease on release from metaphase arrest at fertilization (192). CSF inactivation and MPF inactivation were found to occur by independent pathways, and cyclin B proteolysis preceded disappearance of $p39^{mos}$ (193). Thus MPF inactivation is not simply the result of loss of CSF activity. The physiologically important substrates for the $p39^{mos}$ protein kinase have yet to be identified, though cyclin B can act as a substrate in vitro (194), and cyclin B phosphorylation in oocyte extracts was decreased by prior treatment with *c-mos* antisense oligonucleotides. The available data suggest that $p39^{mos}$ has a dual function firstly in meiotic maturation (191), and secondly in stabilization of metaphase arrest (190), but do not indicate if phosphorylation of a single substrate could explain both activities. There may also be alternative pathways causing metaphase arrest, as activated mutants of the $p21^{ras}$ oncogene product induced meiotic maturation in *Xenopus* oocytes and exhibited CSF-like activity when introduced into early embryos, in each case independently of $p39^{mos}$ (195).

The extent to which these $p39^{mos}$-independent pathways are relevant to maturation and metaphase arrest in vivo remains to be established. Some 20% of the M phase–specific histone H1 kinase of *Xenopus* eggs was found to derive from protein fractions lacking detectable $p34^{cdc2}$ (196). Fractions containing this histone kinase activity were able to accelerate progesterone-

induced oocyte maturation, though unlike MPF they could not induce maturation alone. The H1 kinase activity was bound by the mitosis-specific monoclonal antibody MPM-2, previously shown to inhibit oocyte maturation (197).
This additional H1 kinase activity appears to act as a positive regulator of M
phase entry, but the physiologically relevant substrates for this activity have
yet to be identified.

Cell Cycle Roles for Phosphoprotein Phosphatases

If control of progression through the animal cell cycle can be explained at
least in part at the level of changes in protein phosphorylation, then protein
phosphatases are potentially as important for this control as are protein
kinases such as $p34^{cdc2}$ and $p39^{mos}$. A number of roles for protein phosphatases in cell cycle control have been suggested, of which perhaps the most
completely understood is the role for $p80^{cdc25}$-related phosphatases that
appear to activate pre-MPF by removing phosphate from $p34^{cdc2}$ residue
Tyr15 and probably Thr14 as discussed above (161–163).

Less well-defined roles for the serine/threonine-specific phosphatases types
1 and 2A have also been proposed. Specific inhibition of type 2A phosphatases by okadaic acid was found to cause premature activation of $p34^{cdc2}$/
cyclin B complexes in activated egg extracts (170). Inhibition of the type 2A
enzyme was probably also responsible for the premature mitotic features
observed after treatment of BHK cells with okadaic acid (182). Crude biochemical fractionation of oocyte extracts identified a protein fraction termed
INH that inhibited activation of pre-MPF (148). Using this inhibition as an
assay, INH was purified more than 2000-fold from oocyte extracts (198). The
purest fractions contained approximately six protein species detectable by
silver staining, one of which was the catalytic subunit of phosphatase 2A.

Further evidence for the identity between INH and a type 2A activity came
from the demonstration that INH was itself inhibited by incubation with
okadaic acid (198). The protein phosphate groups targeted by INH, stabilized
by okadaic acid addition, and responsible for premature MPF activation in
egg extracts have not been identified. Nonetheless, the observation of premature mitotic features on inhibition of phosphatase 2A indicates a positive
requirement for this activity to suppress MPF during interphase. This could be
an indirect effect that promotes phosphorylation of $p34^{cdc2}$ on Thr14 and
Tyr15 by influencing their relative rates of phosphorylation or dephosphorylation. Addition of INH to purified active $p34^{cdc2}$/cyclin B (in which $p34^{cdc2}$
residues Thr14 and Tyr15 were largely dephosphorylated but Thr161 was
probably phosphorylated) resulted in dephosphorylation of both $p34^{cdc2}$ and
cyclin B, and loss of histone H1 kinase activity with kinetics similar to those
of $p34^{cdc2}$ dephosphorylation (198). Thus removal of phosphate from Thr161
of $p34^{cdc2}$ may be responsible for inactivation of H1 kinase/MPF activity in

vitro, perhaps by destabilizing the $p34^{cdc2}$/cyclin B heterodimer (Figure 1). It remains to be seen if this mechanism is related to the normal physiological action of INH/type 2A phosphatase in preventing premature activation of pre-MPF.

Inhibition of type 1 phosphatase by the specific peptide inhibitors 1 and 2 did not cause premature MPF activation in mitotic egg extracts (170), but blocked maturation when the inhibitors were injected into *Xenopus* oocytes (199, 200). Genetical experiments using *Aspergillus* (201), yeast (202, 203), and *Drosophila* (204) all show that type 1 phosphatases have additional roles in late mitosis. These functions may well be conserved in vertebrate cells, but again the relevant targets for dephosphorylation have yet to be identified.

CONTROL OF THE ANIMAL CELL CYCLE IN G1

One approach to understanding control of progression through the cell cycle is to examine the ways in which cells reversibly withdraw from or become arrested within the cycle in response to the presence or absence of various stimuli. This approach has been successful in identifying the factors MPF and CSF involved in metaphase establishment and arrest in eggs, respectively, which in turn has shed light on the process of M phase entry generally as discussed above. In principle, analysis of cell cycle arrest other than at the onset of or during metaphase could be correspondingly informative about key regulatory steps in interphase. One problem encountered in this type of analysis is that in animals the physiologically significant cell cycle arrest points are likely to be cell type–specific. Activated B-lymphocytes, for example, were found to require distinct external stimuli in order to progress beyond each of three cell-cycle arrest points: one immediately after M phase, a second in mid-G1, and a third in mid-G2 (45).

In contrast, mouse fibroblasts appear to require the continued presence of extracellular growth factors only during G1, or the first part of G1 (reviewed in 1). During late G1 and the remainder of the cycle, fibroblasts are relatively insensitive to removal of growth factors or the partial inhibition of protein synthesis. Although "commitment to enter S phase" is a phrase frequently used in the literature, for fibroblasts this is usually taken to mean commitment to enter and complete both S phase and the following mitosis, though this would clearly not apply to all cell types. The fibroblast experimental system has been the focus of considerable attention for a number of reasons. Like hematopoietic cells, fibroblasts have growth factor requirements that are increasingly well understood, and essentially homogeneous 3T3 fibroblast populations are relatively easily generated and maintained in culture (205).

The commitment of fibroblasts to enter the cell cycle from G1 is furthermore somewhat analogous to the "start" control point in yeasts, where cells become similarly committed to entering the mitotic cycle (206). This analogy raises the possibility that investigating start control through primarily genetical experiments in yeasts could lead to a greater understanding of the commitment process in fibroblasts, although this potential has yet to be realized. The obvious precedent here is the contribution that yeast genetics has made to understanding M phase onset in animals. Study of the commitment of fibroblasts to cell cycle reentry was further intensified by the discovery that this control step seems to be aberrantly regulated in neoplastically transformed cells (reviewed in 46).

Growth Factor Requirements for Commitment of Fibroblasts to S Phase Entry

Though fibroblasts are probably arrested primarily as a result of contact with adjacent cells in vivo, there is evidence to suggest that this state of contact inhibition is to some extent physiologically equivalent to the quiescent or G0 state elicited in culture by reduced serum growth factor concentrations (207). It has been proposed that cells in close contact have a reduced surface area exposed to growth factors, and it may be possible to mimic this situation in culture by reducing the serum concentration in the medium. Experiments in support of this proposal have shown that cells that cannot spread out because of the nature of the cell support matrix do not enter the cell cycle as readily as cells that do spread out (208, 209). Quiescent or density-arrested fibroblasts leave G0 in response to serum growth factors or to reduction in cell density, and begin to enter S phase after 8 to 20 hours (207). Quiescent fibroblasts, and those that are in cycle and have just completed mitosis, require exposure to specific growth factors before becoming committed to enter a new cell cycle.

After the completion of cell division, growing Swiss mouse 3T3 fibroblasts were found to spend approximately 3.5 hours in a state termed G1pm (postmitotic) in which they were especially sensitive to the removal of serum growth factors (210). During G1pm, removal of serum for as little as one hour caused cells to enter a G0 state, from which they took eight hours to return to the growth factor-independent G1ps (presynthetic) phase. Of the serum growth factors tested, only platelet-derived growth factor (PDGF) was fully able to prevent cells leaving G1pm and becoming quiescent (210). In a separate study, it was concluded that insulinlike growth factor 1 (IGF-1) was the sole growth factor required for cycling mouse BALB/c-3T3 (A31) fibroblasts in G1 to enter S phase (211). If this discrepancy is the result simply of the different strains of mouse 3T3 fibroblasts used in the two studies, this serves to underline the high degree of specificity of growth factor requirements in a given cell type. A role for PDGF in controlling fibroblast entry into and exit from G0 was also demonstrated in experiments using quiescent, serum-deprived BALB/c-3T3 (A31) cells (212). Treatment with PDGF rendered these cells "competent" to respond to other serum components. Transient exposure of competent cells to secondary stimuli such as epidermal growth factor (EGF), and insulin allowed progression to a point termed "V" some six hours before the onset of S phase (212; reviewed in 46). Further progression beyond the V point was found to require only IGF-1.

Experiments carried out by other investigators using growth factors have defined two control points by alternative methods. Most fibroblasts progressing from quiescence towards S phase were found to become independent of serum at a restriction point termed "R" (207) approximately two hours before the onset of S phase (213). The R point was also identified as a unique point in late G1 at which cells escaped their requirement for essential amino acids (207). Using a different approach, in which competent BALB/c-3T3 (A31) fibroblasts were allowed to progress towards S phase for 12 to 15 hours before removal of growth factors, a subpopulation of cells was generated that appeared to be blocked just before S phase and yet still required serum factors for S phase entry (212). This block point, termed "W," does not seem to lie within either of the G1 compartments observed in cycling fibroblasts. Cells in G1pm became quiescent if serum was removed for as little as one hour (210, 214), whereas cells blocked at the W point survived 24 hours of serum deprivation while remaining poised to initiate DNA replication (212). The W point could not be placed in the G1ps compartment either, as cells blocked at W still required the presence of growth factors in order to enter S phase. It remains possible that the W point represents the pm-to-ps transition, as for some cells G1ps is very short and so this transition would coincide with S phase entry (210). If this is the case, then the distinction between the R and W points might simply reflect the variable duration of G1ps. Although some cells entered S directly after leaving G1pm, an average cell spent 3.5 hours in growth factor–independent G1ps before initiating DNA replication (210).

Despite the complexities of nomenclature and inconsistencies regarding specific growth factors that have arisen from the different experimental systems used, this type of approach at the very least identifies a control point in G1 at which fibroblasts become growth factor independent and committed to entering the next cycle, though the time interval between traversal of this control point and the initiation of DNA replication varies from cell to cell over a range of approximately zero to seven hours. During this time the cells enter S phase with apparent first-order kinetics (215) at a rate that is determined by the concentration of growth factors in the surrounding medium (216). This kinetic analysis attracted considerable attention and gave rise to a transition probability model for S phase entry, in which cells acquire a fixed probability of initiating DNA replication on arrival at a late G1 control point (215, 217). Thus commitment in fibroblasts is usually temporally distinct from the initiation of DNA replication itself. Whether or not this control point is a general feature of postembryonic animal cell cycles remains to be established, though the position of the pm-to-ps transition, on average halfway through a seven-hour G1 period (210), is at least superficially similar to the mid-G1 restriction point of activated B-lymphocytes (45).

Protein Synthesis Requirements for Commitment of Fibroblasts to S Phase Entry

Partial (50 to 70%) inhibition of protein synthesis caused fibroblasts to accumulate in G1 (218–220). The requirement for a rapid rate of protein synthesis in G1 is confined to the G1pm or pre-R point period (220), and kinetic experiments suggested that fibroblasts escape from their serum growth factor and essential amino acid requirements at a single R point (207). It was thus proposed that the growth factor requirement between the V and R points reflects a requirement for rapid synthesis of a critical regulatory labile protein (220). The half-life of this putative protein was estimated at two to three hours, and a critical threshold level appeared to be required in order for cells to pass the R point. This provides a further kinetic distinction between R point traversal and the initiation of DNA replication itself, as cells blocked at the G1/S boundary by hydroxyurea in the absence of growth factors lost their ability to enter S phase by a first-order decay with a half-time of approximately five hours (221). These kinetics were shown to be inconsistent with the decay below a critical threshold of a protein such as that proposed to be required for R point traversal.

This kinetic approach is rather unlikely by itself to lead to the identification of key regulatory molecules important for commitment to and initiation of S phase, but has provided data that predict how candidate molecules of this sort should behave, at least in fibroblasts.

The Search for an S Phase–Promoting Factor

If kinetic studies of cell physiology are unlikely to reveal the molecular mechanisms that underlie animal cell cycle commitment in G1, alternative approaches to this problem may be more illuminating. In principle genetical approaches could identify gene products involved in the commitment process. Dominant gain-of-function mutations that cause cells to become committed to a new cell cycle under inappropriate conditions could include those lying in cellular oncogenes known to cause transformation (222). Many of these oncogenes have identified diverse components of growth factor signal transduction pathways (223), and these may be biochemically rather far removed from the molecules responsible for commitment itself. These growth factor signalling pathways lie outside the scope of this review, but their multitudinous components and some of the interactions between them have been reviewed in depth elsewhere (222–224).

An alternative genetical strategy is to look for analogous functions in yeasts, an approach successfully applied to the control of M phase onset as reviewed above. In yeasts, as in fibroblasts, traversal of a major control point in G1 commits cells to entering the next mitotic cell cycle (49–51, 206, 225). Passing this start control point requires the *cdc2* gene function in fission yeast or its functional homologue *CDC28* in budding yeast, an organism as distantly related to fission yeast as either is to mammals (226). Conservation of the *cdc2* start function over this large phylogenetic distance implies that this function is conserved in other eukaryotes. Thus the gene product p34$^{cdc2/28}$ carries the responsibility not only for regulating entry into M phase, but also for commitment to the mitotic cycle in G1 in a number of eukaryotic species. This idea was supported by the cloning of the human *cdc2* homologue by functional complementation of a fission yeast *cdc2* mutant (67), and the subsequent demonstration that this human gene function can fully substitute for the yeast counterpart in both its G1 (start) and G2/M roles. The specific inhibition of new p34^{cdc2} synthesis by antisense oligonucleotides in T lymphocytes stimulated to enter the cell cycle from G0 had no effect on early or mid-G1 events, but significantly inhibited DNA replication (227). This finding supports the proposition that p34^{cdc2} functions in a wide variety of eukaryotic species to commit cells in G1 to enter the mitotic cycle.

Execution of this function could be the biochemical basis for traversal of the R point or G1pm/ps transition in fibroblasts. If this is the case, then what is different about p34^{cdc2} at G1/S compared with G2/M? One possibility is that the activity of p34^{cdc2} is essentially identical at both points in the cycle, but the consequence of this activity could be interpreted by the cell as a signal either for M phase entry or commitment to a new cycle depending on the cell cycle status as defined by other components. Alternatively, the p34^{cdc2} protein may be intrinsically different at the two points in the cell cycle. Support for the latter model came from experiments in fission yeast that suggested that the biochemical distinction between G1 and G2 may reside in the state of p34^{cdc2} itself (228). How, then, could the G1 and G2 forms of p34^{cdc2} be distinguished? Studies of the premature appearance of mitotic features discussed above suggested that p34^{cdc2} becomes activatable for its G2/M function at a point close to the initiation of S phase. The appearance of this form could indicate the transition of p34^{cdc2} from an intrinsic G1 state to a G2 state. In the tsBN2 cell line, this presumptive transition was accompanied by the first appearance of Tyr15 phosphorylation (188), which could perhaps form part of the distinction between the G1 and G2 forms of p34^{cdc2}.

Another possibility is that association between p34^{cdc2} and a B cyclin, which is characteristic of cells preparing to enter M phase, is a means by which the two states of p34^{cdc2} are distinguished. In budding yeast, a family of proteins collectively known as G1 cyclins, which have limited but significant homology to the B cyclins, is involved in determining the cell cycle timing of start, probably at least in part by interacting with the p34^{cdc2} homologue, p34^{CDC28} (229–234). This raises the possibility that binding G1 cyclins rather than cyclin B distinguishes the G1 form of p34$^{cdc2/28}$ from the G2 form.

Genes encoding proteins with some degree of homology to cyclins have been isolated from a variety of organisms, either on the basis of functional complementation of budding yeast G1 cyclin mutants (235–237), or by alternative strategies that do not rely upon cell cycle function (238–241). The extent to which these cyclin homologues are involved in regulating commitment to the cell cycle in animal cells has yet to be established. It is clear from the preceding discussion that the distinction between the G1 and G2 forms of p34^{cdc2} remains an important unsolved question. The distinction could be quite subtly based on, for example, location within the cell, which could itself be determined by progression through some earlier cell cycle phase (242). It is also possible that p34^{cdc2} does not act as a protein kinase at the G1/S transition. Certainly p34^{cdc2} kinase activity appears to be much reduced at this point of the cycle compared with M phase (15, 135, 138, 243), though this could be due to inappropriate substrates being used in the assays in vitro, or to a requirement for a reduced overall level of activity in G1. An alternative explanation is that p34^{cdc2} could perform some other, as-yet-undefined function at G1/S.

In some vertebrate systems, there is evidence indicating that two distinct protein species could perform the G1/S and G2/M roles carried out in yeasts by p34^{cdc2}. This would represent the most extreme means of distinction between the G1 and G2 forms of the protein. The strongest support for this notion has come from studies carried out using *Xenopus* egg extracts that accurately perform DNA replication in vitro. In these extracts, entry into S phase was prevented if p34^{cdc2}-related proteins were removed by either of two affinity reagents (244). It was not possible using these reagents to distinguish between bona fide p34^{cdc2} (the protein that forms the catalytic subunit of MPF) and closely related but functionally distinct proteins known to exist in a variety of species including *Xenopus*. A recent study indicates that a p34^{cdc2}-related 32-kDa protein termed cdc2B or cdk2 rather than p34^{cdc2} itself is required for S phase entry in this system (145). Specific immunodepletion of this 32-kDa species significantly reduced the capacity of the extract to replicate added DNA, whereas depletion of p34^{cdc2} had no such effect. Conversely, depletion of p34^{cdc2} prevented entry into mitosis, but depletion of the 32-kDa species did not. The observations that microinjection of antibodies against p34^{cdc2} into fibroblasts (117) and shifting a mouse *cdc2* mutant cell line to its restrictive temperature (66) both resulted in arrest in G2 have been taken as supportive evidence for a "G2 only" role for p34^{cdc2} in vertebrates.

Assays based on replication of the DNA virus SV40 have revealed that extracts prepared from S phase human cells are 20-fold more active in initiating SV40 DNA replication than those prepared from cells in G1 (245). A factor termed RF-S has been described and partially purified from S phase cells that converts a G1 extract to an S phase extract, in the sense that initiation of SV40 DNA replication is substantially increased (246). Partially purified RF-S contained $p34^{cdc2}$ as well as cyclins A and B, and addition of cyclins of either type to a G1 extract could mimic RF-S action. One interpretation of these data is that $p34^{cdc2}$/cyclin B complexes, implicated elsewhere in the initiation of M phase, are capable in this system of performing a G1/S function analogous to that performed in *Xenopus* egg extracts by the $p34^{cdc2}$-like protein cdc2B/cdk2. Thus it is not yet clear if the apparent allocation of G1 and G2 roles of $p34^{cdc2}$ to two different proteins is a general feature of postembryonic vertebrate cell cycles. At present perhaps an open mind should be kept.

An important unanswered question concerns the relationship between commitment to enter the S phase, as defined by the start control in yeasts or the late G1 controls of fibroblasts, and the initiation of DNA replication itself. In vitro systems for studying initiation of DNA replication such as those described above might ultimately help to clarify this relationship.

Growth Control and Cell Cycle Control

Animal gene functions required for cell cycle commitment in G1 might be identified by recessive loss-of-function mutations, in much the same way that such mutations identified $p34^{cdc2/28}$ as a start function in yeasts. Temperature-sensitive animal cell lines that appear to be conditionally defective in G1 functions have been isolated, and in some cases the corresponding genes have been cloned and sequenced (61, 64). This approach has been limited by the lack of a straightforward screen capable of distinction between growth defects and cell cycle defects. In yeast this distinction was made by selecting for further study only those mutants that exhibited a cell division cycle (cdc) phenotype, that is to say continued to grow while some feature of the nuclear division cycle was blocked. A detailed view of the way in which animal cell growth and the cell cycle are coordinated to produce "balanced growth" is not yet available, but it seems that the two processes are separable, for example in terms of their specific growth factor requirements in fibroblasts (247). Some experiments, although not others, have shown that in fibroblasts as in yeasts, commitment to the cycle depends upon the attainment of a critical cell mass (248–251), though in the case of fibroblasts no further growth was necessary for completion of the S, G2, and M phases (252). The relationship between this critical mass requirement and the G1 controls such as the R point identified in fibroblasts is unclear, though the observed hypersensitivity to CHX in G1pm (220) could reflect a mechanism for sensing growth rate, rather than a requirement for synthesis of a specific labile protein species. Thus animal cell mutants that become arrested at their restrictive temperature with a G1 DNA content could be defective in key cell cycle control functions, but it is equally if not more likely that they are defective in some aspect of cell growth, and that arrest in G1 is a secondary effect.

Negative controls over cell proliferation that operate in vivo are increasingly the focus of experimental attention (253), and studies of this area may offer new insights into the control of cell cycle progression in G1. Here again care must be taken to distinguish between cell cycle controls and controls over cell growth, with attendant cell cycle effects that could be secondary to such growth controls. This is not to say that these controls are not worthy of investigation, but rather that such investigations might reveal aspects of growth control that are not immediately relevant to cell cycle control mechanisms per se.

One negative regulator of proliferation that is thought to be physiologically relevant is distinguished by the similarity of its arrest point to the R point of fibroblasts. This is transforming growth factor $\beta1$ (TGF$\beta1$), a peptide factor that was shown to cause the reversible arrest of a variety of cell types in G1 (254). For mink lung endothelial cells, this arrest point was found to be in late G1, approximately one hour before the onset of S phase (255). Arrest at this point appeared to prevent the accumulation of the presumptive G2/M form of p34^{cdc2} in these cells, as after TGF$\beta1$ treatment a dephosphorylated, histone H1 kinase-negative form of p34^{cdc2} characteristic of G0/early G1 cells (13, 15, 243, 256) was detected (255). This clearly suggests that the arrest point is before the point at which the G2 form of p34^{cdc2} is first generated, but cannot be taken as evidence for a direct link between TGF$\beta1$ action and p34^{cdc2} itself. This peptide factor is mitogenic rather than inhibitory for mesenchymal cells (254), so the signal generated by TGF$\beta1$ at the cell surface can be interpreted differently in different tissues. This further suggests that the connections between TGF$\beta1$ and cell cycle control mechanisms are complex and indirect.

Cellular components of the mechanisms that negatively control proliferation include the products of tumor suppressor genes such as p105^{Rb} (257, 258) and p53 (259). Deletion or mutation of these genes appears to be a widespread feature of tumorigenesis in vivo, and introduction of the wild-type genes into transformed cells suppressed the transformed phenotype. Whether these gene products act principally to suppress cell growth or to block cell cycle progression has yet to be established, however.

PROSPECTS

The biochemical characterization of the G1 controls defined in yeasts as start and in fibroblasts as the R point or G1pm-to-ps transition remains a major challenge to workers in this field. It will be especially interesting to learn something of the relationship between these cell cycle controls and controls of cell growth. The relevance of each of these types of control to positive and negative controls over proliferation such as those mediated by oncogenes and tumor suppressor genes is expected to be the subject of intense investigation.

Characterization of the G1 form of $p34^{cdc2}$ that is required for passing start in yeasts will greatly facilitate investigation of the involvement of $p34^{cdc2}$ or related proteins in the analogous controls in animal cells. If we assume that the G1 form of $p34^{cdc2}$ functions as a protein kinase, it would be expected to phosphorylate substrates other than those phosphorylated at M phase onset. In vitro DNA replication systems may help to clarify the relationship that clearly must exist between G1 controls over cell cycle commitment and the initiation of S phase itself.

The pathway linking completion of S phase to the onset of mitosis provides further challenges to experimental design at the interface between cell biology and biochemistry. At the far end of this pathway dephosphorylation of $p34^{cdc2}$ in its ATP-binding region by a $p80^{cdc25}$-related phosphatase is thought to lead to activation of the protein kinase and the initiation of mitosis, but earlier components of the pathway remain to be discovered. Molecular mechanisms capable of sensing lack of completion of DNA replication promise to be fascinating. Finally, there is still much to be learned about the onset of M phase itself. While activation of MPF/$p34^{cdc2}$ could lead directly to some aspects of M phase–specific nuclear envelope and cytoskeletal changes as reviewed above, little is known of the events leading to chromatin condensation and spindle formation, for example. The identification of a universal mechanism for the onset of M phase might have been the end of a chapter, but it is expected to be only part of rather a long book.

ACKNOWLEDGMENTS

We thank Tamar Enoch and Jacky Hayles for critical reading of the manuscript and helpful discussions during its preparation. Work in our laboratory is supported by the Imperial Cancer Research Fund and the Medical Research Council.

Literature Cited

1. Pardee, A., Dubrow, R., Hamlin, J. L., Kletzien, R. F. 1978. *Annu. Rev. Biochem.* 47:715–50
2. Nurse, P. 1990. *Nature* 344:503–8
3. Smith, L. D., Ecker, R. E. 1969. *Dev. Biol.* 19:281–309
4. Smith, L. D., Ecker, R. E. 1971. *Dev. Biol.* 25:233–47
5. Masui, Y., Markert, C. 1971. *J. Exp. Zool.* 177:129–46
6. Lohka, M. J., Hayes, M. K., Maller, J.

L. 1988. *Proc. Natl. Acad. Sci. USA* 85:3009–13
7. Gautier, J., Norbury, C., Lohka, M., Nurse, P., Maller, J. 1988. *Cell* 54:433–39
8. Dunphy, W. G., Brizuela, L., Beach, D., Newport, J. 1988. *Cell* 54:423–31
9. Nurse, P. 1975. *Nature* 256:547–51
10. Nurse, P., Thuriaux, P. 1980. *Genetics* 96:627–37
11. Labbé, J. C., Capony, J. P., Caput, D.,

Cavadore, J. C., Derancourt, M. K., et al. 1989. *EMBO J.* 8:3053–58

12. Gautier, J., Minshull, J., Lohka, M., Glotzer, M., Hunt, T., et al. 1990. *Cell* 60:487–94

13. Simanis, V., Nurse, P. 1986. *Cell* 45:261–68

14. Labbé, J. C., Picard, A., Peaucellier, G., Cavadore, J. C., Nurse, P., et al. 1989. *Cell* 57:253–63

15. Morla, A. O., Draetta, G., Beach, D., Wang, J. Y. J. 1989. *Cell* 58:193–203

16. Gautier, J., Matsukawa, T., Nurse, P., Maller, J. 1989. *Nature* 339:626–29

17. Gould, K. L., Nurse, P. 1989. *Nature* 342:39–45

18. Krek, W., Nigg, E. A. 1991. *EMBO J.* 10:305–16

19. Norbury, C., Blow, J., Nurse, P. 1991. *EMBO J.* 10: 3321–29

20. Gould, K., Moreno, S., Owen, D., Sazer, S., Nurse, P. 1991. *EMBO J.* 10:3297–309

21. Maller, J. L., Smith, D. S. 1985. *Dev. Biol.* 109:150–56

22. Moreno, S., Nurse, P. 1990. *Cell* 61:549–51

23. Lin, B. T. Y., Gruenwald, S., Morla, A., Lee, W.-H., Wang, J. Y. J. 1991. *EMBO J.* 10:857–64

24. Bellé, R., Derancourt, J., Poulhe, R., Capony, J.-P., Ozon, R., et al. 1989. *FEBS Lett.* 255:101–4

25. Peter, M., Nakagawa, J., Dorée, M., Nigg, E. A. 1990. *Cell* 61:591–602

26. Peter, M., Nakagawa, J., Dorée, M., Labbé, J. C., Nigg, E. A. 1990. *Cell* 60:791–801

27. Shenoy, S., Choi, J. K., Bagrodia, S., Copeland, T. D., Maller, J. L., et al. 1989. *Cell* 57:763–74

28. Belenguer, P., Caizergues, F. M., Labbé, J. C., Dorée, M., Amalric, F. 1990. *Mol. Cell. Biol.* 10:3607–18

29. Kipreos, E. T., Wang, J. Y. J. 1990. *Science* 248:217–20

30. Yamashiro, S., Yamakita, Y., Hosoya, H., Matsumura, F. 1991. *Nature* 349: 169–72

31. Luscher, B., Brizuela, L., Beach, D., Eisenman, R. N. 1991. *EMBO J.* 10: 865–75

32. Bischoff, J. R., Friedman, P. N., Marshak, D. R., Prives, C., Beach, D. 1990. *Proc. Natl. Acad. Sci. USA* 87: 4766–70

33. Cisek, L. J., Corden, J. L. 1989. *Nature* 339:679–84

34. McVey, D., Brizuela, L., Mohr, I., Marshak, D. R., Gluzman, Y., et al. 1989. *Nature* 341:503–7

35. Morgan, D. O., Kaplan, J. M., Bishop,

J. M., Varmus, H. E. 1989. *Cell* 57:775–86

36. Mulner-Lorillon, O., Cormier, P., Labbé, J. C., Dorée, M., Poulhe, R., et al. 1990. *Eur. J. Biochem.* 193:529–34

37. Reeves, R., Langan, T. A., Nissen, M. S. 1991. *Proc. Natl. Acad. Sci. USA* 88:1671–75

38. Ward, G. E., Kirschner, M. W. 1990. *Cell* 61:561–77

39. Heald, R., McKeon, F. 1990. *Cell* 61:579–89

40. Enoch, T., Peter, M., Nurse, P., Nigg, E. 1991. *J. Cell Biol.* 112:797–807

41. Chou, Y.-H., Bischoff, J. R., Beach, D., Goldman, R. D. 1990. *Cell* 62:1063–71

42. Pederson, T., Gelfant, S. 1970. *Exp. Cell Res.* 59:32–36

43. Edgar, B. A., O'Farrell, P. H. 1990. *Cell* 62:469–80

44. Gelfant, S. 1962. *Exp. Cell Res.* 26:395–403

45. Melchers, F., Lernhardt, W. 1985. *Proc. Natl. Acad. Sci. USA* 82:7681–85

46. Pardee, A. B. 1989. *Science* 246:603–8

47. Li, J. J., Kelly, T. J. 1984. *Proc. Natl. Acad. Sci. USA* 81:6973–77

48. Stillman, B. 1989. *Annu. Rev. Cell Biol.* 5:197–245

49. Hartwell, L. H., Mortimer, R. K., Culotti, J., Culotti, M. 1973. *Genetics* 74:267–86

50. Reed, S. 1980. *Genetics* 95:566–77

51. Nurse, P., Bissett, Y. 1981. *Nature* 292:558–60

52. Roth, S. Y., Collini, M. P., Draetta, G., Beach, D., Allis, C. D. 1991. *EMBO J.* 10:2069–75

53. Paris, J., LeGuellec, R., Couturier, A., LeGuellec, K., Omilli, F., et al. 1991. *Proc. Natl. Acad. Sci. USA* 88:1039–43

54. Lehner, C. F., O'Farrell, P. H. 1990. *EMBO J.* 9:3573–81

55. Norbury, C. J., Nurse, P. 1989. *Biochim. Biophys. Acta* 989:85–95

56. Elledge, S. J., Spottswood, M. R. 1991. *EMBO J.* 10:2653–59

57. Siminovich, L. 1976. *Cell* 7:1–11

58. Basilico, C. 1977. *Adv. Cancer Res.* 24:223–66

59. Nishimoto, T., Eilen, E., Basilico, C. 1978. *Cell* 15:475–83

60. Marcus, M., Fainsod, A., Diamond, G. 1985. *Annu. Rev. Genet.* 19:389–421

61. Greco, A., Ittman, M., Basilico, C. 1987. *Proc. Natl. Acad. Sci. USA* 84:1565–69

62. Kai, R., Ohtsubo, M., Sekiguchi, M., Nishimoto, T. 1986. *Mol. Cell. Biol* 6:2027–32

63. Mineo, C., Murakami, Y., Ishimi, Y., Hanaoka, F., Yamada, M. 1986. *Exp. Cell Res.* 167:53–62
64. Nishitani, H., Kobayashi, H., Ohtsubo, M., Nishimoto, T. 1990. *J. Biochem.* 107:228–35
65. Uchida, S., Sekiguchi, T., Nishitani, H., Miyauchi, K., Ohtsubo, M., et al. 1990. *Mol. Cell. Biol.* 10:577–84
66. Th'ng, J. P. H., Wright, P. S., Hamaguchi, J., Lee, M. G., Norbury, C. J., et al. 1990. *Cell* 63:313–24
67. Lee, M. G., Nurse, P. 1987. *Nature* 327:31–35
68. Jimenez, J., Alphey, L., Nurse, P., Glover, D. M. 1990. *EMBO J.* 9:3565–71
69. Ely, S., D'Arcy, A., Jost, E. 1978. *Exp. Cell Res.* 116:325–31
70. Zelligs, J., Wollman, S. 1979. *J. Ultrastruct. Res.* 66:53–77
71. Gerace, L., Blobel, G. 1980. *Cell* 19:277–87
72. Marsden, M., Laemmli, U. K. 1979. *Cell* 17:849–58
73. Earnshaw, W. C. 1988. *BioEssays* 9:147–50
74. Lazarides, E. 1975. *J. Histochem. Cytochem.* 23:507–28
75. Schroeder, T. E. 1976. *Cell Motility*, pp. 265–78. Cold Spring Harbor: Cold Spring Harbor Lab.
76. Sanger, J. W., Sanger, J. M. 1976. See Ref. 75, pp. 1295–316
77. Aubin, J. E., Osborn, M., Franke, W. W., Weber, K. 1980. *Exp. Cell Res.* 129:149–65
78. Hartwell, L., Weinert, T. 1989. *Science* 246:629–34
79. Enoch, T., Nurse, P. 1991. *Cell* 65:921–23
80. Rao, P. N., Johnson, R. T. 1970. *Nature* 225:159–64
81. Wasserman, W., Masui, T. 1975. *Exp. Cell Res.* 91:381–88
82. Newport, J., Kirschner, M. 1984. *Cell* 37:731–42
83. Weintraub, H., Buscaglia, M., Ferrez, M., Weiller, S., Boulet, A., et al. 1982. *C. R. Acad. Sci. Ser. III* 295:787–90
84. Sunkara, P. S., Wright, D. A., Rao, P. N. 1979. *Proc. Natl. Acad. Sci. USA* 76:2799–802
85. Nelkin, B., Nichols, C., Vogelstein, B. 1980. *FEBS Lett.* 109:233–38
86. Sorensen, R. A., Cyert, M. S., Pedersen, R. A. 1985. *J. Cell Biol.* 100:1637–40
87. Hashimoto, N., Kishimoto, T. 1988. *Dev. Biol.* 126:242–52
88. Kishimoto, T., Kanatani, H. 1976. *Nature* 260:321–22
89. Kishimoto, T., Kuriyama, R., Kondo, H., Kanatani, H. 1982. *Exp. Cell Res.* 137:121–26
90. Kishimoto, T., Kondo, H. 1986. *Exp. Cell Res.* 163:445–52
91. Tachibana, K., Yanagishima, N., Kishimoto, T. 1987. *J. Cell Sci.* 88:273–81
92. Wasserman, W. J., Smith, L. D. 1978. *J. Cell Biol.* 78:R15-R22
93. Lake, R. S., Salzman, N. P. 1972. *Biochemistry* 11:4817–25
94. Lake, R. S. 1973. *J. Cell Biol.* 58:317–31
95. Bradbury, E. M., Inglis, R. J., Matthews, H. R. 1974. *Nature* 247:257–61
96. Woodford, T. A., Pardee, A. B. 1986. *J. Biol. Chem.* 261:4669–76
97. Zeilig, C. E., Langan, T. A. 1980. *Biochem. Biophys. Res. Commun.* 95:1372–79
98. Sano, K. 1985. *Dev. Growth Differ.* 27:263–75
99. Labbé, J. C., Picard, A., Karsenti, E., Dorée, M. 1988. *Dev. Biol.* 127:157–69
100. Langan, T. A., Gautier, J., Lohka, M., Hollingsworth, R., Moreno, S., et al. 1989. *Mol. Cell Biol.* 9:3860–68
101. Verde, F., Labbé, J.-C., Dorée, M., Karsenti, E. 1990. *Nature* 343:233–38
102. Lohka, M., Maller, J. 1985. *J. Cell Biol.* 101:518–23
103. Lohka, M., Masui, Y. 1983. *Science* 220:719–21
104. Tuomikoski, T., Félix, M.-A., Dorée, M., Gruenberg, J. 1989. *Nature* 342:942–44
105. Hindley, J., Phear, G. 1984. *Gene* 31:129–34
106. Beach, D., Durkacz, B., Nurse, P. 1982. *Nature* 300:706–9
107. Reed, S. I., Hadwiger, J. A., Lorincz, A. T. 1985. *Proc. Natl. Acad. Sci. USA* 82:4055–59
108. Lorincz, A., Reed, S. 1984. *Nature* 307:183–85
109. Piggott, J. A., Rai, R., Carter, B. L. A. 1982. *Nature* 298:391–94
110. Fantes, P. 1981. *J. Bacteriol.* 146:746–54
111. Krek, W., Nigg, E. A. 1989. *EMBO J.* 8:3071–78
112. Hirt, H., Pay, A., Gyorgyey, J., Bako, L., Nemeth, K., et al. 1991. *Proc. Natl. Acad. Sci. USA* 88:1636–40
113. Ferreira, P. C. G., Hemerly, A. S., Villarroel, R., Van Montagu, M., Inzé, D. 1991. *The Plant Cell* 3:531–40
114. Colasanti, J., Tyers, M., Sundaresan, V. 1991. *Proc. Natl. Acad. Sci. USA* 88:3377–81
115. Draetta, G., Brizuela, L., Potashkin, J., Beach, D. 1987. *Cell* 50:319–25
116. John, P. C., Sek, F. J., Carmichael, J.

P., McCurdy, D. W. 1990. *J. Cell Sci.* 97:627–30

117. Riabowol, K., Draetta, G., Brizuela, L., Vandre, D., Beach, D. 1989. *Cell* 57:393–401

118. John, P. C. L., Sek, F. J., Lee, M. G. 1989. *The Plant Cell* 1:1185–93

119. Evans, T., Rosenthal, E. T., Youngbloom, J., Distel, D., Hunt, T. 1983. *Cell* 33:389–96

120. Swenson, K. I., Farrell, K. M., Ruderman, J. V. 1986. *Cell* 47:861–70

121. Pines, J., Hunt, T. 1987. *EMBO J.* 6:2987–95

122. Pines, J., Hunter, T. 1989. *Cell* 58:833–46

123. Lehner, C. F., O'Farrell, P. H. 1989. *Cell* 56:957–68

124. Lehner, C. F., O'Farrell, P. H. 1990. *Cell* 61:535–47

125. Whitfield, W. G. F., Gonzalez, C., Sanchez-Herrero, E., Glover, D. M. 1989. *Nature* 338:337–40

126. Minshull, J., Blow, J. J., Hunt, T. 1989. *Cell* 56:947–56

127. Westendorf, J. M., Swenson, K. I., Ruderman, J. V. 1989. *J. Cell Biol.* 108:1431–44

128. Pines, J., Hunter, T. 1990. *Nature* 346:760–63

129. Minshull, J., Golsteyn, R., Hill, C. S., Hunt, T. 1990. *EMBO J.* 9:2865–75

130. Booher, R., Beach, D. 1988. *EMBO J.* 8:2321–27

131. Goebl, M., Byers, B. 1988. *Cell* 54:739–40

132. Hagan, I., Hayles, J., Nurse, P. 1988. *J. Cell Sci.* 91:587–95

133. Solomon, M., Booher, R., Kirschner, M., Beach, D. 1988. *Cell* 54:738–39

134. Booher, R., Beach, D. 1987. *EMBO J.* 6:3441–47

135. Moreno, S., Hayles, J., Nurse, P. 1989. *Cell* 58:361–72

136. Nurse, P., Thuriaux, P., Nasmyth, K. 1976. *Mol. Gen. Genet.* 146:167–78

137. Murray, A. W., Kirschner, M. W. 1989. *Nature* 339:275–80

138. Draetta, G., Beach, D. 1988. *Cell* 54:17–26

139. Solomon, M. J., Glotzer, M., Lee, T. H., Phillippe, M., Kirschner, M. 1990. *Cell* 63:1013–24

140. Draetta, G., Luca, F., Westendorf, J., Brizuela, L., Ruderman, J., et al. 1989. *Cell* 56:829–38

141. Booher, R. N., Alfa, C. E., Hyams, J. S., Beach, D. H. 1989. *Cell* 58:485–97

142. Murray, A. W., Solomon, M. J., Kirschner, M. W. 1989. *Nature* 339:280–86

143. Glotzer, M., Murray, A. W., Kirschner, M. W. 1991. *Nature* 349:132–38

144. Norbury, C., Nurse, P. 1991. *Curr. Biol.* 1:23–24

145. Fang, F., Newport, J. W. 1991. *Cell* 66:731–42

146. Mudryj, M., Devoto, S. H., Hiebert, S. W., Hunter, T., Pines, J., et al. 1991. *Cell* 65:1243–53

147. Bandara, L., Adamczewski, J. P., Hunt, T., La Thangue, N. B. 1991. *Nature* 352:249–51

148. Cyert, M., Kirschner, M. 1988. *Cell* 53:185–95

149. Gautier, J., Maller, J. L. 1991. *EMBO J.* 10:177–82

150. Wilt, F. H., Sakai, H., Mazia, D. 1967. *J. Mol. Biol.* 27:1–7

151. Wagenaar, E. B., Mazia, D. 1978. *Cell Reproduction, ICN-UCLA Symp. Mol. Cell. Biol.*, pp. 539–45. New York: Academic

152. Wagenaar, E. B. 1983. *Exp. Cell Res.* 144:393–403

153. Gerhart, J., Wu, M., Kirschner, M. 1984. *J. Cell Biol.* 98:1247–55

154. Russell, P., Nurse, P. 1987. *Cell* 49:559–67

155. Lundgren, K., Walworth, N., Booher, R., Dembski, M., Kirschner, M., et al. 1991. *Cell* 64:1111–22

156. Featherstone, C., Russell, P. 1991. *Nature* 349:808–11

157. Parker, L. L., Atherton-Fessler, S., Lee, M. S., Ogg, S., Falk, J. L., et al. 1991. *EMBO J.* 10:1255–63

158. Fantes, P. 1979. *Nature* 279:428–30

159. Russell, P., Nurse, P. 1986. *Cell* 45:145–53

160. Gould, K. L., Moreno, S., Tonks, N. K., Nurse, P. 1990. *Science* 250:1573–76

161. Kumagai, A., Dunphy, W. G. 1991. *Cell* 64:903–14

162. Strausfeld, U., Labbé, J.-C., Fesquet, D., Cavadore, J. C., Picard, A., et al. 1991. *Nature* 351:242–45

163. Moreno, S., Nurse, P. 1991. *Nature* 351:194

164. Moreno, S., Nurse, P., Russell, P. 1990. *Nature* 344:549–52

165. Ducommun, B., Draetta, G., Young, P., Beach, D. 1990. *Biochem. Biophys. Res. Commun.* 167:301–9

166. Enoch, T., Nurse, P. 1990. *Cell* 60:665–73

167. Sadhu, K., Reed, S. I., Richardson, H., Russell, P. 1990. *Proc. Natl. Acad. Sci. USA* 87:5139–43

168. Edgar, B. A., O'Farrell, P. H. 1989. *Cell* 57:177–87

169. Russell, P., Moreno, S., Reed, S. 1989. *Cell* 57:295–303

170. Félix, M.-A., Cohen, P., Karsenti, E. 1990. *EMBO J.* 9:675–83

171. Ferrell, J. J., Wu, M., Gerhart, J. C., Martin, G. S. 1991. *Mol. Cell. Biol.* 11:1965–71
172. McKeon, F., Kirschner, M., Caput, D. 1986. *Nature* 319:463–68
173. Newport, J. W., Forbes, D. J. 1987. *Annu. Rev. Biochem.* 56:535–65
174. Gerace, L., Burke, B. 1988. *Annu. Rev. Cell Biol.* 4:335–74
175. Bretscher, A. 1986. *Nature* 321:726–27
176. Yamashiro, S., Yamakita, Y., Ishikawa, R., Matsumura, F. 1990. *Nature* 344:675–78
177. Chou, Y.-H., Rosevear, E., Goldman, R. D. 1989. *Proc. Natl. Acad. Sci. USA* 86:1885–89
178. Erikson, E., Maller, J. 1989. *J. Biol. Chem.* 264:13711–17
179. Gotoh, Y., Nishida, E., Matsuda, S., Shiina, N., Kosako, H., et al. 1991. *Nature* 349:251–54
180. Ajiro, K., Nishimoto, T., Takahashi, T. 1983. *J. Biol. Chem.* 258:4534–38
181. Schlegel, R., Pardee, A. 1986. *Science* 232:1264–66
182. Yamashita, K., Yasuda, H., Pines, J., Yasumoto, K., Nishitani, H., et al. 1990. *EMBO J.* 9:4331–38
183. Hara, K., Tydeman, P., Kirschner, M. 1980. *Proc. Natl. Acad. Sci. USA* 77: 462–66
184. Kimelman, D., Kirschner, M., Scherson, T. 1987. *Cell* 48:399–407
185. Raff, J. W., Glover, D. M. 1988. *J. Cell Biol.* 107:2009–19
186. Dasso, M., Newport, J. 1990. *Cell* 61:811–23
187. Enoch, T., Gould, K. L., Nurse, P. 1991. *Cold Spring Harbor Symp. Quant. Biol.* 56: In press
188. Nishitani, H., Ohtsubo, M., Yamashita, K., Iida, H., Pines, J., et al. 1991. *EMBO J.* 10:1555–64
189. Ohtsubo, M., Okazaki, H., Nishimoto, T. 1989. *J. Cell Biol.* 109:1389–97
190. Sagata, N., Watanabe, N., Vande Woude, G. F., Ikawa, Y. 1989. *Nature* 342:512–18
191. Sagata, N., Oskarsson, M., Copeland, T., Brumbaugh, J., Vande Woude, G. F. 1988. *Nature* 335:519–25
192. Watanabe, N., Vande Woude, G. F., Ikawa, Y., Sagata, N. 1989. *Nature* 342:505–11
193. Watanabe, N., Hunt, T., Ikawa, Y., Sagata, N. 1991. *Nature* 352:247–48
194. Roy, L. M., Singh, B., Gautier, J., Arlinghaus, R. B., Nordeen, S. K., et al. 1990. *Cell* 61:825–31
195. Daar, I., Nebreda, A. R., Yew, N., Sass, P., Paules, R., et al. 1991. *Science* 253:74–76
196. Kuang, J., Penkala, J. E., Wright, D.

A., Saunders, G. F., Rao, P. N. 1991. *Dev. Biol.* 144:54–64
197. Kuang, J. K., Zhao, J., Wright, D. A., Saunders, G. F., Rao, P. N. 1989. *Proc. Natl. Acad. Sci. USA* 86:4982–86
198. Lee, T. H., Solomon, M. J., Mumby, M. C., Kirschner, M. W. 1991. *Cell* 64:415–23
199. Huchon, D., Ozon, R., Demaille, J. G. 1981. *Nature* 294:358–59
200. Foulkes, J. G., Maller, J. L. 1982. *FEBS Lett.* 150:155–60
201. Doonan, J. H., Morris, N. R. 1989. *Cell* 57:987–96
202. Booher, R., Beach, D. 1989. *Cell* 57:1009–16
203. Okhura, H., Kinoshita, N., Miyatani, S., Toda, T., Yanagida, M. 1989. *Cell* 57:997–1007
204. Axton, J. M., Dombradi, V., Cohen, P. T., Glover, D. M. 1990. *Cell* 63:33–46
205. Freshney, R. I. 1987. *Culture of Animal Cells: A Manual of Basic Technique.* New York: Liss
206. Bartlett, R., Nurse, P. 1990. *BioEssays* 12:457–63
207. Pardee, A. B. 1974. *Proc. Natl. Acad. Sci. USA* 71:1286–90
208. Folkman, J., Moscona, A. 1978. *Nature* 273:345–49
209. O'Neill, C., Jordan, P., Ireland, G. 1986. *Cell* 44:489–96
210. Zetterberg, A., Larsson, O. 1985. *Proc. Natl. Acad. Sci. USA* 82:5365–69
211. Campisi, J., Pardee, A. B. 1984. *Mol. Cell. Biol.* 4:1807–14
212. Pledger, W. J., Stiles, C. D., Antoniades, H. N., Scher, C. D. 1978. *Proc. Natl. Acad. Sci. USA* 75:2839–43
213. Yen, A., Pardee, A. B. 1978. *Exp. Cell Res.* 116:103–13
214. Campisi, J., Morreo, G., Pardee, A. B. 1984. *Exp. Cell Res.* 152:459–66
215. Smith, J. A., Martin, L. 1973. *Proc. Natl. Acad. Sci. USA* 70:1263–67
216. Brooks, R. F. 1976. *Nature* 260:248–50
217. Shields, R. 1977. *Nature* 267:704–7
218. Schneiderman, M. H., Dewey, W. C., Highfield, D. P. 1971. *Exp. Cell Res.* 67:147–55
219. Brooks, R. F. 1977. *Cell* 12:311–17
220. Rossow, P. W., Riddle, V. G. H., Pardee, A. B. 1979. *Proc. Natl. Acad. Sci. USA* 76:4446–50
221. Das, M. 1981. *Proc. Natl. Acad. Sci. USA* 78:5677–81
222. Bishop, J. M. 1991. *Cell* 64:235–48
223. Cantley, L. C., Auger, K. R., Carpenter, C., Duckworth, B., Graziani, A., et al. 1991. *Cell* 64:281–302
224. Hunter, T. 1991. *Cell* 64:249–70
225. Forsburg, S. L., Nurse, P. 1991. *Annu. Rev. Cell Biol.* 7:227–56

226. Sipiczki, M. 1989. *Molecular Biology of the Fission Yeast,* pp. 431–52. San Diego: Academic
227. Furukawa, Y., Piwnica-Worms, H., Ernst, T. J., Kanakura, Y., Griffin, J. D. 1990. *Science* 250:805–8
228. Broek, D., Bartlett, R., Crawford, K., Nurse, P. 1991. *Nature* 349:388–93
229. Sudbery, P. E., Goodey, A. R., Carter, B. L. A. 1980. *Nature* 288:401–4
230. Cross, F. 1988. *Mol. Cell. Biol.* 8:4675–84
231. Nash, R., Tokiwa, G., Anand, S., Erickson, K., Futcher, A. B. 1988. *EMBO J.* 7:4335–46
232. Hadwiger, J. A., Wittenberg, C., Richardson, H. E., De Barros Lopes, M., Reed, S. I. 1989. *Proc. Natl. Acad. Sci. USA* 86:6255–59
233. Richardson, H. E., Wittenberg, C., Cross, F., Reed, S. I. 1989. *Cell* 59:1127–33
234. Wittenberg, C., Sugimoto, K., Reed, S. I. 1990. *Cell* 62:225–37
235. Forsburg, S. L., Nurse, P. 1991. *Nature* 351:245–48
236. Xiong, Y., Connolly, T., Futcher, B., Beach, D. 1991. *Cell* 65:691–99
237. Lew, D. J., Dulic, V., Reed, S. I. 1991. *Cell* 66:1197–206
238. Motokura, T., Bloom, T., Kim, H. G., Juppner, H., Ruderman, J. V., et al. 1991. *Nature* 350:512–15
239. Bueno, A., Richardson, H., Reed, S. I., Russell, P. 1991. *Cell* 66:149–59
240. Matsushime, H., Roussel, M. F., Ashmun, R. A., Sherr, C. J. 1991. *Cell* 65:701–13
241. Ghiara, J. B., Richardson, H. E., Sugimoto, K., Henze, M., Lew, D. J., et al. 1991. *Cell* 65:163–74
242. Bailly, E., Dorée, M., Nurse, P., Bornens, M. 1989. *EMBO J.* 8:3985–95
243. Norbury, C., Nurse, P. 1990. *Ciba Found. Symp.* 150:168–77
244. Blow, J. J., Nurse, P. 1990. *Cell* 62:855–62
245. Roberts, J. M., D'Urso, G. 1988. *Science* 241:1486–89
246. D'Urso, G., Marraccino, R. L., Marshak, D. R., Roberts, J. M. 1990. *Science* 250:786–91
247. Zetterberg, A., Engstrom, W., Dafgard, E. 1984. *Cytometry* 5:368–75
248. Killander, D., Zetterberg, A. 1965. *Exp. Cell Res.* 40:12–20
249. Paul, D., Brown, K. D., Rupniak, H. T., Ristow, H. J. 1978. *In Vitro* 14:76–85
250. Johnston, G. C., Pringle, J. R., Hartwell, L. H. 1977. *Exp. Cell Res.* 105:79–98
251. Nurse, P., Thuriaux, P. 1977. *Exp. Cell. Res.* 107:365–75
252. Larsson, O., Dafgard, E., Engstrom, W., Zetterberg, A. 1986. *J. Cell. Physiol.* 127:267–73
253. Marshall, C. J. 1991. *Cell* 64:313–26
254. Moses, H. L., Yang, E. Y., Pietenpol, J. A. 1990. *Cell* 63:245–47
255. Howe, P. H., Draetta, G., Leof, E. B. 1991. *Mol. Cell. Biol.* 11:1185–94
256. Lee, M. G., Norbury, C. J., Spurr, N. K., Nurse, P. 1988. *Nature* 333:676–79
257. Lee, W. H. 1989. *Int. Symp. Princess Takamatsu Cancer Res. Fund* 20:159–70
258. Weinberg, R. A. 1990. *Trends Biochem. Sci.* 15:199–202
259. Levine, A. J., Momand, J., Finlay, C. A. 1991. *Nature* 351:453–56
260. Taya, Y., Yasuda, Y., Kamijo, M., Nakaya, K., Nakamura, Y., et al. 1989. *Biochem. Biophys. Res. Commun.* 164:580–86
261. Satterthwaite, L., Cisek, L., Corden, J., Pollard, T. 1989. *J. Cell Biol.* 109:284a

Annu. Rev. Biochem. 1992. 61:471–516

VESICLE-MEDIATED PROTEIN SORTING

Nancy K. Pryer, Linda J. Wuestehube, and Randy Schekman

Department of Molecular and Cell Biology and Howard Hughes Medical Institute, University of California, Berkeley, California 94720

KEY WORDS: vesicular transport, secretion, membrane traffic, protein transport, protein sorting

CONTENTS

PERSPECTIVES

The transport of proteins between membrane-bounded organelles is an immensely complex process. Protein transport within both the secretory and endocytic pathways involves the packaging of proteins into transport vesicles

471

0066-4154/92/0701-0471$02.00

that bud from one membrane and fuse with another. To understand vesicle-mediated protein transport and sorting, two central questions must be addressed. The first question concerns the physical basis of vesicular transport. How are vesicles assembled, released, targeted to the appropriate membrane, and then consumed? The second question considers how protein sorting is achieved by vesicular carriers. More specifically, how are proteins destined for the next membrane compartment efficiently exported, while resident proteins necessary for organelle function are retained?

Recent progress in understanding vesicle-mediated protein sorting has been achieved by the convergence of classical biochemical and genetic approaches. The genetic complexity of vesicular transport, revealed by the isolation and analysis of yeast mutants, is beginning to be translated into the language of protein biochemistry. A number of cell-free assays reconstituting steps of intermembrane vesicular transport have been used to determine the requirements for transport and to purify a few of the many proteins that function in vesicle formation, targeting, and fusion. The development of in vitro assays that require the activity of specific yeast gene products, combined with the discovery that yeast and mammalian gene products are interchangeable in vitro, has opened the door to biochemical analysis of gene products defined by yeast mutants. Such analysis has yielded several factors common to multiple transport steps. In this review we discuss recent advances that have been made in understanding the mechanism of vesicle-mediated protein sorting. These advances include:

1. The interweaving of genetic and biochemical techniques in identifying conserved proteins required for vesicle-mediated transport;
2. The discovery that small GTP-binding proteins participate in vesicle targeting;
3. Analysis of structural proteins composing a coat that assembles on the face of some vesicles during vesicle formation;
4. Identification of sorting signals that mediate the inclusion of cargo proteins into vesicles and the retention or retrieval of organelle resident proteins, and elucidation of the cellular machinery that recognizes these signals.

INTRODUCTION TO VESICULAR TRANSPORT

The flow of membrane and proteins between compartments of a typical eukaryotic cell is diagrammed in Figure 1. Both the biosynthetic (or secretory) pathway and the endocytic pathway employ vesicular intermediates to transport proteins. In the biosynthetic pathway (1), membrane and secretory proteins are cotranslationally inserted into the endoplasmic reticulum (ER). Within the lumen of the ER, proteins may be proteolytically processed, folded, and receive oligosaccharide modifications. Newly synthesized proteins are then packaged into vesicles and transported to the Golgi apparatus. Further maturation of proteins occurs in this compartment, including proteolytic processing and additions and modifications of oligosaccharide residues. The Golgi apparatus is composed of discrete, closely apposed membrane stacks, designated cis-, medial-, and trans-, flanked by cis- and trans- tubular membranous sorting compartments (2, 3). The structural complexity of the Golgi apparatus reflects the multiple functions of protein maturation, recognition, and sorting occurring in this organelle. The cis-Golgi network (4), also designated the salvage compartment (5) or 15 °C intermediate (6), is the site of recognition and recycling of escaped ER resident proteins. At the opposite face of the Golgi apparatus, numerous sorting events take place in the trans-Golgi network (TGN) (7). In the TGN, newly synthesized lysosomal resident proteins are diverted from the constitutive secretory pathway, and in cell types that have a regulated secretory pathway, hormones and neurotransmitters are sorted into secretory vesicles. Constitutive secretory proteins also are packaged into transport vesicles in the TGN. In the endocytic pathway (8–10), receptor-ligand complexes and solutes taken in from the plasma membrane by clathrin-coated vesicles are delivered to the acidic early endosome, where receptor and ligand are uncoupled. Receptors may be recycled directly to the plasma membrane or may recycle via the trans-Golgi network. Ligands and solutes are directed to late endosomes and ultimately degraded in lysosomes.

Given this constant flow of membrane between organelles, how does the cell ensure efficient sorting of exported proteins from resident organelle proteins? The current paradigm for protein sorting by vesicular transport in the secretory pathway is the bulk flow model (1, 11). In this model, constitutive secretion from the ER through the Golgi to the cell surface occurs by default (i.e. no sorting signal is required). All diversions from this pathway to other compartments must be signal mediated, as must be a recycling scheme to recover escaped resident proteins. Evidence for the bulk flow model is derived from the existence of sorting signals for delivery of proteins to lysosomes (12) and to the regulated secretory pathway (13). The recent discovery of a signal-mediated pathway for the recycling of escaped ER resident proteins also supports a bulk flow model (14). It appears that signal sequences are not required for efficient export of soluble proteins. To date, no signals for constitutive export have been detected. Experimental tests of bulk flow have measured the rate of secretion of small, soluble, marker peptides (15, 16). These signal-less tripeptides are secreted from the ER to the cell surface at the same rate as the fastest endogenous secretory protein, indicating that no signal is required for efficient export of soluble proteins. However, the

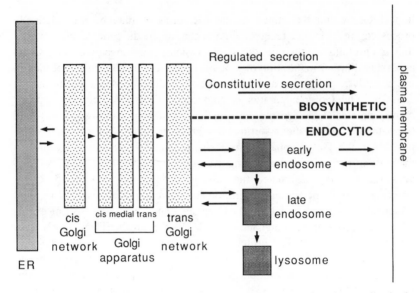

Figure 1 Schematic diagram of transport steps between membrane-bound organelles in the biosynthetic and endocytic pathways. ER, endoplasmic reticulum.

bulk flow model does not explain the measured differences in the rates of export of endogenous secretory proteins from the ER, nor does it account for potential subcompartmentation in the ER (17, 18).

The transfer of material between membranes in the biosynthetic and endocytic pathways has long been believed to occur by vesicular transport (19). The best early evidence for vesicular transport was obtained from pulse-chase studies of pancreatic exocrine cells showing that labeled material moved from the ER to the Golgi apparatus via small vesicles (20). In the past five years, transport vesicles mediating a number of different transport steps have been isolated directly from various cells, usually by monitoring the enrichment of a vesicle cargo protein (21–27), and from cell-free reactions reconstituting a transport step (28–34). The authenticity of isolated vesicular intermediates is best documented in those cases in which the isolated vesicle has been demonstrated to act as a functional intermediate in a reconstituted transport reaction (22, 25, 29, 30). The great majority of the transport steps indicated by arrows in Figure 1 are vesicle mediated, with a few possible exceptions currently subject to scrutiny (35–37). Because so many transport steps involve vesicular intermediates, considerable progress in understanding protein transport will be made by discerning the common mechanisms underlying vesicle formation, targeting, and fusion.

The formation of a transport vesicle requires, at the least, the deformation of membrane from a planar surface into a curved vesicle, and a mechanism to release the complete vesicle from the donor membrane. In some cases, vesicle formation will be preceded by the sorting of included membrane and lumenal proteins, with the concurrent exclusion of organelle resident proteins from the budding vesicle. In cases of nonselective vesicle formation (bulk flow), a recycling step to identify and retrieve runaway resident proteins is expected to be present. As discussed further in the last two sections of this review, sorting mechanisms may exist for both the case of selective inclusion of cargo into vesicles, as well as the case of recycling of resident proteins.

Three models have been proposed to explain the mechanism of vesicle formation. In the first model, vesicle formation is mediated by a protein coat. Coat proteins recruited to the cytoplasmic face of the membrane are thought to initiate or provide the driving force for membrane deformation. Transmembrane cargo proteins may be selectively included in forming vesicles by interaction across the membrane with coat proteins, or alternatively, coat proteins may assemble at pre-formed sites enriched in cargo proteins. An example for this model is provided by the clathrin coat that mediates vesicle formation from the plasma membrane and the trans-Golgi network (38, 39). The structure of the clathrin coat and its proposed role in sorting and vesicle formation is discussed further in section entitled STRUCTURE AND FUNCTION OF VESICLE COATS.

In the second model, vesicle formation is mediated by the cargo molecules. This model is based on studies of enveloped viruses budding from the plasma membrane of infected cells (40, 41). The spherical nucleocapsid of alphaviruses is assembled in the cytosol and associates with the plasma membrane by binding to viral transmembrane glycoproteins. Interactions between the membrane glycoprotein spikes and the capsid shell are thought to provide the force to wrap the plasma membrane around the capsid, forming a vesicle. There are currently no known examples of core-initiated vesicle formation in uninfected cells.

A third model proposes that vesicle formation may be driven by changes in membrane organization distinct from coat or core effects. The bilayer couple hypothesis (42) suggests that membrane deformation could be achieved by changes in membrane protein distribution. Biological membranes are proposed to act as bilayer couples, such that an asymmetric distribution of proteins and/or lipids between the halves of a bilayer results in differential expansion or contraction of one monolayer and a corresponding change in membrane shape. Experiments treating erythrocyte membranes with anionic drugs that intercalate preferentially into one monolayer revealed compensatory changes in membrane shape in the form of crenations or cups in the membrane (42). To date, no studies have been performed on cellular vesicle formation events at this level of molecular detail.

Little is known about the scission event releasing vesicles and re-sealing the membrane. A clue to understanding scission is provided by the *Drosophila shibire* mutant, in which mature clathrin-coated vesicles accumulate attached to the plasma membrane by thin membrane "necks" (43). The *shibire* gene encodes a protein with GTP-binding motif, related to a yeast protein, Vpslp, required for protein sorting to the vacuole (44, 45). Perhaps these proteins couple nucleotide hydrolysis to vesicle scission from the donor membrane.

Transport and targeting of mature vesicles are critical steps for ensuring the efficiency and specificity of vesicular transport. Vesicle transport may occur by the active translocation of vesicles along elements of the cytoskeleton, particularly in polarized cells such as neurons (13, 46, 47). Targeting (and/or docking) of the vesicle to the acceptor membrane is likely to be closely regulated. A group of related GTP-binding proteins, discussed in the section entitled SMALL GTP-BINDING PROTEINS IN VESICULAR TRANSPORT, may be responsible for regulating a targeting step. These proteins are members of a large class of Ras-like GTP-binding proteins, each perhaps localized to a particular membrane compartment (48, 49). By analogy to other GTP-binding proteins, these proteins could determine the directionality and fidelity of vesicle targeting.

Vesicle consumption encompasses all events following vesicle docking to the membrane, including vesicle uncoating and membrane fusion. Cellular membrane bilayer fusion events are highly specific, rapid, and proceed without disrupting the membrane barrier function (50). Membrane fusion has been studied primarily using pure phospholipid liposomes (51), but fusion of protein-containing membranes is apt to be quite different. Information on fusion events in biological membranes has been derived from studies of fusogenic viral proteins (52, 53) and from measurements of membrane fusion during exocytosis (54). Influenza virus fusion with the host cell endosome membrane is achieved by the fusogenic activity of influenza hemagglutinin. Upon exposure to acidic pH in the endosome, hemagglutinin undergoes a conformational change, exposing highly hydrophobic fusion peptides. These peptides are thought to embed in the adjacent endosome membrane, bringing the two membranes close enough to overcome the resisting hydration force and fuse. Fusion of exocytic vesicle membrane with the plasma membrane has been studied in the large mast cells of a mutant mouse (55). Measurements of conductance across the membrane have detected the existence of a "fusion pore" that forms as an early event in exocytosis. These pores open abruptly and then dilate, presumably by the lateral incorporation of lipid into the pores, leading to bilayer fusion (56).

Several vesicle consumption reactions reconstituted in vitro require the same protein activity at a step prior to membrane fusion (57–59). This protein, NSF, discussed further below, is believed to function as part of a multisubunit fusion machine, although NSF itself is lacking the hydrophobic sequences characteristic of viral fusion proteins. An NSF-independent vesicle consumption activity also has been detected in vitro (60). The mechanism by which these proteins function in promoting membrane bilayer fusion awaits further study.

GENETIC AND BIOCHEMICAL APPROACHES TO VESICULAR TRANSPORT

Genetic Studies of Vesicular Transport

A genetic approach has proven fruitful in elucidating the complexity of vesicle-mediated protein transport. Genetic strategies have most successfully been applied in yeast because of the facile genetics and ease of manipulation of genes in this unicellular eukaryote (61). A number of screens and selection schemes have been devised to isolate mutants defective in protein transport and sorting.[1] Yeast (*Saccharomyces cerevisiae*) genes defined by mutants defective in intercompartmental transport are listed in Table 1. Only those genes for which the gene product has been characterized biochemically have been included. Many of these genes were isolated as temperature-sensitive conditional mutants, which are useful in examining the function of essential genes because of the ability to impose the defect under nonpermissive conditions. The first *sec* (secretion) mutants were isolated by screening an unselected collection of temperature-sensitive cells for mutants that accumulate precursor forms of exported proteins (62). Subsequently, 23 complementation groups (*sec1–23*) were isolated in a selection scheme based on the increased density of mutagenized cells upon a shift to the nonpermissive temperature (63). Two additional secretion mutants, *bet1* and *bet2* (blocked early in transport), were isolated by surviving a [³H]mannose suicide selection in which cells with defective protein transport to the Golgi apparatus do not receive lethal mannose additions to their glycoproteins (64). Another large group of mutants defective in sorting and transport to the vacuole (*vps*, for vacuolar protein sorting) were isolated by the mislocalization of a vacuolar resident protein to the cell surface (65–68). Most recently, the *erd* mutants (ER retention defective) also were isolated by mislocalization of a marker protein, in this case an ER resident, to the cell surface (69). An added level of sophistication is contributed by the isolation of genes that suppress conditional mutant defects. *SAR1* was isolated as a suppressor of *sec12* (70), *BOS1* as a suppressor of *bet1* (71), and four *SLY* genes as suppressors of loss of *YPT1* function (72).

A "reverse genetics" strategy was used to identify some of the yeast genes involved in vesicular traffic. In this strategy, a protein of interest in another

---

[1]Yeast nomenclature: *SEC1* indicates wild-type gene, *sec1* indicates mutant gene, and Sec1p indicates protein encoded by *SEC1* gene.

Table 1 Yeast genes that may function in vesicle-mediated protein sorting[e]

gene	phenotype[a,b]	M_r (kDa)	mammalian homologue	sequence features	protein features and localization[b,d]	protein function (in vitro)	Refs.
ER to Golgi, vesicle formation							
SEC12	ER, core-glycosyl. proteins	70	—	—	integral membrane glycoprotein	ER-Golgi transport and ER-derived vesicle formation blocked in vitro	(30, 80, 81, 86, 87, 239, 240)
SAR1	core-glycosyl. proteins	21	—	ras GTP-binding domain	peripheral membrane protein, increased association with membranes from SEC12 multicopy strain	GTP-binding; rescues sec12 ER-Golgi defect in vitro; Sar1p restores vesicle formation activity to reaction inhibited by Sec12p overproduction	(70, 85, 86, 241)
SEC13	ER, core-glycosyl. proteins	34	—	G-protein β subunit internal repeats	peripheral membrane protein, soluble form in heteromeric complex	—	(80, 81, 239, 242, 243)
SEC23	ER, core-glycosyl. proteins	84	anti-Sec23p Ab cross-reacts with mammalian homolog	—	peripheral membrane protein, soluble form in complex with 105-kDa protein	Sec22p-105-kDa complex rescues sec23 ER-Golgi defect in vitro; formation of ER-Golgi vesicles blocked in vitro	(30, 80, 81, 92, 104–106, 108, 109, 239)
ER to Golgi, vesicle consumption							
SEC17	50 nm vesicles, core-glycosyl. proteins	33	α-SNAP	homology with α-SNAP peptides	peripheral membrane protein	mammalian α-SNAP rescues sec17 defect in an NSF-dependent step in vitro	(80, 81, 114; C. Kaiser, personal communication)
SEC18	50 nm vesicles, core-glycosyl. proteins	84	NSF	48% identity with mammalian NSF	peripheral membrane protein	ER-Golgi vesicles accumulate in vitro, SEC18 encodes NSF activity for mammalian intra-Golgi transport	(30, 80, 81, 91, 112)
YPT1	ER, 50 nm vesicles, core-glycosyl. proteins	23.5	Rab1	ras GTP-binding domain, 48% identity with SEC4	peripheral membrane protein, C-terminal prenylation, GTP-binding	ER-Golgi transport blocked in vitro at vesicle targeting step by Ab or mutant	(30, 93, 98, 122, 123, 125, 126, 138, 144, 244, 245)

ER to Golgi, undefined step

Gene	Location	Size		Sequence feature	Protein characteristics	Notes	References
SEC20	ER, 50 nm vesicles, core-glycosyl. proteins	50	—	C-terminal HDEL	integral membrane glycoprotein	—	(80, 81, 236)
SEC21	ER, 50 nm vesicles, core-glycosyl. proteins	105	—	weak homolgy to α-, β-adaptins, β-COP	peripheral membrane protein; soluble complex of approx. 600 kDa	—	(80, 81, 246)
BET2	ER, core-glycosyl. proteins	36.5	—	yeast Ras prenyl-transferase (DPR1/RAM1)		type II geranylgeranyltransferase activity in vitro; required in vivo for membrane association of Sec4p and Ypt1p	(64, 82, 148)
BOS1	ER, 50 nm vesicle cluster	27	—	—	integral membrane protein	—	(71, 247)
SLY2	ER, core-glycosyl. proteins	25	—	restricted homology with synaptobrevin	integral membrane protein	—	(72, 248)

Intra-Golgi transport

Gene	Location	Size		Sequence feature	Protein characteristics	Notes	References
SEC7	Golgi membrane, heterogeneously glycosyl. proteins	230	—	acidic N-terminus	peripheral membrane protein, phosphorylated, punctate localization, partially colocalized with late Golgi marker	anti-Sec7p IgG blocks ER-Golgi transport in vitro	(80, 94, 249–251)
SEC14	Golgi membrane, 80–100 nm vesicles, partially glycosyl. proteins	37	—	identity with yeast PI/PC transferase	cytosolic and peripheral membrane protein; fractionates with late Golgi marker	*sec14* defective in PI/PC transferase activity in vitro	(80, 115, 116, 118)
ARF1/ARF2	partially glycosyl. proteins	21	ADP ribosylation factor	*ras* GTP-binding domain; 2 genes 96% identity	peripheral membrane protein; *ARF1* produces 90% of Arf protein	—	(78, 79, 124, 147)

Table 1 (*Continued*)

gene	phenotype[a,b]	M_r (kDa)	mammalian homologue	sequence features	protein features and localization[b,d]	protein function (in vitro)	Refs.
Golgi apparatus to cell surface							
SEC2	80–100 nm vesicles, mature secretory proteins	105	—	N-terminal coiled-coil domain	soluble, in complex of 500–700 kDa	—	(80, 84)
SEC4	80–100 nm vesicles, mature secretory proteins	23.5	—	*ras* GTP-binding domain, 48% identity with *YPT1*	GTP-binding, peripheral membrane protein, cytoplasmic surface of secretory vesicles and plasma membrane	GTP hydrolysis in vitro	(80, 83, 127, 140)
SEC15	80–100 nm vesicles, mature secretory proteins	116	—	—	partially soluble	—	(80, 152)
Sorting/retention							
CHC1	mislocalizes late Golgi protein to plasma membrane, slow growth	190	mammalian clathrin heavy chain	—	peripheral membrane protein, coats 60 nm vesicles	—	(74, 75, 77, 252)
CLC1	slow growth	38	mammalian clathrin light chain	—	peripheral membrane protein, coats 60 nm vesicles	—	(76)
ERD2	ER-like membrane, secretes lumenal ER proteins	26	50% identity with human KDEL receptor homologue	—	integral membrane protein, punctate localization	—	(69, 229, 230)

Transport and sorting to vacuole

VPS	Phenotype	(kDa)		identity/motifs	Property[d]	GTP-binding	References
VPS1	Golgi membrane, multivesicular bodies; secretes vacuolar proteins	80	Mx protein, dynamin	identity with S.c. SPO15, similarity to D.m. shibire[c]	punctate localization	—	(44, 65, 153, 253, 254)
VPS3	secretes soluble vacuolar proteins	140	—	—	partially soluble protein	—	(65, 255)
VPS5	fragmented vacuole, secretes soluble vacuolar proteins	90	—	—	partially soluble protein, phosphorylated	—	(66; D. Horazdovsky, S. Emr, personal communication)
VPS15	secretes soluble vacuolar proteins; no effect on membrane proteins	166	—	catalytic domain of Ser/Thr kinases, type II phosphatase	peripheral membrane protein, N-terminal myristolation, autophosphorylated, fractionates with late Golgi marker	—	(67, 200, 201)
VPS17	fragmented vacuole, secretes soluble vacuolar proteins	70	—	—	peripheral membrane protein, phosphorylated	—	(67; K. Köhrer, S. Emr, personal communication)
VPS33	vacuole absent, secretes vacuolar proteins	75	—	ATP-binding motifs	soluble protein	—	(67, 260, 261)
VPS34	secretes vacuolar proteins	95	—	—	partially soluble protein, punctate distribution, phosphoprotein, co-immunoprecipitates with Vps15p	—	(67; J. Stack, P. Herman, S. Emr, personal communication)

[a] glycosyl. = glycosylated; core-glycosyl. indicates protein modified by addition of core oligosaccharides in ER.

[b] 50 nm vesicles mediate ER-Golgi transport (81); 60 nm vesicles are purified clathrin-coated vesicles (252); 80–100 nm vesicles mediate Golgi-cell surface transport (63).

[c] S.c. = Saccharomyces cerevisiae; D.m. = Drosophila melanogaster.

[d] Proteins designated peripheral membrane may also exist in soluble, cytoplasmic form; proteins designated partially soluble sediment with a Triton X-100-insoluble fraction.

[e] Gene products not yet characterized include Sec1p, Sec3p, Sec5p, Sec6p, Sec8p, Sec9p, Sec10p, Sec16p, Sec19p, Sec22p (63), Bet1p (64, 71), Sly1p, Sly12p, Sly41p (72, 248), Erd1p (226, 228), Uso1p (264), many Vps proteins (192).

organism is used to identify conserved yeast gene homologues, which can then be analyzed genetically. For example, clathrin heavy and light chains are involved in vesicle formation and membrane traffic in mammalian cells (38, 73). Conserved properties of clathrin heavy and light chains were used to isolate yeast genes encoding conserved clathrin homologues (74–76). Analysis of clathrin heavy chain–deficient yeast revealed that the retention of a Golgi resident membrane protein was altered (77). Similarly, Arf1p, a member of the Ras-like family of small GTP-binding proteins, was first identified in mammalian cells, and then homologues (Arf1p and Arf2p) were detected in yeast by antibody cross-reactivity (78). Arf1p-deficient yeast cells were found to be defective in protein processing and transit through the Golgi apparatus (79).

Once a collection of mutants has been isolated, double mutant analysis can be used to determine the order in which genes function and thereby dissect the pathway of interest. *sec* mutants were initially classified by analyzing the glycosylation and processing state of accumulated secretory protein precursors and by electronmicroscopic observation of aberrant membranes that accumulated at a restrictive temperature. Mutants that accumulated ER membrane and partially glycosylated precursors were categorized as blocked in ER to Golgi transport. Other mutants fell into one of two categories: those blocked in transport within the Golgi and those blocked in transport from the Golgi to the plasma membrane (80). A more recent analysis using improved electronmicroscopic techniques for the resolution of membranes further subdivided the ER to Golgi–blocked *sec* mutants into two classes: those that accumulate primarily ER membrane and those that in addition accumulate small (50 nm) vesicles. Epistasis tests of double mutants between the two groups showed that the ER accumulation block was imposed first, suggesting that these mutants are blocked prior to vesicle formation from the ER and the vesicle accumulation group of mutants is blocked in vesicle consumption (81).

Tests of genetic interactions have been used to identify gene products that interact directly or function in the same pathway. Synthetic lethality, in which two slightly defective conditional mutants are more defective in combination as a double mutant, may indicate a functional relationship between the corresponding proteins. For example, a double mutant of *bet2*, a mutant blocked in ER to Golgi transport, and *sec4*, blocked at a step between the Golgi and the plasma membrane, is incapable of growth under conditions where the single mutants are each able to grow, indicating a synthetic lethal interaction between the mutant genes (82). Further studies have established that *BET2* function is required for the attachment of Sec4p, a Ras-like, GTP-binding protein, to membranes. *BET2* is homologous to a mammalian prenyltransferase that modifies small GTP-binding proteins to facilitate their membrane attachment. It is likely that *BET2* acts to attach a prenyl group to

Sec4p, and this modification permits Sec4p membrane attachment and function. In this case, a genetic interaction between mutants in membrane transport provided a clue to functional protein-protein interactions. Synthetic lethality between a number of ER to Golgi *sec* mutants (71, 81), as well as between many late *sec* mutants (83, 84), suggests that the proteins defined by these mutations may act in concert.

Suppression of the genetic defect of one mutant by another gene also may indicate that the two gene products interact. *SAR1*, a suppressor of *sec12*, encodes a 21-kDa, Ras-like GTP-binding protein. Analysis of Sar1p and Sec12p in vitro has revealed that these proteins are required for ER to Golgi transport in a reconstituted transport assay. The temperature-sensitive *sec12* defect in ER to Golgi transport can be reproduced in semi-intact cells (30, 85). The addition of bacterially expressed Sar1p to this reaction restores activity to *sec12* membranes (85). Furthermore, in a refined assay measuring vesicle formation from the ER, membranes containing elevated levels of Sec12p competitively inhibit vesicle formation, and this inhibition is overcome by the addition of a Sar1p-enriched fraction (86). Finally, the association of Sar1p with a membrane fraction is stimulated by Sec12p (87). Each of these tests indicates functional interaction between the Sar1 and Sec12 proteins, as initially suggested by the genetic interaction between *SAR1* and *sec12*.

Thus, genetic analysis of vesicular transport has been useful in assembling the list of players, defining the order of gene function, and providing clues for protein-protein interactions. Elucidation of the mechanism by which these gene products act, however, requires biochemical studies. Much of the information in Table 1 has been obtained by the use of specific antibody probes, generated against bacterially expressed fusions of the gene of interest with an inducible bacterial gene. These antibodies are commonly used to determine the molecular weight and intracellular localization of each protein, and as tools for the development of functional assays. Advances in understanding the role of some of these genes also have been made by detecting sequence homology with better-characterized mammalian proteins. As discussed further below, the most definitive evidence for the specific function of a gene product has been obtained with reconstitution studies. In vitro assays that require the function of specific gene products have been developed for only a few of the genes that function in protein transport. It is clear that continued biochemical analysis of yeast mutants will be vital in dissecting the mechanism of vesicle-mediated protein sorting.

Reconstituted Reactions of Vesicular Transport

Reconstitution of intercompartmental transport in a cell-free system was first achieved more than 10 years ago for transport between Golgi compartments of CHO cells (88). Since that time, there has been rapid progress in the reconstitution of transport between other membrane compartments in many organisms. A summary of in vitro reactions that measure specific transport steps in both the biosynthetic and endocytic pathways is presented in Table 2. We have included in this table most assays developed since 1984. Striking features of this compilation include the sheer number of transport steps that have been amenable to reconstitution, the diversity of cell-free systems devised, and the many creative ways in which markers have been used to track intercompartmental transport.

Reconstituted transport reactions share a dependence on physiologically relevant parameters such as ATP, physiological temperature, and cytosolic proteins. The requirement of a cytosolic fraction for a membrane-based process reflects the interaction of cytosolic factors with membrane components in transport processes. The reconstituted assays also reflect a trend toward more defined components. Many transport steps were initially reconstituted in semi-intact cells and later refined by the substitution of more purified membrane components. With the further purification of membrane and cytosolic components, the reconstitution of a transport step in a fully resolved system may be possible.

Two advantages that these cell-free assays offer may not be immediately apparent. One is the ability to exploit heterologous mixtures in which a component shown to be essential in one organism is combined with cell fractions from another organism (89, 90). For example, a protein initially identified by a yeast *sec* mutation can replace a mammalian cytosolic protein in supporting transport between isolated mammalian Golgi membranes (91). A second advantage is the ability to accumulate vesicular intermediates in vitro, particularly those intermediates that may be transient in vivo. Such vesicular intermediates may accumulate in vitro in the absence of an active acceptor fraction (29, 31–34), or by the imposition of specific blocks in vesicle consumption by the addition of biochemical inhibitors, antibodies, or the use of mutant fractions (28, 30, 92–94).

Specific inhibitors that block transport reversibly are useful for biochemical "epistasis" tests to rank in temporal order the biochemical requirements for a transport step (30, 95–99). For example, Ca^{2+}, GTP hydrolysis, and physiological temperature all are required for ER to Golgi transport in semi-intact CHO cells (96, 100). Each of these requirements was detected by a specific block: EGTA, GTPγS (guanosine-5'-O-thiotriphosphate), and 15 °C incubation, respectively. Two of these blocks are reversible, permitting experiments to determine the order of requirements for these components. When transport was blocked by incubation at low temperature, and then reversed by shifting to higher temperature in the presence of GTPγS, progression to the cis Golgi was inhibited, indicating that GTP hydrolysis is required after the temperature-dependent step. In contrast, upon reversal of the EGTA block in transport, GTPγS did not affect further transport, indicating that the GTP-requiring step precedes the Ca^{2+}-requiring step (96). The order of the steps requiring these components is thus: physiological temperature $\rightarrow$ GTP hydrolysis $\rightarrow Ca^{2+}$. These studies provide the means to further dissect an individual transport step into subreactions, each defined by specific biochemical requirements.

Specific blocks also can be exploited to devise an activity assay for the purification of a required component in transport. The best example of this strategy is the purification of NSF (NEM-Sensitive Factor) from mammalian Golgi membrane fractions. Treatment of isolated mammalian Golgi membranes with a low concentration of the sulfhydryl alkylating agent N-ethylmaleimide (NEM) selectively inactivates the acceptor membrane fraction for intra-Golgi transport in vitro (101). Activity could be restored by the addition of untreated Golgi membrane peripheral proteins. This observation provided the basis for the purification of NSF, a 76-kDa protein that exists as a homotetrameric complex (57). Electronmicroscopic examination of Golgi membranes blocked by NEM treatment revealed that intermediate transport vesicles accumulate in an uncoated, membrane-associated, pre-fusion state (102). Based on these observations, NSF has been proposed to act as part of a fusion complex that promotes the fusion of the vesicle membrane bilayer with the acceptor membrane.

The observed requirement for NSF in transport between Golgi membranes prompted a search for other NEM-sensitive, NSF-requiring fusion events. NSF was subsequently found to be required for ER to Golgi transport in vitro (58) and for the fusion of endocytic carrier vesicles in vitro (59). The criteria used for both these studies was the ability of purified NSF to restore activity to NEM-treated membranes in vitro, and the inhibition of both transport reactions by anti-NSF antibodies. Thus the importance of reconstituted reactions is exemplified by this use of a specific inhibitor of transport in vitro to discover an activity required in several transport steps. As discussed further in the next section, the role of NSF also is conserved across species boundaries. To date only five proteins, in addition to NSF, have been purified based on their activity in vesicle-mediated transport reactions: α-, β-, γ-SNAP (103), Sec23p (104), and Sar1p (85, 86). Clearly this represents only a small fraction of the proteins that act in each transport step.

Convergence of Biochemical and Genetic Approaches

The advent of in vitro assays to measure ER to Golgi transport in gently lysed yeast cells has provided the means to analyze yeast mutants biochemically (105–107). Semi-intact cells provide both donor and acceptor membranes, and transport is dependent on ATP, cytosol, and physiological temperature. As a marker for transport, radiolabeled yeast prohormone α-factor is syn-

Table 2 Cell-free reactions reconstituting steps in the biosynthetic and endocytic pathways[a]

Transport step	Source	Cell fractions	Measurement	Requirements and inhibitors	Protein activities[b]	Refs.
Biosynthetic pathway: ER to Golgi transport						
ER to cis- and medial-Golgi	CHO cells, 15B mutant	Semi-intact cells, hypotonic lysis or NC stripping; or crude mitotic cell homogenates	oligosaccharide processing intermediates of VSV-G	ATP, cytosol, time, temp.; inhibited by NEM, GTP-γS, EGTA, Rab peptide	NSF	(58, 96, 97, 100, 137, 265, 266)
ER to Golgi	hepatocyte	ER-enriched and Golgi-enriched membrane fractions	Transfer of radioactivity from ER to Golgi membranes adsorbed on NC and fucosylation of dipeptidyl-aminopeptidase IV.	ATP or GTP, cytosol, time, temp.	—	(22, 267)
ER to basolateral plasma membrane	MDCK cells	semi-intact cells lysed by NC stripping	endoglycosidase H resistance of VSV-G; C6-NBD-ceramide glucosylation, movement of fluorescent lipids to cell surface	ATP, time, temp.	—	(268)
ER to Golgi	Saccharomyces cerevisiae	Semi-intact cells, freeze-thaw or hypotonic lysis; or ER-enriched membranes	Golgi-specific glycosylation of prohormone	ATP, cytosol, time, temp.; inhibited by EGTA, NEM, GTP-γS	Sec23p, Ypt1p, Sec18p, Sec12p, Sar1p, Sec7p	(29, 30, 85, 86, 92–94, 98, 104–106, 108–110, 138, 242, 269)
Biosynthetic pathway: intra-Golgi transport						
cis-Golgi to medial-Golgi	CHO cells, wild-type and 15B mutant	Golgi-enriched membrane fractions (donor from 15B; acceptor from wild-type)	incorporation of N-acetyl-glucosamine into VSV-G	ATP, cytosol, time, temp.; inhibited by NEM, ATP-γS, GTP-γS, Rab peptide	NSF (Sec18p), α-SNAP (Sec17p), β-, γ-SNAP, acyl CoA	(28, 57, 91, 95, 101–103, 114, 135, 137, 270–275)
transport to trans-Golgi	CHO cells, wild-type and 15B mutant	Golgi-enriched membrane fractions (1021 donor; wild-type acceptor)	incorporation of sialic acid into VSV-G	ATP, cytosol, time, temp.; stimulated by DTT; inhibited by NEM	acyl CoA	(276)

Biosynthetic pathway: post-Golgi transport

Process	Cell type	Assay system	Readout	Requirements	Proteins	Ref.
apical and basolateral vesicle formation from the TGN	MDCK cells	semi-intact cells lysed by NC stripping	proteolysis of hemagglutinin (HA), Ab accessibility of HA and VSV-G	ATP; inhibited by CaCl$_2$	—	(34, 221, 277)
formation of transport vesicles from the TGN	BHK cells	crude homogenate of SFV- or VSV-infected cells	release of mature, sialylated p62 glycoprotein from the TGN detected by gradient separation	ATP, cytosol, time, temp.	—	(32)
formation of vesicles from the TGN	hepatocyte	immunoisolated stacked Golgi fraction	release of albumin, sialylated transferrin, and mature polymeric IgA receptor in sedimentable form	ATP, cytosol, time, temp.; inhibited by NEM	—	(31)
formation of regulated and constitutive secretory vesicles from the TGN	PC12 cells	crude cell homogenate	release of secretogranin 1 (regulated) and proteoglycan (constitutive) markers from the TGN, detected by gradient separation	ATP, time, temp.; inhibited by GTPγS	—	(33, 278)
Golgi to vacuole	S. cerevisiae	semi-intact cells, freeze-thaw and hypotonic lysis	proteinase A processing of carboxypeptidase Y	ATP, cytosol, time, temp.	Vps15p, Vps33p, Vps34p	(202)
trans-Golgi to plasma membrane	CHO cells	Streptolysin-O-perforated semi-intact cells	movement of VSV-G to cell surface detected by immunofluorescence, release of soluble glycosaminoglycan chains	ATP, cytosol, time, temp.; inhibited by NEM, EGTA, GTPγS	—	(279)
fusion of trans-Golgi vesicles with the plasma membrane	BHK-21 cells	homogenates of influenza-infected cells with neuraminidase and cells with plasma membrane-bound viral proteins bearing N-acetylneuraminic acid)	release of free N-acetylneuraminic acid	ATP, time, temp. trypsin-sensitive factors	—	(280)

Table 2 (*Continued*)

Transport step	Source	Cell fractions	Measurement	Requirements and inhibitors	Protein activities[b]	Refs.
Endocytic pathway						
formation of clathrin-coated pits and vesicles	A431 cells	semi-intact cells, hypotonic lysis	morphometric analysis; accessibility of internalized biotin-S-S-transferrin to Ab and to S-S cleaving agent	ATP, cytosol, time, temp.; inhibited by clathrin mAb and mAb against transferrin receptor cytoplasmic domain	—	(154, 155)
formation of clathrin-coated pits and vesicles	human fibroblasts	plasma membrane fragments (prelabeled with anti-low density lipoprotein receptor IgG-gold) on cover slips	binding of anti-clathrin mAb	ATP, Ca^{2+}, cytosol, time, temp.; inhibited by NEM treatment of membranes	clathrin heavy and light chains, AP-2	(156, 157, 173, 281)
plasma membrane endocytic vesicle-endosome fusion	macrophages	homogenate of cells with internalized anti-DNP Ab (plasma membrane vesicles), and homogenate or endosome-enriched fraction from cells with internalized DNP-β-glucuronidase	formation of immune complex, morphological analysis	ATP, cytosol, time, temp.; inhibited by NEM, or trypsin treatment of membranes	—	(282, 283)
endosome-endosome fusion	macrophages	homogenates of cells with internalized mannosylated anti-DNP IgG and cells with internalized DNP-β-glucuronidase	formation of immune complex	ATP, cytosol, time, temp.; inhibited by NEM and anti-NSF mAb	NSF	(59, 284)
endosome-endosome fusion	BHK-21 cells	homogenates of cells with internalized viral neuraminidase and cells with internalized N-acetylneuraminic acid-bearing viral glycoproteins	release of free N-acetylneuraminic acid	ATP	—	(285)

Process	Cell type	System	Assay	Requirements	Protein activities	Ref.
endosome-endosome fusion	BHK-21 cells	homogenate of cells with internalized lactoperoxidase or biotin-HRP and immunoisolated endosomes with either VSV-G or VSV-G-avidin	morphological analysis, iodination of VSV-G, or formation of biotin-avidin-HRP complex	ATP, time, temp.	—	(35, 286, 287)
endosome-endosome fusion	CHO cells	homogenate of cells with internalized avidin-β-gal or biotinylated mouse Ab	formation of avidin-biotin complex	ATP, cytosol, time, temp.; inhibited by NEM	—	(288)
endosome-endosome fusion	A431 cells	homogenate of cells with internalized radiolabeled transferrin or anti-transferrin receptor Ab	formation of radiolabeled immune complex	ATP, cytosol, time; inhibited by NEM, or trypsin treatment of membranes	—	(289)
early endosome-late endosome fusion	MDCK cells	homogenate of cells with internalized avidin or internalized biotin-HRP	formation of avidin-biotin-HRP complex, morphological analysis	ATP, cytosol, time, temp.; stimulated by MT assembly; inhibited by excess MT, GTPγS, dynein cleavage	MAPs, kinesin	(290)
late endosome to trans-Golgi network	CHO cells (1021 mutant), and rat liver	semi-intact CHO cells, hypotonic lysis (donor) and Golgi-enriched rat liver membranes (acceptor)	sialylation of mannose 6-phosphate receptor	ATP, cytosol, time, temp.; inhibited by GTPγS	—	(60, 218, 219)

[a] Abbreviations/explanations: CHO 15B mutant, lacks the medial-Golgi enzyme GlcNAc transferase 1; CHO 1021 mutant, lacks a permease that allows CMP-sialic acid to enter the Golgi lumen; VSV-G, vesicular stomatitis virus glycoprotein; HRP, horseradish peroxidase; DNP, di-nitro-phenol; DTT, dithiothreitol; NC, nitrocellulose; Ab, antibody; mAb, monoclonal antibody; MT, microtubule; MAP, microtubule-associated protein; SFV, Semliki Forest Virus.

[b] Protein activities designates those proteins that have been shown to be required in vitro either by antibody or mutant blocks or by the addition of purified proteins.

thesized in vitro and posttranslationally translocated into the ER. Transport to the Golgi complex is detected by the addition of Golgi-specific outer chain mannose to α-factor using an antibody that specifically binds this oligosaccharide linkage. The ER to Golgi transport defect of *sec23* was reproduced in vitro by using membranes and cytosol derived from the *sec23* mutant strain at the restrictive temperature (105, 106). Transport activity was restored by the addition of Sec23p-containing wild-type cytosol, providing an assay for the purification of Sec23p by classical biochemical techniques (104, 108). A 105-kDa protein copurified in a complex with Sec23p (109). A "reverse genetics" approach can now be applied to isolate and characterize the gene encoding p105 and determine whether p105 is required for ER to Golgi transport in vivo. In addition, classical biochemical methods can be used to study p105 activity in vitro. Methods to reconstitute ER to Golgi transport have now been developed using more purified yeast membrane components (98, 110). These reconstituted systems hold promise for the purification and characterization of additional Sec proteins, and the potential to analyze the protein interactions suggested by genetic interactions among vesicular transport mutants.

The success of purifying active Sec23p from yeast inspired a search for functional homologues of Sec23p in other organisms. Antibodies raised against purified Sec23p have been used to detect cross-reactive proteins in mammalian cell extracts. Immuno-electronmicroscopy analysis demonstrated that the mammalian Sec23p homologue is localized to the transitional region between the ER and the Golgi of pancreatic exocrine cells (111). This location of the Sec23p homologue in mammalian cells is consistent with the proposed role of *SEC23* in the formation of ER-Golgi transport vesicles in vivo (81). These findings suggest that Sec23p may have an evolutionarily conserved function in ER-derived vesicle formation.

Another advance in the biochemical analysis of yeast mutants has been the development of heterologous in vitro systems in which yeast and mammalian cell fractions are interchangeable. As discussed earlier, mammalian intra-Golgi transport assays require NSF activity. The finding that NSF shares 48% sequence identity with yeast Sec18p (91, 112) led to experiments to test whether yeast Sec18p can functionally replace NSF in vitro. Unfractionated cytosol from wild-type yeast cells, when substituted for mammalian cytosol, was found to provide NSF activity in intra-Golgi transport assays. Furthermore, NSF activity was higher in cytosol prepared from a yeast strain that overproduces Sec18p, and NSF activity was absent from *sec18* mutant cytosol (91). These results demonstrate that NSF (Sec18p) is functionally conserved across evolutionarily divergent species.

An extension of this heterologous transport assay was used to assign a biochemical activity to a second yeast protein. Yeast cytosol was unable to support transport between salt-extracted Golgi membranes in the presence of mammalian NSF, but transport activity was restored by the addition of bovine brain cytosol. Fractionation of bovine brain cytosol yielded the purification of three related 35-kDa proteins, each of which restored the missing activity (103). These three proteins were found to fulfill a previously identified activity required to bind NSF to Golgi membranes (113), and were designated α-, β-, and γ-SNAP (for Soluble NSF Attachment Protein) (114). The substitution of yeast cytosol containing active Sec18p (NSF) for mammalian brain SNAP fractions showed that yeast cytosol contains SNAP activity. Screening cytosol from each of the *sec* mutants defective in ER to Golgi transport revealed that *sec17* was defective in SNAP activity and the deficiency could be rescued by the addition of purified bovine brain α-SNAP (114). It is likely that Sec17p provides SNAP activity required for NSF function in mammalian cells.

The in vitro results obtained with these heterologous systems closely agree with the in vivo functions of *SEC17* and *SEC18*. Both the *sec17* and *sec18* mutant cells are blocked in ER to Golgi transport at a vesicle consumption step. Furthermore, analysis of *sec17,sec18* double mutants indicates that these genes display synthetic lethality, a genetic interaction that may suggest a protein interaction (81). These observations suggest that in yeast, Sec18p and Sec17p might act in concert in ER to Golgi transport in the same way that NSF and α-SNAP act together in mammalian intra-Golgi transport.

Sequence homologies also have provided insights into the possible function of other yeast genes involved in protein transport. Sequencing of the *SEC14* gene, identified as a mutant blocked in protein transport from the Golgi apparatus, revealed that Sec14p is identical to yeast PI/PC transferase, a protein that exchanges phosphatidylinositol or phosphatidylcholine between membranes (115, 116). Phospholipid transferase proteins from mammalian sources have been extensively studied and found to mediate phospholipid transfer in vitro. *sec14* mutant extracts are deficient in a phospholipid transfer assay, while wild-type yeast extracts supported phospholipid exchange. Sec14p has been proposed to regulate the ratio of PI to PC in the Golgi membrane to maintain the proper phospholipid ratio, which may be required for efficient formation of mature secretory vesicles (117, 118). These studies of Sec14p function, initiated by detecting sequence homologies, have provided insight into a previously unknown level of regulation of vesicular transport.

A particularly illustrative example of the interplay between genetic and biochemical approaches to vesicular transport is provided by the small GTP-binding proteins (49). The homology of yeast *SEC4* to the mammalian *ras* gene family (83) directed attention to the possible function of Ras-like GTP-binding proteins in vesicular transport. A large class of yeast and mammalian GTP-binding proteins have now been identified through the use of molecular cloning techniques such as the polymerase chain reaction (PCR), genetic analysis, and biochemical isolation. As discussed in detail in the following section, ultrastructural analysis of mammalian cells, characterization of yeast mutants, and development of in vitro assays have contributed to our knowledge of how these proteins may function in vesicular transport.

SMALL GTP-BINDING PROTEINS IN VESICULAR TRANSPORT

GTP-binding proteins function in a variety of cellular processes and are characterized by a strongly conserved guanine nucleotide–binding domain required for the binding, hydrolysis, and exchange of GTP and GDP. Within this family are a group of small GTP-binding proteins (20–25-kDa), which share homology to the mammalian Ras 21-kDa GTP-binding protein (p21^{ras}) (48, 119, 120). These proteins contain the characteristic GTP-binding domain and a second, more divergent domain thought to mediate the interaction of small GTP-binding proteins with specific effector proteins. Extensive study of two well-characterized GTP-binding proteins, p21^{ras} and elongation factor Tu (EF-Tu), established that the activity of GTP-binding proteins is regulated by the phosphorylation state of the bound guanine nucleotide. GTP-binding proteins undergo a conformational change upon dissociation of GDP and replacement by GTP, which renders the GTP-bound form able to interact with an effector protein. GTP hydrolysis terminates interaction with the effector and the protein reverts to an inactive conformation. The proportion of time spent in the active and inactive states is intricately coordinated by the action of regulatory proteins. The rate of exchange of GDP for GTP can be stimulated by a guanine nucleotide release factor (GNRP) or, conversely, inhibited by a GDP dissociation inhibitor (GDI). The rate of GTP hydrolysis can also be accelerated by the action of a GTPase activating protein (GAP). These factors are likely to vary among the individual GTP-binding proteins and their relative affinities for GTP and GDP.

Based on the known activities of a few well-characterized GTP-binding proteins, two potential mechanisms of action of GTP-binding proteins in secretion have been proposed (48). By analogy to the α subunit of heterotrimeric G-proteins, small GTP-binding proteins may be involved in the amplification or transduction of a signal required for a subsequent event in vesicular transport. A G_α subunit localized to the Golgi complex has been implicated in regulating secretion (121). Alternatively, GTP-binding proteins may confer directionality and fidelity upon vesicular transport in the same manner that EF-Tu "proofreads" protein synthesis.

Several lines of evidence support a role for small GTP-binding proteins in vesicular transport (48, 49, 120). First, mutations in each of four yeast genes encoding small GTP-binding proteins (*ARF1*, *SAR1*, *YPT1*, and *SEC4*) result in defects in various steps of secretion. Second, in mammalian cells, individual Rab proteins (homologues of Ypt1p) are localized to specific organelles of the secretory pathway. Finally, in vitro assays (Table 2) that reconstitute steps of intracellular transport require GTP, are blocked by treatments that inhibit small GTP-binding protein activity, and are defective when yeast mutant fractions are used.

As discussed earlier, *SEC4* and *SAR1* were identified by virtue of their role in secretion (63, 70), whereas *YPT1* and *ARF1* were initially of interest due to their sequence homology with *ras* and later found to function in secretion (78, 122–124). *SEC4* and *YPT1* are similar to one another in sequence and are related to mammalian *rab* genes, while *SAR1* and *ARF1* are more divergent from the other *ras*-like genes (48). The accumulation of secretory precursor proteins and aberrant membranous organelles in these mutants is diagnostic of their transport defect. Depletion of Sar1p by repression of *SAR1* expression results in the accumulation of ER-modified forms of secretory protein precursors (70). *ypt1* mutants also accumulate ER forms of secretory protein precursors, and some alleles exhibit defects in Golgi-specific protein modifications (123, 125, 126). Morphological analysis revealed that *ypt1* mutants accumulate aberrant membranes and 50 nm vesicles (125). *SEC4* is 48% identical in sequence to *YPT1*, but unlike *ypt1*, *sec4* mutants do not block ER to Golgi transport. Rather, *sec4* mutants accumulate the 80–100 nm vesicles involved in transport of proteins from the Golgi apparatus to the cell surface (63). Consistent with a role in post-Golgi transport, Sec4p has been localized to both the plasma membrane and the cytosolic face of mature 80–100 nm secretory vesicles (127). Antibodies to Arf1p and Ypt1p both label the Golgi apparatus of mammalian cells, but because of potential cross-reactivity with a number of mammalian homologues, the exact site of Ypt1p and Arf1p localization in yeast has not been determined.

The well-defined ultrastructure of the secretory pathway in mammalian cells facilitated the subcellular localization of specific GTP-binding proteins. The specific association of small GTP-binding proteins with membranes of the secretory and endocytic pathways, summarized in Table 3, is highly suggestive of a role in the regulation of membrane traffic. At least 12 mammalian *rab* genes have been isolated from cDNA libraries using probes that recognize conserved *ras* sequences (128–130). The *rab* genes are most closely related to yeast *YPT1* and *SEC4* genes; *rab1* is 75% identical to *YPT1*. Rab proteins 1 through 6 have been shown to bind and hydrolyze GTP in vitro (129). Monospecific antibodies raised against the Rab proteins and peptides have been used to determine the location of Rab proteins in the cell using

Table 3 Localization of small GTP-binding proteins to organelles of the biosynthetic and endocytic pathways

protein	localization[a]	reference
Mammalian		
Rab1	nd	—
Rab2	cis Golgi network	(131)
Rab3	synaptic vesicles	(134)
Rab4	early endosomes	(133)
Rab5	plasma membrane, early endosomes	(131)
Rab6	cis and medial Golgi	(132)
Rab7	late endosomes	(131)
Rab8–11	nd	—
Arf1	Golgi membranes	(79)
Yeast		
Sec4p	plasma membrane, post-Golgi vesicles	(127)
Sar1p	endoplasmic reticulum	(241)
Ypt1p	nd	—
Arf1p, Arf2p	nd	—

[a] nd = not determined

subcellular fractionation, immunofluorescence, and immuno-electronmicroscopy (131–134) (Table 3). The highly restricted distribution of each of the Rab proteins suggests that different GTP-binding proteins may act at each step of transport. A test of this hypothesis awaits reconstitution of Rab protein function in vitro.

Cell-free assays have provided the most compelling evidence implicating small GTP-binding proteins in transport. The nonhydrolyzable GTP analogue GTPγS inhibits most in vitro transport reactions (Table 2), probably as a result of a direct effect on one or more GTP-binding proteins (135). In addition to GTPγS, other inhibitors have been used to study the function of GTP-binding proteins in vitro. Antibody inhibition was used as a block of Rab function in an assay reconstituting fusion between early endosomes in vitro (136). Antibodies against Rab5, but not Rab2 or Rab7, blocked the fusion of isolated early endosomes in vitro. Fusion could be restored by the addition of a Rab5-containing cytosol. Endosome fusion was also defective in a reaction containing a point mutation of Rab5 that did not bind GTP, or a deletion of the C-terminal domain of Rab5, thought to mediate Rab5 binding to the membrane. Taken in combination, these data present convincing evidence that Rab5 function is required for early endosome fusion in vitro.

Another approach to study Rab protein function in vitro is to compete for Rab effector binding sites by adding peptides that mimic Rab sequences (137). Synthetic peptides prepared from the Rab3 effector domain sequence blocked ER to Golgi transport in a mammalian in vitro assay, presumably by competing with Rab proteins for binding to a downstream effector. ER to Golgi transport was blocked at a late step, preceding vesicle fusion. Intra-Golgi transport between isolated mammalian Golgi membranes also was inhibited by effector domain peptide. Rab3 itself probably does not function in ER to Golgi or intra-Golgi transport, but instead the effector domains are conserved to a sufficient degree that Rab3 peptides inhibit the action of other Rab proteins in the ER and Golgi membranes.

In vitro transport assays using yeast cell fractions have been used to assess the function of small GTP-binding proteins identified genetically. Antibody fragments against Ypt1p were found to inhibit ER-Golgi transport, and cell fractions prepared from the *ypt1* mutant strain are defective in transport assays (98, 138). The Ypt1p requirement has been further defined to occur at a point after vesicle formation, but preceding vesicle targeting or fusion, since blocking Ypt1p function results in the accumulation of a vesicular intermediate (30, 93). Furthermore, Ypt1p cofractionates with these accumulated vesicles (93).

Although a role for small GTP-binding proteins in secretion has been demonstrated in vitro, the actual molecular mechanism by which these proteins couple GTP hydrolysis to membrane transport is not understood. Mutagenesis of *SEC4* and an examination of mutant phenotypes has led to a model in which Sec4p ensures the proper targeting of secretory vesicles to the plasma membrane (139). GTP-bound Sec4p is proposed to associate with the secretory vesicle membrane. In the GTP-bound form, Sec4p recognizes an effector on the plasma membrane, targeting the secretory vesicle to its appropriate acceptor membrane. GTP hydrolysis promotes the release of Sec4p-GDP from the effector, and Sec4p-GDP is recycled through the cytosol in an inactive form and made available for re-binding to secretory vesicles. This model is supported by the observation that a *SEC4* mutant unable to bind nucleotide and thought to be locked into the active conformation acts as a dominant inhibitor of secretory vesicle fusion (139). By analogy to the role of EF-Tu in the positioning of aminoacyl tRNAs, Sec4p is proposed to ensure the unidirectional delivery of a vesicle to its target.

The regulation of small GTP-binding protein activity by accessory factors is beginning to be investigated. Because purified Sec4p has a low intrinsic rate of GTP hydrolysis and a high affinity for GTP (140), isolated Sec4p is expected to exist predominantly in the active, GTP-bound form. It is likely that Sec4p GTP hydrolysis and exchange are closely regulated in the cell. Recently, a GDI from mammalian cells was found to act upon yeast Sec4p, providing a mechanism whereby the GDP-bound form of Sec4p might be favored (141). The existence of a Sec4p GAP activity to stimulate GTP hydrolysis also has been proposed. A mammalian GAP has been shown to act upon yeast Ypt1p, and this activity is blocked by mutations in the *YPT1* effector domain (125). Another site of regulation of GTP-binding protein activity is suggested by the cell cycle–dependent phosphorylation of Rab1A and Rab4, which might be related to the cessation of membrane traffic during mitosis (142).

Membrane association of the hydrophilic small GTP-binding proteins is achieved by the attachment of a hydrophobic modified lipid group to a C-terminal cysteine. For p21^{ras}, the C-terminal sequence CAAX (A is aliphatic and X is any amino acid) specifies, among other posttranslational modifications, the addition of a 15-carbon farnesyl group by thioether linkage to cysteine [(143); see Clarke in this volume]. Farnesylation is required for p21^{ras} association with membranes. In contrast, Sec4p, Ypt1p, and the Rab proteins all appear to be modified by the addition of a 20-carbon geranylgeranyl group onto a C-terminal cysteine, specified by a conserved CC sequence. Deletion of the C-terminal amino acids prevents prenylation and increases the amount of soluble Sec4p and Ypt1p (139, 144). In fibroblasts starved for mevalonate, a precursor in the pathway of prenyl synthesis, Rab1 shifts from a membrane to cytosolic distribution (145). Attachment to the membrane also is required for function, as demonstrated by the inability of a C-terminal deletion of *SEC4* to replace a *sec4* mutant defect. Several of the Rab proteins have been shown directly to be modified by the addition of geranylgeranyl groups (145, 146). Ypt1p has been reported to be palmitoylated at the C-terminus (144). The mechanism by which Sar1p attaches to the membrane is unknown. Sar1p has only one cysteine, 20 amino acids from the C-terminus, and posttranslational modifications of this residue have not been detected (70). Mammalian Arf1, in contrast to other small GTP-binding proteins, is modified at the N-terminus by a myristate group attached to glycine (147).

The membrane association of Ypt1p and Sec4p has recently been shown to depend on the activity of the *BET2* gene product (82). *BET2*, identified as a secretory mutant blocked in ER to Golgi transport (64), is similar in sequence to *DPR1/RAM1*, which encodes a prenyltransferase that modifies Ras protein in yeast. Furthermore, Bet2p exhibits geranylgeranyltransferase activity in vitro (148), suggesting that Sec4p and Ypt1p function are dependent on a geranylgeranyl modification. Mammalian prenyltransferases are heterodimers composed of a unique β subunit, which recognizes the substrate, and of an α subunit shared between at least two prenyltransferases (149). The product of the yeast *DPR1/RAM1* gene shares homology with the β subunit of mammalian farnesyltransferase (150, 151). *BET2* may encode a specific β subunit of the geranylgeranyltransferase, and function with an α subunit that has yet to be identified.

The attachment of a hydrophobic modified lipid anchor accounts for the strong membrane association of GTP-binding proteins, but does not explain the specific localization of these proteins to different membrane compartments. Additional signals within each small GTP-binding protein and corresponding receptors for those signals may dictate their specific localization. Sec12p is a candidate for the specific recruitment of Sar1p to the ER membrane. While direct Sar1p binding to Sec12p has not been demonstrated, the attachment of Sar1p to membranes is correlated with the amount of Sec12p present (87). Perhaps initial recruitment of Sar1p to the membrane occurs via binding to Sec12p, and subsequent modification of Sar1p with an as-yet-undetected lipid anchor ensures its stable association. Analogous receptors may exist to specify the association of each small GTP-binding protein with the membrane of a compartment of the secretory pathway. Additionally, membrane association must be reversible for proteins such as Sec4p, which has been proposed to recycle via the cytosol from the plasma membrane to secretory vesicles (127).

Future efforts to understand the exact role of GTP-binding proteins in vesicular transport are likely to address how small GTP-binding proteins specifically associate with different membranes, and how that association is regulated. The identity of the downstream effectors of small GTP-binding proteins is of great interest. A potential Sec4p-interacting effector, the product of the *SEC15* gene, has been identified genetically (152). Effector proteins may also be identified biochemically using peptides that mimic the effector domain of GTP-binding proteins. Given the potential number of vesicle-mediated transport steps between various cellular compartments, it is reasonable to expect multiple GTP-binding proteins to participate. What remains to be elucidated is how fidelity, directionality, and control of vesicle-mediated transport are conferred by this family of proteins.

STRUCTURE AND FUNCTION OF VESICLE COATS

A consideration of vesicle formation and protein sorting requires an examination of the protein coats surrounding many vesicles. Two types of coated vesicles of distinct coat compositions have been characterized biochemically and morphologically. Clathrin-coated vesicles function in steps of vesicular transport that appear to require selection of vesicle cargo. Clathrin-coated vesicles mediate endocytic uptake from the plasma membrane and carry protein that is diverted from the constitutive pathway at the trans-Golgi network (38). In contrast, vesicles coated with a distinct nonclathrin coat mediate the transport of proteins between Golgi compartments. This transport step is thought to occur by nonselective capture (bulk flow) of membrane and secretory protein for movement between Golgi cisternae (153).

Clathrin-Coated Vesicles

Clathrin-coated vesicles (38, 39) form by the invagination of a clathrin-coated indentation of membrane called a coated pit. Coated pits concentrate selected vesicle cargo, usually transmembrane receptors bound to a soluble ligand, while excluding resident membrane proteins. The cytoplasmic face of a coated pit is composed of a layer of clathrin-associated proteins (adaptins) bound to the membrane, surrounded by an attached lattice of clathrin sub-units, thought to be assembled by the recruitment of soluble forms of clathrin and adaptins to the membrane. Coated pits are converted into mature coated vesicles and released from their membrane attachment. Once released, mature coated vesicles are uncoated rapidly, releasing clathrin and adaptins for another round of coated vesicle formation, and generating a fusion-competent vesicular intermediate. Several research groups recently reconstituted clathrin-coated vesicle formation in vitro in permeabilized cells (154, 155) and from patches of plasma membrane (156, 157). These assays will permit further functional analyses of coat proteins and investigation of the mechanisms involved in clathrin-coated vesicle formation.

The major structural component of the clathrin-coated vesicle is the clathrin triskelion (158), which is composed of three clathrin heavy chains (180 kDa), associated at their carboxy-termini to form a three-legged structure. Triskelions of heavy chain alone are capable of self assembly in physiological buffers to form the polyhedral lattice characteristic of clathrin-coated pits and vesicles. Each triskelion also contains three clathrin light chains (30–40 kDa), each bound to a heavy chain near the vertex of the triskelion (159).

Clathrin heavy chain is thought to serve a structural role in vesicle formation and regulation of membrane traffic. The self-assembly properties of the clathrin lattice in vitro have led to the proposal that clathrin may effect the change in membrane shape required to produce a curved vesicle. Rearrangement of the interactions between adjacent clathrin triskelions theoretically is sufficient to generate curvature (160). The transition from a shallow coated pit-type lattice to a rounded coated vesicle cage has been reproduced in vitro (157), providing a means to study the mechanics of clathrin-coated vesicle formation.

Clathrin heavy chains also are involved in the retention of resident proteins to the proper organelle. A yeast strain lacking a functional clathrin heavy chain gene grows slowly but has no general secretory or endocytic defects (74). However, this mutant is defective in the processing of a secreted prohormone that is normally proteolytically cleaved in a late Golgi compartment to generate the mature form. This defect is due to a mislocalization of the protease, Kex2p, to the plasma membrane, suggesting that clathrin heavy chain is required either to retain Kex2p in the Golgi apparatus or to retrieve Kex2p that has escaped to the plasma membrane (77).

Clathrin light chains are likely to play a regulatory role in disassembly of the clathrin coat. Disassembly of the clathrin lattice and release from the vesicle surface requires clathrin light chain and is stimulated by the enzymatic action of Hsc70, a member of the heat shock protein family (161). By controlling uncoating, clathrin light chain could regulate the equilibrium between clathrin coat assembly and disassembly (162), and in this manner regulate vesicle consumption.

In addition to the clathrin heavy and light chains, the clathrin coat also contains adaptins, thought to mediate clathrin binding to membranes and to play a role in the selection of vesicle cargo (163, 164). Two major adaptin complexes, designated AP-1 and AP-2 (for assembly or adaptor protein) have been identified biochemically. Each is a complex of approximately 300 kDa, composed of proteins of 100–110 kDa (α-, β-, β'-, and γ-adaptin) and proteins in the range of 50 kDa and of 18–20 kDa. The homologous β- and β'- proteins share identical tryptic peptide fragments and contain a clathrin-binding domain (165, 166). The γ- and α-adaptin genes have some homologous sequences (167, 168), but other adaptin subunits appear to be distinct. AP-2 is the more abundant complex and is made up of α- and β-adaptin, plus proteins of 50 kDa and 17 kDa. Immunolocalization using monoclonal antibodies specific for α-adaptin shows that AP-2 is found only in association with clathrin on the plasma membrane (169). AP-1 is composed of β'- and γ-adaptin, and 47- and 20-kDa proteins, and is associated with clathrin in the trans-Golgi network (165).

Adaptin complexes have been proposed to link the clathrin lattice with the membrane, perhaps by binding both to clathrin and to transmembrane proteins in the coated vesicle membrane. Electronmicrographs of clathrin-coated vesicles show that adaptins are positioned between the vesicle surface and clathrin lattice, in the appropriate position to mediate clathrin binding to the membrane (170). Adaptin complexes have been shown to bind in vitro to the cytoplasmic domains of transmembrane receptors that are packaged into clathrin-coated vesicles: for example, the LDL receptor cytoplasmic domain binds AP-2 (171). The cytoplasmic domain of the mannose 6-phosphate receptor, which is sorted into clathrin-coated vesicles from both the Golgi apparatus and the plasma membrane, binds to both AP-1 and AP-2 (172). AP-2 also binds to plasma membrane fragments stripped of endogenous clathrin in a reaction reconstituting coated pit formation in vitro (156). In this assay, AP-2 binds to an uncharacterized integral membrane protein, perhaps a receptor cytoplasmic domain (173). Adaptin function has not yet been investigated in intact cells, but the recent identification of yeast adaptin homologues may facilitate a genetic approach to clathrin-adaptin interactions (174, 175).

Golgi-Derived (COP-Coated) Vesicles

The second type of vesicle coat, surrounding vesicular intermediates of intra-Golgi transport, is less well defined (4, 153). Golgi-derived coated vesicles were isolated by inhibition of a vesicle targeting/fusion step in an intra-Golgi transport reaction. The addition of GTPγS blocks the uncoating and subsequent consumption of transport vesicles, resulting in a fivefold increase in vesicle number (28, 135). Inhibition of vesicle targeting or fusion by GTPγS is irreversible, thus it has not yet been possible to determine whether these vesicles are functional intermediates. The protein coat surrounding the accumulated vesicles is composed of four major proteins of 160, 110, 98, and 61 kDa, designated α, β, γ, and δ COP (for coat protein), respectively (176). A soluble protein complex containing these four COPs has been isolated from bovine brain cytosol (177), suggesting that, like clathrin, the Golgi vesicle coats may exist in a dynamic equilibrium between soluble and membrane-bound forms. In contrast to clathrin and adaptins, the COP complex is a preassembled protein coat subunit. The soluble COP complex, or "coatomer," also contains proteins of 36, 35, and 20-kDa.

With the exception of β-COP, little is known about the Golgi-derived coated vesicle proteins. The N-terminal half of the β-COP sequence has homology to β-adaptin of clathrin-coated vesicles (178), which suggests the intriguing possibility that the two proteins play analogous roles in both types of vesicles. Immunofluorescence localization of β-COP shows that it is restricted to the Golgi apparatus. β-COP has recently been identified as an early target of the fungal metabolite brefeldin A, which disrupts Golgi function. Upon the addition of brefeldin A to permeabilized cells, β-COP is rapidly released from Golgi membranes into the cytosol (179) and reassembly of coat proteins onto membranes is inhibited (180). The effect of brefeldin A is abrogated by the inclusion of GTPγS (181), suggesting that GTP hydrolysis may regulate COP coat assembly and disassembly.

The cycle of Golgi-derived coated vesicle formation and consumption is similar to that of clathrin-coated vesicles. COP-coated buds have been visualized on Golgi membranes, and these are presumably precursors of coated vesicles. A single round of vesicle formation and consumption was studied in vitro by applying inhibitors of intra-Golgi transport (99). The addition of GTPγS results in fully coated vesicles bound to acceptor membranes, but incapable of fusion. This result contrasts with the situation in clathrin-coated vesicles, which are uncoated immediately after formation, and suggests that the COP complex may play a role in vesicle targeting or binding to acceptor membranes. A later block is imposed by treatment with the alkylating agent NEM, resulting in the accumulation of bound, uncoated vesicles, suggesting that the COP complex, like clathrin, is removed prior to fusion.

The exact function of the COP complex in vesicle formation, protein sorting, and the regulation of vesicle fusion awaits further study. The sequence similarity between β-COP and β-adaptin suggests that the COP coat proteins may serve similar roles to the clathrin-adaptin coat. For example, β-COP may link the COP complex to Golgi membranes. It is not clear whether the COP complex fulfills vesicle coat functions unique to intra-Golgi transport, or also serves in other transport events. A firm answer concerning the role of the COPs in vesicle transport will require the development of a fractionated transport reaction that depends on the addition of the COP complex.

SIGNAL-MEDIATED SORTING INTO VESICLES

The selective packaging of cargo into vesicles is the major opportunity for the active sorting of proteins, and mechanisms for sorting both membrane and soluble proteins into vesicles have been described. For soluble proteins, the best-studied case of signal-mediated sorting is the selective diversion of lysosomal hydrolases from the constitutive secretory pathway in the trans-Golgi network (12). Sorting of lysosomal proteins requires the recognition of sorting signals in both soluble and transmembrane proteins, and the selective activity of clathrin-coated vesicles. The sorting of soluble proteases to the yeast vacuole is signal mediated but does not appear to involve the same recognition process as lysosomal sorting (182). In cells with a regulated pathway of secretion, the sorting of soluble hormones and neurotransmitters into the lumen of regulated secretory vesicles also may be signal mediated (13, 183).

Membrane protein sorting into vesicles relies on signals, usually in the cytoplasmic domain, that are recognized by the cellular sorting machinery and specify inclusion in the membrane of a budding vesicle, as in the clustering of transmembrane receptors into clathrin-coated pits (38, 184). The differential distribution of membrane proteins to the apical and basolateral domains of polarized epithelial cells may also require sorting signals (185–187). Many transmembrane receptors that are subjected to repeated sorting steps in the plasma membrane, endosome, and TGN are likely to have multiple signals, intricately regulated, to direct sorting at each step.

The general strategy for identifying sorting signals involves molecular manipulations of the gene encoding an easily traced, sorted marker protein. Gene deletions, fusions, and site-directed mutagenesis have been used to identify sequences required and sufficient for targeting proteins to the appropriate compartment. A difficulty frequently encountered in this strategy is that changes in sequence may result in the perturbation of protein folding and thus affect sorting. One common theme emerging from these studies of sorting signals is that often there is no consensus sequence shared by proteins sorted in the same pathway. Many sorting signals are probably provided by secondary structure or conformational motifs in addition to aspects of primary sequence.

Signal-Mediated Sorting of Soluble Proteins

SORTING TO HYDROLYTIC COMPARTMENTS   The sorting of soluble
hydrolytic enzymes to lysosomes is achieved by signal-mediated diversion
from the secretory protein pathway [(12, 188, 189); see also Kornfeld, this
volume]. Proteins destined to reside in the lysosome transit the early steps of
the secretory pathway in parallel with secreted proteins. The pathways diverge in the trans-Golgi network, where lysosomal proteins are sorted into the
endocytic system for delivery to the lysosome. Lysosomal protein sorting
requires at least three recognition steps. First, lysosomal proteins are recognized and modified by the addition of a unique mannose 6-phosphate residue.
The selectively modified proteins next are recognized by a receptor specific
for mannose 6-phosphate. Lastly, the receptor-protein complex in turn is
packaged selectively into vesicles and transported to an acidic, prelysosomal
compartment. Upon delivery to this compartment, the mannose 6-phosphate
receptor is uncoupled from the lysosomal proteins and recycled back to the
trans-Golgi network, while the hydrolases are transported further to the
lysosome.

The modification of soluble lysosomal proteins is achieved in an early (cis)
Golgi compartment by the concerted action of a GlcNAc phosphotransferase
and glucosaminidase. The result is the addition of a unique mannose 6-
phosphate residue, marking proteins destined for the lysosome. No consensus
sequence specifying mannose 6-phosphorylation has been detected in lysosomal enzymes, but a conformation-dependent recognition domain recently
was identified (190). Two sorting sequences in the lysosomal protein cathepsin D were identified by their ability to specify mannose 6-phosphorylation
when transplanted into normally unmodified, secreted pepsinogen. The two
putative sorting domains are far apart in the primary sequence of cathepsin D,
but are predicted to be in close contact on a three-dimensional model, hinting
at the potential complexity of sorting signals.

The mannose 6-phosphate–tagged proteins encounter and bind to one of
two related mannose 6-phosphate receptors in the trans-Golgi network. Receptor-protein complexes are sorted into clathrin-coated vesicles, probably by
the recognition of a sorting signal on the cytoplasmic domain of the mannose
6-phosphate receptor. Partial deletions of the cytoplasmic tail of the mannose
6-phosphate receptor results in the impaired sorting of soluble lysosomal
proteins (191). The adaptin complexes are reasonable candidates for the
cellular recognition and sorting component, because both the Golgi-localized
AP-1 and plasma membrane AP-2 complexes bind the mannose 6-phosphate
receptor tail in vitro (172).

The yeast vacuole serves a similar function to mammalian lysosomes, and
as for lysosomes, the sorting of soluble hydrolases to the yeast vacuole is

signal mediated (182, 192). Vacuolar hydrolases transit the secretory pathway and are glycosylated, but unlike for lysosomal enzymes, sorting to the yeast vacuole does not require glycosylation. Diversion to the vacuole occurs in a late Golgi compartment. Overproduction of the soluble vacuolar protein carboxypeptidase Y (CPY) results in mislocalization to the cell surface, suggesting the existence of a saturable recognition system for vacuolar sorting (193). Deletions and mutagenesis of the gene encoding CPY established a sorting sequence in the first 10 amino acids of the propeptide that is necessary and sufficient for CPY localization to the vacuole (194, 195). Site-directed mutagenesis of this region was used to identify a tetrapeptide QRPL (Gln-Arg-Pro-Leu) responsible for sorting of proCPY (196). A similar analysis of a second vacuolar protease, proteinase A, also identified a sorting signal in the N-terminal propeptide region (197). However, no QRPL-like sequence was detected in the propeptide of proteinase A, suggesting that either more than one recognition and sorting pathway exists for targeting to the vacuole, or that features in addition to the primary sequence play a role in sorting. The observation that overproduction of CPY results in its missorting to the cell surface while the localization of other soluble vacuolar proteins is unaffected is consistent with the former possibility (193). Sorting information for the correct localization of a vacuolar membrane protein has been localized to the short cytoplasmic and/or transmembrane domain (198). It is not yet clear whether this same sorting information specifies vacuolar localization for other membrane proteins.

The isolation of an abundance of yeast mutants defective in vacuolar sorting will aid in the identification of both sorting sequences and the cellular vacuolar sorting machinery. Two separate selection schemes for mutants that mislocalize CPY to the cell surface have resulted in the isolation of nearly 50 different *vps* (vacuolar protein sorting) complementation groups (65–68). Many *vps* mutants are defective in vacuolar morphology and vacuolar distribution during cell division. Phenotypic analysis has revealed that a subclass of *vps* mutants mislocalize soluble vacuolar proteins to the cell surface while membrane protein localization to the vacuole is unaffected, suggesting that different sorting mechanisms may operate for soluble and membrane proteins. Some mutants (e.g. *vps18*) also secrete unprocessed α-factor, indicating a defect in the function of a late Golgi compartment containing Kex2p (199). The protein products of several *vps* mutants have been characterized (see Table 1). Vps1p is homologous to a group of proteins with a GTP-binding motif, some of which also bind to microtubules (45). Another Vps protein, Vps15p, is a membrane-associated serine/threonine kinase (200, 201). The recent development of an in vitro assay to measure the delivery of CPY to the vacuole in permeabilized cells should provide the means for an analysis of the *vps* mutant defects in vitro (202).

SORTING TO THE REGULATED SECRETORY PATHWAY   Cells that secrete hormones and neurotransmitters in response to external stimuli store these materials in secretory vesicles, also called secretory granules because of their dense appearance by electronmicroscopy. The fusion of these stored vesicles with the plasma membrane is triggered by an external signal, in contrast to the case of the constitutive pathway, in which vesicle fusion is not temporally regulated. As with the other specific diversions from the constitutive pathway, proteins destined for inclusion in secretory granules are sorted in the trans-Golgi network (183). The distribution of hormones and constitutively secreted marker proteins has been traced by immuno-electronmicroscopy. Regulated and constitutive proteins comingle in the lumen of the Golgi apparatus, but hormones are packaged selectively into clathrin-coated vesicles from which constitutive proteins are excluded (203, 204). The best evidence that sorting to the regulated secretory pathway is signal mediated is derived from the observation that the construction of human growth hormone–VSV-G fusion protein results in the redirection of VSV-G from the constitutive to the regulated pathway (205). Selective aggregation of hormones and other proteins found in secretory granules also has been proposed as a mechanism for sorting regulated secretory proteins. Several secretory granule marker proteins selectively aggregate in vitro in the presence of calcium and low pH (206). This selective condensation is proposed to sort proteins by excluding constitutive secretory proteins from the lumen of a budding secretory granule. The formation of immature secretory granules and a separate population of constitutive secretory vesicles from the trans-Golgi network has been reconstituted in cell homogenates, facilitating an analysis in vitro of secretory vesicle formation (33).

Signal-Mediated Sorting of Membrane Proteins

SORTING OF TRANSMEMBRANE RECEPTORS INTO COATED PITS   The selective inclusion of single-spanning transmembrane receptors into clathrin-coated endocytic vesicles formed at the plasma membrane is dependent on a signal in the receptor cytoplasmic domain. Deletion of the cytoplasmic domain of a number of transmembrane receptors [LDL receptor (207), mannose 6-phosphate receptor (191), polymeric IgG receptor (208), transferrin receptor (209), and Fc receptor (210)] abolishes their endocytic uptake. The mutant receptors are left stranded on the cell surface because of a failure in recruiting these receptors into clathrin-coated pits (211). A clue to the nature of the signal for inclusion in coated pits was provided by a naturally occurring mutation in the LDL receptor. A single point mutation in the cytoplasmic domain replaces tyrosine with cysteine and the receptor is not internalized (212). The function of tyrosine in the cytoplasmic domain was tested further by the introduction of tyrosine into several sites of the cytoplasmic tail of influenza hemagglutinin (HA), a protein normally excluded from coated pits. The addition of a single tyrosine in the appropriate position was sufficient for the internalization of the mutant HA (213). Recent mutagenesis studies of the regions surrounding tyrosine residues in receptor cytoplasmic domains have led several groups to propose internalization signals comprising 4–10 amino acids, varying in sequence, surrounding tyrosine (214–216). The absence of a consensus amino acid sequence for endocytosis of these receptors suggests that the sorting signal is likely to be a structural motif encompassing tyrosine.

The clathrin-adaptin complex is implicated as a likely part of the cellular machinery recognizing sorting signals, since the endocytosis defect of cytoplasmic tail mutant receptors is manifested at the point of receptor association with clathrin- and adaptin-coated pits. This idea is supported by the observed specificity in adaptin complex binding to mutant and wild-type receptors in vitro. Plasma membrane AP-2 binds to the mannose 6-phosphate receptor cytoplasmic tail in vitro, but fails to bind if two tyrosine residues of the receptor are changed to uncharged amino acids (172). In contrast, the Golgi AP-1 binds to both the unaltered and mutant forms of the mannose 6-phosphate receptor, suggesting that the Golgi adaptin complex recognizes a signal distinct from the AP-2 signal. AP-2 also binds influenza HA in which tyrosine has been inserted, but does not bind wild-type influenza HA (171). These studies suggest that, at least for endocytosis, sorting specificity can be achieved by the recognition of sorting signals by clathrin-adaptin complexes and the selective inclusion of recognized proteins into clathrin-coated vesicles.

The regulation of sorting signals is less well understood. Some receptors are endocytosed only when ligand is bound (38), suggesting that signals for inclusion into coated pits can be turned on and off. Once endocytosed, receptors follow a complex recycling pathway through at least one endosomal compartment to the trans-Golgi network and back to the plasma membrane. It is possible to imagine that different signals are switched on and off in each compartment, or that the recognition and sorting machinery vary in each compartment. For example, a partial deletion of the cytoplasmic tail of the mannose 6-phosphate receptor prevents sorting to the lysosome, while a more extensive deletion precludes endocytosis of the receptor from the cell surface (191, 217). This suggests the presence of at least two distinct sorting signals in the mannose 6-phosphate receptor cytoplasmic domain that may be recognized by different sorting factors, such as the adaptin complexes, in different organelles. A method recently has been developed to reconstitute one step in this recycling pathway in vitro (218). The transport of the mannose 6-phosphate receptor from a late endosomal compartment to the trans-Golgi network in permeabilized cells is measured by the Golgi-specific addition of sialic acid residues to the receptor. The reaction does not depend on clathrin, as antibodies that block a cell-free endocytic reaction have no effect on recycling to the Golgi (219). This assay will be useful in defining the biochemical requirements for sorting at this step, and in identifying the cellular sorting machinery.

MEMBRANE PROTEIN SORTING IN POLARIZED EPITHELIA   Polarized epithelial cells maintain distinct apical and basolateral membrane domains in part by sorting newly synthesized membrane proteins as they exit the TGN (220, 220a). Apical and basolateral membrane proteins are segregated into separate secretory vesicles that bud from the TGN and can be accumulated in vitro in semi-intact cells (34, 221). The study of membrane protein sorting has been complicated by variations in the degree to which different cell types use each delivery pathway. Despite these variations, the current model for sorting of apical and basolateral membrane proteins proposes that apical targeting is signal mediated, while delivery to the basolateral domain occurs by default (185). The model is supported by the observation that proteins unique to epithelial cells are found only in the apical domain (185). However, distinct apical targeting signals have not been detected. Some proteins restricted to the apical surface are sorted by a distinct mechanism by which they are anchored to apical surface glycolipids (222). Contrary to the predictions of the model, recent studies on the distribution of Fc receptor isoforms have revealed a potential basolateral sorting signal. Basolateral delivery of Fc receptors requires a cytoplasmic determinant that is also required for efficient endocytosis of the receptor (223). It may be that sorting signals operate in both delivery pathways.

A basolateral sorting signal has also been detected in the polymeric immunoglobulin (pIg) receptor responsible for the transcytosis of IgA and IgM from the basolateral surface to the apical surface. A distinct signal for the delivery of newly made pIg receptor to the basolateral domain resides in the 14 amino acids of the cytoplasmic domain proximal to the membrane (224). Because the pIg receptor mediates transcytosis, it is likely to contain additional signals to specify endocytic uptake from the basolateral membrane, sorting in the endocytic pathway, and delivery to the apical domain. As discussed above for the mannose 6-phosphate receptor, the regulation of these signals and the means by which they are recognized by the cell will be of great interest in the future.

SIGNAL-MEDIATED RETENTION/RETRIEVAL OF RESIDENT PROTEINS

The retention of resident proteins in the endoplasmic reticulum and Golgi cisternae in the face of the inexorable flow of membrane and proteins is a major challenge for the cell. This problem is solved in part by the presence of signals that specify retention of proteins in specific compartments.

A model in which retention of soluble ER resident proteins is achieved by signal-mediated retrieval of escaped proteins from an early Golgi compartment has been proposed (14). In mammalian cells, the C-terminal tetrapeptide, KDEL (Lys-Asp-Glu-Leu), serves as a signal that is required for resident retention in the ER and confers ER localization on proteins normally destined for the cell surface (225). However, no ER protein is sufficiently abundant to act as an anchor for KDEL-containing proteins, suggesting that retention may be due to the retrieval of escaped ER resident proteins from a later compartment. The observation that ER lumenal proteins receive Golgi-specific modifications supports the existence of a recycling pathway. The fusion of KDEL to the lysosomal protein cathepsin D results in its retention in the ER (226). Yet, this chimeric KDEL–cathepsin D is phosphorylated, indicating that it has been exposed to the phosphotransferase that resides in an early Golgi compartment, and perhaps subsequently returned to the ER.

Further evidence for a recycling pathway was obtained in yeast, where ER retention is specified by the sequence HDEL (69). The retention system bears the hallmarks of a receptor-mediated process. Only proteins containing HDEL and not KDEL sequences are recognized in yeast, indicating that the system is highly selective. Also, the system is saturable, since overexpression of HDEL-fusion proteins disrupts the retention of HDEL-bearing ER resident proteins. Upon fusion of the HDEL sequence to the normally secreted proteins invertase and pro-α-factor, these proteins appear to be retained in the ER, yet are modified with Golgi-specific oligosaccharide linkages (227). The modification of these fusion proteins does not occur in an ER to Golgi–blocked *sec18* mutant. These results indicate that retention of resident proteins in the ER is a dynamic process of lumenal ER resident protein escape followed by retrieval from a later compartment.

A genetic selection was devised to isolate mutants that secrete HDEL-invertase fusion protein as a means to identify the potential HDEL-receptor in yeast. This selection yielded two complementation groups of *erd* (ER retention defective) mutants (69). *ERD1* encodes a potential transmembrane protein, which may be required for normal Golgi function, or perhaps facilitates the operation of the salvage compartment. Deletion of *ERD1* is not lethal, but results in the secretion of ER resident proteins and in defective Golgi-specific modifications of proteins (228). *ERD2* encodes an integral membrane protein of 26 kDa that may function as a retention signal receptor.

ERD2 regulates both the capacity and the specificity of the recycling system. First, the level of expression of *ERD2* is correlated with the capacity of the retention system. The secretion of resident ER proteins in cells over-expressing HDEL-fusion proteins is suppressed in strains coordinately overproducing *ERD2* (229). The role of *ERD2* in regulating the specificity of retention was demonstrated by exchanging the *ERD2* gene in *Saccharomyces cerevisiae* with the *ERD2* gene from a related yeast, *Kluveromyces lactis*. Whereas the *S. cerevisiae* retention system only recognizes HDEL, *K. lactis* retains both HDEL- and DDEL-bearing protein in the ER. Replacement of the *S. cerevisiae ERD2* gene with *K. lactis ERD2* resulted in the retention of both HDEL and DDEL fusion proteins, thus changing the specificity of the retention system (230).

A mammalian *ERD2* homologue was identified by using a polymerase chain reaction strategy with probes designed from conserved regions of yeast *ERD2* genes (231). The putative human KDEL receptor isolated in this approach is 50% identical to each of the yeast *ERD2* genes and encodes a protein of about 25 kDa, which is localized to the Golgi apparatus of mammalian cells.

An alternate approach to identification of the mammalian recycling receptor exploited the anti-idiotype antibody technique (5, 232, 233). Antibodies were raised against antibodies recognizing KDEL sequences. These anti-idiotype antibodies should recognize the conformational determinants of a KDEL-binding region. This strategy identified a 72-kDa transmembrane protein that binds KDEL in vitro. This protein is localized to an intermediate compartment between the ER and Golgi. The fact that this protein is quite different from the *ERD2*-homologous recycling receptor raises the possibility that mammalian cells have more than one recycling receptor or that both proteins function together for optimum specificity.

An additional tool for study of a recycling pathway is provided by the observation that brefeldin A perturbs the secretory pathway. Treatment of mammalian cells with brefeldin A blocks the export of newly synthesized proteins from the ER (234). Brefeldin A also induces the rapid disappearance of the Golgi apparatus, and the redistribution of cis and medial Golgi markers to the ER (235). Thus, brefeldin A is thought to perturb the dynamic balance between biosynthetic export from the ER in one direction and recycling or retrograde transport from the Golgi back to the ER in the opposite direction. Inhibition of export from the ER results in the domination of the retrograde pathway and the resorption of the Golgi. A closer inspection revealed that in the presence of brefeldin A, retrograde transport of Golgi markers to the ER is microtubule dependent and occurs by the extension of tubular processes from the Golgi to the ER (37). It is not known whether these unique membrane tubules also mediate recycling in the absence of brefeldin A.

The effects of brefeldin A on membrane traffic can be explained by two models (4). Both models are based on the observation that brefeldin A–induced dissociation of β-COP from Golgi membranes is correlated with an absence of new coated buds and vesicles and the appearance of membrane tubules connecting Golgi membranes, as discussed earlier in the section on COP-coated vesicles. One possibility is that β-COP functions in a vesicle formation step necessary for biosynthetic protein export. In the presence of brefeldin A, vesicle formation is perturbed and the recycling pathway, mediated by membrane tubules, predominates. Alternatively, β-COP is required for the maintenance of Golgi structure and brefeldin A results in the dissolution of the Golgi. Further study of the recycling pathway, specifically the role of β-COP, will be necessary to elucidate how retrograde transport occurs and how the net retention of ER proteins is achieved.

Transmembrane proteins may be retained in the ER and Golgi compartments by a different mechanism. Membrane protein retention appears signal mediated, but as yet there is no evidence for recycling of membrane proteins. Several possible retention signals for membrane proteins have been identified. The yeast membrane protein Sec20p contains an HDEL sequence (D. Sweet, H. R. B. Pelham, personal communication). A different sequence in the cytoplasmic domain of a viral glycoprotein ER resident is necessary for retention in the ER and also confers ER residency when transferred to proteins normally found in the plasma membrane (237). This signal also is found in the cytoplasmic domain of an endogenous ER protein. Deletion experiments have identified a transmembrane domain required for Golgi retention of a Golgi-localized viral glycoprotein (238). The mechanism by which these signals operate is unknown. It is possible that retention signals for membrane proteins interact with a scaffold material to anchor resident proteins in place or to sequester them away from regions of active vesicle budding. Alternatively, membrane proteins may be recycled in a manner similar to that of soluble proteins.

CONCLUDING REMARKS

Our current knowledge of the mechanism of vesicular transport has been built on a synergism of genetic and biochemical techniques and the utilization of different organisms to exploit the particular advantages of each. The biochemical complexity that is emerging as proteins are purified using reconstituted reactions is only partially reflected by the genetic complexity defined by available yeast mutants. Compiling the cast of players will be facilitated both by designing new screens and selections for the isolation of new mutants and by developing fractionation schemes to purify proteins based on their activity in vitro. Refinement of existing transport assays to distinguish the requirements of specific subreactions from the set of reactions that compose overall transport will allow the assignment of precise functions to purified proteins

Progress towards a mechanistic description of vesicle-mediated protein sorting requires an understanding of the cascade of protein-protein interactions that occurs each time a vesicle is assembled and transported. Clathrin-coated vesicles may continue to serve as the paradigm for vesicle biogenesis. Reconstituted assays that measure clathrin-coated pit and vesicle formation will be useful in elucidating functional domains of coat proteins and the role of coats in initiating or stabilizing vesicle formation. The discovery of related proteins in the nonclathrin coat opens the door to functional comparisons of COP proteins with their clathrin-adaptin homologues. Similarly, in vitro reactions that focus on vesicle consumption will aid in describing how fusion proteins function to bring about the final melding of vesicle and acceptor membrane bilayers.

Protein sorting in vesicle-mediated transport is an intricate process requiring at least two levels of regulation. Signal-mediated pathways that select proteins destined for transport or act to retain or recycle resident proteins may serve as a primary level of sorting. Additional specificity is conferred by regulating vesicle targeting to the appropriate acceptor membrane. The multiplicity of GTP-binding proteins implicated in this process provides a glimpse of the precision with which targeting may be regulated. A knowledge of the proteins that interact with sorting signals and with GTP-binding proteins will be required to understand the molecular details of how these proteins function to regulate sorting.

As illustrated by the examples presented in this review, efforts toward reconstituting transport in a fully resolved system coupled with mutant analysis to establish the physiological relevance of individual protein function will lead to a molecular definition of vesicle-mediated protein sorting.

ACKNOWLEDGMENTS

We thank Charles Barlowe, Michael Rexach, and Alex Franzusoff for their careful reviews of this manuscript.

Literature Cited

1. Pfeffer, S. R., Rothman, J. E. 1987. *Annu. Rev. Biochem.* 56:829–52
2. Farquhar, M. G. 1985. *Annu. Rev. Cell Biol.* 1:447–88
3. Dunphy, W. G., Rothman, J. E. 1985. *Cell* 42:13–21
4. Duden, R., Allan, V., Kreis, T. 1991. *Trends Cell Biol.* 1:14–19
5. Warren, G. 1990. *Cell* 62:1–2
6. Saraste, J., Kuismanen, E. 1984. *Cell* 38:535–49
7. Griffiths, G., Simons, K. 1986. *Science* 234:438–43
8. Hubbard, A. L. 1990. *Curr. Opin. Cell Biol.* 1:675–83.
9. Gruenberg, J., Howell, K. E. 1989. *Annu. Rev. Cell Biol.* 5:453–81
10. Rodman, J. S., Mercer, R. W., Stahl, P. D. 1990. *Curr. Opin. Cell Biol.* 2:664–72
11. Rothman, J. E. 1987. *Cell* 50:521–22
12. Kornfeld, S., Mellman, I. 1989. *Annu. Rev. Cell Biol.* 5:483–525
13. Burgess, T. L., Kelly, R. B. 1987. *Annu. Rev. Cell Biol.* 3:243–93
14. Pelham, H. R. B. 1989. *Annu. Rev. Cell Biol.* 5:1–23
15. Wieland, F. T., Gleason, M. L., Serafini, T. A., Rothman, J. E. 1987. *Cell* 50:289–300
16. Helms, J. B., Karrenbauer, A., Wirtz, K. W. A., Rothman, J. E., Wieland, F. T. 1990. *J. Biol. Chem.* 265:20027–32

17. Lodish, H. F. 1988. *J. Biol. Chem.* 263:2107–10
18. Rose, J. K., Doms, R. W. 1988. *Annu. Rev. Cell Biol.* 4:257–88
19. Palade, G. 1975. *Science* 189:347–58
20. Jamieson, J., Palade, G. 1967. *J. Cell Biol.* 34:577–96
21. Lodish, H. F., Kong, N., Hirani, S., Rasmussen, J. 1987. *J. Cell Biol.* 104:221–30
22. Paulik, M., Nowack, D. D., Morré, D. J. 1988. *J. Biol. Chem.* 263:17738–48
23. Holcomb, C. L., Hansen, W. J., Etcheverry, T., Schekman, R. 1988. *J. Cell Biol.* 106:641–48
24. Walworth, N. C., Novick, P. J. 1987. *J. Cell Biol.* 105:163–74
25. Woodman, P. G., Warren, G. 1991. *J. Cell Biol.* 112:1133–41
26. Sztul, E., Kaplin, A., Saucan, L., Palade, G. 1991. *Cell* 64:81–89
27. Cameron, R. S., Cameron, P. L., Castle, J. D. 1986. *J. Cell Biol.* 103:1299–313
28. Malhotra, V., Serafini, T., Orci, L., Shepherd, J. C., Rothman, J. E. 1989. *Cell* 58:329–36
29. Groesch, M. E., Ruohola, H., Bacon, R., Rossi, G., Ferro-Novick, S. 1990. *J. Cell Biol.* 111:45–53
30. Rexach, M. R., Schekman, R. W. 1991. *J. Cell Biol.* 114:219–29
31. Salamero, J., Sztul, E. S., Howell, K. E. 1990. *Proc. Natl. Acad. Sci. USA* 87:7717–21
32. deCurtis, I., Simons, K. 1989. *Cell* 58:719–29
33. Tooze, S. A., Huttner, W. B. 1990. *Cell* 60:837–47
34. Wandinger-Ness, A., Bennett, M. K., Antony, C., Simons, K. 1990. *J. Cell Biol.* 111:987–1000
35. Gruenberg, J., Griffiths, G., Howell, K. E. 1989. *J. Cell Biol.* 108:1301–16
36. Stoorvogel, W., Strous, G. J., Geuze, H. J., Oorschot, V., Schwartz, A. L. 1991. *Cell* 65:417–27
37. Lippincott-Schwartz, J., Donaldson, J. G., Schweizer, A., Berger, E. G., Hauri, H.-P., et al. 1990. *Cell* 60:821–36
38. Brodsky, F. M. 1988. *Science* 242:1396–402
39. Morris, S. A., Ahle, S., Ungewickell, E. 1989. *Curr. Opin. Cell Biol.* 1:684–90
40. Simons, K., Fuller, S. 1987. *Biological Organization: Macromolecular Interactions at High Resolution,* ed. R. M. Burnett, H. J. Vogel, pp. 139–50. Orlando: Academic
41. Roman, L. M., Garoff, H. 1985. *Trends Biochem. Sci.* 10:428–32
42. Sheetz, M. P., Singer, S. J. 1974. *Proc. Natl. Acad. Sci. USA* 71:4457–61
43. Kosaka, T., Ikeda, K. 1983. *J. Cell Biol.* 97:499–507.
44. van der Bliek, A. M., Meyerowitz, E. M. 1991. *Nature* 351:411–14
45. Rothman, J. H., Raymond, C. K., Gilbert, T., O'Hara, P. J., Stevens, T. H. 1990. *Cell* 61:1063–74
46. Kelly, R. B. 1990. *Cell* 61:5–7
47. Vale, R. D. 1987. *Annu. Rev. Cell Biol.* 3:347–78
48. Bourne, H. R., Sanders, D. A., McCormick, F. 1990. *Nature* 348:125–32
49. Balch, W. E. 1990. *Trends Biochem. Sci.* 15:473–77
50. Wilschut, J. 1989. *Curr. Opin. Cell Biol.* 1:639–47
51. Rand, R. P., Parsegian, V. A. 1986. *Annu. Rev. Physiol.* 48:201–12
52. Stegmann, T., Doms, R. W., Helenius, A. 1989. *Annu. Rev. Biophys. Biophys. Chem.* 18:187–211
53. White, J. M. 1990. *Annu. Rev. Physiol.* 52:675–97
54. Almers, W. 1990. *Annu. Rev. Physiol.* 52:607–24
55. Fernandez, J. M., Neher, E., Gomperts, B. 1984. *Nature* 312:453–55
56. Spruce, A. E., Breckenridge, L. J., Lee, A. K., Almers, W. 1990. *Neuron* 4:643–54
57. Block, M. R., Glick, B. S., Wilcox, C. A., Wieland, F. T., Rothman, J. E. 1988. *Proc. Natl. Acad. Sci. USA* 85:7852–56
58. Beckers, C. J. M., Block, M. R., Glick, B. S., Rothman, J. E., Balch, W. E. 1989. *Nature* 339:397–98
59. Diaz, R., Mayorga, L. S., Weidman, P. J., Rothman, J. E., Stahl, P. D. 1989. *Nature* 339:398–400
60. Goda, Y., Pfeffer, S. 1991. *J. Cell Biol.* 112:823–31
61. Guthrie, C., Fink, G. R. 1991. *Methods Enzymol.* 194:1–933
62. Novick, P., Schekman, R. 1979. *Proc. Natl. Acad. Sci. USA* 76:1858–62
63. Novick, P., Field, C., Schekman, R. 1980. *Cell* 21:205–15
64. Newman, A. P., Ferro-Novick, S. 1987. *J. Cell Biol.* 105:1587–94
65. Rothman, J. H., Stevens, T. H. 1986. *Cell* 47:1041–51
66. Bankaitis, V. A., Johnson, L. M., Emr, S. D. 1986. *Proc. Natl. Acad. Sci. USA.* 83:9075–79
67. Robinson, J. S., Klionsky, D. J., Banta, L. M., Emr, S. D. 1988. *Mol. Cell. Biol.* 8:4936–48
68. Rothman, J. H., Howald, I., Stevens, T. H. 1989. *EMBO J.* 8:2065–75
69. Pelham, H. R. B., Hardwick, K. G.,

Lewis, M. J. 1988. *EMBO J.* 6:1757–62

70. Nakano, A., Muramatsu, M. 1989. *J. Cell Biol.* 109:2667–91
71. Newman, A. P., Shim, J., Ferro-Novick, S. 1990. *Mol. Cell Biol.* 10:3405–14
72. Dascher, C., Ossig, R., Gallwitz, D., Schmitt, H. D. 1991. *Mol. Cell Biol.* 11:872–85
73. Payne, G. S. 1990. *J. Membr. Biol.* 116:93–105
74. Payne, G. S., Schekman, R. 1985. *Science* 230:1009–14
75. Lemmon, S. K., Jones, E. W. 1987. *Science* 238:504–9
76. Silveira, L. A., Wong, D. H., Masiarz, F. R., Schekman, R. 1990. *J. Cell Biol.* 111:1437–49
77. Payne, G. S., Schekman, R. 1989. *Science* 245:1358–65
78. Sewell, J. L., Kahn, R. A. 1988. *Proc. Natl. Acad. Sci. USA* 85:4620–24
79. Stearns, T., Willingham, M. C., Botstein, D., Kahn, R. A. 1990. *Proc. Natl. Acad. Sci. USA* 87:1238–42
80. Novick, P., Ferro, S., Schekman, R. 1981. *Cell* 25:461–69
81. Kaiser, C. A., Schekman, R. 1990. *Cell* 61:723–33
82. Rossi, G., Jiang, Y., Newman, A. P., Ferro-Novick, S. 1991. *Nature* 351:158–61
83. Salminen, A., Novick, P. J. 1987. *Cell* 49:527–38
84. Nair, J., Müller, H., Peterson, M., Novick, P. 1990. *J. Cell Biol.* 110:1897–909
85. Oka, T., Nishikawa, S., Nakano, A. 1991. *J. Cell Biol.* 114:671–79
86. d'Enfert, C., Wuestehube, L. J., Lila, T., Schekman, R. 1991. *J. Cell Biol.* 114:663–70.
87. d'Enfert, C., Barlowe, C., Nishikawa, S. I., Nakano, A., Schekman, R. 1991. *Mol. Cell Biol.* 11:5727–34
88. Fries, E., Rothman, J. E. 1980. *Proc. Natl. Acad. Sci. USA* 77:3870–74
89. Dunphy, W. G., Pfeffer, S. R., Clary, D. O., Wattenberg, B. W., Glick, B. S., et al. 1986. *Proc. Natl. Acad. Sci. USA* 83:1622–26
90. Pâquet, M. R., Pfeffer, S. R., Burczaf, J. D., Glick, B. S., Rothman, J. E. 1986. *J. Biol. Chem.* 261:4367–70
91. Wilson, D. W., Wilcox, C. A., Flynn, G. C., Chen, E., Kuang, W., et al. 1989. *Nature* 339:355–59
92. Rexach, M. R., Schekman, R. 1992. *Methods Enzymol.* In press
93. Segev, N. 1991. *Science* 252:1553–56
94. Franzusoff, A., Lauzé, E., Howell, K. 1992. *Nature* 355:173–75

95. Balch, W. E., Dunphy, W. G., Braell, W. A., Rothman, J. E. 1984. *Cell* 39:405–16
96. Beckers, C. J. M., Balch, W. E. 1989. *J. Cell Biol.* 108:1245–56
97. Beckers, C. J. M., Plutner, H., Davidson, H. W., Balch, W. E. 1990. *J. Biol. Chem.* 265:18298–310
98. Baker, D., Wuestehube, L., Schekman, R., Botstein, D., Segev, N. 1990. *Proc. Natl. Acad. Sci. USA* 87:355–59
99. Orci, L., Malhotra, V., Amherdt, M., Serafini, T., Rothman, J. E. 1989. *Cell* 56:357–68
100. Beckers, C. J. M., Keller, D. S., Balch, W. E. 1987. *Cell* 50:523–34
101. Glick, B. S., Rothman, J. E. 1987. *Nature* 326:309–12
102. Malhotra, V., Orci, L., Glick, B. S., Block, M. R., Rothman, J. E. 1988. *Cell* 54:221–27
103. Clary, D. O., Rothman, J. E. 1990. *J. Biol. Chem.* 265:10109–17
104. Hicke, L., Schekman, R. 1989. *EMBO J.* 8:1677–84
105. Baker, D., Hicke, L., Rexach, M., Schleyer, M., Schekman, R. 1988. *Cell* 54:335–44
106. Ruohola, H., Kabcenell, A. K., Ferro-Novick, S. 1988. *J. Cell Biol.* 107:1465–76
107. Hicke, L., Schekman, R. 1990. *BioEssays* 12:253–58
108. Hicke, L., Schekman, R. 1992. *Methods Enzymol.* In press
109. Hicke, L., Yoshihisa, T., Schekman, R. 1992. In preparation
110. Wuestehube, L. J., Schekman, R. 1992. *Methods Enzymol.* In press
111. Orci, L., Ravazzola, M., Meda, P., Holcomb, C., Moore, H., et al. 1991. *Proc. Natl. Acad. Sci. USA* 88:8611–15
112. Eakle, K. A., Bernstein, M., Emr, S. D. 1988. *Mol. Cell Biol.* 8:4098–109
113. Weidman, P. J., Melançon, P., Block, M. R., Rothman, J. E. 1989. *J. Cell Biol.* 108:1589–96
114. Clary, D. O., Griff, I. C., Rothman, J. E. 1990. *Cell* 61:709–21
115. Bankaitis, V. A., Malehorn, D. E., Emr, S. D., Greene, R. 1989. *J. Cell Biol.* 108:1271–81
116. Bankaitis, V. A., Aitken, J. R., Cleves, A. E., Dowhan, W. 1990. *Nature* 347:561–62
117. Cleves, A., McGee, T., Bankaitis, V. 1991. *Trends Cell Biol.* 1:30–34
118. Cleves, A. E., McGee, T. P., Whitters, E. A., Champion, K. M., Aitken, J. R., et al. 1991. *Cell* 64:789–800
119. Bourne, H. R., Sanders, D. A., McCormick, F. 1991. *Nature* 349:117–27
120. Hall, A. 1990. *Science* 249:635–40

121. Stow, J. L., Bruno de Almeida, J., Narula, N., Holtzman, E. J., Ercolani, L., et al. 1991. *J. Cell Biol.* 114:1113–24
122. Gallwitz, D., Donath, C., Sander, C. 1983. *Nature* 306:704–7
123. Segev, N., Mulholland, J., Botstein, D. 1988. *Cell* 52:915–24
124. Stearns, T., Kahn, R. A., Botstein, D., Hoyt, M. A. 1990. *Mol. Cell. Biol.* 10:6690–99
125. Becker, J., Tan, T.-J., Trepte, H.-H., Gallwitz, D. 1991. *EMBO J.* 10:785–92
126. Schmitt, D., Puzicha, M., Gallwitz, D. 1988. *Cell* 53:635–47
127. Goud, B., Salminen, A., Walworth, N. C., Novick, P. J. 1988. *Cell* 53:753–68
128. Touchot, N., Chardin, P., Tavitian, A. 1987. *Proc. Natl. Acad. Sci. USA* 84:8210–14
129. Zahraoui, A., Touchot, N., Chardin, P., Tavitian, A. 1989. *J. Biol. Chem.* 264:12394–401
130. Chavrier, P., Vingron, M., Sander, C., Simons, K., Zerial, M. 1990. *Mol. Cell Biol.* 10:6578–85
131. Chavrier, P., Parton, R. G., Hauri, H.-P., Simons, K., Zerial, M. 1990. *Cell* 62:317–29
132. Goud, B., Zahraoui, A., Tavitian, A., Saraste, J. 1990. *Nature* 345:553–56
133. van der Sluijs, P., Hull, M., Zahraoui, A., Tavitian, A., Goud, B., Mellman, I. 1991. *Proc. Natl. Acad. Sci. USA* 88:6313–17
134. Fischer von Mollard, G., Mignery, G. A., Baumert, M., Perin, M. S., Hanson, T. J., et al. 1990. *Proc. Natl. Acad. Sci. USA* 87:1988–92
135. Melançon, P., Glick, B. S., Malhotra, V., Weidman, P. J., Serafini, T., et al. 1987. *Cell* 51:1053–62
136. Gorvel, J.-P., Chavrier, P., Zerial, M., Gruenberg, J. 1991. *Cell* 64:915–25
137. Plutner, H., Schwaninger, R., Pind, S., Balch, W. E. 1990. *EMBO J.* 9:2375–83
138. Bacon, R. A., Salminen, A., Ruohola, H., Novick, P., Ferro-Novick, S. 1989. *J. Cell Biol.* 109:1015–22
139. Walworth, N. C., Goud, B., Kabcenell, A. K., Novick, P. J. 1989. *EMBO J.* 8:1685–93
140. Kabcenell, A. K., Goud, B., Northup, J. K., Novick, P. J. 1990. *J. Biol. Chem.* 265:9366–72
141. Sasaki, T., Kaibuchi, K., Kabcenell, A. K., Novick, P. J., Takai, Y. 1991. *Mol. Cell Biol.* 11:2909–12
142. Bailly, E., McCaffrey, M., Touchot, N., Zahraoui, A., Goud, B., et al. 1991. *Nature* 350:715–18
143. Lowy, D. R., Willumsen, B. M. 1989. *Nature* 341:384–85
144. Molenaar, C. M. T., Prange, R., Gallwitz, D. 1988. *EMBO J.* 7:971–76
145. Khosravi-Far, R., Lutz, J. R., Cox, A. D., Conroy, L., Bourne, J. R., et al. 1991. *Proc. Natl. Acad. Sci. USA* 88:6264–68
146. Kinsella, B. T., Maltese, W. A. 1991. *J. Biol. Chem.* 266:8540–44
147. Kahn, R. A., Goddard, C., Newkirk, M. 1988. *J. Biol. Chem.* 263:8282–87
148. Kohl, N. E., Diehl, R. E., Schober, M. D., Rands, E., Soderman, D. D., et al. 1991. *J. Biol Chem.* 266:18884–88
149. Seabra, M. C., Reiss, Y., Casey, P. J., Brown, M. S., Goldstein, J. L. 1991. *Cell* 65:429–34
150. Powers, S., Michaelis, S., Broek, D., Anna, S. S., Field, J., et al. 1986. *Cell* 47:413–22
151. Chen, W.-J., Andres, D. A., Goldstein, J. L., Russell, D. W., Brown, M. S. 1991. *Cell* 66:327–34
152. Salminen, A., Novick, P. J. 1989. *J. Cell Biol.* 109:1023–36
153. Rothman, J. E., Orci, L. 1990. *FASEB J.* 4:1460–68
154. Smythe, E., Pypaert, M., Lucocq, J., Warren, G. 1989. *J. Cell Biol.* 108:843–53
155. Schmid, S. L., Smythe, E. 1991. *J. Cell Biol.* 114:869–80
156. Moore, M. S., Mahaffey, D. T., Brodsky, F. M., Anderson, R. G. W. 1987. *Science* 236:558–63
157. Lin, H. C., Moore, M. S., Sanan, D. A., Anderson, R. G. W. 1991. *J. Cell Biol.* 114:881–91
158. Pearse, B. M. F., Crowther, R. A. 1987. *Annu. Rev. Biophys. Biophys. Chem.* 16:49–68
159. Brodsky, F. M., Hill, B. L., Acton, S. L., Nathke, I., Wong, D., et al. 1991. *Trends Biochem. Sci.* 16:208–13
160. Heuser, J. 1980. *J. Cell Biol.* 84:560–83
161. Rothman, J. E., Schmid, S. L. 1986. *Cell* 46:5–9
162. DeLuca-Flaherty, C., McKay, D. B., Parham, P., Hill, B. L. 1990. *Cell* 62:875–87
163. Pearse, B. M. F., Robinson, M. S. 1990. *Annu. Rev. Cell Biol.* 6:151–71
164. Keen, J. H. 1990. *Annu. Rev. Biochem.* 59:415–38
165. Ahle, S., Mann, A., Eichelsbacher, U., Ungewickell, E. 1988. *EMBO J.* 7:919–29
166. Schröder, S., Ungewickell, E. 1991. *J. Biol. Chem.* 266:7910–18
167. Robinson, M. S. 1989. *J. Cell Biol.* 108:833–42
168. Robinson, M. S. 1990. *J. Cell Biol.* 111:2319–26

169. Robinson, M. S. 1987. *J. Cell Biol.* 104:887–95
170. Vigers, G. P. A., Crowther, R. A., Pearse, B. M. F. 1986. *EMBO J.* 5:2079–85
171. Pearse, B. M. F. 1988. *EMBO J.* 7:3331–36
172. Glickman, J. N., Conibear, E., Pearse, B. M. F. 1989. *EMBO J.* 8:1041–47
173. Mahaffey, D. T., Peeler, J. S., Brodsky, F. M., Anderson, R. G. W. 1990. *J. Biol. Chem.* 265:16514–20
174. Kirchausen, T. 1990. *Mol. Cell Biol.* 10:6089–90
175. Kirchausen, T., Davis, A. C., Frucht, S., Greco, B. O., Payne, G. S., et al. 1991. *J. Biol. Chem.* 266:11153–57
176. Serafini, T., Stenbeck, G., Brecht, A., Lottspeich, F., Orci, L., et al. 1991. *Nature* 349:215–20
177. Waters, M. G., Serafini, T., Rothman, J. E. 1991. *Nature* 349:248–51
178. Duden, R., Griffiths, G., Frank, R., Argos, P., Kreis, T. E. 1991. *Cell* 64:649–65
179. Donaldson, J. G., Lippincott-Schwartz, J., Bloom, G. S., Kreis, T. E., Klausner, R. D. 1990. *J. Cell Biol.* 111:2295–306
180. Orci, L., Tagaya, M., Amherdt, M., Perrelet, A., Donaldson, J. G., et al. 1991. *Cell* 64:1183–95
181. Donaldson, J. G., Lippincott-Schwartz, J., Klausner, R. D. 1991. *J. Cell Biol.* 112:579–88
182. Rothman, J. H., Yamashiro, C. T., Kane, P. M., Stevens, T. H. 1989. *Trends Biochem. Sci.* 14:347–50
183. Miller, S. G., Moore, H.-P. H. 1990. *Curr. Opin. Cell Biol.* 2:642–47
184. Goldstein, J. L., Brown, M. S., Anderson, R. G. W., Russell, D. W., Schneider, W. J. 1985. *Annu. Rev. Cell Biol.* 1:1–39
185. Simons, K., Wandinger-Ness, A. 1990. *Cell* 62:207–10
186. Breitfeld, P. P., Casanova, J. E., Simister, N. E., Ross, S. A., McKinnon, W. C., et al. 1989. *Curr. Opin. Cell Biol.* 1:617–23
187. Hopkins, C. R. 1991. *Cell* 66:827–29
188. Pfeffer, S. R. 1988. *J. Membr. Biol.* 103:7–16
189. Dahms, N. M., Lobel, P., Kornfeld, S. 1989. *J. Biol. Chem.* 264:12115–18
190. Baranski, T. J., Faust, T. L., Kornfeld, S. 1990. *Cell* 63:281–91
191. Lobel, P., Fujimoto, K., Ye, R. D., Griffiths, G., Kornfeld, S. 1989. *Cell* 57:787–96
192. Klionsky, D. J., Herman, P. K., Emr, S. D. 1990. *Microbiol. Rev.* 54:266–92
193. Stevens, T. H., Rothman, J. H., Payne, G. S., Schekman, R. 1986. *J. Cell Biol.* 102:1551–57
194. Johnson, L. M., Bankaitis, V. A., Emr, S. D. 1987. *Cell* 48:875–85
195. Valls, L. A., Hunter, C. P., Rothman, J. H., Stevens, T. H. 1987. *Cell* 48:887–97
196. Valls, L. A., Winther, J. R., Stevens, T. H. 1990. *J. Cell Biol.* 111:361–68
197. Klionsky, D. J., Banta, L. M., Emr, S. D. 1988. *Mol. Cell Biol.* 8:2105–16
198. Klionsky, D. J., Emr, S. D. 1990. *J. Biol. Chem.* 265:5349–52
199. Robinson, J. S., Graham, T. R., Emr, S. D. 1991. *Mol. Cell Biol.* 12:5813–24
200. Herman, P. K., Stack, J. H., DeModena, J. A., Emr, S. D. 1991. *Cell* 64:425–37
201. Herman, P. K., Stack, J. H., Emr, S. 1991. *EMBO J.* 10:4049–60
202. Vida, T. A., Graham, T. R., Emr, S. D. 1990. *J. Cell Biol.* 111:2871–84
203. Orci, L., Ravazzola, M., Storch, M. J., Anderson, R. G. W., Vassalli, J., et al. 1987. *Cell* 49:865–68
204. Tooze, J., Tooze, S. A., Fuller, S. D. 1987. *J. Cell Biol.* 105:1215–26
205. Moore, H.-P. H., Kelly, R. B. 1986. *Nature* 321:443–46
206. Huttner, W. B., Gerdes, H. H., Rosa, P. 1991. *Trends Biochem. Sci.* 16:27–30
207. Lehrman, M. A., Goldstein, J. L., Brown, M. S., Russell, D. W., Schneider, W. J. 1985. *Cell* 41:735–43
208. Mostov, K. E., Kops, A. D., Deitcher, D. L. 1986. *Cell* 47:359–64
209. Rothenberger, S., Iacopetta, B. J., Kühn, L. C. 1987. *Cell* 49:423–31
210. Mieteinen, H. M., Rose, J. K., Mellman, I. 1989. *Cell* 58:317–27
211. Anderson, R. G. W., Goldstein, J. L., Brown, M. S. 1977. *Nature* 270:695–99
212. Davis, C. G., Lehrman, M. A., Russell, D. W., Anderson, R. G. W., Brown, M. S., et al. 1986. *Cell* 45:15–24
213. Lazarovits, J., Roth, M. 1988. *Cell* 53:743–52
214. Chen, W.-J., Goldstein, J. L., Brown, M. S. 1990. *J. Biol. Chem.* 265:3116–23
215. Collawn, J. F., Stangel, M., Kuhn, L. A., Esekogwu, V., Jing, S. Q., et al. 1990. *Cell* 63:1061–72
216. Ktistakis, N. T., Thomas, D., Roth, M. G. 1990. *J. Cell Biol.* 111:1393–407
217. Canfield, W. M., Johnson, K. F., Ye, R. D., Gregory, W., Kornfeld, S. 1991. *J. Biol. Chem.* 266:5682–88
218. Goda, Y., Pfeffer, S. R. 1988. *Cell* 55:309–20
219. Draper, R. K., Goda, Y., Brodsky, F. M., Pfeffer, S. R. 1990. *Science* 248:1539–41

220. Simons, K., Fuller, S. D. 1985. *Annu. Rev. Cell Biol.* 1:243–88
220a. Rodriguez-Boulan, E., Sabatini, D. D. 1978. *Proc. Natl. Acad. Sci. USA* 75:5070–75
221. Bennett, M. K., Wandinger-Ness, A., Simons, K. 1988. *EMBO J.* 7:4075–85
222. Lisanti, M. P., Rodriguez-Boulan, E. 1990. *Trends Biochem. Sci.* 15:113–18
223. Hunziker, W., Harter, C., Matter, K., Mellman, I. 1991. *Cell* 66:907–20
224. Casanova, J. E., Apodaca, G., Mostov, K. E. 1991. *Cell* 66:65–75
225. Munro, S., Pelham, H. R. B. 1987. *Cell* 48:899–907
226. Pelham, H. R. B. 1988. *EMBO J.* 7:913–18
227. Dean, N., Pelham, H. R. B. 1990. *J. Cell Biol.* 111:369–77
228. Hardwick, K. G., Lewis, M. J., Semenza, J., Dean, N., Pelham, H. R. B. 1990. *EMBO J.* 9:623–30
229. Semenza, J. C., Hardwick, K. G., Dean, N., Pelham, H. R. B. 1990. *Cell* 61:1349–57
230. Lewis, M. J., Sweet, D. J., Pelham, H. R. B. 1990. *Cell* 61:1359–63
231. Lewis, M. J., Pelham, H. R. B. 1990. *Nature* 348:162–63
232. Vaux, D., Tooze, J., Fuller, S. 1990. *Nature* 345:495–502
233. Kelly, R. B. 1990. *Nature* 345:480–81
234. Misumi, Y., Misumi, Y., Miki, K., Takatsuki, G., Tamura, G., et al. 1986. *J. Biol. Chem.* 261:11398–403
235. Lippincott-Schwartz, J., Yuan, L. C., Bonifacio, J. S., Klausner, R. D. 1989. *Cell* 56:801–13
236. Deleted in proof
237. Nilsson, T., Jackson, M., Peterson, P. A. 1989. *Cell* 58:707–18
238. Machamer, C. E., Rose, J. K. 1987. *J. Cell Biol.* 105:1205–14
239. Stevens, T., Esmon, B., Schekman, R. 1982. *Cell* 30:439–48
240. Nakano, A., Brada, D., Schekman, R. 1988. *J. Cell Biol.* 107:851–63
241. Nishikawa, S., Nakano, A. 1991. *Biochem. Biophys. Acta* 1093:135–43
242. Pryer, N. K., Salama, N. R., Kaiser, C. A., Schekman, R. 1990. *J. Cell Biol.* 111:325a
243. Pryer, N. K., Salama, N. R., Schekman, R., Kaiser, C. A. 1992. In preparation
244. Schmitt, H. D., Wagner, P., Pfaff, E., Gallwitz, D. 1986. *Cell* 47:401–12
245. Segev, N., Botstein, D. 1987. *Mol. Cell Biol.* 7:2367–77
246. Hosobuchi, M., Bernstein, M., Schekman, R. 1990. *J. Cell Biol.* 111:207a
247. Shim, J., Newman, A., Ferro-Novick, S. 1991. *J. Cell Biol.* 113:55–64
248. Ossig, R., Dascher, C., Trepte, H.-H., Schmitt, H. D., Gallwitz, D. 1991. *Mol. Cell Biol.* 11:2980–93
249. Achstetter, T., Franzusoff, A., Field, C., Schekman, R. 1988. *J. Biol. Chem.* 263:11711–17
250. Franzusoff, A., Schekman, R. 1989. *EMBO J.* 8:2695–702
251. Franzusoff, A., Redding, K., Crosby, J., Fuller, R. S., Schekman, R. 1991. *J. Cell Biol.* 112:27–37
252. Mueller, S. C., Branton, D. 1984. *J. Cell Biol.* 98:341–46
253. Obar, R. A., Collins, C. A., Hammarback, J. A., Shpetner, H. S., Vallee, R. B. 1990. *Nature* 347:256–60
254. Yeh, E., Driscoll, R., Coltrera, M., Olins, A., Bloom, K. 1991. *Nature* 349:713–15
255. Raymond, C. K., O'Hara, P. J. O., Eichinger, G., Rothman, J. H., Stevens, T. H. 1990. *J. Cell Biol.* 111:877–92
256. Deleted in proof
257. Deleted in proof
258. Deleted in proof
259. Deleted in proof
260. Banta, L. M., Robinson, J. S., Klionsky, D. J., Emr, S. D. 1988. *J. Cell Biol.* 107:1369–83
261. Banta, L. M., Vida, T. A., Herman, P. K., Emr, S. D. 1990. *Mol. Cell Biol.* 10:4638–49
262. Deleted in proof
263. Deleted in proof
264. Nakajima, H., Hirata, A., Ogawa, Y., Yonehara, T., Yoda, K., et al. 1991. *J. Cell Biol.* 113:245–60
265. Balch, W. E., Wagner, K. W., Keller, D. S. 1987. *J. Cell Biol.* 104:749–60
266. Schwaninger, R., Beckers, C. J. M., Balch, W. E. 1991. *J. Biol. Chem.* 266:13055–63
267. Nowack, D. D., Morré, D. M., Paulik, M., Keenan, T. W., Morré, D. J. 1987. *Proc. Natl. Acad. Sci. USA* 84:6098–102
268. Simons, K., Virta, H. 1987. *EMBO J.* 6:2241–47
269. Baker, D., Schekman, R. 1989. *Methods Cell Biol.* 31:127–41
270. Balch, W. E., Glick, B. S., Rothman, J. E. 1984. *Cell* 39:525–36
271. Braell, W. A., Balch, W. E., Dobbertin, D. C., Rothman, J. E. 1984. *Cell* 39:511–24
272. Balch, W. E., Rothman, J. E. 1985. *Arch. Biochem. Biophys.* 240:413–25
273. Wattenberg, B. W., Rothman, J. E. 1986. *J. Biol. Chem.* 261:2208–13
274. Wattenberg, B. W., Balch, W. E.,

Rothman, J. E. 1986. *J. Biol. Chem.* 261:2202–7

275. Wattenberg, B. W., Hiebsch, R. R., LeCurex, L. W., White, M. P. 1990. *J. Cell Biol.* 110:947–54

276. Rothman, J. E. 1987. *J. Biol. Chem.* 262:12502–10

277. Bennett, M. K., Wandinger-Ness, A., deCurtis, I., Antony, C., Simons, K., et al. 1989. *Methods Cell Biol.* 31:103–26

278. Tooze, S. A., Weiss, U., Huttner, W. B. 1990. *Nature* 347:207–8

279. Miller, S. G., Moore, H.-P. H. 1991. *J. Cell Biol.* 112:39–54

280. Woodman, P. G., Edwardson, J. M. 1986. *J. Cell Biol.* 103:1829–35

281. Mahaffey, D. T., Moore, M. S., Brodsky, F. M., Anderson, R. G. W. 1989. *J. Cell Biol.* 108:1615–24

282. Mayorga, L. S., Diaz, R., Stahl, P. D. 1988. *J. Biol. Chem.* 263:17213–16

283. Mayorga, L. S., Diaz, R., Stahl, P. D. 1989. *J. Biol. Chem.* 264:5392–99

284. Diaz, R., Mayorga, L., Stahl, P. 1988. *J. Biol. Chem.* 263:6093–100

285. Davey, J., Hurtley, S. M., Warren, G. 1985. *Cell* 43:643–52

286. Gruenberg, J., Howell, K. 1986. *EMBO J.* 5:3091–101

287. Gruenberg, J., Howell, K. E. 1987. *Proc. Natl. Acad. Sci. USA* 84:5758–62

288. Braell, W. A. 1987. *Proc. Natl. Acad. Sci. USA* 84:1137–41

289. Woodman, P. G., Warren, G. 1988. *Eur. J. Biochem.* 173:101–8

290. Bomsel, M., Parton, R., Kuznetsov, S. A., Schroer, T. A., Gruenberg, J. 1990. *Cell* 62:719–31

Annu. Rev. Biochem. 1992. 61:517–57

N-(CARBOXYALKYL)AMINO ACIDS:
Occurrence, Synthesis, and Functions[1]

John Thompson and Jacob A. Donkersloot

Laboratory of Microbial Ecology, National Institute of Dental Research, National Institutes of Health, Bethesda, Maryland 20892

KEY WORDS: opines, pyruvate:oxidoreductase, crown gall, NAD(P)-binding domains, amino acid dehydrogenase

CONTENTS

[1]The US Government has the right to retain a nonexclusive royalty-free license in and to any copyright covering this paper

PERSPECTIVES AND SUMMARY

The properties and function of the constituent amino acids of proteins have been presented in classic texts by Greenstein & Winitz (1), Meister (2), and in more recent publications by Bender (3) and Barrett (4). In their treatise of 1961, Greenstein & Winitz described 90 or so amino acids, but in 1985, Hunt (5) tabulated approximately 700 amino and imino acid derivatives. Many of these "new" compounds either are derivatives of the 20 "primary" amino acids, or originate by posttranslational modifications of these residues in proteins. In a comprehensive review, Wold (6) mentions 40 such derivatives of lysine alone!

An unusual modification occurs when the α- or ω-NH_2 group of an amino acid undergoes reductive condensation with the ketone carbonyl of an α-keto acid. The products of these reactions, collectively termed N-(carboxyalkyl)amino acids [N-(CA)amino acids], have been isolated from phylogenetically diverse sources including eukaryotic cells, plant tumors, and muscle tissue of marine invertebrates (Table 1). N-(CA)amino acids have also been identified during studies of certain enzyme-catalyzed reactions. In the pharmaceutical industry, dipeptides containing N^α-(CA)-moieties have been synthesized for therapeutic use.

N-(CA)amino acids have not previously been discussed in *Annual Review of Biochemistry*. The invitation from the Editorial Committee to summarize biochemical aspects of these compounds stems in part from our discovery of N^ω-(carboxyethyl)amino acids in bacteria [*Lactococcus lactis* (7–9; Table 1)]. More importantly, the request provides the opportunity to overview the N-(CA)amino acids as a group, and to highlight the different roles these compounds play in the physiologies of very diverse organisms. Additionally, we reflect on how knowledge gained from the study of these compounds has contributed to developments in drug design, and to the genetic engineering of plants. Finally, we speculate on future developments in this exciting area of amino acid biochemistry.

NOMENCLATURE AND PROPERTIES OF N-(CARBOXYALKYL)AMINO ACIDS

The general structures of the N^α- and N^ω-(CA)amino acids are shown in Figure 1. These compounds may be regarded as two amino acids linked via a common imino nitrogen or as secondary amines. From a biosynthetic viewpoint, the N-(CA)amino acids [which are frequently called *opines* (10)] are the products of the NAD(P)H-dependent reductive condensation between an α-keto acid and the α- or ω-NH_2 group of an amino acid (11). In this review, the terms *synthase* and *dehydrogenase* are used interchangeably with reference to the NAD(P)$^+$-dependent oxidoreductases responsible for the biosynthesis of N-(CA)amino acids (Table 1). These amino acids usually have two asymmetric centers, and in nature may exhibit (L,L) or (D,L) stereochemistry. However, when synthesized chemically, they are often a mixture of these diastereomeric forms. The nomenclature for the N^α- or N^ω-compounds is provided by the examples shown in Figure 1. Thus for A and B ($n=4$), when $R=R_1=CH_3$ the N-(CA)amino acids are N^2-[1-(DL)-carboxyethyl]-L-alanine and N^6-[1-(DL)-carboxyethyl]-L-lysine, respectively. Many N-(CA)amino acids possess both primary and secondary amine groups and the molecules contain two or more carboxyl groups. Although amphoteric, the N-(CA)amino acids are less basic than the parent amino acids, and derivatives of arginine, lysine, and ornithine exhibit neutral or even slightly acidic properties.

CHEMICAL AND ENZYMATIC SYNTHESES OF N-(CARBOXYALKYL)AMINO ACIDS

Two excellent reviews (10, 12) provide details of the physical properties, chemical syntheses, and methods for extraction, detection, and structure determination of many N-(CA)amino acids. The Abderhalden-Haase procedure (13), in which free or N-protected amino acids are reacted with the resolved (D) or (L) enantiomers of α-halo acids under alkaline conditions, has frequently been employed for the stereoselective synthesis of N-(CA)amino acids (14–16). Thus the reaction between L-arginine and α-(L)-bromopropionate yields N^2-(1-D-carboxyethyl)-L-arginine, the naturally occurring isomer of octopine (17, 18). Reductive condensation between an amino acid and the desired α-keto acid using $NaBH_4$ or $NaBH_3CN$ also provides an efficient, but nonstereospecific method of synthesis (18–20). An excellent method for the diastereoselective synthesis of N^α- and N^ω-(CA)amino acids has been reported (21), whereby the trifluoromethanesulfonates of enantiomerically pure lactates, β-phenyllactates and dimethyl α-hydroxyglutarate, are reacted with suitably protected esters of (D)- or (L)-α-

Table 1 Naturally occurring N-(carboxyalkyl)amino acids: distribution and biosynthesis

N-(carboxyalkyl)amino acid	Trivial name	Enzyme	Cofactor	M_r^a	References[b]
Plants (crown gall)					
N²-(1-D-Carboxyethyl)-L-arginine	octopine	D-octopine synthase	NAD(P)H	38,000	109
N²-(1-D-Carboxyethyl)-L-ornithine	octopinic acid	(EC 1.5.1.11)		38,801[c]	110
N²-(1-D-Carboxyethyl)-L-lysine	lysopine			38,733[c]	111
N²-(1-D-Carboxyethyl)-L-histidine	histopine				
N²-(1-D-Carboxyethyl)-L-methionine	methiopine				
N²-(1,3-D-Dicarboxypropyl)-L-arginine	nopaline	D-nopaline synthase	NAD(P)H	158,000	114
N²-(1,3-D-Dicarboxypropyl)-L-ornithine	nopalinic acid (ornaline)	(EC 1.5.1.19)		(40,000)	114
				(45,394)[c]	115, 116
N²-(1,3-D-Dicarboxypropyl)-L-leucine	leucinopine	Enzymes not yet characterized			
N²-(1,3-D-Dicarboxypropyl)-L-asparagine	asparaginopine (succinamopine)				
N²-(1,3-D-Dicarboxypropyl)-L-glutamine	glutaminopine				
	cucumopine				
	mikimopine				
N²-(Carboxymethyl)-L-arginine	acetopine (noroctopine)				

Bacteria

Compound	Trivial name	Enzyme	Cofactor	Molecular weight	Reference
N^5-(Carboxymethyl)-L-ornithine	—	N^5-(Carboxyethyl)-ornithine synthase (EC 1.5.1.24)	NADPH	150,000	38
N^6-(Carboxymethyl)-L-lysine	—			(78,000)	38
N^5-(1-L-Carboxyethyl)-L-ornithine	—			(38,000)	38
N^6-(1-L-Carboxyethyl)-L-lysine	—			(35,323)[c]	[d]
N^2-(1-D-Carboxyethyl)-L-phenylalanine	—	N^2-(Carboxyethyl)-phenylalanine synthase	NADH	70,000 (36,000)	69

Marine invertebrates

Compound	Trivial name	Enzyme	Cofactor	Molecular weight	Reference
N^2-(1-D-Carboxyethyl)-L-arginine	octopine	D-octopine dehydrogenase (EC 1.5.1.11)	NADH	37,000–40,000	150, 158
N-(1-D-Carboxyethyl)-L-alanine	alanopine	alanopine dehydrogenase (EC 1.5.1.17)	NADH	39,000–46,000	174, 176, 177
N-(1-Carboxyethyl)-β-alanine	β-alanopine	β-alanine dehydrogenase (EC 1.5.1.26)	NADH	N.D.[e]	188
N-(Carboxymethyl)-D-alanine	strombine	strombine dehydrogenase (EC 1.5.1.22)	NADH	38,000	176
N-(1-D-Carboxyethyl)-taurine	taurine	tauropine dehydrogenase	NADH	38,000–42,000	184, 187

Eukaryotes

Compound	Trivial name	Enzyme	Cofactor	Molecular weight	Reference
N^6-(1,3-L-Dicarboxypropyl)-L-lysine	saccharopine	saccharopine dehydrogenase (EC 1.5.1.7)	NADH	39,000 40,556[c]	198 201

[a] Molecular weights given in parentheses refer to monomeric or dimeric forms of the enzyme.
[b] References provide details of enzyme purification and characterization.
[c] Molecular weight calculated from the DNA sequence.
[d] J. A. Donkersloot and J. Thompson, unpublished data.
[e] Not determined.

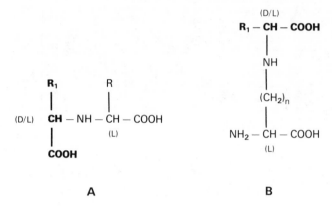

Figure 1   General structures for *A*, N^{α}- and *B*, N^{ω}-(carboxyalkyl)amino acids. Bold type indicates the carboxyalkyl moiety.

amino acids. Of several methods described by Miyazawa (22), the condensation between an amino acid ester and halo acid ester is particularly attractive, because either the diester products are available for peptide synthesis, or after acid hydrolysis, they provide the free N-(CA)amino acids.

In the past decade several opine synthases have been purified from crown-gall tumors, marine invertebrate tissues, yeast, and, more recently, from bacteria (Table 1). These enzymes catalyze the NAD(P)H-dependent reductive condensation between certain α-keto acids and either the N^{α}- or N^{ω}-NH$_2$ group of a variety of L-amino acids. These biosynthetic reactions are rapid, extremely efficient, and importantly, the products are regiochemically and stereochemically defined (Table 1).

N-(CARBOXYALKYL)AMINO ACIDS IN BACTERIA

Agrobacterium tumefaciens is the causative agent of crown gall in plants (23–26), and a variety of N-(CA)amino acids (opines) have been isolated from these tumors (10, 11, 27). The association between *A. tumefaciens* and tumor development in the plant host has been recognized for more than 80 years (28), but it is only in the past two decades that the molecular basis for the etiology of crown gall disease and opine formation has been elucidated (29–33). It is a remarkable fact that although *A. tumefaciens* induces opine synthesis in the plant host, neither the opine synthases nor their products are detectable in the bacterium. The relationship between *A. tumefaciens* and opine formation by crown gall tumors is discussed in the section on N-(CARBOXYALKYL)AMINO ACIDS IN PLANTS (CROWN GALL).

N-(Carboxyalkyl)Amino Acids in Lactococcus lactis

In 1986, during investigations concerning the regulation of sugar transport and metabolism in lactic acid bacteria (34–36), high levels of an unusual ninhydrin-positive compound were detected in the amino acid pool of *Lactococcus lactis* subsp. *lactis* (formerly *Streptococcus lactis*). Purification of this compound (which coeluted with valine from the amino acid analyzer) was achieved by conventional ion-exchange and thin-layer chromatography (7). The structure of this novel compound was determined by ^{1}H and ^{13}C-NMR spectroscopy, and confirmation was provided by chemical synthesis and GC-mass spectrometry (8). The configuration of N^5-(1-L-carboxyethyl)-L-ornithine was established by stereoselective synthesis (8) via alkylation of N^2-Boc-L-ornithine with α-D-bromopropionic acid (Figure 2A). The next higher homolog, N^6-(1-L-carboxyethyl)- L-lysine, was subsequently identified and purified from *L. lactis* (9; Figure 2B). The (L,L) stereochemistry of N^6-(CE)lysine from *L. lactis* was assigned by comparing the ^{13}C-NMR spectrum with those of the chemically prepared (D,L) and (L,L) diastereomers [kindly provided by Dr. Fujioka (37)]. To our knowledge, neither of the N^ω-(CA)amino acids from *L. lactis* has previously been found in either prokaryotic or eukaryotic cells (10).

Biosynthesis of N^5-(Carboxyethyl)ornithine and N^6-(Carboxyethyl)lysine

Two potential routes exist for the biosynthesis of N^5-(CE)ornithine (Figure 3). Since radioactivity from exogenous ^{14}C-labeled arginine, ornithine, and glutamic acid (but not alanine) was incorporated into N^5-(CE)ornithine, route 4 appeared to be operational. Support for this biosynthetic route was provided by the finding that [^{14}C]pyruvic acid was the precursor of the carboxyethyl moiety of [^{14}C]N^5-(CE)ornithine (7).

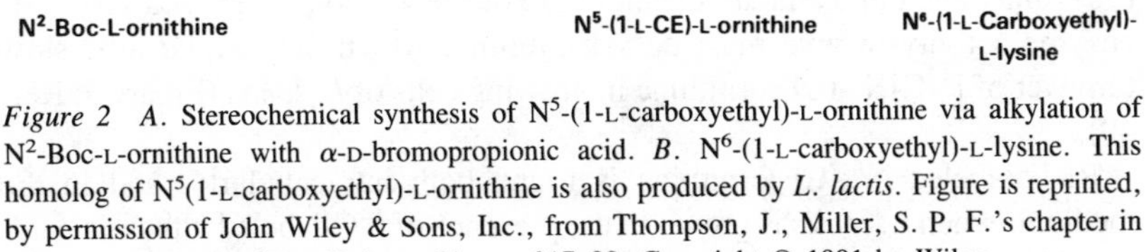

A

N²-Boc-L-ornithine   N⁵-(1-L-CE)-L-ornithine   **B**   Nᵉ-(1-L-Carboxyethyl)-L-lysine

Figure 2 A. Stereochemical synthesis of N^5-(1-L-carboxyethyl)-L-ornithine via alkylation of N^2-Boc-L-ornithine with α-D-bromopropionic acid. *B.* N^6-(1-L-carboxyethyl)-L-lysine. This homolog of N^5-(1-L-carboxyethyl)-L-ornithine is also produced by *L. lactis.* Figure is reprinted, by permission of John Wiley & Sons, Inc., from Thompson, J., Miller, S. P. F.'s chapter in *Advances in Enzymology,* Volume 64, pp. 317–99. Copyright © 1991 by Wiley.

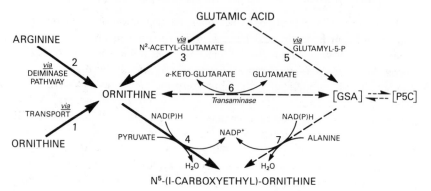

Figure 3 Potential routes for the formation of N^5-(1-carboxyethyl)ornithine by *L. lactis*. Figure is reprinted, by permission of John Wiley & Sons, Inc., from Thompson, J., Miller, S. P. F.'s chapter in *Advances in Enzymology*, Volume 64, pp. 317–99. Copyright © 1991 by Wiley.

N^5-(1-L-Carboxyethyl)-L-ornithine:NADP$^+$ Oxidoreductase

The addition of pyruvic acid, NADPH, and either ornithine or lysine to a cell-free extract of *L. lactis* K1 caused the formation of N^5-(CE)ornithine (Figure 4) and N^6-(CE)lysine, respectively. The enzyme responsible for the synthesis of both N^ω-(CE)amino acids has been purified 8000-fold to homogeneity (38). This enzyme, N^5-(1-L-carboxyethyl)- L-ornithine:NADP$^+$ oxidoreductase [N^5-(CE)ornithine synthase, CEOS; EC 1.5.1.24] requires NADPH as cofactor (K_m ~6.6 μM), and exists as a mixture of dimers and tetramers of 36–38-kDa subunits (38). The gene (*ceo*) encoding this enzyme has been cloned (39) and sequenced (J. A. Donkersloot and J. Thompson, unpublished data).

N^5-(CE)ornithine synthase exhibits a narrow specificity with respect to amino acid and α-keto acid substrates. Only L-ornithine (K_m = 3.3 mM), L-lysine (K_m = 18.2 mM), and to a lesser degree the lysine analog, S-(2-aminoethyl)cysteine, serve as amino acid substrates. From a variety of carbonyl compounds tested, only pyruvate (K_m = 150 μM), 3-fluoropyruvate, and glyoxylate were substrates in the condensation reaction. N^5-(1-CE)ornithine is a potent product inhibitor, and when present at a concentration of 0.1 mM, the compound decreased activity of CEOS by more than 50% (38). Since the intracellular concentration of the amino acid exceeds 10 mM, enzyme activity in vivo must be self-limiting, which may explain the slow turnover of [^{14}C]N^5-(CE)ornithine in growing cells of *L. lactis* (Figure 5; Ref. 40).

Radiolabeled N^5-(CE)ornithine is found both intracellularly, and in the medium, when *L. lactis* is grown in the presence of ^{14}C-labeled glutamic acid, ornithine, or arginine. [*L. lactis* metabolizes arginine via the arginine de-

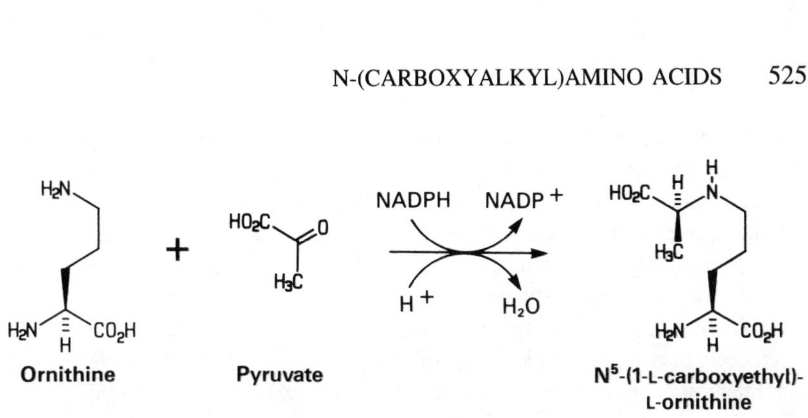

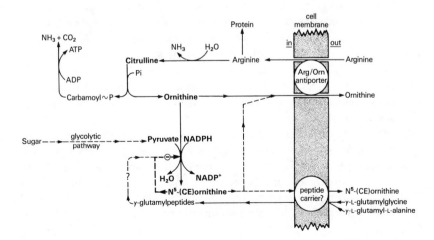

Figure 4 Reaction catalyzed by N^5-(carboxyethyl)ornithine synthase. Figure is reprinted, by permission of John Wiley & Sons, Inc., from Thompson, J., Miller, S. P. F.'s chapter in *Advances in Enzymology*, Volume 64, pp. 317–99. Copyright © 1991 by Wiley.

iminase pathway to ornithine, NH_3, and CO_2 (Figure 5; Ref. 11).] The mechanism of exit of N^5-(CE)ornithine is unknown. However, translocation of the N^ω-(CA)amino acid may occur via the arginine:ornithine antiporter (41, 42), or perhaps via a γ-glutamyl dipeptide carrier.

N^5-(Carboxyethyl)ornithine Synthase: Genetic Locus and Linkage

Experiments designed to locate the gene encoding CEOS (*ceo*) in *L. lactis* have contributed, quite unexpectedly, to the solution of a long-term riddle concerning the locus of the genetic determinants for the synthesis of the

Figure 5 Biosynthesis of N^5-(carboxyethyl)ornithine during operation of the arginine deiminase and glycolytic pathways in *L. lactis*. The scheme also illustrates potential routes for exit of the amino acid, and ⊖ indicates inhibition of N^5-(carboxyethyl)ornithine synthase.

antibiotic nisin (43–48) and for sucrose fermentation in these dairy organisms. It was known that the traits for nisin production and sucrose fermentation are lost simultaneously under certain conditions used for plasmid curing (49–52), and that these properties are jointly transferable by conjugation (50–53). From these data, a plasmid locus had frequently been proposed for these traits (49–52, 54), although with one exception (51), such a plasmid was usually not evident in transconjugants (50, 52, 53). During experiments to ascertain whether *ceo* was located on one of the five plasmids in *L. lactis* K1, a "spontaneous" nisin-sucrose negative mutant was obtained, which surprisingly also lacked CEOS (55). However, comparison of parent and mutant strains showed no difference in their plasmid profiles. Hybridization studies with oligonucleotide probes specific for (*a*) the nisin gene [*spaN* (46, 47)], (*b*) EII^{Scr} (*scrA*), (*c*) sucrose-6-phosphate hydrolase (*scrB*), (*d*) fructokinase (*scrK*), (*e*) *ceo,* and (*f*) IS*904* (56), showed that all these genes were absent in the mutant (39, 55, 57). Based on these results, we proposed that in *L. lactis* K1, *ceo* resides on a large (> 40 kbp) chromosomal transposon (Tn*5306*), which also encodes the genes required for nisin synthesis (and resistance) and sucrose metabolism (39, 55, 57). Another nisin-sucrose transposon (Tn*5301*) has been recently identified in *L. lactis* NCFB 894, and the size of this conjugative element is 70 kbp (57a).

Inhibition of N^5-(Carboxyethyl)ornithine Synthase

N^5-(CE)ornithine may be regarded as a composite of portions of L-ornithine and L-alanine linked via a secondary amine. Alternatively, one may view N^5-(CE)ornithine as an analog of--the dipeptide γ-L-glutamyl-L-alanine, in which the carbonyl moiety of the peptide linkage has been replaced by a methylene group. Interestingly, several dipeptides bearing structural resemblance to N^5-(CE)ornithine, e.g. γ-L-glutamyl-L-alanine and γ-L-glutamylglycine, were effective inhibitors of CEOS. In contrast, the two diastereoisomers N^5-(1-D-carboxyethyl)-L-ornithine and N^6-(1-D-carboxyethyl)-L-lysine, and other dipeptides including β-L-aspartylglycine, γ-D-glutamylglycine, γ-D-glutamyl-L-alanine, and γ-L-glutamyl-D-alanine were not inhibitory (J. Thompson, unpublished information). Clearly, both stereochemical configuration (L,L) and chain length between the two chiral centers are important determinants for inhibitory potency of these dipeptides. Conceivably, compounds similar to the N^5- and N^6-(carboxymethyl) and (carboxyethyl) derivatives of L-ornithine and L-lysine, may find application in the study of γ-glutamyl dipeptide transport systems, or as potential inhibitors of γ-glutamyl cyclotransferase(s) in the γ-glutamyl cycle in eukaryotic cells (58–61).

L-Canaline is the aminooxy analog of L-ornithine (62, 63) in which the δ-carbon atom of the ornithyl backbone has been replaced by an oxygen atom

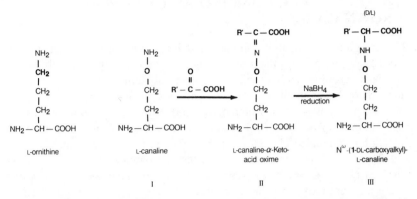

Figure 6   Potential route for synthesis of N^{ω}-(1-DL-carboxyalkyl) derivatives of L-canaline with appropriate keto acids. Syntheses of the glyoxylic-, pyruvic-, α-ketobutyric-, and α-ketoglutaric acid-oximes of L-canaline (R' = H, CH_3-, $CH_3 \cdot CH_2-$, and $CH_3 \cdot CH_2 \cdot CH_2-$, respectively) are described in Ref. 68.

(Figure 6, I). This compound is a potent inhibitor of several ornithine-requiring enzymes, particularly for those in which pyridoxal phosphate is a cofactor for catalysis (64, 65). Inactivation is attributed to the formation of a stable L-canaline:pyridoxal phosphate oxime between the ω-amino group of L-canaline and the aldehydic moiety of pyridoxal phosphate (63, 66, 67). L-Canaline was neither a substrate nor an inhibitor of CEOS. However, the L-canaline-α-keto acid oximes (Figure 6, II; Ref. 68) and their corresponding reduction products [N^{ω}-(1-carboxyalkyl)-L-canaline derivatives, Figure 6, III] are potential inhibitors of the enzyme.

N^2-(1-D-Carboxyethyl)-L-Phenylalanine Dehydrogenase in Arthrobacter

Asano et al (69) have recently described the purification of an opine synthase from *Arthrobacter* sp. strain 1C. This enzyme, N^2-(1-D-carboxyethyl)-L-phenylalanine:NAD^+ oxidoreductase, was induced by growth of the organism on N^2-[1-(DL)-carboxyethyl]-L-phenylalanine. The purified enzyme (M_r = 70,000; subunit $M_r \simeq 36,000$) requires NADH as reductant, and a variety of hydrophobic L-amino acids, including methionine, isoleucine, valine, and phenylalanine, could participate in the reaction. No activity was detectable with L-lysine, L-arginine, or L-ornithine. Pyruvate, oxaloacetate, and to a lesser extent glyoxylate, and α-ketobutyrate, could be used as α-keto sub-strates. All of the N^2-(CA)amino acids formed by the *Arthrobacter* enzyme have the (D,L) configuration.

N-(CARBOXYALKYL)AMINO ACIDS IN PLANTS (CROWN GALL)

The octopine and nopaline families of opines found in crown gall tumors (10–12, 27) are perhaps the most extensively studied of the naturally occurring N-(CA)amino acids (Table 1). Crown gall tumors are widespread on many dicotyledonous plants (23, 24), and these plant hyperplasias (25, 26) were described by Aristotle (referenced in 23). However, it was not until 1907 that Smith & Townsend (28) established the causal relationship between infection of wounded plants by agrobacteria and subsequent tumor etiology (29–31, 70, 71). This relationship is further discussed in the section *Biosynthesis of Crown Gall Opines*. An earlier review by Tempé (10) provides an interesting and informative account of the discovery, isolation, purification, chemical syntheses, detection (72), and properties of the crown gall opines.

Octopine and Nopaline Families

Members of the octopine family are formed by the reductive condensation between pyruvic acid and the α-NH_2 groups of arginine, lysine, ornithine, histidine, and methionine (see Table 1 and Figure 7). The corresponding N^2-(1-D-carboxyethyl) derivatives are octopine (14, 18, 73), lysopine (15, 74, 75), octopinic acid (16, 76, 77), histopine (78–80), and methiopine (81).

Opines of the nopaline family are derived from the reductive condensation between α-ketoglutarate and arginine or ornithine, to yield the corresponding N^2-(1,3-dicarboxypropyl) derivatives nopaline (19, 20, 82, 83) and nopalinic acid (84, 85), respectively (Figure 8). Derivatives of L-leucine, L-asparagine, and L-glutamine have also been described: N^2-(1,3-dicarboxypropyl)-L-leucine [leucinopine (86, 87)], N^2-(1,3-dicarboxypropyl)-L-asparagine [asparaginopine or succinamopine (88–91)], and N^2-(1,3-dicarboxypropyl)-L-glutamine [glutaminopine (88)]. Two diastereomers, cucumopine (92, 93) and mikimopine (94, 95), are formed by condensation and cyclization (rather than by reduction) of L-histidine and α-ketoglutarate. Members of the nopaline family readily cyclize to the corresponding lactams (Figure 8), a property that hindered the initial attempts to purify nopaline (96, 97). Similar lactamizations have been described for leucinopine, asparaginopine, cucumopine, and mikimopine.

Biosynthesis of Crown Gall Opines

The crown gall opines are synthesized by enzymes present in the transformed plant cells. However, the structural genes for octopine synthase (*ocs*) and nopaline synthase (*nos*) are encoded on large (~200 kbp) plasmids resident in virulent strains of *A. tumefaciens* (98, 99). These Ti (tumor-inducing) plasmids are required for crown gall induction, and tumorigenesis involves

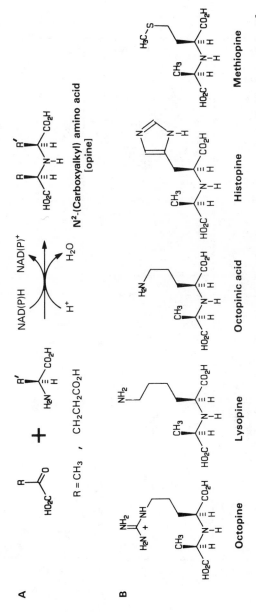

Figure 7 A. General reaction for biosynthesis of the N^2-(carboxyalkyl)amino acids (opines) in crown gall. Members of the N^2-(carboxyethyl) or octopine family are formed from pyruvic acid, $R = CH_3$. The N^2-(1,3-dicarboxypropyl) or nopaline family members are formed from α-ketoglutarate, $R = CH_2 \cdot CH_2 \cdot CO_2H$. *B.* Members of the octopine family of opines.

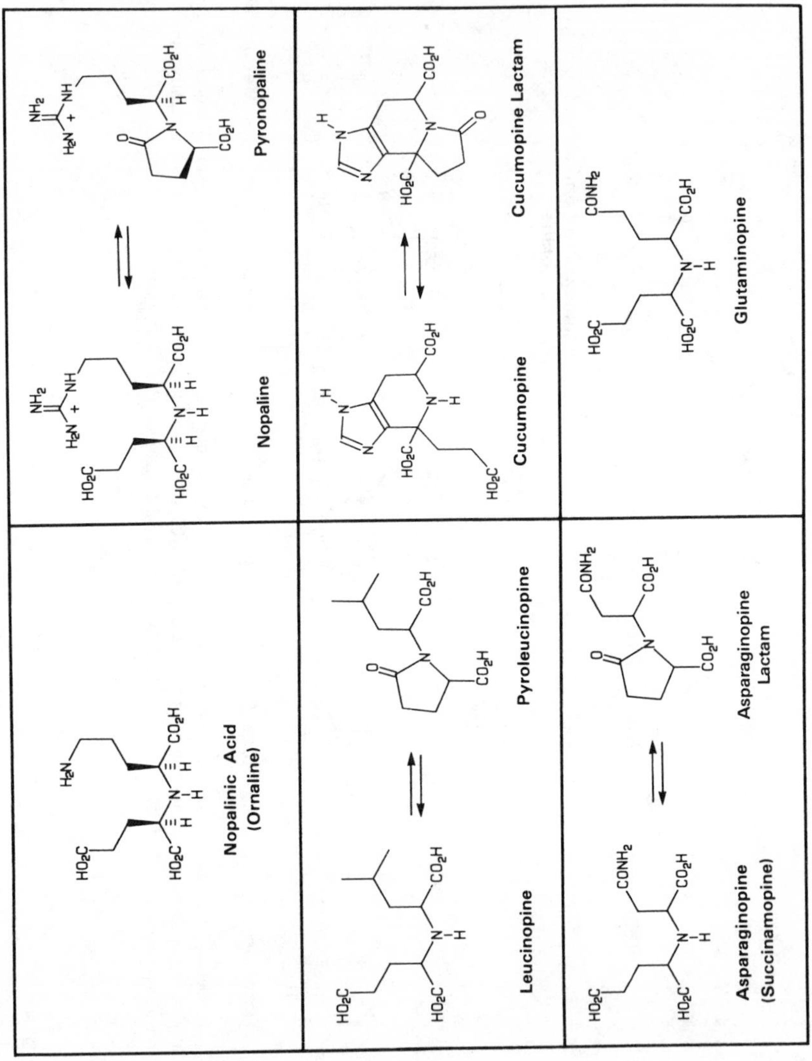

Figure 8   Members of the nopaline family and related opines.

excision of a (~20 kb) segment on which the opine synthase gene is located. After transfer of this T-DNA (in single-strand form) to the plant cell (100–102), and integration into the host genome (30, 103), opine synthase activity is expressed. Whether the octopine or nopaline families of opines are synthesized depends upon the type of synthase encoded by the T-DNA (25, 70, 104, 105).

D-Octopine Synthase

Lysopine was the first of the N-(CA)amino acids identified in crown galls; the absolute (D,L) stereochemistry of this compound was established by Biemann et al (15). Lejeune & Jubier (106) demonstrated the enzymatic synthesis of lysopine upon incubation of crown gall extracts with L-lysine, pyruvate, and NADH. The homologous N^2-(CE)-derivatives of arginine and ornithine were formed by substitution of these amino acids for lysine. Goldmann (107), and Hack & Kemp (108) proposed that a single enzyme mediated the synthesis of all members of the octopine family. Subsequently, N^2-(1-D-carboxyethyl)-L-arginine: NAD^+ oxidoreductase [D-octopine synthase (dehydrogenase) OCS, EC 1.5.1.11] was purified to homogeneity, and this enzyme catalyzed the NAD(P)H- and pyruvate-dependent synthesis of the N^2-(CE)-derivatives of arginine, lysine, ornithine, histidine, glutamic acid, and methionine (109). Kinetic analyses of the monomeric enzyme ($M_r = 38,000$) suggest that the reaction mechanism may depend upon the specific amino acid used as substrate. For octopinic acid synthesis, the order of substrate binding in the Ter-Bi reaction was NAD(P)H first, followed by ornithine and pyruvate in sequential order. For octopine synthesis, NAD(P)H was also bound first, but arginine and pyruvate appeared to bind in random order thereafter. The gene encoding octopine synthase (ocs) has been cloned and sequenced (110, 111). A "lysopine dehydrogenase" has also been purified from crown gall (112). Although classified as a distinct enzyme, it is now believed that this enzyme, N^2-(1-D-carboxyethyl)- L-lysine:$NADP^+$ oxidoreductase (EC 1.5.1.16), and octopine synthase are identical. Interestingly, Otten & Szegedi (113) recently reported that extracts from galls incited by biotype 3 strains of A. tumefaciens catalyzed the synthesis of octopine and lysopine, but that neither histidine nor methionine were substrates for the condensation reaction.

D-Nopaline Synthase

Nopaline and nopalinic acid are synthesized by the NAD(P)H- and α-ketoglutarate-dependent enzyme, N^2-(1,3-D-dicarboxypropyl)-L-arginine: $NADP^+$ oxidoreductase [D-Nopaline synthase (nopaline dehydrogenase) NOS, EC 1.5.1.19]. Nopaline synthase has been purified from tumor tissue of Nicotiana tabacum (114), and the gene (nos) has been sequenced (115, 116) and expressed in Escherichia coli (117). This enzyme is a tetramer ($M_r =$

158,000) comprising four identical subunits. The five novel opines leucino-pine, glutaminopine, asparaginopine (succinamopine), cucumopine, and mikimopine are also derived via condensations with α-ketoglutarate and may, therefore, be included in the nopaline family. However, there is no evidence that NOS catalyzes the formation of these N-(CA)amino acids.

Functions of Opines in Crown Gall

Opines may be regarded as unusual end-products of glycolytic activity in tumor tissue (see Figure 9), but they confer few obvious advantages to the tumor cells. Earlier investigations showed that application of opines to the surface of bean leaves increased the rate of crown gall development (118, 119). Insight into the molecular basis for the earlier observations stems from the recent discovery (120) that various opines, in the presence of wound-released acetosyringone, increase transcription of Ti plasmid–encoded viru-lence (*vir*) genes. Since expression of *vir* is required for excision and transfer of single-stranded T-DNA to the plant cell, an acceleration of the tumorigenic process might be anticipated. In an altruistic sense, the opines ensure the survival of their own biosynthetic genes (23) by promoting the dissemination of the Ti plasmids throughout the general population of agrobacteria. Ream, in an excellent review (26), reiterates this theme as follows (*a*) opines produced by transformed plant cells provide nutrients and a chemical "mes-

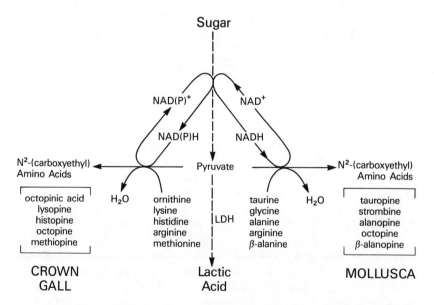

Figure 9   Biosynthesis of N^2-(carboxyethyl)amino acids in crown gall and marine invertebrates. These amino acid derivatives represent alternatives to lactic acid as the end-product of glycolysis.

sage," and (*b*) acetosyringone and other phenolic compounds in the exudate serve as chemoattractants (121) for the phytopathogen, to indicate a host suitable for colonization (29). This unique example of naturally occurring interkingdom genetic engineering between bacteria and plants was formulated earlier by Tempé and collaborators as the *opine concept* (10, 122). As shown in simplified form in Figure 10, this intricate prokaryote-eukaryote interrelationship involves (*a*) attraction of Ti plasmid–containing *A. tumefaciens* by compounds secreted by the wounded plant, (*b*) induction of the *vir* operon, (*c*) transfer and integration of T-DNA, (*d*) expression of opine synthase genes, (*e*) opine biosynthesis and exit to the tumerosphere, and (*f*) proliferation of virulent strains of *A. tumefaciens* due to transport, and utilization of opines as C and N sources.

Catabolism of Opines by Agrobacterium tumefaciens

It has been found that strains of *A. tumefaciens* metabolize only those opines whose synthesis they induce in hyperplasias (104, 105, 122). The genes encoding some of the opine catabolic enzymes are also localized on Ti plasmids (123), but are located outside the T-DNA region. Prior to catabolism, the opines exit the tumor cells via T-DNA-encoded transport protein(s) (124) and are taken up by opine-inducible transport systems in *A. tumefaciens* (125, 126; but see also 127). The high-affinity octopine permease (K_m ~1.5 μM) mediates the active transport of all members of the octopine family, but not nopaline. In contrast, the nopaline-inducible permease exhibits high affinity for octopine (128). After membrane translocation, the intracellular opines are oxidized via cytochrome-linked oxidases (129, 130) to yield lysine and pyruvate from lysopine, arginine plus pyruvate from octopine, and arginine plus α-ketoglutarate from nopaline (128, 131).

In the past decade, the pathway(s) for the intracellular catabolism of nopaline and octopine have been elucidated (123, 130–132). Many of the enzymes required for the metabolism of nopaline, octopine, and arginine to glutamic acid (see Figure 10) have been isolated and characterized. Several genes of the *noc* operon, which encodes the periplasmic nopaline-binding and transport proteins, nopaline oxidase, arginase, and ornithine cyclodeaminase, have been cloned and sequenced (133–135). There is uncertainty concerning the localization of the gene that encodes the enzyme proline dehydrogenase (oxidase), but it seems probable that this gene, and those encoding ornithine-δ-transaminase and Δ^1-pyrroline-5-carboxylate dehydrogenase, are localized on the chromosome. Wabiko et al (136) have provided recent evidence that for some nonpathogenic strains of *A. tumefaciens* that can be grown on nopaline and octopine (e.g. PyON 8), the *noc* genes are chromosomally encoded.

The ability to catabolize opines may confer a selective growth advantage

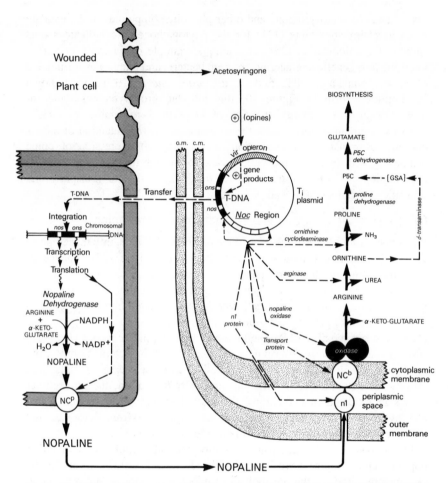

Figure 10   Diagrammatic representation of the opine concept with reference to the biosynthesis of nopaline by crown gall tumor tissues, and catabolism of this N^2-(carboxyalkyl)amino acid by *A. tumefaciens*. This illustration depicts the T-DNA-mediated, interkingdom transfer of genetic information from a prokaryotic to a eukaryotic cell. Genes encoded by the T-DNA include: *nos*, nopaline synthase;*ons*, octopine and nopaline carrier protein (NC^P). The genes for the periplasmic nopaline-binding protein (n1); the nopaline transporter (NC^b), and the *noc* genes for nopaline catabolism are found elsewhere on the Ti-plasmid. Figure is reprinted, by permission of John Wiley & Sons, Inc., from Thompson, J., Miller, S. P. F.'s chapter in *Advances in Enzymology*, Volume 64, pp. 317–99. Copyright © 1991 by Wiley.

upon *A. tumefaciens,* but recent reports of the metabolism of these N-(CA)amino acids by other bacteria indicates that this generalization must be treated with reservation (137–142). Several species when grown either in extracts from crown gall (140), or under opine limitation in chemostats (143, 144), also utilize these N-(CA)amino acids as carbon and nitrogen sources.

These organisms have the potential therefore, to compete with *A. tumefaciens* for environmentally available opines. Information is sparse concerning the transport and catabolism of opines by these other species (145), but arginine is an intermediate in the degradation of both nopaline and octopine. Transposon mutagenesis has been used to isolate mutants of *Pseudomonas aureofaciens* 211 defective in the catabolism of nopaline, and DNA sequences encoding genes for nopaline metabolism have been cloned from this organism (146). However, hybridization studies revealed little homology between these genes and the Ti plasmid–encoded *noc* genes from *A. tumefaciens*.

N-(CARBOXYALKYL)AMINO ACIDS IN MARINE INVERTEBRATES

L-lactic acid is the major product of sugar fermentation by the Embden-Meyerhof-Parnas pathway. In most prokaryotic and eukaryotic cells, lactate dehydrogenase (L-lactate:NAD$^+$ oxidoreductase, EC 1.1.1.27) couples the oxidation of NADH to the reduction of pyruvic acid, thereby regenerating the NAD$^+$ necessary for the continued operation of the glycolytic pathway. However, the tissues of certain marine invertebrates (e.g. *Gastropoda, Pelecypoda,* and *Cephalopoda*) contain little or no lactate dehydrogenase. In these species, NAD$^+$ is regenerated by the NADH-dependent reductive condensation of pyruvate with the α-NH$_2$ group of a variety of amino acids. These reactions (which are analogous to those catalyzed by the crown gall opine synthases) are mediated by a variety of pyruvate:oxidoreductases (147–149). The N^2-(CA)amino acids formed by the *Mollusca* include D-octopine, alanopine, β-alanopine, strombine, and tauropine (Table 1 and Figure 11). These compounds may be considered end-products of glycolysis in the marine invertebrates (Figure 9; 150–153).

Biosynthesis of Opines in Marine Invertebrates

D-Octopine was the first N-(CA)amino acid to be extracted and identified from octopus by Morizawa in 1927 (154). However, the stereochemistry of this opine was only established 50 years later (18). Octopine biosynthesis in marine invertebrates (as in crown gall) occurs via a reductive condensation

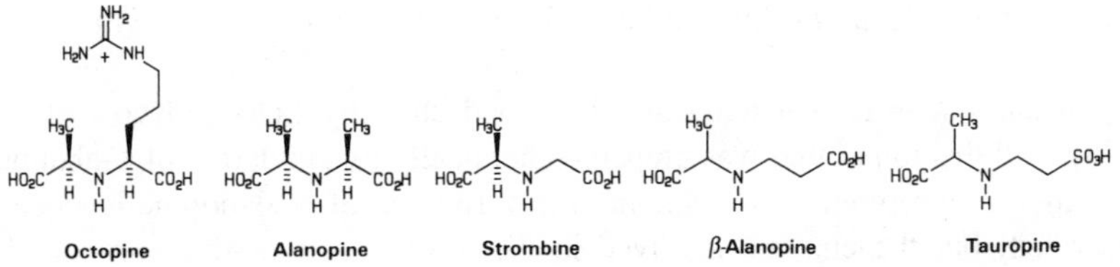

Figure 11   N^2-(Carboxyalkyl)amino acids found in marine invertebrates.

between pyruvic acid and arginine, and the in vitro formation of this compound by an extract from scallops (*Pecten maximus*) was described by Thoai et al (155, 156). Octopine dehydrogenase [N^2-(1-D-carboxyethyl)- L-arginine: NAD^+ oxidoreductase, ODH, EC 1.5.1.11] has been purified (157, 158) and crystallized (159) from *Pecten maximus;* the enzyme ($M_r \sim$ 37,000–40,000) has also been purified from other marine invertebrates (148, 150, 160). The kinetic properties (161, 162), specificity of hydride transfer (163, 164), and the physicochemical characteristics of this enzyme have also been reported (165–167). Isozymes of ODH are present in scallops (168), cuttlefish (169), and squid (170). These forms are apparently tissue-specific, and because of their location and kinetic dissimilarities, different physiological roles have been attributed to these isozymes (171, 172).

Alanopine [*meso*-N-(1-carboxyethyl)-alanine] was isolated from squid muscle tissue by Sato et al (173), and alanopine dehydrogenase [*meso*-N-(1-carboxyethyl)-alanine:NAD^+oxidoreductase, EC 1.5.1.17] was purified approximately 400-fold from adductor muscle of the oyster (174). This monomeric enzyme (M_r = 42,000) has also been partially purified from lugworm (175) and mussel (176). Tissue-specific isozymes of alanopine dehydrogenase also occur (148), and Plaxton & Storey (177) have purified three forms of this enzyme (with slightly different pI and K_m values) from the channeled whelk.

Strombine is the product of a reductive condensation between pyruvic acid and glycine, and this compound was prepared from conch tissue by Sangster et al (178). The structure of this marine opine [N-(1-carboxymethyl)-D-alanine] was determined by ^{1}H-NMR spectroscopy, and confirmed by chemical synthesis from alanine and monochloroacetic acid. Strombine dehydrogenase (EC 1.5.1.22) has been partially purified from the mussel (*Mytilus edulis* L.; 176). Several methods, including GLC (179) and HPLC (180, 181), are available to separate and quantitate alanopine and strombine from marine invertebrate tissue.

Tauropine (or, D-rhodoic acid) is an unusual sulfur-containing opine, first isolated from red algae (182), and subsequently prepared from muscle extracts of the abalone (183). The biosynthesis of tauropine is catalyzed by tauropine dehydrogenase [N-(1-D-carboxyethyl)-taurine:NAD^+oxidoreductase]. This NADH-specific enzyme (M_r =42,000), has been purified to homogeneity from the abalone species *Haliotis lamellosa* (184, 185), and *Haliotis discus hannai* (186), and also from the brachiopod *Glottidia pyrimidata* (187).

β-Alanopine was isolated from the blood shell by Sato and coworkers (188), and the compound was prepared chemically by alkylation of D-alanine with 3-chloropropionic acid. The in vitro formation of β-alanopine has been described, but the enzyme involved in the biosynthesis of this compound, β-alanopine dehydrogenase (EC 1.5.1.26), has only been partially purified.

Immunological studies have shown that antibodies prepared against OCS from crown gall do not cross-react with either of the ODH isozymes from the scallop (189, 190). Conversely, antibodies to the A- and B-isozymes from the scallop do not cross-react with the crown gall enzyme; neither is any cross-reaction detected between the former antibodies and alanopine, strombine, or octopine dehydrogenases prepared from other bivalves (191).

N-(CARBOXYALKYL)AMINO ACIDS IN YEASTS

Few N-(CA)amino acids have been identified in simple eukaryotic cells, and only saccharopine [N^6-(1,3-L-dicarboxypropyl)-L-lysine] has been extensively studied. Saccharopine is an intermediate in lysine biosynthesis in yeasts and fungi (192, 193), and was first prepared from brewer's yeast by Larsen and coworkers (194, 195). Several chemical methods are available for the synthesis of saccharopine (196, 197). In solution, particularly under acidic conditions, saccharopine exists in equilibrium with its lactam, pyrosaccharopine (Figure 12). The biosynthesis of saccharopine is catalyzed by saccharopine dehydrogenase (SDH; NAD^+-L-lysine forming, EC 1.5.1.7) via the NADH-dependent reductive condensation of α-ketoglutarate with L-lysine (Figure 12). SDH ($M_r = 39,000$) was purified to homogeneity from *Saccharomyces cerevisiae* by Ogawa & Fujioka (198), and the physicochemical and kinetic characteristics of the enzyme have been reported in detail [see Fujioka (199) for discussion]. The SDH gene (*LYS5*) from *Yarrowia lipolytica* has recently been cloned (200) and sequenced (201). Of all naturally occurring N-(CA)amino acids, saccharopine most closely resembles the two *L. lactis* metabolites, N^5-(CE)ornithine and N^6-(CE)lysine. All three compounds exhibit (L,L) stereochemistry, and their synthesis involves condensation at the ω-NH_2 group of the amino acid substrates. These properties differentiate them from both the crown gall and the marine invertebrate opines. Furthermore, SDH, when incubated with NADH and abnormally high concentrations of

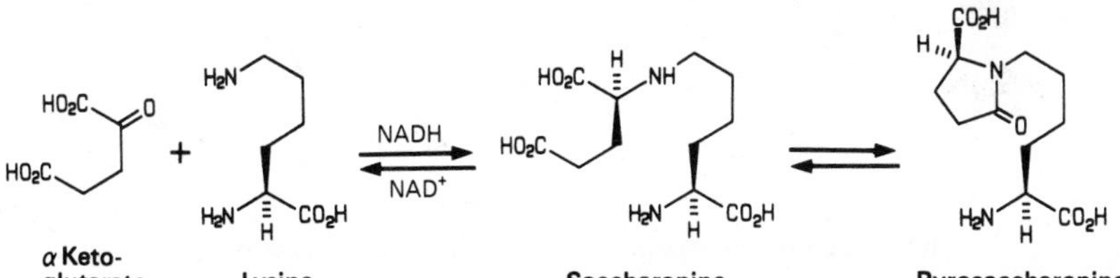

Figure 12 Biosynthesis of saccharopine and cyclization of this compound to its lactam, pyrosaccharopine.

pyruvate and lysine, can also catalyze the synthesis of N^6-(1-L-CE)-L-lysine (37, 202), but this compound (Figure 13) has not been found in yeast. These chemical and enzymatic data suggest that the reactions catalyzed by SDH from yeast and CEOS from *L. lactis* may have much in common.

N-(CARBOXYALKYL)AMINO ACIDS: OTHER SOURCES

N-(CA)amino acids, particularly the carboxyethyl (CE) and carboxymethyl (CM) derivatives (Figure 13) have also been found in healthy (i.e. nontumorous) plants, but in general neither their function(s) nor routes of biosynthesis have been established. N-(CM)-β-alanine and N^2-(CM)-L-serine have been isolated from green gram seeds and asparagus (203, 204), and N^6-(CM)lysine has been detected in an acid hydrolysate of *Sagittaria pygmaea* (205).

In 1975 Wadman and coworkers (206) identified N^6-(CM)lysine in the urine of sick infants, but no correlation was established between pathological condition and excretion of this N^{ω}-(CA)amino acid. Subsequently, Liardon et al (207) detected this compound in the urine of rats that had been fed a diet of alkali-treated whey proteins. From studies of the Maillard (browning) reaction of proteins, it now appears that in both rats and humans, N^6-(CM)lysine is of dietary origin (208–211). N^6-(CM)lysine is a product of the oxidation of N^6-fructoselysine (an Amadori adduct) formed initially via glycation of ϵ-NH_2 groups of lysyl residues in proteins with reducing sugars.

N-(CA)amino acids have also been reported in studies detailing the catalytic mechanisms of aldolases, amino acid oxidases, and pyruvoyl-containing amino acid decarboxylases. For instance, addition of $NaBH_4$ to the reaction catalyzed by 2-keto-4-hydroxyglutarate aldolase caused reduction of azomethines formed between ϵ-NH_2 groups of lysyl residues of the enzyme and the α-keto reaction products (glyoxylate and pyruvate). The reduced, covalently linked derivatives, N^6-(CM)lysine and N^6-(CE)lysine, were recov-

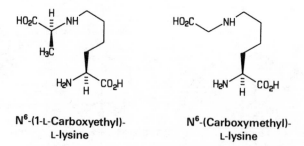

N^6-(1-L-Carboxyethyl)-
L-lysine

N^6-(Carboxymethyl)-
L-lysine

Figure 13   Structures of: *left*, N^6-(1-L-carboxyethyl)- and *right*, N^6-(carboxymethyl)-derivatives of L-lysine.

ered in the acid hydrolysate of the aldolase (212, 213). N^6-(CE)lysine has also been found in the acid hydrolysate of D-amino acid oxidase after quenching of the reaction with $NaBH_4$ (214, 215). These studies, and other investigations (216), established (a) that an α-imino acid was an intermediate in the amino acid oxidase reaction, and (b) that this highly reactive compound could also participate in transimination reactions.

Another route to the formation of N-(CA)amino acids has become apparent from the elegant studies of Snell and collaborators on the catalytic function of the NH_2-terminal pyruvoyl moiety of amino acid decarboxylases (summarized in Refs. 217, 218). Evidence for Schiff base formation between the pyruvoyl moiety of histidine decarboxylase and the substrate (histidine) was obtained by trapping of the intermediate as a covalently linked, secondary amine by addition of $NaBH_4$ to the reaction. N^2-(1-CE)histidine was subsequently detected in the acid hydrolysate of the inactivated enzyme. Furthermore, the pyruvoyl moiety of histidine decarboxylase was recovered in the carboxyethyl group of the N^2-(CA)amino acid (Figure 14).

N-(CARBOXYALKYL) DIPEPTIDES AS ENZYME INHIBITORS

In the early 1970s, it was noted (219–222) that several proline-rich peptides in snake venom were potentiators of bradykinin and inhibitors of angiotensin I-converting enzyme (ACE, dipeptidyl peptide carboxypeptidase, EC 3.4.15.1). This Zn^{2+}-containing enzyme catalyzes the removal of the C-terminal dipeptide (L-His-L-Leu) from the inactive decapeptide angiotensin I to generate the potent vasoconstrictor, angiotensin II (223–225). These reports raised the expectation that di- and tri-peptides containing proline might find pharmaceutical application as ACE-inhibitory, antihypertensive agents (226, 227). Initial results obtained with the dipeptide L-Ala-L-Pro were encouraging and provided the incentive for vigorous programs at the Squibb Institute (226–228) and Merck Sharp & Dohme Laboratories (229, 230), which culminated in the synthesis of the N^α-(carboxyalkanoyl) and N^α-(carboxyalkyl) compounds Captopril and Enalaprilat, respectively (Figure 15). A landmark in this field was the report by Patchett et al (231) in 1980 of a variety of N^α-[carboxyalkyl(CA)]-L-Ala- L-Pro derivatives that were potent inhibitors of ACE. The incorporation of the N^α-(CA) moiety into the L-Ala-L-Pro dipeptide was a major advance in the design (and the potency) of the ACE inhibitors (229, 230). This fact became apparent after the synthesis and testing of approximately 200 N^α-(CA)-dipeptides (231). A correlation between the size and complexity of the carboxyalkyl substituent and extent of ACE inhibition was noted (see Table 1 of Ref. 231). Thus, progression from

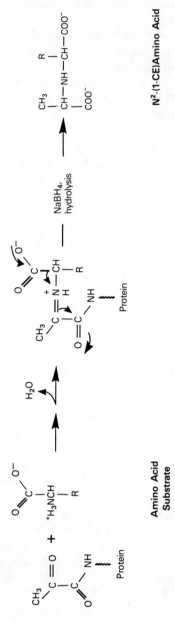

Figure 14 Formation of an N^2-(1-CE)amino acid by trapping of the Schiff base generated during the amino acid decarboxylase reaction [modified from (218)].

the parent N-(carboxymethyl)-L-Ala-L-Pro dipeptide, via N^α-(1-carboxy-ethyl)-L-Ala-L-Pro to N^α-[1-(S)-carboxy-3-phenylpropyl]- L-Ala-L-Pro (Enalaprilat), was accompanied by a 1000-fold increase in potency of the inhibitor. Kinetic analyses (232) of the mechanism of ACE inhibition by Enalaprilat and its analog N^α-[1-(S)-carboxy-3-phenylpropyl]-L-Lys-L-Pro (Lisinopril, Figure 15), showed that Zn^{2+} and the -COOH group of the N-(carboxyalkyl) moiety of the dipeptide were major determinants for the affinity of binding between inhibitor and enzyme. Significantly, the affinity of Lisinopril for the (Zn^{2+}-containing) holoenzyme was 20,000-fold greater than that for the Zn^{2+}-free apoenzyme. Elimination of the -COOH group from the N-(carboxyalkyl) moiety dramatically reduced the potency of Enalaprilat, and the I_{50} concentration for the descarboxy compound was increased by ~4000 fold (see Table XXI of Ref. 229). It seemed reasonable to attribute inhibition of ACE and other metalloendopeptidases (233–236) to carboxylate-mediated chelation or coordination with Zn^{2+} at the active site (237, 238), and a role for the N^α-(CA) dipeptides as transition-state analogs (239) was also proposed. X-ray crystallographic analyses of binding of the inhibitor N^α-(1-carboxy-3-phenylpropyl)-L-Leu-L-Trp to the Zn^{2+}-containing endopeptidase thermolysin have been conducted by Matthews and coworkers (240, 241). Results from these studies, which may also apply to ACE, provided evidence for a dual function of the N^α-carboxylate group in inhibition via (a) bidentate coordination and overall pentacoordination with Zn^{2+}, and (b) promotion of a hydrogen bond network between the inhibitor and neighboring amino acids at the active site.

N-(CARBOXYALKYL) DIPEPTIDES IN ENZYME PURIFICATION

A logical development from the inhibitor studies was the use of N^α-(CA) peptides as ligands for the purification of enzymes such as ACE, thermolysin, and collagenase by affinity chromatography. This novel approach has been remarkably successful in the purification of Zn^{2+}-metalloproteinases. Lisinopril was used as an affinity ligand by El-Dorry et al (242) to purify rabbit testicular dipeptidyl carboxypeptidase. The same compound, coupled to epoxy-activated Sepharose CL-4B, permitted a single-step (1000-fold) purification to homogeneity of ACE from rabbit lung tissue, and also provided a one-step 100,000-fold enrichment of the enzyme from human plasma (243). Stack et al (244) have recently employed the same methodology but a different ligand [N^α-(1-(R,S)-carboxy-n-butyl)-L-Leu-L-Phe- L-Ala-NH$_2$ coupled via the C-terminal amino group to EAH-Sepharose 4B] for the purification of collagenase, gelatinase, and stromelysin from porcine synovial membrane.

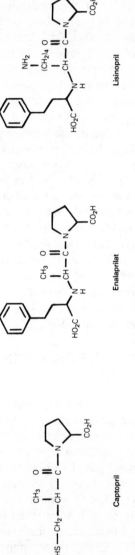

Figure 15 Structures of angiotensin I-converting enzyme (ACE) inhibitors: Captopril, Enalaprilat, and Lisinopril.

SEQUENCE ANALYSIS OF N-(CARBOXYALKYL)AMINO ACID DEHYDROGENASES

Amino acid sequences are currently available for four N-(CA)amino acid dehydrogenases: NOS, OCS, SDH, and CEOS. In this section, we first discuss, and then analyze these sequences for dinucleotide-binding domains. Where appropriate, comparisons are made with other, more distantly related, proteins.

Nopaline and Octopine Synthases

The *nos* gene has been sequenced by two groups of investigators (115, 116). Both groups deduced an open reading frame of 413 amino acids for NOS. A probable transcription initiation site was identified about 35 nucleotides (nt) upstream of the translation initiator codon, and putative transcriptional control elements, resembling those in eukaryotes, were identified. Specifically, the sequence 5'-CATAAA (resembling a "TATA" box) was found 25 nt upstream and the sequence 5'-GGTCACTAT (resembling a "CCAAT" box) was detected 70–80 nt upstream from the transcription initiation site. No intervening sequences were detected, but putative polyadenylation signals (5'-AATAAA/T) were present in the 3'-flanking area.

The *ocs* genes from pTiAch5 and pTi5955 were found to encode proteins of 358 and 359 amino acids, respectively (110, 111). As discussed previously for *nos*, transcriptional control regions resembling those of eukaryotic genes transcribed by RNA polymerase II were identified, and the absence of a typical prokaryotic ribosome-binding site was noted.

In addition to the "classical" TATA and CCAAT sequence boxes, further upstream sequences with enhancerlike activities are required for efficient transcription of both *nos* (245–247) and *ocs* (248–250). Within the plant, *nos* promoter activity is organ specific and developmentally regulated (251). Promoter activity is increased upon wounding of the plant and is also auxin inducible (252). The *ocs* promoter contains a 16-base-pair palindromic sequence that acts as an enhancer (248, 253). Two proteins have recently been identified that bind to this *ocs* element, and which may regulate expression of *ocs* during development of the plant (254, 255).

Saccharopine Dehydrogenase

To date, only the SDH gene from the yeast *Yarrowia lipolytica* has been sequenced (201). This gene (*LYS5*) was found to encode a protein of 369 amino acids with a calculated molecular weight (40,566) similar to that of the enzyme purified from *Saccharomyces cerevisiae*, M_r = 39,000 (198). Transcription was initiated about 50 nt upstream from the translational start codon. Whereas no typical TATA box was evident, putative CAAT/G boxes (with the sequence 5'-CAAT) were identified 183 and 365 nt upstream from the initiation codon. In addition, a 5'-TGACTC sequence was present another 20 nt further upstream, which suggests that expression of SDH might be regulated by the general amino acid control derepression system (previously characterized in *Saccharomyces*). This latter system is mediated by the positive transcription factor GCN4 (256), which binds to TGACTC sequences found upstream of many yeast genes encoding amino acid biosynthetic enzymes.

N^5-(Carboxyethyl)ornithine Synthase

The cloning and sequencing of *ceo* (from *L. lactis* K1) have resulted in the identification of an open reading frame of 313 amino acid residues with a deduced molecular weight of 35,323 (J. A. Donkersloot, J. Thompson, unpublished information). The recombinant *Escherichia coli* cells had fivefold greater specific CEOS activity than *L. lactis,* which is probably due to a gene-dosage effect. A putative ribosome-binding site (5'-AAGGA), centered 8 nt upstream from the initiator codon, has been identified, and sequences resembling consensus "−10" and "−35" regions of lactococcal promoters (257, 258) are also present. Clearly, control of *ceo* expression in *L. lactis* is distinct from that of *nos* and *ocs* in *A. tumefaciens,* which are not expressed because they lack the appropriate prokaryotic expression signals.

"Global" Similarities

Alignment of the four N-(CA)amino acid dehydrogenases (NOS, OCS, SDH, and CEOS) yielded amino acid identity values ranging from 14 to 25%, suggesting that little overall homology exists among these proteins. [The program BestFit (259) was used; J. A. Donkersloot, unpublished information.] NOS and OCS mediate analogous reactions, and while it may seem surprising that these enzymes have only 22% amino acid sequence identity, it should be noted that these proteins are also immunologically distinct (189–191). Previously, Monneuse & Rouzé (who used a different program and a different set of parameters) reported an identity of 26% after aligning NOS and OCS (260). Several regions of homology were apparent upon inspection of the alignment and dot-matrix analysis (260). As shown in Figure 16, one partially homologous segment extends for about 100 residues from the N-terminus of OCS; this segment corresponds to the more highly conserved part of the dinucleotide-binding domain of these two enzymes (see section on *Dinucleotide-Binding Domains*). Homology was also evident in the C-terminal areas, which contain a number of residues presumed to be required for catalysis [see section on REACTION MECHANISM(S) AND CATALYTIC DOMAINS]. A third homology in the middle of the sequence is also apparent, but no function has been associated with this region. Monneuse & Rouzé also noted in their alignment that NOS extends OCS by 20 amino acids at the N-terminus. They speculated that as for LDH, this overlapping "arm" might also be involved in subunit interactions in tetrameric NOS (OCS is monomeric).

SDH and CEOS showed 24% amino acid identity after alignment (J. A. Donkersloot, unpublished information). Inspection of the dot plot (Figure 16) shows that homologous sequences are present in the NH$_2$- and COOH-terminal segments. As is discussed later [see section on REACTION MECHANISM(S) AND CATALYTIC DOMAINS], these sequences may be involved in catalysis.

Thus far, we have compared the four N-(CA)amino acid dehydrogenases currently amenable to sequence analysis. However, we should again emphasize that, whereas the products synthesized by NOS and OCS have the (D,L) configuration, those formed by SDH and CEOS exhibit (L,L) stereochemistry and are also regiochemically distinct. One must envisage differences in the relative juxtaposition of the α-keto and amino acid substrates for these two groups of enzymes. These differences are likely to be reflected in the amino acid sequence, and in the folding of the polypeptide chain to form the appropriate catalytic domains. To exemplify this issue, we note that D-lactate dehydrogenase from *Lactobacillus plantarum* shares no significant homology with the L-lactate dehydrogenase family (261).

Dinucleotide-Binding Domains

Initial X-ray diffraction studies of many dehydrogenases revealed remarkable similarities in their dinucleotide-binding domains (262, 263). Particularly conserved is a βαβ-fold, which binds the ADP moiety of NAD(H). The

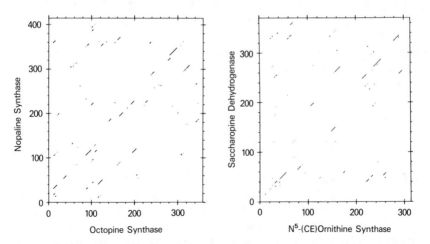

Figure 16 Amino acid sequence comparison among the four N-(CA)amino acid dehydrogenases: NOS and OCS (*left*); SDH and CEOS (*right*). The programs Compare (window size = 20; stringency = 11) and DotPlot were used (259). NOS and OCS have previously been compared (260).

β-strands of this fold are part of a four- or six-stranded parallel sheet, whose C-terminal edge is in close proximity to the pyrophosphate bridge of NAD(H). Several amino acids play important roles in this interaction, and a "core fingerprint" for ADP-binding βαβ-folds has been proposed (262, 264, 265). Within this fingerprint, the motif GXGXXG (single-letter amino acid code) is highly conserved (Figure 17). The first Gly residue marks the sharp turn from the first β-strand (βA) to the succeeding α-helix, and is located close to the pyrophosphate bridge. The second Gly, which is located at the N-terminus of the α-helix, permits a close approach between this terminus and the pyrophosphate moiety, which allows the charge on this group to be stabilized by the dipole originating from the helix (266). The third Gly of the fingerprint facilitates a close interaction between the α-helix and the β-strands. In addition, a conserved acidic amino acid (Asp or Glu) was identified at the end of the second β-strand (βB), which permits H-bonding between the carboxylate group of this amino acid and the 2'-hydroxy of the adenine ribose. Although less is known about NADPH-binding regions, they typically contain the GXGXXA motif (Figure 17; 265, 267, 268). The Gly-to-Ala substitution in the third position of the motif causes the βαβ-fold to be slightly less compact (265). Furthermore, the end of the βB strand of these enzymes does not usually terminate with an acidic residue, presumably to avoid the electrostatic repulsion that would occur with the ribose 2'-phosphate (265, 269). Comparison of the sequences of several NADH- and NADPH-binding flavoprotein disulfide oxidoreductases indicates that in the latter group of enzymes two conserved Arg residues separated by five other amino acids may serve to neutralize the "extra" phosphate group of NADPH (268). Moreover, the three-dimensional structure of glutathione reductase shows that these residues are indeed close to this phosphate (270, 271). Site-directed mutagenesis has confirmed the importance of these two Arg residues in the binding of NADPH by these oxidoreductases (268). In other NADP(H)-linked dehydrogenases (e.g. glutamate dehydrogenase, see Figure 17), amino acid residues located elsewhere in the sequence may serve the same function (272).

Recently, a new type of NADPH-binding domain was discovered in FAD-containing ferredoxin-NADP$^+$ reductases (273). This domain is characterized by a central five-stranded parallel β-sheet and six surrounding helices. Whereas the actual NADPH-binding site is also located at the C-terminal edge of this β-sheet, the fingerprint of the pyrophosphate-binding loop appears to be MXXXGTGXXP.

N^5-(CARBOXYETHYL)ORNITHINE SYNTHASE   Purified CEOS has a strong preference for NADPH (38), and a motif near the center of the sequence closely resembles the core fingerprint for the binding of this cofactor (Figure

NADP(H) Binding

Enzyme	Sequence
N⁵-(CE)Ornithine synthase (*L. lactis*)	166 K I A I L G GNVAQGAFSSISKYSSNI RMYYR KTMSIFKEN
Glutamate dehydrogenase (*E. coli*)	232 G M R V S V S G GNVAQYAIEKAMEFG ARVITASD SSGTVVDES
Glutathione reductase (human)	188 G R S V I V G GYIAVEMAGILSALG KTSLMI RHDKVLRSF
Mercuric reductase (*S. aureus* pI258)	251 Q R L A V I G GYIAAELGQMFHNLG TEVTLMQ RSERLFKTYD
Octopine synthase (*A. tumefaciens*)	1 MAKVAILG GNVALTLAGDLARRLG QVSSIWAP ISNRNSFN
Nopaline synthase (*A. tumefaciens*)	21 PLTVGVLG GHAGTALAAWFASR HVPTALWAP ADHPGSIS
Saccharopine dehydrogenase (*Y. lipolytica*)	200 GALGRCGSGAIDLARKVGIPEENIIRWDMNETKKGGPFQE

NAD(H) Binding

Enzyme	Sequence
Lipoamide dehydrogenase (*E. coli*)	176 E R L L V M G GGIIGLEMGTVYHALG QIDVV EMFDQVIPAAD
Phenylalanine dehydrogenase (*B. sphaericus*)	184 GKTYAIQG LGKVGYKVAEQLLKAG ADLFVD IHENVLNSIK
Leucine dehydrogenase (*B. stearothermophilus*)	173 GKVVAVQG VGNVAYHLCRHLHEEG AKLIVTD INKEVVARAV
Alanine dehydrogenase (*B. sphaericus*)	165 GVHARKVTVIG GIAGTNAAKIAVGMG ADVTVID LSPERLRQLE
Lactate dehydrogenase (*B. stearothermophilus*)	6 GARVVVIG AGFVGASYVFALMNQG IADEIVLID ANESKAIG

Figure 17 Alignment of putative nucleotide-binding folds of N-(CA)amino acid synthases and other dehydrogenases. The three strands that make up the putative βαβ fold are boxed, and conserved amino acid residues are in bold type. The NADP-binding fold of glutathione reductase is derived from a three-dimensional model of the enzyme (271). The putative βαβ folds of mercuric reductase and lipoamide dehydrogenase are based on the structural and sequence homologies of these disulfide oxidoreductases with glutathione reductase (268, 294). The prediction for glutamate dehydrogenase is based on earlier reports (268, 276, 281, 282). The octopine synthase and nopaline synthase predictions have been described previously (260). The alignment shown for the four *Bacillus* enzymes is slightly modified from an earlier prediction (295). References for the sequence data are: N⁵-(CE)ornithine synthase (J. A. Donkersloot and J. Thompson, unpublished information), glutamate dehydrogenase (274, 275), glutathione reductase (271), mercuric reductase (296), octopine synthase (110, 111), nopaline synthase (115, 116), saccharopine dehydrogenase (201), lipoamide dehydrogenase (297), phenylalanine dehydrogenase (298), leucine dehydrogenase (299), alanine dehydrogenase (295), lactate dehydrogenase (300).

17). Interestingly, 8 of the 9 amino acids in this segment are identical to the equivalent sequence of (NADPH-dependent) glutamate dehydrogenase (GDH) from *E. coli* (274, 275), *Salmonella typhimurium* (276), yeast (277, 278), and *Neurospora crassa* (279). No structural data are available for CEOS, and until recently only a relatively low-resolution (0.6 nm) model (280) existed for GDH from *Clostridium symbiosum*. However, in 1974, a $\beta\alpha\beta$-fold for nucleotide-binding was predicted for bovine liver GDH on the basis of sequence homologies to other ADP-binding units (281). Similar predictions have since been made for GDH from *Neurospora,* and *Saccharomyces cerevisiae* (277, 282). The structure of the NAD^+-dependent GDH from *Clostridium symbiosum* was recently solved to 1.96 Å resolution. One of the two domains of this enzyme is indeed structurally similar to the classical dinucleotide-binding fold. Interestingly, however, the direction of the third strand of the (seven-stranded) β-sheet is reversed (282a). Based on these results and on the alignment shown in Figure 17, a nucleotide-binding $\beta\alpha\beta$-fold can be reasonably predicted for CEOS. When aligned in this fashion, Arg195 (or Lys196) and Lys202 of CEOS virtually coincide in relative position with the two Arg residues involved in the binding of NADPH by the flavoprotein disulfide oxidoreductases.

If we assume that the size of the dinucleotide-binding domain in CEOS is similar to that of other dehydrogenases (about 130 residues), then this domain would encompass, approximately, residues 160–290. One might envisage that by suitable folding of the polypeptide chain, amino acids from residue 290 to the COOH-terminus can interact with residues in the NH_2-terminus, to form the catalytic domain of the enzyme.

NOPALINE AND OCTOPINE SYNTHASES   Monneuse & Rouzé reported that the fingerprint for ADP binding is located close to the NH_2-terminus of both proteins, and on this basis a $\beta\alpha\beta$ nucleotide-binding fold was tentatively identified [Figure 17; (260)]. Furthermore, from an alignment of these two enzymes with 18 nucleotide-binding proteins with known structures, four additional β-strands were predicted, thus generating a putative six-stranded nucleotide-binding sheet (260). There seems little doubt about the initial $\beta\alpha\beta$ fold, but the assignment of the four other β-strands should be considered speculative. Clearly, the determination of the three-dimensional structure of one (or both) of these enzymes would be of considerable interest. NOS and OCS are unusual in the sense that both enzymes use either NADPH or NADH as cofactor. In this context, OCS has the fingerprint of NADP(H)-linked dehydrogenases (GXGXXA), but NOS has the GXGXXG motif usually seen in NAD(H)-linked dehydrogenases (Figure 17).

SACCHAROPINE DEHYDROGENASE   Xuan et al (201) recently reported a weak resemblance between the fingerprint of Wierenga et al (264) and the *Y*.

lipolytica SDH segment between residues 195 and 230. Indeed, inspection of the sequence does not reveal the core fingerprint for the binding of NADH. The sequence GSG that is centered around Ser207 may be involved in the binding of the dinucleotide pyrophosphate group, but the characteristics of a typical $\beta\alpha\beta$-binding fold are not readily discernable in this segment (Figure 17). However, it may be noted that if the sequence Gly203-Gly208 is reversed, the sequence GXGXXG is apparent.

REACTION MECHANISM(S) AND CATALYTIC DOMAINS

Attempts have been made to associate specific amino acids, or particular regions, with catalytic function of the N-(CA)amino acid dehydrogenases. Since the dinucleotide-binding domains of the opine dehydrogenases, NOS and OCS, are located near the NH_2-termini, it has been suggested (260) that residues involved in catalysis are located in the C-terminal moieties of these crown gall enzymes. The reaction mechanism of NOS and OCS [and the other N-(CA)amino acid dehydrogenases] involves the binding of an α-keto acid, and it is tempting to draw analogies with the reactions mediated by lactate dehydrogenase (LDH) and malate dehydrogenase (MDH). The active sites of the latter enzymes contain Asp, His, Arg (and other) residues (283, 284). Of these residues, the Asp-His pair has been shown to be part of a proton relay system (285, 286), and Arg stabilizes the α-keto acid by providing a two-point interaction with the carboxylate group of this acid (287). In LDH and MDH, the catalytically relevant Asp and Arg residues are contained in a conserved Asp-Xaa-Xaa-Arg sequence. Monneuse & Rouzé (260) noted that such a sequence is also present in conserved segments in the C-terminal moieties of both NOS (residues 284–287) and OCS (residues 235–238). These authors also suggested that His373 of NOS, and His263 of OCS, could be part of a putative protein relay system in the opine dehydrogenases. However, these residues are not homologous in the alignment proposed by these authors, whereas His373 of NOS and His333 of OCS are contained in homologous sequences. Furthermore, His288 of NOS and His338 of OCS should also be considered as relay candidates, particularly since they are contained in a highly conserved sequence of 11 residues.

In both CEOS and SDH, the nucleotide-binding region is centrally located, and the dot-plots (Figure 16) indicate homology between the N- and C-terminal regions of these proteins. The sequence, Asp-Xaa-Xaa-Arg, is also found in the C-terminus of CEOS (residues 300–303), and two such sequences (at positions 127–130 and 211–214) are present in SDH.

Chemical modification and inhibitor studies indicate participation of Lys, His, and Arg residues in SDH catalysis (Refs. 288, 289, 290, respectively). From these data Fujioka (291) has proposed an attractive mechanism for saccharopine biosynthesis, which may also be applicable to the synthesis of N^5-(CE)ornithine by CEOS. According to this model, the first step in the synthesis of saccharopine involves hydrogen bonding between a lysyl residue in the enzyme and the carbonyl oxygen of α-ketoglutarate. Protonation of the oxygen atom, and nucleophilic attack by the ϵ-NH_2 group of lysine (substrate) on the α-keto carbon, yields a carbinolamine. Migration of a proton from the secondary amine to the -OH group of the carbinolamine causes elimination of water, and the formation of a Schiff base intermediate. The positively charged imine nitrogen facilitates the stereospecific (A-side) transfer of hydride ion, from NADH to C-2 of the 1,3-dicarboxypropyl moiety of the Schiff base, to yield saccharopine. Now that the SDH gene has been cloned, it should be possible to test this model by site-directed mutagenesis.

CONCLUSIONS AND FUTURE DIRECTIONS

Octopine was isolated by Morizawa in 1927, and the first crown gall opines were described in the late 1950s and early 1960s. At the time of their discovery, these unusual N-(CA)amino acids were primarily of academic interest and were considered "curiosity" chemicals. However, the continued study of these compounds has had ramifications for many areas of biochemistry, physiology, agriculture, and medicine.

The "trapping" of N-(CA)amino acids has provided insight into the mechanisms of the reactions catalyzed by aldolases, amino acid oxidases, and pyruvoyl-linked amino acid decarboxylases (214–218). The discovery of these amino acids in marine invertebrates provided new (and unexpected) perspectives concerning the regulation of glycolysis and the anaerobic physiology of these species (150–153). The *A. tumefaciens* opines have been used as "reporter" molecules, and the *ocs* and *nos* promoters have been incorporated into T-DNA vectors for the improvement of crops and plants by genetic engineering (292, 292a, 293). Last but not least, N-(CA)amino acids have been instrumental in the design of drugs for control of hypertension in humans (229–231).

The discovery in the mid-1980s of the N^{ω}-(CA)amino acids in bacteria (*L. lactis*) was accidental, and their functions are presently unknown. One wonders if these compounds are involved in the regulation of gene expression, or whether they are allosteric modifiers of enzyme activity. Are they intermediates of as-yet-unidentified biosynthetic pathways, or precursors to more complex molecules, e.g. bacteriocins? Are they, perhaps, simply detoxification products? From our experience, it seems probable that N-(CA)amino acids have been overlooked in other bacterial species, and careful re-examination may reveal these compounds to be more widespread than currently envisaged. Such studies may also yield clues to the function(s) of these metabolites in *L. lactis,* and previous experience suggests surprise may be in store!

The N-(CA)amino acid synthases (EC 1.5.1-) act on the CH-NH group, and the amino acid dehydrogenases (EC 1.4.1-) on the $CH-NH_2$ group of donors. In the reverse direction, these enzymes use NAD(P)H, α-keto acids, and amino acids (or NH_3) as substrates. One is tempted to speculate on the structural and catalytic similarities of these enzymes, and ponder whether the respective genes were derived from a common progenitor. The future determination of the structures of these proteins by X-ray diffraction, or NMR-spectroscopy, will be of considerable interest and should provide details of the nucleotide-binding and catalytic domains of these (three-substrate) enzymes.

It augurs well for investigators that there is much to learn, and many questions to be answered, in this area of amino acid biochemistry.

ACKNOWLEDGMENTS

Professors Emeriti Frank C. Happold of Leeds University, England, and Robert A. MacLeod of McGill University, Montreal, taught J. T. the "ways" of science. Chuck Wittenberger, Jack Folk, and Alton Meister provided the encouragement to take the "road less-travelled." We acknowledge the excellent technical assistance of Robert Harr and the expert help of Charlette Cureton in the preparation of this review. Finally, we thank Stephen Miller, Stanley Robrish, John Wootton, and our many colleagues at the National Institutes of Health, for their contributions, advice, and criticisms.

Literature Cited

1. Greenstein, J. P., Winitz, M. 1961. *Chemistry of the Amino Acids.* New York: Wiley
2. Meister, A. 1965. *Biochemistry of the Amino Acids.* New York: Academic. 2nd ed.
3. Bender, D. A. 1985. *Amino Acid Metabolism.* New York: Wiley. 2nd ed.
4. Barrett, G. C., ed. 1985. *Chemistry and Biochemistry of the Amino Acids,* ed. G. C. Barrett. London: Chapman & Hall
5. Hunt, S. 1985. See Ref. 4, pp. 55–138
6. Wold, F. 1981. *Annu. Rev. Biochem.* 50:783–814
7. Thompson, J., Curtis, M. A., Miller, S. P. F. 1986. *J. Bacteriol.* 167:522–29
8. Miller, S. P. F., Thompson, J. 1987. *J. Biol. Chem.* 262:16109–15
9. Thompson, J., Miller, S. P. F. 1988. *J. Biol. Chem.* 263:2064–69
10. Tempé, J. 1983. In *Chemistry and Biochemistry of Amino Acids, Peptides, and Proteins,* ed. B. Weinstein, pp. 113–203. New York: Dekker
11. Thompson, J., Miller, S. P. F. 1991. *Adv. Enzymol.* 64:317–99
12. Tempé, J., Goldmann, A. 1982. In *Molecular Biology of Plant Tumors,* ed. G. Kahl, J. S. Schell, pp. 427–49. Orlando: Academic
13. Abderhalden, E., Haase, E. 1931. *Hoppe Seyler's Z. Physiol. Chem.* 202: 49–55
14. Izumiya, N., Wade, R., Winitz, M., Otey, M. C., Birnbaum, S. M., et al. 1957. *J. Am. Chem. Soc.* 79:652–58
15. Biemann, K., Lioret, C., Asselineau, J., Lederer, E., Polonsky, J. 1960. *Bull. Soc. Chim. Biol.* 42:979–91
16. Goto, K., Waki, M., Mitsuyasu, N., Kitajima, Y., Izumiya, N. 1982. *Bull. Chem. Soc. Jpn.* 55:261–65
17. Karrer, P., Appenzeller, R. 1942. *Helv. Chim. Acta* 25: 595–99
18. Biellmann, J. F., Branlant, G., Wallen, L. 1977. *Bioorg. Chem.* 6:89–93
19. Jensen, R. E., Zdybak, W. T., Yasuda,

K., Chilton, W. S. 1977. *Biochem. Biophys. Res. Commun.* 75:1066–70
20. Cooper, D., Firmin, J. L. 1977. *Org. Prep. Proc. Int.* 9:99–101
21. Effenberger, F., Burkard, U. 1986. *Liebigs Ann. Chem.,* pp. 334–58
22. Miyazawa, T. 1980. *Bull. Chem. Soc. Jpn.* 53:2555–65
23. Drummond, M. 1979. *Nature* 281:343–47
24. Nester, E. W., Gordon, M. P., Amasino, R. M., Yanofsky, M. F. 1984. *Annu. Rev. Plant Physiol.* 35:387–413
25. Nester, E. W., Kosuge, T. 1981. *Annu. Rev. Microbiol.* 35:531–65
26. Ream, W. 1989. *Annu. Rev. Phytopathol.* 27:583–618
27. Kemp, J. D., Hack, E., Sutton, D. W., El-Wakil, M. 1978. *Proc. Int. Conf. Plant Pathog. Bacteria* 4:183–88
28. Smith, E. F., Townsend, C. O. 1907. *Science* 25:671–73
29. Schell, J., van Montagu, M., De Beuckeleer, M., De Block, M., Depicker, A., et al. 1979. *Proc. R. Soc. London Ser. B* 204:251–66
30. Chilton, M.-D., Drummond, M. H., Merlo, D. J., Sciaky, D., Montoya, A. L., et al. 1977. *Cell* 11:263–71
31. Hooykaas, P. J. J., Schilperoort, R. A. 1984. *Adv. Genet.* 22:209–83
32. Gheysen, G., Dhaese, P., van Montagu, M., Schell, J. 1985. *Adv. Plant Gene Res.* 2:11–47
33. Binns, A. N., Thomashow, M. F. 1988. *Annu. Rev. Microbiol.* 42:575–606
34. Thompson, J. 1987. In *Sugar Transport and Metabolism in Gram-Positive Bacteria,* ed. J. Reizer, A. Peterkofsky, pp. 13–38. Chichester: Horwood
35. Thompson, J. 1987. *FEMS Microbiol. Rev.* 46:221–31
36. Thompson, J. 1988. *Biochimie* 70:325–36
37. Fujioka, M., Tanaka, M. 1978. *Eur. J. Biochem.* 90:297–300
38. Thompson, J. 1989. *J. Biol. Chem.* 264:9592–601
39. Thompson, J., Nguyen, N. Y., Sackett, D. L., Donkersloot, J. A. 1991. *J. Biol. Chem.* 266:14573–79
40. Thompson, J., Harr, R. J., Donkersloot, J. A. 1990. *Curr. Microbiol.* 20:239–44
41. Poolman, B., Driessen, A. J. M., Konings, W. N. 1987. *J. Bacteriol.* 169:5597–604
42. Thompson, J. 1987. *J. Bacteriol.* 169:4147–53
43. Hurst, A. 1981. *Adv. Appl. Microbiol.* 27:85–123
44. Rayman, K., Hurst, A. 1984. In *Biotechnology of Industrial Antibiotics,*

ed. E. J. Vandamme, pp. 607–28. New York/Basel: Dekker
45. Gross, E., Morell, J. L. 1971. *J. Am. Chem. Soc.* 93:4634–35
46. Buchman, G. W., Banerjee, S., Hansen, J. N. 1988. *J. Biol. Chem.* 263:16260–66
47. Kaletta, C., Entian, K.-D. 1989. *J. Bacteriol.* 171:1597–601
48. Kozak, W., Rajchert-Trzpil, M., Dobrzanski, W. T. 1974. *J. Gen. Microbiol.* 83:295–302
49. LeBlanc, D. J., Crow, V. L., Lee, L. N. 1980. In *Plasmids and Transposons. Environmental Effects and Maintenance Mechanisms,* ed. C. Stuttard, K. R. Rozee, pp. 31–41. New York: Academic
50. Gonzalez, C. F., Kunka, B. S. 1985. *Appl. Environ. Microbiol.* 49:627–33
51. Tsai, H.-J., Sandine, W. E. 1987. *Appl. Environ. Microbiol.* 53:352–57
52. Steele, J. L., McKay, L. L. 1986. *Appl. Environ. Microbiol.* 51:57–61
53. Gasson, M. J. 1984. *FEMS Microbiol. Lett.* 21:7–10
54. Fuchs, P. G., Zajdel, J., Dobrzanski, W. T. 1975. *J. Gen. Microbiol.* 88:189–202
55. Donkersloot, J. A., Thompson, J. 1990. *J. Bacteriol.* 172:4122–26
56. Dodd, H. M., Horn, N., Gasson, M. J. 1990. *J. Gen. Microbiol.* 136:555–66
57. Thompson, J., Sackett, D. L., Donkersloot, J. A. 1991. *J. Biol. Chem.* 266:22626–33
57a. Horn, N., Swindell, S., Dodd, H., Gasson, M. 1991. *Mol. Gen. Genet.* 228:129–35
58. Griffith, O. E., Meister, A. 1977. *Proc. Natl. Acad. Sci. USA* 74:3330–34
59. Taniguchi, N., Meister, A. 1978. *J. Biol. Chem.* 253:1799–806
60. Meister, A., Anderson, M. E. 1983. *Annu. Rev. Biochem.* 52:711–60
61. Anderson, M. E., Meister, A. 1986. *Proc. Natl. Acad. Sci. USA* 83:5029–32
62. Rosenthal, G. A. 1978. *Life Sci.* 23:93–98
63. Williamson, J. D., Archard, L. C. 1974. *Life Sci.* 14:2481–90
64. Rahiala, E-L., Kekomäki, M., Jänne, J., Raina, A., Räihä, N. C. R. 1971. *Biochim. Biophys. Acta* 227:337–43
65. Rosenthal, G. A., Dahlman, D. L. 1990. *J. Biol. Chem.* 265:868–73
66. Rosenthal, G. A. 1981. *Eur. J. Biochem.* 114:301–4
67. Rosenthal, G. A., Berge, M. A., Bleiler, J. A. 1989. *Biochem. System. Ecol.* 17:203–6
68. Cooper, A. J. L. 1984. *Arch. Biochem. Biophys.* 233:603–10

69. Asano, Y., Yamaguchi, K., Kondo, K. 1989. *J. Bacteriol.* 171:4466–71
70. Holsters, M., Hernalsteens, J. P., van Montagu, M., Schell, J. 1982. See Ref. 12, pp. 269–98
71. Petit, A., Tempé, J. 1985. In *Molecular Form and Function of the Plant Genome*, ed. L. van Vloten-Doting, G. S. P. Groot, T. C. Hall, pp. 625–36. New York: Plenum
72. Firmin, J. L. 1990. *J. Chromatogr.* 514:343–47
73. Ménagé, A., Morel, G. 1964. *C. R. Acad. Sci. Paris* 259:4795–96
74. Lioret, C. 1956. *Bull. Soc. Fr. Physiol. Veg.* 2:76
75. Lejeune, B. 1967. *C. R. Acad. Sci. Paris Ser. D* 265:1753–55
76. Ménagé, A., Morel, G. 1964. *C. R. Soc. Biol.* 159:561–62
77. Goldmann-Ménagé, A. 1970. *Ann. Sci. Nat. Bot. Biol. Veg.* 11:223–310
78. Kemp, J. D. 1977. *Biochem. Biophys. Res. Commun.* 74:862–68
79. Kitajima, Y., Waki, M., Izumiya, N. 1982. *Mem. Fac. Sci. Kyushu Univ. Ser. C* 13:349–56
80. Bates, H. A., Kaushal, A., Deng, P.-N., Sciaky, D. 1984. *Biochemistry* 23:3287–90
81. Firmin, J. L., Stewart, I. M., Wilson, K. E. 1985. *Biochem. J.* 232:431–34
82. Goldmann, A., Thomas, D. W., Morel, G. 1969. *C. R. Acad. Sci. Paris Ser. D* 268:852–54
83. Bates, H. A. 1986. *Ann. NY Acad. Sci.* 471:289–90
84. Firmin, J. L., Fenwick, R. G. 1977. *Phytochemistry* 16:761–62
85. Kemp, J. D. 1978. *Plant Physiol.* 62:26–30
86. Chang, C.-C., Chen, C.-M., Adams, B. R., Trost, B. M. 1983. *Proc. Natl. Acad. Sci. USA* 80:3573–76
87. Chilton, W. S., Hood, E., Chilton, M.-D. 1985. *Phytochemistry* 24:221–24
88. Chang, C.-C., Chen, C.-M. 1983. *FEBS Lett.* 162:432–35
89. Chilton, W. S., Tempé, J., Matzke, M., Chilton, M.-D. 1984. *J. Bacteriol.* 157:357–62
90. Chilton, W. S., Rinehart, K. L. Jr., Chilton, M.-D. 1984. *Biochemistry* 23:3290–97
91. Chilton, W. S., Hood, E., Rinehart, K. L. Jr., Chilton, M.-D. 1985. *Phytochemistry* 24:2945–48
92. Davioud, E., Petit, A., Tate, M. E., Ryder, M. H., Tempé, J. 1988. *Phytochemistry* 27:2429–33
93. Davioud, E., Quirion, J-C., Tate, M.

E., Tempé, J., Husson, H-P. 1988. *Heterocycles* 27:2423–29
94. Isogai, A., Fukuchi, N., Hayashi, M., Kamada, H., Harada, H., et al. 1988. *Agric. Biol. Chem.* 52:3235–37
95. Isogai, A., Fukuchi, N., Hayashi, M., Kamada, H., Harada, H., et al. 1990. *Phytochemistry* 29:3131–34
96. Hall, L. M., Schrimsher, J. L., Taylor, K. B. 1983. *J. Biol. Chem.* 258:7276–79
97. Hatanaka, S-I., Atsumi, S., Furukawa, K., Ishida, Y. 1982. *Phytochemistry* 21:225–27
98. DeGreve, H., Decraemer, H., Seurinck, J., van Montagu, M., Schell, J. 1981. *Plasmid* 6:235–48
99. Holsters, M., Silva, B., van Vliet, F., Genetello, C., De Block, M., et al. 1980. *Plasmid* 3:212–30
100. Zambryski, P. 1988. *Annu. Rev. Genet.* 22:1–30
101. Zambryski, P., Tempé, J., Schell, J. 1989. *Cell* 56:193–201
102. Howard, E., Citovsky, V. 1990. *BioEssays* 12:103–8
103. Chilton, M.-D. 1982. See Ref. 12, pp. 299–319
104. Montoya, A. L., Chilton, M.-D., Gordon, M. P., Sciaky, D., Nester, E. W. 1977. *J. Bacteriol.* 129:101–7
105. Bomhoff, G., Klapwijk, P. M., Kester, H. C. M., Schilperoort, R. A., Hernalsteens, J. P., et al. 1976. *Mol. Gen. Genet.* 145:177–81
106. Lejeune, B., Jubier, M. 1970. *Physiol. Veg.* 8:308
107. Goldmann, A. 1977. *Plant Sci. Lett.* 10:49–58
108. Hack, E., Kemp, J. D. 1977. *Biochem. Biophys. Res. Commun.* 78:785–91
109. Hack, E., Kemp, J. D. 1980. *Plant Physiol.* 65:949–55
110. Barker, R. F., Idler, K. B., Thompson, D. V., Kemp, J. D. 1983. *Plant Mol. Biol.* 2:335–50
111. De Greve, H., Dhaese, P., Seurinck, J., Lemmers, M., van Montagu, M., et al. 1983. *J. Mol. Appl. Genet.* 1:499–511
112. Otten, L. A. B. M., Vreugdenhil, D., Schilperoort, R. A. 1977. *Biochim. Biophys. Acta* 485:268–77
113. Otten, L., Szegedi, E. 1985. *Plant Sci.* 40:81–85
114. Kemp, J. D., Sutton, D. W., Hack, E. 1979. *Biochemistry* 18:3755–60
115. Depicker, A., Stachel, S., Dhaese, P., Zambryski, P., Goodman, H. M. 1982. *J. Mol. Appl. Genet.* 1:561–73
116. Bevan, M., Barnes, W. M., Chilton, M.-D. 1983. *Nucleic Acids Res.* 11:369–85

117. Gafni, Y., Chilton, M.-D. 1985. *Gene* 39:141–46
118. Lippincott, J. A., Lippincott, B. B. 1970. *Science* 170:176–77
119. Lippincott, J. A., Lippincott, B. B., Chang, C-C. 1972. *Plant Physiol.* 49:131–37
120. Veluthambi, K., Krishnan, M., Gould, J. H., Smith, R. H., Gelvin, S. B. 1989. *J. Bacteriol.* 171:3696–703
121. Shaw, C. H. 1991. *BioEssays* 13:25–29
122. Petit, A., Dessaux, Y., Tempé, J. 1978. *Proc. Int. Conf. Plant. Pathog. Bacteria* 4:143–52
123. Ellis, J. G., Kerr, A., Tempé, J., Petit, A. 1979. *Mol. Gen. Genet.* 173:263–69
124. Messens, E., Lenaerts, A., van Montagu, M., Hedges, R. W. 1985. *Mol. Gen. Genet.* 199:344–48
125. Klapwijk, P. M., Hooykaas, P. J. J ., Kester, H. C. M., Schilperoort, R. A., Rorsch, A. 1976. *J. Gen. Microbiol.* 96:155–63
126. Klapwijk, P. M., Oudshoorn, M., Schilperoort, R. A. 1977. *J. Gen. Microbiol.* 102:1–11
127. Krishnan, M., Burgner, J. W., Chilton, W. S., Gelvin, S. B. 1991. *J. Bacteriol.* 173:903–5
128. Klapwijk, P. M., Schilperoort, R. A. 1982. See Ref. 12, pp. 475–95
129. Jubier, M-F. 1972. *FEBS Lett.* 28:129–32
130. Schardl, C. L., Kado, C. I. 1983. *Mol. Gen. Genet.* 191:10–16
131. Tempé, J., Petit, A. 1982. See Ref. 12, pp. 451–59
132. Dessaux, Y., Petit, A., Tempé, J., Demarez, M., Legrain, C., et al. 1986. *J. Bacteriol.* 166:44–50
133. Sans, N., Schröder, G., Schröder, J. 1987. *Eur. J. Biochem.* 167:81–87
134. Sans, N., Schindler, U., Schröder, J. 1988. *Eur. J. Biochem.* 173:123–30
135. Schindler, U., Sans, N., Schröder, J. 1989. *J. Bacteriol.* 171:847–54
136. Wabiko, H., Kagaya, M., Sano, H. 1990. *J. Gen. Microbiol.* 136:97–103
137. Rossignol, G., Dion, P. 1985. *Can. J. Microbiol.* 31:68–74
138. Dion, P. 1986. *Can. J. Microbiol.* 32:959–63
139. Boivin, R., Lebeuf, H., Dion, P. 1987. *Can. J. Microbiol.* 33:534–40
140. Saint-Pierre, B., Dion, P. 1988. *Can. J. Microbiol.* 34:793–801
141. Beauchamp, C. J., Kloepper, J. W., Lifshitz, R., Dion, P., Antoun, H. 1990. *Can. J. Microbiol.* 37:158–64
142. Nautiyal, C. S., Dion, P. 1990. *Appl. Environ. Microbiol.* 56:2576–79
143. Bell, C. R. 1990. *Appl. Environ. Microbiol.* 56:1775–81
144. Bell, C. R., Moore, L. W., Canfield, M. L. 1990. *Appl. Environ. Microbiol.* 56:2834–39
145. Bergeron, J., MacLeod, R. A., Dion, P. 1990. *Appl. Environ. Microbiol.* 56: 1453–58
146. Beaulieu, C., Gill, S., Miville, L., Dion, P. 1988. *Can. J. Microbiol.* 34:843–49
147. Gäde, G. 1980. *Mar. Biol. Lett.* 1:121–35
148. Dando, P. R., Storey, K. B., Hochachka, P. W., Storey, J. M. 1981. *Mar. Biol. Lett.* 2:249–57
149. Livingstone, D. R., de Zwaan, A., Leopold, M., Marteijn, E. 1983. *Biochem. System. Ecol.* 11:415–25
150. Gäde, G., Grieshaber, M. K. 1986. *Comp. Biochem. Physiol.* 83B:255–72
151. Fields, J. H. A. 1983. *J. Exp. Zool.* 228:445–57
152. Gäde, G. 1983. *J. Exp. Zool.* 228:415–29
153. de Zwaan, A., Dando, P. R. 1984. *Mol. Physiol.* 5:285–310
154. Morizawa, K. 1927. *Acta Schol. Med. Univ. Imp. Kioto* 9:285
155. Thoai, N. V., Robin, Y. 1959. *Biochim. Biophys. Acta* 35:446–53
156. Thoai, N. V., Robin, Y. 1961. *Biochim. Biophys. Acta* 52:221–33
157. Thoai, N. V., Huc, C., Pho, D. B., Olomucki, A. 1969. *Biochim. Biophys. Acta* 191:46–57
158. Olomucki, A., Huc, C., Lefebure, F., Thoai, N. V. 1972. *Eur. J. Biochem.* 28:261–68
159. Olomucki, A. 1981. *Biochem. Soc. Trans.* 9:278–79
160. Storey, K. B., Dando, P. R. 1982. *Comp. Biochem. Physiol.* 73B:521–28
161. Monneuse-Doublet, M.-O., Olomucki, A. 1981. *Biochem. Soc. Trans.* 9:300–2
162. Carvajal, N., Kessi, E. 1988. *Biochim. Biophys. Acta* 953:14–19
163. Biellmann, J-F., Branlant, G., Olomucki, A. 1973. *FEBS Lett.* 32:254–56
164. Schrimsher, J. L., Taylor, K. B. 1982. *J. Biol. Chem.* 257:8953–56
165. Zettlmeissl, G., Teschner, W., Rudolph, R., Jaenicke, R., Gäde, G. 1984. *Eur. J. Biochem.* 143:401–7
166. Thomé, F., Vachette, P., Dubord, C., Olomucki, A. 1987. *Biochim. Biophys. Acta* 915:342–45
167. Teschner, W., Rudolph, R., Garel, J-R. 1987. *Biochemistry* 26:2791–96
168. Monneuse-Doublet, M.-O., Lefebure, F., Olomucki, A. 1980. *Eur. J. Biochem.* 108:261–69
169. Storey, K. B. 1977. *J. Comp. Physiol. B* 115:159–69

170. Gäde, G. 1980. *Comp. Biochem. Physiol.* 67B:575–82
171. Storey, K. B., Storey, J. M. 1979. *Eur. J. Biochem.* 93:545–52
172. Storey, K. B., Storey, J. M. 1979. *J. Comp. Physiol. B* 131:311–19
173. Sato, M., Sato, Y., Tsuchiya, Y. 1977. *Nippon Suisan Gakkaishi* 43:1077–79
174. Fields, J. H. A., Hochachka, P. W. 1981. *Eur. J. Biochem.* 114:615–21
175. Siegmund, B., Grieshaber, M., Reitze, M., Zebe, E. 1985. *Comp. Biochem. Physiol.* 82B:337–45
176. Dando, P. R. 1981. *Biochem. Soc. Trans.* 9:297–98
177. Plaxton, W. C., Storey, K. B. 1982. *Can. J. Zool.* 60:1568–72
178. Sangster, A. W., Thomas, S. E., Tingling, N. L. 1975. *Tetrahedron* 31:1135–37
179. Storey, K. B., Miller, D. C., Plaxton, W. C., Storey, J. M. 1982. *Anal. Biochem.* 125:50–58
180. Siegmund, B., Grieshaber, M. K. 1983. *Hoppe-Seyler's Z. Physiol. Chem.* 364:807–12
181. Fiore, G. B., Nicchitta, C. V., Ellington, W. R. 1984. *Anal. Biochem.* 139:413–17
182. Kuriyama, M. 1961. *Nature* 192:969
183. Sato, M., Kanno, N., Sato, Y. 1985. *Bull. Jpn. Soc. Fish.* 51:1681–83
184. Sato, M., Gäde, G. 1986. *Naturwissenschaften* 73:207–9
185. Gäde, G. 1986. *Eur. J. Biochem.* 160:311–18
186. Sato, M., Takeuchi, M., Kanno, N., Nagahisa, E., Sato, Y. 1991. *Tohoku J. Agric. Res.* 41:83–95
187. Doumen, C., Ellington, W. R. 1987. *J. Exp. Zool.* 243:25–31
188. Sato, M., Takahara, M., Kanno, N., Sato, Y., Ellington, W. R. 1987. *Comp. Biochem. Physiol.* 88B:803–6
189. Goldmann, A., Moureaux, T., Rouzé, P. 1981. *FEBS Lett.* 130:213–16
190. Baldwin, J. 1982. *Pac. Sci.* 36:357–63
191. Fort, L., Dando, P. R., Rouzé, P., Monneuse, M.-O., Olomucki, A. 1982. *Comp. Biochem. Physiol.* 73B:865–71
192. Bhattacharjee, J. K. 1985. *CRC Crit. Rev. Microbiol.* 12:131–51
193. Broquist, H. P. 1971. *Methods Enzymol.* 17B:112–29
194. Darling, S., Larsen, P. O. 1961. *Acta Chem. Scand.* 15:743–49
195. Kjaer, A., Larsen, P. O. 1961. *Acta Chem. Scand.* 15:750–59
196. Larsen, P. O. 1972. *Acta Chem. Scand.* 26:2562–63
197. Burkard, U., Walther, I., Effenberger, F. 1986. *Liebigs Ann. Chem.*, pp. 1030–43

198. Ogawa, H., Fujioka, M. 1978. *J. Biol. Chem.* 253:3666–70
199. Fujioka, M. 1984. *Arch. Biochem. Biophys.* 230:553–59
200. Xuan, J-W., Fournier, P., Gaillardin, C. 1988. *Curr. Genet.* 14:15–21
201. Xuan, J-W., Fournier, P., Declerck, N., Chasles, M., Gaillardin, C. 1990. *Mol. Cell. Biol.* 10:4795–806
202. Sugimoto, K., Fujioka, M. 1978. *Eur. J. Biochem.* 90:301–7
203. Kasai, T., Sakamura, S., Sakamoto, R. 1971. *Agric. Biol. Chem.* 35:1603–6
204. Kasai, T., Sakamura, S. 1981. *Agric. Biol. Chem.* 45:1483–85
205. Matsutani, H., Kusumoto, S., Koizumi, R., Shiba, T. 1979. *Phytochemistry* 18:661–62
206. Wadman, S. K., de Bree, P.K., van Sprang, F. J., Kamerling, J. P., Haverkamp, J., et al. 1975. *Clin. Chim. Acta* 59:313–20
207. Liardon, R., De Weck-Gaudard, D., Philippossian, G., Finot, P-A. 1987. *J. Agric. Food Chem.* 35:427–31
208. Ahmed, M. U., Thorpe, S. R., Baynes, J. W. 1986. *J. Biol. Chem.* 261:4889–94
209. Büser, W., Erbersdobler, H. F., Liardon, R. 1987. *J. Chromatogr.* 387:515–19
210. Ronca, G., Chiti, R., Lucacchini, A. 1970. *J. Chromatogr.* 47:114–15
211. Gould, B. J., Walsh, D. P., Ah-Sing, E., Wright, J., Stace, B. C. 1989. *Biochem. Soc. Trans.* 17:1087–88
212. Kobes, R. D., Dekker, E. E. 1971. *Biochemistry* 10:388–94
213. Vlahos, C. J., Dekker, E. E. 1986. *J. Biol. Chem.* 261:11049–55
214. Hellerman, L., Coffey, D. S. 1967. *J. Biol. Chem.* 242:582–89
215. Hafner, E. W., Wellner, D. 1971. *Proc. Natl. Acad. Sci. USA* 68:987–91
216. Porter, D. J. T., Bright, H. J. 1972. *Biochem. Biophys. Res. Commun.* 46:571–77
217. Recsei, P. A., Snell, E. E. 1984. *Annu. Rev. Biochem.* 53:357–87
218. van Poelje, P. D., Snell, E. E. 1990. *Annu. Rev. Biochem.* 59:29–59
219. Ferreira, S. H., Bartelt, D. C., Greene, L. J. 1970. *Biochemistry* 9:2583–93
220. Kato, H., Suzuki, T. 1970. *Experientia* 26:1205–6
221. Kato, H., Suzuki, T. 1971. *Biochemistry* 10:972–80
222. Ondetti, M. A., Williams, N. J., Sabo, E. F., Pluščec, J., Weaver, E. R., et al. 1971. *Biochemistry* 10:4033–39
223. Ondetti, M. A., Cushman, D. W. 1982. *Annu. Rev. Biochem.* 51:283–308
224. Soffer, R. L. 1981. In *Biochemical*

Regulation of Blood Pressure, ed. R. L. Soffer, pp. 123–64. New York: Wiley

225. Hubert, C., Houot, A-M., Corvol, P., Soubrier, F. 1991. *J. Biol. Chem.* 266:15377–83

226. Ondetti, M. A., Rubin, B., Cushman, D. W. 1977. *Science* 196:441–44

227. Cushman, D. W., Cheung, H. S., Sabo, E. F., Ondetti, M. A. 1977. *Biochemistry* 16:5484–91

228. Ondetti, M. A., Cushman, D. W. 1981. See Ref. 224, pp.165–204

229. Wyvratt, M. J., Patchett, A. A. 1985. *Med. Res. Rev.* 4:483–531

230. Patchett, A. A., Cordes, E. H. 1985. *Adv. Enzymol.* 57:1–84

231. Patchett, A. A., Harris, E., Tristram, E. W., Wyvratt, M. J., Wu, M. T., et al. 1980. *Nature* 288:280–83

232. Bull, H. G., Thornberry, N. A., Cordes, M. H. J., Patchett, A. A., Cordes, E. H. 1985. *J. Biol. Chem.* 260:2952–62

233. Maycock, A. L., DeSousa, D. M., Payne, L. G., ten Broeke, J., Wu, M. T., et al. 1981. *Biochem. Biophys. Res. Commun.* 102:963–69

234. Fournié-Zaluski, M-C., Soroca-Lucas, E., Waksman, G., Llorens, C., Schwartz, J-C., et al. 1982. *Life Sci.* 31:2947–54

235. Fournié-Zaluski, M-C., Chaillet, P., Soroca-Lucas, E., Marçais-Collado, H., Costentin, J., et al. 1983. *J. Med. Chem.* 26:60–65

236. Gray, R. D., Pierce, W. M. Jr., Harrod, J. W. Jr., Rademacher, J. M. 1987. *Arch. Biochem. Biophys.* 256:692–98

237. Almenoff, J., Orlowski, M. 1983. *Biochemistry* 22:590–99

238. Chu, T. G., Orlowski, M. 1984. *Biochemistry* 23:3598–603

239. Lolis, E., Petsko, G. A. 1990. *Annu. Rev. Biochem.* 59:597–630

240. Monzingo, A. F., Matthews, B. W. 1984. *Biochemistry* 23:5724–29

241. Hangauer, D. G., Monzingo, A. F., Matthews, B. W. 1984. *Biochemistry* 23:5730–41

242. El-Dorry, H. A., Bull, H. G., Iwata, K., Thornberry, N. A., Cordes, E. H., et al. 1982. *J. Biol. Chem.* 257:14128–33

243. Bull, H. G., Thornberry, N. A., Cordes, E. H. 1985. *J. Biol. Chem.* 260:2963–72

244. Stack, M. S., Emberts, C. G., Gray, R. D. 1991. *Arch. Biochem. Biophys.* 287:240–49

245. An, G., Ebert, P. R., Yi, B-Y., Choi, C-H. 1986. *Mol. Gen. Genet.* 203:245–50

246. Ebert, P. R., Ha, S. B., An, G. 1987. *Proc. Natl. Acad. Sci. USA* 84:5745–49

247. Mitra, A., An, G. 1989. *Mol. Gen. Genet.* 215:294–99

248. Ellis, J. G., Llewellyn, D. J., Walker, J. C., Dennis, E. S., Peacock, W. J. 1987. *EMBO J.* 6:3203–8

249. Leisner, S. M., Gelvin, S. B. 1988. *Proc. Natl. Acad. Sci. USA* 85:2553–57

250. Bouchez, D., Tokuhisa, J. G., Llewellyn, D. J., Dennis, E. S., Ellis, J. G. 1989. *EMBO J.* 8:4197–204

251. An, G., Costa, M. A., Mitra, A., Ha, S-B., Marton, L. 1988. *Plant Physiol.* 88:547–52

252. An, G., Costa, M. A., Ha, S-B. 1990. *Plant Cell* 2:225–33

253. Singh, K., Tokuhisa, J. G., Dennis, E. S., Peacock, W. J. 1989. *Proc. Natl. Acad. Sci. USA* 86:3733–37

254. Tokuhisa, J. G., Singh, K., Dennis, E. S., Peacock, W. J. 1990. *Plant Cell* 2:215–24

255. Singh, K., Dennis, E. S., Ellis, J. G., Llewellyn, D. J., Tokuhisa, J. G., et al. 1990. *Plant Cell* 2:891–903

256. Arndt, K., Fink, G. R. 1986. *Proc. Natl. Acad. Sci. USA* 83:8516–20

257. van der Vossen, J. M. B. M., van der Lelie, D., Venema, G. 1987. *Appl. Environ. Microbiol.* 53:2452–57

258. Koivula, T., Sibakov, M., Palva, I. 1991. *Appl. Environ. Microbiol.* 57:333–40

259. Devereux, J., Haeberli, P., Smithies, O. 1984. *Nucleic Acids Res.* 12:387–95

260. Monneuse, M.-O., Rouzé, P. 1987. *J. Mol. Evol.* 25:46–57

261. Taguchi, H., Ohta, T. 1991. *J. Biol. Chem.* 266:12588–94

262. Rossmann, M. G., Liljas, A., Branden, C-I., Banaszak, L. J. 1975. *The Enzymes* 11:61–102

263. Birktoft, J. J., Banaszak, L. J. 1984. *Peptide Protein Rev.* 4:1–46

264. Wierenga, R. K., Terpstra, P., Hol, W. G. J. 1986. *J. Mol. Biol.* 187:101–7

265. Wierenga, R. K., De Maeyer, M. C. H., Hol, W. G. J. 1985. *Biochemistry* 24:1346–57

266. Hol, W. G. J. 1985. *Prog. Biophys. Mol. Biol.* 45:149–95

267. Hanukoglu, I., Gutfinger, T. 1989. *Eur. J. Biochem.* 180:479–84

268. Scrutton, N. S., Berry, A., Perham, R. N. 1990. *Nature* 343:38–43

269. Feeney, R., Clarke, A. R., Holbrook, J. J. 1990. *Biochem. Biophys. Res. Commun.* 166:667–72

270. Pai, E. F., Karplus, P. A., Schulz, G. E. 1988. *Biochemistry* 27:4465–74

271. Karplus, P. A., Schulz, G. E. 1987. *J. Mol. Biol.* 195:701–29

272. Corbier, C., Clermont, S., Billard, P.,

Skarzynski, T., Branlant, C., et al. 1990. *Biochemistry* 29:7101–6

273. Karplus, P. A., Daniels, M. J., Herriot, J. R. 1991. *Science* 251:60–66

274. McPherson, M. J., Wootton, J. C. 1983. *Nucleic Acids Res.* 11:5257–66

275. Valle, F., Becerril, B., Chen, E., Seeburg, P., Heyneker, H., et al. 1984. *Gene* 27:193–99

276. Bansal, A., Dayton, M. A., Zalkin, H., Colman, R. F. 1989. *J. Biol. Chem.* 264:9827–35

277. Moye, W. S., Amuro, N., Rao, J. K. M., Zalkin, H. 1985. *J. Biol. Chem.* 260:8502–8

278. Nagasu, T., Hall, B. D. 1985. *Gene* 37:247–53

279. Kinnaird, J. H., Fincham, J. R. S. 1983. *Gene* 26:253–60

280. Rice, D. W., Baker, P. J., Farrants, G. W., Hornby, D. P. 1987. *Biochem. J.* 242:789–95

281. Rossmann, M. G., Moras, D., Olsen, K. W. 1974. *Nature* 250:194–99

282. Wootton, J. C. 1974. *Nature* 252:542–46

282a. Baker, P. J., Britton, K. L., Engel, P. C., Farrants, G. W., Lilley, K. S., et al. 1992. *Proteins: Struct. Funct. Genet.* 12:75–86

283. Clarke, A. R., Atkinson, T., Holbrook, J. J. 1989. *Trends Biochem. Sci.* 14: 101–5

284. Clarke, A. R., Atkinson, T., Holbrook, J. J. 1989. *Trends Biochem. Sci.* 14: 145–48

285. Birktoft, J. J., Banaszak, L. J. 1983. *J. Biol. Chem.* 258:472-82

286. Clarke, A. R., Wilks, H. M., Barstow, D. A., Atkinson, T., Chia, W. N., et al. 1988. *Biochemistry* 27:1617–22

287. Hart, K. W., Clarke, A. R., Wigley, D.

B., Chia, W. N., Barstow, D. A., et al. 1987. *Biochem. Biophys. Res. Commun.* 146:346–53

288. Ogawa, H., Fujioka, M. 1980. *J. Biol. Chem.* 255:7420–25

289. Fujioka, M., Takata, Y., Ogawa, H., Okamoto, M. 1980. *J. Biol. Chem.* 255:937–42

290. Fujioka, M., Takata, Y. 1981. *Biochemistry* 20:468–72

291. Fujioka, M. 1981. *Biochem. Soc. Trans.* 9:281–82

292. Zambryski, P., Herrera-Estrella, L., De Block, M., van Montagu, M., Schell, J. 1984. In *Genetic Engineering, Principles and Methods*, ed. J. Setlow, A. Hollaender, 6:253–78. New York: Plenum

292a. Schell, J. S. 1987. *Science* 237:1176–83

293. Gasser, C. S., Fraley, R. T. 1989. *Science* 244:1293–99

294. Rice, D. W., Schulz, G. E., Guest, J. R. 1984. *J. Mol. Biol.* 174:483–96

295. Kuroda, S., Tanizawa, K., Sakamoto, Y., Tanaka, H., Soda, K. 1990. *Biochemistry* 29:1009–15

296. Laddaga, R. A., Chu, L., Misra, T. K., Silver, S. 1987. *Proc. Natl. Acad. Sci. USA* 84:5106–10

297. Stephens, P.E., Lewis, H. M., Darlison, M. G., Guest, J. R. 1983. *Eur. J. Biochem.* 135:519–27

298. Okazaki, N., Hibino, Y., Asano, Y., Ohmori, M., Numao, N., Kondo, K. 1988. *Gene* 63:337–41

299. Nagata, S., Tanizawa, K., Esaki, N., Sakamoto, Y., Ohshima, T., et al. 1988. *Biochemistry* 27:9056–62

300. Barstow, D. A., Clarke, A. R., Chia, W. N., Wigley, D., Sharman, A. F., et al. 1986. *Gene* 46:47–55

Annu. Rev. Biochem. 1992. 61:559–601

NEURONAL Ca²⁺/CALMODULIN-DEPENDENT PROTEIN KINASES

Phyllis I. Hanson and Howard Schulman

Department of Pharmacology, Stanford University School of Medicine, Stanford, California 94305-5332

KEY WORDS: protein kinase, calcium, calmodulin, neuronal, autophosphorylation

CONTENTS

0066-4154/92/0701-0559$02.00

INTRODUCTION AND PERSPECTIVES

Most hormones, neurotransmitters, and other signalling molecules stimulate receptors that are coupled to the generation of the second messengers cAMP, diacylglycerol, and Ca^{2+}. These second messenger systems enable coordinated cellular responses to a single stimulus. Studies on cAMP and diacylglyerol have established that their effects are mediated by general or multifunctional protein kinases that phosphorylate numerous substrates and thereby integrate related functions. Protein kinase C (PKC) mediates the actions of diacylglycerol, while cAMP-dependent protein kinase (PKA) is the primary mediator of cAMP.

Ca^{2+} plays a central role as a second messenger with a regulatory involvement in many aspects of cellular signalling. Intracellular Ca^{2+} levels are buffered to a low concentration (ca. 100 nM), and rapidly rise to levels of 1 μM or more in response to incoming signals. In contrast to the few mediators of cAMP and diacylglycerol, a variety of proteins and enzymes mediate intracellular responses to Ca^{2+}. In many cells, calmodulin is the predominant intracellular "Ca^{2+}-receptor" that activates enzymes in response to Ca^{2+}. Protein kinases are prominent among these, and a number of Ca^{2+}/calmodulin-activated protein kinases have been described in both neuronal and non-neuronal tissues. Several, including myosin light chain kinase (MLCK), phosphorylase kinase, and CaM kinase III (an elongation factor-2 kinase), are dedicated to the phosphorylation and regulation of only a single class of substrate. These enzymes have been primarily studied in non-neuronal tissues; the reader is referred to recent reviews of these fields (1–4). Despite this diversity of Ca^{2+} mediators, a general or multifunctional kinase mediating the effects of Ca^{2+} has also been identified, and this enzyme is the major focus of this review.

Recent studies have established that multifunctional Ca^{2+}/calmodulin-dependent protein kinase (CaM kinase)—also referred to as calmodulin-dependent multiprotein kinase, CaM kinase II, and Type II CaM kinase—is the general or multifunctional protein kinase of Ca^{2+} signalling systems. Its physiological niche, along with those of PKA and PKC, is depicted in Figure 1. CaM kinase is widespread in nature, and phosphorylates a large variety of proteins in response to Ca^{2+} signals. It regulates a broad array of functions, including metabolism of carbohydrates, lipids, and amino acids, neurotransmitter release, neurotransmitter synthesis, ion channels, Ca^{2+} homeostasis, cytoskeletal function, and perhaps gene expression. It is activated by the variety of signal transduction pathways that elevate intracellular free Ca^{2+}. CaM kinase has turned out to be as interesting biochemically as any of its substrates. The study of its activation by calmodulin and of the autophosphorylation that converts it to a Ca^{2+}-independent kinase has produced a number of insights into possible physiological functions as well as to the structure/functional design of this and other regulated protein kinases. A sufficient body of knowledge has now established CaM kinase as one of the three major protein kinases orchestrating cellular responses to second messengers.

Three considerations justify a special interest in CaM kinase as an important factor in neuronal signal transduction. (*a*) It is highly concentrated in brain, and in particular in cortical structures and hippocampus. It is localized on both sides of the synapse where events central to neurotransmission are likely to be directly regulated. (*b*) Many of its substrates are involved in neuronal signalling. CaM kinase modulates both neurotransmitter release and neurotransmitter synthesis. Recently identified substrates contribute to regulation of ion flux, Ca^{2+} homeostasis, Ca^{2+}-stimulated gene expression, and the intercellular signalling potentially mediated by nitric oxide synthase. CaM kinase is also a necessary component in the induction of long-term potentia-

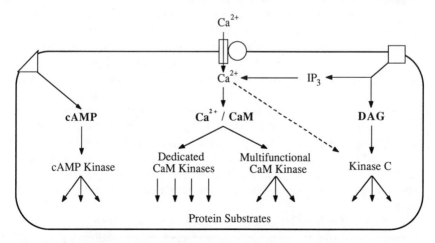

Figure 1 Signal transduction by cAMP, calcium, and diacylglycerol (DAG)

tion, a model of long-term memory. (*c*) Brief Ca^{2+} signals activate CaM kinase, and stimulate an autophosphorylation which allows the kinase to maintain its activated state beyond the duration of a particular Ca^{2+} signal. Furthermore, recently described properties of the autophosphorylation may enable CaM kinase to respond to repetitive Ca^{2+} signals on the basis of their frequency. Potentiation of Ca^{2+} signals by autophosphorylation could play an important role in synaptic plasticity. We focus this review on current understanding of these three aspects of the multifunctional CaM kinase. We also briefly discuss the properties of two other CaM kinases that may have multifunctional roles in transducing Ca^{2+} signals—CaM kinase I and CaM kinase IV (CaM kinase-Gr). Several other reviews on multifunctional CaM kinase provide additional perspectives for the interested reader (5–11).

BIOCHEMICAL AND STRUCTURAL CHARACTERISTICS

Purification and Characterization

Early studies of cellular responses to cAMP led to identification of a central protein kinase responsible for transducing the effects of cAMP within the cell. By analogy to this ubiquitous effector system, it was proposed that other signalling systems such as those involving cGMP and Ca^{2+} might utilize a similar central protein kinase. Ca^{2+} was known to play an important role in regulating a variety of neuronal functions, including exocytosis and excitability. Searches for cAMP-independent protein phosphorylation systems were undertaken, and a Ca^{2+}-sensitive kinase activity was identified in synaptosomal (nerve terminal) preparations. Multifunctional CaM kinase was first specifically described in these synaptosomal membranes (12), and was subsequently shown to be present in membranes from a variety of tissues (13). This kinase phosphorylated the vesicle-associated protein synapsin I as well as several other proteins, two or three of which were later found to be autophosphorylated subunits of the enzyme itself. The soluble form of CaM kinase was described as a novel Ca^{2+}/calmodulin-stimulated protein kinase in rat brain cytosol on the basis of its large size and ability to phosphorylate several substrates that distinguished it from the known dedicated protein kinases (14, 15). Multifunctional CaM kinase was subsequently purified to homogeneity from rat forebrain cytosol using a number of substrates, including myosin light chains (16), synapsin I (17, 18), tryptophan hydroxylase (19), tubulin (20), microtubule-associated protein-2 (MAP-2) (21), and casein (22). All of these kinases share the broad substrate specificity characteristic of multifunctional CaM kinase and are now recognized as being the identical enzyme.

Parallel studies in rabbit skeletal muscle (23) and rabbit liver (24, 25) led to the purification of Ca^{2+}/calmodulin-dependent glycogen synthase kinases that are the non-neuronal isoforms of multifunctional CaM kinase (26, 27). Tissue-specific isoforms of multifunctional CaM kinase have since been purified from dog heart (28, 29), chicken gizzard (30), rat pancreas (31), rat lung (32), and rat spleen (33). Where examined, their properties are similar to the enzyme from brain, although they have slightly different subunit sizes. The kinase prepared from spleen contains subunits of 18, 20, and 21 kDa in addition to two of 50 and 51 kDa, and may include a novel type of subunit (33). Neuronal CaM kinase has also been prepared from species other than rat, including rabbit brain (34), bovine brain (35, 36), mouse brain (37), chicken brain (38), squid synaptosomes (39), and Torpedo electric organ (40).

CaM kinase prepared from rat forebrain typically contains subunits of 50–54 kDa (α subunits) and 58–60 kDa (β subunits) in a ratio of 3–4:1. Ca^{2+}/calmodulin stimulates autophosphorylation of both subunits. Autophosphorylation of α and β will also occur after separation of the subunits on SDS-gels, confirming that both are catalytically active (41, 42). Although the 58–60-kDa subunits in enzyme purified from neuronal tissues are typically described as β subunits, cDNA sequencing described below has identified additional isozymes (β', γ, δ) that are unlikely to be distinguishable from β on the basis of mobility on SDS-gels. We use the convention of referring to the 58–60-kDa subunits as a generic β except when discussing distinct cDNA clones. CaM kinase holoenzyme isolated from rat forebrain has an apparent molecular weight of between 460,000 and 650,000 based on hydrodynamic characterization of the purified enzymes, and is therefore likely a decamer or dodecamer [see (9)]. The cerebellar enzyme also consists of α and β subunits, but in an apparent 1:3–4 ratio (43, 44). In spite of this, cerebellar CaM kinase is similar to forebrain kinase in its holoenzyme size, chromatographic behavior, autophosphorylation, and broad substrate specificity.

Cloning of several CaM kinase isoforms has enabled overexpression and rapid purification of wild-type and mutant CaM kinases for biochemical analysis. Recombinant α-CaM kinase containing only α subunits can be produced by expression in mammalian cells (COS-7, CHO) or by baculovirus infection of insect Sf9 cells (45–47). Such α-CaM kinase displays the substrate phosphoryation, autophosphorylation, calmodulin activation, and molecular size characteristic of the enzyme purified from rat brain (46, 47; P. Hanson and H. Schulman, unpublished observations). Recombinant β-CaM kinase has been similarly produced in CHO cells; this enzyme differs from both rat brain and α-CaM kinase in its apparently monomeric structure (47). α-CaM kinase has also been expressed efficiently in *Escherichia coli*, although solubility and proteolysis of the expressed protein have presented difficulties (48–51). An additional problem with bacterial expression is that it generates a monomeric rather than oligomeric α-CaM kinase (52).

Holoenzyme composition varies in different brain regions. A survey of Ca^{2+}/calmodulin-stimulated autophosphorylation in numerous brain regions demonstrated variable α:β subunit ratios (53, 54). The actual holoenzyme composition in these different regions may be a result of random assembly of available subunits, or may be determined by structural rules governing which subunit combinations are permitted. Immunoprecipitations with antibodies specific to either α or β precipitate both subunits from cerebellum and forebrain, suggesting that both are typically present in individual holoenzymes (44). The cation exchanger S-sepharose has been used to separate holoenzymes on the basis of their β subunit content, and several pools of kinase containing variable α:β ratios (ranging from 6:1 to 1:8) have been isolated from rat forebrain and cerebellar extracts (37, 55, 56). These studies again suggest that CaM kinase holoenzymes contain varying mixtures of isozymic subunits. A recent EM study, however, raises questions about this conclusion (56a). EM images of rat brain CaM kinase suggest that the holoenzymes are primarily octamers and decamers with 8 or 10 peripherally arranged catalytic particles tethered together by a central association domain. Forebrain kinase is relatively enriched in decamers, while cerebellar kinase predominantly contains octamers. Antibody specific for the α subunit binds only to decamers, suggesting that α subunits preferentially form decamers and β subunits octamers. The predominance of decamers in forebrain and octamers in cerebellum is also consistent with a segregation of α into decamers and β into octamers.

Cloning and Analysis of cDNA Clones

cDNAs encoding the gene products for four distinct CaM kinase isoforms have been cloned and sequenced from rat brain libraries. Two brain-specific clones encode α and β subunits (57–60). α-CaM kinase cDNA encodes a protein of 478 amino acids (M_r 54,111) on a 5.0-kilobase (kb) message. The β-CaM kinase cDNA translates to a 542-amino-acid protein (M_r 60,333) contained on a 4.8-kb message. The α and β sequences are closely related, with 91% identical amino acids in the amino-terminal half, and 76% identical in the carboxyl-terminal half (58). A cDNA encoding the α subunit has also been isolated from mouse brain (61). Alternative splicing generates both the β mRNA and a distinct β' isoform (58).

Two other isoforms, designated γ and δ, have been cloned from rat brain (62, 63). Northern blots demonstrate a broad tissue distribution for both subunits, with γ detectable in all tissues examined and δ detectable in all except liver. Each of these is a distinct gene product with approximately 85% homology in amino acids and 75% homology in nucleotides to the α subunit. The γ cDNA encodes a protein of 527 amino acids (M_r 59,038), and δ one of 533 amino acids (M_r 60,080). The primary differences among the isoforms are in variable regions located at the end of the calmodulin-binding domain (Figure 2). Since CaM kinase subunits that do not correspond to any of the cloned isoforms in size have been found in both neuronal and non-neuronal tissues, it is likely that additional isoforms exist that are either distinct gene products or alternatively spliced variants of the known isoforms. For example, autophosphorylated CaM kinase in hypothalamus and midbrain displays a prominent 56-kDa subunit that may be a variant of α (53, 64). This subunit may correspond to the polymerase chain reaction (PCR) product recently identified in monkey brain cDNA in which 11 amino acids are inserted into α in the second variable region (65). Thus far all mammalian isoforms identified, including those from rat heart and human lymphocytes (C. Edman, P. Nghiem, and H. Schulman, unpublished observations), utilize the same structural organization and differ from each other primarily in the size and nature of the inserts at the end of the calmodulin-binding domain. Further analysis is needed to establish whether there are more distantly related isoforms, which may, for example, lack the ability to become Ca^{2+}-independent or are primarily monomeric and not oligomeric. A cDNA encoding an α subunit 77% identical to that from rat brain has recently been isolated from *Drosophila melanogaster* (65a). Like the rat brain α subunit, this cDNA is enriched in the fly head as compared with the body. Related kinases have been cloned from yeast (66, 67) and *Aspergillus* (68).

Comparison with other known protein kinases and subsequent experimental

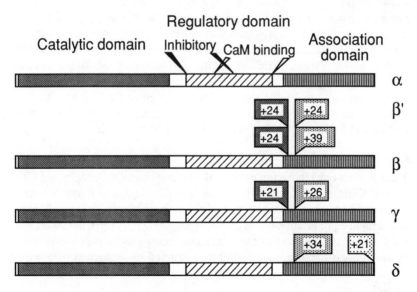

Figure 2 Domain structure of cloned CaM kinase isoforms

confirmation has revealed the overall domain structure of CaM kinase subunits shown in Figure 2 (57, 60). The catalytic moiety of the protein kinase is encoded in the N-terminal half of each subunit (amino acids 1 to 263–274 in α and comparable positions in the others). Partial proteolysis of the kinase, discussed below, generates a catalytically active fragment containing only this domain. The sequence of this catalytic domain is most homologous to that of other calmodulin-regulated kinases (e.g. MLCK and phosphorylase kinase), but retains all of the conserved amino acids characteristic of protein Ser/Thr kinases (69). The region extending from approximately amino acid 275 to 314 is a regulatory domain that contains both autoinhibitory and calmodulin-binding sequences, to be detailed below. The remaining C-terminal domain is an association and oligomerization domain (60), and is required for assembly of the subunits into a holoenzyme (47, 70). Recombinant kinase lacking this domain (e.g. amino acids 1–358 of the α subunit) is fully Ca^{2+}/calmodulin-dependent but monomeric rather than oligomeric.

CATALYTIC ACTIVITY AND ITS REGULATION BY CA^{2+} AND CALMODULIN

Mechanistic studies of multifunctional CaM kinase and its activation by Ca^{2+}/calmodulin have provided evidence for a model in which CaM kinase is kept relatively inactive in its basal state by the presence of an autoinhibitory domain. Binding of Ca^{2+}/calmodulin eliminates this inhibitory constraint and allows the kinase to phosphorylate its substrates, but also itself. This autophosphorylation is extremely interesting, with several consequences for subsequent kinase activity. It significantly slows dissociation of calmodulin, thereby trapping calmodulin even when Ca^{2+} levels are below those initially needed for activation. Once calmodulin dissociates, CaM kinase remains Ca^{2+}-independent until the kinase is dephosphorylated. These effects of autophosphorylation enable the potentiation of brief Ca^{2+} transients. The following sections elaborate on the molecular mechanisms controlling activation of CaM kinase and allowing potentiation of its activity.

Catalytic and Kinetic Properties

Three reports have looked in detail at the kinetic properties of activated CaM kinase. Kuret & Schulman, using casein as a substrate, and Katoh & Fujisawa, using MAP-2 as substrate, studied the native rat brain kinase and identified a rapid equilibrium random bi-bi mechanism (22, 71). Kwiatkowski et al used a catalytic fragment of the kinase (see below) and a peptide substrate, syntide-2, and found a rapid equilibrium ordered bi-bi reaction mechanism (72). The differences may be related to the forms of enzyme used in the studies.

Regulation by Ca^{2+} and Calmodulin

Purified CaM kinase is essentially inactive in the absence of both Ca^{2+} and calmodulin. In the presence of saturating calmodulin, half-maximal activation occurs at 0.5–1 μM free Ca^{2+} (9). Four Ca^{2+} ions bind per calmodulin and induce conformational changes that increase calmodulin's affinity for target enzymes such as CaM kinase [see (73) for reviews]. Half-maximal activity is attained at a calmodulin concentration of between 25 and 100 nM when Ca^{2+} is saturating (9). This is a lower affinity than that of calmodulin for many of its other targets, including MLCK and calcineurin, which are activated at 1 nM or less (4, 74). The affinity of CaM kinase for calmodulin is improved at increased ATP concentrations (71).

Calmodulin activation of CaM kinase displays positive cooperativity. Both autophosphorylation and substrate phosphorylation of a cytoskeletal CaM kinase exhibit cooperativity with Hill coefficients of 1.6 (75). Calmodulin activation of MAP-2 phosphorylation by purified forebrain enzyme shows similar cooperativity with a Hill coefficient of 1.85 (71). Full activation of CaM kinase requires stoichiometric binding of calmodulin (71). Fluorescence emission anisotropy has been used to directly measure calmodulin binding to CaM kinase in solution, allowing confirmation of a one-to-one stoichiometry in calmodulin binding to kinase subunits (T. Meyer, P. Hanson, L. Stryer, and H. Schulman, unpublished observations).

Calmodulin binds to a short basic and hydrophobic domain of the kinase, which has been identified and studied using model peptides. The core calmodulin-binding domain is contained in amino acids 296–309 of the α subunit (76, 77). The last five residues are hydrophobic and particularly important for calmodulin binding. This calmodulin-binding domain peptide has a K_d for calmodulin of 0.1–0.3 nM, which reflects tighter binding than that of calmodulin to CaM kinase. Accessibility of calmodulin to this domain may be constrained in the native enzyme. Studies described below suggest that the intrinsic high affinity of this domain for calmodulin may be realized after autophosphorylation.

Mutant calmodulins enable identification of the features on calmodulin that are important for its interaction with other proteins. CaM kinase is less sensitive to mutations in the calmodulin central helix than other calmodulin targets studied, but is sensitive to carboxyl-terminal calmodulin modifications (78, 79). Determinants of the interaction with calmodulin are indeed encoded in the identified calmodulin-binding domain, since replacement of the calmodulin-binding domain of MLCK with that of CaM kinase (amino acids 296–314) generated a recombinant MLCK with calmodulin affinity and interactions with mutant calmodulins characteristic of CaM kinase (80).

Study of the Autoinhibitory Domain

The regulatory domain of CaM kinase is responsible for autoinhibition as well as calmodulin binding, and maintains basal kinase activity at levels 100- to 1000-fold below the maximum stimulated by Ca^{2+}/calmodulin. The autoinhibitory domain has been explored with synthetic peptides and either native CaM kinase or a catalytically active fragment of the enzyme, as well as by site-directed mutagenesis. A synthetic peptide encoding the autoinhibitory domain (amino acids 281–309 in α) is bifunctional: it both binds calmodulin and inhibits CaM kinase activity (81). Peptides lacking the amino acids critical for binding of calmodulin (amino acids 273–302 in α) effectively inhibit CaM kinase in either the presence or absence of calmodulin (82). Determinants at either end of this regulatory domain contribute significantly to autoinhibition. Replacement of the N-terminal amino acids His282 and Arg283 with Asp by site-directed mutagenesis in α-CaM kinase generated a completely Ca^{2+}-independent kinase (67% Ca^{2+}-independent activity) (83), although replacement of Arg283 alone had little effect (84). At the C-terminal end, deletions in the calmodulin-binding domain generated by site-directed mutagenesis (deletion of 291–315 or of 304–315) also resulted in significant disruption of the autoinhibitory domain and a Ca^{2+}-independent kinase (48).

Kinetic analysis of the bifunctional calmodulin-binding and inhibitory peptide showed that its inhibition of CaM kinase was competitive with ATP and noncompetitive with peptide substrate. The autoinhibitory domain therefore keeps the kinase inactive by inhibiting access to both peptide- and ATP-binding sites (85, 86). Binding of Ca^{2+}/calmodulin to the bifunctional peptide reduces its inhibitory potency by more than 10-fold (81, 85, 86). Binding of Ca^{2+}/calmodulin to CaM kinase itself disables the autoinhibitory domain of the enzyme; this is evident both in the stimulation of substrate phosphorylation as well as in the enhanced rate at which the kinase's ATP-binding site can be modified by reagents such as phenylglyoxal and 5'-p-fluorosulfonylbenzoyl adenosine (FSBA) (87–89). Further analysis with shorter peptides has revealed that the carboxyl-terminal portion of the bifunctional peptide (amino acids 290–309 in α), not only binds calmodulin but also inhibits the kinase. It inhibits phosphorylation competitively with peptide substrate and noncompetitively with ATP (77, 85, 86). These studies, along with the studies of the longer peptides and site-directed mutants described above, suggest that the carboxyl-terminal end of the autoinhibitory domain may function to inhibit binding of peptide substrate, while the amino-terminal end may inhibit binding of ATP.

Catalytic Fragment

The discussion above implies that the catalytic domain of CaM kinase is intrinsically active, and that deinhibition by calmodulin might be simulated by removal of the autoinhibitory domain. Indeed, CaM kinase, like several other calmodulin-dependent enzymes and kinases, can be activated by proteolytically removing its autoinhibitory domain (70, 85, 90–92). CaM kinase is a major constituent of the postsynaptic density (PSD), a thickening of unknown function found on the cytoplasmic face of the postsynaptic membrane in synaptic junctions. Limited proteolysis with chymotrypsin solubilizes a monomeric 30-kDa Ca^{2+}-independent fragment from the PSD-associated CaM kinase (70). Proteolysis of soluble kinase under similar conditions seems to require prior autophosphorylation to produce an active 30-kDa fragment; proteolysis without autophosphorylation generates a 31-kDa inactive fragment (90, 92). Peptide sequencing of these purified fragments suggests that the carboxyl end of the active fragment from a limited chymotryptic digest is Ile271, while that of the inactive fragment is Phe293 or Leu299 (92). A fragment truncated at Ile271 therefore contains the entire catalytic domain of the kinase without autoinhibitory constraints. On the other hand, the fragment extending to Phe293 contains inhibitory determinants but lacks calmodulin-binding sequences to enable activation. Trypsin, chymotrypsin, and calpain can all produce such an active fragment (90), suggesting the presence of a hinge between the catalytic and regulatory domains of the kinase. Ca^{2+}-regulated proteolysis by calpain is a particularly intriguing possibility, since it could provide a mechanism by which activated kinase would be released from the holoenzyme as a constitutive monomeric enzyme.

CAM KINASE REGULATION BY AUTOPHOSPHORYLATION

Ca^{2+}/calmodulin-stimulated autophosphorylation is a prominent characteristic of multifunctional CaM kinase. Autophosphorylation occurs on multiple sites in purified CaM kinase, with perhaps as many as 42 moles of phosphate incorporated per mole of holoenzyme. Autophosphorylation is an intramolecular reaction within holoenzymes, since the reaction rates are not dependent on enzyme concentration (42, 93). In 1985, Saitoh & Schwartz presented the first report of a striking effect of CaM kinase autophosphorylation (94). After autophosphorylation, the *Aplysia* CaM kinase developed activity towards synapsin I that was no longer dependent on Ca^{2+}/calmodulin. Activity in the absence of Ca^{2+}/calmodulin increased from <5% to 74% of the activity in the presence of Ca^{2+}/calmodulin. This dramatic conversion of CaM kinase to a Ca^{2+}-independent species has since been the source of much study and speculation.

Autophosphorylation Converts CaM Kinase to a Ca^{2+}-Independent Kinase

Ca^{2+}/calmodulin stimulates autophosphorylation of CaM kinase, and this autophosphorylation enables the kinase to phosphorylate substrates in a Ca^{2+}-

independent manner. (93–97). In comparison to the Ca^{2+}-stimulated kinase, the autonomous or Ca^{2+}-independent kinase has a two- to five-fold higher K_m that varies with substrate (98, 99), and a somewhat reduced (98) or unchanged V_{max} (99). A decreased K_m for selected substrates has also been reported (100–102). As a result of these kinetic differences, the extent of Ca^{2+}-independent activity is a function of both the substrate in question and its concentration. Typically observed Ca^{2+}-independent or autonomous activities range from 20 to 80% of the full Ca^{2+}-stimulated activity. CaM kinases other than the soluble enzyme from rat forebrain have also been shown to undergo autophosphorylation and develop Ca^{2+}-independent activity, suggesting that this property is conserved among isozymes. These include CaM kinase from rat cerebellum (enriched in β isoform subunits) (103), rat forebrain PSDs (102, 104), pig retinas (105), squid synaptosomes (39), and non-neuronal tissues (97).

Autophosphorylation has also been reported to inhibit maximal stimulation of kinase activity, as noted particularly in earlier studies (22, 95, 100, 106, 107). This loss of activity is more pronounced when autophosphorylation is performed at low concentrations of ATP (95). Autophosphorylation at 5 μM ATP produced a 75% loss in total kinase activity and no coincident autonomy, while at 500 μM ATP there was no decrease in total activity and substantial generation of Ca^{2+}-independent activity (95). Autophosphorylation also appears to make the kinase more sensitive to irreversible thermal inactivation, and ATP may protect against this (93, 108).

Is there a difference in the substrate specificity of the autonomous kinase assayed with and without Ca^{2+}/calmodulin? For most substrates the specificity does not appear to change. However, a small but potentially significant group of substrates has been identified that are phosphorylated only by the Ca^{2+}-independent CaM kinase in the absence of Ca^{2+}/calmodulin. These are other calmodulin-stimulated enzymes that have phosphorylation sites within their own calmodulin-binding domains. The phosphorylation sites are therefore blocked when Ca^{2+} levels are high and calmodulin is bound. Calcineurin (109–111), Ca^{2+}/calmodulin-dependent phosphodiesterase (112), and smooth muscle MLCK (113, 114) are among these. The phosphorylation of a number of other proteins, including myelin basic protein and histone, has also been shown to increase following autophosphorylation of CaM kinase (102).

Analysis of Ca^{2+}/Calmodulin-Stimulated Autophosphorylation

Is autophosphorylation directly responsible for Ca^{2+}-independent CaM kinase activity? Evidence from several studies demonstrates that it is (93, 95–97). Nonhydrolyzable ATP analogs (AdoPP[NH]P, AMP-PNP, AMP-PCP) are unable to substitute for ATP (93, 97), and treatment of autophosphorylated CaM kinase with protein phosphatases 1 or 2A restores complete Ca^{2+}-

dependence to the activity (93, 96, 97). ATPγS induces Ca^{2+}-independence that is resistant to phosphatases (96). Significantly, the degree of Ca^{2+} independence correlates with the extent of autophosphorylation at early points in the reaction (93).

By slowing initial rates of autophosphorylation, a unique P-threonine containing phosphopeptide that coincided with the appearance of Ca^{2+}-independent kinase activity was identified (115–117). Peptide sequencing identified Thr286 as the critical phosphorylation site in the α subunit (116–119). The comparable phosphopeptide generated from the β subunit was also sequenced and shown to contain Thr287 (118). The sites in both α and β are preceded by the consensus phosphorylation sequence of Arg-Gln-Glu-Thr, and a similar site is found in each of the known isoforms. Thr382 is also rapidly phosphorylated in the β subunit, but is not required to maintain Ca^{2+}-independent activity (118). This site is not found in either α or β' subunits (58).

Site-directed mutagenesis studies demonstrate that phosphorylation at Thr286 is necessary and sufficient for the generation of Ca^{2+}-independent activity. Replacement of Thr286 with Leu (46), Ala (52, 120, 121), or Pro (121) abrogates generation of autonomous activity. These mutant enzymes retain normal kinase activity and continue to autophosphorylate at several sites other than Thr286. Replacement of other potentially phosphorylated residues in and around the regulatory region (Thr253, Thr261, Thr276, Ser279) with nonphosphorylatable amino acids has no effect on auto-phosphorylation at Thr286; Ca^{2+}-independent activity is obtained just as in the wild-type enzyme (46). Further replacement of Thr286 with Asp demonstrates that negative charge in place of Thr286 is alone sufficient to make a kinase significantly Ca^{2+}-independent or constitutive when expressed in COS cells (83) or prepared by in vitro translation (120).

Ca^{2+}/calmodulin-stimulated autophosphorylation (of Thr286) occurs largely as an intraholoenzyme reaction (42, 93). However, within the holoenzyme, the phosphorylation could occur as either an intra- or inter-subunit reaction. This has been difficult to address with rat brain CaM kinase, since the subunits are not readily dissociated into monomers. Recombinant α subunits truncated between the calmodulin-binding and association domains have therefore been used to generate monomeric CaM kinase (P. Hanson and H. Schulman, in preparation). In these monomers, Ca^{2+}/calmodulin-stimulated autophosphorylation of Thr286 occurs largely as an intermolecular reaction. Intersubunit autophosphorylation of Thr286 also occurs in holoenzymes (P. Hanson and H. Schulman, in preparation). Holoenzymes containing a mixture of catalytically inactive subunits (mutated at Lys42, a conserved residue important for catalytic activity) and active subunits lacking their own Thr286 readily autophosphorylate Thr286 in the inactive subunits. Intersubunit phosphorylation of Thr286 can therefore occur; the extent of possible intrasubunit phosphorylation in the holoenzymes remains to be addressed.

Is Autophosphorylation a Prerequisite for Substrate Phosphorylation?

Kinetic analysis has been interpreted to suggest that autophosphorylation is a prerequisite for activity toward exogenous substrates (122, 123). In one study, phosphorylation of substrate was shown to lag behind autophosphorylation when the reaction was sufficiently slowed by limiting ATP (123). In another study, a high concentration of substrate was used to compete with autophosphorylation, and this was found to inhibit substrate phosphorylation as well (122). Kinase that was autophosphorylated in a preincubation was not inhibited by subsequent addition of high concentrations of substrate. However, such studies cannot discriminate between a model in which autophosphorylation is a prerequisite for kinase activity and a model in which autophosphorylation competes with substrate phosphorylation so that maximal rates of substrate phosphorylation are not attained until autophosphorylation is completed. In the site-directed mutagenesis experiments with CaM kinase described above, replacement of Thr286 with either Ala, Leu, or Pro did not affect enzyme activity or activation. This is the principal evidence arguing against an essential role of autophosphorylation in activation of the kinase.

Ca^{2+}-Independent Autophosphorylation Inhibits Activation by Calmodulin

Initial autophosphorylation with Ca^{2+}/calmodulin enables the kinase to continue autophosphorylation even after Ca^{2+} (and presumably calmodulin) is removed. After a threshold level of initial autophosphorylation, Ca^{2+}-independent autophosphorylation proceeds to the same (96) or higher (98, 116) final stoichiometry as seen in the continued presence of Ca^{2+}/calmodulin. This phase of autophosphorylation decreases subsequent calmodulin binding to the kinase (100) and thereby inhibits activation of the Ca^{2+}-independent kinase by calmodulin, leaving a kinase with only its Ca^{2+}-independent activity (98). The inhibition is fully reversible upon treatment with protein phosphatases 1 or 2A (98, 116, 124). It has not been established whether this inhibitory autophosphorylation occurs via an intersubunit or an intrasubunit reaction.

Tryptic phosphopeptide mapping demonstrates that Ca^{2+}-independent autophosphorylation occurs at sites largely distinct from those phosphorylated in the presence of Ca^{2+}/calmodulin (100, 116, 124). The sites are located in peptides containing Thr305 and Ser314 in the calmodulin-binding domain (of the α subunit) (124). Phosphatases 1 and 2A were able to remove phosphate from Thr305 selectively, and coincidently reverse the inhibition. Phosphate at Thr305, and not Ser314, is therefore likely to be involved in blocking the interaction between kinase and calmodulin. Studies with synthetic phosphopeptides corresponding to this domain also indicate that affinity for calmodulin is reduced 10-fold when either Thr305 or Thr306 is phosphorylated, whereas phosphate on Ser314 has no inhibitory effect (125). In a parallel approach, site-directed mutagenesis was used to identify the consequences of autophosphorylation at each of these sites (P. Hanson and H. Schulman, submitted). Replacement of Thr305, Thr306, and Ser314 with Ala, both individually and in various combinations, demonstrated that autophosphorylation at either Thr305 or Thr306 is sufficient for complete inhibition of calmodulin-stimulated activity. Autophosphorylation at Ser314 or a minor site, Thr310, had little effect on the subsequent interaction with calmodulin. Calmodulin binding by these enzymes also paralleled the ability of calmodulin to stimulate enzyme activity.

Is Generation of the Ca^{2+}-Independent State a Cooperative Process?

One of the enduring mysteries of CaM kinase is the reason for its unusually large oligomeric size. Cooperativity in calmodulin activation and intersubunit autophosphorylation described above are two features that take advantage of the oligomeric organization of the enzyme. A more exciting rationale was offered by early reports that autophosphorylation of only a few subunits per holoenzyme could switch all subunits to a Ca^{2+}-independent state (93, 96). Approximately 2–4 moles of phosphate per holoenzyme resulted in maximal levels of both Ca^{2+}-independent substrate phosphorylation and continued autophosphorylation (96, 117). When autophosphorylation was slowed by incubation at $0°$ C, total phosphorylation peaked at only 3–4 moles of phosphate per holoenzyme, yet maximal Ca^{2+}-independence was attained (93). Cooperative interaction between subunits after modification of a small fraction of them, perhaps even only one (100), was therefore felt to be sufficient to induce full Ca^{2+} independence.

Given these observations, Miller & Kennedy proposed that CaM kinase might have the characteristics of a Ca^{2+}-sensitive molecular switch (96). Lisman & Goldring extended previous models of molecular switches and simulated conditions for operation of a CaM kinase switch (126–129). Together, these experimental and theoretical studies proposed a mechanism by which CaM kinase could act as a neuronal memory molecule capable of encoding memory at the protein level. In this model, a "threshold" level of autophosphorylation switches the kinase to the "on" state, and as long as this minimal level of phosphate is retained, all subunits in the oligomer are in the Ca^{2+}-independent configuration. Since a switched kinase also continues to autophosphorylate, the kinase should be able to maintain its "on" state. Autophosphorylation can counter the action of cellular phosphatases by enabling refilling of dephosphorylated subunits by neighboring phosphorylated subunits. Problems of protein turnover may be overcome by either interholoenzyme autophosphorylation that would allow nascent holoenzymes to be phosphorylated by preexisting switched kinase molecules or by intersubunit autophosphorylation following insertion of individual nascent subunits into preexisting holoenzymes. The main requirements of this model are: (a) that autophosphorylation convert kinase subunits to a Ca^{2+}-independent state; (b) that there be cooperativity in the effect of autophosphorylation, i.e. a minimal threshold for maximal Ca^{2+}-independent activity; (c) that autophosphorylation of a switched kinase continue in the absence of Ca^{2+}, including autophosphorylation of the autonomy site responsible for the switch. Phosphorylation of this site would be needed to propagate the "on" state in the presence of phosphatases.; (d) that nascent subunits be inserted into preexisting oligomers, since autophosphorylation is an intraholoenzyme reaction (42, 93, 96).

Recent studies have shown that the Ca^{2+}-independent form of CaM kinase is generated in situ, and therefore support the idea that autophosphorylation potentiates the activity of CaM kinase. However, other findings suggest a need for modification of the composite model described above. The question of how many subunits need to be autophosphorylated to "switch" the kinase to maximal levels of Ca^{2+}-independent activity was recently reevaluated by Ikeda et al (99), who found an apparent one-to-one correlation between autophosphorylation of subunits and Ca^{2+}-independent activity, i.e. a switch without a threshold. The extent of Ca^{2+}-independent activity paralleled the number of subunits phosphorylated, with maximal autonomy only obtained at a stoichiometry of 10 per holoenzyme. A lack of a threshold or cooperativity was also reported for liver CaM kinase, which required 1.5 moles of phosphate per subunit to attain full Ca^{2+}- independence (97). The discrepancies in the stoichiometry associated with full autonomy need to be resolved by quantitative measurements, specifically of the level of Thr286 phosphorylation, to assess the presence or absence of a threshold. Additional approaches will be necessary to fully understand the effect of phosphorylated subunits on their neighbors in a holoenzyme.

The critical event in propagation of the "on" state in the model above is continued autophosphorylation at Thr286 in the absence of Ca^{2+}/calmodulin. Although we know that autophosphorylation continues after removal of Ca^{2+}/calmodulin, continued phosphorylation of Thr286 during this phase has not been directly demonstrated. Experiments analyzing Ca^{2+}-independent phosphorylation in kinase after Ca^{2+}-stimulated phosphorylation of only a minimal number of subunits (e.g. four) have not been reported. Studies directed at analysis of the Ca^{2+}-independent phase of autophosphorylation have demonstrated prominent labelling of non-Thr286 sites (100, 116, 124), but could have missed Thr286 if it was completely phosphorylated during the initial Ca^{2+}-stimulated autophosphorylation. Indirect studies bearing on this question include an analysis of autoinhibitory domain peptide phosphorylation in the presence and absence of Ca^{2+}/calmodulin. The bifunctional auto-inhibitory peptide discussed earlier (corresponding to amino acids 281–309) is an inhibitor of the autonomous CaM kinase in the absence of Ca^{2+}/calmodulin. Binding of Ca^{2+}/calmodulin to the peptide not only decreases the peptide's inhibitory potency, but also promotes phosphorylation of the peptide on Thr286. This peptide may simulate the interaction of the auto-inhibitory domain with the kinase, suggesting that Thr286 may only be available as an effective substrate when Ca^{2+}/calmodulin is bound to the kinase.

Autophosphorylation of Thr286 Traps Bound Calmodulin

A recent study suggests that in addition to producing a Ca^{2+}-independent state, autophosphorylation traps calmodulin bound to CaM kinase by markedly reducing its dissociation rate (129a). Trapping may allow the kinase to respond to repetitive Ca^{2+} spikes common in neuronal systems by introducing a threshold frequency at which the kinase becomes highly active as discussed below. Fluorescence emission anisotropy was used to measure dissociation rates of dansylated calmodulin (F-calmodulin) from CaM kinase, taking advantage of the difference in rotational mobility of free F-calmodulin as compared to F-calmodulin bound to the oligomeric kinase. At high Ca^{2+}, bound F-calmodulin exchanges with free calmodulin with a dissociation time of approximately 0.4 seconds. If the kinase is first autophosphorylated, there is a 1000-fold reduction in off-rate, so that F-calmodulin remains bound for hundreds of seconds. The affinity of CaM kinase for F-calmodulin therefore decreases from 15 nM to less than 10 pM. Autophosphorylation converts CaM kinase from an enzyme with one of the weakest affinities for calmodulin to an enzyme with one of the highest affinities for calmodulin. When measured at low Ca^{2+} (100 nM), the dissociation rate of calmodulin from the kinase is reduced from 0.2 seconds to 10–20 seconds. Thus, a F-calmodulin that is bound to an autophosphorylated subunit remains trapped for at least 10–20 seconds after Ca^{2+} is reduced to physiologically basal levels. Trapping is seen in recombinant α-CaM kinase but not in a mutant in which Thr286 is replaced with Ala286 (129a). Phosphorylation of the Thr286 autonomy site is therefore responsible for both trapping and conversion to a Ca^{2+}-independent state. Trapping is similarly seen with cerebellar enzyme enriched in β subunits, suggesting that isoforms other than α are capable of trapping, and can be seen in α-CaM kinase monomers, suggesting that once autophosphorylated, each individual subunit traps calmodulin (129a).

Autophosphorylation and trapping are both cooperatively stimulated by calmodulin; the efficiency of each process increases as the number of kinase subunits with bound calmodulin increases (75; T. Meyer, P. Hanson, L. Stryer, and H. Schulman, unpublished observations). Autophosphorylation and subsequent trapping may involve two neighboring subunits of an oligomer, each bound by calmodulin. If one assumes that during a single isolated Ca^{2+} spike only a few of the 10–12 subunits in an oligomer bind calmodulin, then trapping may provide a mechanism for sequestration of calmodulin by CaM kinase during high-frequency spiking as shown in Figure 4 and described below.

A Model for CaM Kinase Regulation by Autophosphorylation

Our present understanding of the sequence of events leading to autophosphorylation and Ca^{2+}-independent activity in subunits of the CaM kinase holoenzyme is depicted in Figure 3. Activation and deactivation of just two neighboring subunits in a holoenzyme are shown. Prior to activation by Ca^{2+}/calmodulin the subunits are inactive, with the autoinhibitory domain of each subunit inhibiting catalytic function. In the presence of Ca^{2+}, calmodulin binds to both subunits, displacing their autoinhibitory domains, and opening their catalytic sites to protein substrates and ATP. Intersubunit autophosphorylation proceeds only when an activated subunit is presented with Thr286 from an adjacent subunit by the binding of calmodulin to that subunit. A subunit that is not autophosphorylated during the Ca^{2+} transient, such as the lower one shown, rapidly deactivates when Ca^{2+} levels fall and calmodulin dissociates. By contrast, the autophosphorylated subunit remains maximally active for many seconds while calmodulin is trapped, and retains Ca^{2+}-independent activity after calmodulin slowly dissociates. When the calmodulin-binding domain of this activated subunit is vacated, it is rapidly autophosphorylated on Thr305 or Thr306 in a Ca^{2+}-independent reaction and further stimulation of this subunit by calmodulin is blocked.

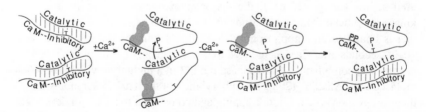

Figure 3 Addition of Ca^{2+}/calmodulin stimulates intersubunit autophosphorylation, resulting in trapping of calmodulin and conversion of the enzyme to a Ca^{2+}-independent state

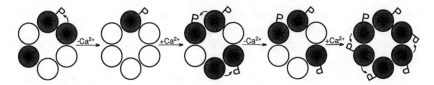

Figure 4 Trapping of calmodulin by a CaM kinase oligomer during successive Ca^{2+} spikes

The role of cooperative trapping of calmodulin in activation of a CaM kinase holoenzyme during high-frequency stimulation is shown in Figure 4. For the sake of simplicity, the kinase is represented as a hexamer, activated calmodulin-bound subunits are shaded, and *P*-Thr286 is indicated with a "P." In this scenario, each Ca^{2+} spike is subsaturating and leads to binding of only a few molecules of calmodulin per holoenzyme. Although all calmodulin-bound subunits are transiently activated, most lack a calmodulin-bound neighbor and are therefore not autophosphorylated. Occasionally, however, calmodulin binds to adjacent subunits, and Thr286 is phosphorylated, as shown in the first panel of the figure. When Ca^{2+} levels decrease, calmodulin dissociates rapidly from all but the autophosphorylated subunit. If Ca^{2+} levels rise again within the next 1–10 seconds, calmodulin again binds to a few subunits. However, because of calmodulin retained from the previous cycle, the probability and rate of intersubunit autophosphorylation are increased. More subunits are phosphorylated on Thr286, and in turn retain calmodulin after Ca^{2+} levels next fall. In a third cycle of Ca^{2+} elevation, the probability and rate of Thr286 phosphorylation are yet further increased. Repetitive Ca^{2+} signals thus lead to the sequestration of calmodulin by CaM kinase when the frequency of Ca^{2+} spiking is sufficiently high. When Ca^{2+} signals instead occur at low frequency, the time between sequential Ca^{2+} spikes is long enough to permit dissociation of trapped calmodulin, and each successive Ca^{2+} spike generates the same CaM kinase response. The cooperativity of autophosphorylation and the subsequent trapping of calmodulin may thus enable a frequency modulation of the kinase with a threshold frequency at which it is rapidly and highly activated.

Comparisons with Other Protein Kinases

The structure/function analysis of CaM kinase described above reveals some striking similarities as well as differences with other second messenger–regulated kinases, such as PKA and PKC. Each of these three kinases is designed with an autoinhibitory domain that maintains the catalytic domain in the basal state (10, 130, 131). In the case of PKA regulatory and catalytic functions are on separate subunits, whereas in CaM kinase and PKC the regulatory domain inhibits the catalytic domain on the same subunit. For all three, the point of interaction of the second messenger, be it cAMP, Ca^{2+}/calmodulin , or Ca^{2+}/phospholipid/diacylglycerol, is at a site that is either next to or overlapping with the inhibitory domain. A deinhibition of kinase activity by second messenger may be a general motif for activation of regulated protein kinases.

There are also some similarities between autophosphorylation of PKA and CaM kinase. Unlike CaM kinase, which autophosphorylates only after being activated, the regulatory subunit of PKA (type II) is phosphorylated in the basal state. The primary effect of this autophosphorylation is a reduced affinity for the catalytic subunit, resulting in more rapid dissociation and activation upon exposure to cAMP, as well as in a slowed deactivation or reassociation of regulatory and catalytic subunits (132, 133). The read-out of this autophosphorylation is therefore catalytic activity that outlasts the actual cAMP signal. Kinetic modelling has suggested that under appropriate circumstances, free catalytic subunits could persist for as long as 20 minutes (134). However, activation of PKA is self-limiting even in the absence of cellular phosphatases, since the subunits eventually reassociate and inactivate. Thus, autophosphorylation is a mechanism for potentiating the "on" state of both CaM kinase and PKA. Important differences between their regulation arise because the multimeric structure of CaM kinase enables cooperative interactions among the enzyme's subunits as discussed above.

CAM KINASE AUTOPHOSPHORYLATION IN CELLS AND TISSUES

The intricate autoregulatory behavior of CaM kinase is interesting if not remarkable, but does it occur in situ? Indeed, conversion of the kinase to a Ca^{2+}-independent species upon elevation of Ca^{2+} has now been amply demonstrated in a number of cell systems. This has been assessed by lysing cells and measuring the extent of Ca^{2+}-independent CaM kinase activity in an in vitro assay. Alternatively, cells or tissues have been labelled in situ with $^{32}P_i$ followed by a determination of Thr286 phosphorylation. Stimulation of Ca^{2+} influx by depolarization with high K^+ has been used to demonstrate a significant increase in Ca^{2+}-independent activity in intact synaptosomes (135), cerebellar granule cells (136), PC12 cells (137), and GH3 cells (138). A depolarization-induced increase in Ca^{2+}-independent activity has also been observed in intact acute hippocampal slices (139). In synaptosomes and cell lines, depolarization converted up to 50% of the CaM kinase to the autonomous state from a basal state in which less than 6% of the molecules were autonomous. In synaptosomes, PC12 cells, and GH3 cells this activation was associated with enhanced labelling of Thr286 (135, 137, 138). Thus, autophosphorylation of Thr286 occurs in situ and produces a Ca^{2+}-independent form of CaM kinase.

Ca^{2+}-independent CaM kinase activity is also induced by neurotransmitters, hormones, and growth factors, thus suggesting that its activation is on the pathway of numerous signalling systems, including those involving phosphatidylinositol (PI) turnover, ligand-gated Ca^{2+} channels, and tyrosine kinase–bearing receptors. For example, CaM kinase responds to activation of the PI signalling system in PC12 cells and GH3 cells. Ca^{2+}-independent kinase activity is enhanced in PC12 cells after stimulation with bradykinin (137) and in GH3 cells after stimulation with thyrotropin-releasing hormone (138). Bradykinin stimulation causes an initial Ca^{2+} spike as intracellular Ca^{2+} stores are released in response to IP_3. Ca^{2+}-independent kinase activity rises from a baseline of 3% to a stimulated level of 22% within 10 seconds, tracking the timecourse of the Ca^{2+} spike. Since the purified kinase is maximally 60–70% Ca^{2+}-independent, this activation involves up to one third of the kinase molecules in the PC12 cells. Following the initial spike, an elevated Ca^{2+}-independent activity is seen for several minutes after Ca^{2+} falls below threshold for activation of the kinase. A similar potentiation is seen in GH3 cells stimulated with thyrotropin-releasing hormone (138). Autophosphorylation may therefore potentiate CaM kinase activity in situ.

CaM kinase also responds to stimulation of ligand-regulated Ca^{2+} channels and to tyrosine kinase–linked receptors. Stimulation of cerebellar granule cells with glutamate, an excitatory amino acid, increased the Ca^{2+}-independent activity of CaM kinase via the N-methyl-D-aspartate subtype of glutamate receptor (140). This glutamate receptor subtype is known to be linked to a Ca^{2+} channel. Activation of the kinase required extracellular Ca^{2+} and correlated nicely with its autophosphorylation. Immunostaining of cultured hippocampal neurons treated with N-methyl-D-aspartate using a phosphoselective P-Thr286 antibody revealed substantially enhanced staining of phosphorylated CaM kinase throughout the cells after stimulation (140a). In PC12 cells, influx of Ca^{2+} can be elicited by stimulation with ATP at a purinergic receptor or with nicotinic agonists at a nicotinic acetylcholine receptor (M. MacNicol and H. Schulman, unpublished observations). Both treatments stimulate autophosphorylation of CaM kinase and convert a significant fraction of it to a Ca^{2+}-independent state. Stimulation of 3Y1 fibroblasts with serum, epidermal growth factor, or platelet-derived growth factor increases phosphorylation of CaM kinase, although neither the site of phosphorylation nor changes in the Ca^{2+}-independent state of the kinase were examined (141). Thus, CaM kinase responds to stimuli that modulate free Ca^{2+} by either influx or intracellular release. The range of receptors and signalling systems subserved by CaM kinase is likely to be broader than that of either PKA or PKC.

CaM kinase in neuronal tissue has a very high basal level of Ca^{2+}-independent activity. Organotypic hippocampal cultures were found to have kinase with 16% Ca^{2+}-independent activity, corresponding to 34% of the kinase molecules, under conditions in which free Ca^{2+} was at a low basal level (142). Thr286 of the kinase was shown to be phosphorylated in unstimulated cultures, consistent with the high Ca^{2+}-independent activity. Acute hippocampal slices from young animals also had high basal levels of activated kinase (20%) (142), while slices from older animals had lower levels (7–9%) (139, 142). Depolarization of acute slices from older animals and incubation in high levels of extracellular Ca^{2+} doubled the Ca^{2+}-independent activity to 12%, accounting for approximately one quarter of CaM kinase molecules (139). How is this high Ca^{2+}-independent activity maintained? The activated state was unaffected by tetrodotoxin or inhibitors of excitatory amino acid receptors in the cultures (142). Inhibition of the kinase in situ by addition of a kinase inhibitor (H-7) or calmodulin antagonist (W-7) significantly lowered the Ca^{2+}-independent activity as did lowering extracellular Ca^{2+} (139, 142). The finding that Ca^{2+} independence, be it in 7% or 34% of all CaM kinase molecules, is maintained in apparently unstimulated cells is consistent with the CaM kinase memory model (129). However, the requirement for Ca^{2+} and calmodulin to maintain this state is inconsistent with the model. Transient activation in both neuronal and non-neuronal systems (135–137), as well as potentiation of the activated state by phosphatase inhibitors (136, 140), suggests that there is a steady state of autophosphorylation and dephosphorylation. The sustained high Ca^{2+}-independent activity at low free Ca^{2+} may be a result of a shift in the Ca^{2+} affinity of the system, e.g. the increase in affinity for Ca^{2+}/calmodulin following autophosphorylation, or an extremely low rate of dephosphorylation, e.g. by activation of a phosphatase inhibitor.

CAM KINASE LOCALIZATION

Regional Distribution

CaM kinase is 20–50-fold more concentrated in brain than in non-neuronal tissues. There are striking regional and developmental variations in its concentration, isoform ratio, and regional and subcellular localization, consistent with a highly regulated role in neuronal functions. Immunocytochemical studies with antibodies recognizing CaM kinase subunits demonstrate high levels of kinase in regions of the telencephalon, including particularly hippocampus, cortex, amygdala, striatum, and lateral septum (64, 143, 144). Antibodies recognizing both subunits also stain the cerebellar cortex (143, 144). Expression levels have been quantified in a radioimmunoassay with an α-specific antibody; α-CaM kinase constitutes 0.74% of total brain protein and 1.4% of hippocampal protein (64). When β subunits are included, CaM kinase accounts for 1% of brain protein and 2% of hippocampal protein.

Within these regions, CaM kinase expression is prominent chiefly in neurons, although it is uncertain how well antibodies react with isoforms in glial cells. High-level expression in particular neurons is striking in hippocampus, cerebral cortex, and cerebellar cortex. Within the hippocampus, the strongest immunoreactivity is observed in molecular and pyramidal cell layers and in the ectal and endal limbs of the dentate gyrus. In the neocortex, the pyramidal neurons of layer V and the dendrites of layer I are most prominently stained, and layer IV exhibits lowest staining. Glial cells exhibit no immunoreactivity with some CaM kinase antibodies (64, 144), and low immunoreactivity with a polyclonal antiserum that efficiently recognized a variety of non-neuronal isozymes (143). CaM kinase has since been purified from cultured astrocytes, but was not found in cultured oligodendrocytes (145, 146). The level of expression in astrocytes is far below levels observed in neurons.

CaM kinase localization in human brain tissues was found to be similar to that of the rat (147, 148). Hippocampal pyramidal neurons of CA1 and the subiculum were particularly intensely stained (148). Sections from Alzheimer's disease brains contained fewer CA1 neurons, but each exhibited far greater concentrations of immunoreactive CaM kinase protein (148). These are the same neurons that form the neurofibrillary tangles characteristic of Alzheimer's disease, and a relationship between overexpression of CaM kinase and abnormal tau protein phosphorylation in the neurofibrillary tangles is plausible. Elevated CaM kinase levels have also been demonstrated in association with the Lewy bodies characteristic of Parkinson's disease (147).

Similar regional distribution has been described for the mRNAs that encode CaM kinase. In situ hybridization with probes specific for α (60, 149) or β (149) subunits demonstrated high levels of α expression in hippocampal and cortical neurons, and high levels of β expression in the Purkinje and granule cells of the cerebellum. Each of the four isozymes (α, β, γ, δ) is expressed in brain, as shown by Northern blot analysis of message levels. α subunit is most abundant in forebrain, less in the brain stem or lower brain, and much less in cerebellum (58, 63). β is highest in cerebellum, and somewhat less abundant in the lower brain and forebrain (58, 63). γ is present at highest levels in forebrain, with slightly lower levels in brain stem and cerebellum (63). δ is most prominent in cerebellum, and slightly lower in forebrain and brain stem (62).

Cerebellum contains lower total levels of CaM kinase activity, with a higher ratio of β:α subunits than forebrain. β subunit is prominently expressed in both Purkinje and granule cells, while α is prominent only in the Purkinje cells (143, 150). Primary cultures of cerebellar granule cells do, however, contain low concentrations of α subunit (56, 151). In retina, a high level of CaM kinase expression is found in synaptic regions (inner and outer plexiform), ganglion cells, and also retinal pigment epithelium (152).

Subcellular Localization

Immunolocalization with monoclonal or polyclonal antibodies demonstrates high concentrations of CaM kinase in neuronal cell bodies and dendritic processes (143, 144). Moderate levels are seen in nerve terminals and spines, and lower levels in the axons of the main fiber tracts. Nuclei are not stained (144) or weakly stained (143). Electron microscopic analysis resolves association with plasma membrane, synaptic vesicles, mitochondria, microtubules, and the PSD. CaM kinase is thus present throughout the neuron, and is available in both the pre- and postsynaptic compartments of a synapse.

Biochemical fractionations demonstrate that CaM kinase is present in the cytosol, as well as in particulate fractions, including plasma membrane, PSD, cytoskeleton, synaptic vesicles, and the nucleus [see (9)]. As assessed by ^{125}I-calmodulin binding, 35% of α subunit is soluble in adult rat forebrain, and 65% is particulate (153). CaM kinase activity is enriched in nerve terminals or synaptosomes, where in fact it was first described (12). Complete subcellular fractionation demonstrated that 80% of CaM kinase activity in cerebral cortex is particulate, with 45% present in nerve terminals or synaptosomes (154). In synaptosomes CaM kinase is found both "inside" the resealed nerve terminals (38%) and "outside" (62%), presumably in association with PSD (155). The presence of CaM kinase in nerve terminals in vivo has been confirmed in a study of lesion-induced degeneration of pre- and postsynaptic neuronal populations in the neostriatum and substantia nigra of the rat (156). Degeneration of nerve terminals was associated with a 30–40% loss of local kinase activity, consistent with a presynaptic localization of CaM kinase.

CaM kinase is particularly concentrated in the postsynaptic density (PSD), and was identified as a major component of the PSD after it was realized that the major PSD protein (mPSDp) is the α subunit of CaM kinase (157–159). In PSD preparations from forebrain, CaM kinase constitutes between 20 and 40% of the total protein. Unlike PSD from rat forebrain, PSDs isolated from cerebellum do not contain much CaM kinase (44). Kinase in PSD preparations has low apparent activity (104, 153, 160). In other respects the PSD-associated kinase is similar to soluble CaM kinase, and is able to autophosphorylate and become Ca^{2+}-independent (104). Factors responsible for the association of kinase with the PSD have not been identified, although the association seems to be correlated with high α subunit content (160). PSD-associated kinase may be covalently bound during preparation, since it is only solubilized by SDS in the presence of sulfhydryl reagents or by proteolysis (91). The significance of the association of CaM kinase with isolated PSDs remains somewhat controversial. Evidence in favor of an association in vivo includes the presence of two pools of CaM kinase; CaM kinase in the PSD pool has a $t_{1/2}$ of about 24 days, while kinase in the soluble fraction has a $t_{1/2}$ of about 12 days (161).

Association with the nucleus and nuclear matrix has been reported (162), and immunostaining with a polyclonal antiserum detected low levels of CaM kinase in the interphase nucleus and in association with the mitotic apparatus (143, 163). Other immunological studies have not detected CaM kinase in the nucleus. Immunostaining of nuclear proteins may be hampered by difficulties in permeabilizing the nuclear membrane or may be artifactually enhanced by association of cytosolic or cytoskeletal proteins with nuclear membranes, and these studies are therefore difficult to interpret. It remains to be determined whether there is a specific CaM kinase isoform localized in nuclei or responsible for regulation of nuclear events.

In situ hybridization studies with probes specific to α and β subunits revealed surprising hybridization of the α probe to layers of hippocampus and cortex that contain very few cell bodies (149). There was substantial expression of α mRNA in hippocampal and cortical neuropil and dendritic processes in which β mRNA was hardly detected. Dendritic localization of mRNA has been described for the microtubule-associated protein MAP-2 (164), and ribosomes have been found distal from the cell soma in neuropil. Some proteins of the postsynaptic structure, including the α subunit of CaM kinase, may therefore be synthesized at dendrites where their transcription could be locally regulated.

Changes in CaM Kinase Localization

Changes in the subcellular localization of CaM kinase following neuronal activity have not been widely observed. However, a few scenarios in which an apparent translocation occurs have been described. Activation and autophosphorylation of CaM kinase in *Aplysia* cytoskeletal preparations are accompanied by release of the cytoskeletal kinase into the soluble fraction (94). Translocation can also be induced by pretreatment of the intact ganglia with serotonin. Long-term visual adaptation of *Drosophila* to blue light affects light/dark choice behavior for many hours after training. CaM kinase in the fly head (brain) becomes increasingly membrane bound in parallel with the blue adaptation (165). Retinal CaM kinase in rats is similarly enhanced in the membrane/cytoskeletal fraction after dark adaptation (166).

Neurons are extremely sensitive to ischemic cytotoxicity, and the biochemical changes accompanying ischemia are complex. A number of studies have now reported decreases in CaM kinase activity as well as changes in its compartmentalization after ischemic insults to neuronal tissues. CaM kinase activity decreases after global ischemia in gerbils (167, 168); similarly, it is reduced in rabbit spinal cord following ischemia (169). Immunoreactivity decreases in rat CA1 neurons following forebrain ischemia (170). CaM kinase activity following ischemia decreases more rapidly in the cytosol than in the particulate fractions (171), and increased concentrations of less active kinase are found in the particulate fractions (172). The molecular bases for the ischemia-induced translocation and reduced activity remain to be elucidated, but could be related to autophosphorylation, proteolysis, or other posttranslational modifications.

REGULATION OF CAM KINASE LEVELS

Developmental Regulation

Developmental changes in CaM kinase expression have been well documented, with changes noted in total CaM kinase activity, subunit ratios, and subcellular distribution during neuronal development [reviewed in (7)]. Early studies of the major PSD protein (the α subunit) demonstrated a 20-fold increase in its level during development (173). Subsequent analyses have confirmed that α subunit levels are sharply regulated, with concentrations in rat forebrain rising linearly 10-fold between day 5 and 25 (153). CaM kinase β subunit is the dominant isozyme at birth and remains at a relatively stable level throughout development, although some decline is evident after the second week (174, 175). The large developmental increase in forebrain α subunit levels changes the overall ratio of $\alpha{:}\beta$ from less than one in the neonate to greater than two in the adult (175, 176). Interestingly, the increase in α subunit levels peaks near the end of the most active period of synaptogenesis (176), and may be coincident with synaptic maturation (7). Cerebellum maintains relatively low levels of α subunit throughout development, and therefore always contains kinase in which β subunits exceed α subunits (174). A cerebellar isoform with an $\alpha{:}\beta$ ratio of 1:8 is dominant at postnatal day 10, whereas one with an $\alpha{:}\beta$ ratio of 1:5 predominates at postnatal day 40 (56).

Message levels are similarly regulated; rat forebrain shows a 10-fold increase in α message between days 1 and 21, and a further 2- to 5-fold increase by day 90 (59). This increase was confirmed by quantitative in situ hybridization to rat brain sections, with observation of a 10-fold increase in α-specific hybridization to frontal cortex between days 4 and 16 (149). In contrast, β message was relatively high at 4 days postnatal, and decreased slightly during development to the adult. In cerebellum the levels of α message never increased substantially, and instead β-specific hybridization increased by twofold in granule and Purkinje cells.

The other major change in CaM kinase observed during postnatal development is a shift in the subcellular distribution from a predominantly cytosolic localization to particulate or membrane fractions (153, 174, 175). CaM kinase is approximately 75% soluble in the neonate and 75% particulate in the 90-day adult (153). As described above, particulate localization includes synaptic plasma membranes, synaptic junctions, and PSD.

CaM kinase expression and localization during differentiation and synaptogenesis in relatively homogeneous cultured hippocampal pyramidal neurons are similar to those observed in postnatal rat forebrain (146). Only 60-kDa subunits were detectable on the first day of culture; by 10 days the 50-kDa subunits predominated. After two days in culture, immunoreactive CaM kinase was half particulate and half cytosolic; after 29 days the particulate exceeded the cytosolic by fivefold. Immunofluorescence demonstrated diffuse CaM kinase expression in cell bodies and processes after four days in culture, concentrations in growth cones and large spots in processes after seven days, and more concentrated spots in processes and at the bottom of the cell bodies after 10 days. These spots exhibiting high concentrations of CaM kinase may reflect a PSD localization. A cultured neuroblastoma/glioma cell line (NG108) also contains CaM kinase, and both the levels of activity and of kinase protein increase when differentiation is induced in culture (177).

Activity-Dependent Changes in CaM Kinase Expression

In adult macaque monkeys, retinal activity has been observed to exert a strong influence on the visual cortex. Input from each eye is targeted to adjacent columns in the primary visual cortex, resulting in alternating columns of ocular dominance from each eye. Visual deprivation leads to changes in the levels of several proteins with putative roles in signalling in the visual cortex. CaM kinase is one of these, with immunoreactivity increasing well above normal levels following monocular deprivation precisely in the columns deprived of their normal visual input (178). This increase in CaM kinase concentration in the deprived columns appears to be a result of corresponding increases in mRNA levels (65). In situ hybridization demonstrated enhanced levels of α CaM kinase mRNA in the deprived eye dominance columns of cortical layers II–VI from as early as 48 hours after monocular deprivation. Similar increases in CaM kinase mRNA have been reported in kitten visual cortex when the animal is deprived of visual stimulation by rearing it in the dark (179).

Neuronal hyperexcitability is associated with decreases in CaM kinase levels. Kindling is a model of synaptic plasticity in which neuronal sensitivity to seizure activity is enhanced by repeated stimulation of particular brain regions with low levels of electrical current or excitatory neurotransmitters. Once established, the enhanced sensitivity to stimulated seizures becomes a stable, long-term modification in the brain. Changes in Ca^{2+}/calmodulin-stimulated phosphorylation of plasma membrane proteins in kindled animals (180, 181) have been directly associated with decreases in CaM kinase immunoreactivity (182). An in situ hybridization study demonstrates that message levels driving CaM kinase expression are decreased in the kindled animals, suggesting that CaM kinase gene expression is modulated by kindling (J. M. Bronstein, P. Micevych, P. Popper, G. Huez, D. B. Farber, C. B. Wasterlain, personal communication).

Another model of neuronal hyperexcitability is status epilepticus, a seizure condition induced by the application of γ-aminobutyric acid (GABA) antagonists. Comparison of particulate CaM kinase activity in seizure rats to that in normals revealed a threefold decrease in the autophosphorylated CaM kinase in cortex, and an eightfold decrease in the hippocampus (183). Exposure of hippocampal slices to magnesium-free medium results in spontaneous epileptiform activity, and was associated with a 38% decrease in CaM kinase activity in homogenates (184).

SUBSTRATES IDENTIFY CELLULAR FUNCTIONS OF CAM KINASE

Although CaM kinase, the enzyme, has turned out to be as interesting as any of its substrates, its primary role is as a kinase that regulates cellular processes by phosphorylating critical proteins in the cell. The task of identifying its substrates is a multistage process that also serves to establish whether it is a general or multifunctional protein kinase that phosphorylates many substrates in situ. As part of this ongoing process, a growing list of putative substrates can be assembled based on in vitro assays in which proteins are found to be phosphorylated by the kinase at a reasonable rate and stoichiometry. The list can be used to identify the consensus amino acid sequence around the phosphorylation sites, and this in turn enables prediction of other possible substrates. Ultimately, it is then essential to demonstrate that CaM kinase phosphorylates substrate proteins in situ. As a minimum, it is necessary to show that in stimulated cells the site(s) of phosphorylation of a putative substrate in situ match the site(s) of phosphorylation obtained when the purified substrate is phosphorylated by the purified kinase in vitro. This approach can be extended with the use of inhibitors to block kinase activity in situ or antisense oligonucleotides and targeted gene inactivation to eliminate expression of the kinase. In addition, introduction of a constitutive form of the kinase can be used to achieve CaM kinase activity selectively in situ without the accompanying increase in Ca^{2+}. The sections that follow describe the putative substrates of CaM kinase and detail how some of these have been demonstrated to be substrates in situ. The results of these studies validate the description of this kinase as multifunctional.

Putative CaM Kinase Substrates

Table 1 provides a list of the proteins that have been found to be good substrates of CaM kinase in vitro. Consistent with being a multifunctional kinase, the enzyme phosphorylates a diverse group of proteins that includes enzymes, cytoskeletal proteins, ion channels, and transcriptional factors. Although some proteins on this list will likely be false leads, others have been and will be found to be substrates in situ. The primary phosphorylation sites for a number of substrates of CaM kinase have been sequenced and are shown in Table 2. Earlier studies identified Arg-Xxx-Yyy-Ser/Thr as the minimal consensus sequence for phosphorylation by CaM kinase (230, 232). Phosphorylation of peptide substrates was found to be diminished by the presence of two rather than one basic residue or by replacement of Arg with Lys. The majority of identified sites still conform to this consensus, but numerous exceptions have been found. In several examples, there are two or more basic residues on the N-terminal side of the phosphorylated residue, and in two substrates the basic residue is Lys and not Arg. A more striking difference is found in six of the listed sites (see acetyl-CoA carboxylase, caldesmon, tau, and vimentin) in which there is no basic residue at the third amino acid on the N-terminal side of the phosphorylated amino acid. There clearly have to be other determinants that enable phosphorylation of these sites. For the anomalous site in vimentin, the acidic residue in the second position on the C-terminal side of the phosphorylation site has been shown to be essential and thus may be a critical determinant of site specificity (198). In all of the substrates there appear to be a disproportionate number of Gln, Glu, and Asp residues within two amino acids of the phosphorylated amino acid, and a hydrophobic amino acid is often the first amino acid on the C-terminal side of the phosphorylation site.

Inhibitors of CaM Kinase

Inhibitors, including nonselective inhibitors used judiciously, can be very helpful in identifying functions and substrates that are regulated by CaM kinase. CaM kinase as well as other calmodulin-dependent enzymes are inhibited by a variety of calmodulin antagonists, including neuroleptics (trifluoperazine), miconazoles (calmidazolium), naphthalenesulfonamides (W-7), and a variety of calmodulin-binding peptides. Inhibitors designed to interact at the ATP-binding site are effective but also not selective for CaM kinase. Among these are the staurosporine family (staurosporine, K-252a, K-252b, and UCN-01) and isoquinolinesulfonamides, such as H-7. These inhibitors typically have higher affinities for PKC than for CaM kinase. Two normal metabolites, sphingosine and 12-hydroperoxyeicosatetraenoic acid (12-HPETE) have been found to be inhibitors of CaM kinase (233, 234). Sphingosine blocks calmodulin-dependent enzymes as well as PKC and is

Table 1 CaM kinase substrates[a]

Protein[b] (size, kDa)	Function	References
Acetyl-CoA carboxylase (260)	fatty acid synthesis	(185, 186)
ATP-citrate lyase (110)	fatty acid synthesis	(185)
C protein (155)	myosin-associated protein	(187)
Calcineurin (58–61)	phosphatase	(110, 111, 188)
Caldesmon (87)	regulates actin-myosin binding	(189)
CRE-binding protein (37)	transcriptional regulation	(190, 191)
Cyclic nucleotide phosphodiesterase (59/63)	cyclic nucleotide metabolism	(112, 192)
GABA-modulin (16.5)	regulation of GABA receptors	(193, 194)
Glycogen synthase (85–90)	carbohydrate metabolism	(23–25, 193)
HMG-CoA reductase (100)	cholesterol biosynthesis	(195, 196)
IP$_3$ receptor (260)	IP$_3$-stimulated Ca^{2+} release	(197)
Intermediate filaments:		
Vimentin (51)	mesenchymal intermediate filament protein	(193, 198, 199)
Desmin (50)	muscle intermediate filament	(199)
Neurofilaments (68, 160, 200)	neuronal intermediate filament	(199)
Glial fibrillary acidic protein (50)	astroglial intermediate filament	(199)
Keratin, type I (acidic, 47)	epithelial cell intermediate filament	(200)
Keratin, type II (basic, 55)	epithelial cell intermediate filament	(200)
Microtubule-associated proteins (MAPs):		
MAP-2 (280)	microtubule assembly	(21, 201–204)
Tau (55–70)	microtubule assembly	(35, 202, 203, 205)

Protein	Function	References
Myelin basic protein (14 & 18)	major myelin protein	(16, 206)
Myosin light chains (20) (smooth muscle)	myosin subunit	(16, 26, 207)
Smooth muscle myosin light chain kinase (130)	initiates muscle contraction	(113, 114, 208)
Lyso-PAF-acetyltransferase	platelet factor synthesis	(209)
Phenylalanine hydroxylase (51)	tyrosine synthesis	(210)
Phosphofructokinase	glycolysis	(211, 212)
(rat liver, 80)		
(sheep heart, 85)		
Phospholamban (11)	Ca^{2+} uptake in SR	(213, 214)
Phospholipase A_2	receptor-stimulated lipid hydrolysis	(215)
Plectin (300)	IF-associated protein	(216)
Pyruvate kinase (61)	carbohydrate metabolism	(212, 217–219)
Ribosomal protein S6 (29)	protein synthesis	(31)
Ryanodine receptor, cardiac (565)	Ca^{2+} release	(220, 221)
Synapsin I (80 & 86)	neurotransmitter release	(222, 223)
Troponin T (31) and I (21) (skeletal muscle)	components of tropomyosin complex	(224)
Troponin I (21) (cardiac muscle)	component of tropomyosin complex	(29)
Tryptophan hydroxylase (59)	serotonin synthesis	(19)
Tyrosine hydroxylase (56)	catecholamine synthesis	(225–227)

[a] Proteins listed are substrates in vitro and potential substrates in vivo. Where possible, references to demonstrated in situ phosphorylation have been included. The references listed are representative, and do not include all substrates or all studies.

[b] Abbreviations used: CoA, coenzyme A; ATP, adenosine triphosphate; CRE, cAMP response element; GABA, γ-aminobutyric acid; HMG, 3-hydroxy-3-methylglutaryl; IP_3, inositol 1,4,5-trisphosphate; and PAF, platelet-activating factor.

therefore limited in usefulness. The selectivity of action of 12-HPETE, a metabolite of arachidonic acid, has not been determined, but it is a potent inhibitor of CaM kinase (IC_{50} of 0.7 μM). It will be interesting to assess whether sphingosine and related metabolites and 12-HPETE are endogenous regulators of CaM kinase. Some of the observed inhibitory effects of 12-HPETE on presynaptic neurotransmitter release may be due to inhibition of CaM kinase (234).

A recently described compound, 1-[N,O-bis(5-isoquinolinesulfonyl)-N-methyl-L-tyrosyl]-4-phenylpiperazine, or KN-62, exhibits apparent selectivity in inhibiting CaM kinase (235, 236). KN-62 binds to CaM kinase and blocks its activation by calmodulin. Potential difficulties in using KN-62 include a low solubility in aqueous solutions and an inability to inhibit the autophosphorylated CaM kinase. Newer generations of this compound should improve on its membrane permeability and solubility while retaining selectivity of action.

Table 2 Sequences of CaM kinase phosphorylation sites[a]

Substrate (reference)	Sequence[b,c]	
Acetyl-CoA carboxylase, rat (186)	I I G S V S* Q D N S E	(25)
Alzheimer's disease amyloid precursor peptide (228)	K K K Q Y T* S* I H H G V	(654, 655)
Calcineurin, bovine (110, 111)	M A R V F S* V L R E E	(197)
Caldesmon, turkey gizzard (189)	V N A Q N S* V A E E E	(73)
	S P K G S S* L K I E R	(587)
	L W E K Q S* V E K P A	(726)
	E A E R L S* Y Q R N D	(26)
Glycogen synthase, rabbit (25, 229, 230)	S K R S N S* V D T S S	(site 1b)
	L S R T L S* V S S L P	(site 2)
Myosin light chain (smooth muscle) (25, 207)	P Q R A T S* N V F S	(19)
Myosin light chain kinase (smooth muscle) (114)	I G R L S S* M A M I S	(512)
Phenylalanine hydroxylase, rat (210)	L S R K L S* D F G Q E	(16)
Phospholamban, dog (213)	I R R A S T* I E M P Q	(17)
Pyruvate kinase, rat (liver) (219)	Y L R R A S* V A Q L T* Q E	(43, 48)
Ryanodine receptor, dog (cardiac) (221)	R T R R I S* Q T S Q V	(2809)
Synapsin Ib (222, 231)	A T R Q T S* V S G Q A	(568, bovine, site 2)
	A T R Q A S* I S G P A	(566, rat)
	P T R Q A S* Q A G P M	(605, bovine, site 3)
	P I R Q A S* Q A G P G	(603, rat)
Tau, bovine (205)	V S S T G S* I D M V D	(405)
Tyrosine hydroxylase, rat (225)	F R R A V S* E Q D A K	(19)
Vimentin, mouse (198)	S T R T Y S* L G S A L	(38)
	R L L Q D S* V D F S L	(82)

[a] Sequences shown are based on sequenced phosphopeptides or comparison to sequenced phosphopeptides in the studies cited following the name of the protein.

[b] The number corresponding to the position of the phosphorylated site in the protein is indicated in parentheses.

[c] The single-letter amino acid code is Ala (A), Arg (R), Asn (N), Asp (D), Asx (B), Cys (C), Gln (Q), Glu (E), Glx (Z), Gly (G), His (H), Ile (I), Leu (L), Lys (K), Met (M), Phe (F), Pro (P), Ser (S), Thr (T), Trp (W), Tyr (Y), Val (V).

Peptide inhibitors based on the autoinhibitory domain of CaM kinase can selectively inhibit CaM kinase (see above), but must be microinjected or bulk loaded into cells. Peptides encompassing both the calmodulin-binding domain and the autoinhibitory domain (e.g. amino acids 281–309) are potent at blocking calmodulin action, and at higher concentrations directly inhibit the kinase (81, 85). Related peptides designed to avoid binding to calmodulin (e.g. amino acids 273–302 or 281–302) inhibit CaM kinase relatively selectively (77, 82, 237). Nonselective inhibition of other kinases poses a potential problem with autoinhibitory domain peptides, since many features of protein kinase catalytic domains are highly conserved (238).

Substrates Phosphorylated in Situ

The role of CaM kinase in phosphorylation of several substrates in situ has been demonstrated by isolating the putative substrate from ^{32}P-labelled cells and showing that stimulation of the cells leads to increased phosphorylation of the sites known to be phosphorylated by CaM kinase in vitro. Such an analysis has been used to implicate synapsin I (239, 240), pyruvate kinase (217), phenylalanine hydroxylase (241), tyrosine hydroxylase (226, 227), phospholamban (214), MLCK (208), and MAP-2 (201) as bona fide in situ substrates of CaM kinase. Comparison of phosphorylation sites cannot be used, of course, to assess the role of CaM kinase in regulating an enzymatic activity or physiological process whose molecular components have not been identified. This can be investigated, however, with the use of selective inhibitors. For example, CaM kinase was found to be essential for inducing long-term potentiation in the hippocampus, a model of neuronal plasticity, by injecting a peptidergic inhibitor based on its autoinhibitory domain (amino acids 273–302) into the postsynaptic cell (82). The same or a related peptide has been used to identify an essential role for CaM kinase in nuclear envelope breakdown (242), in release of glutamate and norepinephrine (237), in mediating the Ca^{2+} pathway for activation of an unidentified Cl^- channel in epithelial cells (243, 244), and in regulating ileal brush border Na^+/H^+ exchange (245). Antibodies to the kinase were also used to show its role in nuclear envelope breakdown. With the exception of the experiments on synaptic release, these studies were facilitated by the ability to microinject the inhibitor and assess CaM kinase function in single cells. Introduction of peptidergic inhibitors by transfection of a minigene encoding the inhibitor should extend the applicability of these inhibitors. The breadth of functions that can be tested is extended by the use of KN-62, which is not limited by microinjection (235, 236). This inhibitor blocks induction of long-term potentiation (246), depolarization-induced phosphorylation of tyrosine hydroxylase in PC12 cells (247), and GABA release into the cerebrospinal fluid of the rat (235).

The use of a constitutive form of CaM kinase will also facilitate identification of proteins and functions that it regulates in situ. A constitutive form of the kinase can be prepared by taking advantage of autophosphorylation with ATP or autothiophosphorylation with ATPγS. Microinjection of such a constitutive CaM kinase into epithelial cells activates a Ca^{2+}-regulated Cl^- channel at low Ca^{2+} (243). Similarly, bulk loading the constitutive kinase into freeze-thawed synaptosomes has been shown to enhance stimulated release of glutamate and norepinephrine (237). A constitutive CaM kinase can also be introduced by microinjecting or transfecting DNA encoding a constitutive construct. The constitutive α-CaM kinase in which Thr286 is replaced with Asp has been used to demonstrate a role of the kinase in maturation of *Xenopus* oocytes (83) and a truncated construct, α-CaM kinase (1-290), was used to show that CaM kinase can increase expression from a rous sarcoma virus promoter (49).

It is anticipated that the reagents described above, as well as other tools, including antisense oligonucleotides to block expression of the kinase, will be utilized to confirm and extend the list of CaM kinase substrates. Studies to date already make a convincing argument that CaM kinase, along with PKA and PKC, is a multifunctional protein kinase that coordinates multiple effects of certain neurotransmitters, hormones, and growth factors.

OTHER (POSSIBLY) MULTIFUNCTIONAL CAM KINASES

CaM Kinase I

Another Ca^{2+}/calmodulin-stimulated kinase activity in neuronal tissues is CaM kinase I (14, 248). A number of proteins have been found to be substrates of this kinase in vitro, including synapsin I, synapsin II, smooth muscle myosin light chain, and the transcriptional activator CREB (191, 248). Its specificity is closer to PKA than to that of multifunctional CaM kinase with regard to synapsin I, since it phosphorylates at the "PKA site," a serine in the globular domain of the protein (site I). This site is distinct from the serines in the vesicle-associating domain phosphorylated by the multifunctional CaM kinase (site II). CaM kinase I has been purified more than 6000-fold from bovine brain (248). It is a monomeric enzyme with a calculated M_r of 42,000 based on hydrodynamic analysis (248). Three polypeptides (37, 39, and 42 kDa) were identified by SDS-PAGE of the purified enzyme, and all three autophosphorylated on threonine residues. The cDNA structure and in situ function of this kinase remain to be described.

CaM Kinase IV (CaM Kinase-Gr)

CaM kinase IV, also referred to as CaM kinase-Gr, was initially isolated in a screen of expression libraries with ^{125}I-calmodulin aimed at finding novel

Ca^{2+}/calmodulin-dependent kinases. Isolation of partial and complete cDNAs encoding the kinase has established that the message for CaM kinase IV encodes a protein of 474 amino acids with a calculated M_r of 53,159 (249–252). The kinase domain maintains the conserved features of protein kinase catalytic domains (69), but is only 46% identical to multifunctional CaM kinase in the catalytic domain and 32% identical overall (249). In rat, the carboxy-terminal sequence of CaM kinase IV was identical to that of calspermin (253), prompting analysis of the genomic structure encoding these two proteins. In fact, one gene with several exons was isolated encoding both calspermin and CaM kinase IV (249, 250). Two transcriptional start sites are utilized to generate the different messages. Exons encoding the calmodulin-binding and acidic domains of the gene product are used in both messages. Calspermin message is only detected in adult male testis, while that encoding CaM kinase IV has been detected in brain, testis, spleen, and thymus. The gene encoding CaM kinase IV is present in the centromeric region of mouse chromosome 18 (254). The human homologue maps to chromosome 5q (255).

CaM kinase IV has been purified from rat cerebellum (250, 255a) and thymus (256). On SDS-gels two subunits of 65 and 67 kDa were purified; hydrodynamic analysis indicates that the enzyme is primarily monomeric with a calculated M_r of 58,000 (255a, 257). CaM kinase IV phosphorylates synapsin I on both sites I and II (251). This differs from multifunctional CaM kinase, which phosphorylates synapsin I only on site II, and from CaM kinase I, which phosphorylates only on site I. CaM kinase IV also phosphorylates the peptide syntide2 (based on the sequence of site 2 in glycogen synthase) (257) and Rap-1b (a Ras-related GTP-binding protein) (258); the latter is not phosphorylated by multifunctional CaM kinase.

CaM kinase IV autophosphorylates on threonine residues in a slow intramolecular reaction to a final stoichiometry of between one and two moles of phosphate per mole of enzyme (257). The effects of autophosphorylation on subsequent activity are complex, and include generation of some Ca^{2+}-independent activity as well as an enhanced affinity of Ca^{2+}/calmodulin-stimulated kinase for ATP and peptide substrate (257). A direct Thr286 homologue is not present in the CaM kinase IV sequence, and the sites and effects of autophosphorylation remain to be further characterized.

CaM kinase IV is expressed at highest levels in cerebellum, forebrain, and thymus, at somewhat lower levels in spleen and testis, and is undetectable in several other tissues (249, 251, 256). The relative abundance of the 65- and 67-kDa isoforms varies. Immunostaining of neurons with antisera against CaM kinase IV revealed prominent staining in nuclei as well as dendritic processes (259). In the adult cerebellum, immunoreactivity was confined to granule cells, with Purkinje cells notably lacking in product (251), hence the name CaM kinase-Gr for granule cells. However, developmental analysis revealed that early in development (embryonic day 22), cerebellar expression was prominent in Purkinje cells instead of granule cells (260). Only after the granule cells had migrated to their final position (by postnatal day 7) did they begin to express CaM kinase IV. During the same period, expression in Purkinje cells declined and was undetectable by postnatal day 14. In contrast, hippocampal pyramidal neurons expressed immunoreactive kinase at relatively constant levels throughout development (260).

What role does CaM kinase IV play in neuronal Ca^{2+} signalling? Is it indeed a multifunctional CaM kinase? Identification of potential functions in situ is important, but will probably have to await development of specific inhibitors. This has been the problem in identifying specific functions for most Ca^{2+}-activated enzymes. Nuclear proteins might be particularly likely CaM kinase IV targets, given the possibility that sequences rich in acidic amino acids may play a role in targetting the kinase to binding sites on nuclear chromatin. Further characterization of effects of autophosphorylation, as well as a detailed characterization of the activation by Ca^{2+}/calmodulin, will enable a greater understanding of signalling systems subserved by this kinase.

SUMMARY

Widespread localization, responsiveness to numerous signal transduction systems, and broad substrate specificity enable the multifunctional CaM kinase to mediate regulation of many cellular functions. The abundance and diversity of CaM kinase substrates attest to its role as a multifunctional kinase. However, expanded identification of its in situ substrates as well as the consequences of their regulation by phosphorylation needs to be accomplished. Recently identified substrates have contributed to the list of potential functions for the CaM kinase. CREB is a hormonally stimulated transcriptional activator, and CaM kinase may lie on the pathway to its activation. This pathway could provide an interface between the potentiation of Ca^{2+} signals by CaM kinase and longer-term modifications of neuronal gene expression. The ryanodine receptor, as well as phospholamban, are involved in cardiac Ca^{2+} homeostasis, and their regulation by CaM kinase phosphorylation suggests the possibility of some feedback control of intracellular Ca^{2+} levels by CaM kinase. Regulation of neuronal plasticity by phosphorylation of synapsin I and of postsynaptic substrates necessary for long-term potentiation is another dynamic area of investigation. The study of substrates and their functions promises to continue providing exciting insights into the control of cellular signalling by Ca^{2+}.

Molecular cloning has enabled structural comparison of neuronal isoforms of the kinase, and has revealed the existence of closely related subunits. Subunits identified to date differ substantially only in two small variable domains, yet their expression in various tissues and during the course of development is precisely controlled. What unique properties do these small variable domains impart to the different isoforms? What directs high concentrations of kinase to a particular subcellular localization, and especially to the PSD? Further molecular cloning will undoubtedly determine whether other multifunctional CaM kinases with unique structures and properties exist.

Finally, studies on the autoregulatory properties of CaM kinase have provided a fascinating picture of how this molecule can alone encode responses to Ca^{2+} signals, potentiating both the duration and magnitude of its activity. Autophosphorylation of the Thr286 autonomy site both traps calmodulin and permits Ca^{2+}-independent activity after calmodulin dissociates. Further analysis of the role of the holoenzyme structure in these modulations will help clarify remaining mechanistic questions. Studies performed during the past few years have clearly established that this Ca^{2+}-independent activity is generated in situ in response to a variety of cell stimuli. Under which conditions and to what extent, however, does this contribute to physiological processes? Does calmodulin trapping enable the kinase to act as a stimulus frequency detector that decodes neuronal information? Does it allow the kinase to maintain its activated and apparently Ca^{2+}-independent state at low Ca^{2+} concentrations? Rigorous analysis of its characteristics in vitro will, no doubt, continue to provide new and surprising properties to be tested in situ and new insights into its regulation of cellular function.

Literature Cited

1. Kamm, K. E., Stull, J. T. 1989. *Annu. Rev. Physiol.* 51:299–313
2. Pickett-Gies, C. A., Walsh, D. A. 1986. *The Enzymes* 17:395–459
3. Ryazanov, A. G., Rudkin, B. B., Spirin, A. S. 1991. *FEBS Lett.* 285:170–75
4. Stull, J. T. 1988. In *Calmodulin,* ed. P. Cohen, C. Klee, pp. 91–122. Amsterdam: Elsevier
5. Cohen, P. 1988. See Ref. 4, pp. 145–94
6. Colbran, R. J., Soderling, T. R. 1990. *Curr. Top. Cell. Regul.* 31:181–221
7. Rostas, J. A. P. 1991. In *Neural and Behavioral Plasticity. The Use of the Domestic Chicken as a Model,* ed. R. J. Andrew, pp. 177–211. Oxford: Oxford Science
8. Rostas, J. A. P., Dunkley, P. 1992. *J. Neurochem.* In press
9. Schulman, H. 1988. *Adv. Second Messenger Phosphoprotein Res.* 22:39–112
10. Schulman, H. 1991. *Curr. Opin. Neurobiol.* 1:43–52
11. Schulman, H., Lou, L. L. 1989. *Trends Biochem. Sci.* 14:62–66
11a. Dunkley, P. R. 1992. *Mol. Neurobiol.* In press
12. Schulman, H., Greengard, P. 1978. *Nature* 271:478–79
13. Schulman, H., Greengard, P. 1978. *Proc. Natl. Acad. Sci. USA* 75:5432–36
14. Kennedy, M. B., Greengard, P. 1981. *Proc. Natl. Acad. Sci. USA* 78:1293–97
15. Yamauchi, T., Fujisawa, H. 1980. *FEBS Lett.* 116:141–44
16. Fukunaga, K., Yamamoto, H., Matsui, K., Higashi, K., Miyamoto, E. 1982. *J. Neurosci.* 39:1607–17
17. Bennett, M. K., Erondu, N. E., Kennedy, M. B. 1983. *J. Biol. Chem.* 258:12735–44
18. Kennedy, M. B., McGuinness, T., Greengard, P. 1983. *J. Neurosci.* 3:818–31
19. Yamauchi, T., Fujisawa, H. 1983. *Eur. J. Biochem.* 132:15–21
20. Goldenring, J. R., Gonzalez, B.,

McGuire, J. S., DeLorenzo, R. J. 1983. *J. Biol. Chem.* 258:12632–40

21. Schulman, H. 1984. *J. Cell Biol.* 99:11–19

22. Kuret, J., Schulman, H. 1984. *Biochemistry* 23:5495–504

23. Woodgett, J. R., Davison, M. T., Cohen, P. 1983. *Eur. J. Biochem.* 136:481–87

24. Ahmad, Z., DePaoli-Roach, A. A., Roach, P. J. 1982. *J. Biol. Chem.* 257:8348–55

25. Payne, M. E., Schworer, C. M., Soderling, T. R. 1983. *J. Biol. Chem.* 258:2376–82

26. McGuinness, T. L., Lai, Y., Greengard, P., Woodgett, J. R., Cohen, P. 1983. *FEBS Lett.* 163:329–34

27. Schworer, C. M., McClure, R. W., Soderling, T. R. 1985. *Arch. Biochem. Biophys.* 242:137–45

28. Gupta, R. C., Kranias, E. G. 1989. *Biochemistry* 28:5909–16

29. Iwasa, T., Inoue, N., Fukunaga, K., Isobe, T., Okuyama, T., et al. 1986. *Arch. Biochem. Biophys.* 248:21–29

30. Ikebe, M., Reardon, S., Scott-Woo, G. C., Zhou, Z., Koda, Y. 1990. *Biochemistry* 29:11242–48

31. Cohn, J. A., Kinder, B., Jamieson, J. D., Delahunt, N. G., Gorelick, F. S. 1987. *Biochim. Biophys. Acta* 928:320–31

32. Nishimura, H., Fukunaga, K., Okamura, H., Miyamoto, E. 1988. *Jpn. J. Pharmacol.* 46:173–82

33. Sato, H., Yamauchi, T., Fujisawa, H. 1990. *J. Biochem.* 107:802–9

34. Yamauchi, T., Fujisawa, H. 1986. *Biochim. Biophys. Acta* 886:57–63

35. Baudier, J., Cole, R. D. 1987. *J. Biol. Chem.* 262:17577–83

36. Nose, P., Schulman, H. 1982. *Biochem. Biophys. Res. Commun.* 107:1082–90

37. Vallano, M. L. 1988. *Biochem. Pharmacol.* 37:2381–88

38. Rostas, J. A. P., Brent, V. A., Seccombe, M., Weinberger, R. P., Dunkley, P. R. 1989. *J. Mol. Neurosci.* 1:93–104

39. Bass, M., Pant, H. C., Gainer, H., Soderling, T. R. 1987. *J. Neurochem.* 49:1116–23

40. Palfrey, H. C., Rothlein, J. E., Greengard, P. 1983. *J. Biol. Chem.* 258:9496–503

41. Kameshita, I., Fujisawa, H. 1989. *Anal. Biochem.* 183:139–43

42. Kuret, J., Schulman, H. 1985. *J. Biol. Chem.* 260:6427–33

43. McGuinness, T. L., Lai, Y., Greengard, P. 1985. *J. Biol. Chem.* 260:1696–704

44. Miller, S. G., Kennedy, M. B. 1985. *J. Biol. Chem.* 260:9039–46

45. Brickey, D. A., Colbran, R. J., Fong, Y.-L., Soderling, T. R. 1990. *Biochem. Biophys. Res. Commun.* 173:578–84

46. Hanson, P. I., Kapiloff, M. S., Lou, L. L., Rosenfeld, M. G., Schulman, H. 1989. *Neuron* 3:59–70

47. Yamauchi, T., Ohsako, S., Deguchi, T. 1989. *J. Biol. Chem.* 264:19108–16

48. Hagiwara, T., Ohsako, S., Yamauchi, T. 1991. *J. Biol. Chem.* 266:16401–8

49. Kapiloff, M. S., Mathis, J. M., Nelson, C. A., Lin, C. R., Rosenfeld, M. G. 1991. *Proc. Natl. Acad. Sci. USA* 88:3710–14

50. Ohsako, S., Watanabe, A., Sekihara, S., Ikai, A., Yamauchi, T. 1990. *Biochem. Biophys. Res. Commun.* 170:705–12

51. Waxham, M. N., Aronowski, J., Kelly, P. T. 1989. *J. Biol. Chem.* 264:7477–82

52. Waxham, M. N., Aronowski, J., Westgate, S. A., Kelly, P. T. 1990. *Proc. Natl. Acad. Sci. USA* 87:1273–77

53. Walaas, S. I., Nairn, A. C., Greengard, P. 1983. *J. Neurosci.* 3:291–301

54. Walaas, S. I., Nairn, A. C., Greengard, P. 1983. *J. Neurosci.* 3:302–11

55. Vallano, M. L. 1989. *J. Neurosci. Methods* 30:1–9

56. Vallano, M. L. 1990. *J. Neurobiol.* 21:1262–73

56a. Kanaseki, T., Ikeuchi, Y., Sugiura, H., Yamauchi, T. 1991. *J. Cell Biol.* 115:1049–60

57. Bennett, M. K., Kennedy, M. B. 1987. *Proc. Natl. Acad. Sci. USA* 84:1794–98

58. Bulleit, R. F., Bennett, M. K., Molloy, S. S., Hurley, J. B., Kennedy, M. B. 1988. *Neuron* 1:63–72

59. Hanley, R. M., Means, A. R., Ono, T., Kemp, B. E., Burgin, K. E., et al. 1987. *Science* 237:293–97

60. Lin, C. R., Kapiloff, M. S., Durgerian, S., Tatemoto, K., Russo, A. F., et al. 1987. *Proc. Natl. Acad. Sci. USA* 84:5962–66

61. Hanley, R. M., Payne, M. E., Cruzalegui, F., Christenson, M. A., Means, A. R. 1989. *Nucleic Acids Res.* 17:3992

62. Tobimatsu, T., Fujisawa, H. 1989. *J. Biol. Chem.* 264:17907–12

63. Tobimatsu, T., Kameshita, I., Fujisawa, H. 1988. *J. Biol. Chem.* 263:16082–86

64. Erondu, N. E., Kennedy, M. B. 1985. *J. Neurosci.* 5:3270–77

65. Benson, D. L., Isackson, P. J., Gall, C. M., Jones, E. G. 1991. *J. Neurosci.* 11:31–47

65a. Cho, K.-O., Wall, J. B., Pugh, P. C., Ito, M., Mueller, S. A., Kennedy, M. B. 1991. *Neuron* 7:439–50

66. Ohya, Y., Kawasaki, H., Suzuki, K., Londesborough, J., Anraku, Y. 1991. *J. Biol. Chem.* 266:12784–94
67. Pausch, M. H., Kaim, D., Kunisawa, R., Admon, A., Thorner, J. 1991. *EMBO J.* 10:1511–22
68. Kornstein, L. B., Gaiso, M. L., Hammell, R. L., Bartelt, D. C. 1992. *Gene.* In press
69. Hanks, S. K., Quinn, A. M., Hunter, T. 1988. *Science* 241:41–52
70. LeVine, H. III, Sahyoun, N. E. 1987. *Eur. J. Biochem.* 168:481–86
71. Katoh, T., Fujisawa, H. 1991. *Biochim. Biophys. Acta* 1091:205–12
72. Kwiatkowski, A. P., Huang, C. Y., King, M. M. 1990. *Biochemistry* 29: 153–59
73. Cohen, P., Klee, C. B., eds. 1988. See Ref. 4, multiple articles
74. Klee, C. B. 1988. See Ref. 4, pp. 225–48
75. LeVine, H. III, Sahyoun, N. E., Cuatrecasas, P. 1986. *Proc. Natl. Acad. Sci. USA* 83:2253–57
76. Hanley, R. M., Means, A. R., Kemp, B. E., Shenolikar, S. 1988. *Biochem. Biophys. Res. Commun.* 152:122–28
77. Payne, M. E., Fong, Y.-L., Ono, T., Colbran, R. J., Kemp, B. E., et al. 1988. *J. Biol. Chem.* 263:7190–95
78. VanBerkum, M. F. A., George, S. E., Means, A. R. 1990. *J. Biol. Chem.* 265:3750–56
79. Weber, P. C., Lukas, T. J., Craig, T. A., Wilson, E., King, M. M., et al. 1989. *Protein:Struct. Funct. Genet.* 6: 70–85
80. Shoemaker, M. O., Lau, W., Shattuck, R. L., Kwiatkowski, A. P., Matrisian, P. E., et al. 1990. *J. Cell Biol.* 111: 1107–25
81. Kelly, P. T., Weinberger, R. P., Waxham, M. N. 1988. *Proc. Natl. Acad. Sci. USA* 85:4991–95
82. Malinow, R., Schulman, H., Tsien, R. W. 1989. *Science* 245:862–66
83. Waldmann, R., Hanson, P. I., Schulman, H. 1990. *Biochemistry* 29:1679–84
84. Fong, Y.-L., Soderling, T. R. 1990. *J. Biol. Chem.* 265:1091–97
85. Colbran, R. J., Fong, Y. L., Schworer, C. M., Soderling, T. R. 1988. *J. Biol. Chem.* 263:18145–51
86. Colbran, R. J., Smith, M. K., Schworer, C. M., Fong, Y.-L., Soderling, T. R. 1989. *J. Biol. Chem.* 264:4800–4
87. Shields, S. M., Vernon, P. J., Kelly, P. T. 1984. *J. Neurochem.* 43:1599–609
88. King, M. M. 1988. *J. Biol. Chem.* 263:4754–57
89. King, M. M., Shell, D. J., Kwiatkowski, A. P. 1988. *Arch. Biochem. Biophys.* 267:467–73
90. Kwiatkowski, A. P., King, M. M. 1989. *Biochemistry* 28:5380–85
91. Rich, D. P., Schworer, C. M., Colbran, R. J., Soderling, T. R. 1990. *Mol. Cell. Neurosci.* 1:107–16
92. Yamagata, Y., Czernik, A. J., Greengard, P. 1991. *J. Biol. Chem.* 266: 15391–97
93. Lai, Y., Nairn, A. C., Greengard, P. 1986. *Proc. Natl. Acad. Sci. USA* 83:4253–57
94. Saitoh, T., Schwartz, J. H. 1985. *J. Cell Biol.* 100:835–42
95. Lou, L. L., Lloyd, S. J., Schulman, H. 1986. *Proc. Natl. Acad. Sci. USA* 83:9497–501
96. Miller, S. G., Kennedy, M. B. 1986. *Cell* 44:861–70
97. Schworer, C. M., Colbran, R. J., Soderling, T. R. 1986. *J. Biol. Chem.* 261:8581–84
98. Hashimoto, Y., Schworer, C. M., Colbran, R. J., Soderling, T. R. 1987. *J. Biol. Chem.* 262:8051–55
99. Ikeda, A., Okuno, S., Fujisawa, H. 1991. *J. Biol. Chem.* 266:11582–88
100. Lickteig, R., Shenolikar, S., Denner, L., Kelly, P. T. 1988. *J. Biol. Chem.* 263:19232–39
101. Yasugawa, S., Fukunaga, K., Yamamoto, H., Miyakawa, T., Miyamoto, E. 1988. *Biomed. Res.* 9:497–502
102. Yasugawa, S., Fukunaga, K., Yamamoto, H., Miyakawa, T., Miyamoto, E. 1991. *Jpn. J. Pharmacol.* 55:263–74
103. Yamauchi, T., Sekihara, S., Ohsako, S. 1991. *Brain Res.* 541:198–205
104. Rich, D. P., Colbran, R. J., Schworer, C. M., Soderling, T. R. 1989. *J. Neurochem.* 53:807–16
105. Bronstein, J. M., Wasterlain, C. G., Farber, D. B. 1988. *J. Neurochem.* 50:1438–46
106. Bronstein, J. M., Farber, D. B., Wasterlain, C. G. 1986. *FEBS Lett.* 196:135–38
107. Yamauchi, T., Fujisawa, H. 1985. *Biochem. Biophys. Res. Commun.* 129:213–19
108. Rostas, J. A. P., Brent, V., Dunkley, P. R. 1987. *Neurosci. Res. Commun.* 1:3–8
109. Hashimoto, Y., King, M. M., Soderling, T. R. 1988. *Proc. Natl. Acad. Sci. USA* 85:7001–5
110. Hashimoto, Y., Soderling, T. R. 1989. *J. Biol. Chem.* 264:16524–29
111. Martensen, T. M., Martin, B. M., Kincaid, R. L. 1989. *Biochemistry* 28:9243–47

112. Hashimoto, Y., Sharma, R. K., Soderling, T. R. 1989. *J. Biol. Chem.* 264:10884–87
113. Hashimoto, Y., Soderling, T. R. 1990. *Arch. Biochem. Biophys.* 278:41–45
114. Ikebe, M., Reardon, S. 1990. *J. Biol. Chem.* 265:8975–78
115. Lai, Y., Nairn, A. C., Gorelick, F., Greengard, P. 1987. *Proc. Natl. Acad. Sci. USA* 84:5710–14
116. Lou, L. L., Schulman, H. 1989. *J. Neurosci.* 9:2020–32
117. Schworer, C. M., Colbran, R. J., Keefer, J. R., Soderling, T. R. 1988. *J. Biol. Chem.* 263:13486–89
118. Miller, S. G., Patton, B. L., Kennedy, M. B. 1988. *Neuron* 1:593–604
119. Thiel, G., Czernik, A. J., Gorelick, F., Nairn, A. C., Greengard, P. 1988. *Proc. Natl. Acad. Sci. USA* 85:6337–41
120. Fong, Y.-L., Taylor, W. L., Means, A. R., Soderling, T. R. 1989. *J. Biol. Chem.* 264:16759–63
121. Ohsako, S., Nakazawa, H., Sekihara, S., Ikai, A., Yamauchi, T. 1991. *J. Biochem.* 109:137–43
122. Katoh, T., Fujisawa, H. 1991. *J. Biol. Chem.* 266:3039–44
123. Kwiatkowski, A. P., Shell, D. J., King, M. M. 1988. *J. Biol. Chem.* 263:6484–86
124. Patton, B. L., Miller, S. G., Kennedy, M. B. 1990. *J. Biol. Chem.* 265:11204–12
125. Colbran, R. J., Soderling, T. R. 1990. *J. Biol. Chem.* 265:11213–19
126. Crick, F. 1984. *Nature* 312:101
127. Lisman, J. E., Goldring, M. 1988. *J. Physiol. (Paris)* 83:187–97
128. Lisman, J. E. 1985. *Proc. Natl. Acad. Sci. USA* 82:3055–57
129. Lisman, J. E., Goldring, M. A. 1988. *Proc. Natl. Acad. Sci. USA* 85:5320–24
129a. Meyer, T., Hanson, P. I., Stryer, L., Schulman, H. 1992. *Science.* In press
130. Hardie, G. 1988. *Nature* 335:592–93
131. Soderling, T. R. 1990. *J. Biol. Chem.* 265:1823–26
132. Rangel-Aldao, R., Rosen, O. M. 1977. *J. Biol. Chem.* 252:7140–45
133. Rosen, O. M., Erlichman, J. 1975. *J. Biol. Chem.* 250:7788–94
134. Buxbaum, J. D., Dudai, Y. 1989. *J. Biol. Chem.* 264:9344–51
135. Gorelick, F. S., Wang, J. K. T., Lai, Y., Nairn, A. C., Greengard, P. 1988. *J. Biol. Chem.* 263:17209–12
136. Fukunaga, K., Rich, D. P., Soderling, T. R. 1989. *J. Biol. Chem.* 264:21830–36
137. MacNicol, M., Jefferson, A. B., Schulman, H. 1990. *J. Biol. Chem.* 265:18055–58
138. Jefferson, A. B., Travis, S. M., Schulman, H. 1991. *J. Biol. Chem.* 266:1484–90
139. Ocorr, K. A., Schulman, H. 1991. *Neuron* 6:907–14
140. Fukunaga, K., Soderling, T. R. 1990. *Mol. Cell Neurosci.* 1:133–38
140a. Suzuki, T., Okumura-Noji, K., Ogura, A., Kudo, Y., Tanaka, R. 1992. *Proc. Natl. Acad. Sci. USA* 89:109–13
141. Ohta, Y., Ohba, T., Fukanaga, K., Miyamoto, E. 1988. *J. Biol. Chem.* 263:11540–47
142. Molloy, S. S., Kennedy, M. B. 1991. *Proc. Natl. Acad. Sci. USA* 88:4756–60
143. Fukunaga, K., Goto, S., Miyamoto, E. 1988. *J. Neurochem.* 51:1070–78
144. Ouimet, C. C., McGuinness, T. L., Greengard, P. 1984. *Proc. Natl. Acad. Sci. USA* 81:5604–8
145. Bronstein, J., Nishimura, R., Lasher, R., Cole, R., de Vellis, J., et al. 1988. *J. Neurochem.* 50:45–49
146. Scholz, W. K., Baitinger, C., Schulman, H., Kelly, P. T. 1988. *J. Neurosci.* 8:1039–51
147. Iwatsubo, T., Nakano, I., Fukunaga, K., Miyamoto, E. 1991. *Acta Neuropathol.* 82:159–63
148. McKee, A. C., Kosik, K. S., Kennedy, M. B., Kowall, N. W. 1990. *J. Neuropathol. Exp. Neurol.* 49:49–63
149. Burgin, K. E., Waxham, M. N., Rickling, S., Westgate, S. A., Mobley, W. C., et al. 1990. *J. Neurosci.* 10:1788–98
150. Walaas, S. I., Lai, Y., Gorelick, F. S., DeCamilli, P., Moretti, M., et al. 1988. *Mol. Brain Res.* 4:233–42
151. Beaman-Hall, C. M., Hozza, M. J., Vallano, M. L. 1992. *J. Neurochem.* In press
152. Bronstein, J. M., Wasterlain, C. G., Bok, D., Lasher, R., Farber, D. B. 1988. *Exp. Eye Res.* 47:391–402
153. Kelly, P. T., Vernon, P. 1985. *Brain Res.* 350:211–24
154. Edelman, A. M., Hunter, D. D., Hendrickson, A. E., Krebs, E. G. 1985. *J. Neurosci.* 5:2609–17
155. Dunkley, P., Cote, A., Harrison, S. M. 1991. *J. Mol. Neurosci.* 2:193–201
156. Walaas, S. I., Gorelick, F. S., Greengard, P. 1989. *Synapse* 3:356–62
157. Goldenring, J. R., McGuire, J. S., DeLorenzo, R. J. 1984. *J. Neurochem.* 42:1077–84
158. Kelly, P. T., McGuinness, T. L., Greengard, P. 1984. *Proc. Natl. Acad. Sci. USA* 81:945–49
159. Kennedy, M. B., Bennett, M. K., Erondu, N. E. 1983. *Proc. Natl. Acad. Sci. USA* 80:7357–61

160. Rostas, J. A. P., Weinberger, R. P., Dunkley, P. R. 1986. *Prog. Brain Res.* 69:355–70
161. Sedman, G. L., Jeffrey, P. L., Austin, L., Rostas, J. A. P. 1986. *Mol. Brain Res.* 1:221–30
162. Sahyoun, N., LeVine, H. III., Bronson, D., Cuatrecasas, P. 1984. *J. Biol. Chem.* 259:9341–44
163. Ohta, Y., Ohba, T., Miyamoto, E. 1990. *Proc. Natl. Acad. Sci. USA* 87:5341- 45
164. Garner, C. C., Tucker, R. P., Matus, A. 1988. *Nature* 336:674–77
165. Willmund, R., Mitschulat, H., Schneider, K. 1986. *Proc. Natl. Acad. Sci. USA* 83:9789–93
166. Bronstein, J., Wasterlain, C. G., Lasher, R., Farber, D. B. 1989. *Brain Res.* 495:83–88
167. Churn, S. B., Taft, W. C., DeLorenzo, R. J. 1990. *Stroke* 21(III):112–16
168. Taft, W. C., Tennes-Rees, K. A., Blair, R. E., Clifton, G. L., DeLorenzo, R. J. 1988. *Brain Res.* 447:159–63
169. Kochhar, A., Saitoh, T., Zivin, J. A. 1991. *Brain Res.* 542:141–46
170. Onodera, H., Hara, H., Kogure, K., Fukunaga, K., Ohta, Y., et al. 1990. *Neurosci. Lett.* 113:134–38
171. Yamamoto, H., Fukunaga, K., Lee, K., Soderling, T. R. 1992. *J. Neurochem.* 58:1110–17
172. Aronowski, J., Grotta, J. C., Waxham, M. N. 1992. *J. Neurochem.* In press
173. Kelly, P. T., Cotman, C. W. 1981. *Brain Res.* 206:251–71
174. Sahyoun, N., LeVine, H. III, Burgess, S. K., Blanchard, S., Chang, K. J., et al. 1985. *Biochem. Biophys. Res. Commun.* 132:878–84
175. Weinberger, R. P., Rostas, J. A. P. 1986. *Dev. Brain Res.* 29:37–50
176. Kelly, P. T., Shields, S., Conway, K., Yip, R., Burgin, K. 1987. *J. Neurochem.* 49:1927–40
177. Vallano, M. L., Beaman-Hall, C. M. 1989. *J. Neurosci.* 9:539–47
178. Hendry, S. H. C., Kennedy, M. B. 1986. *Proc. Natl. Acad. Sci. USA* 83: 1536–40
179. Neve, R. L., Bear, M. F. 1989. *Proc. Natl. Acad. Sci. USA* 86:4781–84
180. Goldenring, J. R., Wasterlain, C. G., Oestreicher, A. B., deGraan, P. E., Farber, D. B., et al. 1986. *Brain Res.* 377:47–53
181. Wasterlain, C. G., Farber, D. B. 1984. *Proc. Natl. Acad. Sci. USA* 81:1253–57
182. Bronstein, J. M., Farber, D. B., Micevych, P. E., Lasher, R., Wasterlain, C. G. 1990. *Brain Res.* 524:49–53
183. Bronstein, J., Farber, D., Wasterlain, C. 1988. *Neurochem. Res.* 13:83–86
184. Churn, S. B., Anderson, W. W., DeLorenzo, R. J. 1992. *Epilepsy Res.* In press
185. Hardie, D. G., Carling, D., Ferrari, S., Guy, P. S., Aitken, A. 1986. *Eur. J. Biochem.* 157:553–61
186. Haystead, T. A. J., Campbell, D. G., Hardie, D. G. 1988. *Eur. J. Biochem.* 175:347–54
187. Schlender, K. K., Bean, L. J. 1991. *J. Biol. Chem.* 266:2811–17
188. Calalb, M. B., Kincaid, R. L., Soderling, T. R. 1990. *Biochem. Biophys. Res. Commun.* 172:551–56
189. Ikebe, M., Reardon, S. 1990. *J. Biol. Chem.* 265:17607–12
190. Dash, P. K., Karl, K. A., Colicos, M. A., Prywes, R., Kandel, E. R. 1991. *Proc. Natl. Acad. Sci. USA* 88:5061–65
191. Sheng, M., Thompson, M. A., Greenberg, M. E. 1991. *Science* 252:1427–30
192. Sharma, J. K., Wang, J. H. 1986. *J. Biol. Chem.* 261:1322–28
193. Schulman, H., Kuret, J., Jefferson, A. B., Nose, P. S., Spitzer, K. 1985. *Biochemistry* 24:5320–27
194. Wise, B. C., Guidotti, A., Costa, E. 1980. *Proc. Natl. Acad. Sci. USA* 80:886–90
195. Beg, Z. H., Stonik, J. A., Brewer, H. B. 1987. *J. Biol. Chem.* 262:13228–40
196. Clarke, P. R., Hardie, D. G. 1990. *FEBS Lett.* 269:213–17
197. Ferris, C. D., Huganir, R. L., Bredt, D. S., Cameron, A. M., Snyder, S. H. 1991. *Proc. Natl. Acad. Sci. USA* 88: 2232–35
198. Ando, S., Tokui, T., Yamauchi, T., Sugiura, H., Tanabe, K., et al. 1991. *Biochem. Biophys. Res. Commun.* 175:955–62
199. Tokui, T., Yamauchi, T., Yano, T., Nishi, Y., Kusagawa, M., et al. 1990. *Biochem. Biophys. Res. Commun.* 169:896–904
200. Yano, T., Tokui, T., Nishi, Y., Nishizawa, K., Shibata, M., et al. 1991. *Eur. J. Biochem.* 197:281–90
201. Jefferson, A. B., Schulman, H. 1991. *J. Biol. Chem.* 266:346–54
202. Schulman, H. 1984. *Mol. Cell. Biol.* 4:1175–78
203. Yamamoto, H., Fukunaga, K., Tanaka, E., Miyamoto, E. 1983. *J. Neurochem.* 41:1119–25
204. Yamauchi, T., Fujisawa, H. 1982. *Biochem. Biophys. Res. Commun.* 109:975–81
205. Steiner, B., Mandelkow, E.-M., Biernat, J., Gustke, N., Meyer, H. E., et al. 1990. *EMBO J.* 9:3539–44

206. Shoji, S., Ohnishi, J., Funakoshi, T., Fukunaga, K., Miyamoto, E., et al. 1987. *J. Biochem.* 102:1113–20
207. Edelman, A. M., Lin, W., Osterhout, D. J., Bennett, M. K., Kennedy, M. B., et al. 1990. *Mol. Cell Biochem.* 97:87–98
208. Stull, J. T., Hsu, L., Tansey, M. G., Kamm, K. E. 1990. *J. Biol. Chem.* 265:16683–90
209. Domenech, C., Domenech, E. M.-D., Soling, H.-D. 1987. *J. Biol. Chem.* 262:5671–76
210. Døskeland, A. P., Schworer, C. M., Døskeland, S. O., Chrisman, T. D., Soderling, T. R., et al. 1984. *Eur. J. Biochem.* 145:31–37
211. Mahrenholz, A. M., Lan, L., Mansour, T. E. 1991. *Biochem. Biophys. Res. Commun.* 174:1255–59
212. Mieskes, G., Kuduz, J., Soling, H. D. 1987. *Eur. J. Biochem.* 167:383–89
213. Simmerman, H. K. B., Collins, J. H., Theibert, J. L., Wegener, A. D., Jones, L. R. 1986. *J. Biol. Chem.* 261:13333–41
214. Wegener, A. D., Simmerman, H. K. B., Lindemann, J. P., Jones, L. R. 1989. *J. Biol. Chem.* 264:11468–74
215. Piomelli, D., Greengard, P. 1991. *Proc. Natl. Acad. Sci. USA* 88:6770–74
216. Herrmann, H., Wiche, G. 1987. *J. Biol. Chem.* 262:1320–25
217. Connelly, P. A., Sisk, R. B., Schulman, H., Garrison, J. C. 1987. *J. Biol. Chem.* 262:10154–63
218. Schworer, C. M., El-Maghrabi, M. R., Pilkis, S. J., Soderling, T. R. 1985. *J. Biol. Chem.* 260:13018–22
219. Soderling, T. R., Schworer, C. M., El-Maghrabi, M. R., Pilkis, S. J. 1986. *Biochem. Biophys. Res. Commun.* 139:1017–23
220. Takasago, T., Imagawa, T., Furukawa, K., Ogurusu, T., Shigekawa, M. 1991. *J. Biochem.* 109:163–70
221. Witcher, D. R., Kovacs, R. J., Schulman, H., Cefali, D. C., Jones, L. R. 1991. *J. Biol. Chem.* 266:11144–52
222. Czernik, A. J., Pang, D. T., Greengard, P. 1987. *Proc. Natl. Acad. Sci. USA* 84:7518–22
223. Huttner, W. B., DeGennaro, L. J., Greengard, P. 1981. *J. Biol. Chem.* 256:1482–88
224. Sato, H., Fukunaga, K., Araki, S., Ohtsuki, I., Miyamoto, E. 1988. *Arch. Biochem. Biophys.* 260:443–51
225. Campbell, D. G., Hardie, D. G., Vulliet, P. R. 1986. *J. Biol. Chem.* 261:10489–92
226. Griffith, L. C., Schulman, H. 1988. *J. Biol. Chem.* 263:9542–49

227. Waymire, J. C., Johnston, J. P., Hummer-Lickteig, K., Lloyd, A., Vigny, A., et al. 1988. *J. Biol. Chem.* 263:12439–47
228. Gandy, S., Czernik, A. J., Greengard, P. 1988. *Proc. Natl. Acad. Sci. USA* 85:6218–21
229. Juhl, H., Sheorain, V. S., Schworer, C. M., Jett, M. F., Soderling, T. R. 1983. *Arch. Biochem. Biophys.* 222:518–26
230. Pearson, R. B., Woodgett, J. R., Cohen, P., Kemp, B. E. 1985. *J. Biol. Chem.* 260:14471–76
231. Südhof, T. C., Czernik, A. J., Kao, H.-T., Takei, K., Johnston, P. A., et al. 1989. *Science* 245:1474–80
232. Kennelly, P. J., Krebs, E. G. 1991. *J. Biol. Chem.* 266:15555–58
233. Jefferson, A. B., Schulman, H. 1988. *J. Biol. Chem.* 263:15241–44
234. Piomelli, D., Wang, J. K. T., Sihra, T. S., Nairn, A. C., Czernik, A. J. 1989. *Proc. Natl. Acad. Sci. USA* 86:8550–54
235. Ishikawa, N., Hashiba, Y., Hidaka, H. 1990. *J. Pharmacol. Exp. Ther.* 254:598–602
236. Tokumitsu, H., Chijiwa, T., Hagiwara, M., Mizutani, A., Terasawa, M., et al. 1990. *J. Biol. Chem.* 265:4315–20
237. Nichols, R. A., Sihra, T. S., Czernik, A. J., Nairn, A. C., Greengard, P. 1990. *Nature* 343:647–51
238. Smith, M. K., Colbran, R. J., Soderling, T. R. 1990. *J. Biol. Chem.* 265:1837–40
239. Nestler, E. J., Greengard, P. 1982. *J. Neurosci.* 2:1011–23
240. Nestler, E. J., Greengard, P. 1982. *Nature* 296:452–54
241. Garrison, J. C., Johnsen, D. E., Campanile, C. P. 1984. *J. Biol. Chem.* 259:3283–92
242. Baitinger, C., Alderton, J., Poenie, M., Schulman, H., Steinhardt, R. A. 1990. *J. Cell Biol.* 111:1763–73
243. Wagner, J. A., Cozens, A. L., Schulman, H., Gruenert, D. C., Stryer, L., et al. 1991. *Nature* 349:793–96
244. Worrell, R. T., Frizzell, R. A. 1991. *Am. J. Physiol.* 260:C877–82
245. Cohen, M. E., Reinlib, L., Watson, A. J. M., Gorelick, F., Rys-Sikora, K., et al. 1990. *Proc. Natl. Acad. Sci. USA* 87:8990–94
246. Ito, I., Hidaka, H., Sugiyama, H. 1991. *Neurosci. Lett.* 121:119–21
247. Ishii, A., Kiuchi, K., Kobayashi, R., Sumi, M., Hidaka, H., et al. 1991. *Biochem. Biophys. Res. Commun.* 176:1051–56
248. Nairn, A. C., Greengard, P. 1987. *J. Biol. Chem.* 262:7273–81

249. Means, A. R., Cruzalegui, F., LeMa-
    gueresse, B., Needleman, D. S.,
    Slaughter, G. R., et al. 1991. *Mol. Cell.
    Biol.* 11:3960–71
250. Ohmstede, C. A., Bland, M. M., Mer-
    rill, B. M., Sahyoun, N. 1991. *Proc.
    Natl. Acad. Sci. USA* 88:5784–88
251. Ohmstede, C. A., Jensen, K. F.,
    Sahyoun, N. E. 1989. *J. Biol. Chem.*
    264:5866–75
252. Sikela, J. M., Hahn, W. E. 1987. *Proc.
    Natl. Acad. Sci. USA* 84:3038–42
253. Ono, T., Slaughter, G. R., Cook, R. G.,
    Means, A. R. 1989. *J. Biol. Chem.*
    264:2081–87
254. Sikela, J. M., Adamson, M. C., Wil-
    son-Shaw, D., Kozak, C. A. 1990.
    Genomics 8:579–82
255. Sikela, J. M., Law, M. L., Kao, F. T.,

    Hartz, J., Wei, Q., et al. 1989. *Ge-
    nomics* 4:21–27
255a. Miyano, O., Kameshita, I., Fujisawa,
    H. 1992. *J. Biol. Chem.* 267:1198–203
256. Frangakis, M. V., Chatila, T., Wood,
    E. R., Sahyoun, N. 1991. *J. Biol.
    Chem.* 266:17592–96
257. Frangakis, M. V., Ohmstede, C.-A.,
    Sahyoun, N. 1991. *J. Biol. Chem.*
    266:11309–16
258. Sahyoun, N., McDonald, O. B., Farrell,
    F., Lapetina, E. G. 1991. *Proc. Natl.
    Acad. Sci. USA* 88:2643–47
259. Jensen, K. F., Ohmstede, C.-A., Fisher,
    R. S., Sahyoun, N. 1991. *Proc. Natl.
    Acad. Sci. USA* 88:2850–53
260. Jensen, K. F., Ohmstede, C.-A., Fisher,
    R. S., Olin, J. K., Sahyoun, N. 1991.
    Proc. Natl. Acad. Sci. USA 88:4050–53

Annu. Rev. Biochem. 1992. 61:603–40

ENZYMES AND MOLECULAR MECHANISMS OF GENETIC RECOMBINATION

Stephen C. West

Imperial Cancer Research Fund, Clare Hall Laboratories, South Mimms, Hertfordshire EN6 3LD, United Kingdom

KEY WORDS: homologous recognition, strand exchange, triplex DNA, RecA protein, Ruv proteins

CONTENTS

0066-4154/92/0701-0603$02.00

PERSPECTIVES AND SUMMARY

Although it is more than 25 years since Clark & Margulies reported the identification of the first recombination-deficient mutants of *Escherichia coli* (1), it is only now that we are beginning to understand some of the intricate biochemical manipulations of DNA necessary for the formation of recombinant molecules.

Much of our current understanding of the mechanism of homologous pairing is derived from in vitro studies of the RecA protein of *E. coli,* and this protein is the focus of the first part of the review. RecA plays a central role in recombination both as a structural protein and as a reaction catalyst. Detailed biochemical studies have shown that RecA is able to pair two homologous DNA molecules and catalyze strand-exchange reactions leading to the formation of heteroduplex DNAs.

RecA is only a small protein (M_r 37,842), but it polymerizes on DNA to form a right-handed helical filament within which the pairing reactions occur. The regularity of these nucleoprotein structures, as seen by electron microscopy, has allowed three-dimensional image reconstruction studies to precede structural studies of RecA by X-ray crystallography. The DNA lies within the deep groove of the RecA filament and is underwound to $1.5\times$ the length of B-form DNA, presumably to fulfil the requirements for a search for homology and the exchange of strands. Current experimental studies on the mechanism of pairing by RecA protein indicate the formation of three-stranded DNA helices within the RecA filament. Our understanding of the mechanism of strand exchange is continually evolving, and until recently it was believed that ATP hydrolysis provided the driving force necessary for strand exchange and the directional movement of the branch point (or Holliday junction). However, this belief has been thrown into doubt by recent work that shows that under some conditions heteroduplex DNA can be formed without ATP hydrolysis. An alternative proposal is that the conversion of ATP to ADP may be required for the release and recycling of RecA itself.

RecA protein is ubiquitous amongst prokaryotic organisms. Proteins that may promote similar reactions in yeast and mammalian cells have been described; however, there have been no reports of these proteins being able to form helical filaments on DNA. The mechanisms of pairing and strand exchange in eukaryotic organisms may therefore differ from the mechanisms catalyzed by RecA.

Until recently, it was not clear which *E. coli* gene(s) encode the enzyme(s) responsible for resolution of the recombination intermediates formed by RecA protein, and efforts to understand this process were limited to in vitro studies of the structure and resolution of synthetic structures that resemble Holliday junctions. These synthetic junctions, made by annealing four single strands, contain an immobile cross-over point at the center of four duplex arms (an X-structure). Particular emphasis has been placed on the structure of the small molecules in solution, and on the way in which they are resolved by endonucleases purified from bacteriophage-infected cells (T4 endonuclease VII and T7 endonuclease I). Whether they represent reasonable models for recombination intermediates remains to be determined. In an alternative approach, recombination reactions catalyzed by RecA protein have been used to detect a resolution activity in fractionated *E. coli* extracts. The development of this in vitro strand exchange/resolution system, as well as the identification of the resolution activity as the *ruvC* gene product, provides new impetus to studies of the resolution process and the development of more complete in vitro recombination reactions.

Studies of the enzymology of general genetic recombination (and in particular RecA protein) have led to a large amount of literature, including many reviews relevant to the work presented here. These include general reviews (2–7), and more specialized reviews on RecA and RecBCD proteins (8–18). With regard to the biochemistry of RecA, the recent reviews by Cox and coworkers (11, 16), Radding (12), and Kowalczykowski (18) are particularly rich in detail, and the short synopsis by Radding (19) is recommended. The only way to keep this review within a reasonable length has been to focus on selected topics, and often to provide a particular viewpoint. This cannot be done without appearing arbitrary and neglecting many excellent papers.

E. COLI RecA PROTEIN

General Activities and In Vitro Systems

The RecA protein of *E. coli* promotes reactions that are important for genetic recombination, DNA repair, and UV-induced mutagenesis. RecA protein plays a central role in the SOS response to DNA damage (20), and catalyzes the autodigestion of LexA repressor (21), UmuD protein (22–24), and λ repressor (21, 25). Cleavage of LexA leads to the derepression of more than 20 genes, including *lexA, recA, recN, recQ, ruvA, ruvB, uvrA, uvrB, uvrD, umuC,* and *umuD*.

In addition to its role in the regulation of recombination, repair, and mutagenesis, RecA protein catalyzes in vitro reactions that are consistent with a direct role in the recombination process. Two popular in vitro systems are illustrated in Figure 1.

REACTIONS INVOLVING THREE DNA STRANDS   To study homologous pairing and strand-exchange reactions catalyzed by RecA protein, most workers have used single-stranded circular and linear duplex DNA substrates, as shown in Figure 1*A* (11, 12, 18).

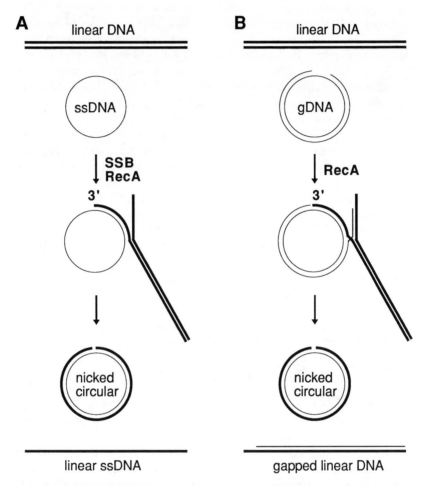

Figure 1   Comparison of the three-stranded (*A*) and four-stranded (*B*) reactions used to demonstrate homologous pairing and strand exchange by RecA protein. The substrates are usually bacteriophage φX174 or M13 DNAs. Intermediates of reaction *B* may be used to study enzymes that resolve recombination intermediates.

The reactions may be divided into a number of steps, the first of which involves the stoichiometric binding of RecA to single-stranded DNA to form a nucleoprotein complex in which the ssDNA lies within the RecA filament (26–31). The presence of a single-stranded DNA-binding protein such as *E. coli* SSB protein facilitates filament formation by removal of secondary structure from the ssDNA (32, 33).

The second step involves the interaction of a nucleoprotein filament with naked duplex DNA, leading to the establishment of homologous contacts. Pairing is thought to occur by multiple random contacts (34–36), which in vitro are facilitated by the formation of large protein-DNA aggregates that serve to concentrate the DNA (37, 38). It is likely that the two interacting DNA molecules become interwound within the RecA filament in the form of a triple-stranded DNA helix (10, 28, 39–43). Pairing may occur over an extensive length, and can take place in the absence of any free ends, i.e. single-stranded circular DNA can pair with covalently closed circular DNA (44–49).

If a free end is present in one of the two interacting molecules (such as with the substrates of Figure 1A), the reaction proceeds into its next phase, in which strands are exchanged to form heteroduplex DNA (44, 50, 51). Strand exchange occurs relatively slowly (2–10 base pairs per second) and with a defined polarity (52–54). The polarity is usually defined relative to the single strand on which filament formation occurs, and is 5'→3'. Thus, transfer is initiated as the 3'-end of the complementary strand of the linear duplex is transferred to the ssDNA. During the course of the reaction, large amounts of ATP are hydrolyzed by RecA (16, 18). At completion, the products are nicked circular duplex and linear single-stranded DNA. Because RecA initiates the reaction by filament formation on the single-stranded DNA, and remains bound to this molecule throughout the reaction, it is subsequently found on the nicked circular product (28, 55, 56). The other product, linear ssDNA, binds free RecA from solution, and no back reaction occurs because two nucleoprotein filaments cannot undergo pairing (13).

Although homologous pairing requires near perfect homology (57), it is remarkable that the RecA-mediated strand exchange process can accommodate significant amounts of base mismatches (58, 59), DNA lesions (60), or heterologous insertions (59). Indeed, the strand-exchange reaction can pass heterologous insertions of up to 1000 nucleotides when placed in the single-stranded DNA, although significant reductions in the amount of product are observed when inserts of 140 base pairs are present within duplex DNA (59). The ability to pass a region of heterology is position dependent, and requires homology beyond the heterologous insert (B. Jwang and C. M. Radding, unpublished). These results are consistent with in vivo observations by Lichten & Fox (61), which indicate that RecA-dependent recombination of bacteriophage λ is capable of including substantial lengths of heterology (>1000 bases) into heteroduplex DNA.

REACTIONS INVOLVING FOUR DNA STRANDS   Although pairing between two fully homologous duplex DNA molecules has not been described, RecA protein can pair two duplexes if one contains a short region of single-stranded DNA (62, 63). The need for ssDNA results in part from the inability of RecA to bind fully duplex DNA at physiological pH (27). Using plasmid DNA molecules that contain short gaps, early studies showed that RecA protein bound stoichiometrically to gapped DNA (26). Binding resulted in filament formation at the gap, which extended into the contiguous duplex DNA, as indicated by a stimulation of the ATPase activity of RecA protein and by electron microscopy (26, 64). Subsequent studies using gapped linear molecules showed that the amount of ATP hydrolysis was directly proportional to the length of the DNA molecule (65). The weak binding of RecA to duplex DNA at pH 7.5 reflects a slow nucleation step in the association pathway rather than an unfavorable binding equilibrium (66–68). The presence of a single-stranded gap (26), low pH (66), intercalating drugs (69), Z-form DNA (70), or other structural perturbations (71, 72) can overcome this barrier to binding. Following nucleation from a single-stranded gap, the binding of RecA to flanking duplex DNA occurs with a 5'→3' polarity (30, 65, 73).

Complexes between gapped DNA and RecA protein, and reactions that occur between gapped and homologous duplex DNA, have been studied in some detail because single-stranded gaps are thought to initiate recombination during postreplication repair (74–76). Gapped DNA-RecA complexes promote pairing and strand-exchange reactions in much the same way as ssDNA-RecA complexes, as indicated in Figure 1B. Again, pairing occurs rapidly and is followed by a processive strand exchange (~3 base pairs/sec), which occurs with a 5'→3' polarity defined relative to the single-stranded DNA on which filament formation is initiated (54, 76). However, in this case, strand exchange is reciprocal, and electron microscopic studies of deproteinized intermediates (77, 78) indicate structures that resemble classical recombination intermediates similar to those proposed by Holliday (79). At the completion of strand exchange, the reaction products are nicked circular and gapped linear DNA. In contrast to reactions involving ssDNA, reactions between gapped and linear duplex DNA do not require SSB to remove secondary structure (76, 80), unless large gaps (>300 nucleotides) are used (81, 82).

The single-stranded DNA in the gap serves a dual purpose. In addition to its role as a nucleation site at which RecA initiates binding, it also provides the site at which strand exchange is initiated. Using a series of linear duplexes, it was shown that strand exchange needed the end of the duplex to be complementary to the single-stranded DNA in the gap (76). Moreover, the length of the overlap into the gap needed to be on the order of 40–100 base pairs for maximal strand exchange efficiency (82, 83).

Strand-exchange reactions between duplex substrates are more sensitive to heterologous insertions than equivalent reactions in which one DNA is single stranded. Using gapped and linear duplex DNA with insertions or deletions of 4 to 38 base pairs, little effect was observed on the efficiency of strand exchange to produce heteroduplex products (84). However, insertions or deletions of 120 or more base pairs reduced the product yield to a level below the threshold of detection. In the same study, the effect of ultraviolet damage to the substrate DNA was tested. Strand exchange was found to pass 30 or more pyrimidine dimers in each duplex (5386 base pairs in length).

A number of variant duplex-duplex reactions are shown in Figure 2. Reactions between gapped circular and covalently closed circular duplex DNA (Figure 2A) lead to the formation of joint molecules in which two duplexes are linked (62, 63). Because only one strand has a free end, complete strand exchange is prohibited and the molecules become linked by a structure that resembles a double Holliday junction that is driven away from the gap by RecA protein. The polarity of this reaction has not been determined, but is likely to be 5'→3' as described above. A surprising apparent lack of polarity was observed when joint molecules (containing Holliday junctions) were formed between two linear duplexes (Figure 2B), one of which contained a 5'- or 3'- gapped tail (77). However, it is possible that a bias in the formation of joints did occur, but the joints formed with the "wrong" polarity were preferentially stable. Such effects have been observed in reactions involving three strands of DNA (43, 85–87).

Although pairing between single-stranded circular and covalently closed duplex DNA is quite easy to demonstrate, homologous contacts between

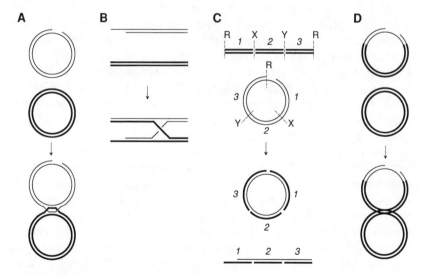

Figure 2 Combinations of duplex DNA substrates from which RecA protein can make joint molecules. In reaction *C*, restriction endonuclease cuts are designated R, X, and Y. In reaction *D*, the gap is located in heterologous sequences (thin lines). The joint formed in this reaction is paranemic in nature, and may involve only transient pairing within the RecA filament (see section on REACTIONS INVOLVING FOUR DNA STRANDS).

regions of duplex DNA have yet to be shown directly. However, indirect evidence suggests that duplex-duplex joints can form at least transiently. The demonstration that RecA protein can promote strand exchange through double-strand breaks provided one of the first forms of evidence for duplex-duplex pairing (88). The substrates used were gapped circular DNA and homologous duplex DNA that had been cut into smaller fragments by a restriction endonuclease (Figure 2C). Strand exchange was initiated by the fragment complementary to the gap, but the second and third fragments that participated in the reactions were homologous only to duplex regions of the gapped DNA. The reaction was independent of any particular end; blunt ends were as efficient as 5'- or 3'-overhangs. The heteroduplex products contained nicks that could be ligated, thereby effecting the repair of double-strand breaks. Whether such a reaction has biological significance in vivo remains to be determined.

To account for the observations just described, it was proposed that each linear duplex fragment was able to pair with homologous duplex sequences within the RecA filament. As strand exchange proceeded towards the end of one linear duplex, a second would already be aligned with its termini in juxtaposition (88). The experiments described above led to the proposal that the RecA filament contains two DNA-binding sites and that paired molecules might take the form of a four-stranded helix within the filament.

More direct evidence for the formation of duplex-duplex joints was provided by experiments in which filaments formed on gapped DNA were reacted with partially homologous superhelical DNA (Figure 2D). Since the molecules shared homology only away from the site of the gap, strand exchange was not permitted. As described in a subsequent section, filament formation results in an underwinding of the DNA by 39.5% relative to B-form DNA. Underwinding of the superhelical DNA, which occurred concomitant with homologous pairing, was detected by Conley & West (81, 89) and Lindsley & Cox (82) using a sensitive topoisomerase assay. The joints were estimated to be on the order of 300–400 base pairs in length, assuming a single joint per molecule (81). The efficiency of formation or stability of these duplex-duplex joints, as measured by the unwinding assay, was low in comparison with joints formed between single- and double-stranded DNA (82). Moreover, they were dependent upon superhelicity and were not observed when interacting molecules were both relaxed (81, 82). In contrast, utilizing similar substrates but using a nitrocellulose filter-binding assay, Chiu et al reported efficient duplex-duplex joint formation (64). Further evidence for duplex-duplex interactions was provided by experiments in which addition of homologous, but not heterologous, competitor DNA blocked potential sites of pairing on the gapped DNA-RecA filament and thereby reduced pairing and strand exchange with a linear duplex (89).

The mechanics of early steps in the reaction of Figure 1B can therefore be summarized in the following way. The gap acts as the initial binding site for RecA protein, which then polymerizes in the $5' \rightarrow 3'$ direction to cover duplex DNA sequences. All DNA sequences within the filament, both single-stranded and duplex, participate in the search for homology with naked duplex DNA, and homologous contacts are established by the formation of paranemic joints. Once in alignment, these nascent intermediates can be converted to stable joint molecules by movement of the joint to the gapped region where strand exchanges may be initiated.

ATPase Activity

Early studies showed that RecA protein is a DNA-dependent ATPase (90, 91). Although only a weak ATPase with a turnover number of ~ 30 min^{-1} (16, 18), the amount of ATP that RecA consumes becomes significant when it is realized that the active species in hydrolysis is a nucleoprotein filament that may be several thousand RecA monomers in length. Hydrolysis of ATP requires filament assembly because oligonucleotides shorter than 50 nucleotides are poor cofactors, and ATP hydrolysis occurs at a uniform level throughout the length of the filament (33, 92). There are no treadmilling effects involving enhanced ATP hydrolysis at one end of the filament compared with the other (92–94).

The ability of RecA to promote strand exchange, coincident with the requirement for ATP, led to the assumption that the energy derived from ATP hydrolysis drives strand exchange. However, it is clear that the two reactions are uncoupled, because the rate of ATP hydrolysis is the same whether the DNA is undergoing strand exchange or not (92). Indeed, there is no reason to invoke an energy requirement at all, as the strand-exchange reaction is isoenergetic (i.e. for every base pair broken, another one reforms). The identification of ATP-independent eukaryotic pairing activities (95–99) therefore led to a reevaluation of the relationship between RecA-mediated strand exchange and ATP hydrolysis. In recent experiments, Menetski et al (100) and Rosselli & Stasiak (101) showed that RecA protein could promote strand exchange in the absence of ATP hydrolysis. Using a nonhydrolyzable analogue of ATP, ATP[γ]S, it was shown that at low Mg^{2+} concentrations RecA could make significant amounts of heteroduplex DNA, as detected following deproteinization of reaction intermediates. Moreover, a mutant RecA protein (RecA72, which contains a single amino acid change at position 72) that binds but can not hydrolyze ATP forms joint molecules (W. M. Rehrauer and S. C. Kowalczykowski, unpublished).

The need for ATP binding, but not its hydrolysis, may be rationalized in the following way (16, 18). In the process of binding ATP, RecA protein undergoes an allosteric transition into a high-affinity DNA-binding state, the filament is assembled on DNA, and DNA molecules undergo pairing reactions. The process of ATP hydrolysis leads to the production of bound ADP and induction of the low-affinity DNA-binding state. Formation of this low-affinity state results in the release of heteroduplex DNA. Experimentally, the need for the transition from high- to low-affinity state can be circumvented by deproteinization using phenol or SDS (100, 101). In simple terms, heteroduplex formation requires dissociation of RecA from the DNA, either by ATP hydrolysis or by artificial means. However, further investigation of the role of ATP hydrolysis is required, since it is clear that hydrolysis and dissociation are not tightly coupled and under some conditions ATP hydrolysis can occur without measurable dissociation (16, 18).

Two interesting points can be made regarding the formation of heteroduplex products in the absence of ATP hydrolysis. Firstly, strand exchange in the presence of ATP[γ]S occurs with the same polarity as that in reactions in which ATP is hydrolyzed (101). This result provides further evidence that the unidirectional nature of RecA-driven strand exchange is a consequence of the polarization of the filament with regard to the single-stranded DNA on which nucleation is initiated (30, 102, 103). This directional polymerization of monomers results from defined protein-protein contacts between high-affinity monomeric subunits (103, 104). Secondly, the role of the high-affinity state is to bring about the formation of a nucleoprotein structure in which two interacting DNA molecules are aligned in homologous register (28, 39). The DNA within the filament could adopt (*a*) a structure in which Watson-Crick base pairs are maintained in a three- or four-stranded DNA structure that is stabilized by additional hydrogen bonds (10, 40–43, 88, 89, 100), (*b*) a transition state in which the DNA strands are prepared for strand exchange (100), or (*c*) a state in which strand exchange has occurred and the products remain bound within the filament. It is possible that the role of ATP hydrolysis by RecA is to convert the triple-helical, or four-stranded, DNA structure into heteroduplex products and to recycle the RecA protein.

While the demonstration of heteroduplex formation in the absence of ATP hydrolysis provides important mechanistic clues as to the way in which pairing and strand exchange occur, it is important to stress that RecA protein is an ATPase, and ATP hydrolysis is required for efficient strand exchange and the continued extension of heteroduplex DNA. In particular, the hydrolysis of ATP may play an important role during DNA repair when DNA lesions or deviations from perfect homology are encountered (16). In support of this role, it has been shown that the formation of heteroduplex DNA in the presence of ATP[γ]S is blocked by very short regions of heterology (104a; J. Kim, R. Inman, and M. Cox, unpublished results).

Structure of RecA-DNA Nucleoprotein Complexes

The view that RecA protein exists in two conformational states, dependent upon the binding and hydrolysis of ATP, is representative of classical allostery and is supported by physical analyses of the structure of RecA filaments. In the absence of a nucleotide cofactor, or in the presence of ADP, RecA binds ssDNA (low-affinity state) to form filamentous complexes with a pitch of approximately 64 Å (13, 105–107). The axial rise per nucleotide is about 2.1 Å, indicating 30 nucleotides per turn of the helix. In contrast, electron microscopy of filaments formed on either single-stranded or duplex DNA in the presence of ATP[γ]S or ATP (i.e. high-affinity state) shows the DNA to be stretched to 150% the length of native B-form duplex DNA, with the axial spacing between base pairs changed from 3.4 Å to 5.1 Å (108–110). The nucleoprotein filament is seen as a right-handed helical structure with a pitch of 95 Å and 18.6 base pairs per turn (106, 107, 111, 112). The average rotation per base pair in the complex is 19.4° (360°/18.6), significantly underwound from the 34.3° observed with B-form DNA (113, 114). The stoichiometry of binding indicates saturation at approximately one RecA per three nucleotides of ssDNA or three base pairs of duplex DNA (112, 115–117), and there are six RecA monomers per turn of the nucleoprotein filament (104, 107, 111). Three-dimensional image reconstructions of filaments formed on duplex DNA in the presence of ATP[γ]S confirm the presence of a right-handed helical filament with a deep groove, and six asymmetric units (RecA monomers) per turn of the helix (104, 118).

Further evidence for the presence of two conformational states comes from proteolytic digestion patterns in the presence of different nucleotide cofactors (119), or from electron microscopy (120). These studies suggest that RecA consists of two distinct regions connected by a flexible linker region (residues 150–180). Two mutant RecA proteins have been identified that have single amino acid substitutions within the proposed linker. These possess ATPase activity but are unable to carry out strand exchange (121, 122). It is likely that the defect lies in their inability to switch from high- to low-affinity states upon hydrolysis of ATP.

Two crystal forms of RecA protein have been obtained at low pH, one hexagonal and the other tetragonal (123), and the X-ray structure of the hexagonal crystal has been solved at a resolution of 2.3 Å (123a, 123b). The crystal structure indicates the formation of a unit cell that contains six protein monomers (space group $P6_1$) arranged with a deep helical groove along the longitudinal axis. The pitch of the helix is 82.7 Å. The crystal structure displays clear protein-protein interactions between monomeric units, as expected for filament formation, and amino acid residues that are highly conserved amongst RecA-like proteins from various bacterial species are located towards the center of the filament (123a).

The DNA on which RecA protein initiates filament formation (either single-stranded or gapped duplex DNA) lies deep within the filament along its longitudinal axis (31). The binding is sequence independent (minor groove binding) and occurs by contacts with the phosphodiester backbone (112, 124, 125), leaving the major groove of the DNA accessible for possible pairing with a second DNA molecule. The bases are oriented perpendicular to the axis of the filament (126, 127).

Three-Stranded DNA Helices as Recombination Intermediates

The primary function of a pairing protein is to bring two DNA molecules into close proximity, with homologous sequences in alignment ready for strand exchange and the formation of heteroduplex DNA. However, since DNA molecules are helical structures, the base contacts necessary for DNA-DNA recognition are effectively on the inside of each DNA molecule. Early models for recombination invoked some form of side-by-side pairing with the DNA helices linked by a Holliday junction (79, 128, 129). However, electron microscopic (28, 39, 49) and biochemical studies (40–43) indicate that pairing occurs within the RecA filament and involves the formation of triple-stranded DNA. Using single-stranded and linear duplex DNA, Rao et al (40) isolated strand-exchange intermediates and showed that the protein-free molecules could serve as substrates for further strand exchange. Upon addition of fresh RecA protein to the isolated joint molecules, strand exchange reinitiated immediately, and progressed at normal rates to form heteroduplex products. Thus, the pairing phase of homologous alignment was bypassed and the structure of these pairing intermediates could be probed biochemically. Although the third strand was expected to be in a displaced single-stranded configuration, it was found to be protected from attack by P1 nuclease. In addition, when intermediates were deproteinized and isolated by gel filtration, 40% resisted melting until the temperature was high enough to melt duplex DNA itself. These results indicate the formation of a triple-stranded DNA structure in which the single strand is hydrogen bonded (paired) into the major groove of the duplex. Joint molecules made by RecA protein have been analyzed by electron microscopy following treatment with chemical crosslinking agents (42, 49). Surprisingly, the DNA molecules were found to be juxtaposed over several kilobases, although it is not clear whether this represents one long joint or a number of smaller ones. For structures of such length to occur it is clear that net helical interwinding is necessary, a requisite that is not available when both DNA molecules are circular. In this case, the joints were short, with an average cross-linked length of only 200 nucleotides.

In related studies, the stability of protein-free joint molecules formed in RecA-mediated reactions between single-stranded DNA and partially homologous duplex DNA has been investigated (43). Interestingly, the stability of the joints was dependent on the position of the homologous DNA sequences, with joints formed at the distal end (Figure 3C; i.e. "wrong" polarity for strand exchange) being much more stable than either medial (Figure 3*B*) or proximal joints (Figure 3*A*). Indeed, joints formed at the distal end were stable up to 80°C, again indicating the presence of three-stranded DNA. These results are consistent with similar observations by Hsieh et al (41), who found that distal joints of 26 nucleotides in length showed exceptional heat stability.

Mechanistically, the pairing of duplex molecules may be similar with two duplexes pairing major-groove to major-groove, in the form of a four-stranded DNA helix (51, 88, 89). Intermediates formed by RecA-mediated strand exchange between gapped circular and linear duplex DNA (reaction of Figure 1*B*) have been isolated following deproteinization, and found to be stable to ~70–75°C (129a). Their stability is therefore similar to that of triple-stranded intermediates formed by RecA.

That genetic recombination could occur by the formation of a structure in which the DNA strands are interwound has been suggested previously. In theoretical studies, McGavin (130, 131) and Wilson (132) proposed that two duplex DNA molecules might pair by the formation of a four-stranded helix in which one duplex lies in the major groove of the other; this concept of pairing was subsequently invoked to explain the way in which RecA protein promotes homologous pairing and strand exchange (51). Indeed, the hypothesis of three- or four-stranded DNA helical complexes is the central feature of the molecular model for recombination proposed by Howard-Flanders et al (10, 133), to which the reader is recommended. However, it is important to stress that while compelling data has been obtained for the formation of three-stranded DNA by RecA protein (40–43, 100), evidence to support the hypothesis of four-stranded DNA remains largely circumstantial (88, 89, 129a) and further studies are required.

The possibility that triple-helical structures exist was suggested some years

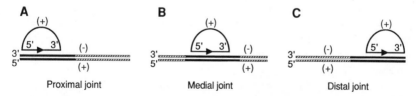

A   Proximal joint   **B**   Medial joint   **C**   Distal joint

Figure 3   Pairing of single-stranded with linear duplex DNA that contains limited homology. Homologous sequences are solid, heterologous are shaded. Joints formed at the proximal end (*A*) undergo strand exchange as the 3'-end of the complementary strand is transferred to the single-strand circle. Medial joints (*B*) are unable to undergo strand exchange and are paranemic in nature. Distal joints (*C*) are unable to strand exchange ("wrong" polarity) and form stable three-stranded DNA structures. For details, see (43) and section on *Three-Stranded DNA Helices as Recombination Intermediates*.

ago (134), and has been supported by more recent observations of three-stranded structures in supercoiled plasmid molecules (135–137), and by the way in which a homopyrimidine oligonucleotide binds into a homopurine-homopyrimidine tract of double-helical DNA (138–142). However, while the formation of triple-helical structures can occur spontaneously given the oligo-purine-oligopyrimidine nature of the DNA, reactions of this type have not been demonstrated with homologous DNA molecules of random sequence. Moreover, these triple-stranded structures differ significantly from structures made by RecA protein, in which the axial spacing between base pairs is stretched to 5.1 Å and strands of like polarity are aligned in parallel fashion. Clearly one role of RecA protein is to overcome some barrier that prohibits DNA of normal sequence from assuming a three- or four-stranded configuration, to form a structure that is stable following deproteinization.

The formation of triple-stranded DNA within the RecA filament has become a central feature of models that attempt to account for the mechanism of homologous pairing and strand exchange (10, 16, 18, 19). In the model presented in Figure 4, it is assumed that three- and four-stranded reactions occur by a similar mechanism in which two DNA molecules are interwound within the deep groove of the RecA filament. This simplistic approach is based on the biochemical similarities between three- and four-stranded reactions as discussed earlier. However, other models or adaptations of the model presented here can also be considered (11, 16, 100). In particular, Cox and coworkers (11, 16) have argued that triple-stranded structures made by RecA have no four-stranded analogue. In this case, it is assumed that all RecA-mediated reactions are initiated as three-stranded intermediates and that strand exchange through regions of four-stranded DNA occurs by a RecA-facilitated branch migration process.

The diagram of Figure 4 is developed from the original proposal of Howard-Flanders et al (10, 133), and accommodates recent data indicating that strand exchange involves rotation of the filament along its longitudinal axis (143). This concept, known as *spooling,* leads to the winding of naked duplex DNA into the filament, the establishment of a region where three- or four-strands are stably paired in the form of a triple- or four-stranded DNA helix, and the release of heteroduplex DNA following unidirectional strand exchange. The rate of movement of the three- or four-stranded DNA (right to left as indicated in Figure 4), and presumably the formation of heteroduplex DNA, is therefore determined by the rate of rotation. The formation of intermediates in which pairing occurs in advance of strand exchange is consistent with the heat stability of protein-free intermediates (40, 41, 43, 129a), electron microscopic observations of on-going strand-exchange reactions (13, 28, 39, 129a), and physical studies of RecA filaments (31, 144). In reactions involving two duplexes (Figure 4, *bottom*), removal of RecA protein

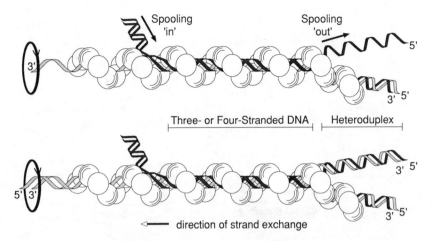

Figure 4 Comparison of hypothetical three-stranded (*top*) and four-stranded (*bottom*) exchange reactions within the RecA filament. The DNA lies within the deep groove of the RecA filament and is stretched (axial rise = 5.1 Å compared with 3.4 Å in B-form duplex DNA; see section on *Structure of RecA-DNA Nucleoprotein Complexes*). Homologous DNA molecules are paired in advance of strand exchange in the form of a three- or four-stranded DNA helix (see section on *Three-Stranded DNA Helices as Recombination Intermediates*). Rotation of the RecA filament along its longitudinal axis causes the naked duplex DNA to be spooled in, such that strand exchange proceeds right to left as drawn (5'→3' relative to the single-stranded (+) circular DNA on which filament formation was initiated). For clarity, the length of the RecA filament is drawn as only a few protein monomers; in reality it may extend over the entire DNA molecule. Similarly, the diameter of the RecA filament (approx 100 Å), and the length of three- or four-stranded DNA are underestimated for diagrammatic purposes. Watson-Crick base pairing, and any additional hydrogen bonds within the three- or four-stranded DNA are not shown. Deproteinization of the lower structure and disruption of the hypothetical region of four-stranded DNA would produce a classical Holliday structure.

and disruption of the hypothetical four-stranded intermediate would lead to the formation of a classical Holliday junction in which two duplex molecules are linked by a single cross-over point.

   To visualize how RecA protein might promote homologous pairing and strand exchange, it is better to view the filament in cross-section (Figure 5). In the following model, it will be assumed that there are two binding sites (I and II) for DNA within the deep groove of the RecA filament (10, 31, 88, 117, 126, 145, 146). Upon binding of ATP, RecA protein assumes the high-affinity DNA-binding state, and a filament is assembled by the binding of ssDNA into site I (Figure 5A, section *a*). Since RecA forms filaments on gapped duplex DNA, it is assumed that the same site can also accept a duplex (Figure 5B, section *a*). The nucleoprotein complex then binds naked duplex DNA into site II and the two DNA molecules become paired within the deep

A. single-strand + duplex

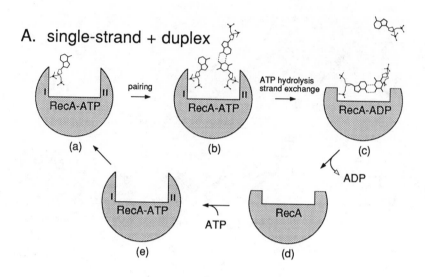

B. duplex + duplex

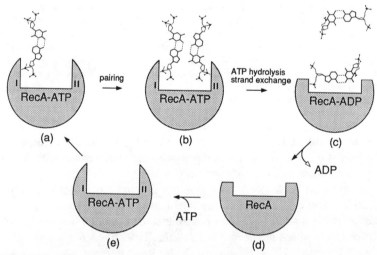

Figure 5 Proposed mechanism for pairing and strand exchange within the filament, and role of ATP hydrolysis. Three-stranded (*A*) and four-stranded (*B*) reactions are proposed to occur by a similar mechanism in which interacting DNA molecules bind into sites I and II of the RecA filament (shown in cross-section). Bases are indicated in axial view with two phosphates. The diameter of the RecA filament (approx 100 Å) is much greater in relation to the DNA than indicated in this diagram.

groove to form triple-stranded (Figure 5A, section b) or four-stranded DNA (Figure 5B, section b). It is likely that homologous pairing involves base-base contacts between the two paired molecules, thus prohibiting the pairing of heterologous DNA sequences.

The nature of the hydrogen bonding required for the establishment of three- and four-stranded DNA helices, and the status of the Watson-Crick base pairing, are unknown. Models in which Watson-Crick bonds are maintained while pairing is accomplished by additional hydrogen bonds (10, 130–132), or the formation of a transition state (or products) in which the substrate Watson-Crick bonds have been disrupted (18, 100), may be considered. Hydrolysis of ATP to ADP leads to the production of the low-affinity DNA-binding state (Figure 5A and B, section c). As described earlier, the high- to low-affinity switch involves an allosteric change of RecA protein, which may be the driving force for strand exchange. In accord with previous models, it is assumed that the conformational switch brings about base rotation and the release of heteroduplex DNA (10, 51, 88, 133). As described by Kowalczykowski (18), RecA protein recycles to the high-affinity DNA-binding state following release of ADP and the rebinding of ATP (Figure 5, sections d and e). Thus, ATP binding is required to provide the free energy required for the establishment and stabilization of the three- or four-stranded DNA intermediate, and ATP hydrolysis may play a role in the release of heteroduplex DNA and the recycling of RecA (100, 101).

Role of RecBCD Protein

In addition to RecA and SSB proteins, a number of other activities are required for genetic recombination and the recombinational repair of damaged

---

(A) Pairing and strand exchange between single-stranded DNA and duplex DNA. RecA protein (with bound ATP, high-affinity state) binds single-stranded DNA into site I to form a nucleoprotein filament (a). The filament binds naked duplex DNA into site II to form a three-stranded DNA helix (b). Watson-Crick bonding may be maintained with additional hydrogen bonds to the third strand, or some form of transition state may occur. RecA undergoes a conformational change (to low-affinity state) upon ATP hydrolysis (c). Heteroduplex formation may occur by base rotation driven by the conformation change. The cycle continues as RecA monomers release ADP (d) and return to the high-affinity state (e) by rebinding ATP.

(B) Pairing and strand exchange between two duplexes. Nucleoprotein filament is assembled as RecA protein (with ATP, high-affinity state) binds duplex DNA following nucleation from a single-stranded gap. One strand of the duplex is bound into site I (a), in readiness for pairing with naked duplex DNA in site II (b). A hypothetical four-stranded DNA helix forms with the twin helices bound major-groove to major-groove. This structure may be stabilized by additional hydrogen bonds. Strand exchange occurs and RecA recycles as described above [for further details, see (10, 18) and sections on reactions involving four DNA strands, ATPase activity, and three-stranded DNA helices as recombination intermediates].

DNA. These include DNA polymerase I, DNA ligase, DNA gyrase, DNA topoisomerase I, and the products of the *recB, recC, recD, recE, recF, recG, recJ, recN, recO, recQ, recR, ruvA, ruvB,* and *ruvC* genes (4–6). Many of these genes have been cloned and their protein products purified. The *recB, recC,* and *recD* gene products form an enzyme (RecBCD) that shows multiple activities, including (*a*) ATP-dependent single- and double-stranded exonuclease, (*b*) a weak ATP-stimulated single-stranded endonuclease, and (*c*) a unidirectional DNA-helicase. Moreover, RecBCD contains a site-specific endonuclease activity that nicks DNA at the sequence 5'-GCTGGTGG-3'. This sequence, designated Chi, specifies a recombination hotspot. Nicking occurs four to six nucleotides to the 3'-side of Chi, and takes place as RecBCD enzyme unwinds the DNA. New results indicate that RecBCD contains a short RNA molecule that is required for Chi cutting activity (S. K. Amundsen, G. R. Smith, unpublished results). The activities of RecBCD have been the subject of a number of recent reviews (11, 14, 15).

Other recombination proteins that have been purified include RecF, which binds linear single-stranded DNA (147), RecJ, an exonuclease active on single-stranded DNA (148), and RecQ, a DNA helicase (149). Attempts to purify the *recG* (150, 151, 151a), *recN* (152), *recO* (153), and *recR* (154) gene products are under way, and new work with the RuvA, RuvB, and RuvC proteins is discussed in a later section.

A mutation in *recB,* or *recC* reduces conjugal recombination to approximately 0.1–1% of wild-type levels (155), indicating that RecBCD enzyme plays a major role in conjugation and transduction. In comparison, *recA* mutants are defective by up to five orders of magnitude, suggesting a redundancy of function or the presence of RecBCD-independent pathways for the formation of recombinant DNA (1, 156). RecBCD has been hypothesized to act at either early or late steps in the recombination process (157). The "initiation" model, suggested by Smith (3, 158), proposes that RecBCD produces a 3'-tail that can be utilized for the initiation of strand exchange by RecA protein. In this model, RecBCD gains entry to duplex DNA and tracks unidirectionally along the DNA, unwinding it as it goes, to form rapidly growing single-stranded loops. When a Chi sequence, in the proper orientation, is encountered, RecBCD cleaves one strand to generate a 3'-tail that is utilized by RecA. Although each individual step had been demonstrated in vitro, only recently was the whole process demonstrated in a coordinated way. Using linear and supercoiled double-stranded DNA substrates, it was shown that the combined action of RecA, RecBCD, and SSB proteins leads to the formation of joint molecules (159). As indicated in the model, RecBCD nicked the DNA at a Chi sequence to produce a free end that was used by RecA for pairing with the superhelical molecule (160). Such a scheme may function during conjugation, since transfer of linear single-stranded DNA leads to its replication to form a linear duplex (161), which may be entered and unwound by RecBCD enzyme.

Alternative models for the action of RecBCD propose that it acts late in recombination and resolves recombination intermediates (162–164). However, attempts to demonstrate RecBCD-mediated resolution of cruciform structures (165) or recombination intermediates made by RecA protein (166) have not provided support for this proposal.

The recombination defects of *recB* or *recC* mutants can be suppressed by secondary mutations in either *sbcA* or *sbcB* (156). Mutations in *sbcA* lead to activation of exonuclease VIII (the product of *recE*), whereas a mutation in *sbcB* inactivates exonuclease I. *sbcB* mutants usually carry an additional mutation, *sbcC*, which suppresses the slow-growth character of *sbcB* mutants (167). If the primary role of RecBCD enzyme is to produce a 3'-end that can be used by RecA, then exonuclease VIII [which digests double-stranded DNA to produce long 3'-tails (168)] could provide a substitute end. Suppression by *sbcB* occurs by a different mechanism. Exonuclease I digests single-stranded DNA from a 3'-end, so its inactivation would leave these ends available for RecA.

Cells with mutations in *recF*, *recG*, *recJ*, *recN*, *recO*, *recQ*, and *recR* were originally isolated as recombination-deficient derivatives of recombination-proficient *recB sbcA* or *recB sbcBC* strains. Historically, each mutation was classified as a member of the RecBCD, RecE, or RecF pathway of recombination, according to its effect on conjugation in wild-type, *recBC sbcA*, or *recBC sbcBC* genetic backgrounds, respectively (4). However, genetic studies indicate that there is a substantial overlap of function. Whereas *recD*, *recJ*, and *recN* single mutants show little or no recombination defect, *recD recJ* double mutants are more defective (169, 170), and *recD recJ recN* triple mutants are quite deficient (171). Moreover, some mutations affect different pathways (or have no effect) according to the type of cross that is measured, while others, such as *recG* (150, 151, 151a), do not fit any of the existing pathway groupings. It is likely that each enzyme plays a specific, but not exclusive, role in recombination, presumably suited to a certain type of DNA substrate (e.g. those provided by conjugation, or those that arise following DNA damage).

EUKARYOTIC PAIRING ACTIVITIES

The relatively simple in vitro assays for homologous pairing and strand exchange developed during studies of the *E. coli* RecA protein have been used to detect and, in some cases, to purify apparently related activities from eukaryotic sources. The first activity of this type was partially purified from the smut fungus *Ustilago maydis* (172). The 70-kDa protein thought to be responsible for activity was named Rec1, since its activity was not observed in extracts from *rec1* mutants (173). However, recent work has shown that Rec1 is not the product of the *rec1* gene, which encodes a 58-kDa protein (174, 175). The properties of the *Ustilago* pairing activity are reviewed elsewhere (176).

Saccharomyces cerevisiae

ACTIVITIES FROM MITOTIC CELLS   The first yeast activity capable of effecting pairing was isolated by Kolodner and coworkers (96). Initially identified as a 132-kDa proteolytic fragment designated Sep1 (strand exchange protein 1), its gene has been cloned and encodes a 175-kDa protein (177). The full-size protein has been overproduced and purified, and initiates the transfer of one strand from a linear duplex to a single-stranded circle (177). Strand exchange occurs with a $5' \rightarrow 3'$ polarity relative to the single strand, but unlike RecA protein, Sep1 was found not to require ATP (96, 177).

The purified Sep1 protein contains an intrinsic exonuclease activity that degrades single-stranded and double-stranded DNA in a $5' \rightarrow 3'$ direction (178). The exonuclease is distinct from the pairing activity. Inhibition of the exonuclease by Ca^{2+} blocked pairing unless the linear ends had been pre-treated with an exogenous exonuclease. These results indicate that the role of the exonuclease is to produce a $3'$-single-stranded tail that can be reannealed with the single-stranded circular DNA. The polarity of the resulting strand-exchange reaction is therefore determined by the polarity of the exonuclease. Insertion mutations in *SEP1* result in a reduced mitotic growth rate and a defect in sporulation (177). Although the mutants are only slightly sensitive to DNA-damaging agents, they are delayed in their ability to recover from DNA damage. *SEP1* mutants show altered meiotic recombination properties, and are defective in the repair of double-stranded breaks.

A second strand transfer protein isolated from mitotic yeast cells (referred to as STPβ, the product of the *DST2* gene) appears to be identical to Sep1, based on its biochemical properties, antibody cross-reaction, and DNA sequence (179, 180).

The stoichiometry of binding of Sep1 to DNA is approximately one monomer per 12 nucleotides of single-stranded DNA, but this was found to be reduced by addition of a 34-kDa yeast SSB protein (181) or the stimulatory factor SF1 (182, 183). Sequence analysis of the cloned SSB gene indicated an open reading frame corresponding to a 70-kDa protein. This gene, now designated *RPA1*, was highly homologous to the 70-kDa subunit of human single-strand binding protein (hRP-A), a cellular protein required for SV40 DNA replication in vitro (184), and was essential for growth. The 33-kDa stimulatory factor SF1 has been purified, and at optimal concentrations (one

SF1 per 20 nucleotides) reduced the need for Sep1 by two orders of magnitude (182). SF1 is thought to play a direct role in the pairing reaction.

Using a similar approach, a 120-kDa DNA pairing activity (DPA) was partially purified from mitotic yeast cells (185). This protein is required stoichiometrically, is ATP independent, and promotes the formation of heteroduplex DNA in a nonpolar way. DPA is unable to promote strand transfer between linear duplex and single-stranded circular DNA unless the linear duplex contains short single-stranded tails. DPA is not thought to be encoded by the *SEP1* gene.

ACTIVITIES FROM MEIOTIC CELLS   The activity of a 38-kDa strand transfer protein (STPα) has been shown to increase 15-fold during meiosis, shortly before the cells are committed to recombination (97, 186). The gene encoding STPα, known as *DST1*, is distinct from *SEP1 (DST2)*. The ATP-independent meiotic protein is required in catalytic amounts and is stimulated by a variety of yeast single-stranded DNA-binding proteins. Exonuclease activity has not been reported. In *dst1-dst1* diploids, induced intragenic recombination was reduced to 10% of the levels seen in wild-type cells (186). Mitotic recombination was not affected. The level of STPα protein appears to be constant during meiosis, suggesting that increased activity may result from posttranslational modification (186).

Genetic studies of meiosis in yeast indicate the participation of a number of gene products, many of which are now the focus of biochemical investigation. Meiotic recombination may be initiated by a double-stranded break (7, 187–189), and a group of genes including *RAD50, RAD51, RAD52, RAD54, RAD55,* and *RAD57* are required for the repair of double-strand breaks in vegetative cells and for recombination during meiosis (190). Other genes specific for meiotic recombination include *SPO11* (191), *MEI4* (192), and *MER1* (193). Mutations in two genes, *HOP1* (194) and *RED1* (195), affect different types of meiotic recombination.

The *RAD50* gene has been sequenced and indicates the presence of an ATP-binding domain (196, 197), and the purified 130-kDa Rad50 protein binds stoichiometrically to duplex DNA dependent upon the presence of ATP (W. Raymond and N. Kleckner, unpublished).

Recent studies show that the products of *RAD51* and the newly identified *DMC1* gene show potentially significant sequence homology with the *E. coli* RecA protein. Rad51 protein, produced from *E. coli*, binds single- and double-stranded DNA in the presence of ATP, and interacts specifically with Rad52 protein (197a). However, attempts to show ATP hydrolysis and DNA renaturation have so far proven negative. The *DMC1* (disrupted meiotic cDNAs) gene is meiosis specific and was isolated from a meiotic cDNA library from which mitotic sequences were subtracted (197b). *dmc1* mutants show reduced levels of recombinant products during meiosis and instead accumulate intermediates containing double-strand breaks. *DMC1* mRNA is abundant during meiosis, and the DMC1 and Rad51 proteins may be functionally related to *E. coli* RecA protein. The amino acid sequences of DMC1 and Rad51 indicate the presence of a consensus ATP-binding site.

Drosophila

A protein designated Rrp1 (recombination repair protein 1) has been purified from *Drosophila melanogaster* embryos on the basis of its apparent ability to promote strand exchange between single-stranded circular and linear duplex DNA (99, 198). Analysis of the gene encoding this 105-kDa protein revealed a 252-amino-acid C-terminal region that was homologous to *E. coli* exonuclease III and *Streptococcus pneumoniae* exonuclease A (199). The N-terminal region did not show homology with any protein in the data base. The studies of Sander et al (199) now show that Rrp1 (like exonuclease III) contains 3'→5' exonuclease and apurinic endonuclease activities, in addition to its ability to promote the renaturation of single-stranded DNA. To determine whether this protein plays a role in cellular recombination, it will be important to characterize *Drosophila* mutants lacking Rrp1 activity.

Mammalian Cells

Initial studies of human cell-free extracts led to reports of ATP-dependent pairing activities (200–203). However, none of these proteins have been purified. One protein, isolated from a T-lymphoblastoid cell line, was initially identified as having an ATP-dependent component (98), but lost its requirement for ATP upon further purification (204). The 120-kDa protein, designated HPP-1 (human pairing protein 1), is required in stoichiometric amounts and promotes strand exchange with a 3'→5' polarity. Although HPP-1-catalyzed strand exchange does not require ATP, a photoaffinity analogue (8-azido ATP) is bound specifically by the protein. With less pure fractions, the rate of strand exchange is faster and occurs with only catalytic amounts of HPP-1. It is possible that an accessory protein is capable of utilizing the ATP bound to HPP-1. Addition of purified human single-strand binding protein (hRP-A) stimulates the rate of pairing 70-fold and reduces the amount of HPP-1 required by more than 10-fold. Stimulation was specific for hRP-A; no similar effects were observed with *E. coli* SSB, bacteriophage gene 32 protein, or *S. cerevisiae* yRP-A (205). Recent work shows that HPP-1 possesses a 3'→5' exonuclease activity (R. Fishel, unpublished observations).

An ATP-independent pairing activity has also been partially purified from human B-lymphoblasts (95) and from HeLa cells (206), and appears similar to that purified from *Drosophila* embryos (99, 207).

At the time of writing, it is not clear whether reactions catalyzed by the majority of eukaryotic pairing proteins are analogous to those promoted by RecA. As yet, the formation of a paranemic joint by a yeast or mammalian protein remains to be demonstrated. The requirement for exonuclease activities may indicate that the pairing proteins facilitate reannealing reactions that lead to the formation of heteroduplex DNA. Such a reaction suggests that recombinant DNA molecules could form via a different mechanism than that catalyzed by RecA, and further work will be required to establish these proteins as bona fide recombination enzymes.

HOLLIDAY JUNCTIONS AND THE RESOLUTION OF BRANCHED DNA STRUCTURES

Intermediates in recombination are thought to be resolved by specific endonucleolytic cleavage. In the case of recombination between two duplex DNA molecules, the intermediate structure can be drawn as a simple cross-over (or Holliday junction), as shown in Figure 6 (79). Resolution of the intermediate, to produce either *patch* (without exchange of flanking markers) or *splice* (with exchange of flanking markers) recombinant products, has been proposed to require free isomerization of the junction between two chemically equivalent forms. Thus, the pair of noncrossed strands in structure I become the pair of crossed strands in structure II. Recognition by an endonuclease that

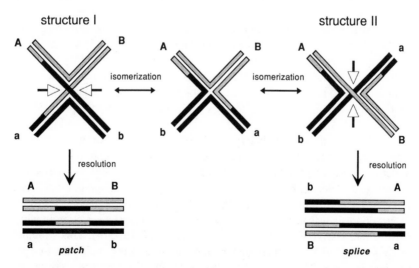

Figure 6 Hypothetical pathway for the resolution of recombination intermediates. Arrows indicate ways in which DNA would need to be cleaved (and religated) to give rise to recombinant products with parental (patch) or recombinant (splice) configurations of the flanking arms.

cuts either pair of crossed strands provides a simple way to imagine how the two types of recombinants might form during genetic crosses.

Evidence for the physical occurrence of recombination intermediates in the form suggested by Holliday comes mainly from electron microscopic visualization of plasmid or viral DNA extracted from bacterial and eukaryotic cells (208–210). Similarly, recombination intermediates made by *E. coli* RecA protein resemble Holliday junctions when viewed by electron microscopy following deproteinization (77, 78, 129a). However, it is important to bear in mind that RecA-driven strand-exchange reactions proceed via quite different structures, in which homologous DNA molecules are interwound within the RecA filament, as described earlier (Figure 4).

A molecular model for a Holliday junction was built by Sigal & Alberts (129). Their studies indicated that complete hydrogen bonding could be maintained on either side of the junction, and that the cross-over point may move by branch migration without any unstacking of the base pairs. However, the proposal that noncrossed strands could be easily converted into crossed strands without base disruption was incomplete, and Sobell (211) demonstrated that conversion of one form to the other (isomerization) would require two 180° rotations as shown in Figure 6. Whether or not isomerization occurs within the cell, and whether it is required for the subsequent processing that gives rise to patch and splice products, remains to be determined.

The Structure of Model Junctions

Attempts to understand the structure of a Holliday junction have recently gained renewed enthusiasm, stimulated in part by the discovery of endonucleases that cleave junctions in duplex DNA. As a model structure for a Holliday junction, Kallenbach, Seeman, and coworkers (212) suggested the use of small synthetic DNA molecules that could be produced by annealing four oligonucleotides to form an X-structure. Since the oligonucleotides differ in their DNA sequence, there is no homology at the point of the cross-over, so prohibiting branch migration. Studies with longer junctions, produced by annealing DNA fragments, showed that the annealed structures had an anomalous mobility through polyacrylamide gels, indicative of a bend at the cross-over point (213). Surprisingly, the junctions were found to be inflexible structures, and the presence of metal ions affected the spatial arrangement of the arms, as detected by changes in gel mobility (214). Since gel mobility is affected by both arm length and the angle between the extended arms, gel electrophoresis was used to show that the junction adopts a structure in which there is pairwise coaxial association of the helices (215, 216). This structure, termed a *stacked X-structure,* may be thought of as being analogous to a pair of open scissors. The two-fold symmetry of the structure was also demonstrated by hydroxylradical probing (217). With regard to helical stacking, the choice of partners is determined by the DNA sequence at the junction point, and single-base-pair changes can result in an exchange of stacking partners (216, 218). In the absence of metal ions, Duckett et al (216) found that junctions of any sequence adopted a square planar structure, in which the four arms were unstacked and the arms maximally extended. However, other gel electrophoretic studies, and measurements of the geometry of a junction in solution by transient electric birefringence, are inconsistent with the adoption of a square planar structure under similar conditions (215, 219).

Surprisingly, the structure adopted in the presence of Mg^{2+} was not a Sigal-Alberts crossed structure, but was a noncrossed structure with the arms aligned in antiparallel fashion (with the small angle being close to 60°). This conclusion is generally supported by gel electrophoretic analyses (215, 216, 220), fluorescence energy transfer spectroscopy (221), transient electric birefringence (219), sensitivity to hydroxylradicals (217), cleavage by T4 endonuclease VII (216, 222, 223), and molecular modelling (224).

However, observations that indicate antiparallel alignment are difficult to reconcile with the biological problems encountered during genetic recombination, since pairing and strand exchange are likely to demand parallel alignment as a way of testing for homology (see section on *Three-Stranded DNA Helices as Recombination Intermediates*). It should also be remembered that the structural studies have so far been limited to junctions that lack homology at the point of the cross-over. It therefore remains to be determined whether their structures are relevant to that of a true recombination intermediate.

Enzymes that Cleave Branched DNA Structures

T4 ENDONUCLEASE VII   Bacteriophage T4 gene *49* encodes an 18-kDa protein that is required during phage maturation. Mutants in gene *49* are defective in packaging, as observed by the formation of highly branched multimeric DNA and abortive infection. The accumulation of branched DNA is caused by the important role that recombination plays in the life-cycle of the T-even phages (225). During growth, DNA replication is initiated at specific origin sequences and results in the formation of single-stranded DNA at the 3'-terminus of the lagging strand. This single-stranded tail is able to invade another chromosome (T4 chromosomes are circularly permuted) or the terminal redundancy of the same molecule, where it acts as a primer for leading-strand DNA synthesis. The resultant branched DNA needs to be resolved to unit length molecules by T4 gene *49* product (endonuclease VII) during packaging. T4 endonuclease VII has been purified from phage-infected cells (226, 227) and resolves branched T4 DNA in vitro (228, 229). The gene has been cloned (230, 231) and the protein purified to homogeneity from an overexpression vector (232).

T4 endonuclease VII assumed the role of the prototypic "junction-resolvase" following important early work by Mizuuchi et al and Lilley & Kemper, in which it was shown to resolve plasmid recombination intermediates and cruciform DNA structures (233–235). Cleavage occurred at the site of the junction by the introduction of nicks in strands of like polarity. The two nicks are introduced in a nonconcerted fashion (235a), and the strand breaks in the DNA products may be sealed by DNA ligase (233). Consistent with its viral role as a debranching enzyme, endonuclease VII cleaves a wide variety of substrates, including X- and Y-structures, heteroduplex loops, and extended single-stranded termini (216, 222, 236–238). As a general rule, cleavage occurs at the 3'-side of the junction or site of secondary structure. The precise sites of cleavage are variable, depend upon DNA sequence, and are usually within 2 to 3 nucleotides of the junction point. Cleavage at the 3'-side of a pyrimidine residue is preferred (234, 235a, 239).

T4 endonuclease VII is dimeric in solution (227, 232) and in the absence of Mg^{2+} binds synthetic X-junctions to form discrete protein-DNA complexes, as detected by band-shift assays and hydroxylradical protection experiments (240). Weak footprints were found in two strands and covered ~5 nucleotides. The regions of protection were symmetrically related and contained the cleavage sites. T4 endonuclease VII can also cleave recombination intermediates made by RecA protein (241). In vivo, it is likely to act upon intermediates made by the T4 UvsX protein, an analogue of RecA (242, 243).

T7 ENDONUCLEASE I   The gene *3* product of bacteriophage T7 has two essential roles during phage growth: (*a*) it degrades cellular host DNA following phage infection, and (*b*) it cleaves junctions in replicating phage DNA to produce unbranched linear duplexes that may subsequently be packaged. Mutations in gene *3* result in mutant phage that are unable to degrade host DNA (244), and are defective in phage maturation (245) and genetic recombination (246–248).

T7 gene *3* has been cloned and complements T4 gene *49⁻* mutants (249–252). The protein product has been overexpressed and purified to homogeneity (252). Endonuclease I has a strong preference for ssDNA (253, 254), and cuts duplex DNA preferentially at X- and Y-junctions (249, 252, 255–257). As observed with T4 endonuclease VII, the efficiency and sites of cleavage are influenced by DNA sequence, and a preference for cleavage at the 5'-side of pyrimidine residues has been observed (239, 255). In studies in which the action of T7 endonuclease I and T4 endonuclease VII were compared on the same synthetic X-junction, a number of differences were observed (240, 256). Whereas endonuclease VII cleaved the junction in one orientation only (at the 3'-side of the junction), endonuclease I promoted cleavage with equal efficiency in the two possible orientations, by the introduction of symmetrical nicks located 1–2 nucleotides to the 5'-side of the junction. T7 endonuclease I is dimeric, composed of two identical 17-kDa subunits, which bind a synthetic X-junction to form a single protein-DNA complex detected by band shift assays (256). Unlike T4 endonuclease VII, hydroxylradical footprinting experiments indicated that all four DNA strands at the junction point were protected. T7 endonuclease I also resolves recombination intermediates made in vitro by RecA protein (258).

EUKARYOTIC PROTEINS   Three apparently distinct activities that cleave cruciforms or synthetic X-structures have been partially purified from *S. cerevisiae*. To distinguish them from each other, the nomenclature suggested by Jensch et al (Endo X1, X2, and X3) is used (259). Endo X1, with a native molecular weight of ~200,000, was partially purified from vegetative yeast cells treated with the DNA-damaging agent mechlorethamine (260, 261). Cleavage of cruciform DNA occurred by the introduction of symmetrically related nicks within the cruciform arms (262). When cruciform junctions that contained heterologous arm sequences were used, cleavage occurred asymmetrically (262, 263). These observations were taken to indicate a need for the homologous alignment of arm sequences in an enzyme-DNA complex as a prerequisite for symmetrical cleavage. In support of this concept, the sites of cleavage in one arm were altered by sequence changes or base methylation in the opposing heterologous arm. Endo X1 activity has been detected in extracts prepared from cells with mutations in the *RAD50, RAD51, RAD52, RAD54, RAD55, RAD56,* and *RAD57* genes (264). Efforts to further purify the nuclease have been unsuccessful, and its gene is unknown.

A second yeast endonuclease, partially purified by its ability to cleave cruciforms, cuts figure-8 DNA molecules to produce circular monomer and dimer products in vitro (265). This protein, Endo X2, binds synthetic X-junctions in which the longest arm is 34 base pairs (266). Although the small junctions were not cleaved, they served as specific inhibitors in cruciform cleavage reactions. Larger X-junctions, produced by annealing DNA fragments containing the *att* sequences of bacteriophage λ, were cut efficiently by Endo X2. These results may indicate a minimum target size that is sufficient for binding but not cleavage, or a requirement for sequence homology. The nuclease did not cleave Y-junctions in DNA (267). The sites of cleavage in the λ *att* junction have been mapped and indicate that resolution occurred by the introduction of symmetrically related nicks. However, studies of a series of mutant junctions showed that changes in base sequence had a dramatic effect on the orientation, rate, and sites of incision (267). Mutations in the gene encoding this nuclease (*CCE1*) fail to affect mitotic or meiotic recombination. However, *cce1* mutants produce a higher than normal frequency of petite cells, suggesting that the CCE1 protein is important for the maintenance of mitochondrial DNA (267a).

The third nuclease, Endo X3, also partially purified from vegetative yeast, has the same native molecular weight as T4 endonuclease VII, as measured by gel filtration, and exhibits remarkably similar enzymatic properties (259). Surprisingly, identical cleavage patterns were produced by Endo X3 and T4 endonuclease VII on cruciform DNA, synthetic X- and Y-junctions, and heteroduplex loop structures. Endo X3 was also found to be inhibited by antibodies raised against T4 endonuclease VII (259).

Until recently, attempts to isolate junction-resolving activities from mammalian cells were largely unsuccessful. Two approaches were attempted using synthetic X-junctions. In the first, gel retardation assays were used to detect proteins that bound junctions specifically. Two proteins have been described, one from rat liver (268) and the other from cultured human cells (269). The rat protein was subsequently identified as the high-mobility group protein, HMG1 (270). Neither protein has been shown to cleave junctions in vitro. In the other approach, it was shown that junctions were processed in some way by HeLa cell-free extracts (271), by an activity from placental tissue (272), and by fractionated calf thymus extracts (273). Elborough & West (273) showed that resolution by the calf thymus nuclease occurred in a manner analogous to that catalyzed by T4 endonuclease VII. Through use of a synthetic X-structure, symmetrically related nicks were observed at sites located 1–2 nucleotides to the 3'-side of the junction.

At the present time, any proposed function for the yeast or calf thymus endonucleases in recombination or repair is purely speculative. However, their in vitro specificities, and similarity to T4 endonuclease VII, are of sufficient interest to hope that continued study will lead to elucidation of their cellular role.

RESOLUTION OF RECOMBINATION INTERMEDIATES IN *E. COLI*

In Vitro Resolution and Genetic Requirements

The formation of recombination intermediates in vitro by RecA protein has been a valuable tool in the detection and characterization of a resolution activity from *E. coli*. In initial studies, covalently closed figure-8 DNA molecules, prepared from two partially homologous plasmid DNA substrates using RecA protein, DNA polymerase I, and DNA ligase, were transformed into *E. coli recA* mutants. Resolution occurred in vivo to produce monomeric and dimeric plasmid progeny (78). Since the plasmids contained genetic markers, both parental and recombinant products could be scored, and crossing-over was found to occur with a 50% frequency. The formation of recombinant products following transformation with a figure-8 recombination intermediate required neither *recA* nor *recB* gene products.

To demonstrate resolution in vitro, Connolly & West (274) used α-structures made by the RecA-catalyzed interaction of gapped circular DNA with linear duplex DNA (as shown in Figure 1*B*). Ongoing strand-exchange reactions were supplemented with fractionated extracts prepared from *E. coli endA recB sbcBC* mutants (defective in endonuclease I, RecBCD enzyme, and exonuclease I), and resolution of the recombination intermediates was observed by the production of (*a*) nicked circular and gapped linear, and (*b*) linear dimer DNA. These products indicate resolution in either of the two possible ways, and correspond to patch and splice recombinant products, respectively. Using partially homologous DNA molecules, in which strand exchange was driven by RecA protein to the point of a heterologous block, the sites of cleavage were mapped to the stalled Holliday junction (274).

Using more purified fractions, it was shown that the activity resided in a small polypeptide that could also resolve synthetic X-junctions (275). Cleavage occurred by the introduction of symmetrically related nicks. Most importantly, this assay was sensitive enough to screen fractionated extracts from a series of recombination-deficient *E. coli*, and Connolly et al showed that resolution activity was absent from extracts prepared from *ruvC* mutants. The involvement of *ruvC* gene product was confirmed by the observation of elevated levels of activity in extracts made from cells containing multicopy plasmids carrying the wild-type *ruvC* gene (275).

Organization of the ruv Locus

Cells carrying *ruv* mutations were originally isolated by their sensitivity to mitomycin C (276) and by their reduced ability to promote recombination between duplicated *gal* genes (277). They are also sensitive to UV light and ionizing irradiation, indicating a role in the recombinational repair of DNA damage. They are recombination deficient in *recBC sbcBC* (278), *recBC sbcA* (279, 280), or *recG* (151) genetic backgrounds, although *ruv* single mutants show only slightly reduced levels of recombinants during conjugal or transductional crosses. That *ruv* gene products might function late in recombination, at the step of resolution, was suggested some time ago by Clark & Low (4) following analysis of unpublished data from the Oliver and Kolodner laboratories. Support for a late role was provided by studies of the formation of F-prime transconjugants, which indicated that the abortive recombination observed in *ruv* mutants was alleviated by a mutation in *recA* (281).

Genetic crosses locate *ruv* at minute 41 on the current linkage map (276). The region is now known to encode three distinct genes designated *ruvA*, *ruvB*, and *ruvC*, and mutations in any of these confer *ruv* phenotype (282). The original *ruv* mutants map within *ruvB*.

The organization of the *ruv* locus is shown in Figure 7. Nucleotide sequencing has shown that *ruvA* and *ruvB* form a single operon that is regulated by

LexA repressor (283–285). Mutations in *ruvA* therefore exert a polar effect on *ruvB*. Sequencing of the *ruvC* gene has revealed the absence of a ribosome-binding site, and *ruvC* forms an operon with *orf-26* (286, 286a), an upstream gene formerly designated *orf-33* (282). The *ruvC* gene product is produced at a very low level, as observed by maxi-cell analysis (286, 286a). Mutations in *orf-26* do not show *ruv* phenotype.

Properties of Ruv Proteins

The *ruvA*, *ruvB*, and *ruvC* genes have been independently cloned by the Shinagawa and Lloyd laboratories, and overexpression vectors have been produced (282, 286–288b). The *ruvA* gene encodes a 22-kDa protein that binds either single- or double-stranded DNA (288). The protein is tetrameric as determined by gel filtration and binds synthetic X-junctions with high affinity (288a, 288c).

The 37-kDa RuvB protein has a weak ATPase activity that is stimulated by the presence of RuvA and DNA, indicating an interaction of the RuvA-DNA complex with RuvB (288). UV-irradiated supercoiled DNA provided the greatest enhancement of the ATPase activity.

The 19-kDa RuvC protein has been purified, and Dunderdale et al (289) showed that it resolves recombination intermediates made by RecA protein. Resolution occurred by endonucleolytic cleavage and gave rise to patch and splice recombinant products. Cleavage was specific for recombination intermediates that contained four strands of DNA; three-stranded intermediates were not resolved. It may be significant that resolution can occur during a reaction in which RecA protein is actively promoting strand exchange, and may indicate that interacting DNA molecules located deep within the RecA filament are accessible to RuvC. It is possible that three stages of the recombination process—homologous pairing, strand exchange, and resolution—all take place within the nucleoprotein filament.

Studies of the interaction of RuvC with small synthetic junctions indicated that resolution was dependent upon the presence of homologous DNA sequences (289). The requirement for homology was demonstrated using synthetic junctions with homologous cores of 12 (junction J12), 6 (J6), 3 (J3), or 0 (J0) base pairs. J12 was cleaved quite efficiently, whereas under the same conditions J6, J3, and J0 were not. These experiments raise the question of

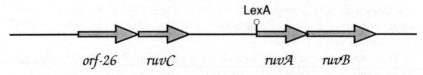

Figure 7   Organization of the *ruv* region of the *E. coli* chromosome.

whether cleavage of a junction by RuvC involves intimate DNA-DNA contacts within the active site. Thus, resolution would be prohibited by a lack of homologous sequences.

The recombination/repair phenotype of *ruvC* mutants, together with the properties of RuvC protein, indicate a direct role in the resolution of recombination intermediates in *E. coli*. The inducible RuvA and RuvB proteins are also likely to play an important role in a late step of recombination, as indicated by their in vivo properties. Recent results show that the RuvA and RuvB proteins act in concert to promote ATP-dependent branch migration of Holliday junctions (290). Addition of RuvA and RuvB to a RecA-mediated recombination reaction (as shown in Figure 1*B*) was found to stimulate the rate of strand exchange and the formation of heteroduplex DNA. Stimulation did not occur via direct interaction with RecA protein; instead, RuvA and RuvB were found to act upon Holliday junctions made by RecA. Individual roles for the RuvA and RuvB proteins have been defined: RuvA protein provides specificity by binding to the Holliday junction (288c), and RuvB (the ATPase) provides the motor force for branch migration (290). Interestingly, RuvAB-mediated branch migration is significantly faster than strand exchange catalyzed by RecA, it occurs at lower energy cost, and can effectively overcome UV-induced DNA damages that severely inhibit RecA-mediated strand exchange (290). These new results may indicate that the primary role of RecA protein is to promote homologous pairing reactions and to initiate strand exchanges leading to the formation of a Holliday junction. This structure is then specifically recognized by RuvA and RuvB proteins, which catalyze more efficient branch migration and the formation of heteroduplex DNA.

CONCLUDING REMARKS

It is clear from the studies presented here that in *E. coli*, and most likely in all organisms, recombination is a complex process that has both structural and catalytic components. The formation of a RecA filament in which DNA helices are interwound is probably the central component of a "protein machine" that carries out the efficient manipulation of DNA. The machine overcomes the kinetic barrier that would normally prevent the pairing of DNA molecules and brings about the close liaison of base sequences in readiness for strand exchange. The exchange reaction involves conformational changes of protein molecules, reactions driven by the binding and release of nucleoside triphosphates. These changes provide order, as the DNA is freed from random motions. Holliday junctions made by RecA may then be recognized by other proteins that promote more efficient movement of the joint along the DNA. Finally, the interacting DNA molecules need to be resolved into mature recombinant products, and this may be catalyzed by the RuvC protein followed by DNA ligase.

Our efforts so far have focussed on a small number of proteins. Although these may be the main players in the game, there is much yet to be learned. The power of bacterial genetics coupled with the relative ease with which important genes can be cloned and purified should lead to many new important mechanistic discoveries. Indeed, the development of a complete in vitro recombination system is beginning to look like a realistic goal for the future.

ACKNOWLEDGMENTS

I am grateful to the members of my laboratory, both past and present, whose efforts have made this review possible. Many people, too numerous to mention, have contributed to this work by sending reprints, preprints, and unpublished data. In particular I would like to thank Mike Cox, Nancy Kleckner, Richard Kolodner, Steve Kowalczykowski, Robert Lloyd, Charles Radding, and Gerry Smith for information and insight. Work in the author's laboratory is supported by the Imperial Cancer Research Fund.

Literature Cited

1. Clark, A. J., Margulies, A. D. 1965. *Proc. Natl. Acad. Sci. USA* 53:451–59
2. Radding, C. M. 1978. *Annu. Rev. Biochem.* 47:847–80
3. Smith, G. R. 1987. *Annu. Rev. Genet.* 21:179–201
4. Clark, A. J., Low, K. B. 1988. In *The Recombination of Genetic Material,* ed. K. B. Low, pp. 155–215. San Diego: Academic
5. Mahajan, S. K. 1988. In *Genetic Recombination,* ed. R. Kucherlapati, G. R. Smith, pp. 87–140. Washington, DC: Am. Soc. Microbiol.
6. Smith, G. R. 1988. *Microbiol. Rev.* 52:1–28
7. Thaler, D. S., Stahl, F. W. 1988. *Annu. Rev. Genet.* 22:169–98
8. Radding, C. M. 1982. *Annu. Rev. Genet.* 16:405–37
9. Dressler, D., Potter, H. 1982. *Annu. Rev. Biochem.* 51:727–62
10. Howard-Flanders, P., West, S. C., Stasiak, A. J. 1984. *Nature* 309:215–20
11. Cox, M. M., Lehman, I. R. 1987. *Annu. Rev. Biochem.* 56:229–62
12. Radding, C. M. 1988. See Ref. 5, pp. 193–230
13. Stasiak, A., Egelman, E. H. 1988. See Ref. 5, pp. 265–308
14. Taylor, A. F. 1988. See Ref. 5, pp. 231–64
15. Smith, G. R. 1990. In *Nucleic Acids and Molecular Biology,* ed. F. Eckstein, D.

M. J. Lilley, 4:78–98. Berlin: Springer-Verlag
16. Roca, A. I., Cox, M. M. 1990. *Crit. Rev. Biochem. Mol. Biol.* 25:415–56
17. Miller, R. V., Kokjohn, T. A. 1990. *Annu. Rev. Microbiol.* 44:365–94
18. Kowalczykowski, S. C. 1991. *Annu. Rev. Biophys. Biophys. Chem.* 20:539–75
19. Radding, C. M. 1991. *J. Biol. Chem.* 266:5355–58
20. Walker, G. C. 1985. *Annu. Rev. Biochem.* 54:425–58
21. Little, J. W. 1984. *Proc. Natl. Acad. Sci. USA* 81:1375–79
22. Burckhardt, S. E., Woodgate, R., Scheuermann, R. H., Echols, H. 1988. *Proc. Natl. Acad. Sci. USA* 85:1811–15
23. Nohmi, T., Battista, J. R., Dodson, L. A., Walker, G. C. 1988. *Proc. Natl. Acad. Sci. USA* 85:1816–20
24. Shinagawa, H., Iwasaki, H., Kato, T., Nakata, A. 1988. *Proc. Natl. Acad. Sci. USA* 85:1806–10
25. Craig, N., Roberts, J. W. 1980. *Nature* 283:26–30
26. West, S. C., Cassuto, E., Mursalim, J., Howard-Flanders, P. 1980. *Proc. Natl. Acad. Sci. USA* 77:2569–73
27. McEntee, K., Weinstock, G. M., Lehman, I. R. 1981. *J. Biol. Chem.* 256:8835–44
28. Stasiak, A., Stasiak, A. Z., Koller, T.

1984. *Cold Spring Harbor Symp. Quant. Biol.* 49:561–70
29. Flory, J., Tsang, S. S., Muniyappa, K. 1984. *Proc. Natl. Acad. Sci. USA* 81:7026–30
30. Register, J. C., Griffith, J. 1985. *J. Biol. Chem.* 260:12308–12
31. Egelman, E. H., Yu, X. 1989. *Science* 245:404–7
32. Muniyappa, K., Shaner, S. L., Tsang, S. S., Radding, C. M. 1984. *Proc. Natl. Acad. Sci. USA* 81:2757–61
33. Kowalczykowski, S. C., Krupp, R. A. 1987. *J. Mol. Biol.* 193:97–113
34. Gonda, D. K., Radding, C. M. 1983. *Cell* 34:647–54
35. Gonda, D. K., Radding, C. M. 1986. *J. Biol. Chem.* 261:13087–96
36. Julin, D. A., Riddles, P. W., Lehman, I. R. 1986. *J. Biol. Chem.* 261:1025–30
37. Tsang, S. S., Chow, S. A., Radding, C. M. 1985. *Biochemistry* 24:3226–32
38. Chow, S. A., Radding, C. M. 1985. *Proc. Natl. Acad. Sci. USA* 82:5646–50
39. Register, J. C., Christiansen, G., Griffith, J. 1987. *J. Biol. Chem.* 262:12812–20
40. Rao, B. J., Jwang, B., Radding, C. M. 1990. *J. Mol. Biol.* 213:789–809
41. Hsieh, P., Camerini-Otero, C. S., Camerini-Otero, R. D. 1990. *Genes Dev.* 4:1951–63
42. Umlauf, S. W., Cox, M. M., Inman, R. B. 1990. *J. Biol. Chem.* 265:16898–912
43. Rao, B. J., Dutreix, M., Radding, C. M. 1991. *Proc. Natl. Acad. Sci. USA* 88:2984–88
44. DasGupta, C., Shibata, T., Cunningham, R. P., Radding, C. M. 1980. *Cell* 22:437–46
45. Bianchi, M., DasGupta, C., Radding, C. M. 1983. *Cell* 34:931–99
46. Riddles, P. W., Lehman, I. R. 1985. *J. Biol. Chem.* 260:165–69
47. Christiansen, G., Griffith, J. 1986. *Proc. Natl. Acad. Sci. USA* 83:2066–70
48. Schutte, B. C., Cox, M. M. 1987. *Biochemistry* 26:5616–25
49. Bortner, C., Griffith, J. 1990. *J. Mol. Biol.* 215:623–34
50. Cox, M. M., Lehman, I. R. 1981. *Proc. Natl. Acad. Sci. USA* 78:3433–37
51. West, S. C., Cassuto, E., Howard-Flanders, P. 1981. *Proc. Natl. Acad. Sci. USA* 78:2100–4
52. Cox, M. M., Lehman, I. R. 1981. *Proc. Natl. Acad. Sci. USA* 78:6018–22
53. Kahn, R., Cunningham, R. P., DasGupta, C., Radding, C. M. 1981. *Proc. Natl. Acad. Sci. USA* 78:4786–90
54. West, S. C., Cassuto, E., Howard-Flanders, P. 1981. *Proc. Natl. Acad. Sci. USA* 78:6149–53
55. Soltis, D. A., Lehman, I. R. 1983. *J. Biol. Chem.* 258:6073–77
56. Pugh, B. F., Cox, M. M. 1987. *J. Biol. Chem.* 262:1337–43
57. DasGupta, C., Radding, C. M. 1982. *Proc. Natl. Acad. Sci. USA* 79:762–66
58. DasGupta, C., Radding, C. M. 1982. *Nature* 295:71–73
59. Bianchi, M. E., Radding, C. M. 1983. *Cell* 35:511–20
60. Livneh, Z., Lehman, I. R. 1982. *Proc. Natl. Acad. Sci. USA* 79:3171–75
61. Lichten, M., Fox, M. S. 1984. *Proc. Natl. Acad. Sci. USA* 81:7180–84
62. Cassuto, E., West, S. C., Mursalim, J., Conlon, S., Howard-Flanders, P. 1980. *Proc. Natl. Acad. Sci. USA* 77:3962–66
63. Cunningham, R. P., DasGupta, C., Shibata, T., Radding, C. M. 1980. *Cell* 20:223–35
64. Chiu, S. K., Wong, B. C., Chow, S. A. 1990. *J. Biol. Chem.* 265:21262–68
65. Shaner, S. L., Flory, J., Radding, C. M. 1987. *J. Biol. Chem.* 262:9220–30
66. Pugh, B. F., Cox, M. M. 1987. *J. Biol. Chem.* 262:1326–36
67. Pugh, B. F., Cox, M. M. 1988. *J. Mol. Biol.* 203:479–93
68. Kowalczykowski, S. C., Clow, J., Krupp, R. A. 1987. *Proc. Natl. Acad. Sci. USA* 84:3127–31
69. Thresher, R. J., Griffith, J. D. 1990. *Proc. Natl. Acad. Sci. USA* 87:5056–60
70. Kim, J.-I., Heuser, J., Cox, M. M. 1989. *J. Biol. Chem.* 264:21848–56
71. Shi, Y. B., Griffith, J., Gamper, H., Hearst, J. E. 1988. *Nucleic Acids Res.* 16:8945–52
72. Rosenberg, M., Echols, H. 1990. *J. Biol. Chem.* 265:20641–45
73. Cassuto, E., Howard-Flanders, P. 1986. *Nucleic Acids Res.* 14:1149–58
74. Rupp, W. D., Wilde, C. E., Reno, D. L., Howard-Flanders, P. 1971. *J. Mol. Biol.* 61:25–44
75. West, S. C., Cassuto, E., Howard-Flanders, P. 1981. *Nature* 294:659–62
76. West, S. C., Cassuto, E., Howard-Flanders, P. 1982. *Mol. Gen. Genet.* 187:209–17
77. DasGupta, C., Wu, A. M., Kahn, R., Cunningham, R. P., Radding, C. M. 1981. *Cell* 25:507–16
78. West, S. C., Countryman, J. K., Howard-Flanders, P. 1983. *Cell* 32:817–29
79. Holliday, R. 1964. *Genet. Res.* 5:282–304
80. West, S. C., Cassuto, E., Howard-Flanders, P. 1982. *Mol. Gen. Genet.* 186:333–38
81. Conley, E. C., West, S. C. 1990. *J. Biol. Chem.* 265:10156–63

82. Lindsley, J. E., Cox, M. M. 1990. *J. Biol. Chem.* 265:10164–71
83. Conley, E. C., Müller, B., West, S. C. 1990. In *Mutation and the Environment.* Part A: *Basic Mechanisms,* ed. M. L. Mendelsohn, R. J. Albertini, 340A:121–33. New York:Wiley-Liss
84. Hahn, T. R., West, S. C., Howard-Flanders, P. 1988. *J. Biol. Chem.* 263:7431–36
85. Dutreix, M., Rao, B. J., Radding, C. M. 1991. *J. Biol. Chem.* 219:645–54
86. Konforti, B. B., Davis, R. W. 1987. *Proc. Natl. Acad. Sci. USA* 84:690–94
87. Konforti, B. B., Davis, R. W. 1990. *J. Biol. Chem.* 265:6916–20
88. West, S. C., Howard-Flanders, P. 1984. *Cell* 37:683–91
89. Conley, E. C., West, S. C. 1989. *Cell* 56:987–95
90. Ogawa, T., Wabiko, H., Tsurimoto, T., Horii, T., Masukata, H., Ogawa, H. 1979. *Cold Spring Harbor Symp. Quant. Biol.* 43:909–15
91. Roberts, J. W., Roberts, C. W., Craig, N. L., Phizicky, E. M. 1979. *Cold Spring Harbor Symp. Quant. Biol.* 43:917–20
92. Brenner, S. L., Mitchell, R. S., Morrical, S. W., Neuendorf, S. K., Schutte, B. C., Cox, M. M. 1987. *J. Biol. Chem.* 262:4011–16
93. Neuendorf, S. K., Cox, M. M. 1986. *J. Biol. Chem.* 261:8276–82
94. Menetski, J. P., Kowalczykowski, S. C. 1987. *J. Biol. Chem.* 262:2093–100
95. Hsieh, P., Meyn, M. S., Camerini-Otero, R. D. 1986. *Cell* 44:885–94
96. Kolodner, R., Evans, D. H., Morrison, P. T. 1987. *Proc. Natl. Acad. Sci. USA* 84:5560–64
97. Sugino, A., Nitiss, J., Resnick, M. A. 1988. *Proc. Natl. Acad. Sci. USA* 85:3683–87
98. Fishel, R. A., Detmar, K., Rich, A. 1988. *Proc. Natl. Acad. Sci. USA* 85:36–40
99. McCarthy, J. G., Sander, M., Lowenhaupt, K., Rich, A. 1988. *Proc. Natl. Acad. Sci. USA* 85:5854–58
100. Menetski, J. P., Bear, D. G., Kowalczykowski, S. C. 1990. *Proc. Natl. Acad. Sci. USA* 87:21–25
101. Rosselli, W., Stasiak, A. 1990. *J. Mol. Biol.* 216:335–52
102. Shaner, S. L., Radding, C. M. 1987. *J. Biol. Chem.* 262:9211–19
103. Stasiak, A., Egelman, E. H., Howard-Flanders, P. 1988. *J. Mol. Biol.* 202:659–62
104. Egelman, E. H., Stasiak, A. 1986. *J. Mol. Biol.* 191:677–98
104a. Rosselli, W., Stasiak, A. 1991. *EMBO J.* 10:4391–96
105. Williams, R. C., Spengler, S. J. 1985. *J. Mol. Biol.* 187:109–18
106. Heuser, J., Griffith, J. 1989. *J. Mol. Biol.* 210:473–84
107. DiCapua, E., Schnarr, M., Ruigrok, R. W. H., Lindner, P., Timmins, P. A. 1990. *J. Mol. Biol.* 214:557–70
108. Stasiak, A., DiCapua, E., Koller, T. 1981. *J. Mol. Biol.* 151:557–64
109. Dunn, K., Chrysogelos, S., Griffith, J. 1982. *Cell* 28:757–65
110. Flory, J., Radding, C. M. 1982. *Cell* 28:747–56
111. DiCapua, E., Engel, A., Stasiak, A., Koller, T. 1982. *J. Mol. Biol.* 157:87–103
112. Dombroski, D. F., Scraba, D. G., Bradley, R. D., Morgan, A. R. 1983. *Nucleic Acids Res.* 11:7487–516
113. Stasiak, A., DiCapua, E. 1982. *Nature* 299:185–86
114. Chrysogelos, S., Register, J. C., Griffith, J. 1983. *J. Biol. Chem.* 258:12624–31
115. Menetski, J. P., Kowalczykowski, S. C. 1989. *Biochemistry* 28:5871–80
116. Takahashi, M., Kubista, M., Norden, B. 1989. *J. Mol. Biol.* 205:137–48
117. Zlotnick, A., Mitchell, R. S., Brenner, S. L. 1990. *J. Biol. Chem.* 265:17050–54
118. Egelman, E. H., Stasiak, A. 1988. *J. Mol. Biol.* 200:329–49
119. Kobayashi, N., Knight, K., McEntee, K. 1987. *Biochemistry* 26:6801–10
120. Yu, X., Egelman, E. H. 1990. *Biophys. J.* 57:555–66
121. Muench, K. A., Bryant, F. B. 1990. *J. Biol. Chem.* 265:11560–66
122. Muench, K. A., Bryant, F. R. 1991. *J. Biol. Chem.* 266:844–50
123. McKay, D., Steitz, T., Weber, I. T., West, S. C., Howard-Flanders, P. 1980. *J. Biol. Chem.* 255:6662
123a. Story, R. M., Weber, I. T., Steitz, T. A. 1991. *Nature* 355:318–24
123b. Story, R. M., Steitz, T. A. 1991. *Nature* 355:374–76
124. Leahy, M. C., Radding, C. M. 1986. *J. Biol. Chem.* 261:6954–60
125. DiCapua, E., Müller, B. 1987. *EMBO J.* 6:2493–98
126. Kubista, M., Takahashi, M., Norden, B. 1990. *J. Biol. Chem.* 265:18891–97
127. Norden, B., Elvingson, C., Eriksson, T., Kubista, M., Sjöberg, B., et al. 1990. *J. Mol. Biol.* 216:223–28
128. Hotchkiss, R. D. 1974. *Annu. Rev. Microbiol.* 28:445–68

129. Sigal, N., Alberts, B. 1972. *J. Mol. Biol.* 71:789–93
129a. Müller, B., Burdett, I., West, S. C. 1992. *EMBO J.* In press
130. McGavin, S. 1971. *J. Mol. Biol.* 55:293–98
131. McGavin, S. 1977. *Heredity* 39:15–25
132. Wilson, J. H. 1979. *Proc. Natl. Acad. Sci. USA* 76:3641–45
133. Howard-Flanders, P., West, S. C., Rusche, J. R., Egelman, E. H. 1984. *Cold Spring Harbor Symp. Quant. Biol.* 49:571–80
134. Felsenfeld, G., Davies, D. R., Rich, A. 1957. *J. Am. Chem. Soc.* 79:2023–24
135. Lyamichev, V. I., Frank-Kamenetskii, M. D., Soyfer, V. N. 1990. *Nature* 344:568–70
136. Kohwi, Y., Kohwi-Shigematsu, T. 1988. *Proc. Natl. Acad. Sci. USA* 85:3781–85
137. Hanvey, J. C., Shimizu, M., Wells, R. D. 1988. *Proc. Natl. Acad. Sci. USA* 85:6292–96
138. Moser, H. E., Dervan, P. B. 1987. *Science* 238:645–50
139. Praseuth, D., Perrouault, L., LeDoan, T., Chassignol, M., Thuong, N., Helene, C. 1988. *Proc. Natl. Acad. Sci. USA* 85:1349–54
140. Griffin, L. C., Dervan, P. B. 1989. *Science* 245:967–71
141. Rajagopal, P., Feigon, J. 1989. *Nature* 339:637–40
142. Strobel, S. A., Dervan, P. B. 1990. *Science* 249:73–75
143. Honigberg, S. M., Radding, C. M. 1988. *Cell* 54:525–32
144. Stasiak, A., Egelman, E. H. 1986. *Biophys. J.* 49:5–7
145. Akaboshi, E., Howard-Flanders, P. 1990. *J. Biochem.* 107:781–86
146. Müller, B., Koller, T., Stasiak, A. 1990. *J. Mol. Biol.* 212:97–112
147. Griffin, T. J., Kolodner, R. D. 1990. *J. Bacteriol.* 172:6291–99
148. Lovett, S. T., Kolodner, R. D. 1989. *Proc. Natl. Acad. Sci. USA* 86:2627–31
149. Umezu, K., Nakayama, K., Nakayama, H. 1990. *Proc. Natl. Acad. Sci. USA* 87:5363–67
150. Lloyd, R. G., Buckman, C. 1991. *J. Bacteriol.* 173:1004–11
151. Lloyd, R. G. 1991. *J. Bacteriol.* 173:5414–18
151a. Lloyd, R. G., Sharples, G. J. 1991. *J. Bacteriol.* 173:6837–43
152. Rostas, K., Morton, S. J., Picksley, S. M., Lloyd, R. G. 1987. *Nucleic Acids Res.* 15:5041–50
153. Morrison, P. T., Lovett, S. T., Gilson,
L. E., Kolodner, R. 1989. *J. Bacteriol.* 171:3641–49
154. Mahdi, A. A., Lloyd, R. G. 1989. *Nucleic Acids Res.* 17:6781–94
155. Amundsen, S. K., Neiman, A. M., Thibodeaux, S., Smith, G. R. 1990. *Genetics* 126:25–40
156. Clark, A. J. 1973. *Annu. Rev. Genet.* 7:67–86
157. Smith, G. R. 1988. See Ref. 4, pp. 115–54
158. Smith, G. R. 1981. *Stadler Genet. Symp.* 13:25–37
159. Roman, L. J., Dixon, D. A., Kowalczykowski, S. C. 1991. *Proc. Natl. Acad. Sci. USA* 88:3367–71
160. Dixon, D. A., Kowalczykowski, S. C. 1991. *Cell* 66:361–71
161. Ippen-Ihler, K. A., Minkley, E. G. 1986. *Annu. Rev. Genet.* 20:593–604
162. Birge, E. A., Low, K. B. 1974. *J. Mol. Biol.* 83:447–57
163. Faulds, D., Dower, N., Stahl, M. M., Stahl, F. W. 1979. *J. Mol. Biol.* 131:681–95
164. Rosenberg, S. M. 1987. *Cell* 48:855–65
165. Taylor, A. F., Smith, G. R. 1990. *J. Mol. Biol.* 211:117–34
166. Müller, B., Boehmer, P. E., Emmerson, P. T., West, S. C. 1991. *J. Biol. Chem.* 266:19028–33
167. Lloyd, R. G., Buckman, C. 1985. *J. Bacteriol.* 164:836–44
168. Joseph, J. W., Kolodner, R. D. 1983. *J. Biol. Chem.* 258:10418–24
169. Lloyd, R. G., Porton, M. C., Buckman, C. 1988. *Mol. Gen. Genet.* 212:317–24
170. Lovett, S. T., Luisi-DeLuca, C., Kolodner, R. D. 1988. *Genetics* 120:37–45
171. Lloyd, R. G., Buckman, C. 1991. *Biochimie* 73:313–20
172. Kmiec, E., Holloman, W. K. 1982. *Cell* 29:367–74
173. Holliday, R., Taylor, S. Y., Kmiec, E. B., Holloman, W. K. 1984. *Cold Spring Harbor Symp. Quant. Biol.* 49:669–73
174. Holden, D. W., Spanos, A., Banks, G. R. 1989. *Nucleic Acids Res.* 17:10489
175. Tsukuda, T., Bauchwitz, R., Holloman, W. K. 1989. *Gene* 85:335–41
176. Eggleston, A. K., Kowalczykowski, S. C. 1991. *Biochimie* 73:163–76
177. Tishkoff, D. X., Johnson, A. W., Kolodner, R. D. 1991. *Mol. Cell. Biol.* 11:2593–608
178. Johnson, A. W., Kolodner, R. D. 1991. *J. Biol. Chem.* 266:14046–54
179. Dykstra, C. C., Hamatake, R. K., Sugino, A. 1990. *J. Biol. Chem.* 265:10968–73
180. Dykstra, C. C., Kitada, K., Clark, A.

B., Hamatake, R. K., Sugino, A. 1991. *Mol. Cell. Biol.* 11:2583–92

181. Heyer, W. D., Kolodner, R. D. 1989. *Biochemistry* 28:2856–62

182. Norris, D., Kolodner, R. 1990. *Biochemistry* 29:7903–10

183. Norris, D., Kolodner, R. 1990. *Biochemistry* 29:7911–16

184. Heyer, W. D., Rao, M. R. S., Erdile, L. F., Kelly, T. J., Kolodner, R. D. 1990. *EMBO J.* 9:2321–30

185. Halbrook, J., McEntee, K. 1989. *J. Biol. Chem.* 264:21403–12

186. Clark, A. B., Dykstra, C. C., Sugino, A. 1991. *Mol. Cell. Biol.* 11:2576–82

187. Nicolas, A., Treco, D., Schultes, N. P., Szostak, J. W. 1989. *Nature* 338:35–39

188. Cao, L., Alani, E., Kleckner, N. 1990. *Cell* 61:1089–101

189. Sun, H., Treco, D., Szostak, J. W. 1991. *Cell* 64:1155–62

190. Resnick, M. A. 1987. In *Meiosis,* ed. P. B. Moens, pp. 157–210. New York: Academic

191. Klapholz, S., Waddell, C. S., Esposito, R. E. 1985. *Genetics* 110:187–216

192. Menees, T. M., Roeder, G. S. 1989. *Genetics* 123:675–82

193. Engebrecht, J., Roeder, G. S. 1990. *Mol. Cell. Biol.* 10:2379–89

194. Hollingsworth, N. M., Goetsch, L., Byers, B. 1990. *Cell* 61:73–84

195. Rockmill, B., Roeder, G. S. 1990. *Genetics* 126:563–74

196. Alani, E., Subbiah, S., Kleckner, N. 1989. *Genetics* 122:47–57

197. Alani, E., Padmore, R., Kleckner, N. 1990. *Cell* 61:419–36

197a. Shinohara, A., Ogawa, T., Ogawa, H. 1992. *Cell.* In press

197b. Bishop, D., Park, D., Xu, L., Kleckner, N. 1992. *Cell.* In press

198. Lowenhaupt, K., Sander, M., Hauser, C., Rich, A. 1989. *J. Biol. Chem.* 264:20568–75

199. Sander, M., Lowenhaupt, K., Rich, A. 1991. *Proc. Natl. Acad. Sci. USA* 88:6780–84

200. Kenne, K., Ljungquist, S. 1984. *Nucleic Acids Res.* 12:3057–68

201. Kucherlapati, R. S., Spencer, J., Moore, P. D. 1985. *Mol. Cell. Biol.* 5:714–20

202. Ganea, D., Moore, P., Chekuri, L., Kucherlapati, R. 1987. *Mol. Cell. Biol.* 7:3124–30

203. Cassuto, E., Lightfoot, L. A., Howard-Flanders, P. 1987. *Mol. Gen. Genet.* 208:10–14

204. Moore, S. P., Fishel, R. 1990. *J. Biol. Chem.* 265:11108–17

205. Moore, S. P., Erdile, L., Kelly, T.,

Fishel, R. 1991. *Proc. Natl. Acad. Sci. USA* 88:9067–71

206. Hseih, P., Camerini-Otero, R. D. 1989. *J. Biol. Chem.* 264:5089–97

207. Eisen, A., Camerini-Otero, R. D. 1988. *Proc. Natl. Acad. Sci. USA* 85:7481–85

208. Benbow, R. M., Zuccarelli, A. J., Sinsheimer, R. L. 1975. *Proc. Natl. Acad. Sci. USA* 72:235–39

209. Thompson, B. J., Escarmis, C., Parker, B., Slater, W. C., Doniger, J., et al. 1975. *J. Mol. Biol.* 91:409–19

210. Bell, L., Byers, B. 1979. *Proc. Natl. Acad. Sci. USA* 76:3445–49

211. Sobell, H. M. 1974. In *Mechanisms in Recombination,* ed. R. F. Grell, pp. 433–38. New York:Plenum

212. Kallenbach, N. R., Ma, R. I., Seeman, N. C. 1983. *Nature* 305:829–31

213. Gough, G. W., Lilley, D. M. J. 1985. *Nature* 313:154–56

214. Diekmann, S., Lilley, D. M. J. 1987. *Nucleic Acids Res.* 15:5765–74

215. Cooper, J. P., Hagerman, P. J. 1987. *J. Mol. Biol.* 198:711–19

216. Duckett, D. R., Murchie, A. I. H., Diekmann, S., Von Kitzing, E., Kemper, B., Lilley, D. M. J. 1988. *Cell* 55:79–89

217. Churchill, M. E. A., Tullius, T. D., Kallenbach, N. R., Seeman, N. C. 1988. *Proc. Natl. Acad. Sci. USA* 85:4653–56

218. Chen, J. H., Churchill, M. E. A., Tullius, T. D., Kallenbach, N. R., Seeman, N. C. 1988. *Biochemistry* 27:6032–38

219. Cooper, J. P., Hagerman, P. J. 1989. *Proc. Natl. Acad. Sci. USA* 86:7336–40

220. Duckett, D. R., Murchie, A. I. H., Lilley, D. M. J. 1990. *EMBO J.* 9:583–90

221. Murchie, A. I. H., Clegg, R. M., von Kitzing, E., Duckett, D. R., Diekmann, S., Lilley, D. M. J. 1989. *Nature* 341:763–66

222. Mueller, J. E., Kemper, B., Cunningham, R. P., Kallenbach, N. R., Seeman, N. C. 1988. *Proc. Natl. Acad. Sci. USA* 85:9441–45

223. Mueller, J. E., Newton, C. J., Jensch, F., Kemper, B., Cunningham, R. P., et al. 1990. *J. Biol. Chem.* 265:13918–24

224. von Kitzing, E., Lilley, D. M. J., Diekmann, S. 1990. *Nucleic Acids Res.* 18:2671–83

225. Mosig, G. 1987. *Annu. Rev. Genet.* 21:347–71

226. Nishimoto, H., Takayama, M., Minagawa, T. 1979. *Eur. J. Biochem.* 100:433–40

227. Kemper, B., Garabett, M. 1981. *Eur. J. Biochem.* 115:123–32

228. Minagawa, T., Ryo, Y. 1978. *Virology* 91:222–33
229. Kemper, B., Garabett, M., Courage, U. 1981. *Eur. J. Biochem.* 115:133–41
230. Tomaschewski, J., Rüger, W. 1987. *Nucleic Acids Res.* 15:3632–33
231. Barth, K. A., Powell, D., Trupin, M., Mosig, G. 1988. *Genetics* 120:329–43
232. Kosak, H. G., Kemper, B. W. 1990. *Eur. J. Biochem.* 194:779–84
233. Mizuuchi, K., Kemper, B., Hays, J., Weisberg, R. A. 1982. *Cell* 29:357–65
234. Kemper, B., Jensch, F., Depka-Prondzynski, M., Fritz, H. J., Borgmeyer, U., Mizuuchi, K. 1984. *Cold Spring Harbor Symp. Quant. Biol.* 49:815–25
235. Lilley, D. M. J., Kemper, B. 1984. *Cell* 36:413–22
235a. Pottmeyer, S., Kemper, B. 1992. *J. Mol. Biol.* In press
236. Jensch, F., Kemper, B. 1986. *EMBO J.* 5:181–89
237. Kleff, S., Kemper, B. 1988. *EMBO J.* 7:1527–35
238. Kemper, B., Pottmeyer, S., Solaro, P., Kosak, H. 1990. In *Structure and Methods. I. Human Genome Initiative and DNA Recombination,* ed. R. H. Sarma, M. H. Sarma, pp. 215–29. New York:Adenine
239. Picksley, S. M., Parsons, C. A., Kemper, B., West, S. C. 1990. *J. Mol. Biol.* 212:723–35
240. Parsons, C. A., Kemper, B., West, S. C. 1990. *J. Biol. Chem.* 265:9285–89
241. Müller, B., Jones, C., Kemper, B., West, S. C. 1990. *Cell* 60:329–36
242. Fugisawa, H., Yonesaka, T., Minagawa, T. 1985. *Nucleic Acids Res.* 13:7473–82
243. Griffith, J., Formosa, T. 1985. *J. Biol. Chem.* 260:4484–91
244. Center, M., Studier, F., Richardson, C. 1970. *Proc. Natl. Acad. Sci. USA* 65:242–48
245. Paetkau, V., Langman, L., Bradley, P., Scraba, D., Miller, R. C. 1977. *J. Virol.* 22:130–41
246. Kerr, C., Sadowski, P. D. 1975. *Virology* 65:281–85
247. Tsujimoto, Y., Ogawa, H. 1978. *J. Mol. Biol.* 125:255–73
248. Lee, D., Sadowski, P. 1981. *J. Virol.* 40:839–47
249. de Massy, B., Studier, F. W., Dorgai, L., Appelbaum, E., Weisberg, R. A. 1984. *Cold Spring Harbor Symp. Quant. Biol.* 49:715–26
250. Pham, T. T., Coleman, J. E. 1985. *Biochemistry* 24:5672–77

251. Panayotatos, N., Fontaine, A. 1985. *J. Biol. Chem.* 260:3173–77
252. de Massy, B., Weisberg, R. A., Studier, F. W. 1987. *J. Mol. Biol.* 193:359–76
253. Center, M., Richardson, C. 1970. *J. Biol. Chem.* 245:6285–92
254. Sadowski, P. D. 1971. *J. Biol. Chem.* 246:209–16
255. Dickie, P., McFadden, G., Morgan, A. R. 1987. *J. Biol. Chem.* 262:14826–36
256. Parsons, C. A., West, S. C. 1990. *Nucleic Acids Res.* 18:4377–84
257. Lu, M., Guo, Q., Studier, F. W., Kallenbach, N. R. 1991. *J. Biol. Chem.* 266:2531–36
258. Müller, B., Jones, C., West, S. C. 1990. *Nucleic Acids Res.* 18:5633–36
259. Jensch, F., Kosak, H., Seeman, N. C., Kemper, B. 1989. *EMBO J.* 8:4325–34
260. West, S. C., Korner, A. 1985. *Proc. Natl. Acad. Sci. USA* 82:6445–49
261. West, S. C., Parsons, C. A., Picksley, S. M. 1987. *J. Biol. Chem.* 262:12752–58
262. Parsons, C. A., West, S. C. 1988. *Cell* 52:621–29
263. Parsons, C. A., Murchie, A. I. H., Lilley, D. M. J., West, S. C. 1989. *EMBO J.* 8:239–46
264. West, S. C., Elborough, K. M., Parsons, C. A., Picksley, S. M. 1989. In *DNA Repair Mechanisms and Their Biological Implications in Mammalian Cells,* ed. M. W. Lambert, J. Laval, 182:233–43. New York:Plenum
265. Symington, L. S., Kolodner, R. 1985. *Proc. Natl. Acad. Sci. USA* 82:7247–51
266. Evans, D. H., Kolodner, R. 1987. *J. Biol. Chem.* 262:9160–65
267. Evans, D. H., Kolodner, R. 1988. *J. Mol. Biol.* 201:69–80
267a. Kleff, S., Kemper, B., Sternglanz, R. 1992. *EMBO J.* 11:699–704
268. Bianchi, M. E. 1988. *EMBO J.* 7:843–50
269. Elborough, K. M., West, S. C. 1988. *Nucleic Acids Res.* 16:3603–14
270. Bianchi, M. E., Beltrame, M., Paonessa, G. 1989. *Science* 243:1056–59
271. Waldman, A. S., Liskay, R. M. 1988. *Nucleic Acids Res.* 16:10249–66
272. Jayaseelan, R., Shanmugam, G. 1988. *Biochem. Biophys. Res. Commun.* 156:1054–59
273. Elborough, K. M., West, S. C. 1990. *EMBO J.* 9:2931–36
274. Connolly, B., West, S. C. 1990. *Proc. Natl. Acad. Sci. USA* 87:8476–80
275. Connolly, B., Parsons, C. A., Benson, F. E., Dunderdale, H. J., Sharples, G. J., et al. 1991. *Proc. Natl. Acad. Sci. USA* 88:6063–67

276. Otsuji, N., Iyehara, H., Hideshima, Y. 1974. *J. Bacteriol.* 117:337–44
277. Stacey, K. A., Lloyd, R. G. 1976. *Mol. Gen. Genet.* 143:223–32
278. Lloyd, R. G., Benson, F. E., Shurvinton, C. E. 1984. *Mol. Gen. Genet.* 194:303–9
279. Lloyd, R. G., Buckman, C., Benson, F. E. 1987. *J. Gen. Microbiol.* 133:2531–38
280. Luisi-DeLuca, C., Lovett, S. T., Kolodner, R. D. 1989. *Genetics* 122:269–78
281. Benson, F., Collier, S., Lloyd, R. G. 1991. *Mol. Gen. Genet.* 225:266–72
282. Sharples, G. J., Benson, F. E., Illing, G. T., Lloyd, R. G. 1990. *Mol. Gen. Genet.* 221:219–26
283. Shurvinton, C. E., Lloyd, R. G. 1982. *Mol. Gen. Genet.* 185:352–55
284. Benson, F. E., Illing, G. T., Sharples, G. J., Lloyd, R. G. 1988. *Nucleic Acids Res.* 16:1541–50
285. Shinagawa, H., Makino, K., Amemura, M., Kimura, S., Iwasaki, H., Nakata, A. 1988. *J. Bacteriol.* 170:4322–29
286. Sharples, G. J., Lloyd, R. G. 1991. *J. Bacteriol.* 173:7711–15
286a. Takahagi, M., Iwasaki, H., Nakata, A., Shinagawa, H. 1991. *J. Bacteriol.* 173:5747–53
287. Iwasaki, H., Shiba, T., Makino, K., Nakata, A., Shinagawa, H. 1989. *J. Bacteriol.* 171:5276–80
288. Shiba, T., Iwasaki, H., Nakata, A., Shinagawa, H. 1991. *Proc. Natl. Acad. Sci. USA* 88:8445–49
288a. Tsaneva, I. R., Illing, G., Lloyd, R. G., West, S. C. 1992. *Mol. Gen. Genet.* In press
288b. Iwasaki, H., Takahagi, M., Shiba, T., Nakata, A., Shinagawa, H. 1992. *EMBO J.* 10:4381–89
288c. Parsons, C. A., Tsaneva, I. R., Lloyd, R. G., West, S. C. 1992. *Proc. Natl. Acad. Sci. USA.* In press
289. Dunderdale, H. J., Benson, F. E., Parsons, C. A., Sharples, G. J., Lloyd, R. G., West, S. C. 1991. *Nature* 354:506–10
290. Tsaneva, I. R., Müller, B. M., West, S. C. 1992. *Cell.* In press

Annu. Rev. Biochem. 1992. 61:641–71

SMALL CATALYTIC RNAs

Robert H. Symons

Department of Plant Science, Waite Institute, University of Adelaide, Glen Osmond, S.A. 5064, Australia

KEY WORDS: ribozymes, catalytic RNAs, hammerhead self-cleavage of RNA, self-cleavage of pathogenic RNAs

CONTENTS

PERSPECTIVES

Catalytic RNAs are those that have the intrinisic ability to break and form covalent bonds (1). Such RNAs were termed *ribozymes* by Cech and his

641

colleagues (1), and this term is used in this review. Ribozymes that act in an intramolecular reaction catalyze only a single turnover and are usually modified during the reaction. By comparison with enzymic reactions, they can, therefore, be considered to be acting in a quasi-catalytic manner (2). However, ribozymes can also act in *trans* in a truly catalytic manner with a turnover greater than one and without being modified; examples are described later in this review.

The first ribozyme to be described was the 413-nucleotide Group I intervening sequence in the nuclear rRNA precursor of the protozoan, *Tetrahymena thermophila* (1). In the presence of a guanosine cofactor, this intervening sequence (IVS) is excised and the two exons ligated in the complete absence of protein (1, 2). Hence, the IVS catalyzes its own excision and is modified during the process with the guanosine covalently coupled to the 5'-end of the excised IVS. A number of IVSs, called Group I, undergo self-splicing by the same basic pathway (3–6).

In 1983, Altman and his colleagues described the first truly catalytic ribozyme by demonstrating that the approximately 400-nucleotide RNA component of bacterial RNase P could cleave its normal substrate of precursor tRNAs (pre-tRNAs) in the absence of its M_r 14,000 protein subunit, with multiple turnover, and without being modified (7). Soon after, Cech and his colleagues showed that a 395-nucleotide fragment of the *Tetrahymena* IVS could catalyze a number of transesterification reactions in a truly catalytic fashion (reviewed in 2, 4, 5).

The self-splicing Group I and the related Group II introns and the bacterial RNases P that have been characterized give cleaved products with 2', 3'-hydroxyl groups and usually 5'-phosphorylated hydroxyls. Such reactions have been extensively reviewed (2, 8–14) and are not considered further here.

The common theme of this review is the self-cleavage reaction of Figure 1 in which an RNA molecule, in the presence of Mg^{2+} or other divalent cation, cleaves to produce fragments containing a 5'-hydroxyl and a 2', 3'-cyclic phosphate. This is a nonhydrolytic reaction in which the same number of phosphodiester bonds is maintained and the transesterification reaction is theoretically reversible. A number of small plant pathogenic RNAs (15–19), one animal viral RNA (20), and an RNA transcript from satellite II DNA of the newt (21, 22) and from a *Neurospora* mitochondrial DNA plasmid (23) undergo this self-cleavage reaction in vitro in a site-specific manner and in the complete absence of protein (Table 1). In the case of the pathogenic RNAs, this reaction is considered to play an essential role in the replication of these RNAs in vivo (see below).

Such reactions are all intramolecular and hence quasi-catalytic with single turnover. However, these RNAs can be manipulated to provide true catalytic cleavage in *trans* in a way similar to that demonstrated for the *Tetrahymena*

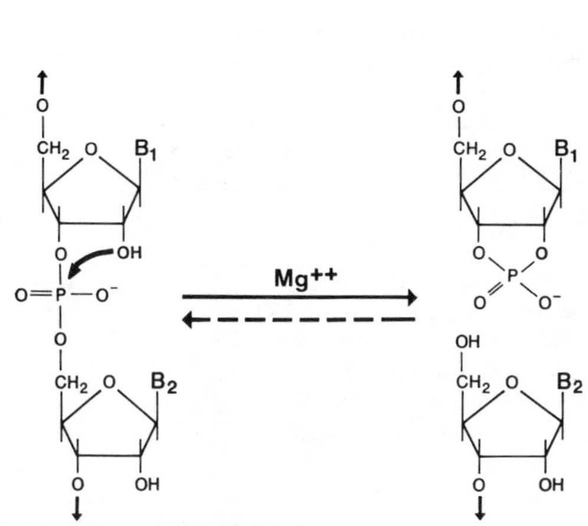

Figure 1   The self-cleavage reaction of RNA catalyzed by Mg^{2+} or other divalent cations. It is a nonhydrolytic, transesterification reaction and is theoretically reversible.

IVS (2–6). It is of interest that the only naturally occurring RNA so far characterized that provides a true catalytic reaction in *trans* without prior modification is the RNA component of bacterial RNase P (8–12).

In this review, the features of all the self-cleavage reactions carried out by the RNAs listed in Table 1 are considered as well as the discussion of two intriguing simple ribozyme reactions that have recently been described. In addition, it is speculated that more self-cleavage reactions are likely to be found and that such reactions may play a role in normal cellular metabolism.

ROLLING CIRCLE REPLICATION REQUIRES SPECIFIC RNA CLEAVAGE

All experimental evidence indicates that circular, plant pathogenic RNAs are replicated by a rolling circle mechanism in vivo (15, 16, 24–28). The two variations of the rolling circle mechanism in Figure 2 are supported by data on the various plus and minus monomeric and multimeric forms of these RNAs found in vivo. In the first variation (Figure 2A), the circular plus strand is copied by an RNA-dependent RNA polymerase to form a concatameric minus strand (step 2). Site-specific cleavage (arrows) of this strand produces a monomer that is circularized by a host RNA ligase (step 3) and then copied by the RNA polymerase to produce a concatameric plus strand. Cleavage of this strand (step 5) produces monomers which, on circularization, produce the progeny circular, plus RNA, the dominant form in vivo.

In the other variation (Figure 2B), the concatameric minus strand of step 1 is not cleaved but is copied directly to give a concatameric plus strand (step

Table 1 Plant and animal pathogenic RNAs and other RNAs that self-cleave in vitro

RNA and self-cleavage structure	Size (nucleotides)	RNA strand cleaved
A. Cleavage by hammerhead structure, reaction not reversible		
Avocado sunblotch viroid (ASBV)	246–251	Plus and minus
Encapsidated linear satellite RNAs of:		
Barley yellow dwarf virus (sBYDV)	322	Plus and minus
Tobacco ringspot virus (sTRSV)	359–360	Plus
Encapsidated circular satellite RNAs (virusoids) of:		
Lucerne transient streak virus (vLTSV)	324	Plus and minus
Solanum nodiflorum mottle virus (vSNMV)	378	Plus only
Subterranean clover mottle virus (vSCMoV)	332 & 388	Plus only
Velvet tobacco mottle virus (vVTMoV)	366	Plus only
RNA transcript of newt satellite II DNA	300	Plus only
B. Cleavage by hairpin/paperclip structure, reaction reversible		
Encapsidated linear satellite RNA of tobacco ringspot virus (sTRSV)	359–360	Minus
C. Cleavage structure and reversibility to be fully defined		
Hepatitis delta virus RNA (HDV RNA)	1700	Plus and minus
D. Cleavage structure and reversibility to be determined		
RNA transcripts of mitochondrial DNA plasmid of *Neurospora*, linear and circular	800	Plus

3), which is cleaved specifically to monomers for ligation to the circular progeny. Those RNAs that self-cleave only in the plus strand in vitro are considered to follow this route.

The rolling circle pathway of replication requires highly specific cleavage within the concatameric RNAs. Since viroids and virusoids appear not to code for any proteins (24), a host enzyme would be required to perform this role if a protein is involved. This is considered unlikely since no known plant ribonuclease has the high specificity required. Further, the very wide host range of some viroids that can replicate in members of many different families (29) would require a universally present enzyme(s) with the appropriate specificity. Another possibility is that these RNAs could usurp specific components of the complexes involved in intron splicing.

Following the pioneering work of Cech & Altman and their colleagues (1, 7), an RNA-catalyzed, self-cleavage event appeared most likely during rolling circle replication, and this was soon demonstrated in vitro for plus and minus sTRSV (30, 31) and plus and minus ASBV (32) and later for the other pathogenic RNAs of Table 1. It seems reasonable that such a self-cleavage reaction is responsible for the cleavage in vivo of concatameric RNAs to monomeric units, although definitive evidence has yet to be provided.

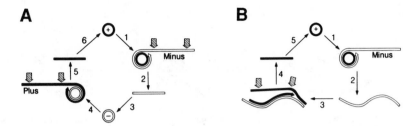

Figure 2 Rolling circle models for the replication of small, circular pathogenic RNAs. *A*. Model in which both plus and minus RNAs self-cleave. *B*. Model in which only the plus RNA self-cleaves. Sites of self-cleavage indicated by arrows. See text for further details. Reprinted with permission from (41).

SELF-CLEAVAGE OF RNA VIA THE HAMMERHEAD STRUCTURE

The history and experimental approaches involved in the discovery and characterization of the hammerhead self-cleavage reaction have been considered in detail elsewhere; readers are referred to a number of reviews (15–19, 30, 33–35). The hammerhead-shaped structures around the self-cleavage sites (arrowed) of (+) vLTSV, and (+) sTRSV are given in Figure 3 with the nucleotides boxed being highly conserved in the RNAs listed in Table 1, part A. The consensus sequence (Figure 3C) is that proposed by Uhlenbeck in 1987 (36, 37) based on the hammerhead sequences of plus and minus ASBV and of the RNA transcript of newt satellite II DNA.

The basic features of the hammerhead structure are the three base-paired stems I, II, and III, surrounding a single-stranded central region, with the 13 conserved bases (boxed). There appear to be few restrictions on the nonconserved nucleotides in the three base-paired stems, but the nucleotide 5' to the self-cleavage site is most commonly C, occasionally A, but never U or G. The base pair on the inside of stem II is usually C:G, whereas the third base pair of this stem is usually G:C. Extensive mutagenesis analysis of the hammerhead structure (see below) has further defined the sequence requirements required for efficient self-cleavage in addition to those determined by a comparative analysis of naturally occurring RNAs.

Double Hammerhead Structure

The single hammerhead structures of (+) ASBV (Figure 4A) and (−) ASBV (32) are different from the other hammerhead structures of naturally occurring RNAs in that stem II contains either two or three base pairs and is enclosed by a three-base hairpin loop. When Epstein & Gall (21) showed that in vitro transcripts of a 330-base-pair tandemly repeated satellite II DNA of the newt

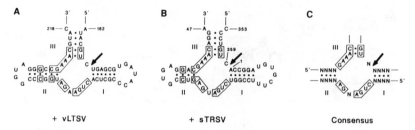

+ vLTSV   + sTRSV   Consensus

Figure 3 Hammerhead structures around the self-cleavage sites of *A*, one virusoid [(+) vLTSV], *B*, a linear satellite RNA [(+) sTRSV], together with *C*, the consensus hammerhead structure based on (+) and (−) ASBV and the RNA transcript of the newt satellite II DNA (21, 24, 36, 41)

could self-cleave in vitro and that a hammerhead structure could be drawn around the self-cleavage site (Figure 4*B*), we considered that there had to be an alternative to this inherently unstable hairpin loop of a two-base-pair stem and a two-base loop.

The alternative, more stable ASBV double hammerhead structure (Figure 4*C*), in which two single hammerhead structures are combined, was demonstrated by transcription of dimeric cDNA clones, which gave dimeric (+) and (−) self-cleaving transcripts (38). By the use of inactivating GAAAC-to-GAAC mutations adjacent to either or both of the self-cleavage sites, the formation of double hammerhead structures during transcription was clearly demonstrated (38). However, with full-length, gel-purified dimeric transcripts, the (+) ASBV still cleaved via a double hammerhead structure while the (−) ASBV cleaved via a single hammerhead structure (39). This was

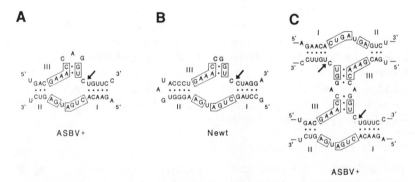

ASBV+   Newt   ASBV+

Figure 4 Single (*A*) and double (*C*) hammerhead structures for (+) ASBV and the single hammerhead structure for the newt satellite DNA II RNA transcript (*B*). The 13 conserved nucleotides are boxed and the sites of self-cleavage are indicated by arrows. Reproduced with permission from (15).

attributed to a more stable stem III of three base pairs and a three-base loop in the (−) ASBV hammerhead structure. These results illustrate the importance of the pathway of folding of the dimeric RNAs in determining which self-cleavage structure is formed.

Indirect evidence that the newt RNA also self-cleaved via a double hammerhead structure was obtained using short (40-mer) RNA transcripts containing the newt hammerhead sequence (40). The self-cleavage reaction showed roughly second order kinetics, indicating a bimolecular reaction by a double hammerhead structure rather than by a single hammerhead or unimolecular reaction, which would give first order kinetics. These results are in contrast with those of Epstein & Pabon-Pena (22), who used purified monomeric and longer transcripts of newt satellite II DNA clones. Monomeric transcripts gave slow cleavage that was independent of concentration, and presumably occurred by a single hammerhead structure that was stabilized by sequences outside the hammerhead domain (22). However, faster self-cleavage was also shown in dimeric RNA transcripts in which partial double hammerhead structures (40) could form that contained one complete and one partial hammerhead structure (22).

Hammerhead Self-Cleavage in Trans

The sequences that make up the (+) and (−) hammerhead domains in ASBV originate from opposite strands in the central regions of the rod-like molecule (Figure 5). Such structure formation differs from that of other naturally occurring RNAs, in which the hammerhead domain is formed from a contiguous set of nucleotides (see Figure 3A, B; Refs. 15–19, 41–44). On the basis of this formation, Uhlenbeck (36) realized the potential for *trans* self-cleavage reactions in which two separate and independent RNA molecules could interact to form an active hammerhead structure and also provide a simple model for the detailed characterization of the hammerhead self-cleavage reaction.

The simple system developed by Uhlenbeck (36, 37) was based on (−) ASBV and consisted (Figure 6A) of a 19-nucleotide ribozyme fragment (R), which caused rapid and specific cleavage of a 24-nucleotide substrate (S) under physiological conditions and had all the properties associated with an RNA enzyme, especially as it catalyzed multiple turnover of substrate. Other examples include the catalytic activity of a 13-mer fragment, the smallest ribozyme described so far, on a 41-mer substrate (Figure 6B; Ref. 45) and a triple in *trans* reaction in which a 19-mer and a 14-mer (R1, R2) combine for cleavage of an 11-mer substrate (Figure 6C; Refs. 47, 48). Larger molecules can also participate in such reactions as shown by the efficient self-cleavage using two large fragments of (−) ASBV (43).

Kinetic data for the three in *trans* reactions of Figure 6 are given in Table 2.

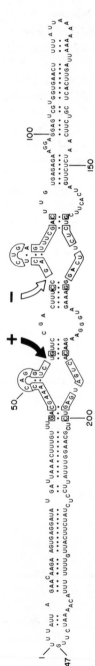

Figure 5 Native structure of ASBV modified to incorporate the plus hammerhead self-cleavage structure and the complement of the minus structure. Sites of self-cleavage indicated by arrows. Reproduced with permission from (32).

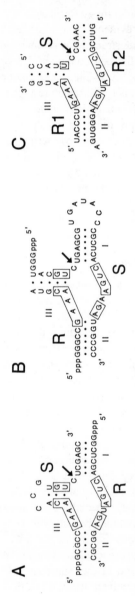

Figure 6 Three examples of self-cleavage in *trans* via a hammerhead structure. R, ribozyme; S, substrate. *A.* Ribozyme of 19 nucleotides and substrate of 24 nucleotides, based on sequence of (−) ASBV (36). *B.* Ribozyme of 13 nucleotides, substrate of 41 nucleotides based on sequence of (+) vLTSV (45). *C.* Three-component structure composed of two ribozyme fragments, R1 and R2, and one substrate fragment (48). Sequence unrelated to any naturally occurring RNA except for 11 conserved nucleotides (boxed). Reproduced with permission from (15).

Although the K_m values for the substrate are in the range found for protein enzymes, the k_{cat} values are much lower than those of 10 to 10^5 min^{-1} for protein-catalyzed reactions. The low k_{cat} value for the triple in *trans* reaction is probably a consequence of the low stability of a hammerhead structure formed from three separate, short oligonucleotides. However, the rate of self-cleavage of RNAs in vivo is likely to much greater as a consequence of the continuous synthesis of large RNA molecules during rolling circle replication, a more favorable ionic environment, and the possible stimulation by host proteins.

For a comparison of catalytic activity, data are also included in Table 2 for in *trans* reactions for the paperclip or hairpin structure of (−) sTRSV considered below. The approximately 400-nucleotide RNA components of two ribonucleases (RNase P) from *Escherichia coli* (M1 RNA) and *Bacillus subtilis* (P RNA) have much lower K_m but similar k_{cat} values to those in Table 2 as do the 395-nucleotide (L-19 IVS) and 393-nucleotide (L-21 ScaI IVS) ribozymes derived from the 413-nucleotide intervening sequence of the nuclear rRNA precursor from *Tetrahymena thermophila* (2–5, 8, 9, 46).

Mutagenesis Analysis of the Hammerhead Reaction

The hammerhead structure lends itself to a detailed mutagenesis analysis, either as a single component, or as two or three components in *trans,* in order to establish the flexibility of the self-cleavage reaction in relation to the minimum consensus structure (Figure 3C) determined from naturally occurring RNAs. The ability to prepare RNA transcripts by the simple and practical T7 RNA transcription method of Milligan & Uhlenbeck (49) makes such an analysis feasible.

The only detailed mutagenesis analysis so far of a single-component hammerhead structure (50) investigated 11 mutations in a 58-mer, the sequence of which corresponded to nucleotides 164 to 216 of (+) vLTSV (Figure 3A) and included five 5′-terminal nucleotides derived from the T7

Table 2   Kinetic data for catalytic self-cleavage in *trans* via hammerhead and hairpin/paperclip structures

System	Kinetic parameter				References
	K_m(mM)	V_{max}(mM.min^{-1})	k_{cat}(min^{-1})	E_{act}(kcal.mol^{-1})	
ASBV- related[a]	0.63	0.046	0.5	13.1	36
vLTSV- related[b]	1.3	0.012	0.5	10.7	45
Triple[c]	0.53	—	0.03	—	48
(−) sTRSV	0.03	—	2.1	19	104

[a] Figure 6A
[b] Figure 6B
[c] Figure 6C

RNA polymerase promoter. The only mutation that effectively eliminated self-cleavage during transcription as well as after purification of the RNA transcript was a C insertion after the C at the self-cleavage site; however, only mutations in two of the conserved nucleotides were tested. The two conserved base pairs in stem I were separately mutated from C:G to U:G and A:U to G:U with some reduction in self-cleavage, especially in the latter case. The only other mutations that caused a significant loss in self-cleavage were the insertion of AA 5' to the conserved GAAAC and the deletion of the A between the conserved CUGA and GA. As described above for (+) and (−) ASBV (38, 39), the mutation of GAAAC to GAAC eliminated self-cleavage activity in dimeric RNA transcripts, either during transcription or after purification.

One limitation of the use of a single oligonucleotide is that a detailed kinetic analysis of the self-cleavage reaction is not possible. Thus, Ruffner et al (51) used their previously well-characterized in *trans* hammerhead system (Figure 6A; Refs. 36, 37) for an extensive mutagenesis study in which they compared cleavage rates of 21 different substrate mutations and 24 different ribozyme mutations. Some of the results are summarized in Figure 7. Only one of the four phylogenetically conserved base pairs (A:U) in Figures 3A and 3B but all nine of the conserved single-stranded residues in the central core are needed for cleavage. The nucleotide 5' to the cleavage site is nearly always C but can be replaced by U, as also in (−) vLTSV (24, 41).

In order to summarize the large amount of mutation data that have been

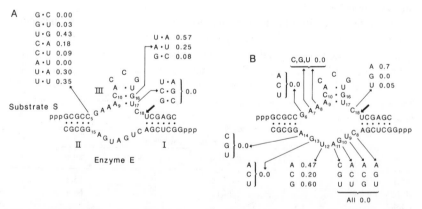

Figure 7 Mutational analysis of a two-component hammerhead structure composed of substrate S and enzyme (ribozyme) E. *A.* Analysis of three base pairs with replacements indicated together with the rates of cleavage relative to the wild-type control. *B.* Analysis of mutations of nucleotides in the central, open region of the hammerhead structure. Mutations indicated together with rates of cleavage relative to the wild-type control. Reproduced with permission from (17).

obtained by Ruffner et al (51) and others (40, 47, 48, 51, 52), two variations of a consensus hammerhead structure are given in Figure 8. Figure 8*A* is the consensus structure for all naturally occurring RNAs with only minor exceptions (17, 18); for example, the C:G base pair on the inside of stem III is U:A in the newt satellite RNA (21, 22, 54; Figure 4*B*) and the third base pair, G:C, is A:U in a satellite RNA of barley yellow dwarf virus (17, 53). Figure 8*B* gives the minimal sequence and structure required for efficient self-cleavage based on both phylogenetic and mutation data (50–52).

What is the Tertiary Structure of the Hammerhead?

The simple two-dimensional hammerhead structure provides little insight into the activated tertiary structure, and the lack of evidence for the reversal of the self-cleavage reaction indicates a relaxation of this structure after cleavage. Presumably, in the presence of a divalent cation, the tertiary structure formed allows the lowering of the activation energy at only one site and the nucleophilic attack of the 2'-hydroxyl on the internucleotide phosphate to provide the phosphoryl transfer reaction and self-cleavage (Figure 1).

Much effort is being spent on the use of proton nuclear magnetic resonance (NMR) spectra to unravel the tertiary structure using RNA prepared by large-scale enzymatic synthesis (49) or by chemical synthesis. However, a requirement for such analyses is a stable active RNA conformation, which is not possible with unmodified RNAs in which the self-cleavage reaction is considered to be very rapid once the active hammerhead structure is formed (42). A further complication is the ability of RNA molecules to form alternative conformations in solution (17, 34, 40–43, 55). However, Heus et al (56) investigated a ribozyme fragment in the absence of its substrate, Pease &

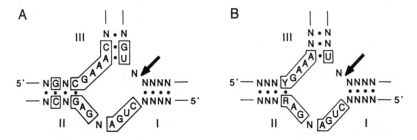

Figure 8   Two consensus hammerhead structures based on conserved sequences in naturally occurring RNAs and on mutational analysis. *A*. Structure containing 17 nucleotides (boxed) conserved in all the self-cleaving RNAs of Table 1*A* except (+) sBYDV (53) and the RNA transcript of newt satellite II DNA (Figure 4). *B*. Revised hammerhead consensus structure of Ruffner et al (51) based on naturally occurring RNAs and mutation data. The only variant is (+) sBYDV, in which the pyrimidine (Y):purine (R) base pair is replaced by a nonpairing A and C (53).

Wemmer (57) examined a single oligonucleotide hammerhead structure after self-cleavage, while Odai et al (58) assigned the resonances of hydrogen-bonded imino protons in a three-RNA oligonucleotide component hammerhead structure in the absence of Mg^{2+}. All the results substantiated some features of the hammerhead structure.

In a more recent proton NMR study, a 34-nucleotide ribozyme was used with a 13-nucleotide, noncleavable, all deoxynucleotide substrate in the presence and absence of Mg^{2+} (59). Data obtained supported the presence of the three helical regions; the minimal spectral changes obtained on the addition of Mg^{2+} indicated the Mg^{2+}-binding site already exists in the Mg^{2+}-free hammerhead structure. The authors concluded that the Mg^{2+} serves primarily in a catalytic, and not a structural, role under the conditions used (59).

A theoretical computational approach was carried out by Mei et al (61) to determine the three-dimensional structure of the minimal self-cleaving hammerhead domain of (+) vLTSV (41, 42). Extensive minimization and dynamics computational studies were performed in vacuum and in the apparent absence of counterions. Although the computational approach and the assumptions required have obvious limitations, the results have provided an interesting working model in which the ribose phosphate backbone takes an abrupt turn such that stems I and III are juxtapositioned. In addition, under this model the structure of the hammerhead per se leads to self-cleavage, and the bases and phosphates do not participate as catalysts although they may be of importance in the coordination of the Mg^{2+} hydrate.

What is the Mechanism of Self-Cleavage?

General acid-base catalysis is most likely involved in self-cleavage in which a basic function within the active hammerhead structure accepts the proton from the 2'-hydroxyl and an acidic function provides a proton for the 5'-hydroxyl at the cleavage site (2, 4). Approaches to investigate the reaction mechanism have included substitution of the phosphate at the self-cleavage site by a phosphorothioate and the replacement of various residues with deoxynucleotides to explore the role of the various 2'-hydroxyl groups.

When RNA polymerases synthesize RNA in the presence of α-phosphorothio-nucleoside triphosphates, only the S_p diastereoisomer is accepted (62, 63) with inversion of the configuration in the phosphodiester bond to give the R_p isomer (64, 65; Figure 9). Incorporation of the R_p phosphorothioate at the cleavage site substantially inhibited self-cleavage of a fragment of (+) sTRSV in the presence of Mg^{2+} (64–66), whereas self-cleavage occurs normally in this and other hammerhead structures in the presence of Mn^{2+} (67). Van Tol

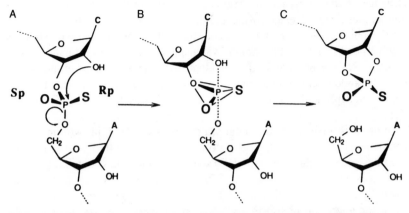

Figure 9 Proposed stereochemical course of hammerhead-mediated phosphodiester cleavage based on phosphorothioate analogues (65, 69). The R_p configuration of the phosphorothioate diester (*A*) proceeds through a postulated trigonal bipyramidal transition state (*B*) to give the cleavage products (*C*) with the endo-isomer of the cyclic phosphorothioate. Adapted with permission from (69).

et al (65) showed that, during the transesterification step of self-cleavage, there is inversion of the configuration of the P-S bond from exo to endo. The results are consistent with transesterification occurring by an in-line $S_N2(P)$ attack by the 2'-hydroxyl group on the phosphorus (19, 65; Figure 9), a result also confirmed by Koizumi & Ohtsuka (68) using a three-component hammerhead.

In another example of phosphorothioate substitution, Slim & Gait (69) prepared oligoribonucleotide substrates with a single phosphorothioate linkage of defined R_p or S_p configuration at the self-cleavage site. On incubation with the appropriate ribozyme, the R_p isomer was cleaved only very slowly, consistent with other results (19, 64, 65, 68), while the rate of cleavage of the S_p isomer was only slightly reduced. The results indicated that the essential Mg^{2+} is bound to the pro-R oxygen in the transition state of the hammerhead cleavage reaction (69; Figure 9). Further support for Mg^{2+} coordination to the nonbridging oxygen is shown by the ability of Mn^{2+} to overcome the inhibition of Mg^{2+}-catalyzed self-cleavage by phosphorothioate substitution at the self-cleavage site, presumably because, unlike Mg^{2+}, Mn^{2+} can coordinate readily with both sulfur and oxygen (67, 69–71).

The presence of phosphorothioate linkages at sites other than the self-cleavage site in the hammerhead domain can also inhibit cleavage. Buzayan et al (72) obtained evidence that phosphorothioate substitution within the highly conserved, single-stranded GpA dinucleotide (Figure 3*B*) inhibited self-cleavage in the presence of Mg^{2+} with their 46-nucleotide (+) sTRSV domain, an effect reversed by the addition of Mn^{2+}. This result suggests that the GA phosphodiester may have a coordination role within the active tertiary hammerhead structure. In a similar way using a two-component hammerhead domain, Ruffner & Uhlenbeck (73) identified four phosphates in the central single-stranded region that play a role in the self-cleavage reaction; the conserved GpA and GpApA, as well as the phosphate at the self-cleavage site.

In order to investigate the binding properties of Mg^{2+} in two- and three-component hammerhead structures, Koizumi & Ohtsuka (68) measured cleavage rates and circular dichroism (CD) spectra for substrates containing inosine or guanosine at the cleavage site. The 2-amino group of the guanine substantially inhibited self-cleavage, consistent with earlier results (50, 51), but the association constants for Mg^{2+} (approximately 2×10^4 M^{-1}) and the number of molecules of Mg^{2+} bound per hammerhead domain (approximately 1) were not affected. Hence, the 2-amino group of guanine interferes with the self-cleavage reaction but not with the binding of Mg^{2+}.

Another approach to explore the role of the 2'-hydroxyl groups in the self-cleavage reaction is to substitute deoxyribo- for ribo-nucleotide residues at various sites as has been done by Cedergren and his colleagues for a 35-nucleotide ribozyme (74) and 14-nucleotide substrate (75), or both (76). The chemical synthesis of these components allowed the insertion of a ribo- or deoxyribo-nucleotide at each site. Various multiple substitutions in the ribozyme in the conserved nucleotides UGA, GA, and GAA present in the central single-stranded region of the hammerhead structure gave reductions in V_{max} which varied from 20- to 200-fold, although there was essentially no change in the K_m of the substrate (74). An all-DNA ribozyme was inactive, confirming an earlier report with a 13-mer ribozyme (45).

Single or multiple deoxynucleotide substitutions in the 14-mer substrate led to a 10-fold decrease in the k_{cat} for self-cleavage and an increase in K_D for the substrate which varied from 8- to 20-fold (75). As the authors concluded (74–76), interpretation of all the data from these various deoxynucleotide substitution experiments is difficult, because such substitutions could result in a conformational change, the removal of one or more 2'-hydroxyls that play a role in catalysis, or a nonspecific relaxation in the three-dimensional structure.

In a similar 34-nucleotide ribozyme and 12-nucleotide substrate, Pieken et al (77) substituted the 2'-hydroxyl group with either a 2'-fluoro or 2'-amino group on certain C, A, or U residues, substitutions that eliminated, as expected, RNase cleavage and alkaline hydrolysis at these sites, as well as self-cleavage when the C residue at the cleavage site was modified. Measurement of k_{cat} and K_m values and the resultant k_{cat}/K_m ratios identified the 2'-hydroxyl groups of the first two A residues in the conserved GAAAC to have a key role in catalytic activity.

Ribozymes Targetting RNA Molecules

In 1986, Zaug et al (78) showed that the L-19 IVS ribozyme derived from the *T. thermophila* precursor rRNA could act as a specific endoribonuclease against defined sequences in α-globin pre-mRNA and in transcripts of the plasmid pBR322. Key sequence variations in the ribozyme led to cleavage at different sites within the pre-mRNA. The L-19 IVS ribozyme and the slightly shortened forms of it have been extensively investigated by Cech and his colleagues (2, 5, 6); although this system is flexible and highly specific, the ribozymes are nearly 400 nucleotides in length. The much smaller hammerhead domain of about 50 nucleotides (36, 42, 43, 79) has provided a unique opportunity to develop simpler ribozyme systems in *trans* to target site-specific cleavage of different RNAs, as first recognized by Uhlenbeck (36). Haseloff & Gerlach (80) extended this approach by the design of small ribozymes based on the hammerhead self-cleaving domain of (+) sTRSV (30, 31, 81, 82; Figure 3*B*). The basic concept was to find GUC sequences in the target RNA to define the cleavage site and then to design the ribozyme around the core sequence of the (+) sTRSV hammerhead with the other nucleotides being determined by the requirements for Watson-Crick base pairs to complete stems I and III (Figure 10). It was recognized that the number of base pairs in stems I and III and their types (A:U, G:C) would affect the specificity, affinity (K_m), and turnover (k_{cat}) of the RNA enzyme (80).

The initial work of Haseloff & Gerlach (80) targetted three sites in a 835-nucleotide transcript prepared in vitro from a cloned bacterial chloramphenicol acetyl transferase gene. They also recognized the potential applications of the use of hammerhead ribozymes in vivo to target specific RNA transcripts as a mechanism for the specific inactivation of host or viral genes. Independently, Koizumi et al (47) prepared a ribozyme by the chemical synthesis of a 13-mer and a 19-mer in order to target a single-stranded region in a 120-nucleotide bacterial 5S RNA. This triple in *trans* reaction was not as efficient as that of the two-component system of Haseloff & Gerlach (80), presumably because of its lower stability.

The Haseloff & Gerlach paper (80) has stimulated others to develop specific ribozymes to target a variety of RNAs in vitro and in vivo. For example, Lamb & Hay (83) prepared short ribozyme transcripts for use in vitro against the coat protein and RNA polymerase genes within the 6-kilobase, single-stranded genomic RNA of potato leafroll virus. Cleavage activity was eliminated in mutant ribozymes in which the conserved GAAAC

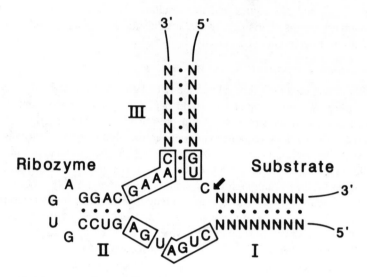

Figure 10 The hammerhead structure of Haseloff & Gerlach (80) derived from *trans*-acting substrate RNA and ribozyme. The nucleotides indicated form part of the (+) sTRSV hammerhead structure, the 13 conserved nucleotides are boxed, and the self-cleavage site is indicated by an arrow. Reproduced with permission from (17).

was mutated to GAAC, consistent with earlier results (38, 39). In a more complex system, Cotten et al (84) targetted the 64-nucleotide U7 snRNA of mouse hybridoma cells by two specific ribozymes to inhibit the 3'-processing of histone pre-mRNA in nuclear extracts in vitro. The ribozyme constructs with stem I and III arms of 8 to 12 nucleotides required a 1000-fold excess for 90% inhibition. In contrast, a full-length antisense RNA required only 6-fold excess, an 80-mer antisense DNA a 60-fold excess, and an 18-mer antisense DNA a 600-fold excess for comparable activity. The authors attributed the poor efficiency of the ribozymes to their instability in the nuclear extracts.

Hammerhead Ribozymes to Manipulate Gene Expression

The ability to switch off or inhibit specific gene expression in prokaryotic or eukaryotic cells has considerable potential as a means of studying gene expression as well as controlling pathogen expression or the suppression of undesirable traits. Although there have been extensive investigations on the use of antisense deoxyribo- and ribo-oligonucleotides, and their chemical variants, as well as antisense RNAs (85–87), the use of hammerhead ribozymes in vivo is in its infancy. There are many aspects to be considered. For example, how efficiently will ribozymes find their target in vivo and cleave their target RNA? Can ribozymes be designed to act in a truly catalytic fashion with multiple turnover? It is this possibility that has been the basis of much of the excitement about the potential of ribozymes in vivo. What will be the half-lives of ribozymes in vivo and what intracellular concentrations are required to be effective? If a ribozyme provides an inhibitory effect on the activity of its target RNA, is the effect due to specific RNA cleavage or is the ribozyme simply acting as an antisense RNA? And finally, will the use of ribozymes offer any practical advantages over antisense DNA and RNA for the specific control of gene expression?

DELIVERY OF RIBOZYMES   Although ribozymes can be delivered to whole cells or to animals, where any transient effect will be determined by the half-lives of the RNAs in vivo (88, 89), the long-term approach is to insert genes for ribozyme constructs with appropriate control and signal regions into target cells. Techniques for the stable insertion of genes into cells for the production of transgenic plants and animals are the subject of intense investigation; they are not considered here.

Two ribozyme-expression strategies are commonly used (88). In the first, one or more ribozyme genes are inserted behind a strong promoter for RNA polymerase II, which may be of viral origin, a retroviral long terminal repeat, or a strong endogenous promoter such as that of the actin gene. A polyadenylation signal is required at the 3'-end of the gene to enable transcription termination and the addition of the poly(A) tail which, together with the m7G cap, should increase RNA stability and transport to the cytoplasm if required. An intron can also be inserted to ensure routing of the transcript through the normal splicing machinery in the nucleus, as such routing may provide an enhanced ribozyme effect.

In the second system, RNA polymerase III promoters derived from tRNA, snRNA, 5S ribosomal RNA, or 7S signal recognition particle RNA gene systems can be used. For example, the ribozyme sequence can be inserted into the anticodon loop of a tRNA gene where the ribozyme sequence would be less likely to be involved in inactive intramolecular interactions with other parts of the transcript (88). RNA polymerase III systems have a number of advantages over those of RNA polymerase II; gene expression appears to occur in all tissues and at an order of magnitude higher level, the gene constructs required are smaller and hence easier to manipulate, and multiple copies of the same or of more than one ribozyme can be inserted.

RIBOZYMES IN VIVO   The in vivo systems developed so far illustrate the great difficulty in proving that it is ribozyme activity, and not antisense activity, that is responsible for any effects observed, and they also emphasize the need for the careful design of controls to indicate what is actually happening. Possibly the best controls are those that contain effective single-base mutations in the ribozyme. Mutation of the conserved GAAAC to GAAC effectively eliminates self-cleavage in vitro (38, 39), but the deletion of one A residue may provide some distortion of the hammerhead structure and hence affect the binding of the stem I and III components of the ribozyme to its target. Such a mutation, therefore, may not provide the best control. The best mutations are likely to be in the conserved CUGA and GA, as for example, the mutation of CUGA to CUUA (51).

The presence of the ubiquitous dsRNA unwinding/modifying activity (90, 91) adds a further complication to interpreting the results of any in vivo ribozyme experiment. This ATP-dependent unwinding activity also modifies adenosine to inosine with the consequent loss of A:U base pairs; in in vitro experiments, at least 40% of A residues in dsRNA formed from mRNA were converted to inosine (92). Hence, this activity in vivo could destroy anti-sense:sense mRNA complexes and could also render the single-stranded regions sensitive to single-strand-specific RNases.

Recent data demonstrate a requirement of at least 15–20 base pairs for the dsRNA unwinding/modifying enzyme to have significant activity with dsRNA formed both intramolecularly and intermolecularly (92a). These results indicate that ribozymes forming short double-stranded stems I and III (Figure 10) would be poor substrates for this modifying activity. However, even one A to I conversion in every 100 nucleotides of an RNA molecule could make the RNA functionally inactive.

The difficulties inherent in in vivo ribozyme experiments are illustrated by several examples. Transient expression in monkey cells of chloramphenicol acetyl transferase mRNA (CAT-mRNA) and its active protein together with a number of ribozymes separately embedded at the same site in the 3'-untranslated region of another mRNA and targetted against three sites in the CAT-mRNA gave up to 60% inhibition of CAT activity, but only when the ribozyme was in 1000-fold molar excess (93). The only control used was a ribozyme inserted in the carrier RNA in the opposite orientation; no inhibition of CAT activity was observed. Neither the CAT mRNA nor the predicted cleavage products could be detected by Northern analysis, presumably because of their rapid degradation (93). In a variation of this system, the same authors (94) coinfected monkey cells with plasmid constructs expressing CAT mRNA and a short sense or antisense 5'-fragment of CAT mRNA. Similar levels of inhibition of CAT activity were found with both the sense and antisense transcripts. Cameron & Jennings (94) consider that one possible effect of the sense RNA is to bind to natural complementarities of sequence in the CAT mRNA, thus providing another mechanism of inhibition.

In another investigation, a ribozyme targetted against the *gag* mRNA of human immunodeficiency virus (HIV) was stably transformed into HeLa cells (95). When these cells were challenged with HIV, there was a significant reduction in *gag* mRNA, proviral DNA, and the *gag*-encoded protein relative to control cells without ribozyme. Polymerase chain reaction analysis of the cellular RNA suggested that specific cleavage had occurred in the *gag* mRNA, but there was no other supporting evidence. In a transient expression system using a ribozyme construct in a RNA polymerase III system, the ribozyme DNA together with its target U7snRNA and a completely unrelated RNA were coinjected into *Xenopus* oocytes (96). The injected U7snRNA had disappeared after 20 hours, but the cleavage products could not be detected; the control RNA was not degraded. However, it was calculated that the ribozyme was at a 500–1000-fold excess over the substrate; an inactive ribozyme control was not used.

In a better-characterized system, Saxena & Ackerman (97) designed ribozymes to target the 28S RNA of *Xenopus* oocytes in regions known to be accessible to the protein cytotoxins, α-sarcin, ricin, Shiga toxin, and Shiga-like toxin II. A ribozyme construct with eight base pairs for stems I and III targetted against a CUC sequence in the α-sarcin domain of 28S RNA cleaved the RNA after injection into *Xenopus* oocytes, although cleavage was much less efficient than that given by α-sarcin. When the conserved CUGA of the ribozyme was mutated to CUUA, no cleavage of the 28S RNA occurred in vivo. However, when protein synthesis was measured by the incorporation of L-^{35}S-methionine, it was decreased to 10% of the control with α-sarcin and 50% of the control by injection of the active ribozyme, the mutated ribozyme, or cycloheximide. Hence, there was no differential effect on protein synthesis between active and inactive ribozyme. Presumably the inhibition of protein synthesis was due to an antisense effect and not to cleavage of the 28S RNA.

Using a prokaryotic system, Chuat & Galibert (98) tested the in vivo activity of ribozymes against *Escherichia coli* β-galactosidase mRNA when the target and ribozyme were on the same or different molecules. Inhibition of enzyme activity was found only when the 39-nucleotide ribozyme was incorporated into the mRNA. The authors concluded that the usual coupled transcription-translation in bacteria prevented any ribozyme action in the intermolecular situation.

MINUS sTRSV RNA SELF-CLEAVES VIA A DIFFERENT MECHANISM THAN PLUS sTRSV

The 359-nucleotide satellite RNA of tobacco ringspot virus (sTRSV or sTobRV) is an intriguing molecule. (+) sTRSV isolated from purified virions is a linear single-stranded molecule, but both circular and linear forms as well as tandem oligomers of plus and minus RNA are found in infected plants (19, 99). As in the case of the viroids and virusoids, replication is considered to follow the rolling circle pathway (19, 35, 100; Figure 2). While (+) sTRSV self-cleaves via the hammerhead structure, (−) sTRSV self-cleaves by an entirely different structure, which has been called a hairpin (101) or paperclip (101a). More recently, a 300-nucleotide satellite RNA of arabis mosaic virus (sArMV), which shows significant sequence homology (50%) to sTRSV, has been described, and it is predicted to show the same two self-cleavage mechanisms (18, 19, 102).

Self-Cleavage of Minus sTRSV

The self-cleavage of both (+) and (−) sTRSV was discovered by Bruening and his colleagues, who showed that the cleavage products contain 5'-hydroxyl and 2', 3'-cyclic phosphate termini (19, 30, 31, 81). As predicted from the sequences at the 5'- and 3'-ends of natural, linear (+) sTRSV, self-cleavage occurs via a hammerhead structure (19, 24, 41) and is not considered further here.

(−) sTRSV transcripts of cDNA clones self-cleave between residues 49 and 48 (18, 19, 31; Figure 11) with the usual practice of the minus numbering being the same as the plus and thus running in the 3'- to 5'-direction. Linker insertion mutagenesis was used to define the two separate regions in (−) sTRSV transcripts that are necessary for self-cleavage (82). These correspond to nucleotides 52 to 43 and 226 to 172 (Figure 11A), whereas deletion analysis by Feldstein et al (103) further reduced the sequence requirements to nucleotides 52 to 43 and 222 to 177. An important difference between the plus and minus self-cleavage events of sTRSV is that (−) sTRSV self-cleavage is readily reversible (19, 31), which contrasts with the negligible religation of (+) sTRSV self-cleavage products and of other RNAs that self-cleave via a hammerhead structure (32, 41, 81).

The transesterification step of self-cleavage of (−) sTRSV when the R_p diastereoisomer of a thiophosphodiester bond was present at the cleavage site led to the inversion of configuration of the P-S bond from exo to endo, as also found for (+) sTRSV (19, 65, 69; Figure 9). Likewise, the results indicate transesterification occurring by an in-line $S_N2(P)$ attack by the 2'-hydroxyl on the phosphorus (19, 65, 69).

A significant difference between (+) and (−) sTRSV is that a phosphorothioate at the self-cleavage site strongly inhibited self-cleavage of the (+) sTRSV, but not of (−) sTRSV, in the presence of Mg^{2+} (66), indicating that the Mg^{2+} in the (−) sTRSV self-cleavage structure is not coordinated to the pro-R oxygen (19). Further, the phosphorothioate in the 2', 3'-cyclic phosphate of the (−) sTRSV self-cleavage fragment showed little inhibition of the religation step (66). This religation reaction for the normal phosphodiester linkage to form either circular or linear molecules had a very low requirement for Mg^{2+}, as it could occur in the presence of the chelator EDTA (66).

Self-Cleavage in Trans *and Mutagenesis*

The nucleotide sequences required for self-cleavage of $(-)$ sTRSV are from two well-separated regions of the RNA molecule (Figure 11A), and the use of these two regions for in *trans* reactions was exploited by Feldstein et al (103) to define the minimum length of 10 nucleotides for the substrate and 46 nucleotides for the ribozyme. A 50-nucleotide ribozyme showed true catalytic activity with a 14-nucleotide substrate with a k_{cat} of 2.1/minute, a K_m for substrate of 0.03 μM, and an energy of activation of 19 kcal/mol (104); these values are of the same order of magnitude as those found for in *trans* reactions via the hammerhead structure (36, 45; Table 2).

The $(-)$ sTRSV self-cleavage domain (Figure 11) lends itself to the same mutational analysis as done for the hammerhead structure (50, 51; Figure 7),

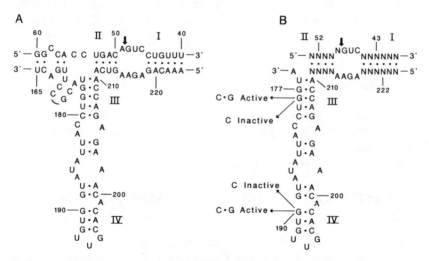

Figure 11   Regions of $(-)$ sTRSV involved in self-cleavage at site indicated by arrow (101, 103, 104). *A*. Predicted two-dimensional structure around self-cleavage site. Residue numbers are derived from those of the plus RNA and hence run in the 3'- to 5'-direction. Four base-paired regions, I to IV, are indicated. *B*. Designated nucleotides considered essential for self-cleavage in *trans* of two fragments as determined by deletion and site-directed mutagenesis. N, any nucleotide, providing base pairing maintained where indicated. Four site-directed mutations, which lead to either active or inactive self-cleaving structures, are indicated. Reproduced with permission from (17).

but so far such analysis has not been as extensive. Elimination of any of the four base pairs in stem II (Figure 11A) gave loss of self-cleavage, but this could be restored by compensatory changes to restore base pairing (19, 105). A more detailed mutagenesis analysis by Hampel et al (101) has provided more data for both the substrate and ribozyme domains; these data are summarized with all other data in Figure 11B. Only the seven nucleotides defined between stems I and II were required in this region for activity, while the only requirement for stems I and II was base pairing. However, large variations in k_{cat} and K_m occurred as a function of the base-paired sequences in stems I and II flanking the single-stranded NGUC.

More limited mutagenesis has been carried out on stem III and its hairpin loop containing stem IV (101). The base pair of G178:C208 could be reversed without loss of ribozyme activity but not when it was removed by conversion of G178 to C, indicating a requirement for stem III. Similar results were obtained for G190:C208 in stem IV. Insertion of a 10-mer linker between G206 and A207 or of a 9-mer linker between U191 and G192, however, did not inhibit self-cleavage (82). More detailed mutagenesis is required to define the sequence and structural requirements in this region.

The mutagenesis data with self-cleavage in *trans* have therefore defined the requirement for two regions, one including stems I and II and the other including stems III and IV and associated nucleotides, with the two regions being joined by two single-stranded nucleotides (Figure 11). Hence, the prediction of an active self-cleavage structure is more difficult than the case of the hammerhead structure (Figures 3, 8). In the model of Hampel et al (101), these two regions are colinear in what they call a hairpin catalytic model. This rod-like structure seems unlikely since interaction between the two regions to form a three-dimensional structure is likely to be necessary for self-cleavage; such a structure would be like a paperclip, a model name introduced by Bruening and his colleagues (101a).

HEPATITIS DELTA VIRUS PLUS AND MINUS RNAs SELF-CLEAVE

Hepatitis delta virus (HDV) is a satellite virus of hepatitis B virus. It contains a covalently closed, single-stranded RNA of about 1700 nucleotides, and it can be folded into an unbranched, rod-like secondary structure similar to that of viroids and virusoids (20, 106–111). A further viroid-like feature is the presence of a sequence highly conserved between sequence variants, which is cross-linked on UV-irradiation (110, 111), as has been shown also for potato spindle tuber viroid (111, 112), and is indicative of an activated tertiary structural element.

HDV can be considered as a negative-strand RNA, satellite virus since the encapsidated RNA does not code for protein; the complementary or plus RNA contains one coding region of 195 amino acids for the HDV antigen within the viral particle (106, 108, 109). All evidence indicates that HDV RNA is replicated by the rolling circle mechanism of Figure 2A (108, 110, 111). Both plus and minus HDV RNAs can self-cleave in vitro to produce products with 5'-hydroxyl and 2', 3'-cyclic phosphate (20, 113-115). There is significant conservation of sequence around the self-cleavage sites, which indicates a similar active tertiary structure for each RNA.

Minimum sequence requirements for self-cleavage of the HDV RNAs and mutational analyses that are being carried out are likely to soon determine the nucleotides in both plus and minus RNA that are necessary for self-cleavage, and to allow the prediction and testing of secondary structure models. Kuo et al (114) have defined a contiguous sequence of 104 nucleotides in the genomic minus RNA that can self-cleave, with 30 nucleotides on the 5'-side and 74-nucleotides on the 3'-side of the cleavage site, while Wu et al (116) have defined a stretch of no more than 117 nucleotides and predicted a secondary structure for this RNA consisting of three hairpin loops in a cloverleaf arrangement. Using an alkaline degradation method (42), Perrotta & Been (117) have reduced the minimum requirement for the same RNA to 82 nucleotides, with evidence that only one to three nucleotides are required 5' to the self-cleavage site as compared to three in the 52-nucleotide hammerhead structure of the plus RNA of the virusoid vLTSV (42; Figure 3A) and four nucleotides for the hairpin/paperclip structure of (−) sTRSV (Figure 11B). Similar pseudoknot-like structures have been proposed for the 80 to 85 nucleotides on the 3'-side of the self-cleavage site of both plus and minus RNAs (118, 118a), which are significantly different from the cloverleaf structure proposed by Wu et al (116) for the minus RNA.

In a study of the conformational requirements for self-cleavage of (−) HDV RNA, Wu & Lai (119) used fragments of 94 to 225 nucleotides long and found that cleavage was usually complete within seconds after the addition of Mg^{2+}, irrespective of the final extent of cleavage. The results indicated that only a small proportion of the RNA molecules were in an active conformation, and these self-cleaved rapidly; the remainder were in an inactive conformation. A similar situation has been extensively characterized for the viroid ASBV and the virusoid vLTSV (15–17, 34, 38, 42, 43). Repeated cycles of denaturation and renaturation increased the extent of self-cleavage of small fragments of (−) HDV RNA (119), as did various treatments with urea and formamide for both the plus and minus RNAs (117, 120–122). All these results indicate that the various RNAs can exist in a variety of conformations and that conformational changes must occur to form an active self-cleavage structure.

Recently, Branch & Robertson (122a) proposed an axehead structural model for the self-cleaving domains of genomic minus and antigenomic plus HDV RNAs (Figure 12). Although these two-dimensional structures are variants of those previously published (116–118, 118a), they have uniquely included the highly conserved sequences between the plus and minus RNAs (boxed in Figure 12) in a simple and continuous structure. The approximately 80% sequence conservation between the two RNAs that are derived from separate areas of the HDV genome provides strong support for the correctness of the axehead model. A similar situation exists for hammerhead self-cleavage domains where there is 75–80% sequence conservation for (+) and (−) vLTSV (41) and approximately 65% for (+) and (−) ASBV (32).

The axehead structures with their three base-paired stems (I, II, and III) around an open region of single-stranded RNA (Figure 12) are reminiscent of a similar arrangement in the hammerhead self-cleavage structures (15–18). The requirement for stem III is uncertain since deletion analyses have indicated that a minimum of only one to three nucleotides are required in addition to those available (117, 118, 118a, 118b, 122) to define precisely the sequence and structural requirements for maximal self-cleavage activity.

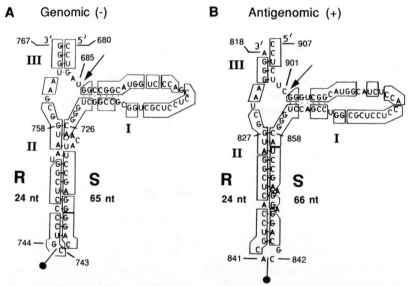

Figure 12   Common structural axehead motifs for the genomic (−) (*A*) and antigenomic (+) (*B*) HDV RNAs (122a). Conserved residues between the plus and minus RNAs are boxed and the site of self-cleavage in each RNA is indicated by an arrow. The three base-paired stems are labeled I, II, and III. Sequences and numbering are from Wang et al (108, 109). For the *trans* reactions, ribozyme R of 24 nt for each RNA and the substrate S of 65 nt or 66 nt are shown with the separation between R and S at the bottom of each structure indicated by knobbed bar. Modified from (122a) and reproduced with permission.

The arrangement of the axehead stems led Branch & Robertson (122a) to develop efficient *trans* cleavage reactions for both plus and minus RNAs by dividing each structure in the hairpin loop of stem II to give 24-nt ribozymes R and either 65-nt or 66-nt substrates S (Figure 12). As in the case of the hammerhead and the hairpin/paperclip structures (Figures 6, 11), such *trans* reactions will allow a detailed kinetic analysis of the cleavage reaction with the systematic variation of sequence in the ribozyme and the substrate.

RNAs FROM *NEUROSPORA* MITOCHONDRIAL DNA PLASMIDS SELF-CLEAVE

The natural Varkud-1c strain of *Neurospora* contains two abundant mitochondrial RNAs, in addition to the normal two mitochondrial ribosomal RNAs, of about 800 nucleotides each (123). These are the circular and linear forms of the same single-stranded RNA, and are derived from circular, double-stranded mitochondrial plasmid DNA, which exists as a series of head-to-tail multimers. The plasmid DNA is called VSDNA and the RNAs, VSRNAs. The multimeric organization for this DNA is typical of that found for other *Neurospora* mitochondrial plasmid DNAs (123).

When plus-strand VSRNA was produced by T7 RNA polymerase transcription of a cloned VSDNA monomer, some of the RNA self-cleaved to two specific fragments with 5'-hydroxyl and 2', 3'-cyclic phosphate termini, as did the full-length transcript when incubated in the presence of Mg^{2+} (23). The site of in vitro self-cleavage was identical to the termini of linear VSRNA isolated from mitchondria. Since there is minimal, if any, sequence homology of this RNA with any of the other self-cleaving RNAs, the VSRNA most likely provides a fourth type of self-cleavage structure (Table 1). Deletion and mutational analyses of the VSRNA are needed to define the minimum sequence requirements for self-cleavage, a possible two-dimensional structure around the self-cleavage site, and the potential for in *trans* reactions.

HOW COMMON ARE SELF-CLEAVAGE REACTIONS IN NATURE?

The 11 naturally occurring RNAs listed in Table 1 self-cleave via four different structures in a reaction that is considered for the plant and animal pathogenic RNAs to be an essential part of their replication pathway (15–20, 27, 111). The limited number of RNAs identified so far that are involved in this simple self-cleavage reaction indicates that many more exist in Nature that have yet to be discovered.

Probably the most prospective area to investigate is that of circular, pathogenic RNAs, since they are likely to be replicated by a rolling circle mechanism, which requires a specific cleavage event to release monomeric units from continuously synthesized RNA (Figure 2). Strong candidates are the plant viroids, among which self-cleavage has only been identified so far in ASBV (Table 1; 32). The 15 other viroids that have been sequenced have a number of sequence and structural features in common (24, 124, 125), and it is feasible that a common self-cleavage structure could be involved. Limited attempts using potato spindle tuber viroid have failed to identify a self-cleavage reaction (126, 127), but a more comprehensive search is clearly warranted.

Lead-Catalyzed Cleavage of Yeast Phe-tRNA

In this reaction, cleavage occurs in the D-loop of yeast phe-tRNA between residues D17 and G18 to give termini with a 5'-hydroxyl and a 2', 3'-cyclic phosphate, a reaction that has been well characterized and reviewed (37, 128–130). The specificity of the reaction is provided by the precise coordination of the lead ion in a pocket formed by eight residues of the tRNA (129). Since the pH optimum for cleavage is about 7, it is considered likely that the hydrated lead ion, $Ph(OH)+$, is the active species, abstracting the proton from the 2'-hydroxyl group of D17, thus allowing the $2'-O^-$ to attack the phosphorus atom in the phosphate between D17 and G18 (129).

Although this reaction is unlikely to be of biological significance, it has been viewed as a model for understanding the self-cleavage reactions of the RNAs considered here, in which similar Mg^{2+}-catalyzed transesterification reactions take place (4, 37, 130). However, for self-cleavage, the proton donor and the proton acceptor in the general acid-base catalyzed reaction have not been identified in any of the four types of reactions described (Table 1).

A Simple Mn^{2+}-Catalyzed Reaction

For a study of divalent cation–catalyzed self-cleavage of RNA, Dange et al (131) incorporated the first 15 nucleotides of the *Tetrahymena* ribosomal RNA intron released during self-splicing (2–6) into the two simple oligonucleotides of Figure 13; the 15 nucleotides correspond to nucleotides G13 to U27. Since these 15 nucleotides are released during the first of two cyclization events of the self-spliced intron (2–6), it was reasoned that its hairpin structure (I in Figure 13) could have special properties. The interest in Mn^{2+} was based on its ability to replace Mg^{2+} in many RNA processing reactions (2, 6, 132). Mn^{2+} can also bind to RNA through phosphate oxygens but, unlike Mg^{2+}, can form complexes with RNA bases (128) and thus contribute to the formation of an active tertiary structure.

In the presence of 10 mM $MnCl_2$ at pH 7.5 for two hours, self-cleavage occurred in both RNAs between G13 and A14 with the production of two fragments with 5'-hydroxyl and 2', 3'-cyclic phosphate termini; a range of other divalent cations were inactive (131). There was a narrow pH range for specific cleavage, removal of hairpin II did not affect cleavage (Figure 13B), and specific cleavage also occurred when the oligonucleotide was incorporated at the 5'-end of a 231-nucleotide RNA transcript. Mutation of A28 to U, but not G or C, eliminated self-cleavage, thus implicating a role for A28 in the reaction. Dange et al (131) considered the reaction to be Mn(OH)+ catalyzed, in a way similar to that described for the Pb(OH)+-catalyzed cleavage of phe-tRNA (128, 129).

This simple reaction emphasizes the potential for many other similar divalent metal ion–catalyzed self-cleavage reactions. As Dange et al (131) suggest, such reactions in cells may be used to control the concentrations of specific RNA substances and to have a role in the control of gene expression at the level of transcription. Unlike Pb^{2+}, Mn^{2+} does play an important role in vivo as a component of many enzymes (133).

Are in Trans Cleavage Reactions Common in Vivo?

All the in *trans* self-cleavage reactions described here have involved in vitro modification and manipulation of naturally occurring RNAs. The high probability that more types of self-cleavage will be discovered, in addition to the four already identified (Table 1), enhances the possibility that in *trans* reactions may play an important role in vivo. For example, modulation of the level of one mRNA may be effected by in *trans* cleavage by another RNA, with the extent of cleavage being determined by its concentration. Another possibility is that the small viroids, which appear incapable of coding for proteins (24, 134), could initiate and maintain the pathogenic state in plants by cleavage in *trans* of one or more host RNAs.

On the basis of the conserved sequence and structural requirements of the

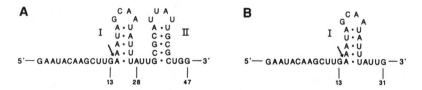

Figure 13   A Mn^{2+}-dependent self-cleavage reaction (131). Sequence and proposed structure of two oligonucleotides A and B, which show a Mn^{2+}-specific self-cleavage at the sites indicated.

hammerhead structure, Ruffner et al (51) calculated that one hammerhead structure would be expected every 11,000 nucleotides, while a search of all *E. coli* sequences in Genebank release 48 identified many potential hammerheads. As more self-cleavage mechanisms are characterized, the possibility increases that these reactions in *trans* may play a role in normal cellular processes and in pathogenic responses in plants and animals.

Literature Cited

1. Kruger, K., Grabowski, P. J., Zaug, A. J., Sands, J., Gottschling, D. E., Cech, T. R. 1982. *Cell* 31:147–57
2. Cech, T. R., Bass, B. L. 1986. *Annu. Rev. Biochem.* 55:599–629
3. Cech, T. R. 1985. *Int. Rev. Cytol.* 93:3–22
4. Cech, T. R. 1987. *Science* 236:1532–39
5. Cech, T. R. 1990. *Annu. Rev. Biochem.* 59:543–68
6. Latham, J. A., Zaug, A. J., Cech, T. R. 1990. *Methods Enzymol.* 181:558–69
7. Guerrier-Takada, C., Gardiner, K., Marsh, T., Pace, N., Altman, S. 1983. *Cell* 35:849–57
8. Altman, S. 1987. *Adv. Enzymol.* 62:1–36
9. Altman, S. 1990. *J. Biol. Chem.* 265:20053–56
10. Forster, A. C., Altman, S. 1990. *Science* 249:783–86
11. Baer, M. F., Arnez, J. G., Guerrier-Takada, C., Vioque, A., Altman, S. 1990. *Methods Enzymol.* 181:569–82
12. Pace, N. R., Smith, D. 1990. *J. Biol. Chem.* 265:3587–90
13. Jacquier, A. 1990. *Trends Biochem. Sci.* 15:351–54
14. Michel, F., Umesono, K., Ozeki, H. 1989. *Gene* 82:5–30
15. Symons, R. H. 1989. *Trends. Biochem. Sci.* 14:445–50
16. Symons, R. H. 1990. *Semin. Virol.* 1:117–26
17. Symons, R. H. 1991. *Crit. Rev. Plant. Sci.* 10:189–234
18. Bruening, G. 1989. *Methods Enzymol.* 180:546–58
19. Bruening, G. 1990. *Semin. Virol.* 1:127–34
20. Taylor, J. 1990. *Semin. Virol.* 1:135–41
21. Epstein, L. M., Gall, J. G. 1987. *Cell* 48:535–43
22. Epstein, L. M., Pabon-Pena, L. M. 1991. *Nucleic Acids Res.* 19:1699–705
23. Saville, B. J., Collins, R. A. 1990. *Cell* 61:685–96
24. Keese, P., Symons, R. H. 1987. In *Viroids and Viroid-like Pathogens*, ed. J. S. Semancik, pp. 1–47. Boca Raton: CRC Press
25. Branch, A. D., Robertson, H. D. 1984. *Science* 223:450–55
26. Branch, A. D., Benenfeld, B. J., Robertson, H. D. 1988. *Proc. Natl. Acad. Sci. USA* 85:9128–32
27. Robertson, H. D., Branch, A. D. 1987. See Ref. 24, pp. 49–70
28. Hutchins, C. J., Keese, P., Visvader, J. E., Rathjen, P. D., McInnes, J. L., Symons, R. H. 1985. *Plant. Mol. Biol.* 4:293–304
29. Diener, T. O. 1979. *Viroids and Viroid Diseases.* New York:Wiley
30. Buzayan, J. M., Hampel, A., Bruening, G. 1986. *Nucleic Acids Res.* 14:9729–43
31. Buzayan, J. M., Gerlach, W. L., Bruening, G. 1986. *Nature* 323:349–53
32. Hutchins, C. J., Rathjen, P. D., Forster, A. C., Symons, R. H. 1986. *Nucleic Acids Res.* 14:3627–40
33. Symons, R. H., Hutchins, C. J., Forster, A. C., Rathjen, P. D., Keese, P., Visvader, J. E. 1987. *J. Cell Sci. Suppl.* 7:303–18
34. Sheldon, C. C., Jeffries, A. C., Davies, C., Symons, R. H. 1990. *Nucleic Acids Mol. Biol.* 4:227–42
35. Bruening, G., Buzayan, J. M., Hampel, A., Gerlach, W. L. 1988. In *RNA Genetics II*, ed. E. Domingo, J. J. Holland, P. Ahlquist, pp. 127–45. Boca Raton: CRC Press
36. Uhlenbeck, O. C. 1987. *Nature* 328:596–600
37. Sampson, J. R., Sullivan, F. X., Behlen, L. S., DiRenzo, A. B., Uhlenbeck, O. C. 1987. *Cold Spring Harbor Symp. Quant. Biol.* 52:267–75
38. Forster, A. C., Davies, C., Sheldon, C. C., Jeffries, A. C., Symons, R. H. 1988. *Nature* 334:265–67

39. Davies, C., Sheldon, C. C., Symons, R. H. 1991. *Nucleic Acids Res.* 19:1893–98
40. Sheldon, C. C., Symons, R. H. 1989. *Nucleic Acids Res.* 17:5665–77
41. Forster, A. C., Symons, R. H. 1987. *Cell* 49:211–20
42. Forster, A. C., Symons, R. H. 1987. *Cell* 50:9–16
43. Forster, A. C., Jeffries, A. C., Sheldon, C. C., Symons, R. H. 1987. *Cold Spring Harbor Symp. Quant. Biol.* 52: 249–59
44. Davies, C., Haseloff, J., Symons, R. H. 1990. *Virology* 177:216–24
45. Jeffries, A. C., Symons, R. H. 1989. *Nucleic Acids Res.* 17:1371–77
46. Herschlag, D., Cech, T. R. 1990. *Biochemistry* 29:10159–71
47. Koizumi, M., Iwai, S., Ohtsuka, E. 1988. *FEBS Lett.* 239:285–88
48. Koizumi, M., Hayase, Y., Iwai, S., Kamiya, H., Inoue, H., Ohtsuka, E. 1989. *Nucleic Acids Res.* 17:7059–71
49. Milligan, J. F., Uhlenbeck, O. C. 1989. *Methods Enzymol.* 180:51–62
50. Sheldon, C. C., Symons, R. H. 1989. *Nucleic Acids Res.* 17:5679–85
51. Ruffner, D. E., Stormo, G. D., Uhlenbeck, O. C. 1990. *Biochemistry* 29: 10695–702
52. Ruffner, D. E., Dahm, S. C., Uhlenbeck, O. C. 1989. *Gene* 82:31–41
53. Miller, W. A., Hercus, T., Waterhouse, P. M., Gerlach, W. L. 1991. *Virology* 183:711–20
54. Epstein, L. M., Gall, J. G. 1987. *Cold Spring Harbor Symp. Quant. Biol.* 52:261–65
55. Fedor, M. J., Uhlenbeck, O. C. 1990. *Proc. Natl. Acad. Sci. USA* 87:1668–72
56. Heus, H. A., Uhlenbeck, O. C., Pardi, A. 1990. *Nucleic Acids Res.* 18:1103–8
57. Pease, A. C., Wemmer, D. E. 1990. *Biochemistry* 29:9039–46
58. Odai, O., Kodama, H., Hiroaki, H., Sakata, T., Tanaka, T., Uesugi, S. 1990. *Nucleic Acids Res.* 18:5955–60
59. Heus, H. A., Pardi, A. 1991. *J. Mol. Biol.* 217:113–24
60. Deleted in proof
61. Mei, H-Y., Kaaret, T. W., Bruice, T. C. 1989. *Proc. Natl. Acad. Sci. USA* 86:9727–31
62. Eckstein, F. 1985. *Annu. Rev. Biochem.* 54:367–402
63. Griffiths, A. D., Potter, B. V. L., Eperon, I. C. 1987. *Nucleic Acids Res.* 15:4145–62
64. Buzayan, J. M., Feldstein, P. A., Segrelles, C., Bruening, G. 1988. *Nucleic Acids Res.* 16:4009–23
65. Van Tol, H., Buzayan, J. M., Feldstein, P. A., Eckstein, F., Bruening, G. 1990. *Nucleic Acids Res.* 18:1971–75
66. Buzayan, J. M., Feldstein, P. A., Bruening, G., Eckstein, F. 1988. *Biochem. Biophys. Res. Commun.* 156: 340–47
67. Dahm, S. C., Uhlenbeck, O. C. 1991. *Biochemistry* 30:9464–69
68. Koizumi, M., Ohtsuka, E. 1991. *Biochemistry* 30:5145–50
69. Slim, G., Gait, M. J. 1991. *Nucleic Acids Res.* 19:1183–88
70. Jaffe, E. K., Cohn, M. 1979. *J. Biol. Chem.* 254:10839–45
71. Pecoraro, V. L., Hermes, J. D., Cleland, W. W. 1984. *Biochemistry* 23:5262–71
72. Buzayan, J. M., Van Tol, H., Feldstein, P. A., Bruening, G. 1990. *Nucleic Acids Res.* 18:4447–51
73. Ruffner, D. E., Uhlenbeck, O. C. 1990. *Nucleic Acids Res.* 18: 6025–29
74. Perreault, J-P., Wu, T., Cousineau, B., Ogilvie, K. K., Cedergren, R. 1990. *Nature* 344:565–67
75. Yang, J-H., Perreault, J-P., Labuda, D., Usman, N., Cedergren, R. 1990. *Biochemistry* 29:11156–60
76. Perreault, J-P., Labuda, D., Usman, N., Yang, J-H., Cedergren, R. 1991. *Biochemistry* 30:4020–25
77. Pieken, W. A., Olsen, D. B., Benseler, F., Aurup, H., Eckstein, F. 1991. *Science* 253:314–17
78. Zaug, A. J., Been, M. D., Cech, T. R. 1986. *Nature* 324:429–33
79. Forster, A. C., Davies, C., Hutchins, C. J., Symons, R. H. 1990. *Methods Enzymol.* 181:583–607
80. Haseloff, J., Gerlach, W. L. 1988. *Nature* 334:585–91
81. Prody, G. A., Bakos, J. T., Buzayan, J. M., Schneider, I. R., Bruening, G. 1986. *Science* 231:1577–80
82. Haseloff, J., Gerlach, W. L. 1989. *Gene* 82:43–52
83. Lamb, J. W., Hay, R. T. 1990. *J. Gen. Virol.* 71:2257–64
84. Cotten, M., Schaffner, G., Birnstiel, M. L. 1989. *Mol. Cell. Biol.* 9:4479–87
85. Helene, C., Toulme, J-J. 1990. *Biochim. Biophys. Acta* 1049:99–125
86. Takayama, K. M., Inouye, M. 1990. *Crit. Rev. Biochem. Mol. Biol.* 25:155–84
87. Uhlmann, E., Peyman, A. 1990. *Chem. Rev.* 90:543–84

88. Cotten, M. 1990. *Trends Biotechnol.* 8:174–78
89. Rossi, J. J., Sarver, N. 1990. *Trends Biotechnol.* 8:179–83
90. Bass, B. L., Weintraub, H. 1987. *Cell* 48:607–13
91. Bass, B. L., Weintraub, H. 1988. *Cell* 55:1089–98
92. Wagner, R. W., Smith, J. E., Cooperman, B. S., Nishikura, K. 1989. *Proc. Natl. Acad. Sci. USA* 86:2647–51
92a. Nishikura, K., Yoo, C., Kim, U., Murray, J. M., Estes, P. A., et al. 1991. *EMBO J.* 10:3523–32
93. Cameron, F. H., Jennings, P. A. 1989. *Proc. Natl. Acad. Sci. USA* 86:9139–43
94. Cameron, F. H., Jennings, P. A. 1991. *Nucleic Acids Res.* 19:469–75
95. Sarver, N., Cantin, E. M., Chang, P. S., Zaia, J. A., Ladne, P. A., et al. 1990. *Science* 247:1222–25
96. Cotten, M., Birnstiel, M. L. 1989. *EMBO J.* 8:3861–66
97. Saxena, S. K., Ackerman, E. J. 1990. *J. Biol. Chem.* 265:17106–09
98. Chuat, J-C., Galibert, F. 1989. *Biochem. Biophys. Res. Commun.* 162:1025–29
99. Buzayan, J. M., Gerlach, W. L., Bruening, G., Keese, P., Gould, A. R. 1986. *Virology* 151:186–99
100. Bruening, G., Passmore, B. K., Van Tol, H., Buzayan, J. M., Feldstein, P. A. 1991. *Mol. Plant Microbe Interact.* 4:219–25
101. Hampel, A., Tritz, R., Hicks, M., Cruz, P. 1990. *Nucleic Acids Res.* 18:299–304
101a. Feldstein, P. A., Buzayan, J. M., van Tol, H., Bruening, G. 1991. *Abstr. Meet. RNA Processing, May 15–19, Cold Spring Harbor Lab., New York*, p. 298.
102. Kaper, J. M., Tousignant, M. E., Steger, G. 1988. *Biochem. Biophys. Res. Commun.* 154:318–25
103. Feldstein, P. A., Buzayan, J. M., Bruening, G. 1989. *Gene.* 83:53–61
104. Hampel, A., Tritz, R. 1989. *Biochemistry* 28:4929–33
105. Feldstein, P. A., Buzayan, J. M., Van Tol, H., deBear, J., Gough, G. R., et al. 1990. *Proc. Natl. Acad. Sci. USA* 87:2623–27
106. Taylor, J. M. 1990. *Cell* 61:371–73
107. Makino, S., Chang, M. F., Shieh, C. K., Kamahora, T., Vannier, D. M., et al. 1987. *Nature* 329:343–46
108. Wang, K. S., Choo, Q. L., Weiner, A. J., Ou, J. H., Najarian, R. C., et al. 1986. *Nature* 328:456
109. Wang, K. S., Choo, Q. L., Weiner, A. J., Ou, J. H., Najarian, R. C., et al. 1986. *Nature* 323:508–14
110. Branch, A. D., Benenfeld, B. J., Baroudy, B. M., Wells, F. V., Gerin, J. L., Robertson, H. D. 1989. *Science* 243:649–52
111. Branch, A. D., Benenfeld, B. J., Baroudy, B. M., Wells, F. V., Gerin, J. L., Robertson, H. D. 1990. *Semin. Virol.* 1:143–52
112. Branch, A. D., Benenfeld, B. J., Robertson, H. D. 1985. *Proc. Natl. Acad. Sci. USA* 82:6590–94
113. Branch, A. D., Benenfeld, B. J., Paul, C. P., Robertson, H. D. 1989. *Methods Enzymol.* 180:418–42
114. Kuo, M. Y.-P., Sharmeen, L., Dinter-Gottlieb, G., Taylor, J. 1988. *J. Virol.* 62:4439–44
115. Sharmeen, L., Kuo, M. Y. P., Dinter-Gottlieb, G., Taylor, J. 1988. *J. Virol.* 62:2674–79
116. Wu, H-N., Lin, Y-J., Lin, F-P., Makino, S., Chang, M-F., Lai, M. M. C. 1989. *Proc. Natl. Acad. Sci. USA* 86:1831–35
117. Perrotta, A. T., Been, M. D. 1990. *Nucleic Acids Res.* 18:6821–27
118. Perrotta, A. T., Been, M. D. 1991. *Nature* 350:434–36
118a. Rosenstein, S. P., Been, M. D. 1991. *Nucleic Acids Res.* 19:5409–16
118b. Thill, G., Blumenfeld, M., Lescure, F., Vasseur, M. 1991. *Nucleic Acids Res.* 19:6519–25
119. Wu, H-N., Lai, M. M. C. 1990. *Mol. Cell. Biol.* 10:5575–79
120. Rosenstein, S. P., Been, M. D. 1990. *Biochemistry* 29:8011–16
121. Belinsky, M. G., Dinter-Gottlieb, G. 1991. *Nucleic Acids Res.* 19:559–64
122. Smith, J. B., Dinter-Gottlieb, G. 1991. *Nucleic Acids Res.* 19:1285–89
122a. Branch, A. D., Robertson, H. D. 1991. *Proc. Natl. Acad. Sci. USA* 88:10163–67
123. Stohl, L. L., Collins, R. A., Cole, M. D., Lambowitz, A. M. 1982. *Nucleic Acids Res.* 10:1439–57
124. Koltunow, A. M., Rezaian, M. A. 1990. *Intervirology* 30:194–201
125. Symons, R. H. 1991. *Mol. Plant Microbe Interact.* 4:111–21
126. Robertson, H. D., Rosen, D. L., Branch, A. D. 1985. *Virology* 142:441–47
127. Tsagris, M., Tabler, M., Sänger, H. L. 1987. *Virology* 157:227–31
128. Jack, A., Ladner, J. E., Rhodes, D.,

Brown, R. S., Klug, A. 1977. *J. Mol. Biol.* 111:315–28
129. Brown, R. S., Dewan, J. C., Klug, A. 1985. *Biochemistry* 24:4785–801
130. Behlen, L. S., Sampson, J. R., DiRenzo, A. B., Uhlenbeck, O. C. 1990. *Biochemistry* 29:2515–23
131. Dange, V., Van Atta, R. B., Hecht, S. M. 1990. *Science* 248:585–88
132. Guerrier-Takada, C., Haydock, K.,

Allen, L., Altman, S. 1986. *Biochemistry* 25:1509–15
133. Scott, T., Eagleson, M., eds. 1988. *Concise Encyclopedia Biochemistry*, pp. 357–58. Berlin/New York: de Gruyter. 2nd ed.
134. Sänger, H. L. 1987. In *The Viroids*, ed. T. O. Diener, pp. 117–66. New York: Plenum

Annu. Rev. Biochem. 1992. 61:673–719

PROKARYOTIC DNA REPLICATION

Kenneth J. Marians

Program in Molecular Biology, Sloan-Kettering Institute, Memorial Sloan-Kettering Cancer Center, New York, NY 10021

KEY WORDS: DNA replication, primosomes, DNA helicases, DNA polymerases, DNA-binding proteins

CONTENTS

PERSPECTIVES AND SUMMARY

Biochemical and genetic analyses have revealed that the action of at least 30 proteins is required to replicate the chromosome of *Escherichia coli*. Much of the effort in this field during the past two decades has involved the identification of genes required during cellular DNA replication, the development of replication systems in vitro that could be used to study the enzymatic mechanisms of replication, and an intensive focus on the enzymology of the replication proteins. These efforts have illuminated the basic activities re-

673

quired and provided detailed information on how individual replication proteins function.

Among the more recent questions asked are: What does the replication fork look like? How do protein-protein interactions between components influence how the replication fork, which in *E. coli* moves at roughly 1000 nucleotides (nt) per second at 37 °C [based on a genome size of 4.7×10^6 base pairs (bp) (1) and 40 min for the replication of one chromosome (2)], functions? How is cellular DNA replication keyed to the cell cycle? And how do replication proteins modulate the secondary and tertiary structure of the DNA to both respond to and send signals?

Prokaryotic DNA replication was last reviewed in these pages 8 years ago (3). Thus, there are many detailed reviews available that should be consulted along with this one. These include specific reviews on DNA helicases (4), DNA topoisomerases (5), single-stranded DNA-binding proteins [SSB(s)] (6, 7), and the *E. coli* DNA polymerase III holoenzyme (Pol III HE) (8), as well as those that consider prokaryotic DNA replication per se (9, 10) and protein-primed DNA replication (11). In addition, 1991 heralded the appearance of a revised *DNA Replication* by Kornberg & Baker (12).

INTRODUCTION TO REPLICATION OPERATIONS

Two problems must be solved during the initiation of DNA replication on a duplex parental template: (*a*) The correct starting point must be identified, and (*b*) since all the enzymatic components of the replication fork act on a single-stranded (ss) DNA template, the origin region must be denatured to allow formation of the replication fork. In this review, two solutions to these problems are considered (Figure 1).

The most general solution to identifying the origin and effecting the subsequent localized denaturation required is via the association of a sequence-specific duplex DNA-binding protein with a series of direct repeats representing its cognate binding sequence to form a nucleoprotein complex at the origin (Figure 1*A*, ii). This scheme is exemplified by the binding of the DnaA protein during the replication of minichromosomes carrying the *E. coli* chromosomal origin (*oriC*) (13) and by the binding of the bacteriophage λO protein to the λ origin (*ori*λ) (14). These nucleoprotein complexes, referred to as specialized nucleoprotein structures (snups) (15), generally consist of 150–250 bp of DNA and multimers of the DNA-binding proteins. Subsequently, either because of the direct action of the bound initiator protein, an inherent instability in the DNA sequence that is exacerbated by the binding of these proteins, or a combination of these events, an A+T-rich region of DNA directly adjacent to the snup becomes denatured (and presumably coated with SSB) (Figure 1*A*, iii). By so marking the origin region and providing available

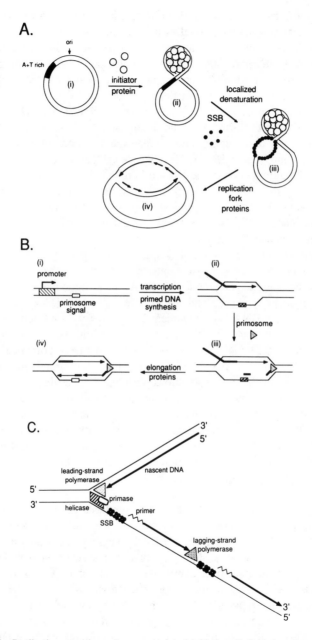

Figure 1   Replication operations. *A*. snup-catalyzed initiation. *B*. Initiation via transcription-primed DNA synthesis. *C*. The replication fork.

single strands, the replication fork proteins can be recruited to the correct initiation point and nascent strand synthesis can ensue (Figure 1A, iv).

In a variation on this theme, during initiation of replication of the small *E. coli* plasmids ColE1 and pBR322 (16), and also presumably during primary initiation of the replication of bacteriophages T7 (17) and T4 (18), replication signals that are functional only in the single-stranded conformation are exposed and activated by a strand displacement reaction (Figure 1B). Nascent DNA synthesis, often primed by the action of RNA polymerase (RNAP), which copies only one template strand, creates, either transiently or in a stable fashion, a D-loop (Figure 1B, ii). The displaced single strand contains signals that allow the introduction to the DNA of a mobile helicase/primase that can simultaneously drive the replication fork forward and synthesize primers for Okazaki fragment synthesis (Figure 1B, iii). A replication fork forms when the enzyme machines that were initially acting separately on the two template strands join together (Figure 1B, iv).

Once formed, a replication fork has four basic components (Figure 1C): the DNA polymerase required for nascent strand synthesis, the DNA helicase required to unwind the parental duplex DNA, the primase required to initiate Okazaki fragment synthesis, and the SSB required to coat exposed single-stranded template DNA. A fundamental asymmetry is introduced to the enzymatic requirements for DNA synthesis at the fork because of the anti-parallel nature of the template strands and the fact that DNA polymerases only synthesize DNA in the $5' \rightarrow 3'$ direction. Thus, one strand—the leading strand—can be made in a continuous fashion. The leading-strand DNA polymerase, if processive enough, need be introduced to the replication fork only once. On the other hand, the other strand—the lagging strand—is made discontinuously in small pieces roughly 2 kb in length (Okazaki fragments). Thus, a replication fork appears to require two distinct DNA polymerases, one highly processive and one moderately processive. As described within, only recently has an appreciation for how this might be accomplished been developed.

DNA polymerases cannot initiate DNA synthesis de novo. Unique enzymes, the primases, have therefore evolved to synthesize short ribonucleotide primers that are then utilized by the lagging-strand polymerase for Okazaki fragment synthesis. In the prokaryotic systems, a new Okazaki fragment must be initiated once every one or two seconds. To ensure that the primase has ready access to the template when and where it is needed, its association with the replication fork is mediated by the DNA helicase acting on the lagging-strand template to unwind the parental duplex in the $5' \rightarrow 3'$ direction. This mobile complex of helicase and primase has been termed a *primosome* (19). While the enzymatic components that form the T4 and T7 primosome on the DNA are clear, those of the *E. coli* primosome are a subject of current investigation.

Since initiation, synthesis, and completion of an Okazaki fragment occur very rapidly, the question arises as to the disposition of the lagging-strand DNA polymerase. Does the same DNA polymerase complex synthesize each and every Okazaki fragment made by the replication fork? This would require mechanisms to cycle the polymerase from completed Okazaki fragment to new primer terminus very rapidly and to ensure that when the polymerase releases the template after completion of the Okazaki fragment, it remains associated with the replication fork. Or is a new polymerase recruited from the pool in the cytoplasm for each new Okazaki fragment? At least in the case of the *E. coli* fork, it now seems clear that the former scenario applies.

What happens after the replication forks have completed nascent DNA synthesis? Our understanding of these events is less clear. Although a specific mechanism exists to arrest replication fork progression on the *E. coli* chromosome (20, 21), the trans-acting gene product required is dispensable (22). However, there does seem to be a defined order at the end of the replication cycle of termination and topological disengagement (segregation of the replicated daughter chromosomes) that triggers the synthesis of a new protein required for subsequent partition (physical separation of the daughter chromosomes at the quarters of the cell) (23, 24) before septum formation and cell division occur. Thus, it seems unlikely that the arrest of replication fork progression would be a haphazard event.

This review considers the initiation and elongation stages of prokaryotic DNA replication as well as the mechanism of replication fork arrest. Elongation is addressed first in order to introduce the replication fork proteins.

ELONGATION

The DNA Polymerase Complexes

At the replication fork, the DNA polymerase complex engaged in leading-strand DNA synthesis must be extremely processive, while its counterpart, engaged in lagging-strand DNA synthesis, is considerably less processive. In *E. coli,* the difference in the degree of processivity required is on the order of 1000-fold. Much of the focus on the replicative DNA polymerase complexes has been on how an enzyme containing the catalytic site for polymerization can be converted from an essentially nonprocessive enzyme (1–50 nt added per primer terminus binding event) to one with a processivity measured in megabases. The answer to the flip side of this question—How is processivity limited on the lagging-strand template?—has only recently begun to be developed. An associated issue addresses how an inherently very processive polymerase complex—one that likes to stay bound to the 3'-OH end of the nascent chain—is induced to leave the 3'-OH end of a completed Okazaki fragment and transit rapidly to the new primer for the next Okazaki fragment.

In *E. coli* and bacteriophages T4 and T7, the solution to generating a very processive polymerase complex is the same. Additional subunits (or accessory proteins) interact with the catalytic polymerase subunit to increase its affinity for the primer terminus. In *E. coli* and T4, energy is required to form the stable structure, whereas this is not the case with T7.

BACTERIOPHAGE T7 GENE 5 PROTEIN AND THIOREDOXIN   The product of gene 5 of bacteriophage T7 (T7 g5P) is a DNA polymerase that also possesses a $3' \rightarrow 5'$ proofreading exonuclease that is the trademark of a replicative polymerase (25). As purified originally from T7-infected *E. coli,* the T7 DNA Pol was shown to be a 1:1 complex of the T7 g5P (80 kDa) and the host thioredoxin (12 kDa, the product of *trxA*) (26, 27). (Unless stated otherwise, in this review the term T7 DNA Pol refers to the complex of the T7 g5P and thioredoxin.) Subsequent cloning and overproduction of the T7 g5P have allowed examination of its properties in the absence of TrxA.

The g5P alone has a limited processivity of only 1–50 nt. On the other hand, the processivity of the g5P-TrxA complex is measured in the thousands (28). Formation of the g5P-TrxA complex requires neither ATP nor a primer-template (28). This increase in processivity is accompanied by a 20- to 50-fold increase in the half-life of a g5P-TrxA complex on a primer-template and a 200-fold increase in the double-stranded (ds) DNA exonuclease activity of the complex compared to that of g5P alone (29). On the other hand, TrxA does not alter significantly the affinity of g5P for ssDNA. Thus, TrxA acts to clamp the g5P to the primer-template. The reducing activity of TrxA is not required for this effect (30, 31). Interestingly, since TrxA decreases the dissociation rate of g5P from the primer-template (29), its presence exacerbates the problem of how rapid cycling of the lagging-strand polymerase is achieved.

BACTERIOPHAGE T4 GENE 43 PROTEIN AND ITS ACCESSORY PROTEINS
The product of gene 43 of bacteriophage T4 (T4 g43P, 110 kDa) is a relatively nonprocessive DNA polymerase that is incapable of either strand displacement synthesis (32, 33) or efficient elongation of primer-templates where there are long stretches of ssDNA (34, 35). This type of primer-template could be replicated when the products of gene 45 (T4 g45P, 27 kDa), gene 44 (T4 g44P, 34 kDa), and gene 62 (T4 g62P, 20 kDa) were added to the reaction (34, 35). This stimulation required ATP hydrolysis (36, 37). The g44P and g62P form a tight complex (g44/62P) that can be purified from infected cells (38, 39). Together with the g45P, the g44/62P complex eliminated pausing by the g43P (39) and decreased the ratio of dNTP $\rightarrow$ dNMP turnover to dNMP incorporation when g43P was replicating a long single-stranded region (40). These observations supported the suggestion of Huang et al (39) that these three proteins, known collectively as the T4 DNA Pol

accessory proteins, form, in an ATP hydrolysis-dependent manner, a sliding clamp that keeps the g43P bound to the 3'-OH terminus of the nascent chain. This sliding clamp increases significantly the processivity of the g43P.

The mechanism of the sliding clamp has received considerable attention in recent years. Initial observations had indicated that the g44/62P complex was a DNA-dependent ATPase that was stimulated by the g45P (36, 37, 41). It is now clear that this ATPase activity is intrinsic to the g44P (42). Interestingly, the g44P ATPase is only stimulated slightly by the g45P (42), suggesting that the g62P mediates this interaction. Jarvis et al (43) have shown that the preferred DNA effector of the accessory protein ATPase is a primer-template. The g45P lowers the K_m of the g44/62P complex for the primer-template by a factor of 20. Since the g44/62P exhibits little effector preference, Jarvis et al (43) have suggested that the g45P is the primary determinant of primer-template recognition. Optimal primer length for ATPase stimulation was 18 nt.

The half-life of g43P on a primer terminus is increased 20-fold in the presence of the accessory proteins (44). It had been suggested previously (39) that the clamp might be renewed by additional ATP hydrolysis (other than that required for assembly). A recent examination of the activity, during single turnover experiments of the holoenzyme (g43P plus the accessory proteins) on primed single-stranded templates that contained regions of significant secondary structure, indicated that additional ATP hydrolysis was not required after assembly of the holoenzyme to maintain a processive enzyme (44). Jarvis et al (44) have therefore suggested that the sliding clamp constitutes a state where the g43P has stabilized the ADP form of the accessory protein complex on the primer terminus and that this stabilization is reciprocal. Consistent with this proposal, Richardson et al (45) demonstrated that a functional, isolable complex of the g44/62P with the g43P could form on a primer-template coated with the T4 SSB (the g32P), but that a similar g45P-g44/62P complex could not be detected. However, a complex that required ATP for formation containing the g45P, the g44/62P, and the g43P could be detected.

A picture of the structure of the elongation complex of the T4 polymerase and accessory proteins on the primer terminus has begun to develop. Capson et al (46) have used primers containing photo-activated cross-linkable residues at specific positions to study how the holoenzyme assembles on the primer template. Variation of the position of the derivatized residue and partial elongation of the primer terminus allowed detection of cross-linked proteins over a primer-template stretch of 30 nt. The use of both ATP and ATPγS during the cross-linking reactions revealed a sequential rearrangement of the accessory proteins in response to the addition of the g43P: (a) Initial binding of the g44/62P and g45P to g32P-coated primer-template resulted, in the absence of ATP hydrolysis (i.e. in the presence of ATPγS), in a structure in which g44P and g62P could be detected near the primer terminus (at positions −4 and −14, respectively, relative to the primer terminus). g45P could be detected behind this complex at position −20. (*b*) ATP hydrolysis apparently caused the g44P to move away from the primer (it could barely be detected), g62P binding remained in the same position, but was reduced. g45P binding was unchanged. (*c*) This rearrangement presumably exposes the 3'-OH end for g43P binding. The g43P covered 5–7 nt of the primer terminus. g62P binding was further decreased and the g45P moved forward and could now be detected at position −14. In the presence of ATP, efficient cross-linking of the g43P required the accessory proteins and the g32P, whereas cross-linking of the g45P required g43P, g44/62P, and g32P. The elongation complex appeared to extend at least 20 bp behind the primer terminus.

Munn & Alberts (47) have made similar observations using a hairpinned primer-template and DNase I footprinting. They found that g32P bound the single-stranded template DNA up to the primer terminus. They could only detect an accessory protein complex, which covered 20 nt of the primer-template duplex and 4–5 nt of the adjacent single-stranded template, in the presence of ATPγS. The presence of the g32P lowered the amount of the g44/62P required to give a footprint by 100-fold, suggesting a direct interaction. Failure to observe a complex in the presence of ATP indicated that ATP hydrolysis must be destabilizing. Capson et al (46) could detect such a complex because of the increased time resolution of the photocross-linking technique.

Although a footprint of an idling g43P could be observed using neocarzinostatin as the cleavage agent, Munn & Alberts (48) could not detect a holoenzyme complex. Surprisingly, DNase I footprinting did show an expanded footprint with the g43P in the presence of the accessory proteins. However, this could be attributed to the independent action of the g45P in the absence of ATP. Competition could be demonstrated between the ATPγS-dependent accessory protein footprint and the ATP-independent g43P-g45P footprint, suggesting these groups of proteins were competing for the same site and supporting the proposal of Capson et al (46) that the accessory proteins rearrange their position when the g43P is added to the complex. Since the g45P binds to a g43P-affinity column but not to DNA, whereas the g44/62P does bind DNA, and since a g43P-g45P footprint could be obtained, Munn & Alberts (48) place the g45P directly behind the g43P on the primer-template in their model of the elongation complex.

Furthermore, since an accessory protein complex could only be observed in the presence of ATPγS and a holoenzyme complex could not be observed in the presence of ATP, even at concentrations of the g44/62P 1000-fold greater than that required to stimulate maximally DNA synthesis by an equivalent amount of g43P, Munn & Alberts (48) argue that a stalled polymerase complex is fundamentally unstable and dissociates rapidly from the primer-terminus, whereas a moving polymerase complex is bound far more tightly. These authors suggest that once formed, the sliding clamp is maintained in a high-energy state where $ADP+P_i$ remains bound to the accessory protein complex aided by either repeated occasional ATP hydrolytic events or by energy transfer to the accessory protein complex from the exothermic process of polymerization. Once the holoenzyme complex is stalled, the inability to recharge results in a rapid decay of the high-energy state and the subsequent dissociation of the holoenzyme complex. However, since there are likely to be several intermediate energy states of the accessory protein complex that are competent for maintaining the clamp, a time lag would be introduced once the holoenzyme stalls before it dissociates. This lag represents a clock mechanism that prevents the polymerase from dissociating until a minimum stalling time is reached. This feature has been incorporated into a model, described in a subsequent section and detailed by Alberts and his colleagues (49), which describes the control of Okazaki fragment synthesis on the lagging-strand template. An alternative view (44), mentioned above, is that an accessory protein-g43P holoenzyme complex actually exists in a potential energy well, and not in a high-energy state. If this is the case, the works of the clock mechanism of Munn & Alberts (48) would require adjustment.

THE *E. COLI* DNA POLYMERASE III HOLOENZYME   DNA polymerase III, the *dnaE* gene product, was clearly identified as the cellular replicative polymerase (50, 51). The original preparations of this enzyme, while active in assays using activated DNA as templates, were essentially unable to copy long stretches of ssDNA (52). A search for a form of enzyme capable of this task led to the identification of either the DNA Pol III HE (53) or of factors that stimulated the action of Pol III core (54, 55). In its current form, the DNA Pol III HE can be purified as a large (>1 MDa) complex of 10 subunits (56): α (132 kDa), ϵ (27 kDa), θ (10 kDa), β (38 kDa), τ (71 kDa), γ (47 kDa), δ (35 kDa), δ' (33 kDa), χ (15 kDa), and ψ (12 kDa). These subunits assort in the following manner: The catalytic Pol III core (57) is composed of a heterotrimer of α [the *dnaE* gene product (58)], ϵ [the *dnaQ* gene product (59)], and θ, and contains the DNA polymerase activity [resident in DnaE (60, 61)] and the proofreading $3' \rightarrow 5'$ exonuclease [resident in DnaQ (62)]. The Pol III core has a very low processivity (63). The addition of τ [the *dnaX* gene product (64, 65)] to the Pol III core causes it to dimerize, generating Pol III' (66). The processivity of Pol III' is increased severalfold over that of the Pol III core (67). Pol III* (68, 69) is a more complex assembly of subunits that contains, in addition to Pol III', the $\gamma\delta$ complex (69), which is composed of γ

[also a product of *dnaX* (70–72)], δ, δ', χ, and ψ (73, 74). Pol III* can be distinguished from Pol III' because it is more processive and can be stimulated by SSB (67). Addition of β [the *dnaN* gene product (75)] to Pol III* generates the functional Pol III HE.

It has recently been demonstrated that τ is the full-length product of *dnaX*, while γ is a shorter gene product produced by a translational frameshift that results in subsequent premature termination (70–72). The key mRNA feature necessary for frameshifting is an A_6 stretch encoding two Lys residues at codons 429 and 430 that allows the ribosome to slip, when it is paused at a stable (−28 kcal) hairpin formed from codons 433–442, into the −1 reading frame. Subsequent translation in the new frame results in termination after the addition of two amino acids, resulting in a 431-amino-acid γ protein. The first 430 amino acids are identical to those of τ.

In addition, looming on the horizon is the final assignment (after some 20 years) of the genetic identity of all the holoenzyme subunits. M. O'Donnell's group (personal communication) have, using the techniques of reverse genetics, identified the genes for θ, δ, δ', χ, and ψ. All five proteins are clearly distinct gene products, based on nucleotide sequence and map position. Their requirement for *E. coli* viability is currently being explored.

Conversion of the catalytic Pol III core to a highly processive form, capable of adding at least 0.5 Mb to a primer terminus in one binding event (75a), requires the ATP-dependent formation of an initiation complex (63, 67, 76, 77) in which the core is clamped onto the primer terminus. The mechanism of assembly of this sliding clamp and its identity have been clarified recently. S. Wickner (78) demonstrated previously that the presence of β (factor I) was essential for formation of an initiation complex capable of subsequent processive DNA replication, and that in the absence of Pol III core, β could be bound to the primer terminus in an activated form by the action of purified γ (DnaZ) and elongation factor III (possibly τδ' or δδ') in a reaction that required ATP hydrolysis. Subsequent addition of the Pol III core resulted in the formation of a processive complex.

Using highly purified holoenzyme subunits, O'Donnell (77) defined these steps precisely, showing that preinitiation complex formation was a result of the γδ complex transferring, in an ATP-dependent fashion, β to the primer-template. This preinitiation complex could be isolated and converted to an initiation complex by the addition of the Pol III core. Initiation complexes are stable in the absence of nucleotide polymerization as long as ATP is present (67, 76). Further resolution of the Pol III core and the γδ complex into their constituent components has facilitated a more detailed understanding of subunit function. The αε combination is as active and processive as αεθ (core) when added to a preinitiation complex, indicating the θ is not required for processive polymerization (79). However, α alone is slower and less processive than αε. Thus, ε, aside from carrying the 3'→5' exonuclease, must also provide a structural function. On the other hand, either combination of $\gamma\delta$ or $\tau\delta'$ was functional in forming a processive polymerase with $\alpha\epsilon$ and β. δ' was not active with γ (80). O'Donnell & Studwell (80) suggest that since one holoenzyme contains β, τ, γ, δ, and δ', the presence of two primer-binding mechanisms might mediate the rapid cycling required of the lag-ging-strand polymerase. This is consistent with O'Donnell's (77) previous observation that cycling of the holoenzyme from a completely replicated primed single-stranded circular [ss(c)] DNA template to another primed ss(c) DNA was accelerated by the presence of a preinitiation complex on the second DNA.

Differential action of τ and γ is clearly indicated in vivo, since purified τ will restore activity to crude soluble extracts that can convert ϕX174 ss(c) DNA to the replicative form and that have been prepared from either a *dnaX* or *dnaZ* strain, whereas purified γ will only complement the extract prepared from the *dnaZ* strain (81). γ lacks a putative DNA-binding site present in the COOH-terminal portion of τ (82). This correlates with the observation that while both τ and γ possess ATP-binding activities, only τ is a DNA-dependent ATPase (70–72, 83). A holoenzyme reconstituted from purified subunits that lacks τ is more prone to dissociate when it encounters secondary structure in its path (84). An initiation complex assembled with core, the $\gamma\delta$ complex, β, and τ contained τ but no γ; however, γ could be detected when τ was omitted from the assembly reaction. These observations have led Maki & Kornberg (84) to suggest that τ might function only in the leading-strand polymerase to potentiate the extraordinary processivity required on that side of the replication fork, while the $\gamma\delta$ complex would function on the lagging-strand side of the fork to mediate the rapid cycling required of that polymerase complex. More about this follows in a subsequent section.

Stukenberg et al (85) have shown that the $\gamma\delta$ complex functions catalytical-ly in preinitation complex formation and cannot be detected (by Western blotting) in an isolated preinitiation complex. This argues that a preinitiation complex is simply a β dimer transferred in active form to the primer-template by the ATP hydrolysis-dependent chaperonin-like action of the $\gamma\delta$ complex. These authors also demonstrated that while β could be isolated by gel filtration bound to a fully replicated form II DNA, linearization of the DNA prior to gel filtration resulted in the inability to recover β in the excluded fractions, suggesting that under the latter circumstances the β slid off the DNA. Additional studies showed that while β could diffuse along duplex DNA in either direction, it could not diffuse over ssDNA, whether it was coated with SSB or not. Clearly the nature of the sliding clamp of a pro-cessively engaged Pol III core is this ability of β to diffuse along duplex DNA. Polymerization adds a vector sense to the diffusion so that one may think of β as being pulled along by the Pol III core while at the same time passively locking the polymerase on the primer-template.

Formation of the sliding clamp requires ATP hydrolysis (63, 67, 76, 77). The amino acid sequence of δ has a match to the consensus nucleotide-binding site (M. O'Donnell, personal communication). Onrust et al (86) have demonstrated that the $\gamma\delta$ complex is a DNA-dependent ATPase that required an SSB-coated primer-template and was stimulated by β. The $\gamma\delta$ and $\gamma\delta'$ combination were active ATPases, even though only the $\gamma\delta$ complex was functional during replication. However, δ' could be shown to stimulate $\gamma\delta$ four- to five-fold during replication. The significance of these observations is not clear at the present time.

As with the T4 system, rearrangements of the subunits of the holoenzyme appear to accompany initiation complex formation. Griep & McHenry (87) could assign different changes in the fluorescent properties of labeled primers in primer-templates upon SSB binding, Pol III* binding, and initiation complex formation in the presence of ATP. In addition, the footprint of a preinitiation complex is larger than the footprint of an initiation complex (30 nt) (C. McHenry, personal communication).

The footprinting studies described above for the T4 and *E. coli* systems imply that the elongating polymerase complex covers 18–25 bp behind the primer terminus. As described in the next section, the primers actually used at the replication fork are 4, 5, and 10 nt long in the T7, T4, and *E. coli* systems, respectively. It remains to be determined how a stable initiation complex forms on these small primers on the lagging-strand side of the replication fork.

Primosomes

During progression of the replication fork, the parental duplex must be unwound and the synthesis of Okazaki fragments initiated every 1 or 2 sec. In the prokaryotic systems that have been well characterized, the enzymes responsible for unwinding (the DNA helicase) and priming (the primase) have been engineered to attract each other with a high affinity, and form complexes (primosomes). These replication fork helicases act processively in the $5'\rightarrow3'$ direction along the lagging-strand template (i.e. they do not dissociate from the template during the lifetime of the fork), which ensures that the primase can localize very rapidly the region where new primer synthesis is required. In the cases to be discussed, the labor is divided between at least two proteins, and in one case, as many as seven may be involved.

T7 GENE 4 PROTEINS   The T7 DNA Pol is incapable of copying a ss(c) DNA template in the absence of a primer; however, DNA synthesis can occur in the presence of the T7 gene 4 protein (T7 g4P) and NTPs (88–90). The g4P is a primase capable of responding to the signals 3'-CTGGG-5' and 3'-CTGGT-5' to synthesize the tetraoligoribonucleotides 5'-pppACCC-3' and 5'-

pppACCA-3', respectively (88, 90). In addition, it was shown that the g4P was a DNA helicase that could unwind DNA in the 5'→3' direction (along the strand to which the enzyme was bound) in a reaction requiring the hydrolysis of dNTPs or NTPs (TTP is preferred) (91–93). Helicase activity required a nonhybridized single-stranded tail on the 3'-end of the fragment to be displaced (91). It could be shown that selection of priming sites along a single-stranded template was also biased in the 5'→3' direction (94), suggesting that even in the absence of duplex DNA to be displaced, the g4P will bind the substrate and move along it in the 5'→3' direction until it encounters a priming site.

The exchange of a primer between a primase and the lagging-strand polymerase complex is a key event during Okazaki fragment synthesis. Thus, one would anticipate the ability to demonstrate mutual interaction between these components. This is quite clear in the case of the T7 DNA Pol and the g4P. Even if prevented from synthesizing its own primer, the g4P will stabilize on the template short oligonucleotides of either the natural sequence or similar ones so that they may be elongated by the polymerase (89). The T7 DNA Pol and g4P form a complex that can be isolated by gel filtration (95). In addition, the simultaneous presence of both the polymerase and the g4P on the template DNA prior to priming and DNA synthesis results in the formation of a complex that is stable to challenge by excess ssDNA (96). The stability of this complex presumably derives from both the independent DNA-binding activities of the two proteins and their interaction.

Preparations of g4P from bacteriophage T7–infected cells are composed of two polypeptides of 63 kDa and 56 kDa (88, 97). The shorter protein is the result of an internal translational start (98). Richardson and his colleagues have been analyzing the independent activities of these proteins using either preparations that are severely deficient in one or the other (99–101) or pure preparations of the 63-kDa g4P or the 56-kDa g4P prepared from overexpression strains driven from plasmids engineered to produce only one or the other polypeptide (102, 103).

The 56-kDa g4P lacks the 7-kDa amino-terminal region of the 63-kDa g4P, a region that contains a putative zinc finger DNA-binding domain (99). This is consistent with the finding that while both forms of g4P are roughly equivalent in their action as a TTPase and DNA helicase, the primase activity of the 63-kDa form is 150-fold greater than that of the 56-kDa form (101). Interestingly, whereas both the 56-kDa form and the native mixture are monomeric in solution (103), the active form of the 56-kDa g4P appears to be a multimer (102). Consistent with this observation is that addition of a sevenfold excess of the 56-kDa form to the 63-kDa form increases the priming activity of the latter by 100-fold (103). Thus, the active species of g4P might be a heterodimer composed of the 63-kDa primase form piggy-backed on the

56-kDa helicase form. This is supported by the finding that during rolling circle DNA synthesis catalyzed by the T7 DNA Pol and g4P, leading-strand synthesis (and therefore g4P-catalyzed helicase action at the fork) is stable both to challenge by ssDNA and dilution, whereas lagging-strand synthesis (and therefore g4P-catalyzed priming) is affected in each case (100). This is manifested by a shift in Okazaki fragment size to larger values, indicating that the frequency of successful initiations on the lagging-strand had decreased. Thus, the priming component of the g4P acts distributively during multiple cycles of Okazaki fragment synthesis.

Additional support for the view that the two forms of the g4P cooperate but act independently is found in the observation of Mendelman & Richardson (103) that even on a 25-nt-long substrate where the priming site was in the center, TTP hydrolysis was required for primer synthesis. The authors propose that the 63-kDa form, piggy-backed on and with its zinc finger domain trailing behind the 56-kDa form, must be dragged in the $5' \rightarrow 3'$ direction across the priming site in order for proper recognition to occur.

T4 GENE 41 AND GENE 61 PROTEINS   The 59-kDa T4 gene 41 protein (T4 g41P) catalytically stimulated strand displacement synthesis by the T4 DNA Pol holoenzyme and low concentrations of the g32P, leading to the suggestion that it was a DNA helicase (39). Venkatesan et al (104) subsequently showed that this was indeed the case. The $5' \rightarrow 3'$ helicase activity (104) of the g41P is supported by its ssDNA-dependent nucleotidase activity (ATP, GTP, dATP, and dGTP) (104–106).

DNA synthesis by the g43P and the accessory proteins on g32P-coated ss(c) DNAs required, in addition to the g41P, the 30-kDa T4 gene 61 protein (T4 g61P) and NTPs (107, 108). It was demonstrated that the combination of the g41P and the g61P was a primase capable of responding to the template sequences 3'-TTG-5' and 3'-TCG-5' to synthesize the pentaoligoribonucleotides 5'-pppACN$_3$-3' and 5'-pppGCN$_3$-3' (107, 108). With the natural template (glycosylated, hydroxymethyl dCMP-containing T4 DNA), only the pppACN$_3$ primers are synthesized (109).

The g61P is the primase (110, 111) and can synthesize dimers in a template-directed fashion (112). However, these dimers cannot be elongated by the T4 DNA Pol holoenzyme. The DNA helicase activity of the g41P, which acts in the reaction as a multimer, is stimulated by the g61P and to a lesser extent by the polymerase accessory proteins (104, 113). Like that of the T7 g4P, the helicase activity of the T4 g41P is stimulated by the presence of a nonhybridized ssDNA tail on the 3'-end of the fragment to be displaced. The efficiency of g61P stimulation of the g41P DNA helicase decreases as the 3'-tail length increases (113). However, efficient unwinding of a long (650-bp) duplex DNA required both proteins, suggesting some stabilization of the g41P on the DNA by the g61P. Indeed, detection, by gel mobility shift, of a stable protein-DNA complex required both g61P and g41P as well as ATP or GTP (113).

A complex series of protein-protein interactions between the proteins present on the lagging-strand template is indicated as a result of a study, by Richardson & Nossal (114), on the activities of a trypsinized product of the g41P. Removal of the COOH-terminal 20 amino acids of the g41P generates a protein (g41TP) with equivalent GTPase and DNA helicase activities. However, in the presence of the g41TP, the normal g32P stimulation of DNA synthesis by the combination of the g43, g44/62, g45, and g61 proteins was not observed. In addition, primer synthesis by either the g41P or g41TP with the g61P was inhibited by coating the ssDNA with g32P. This inhibition could be relieved in the presence of the g41P by the addition of the polymerase accessory proteins. However, the inhibition was maintained when the g41TP was substituted for the g41P. The authors concluded that the COOH-terminal 20 amino acids of the g41P were required to mediate an interaction between the T4 primosome (g41P + g61P) and the polymerase accessory proteins in the presence of g32P. Since this situation is obtained on the lagging-strand template, Richardson & Nossal (114) proposed that this interaction was required to regulate primer synthesis at the replication fork. In a related observation, Cha & Alberts (115) have shown that at a replication fork, the efficiency of primer utilization (the fraction of all primers synthesized that are actually utilized to initiate synthesis of an Okazaki fragment) is high (90%) in the presence of the g32P, whereas it is reduced considerably in the absence of the SSB. The possible role of these protein-protein interactions in controlling Okazaki fragment size is discussed in a subsequent section.

E. COLI—WHAT CONSTITUTES THE REPLICATIVE PRIMOSOME? In E. coli, the DnaB protein (52 kDa) is a 5'→3' DNA helicase that requires a nonhybridized 3' ssDNA tail on the fragment to be displaced (116). DnaB is clearly the replication fork helicase—dnaB ts mutants are of the immediate-stop variety (117), and of all the known dna gene products (DnaA, DnaB, DnaC, DnaG, DnaJ, DnaK, DnaN, DnaQ, DnaT, and DnaX), it alone is capable of unwinding duplex DNA. The DnaB helicase activity is driven by its ssDNA-dependent NTPase activity (118–121).

The DnaG protein (64 kDa) was identified as a primase based on resolution and reconstitution of the enzymes required for the conversion of bacteriophage G4 ss(c) DNA to the replicative form in vitro (122, 123). Temperature-sensitive mutants in dnaG also display an immediate-stop phenotype (117). In addition, some mutant alleles of dnaG have a partition defect at the nonpermissive temperature (124). DnaG is unable to synthesize primers on ssDNA that does not contain the highly specialized G4*ori* sequence unless the DnaB protein is present. In a reaction termed *general priming,* DnaG will, in the presence of DnaB, synthesize primers on any naked ssDNA that can be used to prime subsequent DNA synthesis by the DNA Pol III HE (125, 126). Presumably, DnaG associates with DnaB bound to the DNA by protein-protein interactions, or, as has been suggested previously, DnaB generates a specific DNA structure that promotes binding of DnaG to the template (127).

Both general priming and the helicase activity of DnaB [which is stimulated by DnaG (116)] are inhibited if the DNA is completely coated with the *E. coli* SSB (116, 125, 126). Presumably, this mechanism exists to prevent spurious priming and formation of replication forks on any ssDNA exposed in the cell. Instead, DnaB is loaded onto DNA by mechanisms keyed to the initiation process, thereby ensuring that its access will be limited to the proper regions of the chromosome.

Reconstitution in vitro with purified replication proteins of the rifampicin-resistant conversion of bacteriophage ϕX174 ss(c) DNA to the replicative form led to the discovery of one of these pathways (128, 129). Replication required the DNA Pol III HE, the SSB, and seven other host proteins that were required to effect the synthesis of primers on the DNA. These proteins are the aforementioned DnaB and DnaG proteins, as well as DnaC (29 kDa), DnaT [22 kDa, formerly protein i (130) or factor X (128)], PriA [81 kDa, formerly factor Y (131) or protein n' (132)], PriB [12 kDa, formerly protein n (133)], and PriC [21 kDa, formerly protein n'' (133)]. In the absence of DnaG, these proteins assemble, in an ATP-dependent manner, an isolable prepriming replication intermediate on SSB-coated ϕX174 ss(c) DNA (134, 145) that has been termed the (ϕX-type) preprimosome (19).

Assembly occurs in discrete steps and requires a specific 70-nt sequence in the DNA that has been termed a *primosome assembly site* [PAS (9)]. PriA specifically recognizes and binds to the PAS (136). This activates its ssDNA-dependent ATPase (131, 136), although PriA-catalyzed ATP hydrolysis is not required for primosome assembly (see below). PriB binds to the PriA-DNA complex. DnaT then acts to transfer DnaB from a DnaB-DnaC complex in solution [which requires ATP for formation (137, 138)] to the PriA-PriB-DNA complex, forming the preprimosome. The role of PriC in this process is unclear. Addition of DnaG to the preprimosome completes the assembly of the (ϕX-type) primosome (134, 135).

It is still not clear which proteins remain on the DNA after ϕX-type primosome assembly. In their analysis of a similar priming system utilizing the action of the proteins λO, λP, DnaK, DnaJ, and DnaC to transfer DnaB to SSB-coated DNA, LeBowitz et al (139) suggested that only DnaB remained on the DNA. However, PriA is a potent 3'→5' DNA helicase (140, 141), and Lee & Marians (142) demonstrated that the ϕX-type preprimosome could act as a DNA helicase in both directions along the DNA, driven in the 5'→3' direction by DnaB and in the 3'→5' direction by PriA. Furthermore, it could be demonstrated that the complete primosome could synthesize primers downstream from and on either side of a PAS in a linear DNA template. Moreover, a PriA-PriB-DnaT complex isolated on a linear DNA that could act as a helicase only in the 3'→5' direction could be converted, upon the addition of DnaB and DnaC to complete preprimosome assembly, to a protein complex that maintained its 3'→5' activity and now manifested an equivalent amount of 5'→3' helicase activity. This suggests that both PriA and DnaB are present in the same protein-DNA complex.

On the other hand, the presence of only DnaB on the DNA is supported by a different loading mechanism that has been termed the *ABC primosome* (143). Here, DnaA, the cellular initiator protein, binds to its cognate dsDNA-binding sequence (see the section on INITIATION) present in the form of a hairpin on a SSB-coated ss(c) DNA. Subsequent addition of DnaB and DnaC results in the generation of a mobile prepriming complex that exhibits the helicase properties of DnaB and can be converted to a complete primosome by the addition of DnaG. Presumably, the ABC primosome is identical or very similar to the primosome that assembles at *oriC*. Furthermore, replication forks capable of coordinated leading- and lagging-strand DNA synthesis can be formed with the DNA Pol III HE and either the complete complement of ϕX-type primosomal proteins or only DnaB, DnaC, and DnaG, indicating that neither DnaT, PriA, PriB, nor PriC is required to attract DnaG to a bona fide replication fork (75a).

In an effort to assess the contribution of the ϕX-type primosome to cellular replication, both Lee & Kornberg (144) and Nurse et al (145) have exploited the recent cloning of *priA* (146, 147) to construct *E. coli* strains deficient in PriA activity. Although such mutants can be isolated, cell viability is reduced 10- to 100-fold and the cells exhibit extreme filamentation (144, 145). Nurse et al (145) noted that the SOS response had been induced and that filamentation could be suppressed in a *sulA* [*sfiA,* encoding a SOS-inducible inhibitor of septum formation (148)] background. Interestingly, in the *priA sulA* double mutant, both ColE1-type plasmids [which require the ϕX-type primosome for replication (see section on INITIATION)] and *oriC*-based plasmids could be maintained (145), whereas in the *priA* single mutant they could not (144).

Nurse et al (145) proposed, to account for SOS induction in the *priA* strain, that while replication fork assembly was not dependent on the ϕX-type primosome pathway at *oriC*, completion of the chromosome in a significant fraction of cells could depend on subsequent assembly of ϕX-type primosome-dependent forks if the *oriC* forks were to stall or dissociate. This would account for the observed SOS induction whether PriA was an actual component of the ϕX-type primosome on the DNA or only required for its assembly. In support of PriA function being confined to assembly of a φX-type primosome, Zavitz & Marians (148a) have shown that the gene encoding an ATPase-defective mutant PriA that is not a helicase but still supports φX-type primosome assembly in vitro, will, when provided in trans, suppress the filamentation phenotype of and restore cell viability to a *priA* strain. Similar investigations with the recently cloned *priB* (149, 150) and *priC* (150) should help clarify matters.

The Replication Fork

The action of replication forks has been studied in a similar fashion in the T7 (100, 151), T4 (45, 49, 115, 152, 153), and *E. coli* systems (75a, 154–156, 156a–156d). Generally, in these replication systems, a form II DNA carrying a nonhomologous 5'-single-stranded tail is used as a template to sustain rolling circle DNA replication in the presence of the respective SSB, primosome, and DNA polymerase holoenzyme. The nascent DNA products are a long, continuous leading strand and short (usually 1–5 kb) Okazaki fragments that can be separated easily from each other by alkaline agarose gel electrophoresis. Omission of the primase allows study of leading-strand synthesis in the absence of lagging-strand synthesis (100, 153, 155). The replication forks formed move at rates very similar to those observed in vivo, and can processively synthesize very long (150–500 kb) leading strands (49, 75a, 100, 155). Thus, any particular replication fork will synthesize between 75 and 250 Okazaki fragments, providing an excellent window for the investigation of factors that affect lagging-strand DNA synthesis specifically.

Okazaki fragment synthesis can be viewed as a repetitive cycle that includes: (*a*) synthesis of a new primer, (*b*) transit of the lagging-strand polymerase from the 3'-OH terminus of the just-completed Okazaki fragment to the new primer terminus (an event sometimes called polymerase cycling or travel time), (*c*) nascent lagging-strand DNA synthesis, and (*d*) termination of Okazaki fragment synthesis. Because of this repetition, investigators have asked whether enzymes at the fork act processively—i.e. do they remain bound continuously through multiple cycles of Okazaki fragment synthesis?—or distributively—i.e. do they act once during a cycle of Okazaki fragment synthesis to be replaced by a new molecule from the pool in solution for the next cycle? This is generally examined by diluting active replication forks into reaction mixtures that contain no additional template but in which the concentrations of all the enzyme components except the one under investigation have been kept constant. After dilution, Okazaki fragment size will not change if the enzyme acts processively, whereas the fragments will increase in size if the enzyme acts distributively.

In the T7 (100), T4 (153), and *E. coli* (154) systems, the extraordinary processivity of leading-strand DNA synthesis is manifested through the stable asssociation of both the replication fork helicase (the 56-kDa T7 g4P, the T4 g41P, or DnaB) and the leading-strand polymerase with the fork for its apparent lifetime. Interestingly, there is no requirement for SSB to maintain processive leading-strand synthesis at high rates in any of the three systems (151, 153, 155). Since all of the polymerase complexes are sensitive to secondary structure in the DNA, this clearly indicates that there is little or no free single-stranded leading-strand template DNA between the replication fork helicase and the leading-strand polymerase.

On the other hand, Nakai & Richardson (100) and Wu et al (75a) demonstrated that in the cases of T7 and *E. coli,* respectively, the primase (the 63-kDa T7 g4P or DnaG, respectively) acted distributively. The T4 situation is slightly more complicated, since Richardson et al (45) have shown that the T4 g61P appears to act distributively early in the reaction, but then processively as the replication forks mature (after 5 min). The authors suggested that a protein (possibly the gene 59 product) was missing that was required normally to glue the fork components together.

ASYMMETRIC, DIMERIC DNA POLYMERASES AT THE FORK   Alberts and his colleagues (35, 157) noted early on that during the multiple cycles of Okazaki fragment synthesis at the fork, the lagging-strand polymerase must be released from the template and the nascent DNA after the completion of an Okazaki fragment. Synthesis of the next Okazaki fragment could therefore be accomplished by the same polymerase or a new one from the pool in solution. To maximize the speed and efficiency of targeting a polymerase to the new primer, Sinha et al (157) proposed that the same polymerase could be targeted very efficiently if it remained bound at the fork through protein-protein interactions with the leading-strand polymerase while it transited from the completed Okazaki fragment to the new primer.

Considerable effort has been expended to assess the accuracy of this proposal. There is little evidence that either the T7 or T4 DNA polymerases exist as a dimer in solution, whereas this does seem to be the case for the *E. coli* DNA Pol III HE. In initial studies, Alberts et al (152) demonstrated that the size of Okazaki fragments was insensitive to the concentration of the T4 g43P present during assembly of the forks, suggesting that the leading- and lagging-strand polymerases were coupled. Richardson et al (45), however, showed that this observation depended on the age of the replication forks. As was the case with the T4 g61P and as has been confirmed by dilution experiments, the lagging-strand polymerase appeared to act distributively early and processively late in the reaction. Thus, at the moment, the presence of a dimeric T4 DNA polymerase holoenzyme at the fork is problematic.

On the other hand, considerable evidence exists that the *E. coli* DNA Pol III HE is a dimer of two Pol III cores and associated subunits. The addition of

τ to the core (66) or to $\alpha\epsilon$ (158) results in the formation of Pol III' $[(\alpha\epsilon\theta\tau)_2]$ or $(\alpha\epsilon\tau)_2$, respectively. Stoichiometry studies of various assemblies of the holoenzyme subunits led Maki & Kornberg (56) to suggest that the holoenzyme took the form $(\alpha\epsilon\theta\beta_2)_2\tau_2\gamma_2(\delta\delta'\psi\chi)_2$. In addition, a number of observations have been made indicating that the function of the core-β halves of the polymerase are distinct and possibly modulated by an asymmetric assortment of subunits. Johanson & McHenry (159) showed that it was possible to build two types of initiation complexes of the DNA Pol III HE that showed both differential requirements for and sensitivity to ATP and ATPγS. The independent action of $\tau\delta'$ and $\gamma\delta$ in preinitiation complex formation (80), the stimulation of the primer-template-dependent ATPase activity of $\gamma\delta$ and $\gamma\delta\delta'$, but not $\gamma\delta$ by β (86), the action of τ to move the holoenzyme past secondary structure in the template (84), and the apparent exclusion of γ from initiation complexes formed in the presence of τ (56) have all been described in a prior section of this review.

The functional assortment of subunits and dimeric nature of the *E. coli* DNA Pol III HE at the replication fork has been investigated by Marians and his colleagues (156c, 156d). β and at least some of the components of the $\gamma\delta$ complex acted distributively. τ was not required, under standard reaction conditions, for leading- or lagging-strand synthesis. However, in the presence of moderate (200–400 mM) concentrations of potassium glutamate, τ action at active replication forks was limited to the lagging-strand side, where it appeared to increase the efficiency with which the lagging-strand Pol III core transited from the just-completed Okazaki fragment to the new primer. Thus, there may be some functional asymmetry in holoenzyme subunit action at the fork. It should be noted, however, that active replication forks producing both leading- and lagging-strands could be reconstituted with only the Pol III core and β (156c). These authors have also shown, by dilution experiments, that the lagging-strand Pol III core acts processively during multiple cycles of Okazaki fragment synthesis. Thus, protein-protein interactions, most likely with the leading-strand Pol III core, act to retain the lagging-strand polymerase at the fork.

CONTROL OF OKAZAKI FRAGMENT SYNTHESIS   In order to account for their observation that Okazaki fragment size remained constant at both high and very low concentrations of g43P, Alberts et al (152) proposed that the length of Okazaki fragments synthesized on a particular DNA template might be determined by the position of the first oligonucleotide that primes Okazaki fragment synthesis on the parental strand, relative to the advancing fork. If the rates of elongation of the DNA polymerase complexes acting on both strands were the same, and if the signal for the T4 primosome to initiate primer synthesis was completion of the penultimate Okazaki fragment, then the length of all of the subsequent DNA chains would be templated by the length of the preceding fragment. Experiments by Selick et al (49) dispelled that hypothesis by showing that the length of Okazaki fragments synthesized by a population of active replication forks could be varied during the reaction in response to variation of the concentration of the NTPs, substrates for the g41P-g61P helicase/primase complex.

These authors proposed an alternative model, suggesting that the relative rates of elongation of the two halves of a dimerized assembly of leading- and lagging-strand DNA polymerases and the duration of the pause between termination of elongation and the release of the nascent fragment (as governed by the accessory protein clock mechanism discussed in a prior section), and not a mechanism that measured distance, might determine Okazaki fragment length. In this model, the signal to prime was again conveyed from the lagging-strand polymerase to the T4 primosome following the completion of the nascent DNA chain. The rate of the lagging-strand DNA polymerase, however, was postulated to be faster than the rate of elongation of the leading-strand polymerase. Thus, the placement of primers on the lagging-strand DNA template, and, therefore, the length of the Okazaki fragments, could vary in relation to the expanse of leading-strand DNA template dupli-cated in a given amount of time. This type of model would permit regulatory mechanisms acting on an advancing replication fork to be modified in re-sponse to changing reaction conditions.

Available evidence in support of this model is scanty. Whereas the T4 DNA Pol does synthesize DNA at a rate of 600 nt/sec on g32-coated ssDNA (presumably reflective of lagging-strand synthesis) even though the T4 replication fork moves at 250 nt/sec during rolling circle DNA synthesis (49), there have been, to date, no direct measurements in any system of the rate of the lagging-strand polymerase at an active replication fork. On the other hand, the majority of rolling circle forks examined by electron microscopy showed only one single-stranded gap on the lagging-strand template rather than two, suggesting that the lagging-strand polymerase does spend most of its time stalled and waiting for the next primer (49). In addition, Richardson & Nossal's (114) observation that the g41TP cannot, in conjunction with the g61P and the accessory proteins, participate in primer synthesis in the pres-ence of the g32P, suggests that contact between the accessory proteins in the lagging-strand polymerase complex and the g41P at the fork, which may only occur when Okazaki fragment synthesis is complete, is required to generate a g32P-free region on the lagging-strand template permitting g61P-catalyzed primer synthesis. This supports the proposal (152) that the signal to prime is generated only after completion of nascent lagging-strand synthesis.

Marians and his colleagues (75a, 156a–156d) have developed a somewhat different model to account for the observed changes in Okazaki fragment size at replication forks formed with the *E. coli* DNA Pol III HE and either the ϕX-type primosome or a combination of DnaB, DnaC, and DnaG. They found that all factors that affected Okazaki fragment size did so by perturbing an event during the cycle of Okazaki fragment synthesis that occurred prior to actual nascent chain elongation. Variation in the concentration of NTPs or the distributively acting DnaG affected the frequency of primer synthesis, while variation in the concentration of dNTPs or the β subunit of the DNA Pol III HE affected the efficiency of utilization of primers for the initiation of Okazaki fragment synthesis (156a).

Regulation of primer size [which, when uncoupled from DNA synthesis can reach 70 nt (160, 161)] at 10 nt, the same size observed in vivo (162), could be attributed to an interaction between the Pol III core and DnaG. This interaction was independent of DNA synthesis and preceded preinitiation complex formation on the primer terminus (156b). As in the case of the T4 fork, active forks could respond to changes in the environment by altering the size of the fragments synthesized (75a). Thus, a mechanism where Okazaki fragment size was regulated by templating the size of the penultimate fragment could be excluded. In addition, the dNTP effect could not be explained by variation in the rate of replication fork movement. It was therefore unlikely that variation in Okazaki fragment size could be accounted for solely by differential rates of polymerization of the leading- and lagging-strand DNA polymerases.

Finally, under certain conditions, replication forks could be induced to synthesize aberrantly short Okazaki fragments that were separated on the lagging-strand template by large gaps (156d). Nevertheless, under the same conditions, fragment size could be increased by a factor of 10 when the priming frequency was decreased (by decreasing the DnaG concentration). This indicated that the signal for synthesis of a new primer could not be the stalling of the lagging-strand polymerase when it encountered the penultimate Okazaki fragment; instead it most likely was the association of DnaG with the fork that acted to signal the lagging-strand polymerase to terminate Okazaki fragment synthesis, whether or not all the available template had been copied.

In this view, the cycle of Okazaki fragment synthesis is keyed by the association of DnaG (at a rate governed by its K_a) with DnaB at the fork (Figure 2, i). This occurs while Okazaki fragment synthesis is ongoing. Primer synthesis is initiated (Figure 2, ii), and a DnaG-lagging-strand Pol III core interaction is established (Figure 2, iii) that serves to limit primer synthesis and most likely also exposes the 3'-end of the primer. Okazaki fragment synthesis continues. A preinitiation complex then assembles on the primer-terminus (Figure 2, iv). During this time, under normal circumstances, Okazaki fragment synthesis is completed. An assembled preinitiation complex is likely to be the signal (perhaps mediated by some holoenzyme

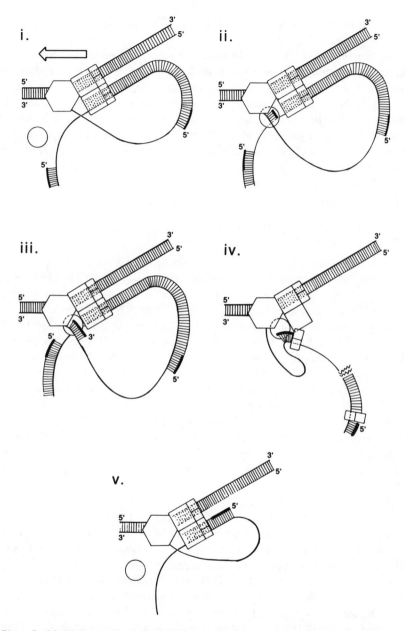

Figure 2 Model for a cycle of Okazaki fragment synthesis at the *E. coli* replication fork. The arrow indicates the direction of fork movement. Parental DNA, heavy lines. Nascent DNA, thin lines. Primers, short black line segments. Elongating Pol III core assemblies, large rectangles. β subunit, small rectangles. Primase, the circle. DnaB, the hexagon. Steps i–v are described in the text.

subunits) for the lagging-strand Pol III core to release and transit to the new primer terminus (Figure 2, v).

Within this context, it can be seen that since lagging-strand template is generated continuously, increased priming frequency (step i) would result in smaller fragments, whereas decreased efficiency of primer utilization, either at step iii because of premature termination of primer synthesis, or step iv because of bungled preinitiation complex formation, would result in larger fragments. The appearance of small fragments separated by large single-stranded gaps could be attributed to an increase in the time required to execute steps iv and v.

INITIATION

At oriC

E. coli chromosomal DNA replication initiates bidirectionally at a specific sequence (*oriC*) located at 84 min on the genetic map (163). The replication forks synthesize roughly 2.3×10^6 bp of DNA and meet in the terminus region (*Ter*) 180° away from *oriC*. The action of RNAP is needed for the initiation of DNA replication (164). The origin region itself was identified originally by both biochemical (165) and genetic (166) gene dosage experiments. Subsequent isolation by cloning exploited either the ability of *oriC* to drive replication of nonreplicating DNA segments that encoded gene products that could be scored easily in vivo (167) or its location near *asn* and *tna* (168, 169). The minimal origin region was isolated by determining the smallest *oriC* segment that could drive replication of a ColE1 plasmid in a host carrying a temperature-sensitive *polA* allele (163). Since DNA polymerase I is required for initiation of ColE1 leading-strand DNA replication (170), propagation of the plasmid at the nonpermissive temperature indicated that *oriC* was active.

The minimal *oriC* is 245 bp (Figure 3) and is highly conserved among enterobacteria (171). *oriC* has several distinctive DNA sequence features. There are four copies of the sequence 5'-TTATC/ACAC/AA-3' in the form of two inverted repeats. The positions of these sequences, denoted R1–R4, are also highly conserved (171). This sequence was shown to be the binding site for the DnaA protein (172, 174), and is referred to as a DnaA box. Directly to the left of the DnaA boxes is a series of three A+T-rich 13mers. This region plays a critical role in the process of localized denaturation required to initiate DNA replication. Finally, *oriC* contains 11 copies of the recognition sequence for the Dam methylase, 5'-GATC-3' (175). Eight of these are in conserved positions. As discussed below, the methylation state of these sequences appears to govern the binding of *oriC* DNA to the outer membrane.

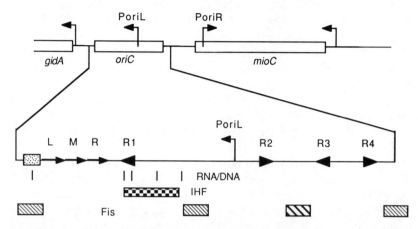

Figure 3   oriC. Arrows above the lines represent the direction of transcription from the indicated promoters. R1–R4 are the DnaA boxes. L, M, and R are the 13mers. The stippled box on the exploded view is the additional A+T-rich region described by Assai et al (189). Vertical lines indicate the leftward-moving RNA/DNA transition points. Also shown is the IHF-binding site and the Fis-binding sites (the one with darkest fill is most likely the only one actually occupied).

A major pathway to understanding the mechanism of initiation at *oriC* was opened by the development in Kornberg's laboratory of a crude soluble enzyme system capable of replicating minichromosomes containing *oriC* in a DnaA- and *oriC*-dependent manner (176). The key event was shown to be the binding of the DnaA protein to the DnaA box sequences (172–174). Subsequent efforts have now outlined the steps necessary for initiation. The central feature of initiation at *oriC* involves the highly cooperative binding of DnaA to the four DnaA boxes (173). DnaA is a weak ATPase that slowly hydrolyzes ATP and can be found either bound to ATP or ADP or free of the nucleotide (177). Release of ADP from DnaA·ADP is a slow process unless stimulated by the presence of acidic phospholipids (178, 179) or cAMP (180). Whereas all forms of DnaA can bind *oriC*, only DnaA·ATP is competent to proceed through subsequent stages.

Initial binding of DnaA to *oriC* requires low concentrations of ATP (30 nM) (177) and can occur at low temperature (181). In the presence of enough of the small dsDNA-binding protein HU (182) to bind only 3% of the template, an initial complex forms that contains 10–20 monomers of DnaA and 200–250 bp of *oriC* (183). The initial complex can be converted, in the presence of high temperature (38 °C) and 5 mM ATP, to an open complex where the three A+T-rich 13mers have become sensitive to cleavage by the single-strand-specific endonuclease P1 (13). Unwinding of the 13mers appears to be sequential, starting with the rightmost 13mer (R-13mer) and spreading to the middle and leftmost ones. DnaA then guides the DnaB protein from a DnaB-DnaC complex in solution to its place between the strands of the denaturation bubble, forming a prepriming complex. If provided with SSB and DNA gyrase, DnaB will then act to unwind the template DNA bidirectionally from *oriC* (184, 185). In addition, if both DnaG and the DNA Pol III HE are provided as well, complete replication forks form and synthesis of the daughter molecules can occur.

Formation of the open complex appears to be the key step in initiation at *oriC*. Kowalski & Eddy (186) demonstrated that the L-13mer exists stably unwound in a supercoiled plasmid in vitro. The specific sequence of the L-13mer was not required for this effect, only the A+T-rich nature of the region. Thermodynamic instability of the 13mer region has also been reported by Gille & Messer (187). However, they showed that in the presence of Mg^{2+} all *oriC* melting was DnaA-dependent. Nevertheless, this suggests an inherent instability in the region. Sequential melting from right to left of the 13mers as reported by Bramhill & Kornberg (13) can now be accounted for based on Yung & Kornberg's (188) observation that DnaA bound to the DnaA boxes in the initial, closed complex interacts specifically with the R-13mer, causing it to unwind. This likely creates a strain in at least the L- and possibly the M-13mer, causing them to denature spontaneously. It has also been suggested, based on the ability of pUC-*oriC* constructs to transform *polA* strains, that an additional A+T-rich region (see Figure 3) directly to the left of the 13mers is required for *oriC* replication (189). Asai et al (189) point out that in the original experiments defining the minimal *oriC* (163), this region was replaced by a similar A+T-rich region from pBR322 DNA. The importance of open complex formation is further underscored by the discovery in Kornberg's laboratory of a 33-kDa protein that inhibits *oriC* replication in vitro by binding to the 13mers and preventing open complex formation (190). Action of this protein and of DnaA are mutually exclusive. Interestingly, null mutants of *E. coli* in *iciA* (for inhibitor of chromosomal initiation), which was shown to encode the 33-kDa protein, are viable and grow at the same rate as wild-type (191).

Open complex formation is stimulated three- to five-fold at HU to DNA ratios sufficient to coat only 3 to 5% of the template (192). Thus, this stimulation is not likely to be related to sequestration of free negative supercoils upon HU binding (193). Instead, it may be related to DNA bending at or near the origin that acts in some manner to facilitate denaturation of the 13mers. Indeed, IHF, a protein with high affinity for curved DNA (194), can substitute in vitro for HU (13). In fact, Polaczek (195) showed that IHF binds *oriC* specifically between R1 and R2 (Figure 3) and bends the DNA. This author notes that this bend allows R1 to face R2–R4 and the 13mers and may be important in the interaction between DnaA and the R-13mer. Filutowicz et al (196) and Gille et al (196a) have also recently shown that another small

DNA-binding protein, Fis, can bind *oriC* at several sites (Figure 3). Fis binding to its highest-affinity site (near R3) also bends the DNA (196a).

Independently, double *hupA hupB* [encoding the two isoforms of HU (197)] null mutants or mutants in either *himA* or *hip* [encoding the two subunits of IHF (198, 199)], can support transformation by *oriC* minichromosomes (200, 201). However, neither the *hupA hupB himA* nor the *hupA hupB hip* triple mutant could maintain these minichromosomes (201). *oriC* minichromosomes can be established but are maintained poorly in *fis* null mutants (196). Caution needs to be exercised here in concluding that these proteins are required for *oriC* replication, since in both the *hupA hupB* double mutant (202) and *hupA hupB himA* triple mutant, chromosomal replication seemed normal (201). However, the distribution of supercoils across the *oriC* region, which is likely to modulate the requirement for these small proteins, can certainly be different in the two contexts.

Superhelicity is required for open complex formation (13, 192, 203). Relaxed templates could form the initial closed complex, but could not be converted to the open complex. In addition, high HU to DNA ratios, conditions that sequester free supercoils (193), inhibit open complex formation (192) and therefore *oriC* replication in vitro (192, 204). This inhibition can be relieved by the action of RNAP. The RNAP activation step could be separated from and was shown to precede DNA replication. RNAP activation was defeated by RNase H, leading to the conclusion that the formation of RNA-DNA hybrids accounted for the stimulation (192). These RNAs were not used as primers and the hybrids activated the DNA even when they were hundreds of nucleotides away from *oriC*. Skarstad et al (205) subsequently showed that activation could be blocked by placing a stretch of G+C-rich DNA between the hybrid and *oriC*, reinforcing the suggestion made originally by Baker & Kornberg (192) that activation was a result of instability created by the RNA-DNA hybrid being transmitted as transient melting through the DNA to the 13mers where it then facilitates their melting during open complex formation. These authors therefore suggested that the RNAP requirement for cellular DNA replication reflected these events rather than a requirement for RNAP transcription to generate a primer terminus. This is consistent with the observations of Ogawa et al (204) that at the low HU to DNA ratio that stimulates open complex formation, only DnaG is required for priming DNA replication, and that even though both RNAP and DnaG are required for DNA replication at the high, inhibitory HU to DNA ratio, DnaG is still responsible for the bulk of the priming.

The role of transcription in initiation of replication at *oriC* is clearly complex. *oriC* is flanked by two active transcription units. Transcripts of *mioC*, which is located to the right, are directed toward the origin, while transcripts of *gidA*, which is located to the left, are directed away from *oriC*

(206) (Figure 3). In addition, there are probably at least two promoters within *oriC* (206, 207, 209) (Figure 3). Lack of transcription from the *gidA* promoter decreased the efficiency of transformation of pBR322-*oriC* chimeras sixfold in a *polA* strain (210). Inactivity of the *mioC* promoter was only inhibitory in the absence of *gidA* transcription (210), although transcription from *mioC* was required in *himA* strains (201). Asai et al (189) showed that *gidA* transcription could suppress the transformation defect of a pUC-*oriC* chimera deleted for the A+T-rich cluster to the left of the 13mers (see above), but only when transcription was oriented away from *oriC* and when *gidA* was on the left. Both Asai et al (189) and Ogawa & Okazaki (210) argue that these effects are best explained as consequences of the Liu & Wang (211) twin domain supercoiling model of transcription, suggesting that *gidA* transcription is necessary to increase negative supercoiling across *oriC*.

RNA/DNA transition points have been mapped at *oriC* in vivo (212, 213). In the minimal *oriC*, these sites occur only in the counterclockwise direction. Zyskind and her colleagues (214, 215) have shown that these leftward-moving transitions are a subset of the positions where RNAP transcribing leftward in vitro from the *mioC* promoter pauses and where 3'-ends of transcripts entering *oriC* from the right can be found. Assignment of these RNA/DNA transitions as initiation points for the counterclockwise leading strand would be consistent with a number of observations indicating that replication initially appears to move in that direction. RNA/DNA transitions in the opposite direction were not detected, leaving the location of the origin of the leading strand of the clockwise moving fork somewhat uncertain.

These data are consistent in indicating the existence of a counterclockwise (to the left) moving fork away from *oriC*, but leave open how the clockwise moving fork is established. Hirose et al (212) and Seufert & Messer (216) have suggested that the first Okazaki fragment synthesized in the counter-clockwise moving fork crosses *oriC* to become the clockwise moving leading strand. However, this is insufficient to account for assembly of the clockwise moving fork. Studies from Kornberg's laboratory demonstrate bidirectional DNA replication on *oriC* templates in the crude soluble system (217), and have established that when unwinding is allowed to proceed from a prepriming complex in the absence of DNA synthesis, there are two foci moving in opposite directions around the template (184). Immunogold antibody labelling was used to demonstrate the presence of DnaB at each unwound branch (185). Thus, the clockwise moving fork may form between a leading strand generated from the counterclockwise-moving lagging strand and the clockwise-moving DnaB loaded at *oriC* via DnaA. The relative contributions of RNAP and DnaG to priming at *oriC* on the chromosome remain problematic. It would be interesting to study with the purified replication system how priming and fork movement are affected by the simultaneous presence of HU,

IHF, and Fis, as well as outer membrane preparations (see below), in the reaction mixture.

It should also be noted that models of initiation of bidirectional replication that invoke a transition of, e.g., the first Okazaki fragment in the counter-clockwise-moving fork becoming the leading strand in the clockwise direction, require the independent action of the two polymerase complexes at the fork and are thus inconsistent with the recent evidence that the leading- and lagging-strand polymerases acting at the fork are physically coupled.

INITIATION AT oriC AND THE CELL CYCLE   E. coli growing with a doubling time between 20 and 60 min exhibits a remarkable constancy in the time [42 min, the C period (2)] required to replicate the chromosome after initiation. Cells divide roughly 20 min [the D period (2)] after termination of replication. Thus, in rapidly growing cells, initiation takes place on nascent chromosomes before their completion, leading to multiforked replication. Under these conditions, any given cell division observed is a consequence of the penultimate or antepenultimate initiation event in that cell. In order to avoid the production of anucleate cells, replication must be completed and the chromosomes partitioned before septum formation completes cell division. Since, under the conditions described, C + D is relatively constant, this implies that the timing of initiation must be precisely regulated during the cell cycle. Donachie (218) proposed that this could occur if initiation was coordinated with mass increase (i.e. cell growth) in such a way that it occurred when a certain mass per origin was achieved (the initiation mass). This could be accomplished by either the dilution, as the cell grows, of an inhibitor whose concentration was high at the time of initiation (219) or by the accumulation of an autoregulated initiator (220).

DnaA has long been a favorite for the key player in this process. Its transcription is both autoregulated (221, 222) and partially under stringent control (222–225) (and therefore linked to growth rate). It is only required for initiation (117, 226, 227), and is dispensable if cells are allowed to replicate by stable DNA replication (228, 229). Overproduction of DnaA abolished the cell cycle specificity of oriC minichromosome replication (230). However, the intracellular concentration of DnaA does not vary appreciably with the cell cycle (231). Recent studies suggest a way out of this cunundrum.

By providing DnaA under control of a lac promoter in a dnaA ts strain and observing cell growth rate, size, DNA content, and timing of initiation as functions of lac promoter induction, Løbner-Olesen et al (232) have demonstrated convincingly that DnaA determines the initiation mass. When DnaA was present at concentrations below wild-type levels, the initiation mass increased, whereas at concentrations higher than wild-type levels, initiation occurred earlier in the cell cycle, i.e. the initiation mass decreased. Autoregulation of *dnaA* transcription would therefore act to keep the initiation mass constant.

How could this be translated into a signal for initiation? Samit et al (233), by examining the footprints of DnaA on *oriC* carried on a minichromosome in vivo, found that R3 seemed to be unprotected during most of the cell cycle. Since the time required for replication of the minichromosome was short, and since all four DnaA boxes are required for replication, they suggested that binding of DnaA to R3 might be the key event required to trigger initiation. This would seem to require the intracellular concentration of DnaA, reported to be constant with the cell cycle (231), to increase just before initiation. This may occur: Even though the total intracellular DnaA concentration is constant, the concentration of free DnaA may vary. Schaefer & Messer (234) have calculated, on the basis of a search of existing sequences in GenBank, that there are roughly 1600 DnaA boxes per genome, enough to sequester all the DnaA in the cell. This analysis used a relaxed DnaA box consensus sequence derived from a study of the ability of bound DnaA to terminate *galK* transcription. A rise in the concentration of free DnaA as the cell cycle approaches the time for initiation may be effected by the cyclical association of the *oriC* region (which includes *dnaA*, mapping only 43 kb away from *oriC*) with the outer membrane, resulting in both an inhibition of immediate reinitiation at *oriC* (235) and a transient 10-fold drop in the steady-state level of *dnaA* transcripts (236).

Association of *oriC* and the outer membrane was predicted on the basis of the replicon hypothesis (237) as the means of ensuring that each daughter cell would receive one completed chromosome. Recent studies have focussed on the mechanism of this binding. Russell & Zinder (238), following up on the observations by Smith et al (239) and Messer et al (240) that *oriC* plasmids fully methylated at the GATC sites transformed *dam* strains with poor efficiency, showed that this efficiency was reversed when unmethylated plasmids were used. Furthermore, it was demonstrated that the unmethylated plasmids accumulated in the hemimethylated state, suggesting that there was a specific block to the replication of hemimethylated DNA. This has now been explained as resulting from a specific binding of hemimethylated DNA to the outer membrane (235). Outer membrane preparations inhibit replication of hemimethylated *oriC* plasmids in the crude soluble system (241). This inhibition can be defeated by allowing DnaA to bind *oriC* before introduction of the membrane fraction. Thus, binding to the membrane prevents DnaA binding to the new daughter *oriC* regions. Furthermore, an elegant study by Campbell & Kleckner (236) demonstrated, by tracking the time required for conversion from the hemimethylated to the fully methylated state, that both *oriC* and the *dnaA* promoter remain sequestered for an average time fourfold longer then that for any other region on the chromosome. This amounted to 30–40% of the cell cycle, regardless of growth rate. These observations are consistent with indications that the normally precise timing between initiation events is lost and becomes random in *dam* mutants of *E. coli* (242, 243).

It therefore seems likely that *oriC* sequestration prevents immediate reinitiation in the cell. Simultaneous sequestration of the hemimethylated *dnaA* promoter, which needs to be fully methylated to be completely active, inhibits *dnaA* transcription and creates a transient drop in the concentration of free DnaA. When finally released from the membrane, there may be insufficient DnaA available to bind *oriC* completely, leading to a time lag before initiation can occur again.

These exciting developments have obviously sparked a search for the proteins that mediate binding of the hemimethylated DNA to the membrane. The involvement of MutH, known to bind hemimethylated GATC sequences during the methyl-directed DNA repair reaction (244), in this process is not yet clear. P. Hughes, A. Malki, J. Herrick, R. Kern, and M. Kohiyama (personal communication) have, using Southwestern blots of outer membrane fractions with hemimethylated *oriC* DNA, identified a 24-kDa protein that may mediate this effect. In addition, J. Campbell & N. Kleckner (personal communication) have isolated a mutation, *seqA*, that increases the transformation frequency of *dam* mutants by methylated *oriC* DNA by 100-fold while leaving that of unmethylated *oriC* DNA unaffected.

Initiation at oriλ

Replication of bacteriophage λ in *E. coli* requires two viral proteins, the products of the O and P genes, and all the *E. coli* DNA replication proteins except DnaA and DnaC (245). Studies have shown that functionally, the λO and λP proteins are DnaA and DnaC analogs, respectively. The origin of replication lies within the O gene and has been shown to consist of a series of four 18mer direct repeats (each of which is an inverted repeat) to the right of which is an A+T-rich region of 35–40 bp (Figure 4) (245). Only two of the repeats and the A+T-rich region are required for replication of plasmids carrying *oriλ* in vitro (246).

Investigation of the mechanisms of *oriλ* replication in vitro began with the cloning and overproduction of the λO and λP proteins and the finding that the same crude soluble extract developed for *oriC* replication could, when supplemented with λO and λP, support the replication of *oriλ*-plasmids in an origin-, λO-, and λP-dependent fashion (247–249). Replication in this system was both RNAP-dependent and inhibited by the λ repressor, as predicted by the requirements for propagation of the bacteriophage (245).

At the current vantage point, it is clear that one can divide the proteins that act at *oriλ* into three groups: (*a*) the viral O and P proteins, (*b*) a group of three heat shock proteins, DnaK, DnaJ, and GrpE, and (*c*) proteins required

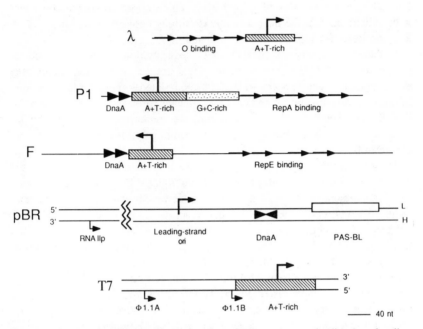

Figure 4 Origins of replication. The arrows above the lines represent the direction of replication. The case illustrated for λ is in the absence of GrpE. The case illustrated for T7 describes the
signals for the rightward-moving fork only.

for replication fork propagation—DnaB, DnaG, the DNA Pol III HE, the
SSB, and DNA gyrase. The initial steps of *ori*λ replication parallel those that
occur at *oriC*. A nucleoprotein complex is formed at the origin that serves to
attract the replication fork helicase and introduce it to the template at the
easily destabilized A+T-rich region.

λO was shown to bind the 18mer direct repeats (250, 251), most likely as a
dimer (252). Thus, eight λO monomers will be bound to at least 95 bp of
DNA. Dodson et al (253) demonstrated, by electron microscopy, that in this
structure the DNA was wrapped around the λO protein aggregate and dubbed
it the O-some. Initial studies showed that additional complexes could be
detected when other proteins involved in *ori*λ replication were added (14,
253). It is now clear that the sequential assembly of a large protein aggregate
at *ori*λ is followed by a heat shock protein-mediated disassembly reaction that
results in the transfer of the DnaB protein to the DNA.

Following formation of the O-some, λP and DnaB add to the complex,
probably as a λP-DnaB complex, to form a λO-λP-DnaB-*ori*λ complex (14,
253–256). This O-P-B complex is functional for subsequent stages of replication (254). Like DnaC, λP forms an isolable complex with DnaB in which

there appears to be three λP protomers to six DnaB protomers (257). Like DnaC, λP suppresses the DNA-dependent ATPase activity of DnaB. However, unlike DnaC, λP does not stimulate the general priming reaction. λP has a higher affinity for DnaB than DnaC does and can displace DnaC from a DnaB-DnaC complex, thereby accounting for its ability to capture DnaB from chromosomal replication after λ infection (257). Addition of λP-DnaB to the O-some is probably mediated by a weak interaction between λP and the O-some (256).

A third-stage complex is formed, which can also be isolated and shown to be functional for subsequent steps, when DnaJ adds to the O-P-B complex (254, 255). DnaB can be retained on a DnaJ column, although it can be washed off at 0.2 M salt (258); thus, this interaction may direct DnaJ to the O-P-B complex and perhaps to the λP-DnaB interface. The addition of DnaK and ATP to the O-P-B-J complex converts it to a partially unwound, open structure (14, 255, 259). Thus, the DnaB protein must be separated from λP by an event catalyzed by DnaK. DnaK is a prototypical Hsp70 protein or molecular chaperone. It possesses a weak DNA-independent ATPase activity (260) that is stimulated by the simultaneous presence of both DnaJ and GrpE (261). Disassembly of the O-P-B-J-K complex (which can be isolated on a linear DNA—conditions where the A+T-rich region cannot become denatured because of lack of superhelicity) requires ATP hydrolysis by DnaK (256, 262). Both Alfano & McMacken (262) and Liberek et al (256, 258) have shown that DnaK and DnaJ act in a concerted manner during the disassembly reaction.

Formation of the O-some is aided by an intrinsic bend in the *ori*λ sequence and induces a structural deformation in the adjacent A+T-rich region that is sensitive to (the single strand-specific) S1 nuclease (263). This instability probably aids the insertion of DnaB between the strands of the template as the disassembly reaction occurs. The primary target for ejection from the O-P-B-J-K complex is clearly λP, although some λO and DnaB can be thrown off into solution as well (255, 259, 262). Demonstration of a DnaK-λP interaction has led to the suggestion that DnaK, which is only loosely associated with the nucleoprotein complex, binds the free λP to prevent its reassociation with DnaB (259).

Under the conditions described above, subsequent unwinding (14) and DNA replication (264) are primarily unidirectional and to the right. This is presumably because, as has been demonstrated (255, 259, 262), disassembly of the O-P-B-J-K complex is incomplete with this set of proteins. Both McMacken's (262, 264) and Georgopoulos's (259) laboratories have reported that a GrpE requirement can be observed in the reconstituted systems when the concentration of DnaK is lowered by a factor of 10; however, the DnaK is still required. It has been suggested (15) that under these conditions, the disassembly reaction is considerably more complete, and bidirectional unwinding can be observed.

Reconstitution of *ori*λ replication in vitro with purified proteins also revealed the role of RNAP in *ori*λ replication. While the crude soluble system was sensitive to rifampicin (247–249), the reconstituted system did not require RNAP and was refractory to inhibition by the λ repressor (264). Mensa-Wilmot et al (265) demonstrated that RNAP was required in the crude soluble system to relieve an inhibition of replication caused by the binding of HU protein to the DNA. HU inhibition could be attributed to a block in the progressive formation of the nucleoprotein complexes necessary for initiation at *ori*λ. When RNAP and HU are included in the reconstituted system, inhibition by the λ repressor is recovered, presumably via its action to prevent transcription from the λ promoters. The authors (265) suggested that this mechanism could account for the transcriptional activation step required during replication of bacteriophage λ (266).

The similarity between initiation at *ori*λ and *oriC* is therefore quite high. Nucleoprotein complexes assemble at a specific site on the DNA and, in a reaction that requires superhelicity (267), direct DnaB to an adjacent, destabilized A+T-rich region so that replication forks can form. One clear difference, however, is that at *ori*λ, activation of the helicase activity of DnaB requires a heat shock protein–mediated disassembly reaction, whereas this is not necessary at *oriC*. This leads to the question of whether DnaJ, DnaK, and GrpE play a role during *oriC* replication. It has been shown recently that DnaA will bind to a DnaK column (268), and that DnaK will both dissociate aggregated DnaA protein (269) and stimulate the activity of a mutant DnaA protein (270). However, it is not clear at this time whether this reflects a specific requirement for DnaK in *oriC* replication or the generalized role of the heat shock proteins in healing bruised proteins.

Origins that Require DnaA, but not for Unwinding—P1 and F

DnaA can play a role in initiation of replication that is different from its mechanism of action at *oriC*. In the replication of the mini-P1 and, most likely, mini-F plasmids, DnaA appears to act only to guide DnaB to a plasmid initiator protein-denatured A+T-rich region.

Bacteriophage P1 exists in the lysogenic state as a plasmid at a stringently controlled copy number of one or two per *E. coli* chromosome (271). Mini-P1 plasmids can be derived that exhibit the same copy number control and replication requirements (272, 273). Replication requires a small, plasmid-encoded protein, RepA, which is also required for copy number control (272). The minimal P1 origin (*oriP1*) of 246 bp (274) lies just to the left of *repA* and consists of (from left to right): two DnaA boxes, a 60-bp A+T-rich region that contains five GATC sites for the Dam methylase, a 40-bp G+C-rich spacer region, and five directly repeated 19-bp sequences that have been shown to be binding sites for RepA (275) (Figure 4). Both DnaA boxes are required for replication in vitro (276), whereas one that matches the consensus sequence precisely suffices in vivo (277). Dam methylation is required for replication (278, 279), which is unidirectional to the left (280). Stable plasmid maintenance requires *dnaA, dnaB, dnaC, dnaJ, dnaK,* and *grpE* (281–284).

Wickner & Chattoraj (280) demonstrated mini-P1 replication in vitro using the crude soluble system developed for *oriC* replication (172) supplemented with purified DnaA and RepA. Replication was sensitive to both rifampicin and novobiocin. Requirements for DnaK, DnaJ, and GrpE were demonstrated subsequently in vitro by complementation with the respective purified proteins of replication-deficient crude extracts prepared from strains mutated in *dnaK, dnaJ,* or *grpE* (285). A 10-fold higher concentration of DnaK than DnaJ was required for complementation; however, unlike in the case of *oriλ* replication, the DnaK requirement for *oriP1* replication could not be reduced in the presence of GrpE.

DnaK and DnaJ have been shown to act in the *oriP1* replication reaction as molecular chaperones that bind and activate RepA for binding to the DNA (286). DnaJ forms an isolable equimolar complex with RepA (285). Interestingly, a 100-fold excess of either RepA or the DnaJ-RepA complex is required to saturate RepA binding in vitro to the 19mers. However, in the presence of DnaK, in a reaction requiring ATP hydrolysis and the DnaJ-RepA complex but not the presence of *oriP1*, RepA binding to *oriP1* could be activated 100-fold. Only RepA was subsequently left on the DNA (286). Since GrpE was not required for this reaction, it presumably acts sometime after RepA is transferred to the DNA.

These data, and inference from the mechanism of initiation at *oriC* and *oriλ*, suggest that the DnaK- and DnaJ-activated binding of RepA to the 19mers results in the denaturation of the A+T-rich region. DnaB is deposited at that site via the interaction of DnaA bound at the DnaA boxes with the DnaB-DnaC complex in solution. The fork thus formed proceeds to the left. It has been suggested that the G+C-rich region in the origin acts to both localize denaturation and prevent rightward fork movement (287). The distinctly different action of DnaA at *oriP1* compared to its action at *oriC*, is supported by the following observations: (*a*) *oriP1* requires a much lower concentration of DnaA (276, 280) than *oriC* and unlike *oriC*, can tolerate changes in the position of the DnaA boxes (277) (because DnaA is not involved in nucleoprotein complex formation) and (*b*) the ADP form of DnaA is sufficient to support *oriP1* replication (276) (because DnaA is not involved in denaturing the A+T-rich region).

While similar detailed enzymological studies have not been reported for mini-F [a self-replicating, stringently controlled plasmid derived from the F factor (288)], the mechanism is likely to be highly analogous. The origin structure is very similar (289) (Figure 4). Replication requires the plasmid-encoded initiation protein, RepE (290), and DnaA (281), DnaJ, DnaK, and GrpE (291). In an interesting twist, transcription of *repE* is controlled by the heat shock sigma factor RpoH (σ^{32}) (292), while *rpoH* expression is controlled by DnaA (293).

Initiation by D- or R-Loop Formation—pBR322

The replication and copy number control of pBR322 DNA, a small 4.6-kb plasmid that propagates via a ColE1-type origin (294), has been studied intensely in vivo and in vitro. Transcription by RNAP is essential to the initiation of pBR322 DNA replication. Here, unlike at *oriC*, the transcript clearly acts as a primer for DNA synthesis and is also necessary for the specific unwinding of the origin region, thereby activating other replication signals.

Transcription of the leading-strand primer precursor (RNA II) is initiated 555 nt upstream of the origin of leading-strand DNA synthesis (295) (Figure 4). RNA II forms a stable RNA-DNA hybrid with the leading-strand template DNA that starts 10–20 nt upstream of the origin. RNase H specifically processes this RNA-DNA hybrid at one of three consecutive A residues providing a 3'-OH primer terminus for the initiation of leading-strand DNA replication (296). This 3'-OH end can only be extended by DNA Pol I, probably because steric hindrance by the specifically folded, upstream, nonhybridized portion of RNA II prevents access of the larger DNA Pol III HE (297). DNA Pol I extends this primer for 200–400 nt, forming the 6sL fragment (170), the first detectable nascent DNA. In addition, this synthetic event results in the formation of a stable D-loop between the nascent 6sL fragment and the plasmid H strand (the leading-strand template) (298). Displacement of the plasmid L strand from a double-stranded conformation to a single-stranded one, as well as the concomitant binding of SSB to the displaced strand, leads to the activation of a PAS [PAS-BL (9)] situated 150 nt downstream of the origin. A ϕX-type primosome assembles on the activated PAS that subsequently migrates in the 5'→3' direction along the lagging-strand template, synthesizing primers for Okazaki fragment synthesis and unwinding the duplex DNA. The requirement for the ϕX-type primosome has been established both in vivo (299) and in vitro, either in a replication system utilizing a crude soluble extract (300) or in one completely reconstituted with purified proteins (16). With the addition of the DNA Pol III HE, a replication fork forms and moves unidirectionally to the right to complete replication of the molecule. It is not clear why nascent leading-strand synthesis by DNA Pol I is limited to the 6sL fragment. However, it is clear that the DNA Pol III HE is responsible for the bulk of nascent strand synthesis (301).

Surprisingly, pBR322 DNA can be propagated in *E. coli* hosts that lack RNase H (302). Under this condition, DNA Pol I is also no longer required (303, 304). Dasgupta et al (304) suggested that this could be accounted for by DnaB binding to a small R-loop just downstream of the normal origin. However, if this region were coated with SSB, DnaB access would be prevented. On the other hand, Parada & Marians (305) showed that in the absence of RNase H, the RNA II-H strand hybrid can extend as far as 3 kb downstream of the origin. Even in the absence of DNA gyrase, the hybrid extended far enough downstream to activate the PAS on the lagging-strand template. These authors suggested that initiation under these conditions would be asymmetric. Normal lagging-strand initation would occur. The leading-strand could initiate via the DNA Pol III HE finding the 3'-OH of the RNA II once the RNAP had dissociated. Since the primer terminus would now be far removed from the upstream, nonhybrdized portion of RNA II, access by the DNA Pol III HE would not be hindered and the DNA Pol I requirement would be obviated.

DnaA AND pBR322 REPLICATION   There are two DnaA boxes in the form of an inverted repeat on pBR322 DNA between the origin of leading-strand DNA synthesis and PAS-BL (Figure 4). One DnaA box fits the consensus sequence precisely, while the other is off by two nucleotides. Two reports indicated that pBR322 DNA replication was affected in a *dnaA* strain (306, 307). The rate of pBR322 DNA synthesis decreased at the nonpermissive temperature and the mode of replication switched from θ-type to a rolling circle mechanism. Subsequently, it was demonstrated that pBR322 DNA replication in crude soluble systems could be stimulated by DnaA (308, 309). Replication still required RNAP and DnaB, but no longer required DnaT, indicating that DnaA was acting in some manner to load an *oriC*-type primosome to the DNA.

Subsequently, Parada & Marians (310) have shown that either R- or D-loop formation in the region of the DnaA boxes is required for DnaA-mediated pBR322 DNA replication. Functional DnaA binding, which resulted in subsequent replication, was shown to be to the double-stranded side of these displacement loops, leading to the conclusion that since replication under these conditions was unidirectional and to the right (Figure 4), DnaA was mediating the loading of DnaB in trans to the displaced single strand. Nurse et al (145) found that the copy number of a pBR322 derivative in which the DnaA boxes had been inactivated was two-thirds that of the wild-type, suggesting that DnaA-mediated replication could account for the maintenance of about one-third of the pBR322 DNA in the cell.

Primary Initiation on Bacteriophages T4 and T7

Early studies showed that initiation of the ~40 kb bacteriophage T7 DNA (at *oriT7*) occurs via the formation of a replication bubble at a position 17% from the lefthand end and proceeds bidirectionally from the bubble (311). Subsequent deletion mapping indicated that a 200-bp region between 14.75 and 15% from the left end was required for this event (312). Mapping of a primary origin in bacteriophage T4 has been more difficult because of the circular permutation of the molecules, the rapid inactivation of the primary pathway, and subsequent initiations via at least two different mechanisms (313). Identification of one particular origin (*oriA*) relied on the isolation of early nascent initiator DNA and subsequent mapping to the chromosome (17, 314).

The mechanisms of initiation at *oriT7* and *oriA* are very similar to the mechanism of initiation of pBR322 DNA replication. *oriT7* consists of, from left to right, two T7 RNAP promoters (ϕ1.1A and ϕ1.1B) followed by a 61-bp A+T-rich region that contains a T7 g4P priming site on the template strand (315) (Figure 4). Initiation in vivo requires the T7 RNAP (17). Studies in Richardson's laboratory using plasmids carrying *oriT7* as templates for replication in vitro with purified T7 replication proteins have indicated the operative mechanism (17, 316, 317). As with pBR322 DNA, leading-strand DNA synthesis to the right is initiated by the extension by the T7 DNA Pol of a 3'-OH primer terminus derived from transcripts originating at ϕ1.1A or ϕ1.1B. The transcripts used to prime DNA synthesis are 10–60 nt long. The combination of transcription and the high A+T (79%) character of the region downstream of the promoters melts the DNA and allows access of the T7 g4P to the nontemplate strand. The g4P translocates in the 5'→3' direction, unwinding the duplex until the first priming site is reached. At this point a rightward-moving fork is established. The first Okazaki fragment, which is being elongated to the left, becomes the leftward-moving leading strand. Subsequent DNA denaturation action of an SSB moves this strand to the left until a T7 g4P priming site is reached on the opposite template strand, allowing access of a g4P to provide helicase action and priming of Okazaki fragment synthesis for the now formed leftward-moving fork. This model is consistent with the positions of RNA/DNA transition points that have been mapped in vivo (318). However, other regions that do not act as origins exist on the T7 genome, in which promoters are followed by A+T-rich stretches. Thus, some unique feature of the primary origin region has yet to be uncovered.

Primary initiation on bacteriophage T4 requires the host RNAP (319). Mosig and her colleagues (18) have proposed that leftward leading-strand initiation occurs at *oriA* via a 3'-OH primer terminus provided by transcripts from one of three promoters, $P_{ori}1$, $P_{ori}3$, or $P_{ori}4$. A specific nascent DNA of ~70 nt can be isolated attached to RNA and may be a result of a pause by the

T4 DNA Pol at a polypurine tract. If so, this would act to displace the nontemplate strand. The region opposite the nascent leading-strand 70mer contains three priming sites for the T4 g61P. Thus, presumably, the T4 primosome can gain access to the DNA at this point, resulting in the formation of the leftward-moving fork. The pathway to formation of the rightward-moving fork is not as obvious. The T4 topoisomerase is required (313). Mosig et al (18) point out that another promoter, $P_{ori}2$, which is on the opposite strand from and overlaps $P_{ori}4$, could act to reiterate the same mechanism in the rightward direction. The bulk of T4 replication, however, appears to be a result of host RNAP-independent pathways that involve recombinational initiation (319–321) and *motA*-dependent, middle-mode T4 promoters (322-324).

ARREST OF REPLICATION FORK PROGRESSION

In *E. coli,* the bidirectionally moving replication forks that form at *oriC* meet 180° around on the chromosome in the *Ter* region. Marker frequency experiments demonstrated that the *Ter* region was, in fact, a replication fork trap of about 450 kb (20, 21). Replication forks from either direction were not prevented from entering the *Ter* region, but they were prohibited from exiting when they progressed across the terminus and reached the other side. Subsequent studies showed this could be attributed to the trans-action of the *tus* gene (22) and the cis-action of six *Ter* sites symmetrically disposed about the terminus region (325–327). Although the requirement in vivo for this termination mechanism is not clear—strains deleted of 340 kb of the *Ter* region are viable, although they produce many nonviable cells and exhibit filamentation (328)—the biochemistry of the reaction has proved interesting.

The 36-kDa Tus protein (329–331) binds *Ter* sequences with high affinity. K_DS range from 3×10^{-13} M for the Tus-*TerB* [(derived from the chromosomal *Ter* region (327)] interaction (with a half-life of 550 min) to 1×10^{-11} for the Tus-*TerR2* [from plasmid R6K (330, 332)] interaction (332a). Arrest of replication fork progression on small plasmids in vivo (325, 326, 330) or in crude extracts in vitro (333) was polar. That is, only one orientiation of the nonpalindromic *Ter* sequences would arrest progression of the approaching replication fork. Using purified Tus and replication systems reconstituted in vitro with purified proteins, Lee et al (334) and Hill & Marians (335) demonstrated a similar polar effect on replication forks formed at *oriC* and the pBR322 origin, respectively.

Khatri et al (336) and Lee et al (334) found that this polar effect could be explained by the action of bound Tus to inhibit the the unwinding catalyzed by DNA helicases in a polar fashion. In experiments that utilized short regions (~30 nt) bound by Tus as the duplex DNA to be displaced, Tus-induced, *Ter* polarity–specific inhibition of DnaB (334, 336), UvrD (334) (helicase II), Rep (334), and PriA [A. Kornberg, personal communication; Hiasa and Marians (336a)] could be demonstrated. Thus, it has been suggested that replication fork arrest is simply a matter of generalized (but polar) inhibition of the replication fork helicase. This is consistent with footprints of Tus on the DNA that show an asymmetric binding of the strands across the *Ter* sequence such that when it is oriented in the active direction, an approaching replication fork would encounter the Tus protein spanning the major grove, bound to both strands; while, when it was oriented in the opposite direction, the approaching fork would encounter Tus bound only to one strand and thus could presumably displace it (332a). The observation that leading-strand synthesis is terminated one nucleotide away from bound Tus also supports this view (335; A. Kornberg, personal communication).

Some recent observations, however, run counter to this view. Hiasa & Marians (336a) have demonstrated that while DnaB is blocked by Tus bound to short oligonucleotides (30–60 nt), its helicase activity is unaffected when the *Ter* site is in the middle of a 250-nt-long fragment. This suggests that while Tus bound to a short fragment may be capable of preventing the translocation of a protein on a single strand, its presence is insufficient to block a helicase actively engaged in unwinding a long stretch of duplex DNA. These authors also observed that, using the same elongated helicase substrate, the $5' \rightarrow 3'$ helicase activity of the ϕX-type primosome was inhibited. Thus, the role of protein-protein interactions in Tus action must be reconsidered.

ACKNOWLEDGMENTS

I would like to thank all those colleagues who provided reprints and preprints. I am grateful that Drs. H. Hiasa, J. Hurwitz, S. Rabkin, S. Shuman, and K. Zavitz took the time to provide critical commentary on the manuscript. I also thank Drs. K. Zavitz and E. Zechner for Figures 1 and 2, respectively. Studies from the author's laboratory were supported by NIH grants GM34557 and GM34558.

Literature Cited

1. Kohara, Y., Akiyama, K., Isono, K. 1987. *Cell* 50:495–508
2. Helmstetter, C. E. 1987. *Escherichia coli and Salmonella typhimurium*, ed. F. C. Neidhardt, J. L. Ingraham, K. B. Low, B. Magasanik, M. Schechter, H. E. Umbarger, pp. 1594–605. Washington, DC: Am. Soc. Microbiol.
3. Nossal, N. G. 1983. *Annu. Rev. Biochem.* 52:581–615
4. Matson, S. W., Kaiser-Rogers, K. A. 1990. *Annu. Rev. Biochem.* 59:289–329

5. Cozzarelli, N. R., Wang, J. C., eds. 1990. *DNA Topology and its Biological Effects*. Cold Spring Harbor, NY: Cold Spring Harbor Lab. 480 pp.
6. Chase, J. W., Williams, K. R. 1986. *Annu. Rev. Biochem.* 55:103–36
7. Meyer, R. R., Laine, P. S. 1990. *Microbiol. Rev.* 54:342–80
8. McHenry, C.S. 1988. *Annu. Rev. Biochem.* 57:519–50
9. Marians, K. J. 1984. *CRC Crit. Rev. Biochem.* 17:153–215

10. McMacken, R., Silver, L., Georgopoulos, C. 1987. See Ref. 2, pp. 564–612
11. Salas, M. 1991. *Annu. Rev. Biochem.* 60:39–71
12. Kornberg, A., Baker, T. 1992. *DNA Replication.* New York: Freeman. 931 pp. 2nd ed.
13. Bramhill, D., Kornberg, A. 1988. *Cell* 52:743–55
14. Dodson, M., Echols, H., Wickner, S., Alfano, C., Mensa-Wilmot, K., et al. 1986. *Proc. Natl. Acad. Sci. USA* 83: 7638–42
15. Echols, H. 1990. *J. Biol. Chem.* 265: 14697–700
16. Minden, J., Marians, K. J. 1985. *J. Biol. Chem.* 260:9316–25
17. Romano, L. J., Tamonoi, F., Richardson, C. C. 1991. *Proc. Natl. Acad. Sci. USA* 78:4107–11
18. Mosig, G., Macdonald, P. M., Powell, D., Trupin, M., Gary, T. 1987. *DNA Replication and Recombination,* ed. R. McMacken, T. J. Kelly, pp. 403–14. New York: Liss
19. Arai, K.-I., Kornberg, A. 1981. *Proc. Natl. Acad. Sci. USA* 78:69–73
20. Hill, T. M., Henson, J. M., Kuempel, P. L. 1987. *Proc. Natl. Acad. Sci. USA* 84:1754–58
21. deMassy, B., Bejar, S., Louarn, J., Louarn, J. M., Bouche, J. P. 1987. *Proc. Natl. Acad. Sci. USA* 84:1759–63
22. Hill, T. M., Kopp, B. J., Kuempel, P. L. 1988. *J. Bacteriol.* 170:662–68
23. Donachie, W. D., Begg, K. J. 1989. *J. Bacteriol.* 171:5405–9
24. Hiraga, S., Ogura, T., Niki, H., Ichinose, C., Mori, H. 1990. *J. Bacteriol.* 172:31–39
25. Grippo, P., Richardson, C. C. 1971. *J. Biol. Chem.* 246:6867–73
26. Modrich, P., Richardson, C. C. 1975. *J. Biol. Chem.* 250:5508–14
27. Modrich, P., Richardson, C. C. 1975. *J. Biol. Chem.* 250:5515–22
28. Tabor, S., Huber, H. E., Richardson, C. C. 1987. *J. Biol. Chem.* 262:16212–23
29. Huber, H. E., Tabor, S., Richardson, C. C. 1987. *J. Biol. Chem.* 262:16224–32
30. Russel, M., Model, P. 1986. *J. Biol. Chem.* 261:14997–5005
31. Huber, H. E., Russel, M., Model, P., Richardson, C. C. 1986. *J. Biol. Chem.* 261:15006–12
32. Goulian, M., Lucas, Z. J., Kornberg, A. 1968. *J. Biol. Chem.* 243:627–38
33. Nossal, N. G. 1974. *J. Biol. Chem.* 249:5668–76
34. Morris, C. F., Sinha, N. K., Alberts, B. M. 1975. *Proc. Natl. Acad. Sci. USA* 72:4800–4
35. Alberts, B. M., Morris, S., Mace, D., Sinha, N., Bittner, M., Moran, L. 1975. *DNA Synthesis and its Regulation,* ed. M. Goulian, P. Hanawalt, pp. 241–69. Menlo Park, Calif: Benjamin
36. Piperno, J. R., Alberts, B. M. 1978. *J. Biol. Chem.* 253:5174–79
37. Piperno, J. R., Kallen, R. G., Alberts, B. M. 1978. *J. Biol. Chem.* 253:5180–85
38. Barry, J., Alberts, B. M. 1972. *Proc. Natl. Acad. Sci. USA* 69:2717–21
39. Huang, C.-C., Hearst, J. E., Alberts, B. M. 1981. *J. Biol. Chem.* 256:4087–94
40. Roth, A. C., Nossal, N. G., Englund, P. T. 1982. *J. Biol. Chem.* 257:1267–73
41. Mace, D. C., Alberts, B. M. 1984. *J. Mol. Biol.* 177:279–93
42. Rush, J., Lin, T.-C., Quinones, M., Spicer, E. K., Douglas, I., et al. 1989. *J. Biol. Chem.* 264:10943–53
43. Jarvis, T. C., Paul, L. S., Hockensmith, J. W., von Hippel, P. H. 1989. *J. Biol. Chem.* 264:12717–29
44. Jarvis, T. C., Newport, J. C., von Hippel, P. H. 1991. *J. Biol. Chem.* 266: 1830–40
45. Richardson, R. W., Ellis, R. L., Nossal, N. G. 1990. *Molecular Mechanisms in DNA Replication and Recombination,* ed. C. C. Richardson, I. R. Lehman, pp. 247–59. New York: Wiley-Liss
46. Capson, T. L., Benkovic, S. J., Nossal, N. G. 1991. *Cell* 65:249–58
47. Munn, M. M., Alberts, B. 1991. *J. Biol. Chem.* 266: 20024–33
48. Munn, M. M., Alberts, B. 1991. *J. Biol. Chem.* 266: 20034–44
49. Selick, H. E., Barry, J., Cha, T.-A., Munn, M., Nakanishi, M., et al. 1987. See Ref. 18, pp. 183–214
50. Nusslein, V., Otto, B., Bonhoeffer, F., Schaller, H. 1971. *Nature New Biol.* 234:285–86
51. Gefter, M. L., Hirota, Y., Kornberg, T., Wechsler, J. A., Barnoux, C. 1971. *Proc. Natl. Acad. Sci USA* 68:3150–53
52. Kornberg, T., Gefter, M. L. 1972. *J. Biol. Chem.* 247:5369–75
53. Wickner, W., Kornberg, A. 1974. *J. Biol. Chem.* 249:6244–49
54. Hurwitz, J., Wickner, S., Wright, M. 1973. *Biochem. Biophys. Res. Commun.* 51:257–67
55. Hurwitz, J., Wickner, S. 1974. *Proc. Natl. Acad. Sci. USA* 71:6–10
56. Maki, H., Maki, S., Kornberg, A. 1988. *J. Biol. Chem.* 263:6570–78
57. McHenry, C. S., Crow, W. 1979. *J. Biol. Chem.* 254:1748–53
58. Welch, M. M., McHenry, C. S. 1982. *J. Bacteriol.* 152:351–56

59. Scheuermann, R., Tam, S., Burgers, P. M. J., Lu, C., Echols, H. 1983. *Proc. Natl. Acad. Sci. USA* 80:7085–89
60. Spanos, A., Sedwick, S. G., Yarronton, G. T., Hübscher, U., Banks, G. R. 1981. *Nucleic Acids Res.* 9:1825–39
61. Maki, H., Horiuchi, T., Kornberg, A. 1985. *J. Biol. Chem.* 260:12982–86
62. Scheuermann, R., Echols, H. 1984. *Proc. Natl. Acad. Sci. USA* 81:7747–51
63. Fay, P. J., Johanson, K. D., McHenry, C. S., Bambara, R. A. 1981. *J. Biol. Chem.* 256:976–83
64. Mullin, D., Woldringh, C., Henson, J., Walker, J. 1983. *Mol. Gen. Genet.* 192: 73–79
65. Kodaira, M., Biswas, S. B., Kornberg, A. 1983. *Mol. Gen. Genet.* 192:80–86
66. McHenry, C. S. 1982. *J. Biol. Chem.* 257:2657–63
67. Fay, P. J., Johanson, K. O., McHenry, C. S., Bambara, R. A. 1982. *J. Biol. Chem.* 257:5692–99
68. Wickner, W., Schekman, R., Geider, K., Kornberg, A. 1973. *Proc. Natl. Acad. Sci. USA* 70:1764–67
69. McHenry, C. S., Kornberg, A. 1977. *J. Biol. Chem.* 252:6478–84
70. Tsuchihashi, Z., Kornberg, A. 1990. *Proc. Natl. Acad. Sci. USA* 87:2516–20
71. Flower, A. M., McHenry, C. S. 1990. *Proc. Natl. Acad. Sci. USA* 87:3713–17
72. Blinkowa, A. L., Walker, J. R. 1990. *Nucleic Acids Res.* 18:1725–29
73. McHenry, C. S., Oberfelder, R., Johanson, K. O., Tomasiewicz, H., Franden, M. 1987. See Ref. 18, pp. 47–61
74. Maki, S., Kornberg, A. 1988. *J. Biol. Chem.* 263:6555–60
75. Burgers, P. M. J., Kornberg, A., Sakakibara, Y. 1981. *Proc. Natl. Acad. Sci. USA* 78:5391–95
75a. Wu, C. A., Zechner, E. L., Marians, K. J. 1992. *J. Biol. Chem.* 267:4030–44
76. Johanson, K. O., McHenry, C. S. 1980. *J. Biol. Chem.* 255:10984–90
77. O'Donnell, M. E. 1987. *J. Biol. Chem.* 262:16558–65
78. Wickner, S. 1976. *Proc. Natl. Acad. Sci. USA* 73:3511–15
79. Studwell, P. S., O'Donnell, M. 1990. *J. Biol. Chem.* 265:1171–78
80. O'Donnell, M., Studwell, P. S. 1990. *J. Biol. Chem.* 265:1179–87
81. Maki, S., Kornberg, A. 1988. *J. Biol. Chem.* 263:6547–54
82. Lee, S. H., Walker, J. 1987. *Proc. Natl. Acad. Sci. USA* 84:2713–17
83. Tsuchihashi, Z., Kornberg, A. 1989. *J. Biol. Chem.* 264:17790–95
84. Maki, S., Kornberg, A. 1988. *J. Biol. Chem.* 263:6561–69
85. Stukenberg, P. T., Studwell-Vaughan, P. S., O'Donnell, M. 1991. *J. Biol. Chem.* 266:11328–34
86. Onrust, R., Stuckenberg, P. T., O'Donnell, M. 1991. *J. Biol. Chem.* 266: 21681–86
87. Griep, M. A., McHenry, C. S. 1990. *J. Biol. Chem.* 265:20356–63
88. Scherzinger, E., Lanka, E., Morelli, G., Seifert, D., Yuki, A. 1977. *Eur. J. Biochem.* 72:543–58
89. Scherzinger, E., Lanka, E., Hillenbrand, G. 1977. *Nucleic Acids Res.* 4:4151–63
90. Romano, L., Richardson, C. C. 1979. *J. Biol. Chem.* 254:10483–89
91. Matson, S. W., Tabor, S., Richardson, C. C. 1983. *J. Biol. Chem.* 258:14017–24
92. Kolodner, R., Richardson, C. C. 1977. *Proc. Natl. Acad. Sci. USA* 74:1525–29
93. Matson, S. W., Richardson, C. C. 1983. *J. Biol. Chem.* 258:14009–16
94. Tabor, S., Richardson, C. C. 1981. *Proc. Natl. Acad. Sci. USA* 78:205–9
95. Nakai, H., Richardson, C. C. 1986. *J. Biol. Chem.* 261:15208–16
96. Nakai, H., Richardson, C. C. 1986. *J. Biol. Chem.* 261:15217–24
97. Hinkle, D. C., Richardson, C. C. 1975. *J. Biol. Chem.* 250:5523–29
98. Dunn, J. J., Studier, F. W. 1981. *J. Mol. Biol.* 148:303–30
99. Bernstein, J. A., Richardson, C. C. 1988. *Proc. Natl. Acad. Sci. USA* 85: 396–400
100. Nakai, H., Richardson, C. C. 1988. *J. Biol. Chem.* 263:9818–30
101. Bernstein, J. A., Richardson, C. C. 1989. *J. Biol. Chem.* 264:13066–73
102. Bernstein, J. A., Richardson, C. C. 1988. *J. Biol. Chem.* 263:14891–99
103. Mendelman, L. V., Richardson, C. C. 1991. *J. Biol. Chem.* 266:23240–50
104. Venkatesan, M., Silver, L. L., Nossal, N. G. 1982. *J. Biol. Chem.* 257:12426–34
105. Morris, C. F., Moran, L. A., Alberts, B. M. 1979. *J. Biol. Chem.* 254:6797–802
106. Liu, C.-C., Alberts, B. M. 1981. *J. Biol. Chem.* 256:2813–20
107. Nossal, N. G. 1980. *J. Biol. Chem.* 255:2176–82
108. Liu, C.-C., Alberts, B. M. 1980. *Proc. Natl. Acad. Sci. USA* 77:5698–702
109. Liu, C.-C., Alberts, B. M. 1981. *J. Biol. Chem.* 256:2821–29
110. Burke, R. L., Munn, M., Barry, J., Alberts, B. M. 1985. *J. Biol. Chem.* 260:1711–22
111. Hinton, D. M., Nossal, N. G. 1985. *J. Biol. Chem.* 260:12858–65

112. Nossal, N. G., Hinton, D. M. 1987. *J. Biol. Chem.* 262:10879–85
113. Richardson, R. W., Nossal, N. G. 1989. *J. Biol. Chem.* 264:4725–31
114. Richardson, R. W., Nossal, N. G. 1989. *J. Biol. Chem.* 264:4732–39
115. Cha, T.-A., Alberts, B. M. 1990. *Biochemistry* 29:1791–98
116. LeBowitz, J. H., McMacken, R. 1986. *J. Biol. Chem.* 261:4738–48
117. Wechsler, J. A., Gross, J. D. 1971. *Mol. Gen. Genet.* 113:273–84
118. Wickner, S., Wright, M., Hurwitz, J. 1974. *Proc. Natl. Acad. Sci. USA* 71:783–87
119. Ueda, K., McMacken, R., Kornberg, A. 1978. *J. Biol. Chem.* 253:261–69
120. Reha-Krantz, L. J., Hurwitz, J. 1978. *J. Biol. Chem.* 253:4051–57
121. Arai, K.-I., Kornberg, A. 1981. *J. Biol. Chem.* 256:5253–59
122. Zechel, K., Bouche, J.-P., Kornberg, A. 1975. *J. Biol. Chem.* 250:4684–89
123. Bouche, J.-P., Zechel, K., Kornberg, A. 1975. *J. Biol. Chem.* 250:5995–6001
124. Grompe, M., Versalovic, J., Koeuth, T., Lupski, J. R. 1991. *J. Bacteriol.* 173:1268–78
125. Arai, K.-I., Kornberg, A. 1979. *Proc. Natl. Acad. Sci. USA* 76:4308–12
126. Arai, K.-I., Kornberg, A. 1981. *J. Biol. Chem.* 256:5267–72
127. Arai, K.-I., Kornberg, A. 1981. *J. Biol. Chem.* 256:5260–66
128. Wickner, S., Hurwitz, J. 1974. *Proc. Natl. Acad. Sci. USA* 71:4120–24
129. Schekman, R., Weiner, J., Weiner, A., Kornberg, A. 1975. *J. Biol. Chem.* 250:5859–65
130. Arai, K.-I., McMacken, R., Yasuda, S.-I., Kornberg, A. 1981. *J. Biol. Chem.* 256:5281–86
131. Wickner, S., Hurwitz, J. 1975. *Proc. Natl. Acad. Sci. USA* 72:3342–46
132. Shlomai, J., Kornberg, A. 1980. *J. Biol. Chem.* 256:6789–93
133. Low, R. L., Shlomai, J., Kornberg, A. 1982. *J. Biol. Chem.* 257:6242–50
134. Weiner, J., McMacken, R., Kornberg, A. 1976. *Proc. Natl. Acad. Sci. USA* 73:752–56
135. Wickner, S. 1978. *The Single-Stranded DNA Phages,* ed. D. T. Denhardt, D. Dressler, D. S. Ray, pp. 255–71. Cold Spring Harbor, NY: Cold Spring Harbor Lab.
136. Shlomai, J., Kornberg, A. 1980. *Proc. Natl. Acad. Sci. USA* 77:799–803
137. Wickner, S., Hurwitz, J. 1975. *Proc. Natl. Acad. Sci. USA* 72:921–25
138. Kabori, J., Kornberg, A. 1982. *J. Biol. Chem.* 257:13770–75
139. LeBowitz, J. H., Zylicz, M., Georgopoulos, C., McMacken, R. 1985. *Proc. Natl. Acad. Sci. USA* 82:3988–92
140. Lee, M. S., Marians, K. J. 1987. *Proc. Natl. Acad. Sci. USA* 84:8345–49
141. Lasken, R. S., Kornberg, A. 1988. *J. Biol. Chem.* 263:5512–18
142. Lee, M. S., Marians, K. J. 1989. *J. Biol. Chem.* 264:14531–42
143. Masai, H., Nomura, N., Arai, K.-I. 1990. *J. Biol. Chem.* 265:15134–44
144. Lee, E. H., Kornberg, A. 1991. *Proc. Natl. Acad. Sci. USA* 88:3029–32
145. Nurse, P., Zavitz, K. H., Marians, K. J., 1991. *J. Bacteriol.* 173:6686–93
146. Nurse, P., DiGate, R. J., Zavitz, K. H., Marians, K. J. 1990. *Proc. Natl. Acad. Sci. USA* 87:4615–19
147. Lee, E. H., Masai, H., Allen, G. C. Jr., Kornberg, A. 1990. *Proc. Natl. Acad. Sci. USA* 87:4620–24
148. Gottesman, S., Halpern, E., Trisler, P. 1981. *J. Bacteriol.* 148:265–73
148a. Zavitz, K., Marians, K. J. 1992. *J. Biol. Chem.* 267:In press
149. Allen, G. C. Jr., Kornberg, A. 1991. *J. Biol. Chem.* 266:11610–13
150. Zavitz, K. H., DiGate, R. J., Marians, K. J. 1991. *J. Biol. Chem.* 266:13988–95
151. Nakai, H., Richardson, C. C. 1988. *J. Biol. Chem.* 263:9831–39
152. Alberts, B., Barry, J., Bedinger, P., Formosa, T., Jongeneel, C. V., Kreuzer, K. N. 1983. *Cold Spring Harbor Symp. Quant. Biol.* 47:655–88
153. Cha, T.-A., Alberts, B. M., 1989. *J. Biol. Chem.* 264:12220–25
154. Mok, M., Marians, K. J. 1987. *J. Biol. Chem.* 262:2304–9
155. Mok, M., Marians, K. J. 1987. *J. Biol. Chem.* 262:16644–54
156. Marians, K. J., DiGate, R. J., Mok, M., Johnson, E., Parada, C., et al. 1990. See Ref. 45, pp. 289–301
156a. Zechner, E. L., Wu, C. A., Marians, K. J. 1992. *J. Biol. Chem.* 267:4045–53
156b. Zechner, E. L., Wu, C. A., Marians, K. J. 1992. *J. Biol. Chem.* 267:4054–63
156c. Wu, C. A., Zechner, E. L., Hughes, A. J. Jr., Frandan, M. A., McHenry, C. S., Marians, K. J. 1992. *J. Biol. Chem.* 267:4064–73
156d. Wu, C. A., Zechner, E. L., Reems, J. A., McHenry, C. S., Marians, K. J. 1992. *J. Biol. Chem.* 267:4074–83
157. Sinha, N. K., Morris, C. F., Alberts, B. M. 1980. *J. Biol. Chem.* 255:4290–303
158. Studwell-Vaughan, P., O'Donnell, M. 1991. *J. Biol. Chem.* 266:19833–41
159. Johanson, K. O., McHenry, C. S. 1984. *J. Biol. Chem.* 259:4589–95

160. McMacken, R., Ueda, K., Kornberg, A. 1977. *Proc. Natl. Acad. Sci. USA* 74:4190–94
161. McMacken, R., Kornberg, A. 1978. *J. Biol. Chem.* 253:3313–19
162. Kitani, T., Yoda, K., Ogawa, T., Okazaki, T. 1985. *J. Mol. Biol.* 184:45–52
163. Oka, A., Sugimoto, K., Takanami, M. M., Hirota, Y. 1980. *Mol. Gen. Genet.* 178:9–20
164. Lark, K. D. 1972. *J. Mol. Biol.* 64:47–60
165. Bird, R. E., Louarn, J. M., Martuscelli, J., Caro, L. 1972. *J. Mol. Biol.* 70:549–66
166. Masters, M., Broda, P. 1971. *Nature New Biol.* 232:137–40
167. Yasuda, S., Hirota, Y. 1977. *Proc. Natl. Acad. Sci. USA* 74:5458–62
168. vonMeyenburg, K., Hansen, F. G., Nielsen, L. D., Riise, E. 1978. *Mol. Gen. Genet.* 160:287–95
169. Miki, T., Hiraga, S., Nagata, T., Yura, T. 1978. *Proc. Natl. Acad. Sci. USA* 75:5099–103
170. Sakakibara, Y., Tomizawa, J.-I. 1974. *Proc. Natl. Acad. Sci. USA* 71:1403–7
171. Zyskind, J. W., Cleary, J. M., Brusilow, W. S., Harding, N. E., Smith, D. W. 1983. *Proc. Natl. Acad. Sci. USA* 80:1164–68
172. Fuller, R. S., Kornberg, A. 1983. *Proc. Natl. Acad. Sci. USA* 80:5817–21
173. Fuller, R. S., Funnell, B. E., Kornberg, A. 1984. *Cell* 38:889–900
174. Matsui, M., Oka, A., Takanami, M., Yasuda, S., Hirota, Y. 1985. *J. Mol. Biol.* 184:529–33
175. Geier, G. G., Modrich, P. 1979. *J. Biol. Chem.* 254:1408–13
176. Fuller, R. S., Kaguni, J. M., Kornberg, A. 1981. *Proc. Natl. Acad. Sci. USA* 78:7370–74
177. Sekimizu, K., Bramhill, D., Kornberg, A. 1987. *Cell* 50:259–65
178. Sekimizu, K., Kornberg, A. 1988. *J. Biol. Chem.* 263:7131–35
179. Yung, B. Y., Kornberg, A. 1988. *Proc. Natl. Acad. Sci. USA* 85:7202–5
180. Hughes, P., Landoulsi, A., Kohiyama, M. 1988. *Cell* 55:343–50
181. Sekimizu, K., Bramhill, D., Kornberg, A. 1988. *J. Biol. Chem.* 263:7124–30
182. Rouvière-Yaniv, J., Gros, F. 1975. *Proc. Natl. Acad. Sci. USA* 72:3428–32
183. Funnell, B. E., Baker, T. A., Kornberg, A. 1987. *J. Biol. Chem.* 262:10327–34
184. Baker, T. A., Sekimizu, K., Funnell, B. E., Kornberg, A. 1986. *Cell* 45:53–64
185. Baker, T. A., Funnell, B. E., Kornberg, A. 1987. *J. Biol. Chem.* 262:6877–85
186. Kowalski, D., Eddy, M. J. 1989. *EMBO J.* 8:4335–44

187. Gille, H., Messer, W. 1991. *EMBO J.* 10:1579–84
188. Yung, B. Y., Kornberg, A. 1989. *J. Biol. Chem.* 264:6146–50
189. Asai, T., Takanami, M., Imai, M. 1990. *EMBO J.* 9:4065–72
190. Hwang, D. S., Kornberg, A. 1990. *Cell* 63:325–31
191. Thöny, B., Hwang, D. S., Fradkin, L., Kornberg, A. 1991. *Proc. Natl. Acad. Sci. USA* 88:4066–70
192. Baker, T. A., Kornberg, A. 1988. *Cell* 55:113–23
193. Broyles, S. S., Pettijohn, D. E. 1986. *J. Mol. Biol.* 187:47–60
194. Stenzel, T. T., Patel, P., Bastia, D. 1987. *Cell* 49:709–17
195. Polaczek, P. 1990. *New Biol.* 2:265–71
196. Filutowicz, M., Ross, W., Wild, J., Gourse, R. L. 1992. *Proc. Natl. Acad. Sci. USA.* In press
196a. Gille, H., Egan, J. B., Roth, A., Messer, W. 1991. *Nucleic Acids Res.* 19:4167–72
197. Wada, M., Kano, Y., Ogawa, T., Okazaki, T., Imamoto, F. 1988. *J. Mol. Biol.* 204:581–91
198. Mechulam, Y., Fayat, G., Blanquet, S. 1985. *J. Bacteriol.* 163:787–91
199. Flamm, E. L., Weisberg, R. A. 1985. *J. Mol. Biol.* 183:117–28
200. Ogura, T., Niki, H., Kono, Y., Imamoto, F., Hiraga, S. 1990. *Mol. Gen. Genet.* 220:197–203
201. Kano, Y., Ogawa, T., Hiraga, S., Okazaki, T., Imamoto, F. 1992. *Gene.* In press
202. Ogawa, T., Wada, M., Kano, Y., Imamoto, F., Okazaki, T. 1989. *J. Bacteriol.* 171:5672–79
203. Funnell, B. E., Baker, T. A., Kornberg, A. 1986. *J. Biol. Chem.* 261: 5616–5624
204. Ogawa, T., Baker, T. A., van der Ende, A., Kornberg, A. 1985. *Proc. Natl. Acad. Sci. USA* 82:3562–66
205. Skarstad, K., Baker, T. A., Kornberg, A. 1990. *EMBO J.* 9:2341–48
206. Morelli, G., Buhk, H., Fisseau, C., Lother, H., Yoshinaga, K., Messer, W. 1981. *Mol. Gen. Genet.* 184:255–59
207. Buhk, H. J., Messer, W. 1983. *Gene* 24:265–79
208. Deleted in proof
209. Lother, H., Messer, W. 1981. *Nature* 294:376–78
210. Ogawa, T., Okazaki, T. 1991. *Mol. Gen. Genet.* 230:193–200
211. Liu, L. F., Wang, J. C. 1987. *Proc. Natl. Acad. Sci. USA* 84:7024–27
212. Hirose, S., Hiraga, S., Okazaki, T. 1983. *Mol. Gen. Genet.* 189:422–31
213. Kohara, Y., Tohdoh, N., Jiang, X. W.,

Okazaki, T. 1985. *Nucleic Acids Res.* 13:6847–66
214. Rokeach, L. A., Zyskind, J. W. 1986. *Cell* 46:763–71
215. Rokeach, L. A., Kassavetis, G. A., Zyskind, J. W. 1987. *J. Biol. Chem.* 262:7264–72
216. Seufert, W., Messer, W. 1986. *EMBO J.* 5:3401–6
217. Kaguni, J., Fuller, R. S., Kornberg, A. 1982. *Nature* 296:623–27
218. Donachie, W. D. 1968. *Nature* 219:1077–79
219. Pritchard, R. H., Barth, P. T., Collins, J. 1969. *Symp. Soc. Gen. Microbiol.* 19:263–97
220. Sompayrac, L., Maaløe, O. 1973. *Nature New Biol.* 241:133–35
221. Braun, R. E., O'Day, K., Wright, A. 1985. *Cell* 40:159–69
222. Atlung, T., Clausen, E. S., Hansen, F. G. 1985. *Mol. Gen. Genet.* 200:442–50
223. Chiaramello, A. E., Zyskind, J. W. 1989. *J. Bacteriol.* 171:4272–80
224. Chiaramello, A. E., Zyskind, J. W. 1990. *J. Bacteriol.* 172:2013–19
225. Polaczek, P., Wright, A. 1990. *New Biol.* 2:574–82
226. Hirota, Y., Mordoh, J., Jacob, F. 1970. *J. Mol. Biol.* 53:369–87
227. Hansen, F. G., Rasmussen, K. V. 1977. *Mol. Gen. Genet.* 155:219–25
228. Kogoma, T., vonMeyenburg, K. 1983. *EMBO J.* 2:463–68
229. Kogoma, T., Subia, N. L., vonMeyenburg, K. 1985. *Mol. Gen. Genet.* 200:103–9
230. Pierucci, O., Rickert, M., Helmstetter, C. E. 1989. *J. Bacteriol.* 171:3760–66
231. Sakakibara, Y., Yuasa, S. 1982. *Mol. Gen. Genet.* 186:87–94
232. Løbner-Olesen, A., Skarstad, K., Hansen, F. G., vonMeyenburg, K., Boye, E. 1989. *Cell* 57:881–89
233. Samitt, C. E., Hansen, F. G., Miller, J. F., Schaechter, M. 1989. *EMBO J.* 8:989–93
234. Schaefer, C., Messer, W. 1991. *Mol. Gen. Genet.* 226:34–40
235. Ogden, G. B., Pratt, M. J., Schaechter, M. 1988. *Cell* 54:127–35
236. Campbell, J. L., Kleckner, N. 1990. *Cell* 62:967–79
237. Jacob, F., Brenner, S., Cuzin, F. 1963. *Cold Spring Harbor Symp. Quant. Biol.* 28:329–48
238. Russell, D. W., Zinder, N. D. 1987. *Cell* 50:1071–79
239. Smith, D. W., Garland, A. M., Herman, G., Enns, R. E., Baker, T., Zyskind, J. W. 1985. *EMBO J.* 4:1319–26
240. Messer, W., Bellekes, U., Lother, H. 1985. *EMBO J.* 4:1327–32

241. Landoulsi, A., Malki, A., Kern, R., Kohiyama, M., Hughes, P. 1990. *Cell* 63:1053–60
242. Bakker, A., Smith, D. W. 1989. *J. Bacteriol.* 171:5738–42
243. Boye, E., Løbner-Olesen, A. 1990. *Cell* 62:981–89
244. Welsh, K. M., Lu, A. L., Clark, S., Modrich, P. 1987. *J. Biol. Chem.* 262:15624–29
245. Furth, M. E., Wickner, S. H. 1983. *Lambda II*, ed. R. W. Hendrix, J. W. Roberts, F. W. Stahl, R. A. Weisberg, pp. 145–73. Cold Spring Harbor, NY: Cold Spring Harbor Lab.
246. Wickner, S., McKenney, K. 1987. *J. Biol. Chem.* 262:13163–67
247. Wold, N., Mallory, J., Roberts, J., LeBowitz, J., McMacken, R. 1982. *Proc. Natl. Acad. Sci. USA* 79:6176–80
248. Tsurimoto, T., Matsubara, K. 1982. *Proc. Natl. Acad. Sci. USA* 79:7639–43
249. Anderl, A., Klein, A. 1982. *Nucleic Acids Res.* 10:1733–40
250. Tsurimoto, T., Matsubara, K. 1981. *Nucleic Acids Res.* 9:1789–99
251. Zahn, K., Blattner, F. R. 1985. *EMBO J.* 13A:3605–16
252. Wickner, S., Zahn, K. 1986. *J. Biol. Chem.* 261:7537–43
253. Dodson, M., Roberts, J., McMacken, R., Echols, H. 1985. *Proc. Natl. Acad. Sci. USA* 82:4678–82
254. Alfano, C., McMacken, R. 1989. *J. Biol. Chem.* 264:10699–708
255. Dodson, M., McMacken, R., Echols, H. 1989. *J. Biol. Chem.* 264:10719–25
256. Liberek, K., Georgopoulos, C., Zylicz, M. 1988. *Proc. Natl. Acad. Sci. USA* 85:6632–36
257. Mallory, J. B., Alfano, C., McMacken, R. 1990. *J. Biol. Chem.* 265:13297–307
258. Liberek, K., Osipuk, J., Zylicz, M., Ang, D., Skorko, J., Georgopoulos, C. 1990. *J. Biol. Chem.* 265:3022–29
259. Zylicz, M., Ang, D., Liberek, K., Georgopoulos, C. 1989. *EMBO J.* 8:1601–8
260. Zylicz, M., LeBowitz, J., McMacken, R., Georgopoulos, C. 1983. *Proc. Natl. Acad. Sci. USA* 80:6431–35
261. Liberek, K., Marszalik, J., Ang, D., Georgopoulos, C., Zylicz, M. 1991. *Proc. Natl. Acad. Sci. USA* 88:2874–78
262. Alfano, C., McMacken, R. 1989. *J. Biol. Chem.* 264:10709–18
263. Schnos, M., Zahn, K., Inman, R. B., Blattner, F. R. 1988. *Cell* 52:385–95
264. Mensa-Wilmot, K., Seaby, R., Alfano, C., Wold, M. S., Gomes, B., McMacken, R. 1989. *J. Biol. Chem.* 264:2853–61
265. Mensa-Wilmot, K., Carroll, K.,

McMacken, R. 1989. *EMBO J.* 8:2393–402

266. Dove, W. F., Inokuchi, H., Stevens, W. F. 1971. *The Bacteriophage Lambda*, ed. A. D. Hershey, pp. 747–71. Cold Spring Harbor NY: Cold Spring Harbor Lab.

267. Alfano, C., McMacken, R. 1988. *Nucleic Acids Res.* 16:9611–30

268. Malki, A., Hughes, P., Kohiyama, M. 1991. *Mol. Gen. Genet.* 225:420–26

269. Hwang, D. S., Crooke, E., Kornberg, A. 1990. *J. Biol. Chem.* 265:19244–48

270. Hwang, D. S., Kaguni, J. 1991. *J. Biol. Chem.* 266:7537–41

271. Prentiki, P., Chandler, M., Caro, L. 1977. *Mol. Gen. Genet.* 152:71–76

272. Abeles, A. L., Snyder, K. M., Chattoraj, D. K. 1984. *J. Mol. Biol.* 173:307–24

273. Pal, S. L., Mason, R. J., Chattoraj, D. K. 1986. *J. Mol. Biol.* 192:275–85

274. Chattoraj, D. K., Snyder, K. M., Abeles, A. L. 1985. *Proc. Natl. Acad. Sci. USA* 82:2588–92

275. Abeles, A. L. 1986. *J. Biol. Chem.* 261:3548–55

276. Wickner, S., Hoskins, J., Chattoraj, D., McKenney, K. 1990. *J. Biol. Chem.* 265:11622–27

277. Abeles, A. L., Reaves, L. D., Austin, S. J. 1990. *J. Bacteriol.* 172:4386–91

278. Austin, S. J., Mural, R. J., Chattoraj, D. K., Abeles, A. L. 1985. *J. Mol. Biol.* 183:195–202

279. Abeles, A. L., Austin, S. J. 1987. *EMBO J.* 6:3185–89

280. Wickner, S. H., Chattoraj, D. K. 1987. *Proc. Natl. Acad. Sci. USA* 84:3668–72

281. Hansen, E. B., Yarmolinsky, M. B. 1986. *Proc. Natl. Acad. Sci. USA* 83:4423–27

282. Scott, J., Vapnek, D. 1980. *Mechanistic Studies of DNA Replication and Genetic Recombination*, ed. B. M. Alberts, pp. 335–45. New York: Academic

283. Tilly, K., Yarmolinsky, M. B. 1989. *J. Bacteriol.* 171:6025–29

284. Bukau, B., Walker, G. C. 1989. *J. Bacteriol.* 171:6030–38

285. Wickner, S. H. 1990. *Proc. Natl. Acad. Sci. USA* 87:2690–94

286. Wickner, S. H., Hoskins, J., McKenney, K. 1991. *Nature* 350:165–67

287. Brendler, T., Abeles, A., Austin, S. 1991. *J. Bacteriol.* 173:3935–42

288. Murotsu, T., Matsubara, K., Sugisaki, H., Takanami, M. 1981. *Gene* 24:235–42

289. Murotsu, T., Tsutsui, H., Matsubara, K. 1984. *Mol. Gen. Genet.* 196:373–78

290. Maki, S., Miki, T., Horiuchi, T. 1984. *Mol. Gen. Genet.* 194:337–39

291. Kawasaki, Y., Wada, C., Yura, T. 1990. *Mol. Gen. Genet.* 220:277–82

292. Wada, C., Imai, M., Yura, T. 1987. *Proc. Natl. Acad. Sci. USA* 84:8849–53

293. Wang, Q. P., Kaguni, J. M. 1989. *J. Biol. Chem.* 264:7338–44

294. Sutcliffe, J. C. 1978. *Cold Spring Harbor Symp. Quant. Biol.* 43:77–90

295. Itoh, T., Tomizawa, J.-I. 1980. *Proc. Natl. Acad. Sci. USA* 77:2450–54

296. Selzer, G., Tomizawa, J.-I. 1982. *Proc. Natl. Acad. Sci. USA* 78:7082–86

297. Masukata, H., Tomizawa, J.-I. 1984. *Cell* 36:513–22

298. Itoh, T., Tomizawa, J.-I. 1978. *Cold Spring Harbor Symp. Quant. Biol.* 43:409–17

299. Masai, H., Arai, K.-I. 1989. *J. Bacteriol.* 171:2975–80

300. Masai, H., Arai, K. 1988. *J. Biol. Chem.* 263:15016–23

301. Staudenbauer, W. 1977. *Mol. Gen. Genet.* 156:27–34

302. Naito, S., Kikanu, T., Ogawa, T., Okazaki, T., Uchida, H. 1984. *Proc. Natl. Acad. Sci. USA* 81:550–54

303. Kogoma, T. 1984. *Proc. Natl. Acad. Sci. USA* 81:7845–49

304. Dasgupta, S., Masukata, H., Tomizawa, J.-I. 1987. *Cell* 51:1113–22

305. Parada, C., Marians, K. J. 1989. *J. Biol. Chem.* 264:15120–29

306. Abe, M. 1980. *J. Bacteriol.* 141:1024–30

307. Polaczek, P., Ciesla, Z. 1984. *Mol. Gen. Genet.* 194:227–31

308. Seufert, W., Messer, W. 1987. *Cell* 48:73–78

309. Ma, D., Campbell, J. L. 1988. *J. Biol. Chem.* 263:15008–15

310. Parada, C., Marians, K. J. 1991. *J. Biol. Chem.* 266:18895–906

311. Dressler, D., Wolfson, J., Magazin, M. 1972. *Proc. Natl. Acad. Sci. USA* 69:998–1002

312. Tamonoi, F., Saito, H., Richardson, C. C. 1980. *Proc. Natl. Acad. Sci. USA* 77:2656–60

313. Mosig, G. 1983. *Bacteriophage T4*, ed. C. K. Mathews, E. M. Kutter, G. Mosig, P. B. Berget, pp. 120–30. Washington, DC: Am. Soc. Microbiol.

314. MacDonald, P. M., Seaby, R. M., Brown, W., Mosig, G. 1983. *Microbiology—1983*, ed. D. Schlessinger, pp. 111–16. Washington, DC: Am. Soc. Microbiol.

315. Saito, H., Tabor, S., Tamonoi, F., Richardson, C. C. 1980. *Proc. Natl. Acad. Sci. USA* 77:3917–21

316. Fuller, C. W., Richardson, C. C. 1985. *J. Biol. Chem.* 260:3185–96

317. Fuller, C. W., Richardson, C. C. 1985. *J. Biol. Chem.* 260:3197–206
318. Fujiyama, A., Kohara, Y., Okazaki, T. 1981. *Proc. Natl. Acad. Sci. USA* 78:903–7
319. Luder, A., Mosig, G. 1982. *Proc. Natl. Acad. Sci. USA* 79:1101–5
320. Dannenberg, R., Mosig, G. 1983. *J. Virol.* 45:813–31
321. Kreuzer, K. N., Yap, W. Y., Menkens, A., Engman, H. W. 1988. *J. Biol. Chem.* 263:11366–73
322. Kreuzer, K. N., Alberts, B. M. 1985. *Proc. Natl. Acad. Sci. USA* 82:3345–49
323. Kreuzer, K. N., Engman, H. W., Yap, W. Y. 1988. *J. Biol. Chem.* 263:11348–57
324. Menkens, A. E., Kreuzer, K. N. 1988. *J. Biol. Chem.* 263:11358–65
325. Hill, T. M., Pelletier, A. J., Tecklenburg, M. L., Kuempel, P. L. 1988. *Cell* 55:459–66
326. Hidaka, M., Akiyama, M., Horiuchi, T. 1988. *Cell* 55:467–75
327. Kuempel, P. L., Pelletier, A. J., Hill, T. M. 1989. *Cell* 59:581–83
328. Henson, J. M., Kuempel, P. L. 1985. *Proc. Natl. Acad. Sci. USA* 82:3766–70
329. Hill, T. M., Tecklenburg, M., Pelletier, A. J., Kuempel, P. L. 1989. *Proc. Natl. Acad. Sci. USA* 86:1593–97
330. Sista, P. R., Mukherjee, S., Patel, P., Khatri, G. S., Bastia, D. 1989. *Proc. Natl. Acad. Sci. USA* 86:3026–30
331. Hidaka, M., Kobayashi, T., Takenaka, S., Takeya, H., Horiuchi, T. 1989. *J. Biol. Chem.* 264:21031–37
332. Germino, J., Bastia, D. 1981. *Cell* 23:681–87
332a. Gottlieb, P. A., Wu, F., Zhang, X., Tecklenburg, M., Kuempel, P., Hill, T. M. 1992. *J. Biol. Chem.* 267: In press
333. MacAllister, T., Khatri, G. S., Bastia, D. 1990. *Proc. Natl. Acad. Sci. USA* 87:2828–32
334. Lee, E. H., Kornberg, A., Hidaka, M., Kobayash, T., Horiuchi, T. 1989. *Proc. Natl. Acad. Sci. USA* 86:9104–8
335. Hill, T. M., Marians, K. J. 1990. *Proc. Natl. Acad. Sci. USA* 87:2481–85
336. Khatri, G. S., MacAllister, T., Sista, P. R., Bastia, D. 1989. *Cell* 59:667–74
336a. Hiasa, H., Marians, K. J. 1992. *J. Biol. Chem.* 267: In press

Annu. Rev. Biochem. 1992. 61:721–59

CONTROL OF NONMUSCLE MYOSINS BY PHOSPHORYLATION

John L. Tan, Shoshana Ravid, and James A. Spudich

Departments of Cell Biology and Developmental Biology, Stanford University School of Medicine, Stanford, California 94305

KEY WORDS: cytoskeleton, cell motility, actin-based motors, protein kinases, protein phosphates

CONTENTS

OVERVIEW AND PERSPECTIVES

Myosins are enzymes capable of utilizing the chemical energy stored in ATP to support translational movement along actin filaments. These enzymes

721

0066-4154/92/0701-0721$02.00

generally comprise two domains, a globular head domain that is common to all myosins and a tail domain that has properties unique to each type of myosin. All elements necessary for force production, including the sites of ATP hydrolysis and actin binding, are contained within the relatively conserved head domain. The structural motifs of the various tail domains presumably designate the specific cellular functions of each type of myosin.

Myosins have been broadly grouped into two general classes. The myosin I class comprises an eclectic assortment of molecules consisting of a single heavy chain with molecular weight ranging from 110,000 to 190,000 and at least one light chain with molecular weight ranging from 14,000 to 27,000. Those myosin Is that have been characterized biochemically consist of a single globular head domain and a tail domain that contains minimal conserved regions and appears to be unique to each myosin I subclass. In contrast, members of the myosin II class have relatively homogeneous structures made up of two heavy chains, each with a molecular weight of approximately 200,000, and two sets of light chains with molecular weights of approximately 16,000–20,000. These hexameric molecules consist of two identical globular head domains and a helical coiled-coil tail domain. The globular heads consist of the amino-terminal parts of the heavy chains and the two sets of light chains. The tail is formed by the intertwining of the carboxyl-terminal portions of the heavy chains. Owing to the inherent properties of the tail domains, myosin IIs are able to associate intermolecularly to form filaments. Myosins isolated from muscle tissues represent prototypical myosin II molecules.

Twenty years ago, most actin-based motility in nonmuscle cells was postulated to involve the two-headed filamentous myosin, now referred to as myosin II. This form of myosin is widely prevalent in eukaryotic cells and had long before been shown to drive the contraction of muscle. However, following the discovery of myosin I molecules in *Acanthamoeba* (1), it has become apparent that there are potentially multiple motors that are capable of generating actin-based movement. Studies on *Dictyostelium* cells genetically engineered such that they express only minor levels of myosin II (2) or a truncated form of myosin II (3) supported this notion. These cells continued to display numerous motile activities including vesicular movement, membrane ruffling, pseudopod extension, and cell locomotion. Subsequent studies on *Dictyostelium* cells in which the single-copy myosin II gene was eliminated by gene replacement, creating amoebae that do not express any myosin II molecules, confirmed that a variety of the cell's motile functions that were widely thought to be actin based do not involve myosin II (4). It is now generally speculated that these various behaviors are driven at least in part by myosin I molecules. The *Dictyostelium* studies also demonstrated that myosin

II is essential to a number of motile functions intrinsic to most nonmuscle cells. These include maintainance of cortical tension, capping of cell surface receptors, morphogenetic changes in shape associated with developmental processes, and, most dramatically, cytokinesis. Unlike skeletal muscle myosins, which are organized into relatively stable sarcomeres, nonmuscle myosin II molecules are thought to undergo a dynamic assembly/disassembly process that permits specific spatial and temporal localization.

How the various actin-based motors are regulated in vivo remains one of the fundamental questions of cell motility. This regulation must function such that the various myosins, all of which utilize tracks of actin filaments, do not interfere with each other's individual functions and may be reversibly organized in a particular cell structure at a particular time. Although the mechanisms that regulate these molecules are obscure, in virtually all myosins that have been biochemically characterized, phosphorylation has been shown to play a central role in regulating enzymatic activity. In addition, phosphorylation appears to control the structural and assembly properties of most myosin II molecules. Much of the work has focused on phosphorylation control of myosin II, although studies on phosphorylation of the myosin I heavy chain from *Acanthamoeba* have provided valuable insights into the regulation of this diverse class of myosins.

This review focuses on the various kinases and phosphorylation sites thought to be involved in the regulation of myosins from nonmuscle cells, and the effects these phosphorylations have on the enzymatic and structural properties of these molecules. Several recent reviews provide excellent summaries of general current knowledge of nonmuscle myosins (5–8). Although a brief summary is presented in this review on phosphorylation of smooth muscle myosin, the reader is referred to other reviews that thoroughly summarize the extensive literature on the regulation of smooth muscle (9–12), as well as recent, short reviews that concentrate on the regulation of smooth muscle and cytoplasmic myosins (13–15).

MYOSIN STRUCTURE AND FUNCTION

The Motor Unit

Myosin from muscle tissues can be proteolytically cleaved into various fragments that distinguish the head and the tail domains. Two proteolytic products, heavy meromyosin (HMM) and subfragment 1 (S1), contain the globular head domain (Figure 1). HMM consists of two globular heads and approximately 1/3 of the α-helical coiled-coil tail domain. S1 consists of a single globular head with no tail sequences. Studies with both HMM and S1

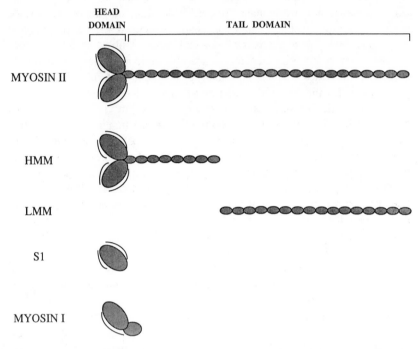

Figure 1 Schematic diagram of myosin domains. The globular head domain comprises the amino-terminal portion of the heavy chain and the light chain(s). Within the head domain, the heavy chain sequences are relatively conserved in both myosin Is and IIs. However, the number and composition of the light chains of the two classes of myosin vary. Myosin II can be proteolytically cleaved into various fragments. Heavy meromyosin (HMM) contains the two globular heads and approximately 1/3 of the α-helical coiled-coil tail domain; light meromyosin (LMM) contains the terminal 2/3 of the tail domain. Subfragment 1 (S1), like myosin I molecules, is composed of a single head domain.

from skeletal muscle tissues have demonstrated that the globular head domain is sufficient to support movement (16–18). Similarly, fragments of the *Dictyostelium* myosin II that are analogous to HMM and S1 have been generated by molecular biological techniques and have also been shown to move actin filaments (19, 20).

Such studies have not been done with the myosin I class of molecules. However, sequence comparisons of both classes of myosins and recent mapping of the brush border myosin I (21, 22) suggest a similar molecular organization of the head domains of both classes of myosin; it is generally assumed that the head domain of myosin I molecules is sufficient to support movement. Numerous conserved regions of sequence in the heavy chains are present in the head domains of all myosins (for review, see 5). Indeed, with the exception of three myosin I molecules studied in *Acanthamoeba* (1, 23–25), the 110-kDa brush border myosin I (26, 27), and a single member of the myosin I class in *Dictyostelium* (28), which were identified by classical biochemical techniques, all other members of the myosin I class have been identified by sequence analysis of conserved regions within the head domain (29–34).

Although the myosin head is composed mainly of the heavy chain, in at least the myosin II class, light chains are an important component of this domain. The head domains of myosin II molecules have two light chains of approximately 16,000–20,000 daltons. One of the two light chains regulates the activity of the myosin head and is thus referred to as the *regulatory light chain*. This chain is also referred to as the *phosphorylatable light chain*, since in almost all systems studied, regulation of myosin activity is mediated by phosphorylation of this chain. The notable exception is molluscan striated muscle, in which myosin activity is regulated by direct binding of Ca^{2+} (35, 36, for review, see 37). Recent genetic studies in *Drosophila* have shown that cells lacking this light chain display a cytokinesis defect (38), suggesting that the regulatory light chain is essential to the in vivo function of myosin II.

The other light chain, historically referred to as the *essential light chain*, does not incorporate phosphates. This light chain has been shown to constitute part of the active site of smooth muscle myosin (39), and may be involved in maintaining the head structure (40) as well as serving as a link between the active site and the regulatory light chain (12). Binding of a monoclonal antibody to the essential light chain has been shown to increase the actin-activated ATPase of smooth muscle myosin (41).

In myosin I molecules, the function of the light chains is unknown. Although the light chains of myosin I molecules are commonly assumed to be an integral part of the head, this has not been established. Indeed, the light chains of the brush border myosin Is from avian and bovine intestine are thought to be carboxyl terminal of the head domain (42). Neither the actin-activated ATPase activity nor the phosphorylation of the heavy chain of an *Acanthamoeba* myosin I requires the presence of the light chain (23).

From the few cases in which the light chains have been characterized in myosin I molecules, it is evident that the number and the composition of the light chains are very variable. The biochemically characterized *Acanthamoeba* myosin Is, myosin IA, IB, and IC, are associated with a single light chain of 15,000, 27,000, and 14,000 daltons, respectively (43). In contrast, the brush border 110-kDa myosin I molecule binds at least three calmodulin molecules, which are the light chains of this myosin (21). Aside from the calmodulin of brush border myosin I, the biochemical properties and the primary structures of other myosin I light chains remain obscure.

The Tail Domain

In skeletal muscle, myosin filaments are arranged in ordered arrays between actin filaments. Together with accessory proteins, these filaments constitute the sarcomeres. Myosin II filaments similar to those present in muscle tissues are also found in nonmuscle cells. In *Dictyostelium,* filaments composed of myosin II molecules have been visualized by immunocytochemical techniques (44, 45). The importance of the assembly portion of the tail domain to the cellular function of myosin II has been addressed by molecular genetic studies. Cells were created that express only a truncated form of myosin II instead of the intact molecule. Cells expressing this truncated form, which corresponds to the proteolytic fragment HMM and thus contains a fully functional head domain, are defective in cytokinesis and in their ability to complete development (3). Further analysis of these mutant cells reveals that although the cells are still motile, their general motility and chemotactic movement are severely impaired (46). These cells also fail to cap cell surface proteins (47). Strikingly, *Dictyostelium* cells that do not express any myosin II exhibit the same phenotype as the HMM mutant (4). This finding indicates that the head domain, which contains all the elements necessary for force generation, is not sufficient for the proper cellular function of myosin II.

Analysis of the tail sequences of muscle and nonmuscle myosin II molecules reveals multiple repeating patterns throughout the entire domain that underlies the α-helical coiled-coil structure of the myosin II tail. The smallest repeating motif contains seven amino acids in which small, generally hydrophobic amino acids are usually found in the first and fourth position. The seven residues in this repeat form two turns of the α-helix with the first and fourth amino acids falling along a single aspect of the helix (48, 49). This configuration provides the hydrophobic core essential to the formation of the α-helical coiled-coil. A repeating pattern of four seven-residue motifs (28 residues) is also evident along the tail. The 28-residue motifs are charged and positioned such that they create alternating bands of positive and negative charges on the surface of the coiled-coil. These charged regions interact with those along the tails of adjacent myosin II molecules to form the myosin filaments.

Early proteolytic studies on muscle myosin demonstrated that the carboxyl-terminal 2/3 of the myosin II rodlike tail domain, termed light meromyosin (LMM), contains all the properties necessary for assembly into filaments (Figure 1). More recent studies indicate that a specific region within the myosin II tail domain is necessary and sufficient for filament formation (Figure 2). Cross & Vandekerckhove (50) showed that the assembly subdomain of smooth muscle myosin is localized to a carboxyl-terminal region comprising approximately 140 residues (20 nm of the tail length). In *Acantha-moeba* myosin II, the carboxyl-terminal 100 amino acids of the tail contain all the elements needed for the assembly process (51) as well as the sites of heavy chain phosphorylation (52, 53). A number of studies have also mapped important regions within the tail domain of the *Dictyostelium* myosin II molecule. Like the *Acanthamoeba* molecule, *Dictyostelium* myosin II possesses an assembly subdomain currently localized to a 34-kDa region (54, 55). This subdomain, located 34 kDa from the carboxyl terminus, is situated within the region found to be the major contact area between the two myosin II molecules in a parallel dimer (56, 57). In addition, carboxyl terminal to this subdomain is a regulatory region that contains sites of heavy chain phosphorylation, which appear to play an important role in disassembling filaments. This was demonstrated by studies in which *Dictyostelium* cells that

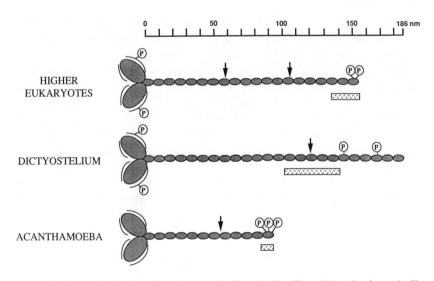

Figure 2 Schematic diagram of various myosin II molecules. The tail length of myosin II molecules varies depending on the source. The tail of smooth muscle myosin is 155 nm long, while those of *Dictyostelium* and *Acanthamoeba* myosin II are 186 and 90 nm, respectively. The positions of bends within the tail domain are indicated by the arrows. Smooth muscle myosin is known to bend in at least two sites approximately 50 and 100 nm from the end of the tail. The *Dictyostelium* myosin II bends at a site 66 nm from the end of the tail. Although the *Acantha-moeba* myosin II is known to bend at a site 40 nm from the end of the tail, this bending, unlike those of higher eukaryotic and *Dictyostelium* myosin IIs, is not dependent on phosphorylation. The hatched boxes represent the assembly subdomains. The sites of phosphorylation are also indicated (by circled Ps). Whereas higher eukaryotic and *Dictyostelium* myosins are phosphory-lated on both the heavy and the light chains, *Acanthamoeba* myosin is phosphorylated only on the heavy chain.

express only a truncated form of myosin II containing the assembly sub-domain but lacking sequences carboxyl terminal to this subdomain were found to exhibit severe abnormalities in their ability to disassemble myosin filaments (58).

All nonmuscle myosin II molecules possess hinge regions within their tail domains. In smooth muscle and vertebrate nonmuscle myosin IIs, two hinge regions are located at approximately 1/3 and 2/3 the length of the tail from the head domain (Figure 2; 59). Bending at these sites is regulated by light chain phosphorylation. In contrast, *Dictyostelium* myosin II possesses a single hinge region, located at a position similar to the second bend of the vertebrate myosin II and regulated by phosphorylation of the heavy chain (Figure 2; 57). The *Acanthamoeba* myosin II also contains a hinge region located at a relatively similar position as the *Dictyostelium* region (Figure 2). It is not known whether phosphorylation regulates bending of the *Acanthamoeba* myosin II.

In contrast to myosin IIs, which all have very similar tail structures, myosin I tail domains are structurally very different from one another. As more myosin I molecules are characterized, this broad class is beginning to be subgrouped based on common motifs within the tail domain as well as sequence conservation within the head domain. For example, the myosin Is can be classified by the presence of recurrent motifs such as the SH3 region or a basic region (8), or by general sequence comparison as is the case with the *Saccharomyces MYO2* gene (33) and the mouse *dilute* gene (34). The functions of the various tail domains remain unclear. However, preliminary studies indicate that the SH3 region may contribute to the ATP-independent actin-binding site within the tail domain or may anchor the myosin I molecule to a membrane-associated protein (8). Similarly, the basic region may serve to bind the myosin I molecule directly to membranes (60).

MYOSIN I HEAVY CHAIN PHOSPHORYLATION

Our current knowledge of phosphorylation of myosin I heavy chains is based primarily on work in *Acanthamoeba*. A single myosin I heavy chain kinase (MIHCK) has been purified from this organism (61). This kinase appears to be a general protein kinase. It is capable of phosphorylating not only all biochemically characterized myosin Is from *Acanthamoeba* (61) as well as a myosin I isolated from *Dictyostelium* (28), but also several other substrates such as smooth muscle myosin light chains and casein (61, 62).

The sites phosphorylated by MIHCK have been identified in a number of the *Acanthamoeba* myosin Is. Unlike the phosphorylation sites on the myosin

II heavy chain, the phosphorylation sites on the myosin I heavy chain are within the head domain. The sites of phosphorylation by MIHCK are serine 315 in myosin IB, serine 311 in myosin IC, and a threonine residue at a corresponding position in myosin IA (63). These residues are located between the putative ATP- and actin-binding sites (44, 65). In all three of these myosin I molecules, there are two to three basic amino acids preceding the phosphorylated residue. Using synthetic peptides, Brzeska et al (66) further defined the MIHCK site and showed that a peptide composed of nine amino acids is sufficient as a substrate for MIHCK. While two basic amino acids at the amino-terminal side of the phosphorylated residue are preferred, at least one is essential. In addition, tyrosine residues situated at the carboxyl terminus of the peptide appear to be essential. This phosphorylation site is conserved in a number of myosin I sequences including *Dictyostelium* myosin IA (30), IB (31), IC (30), and ID (M. A. Titus, personal communication) and bovine (67) and avian (68) brush border myosin I.

Phosphorylation of the heavy chain serves to increase the actin-activated ATPase activity of myosin I molecules from *Acanthamoeba* (61, 69, 70) and from *Dictyostelium* (28). This phosphorylation results in a 20-fold increase in the actin-activated ATPase but has no effect on the binding of myosin I to filamentous actin (25). In addition, using the *Nitella* in vitro movement assay (71, 72), Albanesi et al (73) showed that the movement of beads coated with myosin IA or IB along actin filaments depends on heavy chain phosphorylation.

The *Acanthamoeba* MIHCK is regulated by autophosphorylation and by phospholipid. Incorporation of up to approximately eight moles of phosphate per mole of kinase appears to stimulate the activity of MIHCK (74). MIHCK activity is also enhanced 20-fold by phospholipids (74). The addition of Ca^{2+}, however, has no effect (69, 74), suggesting that the activation of this kinase is distinct from that of PKC. Since myosin Is bind membrane lipids (75–77), the MIHCK appears to be localized such that it may regulate myosin I molecules that are membrane bound.

Several myosin I molecules, including the *Acanthamoeba* High Molecular Weight myosin I (32), the *Saccharomyces MYO2* gene product (33), the mouse *dilute* myosin I (34), and the *Drosophila ninaC* protein (29), do not contain the phosphorylation site for MIHCK. The mechanism of regulating these myosins is not known. The *Drosophila ninaC* protein encodes a protein kinase catalytic domain (29), which may regulate the activity of the putative motor domain. In addition, the *Saccharomyces* MYO2 sequence contains a possible pp34^{cdc2} phosphorylation site (33). However, further biochemical characterizations are necessary to elucidate the regulatory mechanisms of these myosins.

MYOSIN II LIGHT CHAIN PHOSPHORYLATION

Phosphorylation of the regulatory light chain was first demonstrated by Perrie et al (78) in myosin from skeletal muscle and by Adelstein et al (79) in platelet myosin. Since that time, myosin light chain phosphorylation has been shown to occur in every system in which it has been investigated with the exception of *Acanthamoeba*. Early work with platelet myosin demonstrated that phosphorylation of the regulatory myosin II light chain resulted in an increase in the actin-activated ATPase of myosin (80). These works led to the characterization of specific myosin light chain kinases (MLCKs) from a variety of muscle and nonmuscle sources (for review, see 81) including skeletal muscle, smooth muscle, brain, platelets, and more recently, fibroblast (82) and *Dictyostelium* (83, 84). Light chain phosphorylation by the Ca^{2+}/phospholipid-dependent protein kinase, protein kinase C (PKC) also appears to regulate myosin II activity. In addition to MLCK and PKC, a number of other general kinases have been shown to phosphorylate the myosin II regulatory light chains, at least in vitro. These include the Ca^{2+}/calmodulin-dependent protein kinase type II (CaM kinase II), the cell cycle-dependent protein kinase, $pp34^{cdc2}$, the reticulocyte protease-activated protein kinase I, and the *Acanthamoeba* myosin I heavy chain kinase (Table 1). Several other general kinases, including the cAMP-dependent protein kinase (85), epidermal growth factor receptor (86), insulin receptor (87), casein kinase I (88) and II (89, 90), and phosphorylase kinase (88), also phosphorylate isolated myosin II regulatory light chains. However, it is unlikely that these kinases regulate myosin, since they do not phosphorylate the regulatory light chain in the context of an intact myosin II molecule.

Although the majority of studies focused on myosin II light chain phosphorylation have been done on smooth muscle myosin, several studies demonstrate that phosphorylation of nonmuscle myosin II molecules of higher eukaryotes results in effects essentially identical to that observed with smooth muscle myosin. In addition, molecular cloning of the cDNAs of several nonmuscle regulatory light chains from higher eukaryotic organisms (91–93) indicates that the various isoforms of the regulatory light chains are very highly homologous, conserving all characterized phosphorylation sites (Figure 3). Thus it is likely that results seen with the phosphorylation of smooth muscle myosin will hold true for the nonmuscle myosin IIs of higher eukaryotic cells. However, as detailed below, this is not the case with all myosin II molecules from lower eukaryotes.

Phosphorylation by Myosin Light Chain Kinase

PHOSPHORYLATION SITES   There are two sites in the regulatory light chain of myosin II that are phosphorylated by MLCK, at least in vitro. These sites

Table 1   Summary of myosin II regulatory light chain phosphorylation

Site of phosphorylation[a]	Kinase(s)[b]	Effects on myosin in vitro	Evidence of phosphorylation in vivo
Ser19	MLCK CaMK II PAPK I MIHCK	Increase in ATPase Stabilization of 6S Increase in rate of movement	Yes
Ser19 + Thr18	MLCK	Further increase in ATPase Further stabilization of 6S No effect on rate of movement	No
Thr9	PKC	Decrease in ATPase[c] Increase in K_a for actin No effect on V_{max} No effect on rate of movement Decrease in rate of MLCK phosphorylation	No
Thr9 + Ser1 or 2	PKC	Further decrease in ATPase No effect on rate of movement	Yes (Ser1 and 2 only)
Ser1 or 2	pp34[cdc2]	? Decrease in ATPase	—

[a] Sites listed are based on the protein sequence of chicken smooth muscle regulatory light chain.

[b] Myosin light chain kinase (MLCK), Ca^{2+}/calmodulin-dependent protein kinase type II (CaMK II), protease-activated protein kinase I (PAPK I), myosin I heavy chain kinase (MIHCK), protein kinase C (PKC), cell cycle–dependent protein kinase (pp34[cdc2]).

[c] Effects are on myosin previously phosphorylated by MLCK. Decrease in ATPase is not seen with thymus myosin II.

are serine 19 (94) and threonine 18 (95–97). Although phosphorylation of each of these two sites is independent of previous phosphorylation at the other site, the rates of phosphorylation are considerably different, with phosphorylation at serine 19 being approximately 1000-fold greater than that at threonine 18 (95). Phosphorylation at threonine 18 has been demonstrated in both smooth muscle and platelet myosins, and is seen only with high concentrations of MLCK (95, 97).

EFFECTS OF PHOSPHORYLATION ON ENZYMATIC ACTIVITY   Light chain phosphorylation by MLCK has been shown to result in an increase in the actin-activated ATPase activity of myosin IIs in all systems investigated, including platelet (80), smooth muscle (98, 99), and *Dictyostelium* (83). One notable exception is skeletal muscle myosin, which undergoes but does not require light chain phosphorylation for significant actin-activated ATPase activity (100). Light chain phosphorylation in skeletal muscle is thought to be

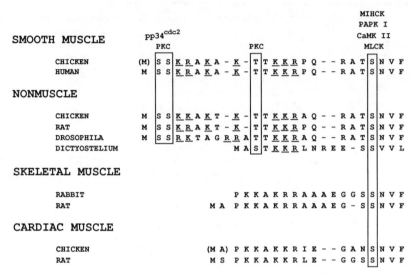

Figure 3 Phosphorylation sites of myosin II light chains. The sequences around the phosphorylation sites of smooth muscle and nonmuscle myosin II of higher eukaryotes are very highly conserved. The phosphorylation sites of MLCK, protein kinase C (PKC), pp34^{cdc2}, Ca^{2+}/calmodulin dependent protein kinase II (CaMK II), protease-activated protein kinase I (PAPK I), and myosin I heavy chain kinase (MIHCK) are boxed. Basic residues thought to be important in PKC phosphorylation are underlined. The *Dictyostelium* myosin II light chain sequence appears more similar to those of skeletal and cardiac myosins. However, while skeletal and cardiac myosins do not contain putative PKC sites, the *Dictyostelium* sequence does. The residues in parentheses are amino acids that are not contained in the mature polypeptide.

associated with increased isometric twitch tension (101). Although the phosphorylation site has not been mapped in all of these systems, it is generally assumed that the site is serine 19, or is analogous to serine 19. In *Dictyostelium*, the MLCK phosphorylates a serine residue (83). This site is postulated to be serine 13 or 14, sites analogous to serine 19 of higher eukaryotes (102).

Studies utilizing in vitro movement assays have demonstrated that phosphorylation at serine 19 in myosins from turkey gizzard, bovine trachea, bovine aorta, and human platelet is required for force production (103–105). Similarly, light chain phosphorylation of the *Dictyostelium* myosin II regulatory light chain is essential to movement, as measured in the *Nitella* in vitro movement assay (83). With this assay, beads coated with myosin II molecules that incorporated approximately one mole phosphate per mole regulatory light chain moved along arrays of *Nitella* actin filaments at a rate of 0.2 μm/s for turkey gizzard myosin, 0.12 μm/s for bovine trachea and aorta myosin, 0.04

μm/s for human platelet myosin II (104), and 1 μm/s for *Dictyostelium* myosin II (83). In addition, studies of smooth muscle myosin reveal an approximately linear relationship between the amount of light chain phosphorylation and the speed of movement at high myosin concentration (103). These results suggest that unphosphorylated myosin produces a negative effect on the velocity of movement and may do so by an increase in the portion of the cycle time spent bound to actin, as suggested by Warshaw et al (106). This phenomenon may be analogous to that seen in smooth muscle fibers, termed the "latch state," in which tension is maintained (presumably by slowly cycling unphosphorylated myosin crossbridges) despite a return to the resting levels of light chain phosphorylation and velocity of shortening (107). Nonmuscle myosin IIs are thought to be similarly controlled.

The mechanism by which light chain phosphorylation affects the enzymatic function of myosin remains controversial. Sellers & Adelstein (108) have suggested that the kinetic step predominately affected by light chain phosphorylation involves the release of phosphate from the actin-bound myosin-ADP-P_i state, the step thought to induce the conformational changes in the myosin molecule that result in force production. This hypothesis is supported by transient kinetic experiments with smooth muscle myosin that indicate that the release of P_i from unphosphorylated HMM is slow with a rate constant of approximately 0.002 per s, and is not activated by the presence of actin (109). However, a series of studies with smooth muscle myosin showed that filaments composed of unphosphorylated myosin exhibited a V_{max} for actin-activated ATPase that was comparable to that of filaments composed of phosphorylated myosin (110, 111), suggesting that the primary effect of phosphorylation is not on the kinetic cycle of the myosin head but rather on the assembly state of the molecule. However, these results may be specific to the high Mg^{2+} concentration used to induce the unphosphorylated myosin into filaments (112). Using a monoclonal antibody that stabilizes filaments composed of unphosphorylated myosin at low Mg^{2+} concentrations, Trybus (112) showed that the ATPase activity of unphosphorylated myosin filaments is similar to that of unphosphorylated HMM and is not activated by actin, and that phosphorylation increases the ATPase activity 60-fold.

Phosphorylation at threonine 18 has also been shown to increase the actin-activated ATPase above that attributable to phosphorylation of serine 19. In studies using smooth muscle HMM, Ikebe et al (113) determined a value of 3.8 per s for the V_{max} of the actin-activated ATPase of HMM phosphorylated at only serine 19, and 8.1 per s for the V_{max} of HMM phosphorylated at both serine 19 and threonine 18, with no significant differences in the K_a for actin. However, no effect on movement was observed by the additional phosphorylation of threonine 18 in the *Nitella* assay (104).

EFFECTS ON MYOSIN CONFORMATION   Myosin II from smooth muscle (114) and a number of nonmuscle cells of higher eukaryotes including brush border (115), thymus (114), erythrocytes (96), and platelets (116), is known to exist in two conformations, which are determined by the phosphorylation state of the regulatory light chains. The folded state, referred to as the 10S form after its sedimentation coefficient, is favored by unphosphorylated molecules and a low-ionic-strength environment. Two bends in the myosin II tail domain, at approximately 50 and 100 nm from the head-tail junction (59), result in the 10S form. The equilibrium between this state and the extended state, the 6S form, is shifted towards the 6S form with light chain phosphorylation. Phosphorylation at serine 19 is sufficient to induce this conformational change (97, 113). However, additional phosphorylation at threonine 18 results in further stabilization of the extended 6S form, as measured by sedimentation velocity measurements and by increased susceptibility of the head-neck junction to proteolysis by papain of HMM phosphorylated at both sites (97, 113).

In contrast to the extended 6S form, the 10S form is able to release the products of ATP hydrolysis at only a very slow rate, less than 0.0005 per s (117, 118), bind actin relatively weakly (119), and is unable to form filaments under physiologic conditions (Figure 4; 114). However, the release of the products of ATP hydrolysis by monomeric unphosphorylated myosin (in the 10S conformation) is approximately 10-fold slower than that of unphosphorylated myosin in the stabilized filaments (112). Unphosphorylated HMM, which cannot assume the folded conformation, has also been shown to release the products of ATP hydrolysis at a rate greater than that of monomeric 10S myosin (118). In addition, the binding of actin to HMM is only slightly affected by phosphorylation (120). Thus, it is likely that the properties of the 10S myosin reflect the steric hindrance imposed by the folded tail onto the head domain rather than an intrinsic enzymatic property of the unphosphorylated molecule.

Several studies suggest that the 10S conformation per se is not required for the inhibition of enzymatic activity. In one study, unphosphorylated myosin, which cannot assume the folded 10S conformation in the context of stabilized filaments, continued to exhibit minimal actin-activated ATPase activity (112). Cross et al (118) have also demonstrated that phosphorylated myosin in the 10S conformation releases the products of ATP hydrolysis at a greater rate than that of unphosphorylated 10S myosin, indicating that phosphorylation can affect the enzymatic activity of myosin without any gross change in conformation.

It also remains quite possible that the conformational isoforms observed for these molecules are reflections of the in vitro milieu. Despite the abundant in

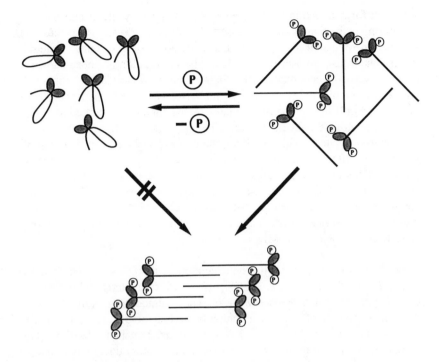

Figure 4 Assembly of higher eukaryotic myosin II. Molecules that are not phosphorylated by MLCK exist mainly in a folded conformation and do not form filaments. MLCK phosphorylation of the regulatory light chains induces a conformational change to an extended form that readily forms filaments.

vitro evidence to support the dependence of the assembly/disassembly process of myosin II on light chain phosphorylation, the presence of the folded conformation has not been observed in intact smooth muscle or in any nonmuscle cells. Indeed, in relaxed smooth muscle, filaments composed of nonphosphorylated myosin molecules are found (121), although experiments with rat smooth muscle indicate that the density of myosin filaments increases with contraction (122). In lower eukaryotic systems, phosphorylation of the myosin II light chain does not result in the conformational changes observed with myosin II of higher eukaryotes. Although bending within the tail domain is evident in some of these systems, conformational changes appear to be dependent on heavy chain and not light chain phosphorylation.

IN VIVO EVIDENCE OF PHOSPHORYLATION   In smooth muscle fibers, phosphorylation of the regulatory light chain by MLCK has been shown to be sufficient for contraction (123). Serine 19 is the primary site of phosphorylation during activation of smooth muscle fibers (124, 125) as well as thrombin stimulation of platelets (126). In almost all nonmuscle systems, including lower eukaryotes, myosin II light chain phosphorylation has been shown to occur in vivo. In endothelial cells, light chain phosphorylation is involved with cell retraction (127). Similarly, in *Dictyostelium* cells, phosphorylation of the myosin II light chain is stimulated by the chemoattractant cAMP (128) with a time course consistent with cell shape changes (129, 130).

Diphosphorylated regulatory light chains have been observed in smooth muscle cells after neural stimulation or treatment with high concentrations of the acetylcholine agonist, carbachol. However, the maximal levels of light chains incorporating two phosphates per molecule are low; 11% for agonist stimulated and 5% for neurally stimulated cells (131). Diphosphorylation, presumably at serine 19 and threonine 18, thus may not be of physiologic significance in smooth muscle.

REGULATION OF MLCK   With the exception of the *Dictyostelium* (83, 84) and possibly the *Physarum* MLCK (132), all characterized MLCKs are Ca^{2+}/calmodulin activated enzymes. The calmodulin-binding domain, thought to consist of a basic amphiphilic α helix (133, 134), is situated carboxy terminal of the protein kinase catalytic domain (135, 136). This domain is postulated to interact with a second regulatory domain that inhibits inherent kinase activity. Upon binding of calmodulin, MLCK is thought to undergo a conformational change in which the autoinhibitory domain is removed from the active site, thus derepressing kinase activity (137–139). Early peptide studies indicated that the calmodulin-binding domain overlaps the autoinhibitory domain (137, 139). More recent studies of smooth muscle MLCK using site-directed mutagenesis define a minimum five-residue overlap of the autoinhibitory and the calmodulin-binding sites (140). These studies are in contrast to deletion and mutagenesis studies done with fibroblast MLCK that suggest that the autoinhibitory and calmodulin-binding sites are distinct (82). Whether these differences reflect the different sources of the kinases remains to be determined.

The regulation of MLCK by calmodulin is itself modulated by phosphorylation. Phosphorylation by cAMP-dependent protein kinase has been shown to occur at two sites (141). One site, phosphorylated only in the absence of Ca^{2+}/calmodulin, is known as the A site. The second site, the B site, is phosphorylated in either the presence or absence of Ca^{2+}/calmodulin. The A site has been identified as serine 512 in smooth muscle MLCK (142). Phosphorylation at this site results in a decreased affinity of MLCK for Ca^{2+}/calmodulin (143). Thus phosphorylation at this site serves to attenuate the signal for activating MLCK. MLCK is similarly phosphorylated by the cGMP-dependent protein kinase, resulting in decreased affinity for Ca^{2+}/ calmodulin (143) and PKC (144, 145). Recently, phosphorylation of smooth muscle MLCK by the multifunctional CaM kinase II has been observed (142, 146). Phosphorylation by this kinase occurs at the same sites as those targeted by the cyclic nucleotide–dependent protein kinases, and as may be suspected, results in identical effects. The role of MLCK phosphorylation in vivo, however, has not been clearly demonstrated, with data either suggesting (147) or disputing (148) an in vivo role for cAMP-dependent protein kinase.

Phosphorylation by Protein Kinase C

PHOSPHORYLATION SITES   Phosphorylation by PKC has been demonstrated in a number of systems, including smooth muscle (149), platelet (126, 150–152), thymus (153), and brain (154). Three major sites of phosphorylation by PKC have been identified in both smooth muscle (155, 156) and platelet (116) myosin II light chains. These sites are threonine 9, serine 1, and serine 2 (Figure 3). Phosphorylation at these sites occurs randomly, although phosphorylation at threonine 9 occurs approximately 5 times faster than that at the other sites (156). Phosphorylation of the second site occurs without preference at either serine residue. However, phosphorylation of the third site occurs only on prolonged incubation (156). Although these sites have not been mapped for other systems, it is generally assumed that the sites are identical or analogous. For example, two-dimensional peptide mapping of the regulatory light chain of thymus myosin phosphorylated by PKC shows a pattern similar to that of smooth muscle myosin (153). In addition, although the major phosphorylation sites are identical in the smooth muscle and platelet systems, PKC phosphorylation of platelet myosin II light chains occurs 10 times faster than PKC phosphorylation of smooth muscle light chains (116).

No data have been collected regarding PKC phosphorylation of myosin II light chains in lower eukaryotes. However, serine 3 or threonine 4 in the *Dictyostelium* regulatory myosin II light chain sequence (102) corresponds to threonine 9 in higher eukaryotic light chains. PKC phosphorylation sites are characterized by basic residues flanking, or on either side of, the phosphory-lated residue (for review, see 157). Serine 3 and threonine 4 are situated amino terminal of three consecutive basic residues (lysine 5, arginine 6 and 7), and thus represent putative PKC phosphorylation sites (Figure 3).

EFFECTS OF PHOSPHORYLATION ON ENZYMATIC ACTIVITY   PKC phosphorylation of the regulatory light chains of myosin II molecules that have not been previously phosphorylated by MLCK appears to have no effects on the enzymatic properties of myosin II. In smooth muscle, platelets, and thymus, PKC phosphorylation of totally dephosphorylated myosin (up to approximately 2 moles of phosphate per mole light chain), unlike phosphorylation by MLCK, does not result in an increase of the actin-activated ATPase activity (143, 153, 156, 158). In addition, phosphorylation by PKC of a variety of smooth muscle and platelet myosin IIs did not result in movement of myosin-coated beads in the *Nitella* assay (104).

However, phosphorylation by PKC of the regulatory light chain that has been previously phosphorylated by MLCK has different effects in various systems. For smooth muscle (143, 156, 158) and platelet (116) myosin II molecules prephosphorylated by MLCK, PKC phosphorylation alters the actin-activated ATPase activity. In experiments with smooth muscle myosin, phosphorylation by PKC to one mole of phosphate per mole light chain results in a 50% decrease in the actin-activated ATPase activity of myosin II that had incorporated two moles of phosphate per mole light chain by MLCK phosphorylation (156). Additional phosphorylation by PKC to two moles of phosphate per mole light chain results in a further decrease of the myosin II ATPase activity to 25% that of myosin phosphorylated by MLCK alone (156). Determinations of the kinetic constants of myosin phosphorylated by MLCK alone or myosin phosphorylated by both MLCK and PKC revealed that the V_{max} of ATPase activity was unaffected by PKC phosphorylation, but that the K_a for actin of myosin phosphorylated by both kinases was higher than that of myosin phosphorylated by MLCK alone [340 μM and 60 μM, respectively for smooth muscle myosin (156); 63 μM and 8.8 μM respectively for platelet myosin (116)]. The effect on the actin-activated ATPase activity by PKC phosphorylation thus appears to reflect a decrease in actin affinity (143).

In contrast, in their study of a unique embryonic smooth muscle myosin isoform, de Lanerolle & Nishikawa (159) found that PKC phosphorylation resulted in an increase in the actin-activated ATPase. The site of phosphorylation by PKC in this system, however, has not been identified, and it has been suggested that PKC may phosphorylate the same site as MLCK in this particular myosin (159). In addition, Carroll & Wagner (153) reported that PKC phosphorylation of thymus myosin II resulted in no changes in the actin-activated ATPase activity of myosin prephosphorylated by MLCK.

Despite these various observed effects on myosin ATPase activity, PKC phosphorylation of smooth muscle and platelet myosin does not seem to affect the rate of myosin-driven movement as measured in an in vitro motility assay. Using the *Nitella* system, Umemoto et al (104) showed that PKC phosphorylation of myosins from turkey gizzard, bovine trachea, bovine aorta, and human platelets that had already been phosphorylated by MLCK to

1 or 2 moles phosphate per mole light chain exhibited similar rates of movement as those phosphorylated by MLCK alone.

Phosphorylation by PKC may affect myosin activity indirectly by modulating MLCK phosphorylation, since in at least the smooth muscle system, PKC phosphorylation appears to affect the ability of MLCK to phosphorylate the regulatory light chain. In experiments in which smooth muscle myosin was sequentially phosphorylated by PKC and then MLCK, Nishikawa et al (143) demonstrated that the initial rate of phosphorylation by MLCK was twofold lower than with myosin that had not been prephosphorylated by PKC. This decrease in initial rate was shown to result from an approximately 10-fold decrease in affinity of MLCK for the substrate; no effect on the V_{max} of MLCK phosphoryation was observed whether unphosphoryated myosin or myosin prephosphorylated by PKC was used. In addition, PKC phosphorylation may alter the site of MLCK phosphorylation (160).

Prephosphorylation by MLCK has also been shown to inhibit subsequent phosphorylation by PKC. Similarly, this inhibition appears to reflect a twofold decrease in the affinity for myosin prephosphorylated by MLCK (143), and thus at least in vitro, these two kinases appear to modulate one another.

EFFECTS ON MYOSIN CONFORMATION   Studies of smooth muscle (161), platelet (116), and thymus (153) myosin indicate that PKC phosphorylation does not induce the conformational changes seen with phosphorylation by MLCK. Myosin phosphorylated by PKC remains predominantly in the 10S conformation, and behaves in a similar fashion to unphosphorylated myosin under various ionic strength conditions. However, smooth muscle myosin that had been phosphorylated by both MLCK and PKC exhibits a sedimentation coefficient of 7.3S, intermediate to that of unphosphorylated myosin or myosin phosphorylated by either PKC or MLCK (161). Analysis of electron micrographs of myosin phosphorylated by both MLCK and PKC revealed that the percent of molecules in the folded conformation (59%) in 0.15–0.08 M ammonium acetate was intermediate to that observed with myosin phosphorylated by MLCK alone (44%) and by PKC alone (93%) (161). Whether this change in the equilibrium between the 10S and the 6S forms is significant remains unclear. Using viscosity measurements to assess the conformation state of platelet myosin, Ikebe & Reardon (116) observed no differences in viscosity of myosin phosphorylated by MLCK alone from that phosphorylated by both MLCK and PKC. However, no direct visualization of myosin conformation was done in this study. The discrepancy between the results seen with smooth muscle and platelet myosins may reflect differences in these systems, but more likely reflects differences in ionic strength conditions and in the assays themselves.

IN VIVO EVIDENCE OF PHOSPHORYLATION   The results of studies of PKC phosphorylation of the myosin II regulatory light chain in intact cells differ in many respects from those obtained from in vitro experiments. Initial studies with smooth muscle tissue from porcine carotid artery suggested that phorbol ester–induced contraction was mediated by low levels, approximately 0.05 mole of phosphate per mole light chain, of PKC phosphorylation of the myosin light chain (162). However, more recent studies with intact and glycerinated smooth muscle tissue from the same source indicate that PKC phosphorylation of myosin light chains, even to relatively high levels (greater than one mole phosphate per mole light chain) is not sufficient to induce contraction or to modulate the isotonic shortening or the isometric force produced by muscle fibers consisting of myosin phosphorylated at the MLCK sites (163). In addition, muscarinic stimulation of intact bovine tracheal smooth muscle failed to reveal any evidence of PKC phosphorylation of the myosin light chains (9). Thus in the smooth muscle system, no compelling evidence exists for the role of PKC phosphorylation in intact cells.

Phosphorylation of the myosin II light chains at the PKC sites in human platelets in response to phorbol ester stimulation (152, 164) and thrombin (126), and in rat basophilic leukemia cells in response to antigen stimulation (165), have also been demonstrated. However, the possible effects of phosphorylation at the PKC sites in intact nonmuscle cells have yet to be elucidated.

In addition, mapping of the PKC sites indicates that only the serine residues are phosphorylated in vivo (126, 152, 165, 166). No evidence exists that the threonine residue that is preferentially phosphorylated by PKC in vitro is targeted by PKC in vivo. The effects of PKC phosphorylation in vivo remain obscure, since in vitro experiments that imply that phosphorylation by PKC inhibits the actin-activated ATPase may reflect the effects of phosphorylation at the threonine residue.

Phosphorylation by Other Kinases

PHOSPHORYLATION BY THE CaM KINASE II   Early studies on the substrates of CaM kinase II indicated that this general protein kinase is able to phosphorylate isolated myosin II regulatory light chain (167, 168) but not intact myosin (169, 170). However, more recent studies have demonstrated that myosin from brain (171), sea urchin egg (172), and smooth muscle (173) are phosphorylated by CaM kinase II. The discrepancy between these results and previous results may reflect the different tissue sources of the kinase and, for smooth muscle myosin, the calmodulin concentrations used to assay kinase activity. In the early studies using smooth muscle myosin, activation of CaM kinase II was achieved by the addition of approximately stoichiometric amounts of calmodulin with respect to kinase. However, since then smooth muscle myosin has been reported to bind Ca^{2+}/calmodulin with high affinity (174). This affinity is apparently higher than that of CaM Kinase II for Ca^{2+}/calmodulin, since phosphorylation of myosin by this kinase requires a calmodulin concentration greater than that of myosin (173). This phenomenon is not observed with the Ca^{2+}/calmodulin-dependent MLCK, since the affinity of MLCK for Ca^{2+}/calmodulin is approximately 100-fold greater than that of CaM kinase II (173).

The site of phosphorylation by CaM kinase II has been determined for smooth muscle myosin and appears to be serine 19. This was inferred from experiments that showed that the phosphorylation site is a serine residue, that two-dimensional mapping of tryptic digests of myosin phosphorylated by CaM kinase II and MLCK are identical, and that phosphorylation by CaM kinase II is not seen after phosphorylation by MLCK (173). As would be expected, phosphorylation by CaM kinase II results in an increase in the actin-activated myosin ATPase activity (171, 173).

What role the CaM kinase II may play in vivo remains to be determined. The greater rate of phosphorylation of the myosin II regulatory light chain by MLCK and the greater affinity for Ca^{2+}/calmodulin of MLCK imply that phosphorylation by CaM kinase II may not be significant in vivo. However, Edelman et al (173) suggest that light chain phosphorylation by CaM kinase II may be important during the phase following the intracellular Ca^{2+} wave since, unlike MLCK, CaM kinase II remains active as a result of an initial Ca^{2+}/calmodulin-dependent autophosphorylation. In addition, in certain nonmuscle tissues such as brain, the concentration of CaM kinase II has been estimated to be at least 1000-fold greater than that of MLCK (173). Thus, the higher concentration of CaM kinase II may offset the lower rate of light chain phosphorylation.

PHOSPHORYLATION BY pp 34^{cdc2} Preliminary studies using smooth muscle myosin indicate that myosin II molecules may be a substrate for the cell cycle–dependent protein kinase pp34^{cdc2} (175). The phosphorylation site is believed to be serine 1 or serine 2, sites that are also phosphorylated by PKC (176). The effects of this phosphorylation, however, have not been conclusively determined. Although the sites are also those targeted by PKC, at least in vitro, PKC phosphorylates threonine 9 with faster kinetics than it phosphorylates either serine residue (156). It remains unclear whether the decrease in the actin-activated ATPase activity seen with PKC phosphorylation of myosin previously phosphorylated by MLCK results from phosphorylation of threonine 9 or whether phosphorylation of serine 1 or serine 2 alone is sufficient to affect the actin-activated ATPase. Thus no evidence exists to indicate that phosphorylation by the pp34^{cdc2} results in the same in vitro effects on myosin activity observed with PKC phosphorylation. It would be of great interest to see what effects phosphorylation of the regulatory light chain by pp34^{cdc2} has on myosin II function in vitro. This should provide some insights into the in vivo effects of phosphorylation not only by this kinase but also by PKC, since the in vivo sites of PKC phosphorylation appear to be only at serine 1 and 2 (126, 152, 166, 167).

PHOSPHORYLATION BY PROTEASE-ACTIVATED PROTEIN KINASE I   The phosphorylation of the regulatory light chain from skeletal muscle (177, 178) and smooth muscle (90) by the protease-activated protein kinase I from skeletal muscle, reticulocytes, and chicken gizzard has been reported. This kinase, which phosphorylates myosin light chain in a Ca^{2+}-independent manner and is activated by limited proteolysis by trypsin, appears to phosphorylate the regulatory light chain at serine 19, as assessed by two-dimensional tryptic and chymotryptic peptide mapping of light chain phosphorylated by this kinase and MLCK, and by phosphoamino acid determination (90). Activation of the myosin II actin-activated ATPase activity has been demonstrated with phosphorylation of the regulatory light chain by the protease-activated protein kinase (90). However, no physiological significance of light chain phosphorylation by this kinase has been established.

PHOSPHORYLATION BY MYOSIN I HEAVY CHAIN KINASE   The myosin I heavy chain kinase isolated from *Acanthamoeba* has been shown to phosphorylate smooth muscle myosin regulatory light chains (61) at the same residues phosphorylated by MLCK and with comparable rates, resulting in an increase in the actin-activated ATPase activity (62). However, this phosphorylation is most likely fortuitous since the *Acanthamoeba* myosin I heavy chain kinase does not phosphorylate the *Acanthamoeba* myosin II regulatory light chain. Thus although it is intriguing to consider the role of a single protein kinase able to activate both the myosin I and the myosin II systems, no evidence exists that such a kinase exists, although several general protein kinases, in particular the membrane-associated PKC, may be candidates.

Dephosphorylation by Myosin Light Chain Phosphatases

A number of general protein phosphatases isolated from skeletal and smooth muscle tissues are known to remove the phosphate from the regulatory myosin light chain (for review, see 179). Protein phosphatase type 1M (PP-1M) appears to be the major enzyme responsible for dephosphorylating myosin light chain in skeletal and cardiac muscle, accounting for approximately 60% and 90% of the total myosin light chain phosphatase activity, respectively (179–181). Two protein phosphatases with high myosin light chain dephosphorylating activity have also been identified from turkey gizzard. These enzymes have been termed smooth muscle phosphatases III and IV (182). Smooth muscle phosphatase IV binds myosin (183, 184), and is thought to be analogous to PP-1M (179). In addition, a phosphatase associated with smooth muscle myosin that appears distinct from the smooth muscle phosphatase IV has been recently identified (185).

Since the general protein phosphatases appear to be ubiquitous, it is presumed that an enzyme equivalent to PP-1M also functions to dephosphorylate the myosin II regulatory light chain in nonmuscle cells. A myosin II light chain phosphatase activity has been partially purified from *Dictyostelium* cells. This phosphatase dephosphorylates the light chain but not the heavy chain of *Dictyostelium* myosin II (83). It would be interesting to determine whether it is a nonmuscle isoform of PP-1M.

MYOSIN II HEAVY CHAIN PHOSPHORYLATION

Phosphorylation of myosin II heavy chain has been found to occur in a variety of nonmuscle cells as well as in the catch muscle of mollusks (Table 2), and appears to be a general mechanism of regulating myosin II function. In most nonmuscle cells, heavy chain phosphorylation occurs in addition to light chain phosphorylation. Much of the work on the molecular details of myosin II heavy chain phosphorylation, including the purification and characterization of specific myosin II heavy chain kinases (MHCKs), has focused on lower eukaryotic systems.

The myosin II heavy chain has been shown to be a substrate for several protein kinases. In addition to substrate-specific MHCKs, the heavy chain can be phosphorylated by a number of general kinases, including PKC as well as casein kinase II (CK II) and a Ca^{2+}/calmodulin dependent protein kinase (CaM kinase). With the exception of myosin II from *Physarum*, phosphorylation of the heavy chain by endogenous kinases results in a decrease of the actin-activated ATPase (Table 2). In all myosin IIs studied, the phosphorylation sites on the heavy chain are located within the tail domain, and in most cases, are in close proximity to the carboxyl terminus (Figure 2). The mechanism by which heavy chain phosphorylation regulates the activity of the head domain is not known. Heavy chain phosphorylation regulates the assembly state of the *Dictyostelium* myosin II molecule, which may in turn affect the measured actin-activated ATPase (186, 187). Heavy chain phosphoryla-

Table 2 Summary of myosin II heavy chain phosphorylation

Myosin source	Kinase[a]	Phosphorylation	Phospho-amino acid	Phosphorylation site	P_i/heavy chain (mole/mole)	Effect on myosin	References
Fibroblast	endogenous	in vitro	—	—	0.3–0.5	—	210
Leukemic myoblast	endogenous	in vivo	—	—	1.4	Decrease in ATPase	244
Lymphocyte	endogenous	in vivo & in vitro	—	—	—	Decrease in ATPase	211
Physarum	endogenous	in vitro	Thr	—	1	Increase in ATPase	212, 245
Pancreatic acinar cell	endogenous	in vivo	Ser & Thr	—	0.05	—	246
Acanthamoeba	MHCK	in vivo & in vitro	Ser	1489, 1494, 1499	3	Decrease in ATPase, Decrease in filament formation	52, 53, 199, 201
Dictyostelium	MHCK	in vivo & in vitro	Ser & Thr	1823, 1833, 2029	2–4	Decrease in ATPase, Decrease in filament formation	128, 186, 192, 197, 198, 207, 226, 247

Brush border	CaMK	in vitro	Thr	—	1	No effect[b]	221–224
Ehrlich ascites tumor cell	CK II	in vivo & in vitro	—	COOH-terminal 10 kDa	—	No effect	219
Macrophage	CK II	in vivo & in vitro	Ser	COOH-terminal 10 kDa	2	No effect	218, 248, 249
Brain	CK II	in vitro	Ser	COOH-terminal 5 kDa	1	No effect	220, 250–252
Human platelet	PKC	in vivo & in vitro	Ser	COOH-terminal 50 amino acids	1	—	165, 214, 215
Chicken epithelium	PKC	in vitro	Ser	1915	1	—	250
Rat basophilic leukemia cells	PKC	in vivo	Ser	last 50 amino acids	1	—	165, 217
Bovine platelet	PKC	in vitro	—	—	1	—	116
Thymus	PKC	in vitro	—	—	1	—	153
Molluscan catch muscle	cAMPK	in vivo & in vitro	—	rod portion	1.3–1.5	Decrease in filament formation	253, 254

[a] Endogenous kinases refer to identified activities that have not been characterized. Myosin heavy chain kinase (MHCK), Ca^{2+}/calmodulin-dependent kinase (CaMK), casein kinase II (CK II), protein kinase C (PKC), cAMP-dependent protein kinase (cAMPK).
[b] No effect on ATPase activity or filament formation was detected.

tion of the *Acanthamoeba* myosin II, however, does not appear to affect the assembly state of the molecule (188) but inhibits the actin-activated ATPase (52).

Phosphorylation by Myosin Heavy Chain Kinases

PHOSPHORYLATION SITES   In the lower eukaryotic systems in which MHCKs have been characterized, there are multiple sites of phosphorylation by MHCK. In *Acanthamoeba*, three sites within the tail domain have been identified (53). These sites lie within the nonhelical carboxyl-terminal region of the tail and have been mapped to serine residues 1489, 1494, and 1499 (189). *Dictyostelium* myosin II also contains multiple sites of phosphorylation near the carboxyl terminus of the tail domain (54, 55, 190–194). Three sites of phosphorylation have been mapped to threonine residues 1823, 1833, and 2029 (54, 195, 196). Serine phosphorylation of the *Dictyostelium* myosin II heavy chain has also been shown (128, 197, 198), but the site(s) of serine phosphorylation remain to be determined.

EFFECTS OF PHOSPHORYLATION   In *Acanthamoeba*, myosin II heavy chain phosphorylation results in a decrease of the actin-activated ATPase (199). Phosphorylation of the three serines near the carboxyl terminus by a partially purified MHCK completely inhibits the myosin ATPase activity (52). Phosphorylation of only serine 1489 appears to be sufficient for this inhibition (51, 200).

Early studies suggested that heavy chain phosphorylation of the *Acanthamoeba* myosin II resulted in destabilization of myosin filaments (201, 202). However, more recent studies indicate that heavy chain phosphorylation has no effect on the assembly properties of *Acanthamoeba* myosin II (188). In addition, studies using a monoclonal antibody that binds near the tip of the *Acanthamoeba* myosin II tail and depolymerizes filaments indicate that disassembly of filaments is associated with an inhibition of the actin-activated ATPase activity (203–205). Thus, although phosphorylation and assembly do not appear to be directly related, both dephosphorylation of the heavy chain and filament formation appear to be necessary for maximal actin-activated ATPase activity.

Other studies suggest that the overall phosphorylation level of the filament rather than an intramolecular change induced by phosphorylation at the tip of the tail is what affects the ATPase activity. These studies showed that heteropolymers of dephosphorylated and phosphorylated myosin (202) and of dephosphorylated myosin and phosphorylated 28-kDa tryptic fragment derived from the carboxyl terminus of the tail (200) exhibited decreased ATPase activity as compared to filaments exclusively composed of dephosphorylated myosin. Atkinson et al (206) have proposed that the state of phosphorylation of the carboxyl-terminal nonhelical region regulates the actin-activated ATPase activity by altering the conformation of the filament in such a way that the myosin heads, although still able to bind filamentous actin, cannot proceed through a normal catalytic cycle.

In *Dictyostelium*, phosphorylation of the myosin II heavy chain also results in a decrease in the actin-activated ATPase of the molecule. In initial studies, phosphorylation by a partially purified MHCK to one mole of phosphate incorporated per mole of heavy chain resulted in an 80% inhibition of the ATPase activity, which is reversible with dephosphorylation (191, 197). In contrast to phosphorylation of the *Acanthamoeba* myosin II, however, phosphorylation of the *Dictyostelium* myosin II heavy chain has a dramatic effect on the ability of the molecule to assemble into filaments, which may be the cause of the apparent increase in the actin-activated ATPase (186, 187, 192, 207).

Although unphosphorylated *Dictyostelium* myosin II will form filaments at salt concentrations less than 200–250 mM, phosphorylated molecules will not assemble at any salt concentration (187, 192). Studies utilizing rotary shadowing techniques to visualize myosin II molecules reveal that phosphorylation of the heavy chain promotes a folding of myosin II at approximately 2/3 the length of the tail from the head domain (57), within the 34-kDa assembly subdomain defined by O'Halloran et al (55). Under conditions that promote filament assembly, molecules in this bent conformation are excluded from the myosin filament. In the bent conformation, the assembly subdomain is presumably sequestered by steric hindrance or by association with an adjacent region as a result of local conformational changes induced by the introduction of negative charges at the nearby phosphorylation sites. Thus, unlike assembly of higher eukaryotic myosin IIs, which is predominately dependent on light chain phosphorylation, assembly of the *Dictyostelium* myosin II appears to be regulated by heavy chain phosphorylation, which favors a folded monomeric conformation that is unable to incorporate into dimers or filaments (Figure 5).

IN VIVO EVIDENCE OF PHOSPHORYLATION   Phosphorylation of the myosin II heavy chain has been demonstrated in a number of cell systems (Table 2). In *Dictyostelium* cells, both serine and threonine phosphorylation of the myosin II heavy chain occurs in vivo (128, 197). Stimulation of chemoattractant-competent cells by cAMP results in an increase in the phosphorylation levels of the myosin II regulatory light chain and heavy chain (128, 208, 209). This increase appears to be predominantly due to an increase in threonine

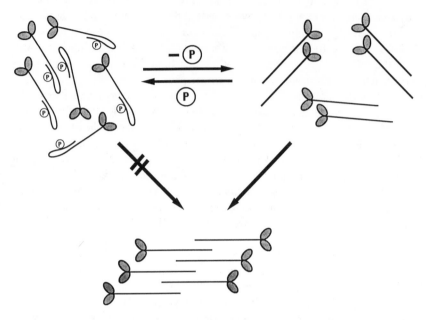

Figure 5   Assembly of *Dictyostelium* myosin II. Such assembly is dependent on heavy chain dephosphorylation in the tail domain. Phosphorylation of the tail induces a bend within the tail. These bent molecules cannot assemble into filaments. Dephosphorylated molecules exist mainly in an extended form and can form filaments.

phosphorylation and parallels myosin translocation to the cytoskeleton and cell shape changes (130). However, little is known about the precise in vivo phosphorylation sites. Thus whether the observed in vivo phosphorylation reflects known MHCK activity or other kinase activity is still unclear. In *Acanthamoeba*, at least two of the three serine sites identified by in vitro phosphorylation by MHCK are known to occur in vivo (52).

REGULATION OF MHCK   MHCK from fibroblasts (210), lymphocytes (211), and *Physarum* (212) was initially described as a protein kinase activity that copurifies with actomyosin. These kinases phosphorylate only the myosin II heavy chain and require only Mg^{2+} and ATP as cofactors. Similarly, a partially purified MHCK from *Acanthamoeba* (52) appears to be specific for the *Acanthamoeba* myosin II heavy chain and also requires only Mg^{2+} and ATP for activity. The *Acanthamoeba* MHCK is not regulated by cAMP or cGMP and is only slightly inhibited by Ca^{2+}/calmodulin.

At least four distinct MHCKs have been identified in *Dictyostelium* (187,

191, 198, 207). At least two have been studied from growing cells. Maruta et al (191) reported the partial purification of a MHCK activity from growing *Dictyostelium* cells that phosphorylates only threonine residues and is not regulated by known kinase effectors such as Ca^{2+}/calmodulin, cAMP, and cGMP. Kuczmarski (198) identified at least two MHCK activities from growing cells. Both are specific for the *Dictyostelium* myosin II heavy chain and require only Mg^{2+} and ATP as cofactors. One activity phosphorylates both serine and threonine residues while the other phosphorylates only threonine residues. In addition, Cote & Bukiejko (207) have purified to near homogeneity a MHCK from the soluble fraction of growing cells. This kinase, which is specific for *Dictyostelium* myosin II heavy chain, phosphorylates only threonine residues. Its activity is unaffected by the presence Ca^{2+}/calmodulin, cAMP, or cGMP. However, this kinase appears to be activated by autophosphorylation (213). It is unclear whether the threonine-specific activity partially purified by Maruta et al (191) and Kuczmarski (198) represents the MHCK characterized by Cote & Bukiejko (207).

Two additional MHCKs have been purified from aggregation-competent *Dictyostelium* cells. One has been shown to phosphorylate *Dictyostelium* as well as *Physarum* myosin II heavy chain on threonine residues and is inhibited by the presence of Ca^{2+}/calmodulin (191). A second MHCK, purified from the membrane-associated fraction, phosphorylates only threonine residues in the *Dictyostelium* myosin II heavy chain and undergoes autophosphorylation (187). What effect, if any, autophosphorylation has on the activity of this kinase is unknown. As with most of the various MHCKs studied in *Dictyostelium*, this MHCK is not affected by Ca^{2+}/calmodulin, cAMP, or cGMP. Recently, a cDNA clone corresponding to this MHCK has been isolated and characterized (S. Ravid and J. A. Spudich, personal communication). The primary structure of this MHCK is very reminiscent of that of PKC. In addition to the highly conserved protein kinase catalytic domain, this MHCK contains the cysteine repeat motif characteristic of the PKC family. It is intriguing to speculate that this MHCK may be a substrate-specific member of the PKC family.

Phosphorylation by Other Kinases

PHOSPHORYLATION BY PKC   The initial observation that PKC phosphorylates myosin II heavy chain came from experiments with rat basophilic leukemia (RBL) cells and human platelets. In RBL cells, aggregation of IgE receptors present on the plasma membrane results in hydrolysis of inositol phospholipids and activation of PKC. This activation correlates with an increase in the level of myosin II heavy chain phosphorylation (166). Similar results were observed in human platelets treated with phorbol esters (214).

Phosphopeptide mapping of myosin II from activated RBL cells and human platelets revealed that the phosphorylated peptides are similar, suggesting that sites of phosphorylation in these systems are analogous if not identical.

Phosphorylation by PKC has also been demonstrated in vitro with myosin II heavy chain from human platelet (214), chicken intestinal epithelial cells (215), bovine platelets (116), and thymus (153). The phosphorylated amino acid is a serine residue, corresponding to serine 1915 of the myosin heavy chain sequence of chicken intestinal epithial cell (216), which is located within the α-helical region about 50 amino acids from the carboxyl terminus (215, 217). Although the role of PKC phosphorylation of myosin II heavy chain is unknown, Adelstein et al (217) have suggested that this phosphorylation destabilizes myosin filaments in RBL cells, a role similar to MHCK phosphorylation of *Dictyostelium* myosin II heavy chain. Changes in myosin localization occur after stimulation of RBL cells with kinetics that are consistent with the myosin phosphorylation (A. Spudich, personal communication). In RBL cells, destabilization of the myosin filaments may also promote disruption of actin filaments and allow histamine granules to approach the cell membrane where their contents can be released (166, 217).

PHOSPHORYLATION BY CK II   Myosin II heavy chains from macrophages (218), Ehrlich ascites tumor cells (219), and bovine brain (220) have been shown to be phosphorylated by CK II. This phosphorylation is known to occur in vivo in at least the macrophage and the Ehrlich ascites tumor cell system. The site of phosphorylation by CK II is located near the tip of the tail but is distinct from the site phosphorylated by PKC (215). In brain myosin II, the site of CK II phosphorylation is a serine residue found in the nonhelical region at the carboxyl terminus of the tail (220). Interestingly, smooth muscle myosins from chicken gizzard and rat aorta contain neither the PKC nor the CK II phosphorylation sites (215). However, phosphorylation by CK II does not appear to have any effect on the actin-activated ATPase activity or the assembly properties of myosin II in vitro. Whether CK II phosphorylation affects some aspect of myosin II function in vivo remains to be determined.

PHOSPHORYLATION BY CaM KINASE Ca^{2+}/calmodulin-dependent phosphorylation of the myosin II heavy chain from brush border by an endogenous kinase (221, 222) has been demonstrated in vitro (223, 224). This phosphorylation occurs on a threonine residue. However, like CK II phosphorylation, no effects on the properties of myosin II have been detected as a consequence of phosphorylation by the CaM kinase.

Dephosphorylation by Myosin Heavy Chain Phosphatases

Phosphatases capable of dephosphorylating myosin II heavy chains have been identified from *Acanthamoeba* and *Dictyostelium*. The phosphatase from *Acanthamoeba* has an apparent molecular weight of 39,000 and has broad substrate specificity (225). This enzyme appears to be a type 2A phosphatase (179) in that it is inhibited by ATP, pyrophosphate, and NaF but is stabilized by Mn^{2+} (225). Dephosphorylation of the *Acanthamoeba* myosin II heavy chain using this phosphatase results in an increase of the actin-activated myosin ATPase.

Two phosphatases have also been partially purified from *Dictyostelium* (226). One phosphatase dephosphorylates the myosin II heavy chain as well as the regulatory light chain. Both also dephosphorylate histone and casein but at a much slower rates. These phosphatases require Mg^{2+} for their activity but are unaffected by Ca^{2+}. Dephosphorylation of *Dictyostelium* myosin II by these phosphatases increases the actin-activated myosin ATPase and induces filament formation.

CONCLUSIONS AND FUTURE DIRECTIONS

Virtually all nonmuscle myosins, with the possible exception of brush border myosin I, are controlled by phosphorylation. The specific mechanisms by which phosphorylation controls the enzymatic and structural properties of these motor proteins vary. In higher eukaryotes, phosphorylation of myosin II molecules on the regulatory light chain appears to be necessary for both the actin-activated myosin ATPase as well as the ability to form filaments. Although phosphorylation of the heavy chain is also known to occur, the specific effects of this phosphorylation on the properties of myosin II are unclear.

In contrast, the properties of *Dictyostelium* myosin II are clearly modulated by regulatory pathways involving phosphorylation of the regulatory light chain and the heavy chain. Distinct kinases phosphorylate these two subunits. In this system, phosphorylation of the regulatory light chain enhances motor activity while that of the heavy chain inhibits filament formation. These phosphorylations thus must be spatially and temporally coordinated in vivo to organize the apparatus essential to the cellular function of myosin II in this cell. In *Physarum*, phosphorylation also appears to increase myosin activity. Although phosphorylation of both polypeptide chains occurs, the specific effects of the individual phosphorylations remain to be determined. In addition, binding of Ca^{2+} to the essential light chain regulates the *Physarum* myosin II by negating the activating influence of phosphorylation (210, 227).

Unlike these myosin IIs, *Acanthamoeba* myosin II is predominantly controlled by phosphorylation of the heavy chain. However, in contrast to the effects of this phosphorylation in *Dictyostelium*, it appears that heavy chain phosphorylation at the tip of the tail domain does not affect the assembly properties of the *Acanthamoeba* myosin II but rather may affect the quaternary structure of the filament to modulate ATPase activity.

Limited data is available on the regulation of the very diverse class of myosin I molecules. Phosphorylation of the heavy chain within the head domain activates amoeboid myosin Is. However, whether this mechanism underlies the regulation of other types of myosin I is unknown. The brush border myosin I binds Ca^{2+}/calmodulin and thus seems to be a Ca^{2+}-regulated enzyme. However, studies on the Ca^{2+} dependence of brush border myosin I–based motility have arrived at inconsistent results. Whereas one study (228) demonstrated that the brush border myosin I is active at low Ca^{2+} concentrations, another study suggested that the enzyme is active at high Ca^{2+} concentrations (229). Although this discrepancy may be explained by a number of possibilities, it is intriguing to speculate that the difference in the two preparations of brush border myosin I may reflect the phosphorylation state of the molecule and that the brush border myosin I, like *Physarum* myosin II, is dually regulated by a direct effect of Ca^{2+} and by phosphorylation. The sequences of brush border myosin I from both bovine and avian sources contain the phosphorylation site recognized by the *Acanthamoeba* MIHCK. It would thus be interesting to determine whether the *Acanthamoeba* MIHCK will phosphorylate brush border myosin I or if an analogous kinase exists in brush border.

Over the past several years, it has become apparent that the molecular basis of cell motility involves a multitude of motor molecules. Novel myosins are being identified not only in animal cells, but also in a number of plant systems including *Nitella* (230), *Egeria* (231), *Lycopersicon* (232), *Heracleum* (233), *Chara* (234, 235), *Pisum* (236), and *Ernodermis* (237). In addition, several reports have suggested the presence of novel myosin molecules in bacterial cells (238–240). However, a number of these newly identified myosins, especially those from eukaryotic sources, have only been characterized at a molecular genetic level. Little is known regarding their cellular functions, their structure, their enzymatic properties, or the regulatory mechanisms of these molecules.

In addition to the myriad of myosins that are now seemingly present in all nonmuscle (and possibly muscle) cells, a number of signalling pathways may control the activity of each motor molecule. In higher eukaryotes and in *Dictyostelium* and *Physarum*, several kinases may regulate at least myosin II activity in vivo by multiple phosphorylations on both the regulatory light chain and the heavy chain. These phosphorylations are presumably coordinated such that the myosin II molecule is appropriately localized and activated for a particular cellular function. For example, phosphorylation of the light chain may regulate the role of myosin II in cell division and cell shape changes. Early studies involving microinjection of antibodies specific to smooth muscle MLCK into sea urchin eggs suggest that MLCK activity and presumably light chain phosphorylation are necessary for cell division (241). More recent studies using antisense oligonucleotides indicate that a decrease in MLCK results in morphological cell shape changes and a possibly slowed cell proliferation (82). In contrast, PKC phosphorylation may modulate myosin II in such a way that vesicular fusion to the plasma membrane and thus secretion may occur. However, the precise relationships between particular phosphorylation events and specific cellular functions of myosin II remain obscure.

The determination of such a relationship can only come about from mapping of in vivo phosphorylation sites coupled with the study of cells that have been specifically mutated such that they no longer express the specific kinases, or that contain myosin that does not possess the phosphorylation sites. One system amenable to such studies is *Dictyostelium*. Both the myosin heavy chain (242) and the regulatory light chain (102) as well as the complementary DNA of the MLCK (243) and the MHCK (S. Ravid and J. A. Spudich, personal communication) have been characterized. The disruption of the *Dictyostelium* myosin heavy chain has already led to new insights into myosin II function. It is hoped that new insights will also be gained from disruption of the myosin kinase genes. Specific mutations that eliminate the phosphorylation site in both the heavy and the light chain can also be made. Such mutated molecules can subsequently be introduced into cells genetically engineered such that they do not express the particular myosin polypeptide chain. This approach would be better suited in determining the role of myosin II phosphorylation by general protein kinases such as PKC or $pp34^{cdc2}$. The availability of gene targeting techniques and the feasibility of biochemical characterizations of myosins in organisms such as *Dictyostelium* as well as in mammalian cells should allow for a better understanding of the role of myosin phosphorylations in controlling these motor proteins.

ACKNOWLEDGMENTS

We are indebted to Kathy Ruppel for critical review of the manuscript and for extensive assistance in its preparation. We would also like to thank Tom Egelhoff, Hans Warrick, Meg Titus, Mitsuo Ikebe, and Janet Smith for many helpful discussions.

Literature Cited

1. Pollard, T. D., Korn, E. D. 1973. *J. Biol. Chem.* 248:4682–90
2. Knecht, D. A., Loomis, W. F. 1987. *Science* 236:1081–86
3. De Lozanne, A., Spudich, J. A. 1987. *Science* 236:1086–91
4. Manstein, D. J., Titus, M. A., De-Lozanne, A., Spudich, J. A. 1989. *EMBO J.* 8:923–32
5. Warrick, H. M., Spudich, J. A. 1987. *Annu. Rev. Cell Biol.* 3:379–421
6. Korn, E. D., Hammer, J. A. III. 1988. *Annu. Rev. Biophys. Biophys. Chem.* 17:23–45
7. Korn, E. D., Hammer, J. A. III. 1990. *Curr. Opin. Cell Biol.* 2:57–61
8. Pollard, T. D., Doberstein, S. K., Zot, H. G. 1991. *Annu. Rev. Physiol.* 53:653–81
9. Kamm, K. E., Stull, J. T. 1989. *Prog. Clin. Biol. Res.* 315:265–78
10. Ito, M., Hartshorne, D. J. 1990. *Prog. Clin. Biol. Res.* 327:57–72
11. Stull, J. T., Bowman, B. F., Gallagher, P. J., Herring, B. P. 1990. *Prog. Clin. Biol. Res.* 327:107–26
12. Trybus, K. M. 1991. *Cell Motil. Cytoskelet.* 18:81–85
13. Kuznicki, J., Barylko, B. 1988. *Int. J. Biochem.* 20:559–68
14. Sellers, J. R. 1991. *Curr. Opin. Cell Biol.* 3:98–104
15. Trybus, K. M. 1991. *Curr. Opin. Cell Biol.* 3:105–11
16. Hynes, T. R., Block, S. M., White, B. T., Spudich, J. A. 1987. *Cell* 48:953–63
17. Harada, Y., Noguchi, A., Kishino, A., Yanagida, T. 1987. *Nature* 326:805–8
18. Toyoshima, Y. Y., Kron, S. J., McNally, E. M., Niebling, K. R., Toyoshima, C., Spudich, J. A. 1987. *Nature* 328:536–39
19. Manstein, D. J., Ruppel, K. M., Spudich, J. A. 1989. *Science* 246:656–58
20. Ruppel, K. M., Egelhoff, T. T., Spudich, J. A. 1990. *Ann. NY Acad. Sci.* 582:147–55
21. Coluccio, L. M., Bretscher, A. 1988. *J. Cell Biol.* 106:495–502
22. Coluccio, L. M., Bretscher, A. 1990. *Biochemistry* 29:11089–94
23. Maruta, H., Gadasi, H., Collins, J. H., Korn, E. D. 1978. *J. Biol. Chem.* 253:6297–300
24. Maruta, H., Gadasi, H., Collins, J. H., Korn, E. D. 1979. *J. Biol. Chem.* 254:3624–30
25. Albanesi, J. P., Hammer, J. A. III,

Korn, E. D. 1983. *J. Biol. Chem.* 258:176–81
26. Collins, J. H., Borysenko, C. W. 1984. *J. Biol. Chem.* 259:4128–35
27. Mooseker, M. S., Coleman, T. R. 1989. *J. Cell Biol.* 108:2395–400
28. Cote, G. P., Albanesi, J. P., Ueno, T., Hammer, J. A. III, Korn, E. D. 1985. *J. Biol. Chem.* 260:4543–46
29. Montell, C., Rubin, G. M. 1988. *Cell* 52:757–72
30. Titus, M. A., Warrick, H. M., Spudich, J. A. 1989. *Cell Regul.* 1:55–63
31. Jung, G., Saxe, C. L. III, Kimmel, A. R., Hammer, J. A. III. 1989. *Proc. Natl. Acad. Sci. USA* 86:6186–90
32. Horowitz, J. A., Hammer, J. A. III. 1990. *J. Biol. Chem.* 265:20646–52
33. Johnstone, G. C., Prendergast, J. A., Singer, R. A. 1991. *J. Cell Biol.* 113:539–51
34. Mercer, J. A., Seperack, P. K., Strobel, M. C., Copeland, N. G., Jenkins, N. A. 1991. *Nature* 349:709–13
35. Kendrick-Jones, J., Lehman, W., Szent-Gyorgyi, A. G. 1970. *J. Mol. Biol.* 54:313–26
36. Jakes, R., Northrop, F., Kendrick-Jones, J. 1976. *FEBS Lett.* 70:229–34
37. Szent-Gyorgyi, A. G., Chantler, P. D. 1986. *Myology*, ed. A. G. Engel, B. Q. Banker, 1:589–612. New York: McGraw-Hill
38. Karess, R. E., Chang, X-J., Edwards, K. A., Kulkarni, S., Aguilera, I., Kiehart, D. P. 1991. *Cell* 65:1177–89
39. Okamoto, Y., Sekine, T., Grammer, J., Yount, R. G. 1986. *Nature* 324:78–80
40. Ochiai, Y., Handa, A., Watabe, S., Hashimoto, K. 1990. *Int. J. Biochem.* 22:1097–103
41. Highashihara, M., Frado, L-L. Y., Craig, R., Ikebe, M. 1989. *J. Biol. Chem.* 264:5218–25
42. Hoshimaru, M., Fujio, Y., Sobue, K., Sugimoto, T., Nakanishi, S. 1989. *J. Biochem.* 106:455–59
43. Lynch, T. J., Brzeska, H., Baines, I. C., Korn, E. D. 1991. *Methods Enzymol.* 196:12–23
44. Yumura, S., Fukui, Y. 1985. *Nature* 314:194–96
45. Fukui, Y. 1990. *Ann. NY Acad. Sci.* 582:156–65
46. Wessels, D., Soll, D. R., Knecht, D., Loomis, W. F., De Lozanne, A., Spudich, J. A. 1988. *Dev. Biol.* 128:164–77
47. Pasternak, C., Spudich, J. A., Elson, E. L. 1989. *Nature* 341:549–51

48. Karn, J., Brenner, S., Barnett, L. 1983. *Proc. Natl. Acad. Sci. USA* 80:4253–57
49. McLachlan, A. D. 1984. *Annu. Rev. Biophys. Bioeng.* 13:167–89
50. Cross, R. A., Vandekerckhove, J. 1986. *FEBS Lett.* 200:355–60
51. Sathyamoorthy, V., Atkinson, M. A. L., Bowers, B., Korn, E. D. 1990. *Biochemistry* 29:3793–97
52. Cote, G. P., Collins, J. H., Korn, E. D. 1981. *J. Biol. Chem.* 256:12811–16
53. Collins, J. H., Cote, G. P., Korn, E. D. 1982. *J. Biol. Chem.* 257:4529–34
54. Pagh, K., Maruta, H., Claviez, M., Gerisch, G. 1984. *EMBO J.* 3:3271–78
55. O'Halloran, T. J., Ravid, S., Spudich, J. A. 1990. *J. Cell Biol.* 110:63–70
56. Pagh, K., Gerisch, G. 1986. *J. Cell Biol.* 103:1527–38
57. Pasternak, C., Flicker, P. F., Ravid, S., Spudich, J. A. 1989. *J. Cell Biol.* 109:203–10
58. Egelhoff, T. T., Brown, S. S., Spudich, J. A. 1991. *J. Cell Biol.* 112:677–88
59. Onishi, H., Wakabayashi, T. 1982. *J. Biochem.* 92:871–79
60. Doberstein, S. K., Pollard, T. D. 1989. *J. Cell Biol.* 109:86a
61. Hammer, J. A. III, Albanesi, J. P., Korn, E. D. 1983. *J. Biol. Chem.* 258:10168–75
62. Hammer, J. A. III, Sellers, J. R., Korn, E. D. 1984. *J. Biol. Chem.* 259:3224–29
63. Brzeska, H., Lynch, T. J., Martin, B., Korn, E. D. 1989. *J. Biol. Chem.* 264:19340–48
64. Brzeska, H., Lynch, T. J., Korn, E. D. 1988. *J. Biol. Chem.* 263:427–35
65. Brzeska, H., Lynch, T. J., Korn, E. D. 1989. *J. Biol. Chem.* 264:243–50
66. Brzeska, H., Lynch, T. J., Martin, B., Corigliano-Murphy, A., Korn, E. D. 1990. *J. Biol. Chem.* 265:16138–44
67. Hoshimaru, M., Nakanishi, S. 1987. *J. Biol. Chem.* 262:14625–32
68. Garcia, A., Coudrier, E., Carboni, J., Anderson, J., Vandekerckhove, J., et al. 1989. *J. Cell Biol.* 109:2895–903
69. Maruta, H., Korn, E. D. 1977. *J. Biol. Chem.* 252:8329–32
70. Lynch, T. J., Brzeska, H., Miyata, H., Korn, E. D. 1989. *J. Biol. Chem.* 264:19333–39
71. Sheetz, M. P., Spudich, J. A. 1983. *Nature* 303:31–35
72. Sheetz, M. P., Chasan, R., Spudich, J. A. 1984. *J. Cell Biol.* 99:1867–71
73. Albanesi, J. P., Fujisaki, H., Hammer, J. A. III, Korn, E. D., Jones, R., Sheetz, M. P. 1985. *J. Biol. Chem.* 260:8649–52

74. Brzeska, H., Lynch, T. J., Korn, E. D. 1990. *J. Biol. Chem.* 265:3591–94
75. Adams, R. J., Pollard, T. D. 1989. *Nature* 340:565–88
76. Adams, R. J., Pollard, T. D. 1989. *Cell Motil. Cytoskelet.* 14:178–82
77. Miyata, H., Bowers, B., Korn, E. D. 1989. *J. Cell Biol.* 109:1519–28
78. Perrie, W. T., Smillie, L. B., Perry, S. V. 1973. *Biochem. J.* 135:151–56
79. Adelstein, R. S., Conti, M. A., Anderson, W. Jr. 1973. *Proc. Natl. Acad. Sci. USA* 70:3115–19
80. Adelstein, R. S., Conti, M. A. 1975. *Nature* 256:597–98
81. Stull, J. T., Nunnally, M. H., Michnoff, C. H. 1986. *The Enzymes* 17:113–66
82. Shoemaker, M. O., Lau, W., Shattuck, R. L., Kwiatkowski, A. P., Matrisian, P. E., et al. 1990. *J. Cell Biol.* 111:1107–25
83. Griffith, L. M., Downs, S. M., Spudich, J. A. 1987. *J. Cell Biol.* 104:1309–23
84. Tan, J. L., Spudich, J. A. 1990. *J. Biol. Chem.* 265:13818–24
85. Noiman, E. S. 1980. *J. Biol. Chem.* 255:11067–70
86. Gallis, B., Edelman, A. M., Casnellie, J. E., Krebs, E. G. 1983. *J. Biol. Chem.* 258:13089–93
87. Pike, L. J., Kuenzel, E. A., Casnellie, J. E., Krebs, E. G. 1984. *J. Biol. Chem.* 259:9913–21
88. Singh, T. J., Akatsuka, A., Huang, K-P. 1983. *FEBS Lett.* 159:217–20
89. Matsumura, S., Murakami, N., Tashiro, Y., Yasuda, S., Kumon, A. 1983. *Arch. Biochem. Biophys.* 227:125–35
90. Tuazon, P. T., Traugh, J. A. 1984. *J. Biol. Chem.* 259:541–46
91. Taubman, M. B., Grant, J. W., Nadal-Ginard, B. 1987. *J. Cell. Biol.* 104:1505–13
92. Zavodny, P. J., Petro, M. E., Lonial, H. K., Dailey, S. H., Narula, S. K., et al. 1990. *Circ. Res.* 67:933–40
93. Grant, J. W., Zhong, R. Q., McEwen, P. M., Church, S. L. 1990. *Nucleic Acids Res.* 18:5892
94. Pearson, R. B., Jakes, R., John, M., Kendrick-Jones, J., Kemp, B. E. 1984. *FEBS Lett.* 168:108–12
95. Ikebe, M., Hartshorne, D. J., Elzinga, M. 1986. *J. Biol. Chem.* 261:36–39
96. Highashihara, M., Hartshorne, D. J., Craig, R., Ikebe, M. 1989. *Biochemistry* 28:1642–49
97. Ikebe, M. 1989. *Biochemistry* 28:8750–55
98. Gorecka, A., Aksoy, M. O., Hart-

shorne, D. J. 1976. *Biochem. Biophys. Res. Commun.* 68:325-31
99. Sobieszek, A. 1977. *The Biochemistry of Smooth Muscle,* ed. N. L. Stephens, pp. 413-43. Baltimore, Md: University Park
100. Morgan, M., Perry, S. V., Ottaway, J. 1976. *Biochem. J.* 157:687-97
101. Pemrick, S. M. 1980. *J. Biol. Chem.* 255:8836-41
102. Tafuri, S. R., Rushforth, A. M., Kuczmarski, E. R., Chisholm, R. L. 1989. *Mol. Cell. Biol.* 9:3073-80
103. Sellers, J. R., Spudich, J. A., Sheetz, M. P. 1985. *J. Cell Biol.* 101:1897-902
104. Umemoto, S., Bengur, A. R., Sellers, J. R. 1989. *J. Biol. Chem.* 264:1431-36
105. Umemoto, S., Sellers, J. R. 1990. *J. Biol. Chem.* 265:14864-69
106. Warshaw, D. M., Desrosiers, J. M., Work, S. S., Trybus, K. M. 1990. *J. Cell Biol.* 111:453-63
107. Murphy, R. A. 1989. *Annu. Rev. Physiol.* 51:275-83
108. Sellers, J. R., Adelstein, R. S. 1985. *Curr. Top. Cell Regul.* 27:51-62
109. Sellers, J. R. 1985. *J. Biol. Chem.* 260:15815-19
110. Wagner, P. D., Vu, N-D. 1986. *J. Biol. Chem.* 261:7778-83
111. Wagner, P. D., Vu, N-D. 1987. *J. Biol. Chem.* 262:15556-62
112. Trybus, K. M. 1989. *J. Cell Biol.* 109:2887-94
113. Ikebe, M., Koretz, J., Hartshorne, D. J. 1988. *J. Biol. Chem.* 263:6432-37
114. Craig, R., Smith, R., Kendrick-Jones, J. 1983. *Nature* 302:436-39
115. Citi, S., Kendrick-Jones, J. 1986. *J. Mol. Biol.* 188:369-82
116. Ikebe, M., Reardon, S. 1990. *Biochemistry* 29:2713-20
117. Cross, R. A., Cross, K. E., Sobieszek, A. 1986. *EMBO J.* 5:2637-41
118. Cross, R. A., Jackson, A. P., Citi, S., Kendrick-Jones, J., Bagshaw, C. R. 1988. *J. Mol. Biol.* 203:173-81
119. Ikebe, M., Hartshorne, D. J. 1986. *Biochemistry* 25:6177-85
120. Sellers, J. R., Adelstein, R. S. 1987. *The Enzymes* 18:382-418
121. Somlyo, A. V., Butler, T. M., Bond, M., Somlyo, A. P. 1981. *Nature* 294:567-69
122. Gillis, J. M., Cao, M. L., Godfraind-de Becker, A. 1988. *J. Muscle Res. Cell Motil.* 9:18-28
123. Itoh, T., Ikebe, M., Kargacin, G. J., Hartshorne, D. J., Kemp, B. E., Fay, F. S. 1989. *Nature* 338:164-67
124. Haeberle, J. R., Sutton, T. A., Trock-

man, B. A. 1988. *J. Biol. Chem.* 263:4424-29
125. Colburn, J. C., Michnoff, C. H., Hsu, L., Slaughter, C. A., Kamm, K. E., Stull, J. T. 1988. *J. Biol. Chem.* 263:19166-73
126. Naka, M., Saitoh, M., Hidaka, H. 1988. *Arch. Biochem. Biophys.* 261:235-40
127. Wysolmerski, R. B., Lagunoff, D. 1990. *Proc. Natl. Acad. Sci. USA* 87:16-20
128. Berlot, C. H., Spudich, J. A., Devreotes, P. N. 1985. *Cell* 43:307-14
129. Chisholm, R. L., Fontana, D., Theibert, A., Lodish, H., Devreotes, P. N. 1985. *Microbial Development,* ed. R. Losick, L. Shapiro, pp. 219-54. Cold Spring Harbor, NY: Cold Spring Harbor Lab.
130. Nachmias, V. T., Fukui, Y., Spudich, J. A. 1989. *Cell Motil. Cytoskelet.* 13:158-69
131. Kamm, K. E., Hsu, L. C., Kubota, Y., Stull, J. T. 1989. *J. Biol. Chem.* 264:21223-29
132. Okagaki, T., Ishikawa, R., Kohama, K. 1991. *Biochem. Biophys. Res. Commun.* 176:564-70
133. Erickson-Viitanen, S., DeGrado, W. F. 1987. *Methods Enzymol.* 139:455-78
134. O'Neil, K. T., DeGrado, W. F. 1990. *Trends Biochem. Sci.* 15:59-64
135. Blumenthal, D. K., Takio, K., Edelman, A. M., Charbonneau, H., Titani, K., et al. 1985. *Proc. Natl. Acad. Sci. USA* 82:3187-91
136. Lukas, T. J., Burgess, W. H., Prendergast, F. G., Lau, W., Watterson, D. M. 1986. *Biochemistry* 25:1458-64
137. Kemp, B. E., Pearson, R. B., Guerriero, V. Jr., Bagchi, I. C., Means, A. R. 1987. *J. Biol. Chem.* 262:2542-48
138. Ikebe, M., Stepinska, M., Kemp, B. E., Means, A. R., Hartshorne, D. J. 1987. *J. Biol. Chem.* 262:13828-34
139. Pearson, R. B., Wettenhall, R. E., Means, A. R., Hartshorne, D. J., Kemp, B. E. 1988. *Science* 241:970-73
140. Ito, M., Guerriero, V. Jr., Chen, X., Hartshorne, D. J. 1991. *Biochemistry* 30:3498-503
141. Conti, M. A., Adelstein, R. S. 1981. *J. Biol. Chem.* 256:3178-81
142. Hashimoto, Y., Soderling, T. R. 1990. *Arch. Biochem. Biophys.* 278:41-45
143. Nishikawa, M., de Lanerolle, P., Lincoln, T. M., Adelstein, R. S. 1984. *J. Biol. Chem.* 259:8429-36
144. Ikebe, M., Inagaki, M., Kanamaru, K., Hidaka, H. 1985. *J. Biol. Chem.* 260:4547-50
145. Nishikawa, M., Shirakawa, S., Adel-

stein, R. S. 1985. *J. Biol. Chem.* 260: 8978–83

146. Ikebe, M., Reardon, S. 1990. *J. Biol. Chem.* 265:8975–78

147. de Lanerolle, P., Nishikawa, M., Yost, D. A., Adelstein, R. S. 1983. *Science* 223:1415–17

148. Miller, J. R., Silver, P. J., Stull, J. T. 1983. *Mol. Pharmacol.* 24:235–42

149. Endo, T., Naka, M., Hidaka, H. 1982. *Biochem. Biophys. Res. Commun.* 105: 942–48

150. Chiang, T. M., Cagen, L. M., Kang, A. N. 1981. *Thromb. Res.* 21:611–22

151. Carroll, R. C., Butler, R. G., Morris, P. A., Gerrard, J. M. 1982. *Cell* 30:385–93

152. Naka, M., Nishikawa, M., Adelstein, R. S., Hidaka, H. 1983. *Nature* 306: 490–92

153. Carroll, A. G., Wagner, P. D. 1989. *J. Muscle Res. Cell Motil.* 10:379–84

154. Ikeda, N., Yasuda, S., Muguruma, M., Matsumura, S. 1990. *Biochem. Biophys. Res. Commun.* 169:1191–97

155. Bengur, A. R., Robinson, E. A., Appella, E., Sellers, J. R. 1987. *J. Biol. Chem.* 262:7613–17

156. Ikebe, M., Hartshorne, D. J., Elizinga, M. 1987. *J. Biol. Chem.* 262:9569–73

157. Kennelly, P. J., Krebs, E. G. 1991. *J. Biol. Chem.* 266:15555–58

158. Nishikawa, M., Hidaka, H., Adelstein, R. S. 1983. *J. Biol. Chem.* 258:14069–72

159. de Lanerolle, P., Nishikawa, M. 1988. *J. Biol. Chem.* 263:9071–74

160. Erdodi, F., Rokolya, B., Barany, M., Barany, K. 1988. *Arch. Biochem. Biophys.* 266:583–91

161. Umekawa, H., Naka, M., Inagaki, M., Onishi, H., Wakabayashi, T., Hidaka, H. 1985. *J. Biol. Chem.* 260:9833–37

162. Adam, L. P., Haeberle, J. R., Hathaway, D. R. 1989. *J. Biol. Chem.* 264:7698–703

163. Sutton, T. A., Haeberle, J. R. 1990. *J. Biol. Chem.* 265:2749–54

164. Castagna, M., Takai, Y., Kaibuchi, K., Sano, K., Kikkawa, U., Nishizuka, Y. 1982. *J. Biol. Chem.* 257:7847–51

165. Ludowyke, R. I., Peleg, I., Beaven, M. A., Adelstein, R. S. 1989. *J. Biol. Chem.* 264:12492–501

166. Nachmias, V. T., Yoshida, K., Glennon, M. C. 1987. *J. Cell Biol.* 105:1761–69

167. Fukunaga, K., Yamamoto, H., Matsui, K., Higashi, K., Miyamoto, E. 1982. *J. Neurochem.* 39:1607–17

168. Bennett, M. K., Erondu, N. E., Kennedy, M. B. 1983. *J. Biol. Chem.* 258:12735–44

169. Payne, M. E., Schworer, C. M., Soderling, T. R. 1983. *J. Biol. Chem.* 258:2376–82

170. Kloepper, R. F., Landt, M. 1984. *Cell Calcium* 5:351–64

171. Tanaka, T., Sobue, K., Owada, M. K., Hakura, A. 1985. *Biochem. Biophys. Res. Commun.* 131:987–93

172. Chou, Y-H., Rebhun, L. I. 1986. *J. Biol. Chem.* 261:5389–95

173. Edelman, A. M., Lin, W-H., Osterhout, D. J., Bennett, M. K., Kennedy, M. B., Krebs, E. G. 1990. *Mol. Cell. Biochem.* 97:87–98

174. Sobieszek, A. 1985. *Biochemistry* 24:1266–74

175. Satterwhite, L., Cisek, L., Corden, J., Pollard, T. 1990. *Ann. NY Acad. Sci.* 582:307

176. Pollard, T. D., Satterwhite, L., Cisek, L., Corden, J., Sato, M., Maupin, P. 1990. *Ann. NY Acad. Sci.* 582:120–30

177. Tuazon, P. T., Stull, J. T., Traugh, J. A. 1982. *Biochem. Biophys. Res. Commun.* 108:910–17

178. Tuazon, P. T., Stull, J. T., Traugh, J. A. 1982. *Eur. J. Biochem.* 129:205–9

179. Cohen, P. 1989. *Annu. Rev. Biochem.* 58:453–508

180. Chisholm, A. A. K., Cohen, P. 1988. *Biochim. Biophys. Acta* 968:392–400

181. Chisholm, A. A. K., Cohen, P. 1988. *Biochim. Biophys. Acta* 971:163–69

182. Pato, M. D., Adelstein, R. S. 1983. *J. Biol. Chem.* 258:7047–54

183. Sellers, J. R., Pato, M. D. 1984. *J. Biol. Chem.* 259:11383–90

184. Pato, M. D., Kerc, E. 1985. *J. Biol. Chem.* 260:12359–66

185. Inagaki, M., Mitsui, T., Ikebe, M. 1990. *Biophys. J.* 57:A145

186. Kuczmarski, E. R., Tafuri, S. R., Parysek, L. M. 1987. *J. Cell Biol.* 105:2989–97

187. Ravid, S., Spudich, J. A. 1989. *J. Biol. Chem.* 264:15144–50

188. Sinard, J. H., Pollard, T. D. 1989. *J. Cell Biol.* 109:1529–35

189. Cote, G. P., Robinson, E. A., Appella, E., Korn, E. D. 1984. *J. Biol. Chem.* 259:12781–87

190. Peltz, G., Kuczmarski, E. R., Spudich, J. A. 1981. *J. Cell Biol.* 89:104–8

191. Maruta, H., Baltes, W., Dieter, P., Marme, D., Gerisch, G. 1983. *EMBO J.* 2:535–42

192. Cote, G. P., McCrea, S. M. 1987. *J. Biol. Chem.* 262:13033–38

193. De Lozanne, A., Berlot, C. H., Leinwand, L. A., Spudich, J. A. 1987. *J. Cell Biol.* 105:2999–3005

194. Wagle, G., Noegel, A., Scheel, J.,

Gerisch, G. 1988. *FEBS Lett.* 227:71–75

195. Vaillancourt, J. P., Lyons, C., Cote, G. P. 1988. *J. Biol. Chem.* 263:10082–87

196. Luck-Vielmetter, D., Schleicher, M., Grabatin, B., Wippler, J., Gerisch, G. 1990. *FEBS Lett.* 269:239–43

197. Kuczmarski, E. R., Spudich, J. A. 1980. *Proc. Natl. Acad. Sci. USA* 77:7292–96

198. Kuczmarski, E. R. 1986. *J. Muscle Res. Cell Motil.* 7:501–9

199. Collins, J. H., Korn, E. D. 1980. *J. Biol. Chem.* 255:8011–14

200. Ganguly, C., Atkinson, M. A. L., Attri, A. K., Sathyamoorthy, V., Bowers, B., Korn, E. D. 1990. *J. Biol. Chem.* 265:9993–98

201. Collins, J. H., Kuznicki, J., Bowers, B., Korn, E. D. 1982. *Biochemistry* 21:6910–15

202. Kuznicki, J., Albanesi, J. P., Cote, G. P., Korn, E. D. 1983. *J. Biol. Chem.* 258:6011–14

203. Kiehart, D. P., Kaiser, D. A., Pollard, T. D. 1984. *J. Cell Biol.* 99:1015–23

204. Atkinson, M. A. L., Appella, E., Corigliano-Murphy, M. A., Korn, E. D. 1988. *FEBS Lett.* 234:435–38

205. Rimm, D. L., Kaiser, D. A., Bhandari, D., Maupin, P., Kiehart, D. P., Pollard, T. D. 1990. *J. Cell Biol.* 111:2405–16

206. Atkinson, M. A. L., Lambooy, P. K., Korn, E. D. 1989. *J. Biol. Chem.* 264:4127–32

207. Cote, G. P., Bukiejko, U. 1987. *J. Biol. Chem.* 262:1065–72

208. Berlot, C. H., Devreotes, P. N., Spudich, J. A. 1987. *J. Biol. Chem.* 262:3918–26

209. Malchow, D., Bohme, R., Rahmsdorf, H. J. 1981. *Eur. J. Biochem.* 117:213–18

210. Muhlrad, A., Oplatka, A. 1977. *FEBS Lett.* 77:37–40

211. Fechheimer, M., Cebra, J. J. 1982. *J. Cell Biol.* 93:261–68

212. Ogihara, S., Ikebe, M., Takahashi, K., Tonomura, Y. 1983. *J. Biochem.* 93:205–23

213. Medley, Q. G., Gariepy, J., Cote, G. P. 1990. *Biochemistry* 29:8992–97

214. Kawamoto, S., Bengur, A. R., Sellers, J. R., Adelstein, R. S. 1989. *J. Biol. Chem.* 264:2258–65

215. Conti, M. A., Sellers, J. R., Adelstein, R. S., Elzinga, M. 1991. *Biochemistry* 30:966–70

216. Shohet, R. V., Conti, M. A., Kawamoto, S., Preston, Y. A., Brill, D. A., Adelstein, R. S. 1989. *Proc. Natl. Acad. Sci. USA* 86:7726–30

217. Adelstein, R. S., Feleg, I., Ludowyke, R., Kawamoto, S., Conti, M. A. 1990. *Adv. Second Messenger Phosphorylation Res.* 24:405–11

218. Atkinson, M. A. L., Korn, E. D., Trotter, J. A. 1986. *J. Cell Biol.* 103:118a

219. Kuznicki, J., Filipek, A. 1988. *Int. J. Biochem.* 20:1203–9

220. Murakami, N., Healy-Louie, G., Elzinga, M. 1990. *J. Biol. Chem.* 265:1041–47

221. Rieker, J. P., Swanljung-Collins, H., Montibeller, J., Collins, J. H. 1987. *Methods Enzymol.* 139:105–14

222. Rieker, J. P., Swanljung-Collins, H., Collins, J. H. 1987. *J. Biol. Chem.* 262:15262–68

223. Keller, T. C. S., Mooseker, M. S. 1982. *J. Cell Biol.* 95:943–59

224. Rieker, J. P., Swanljung-Collins, H., Montibeller, J., Collins, J. H. 1987. *FEBS Lett.* 212:154–58

225. McClure, J. A., Korn, E. D. 1983. *J. Biol. Chem.* 258:14570–75

226. Kuczmarski, E. R., Pagone, J. 1986. *J. Muscle Res. Cell Motil.* 7:510–16

227. Kohama, K., Okagaki, T. 1990. *Adv. Exp. Med. Biol.* 269:181–85

228. Collins, K., Sellers, J. R., Matsudaira, P. 1990. *J. Cell Biol.* 110:1137–47

229. Mooseker, M. S., Conzelman, K. A., Coleman, T. R., Heuser, J. E., Sheetz, M. P. 1989. *J. Cell Biol.* 109:1153–61

230. Kato, T., Tomomura, Y. 1977. *J. Biochem.* 82:777–82

231. Oshuka, K., Inoue, A. 1979. *J. Biochem.* 85:375–78

232. Vahey, M., Titus, M., Trautwein, R., Scordilis, S. 1982. *Cell Motil.* 2:131–47

233. Sokolov, O. I., Bogatyrev, V. A., Turkina, M. V. 1986. *Sov. Plant Physiol.* 33:323–31

234. Grolig, F., Williamson, R. E., Parke, J., Miller, C., Anderton, B. H. 1988. *Eur. J. Cell Biol.* 47:22–31

235. Qiao, L., Grolig, F., Jablonsky, P. P., Williamson, R. E. 1989. *Cell Biol. Int. Rep.* 13:107–17

236. Ma, Y.-Z., Yen, L.-F. 1989. *Plant Physiol.* 89:586–89

237. La Claire, J. W. II. 1991. *Planta* 184:209–17

238. Nakamura, K., Takahashi, K., Watanabe, S. 1978. *J. Biochem.* 84:1453–58

239. Casaregola, S., Norris, V., Goldberg, M., Holland, I. B. 1990. *Mol. Microbiol.* 4:505–11

240. Niki, H. N., Jaffe, A., Imamura, R., Ogura, T., Hiraga, S. 1991. *EMBO J.* 10:183–93

241. Mabuchi, I., Takanoohmuro, H. 1990. *Dev. Growth Differ.* 32:549–56

242. Warrick, H. M., De Lozanne, A., Leinwand, L. A., Spudich, J. A. 1986. *Proc. Natl. Acad. Sci. USA* 83:9433–37
243. Tan, J. L., Spudich, J. A. 1991. *J. Biol. Chem.* 266:16044–49
244. Sagara, J., Nagata, K., Ichikawa, Y. 1983. *Biochem. J.* 214:829–34
245. Takahashi, K., Ogihara, S., Ikebe, M., Tonomura, Y. 1983. *J. Biochem.* 93: 1175–83
246. Watanabe, T. K., Kuczmarski, E. R., Reddy, K. 1987. *Biochem. J.* 247:513–18
247. Kuczmarski, E. R., Routsolias, L., Parysek, L. M. 1988. *Cell Motil. Cytoskelet.* 10:471–81

248. Trotter, J. A. 1982. *Biochem. Biophy. Res. Commun.* 106:1071–77
249. Trotter, J. A., Nixon, C. S., Johnson, M. A. 1985. *J. Biol. Chem.* 260:14374–78
250. Barylko, B., Tooth, P., Kendrick-Jones, J. 1986. *Eur. J. Biochem.* 158:271–82
251. Matsumura, S., Murakami, N., Yasuda, S., Kumon, A. 1982. *Biochem. Biophys. Res. Commun.* 108:1595–600
252. Murakami, N., Matsumura, S., Kumon, A. 1984. *J. Biochem.* 95:651–60
253. Castellani, L., Cohen, C. 1987. *Science* 235:334–37
254. Castellani, L., Cohen, C. 1987. *Proc. Natl. Acad. Sci. USA* 84:4058–62

Annu. Rev. Biochem. 1992. 61:761–807

THE UBIQUITIN SYSTEM FOR PROTEIN DEGRADATION[1]

Avram Hershko and Aaron Ciechanover

Unit of Biochemistry, Faculty of Medicine and the Rappaport Institute for Research in Medical Sciences, Technion-Israel Institute of Technology, Haifa 31096, Israel

KEY WORDS: ubiquitin, protein degradation, protein turnover

CONTENTS

[1] Abbreviations used: E_1, ubiquitin-activating enzyme; E_2, ubiquitin-carrier protein; E_3, ubiquitin-protein ligase; CF-1 to CF-3, conjugate-degrading factors 1 to 3; SDS, sodium dodecyl sulfate; DTT, dithiothreitol; ODC, ornithine decarboxylase; DHFR, dihydrofolate reductase; MPF, maturation promoting factor; ATPγS, adenosine-5'-O(-3-thio)triphosphate.

0066-4154/92/0701-0761$02.00

PERSPECTIVES AND SUMMARY

Intracellular protein degradation has important roles in the modulation of the levels of specific proteins and the elimination of damaged proteins. The process is highly selective: some proteins are degraded within minutes, while others are practically stable. Usually, regulatory proteins or enzymes have fast turnover rates, so that their levels can be rapidly changed in response to appropriate stimuli. In some cases, the rates of degradation of regulatory proteins are controlled with high precision, such as with cyclins, whose levels oscillate at certain stages of the cell cycle and thus control cell cycle progression.

To understand the mechanisms underlying the remarkable selectivity and regulation of intracellular protein breakdown, the identification and characterization of the enzymatic reactions involved are necessary. In this article we describe our current knowledge of the enzymatic steps in the ubiquitin pathway. In this pathway, which seems to be a major system for selective protein degradation in eukaryotic cells, proteins destined for degradation are ligated to the polypeptide ubiquitin, and then are degraded by a specific protease complex that acts on ubiquitinated proteins. Since proteins are selected for degradation at the stage of ubiquitin ligation, we emphasize what we know, and what we would like to know, about the nature of signals in proteins recognized by the ubiquitin ligation system, and how such signals are recognized by ubiquitin ligases. In addition, the mode of action of the multienzyme complex that degrades ubiquitinated proteins is of great interest, since mechanisms must exist to ensure that only appropriately ubiquitinated proteins are translocated to the protease catalytic sites, to avoid proteolytic damage to non-ubiquitinated cellular proteins. We discuss possible mechanisms that may be involved in the regulation of the degradation of specific proteins, and point out future challenges in this field.

INTRODUCTION

The dynamic state of cellular proteins was discovered by Schoenheimer and coworkers more than half a century ago (1). However, the biochemical mechanisms underlying this process remained unknown, and only relatively recently have some of the systems involved begun to be uncovered. It now seems that the selective degradation of many proteins in eukaryotic cells is carried out by the ubiquitin-mediated pathway. The elucidation of this system had begun in 1978, when we resolved a heat-stable polypeptide required for the activity of an ATP-dependent proteolytic system from reticulocytes (2). This polypeptide was subsequently identified as ubiquitin, a 76-amino-acid residue, highly conserved protein present in all eukaryotes. In 1980 we found, in collaboration with Irwin Rose, that ubiquitin is covalently ligated to protein substrates in an ATP-dependent reaction, and proposed that ubiquitin ligation commits proteins for degradation (3, 4). This hypothesis was subsequently supported by a variety of evidence from different laboratories (reviewed in Refs. 5, 6).

Since our last review of the subject in this series (7), a large body of information has become available on the biochemical mechanisms, molecular genetics, and various biological functions of the ubiquitin system. The ubiquitination of proteins has been implicated in a variety of processes such as the heat shock response, DNA repair, cell cycle progression, the modification of histones and of receptors, and the possible pathogenesis of some neurodegenerative diseases. Increasing interest in the ubiquitin field is attested by the large number of recent reviews (5, 8), minireviews (9–12), and monographs (13, 14) on this subject. While this article was being written, three reviews on the ubiquitin system were published, stressing mainly aspects of molecular genetics (15, 16), various biological roles (15), and involvement in pathological states (17). To avoid repetition, we do not review these aspects of the ubiquitin system, or if necessary, we mention them only very briefly. Because of space limitations, we also do not review other pathways of protein degradation, such as lysosomal proteolysis (18) or ATP-dependent (but ubiquitination-independent) proteases in bacterial and eukaryotic cells (19–21). The main focus of our discussion is on recent information on the biochemistry and enzymology of various stages of the ubiquitin pathway.

Presently available information on the enzymatic reactions of the ubiquitin pathway is summarized in Figure 1. This outline is used for the organization of the following detailed discussion of the various stages of the system. Briefly, ligation of ubiquitin to proteins is initiated by the activation of its C-terminal Gly residue, catalyzed by a specific ubiquitin-activating enzyme, E_1 (Step 1). This step consists of the intermediary formation of ubiquitin adenylate (with the displacement of PP_i from ATP) and the transfer of activated ubiquitin to a thiol site of E_1 (with the release of AMP). Next, activated ubiquitin is further transferred by transacylation to thiol groups of a family of ubiquitin-carrier proteins, or E_2s (Step 2). E_2-ubiquitin thiol esters are the donors of ubiquitin for isopeptide bond formation between the C-terminal Gly residue of ubiquitin and ϵ-amino groups of Lys residues of proteins. Ubiquitin-protein ligation may occur by direct transfer from E_2 (Step 3), or by a process in which target proteins are first bound to specific sites of ubiquitin-protein ligases, E_3s (Step 4), and then ubiquitin is transferred from E_2 (Step 5). In vitro substrates of E_3-independent ubiquitin ligation are mainly basic proteins, such as histones. In all cases of E_3-dependent ligation, and in some E_3-independent reactions, multiple ubiquitin units are linked to proteins. Some of these are arranged in polyubiquitin chains (22), in which a major point of linkage was shown to be at Lys48 of ubiquitin (23). A certain species of E_2 can carry out the formation of polyubiquitin chains that are not attached to protein substrate (Ref. 24, Step 3). The function of this reaction is not known. Proteins ligated to multiple ubiquitin units are degraded by a 26S protease complex. The 26S complex is formed by the assembly of three components, designated CF-1 to CF-3, by an ATP-dependent process (Step 6). CF-3 was identified as the 20S "multicatalytic" protease complex, a particle observed in many eukaryotic cells. The assembled 26S protease complex degrades the protein moiety of ubiquitin-protein conjugates to small peptides, in a process that requires the hydrolysis of ATP (Step 7). The roles of ATP in the assembly of the 26S complex and in its proteolytic action remain to be elucidated. Following proteolysis, free and reusable ubiquitin is released, presumably by the action of several ubiquitin-C-terminal hydrolases. One of these is an enzyme that acts on polyubiquitin chains (Step 8).

In this review we describe our current knowledge of the various enzymatic steps outlined above. We also discuss available information on signals in proteins that may be involved in the selection of proteins for ubiquitin ligation and degradation. Large gaps in our knowledge, and areas for future research on the mechanisms, selectivity, and regulation of ubiquitin-mediated protein breakdown, are pointed out.

ENZYMES OF UBIQUITIN ACTIVATION AND LIGATION

Ubiquitin-Activating Enzyme, E_1

The first step in the ubiquitin pathway is mediated by the ubiquitin-activating enzyme, E_1. The activation involves the formation of an E_1-ubiquitin thiol ester via a ubiquitin-adenylate intermediate. Using the "covalent affinity" method (25), the enzyme has been purified from various sources. It appears to be a homodimer with an apparent molecular mass of 210 kDa that is composed of two identical 105-kDa subunits. The mechanism of activation of ubiquitin has been studied thoroughly (26–28), and has been reviewed previously (8, 29).

Genes encoding E_1 have been cloned from yeast (30), wheat (31), and humans (32, 33). As is the case with other enzymes of the ubiquitin system, E_1 is strongly conserved in evolution. The human and the yeast proteins are 53% identical. If conservative amino acid substitutions are included, homology increases to 73%. In wheat, two additional cDNAs encoding E_1 were isolated. One clone was almost identical (~99%) to the first clone isolated, while the second clone displayed 75% homology to the first clone (31). While heterogeneous forms of wheat E_1 have been observed by gel electrophoresis

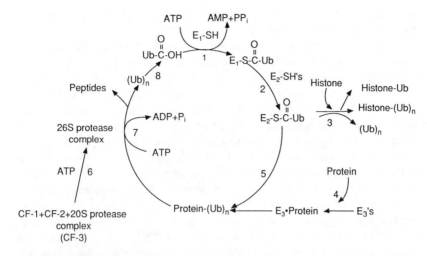

Figure 1 Proposed sequence of events in the ubiquitin proteolytic pathway. See the text. Ub, ubiquitin. Modified from Ref. 6.

(34), the functional significance of multiple E_1 genes observed in plants is not clear. The yeast enzyme contains two Gly-X-Gly-X-X-Gly motifs that are characteristic of nucleotide-binding domains (35). Furthermore, these motifs are followed by a lysine residue and an Asp-X-X-Gly sequence, additional themes that are characteristic of the nucleotide-binding domain of many proteins (36, 37). One such motif is present in both the wheat and the human enzymes (31–33).

The finding of ubiquitin-protein conjugates in both the cytosol and the nucleus suggests that the ubiquitin-ligase system is present in both compartments. Indeed, both the yeast (30) and the human (32) E_1s contain two short stretches of amino acids that resemble the nuclear targeting signal in the yeast MATα2 protein (KIPIK; Ref. 38). Immunocytochemical studies using antibodies raised against purified E_1 show that E_1 is highly concentrated in both the yeast (30) and the mammalian cell nuclei (39). The roles of E_1 in the nucleus are not known. E_1 may be involved in ubiquitination of specific nuclear proteins such as histones (reviewed in Ref. 15) or in the degradation of other nuclear proteins. Comparison of the yeast E_1 sequence with that of other proteins that interact with ubiquitin, such as several of the yeast E_2 enzymes (10, 16), two of the yeast hydrolases (40, 41), or the yeast E_3 protein (42), revealed no sequence similarities. Only Cys600 of the yeast E_1 protein resides in a region with low homology to several of the E_2 enzymes, suggesting that this residue may be the thiol acceptor site of the activated ubiquitin moiety (10, 16, 30). Deletion of *UBA1,* the yeast gene that encodes E_1, is lethal. The indispensability of the E_1 enzyme in yeast, as well as that of its homologs in mouse (43, 44) and humans (33), is consistent with the vital nature of ubiquitin conjugation for cell viability.

A search for conditional, temperature-sensitive mammalian cell cycle arrest mutants has lead to the discovery of several cell lines that bear a single lesion, a mutation in the gene for E_1. Proven E_1 mutants are ts85 from mouse mammary carcinoma cells (43), ts20 from CHO cells (45), and tsA1S9 from mouse L cells (33). In addition, several mutants in which the defect has not been characterized are probable E_1 mutants, since they belong to the same complementation group and have similar phenotypes of cell cycle arrest. The latter mutants include tsFT5, tsFT21, tsFT144, and tsFT173 (46), as well as tsBN75 (47). The high frequency of E_1 mutations is noteworthy. As the E_1 coding genes from these cells have not been cloned and sequenced, it is not known whether the mutations are randomly distributed along the gene, or whether they are clustered in one or more mutational "hot spots." Ohtsubo & Nishimoto (47) have shown that the defect in tsBN75 cells is linked to the gene of hypoxanthine-guanosine phosphoribosyl transferase, suggesting that the mutation is located on the X chromosome. Kudo and colleagues showed by the analysis of somatic hybrids between mouse ts85 and human IMR-90 cells that the human E_1 gene is located on the X chromosome (48); mapping data specify the location as on the short arm of the X chromosome at the region Xp11.2-p11.4 (49). The E_1 mutants are S/G2 or S phase cell cycle arrest mutants, indicating a role for ubiquitin conjugation in progression along the cell cycle.

As many ubiquitin-signalled proteins are targeted for degradation in the cell-free system, it was expected that the E_1 mutants would serve as useful tools for studying the involvement of the ubiquitin system in the degradation of proteins in the intact cell. The ts85, as well the ts20 cells, fail to degrade short-lived and puromycin-containing abnormal proteins (44, 50) following heat inactivation of E_1. It should be noted that E_1 must be inhibited significantly before a reduction of protein degradation can be observed. This can be achieved only by heating the cells to 43.5 °C for 1 h (50, 51). Incubation of the cells at lower temperatures failed to inactivate E_1 to the extent required to reduce protein degradation (50–52). It is interesting to note that cell cycle arrest can be obtained following incubation of the cells at 39.5 °C, conditions under which E_1 is only partially inactivated. It is possible that E_1 is rate-limiting in the conjugation of ubiquitin to the putative cell cycle regulatory proteins, while even residual levels of active enzyme are sufficient to target other proteins for degradation. This might be the case, for example, if E_2s of different functions have markedly different affinities for E_1. Rates of degradation of long-lived proteins were not affected by inactivation of E_1 (50) and were similar to those measured at the permissive temperature. Degradation of this class of proteins may be mediated by an ATP-dependent, but ubiquitin-independent system(s) (see below).

Ubiquitin-Carrier Proteins, E_2s

The ubiquitin-carrier proteins, E_2s (designated also ubiquitin-conjugating enzymes, UBCs;[2] reviewed in Refs. 10, 16), is a family of enzymes with distinct substrate specificities and functions. They mediate the transfer of ubiquitin from E_1 to the protein substrate, with or without the participation of ubiquitin-protein ligases, E_3s.

The structural and functional heterogeneity of the E_2 enzymes was first reported by Pickart & Rose, who isolated from reticulocytes five distinct proteins that were able to generate thiol esters with ubiquitin in the presence of E_1 (53). The enzymes have subunit molecular masses of 32, 24, 20, 17, and 14 kDa. More recently, a 230-kDa E_2 species has been purified from rabbit reticulocytes (54). Only a single species of E_2 ($E_{2\ 14kDa}$) was observed to function in E_3-dependent conjugate formation and protein breakdown (53). The protein substrates used in these studies were specific for $E_3\alpha$ (see below). Haas and colleagues (55) have shown that the $E_{2\ 14kDa}$ can be resolved into two isoforms, and that the two enzymes are $\sim$10-fold more active in E_3-dependent ubiquitin-protein ligation than either the $E_{2\ 20kDa}$ or the $E_{2\ 32kDa}$. The $E_{2\ 14kDa}$, $E_{2\ 20kDa}$, $E_{2\ 32kDa}$, and $E_{2\ 230kDa}$, can catalyze E_3-independent monoubiquitination of histones H1, H2A, H2B, H3, and H4. Interestingly, these four enzymes can also support slow rates of direct polyubiquitination of these histone molecules (54, 56, 57; for polyubiquitination, see below). $E_{2\ 17kDa}$ and $E_{2\ 24kDa}$ do not catalyze either mono or multiple ubiquitinations (56). Recently, Chen & Pickart have purified and characterized a novel 25-kDa E_2 from calf thymus. In the presence of E_1, MgATP, and relatively high concentrations of ubiquitin, the enzyme catalyzes synthesis of polyubiquitin chains that are not attached to any other protein (24, 58). The kinetics of the conjugative process are consistent with diubiquitin acting as a steady-state intermediate. Structural analyses of the conjugates have shown that the $E_{2\ 25kDa}$ ligates successive ubiquitin molecules primarily via lysine 48. It was suggested that the 25-kDa E_2 may function as a polyubiquitinating enzyme in the ubiquitin-dependent proteolytic pathway, though there is no experimental evidence to support this notion.

Ten genes that encode different species of E_2s have been isolated so far from the yeast *Saccharomyces cerevisiae* (reviewed in Refs. 10, 16; see also Ref. 59). Several genes of E_2 enzymes have been described in other eukaryotes (16). The large (and probably growing) number of members of this gene family may reflect the requirement for recognition of different subsets of proteins or the recognition of an equally large number of E_3s that act on specific proteins. Sequence analyses of some of the E_2 genes in yeast have

---

[2]Different authors use different terminologies for the same E_2 enzymes or genes. We use the nomenclature of the author who worked first on the subject, and mention only once the alternative designations.

yielded useful structural and functional information. Comparison of the primary sequences show that all the proteins share a core domain of approximately 150 amino acid residues with 35% sequence homology. The active site cysteine that is required for the formation of thiol ester with ubiquitin is localized to this domain. Some of the E_2 enzymes are small 14–18-kDa proteins that consist almost entirely of this conserved domain (UBC4, 5, and 7 for example; Refs. 10, 16, 60). Others have C-terminal extensions, which are either neutral (such as that of UBC1; 61), or highly acidic (such as those of UBC2 and 3; Refs. 62–64). In order to simplify the discussion on the structure and functions of the various E_2 enzymes and their relationship to protein degradation, we describe separately (a) the RAD6/UBC2, (b) the UBC1, 4, and 5, and (c) the CDC34/UBC3 enzymes.

RAD6/UBC2   Yeast cells contain an abundant 20-kDa E_2 that has been isolated, and its corresponding gene, *UBC2*, has been cloned and sequenced. Sequence analysis revealed that *UBC2* is identical to an already-characterized yeast gene, *RAD6* (63). The gene product plays an important role in DNA repair. Mutations in *RAD6* are pleiotropic. The phenotype demonstrates slow growth rate, severe defects in induced mutagenesis, extreme sensitivity to DNA-damaging agents, defective sporulation, increased mitotic recombination, increased rate of spontaneous mutations, and hypersensitivity to the antifolate drug, trimethoprim (for a review, see Ref. 65). *RAD6* encodes a 20-kDa protein that contains 172 amino acids. A 23-amino-acid residues carboxy-terminal domain contains 20 acidic residues, including an uninterrupted stretch of 13 aspartates (62). Cys88, the single cysteine residue in *RAD6*, is involved in the formation of the thiol ester adduct with ubiquitin. Substitution of this residue with either Ala or Val results in a protein that is completely devoid of E_2 activity (66). Yeast cells harboring the mutated gene have a phenotype similar to that displayed by the *RAD6* null allele (66). These findings suggest that all the biological functions of *RAD6* require its E_2 activity. Further studies have shown that in vitro, the enzyme catalyzes E_3-independent multiple ubiquitination of histones H2A and H2B to yield products that contain seven or more ubiquitin residues. The highly acidic 23-residue carboxy-terminal tail is essential for the histone polyubiquitinating activity: mutant RAD6 proteins that lack all or most of the polyacidic tail fail to polyubiquitinate the histone molecules (67). It appears that the acidic tail is involved in targeting the enzyme to basic protein substrates. Interestingly, the polyacidic tail is required for sporulation as well: a/α diploid strains homozygous for *RAD6* tail-deficient allele are highly defective in sporulation. Addition of the first four residues of the tail partially corrects both the sporulation defect and the histone polyubiquitinating activity (68). Interestingly, the tail-less *RAD6* mutant protein shows almost a wild-type level of resistance to UV radiation and unimpaired ability to convert lesions in DNA to mutations (68). It appears that the repair and mutagenesis functions of the *RAD6*, while completely dependent upon the E_2 function of the protein, are not dependent on its ability to recognize basic proteins (68).

It is interesting to note that a homolog of *RAD6* was isolated from the yeast *Schizosaccharomyces pombe*. The gene product, rhp6$^+$, shares a high degree of structural and functional homology to *RAD6* (77%), but lacks the acidic carboxy-terminal tail (69). Like *RAD6*, null mutations of *rhp6$^+$* gene confer a defect in DNA repair as well as in UV mutagenesis and sporulation. Furthermore, the two genes can substitute for one another, though introduction of the *rhp6$^+$* gene to *RAD6*-deficient *Saccharomyces cerevisiae* did not repair the sporulation defect in this organism (69). It appears, therefore, that the structural requirement for the polyacidic tail may be species specific, and that the tail is not involved in a ubiquitous mechanism.

Recent results from several laboratories indicate that in addition to its E_3-independent action, *RAD6* is involved in E_3-dependent protein breakdown. Sung and colleagues have shown that the RAD6 protein catalyzes multiple ubiquitination of human α-lactalbumin and bovine β-lactoglobulin in the presence of E_1, $E_3\alpha$, ubiquitin, and ATP (for $E_3\alpha$, see below). Addition of the 26S protease complex resulted in degradation of the conjugated proteins (70). *RAD6* was as effective as the reticulocyte E_2 $_{14kDa}$ in catalyzing both the conjugation and degradation reactions. The highly acidic carboxy-terminus of *RAD6* is not essential for conjugation and proteolysis to occur. In agreement with the above results, Dohmen and colleagues have recently demonstrated that the degradation of β-galactosidase species with Arg or Leu residues at the N-terminal position (that is mediated by the $E_3\alpha$ homolog of yeast, see below), was significantly inhibited in a *RAD6* null mutant (71). Mutations in some other yeast E_2 genes had no influence on this process. As in the case of the reticulocyte enzymes (Ref. 72 and see below), the RAD6 protein was found to be physically associated with $E_3\alpha$: immunoprecipitation of the yeast $E_3\alpha$ enzyme yielded not only the ligase protein, but that of *RAD6* as well (71). It thus seems that the *RAD6* protein is the yeast homolog of reticulocyte E_2 $_{14kDa}$ protein, with an extra polyacidic tail. This conclusion is supported by recent sequence data on cDNA of rabbit E_2 $_{14kDa}$, which showed a 69% identity with the *RAD6* enzyme, much higher than with any other yeast E_2 enzyme (S. S. Wing and D. Banville, personal communication). The gene of the rabbit E_2 enzyme is similar to that of a 17-kDa human placental enzyme that was reported previously (73).

It is interesting to note that while the *RAD6* null mutant displays a pleiotropic phenotype, a null mutant of *UBR1* (the yeast homolog of the reticulocyte $E_3\alpha$; see below) displays a very mild phenotype (42). It seems, therefore, that unlike $E_3\alpha$, whose known functions are confined to the recognition of certain proteins with defined N-terminal residues (the "N-end rule" pathway: see below), the *RAD6* enzyme is an essential component of other ubiquitin-dependent pathway(s) as well. Some of these additional pathways may be E_3-independent; in other pathways the *RAD6* enzyme may interact with other species of E_3, such as a recently reported E_3 from yeast of as yet unknown functions (Ref. 74 and see below).

UBC1, UBC4, AND UBC5   The subfamily of these three E_2 enzymes in yeast has been cloned by Jentsch and coworkers and shown to be involved in protein breakdown (reviewed in Refs. 10, 16). UBC4 and UBC5 encode almost identical tail-less proteins (92% identical residues) of 16 kDa that can functionally substitute for one another (60). Null mutations in either *UBC4* or *UBC5* had no significant effect, but the double mutant *UBC4/UBC5* had strongly reduced growth rate. This was accompanied by greatly decreased rates of degradation of abnormal and short-lived proteins, and reduced levels of ubiquitin-protein conjugates (60). Both enzymes are induced by heat and contain heat shock elements in the upstream region of their genes. It thus appears that these two enzymes are involved in some aspects of ubiquitin-mediated protein degradation and are also components of the cellular stress response. In the latter case, it may well be that they are involved in the elimination of heat-induced abnormal and misfolded proteins. Such proteins can induce a heat-shocklike response in the cell (75), and it may well be that one role of the ubiquitin system is to remove the inducing signal.

UBC1 encodes a 24-kDa protein with a C-terminal extension rich in charged amino acids (61). Genetic analysis revealed that the enzyme may be involved in protein degradation. A *UBC1* null mutant was sevenfold more sensitive to canavanine, and showed a slight, but significant decrease in the degradation of canavanyl peptides (61). Also, overexpression of *UBC1* protein in *UBC4/UBC5* double mutant could complement the degradation defect observed in this mutant cell. *UBC1* null mutant displays only a mild slow growth phenotype in the mitotic phase, but is markedly impaired in growth following germination (61). The specific requirement for the *UBC1* enzyme following quiescence suggests that conjugation to ubiquitin, and possibly degradation of certain proteins, may be crucial at the point of spore germination in the yeast development.

CDC34/UBC3   The products of two yeast genes, *CDC4* and *CDC34*, must be present in order to allow replication of chromosomal DNA to occur. Temperature-sensitive mutants of these genes are arrested at late G1, and the spindle pole body fails to separate. Sequence analysis of *CDC34* revealed that it encodes a 34-kDa protein with a C-terminal extension rich in acidic amino acid residues and with a substantial similarity to the yeast RAD6 protein (64).

Overall, 38% of the amino acids in RAD6 are identical to those in CDC34 protein, with 55% homology if conservative replacements are included. A lysate of *Escherichia coli* expressing *CDC34* will not promote conjugation of ubiquitin to any bacterial protein (64). This finding suggests that the spectrum of substrates of this species of E_2 is rather narrow. The enzyme can catalyze monoubiquitination of histones H2B (64) and H2A (76) in the presence of E_1, ubiquitin, and MgATP. Haas and colleagues reported that the enzyme can also polyubiquitinate bovine serum albumin via Lys48 of the ubiquitin molecule. This reaction occurs in an E_3-independent mode (76). Also, the enzyme can conjugate ubiquitin to proteins present in crude reticulocyte extract in an E_3-dependent manner (76). The significance of these reactions, and whether *CDC34* is involved in E_3-dependent protein degradation, are not clear yet. The enzymatic activity of *CDC34* suggests that its cell cycle function is mediated by ubiquitination of specific target proteins. The nature of the target proteins remains to be elucidated.

Ubiquitin-Protein Ligases, E_3s

Two species of E_3, termed $E_3\alpha$ and $E_3\beta$, have been isolated from extracts of reticulocytes. $E_3\alpha$ was more thoroughly studied. It was first discovered as an enzyme component necessary, in addition to E_1 and E_2, for the ligation of ubiquitin to certain proteins and for their degradation (77). It promotes the addition of multiple ubiquitins to the substrate protein as well as the formation of polyubiquitin chains (22). Initially, the enzyme was partially purified by affinity chromatography on ubiquitin-Sepharose (to which it is noncovalently bound) and elution at pH 9 (77). It has an apparent molecular mass of ~350 kDa and a 180-kDa subunit (77, 78); it thus may be composed of two identical subunits. Further studies showed that $E_3\alpha$ contains specific sites for the binding of protein substrates prior to ubiquitination. Initial evidence for this conclusion included chemical cross-linking of ^{125}I-labeled proteins to partially purified $E_3\alpha$, and the functional conversion of E_3-bound labeled proteins to ubiquitin conjugates in pulse-chase experiments (78). There was a general correlation between the binding of different proteins to $E_3\alpha$ and their susceptibility to conjugation and ubiquitin-mediated degradation. It was concluded that the selection of at least some proteins for degradation by the ubiquitin system is carried out by their specific binding to E_3 (78).

Following observations in the laboratories of Varshavsky & Ciechanover on the influence of the N-terminal residues of some proteins on their degradation (see below), Reiss et al (79) showed that the N-terminal recognition is based on the specificities of the binding sites of $E_3\alpha$. The enzyme has two distinct sites that interact with specific N-terminal residues of protein substrates: one site recognizes basic residues (Arg, His, Lys), while the other recognizes bulky-hydrophobic N-terminal residues (Leu, Phe, Trp, Tyr) (79).

Dipeptides, hydroxamates, and methyl esters with a basic N-terminal residue (such as His-Ala, His-hydroxamate, and His-methyl ester), strongly inhibited the binding to $E_3\alpha$, the conjugation, and the degradation of proteins with basic N-termini (such as lysozyme and oxidized RNase A; Lys at the N-terminal position). Likewise, derivatives with a bulky-hydrophobic N-terminal residue (such as Leu-Ala, Phe-Ala, Trp-Ala, and Tyr-Ala) inhibited the binding, conjugation, and degradation of β-lactoglobulin (Leu at the N-terminal position). Dipeptides with a reverse order of residues (such as Ala-His or Ala-Trp) had almost no effect on the tested functions even at relatively high concentrations. Furthermore, the basic N-termini derivatives were specific to the basic N-termini proteins and did not inhibit the binding of bulky-hydrophobic N-termini proteins. Similarly, the bulky-hydrophobic N-termini derivatives affected specifically only proteins with homologous N-termini (79). Taken together, these findings show that the effect of the various derivatives is specific, and is probably mediated via their N-terminal residue. Moreover, these findings demonstrated that the recognition sites for the two classes of proteins are distinct. These investigators designated the basic N-termini proteins as Type I substrates, and the bulky-hydrophobic N-termini proteins as Type II substrates. A third (Type III) class of protein substrates have amino termini that are neither basic nor hydrophobic. The signals responsible for the binding of Type III proteins to $E_3\alpha$ remain unknown.

Based on the above information, Reiss & Hershko developed an affinity technique aimed at purifying this and other species of ligases (80). The enzyme, present in crude reticulocyte extracts, was adsorbed to immobilized human α-lactalbumin (Lys at the N-terminus) or β-lactoglobulin (Leu at the N-terminus). $E_3\alpha$ bound strongly to such protein substrate affinity columns and could not be eluted with high concentrations of salt. However, it was specifically eluted by dipeptides that have NH_2-terminal residues similar to those of the immobilized substrates. Lys-Ala was used to elute the enzyme from human α-lactalbumin, while Phe-Ala eluted the enzyme that was bound to β-lactoglobulin. As expected, Lys-Ala was inactive in eluting the enzyme bound to β-lactoglobulin, and Phe-Ala could not displace the α-lactalbumin-bound enzyme (80). Preparations of $E_3\alpha$ purified either on basic or on bulky-hydrophobic N-termini substrates act on both types of protein substrates, indicating that the separate binding sites for the two different NH_2-terminal residues reside on the same molecule (80).

Our current knowledge of the different binding sites of $E_3\alpha$ is schematically illustrated in Figure 2. As discussed above, the separate "head" sites that bind Type I or Type II N-terminal amino acid residues reside on the same enzyme. In addition, the existence of a "body" site of $E_3\alpha$, which binds regions in proteins other than the N-terminal residue, is suggested by observations such as that $E_3\alpha$ can act on proteins that do not have Type I or II N-termini (80) or on N-α-acetylated proteins (81). The signals in proteins recognized by the putative "body" site of $E_3\alpha$ remain to be elucidated. In other experiments, Reiss and coworkers (72) found that $E_{2\ 14kDa}$ binds strongly to $E_3\alpha$, as indicated by comigration in glycerol density gradient centrifugation. The existence of a specific E_2-binding site of $E_3\alpha$ may facilitate the efficient transfer of activated ubiquitin from E_2 to amino groups of the protein substrate bound to $E_3\alpha$. It was furthermore observed that $E_3\alpha$ binds the ubiquitin moiety of ubiquitin-protein conjugates, as indicated by the observation that ubiquitin-lysozyme conjugates bind more strongly to the enzyme than free lysozyme (72). These ubiquitin-binding site(s) of $E_3\alpha$ may be required for the processive addition of multiple ubiquitin units to the protein substrate. Conjugates containing multiple ubiquitins were not released from $E_3\alpha$ (72). It might be that in the complete proteolytic system, such ubiquitin-protein conjugates are channelled directly from E_3 to the 26S protease complex.

The gene of the yeast homolog of $E_3\alpha$ was cloned by Bartel and colleagues (42). It is required for the in vivo degradation of test proteins with either basic or bulky-hydrophobic N-termini (Ref. 42 and see below). The gene, called *UBR1*, encodes a 225-kDa protein, the sequence of which bears no homology to other known proteins. Like $E_3\alpha$ from reticulocytes, it contains separate binding sites for basic and bulky-hydrophobic N-termini of proteins (42). Its possible physiological functions are discussed below.

Much less is known about ubiquitin-protein ligases other than $E_3\alpha$. Gonda and colleagues have reported that in crude reticulocyte lysates, the degradation of derivatives of β-galactosidase with small uncharged N-termini (Ala, Ser, Thr) can be specifically inhibited by dipeptides with the similar residues at their N-terminal position (82). Heller & Hershko have characterized and partially purified from crude reticulocyte extracts a novel species of E_3,

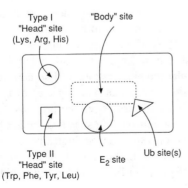

Type I
"Head" site
(Lys, Arg, His)

"Body" site

Type II
"Head" site
(Trp, Phe, Tyr, Leu)

E_2 site

Ub site(s)

Figure 2 Schematic representation of the different binding sites of $E_3\alpha$. See the text. Ub, ubiquitin. Modified from Ref. 80.

designated $E_3\beta$ (83). Following a complete removal of $E_3\alpha$ over immobilized β-lactoglobulin and human α-lactalbumin, these investigators found that the unadsorbed material contains an activity that can conjugate efficiently ubiquitin to certain Type III substrates, mostly with Thr, Ser, and Ala residues at the N-terminal position. Conjugation of multiple ubiquitin molecules to the protein substrates requires E_2, and the conjugated proteins are subsequently degraded. It is not clear yet whether the two enzymes [the one described by Gonda and colleagues (82), and the one purified by Heller & Hershko (83)] are identical, as the signal recognized by $E_3\beta$ appears to involve amino acid residues that reside downstream from the N-terminal residue (83; H. Heller and A. Hershko, unpublished results). $E_3\beta$ has no significant activity on substrates of $E_3\alpha$ that have Type I or Type II NH$_2$-termini. Though they have different substrate specificities, $E_3\alpha$ and $E_3\beta$ share several properties. Both have an apparent molecular mass of approximately 350 kDa as determined by molecular sieving chromatography. They are precipitated at similar concentrations of ammonium sulfate and show similar elution profiles from the hydrophobic matrix phenyl-Sepharose. Like $E_3\alpha$, $E_3\beta$ will ligate multiple ubiquitin molecules to either different lysine residues of the protein substrate or to one another (83). High-molecular-mass conjugates of ^{125}I-S-protein are formed with reductively methylated ubiquitin that cannot form polyubiquitin chains. With ribonuclease T1 that contains a single lysine residue, a single low-molecular-mass conjugate was detected when methylated ubiquitin was present in the reaction mixture. Multiple high-molecular-mass polyubiquitin derivatives of ribonuclease T1 were synthesized when native ubiquitin was present (83).

Recently, Sharon and colleagues have partially purified from yeast extract an enzyme that in the presence of E_1 and of the RAD6 protein promotes the transfer of ubiquitin to several proteins (74). The enzyme ubiquitinates efficiently α- and β-caseins and β-lactoglobulin (Lys, Arg, and Leu at the N-terminal position, respectively). Also, it acts on κ-casein that has a blocked N-terminal residue (pyroglutamate). As activity of this novel E_3 was also present in *UBR1* null mutant, this enzyme is distinct from the yeast $E_3\alpha$ enzyme. The physiological roles of this enzyme remain to be identified. Its possible involvement in functions of RAD6 is of considerable interest.

It is evident from the above discussion how little we know about E_3 enzymes other than $E_3\alpha$. Because the cellular protein substrates of $E_3\alpha$ appear to be very limited (see below), it may be expected that other species of E_3 exist that may recognize other specific features in various cellular proteins. It is hoped that the knowledge gained on the various binding sites and mode of action of $E_3\alpha$ will facilitate future work on other, physiologically more important species of ubiquitin-protein ligases. As in the case of the specific association of $E_3\alpha$ with one species of E_2 ($E_{2\ 14kDa}$ or RAD6 protein), it might be that other species of E_3 pair up with certain E_2s to act on proteins containing specific signals. This possibility, and the alternative possibility of direct action of some E_2s on some cellular proteins, remain to be investigated.

DEGRADATION OF PROTEINS LIGATED TO UBIQUITIN

The 26S Protease Complex and its Three Components

In the original proposal of the ubiquitin marking hypothesis in 1980, it was suggested that proteins ligated to ubiquitin are degraded by a specific proteolytic system that recognizes such conjugates (4). Evidence supporting this suggestion was obtained only several years afterwards, and we still know very little about the mode of action of the conjugate-degrading machinery. In 1984, Rechsteiner and coworkers (84) and Hershko and colleagues (85) independently observed that lysozyme ligated to ubiquitin is rapidly degraded in crude extracts of reticulocytes and that ATP is required for this process. Hershko and coworkers have used a fraction of reticulocyte extract devoid of E_1, E_2, and E_3 to show that this protease system specifically acts on proteins conjugated to ubiquitin, but not on unconjugated proteins (85). In 1986, Rechsteiner and coworkers identified and partially characterized a large ATP-dependent protease complex from rabbit reticulocytes that degraded ubiquitin-ated proteins (86). A similar protease complex was reported by Goldberg and coworkers (87). The sedimentation coefficient of the conjugate-degrading protease complex was estimated at 26S (88), and its molecular mass between 1000 (88) and 1500 (87) kDa. The 26S protease complex was extensively purified by Rechsteiner and coworkers (88), and was shown to contain a characteristic set of subunits with M_r values between 34,000 and 110,000. It was noted that in addition to the above-mentioned subunits, the purified 26S protease complex contained a set of smaller subunits (M_r = 21,000 to 32,000), similar to those of a previously described ATP-independent 20S protease complex (Ref. 88 and see below). Though most of the work on the 26S protease complex was done on complex isolated from reticulocytes, similar protease complexes were observed in other tissues, such as muscle and liver (89), cerebral cortex (90), and cultured leukemia cells (91).

Using a different experimental protocol for the fractionation of reticulocyte extracts, Ganoth et al resolved three separate factors that were all required for the degradation of proteins conjugated to ubiquitin (92). These were termed CF-1, CF-2, and CF-3, for conjugate-degrading factors 1–3. The approximate molecular masses of CF-1, CF-2, and CF-3, estimated by gel filtration, are 600, 250, and 650 kDa, respectively. The three factors were resolved from each other and partially purified. CF-1 and CF-2 contain ATP-binding sites, as indicated by the observation that both factors are protected by ATP against inactivation at 42 °C. The various factors did not seem to be different ATP-dependent proteases that carry out successive cleavage reactions in the conversion of ubiquitin-conjugated proteins to small peptides, as indicated by the lack of accumulation of intermediary cleavage products when ^{125}I-lysozyme-ubiquitin conjugates were incubated with single factors, or with combinations of any two factors. A clue to an interaction between the factors was provided by the observation that when the three factors were incubated with ^{125}I-lysozyme-ubiquitin conjugates and ATP, there was a time lag in the formation of acid-soluble products. This lag could be abolished by the incubation of all three factors with MgATP prior to the addition of the substrate. The time lag was not abolished when any of the three factors was omitted from preincubation (and added later on), or when ATP was replaced with a nonhydrolyzable analog. It was then found that the abolishment of the lag was due to the assembly of the three factors to form an active multienzyme complex (92). All three factors enter the multienzyme complex, as indicated by the observation that the levels of the free forms of all three factors decrease during complex formation. The assembly of the complex requires ATP and Mg^{2+}, and ATP cannot be replaced by nonhydrolyzable analogs. It was concluded that one role of ATP in conjugate breakdown is the formation of the active multienzyme complex (92). Judged by size analysis on glycerol density gradient centrifugation, the multienzyme complex formed by the assembly of the three factors appeared to be identical to the 26S protease complex described by other investigators. It should be noted that the 26S complex was isolated from extracts of untreated reticulocytes (86, 87), whereas the sources for the three free factors were reticulocytes that were treated by ATP depletion (to convert protein-conjugated ubiquitin to free ubiquitin, Ref. 77). It appears reasonable to assume that the 26S complex is dissociated to its three components during ATP depletion.

Is the assembly of the multienzyme complex the sole function of ATP in the degradation of proteins conjugated to ubiquitin? For example, the continuous dissociation and reassembly of the 26S complex may be required for the action of the complex to degrade proteins ligated to ubiquitin. This does not seem so, based on the following observations: (*a*) The nucleotide specificities of complex assembly and of conjugate proteolysis by the (assembled) 26S protease complex are different. Thus, ATP can be replaced only by CTP (and not by GTP or UTP) for complex assembly (92), whereas all four nucleoside triphosphates stimulate the degradation of ubiquitin-conjugated proteins by the 26S complex (86). These observations suggest that different nucleotide-binding sites are involved in these two processes. (*b*) Following the removal of ATP (by the supplementation of hexokinase and glucose), the breakdown of ubiquitin-conjugated proteins by the 26S complex stops immediately. Under similar conditions of ATP-depletion, the decay of assembled active

26S complex is much slower with a half-life time of ~8 min at 37 °C (D. Ganoth and A. Hershko, unpublished results). These findings indicate that ATP is required at two distinct stages in the assembly and function of the 26S protease complex.

Identification of CF-3 as the 20S Protease Complex

One advantage of having a multienzyme complex dissociated into smaller components is that it might be possible to identify the specific function of each separate component. Thus, the CF-3 component of the 26S protease complex has been identified as the previously known 20S protease complex. The subject of the 20S protease complex was reviewed previously (93–95), and here we merely mention briefly some of its main features. It has been observed by many investigators in a variety of eukaryotic cells, and has been called by many names such as the multicatalytic proteinase complex (96, 97), high-molecular-weight protease (98), alkaline protease (99), macropain (100), prosome (101, 102), and proteasome (103). It is a cylindrical particle organized as a stack of four rings of subunits (104). Most of the subunits are nonidentical and are in the molecular mass range of 20 to 32 kDa. Different preparations of the 20S protease complex have variable ATP-independent proteolytic activity. Usually, such protease activity can be increased by mildly denaturing treatments (96, 105, 106), suggesting the exposure of otherwise cryptic catalytic sites. Wilk & Orlowski (96) first showed that this complex possesses three distinct types of endopeptidase activities, cleaving peptide bonds on the carboxyl side of basic, hydrophobic-neutral, and acidic amino acid residues. The genes of several subunits of the 20S complex have been cloned from rat (107, 108, 110–112), humans (113), *Drosophila* (114–116), and yeast (117, 118). The predicted amino acid sequences of these subunits do not resemble any of the known proteases. There is considerable sequence similarity among different subunits of the 20S complex, and some subunits are strongly conserved in evolution, from yeast to mammals. The deletion in yeast of genes of several different subunits is lethal (117–119), indicating essential functions of the 20S protease complex. The functions of this ATP-independent protease complex remained unknown, since intracellular protein degradation requires metabolic energy (7). It seemed possible that ATP-independent proteolytic activity is an in vitro artifact that arises by the "uncoupling" of catalytic sites from regulation.

A relationship between the 26S and 20S protease complexes was first suggested by Rechsteiner and coworkers (88, 93), who noted the presence of similar-sized subunits in both complexes, as well as similarities in substrate specificity and sensitivity to common inhibitors. Those investigators suggested that the 26S and 20S complexes may be related and share common subunits. Subsequently, McGuire et al (120, 121) reported that immunoadsorption of extracts of BHK fibroblasts by antibodies directed against the 20S protease complex abolished ATP-dependent proteolytic activity. Tanaka & Ichihara (122) made similar observations with extracts of rat reticulocytes. In both of the latter two studies, it was not reported whether the ATP-dependent proteolytic activity inhibited by immunoadsorption was also ubiquitin dependent, and it seemed possible that these antibodies cross-react with a related ATP-dependent protease. This was ruled out in a study by Matthews et al (123), who showed that immunoadsorption of extracts of muscle or of reticulocytes by antibodies directed against subunits of the 20S protease complex inhibited ubiquitin-dependent degradation of lysozyme, as well as ATP-dependent degradation of ubiquitinated lysozyme (123). Subsequently, DeMartino and coworkers (124) showed that ATP-dependent proteolysis of BHK cell proteins could be inhibited by treatment with antibodies directed against either the 20S complex or E_1, confirming the involvement of the 20S complex in the ubiquitin system.

Direct evidence for the relationship between the 20S and 26S protease complexes was obtained independently by Eytan et al (125) and Driscoll & Goldberg (126). Using an improved procedure for the purification of the 26S complex, it was well separated from the 20S protease complex (125). With this preparation, the observation of Rechsteiner and coworkers that subunits of the 20S protease complex are contained in the 26S complex was confirmed. It was then asked which component of the 26S complex contains the 20S protease complex. Of the three conjugate-degrading factors, the estimated size of CF-3 ($\sim$650 kDa, Ref. 92) was closest to that reported for the 20S protease complex ($\sim$670–700 kDa, Refs. 88, 89). Furthermore, as was noted previously, preparations of CF-3 contained low levels of ATP-dependent protease activity (92). To examine the possible identity of CF-3 with the 20S protease complex, the observation was utilized that treatment with low concentrations of detergents stimulates ATP-independent activity of this complex (96, 105, 106). Treatment with 0.04% SDS stimulated markedly ATP-independent protease activities in both CF-3 and purified 26S complex (125). SDS-stimulated protease activity coincided exactly with CF-3 in several purification procedures, such as chromatography on DEAE-cellulose and hydroxylapatite and in glycerol density gradient centrifugation. Most conclusively, following incubation of the three conjugate-degrading factors with MgATP, most of SDS-activated protease activity shifted from the region of CF-3 to that of the 26S complex, in parallel to the disappearance of free CF-3 and the formation of the active 26S complex (125). In other experiments, it was found that this shift in size of SDS-activated protease activity of CF-3 required the presence of the other two factors and of MgATP, confirming the notion that it is due to the assembly of the 26S complex. It was concluded that CF-3 is identical to the 20S protease complex and becomes incorporated into the 26S complex in the ATP-dependent assembly process (125).

A similar conclusion was reached by Driscoll & Goldberg (126), based on different experiments. Fraction II from reticulocytes was incubated with or without MgATP and was then precipitated with ammonium sulfate at 38% saturation. This concentration of ammonium sulfate precipitates the 26S complex (87), as well as CF-1 and CF-2 (92), while the 20S protease complex remains in the supernatant (87, 127). It was found that following incubation of Fraction II with MgATP, activities that hydrolyze fluorigenic peptide substrates of the 26S complex (such as succinyl-Leu-Leu-Val-Tyr-4-methylcoumaryl-7-amide) appeared in the 38% ammonium sulfate pellet. This was accompanied by an increase in the size of the complex, assayed by peptide hydrolyzing activity. The higher-molecular-weight form (presumably the 26S protease complex) reacted with a monoclonal antibody directed against one of the subunits of the 20S protease complex (126). A similar ATP-dependent shift of 20S to 26S complex was subsequently observed in extracts of human HL-60 cells by Orino and coworkers (91). By silver staining of immunoprecipitated complexes, these investigators confirmed that the 26S complex contains the subunits of the 20S complex in addition to a set of 35–110-kDa subunits (91). Studies of Heinemeyer et al (119) on yeast mutants defective in subunits of the 20S protease complex provided further evidence for the role of this complex in the degradation of ubiquitinated proteins. One of the mutants, affected in a 23-kDa subunit, has decreased "chymotrypsin-like" activity (cleavage on the carboxyl side of hydrophobic-neutral amino acid residues) of the 20S complex. The mutation is presumably due to an amino acid replacement, since no difference could be detected between mutant and normal subunits on two-dimensional polyacrylamide gel electrophoresis. The rate of degradation of canavanine-containing abnormal proteins was decreased in the mutant. That this was due to impaired break-down of proteins ligated to ubiquitin was suggested by the finding that high-molecular-weight ubiquitin-protein conjugates accumulated in mutant, but not in wild-type yeast cells (119).

The cumulative evidence suggests that the 20S protease complex is the "catalytic core" of the 26S complex that degrades ubiquitinated proteins. It is not yet clear, however, whether the 20S protease complex is the only source of protease catalytic sites for the 26S complex. We did not find any protease activities in preparations of CF-1 and CF-2 from reticulocytes. On the other hand, Estrela & Goldberg have recently reported the partial purification of a 500-kDa proteolytic activity from rabbit skeletal muscle (128, 129). It de-grades both ubiquitinated and free lysozyme, and these activities are stimu-lated by ATP. Peptide substrates for "chymotrypsin-like" proteases are de-graded by this enzyme at a much lower rate than by the 20S protease complex. Activities of this enzyme on other types of protease substrates were not reported. Following the incubation of the 500-kDa enzyme with the 20S protease complex, a high-molecular weight complex, similar in size and several properties to the 26S complex, was formed. It was suggested that the 500-kDa enzyme from muscle is similar to CF-1, but contains some proteolytic activities (128). It should be noted that the time-course of an ATP-dependent degradation of ubiquitinated lysozyme by the 500-kDa enzyme (without the 20S complex) showed a pronounced lag (128). This is reminiscent of the time lag observed in the assembly of the three components of reticulocyte 26S complex (92). It is possible, therefore, that the partially purified preparation of the 500-kDa enzyme contains a mixture of components that associate in the presence of ATP to form an active complex.

Another open question is whether the 20S complex has roles other than those associated with the 26S complex. The level of free CF-3 in reticulocyte extracts is much higher than those of CF-1 and CF-2 (92), and the 20S complex is very abundant in various eukaryotic cells, estimated to constitute 0.5–1% of cellular proteins (95). Thus, additional roles of the 20S complex, or of related particles, are quite possible. It has been reported by several workers that when isolated from certain tissues or under certain conditions, the 20S complex has ATP-dependent (but ubiquitination-independent) protease activity (120, 130). The molecular alterations responsible for ATP dependence of the 20S protease complex have not been defined. In addition, it has been reported that a pre-tRNA 5'-processing endonuclease activity copurifies with a particle that very much resembles the 20S complex by electron microscopy and subunit composition (131). If the particles are similar, an intriguing possibility is that the 20S particle is a "degradosome," containing protease and nuclease catalytic sites. The existence of different populations of 20S complexes, with slightly different subunit compositions, also cannot be ruled out. In this connection, it is noteworthy that several investigators observed that the 20S complex is located both in the cytosol and in the nucleus (131–135). The nuclear localization seems to be especially prominent in some rapidly dividing cells such as sea urchin embryos (133) or malignantly transformed cells (134). It remains to be seen whether nuclear localization of the 20S complex in such cells is related to high rates of RNA processing or to rapid turnover of some nuclear proteins.

Possible Roles of ATP

As mentioned above, ATP has two distinct sites of action in the assembly of the 26S complex and in the action of the assembled complex to degrade ubiquitin-conjugated proteins. In both cases, the mode of action of ATP remains unknown. One possibility is that an ATP-dependent modification reaction of one or more of the components is involved in complex assembly. Since both CF-1 and CF-2 contain ATP-binding sites (92), it seemed possible that one of these factors contains a protein kinase activity responsible for such a reaction. However, copurification of protein kinase activity with the above factors was not observed. In purified preparations of the 20S complex, phosphorylation of a ~30-kDa subunit was found, indicating the association of a protein kinase activity with this complex (D. Ganoth and A. Hershko, unpublished results). It is not clear, however, whether this "autophosphorylation" reaction is relevant to complex assembly.

The action of ATP at its second site (i.e. conjugate breakdown by assembled 26S complex) appears to be analogous to its roles in simpler ATP-dependent bacterial proteases such as the Lon (La) protease or the Clp protease system (reviewed in Refs. 19–21). The Clp protease of *E. coli* may be a simpler counterpart of eukaryotic 26S protease complex, since it has two components: one subunit has protease activity that is stimulated by another subunit that has ATPase activity (136, 137). Armon et al (138) have noted that purified preparations of the 26S complex have considerable ATPase activity. That the ATPase activity was indeed associated with the 26S complex was shown by its exact coincidence with conjugate-degrading activity in glycerol density gradient centrifugation. Partially purified preparations of CF-1, CF-2, and CF-3 had no significant ATP-hydrolyzing activity, but ATPase activity appeared following the assembly of the three factors to form the 26S complex. The nucleotide specificity of complex-associated ATPase was similar to that of the second site of ATP action: GTP and UTP, which can replace ATP in conjugate breakdown (but not in the assembly of the 26S complex), effectively competed with the hydrolysis of $[\gamma\text{-}^{32}P]ATP$ by the 26S complex (138). ATP hydrolysis was not stimulated by the supplementation of ubiquitin-protein conjugates. This observation rules out a mechanism of substrate phosphorylation prior to amide bond hydrolysis, as is the case with 5-oxoprolinase (139, 140).

It is evident from the above discussion how much has yet to be learned about the dual roles of ATP, the functions of CF-1 and CF-2, and the mode of action of the integrated proteolytic machinery. A reasonable working hypothesis is that the energy of ATP hydrolysis may be required for the translocation of ubiquitin-protein conjugates. Since CF-1 and CF-2 confer specificity for ubiquitinated proteins, it may be assumed that one or both of these components contain specific binding site(s) for ubiquitin-protein conjugates. Enzyme-bound conjugates may then have to be transferred to the protease catalytic sites contained in the 20S component of the 26S complex. The dependence of ATPase activity on assembled 26S structure may indicate that the operation of the translocation machinery is possible only in the complete multienzyme complex.

UBIQUITIN-C-TERMINAL HYDROLASES[3]

Essential to the operation of the ubiquitin system is the recycling of free ubiquitin. Several ubiquitin-C-terminal hydrolases (i.e. enzymes that hydrolyze the linkage between the C-terminal glycine residue of ubiquitin and various adducts) have been described, but in most cases, their functions have not been satisfactorily established. Table 1 lists some possible functions of such enzymes in various aspects of ubiquitin metabolism. In the protein degradation pathway, ubiquitin-C-terminal hydrolase(s) is required to release ubiquitin from isopeptide linkage with Lys residues of the protein substrate, during or after the proteolytic degradation of the substrate. In addition, a ubiquitin C-terminal isopeptidase activity may be expected to disassemble polyubiquitin chains linked to the protein substrate, following the degradation of the protein. A "proofreading" function has been proposed by Rose (141) for hydrolases to release free protein from incorrectly ubiquitinated derivatives. This appears to be a plausible expectation, since most highly specific biochemical processes are accompanied by suitable editing systems. Still another possibility is that ubiquitin-C-terminal isopeptidases may be required for the "trimming" of abnormal polyubiquitin chains, such as too long or abnormally branched polyubiquitin chains that may not be suitable for binding of conjugates to the 26S protease complex.

In addition to the above possible functions in the protein breakdown pathway, some ubiquitin-C-terminal hydrolases are certainly required for the processing of precursors in the biosynthesis of ubiquitin, since all ubiquitin genes either are arranged in linear polyubiquitin arrays, or are fused to ribosomal proteins (reviewed in Refs. 15, 16). Therefore, processing functions include the release of ubiquitin from linear polyubiquitin precursors, the cleavage of ubiquitin from ribosomal proteins, and the removal of extra amino acid residues that are at the C-termini of some polyubiquitin genes (Table 1, function IIc).

Two other functions of ubiquitin-C-terminal hydrolases are notable. Rose (141) pointed out that products of high-energy E_1-ubiquitin and E_2-ubiquitin thiol esters reacting with intracellular nucleophiles (such as glutathione or polyamines) would very quickly deplete the pool of free ubiquitin unless such side products are rapidly cleaved by appropriate ubiquitin-C-terminal hydrolases. In addition, in cases when ubiquitination of proteins has nondegradative roles (such as the modification of protein function), deubiquitination by an appropriate hydrolase will reverse the modification. Such appears to be the case with the ubiquitinated histones H2A and H2B, which are rapidly deubiquitinated during mitosis and reubiquitinated shortly afterwards (142).

---

[3]Two terminologies have been proposed previously for such enzymes (141, 151). Both terminologies classified these enzymes according to the size of the leaving group (protein or small compound) and the nature of the amino group in protein (α-NH$_2$ or ϵ-NH$_2$) to which ubiquitin is linked. As shown in this section, several hydrolases are now known to act on different substrates of several of these categories. Therefore, we call all these enzymes collectively *ubiquitin-C-terminal hydrolases*. In cases where it is clear that the bond cleaved in a major substrate is a ubiquitin-ϵ-NH$_2$-protein isopeptide linkage, the term *isopeptidase* is used.

Table 1 Possible functions of ubiquitin-C-terminal hydrolases

---

I.   Protein degradation
     a. Release of ubiquitin from Lys residues of end products
     b. Disassembly of polyubiquitin chains
     c. "Proofreading": release of incorrectly ubiquitinated proteins
     d. "Trimming" of abnormal polyubiquitin structures

II. Processing of precursors in ubiquitin biosynthesis
     a. Cleavage of linear polyubiquitin precursors
     b. Release of ubiquitin from ribosomal proteins
     c. Removal of extra C-terminal amino acid residues

III. Recycling of ubiquitin from side products with thiols and amines

IV. Reversal of nondegradative protein modification by ubiquitination

---

In view of the numerous and important functions of ubiquitin-C-terminal hydrolases, it is not surprising that a great number of such enzymes exist. However, we know very little about their properties and functions. Information available prior to 1988 was reviewed by Rose (141). Briefly, the first ubiquitin-C-terminal hydrolase discovered was an enzyme from rat liver nucleoli that cleaved the isopeptide linkage between ubiquitin and histone H2A in a ubiquitin-H2A conjugate (143, 144). Enzyme(s) with similar activity was detected in a variety of eukaryotic cells (145), and one was partially purified from calf thymus (146). It is not clear whether or not this enzyme is specific for the histone conjugate. A much better characterized enzyme is a ubiquitin-C-terminal hydrolase from reticulocytes and erythrocytes, discovered by Rose and coworkers (147, 148). It was initially detected as a thiol esterase that cleaves ubiquitin-DTT thiol ester that was formed by the reaction of E_1 in the presence of DTT (147). Subsequently, Pickart & Rose found that the enzyme acts on a variety of carboxyl-terminal amide derivatives of ubiquitin with small compounds (148). The enzyme was purified to near homogeneity, and it is a monomer of a 30-kDa subunit (148). Studies on the mechanism of the inhibition of this enzyme by borohydride led to the discovery of ubiquitin-C-terminal aldehyde, a powerful inhibitor of this (149) and of some other (150) ubiquitin-C-terminal hydrolases.

More recently, Mayer & Wilkinson (151) have resolved from calf thymus four ubiquitin-C-terminal hydrolytic activities that act on ubiquitin ethyl ester. Three of these appear to be isoenzymes of ~30 kDa; the fourth hydrolase is larger, estimated at 100–200 kDa. The major 30-kDa isoenzyme from thymus is similar to the previously described ubiquitin-C-terminal hydrolase from reticulocytes. This hydrolase was cloned from a human cDNA library (152). It was found to be 54% identical with PGP 9.5 (152), a protein present at high levels (1–5% of total protein) in neuronal cells (153). PGP 9.5 is apparently similar to one of the minor 30-kDa isoenzymes of ubiquitin-C-terminal hydrolase purified from calf thymus (152). It is not known why this isoenzyme is expressed at such high levels in neurons. This class of ubiquitin-C-terminal hydrolases is apparently strongly conserved in evolution: the sequence of a similar-sized hydrolase from yeast, YUH1 (40), is 32% identical to the major (nonneuronal) mammalian enzyme.

The functions of the 30-kDa class of ubiquitin-C-terminal hydrolases are not completely understood. Since they act strongly on adducts of ubiquitin with small compounds, they may function to rescue ubiquitin from side products, as proposed originally (147). However, it appears likely that these enzymes have additional functions. The 30-kDa enzymes do not act on high-molecular-mass ubiquitin-protein conjugates (148, 151). However, they do act on some ubiquitin-protein adducts. The 30-kDa hydrolase from yeast cleaves the linkage between ubiquitin and the α-NH$_2$ group of small proteins fused to it (40, 153a). Similarly, the 30-kDa hydrolase from erythrocytes also acts strongly on ubiquitin-protein fusions such as ubiquitin-metallothionein or the natural fusion protein of ubiquitin to the 52-amino-acid human ribosomal extension protein (H. Heller, Y. Moskowitz, and A. Hershko, unpublished results). Thus, the 30-kDa class of hydrolases may also have roles in the processing of at least some biosynthetic precursors of ubiquitin. However, Tobias & Varshavsky (41) have recently reported that the yeast 30-kDa hydrolase does not act on engineered fusions of ubiquitin to larger proteins, such as β-galactosidase. The latter fusions are efficiently cleaved by another, 90-kDa yeast hydrolase, the product of gene *UBP1*, cloned by these investigators (41). The 90-kDa enzyme also processes efficiently natural ubiquitin-protein fusions, such as ubiquitin linked to ribosomal proteins or linear polyubiquitin. The latter precursor was processed by the 90-kDa hydrolase only when the polyubiquitin gene was coexpressed with the enzyme in *E. coli*, but not in cell-free extracts, suggesting that the processing of the polyubiquitin precursor may be cotranslational (41). Null mutants in *Saccharomyces cerevisiae* of the 90-kDa hydrolase, of the 30-kDa hydrolase, or of both hydrolases, are phenotypically normal (41). This is presumably due to the presence of at least two other ubiquitin-C-terminal hydrolases of overlapping functions in yeast (cited in Ref. 41). Multiple species of ubiquitin-α-NH$_2$-protein hydrolase activities were also observed in reticulocyte extracts (154). It seems, therefore, that families of ubiquitin-C-terminal hydrolases exist that are involved in the rapid processing of biosynthetic precursors of ubiquitin.

The 90-kDa yeast hydrolase, like the 30-kDa enzyme, does not act on posttranslationally formed, branched polyubiquitin-protein conjugates (41). This does not necessarily mean that the action of these two classes of enzymes is limited to linear ubiquitin-α-NH$_2$-protein linkages. The branched ubiquitin-ϵ-NH$_2$-H2A conjugate is an excellent substrate for the 30-kDa hydrolase purified from erythrocytes (Y. Moskowitz, E. Eytan, and A. Hershko, unpublished results). It may well be, therefore, that these hydrolases have further roles in the removal of ubiquitin from proteins modified by mono-ubiquitination on ϵ-NH_2 groups. The not-too-well characterized ubiquitin-H2A isopeptidases mentioned above (143–146) may all belong to these families of ubiquitin-C-terminal hydrolases.

Another ubiquitin hydrolase of apparently different functions is a 100-kDa enzyme from erythrocytes, characterized recently by Hadari et al (155). It is a very abundant ubiquitin-binding protein in erythrocytes and reticulocytes, and it is easily purified by affinity chromatography on ubiquitin-Sepharose. It acts preferentially on ubiquitin-Lys48-ubiquitin linkages in branched polyubiquitin chains. It might be analogous to a 160-kDa hydrolase from yeast that also acts on conjugates containing polyubiquitin chains (156). The enzyme from erythrocytes, called isopeptidase T, stimulates the breakdown of proteins linked to polyubiquitin chains by the 26S protease complex. Studies on the mechanism of stimulation of protein breakdown by this isopeptidase revealed its following properties: (a) The enzyme converts high-molecular-weight polyubiquitin-protein conjugates to lower-molecular-weight forms, with the release of free ubiquitin. (b) The lower-molecular-weight products are resistant to further action of this enzyme. (c) The hydrolase does not cleave isopeptide linkages between polyubiquitin chains and the protein substrate. The latter two properties of isopeptidase T may reflect a protective mechanism that prevents the premature disassembly of polyubiquitin-protein conjugates by this abundant enzyme, prior to the action of the 26S protease complex. (d) Preincubation of polyubiquitin-protein conjugates with isopeptidase T did not much increase their susceptibility to proteolysis by the 26S complex. This observation indicates that the stimulation of proteolysis by the isopeptidase is not due to "trimming" of the conjugates to forms more susceptible to the action of the 26S protease complex. (e) Preincubation of polyubiquitin-protein conjugates with the 26S protease complex in the presence of ATP allowed the release of large amounts of free ubiquitin upon further incubation with the isopeptidase. This suggests that the action of the 26S protease converts the conjugates to derivatives (presumably, protein-free polyubiquitin chains) more susceptible to the action of isopeptidase T. Thus, a major role of this isopeptidase in protein breakdown may be to remove polyubiquitin chain remnants following the degradation of the protein substrate moiety by the 26S complex. The stimulatory effect of the isopeptidase on the 26S protease complex may be explained by the assumption that protein-free polyubiquitin chains compete on the binding of polyubiquitin-protein conjugates to the 26S complex, and that the activity of the 26S complex is limited by the rate of dissociation of polyubiquitin chain remnants (155). The high levels of isopeptidase T in cells may be required for the recycling of the large amounts of polyubiquitin chain remnants that are produced by ubiquitin-dependent protein degradation.

Among the large gaps in our knowledge on isopeptidase function (cf. Table 1), a major unanswered question is which enzymes are responsible for the cleavage of isopeptide bonds of ubiquitin to the protein substrate at the final stages of protein breakdown, and how is the action of these hydrolases coordinated with that of the 26S protease complex? Here again, it may be assumed that mechanisms exist that prevent the premature action of such enzymes prior to completion of proteolysis. One possible mechanism to prevent premature action of such "terminal" isopeptidases may be by sequential specificity, so that the isopeptidase would specifically recognize only the products of proteolysis. Another possibility is compartmentation in the 26S complex, so that intermediates are channelled in a way that the site of proteolysis is separate from (and precedes that) of the isopeptidase. Rose reported (157) that ATP stimulates the release of free ubiquitin from ubiquitin-H2A conjugate, in the presence of the 26S protease complex. This observation was recently confirmed, using highly purified preparations of 26S complex and several ubiquitin-protein conjugate substrates (E. Eytan, T. Armon, I.A. Rose, and A. Hershko, unpublished results). These findings indicate that the "terminal" isopeptidase is a part of the 26S protease complex. The integration of the action of this isopeptidase with the proteolytic machinery and the mechanisms of ATP stimulation of isopeptidase action remain to be elucidated.

SIGNALS IN PROTEINS FOR UBIQUITIN-MEDIATED DEGRADATION

The N-Terminal Recognition Signal

A centrally important problem is what specific structural features of proteins are recognized by the ubiquitin system. The first type of recognition signal that was elucidated is probably the simplest: it is determined by the amino acid residue at the N-terminal position. Hershko and colleagues first noted that N-α-acetylated proteins are not degraded by a reconstituted ubiquitin proteolytic system from reticulocytes (158; see, however, below). It was found that preferential modification of the N-terminal residue of lysozyme blocks almost completely its degradation. Inhibition of degradation was accompanied by a corresponding decrease in the formation of high-molecular-weight ubiquitin conjugates. Addition of polyalanine side chains (that contain free α-NH$_2$ groups) to the N-terminal blocked protein restored almost completely its sensitivity to ubiquitin conjugation and degradation. It was concluded that the N-terminal α-NH$_2$ group of proteins is required for ubiquitin ligation (158).

Subsequent studies from other laboratories showed that the nature of the N-terminal amino acid residue has a strong influence. Varshavsky and coworkers (159) constructed a ubiquitin-*lacZ* fusion gene in which the cDNA that encodes ubiquitin was fused via a 45-residue amino-terminal extension to the *β-gal* gene of *E. coli*. The bulk of the sequence of the N-terminal extension (residues 8–45, added because of the design of the expression vector) was derived from an internal sequence of the *lac* repressor encoded by the *lacI* gene. *E. coli* transformed with the gene expressed the chimeric protein. In contrast, expression of the fusion gene in yeast resulted in a smaller, cleaved product. Analysis of the N-terminal residues of this product revealed that the fused protein was cleaved at the junction between the C-terminal glycine of the ubiquitin moiety and the N-terminal residue of the extension of the *β*-galactosidase moiety (159; for simplicity, we refer to the junction between ubiquitin and the *lac* repressor extension as ubiquitin-*β*-*gal* junction). The rapid removal of ubiquitin from the fusion protein is presumably due to the action of ubiquitin-C-terminal hydrolases, abundant in yeast. The researchers then converted the ATG codon that specifies the Met residue at the Gly-Met junction into codons that define the 19 other amino acids. Expression of the 20 genes in yeast revealed that with one exception (Ubiquitin-Pro-*β*-Gal), all the chimeric proteins were rapidly de-ubiquitinated. This result made it possible to express in cells proteins that differ solely in their N-terminal position. Measurement of the in vivo half-lives of 20 different species of *β*-Gal proteins expressed in *Saccharomyces cerevisiae* revealed that they vary from more than 20 hr to less than 3 min, depending on the nature of the amino acid residue exposed at their N-terminal position. Proteins with Val, Met, Gly, Ala, Ser, Thr, and Cys residues at their N-terminal position are extremely stable with half-lives longer than 20 hr. Proteins with Leu, Glu, His, Tyr, Gln, Asp, Asn, Phe, Leu, Trp, Lys, and Arg residues at their N-terminal position are extremely unstable with half-lives of less than 30 min. The rate of deubiquitination of Ubiquitin-Pro-*β*-Gal is extremely low (this is probably the only species that is poorly cleaved by the ubiquitin-C-terminal hydrolases). The deubiquitinated product, Pro-*β*-Gal, is long-lived (160). Most of the fused Pro-*β*-Gal protein is degraded prior to its deubiquitination and is extremely unstable ($t_{1/2}$ of approximately 7 min; 160). It is probably targeted for degradation by the ubiquitin molecule to which it is stably bound. The 20 different amino acid residues were classified into "stabilizing" or "destabilizing" with respect to the half-lives they conferred on *β*-Gal when exposed at the N-terminal position of the protein. The resulting rule, designated by the authors the "N-end rule," proposes that the in vivo half-life of a protein is a function of its amino-terminal residue (159, 160). As discussed above, subsequent biochemical work showed that the rule reflects the specificities of the two binding sites of $E_3\alpha$ for the N-terminal residues of proteins.

Some N-terminal amino acid residues of proteins have to be converted to residues bindable to the N-terminal sites of $E_3\alpha$. Ferber & Ciechanover have shown that proteins with acidic N-termini can be degraded only after they are posttranslationally modified by the addition of an arginine residue to their N-terminal residue (161). Further studies have shown that the modification is required in order to allow conjugation of ubiquitin to occur (162). The reaction is catalyzed by arginyl-protein-tRNA transferase, an enzyme investigated previously by Soffer (163) that catalyzes the transfer of an activated arginyl residue from arginyl-tRNA to acidic N-termini of proteins. The enzyme is composed of several ~50-kDa subunits that are associated with several molecules of arginyl-tRNA synthetase to yield a complex with a native molecular mass of approximately 360 kDa (162). Gonda and colleagues have recently shown that in reticulocyte lysates, proteins that have an Asn or a Gln residue at their N-terminal position are also modified by arginine after their N-terminal residues are deamidated by asparginase and glutaminase-like activities (82). Interestingly, proteins that have a cysteine residue at the N-terminal position must also be arginylated at this residue prior to their conjugation and degradation (82). Balzi and colleagues have recently cloned the yeast arginyl-transferase gene, *ATE1*, and found that it encodes a 503-residue protein (164). As expected, proteins with Asp, Asn, Glu, and Gln residues at the N-terminal position are stabilized in *ATE1* null mutant. The fate of proteins with a Cys residue at the N-terminal position has not been studied in this mutant, but it appears to be different in yeast and in mammals (Ref. 82). Elias & Ciechanover have recently shown by direct methods that the tRNA-dependent addition of arginine to acidic amino termini of proteins is necessary for recognition and binding of the substrate to E_3 (165). This is presumably due to the binding of arginylated proteins to the Type I "head" site of $E_3\alpha$.

Is recognition of the N-terminal residue of a protein sufficient to target it for degradation? Bachmair & Varshavsky expressed in yeast 20 chimeric genes between ubiquitin and dihydrofolate reductase (DHFR). The genes differed from each other by the amino acid residue at the N-terminus of DHFR (X at the ubiquitin-X-DHFR construct). All the expressed proteins (except for Pro at the N-terminus of DHFR) were rapidly and efficiently deubiquitinated. The resulting DHFR species were all stable, very much like the wild-type protein (Met at the N-terminus) (160). To understand the mechanistic basis for the difference in stabilities between the X-β-Gal and the X-DHFR proteins, the 38-residue amino-terminal region of X-β-Gal (as described above, the bulk of this region was derived from the *lac* repressor) was positioned upstream to the original N-terminus of DHFR. When these chimeric genes (ubiquitin-X-β-*gal* N-terminal extension-DHFR) were expressed in yeast, the resulting de-ubiquitinated proteins had half-lives similar to those of their

X-β-Gal counterparts. The signal was identified as a pair of lysine residues residing in positions 15 and 17 of the N-terminal extension (160). Either of these two residues is sufficient for targeting the protein for degradation (160).

Further studies have shown that the polyubiquitin chain that targets β-Gal for degradation is attached to one of these particular lysine residues (23). The authors suggested that N-terminal-recognizing E_3 (most probably $E_3\alpha$) must have two substrate recognition sites: one for the N-terminal residue and one for a specific lysine residue on which the polyubiquitin chain is synthesized and anchored. It is also possible that the second site resides on E_2. On the protein substrate, the two motifs must be of sufficient proximity so that they will bind concomitantly to the corresponding sites of E_3 or the $E_3 \cdot E_2$ complex. Interestingly, mapping of ubiquitination sites of the yeast iso-2-cytochrome c revealed that the polyubiquitin chain is synthesized almost exclusively on Lys13 (166). Similarly, in the yeast iso-1-cytochrome c, the polyubiquitin chain is confined to Lys9 (C. W. Sokolik and R. E. Cohen, personal communication).

The requirement for concomitant binding of two motifs does not necessarily mean that they have to be adjacent on the polypeptide chain. Spatial vicinity that is conferred by the three-dimensional structure of the protein or that results from the movement of one segment towards the other may well fulfill the requirement for close association of the two domains. In this context, Johnson and colleagues found that in a multisubunit protein, the two determinants may even reside on different subunits and still target the protein for degradation. In this case (*trans* recognition), only the subunit that contains the available lysine residue is proteolyzed (167). Dunten and colleagues have shown that specific reduction of Cys6-Cys127 disulfide bridge in lysozyme facilitates significantly the ubiquitination of the protein and is a prerequisite to the formation of polyubiquitin chain on at least one of the four chain initiation sites of the molecule (168). Lysozyme has a "destabilizing" residue (Lys) at the N-terminal position and is recognized by $E_3\alpha$ (79, 168). It appears that reduction of the Cys6-Cys127 bond unhinges the N-terminal region of the molecule and facilitates its binding to the N-terminal binding site of E_3. Reduction of this bond may also allow movement of the N-terminal residue to the vicinity of the lysine residues on which the polyubiquitin chains are synthesized and anchored. Thus, the two domains may be able to bind efficiently to their corresponding $E_3\alpha$ or $E_2 \cdot E_3\alpha$ recognition sites. Similarly, while native RNase A is not ubiquitinated and degraded despite having a "destabilizing" N-terminal residue, many modifications that lead to unfolding of the protein will convert it into an efficient substrate (78, 168a).

It should be noted that the formation of a polyubiquitin chain on a certain lysine residue, or on any lysine, is not always absolutely required for protein breakdown. Methylated ubiquitin, in which all amino groups are blocked by reductive methylation (and therefore can be ligated to proteins, but cannot form polyubiquitin chains), significantly stimulates the breakdown of some proteins (22, 169, 170), though at rates generally lower than those obtained with native ubiquitin. In these cases, several molecules of methylated ubiquitin are usually ligated to the substrate protein. Such conjugates of methylated ubiquitin with lysozyme, for example, are good substrates for the 26S protease complex (22, 92). For reasons not yet understood, there is a great variability in the requirement for polyubiquitin chain formation for the degradation of different proteins (170). Thus, the degradation of derivatives of β-galactosidase by reticulocyte extracts requires polyubiquitin chain formation much more strongly (23, 171) than the breakdown of some other test proteins surveyed (170).

An important problem concerns the physiological significance of the N-terminal recognition mechanism. What are the natural substrates of this pathway, and what may be the pathological consequences of inhibiting it? Many lines of evidence indicate that the scope of the N-terminal recognition pathway is rather limited. Following protein synthesis, the removal of initiator methionine by methionine aminopeptidase takes place only if the second amino acid residue has a radius of gyration of less than 1.29 Å (172). As it turns out, all the "destabilizing" "N-end rule" amino acid residues have radii of gyration larger than 1.29 Å, and therefore, such residues will not be exposed at the N-terminus by methionine aminopeptidase (reviewed in Ref. 173). That the population of cellular proteins that may be subjected to recognition and degradation by the "N-end" pathway is rather limited is also shown by the observation that null mutant of *UBR1*, the yeast homolog of $E_3\alpha$, displays only a mild phenotype. The cells are partially defective in sporulation and grow slightly slower than their wild-type counterparts (42). As expected, null mutant of *ATE1*, the gene that encodes arginyl-transferase, displays an even milder phenotype (164). It is possible that the "N-end pathway" proteolyzes selectively compartmentalized proteins in which a destabilizing amino acid residue is exposed by the removal of the signal peptide and that are mislocalized to the cytosol (42, 159, 160). The highly active $E_3\alpha$ in reticulocytes may be involved in the degradation of previously compartmentalized mitochondrial proteins that are broken down in the course of reticulocyte maturation (174).

Signals that are Distinct from the N-Terminal Residue

While the N-terminal residue of a protein plays a role in the recognition of certain proteins, the evidence discussed above indicates that the bulk of the proteins degraded by the ubiquitin system are recognized by different signals.

A great majority of the cellular proteins have acetylated N-termini (175). Initial studies by Hershko and colleagues have shown that these proteins are not degraded by the cell-free ubiquitin system. However, in these studies the researchers used ubiquitin-supplemented Fraction II, a crude fraction devoid of ubiquitin and other neutral or basic proteins (158). A more recent study by Mayer and colleagues has shown, however, that certain N-α-acetylated proteins, such as histone H2A, actin, and α-crystallin, are degraded in whole lysate of reticulocytes in an ATP- and E_1-dependent manner (176). The researchers suggested that a specific factor that is essential for the degradation of these proteins may be removed or inactivated during the fractionation and partial purification of the lysate. Three independent, albeit indirect, lines of evidence suggest that exposure of a free N-terminal residue (by deacetylase, for example) does not play a role in rendering these proteins susceptible to degradation:

1. Preincubation of the proteins in E_1-depleted lysate does not sensitize them to further proteolysis in Fraction II. The working hypothesis behind this experiment was that the deacetylase is the factor that may be missing in Fraction II, but is contained in the whole lysate.
2. The degradation of the substrates was not sensitive to dipeptides with either Type I, Type II, or Type III residues at their N-terminal position.
3. The degradation of actin, a protein with four successive acidic amino acid residues that follow the acetyl group, was not sensitive to ribonucleases. As tRNA^{Arg} is involved in posttranslational modification of acidic N-termini proteins, their degradation is sensitive to ribonucleases (82, 161, 162).

Gonen and colleagues have recently purified from Fraction I a novel protein that seems to be involved in the degradation of at least one N-α-acetylated protein, histone H2A (81). The protein, designated Factor H (FH) is a homodimer composed of two 46-kDa subunits. It is removed from Fraction I during purification of ubiquitin and therefore has not been previously recognized. Surprisingly, FH is not involved in recognition and targeting the substrate for conjugation. Rather, it acts along with the 26S protease complex to promote degradation of high-molecular-mass ubiquitin conjugates of histone H2A. Synthesis of these adducts was mediated by $E_3\alpha$. It is not known, however, whether $E_3\alpha$ plays also a major role in conjugation of N-α-acetylated proteins in the cell.

Using computer analysis of sequences of proteins, Rogers and colleagues have noticed that extremely short-lived proteins contain stretches of amino acids enriched in proline (P), glutamic acid (E), serine (S), and threonine (T) (177). These PEST sequences, which were typically flanked by clusters of positively charged amino acids, led the researchers to propose that they constitute a signal for rapid degradation of these proteins. The statistical evidence that supports the hypothesis is rather strong. Thirty of 32 short-lived proteins with known sequences were found to contain one or more PEST sequences (178). However, neither the mechanistic basis by which the PEST signal targets proteins for degradation, nor the system that recognizes the signal, has been identified. Ghoda and colleagues deleted the carboxy-terminal tail of the mouse ornithine decarboxylase (ODC; residues 423–461) that contains one of the two PEST sequences of the protein (residues 423–449; the first PEST region is confined to residues 293–333) and found that the truncated protein becomes stable (179). On the other hand, Rosenberg-Hasson et al (180) have recently reported that the deletion of residues 447–461 at the C-terminus of ODC, which leaves intact most of the second PEST region, stabilizes this protein. Since deletions in proteins may produce marked alterations in protein structure, deletion analysis cannot conclusively support, or rule out, the identification of putative recognition signals. Loetscher and colleagues have recently constructed chimeric proteins in which the C-terminal PEST-containing region of ODC is fused to the amino or carboxy termini of DHFR. They found that the addition specifically destabilized in vitro the otherwise stable DHFR (181). It should be noted that the degradation of ODC is not mediated by the ubiquitin system as has been shown both in vitro (182) and in vivo (183). On the other hand, the degradation of many other PEST-containing proteins such as phytochrome, certain nuclear oncoproteins, and the tumor-suppressing protein p53 may be mediated, at least in vitro, by the ubiquitin system (see below). It is possible that recognition of PEST sequences is shared by several different proteolytic pathways, or that the PEST sequences play an auxiliary role in different pathways of protein degradation.

DEGRADATION OF SPECIFIC CELLULAR PROTEINS BY THE UBIQUITIN SYSTEM: REGULATORY ASPECTS

For reasons of technical convenience, most of the biochemical work on the ubiquitin proteolytic system was done with nonphysiological protein substrates, such as extracellular proteins. In order to gain a better understanding of the mechanisms of the selectivity and regulation of this system, studies on rapidly degraded specific cellular proteins are obviously required. To date, the possible involvement of the ubiquitin system in the degradation of specific cellular proteins has been suggested in a few cases, reviewed recently (12), and discussed below.

Phytochrome

Phytochrome is a plant regulatory photoreceptor that consists of a linear tetrapyrrole linked to a polypeptide of ~120 kDa (reviewed in 184, 185). It exists in two interconvertible forms: Pr, which absorbs red light, and Pfr, which maximally absorbs in the far red region. Upon the absorption of red light, Pr is converted to Pfr; this reaction initiates, by as yet unknown mechanisms, a series of developmental and morphogenetic responses that facilitate the adaptation of plants to the availability of photosynthetic light. It has been known for many years that the Pr form of phytochrome is quite stable, but the protein is rapidly degraded following photoconversion to Pfr ($t_{1/2}$ ~1–2 h). The significance of this light-induced degradation of phytochrome is not understood; it may serve to down-regulate the active form of the photoreceptor. Shanklin et al (186) examined the possible involvement of the ubiquitin system in phytochrome degradation. Following irradiation of etiolated oat seedlings with red light, a series of higher-molecular-mass derivatives of phytochrome were formed. These were identified as phytochrome-ubiquitin conjugates by immunoprecipitation with anti-phyto-chrome antibodies, followed by immunoblotting with anti-ubiquitin antibodies. It was suggested that phytochrome degradation is carried out by the ubiquitin system (186). Similar accumulation of ubiquitin-phytochrome conjugates during red light–induced degradation of phytochrome was observed in a variety of plant species, including monocots and dicots (187). Quantitation of the levels of ubiquitin-phytochrome conjugates at different irradiation conditions showed a general correlation with phytochrome degradation: when seedlings were transferred from red light back to darkness, phytochrome degradation stopped, and levels of phytochrome-ubiquitin conjugates declined rapidly (188). An unexplained observation on the kinetics of the changes in levels of ubiquitin-phytochrome conjugates is that under conditions of continuous irradiation by red light, levels of these conjugates continued to increase until ~90 min (188). Since rapid degradation of phytochrome is initiated immediately following red light irradiation, an immediate increase in steady-state levels of ubiquitin-phytochrome conjugates would be expected, assuming that irradiation increases the rate of conjugation. It might be that the rate of degradation of phytochrome-ubiquitin conjugates is slowed down by secondary changes at later times of irradiation. It is also possible that the population of ubiquitin-phytochrome conjugates is heterogeneous: the conjugates that accumulate may be poor intermediates in degradation, and therefore be more stable than another population of conjugates that are efficient intermediates and whose steady-state level is very low at all times.

An interesting question is what biochemical alteration(s) causes the rapid degradation of phytochrome following red light irradiation. As in other cases of regulated protein degradation (see below for the case of cyclins), the possibilities are an alteration in the substrate that is recognized by the ligase system, or the activation of a specific ubiquitin ligase that acts on this protein. In the case of phytochrome, substrate alteration obviously occurs, but it is not at all clear that the Pfr form is directly recognized by a ubiquitin ligase. It has been known that following photoconversion, Pfr rapidly becomes "pelletable," i.e. it aggregates or becomes associated with undefined subcellular structures (184). Most of ubiquitin-phytochrome conjugates are found in the "pelletable" fraction, and it was proposed that the processes involved in Pfr aggregation may commit phytochrome to degradation (188). It should be noted, however, that the possible activation of a specific phytochrome-ubiquitin ligase cannot be ruled out at this stage. Since the photoconversion of this receptor elicits a variety of cellular responses, one of them may be the activation of such a specific enzyme.

Oncoproteins

Many nuclear oncoproteins have rapid turnover rates (177), but little is known about the mechanisms of their degradation. The degradation in cultured cells of oncoproteins c-myc and c-myb (189) and of tumor suppressor protein p53 (190) were shown to be nonlysosomal and ATP-dependent. Ubiquitin conjugates of c-myc could not be detected in cells (189); such inability to detect rapidly degraded intermediates does not justify, however, the conclusion (189) that the ubiquitin system is not involved in c-myc degradation. Ciechanover and coworkers (191) have examined the degradation of different oncoproteins by reticulocyte extracts. All oncoproteins examined were synthesized in vitro by translation of the corresponding RNAs in reticulocyte lysates. The degradation of all proteins tested (N-Myc, c-Myc, c-Fos, p53, and E1A) was found to be ATP-dependent and ubiquitin-mediated, as shown by the inhibition of their degradation by an antibody directed against E_1, and by the reversal of inhibition upon the supplementation of E_1. That the degradation signals recognized by the ubiquitin system in reticulocyte lysates may be similar to those recognized in other cells was indicated by the observation that the relative rates of degradation of different derivatives of adenovirus protein E1A are correlated with their rates of degradation in intact *Xenopus* oocytes (192). These results show that the above oncoproteins can be degraded in vitro by the ubiquitin system from reticulocytes. It remains to be shown that this actually occurs in intact nucleated cells. The active reticulocyte cell-free system is derived from a terminally differentiating and highly specialized cell, which may have lost some of the control mechanisms operating in other cells. In addition, the possible interaction of oncoproteins with nuclear components may influence their degradation, and such components may be missing in extracts from anucleate reticulocytes. The same reservations apply to another study on the influence of E6 oncoprotein from papillomaviruses on the degradation of p53 protein in reticulocyte lysates (193). It was previously observed that proteins E6 and p53, translated in reticulocyte lysates, are strongly associated (194). It was noted that this association is followed by the loss of p53 protein. This was shown to be due to the stimulatory effect of E6 protein on the degradation of p53 by reticulocyte lysates (193). Under the experimental conditions employed, p53 was not degraded by reticulocyte lysates unless E6 protein was also present.[4] Using E6 proteins from different papillomaviruses, it was shown that the stimulation of the degradation of p53 is directly related to the affinities of the different E6 proteins to bind p53. E6-stimulated degradation of p53 protein in reticulocyte lysates was ATP-dependent, as shown by its inhibition by the analog ATPγS. It was noted that upon incubation with ATPγS, high-molecular-mass derivatives of p53 accumulated. These were shown to be p53-ubiquitin conjugates, by immunoblotting with anti-ubiquitin antibodies.[5] The formation of ubiquitin conjugates of p53 required the supplementation of protein E6. It was concluded that the association of protein E6 with p53 promotes the degradation of the latter by the ubiquitin system (193). The authors commented that the association of p53 with different proteins may have opposite effects on its degradation. Thus, the formation of complexes of p53 with large T antigen or with E1B 55-kDa protein slow down its degradation and increase the steady-state levels of p53 in transformed cells (cited in Ref. 193). Reduced levels of p53 were observed in some cell lines transformed with papillomaviruses (cited in Ref. 193). It remains to be shown that protein E6 actually stimulates the degradation of tumor suppressor p53 in transformed cells.

MATα2 Repressor

The MATα2 repressor is a transcriptional regulator of mating type switching in *Saccharomyces cerevisiae* (reviewed in Ref. 195). Hochstrasser & Varshavsky (196) have shown that this regulatory protein turns over very rapidly in yeast, with a half-life time of ~5 min. Fusion of the α2 repressor to β-galactosidase caused the rapid degradation of the latter protein. These investigators used deletion analysis of α2-β-galactosidase fusion proteins to search for degradation signals in MATα2 repressor. The α2 repressor has 210 amino acid residues. The deletion of residues 67–210 of the α2 repressor did not affect the rapid degradation of the fusion protein, while the deletion of residues 52–210 produced a stable fusion protein. Furthermore, the deletion of residues 1–136 had no influence on degradation, whereas the deletion of

---

[4]These results are different from the observations of Ciechanover et al (191) on the rapid degradation of p53 in reticulocyte lysates. This is presumably due to differences in incubation conditions and in the mode of preparation of reticulocyte lysates.

[5]ATPγS efficiently carries out ubiquitin-protein ligation (since the scission of α-β bond of ATP takes place in the E₁ reaction), but it cannot support conjugate proteolysis by the 26S complex, which requires β-γ bond hydrolysis of ATP. Therefore, ATPγS and some nonhydrolyzable β-γ ATP analogs (210) may be used as agents for the accumulation of ubiquitin-protein conjugates in crude extracts.

residues 1–141 of the $\alpha 2$ repressor stabilized its fusion protein with β-galactosidase. The investigators concluded that the MAT$\alpha 2$ repressor contains two degradation signals, one requiring residues 53–67 of its amino-terminal domain, and another in residues 136–140 in the carboxyl-terminal domain (196). It should be noted that the deletion of portions of proteins may have drastic effects on protein structure, and thus may affect indirectly their stability and degradation. Therefore, deletion analysis may be used merely for preliminary studies, to be followed by more detailed amino acid replacement analysis or the identification of transferrable signals. It was shown, in fact, in this study that the fusion of residues 136–157 of the $\alpha 2$ repressor to β-galactosidase did not cause its degradation, indicating that this sequence from the carboxyl-terminal domain is not sufficient, by itself, to signal degradation (196). Though the identity of the exact recognition signals in MAT$\alpha 2$ remains to be established, it seems certain that the amino-terminal and carboxyl-terminal domains of this protein are recognized by different mechanisms. This conclusion is supported by the observation that mutations in yeast that inhibit the degradation of fusion proteins containing the amino-terminal domain of $\alpha 2$ repressor do not inhibit the degradation of those containing the carboxyl-terminal domain of the repressor (196).

More recently, Hochstrasser et al (197) have reported evidence suggesting that the degradation of MAT$\alpha 2$ repressor in yeast is carried out, at least in part, by the ubiquitin system. Using derivatives of ubiquitin in which the amino terminus was extended by epitopes from c-myc or ha of influenza virus, they showed that the $\alpha 2$ repressor is multiply ubiquitinated in yeast cells. The addition of these tags was necessary because the low levels of ubiquitin-$\alpha 2$ conjugates in cells could not be detected by anti-ubiquitin antibodies, which have relatively low affinities. That the ubiquitination of MAT$\alpha 2$ is indeed involved in its degradation was shown by the finding that overexpression in yeast of a mutant ubiquitin containing a Lys48 $\rightarrow$ Arg48 substitution (that prevents polyubiquitin chain formation at Lys48) slowed down twofold the rate of the degradation of this protein (197). The incomplete inhibition of $\alpha 2$ degradation by the mutant ubiquitin may be due to competition by native ubiquitin, and also to the fact that the degradation of different proteins has variable requirements for polyubiquitin chain formation (see above).

Cyclins

Cyclins are proteins involved in cell cycle control in eukaryotes (reviewed in 198–200). Several types of cyclins exist, including A-type, B-type, and G1 cyclins. Cyclin B was first discovered in fertilized sea urchin oocytes as a protein that is synthesized during the interphase, and then suddenly degraded at the end of metaphase (201). The levels of the other cyclins rise and fall at other stages of the cell cycle. All cyclins appear to be associated with p34^{cdc2} protein kinase or related kinases, and presumably act by regulating the activity of the kinases. For example, maturation or M-phase promoting factor (MPF) protein kinase is known to consist of cyclin B bound to the p34^{cdc2} protein. The rise in levels of MPF at the onset of mitosis requires the synthesis of cyclin B (and some as yet incompletely defined posttranslational modifications of both proteins), whereas the subsequent degradation of cyclin B is apparently responsible for the inactivation of MPF at the metaphase-anaphase transition. In a cell-free system from *Xenopus* eggs that reproduces cell cycle events, it has been shown that the synthesis of cyclin B is sufficient to drive entry into mitosis (202), whereas cyclin degradation is required to exit from mitosis (203).

The programmed degradation of cyclins at specific stages of the cell cycle is a dramatic example of regulated protein degradation, and raises interesting questions about the mechanisms involved. The elucidation of such mechanisms became possible by the establishment of cell-free systems that carry out regulated cyclin degradation. Luca & Ruderman (204) have studied the regulation of cyclin degradation in a cell-free system from early embryos of the surf clam, *Spisula solidissima*. In this system, fertilized clam oocytes were labeled with [^{35}S]-methionine, and extracts were made at the onset of mitosis. Upon further incubation of concentrated extracts, progress of cell cycle events apparently continued in vitro, and following a short lag period, the degradation of labeled cyclins took place. The following characteristics of cyclin degradation in the cell-free system indicated that the cell-free system faithfully reproduces this process in intact cells: (*a*) The degradation of cyclin A preceded that of cyclin B by several minutes, as in intact cells. (*b*) Only cyclins were specifically degraded in the course of the incubation, while all other labeled proteins were stable. (*c*) The timing of cyclin degradation was determined by the stage of the cell cycle at which extracts were made. Thus, when extracts were made earlier at the interphase, the lag period that preceded the degradation of cyclins A and B was longer than when extracts were made at the beginning of mitosis. (*d*) Cyclin degradation in the cell-free system required ATP, as is the case with most types of protein degradation in intact cells. Further experiments showed that interphase extracts do not contain an inhibitor of cyclin degradation in mitotic extracts, as indicated by incubations in which extracts made at different stages of the cell cycle were combined. In other mixing experiments, in which small amounts of extracts containing labeled cyclins were combined with a great excess of unlabeled extracts made at different stages of the cell cycle, it was found that the timing of cyclin degradation is determined by the cell cycle stage of the enzyme system, and not by the cell cycle stage of the cyclin substrates (204). It could not be ruled out, however, that modification of cyclins does take place, but that modification is reversed by enzymes active in extracts made at other stages of the cell cycle.

To examine the molecular mechanisms of the regulation of cyclin degradation, it had first to be determined whether cyclin degradation is carried out by the ubiquitin system. That this is indeed the case was suggested by independent work of Glotzer and coworkers (205) and of Hershko et al (170). The latter investigators used methylated ubiquitin as a specific inhibitor of the ubiquitin system. As mentioned above, methylated ubiquitin, which cannot form polyubiquitin chains, is a weak stimulator of protein breakdown, and its effectiveness depends on the protein substrate degraded (170). Therefore, in the case of proteins whose degradation depends more strongly on polyubiquitin chain formation, methylated ubiquitin can be used as a specific (though partial) inhibitor of degradation. The addition of methylated ubiquitin to the clam embryo cell-free system described above inhibited markedly (though not completely) the degradation of both cyclin A and cyclin B. That this was due to specific interference with ubiquitin function was indicated by the observation that the supplementation of excess ubiquitin completely overcame the inhibitory action of methylated ubiquitin on cyclin degradation (170). These data indicated that polyubiquitin chain formation is required for cyclin degradation. It seemed reasonable to assume that polyubiquitination of cyclin takes place, but cyclin-ubiquitin conjugates could not be detected in clam embryo extracts under these conditions, possibly due to their rapid degradation. However, Glotzer and coworkers (205) have shown by direct methods that cyclin is ubiquitinated in mitotic extracts of *Xenopus* eggs. These investigators have produced stable mitotic extracts of *Xenopus* eggs by incubation of interphase extracts with a derivative of sea urchin cyclin B that lacks the N-terminal 90-amino-acid part. It had been shown previously in the same laboratory that truncated $\Delta90$ cyclin B can activate MPF without being degraded (203). Interphase extracts incubated with $\Delta90$ cyclin B progress to mitosis, but are arrested at a stable metaphase state (203). When different derivatives of cyclin (see below) were incubated with stable mitotic extracts, a transient accumulation of higher-molecular-mass forms was observed. These were shown to be ubiquitin conjugates, by the incorporation of ^{125}I-ubiquitin into the same high-molecular-mass derivatives (205). In interphase extracts, the level of ubiquitin-cyclin conjugates was ~10-fold lower than in mitotic extracts, and the conjugates formed in interphase extracts were of smaller size. It thus seems that in mitotic extracts, both the ubiquitination of cyclin and the processive addition of multiple ubiquitins are markedly increased.

Another problem addressed by these investigators was which regions of the cyclin molecule are required for ubiquitin ligation and degradation. Fusion of the amino-terminal 13–91 amino acid residues of sea urchin cyclin B to protein A produced a protein that was rapidly degraded in mitotic, but not in interphase, extracts.[6] This finding indicated that the amino-terminal region is not only necessary, but is also sufficient for cyclin degradation. Based on partial mutational analysis, it was proposed that a "destruction box" of the sequence RAALGNISN at residues 42–50 of sea urchin cyclin B is required for its degradation. Homologous sequences are strongly conserved in the amino-terminal regions of cyclins A and B from various species. When the invariant Arg residue in this sequence was replaced by Cys, the resulting derivative was not degraded in mitotic extracts, and only low-molecular-mass derivatives were formed with ubiquitin (205). It is possible that this amino acid replacement decreases the affinity of cyclin to a ubiquitin ligase, which would also interfere with the processive addition of multiple ubiquitins. It should be noted, however, that the Arg → Cys replacement could have secondary effects on protein structure, since a basic residue was changed to a weakly acidic residue. Further mutational analysis by suitable amino acid replacements appears to be required to establish the role of the "destruction box" in cyclin ubiquitination and degradation. Also, a specific inhibitor of the ubiquitin pathway was not used in this study, and therefore it could not be ruled out that ubiquitin-cyclin conjugates are side products, rather than intermediates in cyclin degradation. However, the latter work (205), taken together with the study on the specific inhibition of cyclin degradation by methylated ubiquitin (170), provides quite a strong case for the involvement of the ubiquitin system in cyclin degradation.

Assuming that cyclins are degraded by the ubiquitin system, the next question is what regulatory mechanisms are responsible for its rapid degradation in mitosis. The two main possibilities are that a cyclin-specific ubiquitin ligase may be suddenly activated at mitosis, or that cyclin is converted to a form susceptible to the action of a constitutively active ubiquitin ligase. At present, there is no evidence available to support or rule out either mechanism. Cell cycle–dependent phosphorylation of cyclin B has been observed, but it is correlated with the activation of MPF, rather than with cyclin degradation (206). Since the system is now experimentally approachable, further significant progress on the mechanisms of the regulation of cyclin degradation may be expected in the near future.

DIVERSE FUNCTIONS OF UBIQUITIN CONJUGATION

Conjugation of ubiquitin and the synthesis of naturally occurring ubiquitin-protein fusion products have been implicated in many important cellular

---

[6] This derivative was used for most of the above-mentioned experiments on the identification of ubiquitin-cyclin conjugates formed in mitotic extracts, since fusions with protein A can be readily isolated with IgG-Sepharose.

processes. These include DNA repair, control of cell cycle, response to heat and other stresses, biogenesis of ribosomes, regulation of cell surface receptors, biogenesis of peroxisomes, regulation of transcription, viral infection, programmed cell death, cellular differentiation, and the pathogenesis of certain neurodegenerative diseases. Description of the association between ubiquitination and many of these processes is currently circumstantial, and the underlying mechanisms are still obscure. Conjugation of ubiquitin probably targets some of the proteins involved in these processes to degradation. However, it is possible that attachment of ubiquitin to certain proteins serves a different, yet unknown, nonproteolytic function(s). In particular, monoubiquitination may serve such a role. The involvement of ubiquitin in these processes has been reviewed recently (15, 17).

Recently, another novel function of the ubiquitin system has been described. It was found that the two, apparently distinct, proteolytic systems, the ubiquitin system and the lysosomal-autophagic system, may be functionally associated. Gropper and colleagues have shown that upon shifting the ts85 and the ts20 cells (which harbor a mutated thermolabile E_1; Refs. 43–45, and see above) to the restrictive temperature, there was no change in the rate of degradation of long-lived proteins. In contrast, shifting the wild-type cells to the high temperature is accompanied by a 2–3-fold increase in the rate of degradation of this group of proteins. The heat-induced accelerated degradation can be completely inhibited by the lysosomotropic agents NH_4Cl and chloroquine, indicating that it occurs mostly within lysosomes and autophagic vacuoles (50). Exposure of the cells to starvation (a stimulus that activates the autophagic response) during the heat treatment resulted in even more striking results. While there was no change in the rate of degradation measured in the mutant cells, a fourfold increase was observed in the wild-type cells. Taken together, the results suggest that an active E_1 is required for stress-induced accelerated degradation of cellular proteins in lysosomes.

In considering the role of ubiquitin conjugation in autophagy, two distinct mechanisms can be postulated. One possibility is that cellular proteins are preferentially targeted for degradation in lysosomes only after their conjugation. Because proteins cannot be conjugated to ubiquitin in the mutant cells at the nonpermissive temperature, their degradation is inhibited. Consistent with this hypothesis is the finding that ubiquitin-protein conjugates are selectively enriched in lysosomes of fibroblasts (207, 208). Free ubiquitin can also be found in lysosomes (209). Ubiquitin, which is relatively resistant to the action of proteases, may be released from conjugates during their digestion in the lysosome. A second possibility is that the ubiquitin system is involved directly or indirectly in the formation, maturation, or functions of the autophagic vacuoles. For example, a protein that is essential for the biogenesis of the vacuoles must be first conjugated to ubiquitin in order to become active; or, a short-lived protein that inhibits the formation of these vacuoles may have to be proteolyzed in order to alleviate the inhibition.

CONCLUDING REMARKS

The considerable progress achieved in recent years in the ubiquitin field demonstrates the power of both biochemical and molecular genetic approaches. Most of the events in this pathway were elucidated by the use of biochemical methods, consisting of the resolution of the components of the reticulocyte cell-free system, the purification and characterization of its various enzymes, study of partial reactions, and the reconstitution of the complete system from purified components. The use of molecular genetic methods in yeast not only confirmed the conclusions of the biochemical work, but added further important dimensions by the cloning of the genes of ubiquitin and of various enzymes of the system, the study of their functions by gene disruption and mutation, and the use of site-directed mutagenesis of protein substrates to define the selectivity of the system. It may be hoped that the same two complementing approaches will help in the future to achieve significant progress in the elucidation of major problems in this field. As is often the case, the solving of some questions raises many more new questions. Since it is now clear that the N-terminal signal is not involved in the degradation of most cellular proteins, the identification of other degradation signals in proteins is of obvious importance. Equally important is the identification of enzymes that recognize these signals, such as new species of E_3s, or E_2s, or both. This last objective is certainly approachable by biochemical or genetic searches for specific enzymes required for the breakdown of rapidly degraded specific cellular proteins. Among other major open questions, notable is the elucidation of the mode of action of the 26S protease complex, including the characterization of the functions of CF-1 and CF-2, the identification of the ubiquitin-protein conjugate binding site, and the elucidation of the roles of ATP in complex assembly and proteolysis. The identification of the functions of different ubiquitin C-terminal hydrolases in protein degradation, the integration of their action with that of the 26S protease complex, and the possible editing function of some hydrolases to salvage incorrectly ubiquitinated proteins remain to be explored.

An important direction for future research is the elucidation of the mechanisms of the regulation of the degradation of specific proteins. As in the case of cell-cycle-controlled degradation of cyclins, it may be expected that the programmed destruction of regulatory proteins ensures irreversibility in the progression of temporally controlled processes. To gain an understanding of the regulatory mechanisms involved, the specific ubiquitin ligases that act on these proteins will have to be identified. Possible signals recognized by such enzymes (such as the putative "destruction box" in cyclins) will have to be characterized, and possible regulatory mechanisms, such as the modulation of ligase activity or substrate accessibility, will have to be defined. The integration of the control of the breakdown of such regulatory proteins with the general regulatory networks of temporally controlled processes may thus be elucidated.

NOTE ADDED IN PROOF   Recent work from several laboratories suggests the possible involvement of the 20S protease complex, or of a related particle, in the processing of cytoplasmic protein antigens to peptides for presentation at the cell surface by products of the major histocompatibility complex (MHC). Two genes located at the MHC region have sequence homology to certain subunits of the 20S protease complex (211–213), and the products of these genes are associated with a structure very similar to the 20S complex (214–216). Furthermore, these subunits are inducible by τ-interferon (211, 215, 216), as are the other components of the antigen presentation system. Thus, a subpopulation of 20S complex that contains some MHC-encoded subunits may be involved in antigen processing. While the results are strongly suggestive, there is no direct evidence yet for the above conclusion. It also remains to be seen whether such an antigen-processing 20S particle is a component of a ubiquitin-specific 26S protease complex.

ACKNOWLEDGMENTS

We thank Dr. Irwin A. Rose for comments on the manuscript and Mary Williamson for excellent secretarial assistance. Thanks are due to many colleagues who provided manuscripts prior to publication. Work in the laboratory of A. H. was supported by NIH grant DK-25614 and grants from the United States-Israel Binational Science Foundation. Work of A. H. at the Institute for Cancer Research, Philadelphia, was supported by American Cancer Society Grant BC596 to Irwin A. Rose. Work in the laboratory of A. C. is supported by grants from the United States-Israel Binational Science Foundation (BSF), the German-Israeli Foundation for Scientific Research and Development (GIF), the Israel Cancer Society, the Israel Cancer Research Fund (ICRF), USA, Monsanto, Inc., and the Fund for Basic Research administered by The Israeli Academy of Sciences and Humanities. A. C. is a research career development awardee (RCDA) of the ICRF, USA.

Literature Cited

1. Schoenheimer, R. 1942. *The Dynamic State of Body Constituents*, pp. 25–46. Cambridge, Mass: Harvard Univ. Press
2. Ciechanover, A., Hod, Y., Hershko, A. 1978. *Biochem. Biophys. Res. Commun.* 81:1100–5
3. Ciechanover, A., Heller, H., Elias, S., Haas, A. L., Hershko, A. 1980. *Proc. Natl. Acad. Sci. USA* 77:1365–68
4. Hershko, A., Ciechanover, A., Heller, H., Haas, A. L., Rose, I. A. 1980. *Proc. Natl. Acad. Sci. USA* 77:1783–86
5. Rechsteiner, M. 1987. *Annu. Rev. Cell Biol.* 3:1–30
6. Hershko, A. 1988. *J. Biol. Chem.* 263:15237–40
7. Hershko, A., Ciechanover, A. 1982. *Annu. Rev. Biochem.* 51:335–64
8. Hershko, A., Ciechanover, A. 1986. *Prog. Nucleic Acid Res. Mol. Biol.* 33:19–56
9. Ciechanover, A., Schwartz, A. L. 1989. *Trends Biochem. Sci.* 14:483–88
10. Jentsch, S., Seufert, W., Sommer, T., Reins, H.-A. 1990. *Trends Biochem. Sci.* 15:195–98
11. Hershko, A. 1991. *Trends Biochem. Sci.* 16:265–68
12. Rechsteiner, M. 1991. *Cell* 66:615–18
13. Rechsteiner, M., ed. 1988. *Ubiquitin.* New York: Plenum
14. Schlesinger, M. J., Hershko, A., eds. 1988. *The Ubiquitin System.* Cold Spring Harbor, NY: Cold Spring Harbor Lab.
15. Finley, D., Chau, V. 1991. *Annu. Rev. Cell Biol.* 7:25–69
16. Jentsch, S., Seufert, W., Hauser, H.-P. 1991. *Biochim. Biophys. Acta* 1089:127–39
17. Mayer, R. J., Arnold, J., László, L., Landon, M., Lowe, J. 1991. *Biochim. Biophys. Acta* 1089:141–57
18. Dice, J. F. 1990. *Trends Biochem. Sci.* 15:305–9
19. Gottesman, S. 1989. *Annu. Rev. Genet.* 23:163–98
20. Goldberg, A. L. 1990. *Sem. Cell Biol.* 1:423–32
21. Goldberg, A. L. 1992. *Eur. J. Biochem.* 203:9–23
22. Hershko, A., Heller, H. 1985. *Biochem. Biophys. Res. Commun.* 128:1079–86
23. Chau, V., Tobias, J. W., Bachmair, A., Marriott, D., Ecker, D. J., et al. 1989. *Science* 243:1576–83
24. Chen, Z., Pickart, C. M. 1990. *J. Biol. Chem.* 265:21835–42
25. Ciechanover, A., Elias, S., Heller, H., Hershko, A. 1982. *J. Biol. Chem.* 257:2537–42
26. Ciechanover, A., Heller, H., Katz-Etzion, R., Hershko, A. 1981. *Proc. Natl. Acad. Sci. USA* 78:761–65
27. Haas, A. L., Rose, I. A. 1982. *J. Biol. Chem.* 257:10329–37
28. Haas, A. L., Warms, J. V. B., Rose, I. A. 1983. *Biochemistry* 22:4388–94
29. Pickart, C. M. 1988. See Ref. 13, pp. 77–99
30. McGrath, J. P., Jentsch, S., Varshavsky, A. 1991. *EMBO J.* 10:227–36
31. Hatfield, P. M., Callis, J., Vierstra, R. D. 1990. *J. Biol. Chem.* 265:15813–17
32. Handley, P. M., Mueckler, M., Siegel, N. R., Ciechanover, A., Schwartz, A. L. 1991. *Proc. Natl. Acad. Sci. USA* 88:258–62
33. Zacksenhaus, E., Sheinin, R. 1990. *EMBO J.* 9:2923–29
34. Hatfield, P. M., Vierstra, R. D. 1989. *Biochemistry* 28:735–42
35. Wierenga, R. K., Hol, W. G. J. 1983. *Nature* 302:842–44
36. Hannink, M., Donoghue, D. J. 1985. *Proc. Natl. Acad. Sci. USA* 82:7894–98
37. Dever, T. E., Glynias, M. J., Merrick, W. C. 1987. *Proc. Natl. Acad. Sci. USA* 84:1814–18
38. Hall, M. N., Hereford, L., Herskowitz, I. 1984. *Cell* 36:1057–65
39. Cook, J. C., Chock, P. B. 1991. *Biochem. Biophys. Res. Commun.* 174:564–71
40. Miller, H. I., Henzel, W. J., Ridgway, J. B., Kuang, W.-J., Chisholm, V., Liu, C.-C. 1989. *Bio/Technology* 7:698–704
41. Tobias, J. W., Varshavsky, A. 1991. *J. Biol. Chem.* 266:12021–28
42. Bartel, B., Wünning, I., Varshavsky, A. 1990. *EMBO J.* 9:3179–89
43. Finley, D., Ciechanover, A., Varshavsky, A. 1984. *Cell* 37:43–55
44. Ciechanover, A., Finley, D., Varshavsky, A. 1984. *Cell* 37:57–66
45. Kulka, R. G., Raboy, B., Schuster, R., Parag, H. A., Diamond, G., et al. 1988. *J. Biol. Chem.* 263:15726–31
46. Eki, T., Enomoto, T., Miyajima, A., Miyazawa, H., Murakami, Y., et al. 1990. *J. Biol. Chem.* 265:26–33
47. Ohtsubo, M., Nishimoto, T. 1988. *Biochem. Biophys. Res. Commun.* 153:1173–78

48. Kudo, M., Sugasawa, K., Hori, T.-A., Enomoto, T., Hanaoka, F., Ui, M. 1991. *Exp. Cell Res.* 192:110–17
49. Zacksenhaus, E., Sheinin, R., Wang, H. S. 1990. *Cytogenet. Cell Genet.* 53:20–22
50. Gropper, R., Brandt, R. A., Elias, S., Bearer, C. F., Mayer, A., et al. 1991. *J. Biol. Chem.* 266:3602–10
51. Mayer, A., Gropper, R., Schwartz, A. L., Ciechanover, A. 1989. *J. Biol. Chem.* 264:2060–68
52. Deveraux, Q., Wells, R., Rechsteiner, M. 1990. *J. Biol. Chem.* 265:6323–29
53. Pickart, C. M., Rose, I. A. 1985. *J. Biol. Chem.* 260:1573–81
54. Klemperer, N. S., Berleth, E. S., Pickart, C. M. 1989. *Biochemistry* 28:6035–41
55. Haas, A. L., Bright, P. M. 1988. *J. Biol. Chem.* 263:13258–67
56. Haas, A. L., Bright, P. M., Jackson, V. E. 1988. *J. Biol. Chem.* 263:13268–75
57. Pickart, C. M., Vella, A. T. 1988. *J. Biol. Chem.* 263:15076–82
58. Chen, Z., Niles, E. G., Pickart, C. M. 1991. *J. Biol. Chem.* 266:15698–704
59. Qin, S., Nakajima, B., Nomura, M., Arfin, S. M. 1991. *J. Biol. Chem.* 266:15549–54
60. Seufert, W., Jentsch, S. 1990. *EMBO J.* 9:543–50
61. Seufert, W., McGrath, J. P., Jentsch, S. 1990. *EMBO J.* 9:4535–41
62. Reynolds, P., Weber, S., Prakash, L. 1985. *Proc. Natl. Acad. Sci. USA* 82: 168–72
63. Jentsch, S., McGrath, J. P., Varshavsky, A. 1987. *Nature* 329:131–34
64. Goebl, M. G., Yochem, J., Jentsch, S., McGrath, J. P., Varshavsky, A., Byers, B. 1988. *Science* 241:1331–35
65. Haynes, R. H., Kunz, B. A. 1981. In *The Molecular Biology of the Yeast Saccharomyces cerevisiae: Life Cycle and Inheritance,* ed. J. N. Strathern, E. W. Jones, J. R. Broach, pp. 371–414. Cold Spring Harbor, NY: Cold Spring Harbor Lab.
66. Sung, P., Prakash, S., Prakash, L. 1990. *Proc. Natl. Acad. Sci. USA* 87: 2695–99
67. Sung, P., Prakash, S., Prakash, L. 1988. *Genes Dev.* 2:1476–85
68. Morrison, A., Miller, E. J., Prakash, L. 1988. *Mol. Cell. Biol.* 8:1179–85
69. Reynolds, P., Koken, M. H. M., Hoeijmakers, J. H. J., Prakash, S., Prakash, L. 1990. *EMBO J.* 9:1423–30
70. Sung, P., Berleth, E., Pickart, C., Prakash, S., Prakash, L. 1991. *EMBO J.* 10:2187–93
71. Dohmen, R. J., Madura, K., Bartel, B., Varshavsky, A. 1991. *Proc. Natl. Acad. Sci. USA* 88:7351–55
72. Reiss, Y., Heller, H., Hershko, A. 1989. *J. Biol. Chem.* 264:10378–83
73. Schneider, R., Eckerskorn, C., Lottspeich, F., Schweiger, M. 1990. *EMBO J.* 9:1431–35
74. Sharon, G., Raboy, B., Parag, H. A., Dimitrovsky, D., Kulka, R. G. 1991. *J. Biol. Chem.* 266:15890–94
75. Pelham, H. R. B. 1986. *Cell* 46:959–61
76. Haas, A. L., Reback, P. B., Chau, V. 1991. *J. Biol. Chem.* 266:5104–12
77. Hershko, A., Heller, H., Elias, S., Ciechanover, A. 1983. *J. Biol. Chem.* 258:8206–14
78. Hershko, A., Heller, H., Eytan, E., Reiss, Y. 1986. *J. Biol. Chem.* 261: 11992–99
79. Reiss, Y., Kaim, D., Hershko, A. 1988. *J. Biol. Chem.* 263:2693–98
80. Reiss, Y., Hershko, A. 1990. *J. Biol. Chem.* 265:3685–90
81. Gonen, H., Schwartz, A. L., Ciechanover, A. 1991. *J. Biol. Chem.* 266: 19221–31
82. Gonda, D. K., Bachmair, A., Wünning, I., Tobias, J. W., Lane, W. S., Varshavsky, A. 1989. *J. Biol. Chem.* 264: 16700–12
83. Heller, H., Hershko, A. 1990. *J. Biol. Chem.* 265:6532–35
84. Hough, R., Rechsteiner, M. 1984. *Proc. Natl. Acad. Sci. USA* 81:90–94
85. Hershko, A., Leshinsky, E., Ganoth, D., Heller, H. 1984. *Proc. Natl. Acad. Sci. USA* 81:1619–23
86. Hough, R., Pratt, G., Rechsteiner, M. 1986. *J. Biol. Chem.* 261:2400–8
87. Waxman, L., Fagan, J. M., Goldberg, A. L. 1987. *J. Biol. Chem.* 262:2451–57
88. Hough, R., Pratt, G., Rechsteiner, M. 1987. *J. Biol. Chem.* 262:8303–13
89. Fagan, J. M., Waxman, L., Goldberg, A. L. 1987. *Biochem. J.* 243:335–43
90. Okada, M., Ishikawa, M., Mizushima, Y. 1991. *Biochim. Biophys. Acta* 1073: 514–20
91. Orino, E., Tanaka, K., Tamura, T., Sone, S., Ogura, T., Ichihara, A. 1991. *FEBS Lett.* 284:206–10
92. Ganoth, D., Leshinsky, E., Eytan, E., Hershko, A. 1988. *J. Biol. Chem.* 263:12412–19
93. Hough, R., Pratt, G., Rechsteiner, M. 1988. See Ref. 13, pp. 101–34
94. Rivett, A. J. 1989. *Arch. Biochem. Biophys.* 268:1–8
95. Orlowski, M. 1990. *Biochemistry* 29: 10289–97

96. Wilk, S., Orlowski, M. 1983. *J. Neurochem.* 40:842–49
97. Dahlmann, B., Kuehn, L., Rutschmann, M., Reinauer, H. 1985. *Biochem. J.* 228:161–70
98. Rose, I. A., Warms, J. V. B., Hershko, A. 1979. *J. Biol. Chem.* 254:8135–38
99. DeMartino, G. N., Goldberg, A. L. 1979. *J. Biol. Chem.* 254:3712–15
100. McGuire, M. J., DeMartino, G. N. 1986. *Biochim. Biophys. Acta* 873:279–89
101. de Sa, C. M., de Sa, M.-F. G., Akhayat, O., Broders, F., Scherrer, K., et al. 1986. *J. Mol. Biol.* 187:479–93
102. Falkenburg, P.-E., Haass, C., Kloetzel, P.-M., Niedel, B., Kopp, F., et al. 1988. *Nature* 331:190–92
103. Arrigo, A.-P., Tanaka, K., Goldberg, A. L., Welch, W. J. 1988. *Nature* 331:192–94
104. Kopp, F., Steiner, R., Dahlmann, B., Kuehn, L., Reinauer, H. 1986. *Biochim. Biophys. Acta* 872:253–60
105. Dahlmann, B., Rutschmann, M., Kuehn, L., Reinauer, H. 1985. *Biochem. J.* 228:171–77
106. Tanaka, K., Ii, K., Ichihara, A., Waxman, L., Goldberg, A. L. 1986. *J. Biol. Chem.* 261:15197–203
107. Fujiwara, T., Tanaka, K., Kumatori, A., Shin, S., Yoshimura, T., et al. 1989. *Biochemistry* 28:7332–40
108. Tanaka, K., Fujiwara, T., Kumatori, A., Shin, S., Yoshimura, T., et al. 1990. *Biochemistry* 29:3777–85
109. Deleted in proof
110. Tamura, T., Tanaka, K., Kumatori, A., Yamada, F., Tsurumi, C., et al. 1990. *FEBS Lett.* 264:91–94
111. Kumatori, A., Tanaka, K., Tamura, T., Fujiwara, T., Ichihara, A., et al. 1990. *FEBS Lett.* 264:279–82
112. Tanaka, K., Kanayama, H., Tamura, T., Lee, D. H., Kumatori, A., et al. 1990. *Biochem. Biophys. Res. Commun.* 171:676–83
113. DeMartino, G. N., Orth, K., McCullough, M. L., Lee, L. W., Munn, T. Z., et al. 1991. *Biochim. Biophys. Acta.* 1079:29–38
114. Haass, C., Pesold-Hurt, B., Multhaup, G., Beyreuther, K., Kloetzel, P. M. 1989. *EMBO J.* 8:2373–79
115. Haass, C., Pesold-Hurt, B., Kloetzel, P. M. 1990. *Nucleic Acids Res.* 18:4018
116. Haass, C., Pesold-Hurt, B., Multhaup, G., Beyreuther, K., Kloetzel, P. M. 1990. *Gene* 90:235–41
117. Fujiwara, T., Tanaka, K., Orino, E., Yoshimura, T., Kumatori, A., et al. 1990. *J. Biol. Chem.* 265:16604–13
118. Emori, Y., Tsukahara, T., Kawasaki, H., Ishiura, S., Sugita, H., Suzuki, K. 1991. *Mol. Cell. Biol.* 11:344–53
119. Heinemeyer, W., Kleinschmidt, J. A., Saidowsky, J., Escher, C., Wolf, D. H. 1991. *EMBO J.* 10:555–62
120. McGuire, M. J., Croall, D. E., DeMartino, G. N. 1988. *Arch. Biochem. Biophys.* 262:273–85
121. McGuire, M. J., DeMartino, G. N. 1989. *Biochem. Biophys. Res. Commun.* 160:911–16
122. Tanaka, K., Ichihara, A. 1988. *FEBS Lett.* 236:159–62
123. Matthews, W., Tanaka, K., Driscoll, J., Ichihara, A., Goldberg, A. L. 1989. *Proc. Natl. Acad. Sci. USA* 86:2597–601
124. DeMartino, G. N., McCullough, M. L., Reckelhoff, J. F., Croall, D. E., Ciechanover, A., McGuire, M. J. 1991. *Biochim. Biophys. Acta* 1073:299–308
125. Eytan, E., Ganoth, D., Armon, T., Hershko, A. 1989. *Proc. Natl. Acad. Sci. USA* 86:7751–55
126. Driscoll, J., Goldberg, A. L. 1990. *J. Biol. Chem.* 265:4789–92
127. Eytan, E., Hershko, A. 1984. *Biochem. Biophys. Res. Commun.* 122:116–23
128. Estrela, J. M., Goldberg, A. L. 1992. *J. Biol. Chem.* In press
129. Estrela, J. M., Goldberg, A. L. 1992. *J. Biol. Chem.* In press
130. Driscoll, J., Goldberg, A. L. 1989. *Proc. Natl. Acad. Sci. USA* 86:787–91
131. Castaño, J. G., Ornberg, R., Koster, J. G., Tobian, J. A., Zasloff, M. 1986. *Cell* 46:377–85
132. Domae, N., Harmon, F. R., Busch, R. K., Spohn, W., Subrahmanyam, C. S., Busch, H. 1982. *Life Sci.* 30:469–77
133. Grainger, J. L., Winkler, M. M. 1989. *J. Cell. Biol.* 109:675–83
134. Kumatori, A., Tanaka, K., Inamura, N., Sone, S., Ogura, T., et al. 1990. *Proc. Natl. Acad. Sci. USA* 87:7071–75
135. Tanaka, K., Yoshimura, T., Tamura, T., Fujiwara, T., Kumatori, A., Ichihara, A. 1990. *FEBS Lett.* 271:41–46
136. Katayama, Y., Gottesman, S., Pumphrey, J., Rudikoff, S., Clark, W. P., Maurizi, M. R. 1988. *J. Biol. Chem.* 263:15226–36
137. Woo, K. M., Chung, W. J., Ha, D. B., Goldberg, A. L., Chung, C. H. 1989. *J. Biol. Chem.* 264:2088–91
138. Armon, T., Ganoth, D., Hershko, A. 1990. *J. Biol. Chem.* 265:20723–26
139. Seddon, A. P., Meister, A. 1986. *J. Biol. Chem.* 261:11538–43
140. Li, L., Seddon, A. P., Meister, A. 1987. *J. Biol. Chem.* 262:11020–25

141. Rose, I. A. 1988. See Ref. 13, pp. 135–55
142. Mueller, R. D., Yasuda, H., Hatch, C. L., Bonner, W. M., Bradbury, E. M. 1985. J. Biol. Chem. 260:5147–53
143. Andersen, M. W., Ballal, N. R., Goldknopf, I. L., Busch, H. 1981. Biochemistry 20:1100–4
144. Andersen, M. W., Goldknopf, I. L., Busch, H. 1981. FEBS Lett. 132:210–14
145. Matsui, S.-I., Sandberg, A. A., Negoro, S., Seon, B. K., Goldstein, G. 1982. Proc. Natl. Acad. Sci. USA 79:1535–39
146. Kanda, F., Sykes, D. E., Yasuda, H., Sandberg, A. A., Matsui, S.-I. 1986. Biochim. Biophys. Acta 870:64–75
147. Rose, I. A., Warms, J. V. B. 1983. Biochemistry 22:4234–37
148. Pickart, C. M., Rose, I. A. 1985. J. Biol. Chem. 260:7903–10
149. Pickart, C. M., Rose, I. A. 1986. J. Biol. Chem. 261:10210–217
150. Hershko, A., Rose, I. A. 1987. Proc. Natl. Acad. Sci. USA 84:1829–33
151. Mayer, A. N., Wilkinson, K. D. 1989. Biochemistry 28:166–72
152. Wilkinson, K. D., Lee, K. M., Deshpande, S., Duerksen-Hughes, P., Boss, J. M., Pohl, J. 1989. Science 246:670–73
153. Jackson, P., Thompson, R. J. 1981. J. Neurol. Sci. 49:429–38
153a. Liu, C.-C., Miller, H. I., Kohr, W. J., Silber, J. I. 1989. J. Biol. Chem. 264:20331–38
154. Jonnalagadda, S., Butt, T. R., Monia, B. P., Mirabelli, C. K., Gotlib, L., et al. 1989. J. Biol. Chem. 264:10637–42
155. Hadari, T., Warms, J. V. B., Rose, I. A., Hershko, A. 1992. J. Biol. Chem. 267:719–27
156. Agell, N., Ryan, C., Schlesinger, M. J. 1991. Biochem. J. 273:615–20
157. Rose, I. A. 1988. See Ref. 14, pp. 111–14
158. Hershko, A., Heller, H., Eytan, E., Kaklij, G., Rose, I. A. 1984. Proc. Natl. Acad. Sci. USA 81:7021–25
159. Bachmair, A., Finley, D., Varshavsky, A. 1986. Science 234:179–86
160. Bachmair, A., Varshavsky, A. 1989. Cell 56:1019–32
161. Ferber, S., Ciechanover, A. 1987. Nature 326:808–11
162. Ciechanover, A., Ferber, S., Ganoth, D., Elias, S., Hershko, A., Arfin, S. 1988. J. Biol. Chem. 263:11155–67
163. Soffer, R. L. 1980. In Transfer RNA: Biological Aspects, ed. D. Söll, J. Abelson, P. R. Schimmel, pp. 493–505.

Cold Spring Harbor, NY: Cold Spring Harbor Lab.
164. Balzi, E., Choder, M., Chen, W., Varshavsky, A., Goffeau, A. 1990. J. Biol. Chem. 265:7464–71
165. Elias, S., Ciechanover, A. 1990. J. Biol. Chem. 265:15511–17
166. Sokolik, C. W., Cohen, R. E. 1991. J. Biol. Chem. 266:9100–7
167. Johnson, E. S., Gonda, D. K., Varshavsky, A. 1990. Nature 346:287–91
168. Dunten, R. L., Cohen, R. E., Gregori, L., Chau, V. 1991. J. Biol. Chem. 266:3260–67
168a. Dunten, R. L., Cohen, R. E. 1989. J. Biol. Chem. 264:16739–47
169. Haas, A., Reback, P. M., Pratt, G., Rechsteiner, M. 1990. J. Biol. Chem. 265:21664–69
170. Hershko, A., Ganoth, D., Pehrson, J., Palazzo, R. E., Cohen, L. H. 1991. J. Biol. Chem. 266:16376–79
171. Gregori, L., Poosch, M. S., Cousins, G., Chau, V. 1990. J. Biol. Chem. 265:8354–57
172. Moerschell, R. P., Hosokawa, Y., Tsunasawa, S., Sherman, F. 1990. J. Biol. Chem. 265:19638–43
173. Sherman, F., Moerschell, R. P., Tsunasawa, S., Sternglanz, R., Dumont, M. E. 1992. In Translational Regulation of Gene Expression II, ed. J. Ilan. New York: Plenum. In press
174. Müller, M., Dubiel, W., Rothmann, J., Rapoport, S. 1980. Eur. J. Biochem. 109:405–10
175. Brown, J. L., Roberts, W. K. 1976. J. Biol. Chem. 251:1009–14
176. Mayer, A., Siegel, N. R., Schwartz, A. L., Ciechanover, A. 1989. Science 244:1480–83
177. Rogers, S., Wells, R., Rechsteiner, M. 1986. Science 234:364–68
178. Rechsteiner, M. 1990. Semin. Cell Biol. 1:433–40
179. Ghoda, L., van Daalen Wetters, T., Macrae, M., Ascherman, D., Coffino, P. 1989. Science 243:1493–95
180. Rosenberg-Hasson, Y., Bercovich, Z., Kahana, C. 1991. Eur. J. Biochem. 196:647–51
181. Loetscher, P., Pratt, G., Rechsteiner, M. 1991. J. Biol. Chem. 266:11213–20
182. Bercovich, Z., Rosenberg-Hasson, Y., Ciechanover, A., Kahana, C. 1989. J. Biol. Chem. 264:15949–52
183. Rosenberg-Hasson, Y., Bercovich, Z., Ciechanover, A., Kahana, C. 1990. Eur. J. Biochem. 185:469–74
184. Pratt, L. H. 1978. Photochem. Photobiol. 27:81–105

185. Quail, P. H. 1984. *Trends Biochem. Sci.* 9:450–53
186. Shanklin, J., Jabben, M., Vierstra, R. D. 1987. *Proc. Natl. Acad. Sci. USA* 84:359–63
187. Jabben, M., Shanklin, J., Vierstra, R. D. 1989. *Plant Physiol.* 90:380–84
188. Jabben, M., Shanklin, J., Vierstra, R. D. 1989. *J. Biol. Chem.* 264:4998–5005
189. Lüscher, B., Eisenman, R. N. 1988. *Mol. Cell. Biol.* 8:2504–12
190. Gronostajski, R. M., Goldberg, A. L., Pardee, A. B. 1984. *Mol. Cell. Biol.* 4:442–48
191. Ciechanover, A., DiGiuseppe, J. A., Bercovich, B., Orian, A., Richter, J. D., et al. 1991. *Proc. Natl. Acad. Sci. USA* 88:139–43
192. Slavicek, J. M., Jones, N. C., Richter, J. D. 1988. *EMBO J.* 7:3171–80
193. Scheffner, M., Werness, B. A., Huibregtse, J. M., Levine, A. J., Howley, P. M. 1990. *Cell* 63:1129–36
194. Werness, B. A., Levine, A. J., Howley, P. M. 1990. *Science* 248:76–79
195. Nasmyth, K., Shore, D. 1987. *Science* 237:1162–70
196. Hochstrasser, M., Varshavsky, A. 1990. *Cell* 61:697–708
197. Hochstrasser, M., Ellison, M. J., Chau, V., Varshavsky, A. 1991. *Proc. Natl. Acad. Sci. USA* 88:4606–10
198. Minshull, J., Pines, J., Golsteyn, R., Standart, N., Mackie, S., et al. 1989. *J. Cell. Sci. Suppl.* 12:77–97
199. Lewin, B. 1990. *Cell* 61:743–52
200. Pines, J. 1991. *Cell Growth Differ.* 2:305–10
201. Evans, T., Rosenthal, E. T., Young-blom, J., Distel, D., Hunt, T. 1983. *Cell* 33:389–96
202. Murray, A. W., Kirschner, M. W. 1989. *Nature* 339:275–80
203. Murray, A. W., Solomon, M. J., Kirschner, M. W. 1989. *Nature* 339:280–86
204. Luca, F. C., Ruderman, J. V. 1989. *J. Cell. Biol.* 109:1895–909
205. Glotzer, M., Murray, A. W., Kirschner, M. W. 1991. *Nature* 349:132–38
206. Meijer, L., Arion, D., Golsteyn, R., Pines, J., Brizuela, L., et al. 1989. *EMBO J.* 8:2275–82
207. Doherty, F. J., Osborn, N. U., Wassell, J. A., Heggie, P. E., László, L., Mayer, R. J. 1989. *Biochem. J.* 263:47–55
208. László, L., Doherty, F. J., Osborn, N. U., Mayer, R. J. 1990. *FEBS Lett.* 261:365–68
209. Schwartz, A. L., Ciechanover, A., Brandt, R. A., Geuze, H. J. 1988. *EMBO J.* 7:2961–66
210. Johnston, N. L., Cohen, R. E. 1991. *Biochemistry* 30:7514–22
211. Glynne, R., Powis, S. H., Beck, S., Kelly, A., Kerr, L.-A., Trowsdale, J. 1991. *Nature* 353:357–60
212. Martinez, C. K., Monaco, J. J. 1991. *Nature* 353:664–67
213. Kelly, A., Powis, S. H., Glynne, R., Radley, E., Beck, S., Trowsdale, J. 1991. *Nature* 353:667–68
214. Parham, P. 1990. *Nature* 348:674–75
215. Brown, M. G., Driscoll, J., Monaco, J. J. 1991. *Nature* 353:355–57
216. Ortiz-Navarrete, V., Seelig, A., Gernold, M., Frentzel, S., Kloetzel, P. M., Hammerling, G. J. 1991. *Nature* 353:662–64

Annu. Rev. Biochem. 1992. 61:809–60

myc FUNCTION AND REGULATION

Kenneth B. Marcu[1,2,3], Steven A. Bossone[2], and Amanda J. Patel[1]

Departments of Biochemistry and Cell Biology[1], Pathology[2], and Microbiology[3], State University of New York at Stony Brook, Stony Brook, New York 11794

KEY WORDS: nuclear oncogenes and malignancy, cell growth and differentiation, gene regulation, leucine zipper proteins, helix-loop-helix proteins

CONTENTS

0066-4154/92/0701-0809$02.00

PERSPECTIVES

For an oncogene, *myc* has a rather pivotal spot in the history of the molecular genetics of neoplastic disease. It was only about 10 years ago that *myc* represented the first example of insertional mutagenesis in malignancy with avian leukosis virus (ALV)–induced B cell lymphomagenesis in chickens (1). This seminal discovery, along with the prior identification of the chicken *myc* gene as the captured cellular coding sequence of the acutely transforming avian myelocytomatosis virus (AMV), led to a variety of similar findings on a number of cellular genes with oncogenetic potential (2–4). Soon thereafter, the mammalian homologues of the avian *myc* gene made their way to the scene. An unexpected series of events in a number of laboratories placed the mammalian *c-myc* locus at the breakpoints of nonrandom chromosome translocations with immunoglobulin loci, in Burkitt lymphomas in humans and plasmacytomas in mice (5–9). Eight to nine years ago, *myc* seemed like a neatly wrapped package to all its investigators. The corroborative evidence was overwhelming that *myc* was an oncogene that played a critical if not essential role in the genesis of avian and mammalian B lymphoid malignancies by chromosomal rearrangement mechanisms. However, that was only the beginning of the *myc* saga. A host of research groups have descended upon *c-myc* in the past 10 years, exploring its complex regulation in a variety of cellular contexts, and the biochemical activities and cellular fates of its polypeptide (7–12). These efforts have generated an impressive amount of data, but most importantly, a number of paradigms for the study of cellular genes that regulate themselves and/or other genes.

Studies of *c-myc* have touched the fields of normal and abnormal cell growth and differentiation, in vitro in culture and in vivo in animals. The controversial biochemical properties of the *myc* polypeptide have yielded proposals for its potential roles in the transcriptional and posttranscriptional control of other cellular genes and in DNA replication. The greatest leap in the quest to decipher the "*myc* enigma" arose unexpectedly from the discovery of two classes of transcription factors operationally defined by virtue of one of two common domains of limited structural similarity: the leucine zipper (LZ) (13) and the helix-loop-helix (H-L-H) motif (14). Proteins containing LZ and H-L-H domains have been shown in numerous cases to be sequence-specific DNA-binding factors that regulate the transcription of genes by RNA polymerase II (15). The LZ consists of a 20–30-residue amphipathic α-helix, with a leucine positioned at every seventh residue creating a hydrophobic interface on one side of the helix (13). Zipper proteins also possess a helical stretch of basic amino acids, called the basic region (BR), spaced six residues amino-terminal to their LZ domains. The spatial arrangement of the (BR)LZ domain is related to those of the (BR)H-L-H domains of other transcription factors (15). The H-L-H domain consists of two amphipathic α-helices of about 15 amino acids, separated by a spacer segment of varying length which is preceded by a basic region of about 13 amino acids rich in arginines (14). The amphipathic LZ and H-L-H domains are indirectly required for the stable juxtapositioning of two basic regions that facilitate the formation of a homo or heterodimeric protein complex with a specific DNA sequence (13–15). The *c-myc* polypeptides contain (BR)H-L-H and LZ domains, and strong evidence is accumulating that they mediate their functions as site-specific DNA-binding proteins. The *myc* polypeptide has significant effects on the growth and differentiation of a variety of diverse cell types, and if expressed inappropriately, the normal protein also contributes to neoplasia. Cells keep *myc* under very tight regulation. Here, we have made a valiant effort to review the literature on biological effects mediated directly or indirectly by *myc*, and on the regulation of the gene, its mRNAs, and its polypeptides. The field is vast and at times controversial. We have pointed out where evidence for various phenomena is weak or strong and the direction(s) in which future efforts are likely headed.

myc BELONGS TO A SMALL FAMILY OF HIGHLY RELATED PROTO-ONCOGENES

In 1983 and 1985 reports appeared on the identification of two amplified coding sequences with strong homology to *c-myc* in human neuroblastomas and small cell lung carcinomas, which were dubbed the *N-myc* and *L-myc* genes respectively (12, 16–18). Each of the three mammalian *myc* family genes has the same characteristic three-exon structure with the major polypeptide open reading frame residing in the second and third exons. The first exon of the *myc* genes is not conserved, but possesses regulatory functions, which have been definitively established for the murine and human *c-myc* genes (see below). Highly homologous blocks of amino acids separated by areas of diminished conservation are spread throughout the two coding exons, suggesting early on that *myc* polypeptides may have discreet, independent functional domains (Figure 1).

Other cellular sequences have been identified with comparable homology to some portions of *c-myc*, but available information suggests that they represent pseudogenes (19), truncated genes (20, 21), and in one instance, a developmentally regulated secondary *c-myc* allele (22). *B-myc* contains homology to *c-myc* exon 2, and encodes a 188-amino-acid protein (20). The *B-myc* polypeptide could be analogous to the putative protein encoded by shortened human *L-myc* transcripts that contain only first- and second-exon sequences. *B-myc* is expressed almost ubiquitously, but its functions remain unknown. A fourth gene, *S-myc*, exhibits homology to the second and third

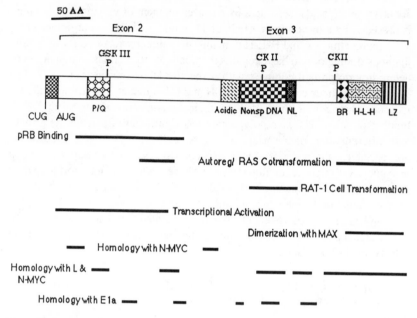

Figure 1 Summary of *c-myc* polypeptide and its functional domains. Abbreviations used: CUG, initiation site within exon; AUG, initiation site defining beginning of exon 2; P/Q, proline/ glutamine-rich region; Nonsp DNA, nonspecific DNA-binding region; NL, nuclear localization signal; BR, basic region; H-L-H, helix-loop-helix protein dimerization domain; LZ, leucine zipper; GSK III, phosphorylation by glycogen synthase kinase; CK II, phosphorylation by casein kinase II.

myc exons, but its protein product is encoded by a single exon (21). Transfection of an *S-myc* gene into a rat tumor line suppressed its malignant phenotype (21). The *Xenopus laevis* genome contains two *c-myc* genes, *c-myc* I and II (22). *c-myc* I and II are extremely similar and mostly differ in their 5' and 3' untranslated regions. More significantly, *c-myc* II is only expressed from the maternal genome during oogenesis, whereas *c-myc* I is the major source of *myc* transcripts during oogenesis and is re-expressed in postgastrula embryos. The significance of *c-myc* II transcripts for *X. laevis* oogenesis and subsequent development remains to be established.

The three bona fide members of the *myc* gene family (*c-*, *N-*, and *L-myc*) are differentially expressed in mammalian development, with *N-myc* and *L-myc* exhibiting a very restricted pattern of tissue and developmental stage specificity (12, 23–28). High levels of *N-myc* and *L-myc* expression have been documented for the early stages of various differentiating cell lineages (24, 25, 27). The amount of *c-myc* expression is by contrast rather generalized, though it can vary considerably (23, 25, 26, 28). In mouse embryos, in situ hybridization has shown *c-myc* to be preferentially expressed in endodermal and mesodermal tissues, while organs developing from ectoderm, including skin, brain, and spinal cord, revealed low levels of *c-myc* RNAs (23, 26). Expression of *c-myc* during development is strongly correlated with proliferation and folding of partially differentiated epithelium (26). *c-myc* expression was also reported to vary with age in mice in a tissue-specific manner with highest levels of expression in newborn and old animals (28). *N-* and *L-myc* expression was only high in a more restricted set of perinatal and newborn tissues (28). Age-associated changes in *c-myc* DNA methylation have also been noted, and are in part correlated with tissue-specific changes in *myc* expression (29). Elevated *c-myc* expression has been associated with a wide variety of malignancies of different cell types (7, 9, 12). On the other hand, *N-myc* is expressed in a restricted set of tumors, which generally display some neural or early differentiated phenotype such as neuroblastomas, retinoblastomas, or embryonal or small cell lung carcinomas (SCLC) (12, 16, 17, 30). *L-myc* expression is restricted to SCLC (18). Elevated levels of *N-* and *L-myc* in malignancies have generally been ascribed to their genomic amplification (12, 16–18, 30). However, this is not always the case, suggesting that *N-* and *L-myc* expression may selectively become deregulated in malignancies of particular cell types, which constitute their normal in vivo sites of expression (12, 31). The three *myc* family genes do not possess distinguishing functional attributes, though their comparisons have been limited in scope (12). Their potencies in fibroblast transformation with cooperating *ras* oncogenes differ, ranked *c-myc* > *N-myc* > *L-myc*, and this is similarly reflected in their potential to elicit lymphoid tumors in transgenic mice (12, 19). Since the manifestation of *myc* functions most likely requires the association of *myc* with "partner proteins" (10), future studies in specific cell types that normally express *L-* and *N-myc* will be essential.

General Structure-Function Properties of c-myc Polypeptides

The human *c-myc* gene encodes two polypeptides (Myc-1 and Myc-2) of 439 and 453 amino acids with apparent molecular weights of 65,000 and 68,000, respectively (10, 32). The *myc* proteins are phosphorylated at multiple sites by casein kinase II (CK II), contain alpha-helical domains and separate areas rich in acidic and basic amino acids, localize to the nucleus, and bind DNA (Figure 1) (10, 33–37). A region of the human *myc* polypeptide encoded in the gene's third exon directs the polypeptide's translocation to the nucleus (Figure 1) (35). This nuclear localization signal has been defined as a nonapeptide, PAAKRVKLD (denoted M1), specified by amino acids 320 to 328 (35). M1 is evolutionarily conserved amongst other *c-myc* genes, and an M1-like peptide is present in *N-myc* but curiously absent in murine and human *L-myc*. M1 resembles the nuclear targeting signal of the polyoma and SV40 large T polypeptides. A second peptide, denoted M2 (RQRRNELKRSF) (amino acids 364 to 374), functions as a surrogate nuclear localization signal in the absence of M1 but is normally not responsible for the protein's cellular distribution (35). M2 like M1 is rich in basic amino acids and is more highly conserved among the *myc* family proteins than is M1. M2 normally functions as a portion of a domain responsible for DNA sequence-specific binding (discussed in more detail below). The human *c-myc* polypeptide also possesses a sequence-nonspecific DNA-binding property specified by amino acids 290 to 318, and this region is conserved among *c-myc* genes of different species (Figure 1) (36, 38). A DNA sequence-specific binding domain is carboxy-terminal to these regions and will be discussed in more detail below. The *myc* polypeptides cofractionate with the nuclear matrix, a nuclear remnant resistant to harsh isolation procedures (39). However, this observation may be only a biologically irrelevant consequence of experimental manipulation (40). Recent work on a *c-myc* protein in the myxomycete physarum polycephalum suggested that *c-myc* transiently associated with the periphery of the nuclear matrix structure during S phase, possibly as a component of the nuclear scaffold (41). The discrepancies in the literature on the association of *myc* with other matrix proteins may be a consequence of cell cycle variation, but nevertheless the biological significance of such associations for *myc* function remains questionable at best. The association of the *myc* polypeptide with matrix components would also be in keeping with a proposed role of *myc* in DNA replication, but this remains controversial (see below). The *myc* polypeptide has been shown to immortalize primary fibroblasts and to cooperate with mutated *ras* oncoproteins to fully transform primary cells to a malignant phenotype (42, 43). The regions of the *myc* polypeptide involved in these and other biological properties (discussed in more detail in subsequent sections) have been mapped and are indicated in Figure 1 (44, 45).

myc-1 and *myc*-2 polypeptides only differ at their amino termini by 14 amino acids, and the human gene's first exon also has the coding potential for a third unrelated protein (32, 51, 52). *myc*-1, the slightly larger variety, is initiated at a CUG codon near the 3' end of the gene's first exon (Figure 1). The remainder of the first exon is a large noncoding sequence, which is remarkably well conserved between the murine and human *myc* genes and has multiple regulatory properties (more details provided in a subsequent section) (7, 9, 11). The smaller, generally more predominant *myc*-2 protein starts at a standard AUG initiation codon at the beginning of the second exon (Figure 1) (32). The *N-myc* gene encodes two phosphoproteins of 58 and 64 kDa, which like *c-myc* differ at their amino termini, but these initiate at two in-frame AUGs at the beginning of the second exon (12, 48, 49). *L-myc* encodes multiple phosphoproteins of 60–66 kDa, which all initiate from one AUG, but phosphatase treatment indicated a single species of 60 kDa (12, 19, 50–52).

Blocks of amino acids are highly conserved between the *myc* family of polypeptides, but these are separated by very divergent regions with *L-myc* the most dissimilar of the three (see Figure 1) (12, 19, 50). Evidence exists for a third protein, *myc*HEX1, of 188 amino acids, encoded by the first exon of the human *c-myc* gene, which unlike the murine gene has an open reading frame (46, 47). Murine cells appear to produce a *myc*HEX1-related protein, but its genomic location remains unknown (53). The *myc*HEX1 reading frame does not overlap with *myc*-1 and *myc*-2 and its function also remains unknown (52, 53). *myc*-1 and *myc*-2 are highly unstable in cells with estimated half-lives of 20–30 minutes (32, 33, 54).

Light at the End of the *myc* Tunnel: A DNA-Binding Site and an Elusive Protein Partner

The *myc* family of oncoproteins possesses a (BR)H-L-H(LZ) chimeric architecture. The (BR)H-L-H domain of the *myc* gene family covers ~55 amino acids positioned about 30 residues from the proteins' carboxy termini; these remaining 30 amino acids constitute their LZ domain (10, 13, 14). Unique functional constraints of the *myc* polypeptides might require an extended second amphipathic helix provided by an appended LZ. Indeed, several recently described transcription factors have (BR)H-L-H(LZ) domains that are spatially similar to that of *myc*: TFE3 (55) and TFEB (56), which both bind to the μE3 sequence motif of immunoglobulin gene enhancers, and USF (57), a cellular factor required for the activity of the adenovirus major late gene promoter. The DNA sequence specificity is not entirely a property of the precise amino acids in the basic region, since protein-protein interactions mediated by the H-L-H and LZ domains can be highly specific with some combinations favored and others highly unfavorable or essentially disallowed (10, 15, 58). For instance, conditions that favor the dimerization of the MyoD and E12 (BR)H-L-H proteins or *fos* and *jun* (BR)LZ proteins do not result in dimers with *myc* (10). Similar negative results have been obtained with USF and *myc* (10). Recombinant *myc* protein did form homodimers and even tetramers, which required an intact LZ, but only under high protein concentrations, implying that such associations may not be physiologically significant (59).

The past year has witnessed great progress, not only in defining the *myc* polypeptide's DNA-binding site, but also in discovering a *myc* partner protein. Blackwell and colleagues crafted a neat polymerase chain reaction (PCR) strategy to identify the DNA-binding site of *myc* homodimers (60). Two assumptions were made. The first, more straightforward assumption was that *myc* homodimers bind a target sequence that contains a CA—TG consensus. These bases are found in the target sequences of many other (BR)H-L-H class transcription factors that bind to the E box motif of transcriptional enhancers (10, 58, 60, 61). The second assumption was more akin to a leap of faith. A novel technique dubbed "selected and amplified binding sequence" (SAAB) imprinting was used to identify the *myc*-binding site amongst a mixture of oligonucleotides that all contained the CA—TG consensus (61). SAAB makes use of the remarkable sensitivity of the PCR to amplify DNA sequences selected by a DNA-binding protein in an electrophoretic mobility shift assay (EMSA), and was initially used to retrieve the previously defined MyoD/E12-binding site out of a partially degenerate mixture (61). However, MyoD/E12 forms a stable, easily detectable protein-DNA complex in an EMSA, but purified *myc* protein is insoluble and forms homodimeric pairs very inefficiently. To solve the latter problem, a carboxy-terminal 92-amino-acid fragment of the human *myc* polypeptide, which contains its (BR)H-L-H(LZ) domain, was fused to the carboxy terminus of glutathione S-transferase (GST). The soluble GST-*c-myc*92 chimera was used in an EMSA with a degenerate E box motif mixture (5'-TCCCCNNNNCANNTGNNNNCTGAT-3'). No band shift was detectable; but after several rounds of PCR, complexes became visible and in sufficient quantity for DNA sequencing to reveal the binding site, 5'-CCCCACCACGTGGTGCCTGA-3'. Interestingly, the core of this SAAB, CACGTG, is identical to the site bound by USF (57, 60). Subsequent to this publication, Prendergast & Ziff reported that a *myc* (BR)-E12 (H-L-H) chimeric protein, which readily formed homodimers, bound the sequence GGCCACGTGACC, which has the same six-base-pair (bp) core revealed in the Blackwell et al study (62). Situations have been independently reported wherein *myc* either directly or indirectly appeared to activate or repress the expression of other cellular genes (discussed in more detail below). Future work will likely be focused towards screening for the CACGTG sequence in the promoters of these putative *myc* target genes.

Blackwood & Eisenman, who collaborated along with Blackwell, Kretzner, and Weintraub to identify the *myc* target sequence, used a radiolabeled GST-*c-myc*92 chimera to screen for a gene encoding a *myc* partner protein in a bacterial expression library of primate cDNAs, and pulled out two overlapping clones (63). These cDNAs encoded a (BR)H-L-H(LZ) protein of just 160 amino acids, which was dubbed *max* for "*myc* Associated X" protein. *myc* and *max* formed stable heterodimers, which required the H-L-H and LZ motifs of each protein and bound about 100-fold more efficiently to the *myc* target sequence (63). The discovery of *max* facilitated the cloning of *myn*, its murine homologue, which was reported by Prendergast et al along with some of the biological properties of *myc/max* heterodimers (64). Cells transfected with *myc* and *myn* displayed a potentiation of the biological properties exhibited by *myc* alone, but this was not observed by transfecting *myn* alone, implying that *myc* is the limiting partner of this functional heterodimer in vivo (64). Other recent observations of Luscher & Eisenmann indicate that (*a*) *max* is considerably more stable than *myc*; (*b*) *myc* is limiting in cells compared to *max;* (*c*) *myc/max* heterodimers are present in cell nuclei (B. Luscher, R. N. Eisenman, unpublished observations). *max* heterodimerized equally well with other *myc* family polypeptides, but not with MyoD (63). It would seem that the unique distances of the H-L-H and LZ domains would determine the specificity of H-L-H(LZ) protein interactions. Other preliminary data have revealed that *myc* transactivated a reporter gene with four tandem copies of its binding site fused upstream of a minimal *HSV-tk* gene promoter, while *max* repressed its basal activity and *myc* + *max* only yielded a modest level of transactivation in fibroblastic cells (L. Kretzner, R. N. Eisenman, personal communication). Will *myc* remain faithful solely to *max*, or will *max* only represent one of *myc*'s promiscuous relations? Future work will likely be directed in part towards searching for other *max*'s in a variety of differentiated cell types where *c-myc* and its immediate family members appear to play important roles in cell fate, and in part towards defining other potential binding sites of *N-myc/max* and *L-myc/max* heterodimers or unique properties of their transcriptional activation domains.

EFFECTS ON AND BY *myc* IN CELL-CYCLE PROGRESSION AND PROLIFERATION

myc belongs to a set of cellular messengers commonly referred to as the "immediate early response" genes because their expression is activated by a variety of mitogenic stimuli, independent of de novo protein synthesis, early during the G0 to G1 transition of cells from a resting to a growing state (65–68). The products of immediate early genes are believed to facilitate a cell's progression through the cycle to eventually achieve DNA synthesis in S phase (69, 70). A number of immediate early genes have been identified and characterized; and like *myc*, some were found to encode nuclear proteins with oncogenic potential that function as transcription factors (e.g. *fos* and *jun*) (66, 67, 71, 72). Modes of *myc* induction have largely been investigated in established fibroblastic cell lines, because their growth stimuli and associated intracellular pathways have been well documented (65–70). Even though *myc* is strongly induced within the first two hours of G1 by most growth stimuli, its expression is not restricted to G1, because its levels are invariant throughout the cycle in continuously proliferating cells (73, 74). Early work elegantly demonstrated *myc* activation in response to platelet-derived growth factor (PDGF), and other substances that mimic the effects of PDGF in fibroblasts including serum, fibroblast growth factor (FGF), and the tumor promoter 12-O-tetradecanoylphorbol-13-acetate (TPA) (66). PDGF receptor occupancy activates phospholipase C, which cleaves phosphatidylinositol to yield the secondary messengers inositol triphosphate and diacylglycerol, which respectively stimulate intracellular calcium release and protein kinase C (PKC) (75–77). PKC is directly activated by TPA, and its desensitization by prolonged TPA exposure was shown to reduce PDGF-mediated *myc* activation by up to 70% in BALB/c-3T3 cells, thereby strongly implicating a PKC pathway in most of the PDGF-mediated effect on *myc* in this particular fibroblastic line (66, 78).

The contribution of the PKC pathway for *myc* activation is not as straightforward in other cell backgrounds or physiological states in which protein kinase A (PKA) or yet other signal transduction pathways are involved. PDGF upregulates *myc* in quiescent Swiss 3T3 cells by increasing intracellular cyclic AMP (cAMP) levels via an indirect mechanism that is also observed with forskolin, a direct activator of adenylate cyclase (79, 80). Additional complexities of *myc* activation routes are evident from recent findings with MG-63 human osteosarcoma cells, wherein PDGF induced *myc* independent of PKC and PKA (81). All of these mitogenic stimuli largely activated *myc* gene transcription, while serum also affected *myc* expression at the posttranscriptional level (delineated in more detail below) (82, 83). Elevation of cAMP alone was sufficient to induce *myc* in the MG-63 line, and this effect was potentiated by the calcium ionophore A23187 (81). Modes of *myc* activation also differed in BALB/c3T3(A31) cells depending on their physiological state prior to mitogenic stimulation. If cells were growth arrested at confluence vs quiescent at subconfluence by serum deprivation prior to receiving mitogens, *myc* and other immediate early genes such as *fos* were induced by stimulators of PKC, while other growth factors required for further progression towards S phase, such as epidermal growth factor (EGF) and insulinlike growth factor 1 (IGF-1), did not activate PKC and poorly induced *myc* and *fos* (82). In contrast, subconfluent, quiescent BALB/c3T3(A31) cells only required signalling provided by EGF and IGF-1 to progress to S phase; and EGF was a more potent *myc* inducer than TPA in these cells (82). Induction of *myc* and *fos* by EGF alone has also been reported for the euploid EL2 rat embryo fibroblast line, which acquired a transformed phenotype under EGF alone (84). The EGF effect in BALB/c3T3 cells is potentiated by substances that elevate intracellular cAMP (cholera toxin and 3-isobutyl-1-methylxanthine), and calcium ionophore A23187 mimicked EGF when provided along with a cAMP inducer (80). However, when administered individually, calcium ionophores and cAMP inducers failed to induce *myc* (80). Furthermore, the EGF route to *myc* activation remains enigmatic given recent observations with C3H10T1/2 mouse fibroblasts (85). In the latter cellular context, activation of *myc* and *fos* by EGF and transforming growth factor alpha (TGF-α), which both employ the EGF receptor, were independent of a transient elevation in intracellular calcium levels (85).

Clearly, differences in cellular physiology must be strongly considered to decipher each intracellular activation pathway, including their immediate nuclear targets leading to *myc* induction.

The induction of *myc* by mitogenic stimuli is not unique to fibroblastic cells, implying that it likely resides at a pivotal point of intracellular growth pathways in most cell types. Early on, *myc* mRNAs were shown to be activated in murine T and B lymphocytes by concanavalin A (Con A) and lipopolysaccharide (LPS), respectively (66), and by phytohemagglutinin (PHA), phorbol 12-myristate 13-acetate (PMA), ionomycin, and OKT3 anti-antigen receptor monoclonal antibody in human peripheral mononuclear cells (PBMC) (66, 86–88). A number of these effects were superinduced by protein synthesis inhibitors, thereby implicating labile repressors in *myc* RNA accumulation (66, 87, 88). Growth factors involved in the progression of human PBMC to S phase, such as interleukin 2 (IL-2), only elevated *myc* expression when provided along with PHA with a 24-hr delay (86, 87). Inhibitors of lymphocyte proliferation, such as cyclosporin A and dexamethasone, suppressed the induction of *myc* by PHA within 3 hr, but anti-IL-2 receptor antibody only inhibited after 24 hr (86). Hydroxyurea and anti-transferrin receptor antibody, which block the transition of cycling cells from G1 to S phase, did not inhibit *myc* but delayed its diminution normally observed just prior to S phase (86). The induction of *myc* in cloned human T lymphocytes was more rapid with IL-2 than Con A and possibly biphasic with IL-2 (89). Since Con A induces T cell lines to produce lymphokines such as IL-2, which induces their proliferation, the transient activation of *myc* was insufficient to induce their growth (89), and this was also the case for human B lymphocytes (90). Stimulation of human B lymphocytes by membrane immunoglobulin (mIg) crosslinking activated *myc*, as did addition of antigen to antigen-specific mouse B cells (90, 91). Interestingly, IgG anti-Ig and Fab2' anti-Ig induced comparable levels of *myc*; but its expression was only maintained by the latter stimulus, which was mitogenic in contrast to the former, providing additional evidence for a *myc* requirement in S phase (91). As opposed to the proliferative effect of mIg crosslinking observed with mature B cells, immature B cell growth was inhibited by similar treatment, reflected in an initial rise followed by a rapid fall of *myc* levels below preinduced levels (91). In keeping with these findings, crosslinking of ex-ogenously introduced mIgD, in contrast to mIgM on immature B cells, delivered a growth signal and also sustained a high level of *myc* activity (92).

Given that competence factors activate *myc* expression, the realization that *myc* functions could replace them was not unexpected, but gratifying never-theless. Armelin and colleagues demonstrated that the activation of an ex-ogenous, inducible *myc* gene imparted to BALB/c-3T3 cells the ability to proliferate in platelet-poor plasma (PPP) (93). After this report, microinjected *myc* polypeptide was directly shown to cooperate with PPP to stimulate DNA synthesis (94); and *myc* was also observed to enhance the responsiveness of the C3H/10T1/2 mouse embryo line to competence (PDGF) and progression (EGF) factors (95) and to enhance EGF mitogenic effects on FR3T3 cells (96). The abrogation of growth factor requirements by *myc* is not unique to fibroblastic cells, because the constitutive overexpression of exogenous *v-myc* and *c-myc* genes differentially increased the viability of interleukin-3 (IL-3)–dependent FDC(P1) myeloid cells in the absence of IL-3, with *v-myc* entirely abrogating their IL-3 requirement (97, 98). Furthermore, FDC(P1) myeloid cells, which have lost their IL-3 dependency, have acquired activated endogenous *c-myc* or *N-myc* genes (98). However, other recent work with an IL-3-dependent, nontumorigenic myeloid line, 32Dc13, revealed that constitutive *c-myc* expression prevented G1 cell cycle arrest upon IL-3 deprivation, but also accelerated cell death instead of improving their viability (J. Cleveland, personal communication). This latter observation suggested that the abrogation of IL-3 dependency is not a direct effect of *c-myc* overexpression. Another line of evidence for a functional association between *myc* activation and cell-cycle progression was recently obtained by noting that heparin, an inhibitor of cellular proliferation, appeared to act at least in part by inhibiting the PKC-mediated induction of *myc* and *fos* distal to PKC activation (99). Cells expressing constitutively high levels of an exogenous *myc* gene cycle more rapidly have a dramatically shortened G1 phase, implying that *myc* levels have a rate-limiting effect on the duration of the G1 phase (100). Indeed, the timed microinjection of *c-myc* antibodies and the induction of an antisense *myc* mRNA after serum stimulation of quiescent Swiss mouse 3T3 cells indicated that the biological effects of the *myc* polypeptide may only be essential during the first six hours of the G1 phase (N. Sullivan, personal communication). Other nuclear oncoproteins, such as *p53*, also replaced the requirement for PDGF, but endogenous *myc* expression remained elevated and actually increased in such cells in response to PPP and insulin (101). This later observation is presumably a consequence of *myc* induction via a progression factor pathway (possibly comparable to EGF and insulin-initiated responses), suggesting that *myc* represented an essential factor for progression to S phase (101). Indeed, recent work using novel chimeras of the *myc* polypeptide and the ligand-binding domain of the human estrogen receptor have elegantly shown that activation of *myc* was sufficient to stimulate DNA synthesis (102, 103). Entry into S phase occurred earlier than in normal serum-stimulated fibroblasts and remarkably, the induction of other immediate early response genes was not observed, suggesting that at least some early events in the cycle were not essential after *myc* activation (103).

EFFECTS ON AND BY *myc* IN DIFFERENTIATION

Just as *myc* was induced in response to mitogenic stimuli, its expression was also affected by a host of differentiation inducers in a variety of cell types (7–9, 12). To mention only a few examples, *myc* was rapidly downmodulated upon induction of differentiation pathways in mouse erythroleukemia (MEL) cells (104), human promyelocytic HL60 (105, 106), monoblastic U-937 (107) and proerythroid K-562 cells (108), murine primary keratinocytes (109), and P19 embryonal carcinoma cells (110). In murine F9 teratocarcinoma and WEHI3B myeloid leukemia cells, the loss of *myc* expression was delayed until the terminal stages of differentiation and growth arrest (111, 112). Similarly, natural induction of murine myeloid M1 cells to differentiate into macrophages and granulocytes resulted in *myc* downregulation during terminal differentiation at their G0/G1 interface, and this effect was associated with the production of an endogenous interferon (IFN) (113). The antimitogenic effects of IFN in this latter system were confirmed upon the addition of antibodies against type I IFN, which abrogated the decline in *myc* and also sustained cell proliferation (113). Another study, using sodium butyrate to induce some differentiated properties in a human colon carcinoma line, revealed that labile factors were involved in the delayed inhibition of *myc* RNA accumulation (114). Nevertheless, the role of cell-type-specific factors in the rapid and delayed inhibitory effects on *c-myc* in different differentiation programs remains largely unknown. The downregulation of *myc* exhibits a biphasic pattern in L6E9 (115), MEL (104), and P19 (110) cells, with an initial transient inhibition followed by a peak of activity that can exceed normal levels by 5- to 20-fold (MEL and P19 respectively), and then a final decline below basal levels with the onset of terminal differentiation. A recent study in MEL cells revealed that enhanced *myc* polypeptide turnover was associated with their postcommitment differentiation program, presumably a reflection of their irreversible loss of proliferative potential (116). This observation suggested that the initial loss of *myc* expression during the MEL commitment period was not associated with their terminal differentiation and may have reflected a nondifferentiated response to various chemical inducers (116). Alternatively, a transient decrease in DNA synthesis may be associated with the initial suppression of *myc*. The correlation between the number of cells undergoing this temporary cell cycle withdrawal with the number that go on to terminally differentiate suggested that the early, temporary cessation of DNA synthesis was a necessary event in their dimethylsulfoxide (DMSO)-induced differentiation (116). Similarly, the downmodulation of *myc* upon initiation of muscle differentiation programs in L6E9 and BC3H1 myoblastic cells does not reflect their absolute proliferative capacity (115, 117), and even more surprisingly *myc* reinduction by growth factors in L6E9 myoblasts failed to downregulate muscle-specific genes (115). In contrast, *myc* was abundant in proliferating embryonic and perinatal cardiac muscle, but *myc* levels were low in adult myocardium which has irreversibly withdrawn from the cell cycle (118). Therefore, it remains somewhat enigmatic whether the early loss of *myc* expression is responsible for or simply a by-product of differentiation. To further complicate matters (superficially at least), *myc* was transiently activated by nerve growth factor (NGF) in PC12 pheochromocytoma cells, but EGF had the same effect without initiating neurite differentiation, indicating that *myc* activation might be a consequence of their differentiation but was insufficient to cause neuronal development (119). In another study, *myc* RNAs were induced upon PMA stimulation of a chronic lymphocytic leukemia (CLL) cell line to differentiate into immunoglobulin-secreting plasmablasts, but these cells only progressed from G0 to G1, never entering S phase (120). However, a more recent report indicated that such cells are heterogeneous in phenotype and that *myc* protein was actually absent in terminally differentiated, nonproliferative human plasma cells (121). In other work, the *N-myc* gene was shown to be downregulated in retinoic acid (RA)–induced differentiation of teratocarcinoma (122) and neuroblastoma (24, 123, 124) cell lines, but in F9 cells downregulation of *N-myc*, unlike that of *c-myc*, was rapid and transient, implying that this effect was associated with early phases of cell-type-specific differentiation pathways (122).

More direct evidence for the importance of downregulating *myc* to establish differentiated phenotypes in vitro was obtained with exogenously introduced sense and antisense *myc* genes. In a number of instances, the inappropriate, deregulated expression of a transfected *c-myc* gene in MEL cells interfered with their terminal differentiation (125–128) but not their commitment period, and *L-myc*-transfected cells behaved in a similar manner (129). Another group reported that *c-myc* antisense transcripts accelerated MEL cell differentiation but also inhibited their G1 progression (130). The constitutive, overexpression of an exogenous *myc* gene had no obvious effect on the early phase of the MEL differentiation program since the endogenous *myc* genes were still normally downregulated in response to differentiation inducers (125–127), but a cell fusion approach revealed that a DMSO-inducible activity involved in triggering their differentiation was blocked (128). Other recent evidence indicated no disruption of cell cycle control or gross alterations in endogenous gene expression early in the MEL cell induction program, suggesting that the initial downregulation of *myc* was not essential for their early transient cessation of DNA synthesis (131). In contrast to these reports, one MEL cell study using an inducible *myc* gene revealed that *myc* might also have a positive influence on their differentiation. Here the induction of an exogenous *myc* gene allowed for a more rapid DMSO-induced recovery of endogenous *myc* expression, and these cells exhibited a shorter commitment period (132). The same study found that an antisense *myc* gene delayed the commitment-associated restoration of *myc* expression and slowed their differentiation (132), which directly contrasts with the antisense experiments of an independent group (130). Furtheremore, the introduction of an antisense *myc* gene in HL60 cells suppressed endogenous *myc* expression, decreased their proliferation, and triggered their differentiation to monocytes (133, 134). Similarly, F9 teratocarcinoma cells transfected with a constitutive antisense *myc* expression vector spontaneously differentiated into cells ressembling those exposed to RA, while overexpression of a sense *myc* construct resulted in cell clones resistant to RA-induced differentiation (135). However, the *myc* plot thickens once again if one considers two reports, both using an inducible *myc* gene in F9 cells, which yielded conflicting results: (*a*) in the first study, the overexpression of exogenous *myc* caused the death of 70–80% of the cells with the remaining 20–30% differentiating in response to RA plus cAMP, suggesting that the effects of enforced *myc* expression are cell cycle phase dependent (136); (*b*) in a second, more recent study, enforced *myc* expression did not significantly alter the F9 differentiation program, but endogenous *myc* depletion achieved with an inducible antisense *myc* vector resulted in the early cessation of DNA synthesis (137). In agreement with most of the findings described above with MEL, HL60, and F9 cells, the PMA-induced differentiation of U-937 monoblastic cells into macrophages was inhibited by the expression of an avian *v-myc* gene (138). A recent follow-up study indicates that interferon-γ (IFN-γ) bypasses the effects of constitutive *v-myc* expression in the U-937 line, and provides the interesting implication that the downregulation of *myc* is not obligatory for U-937 differentiation (F. Oberg, L. G. Larsson, R. Anton, K. Nilsson, submitted for publication). Enforced *myc* expression in 3T3-L1 preadipocyte cells did not prevent their G0/G1 arrest at confluence or their ability to replicate the genome upon initiation of their differentiation program, but did leave intact their responsiveness to mitogens, which prevented their irreversible cell cycle withdrawal and differentiation (139). These findings led to the conclusion that enforced *myc* expression prevented preadipocytes from entering a unique G0 state, operationally defined as GD, which is essential for their mesenchymal differentiation (139). Early work demonstrated that quail embryo myoblasts were transformed by the *v-myc* gene of the MC29 virus, and their differentiation into multinucleated myotubes was irreversibly suppressed, while a transformation-defective MC29 mutant failed to prevent their differentiation (140). Moderate overexpression of an exogenous murine *myc* gene in BC3H1 myogenic cells partially inhibited the expression of muscle-specific genes, but was insufficient to prevent their differentiation (141). Deregulated *myc* expression in the developing heart of a line of transgenic mice resulted in massive cardiac enlargement caused by myocyte hyperplasia (142). However, recent work has revealed that enforced *myc* expression can prevent the initiation of myoblast differentiation by the myogenic regulatory genes *myoD* and *myogenin* by a mechanism independent of Id, a negative regulator of muscle differentiation that sequesters MyoD and Myogenin polypeptides in inactive hetero-oligomers (143). Clearly, more work is needed to dissect the multiple molecular routes whereby deregulated *myc* expression antagonizes the commitment to various cellular differentiation programs.

TUMORIGENESIS

The *myc* genes belong to the category of immortalizing oncogenes such as *E1a* (144), *large T* (145), and *p53* (146) as they are able to partially transform primary cells in culture but not induce a tumorigenic phenotype. It has been proposed that abnormal *myc* expression would alter the regulation of cellular genes, rendering cells more susceptible to transformation to a malignant phenotype. The dominant action of another oncogene or the loss of a tumor suppressor gene would then accelerate or promote tumorigenicity (147). Several groups have looked at the effect of *myc* on primary cells and established cell lines as well as using transgenic animals to extend these studies in vivo. Although it is well established that *c-* (42), *L-* (148), and *N-myc* (19, 149–151) can cooperate with an activated *ras* oncogene to transform primary rat fibroblasts in tissue culture, deregulated *myc* expression is alone insufficient to elicit a malignant phenotype in the absence of secondary genetic events.

Overexpression of exogenous *myc* genes can elicit transformed and malignant phenotypes in various established cell lines. Deregulated expression of *c-myc* did not induce focus formation in early passage (10 generation) FR3T3 cells, but was able to elicit a tumorigenic phenotype in late passage (60 generation) cells, which indicates that a secondary event was needed (152); and other results suggested that such secondary events randomly occurred during prolonged in vitro culture (153). NIH/3T3 or Rat-2 cells, which contained *c-myc* genes linked to a viral promoter, showed a slight increase in refractility, the ability to grow in soft agar, as well as a decreased serum requirement, and also formed tumors in nude mice and syngeneic rats (154). When *c-myc* was transfected in Epstein-Barr Virus (EBV)–infected lymphoblastoid cells, it induced a transformed phenotype and the cells formed tumors in nude mice (155). Rat-1 cells were transformed by either *N-myc* or *c-myc* acting alone when overexpressed (151). Clearly, *myc* has a role in tumorigenesis, but events in addition to *myc* overexpression are needed to elicit a full malignant phenotype. On balance, the potency and range of effects attributed to abnormal *myc* expression in established cell lines should be interpreted in the context of the cellular background.

The cooperation of *myc* with other oncogenes is consistent with the idea that *myc* alone does not cause malignancies, but is one step toward the progression to a malignant state. *myc* cooperates with *ras* to transform rat fibroblasts (156, 157), rat embryo cells (158, 159), human epithelial cells (160), and pre-B cells but not mature B cells (160, 161). Baby mouse kidney cells immortalized by HPV-16 and *c-myc* displayed a reduced need for growth hormones and acquired a malignant phenotype (162). IgM-secreting murine plasmacytomas were induced by a *c-myc* and *v-Ha-ras* containing retrovirus in pristane-primed mice (163). *c-myc* also cooperated with *raf* to produce lymphoid tumors in mice of B and T lineage (164–166), accelerated pristane-primed murine plasmacytogenesis in BALB/C mice (166, 167), and transformed chicken embryo neuroretinal cells (168). *c-myc* and *c-raf*-1 transformed SV40 large T-immortalized bronchial epithelial cells, which then formed large cell carcinomas in nude mice (169). The subsequent acquisition of a *c-myc* translocation in an indolent low-grade B cell lymphoma, also harboring a *bcl*-2 locus rearrangement, was correlated with its progression to a very aggressive lymphocytic leukemia (170).

The use of transgenic animals harboring activated oncogenes has confirmed and extended the in vitro results described above. Early work demonstrated that mice harboring a *myc* transgene driven by a glucocorticoid-responsive murine mammary tumor virus long terminal repeat (MMTV LTR) developed adenocarcinoma of the breast (171, 172), testicular tumors, and B and T lymphoid tumors, along with tumors of mast cell origin (172). Cell proliferation appeared normal in these animals. The fact that in all tissues where *c-myc* was overexpressed, malignancies did not develop, could indicate that indeed a secondary event is needed and that *c-myc* is required but not sufficient for tumor formation (172). Contributing to this idea is the study in which H-2K/*myc* transgenic mice, which showed expression of the transgene in nearly all tissues, developed a lymphoproliferative disorder with no tumor formation seen even after 20 months (173). Thus, in the absence of a so-called "second hit," no tumors develop. When the expression of a *c-myc* transgene is targeted to lymphoid tissues by the action of the Eμ immunoglobulin enhancer, mice developed monoclonal B cell tumors (generally of the pre-B cell stage but occasionally mature surface Ig+ B cells) (174–177). The incidence of these B cell malignancies varied greatly in different genetic backgrounds, arguing that recessive loci in different inbred mouse strains strongly contribute to their *myc*-induced susceptibility towards malignancy. Young mice exhibit proliferation of pre-B cells, which display altered development as decreased amounts of mature B cells are seen in the adult (175). However, pre-B cells of Eμ *myc* mice did not show altered growth requirements in vitro, but acquired a malignant phenotype upon introduction of other activated oncogenes such as *ras* (178, 179). In contrast, Eμ *myc* rabbits developed oligoclonal lymphocytic leukemias within three weeks of birth. Transgene expression seems to be limited to a specific stage of B cell development (180). When c-myc expression was directed to the thymus by inserting c-myc coding regions into the Thy-1 (pan T cell surface marker) transcription unit, thymic tumors are seen, composed of proliferating thymocytes and epithelial cells (181). Eμ-N-myc mice also developed pre-B and B lymphoid malignancies (182, 183). In Eμ-L-myc mice, L-myc expression was confined to thymocytes, and these mice developed T cell lymphomas at reduced incidence and with a longer latency relative to Eμ-N- or c-myc tumors (184). Interestingly, in these tumors generated by deregulated N- and L-myc, no expression of c-myc was observed, which indicated that negative crossregulation may be operating, and that L- and N-myc can substitute for c-myc function in vivo (182, 184).

Transgenic studies on oncogenic cooperation have shown an altered picture of malignancies compared to transgenics harboring individual oncogenes. Matings between myc and bcl-2 transgenics yielded mice with novel, polyclonal malignancies of lymphoid-committed stem cells, in utero, which were not seen with mice singly transgenic with bcl-2 or myc (185). In contrast to v-Ha-ras, v-abl did not cooperate with c-myc in the development of early B cell tumors but did yield a preponderance of end-stage B cell tumors (plasmacytomas) (186). Indeed, singly transgenic Eμ-abl mice developed plasma cell tumors several months after birth, and these murine plasmacytomas (MPCs) contained endogenous myc translocations (186). Positive cooperation between oncogenes can thus either accelerate the development of tumor formation or lead to the formation of novel tumor types. Evidence that stages of developing B cells have different susceptibilities to myc was revealed in other transgenic mice harboring a membrane form immunoglobulin transgene in addition to an IgH enhancer-driven myc gene. These mice did not develop early B cell tumors, because the presence of the immunoglobulin transgene was believed to accelerate B cell maturation and thereby reduce the available pool of early proliferating B cells (187). Some experiments with retroviruses harboring individual or multiple oncogenes in pristane-primed adult mice have revealed a similar phenomenon. Retroviral vectors containing c-myc (188, 189) or v-abl (190) induced the production of myeloid or pre-B cell tumors, respectively, while a vector containing both c-myc and v-abl largely generated MPCs with some occasional pre-B tumors. The myc/abl combination was particularly effective in generating plasma cell tumors, which even appeared without pristane administration required for myc/ras- and myc/raf-induced plasmacytogenesis (F. Mushinski and M. Potter, personal communication). The occasional pre-B tumors seen with the double vector were subsequently found to have a defective c-myc gene (F. Mushinski, personal communication). These results suggested that c-myc overexpression was antagonistic to the oncogenic effects of v-abl in immature B cells.

myc collaborators have been discovered using Mo-MLV infection of Eμ-*myc* transgenic mice. Mo-MuLV accelerated B lymphomagenesis in these mice, and several novel proviral integration sites have been described (191, 192). These have been designated as *pim-1* (191), *pim-2* (192), *emi-1* (192), *bmi-1* (191, 192), *pal-1* (191), and *bla-1* (191). A gene encoding a novel zinc finger protein, *bmi-1* (B lymphoma Mo-MuLV integration region 1), was discovered at one of these sites. A related gene, *mel-18*, has recently been proposed to be in the same family as *bmi-1* (193). *mel-18* is expressed in several tumor cells but poorly expressed in normal tissues. These genes are differentially expressed in various tissues during development, and as such may define a novel family of genes involved in cell proliferation and progression to a malignant state (193).

A ROLE FOR *c-myc* IN DNA REPLICATION: FACT OR FANTASY?

As summarized above, *c-myc* expression is generally associated with the onset and maintenance of cellular proliferation but can be antagonistic for differentiation. Roles for *myc* in RNA transcription and/or DNA replication have been proposed, given that it is a nuclear phosphoprotein with DNA-binding activity (10, 33–37). The wealth of evidence that supports the notion that *c-myc* is a sequence-specific DNA-binding transcriptional regulator (the presence of the basic-H-L-H domain and leucine zipper, nuclear localization, and the ability to transactivate cellular genes) is quite compelling. RNA transcription and DNA replication need not be mutually exclusive (194), as transcriptional control elements have been found in DNA replication origins. The data that support a role for *c-myc* in DNA replication are controversial and mired in conflicting reports and in some cases irreproducible results.

Published reports have shown that in some human lymphoid cell lines, *c-myc* overexpression stimulates replication of an SV40-based vector (195–197). Both *N-myc* and *L-myc* were found to facilitate SV40 replication when overexpressed, but the cell background for the effect of *L-myc* was more restricted (197). Another group reported the ability of anti-*c-myc* antibodies to inhibit DNA replication (198), but this was shown to be an experimental artifact as an inhibitor of DNA replication was identified in the anti-*c-myc* preparation (199). The activity of the inhibitor was separated from the anti-*c-myc* activity, with the resulting purified antibody unable to reproduce the published results (199). Along this same line, it has been reported that *c-myc* can substitute for large T antigen in SV40 replication (196), but another group has not been able to reproduce this observation (200). Contributing to the idea that *c-myc* functions both as a transcriptional regulator and in DNA replication are several papers published over the past four to five years on the existence of a putative binding site for the *c-myc* protein, approximately 2 kilobases (kb) upstream from the human *c-myc* coding sequences (201–203). This sequence spanned about 200 bp, was described as possessing enhancer activity, and was able to function as an origin of DNA replication (201). A seven-base-pair sequence within this region was reported to bind the *myc* protein (202). However, other groups have been unable to confirm various aspects of these data (200). In an autonomously replicating human sequence, the *c-myc* protein was found to promote DNA replication by binding to the origin (204). Since this putative ori 5' of human *c-myc* does not resemble the recently defined *c-myc* binding sequence (60), its ability to bind *c-myc* with specificity in vivo remains questionable.

While it is possible for a protein to possess a dual role as a sequence-specific transcription factor and as a DNA replication factor (194), the data for *c-myc* as a factor directly involved in DNA replication is suggestive but not compelling. *Xenopus* embryogenesis has to date provided the most convincing evidence for a role of *c-myc* in DNA replication, even though the data remain largely correlative at this point. This study concerns the intracellular redistribution of a maternal store of *c-myc* protein immediately following fertilization (205). In contrast to the somatic store of *c-myc* protein, which has a nuclear localization and is rapidly turned over, the *c-myc* protein found in the oocyte is cytoplasmic and extremely stable (206, 207). Upon fertilization, this cytoplasmic store of *c-myc* is exported to the nucleus; this continues throughout the early cleavages, until the mid-blastula transition (205). The pool of *c-myc* protein in the nuclei remained constant, so that if redistribution was ongoing during the early cleavages, then the protein was turning over in the nucleus. Considering that *Xenopus* embryos are transcriptionally silent before the mid-blastula transition, the possibility arises that *c-myc* is playing a role in DNA replication, though further experimentation must directly test this hypothesis. The reports stating that the overexpression of *c-myc* enhances SV40 replication (195–197), and a paper that showed that a *c-myc* antisense oligodeoxynucleotide inhibits entry of cells into S phase (208), suggest a role for *c-myc* in DNA replication, but perhaps through an indirect route by inducing the expression of certain limiting cellular factors that are required for DNA replication.

REGULATION OF *myc* AND BY *myc*

myc Autoregulation

When *c-myc* is involved in chromosomal translocations, as found in human Burkitt lymphoma and MPCs, expression is predominantly from the translocated allele, with the nontranslocated allele being transcriptionally silent (6–9, 11, 209). Although the regulation of *myc* expression in malignancies such as Burkitt lymphoma has been shown to be exceedingly complex (see below), two simple models for the allelic exclusion seen for *myc* expression can be proposed. The *myc* gene could normally be transcriptionally silent at the particular stage of development of the malignancies, and the translocated allele would be expressed by its placement in an active chromatin domain. Alternatively, *myc* could be involved in a negative autoregulatory loop, and the overexpression of the translocated gene would feedback and downregulate the nontranslocated allele (210). Either the normal autoregulatory signal is lost by the translocated allele, or the negative signals controlling expression are overcome by compensatory positive signals.

Early experiments yielded conflicting findings for an autoregulation model. In some studies, the introduction of an exogenous *c-myc* gene effectively repressed the endogenous *c-myc* gene (164, 211), whereas in a larger number of studies performed with various established tissue culture cell lines, no downregulation of endogenous *myc* was seen (125, 139, 141, 154). The transgenic mouse system provided direct evidence for some type of autoregulatory loop. Early B-cell populations of mice expressing a *myc* transgene under the constitutive control of an immunoglobulin gene enhancer were negative for endogenous *myc* expression (212).

Penn et al described a negative autoregulatory loop for *c-myc* in tissue culture cell lines (213). A more recent comprehensive study has shed some light on conditions under which an autoregulatory mechanism for *c-myc* may be operating (214). Using constructs placing *c-myc* under the control of the SV40 promoter/enhancer or using a heavy metal–inducible construct (the metallothionein promoter), Grignani et al showed that exogenous *c-myc*, when overexpressed, could negatively regulate endogenous *c-myc* expression in a dose-dependent manner. This effect was not universal, as some cell lines had lost the ability to regulate their own expression of *c-myc*. After analyzing a great number of cell lines, the authors concluded that the autoregulatory loop was functional for primary cells and some established cell lines (214), but that autoregulation was lost by cells with a transformed phenotype (214). These observations could explain some of the discrepancies in the literature.

The different members of the *myc* family exhibit crossregulation. When *N-myc* is amplified, either in tumors (215, 216), transfected cell lines (217), or in transgenic mice (25), *c-myc* expression is downregulated. Likewise, in some small cell lung carcinomas, where expression of *L-myc* is high (216), *c-myc* expression is again downregulated. However, normal levels of coexpression of *myc* family genes (25) in a number of cell lines indicate that expression of *myc* gene family members is not mutually exclusive and likely depends on other interacting cellular factors.

The domains of the *myc* protein that are critical for autoregulation have been identified. Mutations, in the conserved H-L-H and LZ domains, shown to be essential for *ras* cotransformation, abrogated autoregulation (218). It is very appealing to consider that heterodimeric complexes of *myc* and its recently identified BR-H-L-H-LZ partner, *max*, would bind to a site in the *myc* regulatory region to downregulate *myc* expression as *myc*'s protein dimerization domains were needed for autoregulation (218). Unfortunately, the specific *myc/max*-binding site has not been found in the known regulatory regions for either human or mouse c-*myc*. It is possible that c-*myc* could have more than one partner and perhaps recognizes a different binding site when complexed with a different protein(s). Alternatively, the autoregulatory loop could be indirect with *myc/max* heterodimers regulating other cellular genes whose expression would be directly involved in negatively controlling *myc* transcription.

Regulation of Other Cellular Genes by myc

Architecturally, the c-*myc* protein has all of the necessary attributes of a transcriptional regulatory factor, with protein dimerization, transcriptional activation, and DNA-binding domains. In light of these well-characterized motifs, and c-*myc*'s role in cell growth and proliferation, the pursuit of genes regulated by *myc* could hold the keys to the enigmatic function(s) of the *myc* polypeptide.

Early studies indicated that the c-*myc* protein had the potential to regulate other genes. The activation domains of a variety of mammalian cell transcription factors have been shown to function in yeast cells upon fusion to DNA-binding domains of other transcriptional activators that recognize their target sequences in yeast. Early work revealed that the *myc* polypeptide contained a weak transcriptional activation domain in comparison to that of the *fos* oncoprotein in yeast. This study relied upon the ability of the DNA-binding domain of the bacterial *LexA* repressor to specifically recognize its DNA-binding site linked to a reporter gene in yeast (219). In a more recent study, different portions of the human *myc* gene were fused in frame to the DNA-binding domain of the *GAL4* yeast transcriptional activator protein, and the resultant chimeric polypeptides were assayed for their relative abilities to activate expression of a cotransfected reporter gene linked to *GAL4* target sequences in yeast. Strong evidence for the existence of three independent activation regions between amino acids 1 and 143 of the human *myc* polypeptide was obtained (220). In other studies, *myc* could be shown to positively regulate the *hsp70* promoter (221) and downregulate the metallothionein promoter (222) to variable degrees. Expression of two cellular genes, *mr1* and *mr2*, was enhanced through a posttranscriptional mechanism in c-*myc*-immortalized rodent fibroblasts (223), and *mr1* was later shown to be plasminogen activator inhibitor 1 (224).

Most findings indicated that overexpression of *myc* and/or deregulated *myc* expression was correlated with the downregulation of a variety of different cellular genes. Two variant H1 histone genes were downregulated in MEL cells in which terminal differentiation was blocked by the stable integration of a transfected deregulated *c-myc* gene (225). Differentiation of MEL cells is also blocked by deregulated expression of *c-myb* (225), but the variant histone genes were not affected in *c-myb* blocked cells, suggesting that the effect on the expression of some H1 genes was *myc* specific (225). A variety of cell surface molecules have been shown to be downregulated by augmented *myc* expression. Expression of major histocompatibility complex (MHC) Class I molecules is depressed by both *c-myc* and *N-myc* (226). N-CAM expression is abrogated by *N-myc* expression (227). *LFA-1* was shown by one group to be downregulated by *c-myc* and *N-myc* in EBV-immortalized B cells (228). However, a recent report failed to reproduce these data, as no change in *LFA-1* expression was seen when *c-myc* was overexpressed in EBV-immortalized human B cells (229), suggesting that the levels of the *c-myc* polypeptide and/or the vagaries of cellular context must be considered. *c-myc* has recently been reported to downregulate three collagen genes, the effect of which is dependent upon an intact leucine zipper domain (230). In vitro DNA-binding studies of *c-myc* protein on the pro-alpha 2(I) collagen promoter indicated that *c-myc* was able to bind this promoter, but no sequence specificity was determined (230). *c-myc* was also found to downregulate *neu* expression and abrogate *neu*-induced transformation via a sequence element in the *neu* gene promoter (231).

The most direct evidence for regulation of a cellular gene by *c-myc* was obtained by using steroid hormone-*myc* chimeras (102). *myc* was shown to activate transcription of the α-prothymosin gene in the absence of protein synthesis (103), implying a direct role for *myc* in its regulation; but it remains to be established that *myc* or *myc/max* directly interacts with α-prothymosin regulatory sequences.

Regulation of myc *Transcriptional Initiation*

c-myc is transcribed from multiple, independently regulated transcription initiation sites, which yields a complex picture of gene control. The gene has two major promoters (P1 and P2), which are positioned 161 and 164 bp apart in humans (232, 233) and mice, respectively (234). P2 is the predominant promoter, giving rise to 75–90% of cytoplasmic *c-myc* RNAs, whereas P1 generates 10–25% of them (235–239). A third promoter, P0 (550 to 650 bp 5' of P1), is unique to the human gene, but only generates 5% of effective *myc* transcription (240). Human, mouse, and rat *c-myc* genes contain yet another minor promoter, P3, near the 3' end of their first intron, which yields about 5% of total *c-myc* mRNAs (241). P0 and P3 both lack TATA boxes, unlike P1 and P2. The sizes of the mature mRNAs for P0, P1, P2, and P3 are approximately 2.5–3.1 kb, 2.2 kb, 2.4 kb, and 2.3 kb, respectively (232,

234, 240, 242). Differential promoter usage has been observed in a variety of cell types with ratios of P1:P2 varying from 0.1 to 1.0 (236–239, 243, 244). However, the P1:P2 ratio rarely exceeds 0.2, and atypical cases in which it approaches 1.0 or higher have generally been associated with abnormal or deregulated control (245).

Studies of *c-myc* chromatin have revealed multiple DNase I hypersensitive (DH) sites located approximately 4 kb upstream and 2 kb downstream of P1 (9) as seen in Figure 2. Hypersensitive sites that are shared by the human and murine *c-myc* genes are relative to P1 as follows: I (-1850), II1 (-1380), II2 (-750), III1 (-130), III2 ($+90$), IV ($+800$), and V ($+1800$) (236, 245–49). DH sites III1 and III2 are associated with P1 and P2 promoter activity, while effects attributed to the remaining sites are correlative in nature and not always consistent (236, 249). For instance, DH site I is retained in some but not all cases of *c-myc* repression and was originally hypothesized to reflect some type of negative control (possibly autoregulatory in nature) (245). However, this may also reflect differences in *myc* regulation in differentiated

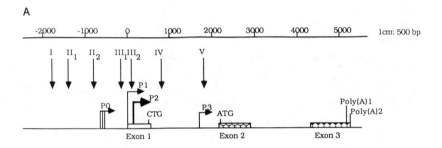

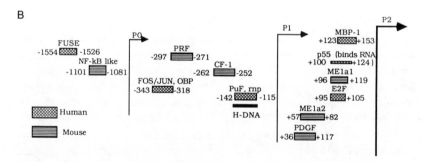

Figure 2   *A.* Arrangement of *c-myc* exons, promoters (P0, P1, P2, P3), major translation start codons (CTG, ATG), polyadenylation sites [poly(A)1, poly(A)2], and DNase I hypersensitive sites. *B.* Summary of some human and murine *c-myc* transcription control elements and factor-binding sites. All coordinates are relative to P1. References are in the text.

cells or transcription from upstream sequences that may or may not have an effect on P1 and P2 initiation. Chen & Allfrey have found direct evidence for changes in nucleosome topography wherein the binding of the *c-myc* nucleosomes to mercury(II) affinity chromatography was directly related to *myc* expression (250).

Despite a large body of work, the molecular requirements for the control of *myc* have remained elusive, because of conflicting or unsubstantiated findings. Some of these controversies arise from primary sequence differences and/or variations in the positions of putative regulatory elements upstream and within the murine and human *c-myc* genes, and also from differences in gene expression assays, chimeric construct boundaries, gene transfer methodologies, and recipient cell backgrounds used by aficionados of *myc* control. The significance of these myriad studies for physiologically relevant modulation of normal and abnormal *myc* expression becomes suspect when the effects of a newly defined putative control element on P1 or P2 remain unproven. In performing gene transfer experiments with mutated and/or chimeric genes, special attention should be given to alterations in the P1:P2 expression of the exogenous construct in comparison to that of the endogenous *myc* gene in the same recipient cells. Most findings have indicated that the closely spaced P1 and P2 promoters are subject to independent regulation, but this picture can easily increase in complexity if the usage of one start site interferes with the optimum use of the other. A summary of *myc* initiation control elements is presented in Table 1. We attempt to simplify matters by separately discussing *cis*-acting transcriptional initiation elements residing outside and inside the *myc* gene.

From the results summarized in Table 1, a variety of putative positive and negative effector sites have been identified upstream of the P1 initiation site, and in some cases controversies have arisen over their existence and locations. Deletion/transfection studies of the human *c-myc* sequence have shown that regions upstream of -101 (P1 at $+1$) have no effect (251), regions upstream of -350 have a negative effect (252), and regions upstream of -353 have a positive effect (253). In vivo competition experiments yield the same kind of perplexing results. Whereas Chung et al reported that -608 to -407 may contain a negative element, Lipp et al observed no evidence for such a negative element in a construct encompassing the same region. However, the 5' and 3' limits of the constructs and the recipient cell types differed in these two studies. The same or overlapping DNA sequences may function in a dominant positive or negative fashion depending on the DNA-binding factor(s) present in different cell types. Accurate transcription of the human *c-myc* gene from P1 and P2 was obtained in *Xenopus* oocytes (254), and the region described by Nishikura (255) (Table 1) as necessary for P1 transcriptional initiation contains a TATA box and a GC-rich sequence at -44

Table 1 Regulatory elements inside and outside the *myc* gene

References	Species	Regions of interest relative to P1	Effect on activity	Cell background
Nishikura (255)	human	−2.3 to +5.9 kb WT[a]		X. *laevis* oocytes
		−2.3 to −0.35 kb	0	
		−60 to −37 bp	+(P1)	
		+95 to +105 bp	+(P2)	
Chung et al (251)	human	−6.8 to +2.2 kb-CAT WT[a]		Ly65 (Burkitt lymphoma), NIH 3T3 (murine fibroblast)
		>−101 bp	0	
		−4.5 to +1.3 kb-CAT WT[b]		
		−608 to −407 bp	−	
		+47 to +513 bp	−	
		+510 to +1180 bp	−	
		−400 to −293 bp	−	
Remmers et al (268)	mouse	−1.1 to +0.6 kb-CAT WT[a]		BJB (human lymphoblastoid line), COS (monkey fibroblast)
		−1.1 to 0.4 kb	−	
Yang et al (274)	mouse	−1.1 to +0.6 kb-CAT WT[a]		BJAB, COS, HeLa (human epithelial line)
		+517 to +563 bp	+(P2)	
Hay et al (253)	human	−2.3 to +0.5 kb-CAT WT[a]		Ltk-(murine fibroblast), C39 (SV40 immortalized human
		−2.3 to −1.3 kb	+	

Reference	Species	Construct	Activity	Cell type
Lipp et al (252)	human	−1.3 to −0.3 kb	+	fibroblasts), COL0320 (human colon carcinoma cells)
		−353 to −293 bp	−	
		−293 to −101 bp	+(P1)	
		−6.7 to +0.5 kb-CAT WT[a]		Ltk-, 143tk- (human fibroblast), BHK (Baby hamster kidney fibroblasts)
		>−350bp	−	
		−353 to −101 bp	+	
		−101 to +66 bp	−(P2)	
		>−350 bp[b]	0	
		−353 to +513 bp	+	
		−353 to −101 bp	+	
		−101 to +513 bp	+	
Thalmeier et al (272)	human	−2.3 to +0.5 kb-CAT WT[a]	+	HeLa, 293 (+Ad E1a)
		+95 to +141 bp		
Asselin et al (263)	mouse	−0.4 to +0.6 kb- CAT WT[a]	+	CV 1 (monkey kidney fibroblasts), HeLa
		+108 to +124 bp	+(P2)	
		>−109 bp	0(P1)	
Postel et al (256)		−142 to +513 bp[c]	+(P2)	HeLa extracts
		−408 to −294 bp	+(P2)	
Hall (271)		+24 to +140 bp	+(P2)	

[a] Deletion/transfection assay
[b] in vivo competition assay
[c] in vitro transcription assay

from P1 (CCGCCC), which is perfectly conserved between human and mouse. Another region of particular interest is -293 to -101, which was originally found essential for P1 usage and necessary for efficient P2 initiation in gene transfer assays (253, 256). More recently, a GGGTGGG sequence motif (-142 to -115) within this region was found to bind a nuclear factor, PuF (256), as well as a ribonucleoprotein complex (257), and was also deemed necessary for P2 activity in an in vitro transcription assay (256). PuF may reside in the ribonucleoprotein complex, or these may represent distinct factors recognizing the same or overlapping sequences. A factor termed nuclease sensitive element protein-1 (NSEP-1), which binds to this region, has recently been cloned (R. Kolluri, T. A. Torrey, A. J. Kinniburgh, personal communication). More recently, Postel et al reported that an antisense oligonucleotide of the PuF-binding site dramatically inhibited P1 activity in vivo but only had a modest effect on P2 activity. However, the disparity between the in vitro and in vivo data was not addressed (258). The lack of in vivo footprinting data and *trans*-activation assays with cloned genes encoding these binding activities make it difficult to assess their relative significance at present. The region between -142 and -115 contains a purine-pyrimidine-rich sequence that can adopt a triple-helical H-DNA conformation, containing S1 nuclease and DH sites (256, 259-61). A similar region in the chicken and mouse *c-myc* genes binds multiple proteins including SP1 (262, 263). Common Factor 1 (CF1), a ubiquitous transcription factor, binds at -262 to -252 relative to P1 start, and plasmacytoma-specific factor PRF binds just 5' of CF1 at -290 (264–266). The PRF site appears to function in a cell-type-specific negative manner, but its role in differential P1 and P2 usage remains to be determined, while a multimerized CF1 site was found to activate transcription of a basal promoter in both B cells and fibroblasts (266). An NFκB-like factor has been identified in WEHI 231 extracts that binds to a sequence -1101 to -1081 upstream of the murine P1 start site, and like the CF1 site has the potential to contribute to positive control, since two or more copies activated a heterologous promoter (267). An earlier study focused on a further upstream region of the murine *myc* gene, between -1188 and -428, which behaved in the manner of a transcriptional "dehancer" of the *myc* promoter or other heterologous promotors (268). This negative region was subdivided into two negative elements that bind multiple nuclear factors. However, the dehancer effect was more pronounced with heterologous promoters in B lymphoid cell lines, making its importance for regulating P1- and P2-initiated transcripts less general and possibly cell-type-specific. These negative elements may function by interfering with transcription factors binding near the P1 and P2 start sites (269). Avigan et al have identified an element far upstream of human P1 (-1554 to -1526) termed FUSE. They demonstrate that when HL60 and U937 leukemia cells are induced to differentiate with DMSO, there is a dramatic decrease in binding to FUSE. The loss of binding correlates closely with a loss of transcription initiation of the endogenous *c-myc* gene and also with commitment to differentiation (270).

Cis-acting elements residing in the *c-myc* first exon and intron have been proposed to regulate P2 transcriptional initiation. Two control elements (ME1a1 and ME1a2) positioned between P1 and P2 were first described by Asselin et al (263). ME1a1 (+96 to +119) and ME1a2 (+57 to +82) were defined as major target sites for the binding of nuclear factors and are well conserved between the mouse and human genes. Two gel mobility shift complexes were observed with an ME1a1 synthetic oligonucleotide; one of these was also observed with the ME1a2 site (S. Bossone and C. Asselin, unpublished results; A. Nepveu, personal communication). Deletion of the ME1a1-binding site resulted in a dramatic reduction of P2 activity in fibroblastic cells and a modest upregulation of P1, but this could be an indirect consequence of reduced P2 function if these nearby start sites normally interfere with each other (263). Recent findings have shown that an ME1a1 site-directed mutant lost virtually all P2 activity with a modest increase in P1, and the same mutant only generated one instead of two gel shift complexes (S. Bossone, C. Asselin, A. Patel, K. Marcu, submitted for publication). An ME1a2 site-directed mutant reduced P2 activity, raising the P1:P2 ratio from 0.2 to ~1.0 (C. Asselin, unpublished results). Recently, a human gene encoding a novel zinc finger protein that binds the ME1a1 sequence was identified (S. Bossone, C. Asselin, A. Patel, K. Marcu, submitted for publication). Preliminary evidence suggests that this factor is involved in P2 regulation because it failed to bind the mutated ME1a1 site described above that inactivated the P2 promoter (S. Bossone, C. Asselin, A. Patel, K. Marcu, submitted for publication). Hall demonstrated that the ME1a1 site was required for transcriptional initiation from P2 in an in vitro transcription system (271). A sequence element (GGCGGGAAAA) located between ME1a1 and ME1a2 and conserved between mouse and human was differentially required for P2 initiation in different cellular backgrounds (255, 272) and found to bind to the E2F transcription factor (272, 273). Interestingly, this region has also been shown to bind as an RNA to a p55 protein found in HeLa and MEL cells (273a). An earlier study identified another positive element downstream of the P2 start site within exon 1, which behaved like a position-dependent positive modulator of P2 activity (274). This positive element functioned when it was placed downstream of P2 or other heterologous promoters. Removal of 60 bp of DNA at the exon 1/intron 1 junction of the mouse gene diminished P2 activity (265, 274); this region was shown to contain a sequence that resembles the binding site of a nuclear factor in the promoter of a mouse ribosomal protein gene (275). A 37-kDa protein, MBP-1, which binds upstream of human P2, has recently been cloned (276). DNase I protection analysis showed that MBP-1 protects a region from +123 to +153 relative to P1 just 5' of the P2 TATA motif, and appears to be a negative regulator of c-myc transcription from cotransfection experiments with myc-CAT constructs. A 20-bp sequence located at the beginning of intron 1 of the human myc gene binds a 138-kDa nuclear phosphoprotein denoted MIF (277). The MIF-binding site is inactivated by mutation(s) and rearrangements in Burkitt lymphoma cells, suggesting that it might play a role in limiting myc activity (278).

myc transcription initiation can either be upregulated or downregulated in response to growth factors and other stimuli. Growth factors such as TGF-α (85), PDGF (81, 66, 71), IL-2 (279, 280), PHA (280), and ionomycin (280) all can potentially upregulate myc transcriptional initiation. Furthermore, a PDGF response element has been identified in the murine c-myc gene, +36 to +117 relative to P1, which stimulates transcription initiation from P1 3–5-fold in the presence of PDGF (281). Elevation of myc initiation was also shown to occur in B cells treated with anti-Ig antibody (282). TNF-α and TGF-β both suppressed c-myc, with TGF-β directly affecting initiation (283), but the mechanism of TNF-α inhibition of myc transcription remains to be elucidated (284, 285). TGF-β appeared to interact with an element near P1 (+100 to −71) (283), possibly via the RB gene product (286). Indeed, a DNA sequence within the c-myc gene (+20 to +50 from P1) resembles the RB control element of the c-fos gene (287). c-myc was also transcriptionally inactivated in dibutyryl cAMP-treated HL60 cells, but this mechanism also remains to be elucidated (288).

Regulation of myc Transcriptional Elongation

Regulation at the level of transcriptional elongation has been described in prokaryotic systems, but only recently in the control of eukaryotic genes. This phenomenon plays an important role in the regulation of a number of proto-oncogenes, including c-myc (289), L-myc (290), c-myb (291), c-fos (292, 293), and c-mos (294) genes, as well as a number of other cellular and viral genes (289) in which a block to transcriptional elongation has been described. Multiple mechanisms that have been ascribed to attenuation of prokaryotic gene transcription may be relevant to c-myc transcriptional blockage (289). These include (a) association of termination or antitermination factors with the RNA polymerase II transcriptional complex at specific nucleotide sequences; (b) regions of dyad symmetry followed by polyuridine stretches in the RNA that may or may not associate with accessory factors and impede the progress of the transcription bubble; (c) steric interference to polymerase progression by DNA-binding factors that might also alter the conformation of the DNA template. Truncated first-exon transcripts have not been observed to accumulate in the nucleus or cytoplasm of mammalian cells and appear to be rapidly wasted in the nucleus. However, Re et al (295) recently identified a 380-nucleotide (nt) truncated human *c-myc* first-exon transcript, which increased in quantity upon enhanced *c-myc* transcriptional blockage. Pausing of polymerase may be involved in this phenomenon, but most evidence points to premature transcriptional termination.

The regions necessary for transcriptional blockage within *c-myc* have been defined by a number of studies using the in vitro nuclear run-on assay. Wright & Bishop specified a 180-bp region at the 3' end of murine exon 1 to be necessary and sufficient to mediate an elongation block within an α-globin gene in HeLa cells (296). Experiments also showed that the *c-fos* attenuator could replace the *c-myc* attenuator and mediate a block (297). This region contains GC-rich sequences that have the potential to form a stem-loop structure followed by a T5 tract, and therefore bears a strong architectural resemblance to the rho-independent terminators of prokaryotes. The T5 tract was also observed to be the site of premature termination when the murine *c-myc* gene was transcribed in *Xenopus* oocytes (298). Miller et al reported murine exon 1 sequences were insufficient to block transcripts initiated from the SV40 early and MHC H-2K promoters (299). The block was restored when *myc* P2 promoter sequences were introduced in their normal location into the construct (299). *myc* P2 promoter deletion mutants assayed in HeLa and CV-1 identified a nuclear factor–binding site, ME1a1 (see Figure 2), required for efficient P2 initiation and blockage (299). Indeed, along with exon 1 downstream sequences, the ME1a1-binding site was alone sufficient to restore transcriptional blockage when inserted in either orientation 3' of the MHC H-2K promoter (300). Interestingly, a sequence bearing a strong resemblance to ME1a1 was recently shown to mediate transcription termination of human complement C2 gene transcription (R. Ashfield, P. Enriquez-Harris, N. Proudfoot, personal communication).

Studies on transcriptional blockage of the human *myc* gene have revealed many common features with the murine gene. Bentley & Groudine were the first to describe the *myc* transcriptional block phenomenon with their work on the human *myc* gene (236). A 95-bp sequence located 35 bp 5' of the exon 1/intron 1 boundary of the human *c-myc* gene was sufficient to effect premature transcriptional termination in *Xenopus* oocytes when cloned downstream of the *HSV tk* promoter but not the adenovirus major late promoter (298). Although these transcripts terminated at one of two thymine-rich sequences, deletion analysis showed that a region with dyad symmetry resembling a rho-independent terminator caused termination 10–20 bp downstream (298). Linker/scanner and deletion mutagenesis studies in *Xenopus* oocytes define sequences that appear to be essential for efficient P1 elongation between the P2 TATA box and CAP site. Furthermore, these elements only influence P1 attenuation when placed downstream of the start site in chimeric constructs. In their normal context, these sequences are not required for efficient elongation of P2-initiated transcripts (T. Meulia, T. Krumm, M. Groudine, personal communication). P2 supports transcription by RNA pol II and III; however, pol III transcripts are always blocked near the 3' end of exon 1, while pol II is only partially blocked (301, 302). An in vitro transcription system using purified RNA polymerase II and polymerase III showed that termination occurred at clusters of 7 Ts in two regions: 20 bp upstream of the exon 1/intron 1 junction with pol III, and 35 bp downstream of exon 1 with pol II and pol III (303). In a recent report, no premature termination of *myc* transcription was observed in vitro with HeLa nuclear extracts under standard conditions, but transcripts initiating at *myc* P2 or the adenovirus MLP prematurely terminated 30 bp downstream of exon 1 in the presence of 400 mM KCl (304). Deletional analysis of the human *c-myc* gene revealed that as in the murine gene (299), the block to elongation was promoter dependent (239). Transcripts that initiated at P1 did not terminate at the 3' end of exon 1, whereas P2-initiated transcripts either terminated or read through the exon 1 block signals (239).

c-myc transcription has been shown to be extremely pliable to various growth factors, mitogens, and differentiating agents that may act by altering elongation. Bentley & Groudine observed a rapid, dramatic increase in *myc* transcriptional blockage upon exposure of HL60 cells to RA (236). The same phenomenon was observed when HL60 cells were treated with DMSO (305), phorbal 12,13-dibutyrate, 1,2-dioconoylglycerol (306), and probably with 1,25-dihydroxyvitamin D3 as well (307, 308). DH site II_2 upstream of P1 decreased in intensity upon block induction, while DH site IV within intron 1 increased (236, 305). Enhanced blockage of the murine *c-myc* gene has been reported for fibroblasts transformed by Abelson murine leukemia virus (A-MuLV) (309, 310), differentiating MEL cells (292, 311, 312), and P19 cells (110). Induction of the elongation block appears to be a common mechanism for the early downregulation of *c-myc* during differentiation. Some growth factors and mitogens act by relieving the block to elongation, as seen in pokeweed mitogen (PWM)-treated human tonsillor mononuclear cells; PMA-treated human T lymphocytes; mouse spleen lymphocytes and antigen-specific T cells exposed to Con A; and EGF-treated mouse fibroblasts (85, 88, 280, 311, 313, 314). Furthermore, EGF induction of *c-myc* appears to be cAMP dependent (80). The block effect also occurred in the absence of de novo protein synthesis, indicating the *trans*-acting factors required for enhancing blockage already reside in cells, but their mechanism(s) of activation remain to be elucidated. It also remains to be determined if the *cis*-acting regulatory elements required for the manifestation of the block are sufficient for its modulation.

Other Oncoproteins Contribute to myc Transcriptional Regulation

Effects on c-myc transcription mediated by a number of oncoproteins have been documented, including v-abl (and possibly other tyrosine kinases), p53, fos/jun, adenovirus E1a, v-erb B, v-raf, early products of polyoma virus, and c-myb. One study demonstrated that the early transforming proteins of polyoma virus induced myc RNA accumulation in BALB/c3T3 cells, but the molecular mechanism was not determined (315). Direct evidence for the induction of constitutive myc transcription was reported in IL-3-dependent FDC-P1 murine myeloid cells infected with a temperature-sensitive (ts) v-abl mutant (316). Functional v-abl protein was associated with IL-3 abrogation and deregulated myc transcription, which was manifested by upregulated initiation and abrogation of the first exon transcriptional block. The use of a ts v-abl mutant ruled out a requirement for protein synthesis after v-abl induction, implying a direct effect of the v-abl tyrosine kinase. Other tyrosine kinases such as fms, src, and trk also abrogated IL-3 and resulted in constitutive myc expression, suggesting that a common phosphorylation pathway was involved (316). In contrast to these observations with a myeloid cell line, several independent clones of NIH 3T3 fibroblasts, infected with wild-type v-abl viruses, amplified their c-myc loci and displayed a greatly enhanced transcriptional block (309, 310). Together, these observations suggested that effects on myc elicited by an active v-abl tyrosine kinase depend on factors that are likely to be cell-type-specific. Ki-MSV v-raf oncoprotein was found to inhibit the PDGF-induced transcription of myc in BALB/c3T3 cells, while other growth factors on myc-like FGF remained unaffected (317). Subsequent to these reports, Evans et al published that the c-myb protein upregulated endogenous myc and an exogenous myc promoter–driven CAT gene in a murine T cell line (318). Recently, the same group of investigators have identified multiple c-myb-binding sites upstream of the murine c-myc gene, and found evidence that (a) these c-myb-binding sites are differentially required for exogenous myc promoter activation in myeloid and T cell lines; (b) c-myb DNA-binding and trans-activation domains are required for myc promoter activation (J. Cogswell et al, personal communication). A recent study by Zobel et al identified multiple v- and c-myb-binding sites 5' of the human c-myc gene, and also observed myb trans-activation of a human myc promoter construct (319). Future work will undoubtedly determine whether these effects are mediated by c-myb in normal or transformed cells, and more importantly, their physiological significance for endogenous myc expression. In other work, trans-activation of the human myc gene's P2 promoter by the adenovirus E1a oncoprotein was shown to require an E2F transcription factor–binding site (GGCGGGAAAA) located between the ME1a1 and

ME1a2 nuclear factor–binding sites 5' of the P2 initiation site (272, 273, 320). However, *E1a* was also found to inhibit *myc* transcription (321, 322), again possibly underlying the important contribution(s) of cellular context. A mutated *p53* protein was recently found to *trans*-activate a human *myc* P2 promoter-driven reporter gene by an indirect mechanism, while wild-type *p53* suppressed a chimeric *myc/CAT* construct, implying that some of the cell growth control attributable to *p53* may be mediated via *c-myc* transcriptional control (323). Other evidence for the products of proto-oncogenes negatively regulating *myc* were reported early on by Hay et al (324), who identified a negative element 318–343 bp 5' of the human *myc* P1 that formed a complex with *fos/jun* heterodimers (324, 325) and more recently also an octamer-binding protein (325). Interestingly, a 13-bp sequence closely resembling an AP1 (*fos/jun* complex) transcription factor–binding site, believed to act as a negative regulator of the adipocyte *aP2* gene, was identified in this *myc* negative element (324). Recently, a novel *myc* RNA species (designated PO1) ~5 kb in size, containing no exon 1 sequences, exon 2, most of exon 3, and new exons at its 5' and 3' ends, was identified in fibroblastic cells transfected with the avian *v-erb* B oncogene (Y. Echelard, K. B. Marcu, E. N. Olson, and A. Nepveu, submitted for publication). Endogenous *c-myc* genes expressing PO1 RNAs were not rearranged, indicating that PO1 production involved dramatic, dominant alterations in *c-myc* gene regulation, possibly operating at both the transcriptional and posttranscriptional levels. PO1 *myc* RNAs were found to encode the 65-kDa *c-myc*-II polypeptide and a novel *myc* protein of 70 kDa. Cells expressing PO1 *myc* RNAs were malignant in mice. Unfortunately, the relationship between PO1 *myc* RNAs and *v-erb* B remains uncertain because sustained *v-erb* B expression was not required to maintain PO1 expression. Considering *myc*'s pivotal role in cellular growth and differentiation pathways, it is not at all unreasonable that regulators of *myc* may themselves represent the products of other oncoproteins, making these findings the tip of a very large iceberg.

Significance of c-myc Antisense Transcription

RNA polymerase II–derived *myc* antisense transcripts, denoted here as "cym," have been observed in various regions of the human and murine *c-myc* genes by nuclear run-on assay. However, "cym" transcripts neither accumulated (236, 309, 311) nor exhibited any modulation associated with cellular growth states or phenotypes. Cells expressing *myc* genes fused in a head-to-head configuration with Ig genes accumulate "cym-Ig" transcripts in addition to aberrant sense *myc* transcripts (326, 327). Such "cym-Ig" RNAs are found in polyribosomes, but their biological significance remains unknown. Therefore, at best, "cym" RNAs remain a molecular oddity awaiting some functional distinction.

Posttranscriptional Control of c-myc Expression

Variations in steady-state levels of *c-myc* polypeptides are only in part attributable to transcriptional modulation. Posttranscriptional control mechanisms play a key role in *myc* gene regulation, because its mRNAs (328–331) and their encoded polypeptides (331, 332) are normally highly unstable ($t_{1/2}$ of 10–30 min) and multiple molecular parameters are responsible. Posttranscriptional phenomena strongly contribute to downregulation of *myc* expression in proliferating and differentiating cells. Posttranscriptional mechanisms have been implicated in the rapid depletion of *c-myc* mRNAs in the later phase of fibroblastic cell responses to growth stimuli; the degree of this effect depends on the cell background and physiological status prior to mitogen addition (310). In several cases, the rapid cessation of de novo protein synthesis has abrogated some of these negative effects, implicating labile factors (328, 338). Remarkably, the overriding means of *myc* negative control in several types of cells—differentiating F9 teratocarcinoma cells (83, 111, 333, 334), P19 embryonal carcinoma cells induced to yield muscle or neuronal cells (110), MEL cells induced to become erythrocytes (292, 311, 312), and BC3H1 myogenic cells converting to myoblasts—is posttranscriptional (A. Nepveu, E. Olsen, K. B. Marcu, unpublished). However, in other cases, posttranscriptional effects are only contributory and at times controversial (335–337, 339, 340). Alterations in *myc* mRNA half-lives are insufficient to explain some of these posttranscriptional phenomena, necessitating the invocation of more "nebulous" nuclear factors. The degree of posttranscriptional regulation of *c-myc* can also be tissue-specific in vivo, as documented for adult mouse, lymphoid, liver, and brain, and fetal mouse liver development (341). *N-myc* transcripts are also strongly subject to posttranscriptional control, given their preferential accumulation in fetal mouse brain, even though the gene is transcribed fairly equivalently in brain, liver, spleen, and placenta (342).

Posttranscriptional phenomena have also been shown to engender positive effects on *myc* expression. When G0-arrested Chinese hamster lung fibroblasts were treated with growth factors, the dramatic increase in *c-myc* RNA levels was not accompanied by an appreciable change in transcription (343). The stability of *c-myc* RNA was altered in EBV-positive BJAB human B lymphoblastoid cells, with the half-life increasing from < 36 min. to > 70 min. (344). Initial increases in *c-myc* RNA levels in anti-Ig-treated WEHI.231 murine B cells were not entirely due to enhanced transcription (345, 346). A transient increase of *c-myc* expression during the commitment phase of P19 and MEL cell differentiation has a significant posttranscriptional component (110, 311). Posttranscriptional mechanisms are largely responsible for the dramatic induction of *c-myc* RNAs in regenerating kidney tissue and mouse liver in vivo (347–349). Similarly, experiments with a human *myc* gene introduced into FR3T3 rat fibroblasts have revealed that most of the inductive effect of serum was posttranscriptional and prior to mRNA stabilization (350). Stabilization of *c-myc* RNA is observed in *X. laevis* oocytes during oogenesis followed by destabilization after fertilization (351). Alterations in posttranscriptional control have also been suggested to contribute to *myc* activation in malignancies given that *myc* transcripts with abnormal structures (absence of first exon and presence of first intron sequences), produced as a consequence of chromosome translocation, have longer half-lives (332, 352–354).

Different portions of *c-myc* transcripts have been implicated as target sites for posttranscriptional control at the RNA level. Pei & Calame (355) reported that *myc* exon 1 was only able to confer rapid turnover in the context of other *myc* RNA sequences. Deletion/transfection studies of *myc* constructs in mouse cells found that exon 1 sequences were insufficient to confer transcript instability, but a 140-bp U-rich sequence in the 3' untranslated region (UTR) was sufficient (356). Many *c-myc* RNAs with truncated or chimeric 3' UTRs have greater stability (357). The responsible target sequences in the *myc* 3' UTR are rich in As and Us (AUUUA/UUAUUUA) and are involved in rapid turnover of a number of cellular RNAs (358, 359). Surprisingly, site-directed mutagenesis of these 3' UTR sequences in murine *c-myc* (AUUUA to AGGGA) showed no change in the half-life of *c-myc* mRNAs, implicating yet other ill-defined sequences in the body of the *myc* transcript (360). Although the *c-myc* gene contains two conserved polyadenylation sites, the second of which is used sixfold more frequently, there is no apparent function for their conservation (361). *c-myc* transcript degradation was recently found to occur step-wise in a 3' to 5' direction: shortening of the poly(A) tail (330, 362, 364–366), followed by removal of the adjacent A+U-rich sequences, and finally degradation of the body of the message. Using a cell-free mRNA decay system, a ribonucleoprotein that binds the A+U-rich sequence was found to contribute to *myc* RNA stability (364). Inhibition of translation-dependent *myc* poly(A) shortening also stabilizes the transcript (366). In apparent contrast to all of these observations, one group reported that de-adenylated *myc* transcripts in cells were more stable (365). Inhibition of translation-dependent poly(A) shortening also stabilizes the transcript, implying that a labile protein and/or translation is required for degradation of the transcript (366). The reasons for multiple sequences possibly providing redundant regulation remain unclear, but the modes of posttranscriptional control could conceiveably differ in various cell types.

Evidence for *myc* translational control has also been reported. The possibility that the noncoding first exon modulates the translational efficiency of *c-myc* RNA was first proposed by Saito et al (367). However, in vivo transfection studies indicated that the presence of 5' UTR sequences did not influence *myc* mRNA translational efficiency (368). In contrast, in vitro translation assays have shown that full-length *c-myc* mRNAs have lower translational efficiencies than those lacking exon 1 (369), and the presence of *myc* exon 1 sequences at the 5' ends of heterologous transcripts inhibited their in vitro translation (370). Some of these translational inhibitory effects may be due to increased secondary structure (371, 372), and *trans*-acting factors that melt other cellular RNAs may relieve them (373–375). Translational effects engendered by *myc* exon 1 sequences have been observed in vivo; modulation of *myc* translational control was reported in *X. laevis* oocytes and fertilized eggs (376). Here, a 360-nt portion of murine exon 1 fused 5' of the *CAT* gene decreased translational efficiency, but only at the oocyte stage and not in mature or in vitro fertilized eggs (376). Discordancies between levels of *myc* RNAs and proteins in tissue culture cells expressing exogenous *myc* genes implied that translational control may normally operate to keep *myc* protein quantities in check (152). Indeed, one study reported that the induced overexpression of *myc* polypeptides was toxic in Chinese hamster ovary cells (377). More recently, one group of investigators has obtained evidence for unusually high levels of *myc* polypeptide in human myeloma cell lines without concomitant changes in *myc* mRNAs or polypeptide half-life, implying that *myc* translational control may somehow be circumvented in some malignancies (N. F. Sullivan, personal communication). Clearly, *myc* translational control is both controversial and in need of mechanistic details, which we hope are just over the horizon.

MODES OF *myc* ACTIVATION IN MALIGNANCIES

Gene Amplification

Amplification of the *myc* family of oncogenes has been observed late in the progression of a number of human tumors and generally associated with an aggressively malignant phenotype (378–380). Elevated *c-myc* expression through gene amplification has been seen in gastric adenocarcinoma cells (381), variant small-cell lung carcinoma (SCLC) (31, 382), glioblastoma (383), carcinoma of the breast (384, 385), colon carcinoma (386), plasma-cell leukemia (387), promyelocytic leukemia (388, 389), and granulocytic leukemia (390). The HL60 promyelocytic leukemia cell line is atypical in that *c-myc* is also amplified in the primary tumor (389). *N-myc* is amplified in several neuroblastoma cell lines (17, 378, 391), retinoblastomas (30), SCLC (216), and in one malignant astrocytoma (393). To date, *L-myc* amplification seems to occur solely in SCLC (18).

Proviral Insertion

Activation by proviral insertion has been observed for both *c-* and *N-myc*. In bursal lymphomagenesis, the *c-myc* locus was found to be a common integration site of avian leucosis virus (ALV), which resulted in 20–100-fold upregulated *myc* expression (394–396). Three configurations of ALV insertions, relative to *c-myc*, have been described: (*a*) integration upstream of *c-myc* coding sequences in the same transcriptional orientation, producing a hybrid mRNA containing proviral sequences fused to *c-myc* 5' sequences, (*b*) insertion upstream of *c-myc* but in the opposite transcriptional orientation, (*c*) insertion downstream, creating a hybrid transcript with proviral sequences fused to the 3' end of *c-myc* mRNA (394). The mechanism(s) by which proviral insertion enhances *c-myc* expression appears complex. In the case where the proviral LTR is inserted upstream of *c-myc* in a positive orientation, *c-myc* is probably being driven by the viral LTR with its normal transcriptional control disrupted. By inserting downstream of *c-myc*, the provirus could disrupt numerous negative regulatory elements, or the effects of a LTR enhancer could be positively regulating *myc* expression. The same argument could be made for proviral insertion upstream of *c-myc* but in opposite orientation.

Insertional activation of *myc* by Moloney murine leukemia virus (Mo-MuLV) in mice has been shown to contribute to T cell lymphomagenesis (397–400). In the great majority of tumors exhibiting proviral activation of the *c-myc* locus, the integration site was upstream of *c-myc* (or more rarely within exon 1), generally in the opposite transcriptional orientation (397, 398). In about 35% of T cell lymphomas *N-myc* was found to be activated by proviral insertion (400). In contrast to *c-myc*, the *N-myc* integration site was limited to a small segment within its 3' noncoding sequence, suggesting that this region contains regulatory elements that normally downregulate *N-myc* expression (400).

Finally, both *c-* (401) and *N-myc* (402) were found to be activated by woodchuck hepatitis virus (WHV) in hepatocellular carcinoma. In one study, *c-myc* was found to be activated by insertion of viral sequences either upstream in the opposite transcriptional orientation, or downstream in the same orientation (401). Another study found that the WHV activated *N-myc* by insertion 3' of coding sequences in manner similar to that seen in T cell lymphomas activated by Mo-MuLV (402). In contrast, human hepatitis B virus integrates into the host genome in a large number of hepatocellular carcinomas, but a single, nonrandom site of integration has not been observed (403).

Chromosomal Translocations

Nonrandom chromosomal translocations involving the *c-myc* locus have been described in a number of tumors. Along with the well-documented *myc* locus rearrangements in BL and MPC (6–9, 11, 209, 404, 405), translocations involving *c-myc* have been described in T cell leukemias (406–408), B cell acute lymphocytic leukemia (409, 410), and rat immunocytomas (411, 412).

In both MPC and BL, the *myc* allele, (residing on chromosomes 15 and 8, respectively) is reciprocally translocated to either the immunoglobulin heavy (IgH) or light chain (IgL) loci. Only the translocated *myc* allele is expressed in these malignancies, with the normal allele being transcriptionally silent (209). This phenomenon strongly implies that *myc* expression in BL and MPC is abnormal. The most common translocation seen in the neoplasms results in a head-to-head (5'-5' orientation) of the *c-myc* and IgH loci. The *c-myc* coding exons are left intact with the break occurring either 5' of exon 1 or within exon 1 or intron 1. Variant translocations have been described that place one of the immunoglobulin light chain loci several hundred kb 3' of *c-myc*. A number of these distant 3' *c-myc* rearrangements reside at a common locus denoted *pvt-1*. Large (> 12 kb) transcripts originate from *pvt-1*, and smaller transcripts representing fusions with IgC kappa sequences have been seen in the variant translocations (413, 414). The normal function of *pvt-1* and its role in malignancy remain important unanswered queries. Evidence for both lambda and kappa translocations has been seen in BL, but in MPC only kappa translocations have been documented (9). In BL, the majority of translocated *myc* genes wherein the IgH locus fusions occurred outside of the *myc* gene accumulated mutations, within the gene's first exon and intron, suggesting that some translocated *myc* alleles are subjected to a similar if not analogous phenomenon (hypermutation) that mutates rearranged Ig-variable-region gene segments (404). Exon 1 mutations have been shown to contribute to the abrogation of RNA polymerase II blockage, thereby resulting in deregulated *myc* transcription and presumably higher levels of cytoplasmic mRNAs (415). However, other recent data have demonstrated that the presence of such exon 1 mutations are alone insufficient to abrogate transcription blockage in other B cell backgrounds (239). In one BL cell line harboring a variant translocation of *c-myc* with the lambda light chain (8:22), a point mutation in intron 1 abrogated binding of a nuclear protein (277). Five of of seven BL alleles tested for binding of this factor showed mutations that altered its binding capacity (277). How this intron-binding factor relates to normal *c-myc* regulation is not known.

c-myc expression in MPC and BL is elevated over that seen in resting B cells, but is not elevated over the level seen in EBV-immortalized B lymphocytes (32, 239). Expression of *myc* protein from a non-AUG codon near the 3' end of exon 1 has been shown to be absent in some BL lines (32). In the Raji BL line, expression from *c-myc*'s upstream promoter, P0, was restricted to the translocated allele, although expression from the nontranslocated allele could be induced by TPA while P1 and P2 remained silent (416). P3-initiated transcripts were seen from both translocated and nontranslocated alleles (416). Taken together, these observations indicated that P0, P1, P2, and P3 were independently regulated and differentially affected in *c-myc* deregulation in BL.

A shift in promoter usage in favor of P1 has been consistently reported for translocated alleles in BL and MPC (237, 242). The P2 start site is preferred over P1 because its promoter is inherently stronger (299; A. Nepveu, personal communication). Translocated *myc* alleles transfected into other cell backgrounds transcriptionally recover their normal P1:P2 ratios along with the exon 1 transcriptional block, suggesting that most of the block originates from P2-initiated mRNAs (239). Differential *myc* mRNA stability does not appear responsible for alterations in P1:P2 ratios in neoplasia, since these transcripts retain their inherent instability as seen in normal cells (243). Both cell background and presence of immunoglobulin sequences seem to contribute to the deregulated *c-myc* transcription. Perhaps immunoglobulin sequences enhance the "accessibility" of the *c-myc* gene to the transcription machinery, or enhancer elements of the immunoglobulin loci have dominant, differential effects on *c-myc* expression. However, the intronic enhancers of the IgH and IgL loci are not directly involved with any consistency, suggesting that other enhancers located elsewhere in the various Ig loci may be more relevant for deregulated *myc* activity. In support of this view, a second IgH locus enhancer positioned some distance 3' of C-alpha would remain associated with the majority of the BL T(8;14) and MPC T(12;15) IgH-*myc* rearrangements (416a). Furthermore, recent findings indicate that the 3' C-kappa enhancer may be responsible for the P2 to P1 promoter shift in BL variant translocations (416b; G. Bornkamm, personal communication).

Since the presence of a translocated *c-myc* allele is a common feature of BL and MPC, a question arises as to the mechanism of this nonrandom translocation. It was proposed early on that the *myc* locus may be a substrate for the lymphoid cell-specific recombination machinery normally responsible for IgH locus rearrangements during B cell ontogeny (404). Therefore, when immunoglobulin loci are actively undergoing intramolecular rearrangements, they may also be prone to rearrangements with other cellular genes, but direct evidence for the *c-myc* locus in this phenomenon has never been found. On the contrary, results of experiments to test directly if *c-myc* was a likely target for Ig CH switch-recombination activity in B cells were negative (404a). It would be interesting to determine whether *RAG1* and *RAG2* (the proposed recombination activating genes) required for IgV gene assembly play a role in *c-myc* rearrangements. However, since their *cis*-acting target sequences are not found near all sites of *c-myc* translocations, this appears unlikely. Piccoli et al reported the presence of a GAGG sequence near *c-myc* rearrangements (404b), but this would seem too simplistic a target site to mediate this phenomenon and again is not found at the site of all *myc* translocations. Why is the *c-myc* gene the only oncogene thus far involved in these translocations in BL and MPC? This is likely to be a selected phenomenon, such that nonproductive translocations not leading to a malignant phenotype are lost. Perhaps *c-myc* resides at a fragile chromosomal site more prone to breakage (404c).

myc INTERACTIONS WITH TUMOR SUPPRESSORS

The discovery of tumor suppressor genes, such as the retinoblastoma susceptibility gene (*pRB*) and *p53*, has led to a great deal of speculation as to the mechanism(s) by which these suppressor genes may be exerting their effect(s). One could theorize that the product of these genes would interact, either directly or indirectly, with normal cellular genes that are involved in cell growth and/or proliferation. The suppressor genes could regulate the activity of these growth-associated cellular genes. The loss of function of the proposed tumor suppressors through either a gross deletion or some other mutational event could perturb the normal regulatory loop of these cell cycle–associated genes, such as *c-myc*, thus leading to inappropriate expression and possible tumor progression. Recently, the interaction of *c-myc* with *pRB* and *p53* has been under scrutiny, and not unexpectedly, the results have proven both titillating and confusing, not to mention controversial.

Pietenpol et al (283) have shown that TGF-β downregulates *c-myc* expression in mouse keratinocytes in a manner that seems to be mediated by *pRB* (286). The effect of TGF-β could be blocked by the viral proteins HPV-16 E7, adenovirus type 5 *E1a*, and SV40 large T antigen, when these transforming proteins had an intact *pRB*-binding domain (283). Adding confusion to this issue is a report showing that *c-myc* expression is induced, rather than downregulated, in mink lung epithelial cells by TGF-β (418). A recent paper has shown that the *pRB* protein can bind directly to the *c*- and *N-myc* proteins, and a rough map of the binding site on *c-myc* was given (419). There seem to be two *pRB*-binding sites in the *c-myc* protein between amino acids numbered 41 to 178, as a deletion of these residues abrogated binding of *c-myc* to *pRB*. Two sites are thought to exist, because *pRB* binds to mutant *c-myc* proteins containing deletions of residues 1–74 or 56–103 or 109–204. The domain of *pRB* that interacts with *myc* overlaps the binding site for the HPV-16 E7 protein (420), as it is able to compete with *c-myc* for binding to *pRB*. *E1a* has been shown to occupy the same binding site on *pRB* (421, 422). Interestingly, *c-myc*, like *E1a*, cooperates with an activated *ras* oncogene in transformation (see tumorigenesis section) and has now been shown to bind *pRB* at a site that is functionally relevant, as this domain in *RB* is often deleted or mutated in various human tumors (423–429). However, in vivo evidence for an interaction between *pRB* and *c-myc* remains to be established.

The interaction of *c-myc* with *p53* is not as clearly defined as that with *pRB*. The *c-myc* protein has not been shown to bind directly to *p53* (419), though wild-type *p53* has been shown to suppress transcription from a *c-myc/CAT* construct, while mutant *p53* actually induces transcription from these constructs (323). The effect of *p53* on *myc* is most likely not due to its direct interaction with the *c-myc* promoter, because a mutant *p53* lacking the protein's DNA-binding domain retained repressor activity. However, the *p53* polypeptide's nuclear localization signal was essential for its effect on the *c-myc* promoter (323).

Mechanistically, the interaction of *myc* with antioncogenes is appealing, as *pRB* could directly bind the *c-myc* protein, and one could envision a model whereby *c-myc* would be sequestered from the available active pool owing to its interaction with *pRB* and its activity effectively neutralized. The loss of this association of *c-myc* with *pRB* may lead to overexpression or inappropriate expression of *c-myc* and contribute to tumor progression. It is critical to determine whether *c-myc* binds to the underphosphorylated, active, growth-suppressing form of *pRB* (430–432)—and whether the association of *pRB* and *c-myc* renders *c-myc* inactive, as the *pRB*-binding domain (419) maps to one region of *c-myc* that is required for transformation but is distinct from *c-myc*'s BR-H-L-H and leucine zipper regions, which are also essential for *c-myc*'s function in transformation (44, 45).

RETROSPECTIVE

In considering an appropriate title for our review, we came down to two choices: (*a*) the one provided, and (*b*) "Everything you wanted to know about *myc* but were afraid to ask." We decided on the former, more conservative choice because *myc* itself is sufficiently provocative. Data base searches revealed about 3000 citations on *myc* since its original discovery about 12 years ago. If we have neglected to direcly reference a particular laboratory's contributions, we hope that these can be found in an earlier review (of which a number have been cited). We conclude that *myc* is a lot of fun to work on, but may be like a puzzle that has one too many solutions. Where does *myc* go from here? Believe it or not, as Ripley aptly put it once too often, many places! In the coming years, we will likely witness the molecular and biochemical dissection of the cellular pathways leading to and away from the *myc* polypeptide. More specifically, we anticipate: (*a*) a proof of a role for *myc* in DNA replication, (*b*) direct demonstrations for *c-myc*, *N-myc*, and *L-myc* heterodimers with *max*-activating synthetic and natural promoters, which we hope will lead to the elucidation of differential functions for the other members of the *myc* family, (*c*) elucidation of the mechanism of *myc* autoregulation and its role in the maintenance of cell growth control, (*d*) identification of the immediate cellular targets of *myc* that determine cell cycle progression and/or a cell's differentiated fate. For diehard *myc* aficionados who also happen to remain optimistic souls, we predict that the best is yet to come!

ACKNOWLEDGMENTS

We thank our many colleagues who provided their unpublished observations (especially R. N. Eisenman, M. Groudine, A. Nepveu, N. Sullivan, A.

Kinniburgh, J. Cogswell, J. F. Mushinski, and K. Nilsson). KBM thanks Margarita Reyes for her heroic efforts in manuscript preparation. SAB and AJP also gratefully acknowledge C. Bossone and I. Stupakoff for their help and patience. We dedicate this review to the memory of Dr. Paul Fahrlander, for all that he was and all that he might have become.

Literature Cited

1. Hayward, W. S., Neil, B. G., Astrin, S. M. 1981. *Nature* 209:475–79
2. Sheiness, D., Bishop, J. M. 1979. *J. Virol.* 31:514–21
3. Bishop, J. M. 1983. *Annu. Rev. Biochem.* 52:301–54
4. Varmus, H. E. 1984. *Annu. Rev. Genet.* 28:553–612
5. Klein, G. 1981. *Nature* 294:313–18
6. Klein, G., Klein, E. 1985. *Immunol. Today* 6:208–15
7. Kelly, K., Siebenlist, U. 1986. *Annu. Rev. Immunol.* 4:317–38
8. Cole, M. D. 1986. *Annu. Rev. Genet.* 20:361–84
9. Marcu, K. B. 1987. *BioEssays* 6:28–32
10. Luscher, B., Eisenman, R. N. 1990. *Genes Dev.* 4:2025–35
11. Spencer, C. A., Groudine, M. 1991. *Adv. Cancer Res.* 56:1–48
12. Zimmerman, K., Alt, F. W. 1990. *Crit. Rev. Oncogen.* 2:75–95
13. Landschulz, W. H., Johnson, P. F., McKnight, S. L. 1988. *Science* 240:1759–64
14. Murre, C., McCaw, P. S., Baltimore, D. 1989. *Cell* 56:777–83
15. Johnson, P. F., McKnight, S. L. 1989. *Annu. Rev. Biochem* 58:799–839
16. Kohl, N. E., Kanda, N., Schreck, R. R., Bruns, G., Latt, S. A., et al. 1983. *Cell* 35:359–67
17. Schwab, M., Alitalo, K., Klempnauer, K. H., Varmus, H. E., Bishop, J. M., et al. 1983. *Nature* 305:245–48
18. Nau, M. M., Brooks, B. J., Battey, J., Sausville, E., Gazdar, A. F., et al. 1985. *Nature* 318:69–73
19. DePinho, R. A., Hatton, K. S., Tesfaye, A., Yancopoulos, G. D., Alt, F. W. 1987. *Genes Dev.* 1:1311–26
20. Ingvarsson, S., Asker, C., Axelson, H., Klein, G., Sumegi, J. 1988. *Mol. Cell. Biol.* 8:3168–74
21. Sugiyama, A., Kume, A., Nemoto, K., Lee, S., Asami, Y., et al. 1989. *Proc. Natl. Acad. Sci. USA* 86:9144–48
22. Vriz, S., Taylor, M., Mechali, M. 1989. *EMBO J.* 8:4091–97
23. Pfeifer-Ohlsson, S., Rydnert, J., Goustin, A. S., Larsson, E., Betsholtz, C., et al. 1985. *Proc. Natl. Acad. Sci. USA* 82:5050–54
24. Jakobovits, A., Schwab, M., Bishop, J. M., Martin, G. R. 1985. *Nature* 318:188–91
25. Zimmerman, K., Yancopoulos, G., Collum, R., Smith, R., Kohl, N., et al. 1986. *Nature* 319:780–83
26. Schmid, P., Schulz, W. A., Hameister, H. 1989. *Science* 243:226–29
27. Downs, K. M., Martin, G. R., Bishop, J. M. 1989. *Genes Dev.* 3:860–69
28. Semsei, I., Ma, S., Cutler, R. G. 1989. *Oncogene* 4:465–71
29. Ono, T., Takahashi, N., Okada, S. 1989. *Mutat. Res.* 219:39–50
30. Lee, W. H., Murphree, A. L., Benedict, W. F. 1984. *Nature* 309:458–60
31. Krystal, G., Birrer, M., Way, J., Nau, M., Sausville, E., et al. 1988. *Mol. Cell. Biol.* 8:3373–81
32. Hann, S. R., King, M. W., Bentley, D. L., Anderson, C. W., Eisenman, R. N. 1988. *Cell* 52:185–95
33. Hann, S. R., Eisenman, R. N. 1984. *Mol. Cell. Biol.* 4:2486–97
34. Persson, H., Leder, P. 1984. *Science* 225:728–31
35. Dang, C. V., Lee, W. M. F. 1988. *Mol. Cell. Biol.* 8:4048–54
36. Dang, C. V., Dam, H. V., Buckmire, M., Lee, W. M. F. 1989. *Mol. Cell. Biol.* 9:2477–86
37. Luscher, B., Kuenzel, E. A., Krebs, E. G., Eisenman, R. N. 1989. *EMBO J.* 8:1111–19
38. Donner, P., Greiser-Wilke, I., Moelling, K. 1982. *Nature* 296:262–64
39. Eisenman, R. N., Tachibana, C. Y., Abrams, H. D., Hann, S. R. 1985. *Mol. Cell. Biol.* 5:114–26
40. Evan, G. I., Hancock, D. C. 1985. *Cell* 43:253–61
41. Waitz, W., Loide, P. 1991. *Oncogene* 6:29–35
42. Land, H., Parada, L. F., Weinberg, R. A. 1983. *Nature* 304:596–602
43. Mougneau, E., Lemieux, L., Rassoulzadegan, M., Cuzin, F. 1984. *Proc. Natl. Acad. Sci. USA* 81:5758–62

44. Sarid, J., Halazonetis, T. D., Murphy, W., Leder, P. 1987. *Proc. Natl. Acad. Sci. USA* 84:170–73
45. Stone, J., de Lange, T., Ramsay, G., Jakobovits, E., Bishop, J. M., et al. 1987. *Mol. Cell. Biol.* 7:1697–709
46. Gazin, C., Dupont de Dinechin, S., Hampe, A., Masson, J., Maring, P., et al. 1984. *EMBO J.* 3:383–87
47. Gazin, C., Rigolet, M., Briand, J. P., Van Regenmortal, M. H. V., Galibert, F. 1986. *EMBO J.* 5:2241–50
48. Ramsay, G., Stanton, L., Schwab, M., Bishop, J. M. 1986. *Mol. Cell. Biol.* 6:4450–57
49. Makela, T. P., Saksela, K., Alitalo, K. 1989. *Mol. Cell. Biol.* 9:1545–52
50. Legouy, E., DePinho, R., Zimmerman, K., Collum, R., Yancopoulos, G., et al. 1987. *EMBO J.* 6:3359–66
51. DeGreve, J., Battey, J., Fedorko, J., Birrer, M., Evan, G., et al. 1988. *Mol. Cell. Biol.* 8:4381–88
52. Ikegaki, N., Minna, J., Kennett, R. H. 1989. *EMBO J.* 8:1793–99
53. Dedieu, J., Gazin, C., Rigolet, M., Galibert, F. 1988. *Oncogene* 3:523–29
54. Persson, H., Gray, H. E., Godeau, F. 1985. *Mol. Cell. Biol.* 5:2903–12
55. Beckmann, H., Su, L-K., Kadesch, T. 1990. *Genes Dev.* 4:167–79
56. Carr, C. S., Sharp, P. A. 1990. *Mol. Cell. Biol.* 10:4384–88
57. Gregor, P. D., Sawadogo, M., Roeder, R. G. 1990. *Genes Dev.* 4:1730–40
58. Murre, C., McCaw, P. S., Vaessin, H., Caudy, M., Jan, L. Y., et al. 1989. *Cell* 58:537–44
59. Dang, C. V., McGuire, M., Buckmire, M., Lee, W. M. F. 1989. *Nature* 337:664–66
60. Blackwell, T. K., Kretzner, L., Blackwood, E. M., Eisenman, E. N., Weintraub, H. 1990. *Science* 250:1149–51
61. Blackwell, T. K., Weintraub, H. 1990. *Science* 250:1104
62. Prendergast, G. C., Ziff, E. B. 1991. *Science* 251:186–89
63. Blackwood, E. M., Eisenman, E. N. 1991. *Science* 251:1211–17
64. Prendergast, G. C., Lawe, D., Ziff, E. B. 1991. *Cell* 65:395–408
65. Cochran, B. H., Reffel, A. C., Stiles, C. D. 1983. *Cell* 33:939–47
66. Kelley, K., Cochran, B. H., Stiles, C. D., Leder, P. 1983. *Cell* 35:603–10
67. Greenberg, M. E., Ziff, E. B. 1984. *Nature* 311:433–38
68. Lau, L. F., Nathans, D. 1987. *Proc. Natl. Acad. Sci. USA* 84:1182–86
69. Pledger, W. J., Stiles, C. D., Antoniades, H. N., Scher, C. D. 1977. *Proc. Natl. Acad. Sci. USA* 74:4481–85
70. Pledger, W. J., Stiles, C. D., Antoniades, H. N., Scher, C. D. 1978. *Proc. Natl. Acad. Sci. USA* 75:2839–43
71. Sassone-Corsi, P., Lamph, W. W., Kamps, M., Verma, I. M. 1988. *Cell* 54:553–60
72. Rauscher, F. J., Cohen, D. R., Curran, T., Bos, T. J., Vogt, P. K., et al. 1988. *Science* 240:1010–16
73. Thompson, C. B., Challoner, P. B., Neiman, P. E., Groudine, M. 1985. *Nature* 314:363–66
74. Hann, S. R., Thompson, C. B., Eisenman, R. N. 1985. *Nature* 314:366–69
75. Habenicht, A. J. R., Glomset, J. A., King, W. C., Nist, C., Mitchell, C. D., Ross, R. 1981. *J. Biol. Chem.* 256: 12329–35
76. Nishizuka, Y. 1984. *Nature* 308:693–98
77. Hokin, L. E. 1985. *Annu. Rev. Biochem.* 54:205–35
78. Coughlin, S. R., Lee, W. M. F., Williams, P. W., Giels, G. M., Williams, L. T. 1985. *Cell* 43:243–51
79. Rozengurt, E., Stroobant, P., Waterfield, M. D., Deuel, T. F., Keehan, M. 1983. *Cell* 34:265–72
80. Ran, W., Dean, M., Levine, R. A., Henkle, C., Campisi, J. 1986. *Proc. Natl. Acad. Sci. USA* 83:8216–20
81. Frick, K. K., Scher, C. D. 1990. *Mol. Cell. Biol.* 10:184–92
82. Dean, M., Levine, R. A., Ran, W., Kindy, M. S., Sonenshein, G. E. 1986. *J. Biol. Chem.* 261:9161–66
83. Nepveu, A., Levine, R. A., Campisi, J., Greenberg, M. E., Ziff, E. B., Marcu, K. B. 1987. *Oncogene* 1:243–50
84. Liboi, E., Pelosi, E., Testa, U., Peschle, C., Rossi, G. B. 1986. *Mol. Cell Biol.* 6:2275–78
85. Cutry, A. F., Kinniburgh, A. J., Krabak, M. J., Hui, S-W., Wenner, C. E. 1989. *J. Biol. Chem.* 264:19700–5
86. Reed, J. C., Nowell, P. C., Hoover, R. G. 1985. *Proc. Natl. Acad. Sci. USA* 82:4221–24
87. Reed, J. C., Alpers, J. D., Nowell, P. C., Hoover, R. G. 1986. *Proc. Natl. Acad. Sci. USA* 83:3982–86
88. Lindsten, T., June, C. H., Thompson, C. B. 1988. *EMBO J.* 7:2787–94
89. Reed, J. C., Alpers, J. D., Scherle, P. A., Hoover, R. G., Nowell, P. C. 1987. *Oncogene* 1:223–28
90. Lacy, J., Sarkar, S. N., Summers, W. C. 1986. *Proc. Natl. Acad. Sci. USA* 83:1458–62
91. Phillips, N. E., Parker, D. C. 1987. *Mol. Immunol.* 24:1199–205
92. Tisch, R., Roifman, C. M., Hozumi, N. 1988. *Proc. Natl. Acad. Sci. USA* 85: 6914–18

93. Armelin, H. A., Armelin, M. C. S., Kelly, K., Steward, T., Leder, P., et al. 1984. *Nature* 310:655–60
94. Kaczmarek, L., Hyland, J. K., Watt, R., Rosenberg, M., Baserga, R. 1985. *Science* 228:1313–15
95. Sorrentino, V., Drozdoff, V., McKinney, M. D., Zeitz, L., Fleissner, E. 1986. *Proc. Natl. Acad. Sci. USA* 83:8167–71
96. Stern, D. F., Roberts, A. B., Roche, N. S., Sporn, M. B., Weinberg, R. A. 1986. *Mol. Cell Biol.* 6:870–77
97. Rapp, U. R., Cleveland, J. L., Brightman, K., Scott, A., Ihle, J. N. 1985. *Nature* 317:434–38
98. Dean, M., Cleveland, J. L., Rapp, U. R., Ihle, J. N. 1987. *Oncogene Res.* 1:279–96
99. Wright, T. C., Pukac, L. A., Castellot, J. J., Karnovsky, M. J., Levine, R. A., et al. 1989. *Proc. Natl. Acad. Sci. USA* 86:3199–203
100. Karn, J., Watson, J. V., Lowe, A. D., Green, S. M., Vedeckis, W. 1989. *Oncogene* 4:773–87
101. Gai, X. X., Rizzo, M-G., Valpreda, S., Baserga, R. 1989. *Oncogene Res.* 5:111–20
102. Eilers, M., Picard, D., Yamamoto, K. R., Bishop, J. M. 1989. *Nature* 340:66–68
103. Eilers, M., Schirm, S., Bishop, J. M. 1991. *EMBO J.* 10:133–41
104. Lachman, H. M., Skoultchi, A. I. 1984. *Nature* 310:592–94
105. Westin, E. H., Wong-Staal, F., Gelmann, E. P., Dalla-Favera, R., Papas, T. S., et al. 1982. *Proc. Natl. Acad. Sci. USA* 79:2490–94
106. Reitsma, P. H., Rothberg, P. G., Astrin, S. M., Trial, J., Bar-Shavit, Z., et al. 1983. *Nature* 306:492–94
107. Einat, M., Resnitzky, D., Kimchi, A. 1985. *Nature* 313:597–600
108. Bianchi-Scarra, G. L., Romani, M., Coviello, D. A., Garre, C., Ravazzolo, R., et al. 1986. *Cancer Res.* 46:6327–32
109. Dotto, G. P., Gilman, M. Z., Maryama, M., Weinberg, R. A. 1986. *EMBO J.* 5:2853–57
110. St-Arnaud, R., Nepveu, A., Marcu, K. B., McBurney, M. W. 1988. *Oncogene* 3:553–59
111. Campisi, J., Gray, H. E., Pardee, A. B., Dean, M., Sonenshein, G. E. 1984. *Cell* 36:241–47
112. Gonda, T. J., Metcalf, D. 1984. *Nature* 310:249–51
113. Resnitzky, D., Yarden, A., Zipori, D., Kimchi, A. 1986. *Cell* 46:31–40
114. Herald, K. M., Rothberg, P. G. 1988. *Oncogene* 3:423–28
115. Endo, T., Nadal-Ginard, B. 1986. *Mol. Cell. Biol.* 6:1412–21
116. Spotts, G. D., Hann, S. R. 1990. *Mol. Cell. Biol.* 10:3952–64
117. Spizz, G., Roman, D., Strauss, A., Olson, E. N. 1986. *J. Biol. Chem.* 261:9483–88
118. Schneider, M. D., Payne, P. A., Ueno, H., Perryman, M. B., Roberts, R. 1986. *Mol. Cell. Biol.* 6:4140–43
119. Greenberg, M. E., Greene, L. A., Ziff, E. B. 1985. *J. Biol. Chem.* 260:14101–110
120. Larsson, L-G., Gray, H. E., Totterman, T., Pettersson, U., Nilsson, K. 1987. *Proc. Natl. Acad. Sci. USA* 84:223–27
121. Larsson, L-G., Schena, M., Carlsson, M., Sallstrom, J., Nilsson, K. 1991. *Blood* 77:1025–32
122. Kato, K., Kanamori, A., Kondoh, H. 1990. *Mol. Cell. Biol.* 10:486–91
123. Thiele, C. J., Reynolds, C. P., Israel, M. A. 1985. *Nature* 313:404–6
124. Larcher, J. C., Vayssiere, J. L., Lossouarn, L., Gros, F., Croizat, B. 1991. *Oncogene* 6:633–38
125. Coppola, J. A., Cole, M. D. 1986. *Nature* 320:760–63
126. Dmitrovsky, E., Kuehl, W. M., Hollis, G. F., Kirsch, I. R., Bender, T. P., et al. 1986. *Nature* 322:748–50
127. Prochownik, E. V., Kukowska, J. 1986. *Nature* 322:848–50
128. Kaneko-Ishino, T., Kume, T. U., Sasaki, H., Obinata, M., Oishi, M. 1988. *Mol. Cell. Biol.* 8:5545–48
129. Birrer, M. J., Raveh, L., Dosaka, H., Segal, S. 1989. *Mol. Cell. Biol.* 9:2734–37
130. Prochownik, E. V., Kukowska, J., Rodgers, C. 1988. *Mol. Cell. Biol.* 8:3683–95
131. Coppola, J. A., Parker, J. M., Schuler, G., Cole, M. D. 1989. *Mol. Cell. Biol.* 9:1714–20
132. Lachman, H., Cheng, G., Skoultchi, A. 1986. *Proc. Natl. Acad. Sci. USA* 83:6480–84
133. Yokoyama, K., Imamoto, F. 1987. *Proc. Natl. Acad. Sci. USA* 84:7363–67
134. Holt, J. T., Redner, R. L., Nienhuis, A. W. 1988. *Mol. Cell. Biol.* 8:963–73
135. Griep, A. E., Westphal, H. 1988. *Proc. Natl. Acad. Sci. USA* 85:6806–10
136. Onclercq, R., Babinet, C., Cremisi, C. 1989. *Oncogene Res.* 4:293–302
137. Nishikura, K., Kim, U., Murray, J. M. 1990. *Oncogene* 5:981–88
138. Larsson, L-G., Ivhed, I., Gidlund, M., Pettersson, U., Vennstrom, B., et al. 1988. *Proc. Natl. Acad. Sci. USA* 85:2638–42

139. Freytag, S. O. 1988. *Mol. Cell. Biol.* 8:1614–24
140. Falcone, G., Tato, F., Alema, S. 1985. *Proc. Natl. Acad. Sci. USA* 82:426–30
141. Schneider, M. D., Perryman, M. B., Payne, P. A., Spizz, G., Roberts, R., et al. 1987. *Mol. Cell. Biol.* 7:1973–77
142. Jackson, T., Allard, M. F., Sreenan, C. M., Doss, L. K., Bishop, S. P., et al. 1990. *Mol. Cell. Biol.* 10:3709–16
143. Miner, J. H., Wold, B. J. 1991. *Mol. Cell. Biol.* 11:2842–51
144. Ruley, H. E. 1983. *Nature* 304:602–6
145. Rassoulzadegan, M. A., Cowie, A., Carr, A., Glaichenhaus, N., Kamen, R., et al. 1982. *Nature* 300:713–18
146. Eliyahu, D., Raz, A., Gruss, P., Givol, D., Oren, M. 1984. *Nature* 312:646–49
147. Hunter, T. 1991. *Cell* 64:249–70
148. Birrer, M. J., Segal, S., DeGreve, J. S., Kaye, F., Sausville, E. A., et al. 1988. *Mol. Cell. Biol.* 8:2668–73
149. Schwab, M., Varmus, H. E., Bishop, J. M. 1985. *Nature* 316:160–62
150. Yancopoulos, G. D., Nisen, P. D., Tesfaye, A., Kohl, N. E., Goldfarb, M. P., et al. 1985. *Proc. Natl. Acad. Sci. USA* 82:5455–59
151. Small, M. B., Hay, N., Schwab, M., Bishop, J. M. 1987. *Mol. Cell. Biol.* 7:1638–45
152. Zerlin, M., Julius, M. A., Cerni, C., Marcu, K. B. 1987. *Oncogene* 1:19–27
153. Mougneau, E., Cerni, C., Tillier, F., Cuzin, F. 1988. *Oncogene Res.* 2:177–88
154. Keath, E. J., Caimi, P. G., Cole, M. D. 1984. *Cell* 39:339–48
155. Lombardi, L., Newcomb, E. W., Dalla-Favera, R. 1987. *Cell* 49:161–70
156. Land, H., Chen, A. C., Morgenstern, J. P., Parada, L. F., Weinberg, R. A. 1986. *Mol. Cell. Biol.* 6:1917–25
157. Kohl, N. E., Ruley, H. E. 1987. *Oncogene* 2:41–48
158. Lee, M. F., Schwab, M., Westaway, D., Varmus, H. E. 1985. *Mol. Cell. Biol.* 5:3345–56
159. Storer, R. D., Allen, H. L., Kaynak, A. R., Bradley, M. O. 1988. *Oncogene* 2:141–47
160. Schwartz, R., Stanton, L. W., Marcu, K. B., Witte, O. 1986. *Mol. Cell. Biol.* 6:3221–31
161. Overall, R. W., Weisser, K. E., Hess, B., Namen, A. E., Grabstein, K. H. 1989. *Oncogene* 4:1425–32
162. Crook, T., Almond, N., Murray, A., Stanley, M., Crawford, L. 1989. *Proc. Natl. Acad. Sci. USA* 86:5713–17
163. Clynes, R., Wax, J., Stanton, L. W., Smith-Gill, S., Potter, M., et al. 1988. *Proc. Natl. Acad. Sci. USA* 85:6067–71
164. Rapp, U. R., Cleveland, J. L., Frederickson, T. N., Holmes, K. L., Morse, H. C., et al. 1985. *J. Virol.* 55:23–33
165. Principato, M., Cleveland, J. L., Rapp, U. R., Holmes, K. L., Pierce, J. H., et al. 1990. *Mol. Cell. Biol.* 10:3562–68
166. Kurie, J. M., Morse, H. C. III, Principato, M. A., Wax, J. S., Troppmair, J., et al. 1990. *Oncogene* 5:577–82
167. Troppmair, J., Potter, M., Wax, J. S., Rapp, U. R. 1989. *Proc. Natl. Acad. Sci. USA* 86:9941–45
168. Bechade, C., Calothy, G., Pessac, B., Martin, P., Coll, J., et al. 1985. *Nature* 316:559–62
169. Pfeifer, A. M. A., Mark, G. E. III, Malan-Shibley, L., Graziano, S., Amstad, P., et al. 1989. *Proc. Natl. Acad. Sci. USA* 86:10075–79
170. Gauwerky, C. E., Huebner, K., Isobe, M., Nowell, P. C., Croce, C. M. 1989. *Proc. Natl. Acad. Sci. USA* 86:8867–71
171. Stewart, T. A., Pattengale, P. K., Leder, P. 1984. *Cell* 38:627–37
172. Leder, A., Pattengale, P. K., Kuo, A., Stewart, T. A., Leder, P. 1986. *Cell* 45:485–95
173. Morello, D., Lavenu, A., Bandeira, A., Portnoi, D., Gaillard, J., et al. 1989. *Oncogene Res.* 4:111–25
174. Adams, J. M., Harris, A. W., Pinkert, C. A., Corcoran, L. M., Alexander, W. S., et al. 1985. *Nature* 318:533–38
175. Langdon, W. Y., Harris, A. W., Cory, S., Adams, J. M. 1986. *Cell* 47:11–18
176. Alexander, W. S., Schrader, J. W., Adams, J. M. 1987. *Mol. Cell. Biol.* 7:1436–44
177. Alexander, W. S., Bernard, O., Cory, S., Adams, J. M. 1989. *Oncogene* 4:575–81
178. Langdon, W. Y., Harris, A. W., Cory, S. 1989. *Oncogene Res.* 4:253–58
179. Alexander, W. S., Adams, J. M., Cory, S. 1989. *Mol. Cell. Biol.* 9:67–73
180. Knight, K. L., Spieker-Polet, H., Kazdin, D. S., Oi, V. T. 1988. *Proc. Natl. Acad. Sci. USA* 85:3130–34
181. Spanopoulou, E., Early, A., Elliott, J., Crispe, N., Ladyman, H., et al. 1989. *Nature* 342:185–89
182. Dildrop, R., Ma, A., Zimmerman, K., Hsu, E., Tesfaye, A., et al. 1989. *EMBO J.* 8:1121–28
183. Zimmerman, K., Legouy, E., Stewart, V., Depinho, R., Alt, F. W. 1990. *Mol. Cell. Biol.* 10:2096–103
184. Moroy, T., Fisher, P., Guidos, C., Ma, A., Zimmerman, K., et al. 1990. *EMBO J.* 9:3659–66
185. Strasser, A., Harris, A. W., Bath, M. L., Cory, S. 1990. *Nature* 348:331–33

186. Rosenbaum, H., Harris, A. W., Bath, M. L., McNeall, J., Webb, E., et al. 1990. *EMBO J.* 9:897–905
187. Nussenzweig M. C., Schmidt, E. V., Shaw, A. C., Sinn, E., Campos-Torres, J., et al. 1988. *Nature* 336:446–50
188. Baumbach, W. R., Keath, E. J., Cole, M. D. 1986. *J. Virol.* 59:276–83
189. Green, S. M., Lowe, A. D., Parrington, J., Karn, J. 1989. *Oncogene* 4:737–51
190. De Klein, A., Hermans, A., Heister-kamp, N., Groffen, J., Grosveld, G. 1988. In *Cellular Oncogene Activation*, ed. G. Klein, pp. 295–312. New York: Decker
191. van Lohuizen, M., Verbeek, S., Schei-jen, B., Wientjens, E., van der Gulden, H., et al. 1991. *Cell* 65:737–52
192. Haupt, Y., Alexander, W. S., Barri, G., Klinken, S. P., Adams, J. M. 1991. *Cell* 65:753–63
193. Goebl, M. G. 1991. *Cell* 66:623
194. DePamphilis, M. L. 1988. *Cell* 52:635–38
195. Classon, M., Henriksson, M., Sumegi, J., Klein, G., Hammaskjold, M.-L. 1987. *Nature* 330:272–74
196. Iguchi-Ariga, S. M. M., Itani, T., Yamaguchi, M., Ariga, H. 1987. *Nucleic Acids Res.* 15:4889–99
197. Classon, M., Henriksson, M., Klein, G., Sumegi, J. 1990. *Oncogene* 5:1371–76
198. Studzinski, G. P., Brelvi, Z. S., Feldman, S. C., Watt, R. A. 1986. *Science* 234:467–70
199. Gutierrez, C., Guo, Z.-S., Farrell-Towt, J., Ju, G., DePamphilis, M. L. 1987. *Mol. Cell. Biol.* 7:4594–98
200. Gutierrez, C., Guo, Z.-S., Burhans, W., DePamphilis, M. L. 1988. *Science* 240:1202–3
201. Iguchi-Ariga, S. M. M., Okazaki, T., Itani, T., Ogata, M., Sato, Y., et al. 1988. *EMBO J.* 7:3135–42
202. Ariga, H., Imamura, Y., Iguchi-Ariga, S. M. M. 1989. *EMBO J.* 8:4273–79
203. McWhinney, C., Leffak, M. 1990. *Nucleic Acids Res.* 18:1233–42
204. Iguchi-Ariga, S. M. M., Itani, T., Kiji, Y., Ariga, H. 1987. *EMBO J.* 6:2365–71
205. Gusse, M., Ghysdael, J., Evan, G., Soussi, T., Mechali, M. 1989. *Mol. Cell. Biol.* 9:5395–403
206. King, M. W., Roberts, J. M., Eisenman, R. N. 1986. *Mol. Cell. Biol.* 6:4499–508
207. Taylor, M. V., Gusse, M., Evan, G. I., Dathan, N., Mechali, M. 1986. *EMBO J.* 5:3563–70
208. Heikkila, R., Schwab, G., Wickstrom, E., Loke, S. L., Pluznik, D. H., et al. 1987. *Nature* 328:445–49
209. Cory, S. 1986. *Adv. Cancer Res.* 47:189–234
210. Leder, P., Battey, J., Lenoir, G., Moulding, C., Murphy, W., et al. 1983. *Science* 222:765–71
211. Cleveland, J. L., Huleihel, M., Bressler, P., Siebenlist, U., Akiyama, L., et al. 1988. *Oncogene Res.* 3:357–75
212. Alexander, W. S., Schrader, J. W., Adams, J. M. 1987. *Mol. Cell. Biol.* 7:1436–44
213. Penn, L. J. Z., Brooks, M. W., Laufer, E. M., Land, H. 1990. *EMBO J.* 9:1113–21
214. Grignani, F., Lombardi, L., Inghirami, G., Sternas, L., Cechova, K., et al. 1990. *EMBO J.* 9:3913–22
215. Nisen, P. D., Zimmerman, K. A., Cotter, S. V., Gilbert, F., Alt, F. W. 1986. *Cancer Res.* 46:6217–22
216. Nau, M. M., Brooks, B. J., Carney, D. N., Gazdar, A. F., Battey, J. F., et al. 1986. *Proc. Natl. Acad. Sci. USA* 83:1092–96
217. Rosenbaum, H., Webb, E., Adams, J. M., Cory, S., Harris, A. W. 1989. *EMBO J.* 8:749–55
218. Penn, L. J. Z., Brooks, M. W., Laufer, E. M., Littlewood, T. D., Morgenstern, J. P., et al. 1990. *Mol. Cell. Biol.* 10:4961–66
219. Lech, K., Anderson, K., Brent, R. 1988. *Cell* 52:179–84
220. Kato, G. J., Barrett, J., Villa-Garcia, M., Dang, C. V. 1990. *Mol. Cell. Biol.* 10:5914–20
221. Kingston, R. E., Baldwin, A. S., Sharp, P. A. 1984. *Nature* 312:280–82
222. Kaddurah-Daouk, R., Greene, J. M., Baldwin, A. S., Kingston, R. E. 1987. *Genes Dev.* 1:347–57
223. Prendergast, G. C., Cole, M. D. 1989. *Mol. Cell. Biol.* 9:124–34
224. Prendergast, G. C., Diamond, L. E., Dahl, D., Cole, M. D. 1990. *Mol. Cell. Biol.* 10:1265–69
225. Cheng, G., Skoultchi, A. I. 1989. *Mol. Cell. Biol.* 9:2332–40
226. Lenardo, M., Rustgi, A. K., Schievella, A. R., Bernards, R. 1989. *EMBO J.* 8:3351–55
227. Akeson, R., Bernards, R. 1990. *Mol. Cell. Biol.* 10:2012–16
228. Inghirami, G., Grignani, F., Sternas, L., Lombardi, L., Knowles, D. M., et al. 1990. *Science* 250:682–86
229. Endo, K., Borer, C. H., Tsujimoto, Y. 1991. *Oncogene* 6:1391–96
230. Yang, B.-S., Geddes, T. J., Pogulis, R. J., De Crombrugghe, B., Freytag, S. O. 1991. *Mol. Cell. Biol.* 11:2291–95

231. Suen, T.-C., Hung, M.-C. 1991. *Mol. Cell. Biol.* 11:354–62
232. Battey, J., Moulding, C., Taub, R., Murphy, W., Stewart, T., et al. 1983. *Cell* 34:779–87
233. Watt, R., Nishikura, K., Sorrentino, J., Ar-Rushdi, A., Croce, C. M., et al. 1983. *Proc. Natl. Acad. Sci. USA* 80:6307–11
234. Bernard, O., Cory, S., Gerondakis, S., Webb, E., Adams, J. M. 1983. *EMBO J.* 2:2375–83
235. Stewart, T. A., Bellve, A. R., Leder, P. 1984. *Science* 226:707–10
236. Bentley, D. L., Groudine, M. 1986. *Nature* 321:702–6
237. Taub, R., Moulding, C., Battey, J., Latt, S., Lenoir, G. M., et al. 1984. *Cell* 36:339–48
238. Taub, R., Kelly, K., Battey, J., Latt, S., Lenoir, G. M., et al. 1984. *Cell* 37:511–20
239. Spencer, C. A., LeStrange, R. C., Hayward, W. S., Novak, U., Groudine, M. 1990. *Genes Dev.* 4:75–88
240. Bentley, D. L., Groudine, M. 1986. *Mol. Cell. Biol.* 6:3481–89
241. Ray, D., Robert-Lézéngés, J. 1989. *Oncogene Res.* 5:73–78
242. Yang, J., Bauer, S. R., Mushinski, J. F., Marcu, K. B. 1985. *EMBO J.* 4:1441–47
243. Nishikura, K., Murray, J. M. 1988. *Oncogene* 2:493–98
244. Broome, H. E., Reed, J. C., Godillot, E. P., Hoover, R. G. 1987. *Mol. Cell. Biol.* 7:2988–93
245. Siebenlist, U., Hennighausen, L., Battey, J., Leder, P. 1984. *Cell* 37:381–91
246. Fahrlander, P. D., Piechaczyk, M., Marcu, K. B. 1985. *EMBO J.* 4:3195–202
247. Dyson, P. J., Littlewood, T. D., Forster, A., Rabbitts, T. H. 1985. *EMBO J.* 4:2885–91
248. Dyson, P. J., Rabbitts, T. H. 1985. *Proc. Natl. Acad. Sci. USA* 82:1984–88
249. Siebenlist, U., Bressler, P., Kelly, K. 1988. *Mol. Cell. Biol.* 8:867–74
250. Chen, T. A., Allfrey, V. G. 1987. *Proc. Natl. Acad. Sci. USA* 84:5252–56
251. Chung, J., Sinn, E., Reed, R. R., Leder, P. 1986. *Proc. Natl. Acad. Sci. USA* 83:7918–22
252. Lipp, M., Schilling, R., Wiest, S., Laux, G., Bornkamm, G. W. 1987. *Mol. Cell. Biol.* 7:1393–400
253. Hay, N., Bishop, J. M., Levens, D. 1987. *Genes Dev.* 1:659–71
254. Nishikura, K., Goldflam, S., Vuocolo, G. A. 1985. *Mol. Cell. Biol.* 5:1434–41
255. Nishikura, K. 1986. *Mol. Cell. Biol.* 6:4093–98

256. Postel, E. H., Mango, S. E., Flint, S. J. 1989. *Mol. Cell. Biol.* 9:5123–31
257. Davis, T. L., Firulli, A. B., Kinniburgh, A. J. 1989. *Proc. Natl. Acad. Sci. USA* 86:9682–86
258. Postel, E. H., Flint, S. J., Kessler, D. J., Hogan, M. E. 1991. *Proc. Natl. Acad. Sci. USA* 88:8227–31
259. Boles, T. C., Hogan, M. E. 1987. *Biochemistry* 198:367–76
260. Cooney, M., Czernuszewicz, G., Postel, E. H., Flint, S. J., Hogan, M. E. 1988. *Science* 241:456–59
261. Kinniburgh, A. J. 1989. *Nucleic Acids Res.* 17:7771–78
262. Lobanenkov, V. V., Nicolas, R. H., Alder, V. V., Paterson, H., Klenova, E. M., et al. 1990. *Oncogene* 5:1743–53
263. Asselin, C., Nepveu, A., Marcu, K. B. 1989. *Oncogene* 4:549–58
264. Kakkis, E., Calame, K. 1987. *Proc. Natl. Acad. Sci. USA* 84:7031–35
265. Kakkis, E., Riggs, K. J., Gillespie, W., Calame, K. 1989. *Mol. Cell. Biol.* 339:718–21
266. Riggs, K. J., Merrell, K. T., Wilson, G., Calame, K. 1991. *Mol. Cell. Biol.* 11:1765–69
267. Duyao, M. P., Buckler, A. J., Sonenshein, G. E. 1990. *Proc. Natl. Acad. Sci. USA* 87:4727–31
268. Remmers, E. F., Yang, J-Q., Marcu, K. B. 1986. *EMBO J.* 5:899–904
269. Weisinger, G., Remmers, E. F., Hearing, P., Marcu, K. B. 1988. *Oncogene* 3:635–46
270. Avigan, M. I., Strober, B., Levens, D. 1990. *J. Biol. Chem.* 265:18538–45
271. Hall, D. J. 1990. *Oncogene* 5:47–54
272. Thalmeier, K., Synovzik, H., Mertz, R., Winnacker, E-L., Lipp, M. 1989. *Genes Dev.* 3:527–36
273. Hiebert, S. W., Lipp, M., Nevins, J. R. 1989. *Proc. Natl. Acad. Sci. USA* 86:3594–98
273a. Parkin, N. T., Sonenberg, N. 1989. *Oncogene* 4:815–22
274. Yang, J-Q., Remmers, E. F., Marcu, K. B. 1986. *EMBO J.* 5:3553–62
275. Atchison, M. L., Meyuhas, O., Perry, R. P. 1989. *Mol. Cell. Biol.* 9:2067–74
276. Ray, R., Miller, D. M. 1991. *Mol. Cell. Biol.* 11:2154–61
277. Zajac-Kaye, M., Gelmann, E. P., Levens, D. 1988. *Science* 240:1776–80
278. Zajac-Kaye, M., Levens, D. 1990. *J. Biol. Chem.* 265:4547–51
279. Kelly, K., Siebenlist, U. 1988. *J. Biol. Chem.* 263:4828–31
280. Heckford, S. E., Gelmann, E. P., Matis, L. A. 1988. *Oncogene* 3:415–21
281. Sacca, R., Cochran, B. H. 1990. *Oncogene* 5:1499–505

282. Buckler, A. J., Rothstein, T. L., Sonenshein, G. E. 1990. *J. Immunol.* 145: 732–36
283. Pietenpol, J. A., Holt, J. T., Stein, R. W., Moses, H. L. 1990. *Proc. Natl. Acad. Sci. USA* 87:3758–62
284. Krönke, M., Schluter, C., Pfizenmaier, K. 1987. *Proc. Natl. Acad. Sci. USA* 84:469–73
285. Schachner, J., Blick, M., Freireich, E., Gutterman, J., Beran, M. 1988. *Leukemia* 2:749–53
286. Pietenpol, J. A., Stein, R. W., Moran, E., Yaciuk, P., Schelgel, R., et al. 1990. *Cell* 61:777–85
287. Robbins, P. D., Horowitz, J. M., Mulligan, R. C. 1990. *Nature* 346:668–71
288. Trepel, J. B., Colamonici, O. R., Kelly, K., Schwab, G., Watt, R. A., et al. 1987. *Mol. Cell. Biol.* 7:2644–48
289. Spencer, C.A., Groudine, M. 1990. *Oncogene* 5:777–85
290. Krystal, G., Birrer, M., Way, J., Nau, M., Sausville, E., et al. 1988. *Mol. Cell. Biol.* 8:3373–81
291. Bender, T. P., Thompson, C. B., Kuehl, W. M. 1987. *Science* 237:1473–76
292. Watson, R. J. 1988. *Mol. Cell. Biol.* 8:3938–42
293. Collart, M. A., Tourkine, N., Belin, D., Vassalli, P., Jeanteur, Ph., et al. 1991. *Mol. Cell. Biol.* 11:2826–31
294. McGeady, M. L., Wood, T. G., Maizel, J. V., Vande Woude, G. F. 1986. *DNA* 5:289–98
295. Re, G. G., Antoun, G. R., Zipf, T. F. 1990. *Oncogene* 5:1247–50
296. Wright, S., Bishop, J. M. 1989. *Proc. Natl. Acad. Sci. USA* 86:505–9
297. Wright, S., Bishop, J. M. 1991. *J. Cell. Biochem.* 15(G):265 (Abstr.)
298. Bentley, D. L., Groudine, M. 1988. *Cell* 53:245–56
299. Miller, H., Asselin, C., Dufort, D., Yang, J-Q., Gupta, K., et al. 1989. *Mol. Cell. Biol.* 9:5340–49
300. Dufort, D., Miller, H., Nepveu, A. 1991. *J. Cell. Biochem.* 15(G):271 (Abstr.)
301. Chung, J., Sussman, D. J., Zeller, R., Leder, P. 1987. *Cell* 51:1001–8
302. Bentley, D. L., Brown, W. L., Groudine, M. 1989. *Genes Dev.* 3: 1179–89
303. Kerppola, T. K., Kane, C. 1988. *Mol. Cell. Biol.* 8:4389–94
304. London, L., Keene, R. G., Landick, R. 1991. *Mol. Cell. Biol.* 11:4599–615
305. Eick, D., Bornkamm, G. W. 1986. *Nucleic Acids Res.* 14:8331–46
306. Salehi, Z., Taylor, J. D., Niedel, J. E. 1988. *J. Biol. Chem.* 263:1898–90

307. Simpson, R. U., Hsu, T., Begely, D. A., Mitchell, B. S., Alizadeh, B. N. 1987. *J. Biol. Chem.* 262:4104–8
308. Simpson, R. U., Hsu, T., Wendt, M. D., Taylor, J. M. 1989. *J. Biol. Chem.* 264:19710–14
309. Nepveu, A., Marcu, K. B. 1986. *EMBO J.* 5:2859–65
310. Nepveu, A., Fahrlander, P. D., Yang, J-Q., Marcu, K. B. 1985. *Nature* 317: 440–43
311. Nepveu, A., Marcu, K. B., Skoultchi, A. I., Lachman, H. M. 1987. *Genes Dev.* 1:938–45
312. Mechti, N., Piechaczyk, M., Blanchard, J. M., Marty, L., Bonnieu, A., et al. 1986. *Nucleic Acids Res.* 14:9653–66
313. Eick, D., Berger, R., Polack, A., Bornkamm, G. W. 1987. *Oncogene* 2:61–65
314. Schneider-Schaulies, J., Schimpl, A., Wecker, E. 1987. *Eur. J. Immunol.* 17:713–18
315. Zullo, J., Stiles, C. D., Garcea, R. L. 1987. *Proc. Natl. Acad. Sci. USA* 84:1210–14
316. Cleveland, J. L., Dean, M., Rosenberg, N., Wang, J. Y. J., Rapp, U. R. 1989. *Mol. Cell. Biol.* 9:5685–95
317. Zullo, J. N., Faller, D. V. 1988. *Mol. Cell. Biol.* 8:5080–85
318. Evans, J. L., Moore, T. L., Kuehl, W. M., Bender, T., Ting, J. P. 1990. *Mol. Cell. Biol.* 10:5747–52
319. Zobel, A., Kalkbrenner, F., Guehmann, S., Nawrath, M., Vorbrueggen, G., et al. 1991. *Oncogene* 6:1397–407
320. Lipp, M., Schilling, R., Bernhart, G. 1989. *Oncogene* 4:535–41
321. Timmers, H. T. M., De Wit, D., Bos, J. L., van der Eb, A. J. 1988. *Oncogene Res.* 3:67–71
322. van Dam, H., Offringa, R., Smits, A. M. M., Bos, J. L., Jones, N. C., van der Eb, A. J. 1989. *Oncogene* 4:1207–12
323. Jenkins, J. 1992. Personal communication
324. Hay, N., Takimoto, M., Bishop, J. M. 1989. *Genes Dev.* 3:293–303
325. Takimoto, M., Quinn, J. P., Farina, A. R., Staudt, L. M., Levens, D. 1989. *J. Biol. Chem.* 264:8992–99
326. Keath, E. J., Kalikan, A., Cole, M. D. 1984. *Cell* 37:521–28
327. Calabi, F., Neuberger, M. S. 1985. *EMBO J.* 4:667–74
328. Dani, C., Blanchard, J-M., Piechaczyk, M., El Sabouty, S., Marty, L., et al. 1984. *Proc. Natl. Acad. Sci. USA* 81: 7046–50
329. Brewer, G., Ross, J. 1989. *Mol. Cell. Biol.* 7:1996–2006
330. Swartwout, S. G., Kinniburgh, A. J. 1989. *Mol. Cell. Biol.* 9:288–95

331. Rabbitts, P. H., Watson, J. V., Lamond, A., Forster, A., Stinson, M. A., et al. 1985. *EMBO J.* 4:2009–15

332. Rabbitts, P. H., Forster, A., Stinson, M. A., Rabbitts, T. H. 1985. *EMBO J.* 4:3727–33

333. Dean, M., Levine, R. A., Campisi, J. 1986. *Mol. Cell. Biol.* 6:518–24

334. Dony, C., Kessel, M., Gruss, P. 1985. *Nature* 317:636–39

335. Jonak, G. J., Knight, E. 1984. *Proc. Natl. Acad. Sci. USA* 81:1747–50

336. Knight, E., Anton, E. D., Fahey, D., Friedland, B. K., Jonak, G. J. 1985. *Proc. Natl. Acad. Sci. USA* 82:1151–54

337. Dani, C., Mechti, N., Piechaczyk, M., Lebleu, B., Jeanteur, Ph., et al. 1985. *Proc. Natl. Acad. Sci. USA* 82:4896–99

338. Linial, M., Gunderson, N., Groudine, M. 1985. *Science* 230:1126–32

339. Einat, M., Resnitzky, D., Kimchi, A. 1985. *Nature* 313:597–600

340. Dron, M., Modjtahedi, N., Brison, O., Tovey, M. G. 1986. *Mol. Cell. Biol.* 6:1374–78

341. Morello, D., Asselin, C., Lavenu, A., Marcu, K. B., Babinet, C. 1989. *Oncogene* 4:955–61

342. Babiss, L. E., Friedman, J. M. 1990. *Mol. Cell. Biol.* 10:6700–8

343. Blanchard, J-M., Piechaczyk, M., Dani, C., Chambard, J-C., Franchi, A., et al. 1985. *Nature* 317:443–45

344. Lacy, J., Summers, W. P., Summers, W. C. 1989. *EMBO J.* 8:1973–80

345. Levine, R. A., McCormack, J. E., Buckler, A., Sonenshein, G. E. 1986. *Mol. Cell. Biol.* 6:4112–16

346. McCormack, J. E., Pepe, V. H., Kent, R. B., Dean, M., Marshak-Rothstein, A., et al. 1984. *Proc. Natl. Acad. Sci. USA* 81:5546–50

347. Asselin, C., Marcu, K. B. 1989. *Oncogene Res.* 5:67–72

348. Sobczak, J., Mechti, N., Tournier, M-F., Blanchard, J-M., Duguet, M. 1989. *Oncogene* 4:1503–8

349. Morello, D., Lavenu, A., Babinet, C. 1990. *Oncogene* 5:1511–19

350. Richman, A., Hayday, A. 1989. *Mol. Cell. Biol.* 9:4962–69

351. Taylor, M. V., Gusse, M., Evan, G. I., Dathan, N., Mechali, M. 1986. *EMBO J.* 5:3563–70

352. Bauer, S. R., Piechaczyk, M., Nordan, R., Owens, J. D., Nepveu, A., et al. 1989. *Oncogene* 4:615–23

353. Piechaczyk, M., Yang, J-Q., Blanchard, J-M., Jeanteur, Ph., Marcu, K. B. 1985. *Cell* 42:589–97

354. Eick, D., Piechaczyk, M., Henglein, B., Blanchard, J-M., Traub, B., Kofler, E., et al. 1985. *EMBO J.* 4:3717–25

355. Pei, R., Calame, K. 1988. *Mol. Cell. Biol.* 8:2860–68

356. Jones, T. R., Cole, M. D. 1987. *Mol. Cell. Biol.* 7:4513–21

357. Hollis, G. F., Gazdar, A. F., Bertness, V., Kirsch, I. R. 1988. *Mol. Cell. Biol.* 8:124–29

358. Shaw, G., Kamen, R. 1986. *Cell* 46:659–67

359. Caput, D., Beutler, B., Hartog, K., Thayer, R., Brown-Shimer, S., Cerami, A. 1986. *Proc. Natl. Acad. Sci. USA* 83:1665–70

360. Bonnieu, A., Roux, P., Marty, L., Jeanteur, Ph., Piechaczyk, M. 1990. *Oncogene* 5:1670–74

361. Laird-Offringa, I. A., Elfferich, P., Knaken, H. J., de Ruiter, J., van der Eb, A. J. 1989. *Nucleic Acids Res.* 17:6499–514

362. Brewer, G., Ross, J. 1988. *Mol. Cell. Biol.* 8:1697–708

363. Deleted in proof

364. Brewer, G. 1991. *Mol. Cell. Biol.* 11:2460–66

365. Swartwout, S. G., Preisler, H., Guan, W., Kinniburgh, A. J. 1987. *Mol. Cell. Biol.* 7:2052–58

366. Laird-Offringa, I. A., De Wit, C. L., Elfferich, P., van der Eb, A. J. 1990. *Mol. Cell. Biol.* 10:6132–40

367. Saito, H., Hayday, A. C., Wiman, K., Hayward, W. S., Tonegawa, S. 1983. *Proc. Natl. Acad. Sci. USA* 80:7476–80

368. Butnick, N. Z., Miyamoto, C., Chizzonite, R., Cullen, B. R., Ju, G., et al. 1985. *Mol. Cell. Biol.* 5:3009–16

369. Darveau, A., Pelletier, J., Sonenberg, N. 1985. *Proc. Natl. Acad. Sci. USA* 82:2315–19

370. Parkin, N., Darveau, A., Nicholson, R., Sonenberg, N. 1988. *Mol. Cell. Biol.* 8:2875–83

371. Baim, S. B., Pietras, D. F., Eustice, D. C., Sherman, F. 1985. *Mol. Cell. Biol.* 5:1839–46

372. Pelletier, J., Sonenberg, N. 1985. *Cell* 40:515–26

373. Bass, B. L., Weintraub, H. 1987. *Cell* 48:607–13

374. Rebagliati, M. R., Melton, D. A. 1987. *Cell* 48:599–606

375. Wagner, R. W., Nishikura, K. 1988. *Mol. Cell. Biol.* 8:770–76

376. Lazarus, P., Parkin, N., Sonenberg, N. 1988. *Oncogene* 3:517–21

377. Wurm, F. M., Gwinn, K. A., Kingston, R. E. 1986. *Proc. Natl. Acad. Sci. USA* 83:5414–18

378. Schwab, M., Ellison, J., Busch, M., Rosenau, W., Varmus, H. E., et al. 1984. *Proc. Natl. Acad. Sci. USA* 81:4940–44

379. Brodeur, G. M., Seeger, R. C., Schwab, M., Varmus, H. E., Bishop, J. M. 1984. *Science* 224:1121–24
380. Yokota, J., Tsunetsugu-Yokota, Y., Battifora, H., LeFevre, C., Cline, M. J. 1986. *Science* 231:261–65
381. Shibuya, M., Yokota, J., Ueyama, Y. 1985. *Mol. Cell. Biol.* 5:414–18
382. Little, C. D., Nau, M. M., Carny, D. N., Gazdar, A. F., Minna, J. D. 1983. *Nature* 306:194–96
383. Trent, J., Meltzer, P., Rosenblum, M., Harsh, G., Kinzler, K., et al. 1986. *Proc. Natl. Acad. Sci. USA* 83:470–73
384. Kozbor, D., Croce, C. M. 1984. *Cancer Res.* 44:438–41
385. Escot, C., Theillet, C., Lidereau, R., Spyratos, F., Champeme, M.-H., et al. 1986. *Proc. Natl. Acad. Sci. USA* 83:4834–38
386. Alitalo, K., Schwab, M., Lin, C. C., Varmus, H. E., Bishop, J. M. 1983. *Proc. Natl. Acad. Sci. USA* 80:1707–11
387. Sumegi, J., Hedberg, T., Bjorkholm, M., Godal, T., Mellstedt, H., et al. 1985. *Int. J. Cancer* 36:367–71
388. Collins, S., Groudine, M. 1982. *Nature* 298:679–81
389. DallaFavera, R., Wong-Staal, F., Gallo, R. C. 1982. *Nature* 299:61–63
390. McCarthy, D. M., Rassool, F. V., Goldman, J. M., Graham, S. V., Birnie, G. D. 1984. *Lancet* 2:1362–65
391. Michitsch, R. W., Montgomery, K. T., Melera, P. W. 1984. *Mol. Cell. Biol.* 4:2370–80
392. Deleted in proof
393. Garson, J. A., McIntyre, P. G., Kemshead, J. T. 1985. *Lancet 2:718–19*
394. Payne, G. S., Bishop, J. M., Varmus, H. E. 1982. *Nature* 295:209–14
395. Westaway, D., Payne, G., Varmus, H. E. 1984. *Proc. Natl. Acad. Sci. USA* 81:843–47
396. Linial, M., Groudine, M. 1985. *Proc. Natl. Acad. Sci. USA* 82:53–57
397. Selten, G., Cuypers, H. T., Zijlstra, M., Melief, C., Berns, A. 1984. *EMBO J.* 3:3215–22
398. Corcoran, L. M., Adams, J. M., Dunn, A. R., Cory, S. 1984. *Cell* 37:113–22
399. Lazo, P. A., Lee, J. S., Tsichlis, P. N. 1990. *Proc. Natl. Acad. Sci. USA* 87:170–73
400. van Lohuizen, M., Breuer, M., Berns, A. 1989. *EMBO J.* 8:133–36
401. Hsu, T. Y., Moroy, T., Etiemble, J., Louise, A., Trepo, C., et al. 1988. *Cell* 55:627–35
402. Fourel, G., Trepo, C., Bougueleret, L., Henglein, B., Ponzetto, A., et al. 1990. *Nature* 347:294–98
403. Tiollais, P., Pourcel, C., Dejean, A. 1985. *Nature* 317:489–95
404. Showe, L. C., Croce, C. M. 1987. *Annu. Rev. Immunol.* 5:253–77
404a. Ott, D. E., Marcu, K. B. 1989. *Int. Immunol.* 1:582–91
404b. Piccoli, S. P., Caimi, P. G., Cole, M. D. 1984. *Nature* 310:327–30
404c. Yunis, J. J., Soreng, A. L. 1984. *Science* 226:1199–204
405. Mushinski, F. M. 1988. See Ref. 190, pp. 181–222
406. Erikson, J., Finger, L., Sun, L., Ar-Rushdi, A., Nishikura, K., et al. 1986. *Science* 232:884–86
407. Finver, S. N., Nishikura, K., Finger, L. R., Haluska, F. G., Finan, J., et al. 1988. *Proc. Natl. Acad. Sci. USA* 85:3052–56
408. Finger, L. R., Huebner, K., Cannizzaro, L. A., McLeod, K., Nowell, P. C., et al. 1988. *Proc. Natl. Acad. Sci. USA* 85:9158–62
409. Peschle, C., Mavilio, F., Sposi, N. M., Giampaolo, A., Care, A., et al. 1984. *Proc. Natl. Acad. Sci. USA* 81:5514–18
410. Care, A., Cianetti, L., Giampaola, A., Sposi, N. M., Zappavigna, V., et al. 1986. *EMBO J.* 5:905–11
411. Sumegi, J., Spira, J., Bazin, H., Szpirer, J., Levan, G., et al. 1983. *Nature* 306:497–98
412. Pear, W. S., Ingvarsson, S., Steffen, D., Munke, M., Francke, U., et al. 1986. *Proc. Natl. Acad. Sci. USA* 83:7376–80
413. Shtivelman, E., Bishop, J. M. 1990. *Mol. Cell. Biol.* 10:1835–39
414. Huppi, K., Siwarski, D., Skurla, R., Klinman, D., Mushinski, J. F. 1990. *Proc. Natl. Acad. Sci. USA* 87:6964–68
415. Cesarman, E., Dalla-Favera, R., Bentley, D., Groudine, M. 1987. *Science* 238:1272–75
416. Eick, D., Polack, A., Kofler, E., Lenoir, G. M., Rickinson, A. B., et al. 1990. *Oncogene* 5:1397–402
416a. Petterson, S., Cook, G. P., Brüggerman, M., Williams, G. T., Neuberger, M. S. 1990. *Nature* 344:165–68
416b. Meyer, K. B., Neuberger, M. S. 1989. *EMBO J.* 8:1959–64
417. Deleted in proof
418. Kim, S.-J., Lee, H.-D., Robbins, P. D., Busam, K., Sporn, M. B., et al. 1991. *Proc. Natl. Acad. Sci. USA* 88:052–56
419. Rustgi, A. K., Dyson, N., Bernards, R. 1991. *Nature* 352:541–44
420. Dyson, N., Howley, P. M., Munger, K., Harlow, E. 1989. *Science* 243:934–37

421. Egan, C., Jelsma, T. N., Howe, J. A., Bayley, S. T., Ferguson, B., et al. 1988. *Mol. Cell. Biol.* 8:3955–59
422. Egan, C., Bayley, S. T., Branton, P. E. 1989. *Oncogene* 4:383–88
423. Bookstein, R., Lee, E. Y.-H. P., Peccei, A., Lee, W.-H. 1989. *Mol. Cell. Biol.* 9:1628–34
424. Friend, S. H., Bernards, R., Rogelj, S., Weinberg, R. A., Rapaport, J. M., et al. 1986. *Nature* 323:643–46
425. Hansen, M. F., Koufos, A., Gallie, B. L., Phillips, R. A., Fodstad, O., et al. 1985. *Proc. Natl. Acad. Sci. USA* 82:6216–20
426. Harbour, J. W., Lai, S.-L., Whang-Peng, J., Gazdar, A. F., Minna, J. D., et al. 1988. *Science* 241:353–56

427. Horowitz, J. M., Yandell, D. W., Park, S.-H., Canning, S., Whyte, P., et al. 1989. *Science* 243:937–40
428. T'Ang, A., Varley, J. ? ., Chakraborty, S., Murphree, A. L., Fung, Y. K. T. 1988. *Science* 242:263–66
429. Toguchida, J., Ishizaki, K., Sasaki, M. S., Nakamura, Y., Ikenaga, M., et al. 1989. *Nature* 338:156–58
430. Buchkovich, K., Duffy, L. A., Harlow, E. 1989. *Cell* 58:1097–105
431. Furukawa, Y., DeCaprio, J. A., Freedman, A., Kanakura, Y., Nakamura, M., et al. 1990. *Proc. Natl. Acad. Sci. USA* 87:2770–74
432. Mihara, K., Cao, X.-R., Yen, A., Chandler, S., Driscoll, B., et al. 1989. *Science* 246:1300–3

Annu. Rev. Biochem. 1992. 61:861–96

PROTON TRANSFER IN REACTION CENTERS FROM PHOTOSYNTHETIC BACTERIA

M. Y. Okamura and G. Feher

Department of Physics, University of California—San Diego, La Jolla, California 92093-0319

KEY WORDS: quinone, electron transfer, site-directed mutagenesis, protonation, photosynthesis

CONTENTS

INTRODUCTION AND PERSPECTIVE

In purple photosynthetic bacteria light energy is transformed into chemical energy by the action of a light-driven proton pump coupled to electron

0066-4154/92/0701-0861$02.00

transfer. A central role in this process is played by the bacterial reaction center (RC), a membrane protein that absorbs photons, performs the initial rapid electron transfer reactions and is the site for the initial proton uptake reactions of the proton pump (reviewed in references 1–4). The key reactions that occur in the RC involve the two-electron reduction and concomitant binding of two protons from the cytoplasm side of the membrane by a bound quinone, Q_B, on the RC (Figure 1). The doubly reduced quinone subsequently dissociates from the RC and is reoxidized by the cytochrome b/c_1 complex, releasing protons on the periplasmic side of the membrane. The net result of these reactions is the vectorial transport of protons across the membrane, driven by electron transfer, as proposed by Mitchell (5). This proton transport produces a pH gradient that drives ATP synthesis (reviewed in reference 6).

The transmembrane proton-pumping function of the RC has been known since the initial observation by Baltscheffsky & von Stedingk (8) of proton uptake by chromatophores from *Rhodospirillum rubrum*. Further studies of this proton uptake in intact membranes (9–11) established that it was associated with electron transfer to the secondary acceptor species (now known to be Q_B). An important advance was the isolation of RCs by Reed & Clayton (12) and subsequent further purification (13) and characterization of its electron transfer components (reviewed in 14). Proton uptake measurements on isolated RCs led Wraight to propose the basic ideas for proton transport in RCs, i.e. proton uptake by the doubly reduced quinone and the involvement of acidic residues from the protein in this process (15). However, the molecu-

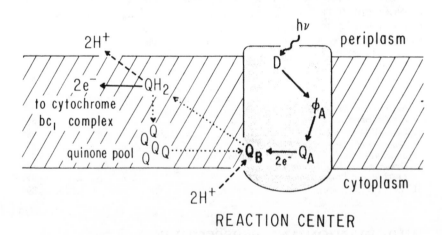

Figure 1   Schematic representation of electron and proton transfer in bacterial photosynthesis. Electron transfer steps are indicated by solid lines, proton transfer steps by dashed lines, and diffusion steps by dotted lines. Proton transfer is coupled to electron transfer via the protonation of Q_B^{2-} in the RC. Reproduced from reference 7 with permission.

lar details of these mechanisms have only recently begun to be elucidated. Two major developments have contributed to our present understanding: (*a*) the determination of the X-ray crystal structure of the RCs from *Rhodopseudomonas viridis* by Deisenhofer et al (16) and from *Rhodobacter sphaeroides* by Allen et al (17) and Chang et al (18), and (*b*) site-directed mutagenesis on RCs from *Rb. sphaeroides* by Paddock et al (7, 19) and Takahashi & Wright (20), which was guided by these structures. The X-ray crystal structure revealed that Q_B is located in the interior of the protein, out of contact with the aqueous solution, and suggested the possibility that protonatable amino acid side chains from the protein were responsible for proton transport to Q_B. Site-directed mutagenesis of several of these residues to nonprotonatable groups resulted in loss of proton transport to Q_B and conclusively demonstrated that the protein plays an important role in proton transport. These findings have led to renewed interest in the mechanistic and structural interpretation of earlier measurements of light-induced proton uptake and the pH dependence of electron transfer rates and equilibria, and they have encouraged new measurements that are sensitive to proton transfer, e.g. light-induced electrogenicity and infrared (IR) spectroscopy. In this review we discuss the results of biophysical measurements made on native and mutant RCs, which address the following questions: (*a*) What proton transfer steps are involved? (*b*) Which amino acid residues are involved in these steps? (*c*) What are the rates and energetics for these steps? (*d*) What are the mechanisms of proton transfer? These questions are similar to those posed for other proton transport proteins in biological membranes, e.g. bacteriorhodopsin (21), F_1F_0 ATPase (22), *lac* permease (23), and cytochrome oxidase (24). Recent reviews have been published on the RC structure (1, 25, 26), electron transfer reactions (27), and quinone chemistry (28), as well as on proton transfer processes in biology (29, 30). The electron and proton transfer in bacterial RCs has been recently discussed (31, 32).

STRUCTURE OF BACTERIAL REACTION CENTER

The RC from *Rb. sphaeroides* (shown in Figure 2) contains three protein subunits, L, M, and H (33, 34). The L and M subunits each have five membrane-spanning helices, which form the core of the RC. The H subunit contains only one membrane-spanning helix. Most of the H subunit consists of a globular protein localized on the cytoplasmic side of the membrane covering the quinone Fe complex. The H subunit is asymmetrically located in RCs from both *Rps. viridis* (16) and *Rb. sphaeroides* (33), forming a cap over the Q_A region of the RC. This cap may serve to isolate Q_A from the exterior aqueous environment.

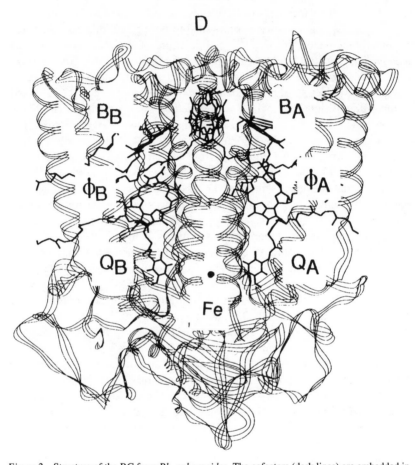

Figure 2   Structure of the RC from *Rb. sphaeroides*. The cofactors (dark lines) are embedded in the protein, composed of three subunits (L, M, H) containing 11 membrane-spanning α-helices and a globular part on the cytoplasmic side. Electron transfer proceeds from the primary donor D across the membrane along the A branch via Q_A to Q_B. Proton transfer proceeds from the external aqueous phase to Q_B, which is buried in the protein. The doubly reduced $Q_B H_2$ leaves the pocket presumably by a pathway along the isoprenoid tail. Modified from reference 33 with permission.

Associated with the bacterial RC from *Rb. sphaeroides* are the following cofactors: four bacteriochlorophylls, two bacteriopheophytins, two ubiquinone molecules, and one nonheme Fe^{2+}. The arrangement of these cofactors (35, 36) in the RC from *Rb. sphaeroides* is shown in Figure 2. Two closely associated bacteriochlorophyll molecules near the periplasmic side of the RC form a bacteriochlorophyll dimer that serves as the primary electron donor (D). The other cofactors, bacteriochlorophyll (B), bacteriophenophytin

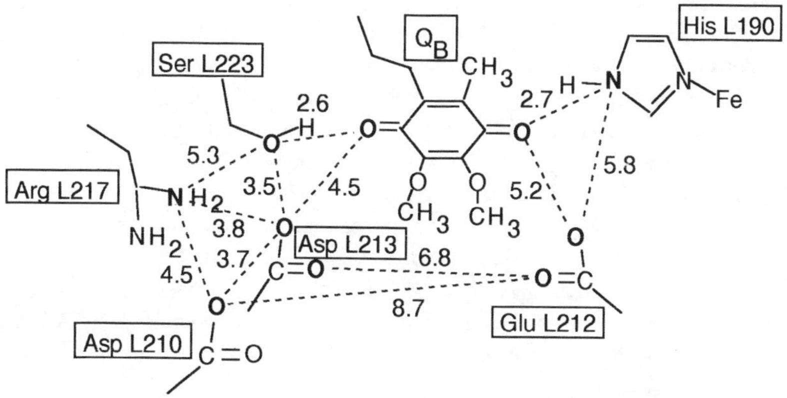

Figure 3 Schematic representation of the structure near the Q_B-binding site. The closest distances (in angstroms) between neighboring residues are indicated. Modified from reference 38 with permission.

(ϕ), and quinone (Q), are symmetrically arranged in pairs around an approximately twofold axis of symmetry connecting the dimer and the Fe^{2+} ion in a direction perpendicular to the plane of the membrane. Two quinone molecules, Q_A and Q_B, are associated with the nonheme Fe^{2+} by making hydrogen bonds to histidine ligands of the Fe^{2+}, thereby forming a quinone-Fe^{2+} complex near the cytoplasmic side of the membrane. Despite the symmetry of the RC, electron transfer proceeds along only one side from D to ϕ_A to Q_A to Q_B. Furthermore, the properties of Q_A and Q_B are very different. Whereas Q_A can only be singly reduced to Q_A^-, Q_B can be further reduced to Q_BH_2. This difference may be due to the relative accessibility of Q_B to protons from the solvent, in contrast to the inaccessibility of Q_A to protons.[1]

Proton transfer in the bacterial RC is determined largely by the protein structure near the Q_B-binding site, shown schematically in Figure 3 (38). The

---

[1] The differences in the electron transfer rates to Q_A and Q_B may be understood in terms of their accessibility to protons. The electron transfer to Q_A must be fast to trap the energy of the charge separation and to prevent the back reaction between D^+ and ϕ^-, which occurs in nanoseconds. The forward rate is maximized by matching the reorganization energy with the energy gap for the reaction (37). The reorganization energy, however, will be increased by contributions from polar groups necessary for proton conduction to the quinone. Consequently, the presence of a proton transfer chain may necessitate a lower quinone reduction rate. Thus, optimal energy trapping and proton transfer require different environments. This may be the reason why bacterial RCs (and PSII RCs in plants) contain two different quinones, Q_A and Q_B, to perform the energy-trapping and proton transport functions, respectively. The requirement for two quinones may be responsible for the apparent twofold symmetry of bacterial RCs.

distances between selected nearby residues are indicated. The binding site consists of a loop of amino acid residues from the section of residues between the D and E transmembrane helices of the L subunit. The quinone carbonyl oxygens are hydrogen bonded on one side to HisL190 (which also is a ligand to Fe) and on the other side to the hydroxyl group of SerL223. The quinone ring is near the hydrophobic amino acid residues IleL229, ValL220, and PheL216. Also nearby are the acidic residues GluL212, and AspL213, AspL210, and the basic residue ArgL217. These acidic and basic residues could serve to modify the electrostatic potential at Q_B and/or serve as proton donor groups in a proton transfer pathway to Q_B. In addition, there are other residues from the M and H subunits between Q_B and the exterior aqueous solvent. Two chains of protonatable residues have been identified by Allen et al (38) as possible paths for proton transfer to Q_B. No similar pathways were seen near Q_A.

QUINONE CHEMISTRY

The coupling between electron transfer and proton transfer in the RC results from the acid-base properties of the quinone molecule in its different redox states: quinone (Q), semiquinone (Q^-), and dihydroquinone (Q^{2-}) (28, 39, 40). Proton binding can accompany electron binding as shown below.

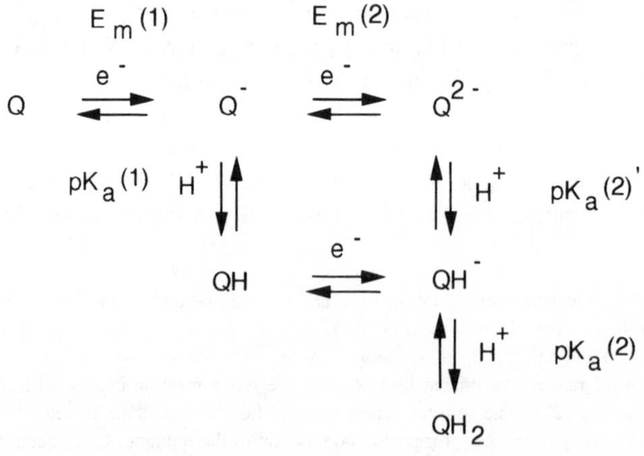

$$
\begin{array}{ccccc}
& E_m(1) & & E_m(2) & \\
& \xrightarrow{e^-} & & \xrightarrow{e^-} & \\
Q & \rightleftharpoons & Q^- & \rightleftharpoons & Q^{2-} \\
pK_a(1)\ H^+ \updownarrow & & & \updownarrow H^+ & pK_a(2)' \\
& & \xrightarrow{e^-} & & \\
QH & \rightleftharpoons & QH^- & & \\
& & & \updownarrow H^+ & pK_a(2) \\
& & QH_2 & &
\end{array}
\qquad 1.
$$

In aqueous solution, e.g. 80% ethanol, quinone reduction proceeds through two one-electron steps to the fully reduced state, QH_2. The midpoint potential for ubiquinone measured in this case, E_m (Q/QH$_2$) = 110 mV (28), is the average of the two one-electron potentials $E_m(1)$ and $E_m(2)$ [all E_m given with respect to standard hydrogen electrodes (pH 7)]. Swallow (39) has estimted the one-electron potential (pH 7) for ubiquinone in aqueous solution to be about $E_{m,7}(1) = -230$ mV. From this value one can estimate that E_m in aqueous ethanol is approximately equal to the value in the aqueous solution. $E_{m,7}(2) \cong 450$ mV. The second electron reduction occurs more readily (at higher potential) than the first. For quinones in aprotic solvents, e.g. acetonitrile, the intermediate semiquinone state is stable. The values for the midpoint potentials are $E_m(1) = -0.43$ V and $E_m(2) = -1.15$ V (28). The second electron has a midpoint potential, $E_m(2)$, about 720 mV more negative than the first (28) owing to the accumulation of negative charge. Thus, the ease of the second electron reduction in aqueous solvents is aided by proton binding, or equivalently the proton binding is driven by quinone reduction. The pK_a for the ubisemiquinone was determined to be about 6 (39), whereas that for the dihydroquinone was determined to be 13.3 [with $pK_a(2)'$ for the second proton QH^- being somewhat higher than $pK_a(2)$] (28). Since these measurements were made in organic solvents, the pK_as in water should be somewhat lower, e.g. a pK_a of 5 for the semiquinone (39) and about 12 for the dihydroquinone.

In the RC the midpoint potential $E_{m,7}(1)$ for Q_A in chromatophores from *Rb. sphaeroides* has been determined to be about -20 mV at pH 7 (41) and between -50 and $+20$ mV in isolated RCs of *Rb. sphaeroides* (42, 43). Lower values of $E_{m,7}(2)$ have been observed at high pH in both chromatophores (-180 mV, pH 10) (41) and isolated RCs (43). The midpoint potential for the second electron reduction $E_{m,7}(2)$ of Q_A has not been measured, possibly because it may be too negative, as expected for a quinone in an aprotic environment. This may be due to the inaccessibility of Q_A to protons.

The midpoint potentials for $E_{m,8}(1)$ and $E_{m,8}(2)$ for Q_B in chromatophores of *Rb. sphaeroides* are 40 and -40 mV, respectively (44). The small difference between the midpoint potentials for the first and second electron reductions of Q_B is an important feature in the design of the RC, allowing the efficient reduction of Q_B by electrons from Q_A^-. This property is not found for ubiquinone either in aqueous solution or in aprotic solvents, in which the differences between $E_m(1)$ and $E_m(2)$ are large ($+680$ and -720 mV for aqueous solution at pH 7 and for the aprotic solvent acetonitrile, respectively). The midpoint potential for the semiquinone formation (Q_B^-) is raised in the RC (making it easier to reduce) to match the average potential for the two-electron reduction of the quinone. This may be accomplished by hydrogen bonding and electrostatic interactions with residues near Q_B. Wraight has discussed the electrostatic stabilization of the semiquinone of Q_B by coupling with external acidic residues (40).

QUINONE REDUCTION CYCLE

The sequence of light-induced electron and proton transfer reactions that occur within the RC in the process of reducing Q_B to Q_BH_2 is represented by the quinone reduction cycle shown in Figure 4 (15, 45, 46). In this cycle two electrons are transferred and two protons are bound to Q_B as a result of two separate photochemical events that result in the oxidation of two cytochrome molecules. Each step in the cycle represents a change in the state of the quinone complex. The most important reactions as far as proton transfer is concerned are reactions 4 and 5, in which proton uptake is coupled to second electron transfer to Q_B.

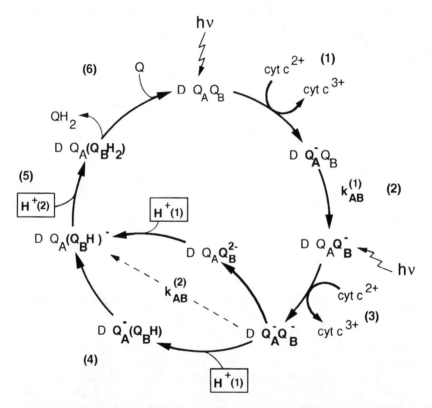

Figure 4   Quinone reduction cycle. The uptake of protons (rectangles) occurs in two reactions (steps 4 and 5) that are coupled to photon absorption, cytochrome oxidation (steps 1 and 3), and electron transfer (steps 1 through 4) followed by quinone exchange (step 6). Steps 2, 4, 5, and 6 are reversible. Two possible pathways are indicated for step 4. In the upper, electron transfer precedes protonation; in the lower, proton transfer precedes electron transfer. In native RCs the cycling time is ca. 1 ms.

The cycle starts with the RC in the initial state DQ_AQ_B. The first step in the cycle, $DQ_AQ_B \rightarrow DQ_A^-Q_B$, involves three rapid electron transfer steps not explicitly shown in Figure 4 (reviewed in 27). These are (a) electron transfer from the primary donor to the pheophytin molecule along the A branch, which occurs in 3.5 ps (the bacteriochlorophyll molecule has been proposed as a mediator in this reaction, either as a real electron acceptor or by a superexchange mechanism); (b) electron transfer from bacteriopheophytin to Q_A, which occurs in ca. 200 ps; and (c) electron transfer from a cytochrome c_2 molecule to D^+, which occurs in ca. 1 μs. The second reaction in the cycle, $DQ_A^-Q_B \rightarrow DQ_AQ_B^-$, involves the transfer of the first electron [$k^{(1)}_{AB}$] to Q_B in ca. 100 μs (pH 7.5) (15, 47, 48). This reaction is reversible and gives rise to an equilibrium between the states $Q_A^-Q_B$ and $Q_AQ_B^-$. The third step is another photochemical cytochrome oxidation that gives rise to the diradical state $DQ_A^-Q_B^-$. The fourth reaction, which gives rise to the singly protonated, doubly reduced state $DQ_A(Q_BH)^-$, could proceed through two possible paths: proton uptake followed by electron transfer or electron transfer followed by proton uptake. These two mechanisms are shown as parallel paths in Figure 4 (step 4) and will be discussed in more detail in a later section. Following the formation of the $DQ_A(Q_BH)^-$ state, Q_B takes up the second proton in the fifth reaction to form the dihydroquinone state, $DQ_AQ_BH_2$. The rates for the individual steps in reactions 4 and 5 have not been resolved, but the overall time for the two reactions is about 1 ms (pH 7.5) (15, 49). Results from site-directed mutagenesis suggest that the two protons, $H^+(1)$ and $H^+(2)$, are taken up via two separate pathways (see below). After reduction, the dihydroquinone dissociates from the RC (reaction 6) and is replaced by an oxidized quinone (50), thereby completing the cycle. The exchange times for the quinone depend on the nature of the quinone and detergent composition, but are generally quite short, i.e. in the millisecond range. The cyclic electron transfer rate for this cycle, which is the turnover rate for cytochrome photooxidation by RCs under saturating light intensity in the presence of excess cytochrome and quinone, is about 1000 s^{-1}.

EXPERIMENTAL APPROACHES TO THE PROBLEM OF PROTON TRANSFER

The photochemical charge separation process results in the creation of negatively charged quinone species inside the protein molecule. The proton transport processes that occur in the RC are driven by the electric fields due to these charges. In principle two types of proton transfer steps are possible: (a) the protonation of the reduced quinone, and (b) the protonation of acidic residues in the protein near the quinone whose pK$_a$s are changed as a result of the negative charge on the quinone. The former process is the basis for the proton pump. The latter process can give important information about nearby acidic residues, some of which may be involved in the proton transport pathway. Evidence for the proton transport into the RC upon quinone reduction comes from spectroscopic measurements, direct measurements of proton uptake, measurements of electrogenic events across membranes, and the influence of pH on electron transfer rates and equilibria. The proton uptake that results from the formation of the one-electron states, Q_A^- and Q_B^-, and the proton uptake that results from the Q_B^{2-} state in the quinone reduction cycle are discussed below.

Spectroscopy

OPTICAL   The protonation state of the components of the RC have been investigated by several types of spectroscopic measurements. Optical and electron paramagnetic resonance (EPR) studies (see below) of bacterial RCs indicated that the singly reduced semiquinones Q_A^- (45, 46, 51, 52) and Q_B^- (45, 46) had characteristic absorption peaks at 450 nm that are indicative of the anionic semiquinone (39). The characteristic absorption spectrum of a protonated semiquinone [i.e. a shift of the absorption peak to 425 nm (39)] has not been reported (15, 45) for pH as low as 5 (P. H. McPherson, unpublished data). Morrison et al (28) have compared the semiquinone spectra measured in different solvents and concluded that the optical spectrum of Q_A^- observed in RCs was that of an anionic semiquinone in a protic environment (28), as expected for quinones with strong H bonds. The doubly reduced Q_B picks up two protons as determined from the optical spectra, which show the doubly reduced state to be the dihydroquinone Q_BH_2 (45, 53).

ELECTRON PARAMAGNETIC RESONANCE   The EPR spectra of the Q_A^- and Q_B^- states are broad (ca. 300 G) lines with g values near 1.8 observable only at cryogenic temperatures. The lines have been interpreted in terms of a magnetically coupled Fe^{2+} Q^- interacting spin system (54). The EPR spectra of Q_A^- were studied in RCs from which the Fe^{2+} has been removed (55–57) or replaced by Zn^{2+} (57a). The line widths of the EPR signals were found to be close to 8.0 G, characteristic of the ubisemiquinone anion radical: this is considerably narrower than the 12 G expected for the protonated semiquinone (57). The electron nuclear double resonance (ENDOR) spectrum of Q_A^- and Q_B^- in RCs in which Fe^{2+} was replaced by Zn^{2+} (58, 59) showed strong couplings which were attributed to two exchangeable protons. These couplings were assigned to H bonds to the carbonyl oxygens of the quinone. By using a model for dipolar coupling, the distances from the quinone oxygen to the proton were estimated to be 1.55 and 1.78 Å for Q_A and 1.69 and 1.97

(or both 2 Å) for Q_B. These distances are in the range expected for hydrogen bonds (1.5 to 2 Å) and are larger than those for a protonated quinone (1 Å).

INFRARED   IR spectral changes in bacterial RCs have been observed by using a variety of procedures to identify changes due to quinone reduction. Several IR bands due to quinone reduction were observed that could be caused by either quinone vibrations or vibrational bands of protein residues perturbed by reduction (60–62). Recently, time-resolved IR spectroscopy showed IR absorption changes due to electron transfer from $Q_A^- Q_B$ to $Q_A Q_B^-$ with a half time of 150 μs (63). In addition to those changes, slower absorption changes with a half time of about 1 ms were seen at 1617 and 1750 cm^{-1} (W. Mäntele, personal communication). These changes are at absorption frequencies characteristic of carboxylate and carboxylic acid and could be due to proton uptake following electron transfer.

Stoichiometry of Proton Uptake

Proton uptake was measured with pH-sensitive dyes (15, 64), pH electrodes (65), and conductance measurements (64). Results with isolated RCs showed that proton uptake occurs upon formation of either Q_A^- or Q_B^-. However, the stoichiometry for proton uptake was less than one proton per RC (Figure 5) and can be accounted for by pK_a shifts of nearby acidic residues in the protein. This is in accord with the spectroscopic observations that indicated that the semiquinones are not protonated. Proton uptake by the doubly reduced Q_B has been found to involve two protons per Q_B, in accord with the formation of the dihydroquinone state $Q_B H_2$ (discussed in a later section).

The pH dependence of proton uptake after one-electron reduction of the quinones provides information about interaction between quinones and titrable residues. The proton uptakes for both DQ_A^- and DQ_B^- showed similar pH dependencies, being smaller at low pH and having a maximum value at pH 9 to 10 (43, 65). The proton uptake of DQ_A^- has been modeled by using an interaction between Q_A^- and four acidic residues with pK_as in the range from 5 to 10 (43, 65). The pK_as of these residues are increased (by an amount ΔpK) after quinone reduction. The major interaction is with a residue with a pK_a of ca. 10 ($\Delta pK \approx 0.8$). A smaller interaction is with a residue with a pK_a of ca. 6 ($\Delta pK \approx 0.6$). Proton uptake due to DQ_B^- ia larger than that due to DQ_A^- and was also modeled by interaction of the reduced quinone with four titrable residues. The interacting groups need not be the same as those that interact with Q_A^-. The difference in the proton uptake between Q_A^- and Q_B^- is related to the pH dependence of the equilibrium between $Q_A^- Q_B$ and $Q_A Q_B^-$. The proton uptake by DQ_B^- is greater than that by DQ_A^- in the region above pH 9 and below pH 6.5 (Figure 5). The difference

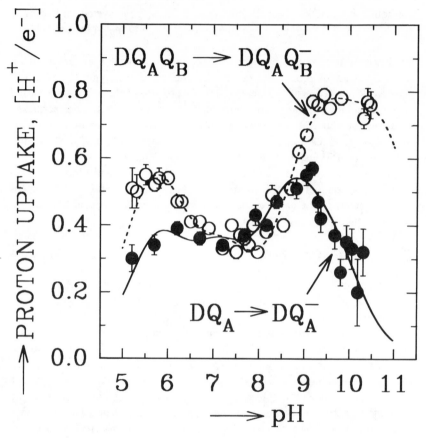

Figure 5 Proton uptake due to the formation of the semiquinone states DQ_A^- and DQ_B^-. Following light-induced charge separation D^+Q^-, the oxidized donor was reduced by cytochrome c. The larger proton uptake by DQ_B^- compared with DQ_A^- near pH 6 and 10 is due to a few specific protonatable residues near Q_B. Similar results were obtained by Maroti & Wraight (43). Modified from reference 65 with permission.

in proton uptake is related to the change in free energy $\Delta G(\text{pH})$ for the reaction $Q_A^- Q_B \rightarrow Q_A Q_B^-$ by the following equation:

$$\Delta G(\text{pH2}) - \Delta G(\text{pH1}) = \ln 10 \, RT \int \Delta H d(\text{pH}), \qquad 2.$$

where ΔH is the difference in proton uptake between the $Q_A Q_B^-$ and $Q_A^- Q_B$ states as a function of pH and the integral is taken from pH1 to pH2. By using this equation, the proton uptake and semiquinone equilibria were found to be in agreement (65).

The identity of the acidic residues responsible for proton uptake is of importance since some of them may be involved in the proton transfer chain to

Q_B. In principle each proton uptake peak may be due to contributions from many different residues with effective pK_as within the observed range. Electrostatic calculations (P. Beroza, unpublished results) have shown that the proton uptake will tend to be dominated by nearby buried residues since surface residues are shielded by the aqueous solvent. However, distant residues still contribute to the proton uptake because of their relatively large abundance. In addition, the assignment of observed pK_as to specific residues is complicated by electrostatic interactions between titrable residues (66). The identity of some of the strongly interacting residues will be discussed below.

Rates of Electron Transfer and Proton Uptake

A comparison between the rates of electron transfer and proton uptake is necessary to understand the coupling between electron and proton transfer. The rates of proton uptake associated with quinone reduction have been measured with pH indicator dyes (15, 32, 43, 64). The following results were obtained for the different transfer steps.

For the reduction of Q_A by reduced bacteriophenophytin,

$$\phi_A^-Q_A \xrightarrow{k_{\phi_A Q_A}} \phi_A Q_A^-,$$ 3.

the electron transfer rate, $k_{\phi_A Q_A}$, is 5×10^9 s^{-1} and is relatively independent of temperature and pH (27). The proton uptake rate associated with this reaction was found to be much lower, 10^4 s^{-1} (pH 6). The rate depends on pH, decreasing at higher pH with a slope $d(\log k)/d(pH) = -0.3$ (15). From the temperature dependence of the rate, an activation energy of >40 kJ/mol (0.4 eV) was found (32). Takahashi et al have explained the proton uptake rate, the relatively high activation energy, and the low viscosity dependence for the reaction in terms of a collisional reaction between pH indicator molecules and proton acceptor groups that reside on the protein and whose accessibility may be determined by conformational equilibria of the protein (32). Supporting evidence for a conformational change at this step is the observation by Brzezinski et al. (67) of an electrogenic event upon reduction of Q_A to Q_A^- that has the opposite sign from that observed for proton uptake. This electrogenic event was suggested to be due to a conformational change upon Q_A reduction.

The reduction of Q_A^- to the doubly reduced Q_A^{2-} state occurs much more slowly (10 ms) (68) than the first reduction (200 ps) and is observed only under extreme conditions of high illumination. The reduced rate of the second electron transfer to Q_A may be related to the absence of an efficient proton transport chain to Q_A from the solvent. The double reduction of Q_A is enhanced in RCs from which Fe^{2+} has been removed. This is probably due to an increased proton accessibility of the Q_A site (69), brought about by possible structural changes associated with the removal of Fe^{2+}. In these RCs the electron transfer rate $k_{\phi_A Q_A}$ is lowered considerably (70). Thus, there seems to be a correlation between fast electron transfer and inaccessibility to protons (see footnote 1).

The rates of proton transfer and electron transfer due to the reaction

$$Q_A^- Q_B \xrightarrow{k_{AB}^{(1)}} Q_A Q_B^- \qquad\qquad 4.$$

were found to be very similar. Both the electron transfer and proton transfer rates were about 10^4 s^{-1} at low pH and decreased with increasing pH. The reason for the coincidence between these rates may be that proton transfer is limited by electron transfer (or electron transfer is limited by proton transfer).

In RCs from *Rb. sphaeroides* different values for $k_{AB}^{(1)}$ have been found by different workers. Wraight (15) and Vermeglio & Clayton (47) found a relatively constant decrease in the log of the rate with increasing pH with a slope [d(log k)/d(pH)] of about -0.3. Kleinfeld et al (48) reported a pH-independent rate at low pH that decreases above pH 9. Takahashi et al (32) have reported that the rates are biphasic, with the fast phase agreeing with the results of Kleinfeld but with an additional slower phase.

The decrease in $k_{AB}^{(1)}$ with increasing pH is consistent with the presence of an acidic residue near Q_B that must be protonated before electron transfer can occur (15). The data of Kleinfeld et al could be fitted to a simple model in which the protonation state of a residue with pK_a of ca. 9 determined the rate of electron transfer (48). This residue was shown by site-directed mutagenesis to be GluL212 (7, 71). In mutant RCs containing Gln in place of Glu at position L212 the rate of electron transfer remained constant at high pH, showing the absence of an interacting, titratable residue.

At high pH, the negatively charged GluL212 must become protonated before fast electron transfer can occur. Thus, the kinetics of electron transfer can give information about the rates of proton transfer to GluL212. The kinetics of electron transfer seen by Kleinfeld et al (48) at higher pH are monophasic, suggesting that the rates of protonation and deprotonation of GluL212 are high compared with those of electron transfer; e.g. at pH 9 the proton transfer rate should be higher than 10^4 s^{-1}. The biphasic kinetics for $k_{AB}^{(1)}$ observed by Wraight and coworkers (32) could be due to sample heterogeneity or to equilibrium mixtures of conformational or protonation states that relax more slowly than the measured rate. Takahashi et al (32) have argued that the biphasic kinetics do not arise from sample heterogeneity since the recombination kinetics are monophasic; they have explained their results by a series scheme in which the rapid phase of electron transfer does not immediately go to completion but goes to a steady-state equilibrium between $Q_A^- Q_B$ and $Q_A Q_B^-$. A slower proton uptake reaction corresponding to the slower phase then occurs to drive the reaction to equilibrium. This could account for both a high rate ($>10^4$ s^{-1}; pH 9) and a low rate (200 s^{-1}; pH 9) of proton transfer. The differences between the results seen by Kleinfeld et al (48) and Wraight (15) are not understood at present but may be due to differences in sample preparation or detergent. Biphasic rates of other reactions have also been observed: for $k_{AB}^{(1)}$ in RCs from a mutant in which AspL213 was changed to Asn (72), for k_{AD} in RCs in which ubiquinone was replaced by anthraquinone (73), and in the recombination reactions of *R. viridis* (74).

The proton uptake rate associated with the second electron transfer reaction,

$$2H^+ + Q_A^-Q_B^- \xrightarrow{k_{AB}^{(2)}} Q_AQ_BH_2, \qquad\qquad 5.$$

has been found to be similar to the electron transfer rate $k_{AB}^{(2)}$ (15, 32, 53). Both the electron transfer and proton uptake rates associated with the second-electron $k_{AB}^{(2)}$ are about 10^4 s^{-1} at pH 5 and decrease with increasing pH. The pH dependence is complex, exhibiting a slope d(log k)/d(pH) $= -0.3$ below pH 8. The slope decreases to about -1 above pH 8. The rates were explained by a model in which proton uptake occurs before electron transfer (15, 19), but are also consistent with a model in which electron transfer can occur before proton transfer (53) (see below).

Effects of Site-Directed Mutagenesis on Electron and Proton Transfer Rates

The involvement of specific amino acid residues in proton transfer has been investigated experimentally by site-directed mutagenesis of amino acid residues in the Q_B-binding site. Protonatable residues near Q_B were changed to nonprotonable residues. RCs containing the altered residues were isolated and characterized by measuring the cytochrome turnover rate (i.e. the rate of cytochrome oxidation during continuous illumination in the presence of excess cytochrome c and quinone), as well as by electron and proton transfer measurements. Mutants that were deficient in proton transfer were identified by having one or more of the following characteristics: (*a*) reduced rates of cytochrome turnover, (*b*) reduced rates of proton uptake, and (*c*) reduced rates of electron transfer, $k_{AB}^{(2)}$.

Changes in these rates can result from a decreased rate of proton uptake by Q_B, which can become the rate-limiting step for these rates (see Figure 4). The rate of cytochrome photooxidation can be limited by proton transfer if the uptake of either of the two protons $H^+(1)$ or $H^+(2)$ becomes the rate-limiting step. If the uptake of $H^+(1)$ is inhibited, slow cytochrome oxidation will be observed after a fast oxidation of two cytochromes (steps (1) and (3) in Figure 4). If uptake of $H^+(2)$ is inhibited, a fast oxidation of three cytochromes will be observed owing to the additional reduction of the quinone complex (step 4

Table 1 RC mutants from *Rb. sphaeroides* used to investigate proton transfer (pH 7.5)

RC		$k_{AB}^{(1)}$ (s^{-1})	$k_{AB}^{(2)}$ (s^{-1})	k_H (slow) (s^{-1})	Turnover (cyt/RC) (s^{-1})	Ref.
Native		6000	1300	1200	>500	7, 19
GluL212	→ Gln	3500	750	6	12	7, 71
	→ Asp	1750	>500	>500	~200	76
SerL223	→ Ala	15,000	4	4	8	19
	→ Asn	>300	~10	—	~10	74a
	→ Thr	1000	>1000	—	>150	19
	→ Asp	>500	>500	—	>150	74a
AspL213	→ Asn	350	0.4	0.3	0.8	20, 78
	→ Leu	300	—	—	0.6	M. Paddock, unpublished
AspL210	→ Asn	800	600	—	290	M. Paddock, unpublished
ArgL217	→ Gln	4000	—	>200	>200	M. Paddock, unpublished
ArgL217	→ Leu	3200	—	—	>150	M. Paddock, unpublished
HisL190	→ Gln	1500	—	—	200	J. Williams, unpublished
AspL213	→ Asn,	4000	0.4		0.6	71; M. Paddock, unpublished
GluL212	→ Gln					

in Figure 4). When proton transfer is the rate-limiting step in the quinone reduction cycle, the rate of cytochrome turnover equals twice the proton transfer rate since two cytochrome molecules are oxidized in one cycle.

A reduced proton uptake rate k_H can result from a mutation that changes the pathway to Q_B. A slow protonation would be observed when the proton uptake associated with the second electron transfer is measured either after 2 single turnover flashes (< 1 μs duration) or after a single multiturnover flash (> 1 ms duration). In these cases the reaction center would be rapidly driven to the rate-limiting proton uptake step (of either H$^+$(1) or H$^+$(2), see Figure 4).

A slow rate of electron transfer would occur in a mutant in which proton uptake was inhibited and in which proton uptake was required before electron transfer. Thus, in the model shown in Figure 4 transfer of the second electron would be slow if H$^+$(1) were blocked but would be unaffected if H$^+$(2) were blocked. (The state $DQ_AQ_B^{2-}$ in Figure 4 is expected to be a transient intermediate state with a low occupancy.)

A list of mutants constructed to study proton transfer is shown in Table 1.

In the first proton uptake mutant constructed, GluL212, an acidic group close to the Q_B site, was changed to Gln by Paddock et al (7). The cytochrome turnover rate in this mutant was slowed by a factor of >40 following the fast oxidation of three cytochromes (Figure 6). This can be explained by a bottleneck in the transfer of the second proton, H$^+$(2) (see step 5 in Figure 4),

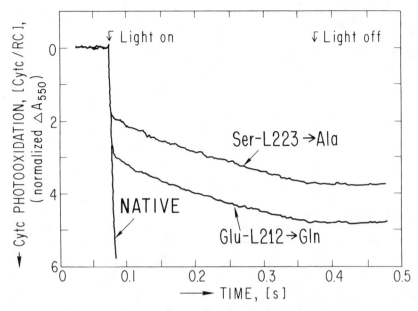

Figure 6 Cytochrome turnover rates in native and mutant RCs. RCs were illuminated in the presence of excess cytochrome *c* and Q_{10}. The mutants SerL233 → Ala and GluL212 → Gln show a fast oxidation of two and three cytochromes, respectively, followed by a reduced cytochrome oxidation rate that indicates a bottleneck in the proton transfer rate. Modified from references 7 and 19.

which slows the turnover of the quinone reduction cycle. Since this block occurs after electron transfer of the second electron, the RC can oxidize another cytochrome (three altogether) to form the fully reduced quinone complex $Q_A^-(Q_BH)^-$. This suggested that electron transfer of the first and second electrons is unimpeded and that the rate limitation is due to a bottleneck in proton transfer (7). This proposal was verified by direct measurements of the electron and proton transfer rates that showed that electron transfer rates for both $k_{AB}^{(1)}$ and $k_{AB}^{(2)}$ were not appreciably changed. However, when RCs were illuminated with a multiturnover flash, biphasic proton uptake was observed with fast uptake of ca. one proton and slow uptake of a second proton (Figure 7) (31, 75). The proton uptake rate was slowed by a factor of about 200 to 6 s^{-1}, which limits the turnover of the quinone reduction cycle (12 cytochrome RC^{-1}s^{-1}) (i.e. proton transfer was half the cytochrome turnover rate). This mutation was later also constructed by Takahashi & Wraight (71), who found a biphasic rate of proton and electron transfer after two flashes. This indicates a rapid transfer of the second electron to a steady-state level (step 4 in Figure 4), which subsequently decays as a result of slow proton uptake (step 5 in Figure 4). This slow proton uptake

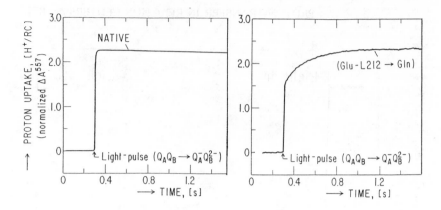

Figure 7 Proton uptake in native *(left)* and mutant *(right)* RCs. RCs in the presence of excess cytochrome *c* were given a saturating light pulse ($\tau \approx 1$ ms) driving the quinone reduction cycle to the rate-limiting step. The slow proton uptake in the GluL212 → Gln mutant indicates a proton transfer bottleneck. The deviation from an integer (one or two) proton uptake is due to the partial proton uptake associated with the semiquinone state (see Figure 5). Modified from reference 31 with permission.

supports the conclusion that GluL212 is required for proton uptake in the quinone reduction cycle.

The change of GluL212 to Gln is structurally conservative since the size and shape of the residue are retained. Kinetic measurements ($k_{AB}^{(1)}$, k_{BD}) have indicated that the pK_a for GluL212 is 9.5 (7, 71). Thus, no changes due to electrostatic effects are expected near neutral pH. Preliminary X-ray crystallography measurements have indicated that no large changes (<0.5 Å) in the structure of the mutant occurred (A. Chirino & D. C. Rees, unpublished results). The main change is in the activity of the protonatable group. Control experiments in which GluL212 was changed to Asp gave RCs with normal turnover rates (7, 76). These experiments showed that proton uptake is an important part of the photochemical cycle and that under some conditions can be the rate-limiting step. They also point to the important role that the protein plays in the proton conduction pathway to Q_B.

A second mutation, in which SerL223, which provides a hydrogen bond to Q_B, was changed to Ala was constructed (19). In this mutant the cytochrome turnover rate slowed by a factor of 60 after an initial fast oxidation of two cytochromes (Figure 6). This was explained by a proton transfer bottleneck in the transfer of the first proton, H(1)$^+$, in the quinone reduction cycle (step 4 in Figure 4). A block in the uptake of this proton would reduce the rate of the second electron transfer, $k_{AB}^{(2)}$, and lead to the accumulation of the state $Q_A^- Q_B^-$ after oxidation of only two cytochromes. This suggestion is supported by the result that both proton uptake and electron transfer rates, $k_{AB}^{(2)}$, were decreased by a factor of ca. 300 (pH 7.5) (Table 1). The change of SerL223 to Ala does not lead to large structural changes, as indicated by the lack of change in the electron transfer rates for the first electron, $k_{AB}^{(1)}$, and charge recombination, k_{BD} (see Eq. 6), and by an unchanged EPR spectrum of $Q_B^- Fe^{2+}$ (19). Control experiments in which SerL223 was changed to protonable residues Thr and Asp gave RCs with normal turnover rates, while substitution with Asn resulted in a slow turnover (Table 1). Thus, this mutation shows that the proton transfer of the first proton, $H^+(1)$, involves SerL223.

A third mutation, in which AspL213, an acidic residue near Q_B, was changed to Asn, was constructed in two laboratories (19, 20, 71, 78). The findings of the two groups are in good agreement. The cytochrome turnover rate in this mutant was decreased by at least a factor of 600 after a fast oxidation of two cytochromes similar to the effect of the SerL233 $\rightarrow$ Ala mutation, indicating a block in proton and electron transfer associated with the second electron (78). The proton uptake and electron transfer rates, $k_{AB}^{(2)}$, were reduced by a factor of 4000 to a rate of ca. $0.3\ s^{-1}$ (at pH 7.5) (71, 78). Evidence that the reduction in proton uptake and electron transfer rates is due to a proton transfer bottleneck comes from the observation (79) that these rates are increased by addition of azide and other weak acids that are presumed to act as proton carriers. A similar increase in proton transfer rate was observed by addition of azide to proton transfer mutants of bacteriorhodopsin (80, 81). These reductions in proton and electron transfer rates can be explained if AspL213, along with SerL223, is involved in the proton transfer pathway of the first proton, $H^+(1)$, which is necessary for the transfer of the second electron to occur (step 4 in Figure 4).

There is some indication that AspL213 is also involved in the pathway for the second proton, $H^+(2)$, possibly as a donor to GluL212. McPherson et al base their suggestion on the kinetics of the first electron transfer, $k_{AB}^{(1)}$, in the AspL213 $\rightarrow$ Asn mutant (72). Although this reaction is thermodynamically more favorable in mutant RCs than in native RCs (20), the rate is lowered by a factor of about 10 at pH 7 and 100 at pH 9 (71, 72). One explanation for this is that in the mutant a new rate-limiting step, due to slow proton transfer, is involved. In this mutant the pK_a for GluL212 is lowered to <7 (32, 71; see below). Since the protonation of GluL212 was shown to be necessary for $k_{AB}^{(1)}$ to proceed in native RCs (7, 71), the protonation of GluL212 may be the rate-limiting step in the mutant. The association of the low rate with the ionization of GluL212 is consistent with the finding that $k_{AB}^{(1)}$ is fast in a double mutant in which GluL212 is changed to Asn, in addition to the AspL213 $\rightarrow$ Asn mutation (71; M. Paddock, unpublished data). The idea that proton transfer to GluL212 is the rate-limiting step for $k_{AB}^{(1)}$ in the AspL213 $\rightarrow$ Asn mutant is supported by the following observations: (a) The rate is biphasic in the pH range 5.5 to 6.5. The high (pH independent) rate is attributed to RCs in which GluL212 is protonated (72), and the low (pH dependent)-rate is attributed to RCs in which GluL212 is unprotonated (if proton transfer were fast, one should see monophasic kinetics). (b) The low rate is accelerated by the addition of azide (P. McPherson, unpublished data) which may act as a protonophore. A low rate of proton transfer in the mutant suggests that AspL213 is involved in the proton transfer pathway to GluL212. Takahashi et al have also suggested that involvement of AspL213 in the $H^+(2)$ pathway (32). Their argument is based on the low rate of the second electron transfer, $k_{AB}^{(2)}$, in the AspL213 → Asn mutant. They argue that if either of the $H^+(1)$ or $H^+(2)$ pathways are operative, $k_{AB}^{(2)}$ should proceed normally. The absence of electron transfer in the AspL213 → Asn mutant then indicates that AspL213 is involved in the pathway for both protons. This argument seems inconsistent with the slow $k_{AB}^{(2)}$ in the SerL223 → Ala mutant, since SerL223 is unlikely to be involved in both pathways (see Figures 3 and 9).

Although it is clear that the mutation of AspL213 to Asn has large effects on the rates of proton and electron transfer, the interpretation of these effects in terms of proton transfer mechanisms is complicated by electrostatic effects due to the removal of a negative charge. This change stabilizes the $Q_A Q_B^-$ state relative to the $Q_A^- Q_B$ state and changes the pK_a of neighboring residues. Takahashi & Wraight (71) have pointed out that the stabilization of Q_B^- could make $Q_A^- Q_B^-$ more stable than $Q_A Q_B H_2$ and that therefore the second electron would not transfer even if proton transfer were fast. In this case the observed slow proton and electron transfer could be due to the oxidation of the semiquinones by an external oxidant and might not provide information about the pathway for proton transfer.

Mutations that did not produce large changes in the proton transfer rate include several controls that preserved the proton-donating ability, e.g. GluL212 → Asp (76), SerL223 → Thr (19), and SerL223 → Asp (M. Paddock, unpublished data). Several other mutations that did not result in a significant loss of activity include ArgL217 → Gln and AspL210 → Asn (M. Paddock, unpublished data). Preliminary results on a mutant in which the residue HisL190, which provides the second H bond to Q_B, was changed to Gln showed a near-normal cytochrome turnover rate (J. Williams, unpublished data). This indicates that the nature of this hydrogen bond is not essential for proton transfer to Q_B.

The AspL213 → Asn and SerL223 → Ala mutants are not able to grow photosynthetically. Several second-site suppressor mutations that restore photosynthetic growth have been isolated from these mutants (S. Rongey, unpublished data). These mutants should help to determine the requirements for efficient proton transfer.

Electron and Proton Transfer Equilibria

The pH dependence of the equilibrium between electron transfer states can be used to obtain information about the energetics of proton and electron transfer and the pK_as of residues in the RC. The equilibrium constant between the states $Q_A^-Q_B$ and $Q_AQ_B^-$ can be determined by measuring the back reaction rate, k_{BD}, defined in Eq. 6 (48, 82).

$$DQ_AQ_B \underset{k_{AD}}{\overset{h\nu}{\rightleftarrows}} D^+Q_A^-Q_B \underset{k_{BA}}{\overset{k_{AB}}{\rightleftarrows}} D^+Q_AQ_B^- \qquad\qquad 6.$$

with k_{BD} indicated by the dashed arrow returning to DQ_AQ_B.

The recombination reaction (k_{BD}) has been shown in native RCs to proceed through the intermediate $D^+Q_A^-Q_B$ state, which is in equilibrium with the $D^+Q_AQ_B^-$ state. Consequently, the recombination rate k_{BD} is proportional to the fraction, α, of RCs in the $D^+Q_A^-Q_B$ state and depends on the equilibrium free energy ΔG for the reaction $Q_A^-Q_B \Leftrightarrow Q_AQ_B^-$, i.e.

$$k_{BD} = \alpha k_{AD} = k_{AD}/(1 + e^{-\Delta G/kT}) \qquad\qquad 7.$$

From the pH dependence of k_{BD}, the pK_as of groups interacting with Q_B (and Q_A) may be obtained from the equation (48)

$$\Delta G = \Delta G^0 - \sum_i kT \ln[(1 + 10^{[pH - pK_i(Q_B^-)]})/(1 + 10^{[pH - pK_i(Q_B)]})] \qquad 8.$$

where ΔG^0 is the standard free energy change for the reaction and $pK_i(Q_B)$ and $pK_i(Q_B^-)$ are the pK_as of acidic residue i in RCs containing Q_B and Q_B^-, respectively. The increase in pK_a due to quinone reduction [$\Delta pK = pK(Q_B^-) - pK(Q_B)$] is attributed to an electrostatic interaction.

The back reaction rate as a function of pH has been measured for native RCs and for RCs (7, 20, 32, 71, 77, 83) containing mutations to acidic groups near the Q_B-binding site (Figure 8). These rates exhibit a complex dependence on pH, indicating interaction of titratable residues with the charged semi-quinone states. The pH dependence is altered in the mutants in which GluL212, AspL213, or AspL210 has been changed to a nontitratable residue. The values for the pK_as of the various acidic residues are summarized in Table 2.

In native RCs the back reaction has two pH-dependent regions, indicating interactions with at least two acidic residues (Figure 8). Fitting it with two residues gives $pK_a(Q_B)$ and $pK_a(Q_B^-)$ 5.5 and 6.1, for one residue and 9.3

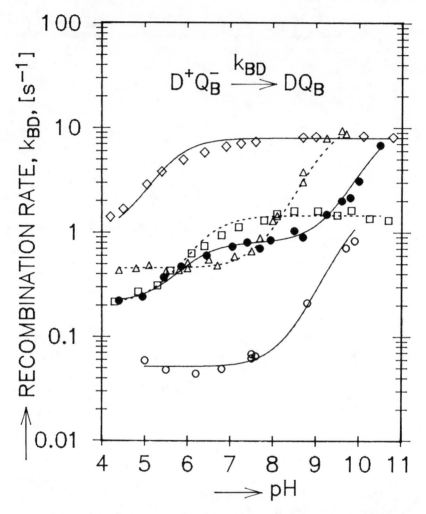

Figure 8   Recombination rate k_{BD} as a function of pH for native and mutant RCs. Symbols: •, native; ○, AspL213 → Asn (83); △ AspL210 → Asn (77); □, GluL212 → Gln (7); ◇, GluL212 → Asp (76). The pH dependence gives information about the protonation state of residues near Q_B. Similar results for the AspL213 → Asn and GluL212 → Gln mutants were obtained by Takahashi & Wraight (71).

and >11, for the other (Table 2). Tentative assignments of the groups responsible for the pH dependence of the rates have been made by studying RCs modified by site-directed mutagenesis.

The pH dependence above pH 9.3 in native RCs is eliminated in RCs from the GluL212 → Gln mutant (7, 71). This is consistent with the assignment of a pK$_a$ of 9.3 to GluL212. In RCs in which AspL213 (32, 71, 76) and

Table 2 pK$_a$s for acidic residues obtained from k_{BD} measurements[a]

RC	pK$_1$ (Q$_B$)	pK$_1$ (Q$_B^-$)	pK$_2$ (Q$_B$)	pK$_2$ (Q$_B^-$)
Native	5.5[b]	6.1	9.3	>11
GluL212 → Gln	5.5[b]	6.4	—	—
GluL212 → Asp	4.7	6.4	—	—
AspL213 → Asn	—	—	8.3[b]	>10
AspL210 → Asn	—	—	7.8	>9

[a] These pK$_a$ values were obtained by using Eq. 7 and 8 for residues labeled 1 or 2 in RCs containing Q$_B$ or Q$_B^-$.
[b] This represents a maximum value. If the pH-independent region is due to a direct back reaction (71), the actual pK$_a$ would be lower.

AspL210 were replaced by Asn (77), the transition at 9.3 is shifted to lower pH (8.3 and 7.8 for AspL213 and AspL210, respectively). Takahashi & Wraight (71) have suggested that the pH-independent region in the AspL213 → Asn mutant is due to the direct electron transfer from $D^+Q_AQ_B^-$ to DQ_AQ_B. (In this case the pK$_a$ for GluL212 in this mutant may be at a lower pH.) In the double mutant with AspL213 → Asn and GluL212 → Gln the value of k_{BD} is ca. 0.1 s^{-1} and is pH independent (71), showing that the pH dependence in the single AspL213 → Asn mutation was due to the shifted pK$_a$ of Glu-L212.

The changes in pK$_a$ can be understood in terms of electrostatic interactions between titrating residues (for a good discussion see reference 84). The shift in the pK$_a$ of GluL212 from 9.3 to 7.8 in the AspL210 mutant indicates an interaction between GluL212 and AspL210 of ca. 1.5 pH units (90 meV). From this interaction an effective dielectric constant of ca. 16 can be estimated if the energy is calculated by using Coulomb's law and the known distance of 9.8 Å (average distance between carboxyl oxygens).

The electrostatic interactions between the two Asp residues and GluL212 is partially responsible for the apparent high pK$_a$ of 9.3 for GluL212. The two Asp residues must have intrinsic pK$_a$s (owing to interaction with ArgL217, and possibly more effective dielectric screening) lower than that of GluL212 and hence ionize at a lower pH. The pK$_a$ of GluL212 is raised by the sum of the electrostatic interactions with the Asp residues (ca. 3 to 5 pH units).

In RCs from the GluL212 → Asp mutant the high-pH transition changes from pK$_a$ of 9.3 to pK$_a$ of <5. This large change in the apparent pK$_a$ of an acidic residue as a result of the removal of a methylene group is unusual. However, it may be explained if the mutation results in a change in the order of ionization of strongly interacting residues; i.e. the intrinsic pK$_a$ of AspL212 has been lowered below the intrinsic pK$_a$s of AspL210 and AspL213.

The pH dependence in the region from pH 5 to 7 in native RCs was eliminated in both the AspL213 → Asn (20, 83) and AspL210 → Asn (77) mutants. This may be explained if in these mutants the pK_a of the remaining carboxylic acid residue besides GluL212 (AspL210 or AspL213) was shifted to lower values (ca. 3). Electrostatic calculations have indicated that the AspL213 has a lower $pK_a(Q_B)$ (<4) than AspL210 (P. Beroza, unpublished data) owing to interaction with ArgL217. The $pK_a(Q_B)$ of AspL210 is then assigned a value of ca. 5. The $pK_a(Q_B^-)$ values of both AspL210 and AspL213 are estimated to be similar and are consistent with the value of ca. 6 observed experimentally (Figure 8; Table 2). It seems, therefore, that both are implicated in the 5 < pH < 7 dependence of k_{AB}. Why, then, is the pH dependence removed when either of the two Asp residues is mutated? An explanation lies in the interaction between the two acidic groups; i.e. in the mutant the remaining carboxylic acid residue is shifted to a lower pK_a value (below the observed range) owing to loss of interaction with the acidic group that has been removed.

Another way to obtain information about proton and electron transfer equilibria is to measure the pH dependence of the midpoint potential (E_m) for quinone reduction in the RC. The E_m for the reduction of Q_A has been found by many workers to decrease by 60 mV per pH unit, as expected for a reduction in which one proton is taken up per electron (41). This pH dependence is inconsistent with the measured proton uptake of less than one proton per electron discussed earlier. A possible reason for this discrepancy is the difference in the time scale of the measurements of midpoint potential and proton uptake or the difficulty in achieving redox equilibrium between the RC and the monitoring electrodes. An indirect method of measuring the pH dependence of the free energy change due to Q_A^- formation, based on the intensity of delayed fluorescence, agrees with the lower values for proton uptake (85).

Electrogenicity

The proton transfer events associated with quinone reduction have also been studied by measuring the photoinduced voltage across membranes (e.g. lipid monolayers or bilayers) containing oriented RCs (86–90). This photo voltage (electrogenicity) is a measure of the electrical work performed in going from one state of the RC to another and is related to the dielectrically weighted charge displacement across the membrane. The electron transfer $k_{AB}^{(1)}$ from Q_A^- to Q_B ws shown to be nonelectrogenic at pH 7.5 in RCs (89, 90) and in chromatophores (91). This is expected from the structure of the RC since the charge displacement vector between Q_A and Q_B is parallel to the plane of the membrane (i.e. along an equipotential surface). However, it was found (92) that at pH < 7 and pH > 8 the electron transfer was electrogenic, indicating a charge displacement perpendicular to the membrane. The displacements were interpreted to arise from proton uptake by acidic residues driven by pK_a shifts due to Q_B^- formation. The pH dependence of the amplitudes of the displacement currents agreed with the pK shifts expected from the pH dependence of the back-reaction kinetics, k_{BD}.

The electrogenicity associated with the second electron transfer reaction, $k^{(2)}_{AB}$, which occurred in oriented RCs after a second flash, had a characteristic time of ca. 1 ms (pH 7.5) (89, 91). This electrogenicity is most probably due to proton transfer since electron transfer is parallel to the membrane. The amplitude of the electrogenic signal associated with this proton uptake decreased approximately by one-half (pH 7) in the GluL212 → Gln mutant (93). This result supports the model in which GluL212 is responsible for the transport of one of the two protons taken up in this step.

MECHANISM OF PROTON TRANSFER

Proton transfer in bacterial RCs is a complex phenomenon, whose details have so far not been worked out in final form. We shall summarize here the present state of knowledge about four aspects of the reaction mechanism: (a) the sequence of electron and proton transfer steps, (b) the pathway of proton transport, (c) the dynamic model of proton transport, and (d) the energetics of proton transport.

Sequence of Electron and Proton Transfer Steps

The direct protonation of Q_B is associated with the transfer of the second electron to Q_B as shown by spectroscopic and proton uptake measurements. The proposed mechanism for this step involves the uptake of two protons, $H^+(1)$ and $H^+(2)$. The first proton uptake step could occur either before or after electron transfer as shown below (also see Figure 4) (53):

$$9.$$

The experimental data are consistent with either mechanism. The upper path (proton transfer before electron transfer) would be expected to dominate if $Q_A Q_B^{2-}$ is higher in energy than $Q_A Q_B H$; the lower path (electron transfer before proton transfer) would dominate if $Q_A Q_B H$ is higher in energy than $Q_A Q_B^{2-}$. A difficulty in deciding experimentally between the two mechanisms is that both $Q_A Q_B H$ and $Q_A Q_B^-$ are high-energy intermediates that are not formed in sufficient quantity to be observed. Furthermore, we do not know from experimental measurements whether electron transfer precedes or follows proton transfer, since these individual steps have not yet been resolved kinetically. Thus, in the absence of reliable calculations of the energies of the intermediate states, we can only speculate under what conditions one or the other pathway would dominate.

The energy required to reduce Q_B^- to Q_B^{2-} (without protonation) depends on the environment of the quinone and is higher in aprotic solvents than in aqueous solutions. For instance, for ubiquinone the midpoint potential E_m between Q^- and Q^{2-} (unprotonated) is at -1.15 V in acetonitrile (28) but can be estimated to be about -0.3 V in aqueous ethanol [calculated by assuming $E_m(2) = 0.45$ V, $pK_a(2) = 13$, and $pK_a(2)' = 14$]. If the E_m for Q_B^{2-} in RCs is close to that in acetonitrile, the proton transfer (upper) path should dominate. However, if E_m is similar to that in aqueous ethanol, the reaction (lower path) could have an activation energy less than 300 mV, making it favorable for electron transfer to precede proton transfer.

The energy required to form the protonated semiquinone state QH depends on the difference between the pH and the pK_a. The protonated semiquinone state has never been observed by optical spectroscopy even at low pH (down to pH 5) despite the fact that there are negatively charged residues, e.g. AspL213, near Q_B. Thus its pK_a might be very low (<4). In this case the reduced Q_B^- and Q_B^{2-} states may be stabilized with respect to the protonated state, for instance by hydrogen bonding, permitting the electron transfer to proceed before protonation. However, an alternative explanation for the lack of observation of the protonated semiquinone invokes interactions with nearby acid residues.[2] These interactions could lower the energy needed to protonate the semiquinone at pH 7 to a low value (e.g. 1 pH unit, i.e. 60 meV) and proton transfer could precede electron transfer. It is also possible that proton transfer and electron transfer are highly cooperative and closely coupled. In this case electron transfer and proton transfer may proceed simultaneously.

Structural Models for the Proton Transfer Pathway

A structural model for the proton transfer pathway in RCs from *Rb. sphaeroides* has been based on site-directed mutagenesis studies in which the residues that significantly affect proton transfer were identified (19, 71) (see above) and on inspection of the X-ray crystal structure of the RC. This model is shown in Figure 9. In this model two protons, $H^+(1)$ and $H^+(2)$, are transferred from the external solvent to Q_B along two pathways. The first proton, $H^+(1)$, is transferred along a pathway involving AspL213 and SerL223. The second proton is transferred along a pathway involving AspL213 and GluL212. Inspection of the X-ray crystal structure of the protein shows that protons from solution can find access to the region near AspL210

---

[2]The lack of observation of the protonated semiquinone at lower pH may be due to a pH-dependent pK_a in which the pK_a for the protonated semiquinone decreases at lower pH owing to the protonation of nearby acid groups, e.g. AspL213.

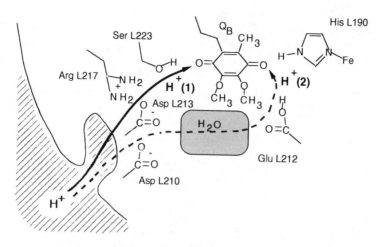

Figure 9 Proposed pathways for proton transfer in RCs from *Rb. sphaeroides*. Protons from the aqueous phase (hatched area) can approach AspL213 via aqueous channels in the protein. The first proton, $H^+(1)$ (solid line), taken up by the quinone carbonyl H-bonded to SerL223, was shown to transfer via a pathway involving AspL213 and SerL223. The second proton, $H^+(2)$ (dashed line), taken up by the carbonyl H-bonded to HisL190, was shown to transfer via a pathway involving GluL212 and possibly AspL213. A cavity (shaded) near the methoxy groups of Q_B, presumably containing internal water, is likely to play a role in the proton transfer.

or ArgL217, although evidence for the direct involvement of these residues is lacking (Table 1). In addition, a void in the protein structure large enough to accommodate five or six water molecules can be visualized in the X-ray crystal structure in a region bordered by the methoxy groups from Q_B, AspL213, and GluL212 (P. Beroza, unpublished results). This pocket is likely to contain disordered water molecules, which undoubtedly play an important role in the proton transfer to Q_B. Bound water molecules have been observed in this region in the X-ray crystal structure of the RC from *Rps. viridis* (25).

The large changes in proton transfer rates that accompany the structurally conservative mutations of proton donor residues suggest that proton transport in the *Rb. sphaeroides* RC is a specific process involving defined pathways (at least for residues close to the quinone). This would be expected since proton transfer can occur only if appropriate donor and acceptor groups are positioned within the proton transfer distance of a few angstroms of each other (94). The difficulty in observing changes in proton transfer rates as a result of mutations of residues farther away from the quinone site (e.g. AspL210, ArgL217) may be due to multiple alternative parallel pathways that circumvent these residues.

The pathway shown in Figure 9 is based on the assumption that the large changes in proton transfer rate due to mutation of GluL212, SerL223, and

AspL213 result from local changes in the proton accessibility near the mutated residue and not from conformational changes. In addition, the model makes the simplest assumption, i.e. that the loss of activity due to mutation of a protonatable residue indicates its role as a proton donor in a proton transfer chain. Other interpretations of the results from site-directed mutagenesis are possible; these include (a) short-range steric effects or long-range conformational effects not related to the role of the mutated residue as a proton donor and (b) electrostatic effects on either the rate or the equilibrium in the proton transfer reaction. Mechanisms for proton transfer consistent with the observed changes are discussed in the next section.

There is evidence that in RCs from other bacterial species different pathways are operative. In RCs from Rps. viridis and Rsp. rubrum AspL213 is not conserved but is changed to Asn (95). However, these RCs have a compensating charge modification. The residue AsnM44 close to Q_B in Rb. sphaeroides is found to be changed to Asp at the homologous M43 position in RCs from both Rps. viridis and Rsp. rubrum. Recently a double mutant containing the AspL213 → Asn and Asn M44 → Asp mutations was constructed (S. Rongey, unpublished results). The second mutation (AsnM44 → Asp) restored high turnover and proton transfer rates and photosynthetic competence that was lost in the AspL213 → Asn single mutant. This supports the idea that proton transfer in Rps. viridis and Rsp. rubrum involves AspM44. Thus, the proton transfer pathways may be quite different in different RCs.

The RC of photosystem II (PSII) from oxygenic photosynthetic organisms contains a Q_B site with properties similar to those in the RC from purple bacteria (reviewed in reference 96). In the PSII RCs from Aramanthus hybridus, mutation of the residue Ser264 on the D1 subunit (homologous to the SerL223 in Rb. sphaeroides) to Gly does not alter the rate $k_{AB}^{(2)}$ (97), as it does in the RC from Rb. sphaeroides. This result may be due to a different pathway for proton transfer to Q_B in PSII RCs. Another interpretation is that a water molecule substitutes for the Ser OH group in the Gly-containing mutant. Takahashi & Wraight (79) have found that weak acids such as azide will facilitate the transfer of protons to Q_B in RCs from Rb. sphaeroides from mutants whose proton transfer rates were blocked. These workers have proposed that bicarbonate plays a similar role in RCs from PSII. This proposal would explain many of the important regulatory effects of bicarbonate on the functioning of the PSII RC (98). Thus, bicarbonate regulates proton transfer, which in turn may regulate photosynthetic growth of plants. A model in which bicarbonate binds to the nonheme iron in the PSII RC has been discussed by Diner et al (96).

Dynamic Models of Proton Transfer

A mechanism of proton transfer must ultimately describe the molecular events that lead to proton transport. Several molecular mechanisms for proton transfer in proteins have been discussed (99–101). One model involves proton transfer through a hydrogen-bonded chain of proton donor residues (99). An examination of the RC structure from *Rb. sphaeroides* did not reveal a hydrogen-bonded chain between protein residues from the exterior of the protein to Q_B. However, a hydrogen-bonded chain could be present if bound water molecules that are unresolved in the X-ray crystal structure are present. For instance, the proposed proton donor residue to GluL212 is AspL213 at a distance of 7 Å. There is a void between these two residues that probably contains water molecules which could form hydrogen bridges between the residues. Thus, water molecules undoubtedly play an important role in transporting protons to Q_B.

Several different dynamic models for proton transfer may be invoked to explain the results obtained by site-directed mutagenesis. These are shown on the left of Figure 10 for the transfer of the first proton and involve SerL223 and AspL213. The series of steps 1, 2, and 3 (sequential mechanism) involve the sequential protonation of AspL213, SerL223, and Q_B^-. A similar mechanism (concerted mechanism) involves steps 1 and 4, in which proton transfer from AspL213 and SerL223 occurs in a concerted manner [a similar mechanism, proposed for serine proteases, has been extensively investigated (102)]. Whether the sequential or concerted mechanism dominates depends on the energies of the charged states. Steps 1 and 5 constitute a direct mechanism involving the direct transfer of a proton from AspL213 to Q_B^-. In this mechanism the reaction product, the protonated quinone, initially has a different structure, since the transferred proton does not come from the hydrogen bond with the SerL223. Takahashi & Wraight favor the direct mechanism (71), in which case the effects of the Ser → Ala mutation at L223 is attributed to a steric effect. This assignment does not explain the effects of the Ser → Thr, Ser → Asp, and Ser → Asn mutations in which only the mutants with protontable residues exhibit normal proton transfer although all may have similar steric effects. Another, and possibly the most general, mechanism (hydronium ion mechanism) involves steps 6 and 7 (or 6, 8, and 3), in which a protonated internal water molecule, H_3O^+, acts as the proton donor group and the role of AspL213 (negatively charged) is to stabilize the positively charged hydronium ion. The hydronium ion mechanism would explain the proton transport in *Rsp. rubrum* and *Rps. viridis* RCs, and in the double mutant (AspL213 → Asn, AsnM44 → Asp) in *Rb. sphaeroides* RCs, where the residue at AspL213 is replaced by Asn with a compensating modification of Asn to Asp at the position homologous to M44 in *Rb. sphaeroides*. This modification conserves the electrostatic stabilizing properties of a negatively charged carboxyl group without having to invoke an entirely new pathway for proton transfer.

Several mechanisms for proton transport of the second proton are shown on the right of Figure 10. The doubly reduced, singly protonated quinone can

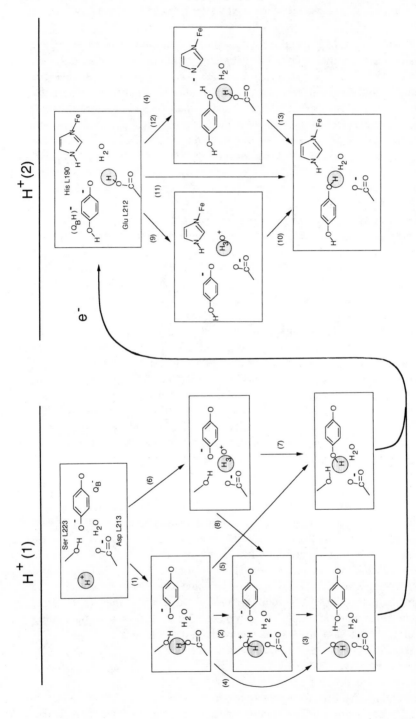

Figure 10 Possible mechanisms for the transfer of the first proton, $H^+(1)$ (*left*), to Q_B^- and the second proton, $H^+(2)$ (*right*), to $(Q_BH)^-$. The circled H represents the proton originating in the outside solvent.

accept a proton by different mechanisms involving GluL212. The proton can be transferred from GluL212 to the quinone directly (step 11). Alternatively, a protonated water molecule may be involved (steps 9 and 10). Another mechanism might involve the stabilization of the $Q_B{}^{2-}$ state by deprotonation and reprotonation of HisL190 (steps 12 and 13) (this proton transfer might also occur before electron transfer). After protonation, the doubly reduced quinone dissociates from the Q_B-binding site and is replaced by an oxidized quinone molecule (50). The driving force for the dissociation of Q_BH_2 is the lowered binding affinity of the dihydroquinone, probably as a result of the loss of H-bonds from the amino acids.

Energetics of Proton Transport

In the previous section we discussed different mechanisms of proton transfer. Which of these mechanisms dominates will depend critically on the energy of the different states involved in the process (101). Thus, an understanding of the energetics is of the utmost importance. A critical point in the energetics is the calculation of the energy required for transferring charges inside proteins (102–104). These calculations present a difficult task since they require the calculation of small differences between large energies. An additional complication is the electrostatic interaction between the many charged residues of the protein (66). Work in this area is currently in progress in several laboratories, and the results will be important for the full understanding of the mechanism of proton transfer.

In the absence of accurate calculations, a tentative working model for the energetics of proton transfer for the sequential mechanism based on ex-

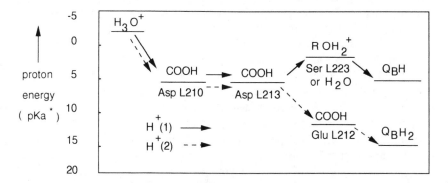

Figure 11 Estimated energies of the protonated donor groups in the RC from *R. sphaeroides* that are thought to be involved in the transfer of $H^+(1)$ (solid arrows) and $H^+(2)$ (dashed arrows). The $pK_a{}^*$ levels in the diagram represent energy levels of the protonated states in the proton transfer pathway for a single proton as it proceeds from solution to a $Q_B{}^-$ molecule with a single negative charge [either $Q_B{}^-$ or $(Q_BH)^-$] at pH 7. At pH 7 the states with $pK_a{}^*$s less than 7 will be unprotonated and those with $pK_a{}^*$s greater than 7 will be protonated.

perimental results and rough estimates is presented in Figure 11. It shows the energy levels (pK_a^*) of the different groups proposed to be involved in the transfer of the first and second proton. The pK_a^* values are the energies required to protonate residues in the RCs when the protonation states of the acidic residues are in equilibrium at pH 7 and Q_B has a single negative charge, i.e. the state of the RC during the uptake of the first and second proton, assuming that proton transfer occurs before the second electron transfer. The pK_a^* values will generally not be the pK_as that would be measured by titration since the protonation state of interacting residues would change. The estimates are based on interaction energies of 2 to 4 pH units for nearby charged residues calculated from Coulomb's Law and assuming an effective dielectric constant of 10 to 20. The value of the intrinsic pK_as were those in aqueous solution. No attempt was made to account for the change in the intrinsic pK_a value due to the protein environment.

The pK_a^* for the protonated semiquinone Q_BH in an environment containing several negative charges (AspL213, AspL210) is partially compensated by the positive charges of ArgL217 and Fe^{2+}. Its value is estimated to be ca. 6, increased by ca. 1 pH unit over the value in aqueous solution (39) but less than 7 to explain the lack of observation of the protonated semiquinone by optical spectroscopy at pH 7. The pK_a^* for the protonated Ser hydroxyl group ROH_2^+ is estimated to be ca. 2 from the solution value of -2 (105) shifted by about $+4$ pK units, mainly owing to strong interactions with the nearby negative charges on Q_B and AspL213. A hydronium ion (H_3O^+) in the same region would have approximately the same pK_a. The pK_a^* for the carboxyl group of AspL213 is estimated to have a value of ca. 6 owing to interactions with ArgL217, Asp210, and Q_B^-. The values for protonated Ser, the hydronium ion, and AspL213 are all close enough to that of the protonated semiquinone to indicate that they can serve as efficient proton donors to Q_B^-. In contrast, the pK_a^* of the carboxyl group of GluL212 (when Q_B is negatively charged) is ca. 12, which may be too high for it to serve as an efficient proton donor to Q_B^-. This may explain why GluL212 does not replace SerL223 as the donor of the first proton in the SerL223 $\rightarrow$ Ala mutant. The pK_as of the neutral Ser hydroxyl group and water molecule are ca. 15 in solution (ca. 19 in the region near Q_B^-), making proton transfer to Q_B^- by these neutral species unlikely. The pK_a^* of the carboxyl group on AspL210 is estimated from the pH dependence of k_{BD} to be ca. 6.

The proton transfer chain for the second proton is shown in the lower path of Figure 11. The pK_a^* of the dihydroquinone Q_BH_2 is ca. 12 in aqueous solution (28); the value in the RC should be somewhat higher, ca. 13, again, increased by ca. 1 as was the pK_a^* of the semiquinone[3] (see above). This

---

[3]The pK_a^* for Q_BH_2 after proton transfer should also be increased by interaction with the ionized GluL212 if this group is its immediate proton donor. However, the ionized GluL212 should become rapidly reprotonated.

large change in pK_a upon transfer of the second electron is the driving force for the uptake of both protons. The proton transfer path is proposed to be the same as that for the first proton up to the branch point where a proton is transferred to GluL212. The pK_a of the carboxyl group of GluL212 (in the presence of a single charge on Q_B) of 12 makes GluL212 a good donor to Q_BH^- but not to Q_B^-. Although the estimates are very rough and will likely require revision, they serve as the basis for discussion and further experimentation. The energetics of proton transfer appear to avoid large increases in the free energy of proton transfer between members of the chain. For a proton to be transported to Q_B^- it must surmount a barrier imposed by the protonation of groups with low pK_as, e.g. Ser, hydroxyl group, or internal water. Although the energies of the protonated states of these groups are lowered by negative charges, there still remains an activation barrier of approximately 4 pH units (240 meV), assuming that a proton donor with a pK_a^* of 6 (e.g. AspL210) is involved in the chain. This barrier is not unreasonably high and should appear as an activation energy for proton transfer.

SUMMARY

Proton transfer in the bacterial RC associated with the reduction of the bound Q_B to the dihydroquinone is an important step in the energetics of photosynthetic bacteria. The binding of two protons by the quinone is associated with the transfer of the second electron to Q_B at a rate of ca. 10^3 s^{-1} (pH 7). Mutation of three protonatable residues, GluL212, SerL223, and AspL213, located near Q_B to nonprotonatable residues (Gln, Ala, and Asn, respectively) resulted in large reductions (by 2 to 3 orders of magnitude) in the rate or proton transfer to Q_B. These mutations can be grouped into two classes: those that blocked both proton transfer and electron transfer (SerL223, and AspL213) and those that blocked only proton transfer (GluL212). These results were interpreted in terms of a pathway for proton transport in which uptake of the first proton, required for the transfer of the second electron, occurs through a pathway involving AspL213 and SerL223. Uptake of the second proton, which follows electron transfer, occurs through a pathway involving GluL212 and possibly AspL213.

Acidic residues near Q_B affect electron transfer rates via electrostatic interactions. One residue, with a pK_a of ca. 10 interacting strongly with the charge on Q_B ($\Delta pK_a > 2$), was shown to be GluL212. A second residue with a pK_a of ca. 6, which interacts more weakly with the charge on Q_B ($\Delta pK \approx$ 1), could be either AspL210 or AspL213.

Several possible mechanisms for proton transfer are consistent with the observed experimental results and proposed proton pathways. These involve proton transfers from individual amino acid residues or internal water molecules either as single steps or in a concerted fashion. The determination of the dominant mechanism will require evaluation of the energetics of the various steps.

ACKNOWLEDGMENTS

We thank our collaborators, especially Paul McPherson and Mark Paddock, whose PhD theses formed the basis for much of this work; Paul Beroza, Peter Brzezinski, Art Chirino, Don Fredkin, Adam Messinger, Doug Rees, Scott Rongey, and JoAnn Williams for helpful discussions and permission to cite their unpublished work; and Colin Wright for sending copies of manuscripts prior to publication. The work from our laboratory was supported by grants from NSF and NIH.

Literature Cited

1. Feher, G., Allen, J. P., Okamura, M. Y., Rees, D. C. 1989. *Nature* (London) 339:111–16
2. Breton, J., Vermeglio, A. 1988. *The Photosynthetic Bacterial Reaction Center: Structure and Dynamics.* New York: Plenum. 443 pp.
3. Norris, J. R., Schiffer, M. 1990. *Chem. Eng. News* 68(31):22–28
4. Michel-Beyerle, M. E., ed. 1990. *Reaction Centers of Photosynthetic Bacteria.* New York: Springer-Verlag. 469 pp.
5. Mitchell, P. 1961. *Nature* (London) 191:144–48
6. Cramer, W. A., Knaff, D. B. 1990. *Energy Transduction in Biological Membranes.* New York: Springer-Verlag. 545 pp.
7. Paddock, M. L., Rongey, S. H., Feher, G., Okamura, M. Y. 1989. *Proc. Natl. Acad. Sci. USA* 86:6602–6
8. Baltscheffsky, H., von Stedingk, L.-V. 1966. In *Currents in Photosynthesis,* ed. J. B. Thomas, J. C. Goedheer, pp. 253–61. Rotterdam: Ad. Donker.
9. Cogdell, R. J., Jackson, J. B., Crofts, A. R. 1973. *J. Bioenerg.* 4:211–27
10. Halsey, Y. D., Parson, W. W. 1974. *Biochim. Biophys. Acta* 347:404–16
11. Petty, K. M., Dutton, P. L. 1976. *Arch. Biochem. Biophys.* 172:335–45
12. Reed, D. W., Clayton, R. K. 1968. *Biochem. Biophys. Res. Commun.* 30:471–75
13. Feher, G. 1971. *Photochem. Photobiol.* 14:373–87
14. Feher, G., Okamura, M. Y. 1978. See Ref. 108, pp. 349–86
15. Wraight, C. A. 1979. *Biochem. Biophys. Acta.* 548:309–27
16. Deisenhofer, J., Epp, O., Miki, K., Huber, R., Michel, H. 1984. *J. Mol. Biol.* 180:385–98
17. Allen, J. P., Feher, G., Yeates, T. O., Rees, D. C., Deisenhofer, J., et al. 1986. *Proc. Natl. Acad. Sci. USA* 83:8589–93
18. Chang, C.-H., Tiede, D., Tang, J., Smith, U., Norris, J., Schiffer, M. 1986. *FEBS Lett.* 205:82–86
19. Paddock, M. L., McPherson, P. H., Feher, G., Okamura, M. Y. 1990. *Proc. Natl. Acad. Sci. USA* 87:6803–7
20. Takahashi, E., Wraight, C. A. 1990. *Biochim. Biophys. Acta* 1020:107–11
21. Khorana, H. G. 1988. *J. Biol. Chem.* 263:7439–42
22. Senior, A. E. 1990. *Annu. Rev. Biophys. Biophys. Chem.* 19:7–41
23. Kaback, H. R. 1990. *Philos. Trans. R. Soc. London* Ser. B 326:425–36
24. Malmstrom, B. G. 1989. *FEBS Lett.* 250:9–21
25. Deisenhofer, J., Michel, H. 1989. *EMBO J.* 8:2149–70
26. Rees, D. C., Komiya, H., Yeates, T. O., Allen, J. P., Feher, G. 1989. *Annu. Rev. Biochem.* 58:607–33
27. Kirmaier, C., Holten, D. 1987. *Photosynth. Res.* 13:225–60
28. Morrison, L. E., Schelhorn, J. E., Cotton, T. M., Bering, C. L., Loach, P. A. 1982. In *Function of Quinones in Energy Conserving Systems,* ed. B. L. Trumpower, pp. 35–58. New York: Academic
29. Williams, R. J. P. 1988. *Annu. Rev. Biophys. Biophys. Chem.* 17:71–97
30. Copeland, R. A., Chan, S. I. 1989. *Annu. Rev. Phys. Chem.* 40:671–98
31. Feher, G., McPherson, P. H., Paddock, M., Rongey, S., Schönfeld, M., Oka-

mura, M. Y. 1990. See Ref. 107, pp. I.1.39–46
32. Takahashi, E., Maroti, P., Wraight, C. A. 1991. In *Electron and Proton Transfer in Chemistry and Biology*, ed. E. Diemann, W. Junge, A. Muller, H. Ratajczaks. Amsterdam: Elsevier. In press
33. Allen, J. P., Feher, G., Yeates, T. O., Komiya, H., Rees, D. C. 1987. *Proc. Natl. Acad. Sci. USA* 84:6162–66
34. Chang, C.-H., El-Kabbani, O., Tiede, D., Norris, J., Schiffer, M. 1991. *Biochemistry* 30:5352–60
35. Allen, J. P., Feher, G., Yeates, T. O., Komiya, H., Rees, D. C. 1987. *Proc. Natl. Acad. Sci. USA* 84:5730–34
36. El-Kabbani, O., Chang, C.-H., Tiede, D., Norris, J., Schiffer, M. 1991. *Biochemistry* 30:5361–69
37. Marcus, R. A., Sutin, N. 1985. *Biochim. Biophys. Acta* 811:265–322
38. Allen, J. P., Feher, G., Yeates, T. O., Komiya, H., Rees, D. C. 1988. *Proc. Natl. Acad. Sci. USA* 85:8487–91
39. Swallow, A. J. 1982. See Ref. 106, pp. 59–72
40. Wraight, C. A. 1982. See Ref. 106, pp. 181–98
41. Prince, R. C., Dutton, P. L. 1978. See Ref. 108, pp. 439–53
42. Dutton, P. L., Leigh, J. S., Wraight, C. A. 1973. *FEBS Lett.* 36:169–73
43. Maroti, P., Wraight, C. A. 1988. *Biochim. Biophys. Acta* 934:329–47
44. Rutherford, A. W., Evans, M. C. W. 1980. *FEBS Lett.* 110:257–61
45. Vermeglio, A. 1977. *Biochim. Biophys. Acta* 459:516–24
46. Wraight, C. A. 1977. *Biochim. Biophys. Acta* 459:525–31
47. Vermeglio, A., Clayton, R. K. 1977. *Biochim. Biophys. Acta* 461:159–65
48. Kleinfeld, D., Okamura, M. Y., Feher, G. 1984. *Biochim. Biophys. Acta* 766:126–40
49. Kleinfeld, D., Okamura, M. Y., Feher, G. 1985. *Biochim. Biophys. Acta* 809:291–310
50. McPherson, P. H., Okamura, M. Y., Feher, G. 1990. *Biochim. Biophys. Acta* 1016:289–92
51. Clayton, R. K., Straley, S. C. 1972. *Biophys. J.* 12:1221–34
52. Slooten, L. 1972. *Biochim. Biophys. Acta* 275:208–18
53. Maroti, P., Wraight, C. A. 1990. See Ref. 107, pp. I.165–68
54. Butler, W. F., Calvo, R., Fredkin, D. R., Isaacson, R. A., Okamura, M. Y., Feher, G. 1984. *Biophys. J.* 45:947–73
55. Loach, P. A., Hall, R. L. 1972. *Proc. Natl. Acad. Sci. USA* 69:786–90
56. Feher, G., Okamura, M. Y., McElroy, J. D. 1972. *Biochim. Biophys. Acta* 267:222–26
57. Hales, B. J., Case, E. E. 1981. *Biochim. Biophys. Acta* 637:291–302
57a. Kleinfeld, D. 1984. *On the dynamics of electron transfer in photosynthetic reaction centers*. PhD thesis. Univ. Calif., San Diego. 317 pp.
58. Lubitz, W., Abresch, E. C., Debus, R. J., Isaacson, R. A., Okamura, M. Y., Feher, G. 1985. *Biochim. Biophys. Acta* 808:464–69
59. Feher, G., Isaacson, R. A., Okamura, M. Y., Lubitz, W. 1985. In *Antennas and Reaction Centers of Photosynthetic Bacteria*, ed. M. E. Michel-Beyerle, pp. 174–89. Berlin: Springer-Verlag
60. Bagley, K. A., Abresch, E., Okamura, M. Y., Feher, G., Bauscher, M., et al. 1990. See Ref. 107, pp. I.77–80
61. Buchanan, S., Michel, H., Gerwert, K. 1990. See Ref. 107, pp. I.69–72
62. Breton, J., Thibodeau, D. L., Berthomieu, C., Mäntele, W., Vermeglio, A., Nabedryk, E. 1991. *FEBS Lett.* 278:257–60
63. Hienerwadel, R., Thibodeau, D., Lenz, F., Breton, J., Nabedryk, E., et al. 1991. *Fifth Int. Conf. Time Resolved Vibrational Spectrosc.* Berlin: Springer-Verlag. In press
64. Maroti, P., Wraight, C. A. 1988. *Biochim. Biophys. Acta* 934:314–28
65. McPherson, P. H., Okamura, M. Y., Feher, G. 1988. *Biochim. Biophys. Acta* 934:348–68
66. Beroza, P., Fredkin, D. R., Okamura, M. Y., Feher, G. 1991. *Proc. Natl. Acad. Sci. USA* 88:5804–8
67. Brzezinski, P., Paddock, M. L., Messinger, A., Okamura, M. Y., Feher, G. 1992. *Biophys. J.* Manuscript in preparation
68. Okamura, M. Y., Isaacson, R. A., Feher, G. 1979. *Biochem. Biophys. Acta* 546:394–417
69. Debus, R. J., Feher, G., Okamura, M. Y. 1986. *Biochemistry* 25:2276–87
70. Kirmaier, C., Holten, D., Debus, R. J., Feher, G., Okamura, M. Y. 1986. *Proc. Natl. Acad. Sci. USA* 83:6407–11
71. Takahashi, E., Wraight, C. A. 1991. *Biochemistry* 31:855–65
72. McPherson, P. H., Rongey, S. H., Paddock, M. L., Feher, G., Okamura, M. Y. 1991. *Biophys. J.* 59:142a
73. Sebban, P. 1988. *Biochem. Biophys. Acta* 936:124–32
74. Gao, J.-L., Shopes, R. J., Wraight, C.

A. 1991. *Biochim. Biophys. Acta* 1056:259–72

74a. Paddock, M. 1991. *Characterization of mutant photosynthetic reaction centers from* Rhodobacter sphaeroides. PhD thesis. Univ. Calif., San Diego. 155 pp.

75. McPherson, P. H., Schönfeld, M., Paddock, M. L., Feher, G., Okamura, M. Y. 1990. *Biophys. J.* 57:404a

76. Paddock, M. L., Feher, G., Okamura, M. Y. 1990. *Biophys. J.* 57:569a

77. Paddock, M. L., Juth, A., Feher, G., Okamura, M. Y. 1992. *Biophys. J.* In 61:153a

78. Rongey, S. H., Paddock, M. L., Juth, A. L., McPherson, P. H., Feher, G., Okamura, M. Y. 1991. *Biophys. J.* 59:142a

79. Takahashi, E., Wraight, C. A. 1990. *FEBS Lett.* 283:140–44

80. Tittor, J., Soell, C., Oesterhelt, D., Butt, H-J., Bamberg, E. 1989. *EMBO J.* 8:3477–82

81. Otto, H., Marti, T., Holz, M., Mogi, T., Lindau, M., et al. 1989. *Proc. Natl. Acad. Sci. USA* 86:9228–32

82. Wraight, C. A., Stein, R. R. 1980. *FEBS Lett.* 113:73–77

83. Paddock, M. L., Rongey, S. H., McPherson, P. H., Feher, G., Okamura, M. Y. 1991. *Biophys. J.* 59:142a

84. Edsall, J. T., Wyman, J. 1958. *Biophysical Chemistry.* New York: Academic. 699 pp.

85. McPherson, P. H., Nagarajan, V., Parson, W. W., Okamura, M. Y., Feher, G. 1990. *Biochem. Biophys. Acta* 1019:91–94

86. Schönfeld, M., Montal, M., Feher, G. 1979. *Proc. Natl. Acad. Sci. USA* 76:6351–55

87. Packham, N. K., Mueller, P., Tiede, D. M., Dutton, P. L. 1980. *FEBS Lett.* 110:101–6

88. Trissl, H. 1983. *Proc. Natl. Acad. Sci. USA* 80:7173–77

89. Feher, G., Okamura, M. Y. 1984. In *Advances in Photosynthesis Research*, ed. C. Sybesma, 2:155–64. Dordrech: Martinus Nijhoff

90. Blatt, Y., Gopher, A., Montal, M., Feher, G. 1983. *Biophys. J.* 41:121a

91. Kaminskaya, O. P., Drachev, L. A., Konstantinov, A. A., Semenov, A. Y., Skulachev, V. P. 1986. *FEBS Lett.* 202:224–28

92. Brzezinski, P., Paddock, M. L., Rongey, S. H., Okamura, M. Y., Feher, G. 1991. *Biophys. J.* 59:143a

93. Brzezinski, P., Paddock, M. L., Okamura, M. Y., Feher, G. 1991. *Biophys. J.* 59:143a

94. Scheiner, S. 1986. *Methods Enzymol.* 127:456–65

95. Komiya, H., Yeates, T. O., Rees, D. C., Allen, J. P., Feher, G. 1988. *Proc. Natl. Acad. Sci. USA* 85:9012–16

96. Diner, B. A., Petrouleas, V., Wendoloski, J. J. 1991. *Physiol. Plant.* 81:423–36

97. Taoka, S., Crofts, A. R. 1990. See Ref. 107, pp. I.547–50

98. Blubaugh, D. J., Govindjee. 1988. *Photosynth. Res.* 19:85–128

99. Nagle, J. F., Tristam-Nagle, S. 1983. *J. Membr. Biol.* 74:1–14

100. Schulten, Z., Shulten, K. 1986. *Methods Enzymol.* 127:419–38

101. Warshel, A. 1986. *Methods Enzymol.* 127:578–87

102. Warshel, A., Naray-Szabo, G., Sussman, F., Hwang, J.-K. 1989. *Biochemistry* 28:3629–37

103. Warshel, A., Åqvist, J. 1991. *Annu. Rev. Biophys. Biophys. Chem.* 20:267–98

104. Honig, B. H., Hubbell, W. L., Flewelling, R. F. 1986. *Annu. Rev. Biophys. Biophys. Chem.* 15:163–93

105. Stewart, R. 1985. *The Proton: Applications to Organic Chemistry*, Vol. 46. New York: Academic. 313 pp.

106. Trumpower, B. L., ed. 1982. *Function of Quinones in Energy Conserving Systems.* New York: Academic

107. Baltscheffsky, M., ed. 1990. *Current Research in Photosynthesis*, Vol. 1. Dordrecht: Kluwer

108. Clayton, R. K., Sistrom, W. R., eds. 1978. *The Photosynthetic Bacteria.* New York: Plenum. 946 pp.

Annu. Rev. Biochem. 1992. 61:897–946

ZINC PROTEINS: Enzymes, Storage Proteins, Transcription Factors, and Replication Proteins

*Joseph E. Coleman**

Department of Molecular Biophysics and Biochemistry, Yale University, New Haven, Connecticut 06510

KEY WORDS: mechanism of action of zinc enzymes, zinc fingers, hormone receptors, metallothionein, DNA-binding proteins

CONTENTS

*Original work in the author's laboratory was supported by NIH grants DK09070 and GM21919.

897

INTRODUCTION

Since 1869 zinc has been known to be an esssential trace element for eukaryotes (1). The presence of zinc in an enzyme was recognized for the first time in 1940 by Keilin & Mann, who discovered zinc in carbonic anhydrase (2). This remained the sole example of zinc functioning in a biologically important macromolecule until Vallee and coworkers described several other zinc metalloenzymes in the 1950s; these included carboxypeptidase A (3), alcohol dehydrogenase (4, 5), and alkaline phosphatase (6, 7). There are now more than 300 known zinc enzymes (8). More recently zinc has been discovered to play structural and functional roles in an entirely new class of protein molecules, namely a variety of eukaryotic transcription factors. Several proteins involved in DNA replication and reverse transcription also have been discovered to be zinc proteins. This large increase in the number of proteins known to contain zinc with functions extending from catalysis of metabolic pathways and macromolecular synthesis to the regulation of gene expression bring the known functions of zinc more into line with the fact that the average adult human body contains 2.3 g of zinc compared with 4 g of iron. Thus zinc is the second most abundant trace metal in most higher animals, and its functions are beginning to catch up with its abundance.

ZINC ENZYMES

Summary

The earliest zinc enzymes to be discovered contained a single zinc atom at the active site, e.g. carbonic anhydrase and carboxypeptidase A. The ligands to the zinc ions in these two proteins consist of three H nitrogen atoms in carbonic anhydrase (9) and two H nitrogen atoms plus both oxygens of the γ-carboxylate of a glutamate residue in carboxypeptidase (10, 11). The coordination sphere is completed in both enzymes by a water molecule to form a tetrahedral complex in the first instance and a five-coordinate complex in the other. A large majority of the zinc enzymes discovered follow this pattern, in that they have single zinc sites consisting of a combination of N and O ligands with a solvent water molecule completing the coordination sphere. The role of zinc in the structure and function of enzyme molecules is not limited to a single site. Alcohol dehydrogenase contains two zinc ions per subunit, only one of which is present at the active center (12, 13). The other zinc site is located ca. 14 Å away from the active center and has been termed a "structural site." The zinc at the latter site is tetrahedrally coordinated to the -S$^-$ donors from four Cys residues and does not have a coordinated solvent water. The active-center zinc, although following the typical pattern of three protein ligands and a solvent water molecule, is ligated by the -S$^-$ donors from two Cys residues and the N-3 of a His residue (13). Thus sulfur coordination, expected for IIB metal ions, does occur in zinc enzymes, but not as frequently as might have been expected. The great majority of catalytic zinc sites are mixtures of nitrogen and oxygen ligands.

Asparate transcarbamylase is an enzyme that contains only a structural zinc ion. This zinc, one in each of the six regulatory subunits, is coordinated by the -S$^-$ donors from four Cys residues and stabilizes the conformation of the peptide loops which form the interface between the regulatory (R) and catalytic (C) subunits (14). The R and C subunits dissociate when Zn is removed. The constellation of amino acid side chains participating in zinc coordination at enzyme active sites has been reviewed by Vallee and Auld (8). The ligands and their spacing in the polypeptide chains have been used to provide a significant guide to the molecular topology of these enzyme molecules. This brief introduction to the zinc enzymes that contain isolated zinc sites serves to put in perspective the discovery in the last several years of zinc enzymes that contain three closely spaced zinc ions, two of which form a binuclear pair via a bridging carboxylate ligand. There are now available crystal structures of three zinc enzymes, all involved in various aspects of phosphate ester transformation, in which these sites have been found: alkaline phosphatase (15), P1 nuclease (16), and phospholipase C (17). These multi–zinc sites add a new dimension to the participation of zinc in complex reaction mechanisms.

RNA POLYMERASES   Before leaving this brief overview of zinc proteins, it should be recalled that very solid data now exist showing that the bacterial multisubunit RNA polymerases, as well as the eukaryotic RNA polmerases II and III, contain two zinc ions associated with their largest subunits, β and β', following the nomenclature applied to the bacterial enzymes (18–20). The two largest subunits of the eukaryotic enzymes show extensive homology to β and β' of the *Escherichia coli* enzyme. Thus zinc assumes a central role in transcription. One of the zinc sites is almost certainly located in the β' subunit and is a tetrahedral sulfur-containing site based on the presence of d-d absorption bands for the Co(II) derivative typical of a tetrahedral complex along with -S$^-$ $\rightarrow$ Co charge transfer bands (20). A cluster of C residues containing two -C-X$_2$-C- sequences located near the N terminus of the β' polypeptide and preserved in the largest subunit of the eukaryotic polymerases is likely to contribute the ligands for this site (21).

The nature of the second zinc site is less clear. The ready displacement of this zinc by mercurial groups suggests the presence of sulfur coordination; however, the site does not appear to be tetrahedral since no typical tetrahedral d-d absorption spectrum is observed in the Co(II) derivative (21). The zinc at the tetrahedral sulfur site in the *E. coli* enzyme cannot be removed without denaturation of the protein. In contrast, the second zinc is easily removed by treatment with mercurial compounds followed by dialysis against thiols to remove the mercurial compound. Removal of this zinc has no effect on the standard transcription assays carried out in the test tube (21). There are simple one-subunit RNA polymerases which clearly do not contain zinc. The enzyme from bacteriophage T7 shows no zinc dependence (22), and the enzymes from T3 and SP6 are probably similar. This finding suggests, but does not prove, that the zinc in the multisubunit RNA polymerases may not participate directly in the polymerization reaction, but may be required for the folding of certain subdomains of the enzyme or for their interactions with other subunits or proteins of the transcription complex. Exactly such functions are now known to characterize the role of zinc in a variety of eukaryotic transcription factors, as outlined later in this review.

Catalytic Functions of Zinc-Zinc Coordinated Water as a Nucleophile

CARBONIC ANHYDRASE   When carbonic anhydrase was the only zinc enzyme known, several proposals were put forward that the function of the Zn ion could be to generate a Zn hydroxide at neutral pH which would attack the adjacent CO_2 molecule. This was an attractive mechanism, since the attack of ^-OH on CO_2 proceeds rapidly at pH > 10 ($k = 8.5 \times 10^3$ s^{-1} M^{-1}). Variants of this mechanism were suggested in 1949 by Smith (23), in 1959 by Davis (24), and in 1961 by DeVoe and Kistiakowsky (25).

Following these early suggestions, it was thought more likely that zinc enzymes used amino acid side chains in concerted reactions in which the Zn was acting as a Lewis acid directly coordinating a group on the substrate and withdrawing electrons, thus labilizing a bond. While at least one open coordination site on the enzyme-bound zinc would be required for such a function, the solvent water, expected to be present in the absence of the substrate, was thought likely to be displaced by substrate. Thus an amino acid side chain, probably the imidazole ring of H rather than a Zn-coordinated H_2O molecule, was favored as being responsible for the sigmoid pH activity profile of carbonic anhydrase (apparent pK_a of 6.5 to 8 depending on conditions) (25–27).

Two general mechanisms for carbonic anhydrase can be formulated: either the attack of H_2O on CO_2 to produce the netural carbonic acid or the attack of ^-OH on CO_2 to produce the bicarbonate anion. Carbonic acid appears highly unlikely as a substrate in the reverse reaction because its concentration is so low in the pH range in which the enzyme is active that protonation of the predominant species, HCO_3^-, would not supply the substrate fast enough (27, 28). The zinc ion at the active center of carbonic anhydrase is tetrahedrally coordinated at the bottom of a 12-Å-deep cavity as shown in Figure 1, taken from the 2.0-Å crystal structure of human carbonic anhydrase II (29,

30). The amino acid side chains of closest approach to the zinc ion are H64, Y7, E106, and T199. The cavity has a hydrophobic and a hydrophilic side, the former centered around L198 and the latter around H64. Extensive studies have ruled out the pK_a of H64, as well as that of other amino acid side chains, as being responsible for the apparent pK_a of the activity (for complete coverage, see reference 31). It is now generally accepted that the pK_a reflected in the activity profile is that of a Zn-coordinated water molecule, $Zn\text{-}OH_2 \rightleftharpoons Zn\text{-}^-OH + H^+$. Thus the carbonic anhydrase mechanism can be written as in Eq. 1.

$$\overset{-H^+}{\underset{+H^+}{Zn\text{—}OH_2 \rightleftharpoons Zn\text{—}^-OH}} + CO_2 \rightleftharpoons Zn\text{—}HCO_3{}^- + H_2O \rightleftharpoons Zn\text{—}OH_2 + HCO_3{}^- \qquad 1.$$

      Proton transfer   Ligand exchange

Several metal-centered phenomena suggested early that the $Zn\text{-}OH_2 \rightleftharpoons Zn\text{-}^-OH + H^+$ equilibrium accounted for the $EH \rightarrow E + H^+$ transition of carbonic anhydrase. When the anions CN^- and ^-SH or their respective acids,

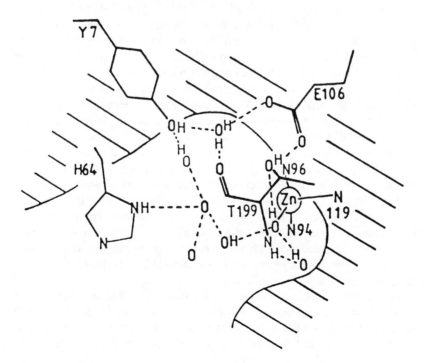

Figure 1 Structure at the active center of carbonic anhydrase (29).

HCN and H_2S, bind to the active-site Zn ion, protons are taken up or released, respectively (32). Fitting of the resulting biphasic release or uptake or protons as a function of pH requires a combination of the pK_as of HCN and H_2S (9.3 and 6.9, respectively) and a single pK_a on the enzyme. The enzyme pK_a is the same for both agents, 7.5 to 8 depending on the isozyme. A protonatable group whose pH equilibrium is directly coupled to formation of a complex between the active-center metal ion and a monodentate anion can best be explained in terms of the proton dissociation from a coordinated water molecule (32).

Second, carbonic anhydrase displays esterase activity against paranitro-phenyl acetate with a pH–activity curve similar to that for CO_2 hydration (33). A substitution of Cd for Zn in the enzyme shifts the apparent pK_a of activity from pH ca. 7.5 to ca. 9.3 (34). An alkaline shift in a proton equilibrium of this magnitude induced by a metal ion substitution seems only satisfactorily explained by the $Zn-OH_2 \rightleftharpoons Zn-{}^-OH + H^+$ hypothesis. The much softer metal, Cd, would be expected to be much less effective in lowering the pK_a of a coordinated solvent.

If the proton acceptor were bulk water in the proton transfer step, the reaction, dependent on a group with a pK_a of 6.5 to 7.5, should not proceed faster than ca. 10^3 s^{-1} (28). In fact, hydration reactions as fast as 10^5 s^{-1} have been observed for the isozyme I and 10^6 s^{-1} for isozyme II of human carbonic anhydrase. These high rates have been explained by the fact that proton transfers during initial rate measurements are transfers to buffer with a pK_a near neutral pH, rather than to the H_2O. The latter involves the H_3O^+ species with a pK_a of ca. -1.7, thus limiting the transfer rate to ca. 10^3 s^{-1}.

Using ^{13}C and ^{18}O isotope exchange methods, Tu and Silverman (35, 36) have measured carbonic anhydrase rate constants at equilibrium and have shown that as buffer is removed, k_{cat} falls toward the theoretical proton transfer limit of 10^3 to 10^4 s^{-1}. In the presence of buffer, some step in ligand exchange may become rate limiting. Ligand exchanges such as that of HCO_3^- for H_2O (Eq. 1) most probably proceed through expansion of the coordination sphere. Recent high-resolution crystal structures of the bicarbonate complexes of both Zn and Co carbonic anhydrase have shown that bicarbonate is a monodentate ligand (A. Liljas, personal communication). The Zn enzyme–bicarbonate complex is tetrahedral, whereas the Co enzyme–bicarbonate complex is five coordinate with a solvent water occupying the fifth position. Thus a five-coordinate intermediate is possible and Co has shifted the equilibrium in favor of a five-coordinate species.

The crystal structures of carbonic anhydrase show that the active-center cavity of carbonic anhydrase contains organized water molecules that form a network of hydrogen bonds linking the side chains of H64, Y6, E106, and T199 and the $Zn-{}^-OH$ (Figure 1). Several effects of this chain of hydrogen bonds on the reaction mechanism have been postulated (37–39). (*a*) The network could serve to orient the nucleophilic hydroxide so that it is properly positioned relative to the carbon of the linear CO_2 molecule in the bottom of the cavity. (*b*) The network may provide a pathway for proton transfer to solvent or buffer via H64. Mutants of H64 show a decreased k_{cat} that is adequately explained by a less facile proton transfer step; e.g. the mutation H64A in human carbonic anhydrase II (HCAII) reduces the k_{cat} to ca. $10^4 \, s^{-1}$. (*c*) The precise hydrogen bond stereochemistry as well as the length of hydrogen bonds in a water network connected to the $Zn\text{-}^-OH$ or $Zn\text{-}OH_2$ can affect the rate of proton transfer from the Zn-bound water and also its pK_a, e.g. by stabilizing $Zn\text{-}OH_2$ over $Zn\text{-}^-OH$ (39).

The discovery of carbonic anhydrase III (CAIII) in skeletal muscle, representing as much as 20% of cytosolic protein, has given important insights into the features around the $Zn\text{-}OH_2$ at the active center of carbonic anhydrase that influence the rate of reaction (40, 41). Although similar in structure to other isozymes of carbonic anhydrase, the CO_2 hydration activity of CAIII is 500-fold lower ($2 \times 10^3 \, s^{-1}$) than that of isozyme II. Various physicochemical parameters including anion binding and d-d absorption spectra of the Co(II) derivative of isozyme III show that the enzyme remains in the alkaline (E) or active form to pH values as low as 5.5 (42). Yet all evidence still suggests that the E form of the enzyme is a $Zn\text{-}^-OH$ species. The crystal structure of carbonic anhydrase III has now been solved, and it is indeed very similar in general structure to that of isozyme II (43). The two polypeptide chains have 58% amino acid identity, and the root mean square (rms) deviation of all main-chain atoms in the two enzymes is 0.92 Å. Three striking amino acid substitutions were found in the three residues near the zinc ion, H64 → K64, N67 → R67, and L198 → F198. Two of these residues, K64 and R67, are on the hydrophilic side of the active center cavity, and one, F198, is on the hydrophobic side.

Recently these three residues in CAIII have been mutated to those that occur in CAII, with surprising results (39). The K64 → H change has essentially no effect on k_{cat}/K_m. In contrast, the F198L mutation increased k_{cat}/K_m for the hydration of CO_2 by ca. 20-fold, while the double mutant (F198L, R67L) resulted in a 70-fold increase in k_{cat}/K_m ($k_{cat} = 2 \times 10^4 \, s^{-1}$). Although k_{cat}/K_m for the double mutant shows little change on the additional mutation, K64H, to produce the triple mutant, there is an additional increase in k_{cat} to $2 \times 10^6 \, s^{-1}$, thus moving k_{cat} for the triple mutant to that observed for CAII. The most surprising finding was that all mutants containing the F198L substitution showed a pH activity profile with an apparent pK_a of 6.8 to 6.9, very near to that of CAII (39). Thus, contrary to what might have been predicted, neither of the two charged side chains in CAIII significantly affects (i.e. lowers) the pK_a of the coordinated solvent. Rather, it is the substitution of one hydrophobic side chain (leucine) for another (phenylalanine) that affects both the pK_a of the coordinated solvent and the rate of the reaction (39).

The best guess as to the source of these effects is that the nature of the hydrophobic side chain in the cavity affects the precise arrangement of the hydrogen bond network connected to the $Zn-OH_2$ or $Zn-^-OH$. As with CAII, the crystal structure of CAIII shows the $Zn-^-OH$ or $Zn-OH_2$ to be involved with the side chain of T199 and two H_2O molecules which form a partially ordered network of at least nine hydrogen-bonded water molecules (39, 43). In CAIII one of these water molecules is observed to form a hydrogen bond with the π electron cloud of the F198 ring. Thus the F198L substitution may alter a hydrogen-bonded structure that is closely coupled to steps in the carbonic anhydrase mechanism and is primarily responsible for adjusting the pK_a of the coordinated solvent molecule.

CARBOXYPEPTIDASE A   In the past several years attention has been focused on the possibility that the solvent water molecule coordinated at the fourth position in many zinc metalloenzymes is a direct participant in their mechanisms as well. Carboxypeptidase A is one of the most interesting of these enzymes because it shows that although Zn may promote the attack of a coordinated solvent on a polarized carbon atom, there can be adjacent protein side chains that participate in both the polarization of the carbon atom and the formation of the incipient coordinated hydroxide species.

The early crystal structure studies of carboxypeptidase A included the study of the structure of a complex with the slowly hydrolyzed dipeptide substrate, Gly-L-Tyr (44–46). Significant features of the active center found at that time included the coordination of Zn to the N3 of H196, the N3 of H69, and the carboxylate of E72, since shown to be bidentate (10, 47). A single coordinated water molecule completed the zinc coordination sphere. The Gly-L-Tyr complex showed the presence of a hydrophobic pocket enclosing the aromatic side chain of the C-terminal residue, while its carboxyl group formed a salt bridge with R145. These features are included in Figure 2, which is not of the Gly-L-Tyr complex, but indicates how a more normal substrate is thought to bind on the basis of recent crystallographic studies of transition-state analogs (reviewed in reference 48).

The structure of the Gly-L-Tyr complex showed the carbonyl oxygen of the peptide bond to be coordinated to the Zn ion, displacing the coordinated H_2O. The placement of Gly-L-Tyr at the active site suggested that the carboxyl side chain of E270 participated either in a direct nucleophilic attack on the carbonyl carbon of the peptide bond or as a general base promoting the attack of an intervening water molecule on the carbonyl carbon (46). The direct nucleophile alternative was favored and requires the formation of an anhydride between the carboxyl of E270 and the new carboxyl of the hydrolysis product as an intermediate. More recent studies have shown that in the Gly-L-Tyr complex the free amino group and the carbonyl oxygen of the peptide group both coordinate the Zn, forming a chelate (48, 49). This nonproductive mode of binding is not available to typical substrates, in which the penultimate amino group is substituted. Coordination of the carbonyl oxygen adjacent to the coordinated H_2O would probably hinder a hydrolysis reaction involving attack of the coordinated water on the carbonyl carbon, since the pK_a of the former would be raised by the additional interaction of Zn with the carbonyl oxygen. An extensive series of cryogenic experiments demonstrated the existence of a series of metastable enzyme-bound intermediates during the hydrolysis of both esters and peptides by carboxypeptidase A (50–52). None of these intermediates react with trapping reagents, suggesting that none of the carboxypeptidase A reaction intermediates have the characteristics expected of the anhydride (50). The absence of an acyl enzyme intermediate, at least in the hydrolysis of peptides, was also suggested by ^{18}O-labeling experiments (53).

Recently there have been a series of crystallographic studies of the complexes between the active center of craboxypeptidase A and substrate analog or inhibitor molecules, which appear to mimic the true substrate-binding mode or the transition state in a convincing manner (48, 54). For example the inhibitor N-(*tert*-butoxycarbonyl)-5-amino-2-benzyl-4-oxo-6-phenylhexanoic acid is an extremely potent inhibitor of carboxypeptidase A ($K_i = 6.7 \times 10^{-7}$ M). This inhibitor is the ketonic analog of the substrate N-tert-boc-L-phenylalanyl-L-phenylalanine. Like other nonactivated ketones, this compound is hydrated at the carbonyl carbon at a level of less than 1% in aqueous solution. Yet the crystal structure of the enzyme–inhibitor complex shows that the compound bound at the active center of carboxypeptidase A is present 100% as the hydrate. The enzyme appears to have favored the hydration reaction to achieve what is essentially the gem-diolate analog of the proteolytic tetrahedral intermediate (54). The polarization of one of the hydrate oxygens by both Zn^{2+} and $Arg127^+$ stabilize it as the anion, the basis for proposing the intermediate pictured in Figure 2. The highest-resolution crystal structures of carboxypeptidase A show that upon the binding of inhibitors that coordinate the metal ion, the Zn moves slightly toward R127 along with a movement of E72 that alters its coordination toward a monodentate form (55, 56). As the pH of the crystalline enzyme is raised from pH 7.5 to 9.5, the Zn–solvent bond decreases in length, possibly indicating an increase in the amount of the zinc hydroxide species (57). All of these rearrangements may be important aspects of substrate binding and catalysis.

The compound N[[(benzyloxycarbonylamino]methyl]hydroxyphosphinyl-L-Phe is the phosphonamidate analog of Cbz-Gly-Phe, one of the most rapidly hydrolyzed peptide substrates for carboxypeptidase A (58). At pH 8.5 this phosphonamidate analog binds to carboxypeptidase A with the anionic moiety asymmetrically straddling the zinc ion (59). If the pH is lowered to 7.5, the extra electron density found at the active center corresponds to the hydrolysis products of the phosphonamidate (60). Whether the enzyme actually participates in the hydrolysis is not clear, since the phosphonamidate linkage is already relatively unstable. Nevertheless, this mode of binding of a transition state analog also supports the notion that productive binding of a substrate is as illustrated in Figure 2. These crystal structures of reasonable transition state analogs support the mechanism of hydrolysis, as illustrated in Figures 2*B* and *C*. The Zn-coordinated water molecule is the nucleophile attacking the carbonyl carbon. The carboxyl of E270 not only activates this water by abstracting a proton, but also is pictured as transferring a proton to the product in the final step. There has been discussion about whether Y248 or E270 is the proton transfer agent, and original references should be consulted.

THERMOLYSIN   Thermolysin is a zinc peptidase from a thermophilic bacterium; it has the same side chain specificity as carboxypeptidase A, but is an endopeptidase rather than a carboxy-exopeptidase. A complete summary of the extensive solution and crystal structure studies on this enzyme has been published by Matthews (61). The fact that the active centers of the two enzymes have similar coordination around the Zn ion [the N of H142, the N of H146, the carboxylate of E166, and a coordinated H_2O molecule (61–64)] is believed to indicate convergent evolution, since the amino acid sequences of the two enzymes are not related (65). The carboxylate of E166 is a monodentate ligand in thermolysin, not a bidentate carboxylate like E72 of carboxypeptidase A (66). The positions and nature of the amino acid side chains at the active centers of carboxypeptidase A and thermolysin are so similar that their reaction mechanisms must be very similar if not identical (61). E143 of thermolysin is adjacent to the zinc site in a position to activate the coordinated water in the same manner as E270 does for carboxypeptidase A. Crystal structures of thermolysin–inhibitor complexes selected to represent a spectrum of analogs of the tetrahedral intermediate of peptide hydrolysis, as well as the reaction product carrying the free amino group, have been completed (summarized in reference 61). From the structure of these complexes, the solution kinetics, kinetic isotope effects, and ^{18}O exchange data, the reaction mechanism pictured in Figure 3 has been proposed for thermolysin (67).

As with carboxypeptidase A, phosphonamidate inhibitors appear to provide reasonable transition state analogs (68). In thermolysin there is an additional stereochemical restriction on E143, which appears to preclude a direct nucleophilic attack of the carboxylate on the carbonyl carbon of a bound

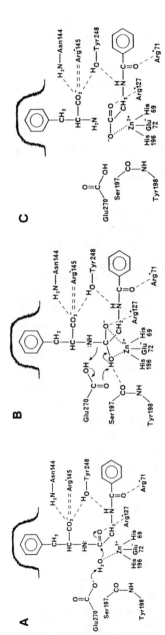

Figure 2   Mechanism of peptide hydrolysis proposed for the zinc exopeptidase carboxypeptidase A. (From reference 48.)

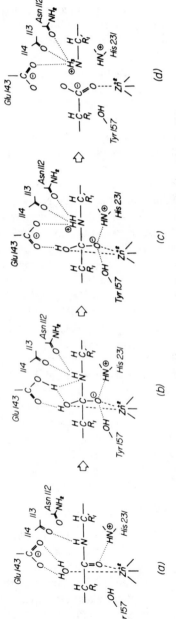

Figure 3   Mechanism of peptide hydrolysis proposed for the zinc endopeptidase thermolysin. (From reference 61.)

peptide (69). Hence the mixed anhydride mechanism was ruled out early. In the mechanism pictured in Figure 3 the Zn-coordinated water, activated by hydrogen bonding to E143, is the nucleophile attacking the carbonyl carbon of the scissile peptide bond. The proton abstracted by E143 could then be easily transferred to the leaving nitrogen. As indicated above, the disadvantage of the additional Zn–carbonyl oxygen interaction is a matter of discussion.

ALCOHOL DEHYDROGENASE   The mechanism of this enzyme, also with three protein ligands to the active site zinc and one solvent H_2O molecule, illustrates another variation on zinc coordination chemistry involving monodentate ligands at the active centers (70). The oxidation–reduction catalyzed by alcohol dehydrogenase involves a typical hydride transfer to the nicotinamide ring of the NAD coenzyme. The precise role of the Zn in the mechanism of hydride transfer from the alcohol has been a matter of debate (70). Some proposals favor a mechanism in which the substrate binds to the coordinated water molecule rather than displacing the water, the latter acting as a general base. The alternative mechanism has the alcohol displace the water to coordinate directly as the alcoxide or at least significantly lower the pK_a of the alcoholic -OH, facilitating the hydride transfer. There are a number of structural arguments, based on the crystal structure of the ternary complex with the substrate, which favor the alcoxide (71). The oxygen of the substrate is coordinated to the zinc, and the coordination sphere is hindered by the sulfur atoms such that water is excluded (see references 70 and 71 for discussion). With regard to the induction of an alcoxide by zinc coordination, several lines of evidence now suggest that the phosphotransferase reaction catalyzed by alkaline phosphatase uses a similar species (see below). The alcohol or phenol acceptor, R_2OH, in the $R_2OH + R_1OPO_3^{2-} \rightarrow R_2OPO_3^{2-} + R_1OH$ transferase reaction appears to coordinate the A-site zinc ion as the alcoxide, thus becoming the nucleophile instead of the ^-OH. The pH dependence of the phosphotransferase activity relative to that of the phosphohydrolase activity supports this conclusion (see below).

Multi-Zinc Sites in Enzymes

ALKALINE PHOSPHATASE   Alkaline phosphatase is a dimeric zinc-containing phosphomonoesterase that is maximally active at alkaline pH and hydrolyzes phosphate monoesters, $ROPO_3^{2-}$. Alkaline phosphatase is widely distributed in mammalian tissues including intestines, kidneys, placenta, and bone. *E. coli* contains a similar enzyme inducible by phosphate starvation (72–74). With the proper adjustments for insertions or deletions, the amino acid sequences of several of the mammalian phosphatases can be fit to that of the *E. coli* enzyme (75). Such comparisons show that the essential active-center residues identified in the crystal structure of the bacterial enzyme are preserved in the eukaryotic enzymes. All alkaline phosphatases form a covalent phosphoseryl intermediate on the reaction pathways, known as E-P (see below). The noncovalent complex with the product phosphate is termed E·P.

In 1962 the enzyme from *E. coli* was shown to contain two zinc atoms per dimer (6). The presence of zinc in the mammalian alkaline phosphatases has since been confirmed in a number of studies (76). Following the initial report, the zinc:protein stoichiometry of the *E. coli* enzyme was revised upward to four or under some circumstances even six zinc atoms per dimer. The original lower stoichiometry occurred because the ammonium sulfate treatment used in the preparation readily removes zinc from one or two of the three metal-binding sites now known to be present in each active center (77). If zinc is added to the preparation buffers, full stoichiometry can be achieved. In the presence of 1 to 10 mM Mg, two Zn ions and one Mg ion, rather than three Zn ions, are bound at each active center.

Coordination chemistry at the active center of alkaline phosphatase   The crystal structure of native zinc alkaline phosphatase (AP) from *E. coli* at 2-Å resolution shows that the active centers on each monomer contain three closely spaced metal-binding sites, two occupied by Zn ions and a third by an Mg ion (78). ^{113}Cd nuclear magnetic resonance (NMR) of the ^{113}Cd$_6$ alkaline phosphatase dimer originally identified the presence of three Cd-binding sites in each monomer (79). Multinuclear NMR investigations of the enzyme have used the designations A, B, and C for the three sites; the crystal structure uses the designation 1, 2, and 3. The "native" enzyme dimer has the metal composition (Zn$_A$Zn$_B$Mg$_C$)$_2$AP.

The general polypeptide fold of the alkaline phosphatase monomer is shown in Figure 4. The active center is located at the carboxyl end of the central β-sheet, and all the ligands to the three metal ions are provided from one monomer. A computer graphics representation of the E·P complex of the native Zn$_4$Mg$_2$ enzyme, including the metal–ligand bonds, the slowly exchanging water molecules, the hydrogen bonds, and the amino acid side chains located within the immediate region of the active center, is shown in Figure 5. The Zn and Mg ions form a cluster in which the metal-to-metal distances trace a triangle of 3.94 by 4.88 by 7.09 Å (79). Despite the close packing of the metal centers, there is only one bridging ligand, the carboxyl of Asp51, which bridges between Zn2 and Mg3 or Cd2 and Cd3 in the Cd$_6$ enzyme.

Zn$_1$(A) coordination   The A-site Zn has four ligands from the protein, which include both carboxyl oxygens of D327, the N3 of H331 and the N3 of H412. In the absence of HPO$_4$$^{2-}$, water relaxation data (80) and ^{35}Cl$^-$ NMR data

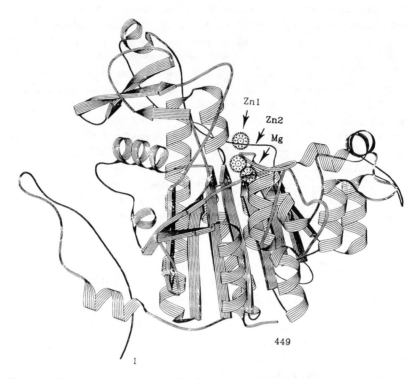

Figure 4   Structure of the monomer of alkaline phosphatase. The monomer-monomer interface of the dimer is on the left. The protein consists of a central 10-stranded β-sheet flanked by 15 helices and another 3-stranded β-sheet with a small helix at the top. (From reference 78.)

(81) suggest that the coordination sphere of the A-site Zn is completed by two H_2O molecules. In the phosphate complex E·P, one of the phosphate oxygens forms a typical coordinate bond with Zn1, with a Zn1-O bond length of 1.97 Å and a Zn1-O-P bond angle of 120° (78). H372, which was originally thought to coordinate Zn1, is not a direct ligand (3.8 Å away from Zn1), but the N3 of H372 is hydrogen bonded to one of the coordinated carboxyl oxygens of D327.

Zn2(B) coordination   In the E·P complex Zn2 is coordinated tetrahedrally by the N3 of H370, one of the carboxyl oxygens of the bridging D51, and one of the carboxyl oxygens of D369. The tetrahedral coordination is completed by a second phosphate oxygen, forming a phosphate "bridge" between Zn1 and Zn2. Although the Zn2-O bond length is 1.97 Å, identical to that for the Zn1-O bond, the Zn2-O-P bond angle at site B is nearly linear (175°).

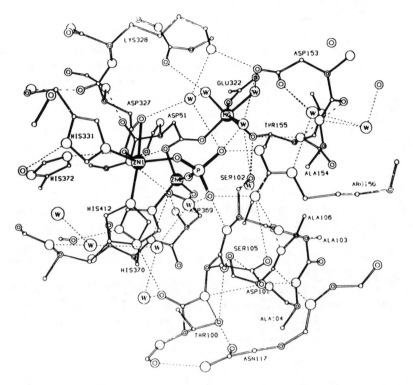

Figure 5   The active-site region as it occurs in the E·P complex of alkaline phosphatase *(E. coli)* including all atoms within 10 Å of the bound phosphate. Water molecules are indicated by W, and hydrogen bonds are shown by the dashed lines. (From reference 78.)

Mg3(C) coordination   The Mg or third Zn site can be described as a slightly distorted octahedron consisting of the second carboxyl oxygen of the bridging D51, one of the carboxyl oxygens of E322, and the hydroxyl of T155, with the rest of the coordination sites filled by three slowly exchanging water molecules. D153 is not a direct ligand as originally believed, but is an indirect ligand in that its carboxyl group forms hydrogen bonds with two coordinated water molecules, which are the direct ligands to Mg (Figure 5). The Mg site does not appear to be close enough to participate directly in the hydrolysis mechanism, but could of course make a contribution to the shape of the electrostatic potential around the active center.

Enzyme-bound phosphate in the E·P imtermediate   The structure of the E·P phosphoenzyme shows that two of the phosphate oxygens form oxygen–metal bonds to Zn1 and Zn2, while the other two form hydrogen bonds to the guanidinium group of R166 (Figure 5). The ^{31}P NMR signal of E·P has provided the demonstration by saturation and inversion transfer that the dissociation of inorganic phosphate, $k_d \approx 35$ s^{-1}, is the rate-limiting step in the mechanism at alkaline pH (82, 83). Likewise the ^{31}P NMR signal of E-P has made possible the demonstration that the phosphoseryl residue in the apoenzyme is stable from pH 2 to 22 (84). Thus both phosphate binding and the dephosphorylation of the phosphoseryl intermediate are Zn dependent. E·P formed by the Zn$_4$Mg$_2$ enzyme has a ^{31}P resonance at 4 ppm, whereas E-P resonates 8 ppm, downfield of phosphoric acid (77, 85). The chemical shift of the E-P signal is relatively insensitive to the substitution of the various metal ion species at sites A and/or B, whereas the chemical shift of the E·P signal is highly sensitive to the nature of the metal ion in both sites: 3.4 ppm in Zn$_4$Mg$_2$ AP and 13.0 ppm in Cd$_6$AP. For the Cd$_6$AP the ^{31}P NMR signal suggested that the phosphate of E·P was coordinated to only one of the active-center Cd ions, since the ^{31}P signal for E·P is a doublet showing a single 30-Hz ^{31}P–^{113}Cd coupling (77). Heteronuclear decoupling shows that this coupling comes from the A-site ^{113}Cd(II), a conclusion supported by the disappearance of this coupling in the Zn$_A$Cd$_B$ hybrid enzyme (86). Although the ^{31}P coupling disappears upon Zn substitution at the A site, the unusual downfield chemical shift of the phosphorus resonance, 12.6 ppm, is maintained in the Zn$_A$Cd$_B$ hybrid, leading to the unexpected conclusion that it is Cd at the B site rather than the A site that induces the unusual downfield shift of the E·^{31}P signal. This became less surprising when the crystal structure of E·P showed the bound phosphate to bridge the two metal ions in sites A and B. The unusual bonding to ^{113}Cd$_B$ may be responsible for the absence of ^{31}P–^{113}Cd coupling.

The electron density map of the active center in the absence of phosphate indicates a bond between the oxygen of S102 and Zn2, supporting the notion that one of the functions of Zn2(B) is to deprotonate the Ser hydroxyl to SerO$^-$. Although the 2-Å map of the E·P complex shows this coordination position to be occupied by one of the phosphate oxygens and the side chain of S102 appears free in the cavity (disordered), S102 may not be accessible to protons in the E·P complex.

It is the metal ion in site A that controls the pH at which [E·P] = [E-P], shifting from pH 5 for Zn in A site to pH 8.7 for Cd in A site (85, 87). This shift of ca. 3 pH units in the pK$_a$ controlling this equilibrium has suggested that this equilibrium, which must reflect the apparent pK$_a$ of the activity, represents the proton dissociation from a coordinated water molecule at the A site (83, 86). This is the most likely explanation for the metal-dependent shift of ca. 3 pH units in this pK$_a$ as discussed above for carbonic anhydrase. If this is correct, Zn-$^-$OH becomes the nucleophile in the second step of the hydrolysis mechanism attacking the phosphorus of E-P.

Details of the many multinuclear NMR investigations of alkaline phosphatase have been reviewed previously (87, 88).

Michaelis complex of alkaline phosphatase with a phosphate monoester   The binding mode of the phosphate to the zinc ions is such that the oxygen of S102 is in the required apical position to initiate a nucleophilic attack on the phosphorus nucleus (Figure 5). If this E·P structure is extrapolated to that of E·ROP, with which it must at least bear some features in common, the oxygen coordinated to Zn_A must be the ester oxygen. None of the other positions would allow space for the R group of the substrate. This suggests that in the normal hydrolysis reaction, Zn_A is an electrophile activating the leaving group, much as protonation of the ester oxygen does in the model systems. Thus it is possible that the alkaline phosphatase mechanism has as significant a dissociative component provided by the Zn_A as it does an associative component provided by the attack of Ser102, the latter being activated by the second Zn ion.

A dissociative character of the alkaline phosphatase reaction has been suggested by the low magnitude of secondary isotope effects when the nonbridging oxygens of the substrate are labeled with ^{18}O (90). An electrophilic activation by the Zn would explain why there is not a significant difference in the β values for k_{cat}/K_m observed for substrates that cannot protonate the leaving group (phosphopyridines) compared with those that can (oxyesters) (89); both β values are equally small (89).

Phosphoseryl intermediate   Phosphorylation of the Cd_6 enzyme in solution and the crystal at pH 7.5 results in exclusive formation of the E-P intermediate (79). The electron density map of the resulting complex shows that a normal ester bond has formed with S102 and that the phosphoryl group is slightly deeper in the active-center cavity than in the E·P complex. The phosphate oxygens remain very close to the metal ions, but the phosphate appears to have moved far enough from Cd_A to provide for the positioning of a coordinated H_2O (^-OH) at Cd_A such that it can be the nucleophile attacking the phosphorus from the apical position opposite the seryl oxygen. There is a bond between the new ester oxygen of the phosphoseryl group and Cd2(B); thus the metal ion at site B appears to activate the leaving group in the second step of the mechanism in the same manner as the metal ion at site A does in the first step. Dissociation of inorganic phosphate from the positive center is then the final and slow step of the mechanism (82, 83). The two major hydrolysis steps of the mechanism are illustrated in Figure 6.

P1 NUCLEASE   The second zinc enzyme containing three zinc-binding sites in its active center is the endonuclease isolated from *Penicillium citrinum*,

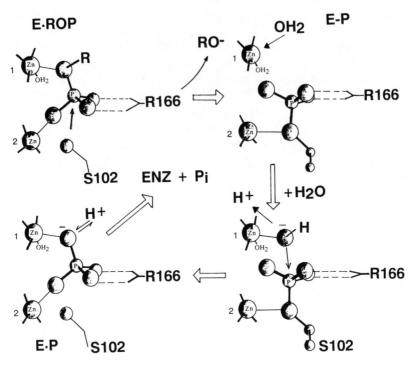

Figure 6 Mechanism of action proposed for alkaline phosphatase. The two half reactions are phosphorylation of Ser102 and the dephosphorylation of this covalent intermediate (E-P).

known as P1 nuclease (92, 93). The crystal structure of P1 nuclease at 2.8-Å resolution shows that each active center contains three closely spaced Zn ions with a bridging aspartate carboxyl between Zn1 and Zn3, as well as a bridging water molecule or hydroxide ion (93). The enzyme is a phosphodiesterase which cleaves between the 3'-hydroxyl and the 5'-phosphoryl group of adjacent nucleotides. It also can act as a phosphomonoesterase and remove the 3'-terminal phosphate group from nucleotides (94). Both activities are zinc dependent (95, 96). The enzyme belongs to a family of zinc-dependent nucleases that are isolated from several sources and show a preference for single-stranded nucleotides. Hydrolysis of the duplex nucleotides may be associated with local melting of the substrate (93). Both DNA and RNA are hydrolyzed, although the phosphomonoesterase activity appears to be significantly greater with RNA substrates.

P1 nuclease contains 270 amino acid residues (97). The amino acid sequence shows the enzyme to be 50% homologous to the S1 nuclease from *Aspergillus oryzae* (93), which has also been reported to require three zinc ions for activity (98). P1 nuclease is largely helical, being made up of no less than 14 α-helices (Figure 7A). The three zinc ions of the active center are bound near the bottom of a large cleft formed between six of the helices.

The Zn ions form a trinuclear site (Figure 7B). Zn1 and Zn3 are ca. 3.2 Å apart and are linked by a bridging carboxylate from D120 and by the oxygen of a water molecule or hydroxide ion (O1, Figure 7B). The third site, Zn2, is ca. 5.8 Å from Zn1 and 4.7 Å from Zn3. Zn1 is coordinated by the N-1 of H60, the N-3 of H116, and the O-δ1 of D45. Zn3 is coordinated by the N-1 of H6, the nitrogen of the α-NH$_2$ group of W1, the carbonyl oxygen of W1, and the other carboxylate oxygen of the bridging D120. Zn2 is coordinated by the N-3s of H126 and H149, the carboxylate of D153, and the oxygens of two solvent water molecules (O-2 and O-3, Figure 7B).

Substrate interactions with the active center of P1 nuclease   Since the *R*-diastereomer of the thiophosphorylated dinucleotide, dA·P(S)·dA, in contrast to the *S*-isomer, is not hydrolyzed by P1 nuclease, it was possible to soak the P1 crystals in the *R*-isomer. The Fourier difference map shows two well-defined binding sites for the nucleotide ca. 20 Å apart. Only one of these is near the trinuclear zinc complex. At the nucleotide-binding site near the Zn ions, one adenine ring is intercalated between the side chains of F61 and V32 (Figure 8). The carboxylate of D63 and the carbonyl oxygens of L25 and E128 are all in positions that could allow hydrogen bonds to form with the N6 of the bound adenine ring. The N1 of the adenine could form a hydrogen bond with the carboxylate of D63, but would require protonation of this side chain. The 3' half of the bound nucleotide is not visible in the difference map and is apparently disordered. The lack of density is not due to hydrolysis of the nucleotide, since no hydrolysis was detected over a period of weeks (93). A high peak of electron density near Zn2 has been interpreted as the phosphate ester group. The 5' ester oxygen and one of the unesterified phosphate oxygens are pictured as forming hydrogen bonds with the guanidinium group of R48. The other unesterified phosphate oxygen is within coordination distance of Zn2 and its two coordinated solvent molecules, O2 and O3.

At the current stage of resolution of this "pseudosubstrate" complex there are several mechanistic possibilities. The phosphate oxygen could coordinate Zn2, displacing the O3 water ligand, with O2 becoming a Zn-coordinated nucleophile in the hydrolysis reaction. Alternatively, O3 could be the direct nucleophile attacking the adjacent phosphorus in the apical position to form the five-coordinate intermediate. This could be a Zn-OH$_2$ or Zn-$^-$OH. Volbeda et al (93) point out that the pH optimum for P1 nuclease is between 4.5 and 6; they invoke several explanations for this, including the presence of two Asp-Glu pairs at the lip of the binding cavity, which would require the low pH for protonation. Otherwise the pairs would split apart, causing large con-

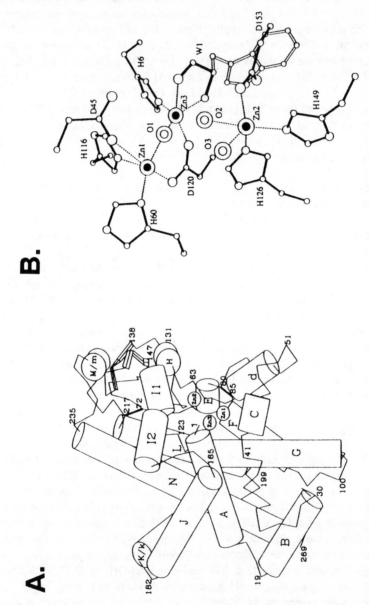

Figure 7 (*A*) Crystal structure of P1 nuclease (*P. citrinum*). (*B*) Ligand structure of the three zinc complexes at the active center of P1 nuclease. (From reference 93.)

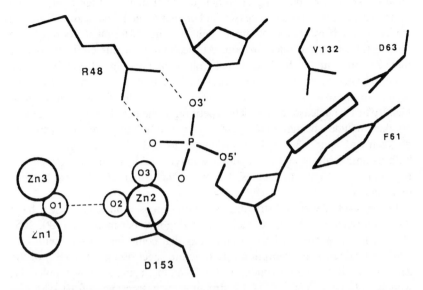

Figure 8 Diagram of the relationships of the thiophosphorylated dinucleotide dA·P(S)·dA bound at the active site of P1 nuclease to the surrounding amino acids. This is the unhydrolyzed *R* isomer. (From reference 93.)

formational changes which might reduce activity. Muscle carbonic anhydrase (CAIII), however, has a pK_a for the transition from the protonated (inactive) to the unprotonated (active) form of the enzyme below pH 5, yet this pK_a represents the Zn-OH$_2$ $\rightleftharpoons$ Zn-$^-$OH + H equilibrium (42) (see above). Therefore an acidic pH optimum does not rule out the possibility that the pH activity profile reflects the ionization of a Zn-coordinated water molecule.

PHOSPHOLIPASE C   The third enzyme discovered to possess a triad of Zn ions at its active center is phospholipase C from *Bacillius cereus* (99, 100). The mammalian phospholipases C, which hydrolyze phosphatidylinositol and phosphatidylcholine, are catalysts in the pathways to the generation of second messengers which include inositol triphosphate (IP$_3$) and diacylglycerol (101, 102). It was therefore of considerable interest to find in *B. cereus* an enzyme of 245 amino acids that catalyzed the hydrolysis of phosphatidylcholine and was similar to the mammalian phospholipases C (103). The functional similarity was suggested when the *B. cereus* enzyme was found to enhance prostaglandin biosynthesis when added to the appropriate cell extract (104). Thus it may be a model for the poorly characterized mammalian enzymes.

The crystal structure of the bacterial enzyme is available at 1.5-Å resolution as refined from 1.9-Å data (99). This enzyme, like P1 nuclease, is essentially an all-helix protein consisting of 10 helices rather than 14 (Figure 9A). Much of its structure, however, can be overlaid on that of the P1 nuclease. There is a cavity formed by parts of three helices, the N-terminal loop of the polypeptide, and one internal loop. At the bottom of this cleft are located three Zn ions with an almost identical arrangement and ligands to those of the three Zn sites found in P1 nuclease. The structure of the Zn triad and its ligands in phospholipase C is summarized in Figure 9B. Because of the near identity to P1 nuclease, a complete repetition of the details will not be given, but they include a bridging D122 carboxylate and bridging water or hydroxide between Zn1 and Zn3 and the coordination to Zn3 of the N-terminal W via both its amino group and its peptide carbonyl. The isolated Zn2 also carries two coordinated water molecules. The rest of the ligands are identical to those of P1 nuclease.

In contrast to the crystal structure, solution studies of phospholipase C have consistently indicated the presence of two zinc ions per molecule (100, 105, 106). There appeared to be two zinc sites that could be replaced with Cd, while extended X-ray absorption fine structure spectroscopy (EXAFS) of the Zn protein indicated a maximum of 2.3 Zn atoms, using a revised molecular weight (99, 107). Yet the crystal structure in the presence of 10 μM free Zn(II) shows three bound Zn ions. This may be similar to alkaline phosphatase, for which preparation conditions in the absence of Zn buffers removed the metal ion from all but the highest-affinity binding site (see above).

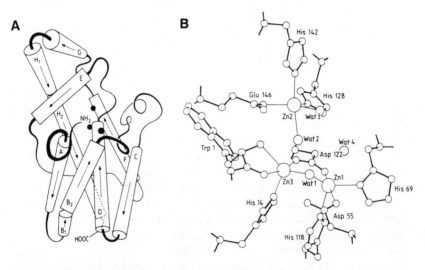

Figure 9 (A) Crystal structure of phospholipase C. The positions of the three zinc ions are represented by the black balls. (B) Ligand structure surrounding the three zinc ions of phospholipase C. (From reference 99.)

Studies with bound nonhydrolyzable analogs of the phosphodiester substrate have not been done for phospholipase C, but a significant difference electrondensity map has been observed for the crystals soaked in phosphate (99). Phosphate oxygens are coordinated to all three Zn ions, displacing the bridging water ($^-$OH) between Zn1 and Zn3 and one of the water molecules from Zn2. This is quite different from the relationships shown by the diester phosphate of the bound nucleotide in P1 nuclease. Preliminary data, however, suggest that a similar complex is formed by the P1 nuclease with inorganic phosphate (93).

COMPARATIVE STRUCTURES OF THE ENZYMES CONTAINING MULTIPLE ZINC SITES   Although native alkaline phosphatase contains two Zn sites and one Mg site, the enzyme is functional with three Zn ions and could be included among those with trinuclear Zn sites. Are the three enzymes with trinuclear Zn sites related? All three metal sites in P1 nuclease and phospholipase C are typical Zn-binding sites with mixed oxygen and nitrogen donors, whereas the third site in alkaline phosphatase is made up of only oxygen donors, which may explain its preference for Mg. The aspartate carboxylate bridging two of the Zn ions appears to define a binuclear zinc site in the two diesterases. In contrast, the bridging aspartate carboxylate defines the Zn–Mg pair in native alkaline phosphatase.

Although all three enzymes are involved in phosphate ester hydrolysis, alkaline phosphatase has appeared to be exclusively a monoesterase, although one recent report has detected weak diesterase activity (J. F. Chlebowski, personal communication). There is only 18% amino acid sequence identity between the polypeptide chains of P1 nuclease and phospholipase C, yet the crystal structures of the two enzymes show that 145 of their Cα atoms can be superimposed with a positional rms deviation of only 1.8 Å. Thus P1 nuclease and phospholipase C may well be evolutionarily related (93). The general polypeptide fold of the alkaline phosphatase monomer is clearly different from that of the other two enzymes (Figure 4). In addition, alkaline phosphatase forms a phosphoseryl intermediate (Figure 6). Both P1 nuclease and phosphlipase C appear to catalyze single-step hydrolyses with inversion of chirality around phosphate. For alkaline phosphatase the primary interaction of the monoester substrate is with the two Zn ions not linked by a bridging ligand. A second feature is the finding that the Zn ions coordinate the ester oxygens, activating the leaving groups in the initial scission of the R-O-P bond as well as in the hydrolysis of the phosphoseryl intermediate. The function of the isolated Mg or Zn site in the mechanism of action of alkaline phosphatase is less clear. The functions of the two Zn sites 3.9 Å apart are obviously closely connected with the formation of the covalent intermediate.

In contrast to alkaline phosphatase, the current information on substrate interactions with the diesterases P1 nuclease and phospholipase C suggests that the isolated Zn site, not the bridged binuclear site, is the site of primary interaction with the diester at least. This is also the site that appears to carry the potential $Zn-OH_2$ or $Zn-{}^-OH$ nucleophile (Figure 8). The lack of significant interactions between the bound nucleotide and the binuclear Zn complex in P1 nuclease has led to the suggestion that the binuclear cluster might serve as an important structural element of the protein (93). The distribution within the polypeptide chain of the residues contributing the ligands to the zincs at the binuclear site is such that its formation ties together rather distant regions of the molecule. This may serve to stabilize the protein.

Studies of the complexes of substrate analogs have not progressed as far for either the P1 nuclease or the phospholipase C as they have for alkaline phosphatase, and the difference maps for phosphate bound to both P1 nuclease and phospholipase C, which show that an interaction with inorganic phosphate can take place with all three zinc ions, raises the possibility that all three Zn ions can be catalytically involved at some stage of a complete mechanism. The simultaneous coordination of phosphate to more than one zinc relates these sites perhaps a little more closely to that of alkaline phosphatase, in which the phosphate of E·P bridges Zn1 and Zn2. P1 nuclease does have a phosphomonoesterase activity, and the mechanistic role of the three Zn ions could be different in the monoesterase and diesterase reactions. It may be premature to postulate exactly how similar or different the functions of the triad of metal ions at the active centers are in these three enzymes.

ZINC STORAGE AND TRANSPORT: ROLE OF METALLOTHIONEIN

We know relatively little about the equilibria that must exist between storage sites in cells, free zinc in cell cytoplasm and body fluids, and the proteins in which zinc has a functional role. Despite the abundance of zinc in animal cells, little information is available on the true free zinc concentration in cells, i.e. the concentration of zinc not complexed to a macromolecule. Similarly, we do not know whether the free zinc concentration varies with time or whether significant zinc gradients develop between cell compartments or between cells. Large varations in the zinc content in different tissues are observed (109). This must mean at the very least that cell contents of zinc-containing macromolecules must vary. These differences are likely to represent other aspects of zinc metabolism as well, such as variation in the transport of the ion.

In the area of zinc storage and transport, some plasma proteins binding zinc have been detected (110) and the metal cluster protein metallothionein has been postulated to have a role in the storage and therefore the delivery of zinc to other macromolecules (see reference 111 for an extensive discussion of these aspects). Solid physiological evidence for this supposition has not been obtained. Metallothionein was first discovered in horse kidney by Margoshes & Vallee in 1957 (112) as the only naturally occurring cadmium protein. The protein also contained significant amounts of Zn (113), and it was appreciated early on that its Cd content increased with the age of the animal. Copper can be bound in vivo by mammalian metallothioneins as well (114). Metallothionein was then discovered to be induced by Cd injection, and one of its roles was postulated to be that of detoxifying heavy metals.

The metal sites in metallothionein were discovered by ^{113}Cd NMR in 1979 to consist of two Cd-S clusters, a four-metal cluster and a three-metal cluster involving 11 and 9 C residues, respectively (115). There are three bridging -S$^-$ ions in the three-metal cluster and five bridging -S$^-$ ions in the four-metal cluster. The solution structure of the entire protein has been determined by two-dimensional (2D) NMR methods (116–118), and recently a crystal structure has been solved at 1.8-Å resolution (119). An earlier low-resolution crystal structure was found to be in error because of a misassignment of the sequential order of the C ligands around the metal ions. The 2D NMR structures of both rabbit (116) and rat (117) metallothionein (the crystal species) showed a different cysteine–metal connectivity, and a revised higher-resolution crystal structure now confirms solution structure as being correct.

Although the protein can be prepared as the Cd$_7$ protein, the "native" protein contains five Cd ions and two Zn ions. The metals are asymmetrically distributed, four Cd ions in the four-metal cluster and two Zn ions and one Cd ion in the three-metal cluster. A Zn$_7$ protein can be made by exchange, and NMR data indicate that its structure is not significantly different from that of the Cd$_7$ derivative (120). The three-metal cluster is less stable and exchanges the metal ions more rapidly. Major compilations of current metallothionein research have been published periodically and the latest, in 1987, covers many structural and functional aspects of this interesting protein (111).

The understanding of the control of expression of the metallothionein gene has advanced rapidly in the last few years. The 5' upstream region of the mammalian gene for this protein contains several regulatory DNA sequences (121, 122). One is a steroid response element, one is a site for the binding of the "housekeeping" transcription factor Sp1, and a third is the so-called metal-responsive element (MRE). The MRE is responsible for the activation of transcription of the gene in the presence of heavy metal ions. The nature of the transcription factor responsible for recognizing the MRE has been identified in the last several years.

Metallothionein is widely distributed in eukaryotic organisms and is not always found as a Zn-Cd protein. In yeasts it is a copper protein with the amino acid sequence shown in Figure 10A (123–125). The structure of the metal cluster is currently under investigation (S. Narula and I. M. Armitage,

A.   **METALLOTHIONEIN**   B.   **TRANSCRIPTION FACTOR**

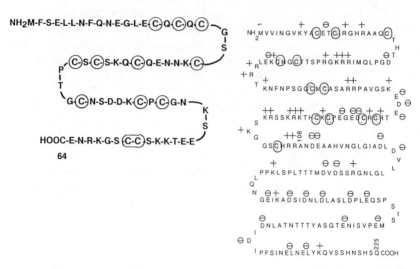

Figure 10 Primary structure of yeast metallothionein (*A*) and the transcription factor (*B*) binding to the metalloregulatory element, a DNA sequence upstream of the transcriptional start signal of the yeast metallothionein gene. (Panel *B* reproduced from reference 127.)

personal communication). It was originally believed to be a single cluster of eight Cu ions, but work with the Ag derivative suggests that it may contain seven metal ions in separate four-metal and three-metal clusters. The folding of the polypeptide which dictates preferences of a given metallothionein for one IIB metal ion or another (considering CuI to be a IIB metal ion) is not understood at present.

Since yeast cells are more easily subject to mapping and isolation of specific genes than are mammalian cells, it has been possible to isolate and sequence a gene whose mutation is associated with copper sensitivity in yeasts (127). The product of this gene, known as ACE1, is the transcription factor binding to the DNA MRE, thus activating transcription of the metallothionein gene (127, 128). The gene codes for a protein of about twice the size of yeast metallothionein (Figure 10*B*) whose N-terminal amino acid sequence reveals a domain containing 12 C residues spaced similarly to the C residues in yeast metallothionein itself. In contrast to metallothionein, this protein contains 10 K and 15 R residues in the sequences between the C residues, suggesting that the N-terminal half of the molecule is the DNA-binding domain. The transcription factor has an equally large C-terminal domain which carries a net negative charge, a characteristic of the transactivating domains of many transcription factors. Thus the metallothionein gene responds to a transcription factor whose DNA-binding domain itself contains a metal cluster. The folding and therefore the affinity of the binding domain for DNA must be controlled by the copper concentration in the cell (129). The existence of a metal-dependent mechanism for the control of specific transcription is in itself intriguing and may be of relevance in studies of other metal-containing transcription factors. It seems reasonable to speculate that an analogous transcription factor exists in mammalian cells responding to Zn and Cd.

ZINC-REQUIRING TRANSCRIPTION FACTORS

Classes of Zinc-Containing Transcription Factors

The first discovery of a zinc metalloprotein that controlled the transcription of a specific gene was made in 1983 by Hanas et al, who showed that transcription factor TFIIIA from *Xenopus* oocytes, necessary for the transcription of the 5S RNA gene by RNA polymerase III, was a zinc protein (130). The interactions of TFIIIA with nucleic acid are somewhat unusual in that the protein not only recognizes the internal control region of ca. 35 bp in the center of the 5S RNA gene but also itself is bound to the product, 5S RNA, to form the 7S RNA-protein particle. Both the 7S particle and the RNA-free TFIIIA protein were reported to contain two Zn ions per molecule (130, 131). The zinc was resistant to removal by chelating agents in the 7S particle, but could be readily removed from the free protein (131). It was necessary to add zinc to the apoTFIIIA to reconstitute the 7S particle. It was also demonstrated by DNase footprinting that only the Zn-containing TFIIIA bound specifically to the 5S RNA gene; removal of zinc from the protein abolished specific binding to the gene.

When the nucleotide sequence of the gene became available in 1985, it was noted by Miller et al (132) that the amino acid sequence of the translated product could be arranged so that a pair of conserved C residues and a pair of conserved H residues separated by a 12- to 13-residue spacer defined a series of 9 approximate repeat amino acid sequences of ca. 30 residues within the TFIIIA protein as follows: $-C-X_{2-5}-C-X_{12-13}-H-X_{3-4}-H-$. Within the X_{12-13} spacers there was an additional pair of conserved residues, an aromatic amino acid, F or Y, and a branched aliphatic amino acid, usually L, so that the conserved sequence becomes $-C-X_3-F-X_5-L-X_{2-3}-H-$. The pairs of C and H residues appeared to be excellent candidates for ligands to form a tetrahedral Zn complex. Miller et al (132) proposed that each of these repeats formed a structure dubbed the "zinc finger," since it was suggested that the 12- to 13-residue loop or "finger" portion formed the DNA-binding surface. EXAFS showed that the Zn absorption edge in TFIIIA was fit by a tetrahedral model with two S and two N atoms as ligands (133). Zinc analysis of the 7S particle showed that the particle contained 7 to 11 zinc atoms, more in agreement with the 9 fingers present than the original report of 2 zinc ions per molecule (132). The precise stoichiometry or the reasons for variable metal ion stoichiometry in different TFIIIA preparations remains a matter of debate, but more recent analytical data still show significant differences in zinc : protein stoichiometry from different laboratories (134).

The number of transcription factors reported to contain the zinc finger motif has increased rapidly and is currently around 45 different proteins. Many of these proteins contain the C_2H_2-type pairs found in TFIIIA; however, many fit similar structures but with C_2C_2 ligand pairs. With the accumulation of significant structural information three rather different structural motifs have emerged. One, represented by the now classical TFIIIA, is characterized by the C_2H_2 ligand motif and represents the structure found in 18 of these proteins (135–139). The transcription factors that use only C residues as Zn ligands have been divided into two groups. One is a large family of hormone receptors that contain two isolated Zn domains at either end of a DNA-binding helix loop (140, 141), and the second is a family of yeast transcription factors in which six C residues form a Zn_2Cys_6 binuclear cluster within their DNA-binding domains (142–145).

Zinc-containing DNA-binding domains compared with transactivating domains   Investigations of the zinc-containing transcription factors naturally focused on the structure of the zinc domains, since these domains coincide with their DNA-binding and nucleotide base recognition functions. The structural studies have made rapid progress by using relatively small subdomains of these transcription factors that contain little more than the intact DNA-binding domain. For example, all of the 2D NMR determinations of the zinc finger structure in solution are of single zinc fingers induced to fold by relatively large Zn : peptide ratios (135–137). These single fingers do not recognize DNA sequences specifically (135). In contrast, the native transcription factors are large proteins, often of 700 to 800 amino acids, in which the majority of the polypeptide chain is involved in interactions with other proteins, most notably some component of the transcription complex. To place the structural work on the DNA-binding and sequence recognition domains in its appropriate context, a brief summary of the functional components of representative examples of complete zinc-containing transcription factors is given in Figure 11.

Solution Structures of the Zinc-Containing DNA-Binding Domains of Transcription Factors as Determined by 2D NMR Methods

There are a number of solution structures of single C_2H_2 zinc fingers that were determined by 2D NMR methods (135–137, 146, 147), as well as two

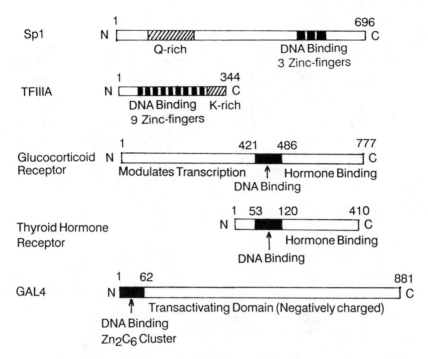

Figure 11 Diagrams of the functional domains of eukaryotic zinc-containing transcription factors; their placement within the primary structure.

examples of solution structures of the DNA-binding domains from the hormone receptor superfamily determined by 2D NMR methods, an 86-residue fragment from the glucocorticoid receptor (148) and an 84-residue fragment from the estrogen receptor (149). The general structure of the binuclear metal cluster found in the 62-residue N-terminal DNA-binding domain of GAL4 has been determined by 2D proton and ^{1}H-^{113}Cd heteronuclear NMR studies (164–168). The motifs characterizing the zinc domains of each of the three classes of transcription factor are compared in Figure 12.

The structure for the single C_2H_2 zinc finger illustrated in Figure 12A is a general one and is close to that found for all the single fingers by 2D NMR methods. These include single fingers from ADR1 (135), Xfin (136), Zfy (147), and human enhancer-binding protein (137). Transcription factors now believed to contain one or more copies of this motif are listed in Table 1. The finger consists of an N-terminal short antiparallel β-sheet of 9 to 11 residues including the β-turn encompassing the C-X_{2-4}-C group and a second turn at the top of the finger followed by ca. 3 turns of α-helix (9 residues), although some of this helix is a 3_{10} helix in some zinc fingers (see the discussion of the

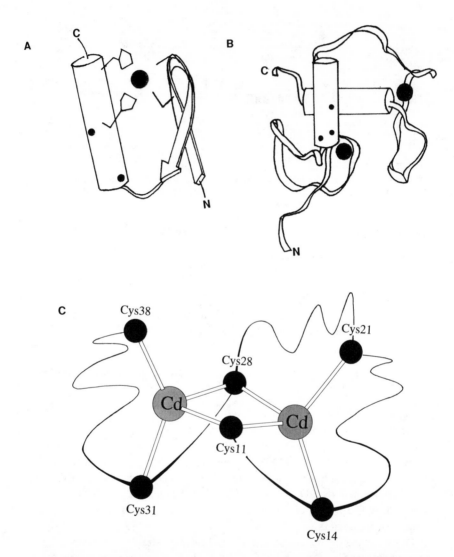

Figure 12 Structural motifs found in the zinc-containing DNA-binding domains of eukaryotic transcription factors. (*A*) Zinc finger. This sketch is modeled from the fingers in the crystal structure of the Zif 268–DNA complex. The orientation has been selected to indicate that the N-terminal end of the helix sits in the major groove. The black dots indicate the approximate positions of the two Arg residues making hydrogen bonds with G residues. (*B*) Steroid receptor DNA-binding domain. The sketch is of the estrogen receptor structure as determined in solution by 2D NMR methods. The small dots on the recognition helix indicate the three residues on the recognition helix making specific base contacts (see Table 3). (*C*) Binuclear metal cluster in the GAL4 DNA-binding domain. The metal can be either Zn or Cd. (Panel *C* reproduced from reference 168.)

Table 1 Zinc-containing transcription factors of the TFIIIA type

Transcription factor (source)	No. of Zn domains	Reference
Sp1 (human)	3	139
Zfy (human)	13	150
Zif 268 (mouse)	3	151
Mkrl (mouse)	7	152
Mkr2 (mouse)	9	152
TFIIIA *(Xenopus)*	9	138
Xfin *(Xenopus)*	37	153
p43 (5S RNA-binding protein) *(Xenopus)*	9	154
Serendipity β *(Drosophila)*	6	155
Serendipity δ *(Drosophila)*	7	155
Kruppel *(Drosophila)*	5	156
Snail *(Drosophila)*	5	157
Krh *(Drosophila)*	6	158
Hunchback *(Drosophila)*	6	159
Terminus *(Drosophila)*	1	160
TRS-1 (trypanosome)	5	161
ADR1 (yeast)	2	162
SWI5 (yeast)	3	163

X-ray structure below). The Zn is held in a tetrahedral complex by the two -S^- ligands from the C residues and the N3 nitrogens of the imidazole side chains of the H-X_{2-5}-H sequence toward the C-terminal end of the helix. The two conserved aromatic residues, Y (or F) and L, were discovered to be in contact, forming a hydrophobic interaction or type of strut, important for stabilizing the distal portion of the finger. This fold of the zinc finger was predicted on theoretical grounds by Berg (169) prior to completion of the solution structures. The amino acid sequences of several zinc fingers found in native proteins chosen to illustrate the spectrum of variation in the amino acid sequences of the C_2H_2 type of structure are collected in Table 2.

The 2D NMR structures of the DNA-binding domains of the glucocorticoid receptor (147) and the estrogen receptor (148, 170) are similar and can be represented by the ribbon diagram in Figure 12*B*. They consist of two separate zinc-binding subdomains folded together to form a compact domain. The two helices found in the structure, one emerging from the N-terminal zinc subdomain and one from the C-terminal zinc subdomain, are packed against each other at right angles. The helices are amphipathic, with the side chains of their hydrophobic faces packing the core of the domain while the exposed surface of the vertical helix facing the reader carries two lysines and an arginine involved in DNA binding (170).

Table 2   Variation in amino acid sequences of zinc fingers

Transcription factor	Finger	Sequence[a]
TFIIIA	Finger 1	C̲SFAD C̲GAAYNKNWKLQ–A H̲L C–K H̲
	Finger 6	C̲KKDD S̲ C̲SFVGKTWTLYLK H̲VAEC H̲
	Finger 7	C̲––DV C̲NRKFRHKDYLR–D H̲QK–T H̲
Zfy	Finger 1 (odd)	C̲ M I C̲GKKFKSRGFLKR H̲MKN H̲
	Finger 2 (even)	C̲KF C̲EYETAEQGLLNR H̲LLAV H̲
SWI5 (yeast)	Finger 1	C̲LFPG C̲TKTFKRRYNIRS H̲IQT H̲
	Finger 3	C̲P–––C̲GKKFNREDALVV H̲RSRMI C̲

[a] The dashes are deletions. These fingers illustrate the spectrum of length of spacers between ligands as well as a possible change in ligand type (finger 3, SWI5) found in "classical" zinc fingers. Underlining indicates proposed ligands.

The four C residues of the N-terminal half of the estrogen receptor (ER) domain and four of the five C residues in the N-terminal half of the glucocorticoid receptor (GR) domain form one tetrahedral Zn complex. Four of the five C residues in the C-terminal half of both the ER and GR domains form the second tetrahedral Zn complex. The presence within these sequences of C residues that are not ligands suggests that caution be exercised in assigning the ligand arrangement from sequence data alone. Even in the NMR structures there can be potentially some doubt, since the NMR data do not directly supply information on the Zn coordination. The possible alternative structures must be entered in the structure determination and examined to see whether one converges better than the other, as was done for the GR fragment (147). ^{113}Cd substitution can give direct information on the coordination arrangement by the detecting ^{113}Cd–^{1}H coupling (see GAL4 studies below).

In both halves of the hormone receptor DNA-binding domains, the α-helix begins with the residue following the third C ligand and continues for 11 to 13 residues. The loops of peptide between the pairs of C ligands, originally thought to be fingers analogous to those of the TFIIIA type, form irregular folded structures on either side of the helix-containing central structure (Figure 12*B*). In the ER domain, the N-terminal loop is 13 residues long and well structured (148, 170), while the analogous C-terminal loop consists of only 9 residues and is less well structured. As the crystal structure of the DNA complex of the GR domain reveals, these loops, in contrast to those of the C_2H_2-type zinc fingers, do not form the major DNA-binding surface. This conclusion was indicated earlier by site-directed mutagenesis.

Site mutation also showed that residues E25, G26, and A29 of the ER domain (indicated in Figure 12*B*) and an analogous triple in the GR domain

(G, S, and V) determine whether the receptor will recognize the palindromic estrogen-responsive element (ERE) or the glucocorticoid-responsive element (GRE) DNA sequence (171–173). These residues are in the so-called "knuckle" of the first zinc complex (finger), and all three fall within the N-terminal helix. The knuckle region of the second zinc complex, residues P44 to Q48 in the ER domain, were found to be important in enabling the protein to discriminate between the different half-site spacing in the ERE and the thyroxine-responsive element (TRE) (172). This is probably related to the fact that this region contributes to the dimerization domain (see the discussion of the crystal structure below). Swapping of this second finger sequence between the GR and ER, however, does not result in loss of the original specificity controlled by the N-terminal specificity helix (174). DNA footprinting studies of the bound ER and GR domains, either as the entire receptors or as their isolated DNA-binding domains, showed that dimers of the ER and GR domains contacted two adjacent major grooves on one face of the DNA helix and did so with exact rotational symmetry (175, 176). The above findings suggested that the N-terminal helix was a DNA recognition helix likely to occupy the major groove of the DNA in the complex. This conclusion, modeled by Hard et al (147) with their NMR-determined solution structure of the GR domain, was confirmed by the crystal structure of the GR-DNA complex (see below). The structural motif found in the hormone receptor has been termed a "zinc twist" (177) to distinguish it from the zinc finger, since its DNA-binding surface and the organization and functions of the loops surrounding the separate zinc domains are rather different from the classical TFIIIA-type zinc finger. The amino acid sequences of the immediate N- and C-terminal Zn domains from the superfamily of hormone receptors are collected in Table 3.

The third distinct type of zinc-containing transcription factor is represented by a set of 14 fungal transcription factors, for which amino acid sequences of the DNA-binding domains are given in Table 4. These proteins bind as dimers to palindromic DNA sequences located in the 5' upstream region of the genes whose transcription they control. The presence in the amino acid sequences of the DNA-binding domains (as determined by deletion mutagenesis) of two pairs of $C-X_2-C$ sequences separated by 13 residues initially led to the assumption that these were zinc finger proteins of the TFIIIA class containing a single zinc finger (204). DNA binding of a cloned fragment consisting of the 149 N-terminal residues of the prototype of this group, GAL4, showed the binding of this fragment to the 17-bp palindromic DNA sequence (UAS_G) recognized by GAL4 to be metal ion dependent (164). Either Zn or Cd restored specific DNA recognition to the apoprotein (164). Analysis of homogeneous samples of the 149-mer [GAL4(149)] dialyzed against metal-free buffers showed zinc contents from 1.2 to 1.6 zinc ions per mol of protein

Table 3 Amino acid sequences between the zinc ligands in the hormone receptor proteins

Hormone receptor[a]	Amino acid sequence[b]								Reference
GR	C LV C SDEASGCHYGVLT C GS C	KV-//-C AGRND C	IIDKIRRKN C	PA C	178
MR	C -- C ------V-- C -- C	KV-//-C ----- C	-V----- C	-- C	179
PR	C -I C --------- C -- C	KV-//-C ----- C	------- C	-- C	180, 181
AR	C -I C -------A-- C -- C	KV-//-C -S--- C	T---F---- C	-S C	182, 183
ER	C A- C N-Y---Y----WS C EG C	KA-//-C PAT-Q C	T---N---- C	Q-- C	184
TR	C V- C G-K-T-Y--RCI- C EG C	KS-//-C TYDGC C	V----T-NQ C	QL C	185, 186
VitD	C G- C G-R-T-F-FNAM- C EG C	KG-//-C PFDG- C	K-T-DN-RH C	E-- C	187
RAR	C FV C Q-KS--Y----SA C EG C	KG-//-C HRDKN C	--N-VT-NR C	QY C	188, 189

[a] Abbreviations: GR, glucocorticoid receptor; MR, mineralocorticoid receptor; PR, progesterone receptor; AR, androgen receptor; ER, estrogen receptor; TR, thyroid hormone receptor; VitD, vitamin D receptor; RAR, retinoic acid receptor.

[b] The sequences do not include the amino acids outside the zinc domains except for the two residues just distal to the first finger; KV, KA, KS, or KG. The second of these, adjacent to the lysine conserved in all the receptors, along with the two residues between the second pair of cysteine ligands in the first domain, either GS or EG, are the three amino acid side chains within the recognition helix which are critical to sequence recognition as found by site mutagenesis and found to make specific interactions in the crystal structure of the GR-DNA complex. Dashes and rules are as in Table 2.

Table 4 Amino acid sequence in the DNA-binding domains of fungal transcription factors containing a cluster of six cysteine residues

Protein	Initial residue no.	Sequence of DNA-binding domain[a]									Reference
GAL4	10	A C	DI C	RLKKLK C	SK--EK-P-K C	AK C	LKNNWEC- C				190
LAC9	94	A C	DA C	RKKKWK C	SK--TV-P-T C	TN C	LKYNLDC- C				191
PDR1	45	S C	DN C	RKRKIK C	NG--KF-P-- C	AS C	EIYSCEC- C				192
PUT3	33	A C	LS C	RKRHIK C	P---GGNP-- C	QK C	VTSNAI- C				193
QUTA	48	A C	DS C	RSKKCK C	DG--AQ-P-I C	ST C	ASLSRP- C				194
qa-1F	75	A C	DQ C	RAAREK C	DG--IQ-P-A C	FP C	VSQGRS- C				195
LEU3	36	A C	VE C	RQQKSK C	DAH-ERAPEP C	TK C	AKKNVP- C				196, 197
amdR	19	A C	VH C	HRRKKR C	DARLVGLP-- C	SN C	RSAGKTD C				198
UGA3	16	G C	IT C	KIRKKR C	SE--DK-P-V C	RD C	RRLSFP- C				199
ARGRII	20	G C	WT C	RGRKVK C	DL--RH-P-H C	QR C	EKSNLP- C				200
HAP1	63	S C	TI C	RKRKVK C	DK--LR-P-H C	QQ C	TKTGVA- C	CLCH			201
MAL63	7	S C	DC C	RVRRVK C	DR--NK-P-- C	NR C	IQRNLN- C				202
MAL6R	7	S C	DC C	RVRRVK C	DR--NK-P-- C	NR C	IQRNLN- C				203

[a] Dashes and rules are as in Table 2.

(164). In contrast, the CdGAL4(149) consistently bound two Cd ions per mol when all Zn was removed (164–166). The presence of two metal-binding sites on GAL4(149) was confirmed by ^{113}Cd NMR of CdGAL4(149), which showed two ^{113}Cd NMR signals at 669 and 707 ppm downfield of the signal for $Cd(ClO_4)_2$ (164). Each signal integrated to one Cd nucleus. The magnitudes of these chemical shifts imply coordination of each ^{113}Cd(II) to four donor sulfur atoms.

The six C residues in the short sequence containing the two C-X_2-C pairs (Table 4) are the only sulfur donors within the DNA-binding domain. Thus the presence of a binuclear cluster, Cd_2Cys_6, involving all six cysteine residues with two bridging -S^- ligands was postulated to account for the cadmium data (164). Cloning of the first 62 residues of the protein, GAL4(62), resulted in a highly soluble monomer with the same ^{113}Cd NMR spectrum as ^{113}Cd$_2$GAL4(149) had. GAL4(62) binds to the UAS$_G$ DNA sequence just as tightly as GAL4(149) does when the monomers are cross-linked (165). A number of ^{1}H-^{113}Cd heteronuclear 2D NMR experiments on ^{113}Cd$_2$GAL4 (62) have now clearly established the presence in GAL4 (62) of the Cd_2Cys_6 binuclear cluster shown in Figure 12C.

The binding of the two Cd ions is highly cooperative. A precisely similar coordination chemistry for zinc cannot be automatically assumed. Although it is possible to form Zn_1-Cd_1 hybrids from the DNA-binding domains cloned from both the GAL4 and LAC9 transcription factors (205), the homogeneous zinc derivative of GAL4 shows a facile equilibrium between Zn_2 and Zn_1 forms of the protein (168). The Zn_2GAL4(62) or Zn_2GAL4(149) can be formed in the presence of free zinc concentrations of $\geq 10\mu$M; however, dialysis against the standard metal-free buffers results in the loss of one of these zinc ions. This accounts for the initial analytical data on the zinc content for both zinc GAL4(63) and GAL4(149), which showed fewer than two zinc ions per molecule (164, 165). Both Zn_1 and Zn_2 derivatives of GAL4(62) bind to the UAS$_G$ DNA sequence. Differences in conformation or DNA-binding affinity are under study. It is pure speculation at present to suggest that the metal-induced folding of the specific DNA-binding surface of these proteins provides a mechanism in which the metal ion concentration could regulate transcription. This mechanism appears to apply for metallothionein (discussed above), but generalizing that to the control of transcription of nonmetalloprotein genes requires more data.

Crystal Structure of the Three Zinc Fingers of the Zif 268–DNA Complex

The three C_2H_2-type zinc fingers of the mouse transcription factor Zif 268 (73 amino acids corresponding to residues 349 to 421 of the native protein) have been cloned as a single fragment, extracted with 6.4 M guanidinium-HCl, and refolded in the presence of Zn or Co. The refolded protein domain was crystallized in the presence of a 12-bp DNA containing the specific 9-bp sequence recognized by Zif 268 (206). The structure of the complex has been solved at 2.1-Å resolution (Figure 13A). The three zinc fingers sequentially "march" around the major groove of the DNA such that their projection down the axis of the DNA double helix appears as a C-shaped clamp surrounding the DNA. The N-terminal end of the α-helix projects into the major groove, but the axis of the α-helix is only approximately aligned with the major groove and is tipped at ca. 45° with respect to the plane of the base pairs. The result is that the zinc ions and their ligands are located outside the 20-Å perimeter of the DNA. The N-terminal end of each α-helix makes the specific contacts with the base pairs, each helix interacting with a 3-bp subset of the 9-bp DNA sequence. The short β-sheet of each zinc finger is folded on the back of the helix away from the base pairs. The N-terminal strand of each β-sheet makes no contact with the DNA, whereas the more C-terminal strand contacts the sugar–phosphate backbone along one DNA strand.

The consensus nucleotide-binding site for Zif 268 is 5'-GCGTGGGCG-3'. The three fingers of the protein make a total of 11 hydrogen bonds with the bases in the major groove. All contacts are made with the G-rich DNA strand shown above. The contacts between the protein side chains and the bases are diagrammed in Figure 13B. On each finger the residue immediately preceding the start of the α-helix and the second, third, and sixth residues of the α-helix are critical to the interaction with the DNA bases. Fingers 1 and 3 have identical residues at these positions (R, D, E, R), whereas finger 2 has R, D, H, T at the analogous positions. Each finger has two primary interactions, R18 and R24, of finger 1 with the GCG subsite, R46 and H49 of finger 2 with the GGT subsite, and R74 and R80 with the GCG subsite. In all three fingers the guanidino group of the arginine just preceding the start of the α-helix, R18, R46, and R74, hydrogen bonds with the N7 and O6 of the G at the 3' end of the subsite. The second residue within each α-helix, the conserved aspartic acid, is placed such that the oxygens of the carboxylate make hydrogen bond–salt bridges with the N_ϵ and N_η of the guanidinum group already hydrogen bonded to the first G of each subsite, thus stabilizing the long R side chain interaction with the base.

The third residue of the α-helix in fingers 1 and 3 is a glutamic acid. Although this residue may be important for conformation, it does not contact the DNA. In contrast, the third residue in finger 2 is a histidine (not a zinc ligand), and the N3 of the imidazole side chain forms a hydrogen bond with N7 of the second G in subsite 2 (Figure 13B). The imidazole ring is coplanar with the guanine ring and stacks against the T of the adjacent base pair. The sixth residue of the α-helix in fingers 1 and 3 is an arginine whose guanidino group again makes a typical hydrogen bond structure with N7 and O6 of the G

A

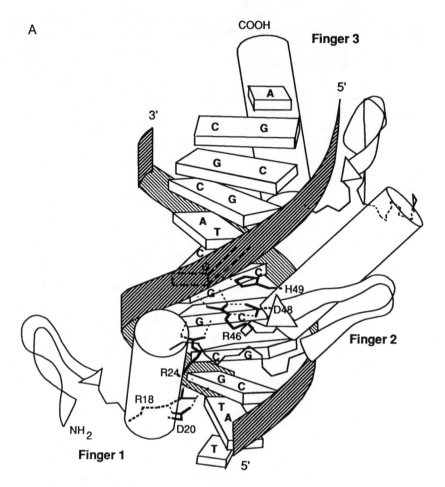

residues at the 5' ends of the first and third subsites. The analogous residue in finger 2 is a threonine, which does not appear to participate directly in DNA binding.

The contacts between Zif 268 and the phosphate backbone include one unusual structure in which the imidazole ring of the first His ligand in each finger, coordinating to Zn through N3, contacts the phosphodiester oxygens through N1. H25 (first finger) contacts the 5' phosphate of bp 7, while H53 (second finger) contacts that of bp 4. These contacts actually overlap into the following subdomains, and therefore the same contact for H81 (third finger) is outside the short DNA in the crystal. Thus the zinc complex is modulating the DNA interaction by a very direct contribution to the structure of the DNA-binding surface of the protein. Since this H ligand is conserved in all zinc

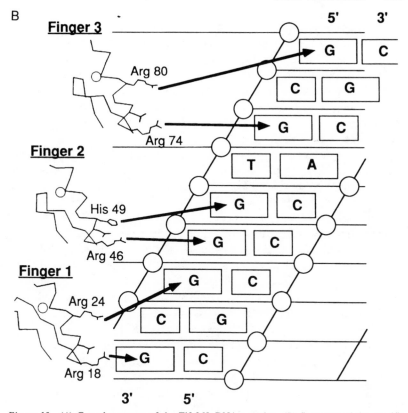

Figure 13 (A) Crystal structure of the Zif 268–DNA complex. (B) Summary of the specific amino acid side chain–base contacts found in the three-finger Zif 268–DNA complex. (From reference 206.)

fingers, they all probably use this unique DNA–protein interaction. The rest of the interactions between the phosphate backbone and the protein involve R and S residues (206).

Crystal Structure of the DNA-Binding Domain of the Glucocorticoid Receptor Complexed with the DNA of the Glucocorticoid-Responsive Element (GRE)

Crystals have been obtained of the same 86-amino-acid residue fragment of the GR domain used for the 2D NMR studies complexed to two different oligonucleotides containing a palindromic sequence corresponding to the GRE (207). The two oligonucleotides differ in that one (GRE$_{S3}$) contains a 3-bp spacer between the half-sites whereas the other (GRE$_{S4}$) contains a 4-bp spacer. The latter provides a target nucleotide with exact twofold symmetry and was expected not to influence the binding. This was based on the observation that the 86-mer is a monomer in solution as proven by the 2D NMR studies (148). Thus two monomers binding noncooperatively were expected to simply bind slightly farther apart. The GRE$_{S4}$ complex has been solved to 2.9-Å resolution, whereas the GRE$_{S3}$ complex is currently solved to 4.0-Å resolution (207). Comparison of the two structures gives a surprising result. Although each of the oligonucleotides binds two molecules of the GR fragment, only the GRE$_{S3}$ complex shows a strictly symmetrical interaction of the protein dimer with the palindrome. In the complex of GRE$_{S4}$, for which the resolution of 2.9 Å provides significant detail, only one monomer of the protein dimer interacts with the half-site in a highly specific manner. Because of the 4-bp rather than the 3-bp spacing found in normal GREs, the other protein monomer is "out of register" and is forced to dock with the other half of the palidrome as if there were a 3-bp spacer rather than a 4-bp spacer. Thus, despite the failure to dimerize in solution, when at least one monomer interacts in the correct way with a specific DNA half-site, a "tight" protein dimer is induced to form and dictates a 3-bp spacing between half sites. Structural analysis of the GRE$_{S3}$ complex shows it to be arranged symmetrically such that the protein has assumed in both sites a binding mode characteristic of only one of the two havles of the GRE$_{S4}$ complex. The half-site interaction common to both complexes is assumed to represent the specific interaction and clearly shows specific base–protein interactions.

The structure of the complex between the 86-mer dimer and the GRE$_{S4}$ is shown by the ribbon diagram in Figure 14. The N-terminal helices of the two monomers dock from one side of the DNA double helix, and lie in adjacent major grooves. At the site representing specific binding (the upper monomer in Figure 14), the N-terminal helix lies deeper in the major groove, which has widened by ca. 2 Å. The N-terminal rather irregular zinc finger–like structure can be seen best at the lower left of Figure 14 and is similar to that found in the NMR structure. The zinc is tetrahedrally coordinated by the C residues 440, 442, 457, and 460. This amino-terminal finger has a short segment of antiparallel β-sheet, which helps to orient the residues that make contacts with the phosphate backbone (207). The N-terminal helix begins with the second pair of C ligands and contributes two major functions in agreement with the general conclusions of the NMR studies. The hydrophobic face opposite the DNA contributes the side chains of F163 and F464 to the hydrophobic core of the molecule, which also receives side chains from the C-terminal helix.

The second zinc finger–like domain has a much better defined structure than found in the NMR solution structure of the monomers of either the GR or ER fragments. It consists of a short β-strand forming the proximal half and a distorted α-helix forming the distal half. The ligands are C476, C482, C492,

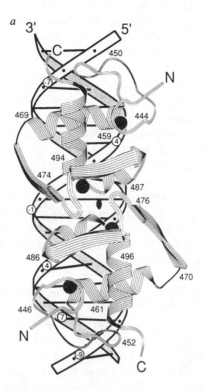

Figure 14   Crystal structure of an 86-amino-acid residue fragment of the glucocorticoid receptor complexed with its hormone receptor element. The zinc ions are indicated by the black balls. (From reference 207.)

and C495, not C500. The C-terminal zinc finger–like structure forms the entire interface responsible for the dimerization of this GR subdomain, largely through contacts of the residues between C476 and C482, the loop referred to as the "D box" (172). Like the first zinc subdomain, it also provides some phosphate backbone contacts.

The individual zinc subdomains in the GR domain have opposite chirality surrounding the zinc, a structural feature of some interest. The N-terminal zinc–sulfur coordination has the *S* configuration like the zinc fingers of the TFIIIA type. In contrast, the C-terminal zinc–sulfur coordination complex has the *R* configuration. This chirality is associated with the fact that the D-box assumes a reverse *β*-turn configuration. Together, this structure provides all the known subunit-subunit interactions. The *R* stereochemistry of the zinc complex is an integral part of this configuration.

The contacts between this GR subdomain and the bases of the HRE are all made with three acid side chains (R466, V462, K461) on the face of the N-terminal α-helix, and they interact principally with bases G4, T5, T6, and G7 of the HRE. R166 forms typical hydrogen bonds between its guanidino group and the N7 and O6 of G4. The methyl group of T5 (invariant in GREs) makes a "glove fit" van der Waals contact with the side chain of V462. G7 is also a constant base in most HREs, and K461 makes one direct hydrogen bond with G7 and one water-mediated hydrogen bond. The same water molecule makes a second hydrogen bond to the O4 of T6. The latter is the second thymine base important in specifying a GR-specific element.

The above specific amino acid side chain base interactions are missing from the other monomer–DNA interaction of the GRE$_{S4}$ complex, supporting the interpretation that it is a nonspecific DNA interaction forced because the dimer of the GR DNA-binding domain is designed to recognize a 3-bp rather than a 4-bp spacer. This finding would correlate well with the observation that the spacer length in the center of the palindromic HRE recognition sequences seems as important as specific base recognition in controlling the discrimination between different DNA sequences.

ANALYTICAL DATA ON ZINC CONTENT OF TRANSCRIPTION FACTORS   In large measure the zinc stoichiometry expected for zinc finger transcription factors has been inferred from the amino acid sequence. Among the zinc finger transcription factors, analytical data have been published for TFIIIA (130, 132, 134), Spl (208), and Zif 268 (206). The Zif 268 data come from the crystal structure of the refolded three fingers (206). There are analytical and EXAFS data on a 150-residue DNA-binding fragment of the glucocorticoid receptor that show 2.3 Zn ions per molecule (209). The 2D NMR structure of the estrogen receptor fragment solidly infers the presence of two zinc ions, but does not directly detect them. For two of the zinc cluster transcription factors, GAL4 (164–168) and LAC9 (205, 210), analytical data show two zinc ions per mol of cloned DNA-binding fragment. However, for GAL4, proteins containing either one or two zinc ions have been prepared (168). The glucocorticoid receptor binds three metal ions under some conditions, probably involving the extra C residues (211). As more of these transcription factors are cloned and overproduced, additional analytical data on this group of proteins will increase our understanding of the zinc $\rightleftharpoons$ protein equilibria that apply to these proteins in solution.

ZINC PROTEINS BINDING TO SINGLE-STRANDED NUCLEIC ACIDS

Gene 32 Protein from Bacteriophage T4

In 1986 the classic accessory protein in DNA replication, gene 32 protein from bacteriophage T4, was shown to contain one atom of Zn (212). This protein is essential for DNA replication, recombination, and repair in T4 (213). The protein binds cooperatively and nonspecifically to the single-stranded DNA at the replication fork, protects it from nuclease attack, and helps assemble other proteins of the replication complex, most notably the T4 DNA polymerase (213). The protein contains three domains. The positively charged N-terminal domain, residues 1 to 21, is required for cooperative binding to DNA, while the C-terminal domain, residues 253 to 275, is negatively charged and interacts with the DNA polymerase. The core region, residues 22 to 252, contains the single-stranded DNA-binding site, and the zinc ion coordinated to C77, H81, C87, and C90 in the sequence shown below (212).

Gene 32	$-C-S-S-T-H-G-D-Y-D-S-C-P-V-C-$
HIV NC protein 1	$-C-F-N-C-G-K-E-G-H-T-A-R-N-C-$
HIV NC protein 2	$-C-W-K-C-G-K-E-G-H-Q-M-K-D-C-$

Spectral and NMR data support this assignment of the zinc ligands (214, 215). Studies comparing the binding of the zinc and zinc-free proteins to DNA show that the apoprotein binds less tightly to DNA and protects the DNA less well from nuclease digestion and that the cooperativity of binding is significantly reduced (216, 217). Gene 32 protein also functions at the translational level in that it represses the translation of its own mRNA. It does so by binding to a secondary structure present near the 5' end of the mRNA (218). It has recently been shown that this binding and translation repression depend critically on the presence of the Zn (219).

Retroviral Nucleocapsid Proteins

At the time of the discovery of Zn in gene 32 protein, it was noted that the nucleocapsid (NC) proteins of several retroviruses contained two repeats of an amino acid sequence similar to the Zn-binding sequence in gene 32 protein (212, 219, 220). These repeats are shown above for the NC protein from human immunodeficiency virus (HIV). Similar sequences are found in the *GAG* gene products from other retroviruses, although the NC protein from bovine leukemia virus contains only one copy of the sequence (212). The HIV NC protein, also known as p15, is a protein of 149 amino acids and is coded for by the 3' end of the *GAG* gene of HIV (221, 222). The NC protein is a single-stranded nucleic acid–binding protein, and 2000 to 3000 copies are incorporated into the virus particle (223).

Several functions have been reported for the retroviral NC proteins. The protein nonspecifically coats the single-stranded RNA genome (223, 224), and mutations in the C residues of the putative metal-binding sequence show these residues to be essential for the specific packaging of viral RNA. The mutant NC proteins lead to viral particles that either do not contain the viral RNA or contain cellular RNA (225–228). The NC proteins appear to bind specifically to a region near the 5' end of the retroviral genome, designated the psi sequence. This sequence, located between a 5' splice donor site and the ATG translational start signal for the *GAG* gene, is also necessary for the correct packaging of viral RNA (229, 230). Synthesis of the double-stranded DNA copy of the viral RNA, following endocytosis of the viral particle, is catalyzed by the reverse transcriptase and appears to involve transcription from a genome still contained in a macromolecular aggregate consisting of the viral RNA and various *GAG* gene products. It has therefore been postulated that the NC protein has a role in assisting the replication process carried out by the reverse transcriptase (231, 232).

Since it has been proposed by a number of investigators that the CCHC-containing sequences in the retroviral NC proteins bind Zn, it has been suggested that the metal binding is interrupted by point mutations in the CCHC residues. This proposal remains controversial because the NC proteins isolated by the standard techniques do not seem to contain stoichiometric amounts of Zn(II) (233), although studies in vitro show that it is possible under controlled conditions to bind metal ions to these proteins or to peptides containing the proposed metal-binding sequence from the NC proteins (234–236).

When the severe oxidation problems accompanying isolation of a native cloned NC protein of HIV were solved, it became possible to isolate the protein containing two Zn, two Cd, or two Co ions per molecule (237). Simple titrations of RNA with the NC protein do not reveal large differences in binding affinity between the metallo and apo forms of the protein. However, the effect of the metal ions on the more specific functions postulated for the NC protein remain to be tested.

SUMMARY

In the past five years there has been a great expansion in our knowledge of the role of zinc in the structure and function of proteins. Not only is zinc required for essential catalytic functions in enzymes (more than 300 are known at present), but also it stabilizes and even induces the folding of protein sub-domains. The latter functions have been most dramatically illustrated by the discovery of the essential role of zinc in the folding of the DNA-binding domains of eukaryotic transcription factors, including the zinc finger transcription factors, the large family of hormone receptor proteins, and the zinc cluster transcription factors from yeasts. Similar functions are highly probable for the zinc found in the RNA polymerases and the zinc-containing accessory proteins involved in nucleic acid replication.

The rapid increase in the number and nature of the proteins in which zinc functions is not unexpected since zinc is the second most abundant trace metal found in eukaryotic organisms, second only to iron. If one substracts the amount of iron found in hemoglobin, zinc becomes the most abundant trace metal found in the human body.

Literature Cited

1. Raulin, J. 1869. *Ann. Sci. Bot. Biol. Veg.* 11:93
2. Keilin, D., Mann, T. 1940. *Biochem. J.* 34:1163–76
3. Vallee, B. L., Neurath, H. 1955. *J. Biol. Chem.* 217:253–61
4. Vallee, B. L., Hoch, F. L. 1957. *J. Biol. Chem.* 225:185–95
5. Vallee, B. L., Hoch, F. L. 1955. *Proc. Natl. Acad. Sci. USA* 41:327–38
6. Plocke, D. J., Levinthal, C., Vallee, B. L. 1962. *Biochemistry* 1:373–78
7. Vallee, B. L. 1955. *Adv. Protein Chem.* 10:317–84
8. Vallee, B. L., Auld, D. S. 1990. *Biochemistry* 29:5647–59
9. Liljas, A., Kannan, K. K., Bergsten, P.-C., Waara, B. I., Fridborg, K., et al. 1972. *Nature* (London) *New Biol.* 235:131–37
10. Lipscomb, W. N., Hartsuck, J. A., Quiocho, F. A., Reeke, G. N. 1969. *Proc. Natl. Acad. Sci. USA* 64:28–35
11. Rees, D. C., Lewis, M., Lipscomb, W. N. 1983. *J. Mol. Biol.* 168:367–87
12. Bränden, C.-I., Eklund, H., Nordström, B., Boiwe, T., Söderlund, G., et al. 1973. *Proc. Natl. Acad. Sci. USA* 70: 2439–42
13. Eklund, H., Branden, C.-I. 1983. In *Zinc Enzymes*, ed. T. G. Spiro, pp. 123–52. New York: Wiley-Interscience.
14. Honzatko, R. B., Crawford, J. L., Monaco, H. L., Ladner, J. E., Edwards, B. F. P., et al. 1982. *J. Mol. Biol.* 160:219–63
15. Kim, E., Wyckoff, H. W. 1991. *J. Mol. Biol.* 218:449–64
16. Volbeda, A., Lahm, A., Sakiyama, F., Suck, D. 1991. *EMBO J.* 10:1607–18
17. Hough, E., Hansen, L. K., Birknes, B., Jynge, K., Hansen, S., et al. 1989. *Nature* (London) 338:357–60
18. Scruton, M. C., Wu, C. W., Goldthwait, D. A. 1971. *Proc. Natl. Acad. Sci. USA* 68:2497–501
19. Halling, S. M., Sanchez-Anzaldo, F. J., Fukuda, R., Doii, R. H., Meares, C. F. 1977. *Biochemistry* 16:2880–84
20. Speckhard, D. C., Wu, F. Y.-H., Wu, C.-W. 1977. *Biochemistry* 16:5228–34
21. Giedroc, D. P., Coleman, J. E. 1986. *Biochemistry* 25:4969–78
22. King, G. C., Martin, C. T., Pham, T. T., Coleman, J. E. 1986. *Biochemistry* 25:36–40
23. Smith, E. L. 1949. *Proc. Natl. Acad. Sci. USA* 35:80–90
24. Davis, R. P. 1959. *J. Am. Chem. Soc.* 81:5674–78
25. DeVoe, H., Kistiakowsky, G. B. 1961. *J. Am. Chem. Soc.* 83:274–79
26. Kernohan, J. C. 1964. *Biochim. Biophys. Acta* 81:346–56
27. Lindskog, S., Coleman, J. E. 1973. *Proc. Natl. Acad. Sci. USA* 70:2505–8
28. Coleman, J. E. 1984. *Ann. N.Y. Acad. Sci.* 249:26–48
29. Eriksson, A. E., Jones, T. A., Liljas, A. In *Zinc Enzymes*, ed. I. Bertini, C. Luchinat, W. Maret, M. Zeppezauer, pp. 317–28. Boston: Birkhauser
30. Eriksson, A. E., Kylsten, P. M., Jones, T. A., Liljas, A. 1988. *Proteins Struct. Funct. Genet.* 4:274–82
31. Tashian, R. E., Hewett-Emmett, D., eds. 1984. *Ann. N.Y. Acad. Sci.* 429:1–640
32. Coleman, J. E. 1964. *J. Biol. Chem.* 242:5212–19
33. Pocker, Y., Stone, J. T. 1968. *Biochemistry* 7:2936–45
34. Bauer, R., Limkilde, P., Johansen, J. T. 1976. *Biochemistry* 15:334–42
35. Tu, C. K., Silverman, D. N. 1975. *J. Am. Chem. Soc.* 97:5935–36
36. Tu, C. K., Silverman, D. N. 1985. *Biochemistry* 24:5881–87
37. Eriksson, A. E., Kylsten, P. M., Jones, T. A., Liljas, A. 1988. *Proteins Struct. Funct. Genet.* 4:283–93
38. Tu, C. K., Paranawithana, S. R., Jewell, D. A., Tanhauser, S. M., LoGrasso, P. V., et al. 1990. *Biochemistry* 29:6400–5
39. LoGrasso, P. V., Tu, C. K., Jewell, D. A., Wynns, G. C., Laipis, P. J., Silverman, D. N. 1991. *Biochemistry* 30: 8463–70
40. Gros, G., Dodgson, S. J. 1988. *Annu. Rev. Physiol.* 50:669–94
41. Tu, C. K., Sanyal, G., Wynns, G. C., Silverman, D. N. 1983. *J. Biol. Chem.* 258:8867–71

42. Engberg, P., Lindskog, S. 1984. *FEBS Lett.* 170:326–30
43. Eriksson, A. E. 1988. PhD dissertation, Uppsala Univ.
44. Lipscomb, W. N., Hartsuck, J. A., Quiocho, F. A., Reeke, G. N. 1969. *Proc. Natl. Acad. Sci. USA* 64:28–35
45. Hartsuck, J. A., Lipscomb, W. N. 1971. In *The Enzymes*, ed. P. Boyer, pp. 16–56. New York: Academic.
46. Lipscomb, W. N., Hartsuck, J. A., Reeke, G. N., Quiocho, F. A., Bethge, P. H., et al. 1968. *Brookhaven Symp. Biol.* 21:24–90
47. Rees, D. C., Lewis, M., Lipscomb, W. N. 1983. *J. Mol. Biol.* 168:367–87
48. Christianson, D. W., Lipscomb, W. N. 1989. *Acc. Chem. Res.* 22:62–69
49. Christianson, D. W., Lipscomb, W. N. 1986. *Proc. Natl. Acad. Sci. USA* 83: 7568–72
50. Auld, D. S., Baldes, A., Geoghegan, K. F., Holmquist, B., Martinelli, R. A., Vallee, B. L. 1984. *Proc. Natl. Acad. Sci. USA* 81:5041–45
51. Geoghegan, K. F., Galdes, A., Martinelli, R. A., Holmquist, B., Auld, D. S., Vallee, B. L. 1983. *Biochemistry* 22:2255–62
52. Geoghegan, K. F., Galdes, A., Hanson, G., Holmquist, B., Auld, D. S., Vallee, B. L. 1986. *Biochemistry* 25:4669–74
53. Breslow, R., Wernick, D. L. 1977. *Proc. Natl. Acad. Sci. USA* 74:1303–7
54. Shoham, G., Christianson, D. W., Oren, D. A. 1988. *Proc. Natl. Acad. Sci. USA* 85:684–88
55. Rees, D. C., Lipscomb, W. N. 1982. *J. Mol. Biol.* 160:475–98
56. Christianson, D. W., Lipscomb, W. N. 1985. *Proc. Natl. Acad. Sci. USA* 82:6840–44
57. Shoham, G., Rees, D. C., Lipscomb, W. N. 1984. *Proc. Natl. Acad. Sci. USA* 81:7767–71
58. Jacobsen, N. E., Bartlett, P. A. 1981. *J. Am. Chem. Soc.* 103:654–57
59. Christianson, D. W., Lipscomb, W. N. 1988. *J. Am. Chem. Soc.* 110:5560–65
60. Christianson, D. W., Lipscomb, W. N. 1988. *J. Am. Chem. Soc.* 108:545–46
61. Matthews, B. W. 1988. *Acc. Chem. Res.* 21:333–40
62. Matthews, B. W., Jansonius, J. N., Colman, P. M., Schoenborn, B. P., Dupourque, D. 1972. *Nature* (London) *New Biol.* 238:37–41
63. Matthews, B. W., Colman, P. M., Jansonius, J. N., Titani, K., Walsh, K. A., Neurath, H. 1972. *Nature* (London) *New Biol.* 238:41–43
64. Colman, P. M., Jansonius, N. N., Matthews, B. W. 1972. *J. Mol. Biol.* 70:701–24
65. Kester, W. R., Matthews, B. W. 1977. *J. Biol. Chem.* 252:7704–10
66. Holmes, M. A., Matthews, B. W. 1982. *J. Mol. Biol.* 160:623–639
67. Hanauer, D. G., Monzingo, A. F., Matthews, B. W. 1984. *Biochemistry* 23:5730
68. Weaver, L. H., Kester, W. R., Matthews, B. W. 1977. *J. Mol. Biol.* 114:119–32
69. Kester, W. R., Matthews, B. W. 1977. *Biochemistry* 16:2506–16
70. Eklund, H., Jones, A., Schneider, G. 1986. In *Zinc Enzymes*, ed. I. Bertini, C. Luchinat, W. Maret, M. Zeppezauer, pp. 377–392. Boston: Birkhauser.
71. Zeppezauer, M. 1986. See Ref. 70, pp. 417–434
72. Horiuchi, T., Horiuchi, S., Mizuno, D. 1959. *Nature* (London) 183:1529–30
73. Torriani, A. 1960. *Biochim. Biophys. Acta* 38:460–69
74. Garen, A., Levinthal, C. 1960. *Biochim. Biophys. Acta* 38:470–83
75. Kim, E. E., Wyckoff, H. W. 1989. *Clin. Chim. Acta* 186:175–88
76. Fosset, M., Chappelet-Tordo, D., Lazdunski, M. 1974. *Biochemistry* 13: 1783–88
77. Gettins, P., Coleman, J. E. 1983. *J. Biol. Chem.* 258:408–16
78. Kim, E. E., Wyckoff, H. W. 1991. *J. Mol. Biol.* 218:449–64
79. Gettins, P., Coleman, J. E. 1983. *J. Biol. Chem.* 258:396–407
80. Schulz, C., Bertini, I., Viezzoli, M. S., Brown, R. D., Koenig, S. H., Coleman, J. E. 1989. *Inorg. Chem.* 28:1490–96
81. Gettins, P., Coleman, J. E. 1984. *J. Biol. Chem.* 259:11036–40
82. Hull, W. E., Halford, S. E., Gutfreund, H., Sykes, B. D. 1976. *Biochemistry* 15:1547–61
83. Gettins, P., Metzler, M., Coleman, J. E. 1985. *J. Biol. Chem.* 260:2875–83
84. Chlebowski, J. F., Armitage, I. M., Tusa, P. P., Coleman, J. E. 1976. *J. Biol. Chem.* 251:1207–16
85. Coleman, J. E., Gettins, P. 1986. See Ref. 70, pp. 77–99.
86. Gettins, P., Coleman, J. E. 1984. *J. Biol. Chem.* 259:4991–97
87. Gettins, P., Coleman, J. E. 1983. *Adv. Enzymol.* 55:381–452
88. Coleman, J. E., Armitage, I. M., Chlebowski, J. F., Otvos, J. D., Schoot Uiterkamp, A. J. M. 1979. In *Biological Applications in Magnetic Resonance*, ed. R. G. Shulman, pp. 345–95. New York: Academic.

89. Labow, B. I. 1989. *Mechanism of phosphoryl transfer by alkaline phosphatase from* E. coli. MS thesis. Brandeis Univ.
90. Weiss, P. M., Cleland, W. W. 1989. *J. Am. Chem. Soc.* 111:1928–29
91. Deleted in proof
92. Shishido, K., Ando, T. 1982. In *Nucleases*, ed. S. M. Linn, R. J. Roberts, pp. 155–185. Cold Spring Harbor: Cold Spring Harbor Laboratory.
93. Volbeda, A., Lahm, A., Sakiyama, F., Suck, D. 1991. *EMBO J.* 10:1607–18
94. Fujimoto, M., Kuninka, A., Yoshino, H. 1974. *Agric. Biol. Chem.* 38:785–90
95. Fujimoto, M., Kuninka, A., Yoshino, H. 1974. *Agric. Biol. Chem.* 38:1555–61
96. Fujimoto, M., Kuninka, A., Yoshino, H. 1974. *Agric. Biol. Chem.* 38:1555–61; 2141–47
97. Maekawa, K., Tsunawa, S., Dibo, G., Sakiyama, F. Reported in Ref. 93.
98. Shishido, K., Habuka, N. 1986. *Biochim. Biophys. Acta* 884:215–18
99. Hough, E., Hansen, L. K., Birknes, B., Jynge, K., Hansen, S., et al. 1989. *Nature* (London) 338:357–60
100. Little, C. 1981. *Acta Chem. Scand.* B35:39–44
101. Berridge, M. J. 1984. *Biochem. J.* 220:345–60
102. Besteman, J. M., Duronio, V., Cuatracosa, P. 1986. *Proc. Natl. Acad. Sci. USA* 83:6785–89
103. Johansen, T., Holm, T., Guddal, P. H., Sletten, K., Haugli, F. B., Little, C. 1988. *Gene* 65:293–304
104. Levine, L., Xiao, D. M., Little, C. 1988. *Prostaglandins* 34:633–42
105. Little, C., Otnaess, A. B. 1975. *Biochim. Biophys. Acta* 391:326–33
106. Bicknell, R., Hanson, G. R., Holmquist, B., Little, C. 1986. *Biochemistry* 25:4219–33
107. Feiters, M. C., Little, C., Waley, S. G. 1986. *J. Phys.* (Paris) 47:1169–72
108. Deleted in proof.
109. Subcommittee on Zinc, National Research Council. 1979. *Zinc. Report of the Subcommittee on Zinc*, pp. 63–172. Baltimore: University Park Press.
110. See Ref. 109, pp. 123–72
111. Kägi, J. H. R., Kojima, Y., ed. 1957. *Metallothionein II. Experientia Supplementum*, Vol. 52. Basel: Berkhäuser.
112. Margoshes, M., Vallee, B. L. 1957. *J. Am. Chem. Soc.* 79:4813–14
113. Kägi, J. H. R., Vallee, B. L. 1960. *J. Biol. Chem.* 235:3460–65
114. Bremner, I. 1957. See Ref. 111, pp. 81–107
115. Otvos, J. D., Armitage, I. M. 1979. *J. Am. Chem. Soc.* 101:7734–36
116. Arseniev, A., Schultze, P., Wörgötter, E., Braun, W., Wagner, G., et al. 1988. *J. Mol. Biol.* 201:637–57
117. Schultze, P., Wörgötter, E., Braun, W., Wagner, G., Vasak, M., et al. 1988. *J. Mol. Biol.* 203:251–68
118. Braun, W., Wagner, G., Wörgötter, E., Vasak, M., Kägi, J. H. R., Wüthrich, K. 1986. *J. Mol. Biol.* 187:125–29
119. Robbins, A. H., McRee, D. E., Williamson, M., Collett, S. A., Xuong, N. H., et al. 1991. *J. Mol. Biol.* 221:1269–93
120. Neuhaus, D., Wagner, G., Vasak, M., Kägi, J. H. R., Wüthrich, K. 1984. *Eur J. Biochem.* 143:659–67
121. Palmiter, R. D. 1987. See Ref. 111, pp. 63–80
122. Hamer, D. H. 1986. *Annu. Rev. Biochem.* 55:913–51
123. Karin, M., Najarian, R., Haslinger, A., Valenzuela, P., Welsh, J., Fogel, S. 1984. *Proc. Natl. Acad. Sci. USA* 81:337–41
124. Butt, T. R., Sternberg, E. J., Gorman, J. A., Clark, P., Hamer, D., et al. 1984. *Proc. Natl. Acad. Sci. USA* 81:3332–36
125. Winge, D. R., Nielson, K. B., Gray, N. R., Hamer, D. H. 1985. *J. Biol. Chem.* 260:14464–70
126. Deleted in proof
127. Furst, P., Hu, S., Hackett, R., Hamer, D. H. 1988. *Cell* 55:705–17
128. Thiele, D. J. 1988. *Mol. Cell. Biol.* 8:2745–52
129. Furst, P., Hamer, D. H. 1989. *Proc. Natl. Acad. Sci. USA* 86:5267–71
130. Hanas, J. S., Hazuda, D. J., Bogenhagen, D. F., Wu, F. Y.-H., Wu, C.-W. 1983. *J. Biol. Chem.* 258:14120–14125
131. Wu, C.-W. 1986. See Ref. 70, pp. 563–78
132. Miller, J., MacLachlan, A. D., Klug, A. 1985. *EMBO J.* 4:1609–14
133. Diakun, G. P., Fairall, L., Klug, A. 1986. *Nature* (London) 324:698–99
134. Shang, Z., Liao, Y.-D., Wu, F. Y.-H., Wu, C.-W. 1989. *Biochemistry* 28:9790–95
135. Parraga, G., Horvath, S. J., Eisen, H., Taylor, W. E., Hood, L., et al. 1988. *Science* 241:1489–92
136. Lee, M. S., Gippert, G. P., Soman, K. V., Case, D. A., Wright, P. E. 1989. *Science* 245:635–37
137. Omichinski, J. G., Clore, G. M., Appella, E., Sakaguchi, K., Gronenborn,

A. M. 1990. *Biochemistry* 29:9324–34

138. Ginsberg, A. M., King, B. O., Roeder, R. G. 1984. *Cell* 39:479–89
139. Kadonaga, J. T., Carner, K. R., Masiarz, F. R., Tjian, R. 1987. *Cell* 51:1079–90
140. Evans, R. M. 1988. *Science* 240:889–95
141. Beato, M. 1989. *Cell* 56:335–44
142. Giniger, E., Varnum, S. M., Ptashne, M. 1985. *Cell* 40:767–74
143. Keegan, L., Gill, G., Ptashne, M. 1986. *Science* 231:699–704
144. Johnston, M. 1987. *Microbiol. Rev.* 51:458–76
145. Oshima, Y. 1982. In *Molecular Biology of the Yeast Saccharomyces*, Vol. 1, ed. J. Strathern, E. Jones, J. K. Broach, pp. 159–180. Cold Spring Harbor: Cold Spring Harbor Laboratory.
146. Neuhaus, D., Nakaseko, Y., Nagai, K., Klug, A. 1990. *FEBS Lett.* 262:179–84
147. Kochoyan, M., Havel, T. F., Nguyen, D. T., Dahl, C. E., Keutmann, H. T., Weiss, M. A. 1991. *Biochemistry* 30:3371–86
148. Hard, T., Kellenbach, E., Boelens, R., Maler, B. A., Dahlman, K. 1990. *Science* 249:157–60
149. Schwabe, J. W. R., Neuhaus, D., Rhodes, D. 1990. *Nature* (London) 348:458–61
150. Page, D. C., Mosher, R., Simpson, E. M., Fisher, E. M. C., Mardon, G., et al. 1987. *Cell* 51:1091–1104
151. Milbrandt, J. 1987. *Science* 238:797–99
152. Chowdhury, K., Deutsch, U., Gruss, P. 1987. *Cell* 48:771–78
153. Ruis, I., Altaba, A., Perry-O'Keefe, H., Melton, D. A. 1987. *EMBO J* 6:3065–70
154. Joho, K. E., Darby, M. K., Crawford, E. T., Brown, D. D. 1990. *Cell* 61:293–300
155. Vincent, A., Colot, H. V., Rosbash, M. 1985. *J. Mol. Biol.* 186:149–66
156. Rosenberg, U. B., Schröder, C., Preiss, A., Kienlin, A., Cote, S., et al. 1986. *Nature* (London) 319:336–39
157. Boulay, J. L., Dennefeld, C., Alberga, A. 1987. *Nature* (London) 330:395–98
158. Schuh, R., Aicher, W., Gaul, U., Cote, S., Preiss, A., et al. 1986. *Cell* 47:1025–32
159. Tautz, D., Lehmann, R., Schnürch, H., Schuh, R., Seifert, E., et al. 1987. *Nature* (London) 327:383–89
160. Baldarelli, R. M., Mahoney, P. A., Salas, F., Gustavson, E., Boyer, P. D., et al. 1988. *Dev. Biol.* 125:85–95
161. Pays, E., Murphy, N. B. 1987. *J. Mol. Biol.* 197:147–48

162. Hartshorne, T. A., Blumberg, H., Young, E. T. 1986. *Nature* (London) 320:283–87
163. Stillman, D. J., Banktier, A. T., Seddon, A., Groenhout, E. G., Nasmyth, K. A. 1988. *EMBO J.* 7:485–94
164. Pan, T., Coleman, J. E. 1989. *Proc. Natl. Acad. Sci. USA* 86:3145–49
165. Pan, T., Coleman, J. E. 1990. *Biochemistry* 29:3023–29
166. Pan, T., Coleman, J. E. 1990. *Proc. Natl. Acad. Sci. USA* 87:2077–81
167. Pan, T., Coleman, J. E. 1991. *Biochemistry* 30:4212–22
168. Gardner, K. H., Pan, T., Narula, S., Rivera, E., Coleman, J. E. 1991. *Biochemistry* 30:11292–302
169. Berg, J. M. 1988. *Proc. Natl. Acad. Sci. USA* 85:99–102
170. Schwabe, J. W. R., Rhodes, D. 1991. *Trends Biochem. Sci.* 16:291–96
171. Mader, S., Kumar, V., de Verneuil, H., Chambon, P. 1989. *Nature* (London) 338:271–74
172. Umesono, K., Evans, R. M. 1989. *Cell* 57:1139–46
173. Danielson, M., Hinck, L., Ringold, G. M. 1989. *Cell* 57:1131–38
174. Green, S., Kumar, V., Theulaz, I., Wahli, W., Chambon, P. 1988. *EMBO J.* 7:3037–44
175. Tsai, S. Y., Carlstedt-Duke, J., Weigel, N. L., Dahlman, K., Gustafsson, J.-A., et al. 1988. *Cell* 55:361–69
176. Klein-Hitpass, L. 1989. *Mol. Cell. Biol.* 9:43–49
177. Vallee, B. L., Coleman, J. E., Auld, D. S. 1991. *Proc. Natl. Acad. Sci. USA* 88:999–1003
178. Hollenberg, S. M., Giguere, V., Segui, P., Evans, R. M. 1987. *Cell* 49:39–46
179. Arriza, J. L., Weinberger, C., Cerelli, G., Glaser, T. M., Handelin, B. L., Housman, D. E., Evans, R. M. 1987. *Science* 237:340–75
180. Misrahi, M., Atger, M., D'Auriol, L., Loosfelt, H., Meriel, C., et al. 1987. *Biochem. Biophys. Res. Commun.* 143:740–48
181. Huckaby, C. S., Connelly, O. M., Beattie, W. G., Dobson, A. D. W., Tsai, M. J., O'Malley, B. W. 1987. *Proc. Natl. Acad. Sci. USA* 84:8380–84
182. Chang, C., Kokonitis, J., Liao, S. 1988. *Science* 240:324–26
183. Lubahn, D. B., Joseph, D. R., Sullivan, P. M., Willard, H. F., French, F. S., Wilson, E. M. 1988. *Science* 340:327–30
184. Walter, P., Green, S., Greene, G.,

Krust, A., Bornert, J.-M., et al. 1985. *Proc. Natl. Acad. Sci.* USA 82:7889–93

185. Sap, J., Murioz, A., Damm, K., Goldberg, Y., Ghysdael, J., et al. 1986. *Nature* (London) 324:635–40

186. Weinberger, C., Thompson, C. C., Ong, E. S., Lebo, R., Gruol, D. J., Evans, R. M. 1986. *Nature* (London) 324:641–46

187. McDonnell, D. P., Mangelsdorf, D. J., Pike, J. W., Haussler, M. R., O'Malley, B. W. 1987. *Science* 235:1214–17

188. Giguere, V., Ong, E. S., Segui, P., Evans, R. M. 1987. *Nature* (London) 330:624–29

189. Petkovich, M., Brand, N. J., Krust, A., Chambon, P. 1987. *Nature* (London) 330:444–50

190. Laughon, A., Gesteland, R. F. 1984. *Mol. Cell. Biol.* 4:260–67

191. Salmeron, J. M., Johnston, S. A. 1986. *Nucleic Acids Res.* 14:7767–81

192. Balzi, E., Chen, W., Ulaszewski, S., Capieaus, E., Goffeau, A. 1987. *J. Biol. Chem.* 262:16871–79

193. Marczak, J. E., Brandriss, M. C. 1991. *Mol. Cell. Biol.* 11:2609–19

194. Beri, R. K., Whittington, H., Roberts, C. R., Hawkins, A. R. 1987. *Nucleic Acids Res.* 15:7991–8001

195. Baum, J. A., Geever, R., Giles, N. H. 1987. *Mol. Cell. Biol.* 7:1256–66

196. Friden, P., Schimmel, P. 1987. *Mol. Cell. Biol.* 7:2708–17

197. Zhou, K., Brisco, P. R. G., Hinkkanen, A. E., Kohlhaw, G. B. 1987. *Gene* 15:5261–73

198. Andrianopoulos, A., Hynes, M. J. 1990. *Mol. Cell. Biol.* 10:3194–3203

199. Andre, B. 1990. *Mol. Gen. Genet.* 220:269–76

200. Messenguy, R. F., Dubois, E., Descamps, F. 1986. *Eur. J. Biochem.* 157:77–81

201. Pfeifer, K., Kim, K.-S., Kogan, S., Guarente, L. 1989. *Cell* 56:291–301

202. Kim, J., Michels, C. A. 1988. *Curr. Genet.* 14:319–23

203. Sollitti, P., Marmur, J. 1988. *Mol. Gen. Genet.* 213:56–62

204. Johnston, M. 1987. *Nature* (London) 328:353–55

205. Pan, T., Halvorsen, Y.-D., Dickson, R. C., Coleman, J. E. 1990. *J. Biol. Chem.* 265:21427–29

206. Pavletich, N. P., Pabo, C. O. 1991. *Science* 252:809–17

207. Luisi, B. F., Xu, W. X., Otwinowski, Z., Freedman, L. P., Yamamoto, K. R., Sigler, P. B. 1991. *Nature* (London) 352:497–505

208. Kuwahara, J., Coleman, J. E. 1990. *Biochemistry* 29:8628–31

209. Freedman, L. P., Luisi, B. F., Korszun, Z. R., Basavappa, R., Sigler, P. B., Yamamoto, K. 1988. *Nature* (London) 334:543–46

210. Halvorsen, Y. D. C., Nandabalan, K., Dickson, R. C. 1990. *J. Biol. Chem.* 265:13283–89

211. Pan, T., Freedman, L. P., Coleman, J. E. 1990. *Biochemistry* 29:9218–25

212. Giedroc, D. P., Keating, K. M., Williams, K. R., Konigsberg, W. H., Coleman, J. E. 1986. *Proc. Natl. Acad. Sci. USA* 83:8452–56

213. Chase, J. W., Williams, K. R. 1986. *Annu. Rev. Biochem.* 55:103–36

214. Giedroc, D. P., Johnson, B. A., Armitage, I. M., Coleman, J. E. 1989. *Biochemistry* 28:2410–18

215. Pan, T., Giedroc, D. P., Coleman, J. E. 1989. *Biochemistry* 28:8828–32

216. Giedroc, D. P., Keating, K. M., Williams, K. R., Coleman, J. E. 1987. *Biochemistry* 26:5251–59

217. Nadler, S. G., Roberts, W. J., Shamoo, Y., Williams, K. R. 1990. *J. Biol. Chem.* 265:10389–94

218. McPheeters, D. S., Stormo, G. D., Gold, L. 1988. *J. Mol. Biol.* 201:517–35

219. Shamoo, Y., Webster, K. R., Williams, K. R., Konigsberg, W. H. 1991. *J. Biol. Chem.* 266:7967–70

220. Berg, J. M. 1986. *Science* 232:485–87

221. Ratner, L., Haseltine, W., Patarca, R., Livak, K. J., Starcich, B., et al. 1985. *Nature* (London) 313:277–83

222. Guyader, M., Emerman, M., Sonigo, P., Clavel, F., Montagnier, L., Alizon, M. 1987. *Nature* (London) 326:662–69

223. Meric, C., Darlix, J. L., Spahr, P. F. 1984. *J. Mol. Biol.* 174:531–38

224. Fleissner, E., Tress, E. 1973. *J. Virol.* 12:1612–15

225. Meric, C., Goff, S. 1989. *J. Virol.* 63:1558–68

226. Gorelick, R. J., Henderson, L. E., Hanser, J. P., Rein, A. 1988. *Proc. Natl. Acad. Sci. USA* 85:8420–24

227. Aldovini, A., Young, R. A. 1990. *J. Virol.* 64:1920–26

228. Gorelick, R. J., Nigida, S. M., Jr., Bess, J. W., Jr., Arthur, L. O., Henderson, L. E., Rein, A. 1990. *J. Virol.* 64:3207–11

229. Mann, R., Mulligan, R. C., Baltimore, D. 1983. *Cell* 33:401–97

230. Lever, A., Gottlinger, H., Haseltine,

W., Sodorski, J. 1989. *J. Virol.*
63:4085–87

231. Sykora, K. W., Moelling, K. 1981. *J. Gen. Virol.* 55:379–91

232. Prats, A. C., Sarik, L., Gabus, C., Litvak, S., Keith, G., Darlix, J. L. 1988. *EMBO J.* 7:1777–83

233. Jentoft, J. E., Smith, L. M., Fu, X., Johnson, M., Leis, J. 1988. *Proc. Natl. Acad. Sci. USA* 85:7094–98

234. Roberts, W. J., Pan, T., Elliott, J. I., Coleman, J. E., Williams, K. R. 1989. *Biochemistry* 28:10043–47

235. South, T. L., Kim, B., Summers, M. F. 1989. *J. Am. Chem. Soc.* 111:395–96

236. South, T. L., Blake, P. R., Sowder, R. C., Arthur, L. O., Henderson, L. E., Summers, M. F. 1990. *Biochemistry* 29:7786–89

237. Fitzgerald, D. W., Coleman, J. E. 1991. *Biochemistry* 30:5196–5201

Annu. Rev. Biochem. 1992. 61:947–75

AMYLOIDOSIS

Jean D. Sipe

Department of Biochemistry, Boston University School of Medicine, Boston, Massachusetts 02118

KEY WORDS: serum amyloid A, amyloid fibrils, amyloid precursor protein, amyloid-P component, transthyretin (prealbumin)

CONTENTS

PERSPECTIVES

Amyloidosis is defined as a group of biochemically diverse conditions in which normally innocuous, soluble proteins polymerize to form insoluble fibrils. The growing mass of *amyloid fibrils* associates with plasma and extracellular matrix proteins and proteoglycans to form *amyloid deposits,*

947

0066-4154/92/0701-0947$02.00

which invade the extracellular spaces of organs destroying normal tissue architecture and function. Amyloidosis accompanies numerous and wide-ranging medical conditions and disorders including cancer, rheumatoid arthritis, Alzheimer's disease, chronic renal dialysis, familial amyloid polyneuropathy, and metabolic diseases such as diabetes (1–9).

The term *amyloid,* meaning starch-like, was introduced by Virchow in 1851 (10); despite demonstration by Friedrich & Kekule within the decade (11) that the major component of amyloid deposits is protein, the name has endured. Although not cellulose (now known to be exclusively a plant carbohydrate) as Virchow had initially thought, carbohydrates of a different biochemical nature, heparan sulfate proteoglycans, probably cause the positive iodine staining reaction exhibited by amyloid deposits of diverse origin (3, 12, 13).

To date, 15 normally nonfibrillar proteins have been identified as amyloid fibrils in clinically diverse conditions (Table 1). In some cases, only a single organ of the body is affected, such as the pancreas in diabetes or the brain in Alzheimer's disease; in other cases amyloid fibrils are deposited in multiple organs of the body. The latter systemic amyloidoses (Figure 1) are derived from circulating plasma protein precursors (1–9, 12, 14). All amyloid fibrils in tissue deposits are resistant to proteolytic digestion (15, 16), insoluble under physiologic conditions (17), and with the exception of neurofibrillary tangles (NFT) in the brain, are morphologically similar (1–4). The NFTs are bundles of paired helical filaments (PHF) in the perinuclear cytoplasm of affected neuronal cell bodies; the PHFs are larger in diameter and arranged in a more pronounced twisted beta-sheet conformation than other amyloid fibrils.

The field of amyloidosis expanded rapidly with the advent of biochemical studies, beginning with the amino acid sequence analysis of immunoglobulin light chain amyloid fibril proteins (18). Since then, amyloid research has uncovered numerous proteins and peptides linked to metabolic processes, including the injury-specific, cytokine-regulated serum amyloid A (SAA) family of apolipoproteins, more than 20 transthyretin (prealbumin) (TTR) variants, the previously unrecognized diabetes-associated islet amyloid polypeptide (IAPP), and the cerebral amyloid beta protein (AB) and prion proteins that cause scrapie and other neurodegenerative diseases (AScr). Now amyloid research encompasses proteolysis and aging; lipid metabolism and injury and infection; thyroxine, transthyretin, and peripheral nerve function; beta protein, amyloid precursor protein (APP), and neurodegenerative disorders; and prions and nonviral, nonbacterial, nonclassical infectious agents. The mononuclear phagocyte system plays a central role in amyloidosis, in that it provides cytokine signals that regulate expression of some amyloid pre-

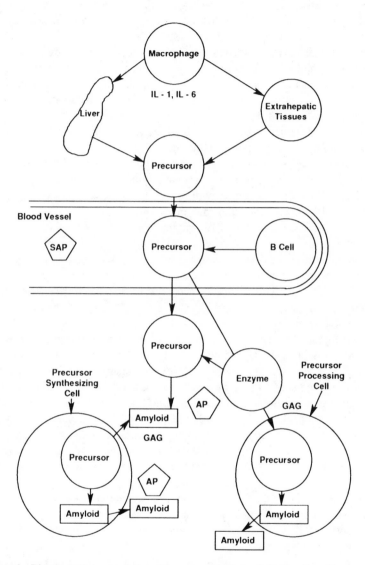

Figure 1   Diagrammatic representation of the many routes to amyloid deposition. The two major pathways are distinguished by local (*bottom left*) vs distal (*upper half*) synthesis of the precursor. Any of the routes may involve either intra- or extracellular formation of amyloid fibrils at the site of amyloid deposition (*bottom half*).

cursors as well as the proteolytic enzymes that convert precursors to fibril proteins. Furthermore, degradation and resorption of fibrils from amyloid deposits can be achieved by phagocytosis.

The cause of amyloid fibril formation and deposition has remained unknown since the time of Virchow, and seems to grow more complex with the identification of each additional fibril protein. In view of the clinical and biochemical diversity of amyloidosis, common pathogenetic mechanism(s) are not evident, except for the nidus theory in which it is conjectured that amyloid fibril fragments, once formed, serve as templates for fibril growth and scaffolding for fibril polymerization.

There are many common features among all of the amyloidoses, but most have at least one exception. For example, amyloidosis is generally a disease of *aging*, yet can (rarely) occur with juvenile rheumatoid arthritis in young children (19). Often amyloidogenic precursors are *structural variants* of precursor proteins, yet normal TTR can be deposited as amyloid fibrils. The *extracellular accumulation* of amyloid fibrils is invariant with the exception of intraneuronal NFT. Amyloidosis is almost always associated with a *poor clinical prognosis,* yet the pathological implications of isolated atrial amyloidosis involving atrial natriuretic factor are unclear. *Defective proteolytic processing* is implicated in many forms of amyloidosis, yet in hemodialysis-associated amyloidosis and in senile cardiac amyloid deposits the fibril proteins are frequently intact precursors. *High local concentrations* of amyloid precursor proteins have been implicated in localized forms of amyloidosis such as pancreatic islet amyloidosis; however, systemic amyloidoses involve circulating precursors ranging in concentration from μg/ml to mg/ml (apolipoproteins, immunoglobulin light chains, and transthyretin) and affect multiple organs of the body with varying degrees of extravascular space. An *amyloidogenic factor,* amyloid enhancing factor (AEF), is associated with several forms of amyloidosis, but thus far AEF can be described only in functional terms since there are conflicting reports as to its biochemical nature.

All amyloid fibrils do, however, have the common feature of the beta-pleated sheet conformation, which is not normally present in nonpathogenic fibril polymers. The amyloid fibril monomers are all small in size, 3–30 kDa, are polyanionic, with a high content of short-chain dicarboxylic amino acid residues and, in tissues, are always associated with two extracellular matrix constituents, amyloid-P component and glycosaminoglycans. Amyloidosis occurs naturally and is widespread throughout the animal kingdom. This review is primarily concerned with the biochemistry of human amyloidosis. Our biochemical knowledge of amyloidosis is derived in large part from human disease, and, with the exception of reactive, systemic amyloidosis, well-developed experimental animal models are not yet available.

AMYLOID DEPOSITS

Properties

Amyloidoses are complex diseases of pathological conformational change and polymerization, in which insoluble fibrils are derived from normally soluble precursors. There have been infrequent reports of more than one biochemical type of fibril present within a single amyloid-laden organ or in different tissues from a single patient (20–22).

Amyloid deposits were identified by the iodine and sulfuric acid technique introduced by Virchow (10) until 1875, when metachromatic staining with aniline dyes such as methyl violet and crystal violet was adopted (23–26). The use of Congo red as a histochemical stain for amyloid was introduced by Bennhold in 1922 (27), and polarization microscopy was added by Divry & Florkin (28). Together the techniques constitute a simple, definitive histologic test for the presence of amyloid that is in widespread use today. Staining with fluorescent dyes such as thioflavin is a sensitive screening technique for amyloid deposits in brain and other tissues, but the presence of amyloid fibrils must be confirmed by another technique, such as Congo red staining (12, 29).

Without polarizing optics, amyloid deposits in tissue sections appear homogenous, eosinophilic, and amorphous when analyzed by light microscopy (12). With polarizing optics, unstained amyloid deposits exhibit birefringence, indicating the presence of highly ordered structures (28). Upon staining with Congo red dye, the birefringence of amyloid deposits is intensely increased in a manner that is not seen with other fibrillar structures such as collagen, elastin, and fibrin. This indicates that the linear dye molecules are arranged in parallel along the amyloid fibril axis (12, 30). Amyloid was described as fibrillar by the electron microscopic studies of Cohen & Calkins (31). Eanes & Glenner (32) determined from the X-ray diffraction patterns of isolated amyloid fibrils (33) that the proteins are arranged in a cross-beta-pleated sheet conformation, with the polypeptide backbones oriented perpendicular to the long axis of the fibril; infrared spectroscopic analysis revealed that the orientation of the polymerized polypeptides is antiparallel (34).

Formation

Amyloidosis is the end result of a dynamic process, in which fibrils are first assembled from filaments, and then together with other constituents accumulate as bundles of fibrils and then as fibrous deposits in tissues. Lateral aggregates of fibrils and filaments have been observed in proximity to cells, oriented perpendicularly to, and merging with, the cell membrane. Amyloid fibrils are observed to be in bundles near cell membranes and to be more dispersed at greater distances from cells (reviewed in 7).

Histological and experimental evidence points to the importance of phagocytic cells in amyloid fibril formation, particularly in inflammation-associated reactive systemic amyloidosis, in which mononuclear phagocytes play a role in the regulation of precursor production, formation of fibrils, and intraphagosomal digestion of fibrils (12, 35–37; Figures 1, 2). Experimental murine reactive amyloidosis is characterized by the appearance of amyloid fibrils within lysosomes in the early stages of amyloid deposition, suggesting that amyloid A (AA) fibrils arise from the intralysosomal cleavage of the precursor SAA (36). In the case of long-standing murine amyloidosis, slow absorption of amyloid fibrils by intraphagosomal digestion results in regression of long-standing amyloid deposits (12, 37–39).

Amyloid enhancing factor (AEF) is a noninflammatory, nonamyloidogenic substance that, when coadministered with a single inflammatory stimulus, synchronizes experimental animals such as mice to deposit AA amyloid within 24–36 hours (40, 41). Despite numerous biochemical studies, AEF has not yet been purified to homogeneity; it is possible that biochemically distinct entities can elicit a specific macrophage response to initiate amyloid fibril formation.

Catabolism of circulating (extracellular) proteins generally takes place by digestion within lysosomes of mononuclear phagocytes; nonlysosomal degradation of intracellular proteins is mediated by ubiquitin (42). Ali-Khan and coworkers report that ubiquitin preparations demonstrate AEF activity (43); ubiquitin is present in neurofibrillary tangles and neuritic plaques of Alzheimer's disease (44). Ubiquitin is covalently attached to lysine residues of target molecules intrinsic to membranes, and has been described as possessing proteolytic activity over a broad pH range (42–44). Zucker-Franklin (42) has proposed that ubiquitin may target aberrant forms of TTR, gelsolin, and other amyloid fibril precursors, resulting in accumulation of crosslinked proteins that would escape degradation.

Amyloidosis, although still virtually untreatable, is, in one case, dramatically preventable. AA amyloidosis in more than 2000 familial Mediterranean fever (FMF) patients has been prevented by chronic prophylactic colchicine treatment (45). In experimental murine amyloidosis, colchicine inhibits induction of AA amyloidosis by blocking AEF formation, but does not inhibit the action of AEF (46, 47). The experimental murine AEF model provides a potential method for screening potential therapeutic intervention in amyloidosis, since AEF activity has been isolated from several types of human amyloid deposits in addition to AA, including brain (48).

The fact that the beta-pleated sheet conformation is a predicted stable conformation for proteins, together with the demonstration that fibrils can be formed in vitro in the absence of amyloid-P component and proteoglycans, suggests that a fibril permissive sequence and appropriate environmental

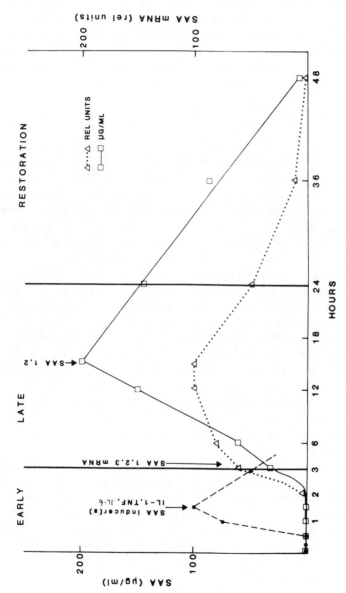

Figure 2 Summary of the acute phase SAA response in mice. Following inflammatory stimulation, SAA-inducing cytokines are produced, followed by expression of SAA genes in liver, synthesis, release of some of the gene products into the blood stream, and finally, deinduction of SAA expression. Reprinted with permission from *Immunophysiology*, ed. J. J. Oppenheim, E. M. Shevach, 1990, New York: Oxford University Press. (TNF is a quantitatively minor SAA-inducing cytokine.)

conditions are the only requirements for fibril formation (1, 12, 49). In vivo, however, amyloid-P component (AP) and heparan sulfate proteoglycans (HSPG) are universal constituents of amyloid deposits. This association may be adventitious, or the presence of AP and HSPG may be essential to establish or maintain appropriate conditions for fibril formation or prevention of immediate resorption of fibrils before formation of bundles of fibrils and accumulation of amyloid deposits (50).

AMYLOID FIBRILS

Properties

Structural studies of amyloid fibril proteins (Table 1) were facilitated by a water extraction fibril isolation method introduced in 1968 by Pras and coworkers (33). Application of the Pras method to various types of amyloid-laden tissues has been reviewed by Skinner & Cohen (51, 52). Briefly, after soluble proteins have been removed from tissue homogenates by extraction with saline, amyloid fibrils are separated from other insoluble proteins by suspension in water followed by differential centrifugation.

Isolation of amyloid fibrils from localized deposits in brain is more difficult (53–58). Glenner & Wong (53) used a modification of the water extraction method to isolate and determine the amino acid sequence of the major fibril protein from cerebrovascular amyloid deposits from a patient with Alzheimer's disease. Isolation of the intraneuronal NFT fraction (consisting of pairs of single 10-nm helical filaments (PHFs) and also some straight, unpaired filaments) and the plaque cores required a different approach. Selkoe & Abraham (54) isolated PHFs and cores after boiling enriched fractions in sodium dodecyl sulfate (SDS) and beta mercaptoethanol, in which the amyloid remains insoluble. The NFTs were further purified by sucrose density gradient centrifugation and the cores by fluorescence-activated cell sorter (FACS) separation.

Fibrils isolated from clinically and biochemically diverse amyloid deposits share a common ultrastructure (7–10 nm wide, rigid, nonbranching fibrils of variable length), except for NFTs, which are bundles of twisted PHFs and are 18 nm in diameter (12, 59, 60). An amyloid fibril consists of two (or more) filamentous subunits of about 3 nm diameter, running in parallel and twisting about each other. Polypeptide subunits in the beta-pleated sheet conformation are polymerized within each filament so that adjacent polypeptide chains are folded with antiparallel orientation, i.e. the carboxyl and amino termini are out of register (49, 61). The polypeptide backbones are directed perpendicularly to an axis of symmetry running parallel to the filaments, the fibril axis, as deduced from their cross-beta X-ray diffraction pattern (49, 62, 63). The beta-pleated sheet conformation is one of the basic secondary

Table 1 Precursors of human amyloid fibril proteins

Distribution	Protein	Size (kDa)	
		Fibril	Precursor
Systemic	Immunoglobulin	5–23	23
Systemic	Lipoproteins		
	Apo-SAA	8	12
	Apo-AI	9–11	26
	Apo-AII (mouse only)	9	9
Systemic	TTR/Prealbumin	5–14	14 (monomer)
Pancreas	Islet amyloid polypeptide (IAPP)	4	9
	Insulin (*Octodon degu* only)	6	6
Thyroid	Calcitonin	6	14.5
Heart	Atrial natriuretic factor	3–4	13
Musculoskeletal	Beta-2-microglobulin	12	12
Brain	Beta protein	4–5	110–135
	Cystatin C	12	13
	Prion	27–30	30–35
Systemic	Gelsolin	7	90–93
Skin	Keratin	?	?

structure configurations for polypeptides predicted by Pauling & Corey (64), and is a structural component of amyloid precursor proteins such as immunoglobulins, TTR/prealbumin, beta-2-microglobulin, and keratin (Table 1). Amyloid deposit formation is probably favored by the stability of the beta-pleated sheet conformation, once amyloid fibril formation is initiated. Although the primary structures of amyloid precursors and fibrils differ widely, the polypeptide backbones fold and polymerize in the same way to result in similar secondary structures (1–9).

Amyloid fibrils have been prepared in vitro from a number of natural polypeptides, including insulin, glucagon, calcitonin, immunoglobulin light chains, and beta 2 microglobulin, and from the synthetic polypeptide poly-L-lysine (12). Synthetic polypeptides can be influenced by heparin to form beta structure rather than alpha-helix, suggesting a causative role for the ubiquitous HSPGs in amyloid deposition (reviewed in 3).

Amyloid fibrils formed in vitro are morphologically indistinguishable from naturally occurring fibrils. Nearly all synthetic and natural amyloid fibrils are antiparallel beta-pleated sheets; however, insulin fibrils formed in vitro are parallel beta-pleated sheets (12, 49). In vitro amyloid formation is influenced by alterations in pH, salt concentration, protein concentration, and proteolysis (49).

The X-ray diffraction patterns of lyophilized in vitro and isolated natural amyloid fibril preparations are similar. Analysis of the X-ray scattering pattern from wet gels of amyloid A (AA) fibrils indicates the presence of alpha-helix/coil structure in addition to beta-pleated conformation (66, 67). It has been suggested that the intrafibril protein is primarily globular and that the X-ray diffraction patterns result from the interfibril alignment of the beta sheets. Studies of both native and reaggregated bovine AA also indicate the presence of alpha-helix/coil in AA fibrils (68). The extent to which nonfibrillar contaminants contribute to these physical measurements with isolated natural fibrils is difficult to determine. In vitro formation of amyloid fibrils from recombinant-generated SAA precursor proteins may enhance structural studies of intra- and interfibrillar protein structure.

Relationship to Other Fibrils

Fibrin deposits, which result from the proteolytic conversion by thrombin of soluble fibrinogen to insoluble fibrin gels, can have an amorphous appearance similar to amyloid deposits in hematoxylin and eosin-stained tissue sections. The fibrin monomer polymerizes side-to-side and end-to-end. The resultant polymers are relatively soluble in dilute acid and 6M urea and susceptible to proteolysis by plasmin until they are crosslinked in the presence of coagulation factor XIIIa to form a tough insoluble clot (69, 70). The stabilization by crosslinking is a feature that distinguishes fibrin, which, although Congophilic, does not exhibit apple green birefringence, from amyloid fibrils (49, 71).

  Collagen fibers occur in bundles of several hundred microns in diameter; they can be separated into fibrils with a periodic banded structure. Congo red staining does not affect the intensity of collagen birefringence, nor does it render collagen birefringence dichroic (72). The presence of glycosaminoglycans (GAGs) may control the rate and size of collagen fibrils in the early stages of collagen formation (73, 74). The effect of GAGs on amyloid fibril polymerization is unknown, although it has been shown that synthetic peptides are influenced by heparin to form beta structures rather than alpha-helix (75).

  Unlike amyloid fibrils and collagen fibers, *elastin fibers* do not appear as arrangements of periodic or ordered subunits when examined by light microscopy. Elastin fibers are Congophilic, but show no birefringence (49, 76–78). Amyloid-P component (AP), which is universally associated with amyloid deposits, is a normal component of basement membrane (79). AP is a major constituent of the microfibrillar network of elastic connective tissue, and is associated with amorphous elastin at the periphery of elastic fibers (80).

Nomenclature

The 1990 guidelines for nomenclature and classification of amyloid and amyloidosis recommended by the International Nomenclature Committee for Amyloidosis state that when possible, amyloids and amyloidoses should be classified by the fibril protein (81). The capital letter A is used as a first letter of designation for all amyloid proteins followed by the protein designation without any open space. The fibril protein must have been characterized by amino acid sequence determination and must be distinguishable from extrafibrillar material such as amyloid-P component (AP). The complete guidelines are published in the proceedings of the VIth International Symposium on Amyloidosis (14).

AMYLOID FIBRIL PRECURSORS

Immunoglobulin Chains

The first amyloid fibril protein to be characterized by amino acid sequence analysis was from the liver of a patient with AL amyloidosis, an uncommon disease in which monoclonal light chains are produced as a result of a plasma cell dyscrasia. After solubilization of fibrils in guanidinium hydrochloride, the major protein isolated by gel filtration was found to be identical to a portion of the variable region of a monoclonal immunoglobulin (Ig) light chain (82). Subsequent studies identified both intact and amino-terminal fragments of monoclonal immunoglobulin light chains as the main protein components of AL amyloid deposits infiltrating heart, kidneys, liver, spleen, and other organs of a number of patients (1).

AL is a disease of aging that is influenced by sex; two thirds of patients are male with a median age of 62 (83). Generally, AL fibrils are deposited systemically; occasionally, localized deposits are observed in lung and skin (1, 12, 84, 85). AL fibrils may originate as independently synthesized light chain fragments rather than as degradation products (86, 87). There are other immunoglobulin deposition diseases related to, but distinct from, AL amyloidosis in which light and heavy chains are deposited to the extent that organ function is compromised, but in which the deposited protein is not Congophilic. Cultured bone marrow cells from all deposition disease patients produce, in addition to intact light chains, fragments that appear to result from truncated synthesis rather than proteolysis.

Given the presence of a monoclonal population of fragmented or truncated immunoglobulin chains, the underlying basis of AL amyloid fibril formation is thought to be the specific amino acid sequence of the variable region, which predisposes the adoption of secondary and tertiary structures necessary for polymerization into fibrils (88). However, even though there are more than 40 partial or complete AL amyloid and Bence-Jones amino acid sequence determinations in the literature (89–91), the distinguishing structural features of AL amyloid fibril formation are difficult to relate to primary structure. The lambda class of light chains has been found in AL deposits more frequently than the kappa class, by a ratio of 2:1. That ratio is the converse of that seen in multiple myeloma without amyloidosis and other monoclonal gammo-pathies (83). The AL lambda/kappa ratio may change as more structural information becomes available (89). Definition of differences between amyloid- and non-amyloid-forming proteins will probably require tertiary structure analysis.

Both fibrillar and nonfibrillar (non-Congophilic) deposits of light chains exhibiting the same antigenic determinants have been described (92); the nonfibrillar deposits also lacked amyloid-P component. This observation suggests that factors other than light chain structure are important in fibril formation, such as local environment and deficiencies in the degradative process (93). The importance of factors other than light chain structure is also supported by early studies in which amyloid fibrils could be formed in vitro by proteolysis of native Bence-Jones proteins from nonamyloid, but not amyloid patients (94–97). More recently, specific proteolytic fragments of reduced and alkylated kappa I Bence-Jones proteins from three amyloidotic patients formed insoluble protein with the tinctorial properties of amyloid (98). The structures of the hydrophobic fibril-forming regions of the three light chains were similar, suggesting that the capacity for AL fibril formation resides in the juxtaposition of specific hydrophobic regions of light chains. The role of carbohydrate in AL fibril formation is unclear. There is a higher rate of glycosylation among many AL proteins (3–5); however, several AL proteins lack carbohydrate (99).

AL research was, until recently, hampered by the lack of animal models. An amyloid fibril protein similar to human lambda II AL fibril proteins has been isolated from cats (100). AL amyloidosis in mice was observed following repeated injection of human Bence-Jones proteins, ~10 mg/gm body weight, from amyloidotic, but not nonamyloidotic patients; detection of human AL fibrils was facilitated by dehydration 24 hours prior to injection (101, 102). Recently a horse lambda light chain amyloid protein was isolated from cutaneous nodule deposits (103). The development of these and other experimental animal models together with more complete structure determination of AL fibril proteins and precursors may facilitate explanation of some cases of plasma cell dyscrasia in which the monoclonal protein product is predisposed to amyloid fibril formation.

High-Density Lipoproteins (HDL)

A novel protein, amyloid A (AA), is associated with the reactive systemic amyloidosis that occurs in a fraction of cases of chronic or recurrent acute inflammatory disease. AA fibrils, which accumulate predominantly in spleen, liver, and kidney, are derived from a circulating precursor, *serum amyloid A (SAA),* by proteolytic removal of one third of the SAA polypeptide from the carboxyl terminus (104–106). SAA was identified in patients with and without amyloidosis as a serum protein having antigenic determinants in common with the fibrillar tissue AA protein (105). The pathogenesis of AA amyloidosis is better understood than that of other forms because of the availability of well-characterized animal models, particularly the mouse.

SAA proteins are injury-specific apolipoproteins, synthesized by many species in response to products of activated macrophages and released into the bloodstream during disturbances of homeostasis known collectively as the *acute phase response* (107–111) (Figure 2). Human plasma contains from two to six SAA isoforms, the expression of which in liver is regulated by products of activated macrophages, predominant among which are interleukin-1 (IL-1) and interleukin-6 (IL-6) (112–115). Apo-SAA is carried by plasma high-density lipoproteins (HDL) at concentrations of 1 μg/ml or less during homeostasis (107, 115, 116); SAA expression may increase by as much as 1000-fold, however, following tissue injury and cell necrosis. Therefore, at points during the acute phase response, apo-SAA can constitute as much as 80% of the total HDL protein, presumably by displacement of apo-AI and apo-AII (117–120).

The magnitude of apo-SAA synthesis appears to be controlled by cytokines acting synergistically (112, 114). IL-1 and IL-6 together stimulate SAA synthesis by human hepatoma cells to a magnitude sufficient to account for SAA concentrations measured during the acute phase response. The IL-1 receptor antagonist protein abolishes >90% IL-1- and IL-6-stimulated SAA production (114). Both NFκB and IL-6 response elements have been identified in the SAA promoter region; nuclear transcription factors induced by IL-1 and IL-6 have different time courses (121). The mechanism of synergistic stimulation of SAA production appears to involve enhanced mRNA transcription and subsequent SAA mRNA stabilization (H. Rokita, J. Sipe, unpublished observations).

Apo-SAA, although not well understood at the functional level, is thought to modulate HDL lipid metabolism. HDL3 particles from plasma of individuals undergoing an acute phase response are larger and have reduced electrophoretic mobility compared with HDL3 particles isolated from individuals during homeostasis (120). HDL charge alteration may occur as a result of association of apo-SAA with HDL particles. A disruption in lipoprotein metabolism occurs during the acute phase response in Syrian hamsters, with a decrease in HDL-cholesterol and retarded electrophoretic mobility of HDL that coincide with increased content of SAA in HDL (122).

The first 11 residues from the amino terminus of apo-SAA are all strongly hydrophobic, and this region has been proposed as a lipid-binding domain (66, 67). Protein and casein kinase phosphorylation sites are present in the carboxyl portion of the apo-SAA proteins (123), and apo-SAA has been phosphorylated by phosphokinase C (124). Exogenous apo-SAA has been shown to modulate in vitro assays of immune and coagulation system function (125, 126).

Structural studies of human apo-SAA genes and proteins have identified six major isoforms of apo-SAA in plasma as the products of three genes (113, 118, 127–131); recently additional allelic and nonallelic variants have been reported (132–134). There is evidence for the existence of molecules related to, but distinct from, "classical" apo-SAA (135). Rabbit synovial fibroblasts produce an SAA-like protein with capacity for autocrine modulation of collagenase production (136). A similar human protein GSAA1 has been described at the gene level (137).

The interlaboratory and interspecies SAA nomenclature is sometimes confusing, because the total number of members of the SAA gene family has not yet been determined. Apo-SAA isoforms can be identified unequivocally by assigning the major isoforms separated by electrofocusing to published gene sequences (130). Direct and derived amino acid sequence data are available for AA and apo-SAA families of numerous species in addition to human, including cat, cow, dog, duck, hamster, horse, mink, mouse, rabbit, rat, and sheep (123). SAA genes are 3–4 kb in size and contain four exons; nearly all genes share a constant region in exon 3 and most of the structural differences reside in exon 4. Two separate branches in the evolution of SAA proteins have been described. Several species, including cat, cow, dog, horse, and mink, have substitutions at positions 72–81 that are not present in the presently characterized human and mouse genes. Rat SAA gene(s) lack exon 3 and a corresponding protein has not been isolated with HDL, although a 1000-fold increase in SAA mRNA is observed during inflammation (138–141). Recent studies show Armenian hamsters do not express exon 3 and do not contain an acute phase–inducible HDL apolipoprotein (J. Sipe, F. deBeer, unpublished observations).

Despite two decades of clinical research and study with animal models, the mechanism by which soluble apo-SAA isoforms are converted and deposited as insoluble AA fibrils in the extracellular spaces of tissues remains undefined. Selective deposition of single murine and Syrian hamster isoforms has been observed, although apparently multiple isoforms are deposited in human AA amyloidosis (142–144). AA fibril formation is an aberrant outcome of the acute phase response, a series of normal events in host defense. The sequential host responses to tissue injury and cell necrosis (Figure 2) include (*a*) macrophage activation and cytokine production, (*b*) transient SAA expression in liver, (*c*) apo-SAA association with HDL, and (*d*) apo-SAA clearance in peripheral microvasculature. SAA production and structure are critical factors in AA amyloidogenesis, but impaired clearance is probably the proximal step to fibril formation.

Apolipoprotein AI (apo-AI), a nonglycosylated 28-kDa protein of 243 amino acids, accounts for nearly 65% of the protein in HDL during homeostasis (145). An autosomal dominant form of hereditary amyloidosis results from deposition of an apo-AI variant in which arginine is substituted for glycine at position 26 (146, 147). Gene carriers exhibit low levels of HDL and exhibit peripheral neuropathy that is clinically similar to that observed in familial amyloidosis caused by variants of TTR. The residence time of the Arg26 apo-AI in plasma is approximately one-half that of the normal protein; however the variant, while removed rapidly from plasma, is not excreted in urine at the expected rate, suggesting that the mutant protein spends more time in the extravascular space where amyloid deposition occurs (148). Jones and coworkers have studied a family carrying the Arg26 variant of apo-AI that presents clinically with renal failure, but no neurological manifestations (147). This is strong evidence for factors other than structure in the anatomic localization of amyloid deposits. Schonfeld (149) reviewed the clinical manifestations of 8–13 apo-AI variants; only 3 exhibited lowered HDL levels, and only the Arg26 variant is associated with amyloidosis.

Another protein associated with HDL, *apolipoprotein AII (apo-AII)*, is deposited intact as fibrils in a number of mouse strains. SAM-P, LLC, SJL/J, and A/J mice develop amyloidosis of the non-AA type spontaneously with aging (150–154). A corresponding human amyloidosis involving apo-AII has not yet been described. Although apo-AII copurifies with apo-SAA on HDL, aged amyloidotic mice with apo-AII deposits do not exhibit apo-SAA in amyloid deposits, although SAA in hepatocytes is evident by immunohistochemistry (153). The affected mouse strains have a mutation in the apo-AII gene, substituting glutamine for proline in position 5. The presence of proline in an amyloidogenic region is thought to prevent amyloidosis by interfering with folding of the polypeptide backbone into the beta-pleated sheet conformation.

The formation of amyloid fibrils from three different HDL apolipoproteins suggests that the local environment in the microvasculature, where HDL is catabolized, may selectively predispose to amyloid fibril formation from apolipoproteins.

Transthyretin (Prealbumin), TTR

TTR is a 55-kDa nonglycosylated plasma protein consisting of four identical 127-amino-acid subunits. TTR was originally named prealbumin because its electrophoretic migration is more rapid than albumin, and was later renamed transthyretin because of its function as a carrier protein for the thyroid hormones (155–157). TTR also serves as a carrier for the retinol (vitamin A alcohol):retinol-binding protein (RBP) complex (158). TTR is encoded by a single locus of ~9 kb; the gene consists of four introns, and multiple nuclear factors interact within the enhancer and promoter regions of TTR, including

HNF 1,3,4, C/EBP, and AP1 (159–161). Both the primary and X-ray crystallographic structures of TTR were determined (162, 163) before its association with amyloidosis was recognized (164).

Costa and coworkers determined that amyloid fibril proteins in kidney, thyroid, and other tissues of patients with familial amyloid polyneuropathy (FAP) were fragments of a protein identical to plasma TTR, but with methionine substituted for valine at position 30 (164). FAP is an autosomal dominant condition, with delayed onset of disease symptoms of from three to seven decades, although the mutant protein is present from birth (157).

To date, 25 TTR variants have been identified by protein and DNA screening studies (165–170; M. Saraiva, personal communication). These include substitutions at positions 6, 30 (2), 33 (2), 36, 42, 45, 49, 50, 58, 60, 64, 68, 70, 77, 84 (2), 90, 102, 109, 111, 114, 119, and 122 of the TTR subunit; at positions 30, 33, and 84, two different substitutions have been found. Variant TTR gene carriers exhibit clinically heterogenous amyloidoses; some TTR variants are not associated with amyloidosis, e.g. Arg102 and Met119, and some affect thyroxine function, e.g. Ser6 and Thr109 (167, 168). Several homozygous Met30 patients, as well as compound heterozygotes with two mutant alleles, Met30 and His90, Met30 and Asn90, or Met119 and Asn90, have been described. Amyloidosis is not always associated with variant alleles in these carriers (157, 165, 170). Fibrils are made up of both full-length TTR and fragments of TTR monomers and of variable quantities of both mutant and normal TTRs (171).

X-ray diffraction studies of Met30 FAP fibrils (172) have shown the TTR tetramer interactions within the fibril to be similar to those of normal TTR within the crystal (163). The TTR molecules in the fiber are stacked so that the thyroxine-binding channel is parallel to the fibril axis. The results suggest that the strength of hydrophobic interactions between TTR tetramers is held in delicate balance. Hamilton and coworkers crystallized TTR from a patient homozygous for the Met30 variation (173). Their X-ray diffraction data demonstrated increases in volume of unit cell and structural differences that were described as not large, but extensive. It has been suggested that the Met30 mutation results in a less avid binding site for thyroxine.

Migita and coworkers have suggested that mutation-specific conformational changes in TTR could permit formation of intersubunit disulfide bonds between the single cysteine at residue 10 in TTR monomers, thus facilitating polymerization into fibrils (174). The presence of a disulfide bond in vitreous Met30 TTR amyloid was demonstrated by SDS-PAGE under reducing and nonreducing conditions (167). Synthetic peptide fragments of TTR have been converted to morphologically similar, but not identical, fibrils, as have normal TTR under acidic conditions (175).

The pathogenesis of ATTR is distinct from AL and AA amyloidoses in that the ATTR fibril precursor is a constitutive protein present from birth, whereas AL requires the production of an amyloid-susceptible monoclonal immunoglobulin light chain and AA requires inflammatory stimulation of the host to elicit the SAA precursor. ATTR is a negative acute phase protein, but there is no evidence for a role of proinflammatory cytokines in ATTR amyloidogenesis. ATTR is similar to Apo-AI in that the same genetic defect can give rise to different phenotypic expressions, suggesting that posttranslational events affect the amyloidogenicity of the protein (166). Amyloidogenic factors are thought to include both TTR conformation, altered either by an amino acid substitution or posttranslational modification of normal or mutant TTR, and proteolysis. TTR fragments, in which positions of cleavage are constant, are found in fibrils. Changes in proteolysis with aging could explain how normal TTR is deposited in heart in senile systemic amyloidosis if proteolytic fragments of normal TTR susceptible to amyloid fibril formation are generated.

Endocrine Amyloid

The concept that amyloid in endocrine tissues is hormonal in origin was introduced by Pearse, who coined the term APUD (Amine Precursor Uptake and Decarboxylation) amyloid because of the common origin of the precursor cells from neuroectoderm (176). Peptide hormones readily form amyloid fibrils in vitro (12, 177), and may be disposed to localized amyloid fibril formation and deposition in vivo by several factors (178). Peptide hormones are small in size and have conformationally flexible structures. Polypeptide hormones such as atrial natriuretic factor (ANF) and insulin and glucagon have fibril-permissive sequences and a high degree of conformational flexibility, and can readily assume the beta-pleated sheet conformation when bound to lipid or under acidic conditions (74, 179, 180). High protein concentration favors amyloid fibril formation. Polypeptide hormones are highly concentrated in secretory granules and extracellularly at the site of their release, and high concentrations may result from overproduction or defects in processing.

Amyloid deposits that occur in the islets of Langerhans in >90% of individuals with type II diabetes (181) and as insulinoma endocrine tumors were initially thought to be related to insulin. However, in 1986, the amyloid fibril protein in the endocrine pancreas was determined to be a novel polypeptide related to calcitonin (182), now identified as IAPP (islet amyloid polypeptide) (178). The human IAPP gene product is an 89-amino-acid residue prepropeptide consisting of a signal peptide and two short propeptides on the amino and carboxyl portions of mature IAPP (183–185). IAPP is expressed by beta cells in the pancreas, stored and released together with insulin, and is also present in plasma (186–188).

Amyloidogenic regions of human IAPP have been identified from in vitro studies with synthetic peptides; the amyloidogenic region has been localized to residues 20–30 of the 37-amino-acid IAPP molecule (184, 189). The central portion is the least conserved region of IAPP, and interspecies variation at position 20–29 may explain why islet amyloid only occurs in some species, such as humans, cats, and monkeys (180, 190). The tetrapeptide corresponding to residues 25–28, -Ala-Ile-Leu-Ser-, is most important to amyloid fibril formation (191).

Medullary carcinoma of the thyroid (MCT) is frequently accompanied by amyloid deposition. A fragment of the procalcitonin was identified as the major amyloid fibril protein in 1976 (192). MCT amyloid deposits contain both procalcitonin and mature calcitonin (192, 193). The fact that both procalcitonin and calcitonin can polymerize as amyloid fibrils in the thyroid suggests a processing defect is involved in this form of amyloidosis (192–194).

Insulin is readily converted to amyloid fibrils in vitro (12). Insulin-derived amyloid in diabetes patients at the insulin injection site was, in one case, purified, analyzed, and found to consist of uncleaved porcine insulin (195, 196). Recent biochemical studies established that islet amyloid in the *Octodon degu* (South American rodent) is unmodified insulin A and B chains. *Octogon degu* insulin differs in structure from the insulin of most other species (197).

In addition to TTR-derived senile cardiac amyloid, isolated atrial amyloid (AANF) is very common in aged individuals and affects more than 50% of persons over 70 years of age (198). Atrial myocytes have an endocrine function and cleave the carboxyl 26-amino-acid part from the 126-amino-acid precursor pro-atrial natriuretic factor (ANF) (199). The resultant ANF can form amyloid fibrils in close association with the sarcolemma, apparently without extensive proteolytic cleavage (177, 200, 201).

Beta-2 Microglobulin

Beta-2-microglobulin amyloid (AB2M) deposition in the musculoskeletal system is a serious and frequent complication for patients with renal failure who have received long-term hemodialysis using cuprophane membranes (22, 206). B2M is a small protein (11.8-kDa, consisting of 99 amino acids) that contains a significant amount of beta-pleated sheet structure and readily crystallizes or precipitates under conditions of high protein or low salt concentration (202–205). The local concentration of B2M is thought to be a key factor in B2M amyloidogenesis, since circulating concentrations in groups of patients with and without amyloidosis do not differ (206). Amyloid fibrils were formed in vitro without proteolysis, by concentration of the protein in low salt (207). The role of proteolytic alteration in B2M amyloidogenesis requires clarification, since B2M fragments are isolated from in vivo AB2M deposits (208).

AB2M may provide insights into anatomic factors specifying site of amyloid deposition. Calcium plays a universal role in amyloidogenesis by bridging AP binding to amyloid fibrils; calcium may have a more specific role in B2M amyloidogenesis because B2M fibrils are closely associated with deposits of calcium compounds in kidneys, heart, and synovium (22). Many systemic amyloid fibrils accumulate in abdominal fat tissue; aspiration of subcutaneous fat followed by Congo red staining is a readily available screening biopsy for amyloidosis (209). AB2M is not detectable by abdominal fat biopsy in B2M amyloid patients, suggesting that fat tissue does not have the necessary extracellular matrix to serve as scaffolding for B2M amyloid fibril formation (210).

Brain Amyloid (APP, Cystatin C, Prions)

Alzheimer's disease (AD) and other dementias are characterized by amyloid deposition in the brain; the amyloid fibrils are considered to be a possible causative lesion. A novel protein, beta amyloid protein (AB), has been identified in three anatomically distinct amyloid deposits, the cerebrovascular walls, the cores of neuritic plaques, and intracellular neurofibrillary tangles (NFT) composed of paired helical filaments (PHF). Amyloid-laden vessels are found in >92% of cases of Alzheimer's disease (211, 212). Glenner & Wong (55) isolated a 28-amino-acid residue protein, which they called beta protein, from amyloid-laden leptomeningeal tissues of a patient who died with Alzheimer's disease. Subsequently, the subunit protein from senile plaque core amyloid was found to be a larger form of AB (4-kDa, 39–42 amino acids extended at the carboxyl terminus) (213, 214). The entire AB consists of 43 amino acids (211).

AB is derived from a large transmembrane glycoprotein called amyloid precursor protein (APP). Four mRNA species are generated from the APP gene by alternative splicing, APP695, 714, 751, and 770, all of which encode AB; the two larger transcripts contain sequences homologous to a Kunitz-type proteinase inhibitor (protease nexin II) (215). APP695 is a transmembrane glycoprotein in which the AB sequence spans residues 597–638, with the carboxyl-terminal half of AB buried within the membrane. Normal processing of APP results in cleavage at Gln15 of the AB protein, such that the first 15 amino acids of ABP form the carboxyl terminus of a large soluble protein and the amino terminus of the ABP is anchored in the cell membrane. A heterogenous group of 110–135-kDa membrane-associated APP-related glycoproteins is found in brain, non-neural tissues, and cultured cells (216).

APP processing is thought to be defective in AD; as a result, the AB portion of APP is not clipped in the middle, but on either side, thereby generating fibril-forming fragments (217). Abraham and coworkers (218) have isolated a protease that may play a role in processing APP to form AB; the cathepsin G–like enzyme hydrolyzes bonds between methionine and aspartic acid, which is the amino-terminal residue of ABP. The protease is inhibited by protease nexin II and alpha-1-antichymotrypsin. Recently, an AB clipping enzyme was identified in brain and liver using synthetic substrates and a synthetic fragment of APP that begins at Met596 of APP695 and contains the first 26 residues of AB (219). The enzyme is biochemically similar to the lysosomal enzyme cathepsin B. Further work will be required to determine the roles of cathepsins B and G and lysosomes and lysosomal enzymes in processing of the four APP proteins and in ABP fibril formation.

Abraham and coworkers (220) found the serine protease inhibitor alpha 1-antichymotrypsin to be tightly associated specifically with the ABP fibril. This suggests that an imbalance in proteases and inhibitors may cause the defective proteolytic processing underlying ABP formation. Others have extended the suggestion that alpha 1-antichymotrypsin and APP expression in the brain is part of a cerebral acute phase response (220) to propose that ABP amyloidogenesis is due to an IL-1- and IL-6-mediated acute phase response in the brain (221). Microglial cells and astrocytes secrete IL-1 and IL-6 respectively (222, 223). IL-1 has been shown to increase expression of APP mRNA in cultured human endothelial cells (224).

Intense investigation of the origin of AB and other proteins essential for amyloid fibril formation in AD—to determine if the proteins are produced locally or at a distance from the site of processing and/or formation—is under way (58, 212, 225). There is evidence for a vascular origin in which BP formation occurs via an alternative pathway for proteolytic processing.

Insights into AB fibrillogenesis may be gained from other cerebral disorders with clinical manifestations different from AD and Down's syndrome, but in which AB fibrils are also deposited. Two such variant conditions are sporadic Congophilic angiopathy, in which amyloid deposition occurs in the walls of leptomeningeal vessels with very few plaques and tangles, and Dutch hereditary cerebral hemorrhage with amyloidosis (HCHWA). In HCHWA, ABP is deposited as amyloid fibrils primarily in the walls of cerebral blood vessels, leading to fatal strokes at a younger age than symptoms of AD would normally appear. In HCHWA patients, AB has a Gln substitution at residue 22 and there is rupture of amyloid-laden vessels, atypical plaques, and no tangles (226). HCHWA may represent an early stage of AD in which there is predominantly vascular deposition.

Human AB deposition has been observed in the brains of transgenic mice; in one case, a chimeric gene was constructed with human APP751 cDNA under control of a rat neural-specific promoter (227); in the other, with the AB cDNA under control of the human APP promoter (228). In both cases AB deposition in the CNS was observed immunohistochemically, and studies are in progress to determine whether amyloid deposition occurs with aging and if the transgenic mice will display evidence of neuronal degeneration and other features of AD.

Hereditary cystatin C amyloid angiopathy (HCCAA) is characterized by deposition of cystatin C amyloid fibrils derived from the protease inhibitor cystatin C in the CNS and in other tissues as well. HCCAA is associated with a point mutation in the codon for leucine at position 68 in exon 2 of the cystatin C gene, resulting in a substitution of glutamine for leucine (229). There are several large families in Iceland afflicted with this disorder in which affected individuals die before age 30 (230). Cystatin C amyloid fibrils differing from the Icelandic variant have been implicated as a causative factor in cerebral hemorrhage afflicting elderly persons in Japan (231).

The pathogenetic mechanism of HCCAA is under investigation using cystatin C expression in monocytes from patients and asymptomatic carriers. Cystatin C secretion from patients' monocytes is reduced, suggesting that amyloid fibril formation occurs because of a generalized secretory defect (232).

The prion protein (PrP), 27–30 kDa, is associated with a group of transmissable neurodegenerative diseases, the spongiform encephalopathies, including scrapie, kuru, Creutzfeldt-Jakob disease, Gerstmann-Straussler-Scheinker syndrome, and bovine spongiform encephalopathy (mad cow disease) (233). The spongiform encephalopathies are characterized by the appearance of prion-derived amyloid plaques (designated AScr for the scrapie form of the disease), neuronal degeneration and vacuolation, and astroglial proliferation, and appear to have both an infectious and a genetic origin. The PrP protein, which is unrelated to AB, is thought to be both an infectious agent and an amyloid fibril protein. The prion gene product is encoded in a single exon of a single-copy gene; the 27–30-kDa PrP neuron surface protein is derived by proteolytic processing from a 30–35-kDa precursor cell membrane protein (reviewed in 233).

Prions are infectious pathogens that differ from viroids and viruses both with respect to their structure and with respect to the diseases that they cause (233). The infectious nature and mechanism of replication of prion particles are controversial and not yet understood. Some models of scrapie infectivity involve nucleic acid encapsulated by either a virus or a host-encoded protein. Prusiner and coworkers suggest that mutant PrPs do not actually replicate themselves, but induce normal host prion protein to adopt abnormal conformations (233). However, it is difficult to explain either the existence of multiple strains of scrapie that can be propagated in a single strain of inbred mice or the mutability of PrP without an obligate nucleic acid genome (234, 235). Weissman has recently proposed a unified theory of prion propagation in which PrP can initiate prion disease, but the phenotypic properties defining strain differences are due to a nucleic acid usually associated with PrP, but that can be recruited from the host (234).

The protein forming the AScr plaques in Gerstmann-Straussler-Scheinker syndrome has been identified as a PrP variant with leucine substituted for proline at residue 102. When the codon 102 point mutation was introduced into the mouse PrP gene in transgenic mice, CNS degeneration occurred; the transgenic mice exhibited a scrapie-like spongiform encephalopathy with vacuoles in brain tissue (236).

Amyloid plaques in prion diseases contain PrP as evidenced by immunohistochemistry and amino acid sequence studies of purified proteins (235–239). Tissue fractions rich with the scrapie infectious agent were found to be enriched for rod-shaped particles resembling amyloid (233, 240). It is questionable whether prion rods are actually amyloid since the in vitro production of amyloid has been considered an artifact of proteolysis used in the isolation procedures (241); however, deposition of prion amyloid is controlled by the PrP sequence (242).

Gelsolin

Gelsolin is a calcium-dependent regulatory protein of actin gel-sol transformation that severs actin molecules and, under appropriate ionic conditions or calcium ion concentration, can laterally associate with other cytoskeletal proteins (243–245). Gelsolin modulates actin by binding actin monomers, nucleating actin filament growth, and severing actin filaments (246, 247). Plasma and cytoplasmic gelsolin are derived from a single gene by alternative transcriptional initiation sites and mRNA processing (246, 247). The potential of gelsolin to form amyloid fibrils was only recently recognized when amyloid fibrils of gelsolin fragments were found in the blood vessels and basement membranes of many tissues throughout the body of patients with Finnish hereditary systemic amyloidosis (248). The major fibril protein is a 71-amino-acid fragment of gelsolin containing the actin-binding domain, 173–243, of secreted plasma gelsolin. There is a single amino acid substitution at residue 187, asparagine for aspartic acid.

Cutaneous Amyloid (Keratin)

Of the many clinical manifestations of amyloidosis, numerous amyloidotic skin lesions can occur, including AL, AA, and ATTR. In primary localized cutaneous amyloidosis, there is histochemical evidence that amyloid originates in the keratogenous epithelial cells (249). Biochemical information is limited, although keratin is a fibrous protein, which in the beta form has a parallel beta-pleated sheet structure (250, 251).

CONCLUSION

The number of proteins implicated in amyloidosis has increased from one to fifteen in the past 20 years; this increase may be expected to continue, particularly in the area of polypeptide endocrine hormones and lipoproteins. The mechanism of amyloid fibril formation remains elusive, and many routes to the same endpoint appear to exist. In view of the diversity of precursor proteins and anatomic sites of deposition, the common factor in all of the amyloidoses seems to be protein-protein interactions stabilized both by the structure of the protein and the microenvironment. The role of regulated proteolysis in amyloidogenesis is only beginning to be addressed. The recent development of transgenic models may be expected to facilitate identification of proteolytic enzymes and further define the role of macrophages in the complex sequence of events leading to amyloid deposition in the central and peripheral nervous systems. Analysis of amyloid precursor processing enzymes in vitro with their substrates and other components of amyloid deposits may provide insights into the sequence of events leading to amyloidosis—from conformational changes, to polymerization into fibrils, to assembly of fibrils into bundles, and finally to amyloid deposition.

ACKNOWLEDGMENTS

I thank Drs. Maria Saraiva and Edgar Cathcart for allowing me to use information prior to publication; Drs. Hanna Rokita and Wayne Gonnerman for critical reading of the text; Ms. Lien Tran and Mr. Peter Sipe for help with literature searching and preparation of the manuscript; and the National Institute on Aging (AG 9006) for support.

Literature Cited

1. Glenner, G. G. 1980. *N. Engl. J. Med.* 302:1283–92, 1333–43
2. Husby, G., Sletten, K. 1986. *Scand. J. Immunol.* 23:253–65
3. Kisilevsky, R. 1987. *Can. J. Physiol. Pharmacol.* 65:1805–15
4. Cohen, A. S., Connors, L. H. 1987. *J. Pathol.* 151:1–10
5. Benson, M. D., Wallace, M. R. 1988. In *The Metabolic Basis of Inherited Disease,* ed. C. R. Scriver, A. L. Beaudet, W. S. Sly, D. Valle, pp. 2439–60. New York: McGraw-Hill. 6th ed.
6. Castano, E. M., Frangione, B. 1988. *Lab. Invest.* 58:122–32
7. Cathcart, E. S. 1992. In *Textbook of Rheumatology,* ed. W. N. Kelly, E. D. Harris, S. Ruddy, C. B. Sledge. Philadelphia: Saunders. In press

8. Stone, M. J. 1990. *Blood* 75:531–45
9. Benditt, E. P. 1986. In *Amyloidosis,* ed. J. Marrink, M. H. van Rijswijk, pp. 101–6. Dordrecht:Nijhoff
10. Virchow, R. 1851. *Verh. Phys. Med. Ges. Wurzburg* 2:51–54
11. Friedrich, N., Kekule, A. 1859. *Arch. Pathol. Anat. Physiol. Klin. Med.* 16: 50–65
12. Glenner, G. G., Page, D. L. 1976. *Int. Rev. Exp. Pathol.* 15:1–92
13. Snow, A. D., Willmer, J., Kisilevsky, R. 1987. *Lab. Invest.* 18:506–8
14. Natvig, J. B., Forre, O., Husby, G., Husebekk, A., Skogen, B., et al. 1991. *Amyloid and Amyloidosis, 1990,* pp. 1–922. Kluwer:Dordrecht
15. Sorenson, G. D., Shimamura, T. 1964. *Lab. Invest.* 13:1409–17

16. Husebekk, A., Skogen, B. 1991. See Ref. 14, pp. 107–10
17. Glenner, G. G., Keiser, H. R., Bladen, H. A., Cuatrecasas, P., Eanes, E. D., et al. 1968. *J. Histochem. Cytochem.* 16: 633–44
18. Glenner, G. G., Harada, M., Isersky, C., Cuatrecasas, P., Page, D., et al. 1970. *Biochem. Biophys. Res. Commun.* 41:1013–19
19. Trainin, E. B., Spitzer, A., Greifer, I. 1978. *NY State J. Med.* 78:72–77
20. Westermark, P., Natvig, J. B., Anders, R. F., Sletten, K., Husby, G. 1976. *Scand. J. Immunol.* 5:31–36
21. Lathi, D., Cathcart, E. S., Sipe, J. D. 1991. See Ref. 14, pp. 745–48
22. Gejyo, F., Maruyama, S., Maruyama, H., Homma, N., Aoyagi, R., et al. 1991. See Ref. 14, pp. 377–80
23. Cohen, A. S. 1986. See Ref. 9, pp.1–19
24. Cornil, A. V. 1875. *C. R. Acad. Sci (Paris)* 80:1288–91
25. Jurgens, R. 1875. *Virchows Arch. Pathol. Anat. Physiol.* 65:189–96
26. Hesell, R. 1875. *Wien Med. Wochenschr.* 25:715–16
27. Bennhold, H. 1922. *Muench. Med. Wochenschr.* 69:1537–38
28. Divry, P., Florkin, M. 1927. *C. R. Soc. Biol.* 97:1808–10
29. Schwartz, P. 1967. In *Amyloidosis. Proc. Symp. Amylodoisis,* ed. E. Mandama, L. Ruinen, J. H. Schollen, A. S. Cohen, pp. 400–17. Amsterdam: Excepta Med.
30. Romhanyi, G. 1949. *Schweiz. Z. Pathol. Bakteriol.* 12:253–62
31. Cohen, A. S., Calkins, E. 1959. *Nature* 183:1202–3
32. Eanes, E. D., Glenner, G. G. 1968. *J. Histochem. Cytochem.* 16:673–77
33. Pras, M., Schubert, M., Zucker-Franklin, D., Rimon, A., Franklin, E. C. 1968. *J. Clin. Invest.* 47:924–33
34. Termine, J. D., Eanes, E. D., Ein, D., Glenner, G. G. 1972. *Biopolymers* 11: 1103–13
35. Teilum, G. 1952. *Ann. Rheum. Dis.* 11:119–36
36. Shirahama, T. 1975. *Am. J. Pathol.* 81:101–16
37. Shirahama, T., Cohen, A. S. 1971. *Am. J. Pathol.* 63:463–85
38. Richter, G. W. 1954. *Am. J. Pathol.* 30:239–51
39. Williams, G. 1967. *J. Pathol. Bacteriol.* 94:331–36
40. Kisilevsky, R. 1983. *Lab. Invest.* 9: 381–90
41. Deal, C. L., Sipe, J. D., Tatsuta, E.,

Skinner, M., Cohen, A. S. 1981. *Ann. NY Acad. Sci.* 389:439–41
42. Zucker-Franklin, D. 1991. See Ref. 14, pp. 917–20
43. Chronopoulos, S., Alizadeh-Khiavi, K., Normand, J., Ali-Khan, Z. 1991. *J. Pathol.* 163:199–203
44. Perry, G., Friedman, R., Shaw, G., Chau, V. 1987. *Proc. Natl. Acad. Sci. USA* 84:3033–36
45. Zemer, D., Sohar, E., Pras, M. 1991. See Ref. 14, pp. 859–62
46. Brandwein, S. R., Sipe, J. D., Tatsuta, E., Skinner, M., Cohen, A. S. 1984. *J. Rheumatol.* 11:597–601
47. Brandwein, S. R., Sipe, J. D., Skinner, M., Cohen, A. S. 1985. *Lab. Invest.* 52:319–25
48. Varga, J., Flinn, M. S. M., Shirahama, T., Rodgers, O. G., Cohen, A. S. 1986. *Virchows Arch.(B)* 51:177–85
49. Glenner, G. G., Eanes, E. D., Bladen, H. A., Linke, R. P., Termine, J. D. 1974. *J. Histochem. Cytochem.* 22:1141–58
50. Kisilevsky, R. 1989. *Neurobiol. Aging* 10:499–500
51. Skinner, M., Shirahama, T., Cohen, A. S. 1986. See Ref. 9, pp. 91–98
52. Skinner, M., Shirahama, T., Cohen, A. S., Deal, C. L. 1983. *Prep. Biochem.* 12:461–76
53. Glenner, G. G., Wong, C. W. 1984. *Biochem. Biophys. Res. Commun.* 120: 855–90
54. Selkoe, D., Abraham, C. 1986. *Methods Enzymol.* 134:388–404
55. Wong, C. W., Quaranta, V., Glenner, G. G. 1985. *Proc. Natl. Acad. Sci. USA* 82:8729–32
56. Prelli,F., Castano, E., Glenner, G. G., Frangione, B. 1988. *J. Neurochem.* 51: 648–51
57. Kang, J., Lemaire, H. G., Unterbeck, A., Grzeschik, K. H., Multhaup, G., et al. 1987. *Nature* 235:733–36
58. Selkoe, D. 1991. See Ref. 14, pp. 713–17
59. Kidd, M. 1963. *Nature* 197:192–93
60. Gorevic, P. D., Goni, F., Pons-Estel, B., Alvarez, F., Peress, N. S., et al. 1986. *J. Neuropathol. Exp. Neurol.* 45:647–64
61. Glenner, G. G., Eanes, E. D., Page, D. L. 1972. *J. Histochem. Cytochem.* 20: 821–26
62. Bonar, L., Cohen, A. S., Skinner, M. 1969. *Proc. Soc. Exp. Biol. Med.* 131:1371–73
63. Keizman, I. K., Ravid, M., Sohar, E. 1976. *Isr. J. Med. Sci.* 12:1137–40
64. Pauling, L., Corey, R. B. 1953. *Proc. Natl. Acad. Sci. USA* 39:253–56

65. Snow, A. D., Wight, T. N. 1989. *Neurobiol. Aging* 10:481–97
66. Turnell, W. G., Pepys, M. B. 1986. See Ref. 9, pp. 127–33
67. Turnell, W. G., Sarra, R., Glover, I. D., Baum, J. O., Caspi, D., et al. 1986. *Mol. Biol. Med.* 3:387–407
68. Van Andel, A. C. J., Niewold, T. A., Lutz, B. T. G., Messing, M. W. J., Gruys, E. 1986. See Ref. 9, pp. 169–76
69. Doolittle, R. F. 1975. In *Plasma Proteins,* ed. F. W. Putnam, 2:110–62. New York: Academic
70. Davie, E. W., Hanahan, D. J. 1975. See Ref. 69, 3:422–44
71. Elghetany, M. T., Saleem, A. 1988. *Stain Technol.* 63:201–12
72. Wolman, M. 1971. *Lab. Invest.* 25:104–10
73. Mathews, M. B., Decker, L. 1968. *Biochem. J.* 109:517–26
74. Obrink, B. 1973. *Eur. J. Biochem.* 34:129–37
75. Gelman, R. A., Blackwell, J., Mathews, M. B. 1974. *Biochem. J.* 141:445–54
76. Puchtler, H., Sweat, F., Levine, M. 1962. *J. Histochem. Cytochem.* 10:355–64
77. Cooper, J. H. 1969. *J. Clin. Pathol.* 22:410–13
78. DeLellis, R. L., Bowling, M. C. 1970. *Hum. Pathol.* 1:655–59
79. Dyck, R. F., Lockwood, C. M., Kershaw, M., McHugh, N., Duance, V. C., et al. 1980. *J. Exp. Med.* 152:1162–74
80. Breathnach, S. M., Melrose, S. M., Bhogal, B., de Beer, F. C., Dyck, R. F., et al. 1981. *Nature* 293:652–54
81. Husby, G., Araki, A., Benditt, E. P., Benson, M. D., Cohen, A. S., et al. 1991. See Ref. 14, pp. 7–11
82. Glenner, G. G., Harbaugh, J., Ohms, J. I., Harada, M., Cuatrecasas, P. 1970. *Biochem. Biophys. Res. Commun.* 41:1287–89
83. Kyle, R. A. 1991. See Ref. 14, pp. 147–52
84. Page, D. L., Isersky, C., Harada, M., Glenner, G. G. 1972. *Res. Exp. Med.* 159:75–86
85. Husby, G., Sletten, K., Blumenkrantz, N., Danielsen, L. 1981. *Clin. Exp. Immunol.* 45:90–96
86. Eulitz, M., Linke, R. 1985. *Hoppe Seylers Z. Biol. Chem.* 366:907–15
87. Buxbaum, J., Caron, D., Gallo, G. 1991. See Ref. 14, pp. 197–200
88. Solomon, A., Frangione, B., Franklin, E. C. 1982. *J. Clin. Invest.* 70:453–60
89. Cornwell, G. G. III, Thomas, B., Kyle, R. A., Slabinski, K., Sletten, K., et al. 1991. See Ref. 14, pp. 203–6
90. Liepnieks, J. J., Benson, M. D., Dwulet, F. E. 1991. See Ref. 14, pp. 153–56
91. Kabat, E. A., Wu, T. T., Reid-Miller, M., Perry, H. M., Gottesman, K. S. 1987. *Sequences of Proteins of Immunological Interest,* pp. 63–77, 287–92. Bethesda:NIH
92. Gallo, G., Picken, M., Frangione, B., Buxbaum, J. 1988. *Mod. Pathol.* 1:453–56
93. Solomon, A., Kyle, R. A., Frangione, B. 1986. In *Amyloidosis,* ed. G. G. Glenner, E. F. Osserman, E. P. Benditt, E. Calkins, A. S. Cohen, D. Zucker-Franklin, pp. 449–62. New York: Plenum
94. Glenner, G. G., Ein, D., Eanes, E. D., Bladen, H. A., Terry, W., et al. 1971. *Science* 174:712–14
95. Linke, R. P., Tischendorf, F. W., Zucker-Franklin, D., Franklin, E. C. 1973. *J. Immunol.* 111:24–26
96. Epstein, W. V., Tan, M., Wood, I. S. 1974. *J. Lab. Clin. Med.* 84:107–12
97. Linke, R. P., Zucker-Franklin, D., Franklin, E. C. 1973. *J. Immunol.* 111:10–23
98. Eulitz, M., Breuer, M., Eblen, A., Weiss, D. T., Solomon, A. 1991. See Ref. 14, pp. 505–10
99. Toft, K. G., Olstad, O. K., Sletten, K., Westermark, P. 1991. See Ref. 14, pp. 169–72
100. Liepnieks, J. J., Benson, M. D., DiBartola, S. P. 1991. See Ref. 14, pp. 189–92
101. Koss, M. N., Pirani, L. L., Osserman, E. F. 1976. *Lab. Invest.* 34:579–91
102. Solomon, A., Weiss, D. 1991. See Ref. 14, pp.193–96
103. Linke, R. P., Geisel, O., Mann, D. 1991. See Ref. 14, p. 247
104. Benditt, E. P., Eriksen, N., Hermodson, M. A., Ericsson, L. H. 1971. *FEBS Lett.* 19:169–73
105. Levin, M., Pras, M., Franklin, E. C. 1973. *J. Exp. Med.* 138:373–80
106. Benditt, E. P., Eriksen, N., Meek, R. L. 1988. *Methods Enzymol.* 163:510–23
107. McAdam, K. P. W. J., Sipe, J. D. 1976. *J. Exp. Med.* 144:1121–27
108. Sipe, J. D. 1978. *Br. J. Exp. Pathol.* 59:305–10
109. Morrow, J. F., Stearman, R. S., Peltzman, C. J., Potter, D. A. 1981. *Proc. Natl. Acad. Sci. USA* 78:4718–22
110. Tape, C., Tan, R., Nesheim, M., Kisilevsky, R. 1988. *Scand. J. Immunol.* 28:317–24

111. Husebekk, A., Skogen, B., Husby, G., Marhaug, G. 1985. *Scand. J. Immunol.* 21:283–87
112. Ganapathi, M. K., Rzewnicki, D., Samols, D., Jiang, S. L., Kushner, I. 1991. *J. Immunol.* 147:1261–65
113. Woo, P., Sipe, J. D., Dinarello, C. A., Colten, H. R. 1988. *J. Biol. Chem.* 262:15790–97
114. Sipe, J. D., Rokita, H., Bartle, L. M., Loose, L. D., Neta, R. 1991. *Cytokine* 3:497
115. Lowell, C. A., Potter, D. A., Stearman, R. S., Morrow, J. F. 1986. *J. Biol. Chem.* 261:8442–52
116. Dubois, D. Y., Malmendier, C. L. 1988. *J. Immunol. Methods* 112:71–75
117. Hoffman, J., Benditt, E. P. 1982. *J. Biol. Chem.* 257:10510–17
118. Coetzee, G. A., Strachan, A. F., van der Westhuyzen, D. R., Hoppe, H. C., Jeenah, M. S., et al. 1986. *J. Biol. Chem.* 261:9644–51
119. Husebekk, A., Skogen, B., Husby, G. 1981. *Scand. J. Immunol.* 25:375–81
120. Strachan, A. F., de Beer, F. C., Coetzee, G. A., Hoppe, H. C., Jeenah, M. S., et al. 1986. *Coll. Prot. Biol. Fluids* 34:359–62
121. Woo, P., Betts, J., Edbrooke, M. 1991. See Ref. 14, pp. 13–19
122. Hayes, K. C., Lim, M., Pronczuk, A., Sipe, J. D. 1991. *FASEB J.* 5:A1287
123. Syversson, P. V., Sletten, K., Husby, G. 1991. See Ref. 14, pp. 111–14
124. Nel, A. E., de Beer, M. C., Shepard, E. G., Strachan, A. F., Vandenplas, M. L., et al. 1988. *Biochem. J.* 255:29–34
125. Benson, M. D., Aldo-Benson, M. 1979. *J. Immunol.* 122:2077–82
126. Zimlichman, S., Danon, A., Nathan, I., Mozes, G., Shainkin-Kestenbaum, R. 1990. *J. Lab. Clin. Med.* 116:180–86
127. Sack, G. 1983. *Gene* 21:19–24
128. Sipe, J. D., Colten, H. R., Goldberger, G., Edge, M., Tack, B. F., et al. 1985. *Biochemistry* 24:2931–36
129. Kluve-Beckerman, B., Long, G., Benson, M. D. 1986. *Biochem. Genet.* 24:795–803
130. Strachan, A., Brandt, W. F., Woo, P., van der Westhuyzen, D. R., Coetzee, G. A., et al. 1989. *J. Biol. Chem.* 264: 18368–73
131. Dwulet, F., Wallace, D. K., Benson, M. D. 1988. *Biochemistry* 27:1677–82
132. Steinkasserer, A., Weiss, E. H., Schwaeble, W., Linke, R. P. 1990. *Biochem. J.* 268:187–93
133. Beach, C. M., de Beer, M. C., Sipe, J. D., Loose, L. D., de Beer, F. C. 1992. *Biochem. J.* 282:615–20
134. Steinmetz, A., Vitt, H., Motzny, S., Kaffaarnik, H. 1991. See Ref. 14, pp. 20–23
135. De Beer, F. C., de Beer, M. C., Sipe, J. D. 1991. See Ref. 14, pp. 890–93
136. Brinckerhoff, C. E., Mitchell, T. I., Karmilowicz, M. J., Kluve-Beckerman, B., Benson, M. D. 1989. *Science* 243:655–57
137. Sack, G. S., Talbot, C. L. Jr. 1989. *Gene* 84:509–15
138. Li, X., Liao, W. S.-L. 1991. *J. Biol. Chem.* 23:15192–201
139. Sipe, J. D., Rokita, H., Koj, A. 1986. *Coll. Prot. Biol. Fluids* 34:331–34
140. Baltz, M. L., Rowe, I. F., Caspi, D., Turnell, W. G., Pepys, M. B. 1987. *Biochem. J.* 242:301–3
141. Meek, R. L., Benditt, E. P. 1989. *Proc. Natl. Acad. Sci. USA* 86:1890–94
142. Hoffman, J. S., Ericsson, L. H., Eriksen, N., Walsh, K. A., Benditt, E. P. 1984. *J. Exp. Med.* 159:641–46
143. Niewold, T. A., Tooten, P. C. J. 1990. *Scand. J. Immunol.* 31:389–96
144. Kluve-Beckerman, B., Liepnieks, J., Benson, M. D. 1991. See Ref. 14, pp. 125–28
145. Osborne, J. C., Brewer, H. B. 1977. *Adv. Protein Chem.* 31:253–337
146. Nichols, W. C., Gregg, R. E., Brewer, H. B., Benson, M. D. 1988. *Biochem. Biophys. Res. Commun.* 156:762–68
147. Jones, L. A., Harding, J. A., Cohen, A. S., Skinner, M. 1991. See Ref. 14, pp. 385–88
148. Benson, M. D., Rader, D. J., Schaefer, J. R., Gregg, R. E., Fairwell, T., et al. 1991. See Ref. 14, pp. 381–84
149. Schonfeld, G. 1990. *Atherosclerosis* 81:81–93
150. Takeda, T., Higuchi, K., Hosokawa, M. 1986. See Ref. 93, pp. 685–90
151. Scheinberg, M. A., Cathcart, E. S., Eastcott, J. W., Skinner, M., Benson, M., et al. 1976. *Lab. Invest.* 35:47–54
152. Chai, C. K. 1976. *Am. J. Pathol.* 85:49–72
153. Warden, C., Bee, L., Lusis, A. J., Lerner, C., Chai, C. K., et al. 1991. See Ref. 14, pp. 397–401
154. Higuchi, K., Naiki, H., Kitagawa, K., Hosokawa, M., Takeda, T. 1991. *Virchows Arch. B* 60:231 38
155. Blake, C. C. F. 1981. *Proc. R. Soc. London Ser. B* 211:413–31
156. Cornwell, G. G. III, Sletten, K., Olofsson, B. O., Johansson, B., Westermark, P. 1987. *J. Clin. Pathol.* 40:226–31
157. Saraiva, M. J. M. 1991. *Neuromuscular Disord.* 1:3–6

158. Smith, F. R., Goodman, D. S. 1971. *J. Clin. Invest.* 50:2426–36
159. Mendel, D. B., Crabtree, G. R. 1991. *J. Biol. Chem.* 266:677–80
160. Mita, S., Maeda, S., Shimada, K., Araki, S. 1984. *Biochem. Biophys. Res. Commun.* 124:558–64
161. Sasaki, H., Yoshioka, N., Tagaki, Y., Sakaki, Y. 1985. *Gene* 37:191–97
162. Kanda, Y., Goodman, D. S., Canfield, R. E., Morgan, F. J. 1974. *J. Biol. Chem.* 249:6796–805
163. Blake, C. C. F., Geisow, J. J., Oatley, S. J., Rerat, B., Rerat, C. 1978. *J. Mol. Biol.* 121:339–56
164. Costa, P. P., Figueira, A. S., Bravo, F. R. 1978. *Proc. Natl. Acad. Sci. USA* 75:4499–503
165. Saraiva, M. J. M., Costa, P. P., Goodman, D. S. 1992. *The Molecular and Genetic Basis of Neurological Disease,* ed. R. N. Rosenberg, S. B. Prusiner, S. DiMauro, R. L. Barchi, L. M. Kunkel. London: Butterworths. In press
166. Almeida, M. R., Altland, K., Rauh, S., Gawinowicz, M. A., Moreira, P., et al. 1991. *Biochim. Biophys. Acta* 1097:224–26
167. Wahlquist, J., Thylen, C., Haettner, E., Sandgren, O., Holmgren, G., et al. 1991. See Ref. 14, pp. 587–90
168. Fitch, N. J. S., Akbari, M. T., Ramsden, D. B. 1991. *J. Endocrinol.* 129:309–13
169. Moses, A. C., Rosen, H. N., Moller, D. E., Tzuzaki, S., Haddow, J. E., et al. 1990. *J. Clin. Invest.* 86:2025–33
170. Jacobson, D. R., Buxbaum, J. N. 1991. *Adv. Human Genet.* 20:69–123
171. Felding, P., Fex, G., Westermark, P., Olofsson, B. O., Pitkanen, P., Benson, L. 1985. *Scand. J. Immunol.* 21:133–40
172. Terry, C. J., Blake, C. C. F. 1991. See Ref. 14, pp. 575–78
173. Hamilton, J. A., Steinrauf, L. K., Liepnieks, J., Benson, M. 1991. See Ref. 14, pp. 579–82
174. Migita, S., Takegami, M., Benson, M. D. 1991. See Ref. 14, pp. 583–86
175. Gustavsson, A., Engstrom, E., Westermark, P. 1991. See Ref. 14, pp. 591–94
176. Pearse, A. G. E. 1980. *Mikroskopie (Wien)* 36:257–69
177. Johansson, B., Westermark, P. 1990. *Exp. Mol. Pathol.* 52:266–78
178. Westermark, P., Johnson, K. H. 1991. See Ref. 14, pp. 427–32
179. Surewicz, W. K., Mantsch, H. H., Stahl, G. L., Epand, R. M. 1987. *Proc. Natl. Acad. Sci. USA* 84:7028–30
180. Westermark, P., Sletten, K., Johansson,

B., Cornwell, G. G. III. 1990. *Proc. Natl. Acad. Sci. USA* 87:5036–40
181. Westermark, P., Wilander, E. 1978. *Diabetologia* 15:417–21
182. Westermark, P., Wernstedt, C., Wilander, E., Sletten, K. 1986. *Biochem. Biophys. Res. Commun.* 140:827–31
183. Sanke, T., Bell, G., Sample, C., Rubenstein, A. H., Steiner, D. F. 1988. *J. Biol. Chem.* 263:17243–46
184. Betsholtz, C., Svensson, V., Rorsman, F., Engstrom, U., Westermark, G. T., et al. 1989. *Exp. Cell. Res.* 183:484–93
185. Mosselman, S., Hoppener, J. W. M., Lips, C. J. M., Jansz, H. S. 1989. *FEBS Lett.* 247:154–58
186. Johnson, K. H., O'Brien, T. D., Hayden, D. W., Jordan, K., Ghobrial, H. K. G., et al. 1988. *Am. J. Pathol.* 130:1–8
187. Lukinius, A., Wilander, E., Westermark, G. T., Engstrom, U., Westermark, P. 1989. *Diabetologia* 32:240–44
188. Clark, A., Edwards, C. A., Ostle, L. R., Sutton, R., Rothbard, J. B., et al. 1989. *Cell Tissue Res.* 257:179–85
189. Glenner, G. G., Eanes, E. D., Wiley, C. A. 1988. *Biochem. Biophys. Res. Commun.* 155:608–14
190. Betsholtz, C., Christmansson, L., Engstrom, U., Rorsman, F., Jordan, K., et al. 1990. *Diabetes* 39:118–22
191. Westermark, P., Johnson, K. J., Engstrom, U., Westermark, G. T., Dominguez, H., et al. 1991. See Ref. 14, pp. 449–52
192. Sletten, K., Westermark, P., Natvig, J. B. 1976. *J. Exp. Med.* 143:993–98
193. Sletten, K., Natvig, J. B., Westermark, P. 1991. See Ref. 14, pp. 477–80
194. Le Mouellec, J. M., Julienne, A., Chenais, J., Lasmoles, G., Guliana, J. M., et al. 1984. *FEBS Lett.* 167:93–97
195. Storkel, S., Schneider, H.-M., Munterfering, H., Kashiwagi, S. 1983. *Lab. Invest.* 48:108–11
196. Dische, F. E., Wernstedt, C., Westermark, G. T., Westermark, P., Pepys, M. B., et al. 1988. *Diabetologia* 31:158–61
197. Hellman, U., Wernstedt, C., Westermark, P., O'Brien, T. D., Rathbun, W. B., Johnson, K. H. 1990. *Biochem. Biophys. Res. Commun.* 169:571–77
198. Westermark, P., Johansson, B., Natvig, J. B. 1979. *Scand. J. Immunol.* 10:303–8
199. Oikawa, S., Imai, M., Ueno, A., Tanaka, S., Noguchi, T. 1984. *Nature* 309:724–26
200. Johansson, B., Wernstedt, C., Wester-

mark, P. 1987. *Biochem. Biophys. Res. Commun.* 148:1087–92

201. Linke, R. P., Voigt, C., Storkel, F. S., Eulitz, M. 1988. *Virchows Arch. B* 55:125–27

202. Berggard, I., Bearn, A. G. 1968. *J. Biol. Chem.* 243:4095–103

203. Hall, P. W., Vasiljevic, M. 1973. *J. Lab. Clin. Med.* 81:897–904

204. Peterson, P. A., Cunningham, B. A., Berggard, I., Edelman, G. M. 1972. *Proc. Natl. Acad. Sci. USA* 60:1697–701

205. Becker, J. W., Reeke, G. N. Jr. 1985. *Proc. Natl. Acad. Sci. USA* 82:4225–29

206. Gejyo, F. F., Homma, N., Suzuki, Y., Arakawa, M. 1986. *N. Engl. J. Med.* 314:585–86

207. Connors, L. H., Shirahama, T., Skinner, M., Cohen, A. S. 1985. *Biochem. Biophys. Res. Commun.* 131:1063–68

208. Linke, R. P., Floege, J., Lottspeich, F., Deutzmann, R. 1991. See Ref. 14, pp. 369–72

209. Westermark, P., Stenkvist, B. 1973. *Arch. Intern. Med.* 132:522–23

210. Merlini, G., Limido, D., Bellotti, V., Bucciarelli, E., Berretta, P., et al. 1991. See Ref. 14, pp. 801–4

211. Glenner, G. G. 1988. *Cell* 52:307–8

212. Glenner, G. G., Mehlhaff, P. M., Kawano, H. 1991. See Ref. 14, pp. 707–12

213. Masters, C. L., Simms, G., Weinman, N. A., Multhaup, G., McDonald, B. L., et al. 1985. *Proc. Natl. Acad. Sci. USA* 82:4245–47

214. Selkoe, D. J., Abraham, C. R., Podlisny, M. B., Duffy, L. K. 1986. *J. Neurochem.* 146:1820–34

215. Van Nostrand, W. E., Wagner, S. L., Suzuki, M., Choi, B. H., Farrow, J. S., et al. 1989. *Nature* 341:546–49

216. Selkoe, D. J., Podlisny, M. B., Joachim, C. L., Vickers, E. A., Lee, G., et al. 1988. *Proc. Natl. Acad. Sci.* 85:7341–45

217. Esch, F. S., Keim, P. S., Beattie, E. C., Blacker, R. W., Culwell, A. R., et al. 1990. *Science* 248:1122–24

218. Abraham, C. R., Razzaboni, B. L., Ben-Meir, A., Papastoitsis, G. 1991. See Ref. 14, pp. 718–21

219. Tagawa, K., Kunishita, T., Maruyama, K., Yoshikawa, K., Kominami, E., et al. 1991. *Biochem. Biophys. Res. Commun.* 177:377–87

220. Abraham, C. R., Selkoe, D. J., Potter, H. 1988. *Cell* 52:487–501

221. Vandenabeele, P., Fiers, W. 1991. *Immunol. Today* 12:217–19

222. Guilian, D., Baker, T. J., Shih, L.-C. N., Lachman, L. B. 1986. *J. Exp. Med.* 164:594–604

223. Griffin, W. S. T., Stanley, L. C., Ling, C., White, L., MacLeod, V., et al. 1989. *Proc. Natl. Acad. Sci. USA* 86:7611–15

224. Goldgaber, D., Harris, H. W., Hla, T., Maciag, T., Donnelly, R. J., et al. 1989. *Proc. Natl. Acad. Sci. USA* 86:7606–10

225. Wiesniewski, H. M., Wegil, J., Kida, E., Burrage, T., Currie, J. 1991. See Ref. 14, pp. 726–31

226. Levy, E., Carman, M. D., Fernandez-Madrid, I., Lieberburg, I., Power, M. D., et al. 1990. *Science* 248:1124–26

227. Wirak, D. O., Bayney, R., Ramabhadran, T. V., Fracasso, R. P., Hart, J. T., et al. 1991. *Science* 253:323–25

228. Quon, D., Wang, Y., Catalano, R., Scardina, J. M., Murakami, K., et al. 1991. *Nature* 352:239–41

229. Palsdottir, A., Abrahamson, M., Thorsteinsson, L., Arnason, A., Olafsson, I., et al. 1988. *Lancet* 2:603–4

230. Jensson, O., Gudmundsson, G., Amason, A., Blondal, H., Petursdottir, I., et al. 1987. *Acta Neurol. Scand.* 76:102–14

231. Fujihara, S., Shimode, K., Nakamura, M., Kobayashi, S., Tsunematsu, T., et al. 1991. See Ref. 14, pp. 365–68

232. Thorsteinsson, L., Georgsson, G., Asgeirsson, B., Bjarnadottir, M., Olafsson, I. O., et al. 1991. See Ref. 14, pp. 357–60

233. Prusiner, S. B. 1991. *Science* 252:1515–22

234. Weissmann, C. 1991. *Nature* 352:679–82

235. Bruce, M. E., Dickinson, A. G. 1987. *J. Gen. Virol.* 68:79–89

236. Hsiao, K. K., Scott, M., Foster, D., Groth, D. F., DeArmond, S. J., et al. 1990. *Science* 250:1587–90

237. Bendheim, P. E., Barry, R. A., DeArmond, S. J., Stites, D. P., Prusiner, S. B. 1984. *Nature* 310:418–21

238. DeArmond, S. J., McKinley, M. P., Barry, R. A., Braunfeld, M. B., McColloch, J. R., et al. 1985. *Cell* 41:221–35

239. Roberts, G. W., Lofthouse, R., Allsop, D., Landon, M., Kidd, M., et al. 1988. *Neurology* 38:1534–40

240. Prusiner, S. B., McKinley, M. P., Bowman, K. A., Bolton, D. C., Bendheim, P. E., et al. 1983. *Cell* 35:349–58

241. Gabizon, R., McKinley, M. P., Prusiner, S. B. 1987. *Proc. Natl. Acad. Sci. USA* 84:4017–21

242. Prusiner, S. B., Scott, M., Foster, D.,

Pan, K -M., Groth, D., et al. 1990. *Cell* 63:673–86

243. Yin, H. L., Stossel, T. P. 1979. *Nature* 281:583–86
244. Yin, H. L., Albrecht, J., Fattoum, A. 1981. *J. Cell. Biol.* 91:901–6
245. Kwiatkowski, D. J., Mehl, R., Yin, H. L. 1988. *J. Cell Biol.* 106:375–84
246. Kwiatkowski, D. J., Stossel, T. P., Orkin, S. H., Mole, J. H., Colten, H. R., et al. 1986. *Nature* 323:455–58
247. Kwiatkowski, D. J., Westbrook, C. A., Bruns, G. A. P., Morton, C. C. 1988. *Am. J. Hum. Genet.* 42:565–72
248. Maury, C. P. J., Alli, K., Baumann, M. 1990. *FEBS Lett.* 260:85–87, 1990
249. Yoneda, K., Watanabe, H., Yanagihara, M., Mori, S. 1989. *J. Cutaneous Pathol.* 16:133–36
250. Gueft, B. 1972. *Mt. Sinai J. Med.* 39:91–102
251. Haurowitz, F. 1963. *The Chemistry and Function of Proteins,* pp. 217–21. New York: Academic

Annu. Rev. Biochem. 1992. 61:977–1010

MASS SPECTROMETRY OF PEPTIDES AND PROTEINS

K. Biemann

Department of Chemistry, Massachusetts Institute of Technology, Cambridge, Massachusetts 02139

KEY WORDS: fast atom bombardment ionization, laser desorption ionization, electrospray ionization, molecular weight measurements, tandem mass spectrometry

CONTENTS

PERSPECTIVES AND SUMMARY

During the 1980s, methods were devised that made it possible to ionize large and polar molecules in a mass spectrometer. These developments triggered

977

0066-4154/92/0701-0977$02.00

the design of commercial mass spectrometers with a mass range commensurate with the molecular weights of the ionized materials. The combination of these two factors suddenly made it possible to apply mass spectrometry (MS) to the solution of many biochemical and biological problems. The relatively high sensitivity (nanomole to picomole; under certain conditions even less) of these instruments was another important feature.

For various reasons, peptides and proteins are particularly suited for these novel ionization techniques, most of which were developed and first demonstrated with molecules of this type. Once it was shown that an underivatized peptide of $M_r > 1000$ could be ionized by "fast atom bombardment" (FAB) (1) and measured accurately to a fraction of a mass unit, the race was on to measure larger and larger molecules and, at this writing, the record stands at $\sim$400 kDa $\pm$ $\sim$4 kDa using "matrix-assisted laser desorption" (maLD) (2). This is a much higher accuracy than is possible with conventional chromatographic or electrophoretic methods which, in addition, generally require calibration standards of similar shape, whereas ionization in the mass spectrometer is quite independent of this parameter.

It should be duly noted that the demonstration in 1976 by Macfarlane & Torgerson (3), that relatively large proteins could not only be ionized upon impact of MeV particles but also then detected in a mass spectrometer, provided considerable encouragement in the search for the ionization methods alluded to above and described in detail later. The drawback of Macfarlane's method was the need for ^{232}Cf, the radioactive decay of which provided the MeV particles, and it was not until 1982 that a commercial instrument became available.

The most recent method for measuring the molecular weights of peptides and proteins is based on their ability to become polyprotonated under certain conditions and, therefore, multiply charged (4). Because a mass spectrometer measures the mass-to-charge ratio, m/z, of a particle, where z is the number of charges, even very large molecules, when highly protonated, form ions that can be detected with relatively small mass spectrometers.

All these ionization methods generate protonated peptide and protein molecules that have very little excess energy and are thus quite stable. For this reason, their molecular weights can be readily measured, even when they are present in mixtures, but the accompanying lack of specific fragmentation makes it impossible to obtain any structure information. If that is desired, more energy must be imparted onto the ion molecule, which is generally accomplished by collision with a small neutral molecule. The resulting structure-specific fragment ions are analyzed in a tandem mass spectrometer. This principle has been used for more than two decades in the study of gaseous ion chemistry, but it is only in the past few years that it has become widely used—in fact, indispensable—in analyzing the structure of peptides and proteins.

Because mass spectrometry is entirely different from the Edman degradation, it overcomes many of the disadvantages of the latter, and the two methodologies complement each other. It is hoped that this review will increase the biochemical community's awareness of the particular advantages of mass spectrometry.

SCOPE OF REVIEW

It is the intent to outline the basic principles of the methodology and demonstrate the usefulness of mass spectrometric strategies in the solution of peptide and protein structure problems. The field has been reviewed previously (5) for the mass spectrometry community. This review concentrates on the more recent developments, particularly the solutions of problems arising in the field of biotechnology. The large number of purely mass spectrometric studies, even though they may utilize peptides or proteins, are not discussed unless they are particularly pertinent to the topics touched upon here. For the others, the reader is referred to the quite exhaustive reviews of mass spectrometry that appear in *Analytical Chemistry* every even-numbered year, the last one by Burlingame et al (6).

Some material from books devoted to the use of mass spectrometry in biology and biochemistry, however, is discussed. Such books are either the results of conferences (7, 8) or multi-authored collections (9–11) containing a wealth of information. Space does not permit discussing the instrumentation used for mass spectrometry and the physical principles of the various methods for mass separation. An excellent description of these is readily accessible to the biochemist (12). The conventional ionization techniques, electron ionization and chemical ionization, which require a sample in the gas phase, are not applicable to underivatized peptides and are therefore not covered, although they played a useful role prior to 1980 (5). The review concentrates on naturally occurring or recombinant proteins. Space limitations do not permit discussing the important role that mass spectrometry has begun to play in the structure verification of chemically synthesized peptides and in the demonstration of their homogeneity (13).

MOLECULAR WEIGHT DETERMINATIONS OF PEPTIDES AND PROTEINS

Fast Atom (Ion) Bombardment Ionization

In 1981, Barber et al (1) reported the mass spectrum of an underivatized undecapeptide, Met-Lys-bradykinin of $M_r = 1318$, which exhibited a strong signal at m/z 1319 due to the protonated molecule $(M + H)^+$. The spectrum was obtained by bombarding a small drop of glycerol containing a few micrograms of the peptide with a beam of argon atoms of a few keV kinetic energy. In this process, protonated (and deprotonated) peptide, glycerol, and glycerol cluster ions are ejected. It takes place in the ion source of the mass spectrometer that separates the resulting ions according to mass, or more precisely, the m/z ratio, but in this case, mostly singly charged ions ($z = 1$) are produced. Depending on the polarity of the electric and magnetic fields of the spectrometer, either positive or negative ions are recorded. Because peptides and proteins contain many sites that can easily be protonated, positive ion spectra are generally more useful.

A typical example is shown in Figure 1, which is the FAB mass spectrum from a fraction of a tryptic digest of a large protein (14). The large peaks numbered 1–11 are certainly due to $(M + H)^+$ ions (the smaller ones may be fragments), indicating the presence of at least 11 individual peptides in this fraction. Peak 7 is shown expanded in the upper right, revealing the isotopic multiplet of the $(M + H)^+$ ions, caused primarily by the natural abundance of ^{13}C. The most abundant peak is due to the species containing only ^{12}C, ^{1}H, ^{14}N, and ^{16}O, and therefore termed "monoisotopic." This spectrum was obtained with a double-focusing magnetic deflection mass spectrometer that easily resolves these isotopic clusters, the components of which differ by one dalton. The accuracy of the mass assignment is routinely within ± 0.3 dalton or better. This resolution and accuracy can be achieved even at $m/z \sim 6000$, as demonstrated in Figure 2, the $(M + H)^+$ ion cluster of human insulin. The large number of carbon atoms present in this molecule makes it unlikely that all of them are ^{12}C, and the monoisotopic peak (m/z 5804.6) is therefore quite small, while the most abundant species contains three ^{13}C atoms. For an unknown, therefore, it is quite difficult to decide which peak is the monoisotopic signal. Furthermore, it is not easy to achieve this resolution at high mass, and it is common practice to record such spectra at much lower resolution (by opening the resolving slits), which causes the multiplet to collapse to a broad peak (Figure 2) and the signal intensity to increase significantly. The centroid of this peak is the "average mass," i.e. the abundance-weighted sum of the isotopes of all the elements present. The average mass is the sum of the "chemical" atomic weights (e.g. 12.011 for carbon) of all the elements present, while the monoisotopic mass is the sum of all their lightest isotopes (e.g. 12.000000 for ^{12}C). As is evident from Figure 2, these two values differ significantly and it is, thus, important to indicate what scale is used, which is practically dictated by the resolution of the mass spectrometer and the mass of the ion. In this chapter, which deals mainly with rather large molecules, the values cited are average masses unless followed by "(monoisotopic)".

In the original FAB experiment (1), a beam of argon atoms was used as the bombarding particle beam and glycerol as the solvent (matrix), but neither of these are unique. Xenon atoms ($Xe^°$) are more efficient than the lighter $Ar^°$

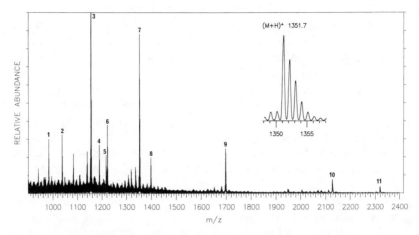

Figure 1   The FAB mass spectrum of an HPLC fraction of a complex tryptic digest. It indicates the presence of at least 11 peptides. The region around peak 7 is shown expanded in the inset. (From Ref. 14, reproduced with permission from *Science*, copyright 1987 by the AAAS.)

atoms (15), and a beam of cesium ions (Cs^+) also has some advantages (16). Quite a few substances are used as matrices (17). The latter must be a liquid of low vapor pressure at room temperature (because the ionization takes place in the vacuum of the mass spectrometer), a reasonably good solvent for polar molecules, and unreactive with the compounds to be analyzed. For peptides and proteins, matrices of glycerol, thioglycerol, or an eutectic mixture of dithiothreitol with dithioerythritol (DTT/DTE) are widely used, often with the

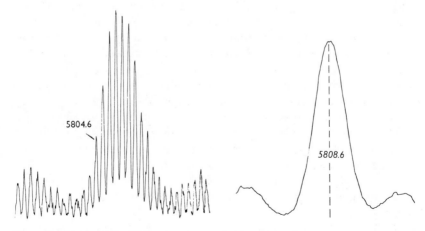

Figure 2   Section of the FAB mass spectrum of human insulin ($C_{257}H_{383}N_{65}O_{77}S_6$) at a resolution of 1:6000 (*left*) and 1:500 (*right*).

addition of acid to facilitate formation of protonated molecules. For hydrophobic peptides, less polar matrices, such as 3-nitrobenzyl alcohol, its octylether, or N-formyl-2-aminoethanol (18), are preferable. For negative ion FAB, basic liquids, such as triethanolamine, are appropriate because they aid in the deprotonation of molecules with acidic sites.

Although it was stated earlier and demonstrated by Figure 1 that mixtures of peptides can be analyzed by FAB-MS, one must be careful in interpreting such data. This is because certain peptides, generally the more hydrophobic ones, seem to be preferentially ionized over hydrophilic peptides. The signal of the latter may be completely suppressed and the peptide may escape detection, even though it ionizes well in the absence of the others. This is one reason why it is advisable to fractionate complex peptide mixtures, such as proteolytic digests of proteins, by reversed-phase high-performance liquid chromatography (RP-HPLC) prior to FAB-MS because such fractions, while often still mixtures, will not contain hydrophobic and hydrophilic peptides together (at least not those of similar size).

A hydrophobicity index has been proposed for the prediction of this suppression (19), along with the suggestion to alleviate it by converting all the carboxyl groups in a peptide to their isopropyl esters. Hexyl or benzyl esters have also been suggested (20), both to reduce the suppression effect and to increase signal intensity, because the hydrophobic groups keep the molecule at the surface of the glycerol drop where it is more accessible to the ionizing beam.

Peptides and small proteins of $M_r < 6000$ are well suited for ionization by FAB, but with increasing molecular weight, it becomes more difficult to obtain a good signal. Molecular weight determinations in the 10–25 kDa range have been reported (21–23), but for that purpose, one now uses other methods (see below). Fast atom (or ion) bombardment is thus most suitable for the molecular weight determination or sequencing (by tandem mass spectrometry) of peptides that are isolated as such, produced by proteolysis or chemical cleavage of proteins, or prepared by chemical synthesis.

Another characteristic of FAB ionization is its compatibility with chromatographic separation techniques. A flow rate of 2–10 μL of column effluent can be tolerated by the vacuum system of the mass spectrometer. The bombarding Xe° or Cs$^+$ beam is directed at a fritted disc covering the exit of a thin glass capillary that connects the chromatograph with the ion source (24), or simply at the open end of the capillary (25, 26). A few percent of glycerol is added, either directly to the solvents used for the chromatograph or later, after the column. A flow of column effluent is thus continuously subjected to ionization under FAB conditions because the volatile solvent immediately evaporates and leaves a film of glycerol behind. Whenever a compound elutes, it is ionized and its spectrum is recorded by continuously scanning the mass spectrometer. This approach is particularly suited to the characterization of proteolytic digests (27–30), which can be injected directly into the HPLC. Both the retention time and the molecular weight of each eluting peptide are obtained in a single experiment. To obtain sequence information, one can couple the chromatograph with a tandem mass spectrometer, but this mode of operation is more complex (31). Capillary zone electrophoresis can also be combined with FAB-MS but, because of the very low flow rates, make-up solvent containing glycerol must be added to the eluent (32, 33).

Continuous flow (CF) FAB-MS has the added advantages that the matrix background is reduced, leading to a lower sample requirement (34), and that the ionization suppression of hydrophilic peptides by hydrophobic ones is greatly diminished (35, 36). For these reasons, the CF method is suggested for the introduction of the sample into the FAB-MS (37).

Plasma Desorption Ionization

The FAB ionization process discussed above utilizes particles of 8–35 keV kinetic energy. Spontaneous fission of ^{252}Cf produces two particles of much higher kinetic energy (e.g. 80 MeV ^{144}Cs and ^{108}Tc) traveling in opposite directions. Because of their high energy, these particles can pass through a thin metal foil and ionize organic material deposited on the other side of the foil. Fission is a discrete event occurring at a rate of $1–5 \times 10^3$ sec^{-1} for a typical ^{252}Cf source and resulting in an ion beam pulsed at that frequency. For this reason, a time-of-flight (TOF) mass spectrometer is used to determine the m/z ratio of the ions. As its name implies, the time is measured that it takes for the ion, after acceleration by a potential gradient of 10–20 kV, to pass to a detector located at the other end of a 0.2–2 m long tube. The flight time, which can be measured quite accurately with the appropriate circuitry, is proportional to the square root of m/z, thus permitting calculation of the m/z ratio of the ion. A TOF mass spectrometer is a relatively simple device (compared to the magnetic deflection or quadrupole mass spectrometers used for FAB) and has a very high mass range. Its disadvantage is the relatively low resolution.

The peptide or protein sample is adsorbed on a layer of nitrocellulose covering the target foil, and rinsed with water to remove salts that might otherwise impede ionization or lead to adducts of Na^+ or other ions, causing a shift in the molecular weight. Plasma desorption (PD) MS has been used in a limited number of laboratories to determine the molecular weights of peptides and proteins. The largest recorded is porcine pepsin of M_r 36,688, determined 34,630 (38), which is within the accuracy range of ±0.1% to ±1% of this method.

As with FAB ionization, PD discriminates among different peptides, but here it is the charge rather than hydrophobicity that causes suppression. A net positive charge favors ejection of positive ions, a net negative charge favors formation of negative ions. Thus, operating the TOF mass spectrometer consecutively in both modes results in more complete information (39).

Although under normal PD-MS conditions mainly singly and multiply charged molecular ions are formed, fragmentation along the peptide bonds has been observed when a large amount of sample (nanomoles instead of picomoles) is applied to the target and the signal is integrated for a long time (up to a few hours). Fragment ions from peptides ranging in molecular weight up to 3800 (40), 4600 (41), and for the 20 N-terminal amino acids of an 8.9-kDa protein (42) were observed, and could be correlated with the known structure of the peptides. Because of the limited resolution and low mass accuracy, this approach has little utility for unknowns.

Matrix-Assisted Laser Desorption Ionization

In 1988, Hillenkamp reported (2) that proteins up to a molecular weight of 60,000 can be ionized if embedded in a large molar excess of an UV-absorbing matrix (e.g. nicotinic acid) and irradiated with a 266-nm laser beam. Because UV lasers are pulsed, a TOF mass spectrometer was used for the analysis. Two years later, the molecular ion of a monoclonal antibody was recorded at m/z 149,190 along with a signal at ~300 kDa, the ionization-induced dimer (43). The mass accuracy at high mass was at first about ±0.5% (44) and improved to ±0.1–0.2% (45). Beavis & Chait (46, 47) use 355-nm radiation and a matrix (sinapinic acid, 3,5-dimethoxy-4-hydroxycinnamic acid) that absorbs at that wave length. With this system, a mass measurement accuracy of ±0.01% with 1 pmol of protein is reported for substances of $M_r <$ 30,000, when using a protein of well-defined molecular weight as an internal calibration standard (48). In the short time since its inception, maLD has outperformed PDMS in the ease of operation, high mass range, sensitivity, and the time required to record a spectrum (< 1 min). Figure 3 is an example of present performance at high mass.

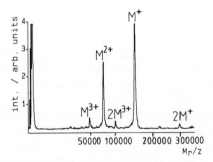

Figure 3 Matrix-assisted LD mass spectrum of an IgG monoclonal antibody, $M^+ = m/z$ 147,430 ± 90. The clearly visible peak just below m/z 300,000 is the ionization-induced dimer. Matrix: nicotinic acid; laser wavelength 266 nm (courtesy F. Hillenkamp).

Recently, Hillenkamp's group has shown that infrared (IR) radiation of 2.94 μm (49) or 10.6 μm (50) can also be used for maLD with about the same results as obtained in the UV. There is even a hint that IR is more effective for larger molecules.

Although at first the emphasis was on obtaining good signals from ever larger proteins, the maLD method is also very useful for much smaller peptides. The lower limit is caused by the background generated by clusters of the matrix, about m/z 1000 for sinapinic acid and m/z 500 for 2,5-dihydroxybenzoic acid (51). Therefore, maLD-MS is also suitable for the analysis of peptide mixtures, such as enzyme digests (52). The advantages over FAB are the much higher sensitivity, simpler operation, and less suppression of hydrophilic peptides, at the expense of lower resolution (isotopic multiplets cannot be resolved) and somewhat lower mass accuracy. These drawbacks may be eliminated by performing maLD in a double-focusing magnetic mass spectrometer (53).

Electrospray Ionization

In 1988, Fenn reported (4) the successful use of the "electrospray" (ES) principle to ionize large proteins, by spraying a very dilute solution thereof from the tip of a needle across an electrostatic field gradient of a few kV. In this process, multiprotonated protein molecules $(M + nH)^{n+}$ are formed that, when mass analyzed, give rise to a series of consecutive peaks at $(M + n)/n$ along the m/z scale. An example is shown in Figure 4, which demonstrates the major advantage of ES: because a protein can easily be multiprotonated, the m/z ratios of the resulting ions appear as a proportional fraction of its molecular weight. A simple mass spectrometer of limited mass range (up to m/z 1500–2000, where $z = 1$) suffices to measure molecular weights up to 50 times larger. For a single compound, each adjacent pair of peaks must differ by one charge. The unknown charge state of the ion can be derived from any two of such pairs because they are all derived from a single molecular mass. These calculations can be carried out by hand or by using a computer algorithm, particularly when the spectrum of a mixture needs to be de-convoluted (54). The relatively high accuracy or, more precisely speaking, the precision ($\pm 0.01\%$) of the mass measurement by ES ionization (ESI) is chiefly the result of averaging individual measurements in the same spectrum. For example, each of the 12 m/z values in Figure 4 contributes to the final result, the molecular weight of the polyprotonated solute. Another factor that may affect the accuracy is the care with which the mass scale of the spectro-meter is calibrated and the stability of the calibration.

For practical purposes, such as low accelerating voltages and tolerance for relatively high pressures in the vacuum envelope of the mass spectrometer, quadrupole instruments were formerly used almost exclusively, and only recently have the magnetic deflection instruments entered this field (55, 56).

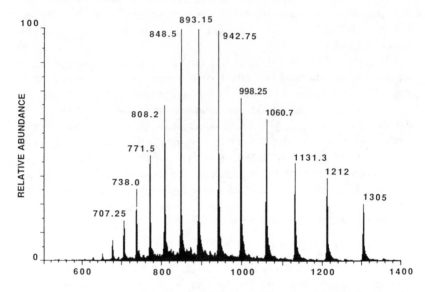

Figure 4   Electrospray mass spectrum of equine myoglobin, $M_r = 16,951.5$. The ion at m/z 893.15 corresponds to the molecule with 19 protons attached. From the 12 ions ranging from 24^+ (m/z 705.25) to 13^+ (m/z 1305), a molecular weight of $16,951.9 \pm 1.7$ is calculated (courtesy R. D. Smith).

One advantage of the magnetic instruments is the higher resolving power that makes it possible to assign the charge state directly, at least for low values of z. For example, if the components of an isotopic multiplet are spaced 1/5 of a dalton apart, then it must be due to an ion with five charges (57). High resolution can also be obtained with a Fourier transform mass spectrometer (FTMS) coupled to an external ES ion source (58), but the resolution decreases dramatically above m/z 1000 (59). At least with quadrupole mass spectrometers, the useful range for ESI-MS is probably limited to $M_r <$ 100,000 because the ion clusters differing by one charge become unresolvable, resulting in a spectrum that resembles a broad hump (60, 61).

In general, the maximum number of protons that attach to a peptide or protein under ES conditions correlates well with the total number of basic amino acids (Arg, Lys, His) plus the N-terminal amino group, unless it is acylated (62). However, the accessibility of these basic sites is an important factor. The distribution of charge states, thus, depends on pH, temperature, and any denaturing agent present in the solution being electrosprayed into the mass spectrometer. This information can be used to probe conformational changes in the protein. For example, for bovine cytochrome c, the most abundant ion has 10 positive charges when electrospraying a solution at pH

5.2, but 16 charges at pH 2.6, and an intermediate bimodal distribution at pH 3.0 (63). A similar effect is observed upon reduction of disulfide bonds. Hen egg white lysozyme with four disulfide bonds shows a charge distribution centering at 12^+ but, upon reduction with DTT, a new cluster centering around 15^+ appears (61).

As ES ionization requires a solution of the sample to be analyzed, it is ideally suited for the direct coupling with a chromatograph. Such combinations have been demonstrated with normal bore (4.6 mm) columns, which require postcolumn stream splitting or direct coupling with a microbore (1 mm) column (64), and with packed capillary HPLC (65). In both cases, tryptic digests were used to evaluate the methodology. A careful comparison was made of the performance of on-line coupling of microbore HPLC with CF-FAB and ES, respectively, using enzymatic digests of rs-CD4 glycoprotein and r-hepatitis B surface antigen as real-life tests (66). The two methods gave comparable results, but the authors concluded that the HPLC-ES-MS combination gave more reliable data.

The potentially high resolving power of capillary zone electrophoresis (CZE) makes it attractive for combination with ES-MS (67). Preliminary experiments using horse and whale myoglobins have been reported (68).

PROBLEMS THAT CAN BE SOLVED BY ACCURATE ($\sim 1{:}10^4$) MOLECULAR WEIGHT INFORMATION ALONE

In many situations, measurement of the molecular weight of a peptide, protein, or modified product and comparison with the predicted value can confirm an hypothesis or differentiate between one or more structural possibilities. The mass accuracy required depends on the questions asked and may range from $\pm < 0.3$ dalton if one needs to differentiate the presence of Asn vs Asp or Gln vs Glu, which differ only by 1 dalton, to ± 50 daltons if the question is the number of hexoses attached to a glycopeptide or -protein. In other cases, even lower mass accuracy may suffice. The ionization methods discussed in the previous sections cover that range very adequately.

Check Correctness of an Amino Acid Sequence

Many recent amino acid sequences of proteins were derived indirectly from the DNA sequence of the corresponding gene. Once the open reading frame of a DNA sequence is translated into the amino acid sequence using the genetic code, one can calculate the molecular weights of all peptides that would be generated by digestion of the gene product with a specific enzyme. The molecular weights of the peptides obtained upon such a digestion of the corresponding protein must match, within experimental error, those predicted. The first example of this approach was the verification of the cDNA sequence coding for glycyl-tRNA synthetase from *Escherichia coli* (69). In this case, the molecular weights of the tryptic peptides were determined by FAB-MS, at the same time as the DNA sequencing work was proceeding. Thus, it was possible to spot errors in the DNA sequence in real time and correct them immediately (70), as was done in the cases of endo-β-N-acetylglucosaminidase H (71), histidine t-RNA synthetase (72), rabbit muscle creatine phosphokinase (73), and glutamyl-tRNA-synthetase (74).

This method is now often used when a published DNA sequence becomes suspect. That of protein S from *Myxococcus xanthus* (75) turned out to be correct. On the other hand, the DNA sequence reported (76) for the gene coding for the widely used protease Endo Glu-C did not correspond to the previously published protein sequence (77) at 12 points. This discrepancy was recently settled by FAB-MS and tandem-MS (78). It turned out that neither the DNA sequence nor the amino acid sequence was entirely correct.

More recently, this strategy, which is often called "FAB-mapping" (79, 80), has been applied to the structure verification of proteins produced by recombinant technology. In these cases, it is not only necessary to demonstrate that the primary sequence is correct but also whether or not any posttranslational modifications have taken place. For example, it could be shown that recombinant human leucocyte interferon A expressed in large quantities in *E. coli* had the expected structure. However, two minor products were found, in which the N-terminal amino acid was acetylated and the two cysteines had been oxidized to cysteic acid (81). Similarly, rhα-2 interferon was found to be mainly the correct product, but it also contained a by-product with an acetylated N-terminus (82). Recombinant human interleukin-1α (22), as well as the rs-CD4 receptor (83) were found to be correct, as was recombinant hepatitis B Surface Antigen Protein, except that 70% of it was N-acetylated (84). However, rat prostatic spermine-binding protein produced by recombinant techniques was found to be missing a stretch of 35 amino acids (85), apparently because the cDNA clone was missing one exon. Computer programs exist (86, 87) that greatly facilitate the matching of experimental data with known or predicted protein sequences, predict proteolytic peptides derived thereof, and many other operations, such as the location of sequons (see below), etc.

An important case where FAB-MS played a crucial role was the characterization of human relaxin, a peptide hormone involved in childbirth (88). It is formed from prerelaxin in a manner analogous to the conversion of preinsulin to insulin, with which it shares the disulfide linked A–B chain structure but exhibits very little homology. The amino acid sequence of human prerelaxin is known from the DNA sequence of the gene, but the point of cleavage to produce the B-chain had to be inferred from the known sequences of the pig and rat analogues, which implicated position 33. In order to produce the correct recombinant human relaxin, this had to be verified. Only a few picomoles of the natural product could be obtained. Since the question related to the C-terminus, Edman degradation was ruled out. A 0.5 pmol sample of relaxin isolated from corpus luteum was subjected to FAB-MS in a reducing matrix (see below) to cleave the disulfide bonds. Two peaks at m/z 2656.6 and m/z 3313.3 were observed. The former was the expected $(M + H)^+$ ion of the A-chain, while the latter indicated that the B-chain was only 29 amino acids long, i.e. four amino acids shorter than predicted. Thus, it was possible to make the correct cDNA for production of recombinant human relaxin. The disulfide bridges were also determined by mass spectrometry, as outlined in a later section. The recombinant product was also fully characterized by mass spectrometric methods (88).

The studies outlined above all utlized FAB-MS, but ES and maLD can be used similarly. ES made it possible to identify the primary gene product of the α-crystallins from bovine lenses, by measuring the molecular weight of the intact protein and also detecting phosphorylated and truncated forms. The location of the phosphate and the extent of truncation were determined from the ES spectra of proteolytic digests (89). The ES spectrum of bovine serum albumin (BSA) gave a value of 66,436 for the molecular weight, as derived from the large number of multiply charged peaks centering around $(M + 45H)^{45+}$ (90). This value was in disagreement with that calculated (66,261.1) for the published 582-amino-acid sequence of BSA (91). This difference of -0.3% is much larger than the average error of such measurements, and suggested that the published structure was incorrect. In the tryptic digest of BSA, 66 peptides were identified by on-line HPLC-CF-FAB. The molecular weights of all but one matched those predicted and covered 93% of the sequence. The molecular weight of the one that did not agree was 163 daltons too high and implied an additional tyrosine at position 156, increasing the total number of amino acids to 583 (90).

Direct molecular weight measurements of intact proteins also allow the rapid, preliminary verification of the fidelity and homogeneity of recombinant proteins. A number of recombinant variants of structural proteins of human immunodeficiency virus (HIV) types 1 and 2, and of α-antitrypsin, ranging in molecular weight from 14,000 to 44,000, were subjected to ES-MS (92). Most measurements were within ± 1 to ± 8 daltons of the expected value, and thus correct. However, one of the HIV-2 proteins turned out to be a mixture of the expected protein, M_r 25,748 $\pm$ 6, derived from the polyprotonated ion cluster centering around $(M + 28H)^{28+}$, calculated M_r 25,745, and a smaller one, M_r 25,497 $\pm$ 8. This result indicated partial C-terminal truncation, causing the loss of Leu-Met and reducing the molecular weight by 244 daltons. This difference of 1% led to a set of well-resolved peaks in the 24^+ to 32^+ ion cluster centered around m/z 900. Another of the recombinant structural proteins of HIV-2 exhibited a very complex ES spectrum consisting of groups of at least nine peaks for each of the 11^+ to 16^+ ions. The measured masses clearly indicated that the original 130-amino-acid long protein had been severely truncated at the C-terminus to nine products, the shortest ending at position 111 and the longest at 119. From the relative intensity of the 13^+ ion signals, the authors concluded that the major component represented 29% and the two least abundant components 2% each.

There is also a report (93) of the ES spectrum of the unfractionated tryptic digest of human apolipoprotein A-I, in which 32 of the 37 predicted proteolysis products were detected in a single spectrum (average of five scans). Their molecular weights were measured with an accuracy sufficient to determine them to within one dalton. The five products not found were single amino acids, dipeptides, and one pentapeptide, but they were accounted for as part of some of seven additional peptides present due to incomplete cleavage. The spectrum mainly consisted of $(M + 2H)^{2+}$ ions because tryptic peptides have two basic sites, unless they also contain histidine, in which case triply charged ions are also formed. While it would be difficult to interpret such a spectrum for a protein of unknown sequence, it is probably adequate for checking the correctness of a proposed sequence.

A word of caution at the end of this section: The molecular weight of a peptide or protein reveals, at best, the sum of the amino acids present but not, of course, the sequence. For example, a reversal of two amino acids cannot be detected in this manner. Such an error is unlikely to arise during DNA sequencing, but is more probable during Edman sequencing (by mixing up cycles or an incorrect interpretation of the data).

Detect Natural or Biosynthetic Mutations

The detection of mutations is akin to the detection of human error in a sequence, but here the error is nature's. Human hemoglobin (Hb) is probably the protein to which mass spectrometry has most frequently been applied for the detection of mutants (94, 95). Many mutants detected have been quickly recognized as already known; others were identified as new. The strategy again involves comparing the molecular weights of proteolytic peptides obtained with various specific enzymes with those known to be formed from the native α, β, γ, and δ chains of Hb. To simplify the experiment, the chains are digested separately. A change in molecular weights indicates the region of the molecule where the mutation occurred, and the mass difference between the wild-type and the mutant peptide suffices to identify the amino acid that has mutated if there is only one possibility. If there are more, they can frequently be sorted out by comparing the data from different digests. If questions still remain, one has to resort to tandem mass spectrometry (see below) or, if not available, to the Edman degradation.

In one specific case, the C-terminal tryptic dipeptide of the β-chain Tyr-His, $(M + H)^+ = m/z$ 319.1 (monoisotopic), was not detected, but it could have been obscured by the high matrix background at low mass. The much larger C-terminal Endo Glu-C peptide was found to have a molecular weight 300 daltons less than that of the corresponding peptide, $(M + H)^+ = m/z$ 2679, of normal Hb (96). This previously known mutation is due to a change of the UAU codon for tyrosine to one of the termination codons UAA or UAG. Another example reported in the same paper revealed two mutations in the same tryptic peptide β41–59, Asp $\rightarrow$ Asn and Pro $\rightarrow$ Ser, deduced from the decrease in the molecular weight by 11 daltons. Unfortunately, there are two aspartic acids and two prolines present in this peptide. A tandem-MS measurement revealed the mutation of Asp-52 $\rightarrow$ Asn and Pro-51 $\rightarrow$ Ser. This new mutant was termed Hb-Grenoble.

Mutants generated by recombinant DNA technology represent a somewhat different problem. In these cases the attempted modification is known, and it is merely necessary to check whether it was correctly carried out. This can be done by measuring the molecular weight with sufficient accuracy. For example, in recombinant bovine somatotropin (M_r 21,816) Asn99 was replaced by Gly, Pro, Ser, Glu, and Asp. The ES spectrum for each of these was found to be within $+3.5$ and $+7.3$ daltons, indicating a systematic positive error, but clearly proving that replacement had taken place in the first four cases. However, the Asn $\rightarrow$ Asp change (one dalton) could not be verified (97). In such a case where the protein is large and the mass difference of the replacing amino acid is small, one must cleave the protein into sufficiently smaller segments. For this purpose, cleavage with cyanogen bromide (CNBr) is sometimes the method of choice, because the low occurrence of methionine gives a small number of fragments in a mass range for which the accuracy of maLD (98) or ES is sufficient.

Identify Chemical Modifications

Chemical modifications of peptides lead to a generally predictable change in molecular weight and are thus subject to confirmation or identification by mass spectrometry. For example, from the increase by 126 daltons, or lack thereof, in the molecular weights of the tryptic peptides obtained from horse cytochrome c, which had been incubated with iodine, it was deduced that only one of the four tyrosines (position 74) was reactive under these conditions (99). Similarly, incubation of hemoglobin with styreneoxide resulted in an increase by 120 daltons in the molecular weight of tryptic peptide β133–144, indicating incorporation of one phenyl-hydroxyethyl group, which was then shown by tandem-MS to be attached to His143 (100).

Photoaffinity labeling is another area in which the change in molecular weight of the peptides obtained from the modified protein, upon proteolysis with one or more specific enzymes, pinpoints the region where the reactive molecule is attached to one or more amino acids. If the exact position must be known, it can again be determined by tandem-MS (101). In another investigation, a synthetic heptadecapeptide containing 4-benzoylphenylalanine was photolyzed in the presence of calmodulin to learn more about the binding of the latter with amphiphilic α-helical peptides. The FAB spectrum of the product demonstrated that it was a 1:1 adduct, and the molecular weights of peptides obtained with Endo Glu-C and trypsin, respectively, narrowed the region where the reaction took place in calmodulin. The final site of attachment was, in this case, determined by conventional Edman degradation (102).

Determine Extent of Glycosylation of Proteins

LOCATE GLYCOSYLATION SITES   The determination of the primary structure of a glycoprotein involves three quite different questions: (*a*) the amino acid sequence; (*b*) the structure or structures of the carbohydrate moieties; and (*c*) the identification of the points of attachment. This section relates to (*c*). In general, the amino acid sequence of the deglycosylated protein is determined first or, at least, in parallel, or it may be known from the corresponding DNA sequence.

Carbohydrates in glycoproteins are either N-linked to asparagine or O-linked to serine or threonine. For the former, there exists a consensus sequence (sequon) Asn-X-Y, where Y is Thr or Ser and X can be any amino acid. This requirement severely limits the positions where N-linked sugars can be found.

For the determination of the site of the carbohydrate attachment by mass spectrometry, two strategies have been used. In one, the glycoprotein is treated with endo-β-N-acetylglucosaminidase H (Endo H) that cleaves the first glycosidic bond between the carbohydrate and N-acetylglucosamine, which remains attached to Asn. After proteolysis, peptides 203 daltons heavier than the corresponding unglycosylated analogue are obtained. Glycosylated peptides, thus, can be easily recognized in the mass spectrum of the digest, which, if complex, should first be fractionated by HPLC. If the glycosylation of a site is not complete, as is sometimes the case, both molecular weights M_r and $M_r + 203$ are observed. The choice of proteolytic enzyme is dictated by the location of the sequons and the desire to produce peptides that contain only one potential glycosylation site.

The Endo H FAB-MS strategy was used to determine the glycosylation sites in yeast external invertase, which contains 14 sequons. In this particular case, there exists a related protein, internal invertase, which was believed to have the same amino acid sequence but not to be glycosylated. This was confirmed by the comparison of the FAB-MS data of two proteolytic digests of internal invertase with those of external invertase, which had first been treated with Endo H. The $(M + H)^+$ ions were either found at the same mass or differed by 203 daltons. The latter was the case for all 13 peptides containing the 14 sequons, because one chymotryptic peptide had the sequence Asn-Asn-Thr-Ser-Gly-Phe-Phe and thus contained two overlapping sequons. Two manual Edman steps performed on the HPLC fraction containing this peptide, and remeasuring the FAB spectrum after each step, revealed that only the first but not the second Asn had an N-acetyl glucosamine group attached (103).

Another strategy, developed by Carr & Roberts (104), makes use of the ability of peptide-N^4-[N-acetyl-β-glucosaminyl] asparagine amidase (PNGase F) to hydrolyze all of the commonly encountered N-linked oligosaccharides. As a result, the glycosylated Asn is converted to Asp, which differs by only one dalton. When the proteolytic digests obtained before and after treatment of the glycoprotein with PNGase F are fractionated by HPLC and analyzed by FAB-MS, the second digest exhibits $(M + H)^+$ ions that are absent in the first. This is because peptides containing large carbohydrate moieties do not ionize well by FAB or, if they do, their molecular weights are much higher than that of the deglycosylated peptide. These newly appearing $(M + H)^+$ ions indicate that the corresponding peptides had previously been glycosylated. They can be located in the amino acid sequence, based on the specificity of the proteolytic enzyme that was used in the experiment and taking into account that Asn was converted to Asp. A recent review details these approaches and discusses various applications (105).

This methodology showed that in tissue plasminogen activator, three of the four potential N-glycosylation sites are occupied, but the fourth is not (106). Recombinant soluble CD4 receptor is a 369-amino-acid long glycoprotein of M_r ~55,000 that contains two Asn sequons. The tryptic digest, after PNGase F treatment, revealed four new peptides detectable by FAB-MS, two of each of which encompassed the two glycosylation sites Asn271 and Asn300, respectively. From the absence of any signal for the corresponding $(M + H)^+$ ions (1 dalton less) in the digest of the intact glycoprotein, the authors concluded that both sites are at least 90% glycosylated (83). The redundant molecular weight information on all the tryptic peptides generated in these experiments, as well as data from digests produced with Endo Glu-C and CNBr, established that the recombinant protein had the sequence corresponding to the cDNA cloned and that it was not modified at either the N- or the C-terminus. The two carbohydrate moieties are large biantennary sugars, heterogeneous in neuraminic acid and fucose content.

In contrast to N-glycosylation sites, there is no consensus sequence for O-glycosylations on serine or threonine. There is also no enzyme that specifically removes O-linked sugars. Fortunately, the carbohydrate groups bound to serine or threonine are generally small, and the $(M + H)^+$ ion of the glycosylated peptide can be observed in the FAB or maLD mass spectrum of a proteolytic digest of the glycoprotein. The mass of this ion does not correspond to any of the predicted unglycosylated peptides (except by coincidence), but is higher by the mass of a single carbohydrate or a combination. Unless there is only one serine or threonine present in the peptide, the specific O-glycosylation sites must be determined by tandem mass spectrometry (107, 108).

DETERMINE CARBOHYDRATE CONTENT OF GLYCOPROTEINS   As implied by the foregoing example, glycoproteins are, in general, homogeneous in the protein portion but heterogeneous with respect to the carbohydrate part. Their molecular weight is, therefore, the abundance-weighted average of the various components, which often cannot be resolved by either maLD or ES. It is, however, possible to determine the total carbohydrate content by measuring the average molecular weight by maLD and subtracting the known mass of the protein. For endoglucanase I from *Trichoderma reesei,* a molecular weight of $52,100 \pm 130$ was determined by maLD. The difference between this value and the protein molecular weight calculated from the DNA-derived amino acid sequence gave 6280 ± 130 daltons for the carbohydrate moiety (109). Alternatively, one can determine the molecular weight of the glycoprotein before and after enzymatic deglycosylation. The maLD spectrum of violet phosphatase gave a molecular weight of 35,050, which shifted downwards by 2400 daltons upon treatment with Endo H (110). The shift represents the mass of the carbohydrate moiety, except the penultimate N-acetylglucosamines. Treatment with PNGase F would perhaps be a better choice.

The former strategy (subtraction of calculated protein mass) was used in an extensive study of monoclonal antibodies. From the molecular weights determined by maLD of intact antibody (Figure 3 is an example), as well as reduced light and heavy chains, the DNA-derived protein molecular weights were subtracted to arrive at the average carbohydrate mass. For a number of antibodies, this value ranged from 2680 ± 200 to 3620 ± 540 daltons. The large uncertainty is due to the error in the molecular weight measurements of these large proteins. In the same work, maLD was used to determine the average number of certain chelators and anticancer drugs covalently bound to a monoclonal antibody (111).

Locate Disulfide Bonds

The linear sequence of a protein is practically always known when the question of the location of -S-S- bridges arises. A mass spectrometric strategy had first been proposed by Morris & Pucci (112) using insulin as a model. In principle, the oxidized form of the protein is digested with a suitable enzyme and the digest analyzed by FAB-MS. The same digest is then reanalyzed after reduction of all -S-S- bonds. The $(M + H)^+$ ions of peptides not involved in disulfide bridges are the same in both sets of mass spectra; those with an internal -S-S- bond are found two daltons higher after reduction, the signal of those that represented two peptides linked by an -S-S- bridge disappears after reduction, and two new $(M + H)^+$ ions are observed that represent the two cysteine-containing peptides. From the specificity of the enzyme used and the known sequence of the protein, the disulfide bonds can be assigned (113, 114).

There are a few problems with this simplistic approach. For the assignments to be correct, disulfide bond interchange, which occurs at high pH, has to be strictly avoided. Often the unreduced protein is resistant to enzymatic cleavage. To overcome both these problems, partial acid hydrolysis has recently been suggested (115). Finally, if two cysteines involved in two different disulfide bonds are located close together in the primary sequence, it may be difficult to obtain peptides containing only one -S-S- bridge. In that case, manual Edman degradations must be performed until the first Cys is reached. This approach has been used in the assignment of the three disulfide bonds in recombinant insulinlike growth factor I and a biologically less active by-product, which turned out to differ by the interchange of two -S-S- bridges involving two neighboring cysteines (116).

Using either Endo Glu-C, trypsin, or both for proteolysis and DTT for reduction, it was possible to determine the three disulfide bridges in rs-CD4 receptor (83), the two bridges in recombinant human growth hormone and its methionine analogue (117), and the three bridges in antithrombin III (118). The two disulfide bonds in the amylase inhibitor PAIM I were also deduced from the $(M + H)^+$ ions of the disulfide-bonded and reduced peptides observable in the FAB spectra, except that a separate reductive step was not necessary (119). Reduction may have occurred accidentally or during FAB ionization. It had been noted previously that the Xe° beam slowly induces reduction if thioglycerol or a DTT/DTE mixture is used as the matrix, an observation that was the basis of an earlier method for FAB-MS-based disulfide bridge assignment (120). Recently, it was reported that the addition of 5% trifluoroacetic acid to the matrix prevents the reduction (121, 122).

Instead of cleaving the -S-S- bond reductively to two cysteines, it can also be oxidized with performic acid to two cysteic acids (123). This approach has the disadvantage that methionine and tryptophan are also oxidized. However, this fact can be used to recognize the presence of these amino acids, based on the shift in molecular weight by 16 or 32 daltons.

Partially Sequence Peptides by Chemical or Enzymatic Truncation

As has been pointed out, FAB ionization and the other "soft" ionization methods described above produce mainly ionized molecules, which undergo little fragmentation. This is an advantage when one is working with mixtures, but a disadvantage if one wishes to obtain structural information. In the first few years after the FAB technique was discovered, peptide spectra were interpreted in terms of their amino acid sequence. When using large samples (2–50 nmol), sufficient fragment ions were produced that could be correlated with mostly known sequences (124), but it is now generally accepted that, at best, a partial sequence can be obtained (125). This is sometimes enough to solve a given problem, for example, when there is a specific question about a published sequence. Such was the case for human myelin basic protein in the region from position 45 to 89. A tryptic digest of 20 nmol of that 45-amino-acid peptide, fractionated by HPLC, gave FAB spectra that exhibited sufficient fragmentation to assign His-Gly and Glu-Asn to positions 77–78 and 83–84, respectively (126), which previously had been in doubt.

Sometimes an attempt is made to overcome the lack of mass spectrometric fragmentation by the use of chemical or enzymatic stepwise degradation, and redetermination of the molecular weight after each step. Manual Edman degradation is used for N-terminal and carboxypeptidases for C-terminal truncation (127). With immobilized exopeptidases, cleaner truncation products are obtained (128). The major difficulty arises from the different rate at which carboxypeptidases cleave peptide bonds, depending on the amino acid encountered. Even in careful time-dependent studies, a certain truncated peptide may be missed because a fast-cleaving amino acid is followed by a slow one. With a mixture of carboxypeptidases Y and B, it was possible to identify the 14 C-terminal amino acids of synthetic human parathyroid hormone, which is 34 amino acids long (129). In this work, maLD was used to measure the molecular weights of the consecutively truncated peptides, but the same could have been accomplished by FAB-MS.

Stepwise removal of successive N-terminal amino acids by the Edman degradation and measuring the molecular weight of the truncated peptide are most useful if the assignment of a peptide to a specific region of a known protein sequence is in doubt for one reason or another, most often because that particular mass would fit two or more places (70). For actual sequencing, it is too tedious compared with tandem mass spectrometry. The major problem with these subtractive methods is that with each additional step, the signal becomes weaker and the background rises, and soon the point is reached at which the assignment of the new molecular ion becomes ambiguous.

Identify Noncovalent Protein Complexes

During the ionization processes discussed above, noncovalently bound complexes almost always dissociate and the $(M + H)^+$ ions of the individual components are observed. For example, both maLD and ES spectra of hemoglobin exhibit separate signals for α-chain, β-chain, and heme. In a few cases, undissociated subunit oligomers have been observed by maLD. For example, streptavidin gives an abundant signal at m/z 52,510 corresponding to the tetramer, when measured with nicotinic acid as the matrix and applied in 10% aqueous ethanol solution. Apparently, the organic solvent has a strong effect and causes more and more dissociation into the subunits with increasing concentration (43).

An important observation is the recent finding that very tight complexes of proteins with other molecules can be detected by ES-MS. The ES spectrum of the cytoplasmic receptor for cyclosporin exhibits an abundant $(M + 7H)^{7+}$ ion at m/z 1688.7. Upon addition of the macrolide FK 506 (an immunosuppressive drug of M_r 804), a new peak appears at m/z 1803.1 corresponding to the 1:1 complex. The same effect is observed with another macrolide rapamycin (M_r 913), and it is even possible to estimate the relative ratio of their binding constants from the relative peak height when adding a 1:1 mixture of the macrolides (130). A potentially even more significant finding is the observation that one can monitor the hydrolysis of hexa-N-acetylglucosamine by hen egg white lysozyme, by infusing the reaction mixture into an ES-MS. The spectrum first exhibits multiply charged ions (maximum of 8^+) for lysozyme, as well as the adduct with the hexasaccharide. Over a 60 min period, partial degradation down to the trisaccharide is observed by the appearance of further adduct ions, the relative abundances of which correlate with the known association constants of the enzyme and oligo-N-acetylglucosamines (131). If generally applicable, this methodology could become an important tool in enzyme-substrate studies.

SEQUENCING OF PEPTIDES AND PROTEINS BY TANDEM MASS SPECTROMETRY

To obtain structural information by mass spectrometry, the molecule to be studied must undergo fragmentation of one or more bonds in a manner that ions are formed, the m/z ratio of which can be related to the structural features. Protein-derived peptides are linear molecules whose structures are constrained by two facts: the linear backbone consists of repetitive α-amino acid amide bonds, and the substituents on each α-carbon are the side chains of one of the 20 naturally occurring amino acids, unless they are posttranslationally modified. As has been mentioned, "soft" ionization methods generate singly or polyprotonated molecular ions that contain insufficient excess energy to fragment. However, by converting some of the kinetic energy from the peptide ion into vibrational energy, fragmentation can be achieved. This can be accomplished in a tandem mass spectrometer.

Principles of Tandem Mass Spectrometry

As outlined earlier, when a peptide is ionized by FAB, it forms an $(M + H)^+$ ion that is then mass analyzed to determine its molecular weight. If structural information is needed, the $(M + H)^+$ ion exiting the first mass spectrometer (MS1) is passed into a region (collision cell) containing a neutral gas (He, Ar, or Xe) at $\sim 10^{-3}$ torr. Upon collision with an atom, the $(M + H)^+$ ion ("precursor ion") fragments to a considerable extent and the resulting "product ions" are then mass analyzed in the second mass spectrometer (MS2) of the tandem instrument.

Figure 5 schematically depicts the ionization of a mixture of five peptides in MS1, which is scanned to produce the spectrum (like the one shown in Figure 1) exhibiting the isotope clusters of the five $(M + H)^+$ ions. After the scan, MS1 is then set for the ^{12}C-only component of the cluster for the peptide of interest, so that this selected ion beam passes into the collision cell. The fragments (F_1-F_n) produced upon collision-induced decomposition (CID) of the precursor ion (part of which remains intact) are then mass analyzed by scanning MS2 to record the product ion (or CID) spectrum. The latter is monoisotopic, which has the advantage that MS2 can be operated at somewhat lower than unit mass resolution (132). Since it takes only 1–2 min to record the spectrum, one can then set MS1 for the next precursor ion and obtain its CID spectrum, and so on. However, as the sample is continuously ionizing, the $(M + H)^+$ ions of the components not being analyzed at the time (P_1-P_3 and P_5 in Figure 5) are lost. Thus, it is best to keep the mixture simple (five precursors of interest, or fewer). More complex mixtures are, therefore, fractionated by HPLC, but it is by no means necessary to purify each peptide, and this is one of the major advantages of tandem mass spectrometry.

The CID spectra can be produced with a variety of instruments. The principle illustrated in Figure 5 involves either two double-focusing magnetic mass spectrometers, three quadrupoles (the second of which functions as the collision cell), or a hybrid of the two, a magnetic spectrometer followed by two quadrupoles. From a practical point of view, the major difference is that in a magnetic instrument the precursor ion has a kinetic energy of 5–10 keV (producing single high-energy collisions), but in the triple quadrupole or hybrid it has 100 eV or less (resulting in multiple low-energy collisions). A

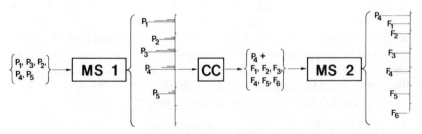

Figure 5 Principle of tandem mass spectrometry. See text for details. (From Ref. 132, reproduced with permission from John Wiley & Sons, Ltd., copyright 1988.)

careful comparison of the performance of a magnetic instrument with a hybrid found that data obtained with the former are generally more complete and reliable (133).

Characteristics and Interpretation of CID Mass Spectra of Peptides

The fragmentation processes of protonated peptides by CID are now quite well understood (132) and are summarized in Figure 6. The bond cleavages indicated at the top of the figure lead to fragments of type a_n, b_n, and c_n, if the charge is retained on the N-terminal piece. The exception is that for a c_n ion, two daltons have to be added because it retains the original protonating hydrogen and picks up one more from the other side of the peptide. The x_n, y_n, and z_n ions are formed when the charge is retained on the C-terminal, with the y_n ions having added two hydrogens analogous to the c_n ions. It is obvious that ions of the same type and incremented n will differ by the mass of the amino acid minus H_2O ("residue mass", see Table 1) at that place in the sequence. From a single, complete ion series, therefore, one can deduce the amino acid sequence of a peptide, except that leucine and isoleucine, and lysine and glutamine, cannot be differentiated at this point. A single, complete ion series is not generally observed, and overlapping N- and C-terminal ions are used to arrive at the sequence.

These fragmentation processes occur both with high- and low-energy CID spectra, except that in the latter the fragments often undergo further loss of H_2O or NH_3. The d_n and w_n ions (middle of Figure 6) require cleavage of the

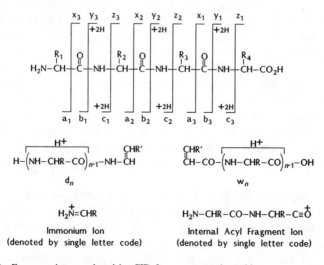

Figure 6   Fragment ions produced by CID from protonated peptides.

Table 1   Residue mass [-NH-CHR-CO-] of common amino acids

Amino acid	Letter code three	single	Mass	Amino acid	Letter code three	single	Mass
Glycine	Gly	G	57	Aspartic acid	Asp	D	115
Alanine	Ala	A	71	Glutamine	Gln	Q	128
Serine	Ser	S	87	Lysine	Lys	K	128
Proline	Pro	P	97	Glutamic acid	Glu	E	129
Valine	Val	V	99	Methionine	Met	M	131
Threonine	Thr	T	101	Histidine	His	H	137
Cysteine	Cys	C	103	Phenylalanine	Phe	F	147
Isoleucine	Ile	I	113	Arginine	Arg	R	156
Leucine	Leu	L	113	Tyrosine	Tyr	Y	163
Aspargine	Asn	N	114	Tryptophan	Trp	W	186

backbone and the side chain, and are found only in high-energy CID spectra. As can be noticed from the structures, both d_n (134) and w_n (135) ions differ by 14 daltons for leucine vs isoleucine, which therefore can be differentiated when at least one of these two fragments is observed.

The ions shown at the bottom of Figure 6 have lost fragments from the N-terminal and the C-terminal portion of the peptide and carry little or no sequence information. Both types are found in high- and low-energy CID spectra, and the immonium ion can be used to identify some or all of the amino acids present in the peptide.

A typical high-energy CID spectrum is shown in Figure 7. It was obtained from a tryptic peptide for which the MS1 scan indicated an $(M + H)^+$ ion of m/z 1001.64. The labeling of the peaks is based on the interpretation of their m/z values. The spectrum exhibits an almost complete set of b_n ions, revealing the sequence from the N-terminus towards the C-terminus and an overlapping y_n series in the opposite direction. The m/z values of the w_3, w_4, w_5, and w_8 ions require that the third, fourth, and fifth amino acids from the C-terminus are all leucines, while the eighth is isoleucine. The presence and position of basic amino acids influence fragmentation. The spectrum shown in Figure 7 exhibits many y_n and w_n ions because the Lys at the C-terminus retains the proton efficiently. The more basic Arg has an even stronger effect. Conversely, a basic amino acid at the N-terminus favors a_n and d_n ions, and a fixed charge (quaternary ammonium ion) has an even more dramatic effect (132).

It may appear to be difficult to recognize a particular ion series in a spectrum seemingly as complex as that shown in Figure 7, but one has to keep in mind that peaks belonging to any one of the backbone fragmentations (a_n, b_n, c_n, x_n, y_n, and z_n) can differ in m/z by only one of the 18 mass values listed in Table 1. Furthermore, fragments a_n, b_n, and y_n generally form a continuous

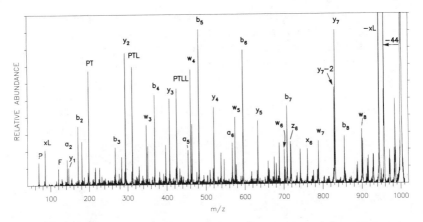

Figure 7 High-energy CID spectrum of the $(M + H)^+$ ion, m/z 1001.64 (monoisotopic), of a tryptic peptide (Gly-Ile-Pro-Thr-Leu-Leu-Leu-Phe-Lys) from *E. coli* thioredoxin. See text for interpretation.

series. For example, the mass difference between the peaks labeled b_2 and b_3 is 97 daltons, indicating Pro in position 3. Peak b_4 is 101 daltons higher, placing Thr in position 4, etc (see Table 1). That Ile rather than Leu is in position 2 is revealed by the w_8 ion. A y_n-2 ion is characteristic of proline, which is at the seventh position from the C-terminus. The internal acyl fragment ions PT, PTL and PTLL are recognized once the sequence is established and lend support to the correct interpretation of the spectrum. The peaks labeled P, xL, and F are the immonium ions, indicating the presence of Pro, Leu and/or Ile, and Phe. The interpretation of these spectra can be carried out manually or with the aid of a computer program (136).

For reasons of simplicity, this example represents only a nonapeptide, but the same type of information can be obtained for peptides with molecular weights up to 2000–2500 daltons (137), and even larger ones reveal at least partial sequences that often suffice to solve a particular problem (14). One also has to keep in mind in the sequencing of a protein that more than one proteolytic digest is generated, and that the CID spectra of the peptides of one provides redundancy with the other set and can be used to resolve remaining ambiguities (see below). This task can also be carried out with the aid of a computer program (138).

The above discussion dealt exclusively with the CID of singly charged precursor ions. The advent of ES made it possible to dissociate multiply charged ions, which is a promising approach, even though the data may be more difficult to interpret. The simplest cases are tryptic peptides, because they produce precursor ions with one charge on either end and thus chiefly b_n and y_n ions (64). The CID spectra of more highly charged precursor ions become more difficult to interpret and provide incomplete sequence information. Because a high charge state increases the kinetic energy of the ion and provides repulsive forces within the multiply charged precursor ion, large molecules still dissociate to a considerable extent (139). Even proteins as large as serum albumin fragment in such a way that significant b_n ions with 3–5 charges are formed from the N-terminal 15–30 amino acids (140). In this case, fragmentation is caused by the nozzle-skimmer potential. There is no specific precursor selection and the charge state of the fragments is not known. Still, it was possible to compare the spectra of serum albumin of eight species of known sequence and extend the short, known N-terminal sequence (1–24) in the dog protein to position 31 (140).

Strategies for Mass Spectrometric Sequencing of Proteins

With the availability of methods, such as maLD and ES, that permit the accurate molecular weight determination of a protein, this measurement is now the first step. The resulting data must not only agree with the final results of the sequence determination but also reveal the degree of homogeneity of the protein preparation at hand. Needless to say, simply knowing how large the protein is influences the detailed strategy.

Unless it is known that there is no cysteine present, the protein is reduced and alkylated prior to digestion with proteolytic enzymes. Redetermination of the molecular weight reveals the total number of cysteines present, because of the increase in mass by 57, 58, or 105 daltons per Cys upon carbamidomethylation, carboxymethylation, or ethylpyridylation, respectively, of the reduced protein.

In order to generate two sets of very different and overlapping sets of proteolytic peptides, the protein is then digested with trypsin (or Endo Lys-C), which generates peptides with a basic amino acid at the C-terminus, and with Endo Glu-C, which provides peptides ending in glutamic acid (except the original C-terminus of the protein). Experimental conditions for small-scale digests for mass spectrometry have been summarized by Lee & Shively (141). About 0.5–2 nmol of protein per digest should suffice to deduce the complete sequence by tandem mass spectrometry. After fractionation by HPLC into simple mixtures, the molecular weight of each peptide present is determined, and those exhibiting sufficiently abundant $(M + H)^+$ ions below m/z 3000 are subjected to CID in the tandem mass spectrometer. Sequences derived from these CID spectra are then assembled, making use of the overlapping peptides from the two digests, as well as the molecular weights of larger peptides for which no sequences are available but that encompass two or more sequenced peptides. Frequently, some ambiguities remain, such as two or more ways to arrange the order of the peptides. In such cases, a third enzyme that has the specificity that would distinguish these remaining possibilities (or CNBr cleavage) may be chosen.

A number of thioredoxins and glutaredoxins have been sequenced by this approach in the author's laboratory. In Figure 8, the assembly of such a set of data to the final structure of the glutaredoxin isolated from rabbit bone marrow is summarized (142). The underlined proteolytic peptides were used in deriving the sequence. Heavy underlining indicates the assignment of the amino acid from CID spectra, while light underlining represents peptides that were not sequenced but whose molecular weights were determined, or sections where the CID spectrum did not reveal fragmentation of the peptide bond. Asterisks indicate positions where mass differences alone did not permit unambiguous assignments. The Lys/Gln mass equivalence could be decided in many cases based on cleavage of trypsin, except for positions 10 and 19 because they are followed by Pro, which prevents cleavage at the preceding Lys. The absence of tryptic cleavage should not be taken as proof that the amino acid of residue mass 128 is Gln because the digest may not have been complete. Those HPLC fractions containing these questionable assignments were, therefore, reacted with acetic anhydride, which acetylates the free N-terminus and the ϵ-amino group of Lys and adds 42 daltons for each of these groups present. For the differentiation of Leu from Ile, a charge-directing group must be present at either the N-terminus (to promote d_n ions) or the C-terminus (to promote w_n ions). The fractions containing the peptides encompassing amino acids 47, 56, 60, and 63 were derivatized in a manner that places a fixed positive charge at the N-terminus, which forces the generation of d_n ions for the differentiation of Leu and Ile (143).

It should be noted that this glutaredoxin has a blocked N-terminus, but this did not, of course, interfere with sequencing it. That structural feature raised doubts about the published structure of the glutaredoxin from calf thymus (144), which showed an N-terminal pyroglutamic acid and missed four amino acids. The error was easily corrected (145) by a mass spectrometric experiment on a chymotryptic digest, which revealed the correct N-ac-Ala N-terminus and the presence of Thr68-Val-Pro-Arg71, a tryptic peptide that had been missed in the earlier work. This is another example of the efficiency of MS and tandem-MS.

Using the strategy outlined for rabbit bone marrow glutaredoxin, the primary structures of the thioredoxins from *Chlorobium thiosulfatophilum* (146), *Chromatium vinosum* (21), *Rhodosporillum rubrum* (147), and from rabbit bone marrow (148) were determined. For the determination of the primary structure of a protease inhibitor from *Sarcophaga bullata*, a small cysteine-rich protein, specific cleavage at the N-terminus of that amino acid by the Stark reaction (149) was used to obtain peptides that could be sequenced by CID-MS (150). This is a useful reaction, which had been rarely employed previously, because the resulting peptides have a blocked N-terminus and are thus not suitable for the Edman procedure.

Hunt et al (151) are taking a similar approach, using either a triple

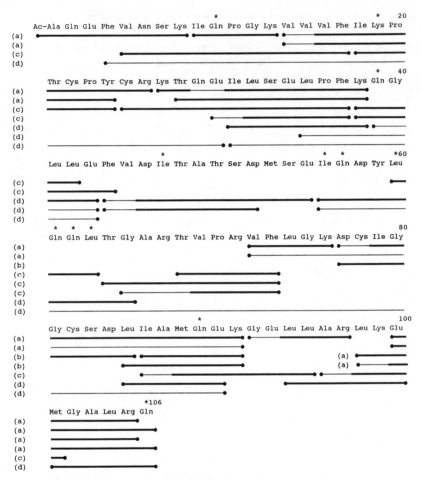

Figure 8 The amino acid sequence of glutaredoxin from rabbit bone marrow, as deduced from the CID spectra of a set of proteolytic digests. For details, see text. (From Ref. 142, reproduced with permission.)

quadrupole or quadrupole Fourier transform mass spectrometer (Q-FTMS) (152, 153), which employs low-energy collisions, and therefore does not produce the d_n and w_n ions necessary to differentiate Leu from Ile. Also, because of the lower resolution of quadrupole mass spectrometers and to gain sensitivity, the entire isotope cluster of the $(M + H)^+$ ion is transmitted and subjected to CID. Often the CID spectra of the peptides are measured before and after methyl ester formation to give a count of the COOH groups present and to aid in the interpretation, because each such group adds 14 daltons (151).

This methodology has been used to determine the amino acid sequence of two murine 9-kDa calbidin isoforms. Both the major and the minor components have a blocked N-terminus and differ only by an additional Gln in the minor isoform inserted after position 43 (154). The location of this Gln, along with other genetic evidence, makes it likely that it is due to alternative gene splicing.

For practical reasons, combinations of Edman results and CID mass spectra are often used to determine the amino acid sequence of proteins. This was the case for the highly homologous purple acid phosphatase and uteroferrin (155), bovine retinaldehyde-binding protein (156), steroid-binding protein from rabbit serum (157), the acidic subunit of Mojave toxin (158), and parvalbumin from chicken muscle and thymus (159).

An interesting application is the exploration of receptor topography by limited proteolysis of exposed sections of a protein. When nicotinic acetylcholine receptor embedded in reconstituted membrane was briefly exposed to either trypsin or Endo Glu-C, a number of peptides were released. Since the sequence of the receptor is known, their molecular weights, as determined by FAB-MS, revealed that they mainly came from the C-terminus of the α-subunit, which extends into the cytoplasmic space, and the N-terminus on the other side of the membrane (160).

Posttranslational Modifications

The protein chains produced along the RNA template in the ribosome are rarely functional as such, and undergo a variety of structural changes, either immediately or later. These modifications, therefore, cannot be deduced from the DNA of the gene coding for the protein. Often, they also cannot be determined by the Edman degradation, either because they render it inapplicable (such as an acylated N-terminus), or they revert to the unmodified amino acid under the Edman conditions (such as γ-carboxy Glu), or they are too hydrophilic to be successfully extracted or chromatographed after release as the phenylthiohydantoin derivative (such as those of glycosylated amino acids). In these cases, therefore, mass spectrometry is the method of choice for the direct identification of the modification.

Of these, N-glycosylation has already been discussed, because it can generally be detected by molecular weight determination of peptides containing a single sequon. For O-glycosylation, tandem mass spectrometry is necessary to locate the site if there is more than one Ser or Thr in the peptide (107).

Phosphorylation of Ser and Thr is another important modification of proteins, the function of which is often regulated by this structural change. It is sometimes possible to determine the location of a phosphate moiety by the judicious choice of proteolytic enzymes, which produce a peptide with only a single Ser or Thr, the molecular weight of which indicates whether or not it is

80 daltons heavier than the native peptide (161). If that is not easily possible, one has to take recourse to CID spectra of the $(M + H)^+$ ion, which are characterized by an abundant $(M + H - H_3PO_4)^+$ ion, and the same holds for fragment ions containing the phosphorylation site. Examples are Troponin T, which has a phosphorylated N-acetyl-Ser at the N-terminus (14), and Photosystem II proteins, which have N-acetyl-phospho-Thr at that position (162). A more detailed review of the mass spectrometry of phosphorylated and sulfated peptides is available (163).

Bovine class III β-tubulin is modified in two ways. The FAB spectrum of the C-terminal CNBr peptide, which had been esterified with methanol, exhibited five peaks spaced 143 daltons apart, indicating the addition of 1–4 Glu residues. The CID spectrum revealed that these are attached to the γ-carboxyl of the first Glu in this peptide. Peaks in the FAB spectrum that were 98 daltons lower than the major peaks indicated phosphorylation at the only hydroxyamino acid (Ser) present in these peptides (164).

An unusual posttranslational modification has been found by CID-MS of a tryptic decapeptide from ribulose bisphosphate carboxylase (165). The spectrum showed only fragments that retained at least half of the molecule, due to the presence (in position 6) of Nϵ-trimethyl Lys, the quaternary ammonium cation of which controlled the fragmentation.

The action of exopeptidases in vivo generates proteins with "ragged ends" either at the N-terminus or the C-terminus. They manifest themselves in the mass spectra of proteolytic digests by the presence of peptides that differ by the residue mass (Table 1) of one or more amino acids. The correctness of this assumption is easily established by tandem mass spectrometry. Two examples have been mentioned earlier (89, 92).

Literature Cited

1. Barber, M., Bordoli, R. S., Sedgwick, R. D., Tyler, A. N. 1981. *J. Chem. Soc. Chem. Comm.* 1981:325–27
2. Karas, M., Hillenkamp, F. 1988. *Anal. Chem.* 60:2299–301
3. Macfarlane, R. D., Torgerson, D. F. 1976. *Science* 191:920–25
4. Meng, C. K., Mann, M., Fenn, J. B. 1988. *Z. Phys. D* 10:361–68
5. Biemann, K., Martin, S. A. 1987. *Mass Spectrom. Rev.* 6:1–76
6. Burlingame, A. L., Millington, D. S., Norwood, D. L., Russell, D. H. 1990. *Anal. Chem.* 62:268R–303R
7. Burlingame, A. L., Castagnoli, N. Jr., eds. 1985. *Mass Spectrometry in the Health and Life Sciences.* Amsterdam: Elsevier. 638 pp.
8. Burlingame, A. L., McCloskey, J. A., eds. 1990. *Biological Mass Spectrometry.* Amsterdam: Elsevier. 700 pp.
9. McEwen, C. N., Larsen, B. S., eds. 1990. *Mass Spectrometry of Biological Materials* New York: Dekker. 515 pp.
10. McCloskey, J. A., ed. 1990. *Methods in Enzymology. Mass Spectrometry,* Vol. 193. San Diego: Academic. 960 pp.
11. Desiderio, D. M., ed. 1991. *Mass Spectrometry of Peptides.* Boca Raton: CRC Press. 421 pp.
12. Jennings, K. R., Dolnikowski, G. G. 1990. See Ref. 10, pp. 37–61
13. Papayannopoulos, I. A., Biemann, K. 1992. *Peptide Res.* 5:83–90
14. Biemann, K., Scoble, H. A. 1987. *Science* 237:992–98
15. Martin, S. A., Costello, C. E., Biemann, K. 1982. *Anal. Chem.* 54: 2362–68
16. Aberth, W., Straub, K. M., Burlingame, A. L. 1982. *Anal. Chem.* 54:2029–34

17. De Pauw, E. 1986. *Mass Spectrom. Rev.* 5:191–212
18. Tip, L., Versluis, C., Dallinga, J. W., Heerma, W. 1990. *Anal. Chim. Acta* 241:219–25
19. Naylor, S., Findeis, A. F., Gibson, B. W., Williams, D. H. 1986. *J. Am. Chem. Soc.* 108:6359–63
20. Falick, A. M., Maltby, D. A. 1989. *Anal. Biochem.* 182:165–66
21. Johnson, R. S., Biemann, K. 1987. *Biochemistry* 26:1209–14
22. Hassell, A. M., Johanson, K. O., Goodhart, P., Young, P. R., Holskin, B. P., et al. 1989. *J. Biol. Chem.* 264:4948–52
23. Siegel, M. M., Tsao, R., Doellins, V. W., Hollander, I. J. 1990. *Anal. Chem.* 62:1536–42
24. Ito, Y., Takeuchi, T., Ishi, D., Goto, M. 1985. *J. Chromatogr.* 346:161–66
25. Caprioli, R. M., Moore, W. T. 1990. See Ref. 10, pp. 214–37
26. Caprioli, R. M., Fan, T., Cottrell, J. S. 1986. *Anal. Chem.* 58:2949–54
27. Caprioli, R. M., Moore, W. T., DaGue, B., Martin, M. 1988. *J. Chromatogr.* 443:355–62
28. Bell, D. J., Brightwell, M. D., Neville, W. A., West, A. 1990. *Rapid Commun. Mass Spectrom.* 4:88–91
29. Henzel, W. J., Bourell, J. H., Stults, J. T. 1990. *Anal. Biochem.* 187:228–33
30. Jones, D. S., Heerma, W., Van Wassenaar, P. D., Haverkamp, J. 1991. *Rapid Commun. Mass Spectrom.* 5:192–95
31. Cappiello, A., Palma, P., Papayannopoulos, I. A., Biemann, K. 1990. *Chromatographia* 30:477–83
32. Moseley, M. A., Deterding, L. J., Tomer, K. B., Jorgenson, J. W. 1989. *J. Chromatogr.* 480:197–209
33. Caprioli, R. M., Moore, W. T., Marlin, M., DaGue, B. B., Wilson, K., et al. 1989. *J. Chromatogr.* 480:247–57
34. Caprioli, R. M., Fan, T. 1986. *Biochem. Biophys. Res. Commun.* 141:1058–65
35. Caprioli, R. M., Moore, W. T., Fan, T. 1987. *Rapid Commun. Mass Spectrom.* 1:15–18
36. Caprioli, R. M., Moore, W. T., Petrie, G., Wilson, K. 1988. *Int. J. Mass Spectrom. Ion Processes* 86:187–99
37. Caprioli, R. M. 1990. *Anal. Chem.* 62:477A–85A
38. Roepstorff, P. 1989. *Acc. Chem. Res.* 22:421–27
39. Nielsen, P. F., Roepstorff, P. 1989. *Biomed. Environ. Mass Spectrom.* 18:131–37
40. Buko, A. M., Sarin, V. K. 1990. *Rapid Commun. Mass Spectrom.* 4:541–45
41. Vorst, H. J., Van Tilborg, M. W. E.

M., Van Veelen, P. A., Tjaden, U. R., Van der Greef, J. 1990. *Rapid Commun. Mass Spectrom.* 4:202–5
42. Van Veelen, P. A., Tjaden, U. R., Van der Greef, J., De With, N. D. 1991. *Org. Mass Spectrom.* 26:345–46
43. Karas, M., Bahr, U., Ingendoh, A., Nordhoff, E., Stahl, B., et al. 1990. *Anal. Chim. Acta* 241:175–85
44. Karas, M., Bahr, U., Hillenkamp, F. 1989. *Int. J. Mass Spectrom. Ion Processes* 92:231–42
45. Karas, M., Ingendoh, A., Bahr, U., Hillenkamp, F. 1989. *Biomed. Environ. Mass Spectrom.* 18:841–43
46. Beavis, R. C., Chait, B. T. 1989. *Rapid Commun. Mass Spectrom.* 3:436–39
47. Beavis, R. C., Chait, B. T. 1989. *Rapid Commun. Mass Spectrom.* 3:432–35
48. Beavis, R. C., Chait, B. T. 1990. *Anal. Chem.* 62:1836–40
49. Overberg, A., Karas, M., Bahr, U., Kaufmann, R., Hillenkamp, F. 1990. *Rapid Commun. Mass Spectrom.* 4:293–96
50. Overberg, A., Karas, M., Hillenkamp, F. 1991. *Rapid Commun. Mass Spectrom.* 5:128–31
51. Strupat, K., Karas, M., Hillenkamp, F. 1991. *Int. J. Mass Spectrom. Ion Processes.* 111:89–102
52. Annan, R. S., Köchling, H. J., Hill, J. A., Biemann, K. 1992. *Rapid Commun. Mass Spectrom.* 6:298–302
53. Hill, J. A., Annan, R. S., Biemann, K. 1991. *Rapid Commun. Mass Spectrom.* 5:395–99
54. Mann, M., Meng, C. K., Fenn, J. B. 1989. *Anal. Chem.* 61:1702–8
55. Meng, C. K., McEwen, C. N., Larsen, B. S. 1990. *Rapid Commun. Mass Spectrom.* 4:147–50
56. Meng, C. K., McEwen, C. N., Larsen, B. S. 1990. *Rapid Commun. Mass Spectrom.* 4:151–55
57. Larsen, B. S., McEwen, C. N. 1991. *J. Am. Soc. Mass Spectrom.* 2:205–11
58. Henry, K. D., Williams, E. R., Wang, B. H., McLafferty, F. W., Shabanowitz, J., et al. 1989. *Proc. Natl. Acad. Sci. USA* 86:9075–78
59. Henry, K. D., McLafferty, F. W. 1990. *Org. Mass Spectrom.* 25:490–92
60. Fenn, J. B., Mann, M., Meng, C. K., Wong, S. F., Whitehouse, C. M. 1990. *Mass Spectrom. Rev.* 9:37–70
61. Loo, J. A., Edmonds, C. G., Udseth, H. R., Smith, R. D. 1990. *Anal. Chem.* 62:693–98
62. Smith, R. D., Loo, J. A., Edmonds, C. G., Barinaga, C. J., Udseth, H. R. 1990. *Anal. Chem.* 62:882–99
63. Chowdhury, S. K., Katta, V., Chait, B.

T. 1990. *J. Am. Chem. Soc.* 112:9012–13

64. Covey, T. R., Huang, E. C., Henion, J. D. 1991. *Anal. Chem.* 63:1193–200
65. Huang, E. C., Henion, J. D. 1991. *Anal. Chem.* 63:732–39
66. Hemling, M. E., Roberts, G. D., Johnson, W., Carr, S. A., Covey, T. R. 1990. *Biomed. Environ. Mass. Spectrom.* 19:677–91
67. Smith, R. D., Olivares, J. A., Nguyen, N. T., Udseth, H. R. 1988. *Anal. Chem.* 60:436–41
68. Loo, J. A., Udseth, H. R., Smith, R. D. 1989. *Anal. Biochem.* 179:404–12
69. Webster, T. A., Gibson, B. W., Keng, T., Biemann, K., Schimmel, P. 1983. *J. Biol. Chem.* 258:10637–41
70. Gibson, B. W., Biemann, K. 1984. *Proc. Natl. Acad. Sci. USA* 81:1956–60
71. Robbins, P. W., Trimble, R. B., Wirth, D. F., Hering, C., Maley, F., et al. 1984. *J. Biol. Chem.* 259:7577–83
72. Freedman, R., Gibson, B., Donovan, D., Biemann, K., Eisenbeis, S., et al. 1985. *J. Biol. Chem.* 260:10063–68
73. Putney, S., Herlihy, W., Royal, N., Pang, H., Aposhian, H. V., et al. 1984. *J. Biol. Chem.* 259:14317–20
74. Breton, R., Sanfacon, H., Papayannopoulos, I., Biemann, K., Lapointe, J. 1986. *J. Biol. Chem.* 261:10610–17
75. Takao, T., Hitouji, T., Shimonishi, Y., Tanabe, T., Inouye, S., et al. 1984. *J. Biol. Chem.* 259:6105–9
76. Carmona, C., Gray, G. L. 1987. *Nucleic Acids Res.* 15:6757
77. Drapeau, G. R. 1978. *J. Biol. Chem.* 253:5899–901
78. Vestling, M., Hua, S., Murphy, C., Smith, P., Fenselau, C. 1990. *J. Prot. Chem.* 9:320–21
79. Biemann, K. 1982. *Methods in Protein Sequence Analysis*, ed. M. Elzinga, pp. 279–88. Clifton, NJ: Humana
80. Morris, H. R., Panico, M., Taylor, G. W. 1983. *Biochem. Biophys. Res. Commun.* 117:299–305
81. Takao, T., Kobayashi, M., Nishimura, O., Shimonishi, Y. 1987. *J. Biol. Chem.* 262:3541–47
82. Padrom, G., Besada, V., Agraz, A., Quinones, Y., Herrera, L., et al. 1989. *Anal. Chim. Acta* 223:361–69
83. Carr, S. A., Hemling, M. E., Folena-Wasserman, G., Sweet, R. W., Anumula, K., et al. 1989. *J. Biol. Chem.* 264:21286–95
84. Hemling, M. E., Carr, S. A., Capiau, C., Petre, J. 1988. *Biochemistry* 27:699–705
85. Anderegg, R. J., Carr, S. A., Huang,

Y., Hiipakka, R. A., Chang, C., et al. 1988. *Biochemistry* 27:4214–21
86. Lee, T. D., Vemuri, S. 1990. *Biomed. Environ. Mass Spectrom.* 19:639–45
87. Papayannopoulos, I. A., Biemann, K. 1991. *J. Am. Soc. Mass Spectrom.* 2:174–77
88. Stults, J. T., Bourell, J. H., Canova-Davis, E., Ling, V. T., Laramee, G. R., et al. 1990. *Biomed. Environ. Mass Spectrom.* 19:655–64
89. Smith, J. B., Thevenon-Emeric, G., Smith, D. L., Green, B. 1991. *Anal. Biochem.* 193:118–24
90. Hirayama, K., Akashi, S., Furuya, M., Fukuhara, K. 1990. *Biochem. Biophys. Res. Commun.* 173:639–46
91. Reed, R. G., Putnam, F. W., Peters, T. Jr. 1980. *Biochem. J.* 191:867–68
92. Van Dorsselaer, A., Bitsch, F., Green, B., Jarvis, S., Lepage, P., et al. 1990. *Biomed. Environ. Mass Spectrom.* 19:692–704
93. Chowdhury, S. K., Katta, V., Chait, B. T. 1990. *Biochem. Biophys. Res. Commun.* 167:686–92
94. Wada, Y., Matsuo, T., Sakurai, T. 1989. *Mass Spectrom. Rev.* 8:379–434
95. Lee, T. D., Rahbar, S. 1991. See Ref. 11, pp. 257–74
96. Prome, D., Prome, J. C., Pratbernou, F. 1988. *Biomed. Environ. Mass Spectrom.* 16:41–44
97. Baczynskyj, L., Bronson, G. E. 1990. *Rapid Commun. Mass Spectrom.* 4:533–35
98. Beavis, R. C., Chait, B. T. 1987. *Proc. Natl. Acad. Sci. USA* 87:6873–77
99. Amico, V., Foti, S., Saletti, R., Cambria, A., Petrone, G. 1988. *Biomed. Environ. Mass Spectrom.* 16:431–37
100. Kaur, S., Medzihradszky, K. F., Yu, Z., Baldwin, M. A., Gillece-Castro, B. L., et al. 1990. See Ref. 8, pp. 285–313
101. Brinegar, A. C., Cooper, G., Stevens, A., Hauer, C. R., Shabanowitz, J., et al. 1988. *Proc. Natl. Acad. Sci. USA* 85:5927–31
102. Larsen, B. S. 1991. *Biol. Mass Spectrom.* 20:139–41
103. Reddy, V. A., Johnson, R. S., Biemann, K., Williams, R. S., Ziegler, F. D., et al. 1988. *J. Biol. Chem.* 263:6978–85
104. Carr, S. A., Roberts, G. D. 1986. *Anal. Biochem.* 157:396–406
105. Carr, S. A., Barr, J. R., Roberts, G. D., Anumula, K. R., Taylor, P. B. 1990. See Ref. 10, pp. 501–18
106. Carr, S. A., Roberts, G. D., Jurewicz, A., Frederick, B. 1988. *Biochimie* 70:1445–54
107. Medzihradszky, K. F., Gillece-Castro,

B. L., Settineri, C. A., Townsend, R. R., Masiarz, F. R., et al. 1990. *Biomed. Environ. Mass Spectrom.* 19:777–81

108. Vath, J. E., Jankowski, M. A., Martin, S. A., Scoble, H. A. 1990. *Proc. ASMS Conf. Mass Spectrom. Allied Top., 38th, Tucson,* pp. 351–52

109. Allmaier, G., Schmid, E. R., Hagspiel, K., Kubicek, C. P., Karas, M., et al. 1990. *Anal. Chim. Acta* 241:321–27

110. Hillenkamp, F., Karas, M. 1989. *Proc. ASMS Conf. Mass Spectrom. Allied Top., 37th, Miami Beach,* pp. 1168–69

111. Siegel, M. M., Hollander, I. J., Hamann, P. R., James, J. P., Hinman, L., et al. 1991. *Anal. Chem.* 63:2470–81

112. Morris, H. R., Pucci, P. 1985. *Biochem. Biophys. Res. Commun.* 126:1122–28

113. Yazdanparast, R., Andrews, P. C., Smith, D. L., Dixon, J. E. 1987. *J. Biol. Chem.* 262:2507–13

114. Smith, D. L., Zhou, Z. 1990. See Ref. 10, pp. 374–89

115. Zhou, Z., Smith, D. L. 1990. *J. Protein Chem.* 9:523–32

116. Raschdorf, F., Dahinden, R., Maerki, W., Richter, W. J., Merryweather, J. P. 1988. *Biomed. Environ. Mass Spectrom.* 16:3–8

117. Nakazawa, H. 1988. *Chem. Pharm. Bull.* 36:988–93

118. Zhou, Z., Smith, D. L. 1990. *Biomed. Environ. Mass. Spectrom.* 19:782–86

119. Akashi, S., Hirayana, K., Seino, T., Ozawa, S., Fukuhara, K., et al. 1988. *Biomed. Environ. Mass Spectrom.* 15:541–46

120. Yazdanparast, R., Andrews, P., Smith, D. L., Dixon, J. E. 1986. *Anal. Biochem.* 153:348–53

121. Visentini, J., Gauthier, J., Bertrand, M. J. 1989. *Rapid Commun. Mass Spectrom.* 3:390–95

122. Mancini, M. L. 1989. *Biomed. Environ. Mass Spectrom.* 18:1102–4

123. Sun, Y., Smith, D. L. 1988. *Anal. Biochem.* 172:130–38

124. Williams, D. H., Bradley, C. V., Santikarn, S., Bojesen, G. 1982. *Biochem. J.* 201:105–17

125. Roepstorff, P., Hoejrup, P., Moeller, J. 1985. *Biomed. Mass Spectrom.* 12:181–89

126. Gibson, B. W., Gilliom, R. D., Whitaker, J. N., Biemann, K. 1984. *J. Biol. Chem.* 259:5028–31

127. Bradley, C. V., Williams, D. H., Hanley, M. R. 1982. *Biochem. Biophys. Res. Commun.* 104:1223–30

128. Wagner, R. M., Fraser, B. A. 1987. *Biomed. Environ. Mass Spectrom.* 14:235–39

129. Schaer, M., Boernsen, K. O., Gassmann, E. 1991. *Rapid Commun. Mass Spectrom.* 5:319–26

130. Ganem, B., Li, Y.-T., Henion, J. D. 1991. *J. Am. Chem. Soc.* 113:6294–96

131. Ganem, B., Li, Y-T., Henion, J. D. 1991. *J. Am. Chem. Soc.* 113:7818–19

132. Biemann, K. 1988. *Biomed. Environ. Mass Spectrom.* 16:99–111

133. Bean, M. F., Carr, S. A., Thorne, G. C., Reilly, M. H., Gaskell, S. J. 1991. *Anal. Chem.* 63:1473–81

134. Johnson, R. S., Martin, S. A., Biemann, K. 1988. *Int. J. Mass Spectrom. Ion Processes* 86:137–54

135. Johnson, R. S., Martin, S. A., Biemann, K., Stults, J. T., Watson, J. T. 1987. *Anal. Chem.* 59:2621–25

136. Johnson, R. S., Biemann, K. 1989. *Biomed. Environ. Mass Spectrom.* 18:945–57

137. Biemann, K. 1990. See Ref. 10, pp. 455–79

138. Ghosh, A., Biemann, K. 1991. *Proc. Am. Soc. Mass Spectrom. Conf. Mass Spectrom & Allied Top., 39th,* Nashville, pp. 1225–26

139. Smith, R. D., Loo, J. A., Barinaga, C. J., Edmonds, C. G., Udseth, H. R. 1990. *J. Am. Soc. Mass Spectrom.* 1:53–65

140. Loo, J. A., Edmonds, C. G., Smith, R. D. 1991. *Anal. Chem.* 63:2488–99

141. Lee, T. D., Shively, J. E. 1990. See Ref. 10, pp. 361–74

142. Hopper, S., Johnson, R. S., Vath, J. E., Biemann, K. 1989. *J. Biol. Chem.* 264:20438–47

143. Vath, J. E., Biemann, K. 1990. *Int. J. Mass Spectrom. Ion Processes* 100:287–99

144. Klintrot, L.-M., Hoog, J.-O., Jornvall, H., Holmgren, A., Luthman, M. 1984. *Eur. J. Biochem.* 144:417–23

145. Papayannopoulos, I. A., Gan, Z. R., Wells, W. W., Biemann, K. 1989. *Biochem. Biophys. Res. Commun.* 159:1448–54

146. Mathews, W. R., Johnson, R. S., Cornwell, K. L., Johnson, T. C., Buchanan, B. B., et al. 1987. *J. Biol. Chem.* 262:7537–45

147. Johnson, T. C., Yee, B. C., Carlson, D. E., Buchanan, B. B., Johnson, R. S., et al. 1988. *J. Bacteriol.* 170:2406–8

148. Johnson, R. S., Mathews, W. R., Biemann, K., Hopper, S. 1988. *J. Biol. Chem.* 263:9589–97

149. Stark, G. R. 1977. *Methods Enzymol.* 47:129–32

150. Papayannopoulos, I. A., Biemann, K. 1992. *Protein Sci.* 1:278–88
151. Hunt, D. F., Yates, J. R. III, Shabano- witz, J., Winston, S., Hauer, C. R. 1986. *Proc. Natl. Acad. Sci. USA* 83: 6233–37
152. Hunt, D. F., Shabanowitz, J., Yates, J. R. III, Zhu, N. Z., Russell, D. H., et al. 1987. *Proc. Natl. Acad. Sci. USA* 84: 620–23
153. Hunt, D. F., Shabanowitz, J., Yates, J. R. III, Griffin, P. R., Zhu, N. Z. 1989. *Anal. Chim. Acta* 225:1–10
154. Hunt, D. F., Yates, J. R. III, Shabano- witz, J., Bruns, M. E., Bruns, D. E. 1989. *J. Biol. Chem.* 264:6580–86
155. Hunt, D. F., Yates, J. R. III, Shabano- witz, J., Zhu, N. Z., Zirino, T., et al. 1987. *Biochem. Biophys. Res. Commun.* 144:1154–60
156. Crabb, J. W., Johnson, C. M., Carr, S. A., Armes, L. G., Sarri, J. C. 1988. *J. Biol. Chem.* 263:18678–87
157. Griffin, P. R., Kumar, S., Shabanowitz, J., Charbonneau, H., Namkung, P. C., et al. 1989. *J. Biol. Chem.* 264:19066– 75
158. Bieber, A. L., Becker, R. R., McPar- land, R., Hunt, D. F., Shabanowitz, J., et al. 1990. *Biochem. Biophys. Acta* 1037:413–21
159. Kuster, T., Staudenmann, W., Hughes, G. J., Heizmann, C. W. 1991. *Biochem- istry* 30:8812–16
160. Moore, C. R., Yates, J. R. III, Griffin, P. R., Shabanowitz, J., Martino, P. A., et al. 1989. *Biochemistry* 28:9184–91
161. Juhasz, P., Papayannopoulos, I. A., Biemann, K. 1992. *Peptides: Chemistry and Biology (Proc. 12th Am. Peptide Symp.),* ed. J. A. Smith, J. E. Rivier, pp. 558–59. Leiden: ESCOM
162. Michel, H., Hunt, D. F., Shabanowitz, J., Bennett, J. 1988. *J. Biol. Chem.* 263:1123–30
163. Gibson, B. W., Cohen, P. 1990. See Ref. 10, pp. 481–501
164. Alexander, J. E., Hunt, D. F., Lee, M. K., Shabanowitz, J., Michel, H., et al. 1991. *Proc. Natl. Acad. Sci. USA* 88:4685–89
165. Houtz, R. L., Stults, J. T., Mulligan, R. M., Tolbert, N. E. 1989. *Proc. Natl. Acad. Sci. USA* 86:1855–59

Annu. Rev. Biochem. 1992. 61:1011–51

TRANSPOSITIONAL RECOMBINATION: Mechanistic Insights from Studies of Mu and Other Elements[1]

Kiyoshi Mizuuchi

Laboratory of Molecular Biology, National Institute of Diabetes, and Digestive and Kidney Diseases, National Institutes of Health, Bethesda, Maryland 20892

KEY WORDS: genome rearrangement, DNA transposition, retroviral DNA integration, Tn10, Tn7

CONTENTS

1011

INTRODUCTION

Genome rearrangements are frequent events in many organisms. Many of these events involve specific protein factors and specific DNA sequences. A segment of DNA that has the ability to translocate or reconfigure is generally called a *mobile genetic element*. While some rearrangements appear to take place by a mechanism unique to the particular reaction, many DNA elements appear to share a similar mechanism.

The group of mobile genetic elements considered here may be classified as the transposon family. They are abundant and found within the genomes of a wide variety of organisms. *Transposons* are characterized by their ability to translocate to a variety of sites on the chromosome of the host organism. Transposons (in this review this term is used to represent all members of the family) can be divided into subgroups based on their transposition mechanism or the sequence similarities between their transposition proteins (1, 2). For example, the amino acid sequences of the integrases or IN proteins of retrotransposons and retroviruses are not only similar to each other, but also similar to that of the transposase of the IS3 family of prokaryotic insertion sequences (2a, 3, 4). Other subgroups of bacterial elements include the Tn3 family, the IS10 family, the IS1 family, and the Mu family of transposing phages. There also are many elements whose phylogenic relationship with other elements has not been established.

Currently, only a handful of transposons have been studied in mechanistic detail, and it would be premature to judge whether most of these elements share a common mechanism of transposition. This review focuses on the mechanisms of the early steps in transposition of a few diverse well-studied elements, which display a remarkable similarity in the basic mechanism, while also showing important differences that result in critical changes in the biological outcome of the transposition process.

These transposition reactions apparently are different from, but possibly related to, another group of reactions classified as conservative site-specific recombinations. These reciprocal recombinations between pairs of unique DNA sites achieve integration, excision, or inversion of a DNA segment (5–7; see more discussions in PERSPECTIVES).

The elements discussed here all utilize a higher order protein-DNA architecture to fine tune the reaction pathway. Regulation of this pathway achieves a physiologically sensible outcome, while avoiding reactions that would be damaging for the organism. The use of complex macromolecular interactions to control the reaction pathway is also likely in other biological systems, especially those with nucleic acid reaction components. The knowledge gained from the study of the DNA transposition mechanisms and their control should thus prove useful in the consideration of other complex biological reactions.

While this review concentrates on the mechanistic aspects of transposition of a few well-studied elements, broader biological aspects of these elements, as well as a wide variety of other elements, are discussed in excellent reviews compiled in a recent book, *Mobile DNA,* edited by D. E. Berg and M. M. Howe (8). Another book, *Genetic Recombination,* edited by R. Kucherlapati and G. R. Smith (9), also contains useful information on variety of recombination reactions. Earlier reviews on related subjects can be found in the book edited by Shapiro (10). Details of some of the elements discussed here are also reviewed in (11–13).

OUTLINE OF THE REACTION

General Structure of the Elements

Transposons carry sequences required for transposition at each end of the element. These end-sequences are unique to each type of element. The two end-sequences of many elements are identical or related by a consensus sequence and form a pair of terminal inverted repeats. The transposon end-sequences play two functional roles: (*a*) to act as the sequence-specific binding site for the transposase protein, and (*b*) to fit into the active site of the transposase during the DNA cleavage and joining reactions.

Structurally simple transposons contain only a copy of transposase-binding sequence at or near each end of the element where the transposase carries out the cleavage and joining of the DNA strands. Many IS elements, such as IS10 (and Tn10 which is made up of two copies of IS10), and the Tn3 family of elements are examples (1, 14, 19). The sequences required for retroviral DNA integration also appear to be simple (15). In more complex examples, additional protein-binding sequences on the transposon are required for efficient transposition (11, 16, 17).

Although most transposons demonstrate some degree of target sequence preference (11, 18), they generally can transpose to many target sites in the chromosome of their host organism, indicating that they have little target sequence specificity. One exception is Tn7, which transposes at high efficiency to a specific site on the bacterial chromosome (17).

Brief Description of the Reaction Steps

The transposon family of genetic elements recombine via a three-step process. Although in detail, there are important variations in the reaction pathway from one element to another, the basic mechanistic aspects appear to be shared by a wide variety of transposons. The first two steps, the nuclease step and strand transfer step, can be considered as a special form of site-specific recombination that generates a transposition intermediate (or the strand transfer product). This process is referred to here as the *transpositional recombination.* In the third step, the intermediate is processed by DNA repair or replication.

In the first step, the transposon donor DNA is specifically cleaved at the two transposon ends to generate a pair of 3'-OH termini of the transposon sequence. On some elements, such as phage Mu and retroviruses, only a single strand is cut, leaving the other strand uncleaved (20–22). On other elements, such as Tn7 and Tn10, which transpose nonreplicatively, double-strand cuts are generated at the ends of the elements (23–25). For all the elements studied, the 3'-OH termini of the element exposed by the endonuclease is utilized for the next step, strand transfer (21–27).

Site-specific cleavages to expose the 3'-ends of the element are bypassed in some retro-elements. Retro-elements transpose via an RNA intermediate that is converted into the double-stranded linear DNA, which functions as the direct precursor for integration into a new location in the host chromosome (15, 28). It is this integration process that is related to other transposition reactions. Most retroviral integration precursor DNAs are two base pairs longer at each end than the sequence that integrates into host chromosome (15). These extra nucleotides at the 3'-ends therefore must be removed by nucleolytic cleavage. However, the integration precursor of most yeast Ty elements appears to be exactly the same length as the sequence that integrates (28). Thus, no nucleolytic cleavage is necessary for these elements.

Strand transfer is the concerted cleavage and joining of phosphoester bonds that results in connection of the donor and target DNA. The target DNA cleavage and joining to the donor DNA ends appear energetically coupled, since for several reactions studied in detail, external energy sources such as ATP are not required (29–34). Further discussion of this subject is presented in MECHANISM OF THE CHEMICAL STEPS.

The cleavage sites on the two strands of the target DNA are generally staggered by a few nucleotides. This has been directly demonstrated for Tn10 (25). For other elements, staggered cleavage is inferred from the structure of the transposition products. This staggered cut of the target DNA is the source of the short target sequence duplication that flanks the transposed elements. The size of the duplication, and thus the size of the stagger, is unique to each element. All the elements studied so far join the 5'-ends of the cut target DNA to the 3'-OH ends of donor DNA, leaving the 3'-ends of the target strands unjoined to the donor (21–23, 25, 27).

The structure of the strand transfer product, and thus the subsequent processing required, depends on the nature of the unjoined donor DNA ends. For elements such as Tn7 and Tn10 that transpose by the "cut and paste" mechanism (23, 35), gap repair completes the transposition process. The strand transfer product of phage Mu carries a pair of branched DNA structures, owing to the remaining connections to the original donor DNA flanking the 5'-ends of the transposon (27). Such an intermediate can be processed by replicating the entire transposon sequence and the short target site sequence if replication proteins assemble at the DNA branch.

In the general outline of the reaction described above, attention centered on the structure of the participating DNA. For reactions studied in detail, evidence is accumulating that the transpositional recombination leading to the strand transfer products takes place through formation of a stable protein-DNA architecture. Details of each reaction step and the current knowledge on the protein-DNA architecture are discussed in subsequent sections. In this review, I do not discuss the mechanisms of the repair/replication step in further detail, since biochemical study has not yet addressed the mechanism.

Protein Participants

The DNA cleavages and joining steps involved in the DNA transposition are well coordinated and controlled by highly specific protein-DNA complexes. The proteins that carry out transposition not only mediate the chemical reactions, but also must play structural roles. In some systems, one protein carries out many of the necessary functions, while in other cases, these functions are distributed among several proteins.

Autonomous transposable DNA elements code for a protein or proteins required for their own movement. These proteins are called the "transposase" for regular transposons, or the "integrase" (or IN protein) for retro-elements. The term "transposase" should not be interpreted to imply that the protein is designed to function in a catalytic way with multiple turnovers. Because of their important structural roles, transposases do not appear to be designed to catalyze multiple events. Some transposons, in addition, require accessory proteins for maximum efficiency. The proteins involved in the transposition of several well-studied elements are described in THE DNA SITES AND PROTEINS.

MECHANISM OF THE CHEMICAL STEPS

Transpositional recombination involves two chemical steps: (a) the site-specific endonucleolytic cleavage of the donor DNA to expose the 3'-OH termini at the transposon ends, and (b) the DNA strand transfer step that joins these donor 3'-OH ends to a target DNA. These two reactions are chemically uncoupled and independent. Donor cleavage can be separated from strand transfer mutationally (36), kinetically (35), or by modifications of the reaction conditions or components (20, 29). Strand transfer can be carried out starting with preprocessed donor DNA substrates provided they have the proper 3'-OH ends. Strand transfer with precut donor DNA has been demonstrated in the Mu, several retroviral, and Tn7 in vitro systems (20, 23, 32–34, 37, 38). In the cases of Mu, Tn10, and retroviruses, both chemical steps are mediated by the same protein, the transposase, or the IN protein (20, 29, 32; H. Benjamin, R. Chalmers, N. Kleckner, personal communication). For Tn7, the assignment of functions to the different proteins has not been completed.

Although structural information on the active site-substrate DNA interaction, which is necessary to describe in detail the mechanisms of the chemical steps, is not yet available, experiments that address one aspect of the mechanism have recently been completed (39, 40). These experiments addressed the number of transesterification steps involved in the strand transfer mechanism. The simplest possible mechanism for each of the two chemical steps is a one-step reaction. A possible one-step mechanism for endonucleolytic cleavage is that H_2O acts as the nucleophile that attacks the phosphoester bond at the site of cleavage; the transposase or IN protein would render the ester bond susceptible to nucleophilic attack. Alternatively, cleavage could require two steps if a reactive amino acid side chain in the transposase or IN protein acts as the nucleophile to cut the DNA strand; the initial product would be a protein-DNA covalent intermediate, with a DNA 5'-phosphoryl group joined to the protein, for example, by a phosphoester bond to a tyrosine or serine side chain. The covalent intermediate must then be attacked by H_2O to generate the final cleaved DNA. More complex models involving more transfer steps, although theoretically possible, seem unlikely. Thus, establishing whether one or two phosphoryl transfer steps are involved in DNA cleavage provides evidence that supports either direct hydrolysis or the two-step model.

The DNA strand transfer reaction may also be achieved either with or without a protein-bound covalent intermediate. In a one-step transesterification model, the OH group of the donor 3'-end is the nucleophile that attacks a phosphoester bond in the target DNA. Cleavage of the target DNA and joining of the 5'-phosphoryl group to the donor 3'-OH end are thus two outcomes of one chemical step. In an alternative two-step reaction, the target DNA may be cut by the transposase with formation of a protein-target DNA covalent intermediate. The protein-bound target DNA must then be transferred to the donor 3'-ends by a second transesterification step. Both of the models are consistent with the observed independence of strand transfer from high-energy cofactors (29, 30). Again, more complex models for the strand transfer mechanism are possible, but seem unlikely. The one-step transesterification model is similar to the mechanism of steps in RNA splicing (41). The two-step model well describes the mechanism of conservative site-specific recombination (5, 39, 42, 42a).

The experimental approach, which was used to address whether one- or two-step mechanisms were involved in DNA cleavage and strand-transfer, utilized chiral phosphorothioates at the reaction center of the substrate DNA (39, 40). The phosphate in regular phosphoester bond is achiral, but can be made chiral by substituting one of the nonbridging oxygens with a sulfur. The chirality inverts with each phosphoryl transfer step that takes place by an in-line nucleophilic substitution, the mechanism almost certainly used during enzymatic phosphoryl transfer (43). Thus, a model that involves an odd number of (e.g. one) transfer steps predicts inversion of the chirality in the course of the reaction. Overall retention of chirality would support models involving even numbers of (e.g. two) transfer steps.

Chemical Mechanism of the Nuclease Reaction

Retroviral IN proteins specifically cleave double-stranded DNA oligonucleotides that match the terminal sequence of the linear viral DNA (see below for details). When the strand to be cleaved is radioactively labeled at the 5'-end, a product shortened by two nucleotides at the 3'-end is observed (32, 44, 45). When label is placed within the dinucleotide portion that is to be cut off, in addition to the expected simple dinucleotide, two other cleavage products are observed (40, 46). One of these products carries glycerol, which was present in the reaction, attached to the 5'-phosphate of the cleaved dinucleotide (46). The other product is a cyclic dinucleotide in which the 3'-OH is linked to the cleaved 5'-phosphoryl group (40).

These unusual products could be explained if the IN protein utilizes a variety of nucleophiles other than H_2O to attack the phosphoester bond to be cleaved. In this way, the dinucleotide could be generated in one step with either H_2O, glycerol (or other related compounds included in the reaction mixture), or the 3'-OH of the dinucleotide itself as the nucleophile. Alternatively, the dinucleotide may be initially cleaved by a nucleophilic residue on the IN protein to form a protein-bound covalent intermediate, and the observed products could result from the secondary attack by another accessible nucleophile on the protein-DNA linkage.

The cyclic dinucleotide product offered a convenient way to ask whether the one-step or two-step model is more likely. A phosphorothioate linkage of Rp configuration was incorporated at the cleavage site in the substrate DNA. The circular dinucleotide containing the phosphorothioate linkage was isolated after cleavage by IN, and the stereochemical configuration of this bond was determined to be the Sp-isomer (40). Inversion of the phosphorothioate chirality during cleavage supports the one-step model, while eliminating the two-step protein-bound covalent intermediate mechanism for formation of the cyclic product.

It is simplest to assume that the other cleavage products are made by the same mechanism. However, this result does not formally rule out the unlikely possibility that the simple dinucleotide can be made by a different mechanism.

Avian sarcoma leukosis virus (ASLV) IN protein has been reported to form a phosphoester bond with cleaved DNA via serine and threonine residues, an observation supporting models involving covalent intermediates (4). The existence of a covalent complex was supported by the detection of phosphoserine and phosphothreonine after acid hydrolysis of the reaction products. The results described above suggest an alternative interpretation. The fact that IN protein utilizes a variety of nucleophiles for the nuclease reaction step suggests that an exposed serine or threonine may also be adventitiously used as a nucleophile. In fact, when human immunodeficiency virus (HIV) IN protein is incubated with substrate DNA in the presence of serine or threonine, a product is formed that is consistent with a structure in which the dinucleotide is attached via the 5'-phosphate to the amino acid (46). However, a three-step model, involving two obligatory protein-DNA covalent intermediates, cannot yet be ruled out.

Chemical Mechanism of DNA Strand Transfer

The stereochemical course of the DNA strand transfer has been determined for both the Mu reaction and HIV DNA integration (39, 40). In both cases, target DNAs containing phosphorothioate linkages were used, and the configuration of the phosphorothioate that bridges the donor and target segment in the strand transfer product was determined. The Rp configuration of the phosphorothioate at the target site inverted to the Sp configuration in the strand transfer product in both systems (39, 40). These results support a one-step transesterification model for DNA strand transfer. Two-step transesterification involving a protein-DNA covalent intermediate is ruled out if the in-line displacement mechanism of phosphoryl transfer is assumed.

Currently, the mechanism of the chemical steps of transpositional recombination has been studied for only a limited number of elements. Nevertheless, considering the basic similarities in the reaction mechanisms used by Mu and retroviruses, this mechanism for the chemical steps may be expected to be shared by essentially all the elements in the group. On the other hand, the surprising flexibility in the nucleophiles that may participate in the nucleolytic cleavage by the HIV IN protein points to possible evolutionary flexibility in the reaction mechanism. This point is discussed in PERSPECTIVES.

THE DNA SITES AND PROTEINS

In this section, features of the *cis*-acting DNA sequences and the proteins involved in transpositional recombination of several elements are described. The elements selected for discussion have been studied biochemically and represent different classes within the transposon family. The complexity of both the protein components and the DNA substrates differ substantially from element to element. In the subsequent sections, the functional interactions among the reaction participants described here are discussed.

Bacteriophage Mu

MU DONOR DNA SEQUENCES   Bacteriophage Mu carries three MuA-binding sites at each end of the Mu genome (47). These sites are named, from outermost to innermost, L1, L2, and L3 at the left end, and R1, R2, and R3 at the right end. The six sites are related by a 22-nucleotide consensus sequence YGTTTCAYNNRAARYRCGAAAR, which shows no obvious internal symmetry (47). The L1 and R1 are similarly arranged at the two ends at position 6–27 counting from the terminal nucleotide of the Mu sequence. The arrangement of other sites at the two ends is different. At the left end, the internal two sites, L2 (position 111–132) and L3 (position 151–172), are close to each other but separated from L1. All three L-end sites are oriented similarly. L2, the weakest MuA-binding site, lacks part of the consensus sequence. At the right end, all three sites are adjacent to each other (R2, 28–49; R3, 56–77), and while R2 is oriented similarly to R1, R3 is oriented in the opposite direction.

The 22-nucleotide consensus sequence is sufficient for the site-specific binding by MuA, and the DNaseI footprint of MuA does not extend much beyond the consensus sequence in a simple MuA–Mu end DNA complex. However, changes in the terminal 5'-CA-3' sequence, (especially the terminal A) located outside of the consensus and common between the two ends, have profound effects in both in vivo and in vitro reactions (48, 49). Thus, the terminal nucleotides seem to be required for reaction steps beyond the initial binding of MuA to DNA, possibly for synapsis of the two ends or the chemical reactions.

The six MuA-binding sites are not equally important for transposition. An earlier in vivo study indicated that R3 can be removed without significantly affecting the transposition efficiency, while the other five sites are all important (50, 51). However, the mini-Mu used in this study lacked the internal activation sequence (IAS, see below). Thus, only the low-efficiency events that do not depend on the IAS were detected. According to in vitro experiments with a set of deletion/substitution mutants of a mini-Mu that carried the IAS, any one site among L2, L3, and R3 could be removed without significantly affecting the mini-Mu donor activity if IHF is present, while removal of two sites is detrimental (52; R. G. Allison, G. Chaconas, personal communication). The functional distinctions between the two sets of Mu-end MuA-binding sites—the end proximal sites L1, R1, and R2, and the internal sites L2, L3, and R3—are discussed in the next section.

In addition to their structural differences, the two Mu ends also appear to be functionally distinct. Both in vivo, and in vitro under normal reaction conditions, a mini-Mu donor must carry the L- and R-ends of Mu sequence in order to be active; a mini-Mu carrying either a pair of L-ends or a pair of R-ends transposes very poorly (M. Mizuuchi, K. Mizuuchi, unpublished).

Among transposons, Mu is unique in having a DNA sequence element required for efficient transposition distantly located from the two end sequences. This *internal activation sequence* (the IAS) located near the Mu left end at position 889–997 in the phage genome is also referred to as a transpositional enhancer (53–55). In vivo transposition is reduced about 100-fold in the absence of the IAS, and under standard reaction conditions, in vitro Mu donor DNA cleavage and strand transfer are greatly diminished in its absence. The exact location of the IAS is not critical; any location between L3 and R3 is acceptable. However, the relative orientation of the IAS is important (53–55).

The IAS overlaps with the Mu operator, which controls early gene transcription. Within the IAS lies a centrally positioned IHF-binding site, flanked by two operator sequences (O1 and O2) to which both MuA and Mu repressor bind (see below; 47, 54). Helical phasing between O1 and O2 is important for the IAS function (54). IHF (see subsections below) is important for IAS function when the donor DNA is supercoiled to the extent considered to be physiological (54, 56). Use of a highly supercoiled donor DNA in an in vitro reaction eliminates the need for IHF (56).

MU TARGET SEQUENCES   While Mu can transpose into many sites within a segment of DNA, close examination reveals that some target sites are clearly preferred (57). DNA bound by MuB protein is preferentially used as the target, and target site selection by Mu is accomplished at three levels. First, Mu displays transposition target immunity, allowing it to avoid transposition near a copy of Mu DNA (58, 59). This aspect of target selection is discussed separately. Second, MuB, which demonstrates weak but recognizable binding site preference, seems to direct Mu transposition to certain areas several hundred base pairs in length. Large oligomers of MuB may be the active form for target activation, thus explaining the regional rather than specific site preference. These "hot" regions are relatively frequent. For example, pBR322 DNA has two such regions, one located near the EcoR1 site, the other close to the end of the *bla* gene (M. Mizuuchi, K. Mizuuchi, unpublished). Preferred MuB binding to the middle of the *bla* "hot" region has been demonstrated by DNaseI protection (M. Mizuuchi, K. Mizuuchi, unpublished). The final level of target sequence selection lies within MuA protein itself. Target sites with the sequence NYSRN are preferred. This same consensus target sequence is used whether or not MuB is present (M. Mizuuchi, K. Mizuuchi, unpublished).

MU TRANSPOSASE   MuA protein (MuA) is the transposase of Mu phage. MuA has (*a*) two separate sequence-specific DNA-binding activities that recognize the Mu donor DNA, (*b*) site-specific single-strand endonuclease activity that exposes the 3'-OH ends of the Mu DNA sequence, and (*c*) DNA strand transfer activity that cuts and joins a target DNA to the donor 3'-ends. 75-kDa MuA, according to the sequence of the gene, consists of 663 amino acids (60).

Partial proteolytic fragments of MuA have begun to give a picture of the functional organization of the protein. A truncated form of MuA, missing a C-terminal portion, is generated during incubation of crude extracts containing MuA (61). Partial proteolysis of MuA by a variety of enzymes in vitro produces discrete peptide fragments (62). The in vitro activities of these polypeptides have helped in the assignment of multiple MuA functions to parts of the protein (53, 55, 62, 63).

The N-terminal portion from aa1 (amino acid position 1) to aa64 or aa75 contains a sequence-specific DNA-binding domain that binds to the operator sequences in the IAS (53, 55). This binding specificity is consistent with the presence of amino acid sequence homology between the N-terminal domains of MuA and Mu repressor (60, 64). The N-terminal sequence homology between transposase and repressor is also found in the closely related phage D108, which shares mechanistic and regulatory strategies with Mu (64). MuA that is missing this domain functions as well as the wild-type protein under reaction conditions that bypass the need for the IAS (53).

From aa77 to aa177 is a second sequence-specific DNA-binding domain that recognizes the 22-base-pair (bp) MuA-binding sites within the Mu L- and R-ends (55, 62). Binding of intact MuA to these sites, either as individual sites or in the cluster, induces a substantial bend in the DNA (65, 66).

Proteolytic cleavage around aa243 to aa247 separates the two site-specific DNA-binding domains from the rest of the protein (62). A cleavage around aa574 further separates a C-terminal peptide (62). The resulting internal 35-kDa peptide (aa248 to aa574) exhibits nonspecific DNA binding activity (62). The C-terminal part (beyond aa575) can be functionally divided into two segments; the endmost segment (aa617 to aa663) is essential for interaction with MuB protein. MuA missing this C-terminal fragment does not respond to the presence of MuB either in vivo or in vitro (63, 67, 68). However, this C-terminal deletion protein retains all other activities of MuA and simply behaves as wild type protein does in the absence of MuB (63, 67, 68); its activities include DNA binding, donor DNA cleavage, and DNA strand transfer into intramolecular target sites. In contrast, a larger C-terminal deletion protein missing aa574 to aa663 loses both donor DNA cleavage and DNA strand transfer activities. Thus, the part of MuA around residue 574 appears to participate in donor DNA cleavage and target DNA interactions. A missense mutant of MuA that substitutes I for T at aa547 has normal activity in Mu DNA binding and donor DNA cleavage, but is temperature sensitive for strand transfer, suggesting the importance of this region of the protein in the later reaction steps (68).

While MuA possesses all the fundamental activities necessary to make the Mu transposition intermediate, several accessory proteins assist in the reactions beyond simple DNA binding. Two *Escherichia coli* DNA-binding proteins, HU and IHF, are involved early before donor DNA cleavage, and

MuB activates strand transfer by MuA and facilitates sensible target site selection.

MU TRANSPOSITION ACCESSORY PROTEINS

MuB protein   MuB is the second Mu-encoded protein directly involved in Mu transposition. In the absence of MuB, (*a*) the in vivo frequency of Mu transposition is greatly reduced (69, 70), and the residual transposition events utilize target sites near or within the Mu donor DNA (71); (*b*) Mu DNA replication is undetectable (72); and (*c*) in a cell-free reaction, strand transfer products utilizing target sites on separate molecules from the Mu donor DNA are virtually undetectable (29). Omission of ATP from the reaction has the same effects, indicating that MuB function requires ATP (29). In the presence of ATP, MuB has a strong sequence-nonspecific DNA-binding activity, and DNA bound by MuB is the preferred target for Mu transposition (58).

According to the gene sequence, MuB contains 312 amino acids; its predicted molecular weight is 35,000 (73). A likely nucleotide-binding site can be found in the sequence (73–75). Partial proteolysis of MuB indicates a protease-sensitive region around amino acid 222; a 25-kDa N-terminal polypeptide (aa1–aa227) binds ATP (75).

MuB possesses a weak ATPase, which is modulated by DNA and MuA (29, 76). MuB free in solution behaves as a monomer; it oligomerizes in the presence of ATP (76). Oligomeric forms, rather than the monomer, appear to be the active state for both the stable DNA-binding and MuA interactions. Further discussion on the MuB ATPase activity and its role in target DNA selection is given later.

IHF   IHF was originally identified as a host protein required for integrative recombination of bacteriophage lambda DNA (77), and is involved in a variety of DNA transactions that are outside the scope of this review. IHF is a member of a family of small heat-stable DNA-binding proteins and is structurally related to HU protein. IHF is a heterodimeric protein composed of two related but distinct subunits with monomer molecular weights of 11,000 (77–80). DNA sites with a consensus sequence C/T AANNNNTTGAT A/T are specifically bound and bent by IHF (81, 82, 82a, 82b). The bend angle has been estimated to be about 140 degrees toward the bound protein (83). The protein has been postulated to interact with DNA primarily through the minor groove, perhaps using a pair of protruding arms that is composed of two-stranded β-sheet (84; see below).

IHF is involved in transposition of several bacterial elements. In each case, specific protein-binding sites on the transposon DNA have been recognized (1). Phage Mu growth requires IHF (85). A mutant phage that can grow on *E. coli* strains missing IHF carries a mutation within the early promotor (86).

This observation suggested that IHF acts on Mu growth by influencing its transcription. Later, it was realized that a part of the promotor and accompanying operator sequence is directly involved in transposition (53–55; see the IAS above).

HU   HU is structurally related to IHF and other type 2 DNA-binding proteins (87). *E. coli* HU, like IHF, is composed of two closely related but distinctive subunits. Each subunit of the heterodimeric protein has a molecular weight of 9500 (88). An analogous protein from *Bacillus stearothermophilus,* HBs, has been crystallized and the structure determined (89). Sequence similarity among the members of this protein family suggests that they share a common structure, which can be divided into two parts. Contacts between the two subunits are made within a globular core of the protein. A pair of antiparallel two-stranded β-sheets that stick out from this core are considered to be involved in DNA binding. HU, unlike IHF, binds sequence nonspecifically to DNA, and when sufficient HU is bound, a condensed structure resembling a nucleosome filament is formed (90).

HU, which is also involved in Tn10 transposition, was first recognized as a host protein required for formation of the Mu transposition intermediate (91). More specifically, it is required prior to the donor DNA cleavage, probably during the synaptic process that brings the two Mu DNA ends into the stable protein DNA complex (91–93).

Mu repressor   The Mu repressor regulates Mu transposition at two levels. The primary function of the Mu repressor is negative regulation of early transcription, i.e. shutting down expression of the MuA and MuB genes, and regulating its own synthesis (94, 95). Repressor also directly inhibits Mu transposition (47, 53). Both inhibitory activities are mediated by its sequence-specific binding to the Mu operator sequences around the early promoter. There are three binding sites in the operator area called O1, O2, and O3. O2 overlaps the early promoter and O3 is adjacent to the repressor promoter (94). O1 and O2 are the essential components of the IAS (54, 55).

The repressor has a molecular weight of 21,500; an earlier estimate of 19,000 from the DNA sequence did not take into account the unexpected use of UUG as the initiator codon, which places the N-terminal methionine 22 amino acids further upstream of the initial prediction (53). The N-terminal half of the repressor shares amino acid sequence similarity with the N-terminal domain of MuA (60, 64). Thus, this domain must be responsible for binding to the same DNA sequence. The repressor most likely prevents the necessary interaction between MuA and IAS by sequestering the operator sequence and thus inhibits transposition at an early stage.

Physiological need for such an elaborate regulation can be rationalized as follows. As the phage infects a sensitive host cell, it must decide between the lysogenic and lytic pathways. However, certain amounts of MuA and MuB protein must be produced for both pathways, since Mu uses transposition for both lysogenic integration and lytic DNA replication. When Mu chooses the lysogenic pathway, by synthesizing the repressor protein, more than one round of transposition should be avoided; such secondary transposition events would likely be detrimental for the survival of the resulting lysogenic cell. Thus, cells that produce a significant amount of the repressor need to inhibit further action of MuA quickly, rather than waiting for dilution or degradation of the MuA already made. This "quick response" is achieved by using the repressor to sequester the IAS from interaction with MuA. Mu repressor would also prevent fortuitous gene rearrangements by Mu transposition during growth of lysogenic cells by inhibiting the action of accidentally expressed MuA.

Many other transposons also possess mechanisms to negatively regulate their transposition; many elements control transposase expression at the transcriptional or translational levels, while some elements also produce direct inhibitor proteins. This subject is covered comprehensively by other reviews (96), and is not discussed further here.

Tn7

Tn7 DONOR DNA SEQUENCES   The end sequences of Tn7 that are required for transposition are complex, and their organization is reminiscent of the phage Mu end sequences. The Tn7 ends are composed of short inverted repeats and multiple binding consensus sequences for the TnsB protein (97–100). The terminal eight nucleotides are identical at the two ends and are presumably required for Tn7 transposition. This 8-bp sequence is outside of the endmost TnsB-binding consensus sequence, and which protein recognizes this sequence is not yet known. TnsB may extend its DNA interactions to involve this sequence. Alternatively, another Tns protein, perhaps TnsA, interacts with this sequence. In either case, this end-proximal sequence is likely to be recognized by the active site for the chemical steps of transpositional recombination.

In addition to the TnsB sites close to the two ends of the element (between positions 9 to 30), five additional 22-nucleotide consensus sequences for TnsB binding (GACAAWAWAGTYBKRAACTRRR) exist further inside. On the left end, two sequences are located at positions 73–94 and positions 129–150. On the right end, the three sequences are located at positions 27–50, 50–71, and 70–91. The TnsB-binding sequences at each end are oriented in the same direction; the two ends are in inverted orientation with respect to each other. A functional left end requires the original sequence from position 1 to 149, indicating that all three TnsB sites are necessary (100). On the right end, deletion of the innermost TnsB-binding site causes only a severalfold reduction of the transposition frequency. Further deletion of the right end sequence is detrimental to transposition (100). The innermost TnsB site at the right end is most likely involved in regulating expression of the Tns genes, since it overlaps the promoter sequence. While the two end sequences of Tn7 are asymmetric, an artifically constructed transposon carrying two right-end sequences can transpose efficiently, while one with two copies of the left end cannot (100).

Tn7 TARGET SEQUENCES   Tn7 is unique among transposons in that it transposes with high frequency into a specific target site (attTn7) on the bacterial chromosome (97, 98, 101–104). By a separate pathway, it also transposes at a lower frequency into nonspecific target sites (97, 98, 101–104).

The sequence determinants of the *E. coli* attTn7 are located near the 3'-end of the coding sequence of the *glmS* gene (97, 102). However, insertion of Tn7 occurs into a site outside of the *glmS* coding sequence, a location some 30 bp distant from the critical sequence determinant (97, 102, 105, 106). This clever arrangement allows conservation of the attTn7 sequence through evolution due to selective pressure that maintains a functional *glmS* gene while avoiding destruction of an important gene by high-frequency transposition into an essential region. Insertion of Tn7 to attTn7 takes place in one orientation with the R-end proximal to *glmS* (101). In the absence of attTn7, a low-frequency transposition into sites with sequences similar to attTn7 is observed (104).

In the second pathway, low-frequency transposition of Tn7 into many target sites occurs, with no apparent sequence specificity (98, 103, 104).

Tn7 TRANSPOSITION PROTEINS   Tn7 encodes five proteins for transposition: TnsA, TnsB, TnsC, TnsD, and TnsE (98, 103, 106a). The first three are required for both the attTn7 and the nonspecific pathways described above (98, 103, 104). Transposition into the attTn7 requires, in addition, TnsD (98, 103, 104). Transposition into nonspecific sites requires TnsE but not TnsD (98, 103, 104).

TnsA has a predicted molecular weight of 31,000 (17, 106a, 107). Other than its requirement for all the Tn7 transposition events, little is known about its function. In an in vitro reaction system, no transposition intermediates are generated if TnsA is omitted (23). Thus, it is likely that this protein directly participates in an early stage of transposition.

TnsB has a predicted molecular weight of 81,000 (106a). As described above, this protein binds specifically to multiple sites at the two ends of Tn7 (99).

TnsD, a protein with predicted molecular weight of 59,000 (106a), is required for the high-frequency transposition into attTn7. TnsD binds sequence specifically to attTn7 DNA and protects the attTn7 determinant sequence around positions 25 to 55 from nuclease digestion (108; K. Kubo, N. Craig, personal communication). Thus, this protein seems to guide other parts of the transposition machinery, including the donor DNA, to the specific target site; a role for TnsC in this process is implicated (see below).

TnsE has a predicted molecular weight of 61,000 (106a, 109). In vivo studies have demonstrated that TnsE serves as an alternative to TnsD, directing transposition to nonspecific target sites (98, 103, 104). The transposition frequency via this pathway is lower than that using TnsD. TnsE, while it does not share sequence similarity with TnsD (106a), may share a common mechanism of action.

TnsC has a predicted molecular weight of 63,000 (106a), and binds to ATP and to DNA (P. Gamas, N. Craig, personal communication). In an in vitro reaction containing TnsA, B, C, and D proteins, ATP (or dATP) is required for the production of Tn7 transposition intermediates (23). In the presence of ATP, TnsC preferentially binds adjacent to TnsD bound to attTn7, extending the nuclease protection towards but not reaching to the insertion site (K. Kubo, N. Craig, personal communication). When ATP is replaced by a nonhydrolyzable analog, TnsD is no longer required for transpositional recombination and random target sites are used for strand transfer (R. Bainton, K. Kubo, N. Craig, personal communication). These observations strongly suggest that the role of TnsC in part resembles that of MuB in Mu transposition. However, TnsC cannot be equivalent to MuB, since another protein, TnsD or TnsE, is required to help select a proper target DNA under normal reaction conditions.

It is unclear which of the five Tns proteins possesses the DNA cleavage and joining activities necessary for transposition.

IS10/Tn10

IS10/Tn10 DONOR DNA SEQUENCES   Tn10 is composed of two inverted copies of IS10, surrounding genes for tetracycline resistance. It transposes by an exclusively nonreplicative mechanism (110). The two ends of each IS10 are called "inside" and "outside" ends according to their position in Tn10; all four end sequences are functional, and artificial transposons with various combinations of these ends have been constructed and used as transposition donor DNAs. The two ends share a 23-nucleotide imperfect inverted repeat sequence (CTGAKRRATCCCCTMATRATTTY). Each part within this sequence appears to have a distinct function. Positions 6–13 seem to be required for binding by the transposase; mutations or chemical modifications at these positions interfere with all the reactions tested (111; J. Bender, H. Benjamin, N. Kleckner, personal communication). Changes at positions 4 and 5 have no significant effect (111). Changes at positions 1–3 inhibit strand transfer without affecting the donor DNA cleavages (111; D. Haniford, H. Benjamin, N. Kleckner, personal communication); this suggests that the interaction (either direct or indirect) of these nucleotides with transposase changes between the two chemical steps.

This terminal sequence is necessary and sufficient for the function of an inside end. Reactions involving only the inside ends do not require host proteins (31; H. Benjamin, N. Kleckner, personal communication). The inside end contains a GATC sequence, which is methylated by the host cell *dam* methylase, and transposition involving this end is controlled by its methylation state (112). The unmethylated end or one of the two possible forms of the hemimethylated end is more active for transposition than the fully methylated form (112). Combined with the increased expression of Tn10 transposase when the promoter is unmethylated or hemimethylated (112), IS10 is likely to transpose only immediately after the passage of the chromosomal replication fork. The activation of only one of the two sister copies (19, 96, 112) when the DNA is hemimethylated would be important. This regulation allows transposons, which leave a double-strand break at the donor site, to avoid killing of the host cell by leaving an intact copy that can be used as the template for gap repair. Tn5 and Tn903, as well as Tn7, all of which transpose nonreplicatively, also exhibit *dam* methylation control over their transposition (113, 114; O. Hughes, N. Craig, personal communication).

The outside end carries, in addition to the terminal inverted repeat, an IHF-binding consensus sequence (positions 30 to 42) that modulates the activity of this end (115). One function of IHF binding to this site may be negative control of transposase expression from the adjacent promoter (111). It also directly affects reactions involving the outside end (31). The terminal inverted repeat at this end differs from the inside end at five positions, and these differences confer the dependence on host proteins. Either HU or IHF can fill the role; action by IHF requires its binding site (111). In addition, IHF appears to stimulate use of a target site within the donor transposon (L. Signon, N. Kleckner, personal communication), possibly by introducing DNA bends that bring the nearby target into easy reach of the donor ends.

IS10/TN10 TARGET SEQUENCES   Tn10 prefers a target sequence of NGCTNAGCN (18). A short region that is present on both sides of this target sequence with no apparent consensus also influences the target site preference (J. Bender, N. Kleckner, personal communication). Tn10 transposase is responsible for this target sequence preference. In an in vitro reaction using highly purified transposase without any host proteins, transposition exhibits the same target site preference as seen in vivo (H. Benjamin, N. Kleckner, personal communication; also see below).

IS10/Tn10 TRANSPOSASE   The Tn10 transposase is the only Tn10-encoded protein required for transposition. It is the only protein essential for the production of the transposition intermediate involving target sites within the transposon in vitro with a donor DNA carrying a pair of inside IS10 ends (31; R. Chalmers, H. Benjamin, N. Kleckner, personal communication). Unlike MuA protein, which cuts only one strand, Tn10 transposase produces double-strand cuts at the donor ends (24, 25; H. Benjamin, N. Kleckner, personal communication). Thus, the transposase must recognize the transposon end sequence, cut both strands of the donor DNA at the transposon ends, and promote strand transfer into a target DNA.

According to the DNA sequence of its gene, Tn10 transposase consists of 402 amino acids with a predicted molecular weight of 46,000 (116). Inside an *E. coli* cell, Tn10 transposase acts preferentially *in cis;* transposon ends adjacent to the gene from which the transposase is produced are preferentially utilized (117). This feature, while unique to prokaryotic systems, is shared by several prokaryotic transposases, including Tn5, IS903, and to an extent, MuA protein (118–121).

A number of transposase mutants have been isolated that render the protein incapable of accomplishing strand transfer but do not affect the donor DNA cutting (36). The mutations lie in several clusters, suggesting the parts of the protein responsible for interacting with target DNA and promoting strand transfer (36). Consistent with this notion, other mutations affecting the target sequence preference also map very close to these sites (122).

Close to the C-terminus of the protein, there is a region in which the amino acid sequence exhibits limited, yet recognizable similarity to transposases of several prokaryotic elements (123, 124). The class of elements that exhibit this similarity includes, in addition to Tn10/IS10, Tn5/IS50, IS4, IS186, Tn4430/IS231, and ISH1 (2).

Retroviruses

RETROVIRAL DNA SEQUENCES   To integrate into the host chromosome, retroviral DNA must carry a pair of the sequence at the end of the so-called "long terminal repeats" (LTRs) (15). The required sequence is relatively short, less than 15 base pairs for many viruses (15). Mutation analysis, both in vivo and in vitro, and modification interference experiments of the terminal sequences indicate that many base changes and chemical modifications within the short terminal sequences are well tolerated (34, 125–129). On the other hand, it is important that the critical parts of the sequence be located near the ends of the linear double-stranded DNA. Normally, the 5'-CA-3' di-nucleotide that marks the end of the integrated proviral DNA is located two base pairs away from the ends of precursor linear viral DNA. This distance can be shortened to zero or one base pair or elongated by a few base pairs, but further modifications are detrimental for the sequence-specific cleavage and strand transfer (126–130).

RETROVIRAL IN PROTEINS   The primary actors in the retroviral DNA integration processes are the viral IN proteins. The IN proteins from several sources have been purified and demonstrated to carry both sequence-specific endonuclease activity and DNA strand transfer activity (32–34, 37, 38, 44, 45, 131).

The IN proteins of all the retro-elements studied are produced by specific proteolytic cleavage of a gag-pol polyprotein. The precursor *pol* gene product generally includes the protease responsible for maturation of the components: reverse transcriptase (RT), RNase H, and integrase (IN) (15). For most viruses, IN is processed into an independent polypeptide of 30 to 46 kDa. In the case of ASLV, the IN domain also exists as a domain of the larger subunit (RT-IN polypeptide) of the dimeric RT protein (132).

A zinc finger–like structural motif is found in the N-terminal part of IN proteins. The C-terminal part contains a domain of conserved amino acid sequence that shares similarities with the transposases of the IS3 family of bacterial elements (2a, 3, 4).

PROTEIN-DNA COMPLEXES AND TRANSPOSITION STEPS

Starting from an ensemble of reversible interactions among the participating proteins and DNA sites, transposition proceeds through formation of progressively more stable protein-DNA complexes. While certain events take place in a compulsory order, other events can occur in alternative orders, thus forming a network of possible reaction pathways. While the details of the reaction pathways and their physiological implications are the subject of ongoing studies, the complex design of this system seems to accommodate a number of important control points.

One important attribute of the assembly process of the higher order protein-DNA complex is discrimination of the relative orientation of the element ends. Recognition of the relative orientation of sequence segments located at a distance on a DNA molecule is often a physiologically important, but mechanistically nontrivial biological problem. If transposition is attempted with a pair of element ends that are on two separate DNA molecules, or that are on the same DNA molecule but in the wrong relative orientation, the result will be nonproductive and may lead to death of the host cell. For a transposon whose copy number per host genome is low, the selection of the two ends to be paired may seem to present no problem. However, even in this case, immediately after replication of the element, or during chromosome pairing, the two sister copies of the element are located close to each other; a mechanism to guard against selection of the wrong pair of ends would thus be important.

All the prokaryotic elements tested exhibit some end-pair selection mechanism. It is noteworthy that all these elements require supercoiled DNA substrates for an efficient reaction. For Mu, the energy of DNA supercoiling and the structure of the high order protein-DNA complex have been directly implicated in the mechanism of proper end pairing (133).

The ability to recognize the relative orientation of two DNA segments in recombination is not unique to transposition. The invertase and resolvase families of site-specific recombination reactions also demonstrate remarkable site orientation specificity (6, 7, 14). In these cases, the special geometrical structure of an essential synaptic protein-DNA complex appears to be designed to utilize the energy of DNA supercoiling such that only a properly oriented pair of DNA sites can efficiently form the productive conformation. In the invertase and resolvase reactions, sites with incorrect relative orientations pair to form a stable synaptic complex that resembles the proper complex but fails to recombine (134–137). This is in contrast to the Mu transposition reaction in which two ends in the wrong relative orientation do not even stably synapse (see below).

The reactions that precede formation of the Mu strand transfer products can be viewed as a macromolecular architecture assembly process. In such a process, not only the substrate DNA molecules, but also the protein molecules make up the product structure. The free energies that govern the reaction outcome are heavily influenced by the protein components as well as by the changes in the DNA parts. Such a process allows protein and DNA conformational transitions to function as possible kinetic control points, thus coordinating separate events so that the overall outcome is physiologically meaningful.

The strand transfer product, which exists as a stable protein-DNA complex, is an intermediate of the DNA transposition process. The stable protein-DNA architecture must be disassembled in such a way as to assist the processing of the DNA to form a simple DNA structure. While this disassembly process and processing of the strand transfer product have not been studied in detail, they are also expected to be coordinated in an orderly fashion.

Current understanding of the structure and function of the protein-DNA complexes involved in the Mu transposition process are discussed first. The protein-DNA complexes made by other elements are then described for comparison.

Protein-DNA Complexes in Mu Transposition

REVERSIBLE INTERACTION BETWEEN PROTEINS AND DNA SITES   The initial interactions between the proteins and DNA sites are all quickly reversible. The six MuA-binding sites at the two Mu ends show different affinities for MuA (47). L1, L3, and R3 bind MuA with a K_d of about 10^{-8}M (K. Adzuma, K. Mizuuchi, unpublished). L2 lacks half of the consensus binding sequence and has a lower affinity for MuA. Binding of MuA to the Mu-end DNA sites does not require divalent metal ions (T. A. Baker, K. Mizuuchi, unpublished). Judged from the stoichiometry of protein to DNA in protein-DNA complexes separated in acrylamide gels, one monomer of MuA binds to each site (66). MuA protein bends DNA upon binding to the Mu-end sites (65, 66). Thus, when all three sites at each end are occupied by MuA, each end is expected to take on a specific geometrical configuration.

The binding of MuA to the IAS is weaker than the MuA–Mu end-site interaction. The DNA-binding domains on MuA for the two DNA sequences, the Mu ends and the IAS, are separate. Thus, a single MuA molecule can, in principle, bind to both sequences simultaneously. However, it has been suggested that there may be a negative cooperative interaction between the two binding sites on MuA, and the tertiary complex may be unstable (55).

PROCESS THAT LEADS TO STABLE SYNAPSIS OF TWO MU ENDS   To participate in the Mu transpositional recombination, the Mu donor DNA must carry the two Mu end-sequences in their natural relative orientation on a single negatively supercoiled DNA molecule (138).

For Mu transposition, the critical step for the end orientation recognition occurs prior to the stable synapsis of the two Mu ends; the stable synapsis does not take place if the relative end orientation is incorrect (M. Mizuuchi, T. A. Baker, K. Mizuuchi, in preparation). Based on experiments that involved topological analogs of the mini-Mu donor DNA, a transient synaptic protein-DNA complex called the plectosome was proposed to explain the end orientation selectivity (133). This model states that the special plectonemic geometry of the two Mu end-segments within this critical reaction intermediate is such that either its formation or its stability requires the negative superhelicity of the DNA and the proper orientation of the two Mu end-sequences.

The plectosome, which has not been isolated as a stable intermediate, is considered to be the obligatory precursor of the stable synaptic complex (SSC, see below) under normal conditions. Stable synapsis of two Mu ends requires, in addition to the proper end orientation, the IAS to be present (M. Mizuuchi, T. A. Baker, K. Mizuuchi, in preparation). Thus, it is most convenient to consider that three DNA segments, two Mu ends and the IAS must cooperate through their interaction with bound MuA molecules prior to stable Mu end synapsis. M. G. Surette and G. Chaconas recently found that the IAS is capable of functioning in *trans* at this stage of the reaction (138a). Protein components that uniquely associate with the IAS, IHF, and the N-terminal domain of MuA are also considered to be specifically required at this stage. Some of the MuA-binding sites at the two Mu end-segments are also required only at the early stage of the reaction as described below.

There are two different states in the interaction between MuA and its binding sites at the Mu ends. The primary interaction between MuA and Mu end binding sites is reversible. On the other hand, once the SSC is formed, interaction between a subset of the Mu end binding sites and MuA molecules within the complex is stable for many hours (M. Mizuuchi, T. A. Baker, K. Mizuuchi, in preparation). Thus, the SSC must be energetically more favorable than the reversibly interacting MuA-DNA mixture. Nevertheless, the SSC cannot be made by simply mixing the MuA molecules and the Mu-end DNA segments together under typical reaction conditions. The difference in stability also suggests significant conformational changes in the participating macromolecules. The slow transition between the states of the protein-DNA complexes and temperature dependence of the transition suggest a high activation energy for the transition (T. A. Baker, M. Mizuuchi, K. Mizuuchi, in preparation). Under normal circumstances, the accessible kinetic path between the two states requires the participation of other factors. These factors, the IAS, IHF, the N-terminal domain of MuA, several MuA-binding sites at the Mu ends, and probably HU, with the superhelical donor DNA as the substrate, can be considered as the intrastructural catalyst of a structural transition. The physiological outcome is at least twofold. First, it allows exclusive usage of the correct pair of Mu ends for transposition; no irreversible step takes place utilizing a wrong pair of ends. Second, it allows direct regulation of Mu transposition by the Mu repressor protein through competitive binding to the IAS, again at an early stage of the reaction prior to any irreversible step.

In an in vitro reaction, this regulatory system can be bypassed by changing the reaction conditions. High superhelicity eliminates the need for IHF (54, 56). The presence of high concentrations of glycerol and lowered salt concentrations in the reaction drastically relax the requirements for the formation of the stable protein-DNA complexes (53). Inclusion of dimethyl sulfoxide (DMSO) in the reaction has even more dramatic effects on the requirements of the reaction. Essentially all of the corequirements for the formation of the stable protein-DNA complexes can be eliminated from the reaction without seriously impeding the transpositional recombination in the presence of DMSO (53, 133).

Finally, precut donor ends can be used for the DNA strand transfer reaction, bypassing the requirements for stable synapsis. A DNA fragment carrying the precut 5'-CA-3'-OH-end with the Mu-R end-sequence can participate in a pairwise attack on a target DNA to accomplish DNA strand transfer (20). The pair of ends do not need to belong to the same DNA molecule (20). Only MuA is required as a protein factor (20).

These reactions, either with special reaction conditions or with the precut DNA substrate, can be considered either to lower the activation energy for the formation of the stable protein-DNA complex, while still using essentially the same kinetic path via the plectosome, or to generate a new kinetic circuit for the reaction.

STABLE SYNAPTIC COMPLEX   Successful synapsis of the two Mu end-segments through formation of the plectosome results in the SSC. In the SSC, the two Mu ends are held together through stably bound MuA molecules, but the endonucleolytic cleavage at the ends of the Mu sequences is yet to take place.

In the absence of any divalent metal ion, while MuA binds to the Mu end-sequences, a stable protein-DNA complex is not formed (M. Mizuuchi, T. A. Baker, K. Mizuuchi, in preparation). In the presence of Mg^{2+}, the reaction efficiently undergoes the subsequent DNA cleavage and strand transfer steps. In contrast, Ca^{2+} does not allow the donor DNA cleavage to take place (M. Mizuuchi, T. A. Baker, K. Mizuuchi, in preparation). Nonetheless, a stable protein-DNA complex (the SSC) containing the two synapsed Mu-end DNA segments is generated during incubation of the protein and DNA substrates in the presence of Ca^{2+} (M. Mizuuchi, T. A. Baker, K. Mizuuchi, in preparation). The SSC is resistant to challenge by a large excess of competitor DNA carrying the Mu end-sequence, and it undergoes donor cleavage and strand transfer upon addition of Mg^{2+} (M. Mizuuchi, T. A. Baker, K. Mizuuchi, in preparation).

The IAS has been shown to be required for the formation of the SSC but not for the subsequent reaction steps (138a; M. Mizuuchi, T. A. Baker, K. Mizuuchi, in preparation). Most of the reaction components that have previously been shown to be dispensable after formation of the cleaved donor complex (CDC, see below) may become dispensable once the SSC is formed. This notion is supported by a nuclease protection experiment that showed that, as in the CDC, the L1, R1, and R2 sites are stably occupied in the SSC, but the L2, L3, and R3 sites become accessible to nucleases in the presence of competitor DNA (M. Mizuuchi, T. A. Baker, K. Mizuuchi, in preparation). While the activity of the SSC is resistant to challenge by competitor DNA, it is less stable than the CDC. It is inactivated by preincubation at a lower temperature than is required to inactivate the CDC (T. A. Baker, M. Mizuuchi, K. Mizuuchi, in preparation). Activity is also gradually lost when the donor DNA is relaxed or when the divalent metal ion is removed (T. A. Baker, M. Mizuuchi, K. Mizuuchi, in preparation).

Protein crosslinking experiments have shown that, as in the CDC (see below), a stable tetrameric unit of MuA exists in the SSC (M. Mizuuchi, T. A. Baker, K. Mizuuchi, in preparation). In contrast, MuA failed to form such a crosslinkable tetramer prior to formation of the SSC (M. Mizuuchi, T. A. Baker, K. Mizuuchi, in preparation; 52).

CLEAVED DONOR COMPLEX   In the presence of Mg^{2+}, the SSC is readily converted to the CDC (20, 139). The CDC, which is also called the type I transpososome (139), accumulates in the presence of Mg^{2+} when MuB is absent from the reaction. In the CDC, the two Mu ends are cleaved to expose the 3'-OH ends of the Mu sequence (20). The tightly bound MuA molecules hold the two Mu ends together and the Mu part of the donor DNA remains negatively supercoiled, while the flanking outside part is relaxed due to the single-strand cuts at the Mu ends (20, 139). The CDC efficiently accomplishes strand transfer upon addition of a target DNA bound by MuB protein (20, 139). Thus the CDC is an immediate precursor of the strand transfer reaction.

The CDC is stable to a variety of treatments: it is stable in the presence of 2M NaCl or 4M urea; 5 min incubation at 60 °C results in 50% loss of the complex; it is stable in the presence of ethylenediaminetetraacetic acid (EDTA) to chelate Mg^{2+} (139). Removal of the superhelicity of the Mu DNA segment does not destabilize the complex (20, 139). The HU protein, which stays associated with the CDC in a low salt buffer, can also be removed from the complex without affecting its ability to carry out DNA strand transfer (92, 93).

Nuclease footprinting experiments on the CDC have been carried out by several laboratories (52, 66, 93). Removal of loosely bound MuA molecules, either by high salt or by addition of competitor DNA, renders L2, L3, and R3 sites susceptible to attack by DNaseI (52, 93). The IAS is also not stably bound by MuA in the CDC (93). In contrast, L1, R1, and R2 sites are occupied by tightly bound MuA molecules (52, 93). In addition to the MuA-binding consensus sequence, the MuA footprints at the L1 and R1 sites extend to the very ends of the Mu sequence and beyond. The extent of the outside DNaseI protection is 7 to 13 nucleotides, depending on the strand (52, 93). This is in contrast to the footprint of reversibly bound MuA at these sites, which does not extend much beyond the end of the consensus sequence that terminates 6 nucleotides short of the Mu ends (47, 52, 93). Use of hydroxyl radicals as the footprinting probe has revealed a hypersensitive site immediately outside of the Mu end-sequence in the CDC, suggesting a distorted DNA conformation (52).

Not only can the loosely bound MuA be removed from the CDC, but also the IAS as well as L2 and L3 sites (and probably the R3 site, although this has not been directly tested) can be cut away from the complex without destroying its ability to carry out DNA strand transfer (93).

Protein crosslinking carried out on the CDC purified away from loosely bound MuA molecules yields a tetrameric form of the MuA protein, suggesting that four monomers of MuA molecules participate in a tight complex (52). The crosslinkable tetramer form is normally not produced in the absence of a functional Mu donor DNA (52). Since there are only three MuA-binding sites stably occupied by MuA in the CDC (or STC, see below), a question arises as to the location of the fourth MuA molecule. One possibility is that it is located outside of the Mu ends and is responsible for the protection there. However, symmetry considerations make this model less attractive. Another possibility is that the complex has a pseudo two-fold axis and the fourth MuA is positioned symmetrically to the MuA bound to R2. However, the corresponding position of the L-end DNA adjacent to the L1 site is not protected as one might expect if this part of the L-end is nonspecifically bound to the fourth MuA. It is possible that the arrangement of the MuA tetramer is such that DNA binding to the fourth MuA requires a sharp DNA bend that does not take place in the absence of a properly located binding consensus sequence adjacent to the L1 site. This possibility is supported by the following observation. In the strand transfer complex made with a pair of precut R-end fragments, four sites (two R1 sites and two R2 sites) are stably occupied by MuA (93). Formation of the crosslinkable MuA tetramer is also observed in the presence of the precut R-end fragments (T. A. Baker, M. Mizuuchi, K. Mizuuchi, in preparation).

STRAND TRANSFER COMPLEX   When the CDC encounters a target DNA bound by MuB, coordinated strand transfer at the two ends of the Mu sequence makes the strand transfer product that is stably complexed with participating proteins (139). This complex is called the strand transfer complex (STC), or type II transpososome (139).

MuB and HU proteins associated with the complex can be removed by a high salt treatment without dissociation of the Mu ends–MuA complex (92). The STC is even more stable than the CDC. It withstands treatment at 65 °C or with 6M urea (139).

The MuA molecules that hold the CDC together stay tightly bound to the two Mu ends in the STC (139). Like the CDC, the MuA molecules bound to L2, L3, and R3 sites can be removed by a competitor DNA, while the L1, R1, and R2 sites are stably occupied (93). The extended protection from DNaseI at L1 and R1 sites expands to about 15 nucleotides outside of uncut strand of the donor flanking DNA (93). The target DNA strands that are joined to the donor 3'-ends are also protected from DNaseI for about 20 nucleotides adjacent to the junction (93).

THE EFFECTS OF MuB ON EVENTS INVOLVING THE DONOR DNA   In the above description of the reactions involving the donor DNA, the effect of interactions with MuB or MuB-bound target DNA was ignored, except for the participation of MuB in the strand transfer reaction to form the STC. However, MuB with or without a bound target DNA can also interact with the MuA-donor DNA complexes, even prior to the formation of the CDC, with significant consequences.

MuB oligomerizes in the presence of ATP and binds nonspecifically to DNA to make it an efficient target for Mu transposition (58, 76). A functional interaction of MuB with any of the other reaction components appears to require ATP. The subject of DNA-MuB-MuA interaction is discussed further in the next section. Here, the effects of MuB on the MuA-donor DNA complexes are described.

The notion that MuB can influence MuA-donor DNA interactions at earlier stages is supported by a variety of observations. For example, if a donor DNA with one of the Mu end-sequences mutated at the terminal A residue is reacted in the absence of MuB, donor DNA cleavage does not take place (49). However, if MuB and ATP are included in the reaction, strand transfer product is generated with significant efficiency (49). Thus MuB must be able to interact with the MuA-donor DNA complex prior to the donor DNA cleavage to enable the otherwise blocked donor cleavage to take place.

MuB can interact with MuA-donor DNA complexes without binding to a potential target DNA. When an excess concentration of MuA is added to a reaction together with all other components, the otherwise efficient intermolecular strand transfer is severely inhibited (63). Instead, DNA sites on the donor DNA itself are used as the strand transfer target (63). This intramolecular reaction was found to be due to the presence of a pool of MuB not bound to DNA (63). Free MuB can interact with MuA-donor DNA complexes at the stage of SSC as well as CDC, causing efficient intramolecular strand transfer (63).

MuB that has lost the ability to stably bind DNA can efficiently stimulate intramolecular, but not intermolecular, DNA strand transfer (63, 139a). Chemical alteration of MuB by the sulfhydryl modifying agent N-ethylmaleimide (NEM) causes loss of the stable DNA-binding activity of MuB in the presence of ATP (63, 139a). The NEM-modified MuB can no longer promote intermolecular Mu DNA strand transfer (63, 139a). However, it retains full activity to stimulate intramolecular strand transfer (63, 139a). Furthermore, in the presence of Mn^{2+}, in place of Mg^{2+}, efficient intramolecular strand transfer takes place without MuB (63). These observations indicate that the CDC possesses the intrinsic ability to carry out efficient DNA strand transfer, an ability that is suppressed under normal conditions until the CDC is allosterically activated by the contact with MuB (63). According to this model, the ability of MuB to stimulate intermolecular strand transfer is based on the fact that MuB is normally localized on intermolecular target DNA when it activates strand transfer by the CDC.

As MuB not bound to DNA can interact with MuA-donor DNA complexes at different reaction stages, DNA-bound MuB can interact not only with the CDC but with the SSC as well. When the SSC that accumulates in the presence of Ca^{2+} is preincubated with MuB and target DNA, intermolecular strand transfer takes place quickly upon addition of Mg^{2+} (M. Mizuuchi, T. A. Baker, K. Mizuuchi, in preparation). In comparison, if MuB-bound target DNA is added together with Mg^{2+}, strand transfer is distinctly slower (M. Mizuuchi, T. A. Baker, K. Mizuuchi, in preparation). This observation indicates that during the preincubation with Ca^{2+}, a complex between the SSC and the target-bound MuB is formed. The association between the SSC and the target-bound MuB depends on a divalent cation; it dissociates quickly upon addition of ethyleneglycol-bis-(β-aminoethylether)N,N, N',N'-tetraacetate (EGTA) (M. Mizuuchi, T. A. Baker, K. Mizuuchi, in preparation).

These observations indicate that the assembly of the reaction components can take place in different orders, resulting in different outcomes of the reaction. The order of the assembly in turn is influenced by the relative concentrations of the reaction components and controlled by the balance of kinetic parameters of each reaction path. The MuB bound to a potential target DNA can interact with MuA at even earlier stages of the reaction, prior to the formation of the SSC. This interaction strongly influences the choice of the target DNA sites, and is discussed separately. Currently, there is no direct evidence of interaction between MuA and MuB in the absence of DNA or ATP.

Tn7 Protein-DNA Complexes

The higher order protein-DNA complexes involved in Tn7 transposition have not yet been directly studied. Nevertheless, current observations point to their critical role in the transposition process. The large number of TnsB-binding sites at the two Tn7 ends is reminiscent of the Mu system. Compared to the Mu reaction, events on the Tn7 donor DNA and those on the target DNA seem to be much more stringently coordinated. In the Tn7 reaction, omission of any of the reaction components, including the attTn7 target DNA, blocks the donor cleavage step (23). Thus, the formation of the protein-DNA complex competent for the donor DNA cleavage requires participation of all four proteins, TnsA, TnsB, TnsC, and TnsD, and proper donor and target DNAs.

Efficient Tn7 transpositional recombination requires the donor DNA to be supercoiled for the TnsD pathway. A supercoiled attTn7 target DNA also appears to be preferred (R. Bainton, N. Craig, personal communication). Tn7 transpositional recombination presumably also requires a donor DNA molecule with two properly oriented Tn7 ends. The mechanism of recognition of the end-sequence orientation may be basically similar to the Mu case, although an IAS or enhancer element has not been reported for Tn7.

Stringent control of the donor DNA cleavage by the availability of the target DNA for Tn7 transposition is understandable considering the limited copies of the potential target site in a cell for high-frequency transposition. Donor cleavage in the absence of an eligible target site would be disadvantageous to inheritance of the element.

IS10/Tn10 Protein-DNA Complexes

Tn10 also requires a supercoiled DNA substrate for efficient donor DNA cleavage and strand transfer, although the superhelical density can be substantially reduced (H. Benjamin, D. Haniford, N. Kleckner, personal communication). Since in the in vitro reaction for Tn10, strand transfer predominantly utilizes intramolecular target sites, it is currently difficult to differentiate the requirement and role of superhelicity of the donor DNA from those of the target DNA. Like Mu, donor DNA cleavage does not take place when the two Tn10 ends are in their wrong relative orientation (24; H. Benjamin, N. Kleckner, personal communication).

Two stable protein-DNA complexes have been identified as Tn10 transposition intermediates. They correspond to the CDC and the STC of Mu.

The CDC of Tn10 contains, as the DNA component, an excised transposon fragment (ETF) (35, 36). The ETF is the Tn10 DNA precisely cleaved at the ends of the Tn10 sequence on both strands (H. Benjamin, N. Kleckner, personal communication). In the CDC, the two ends of the ETF are tightly held together by the bound transposase protein (35). The donor DNA that flanked the Tn10, which is cut away by double-strand cleavages, is not stably retained with the complex (35). The ETF within the CDC is supercoiled; the transposase molecules that bridge the two ends prevent the DNA ends from swivelling (35). The Tn10 transposase is considered to be the only critical protein component of the CDC, since the Tn10 strand transfer products can be made by a substantially purified preparation of the protein without the addition of other protein components (H. Benjamin, N. Kleckner, personal communication). The CDC accumulates in reactions with a special class of the mutant Tn10 transposase that is impaired in the DNA strand transfer step (36). Kinetic experiments with wild-type transposase both in vivo and in vitro indicate that the CDC is a competent intermediate in the Tn10 strand transfer process. It initially accumulates and is subsequently consumed as the strand transfer products accumulate (35).

The STC of Tn10 is produced in vitro mainly by utilizing a target sequence within the ETF (31, 35). The favored use of the target orientation is such that the most abundant strand transfer product is a transposon circle in which a segment of the transposon sequences between one end of the element and the target site is inverted (25, 31, 35). The 3'-ends of the element are joined to the 5' protruding ends of the target site cut by a nine-base-pair stagger (25). The two Tn10 ends within the STC are likely held together, judged by its mobility in gels (35). Like the Mu STC, the Tn10 STC is more stable than the CDC against high temperature or urea treatment (35).

The preferred use of a target site within the donor transposon sequence may be due to the release of the flanking DNA from the CDC. If this were the case, unlike Tn7, Tn10 may form the initial synaptic complex and the CDC without prior commitment to a target site. This would allow relatively random selection of target sites located outside of the donor Tn10, but the system would suffer from a high probability of intramolecular events. Unlike replicative transposons such as Mu, self-destruction of the donor Tn10 would be less damaging to the host cell, which can efficiently repair the double strand gap left by the excised transposon. Recent observations suggest that IHF strongly influences the target site selection of Tn10; it appears to promote use of intra-transposon target sites at the expense of other sites (L. Signon, N. Kleckner, personal communication).

Retroviral Protein-DNA Complexes

The purified IN protein of retroviruses can promote both the endonuclease and the strand transfer steps of the integration reaction. However, the reactions carried out with purified IN protein and DNA substrate by themselves are yet to reproduce efficiently an important aspect of the in vivo reaction. Efficient integration of the viral DNA is achieved by coordination of the events at the two ends of a viral DNA. With Moloney murine leukemia virus (Mo-MLV), coordinated insertion of pairs of DNA substrate molecules matching the ends of Mo-MLV DNA has been observed using semipurified IN protein, but this reaction is inefficient; many events involve insertion of only one copy of the viral DNA substrate into one strand of the target DNA (32, 37). With purified HIV IN protein, the coordinated reaction is even less efficient and has been detected only in experiments that involve strong selection for coordinated events (38).

In contrast, the stable protein-DNA complexes containing linear double-stranded viral DNA isolated from Mo-MLV-, HIV-, or ASLV-infected cells efficiently carry out a coordinate attack by the two ends of the viral DNA (21, 22, 30, 140–144). The protein-viral DNA complex of Mo-MLV, as isolated from infected cells, is fast sedimenting (140). Detailed analysis of the essential structural features necessary for the coordinated insertion of pairs of viral DNA ends has not been reported. The protein-viral DNA complex purified from HIV-infected cells is essentially free from major structural proteins as judged by immunoprecipitation experiments (143). This complex is capable of efficient insertion of the viral DNA into exogenously added target DNA. The essential components of these complexes for the coordinated reaction and the pathway for the assembly of the complex remain to be studied.

Curiously, although the reactions with Mo-MLV viral DNA-protein complexes isolated from infected cells can take place efficiently in the presence of either Mg^{2+} or Mn^{2+}, reactions with purified IN proteins and short DNA substrates matching the ends of the viral DNA strongly prefer Mn^{2+} as the divalent metal ion (30, 32).

Another intriguing question that remains to be addressed concerns the mechanism to avoid "autointegration," or intramolecular target events. In the absence of a special mechanism, essentially all the reactions would be likely to take place intramoleculary. In fact, the ASLV DNA-protein complexes isolated from infected cells carry out predominately "autointegration" events (144). On the other hand, Mo-MLV and HIV DNA-protein complexes can efficiently utilize exogenously added intermolecular target DNA (21, 22, 141, 142). Mo-MLV and HIV complexes may have a mechanism that allows them to avoid destruction of their own DNA, while the essential component(s) of such a mechanism may be easily lost from the ASLV DNA-protein complex preparation.

SELECTION OF THE TARGET SITES AND TRANSPOSITION TARGET IMMUNITY

Transposition target immunity is a phenomenon by which the presence of a resident transposon renders the nearby DNA region a poor target for additional insertion by the same kind of element. Immunity is displayed by a number of transposons that transpose at relatively high frequency, such as the Tn3 family of elements, Tn7, and Mu. The phenomenon was first discovered for Tn3; a single copy of the terminal inverted repeat sequence of Tn3 confers target immunity on a plasmid DNA (145, 146). Similarly, a plasmid carrying at least one Mu end-sequence is a poor target for Mu transposition both in vivo and in vitro (58, 59, 146a). Incomplete end sequences carrying less than two MuA-binding sites are inefficient in conferring target immunity (58, 59). In the case of Tn7, a copy of the right end-sequence efficiently confers target immunity on a plasmid DNA, but the left end-sequence is inefficient in this respect (100).

For high-transposition-frequency elements such as Mu, which undergo multiple rounds of replicative transposition during lytic growth, transposition target immunity can be viewed as a means of preventing self-destruction, by avoiding integrating into itself or into another copy of the element. Immunity can also prevent the genome instability that may result from having multiple copies of the same element in close proximity within a chromosome; potentially detrimental gene rearrangements may occur through homologous recombination.

Mu Transposition Target Immunity

Target immunity is demonstrable in the Mu in vitro reaction, which has allowed us to understand a critical part of the mechanism by which a transposon end-sequence makes neighboring DNA sites poor targets. The basic mechanism involves the following interactions: (*a*) DNA bound by MuB is a preferred target, (*b*) MuA binds specifically to Mu end-sequences, (*c*) MuA speeds up the dissociation of MuB from DNA. The details of how this system functions depend on several complexities involving ATP binding and hydrolysis by MuB, as well as MuB protein oligomerization and MuA-MuB interactions.

In the absence of MuB, but in the presence of HU, MuA can efficiently cleave the Mu ends, forming the CDC (20, 29, 139). DNA strand transfer also takes place in the absence of MuB, but only inefficiently (29). MuB must activate MuA for efficient strand transfer; in the absence of MuB, essentially all strand transfer events utilize intramolecular DNA sites, near the donor Mu, as a target (29). Thus, in the absence of MuB the target site may simply be chosen by its proximity to the Mu ends.

MuB requires ATP to participate in the strand transfer reaction (29). In the absence of ATP, MuB weakly binds DNA without strong sequence specificity (147). The DNA binding is much tighter in the presence of ATP, and with typical concentrations of reaction components, essentially all the MuB in the reaction is bound to DNA (58). Thus, when MuB activates MuA in the CDC for strand transfer, DNA sites near the bound MuB are efficiently used as targets. The target sites for Mu transposition are generally determined by the distribution of MuB among potential DNA sites. While MuB has some DNA sequence preference, it is weak enough to be ignored for the purpose of this discussion. MuB is, therefore, expected to be relatively evenly distributed along all DNAs. This is in fact the case in the presence of a nonhydrolyzable ATP analog, ATPγS (58). Experiments involving ATPγS indicate that the ATP-MuB-DNA complex is very stable, and the MuB in the complex does not move from one DNA site to another in the absence of ATP hydrolysis (58). Even when ATP hydrolysis is occurring, MuB does not move readily from one DNA to another, provided that a high enough concentration of ATP is present to ensure quick binding of a new ATP molecule after release of the hydrolyzed product (58). Thus, practically speaking, MuB does not dissociate from DNA during each cycle of ATP turnover. Only when the ATP site remains unoccupied does the protein dissociate (K. Adzuma, K. Mizuuchi, unpublished).

In the presence of ATP hydrolysis, MuA drastically affects the distribution of MuB among DNA molecules, making DNA sites more or less preferred as targets for attack by the CDC. While relocation of MuB among DNA molecules is stimulated by MuA, if there are no specific MuA-binding sites anywhere, the MuB distribution is not strongly influenced by MuA. MuA simply accelerates the establishment of an "even" distribution of MuB among DNA sites (58). On the other hand, if some of the DNA molecules carry a Mu end-sequence, MuA preferentially stimulates removal of MuB from this DNA molecule (58). Thus, in a mixed population of relatively small DNA molecules (such as plasmids), with and without a Mu end-sequence, MuB accumulates on the molecules lacking a Mu end-sequence (58). On a large DNA molecule (such as the host chromosome) with a Mu end-sequence, MuB presumably accumulates at sites distant from the Mu end-sequence. The MuA bound to the Mu end-sequence raises the effective local concentration of MuA that mediates removal of MuB from nearby DNA sites (58, 148). MuA-mediated removal of MuB from DNA requires ATP hydrolysis (58).

MuB oligomerizes as it binds ATP, and kinetic experiments indicate that this oligomerization is critical for ATP hydrolysis (76). The effects of DNA on the MuB ATPase indicate that, in the absence of MuA, MuB generally stays on the DNA, thus avoiding the oligomer dissociation-association steps during the ATP hydrolysis cycle (76). However, in the presence of DNA, MuA strongly stimulates MuB ATPase and the oligomer dissociation-association appears to occur (29, 76). These observations support the view that MuA acts through the ATP hydrolysis cycle to influence the oligomeric state of MuB and thus to modulate its binding to DNA.

Simple release of MuB from DNA is not what happens when DNA-bound MuB encounters the MuA within the CDC. The MuB-bound DNA is instead efficiently captured as the target for strand transfer and the STC is formed. The MuB remains with the STC unless it is removed, for example, by high salt (92). The MuA molecules associated with the STC may still be able to remove MuB molecules bound elsewhere on the strand transfer product. Otherwise generation of a target DNA that is joined to multiple donor molecules might be expected. Such products are rare in reactions carried out with ATP, while they are abundant in the presence of ATPγS.

The SSC also seems to be capable of removing MuB from nearby DNA sites. Otherwise, the donor DNA that is converted to the SSC in a reaction containing Ca^{2+} should become nonimmune and therefore be used efficiently as a target for insertion by other donor DNA molecules. In contrast, the donor DNA that forms the SSC remains a poor target (M. Mizuuchi, T. A. Baker, K. Mizuuchi, in preparation).

If a simple local effective concentration principle is responsible for removal of MuB from DNA sites near a donor sequence, the final effective concentration of MuB among DNA sites should appear even, when viewed from the donor end (or MuA molecules bound to it). Therefore, some additional mechanism must differentiate the two types of reactions, the removal of MuB from the bound DNA and the capture of MuB-bound DNA as the target by a donor complex. Whether this mechanism involves different sizes of the MuB oligomer needed for the two reactions, or an additional kinetic device such as that discussed below operates in the Mu target immunity, remains to be investigated.

Interaction of free MuB with donor DNA-MuA complexes was discussed above. This interaction stimulates the use of intramolecular target sites near or within the donor DNA, increasing the risk of donor self-destruction. Normally, in the presence of ATP, MuB is tightly bound to DNA and the free MuB concentration is low (58, 63). However, too high concentration of MuA generates a steady-state pool of free MuB (63). Also, when all the potential target DNA in the reaction has a nearby Mu end-sequence, intramolecular transposition becomes frequent either because MuB has no choice but to bind to the immune DNA or because of the raised pool of the free MuB. Thus, it is critical to control the MuA concentration inside the cell to achieve efficient replication of the phage genome.

Several questions remain to be addressed concerning Mu transposition target immunity. How does Mu protect its entire 37-kbp genome from self-destruction? The average distance of the target site from the donor Mu in vivo may be about 20 kbp, and targets as close as several kbp are also used (149). Is any distinction made between the inside and the outside of the Mu sequence? Are there many Mu end–like sequences embedded throughout the Mu genome? Does the IAS also confer target immunity to nearby DNA? Is it possible that the sequence in the middle of the Mu genome that strongly interacts with DNA gyrase and is required for efficient Mu replication (149a) somehow protects the Mu sequence from self-destruction? One should also be reminded of the notion that the effective concentration between two sites on an interwound supercoil is practically independent of their distance within the circle (N. Cozzarelli, personal communication). Additional protective mechanisms yet to be discovered may operate during Mu replication.

Transposition Target Immunity of Other Transposons

The Tn3 family of transposons exhibits strong target immunity. A single copy of their terminal inverted repeat sequence confers immunity to a large stretch of DNA (146). Tn7 also exhibits target immunity. Transposition to the attTn7 site by the TnsD pathway, as well as to nonspecific targets by the TnsE pathway, is inhibited by the presence of a copy of Tn7 on the same DNA (100, 150). Transposition into a 50-kbp plasmid can be inhibited more than 100-fold by a resident element. While an attTn7 plasmid exhibits target immunity when a copy of Tn7 is present at a distance from attTn7, the use of attTn7 is even more severely inhibited when the site is directly occupied by a resident element, even though the attTn7 determinant sequence remains unaltered (100). Tn7 target immunity is also observed in a cell-free reaction for the TnsD pathway (23). Inverted repeat sequences of Tn1000 ($\gamma\delta$) with mutations at the very end that inactivate transposition function, without loss of transposase binding activity, possess the ability to confer the target immunity (151). Thus, it is unlikely that the immune target DNA is involved in chemical steps such as the transposon-end DNA cleavage.

The mechanism postulated for Mu target immunity may require modification in order to explain the immunity mechanism of other elements. The majority of the Tn3 family of elements encode a single polypeptide for transposition (14). Mu's strategy, the use of a separate activator protein and a device to control its distribution, may not apply here unless an unidentified host protein plays the role of MuB. Alternatively, each transposase molecule may play one of the two mutually exclusive roles, interaction with the donor DNA or interaction with the target DNA. More importantly, the target immunity of these elements is very strong, as it functions over much longer distances along the DNA than does the Mu system. Since Mu must replicatively transpose close to 100 times within a single cell during less than one hour of lytic growth, very stringent target immunity would be detrimental for Mu growth. In contrast, Tn3 and Tn7 transpose relatively rarely and a stringent immunity may be advantageous.

Intermolecular collisions with all the element end-sequences take place equally to all possible target sites. Immune target sites experience additional collisions, with the element end-sequence present on the same DNA molecule. The frequency of these intramolecular collisions depend on the effective local concentration between the two DNA sites, and it decreases as the distance between the two sites increase. Thus, the difference in the frequency of collisions with an element end, between an immune site distant from the element end and nonimmune DNA sites, is small. The puzzle is how this small difference can be amplified to a large transposition frequency difference between the immune and nonimmune targets. Let us assume that formation of the target DNA-activator protein complex takes a substantial length of time to be built, for example, owing to sequential addition of a large number of the protomers. On the other hand, destruction of the structure can take place at any time during the process by interaction with the element-end DNA-transposase complex. Then, the completion frequency of the target-activator complex is determined by the probability that the DNA site does not interact with the element-end DNA-transposase complex during the entire building process. By this mechanism, a small difference in the rate of interference in the complex building process can be amplified to a large difference in the probability of successful completion of the active complex. The Tn7 reaction is strongly stimulated by preincubation of the target DNA with the proteins that bind to it, TnsC and TnsD, in the presence of ATP (R. Bainton, N. Craig, personal communication). This finding encourages the view that critical early steps in protein-DNA complex formation start on the target DNA.

Whatever the actual mechanism of target immunity, it is likely to utilize the interplay between the ATPase reaction and macromolecular structural transitions involving higher order protein-DNA complexes. Transposition target immunity is a supreme example of the use of a bioenergetic device for delicate control over reactions so as to achieve a physiologically desirable outcome.

PERSPECTIVES

What is the evolutionary relationship between the large number of subgroups of elements in the transposon family? It is not at present clear whether most subgroups are evolutionarily related or have separate origins. However, it is easy to imagine that transposons may evolve very rapidly because of their parasitic nature. Inactive elements can stay dormant within the host genome and accumulate mutations without facing immediate negative selection. Once a new active form is generated, it can spread rapidly not only within the population of the original organism but also horizontally to different organisms. Thus, it is reasonable to suspect that many elements have evolved from a common ancestor.

The above view is strengthened by the similarity of the fundamental mechanism of transposition of several elements belonging to different subgroups. The polarity of the strand transfer reaction is the same for all reactions studied. In all cases where a functional assignment has been made, a single protein mediates both of the chemical steps, donor DNA cleavage and strand transfer. This is probably true for many, if not all, of the elements in the transposon family; many elements appear to encode only one protein as their putative transposase (1).

If many elements within the transposon family are indeed evolutionarily related and their reaction mechanism is conserved, how does this mechanism relate to the mechanism of other types of site-specific recombination reactions? One group of reactions that requires comment is conservative site-specific recombination and DNA topoisomerization. These reactions utilize protein-DNA covalent intermediates to conserve the phosphoester bond energy between the cleavage and joining steps (5). When one-step transesterification was first proposed as a model for the DNA strand transfer step of Mu transposition, it was pointed out that while the one-step model is possible for the Mu-type strand transfer, the two-step mechanism with a protein-DNA covalent intermediate is a necessity for the topoisomerase and conservative site-specific recombination reactions (11, 20). This is because of the strand passage or strand rotations that must take place between the cleavage and joining steps.

Although these two types of mechanism may appear fundamentally different, close examination of the nuclease reaction of the HIV IN protein offers a possible alternative view. Important for this discussion is the fact that a large variety of nucleophiles can be used by the IN protein for strand cleavage (40, 46).

Two proposals were made based on the above observation, one mechanistic and the other evolutionary (40). First, the same active site on the IN protein may be responsible for both the nuclease reaction and the strand transfer reaction. According to this model, the primary action of the IN protein in both steps is destabilization of a pair of phosphoester bonds (40). In the nuclease step, a water molecule (or other potential nucleophile) cleaves the destabilized bonds in the donor DNA to expose the donor 3'-OH ends. In the second step, a target DNA is bound and the two target phosphoester bonds are similarly activated by the same active sites on possibly a second set of IN protein molecules (40). This takes place presumably only in the presence of the cleaved donor DNA-IN protein complex, which places the donor 3'-OH ends close to the activated target bonds to facilitate the strand transfer. This model would explain why the activity for the two chemical steps may reside in one protein for this class of elements. Further, if two pairs of IN molecules, each monomer activating one phosphoester bond, are indeed involved in the overall reaction, a similarity with the symmetric arrangement in conservative site-specific recombination becomes apparent. Second, it now seems very easy to imagine how the two types of recombination mechanism, the one-step transesterification and the two-step transesterification, can interconvert through evolution. The only critical change in such evolution may be the gain or loss of a properly positioned nucleophilic amino acid residue at the active site of the protein (40).

The above considerations point to a possible evolutionary connection between transposition and conservative site-specific recombination and DNA topoisomerase reactions. Typical conservative site-specific recombination uses a pair of specific sites recognized by a specific recombinase. The critical components of each DNA site are the central core, or the crossover region, and a pair of recombinase-binding sequences symmetrically flanking the core. The recombinase molecules cut each site at one end of the core on each strand and form a covalent bond with the DNA phosphoryl end. After rearrangement of the strands, the process is reversed to form new DNA phosphoester bonds. Four strands are cut and four new phosphoester bonds are made. Overall chemical energy is conserved in this reciprocal process. The sequence of the core is not critical, but it must be homologous between the recombination partners. In contrast, the mechanism of translocation of a group of elements called conjugative transposons appears to deviate from this typical mechanism. Their recombinase sequence resembles that of the lambda Int protein, suggesting a fundamentally similar mechanism (152). However, some of these elements (such as Tn916 and Tn1545) can transpose to many different target sites that do not show strong similarity to the recombinase-binding sequence or even to the core sequence (152). Thus, the reaction on the surface looks more like a typical transposon-type reaction. This leaves open the question of whether or not the recombinases of these elements act in a conservative and reciprocal manner. It may be that the recombinase not only can recognize asymmetric sites, it also does not join all four new phosphoester bonds; instead it may act only as a nuclease at certain steps.

Even more intriguing examples that may eventually bridge the gap between the typical transposon-type mechanism and the typical conservative site-specific recombination mechanism are some of the plant transposable elements and the immunoglobulin VDJ recombination reaction. Of particular interest in the context of this discussion is the structure of the junction left behind after excision of the transposable element (one can consider the VDJ recombination reaction equivalent to a transposition reaction by envisaging that the recombination signal sequences correspond to the transposable element end sequences). The repaired junctions after element excision often contain short palindromes derived from sequence that abutted the element prior to excision (153, 154). Apparently similar short palindromes generated in the VDJ recombination have been termed P-nucleotides (155). One possible way the P-nucleotides might be generated is the following. When the element sequence is cut away by double-strand cleavage, the remaining ends are made into hairpins by joining the 3'- and 5'-ends. A later repair step may involve a single-strand cut internal to the tip of the hairpin, opening the hairpin to form the palindrome. Gap filling and end joining would then produce a junction containing the P-nucleotides.

Although such a hairpin can be made by a one-step mechanism, some proteins such as lambda Int and yeast Flp recombinase efficiently generate DNA hairpins under certain circumstances, presumably by a two-step mechanism (156; M. M. Cox, personal communication). Thus, it seems not unreasonable to entertain the possibility that these plant transposable elements and the VDJ recombination reaction involve proteins that can form a covalent intermediate during the DNA cleavage step. To further evaluate such a model and to assess the relationship among different types of DNA rearrangement reactions we must await future studies.

Whatever the actual mechanism of the above-mentioned reactions, comparative mechanistic studies of a variety of site-specific recombination reactions are expected to yield a wealth of information. This information may give us a glimpse into how one type of reaction system may diverge and evolve into a variety of reaction mechanisms that may be used for different biological purposes.

ACKNOWLEDGMENTS

I am grateful to the colleagues who furnished me with reprints, preprints, and unpublished information. I thank my coworkers for helpful discussions. Special thanks go to Tania A. Baker and Bob Craigie who helped in this writing by their useful discussions, careful reading, and numerous comments on the manuscript. I also thank Howard A. Nash and Nancy Kleckner for their useful comments. The work in my laboratory has been supported in part by the National Institutes of Health Intramural AIDS Targeted Antiviral Program.

Literature Cited

1. Galas, D. J., Chandler, M. 1989. See Ref. 8, pp. 109–62
2. Syvanen, M. 1988. See Ref. 9, pp. 331–56
2a. Fayet, O., Ramond, P., Polard, P., Prere, M. F., Chandler, M. 1990. *Mol. Microbiol.* 4:1771–77
3. Khan, E., Mack, J. P. G., Katz, R. A., Kulkosky, J., Skalka, A. M. 1991. *Nucleic Acids Res.* 19:851–60
4. Katzman, M., Mack, J. P. G., Skalka, A. M., Leis, J. 1991. *Proc. Natl. Acad. Sci. USA* 88:4695–99
5. Craig, N. L. 1988. *Annu. Rev. Genet.* 22:77–105
6. Hatfull, G. F., Grindley, N. D. F. 1988. See Ref. 9, pp. 357–96
7. Glasgow, A. C., Hughes, K. T., Simon, M. I. 1989. See Ref. 8, pp. 637–60
8. Berg, D. E., Howe, M. M., eds. 1989. *Mobile DNA.* Washington, DC: Am. Soc. Microbiol. 972 pp.
9. Kucherlapati, R., Smith, G. R., eds. 1988. *Genetic Recombination.* Washington, DC: Am. Soc. Microbiol. 731 pp.
10. Shapiro, J. A., ed. 1983. *Mobile Genetic Elements.* New York: Academic. 688 pp.
11. Mizuuchi, K., Craigie, R. 1986. *Annu. Rev. Genet.* 20:385–429
12. Grindley, N. D. F., Reed, R. R. 1985. *Annu. Rev. Biochem.* 54:863–96
13. Symonds, N., Toussaint, A., van de Putte, P., Howe, M. M. 1987. *Phage Mu.* Cold Spring Harbor, NY: Cold Spring Harbor Lab. 354 pp.
14. Sherratt, D. 1989. See Ref. 8, pp. 163–84
15. Varmus, H., Brown, P. 1989. See Ref. 8, pp. 53–108
16. Berg, D. E. 1989. See Ref. 8, pp. 185–210
17. Craig, N. L. 1989. See Ref. 8, pp. 211–26
18. Halling, S. M., Kleckner, N. 1982. *Cell* 28:155–63
19. Kleckner, N. 1989. See Ref. 8, pp. 227–68
20. Craigie, R., Mizuuchi, K. 1987. *Cell* 51:493–501
21. Fujiwara, T., Mizuuchi, K. 1988. *Cell* 54:497–504
22. Brown, P. O., Bowerman, B., Varmus, H. E., Bishop, J. M. 1989. *Proc. Natl. Acad. Sci. USA* 86:2525–29
23. Bainton, R., Gamas, P., Craig, N. L. 1991. *Cell* 65:805–16
24. Morisato, D., Kleckner, N. 1984. *Cell* 39:181–90
25. Benjamin, H. W., Kleckner, N. 1989. *Cell* 59:373–83
26. Mizuuchi, K. 1984. *Cell* 39:395–404
27. Craigie, R., Mizuuchi, K. 1985. *Cell* 41:867–76
28. Boeke, J. D. 1989. See Ref. 8, pp. 335–74
29. Maxwell, A., Craigie, R., Mizuuchi, K. 1987. *Proc. Natl. Acad. Sci. USA* 84:699–703
30. Brown, P. O., Bowerman, B., Varmus, H. E., Bishop, J. M. 1987. *Cell* 49:347–56
31. Morisato, D., Kleckner, N. 1987. *Cell* 51:101–11
32. Craigie, R., Fujiwara, T., Bushman, F. 1990. *Cell* 62:829–37
33. Katz, R. A., Merkel, G., Kulkosky, J., Leis, J., Skalka, A. M. 1990. *Cell* 63:87–95
34. Bushman, F. D., Craigie, R. 1991. *Proc. Natl. Acad. Sci. USA* 88:1339–43
35. Haniford, D. B., Benjamin, H. W., Kleckner, N. 1991. *Cell* 64:171–79
36. Haniford, D. B., Chelouche, A. R., Kleckner, N. 1989. *Cell* 59:385–94
37. Fujiwara, T., Craigie, R. 1989. *Proc. Natl. Acad. Sci. USA* 86:3065–69
38. Bushman, F. D., Fujiwara, T., Craigie, R. 1990. *Science* 249:1555–58

39. Mizuuchi, K., Adzuma, K. 1991. *Cell* 66:129–40
40. Engelman, A., Mizuuchi, K., Craigie, R. 1991. *Cell* 67:1211–21
41. Cech, T. R. 1990. *Annu. Rev. Biochem.* 59:543–68
42. Craig, N. L., Nash, H. A. 1983. *Cell* 35:795–803
42a. Pargellis, C. A., Nunes-Duby, S. E., de Vargas, L. M., Landy, A. 1988. *J. Biol. Chem.* 263:7678–85
43. Knowles, J. R. 1980. *Annu. Rev. Biochem.* 49:877–919
44. Katzman, M., Katz, R. A., Skalka, A. M., Leis, J. 1989. *J. Virol* 63:5319–27
45. Sherman, P. A., Fyfe, J. A. 1990. *Proc. Natl. Acad. Sci. USA* 87:5119–23
46. Vink, C., Yeheskiely, E., van der Marel, G. A., van Boom, J. H., Plasterk, R. H. 1991. *Nucleic Acids Res.* 19:6691–98
47. Craigie, R., Mizuuchi, M., Mizuuchi, K. 1984. *Cell* 39:387–94
48. Burlingame, R. P., Obukowicz, D. L., Lynn, D. L., Howe, M. M. 1986. *Proc. Natl. Acad. Sci. USA* 83:6012–16
49. Surette, M. G., Harkness, T., Chaconas, G. 1991. *J. Biol. Chem.* 266:3118–24
50. Groenen, M. A. M., Timmers, E., van de Putte, P. 1985. *Proc. Natl. Acad. Sci. USA* 82:2087–91
51. Groenen, M. A. M., van de Putte, P. 1986. *J. Mol. Biol.* 189:597–602
52. Lavoie, B. D., Chan, B. S., Allison, R. G., Chaconas, G. 1991. *EMBO J.* 10:3051–59
53. Mizuuchi, M., Mizuuchi, K. 1989. *Cell* 58:399–408
54. Surette, M. G., Lavoie, B. D., Chaconas, G. 1989. *EMBO J.* 8:3483–89
55. Leung, P. C., Teplow, D. B., Harshey, R. M. 1989. *Nature* 338:656–58
56. Surette, M. G., Chaconas, G. 1989. *J. Biol. Chem.* 264:3028–34
57. Castilho, B. A., Casadaban, M. J. 1991. *J. Bacteriol.* 173:1339–43
58. Adzuma, K., Mizuuchi, K. 1988. *Cell* 53:257–66
59. Darzins, A., Kent, N. E., Buckwalter, M. S., Casadaban, M. J. 1988. *Proc. Natl. Acad. Sci. USA* 85:6826–30
60. Harshey, R. M., Getzoff, E. D., Baldwin, D. L., Miller, J. L., Chaconas, G. 1985. *Proc. Natl. Acad. Sci. USA* 82:7676–80
61. Betermier, M., Alazard, R., Raguet, F., Roulet, E., Toussaint, A., Chandler, M. 1987. *Mol. Gen. Genet.* 210:77–85
62. Nakayama, C., Teplow, D. B., Harshey, R. M. 1987. *Proc. Natl. Acad. Sci. USA* 84:1809–13
63. Baker, T. A., Mizuuchi, M., Mizuuchi, K. 1991. *Cell* 65:1003–13
64. Mizuuchi, M., Weisberg, R. A., Mizuuchi, K. 1986. *Nucleic Acids Res.* 14:3813–25
65. Adzuma, K., Mizuuchi, K. 1987. *Structure and Expression,* 3:97–104. New York: Adenine
66. Kuo, C.-F., Zou, A., Jayaram, M., Getzoff, E., Harshey, R. 1991. *EMBO J.* 10:1585–91
67. Harshey, R. M., Cuneo, S. D. 1986. *J. Genet.* 65:159–74
68. Leung, P. C., Harshey, R. M. 1991. *J. Mol. Biol.* 219:189–99
69. Faeren, M., Huisman, O., Toussaint, A. 1978. *Nature* 271:580–82
70. O'Day, K. J., Schultz, D. W., Howe, M. M. 1978. *Microbiology—1978,* pp. 48–51. Washington, DC: Am. Soc. Microbiol.
71. Coelho, A., Maynard-Smith, S., Symonds, N. 1982. *Mol. Gen. Genet.* 185:356–62
72. Wijffelman, C., Lotterman, B. 1977. *Mol. Gen. Genet.* 151:169–74
73. Miller, J. L., Anderson, S. K., Fujita, D. J., Chaconas, G., Baldwin, D., Harshey, R. M. 1984. *Nucleic Acids Res.* 12:8627–38
74. Chaconas, G. 1987. See Ref. 13, pp. 137–57
75. Teplow, D. B., Nakayama, C., Leung, P. C., Harshey, R. M. 1988. *J. Biol. Chem.* 263:10851–57
76. Adzuma, K., Mizuuchi, K. 1991. *J. Biol. Chem.* 266:6159–67
77. Nash, H. A., Robertson, C. A. 1981. *J. Biol. Chem.* 256:9246–53
78. Flamm, E. L., Weisberg, R. A. 1985. *J. Mol. Biol.* 183:117–28
79. Miller, H. I. 1984. *Cold Spring Harbor Symp. Quant. Biol.* 49:691–98
80. Mechulam, Y., Fayat, G., Blanquet, S. 1985. *J. Bacteriol.* 163:787–91
81. Craig, N. L., Nash, H. A. 1984. *Cell* 39:707–16
82. Robertson, C. A., Nash, H. A. 1988. *J. Biol. Chem.* 263:3554–57
82a. Goodrich, J. A., Schwartz, M. L., McClure, W. R. 1990. *Nucleic Acids Res.* 18:4993–5000
82b. Nash, H. A. 1990. *Trends Biochem. Sci.* 15:222–27
83. Thompson, J. F., Landy, A. 1988. *Nucleic Acids Res.* 16:9687–705
84. Yang, C. C., Nash, H. A. 1989. *Cell* 57:869–80
85. Miller, H. I., Friedman, D. I. 1977. *Abstr. Annu. Meet. Am. Soc. Microbiol.* 77:137–43
86. Goosen, N., van de Putte, P. 1984. *Gene* 30:41–46

87. Drlica, K., Rouviere-Yaniv, J. 1987. *Microbiol. Rev.* 51:301–19
88. Suryanarayana, T., Subramanian, A. R. 1978. *Biochem. Biophys. Acta* 520:342–57
89. Tanaka, I., Appelt, K., Dijk, J., White, S. W., Wilson, K. S. 1984. *Nature* 310:376–81
90. Rouviere-Yaniv, J., Yaniv, M., Germond, J. E. 1979. *Cell* 17:265–74
91. Craigie, R., Arndt-Jovin, D. J., Mizuuchi, K. 1985. *Proc. Natl. Acad. Sci. USA* 82:7570–74
92. Lavoie, B. D., Chaconas, G. 1990. *J. Biol. Chem.* 265:1623–27
93. Mizuuchi, M., Baker, T. A., Mizuuchi, K. 1991. *Proc. Natl. Acad. Sci. USA* 88:9031–35
94. Goosen, N., van de Putte, P. 1987. See Ref. 13, pp. 41–52
95. Pato, M. L. 1989. See Ref. 8, pp. 23–52
96. Kleckner, N. 1990. *Annu. Rev. Cell Biol.* 6:297–327
97. Lichtenstein, C., Brenner, S. 1982. *Nature* 297:601–3
98. Rogers, M., Ekaterinaki, N., Nimmo, E., Sherratt, D. 1986. *Mol. Gen. Genet.* 205:550–56
99. McKown, R. L., Waddell, C. S., Arciszewska, L. K., Craig, N. L. 1987. *Proc. Natl. Acad. Sci. USA* 84:7807–11
100. Arciszewska, L. K., Drake, D., Craig, N. L. 1989. *J. Mol. Biol.* 207:35–52
101. Lichtenstein, C., Brenner, S. 1981. *Mol. Gen. Genet.* 183:380–87
102. McKown, R. L., Orle, K. A., Chen, T., Craig, N. L. 1988. *J. Bacteriol.* 170:352–58
103. Waddell, C. S., Craig, N. L. 1988. *Genes Dev.* 2:137–49
104. Kubo, K. M., Craig, N. L. 1990. *J. Bacteriol.* 172:2774–78
105. Gay, N. J., Tybulewicz, V. L., Walker, J. E. 1986. *Biochem. J.* 234:111–17
106. Gringauz, E., Orle, K. A., Waddell, C. S., Craig, N. L. 1988. *J. Bacteriol.* 170:2832–40
106a. Flores, C., Qadri, M. I., Lichtenstein, C. 1990. *Nucleic Acids Res.* 18:901–11
107. Orle, K. A., Craig, N. L. 1990. *Gene* 96:1–7
108. Waddell, C. S., Craig, N. L. 1989. *Proc. Natl. Acad. Sci. USA* 86:3958–62
109. Smith, G. L. F., Jones, P. 1986. *Nucleic Acids Res.* 14:7915–27
110. Bender, J., Kleckner, N. 1986. *Cell* 45:801–15
111. Huisman, O., Errada, P. R., Signon, L., Kleckner, N. 1989. *EMBO J.* 8:2101–9
112. Roberts, D. E., Hoopes, B. C., McClure, W. R., Kleckner, N. 1985. *Cell* 43:117–30
113. McCommas, S. A., Syvanen, M. 1988. *J. Bacteriol.* 170:889–94
114. Yin, J. C. P., Krebs, M. P., Reznikoff, W. S. 1988. *J. Mol. Biol.* 199:35–45
115. Way, J. C., Kleckner, N. 1984. *Proc. Natl. Acad. Sci. USA* 81:3452–56
116. Halling, S. M., Simons, R. W., Way, J. C., Walsh, R. B., Kleckner, N. 1982. *Proc. Natl. Acad. Sci. USA* 79:2608–12
117. Morisato, D., Way, J. C., Kim, H. J., Kleckner, N. 1983. *Cell* 32:799–807
118. Derbyshire, K. M., Hwang, L., Grindley, N. D. 1987. *Proc. Natl. Acad. Sci. USA* 84:8049–53
119. Isberg, R. R., Syvanen, M. 1981. *J. Mol. Biol.* 150:15–32
120. Phadnis, S. H., Sasakawa, C., Berg, D. E. 1986. *Genetics* 112:421–27
121. Pato, M. L., Reich, C. 1984. *Cell* 36:197–202
122. Bender, J., Kleckner, N. 1991. *EMBO J.* In press
123. Kathary, R., Jones, D., Candido, E. 1985. *J. Bacteriol.* 164:957–59
124. Mahillon, J., Seurinck, J., Van Rompuy, L., Delcour, J., Zabeau, M. 1985. *EMBO J.* 4:3895–99
125. Bushman, F. D., Craigie, R. 1990. *J. Virol.* 64:5645–48
126. Bushman, F. D., Craigie, R. 1991. *Proc. Natl. Acad. Sci. USA.* In press
127. Vink, C., van Gent, D. C., Elgersma, Y., Plasterk, R. H. A. 1991. *J. Virol.* 65:4636–44
128. LaFemina, R. L., Callahan, P. L., Cordingley, M. G. 1991. *J. Virol.* 65:5624–30
129. Colicelli, J., Goff, S. P. 1988. *J. Mol. Biol.* 199:47–59
130. Ross, M. J., Schwartzberg, P. L., Goff, S. P. 1989. *Cell* 58:47–54
131. Vora, A. C., Fitzgerald, M. L., Grandgenett, D. P. 1990. *J. Virol.* 64:5656–59
132. Skalka, A. M. 1988. See Ref. 9, pp. 701–24
133. Craigie, R., Mizuuchi, K. 1986. *Cell* 45:793–800
134. Parker, C. N., Halford, S. E. 1991. *Cell* 66:781–91
135. Kanaar, R., Klippel, A., Shekhtman, E., Dungan, J. M., Kahmann, R., Cozzarelli, N. R. 1990. *Cell* 62:353–66
136. Heichman, K. A., Moskowitz, I. P., Johnson, R. C. 1991. *Genes Dev.* 5:1622–34
137. Moskowitz, I. P., Heichman, K. A., Johnson, R. C. 1991. *Genes Dev.* 5:1635–45
138. Mizuuchi, K. 1983. *Cell* 35:785–94
138a. Surette, M. G., Chaconas, G. 1992. *Cell.* In press

139. Surette, M. G., Buch, S. J., Chaconas, G. 1987. *Cell* 49:253–62
139a. Surette, M. G., Chaconas, G. 1991. *J. Biol. Chem.* 266:17306–13
140. Bowerman, B., Brown, P. O., Bishop, J. M., Varmus, H. E. 1989. *Genes Dev.* 3:469–78
141. Ellison, V., Abrams, H., Roe, T. Y., Lifson, J., Brown, P. 1990. *J. Virol.* 64:2711–15
142. Farnet, C. M., Haseltine, W. A. 1990. *Proc. Natl. Acad. Sci. USA* 87:4164–68
143. Farnet, C. M., Haseltine, W. A. 1991. *J. Virol.* 65:1910–15
144. Lee, Y. M. H., Coffin, J. M. 1990. *J. Virol.* 64:5958–65
145. Robinson, M. K., Bennett, P. M., Richmond, M. H. 1977. *J. Bacteriol.* 129:407–14
146. Lee, C.-H., Bhagwat, A., Heffron, F. 1983. *Proc. Natl. Acad. Sci. USA* 80:6765–69
146a. Reyes, O., Beyou, A., Mignotte-Vieux, C., Richaud, F. 1987. *Plasmid* 18:183–92

147. Chaconas, G., Gloor, G., Miller, J. L. 1985. *J. Biol. Chem.* 260:2662–69
148. Adzuma, K., Mizuuchi, K. 1989. *Cell* 57:41–47
149. Resibois, A., Pato, M., Higgins, P., Toussaint, A. 1984. *Proteins Involved in DNA Replication,* pp. 69–76. New York: Plenum
149a. Pato, M. L., Howe, M. M., Higgins, N. P. 1990. *Proc. Natl. Acad. Sci. USA* 87:8716–20
150. Hauer, B., Shapiro, J. 1984. *Mol. Gen. Genet.* 194:149–58
151. Wiater, L. A., Grindley, N. D. 1990. *J. Bacteriol.* 172:4959–63
152. Murphy, E. 1989. See Ref. 8, pp. 269–88
153. Fedoroff, N. V. 1989. See Ref. 8, pp. 375–412
154. Coen, E. S., Robbins, T. P., Almeida, J., Hadson, A., Carpenter, R. 1989. See Ref. 8, pp. 413–36
155. Lafaille, J. J., DeCloux, A., Bonneville, M., Takagaki, Y., Tonegawa, S. 1989. *Cell* 59:859–70
156. Nash, H. A., Robertson, C. A. 1989. *EMBO J.* 8:3523–33

Annu. Rev. Biochem. 1992. 61:1053–95

TRANSCRIPTION FACTORS:
Structural Families and Principles of DNA Recognition

Carl O. Pabo

Howard Hughes Medical Institute, Department of Biology, Massachusetts Institute of Technology, Cambridge, Massachusetts 02139

Robert T. Sauer

Department of Biology, Massachusetts Institute of Technology, Cambridge, Massachusetts 02139

KEY WORDS: protein-DNA recognition, DNA-binding protein, helix-turn-helix, homeodomain, zinc finger

CONTENTS

1053

0066-4154/92/0701-1053$02.00

INTRODUCTION

DNA-binding proteins play central roles in biology. Among other activities, they are responsible for replicating the genome, for transcribing active genes, and for repairing damaged DNA. One of the largest and most diverse classes of DNA-binding proteins are the transcription factors that regulate gene expression. In this review, we focus on structural studies of the DNA-binding domains from these transcription factors. A more general review of protein-nucleic acid interactions, which also includes a discussion of restriction enzymes, polymerases, and RNA-binding proteins, can be found in Steitz (1).

Transcription factors regulate cell development, differentiation, and cell growth by binding to a specific DNA site (or set of sites) and regulating gene expression. One cannot fully understand how genetic information is utilized without understanding the structure and DNA-binding properties of these transcription factors. In this review, we address progress in understanding the structural basis for sequence-specific binding. What secondary structures can provide a surface that is complementary to the structure of double-helical DNA? What contacts with the bases and the DNA backbone allow site-specific recognition? How does understanding these structural details enhance our understanding of the molecular mechanisms involved in repression and activation of gene expression?

There have been several recent reviews of transcription factors and DNA-binding proteins (1–7). In addition, there have been reviews focusing on specific families of transcription factors such as the helix-turn-helix proteins (8–10); zinc finger, steroid receptor, and other metal-binding DNA-binding proteins (11–17); leucine zipper proteins (18); homeodomains (19–22); and β-sheet DNA-binding proteins (23). There also have been separate reviews covering the Trp repressor (24) and λ repressor (25). When we wrote our last general review (26), structures were known for only three DNA-binding proteins. Now structures have been reported for more than 10 protein-DNA complexes and for more than 20 DNA-binding proteins. This wealth of new data allows a much broader perspective on the general problem of protein-DNA recognition. Since an exhaustive review of the literature would be overwhelming, we have chosen to focus on a few well-studied systems that illustrate the main biological and structural issues.

FAMILIES OF DNA-BINDING PROTEINS

One of the central observations emerging from structural studies and sequence comparisons is that many DNA-binding proteins can be grouped into classes that use related structural motifs for recognition. Some families, such as the helix-turn-helix proteins, were first recognized because of structural similarities. Other families were first identified by sequence comparisons and later characterized by structural studies. Large, well-established families include the helix-turn-helix (HTH) proteins, the homeodomains, zinc finger proteins, the steroid receptors, leucine zipper proteins, and the helix-loop-helix proteins. Two smaller families have been identified that use β-sheets for DNA binding. Sequence comparisons indicate that there are a number of additional families of DNA-binding proteins, but fewer structural data are available for the families that have been characterized most recently.

Familial relationships are a powerful unifying theme in the study of DNA-binding proteins since they relate evolution, structure, recognition, gene regulation, and design. Thinking about these motifs can—at least in a schematic fashion—give a satisfying general picture of protein-DNA recognition. First, the existence of distinct families shows that there are multiple solutions to the structural problem of designing DNA-binding proteins. There is no single pattern or simple code—evolution has solved the problem in several different ways. However, each of the motifs that has been characterized in detail involves a simple secondary structure (usually an α-helix) that is complementary to the structure of B-DNA in a straightforward way. Side chain contacts play a major role in site-specific binding and often allow the same motif to be used at a set of different sites. In some way, the number of members in a particular family of DNA-binding proteins may measure the relative "evolutionary success" of a particular DNA-binding motif. These major families are nature's most successful designs for DNA-binding proteins.

Although thinking about DNA-binding proteins in terms of families has many attractive aspects, it is important to realize that there are many interesting and important DNA-binding proteins that do not belong to any of the known families. The SV40 large T antigen (27) and the human p53 tumor suppressor gene (28) are two such examples. Of course, there is no reason to believe that we have discovered all the major families of DNA-binding proteins, or to believe that all DNA-binding proteins must belong to clearly identified families. In spite of these cautions, it seems worthwhile to focus on the major families of DNA-binding proteins because information about a single member provides a framework for thinking about the whole family. In addition, comparisons within families and between families of DNA-binding proteins may help us understand the general principles of protein-DNA recognition. Finally, since these motifs have been so successful in proliferating and adopting new roles during evolution, it seems reasonable to imagine that they may also provide the most convenient scaffolds to use in attempting to design new DNA-binding proteins.

Helix-Turn-Helix

The helix-turn-helix (HTH) structure was the first DNA-recognition motif discovered, and structures have now been determined for many HTH proteins and protein-DNA complexes. Since the HTH proteins have been analyzed in such detail, studying these proteins provides an excellent background for thinking about other motifs that are used in site-specific recognition. The first crystal structures of DNA-binding proteins were those of the λ Cro protein (29, 30), the *Escherichia coli* CAP protein (31), and the DNA-binding domain of λ repressor (32). Comparisons of these three proteins revealed a conserved recognition motif consisting of an α-helix, a turn, and a second α-helix (33, 34). Structures are now available for several other HTH proteins, including Lac repressor (35), 434 repressor (36, 37), 434 Cro (38, 39), Trp repressor (40), and Fis protein (41). The DNA-binding domain of LexA contains a related two-helix motif, but the turn region is more extended than in the other HTH proteins (42).

Although structural studies first led to its identification, sequence comparisons suggested that the HTH motif occurred in a large family of prokaryotic DNA-binding proteins (43–45). The most highly conserved residues in the HTH motif include a glycine in the turn and also include several hydrophobic residues (see Figure 1). These conserved residues help to stabilize the arrangement of the two helices in the HTH unit and help this unit to pack against the rest of the protein.

It is worth emphasizing that the HTH motif (unlike many of the other motifs that we discuss) is not a separate, stable domain. The HTH motif cannot fold or function by itself but always occurs as part of a larger

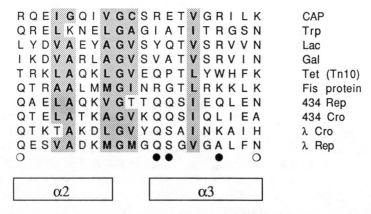

Figure 1 Alignment of the helix-turn-helix sequences of phage and bacterial regulatory proteins. Sequence positions that are generally hydrophobic (A, C, V, I, L, M, F, Y) or small (G, A) are shaded. The open and closed circles represent residue positions that make backbone contacts (○) or that mediate hydrogen bonds with bases in the major groove (●) in the λ and/or 434 repressor-complexes (47, 48). For sequence references, see (26) and references therein and (46).

DNA-binding domain. Comparing the first HTH protein structures immediately revealed that the HTH motif occurred in a number of different structural environments. Thus, the λ Cro protein and the *E. coli* CAP protein contain a number of β-sheets in their DNA-binding domains, while the corresponding domain of λ repressor is entirely α-helical. Studies of complexes involving HTH proteins (discussed below) have revealed that other regions of the DNA-binding domains (outside of the HTH units) can also have significant roles in recognition. Although there is a natural tendency to focus on contacts made by the HTH unit, it is wrong to assume that recognition involves only such contacts. It is even more misleading to focus exclusively on the contacts made by the second helix—often called the "recognition" helix—of the HTH motif.

The structures of several repressor-operator complexes have been determined at high resolution, yielding a wealth of information about how the HTH motif is used in site-specific recognition. The structure of the λ repressor-operator complex (47) is used to illustrate our discussion (Figure 2). The N-terminal domain of λ repressor forms a dimer, and each subunit interacts with one-half of the operator site. In each half-site, side chains from helices 2 and 3 of λ repressor (the HTH unit) make critical contacts with the DNA. For example, Gln44, the first residue of helix 3, makes two hydrogen bonds with an adenine near the outer edge of the operator; Ser45 makes a single hydrogen bond with a guanine; Gly46 and Gly48 make hydrophobic contacts with thymine methyl groups; and Asn52 makes a hydrogen bond to a phosphodiester oxygen. Many of these contacts are shown in Figures 2*B* and 2*C*. Gln33, in helix 2, contacts a phosphodiester oxygen and also hydrogen bonds to the Gln44 side chain. Presumably, this side chain–side chain interaction (Figure 2*C*), which illustrates the structural complexity of the recognition process, helps to stabilize both the backbone contact made by Gln33 and the base contacts made by Gln44. It is noteworthy that 434 repressor has a pair of glutamines at corresponding positions, which have similar roles in recognition of the 434 operator (48, 49).

As mentioned above, residues outside of the HTH unit also are important for DNA recognition. In the λ complex, Lys19 and Tyr22 illustrate how residues from neighboring regions can make contacts with the DNA backbone (Figure 3). λ repressor also has a distinct structural motif—an extended N-terminal arm—that makes critical contacts in the major groove. Lys4, in the N-terminal arm, cooperates with Asn55 (in the loop region following helix 3) to make two hydrogen bonds with a guanine in the major groove (Figure 2*C*). The rest of the N-terminal arm wraps around the center of the operator and makes several additional contacts (47, 50). The N-terminal arms of λ repressor are flexible in solution (32, 51, 52) and only assume a fixed position upon operator binding (47, 50). As we will see, similar disorder → order

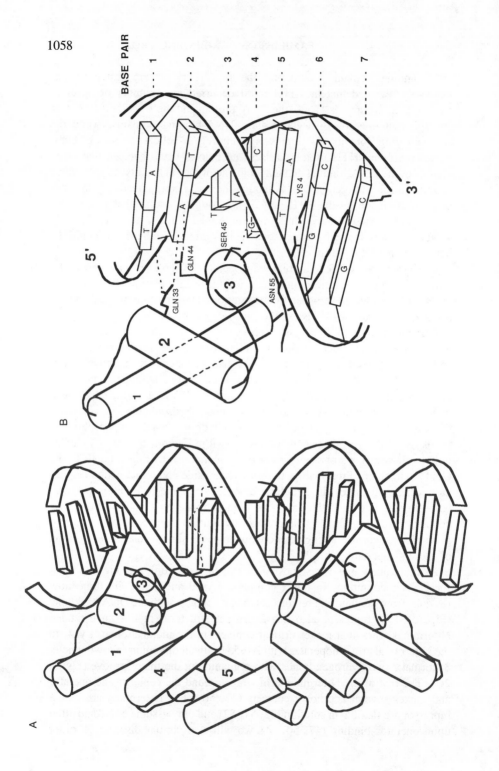

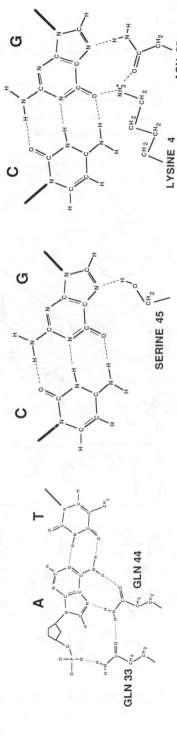

Figure 2 A. Sketch showing overall arrangement of complex containing the N-terminal domain of λ repressor (47). Cylinders are used to represent α-helices, and the helices in one monomer are numbered. Helices 2 and 3 correspond to the conserved HTH unit. *B.* Sketch emphasizing the HTH unit and hydrogen bonds with the bases in one-half of the λ operator site. *C.* Sketches showing hydrogen bonds with bases 2, 4, and 6 in one-half of the λ operator site. Figures reprinted with permission from *Science.*

transitions occur in other DNA-binding motifs (such as the basic region of the leucine zipper proteins).

Crystal structures of complexes have also been reported for the 434 repressor (48), 434 Cro (38, 53), Trp repressor (54), λ Cro (55), and *E. coli* CAP protein (56). These cocrystal structures show that all of the HTH protein-DNA complexes have a number of common features: (*a*) The repressors bind

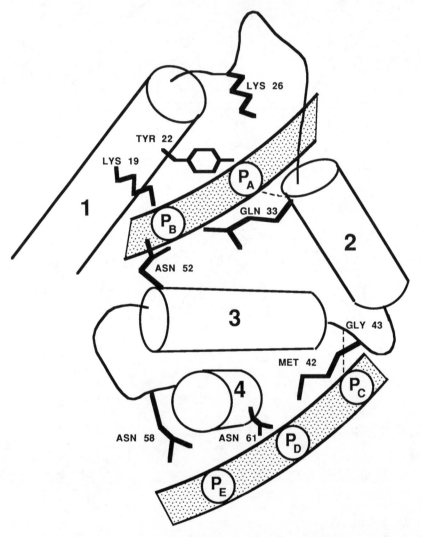

Figure 3 Sketch showing side chains that interact with the sugar-phosphate backbone in one-half of the λ repressor-operator complex (47). Reprinted with permission from *Science*.

as dimers. Each monomer recognizes a half-site, and the approximate symmetry of the DNA-binding site is reflected in the approximate symmetry of the complexes. (b) The conserved HTH unit contacts the DNA in each half of the operator site. The first helix of the HTH unit is somewhat "above" the major groove, but the N-terminus of this helix contacts the DNA backbone. The second helix of the HTH unit fits into the major groove, and the N-terminal portion of this helix is closest to the edges of the base pairs.[1] (c) The operator sites are B-form DNA. (d) Side chains from the HTH units make site-specific contacts with groups in the major groove. (e) Each complex has an extensive network of hydrogen bonds between the protein and the DNA backbone. Some of these contacts are made by lysines or arginines, but many of them are made by short, polar side chains or even by -NH groups from the polypeptide backbone.

Since a set of high-resolution structures is available, the HTH complexes can be used to illustrate principles that may be used by other families of DNA-binding proteins. As mentioned above, λ repressor uses an extended region of peptide chain to wrap around the DNA and augment the contacts made by residues in the HTH region. This reveals another motif that may be important in protein-DNA recognition and illustrates how several distinct structural motifs can contribute to site-specific binding. The 434 repressor is sensitive to base substitutions in the center of its operator site, even though it does not directly contact any of these bases (36, 48). The central bases in the 434 operator are thought to influence binding indirectly via effects on DNA conformation (57, 58). Similar effects may play a role in protein-DNA recognition in a variety of systems. The Trp repressor appears to use several water molecules to provide critical contacts (54). These waters, which are tightly bound at the N-terminal end of an α-helix, show that hydration can have important effects on recognition, and further illustrate the structural complexity of protein-DNA interactions.

Although we have focused on the DNA-binding domains, the HTH proteins often contain additional domains that have important roles in regulating activity. For example, the N-terminal domain of CAP allows dimer formation and also binds cAMP, an allosteric effector of DNA binding (31). In the λ, 434, and LexA repressors, the C-terminal domains allow stable dimer formation, and the process of induction involves proteolytic cleavage between the N-terminal and C-terminal domains (for review, see 25). This modular arrangement, with different functions in different domains, also is a common theme in eukaryotic transcription factors, where the DNA-binding domains represent only part of the intact proteins (2–6).

---

[1]Although each of these proteins uses the HTH unit in a generally similar way, there are some interesting differences in the precise arrangement of the HTH unit with respect to the DNA [for review, see Harrison & Aggarwal (9)].

Homeodomain

The homeodomain is a DNA-binding motif that is present in a large family of eukaryotic regulatory proteins (19, 20). Although the conserved sequences were first recognized in proteins that regulate *Drosophila* development (see Figure 4), it now is clear that the homeodomain has a broader role in eukaryotic gene regulation. A comparison of amino acid sequences suggested that the homeodomain would contain a HTH motif (59, 60), and this has now been confirmed by structural studies. However, unlike the isolated HTH unit, the 60-residue homeodomain forms a stable, folded structure and can bind DNA by itself (61–63).

The structure of the homeodomain was first determined by 2D NMR studies of the *Drosophila* Antennapedia (Antp) homeodomain (63, 64), and there have now been several studies of homeodomains (64a) and homeodomain-DNA complexes (65–67). These studies confirmed that the homeodomain contains a HTH motif. Distance constraints, determined from 2D NMR studies of the Antp complex, were used to dock the homeodomain against a B-form DNA site (65). The crystal structure of the *Drosophila* engrailed homeodomain-DNA complex has also been solved (66), and the structure of a complex containing the yeast MAT α2 homeodomain has recently been determined (67).

The engrailed homeodomain contains an extended N-terminal arm and three α-helices (Figure 5). The overall structure of the protein is quite easy to visualize: helix 1 and helix 2 pack against each other in an antiparallel arrangement. Helix 3 is roughly perpendicular to the first two helices, and the hydrophobic face of this extended helix packs against helices 1 and 2 to form the interior of the protein. The C-terminal residues of the isolated Antp homeodomain have been described as a fourth helix, separated from helix 3 by a kink (63). However, in the engrailed cocrystal and α2 cocrystal these C-terminal residues form a continuous, helical extension of helix 3 (66, 67).

The main contacts in the engrailed complex are made by residues in helix 3, which fits into the major groove, and by residues in the extended N-terminal arm, which fits into the minor groove (Figures 5 and 6). Most of the contacts are made by helix 3. The exposed face of helix 3 fits directly into the major groove, allowing side chains to make extensive contacts with the bases and with phosphodiester oxygens along the edge of the major groove. Residues near the middle of helix 3 are closest to the bases, but residues near the C-terminal end make a number of contacts with the DNA backbone (Figures 4–6). Helices 1 and 2 span the major groove but are much farther from the DNA. There only are two DNA contacts made by this region of the protein. Both are with the DNA backbone. The other critical contacts come from an N-terminal arm that, as predicted by Garcia-Blanco et al (68), has some similarities with λ repressor's arm. Residues 3–9 of engrailed form an ex-

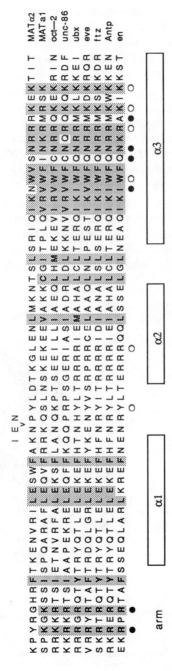

Figure 4 Alignment of homeodomain sequences. Positions that are homologous, generally basic (R, K), or generally hydrophobic (A, C, V, I, L, M, F, Y) are shaded. The positions of the α-helices, backbone contacts (○), and major groove contacts (●) are from the engrailed and/or α2 complexes (66, 67). For sequence references and an excellent discussion of sequence similarities among a larger set of homeodomains, see Scott et al (19).

tended N-terminal arm that fits into the minor groove and supplements the contacts made by helix 3. (Since the homeodomain is just a small part of the intact engrailed protein, this extended N-terminal arm probably would form a loop or linker region in the intact protein.)

The crystal structure of the α2 homeodomain-DNA complex has recently been determined (67), and turns out to be very similar to the engrailed complex. This is particularly interesting because the α2 sequence is one of the most divergent within the homeodomain family. (Some scientists had refused to recognize α2 as a legitimate member of the family.) However, α2 and engrailed fold in very similar ways: the C_α backbone of α2 superimposes quite well on the C_α backbone of engrailed (67). The α2 homeodomain has a three-residue "insertion" in the loop between helices 1 and 2, but this is accommodated without any major changes in the overall fold of the protein. Since α2 differs substantially from engrailed in sequence but has a similar structure, it will be interesting to see whether members of different homeodomain subfamilies, which have been defined by grouping proteins with the most similar sequences (19), have any distinctive structural or functional properties.

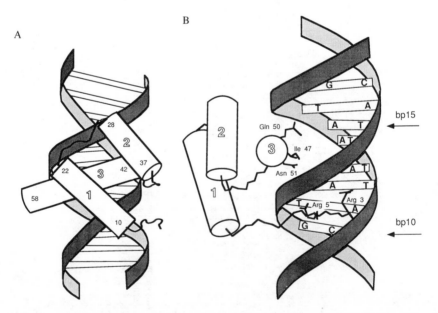

Figure 5   *A*. Sketch of the engrailed homeodomain-DNA complex (66), summarizing the overall relationship of the α-helices and the N-terminal arm with respect to the DNA. Helices 2 and 3 form the conserved HTH unit. *B*. View, at right angles to that shown in panel A, emphasizes the minor groove contacts made by the N-terminal arm and the major groove contacts made by helix 3. Reprinted with permission from *Cell*.

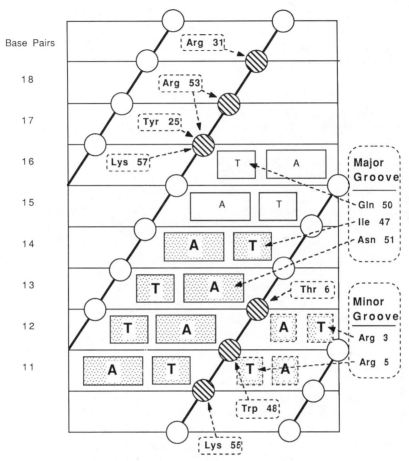

Figure 6   Sketch summarizing all the contacts in the engrailed homeodomain-DNA complex (66). The DNA is represented as a cylindrical projection, and the shading emphasizes the TAAT subsite that occurs in many homeodomain-binding sites. Phosphates are represented by circles; hatched circles show phosphates that are contacted by the engrailed homeodomain. Reprinted with permission from *Cell*.

A comparison of the engrailed and α2 complexes also reveals that these two homeodomains dock against the DNA in very similar ways (67). As observed with the engrailed complex (66) and with the Antp complex (65), the N-terminal arm makes contacts in the minor groove, and helix 3 makes an extensive set of contacts in the major groove. Superimposing the engrailed and α2 complexes by superimposing helix 3 of each protein shows that there are a set of common contacts, which are mediated by conserved residues at the protein-DNA interface. There are two conserved hydrogen bonds with an adenine in the major groove. These are made by Asn51, which is one of the four "invariant" residues found in every homeodomain (19). Six other conserved residues make conserved contacts with phosphodiester oxygens that flank the major groove. Presumably, these sets of conserved contacts play a critical role in determining the precise orientation of helix 3.

Since the α2 and engrailed complexes are so similar, they may provide a basis for modeling other homeodomain-DNA complexes. Further studies will be needed before we can fully explain the different sequence preferences of the homeodomains, but several residues are appropriately positioned to make base contacts and are likely to influence DNA specificity. Residues 47, 50, 51, and 54 appear to be especially important. Genetic and biochemical analyses have shown that residue 50 is important for controlling the differential specificity of homeodomain-DNA interactions (69–71), and the structures show that the side chain of residue 50 points into the major groove and is in an excellent position to contribute to the specificity of binding (Figure 5B). The structures also show that that the side chains of residues 47 and 54 can make base contacts (although the Ala54 side chain of engrailed is too short to reach the bases). As noted above, Asn51 hydrogen bonds to an adenine in each complex. Presumably, this interaction is important for binding, but Asn51 cannot determine differential specificity since it is conserved throughout the homeodomain superfamily. Differences in the sequence and in the precise orientation of the N-terminal arm may also help determine binding specificity. Since engrailed and α2 use slightly different regions of the arm, we will need more structures and more biochemical data before we can understand how these contacts contribute to specificity.

Overall, the structural data agree very well with biochemical and genetic data about the homeodomains and about homeodomain-DNA interactions. For example, Trp48, Phe49, Asn51, and Arg53 occur in every one of the higher eukaryotic homeodomains compiled by Scott et al (19). These invariant residues occur right in the middle of helix 3, in the region that is closest to the major groove. Trp48 and Phe49 form part of the hydrophobic core and must play a key role in stabilizing the correct folded structure (Trp48 also makes a backbone contact). The invariant hydrophilic residues—Asn51 and Arg53—make critical contacts with the DNA. Asn51 makes a pair of hydrogen bonds with an adenine (Figure 7), and Arg53 hydrogen bonds with two phosphate groups on the DNA backbone (Figure 6). The structure also explains the roles of other highly conserved residues and regions of the homeodomain, including residues in the N-terminal arm, other residues in the hydrophobic core, and the set of residues that contact the phosphodiester backbone.

As first predicted from sequence comparisons (59, 60), the homeodomain contains a HTH unit (63) that is similar to those observed in the prokaryotic

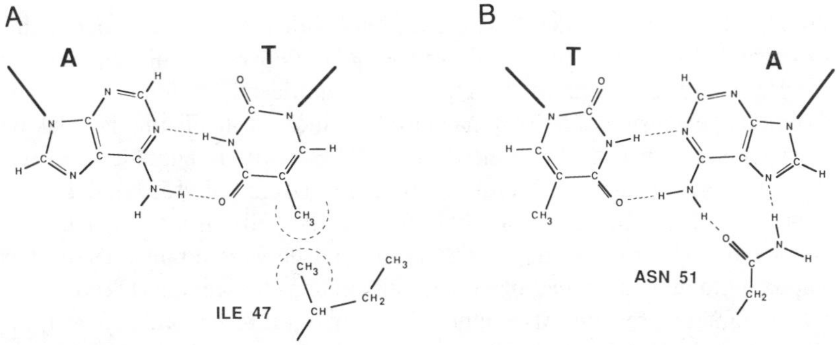

A

B

Figure 7 Sketch showing the major groove contacts made by *A*. Ile47 and *B*. Asn51 in the engrailed homeodomain-DNA complex (66). Reprinted with permission from *Cell*.

repressors. However, structural studies of the complexes (65–67) have shown that the homeodomain uses the HTH unit in a novel way. These differences can be illustrated by comparing the λ repressor-operator and engrailed homeodomain-DNA complexes (66). Superimposing the λ and engrailed HTH units (to provide a common frame of reference) leaves the DNA sites in very different positions, showing that these HTH units dock against the DNA in significantly different ways. In the λ complex, the N-terminal portion of helix 3 (the "recognition helix") is closest to the bases. In the engrailed complex, the center of a much longer helix 3 (again, the "recognition helix") is closest to the bases. Given these differences, it is not surprising that the critical base contacts are made by residues from different portions of the HTH units. In the λ and 434 complexes, the critical base contacts are made by residues in the first turn (i.e. at the N-terminal end) of the recognition helix. In the homeodomain complexes, residues in the second and third turns of the recognition helix make critical contacts with the bases. There also are significant differences in the way that helix 2 (the first helix of the HTH unit) is used in the two complexes. In the λ and 434 complexes, the N-terminal end of helix 2 fits partway into the major groove and the peptide -NH of the first residue hydrogen bonds to a phosphodiester oxygen on the DNA backbone. In contrast, helix 2 of the homeodomain lies entirely above the major groove and cannot use a peptide -NH to hydrogen bond to the DNA.

The different docking arrangements seen for the prokaryotic and eukaryotic HTH units pose some challenging questions about the the evolution and structural significance of the HTH unit. However, it is important to recognize that the helices in the HTH unit of the homeodomain are significantly longer than the corresponding helices in many of the prokaryotic HTH proteins.[2] These differences in helix length presumably are responsible for some of the differences in the docking arrangement. Thus lengthening the first helix of the HTH unit probably prevents this helix from tucking partway into the major groove. To avoid van der Waals collisions with the DNA backbone, this extended helix must lie entirely above the major groove, and this requires that the HTH unit dock against the DNA in a significantly different way. The C-terminal extension of helix 3 may also be critical in stabilizing the docking arrangement seen with the homeodomain HTH motif. In engrailed and $\alpha2$, three residues from this C-terminal extension (Arg53, Lys55, and Lys57) contact the DNA backbone and help to hold the helix in the major groove (Figure 6). When comparing HTH motifs and docking arrangements, it is important to remember that the prokaryotic proteins present the HTH unit in a variety of structural contexts. Future studies may yet reveal prokaryotic HTH motifs that dock as the homeodomain does (72) or may reveal other docking arrangements for the homeodomain, and thus provide a "missing link" that helps us understand the evolutionary and structural relationships between these families.

Although an isolated homeodomain can fold correctly and bind DNA with a specificity similar to that of the intact protein, it seems likely that the precise DNA-binding specificity is modulated by other regions of the protein. Many homeodomain proteins contain other sequence motifs that flank the homeodomain and are conserved within specific subfamilies (19). As noted by Kissinger et al (66), both the N-terminus and C-terminus of the homeodomain are near the DNA, and neighboring residues from the intact protein could easily contact flanking regions of the DNA. The POU proteins (73, 74) seem to provide a particularly clear example of how this may occur. The POU-specific domain, which was first observed in the pit-1, oct-2, and unc-86 proteins, is a conserved 65–75-residue segment just on the N-terminal side of the homeodomain (75). This POU-specific domain appears to make DNA contacts with a set of bases adjacent to those contacted by the homeodomain (73, 74). It is possible that other conserved sequences found in the intact homeodomain proteins have other important roles in DNA recognition. They might influence contacts made by the N-terminal arm or the C-terminal helix, and they could provide "attachment" or "targeting" sites for other proteins that modulate the specificity or affinity of DNA binding.

Protein-protein interactions may also have a role in modulating many homeodomain-DNA interactions. For example, the yeast $\alpha2$ protein forms homodimers, but also forms complexes with a related homeodomain protein a1 and with the general transcription factor Mcm1 (76, 77). Each of these complexes has different site preferences, and it is possible that these acces-

---

[2]It is interesting to consider how our views might be influenced by the fact that the bacterial HTH proteins were solved first. Had the homeodomains been solved first, their extended helices would have been used to define the HTH unit, and subsequent structural comparisons with the prokaryotic HTH proteins would have yielded a far less impressive degree of structural homology.

sory proteins (in addition to adding new DNA contacts) actually alter alter the way that $\alpha2$ interacts with DNA. Recent studies have shown that mutations in the $\alpha2$ HTH motif have different effects on the binding of $\alpha2$ and on the binding of the a1/$\alpha2$ complex (A. Vershon and A. Johnson, unpublished). Studies of the human oct-1 homeodomain protein have also emphasized the importance of accessory proteins in recognition and regulation (78), but additional studies will be needed to understand precisely how these proteins influence recognition. Do they provide additional contacts, or do they affect the way that the homeodomain docks against the DNA? Can they change the orientation of the homeodomain recognition helix or shift the position of the N-terminal arm?

Zinc Finger

Zinc fingers, of the type first discovered in the *Xenopus* transcription factor IIIA (TFIIIA) (79, 80), are another one of the major structural motifs involved in protein-DNA interactions (11–14). Zinc finger proteins are involved in many aspects of eukaryotic gene regulation. Homologous zinc fingers occur in proteins induced by differentiation and growth signals, in proto-oncogenes, in general transcription factors, in genes that regulate *Drosophila* development, and in regulatory genes of eukaryotic organisms (11–14 and references therein). Proteins in this family usually contain tandem repeats of the 30-residue zinc finger motif, with each motif containing the sequence pattern $Cys-X_{2or4}-Cys-X_{12}-His-X_{3-5}-His$ (Figure 8). Unfortunately, the term "zinc finger" has acquired a loose—almost topological—definition and has been used when referring to almost any sequence that has a set of cysteines and/or histidines within a short region of polypeptide chain. Here, we focus on fingers that are structurally homologous to the fingers in TFIIIA. Other cysteine-rich and histidine-rich motifs, such as those that occur in the steroid receptors, in the yeast transcription factor GAL4, and in certain retroviral proteins, are discussed later in this review.

Model building predicted (84, 85) and 2D NMR studies confirmed (86, 87) that the TFIIIA-like zinc fingers contain an antiparallel β-sheet and an α-helix. Two cysteines, which are near the turn in the β-sheet region, and two histidines, which are in the α-helix, coordinate a central zinc ion and hold these secondary structures together to form a compact globular domain. The crystal structure of a zinc finger–DNA (83) complex containing three fingers from zif268 (88) and a consensus zif-binding site has been reported. The crystal structure shows that the zinc fingers bind in the major groove of B-DNA and wrap partway around the double helix (Figure 9). Each finger has a similar way of docking against the DNA and makes base contacts with a three-base-pair subsite. [Sequence analyses and mutational studies had correctly predicted many features of this complex (89).] Neighboring fingers are

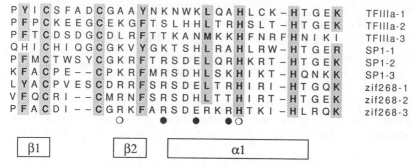

Figure 8 Alignment of zinc finger sequences. Positions of the conserved cysteines (C) and histidines (H) are shaded, as are positions that are generally basic (R, K) or hydrophobic (A, C, V, I, L, M, F, Y). The positions of secondary structure, conserved backbone contacts (○), and major groove contacts (●) are from finger 1, 2, and/or 3 of the zif268 complex. For sequence references see (81, 83, 88, 90). The sequence alignment is based upon that of Miller et al (79). An excellent discussion of sequence relationships between a much larger set of zinc finger proteins can be found in Krizek et al (82).

arranged in a way that reflects the helical pitch of the DNA and the three-base-pair periodicity of the subsites. Thus, a rotation of approximately 96 degrees (3 × 32 degrees/bp) around the DNA axis and a translation of approximately 10 Å (3 × 3.4 Å/bp) along the DNA axis move one finger onto the next. In the zif complex, the antiparallel β-sheet is on the back of the α-helix—away from the base pairs—and the β-sheet is shifted towards one side of the major groove. The first β-strand does not make any contacts with the DNA, whereas the second β-strand contacts the sugar phosphate backbone.

The zif268 fingers make a set of hydrogen bonds with bases in the major groove (see Figures 10 and 11). In each finger, critical base contacts are made by an arginine that immediately precedes the α-helix. This arginine also interacts with an aspartic acid that is the second residue in the α-helix. In finger 2, a histidine (the third residue in this helix) makes an additional base contact. In fingers 1 and 3, an arginine (the sixth residue in these helices) also makes a base contact. All of these contacts involve hydrogen bonds with guanines on the G-rich strand of the consensus binding site (5'-GCGTGGGCG-3'). The peptide binds with finger 1 near the 3' end (and with finger 3 near the 5' end) of this primary strand. The structure also reveals a set of contacts with the DNA backbone, and most of these also involve the primary, guanine-rich, strand of the DNA. In each finger, an arginine in the second β-strand makes a contact to a phosphodiester oxygen, and the first zinc-binding histidine in the α-helix contacts another phosphodiester oxygen (these contacts are indicated schematically by open circles in Figure 8).

The conserved arrangement of the fingers, and the pattern of side chain–

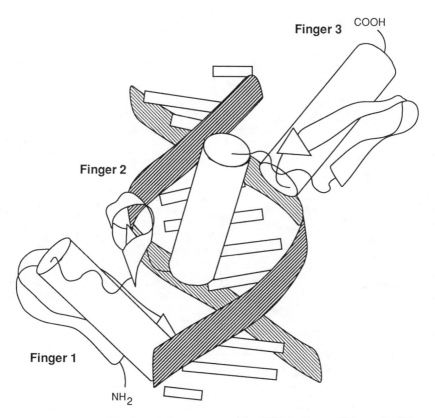

Figure 9   Sketch showing the overall arrangement of the zif268 zinc finger–DNA complex (83). The α-helices are shown as cylinders; β-sheets are shown as ribbons; and the zinc ions have been omitted from this sketch. The three fingers fit in the major groove, and each has a similar relationship to the DNA. Each finger makes base contacts in a three-base-pair subsite. Reprinted with permission from *Science*.

base interactions, are so regular that it may eventually be possible to describe a "code" for zinc finger–DNA interactions. The zinc finger uses the arginine that immediately precedes the α-helix, as well as the second, third, and sixth residues of the α-helix to contact the base pairs. (In the zif complex, the second residue of each helix is an aspartic acid, but a longer side chain at this position might be able to contact the bases.) Overall, there is a relatively simple pattern to the contacts (83). None of the zif fingers contacts all three bases, but the residue immediately preceding the α-helix contacts the third base on the primary strand of the subsite (5' — — G), the third residue in the α-helix can contact the second base on the primary strand (5' — G —), and

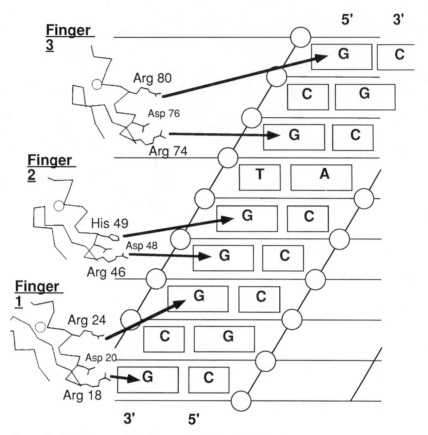

Figure 10   Sketch summarizing base contacts in the zif268 zinc finger–DNA complex (83). The DNA is represented as a cylindrical projection, and arrows indicate contacts in the major groove. The base contacts involve the guanine-rich strand of the binding site. Reprinted with permission from *Science*.

the sixth residue in the α-helix can contact the first base (5' G — —) of the subsite. These simple patterns reflect the fact that each of the three fingers docks against the DNA in a very similar way and is related to the next by a simple helical motion. This appears to be the first instance where the periodicity of a protein structure has such a simple relationship to the periodicity of double-helical DNA. Recognition is based on a simple modular system that can be used to recognize extended, asymmetric sites.

The structure of the zif268 complex should provide a useful guide for modeling complexes, such as Sp1 (90), with closely related fingers. However, there is no reason to believe that all fingers bind in exactly the same way. As we have discussed, the structurally conserved HTH unit can dock against

A B C

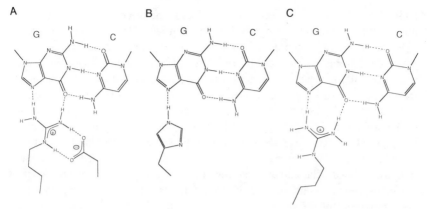

Figure 11 Sketch showing details of the base contacts in the zif268 zinc finger–DNA complex (83). *A*. Conserved contacts made by Arg18, Arg46, and Arg74. *B*. Contact made by His49 in finger 2. *C*. Conserved contacts made by Arg24 in finger 1 and Arg80 in finger 3. Reprinted with permission from *Science*.

the DNA in a number of different ways, and it seems plausible that the zinc finger motif may also have a set of different binding modes. TFIIIA-like zinc fingers are extremely common, and it seems that evolution has had ample opportunity to try different ways of using these fingers. Subtle sequence variants exist, such as the patterns found in the alternating fingers of ZFY (91), and these distinctive sequence patterns may reflect differences in the structure or docking arrangements of the fingers (92). Other distinctive sequence patterns have been noted. For instance, whenever there are five residues between the histidines, this finger usually is the last one in a tandem array (93) or occurs as a separated finger (94, 95). It will be necessary to solve the structures of other zinc finger–DNA complexes to see if other stable docking arrangements exist, to determine whether zinc fingers bind to AT-rich sequences in fundamentally different ways than they bind to GC-rich sequences, and to understand how zinc finger proteins interact with RNA [as in the 7S particle formed by TFIIIA (79 and references therein)].

Steroid Receptor

The steroid receptors are an important family of regulatory proteins that include receptors for the steroid hormones, retinoids, vitamin D, thyroid hormones, and a number of other important compounds. Genetic and bio-chemical studies done in a number of different laboratories revealed that these proteins contain separate domains for hormone binding, DNA binding, and for transcriptional activation (16, 17). The DNA-binding domains, which contain about 70 residues, have eight conserved cysteine residues (Figure 12). Biochemical studies showed that a peptide from the DNA-binding domain of the glucocorticoid receptor could fold in the presence of zinc, and that this peptide specifically recognized the appropriate DNA-binding site (99).

Since the steroid receptors contain two sets of four cysteines, it had been proposed that this region would form a pair of "zinc fingers." However, the sequence patterns seen in the steroid receptors are very different than those found in the TFIIIA-like zinc fingers (cf. Figures 8 and 12), and it was clear that the steroid receptors formed a distinct structural motif (100). NMR studies of the DNA-binding domains from the glucocorticoid and estrogen receptors (101, 102) revealed that each of these peptides folds into a single globular domain with a pair of α-helices. The two extended helices are roughly perpendicular and are held together by hydrophobic contacts. A zinc ion binds near the start of each helix and holds a peptide loop against the N-terminal end of the helix.

Recently, a crystal structure for a complex of the glucocorticoid receptor has been reported (103). This structure shows that the peptide binds as a dimer, even though NMR studies had shown that the peptide exists as a monomer in solution (101). The first helix of each subunit fits into the major groove, and side chains from this helix make contacts with the edges of the base pairs. The second major helix provides phosphate contacts with the DNA backbone and provides the dimerization interface. The crystal structure also gives more information about the loop regions that precede the major helices. The loop of the N-terminal finger contains a short segment of antiparallel β structure, while the C-terminal finger contains a distorted α-helix.

Leucine Zipper and Helix-Loop-Helix

The leucine zipper (104) and helix-loop-helix proteins (105) have important roles in differentiation and development. They also are interesting because they illustrate the important roles that heterodimer formation can play in the regulation of gene expression.

The leucine zipper motif was first discovered as a conserved sequence pattern in several eukaryotic transcription factors (104), and it now is clear that this motif appears in a wide variety of transcription factors from fungi, plants, and animals (see references in Ref. 18). The DNA-binding domains of these leucine zipper proteins generally contain 60–80 residues (104, 106) and contain two distinct subdomains: the leucine zipper region mediates dimerization, while a basic region contacts the DNA.

Leucine zipper sequences are characterized by a heptad repeat of leucines over a region of 30–40 residues (104; see Figure 13). There also tends to be a conserved repeat of hydrophobic residues (often Val or Ile) occurring three residues to the N-terminal side of the leucines (104). Biochemical experiments suggested that the leucine zipper region forms two parallel α-helices in a coiled-coil arrangement (111), and a high-resolution structure of

Figure 12 Alignment of steroid receptor sequences. Positions of the conserved cysteines (C) are shaded, as are positions that are generally acidic (D, E), basic (H, R, K), hydrophobic (A, C, V, I, L, M, F, Y), or small (G, A). The positions of secondary structure, backbone contacts (○), and major groove contacts (●) are from the structure of the glucocorticoid complex (103). The alignment is based upon that of Evans (16). For sequence references see Evans (16) and references therein, and (96–98).

Figure 13 Alignment of basic region-leucine zipper sequences. Positions that are generally conserved, basic (R, K), or hydrophobic (A, C, V, I, L, M, F, Y) are shaded. The position of the leucine zipper α-helix is from O'Shea et al (112). The alignment is based upon that of Kerppola & Curran (18). For sequence references, see (104, 106–110) and references therein. The sis-A sequence was determined by J. Erickson and T. Cline (unpublished).

this region from GCN4 has recently been reported (112). The coiled-coil arrangement readily explains the heptad repeat of the leucines and the offset heptad repeat of hydrophobic residues. These residues form the buried subunit interface of the coiled-coil dimer. Coiled-coils have a periodicity of 3.5 residues per helical turn, and thus every seventh residue is in the same structural environment.

The basic region (which contains about 30 residues) is rich in arginines and lysines, but also contains other residues that are conserved throughout the family or in particular subfamilies (18; Figure 13). Swapping basic regions and zipper regions from different proteins shows that the basic region is primarily responsible for the sequence preferences of the leucine zipper proteins (113). In fact, the basic region of GCN4 (without the leucine zipper!) can bind to its DNA site specifically as long as a disulfide bond is added to allow dimer formation (114). No crystal structure or NMR structure is available for the basic region, but it appears that this region is disordered in solution and α-helical in the complex. Peptides corresponding to this region are only about 25% helical in solution but are nearly 100% helical when bound to DNA (114–117). "Affinity cleavage" and protection experiments suggest that the basic regions are symmetrically positioned in the major groove (118, 120).

No direct structural data are available, but two closely related models for the complex have been proposed (115, 119). In these models, the peptide is shaped like a Y. The parallel leucine zipper region forms the stem of the Y, and the basic region forms α-helices that extend off like the arms of the Y. In each model, the N-terminal end of the two-fold symmetric leucine zipper fits over the center of the binding site, and the helices from the basic region extend in opposite directions along the major grooves of the DNA. In the "scissors grip" model, the α-helix from the basic region is kinked so that it can follow the curve of the major groove (119). In the "induced helical fork" model, this helix is straight and thus extends away from the DNA after contacting three or four base-pairs (115).

Leucine zipper proteins can form heterodimers, and these mixed dimers have important roles in regulating the biological activity of the bZIP proteins. For example, the active transcription factor AP-1 consists of one Fos protein and one Jun protein (121, 122). Heterodimer formation has several different roles in the leucine zipper family. Heterodimers can limit activity: CREB, a cAMP response regulator, is antagonized by formation of heterodimers with CREM (123). Heterodimers may also acquire new DNA-binding specificities (124) and thus be targeted to different sites than homodimers. In other cases, heterodimer formation may allow for different combinations of activation and/or repression domains and thus change the regulatory properties of a molecule bound at a fixed DNA site.

The helix-loop-helix (HLH) proteins (105, 125) appear to have some similarities with the leucine zipper family. Like the leucine zipper proteins, the HLH proteins have a basic region that contacts the DNA and a neighboring region that mediates dimer formation (126). Based upon sequence patterns, it has been proposed that this dimerization region forms an α-helix, a loop, and a second α-helix (105, Figure 14). The sequence of the basic region has some similarities with that of the basic region of the leucine zipper proteins (127), but it is not known whether these regions have similar three-dimensional structures.

Like the leucine zipper proteins, the HLH proteins have many important roles in differentiation and development, and their activity is modulated by heterodimer formation. Thus the MyoD protein, which appears to be the primary signal for differentiation of muscle cells, binds DNA most tightly when it forms a heterodimer with the ubiquitously expressed E2A protein (128). There also is a cellular protein that contains an HLH region—without the basic region—and apparently acts as a negative regulator by forming inactive heterodimers with MyoD (129). Heterodimer formation is used in many different ways in this family of proteins, mixing positive activators, ubiquitously expressed proteins, and negative regulators to modulate gene activity (130). There may be yet other families of proteins that use similar themes of homodimer and heterodimer formation to regulate activity. For example, studies of AP-2 led to proposals about a helix-span-helix class of proteins (131).

β-Sheet Motifs

Most of the major families of DNA-binding proteins that have been structurally characterized bind with an α-helix in the major groove. The MetJ, Arc, and Mnt repressors are interesting because they belong to a family of prokaryotic regulatory proteins that uses an antiparallel β-sheet for DNA binding (132, 133). The crystal structure of MetJ is known, and a preliminary description of the DNA complex has been reported (133, 134). The structure of Arc, both in solution (132) and in a crystal (U. Obeysekare, C. Kissinger, R. Sauer, and C. Pabo, unpublished), has also been determined. MetJ (which contains 104 residues) is significantly larger than Arc (which contains 53 residues), but both proteins form dimers in solution and the structures are homologous in a region that contains a β-strand and two α-helices. In the protein dimers, the β-strands pair to form an antiparallel β-sheet, while the α-helical regions pack against the sheet and against each other to stabilize the dimer.

MetJ binds as a tetramer to an 18-base-pair DNA site (134). In this complex, a dimer binds to each half-site, and each half-site contains a two-fold symmetry axis that is coincident with the two-fold axis of the β-sheet. The β-sheet fills the major groove, and side chains on the exposed

0-15 residues

P S V I R R N A R E R N R R V K Q V N N G F S Q L R Q H I P A A V I A D L S −15−**K V S T L K M A V E Y I** R R L Q	T5 achaete-scute
Q S V Q R R N A R E R N R R V K Q V N N S F A R L R Q H I P Q S I − I T D L T −12−**K V D T L R I A V E Y I** R S L Q	T4 achaete-scute
E R R M A N N A R E R V R R D I N E A F R E L G R M C Q M H L K S D K A − 2 −**K L L I Q Q A V Q V I** L G L E	e47
E R R V A N N A R E R L R V R D I N E A F K E L G R M C Q L H L N S E K P − 2 −**K L L I L H Q A V S V I** L N L E	e12
E R R Q A N N A R E R I R D I N E A L K E L G R M C M T H L K S D K P − 2 −**K L G I N A V E V I** M T L E	daughterless
E R R R H N I L E R Q R N D L R S S F L T L R D H V P E L V K N E K A − 1 −**K V I L K K A T E Y V I** H S L Q	n-myc
V K R R T H N V L E R Q R N E L K R S F F A L R D Q I P E L E N N E K A − 1 −**K V V I L K K A T A Y I** L S V Q	c-myc
D R R K A A T M R E R R R L S K V N E A F E T L K R C T S S N P N Q R L P − 0 −**K V E L L R N A I R Y I** E G L Q	myoD

basic region   helix   loop   helix

Figure 14   Alignment of basic region helix-loop-helix sequences. Positions that are generally conserved, acidic (D, E), basic (R, K), or hydrophobic (A, C, V, I, L, M, F, Y) are shaded. The sequences, alignment, and positions of potential secondary structure are from Murre et al (105) and references therein.

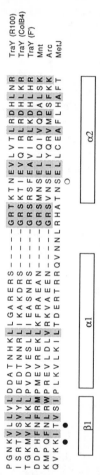

P G N K V **L V L** D D A T N H K **L L** G A R E R S − − − − − **G R T K T N E V L V T L R D H L N R**	TraY (R100)
I S R T V **S Y L** D E D T N N R **L I** K A K D R S − − − − − **G R S K T I E V Q I R L R D H L K R**	TraY (ColB4)
T G K M V **K L K L** P V D V E S L **L I** E A S N R S − − − − − **G R S R S F E A V I R L K D H L H R**	TraY (F')
D D P H F N **F R M P** M E V R E K L **K F** R A E A N − − − − − **G R S M N S E L L Q I V Q D A L S K**	Mnt
K M P Q F N **L R W** P R E V L D L **V** R K V A E E N − − − − **G R S V N S E I Y Q R V M E S F K K**	Arc
Q V K K I T **V S I** P L K V L K I **L** T D E R T R R Q V N N L R H A T N S E L L C E A F L H A F T	MetJ

β1   α1   α2

Figure 15   Alignment of β-sheet DNA-binding protein sequences. Positions that are generally conserved, acidic (D, E), basic (R, K), or hydrophobic (A, C, V, I, L, M, F, Y, W) are shaded. The secondary structure is that found in MetJ (134). The positions of major groove contacts (●) and backbone contacts (○) are from the MetJ complex (133). The alignment is from Breg et al (132). For sequence references, see (132) and (141) and references therein.

face of the sheet contact the base pairs (133). One lysine from each β-strand contacts a guanine, and a neighboring threonine on each strand contacts an adenine. Residues from the N-terminal end of α-helix 2 make backbone contacts. The MetJ tetramer is stabilized by interactions between α-helix 1 on one dimer and α-helix 1 on the other dimer.

It seems likely that Arc and Mnt interact with DNA in a way that is fundamentally similar to MetJ's interaction with DNA. Both proteins bind their operators as tetramers (136, 137), and genetic experiments have implicated the β-sheets in DNA recognition (135, 138, 139). However, the half-sites of the *arc* and *mnt* operators are not two-fold symmetric, and thus some breakdown of symmetry (in comparison with the fully symmetric arrangement in the MetJ complex) would need to occur. Biochemical experiments also show that Mnt uses an N-terminal arm to wrap around and contact the center of its operator (140). These interactions, which supplement the β-sheet contacts made by Mnt, seem generally similar to those made by the N-terminal arm of λ repressor.

Although Arc and Mnt show significant sequence similarities, their relationship with MetJ was only clearly established after structural studies of MetJ and Arc (132, 134). Even when the structures are used to align the sequences, there are relatively few positions at which a specific residue is conserved (Figure 15). Searches for additional members of this family, based upon hydrophobicity patterns, have identified the TraY proteins as probable relatives (141). However, given the limited sequence similarity among known members of this family, there could be other members that are simply too difficult to detect based upon sequence searches.

There appears to be at least one other family of regulatory proteins that uses β-sheets for DNA recognition. Although no structure is available, several arguments suggest that the *E. coli* IHF protein uses β-sheets for site-specific recognition (142). IHF is homologous with the bacterial HU protein, and crystallographic analysis of HU shows that it contains a pair of antiparallel β-arms that might participate in DNA recognition (143, 144). Based on chemical protection, sequence homology, and other biochemical data, a model has been proposed with β-sheet regions from IHF fitting in the minor groove of a sharply bent DNA site (142).

Other Families

There are a number of additional families of DNA-binding proteins that have been identified in studies of transcription factors, and presumably more families will be discovered in the future. As mentioned above, there are several other families (distinct from the TFIIIA-like fingers and the steroid receptors) that use zinc or other metal ions to stabilize their structures. For example, the cysteine-rich motif in the yeast GAL4 repressor has a distinct sequence pattern (Cys-X2-Cys-X6-Cys-X6-Cys-X2-Cys-X6-Cys) and is representative of a separate family of "fungal fingers" (145). NMR studies have shown that GAL4 contains a distinctive binuclear metal cluster (146), and this set of proteins has recently been referred to as the "zinc cluster" family (15). The "CCHC" motif, which has a sequence pattern of the form Cys-X2-Cys-X4-His-X4-Cys, is yet another metal-binding motif involved in nucleic acid recognition (147). This motif occurs in a set of retroviral gag proteins, and the structure, as determined by 2D NMR for a peptide from the HIV gag protein, is similar to the iron-binding region of rubredoxin (148).

There are several other families of DNA-binding proteins with cysteine-rich motifs that may serve as metal-binding domains. For example, a cysteine-rich motif (with the general form Cys-X-Asn-Cys-X17-Cys-Asn-X-Cys) was discovered in the GATA factor, which is specifically expressed in erythroid cells (149, 150). RAG-1 and RAD18 are prototypical members of a distinctive family of DNA-binding proteins that contain another cysteine-rich motif (151). The LIM motif is a distinctive cysteine-rich region that occurs on the N-terminal side of the homeodomain in several regulatory proteins from *Caenorhabditis elegans* (152). The ACE1 transcription factor of yeast (which may be related to metallothionein) has a cysteine-rich region that appears to form a cluster with 6–7 copper ions (153).

Although cysteine-rich and histidine-rich motifs are easily detected in sequence comparisons (these residues occur relatively rarely in most proteins) and have received considerable attention, several other families of DNA-binding proteins have been characterized recently. One prominent family (that was mentioned previously) is the set of POU proteins, which contain a homeodomain and a POU-specific domain that also binds DNA (73–75). There also are families of transcriptional regulatory proteins related to c-ets (154), to the nonhistone high mobility group proteins (155 and references therein), to the serum response factor (156 and references therein), to c-myb (157 and references therein), to NF-κB, and to the *rel* oncogene (158–160). Sequence comparisons also suggest that there are groups of DNA-binding proteins related to the paired domain from *Drosophila* (161, 162), to the homeotic gene fork head (163, 164), and to a set designated as the TEA domain (165). At present, no detailed structural information is available about these families. Obviously, a thorough analysis will require information on the three-dimensional structure, biological distribution, and regulatory roles of each of these families.

PRINCIPLES OF RECOGNITION

Helices in Recognition

Now that a set of protein-DNA complexes has been solved, we can take a broader look at the problem of protein-DNA recognition and search for useful generalizations. Most of the well-characterized families of DNA-binding proteins use α-helices to make base contacts in the major groove. Although β-sheets or regions of extended polypeptide chain can also make contacts, α-helices are used much more frequently. In fact, the use of α-helices in site-specific recognition is so widespread that there is some danger of over-emphasizing their role or of misunderstanding the different ways in which an α-helix can be used. As mentioned above, referring to a particular helix as a "recognition helix" can easily be misleading. There is no evidence that an isolated helix from any of the known motifs can bind DNA in a site-specific fashion, and no regulatory system has been discovered that uses an isolated helical peptide (i.e. a single α-helix) as a site-specific DNA-binding protein. In every reported complex involving α-helical motifs, it appears that the overall binding specificity results from a set of interactions, with some contacts from a "recognition" helix and some from surrounding regions of the protein.

Although the orientation of α-helices with respect to the major groove may be conserved within a given family or subfamily of DNA-binding proteins, there is no unique way of placing an α-helix in the major groove. Thus, the orientations of the α-helices are significantly different when complexes from different DNA-binding families are compared (e.g. λ repressor, the engrailed homeodomain, and the zif268 zinc fingers). Even within a given family, the orientations of the α-helices can be significantly different (as with λ repressor and Trp repressor). The main point is that the overall shape and dimensions of an α-helix allow it to fit into the major groove in a number of related, but significantly different ways. Some helices lie in the middle of the major groove and have the axis of the α-helix approximately tangent to the local direction of the major groove (i.e. at an angle of 32° with respect to the plane that is perpendicular to the helical axis of the DNA). Other α-helices are tipped at different angles, varying at least 15° in each direction, and some are arranged so that only the N-terminal portion of the α-helix fits completely into the major groove. It appears that surrounding regions of the proteins (and not just the sequences of the "recognition helices") help to determine how these α-helices are positioned in the major groove.

Interactions with Bases

Contacts with the bases play a crucial role in site-specific binding, and the known complexes contain a rather diverse set of base contacts. Structural studies have revealed (a) direct hydrogen bonds between the protein side chains and the bases, (b) occasional hydrogen bonds between the polypeptide backbone and the bases, (c) hydrogen bonds mediated by water molecules, and (d) hydrophobic contacts. Although a few base contacts occur in the minor groove (such as those made by the N-terminal arm of the homeodomain), the major groove appears to be more important for site-specific recognition. This is not surprising, since the major groove is larger and more accessible in B-DNA and has more potential sites for hydrogen bonding and hydrophobic contacts. Analyzing the pattern of hydrogen-bonding sites along the edge of the base pairs also suggested that major groove contacts would provide a more reliable basis for sequence-specific recognition (166). In the minor groove, the O2 of thymine and the N3 of adenine occupy very similar positions, and may be difficult to distinguish since both are hydrogen bond acceptors.

A comparison of the base contacts that have been observed in various complexes shows that there is no simple "recognition code"; i.e. there is no one-to-one correspondence between the amino acid side chains and the bases they contact. Figure 16 shows some of the hydrogen-bonding interactions that have been observed between side chains and bases in the known complexes. Some side chains are used to contact more than one kind of base, and some bases are contacted by a variety of side chains. No simple pattern or rule describes all of these contacts, although the table suggests that hydrogen bonds with purines (particularly with guanine!) may have an especially important role in recognition. What accounts for this diversity of contacts? Even when the same secondary structure, such as an α-helix, is used for base recognition, the precise position and orientation of the α-helix may determine what set of side chain–base interactions are possible for a particular residue. Further levels of complexity are added when we consider side chain–side chain interactions, residues that contact more than one base pair, and contacts that involve bridging water molecules or ions. There also are cases—as observed with λ repressor's N-terminal arm—in which the polypeptide backbone of the protein hydrogen bonds with the edge of the base pairs (50). Overall, the variety and structural complexity of the contacts involved in protein-DNA interactions rival those of the contacts involved in protein structure and folding.

In spite of these complexities, some organizing principles emerge when we compare closely related structures, such as the λ and 434 repressors or the engrailed and α2 homeodomains. In these cases, we find that conserved residues at the protein-DNA interface make conserved contacts with the DNA. This was first noted when comparing the λ and 434 repressors, which have three conserved residues on the DNA-binding face of the HTH unit (49). Each of these conserved residues makes conserved contacts with the DNA. A glutamine at the start of the HTH unit makes conserved contacts with the DNA backbone, a glutamine at the start of the "recognition helix" makes conserved hydrogen bonds with an adenine, and an asparagine at the end of the HTH unit makes a conserved contact with the DNA backbone. Comparing the engrailed and α2 homeodomain–DNA complexes shows that conserved residues at the protein-DNA interface also make conserved contacts with the

	A	T	C	G
Arg		CAP		GR Trp zif CAP
Asn	en			lambda
Gln	lambda 434	434		434
Glu			CAP	
His				zif
Lys				GR lambda MetJ
Ser				lambda
Thr	MetJ			

Figure 16 Chart summarizing hydrogen bonds between side chains and bases in the major groove for representative protein-DNA complexes. Contacts are summarized for the *E. coli* CAP protein (56), the *Drosophila* engrailed homeodomain [en, (66)], the glucocorticoid receptor [GR, (103)], the 434 repressor (48), the λ repressor (47), the *E. coli* MetJ repressor (133), the *E. coli* Trp repressor (54), and the zif268 zinc fingers (83).

DNA (66, 67). The amino acid sequences of these two homeodomains are only 27% identical, but six conserved residues—Phe8, Tyr25, Trp48, Arg53, Lys55, and Lys57—make conserved contacts with the DNA backbone. Asn51, which is invariant in the homeodomain family, makes a conserved set of hydrogen bonds with an adenine. Comparing fingers 1, 2, and 3 of the zif268 complex (83) again shows that conserved residues at the protein-DNA interface make conserved contacts with the DNA. There are several conserved contacts with the DNA backbone and several conserved contacts with the bases.

These comparisons suggest that a given family or subfamily of DNA-binding proteins may have—in addition to the conserved folding motif that characterizes the family—a conserved "docking mechanism" and a conserved set of contacts. This may provide a useful simplifying principle when thinking about recognition. Contacts may be more predictable if we focus attention on a particular position within a particular structural motif. In some sense, the position and orientation of the polypeptide backbone with respect to the DNA determine the "meaning" of a particular side chain (167). The fact that a side chain can be presented in different ways by different structural motifs (or even by different positions within a given motif) may be one of the reasons that there is no simple "recognition code" (Figure 16). Since structure is central to recognition, the folding of the polypeptide and the overall docking arrangement help to determine which base contacts are possible in any given situation. In this context, it is intriguing that the conserved base contacts that have been observed all involve the types of side chain–base interactions that had been predicted by Seeman, Rosenberg, and Rich (166). These include the interaction of arginine with guanine, and the interaction of glutamine or asparagine with adenine. Since each of these side chain–base contacts involves a pair of hydrogen bonds, they may play an especially important role in site-specific recognition.

Contacts with the DNA Backbone

Even a cursory examination of protein-DNA complexes suggests that contacts with the DNA backbone play an integral role in site-specific recognition. In the known complexes, roughly half of all the hydrogen bonds involve contacts with the DNA backbone. Almost all of these contacts involve the phosphodiester oxygens. A few hydrophobic interactions with sugar rings in the DNA backbone have been reported, but these are much less common and presumably far less important for recognition.

As observed with side chain–base interactions, there does not appear to be any simple "rule" or pattern describing which residues are used for backbone contacts. Examining the published complexes shows that many different side chains, and even the -NH of the polypeptide backbone, can hydrogen bond to the phosphodiester oxygens. Given the variety of contacts that have been observed, it seems likely that any basic or neutral hydrogen-bonding side chain can be used to contact the phosphodiester oxygens. We assume that aspartic acid and glutamic acid are used less frequently because of the unfavorable electrostatic interactions, but even these side chains could contact the phosphates via bridging divalent cations such as Mg^{2+} or Ca^{2+}. Although lysine and arginine are used in a number of cases, their favorable electrostatic interactions with the phosphates may be partially offset by the inherent flexibility of their long side chains. In some complexes (e.g. λ repressor), short polar side chains and the peptide -NH seem to play a more important role in contacting the backbone. These contacts may provide more stereospecificity than those mediated by arginines or lysines.

The sheer number of hydrogen bonds with the phosphodiester oxygens, and the exquisite hydrogen-bonding networks that have been observed (47, 48, 54), suggest that these contacts are very important for site-specific recognition. We still do not understand the exact role of backbone contacts, but two major roles appear possible: (a) Backbone contacts may serve as "fiducial marks" that help hold the protein against the bases in a fixed arrangement and thereby enhance the specificity of the side chain–base interactions. Without these backbone contacts to orient the protein in the major groove and help establish a fixed register for the interactions, the "recognition helix" might slip or shift in a way that would allow it to hydrogen bond with inappropriate sites. (b) Sequence-dependent variations in the DNA structure may also be important (as discussed in the next section). To the extent that base sequence determines the structure of the DNA—and therefore the most favorable positions for the phosphodiester oxygens—contacts with the DNA backbone may allow indirect recognition of the sequence.

Role of DNA Structure in Recognition

Any thorough discussion of binding specificity requires consideration of variations in DNA structure and/or flexibility. There are several ways in which the local structure of the DNA could influence specificity: (a) There could be sequence-dependent influences on the structure of a given binding site, causing it to have an average structure that deviates from canonical B-form DNA (54, 168). Specificity would be enhanced if a protein were designed to recognize a stable local structure such as a DNA bend or kink. (b) There could be sequence-dependent effects on the flexibility of the DNA, which could contribute to specificity if the protein distorts the DNA (possibly by bending) as it binds. Even if the average solution structure of the DNA were linear, bending could contribute to recognition if the correct site were more flexible (i.e. more easily bent) than canonical B-form DNA.

There now are a number of examples in which repressors bind to bent sites on the DNA, but it still is difficult to evaluate the contributions of these effects to site-specific recognition. Thus significant bending of the DNA occurs in complexes with the Trp repressor (54), 434 repressor (48), 434 Cro protein (53), λ repressor (47), and λ Cro protein (55). In each case, the DNA bends as if it were beginning to "wrap around" the HTH units of the proteins. An even more dramatic bend of 90° occurs in the *E. coli* CAP complex (56). It seems plausible that these bends (especially the dramatic bend observed in the CAP complex) make some contribution to specificity, but it still is difficult to evaluate the magnitude of these effects. The main problems stem from the difficulty of determining the average structure of the free DNA and the energetic cost of bending the DNA. It would be useful to have structures of the free DNA-binding sites, but solution structures determined by 2D NMR do not provide sufficient structural detail (169), and crystal structures of the free DNA may be seriously perturbed by crystal packing forces (170). Energetic analyses present an even more serious problem. To understand the energetic contributions to recognition, we really would need to compare the energetic costs of distorting the correct site (which could be zero if this site is bent in solution) with the energetic costs of distorting other sites. At present, there does not appear to be any reliable way of determining these energies.

Trying to understand the role of contacts with the DNA backbone raises several related problems about recognition. One involves the structural relationship between specific and nonspecific binding. In particular, do DNA-binding proteins make the same backbone contacts when they bind to nonspecific DNA? This is crucial to understanding the role of backbone contacts in recognition, but unfortunately there is very little structural information about nonspecific complexes. [A symmetric DNA site, with an altered spacing between the half-sites, was used when cocrystallizing the DNA-binding domain of the glucocorticoid receptor, and one-half of this complex provides us with our first glimpse of a nonspecific complex (103).] Since biological specificity only results from the free energy differences between specific and nonspecific binding, we can never really understand site-specific binding without understanding the nonspecific "reference state."

Considering the role of contacts with the phosphodiester oxygens raises broader problems involved in any attempts to assign separate energetic contributions to particular contacts or groups within protein-DNA complexes. One can easily study the binding of a mutant protein or of a DNA molecule in which a particular functional group has been removed, but it is not correct to assume that the $\Delta\Delta G$ measured for such a variant only reflects the contribution of this particular contact. First, one would need to know that the mutation had not altered the structure of the complex. Changing even one contact could allow subtle changes throughout the complex. Second, it seems likely that individual contacts—including base contacts and backbone contacts—are intimately coupled. One would need to study many different variants and test various sets of changes in any attempt to assess properly the intrinsic and cooperative energies of the system. Careful studies of the structures and energies of mutant complexes are sure to yield useful insights, but the reductionist approach can only be taken so far. Ultimately, recognition involves a set of contacts with a set of sites on the DNA, and attempting to describe the energy of the complex as a sum of separate, discrete contributions may be meaningless.

General Principles of Site-Specific Recognition

Although the diversity of known DNA-binding motifs and contacts suggests that there are no simple rules or patterns describing site-specific recognition, comparing the known complexes allows us to make a few broad generalizations.

1. Site-specific recognition always involves a set of contacts with the bases and with the DNA backbone.
2. Hydrogen bonding is critical for recognition (although hydrophobic interactions also occur). A complex typically has on the order of 1–2 dozen hydrogen bonds at the protein/DNA interface.
3. Side chains are critical for site-specific recognition. There are instances in which the peptide backbone makes hydrogen bonds with bases or the DNA backbone, but side chains make most of the critical contacts.
4. There is no simple one-to-one correspondence between side chains and the bases they contact. It appears that the folding and docking of the entire protein help to control the "meaning" that any particular side chain has in site-specific recognition.
5. Most of the base contacts are in the major groove. Contacts with purines (which are larger and offer more hydrogen-bonding sites in the major groove) seem to be especially important.
6. Most of the major motifs contain an α-helical region that fits into the major groove of B-form DNA. There are examples of β-sheets and/or extended regions of polypeptide chain that play critical roles in certain proteins, but base contacts from these regions appear to be less common.
7. Contacts with the DNA backbone usually involve hydrogen bonds and/or salt bridges with the phosphodiester oxygens.
8. Multiple DNA-binding domains usually are required for site-specific recognition. The same motif may be used more than once, as occurs when the active binding species is a homodimer or heterodimer, or when a single polypeptide contains tandem recognition motifs. Different motifs (an extended arm and a HTH unit; a homeodomain and POU-specific domain, etc) may also be used in the same complex.

9. Recognition is a detailed structural process. Hydration can play a critical role in recognition; sequence-dependent aspects of the DNA structure may also be important.

DIRECTIONS FOR FUTURE RESEARCH

The past few years have brought exciting progress in the study of protein-DNA recognition, but important questions remain. Although many of the problems are interrelated, some of the most critical questions involve issues of structure, energy, evolution, gene regulation, and protein design.

Structure   Recent structural studies have given us a much better understanding of protein-DNA interactions, but much work remains. Obviously, it will be important to solve the structures of complexes that contain some of the other major DNA-binding motifs (leucine zippers, helix-loop-helix proteins, etc). It also will be important to determine the structures of some intact regulatory proteins. We need to know the structures of other domains to understand transcriptional activation and other activities of the intact proteins. It also is possible that neighboring domains affect the specificity, affinity, or accessibility of the DNA-binding domain. Finally, there is a continuing need to examine more examples of the "known" DNA-binding motifs. The fact that the prokaryotic HTH unit and the eukaryotic homeodomain bind in such different ways should serve as something of a warning. Other surprises may await us, and it certainly is possible that there are multiple ways to use a zinc finger or a homeodomain in DNA recognition.

Modeling presents additional challenges for future structural work. Obviously, it would be a tremendous advance if we could reliably model protein-DNA interactions. Does the wealth of new structural data allow us to improve our methods for modeling? Is it now possible to model reliably homologous complexes? Will the development of improved potential functions and/or the development of new computational strategies eventually allow us to model other protein-DNA complexes?

Studies of DNA-binding proteins certainly will benefit from any improvements that can be made in the methods for correlating amino acid sequence and protein structure. Bowie & Eisenberg (171) have recently presented a method that uses patterns derived from protein crystal structures to search for sequences that could reasonably be expected to fold in the same manner. It will be interesting to see if this method can identify new members for any of the major DNA-binding families for which structures are available.

Energy   A real chemical and physical understanding of recognition also will require a better understanding of the energetics of protein-DNA interactions.

Analyzing the energies either experimentally or theoretically is complicated because recognition always involves a set of contacts. It is difficult to "dissect" these interactions in a way that assigns specific energetic contributions to individual contacts. These problems are further complicated by the fact that we usually are interested in the energetic differences between contacts that occur in a specific complex and contacts that occur with nonspecific DNA. (These difference in the binding energy account for the biological specificity.) Modern computational methods, such as those involving free energy perturbation (172), may be useful, but these studies are difficult and computationally expensive. These calculations also require very high-resolution crystal structures as a starting point, and we need more detailed structural information about nonspecific complexes before we have any chance of understanding the energetic differences that are responsible for site-specific recognition. Obviously, these energetic problems are complicated. At some point, they overlap with more general and fundamental questions about protein folding and macromolecular recognition.

Other important questions—such as trying to understand why so many proteins use α-helices for recognition—involve a combination of structural and energetic issues. Is there some inherent advantage to using an α-helix in recognition? Clearly, the size and shape of an α-helix make it complementary to the major groove, but a β-sheet can also fit in the major groove. Why do relatively few proteins use β-sheets for site-specific binding? Is the α-helix more "rigid" in some way that helps to enhance the specificity of the contacts? Are there more distinct ways of arranging an α-helix in the major groove and thus more sequences that helical proteins can recognize?

Trying to understand the role of flexible regions in site-specific recognition also involves structural and energetic issues. Why are flexible or disordered regions (like λ repressor's N-terminal arm and the basic region of the leucine zipper proteins) sometimes used for DNA-binding? Do these regions of the polypeptide need to be flexible to allow rapid binding? Would a rigid protein bind too slowly and be released too slowly? Alternatively, do these disorder → order transitions provide an entropic way of limiting the overall binding energy while retaining a set of protein-DNA contacts that are essential for specificity?

Evolution   Although the known families provide a plausible way of grouping DNA-binding proteins, there still are many interesting questions about subfamily and superfamily relationships. We cannot yet construct an accurate genealogy for the families of DNA-binding proteins or understand the structural significance of all of the conserved sequence patterns. For instance, are the prokaryotic HTH motif and the eukaryotic homeodomain related by divergent or convergent evolution? Why are these structures so similar and yet used in different ways when binding to DNA? What is the significance of the various homeodomain subfamilies that have been noted? Are the various families of zinc finger proteins related to each other in any meaningful way? Are the leucine zipper and helix-loop-helix proteins structurally related? As we look to the more distant past and try to visualize the earliest stages in the evolution of modern gene regulation, we also wonder: What are the smallest structural units that could plausibly have given some selective advantage during evolution? Could an isolated HTH unit or a single zinc finger have any meaningful regulatory role?

Gene Regulation   Studies of protein-DNA recognition are closely linked to problems of gene regulation, but additional work is needed to really understand the connections between recognition and regulation. A careful analysis of the binding energies is important, since we would like to understand the physical basis for the differential regulation of gene expression. How do homeodomains distinguish among closely related binding sites? What role do accessory proteins play in site-specific binding? How much energy is contributed by protein-protein interactions? How does competition with other regulatory proteins and competition with nucleosomes affect binding? Clearly, additional structural work will be needed to help us understand the structural basis for positive and negative control. As mentioned above, it becomes extremely important to have structural information about the intact proteins so that we can see the domains responsible for positive and negative regulation.

Another question involves the apparent association between certain structural motifs and specific biological roles. Are certain motifs, such as the helix-loop-helix motif, particularly well adapted for regulating differentiation and development? Could other families of proteins, such as zinc fingers, have been used as effectively in these roles? It also seems possible that certain motifs are particularly well suited for recognizing certain base sequences. Does each motif have characteristic DNA sequences that are most suitable for the binding site? Are some motifs more "adaptable" than others and thus suited for recognizing a wider range of binding sites?

Design   Attempts to design DNA-binding proteins provide another exciting challenge for future research. This will provide a rigorous test of our understanding, and could have tremendous practical benefits since new proteins might be used for research, diagnosis, or even gene therapy. When working on design, it may make sense to use known motifs from the major families of DNA-binding proteins. These provide an attractive starting point because each of these motifs is based on a stable secondary structure and has a good way of packing against the DNA. Moreover, the existence of large families of DNA-binding proteins suggests that these motifs offer highly successful and adaptable frameworks for protein-DNA recognition. Pavletich & Pabo (83) have suggested that the zinc finger motif may provide a particularly attractive framework for the design of new DNA-binding proteins. Since the zinc fingers recognize an asymmetrical DNA sequence, there is no need to restrict the design process to work with symmetric target sites. It also is possible that the modular nature of the zinc finger–binding units will allow simplification of the design problem. Since each finger makes its primary contacts in a three-base-pair region, it might be possible to design or select fingers that recognize each of the 64 possible base triplets. If one can "mix and match" these fingers, it should be possible to design proteins that recognize any desired base sequence.

Ultimately, the problems of structure, energy, evolution, gene regulation, and design are closely related. We have made significant progress, but much work is needed before we have a real understanding of protein-DNA interactions. We should not be content with a superficial understanding that merely allows us to rationalize a few key observations. We need a deeper understanding with real explanatory power—an understanding that will let us predict how other proteins bind and that will allow us to design new DNA-binding proteins. Exciting challenges lie ahead.

ACKNOWLEDGMENTS

We thank our colleagues for helpful discussions and for permission to cite unpublished information, and thank Kristine Kelly for assistance in preparation of this manuscript. Work in our laboratories was supported NIH grant AI-16892 (R.T.S.), by NIH grant GM-31471 (C.O.P.), and by the Howard Hughes Medical Institute (C.O.P.).

Literature Cited

1. Steitz, T. A. 1990. *Q. Rev. Biophys.* 23:205–80
2. Johnson, P. F., McKnight, S. L. 1989. *Annu. Rev. Biochem.* 58:799–839
3. Mitchell, P. J., Tjian, R. 1989. *Science* 245:371–78
4. Struhl, K. 1989. *Trends Biol. Sci.* 14:137–40
5. Frankel, A. D., Kim, P. S. 1991. *Cell* 65:717–19
6. He, X., Rosenfeld, M. G. 1991. *Neuron* 7:183–96
7. Harrison, S. C. 1991. *Nature.* 355:715–19
8. Brennan, R. G., Matthews, B. W. 1989. *J. Biol. Chem.* 264:1903–6
9. Harrison, S. C., Aggarwal, A. K. 1990. *Annu. Rev. Biochem.* 59:933–69
10. Brennan, R. G. 1991. *Curr. Opin. Struc. Biol.* 1:80–88
11. Berg, J. M. 1986. *Science* 232:485–87
12. Klug, A., Rhodes, D. 1987. *Trends Biochem. Sci.* 12:464–69
13. Kaptein, R. 1991. *Curr. Opin. Struc. Biol.* 1:63–70
14. Berg, J. M. 1990. *Annu. Rev. Biophys. Biophys. Chem.* 19:405–21
15. Vallee, B. L., Coleman, J. E., Auld, D. S. 1991. *Proc. Natl. Acad. Sci. USA* 88:999–1003
16. Evans, R. M. 1988. *Science* 240:889–95
17. Beato, M. 1989. *Cell* 56:335–44
18. Kerppola, T. K., Curran, T. 1991. *Curr. Opin. Struc. Biol.* 1:71–79
19. Scott, M. P., Tamkun, J. W., Hartzell, G. W. 1989. *Biochim. Biophys. Acta* 989:25–48
20. Gehring, W. J., Muller, M., Affolter, M., Percival-Smith, A., Billeter, M., et al. 1990. *Trends Genet.* 6:323–29

21. Wright, C. V., Cho, K. W., Oliver, G., De Robertis, E. M. 1989. *Trends Biochem. Sci.* 14:52–56
22. Hayashi, S., Scott, M. P. 1990. *Cell* 63:883–94
23. Phillips, S. E. V. 1991. *Curr. Opin. Struc. Biol.* 1:89–98
24. Luisi, B. F., Sigler, P. B. 1990. *Biochim. Biophys. Acta* 1048:113–26
25. Sauer, R. T., Jordan, S. R., Pabo, C. O. 1990. *Adv. Prot. Chem.* 40:1–61
26. Pabo, C. O., Sauer, R. T. 1984. *Annu. Rev. Biochem.* 53:293–321
27. Arthur, A. K., Hoss, A., Fanning, E. 1988. *J. Virol.* 62:1999–2006
28. Kern, S. E., Kinzler, K. W., Bruskin, A., Jarosz, D., Friedman, P., Prives, C., et al. 1991. *Science* 252:1708–11
29. Anderson, W. F., Ohlendorf, D. H., Takeda, Y., Matthews, B. W. 1981. *Nature* 290:754–58
30. Ohlendorf, D. H., Anderson, W. F., Fisher, R. G., Takeda, Y., Matthews, B. W. 1982. *Nature* 298:718–23
31. McKay, D. B., Steitz, T. A. 1981. *Nature* 290:744–49
32. Pabo, C. O., Lewis, M. 1982. *Nature* 298:443–47
33. Steitz, T. A., Ohlendorf, D. H., McKay, D. B., Anderson, W. F., Matthews, B. W. 1982. *Proc. Natl. Acad. Sci. USA* 79:3097–100
34. Ohlendorf, D. H., Anderson, W. F., Lewis, M., Pabo, C. O., Matthews, B. W. 1983. *J. Mol. Biol.* 169:757–69
35. Kaptein, R., Zuiderweig, E. R. P., Sheek, R. M., Boelens, R., van Gunsteren, W. F. 1985. *J. Mol. Biol.* 182:179–82
36. Anderson, J. E., Ptashne, M., Harrison, S. C. 1987. *Nature* 326:846–52
37. Mondragon, A., Subbiah, S., Almo, S. C., Drottar, M., Harrison, S. C. 1989. *J. Mol. Biol.* 205:189–200
38. Wolberger, C., Dong, Y. C., Ptashne, M., Harrison, S. C. 1988. *Nature* 335:789–95
39. Mondragon, A., Wolberger, C., Harrison, S. C. 1989. *J. Mol. Biol.* 205:179–88
40. Schevitz, R. W., Otwinowski, Z., Joachimiak, A., Lawson, C. L., Sigler, P. B. 1985. *Nature* 317:782–86
41. Kostrewa, D., Granzin, J., Koch, C., Choe, H. W., Raghunathan, S., et al. 1990. *Nature* 349:178–80
42. Lamerichs, R. M., Padilla, A., Boelens, R., Kaptein, R., Ottleben, G., et al. 1990. *Proc. Natl. Acad. Sci. USA* 86:6863–67
43. Matthews, B. W., Ohlendorf, D. H., Anderson, W. F., Takeda, Y. 1982. *Proc. Natl. Acad. Sci. USA* 79:1428–32

44. Sauer, R. T., Yocum, R. R., Doolittle, R. F., Lewis, M., Pabo, C. O. 1982. *Nature* 298:447–51
45. Weber, I. T., McKay, D. B., Steitz, T. A. 1982. *Nucleic Acids Res.* 10:5085–102
46. Koch, C., Vandekerckhove, J., Kahmann, R. 1988. *Proc. Natl. Acad. Sci. USA* 85:4237–41
47. Jordan, S. R., Pabo, C. O. 1988. *Science* 242:893–99
48. Aggarwal, A. K., Rodgers, D. W., Drottar, M., Ptashne, M., Harrison, S. C. 1988. *Science* 242:899–907
49. Pabo, C. O., Aggarwal, A. K., Jordan, S. R., Beamer, L. J., Obeysekare, U. R., et al. 1990. *Science* 247:1210–13
50. Clarke, N. D., Beamer, L. J., Goldberg, H. R., Berkower, C., Pabo, C. O. 1991. *Science* 254:267–70
51. Pabo, C. O., Krovatin, W., Jeffrey, A., Sauer, R. T. 1982. *Nature* 298:441–43
52. Weiss, M. A., Sauer, R. T., Patel, D. J., Karplus, M. 1984. *Biochemistry* 23:5090–95
53. Mondragon, A., Harrison, S. C. 1991. *J. Mol. Biol.* 219:321–34
54. Otwinowski, Z., Schevitz, R. W., Zhang, R-G., Lawson, C. L., Joachimiak, A., et al. 1988. *Nature* 335:321–29
55. Brennan, R. G., Roderick, S. L., Takeda, Y., Matthews, B. W. 1990. *Proc. Natl. Acad. Sci. USA* 87:8165–69
56. Schulz, S. C., Shields, G. C., Steitz, T. A. 1991. *Science* 253:1001–7
57. Koudelka, G. B., Harrison, S. C., Ptashne, M. 1987. *Nature* 326:886–88
58. Koudelka, G. B., Harbury, P., Harrison, S. C., Ptashne, M. 1988. *Proc. Natl. Acad. Sci. USA* 85:4633–37
59. Laughon, A., Scott, M. P. 1984. *Nature* 320:25–31
60. Shepherd, J. C. W., McGinnis, W., Carrasco, A. E., De Robertis, E. M., Gehring, W. J. 1984. *Nature* 310:70–71
61. Sauer, R. T., Smith, D. L., Johnson, A. D. 1988. *Genes Dev.* 2:807–16
62. Affolter, M., Percival-Smith, A., Muller, M., Leupin, W., Gehring, W. J. 1990. *Proc. Natl. Acad. Sci. USA* 87:4093–97
63. Qian, Y. Q., Billeter, M., Otting, G., Muller, M., Gehring, W. J., et al. 1989. *Cell* 59:573–80
64. Billeter, M., Qian, Y., Otting, G., Muller, M., Gehring, W. J., et al. 1990. *J. Mol. Biol.* 214:183–97
64a. Phillips, C., Vershon, A., Johnson, A., Dahlquist, F. 1991. *Genes Dev.* 5:764–72
65. Otting, G., Qian, Y. Q., Billeter, M.,

Muller, M., Affolter, M., et al. 1990. *EMBO J.* 9:3085–92

66. Kissinger, C. R., Liu, B. S., Martin-Blanco, E., Kornberg, T. B., Pabo, C. O. 1990. *Cell* 63:579–90

67. Wolberger, C., Vershon, A. K., Liu, B., Johnson, A. D., Pabo, C. O. 1991. *Cell* 67:517–28

68. Garcia-Blanco, M. A., Clerc, R. G., Sharp, P. A. 1989. *Genes Dev.* 3:739–45

69. Hanes, S. D., Brent, R. 1989. *Cell* 57:1275–83

70. Treisman, J., Gonczy, P., Vashishtha, M., Harris, E., Desplan, C. 1989. *Cell* 59:553–62

71. Hanes, S. D., Brent, R. 1991. *Science* 251:426–30

72. Affolter, M., Percival-Smith, A., Muller, M., Billeter, M., Qian, Y. Q., et al. 1991. *Cell* 64:879–80

73. Rosenfeld, M. G. 1991. *Genes Dev.* 5:897–907

74. Ruvkun, G., Finney, M. 1991. *Cell* 64:475–78

75. Herr, W., Sturm, R. A., Clerc, R. G., Corcoran, L. M., Baltimore, D., et al. 1988. *Genes Dev.* 2:1513–16

76. Goutte, C., Johnson, A. D. 1988. *Cell* 52:875–82

77. Keleher, C. A., Passmore, S., Johnson, A. D. 1989. *Mol. Cell. Biol.* 9:5228–30

78. Kristie, T., Sharp, P. 1990. *Genes Dev.* 4:2383–96

79. Miller, J., McLachlan, A. D., Klug, A. 1985. *EMBO J.* 4:1609–14

80. Brown, R. S., Sander, C., Argos, P. 1985. *FEBS Lett.* 186:271–74

81. Ginsberg, A. M., King, B. O., Roeder, R. G. 1984. *Cell* 39:479–89

82. Krizek, B. A., Amann, B. T., Kilfoil, V. J., Merkle, D. L., Berg, J. M. 1991. *J. Am. Chem. Soc.* 113:4518–23

83. Pavletich, N. P., Pabo, C. O. 1991. *Science* 252:809–17

84. Berg, J. M. 1988. *Proc. Natl. Acad. Sci. USA* 85:99–102

85. Gibson, T. J., Postma, J. P. M., Brown, R. S., Argos, P. 1988. *Prot. Eng.* 2:209–18

86. Parraga, G., Horvath, S. J., Eisen, A., Taylor, W. E., Hood, L., et al. 1988. *Science* 241:1489–92

87. Lee, M. S., Gippert, G. P., Soman, K. V., Case, D. A., Wright, P. E. 1989. *Science* 245:635–37

88. Christy, B. A., Lau, L. F., Nathans, D. 1988. *Proc. Natl. Acad. Sci. USA* 85:7857–61

89. Nardelli, J., Gibson, T., Vesque, C., Charnay, P. 1991. *Nature* 349:175–78

90. Kadonaga, J. T., Carner, K. R.,

Masiarz, F. R., Tjian, R. 1987. *Cell* 51:1079–90

91. Mardon, G., Page, D. C. 1989. *Cell* 56:765–70

92. Kochoyan, M., Havel, T., Nguyen, D., Dahl, C., Keutmann, H., et al. 1991. *Biochemistry* 30:3371–86

93. Baldwin, A. Jr., LeClair, K., Singh, H., Sharp, P. 1990. *Mol. Cell Biol.* 10:1406–14

94. Reuter, G., Giarre, M., Farah, J., Gausz, J., Spierer, A., Spierer, P. 1990. *Nature* 344:219–23

95. Fasano, L., Roder, L., Core, N., Alexandre, E., Vola, C., et al. 1991. *Cell* 64:63–79

96. Nauber, U., Pankratz, M. J., Kienlin, A., Seifert, E., Klemm, U., Jaekle, H. 1988. *Nature* 336:489–92

97. Hazel, T. G., Nathans, D., Lau, L. F. 1988. *Proc. Natl. Acad. Sci. USA* 85:8444–48

98. Tilley, W. D., Marcelli, M., Wilson, J. D., McPhual, M. J. 1989. *Proc. Natl. Acad. Sci. USA* 86:327–31

99. Freedman, L. P., Luisi, B. F., Korszun, Z. R., Basavappa, R., Sigler, P. B., et al. 1988. *Nature* 334:543–46

100. Frankel, A. D., Pabo, C. O. 1988. *Cell* 53:675

101. Hard, T., Kellenbach, E., Boelens, R., Maler, B. A., Dahlman, K., et al. 1990. *Science* 249:157–60

102. Schwabe, J., Neuhaus, D., Rhodes, D. 1990. *Nature* 348:458–61

103. Luisi, B. F., Xu, W. X., Otwinowski, Z., Freedman, L. P., Yamamoto, K. R., et al. 1991. *Nature* 352:497–505

104. Landschulz, W. H., Johnson, P. F., McKnight, S. L. 1988. *Science* 240:1759–64

105. Murre, C., McCaw, P. S., Baltimore, D. 1989. *Cell* 56:777–83

106. Hope, I. A., Struhl, K. 1986. *Cell* 46:885–94

107. Roman, C., Platero, J. S., Shuman, J., Calame, K. 1990. *Genes Dev.* 4:1404–15

108. Maekawa, T., Sakura, H., Kanei, I. C., Sudo, T., Yoshimura, T., et al. 1989. *EMBO J.* 8:2023–28

109. Hartings, H., Maddaloni, M., Lazzaroni, N., DiFonzo, N., Motto, M., et al. 1989. *EMBO J.* 8:2795–801

110. Fu, Y. H., Paietta, J. V., Mannix, D. G., Marzluf, G. A. 1989. *Mol. Cell. Biol.* 9:1120–27

111. O'Shea, E. K., Rutkowski, R., Kim, P. S. 1989. *Science* 243:538–42

112. O'Shea, E. K., Klemm, J. D., Kim, P. S., Alber, T. 1991. *Science* 254:539–44

113. Agre, P., Johnson, P. F., McKnight, S. L. 1989. *Science* 246:922–25

114. Talanian, R. V., McKnight, C. J., Kim, P. S. 1990. *Science* 249:769–71
115. O'Neil, K. T., Hoess, R. H., DeGrado, W. F. 1990. *Science* 249:774–78
116. Weiss, M. A., Ellenberger, T., Wobbe, C. R., Lee, J. P., Harrison, S. C., et al. 1990. *Nature* 347:575–78
117. Patel, L., Abate, C., Curran, T. 1990. *Nature* 347:572–74
118. Oakley, M. G., Dervan, P. B. 1990. *Science* 248:847–49
119. Vinson, C. R., Sigler, P. B., McKnight, S. L. 1989. *Science* 246:911–16
120. Nye, J. A., Graves, B. J. 1990. *Proc. Natl. Acad. Sci. USA* 87:3992–96
121. Rauscher, F. J. III, Cohen, D. R., Curran, T., Bos, T. J., Vogt, P. K., et al. 1988. *Science* 240:1010–16
122. Curran, T., Franza, B. R. J. 1988. *Cell* 55:395–97
123. Foulkes, N. S., Borrelli, E., Sassone-Corsi, P. 1991. *Cell* 64:739–49
124. Hai, T., Curran, T. 1991. *Proc. Natl. Acad. Sci. USA* 88:3720–24
125. Murre, C., McCaw, P. S., Vaessin, H., Caudy, M., Jan, L. Y., et al. 1989. *Cell* 58:537–44
126. Voronova, A., Baltimore, D. 1990. *Proc. Natl. Acad. Sci. USA* 87:4722–26
127. Prendergast, G. C., Ziff, E. B. 1989. *Nature* 341:392
128. Weintraub, H., Davis, R., Tapscott, S., Thayer, M., Krause, M., et al. 1991. *Science* 251:761–66
129. Benezra, R., Davis, R. L., Lockshon, D., Turner, D. L., Weintraub, H. 1990. *Cell* 61:49–59
130. Barinaga, M. 1991. *Science* 251:1176–77
131. Williams, T., Tjian, R. 1991. *Science* 251:1067–71
132. Breg, J. N., van Opheusden, J. H. J., Burgering, M. J. M., Boelens, R., Kaptein, R. 1990. *Nature* 346:586–89
133. Phillips, S. 1991. *Curr. Opin. Struc. Biol.* 1:89–98
134. Rafferty, J. B., Somers, W. S., Saint-Girons, I., Phillips, S. E. V. 1989. *Nature* 341:705–10
135. Vershon, A. K., Bowie, J. U., Karplus, T. M., Sauer, R. T. 1986. *Proteins* 1:302–11
136. Vershon, A. K., Liao, S-M., McClure, W. R., Sauer, R. T. 1987. *J. Mol. Biol.* 195:311–22
137. Brown, B. M., Bowie, J. U., Sauer, R. T. 1990. *Biochemistry* 29:11189–95
138. Youderian, P., Vershon, A., Bouvier, S., Sauer, R. T., Susskind, M. M. 1983. *Cell* 35:777–83
139. Knight, K. L., Sauer, R. T. 1989. *Proc. Natl. Acad. Sci. USA* 86:797–801

140. Knight, K. L., Sauer, R. T. 1989. *J. Biol. Chem.* 264:13706–10
141. Bowie, J. U., Sauer, R. T. 1990. *J. Mol. Biol.* 211:5–6
142. Yang, C-C., Nash, H. A. 1989. *Cell* 57:869–80
143. Tanaka, I., Appelt, K., Dijk, J., White, S. W., Wilson, K. S. 1984. *Nature* 310:374–81
144. White, S. W., Appelt, K., Wilson, K. S., Tanaka, I. 1989. *Proteins* 5:281–88
145. Pfeifer, K., Kim, K-S., Kogan, S., Guarente, L. 1989. *Cell* 56:291–301
146. Pan, T., Coleman, J. E. 1990. *Proc. Natl. Acad. Sci. USA* 87:2077–81
147. Green, L. M., Berg, J. M. 1989. *Proc. Natl. Acad. Sci. USA* 86:4047–51
148. Summers, M. F., South, T. L., Kim, B., Hare, D. R. 1990. *Biochemistry* 29:329–40
149. Joulin, V., Bories, D., Eleouet, J. F., Labastie, M-C., Chretien, S., et al. 1991. *EMBO J.* 10:1809–16
150. Tsai, S-F., Martin, D. I. K., Zon, L. I., D'Andrea, A. D., Wong, G. G., et al. 1989. *Nature* 339:446–51
151. Freemont, P. S., Hanson, I. M., Trowsdale, J. 1991. *Cell* 64:483–84
152. Freyd, G., Kim, S. K., Horvitz, H. R. 1990. *Nature* 344:876–79
153. Dameron, C. T., Winge, D. R., George, G. N., Sansone, M., Hu, S., et al. 1991. *Proc. Natl. Acad. Sci. USA* 88:6127–31
154. Karim, F. D., Urness, L. D., Thummel, C. S., Klemsz, M. J., McKercher, S. R., et al. 1991. *Genes Dev.* 4:1451–53
155. Travis, A., Amsterdam, A., Belanger, C., Grosschedl, R. 1991. *Genes Dev.* 5:880–94
156. Christ, C., Tye, B. 1991. *Genes Dev.* 5:751–63
157. Grotewold, E., Athma, P., Peterson, T. 1991. *Proc. Natl. Acad. Sci. USA* 88:4587–91
158. Bours, V., Villalobos, J., Burd, P. R., Kelly, K., Siebenlist, U. 1990. *Nature* 348:76–80
159. Ghosh, S., Gifford, A. M., Riviere, L. R., Tempst, P., Nolan, G. P., Baltimore, D. 1990. *Cell* 62:1019–29
160. Kieran, M., Blank, V., Logeat, F., Vandekerckhove, J., Lottspeich, F., et al. 1990. *Cell* 62:1007–18
161. Treisman, J., Harris, E., Desplan, C. 1991. *Genes Dev.* 5:594–604
162. Chalepakis, G., Fritsch, R., Fickenscher, H., Deutsch, U., Goulding, M., Gruss, P. 1991. *Cell* 66:873–84
163. Lai, E., Prezioso, V. R., Tao, W., Chen, W. S., Darnell, J. E. 1991. *Genes Dev.* 5:416–27

164. Weigel, D., Jackle, H. 1990. *Cell* 63:455–56
165. Burglin, F. 1991. *Cell* 66:11–12
166. Seeman, N. C., Rosenberg, J. M., Rich, A. 1976. *Proc. Natl. Acad. Sci. USA* 73:804–8
167. Pabo, C. O. 1984. *Specificity in protein-DNA interactions. Proc. Robert A. Welch Found. Conf. Chem. Res. XXVII, Stereospecificity in Chemistry and Biochemistry,* pp. 222–55
168. Dickerson, R. E., Drew, H. R. 1981. *J. Mol. Biol.* 149:761–86
169. Wemmer, D. E. 1991. *Curr. Opin. Struc. Biol.* 1:452–58
170. Shakked, Z. 1991. *Curr. Opin. Struc. Biol.* 1:446–51
171. Bowie, J. U., Luthy, R., Eisenberg, D. 1991. *Science* 253:164–70
172. Karplus, M., Petsko, G. A. 1990. *Nature* 347:631–39

Annu. Rev. Biochem. 61:1097–1129
Copyright © 1992 by Annual Reviews Inc. All rights reserved

PHEROMONE RESPONSE IN YEAST

Janet Kurjan

Department of Microbiology and Molecular Genetics, College of Medicine and College of Agriculture and Life Sciences, University of Vermont, Burlington, Vermont 05405

KEY WORDS: mating, G protein, α-factor response

CONTENTS

INTRODUCTION

Over the past several years, G proteins have been found to be highly conserved components of signalling systems. In mammalian systems, a vast array of receptors act through G proteins (1), and the large number of genes encoding the various G protein subunits suggests a degree of complexity that was previously unexpected (2–4). G protein genes have now been identified in a wide variety of eukaryotes, including plants (5), nematodes (6, 7),

1097

Drosophila (8–10), *Dictyostelium* (11, 12), and yeast (13–15), suggesting that all eukaryotes are likely to utilize G protein–mediated signalling systems.

The discovery of a G protein involved in response to pheromone in the yeast *Saccharomyces cerevisiae* provided an opportunity to study a G protein–mediated signalling pathway utilizing the powerful genetics available in this system. Many components involved in pheromone response were identified almost two decades ago in screens for sterile mutants (16, 17), but their roles had not been elucidated. In recent years, many of these genes have been shown to encode pheromone receptors, G protein subunits, and downstream components, including kinases and a transcription factor. In this system, therefore, it should be possible to elucidate the mechanisms and interactions of components of the entire signalling pathway.

There are a number of recent reviews on aspects of yeast mating, including pheromone production and response (18–23). In this review, I give an overview of these processes. I give a more detailed discussion of areas for which there are recent results or for which there are controversies, including receptor topology, G protein mutants, the Fus3 and Kss1 kinases, and desensitization.

STERILE MUTANTS

The isolation of sterile (*ste*) mutants was an important early step that has allowed extraordinary elucidation of components involved in mating (16, 17). Other than mutants defective at *MAT,* the regulatory locus that determines whether a cell is of **a**, α or **a**/α cell type (24), most of the sterile mutants show defects either in pheromone production or pheromone response, indicating the importance of these processes for mating. These mutants have allowed the extensive analysis of the pathways involved in these processes. In some cases, functional redundancy of genes, e.g. the structural genes for the pheromones themselves, prevented their isolation as sterile mutants because mutations in two genes are necessary to cause sterility. A variety of other approaches have allowed the identification of such genes.

PHEROMONES

The existence of the yeast pheromones was initially determined by the ability of vegetatively grown α cells to induce morphological changes in **a** cells without cell-cell contact, suggesting that α cells secrete one or more pheromones (Figure 1) (25). Cell culture supernatants from both **a** and α cells were found to induce cell-cycle arrest and agglutination as well as these morphological changes (26–31). It was unclear whether single or multiple factors were responsible for the multiple responses.

α-Factor

The activity secreted by α cells, called α-factor, was the first to be characterized (27–30). Separate groups investigating agglutination and the cell cycle and morphological responses purified the activating factor and in both cases found it to be the identical 13-amino-acid peptide (32, 33). The ability of synthetic α-factor to elicit the same responses indicates that this peptide alone is sufficient for induction of all of the responses in **a** cells (34, 35).

Two genes, *MFα1* and *MFα2*, encode precursors of 165 and 120 amino acids, respectively, with tandem α-factor or α-factor-like repeats (36, 37). Null mutations in both genes in α cells result in undetectable α-factor production and a mating defect, indicating that there are no other functional α-factor genes (38). The α-factor precursors are translocated into the classical secretory pathway, where they are glycosylated and processed to produce mature α-factor. The steps involved in processing of the *MFα1* precursor, which produces the majority of α-factor, have been extensively characterized, including the identification of genes involved in the proteolytic processing steps (reviewed in 22, 23).

a-Factor

Determination of the molecular nature of a-factor was impeded by its hydrophobicity, which has since been shown to result from lipid modification. Recent characterization has determined that it consists of two 12-amino-acid peptides that differ by a single amino acid substitution; both peptides have a C-terminal cysteine residue that is modified by farnesylation and carboxymethylation (39, 40). Two genes, *MFA1* and *MFA2*, encode precursors of 34 and 36 amino acids, respectively, each of which contains a single a-factor sequence that corresponds to one of the two peptide sequences (41). The precursors do not show features common to most precursors of secreted proteins, and *sec* mutants are not defective in a-factor secretion (42). These results suggest that a-factor is secreted by a novel mechanism that does not involve the classical secretory pathway. The a-factor precursors appear to be synthesized in the cytoplasm, where they undergo proteolytic processing and modification of the cysteine residue (43, 44). The *STE6* gene, which is essential for a-factor production, encodes a protein with interesting and suggestive features similar to proteins with multiple membrane-spanning domains, including the mammalian multidrug resistance protein Mdr, that are involved in ATP-dependent pumping of compounds out of cells (42, 45). This discovery led to the idea that farnesylation of a-factor results in its localization to the cytoplasmic face of the plasma membrane where it can interact with the *STE6* product, which then acts to transport it across the plasma membrane. It will be interesting to see whether this novel form of secretion of lipopeptides also occurs in higher eukaryotes.

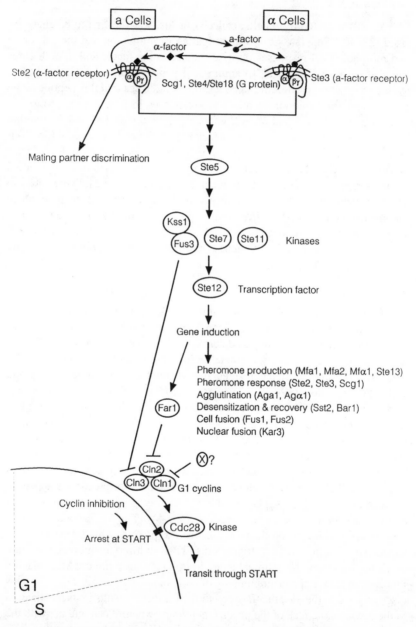

Figure 1 Yeast pheromone response. **a** and α cells secrete peptide pheromones, called **a**- and α-factor, respectively, which bind to receptors on the opposite mating type (Ste3 and Ste2, respectively). A G protein composed of α (Scg1), β (Ste4), and γ (Ste18) subunits is proposed to interact with both receptors. Downstream components necessary to activate the pathway include

PHEROMONE RESPONSE

The initially observed responses to pheromone included arrest of cells as unbudded cells in the G1 phase of the cell cycle, morphological changes called shmoo formation, and induction of agglutinability (26–30). As genes involved in various aspects of mating were isolated, it became clear that many of them are induced by pheromone. These genes play roles in many aspects of pheromone response and mating, including increased pheromone production and response itself (46–50), the ability to recover from pheromone-induced arrest (51, 52, 56), cell fusion (53, 54), and nuclear fusion (55). Many of the original sterile mutants identified components of the pheromone response pathway. In recent years, other components have been identified by other genetic and molecular approaches. Whereas sterile mutations in components of the pheromone response pathway result in an inability to activate the pathway (16, 17), null mutations in some of the components lead to the opposite effect, constitutive activation of the pathway, which results in a growth defect due to arrest in G1, and morphological alterations (13, 57, 58). For null mutants, these opposing phenotypes provide information as to whether the wild-type component plays a positive or a negative role in the pathway. If elimination of a component leads to constitutive activation, this component must play a negative role, i.e. it must act to keep the pathway off in the absence of pheromone. If elimination of a component leads to a defect in activation, this component must play a positive role, i.e. it must act to turn the pathway on in the presence of pheromone. Diploid (a/α) cells do not produce or respond to pheromone and do not mate; therefore the sterile mutations do not result in a phenotype in a/α diploids. Similarly, for a component of the pheromone response pathway with a negative role, the growth defect due to a null mutation is haploid-specific, i.e. a and α haploids show the growth defect, but a/α strains homozygous for the null mutation show normal growth. Such a haploid-specific effect has been useful in identifying negative components of the pathway (13, 57, 58).

Other genes have been identified that are likely to be modifiers of components of the pheromone response pathway (59, 60). Often such genes are also involved in modification of cellular functions not involved in pheromone response; therefore, mutations in such modifiers often lead to a pleiotropic

---

kinases and a transcription factor involved in induction of many aspects of mating and pheromone response. Cell cycle arrest is proposed to occur by inhibition of cell cycle components necessary to transverse START, the G1 cyclins, and Cdc28 kinase. Another aspect of mating is the ability of a cell to discriminate between potential mating partners based on their ability to produce pheromone. In a cells, the discrimination mechanism has been shown to act through the receptor Ste2 but can occur independently of the G protein–mediated pathway.

phenotype. Such pleiotropy can mask the role of a component in pheromone response. In some cases, differences in the phenotype in haploids versus diploids has suggested a role in pheromone response.

Pheromone Receptors

α-FACTOR RECEPTOR, Ste2   The α-factor receptor gene was identified among the original sterile mutants; the *ste2* mutation is **a**-specific, i.e. results in sterility only in **a** cells (16). Production of **a**-factor is normal, but response to α-factor is defective. These phenotypes indicate that *STE2* encodes a product that is specifically involved in pheromone response in **a** cells, consistent with a defect in the α-factor receptor. Labelled α-factor binds only to **a** cells, and binding is temperature sensitive in a *ste2-ts* mutant (61). In addition, the size of the α-factor binding protein is decreased in a *ste2* truncation mutant (62). These results suggest that Ste2 is the receptor, although it could also be one component of a multicomponent receptor. Expression of Ste2 in *Xenopus* oocytes allows binding of α-factor, indicating that Ste2 alone is sufficient for α-factor binding (63). Ste2 also determines specificity, because expression in *S. cerevisiae* of the *STE2* gene from the *S. kluyveri* allows response to *S. kluyveri* α-factor (64). Therefore, Ste2 is responsible for α-factor binding and for specificity, the critical functions of the α-factor receptor.

a-FACTOR RECEPTOR, Ste3   The absence of information on the chemical nature of **a**-factor until recently has impeded the types of biochemical experiments that allowed the identification of the α-factor receptor. However, genetic and molecular results strongly indicate that Ste3 is the **a**-factor receptor. *ste3* mutations result in the complementary set of phenotypes as found for *ste2* mutations; *ste3* mutants show α-specific sterility, produce normal levels of α-factor, but are defective in response to **a**-factor (16, 65). These phenotypes are consistent with a defect in the **a**-factor receptor. A fusion of Ste3 sequences to *LacZ* results in localization of β-galactosidase to the membrane fraction (65). The predicted topology of Ste3, based on DNA sequencing, is similar to those of Ste2 and other receptors (see below). Together these results suggest that Ste3 corresponds to the **a**-factor receptor.

RECEPTOR TOPOLOGY   Sequencing of the *STE2* and *STE3* genes indicated that both could encode proteins with seven putative membrane-spanning domains (Figures 1 and 2) (65–67). This pattern resembles G-protein linked receptors from many systems (68, 69). Based on other receptors in this class, it is predicted that the N-termini of the pheromone receptors are extracellular whereas the C-termini are cytoplasmic. Experiments on both receptors are consistent with these predicted topologies.

To test the predicted topology of Ste3, gene fusions to invertase were made (Figure 2) (70). Ste3 itself is unglycosylated, but the C-terminal invertase sequences should become glycosylated when translocated into the lumen of the secretory pathway. If the C-terminal invertase sequences are not translocated, they should remain unglycosylated. A fusion in the third proposed extracellular loop (amino acid 241) results in glycosylation of invertase. Two fusions to the proposed large cytoplasmic C-terminal domain (amino acids 287 and 467) do not allow glycosylation of invertase. These results indicate that the proposed third loop is extracellular and that the seventh hydrophobic region does act as a transmembrane domain, consistent with the proposed topology (Figure 2A).

An extensive analysis of Ste2 topology was reported recently (71). In this study, tripartite fusions were made with N-terminal domains of Ste2 fused to a Kex2 proteolytic processing site from the precursor of the peptide killer factor (Figure 2B). Kex2 processing sites in the killer and α-factor precursors are cleaved at the Golgi stage of the secretory pathway (22). This processing domain is in turn fused to the *bla* gene of *E. coli* to produce tripartite Ste2-Kex2 site-*bla* fusions (71). If the fusion results in exposure of the Kex2 site within the secretory pathway, it should be cleaved, allowing secretion of β-lactamase, whereas a cytoplasmic localization of the Kex2 processing site should not allow cleavage and secretion. Several fusions at positions predicted to be extracellular or close to the extracellular face of the membrane allow β-lactamase secretion, whereas two fusions predicted to be cytoplasmic do not allow secretion.

A further test involved insertion of a signal sequence between the Ste2 and Kex2 processing sequences, which would be predicted to reverse the orientation of the C-terminal processing and *bla* sequences (Figure 2B) (71). Insertion of the signal sequence at two predicted intracellular positions allows β-lactamase secretion, confirming the intracellular localization of the original fusion point. Insertion of the signal sequence at a predicted extracellular position gives a less clear result, a reduction but not elimination of β-lactamase secretion. For this insertion to reverse the orientation of the C-terminal processing and *bla* sequences, the signal sequence would need to transverse the membrane in the opposite orientation than usual; perhaps this process is inefficient, resulting in the mild effect of the signal sequence insertion. Overall, this study confirmed the extracellular localization of the three proposed extracellular loops and the intracellular localization of the proposed third cytoplasmic loop and the cytoplasmic C-terminus.

SPECIFICITY OF RESPONSE   The majority of the sterile mutations are nonspecific, i.e. they result in a mating defect in both **a** and α cells (16, 17). Most of the **a**- and α-specific mutants are defective in pheromone production

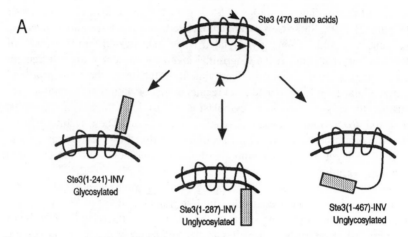

Figure 2   Receptor topology. The pheromone receptors have seven hydrophobic domains that are proposed to represent transmembrane domains, a pattern typical of receptors in G protein–mediated pathways. In-frame fusions to various positions (indicated by arrowheads) of Ste3 (*A*) and Ste2 (*B*) have been constructed to test this proposed topology. If the position of the fusion is extracellular (shown above the membrane), the C-terminus of the hybrid should be translocated into the lumen of the secretory pathway where N-glycosylation and proteolytic processing occur. If the position of the fusion is cytoplasmic (shown below the membrane), the C-terminus of the hybrid should remain unmodified. In (*A*), fusions of Ste3 to invertase (hatched rectangle) were constructed and assayed for glycosylation of invertase (70).

(42, 44, 45, 72, 73). The only **a**- and α-specific mutants that affect pheromone response are the receptor mutants, *ste2* and *ste3* (16, 17). The nonspecific steriles all affect pheromone response, suggesting that these genes encode components of the pathway that act in both **a** and α cells. The receptors are therefore the only identified components of the pheromone response pathway that are specific to one or the other mating type. In cells that do not express any other **a**- or α-specific products, expression of Ste2 is sufficient for response to α-factor and expression of Ste3 is sufficient for response to **a**-factor (74, 75). The pheromone receptors are therefore the only components of the pheromone response pathway that determine the specificity of the response, and the other components of the pathway are identical in **a** and α cells (Figure 1).

G Protein

G PROTEIN SUBUNITS   The *SCG1* gene product (also called *GPA1*) shows significant sequence similarity to the α subunits of G proteins (13, 14). *scg1* null mutants exhibit cell-cycle arrest and morphological changes (Figure 4*B*) (13, 57, 76); the haploid-specific nature of this effect suggested that the *scg1* null phenotype represents constitutive activation of the pheromone response

B

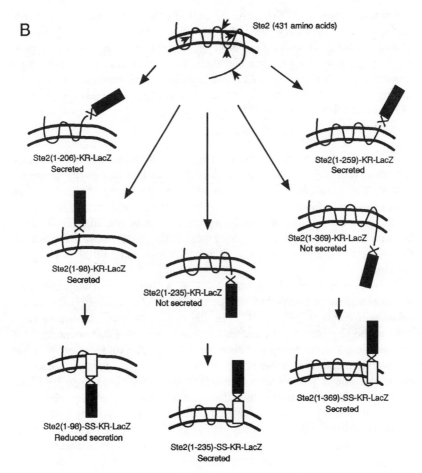

Ste2 (431 amino acids)

Ste2(1-206)-KR-LacZ
Secreted

Ste2(1-259)-KR-LacZ
Secreted

Ste2(1-98)-KR-LacZ
Secreted

Ste2(1-369)-KR-LacZ
Not secreted

Ste2(1-235)-KR-LacZ
Not secreted

Ste2(1-98)-SS-KR-LacZ
Reduced secretion

Ste2(1-369)-SS-KR-LacZ
Secreted

Ste2(1-235)-SS-KR-LacZ
Secreted

In (B), tripartite fusions of Ste2 to a proteolytic processing site (X) followed by the *E. coli bla* gene (solid rectangles) were made and secretion of β-lactamase was assayed (71). In several of these constructs, a signal sequence (open rectangles) was inserted before the proteolytic process- ing site, which would be predicted to transverse the membrane and reverse the orientation of the C-terminal domain.

pathway. More definitive proof was obtained by conditional expression of Scg1 (58, 77). Inhibition of Scg1 activity results in effects virtually identical to the effects of exposure of wild-type cells to pheromone, including cell-cycle arrest, morphological changes, and induction of pheromone-inducible genes. In addition, overexpression of *SCG1* inhibits pheromone response. These results indicate that the wild-type *SCG1* product acts to inhibit the pheromone response pathway, i.e. that it plays a negative role.

The *STE4* gene product shows sequence similarity to mammalian G protein

β subunits, and the *STE18* gene product shows more moderate sequence similarity to G protein γ subunits (15). Some rare *ste4* and *ste18* mutants show an unusual pheromone response phenotype (described below), suggesting that the two gene products are likely to act at the same step (78). Null mutations in either *ste4* or *ste18* result in defects in pheromone response and mating (Figure 4H,J). This phenotype indicates that the wild-type products are required for transmission of the response, i.e. that they play positive roles in the pathway.

MODEL FOR G PROTEIN FUNCTION   In two well-characterized mammalian responses mediated by G proteins, phototransduction and β-adrenergic response, the α subunits play positive roles by activating the downstream effector in the pathway (reviewed in 1). By binding α and thereby preventing its ability to activate the effector, $\beta\gamma$ plays a negative role in these systems. Because these roles are the opposite of those in the yeast pathway, it was clear that aspects of the mechanism of G protein action must differ in the yeast system.

A model was proposed that was based on the biochemical mechanism for the mammalian pathways but that accounted for the opposite roles of the subunits in yeast (Figure 3) (13). In this model, before exposure to pheromone, the G protein is in its heterotrimeric form with GDP bound to the α subunit (Scg1). After binding of pheromone to the receptor (Ste2 or Ste3), the G protein interacts with the receptor, which promotes guanine nucleotide exchange. The α subunit with bound GTP then dissociates from $\beta\gamma$ (Ste4/Ste18), and free $\beta\gamma$ activates the effector, which has not yet been identified. The α subunit is predicted to have an intrinsic GTPase activity that acts to return the system to the basal state. In this model, the activation of the effector by $\beta\gamma$ accounts for its positive role. The negative role of the α subunit occurs through binding to $\beta\gamma$, preventing its activity.

The only differences between the mammalian and yeast models come after dissociation of α-GTP from $\beta\gamma$. Whereas in the mammalian systems α-GTP activates the effector, in the yeast system $\beta\gamma$ activates the effector. Also, a desensitization step is predicted to act at the β subunit to promote its dissociation from the effector and allow the return to the basal state (79), as described below.

Epistasis tests, in which the phenotype of a double mutant containing mutations in two components of a pathway is compared to the single-mutant phenotypes, can provide information about the order of action of the two components in the pathway. The epistatic relationships for mutations in the receptors and the G protein subunits fit the model (Figure 3). Constitutive activation due to an *scg1* mutation in the α subunit occurs even in the absence of pheromone and therefore should be independent of the receptors, i.e. a null mutation in the downstream component, Scg1, should be epistatic to a null

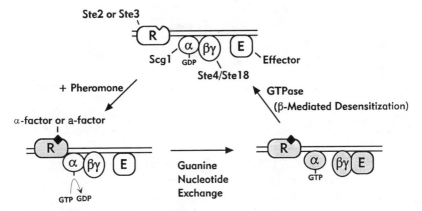

Figure 3   Model for G protein action. The G protein is proposed to exist predominantly in the heterotrimeric state with GDP bound in the absence of pheromone. Activation of the receptor by binding of pheromone allows association with the G protein, which results in guanine nucleotide exchange and dissociation of α-GTP from $\beta\gamma$. Free $\beta\gamma$ is proposed to activate an effector, which is currently unidentified. The system is proposed to return to the basal state by the action of a GTPase intrinsic to the α subunit and phosphorylation of the β subunit, which would promote dissociation of $\beta\gamma$ from the effector. Activated forms of the components are dotted.

mutation in the upstream component, the receptor. *scg1 ste2* double mutants show the *scg1* growth defect (13, 77, 80), indicating that the α subunit acts downstream of the receptor as would be predicted for any G protein–mediated pathway. Activation of the pathway by an *scg1* mutation should require the function of all downstream components necessary for activation of the pathway; therefore a null mutation in any downstream *ste* mutant should prevent the constitutive activation resulting from an *scg1* mutation. Combination of an *scg1* mutation with a *ste4* or *ste18* mutation eliminates the *scg1* growth defect and results in a defect in activation of the pathway, i.e. the sterile phenotype (Figure 4*K,L*) (15). Therefore, both *ste4* and *ste18* are epistatic to *scg1*, indicating that $\beta\gamma$ acts downstream of α.

The results of overexpression of the α and β subunits are also consistent with the proposed model (Figure 3). Overexpression of *SCG1* inhibits pheromone response (13, 76), whereas overexpression of *STE4* constitutively activates pheromone response (81–83). These effects are consistent with the positive and negative roles of *STE4* and *SCG1*, respectively. Constitutive activation resulting from Ste4 overexpression requires Ste18. The role of Ste18 may be only to localize Ste4 to its site of action, or Ste4 and Ste18 may act together in activating the effector. Overexpression of *SCG1* prevents the constitutive activation resulting from overexpression of *STE4*, consistent with the negative action of *SCG1* on *STE4* activity as proposed in the model. As well as providing support for the model, these results indicate the importance of the proper stoichiometry of these components of the pathway.

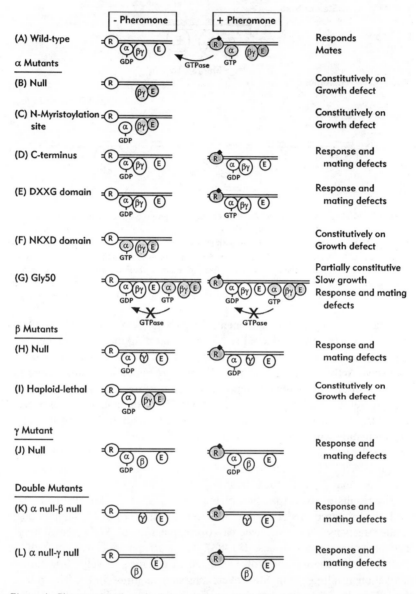

Figure 4 Phenotypes and mechanistic models for G protein mutants. The phenotypes of G protein mutants (shown at the right) are interpreted with respect to the model for G protein action (Figure 3). Diamond, **a**- or α-factor; R, **a**-factor receptor Ste3 or α-factor receptor Ste2; α, Scg1; β, Ste4; γ, Ste18; E, effector. Active forms of the proteins are shaded. (*A*) In the wild-type strain, the G protein is heterotrimeric with bound GDP in the absence of pheromone. In the presence of pheromone, guanine nucleotide exchange occurs (not shown), α-GTP dissociates from $\beta\gamma$, and

G PROTEIN MUTANTS   Each of the G protein subunits is proposed to have multiple functions (Figure 3). The α subunit must interact with the pheromone receptors as well as with the $\beta\gamma$ complex. The β and γ subunits must interact with one another as well as with the α subunit and with the effector. The α subunit is predicted to have guanine nucleotide–binding and GTPase activities. Various modifications of the G protein may affect membrane localization and desensitization, as described below. Given this wide variety of functions, it is not surprising that mutations in the G protein subunits result in quite a variety of phenotypes and that these phenotypes are sometimes difficult to interpret. Although it will be necessary to use biochemical approaches to test the current model for the action of the G protein and to determine the biochemical effect of mutations, it is useful to try to interpret mutant phenotypes with respect to the model (Figure 4).

←——

$\beta\gamma$ activates the effector. The GTPase is necessary to return the system to the basal state. (*B*) In the α null mutant, $\beta\gamma$ is free and activates the effector even in the absence of pheromone. (*C*) The mutation of the N-myristoylation consensus sequence (Figure 5) results in the null phenotype, indicating that $\beta\gamma$ is active in the absence of pheromone. This effect could result from a defect in proper localization of α or in its inability to interact with $\beta\gamma$. (*D*) Several mutations near the C-terminus of α (Figure 5) result in defects in pheromone response and mating. This effect is proposed to be due to an inability of the G protein to interact with the pheromone receptors. (*E*) Substitutions at amino acid 322 (Figure 5) result in defects in pheromone response and mating, indicating that $\beta\gamma$ is not active even in the presence of pheromone. Based on the biochemical effects of the analogous α_s mutation, it is proposed that guanine nucleotide exchange occurs upon exposure to pheromone but the conformational change does not occur, resulting in defects in dissociation of α-GTP from $\beta\gamma$ and in activation of the effector. (*F*) Asn to Lys (amino acid 388) and Asp to Ala (amino acid 391) substitutions (Figure 5) result in the null phenotype, indicating that $\beta\gamma$ is active in the absence of pheromone. This phenotype is proposed to result from an increased dissociation rate for guanine nucleotide, which would shift the equilibrium towards to GTP-bound form due to the higher level of GTP than GDP in the cell. (*G*) Substitutions at amino acid 50 (Figure 5) result in several phenotypes. Phenotypes that indicate increased activity of the pathway even in the absence of pheromone are proposed to result from a GTPase defect. Because the growth defect is not as severe as for null mutants, only a portion of the G protein is proposed to be in the activated form. Alternative explanations for the reduced mating and altered response to pheromone have been proposed as described in the text. (*H*) Null mutations in β result in defects in pheromone response and mating due to an inability to activate the effector. (*I*) A dominant mutation in β results in activation of the pathway in the absence of pheromone. This effect may result from an inability of $\beta\gamma$ to interact with α, therefore leaving it free to activate the effector. (*J*) Null mutations in γ result in defects in response to pheromone and mating due to an inability to activate the effector. These mutations are likely to result in a defect in localization of β. (*K* and *L*) Double mutants with a null mutation in either β or γ as well as a null mutation in α show defects in pheromone response and mating, i.e. the β and γ mutations eliminate or are epistatic to the effect of the α mutation. Because a null mutation in α acts to constitutively activate the pathway, this result indicates that β and γ mutations eliminate the activation of the pathway at a step downstream of α.

Dominant mutants Dominant (or partially dominant) mutations have been isolated in the α and β subunits. These mutations have the opposite effect as the null mutations in the respective genes. The dominant β mutation results in growth and morphological defects in **a** and in α cells but not in **a**/α cells, and therefore has been called a haploid-lethal mutation (80). This constitutive activation indicates that the positive role of $\beta\gamma$ in activation of the pathway has not been eliminated by this mutation. A defect that can explain this phenotype is an inability of $\beta\gamma$ to interact with α; therefore free $\beta\gamma$ would be present even in the absence of pheromone, leading to constitutive activation (Figure 4*I*). Such a mutation would be predicted to be dominant, because even in the presence of wild-type $\beta\gamma$, mutant $\beta\gamma$ would be unable to interact with α and would thus be able to activate the pathway.

The partially dominant α mutations result in defects in pheromone response and mating (Figure 4*D,E*) (84, 85). Based on the model, such a mutant protein would be predicted to be defective in dissociation from $\beta\gamma$; the resulting absence of free $\beta\gamma$ would prevent activation of the pathway. An inability of α to dissociate from $\beta\gamma$ could result from several different mechanistic defects. Dissociation from $\beta\gamma$ is normally preceded by a series of events: an interaction with the activated receptor, guanine nucleotide exchange, and an associated conformational change (86–88). Mutant α subunits defective in any of these processes would be unable to dissociate from $\beta\gamma$, resulting in defects in pheromone response and mating. In a strain containing equivalent levels of wild-type and mutant α, a portion of $\beta\gamma$ would be permanently associated with mutant α and therefore unable to activate the pathway. The remainder of $\beta\gamma$ would be associated with wild-type α; this portion would dissociate from α upon activation of the receptor and activate the pathway. α mutants of this type would be partially dominant because dissociation of only a portion of $\beta\gamma$ should allow activation of the pathway but to a lesser extent than in wild-type cells.

Modification and localization mutants The γ subunit gene *STE18* terminates with a CAAX box sequence (C=cysteine, A=an aliphatic amino acid, X=any uncharged amino acid) (15, 89). Proteins encoded by other genes terminating in a CAAX box, such as the **a**-factor genes and *ras*, undergo proteolytic processing of the AAX residues and isoprenylation and carboxymethylation of the cysteine residue (43, 89, 90). These C-terminal modifications are implicated in membrane localization. The *STE18* CAAX box therefore suggests that localization of $\beta\gamma$ to the membrane involves similar processing and modification events.

One set of β and γ mutants is thought to affect membrane localization of $\beta\gamma$ or the interaction between β and γ (91; M. Whiteway, personal communication). These mutants show response to low levels but not high levels of pheromone (78). Some of these γ mutants have alterations in the CAAX box that would be predicted to prevent modification. Mutations in *DPR1,* which has been implicated in the isoprenylation of ras, **a**-factor, and γ (43), result in the same phenotype (M. Whiteway, personal communication), suggesting that this phenotype results from a defect in modification and a resulting defect in membrane localization of γ. If localization of β to the membrane occurs only through its interaction with γ, mutations that prevent the β-γ interaction would have the same phenotype as mutants defective in membrane localization of γ; therefore, mutants with this unusual phenotype in β and at positions other than the CAAX box of γ are proposed to be defective in the interaction between β and γ.

Modification of mammalian α subunits by myristoylation has been implicated in membrane localization (89). The *SCG1* α subunit has a consensus sequence for N-myristoylation and has been shown to be myristoylated (91a). The *CDC72* gene is identical to the previously identified *NMT1* gene, which encodes the N-myristoyl transferase responsible for this modification (91a, 92). Temperature-sensitive *cdc72* mutants arrest at the same point as pheromone-arrested cells, and an N-terminal Gly-to-Ala mutation, which eliminates the N-myristoylation site in *SCG1,* results in a null phenotype (Figure 4C, Figure 5) (91a). These results indicate that this modification is necessary for α function, possibly by mediating its localization to the membrane or its interaction with βγ.

Receptor interaction mutants In mammalian systems, the C-termini of G protein α subunits have been implicated in receptor interactions (93–96), and a single amino acid substitution near the α$_s$ C-terminus eliminates its interaction with the β-adrenergic receptor (86, 97, 98). Several mutations near the Scg1 C-terminus result in decreased or undetectable mating, a phenotype that would be predicted for mutations that reduce or eliminate the receptor interactions (Figures 4D, 5) (85). A Lys-to-Pro substitution five amino acids from the C-terminus results in severe defects in pheromone response and mating in **a** cells but only reduces mating twofold in α cells. A mutation analogous to the α$_s$ mutation, a Lys-to-Pro substitution six amino acids from the C-terminus, has severe effects in both mating types, but the effect is somewhat greater in α than in **a** cells. These results suggest that particular C-terminal mutations can have a greater effect on the interaction with one of the pheromone receptors than with the other. The two receptors are predicted to have similar topologies, as discussed above, but do not show sequence similarity (65–67). Therefore, the specificity for the interactions with the two receptors might differ. The differential phenotypes of the C-terminal α mutants are consistent with such a difference in specificity.

Guanine nucleotide–binding domain mutants   Although binding of guanine nucleotides to Scg1 has not been shown directly, the increased affinity of α-factor for its receptor, Ste2, upon addition of nonhydrolyzable GTP analogs has been observed (99). This change requires all three subunits of the G protein. These results suggest that Scg1 is a guanine nucleotide–binding protein.

Two mutations in the conserved NKXD domain of Scg1 (Figure 5) result in a phenotype similar to the null phenotype, growth and morphological defects that represent constitutive activation of the pathway (84). Similar mutations in ras result in a transforming phenotype, which should result from a shift in the equilibrium towards the GTP-bound form of ras (100). The biochemical effect of these ras mutations is an increased dissociation rate of guanine nucleotides from the ras protein. Due to the much higher concentration of GTP than GDP within the cell (101, 102), an increased dissociation rate is predicted to result in a shift in the equilibrium such that a larger proportion of ras has GTP bound. A similar equilibrium shift for Scg1 should result in dissociation from βγ even in the absence of pheromone, resulting in the constitutive activation seen for these mutants (Figure 4F).

The glycine residue of the conserved DXXG domain of guanine nucleotide–binding proteins has been proposed to act as a hinge involved in the conformational change that accompanies guanine nucleotide exchange (88, 103–105). A Gly-to-Ala substitution in rat α_s eliminates its ability to activate the effector adenylate cyclase (87, 106, 107). This mutant α_s protein

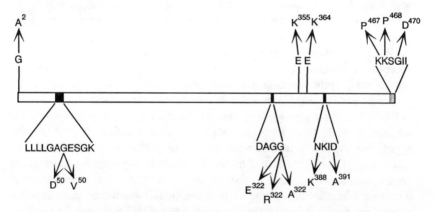

Figure 5 Scg1 mutations. The Scg1 open reading frame is diagrammed as an open rectangle. The three domains that are conserved across families of guanine nucleotide–binding proteins and that form the guanine nucleotide–binding domain are indicated in black and the C-terminal domain is indicated by hatching. Amino acid changes in conserved domains are shown below and changes in nonconserved regions are shown above the rectangle.

does undergo guanine nucleotide exchange, but does not undergo the conformational change, resulting in defects in dissociation from $\beta\gamma$ and in activation of adenylate cyclase. The analogous substitution in Scg1 results in defects in pheromone response and mating (Figure 5) (84). Two other substitutions at this position also resulted in pheromone response defects (108); their effect on mating was not reported. The phenotype of these Scg1 mutants is consistent with a defect in the conformational change necessary for dissociation from $\beta\gamma$ as seen for the analogous α_s mutant (Figure 4E).

Mutations at the Gly12 residue of mammalian ras result in reduced GTPase activity and a transforming phenotype (100), and the analogous mutation in yeast Ras has a similar effect (109). The analogous mutation in α_s shows the predicted GTPase defect but does not show the predicted constitutive activation of adenylate cyclase when tested in vivo, although such an effect is seen in vitro (110, 111). These results suggest that the effect of this mutation may be more complicated in Gα subunits than in ras.

Mutations at the analogous position of Scg1 (amino acid 50; Figure 5) have complex effects and are somewhat difficult to interpret (84, 108, 112). Also, some of the phenotypes differ depending on whether the mutant gene is expressed from a plasmid (108, 112) or as a genomic replacement (84). A GTPase defect in Scg1 would be predicted to result in constitutive activation of the pheromone response pathway by shifting the equilibrium towards the GTP-bound form. When examined on a plasmid, such an effect is not seen (108, 112). When examined as a genomic replacement, however, phenotypes predicted for constitutive activation are observed, including a growth defect and high basal expression of a pheromone-inducible reporter gene (84). The growth defect is not as severe as found for the null or NKXD domain mutants, described above, suggesting that some Scg1 must remain bound to $\beta\gamma$, i.e. that the equilibrium is not shifted completely towards the GTP-bound form (Figure 4G).

Both plasmid-borne and genomic forms of the mutant show greatly decreased mating (84, 112). The mutant with the genomic replacement shows barely detectable induction of the pheromone-inducible reporter gene above its high basal expression level. In a plate assay of pheromone response, a growth pattern is seen that has alternatively been interpreted as increased recovery from pheromone arrest or an inability to mount a sustained response to pheromone (108, 112–114). One interpretation of these results has been that the GTP-bound form of Scg1 turns on desensitization, resulting in reduced mating, as discussed below. My colleagues and I have proposed an alternative explanation, that in addition to a shift in the equilibrium towards the GTP-bound form of Scg1, the portion of Scg1 that remains GDP bound is activated inefficiently upon exposure to pheromone. The latter explanation is consistent with in vivo results for the analogous α_s mutation, for which the predicted decrease in GTPase activity was seen but the predicted increase in adenylate cyclase activity was not seen (110). Any interpretation of this complex phenotype can only be speculative without additional information.

Two additional mutations in *SCG1* result in a similar phenotype as the mutations at amino acid 50 (108). The positions of these mutations (amino acids 355 and 364; Figure 5) are conserved in Gα subunits but not in other families of guanine nucleotide–binding proteins. Because these mutations are not in the highly conserved domains that form the guanine nucleotide–binding pocket (104, 105), the region of these mutations is likely to play a role specific to Gα subunits.

COMPONENTS THAT ACT AT THE LEVEL OF THE G PROTEIN   Several components that act at the level of the G protein have been identified. Mutations in these components result in pleiotropic phenotypes, indicating that they are not specific for the pheromone response pathway.

A *dpr1/ram1* mutant was isolated by suppression of the growth defect of *scg1* mutants (115). This result indicated that Dpr1 modification is necessary for the activity of a component of the pathway that plays a positive role downstream of the α subunit. The role of Dpr1 in farnesylation and localization of γ as well as ras and **a**-factor can account for this effect (116).

cdc72 haploids arrest at the α-factor arrest point, whereas **a**/α diploids homozygous for *cdc72* also show a growth defect, but arrest does not occur at a particular stage of the cell cycle (58). This effect differs from mutants that show constitutive activation of the pheromone response pathway due to mutations in either α or β, in which no effect is seen in diploids (13, 57, 80). The *cdc72* phenotype could be explained by a gene necessary for an essential modification of a component of the pheromone response pathway, explaining the phenotype in haploid cells, but also necessary for the activity of some other essential cellular component that is expressed in diploids. The finding that *CDC72* encodes N-myristoyl transferase and that modification of Scg1 by this enzyme is necessary for its function explains the haploid phenotype (91a). The component(s) modified by Cdc72 in diploids is unknown.

Two other mutants, *cdc36* and *cdc39,* show a similar phenotype as the *cdc72* mutant, cell-cycle arrest at the pheromone arrest point in haploids, and non-cell-cycle-specific arrest in diploids (59, 60, 117). The haploid growth defect is not suppressed by receptor mutations but is suppressed by null mutations in β, indicating that Cdc36 and Cdc39 act between the receptor and the β subunit of the G protein. Overexpression of α was reported to suppress the haploid growth arrest, suggesting that Cdc36 and Cdc39 might both act at the level of the α subunit (60). However, one of the C-terminal *SCG1* mutations that results in sterility (85) suppresses the haploid growth arrest of *cdc39* but not *cdc36*, suggesting that Cdc36 and Cdc39 act at different levels (59). An earlier result led to a different prediction for the level of action of

Cdc36 and Cdc39; *cdc36* and *cdc39* mutations were shown to suppress the sterility of temperature-sensitive *ste4* and *ste5* mutants but not of sterile mutants further downstream in the pathway (118), suggesting that Cdc36 and Cdc39 act just downstream of the G protein in the pathway (Figure 1). Residual activity of the mutant Ste4 and Ste5 proteins in the latter study may have led to an incorrect determination of the order of action of these components.

The differences observed in these studies stress that although epistasis analysis is a powerful tool, it must be interpreted carefully. In this particular set of experiments, different results were obtained between null and conditional mutants and between a dominant inhibitory mutation and overexpression of the inhibitory component. However, the epistasis experiments certainly implicate Cdc36 and Cdc39 as playing a role in G protein function, which should help to direct experiments aimed at determining their mechanism of action.

Downstream Components

A number of genes that play roles in mating, many of which were identified among the original nonspecific sterile mutants, encode components of the pheromone response pathway. Null mutations in most of these genes suppress the growth defects of *SCG1* and *STE4* mutations that result in constitutive activation of the pathway, suggesting that they act downstream of the G protein (Figure 1) (77, 80).

Ste5   The ability of *ste5* mutations to suppress the growth defects of the α and β mutations that lead to growth arrest suggests that Ste5 acts downstream of the G protein, although these results do not eliminate the possibility that Ste5 acts at the same level as βγ (77, 80). No additional information on the mechanism of action of *STE5* is available.

KINASES   Several genes implicated in the pheromone response pathway (*STE7*, *STE11*, *FUS3*, and *KSS1*) encode proteins with sequence similarity to kinases (113, 119–121). For Ste11, kinase activity has been demonstrated (121). Null mutations in either *ste7* or *ste11* result in defects in pheromone response and mating, indicating that both kinases are necessary for mating, and thus suggesting that the two kinases have independent substrates that must be phosphorylated to activate the pathway.

The Kss1 and Fus3 kinases show a high degree of similarity to one another and have overlapping functions (113, 120). In the original report on *FUS3*, the *fus3* null mutant was reported to be sterile (120), but a more recent analysis has shown that the original strain also contains a *kss1* mutation (135). Examination of strains with single null mutations indicate that *fus3* mutations reduce but do not eliminate pheromone response and mating, and *kss1* mutations have little or no effect on pheromone response and mating. The double *fus3 kss1* mutants, however, are defective in pheromone response and mating. These results suggest that the two kinases have a common substrate(s) that must be phosphorylated to activate the response pathway. Other phenotypes indicate that these kinases have independent functions in cell-cycle arrest and desensitization, as described below.

One potential Ste11 substrate has been identified, although its role is not known (121), and no other in vivo substrates for the four kinases have been identified. Phosphorylation of Ste12 (described below), which acts downstream of the kinases, is necessary for activity (122), making it a likely substrate for one or more of the kinases. Several components upstream of Ste12 are also phosphorylated, including the kinases themselves (79, 121). Potentially, therefore, phosphorylation by the kinases could act at multiple levels, i.e. could increase the activity of one another and of both upstream and downstream components. Such a phosphorylation pattern could complicate analysis of the order of action of the kinases by epistasis analysis.

Ste12 TRANSCRIPTION FACTOR   Several results indicate that Ste12 is a transcription factor that induces genes in response to pheromone. Overexpression of Ste12 results in gene induction, morphological changes, and a low level of mating even in *ste7* and *ste11* strains (123). These results indicate that Ste12 acts downstream of Ste7 and Ste11. Ste12 binds to the consensus sequence that has been determined to be responsible for gene induction by pheromone (124, 125). Fusion of *STE12* sequences to the DNA-binding domain of the Gal4 transcription factor allows pheromone-inducible transcription via binding to the Gal4 UAS (upstream activating sequence) (122). The domain of Ste12 involved in this induction correlates well with the region that is phosphorylated in response to pheromone. Exposure to pheromone does not affect the DNA-binding activity of Ste12; therefore, some other Ste12 activity, such as gene activation, is likely to be altered by exposure to pheromone and the resulting phosphorylation of Ste12.

A large number of genes have been shown to be induced by pheromone. In some cases, there is basal expression of the genes and a fewfold increase in expression in the presence of pheromone. In other cases, basal expression is very low or undetectable and induction results in high-level expression. The pheromone and pheromone receptor genes, as well as the *STE12* gene itself, are induced by pheromone (46–50, 122). Other components induced by pheromone are involved in cell adhesion (126–128), cell-cycle arrest (129), cell and nuclear fusion (53–55), and desensitization and recovery of cells from arrest (51, 52). Induction of genes by pheromone therefore promotes many of the steps involved in pheromone response and mating.

CELL-CYCLE ARREST   The discovery of cyclin mutants that are resistant to cell-cycle arrest by pheromone led to the idea that pheromone arrest occurs by inhibition of components of the vegetative cell cycle that are necessary to transverse START (Figure 1) (130). The G1 cyclins are necessary for the activity of the Cdc28 kinase, whose activity is necessary to pass the START step in the G1 phase of the cell cycle (117, 131–133). A *cdc28* mutation results in arrest at this stage, which is the same point at which pheromone arrest occurs. All three genes that encode G1 cyclins (*CLN1, CLN2,* and *CLN3/DAF1/WHI1*) must be mutated to lead to cell-cycle arrest at START, indicating that any one of the three G1 cyclins is sufficient to transverse START, i.e. that they are functionally redundant (131, 132).

The cyclin mutations that cause resistance to pheromone arrest are predicted to result in increased cyclin stability (130, 134). Pheromone arrest was therefore proposed to occur by inactivation of the G1 cyclins, possibly by degradation, although it could alternatively occur at some other level, such as gene repression. To eliminate Cdc28 activity and give rise to arrest, all three cyclins would have to be inactivated (Figure 1) (131). These results therefore suggest that cell-cycle arrest by pheromone occurs by interfering with the activity of components that act during vegetative growth to allow cells to transverse the START step in the G1 phase of the cell cycle.

More specific information on the mechanism by which pheromone response leads to cell-cycle arrest has been elucidated recently by the isolation of mutants that separate gene induction and morphological changes from cell-cycle arrest. *far1* mutants do not undergo cell-cycle arrest in response to pheromone but do show pheromone-induced gene induction and morphological changes (129). The *FAR1* gene is itself inducible by pheromone. Based on these results, it was suggested that exposure to pheromone leads to Far1 production, and that Far1 acts to prevent transit through the G1 phase of the cell-cycle. A *far1 cln2* double mutant shows normal cell-cycle arrest in response to pheromone, whereas *far1 cln1* and *far1 cln3* double mutants show the cell-cycle arrest defect of *far1* single mutants. These results suggest that Far1 acts specifically through Cln2 (Figure 1). In a wild-type strain, all three cyclins are proposed to be inactivated upon exposure to pheromone, whereas in a *far1* strain Cln1 and Cln3 are inactivated by exposure to pheromone, but Cln2 remains active. This Cln2 activity is sufficient to activate Cdc28, thus allowing cell division to continue. In the absence of both Far1 and Cln2, none of the G1 cyclins is active after exposure to pheromone, resulting in a lack of Cdc28 activity and cell-cycle arrest. Far1 could inhibit Cln2 either by preventing its expression or by inactivating the protein.

far1 mutants show decreased mating as well as the defect in cell-cycle arrest (129). The mating defect does not result solely from the defect in cell-cycle arrest, because the *far1 cln2* double mutants arrest but still show the mating defect and *far1* truncation mutants have been isolated that also arrest but show the mating defect. These results suggest that Far1 plays a role in mating in addition to its role in cell-cycle arrest.

The *FUS3* gene plays a specific role in cell-cycle arrest as well as its role in gene induction described above (120, 135). Whereas *fus3 kss1* double mutants show defects in gene induction, cell-cycle arrest, and mating, *fus3 KSS1* mutants show a phenotype similar to that of *far1* mutants, approximately wild-type gene induction and morphological changes, a decrease but not complete defect in mating, and no cell-cycle arrest. Fus3 therefore plays a dual role in pheromone response—a role in gene induction, which can also be fulfilled by Kss1 due to their overlapping function, and a role in cell-cycle arrest, which is not shared by Kss1 (Figure 1). In addition, Fus3 is reported to play a role in vegetative growth. Like Far1, Fus3 is thought to act through the G1 cyclins. *fus3 cln3* double mutants are able to arrest, whereas *fus3 cln1* and *fus3 cln2* double mutants retain the *fus3* defect in cell-cycle arrest, and *fus3 cln1 cln2* triple mutants show only slight arrest in response to pheromone. Cln3 is therefore the main cyclin that retains activity in *fus3* cells and produces the defect in cell-cycle arrest upon exposure to pheromone (Figure 1).

Other results seem somewhat surprising in combination with this result. The transcription of the *CLN1* and *CLN2* genes is repressed upon exposure to pheromone (133), suggesting that elimination of Cln1 and Cln2 activity occurs by transcriptional repression. *fus3* mutants are defective in pheromone-induced repression of *CLN1* and *CLN2* RNAs (135). This result might suggest that Fus3 acts to induce cell-cycle arrest in response to α-factor by repressing the transcription of the *CLN1* and *CLN2* genes. However, the finding that the *fus3 cln3* double mutant arrests in response to pheromone indicates that there is not significant Cln1 or Cln2 activity in these cells even though their transcripts are present, and indicates that the major role of Fus3 in cell-cycle arrest is through Cln3. Some posttranscriptional inactivation of Cln1 and Cln2 must therefore occur upon pheromone exposure in *fus3 cln3* double mutants. For Cln2, this role may be played by Far1, suggesting that Far1 inhibition occurs posttranscriptionally (129). A currently unknown component has been postulated to act to inhibit Cln1.

The multiple responses to pheromone suggests that the pheromone response pathway must branch at some point to produce the different responses, in particular gene induction and cell-cycle arrest (Figure 1). The discovery of *FAR1* suggested that branch is at a very late step, with cell-cycle arrest occurring through gene induction (129). The role of Fus3 in cell-cycle arrest, however, may indicate a more complicated branching pattern (120, 135). Fus3, along with Kss1, acts in gene induction at a stage downstream of the G protein and upstream of the Ste12 transcription factor. The stage at which its role in cell-cycle arrest occurs is less clear. Whereas increased expression of Ste12 results in cell-cycle arrest in a wild-type strain, it does not result in cell-cycle arrest in a *fus3* mutant. This result indicates that the role of Fus3 in cell-cycle arrest must be either downstream of Ste12 or in combination with Ste12. One possible explanation for this result that retains a simple branched pathway is that there is a specific Fus3-dependent phosphorylation of Ste12 that is not necessary for induction of most genes but is necessary for induction of a particular gene(s) that encodes a component that inhibits Cln3.

MORPHOLOGICAL CHANGES   Another aspect of the response to pheromone are the morphological changes that occur upon exposure to high levels of pheromone (25, 26). The cells continue to grow and elongate to form pear-shaped structures, called shmoos. Secretion occurs through the extension or shmoo tip (136, 137), and products implicated in cell-cell interactions and fusion during mating also localize to the shmoo tip (53, 138). It is therefore thought that a pair of mating cells elongate towards one another, localize products involved in fusion to the shmoo tip, and fuse at the shmoo tips to produce the dumbbell-shaped zygotes. Because mutations in G protein sub-units and overexpression of Ste12 lead to morphological changes (13, 57, 58, 80, 123), it seems that the same pathway that leads to gene induction also acts in the morphological alterations. However, conditional *cdc28* mutants and *cln1 cln2 cln3* triple mutants show similar morphological changes when arrested by the restrictive condition (117, 131). Thus cell-cycle arrest at START without exposure to pheromone leads to a similar morphological phenotype, but does not lead to other responses to pheromone such as gene induction. Whether the similar morphological changes induced by these alternative branches of the pathway actually represent different events or whether there are multiple controls on morphological changes is not known.

Desensitization and Recovery

After a period of cell-cycle arrest due to exposure to pheromone, cells are able to recover and resume growth (139, 140). This recovery involves both degradation of pheromone and a cell-intrinsic process of desensitization. Desensitized cells can bind pheromone but do not activate the pathway upon pheromone binding (48). Desensitization therefore uncouples the pathway at some point. In mammalian G protein–mediated signal transduction pathways, receptor-mediated events are involved in desensitization (68). Upon activation of the phototransduction and β-adrenergic pathways, the C-terminal domains of rhodopsin and the β-adrenergic receptor become phosphorylated; arrestin or β-arrestin respectively can bind to the phosphorylated C-terminus and prevent binding of the G protein, thus uncoupling the pathway. Although the mechanism of desensitization in yeast has not been fully characterized, there is evidence for multiple mechanisms acting at different points along the pheromone response pathway. Some of these mechanisms may also occur in higher eukaryotes, whereas others may be specific to the yeast pathway.

PHEROMONE DEGRADATION   Cells of **a** mating type secrete an activity, called barrier, that inactivates α-factor (141, 142). This activity is induced by exposure to α-factor (51, 56). Mutations in the gene that encodes this activity (*BAR1/SST1*) result in increased sensitivity to α-factor and an increase in the lag before recovery occurs; this effect is **a**-specific, i.e. *sst1* mutations do not result in a phenotype in α cells (139, 143). *BAR1* encodes a pepsinlike protease that cleaves α-factor between the sixth and seventh amino acids (141, 142, 144). Thus, the barrier system provides a feedback mechanism in which exposure to α-factor induces an activity that degrades α-factor, thus inhibiting further response to pheromone and promoting recovery from pheromone arrest.

Until recently, no α-specific activity had been found that degrades **a**-factor. An extracellular cell-associated activity that inactivates **a**-factor has now been reported (145). This activity is induced by exposure to pheromone. The predicted phenotype for a mutation in the gene that encodes this activity would be α-specific supersensitivity to **a**-factor. Such a mutant (*ssl1*) has been isolated (146), but surprisingly, it does not result in a defect in the **a**-factor degradative activity, indicating that it does not identify the structural gene (145). The role of *SSL1* remains unknown.

RECEPTOR-MEDIATED DESENSITIZATION   Truncations of the C-terminal cytoplasmic domain of the α-factor receptor, Ste2, lead to increased sensitivity to pheromone, indicating that this domain is not required for the response itself but is likely to play a role in desensitization (147, 148). Like the mammalian G protein–linked receptors, this domain of Ste2 is phosphorylated on Ser and Thr residues, and phosphorylation increases upon exposure to pheromone (148). No arrestinlike analog that could bind this phosphorylated domain and uncouple the pathway has been identified.

G PROTEIN β SUBUNIT–MEDIATED DESENSITIZATION   In the well-characterized mammalian systems, the intrinsic GTPase of the α subunit returns the protein to the GDP-bound form, resulting in its dissociation from the effector and allowing its reassociation with free $\beta\gamma$. In the yeast system, $\beta\gamma$ is proposed to be bound to the effector after dissociation from α (Figure 3). Return to the basal heterotrimeric state therefore requires both a GTPase to return α to the GDP-bound form and dissociation of $\beta\gamma$ from the effector. After both of these events have occurred, free $\beta\gamma$ should bind to α-GDP. A desensitization step that causes $\beta\gamma$ to dissociate from the effector would therefore facilitate return to the basal heterotrimeric state. Interestingly, the yeast β gene, *STE4*, contains a domain not present in the mammalian β genes that is implicated in desensitization (79). A *STE4* mutant with this nonconserved domain deleted can function in activation of the pathway but is supersensitive to pheromone, a phenotype associated with desensitization mutants. In addition, β becomes phosphorylated upon exposure to pheromone, and the phosphorylation sites are within the conserved domain. These results suggest that response to pheromone induces phosphorylation of β, which causes it to dissociate from the effector and reassociate with α-GDP, thus promoting recovery from pheromone arrest.

DO THE G PROTEIN α SUBUNIT AND Sgv1 PLAY ROLES IN DESENSITIZATION?   A role for the α subunit in desensitization has been suggested based on some of the phenotypes of the $scg1^{Val50}$ mutant, described above (108, 112). Because the analogous mutations in ras and α_s result in GTPase defects, it was specifically proposed that the GTP-bound form of α promotes desensitization. One particular phenotype was critical in this interpretation. A plate pheromone response assay was interpreted as indicating increased recovery from pheromone-induced arrest. Dominance tests with the mutant gene on a plasmid suggested that this mutation is dominant (112). It therefore was concluded that α-GTP must promote recovery even in the presence of wild-type Scg1. However, when examined as a genomic replacement, this mutation is recessive, arguing against a specific role for Scg1-GTP (84). In addition, the plate assay is difficult to interpret. Whereas some authors have interpreted this particular phenotype as increased recovery (108, 112, 113), others have interpreted it as an attenuated response to pheromone or inhibition of pheromone response (76, 114). In this assay, many processes are occurring, including pheromone induction of cell-cycle arrest, induction of pheromone degradation, and desensitization. Even if the observed phenotype does represent increased recovery, the mechanism inducing this effect is unclear; e.g. it could result from increased induction of the barrier protease and the resulting degradation of α-factor. There is no direct evidence that Scg1-GTP is acting to promote desensitization from the current results. Given all of these complications, the current information is inadequate to conclude a role for the α subunit and in particular for the GTP-bound form of the α subunit in desensitization.

A new mutant (*sgv1*) was isolated recently that increases the sensitivity of the $scg1^{Val50}$ mutant to pheromone (149). The *SGV1* gene shows kinase homology. Based on the conclusion that $scg1^{Val50}$ promotes desensitization, the interpretation of the *sgv1* phenotype was that it is defective in promotion of recovery by $scg1^{Val50}$. Given the questions as to whether $scg1^{Val50}$ phenotypes represent promotion of desensitization, it seems best to try to interpret the *sgv1* phenotypes in a wild-type strain for evidence for a role in desensitization. The original *sgv1* allele causes cold-sensitivity (cs) and temperature-sensitivity (ts). This phenotype is seen in **a**/α *sgv1*/*sgv1* diploids as well as in haploids. At the permissive temperature, there is a high basal level of unbudded cells, indicating that there is a partial defect even at this temperature. An *sgv1* disruption mutation is lethal. These phenotypes suggest that Sgv1 plays a role in growth in both haploids and diploids, and therefore is not specific to pheromone response or recovery. A *CLN3* allele that is proposed to result in increased Cln3 stability and that promotes transit through START under conditions not seen in the wild-type (130) partially suppresses the cs and ts phenotypes of the original *sgv1* mutant (149). These results indicate that the role of Sgv1 in cell growth is likely to occur upstream of the cyclins.

Does Sgv1 play a role in desensitization to pheromone? The basal and induced expression levels of a pheromone-inducible reporter gene in the *sgv1* mutant are similar to the wild-type (149). Arrest by pheromone occurs at lower pheromone levels than in the wild-type. However, even in the absence of pheromone the *sgv1* cells are partially arrested at the permissive temperature as indicated by the high basal level of unbudded cells. Therefore, the greater sensitivity to pheromone may simply be a synergistic effect of pheromone action on partially arrested cells. The increased pheromone sensitivity of *scg1*^{Val50} *sgv1* may also reflect this synergistic effect.

Because pheromone response seems to occur through inhibition of the G1 cyclins and therefore of Cdc28 activity, it is likely that recovery from pheromone arrest occurs by resumption of G1 cyclin and Cdc28 activity. The results with the *sgv1* mutants (149) implicate Sgv1 as a new component that acts to promote transit through the G1 phase of the cell cycle, although its exact role remains unclear. Given the growth defects associated with *sgv1* mutants, it seems unlikely that Sgv1 specifically plays a role in desensitization. Sgv1 may play a role in the mechanism by which either pheromone response itself or recovery from pheromone arrest feeds into regulation of the G1 cyclins and Cdc28. It will be interesting to determine the mechanism by which Sgv1 acts in both vegetative growth and in pheromone response or recovery.

Sst2-MEDIATED DESENSITIZATION   The *SST2* gene was identified by the isolation of mutants that are supersensitive to pheromone (139). These mutants also show a defect in recovery from pheromone arrest. The effect of *sst2* mutations is similar in **a** and α cells. These phenotypes suggest that Sst2 is involved in desensitization. The *SST2* gene is expressed only in haploid cells and is induced by exposure to pheromone (52). This result suggests that exposure to pheromone increases the level of Sst2, which then acts on some component of the pathway to allow desensitization and recovery.

To analyze components that have been suggested to play a role in desensitization, double mutants with *sst2* and a supersensitive mutant in another component have been constructed. If Sst2 and the second component act independently, the double mutant should be more supersensitive than either single mutant, i.e. the mutations should be synergistic. If Sst2 acts on the second component, the phenotype should be similar to one of the single mutants, i.e. a mutation in the second component should be epistatic to the *sst2* mutation.

Ste2 truncations and *sst2* mutations are synergistic, indicating that Sst2 does not act through the receptor C-terminal domain (147, 148). This result is inconsistent with the attractive model that Sst2 might act as arrestin does by binding phosphorylated receptor and uncoupling the receptor from the G protein (52).

Phosphorylation of β (Ste4) in response to pheromone occurs in *sst2* mutants, indicating that the aspect of desensitization that occurs by β phosphorylation is independent of Sst2 (79). Also, double mutants with the internal β deletion that results in supersensitivity and an *sst2* mutation are inviable. This result was interpreted as indicating that the double mutant cannot even adapt to the low basal activity of the pheromone response pathway, therefore preventing growth. The synergistic effect of these two mutations indicates that Sst2 is not likely to act through β.

Although the results differ with *scg1*$^{\text{Val50}}$ expressed from a plasmid or as a genomic replacement, a double mutant with the genomic replacement of *scg1*$^{\text{Val50}}$ and an *sst2* mutation has the same phenotype as the single *scg1*$^{\text{Val50}}$ mutant, i.e. the *scg1*$^{\text{Val50}}$ mutation is epistatic to *sst2* (84). If Scg1 does play a role in desensitization, then Sst2 could act at the level of the α subunit.

Kss1-MEDIATED DESENSITIZATION   The *KSS1* gene was isolated by virtue of its ability when present on a high-copy plasmid to suppress the supersensitivity of *sst2* mutants (113). Based on a plate assay, this suppression was concluded to occur by promotion of recovery. Expression of *KSS1* under the control of a galactose reporter was also reported to promote recovery in a wild-type strain. It was therefore surprising that *KSS1* has been implicated as a component of the pheromone response pathway itself. As described above, the *KSS1* gene shows kinase homology, with the highest degree of similarity to *FUS3* (113, 120). Either *KSS1* or *FUS3* is necessary to activate the pheromone response pathway, i.e. double *kss1 fus3* mutants are defective in activation of the pheromone response pathway (135). The positive role for Kss1 suggested by this latter experiment seems to contradict its proposed role in turning off pheromone response. A hypothesis that reconciles these results is that phosphorylation of some substrate(s) necessary to activate the response pathway can be catalyzed by either Fus3 or Kss1, whereas phosphorylation of a Kss1-specific substrate promotes desensitization. An appealing candidate for the latter substrate(s) is the G protein β subunit (79).

ZYGOTE FORMATION

After response to pheromone, **a** and α cells can fuse with one another to form a zygote that produces **a**/α diploid buds. Response to pheromone induces components involved in many aspects of this process. Secretion occurs through the shmoo tip in cells responding to pheromone (136), which may act as a mechanism to localize components involved in pheromone response and cell fusion to the position at which cells will eventually fuse.

For the plasma membranes of two cells to fuse, the cell wall must be altered or degraded. Exposure to pheromone results in changes in the cell wall, many of which occur specifically at the shmoo tip (150–153). The *FUS1* and *FUS2* genes are induced by pheromone and play roles in the cell fusion process (53, 54). Fus1 is a membrane protein that localizes to the shmoo tip after pheromone induction and to the neck of the zygote after fusion (53, 154). Genetic results indicate that Fus1 and Fus2 are functionally redundant and therefore are likely to play similar roles. The mechanism by which Fus1 and Fus2 promote cell fusion remains to be determined.

After cells fuse to form the zygote, the two nuclei must fuse to form a single **a**/α nucleus. During pheromone response, the spindle pole body (SPB) orients towards the shmoo tip (155). Connection of microtubules between the two SPBs is likely to promote migration of the nuclei towards one another, and fusion to produce the single nucleus occurs at the SPBs. Efficient nuclear fusion is promoted by exposure to pheromone (156). The *KAR3* gene, which is required for SPB structure and function in vegetative cells, is also required for nuclear fusion within the zygote (55). The *KAR3* gene is induced by pheromone; therefore nuclear fusion may occur by increasing and altering the action of nuclear components also involved in vegetative cell division.

MATING PARTNER DISCRIMINATION

Recently, an aspect of the mating process has been characterized that has been called courtship (157, 158). In this process, a cell is able to discriminate among possible mating partners. Cells choose a mating partner that produces the highest level of pheromone, and as predicted, discrimination by **a** cells requires the α-factor receptor Ste2 (159). Surprisingly, this action of the receptor can occur independently of the G protein–mediated pathway (Figure 1). Discrimination is proposed to act in reorganization of the cytoskeleton and orientation of the nucleus and secretory pathway towards the mating partner. Intracellular components that mediate this novel action of the receptor have not been identified.

PERSPECTIVES

Over the past several years, an extensive amount of information has been learned about the processes involved in pheromone production and response in yeast. However, many interesting questions remain. The mechanism by which Ste6 mediates α-factor secretion needs to be elucidated. It is hoped that information gained in this system will facilitate determination of the mechanism of action of Mdr and other mammalian proteins with a similar structure. The determination of the effector in the pheromone response pathway, i.e. the component immediately downstream of βγ, has been surprisingly elusive. It is possible that one or more of the currently identified components of the pathway is actually the effector. Alternatively, the effector may be encoded by functionally redundant genes, or the phenotypes of effector mutants might be pleiotropic. Either of these possibilities could account for the difficulty in identifying the effector. The identification of multiple kinases that act in the pathway suggests that a complicated phosphorylation pattern may be involved in activation of the pathway and possibly also in desensitization. Identification of the critical substrates for each of the kinases is necessary to elucidate this aspect of the mechanism. Much work needs to be done to determine the number of steps at which desensitization acts and the mechanisms by which it occurs. The discovery of components that act to inhibit regulatory components of the START step in the G1 phase of the cell cycle has provided important information as to how exposure to pheromone induces cell-cycle arrest. However, more components involved in this process remain to be identified, and the mechanism by which the known components, Fus3 and Far1, inhibit the G1 cyclins has not been determined. The recent finding that the choice of a mating partner occurs through the pheromone receptors but can act independently of the G protein pathway is surprising and fascinating. These results indicate that there is a previously unexpected pathway that needs to be elucidated.

The results in this system as well as in other aspects of growth control in yeast have indicated the extensive degree of conservation of many fundamental cellular processes. The ability to use the powerful genetics available in yeast provides an important tool to study these processes. Yeast researchers hope that elucidation of new aspects of these processes will allow the discovery of similar aspects in higher systems that might be difficult to discover otherwise. Clearly not all aspects of particular systems will be conserved and even conserved aspects may not be identical. Examples of differences in characterized systems include the similar mechanism of action of yeast and mammalian ras, but different effectors on which ras acts, and the differential roles of the α versus βγ subunits of G proteins in the yeast mating system versus the best-characterized mammalian G protein–mediated systems. In studying these pathways, therefore, it is important to consider possible differences between systems in the interpretation of results. The variations between systems are as interesting as the extensive similarities.

ACKNOWLEDGMENTS

I thank Doug Johnson, Hans de Nobel, Rinji Akada, and Lorena Kallal for comments on the manuscript.

Literature Cited

1. Iyengar, R., Birnbaumer, L. 1990. In *G Proteins*, ed. R. Iyengar, L. Birnbaumer, pp. 1–14. San Diego:Academic
2. Strathmann, M., Wilkie, T. M., Simon, M. I. 1989. *Proc. Natl. Acad. Sci. USA* 86:7407–9
3. Gautam, N., Northup, J., Tamir, H., Simon, M. I. 1990. *Proc. Natl. Acad. Sci. USA* 87:7973–77
4. Levine, M. A., Smallwood, P. M., Moen, P. T. Jr., Helman, L. J., Ahn, T. G. 1990. *Proc. Natl. Acad. Sci. USA* 87:2329–33
5. Ma, H., Yanofsky, M. F., Meyerowitz, E. M. 1990. *Proc. Natl. Acad. Sci. USA* 87:3821–25
6. Fino-Silva, I., Plasterk, R. H. 1990. *J. Mol. Biol.* 215:483–87
7. van der Voorn, L., Gebbink, M., Plasterk, R. H., Ploegh, H. L. 1990. *J. Mol. Biol.* 213:17–26
8. Yarfitz, S., Provost, N. M., Hurley, J. B. 1988. *Proc. Natl. Acad. Sci. USA* 85:7134–38
9. Yoon, J., Shortridge, R. D., Bloomquist, B. T., Schneuwly, S., Perdew, M. H., Pak, W. L. 1989. *J. Biol. Chem.* 264:18536–43
10. Quan, F., Forte, M. A. 1990. *Mol. Cell. Biol.* 10:910–17
11. Kumagai, A., Pupillo, M., Gundersen, R., Miake-Lye, R., Devreotes, P. N., Firtel, R. A. 1989. *Cell* 57:265–75
12. Pupillo, M., Kumagai, A., Pitt, G. S., Firtel, R. A., Devreotes, P. N. 1989. *Proc. Natl. Acad. Sci. USA* 86:4892–96
13. Dietzel, C., Kurjan, J. 1987. *Cell* 50:1001–10
14. Nakafuku, M., Itoh, H., Nakamura, S., Kaziro, Y. 1987. *Proc. Natl. Acad. Sci. USA* 84:2140–44
15. Whiteway, M., Hougan, L., Dignard, D., Thomas, D. Y., Bell, L., et al. 1989. *Cell* 56:467–77
16. MacKay, V. L., Manney, T. R. 1974. *Genetics* 76:255–71
17. Hartwell, L. H. 1980. *J. Cell Biol.* 85:811–22
18. Cross, F., Hartwell, L. H., Jackson, C., Konopka, J. B. 1988. *Annu. Rev. Cell Biol.* 4:429–57

19. Fields, S. 1990. *Trends Biochem. Sci.* 15:270–73
20. Marsh, L., Neiman, A. M., Herskowitz, I. 1991. *Annu. Rev. Cell Biol.* 7:699–728
21. Sprague, G. F. Jr. 1991. *Trends Genet.* 7:393–98
22. Fuller, R. S., Sterne, R. E., Thorner, J. 1988. *Annu. Rev. Physiol.* 50:345–62
23. Kurjan, J. 1991. In *Microbial Cell-Cell Interactions*, ed. M. Dworkin, pp. 113–44. Washington, DC: Am. Soc. Microbiol.
24. Herskowitz, I. 1989. *Nature* 342:749–57
25. Levi, J. D. 1956. *Nature* 177:753–54
26. Duntze, W., MacKay, V., Manney, T. R. 1970. *Science* 168:1472–73
27. Bucking-Throm, E., Duntze, W., Hartwell, L. H., Manney, T. R. 1973. *Exp. Cell Res.* 76:99–110
28. Shimoda, C., Yanagishima, N. 1975. *Antonie van Leeuwenhoek* 41:521–32
29. Hagiya, M., Yoshida, K., Yanagishima, N. 1977. *Exp. Cell Res.* 104:263–72
30. Betz, R., Duntze, W., Manney, T. R. 1978. *FEMS Lett.* 4:107–10
31. Wilkinson, L. E., Pringle, J. R. 1974. *Exp. Cell Res.* 89:175–87
32. Sakurai, A., Sakata, K., Tamura, S., Aizawa, K., Yanagishima, N., Shimoda, C. 1976. *Agric. Biol. Chem.* 40:1057–58
33. Stotzler, D., Kiltz, H. H., Duntze, W. 1976. *Eur. J. Biochem.* 69:397–400
34. Masui, Y., Chino, N., Sakakibara, S., Tanaka, T., Murakami, T., Kita, H. 1977. *Biochem. Biophys. Res. Commun.* 78:534–38
35. Ciejek, E., Thorner, J., Geier, M. 1977. *Biochem. Biophys. Res. Commun.* 78:952–61
36. Kurjan, J., Herskowitz, I. 1982. *Cell* 30:933–43
37. Singh, A., Chen, E. Y., Lugovoy, J. M., Chang, C. N., Hitzman, R. A., Seeburg, P. H. 1983. *Nucleic Acids Res.* 11:4049–63
38. Kurjan, J. 1985. *Mol. Cell. Biol.* 5:787–96
39. Betz, R., Crabb, J. W., Meyer, H. E.,

Wittig, R., Duntze, W. 1987. *J. Biol. Chem.* 262:546–48

40. Anderegg, R. J., Betz, R., Carr, S. A., Crabb, J. W., Duntze, W. 1988. *J. Biol. Chem.* 263:18236–40

41. Brake, A., Brenner, C., Najarian, R., Laybourn, P., Merryweather, J. 1985. In *Current Communications in Molecular Biology:Protein Transport and Secretion*, ed. M.-J. Gething, pp. 103–8. New York:Cold Spring Harbor Lab.

42. Kuchler, K., Sterne, R. E., Thorner, J. 1989. *EMBO J.* 8:3973–84

43. Schafer, W. R., Trueblood, C. E., Yang, C.-C., Mayer, M. P., Rosenberg, S., et al. 1990. *Science* 249:1133–39

44. Hrycyna, C., Sapperstein, S. K., Clarke, S., Michaelis, S. 1991. *EMBO J.* 7:1699–709

45. McGrath, J. P., Varshavsky, A. 1989. *Nature* 340:400–4

46. Hagen, D. C., Sprague, G. F. Jr. 1984. *J. Mol. Biol.* 178:835–52

47. Hartig, A., Holly, J., Saari, G., MacKay, V. L. 1986. *Mol. Cell. Biol.* 6:2106–14

48. Jenness, D. D., Spatrick, P. 1986. *Cell* 46:345–53

49. Strazdis, J. R., MacKay, V. L. 1983. *Nature* 305:543–45

50. Achstetter, T. 1989. *Mol. Cell. Biol.* 9:4507–14

51. Manney, T. 1983. *J. Bacteriol.* 155:291–301

52. Dietzel, C., Kurjan, J. 1987. *Mol. Cell. Biol.* 7:4169–77

53. Trueheart, J., Boeke, J. D., Fink, G. R. 1987. *Mol. Cell. Biol.* 7:2316–28

54. McCaffrey, G., Clay, F. J., Kelsay, K., Sprague, G. F. Jr. 1987. *Mol. Cell. Biol.* 7:2680–90

55. Meluh, P. B., Rose, M. D. 1990. *Cell* 60:1029–41

56. Kronstad, J. W., Holly, J. A., MacKay, V. L. 1987. *Cell* 50:369–77

57. Miyajima, I., Nakafuku, M., Nakayama, N., Brenner, C., Miyajima, A., et al. 1987. *Cell* 50:1011–19

58. Jahng, K.-Y., Ferguson, J., Reed, S. I. 1988. *Mol. Cell. Biol.* 8:2484–93

59. Neiman, A. M., Chang, F., Komachi, K., Herskowitz, I. 1990. *Cell Regul.* 1:391–401

60. de Barros Lopes, M., Ho, J.-Y., Reed, S. I. 1990. *Mol. Cell. Biol.* 10:2966–72

61. Jenness, D. D., Burkholder, A. C., Hartwell, L. H. 1983. *Cell* 35:521–29

62. Blumer, K. J., Reneke, J. E., Thorner, J. 1988. *J. Biol. Chem.* 263:10836–42

63. Yu, L., Blumer, K. J., Davidson, N., Lester, H. A., Thorner, J. 1989. *J. Biol. Chem.* 264:20847–50

64. Marsh, L., Herskowitz, I. 1988. *Proc. Natl. Acad. Sci. USA* 85:3855–59

65. Hagen, D. C., McCaffrey, G., Sprague, G. F. Jr. 1986. *Proc. Natl. Acad. Sci. USA* 83:1418–22

66. Nakayama, N., Miyajima, A., Arai, K. 1985. *EMBO J.* 4:26432648

67. Burkholder, A. C., Hartwell, L. H. 1985. *Nucleic Acids Res.* 13:8463–75

68. Sibley, D. R., Benovic, J. L., Caron, M. G., Lefkowitz, R. J. 1987. *Cell* 48:913–22

69. Marsh, L., Herskowitz, I. 1987. *Cell* 50:995–96

70. Clark, K. L., Davis, N. G., Wiest, D. K., Hwang-Shum, J.-J., Sprague, G. F. Jr. 1988. *Cold Spring Harbor Symp. Quant. Biol.* 53:611–20

71. Cartwright, C. P., Tipper, D. J. 1991. *Mol. Cell. Biol.* 11:2620–28

72. Hartwell, L. H. 1980. *J. Cell Biol.* 85:811–22

73. Marr, R. S., Blair, L. C., Thorner, J. 1990. *J. Biol. Chem.* 265:20057–60

74. Bender, A., Sprague, G. F. Jr. 1986. *Cell* 47:929–37

75. Nakayama, N., Miyajima, A., Arai, K. 1987. *EMBO J.* 6:249–54

76. Kang, Y.-S., Kane, J., Kurjan, J., Stadel, J. M., Tipper, D. J. 1990. *Mol. Cell. Biol.* 10:2582–90

77. Nakayama, N., Kaziro, Y., Arai, K.-I., Matsumoto, K. 1988. *Mol. Cell. Biol.* 8:3777–83

78. Whiteway, M., Hougan, L., Thomas, D. Y. 1988. *Mol. Gen. Genet.* 214:85–88

79. Cole, G. M., Reed, S. I. 1991. *Cell* 64:703–16

80. Blinder, D., Bouvier, S., Jenness, D. D. 1989. *Cell* 56:479–86

81. Whiteway, M., Hougan, L., Thomas, D. Y. 1990. *Mol. Cell. Biol.* 10:217–22

82. Nomoto, S., Nakayama, N., Arai, K.-I., Matsumoto, K. 1990. *EMBO J.* 9:691–96

83. Cole, G. M., Stone, D. E., Reed, S. I. 1990. *Mol. Cell. Biol.* 10:510–17

84. Kurjan, J., Hirsch, J. P., Dietzel, C. 1991. *Genes Dev.* 5:475–83

85. Hirsch, J. P., Dietzel, C., Kurjan, J. 1991. *Genes Dev.* 5:467–74

86. Sullivan, K. A., Miller, R. T., Masters, S. B., Beiderman, B., Heideman, W., Bourne, H. R. 1987. *Nature* 330:758–60

87. Miller, R. T., Masters, S. B., Sullivan, K. A., Beiderman, B., Bourne, H. R. 1988. *Nature* 334:712–15

88. Hingorani, V. N., Ho, Y.-K. 1987. *FEBS Lett.* 220:15–22

89. Spiegel, A. M., Backlund, P. S. Jr., Butrynski, J. E., Jones, T. L. Z., Sim-

onds, W. F. 1991. *Trends Biochem. Sci.* 16:338–41

90. Magee, T., Hanley, M. 1988. *Nature* 335:114–15

91. Whiteway, M., Hougan, L., Dignard, D., Bell, L., Saari, G., et al. 1988. *Cold Spring Harbor Symp. Quant. Biol.* 53:585–90

91a. Stone, D. E., Cole, G. M., de Barros Lopes, M., Goebl, M., Reed, S. I. 1991. *Genes Dev.* 5:1969–81

92. Duronio, R. J., Towler, D. A., Heuckeroth, R. O., Gordon, J. I. 1989. *Science* 243:796–800

93. Van Dop, C., Yamanaka, G., Steinberg, F., Sekura, R. D., Manclark, C. R., et al. 1984. *J. Biol. Chem.* 259:23–26

94. Cerione, R. A., Kroll, S., Rajaram, R., Unson, C., Goldsmith, P., Spiegel, A. M. 1988. *J. Biol. Chem.* 263:9345–52

95. Hamm, H. E., Deretic, D., Hofmann, K. P., Schleicher, A., Kohl, B. 1987. *J. Biol. Chem.* 262:10831–38

96. Hamm, H. E., Deretic, D., Arendt, A., Hargrave, P. A., Koenig, B., Hofmann, K. P. 1988. *Science* 241:832–35

97. Haga, T., Ross, E. M., Anderson, H. J., Gilman, A. G. 1977. *Proc. Natl. Acad. Sci. USA* 74:2016–20

98. Rall, T., Harris, B. A. 1987. *FEBS Lett.* 224:365–71

99. Blumer, K. J., Thorner, J. 1990. *Proc. Natl. Acad. Sci. USA* 87:4363–67

100. Barbacid, M. 1987. *Annu. Rev. Biochem.* 56:779–827

101. Ditzelmuller, G., Wohrer, W., Kubicek, C. P., Rohr, M. 1983. *Arch. Microbiol.* 135:63–67

102. Feig, L. A., Cooper, G. M. 1988. *Mol. Cell. Biol.* 8:2472–78

103. Bourne, H. R. 1988. *Cell* 53:669–71

104. Jurnak, F., Heffron, S., Bergmann, E. 1990. *Cell* 60:525–28

105. Milburn, M. V., Tong, L., deVos, A. M., Brunger, A., Yamaizumi, Z., et al. 1990. *Science* 247:939–45

106. Bourne, H. R., Kaslow, D., Kaslow, H. R., Salomon, M. R., Licko, V. 1981. *Mol. Pharmacol.* 20:435–41

107. Salomon, M. R., Bourne, H. R. 1981. *Mol. Pharmacol.* 19:109–16

108. Stone, D. E., Reed, S. I. 1990. *Mol. Cell. Biol.* 10:4439–46

109. Broek, D., Samiy, N., Fasano, O., Fujiyama, A., Tamanoi, F., et al. 1985. *Cell* 41:763–69

110. Graziano, M. P., Gilman, A. G. 1989. *J. Biol. Chem.* 264:15475–82

111. Masters, S. B., Miller, R. T., Chi, M.-H., Chang, F.-H., Beiderman, B., et al. 1989. *J. Biol. Chem.* 264:15467–74

112. Miyajima, I., Arai, K.-I., Matsumoto, K. 1989. *Mol. Cell. Biol.* 9:2289–97

113. Courchesne, W. E., Kunisawa, R., Thorner, J. 1989. *Cell* 58:1107–19

114. Konopka, J. B., Jenness, D. D. 1991. *Cell Regul.* 2:439–52

115. Nakayama, N., Arai, K., Matsumoto, K. 1988. *Mol. Cell. Biol.* 8:5410–16

116. Finegold, A. A., Schafer, W. R., Rine, J., Whiteway, M., Tamanoi, F. 1990. *Science* 249:165–69

117. Reed, S. I. 1980. *Genetics* 95:561–77

118. Shuster, J. R. 1982. *Mol. Cell. Biol.* 2:1052–63

119. Teague, M. A., Chaleff, D. T., Errede, B. 1986. *Proc. Natl. Acad. Sci. USA* 83:7371–75

120. Elion, E. A., Grisafi, P. L., Fink, G. R. 1990. *Cell* 60:649–64

121. Rhodes, N., Connell, L., Errede, B. 1990. *Genes Dev.* 4:1862–74

122. Song, O.-K., Dolan, J. W., Yuan, Y.-L. O., Fields, S. 1991. *Genes Dev.* 5:741–50

123. Dolan, J. W., Fields, S. 1990. *Genes Dev.* 4:492–502

124. Dolan, J. W., Kirkman, C., Fields, S. 1989. *Proc. Natl. Acad. Sci. USA* 86:5703–7

125. Errede, B., Ammerer, G. 1989. *Genes Dev.* 3:1349–61

126. Terrance, K., Lipke, P. N. 1987. *J. Bacteriol.* 169:4811–15

127. Lipke, P. N., Wojciechowicz, D., Kurjan, J. 1989. *Mol. Cell. Biol.* 9:3155–65

128. Roy, A., Lu, C. F., Lipke, P. N., Marykwas, D., Kurjan, J. 1991. *Mol. Cell. Biol.* 11:4196–206

129. Chang, F., Herskowitz, I. 1990. *Cell* 63:999–1011

130. Cross, F. 1988. *Mol. Cell. Biol.* 8:4675–84

131. Richardson, H. E., Wittenberg, C., Cross, F., Reed, S. I. 1989. *Cell* 59:1127–33

132. Cross, F. 1990. *Mol. Cell. Biol.* 10:6482–90

133. Wittenberg, C., Sugimoto, K., Reed, S. I. 1990. *Cell* 62:225–37

134. Nash, R., Tokiwa, G., Anand, S., Erickson, K., Futcher, A. B. 1988. *EMBO J.* 7:4335–46

135. Elion, E. A., Brill, J. A., Fink, G. R. 1991. *Proc. Natl. Acad. Sci. USA* 88:9392–96

136. Field, C., Schekman, R. 1980. *J. Cell Biol.* 86:123–28

137. Baba, M., Baba, N., Ohsumi, Y., Kanaya, K., Osumi, M. 1989. *J. Cell Sci.* 94:207–16

138. Watzele, M., Klis, F., Tanner, W. 1988. *EMBO J.* 7:1483–88

139. Chan, R. K., Otte, C. A. 1982. *Mol. Cell Biol.* 2:11–20
140. Moore, S. A. 1984. *J. Biol. Chem.* 259:1004–10
141. Tanaka, T., Kita, H. 1977. *J. Biochem.* 82:1689–93
142. Ciejek, E., Thorner, J. 1979. *Cell* 18:623–35
143. Sprague, G. F. Jr., Herskowitz, I. 1981. *J. Mol. Biol.* 153:357–72
144. MacKay, V. L., Welch, S. K., Insley, M. Y., Manney, T. R., Holly, J., et al. 1988. *Proc. Natl. Acad. Sci. USA* 85:55–59
145. Marcus, S., Xue, C.-B., Naider, F., Becker, J. M. 1991. *Mol. Cell. Biol.* 11:1030–39
146. Steden, M., Betz, R., Duntze, W. 1989. *Mol. Gen. Genet.* 219:439–44
147. Konopka, J. B., Jenness, D. D., Hartwell, L. H. 1988. *Cell* 54:609–18
148. Reneke, J. E., Blumer, K. J., Courchesne, W. E., Thorner, J. 1988. *Cell* 55:221–34
149. Irie, K., Nomoto, S., Miyajima, I., Matsumoto, K. 1991. *Cell* 65:785–95
150. Lipke, P. N., Taylor, A., Ballou, C. E. 1976. *J. Bacteriol.* 127:610–18
151. Lipke, P. N., Ballou, C. E. 1980. *J. Bacteriol.* 141:1170–77
152. de Nobel, J. G., Klis, F. M., Prem, J., Munnik, T., van den Ende, H. 1990. *Yeast* 6:491–99
153. Osumi, M., Shimoda, C., Yanagishima, N. 1974. *Arch. Microbiol.* 97:27–38
154. Trueheart, J., Fink, G. R. 1989. *Proc. Natl. Acad. Sci. USA* 86:9916–20
155. Byers, B., Goetsch, L. 1975. *J. Bacteriol.* 124:511–23
156. Rose, M. D., Price, B. R., Fink, G. R. 1986. *Mol. Cell. Biol.* 6:3490–97
157. Jackson, C. L., Hartwell, L. H. 1990. *Mol. Cell. Biol.* 10:2202–13
158. Jackson, C. L., Hartwell, L. H. 1990. *Cell* 63:1039–51
159. Jackson, C. L., Konopka, J. B., Hartwell, L. H. 1991. *Cell* 67:389–402

Annu. Rev. Biochem. 1992. 61:1131–73
Copyright © 1992 by Annual Reviews Inc. All rights reserved

HORMONE RESPONSE DOMAINS IN GENE TRANSCRIPTION

Peter C. Lucas and Daryl K. Granner

Department of Molecular Physiology and Biophysics, Vanderbilt University Medical School, Nashville, Tennessee 37232-0615

KEY WORDS: hormone response elements, regulation of gene expression, steroid/thyroid/ retinoid receptor family, complex response units, protein-DNA interactions

CONTENTS

PERSPECTIVES AND SUMMARY

During the past several years, studies of the mechanism of hormonal regulation of gene transcription have concentrated on the identification of the DNA sequences, usually located in the 5'-flanking regions of regulated genes, that mediate these effects. Such DNA sequences, termed hormone response elements (HREs), have been defined for several hormones. Paralleling the identification of HREs have been the isolation, characterization, and cloning of proteins that, upon activation by hormone, function both to bind HREs and to mediate transcriptional effects. Recent work suggests that these HRE-bound proteins interact, either directly or indirectly through intermediary protein-protein contacts, with components of the basal transcription apparatus (for reviews, see 1, 2).

The paradigm for hormonal regulation of gene transcription is the regulation of mouse mammary tumor virus (MMTV) DNA by glucocorticoids. Soon after the discovery that gene expression from cloned MMTV sequences is induced by glucocorticoids in transfected mouse cells (3–6), it was found that the glucocorticoid receptor binds, in a sequence-specific manner, to several DNA sites located in the MMTV long terminal repeat (LTR) upstream of the transcription initiation site (7–10). Furthermore, these receptor-binding sites are capable of conferring hormonal responsiveness to a heterologous promoter that is not normally transcriptionally regulated by glucocorticoids (11). Thus, these DNA sequences were referred to as "glucocorticoid response elements" (GREs) (11, 12).

Consensus HREs were soon derived for many hormones based on the comparison of individual HRE sequences from diverse genes, identified using the approach employed to define the MMTV GREs. Rather surprising, in view of their remarkably different physiologic effects, was the observation that glucocorticoids, androgens, mineralocorticoids, and progestins could all regulate transcription through an identical consensus HRE (Figure 1). In the presence of the cognate receptor, all these agents strongly stimulate transcription through this DNA element when it is ligated to hormone-insensitive promoter-reporter gene constructs. This raised the possibilities that: (a) there were subtle but significant functional differences in the sequences of HREs from individual genes; (b) receptor was rate-limiting in certain cells; (c) ligand was limiting in certain cells; or (d) other cis-elements and/or trans-acting factors were involved in determining hormone specificity. As will become apparent, all of these possibilities have subsequently been demonstrated. Of most interest, in the context of this review, is the evolution of thinking regarding the complexity of HREs. While there is no doubt that specific, short segments of DNA can serve as HREs in synthetic constructs (when ligated to a heterologous promoter and a reporter gene), it is not clear how many simple HREs actually exist in naturally occurring genes. It is apparent that receptor-binding elements (the simple HREs) often function optimally only when they exist in multiple copies, or when associated with other cis-elements. In some cases the latter augment the response; in others they are obligatory components. These assemblies, which mediate the effect of a single hormone, are more complex than simple HREs and are designated hormone response units (HRUs) to connote their multicomponent nature. Diversity of hormone action can also be accomplished through protein-protein interactions at a common HRE. These interactions can be between heterologous receptors within the steroid/thyroid/retinoid superfamily, or between members of different superfamilies of transcription factors. In the past year it has also become apparent that DNA elements involved in the developmental, tissue-specific, and hormone-regulated expression of a gene, may be collected, even superimposed, in a region of DNA that acts as a multifunctional regulatory domain. This type of regulation involves combinations of both response elements and protein-protein interactions.

The progressive accumulation of information that led to these formulations is the subject of this review. We apologize for having to omit the citation of so much important work, but it is impossible, in limited space, to provide a comprehensive review of this topic. We have opted to discuss certain examples that illustrate the major point of this review, namely that HREs are similar mainly in their complexity and diversity of structure.

CHARACTERIZATION OF SIMPLE HORMONE RESPONSE ELEMENTS

Hormonal regulation of gene transcription is based on the cis-element/trans-acting factor paradigm. The hormone response elements (HREs) identified in the earliest studies were the cis-elements. Studies of the steroid/thyroid/

GRE (glucocorticoids, androgens mineralocorticoids, progestins)	←——— ———→ AGAACA NNN TGTTCT
ERE (estrogen)	←——— ———→ AGGTCA NNN TGACCT
TRE (thyroid hormone, retinoic acid)	←——— ———→ AGGTCA – – – TGACCT
CRE (cAMP)	←———→ TGACGTCA
AP-1 SITE (phorbol esters)	←— —→ TGA (G/C) TCA

Figure 1   Consensus HREs for a variety of hormones.

retinoid family of hormones proceeded rapidly because the major candidate for the trans-acting factor, the receptor, either existed in highly purified form, or became available after the identification of this superfamily of proteins. The characterization of peptide hormone response elements proceeded much more slowly because the trans-acting factors were not known and, in many instances, that is still the case.

Hormones Acting Through Intracellular Receptors

In general, HREs for hormones of the steroid family are palindromic in structure, often consisting of an inverted repeat sequence separated by a three-nucleotide gap. For example, the sequence AGAACAnnnTGTTCT binds the glucocorticoid hormone-receptor complex, but it also binds several other ligand-receptor complexes (Figure 1). All of these complexes result in transactivation of reporter genes. The consensus estrogen response element (ERE) differs substantially (Figure 1), but it also consists of an inverted repeat separated by three base pairs. Thyroid hormone and retinoic acid enhance gene transcription through a response element that contains palindromic arms identical to those of the ERE, but that are not separated by a three-nucleotide gap (Figure 1) (13). It should be noted that, although these are consensus elements, there is considerable variability in naturally occurring elements. For example, the TGTTCT sequence is highly conserved in GREs, but the leftmost hexanucleotide (AGAACA) can be quite variable. As we shall see in later sections of this review, slight alterations in the primary sequence or structure of an HRE may have profound effects on its activity. It should also be stated that the presence of such a sequence in a hormonally regulated promoter is not sufficient evidence of function; the identification of an HRE requires the demonstration of both receptor binding and transactivation, although as will be seen, the latter may require additional components.

In recent years, specific intracellular receptors for each of the hormones mentioned above have been cloned. All show considerable similarity in their structural organization, and several distinct functional domains reside within each receptor molecule. Most notably, each possesses a DNA-binding domain of approximately 70 amino acids that is thought to fold into two zinc finger motifs similar to the zinc-stabilized DNA-binding domains of *Xenopus laevis* transcription factor IIIA (14). It appears that discrimination in DNA binding, for members of the steroid/thyroid/retinoid superfamily of receptors, is determined primarily by three amino acids at the base of the first finger. Thus, these amino acids are conserved among receptors that recognize and bind to identical DNA elements. A second functional domain common to all receptors is a C-terminal hormone-binding domain. As would be expected, the primary amino acid sequence of this domain varies considerably from receptor to receptor as each recognizes distinct hormonal ligands. A dimerization domain comprises yet a third functional motif in this family of receptors. In at least some cases, this domain consists of nine heptad repeats of hydrophobic amino acids, which could potentially form a hydrophobic dimerization interface along one side of a coiled-coil α-helix (15, 16) in a structure similar to the "leucine zipper" dimerization motif present in other transcription factors (17). Mounting evidence now suggests that hormone receptors form dimers in solution prior to DNA binding (18–21). The ability to dimerize appears to be related to the palindromic structure of consensus HREs; it is now thought that receptors interact with HREs as dimers in a head-to-head arrangement, with each arm of the palindromic DNA elements contacting a single receptor molecule (18, 22, 23). This notion was recently supported by the use of nuclear magnetic resonance and X-ray crystallography to study the interaction of the DNA-binding domain of the glucocorticoid receptor with the GRE (24, 25).

The structural characterization of this family of hormone receptors and their corresponding consensus HREs has been the subject of several recent reviews (16, 26–29), wherein the reader can find a more comprehensive treatment.

Hormones Acting Through Cell Surface Receptors

A variety of other hormones influence the transcription of target genes through distinctly different intracellular mechanisms. For example, unlike the hydrophobic steroid hormones, which are capable of diffusing through the plasma membrane and interacting directly with the intracellular receptors/ transcription factors discussed above, peptide hormones bind to cell surface receptors and set in motion complex intracellular second messenger cascades. One consequence of the initiation of these signaling pathways is the activation of one or more transcription factors. Unfortunately, for some of these peptide hormones, little is known about both the signaling pathways and the transcription factors involved in the regulatory schemes.

One of the best-understood examples of transcriptional regulation by peptide hormones involves the adenylate cyclase system. The binding of many peptide hormones to cell surface receptors results in the elevation of intracellular cAMP levels which, in turn, activates protein kinase A. In characterizing the effects of such agents, it was found that, in a number of cAMP-regulated genes, a short palindromic core sequence was capable of mediating cAMP-induced gene transcription (Figure 1) (30–36; for a review, see Ref. 37). This observation was quickly followed by the cloning of a protein factor that binds to this cAMP response element (CRE), probably in the form of a dimer (38, 39). Functional analyses demonstrated that, like the steroid/thyroid/retinoid HREs, the CRE conferred cAMP-responsiveness to heterologous gene promoters and functioned in a relatively position- and orientation-independent manner. The structure of the CRE-binding protein (CREB) differs substantially from that of the steroid-thyroid hormone class of receptors. CREB dimerizes through a structurally distinct leucine-zipper motif and binds DNA through an adjacent "basic region" (for a review, see Ref. 40). Although specific protein kinase A phosphorylation sites have been identified on CREB molecules (41), the effects of such phosphorylation events are somewhat unclear. Some evidence demonstrates an increased efficacy of DNA binding for phosphorylated CREB (42), while other studies suggest that only the capacity of CREB to activate transcription is affected (43). These different results may, in part, be explained by the finding that there are numerous CREB isoforms, which are all members of an ever-expanding family of transcription factors (40). Further complicating the study of cAMP-dependent transcription is the observation that, for at least some genes, cAMP regulation occurs through DNA sequences distinct from the consensus CRE and requires proteins other than CREB (44–47).

A variety of other agents, including some growth factors, control gene transcription through the diacylglycerol-dependent protein kinase C pathway. Phorbol esters mimic these actions by directly activating protein kinase C. The products of the nuclear proto-oncogenes c-*fos* and c-*jun* mediate the stimulation of transcription initiated by the activation of protein kinase C. Because Fos and Jun share several structural features with one another, and with the CREB proteins, all are members of a superfamily of transcription factors (48). The heterodimeric complex of Fos and Jun, known as the AP-1 transcription factor, binds to a short DNA sequence present in a number of phorbol ester–regulated genes (Figure 1). However, phorbol esters regulate the transcription of several genes through elements that do not conform to the consensus AP-1 site (for a review, see Ref. 49). It is therefore possible that many transcription factors are involved in controlling gene expression in response to activation of the protein kinase C signaling pathway.

The regulation of transcription by a variety of other polypeptide hormones is only poorly understood. For example, although insulin regulates the transcription of numerous genes, only a few DNA sequences responsible for mediating effects of insulin on target gene expression have been identified (50–53). Furthermore, these sequences appear to share no significant homology and may interact with distinct protein factors (for a review, see Ref. 54). Activation of genes by interferons occurs through discrete DNA elements as well (55). However, as is the case with insulin, the intracellular signal transduction pathways involved are still unclear.

FUNCTIONAL DIVERSITY OF SIMPLE HORMONE RESPONSE ELEMENTS

Following the identification of consensus HREs for a number of hormones, several important questions arose: (*a*) What mechanisms are involved in modulating the magnitude of transcriptional regulation of a target gene in response to the hormone? (*b*) Since some consensus HREs mediate the effects of a variety of hormones (Figure 1), what mechanisms provide specificity in hormonal regulation of gene transcription? (*c*) How is negative regulation of transcription by hormones achieved? That is, what are the mechanisms responsible for determining the direction of transcriptional regulation?

Mechanisms for Modulating the Magnitude of Hormone Response

It was first thought that HREs function as independent entities, capable of mediating hormonal effects on transcription through basal promoters but in the absence of other regulatory elements and their corresponding binding proteins. Indeed, the criteria for rigidly identifying an HRE included the demonstration that a DNA element was capable of mediating hormone-dependent regulation of transcription when placed, in isolation, in the context of a heterologous promoter, and that this was often position and orientation independent (4, 11, 12). Considerable evidence indicates that, in fact, many HREs can function as independent elements. Bradshaw et al demonstrated that, in transfected cells, a short GRE/PRE sequence from the tyrosine aminotransferase gene mediates either glucocorticoid or progesterone-induced gene expression when inserted immediately upstream of a chimeric reporter construct consisting only of a minimal promoter/TATA box sequence linked to the chloramphenicol acetyltransferase (CAT) reporter gene (56). Hormonal regulation was dependent only on the presence of coexpressed glucocorticoid or progesterone receptor, respectively. Similar findings have been reported by others for both the GRE and ERE (57, 58). In vitro transcription systems have also been established in which HREs function in the context of just a minimal promoter (59–61). These findings suggest that HRE-receptor complexes can directly influence the binding or activity of components of the basic transcription machinery, and much effort is now directed toward understanding the mechanistic details of this phenomenon.

The tandem insertion of HREs into test promoters often results in a synergistic response to hormones. This appears to be a phenomenon common to HREs for a variety of hormones, including estrogens (62–64), progestins (56), thyroid hormones (65), glucocorticoids (56, 57), and hormones that act through cAMP (66). Multiple copies of HREs exist in some genes. For example, tandem EREs function synergistically in at least one of the vitellogenin genes (62, 67). In some of these cases, cooperative binding of transcription factors to the tandem HREs appears to explain the functional synergism. Tsai et al demonstrated that the binding of progesterone receptor dimers to each of two closely linked PREs produces a protein-DNA complex 100 times more stable than that produced by a single receptor dimer interacting with a lone PRE (68). Cooperative binding of glucocorticoid receptors to tandem GREs has also been demonstrated (69). In other cases, however, cooperative binding of factors to adjacent HREs may not explain functional synergism. While studying estrogen regulation of the *Xenopus* vitellogenin B1 gene, Martinez & Wahli (70) demonstrated cooperative binding of estrogen receptors to the two imperfectly palindromic EREs that constitute a functional "Estrogen-Responsive Unit" (63) in the promoter of this gene. These results, however, were contested more recently; Ponglikitmongkol et al found no evidence for such cooperative binding to either perfect or imperfect tandem EREs, including the EREs of the vitellogenin B1 gene (64). Explanations for this discrepancy are not readily apparent, as both groups used the same source of receptor and similar binding conditions and analyses.

The magnitude of transcriptional regulation may also be affected by the location of HREs relative to the transcription initiation site. Although HREs typically display enhancerlike properties, in that they may function in a position-independent fashion in test gene promoters, there are certainly limitations to this flexibility. In particular, HREs present in chimeric reporter constructs containing only a minimal promoter/TATA box sequence may exhibit profound position-dependency. GREs located immediately upstream of the TATA box in such constructs mediate efficient glucocorticoid or progesterone induction of gene transcription but are almost totally inert when located far upstream of the transcription initiation site (56, 58). When two or more HREs function synergistically to mediate an effect of hormone, the spacing between the elements themselves, as well as their stereoalignment, may critically affect the degree of synergism (62–64). A decrease in cooperative binding of factors to HREs that have been distanced from one another appears to explain the reduced functional cooperativity seen for at least some of these HREs (69). Taken together, these results indicate that very specific protein-protein contacts, both between DNA-bound receptors and between receptors and components of the basic transcription apparatus, may be critical for establishing efficient hormonal regulation of gene transcription. These observations underscore the importance of demonstrating that a candidate HRE mediates hormonal effects on transcription from its natural location and not simply when inserted at an ideal location in a heterologous context.

Finally, the magnitude of transcriptional regulation mediated by an HRE can be affected by slight sequence alterations, the systematic mutation of each base in the consensus GRE being a case in point (71). Therefore, a combination of factors (the primary DNA sequence, multiplicity, spacing, and placement of HREs within the context of a gene) dramatically influence the degree to which a particular hormone can regulate the expression of a target gene.

Mechanisms for Modulating the Specificity of Hormone Response

While certain HREs are capable of mediating the effects of several hormones on transcription, slight alterations in the structure or organization of these elements can create new elements, highly specific for a particular hormone. For example, although retinoic acid (RA) and thyroid hormone (T_3) can induce transcription through a palindromic thyroid response element (TRE) derived from the growth hormone gene (13), numerous genes are responsive to only one or the other of these agents. An intensive effort has been made to identify naturally occurring response elements that specifically mediate effects of RA on gene transcription. Such elements have been identified in the promoters of the laminin B1 (72, 73), RA receptor β (74–76), and phosphoenolpyruvate carboxykinase (PEPCK) (77, 78) genes. Similar elements, also capable of mediating an effect of RA, have been characterized in the alcohol dehydrogenase (ADH3) and cellular retinol-binding protein type I (CRBPI) gene promoters, although the ability of T_3 to regulate transcription through these elements has not yet been addressed (79, 80). A striking feature, common to all these RA response elements (RAREs), is that each consists of two or more direct repeats of a TGACC motif. This is in contrast to the consensus TRE (Figure 1), which contains the TGACC sequence repeated in a palindromic arrangement. Response elements specific for T_3, but not RA, have also been observed in the chicken lysozyme (81, 82) and α-myosin heavy chain genes (83, 84).

Two groups have systematically analyzed the DNA sequence determinants that specify responsiveness to RA, T_3, and also vitamin D and estrogen. By synthesizing a series of artificial response elements, Umesono et al (85) demonstrated that direct repeats of the palindromic TRE half-site (referred to here as an AGGTCA motif) are capable of mediating effects of RA, T_3, and vitamin D on transcription. Interestingly, specificity in responsiveness was strongly influenced by the spacing of the direct repeats. Vitamin D activated transcription only through a direct repeat separated by three base pairs. Constructs with a four-base-pair spacer mediated responsiveness to T_3, and those with a five-base-pair gap conferred responsiveness to RA. In good agreement with these functional studies, each element represented a relatively specific DNA-binding site for the vitamin D, T_3, and RA receptors, respectively.

This finding is complicated by the observation that different receptor subtypes, for one particular hormone, may also recognize distinct arrangements of the AGGTCA motif. For example, Mangelsdorf et al (86) characterized a RARE in the promoter of the cellular retinol-binding protein type II (CRBPII) gene that responds only to a unique receptor, the retinoid X receptor (RXR) (87). This element consists of five tandem copies of the AGGTCA motif, each separated by a single base pair. Although the α subtype of the RA receptor (RAR-α) bound to the DNA element in vitro, this form of the receptor was not capable of mediating a RA response on CRBPII gene transcription. Similarly, Rottman et al reported that a site in the apolipoprotein AI (apoAI) promoter mediates RA induction of this gene, and in an RXR-dependent fashion (88). Here again, RAR-α was seen to bind the apoAI RARE but was ineffective in mediating a RA effect on transcription. Although it is not yet clear which DNA bases within this RARE make contact with either RXR or RAR-α, the element from the human apoAI gene promoter does contain a direct repeat of an AGGTCA-like motif separated by a single base pair (89). By interacting with distinct sequence motifs, different hormone receptor subtypes may then be responsible for regulating specific gene networks. Because apoAI is a major constituent of chylomicrons, which carry derivatives of vitamin A (retinyl esters) through the lymph from its site of absorption in the intestine, it is intriguing to speculate that RXR might regulate a network of genes whose products are involved in vitamin A metabolism. CRBPII also plays an important role in vitamin A metabolism; in the absorptive cells of the intestine, retinol bound to CRBPII represents a substrate for a microsomal enzyme, lecithin-retinol acyltransferase (LRAT), which converts retinol into its esterified form so that it can be packaged into chylomicrons (90, 91). Ong et al recently demonstrated that the appearance of LRAT activity in rat intestine closely follows that of CRBPII activity (92). It will therefore be of interest to determine whether RXR mediates transcriptional effects on the LRAT gene and other genes encoding enzymes involved in retinoid metabolism.

Näär et al also systematically analyzed the sequence determinants that specify responsiveness to particular hormones (82). Consistent with the findings of Umesono et al (85), they showed that T_3 and RA both enhanced transcription through elements consisting of direct repeats of the palindromic TRE half-site; specificity in responsiveness was again determined by the spacing between these direct repeats. In addition, Näär et al characterized a novel "inverted palindromic" arrangement of the core motif. This element, which is quite similar to a naturally occurring TRE in the chicken lysozyme gene (81), selectively mediated induction of gene transcription by T_3. Despite the close agreement of these two groups on a number of experimental findings, there were also some notable differences. Most significantly, Umesono et al demonstrated a general correlation between function and binding specificity (85). That is, DNA elements capable of mediating effects of only one or the other hormone on transcription were also somewhat selective in binding the corresponding hormone receptor. In contrast, Näär et al found that, while the estrogen receptor bound effectively only to the consensus ERE (Figure 1), through which it positively regulates transcription, both the T_3

and RA receptors displayed relatively promiscuous binding characteristics, recognizing a variety of arrangements of the duplicated TRE half-sites (82). These results suggest that specificity in responsiveness may depend more on the conformation of a bound receptor, perhaps determined by a specific DNA sequence–protein interaction, than on selectivity of the binding reaction itself. Alternatively, other elements/factors may be involved. Whatever the resolution to these differences, it is becoming clear that slight alterations in the structure or organization of "consensus" HREs can result in the creation of highly specific response elements.

Selective hormonal responsiveness can be determined not only by the characteristics of an HRE, but also by the abundance or activity of the transcription factors that interact with an element. Although the consensus GRE can mediate positive effects of both glucocorticoids and progesterone on transcription, certain genes carrying elements closely related to the GRE may be regulated by only one of these hormones. Strähle et al demonstrated that the inability of progesterone to induce the expression of several genes ordinarily responsive to glucocorticoids in FTO2B-3 rat hepatoma cells could be overcome simply by introducing into these cells an expression plasmid encoding the progesterone receptor (93). Therefore, although some HREs are capable of mediating the effects of several hormones on transcription, a selective response may be established through differential expression of the hormone-sensitive transcription factors that bind these elements. A higher level of complexity may be achieved when one hormone is able to affect the signal transduction pathway of a second hormone. As is discussed in more depth later, RA and vitamin D both regulate the transcription of the osteocalcin gene, in osteoblastlike cells, through overlapping DNA elements. Interestingly, RA may influence the ability of vitamin D to regulate gene expression in these cells by altering the level of vitamin D receptors (94). Presumably, RA could affect the ability of a variety of other agents to regulate transcription. In the T-47D human breast cancer cell line, RA downregulates the expression of progesterone receptors (95), while in several other cell lines, RA treatment enhances the activity or abundance of both protein kinases A and C (96–100). There are certainly many other examples of signal transduction crosstalk, too numerous to discuss here. In any case, it is likely that the ability of an HRE to mediate selectively the transcriptional effects of one hormone over another, in many instances, reflects a complex interplay between the signal transduction pathways involved.

Mechanisms for Modulating the Direction of Hormone Response

Although most of the HREs that have been characterized mediate positive effects of hormones on transcription, there are numerous examples in which hormones inhibit transcription. Endocrine physiology suggests that a single hormone can have positive and negative effects on transcription of genes within the same cell. Of particular interest in this regard are the effects of hormones such as insulin on metabolic pathways. It remains to be seen whether the same second messenger cascades, transcription factors, or HREs are commonly used to mediate bidirectional effects.

Although insulin induces the expression of many genes, the hormone is also capable of strongly inhibiting others (for a review, see Ref. 54). In the liver, and liver-derived cell lines, insulin plays a dominant role in the hormonal regulation of the PEPCK gene in that it fully inhibits both glucocorticoid and cAMP-induced expression of the gene (101, 102). Inhibition of PEPCK gene transcription occurs while insulin simultaneously stimulates transcription of the glucokinase gene (103). This, in part, accounts for the ability of the hormone to inhibit gluconeogenesis while promoting glycolysis in the hepatocyte. We recently identified a short sequence in the PEPCK promoter that mediates a negative effect of insulin on transcription in H4IIE rat hepatoma cells (50, 104). The element functions in a heterologous context to mediate insulin-responsive inhibition of the thymidine kinase (TK) gene promoter; multimerization of the element allows for nearly complete inhibition of basal transcription of a TK-CAT reporter construct upon addition of insulin (105). Comparison of this insulin response sequence (IRS) with DNA elements that mediate positive effects of insulin on transcription (51–53) reveals little similarity. Efforts are now being made to identify and characterize the proteins that interact with the few IRSs that have been identified.

Negative hormone response elements (nHREs) have been described for several other hormones. Ligand-activated T_3 and RA receptors both inhibit transcription of the epidermal growth factor (EGF) receptor gene promoter through a 36-base-pair element (106). Although DNA-binding studies have demonstrated the specific interaction of these receptors with a fragment of the EGF promoter containing this element, the actual boundaries of the receptor-binding site have not yet been delineated. It is difficult, therefore, to compare the characteristics of this element with those of known positive elements for T_3 and RA. An nTRE has been identified near the transcription start site of both the rat and human β thyroid-stimulating hormone (TSHβ) gene (107–109). It was initially thought that the binding of T_3 receptors to this element might disrupt the formation of an active basal transcription complex, owing to competition for DNA binding with components of the transcription machinery. Recent evidence, however, suggests that this nTRE may function in a position-independent fashion. When inserted in two different locations in a truncated rat growth hormone gene promoter, the TSHβ nTRE mediated T_3-dependent inhibition of transcription (110). It should be noted that both insertion sites were quite close to the transcription start site so that T_3

receptor binding could conceivably inhibit RNA polymerase attachment, or progression, in this heterologous context as well. Arguing against this possibility was the observation that T_3 induction was observed when a positive TRE was inserted into the same sites. More recently, Näär et al demonstrated that an nTRE from the mouse TSHβ gene, which is located at a position similar to the nTRE of the rat gene but is somewhat different in primary sequence, could mediate a negative transcriptional effect of T_3 when ligated to the 5' end of the TK promoter (82). In this case the insertion site, at position -107, was well upstream of the transcription start site. Surprisingly, in these experiments coexpression of T_3 receptor resulted in a 3–4-fold enhancement of basal reporter gene activity; addition of T_3 simply reversed this effect.

Unlike glucocorticoid receptors, T_3 receptors are known to reside in the nucleus (111, 112) and to bind DNA even in the absence of hormone (113–116). It is therefore possible that ligand-free receptors bind certain elements, such as the mouse TSHβ nTRE, in a conformation that allows them to constitutively activate transcription. Interaction of T_3 with these receptors may then abolish this effect. This would imply that HREs may serve, not only to position receptors at appropriate locations in gene promoters, but also to influence their transcriptional-regulatory activity. Examination of the sequences of the rat and mouse TSHβ nTREs reveals that both contain direct repeats of the AGGTCA motif, although the spacing between the repeats differs for the two species. Further work is needed to determine how these sequences mediate negative effects of T_3 on transcription while quite similar elements mediate positive effects of the ligand-bound T_3 receptor.

An nTRE has also been characterized in the promoter of the glycoprotein hormone α-subunit gene (117, 117a). This element is, again, centered closely around the transcription start site. For this reason, it is believed that T_3 receptor binding precludes the binding of the TATA box factor or other associated factors and thereby prevents the assembly of an active transcription complex. It has been suggested that a glucocorticoid receptor-binding site, which overlaps the TATA box of the osteocalcin gene promoter, mediates negative glucocorticoid regulation of this gene through a mechanism similar to that employed by the nTREs (117b).

A slightly different nGRE was described in the bovine prolactin gene (118). Here, DNA fragments containing glucocorticoid receptor-binding sites from the prolactin promoter conferred glucocorticoid-dependent repression of transcription when ligated, in either orientation, upstream or far downstream of the promoter of a TK-CAT fusion gene. Likewise, these sequences functioned as nGREs in several other heterologous promoters and in several transfected cell lines. The experiments also demonstrated that the prolactin nGRE serves as a basal enhancer, as it strongly increased the efficiency of basal transcription when fused to the TK-CAT reporter. Glucocorticoid receptors may bind to the prolactin nGRE and either block the binding or, in some way, neutralize the actions of a distinct enhancer-activating protein that functions through an overlapping site (118). Therefore, in contrast to what has been suggested for several nHREs that overlap the TATA box or RNA polymerase attachment sites, receptor interaction with the prolactin nGRE appears to result in an inhibition of enhancer activity rather than of the activity of the basic transcription machinery.

Glucocorticoids inhibit transcription of the proopiomelanocortin (POMC) gene in the anterior pituitary. Mutation of a region thought to be involved in basal expression of the POMC promoter (from about -75 to -51) reduced glucocorticoid receptor binding to this element and abolished glucocorticoid repression of reporter genes (119). This segment contains elements analogous to a CCAAT box and to the COUP transcription factor recognition site. Nuclear extracts of AtT-20 pituitary cells contain at least three different activities that bind to this POMC nGRE. A CCAAT box oligonucleotide failed to compete for binding of nuclear proteins to the wild-type sequence, but a COUP oligonucleotide competed very efficiently. It was then hypothesized that glucocorticoid receptors, when bound to the nGRE, precluded the binding of a critical component (COUP) of the basal promoter (119) in a manner analogous to the examples cited above. However, in a recent study, a substitution mutation that destroyed the COUP-binding site had no effect on basal promoter activity (119a). Therefore, COUP factor displacement by the glucocorticoid receptor does not appear to be the mechanism for action of the nGRE in the POMC promoter.

For all the cases where hormone receptor binding is thought to preclude the binding or activity of some other positively acting factor(s), one must still explain why the hormone-activated receptor is, itself, unable to enhance gene transcription. Again, it is possible that HREs may act somewhat like allosteric effectors, altering the conformation of bound receptors in a way that either allows or prevents the positive regulatory potential of the receptors to be manifested. While this may at least partially explain the negative activity of some of the nHREs described above, it is clear that inter-element and protein-protein interactions (context relationships) are often crucial in establishing negative regulation.

ROLE OF RECEPTOR HETERODIMERIZATION AND COMPETITIVE DNA BINDING   In recent years, several studies have demonstrated that some hormone-sensitive transcription factors compete for DNA binding with one another, or bind DNA as heterocomplexes, with dramatic functional consequences. Research into the mechanisms of T_3-dependent transcriptional regulation has been complicated by the discovery that two separate genes encode T_3 receptors. Furthermore, multiple splicing variants have been characterized for each receptor subtype. The two fully functional receptors, termed T_3 receptor α (TRα; 120–124) and T_3 receptor β (TRβ; 122, 125, 126), show considerable homology with one another and with the product of the viral oncogene, v-*erbA*, carried by the avian erythroblastosis virus. A variant of the β receptor (TRβ-2; 127) is expressed exclusively in the pituitary and also acts as a functioning T_3 receptor. Interestingly, both the v-*erbA* oncogene product and a naturally occurring splicing product of the TRα gene, TRα-2 (124, 128–132), can bind TREs but not T_3 (130–134). As a result, neither factor is capable of acting as a T_3-responsive transcription factor. In fact, when coexpressed along with the functional T_3 receptors (TRα or TRβ), both are capable of inhibiting T_3 induction of reporter genes harboring a TRE (135–137).

The unliganded TRα and TRβ are also capable of inhibiting such reporter genes. As mentioned earlier, both T_3 and RA induce gene transcription through a perfectly palindromic TRE (13). However, when TRs are coexpressed with RARs, the ability of RA to induce gene expression through a TRE is dramatically blunted (138, 139). The converse does not appear to be true; unliganded RARs do not inhibit the ability of T_3 to trans-activate through the TRE (138). When T_3 is added along with RA, the inhibitory effect is relieved and the two hormones can actually cooperate in enhancing transcription (84, 138). Along the same lines, it is known that TRs inhibit estrogen receptor (ER) activation of transcription, mediated by the palindromic ERE consensus sequence (see Figure 1) (65). Although the effect was initially demonstrated to occur in the presence of T_3, it is now known that inhibition can occur when the TR is in its unliganded form (140).

There are two principle hypotheses to explain the inhibitory effects of the receptor variants (TRα-2 and v-erbA) and the unliganded functional TRs. First, inhibition may occur as the result of competition between these receptor forms and other transcriptionally active receptors for binding response elements in reporter gene promoters. Second, heterodimers may form between receptor species, which are then unable to function as positive transcriptional regulators. These explanations are not mutually exclusive; evidence exists supporting both mechanisms of inhibition, depending upon which receptors are being studied. For example, the data suggest that heterodimers of the TR and ER do not form, despite many attempts to show such an interaction (140, 141). Since TRs are capable of binding alone to the consensus ERE, it is thought that TR inhibition of estrogen responsiveness, mediated by this element, is due to simple competition between TRs and ERs for DNA binding (65). Similarly, TRα-2 and v-erbA may inhibit functional TRs by competing for DNA binding at TREs. Samuels and coworkers, however, have provided convincing evidence that, in some cases, heterodimer formation is responsible for negative transcriptional effects. In transfection experiments, they demonstrated that mutants of the chick TRα that lacked the DNA-binding domain were capable of blocking gene activation by wild-type TRα, TRβ, and RAR (15). This "dominant negative" effect mapped to the series of heptad repeats in the chick TR thought to be responsible for receptor dimerization. These experiments were followed by an analysis of the mechanism of v-erbA inhibition of transcription (142). In striking contrast, v-erbA mutants that lacked the DNA-binding domain were incapable of inhibiting either TRα- or RAR-dependent gene activation. This difference in dominant negative activity mapped to a single amino acid change in the v-erbA oncoprotein, relative to TRα, near one of the heptad repeats (142). Taken together, these results suggest that the unliganded TR may inhibit gene activation through dimerization (e.g. formation of nonfunctional, unliganded TR/liganded RAR heterodimers), while v-erbA appears unable to accomplish inhibition in this way. Instead, v-erbA inhibits gene activation by competing with transcriptionally active receptors for DNA binding.

As described above, the ability of fully functional TRs to inhibit RAR activation of gene transcription through the consensus TRE occurs only in the absence of T_3. There appear to be cases, however, where TRs inhibit RAR function regardless of whether or not T_3 is present. We have recently identified a RA response element (RARE) in the PEPCK gene promoter that mediates a stimulatory effect of RA on transcription (77). Curiously, this element binds both RARs and TRs but fails to mediate a positive transcriptional response to T_3 (78). When both receptors are expressed in transfected cells, the ability of RA to induce transcription through the PEPCK-RARE is significantly hampered; in this case, the inhibitory effect of TRs occurs both in the presence and absence of T_3 (78), reminiscent of the inhibitory actions of TRα-2 and v-erbA. Therefore, it is possible that TRs inhibit RA action by competing for DNA binding at the RARE. Once bound to the PEPCK element, TRs fail to mediate a positive transcriptional response. Alternatively, because DNA mobility shift experiments provided evidence for the binding of TR/RAR heterodimers to the PEPCK-RARE (78), it is possible that inhibition occurs not through simple competition for DNA binding but through the formation of transcriptionally inactive heterodimers. Glass et al made the opposite but analogous observation in studying the regulation of the α-myosin heavy chain gene (84). In that case, the authors characterized a response element (MHC-TRE) that specifically mediated a positive effect of T_3, but not RA, on transcription. Coexpression of both TRs and RARs, however, blunted T_3 responsiveness, both in the presence and absence of RA. DNA-binding data demonstrated that TR/RAR heterodimers could bind to the MHC-TRE as well. Since RARs alone displayed only very weak binding to the MHC-TRE, the inhibitory effect on T_3 action, in this case, was thought to be primarily the result of heterodimerization. Figure 2 summarizes the com-

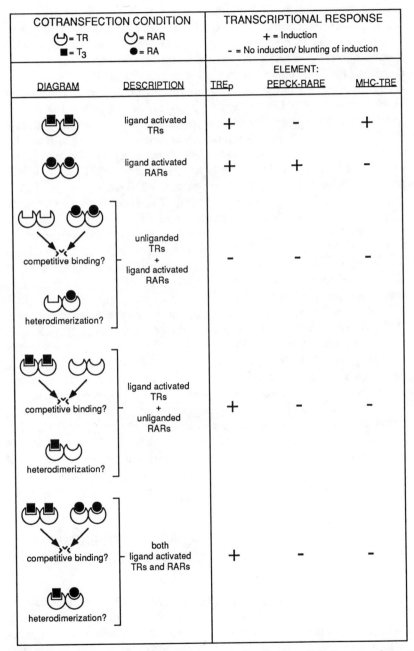

Figure 2   A summary of the transcriptional effects mediated by three distinct response elements under a variety of receptor cotransfection and hormone conditions.

plex inhibitory actions of both the TR and RAR on all three elements discussed—the palindromic TRE, the PEPCK-RARE, and the MHC-TRE.

It is important to note that there are many other examples in which a specific transcription factor inhibits gene expression by either dimerizing or competing for DNA binding with another structurally similar, but positively acting, transcription factor. For example, positive transcriptional regulation by Jun can be inhibited by overexpression of the related factor, Jun-B (143, 144). Although the precise mechanism responsible for this inhibitory effect of Jun-B is not known, it is likely to involve competition between Jun-B and Jun for heterodimerization with Fos and/or DNA binding at phorbol ester response elements. Likewise, Foulkes et al have identified a nuclear factor, termed CRE modulator (CREM), which shows striking similarity to the cAMP-responsive transcription factor, CREB (145). When coexpressed with CREB, CREM inhibits cAMP induction of gene expression mediated by the CRE (145, 146). Again, evidence suggests that CREM acts as a downregulator of cAMP-induced transcription by binding DNA as nonfunctional homodimers or as heterodimers with CREB. The regulation of type I interferon genes involves competition between positive and negative factors as well. Two structurally similar, virus- and interferon-inducible proteins, IRF-1 and IRF-2, bind the identical regulatory elements in the promoters of these genes (147). However, while IRF-1 enhances transcription through this element, IRF-2 does not. Moreover, IRF-2 inhibits IRF-1 activity, presumably by competing for DNA binding. Finally, the t(15;17) chromosomal transloca-tion, which occurs exclusively in cases of acute promyelocytic leukemia, results in the fusion of the RARα gene to a gene encoding a protein of unknown function (148–150). The protein produced by this fusion gene, termed PML-RARα, has recently been cloned and characterized (151, 152). Evidence suggests that PML-RARα may inhibit some of the functions of wild-type RARs, possibly in a manner analogous to the inhibitory actions of v-erbA on TR function. Interactions similar to those discussed in this para-graph are described in more detail below, in the section on the interaction of AP-1 with steroid receptors.

MULTIPLE ELEMENT/FACTOR INTERACTIONS PROVIDE REGULATION THROUGH HORMONE RESPONSE UNITS

As we have discussed, many of the HREs analyzed consist of a single binding site for a cognate transcription factor, which may suffice to regulate the expression of a target gene. Despite the apparent simplicity of this arrange-ment, a considerable diversity of hormonal effects may be mediated by such elements. Variations in their number, spacing, orientation, and primary DNA sequence can all effect the magnitude, direction, and specificity of hormonal responsiveness. Much of this information has been obtained from the analysis of synthetic constructs. In many (perhaps most) natural genes, the hormone responsive regulatory regions are considerably more complex and involve multiple DNA-binding sites for a variety of protein factors. The participation of multiple factors in mediating the response to a hormonal signal, of course, provides the opportunity for even greater flexibility. In the next several sections, we discuss examples of such complex regulatory regions. Although we focus on regulation by hormones of the steroid/thyroid class, similarly complex examples will likely emerge for other hormones such as insulin, or those that act through cAMP.

Six Examples of Naturally Occurring Hormone Response Units

THE MMTV PROMOTER   Although originally viewed as the prototypical simple HRE, an excellent example of multifactorial complexity comes from studies in which the hormonally responsive domain of the MMTV gene was characterized. As noted earlier, the MMTV-LTR contains several DNA sequences that bind steroid hormone receptors (designated GR1, 2, 3, 4 in Figure 3). These sequences, when placed in isolation in a heterologous context, are capable of mediating glucocorticoid, androgen, and progesterone induction of gene transcription. As such, they can be viewed as "simple" HREs. Nevertheless, accumulated evidence indicates that, in their natural setting, these HREs are dependent upon other DNA sequences, and their corresponding binding proteins, in order to confer full hormonal responsiveness upon the MMTV promoter. In particular, a 3' deletion analysis demonstrated that the removal of DNA sequences immediately downstream of the most promoter-proximal glucocorticoid receptor-binding site in the MMTV-LTR nearly abolished responsiveness of the MMTV promoter to glucocorticoids (153). It was subsequently recognized that these deleted sequences included the binding site for the common transcription factor, nuclear factor 1 (NF1) (Figure 3) (154). Furthermore, mutations that disrupted binding of NF1 to this region of the MMTV promoter also dramatically impaired the glucocorticoid induction (155). Finally, complementation studies were performed to establish firmly the role of NF1 in mediating glucocorticoid responsiveness. The MMTV promoter is poorly expressed in JEG-3 human choriocarcinoma cells, which are deficient in NF1, but cotransfection of MMTV promoter constructs with an expression vector encoding NF1 provided a robust response to glucocorticoids (156).

In studying the mechanism of glucocorticoid action, Cordingley et al asked whether hormone treatment would result in enhanced binding of factors other

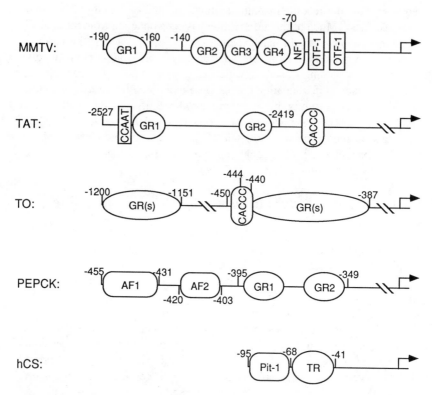

Figure 3 Schematic diagrams of the complex HRUs of several genes.

than the glucocorticoid receptor to the MMTV promoter (154). Using an exonuclease III digestion assay, these investigators demonstrated that NF1 binding to MMTV DNA in vivo closely correlates with hormone activation. It is important to note that because these studies used cells transformed with minichromosomes containing the MMTV promoter, the transforming DNA was packaged in the form of chromatin and was not present in the cells as naked DNA. These results suggested that the binding of glucocorticoid receptors to GRE sequences facilitates the binding of NF1. Surprisingly, however, in vitro DNA-binding experiments demonstrated a lack of cooperativity in the binding of glucocorticoid receptors and NF1 to their neighboring sites on naked fragments of the MMTV promoter (156, 157). Rather, these factors actually competed for DNA binding. To resolve the apparent contradiction between the in vitro and in vivo binding data, several groups examined the effects of chromatin structure on the interaction of these factors with DNA. Nucleosome reconstitution experiments demonstrated that a nucleosome is positioned in a sequence-specific manner over a portion of the MMTV promoter that includes the NF1-binding site (158–161). Furthermore, the nucleosome boundaries detected with these in vitro studies appeared to be consistent with those identified in vivo by Cordingley et al (154). Interestingly, while glucocorticoid receptors bound to such reconstituted nucleosomes and naked DNA with similar affinities, NF1 did not bind to the nucleosomally organized promoter (159–161). This result suggested that the hormone-induced binding of NF1 seen in vivo may result from a disruption or displacement of nucleosomes from the MMTV promoter in response to glucocorticoid receptor binding. Indeed, it has been known for some time that chromatin remodeling does occur over this region of the promoter following hormone treatment (159, 160, 162). Thus, the primary mechanism of action of the glucocorticoid receptor in this system may be to unmask the binding site for the strong transcriptional activator, NF1.

Although it is now well established that NF1 participates in the glucocorticoid regulation of MMTV gene transcription, several studies have implicated yet another factor in the regulation of this promoter by progesterone. While mutations of the NF1-binding site dramatically reduce responsiveness of the MMTV promoter to glucocorticoids, sensitivity to progesterone remains largely intact (163, 164). In a recent study, two binding sites for the ubiquitous transcription factor OTF-1 were identified just downstream of the NF1 site in the MMTV promoter (Figure 3) (165). Mutational analyses demonstrated that the most distal of these OTF-1 binding sites is required for the residual hormonal responsiveness seen following the deletion of NF1-binding sequences. Interestingly, DNA-binding studies revealed that both progesterone and glucocorticoid receptors bind cooperatively with OTF-1 to the MMTV promoter, suggesting that the bound receptors either facilitate or stabilize the interaction of OTF-1 with its cognate binding sites. These findings provide the first example of cooperative binding between a hormone receptor and another transcription factor as a mechanism for establishing synergistic transactivation of a natural promoter. Other investigators, however, using a synthetic oligonucleotide probe containing both receptor- and OTF-1-binding sites, have been unable to demonstrate such cooperative binding (166). These differences may stem from the use of slightly different DNA probes and binding assays (gel retardation versus DNase I footprinting).

It is now clear that the receptor-binding sites alone are insufficient to confer full hormone responsiveness to the MMTV promoter; the role of "accessory factors" has been firmly established. As a result, the receptor-binding sites (GR1, 2, 3, 4) may be viewed as subcomponents of a larger hormone response unit (HRU) that also contains other recognition sequences such as those for NF1 and OTF-1 (Figure 3). It should be noted that yet additional accessory elements and factors may be involved. Cato et al described several mutations, outside of the receptor-binding sites, which differentially affect hormone-responsive transcription depending upon which steroid is being analyzed (163). For example, a mutation some 150 base pairs upstream of the MMTV transcription start site (between GR1 and GR2; see Figure 3) drastically reduced androgen responsiveness in T470 cells, but left glucocorticoid and progestin responsiveness fully intact (163). Therefore, as NF1 and OTF-1 appear to enhance the glucocorticoid and progesterone effects, respectively, another uncharacterized factor may act as an accessory factor specifically for the androgen response.

THE TYROSINE AMINOTRANSFERASE GENE PROMOTER   Expression of the enzyme tyrosine aminotransferase (TAT) occurs primarily in the parenchymal cells of the liver and is regulated, at the transcriptional level, by both glucocorticoid hormones and cAMP (167–169). In characterizing the glucocorticoid induction of the TAT gene, Schütz and coworkers identified a complex glucocortoid response unit (GRU) located some 2500 base pairs upstream of the transcription start site. As in the MMTV system, the administration of glucocorticoids to cultured FTO-2B rat hepatoma cells resulted in a rapid, but reversible, appearance of a DNase I hypersensitive site between positions −2600 and −2300 in the endogenous TAT gene (170–172). The precise boundaries of this hypersensitive site were delineated using a restriction enzyme digestion assay, which demonstrated that the hormone-induced alteration in chromatin structure likely involved the disruption or displacement of two nucleosomes from this location (172). By analyzing glucocorticoid-responsiveness of various 5' and 3' deletion mutants of a transfected TAT-CAT reporter gene, they determined that the DNA sequences responsible for conferring hormone induction mapped between positions −2527 and −2419, fully within the region of hormone-induced DNase I hypersensitivity (171). DNA-binding studies revealed that partially purified glucocorticoid receptor bound to two adjacent sites within this hormone-responsive region (Figure 3), contacting sequences highly similar to the consensus GRE illustrated in Figure 1. Curiously though, the most promoter-proximal receptor-binding site was totally inactive when ligated in isolation to either truncated TAT-CAT or TK-CAT reporter constructs (171). When analyzed in a similar way, the second receptor-binding site displayed some hormone-sensitive enhancer activity but was substantially less effective than the entire −2527/−2419 fragment (57, 171). In contrast, if a segment of the TAT promoter containing both the distal receptor-binding site and sequences immediately upstream, stretching to the −2527 boundary, was ligated to a reporter construct, a strong glucocorticoid effect on CAT expression was observed (57). Inspection of these upstream sequences revealed the presence of a CCAAT motif (Figure 3), which is often required for full promoter activity of other genes (173). The functional significance of the CCAAT motif was confirmed by selectively destroying the CCAAT sequence with a cluster of five point mutations. These mutations abolished the glucocorticoid induction of transcription mediated by the distal receptor-binding site in conjunction with its upstream flanking sequence (57). Finally, transfection studies revealed that when both receptor-binding sites were included along with the CCAAT motif in a test reporter construct, an even greater hormone response could be achieved (171). These results suggested that glucocorticoid receptors must cooperate with one another, and with a distinct CCAAT-binding factor, to mediate a strong inductive effect on TAT gene transcription.

In a separate analysis, Schütz and coworkers used an in vivo footprinting technique to demonstrate the binding of factors, presumably hormone-activated receptors, to the glucocorticoid receptor-binding sites of the TAT gene in hormone-treated cells (172, 174). Interestingly, the data also suggested that one or more additional factors bound a nearby, downstream CACCC-box sequence (Figure 3) in a hormone-dependent fashion. The functional significance of the CACCC-box in mediating glucocorticoid responsiveness, however, is still unclear. Removal of the site by 3' deletion analysis of the TAT promoter failed to dramatically alter hormonal responsiveness (171), but this deletion analysis was performed in Ltk⁻ cells, which do not appear to express or utilize the CACCC-box binding factor as efficiently as FTO-2B cells, in which the in vivo footprinting studies were performed.

Taken together, the above-mentioned studies suggest a model for understanding the glucocorticoid regulation of TAT gene transcription that is reminiscent of that for the transcription of MMTV. Hormone-activated glucocorticoid receptors appear to bind two distinct sites far upstream of the TAT gene transcription start site. The binding of factors to the neighboring CCAAT, and probably the CACCC-box motifs, occurs simultaneously, and the cooperative actions of all these factors are necessary for the full glucocorticoid effect. As in the MMTV system, the binding of these factors occurs with the simultaneous disruption of nucleosomal structure.

THE TRYPTOPHAN OXYGENASE GENE PROMOTER   As with TAT, the expression of the liver enzyme tryptophan oxygenase (TO) is induced by glucocorticoids at the transcriptional level (175). To map the glucocorticoid-responsive region(s) of the TO promoter, Danesch et al analyzed a series of 5' deletion mutants of a TO-CAT reporter construct in transfected culture cells (176). Deletion of promoter sequences to position −444 completely abolished glucocorticoid-dependent enhancement of CAT expression. Surprisingly, footprinting experiments revealed that partially purified glucocorticoid receptors bound to sequences between positions −440 and −387 (Figure 3), which were completely preserved in the −444 5' deletion mutant. Although the exact stoichiometry of receptor binding is not known, several putative GRE half-sites are located between −440 and −387, suggesting that more than one glucocorticoid receptor molecule interacts with DNA in this region of the promoter. In any case, these findings indicated that the receptor-binding sites, themselves, were inert in the absence of upstream flanking sequences.

An analysis of the sequences that were removed to construct the −444 5' deletion mutant revealed the presence of a nearby CACCC-box element (Figure 3), which was identical to that originally found in the β-globin promoter (177, 178). In a later study, this group demonstrated the specific binding of nuclear proteins to this CACCC-box motif and showed that this element cooperated with GRE sequences, placed in a variety of neighboring positions, to mediate glucocorticoid-dependent enhancement of gene transcription (179). Interestingly, footprinting experiments revealed that glucocorticoid receptors also bind far upstream in the TO promoter, between positions −1200 and −1151 (Figure 3) (176). Deletion of these binding sites results in a drop in glucocorticoid-responsiveness equivalent to that seen when the more proximal receptor-binding sites and CACCC-box are deleted. In summary, the distal receptor-binding sites appear to function in an additive manner with the more complex proximal domain in conferring complete hormone-responsiveness to the TO gene promoter.

THE PHOSPHOENOLPYRUVATE CARBOXYKINASE GENE PROMOTER
Several hormones are involved in the regulation of phosphoenolpyruvate carboxykinase (PEPCK) gene transcription. As discussed in a later section, the effects of a number of these hormones are integrated through a relatively small multihormonal control domain upstream of the PEPCK gene transcription start site. However, at this point, it is instructive to compare the complex GRU of the PEPCK gene with the GRUs of the above-mentioned genes.

As with the TAT and TO genes, fusion constructs consisting of the PEPCK promoter ligated to the CAT reporter gene have been used to study the mechanism of glucocorticoid-dependent enhancement of PEPCK gene expression. In transfected H4IIE rat hepatoma cells, a series of 5' deletion mutations of the PEPCK promoter were used to locate the 5' boundary of the glucocorticoid response unit (180). While plasmids with promoter endpoints extending to position −467 or beyond all responded equally to glucocorticoids, a prominent decrease in the glucocorticoid response (from 11- to 2-fold induction of CAT expression) occurred when the promoter was truncated to include only 402 base pairs of promoter sequence. DNase I footprinting experiments were then performed with highly purified glucocorticoid receptor to identify precisely the DNA sequences that interact with the receptor, the expectation being that at least some portion of the footprint would be found between −467 and −402. Surprisingly, there was no evidence of interaction within this region, but the receptor did protect sequences between −395 and −349 (Figure 3) (180). This region contains two putative core sequences, CACACAnnnTGTGCA (GR1) and AGCATAnnnAGTCCA (GR2) required for receptor binding, which match the consensus GRE at 6/12 and 7/12 positions, respectively (180).

The discrepancy between the boundary of the functional GRU and the glucocorticoid receptor-binding sites raised the possibility that accessory elements and associated binding factors were an integral component of the GRU. To ascertain whether accessory factors bind to the region between −467 and −402, evidence for protein-DNA interactions was sought using crude rat liver nuclear extract and the DNase I footprint assay. The results demonstrated that two DNA sequences were protected (180). The first region, from −455 to −431, was denoted accessory factor site 1 (AF1) and the second, from −420 to −403, was denoted accessory factor site 2 (AF2) (Figure 3). The relative functional contributions of the accessory factor–binding sites and the glucocorticoid receptor-binding sites were determined by analyzing the regulation of CAT expression from plasmids in which either one or another of these elements was deleted (180). The promoter containing the entire GRU (all four binding sites shown in Figure 3) gave an 11-fold induction in response to glucocorticoids, whereas promoters lacking either both accessory elements or both receptor-binding sites were almost totally unresponsive. The deletion of either AF1 or AF2 alone caused an approximate 50% reduction in hormone responsiveness as did deletion of either GR1 or GR2 alone. These results demonstrated that the receptor-binding sites themselves are inert, and that each functions independently, accounting for half of the full response, provided both accessory factor sites are present. Our most recent studies have focused on characterizing the proteins that interact with the accessory elements. We now know that multiple proteins interact at these sites and since, as we discuss later, several hormones affect PEPCK expression through this region, it will be a challenge to determine which factors function in the accessory role for the glucocorticoid response and which mediate the response to other hormones.

As with glucocorticoids, agents that increase intracellular cAMP levels in hepatocytes positively affect PEPCK gene transcription. The segment of the PEPCK promoter primarily responsible for mediating this cAMP effect has been mapped to a region centered around position −90 (32, 36, 181). These sequences contain a motif nearly identical to the consensus CRE (Figure 1), and are known to bind the CREB transcription factor (182). Nevertheless, more recent experiments have demonstrated that sequences upstream of the PEPCK CRE are also involved in conferring full cAMP-responsiveness to the PEPCK gene (183). These sequences, spanning the region from about

−290 to −230, appear to contain the binding sites for several distinct proteins, including the C/EBP transcription factor. Together with the CRE, these sequences may then be considered part of a complex cAMP response unit.

THE α_1-ACID GLYCOPROTEIN GENE PROMOTER   The acute phase reactants are a group of plasma proteins released by the liver in response to tissue injury. Although their various functions are not well understood, they presumably help counteract the consequences of trauma. One of these proteins, α_1-acid glycoprotein (AGP), has been studied extensively; expression of the protein is known to be induced by a variety of interleukins and interferons, and also by glucocorticoids. Although glucocorticoids have been shown conclusively to regulate the genes encoding TAT, TO, and PEPCK at the transcriptional level using the nuclear run-on transcription assay, the mode of regulation for AGP has been the subject of conflicting reports. In one report, run-on assays using liver nuclei isolated from rats treated with glucocorticoids demonstrated a clear induction of AGP gene transcription (184). However, similar assays have shown that transcription of the gene in cultured HTC rat hepatoma cells is unaffected by hormone treatment (185). Because AGP mRNA levels were dramatically induced in these cultured cells, it was suggested that glucocorticoids act primarily at a posttranscriptional level to enhance expression of the AGP protein. Despite these conflicting results, several groups have demonstrated that transfected reporter constructs, containing AGP promoter sequences directing the expression of CAT, respond very well to glucocorticoids in HTC and other cell lines (186–188). The consensus, therefore, is that at least part of the glucocorticoid response occurs at the level of transcription, and that AGP promoter sequences mediate this effect.

An interesting observation made in the early studies of AGP regulation was that efficient induction of AGP mRNA by glucocorticoids could be blocked by cycloheximide, indicating that on-going protein synthesis was required for the hormone to be effective (189, 190). In contrast, the induction of MMTV mRNA in the same cells was seen to be unaffected by inhibitors of protein synthesis. Later studies demonstrated the same cycloheximide sensitivity for transfected AGP-CAT reporter genes (187). These results suggested that factors other than the glucocorticoid receptor are necessary for glucocorticoid-dependent transcription of the AGP gene. A careful analysis of the time course of the cycloheximide effect has provided clues as to the nature of these accessory factors. Incubation of cells for relatively short periods (1–2 hours) with both cycloheximide and glucocorticoids resulted in an induction of AGP mRNA comparable to that seen in cells not treated at all with inhibitors of protein synthesis (187). Inhibition was only obvious following longer periods (>4 hours) of exposure to cycloheximide. Moreover, pretreatment of cells with cycloheximide led to an even more pronounced impairment of hormone responsiveness. The same time-dependent effects of cycloheximide were observed when transfected AGP promoter-driven reporter constructs were studied. These results led the authors to conclude that one or more pre-existing but labile proteins are involved in mediating the direct glucocorticoid induction of AGP gene transcription. This conclusion is in contrast to a mechanism whereby glucocorticoid treatment would result in the de novo synthesis of a factor that would then act, somewhat as a hormone-responsive second messenger, to alter expression of the AGP gene.

Subsequent analysis of various 5' and 3' mutants of AGP-CAT constructs revealed that sequences between -120 and -42 in the AGP promoter are responsible for mediating the effect of glucocorticoids on transcription of these chimeric genes (186). Klein et al used DNase I protection and methylation interference assays to demonstrate the binding of partially purified glucocorticoid receptors to an imperfect palindrome, centered between positions -121 and -107, that resembles the consensus GRE (191). Insertion of this sequence into a heterologous reporter construct generated a hybrid promoter that was quite responsive to glucocorticoids. Interestingly, hormone regulation of this hybrid construct was unaffected by cycloheximide. However, when a larger fragment of the AGP promoter, containing the GRE homology as well as some 70 base pairs of downstream sequence, was inserted into the same heterologous reporter construct, two differences were observed. First, the larger fragment conferred severalfold greater hormone-responsiveness to the heterologous promoter than did the short GRE-like sequence. Second, the glucocorticoid effect mediated by the larger fragment displayed cycloheximide sensitivity. Taken together, these results suggested that sequences immediately downstream of the glucocorticoid receptor-binding site in the AGP promoter interact with potentially labile factors that cooperate with bound receptors to enhance transcription of the gene. In a more recent study, DiLorenzo et al characterized the binding of two proteins, or protein complexes, that interact with a total of three distinct sites downstream of the GRE homology (192). An understanding of the relevance of these proteins to the glucocorticoid response, however, awaits the functional analysis of promoter mutations that selectively prevent the binding of these factors.

THE PLACENTAL LACTOGEN GENE PROMOTER   Although our discussion of multifactorial HRUs has focused mainly on transcriptional regulation by glucocorticoids, it is likely that similarly complex response units will be discovered for a variety of hormones. For example, recently Voz et al characterized a T_3 response unit (TRU) in the promoter of the human placental lactogen B (chorionic somatomammotropin; hCS-B) gene (193). Here again, a discrepancy was revealed between the functional analysis of transfected 5' and 3' deletion mutants of hCS-CAT reporter constructs and receptor-binding experiments. While a larger fragment of the hCS promoter, from −97 to −31, was required for T_3 induction of CAT expression, binding studies demonstrated receptor interaction only within a region from −67 to −41 (Figure 3). The possible involvement of a Pit-1 (GHF-1) transcription factor–binding site (194, 195), located just upstream of the receptor-binding site between −95 and −68 (Figure 3), was investigated by introducing a 5-base-pair deletion in the hCS promoter known to abolish Pit-1 recognition. This mutation completely destroyed T_3-responsiveness. Because neither the Pit-1 nor the T_3 receptor-binding sites alone were able to mediate the hormone's effect on transcription, the authors concluded that both sites function cooperatively as a single TRU. Because the hCS gene is normally expressed only in the placenta, while the Pit-1 factor is specific to pituitary cells such as the GC rat pituitary cells in which the functional studies were performed, the relevance of this TRU to the physiological expression of the hCS gene is uncertain.

Synthetic Hormone Response Units

Several groups have systematically analyzed the ability of hormone receptors to cooperate with a variety of general basal transcription factors. Schüle et al constructed a series of reporter constructs, each containing a consensus GRE upstream of TK-CAT, but with different neighboring binding sites for other diverse factors including the CCAAT, Sp1, OTF, NF1, and CACCC-box factors (196). Transfection studies revealed that the inducibility of these constructs by glucocorticoids or progestins was synergistically enhanced by the presence of any of the factor-binding sites. However, there may be some specificity to this phenomenon. Allan et al have demonstrated, using similar constructs in an in vitro transcription system, that the COUP transcription factor is unable to substitute for NF1 in this synergistic capacity (197).

The mechanisms by which steroid receptors cooperate with other general transcription factors to mediate hormone induction of gene transcription are still unclear. Although cooperative DNA binding has been suggested in some cases, this explanation does not seem to be generally applicable. A second model for functional synergism suggests that DNA-bound factors interact with a common, non-DNA-binding protein, which may then directly influence components of the basal transcription apparatus. The ability of this intermediary factor to alter transcriptional activity may depend, in a nonlinear fashion, on the number or strength of its interactions with DNA-bound factors. There may, in fact, be several different mechanisms simultaneously operating in different HRUs. For example, a comparison of the naturally occurring HRUs described above reveals some subtle but important differences, which may reflect on their mechanisms of action. While in some cases accessory elements and factors are obligatory for hormone action (e.g. the PEPCK GRU), in other cases the receptor-binding sites are somewhat functional on their own, and accessory factors are required only for optimal hormone responsiveness. At this point one can only hypothesize in a general way about the mechanisms responsible for hormone induction in each of these cases, but the detailed molecular interactions that link hormone receptors to the transcription initiation complex are currently under intense study.

Role of Auxiliary Factors in Receptor-DNA Interactions

In each of the examples of complex HRUs described above, hormone receptors are presumed to bind, unassisted, to DNA at discrete sites. Indeed, many investigators have successfully used highly purified glucocorticoid, RA, and T_3 receptors to demonstrate sequence-specific DNA binding in both gel mobility shift and DNase I footprinting experiments. However, there is a growing awareness that additional protein factors may, in some cases, be necessary for specific, high-affinity binding of receptors to chromatin.

Recently, several groups have demonstrated that, while T_3 receptors produced by in vitro translation systems bind to a TRE, the addition of crude nuclear extract to the binding reactions results in a markedly enhanced receptor-DNA interaction (198–203). The binding enhancement is not a general effect of added protein, but appears to be mediated by one or more specific proteins, which have been termed T_3 receptor auxiliary proteins (TRAPs). Burnside et al fractionated nuclear proteins with gel filtration chromatography and demonstrated that peak TRAP activity from GH_3 cell nuclei elutes with an estimated molecular weight of 65,000 (199). A protein of similar size, from JEG-3 cell extracts, was crosslinked to T_3 receptors bound to DNA (203). Additional insights into the characteristics of the proteins exhibiting TRAP activity have come from experiments using DNA affinity chromatography; Darling et al demonstrated that TRAP binds tightly to affinity columns composed of concatamers of a TRE oligonucleotide, therefore suggesting that TRAP proteins may bind DNA in the absence of ligand-activated thyroid hormone receptors (201). Although several models have been proposed to explain the mechanism by which TRAPs increase the affinity of nuclear receptors for their response elements (204), it is now thought that these proteins interact with both DNA and receptor to form a more stable ternary complex. Certainly the details of the molecular actions of TRAPs must await their purification and cloning. Although perhaps best described for T_3 receptors, the enhancement of receptor DNA binding by specific nuclear proteins may be a general phenomenon. TRAP-like activities in nuclear extracts have been shown to facilitate the interaction of receptors for RA (201, 205, 206), progesterone (207), estrogen (208), and vitamin D (201, 209) with their cognate response elements.

Spelsberg and coworkers have used a different approach to study receptor-chromatin interactions but have arrived at somewhat similar conclusions. A number of years ago, this group began studying aspects of progesterone action on avian oviduct gene expression by analyzing the binding of radiolabeled progesterone receptor ([^{3}H]PR) to isolated chromatin (210, 211, and references therein). The binding observed in this cell-free system was both saturable and of high affinity. Next, by sequentially removing various components from the intact chromatin with mild salt treatments, they were able to identify a class of nonhistone proteins that was essential for this binding; when these proteins were stripped from the chromatin, [^{3}H]PR binding was dramatically reduced and was nonsaturable. More recently, a 10-kDa candidate "acceptor protein," termed receptor-binding factor 1 (RBF-1), was purified to homogeneity from this source (212, 213). Several lines of evidence support a role for this protein in mediating progesterone receptor binding to DNA. Most significantly, high-affinity binding could be reconstituted by complexing purified RBF-1 with naked genomic DNA (213). Also, monoclonal antibodies raised against RBF-1 dramatically inhibited [^{3}H]PR binding to intact chromatin while not affecting the ability of estrogen receptors to do so under the same conditions (214).

The approach taken by the Spelsberg laboratory has been used to study acceptor proteins for a variety of other receptors, including those for estrogens, androgens, and glucocorticoids (for reviews, see Refs. 210, 211). Among the future challenges will be the determination of how these acceptor proteins interact with both receptors and DNA. It will be particularly interesting to see whether the acceptor proteins display sequence-specific DNA binding to sites in or around HREs that are known to interact with purified receptors. Finally, it will be intriguing to explore the possible relationship between acceptor proteins and TRAP-like factors.

INTEGRATION OF PLEIOTROPIC SIGNALS THROUGH MULTIFUNCTIONAL REGULATORY DOMAINS

Many genes are regulated by combinations of hormones. In some cases, two or more hormones may provide additive responses, or act cooperatively to give synergistic regulation in one direction. Other hormones may exert dominant effects in the opposite direction. How is such complex positive and negative signaling integrated, and what are the mechanisms that provide for dominance, additivity, and synergism within the context of a single promoter? The answers to these questions remain largely unknown, but a common theme is emerging from studies on genes subjected to complex regulation. Instead of operating through distinct, separable response elements, hormone signals sometimes converge on a single regulatory domain in a target gene promoter.

In this way, intricate protein-protein and protein-DNA interactions may play a role in determining the ultimate transcriptional outcome under a particular hormonal condition. We describe below three examples of multihormonal regulation that illustrate this emerging theme.

Three Examples of Hormone Signal Convergence

THE PHOSPHOENOLPYRUVATE CARBOXYKINASE GENE   PEPCK catalyzes the rate-limiting step in gluconeogenesis. Positive and negative regulation of this process is primarily accomplished by hormones that influence the rate of transcription of the PEPCK gene. Glucocorticoids, cAMP, and RA increase the rate of PEPCK gene transcription, whereas insulin and phorbol esters exert dominant, negative effects (77, 101, 102, 215, 216). We have already described, in separate sections of this essay, the elements that mediate responsiveness to glucocorticoids, RA, and insulin. Not discussed was the fact that these coexist within a single regulatory domain. The GRU spans the entire domain ($\sim$110 bp) and consists of two glucocorticoid receptor-binding sites (GR1 and GR2) and two accessory factor–binding sites (AF1 and AF2) (Figure 4A). Interestingly, the AF1 site coincides with a RARE that is largely responsible for mediating the positive effect of RA on transcription (Figure 4A). As described earlier, this RARE also binds T_3 receptors that block the inductive effects of RA. In addition, we now know that the liver protein HNF4 (217), a member of the steroid/thyroid/retinoid receptor superfamily, interacts with the AF1 element and may function both as a tissue-specific transcription activator and as an accessory factor for the glucocorticoid response (218). The observation that AF1 is the site of the first protein-DNA interaction in the developmental expression of the PEPCK gene in liver further establishes the pleiotropic nature of this domain (218a).

The AF2 element also serves more than one function, as it coincides both with an insulin response sequence (IRS) and a phorbol ester response sequence (PRS) (Figure 4A) (50, 105). Mutations that abolish the insulin and phorbol ester effects also disable AF2 as an accessory element for the glucocorticoid response. This coincident location of "elements", which are obviously not exclusive HREs, may explain the dominant action insulin and phorbol esters have over glucocorticoids in regulating expression of the PEPCK gene. We are currently characterizing the proteins that bind the AF2 site in order to better understand this convergence of hormonal signals. It is conceivable that a single protein functions as a glucocorticoid receptor accessory factor but is negatively regulated, perhaps through a phosphorylation/dephosphorylation mechanism, by second messengers for insulin and phorbol esters. It is equally possible, however, that separate proteins mediate stimulatory and inhibitory functions through the AF2 element and compete for DNA binding.

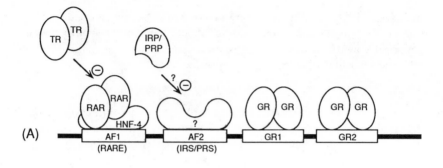

(A)

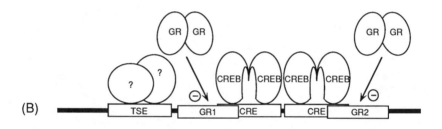

(B)

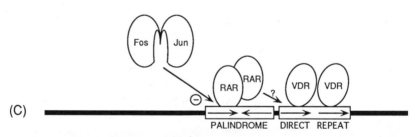

(C)

Figure 4   Schematic diagrams of several multifunctional regulatory domains are shown. Factors depicted as bound to DNA elements all mediate stimulatory effects on gene transcription. Other factors, shown above each promoter diagram, mediate negative effects by either displacing the bound factors or by interfering with the activity of the bound factors solely through protein-protein interactions. (*A*) The PEPCK gene. (*B*) The glycoprotein hormone α-subunit gene. (*C*) The osteocalcin gene. GR, glucocorticoid receptor; VDR, vitamin D receptor; IRP/PRP, insulin response protein/phorbol ester response protein.

THE GLYCOPROTEIN HORMONE α-SUBUNIT GENE   The glycoprotein hormones [luteinizing hormone (LH), follicle stimulating hormone (FSH), chorionic gonadotropin (CG), and thyroid stimulating hormone (TSH)] are heterodimeric proteins consisting of a common α-subunit and distinct β-subunits. Because these hormones are involved in modulating various metabolic and reproductive processes, it is not surprising that their synthesis is subject to complex hormonal feedback regulation. Transcription of the gene encoding the α-subunit has been studied in considerable detail in the JEG-3 human choriocarcinoma cell line. This line, of placental origin, produces CG, and transcription of the α-subunit of this hormone is stimulated by cAMP (219). Promoter sequences responsible for this effect have been mapped to a region between -146 and -111 relative to the transcription start site (34, 35, 220). This region is composed of two 18-base-pair repeats, each of which contains a binding site for the transcription factor CREB (Figure 4B). Located immediately upstream of these CREs is a tissue-specific enhancer (TSE) that binds at least two distinct nuclear factors (220, 221) (Figure 4B). Placental-specific expression, however, requires both the TSE and the CREs (220). As a result, CREB is thought to play a dual role in that it appears to be required for both tissue-specific and hormone-induced gene expression.

In contrast, glucocorticoids negatively regulate α-subunit gene transcription in placental cells, an effect that is receptor dependent (222). Using DNase I protection assays, the binding of glucocorticoid receptors to three sites in the α-subnit promoter was demonstrated (222) (Figure 4B). Two of these binding sites overlap the CREs, while the third (not shown in Figure 4B) overlaps the CCAAT-box sequence in the basal promoter. Based on these findings, the authors hypothesized that glucocorticoid receptors interfere with high-level transcription by directly competing with positive factors, primarily CREB, for DNA binding. However, because competition has not yet been demonstrated using either in vitro or in vivo binding assays, the possibility remains that both the receptors and CREB proteins can bind simultaneously and that inhibition occurs through some undefined protein-protein interaction. In support of their model, they and others have shown that an intact DNA-binding domain in the glucocorticoid receptor is essential for the inhibition to be observed (222, 223). It should be noted, however, that this model has been challenged recently by Chatterjee et al (224), who were unable to demonstrate high-affinity binding of glucocorticoid receptors to the α-subunit gene promoter. This group suggested that receptors interfere with the transactivating potential of enhancer-binding proteins or other associated factors solely through protein-protein interactions.

THE OSTEOCALCIN GENE   Osteocalcin is an abundant noncollagenous protein synthesized by the osteoblast cells of bone. The protein functions in the process of bone mineralization, and is under the direct transcriptional control of vitamin D [1,25(OH)$_2$D$_3$], which has effects on bone remodeling (225). Because the vitamin D receptor is structurally similar to members of the steroid/thyroid/retinoid receptor superfamily, it has been assumed for some time that vitamin D receptors bind to, and regulate transcription through, discrete DNA elements. Indeed, as we discussed earlier, Umesono et al demonstrated transcriptional regulation by vitamin D through a synthetic HRE composed of a direct repeat of an AGGTCA motif separated by three base pairs (85). Until the study of osteocalcin gene expression, however, no naturally occurring vitamin D response elements (VDREs) were known. Using osteoblastlike ROS 17/2.8 cells transfected with osteocalcin-CAT reporter constructs, two groups identified a VDRE within a relatively short region located approximately 500 base pairs upstream of the human osteocalcin gene transcription start site (226, 227). Despite this early success, fine-mapping of the element has proven controversial. It was initially suggested that a palindromic sequence (GGTGACTCACC; see also Figure 4C) was largely responsible for mediating the hormone's effect on transcription (227, 228). More recently, however, a detailed study has implicated sequences immediately downstream of this palindrome in both receptor-binding and in hormone-dependent transactivation (229). These sequences contain a direct repeat of a GGGTCA-like motif separated by three base pairs (Figure 4C). Characterization of the VDRE in the rat osteocalcin gene promoter has provided additional insight into this matter. In both the rat and human, osteocalcin promoter sequences are highly conserved (14/17-base-pair match) through the direct repeat domain, but appear to diverge upstream, where the palindromic sequence is located in the human gene. As in the human gene promoter, the direct repeats of the rat gene appear to both bind vitamin D receptors and mediate the hormone's effect on transcription (230–232). Nevertheless, the fact remains that mutation or deletion of bases within the palindromic region do affect vitamin D responsiveness to some degree (228, 229). It is therefore possible that factors other than the receptor, which binds to the direct repeats, interact with upstream sequences and augment hormone responsiveness.

Schüle et al extended studies of the human osteocalcin VDRE by examining the ability of the element to mediate effects of other hormones (228). Interestingly, RA was equally as effective as vitamin D at inducing transcription from an osteocalcin-CAT reporter construct, an effect that was entirely abolished upon site-directed deletion of the VDRE (228). Furthermore, when the VDRE was transferred to a heterologous, MMTV-based reporter, RA induced expression of this construct as well, although the extent of induction was not reported. It should be noted that these results were in stark contrast to those of Morrison et al, who also observed RA regulation of osteocalcin-CAT constructs but were unable to achieve regulation by ligating the VDRE to a TK-CAT reporter construct (227). These differences may be explained by the fact that, when examining responsiveness of the VDRE-MMTV-CAT construct, Schüle et al carried out their studies by cotransfecting an RAR expression vector to boost receptor levels in transfected cells (228). In further support of the notion that the VDRE also represents an RARE, Schüle et al demonstrated binding of RARs to a short oligonucleotide containing the VDRE. It is not clear, however, exactly where the receptor binds within the element, although a limited analysis with mutant competitor oligonucleotides suggested interaction with the palindromic motif (Figure 4C).

Finally, the central portion of the palindromic motif is identical to the AP-1 element consensus sequence, TGACTCA (see Figure 1). As a result, Schüle et al tested the ability of coexpressed Jun and Fos proteins to regulate transcription from the osteocalcin promoter (228). They demonstrated profound inhibition of osteocalcin-CAT expression by these proteins under either basal conditions, or following induction by vitamin D or RA. Furthermore, the effect was totally dependent on the presence of the AP-1 site, and in vitro binding experiments conclusively demonstrated the interaction of the Fos-Jun complex with DNA at this location (Figure 4C). Other examples of the involvement of AP-1 with hormone receptors are discussed in the next section.

To summarize, a relatively small domain in the human osteocalcin gene promoter appears to integrate signals involved in pleiotropic responses. Owen et al have proposed a model for understanding the role of this multifunctional response domain in the physiological expression of osteocalcin (233). They suggest that elevated AP-1 activity, present in proliferating, immature bone-forming cells, inhibits the production of osteocalcin, thereby preventing the appearance of a differentiated cell phenotype. When proliferation slows and AP-1 activity decreases, both RA and vitamin D can stimulate osteocalcin gene transcription through the gene's regulatory domain, unhindered by the interaction of Fos and Jun with neighboring or overlapping DNA sequences. In this way, osteocalcin is expressed and hormonally regulated only in mature, differentiated cells.

Crosstalk Between Steroid Receptors and AP-1

The osteocalcin gene is only one example of signal transduction convergence involving factors sensitive to steroidlike hormones and to mitogenic stimulation. In the past year, several groups have explored the functional interaction of steroid receptors with the AP-1 complex in the context of a few transcriptionally regulated genes. Although the studies yielded some common conclusions, there are also some striking differences that have yet to be resolved.

Diamond et al (234) characterized a "composite response element" that was known to be required for both phorbol ester stimulation and glucocorticoid repression of proliferin gene transcription (235). They demonstrated that, although this element mediates an inhibitory effect of glucocorticoids in a number of cell types, in others, such as the human HeLa cells, glucocorticoids actually induce gene expression through the element (234). Furthermore, within a single cell type, the element could mediate both repressive and stimulatory effects of the hormone depending upon the intracellular ratio of the Jun and Fos proteins. Glucocorticoids enhanced transcription when Jun levels were elevated by transfection with a c-*jun* expression plasmid, whereas when Fos levels were elevated in a similar way, glucocorticoids repressed transcription (234). These functional studies were followed by an analysis of protein-DNA and protein-protein interactions. DNA-binding studies revealed that the composite element served as a binding site for the AP-1 complex and for the glucocorticoid receptor, both of which can evidently occupy the site simultaneously. Furthermore, crosslinking studies demonstrated that the glucocorticoid receptor and Jun specifically interact with one another, while Fos can bind the receptor only indirectly, through its interaction with Jun. These findings implicate a scheme of regulation whereby the composition of the AP-1 complex (i.e. whether it is composed predominantly of Jun-Jun homodimers or of Fos-Jun heterodimers) affects the ability of the glucocorticoid receptor to mediate a positive or negative transcriptional effect, but only when all proteins are clustered on the composite DNA element. More recently, a similar mechanism has been proposed for regulation of the α-fetoprotein gene (236).

In contrast, somewhat different models have been proposed by groups studying the collagenase gene promoter and synthetic promoters that are repressed by glucocorticoids and stimulated by AP-1 (237–240). In these studies, glucocorticoid repression is mediated through an AP-1 element but is thought to occur without direct binding of receptor to DNA. Rather, some have suggested that the receptor becomes "tethered" to the promoter through protein-protein interaction with the AP-1 complex (237), while others have proposed that receptor/AP-1 interaction occurs in the absence of DNA and results in the mutual inhibition of DNA binding (238, 239). This latter model appears to explain the ability of AP-1 to inhibit glucocorticoid receptor transactivation through a consensus GRE. Perhaps one of the most striking differences among all these studies involves the characterization of the protein interactions responsible for complex formation between the glucocorticoid receptor and AP-1. While some have detected a specific interaction between only Jun and the receptor (234), others have observed both receptor-Jun and receptor-Fos interactions (238), while still others have been unable to detect any such crossfamily, protein-protein binding (239, 240). These differences, as well as others, have been elegantly addressed in a recent review by Miner et al (241). Very recently, numerous studies have implicated similar interactions between AP-1 and various other hormone receptors (242–246). Undoubtedly, future research into this phenomenon of signal transduction crosstalk will help resolve some of the conflicting notions that presently exist in the literature.

CONCLUSIONS

The identification of hormone response elements (HREs), short segments of DNA that mediate effects of hormones on transcription, was a major advance. The interaction of these elements with hormone receptors, or other cognate trans-acting proteins, can alter the expression of reporter gene constructs. As studies progressed, it became apparent that maximal activity of many HREs, particularly in the context of the natural promoter, requires additional elements/factors. In some instances this involves local assemblies, but in other cases components can be quite widely separated. These assemblies may consist of multimers of an HRE, all binding the same trans-acting factor, or of HREs in combination with different cis-elements, each binding a distinct trans-acting factor. A variety of cis-elements can fulfill this role, including those typically associated with promoting basal or tissue-specific activity as well as those responsive to other signal transduction pathways. Of particular interest in this regard is the involvement of AP-1 in hormone signaling. The modulation of several steroid hormone responses by members of the AP-1 family has already been described, and more examples will likely follow. As a result, the interactions between the AP-1 complex and steroid hormone receptors may be considered the prototypic example of crosstalk between divergent classes or families of transcription factors. The multiplicity of intrafamily and interfamily protein-protein interactions, which may even extend to involve chromatin and the nuclear matrix, creates the potential for great flexibility in the hormonal regulation of gene transcription.

The degree of complexity of a hormone-responsive promoter may be directly related to the complexity of the process being regulated by the protein product of the gene. If so, one might expect great diversity in the structure of hormone-responsive promoters, and in the extreme view, each may be unique. In this context the term HRE is too restrictive, since it defines only a hormone receptor-binding site, and it is now apparent that few, if any, such elements function by themselves. The term HRU, meant to describe a complex of elements concerned with regulation by a single hormone, may be applicable in certain instances, but it is too restrictive to explain multihormonal, additive, synergistic, negative, and dominant types of regulation, in addition to tissue-specific and developmentally timed expression. The constellation of elements and factors responsible for combinations of these effects can best be thought of as existing in response domains. A response domain indicates a sphere of influence or control without readily identifiable boundaries. As evidence accumulates, it will be interesting to see whether common features of domains are established, and whether the components of domains might be predictable a priori.

ACKNOWLEDGMENTS

We thank Herbert H. Samuels, Keith R. Yamamoto, Richard M. O'Brien, and Robert K. Hall for their many helpful comments and suggestions, and Deborah Caplenor for her help in preparing the manuscript. Our research on this subject has been supported, in part, by NIH Grant DK35107, the Vanderbilt Diabetes and Endocrinology Research Center (DK20593), and an American Diabetes Association Mentor-Based Training Grant. P.C.L. is supported by the Vanderbilt/NIH Medical Scientist Training Program (GM07347).

Literature Cited

1. O'Malley, B. 1990. *Mol. Endocrinol.* 4:363–69
2. Lewin, B. 1990. *Cell* 61:1161–64
3. Yamamoto, K. R., Chandler, V. L., Ross, S. R., Ucker, D. S., Ring, J. C., Feinstein, S. C. 1981. *Cold Spring Harbor Symp. Quant. Biol.* 45:687–97
4. Buetti, E., Diggelmann, H. 1981. *Cell* 23:335–45
5. Lee, F., Mulligan, R., Berg, P., Ringold, G. 1981. *Nature* 294:228–32
6. Huang, A. L., Ostrowski, M. C., Berard, D., Hager, G. L. 1981. *Cell* 27:245–55
7. Payvar, F., Wrange, Ö., Carlstedt-Duke, J., Okret, S., Gustafsson, J.-Å., Yamamoto, K. R. 1981. *Proc. Natl. Acad. Sci. USA* 78:6628–32
8. Govindan, M. V., Spiess, E., Majors, J. 1982. *Proc. Natl. Acad. Sci. USA* 79:5157–61
9. Payvar, F., Firestone, G. L., Ross, S. R., Chandler, V. L., Wrange, Ö., et al. 1982. *J. Cell. Biochem.* 19:241–47
10. Pfahl, M. 1982. *Cell* 31:475–82
11. Chandler, V. L., Maler, B. A., Yamamoto, K. R. 1983. *Cell* 33:489–99
12. Yamamoto, K. R. 1985. *Annu. Rev. Genet.* 19:209–52
13. Umesono, K., Giguere, V., Glass, C. K., Rosenfeld, M. G., Evans, R. M. 1988. *Nature* 336:262–65
14. Klug, A., Rhodes, D. 1987. *Trends Biochem. Sci.* 12:464–69
15. Forman, B. M., Yang, C.-R., Au, M., Casanova, J., Ghysdael, J., Samuels, H. H. 1989. *Mol. Endocrinol.* 3:1610–26
16. Forman, B. M., Samuels, H. H. 1990. *Mol. Endocrinol.* 4:1293–301
17. Landschulz, W. H., Johnson, P. F., McKnight, S. L. 1988. *Science* 240:1759–64
18. Kumar, V., Chambon, P. 1988. *Cell* 55:145–56
19. Wrange, Ö., Eriksson, P., Perlmann, T. 1989. *J. Biol. Chem.* 264:5253–59
20. DeMarzo, A. M., Beck, C. A., Oñate, S. A., Edwards, D. P. 1991. *Proc. Natl. Acad. Sci. USA* 88:72–76
21. Cairns, W., Cairns, C., Pongratz, I., Poellinger, L., Okret, S. 1991. *J. Biol. Chem.* 266:11221–26
22. Tsai, S. Y., Carlstedt-Duke, J., Weigel, N. L., Dahlman, K., Gustafsson, J.-Å., Tsai, M.-J., O'Malley, B. W. 1988. *Cell* 55:361–69
23. Dahlman-Wright, K., Siltala-Roos, H., Carlstedt-Duke, J., Gustafsson, J.-Å. 1990. *J. Biol. Chem.* 265:14030–35
24. Härd, T., Kellenbach, E., Boelens, R., Maler, B. A., Dahlman, K., et al. 1990. *Science* 249:157–60
25. Luisi, B. F., Xu, W. X., Otwinowski, Z., Freedman, L. P., Yamamoto, K. R., Sigler, P. B. 1991. *Nature* 352:497–505
26. Evans, R. M. 1988. *Science* 240:889–95
27. Beato, M. 1989. *Cell* 56:335–44
28. Tsai, S. Y., Tsai, M.-J., O'Malley, B. W. 1991. In *Nuclear Hormone Receptors,* ed. M. G. Parker, pp. 103–24. London: Academic
29. Martinez, E., Wahli, W. 1991. See Ref. 28, pp. 125–53
30. Montminy, M. R., Sevarino, K. A., Wagner, J. A., Mandel, G., Goodman, R. H. 1986. *Proc. Natl. Acad. Sci. USA* 83:6682–86
31. Comb, M., Birnberg, N. C., Seasholtz, A., Herbert, E., Goodman, H. M. 1986. *Nature* 323:353–56
32. Short, J. M., Wynshaw-Boris, A., Short, H. P., Hanson, R. W. 1986. *J. Biol. Chem.* 261:9721–26
33. Tsukada, T., Fink, J. S., Mandel, G., Goodman, R. H. 1987. *J. Biol. Chem.* 262:8743–47
34. Deutsch, P. J., Jameson, J. L., Habener, J. F. 1987. *J. Biol. Chem.* 262:12169–74
35. Silver, B. J., Bokar, J. A., Virgin, J. B., Valen, E. A., Milsted, A., Nilson, J. H. 1987. *Proc. Natl. Acad. Sci. USA* 84:2198–202

36. Quinn, P. G., Wong, T. W., Magnuson, M. A., Shabb, J. B., Granner, D. K. 1988. *Mol. Cell. Biol.* 8:3467–75
37. Montminy, M. R., Gonzalez, G. A., Yamamoto, K. K. 1990. *Recent Prog. Horm. Res.* 46:219–30
38. Hoeffler, J. P., Meyer, T. E., Yungdae, Y., Jameson, J. L., Habener, J. F. 1988. *Science* 242:1430–33
39. Gonzalez, G. A., Yamamoto, K. K., Fischer, W. H., Karr, D., Menzel, P., et al. 1989. *Nature* 337:749–52
40. Habener, J. F. 1990. *Mol. Endocrinol.* 4:1087–94
41. Gonzalez, G. A., Montminy, M. R. 1989. *Cell* 59:675–80
42. Yamamoto, K. K., Gonzalez, G. A., Biggs, W. H. III, Montminy, M. R. 1988. *Nature* 334:494–98
43. Quinn, P. G., Granner, D. K. 1990. *Mol. Cell. Biol.* 10:3357–64
44. Imagawa, M., Chiu, R., Karin, M. 1987. *Cell* 51:251–60
45. Kagawa, N., Waterman, M. R. 1990. *J. Biol. Chem.* 265:11299–305
46. Kagawa, N., Waterman, M. R. 1991. *J. Biol. Chem.* 266:11199–204
47. Metz, R., Ziff, E. 1991. *Genes Dev.* 5:1754–66
48. Ransone, L. J., Verma, I. M. 1990. *Annu. Rev. Cell Biol.* 6:539–57
49. Rahmsdorf, H. J., Herrlich, P. 1990. *Pharmacol. Ther.* 48:157–88
50. O'Brien, R. M., Lucas, P. C., Forest, C. D., Magnuson, M. A., Granner, D. K. 1990. *Science* 249:533–37
51. Keller, S. A., Rosenberg, M. P., Johnson, T. M., Howard, G., Meisler, M. H. 1990. *Genes Dev.* 4:1316–21
52. Nasrin, N., Ercolani, L., Denaro, M., Kong, X. F., Kang, I., Alexander, M. 1990. *Proc. Natl. Acad. Sci. USA* 87:5273–77
53. Philippe, J. 1991. *Proc. Natl. Acad. Sci. USA* 88:7224–27
54. O'Brien, R. M., Granner, D. K. 1991. *Biochem. J.* 278:609–19
55. Levy, D., Darnell, J. E. Jr. 1990. *New Biol.* 2:923–28
56. Bradshaw, M. S., Tsai, M.-J., O'Malley, B. W. 1988. *Mol. Endocrinol.* 2:1286–93
57. Strähle, U., Schmid, W., Schütz, G. 1988. *EMBO J.* 7:3389–95
58. Schatt, M. D., Rusconi, S., Schaffner, W. 1990. *EMBO J.* 9:481–87
59. Freedman, L. P., Yoshinaga, S. K., Vanderbilt, J. N., Yamamoto, K. R. 1989. *Science* 245:298–301
60. Bagchi, M. K.,Tsai, S. Y., Weigel, N. L., Tsai, M.-J., O'Malley, B. W. 1990. *J. Biol. Chem.* 265:5129–34
61. Tsai, S. Y., Srinivasan, G., Allan, G. F., Thompson, E. B., O'Malley, B. W.,

Tsai, M.-J. 1990. *J. Biol. Chem.* 265:17055–61
62. Martinez, E., Givel, F., Wahli, W. 1987. *EMBO J.* 6:3719–27
63. Klein-Hitpass, L., Kaling, M., Ryffel, G. U. 1988. *J. Mol. Biol.* 201:537–44
64. Ponglikitmongkol, M., White, J. H., Chambon, P. 1990. *EMBO J.* 9:2221–31
65. Glass, C. K., Holloway, J. M., Devary, O. V., Rosenfeld, M. G. 1988. *Cell* 54:313–23
66. Weih, F., Stewart, A. F., Boshart, M., Nitsch, D., Schütz, G. 1990. *Genes Dev.* 4:1437–49
67. Seiler-Tuyns, A., Walker, P., Martinez, E., Mérillat, A.-M., Givel, F., Wahli, W. 1986. *Nucleic Acids Res.* 14:8755–70
68. Tsai, S. Y., Tsai, M.-J., O'Malley, B. W. 1989. *Cell* 57:443–48
69. Schmid, W., Strähle, U., Schütz, G., Schmitt, J., Stunnenberg, H. 1989. *EMBO J.* 8:2257–63
70. Martinez, E., Wahli, W. 1989. *EMBO J.* 8:3781–91
71. Nordeen, S. K., Suh, B. J., Kühnel, B., Hutchison, C. A. III. 1990. *Mol. Endocrinol.* 4:1866–73
72. Vasios, G. W., Gold, J. D., Petkovich, M., Chambon, P., Gudas, L. J. 1989. *Proc. Natl. Acad. Sci. USA* 86:9099–103
73. Vasios, G., Mader, S., Gold, J. D., Leid, M., Lutz, Y., et al. 1991. *EMBO J.* 10:1149–58
74. de Thé, H., Vivanco-Ruiz, M. del M., Tiollais, P., Stunnenberg, H., Dejean, A. 1990. *Nature* 343:177–80
75. Sucov, H. M., Murakami, K. K., Evans, R. M. 1990. *Proc. Natl. Acad. Sci. USA* 87:5392–96
76. Hoffmann, B., Lehmann, J. M., Zhang, X.-K., Hermann, T., Husmann, M., et al. 1990. *Mol. Endocrinol.* 4:1727–36
77. Lucas, P. C., O'Brien, R. M., Mitchell, J. A., Davis, C. M., Imai, E., et al. 1991. *Proc. Natl. Acad. Sci. USA* 88:2184–88
78. Lucas, P. C., Forman, B. M., Samuels, H. H., Granner, D. K. 1991. *Mol. Cell. Biol.* 11:5164–70
79. Duester, G., Shean, M. L., McBride, M. S., Stewart, M. J. 1991. *Mol. Cell. Biol.* 11:1638–46
80. Smith, W. C., Nakshatri, H., Leroy, P., Rees, J., Chambon, P. 1991. *EMBO J.* 10:2223–30
81. Baniahmad, A., Steiner, C., Köhne, A. C., Renkawitz, R. 1990. *Cell* 61:505–14
82. Näär, A. M., Boutin, J.-M., Lipkin, S. M., Yu, V. C., Holloway, J. M., et al. 1991. *Cell* 65:1267–79

83. Izumo, S., Mahdavi, V. 1988. *Nature* 334:539–42
84. Glass, C. K., Lipkin, S. M., Devary, O. V., Rosenfeld, M. G. 1989. *Cell* 59:697–708
85. Umesono, K., Murakami, K. K., Thompson, C. C., Evans, R. M. 1991. *Cell* 65:1255–66
86. Mangelsdorf, D. J., Umesono, K., Kliewer, S. A., Borgmeyer, U., Ong, E. S., Evans, R. M. 1991. *Cell* 66:555–61
87. Mangelsdorf, D. J., Ong, E. S., Dyck, J. A., Evans, R. M. 1990. *Nature* 345:224–29
88. Rottman, J. N., Widom, R. L., Nadal-Ginard, B., Mahdavi, V., Karathanasis, S. K. 1991. *Mol. Cell. Biol.* 11:3814–20
89. Widom, R. L., Ladias, J. A. A., Kouidou, S., Karathanasis, S. K. 1991. *Mol. Cell. Biol.* 11:677–87
90. Ong, D. E., Kakkad, B., MacDonald, P. N. 1987. *J. Biol. Chem.* 262:2729–36
91. MacDonald, P. N., Ong, D. E. 1988. *J. Biol. Chem.* 263:12478–82
92. Ong, D. E., Lucas, P. C., Kakkad, B., Quick, T. C. 1991. *J. Lipid Res.* 32:1521–27
93. Strähle, U., Boshart, M., Klock, G., Stewart, F., Schütz, G. 1989. *Nature* 339:629–32
94. Petkovich, P. M., Heersche, J. N. M., Tinker, D. O., Jones, G. 1984. *J. Biol. Chem.* 259:8274–80
95. Clarke, C. L., Roman, S. D., Graham, J., Koga, M., Sutherland, R. L. 1990. *J. Biol. Chem.* 265:12694–700
96. Ludwig, K. W., Lowey, B., Niles, R. M. 1980. *J. Biol. Chem.* 255:5999–6002
97. Plet, A., Evain, D., Anderson, W. B. 1982. *J. Biol. Chem.* 257:889–93
98. Durham, J. P., Emler, C. A., Butcher, F. R., Fontana, J. A. 1985. *FEBS Lett.* 185:157–61
99. Zylber-Katz, E., Glazer, R. I. 1985. *Cancer Res.* 45:5159–64
100. Makowske, M., Ballester, R., Cayre, Y., Rosen, O. M. 1988. *J. Biol. Chem.* 263:3402–10
101. Granner, D. K., Andreone, T., Sasaki, K., Beale, E. 1983. *Nature* 305:549–51
102. Sasaki, K., Cripe, T. P., Koch, S. R., Andreone, T. L., Petersen, D. D., et al. 1984. *J. Biol. Chem.* 259:15242–51
103. Magnuson, M. A., Andreone, T. L., Printz, R. L., Koch, S., Granner, D. K. 1989. *Proc. Natl. Acad. Sci. USA* 86:4838–42
104. Forest, C. D., O'Brien, R. M., Lucas, P. C., Magnuson, M. A., Granner, D. K. 1990. *Mol. Endocrinol.* 4:1302–10
105. O'Brien, R. M., Bonovich, M. T., For-

est, C. D., Granner, D. K. 1991. *Proc. Natl. Acad. Sci. USA* 88:6580–84
106. Hudson, L. G., Santon, J. B., Glass, C. K., Gill, G. N. 1990. *Cell* 62:1165–75
107. Carr, F. E., Burnside, J., Chin, W. W. 1989. *Mol. Endocrinol.* 3:709–16
108. Wondisford, F. E., Farr, E. A., Radovick, S., Steinfelder, H. J., Moates, J. M., et al. 1989. *J. Biol. Chem.* 264:14601–4
109. Darling, D. S., Burnside, J., Chin, W. W. 1989. *Mol. Endocrinol.* 3:1359–68
110. Brent, G. A., Williams, G. R., Harney, J. W., Forman, B. M., Samuels, H. H., et al. 1991. *Mol. Endocrinol.* 5:542–48
111. Oppenheimer, J. H., Schwartz, H. L., Mariash, C. N., Kinlaw, W. B., Wong, N. C.W., Freake, H. C. 1987. *Endocr. Rev.* 8:288–308
112. Spindler, B. J., MacLeod, K. M., Ring, J., Baxter, J. D. 1975. *J. Biol. Chem.* 250:4113–19
113. MacLeod, K. M., Baxter, J. D. 1976. *J. Biol. Chem.* 251:7380–87
114. Groul, D. J. 1980. *Endocrinology* 107:994–99
115. Jump, D. B., Seelig, S., Schwartz, H. L., Oppenheimer, J. H. 1981. *Biochemistry* 20:6781–89
116. Perlman, A. J., Stanley, F., Samuels, H. H. 1982. *J. Biol. Chem.* 257:930–38
117. Burnside, J., Darling, D. S., Carr, F. E., Chin, W. W. 1989. *J. Biol. Chem.* 264:6886–91
117a. Chatterjee, V. K. K., Lee, J.-K., Rentoumis, A., Jameson, J. L. 1989. *Proc. Natl. Acad. Sci. USA* 86:9114–18
117b. Strömstedt, P.-E., Poellinger, L., Gustafsson, J.-Å., Carlstedt-Duke, J. 1991. *Mol. Cell. Biol.* 11:3379–83
118. Sakai, D. D., Helms, S., Carlstedt-Duke, J., Gustafsson, J.-Å., Rottman, F. M., Yamamoto, K. R. 1988. *Genes Dev.* 2:1144–54
119. Drouin, J., Sun, Y. L., Nemer, M. 1990. *Trends Endocrinol. Metab.* 2:219–25
119a. Therrien, M., Drouin, J. 1991. *Mol. Cell. Biol.* 11:3492–503
120. Sap, J., Muñoz, A., Damm, K., Goldberg, Y., Ghysdael, J., et al. 1986. *Nature* 324:635–40
121. Thompson, C. C., Weinberger, C., Lebo, R., Evans, R. M. 1987. *Science* 237:1610–14
122. Murray, M. B., Zilz, N. D., McCreary, N. L., MacDonald, M. J., Towle, H. C. 1988. *J. Biol. Chem.* 263:12770–77
123. Nakai, A., Sakurai, A., Bell, G. I., DeGroot, L. J. 1988. *Mol. Endocrinol.* 2:1087–92
124. Prost, E., Koenig, R. J., Moore, D. D.,

Larsen, P. R., Whalen, R. G. 1988. *Nucleic Acids Res.* 16:6248
125. Weinberger, C., Thompson, C. C., Ong, E. S., Lebo, R., Gruol, D. J., Evans, R. M. 1986. *Nature* 324:641–46
126. Koenig, R. J., Warne, R. L., Brent, G. A., Harney, J. W., Larsen, P. R., Moore, D. D. 1988. *Proc. Natl. Acad. Sci. USA* 85:5031–35
127. Hodin, R. A., Lazar, M. A., Wintman, B. I., Darling, D. S., Koenig, R. J., et al. 1989. *Science* 244:76–79
128. Benbrook, D., Pfahl, M. 1987. *Science* 238:788–91
129. Nakai, A., Seino, S., Sakurai, A., Szilak, I., Bell, G. I., DeGroot, L. J. 1988. *Proc. Natl. Acad. Sci. USA* 85:2781–85
130. Izumo, S., Mahdavi, V. 1988. *Nature* 334:539–42
131. Mitsuhashi, T., Tennyson, G. E., Nikodem, V. M. 1988. *Proc. Natl. Acad. Sci. USA* 85:5804–8
132. Lazar, M. A., Hodin, R. A., Darling, D. S., Chin, W. W. 1988. *Mol. Endocrinol.* 2:893–901
133. Lazar, M. A., Hodin, R. A., Chin, W. W. 1989. *Proc. Natl. Acad. Sci. USA* 86:7771–74
134. Schueler, P. A., Schwartz, H. L., Strait, K. A., Mariash, C. N., Oppenheimer, J. H. 1990. *Mol. Endocrinol.* 4:227–34
135. Koenig, R. J., Lazar, M. A., Hodin, R. A., Brent, G. A., Larsen, P. R., et al. 1989. *Nature* 337:659–61
136. Damm, K., Thompson, C. C., Evans, R. M. 1989. *Nature* 339:593–97
137. Sap, J., Muñoz, A., Schmitt, J., Stunnenberg, H., Vennström, B. 1989. *Nature* 340:242–44
138. Graupner, G., Wills, K. N., Tzukerman, M., Zhang, X.-K., Pfahl, M. 1989. *Nature* 340:653–56
139. Brent, G. A., Dunn, M. K., Harney, J. W., Gulick, T., Larsen, P. R., Moore, D. D. 1989. *New Biol.* 1:329–36
140. Holloway, J. M., Glass, C. K., Adler, S., Nelson, C. A., Rosenfeld, M. G. 1990. *Proc. Natl. Acad. Sci. USA* 87:8160–64
141. Graupner, G., Zhang, X.-K., Tzukerman, M., Wills, K., Hermann, T., Pfahl, M. 1991. *Mol. Endocrinol.* 5:365–72
142. Selmi, S., Samuels, H. H. 1991. *J. Biol. Chem.* 266:11589–93
143. Chiu, R., Angel, P., Karin, M. 1989. *Cell* 59:979–86
144. Schütte, J., Viallet, J., Nau, M., Segal, S., Fedorko, J., Minna, J. 1989. *Cell* 59:987–97
145. Foulkes, N. S., Borrelli, E., Sassone-Corsi, P. 1991. *Cell* 64:739–49
146. Foulkes, N. S., Laoide, B. M., Schlot-

147. Harada, H., Fujita, T., Miyamoto, M., Kimura, Y., Maruyama, M., et al. 1989. *Cell* 58:729–39
148. de Thé, H., Chomienne, C., Lanotte, M., Degos, L., Dejean, A. 1990. *Nature* 347:558–61
149. Borrow, J., Goddard, A. D., Sheer, D., Solomon, E. 1990. *Science* 249:1577–80
150. Alcalay, M., Zangrilli, D., Pandolfi, P. P., Longo, L., Mencarelli, A., et al. 1991. *Proc. Natl. Acad. Sci. USA* 88:1977–81
151. Kakizuka, A., Miller, W. H. Jr., Umesono, K., Warrell, R. P. Jr., Frankel, S. R., et al. 1991. *Cell* 66:663–74
152. de Thé, H., Lavau, C., Marchio, A., Chomienne, C., Degos, L., Dejean, A. 1991. *Cell* 66:675–84
153. Ponta, H., Kennedy, N., Skroch, P., Hynes, N. E., Groner, B. 1985. *Proc. Natl. Acad. Sci. USA* 82:1020–24
154. Cordingley, M. G., Riegel, A. T., Hager, G. L. 1987. *Cell* 48:261–70
155. Miksicek, R., Borgmeyer, U., Nowock, J. 1987. *EMBO J.* 6:1355–60
156. Brüggemeier, U., Rogge, L., Winnacker, E.-L., Beato, M. 1990. *EMBO J.* 9:2233–39
157. Perlmann, T., Eriksson, P., Wrange, Ö. 1990. *J. Biol. Chem.* 265:17222–29
158. Richard-Foy, H., Hager, G. L. 1987. *EMBO J.* 6:2321–28
159. Perlmann, T., Wrange, Ö. 1988. *EMBO J.* 7:3073–79
160. Piña, B., Brüggemeier, U., Beato, M. 1990. *Cell* 60:719–31
161. Archer, T. K., Cordingley, M. G., Wolford, R. G., Hager, G. L. 1991. *Mol. Cell. Biol.* 11:688–98
162. Zaret, K. S., Yamamoto, K. R. 1984. *Cell* 38:29–38
163. Cato, A. C. B., Skroch, P., Weinmann, J., Butkeraitis, P., Ponta, H. 1988. *EMBO J.* 7:1403–10
164. Kalff, M., Gross, B., Beato, M. 1990. *Nature* 344:360–62
165. Brüggemeier, U., Kalff, M., Franke, S., Scheidereit, C., Beato, M. 1991. *Cell* 64:565–72
166. Muller, M., Baniahmad, C., Kaltschmidt, C., Schüle, R., Renkawitz, R. 1991. See Ref. 28, pp. 155–74
167. Granner, D. K., Beale, E. G. 1985. In *Biochemical Actions of Hormones*, ed. G. Litwack, pp. 89–138. Orlando: Academic
168. Hashimoto, S., Schmid, W., Schütz, G. 1984. *Proc. Natl. Acad. Sci. USA* 81:6637–41
169. Schmid, E., Schmid, W., Jantzen, M.,

ter, F., Sassone-Corsi, P. 1991. *Proc. Natl. Acad. Sci. USA* 88:5448–52

Mayer, D., Jastorff, B., Schütz, G.
1987. *Eur. J. Biochem.* 165:499–506
170. Becker, P., Renkawitz, R., Schütz, G.
1984. *EMBO J.* 3:2015–20
171. Jantzen, H.-M., Strähle, U., Gloss, B.,
Stewart, F., Schmid, W., et al. 1987.
Cell 49:29–38
172. Reik, A., Schütz, G., Stewart, A. F.
1991. *EMBO J.* 10:2569–76
173. Jones, K. A., Yamamoto, K. R., Tjian,
R. 1985. *Cell* 42:559–72
174. Becker, P. B., Gloss, B., Schmid, W.,
Strähle, U., Schütz, G. 1986. *Nature*
324:686–88
175. Danesch, U., Hashimoto, S., Renkawitz, R., Schütz, G. 1983. *J. Biol.
Chem.* 258:4750–53
176. Danesch, U., Gloss, B., Schmid, W.,
Schütz, G., Schüle, R., Renkawitz, R.
1987. *EMBO J.* 6:625–30
177. Dierks, P., van Ooyen, A., Cochran, M.
D., Dobkin, C., Reiser, J., Weissmann,
C. 1983. *Cell* 32:695–706
178. Myers, R. M., Tilly, K., Maniatis, T.
1986. *Science* 232:613–18
179. Schüle, R., Muller, M., Otsuka-
Murakami, H., Renkawitz, R. 1988.
Nature 332:87–90
180. Imai, E., Stromstedt, P.-E., Quinn, P.
G., Carlstedt-Duke, J., Gustafsson,
J.-Å., Granner, D. K. 1990. *Mol. Cell.
Biol.* 10:4712–19
181. Bokar, J. A., Roesler, W. J., Vanden-
bark, G. R., Kaetzel, D. M., Hanson,
R. W., Nilson, J. H. 1988. *J. Biol.
Chem.* 263:19740–47
182. Park, E. A., Roesler, W. J., Liu, J.,
Klemm, D. J., Gurney, A. L., et al.
1990. *Mol. Cell. Biol.* 10:6264–72
183. Liu, J., Park, E. A., Gurney, A. L.,
Roesler, W. J., Hanson, R. W. 1991. *J.
Biol. Chem.* 266:19095–102
184. Kulkarni, A. B., Reinke, R., Feigelson,
P. 1985. *J. Biol. Chem.* 260:15386–89
185. Vannice, J. L., Taylor, J. M., Ringold,
G. M. 1984. *Proc. Natl. Acad. Sci. USA*
81:4241–45
186. Baumann, H., Maquat, L. E. 1986.
Mol. Cell. Biol. 6:2551–61
187. Klein, E. S., Reinke, R., Feigelson, P.,
Ringold, G. M. 1987. *J. Biol. Chem.*
262:520–23
188. Baumann, H., Jahreis, G. P., Morella,
K. K. 1990. *J. Biol. Chem.* 265:22275–
81
189. Baumann, H., Firestone, G. L.,
Burgess, T. L., Gross, K. W., Yamamoto, K. R., Held, W. A. 1983. *J. Biol.
Chem.* 258:563–70
190. Vannice, J. L., Ringold, G. M.,
McLean, J. W., Taylor, J. M. 1983.
DNA 2:205–12

191. Klein, E. S., DiLorenzo, D., Posseck-
ert, G., Beato, M., Ringold, G. M.
1988. *Mol. Endocrinol.* 2:1343–51
192. DiLorenzo, D., Williams, P., Ringold,
G. 1991. *Biochem. Biophys. Res. Commun.* 176:1326–32
193. Voz, M. L., Peers, B., Belayew, A.,
Martial, J. A. 1991. *J. Biol. Chem.*
266:13397–408
194. Catanzaro, D. F., West, B. L., Baxter,
J. D., Reudelhuber, T. L. 1987. *Mol.
Endocrinol.* 1:90–96
195. West, B. L., Catanzaro, D. F., Mellon,
S. H., Cattini, P. A., Baxter, J. D.,
Reudelhuber, T. L. 1987. *Mol. Cell.
Biol.* 7:1193–97
196. Schüle, R., Muller, M., Kaltschmidt,
C., Renkawitz, R. 1988. *Science*
242:1418–20
197. Allan, G. F., Ing, N. H., Tsai, S. Y.,
Srinivasan, G., Weigel, N. L., et al.
1991. *J. Biol. Chem.* 266:5905–10
198. Murray, M. B., Towle, H. C. 1989.
Mol. Endocrinol. 3:1434–42
199. Burnside, J., Darling, D. S., Chin, W.
W. 1990. *J. Biol. Chem.* 265:2500–4
200. Lazar, M. A., Berrodin, T. J. 1990.
Mol. Endocrinol. 4:1627–35
201. Darling, D. S., Beebe, J. S., Burnside,
J., Winslow, E. R., Chin, W. W. 1991.
Mol. Endocrinol. 5:73–84
202. Beebe, J. S., Darling, D. S., Chin, W.
W. 1991. *Mol. Endocrinol.* 5:85–93
203. O'Donnell, A. L., Rosen, E. D., Darling, D. S., Koenig, R. J. 1991. *Mol.
Endocrinol.* 5:94–99
204. Rosen, E. D., O'Donnell, A. L.,
Koenig, R. J. 1991. *Mol. Cell. Endocrinol.* 78:C83–C88
205. Glass, C. K., Devary, O. V., Rosenfeld, M. G. 1990. *Cell* 63:729–38
206. Yang, N., Schüle, R., Mangelsdorf, D.
J., Evans, R. M. 1991. *Proc. Natl.
Acad. Sci. USA* 88:3559–63
207. Edwards, D. P., Kühnel, B., Estes, P.
A., Nordeen, S. K. 1989. *Mol. Endocrinol.* 3:381–91
208. Feavers, I. M., Jiricny, J., Monchar-
mont, B., Saluz, H. P., Jost, J. P. 1987.
Proc. Natl. Acad. Sci. USA 84:7453–57
209. Liao, J., Ozono, K., Sone, T., McDonnell, D. P., Pike, J. W. 1990. *Proc.
Natl. Acad. Sci. USA* 87:9751–55
210. Rories, C., Spelsberg, T. C. 1989.
Annu. Rev. Physiol. 51:653–81
211. Spelsberg, T. C., Rories, C., Rejman, J.
J., Goldberger, A., Fink, K., et al.
1989. *Biol. Reprod.* 40:54–69
212. Goldberger, A., Spelsberg, T. C. 1988.
Biochemistry 27:2103–9
213. Schuchard, M., Rejman, J. J., McCormick, D. J., Gosse, B., Ruesink, T.,

Spelsberg, T. C. 1991. *Biochemistry* 30:4535–42

214. Goldberger, A., Horton, M., Katzmann, J., Spelsberg, T. C. 1987. *Biochemistry* 26:5811–16

215. Lamers, W. H., Hanson, R. W., Meisner, H. M. 1982. *Proc. Natl. Acad. Sci. USA* 79:5137–41

216. Chu, D. T. W., Granner, D. K. 1986. *J. Biol. Chem.* 261:16848–53

217. Sladek, F. M., Zhong, W., Lai, E., Darnell, J. E. Jr. 1990. *Genes Dev.* 4:2352–65

218. Hall, R. K., Mitchell, J., Lucas, P. C., Sladek, F., Granner, D. K. 1991. In *Abstr. Cold Spring Harbor Meet. Cancer Cells: Regulation of Eukaryotic mRNA Transcription,* p. 76. New York: Cold Spring Harbor Lab.

218a. Trus, M., Benvenisty, N., Cohen, H., Reshef, L. 1990. *Mol. Cell. Biol.* 10: 2418–22

219. Jameson, J. L., Jaffe, R. C., Gleason, S. L., Habener, J. F. 1986. *Endocrinology* 119:2560–67

220. Delegeane, A. M., Ferland, L. H., Mellon, P. L. 1987. *Mol. Cell. Biol.* 7:3994–4002

221. Jameson, J. L., Albanese, C., Habener, J. F. 1989. *J. Biol. Chem.* 264:16190–96

222. Akerblom, I. E., Slater, E. P., Beato, M., Baxter, J. D., Mellon, P. L. 1988. *Science* 241:350–53

223. Oro, A. E., Hollenberg, S. M., Evans, R. M. 1988. *Cell* 55:1109–14

224. Chatterjee, V. K. K., Madison, L. D., Mayo, S., Jameson, J. L. 1991. *Mol. Endocrinol.* 5:100–10

225. Haussler, M. R., Mangelsdorf, D. J., Komm, B. S., Terpening, C. M., Yamaoka, K., et al. 1988. *Rec. Prog. Horm. Res.* 44:263–305

226. Kerner, S. A., Scott, R. A., Pike, J. W. 1989. *Proc. Natl. Acad. Sci. USA* 86:4455–59

227. Morrison, N. A., Shine, J., Fragonas, J.-C., Verkest, V., McMenemy, M. L., Eisman, J. A. 1989. *Science* 246:1158–61

228. Schüle, R., Umesono, K., Mangelsdorf, D. J., Bolado, J., Pike, J. W., Evans, R. M. 1990. *Cell* 61:497–504

229. Ozono, K., Liao, J., Kerner, S. A., Scott, R. A., Pike, J. W. 1990. *J. Biol. Chem.* 265:21881–88

230. Demay, M. B., Gerardi, J. M., DeLuca, H. F., Kronenberg, H. M. 1990. *Proc. Natl. Acad. Sci. USA* 87:369–73

231. Markose, E. R., Stein, J. L., Stein, G. S., Lian, J. B. 1990. *Proc. Natl. Acad. Sci. USA* 87:1701-5

232. Terpening, C. M., Haussler, C. A., Jurutka, P. W., Galligan, M. A., Komm, B. S., Haussler, M. R. 1991. *Mol. Endocrinol.* 5:373–85

233. Owen, T. A., Bortell, R., Yocum, S. A., Smock, S. L., Zhang, M., et al. 1990. *Proc. Natl. Acad. Sci. USA* 87:9990–94

234. Diamond, M. I., Miner, J. N., Yoshinaga, S. K., Yamamoto, K. R. 1990. *Science* 249:1266–72

235. Mordacq, J. C., Linzer, D. I. H. 1989. *Genes Dev.* 3:760–69

236. Zhang, X.-K., Dong, J.-M., Chiu, J.-F. 1991. *J. Biol. Chem.* 266:8248–54

237. Jonat, C., Rahmsdorf, H. J., Park, K.-K., Cato, A. C. B., Gebel, S., et al. 1990. *Cell* 62:1189–204

238. Yang-Yen, H.-F., Chambard, J.-C., Sun, Y.-L., Smeal, T., Schmidt, T. J., et al. 1990. *Cell* 62:1205–15

239. Schüle, R., Rangarajan, P., Kliewer, S., Ransone, L. J., Bolado, J., et al. 1990. *Cell* 62:1217–26

240. Lucibello, F. C., Slater, E. P., Jooss, K. U., Beato, M., Müller, R. 1990. *EMBO J.* 9:2827–34

241. Miner, J. N., Diamond, M. I., Yamamoto, K. R. 1991. *Cell Growth Differ.* 2:525–30

242. Weisz, A., Rosales, R. 1990. *Nucleic Acids Res.* 18:5097–106

243. Nicholson, R. C., Mader, S., Nagpal, S., Leid, M., Rochette-Egly, C., Chambon, P. 1990. *EMBO J.* 9:4443–54

244. Gaub, M.-P., Bellard, M., Scheuer, I., Chambon, P., Sassone-Corsi, P. 1990. *Cell* 63:1267–76

245. Schüle, R., Rangarajan, P., Yang, N., Kliewer, S., Ransone, L. J., et al. 1991. *Proc. Natl. Acad. Sci. USA* 88:6092–96

246. Doucas, V., Spyrou, G., Yaniv, M. 1991. *EMBO J.* 10:2237–45

Annu. Rev. Biochem. 1992. 61:1175–1212

DISEASES OF THE MITOCHONDRIAL DNA

Douglas C. Wallace

Department of Genetics and Molecular Medicine, Emory University School of Medicine, 1462 Clifton Road, Atlanta, Georgia 30322

KEY WORDS: mitochondrial diseases, oxidative phosphorylation, maternal inheritance, degenerative diseases, aging

CONTENTS

INTRODUCTION

Over the past three years, a number of mitochondrial DNA (mtDNA) mutations that cause human disease have been identified. These are associated with

1175

0066-4154/92/0701-1175$02.00

a broad spectrum of clinical manifestations, including blindness, deafness, dementia, movement disorders, weakness, cardiac failure, diabetes, renal dysfunction, and liver disease. The molecular and biochemical characterization of mtDNA diseases has provided new insights into the nature of human degenerative diseases and has raised the possibility that deleterious mtDNA mutations may be much more common than previously thought.

THE HUMAN MITOCHONDRIAL DNA AND OXIDATIVE PHOSPHORYLATION

The human mtDNA is a 16,569-nucleotide pair (np) closed circular molecule that codes for 13 essential genes of oxidative phosphorylation (OXPHOS) plus the structural rRNAs and tRNAs necessary for their expression (Figure 1) (1–3). As a consequence of its endosymbiotic origin, the mitochondrion has an independent replication, transcription, and translation system, which combines the features of prokaryotic and eukaryotic cells (4).

Mitochondrial Biogenesis

The two strands of the mtDNA differ markedly in their base composition. The heavy (H)-strand harbors most of the guanine (G) residues, while the light (L)-strand contains most of the cytosine (C) residues. The H-strand functions as the template for the small (12S) and large (16S) rRNAs, 12 of the polypeptides, and 14 of the tRNAs. The L-strand is the template for one polypeptide (ND6) and 8 tRNAs (Figure 1) (2, 5, 6).

Mitochondrial DNA replication uses two origins, one for each strand, which are separated by two thirds of the genome. The H-strand origin is located in the major control region delineated by the displacement (D)-loop. The D-loop contains the two mtDNA promoters, one for the H-strand (P_H, np 545–567) and the other for the L-strand (P_L, np 392–445); three conserved sequence blocks, (CSBI at np 213–235, CSBII at np 299–315, and CSBII at np 346–363), and the termination-associated sequences (TAS) at np 16147–16172. The D-loop is created by a triple-stranded region containing a new H-strand segment, the 7S DNA, base paired to the parental L-strand. The 7S DNA is synthesized using an RNA primer transcribed from P_L and terminated adjacent to CSBII through cleavage of the L-strand transcript by RNAse mitochondrial RNA processing (MRP) (7–9). Synthesis of 7S DNA then begins at CSBII and ends at TAS (10), encompassing about 700 np. H-strand synthesis starts at the 7S DNA and extends clockwise around the mtDNA, in order of decreasing nucleotide numbers, displacing the parental H-strand as a single-strand loop. After traversing two thirds of the genome, the L-strand origin of replication (O_L, np 5721–5798) is exposed and L-strand synthesis is initiated by a specific primase containing the cytosol 5.8S rRNA (11).

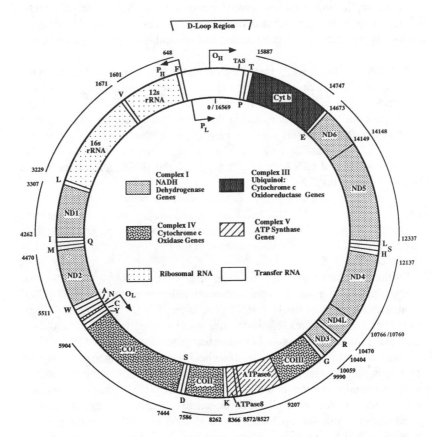

Figure 1 The human mtDNA map. The mtDNA encompasses 16,569 nps with numbering starting at O_H and proceeding counterclockwise around the circular map. The function of each gene is identified by shading according to the designations within the circle. The first and last nucleotides of the rRNAs and mRNA genes are given on the external arcs. The tRNA genes are indicated by the letter of their cognate amino acid. The names for the genes (2) are O_H = MTOH, CSBI-III = MTCSB-3, P_L = MTLSP, P_H = MTHSP1, F = tRNA^{Phe} = MTTF, 12S rRNA = MTRNR1, V = tRNA^{Val} = MTTV, 16S rRNA = MTRNR2, L = tRNA$^{Leu(UUR)}$ = MTTL1, which also contains the transcription terminator, ND1 = MTND1, I = tRNA^{Ile} = MTTI, Q = tRNA^{Gly} = MTTQ, M = tRNA^{Met} = MTTM, ND2 = MTND2, N = tRNA^{Trp} = MTTW, A = tRNA^{Ala} = MTTA, N = tRNA^{Asp} = MTTN, O_L = MTOLR, C = tRNA^{Cys} = MTTC, Y = tRNA^{Tyr} = MTTY, COI = MTCO1, S = tRNA$^{Ser(UCN)}$ = MTTS1, D = tRNA^{Asp} = MTTD, COII = MTCO2, K = tRNA^{Lys} = MTTK, ATPase8 = MTATP8, ATPase6 = MTATP6, COIII = MTCO3, G = tRNA^{Gly} = MTTG, ND3 = MTND3, R = tRNA^{Arg} = MTTR, ND4 = MTND4L, ND4 = MTND4, H = tRNA^{His} = MTTH, S = tRNA$^{Ser(AGY)}$ = MTTS2, L = tRNA$^{Leu(CUN)}$ = MTTL2, ND5 = MTNDS, ND6 = MTND6, E = tRNA^{Glu} = MTTE, Cytb = MTCYB, T = tRNA^{Thr} = MTTT, P = tRNA^{Pro} = MTTP, TAS = termination-associated sequence = MTTAS.

L-strand replication then proceeds back along the free H-strand. Thus, replication is bidirectional but asynchronous. Because the mtDNA is super-coiled, its replication and transcription can be inhibited by ethidium bromide (3).

MtDNA is symmetrically transcribed from the two promoters located in the D-loop, one for each strand (Figure 1). P_H reads counterclockwise and transcribes most of the genes. P_L reads counterclockwise. Both transcripts encompass the entire genome, and as the tRNAs are read, they fold into their three-dimensional structures and are cleaved out of the transcript by process-ing enzymes comparable to RNAseP (12). Since the tRNAs punctuate the protein coding and rRNA genes, cleavage of the tRNAs also releases the rRNAs and the mRNAs (13, 14).

The mtDNA genome is very compact. Consequently, the genome lacks introns, and most of the mitochondrial mRNAs lack 5' and 3' nontranslated sequences, starting with the initiation codon and ending with the termination codon (1, 13, 14). Poly-A tails are added to the mRNAs posttranscriptionally (1, 5, 14).

H-strand transcription generates an excess of rRNA transcripts over mRNAs owing to a bidirectional transcriptional terminator (nps 3237–3249) located downstream of the 16S rRNA gene in the tRNA$^{Leu(UUR)}$ gene. Hence, most H-strand transcripts terminate after reading the rRNA genes, and most L-strand transcripts terminate before reading the rRNA genes (15, 16).

Like bacterial protein synthesis, mitochondrial protein synthesis is initiated with formylmethionine, uses bacterialike elongation factors, and is sensitive to the bacterial ribosome inhibitor chloramphenicol (CAP) (4, 17). However, mitochondria differ from all other organisms in that they use a different genetic code, with the opal stop codon (UGA) as well as UGG used for tryptophan and the arginine codons AGA and AGG for stop codons (1, 4). Since most mitochondrial mRNAs contain multiple UGA codons, these mRNAs can not be translated in the nuclear-cytoplasmic compartment. Hence, the special mtDNA genetic code confines the mtDNA genes to expression within the mitochondrion (4).

Mitochondrial OXPHOS Complexes and their Synthesis

The mitochondrial energy-generating pathway, oxidative phosphorylation (OXPHOS) is composed of five enzyme complexes assembled from subunits derived from both the mtDNA and nDNA plus the adenine nucleotide trans-locator (ANT) (3, 5, 6) (Figure 2). Complexes I to IV make up the electron transport chain, which receives electrons from $NADH + H^+$ and $FADH_2$ and uses oxygen as a terminal electron acceptor. NADH donates electrons to NADH dehydrogenase (NADH:ubiquinone oxidoreductase or Complex I). The electrons traverse the flavin mononucleotide (FMN) and multiple iron-sulfur centers of Complex I and are then passed to coenzyme Q (ubiquinone or CoQ). CoQ can also be reduced by electrons donated from multiple $FADH_2$-containing dehydrogenases, including succinate dehydrogenase (succinate: ubiquinone oxidoreductase or Complex II), electron-transfer-flavoprotein (ETF) dehydrogenase, and glycerol-3-phosphate dehydrogenase. Electrons from reduced CoQ (ubiquinol) are then transferred to ubiquinol: cytochrome c oxidoreductase (Complex III), which encompasses cytochromes b and c_1 and the Rieske-Iron Sulfur protein. Then the electrons pass through cytochrome c, and cytochrome c oxidase (Complex IV) to oxygen. Cytochrome c oxidase contains cytochromes a + a_3 and two copper atoms. As electrons traverse Complexes I, III, and IV, protons are pumped out of the mitochondrial matrix across the mitochondrial inner membrane. This creates an electrochemical gradient that is utilized by the ATP synthase (Complex V) to synthesize ATP from ADP + Pi. The ATP is then exchanged for cytosolic ATP by the ANT (3, 5, 6).

All of the mitochondrial DNA genes encode subunits of the OXPHOS enzymes. Complex I consists of approximately 39 polypeptides, seven (ND1, ND2, ND3, ND4L, ND4, ND5, ND6) encoded by the mtDNA; Complex II consists of four polypeptides, all nuclear; Complex III of about 10 polypeptides, one (cytochrome b, cytb) encoded by the mtDNA, Complex IV of 13 polypeptides, three (COI, COII, and COIII) encoded by the mtDNA; Complex V of 12 polypeptides, two (ATPase6 and 8) encoded by the mtDNA; and ANT is a homodimer of nuclear subunits (1–3, 5, 6) (Figures 1 and 2).

Developmental Regulation of Nuclear OXPHOS Genes

The nuclear location of most OXPHOS genes permits tissue-specific regulation of energy metabolism. One method for accomplishing this is the differential expression of tissue-specific isoforms. Isoforms have been identified for the systemic and testis-specific forms of cytochrome c (18–20); for heart and systemic (liver) forms of the cytochrome c oxidase subunits (21) VIa (22, 23), VIIa (24–26), and VIII (25, 27–30); for the two differentially expressed ATP synthase subunit 9 genes, which have different presequences but the same mature polypeptide (31–33); for the heart and systemic genes for the α subunit of the ATP synthase (34, 35); and for the heart (36–40), fibroblast (41–45), and systemic (36, 39, 43, 45) isoforms of ANT (ANT1, ANT2, and ANT3, respectively). A second regulatory mechanism is the variable expression of single-copy OXPHOS genes. For example, the ATP synthase β subunit gene is expressed at much higher levels in heart and muscle than in other tissues, and the level of expression of both nuclear DNA and mtDNA OXPHOS genes changes during development, cell growth, and neoplastic transformation (46–48).

The differential expression of nuclear OXPHOS genes is regulated, in part,

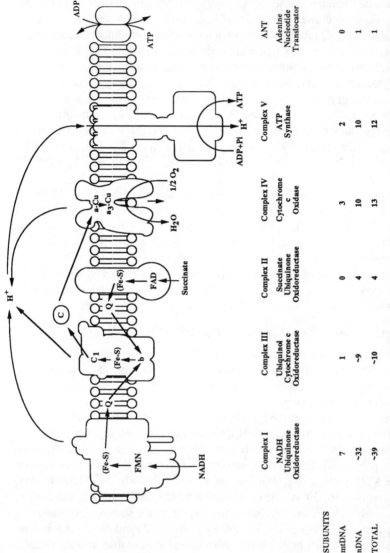

SUBUNITS	Complex I NADH Ubiquinone Oxidoreductase	Complex III Ubiquinol Cytochrome c Oxidoreductase	Complex II Succinate Ubiquinone Oxidoreductase	Complex IV Cytochrome c Oxidase	Complex V ATP Synthase	ANT Adenine Nucleotide Translocator
mtDNA	7	1	0	3	2	0
nDNA	~32	~9	4	10	10	1
TOTAL	~39	~10	4	13	12	1

Figure 2 Diagram of OXPHOS. The five respiratory complexes plus the ANT are shown within the mitochondrial inner membrane. Above is the cytosol and below is the mitochondrial matrix. The subunit composition of each complex is presented in the table below. Abbreviations used: NADH, nicotinamide adenine dinucleotide—reduced; FMN, flavin mononucleotide; FAD, flavin adenine dinucleotide; Fe–S, iron-sulfur center; Q, CoQ; b, cytochrome b; c_1, cytochrome c_1; a and a_3, cytochromes a and a_3.

at the transcriptional level. Ubiquitous *cis*-control elements such as Sp1 and CCAAT have been found in all OXPHOS genes examined (36, 37, 44, 46, 49–52). OXPHOS-specific elements have also been observed, and include the NRF-1 element in cytochrome c (49); the Mt1, Mt3, and Mt4 elements found in several nuclear OXPHOS genes as well as in the mtDNA D-loop (50, 53–55), and tissue-specific OXPHOS gene elements, such as the OXBOX, which is associated with the increased heart and muscle expression of the ANT1 and the ATP synthase β subunit genes (37, 46, 56).

Mitochondrial DNA Genetics

The genetics of the mtDNA differs from that of the nuclear DNA in five important ways (3, 5, 6). First, the mtDNA is semiautonomous, consistent with its endosymbiotic origin. This autonomy can be demonstrated by physically transferring the mitochondria and mtDNAs from one cultured human cell to another by cytoplast fusion or microinjection. Cultured human cells can be isolated that are resistant to CAP. This resistance results from single nucleotide mutations near the 3' end of the large rRNA gene (17, 57, 58). CAP-resistant (CAP-R) mtDNAs can be transferred between cells by enucleating the CAP-R cell and fusing the cytoplasmic fragment containing the mitochondria (cytoplast) to a CAP-sensitive (CAP-S) cell. Viable cells having the recipient cell nucleus and donor cell mtDNAs (cybrids) are isolated by selection in CAP (17, 59). This method has been refined by isolating cells devoid of mtDNA (ρ° cells) through growth in ethidium bromide. Such ρ° cells can be repopulated with mtDNAs by either cybrid fusion or microinjection of isolated mitochondria (60, 61).

Somatic cell genetic studies have revealed that mtDNAs within the cell readily interact, presumably though repeated mitochondrial fusion and fission. Hybrids and cybrids prepared between CAP-R and CAP-S cells that differed in an electrophoretic variant of ND3 revealed that both ND3 variants were expressed when the cell lines were labeled in CAP (62, 63). This demonstrated that the ribosomes of the CAP-R mtDNA are able to translate the mRNAs from the CAP-S mtDNA, showing that the two mtDNAs can complement each other in *trans* and thus must occupy the same compartment (17, 62). In some cells, mitochondrial fusion progresses to the extreme of forming one interconnected mitochondrial network (64).

While mtDNA mixing and complementation appear to be common in human cells, mtDNA recombination is not. Mitochondrial DNA recombinants could not be demonstrated in interspecific somatic cell hybrids (65–67), and repeated crosses within Native American populations between individuals with distinct mtDNA haplotypes have not generated recombinant haplotypes in the past 20,000 or so years (68–70), even though paternal input of mtDNAs into the mammalian egg has been demonstrated (71).

Even though mitochondria are more autonomous than any other mammalian organelle, the transfer of genes from the endosymbiont genome to the nucleus over the past 1.5 billion years has severely curtailed that autonomy (4). The nucleus now contains the genes for 80% of the OXPHOS pathway subunits; all of the genes for mitochondrial intermediary metabolism, including those for the Krebs Cycle, fatty acid oxidation, amino acid metabolism, vitamin biogenesis, etc; and all of the protein genes for mitochondrial biogenesis, including replication proteins, transcription proteins, translation factors, and ribosomal proteins (5, 6). The mRNAs for these genes are translated in the cytosol and the proteins are selectively transported into the mitochondrion. Mitochondrial targeting is generally accomplished via an amino-terminal, positively charged, targetting sequence (72). This is added to the mature protein via one or more 5' exons, as seen in the human ATP synthase β subunit gene (46, 52). The positively charged sequence is attracted to the mitochondrion, binds to a specific receptor system, and is transported through the membrane. The targetting sequence is then proteolytically removed, and the subunit is assembled into the respiratory complexes (72, 73).

The second unique genetic characteristic of the mtDNA is maternal inheritance. The mother transmits her mtDNAs to all of her offspring and her daughters transmit their mtDNAs to the next generation (3, 5, 74, 75). This is the direct consequence of the fact that the egg harbors several hundred thousand mtDNAs, while the sperm has only a few hundred mtDNAs (76). Consequently, the few sperm mtDNAs that enter the egg (71) have little effect on the genotype.

The third feature of mtDNA genetics is that it undergoes replicative segregation during both mitosis and meiosis (3, 5, 77). Each human cell has hundreds of mitochondria and thousands of mtDNAs (78). This means that cells and human lineages can harbor mixtures of mutant and normal (wild-type) mtDNAs (heteroplasmy). As heteroplasmic cells undergo mitotic or meiotic cytokinesis, the proportion of mutant and wild-type mtDNAs that are distributed to daughter cells fluctuates. Consequently, over repeated cell divisions, the mtDNA genotype can progressively drift toward either pure mutant or wild-type mtDNAs (homoplasmy) (77, 79–81). Since a cell's OXPHOS defect is directly proportional to the percentage of mutant mtDNAs (82–84), OXPHOS defects due to mtDNA mutations can range continuously from 0% to 100% of normal activity. Thus, heteroplasmy and replicative segregation mean that cells or individuals with identical nuclear genotypes, such as identical twins, can have different cytoplasmic genotypes and hence different phenotypes.

A fourth feature of mtDNA genetics is threshold expression (3, 5, 77). In patients harboring mtDNA mutations, the phenotype is a product of the severity of the OXPHOS defect (i.e. the nature of the mutation and the percentage of mutant mtDNAs) and the relative reliance of each organ system on mitochondrial energy production. Toxicological studies have shown that different organ systems rely on mitochondrial energy to different extents. Chronic OXPHOS poisoning with cyanide, sodium azide, or dinitrophenol results in optic atrophy, deafness, ataxia, seizures, myoclonus, basal ganglia degeneration, and movement disorders (5). OXPHOS inhibition by dinitrophenol or bonkrekic acid inhibits insulin secretion by pancreatic islets (84a), and damage of islet mitochondria via streptozotocin treatment induces diabetes in animals (84b, 84c). Hypoxia resulting from asphyxiation or ischemia preferentially affects the brain, kidney, heart, and liver, in this order. Ischemia also produces a characteristic muscle pathology, mitochondrial myopathy, in which the oxidative Type I muscle fibers degenerate and accumulate aggregates of abnormal mitochondria. These mitochondrial aggregates stain red by Gomori Modified Trichrome, and hence the affected muscle fibers are called ragged-red fibers (RRF) (5). The differential reliance of organs and tissues on OXPHOS is at least in part a result of the developmental regulation of nuclear OXPHOS gene expression. Therefore, as mitochondrial ATP production declines, it successively falls below the minimum energetic levels necessary for each organ to function normally (energetic thresholds). This results in a progressive increase in the number and severity of clinical symptoms, with the visual pathway, the central nervous system, muscle, heart, pancreatic islets, kidney, and liver being preferentially affected (3, 5, 6, 82, 83, 84a–84c).

The fifth unique aspect of mtDNA genetics is its high sequence evolution rate, about 10 to 20 times higher than the nuclear DNA. This means that gene for gene, deleterious OXPHOS mutations are far more probable in mtDNA than in nuclear DNA (3, 5, 6, 38, 85, 85a). The high mtDNA evolution rate has created extensive sequence variation among individuals and populations (2). Any two human mtDNAs will differ on average by about 4 nucleotides in 1000, or about 70 base substitutions, with a maximum difference of about 3% (2, 68–70, 85a, 86–89). This means that a wide variety of neutral to mildly deleterious mutations have been fixed in various human populations, and that highly deleterious mutations arise frequently. Since deleterious mutations are eliminated by selection, mitochondrial genetic disease must be a common phenomenon.

MITOCHONDRIAL DNA MUTATIONS AND DISEASE

The mtDNA disease mutations that have been described to date fall into four groups: missense mutations, protein synthesis mutations, insertion-deletion mutations, and copy number mutations.

Missense Mutations

Amino acid substitution (missense) mutations have been associated with ophthalmological and neurological diseases. Two classes of phenotypes are known: Leber's Hereditary Optic Neuropathy (LHON) and Neurogenic muscle weakness, Ataxia, and Retinitis Pigmentosa (NARP).

LHON is a maternally inherited form of acute or subacute adult-onset blindness due to death of the optic nerve. Central vision is rapidly lost in both eyes but peripheral vision is retained. Blindness frequently occurs in mid-life, with a mean age of onset of about 27 years, but onset can range from 6 to 60 years. There is a substantial sex bias, with males outnumbering females 4 to 1. Affected individuals may also have cardiac conduction defects and behavior abnormalities (90, 91).

A number of mtDNA mutations are associated with LHON (Table 1). The first mutation was located at np 11778 and changes the 340th amino acid of the ND4 protein from an arginine to a histidine (102). This mutation accounts for about 50% of LHON cases, has not been found in 507 controls, and is readily detected because it eliminates an SfaNI restriction site (88, 102) and creates an MaeI restriction site (100). The mutation changes a highly conserved amino acid, which indicates its biochemical importance (102, 104). Haplotype analysis suggests that the mutation has occurred multiple times in association with LHON (88), consistent with its causing the disease.

In large LHON pedigrees, the np 11778 mutation is essentially homoplasmic (102). However, about 58% of np 11778 LHON pedigrees are singleton cases, and 14% are heteroplasmic with blindness appearing when meiotic segregation increases the proportion of mutant mtDNAs towards 100% (80, 81). Meiotic segregation of the np 11778 mutation can occur quite rapidly, with the proportion of mutant mtDNAs in some families shifting from about 50% to essentially pure mutant in one generation. Moreover, heteroplasmic individuals can have different percentages of mutant mtDNAs in different tissues (81). Hence, mitotic and meiotic replicative segregation of heteroplasmic individuals and lineages generates extensive genetic heterogeneity in LHON.

Eight additional missense mutations have been implicated in causing LHON for patients who lack the 11778 mutation (non-11778 cases) (Table 1). These have been localized in three additional Complex I genes plus the cytochrome b gene of Complex III. Three criteria have been used to implicate these mutations in LHON: the presence of the mutation at significantly higher frequencies in patients than in controls, alteration of an evolutionarily conserved amino acid, and heteroplasmy. While some mutations appear to be sufficient by themselves to cause the disease, others act synergistically to increase or decrease the probability of clinical expression.

Three mutations other than np 11778 appear to be sufficient to cause

Table 1 Disease-associated point mutations: coding regions[a]

Locus*Allele	NP	NT	AA	Cons (H/M/L)	Population frequency 11778+ (%)	11778− (%)	Control (%)	Homoplasmy	Heteroplasmy	References
MTND1*LHON3460	3460	G-A	A-T	M	0	45	<1	+	−	92, 93
MTND1*LHON4136	4136	A-G	Y-C	M	ND	rare	<6	+	−	94
MTND1*LHON4160	4160	T-C	L-P	H	ND	rare	<6	+	−	94
MTND1*LHON4216	4216	T-C	Y-H	L	14	68	7.3	+	−	95
MTND2*LHON4917	4917	A-G	D-N	H	3.2	3	36	+	−	95
MTND2*LHON5244	5244	G-A	G-S	H	0	3.5	<0.05	−	+	96
MTND4*LHON11778	11778	G-A	R-H	H	100	0	<0.2	+	+	80, 81, 88, 90, 91, 95, 97–103
MTND5*LHON13708	13708	G-A	A-T	M	17 / 22	43 / 24	4.2	+	−	95 / 96
MTCYB*LHON15257	15257	G-A	D-N	H	0	17	0.3	+	−	95, 96
MTCYB*LHON15812	15812	G-A	V-M	M	0	8.7+	0.1	+	−	95, 96
MTATP6*NARP8993	8993	T-G	L-R	H	ND	ND	ND	−	+	79

[a] Allele classifications: LHON, Leber Hereditary Optic Neuropathy; NARP, Neurogenic muscle weakness, Ataxia, and Retinitis Pigmentosa; NP, nucleotide position; NT, Nucleotide substitution; AA, amino acid replacement; Cons, interspecies conservation of the substituted nucleotide; H, high; M, medium; L, low; Homoplasmy, pure mutant mtDNAs; Heteroplasmy, mixture of mutant and normal mtDNAs; ND, not determined; < indicates that none of the controls tested was positive for the mutation (e.g. <2 is equivalent to 0 out of 50 controls). Additional data will be necessary to confirm the pathological significance of some of these mutations.

blindness. They occur at nps 3460, 4160, and 15257 (Table 1). The np 3460 mutation in ND1 has been found in nine independent pedigrees but not in 107 controls. It changes a moderately conserved alanine to a threonine and has been found to be homoplasmic (92, 93). The np 4160 mutation in ND1 was identified in a single large LHON pedigree but not in 18 controls. This mutation converts a highly conserved leucine to a proline and was homoplasmic (94). The np 15257 mutation in cytochrome b was found in four independent pedigrees from North America and Europe and in only one of 362 controls. It changes a highly conserved aspartate to an asparagine near one of the heme-binding sites and was also homoplasmic (96).

The remaining five mutations appear to contribute to LHON by interacting with other mutations. These occur at nps 4216, 4917, 5244, 13708, and 15812 (Table 1). Several of these mutations have been found in np 11778 patients as well as in controls. The np 4917 mutation in ND2 has been found in 10 of 28 non-11778 patients (36%), one of 36 np 11778 patients (2.8%), and 4 of 124 controls (3.2%). It alters a highly conserved amino acid and is homoplasmic (95). This mutation has consistently been associated with a second mutation at np 4216 in ND2. The np 4216 mutation was found in 19 of 28 non-11778 patients (68%), 5 of 36 np 11778 patients (14%), and 9 of 124 controls (7.3%). Of 24 patients with the np 4216 mutation, 10 also had the np 4917 mutation (42%), 9 the np 13708 mutation (37%), 4 the np 11778 + np 13708 mutations (16%), and 1 the np 11778 mutation (4%) (95).

The np 13708 mutation has been found in significant frequencies in both np 11778 patients and controls. It converts a moderately conserved alanine to threonine in ND5 (Table 1). In the first of two studies, it was found in 12 of 28 non-11778 patients (43%), 6 of 36 np 11778 patients (17%), and 5 of 117 controls (4.3%). Of the 18 patients with the np 13708 mutation, 9 also harbored the np 4216 mutation (50%), 4 the np 11778 + np 4216 mutations (22%), 2 the np 11778 mutation (11%), and 3 had no other mutation (17%) (95). In the second study, the np 13708 mutation was found in 6 of 25 non-11778 patients (24%), 5 of 23 np 11778 patients (22%), and 14 of 278 controls (5%) (96).

A striking association was also found between the np 13708 mutation, the np 15257 mutation, and two other mutations at nps 15812 and 5244 (96). The np 15812 mutation converts a moderately conserved leucine in cytb to a methionine and is homoplasmic. The np 5244 mutation changes a highly conserved glycine in ND2 to a serine and is heteroplasmic. Furthermore, the four mutations accumulated sequentially along a single mtDNA lineage with the probability of blindness increasing with each new mutation (Figure 3). Six non-11778 patients had the np 13708 mutation. Of these, four also had the cytochrome b mutation at np 15257 and encompassed all known np 15257 patients. Of the four np 15257 patients, two also had the cytochrome b mutation at np 15812, and of these one harbored the ND2 mutation at np 5244

(Figure 3). Haplotype analysis showed that all of these patients belonged to the same mtDNA lineage, and population studies revealed that as each new mutation was added, the probability of vision loss increased. Five percent of the unblind population harbored the np 13708 mutation; 0.3% the nps 13708 and 15257 mutations; 0.1% the nps 13708, 15257, and 15812 mutations; and less than 1 in 2103 the nps 13708, 15257, 15812, and 5244 mutations (Figure 3). Hence, these mtDNA mutations interact synergistically, such that as each new mutation is added, the mitochondrial energy-generating capacity declines and the probability of optic nerve death increases (96).

One mtDNA missense mutation (np 4136) has been proposed to be advantageous, when associated with the np 4160 mutation. The np4160 mutation appears to be more severe than other LHON mutations; with a greater proportion of the maternal relatives going blind, males and females being equally likely to be affected, and affected individuals having additional neurological and psychiatric symptoms (94). One branch of the np 4160 family has a milder phenotype and harbors the np 4136 mutation. Therefore, it was proposed that the np 4136 mutation ameliorates the OXPHOS defect of the np 4160 mutation (94).

As more LHON pedigrees are studied, additional severe mtDNA mutations are likely to be found. This could be the case for maternally inherited LHON associated with early-onset basal ganglia degeneration (Infantile Bilateral Striatal Neurosis, IBSN) and dystonia (105).

The identification of LHON mutations in both Complex I (ND1, ND2, ND4, and ND5) and Complex III (cytb) genes, as well as the demonstration that the mutations can be cumulative, suggests that the probability of vision loss is more a function of the extent of inhibition of the electron transport chain, than due to the loss of a particular enzyme. Since the polypeptides of the election transport chain come from both mtDNA and nDNA genes,

MUTATION	GENE	AMINO ACID SUBSTITUTION	CONSERVATION H / C / M / X	% PATIENTS	% CONTROLS	GENOTYPE
np 13708	ND5	Ala -> Thr	A / L / A / A	23	5	Homoplasmic
np 15257	Cytb	Asp -> Asn	D / D / D / D	9	0.3	Homoplasmic
np 15812	Cytb	Val -> Met	V / V / I / V	4	0.1	Homoplasmic
np 5244	ND2	Gly -> Ser	G / G / G / G	2	0	Heteroplasmic

Figure 3 Synergistic interaction of LHON mutations. The mutations are listed from top to bottom as they accumulated along a mtDNA LHON lineage. The genes affected and amino acid substitutions are given. "Conservation" shows the amino acid in this position for H, human; C, cow; M, mouse; and X, *Xenopus*. Abbreviations used: A, alanine; L, leucine; D, aspartate; V, valine; I, isoleucine; G, glycine; % patients, the percentage of LHON patients with the genotype: np 13708, np 13708 + 15257, np 13708 +15257 + 15812, and np 13708 + 15257 + 15812 + 5244; % control, percentage of the general population with each of the genotypes.

Mendelian genes could also increase the probability of LHON expression. A deleterious X-linked gene has been proposed to be the cause of the male bias in LHON patients (106), but LHON could also be a sex-limited trait. For example, men might have a higher optic nerve metabolic rate then women and thus a lower expression threshold. The probability of blindness might also be increased by environmental intoxication with electron transport chain inhibitors such as cyanide and carbon monoxide. Hence, the factors leading to blindness of a particular LHON patient could be quite complex.

While mutations in electron transport genes have correlated with optic nerve death, a missense mutation in the mitochondrial ATPase6 gene has been found to be associated with retinitis pigmentosa plus other neurological symptoms (NARP). NARP is maternally inherited, with patients having a spectrum of symptoms including bone-spicular retinitis pigmentosa, ataxia, seizures, dementia, proximal neurogenic muscle weakness, sensory neuropathy, and developmental delay. This disease is the result of a T-to-G transversion at np 8993, which changes amino acid 156 of ATPase6 from leucine to arginine. Individuals in these maternal lineages are heteroplasmic, and the severity of the symptoms is related to the percentage of mutant mtDNAs (79).

Missense mutations have thus been identified in genes of most of the respiratory complexes. So far these have preferentially affected the visual pathway and been confined to nervous and excitatory tissues.

Biogenesis Mutations

Nucleotide substitutions in the mtDNA biosynthetic genes generally have more systemic phenotypic consequences than do missense mutations. So far, all biogenesis mutations have been tRNA mutations and associated with mitochondrial myopathy: ragged-red muscle fibers and abnormal mitochondria. In severe cases, the nervous system, muscle, heart, and kidney can also be involved. Transfer RNA mutations are now known for Myoclonic Epilepsy and Ragged-Red Fiber Disease (MERRF); Mitochondrial Encephalomyopathy, Lactic Acidosis, and Strokelike Symptoms (MELAS); and Maternally Inherited Myopathy and Cardiomyopathy (MMC). Preliminary evidence suggests that some cases of Lethal Infantile Mitochondrial Myopathy (LIMM) and of ocular myopathy may also be caused by point mutations (Table 2).

MERRF is a maternally inherited disease in which severely affected family members have an uncontrolled myoclonic epilepsy (periodic jerking) and mitochondrial myopathy (RRFs). The most severely affected maternal relatives also have pronounced aberrations in brain electrophysiology, neurosensory hearing loss, myoclonus, dementia, respiratory failure, dilated cardiomyopathy, and renal dysfunction. However, most maternal relatives of MERRF patients have less severe disease, with clinical manifestations in-

Table 2 Disease-associated point mutations: noncoding regions[a]

Locus*Allele	NP	NT	tRNA	Cons (H/M/L)	Control (%)	Homoplasmy	Heteroplasmy	References
MTTL1*MELAS3243	3243	A-G	Leu(UUR)	H	<2	–	+	107–114
MTTL1*MMC3260	3260	A-G	Leu(UUR)	H	ND	–	+	84
MTTI*FICP4317	4317	A-G	Ile	ND	<3.6	ND	ND	118
MTTK*MERRF8344	8344	A-G	Lys	H	<0.5	–	+	82, 83, 108, 115–117
MTTG*CIPO10006	10006	A-G	Gly	H	ND	+	–	115
MTTS2*CIPO12246	12246	C-G	Ser (AGY)	H	ND	+	–	115
MTTL2*CIPO12308	12308	A-G	Leu(CUN)	H	ND	+	–	115
MTTL2*CPEO12308	12308	A-G	Leu(CUN)	H	ND	+	+	115
MTTT*LIMM15923	15923	A-G	Thr	ND	ND	ND	ND	119
MTTT*LIMM15924	15924	A-G	Thr	ND	ND	ND	ND	119

[a] Allele classifications: MELAS, Mitochondrial Encephalomyopathy, Lactic Acidosis, and Strokelike episodes; MERRF, Myoclonic Epilepsy and Ragged Red Muscle Fibers; CIPO, Chronic Intestinal Pseudoobstruction with myopathy and ophthalmoplegia; CPEO, Progressive Chronic External Ophthalmoplegia; MMC, Maternal Myopathy and Cardiomyopathy; FICP, Fatal Infantile Cardiomyopathy Plus, a MELAS-associated cardiomyopathy; LIMM, Lethal Infantile Mitochondrial Myopathy; Cons, interspecies conservation of the substituted nucleotide; H, high; M, medium; L, low; Homoplasmy, pure mutant mtDNAs; Heteroplasmy, mixture of mutant and normal mtDNAs; ND, not determined. < indicates that none of those tested was positive for the mutation (e.g., <2 is equivalent to 0 out of 50 individuals).

creasing sequentially from electrophysiological aberration, to include mitochondrial myopathy, deafness, myoclonus, dementia, respiratory failure, and cardiomyopathy (Figure 4, right to left) (82, 83).

MERRF pedigrees have been associated with a pleiotropic OXPHOS defect primarily affecting Complexes I and IV. Moreover, the severity of the OXPHOS enzyme defect is directly proportional to severity of the clinical manifestations (Figure 4), consistent with the threshold theory (83). In most

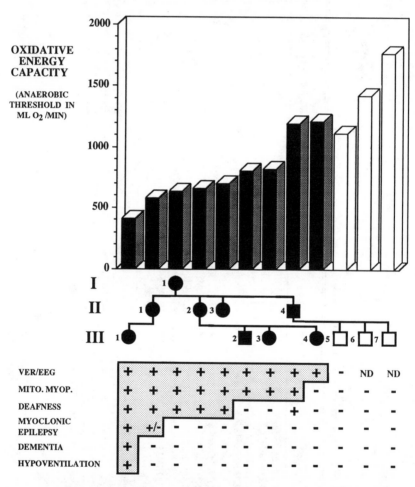

Figure 4 The association between phenotype and OXPHOS defect in a MERFF pedigree. Oxidative Energy Capacity (anaerobic threshold) was determined by an exercise stress test. The values are directly proportional to OXPHOS enzyme levels (83). Abbreviations used: VER/EEG, visual evoked reponse/electroencephalography analysis; Mito. Myop., mitochondrial myopathy; ND, not determined.

pedigrees, MERRF is caused by a mutation in the mtDNA at np 8344, which alters the TψC loop of tRNA^{Lys} (82). The resulting reduction in mitochondrial protein synthesis (120, 121) accounts for the preferential reduction in Complexes I and IV, the complexes with the most mitochondrially encoded subunits.

Maternal relatives of MERRF patients are generally heteroplasmic for the np 8344 mutation, but OXPHOS levels only partially correlate with percentage of mutant mtDNAs. This is because OXPHOS enzyme levels in MERRF family members decline dramatically with age. Consequently, young individuals (those under 20) generally require more than 95% mutant mtDNAs to express the complete MERRF phenotype, while individuals with 85% mutant mtDNAs are clinically normal. Older individuals (those between 60 to 70 years) can be severely affected with 85% mutant mtDNAs or be mildly affected with 63% mutant mtDNAs (82, 122). These observations suggest that MERRF pedigree members are born with normal phenotypes, because their initial OXPHOS defects, determined by their inherited percentage of mutant mtDNAs, are above the tissue expression thresholds. However, as they age, their OXPHOS capacity further declines, ultimately falling below expression thresholds and causing clinical disease. This can account for why many individuals with mtDNA mutations are normal early in life, subsequently acquire symptoms, and then progressively worsen.

MELAS patients have periodic strokelike episodes together with mitochondrial myopathy. The strokes are reversible, affecting cortex and white matter, and can be visualized by computerized tomography or magnetic resonance imaging. MELAS is frequently associated with Complex I defects (123, 124), and only occasional family members have the complete syndrome. However, careful examination of other maternal relatives frequently reveals neurological abnormalities, and individuals harboring the MELAS mutantion frequently get progressively worse with age.

About 80% of MELAS cases result from a heteroplasmic mutation at np 3243 that alters the dihydrouridine loop in tRNA$^{Leu(UUR)}$. This mutation also inactivates the transcriptional terminator encompassed within the tRNA$^{Leu(UUR)}$ gene and downstream from the rRNA genes. Hence, the MELAS mutation may reduce the efficiency of translation and alter the ratio of mitochondrial rRNA and mRNA transcripts (107, 109, 112).

MMC is a maternally inherited disease associated with mitochondrial myopathy, hypertrophic cardiomyopathy, and a combined Complex I and IV OXPHOS defect. Like MELAS, MMC is caused by a mutation in the tRNA$^{Leu(UUR)}$ gene. However, the MMC mutation is at np 3260, which is in the stem of the anticodon loop and outside the transcription terminator sequence. Family members are heteroplasmic for the mutation, and the severity of the Complex I and IV defect is proportional to the percentage of mutant mtDNAs. However, no age effect was observed in the one family studied (84).

LIMM is a heterogeneous group of diseases involving severe neonatal lactic acidosis frequently associated with failure to thrive, hypotonia, neurological deficit, RRFs, cardiomyopathy, and death within months of birth due to respiratory failure. One case has been associated with a mutation in the tRNA^{Ile} gene at np 4317, a mutation not found in 28 controls (118). Two other cases have been been reported with homoplasmic mutations at nps 15923 and 15924 in the anticodon loops of their mtDNA tRNA^{Thr} genes. However, no controls were examined in this study (119). A subsequent survey revealed that the np 15924 variant is a common polymorphism found in 11% of controls (125), and these nucleotides are prone to mutation during cloning (126). Hence the clinical significance of the tRNA^{Thr} mutations needs further investigation.

Ocular myopathy is associated with ptosis (droopy eye lids), ophthalmoplegia (paralysis of eye muscle), and mitochondrial myopathy. Occasional ocular myopathy patients have been found to harbor the tRNA^{Lys} mutation at np 8334 (115) and the tRNA$^{Leu(UUR)}$ mutation at np 3242 (107, 108, 111). In one case, ocular myopathy was associated with chronic intestinal pseudo-obstruction and a combination of mutations in the tRNA$^{Ser(GCU)}$ gene at np 12246, the tRNA$^{Leu(UAG)}$ gene at 12308, and the tRNA^{Gly} gene at 10006. The tRNA$^{Leu(UAG)}$ mutation has also been observed alone in one ocular myopathy patient and one general myopathy patient. None of the new tRNA mutations were found in 17 controls (115); hence, these mutations may cause these phenotypes.

Insertion-Deletion Mutations

Mitochondrial DNA deletions have been found to cause the majority of cases of ocular myopathy and Pearson Syndrome (2, 127–129). Ocular myopathy patients manifest a continuous range of symptoms—from only ophthalmoplegia, ptosis, and mitochondrial myopathy (chronic external ophthalmoplegia plus, CEOP), to retinitis pigmentosa, lactic acidosis, neurosensory hearing loss, ataxia, heart conduction defects, elevated CSF protein, and dementia (Kearn-Sayre Syndrome, KSS). Pearson Marrow/Pancreas Syndrome is a generally fatal childhood disorder associated with pancytopenia (the loss of all blood cells), pancreatic fibrosis, and splenic atrophy (3). Individuals who do survive often progress to a KSS phenotype (130).

Ocular myopathy and Pearson Syndrome patients generally harbor a single mtDNA deletion, but the size and position of the deletion differs markedly among patients (2, 104). More than 100 mtDNA deletions have been characterized, and most of those that have been sequenced are flanked by direct repeats (2). To date, all of the ocular myopathy and Pearson Syndrome deletions have been confined to two sectors delineated by the H-strand and L-strand origins of replications (Figure 5) (2). Ninety-five percent of the deletions occur between O_H and O_L, the righthand sector of Figure 5. The most common deletion, found in 30–50% of patients, is flanked by 13-np direct repeats (5'-ACCTCCCTCACCA) at nps 8468 and 13446 and removes 4997 bp of sequence (Figure 5) (131, 132). The maximum extent of ocular myopathy and Pearson Syndrome deletions in the O_H to O_L sector ranges from np 15945 to 5786 (right side Figure 5), and in the O_L to O_H sector from np 5448 to 470 (left side Figure 5) (2). Thus the deletions extend to, but do not include, the sequences required to initiate replication: the L-strand origin (O_L), including nps 5721–5781; and the H-strand origin (O_H), including the L-strand promoter (P_L) at nps 392–445 through the termination-associated sequences (TAS) at nps 16157–16172. Four deletions remove the H-strand

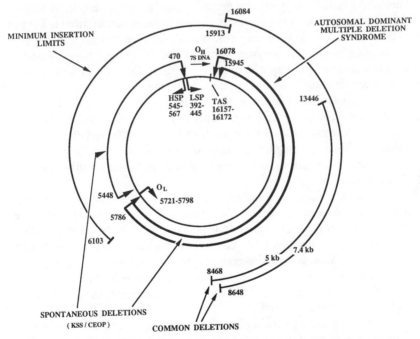

Figure 5   Regions of mtDNA affected by insertion-deletion mutations. The inner circle shows the position of key elements for mtDNA replication (Figure 1). The next arcs outward, reading clockwise from O_H to O_L and O_L to O_H, show the maximum extent of deletions mapped in CEOP, KSS, and Pearson Syndrome patients and in autosomal dominant KSS pedigrees. The outer O_H to O_L arcs show the positions of the common 4997 np (5 kb) and 7436 np (7.4 kb) deletions. The outer O_L to O_H arc shows the minimum area encompassed by the insertion mutations.

promoter (P_H) but spare the L-strand promoter (P_L), thus confirming the critical nature of P_L in DNA replication and demonstrating that transcription of the H-strand genes is not required for maintenance of the mtDNA (2).

Several mechanisms have been proposed for the origin of these deletions: homologous recombination (131, 133), topoisomerase cleavage (134), and slip-replication (132). Slip-replication probably accounts for at least some cases, since the DNA breaks occasionally occur a few nucleotides downstream from the direct repeat, revealing that the upstream repeat is retained and the downstream repeat is lost (132, 135). This is consistent with the upstream repeat on the displaced H-strand base-pairing with the downstream repeat on the L-stand as it is exposed by the replicating fork. Breakage of the H-strand single-strand loop downstream from the direct repeat would create a 3' OH, which could act as a primer for continued H-strand synthesis thus skipping the intervening genes (132).

The great majority of mtDNA deletion cases are spontaneous with no family history. This suggests that most deletions are new mutations that occur during development. If this is the case, tissue lineages derived from cells prior to the deletion should have only normal mtDNAs, while those initiated after the deletion would have varying proportions of deleted mtDNAs. Consistent with this concept, patients with deletions differ widely in the tissue distribution of the deletion, possibly accounting for the extensive variation in the ocular myopathy and Pearson Syndrome phenotypes (133, 136, 137).

Occasional pedigrees have been reported in which mtDNA deletions are present in close maternal relatives, for both identical (138) and different (139) deletions. While this suggests that maternal transmission of deleted molecules that retain both origins occurs, it also indicates that their germline transmission is limited.

Diseases associated with mtDNA deletions progress with age. However, unlike mtDNA point mutations, their progression is associated with an increasing proportion of deleted molecules (140). This implies that the deleted mtDNAs have a replicative advantage over normal molecules. Since all of the mtDNA replication enzymes are encoded by the nucleus, the rate of replication of a mtDNA molecule is directly proportional to its length. Therefore, as the deleted and normal mtDNAs turn over in stable tissues, the deleted molecules are enriched (104).

Staining of muscle biopsies of ocular myopathy patients for cytochrome c oxidase activity has revealed that individual muscle fibers vary from positive to negative along their length. In situ hybridization of mtDNA probes and direct PCR amplification from dissected muscle fibers have revealed that this variation corresponds to variation in the proportion of deleted mtDNAs. Cytochrome c oxidase negative regions have high concentrations of deleted mtDNAs, high levels of mRNAs from the retained genes, and low levels of mRNAs for the deleted genes (141, 142). Cell fusion studies have shown that nuclei within the muscle fiber serve specific metabolic domains (143). Therefore, the regional variation in deleted mtDNAs could be due to the efforts of individual nuclei to compensate for local energy deficiency by stimulating mtDNA proliferation.

While the great majority of ocular myopathy and Pearson Syndrome patients harbor deletions, three cases have been reported with mtDNA duplications. In two cases, ocular myopathy occurred together with diabetes mellitus. The mtDNA duplications of these patients extended from COII to cytochrome b and incorporate both O_H and O_L (144, 145). In one Pearson Syndrome patient, the duplication encompassed about two thirds of the genome, including both origins (133). The progressive enrichment of the larger duplicated molecules over the shorter wild-type molecules could be explained by their two sets of origins. These may permit the duplicated mtDNAs to replicate twice as often as the wild-type (104).

Copy Number Mutations

Certain cases of lethal infantile respiratory failure, lactic acidosis, and muscle, liver, or kidney failure have been found to result from mtDNA depletion. In one family, two siblings died of lethal mitochondrial myopathy, while a cousin related through a maternal great uncle died of hepatic failure. The muscle of one of the siblings had only 2% of the normal mtDNA level, while the liver of her second cousin had only 12% of normal mtDNA levels. Similarly, a child who died of muscle and kidney failure had only 3% of the normal muscle mtDNA, and another child had 17% of the normal muscle and kidney mtDNAs. Deficiencies in mtDNA levels have been associated with reductions in mtDNA gene products but not nuclear OXPHOS gene products, demonstrating an association between the copy number defect and the biochemical defect (146).

A phenocopy of the mtDNA depletion syndrome has been observed in AIDS patients treated with Zidovudine, which blocks both viral and mtDNA replication (147). The Zidovudine-treated patients develop a ragged-red fiber myopathy associated with a reduction in mtDNA levels of 22 to 78% (148).

DEFECTS IN NUCLEAR-CYTOPLASMIC INTERACTIONS

While most cases of ocular myopathy due to mtDNA deletion appear to be spontaneous somatic mutations, some cases are clearly associated with autosomal dominant mutations. Affected individuals in these pedigrees harbor multiple mtDNA deletions instead of only one, and all of the deletions are flanked by direct repeats, suggesting an increased frequency of slip-replication (149–152). Since all of the proteins required for mitochondrial biogenesis are encoded by the nucleus, these pedigrees are probably the result of nuclear mutations in mtDNA replication enzymes.

One KSS patient with multiple deletions also had mtDNA depletion, with muscle mtDNA levels of only 15% normal (153). The relationship between these two phenomena is presently unclear.

Like the spontaneous ocular myopathy deletions, the autosomal dominant patient deletions spare the core elements of the mtDNA O_H and O_L origins (Figure 5). This further supports their importance for mtDNA replication.

THE PROGRESSION OF OXPHOS DISEASES AND AGING

Symptoms in most mtDNA diseases do not appear until later in life, after which they progress. Studies on MERRF suggests that this results from the age-related decline of OXPHOS (82, 83, 122). Therefore, to understand the natural history of mtDNA disease, it is necessary to understand why OX-PHOS declines with age.

In the mtDNA deletion syndromes, the age-related decline in OXPHOS is associated with the progressive enrichment of the deleted mtDNAs (104, 140). This is probably not the case in diseases due to mtDNA point mutations, since these mtDNAs would not have a replicative advantage. Hence, to understand the progression of these diseases, we must identify a second variable that could create an age-related decline in OXPHOS.

Recent morphological, biochemical, and molecular studies have indicated that OXPHOS normally declines with age. The number of cytochrome c oxidase negative skeletal muscle and heart fibers increases with age (154, 155), and both the respiration rate and OXPHOS enzyme activities of Complex I and IV decline progressively with age in human skeletal muscle (156) and liver (157). This age-related decline in OXPHOS is associated with an increase in mtDNA damage. Senescent rat mtDNAs accumulate a variety of low-frequency insertion-deletion mutations (158), and aging human hearts have been found to progressively accumulate both the common 4997 np (159) and the less common 7436 np (160) mtDNA deletions after age 35 (161) (Figure 6). These observations suggest that aging is associated with the progressive accumulation of mtDNA damage in stable tissues, which in turn leads to the decline in OXPHOS capacity (Figure 7) (162).

The most likely cause of age-related mtDNA damage is oxidation by superoxide anions and H_2O_2 (oxygen radicals). These react with lipids, proteins, and nucleic acids, and create thymine glycols and 8-hydroxy-guanosine in DNA, which inhibit replication and transcription. Oxygen radicals are natural by-products of OXPHOS, accounting for between 1% and 4% of oxygen uptake, and they have been shown to be a major DNA-

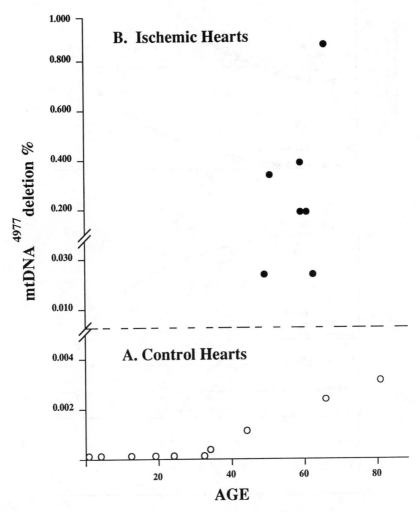

Figure 6   The percentage of cardiac mtDNAs with the 4977 np deletion in normal and ischemic hearts as a function of age. Open circles are determinations for hearts without coronary artery disease. The fill circles are determinations for hearts with substantial coronary artery occlusion. Reprinted by permission from (16).

damaging agent in humans and mammals. Large quantities of DNA oxidation products are excreted by mammals, with each human cell experiencing more than 10^4 oxidative hits per day, resulting in about 7×10^{12} hits per second throughout the body. Evidence indicating that oxidative damage is important in aging comes from the fact that longer-lived animals exhibit a variety of oxygen radical scavenging systems. Evidence that the mitochondria

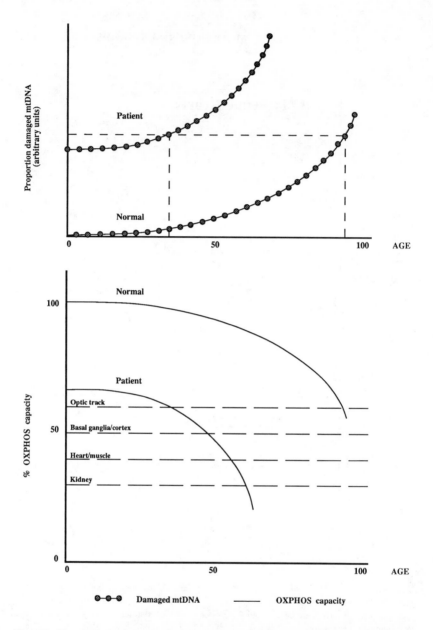

Figure 7 Hypothesis for the mechanism of age-related progression of OXPHOS diseases. Upper panel shows the proposed accumulation of somatic mtDNA mutations with age. Patients are born with a certain percentage of mutant mtDNAs. The dashed lines indicate the relative ages when sufficient mutations accumulate to cause disease. Lower panel shows the decline of OXPHOS capacity of normal individuals and patients. Different tissues have different functional energy minima, shown by dashed lines. OXPHOS declines for both normal individuals and patients, consistent with the accumulation of somatic mtDNA damage. However, because of the inherited mtDNA mutation, the patients start with a lower initial OXPHOS capacity and thus drop below the expression thresholds much earlier than normal individuals.

are important in the aging process comes from the inverse relationship between mammalian life span and basal metabolic rate and oxygen consumption, and the fact that the mtDNA is 16 times more prone to oxidative damage than is nuclear DNA. The extreme sensitivity of mtDNA to oxidative damage may stem from its attachment to the mitochondrial inner membrane, the source of oxygen radicals; its lack of protective histones; and its limited DNA repair systems (163, 164).

Reactive oxygens are generated at two major sites in OXPHOS, the interface between the flavin dehydrogenases and coenzyme Q and at respiratory Complex III. NADH dehydrogenase is the most important dehydrogenase in this process, but succinate dehydrogenase, dihydroorotate dehydrogenase, ETF-dehydrogenase, glycerol-3-phosphate dehydrogenase, and the acyl CoA dehydrogenases may also be involved. Inhibition of Complex III with Antimycin A and Complex IV with cyanide or reduction in electron transport rates due to limiting ADP (state IV respiration) all increase mitochondrial production of superoxide anion and H_2O_2. Presumably, the inhibition of electron flux reduces the flavins and quinones prior to the block stimulating their autooxidation to yield oxygen radicals (164).

The decline of OXPHOS with age may thus reflect the accumulation of oxygen radical damage to the mtDNA (Figure 7). In normal individuals, many mtDNA oxidative hits would be required before mitochondrial OXPHOS would decline significantly. However, once OXPHOS began to decline, electron flow would decrease in the presence of excess substrate, increasing the electronegativity of the flavins, coenzymes Q, and cytochrome b, and stimulating oxygen radical generation and mtDNA damage. As mtDNA oxidative damage increases, replication would be inhibited, thus prolonging the time the mtDNA was triple stranded. This, in turn, would foster mtDNA deletions, which would subsequently become enriched due to their replicative advantage. Individuals who lived long enough would accumulate sufficient oxidative damage to decrease their OXPHOS capacity to below organ-specific energetic thresholds. Individuals born with mtDNA mutations would start with a lower OXPHOS capacity and increased oxygen radical production and thus would fall below OXPHOS expression thresholds much earlier in life (Figure 7).

By this hypothesis, tissues experiencing chronic OXPHOS inhibition should have elevated mtDNA damage. Moreover, to compensate for the concomitant reduction in ATP production, OXPHOS gene expression should also be increased. To test these predictions, mtDNA damage and OXPHOS mRNA levels were compared in normal and ischemic hearts. In ischemic heart disease, coronary artery occlusion due to atherosclerotic plaques starves the heart of nutrients and oxygen essential for mitochondrial function. The ensuing chronic cycle of ischemic and reperfusion would reduce ATP production and increase oxygen radical production. The extent of mtDNA damage in control and ischemic hearts was assessed by determining the proportion of mtDNAs that contained the common 4997 np deletion. DNA was purified from heart tissue and serially diluted. Each dilution was tested for the presence of all mtDNAs and the deleted mtDNAs using the polymerase chain reaction. The ratio of the dilutions at which the deleted molecules and total molecules could no longer be detected permitted calculation of the proportion of deleted molecules. Analysis of normal hearts revealed that deleted molecules were virtually undetectable up to age 35. Then the proportion increased with age, reaching 0.0035% in an 85 year old (Figure 6). Ischemic hearts, by contrast, harbored 8 to 2200 times more deleted mtDNA (0.02 to 0.85%) than the highest control irrespective of age (Figure 6). Hence, hypoxemic inhibition of OXPHOS was associated with a dramatic increase in mtDNA damage (161).

The association between OXPHOS defects and increased mtDNA damage was further supported by testing for mtDNA deletions in the heart of a 16-year-old MELAS patient who harbored the np 3243 mutation and died of hypertrophic cardiomyopathy. While this patient's heart lacked the 4997 np deletion, it did harbor about 0.03% of a 7436-bp deletion (Figure 5) (161). This is significant since normal individuals of this age lack this deletion (160, 165).

Elevated nuclear and mitochondrial OXPHOS mRNA levels were also found to be associated with chronic cardiac overload. Relative to controls, ischemic hearts showed a coordinate induction of the nuclear OXPHOS genes *ANT1, ANT3,* and ATPsyn β, and of the mtDNA genes, cytochrome b, 12S rRNA, and 16S rDNA (161). Therefore, at least in heart, chronic OXPHOS deficiency is associated with increased mtDNA damage and compensatory OXPHOS gene induction.

While the proportion of mtDNAs with the 4997 np deletion was dramatically increased in OXPHOS-deficient heart, the absolute number of these molecules was still low. Even so, this probably reflects a physiologically important phenomenon, since a variety of other mtDNA deletions have also been detected in these hearts and mtDNA point mutations have not yet been examined. Therefore, the cumulative mtDNA damage is likely to be substantial. Furthermore, the effect of damaged mitochondria and mtDNAs could be amplified by the interactive nature of mitochondria. This hypothesis is supported by a study in which mitochondria from the livers of young and old rats were isolated and between 10 and 20 mitochondria microinjected into human diploid fibroblast WI-38 cells, which contain 300 mitochondria. Injection of mitochondria from young animals did not affect significantly the health or longevity of the recipient cells, but injection of mitochondria from aged rats or of mitochondria that had been partially uncoupled resulted in rapid cell loss (166). Thus, even a few damaged mitochondria may have severe deleterious effects on cellular OXPHOS, possibly because mitochondria readily fuse and in some cases form interconnected networks (62, 64). Assuming that oxygen radical damage to the mitochondria and mtDNAs causes the inner membrane to become leaky for protons (uncoupled), then fusion of these mitochondria with normal mitochondria could collapse the electrochemical gradient of the interconnected mitochondria resulting in cell death.

Thus, OXPHOS diseases resulting from mtDNA point mutations may progress because of the secondary accumulation of mtDNA mutations in the patient's somatic cells. Moreover, inherited OXPHOS defects may increase the rate of these somatic mutations by increasing free radical generation. Therefore, somatic mutations in the mtDNA could explain many aspects of the aging process and also the progression of mtDNA diseases (Figure 7).

EVIDENCE FOR OXPHOS DEFECTS IN COMMON DEGENERATIVE DISEASES

The study of mtDNA mutations has shown that OXPHOS diseases have four distinctive characteristics: they are associated with OXPHOS enzyme defects in clinically affected patients, they involve those tissues most reliant on mitochondrial energy (e.g. the central nervous system, heart, skeletal muscle, pancreatic islets, kidney, and liver), they have a complex genetics in which multiple nuclear or mtDNA mutations can give similar phenotypes, and they are frequently expressed later in life and progress with age (Table 3). One or more of these characteristics is found in a variety of common degenerative diseases, including epilepsy, cardiac disease, Parkinson's Disease, Alzheimer's Disease, and maturity-onset diabetes mellitus. This suggests that OXPHOS diseases may be much more common than previously thought.

Epilepsy and seizures are common clinical manifestations of known mtDNA mutations, including the MERRF mutation at np 8344, the MELAS mutation at np 3243, and the KSS deletions. Moreover, the only familial epilepsy for which a gene defect has been identified is MERRF (82, 83). Epidemiological studies of seizure disorders have shown that probands are more likely to have affected mothers than fathers (167). A detailed study of one population revealed that for individuals with seizures up to age 25 years, 8.7% had affected mothers while only 2.4% had affected fathers. Moreover, the earlier the seizures occurred in the parent, the more likely that seizures would occur in the child. While exposure in vitro to maternal seizures or anticonvulsants increased the risk of seizures in the children, these factors could not account for all of the maternal effect (168).

While maternal transmission has been established, maternal inheritance through multiple generations has not. However, this could be explained by

Table 3 OXPHOS defects and age-related diseases

	OXPHOS	Mitochondrial diseases[b]	Degenerative diseases[c]
ATP levels	Maintains	Reduces	Reduces
Tissues[a]	B>M>H>K>L	B>M>H>K>L	B,M,H,K
Genetics	MtDNA + nDNA	MtDNA or nDNA	Complex
Age effect	Declines	Progresses (thresholds)	Progresses

[a] B,M,H,K,L: Brain, Muscle, Heart, Kidney, Lung
[b] possible examples of mitochondrial diseases: LHON, NARP, MERRF, MELAS, MMC
[c] possible examples of degenerative diseases: epilepsy, cardiac disease, Parkinson's, Alzheimer's, diabetes, aging.

replicative segregation and threshold expression. In MERRF and MELAS, clinically overt seizures only occur in individuals with the highest percentage of mutant mtDNAs. Since low percentages of mutant mtDNAs segregate rapidly, but high percentages tend to be inherited, mothers who through the vagaries of replicative segregation inherit a high percentage of mutant mtDNAs and the associated seizures would be more likely to transmit predominantly mutant mtDNAs and seizures to their offspring, even though their mothers were unaffected. Thus seizures due to mtDNA mutations would appear spontaneously and frequently be limited to mother and offspring (82, 83). Consequently, mtDNA mutations might be one cause of seizures.

Somatic and germline mutations in the mtDNA may also be responsible for a number of forms of chronic degenerative heart disease. Ischemic heart disease has already been associated with the accumulation of mtDNA mutations in somatic tissue (161), and mtDNA tRNA$^{Leu(UUR)}$ mutations at np 3243 (161) and np 3260 (84) have been identified in patients with hypertrophic cardiomyopathy. Additional evidence that mtDNA mutations play a role in cardiac disease comes from the detection of a variety of mtDNA point mutations (169) and the 7436 np deletion (160) in a variety of failing hearts. The 7436 np deletion extends from np 8637 to np 16073, and is flanked by 12-bp direct repeats (5'-CATCAACAACCG) (Figure 5) (160, 165). While the 7436 np deletion becomes progressively more common in normal hearts with age, it is much more common in hearts with chronic degenerative disease (160). This observation has been extended by quantitating the levels of the 4997 np deletion in 10 hearts that failed for reasons other than ischemia. Three of these hearts (30%) had elevated levels of this deletion, comparable to those found in ischemic hearts. Two of these were the only cases of idiopathic dilated cardiomyopathy. Moreover, all cardiomyopathy hearts had elevated nuclear and mitochondrial OXPHOS transcript levels, irrespective of their deletion levels (161). Thus, mitochondrial OXPHOS defects may be involved in several forms of non-ischemic heart disease, and most failing hearts may experience ATP deficiency leading to compensatory OXPHOS gene induction.

Parkinson's disease also exhibits many of the characteristics expected for an OXPHOS disease. It is a late-onset movement disorder associated with tremor, slowed and impaired movements, and occasionally dementia. Primary neuronal loss occurs in the substantia nigra, and pathological analysis reveals neuronal Lewy bodies. Parkinsonism can be induced by the Complex I inhibitor MPTP (1-methyl-4-phenyl-1,2,3,6-tetrahydrophyrine) (170–172), demonstrating that the disease can result from OXPHOS deficiency. OX-PHOS enzyme assays in Parkinson patient brains have revealed Complex I defects (173–175), which have been shown to be systemic by assaying for Complex I in blood platelet (176) and skeletal muscle mitochondria (177, 178). The muscle studies further demonstrate that the OXPHOS defects among Parkinson's patients are heterogeneous. Three of six patients were found to have highly statistically significant Complex I defects. One of these also had a Complex III defect and a greatly increased mitochondrial fragility. A fourth patient had a highly statistically significant Complex IV defect and one of the remaining patients had normal muscle OXPHOS (178). Consistent with these variable biochemical findings, the genetics of Parkinson's Disease is complex. Most cases are spontaneous or cluster in families without showing classical Mendelian inheritance. Only rare pedigrees exhibit autosomal dominant transmission (179). Parkinson's patients have been suggested to have distinctive mtDNA haplotypes (113, 180), and since mtDNA deletions accumulate in human brains with age (159, 180), an inborn error in OXPHOS would progress. Hence, Parkinsonism might result from a variety of factors that inhibit OXPHOS: deleterious mutations in mitochondrial or nuclear OXPHOS genes, environmental toxins, or a combination of factors.

Alzheimer's Disease is a late-onset progressive dementia with three pathological hallmarks: neurofibrillary tangles, senile plaques, and cerebral amyloid deposition. In addition to dementia, 10% of Alzheimer's patients develop myoclonus and 30% to 40% have features of Parkinson's disease including tremor, gait impairment, and substantia nigra Lewy bodies. A variety of biochemical studies have associated Alzheimer's disease with defects in mitochondrial respiration. Respiration studies on homogenates of patient neocortex have shown that OXPHOS is uncoupled (181), and enzymatic and immunochemical studies have indicated that patient frontal and occipital cortex shows a 70% reduction in pyruvate dehydrogenase (182). These mitochondrial respiratory defects appear to be systemic, since OXPHOS enzyme assays on blood platelet mitochondria have revealed a threefold reduction in Complex I activity in five of six patients (183), and patient skin fibroblasts have a 21% reduction in mitochondrial glucose and glutamate oxidation and a 145% increase in bound calcium, an ion sequestered by mitochondria (184). OXPHOS defects have been linked to Alzheimer's pathology by showing that treatment of normal human skin fibroblasts with the mitochondrial uncoupler CCCP (carbonyl cyanide m-chlorophenylhydrazone) results in a 10-fold increase in the proportion of cells reacting with an antibody to paired helical filaments and a 157-fold increase in cells reacting to Alzheimer's monoclonal antibody-50 (185). Hence, the development of Alzheimer's pathology and disease could be envisioned as a pathway involving metabolic components of mitochondrial energy production, synaptosomal degeneration due to mitochondrial energetic failure, and the induction and precipitation of toxic cellular proteins. Consistent with this multicomponent model, Alzheimer's Disease has been found to be genetically heterogeneous (186). Segregation analysis indicates that Alzheimer's Disease involves a major autosomal dominant allele with an additional multifactorial component, and that the major locus accounts for about 24% of the transmission variance (187). Linkage studies have identified two dominant loci, one for families with a mean age of onset of greater that 60 years (late onset), and another for families with onset of less than 50 years (early onset) (186). The late onset locus is located in the 19q 12-13 region (188), while the early onset locus maps to the 21q 21 region (189). A small percentage of the chromosome 21q cases have been associated with a valine-to-isoleucine polymorphism at amino acid 717 of the β-amyloid peptide (190), while other cases are not linked to this locus (191, 192). Hence, Alzheimer's can result from of a variety of genetic and environmental factors that promote the death of the highly metabolically active cortical synaptosomes, some associated with defects in structural proteins, but others that may impair the energy production of synaptosomal mitochondria.

Huntington's Disease might also be an OXPHOS disease. Huntington's patients manifest basal ganglia degeneration and associated movement disorders similar to those seen in the maternally inherited LHON with IBSN (105). Biochemical analysis of Huntington's patient brains has revealed a Complex IV defect in the caudate, but not the cortex (193), and analysis of patient blood platelet mitochondria has revealed a Complex I defect (194). This biochemical discrepancy could be more apparent than real, since OXPHOS studies on patients with the same mtDNA mutation have often given variable enzyme defects. The genetics of Huntington's Disease is also unusual. While the disease is caused by an autosomal dominant, chromosome 4 mutation (195), the age of onset of the disease can vary according to the sex of the affected parent. Affected fathers can have either early- or late-onset offspring, whereas affected mothers have only late-onset progeny. This might be explained if a mtDNA gene modulates the expression of the nuclear locus.

Since only late-onset individuals reproduce, affected females would transmit both their mtDNA and nuclear Huntington's gene to their affected offspring and hence would consistently have late-onset children. However, males would transmit only their nuclear Huntington's gene, while their mates would contribute the mtDNA. Assuming that there are two types of mtDNA in the population, one that suppresses the phenotype and the other that exacerbates it, then the mtDNA genotype of the mother would determine the age (early or late) of onset in the children (5, 196).

Myotonic dystrophy is an autosomal dominant disease involving myotonia, muscle wasting, cataracts, hypogonadism, frontal balding, and cardiac dysrhythmias. The disease is associated with the amplification of a CTG repeat in the 3'-nontranslated region of a chromosome 19 protein kinase gene (197, 198). Recently, some patients with myotonic dystrophy have also been found to harbor RRFs and mtDNA deletions (199, 200). Thus, mtDNA mutations may contribute to this phenotype in some cases.

Diabetes mellitus is a common, though genetically heterogeneous, group of disorders that share the feature of glucose intolerance. Two forms of diabetes are generally recognized: juvenile-onset diabetes or Type I, and maturity-onset diabetes or Type II. Type I affects children, results in insulin dependence, and is associated with the HLA (Human Leucocyte Antigens) DR3 and/or DR4 antigens. Type II appears later in life, is associated with normal to low insulin levels, and is frequently associated with obesity (198, 201).

Many cases of diabetes are familial, but the inheritance patterns and ancillary clinical manifestations vary. Epidemiological studies have revealed that as the age-of-onset of the proband increases, the probability that the mother will be the affected parent also increases, reaching a ratio of 3:1 for mean age-of-onset of 46. Moreover, the maternal transmission can be sustained for several generations (201–205). While the maternal transmission of maturity-onset diabetes has generally been attributed to gestational teratological effects (206, 207), diabetes is frequently observed in mtDNA diseases, and it has been reported to be a distinctive feature of some patients with mtDNA duplications (144). Thus, some late-onset diabetes may be due to mtDNA mutations. This possibility has been confirmed for one large pedigree with maternally inherited maturity-onset diabetes and deafness. The cells of the proband had a generalized defect in mitochondrial protein synthesis that was associated with a marked deficiency in all of the proband's muscle mitochondrial OXPHOS enzymes. Yet neither the proband nor any of her maternal relatives exhibited the clinical manifestations commonly associated with mitochondrial disease, including ophthalmolplegia, ptosis, and mitochondrial myopathy with RRFs. Molecular analysis revealed that all maternal relatives harbored a unique 10.4-kb mtDNA deletion, which represented 40 to 70% of the mtDNA molecules in their muscle and/or blood cells. The deletion was flanked by 10-bp direct repeats (5'-CACCCCATCC) at nps 14812 and 4389, and removed O_L as well as all of the mtDNA genes except the rRNAs, ND1, part of cytb, and the adjacent tRNAs (207a). The maternal inheritance and relatively uniform distribution of this deletion among tissues indicated that it does not have the replicative advantage that characterizes CEOP, KSS, and Pearson Syndrome deletions. Probably, the loss of O_L offsets the shorter molecule's replicative advantage, giving it a replication rate comparable to that of the normal mtDNAs. This would mean that it would not undergo progressive enrichment in myotubes leading to RRFs, or in the germline causing the death of the lineage. Rather, it would create a relatively stable and chronic OXPHOS defiency ultimately limiting the metabolic activity of the pancreatic islet cells and causing diabetes. Thus, at least some cases of maturity-onset diabetes appear to be due to mtDNA mutations.

While these observations support the concept that mtDNA mutations can play a role in common degenerative diseases, proof that OXPHOS genes are involved requires the identification of specific OXPHOS gene mutations. Whatever the outcome of these studies, it is clear that the mitochondrial paradigm does provide a useful new perspective from which to examine the common degenerative diseases.

METABOLIC AND GENETIC THERAPY OF OXPHOS DEFECTS

As evidence accumulates that OXPHOS defects contribute to common de-generative diseases, interest in mitochondrial metabolic and genetic therapy is increasing. Current studies emphasize metabolic treatment of existing clinical symptoms. However, as OXPHOS gene mutations are discovered for common diseases, it may become possible to screen the population for individuals at risk for these diseases and to place them on preventative therapy, thus postponing or even eliminating the diseases.

A variety of metabolic therapies have already been developed for treating patients with OXPHOS disease (3). These involve supplementation with common OXPHOS cofactors and oxidizable substrates, stimulation of pyru-vate dehydrogenase, and prevention of oxygen radical damage to mitochon-drial membranes. Examples of initial therapeutic successes include (a) treat-ment of a mitochondrial myopathy patient with a Complex III defect with ascorbate and menadione, resulting in improved high-energy phosphate metabolism detected by ^{31}P-NMR (208, 209), (b) treatment of ocular myopathy patients with CoQ (210, 211) and with succinate and CoQ (132) and observing reversal of respiratory failure, (c) treatment of cardiac patients with CoQ and showing improved prognosis (212), and (d) treatment of lactic acidosis with dichloroacetate and noting reduced acidosis (213). As successful interventional therapies are perfected, they will form the basis for developing new preventative therapies.

While metabolic therapy holds the most promise for rapid development of effective regimes, such treatments are cumbersome and transient. Somatic gene therapy has the potential to provide more permanent solutions to OX-PHOS diseases. One approach for treating mitochondrial myopathies might be myoblast transplantations. Myoblasts of mitochondrial myopathy patients could be explanted and their mutant mtDNAs replaced with normal mtDNA by either replicative segregation or cybrid transfer (17, 59, 77, 120, 214). The genotypically normal muscle cells would then be expanded and injected back into the patients' muscle. There, they could fuse to existing myotubes, contributing more normal mtDNA (215) and supplementing mitochondrial energy production. Alternatively, diseases caused by mtDNA missense mutations, such as LHON and NARP, might be treated by transfer of normal mtDNA genes into the patient's nucleus. The normal mtDNA gene (e.g. ND4 or ATPase 6/8) could be cloned and in vitro mutagenesis used to correct the genetic code. An amino-terminal mitochondrial targeting sequence and an appropriate promoter could then be added by fusing the first exon and upstream sequences of a nuclear-encoded OXPHOS gene (e.g. ATP synthase) to the mtDNA gene (216, 217). This cassette could be inserted into a virus with the appropriate tissue trophism and injected into the patient. Once the construct had been delivered to the appropriate nucleus, the gene would be expressed, and the protein made in the cytosol, transferred to the mitochondria, processed into its mature form, and assembled into the appropriate respiratory complex (218). Since most mtDNA missense mutations require a high percentage of mutant mtDNAs to yield a clinical phenotype, introduction of even 25% normal polypeptide into the mitochondria might be sufficient to increase energy production above the expression threshold. Thus, while the unique genetic characteristics of the mtDNA create perplexing new problems in diagnosis and counseling, they also offer novel new possibilities for the treatment and prevention of disease.

ACKNOWLEDGMENTS

The author thanks Ms. Marie T. Lott and Drs. John M. Shoffner and Antonio Torroni for their important intellectual contributions and Ms. Rhonda E. Burke for assistance in preparing this manuscript. This work was supported by NIH grants NS21328, HL45572, GM46915 and AG10139, and by a Muscular Dystrophy Clinical Research Grant.

Literature Cited

1. Anderson, S., Bankier, A. T., Barrell, B. G., de Bruijn, M. H. L., Coulson, A. R., Drouin, J., Eperon, I. C., Nierlich, D. P., Roe, B. A., Sanger, F., Schreier, P. H., Smith, A. J.H., Staden, R., Young, I. G. 1981. *Nature* 290:457–65

2. Wallace, D. C., Lott, M., Torroni, A., Shoffner, J. 1991. *Cytogenet. Cell. Genet.* 58:1103–23

3. Shoffner, J. M., Wallace, D. C. 1990. *Adv. Hum. Genet.* 19:267–330

4. Wallace, D. C. 1982. *Microbiol. Rev.* 46:208–40

5. Wallace, D. C. 1987. *Birth Defects: Orig. Artic. Ser.* 23:137–90

6. Wallace, D. C. 1986. *Hospital Practice* 21:77–92

7. Chang, D. D., Clayton, D. A. 1987. *Science* 235:1178–84

8. Chang, D. D., Clayton, D. A. 1987. *EMBO J.* 6:409–17

9. Chang, D. D., Clayton, D. A. 1989. *Cell* 56:131–39

10. Doda, J. N., Wright, C. T., Clayton, D. A. 1981. *Proc. Natl. Acad. Sci. USA* 78:6116–20

11. Wong, T. W., Clayton, D. A. 1986. *Cell* 45:817–25

12. Doersen, C-J., Guerrier-Takada, C., Altman, S., Attardi, G. 1985. *J. Biol. Chem.* 260:5942–49

13. Montoya, J., Ojala, D., Attardi, G. 1981. *Nature* 290:465–70

14. Ojala, D., Montoya, J., Attardi, G. 1981. *Nature* 290:470–74

15. Christianson, T. W., Clayton, D. A. 1986. *Proc. Natl. Acad. Sci. USA* 83:6277–81

16. Christianson, T. W., Clayton, D. A. 1988. *Mol. Cell. Biol.* 8:4502–9

17. Wallace, D. C. 1982. *Techniques in Somatic Cell Genetics*, ed. J. W. Shay, pp. 159–87. New York: Plenum

18. Hennig, B. 1975. *Eur. J. Biochem.* 55:167–83

19. Scarpulla, R. C., Agne, K. M., Wu, R. 1981. *J. Biol. Chem.* 256:6480–86

20. Virbasius, J. V., Scarpulla, R. C. 1988. *J. Biol. Chem.* 263:6791–96

21. Lomax, M. I., Grossman, L. I. 1989. *Trends Biochem. Sci.* 14:501–3

22. Schlerf, A., Droste, M., Winter, M., Kadenbach, B. 1988. *EMBO J.* 7:2387–91

23. Smith, E. O., BeMent, D. M., Grossman, L. I., Lomax, M. I. 1991. *Biochim. Biophys. Acta* 1089:266–68

24. Fabrizi, G. M., Rizzuto, R., Nakase, H., Mita, S., Lomax, M. I., et al. 1989. *Nucleic Acids Res.* 17:7107

25. Kadenbach, B., Stroh, A., Becker, A., Eckerskorn, C., Lottspeich, F. 1990. *Biochim. Biophys. Acta* 1015:368–72

26. Seelan, R. S., Scheuner, D., Lomax, M. I., Grossman, L. I. 1989. *Nucleic Acids Res.* 17:6410

27. Lightowlers, R., Ewart, G., Aggeler, R., Zhang, Y-Z., Calavetta, L., Capaldi, R. A. 1990. *J. Biol. Chem.* 265:2677–81

28. Suske, G., Mengel, T., Cordingley, M., Kadenbach, B. 1987. *Eur. J. Biochem.* 168:233–37

29. Rizzuto, R., Nakase, H., Darras, B., Francke, U., Fabrizi, G. M., et al. 1989. *J. Biol. Chem.* 264:10595–600

30. Van Kuilenburg, A. B. P., Muijsers, A. O., Demol, H., Dekker, H. L., Van Beeumen, J. J. 1988. *FEBS Lett.* 240:127–32

31. Dyer, M. R., Gay, N. J., Walker, J. E. 1989. *Biochem. J.* 260:249–58

32. Farrell, L. B., Nagley, P. 1987. *Biochem. Biophys. Res. Commun.* 144:1257–64

33. Gay, N. J., Walker, J. E. 1985. *EMBO J.* 4:3519–24

34. Breen, G. A. M. 1988. *Biochem. Biophys. Res. Commun.* 152:264–69

35. Walker, J. E., Powell, S. J., Vinas, O., Runswick, M. J. 1989. *Biochemistry* 28:4702–8

36. Cozens, A. L., Runswick, M. J., Walker, J. E. 1989. *J. Mol. Biol.* 206:261–80

37. Li, K., Warner, C. K., Hodge, J. A., Minoshima, S., Kudoh, J., et al. 1989. *J. Biol. Chem.* 264:13998–4004

38. Neckelmann, N., Li, K., Wade, R. P., Shuster, R., Wallace, D. C. 1987. *Proc. Natl. Acad. Sci. USA* 84:7580–84

39. Powell, S. J., Medd, S. M., Runswick, M. J., Walker, J. E. 1989. *Biochemistry* 28:866–73

40. Rasmussen, U. B., Wohlrab, H. 1986. *Biochem. Biophys. Res. Commun.* 138:850–57

41. Battini, R., Ferrari, S., Kaczmarek, L., Calabretta, B., Chen, S-T., Baserga, R. 1987. *J. Biol. Chem.* 262:4355–59

42. Chen, S-T., Chang, C-D., Huebner, K., Ku, D-H., McFarland, M., et al. 1990. *Somat. Cell Mol. Gen.* 16:143–49

43. Houldsworth, J., Attardi, G. 1988. *Biochemistry* 85:377–81

44. Ku, D-H., Kagan, J., Chen, S-T., Chang, C-D., Baserga, R., Wurzel, J. 1990. *J. Biol. Chem.* 265:16060–63

45. Lunardi, J., Attardi, G. 1991. *J. Biol. Chem.* 266:16534–40

46. Neckelmann, N., Warner, C. K.,

Chung, A., Kudoh, J., Minoshima, S., et al. 1989. *Genomics* 5:829–43

47. Torroni, A., Stepien, G., Hodge, J. A., Wallace, D. C. 1990. *J. Biol. Chem.* 265:20589–93

48. Webster, K. A., Gunning, P., Hardeman, E., Wallace, D. C., Kedes, L. 1990. *J. Cell. Physiol.* 142:566–73

49. Evans, M. J., Scarpulla, R. C. 1989. *J. Biol. Chem.* 264:14361–68

50. Kagawa, Y., Ohta, S. 1990. *Int. J. Biochem.* 22:219–29

51. Lomax, M. I., Hsieh, C-L., Darras, B. T., Francke, U. 1991. *Genomics* 9:1–9

52. Ohta, S., Tomura, H., Matsuda, K., Kagawa, Y. 1988. *J. Biol. Chem.* 263:11257–62

53. Suzuki, H., Hosokawa, Y., Toda, H., Nishikimi, M., Ozawa, T. 1990. *J. Biol. Chem.* 265:8159–63

54. Suzuki, H., Hosokawa, Y., Nishikimi, M., Ozawa, T. 1991. *J. Biol. Chem.* 266:2333–38

55. Tomura, H., Endo, H., Kagawa, Y., Ohta, S. 1990. *J. Biol. Chem.* 265: 6525–27

56. Li, K., Hodge, J. A., Wallace, D. C. 1990. *J. Biol. Chem.* 265:20585–88

57. Blanc, H., Adams, C. W., Wallace, D. C. 1981. *Nucleic Acids Res.* 9:5785–95

58. Kearsey, S. E., Craig, I. W. 1981. *Nature* 290:607–8

59. Wallace, D. C., Bunn, C. L., Eisenstadt, J. M. 1975. *J. Cell Biol.* 67:174–88

60. King, M. P., Attardi, G. 1988. *Cell* 52:811–19

61. King, M. P., Attardi, C. 1989. *Science* 246:500–3

62. Oliver, N. A., Wallace, D. C. 1982. *Mol. Cell. Biol.* 2:30–41

63. Oliver, N. A., Greenberg, B. D., Wallace, D. C. 1983. *J. Biol. Chem.* 258:5834–39

64. Ruiters, M. H. J., Van Spronsen, E. A., Skjeldal, O. H., Stromme, P., Scholte, H. R., et al. 1991. *J. Inherit. Metab. Dis.* 14:45–48

65. Giles, R. E., Stroynowski, I., Wallace, D. C. 1980. *Somat. Cell Genet.* 6:543–54

66. Wallace, D. C., Eisenstadt, J. M. 1979. *Somat. Cell Genet.* 5:373–96

67. Zuckerman, S. H., Solus, J. F., Gillespie, F. P., Eisenstadt, J. M. 1984. *Somat. Cell Mol. Genet.* 10:85–92

68. Ballinger, S. W., Schurr, T. G., Torroni, A., Gan, Y. Y., Hodge, J. A., et al. 1992. *Genetics.* 130:139–52

69. Schurr, T. G., Ballinger, S. W., Gan, Y.-Y., Hodge, J. A., Merriwether, D. A., et al. 1990. *Am. J. Hum. Genet.* 46:613–23

70. Torroni, A., Schurr, T. G., Yang, C.-C., Szathmary, E. J.E., Williams, R. C., et al. 1992. *Genetics* 130:153–62

71. Gyllensten, U., Wharton, D., Josefsson, A., Wilson, A. C. 1991. *Nature* 352: 255–57

72. Pfanner, N., Neupert, W. 1990. *Annu. Rev. Biochem.* 59:331–53

73. Hendrick, J. P., Hodges, P. E., Rosenberg, L. E. 1989. *Proc. Natl. Acad. Sci. USA* 86:4056–60

74. Case, J. T., Wallace, D. C. 1981. *Somat. Cell Genet.* 7:103–8

75. Giles, R. E., Blanc, H., Cann, H. M., Wallace, D. C. 1980. *Proc. Natl. Acad. Sci. USA* 77:6715–19

76. Michaels, G. S., Hauswirth, W. W., Laipis, P. J. 1982. *Dev. Biol.* 94:246–51

77. Wallace, D. C. 1986. *Somat. Cell Mol. Genet.* 12:41–49

78. Shuster, R. C., Rubenstein, A. J., Wallace, D. C. 1988. *Biochem. Biophys. Res. Commun.* 155:1360–65

79. Holt, I. J., Harding, A. E., Petty, R. K., Morgan-Hughes, J. A. 1990. *Am. J. Hum. Genet.* 46:428–33

80. Holt, I. J., Miller, D. H., Harding, A. E. 1989. *J. Med. Genet.* 26:739- 43

81. Lott, M. T., Voljavec, A. S., Wallace, D. C. 1990. *Am. J. Ophthalmol.* 109:625–31

82. Shoffner, J. M., Lott, M. T., Lezza, A. M., Seibel, P., Ballinger, S. W., Wallace, D. C. 1990. *Cell* 61:931–37

83. Wallace, D. C., Zheng, X., Lott, M. T., Shoffner, J. M., Hodge, J. A., et al. 1988. *Cell* 55:601–10

84. Zeviani, M., Gellera, C., Antozzi, C., Rimoldi, M., Morandi, L., et al. 1991. *Lancet* 338:143–47

84a. Yousufzai, S. Y. K., Bradford, M. W., Shrago, E., Ewart, R. B. L. 1982. *FEBS Lett.* 137:205–8

84b. Welsh, N., Pääbo, S., Welsh, M. 1991. *Diabetologica* 34:626–31

84c. Svensson, C., Welsh, N., Krawetz, S. A., Welsh, M. 1991. *Diabetes* 40:771–76

85. Wallace, D. C., Ye, J., Neckelmann, S. N., Singh, G., Webster, K., Greenberg, B. D. 1987. *Current Genet.* 12:81–90

85a. Merriwether, D. A., Clark, A. G., Ballinger, S. W., Schurr, T. G., Soodyall, H., et al. 1991. *J. Mol. Evol.* 33:543–55

86. Cann, R. L., Stoneking, M., Wilson, A. C. 1987. *Nature* 325:31–36

87. Johnson, M. J., Wallace, D. C., Ferris, S. D., Rattazzi, M. C., Cavalli-Sforza, L. L. 1983. *J. Mol. Evol.* 19:255–71

88. Singh, G., Lott, M. T., Wallace, D. C. 1989. *N. Engl. J. Med.* 320:1300–5

89. Singh, G., Neckelmann, N., Wallace, D. C. 1987. *Nature* 329:270–72

90. Newman, N. J., Lott, M. T., Wallace, D. C. 1991. *Am. J. Ophthalmol.* 111:750–62

91. Newman, N. J., Wallace, D. C. 1990. *Am. J. Ophthalmol.* 109:726–30

92. Howell, N., Bindoff, L. A., McCullough, D. A., Kubacka, I., Poulton, J., et al. 1991. *Am. J. Hum. Genet.* 49:939–50

93. Huoponen, K., Vilkki, J., Aula, P., Nikoskelainen, E. K., Savontaus, M. L. 1991. *Am. J. Hum. Genet.* 48:1147–53

94. Howell, N., Kubacka, I., Xu, M., McCullough, D. A. 1991. *Am. J. Hum. Genet.* 48:935–42

95. Johns, D. R., Berman, J. 1991. *Biochem. Biophys. Res. Commun.* 174:1324–30

96. Brown, M. D., Voljavec, A. S., Lott, M. T., Torroni, A., Yang, C., Wallace, D. C. 1992. *Genetics* 130:163–73

97. Bolhuis, P. A., Bleeker-Wagemakers, E. M., Ponne, N. J., Van Schooneveld, M. J., Westerveld, A., et al. 1990. *Biochem. Biophys. Res. Commun.* 170:994–97

98. Hotta, Y., Hayakawa, M., Saito, K., Kanai, A., Nakajima, A., Fujiki, K. 1989. *Am. J. Ophthalmol.* 108:601–2

99. Huoponen, K., Vilkki, J., Savontaus, M. L., Aula, P., Nikoskelainen, E. K. 1990. *Genomics* 8:583–85

100. Stone, E. M., Copppinger, J. M., Kardon, R. H., Donelson, J. 1990. *Arch. Ophthalmol.* 108:1417–20

101. Vilkki, J., Savontaus, M. L., Nikoskelainen, E. K. 1989. *Am. J. Hum. Genet.* 45:206–11

102. Wallace, D. C., Singh, G., Lott, M. T., Hodge, J. A., Schurr, T. G., et al. 1988. *Science* 242:1427–30

103. Yoneda, M., Tsuji, S., Yamauchi, T., Inuzuka, T., Miyatake, T., et al. 1989. *Lancet* 1:1076–77

104. Wallace, D. C. 1989. *Trends Genet.* 5:9–13

105. Novotny, E. J. Jr., Singh, G., Wallace, D. C., Dorfman, L. J., Louis, A., et al. 1986. *Neurology* 36:1053–60

106. Vilkki, J., Ott, J., Savontaus, M. L., Aula, P., Nikoskelainen, E. K. 1991. *Am. J. Hum. Genet.* 48:486–91

107. Goto, Y., Nonaka, I., Horai, S. 1990. *Nature* 348:651–53

108. Hammans, S. R., Sweeney, M. G., Brockington, M., Morgan-Hughes, J. A., Harding, A. E. 1991. *Lancet* 337:1311–13

109. Hess, J. F., Parisi, M. A., Bennett, J. L., Clayton, D. A. 1991. *Nature* 351:236–39

110. Ino, H., Tanaka, M., Ohno, K., Hattori, K., Ikebe, S., et al. 1991. *Lancet* 337:234–35

111. Johns, D. R., Hurko, O. 1991. *Lancet* 337:927–28

112. Kobayashi, Y., Momoi, M. Y., Tominaga, K., Momoi, T., Nihei, K., et al. 1990. *Biochem. Biophys. Res. Commun.* 173:816–22

113. Ozawa, T., Tanaka, M., Ino, H., Ohno, K., Sano, T., et al. 1991. *Biochem. Biophys. Res. Commun.* 176:938–46

114. Tanaka, M., Ino, H., Ohno, K., Ohbayashi, T., Ikebe, S., et al. 1991. *Biochem. Biophys. Res. Commun.* 174:861–68

115. Lauber, J., Marsac, C., Kadenbach, B., Seibel, P. 1991. *Nucleic Acids Res.* 19:1393–97

116. Yoneda, M., Tanno, Y., Horai, S., Ozawa, T., Miyatake, T., Tsuji, S. 1990. *Biochem. Int.* 21:789–96

117. Zeviani, M., Amati, P., Bresolin, N., Antozzi, C., Piccolo, G., et al. 1991. *Am. J. Hum. Genet.* 48:203–11

118. Tanaka, M., Ino, H., Ohno, K., Hattori, K., Sato, W., et al. 1990. *Lancet* 336:1452

119. Yoon, K. L., Aprille, J. R., Ernst, S. G. 1991. *Biochem. Biophys. Res. Commun.* 176:1112–15

120. Chomyn, A., Meola, G., Bresolin, N., Lai, S. T., Scarlato, G., Attardi, G. 1991. *Mol. Cell Biol.* 11:2236–44

121. Wallace, D. C., Yang, J., Ye, J., Lott, M. T., Oliver, N. A., McCarthy, J. 1986. *Am. J. Hum. Genet.* 38:461–81

122. Shoffner, J. M., Lott, M. T., Wallace, D. C. 1991. *Rev. Neurol.* 147:431–35

123. Ichiki, T., Tanaka, M., Nishikimi, M., Suzuki, H., Ozawa, T., et al. 1988. *Ann. Neurol.* 23:287–94

124. Kobayashi, M., Morishita, H., Sugiyama, N., Yokochi, K., Nakano, M., et al. 1987. *J. Pediatr.* 110:223–27

125. Brown, M. D., Torroni, A., Shoffner, J. M., Wallace, D. C. 1992. *Am. J. Hum. Genet.* In press

126. Mita, S., Monnat, R. J. Jr., Loeb, L. A. 1988. *Mutat. Res.* 199:183–90

127. Holt, I. J., Harding, A. E., Morgan-Hughes, J. A. 1988. *Nature* 331:717–19

128. Lestienne, P., Ponsot, G. 1988. *Lancet* 1:885

129. Rötig, A., Colonna, M., Blanche, S., Fischer, A., Le Deist, F., et al. 1988. *Lancet* 2:567–68

130. McShane, M. A., Hammans, M., Sweeney, I., Holt, I. J., Beattie, T. J., et al. 1991. *Am. J. Hum. Genet.* 48:39–42

131. Schon, E. A., Rizzuto, R., Moraes, C. T., Nakase, H., Zeviani, M., DiMauro, S. 1989. *Science* 244:346–49

132. Shoffner, J. M., Lott, M. T., Voljavec,

A. S., Soueidan, S. A., Costigan, D.
A., Wallace, D. C. 1989. *Proc. Natl.
Acad. Sci. USA* 86:7952–56

133. Rötig, A., Cormier, V., Blanche, S.,
Bonnefont, J.-P., Ledeist, F., et al.
1990. *J. Clin. Invest.* 86:1601–8

134. Nelson, I., d'Auriol, L., Galibert, F.,
Ponsot, G., Lestienne, P. 1989. *C. R.
Acad. Sci.* 309:403–7

135. Mita, S., Rizzuto, R., Moraes, C. T.,
Shanske, S., Arnaudo, E., et al. 1990.
Nucleic Acids Res. 18:561–67

136. Shanske, S., Moraes, C. T., Lombes,
A., Miranda, A. F., Bonilla, E., et al.
1990. *Neurology* 40:24–28

137. Cormier, V., Rotig, A., Quartino, A.
R., Forni, G. L., Cerone, R., et al.
1990. *J. Pediatr.* 117:599–602

138. Poulton, J., Deadman, M. E.,
Ramacharan, S., Gardiner, R. M. 1991.
Am. J. Hum. Genet. 48:649–53

139. Ozawa, T., Yoneda, M., Tanaka, M.,
Ohno, K., Sato, W., et al. 1988.
Biochem. Biophys. Res. Commun. 154:
1240–47

140. Larsson, N. G., Holme, E., Kristians-
son, B., Oldfors, A., Tulinius, M.
1990. *Pediatr. Res.* 28:131–36

141. Mita, S., Schmidt, B., Schon, E. A.,
DiMauro, S., Bonilla, E. 1989. *Proc.
Natl. Acad. Sci. USA* 86:9509–13

142. Shoubridge, E. A., Karpati, G., Hastings, K. E. 1990. *Cell* 62:43–49

143. Pavlath, G. K., Rich, K., Webster, S.
G., Blau, H. M. 1989. *Nature* 337:570–
73

144. Poulton, J., Deadman, M. E., Gardiner,
R. M. 1989. *Lancet* 1:236–40

145. Poulton, J., Deadman, M. E., Gardiner,
R. M. 1989. *Nucleic Acids Res.*
17:10223–29

146. Moraes, C. T., Shanske, S., Trilschler,
H. J., Aprille, J. R., Andreetta, F., et
al. 1991. *Am. J. Hum. Genet.* 48:492–
501

147. Simpson, M. V., Chin, C. D., Keil-
baugh, S. A., Lin, T.-S., Prusoff, W.
H. 1989. *Biochem. Pharmacol.* 38:
1033–36

148. Arnaudo, E., Dalakas, M., Shanske, S.,
Moraes, C. T., DiMauro, S., Schon, E.
A. 1991. *Lancet* 337:508–10

149. Cormier, V., Rotig, A., Tardieu, M.,
Colonna, M., Saudubray, J.-M., et al.
1991. *Am. J. Hum. Genet.* 48:643–48

150. Yuzaki, M., Ohkoshi, N., Kanazawa,
I., Kagawa, Y., Ohta, S. 1989.
Biochem. Biophys. Res. Commun. 164:
1352–57

151. Zeviani, M., Bresolin, N., Gellera, C.,
Bordoni, A., Pannacci, M., et al. 1990.
Am. J. Hum. Genet. 47:904–14

152. Zeviani, M., Servidei, S., Gellera, C.,
Bertini, E., DiMauro, S., DiDonato, S.
1989. *Nature* 339:309–11

153. Otsuka, M., Niijima, K., Mizuno, Y.,
Yoshida, M., Kagawa, Y., Ohta, S.
1990. *Biochem. Biophys. Res. Commun.*
167:680–85

154. Muller-Hocker, J. 1989. *Am. J. Pathol.*
134:1167–73

155. Muller-Hocker, J. 1990. *J. Neurol. Sci.*
100:14–21

156. Trounce, I., Byrne, E., Marzuki, S.
1989. *Lancet* 1:637–39

157. Yen, T. C., Chen, Y. S., King, K. L.,
Yeh, S. H., Wei, Y. H. 1989. *Biochem.
Biophys. Res. Commun.* 165:944–1003

158. Piko, L., Hougham, A. J., Bulpitt, K. J.
1988. *Mech. Ageing Dev.* 43:279–93

159. Cortopassi, G. A., Arnheim, N. 1990.
Nucleic Acids Res. 18:6927–33

160. Hattori, K., Tanaka, M., Sugiyama, S.,
Obayashi, T., Ito, T., et al. 1991. *Am.
Heart J.* 121:1735–42

161. Corral-Debrinski, M., Stepien, G.,
Shoffner, J. M., Lott, M. T., Kanter,
K., Wallace, D. C. 1991. *J. Am. Med.
Assoc.* 266:1812–16

162. Linnane, A. W., Marzuki, S., Ozawa,
T., Tanaka, M. 1989. *Lancet* 1:642–45

163. Ames, B. N. 1989. *Mutat. Res.* 214:41–
46

164. Bandy, B., Davison, A. J. 1990. *Mutat.
Res.* 8:523–39

165. Ozawa, T., Tanaka, M., Sugiyama, S.,
Hattori, K., Ito, T., et al. 1990. *Biochem.
Biophys. Res. Commun.* 170:830–36

166. Corbisier, P., Remacle, J. 1990. *Eur. J.
Cell Biol.* 51:173–82

167. Ottman, R., Hauser, W. A., Susser, M.
1985. *Am. J. Epidemiol.* 122:923–39

168. Ottman, R., Annegers, J. F., Hauser,
W. A., Kurland, L. T. 1988. *Am. J.
Hum. Genet.* 43:257–64

169. Ozawa, T., Tanaka, M., Sugiyama, S.,
Ino, H., Ohno, K., et al. 1991.
Biochem. Biophys. Res. Commun. 177:
518–25

170. Langston, J. W., Ballard, P., Tetrud, J.
W., Irwin, I. 1983. *Science* 219:979–
80

171. Nicklas, W. J., Vyas, I., Heikkila, R.
E. 1985. *Life Sci.* 36:2503–8

172. Singer, T. P., Castagnoli, N. Jr., Ramsay, R. R., Trevor, A. J. 1987. *J.
Neurochem.* 49:1–8

173. Mizuno, Y., Ohta, S., Tanaka, M.,
Takamiya, S., Suzuki, K., et al. 1989.
Biochem. Biophys. Res. Commun. 163:
1450–55

174. Schapira, A. H. V., Cooper, J. M., Dexter, D., Jenner, P., Clark, J. B., Marsden, C. D. 1989. *Lancet* 1:1269

175. Schapira, A. H. V., Cooper, J. M.,
Dexter, D., Clark, J. B., Jenner, P.,
Marsden, C. D. 1990. *J. Neurochem.*
54:823–27

176. Parker, W. D. Jr., Boyson, S. J., Parks,
J. K. 1989. *Ann. Neurol.* 26:719–23

177. Bindoff, L. A., Birch-Machin, M. A., Cartlidge, N. E. F., Parker, W. D. Jr., Turnbull, D. M. 1991. *J. Neurol. Sci.* 104:203–8

178. Shoffner, J. M., Watts, R. L., Juncos, J. L., Torroni, A., Wallace, D. C. 1991. *Ann. Neurol.* 30:332–39

179. Johnson, W. C. 1991. *Neurology* 41:82–87

180. Ikebe, S., Tanaka, M., Ohno, K., Sato, W., Hattori, K., et al. 1990. *Biochem. Biophys. Res. Commun.* 170:1044–49

181. Sims, N. R., Finegan, J. M., Blass, J. P., Bowen, D. M., Neary, D. 1987. *Brain Res.* 436:30–38

182. Sheu, K.-F. R., Kim, Y.-T., Blass, J. P., Weksler, M. E. 1985. *Ann. Neurol.* 17:444–49

183. Parker, W. D. Jr., Filley, C. M., Parks, J. K. 1990. *Neurology* 40:1302–3

184. Peterson, C., Goldman, J. E. 1986. *Proc. Natl. Acad. Sci. USA* 83:2758–62

185. Blass, J. P., Baker, A. C., Ko, L-W., Black, R. S. 1990. *Arch. Neurol.* 47:864–69

186. Haines, J. L. 1991. *Am. J. Hum. Genet.* 48:1021–25

187. Farrer, L. A., Myers, R. H., Connor, L., Cupples, L. A., Growdon, J. H. 1991. *Am. J. Hum. Genet.* 48:1026–33

188. Pericak-Vance, M. A., Bebout, J. L., Gaskell, P. C. Jr., Yamaoka, L. H., Hung, W.-Y., et al. 1991. *Am. J. Hum. Genet.* 48:1034–50

189. St. George-Hyslop, P. H., Tanzi, R. E., Polinsky, R. J., Haines, J. L., Nee, L., et al. 1987. *Science* 235:885–90

190. Goate, A., Chartier-Harlin, M-C., Mullan, M., Brown, J., Crawford, F., et al. 1991. *Nature* 349:704–6

191. Tanzi, R. E., St. George-Hyslop, P. H., Haines, J. L., Polinsky, R. J., Nee, L., et al. 1987. *Nature* 329:156–57

192. Van Broeckhoven, C., Genthe, A. M., Vandenberghe, A., Horsthemke, B., Backhovens, H., et al. 1987. *Nature* 329:153–55

193. Brennan, W. A. Jr., Bird, E. D., Aprille, J. R. 1985. *J. Neurochem.* 44:1948–50

194. Parker, W. D. Jr., Boyson, S. J., Luder, A. S., Parks, J. K. 1990. *Neurology* 40:1231–34

195. Gusella, J. F., Wexler, N. S., Conneally, P. M., Naylor, S. L., Anderson, M. A., et al. 1983. *Nature* 306:234–38

196. Myers, R. H., Goldman, D., Bird, E. D., Sax, D. S., Merril, C. R., Schoenfeld, M., Wolf, P. A. 1983. *Lancet* 1:208–10

197. Brook, J. D., McCurrach, M. E., Harley, H. G., Buckler, A. J., Church, D., et al. 1992. *Cell* 68:799–808

198. McKusick, V. A. 1990. *Mendelian Inheritance in Man*, ed. V. A. McKusick, C. A. Francomano, S. E. Antonarakis. Baltimore/London: Johns Hopkins Univ. Press. 9th ed.

199. Arnaudo, E., Mita, S., Koga, Y., Tritschler, H. J., Shanske, S., DiMauro, S., Karpati, G. 1990. *Neurology* 40(Suppl. 1):376

200. Brais, B., Karpati, G. 1990. *Neurology* 40:376

201. Pimentel, E. 1979. *Acta Diabetol.* 16:193–201

202. Dorner, G., Mohnike, A. 1976. *Endokrinologie* 68:121–24

203. Dorner, G., Mohnike, A., Steindel, E. 1975. *Endokrinologie* 66:225–27

204. Dorner, G., Plagemann, A., Reinagel, H. 1987. *Exp. Clin. Endocrinol.* 89:84–90

205. Freinkel, N., Metzger, B. E., Phelps, R. L., Simpson, J. L., Martin, A. O., et al. 1986. *Horm. Metab. Res.* 18:427–30

206. Dorner, G. 1983. *Psychoneuroendocrinology* 8:205–12

207. Dorner, G., Plagemann, A., Ruckert, J., Gotz, F., Rohde, W., et al. 1988. *Endocrinology* 91:247–58

207a. Ballinger, S. W., Shoffner, J. M., Hedaya, E. V., Trounce, I., Polak, M. A., et al. 1992. *Nature Genet.* 1:11–15

208. Argov, Z., Bank, W. J., Maris, J., Eleff, S., Kennaway, N. G., et al. 1986. *Ann. Neurol.* 19:598–602

209. Eleff, S., Kennaway, N. G., Buist, N. R. M., Darley-Usmar, V. M., Capaldi, R. A., et al. 1984. *Proc. Natl. Acad. Sci. USA* 81:3529–33

210. Bresolin, N., Bet, L., Binda, A., Moggio, M., Comi, G., Nador, F., et al. 1988. *Neurology* 38:892–99

211. Ogasahara, S., Nishikawa, Y., Yorifuji, S., Soga, F., Nakamura, Y., et al. 1986. *Neurology* 36:45–53

212. Langsjoen, P. H., Vadhanavikit, S., Folkers, K. 1985. *Proc. Natl. Acad. Sci. USA* 82:4240–44

213. Stacpoole, P. W., Harman, E. M., Curry, S. H., Baumgartner, T. G., Misbin, R. I. 1983. *N. Engl. J. Med.* 309:390–96

214. Moraes, C. T., Schon, E. A., DiMauro, S., Miranda, A. F. 1989. *Biochem Biophys. Res. Commun.* 160:765–71

215. Partridge, T. A. 1991. *Muscle Nerve* 14:197–212

216. Law, R. H. P., Farrell, L. B., Nero, D., Devenish, R. J., Nagley, P. 1988. *FEBS Lett.* 236:501–5

217. Nagley, P., Devenish, R. J. 1989. *Trends. Biochem. Sci.* 14:31–35

218. Lander, E. S., Lodish, H. 1990. *Cell* 61:925–26

Annu. Rev. Biochem. 1992. 61:1213–30

DIFFERENTIATION REQUIRES CONTINUOUS ACTIVE CONTROL

Helen M. Blau

Department of Pharmacology, Stanford University School of Medicine, Stanford, California 94305

KEY WORDS: gene expression, negative regulation, tissue-specific gene expression

CONTENTS

INTRODUCTION

The vast number of genes involved in the development of a multicellular organism presents a substantial regulatory problem. Complex controls are required to ensure that the right genes are expressed at the correct level in hundreds of different cell types. Since only a small fraction of the total genes

1213

0066-4154/92/0701-1213$02.00

of a differentiated cell are expressed at any given time, negative regulatory mechanisms for preventing inappropriate genes from being expressed are particularly important. Two types of molecular mechanism that could silence genes in differentiated cells are (*a*) "passive control," a mechanism for closing down unneeded genes in a given cell lineage so that they do not need active consideration for the life of the organism, and (*b*) "active control," continual regulation of the expression state of each gene. Under passive control, the commitment to differentiate, like Lyonization of the X chromosome, would result in the permanent inactivation of many unnecessary genes. Under active control, an easily reversible regulatory decision would be made for each gene in each differentiated cell, a decision determined by the protein composition of the cell at any given time. An active mechanism would require that gene expression in the differentiated cells of eukaryotes, like the cells of pro-karyotes (1a), be dynamic and subject to continuous regulation (1b).

The control of differentiation by a passive mechanism appears more likely than control by an active mechanism for the following reasons. First, the complexity of chromosome and DNA structure in eukaryotes vastly exceeds that in prokaryotes, making a simple active mechanism seem untenable. Second, plasticity of gene expression in "terminally" differentiated cells appears unnecessary, even risky. Third, active control seems unduly cumbersome. Indeed, the investment in regulators, especially negative regulators, required to maintain most genes at most times in a silent state appears disproportionately large. Fourth, it is unclear how memory and stability could be achieved by an active control mechanism. Although the rationale in favor of passive control appears strong, accumulating experimental evidence in a number of species suggests that the differentiated state is maintained by active continuous regulation, both by positive and by negative regulators.

In this review I first discuss passive control and evidence that this mechanism may be used to regulate gene expression in some settings during development. I then trace historically the key experiments that led to the hypothesis that the development of differentiated cells in tissues and organs is not passively, but actively, controlled. Finally, I discuss the implications of an active control mechanism for differentiation.

EVIDENCE FOR PASSIVE CONTROL

As a counterpoint to active control, I have selected two examples that currently provide the strongest cases for passive control. The silencing of genes that results from X-chromosome inactivation or as a consequence of imprinting suggests a passive control mechanism. These two events play a critical role in early development, apparently ensuring the balanced contribution of male and female genomes (2). For X chromosomes, it is clear that only one of a pair of genes is active in the same nucleus, an expression state that is established early in embryogenesis and stably transmitted to all progeny cells as methylated heterochromatin. The molecular mechanisms that target genes for this type of repression are not well understood. Once induced, however, this expression state is not easily disrupted and does not appear to be subject to change in the course of normal development; it is altered only in the germline. Imprinting shares several of these properties and, although less well characterized, may share a similar mechanism. X-chromosome inactivation and imprinting both appear to be negative regulatory decisions that persist for the life of the organism.

EVIDENCE FOR ACTIVE CONTROL

It might seem most expedient to silence genes in differentiated cell types by a passive mechanism, such as that used in X-chromosome inactivation. Indeed, why allow muscle genes to be accessible in a liver cell? However, as described below, several lines of experimental evidence suggest that differentiation is actively controlled and requires continuous regulation.

Nuclear Transplantation

Gurdon's classic nuclear transplantation experiments (3) first began to test whether gene expression in differentiation is controlled by an active or a passive mechanism. Gurdon found that when the nuclei of amphibian intestinal cells were transplanted into enucleated eggs, entire feeding tadpoles developed. Nuclei from keratinocytes and noncycling erythrocytes also displayed this potential for dramatic change in nuclear function (4; for a review, see reference 5). These experiments were interpreted as providing strong evidence that genetic material was generally not lost during vertebrate differentiation. In the present context, these experiments also provide support for an active mechanism of gene regulation, because they show that tissue-specific genes, such as globin genes in intestinal cells, are reversibly inactivated. When passed back through egg cytoplasm, the globin genes are resuscitated and are able to function once red blood cells differentiate among the progeny of the reconstituted egg.

Gurdon's data, however, do not prove active control, because passage of the differentiated nuclei through the egg might have stripped the DNA of all passive regulatory influences and allowed reprogramming. Indeed, Di-Berardino et al (6) showed that the frequency of obtaining feeding tadpoles was increased from approximately 2 to 75% if the nuclei were initially injected into maturing oocytes, conditioned with oocyte cytoplasm, and allowed to progress to the blastula stage before being transplanted into enucleated eggs. The high frequency observed in these experiments ruled out the possibility that the results reflected the presence of a small subpopulation of residual stem cells. However, it raised the possibility that the incubation step was necessary to alter gene structure and permit access of regulators. Thus, the environment of the maturing oocyte might be a prerequisite for alleviating passive control and allowing resuscitation of genes inactivated during differentiation.

Transdetermination and Transdifferentiation

Other lines of evidence show that exposure to ooplasm may not be a requirement for the activation of genes from differentiated nuclei that have no need of those genes. For instance, following serial transplantation of imaginal disc cells into the abdomen of an adult *Drosophila* fly, "transdetermination," or a change in determined state, often occurs so that cells originally destined for genital structures first give rise to leg or head structures and eventually to wings (7, 8). A similar developmental transformation of one body part into another occurs when *Drosophila* homeotic genes are mutated (9).

A change termed "transdifferentiation" has also been observed at sites of tissue regeneration following injury. When the iris of an amphibian, chicken, or human eye is damaged, for instance, some of the cells dedifferentiate, lose their melanin pigment, divide, and give rise to lens cells that synthesize characteristic crystallins (10, 11). Similarly, striated muscle tissue isolated from anthomedusae is capable of generating six diverse cell types including sensory neural cells (12, 13). In some cases, these changes in differentiated state cross lineages, so that mesoderm gives rise to ectodermal cell types.

In transdetermination and transdifferentiation experiments, the fate of the cell is altered without recourse to the regulatory hierarchy characteristic of development from the fertilized egg to differentiated tissue. However, these experiments may not rule out a passive control mechanism. A number of cell divisions typically occur between the two differentiated states, and changes critical to reprogramming and activating silent genes could be associated with DNA replication in the course of those divisions, as suggested by the models of Holtzer et al (14), Brown (15), and Weintraub (16). To address this possibility, experiments that test the need for DNA replication as a prerequisite for activating silent tissue-specific genes are required.

Heterokaryons

Strong support for the active-control hypothesis derives from somatic cell hybridization experiments. Upon cell fusion, the influence of one cell type on the function of another can be studied. Experiments with hybrids first showed that gene expression could be altered in transformed cells (17a–c) and in differentiated mammalian somatic cells (17d–h). In some cases, novel gene expression was induced (18a–c). However, since extensive proliferation was required to select the hybrids, a requirement for DNA replication to alleviate repression by a passive mechanism could not be ruled out. Moreover, in dividing cell hybrids, or synkaryons, nuclei were combined and chromosome loss and rearrangement were the norm. Experiments with nondividing cell hybrids, or heterokaryons (19, 20) in which nuclei remained separate and distinct, demonstrated that this resuscitation of genes could occur without cell division or DNA replication (21–23). Upon fusion of muscle cells with primary cell types representing all three embryonic lineages (endoderm, ectoderm, and mesoderm), genes that encoded a wide range of products including enzymes, membrane components, and contractile proteins were activated. Gene dosage, or the balance of regulators contributed by the two fused cell types, was critical. When the number of muscle cell nuclei exceeded the number of liver cell nuclei, extinction of liver genes and activation of muscle genes were observed; conversely, when the number of liver cell nuclei exceeded the number of muscle cell nuclei, muscle gene expression was repressed. By using a similar heterokaryon approach, these results were corroborated for muscle (24, 25) and extended to a variety of tissue-specific genes including hematopoietic (26), hepatic (27), and pancreatic (28) genes, which were activated in fibroblasts after fusion with cell types derived from blood, liver, and pancreas, respectively.

Several aspects of these cell fusion experiments are consistent with an active control mechanism. First, they suggest that genes are available for expression in cells that normally never express them. Thus, genes typical of mesoderm are readily activated in ectodermal cell types. Second, genes are activated in the absence of DNA replication. The frequency of gene activation did not differ when cells were continuously exposed to an inhibitor of DNA synthesis before and after fusion. Therefore if activation of silent genes requires changes in chromatin structure, these changes are mediated by mechanisms independent of DNA synthesis. Third, cells in which differentiation is well under way are as capable of inducing the expression of previously silent tissue-specific genes as are cells initiating differentiation. From this finding, it appears that the activity of *trans*-acting regulators is not required transiently at the onset of differentiation, but continuously to maintain it. Fourth, gene dosage, or the balance of regulators contributed by the two fused cell types, is a critical determinant of whether genes are repressed or activated. Fifth, differences among cell types are observed in the kinetics, frequency, and effects of gene dosage on gene activation or gene repression in heterokaryons. This result is likely to reflect differences in the combination of proteins that interact in each type of heterokaryon. Taken together, these observations provide strong support for the active-control hypothesis.

The Helix-Loop-Helix Family of Myogenic Regulators

Compelling evidence that changes in gene expression that accompany differentiation are actively regulated is also provided by the discovery of MyoD. Weintraub and colleagues (29) showed that the constitutive expression of a cDNA encoding a single *trans*-acting regulator could activate silent muscle genes such as myosin heavy chain and desmin in a range of nonmuscle cell types including fibroblasts, melanocytes, and neuroblastoma cells (30). Indeed, in mesodermal cell types, a complete phenotypic conversion was induced: a muscle-specific distribution of organelles and pattern of gene expression were stably inherited (31). Three additional regulators have now been identified, all members of a helix-loop-helix family of transcription factors (32) that appear to have similar properties (33–37). When constitutively expressed in fibroblasts, each of these myogenic regulators alone is capable of inducing myogenesis, presumably by binding to the consensus E-box sequence found in a large number of muscle-specific genes. These experiments indicate that a single protein, when present at relatively high concentrations, is capable of gaining access to and activating the expression of genes resident in cells that would normally never express them.

Drosophila *and* C. elegans

Further support for the active-control hypothesis derives from elegant experiments showing that in the course of normal development the continuous activity of positive and negative regulators is required to maintain the differentiated state. Most compelling are the results of experiments in which disruption in expression of a nodal, or key, regulatory gene alters the fate of cells. One might expect that developmental decisions such as segment identity or sex would be made early and locked in place by stable, heritable mechanisms. There would seem to be little advantage to allowing for plasticity in the regulation of these decisions, because the welfare of the organism requires that these decisions remain constant in each of its cells throughout its lifetime. However, experiments with *Drosophila* and *Caenorhabditis elegans* indicate that this is often not the case.

The clearest evidence for active control derives from two types of experiments that examine the temporal window during which a particular gene product is required: experiments using temperature-sensitive mutants or somatic mosaics. In the former, gene expression is altered by a shift in temperature. In the latter, mosaic individuals with patches of homozygous mutant cells are generated in heterozygous (normal) individuals by X-ray-induced mitotic recombination or chromosome loss. Studies of this type have shown that the expression of a gene that controls segment identity, such as *Ultrabithorax,* is required throughout development (38–40). Similarly, both positive and negative regulators of sex determination must be expressed continuously or else the sexual characteristics of the cells will change, even in adulthood (41–44). Ongoing gene expression is also required to maintain the identity of neurosensory cells (45). Pattern formation, even at late larval stages, is altered if the expression of critical genes, including the large Polycomb family that encodes negative regulators, is disrupted (46). Perhaps the most striking example of plasticity is found in the *Drosophila* central nervous system: an adult female will engage in a complex courtship behavior typical of an adult male if exposed to a shift in temperature that disrupts the expression of the *tra-2* gene (47).

The results of such classic genetic experiments are corroborated by molecular experiments examining the expression pattern of transcripts by in situ hybridization or of proteins by immunofluorescence. They indicate that the products of homeotic selector genes and sex determination genes are present continuously throughout development (48, 49). These and many other experiments show that the uninterrupted expression of certain nodal negative and positive regulators is essential to the expression of the differentiated state in vivo.

IMPLICATIONS OF ACTIVE CONTROL FOR DIFFERENTIATION

Stoichiometry of Positive and Negative Regulators

The active-control hypothesis suggests that the stoichiometry, or relative concentration, of regulators plays a critical role in the expression of the differentiated state. The effective concentration of a regulator is altered not only when its rate of synthesis or degradation is changed, but also when the concentration of the proteins with which it interacts is altered. Recent evidence indicates that many regulatory proteins form complexes, for example heterodimers via leucine zipper or helix-loop-helix motifs (50, 51). Such interactions can either promote or inhibit the function of a regulator. For instance, the transcription factor MyoD requires the protein E12 to bind DNA efficiently (32), but is prevented from binding DNA when complexed to the protein Id (52). The transcription factor NF-κB is inhibited from entering the nucleus and is therefore inactive when it is complexed to I-κB in the cytoplasm (53, 54). The complexity of these interactions increases as the number of different partners with which a protein can associate increases, as is the case in intact cells (31, 55). Clearly, in addition to abundance, the relative affinity and cooperative interactions of regulators at DNA-binding sites will have a profound impact on gene expression. Recent evidence that synergism among diverse transcriptional regulators occurs even at concentrations at which their DNA-binding sites are saturated suggests that regulators have cooperative effects not just as heterodimers but also as multimeric complexes

(56, 57). Because proteins act in combinations, small changes in the relative concentrations of a single regulator can have large effects on the expression of the differentiated state of the cell, by shifting a critical balance, reaching a threshold, and setting off a cascade of events. These predictions are borne out in vivo. The dosage of genes encoding the helix-loop-helix proteins *daughterless*, *hairy*, and *achaete-scute* determines sex in *Drosophila* (58). Gene dosage also determines sex in *C. elegans* (59) and the phenotype of neurosensory cells in *Drosophila* (60). Gene dosage is also responsible for several hereditary developmental disorders in humans (61).

Negative Control of Gene Expression

As discussed at the outset, a major regulatory dilemma in differentiation is maintaining most genes in an inactive state at any given time. Several lines of evidence suggest that negative *trans*-acting regulators are continuously present. Following cell fusion, many previously expressed tissue-specific genes are shut off (62–66). In *Drosophila*, loss-of-function mutants have revealed loci such as *hairy* and *extramacrochaete* that encode negative regulators of the *achaete-scute* complex (60). These gene products are required continuously to repress the differentiation of sensory organs (67). Similarly, the continuous expression of the homeotic *Polycomb* genes prevents the expression of the *Drosophila Ultrabithorax* gene. If *Polycomb* gene expression is disrupted, segment identity is altered (9, 68). Thus, mechanisms that mediate gene repression are of particular importance.

Negative *cis*-regulatory DNA elements have been revealed in the control regions of many genes, and, like positive *cis*-regulatory elements, they may eventually prove to be present in all genes. This would lend strong support to the active-control hypothesis. Negative elements are used for spatial control of gene expression, ensuring the repression of tissue-specific genes in a number of inappropriate tissues. Many such sites have been identified, for example in genes encoding growth hormone (69), insulin (70–72), renin (73), interleukin-2 (IL-2) receptor α chain (74), immunoglobulin kappa light chain (75), immunoglobulin heavy chain (76), T-cell receptor α chain (77), urokinase plasminogen activator (78), α-fetoprotein (79), vimentin (80), collagen II (81) and ϵ-globin (82). Indeed, subsets of differentiated cell types such as T lymphocytes can be distinguished on the basis of the activity of negative *cis*-regulatory elements: an enhancer controlling the α gene is active in $\alpha\beta$ but not $\gamma\delta$ cells in which a series of negative regulatory elements is active (77).

Negative regulatory sites are also used for temporal control by preventing gene expression at inappropriate times. These sites are used to silence genes until an extracellular signal induces an intracellular signal transduction pathway, leading to derepression. The beta-interferon gene (87, 88) and the yeast heat shock genes (89, 90) are cases in point. Extrinsic signals are also likely to mediate the derepression of genes that occurs during differentiation. Negative regulatory sites appear to mediate lysozyme gene repression prior to macrophage differentiation (91), prevent immunoglobulin kappa genes from being expressed until pre-B cells mature to B cells (75, 92), and suppress the expression of major histocompatibility complex class I genes until embryonic stem cells are induced to differentiate by retinoic acid (93).

In many of the cases cited above, negative regulators do not interact with positive regulators directly. Instead, they act at a distance by a mechanism known as "silencing," which provides a means of shutting off gene expression, even in the presence of positive regulators. First described in the yeast *Saccharomyces cerevisiae* (83, 84), silencers have now been identified in a number of tissue-specific genes in a wide variety of species. Although it has been postulated that they may inhibit transcription (84, 85) or define chromatin loops (86), the mechanism by which silencers act in mammalian cells and the extent to which it parallels that described in yeast remains to be determined.

Negative *cis*-regulatory DNA elements need not be gene specific. A negative element that appears to play a role in repressing nearby genes has been defined in repetitive *alu* sequences (92, 94, 95). Other repetitive elements in the vicinity of several chicken genes, including lysozyme, ovalbumin, calmodulin, and UI genes (91), or near rat genes such as the insulin gene that contain the highly reiterated LINES sequence (71), mediate negative regulation. Such repetitive elements constitute putative "global repression-binding sites," that could provide a means of silencing a number of genes with relatively few negative regulators.

Negative regulators need not act directly by binding to repressive sites in DNA. They can affect the expression of specific genes indirectly by preventing the expression of genes encoding positive regulators (96) or by competing with positive regulators for DNA-binding sites (97–99). Alternatively, negative regulators can complex with positive regulators and prevent DNA binding (53, 54, 100) or render activators nonfunctional once bound to DNA (101–103).

In most cases, the derepression of genes, or relief from negative regulation, is accompanied by an induction of expression of positive regulators. The positive regulators may bind at sites distant from the sites occupied by negative regulators and act indirectly (91), or the binding sites of positive and negative regulators may overlap (87, 88, 104). It is generally not known whether positive effects override negative ones by direct or indirect interference with binding, but both mechanisms are likely to operate, and the stoichiometry, or relative concentration of each type of regulator, is likely to play an important role in the outcome.

Establishment and Maintenance of Differentiation

The active-control hypothesis poses problems for differentiation. Do all regulators have to be active at all times to maintain the differentiated state? How can the requisite number of regulators be continuously produced? How are stability and memory ensured? These questions are best addressed by first examining the origin of the differentiated state.

THE FIRST DIFFERENTIATION

The first cell difference, or differentiation, is likely to result from asymmetry. This is readily apparent in *Drosophila* or *Xenopus* development, in which the components of the egg are unequally distributed. The manifestation of asymmetry in mammalian development occurs later and is more subtle. Cells of the early mouse embryo are totipotent: their function is not restricted. This has been most elegantly demonstrated in experiments in which each of the cells at the four-cell stage was isolated and shown to be capable of contributing to both embryonic and extraembryonic tissues in chimeras formed by juxtaposing genetically distinct blastula-derived cells (105). Asymmetry is cytologically apparent within mouse cells as early as the eight-cell stage: organelles, cytoskeletal elements, and cell surface antigens assume a polar distribution that differs among neighbors (106–110). Although chimera experiments show that the cells are still totipotent at this stage (105), the subsequent fate of their progeny can be reproducibly influenced by their position within the reconstituted eight-cell embryo. Cells on the inside are destined for embryonic tissues, whereas cells on the outside contribute to extraembryonic tissues, presumably as a result of exposure to different signals. These signals induce the cells to elaborate novel protein products that, in turn, allow them to respond to novel signals. This signal–response cycle constitutes a hierarchy of regulatory steps that leads to the generation of neighboring cells that are increasingly differentiated from one another, expressing hundreds of different RNA and protein products.

MEMORY AND STABILITY

If differentiation is actively controlled, how is memory of the differentiated state established, maintained, and propagated to progeny? Is the entire hierarchy of regulators that led to the establishment of each distinct differentiated state continuously required to maintain it? If so, this almost infinite regress of regulators would, indeed, seem prohibitive.

The answer to these questions is that cells have developed mechanisms for circumventing the regulatory hierarchy. Thus, past events can be remembered without being continuously repeated and without recourse to a passive control mechanism. Autoregulatory loops, by which a protein product induces transcription of the gene that encodes it, constitute one active control mechanism that could maintain protein levels in the absence of early steps. This mechanism appears to be used by the bacteriophage lambda repressor, some of the *Drosophila* homeotic selector gene products, the signal transducer c-*jun*, and the helix-loop-helix family of myogenic regulators (111–114). Once genes encoding nodal regulators are activated, autoregulation serves to maintain these regulators at a critical threshold concentration, providing both stability and memory. Another cellular memory mechanism involves auto-catalytic calcium/calmodulin-dependent protein kinases, which have been proposed as effective mediators of long-term storage by virtue of their multi-subunit holoenzyme structure (115). The extracellular matrix components secreted by differentiated cell types help to maintain those differentiated states and can instruct other cells to assume novel fates (116a,b). Additional, as yet unidentified, feedback loops are probably required to maintain the differentiated state by preventing proliferation. By circumventing the regulatory hierarchy, such feedback mechanisms limit the number of regulators required and play a central role in maintaining the differentiated state.

Although actively controlled, the differentiated state is stable. Secreted regulators have limited effects due to differences among cells in membrane components, such as the presence or absence of specific receptors. Cells are also relatively impervious to the effects of a single regulator introduced by injection or transfection. The introduction of tissue-specific transcription factors such as pituitary Pit-I (GHF-1) (117, 118), liver HNF-1 (119), and neural MASH-1 (120) into a variety of cell types causes no phenotypic changes and, in general, does not lead to activation of endogenous target genes. MyoD (see above) is the exception to the rule and is remarkable in its pleiotropic effects. However, even in the case of MyoD, high constitutive expression has diverse effects in different cell types. In some cells MyoD induces a heritable phenotypic conversion; in others it activates the transcription of certain muscle target genes; and in a third category of cells it has no detectable effects at all (30, 31, 121). Experiments examining the ectopic expression or misexpression of single regulators in *Drosophila* corroborrate these findings. When *Antennapedia* and *Deformed*, proteins that control the identity of thoracic and head segments, respectively, are expressed at aberrant times and developmental stages under the control of the heat shock promoter, a range of effects is observed (114, 122). A homeotic transformation, such as antenna to leg, is seen only at the highest concentrations of *Antennapedia* protein. The partial response or lack of response may be because regulator concentrations are insufficient, essential endogenous cell proteins are lacking, or proteins that interfere with regulator activity are present. Such proteins may complex with, modify, or compete directly with the foreign regulator, thereby altering its activity. Taken together, these experiments suggest that the effects of a regulator on the differentiated state are buffered by the protein composition of the cell, which in turn is determined by the cell's heritage or cumulative responses to cues in the course of development. Thus, the differentiated state is stable and is not easily disrupted.

CONCLUSION

Four reasons were presented at the outset that argued in favor of passive control: DNA complexity, the number of regulators, the need for stability and memory, and the risk of unnecessary plasticity. However, as summarized below, active control can satisfy each of these. In addition, active control is advantageous and possibly essential for differentiation.

It appears that we have come full circle in our views of differentiation. Three decades ago Jacob and Monod (1) showed that a balance of positive and negative regulators controlled the expression of structural genes in bacteria. The possibility that similar mechanisms operated to control gene expression in the differentiated cells of higher organisms was questioned when it became apparent that DNA could exist as tightly coiled fibers coated with histones. The discovery of chromatin structure created the expectation that genes were regulated by a different, possibly passive mechanism. In addition, the correlation of newly induced DNase I–hypersensitive sites and of reduced levels of DNA methylation with cell commitment and tissue-specific gene expression was interpreted as indicating that heritable *cis*-acting regulatory mechanisms played a central role in controlling differential gene expression. Finally, it seemed improbable that large numbers of *trans*-acting regulators could efficiently regulate gene expression, because the concentration of DNA in eukaryotic cells was so much greater than in prokaryotes. However, as discussed by Ptashne (111) this is unlikely to present a problem. Therefore, the accumulating evidence suggests that the dynamic mechanism proposed by Jacob and Monod (1) for prokaryotic gene regulation may well be the prevalent mechanism used by eukaryotes to control the expression of differentiation-specific genes.

What is the role of chromatin? Although both active and passive forms of control are associated with changes in "chromatin," the passive forms of gene silencing established early in development are likely to differ at the molecular level from the active forms involved later in the course of differentiation. The repression of genes that accompanies X-chromosome inactivation or imprinting is relatively fixed, or permanent. By contrast, the repression of genes typical of differentiation appears plastic and dynamic. Indeed, contrary to previous models (15, 16), it is now clear that in the absence of DNA replication, inactive genes become hypomethylated, nucleosomes are displaced, and DNase-hypersensitive sites are induced (123, 124). These changes in "chromatin," which alter the expression state of tissue-specific genes, are readily reversible and can all be accounted for by a change in the stoichiometry of *trans*-acting factors (125). Therefore, the molecular mechanisms underlying the stable, heritable gene repression by "passive chromatin" are unlikely to be the same as those responsible for readily altered gene repression by "active chromatin." Further distinctions await an elucidation of the underlying molecular mechanisms.

Active control provides both memory and stability. The sequential activity of a series, or hierarchy, of regulators leads to the establishment of each distinct differentiated state. However, this progression from totipotent to differentiated cell does not, in general, restrict the range of possible genes that a cell can express, as concepts such as "committed stem cell," "terminally differentiated cell," and "fate maps" might suggest. Therefore, regulators in the hierarchy that act for short periods to establish a differentiated state, like those that act continuously to maintain it, need not lead to permanent changes. Cells do not have to establish passive control in order to remember the effects of a transiently expressed regulator. Feedback loops can achieve the same end (see above).

A major problem of the active-control hypothesis appears to be the large number of regulators required, in particular negative regulators, since at any given time only a small fraction of the total genes of a cell are expressed. A solution to this problem may be the discovery of negative regulatory elements, identified in repetitive DNA sequences, that silence several different genes (see above). Although to date relatively few such repetitive sequences have been identified, they could, in theory, constitute highly effective "global repression-binding sites," through which a number of related tissue-specific genes could be silenced by relatively few negative regulators. For example, it seems likely that related DNA sequences mediate the effects of the Polycomb class of negative regulators, which bind to at least 60 sites within polytene chromosomes at which diverse homeotic or other Polycomb target genes reside (126).

Plasticity of gene expression in "terminally" differentiated cells appears unnecessary, even risky. Why should the differentiated state be controlled by mechanisms that are dynamic and reversible? Perhaps active control is an evolutionary vestige: a single jellyfish cell can generate numerous different cell types and axolotls can regenerate entire limbs. On the other hand, active control may provide essential plasticity. The same regulatory genes are used at different times in development to specify different processes and therefore must be accessible and activatable. An example is provided by the segment polarity gene, *engrailed,* in *Drosophila: engrailed* is expressed at two different developmental stages and the regulatory network governing its expression at these two stages differs (127). In addition, differentiation may not be as rigidly determined as it appears. Upon serial transplantation, *Drosophila* cells undergo a transdetermination from leg- to head-type cells and injury to amphibian, chicken, and even mammalian tissues can cause a transdifferentiation, or conversion of melanin-producing iris cells to crystallin-producing lens cells (see above). Moreover, in the course of normal development, cells such as those of the neural crest give rise to a multiplicity of unexpected cell types, including representatives of different embryonic germ layers (128). Diverse postnatal myogenic precursor cells randomly fuse with the entire spectrum of fiber types in their vicinity and, once incorporated, adopt the pattern of gene expression characteristic of the host fiber (129). Finally, neural cells implanted into diverse sites within the brain assume the phenotype of their neighbors (130). Experiments of this type, which use sensitive single-cell markers to monitor the fate of the cells following implantation into novel sites, are revealing unexpected degrees of plasticity of cell function.

Possibly all gene expression is actively controlled. An analysis of position effect variegation raises the possibility that even in cases in which gene expression is generally regarded as passively controlled, the underlying mechanisms may be active. Position effect variegation shares properties with X-chromosome inactivation: gene inactivity is associated with heterochromatin, a region of the chromosome that is permanently condensed (131a,b). Translocations lead to the inactivation by heterochromatin of genes not normally subject to this type of regulation in species ranging from *Drosophila* to humans (131). Recent studies have suggested a link between the spreading of heterochromatin in position effect variegation and the expression of the Polycomb family of genes that encodes *trans*-acting negative regulators. First, the heterochromatin-associated protein HP1 encoded by a member of the Su(var) family responsible for position effect variegation, shares homology with a protein encoded by a member of the Polycomb family (132). Second, the effects of both types of regulation are dose dependent: the size of the domain encompassed in heterochromatin is altered by Su(var) gene dosage (131), and negative regulation of the bithorax complex by the Polycomb gene products depends on the number of copies of that complex (46). These findings raise the possibility that gene regulation in development is entirely active, and apparent differences reflect a continuum, or spectrum, of more or less perturbatable gene states, controlled by different molecular mechanisms.

SUMMARY

The problems posed by differentiation that appear most soluble by a passive control mechanism can readily be solved by an active mechanism. Given the need for plasticity in gene expression in different cell types at different stages, an active mechanism may be advantageous, even essential. It is striking how few changes during differentiation are completely irreversible, the gene rearrangements leading to immunoglobulin expression being one clear exception. Indeed, a prediction of the active-control hypothesis is that any nucleus exposed to the appropriate constellation of proteins at the appropriate concentration should be able to perform functions typical of any given differentiated cell type. In the next decade, the elucidation of novel memory mechanisms, or feedback loops, will substantially increase our understanding of how stable differentiated states can be maintained by continuous active control.

ACKNOWLEDGMENTS

I wish to thank Dr. David Baltimore for numerous stimulating discussions and critique. Special thanks are also due to Dr. Juan Botas for insights that contributed greatly to the current cross-species treatment of the subject. Finally, I am grateful to the many colleagues who generously took time to discuss and dispute the views set forth here.

Literature Cited

1a. Jacob, F., Monod, J. 1961. *J. Mol. Biol.* 3:318–56
1b. Blau, H. M., Baltimore, D. 1991. *J. Cell Biol.* 112:781–83
2. Thomson, J. A., Solter, D. 1988. *Genes Dev.* 2:1344–51
3. Gurdon, J. B. 1962. *J. Embryol. Exp. Morphol.* 10:622–40
4. Gurdon, J. B., Laskey, R. A., Reeves, O. R. 1975. *J. Embryol. Exp. Morphol.* 34:93–112
5. DiBerardino, M. A. 1988. In *Regulatory Mechanisms in Developmental Processes,* ed. G. Eguchi, T. S. Okada, L. Saxén, pp. 129–36. New York: Elsevier
6. DiBerardino, M. A., Orr, N. H., McKinnell, R. G. 1986. *Proc. Natl. Acad. Sci. USA* 83:8231–34
7. Hadorn, E. 1963. *Dev. Biol.* 7:617–29
8. Hadorn, E. 1976. In *The Genetics and Biology of Drosophila,* ed. M. Ashburner, 2C:556–617. San Diego: Academic
9. Lewis, E. B. 1978. *Nature* 276:565–70
10. Yamada, T., McDevitt, D. S. 1974. *Dev. Biol.* 38:104–18
11. Eguchi, G. 1988. See Ref. 5, pp. 147–58
12. Schmid, V., Alder, H. 1984. *Cell* 38:801–9
13. Schmid, V., Alder, H., Plickert, G., Weber, C. 1988. See Ref. 5, pp. 137–46
14. Holtzer, H., Biehl, J., Antin, P., Tokunaka, S., Sasse, J., Pacifici, M., et al. 1983. In *Globin Gene Expression and Hematopoietic Differentiation,* ed. G. Stamatoyannopoulos, A. Niehuis, pp. 213–70. New York: Liss

15. Brown, D. D. 1984. *Cell* 37:359–65
16. Weintraub, H. 1985. *Cell* 42:705–11
17a. Harris, H., Klein, G. 1969. *Nature* 224:1314–16
17b. Harris, H. 1988. *Cancer Res.* 48:3302–6
17c. Peehl, D. M., Stanbridge, E. J. 1982. *Int. J. Cancer* 30:113–20
17d. Carlsson, S. A., Luger, O., Ringertz, N. R., Savage, R. E. 1974. *Exp. Cell Res.* 84:47–55
17e. Konieczny, S. F., Lawrence, J. B., Coleman, J. R. 1983. *J. Cell Biol.* 97:1348–55
17f. Wright, W. E., Aronoff, J. 1983. *Cell Differ.* 12:299–306
17g. Lawrence, J. B., Coleman, J. R. 1984. *Dev. Biol.* 101:463–76
17h. Petit, C., Levilliers, J., Ott, M. O., Weiss, M. C. 1986. *Proc. Natl. Acad. Sci. USA* 83:2561–65
18a. Peterson, J. A., Weiss, M. C. 1972. *Proc. Natl. Acad. Sci. USA* 69:571–75
18b. Davidson, R. L. 1972. *Proc. Natl. Acad. Sci. USA* 69:951–55
18c. Rankin, J. K., Darlington, G. J. 1979. *Somat. Cell Genet.* 5:1–10
19. Harris, H., Sidebottom, E., Grace, D. M., Bramwell, M. E. 1969. *J. Cell Sci.* 4:499–525
20. Ringertz, N. R., Carlsson, S. A., Ege, T., Bolund, L. 1971. *Proc. Natl. Acad. Sci. USA* 68:3228–32
21. Blau, H. M., Chiu, C. P., Webster, C. 1983. *Cell* 32:1171–80
22. Blau, H. M., Pavlath, G. K., Harde-

man, E. C., Chiu, C. P., Silberstein, L., et al. 1985. *Science* 230:758–66
23. Chiu, C. P., Blau, H. M. 1984. *Cell* 37:879–87
24. Wright, W. E. 1984. *J. Cell Biol.* 98:427–35
25. Wright, W. E. 1984. *Exp. Cell Res.* 151:55–69
26. Baron, M. H., Maniatis, T. 1986. *Cell* 46:591–602
27. Spear, B. T., Tilghman, S. M. 1990. *Mol. Cell. Biol.* 10:5047–54
28. Wu, K. J., Samuelson, L. C., Howard, G., Meisler, M. H., Darlington, G. J. 1991. *Mol Cell. Biol.* 11:4423–30
29. Davis, R. L., Weintraub, H., Lassar, A. B. 1987. *Cell* 51:987–1000
30. Weintraub, H., Tapscott, S. J., Davis, R. L., Thayer, M. J., Adam, M. A., et al. 1989. *Proc. Natl. Acad. Sci. USA* 86:5434–38
31. Schaefer, B. W., Blakely, B. T., Darlington, G., Blau, H. M. 1990. *Nature* 344:454–58
32. Murre, C., McCaw, P. S., Vaessin, H., Caudy, M., Jan, L. Y., et al. 1989. *Cell* 58:537–44
33. Wright, W. E., Sasson, D. A., Lin, V. K. 1989. *Cell* 56:607–17
34. Edmondson, D. G., Olson, E. N. 1989. *Genes Dev.* 3:628–40
35. Braun, T., Buschhausen-Denker, G., Bober, E., Tannish, E., Arnold, H. H. 1989. *EMBO J.* 8:701–9
36. Rhodes, S. J., Konieczny, S. F. 1989. *Genes Dev.* 3:2050–61
37. Miner, J. H., Wold, B. 1990. *Proc. Natl. Acad. Sci. USA* 87:1089–93
38. Vogt, V. M. 1946. *Z. Naturforsch.* 1:469–75
39. Lewis, E. B. 1964. In *The Role of Chromosomes in Development,* ed. M. Locke, pp. 231–520. New York: Academic
40. Morata, G., Garcia-Bellido, A. 1976. *Roux's Arch. Dev. Biol.* 179:125–43
41. Wieschaus, E., Nöthiger, R. 1982. *Dev. Biol.* 90:320–34
42. Kimble, J., Edgar, L., Hirsch, D. 1984. *Dev. Biol.* 105:234–39
43. Belote, J. M., Handler, A. M., Wolfner, M. F., Livak, K. J., Baker, B. S. 1985. *Cell* 40:339–48
44. Schedin, P. 1988. PhD thesis. Univ. Colo.
45. Way, J. C., Chalfie, M. 1989. *Genes Dev.* 3:1823–33
46. Duncan, I., Lewis, E. B. 1982. In *Developmental Order: Its Origin and Regulation,* ed. S. Subtelny, P. Green, pp. 533–54. New York: Liss
47. Belote, J. M., Baker, B. S. 1987. *Proc. Natl. Acad. Sci. USA* 84:8026–30
48. Belote, J. M., McKeown, M. B., Andrew, D. J., Scott, T. N., Wolfner, M. F., Baker, B. S. 1985. *Cold Spring Harbor Symp. Quant. Biol.* 50:605–14
49. Duncan, I. 1987. *Annu. Rev. Genet.* 21:285–319
50. Landschulz, W. H., Johnson, P. F., McKnight, S. L. 1989. *Science* 243:1681–88
51. Murre, C., McCaw, P. S., Baltimore, D. 1989. *Cell* 56:777–83
52. Benezra, R., Davis, R. L., Lockshon, D., Turner, D. L., Weintraub, H. 1990. *Cell* 61:49–59
53. Baeuerle, P. A., Baltimore, D. 1988. *Science* 242:540–46
54. Baeuerle, P. A., Baltimore, D. 1988. *Cell* 53:211–17
55. Peterson, C. A., Gordon, H., Hall, Z. W., Paterson, B. M., Blau, H. M. 1990. *Cell* 62:493–502
56. Lin, Y. S., Carey, M., Ptashne, M., Green, M. R. 1990. *Nature* 345:359–61
57. Carey, M., Lin, Y. S., Green, M. R., Ptashne, M. 1990. *Nature* 345:361–64
58. Parkhurst, S. M., Bopp, D., Ish-Horowicz, D. 1990. *Cell* 63:1179–91
59. Hodgkin, J. 1990. *Nature* 344:721–28
60. Botas, J., Moscoso del Prado, J., Garcia-Bellido, A. 1982. *EMBO J.* 1:307–11
61. Epstein, C. J. 1986. In *The Consequences of Chromosome Imbalance: Principles, Mechanisms, and Models,* pp. 65–79. Cambridge: Cambridge Univ. Press
62. Ringertz, N. R., Savage, R. E. 1976. In *Cell Hybrids.* New York: Academic
63. Weiss, M. C. 1982. In *Somatic Cell Genetics,* ed. C. T. Casky, D. C. Robins, pp. 169–820. New York: Plenum
64. Killary, A. M., Fournier, R. E. K. 1984. *Cell* 38:523–34
65. Boshart, M., Weih, F., Schmidt, A., Fournier, R. E. K., Schütz, G. 1990. *Cell* 61:905–16
66. Ruppert, S., Boshart, M., Bosch, F. X., Schmid, W., Fournier, R. E. K., Schütz, G. 1990. *Cell* 61:895–904
67. Moscoso del Prado, J., Garcia-Bellido, A. 1984. *Roux's Arch. Dev. Biol.* 193:242–45
68. Struhl, G., Akam, M. 1985. *EMBO J.* 4:3259–64
69. Larsen, P. R., Harney, J. W., Moore, D. D. 1986. *Proc. Natl. Acad. Sci. USA* 83:8283–87
70. Nir, U., Walker, M. D., Rutter, W. J. 1986. *Proc. Natl. Acad. Sci. USA* 83:3180–84
71. Laimonis, L., Holmgren-Konig, M.,

Khoury, G. 1986. *Proc. Natl. Acad. Sci. USA* 83:3151–55
72. Cordle, S. R., Whelan, J., Henderson, E., Masoska, H., Weil, P. A., Stein, R. 1991. *Cell. Biol.* 11:2881–86
73. Burt, D. W., Nakamura, N., Kelley, P., Dzau, V. J. 1989. *J. Biol. Chem.* 264:7357–62
74. Smith, M. R., Greene, W. C. 1989. *Proc. Natl. Acad. Sci. USA* 86:8526–30
75. Pierce, J. W., Gifford, A. M., Baltimore, D. 1991. *Mol. Cell. Biol.* 11: 1431–37
76. Imler, J. L., Lemaire, C., Wasylyk, C., Wasylyk, B. 1987. *Mol. Cell. Biol.* 7:2558–67
77. Winoto, A., Baltimore, D. 1989. *Cell* 59:649–55
78. Cannio, R., Rennie, P. S., Blasi, F. 1991. *Nucleic Acids Res.* 19:2303
79. Muglia, L., Rothman-Denes, L. B. 1986. *Proc. Natl. Acad. Sci. USA* 83:7653–57
80. Farrell, F. X., Sax, C. M., Zehner, Z. E. 1990. *Mol. Cell. Biol.* 10:2349–58
81. Savagner, P., Miyashita, T., Yamada, Y, 1990. *J. Biol. Chem.* 365:6669–74
82. Cao, S. X., Gutman, P. D., Dave, H. P. G., Schechter, A. N. 1989. *Proc. Natl. Acad. Sci. USA* 86:5306–9
83. Brand, A. H., Breeden, L., Abraham, J., Sternglanz, R., Nasmyth, K. 1985. *Cell* 41:41–48
84. Johnson, A. D., Herskowitz, I. 1985. *Cell* 42:237–47
85. Yu, H., Porton, B., Shen, L., Eckhardt, L. A. 1989. *Cell* 58:441–48
86. Hofmann, J. F. X., Laroche, T., Brand, A. H., Gasser, S. M. 1989. *Cell* 57:725–37
87. Boodbourn, S., Maniatis, T. 1988. *Proc. Natl. Acad. Sci. USA* 85:1447–51
88. Harada, H., Fujita, T., Miyamoto, M., Kimura, Y., Maruyama, M., et al. 1989. *Cell* 58:729–39
89. Park, H. O., Craig, E. A. 1989. *Mol. Cell. Biol.* 9:2025–33
90. Park, H. O., Craig, E. A. 1991. *Genes Dev.* 5:1299–308
91. Baniahmad, A., Muller, M., Steiner, C., Renkawitz, R. 1987. *EMBO J.* 6:2297–303
92. Park, K., Atchison, M. L. 1991. *Proc. Natl. Acad. Sci. USA* 88:9804–8
93. Miyazaki, J.-I., Appella, E., Ozato, K. 1986. *Proc. Natl. Acad. Sci. USA* 83:9537–41
94. Tomilin, N. V., Sanai, M. M., Iguchi-Ariga S. M., Ariga, H. 1990. *FEBS Lett.* 263:69–72
95. Saffer, J. D., Thurston, S. J. 1989. *Mol. Cell. Biol.* 9:355–64
96. McCormick, A., Wu, D., Castrillo, J.

L., Dana, S., Strobl, J., et al. 1988. *Cell* 55:379–89
97. Jaynes, J. B., O'Farrell, P. H. 1988. *Nature* 336:744–49
98. Han, K., Levine, M. S., Manley, J. L. 1989. *Cell* 56:573–83
99. Glass, C. K., Holloway, J. M., Devary, O. V., Rosenfeld, M. G. 1988. *Cell* 54:313–23
100. Mitchell, P. J., Wang, C., Tjian, R. 1987. *Cell* 50:847–61
101. Ma, J., Ptashne, M. 1987. *Cell* 50:137–42
102. Johnston, S. A., Salmeron, J. M. Jr., Dincher, S. S. 1987. *Cell* 50:143–46
103. Levine, M., Manley, J. L. 1989. *Cell* 59:405–8
104. Weissman, J. D., Singer, D. S. 1991. *Mol. Cell. Biol.* 11:4217–27
105. Kelly, S. J. 1977. *J. Exp. Zool.* 200:365–76
106. Maro, B., Johnson, M. H., Pickering, S. J., Louvard, D. 1985. *J. Embryol. Exp. Morphol.* 90:287–309
107. Johnson, M. H., Maro, B. 1985. *J. Embryol. Exp. Morphol.* 90.311–34
108. Fleming, T. P., Pickering, S. J. 1985. *J. Embryol. Exp. Morphol.* 89:175–208
109. Reeve, W. J. D. 1981. *J. Embryol. Exp. Morphol.* 62:351–67
110. Handyside, A. H. 1980. *J. Embryol. Exp. Morphol.* 60:99–116
111. Ptashne, M. 1986. In *A Genetic Switch: Gene Control and Phage λ.* Cambridge, Mass: Cell Press/Blackwell Sci.
112. Angel, P., Hattori, K., Smeal, T., Karin, M. 1988. *Cell* 55:875–85
113. Thayer, M. J., Tapscott, S. J., Davis, R. L., Wright, W. E., Lassar, A. B., Weintraub, H. 1989. *Cell* 58:241–48
114. Kuziora, M. A., McGinnis, W. 1988. *Cell* 55:477–85
115. Lisman, J. E., Goldring, M. A. 1988. *Proc. Natl. Acad. Sci. USA* 85:5320–24
116a. Greenburg, G., Hay, E. D. 1988. *Development* 102:605–22
116b. Streuli, C. H., Bailey, N., Bissell, M. J. 1991. *J. Cell Biol.* 115:1383–95
117. Ingraham, H. A., Chen, R., Mangalam, H. J., Elsholtz, H. P., Flynn, S. E., et al. 1988. *Cell* 55:519–29
118. Bodner M., Castrillo, J. L., Theill, L. E., Deerinck, T., Ellisman, M., Karin, M. 1988. *Cell* 55:505–18
119. Baumhueter, S., Mendel, D. B., Conley, P. B., Kuo, C. J., Turk, C., et al. 1990. *Genes Dev.* 4:372–79
120 Johnston, J. E., Birren, S. J., Anderson, D. J. 1990. *Nature* 346:858–61
121. Hopwood, N. D., Gurdon, J. B. 1990. *Nature* 347:197–200
122. Gibson, G., Gehring, W. J. 1988. *Development* 102:657–75

123. Sullivan, C. H., Grainger, R. M. 1986. *Proc. Natl. Acad. Sci. USA* 83:329–33

124. Bresnick, E. H., John, S., Berard, D. S., LeFebvre, P., Hager, G. L. 1990. *Proc. Natl. Acad. Sci. USA* 87:3977–81

125. Grunstein, M. 1990. *Annu. Rev. Cell Biol.* 6:643–78

126. Zink, B., Paro, R. 1989. *Nature* 337:468–71

127. DiNardo, S., Sher, E., Heemskerk-Jongens, J., Kassis, J. A., O'Farrell, P. H. 1988. *Nature* 332:604–9

128. Le Douarin, N. M. 1986. *Science* 231:1515–22

129. Hughes, S. M., Blau, H. M. 1992. *Cell* 68:659–71

130. Rentranz, P. J., Cunningham, M. G., McKay, R. D. G. 1990. *Cell* 66:713–29

131a. Reuter, G., Giarre, M., Farah, J., Gausz, J., Spierer, A., Spierer, P. 1990. *Nature* 344:219–23

131b. Alberts, B., Sternglanz, R. 1990. *Nature* 344:193–94

132. Paro, R., Hogness, D. S. 1991. *Proc. Natl. Acad. Sci. USA* 88:263–67

AUTHOR INDEX

SUBJECT INDEX

A

ABC primosome
  prokaryotic DNA replication
    and, 689
Abelson murine leukemia virus
    (A-MuLV)
  myc oncogene and, 840
Acanthamoeba spp.
  myosin in, 722-23, 725, 727-
    30, 742, 744, 746-48,
    751-52
Acatalasemia
  peroxisomes and, 176
Accessory factor sites
  hormone response elements
    and, 1155, 1161-62
ACE1 transcription factor
  cysteine-rich motif and, 1080
  metallothionein gene and, 922
AceE gene
  identification of, 295
Acetate
  fatty acid oxidation and, 10
  peroxisomes and, 184
Acetoacetate
  as end-product of fatty acid
    oxidation, 10-11
Acetoacetyl CoA
  peroxisomes and, 170
Acetonitrile solvent
  Diels–Alder reactions and, 38
Acetopine
  distribution and biosynthesis
    of, 520
Acetyl CoA
  isolation of, 14
  neuronal calmodulin-
    dependent protein kinases
    and, 587
  peroxisomes and, 168, 170
Acetyl CoA carboxylase
  calmodulin-dependent protein
    kinase phosphorylation
    sites and, 588, 590
N-Acetylglucosamine
  DNA looping and, 217
Acetyl phosphate
  fatty acid oxidation and, 10-
    11, 13
  lability of, 14
N-Acetylphosphotransferase
  lysosomal proteins and, 502

Acetyltransferase
  peroxisomes and, 168
Acheta domesticus
  apolipoproteins in, 102, 104
α_1-Acid glycoprotein
  gene promoter
    hormone response elements
      and, 1156-57
Acid hydrolases
  mannose 6-phosphate/
    insulinlike growth factor
    II receptor and, 308-9,
    314, 317-18
Acid phosphatase
  cytoplasmic internalization
    signals of, 321
  mass spectrometry and, 1005
  SV40 T antigen and, 70
Acquired immunodeficiency syn-
    drome (AIDS)
  mitochondrial DNA depletion
    syndrome and, 1195
  recombinant immunotoxins
    against, 350
Actin
  gene for, 657
  inositol phosphates and, 244-
    45
  myosin and, 721-23, 726,
    728-29, 732-33, 738,
    741, 743, 746- 47, 750-
    51
  ubiquitin and, 791
Acute lymphoblastic leukemia
    (ALL)
  DNA ligases and, 277
Acylamidoepindolidiones
  β-sheets in, 407
Acyl carrier protein
  membrane-derived oligosac-
    charide biosynthesis and,
    26
Acyl CoA
  acylation of glycerophosphate
    and, 17
  peroxisomes and, 161-62,
    164-68, 171-74, 176-77,
    181-82, 185-87
Acyl CoA dehydrogenase
  mitochondrial DNA and, 1199
Adaptins
  clathrin-coated pits and, 498-
    500, 502

signal-mediated sorting and,
    505
ade8 mutation
  polyadenylation of mRNA
    precursors and, 434
Adenine nucleotide translocator
  mitochondrial oxidative
    phosphorylation com-
    plexes and, 1178-80
Adenosine
  polymerase chain reaction
    and, 143
  SV40 T antigen and, 59, 61-
    62
  telomerases and, 115-20, 122,
    124-125
Adenosine deaminase
  rate acceleration of, 50
Adenosine diphosphate (ADP)
  chromosome partitioning and,
    288, 299
  diphtheria toxin and, 348
  mitochondrial DNA and, 1199
  platelet aggregation in insects
    and, 91-92, 94
  Pseudomonas exotoxin and,
    333, 335-37
  SV40 T antigen and, 61
Adenosine monophosphate
    (AMP)
  DNA ligases and, 253-54,
    259-60, 265-66, 271-72,
    275
  polyadenylation of mRNA
    precursors and, 426-27,
    430
  ubiquitin and, 763
Adenosine triphosphate (ATP)
  acetyl phosphate as precursor
    of, 13
  animal cell cycle control and,
    442, 448-49, 465
  catalytic RNAs and, 658
  chromosome partitioning and,
    288, 293, 299
  cyclin and, 797
  DNA ligases and, 253-54,
    259, 264-67, 269, 272
  heteroduplex DNA and, 604,
    607, 611-13, 617-19,
    623
  inositol phosphates and, 232,
    240

1305

CUMULATIVE INDEXES

CONTRIBUTING AUTHORS, VOLUMES 57–61

1347

CHAPTER TITLES, VOLUMES 57–61

1350

ANNUAL REVIEWS INC.

a nonprofit scientific publisher
4139 El Camino Way
P. O. Box 10139
Palo Alto, CA 94303-0897 • USA

ORDER FORM
ORDER TOLL FREE **1-800-523-8635** (except California)
FAX: 415-855-9815

Annual Reviews Inc. publications may be ordered directly from our office; through booksellers and subscription agents, worldwide; and through participating professional societies.

Prices are subject to change without notice. ARI Federal I.D. #94-1156476

- **Individuals:** Prepayment required on new accounts by check or money order (in U.S. dollars, check drawn on U.S. bank) or charge to MasterCard, VISA, or American Express.

- **Institutional Buyers:** Please include purchase order.

- **Students:** $10.00 discount from retail price, per volume. Prepayment required. Proof of student status must be provided. (Photocopy of Student I.D. is acceptable.) Student must be a degree candidate at an accredited institution. Order direct from Annual Reviews. Orders received through bookstores and institutions requesting student rates will be returned.

- **Professional Society Members:** Societies who have a contractual arrangement with Annual Reviews offer our books at reduced rates to members. Contact your society for information.

- **California orders** must add applicable sales tax.

- **CANADIAN ORDERS:** We must now collect 7% General Sales Tax on orders shipped to Canada. Canadian orders will not be accepted unless this tax has been added. Tax Registration # R 121 449-029. **Note:** Effective 1-1-92 Canadian prices increase from USA level to "other countries" level. See below.

- **Telephone orders,** paid by credit card, welcomed. Call Toll Free **1-800-523-8635** (except in California). California customers use 1-415-493-4400 (not toll free). M-F, 8:00 am - 4:00 pm, Pacific Time. Students ordering by telephone must supply (by FAX or mail) proof of student status if proof from current academic year is not on file at Annual Reviews. Purchase orders from universities require written confirmation before shipment.

- **FAX:** 415-855-9815   **Telex:** 910-290-0275

- **Postage paid by Annual Reviews** (4th class bookrate). UPS domestic ground service (except to AK and HI) available at $2.00 extra per book. UPS air service or Airmail also available at cost. UPS requires street address. P.O. Box, APO, FPO, not acceptable.

- **Regular Orders:** Please list below the volumes you wish to order by volume number.

- **Standing Orders:** New volume in the series will be sent to you automatically each year upon publication. Cancellation may be made at any time. Please indicate volume number to begin standing order.

- **Prepublication Orders:** Volumes not yet published will be shipped in month and year indicated.

- **We do not ship on approval.**

ANNUAL REVIEWS SERIES *Volumes not listed are no longer in print*	Prices, postpaid, per volume		Regular Order Please send Volume(s):	Standing Order Begin with Volume:
	Until 12-31-91 USA & Canada / elsewhere	After 1-1-92 USA / other countries (incl. Canada)		
Annual Review of ANTHROPOLOGY				
Vols. 1-16 (1972-1987)	$33.00/$38.00 ⎫			
Vols. 17-18 (1988-1989)	$37.00/$42.00 ⎬ $41.00/$46.00			
Vols. 19-20 (1990-1991)	$41.00/$46.00 ⎭			
Vol. 21 (avail. Oct. 1992)	$44.00/$49.00	$44.00/$49.00	Vol(s)._____	Vol.____
Annual Review of ASTRONOMY AND ASTROPHYSICS				
Vols. 1, 5-14, (1963, 1967-1976)				
16-20 (1978-1982)	$33.00/$38.00 ⎫			
Vols. 21-27 (1983-1989)	$49.00/$54.00 ⎬ $53.00/$58.00			
Vols. 28-29 (1990-1991)	$53.00/$58.00 ⎭			
Vol. 30 (avail. Sept. 1992)	$57.00/$62.00	$57.00/$62.00	Vol(s)._____	Vol.____
Annual Review of BIOCHEMISTRY				
Vols. 30-34, 36-56 (1961-1965, 1967-1987)	$35.00/$40.00 ⎫			
Vols. 57-58 (1988-1989)	$37.00/$42.00 ⎬ $41.00/$47.00			
Vols. 59-60 (1990-1991)	$41.00/$47.00 ⎭			
Vol. 61 (avail. July 1992)	$46.00/$52.00	$46.00/$52.00	Vol(s)._____	Vol.____

ANNUAL REVIEWS SERIES *Volumes not listed are no longer in print*	Prices, postpaid, per volume		Regular Order Please send Volume(s):	Standing Order Begin with Volume:
	Until 12-31-91 USA & Canada / elsewhere	After 1-1-92 USA / other countries (incl. Canada)		

Annual Review of BIOPHYSICS AND BIOMOLECULAR STRUCTURE

		Until 12-31-91	After 1-1-92	Regular	Standing
Vols. 1-11	(1972-1982)	$33.00/$38.00 ⎤			
Vols. 12-18	(1983-1989)	$51.00/$56.00 ⎬ $55.00/$60.00			
Vols. 19-20	(1990-1991)	$55.00/$60.00 ⎦			
Vol. 21	(avail. June 1992)	$59.00/$64.00	$59.00/$64.00	Vol(s).____	Vol.____

Annual Review of CELL BIOLOGY

Vols. 1-3	(1985-1987)	$33.00/$38.00 ⎤			
Vols. 4-5	(1988-1989)	$37.00/$42.00 ⎬ $41.00/$46.00			
Vols. 6-7	(1990-1991)	$41.00/$46.00 ⎦			
Vol. 8	(avail. Nov. 1992)	$46.00/$51.00	$46.00/$51.00	Vol(s).____	Vol.____

Annual Review of COMPUTER SCIENCE

Vols. 1-2	(1986-1987)	$41.00/$46.00	$41.00/$46.00		
Vols. 3-4	(1988, 1989)	$47.00/$52.00	$47.00/$52.00	Vol(s).____	Vol.____

Series suspended until further notice. Volumes 1-4 are still available at the special promotional price of $100.00 USA /$115.00 other countries, when all 4 volumes are purchased at one time. Orders at the special price must be prepaid.

Annual Review of EARTH AND PLANETARY SCIENCES

Vols. 1-10	(1973-1982)	$33.00/$38.00 ⎤			
Vols. 11-17	(1983-1989)	$51.00/$56.00 ⎬ $55.00/$60.00			
Vols. 18-19	(1990-1991)	$55.00/$60.00 ⎦			
Vol. 20	(avail. May 1992)	$59.00/$64.00	$59.00/$64.00	Vol(s).____	Vol.____

Annual Review of ECOLOGY AND SYSTEMATICS

Vols. 2-18	(1971-1987)	$33.00/$38.00 ⎤			
Vols. 19-20	(1988-1989)	$36.00/$41.00 ⎬ $40.00/$45.00			
Vols. 21-22	(1990-1991)	$40.00/$45.00 ⎦			
Vol. 23	(avail. Nov. 1992)	$44.00/$49.00	$44.00/$49.00	Vol(s).____	Vol.____

Annual Review of ENERGY AND THE ENVIRONMENT

Vols. 1-7	(1976-1982)	$33.00/$38.00 ⎤			
Vols. 8-14	(1983-1989)	$60.00/$65.00 ⎬ $64.00/$69.00			
Vols. 15-16	(1990-1991)	$64.00/$69.00 ⎦			
Vol. 17	(avail. Oct. 1992)	$68.00/$73.00	$68.00/$73.00	Vol(s).____	Vol.____

Annual Review of ENTOMOLOGY

Vols. 10-16, 18	(1965-1971, 1973)				
20-32	(1975-1987)	$33.00/$38.00 ⎤			
Vols. 33-34	(1988-1989)	$36.00/$41.00 ⎬ $40.00/$45.00			
Vols. 35-36	(1990-1991)	$40.00/$45.00 ⎦			
Vol. 37	(avail. Jan. 1992)	$44.00/$49.00	$44.00/$49.00	Vol(s).____	Vol.____

Annual Review of FLUID MECHANICS

Vols. 2-4, 7	(1970-1972, 1975)				
9-19	(1977-1987)	$34.00/$39.00 ⎤			
Vols. 20-21	(1988-1989)	$36.00/$41.00 ⎬ $40.00/$45.00			
Vols. 22-23	(1990-1991)	$40.00/$45.00 ⎦			
Vol. 24	(avail. Jan. 1992)	$44.00/$49.00	$44.00/$49.00	Vol(s).____	Vol.____

Annual Review of GENETICS

Vols. 1-12, 14-21	(1967-1978, 1980-1987)	$33.00/$38.00 ⎤			
Vols. 22-23	(1988-1989)	$36.00/$41.00 ⎬ $40.00/$45.00			
Vols. 24-25	(1990-1991)	$40.00/$45.00 ⎦			
Vol. 26	(avail. Dec. 1992)	$44.00/$49.00	$44.00/$49.00	Vol(s).____	Vol.____

Annual Review of IMMUNOLOGY

Vols. 1-5	(1983-1987)	$33.00/$38.00 ⎤			
Vols. 6-7	(1988-1989)	$36.00/$41.00 ⎬ $41.00/$46.00			
Vol. 8	(1990)	$40.00/$45.00 ⎦			
Vol. 9	(1991)	$41.00/$46.00	$41.00/$46.00		
Vol. 10	(avail. April 1992)	$45.00/$50.00	$45.00/$50.00	Vol(s).____	Vol.____